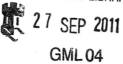

FUNDAMENTAL
NEUROSCIENCE

THIRD EDITION

FUNDAMENTAL NEUROSCIENCE

THIRD EDITION

Edited by

LARRY SQUIRE

VA Medical Center San Diego, California
University of California, San Diego, La Jolla, California

DARWIN BERG

University of California, San Diego
La Jolla, California

FLOYD BLOOM

The Scripps Research Institute
La Jolla, California

SASCHA DU LAC

The Salk Institute
La Jolla, California

ANIRVAN GHOSH

University of California, San Diego
La Jolla, California

NICHOLAS SPITZER

University of California, San Diego
La Jolla, California

AMSTERDAM • BOSTON • HEIDELBERG • LONDON
NEW YORK • OXFORD • PARIS • SAN DIEGO
SAN FRANCISCO • SINGAPORE • SYDNEY • TOKYO

ELSEVIER

Academic Press is an imprint of Elsevier

Academic Press is an imprint of Elsevier

30 Corporate Drive, Suite 400, Burlington, MA 01803, USA
525 B Street, Suite 1900, San Diego, California 92101-4495, USA
84 Theobald's Road, London WC1X 8RR, UK

This book is printed on acid-free paper. ∞ /oo 5442575

Library of Congress Cataloging-in-Publication Data
Fundamental neuroscience / edited by Larry Squire . . . [et al.].—3rd ed.
    p. ; cm.
  Includes bibliographical references and index.
  ISBN 978-0-12-374019-9 (alk. paper)
  1. Neurosciences.   I. Squire, Larry R.
  [DNLM: 1. Nervous System Physiology.   2. Neurosciences.   WL 102 F981 2008]
  QP355.2.F862 2008
  612.8—dc22

                                                          2008001747

British Library Cataloguing-in-Publication Data
A catalogue record for this book is available from the British Library.

ISBN: 978-0-12-374019-9

For information on all Academic Press publications
visit our Web site at www.books.elsevier.com

Printed in Canada
08 09 10 9 8 7 6 5 4 3 2 1

Short Contents

VII
BEHAVIORAL AND COGNITIVE NEUROSCIENCE

Full Contents

III
NERVOUS SYSTEM DEVELOPMENT

IV
SENSORY SYSTEMS

23. Fundamentals of Sensory Systems
STEWART H. HENDRY, STEVEN S. HSIAO, AND
M. CHRISTIAN BROWN

24. Chemical Senses: Taste and Olfaction
KRISTIN SCOTT

25. Somatosensory System
STEWART HENDRY AND STEVEN HSIAO

26. Audition
M. CHRISTIAN BROWN AND JOSEPH SANTOS-SACCHI

27. Vision
R. CLAY REID AND W. MARTIN USREY

V
MOTOR SYSTEMS

VI
REGULATORY SYSTEMS

51. Language and Communication

DAVID N. CAPLAN AND JAMES L. GOULD

52. The Prefrontal Cortex and Executive Brain Functions

EARL MILLER AND JONATHAN WALLIS

53. Consciousness

CHRISTOF KOCH

Preface to the Third Edition

In this third edition of *Fundamental Neuroscience*, we have tried to improve on the second edition with a volume that effectively introduces students to the full range of contemporary neuroscience. Neuroscience is a large field founded on the premise that all of behavior and all of mental life have their origin in the structure and function of the nervous system. Today, the need for a single-volume introduction to neuroscience is greater than ever. Towards the end of the 20th century, the study of the brain moved from a peripheral position within both the biological and psychological sciences to become an interdisciplinary field that is now central within each discipline. The maturation of neuroscience has meant that individuals from diverse backgrounds—including molecular biologists, computer scientists, and psychologists—are interested in learning about the structure and function of the brain and about how the brain works. In addition, new techniques and tools have become available to study the brain in increasing detail. In the last 15 years new genetic methods have been introduced to delete or over-express single genes with spatial and temporal specificity. Neuroimaging techniques such as functional magnetic resonance imaging (fMRI) have been developed that allow study of the living human brain while it is engaged in cognition.

This third edition attempts to capture the promise and excitement of this fast-moving discipline. All the chapters have been rewritten to make them more concise. As a result the new edition is about 30% shorter than previous editions but still covers the same comprehensive range of topics. The volume begins with an opening chapter that provides an overview of the discipline. A second chapter presents fundamental information about the architecture and anatomy of

nervous systems. The remainder of the volume (Sections II–VII) presents the major topics of neuroscience. The second section (Cellular and Molecular Neuroscience) considers the cellular and subcellular organization of neurons, the physiology of nerve cells, and how signaling occurs between neurons. The third section (Nervous System Development) includes discussion of neural induction, cell fate, migration, process outgrowth, development of dendrites, synapse formation, programmed cell death, synapse elimination, and early experience including critical periods. The fourth and fifth sections (Sensory Systems and Motor Systems) describe the neural organization of each sensory modality and the organization of the brain pathways and systems important for locomotion, voluntary action, and eye movements. The sixth section (Regulatory Systems) describes the variety of hypothalamic and extra-hypothalamic systems that support motivation, reward, and internal regulation, including cardiovascular function, respiration, food and water intake, neuroendocrine function, circadian rhythms, and sleep and dreaming. The final section (Behavioral and Cognitive Neuroscience) describes the neural foundations of the so-called higher mental functions including perception, attention, memory, language, spatial cognition, and executive function. Additional chapters cover human brain evolution, cognitive development and aging, and consciousness. The volume will be accompanied by an easily accessible companion website, which will present all the figures and increase the flexibility with which the material can be used.

The authors listed at the ends of the chapters and boxes are working scientists, experts in the topics they cover. The Editors edited the chapters to achieve consistency of style and content. At Academic Press/Elsevier Science, the project was coordinated by Hilary Rowe and Nikki Levy (Publishing Editors), and we are grateful to them for their leadership and advice throughout the project. In addition, Meg Day (Developmental Editor) very capably coordinated the production of the book with the help of Sarah Hajduk (Publishing Services Manager) and Christie Jozwiak (Project Manager).

The Editors of *Fundamental Neuroscience* hope that users of this book, and especially the students who will become the next generation of neuroscientists, find the subject matter of neuroscience as interesting and exciting as we do.

The Editors

About The Editors

Larry R. Squire is Distinguished Professor of Psychiatry, Neurosciences, and Psychology at the University of California School of Medicine, San Diego, and Research Career Scientist at the Veterans Affairs Medical Center, San Diego. He investigates the organization and neurological foundations of memory. He is a former President of the Society for Neuroscience and is a member of the National Academy of Sciences and the Institute of Medicine.

Darwin K. Berg is Distinguished Professor in the Division of Biological Sciences at the University of California, San Diego. He has been chairman of the Biology Department and currently serves as Councilor of the Society for Neuroscience and as a Board member of the Kavli Institute for Brain and Mind. His research is focused on the roles of nicotinic cholinergic signaling in the vertebrate nervous system.

Floyd Bloom is Professor Emeritus in the Molecular and Integrative Neuroscience Department (MIND) at The Scripps Research Institute. His recent awards include the Sarnat Award from the Institute of Medicine and the Salmon Medal of the New York Academy of Medicine. He is a former President of the Society for Neuroscience and is a member of the National Academy of Sciences and the Institute of Medicine.

Sascha du Lac is an Investigator of the Howard Hughes Medical Institute and an Associate Professor of Systems Neurobiology at the Salk Institute for Biological Studies. Her research interests are in the neurobiology of resilience and learning, and her laboratory investigates behavioral, circuit, cellular, and molecular mechanisms in the sense of balance.

Anirvan Ghosh is Stephen Kuffler Professor in the Division of Biological Sciences at the University of California, San Diego and Director of the graduate program in Neurosciences. His research interests include the development of synaptic connections in the central nervous system and the role of activity-dependent gene expression in the cortical development. He is recipient of the Presidential Early Career Award for Scientists and Engineers and the Society for Neuroscience Young Investigator Award.

Nicholas C. Spitzer is Distinguished Professor in the Division of Biological Sciences at the University of California, San Diego. His research is focused on neuronal differentiation and the role of electrical activity and calcium signaling in the assembly of the nervous system. He has been chairman of the Biology Department and the Neurobiology Section, a trustee of the Grass Foundation, and served as Councilor of the Society for Neuroscience. He is a member of the American Academy of Arts and Sciences and Co-Director of the Kavli Institute for Brain and Mind.

Contributors

Jocelyne Bachevalier Emory University, Atlanta, GA

James F. Baker Northwestern University Medical School, Chicago, IL

Floyd E. Bloom The Scripps Research Institute, La Jolla, CA

Scott T. Brady University of Illinois at Chicago, Chicago, IL

Marianne Bronner-Fraser Caltech, Pasadena, CA

Peter J. Brophy University of Edinburgh, Edinburgh, Scotland

M. Christian Brown Harvard Medical School, Boston, MA

Steven J. Burden NYU Medical Center, New York, NY

Ania Busza University of Massachusetts Medical School, Worcester, MA

John H. Byrne University of Texas Medical School at Houston, Houston, TX

David N. Caplan Massachusetts General Hospital, Boston, MA

J. Patrick Card University of Pittsburgh, Pittsburgh, PA

Luz Claudio Mount Sinai School of Medicine, New York, NY

Hollis Cline Cold Spring Harbor Laboratory, Cold Spring Harbor, NY

Carol L. Colby University of Pittsburgh, Pittsburgh, PA

David R. Colman Montreal Neurological Institute, Montreal, Quebec, Canada

Ariel Y. Deutch Vanderbilt University Medical Center, Nashville, TN

Howard B. Eichenbaum Boston University, Boston, MA

Patrick Emery University of Massachusetts Medical School, Worcester, MA

Barry J. Everitt University of Cambridge, Cambridge, United Kingdom

Jack L. Feldman David Geffen School of Medicine at UCLA, Los Angeles, CA

Mary Kay Floeter National Institute of Neurological Disorders and Stroke, Bethesda, MD

Anirvan Ghosh University of California, San Diego, La Jolla, CA

Andrea C. Gore University of Texas at Austin, Austin, TX

Jacqueline P. Gottlieb Columbia University, New York, NY

James L. Gould Princeton University, Princeton, NJ

Sten Grillner Karolinska Institute, Stockholm, Sweden

William A. Harris University of Cambridge, Cambridge, United Kingdom

Volker Hartenstein University of California, Los Angeles, CA

Mary E. Hatten The Rockefeller University, New York, NY

Stewart H. Hendry Johns Hopkins University, Baltimore, MD

J. Allan Hobson Harvard Medical School, Boston, MA

Patrick R. Hof Mount Sinai School of Medicine, New York, NY

Steven S. Hsiao Johns Hopkins University, Baltimore, MD

Yuh-Nung Jan University of California, San Francisco, San Francisco, CA

Jon H. Kaas Vanderbilt University, Nashville, TN

Sabine Kastner Princeton University, Princeton, NJ

Grahame Kidd Lerner Research Institute, Cleveland Clinic, Cleveland, OH

Chris Kintner The Salk Institute for Biological Studies, San Diego, CA

Christof Koch California Institute of Technology, Pasadena, CA

Alex Kolodkin Johns Hopkins University School of Medicine, Baltimore, MD

Eric I. Knudsen Stanford University School of Medicine, Stanford, CA

George F. Koob The Scripps Research Institute, La Jolla, CA

Richard J. Krauzlis The Salk Institute, La Jolla, CA

Jeff W. Lichtman Molecular and Cellular Biology, Harvard University, Cambridge, MA

John C. Longhurst University of California, Irvine, CA

Andrew Lumsden MRC Centre for Developmental Neurobiology, King's College London, U.K.

Pierre J. Magistretti University of Lausanne, Lausanne, Switzerland

Joseph R. Manns Emory University, Altanta, GA

Michael D. Mauk University of Texas Health Science Center at Houston, Houston, TX

David A. McCormick Yale University School of Medicine, New Haven, CT

Donald R. McCrimmon Feinberg School of Medicine, Northwestern University, Chicago, IL

George Z. Mentis The Porter Neuroscience Center, NINDS, NIH, Bethesda, MD

Earl K. Miller Massachusetts Institute of Technology, Cambridge, MA

Jonathan W. Mink University of Rochester School of Medicine and Dentistry, Rochester, NY

Robert Y. Moore University of Pittsburgh School of Medicine, Pittsburgh, PA

Esther A. Nimchinsky Rutgers University, Newark, NJ

Dennis D. M. O'Leary The Salk Institute, La Jolla, CA

Carl R. Olson Carnegie Mellon University, Pittsburgh, PA

Ronald W. Oppenheim Wake Forest University School of Medicine, Winston-Salem, NC

Edward F. Pace-Schott Harvard Medical School, Boston, MA

Luiz Pessoa Department of Psychological and Brain Sciences Indiana University, Bloomington, Bloomington, IN

Terry L. Powley Purdue University, West Lafayette, IN

Todd M. Preuss University of Louisiana at Lafayette, New Iberia, LA

Peter R. Rapp Mount Sinai School of Medicine, New York, NY

R. Clay Reid Harvard Medical School, Boston, MA

Steven M. Reppert University of Massachusetts Medical School, Worcester, MA

John H. Reynolds The Salk Institute, La Jolla, CA

Trevor W. Robbins University of Cambridge, Cambridge, United Kingdom

Robert H. Roth Yale University School of Medicine, New Haven, CT

Joseph Santos-Sacchi Yale University School of Medicine, New Haven, CT

Peter Scheiffele Columbia University, New York, NY

Marc H. Schieber University of Rochester School of Medicine and Dentistry, Rochester, NY

Howard Schulman Stanford University Medical Center, Stanford, CA

Thomas L. Schwarz Children's Hospital, Boston, MA

Kristin Scott University of California, Berkeley, Berkeley, CA

Gordon M. Shepherd Yale University School of Medicine, New Haven, CT

Robert Stickgold Harvard Medical School, Boston, MA

Edward M. Stricker University of Pittsburg, Pittsburg, PA

Larry W. Swanson University of Southern California, Los Angeles, CA

Juan C. Tapia Molecular and Cellular Biology, Harvard University, Cambridge, MA

Marc Tessier-Lavigne Genentech, Inc., South San Francisco, CA

W. Thomas Thach Washington University School of Medicine, St. Louis, MO

Roger B.H. Tootell Martinos Center for Biomedical Imaging, Massachusetts General Hospital, Charlestown, MA

Bruce D. Trapp Cleveland Clinic Foundation, Cleveland Clinic, Cleveland, OH

Leslie G. Ungerlieder National Institute of Mental Health, Bethesda, MD

W. Martin Usrey University of California, Davis, CA

Jean de Vellis University of California, Los Angeles, CA

Joseph G. Verbalis Georgetown University Medical Center, Washington, DC

Christopher S. von Bartheld University of Nevada School of Medicine, Reno, NV

Jonathan D. Wallis University of California at Berkeley, Berkeley, CA

M. Neal Waxham University of Texas Health Science Center, Houston, TX

David R. Weaver University of Massachusetts Medical School, Worcester, MA

Stephen C. Woods University of Cincinnati Medical Center, Cincinnati, OH

SECTION I

NEUROSCIENCE

Fundamentals of Neuroscience

A BRIEF HISTORY OF NEUROSCIENCE

The field of knowledge described in this book is *neuroscience*, the multidisciplinary sciences that analyze the nervous system to understand the biological basis for behavior. Modern studies of the nervous system have been ongoing since the middle of the nineteenth century. Neuroanatomists studied the brain's shape, its cellular structure, and its circuitry; neurochemists studied the brain's chemical composition, its lipids and proteins; neurophysiologists studied the brain's bioelectric properties; and psychologists and neuropsychologists investigated the organization and neural substrates of behavior and cognition.

The term neuroscience was introduced in the mid-1960s, to signal the beginning of an era in which each of these disciplines would work together cooperatively, sharing a common language, common concepts, and a common goal—to understand the structure and function of the normal and abnormal brain. Neuroscience today spans a wide range of research endeavors from the molecular biology of nerve cells (i.e., the genes encoding the proteins needed for nervous system function) to the biological basis of normal and disordered behavior, emotion, and cognition (i.e., the mental properties by which individuals interact with each other and with their environments). For a more complete, but concise, history of the neurosciences see Kandel and Squire (2000).

Neuroscience is currently one of the most rapidly growing areas of science. Indeed, the brain is sometimes referred to as the last frontier of biology. In 1971, 1100 scientists convened at the first annual meeting of the Society for Neuroscience. In 2006, 25,785 scientists participated at the society's 36th annual meeting at which 14,268 research presentations were made.

THE TERMINOLOGY OF NERVOUS SYSTEMS IS HIERARCHICAL, DISTRIBUTED, DESCRIPTIVE, AND HISTORICALLY BASED

Beginning students of neuroscience justifiably could find themselves confused. Nervous systems of many organisms have their cell assemblies and macroscopically visible components named by multiple overlapping and often synonymous terms. With a necessarily gracious view to the past, this confusing terminology could be viewed as the intellectual cost of focused discourse with predecessors in the enterprise. The nervous systems of invertebrate organisms often are designated for their spatially directed collections of neurons responsible for local control of operations, such as the thoracic or abdominal ganglia, which receive sensations and direct motoric responses for specific body segments, all under the general control of a cephalic ganglion whose role includes sensing the external environment.

In vertebrates, the components of the nervous system were named for both their appearance and their location. As noted by Swanson, and expanded upon in Chapter 2 of this volume, the names of the major parts of the brain were based on creative interpretations of early dissectors of the brain, attributing names to brain segments based on their appearance in the freshly dissected state: hippocampus (shaped like the sea horse) or amygdala (shaped like the almond), cerebrum (the main brain), and cerebellum (a small brain).

NEURONS AND GLIA ARE CELLULAR BUILDING BLOCKS OF THE NERVOUS SYSTEM

This book lays out our current understanding in each of the important domains that together define the full scope of modern neuroscience. The structure and function of the brain and spinal cord are most appropriately understood from the perspective of their highly specialized cells: the *neurons*, the interconnected, highly differentiated, bioelectrically driven, cellular units of the nervous system; and their more numerous support cells, the *glia*. Given the importance of these cellular building blocks in all that follows, a brief overview of their properties may be helpful.

Neurons Are Heterogeneously Shaped, Highly Active Secretory Cells

Neurons are classified in many different ways, according to function (sensory, motor, or interneuron), location (cortical, spinal, etc.), the identity of the transmitter they synthesize and release (glutamatergic, cholinergic, etc.), and their shape (pyramidal, granule, mitral, etc.). Microscopic analysis focuses on their general shape and, in particular, the number of extensions from the cell body. Most neurons have one *axon*, often branched, to transmit signals to interconnected target neurons. Other processes, termed *dendrites*, extend from the nerve cell body (also termed the perikaryon—the cytoplasm surrounding the nucleus of the neuron) to receive synaptic contacts from other neurons; dendrites may branch in extremely complex patterns, and may possess multiple short protrusions called dendritic *spines*. Neurons exhibit the cytological characteristics of highly active secretory cells with large nuclei; large amounts of smooth and rough endoplasmic reticulum; and frequent clusters of specialized smooth endoplasmic reticulum (Golgi apparatus), in which secretory products of the cell are packaged into membrane-bound organelles for transport out of the cell body proper to the axon or dendrites. Neurons and their cellular extensions are rich in microtubules—elongated tubules approximately 24 nm in diameter. Microtubules support the elongated axons and dendrites and assist in the reciprocal transport of essential macromolecules and organelles between the cell body and the distant axon or dendrites.

Neurons Communicate Chemically through Specialized Contact Zones

The sites of interneuronal communication in the central nervous system (CNS) are termed *synapses* in the CNS and *junctions* in somatic, motor, and autonomic nervous systems. Paramembranous deposits of specific proteins essential for transmitter release, response, and catabolism characterize synapses and junctions morphologically. These specialized sites are presumed to be the active zone for transmitter release and response. Paramembranous proteins constitute a specialized junctional adherence zone, termed the synaptolemma. Like peripheral junctions, central synapses also are denoted by accumulations of tiny (500 to 1500 Å) organelles, termed synaptic vesicles. The proteins of these vesicles have been shown to have specific roles in transmitter storage; vesicle docking onto presynaptic membranes, voltage- and Ca^{2+}-dependent secretion, and the recycling and restorage of previously released transmitter molecules.

Synaptic Relationships Fall into Several Structural Categories

Synaptic arrangements in the CNS fall into a wide variety of morphological and functional forms that are specific for the neurons involved. The most common arrangement, typical of hierarchical pathways, is either the axodendritic or the axosomatic synapse, in which the axons of the cell of origin make their functional contact with the dendrites or cell body of the target neuron, respectively. A second category of synaptic arrangement is more rare–forms of functional contact between adjacent cell bodies (somasomatic) and overlapping dendrites (dendrodendritic). Within the spinal cord and some other fields of neuropil (relatively acellular areas of synaptic connections), serial axoaxonic synapses are relatively frequent. Here, the axon of an interneuron ends on the terminal of a long-distance neuron as that terminal contacts a dendrite, or on the segment of the axon that is immediately distal to the soma, termed the initial segment, where action potentials arise. Many presynaptic axons contain local collections of typical synaptic vesicles with no opposed specialized synaptolemma. These are termed *boutons en passant*. The release of a transmitter may not always occur at such sites.

Synaptic Relationships Also Belong to Diverse Functional Categories

As with their structural representations, the qualities of synaptic transmission can also be functionally categorized in terms of the nature of the neurotransmitter that provides the signaling; the nature of the receptor molecule on the postsynaptic neuron, gland, or muscle; and the mechanisms by which the postsynaptic cell transduces the neurotransmitter signal into transmembrane changes. So-called "fast" or "classical" neurotransmission is the functional variety seen at the vast majority of synaptic and junctional sites, with a rapid onset and a rapid ending, generally employing excitatory amino acids (glutamate or aspartate) or inhibitory amino acids (γ-aminobutyrate, *GABA*, or glycine) as the transmitter. The effects of those signals are largely attributable to changes in postsynaptic membrane permeability to specific cations or anions and the resulting depolarization or hyperpolarization, respectively. Other neurotransmitters, such as the monoamines (dopamine, norepinephrine, serotonin) and many neuropeptides, produce changes in excitability that are much more enduring. Here the receptors activate metabolic processes within the postsynaptic cells—frequently to add or remove phosphate groups from key intracellular proteins; multiple complex forms of enduring postsynaptic metabolic actions are under investigation. The brain's richness of signaling possibilities comes from the interplay on common postsynaptic neurons of these multiple chemical signals.

THE OPERATIVE PROCESSES OF NERVOUS SYSTEMS ARE ALSO HIERARCHICAL

As research progressed, it became clear that neuronal functions could best be fitted into nervous system function by considering their operations at four fundamental hierarchical levels: molecular, cellular, systems, and behavioral. These levels rest on the fundamental principle that neurons communicate chemically, by the activity-dependent secretion of *neurotransmitters*, at specialized points of contact named *synapses*.

At the *molecular level* of operations, the emphasis is on the interaction of molecules—typically proteins that regulate transcription of genes, their translation into proteins, and their posttranslational processing. Proteins that mediate the intracellular processes of transmitter synthesis, storage, and release, or the intracellular consequences of intercellular synaptic signaling are essential neuronal molecular functions. Such transductive molecular mechanisms include the neurotransmitters' receptors, as well as the auxiliary molecules that allow these receptors to influence the short-term biology of responsive neurons (through regulation of ion channels) and their longer-term regu-

lation (through alterations in gene expression). Completion of the human, chimpanzee, rat, and mouse genomes can be viewed as an extensive inventory of these molecular elements, more than half of which are thought to be either highly enriched in the brain or even exclusively expressed there.

At the *cellular level* of neuroscience, the emphasis is on interactions between neurons through their synaptic transactions and between neurons and glia. Much current cellular level research focuses on the biochemical systems within specific cells that mediate such phenomena as pacemakers for the generation of circadian rhythms or that can account for activity-dependent adaptation. Research at the cellular level strives to determine which specific neurons and which of their most proximate synaptic connections may mediate a behavior or the behavioral effects of a given experimental perturbation.

At the *systems level*, emphasis is on the spatially distributed sensors and effectors that integrate the body's response to environmental challenges. There are sensory systems, which include specialized senses for hearing, seeing, feeling, tasting, and balancing the body. Similarly, there are motor systems for trunk, limb, and fine finger motions and internal regulatory systems for visceral regulation (e.g., control of body temperature, cardiovascular function, appetite, salt and water balance). These systems operate through relatively sequential linkages, and interruption of any link can destroy the function of the system.

Systems level research also includes research into cellular systems that innervate the widely distributed neuronal elements of the sensory, motor, or visceral systems, such as the pontine neurons with highly branched axons that innervate diencephalic, cortical, and spinal neurons. Among the best studied of these divergent systems are the monoaminergic neurons, which have been linked to the regulation of many behavioral outputs of the brain, ranging from feeding, drinking, thermoregulation, and sexual behavior. Monoaminergic neurons also have been linked to such higher functions as pleasure, reinforcement, attention, motivation, memory, and learning. Dysfunctions of these systems have been hypothesized as the basis for some psychiatric and neurological diseases, supported by evidence that medications aimed at presumed monoamine regulation provide useful therapy.

At the *behavioral* level of neuroscience research, emphasis is on the interactions between individuals and their collective environment. Research at the behavioral level centers on the integrative phenomena that link populations of neurons (often operationally or empirically defined) into extended specialized circuits, ensembles, or more pervasively distributed "systems" that integrate the physiological expression of a learned, reflexive, or spontaneously generated behavioral response. Behavioral research also includes the operations of higher mental activity, such as memory, learning, speech, abstract reasoning, and consciousness. Conceptually, "animal models" of human psychiatric diseases are based on the assumption that scientists can appropriately infer from observations of behavior and physiology (heart rate, respiration, locomotion, etc.) that the states experienced by animals are equivalent to the emotional states experienced by humans expressing these same sorts of physiological changes.

As the neuroscientific bases for some elemental behaviors have become better understood, new aspects of neuroscience applied to problems of daily life have begun to emerge. Methods for the noninvasive detection of activity in certain small brain regions have improved such that it is now possible to link these changes in activity with discrete forms of mental activity. These advances have given rise to the concept that it is possible to understand where in the brain the decision-making process occurs, or to identify the kinds of information necessary to decide whether to act or not. The detailed quantitative data that now exist on the details of neuronal structure, function, and behavior have driven the development of computational neurosciences. This new branch of neuroscience research seeks to predict the performance of neurons, neuronal properties, and neural networks based on their discernible quantitative properties.

Some Principles of Brain Organization and Function

The central nervous system is most commonly divided into major structural units, consisting of the major physical subdivisions of the brain. Thus, mammalian neuroscientists divide the central nervous system into the brain and spinal cord and further divide the brain into regions readily seen by the simplest of dissections. Based on research that has demonstrated that these large spatial elements derive from independent structures in the developing brain, these subdivisions are well accepted. Mammalian brain thusly is divided into hindbrain, midbrain, and forebrain, each of which has multiple highly specialized regions within it. In deference to the major differences in body structure, invertebrate nervous systems most often are organized by body segment (cephalic, thoracic, abdominal) and by anterior–posterior placement.

Neurons within the vertebrate CNS operate either within layered structures (such as the olfactory bulb,

cerebral cortex, hippocampal formation, and cerebellum) or in clustered groupings (the defined collections of central neurons, which aggregate into "nuclei" in the central nervous system and into "ganglia" in the peripheral nervous system, and in invertebrate nervous systems). The specific connections between neurons within or across the macro-divisions of the brain are essential to the brain's functions. It is through their patterns of neuronal circuitry that individual neurons form functional ensembles to regulate the flow of information within and between the regions of the brain.

CELLULAR ORGANIZATION OF THE BRAIN

Present understanding of the cellular organization of the CNS can be viewed simplistically according to three main patterns of neuronal connectivity.

Three Basic Patterns of Neuronal Circuitry Exist

Long hierarchical neuronal connections typically are found in the primary sensory and motor pathways. Here the transmission of information is highly sequential, and interconnected neurons are related to each other in a hierarchical fashion. Primary receptors (in the retina, inner ear, olfactory epithelium, tongue, or skin) transmit first to primary relay cells, then to secondary relay cells, and finally to the primary sensory fields of the cerebral cortex. For motor output systems, the reverse sequence exists with impulses descending hierarchically from the motor cortex to the spinal motoneuron. It is at the level of the motor and sensory systems that beginning scholars of the nervous system will begin to appreciate the complexities of neuronal circuitry by which widely separated neurons communicate selectively. This hierarchical scheme of organization provides for a precise flow of information, but such organization suffers the disadvantage that destruction of any link incapacitates the entire system.

Local circuit neurons establish their connections mainly within their immediate vicinity. Such local circuit neurons frequently are small and may have relatively few processes. Interneurons expand or constrain the flow of information within their small spatial domain and may do so without generating action potentials, given their short axons.

Single-source divergent circuitry is utilized by certain neurons of the hypothalamus, pons, and medulla. From their clustered anatomical location, these neurons extend multiple branched and divergent connections to many target cells, almost all of which lie outside the brain region in which the neurons of origin are located. Neurons with divergent circuitry could be considered more as interregional interneurons rather than as sequential elements within any known hierarchical system. For example, different neurons of the noradrenergic nucleus, the locus coeruleus (named for its blue pigmented color in primate brains) project from the pons to either the cerebellum, spinal cord, hypothalamus, or several cortical zones to modulate synaptic operations within those regions.

Glia Are Supportive Cells to Neurons

Neurons are not the only cells in the CNS. According to most estimates, neurons are outnumbered, perhaps by an order of magnitude, by the various nonneuronal supportive cellular elements. Nonneuronal cells include macroglia, microglia, and cells of the brain's blood vessels, cells of the choroid plexus that secrete the cerebrospinal fluid, and *meninges*, sheets of connective tissue that cover the surface of the brain and comprise the cerebrospinal fluid-containing envelope that protects the brain within the skull.

Macroglia are the most abundant supportive cells; some are categorized as astrocytes (nonneuronal cells interposed between the vasculature and the neurons, often surrounding individual compartments of synaptic complexes). Astrocytes play a variety of metabolic support roles, including furnishing energy intermediates and providing for the supplementary removal of excessive extracellular neurotransmitter secretions (see Chapter 13). A second prominent category of macroglia is the myelin-producing cells, the oligodendroglia. Myelin, made up of multiple layers of their compacted membranes, insulates segments of long axons bioelectrically and accelerates action potential conduction velocity. Microglia are relatively uncharacterized supportive cells believed to be of mesodermal origin and related to the macrophage/monocyte lineage. Some microglia reside quiescently within the brain. During periods of intracerebral inflammation (e.g., infection, certain degenerative diseases, or traumatic injury), circulating macrophages and other white blood cells are recruited into the brain by endothelial signals to remove necrotic tissue or to defend against the microbial infection.

The Blood–Brain Barrier Protects Against Inappropriate Signals

The blood–brain barrier is an important permeability barrier to selected molecules between the bloodstream and the CNS. Evidence of a barrier is provided by the greatly diminished rate of access of most lipophobic chemicals between plasma and brain; specific energy-dependent transporter systems permit selected access. Diffusional barriers retard the movement of substances from brain to blood as well as from blood to brain. The brain clears metabolites of transmitters into the cerebrospinal fluid by excretion through the acid transport system of the choroid plexus. The blood–brain barrier is much less prominent in the hypothalamus and in several small, specialized organs (termed *circumventricular organs*; see Chapters 34 and 39) lining the third and fourth ventricles of the brain: the median eminence, area postrema, pineal gland, subfornical organ, and subcommissural organ. The peripheral nervous system (e.g., sensory and autonomic nerves and ganglia) has no such diffusional barrier.

The Central Nervous System Can Initiate Limited Responses to Damage

Because neurons of the CNS are terminally differentiated cells, they cannot undergo proliferative responses to damage, as can cells of skin, muscle, bone, and blood vessels. Nevertheless, previously unrecognized neural stem cells can undergo regulated proliferation and provide a natural means for selected neuronal replacement in some regions of the nervous system. As a result, neurons have evolved other adaptive mechanisms to provide for the maintenance of function following injury. These adaptive mechanisms range from activity dependent regulation of gene expression, to modification of synaptic structure, function, and can include actual localized axonal sprouting and new synapse creation. These adaptive mechanisms endow the brain with considerable capacity for structural and functional modification well into adulthood. This plasticity is not only considered to be activity dependent, but also to be reversible with disuse. Plasticity is pronounced within the sensory systems (see Chapter 23), and is quite prominent in the motor systems as well. The molecular mechanisms employed in memory and learning may rely upon very similar processes as those involved in structural and functional plasticity.

ORGANIZATION OF THIS TEXT

With these overview principles in place, which are detailed more extensively in Section II, we can resume our preview of this book. Another major domain of our field is nervous system development (Section III).

How does a simple epithelium differentiate into specialized collections of cells and ultimately into distinct brain structures? How do neurons grow processes that find appropriate targets some distance away? How do nascent neuronal activity and embryonic experience shape activity?

Sensory systems and motor systems (Sections IV and V) encompass how the nervous system receives information from the external world and how movements and actions are produced (e.g., eye movements and limb movements). These questions range from the molecular level (how are odorants, photons, and sounds transduced into informative patterns of neural activity?) to the systems and behavioral level (which brain structures control eye movements and what are the computations required by each structure?).

An evolutionarily old function of the nervous system is to regulate respiration, heart rate, sleep and waking cycles, food and water intake, and hormones to maintain internal homeostasis and to permit daily and longer reproductive cycles. In this area of regulatory systems (Section VI), we explore how organisms remain in balance with their environment, ensuring that they obtain the energy resources needed to survive and reproduce. At the level of cells and molecules, the study of regulatory systems concerns the receptors and signaling pathways by which particular hormones or neurotransmitters prepare the organism to sleep, to cope with acute stress, or to seek food or reproduce. At the level of brain systems, we ask such questions as what occurs in brain circuitry to produce thirst or to create a self-destructive problem such as drug abuse?

In recent years, the disciplines of psychology and biology have increasingly found common ground, and this convergence of psychology and biology defines the modern topics of behavioral and cognitive neuroscience (Section VII). These topics concern the so-called higher mental functions: perception, attention, language, memory, thinking, and the ability to navigate in space. Work on these problems traditionally has drawn on the techniques of neuroanatomy, neurophysiology, neuropharmacology, and behavioral analysis. More recently, behavioral and cognitive neuroscience has benefited from several new approaches: the use of computers to perform detailed formal analyses of how brain systems operate and how cognition is organized; noninvasive neuroimaging techniques, such as positron emission tomography and functional magnetic resonance imaging, to obtain pictures of the living human brain in action; and molecular biological methods, such as single gene knockouts in mice, which can relate genes to brain systems and to behavior.

THIS BOOK IS INTENDED FOR A BROAD RANGE OF SCHOLARS OF THE NEUROSCIENCES

This textbook is for anyone interested in neuroscience. In preparing it we have focused primarily on graduate students just entering the field, understanding that some of you will have majored in biology, some in psychology, some in mathematics or engineering, and even some like me, in German literature. It is hoped that through the text, the explanatory boxes, and, in some cases, the supplementary readings, you will find the book to be both understandable and enlightening. In many cases, advanced undergraduate students will find this book useful as well.

Medical students may find that they need additional clinical correlations that are not provided here. However, it is hoped that most medical scholars at least will be able to use our textbook in conjunction with more clinically oriented material. Finally, to those who have completed their formal education, it is hoped that this text can provide you with some useful information and challenging perspectives, whether you are active neuroscientists wishing to learn about areas of the field other than your own or individuals who wish to enter neuroscience from a different area of inquiry. We invite all of you to join us in the adventure of studying the nervous system.

CLINICAL ISSUES IN THE NEUROSCIENCES

Many fields of clinical medicine are directly concerned with the brain. The branches of medicine tied most closely to neuroscience are neurology (the study of the diseases of the brain), neurosurgery (the study of the surgical treatment of neurological disease), and psychiatry (the study of behavioral, emotional, and mental diseases). Other fields of medicine also make important contributions, including radiology (the use of radiation for such purposes as imaging the brain—initially with X rays and, more recently, with positron emitters and magnetic waves) and pathology (the study of pathological tissue). To make connections to the many facets of medicine that are relevant to neuroscience, this book includes discussion of a limited number of clinical conditions in the context of basic knowledge in neuroscience.

THE SPIRIT OF EXPLORATION CONTINUES

Less than a decade into the twenty-first century, the Hubble space telescope continues to transmit information about the uncharted regions of the universe and clues to the origin of the cosmos. This same spirit of adventure also is being directed to the most complex structure that exists in the universe—the human brain. The complexity of the human brain is enormous, describable only in astronomical terms. For example, the number of neurons in the human brain (about 10^{12} or 1000 billion) is approximately equal to the number of stars in our Milky Way galaxy. Whereas the possibility of understanding such a complex device is certainly daunting, it is nevertheless true that an enormous amount has already been learned. The promise and excitement of research on the nervous system have captured the attention of thousands of students and working scientists. What is at stake is not only the possibility of discovering how the brain works. It is estimated that diseases of the brain, including both neurological and psychiatric illnesses, affect as many as 50 million individuals annually in the United States alone, at an estimated societal cost of hundreds of billions of dollars in clinical care and lost productivity. The prevention, treatment, and cure of these diseases will ultimately be found through neuroscience research. Moreover, many of the issues currently challenging societies globally—instability within the family, illiteracy, poverty, and violence, as well as improved individualized programs of education—could be illuminated by a better understanding of the brain.

THE GENOMIC INVENTORY IS A GIANT STEP FORWARD

Possibly the single largest event in the history of biomedical research was publicly proclaimed in June 2000 and was presented in published form in February 2001: the initial inventory of the human genome. By using advanced versions of the powerful methods of molecular biology, several large scientific teams have been able to take apart all of an individual's human DNA in very refined ways, amplify the amounts of the pieces, determine the order of the nucleic acid bases in each of the fragments, and then put those fragments back together again across the 23 pairs of human chromosomes.

Having determined the sequences of the nucleic acids, it was possible to train computers to read the sequence information and spot the specific signals that identify the beginning and ending of sequences likely to encode proteins. Furthermore, the computer systems could then sort those proteins by similarity of sequences (motifs) within their amino acid building blocks. After sorting, the computers could next assign the genes and gene products to families of similar proteins whose functions had already been established. In this way, scientists were rapidly able to predict approximately how many proteins could be encoded by the genome (all of the genes an individual has). Whole genome data are now available for humans, for some non-human primates, for rats, and for mice.

Scientifically, this state of information has been termed a "draft" because it is based on a very dense, but not quite complete, sample of the whole genome. What has been determined still contains a very large number of interruptions and gaps. Some of the smaller genes, whose beginning and ending are most certain, could be thought of as parts in a reassembled Greek urn, held in place by bits of blank clay until further excavation is done. However, having even this draft has provided some important realities.

Similar routines allowed these genomic scholars to determine how many of those mammalian genes were like genes we have already recognized in the smaller genomes of other organisms mapped out previously (yeast, worm (*Caenorhabditis elegans*), and fruitfly (*Drosophila melanogaster*)) and how many other gene forms may not have been encountered previously. Based on current estimates, it would appear that despite the very large number of nucleotides in the human and other mammalian genomes, about 30 times the length of the worms and more like 15 times the fruitfly, mammals may have only twice as many genes—perhaps some 30,000 to 40,000 altogether. Compared to other completed genomes, the human genome has greatly increased its representation of genes related to nervous system function, tissue-specific developmental mechanisms, and immune function and blood coagulation. Importantly for diseases of the nervous system that are characterized by the premature death of neurons, there appears to have been a major expansion in the numbers of genes related to initiating the process of intentional cell death, or *apoptosis*. Although still controversial, genes regulating primate brain size have been reported, but links to intellectual capacity remain unproven.

Two major future vistas can be imagined. To create organisms as complex as humans from relatively so few genes probably means that the richness of the required proteins is based on their modifications, either during transcription of the gene or after translation of the intermediate messenger RNA into the protein. These essential aspects of certain proteins account for a small number of brain diseases that can be linked to mutations in a single gene, such as Huntington's Disease (see Chapter 31). Second, though compiling this draft inventory represents a stunning technical achievement, there remains the enormously daunting task of determining, for example, where in the brain's circuits specific genes normally are expressed, and how that expression pattern may be altered by the demands of illness or an unfriendly environment. That task, at present, is one for which there are as yet no tools equivalently as powerful as those used to acquire the flood of sequence data with which we are now faced. This stage has been referred to as the end of "naïve reductionism."

In the fall of 2005, a six-nation consortium of molecular biologists announced the next phase of genomic research. The new focus will be toward refining the initial inventories to compare whole genomes of healthy and affected individuals for a variety of complex genetic illnesses (the HapMap project). Complex genetic diseases, such as diabetes mellitus, hypertension, asthma, depression, schizophrenia, and alcoholism, arise through the interactions of multiple short gene mutations that can increase or decrease one's vulnerability to a specific disease depending on individual life experiences. Ultimately, as the speeds of genome sequencing improve still further and the cost is reduced, it may be possible to predict what diseases will be more likely to affect a given person, and predict the lifestyle changes that person could undertake to improve his or her opportunities to remain healthy.

In order to benefit from the enormously rich potential mother lode of genetic information, next we must determine where these genes are expressed, what functions they can control, and what sorts of controls other gene products can exert over them. In the nervous system, where cell–cell interaction is the main operating system in relating molecular events to functional behavioral events, discovering the still-murky properties of activity-dependent gene expression will require enormous investment.

sented in this book is the culmination of hundreds of years of research. To help acquaint you with some of this work, we have described many of the key experiments of neuroscience throughout the book. We also have listed some of the classic papers of neuroscience and related fields at the end of each chapter, and invite you to read some of them for yourselves.

The pursuit of science has not always been a communal endeavor. Initially, research was conducted in relative isolation. The scientific "community" that existed at the time consisted of intellectuals who shared the same general interests, terminology, and paradigms. For the most part, scientists were reluctant to collaborate or share their ideas broadly, because an adequate system for establishing priority for discoveries did not exist. However, with the emergence of scientific journals in 1665, scientists began disseminating their results and ideas more broadly because the publication record could be used as proof of priority. Science then began to progress much more rapidly, as each layer of new information provided a higher foundation on which new studies could be built.

Gradually, an interactive community of scientists evolved, providing many of the benefits that contemporary scientists enjoy: Working as part of a community allows for greater specialization and efficiency of effort. This not only allows scientists to study a topic in greater depth but also enables teams of researchers to attack problems from multidisciplinary perspectives. The rapid feedback and support provided by the community help scientists refine their ideas and maintain their motivation. It is this interdependence across space and time that gives science much of its power.

With interdependence, however, comes vulnerability. In science, as in most communities, codes of acceptable conduct have evolved in an attempt to protect the rights of individuals while maximizing the benefits they receive. Some of these guidelines are concerned with the manner in which research is conducted, and other guidelines refer to the conduct of scientists and their interactions within the scientific community. Let us begin by examining how new knowledge is created.

NEUROSCIENCE TODAY: A COMMUNAL ENDEAVOR

As scientists, we draw from the work of those who came before us, using other scientists' work as a foundation for our own. We build on and extend previous observations and, it is hoped, contribute something to those who will come after us. The information pre-

THE CREATION OF KNOWLEDGE

Over the years, a generally accepted procedure for conducting research has evolved. This process involves examining the existing literature, identifying an important question, and formulating a research plan. Often, new experimental pathways are launched when one scientist reads with skepticism the observations and interpretations of another and decides to test their validity. Sometimes, especially at the beginning of a new series of experiments, the research plan is purely "descriptive," for example, determining the structure of a protein or the distribution of a neurotransmitter in brain. Descriptive initial research is essential to the subsequent inductive phase of experimentation, the movement from observations to theory, seasoned with wisdom and curiosity. Descriptive experiments are valuable both because of the questions that they attempt to answer and because of the questions that their results allow us to ask. Information obtained from descriptive experiments provides a base of knowledge on which a scientist may draw to develop hypotheses about cause and effect in the phenomenon under investigation. For example, once we identify the distribution of a particular transmitter within the brain or the course of a pathway of connections through descriptive work, we may then be able to develop a theory about what function that transmitter or pathway serves.

Once a hypothesis has been developed, the researcher then has the task of designing and performing experiments that are likely to disprove that hypothesis if it is incorrect. This is referred to as the deductive phase of experimentation, the movement from theory to observation. Through this paradigm the neuroscientist seeks to narrow down the vast range of alternative explanations for a given phenomenon. Only after attempting to disprove the hypothesis as thoroughly as possible may scientists be adequately assured that their hypothesis is a plausible explanation for the phenomenon under investigation.

A key point in this argument is that data may only lend support to a hypothesis rather than provide absolute proof of its validity. In part, this is because the constraints of time, money, and technology allow a scientist to test a particular hypothesis only under a limited set of conditions. Variability and random chance may also contribute to the experimental results. Consequently, at the end of an experiment, scientists generally report only that there is a statistical probability that the effect measured was due to intervention rather than to chance or variability.

Given that one can never prove a hypothesis, how do "facts" arise? At the conclusion of their experiments, the researchers' first task is to report their findings to the scientific community. The dissemination of research findings often begins with an informal presentation at a laboratory or departmental meeting, eventually followed by presentation at a scientific meeting that permits the rapid exchange of information more broadly. One or more research articles published in peer-reviewed journals ultimately follow the verbal communications. Such publications are not simply a means to allow the authors to advance as professionals (although they are important in that respect as well). Publication is an essential component of the advancement of science. As we have already stated, science depends on sharing information, replicating and thereby validating experiments, and then moving forward to solve the next problem. Indeed, a scientific experiment, no matter how spectacular the results, is not completed until the results are published. More likely, publication of "spectacular" results will provoke a skeptical scientist into doing an even more telling experiment, and knowledge will evolve.

RESPONSIBLE CONDUCT

Although individuals or small groups may perform experiments, new knowledge is ultimately the product of the larger community. Inherent in such a system is the need to be able to trust the work of other scientists—to trust their integrity in conducting and reporting research. Thus, it is not surprising that much emphasis is placed on the responsible conduct of research.

Research ethics encompasses a broad spectrum of behaviors. Where one draws the line between sloppy science and unethical conduct is a source of much debate within the scientific community. Some acts are considered to be so egregious that despite personal differences in defining what constitutes ethical behavior, the community generally recognizes certain research practices as behaviors that are unethical. These unambiguously improper activities consist of *fabrication*, *falsification*, and *plagiarism*: Fabrication refers to making up data, falsification is defined as altering data, and plagiarism consists of using another person's ideas, words, or data without attribution. Each of these acts significantly harms the scientific community.

Fabrication and falsification in a research paper taint the published literature by undermining its integrity. Not only is the information contained in such papers misleading in itself, but other scientists may unwittingly use that information as the foundation for new research. If, when reported, these subsequent studies cite the previous, fraudulent publication, the literature is further corrupted. Thus, through a domino-like effect one paper may have a broad negative impact on the scientific literature. Moreover, when fraud is discovered, a retraction of the paper provides only a limited solution, as there is no guarantee that individuals who read the original article will see the retraction. Given the impact that just one fraudulent paper may have, it is not surprising that the integrity of published literature is a primary ethical concern for scientists.

Plagiarism is also a major ethical infraction. Scientific publications provide a mechanism for establishing priority for a discovery. As such, they form the currency by which scientists earn academic positions, gain research grants to support their research, attract students, and receive promotions. Plagiarism denies the original author of credit for his or her work. This hurts everyone: The creative scientist is robbed of credit, the scientific community is hurt by the disincentive to share ideas and research results, and the individual who has plagiarized—like the person who has fabricated or falsified data—may well find his or her career ruined.

In addition to the serious improprieties just described, which are in fact extremely rare, a variety of much more frequently committed "misdemeanors" in the conduct of research can also affect the scientific community. Like fabrication, falsification, and plagiarism, some of these actions are considered to be unethical because they violate a fundamental value, such as honesty. For example, most active scientists believe that honorary authorship—listing as an author someone who did not make an intellectual contribution to the work—is unethical because it misrepresents the origin of the research. In contrast, other unethical behaviors violate standards that the scientific community has adopted. For example, although it is generally understood that material submitted to a peer-reviewed journal as part of a research manuscript has never been published previously and is not under consideration by another journal, instances of retraction for dual publications can be found on occasion.

Scientific Misconduct Has Been Formally Defined by U.S. Governmental Agencies

The serious misdeeds of fabrication, falsification, and plagiarism generally are recognized throughout the scientific community. These were broadly recognized by federal regulations in 1999 as a uniform standard of scientific misconduct by all agencies funding research. What constitutes a misdemeanor is less clear, however, because variations in the definitions of accepted practices are common. There are several sources of this variation. Because responsible conduct is based in part on conventions adopted by a field, it follows that there are differences among disciplines with regard to what is considered to be appropriate behavior. For example, students in neuroscience usually coauthor papers with their advisor, who typically works closely with them on their research. In contrast, students in the humanities often publish papers on their own even if their advisor has made a substantial intellectual contribution to the work reported. Within a discipline, the definition of acceptable practices may also vary from country to country. Because of animal use regulations, neuroscientists in the United Kingdom do relatively little experimental work with animals on the important topic of stress, whereas in the United States this topic is seen as an appropriate area of study so long as guidelines are followed to ensure that discomfort to the animals is minimized.

The definition of responsible conduct may change over time. For example, some protocols that were once performed on human and animal subjects may no longer be considered ethical. Indeed, ethics evolve alongside knowledge. We may not currently be able to know all of the risks involved in a procedure, but as new risks are identified (or previously identified risks refuted), we must be willing to reconsider the facts and adjust our policies as necessary. In sum, what is considered to be ethical behavior may not always be obvious, and therefore we must actively examine what is expected of us as scientists.

Having determined what is acceptable practice, we then must be vigilant. Each day neuroscientists are faced with a number of decisions having ethical implications, most of them at the level of misdemeanors: Should a data point be excluded because the apparatus might have malfunctioned? Have all the appropriate references been cited and are all the authors appropriate? Might the graphic representation of data mislead the viewer? Are research funds being used efficiently? Although individually these decisions may not significantly affect the practice of science, cumulatively they can exert a great effect.

In addition to being concerned about the integrity of the published literature, we must be concerned with our public image. Despite concerns over the level of federal funding for research, neuroscientists are among the privileged few who have much of their work funded by taxpayer dollars. Highly publicized scandals damage the public image of our profession and hurt all of us who are dependent on continued public support for our work. They also reduce the public credibility of science and thereby lessen the impact that we can expect our findings to have. Thus, for our own good and that of our colleagues, the scientific community, and the public at large, we must strive to act with integrity.

SUMMARY

You are about to embark on a tour of fundamental neuroscience. Enjoy the descriptions of the current state of knowledge, read the summaries of some of the classic experiments on which that information is based, and consult the references that the authors have drawn on to prepare their chapters. Think also about the ethical dimensions of the science you are studying—your success as a professional and the future of our field depend on it.

References

Aston-Jones, G., Cohen, J. D. (2005). An integrative theory of locus coeruleus-norepinephrine function: Adaptive gain and optimal performance. *Annu. Rev. Neurosci.* **28**, 403–45050.

Boorstin, D. J. (1983). "The Discoverers." Random House, New York.

Cherniak, C. (1990). The bounded brain: Toward quantitative neuroanatomy. *J. Cog. Neurosci.* **2**, 58–68.

Committee on the Conduct of Science (1995). "On Being a Scientist," 2nd Ed. National Academy Press, National Academy of Sciences, Washington, DC.

Cowan, W. M. and Kandel, E. R. (2001). Prospects for neurology and psychiatry. *JAMA* **285**, 594–600.

Day, R. A. (1994). "How to Write and Publish a Scientific Paper," 4th Ed. Oryx Press, Phoenix, AZ.

Greengard, P. (2001). The neurobiology of slow synaptic transmission. *Science* **294**, 1024–1030.

Kandel, E. R. and Squire, L. R. (2000). Neuroscience: Breaking down scientific barriers to the study of brain and mind. *Science* **290**, 1113–1120.

Kuhn, T. S. (1996). "The Structure of Scientific Revolutions," 3rd Ed. Univ. of Chicago Press, Chicago.

Popper, K. R. (1969). "Conjectures and Refutations: The Growth of Scientific Knowledge," 3rd Ed. Routledge and K. Paul, London.

Shepherd, G. M. (2003). "The Synaptic Organization of the Brain." Oxford Uni. Press, New York.

Swanson, L. (2000). What is the brain? *Trends Neurosci.* **23**, 519–527.

Floyd E. Bloom

2

Basic Plan of the Nervous System

INTRODUCTION

The brain often is compared with a computer these days. True, the brain is a computer, but it is a very special kind of computer—a biological computer that has evolved by natural selection over hundreds of millions of years and countless generations. Furthermore, it has no obvious design features in common with human-engineered computers. Instead, the brain is a unique organ that thinks and feels, generates behavioral interactions with the environment, keeps bodily physiology relatively stable, and enables reproduction of the species—its most important role from evolution's grand perspective. And for strictly personal reasons the brain is the most precious thing we have simply because it is the organ of consciousness, as reflected in René Descartes's famous seventeenth century aphorism, "I think therefore I am."

Aristotle first emphasized that structure and function are inextricably intertwined, two sides of the same coin, with structure providing obvious physical constraints on function. Just think about the difference between a hammer and a saw. Unfortunately, as knowledge becomes more and more specialized, there is a tendency to analyze the structure, function, and chemistry of the nervous system from different, sometimes even isolated, perspectives. The main theme of this chapter is the basic structure-function organization of the nervous system: what are the parts and how are they interconnected into functional systems? In other words, what are the organizing principles—the basic design features—of its circuitry?

Evolution and development are two approaches often used to understand biological complexity—

because they start with the simplest condition—and the human brain is far and away the most complex object we know of, with its roughly 100 billion neurons and 100 trillion axonal connections between them. One remarkable conclusion emerging from these two perspectives is that nerve cells in all animals—from jellyfish to humans—are basically the same in terms of cell biology; what changes most during ontogeny and phylogeny is the *arrangement* of nerve cells into functional circuits: the architecture of the nervous system. The "bricks" or "legos" are similar, but the "buildings" constructed with them can vary tremendously in size and functionality. An ultimate goal of neuroscience may be to understand the human brain, but remarkable progress can nevertheless be made through analyzing "lower" animals and early embryos. The other equally remarkable conclusion is that all vertebrates, from fish to humans, share a common basic plan of the nervous system, with the same major parts and functional systems.

EVOLUTION HIGHLIGHTS: GENERAL ORGANIZING PRINCIPLES

Protists and the simplest multicellular animals (sponges) display ingestive, defensive, reproductive, and other behaviors without any nervous system whatsoever, raising the question: what is the adaptive value of adding a nervous system to an organism? We will now examine key structure-function correlates of nervous system organization in animals with relatively simple body plans and behaviors.

The Nerve Net Is the Simplest Type of Nervous System

In his provocative 1919 book, *The Elementary Nervous System*, George Parker outlined a reasonable scenario for how nervous systems evolved. An updated version begins with the first multicellular animals—similar to modern-day sponges—that emerged over half a billion years ago. They are seemingly amorphous animals that spend their adult lives immobile, submerged in water. Their relatively simple behavior is mediated largely by a set of primitive smooth muscle cell (myocyte) sphincters allowing water flow through body wall pores. These specialized cells are called *independent effectors* because their contraction is evoked by stimuli like stretch or environmental chemicals acting directly on the plasma membrane of individual cells.

The first animal phylum with a nervous system was the Cnidaria, which includes jellyfish, corals, anemones, and the elegantly simple hydra. In contrast to sponges, hydra locomote and show active feeding behavior (Fig. 2.1). These behaviors are coordinated and mediated by a nervous system, a network of specialized units or cells called nerve cells or *neurons* (Box 2.1).

FIGURE 2.1 Locomotor behavior in hydra resembles a series of somersaults, as shown in the sequence beginning on the left. The tiny black dot in the region between the tentacles in the figure at the far right is the animal's mouth. Ingestive (feeding) behavior involves guiding food particles into the mouth with coordinated tentacle movements.

BOX 2.1

THE NEURON DOCTRINE

The cell theory, which states that all organisms are composed of individual cells, was developed around the middle of the nineteenth century by Mattias Schleiden and Theodor Schwann. However, this unitary vision of the cellular nature of life was not immediately applied to the nervous system, as most biologists at the time believed in the cytoplasmic continuity of nervous system cells. Later in the century the most prominent advocate of this **reticularist** view was Camillo Golgi, who proposed that axons entering the spinal cord actually fuse with other axons (Fig. 2.2A). The reticularist view was challenged most thoroughly by Santiago Ramón y Cajal, a founder of contemporary neuroscience and without doubt the greatest observer of neuronal architecture. In beautifully written and carefully reasoned deductive arguments, Cajal presented us with what is now known as the **neuron doctrine**. This great concept in essence states that the cell theory applies to the nervous system: each neuron is an individual entity, the basic unit of neural circuitry (Fig. 2.2B). The acrimonious debate between reticularists and proponents of the neuron doctrine raged for decades. Over the years, the validity of the neuron doctrine has been supported by a wealth of accumulated data. Nevertheless, the reticularist view is not entirely incorrect, because some neurons do act syncytially via specialized intercellular gap junctions, a feature that is more prominent during embryogenesis.

In 1897, Charles Sherrington postulated that neurons establish functional contact with each other and with other cell types via a theoretical structure he called the synapse (Greek *synaptein*, to fasten together). It was not until 50 years later that the structural existence of synapses was demonstrated by electron microscopy (see Fig. 3.3). The synaptic complex is built around an *adhesive* junction, and in this and other respects the complex is quite similar to the desmosome and the adherens junctions of epithelia. In fact, similarities in ultrastructure between the adherens junction and the synaptic complex of central nervous tissue were noted even in early electron microscopic studies (see Peters *et al.*, 1991).

Sensory Neurons

Hydra's body wall is simple, with an outer *ectoderm* layer contacting the external environment, an inner *endoderm* layer facing the body cavity's internal environment and promoting digestion and waste elimination, and a vague middle or *meso* layer in between. Neurons probably differentiated initially from the

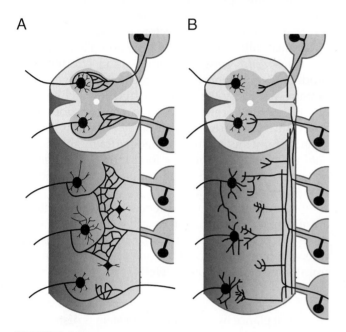

FIGURE 2.2 Two competing views: The nervous system as a reticulum or the neuron doctrine. (A) Proponents of the reticular theory believed that neurons are physically continuous with one another, forming an uninterrupted network. (B) In contrast, the neuron doctrine regards each neuron an individual entity communicating with target cells by way of contiguity rather than continuity, across an appropriate intercellular gap. Adapted from Cajal (1909–1911).

ectoderm, and perhaps the first to evolve were *sensory neurons*. One cytoplasmic extension of these bipolar cells facing the external environment became specialized to detect stimuli much weaker than those activating independent effectors, whereas the other pole became specialized to transmit information about these stimuli to a *group* of independent effectors (Fig. 2.3). Experimental evidence indicates that sensory neurons provide four major selective advantages in evolution:

- Increased stimulus sensitivity
- Faster effector cell responses
- Stronger behavioral responses because multiple effector cells are influenced
- Sensory neurons responding to different stimulus modalities can be distributed strategically in different body regions

The bipolar shape of sensory neurons is fundamentally important. The prototypical theory about neural circuit organization was presented by Santiago Ramón y Cajal in his classic "bible" of structural neuroscience, *The Histology of the Nervous System in Man and Vertebrates* (1909–1911). According to the cornerstone *functional polarity theory*, information normally flows in one direction through most neurons, and thus through most neural circuits—from *dendrites* and *cell body*, the input or receptive parts of the neuron, to a single *axon*, the output or effector part. In other words, most neurons have two classes of processes: one or more dendrites detecting inputs, and a single axon conducting an output that can influence multiple cells through branching or *collateralization*. At least in early developmental stages, all sensory neurons have this fundamental bipolar shape, and over the course of evolution they have become specialized to detect a remarkable

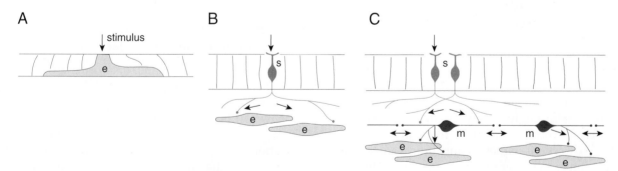

FIGURE 2.3 Activation of effector cells (e) in simple animals. (A) Sponges lack a nervous system; stimuli act directly on effector cells, which are thus called independent effectors. (B) In cnidarians, bipolar sensory neurons (s) differentiate in the ectoderm. The sensory neuron outer process detects stimuli and is thus a dendrite. The inner process of some sensory neurons transmits information directly to effector cells and is thus an axon. Because this type of sensory neuron innervates effector cells directly, it is actually a sensorimotor neuron. (C) Most cnidarian sensory neurons send their axon to motoneurons (m), which in turn send an axon to effector cells. Cnidarian motoneurons may also have lateral processes with other motoneurons, and these processes typically conduct information in either direction (and are thus amacrine processes). Arrows show the direction of information flow.

variety of stimuli from light, temperature, and a wide range of chemicals and ions, to vibration and other mechanical deformations.

Motor Neurons

A second stage of differentiation or complexity in hydra's nervous system was the addition of neurons between sensory and effector. They are defined as motor neurons (*motoneurons*) because they directly innervate effector cells (usually muscle or gland cells), which in turn receive their inputs from sensory neurons (Figs. 2.2B, 2.3C). Conceptually, this provides a *two-layered nervous system*: the first or top layer having sensory neurons and the second or bottom layer having motor neurons. In this prototypical network sensory neurons project (send axon collaterals) to multiple motoneurons, and then each motoneuron innervates a set of effector cells (with a motoneuron and its effector cell set defined as a *motor unit*). During an animal's normal behavior, information flow is unidirectional or polarized from one cell type, sensory neuron, to another cell type, motoneuron, to a third cell type, effector. This is the basic definition of a simple reflex, as defined by Charles Sherrington in his cornerstone of systems neuroscience, *The Integrative Action of the Nervous System* (1906).

In this hypothetical scenario (Fig. 2.3) an environmental stimulus detected by a sensory neuron's dendrite is transmitted by its axon to the dendrites of a motoneuron population. Then the axon of each motoneuron innervates an effector cell population. This is the functional polarity rule applied to a simple two-layer, sensory-motor network mediating reflex behavior.

Another general feature of the hydra two-layered nervous system has been observed: sensory neurons do not innervate each other, whereas motoneurons do interact directly. Here, motoneurons have two projection classes: one to effector cells and another to other motoneurons. Structurally and functionally, many of these hydra motoneurons also have two types of output processes. One is a typical axon innervating an effector cell population. However, the other is a process that contacts homologous processes from other motoneurons. Interestingly, many of these "horizontal" processes between motoneurons transmit information in either direction—either motoneuron can transmit information to the other via these processes. This is an exception to the functional polarity rule and is mediated by reciprocal rather than the more common unidirectional synapses. Cajal (1909–1911) described several examples of neurons that lack a clear axon (in retina, olfactory bulb, and intestine) and called them amacrine cells. As an extension of this it is useful to divide neuronal processes into three types: dendritic (input), axonal (output), or amacrine (bidirectional).

Adding a second layer to the nervous system has obvious adaptive advantages related to increased capacity for response complexity and integration. Consider a stimulus to one specific part of the animal or even one sensory neuron. Its influence may radiate to distant parts of the animal because one sensory neuron innervates multiple motoneurons, those motoneurons innervate additional motoneurons, and each motoneuron innervates multiple effector cells—an example of what Cajal called avalanche conduction. There may be great *divergence* between stimulus and effector cells producing a response, with the actual divergence pattern shaped by the structure–function architecture of the nervous system: how the neurons and their interconnections are arranged in the body. It is easy to imagine how this arrangement in hydra might coordinate the tentacles to bring a food morsel detected by just one of them to the mouth, or how it might coordinate locomotion (Fig. 2.1).

A second basic consequence of this structural arrangement is information *convergence* in the nervous system. Just consider a particular motoneuron: it can receive inputs from more than one sensory neuron and from other motoneurons as well.

Nerve Nets

At first glance hydra's nervous system is distributed fairly uniformly around the radially symmetrical body wall and tentacles (Fig. 2.4). Its essentially double-layered arrangement of distributed sensory and motor neurons is called a *nerve net*. However, in certain regions of the body with specialized function, like around the mouth and base of the tentacles, neurons tend to aggregate—a tendency toward *centralization* that will now be examined more carefully.

Bilateral Symmetry, Centralization, and Cephalization Emerge in Flatworms

In contrast to cnidarians, flatworms are bilaterally symmetrical predators with rostral (head) and caudal (tail) ends, and dorsal and ventral surfaces. These changes in body plan and behavior are accompanied by equally important changes in nervous system organization. Many flatworm neurons are clustered into distinct ganglia interconnected by longitudinal and transverse axon bundles called *nerve cords* (Fig. 2.5). This condensation of neural elements, or centralization, allows faster and thus more efficient communication between neurons because cellular material is conserved and conduction times are reduced. The largest, most complex ganglia (*cephalic ganglia*) are

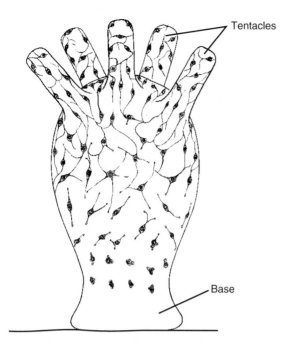

FIGURE 2.4 The nerve net of hydra, a simple cnidarian, is spread diffusely throughout the body wall of the animal. This drawing shows maturation of the nerve net in a hydra bud, starting near the base and finishing near the tentacles. Refer to McConnell (1932) and Koizumi (2002).

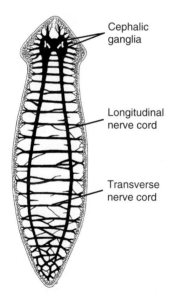

FIGURE 2.5 The nervous system of the planarian, a flatworm, includes longitudinal and transverse nerve cords associated with centralization, and two fused cephalic ganglia in the rostral end associated with cephalization. Centralization and cephalization probably are related to the flatworm's bilateral symmetry and ability to swim forward rapidly. Refer to Lentz (1968). Reproduced with permission from Yale University Press.

localized rostrally where they receive information from specialized sensory receptors in the front of the animal as it swims. Bilateral symmetry, centralization, and cephalization are three cardinal organizational trends in nervous system evolution.

Interneurons

Flatworms are the simplest animals with an abundant, clearly distinct third neuron division, *interneurons*, which are interpolated between sensory and motor neurons (Fig. 2.6). As already noted, Cajal recognized some atypical interneurons that apparently lack distinguishable dendrites and axon (amacrine neurons, or more precisely, amacrine interneurons). However, most interneurons have recognizable dendrites and axon and so presumably transmit information down the axon in only one direction, toward its terminals. They are *typical neurons* conforming to the functional polarity rule.

One consequence of adding a third "layer" of neurons to the nervous system is simply to increase convergence and divergence of information processing, and thus the capacity for response complexity. There are, however, three other critical functions interneurons subserve. They can act as *excitatory* or *inhibitory* "switches" in neuronal networks, assemblies of them can act as *pattern detectors* and *generators* between sensory and motor neurons, and they can be *pacemakers* if they generate intrinsic rhythmical activity patterns.

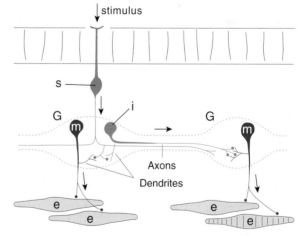

FIGURE 2.6 Invertebrate ganglia (G) usually display two neuron classes: motor neurons (m) and interneurons (i), both typically unipolar, with dendrites arising from a single axon. Here neuronal cell bodies are arranged peripherally and synapses occur in a central region called the neuropil. Sensory neurons (s) usually innervate motoneurons and interneurons but not effectors (e). Arrows show the usual direction of information flow.

BOX 2.2

CAJAL: ICONOCLAST TO ICON

Santiago Ramón y Cajal (1852–1934) is considered by many people to be the founder of modern neuroscience—a peer of Darwin and Pasteur in nineteenth-century biology. He was born in the tiny Spanish village of Petilla de Aragon on May 1, 1852, and as related in his delightful autobiography, he was somewhat mischievous as a child and determined to become an artist, much to the consternation of his father, a respected local physician. However, he eventually entered the University of Zaragoza and received a medical degree in 1873. As a professor of anatomy at Zaragoza his interests were mostly in bacteriology (the nineteenth-century equivalent of molecular biology today in terms of an exciting biological frontier) until 1887, when he visited Madrid at age 35 and first saw through the microscope histological sections of brain tissue treated with the Golgi method, which had been introduced in 1873. Although very few workers had used this technique, Cajal saw immediately that it offered great hope in solving the most vexing problem of nineteenth-century neuroscience: how do nerve cell interact with each other? This realization galvanized and directed the rest of his scientific life, which was extremely productive in terms of originality, scope, and accuracy.

Shortly after Jacob Schleiden, Theodor Schwann, and Rudolf Virchow proposed the cell theory in the late 1830s, Joseph von Gerlach, Sr. and Otto Deiters suggested that nerve tissue was special in the sense that nerve cells are not independent units but instead form a continuous syncytium or reticular net (Fig. 2.2A). This concept was later refined by Camillo Golgi who, based on the use of his silver chromate method, concluded that axons of nerve cells form a continuous reticular net, whereas in contrast dendrites do not anastomose but instead serve a nutritive role, much like the roots of a tree. Using the same technique, Cajal almost immediately arrived at the opposite conclusion, based first on his examination of the cerebellum, and later of virtually all other parts of the nervous system. In short, he proposed that neurons interact by way of contact or contiguity rather than by continuity,

and are thus structurally independent units, which was finally proven when the electron microscope was used in the 1950s. This concept became known as the **neuron doctrine**.

Cajal's second major conceptual achievement was the theory of **functional polarity**, which stated that the dendrites and cell bodies of neurons receive information, whereas the single axon with its collaterals transmits information to the other cells. This rule allows prediction of information flow direction through neural circuits based on the morphology or shape of individual neurons forming them, and it was the cornerstone of Charles Sherrington's (1906) revolutionary physiological analysis of mammalian reflex organization. Recent evidence that many dendrites transmit an action potential or graded potential in the retrograde direction would not violate the tenants of the functional polarity theory unless the potential led to altered membrane potentials in the associated presynaptic axon—and if this were the case the "dendrite" would be classed instead as an amacrine process (see text).

Around the close of the nineteenth century, Cajal made a remarkable series of discoveries at the cellular level. In addition to the two concepts outlined earlier, they include (1) the mode of axon termination in the adult CNS (1888), (2) the dendritic spine (1888), (3) the first diagrams of reflex pathways based on the neuron doctrine and functional polarity (1890), (4) the axonal growth cone (1890), (5) the chemotactic theory of synapse specificity (1892), and (6) the hypothesis that learning could be based on the selective strengthening of synapses (1895).

In one of the great ironies in the history of neuroscience, Cajal and Golgi shared the Nobel Prize for Medicine in 1906 though they had used the same technique to elaborate fundamentally different views on nervous system organization! The meeting in Stockholm may not have diminished the great personal friction between them. In 1931 Cajal wrote: "What a cruel irony of fate to pair like Siamese twins united by the shoulders, scientific adversaries of such contrasting characters."

By this definition the vast majority of vertebrate brain neurons are interneurons. So it is useful at the outset to recognize two broad interneuron categories: local and projection. *Local interneurons*, or *local circuit neurons*, have an axon that remains confined to a distinguishable gray matter region or ganglion, whereas

projection interneurons send a longer axon to a different gray matter region or ganglion, although it may also generate local axon collaterals.

The omnidirectional information flow typical of cnidarian nerve nets is unusual in the rest of the animal kingdom, where most neurons are functionally polarized with information flowing through neural circuits sequentially from dendrites and cell body to axon and axon terminals. However, most invertebrate motoneurons and interneurons are unipolar: a single process, the axon, extends from the cell body. Dendrites branch *from* the axons in the center of a ganglion—entering the *neuropil*—where most synapses are formed (Fig. 2.6). In vertebrates most neurons are multipolar, with several dendrites, plus an axon extending from the cell body or a dendrite.

Features of simple nervous systems are preserved throughout evolution. For example, the part of nervous system in the wall of the human gastrointestinal tract (the enteric nervous system) has many features of a highly refined nerve net, and a "layer" of amacrine interneurons is found in the human retina and olfactory bulb.

A Segmented Ventral Nerve Cord Typifies Annelids and Arthropods

Annelid worms and arthropods have even more complex body plans and behaviors than flatworms, partly because of *segmentation*. Body segments (*metameres*) are repeated serially along the body's rostrocaudal axis, and presumably share a common underlying genetic developmental program, although terminal differentiation (adult structure) may vary. This strategy allows for more complex body plans (including the nervous system) to evolve without a linear or exponential increase in genetic material. Annelids and all the more complex invertebrates share another characteristic feature, a *ventral nerve cord* with a pair of ganglia (or a single fused ganglion) in each segment, and longitudinal axon bundles between ganglia in adjacent segments (Fig. 2.7). Transverse *nerves* also extend from each ganglion to sensory structures and muscles in the same segment.

The Basic Plan of the Vertebrate Nervous System Is Found in Lancelets

Vertebrates are a subphylum of the Chordates and are the most complex of all animals in terms of structure and behavior. They share a basic body plan where common organ systems are arranged in a relatively strict anatomical relationship with one another (Box 2.3 and Figs. 2.8, 2.9). Like other chordates, vertebrates display two key features during some part of their life: a cartilaginous rod, the *notochord*, extending dorsally along the body, and above it a hollow *dorsal nerve cord*. In most vertebrates the notochord's body stiffening and protective functions are supplanted by the verte-

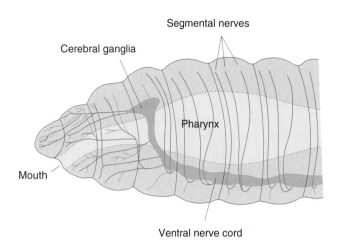

FIGURE 2.7 Nervous system organization in the rostral end of an annelid worm. A ventral nerve cord with more or less distinct ganglia connects with a fused pair of cerebral ganglia (brain) dorsal to the pharynx (part of the digestive tract). Note nerves extending from ventral nerve cord and cerebral ganglia. Refer to Brusca and Brusca (1990).

bral column and bony skull, with the notochord reduced to a series of cartilaginous cushions (discs) between or within the vertebrae. The vertebrate nerve cord is tremendously expanded, thickened, and folded to form the brain and spinal cord (the central nervous system, CNS).

The vertebrate nervous system's basic parts are revealed in the lancelet (amphioxus), a simple, nonvertebrate chordate (subphylum Cephalochordata). The lancelet is a slender, fish-like filter-feeder living half buried in the sand of shallow, tropical marine waters (Fig. 2.9). The body is stiffened by a notochord, and a dorsal nerve cord runs the length of the body, generating segmental nerves innervating muscles and organs. Locomotor behavior (swimming) is produced by alternately contracting right and left segmental muscles (myotomes). Without a notochord these contractions would shorten the animal rather than generate forward propulsive force.

Although typical vertebrate brain regions are not obvious rostrally in the lancelet nerve cord, genes specifying early vertebrate head embryogenesis also are expressed rostrally in the lancelet body. Thus, some components of the molecular program specifying modern vertebrate head development apparently were present early in chordate evolution (Holland and Takahashi, 2005).

Summary

The cniderian nerve net displays most of the basic cellular features of nervous system organization, including convergence and divergence of sensory and motor information. In more complex bilaterally symmetrical invertebrates, neurons and axons tend to aggregate in ganglia, nerve cords, and nerves (centralization), and there is a greater concentration of neurons and sensory organs in the body's rostral end (cephalization). Segmented invertebrates have a ventral nerve cord that includes a bilateral pair of ganglia (or single fused ganglion) in each segment. The primitive chordate, lancelet, displays the basic nervous system organization characteristic of vertebrates, including mammals and humans.

DEVELOPMENT REVEALS BASIC VERTEBRATE PARTS

One nineteenth century biology triumph was the demonstration that early stages of embryogenesis are fundamentally the same in all vertebrates. The CNS and heart are the first organs to differentiate in the embryo, and the basic CNS divisions differentiating early in development are also common to all vertebrates. The names and arrangement of these divisions are the starting point for regional or topo-graphic neuroanatomical nomenclature (Swanson, 2000a).

Nervous System Regionalization Begins in the Neural Plate

During embryogenesis the CNS develops as a hollow cylinder (*neural tube*) from a topologically flat sheet of cells (*neural plate*), by a process of *neurulation* (Chapter 14). Here we simply consider macroscopic structural changes during the transformation.

The neural plate is a spoon-shaped differentiation of the trilaminar embryonic disc's one-cell-thick ectodermal layer (Fig. 2.10). Its wide end lies rostrally and becomes the brain, whereas the narrow end lies caudally and becomes the spinal cord—the two major CNS divisions. A midline *neural groove* divides the neural plate into right and left halves, so the plate displays three cardinal morphogenetic features: *polarity*, *bilateral symmetry*, and *regionalization*. Furthermore, the neural plate differentiates from rostral to caudal, so the brain plate regionalizes first. Signs of this include appearance of the *optic vesicles*, evaginating near the rostral end of the neural plate (in the presumptive hypothalamus); a midline *infundibulum* evaginating

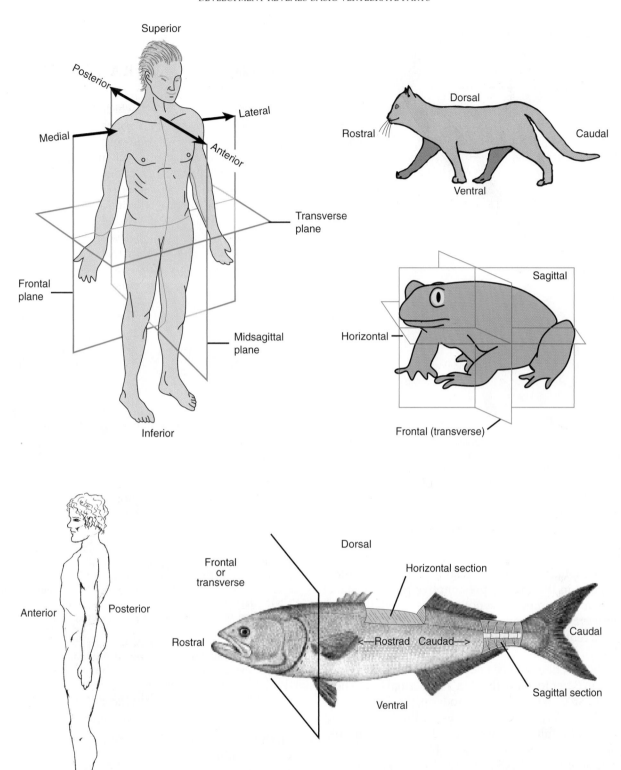

FIGURE 2.8 Orientation of the vertebrate body. Orientation planes for fish, quadrupeds, and bipeds are depicted. Associated with the three cardinal planes (rostrocaudal, dorsoventral, and mediolateral) are three orthogonal planes: horizontal, sagittal, and transverse (or frontal), which are the same in all early vertebrate embryos. For more explanation, see Williams (1995).

A

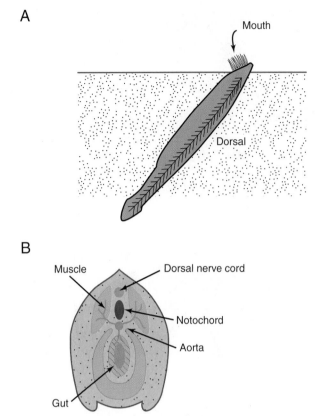

B

FIGURE 2.9 The lancelet (amphioxus) is a forerunner of the vertebrates. (A) Lateral view of the animal in its native habitat under the ocean floor, with its mouth protruding above the sand. (B) A cross-section of the lancelet body showing relationships between gut, aorta, notochord, and dorsal nerve cord. Adapted from Cartmill *et al.* (1987).

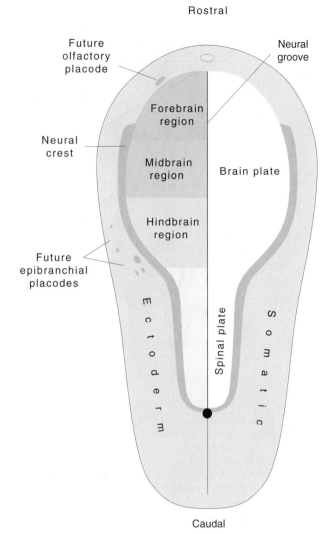

FIGURE 2.10 The neural plate is a spoon-shaped region of ectoderm (neural ectoderm) forming the CNS; surrounding it is somatic ectoderm. The neural plate is polarized (wider rostrally than caudally), bilaterally symmetrical (divided by the midline neural groove), and regionalized (brain plate rostrally, spinal plate caudally). The neural crest is a zone between neural and somatic ectoderm, and a series of placodes develops as "islands" within the somatic ectoderm. The neural crest and placodes generate PNS neurons. The approximate location of future CNS divisions in the neural plate is shown in color on the left. The same color scheme is used in Figs. 2.11, 2.12, and 2.14. Refer to Swanson (1992).

between optic vesicles and rostral end of the notochord, and indicating the presumptive pituitary stalk (again in presumptive hypothalamus); and the *otic rhombomere* (presumptive rhombomere 4), a swelling near the center of what will become the hindbrain (Fig. 2.11, left).

At the junction between neural plate and remaining ectoderm (later forming the skin's epidermal layer) lies a narrow strip of transitional ectoderm, the *neural crest*, a distinctive vertebrate feature (Fig. 2.10). It generates a variety of adult structures, including most neurons of the peripheral nervous system (PNS).

In summary, the CNS and PNS divisions are represented by the neural plate and neural crest, respectively, during the neural plate stage of vertebrate development. The two major CNS divisions, brain and spinal cord, are also indicated in the neural plate, which at this developmental stage is topologically simple: a bilaterally symmetrical, flat sheet that is one cell thick.

Further Regionalization Occurs in the Neural Tube

As neurulation progresses, the neural plate becomes U-shaped as the two halves (*neural folds*) become vertically oriented (Fig. 2.11, right). Then the dorsal tips of the folds fuse, forming an open tube—and finally the tube's ends (*neuropores*) also fuse, producing a completely closed neural tube with a one-cell-thick wall (*neuroepithelium*).

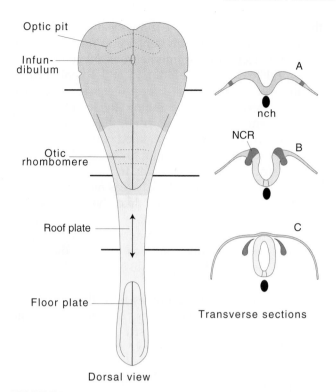

Optic pit

Infun-
dibulum

A

nch

NCR

B

Otic
rhombomere

C

Roof plate

Floor plate

Transverse sections

Dorsal view

FIGURE 2.11 Optic pits, infundibulum, and otic rhombomere (dorsal view on left) are the earliest clear structural differentiations of the neural plate, other than the neural groove. The neural tube forms by neuroectoderm invagination (transverse sections A and B), followed by fusion of the lateral edges of the neural plate (roughly in the neck region of humans), and proceeds both rostrally and caudally (double arrows in roof plate). Note how the neural crest (NCR) pinches off in the process. Also observe notochord (nch) position ventral to neural groove. Refer to Swanson (1992).

Marcello Malpighi, the great seventeenth century founder of histology who also discovered the capillary network between arteries and veins postulated by William Harvey in 1628, recognized that the early chick neural tube displays three rostrocaudally arranged swellings now called primary brain vesicles. They include the *forebrain* (prosencephalic) vesicle, with the optic stalks and infundibulum; the *midbrain* (mesencephalic) vesicle; and the *hindbrain* (rhomben-cephalic) vesicle, with the otic rhombomere (Fig. 2.12A).

These vesicles are the fundamental structural or regional brain divisions. Transitory rhombomeres are the most characteristic hindbrain vesicle feature at this stage, and they develop in association with the pha-ryngeal pouches (Chapter 14). As embryogenesis continues, the forebrain vesicle divides into *endbrain* (telencephalic) and *interbrain* (diencephalic) vesicles, whereas the hindbrain vesicle differentiates vaguely into rostral *pontine* (metencephalic) and caudal *medullary* (myencephalic) regions (Fig. 2.12B). These divi-

sions transform the "three primary vesicle stage" into the "five secondary vesicle stage."

The neural tube lumen becomes the adult CNS *ventricular system* (Fig. 2.12B, left), and its adult shape conforms to extensive differential regionalization of the neural tube wall. Each endbrain vesicle contains a *lateral ventricle*, which communicates through an *interventricular foramen* with the *third ventricle* in the interbrain vesicle's center. The third ventricle continues into the midbrain's *cerebral aqueduct*, which becomes the hindbrain's *fourth ventricle* and then the spinal cord's *central canal*. In older embryos and adults, the ventricular system contains *cerebrospinal fluid* (CSF), much of which is elaborated by specialized, highly vascular regions of *choroid plexus* in the roof of the lateral, third, and fourth ventricles.

Migrating Neurons Form the Mantle Layer's Gray Matter

In the early five secondary vesicle neural tube, cells divide repeatedly although the neural tube remains one cell thick, a *pseudostratified epithelium* of stem cells for neurons and glia. Shortly thereafter many of these cells begin a terminal differentiation into young neurons migrating from the luminal proliferation zone to form a new, more superficial *mantle layer* (Chapter 16). In some CNS regions mantle layer neurons segregate into layers parallel to the surface, whereas in others neurons cluster in nuclei, relatively uniform neuron populations (usually multiple types) that are structurally distinct from surrounding nuclear or layered regions.

Mantle layer formation leads to further CNS regionalization (Fig. 2.12B). In hindbrain and spinal cord, it emerges because motoneurons are generated earliest and ventrally (corresponding to medial neural plate regions). This correlates with observations that gross, relatively uncoordinated embryonic motor behavior starts before reflex pathways become functional—and implying that such behavior is generated endogenously in the CNS itself (Hamburger, 1975).

Ventral mantle layer formation accompanies the transient appearance of a longitudinal groove (*limiting sulcus*) on the neural tube's inner surface. The leading nineteenth century Swiss embryologist Wilhelm His noted that the limiting sulcus divides much of the neural tube into dorsal or *alar plate* and ventral or *basal plate*, with predominantly sensory and motor functions, respectively (Fig. 2.13). This observation complemented the earlier fundamental discovery of François Magendie that spinal sensory and motor fibers are completely segregated in spinal roots: sensory axons enter through dorsal roots whereas motor axons leave

A

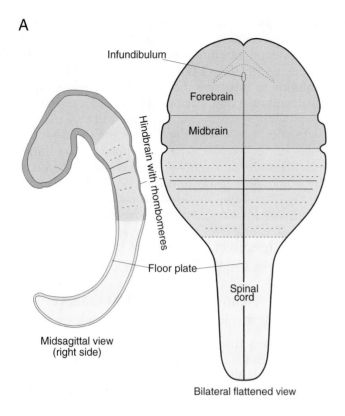

Midsagittal view
(right side)

Bilateral flattened view

B

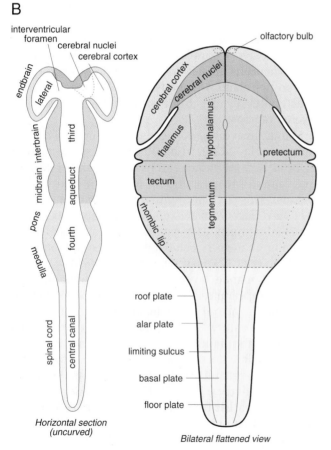

Horizontal section
(uncurved)

Bilateral flattened view

through ventral roots. It is now clear that alar and basal plates are not purely sensory or motor because each contains projection interneurons. Nevertheless, it is helpful to view the hindbrain and spinal cord as having three longitudinal zones: sensory, integrative (reticular formation), and motor. Regionalization of midbrain and forebrain does not fit this scheme neatly and is relatively poorly understood conceptually.

Dorsal regions of the hindbrain's alar plate form a unique structure, the *rhombic lip*. In the pons it generates cerebellar granule cells, whereas more caudally it produces neuron populations like the precerebellar and vestibulocochlear nuclei. Many rhombic lip neuron populations are interesting because they migrate parallel to the neural tube's surface to reach their final destinations, instead of radially like most CNS neurons (Chapter 16).

This differentiation continues until the adult CNS configuration is achieved (Figs. 2.14 and 2.15). The most obvious late-developing structures are the cerebral cortex and cerebellar cortex.

Summary

The vertebrate CNS develops from a sheet of cells called the neural plate that invaginates to form the neural tube. The tube's rostral end differentiates a series of vesicles that constitute the major brain regions, and the caudal end forms the simpler spinal cord. Most PNS neurons differentiate from the neural crest, with the rest arising from nearby somatic ectodermal placodes.

FIGURE 2.12 Formation and regionalization of the neural tube. (A) The early neural tube brain region develops three swellings: forebrain, midbrain, and hindbrain vesicles. The hindbrain vesicle then differentiates a series of transverse swellings called rhombomeres. (B) As differentiation continues, the forebrain vesicle displays right and left endbrain (cerebral hemisphere) vesicles and a medial interbrain vesicle, and the hindbrain vesicle shows vague pontine and medullary regions. This is the five-vesicle stage of neural tube transverse regionalization. Then longitudinal, dorsoventral, regionalization begins. The endbrain vesicle divides into cerebral cortex (including olfactory bulb) and cerebral nuclei (basal ganglia), the interbrain vesicle divides into thalamus and hypothalamus, the midbrain vesicle divides into tectum and tegmentum, the hindbrain vesicle divides into rhombic lip, alar plate, and basal plate, and the spinal cord divides into alar and basal plates. Whether the pretectal region (sometimes called synencephalon) is part of interbrain or midbrain is controversial. At this developmental stage major components of the adult ventricular system are seen in the neural tube lumen. Refer to Swanson (1992) and Alvarez-Bolado and Swanson (1996).

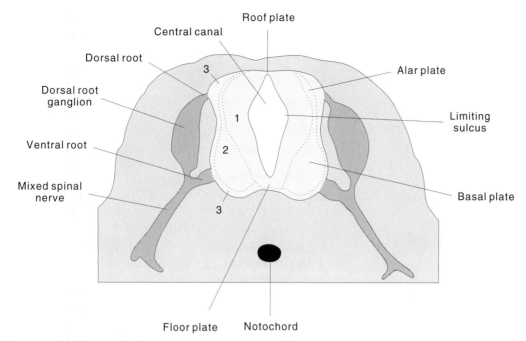

FIGURE 2.13 The early spinal cord and hindbrain are divided into dorsal (alar) and ventral (basal) plates by the limiting sulcus. This morphology reflects earlier ventral differentiation of the mantle layer (2), accompanied by earlier ventral thinning of the neuroepithelial or ventricular layer (1) of the neural tube, which remains as the adult ependymal lining of the ventricular system. The mantle layer develops into adult gray matter. This schematic drawing of a transverse spinal cord histological section also shows dorsal (sensory) and ventral (motor) spinal cord roots, dorsal root ganglia containing sensory neurons derived from the neural crest, and mixed (sensory and motor) spinal nerves distal to the ganglia. The peripheral area (3) is called the marginal zone and develops into the spinal cord white matter or funiculi containing ascending and descending axonal fiber tracts.

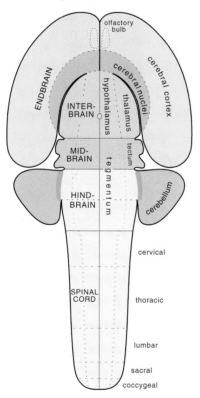

FIGURE 2.14 Major divisions of the adult mammalian CNS are derived from neural plate and neural tube regionalization illustrated in Figs. 2.10–2.12. Modified from Swanson (1992).

THE BASIC PLAN OF NERVOUS SYSTEM CONNECTIVITY

Functional Systems Consist of Interconnected Gray Matter Regions

The nervous system's wiring diagram can be described in terms of the neuronal cell types in each of its distinct gray matter regions and their stereotyped pattern of axonal projections to cell types both locally (within the region) and in other gray matter regions or other tissues (like muscle or gland). A long-term goal of systems neuroscience is to provide a global wiring diagram for the nervous system that systematically accounts for its various functional subsystems—analogous to the circulatory system model provided by Harvey. Little work has been done on this synthetic problem, although interest is accelerating with the development of online neuroinformatics workbenches for connectional information (Bota and Swanson, 2007).

The high-level model of nervous system information processing shown in Figure 2.16 synthesizes basic neurobiological concepts pioneered by Cajal and Sherrington with basic cybernetic principles pioneered by

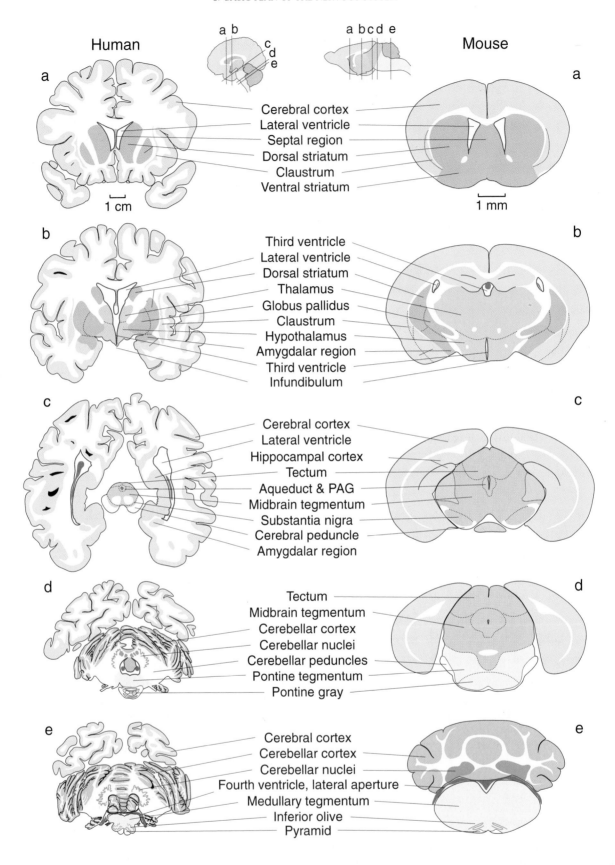

Human

Mouse

a

Cerebral cortex
Lateral ventricle
Septal region
Dorsal striatum
Claustrum
Ventral striatum

1 cm

1 mm

b

Third ventricle
Lateral ventricle
Dorsal striatum
Thalamus
Globus pallidus
Claustrum
Hypothalamus
Amygdalar region
Third ventricle
Infundibulum

c

Cerebral cortex
Lateral ventricle
Hippocampal cortex
Tectum
Aqueduct & PAG
Midbrain tegmentum
Substantia nigra
Cerebral peduncle
Amygdalar region

d

Tectum
Midbrain tegmentum
Cerebellar cortex
Cerebellar nuclei
Cerebellar peduncles
Pontine tegmentum
Pontine gray

e

Cerebral cortex
Cerebellar cortex
Cerebellar nuclei
Fourth ventricle, lateral aperture
Medullary tegmentum
Inferior olive
Pyramid

FIGURE 2.15 Mini atlases to compare major adult brain regions in humans and mice. The brains are cut approximately transversely to the CNS longitudinal axis and illustrate five major levels, arranged from rostral to caudal: a, endbrain; b, interbrain; c, midbrain; d, pons; and e, medulla. The color scheme follows that in Figs. 2.10–2.12 and 2.14, with the choroid plexus of the lateral, third, and fourth ventricles shown in red. Adapted from Nieuwenhuys *et al.* (1988) and Sidman *et al.* (1971).

Norbert Wiener (1948) and John von Neumann (1958). In essence, the model postulates that behavior is determined by CNS motor system output, and that this output is a function of three inputs: sensory system (reflexive), cognitive system (voluntary), and intrinsic behavioral state system. The relative importance of each input in controlling motor output (behavior) varies qualitatively in different species and quantitatively in different individuals. Note that behavior elicits sensory feedback from the external and internal environments that helps determine future motor activity and thus behavior. Each component is now considered further without trying to place all nervous system parts within the global model.

Motor Systems Are Organized Hierarchically

There are three different motor systems: skeletal, autonomic, and neuroendocrine. The first controls striated muscles responsible for voluntary behavior; the second controls smooth and cardiac muscle, and many glands; and the third controls pituitary gland hormone secretion. The skeletal motor system is understood best and thus serves as a prototype for examining basic organizing principles presumably similar for all three.

The skeletal motor system is arranged hierarchically (Fig. 2.17), the lowest level consisting of brainstem-spinal cord α-motoneurons whose axons synapse directly on striated muscle fibers. The next higher level consists of motor pattern generators (MPGs), and the highest level has motor pattern initiators (MPIs) that "recognize" or alter their output in response to specific input patterns, and project to unique sets of MPGs. Ethologists refer to MPIs as "innate releasing mechanisms." One reason central neural circuitry is so complex is that each of the three input types (sensory, intrinsic, cognitive) may go directly to each general level of the motor system hierarchy.

The MPGs and MPIs themselves are hierarchically arranged. This organization is particularly easy to see conceptually for the MPGs subserving locomotor behavior. In the spinal cord, simple MPGs coordinate

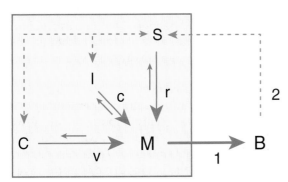

FIGURE 2.16 A model of the nervous system's basic wiring diagram. The model of information flow through the nervous system (yellow box) postulates that behavior (B) is determined by the motor system (M), which is influenced by three classes of neural input: sensory (S), intrinsic behavioral state (I), and cognitive (C). Sensory inputs lead directly to reflex responses (r), cognitive inputs mediate voluntary responses (v), and intrinsic inputs act as control signals (c) to regulate behavioral state. Motor system inputs (1) produce behaviors whose consequences are monitored by sensory feedback (2). Sensory feedback may be used by the cognitive system for perception and by the intrinsic system to generate affect (e.g., positive and negative reinforcement/pleasure and pain). The cognitive, sensory, and intrinsic systems are all interconnected, hence the arrowheads at the ends of each dashed line within the nervous system box. Refer to Swanson (2003).

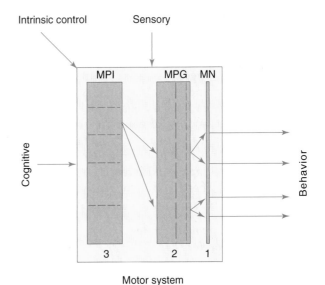

FIGURE 2.17 Hierarchical organization of the skeletal motor system. At the simplest level (1), motoneuron pools (MN) innervate individual muscles generating individual components of behavior. At the next higher level (2), additional interconnected interneuron pools, called motor pattern generators (MPG), innervate specific motoneuron pool sets. At the highest level (3), additional interconnected interneuron pools, called motor pattern initiators (MPI), innervate specific MPG sets. MPIs can activate complex, stereotyped behaviors when activated (or inhibited) by specific patterns of sensory, intrinsic, and/or cognitive inputs. Note that MPGs and MPIs themselves may be organized hierarchically (dashed lines) and that sensory, intrinsic, and cognitive inputs may go directly to any level of the motor system hierarchy. Refer to Swanson (2003).

the reciprocal innervation of muscle pair antagonists across individual joints, more complex MPGs coordinate activity in the set of simpler MPGs for all the joints in a limb, and still more complex MPGs coordinate activity in MPGs for all four limbs. At the next higher level there is a brain hierarchy of MPIs for locomotion that is activated by specific input patterns and projects to the spinal locomotor pattern generator network.

Multiple Sensory Systems Function in Parallel

A set of sensory systems provides information to the CNS from various receptor types, and all the systems can function simultaneously. Cajal noted that unimodal sensory pathways generally branch with some information going directly to the motor system and some going to the cerebral cortex for sensation and perception. The former typically evokes reflex behavior and the latter potentially reaches consciousness and plays an important role in cognition.

Several general features characterize the sensory system (Section IV covers subsystems in detail). First, the CNS receives a wide range of information about the external environment and about the body's internal state. Thus, sensory receptors lie near the body's surface (e.g., touch and olfactory receptors), deep within the body (e.g., aortic stretch receptors), and even within the brain itself (e.g., hypothalamic insulin receptors). Second, each of the three motor systems receives a broad range of sensory inputs. Third, the range of sensory modalities is remarkably similar (though not identical) across vertebrate classes, and information about specific modalities enters the CNS through homologous cranial and spinal nerves in all vertebrates. And fourth, the number of synapses between sensory receptor and cerebral cortex varies in different systems. There is one synapse in the olfactory system, and at least four in the visual system.

The Cognitive System Generates Anticipatory Behavior

It is very likely that the cerebral cortex—along with its cerebral nuclei (basal ganglia)—is the most important, if not sole, part of the cognitive system and that the cerebral cortex is responsible for planning, prioritizing, initiating, and evaluating the consequences of voluntary behavior (Section VII). The fundamental nature of voluntary behavior is obviously a difficult problem to address, but one useful approach is simply to compare it with reflexive behavior. Interestingly, most if not all behaviors mediated by skeletal muscle can be initiated either reflexively or voluntarily, as Descartes pointed out long ago. What seems to distinguish reflexive and voluntary behaviors most clearly

is that the former involves a stereotyped response to a defined stimulus, whereas the latter is anticipatory, with a duration and content impossible to predict with anywhere near the same degree of certainty.

Intrinsic Systems Control Behavioral State

The CNS generates considerable endogenous activity (action potential patterns)—it is definitely not just a passive system waiting to respond to sensory input, as the behaviorist approach a century ago assumed. All CNS parts apparently have a basal activity level that can be either increased or decreased. In many cases, it is still not established whether particular neuronal cell types generate intrinsic activity patterns. It is clear, however, that motoneurons and related MPGs do generate intrinsic activity; as already noted, the embryonic spinal cord produces motor output before sensory circuits develop. Thus, in addition to the three extrinsic input types to the motor system illustrated in Figure 2.16, intrinsic activity within the motor system itself can produce behavior that is neither reflexive nor voluntary.

Certain CNS regions generate intrinsic rhythmic activity patterns. The most important rhythmic behavioral pattern is the sleep–wake cycle that is entrained to the day-night cycle by an endogenous circadian clock, the hypothalamic suprachiasmatic nucleus (Chapters 41 and 42). The sleep–wake cycle is profoundly important because during sleep the body is maintained entirely by ongoing intrinsic and reflexive systems controlling behaviors like respiration and sustained sphincter contractions. In contrast, voluntary mechanisms dominate in wakefulness though reflexive and intrinsic mechanisms are also vitally important then.

Behavioral state control is thus a fundamental intrinsic brain activity. Another aspect of behavioral state—arousal—is especially important during wakefulness. Arousal level generally is correlated with an animal's motivational state or drive level (Chapter 43). The neural system mediating drive is not fully elucidated but is critically dependent on the hypothalamus, and attainment of specific goal objects (foraging behavior) depends on the cognitive system. Arousal and drive may be controlled by subcortical systems but behavior's actual direction and prioritization mainly is determined cortically.

The full identity of neural systems elaborating pleasure and pain is one of neuroscience's deep mysteries. Many regard pleasure and pain as conscious expressions of positive and negative reinforcement, influencing how likely a particular voluntary behavior will be repeated or avoided in the future. Here, reinforcement depends on sensory feedback about a particular behavior's consequences (Fig. 2.16), and one suggestion is that pleasurable and painful sensations, like those associated with drive, are elaborated subcortically within intrinsic control systems. According to this view, thinking or cognition arises in cerebral cortex whereas feeling or affect arises subcortically. It is also possible that all aspects of consciousness (thinking and feeling) arise only from cortical neural activity (Chapter 53).

How Pharmacological and Genetic Networks Relate to Functional Systems

Specific neurotransmitter systems have been incorporated into models of CNS function since the 1950s. Two examples are cholinergic and noradrenergic systems, defined as the total sets of CNS neurons releasing acetylcholine or noradrenalin, respectively, as a neurotransmitter (Chapter 7). In general, these systems are not obviously correlated with traditional CNS functional systems or major topographic parts—typically they are not restricted to one functional system or one major CNS division, though some exceptions may exist. Thus, neurotransmitter systems are not functional systems in the traditional sense. However, they are conceptually or operationally important in helping define circuits or functional systems influenced by particular drug actions. For example, administering centrally acting acetylcholine receptor agonists influences synapses in a variety of traditional *functional systems*, and the set of these functional systems could be defined as a *pharmacological system* with a specific set of behavioral and other responses. If a drug targeted for therapeutic reasons to a specific neural system (e.g., a cholinergic agonist targeted to the cerebral cholinergic system in Alzheimer's disease; Chapter 45), it will also act on other functional systems with appropriate cholinergic receptors (e.g., in the thalamus and lower brainstem). Responses in these other systems produce "side effects" that may be good or bad.

Likewise, any gene product's distribution pattern can also be used to define a chemical, molecular, or neural *gene expression system*. For example, a system could be defined in terms of all neurons expressing the calbindin or μ-opioid receptor gene, and expression of the corresponding gene might be prevented or altered in experimental knockout mice or natural mutations in genetic diseases. These alterations may produce an obvious and stereotyped phenotype or syndrome, but in most instances the gene normally is expressed in multiple functional systems and has complex (even if subtle) physiological and behavioral effects.

Finally, it is important to remember that a genetic program constructs the nervous system's basic *macrocircuitry* during embryogenesis. Determining the correspondence between gene expression networks and neural networks may be the ultimate achievement of systems neuroscience. The nervous system's *microcircuitry*—quantitative aspects of synapse number and strength associated with individual neurons—may be sculpted by experience throughout life.

Summary

There is no simple relationship between the CNS's topographic or regional differentiation and its functional organization. So it is mistaken to assume *a priori* that CNS information simply is processed hierarchically with the spinal cord at the lowest level and the cerebral cortex at the top. An alternative view is that the CNS displays a network rather than hierarchical organization scheme—a circuit where the motor system is driven by sensory, cognitive, and intrinsic behavioral state inputs, and future motor activity is determined partly by sensory feedback about the initial behavior's consequences.

Two major features complicate this simple network model. First, the motor system itself is organized hierarchically, whereas the sensory system transmits multiple modalities in parallel, and this sensory information can reach directly each level of the motor system hierarchy. And second, sensory information also reaches the intrinsic and cognitive systems. In fact, all three input systems are interconnected bidirectionally. The basic plan of neural circuit architecture must be understood on its own terms, not through simple preconceived ideas or superficial analogies with computers, the Internet, irrigation systems, or complicated robots. How traditional CNS functional systems relate to pharmacological systems and genetic networks remains to be determined.

OVERVIEW OF THE ADULT MAMMALIAN NERVOUS SYSTEM

This section reviews structural neuroscience methods used to achieve our current—still very incomplete—understanding of nervous system architectural principles, and introduces the major nervous system components. Long experience teaches that nothing approaches actual dissection for gaining an appreciation of overall brain structure.

where nerves converge and redistribute axons to their target organs.

The ANS has anatomically and functionally *sympathetic* and *parasympathetic* divisions (Chapter 35). The two divisions function in a kind of push-pull relationship with each other. One or the other is never completely on or off. Instead, there are degrees of sympathetic and parasympathetic tone. During sleep, certain involuntary functions like digestion are accelerated. Glands participating in digestion are activated parasympathetically and sympathetic tone is correspondingly decreased. In contrast, Walter B. Cannon noted almost a century ago that during the "fight or flight" reaction characterizing defensive behavior, sympathetic tone is markedly enhanced and parasympathetic tone is reduced sharply. Sympathetic outflow is vastly amplified and coordinated through a set of ganglia and the adrenal medulla, so that sympathetic function occurs relatively synchronously throughout the body. In contrast, parasympathetic system is relatively finely tuned.

The Cerebrospinal Trunk Generates Cranial and Spinal Nerves

From a more systematic perspective on nervous system organization, the spinal cord and brainstem (together the cerebrospinal trunk) generate a continuous series of spinal and cranial nerves, respectively. The human spinal cord, roughly as thick as an adult's little finger, is ultimately surrounded and protected by the vertebral column, whereas the skull protects the brain. In cross-section, the spinal cord's two basic types of nervous tissue are obvious: *gray matter* and *white matter*. Gray matter forms an H-shaped region surrounding the central canal (the ventricular system's spinal segment) and consists mainly of neuronal cell bodies and neuropil. White matter surrounds gray matter in the spinal cord and consists mostly of axons collected into overlapping fiber bundles. Many axons have a myelin sheath, a uniquely vertebrate feature allowing rapid nerve impulse conduction (Chapter 6) and giving white matter its pale appearance.

The spinal cord looks segmented because bilateral pairs of dorsal and ventral roots emerge regularly along its length. These pairs form five sets: cervical (in the neck above the rib cage), thoracic (associated with the rib cage), lumbar (near the abdomen), sacral (near the pelvis), and coccygeal (associated with tail vertebrae). In humans there are typically 31 spinal nerve pairs (8 cervical, 12 thoracic, 5 lumbar, 5 sacral, and 1 coccygeal) that are named according to the intervertebral foramen they pass through. This enumeration varies between species.

Based on human brain macroscopic dissection, Samuel Thomas von Sömmerring in 1778 recognized a sequence of 12 cranial nerve pairs, and this classification scheme remains traditional for vertebrates in general, although it is problematic in terms of completeness (e.g., not including the nasal cavity's terminal nerve) and nonconformance with contemporary fate maps of cranial nerve nucleus development (e.g., motoneurons for nerve VII are generated rostral to those for nerve VI). In any event, cranial nerves are more heterogeneous functionally than spinal nerves, and indeed most cranial nerve pairs have distinct compositions in terms of fiber types.

In humans, seven cranial nerves transmit information about the so-called special senses associated with the head: olfaction (I, olfactory nerve—purely sensory, arising in nasal olfactory epithelium), vision (II, optic nerve from retina), hearing and balance (VIII, vestibuloacoustic nerve from inner ear), and taste (V, VII, IX, and X; parts of the trigeminal, facial, glossopharyngeal, and vagus nerves, respectively). Nerves III (oculomotor), IV (trochlear), and VI (abducens) primarily control conjugate eye movements, although the third nerve also mediates autonomic control of the pupillary light reflex and lens accommodation. Major parts of the trigeminal (V) nerve carry sensory axons from the face (a rostral extension of the spinal somatosensory system) and motor axons innervating the muscles of mastication (chewing). The facial (VII) nerve controls the muscles of facial expression and also innervates the salivary and lacrimal glands—its role in emotional expression is obvious. The glossopharyngeal (IX) nerve innervates the pharynx and mediates the swallowing reflex. The vagus ("wandering," X) nerve has an exceptionally complex and widespread innervation pattern, including laryngeal muscles producing speech, and the parasympathetic innervation of most thoracic and abdominal viscera. The spinal accessory (XI) nerve innervates several muscles that stabilize the head and neck and the hypoglossal (XII) nerve innervates the tongue musculature.

Cerebral Hemispheres and Cerebellum Are Divided into Cortex and Nuclei

Macroscopically the mammalian cerebrospinal trunk has two great expansions—the cerebral hemispheres and cerebellum—and both have an outer laminated cortex surrounding deep nonlaminated nuclei. The most extraordinary growth of the mammalian brain occurs in the endbrain or cerebral hemispheres (Fig. 2.19), which develop more or less as mirror images of one another and are separated in the dorsal midline by a deep interhemispheric (longitudinal) fissure. In

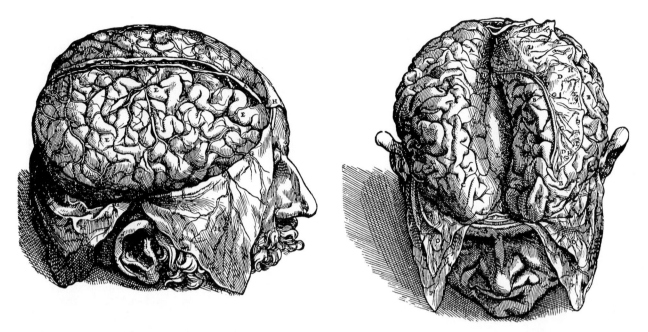

FIGURE 2.19 Surface features of the human cerebral cortex, which is thrown into gyri separated by sulci. In the drawing on the right, the right and left hemispheres have been pulled apart at the interhemispheric or longitudinal fissure to reveal the corpus callosum (L) interconnecting the two hemispheres. The drawings are from perhaps the most important book in the history of medicine by Andreas Vesalius, *Fabric of the Human Body*, published in 1543. The drawings were probably executed by an artist from Titian's studio.

humans, the sulcal pattern is, however, asymmetric and unique in each person, and there are functional asymmetries as well; for example, the speech centers typically are lateralized (Chapter 51). Hemisphere volume is restricted by skull capacity, so as the hemispheres grow during embryogenesis they develop folds (*gyri*) separated by invaginations (*sulci*, and when deeper, *fissures*). This corrugation allows cerebral (and cerebellar) cortex to have a larger surface area. The extent and pattern of folding vary stereotypically with species, although like any trait there are quantitative differences between individuals of a particular species. Two major grooves, the central sulcus and lateral (Sylvian) fissure, are used as anatomical landmarks in the human cerebrum. The central sulcus extends more or less vertically along the hemisphere's lateral surface where it approaches the horizontally oriented lateral fissure. Together they divide arbitrarily the outer cerebral cortical surface into four lobes (*frontal, parietal, occipital,* and *temporal*), named for the overlying cranial bones. In addition, the *insular lobe* is folded completely inside the hemisphere, deep to the lateral fissure (actually about two-thirds of the folded cortical surface lies buried and unexposed to the outer hemisphere surface), and the *limbic lobe* forms the hemisphere's medial border along the interhemispheric fissure.

These lobes are only crude guides to the cerebrum's functional organization. Over the last 150 years pro-

gressively better analysis has parceled the cortical mantle into a mosaic of roughly 50 to 100 areas with more or less distinct structural and functional characteristics. The most famous and enduring cortical *regionalization maps* were generated by Korbinian Brodmann a century ago (Fig. 2.20), although refinements and alternative interpretations abound. Nevertheless, cortical regionalization maps are fundamentally important guides for understanding CNS architecture. Just as one example, virtually the entire thalamus projects topographically on the cortical mantle, which in turn projects topographically on the entire cerebral nuclei (basal ganglia). Information from every sensory modality reaches the cerebral cortex and it in turn sends inputs to virtually the entire motor system.

Most cerebral cortical areas directly modulate activity on the opposite (contralateral) side of the body through descending pathways that cross the midline to reach motor system parts in the contralateral CNS. Furthermore, axon bundles called *commissures* connect cerebral cortical areas of one hemisphere with the same or related areas of the opposite hemisphere—and different areas in the same hemisphere are interconnected through complex *association pathways*. Thus, commissural and association pathways allow comparison and integration of information between cortical areas within and between the cerebral hemispheres.

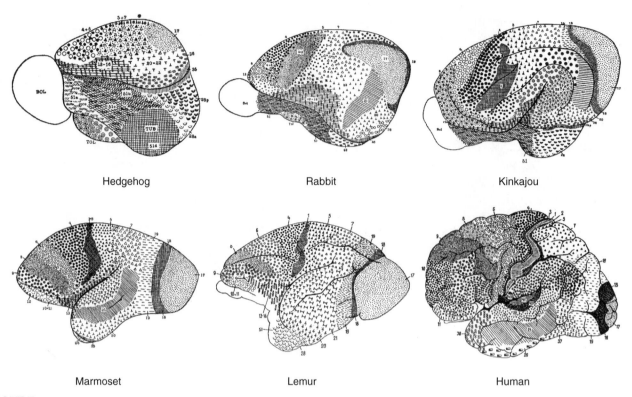

Hedgehog   Rabbit   Kinkajou

Marmoset   Lemur   Human

FIGURE 2.20 A similar cerebral cortical regionalization plan for mammals was proposed by Korbinian Brodmann in 1909. His cortical parceling was based on regional differences in how neuronal cell bodies tend to distribute in layers, an approach referred to as cytoarchitectonics. This figure illustrates his findings in six species, with different regions, or "areas" as he called them, indicated with different symbols and numbers. He distinguished 47 areas in the human cerebral cortex and showed that generally similar patterns applied to all nine species he analyzed.

Communication between hemispheres is eliminated by commissurotomy, the surgical division of all cerebral commissures (including the hippocampal and anterior commissures). This procedure sometimes is used to treat otherwise intractable epilepsy cases, preventing spread of severe epileptic activity from one hemisphere to the other. Incredibly, commissurotomy patients function very well most of the time, and behavioral studies on such "split brain" patients have yielded remarkable information about cerebral cortical organization (Gazzaniga, 2005). Axon bundles (tracts or pathways) connecting very different structures on the two sides of the CNS usually are called *decussations* to distinguish them from commissures.

The Nervous System Is Protected by Membranous Coverings

The CNS is completely surrounded by three concentric connective tissue membranes: *pia*, *arachnoid*, and *dura*. The pia ("faithful") is a very thin, vascular membrane. As the name suggests, it adheres closely to the CNS's surface, even where there are deep invaginations, as in the cerebral and cerebellar cortex. Then comes the arachnoid ("spidery"), which has a tenuous, web-like structure but is histologically similar to pia. Finally, the dura ("tough") is a thick, inelastic covering apposed to the skull and vertebral canal's inner surface. Membranes covering the CNS are continuous with similar coverings of the PNS, where the terminology differs.

In certain CNS regions neural tissue is absent but meninges persist. Here, ependymal cells lining the ventricular system (the monolayer vestige of the embryonic neuroepithelial layer that ends up lining the adult ventricular system) fuse with the pia and arachnoid layers to form structures known as *choroid plexus*, which contains abundant blood vessels and serves as a component of the blood–brain barrier (the blood–CSF barrier; see Chapter 3). The choroid plexus produces CSF in the roof of the lateral, third, and fourth ventricles, and this fluid fills the brain ventricles, and perhaps at least part of the spinal central canal. Under positive hydrostatic pressure, CSF passes out of the brain's interior through three foramina (holes) in the fourth ventricle's roof, to fill the subarachnoid space between pia and arachnoid.

The Brain Is Highly Vascular

The human brain consumes about 20% of the body's oxygen supply at rest, even though it usually weighs only about a kilogram. Thus, the brain must continuously receive a voluminous blood supply, on the order of a liter per minute. Blood reaches the brain through two arterial roots—*vertebral* and *internal carotid arteries*—that anastomose in the circle of Willis, which essentially surrounds the base of the hypothalamus and pituitary gland's stalk. The functional importance of this arterial circle cannot be overemphasized because afterward there is a drastic reduction in anastomoses between brain arteries and arterioles within brain tissue itself. As a result, blockage or rupture of even a small artery or arteriole rapidly deprives the supplied brain region of oxygen, causing a *stroke* or brain attack.

After entering the skull through the foramen magnum with the spinal cord, the paired vertebral arteries fuse into a single *basilar artery*, generating the cerebellar arteries and the posterior cerebral arteries, which supply occipital cerebral regions. The internal carotid arteries divide to form the anterior and middle cerebral arteries; the former supplies each hemisphere's medial surface (especially the limbic lobe), and the latter supplies the rest of the hemisphere (including the speech and somatic sensorimotor areas). By and large, the major arteries course along the cerebral surface and branches dive abruptly into the brain and proliferate into arterioles and capillaries.

A system of large *venous sinuses* collect blood from brain capillaries and return it to the heart, mostly via the paired *internal jugular veins*. The major venous sinuses lie within the dura, whose inelasticity essentially holds the sinuses open. Blood flow through sinuses is slow and under low hydrostatic pressure. Thin-walled venous sinuses surrounded by the tough and immovable dura sets the stage for serious injury when the head is subjected to physical trauma. A well-known example is traumatic injury to the great cerebral vein (of Galen) in the midline that can occur when a boxer is struck in the head. The blow's impact causes the brain to recoil in its CSF cushion, exerting a shearing force against the dura, which remains attached to the skull. This force effectively ruptures the great cerebral vein, leading to serious hemorrhage of venous blood into the subdural space between dura and arachnoid.

Summary

This chapter reviews common approaches to the problem of understanding the nervous system's fundamental structure and wiring diagram—the basic plan or architecture. One approach examines a series of increasingly complex animals from an evolutionary perspective to gain insight into basic organizing principles. It reveals trends toward centralization, cephalization, bilateral symmetry, and regionalization of the nervous system. It also suggests that basic molecular and cellular mechanisms of neuronal function, including electrical signal propagation and neurotransmitter release, have changed little since the appearance of the simplest nervous systems in hydra, jellyfish, and other cnidarians.

Another approach follows the vertebrate nervous system's development from embryo to adult. At early developmental stages the CNS of all vertebrates has the same basic structure. A polarized, bilaterally symmetrical, regionalized neural plate of ectodermal origin invaginates to form a neural tube whose rostral half presents three swellings (forebrain, midbrain, and then hindbrain), followed by a caudal presumptive spinal cord. These four basic CNS divisions, arranged from rostral to caudal, go on to subdivide repeatedly until all laminated and nuclear neuron groups of the adult CNS are formed. A topographic, "geographic," or regional account of the CNS emerges from this developmental approach.

How the CNS's functional systems or circuitry are arranged into a unified whole is a tantalizing, deep, unsolved problem. The model discussed here equates behavior with motor output, which is driven by a combination of sensory, intrinsic behavioral state, and cognitive inputs—as well as by endogenous neuronal activity within each system. Future behavior is determined partly by sensory feedback related to the original behavior's consequences. As this is written, the relationship between CNS macroregionalization (Figs. 2.14 and 2.15) and functional systems (Fig. 2.16) is not obvious. The correspondence between functional neural systems and gene expression networks is even more obscure, although promising results are beginning to emerge in the embryonic spinal cord and brainstem cranial nerve nuclei.

References

Alvarez-Bolado, G. and Swanson, L. W. (1996). "Developmental Brain Maps: Structure of the Embryonic Rat Brain." Elsevier, Amsterdam.

Bota, M. and Swanson, L. W. (2007). Online workbenches for neural network connections. *J. Comp. Neurol.* **500**, 807–814.

Brodmann, K. (1909). *Vergleichende Lokalisationslehre der Grosshirnrinde in ihren Prinzipien dargestellt auf Grund des Zellenbaues.* Barth,

Leipzig. Translated as Brodmann's "Localisation in the Cerebral Cortex" by L. J. Garey. Gordon-Smith, London, 1994.

Brusca, R. C. and Brusca, G. J. (1990). "Invertebrates." Sinauer Associates, Sunderland.

Cajal, S. Ramón y (1909–1911). Histologie du système nerveux de l'homme et des vertébrés, in 2 vols., Maloine, Paris. Translated as Histology of the Nervous System of Man and Vertebrates by N. Swanson and L. W. Swanson. Oxford University Press, New York, 1995.

Cartmill, M., Hylander, W. L., and Shafland, J. (1987). "Human Structure." Harvard University Press, Cambridge.

Furness, J. B. (2006). "The Enteric Nervous System." Blackwell, Malden, MA.

Gazzaniga, M. S. (2005). Forty-five years of split-brain research and still going strong. Nat. Rev. Neurosci. 6, 653–659.

Hamburger, V. (1973). Anatomical and physiological basis of embryonic motility in birds and mammals. In "Studies on the Development of Behavior and the Nervous System" (G. Gottlieb, ed.), Vol. 1, pp. 51076. Academic Press, New York.

Holland, P. W. and Takahashi, T. (2005). The evolution of homeobox genes: Implications for the study of brain development. Brain Res. Bull. 66, 484–490.

Koizumi, O. (2002). Developmental neurobiology of hydra, a model animal of cnidarians. Can. J. Zool. 80, 1678–1689.

Lentz, T. L. (1968). "Primitive Nervous Systems." Yale University Press, New Haven.

McConnell, C. H. (1932). Development of the ectodermal nerve net in the buds of Hydra. Quart. J. Micr. Sci. 75, 495–509.

Nieuwenhuys, R., Voogd, J., and van Huijzen, C. (1988). "The Human Central Nervous System: A Synopsis and Atlas," 3rd ed. Springer-Verlag, Berlin.

Parker, G. H. (1919). "The Elementary Nervous System." Lippincott, Philadelphia.

Peters, A., Palay, S. L., and Webster, H. deF. (1991). "The Fine Structure of the Nervous System: Neurons and Their Supporting Cells," 3rd ed. Oxford University Press, New York.

Sherrington, C. S. (1906). "The Integrative Action of the Nervous System." Scribner's, New York. (Reprinted, Yale University Press, New Haven, 1947).

Sidman, R. L., Angevine, J. B. Jr., and Taber Pierce, E. (1971). "Atlas of the Mouse Brain and Spinal Cord." Harvard University Press, Cambridge, MA.

Singer, C. (1952). "Vesalius on the Human Brain: Introduction, Translation of the Text, Translation of Descriptions of Figures, Notes to the Translations, Figures." Oxford University Press, Oxford.

Swanson, L. W. (1992). "Brain Maps: Structure of the Rat Brain." Elsevier, Amsterdam.

Swanson, L. W. (2000a). What is the brain? Trends Neurosci. 23, 519–527.

Swanson, L. W. (2000b). A history of neuroanatomical mapping. In "Brain Mapping: The Applications," A. W. Toga and J. C. Mazziotta (eds.), Academic Press, San Diego, pp. 77–109.

Swanson, L. W. (2003). "Brain Architecture: Understanding the Basic Plan." Oxford University Press, Oxford.

Von Neumann, J. (1958). "The Computer and the Brain." Yale University Press, New Haven.

Wiener, N. (1948). "Cybernetics, or Control and Communication in the Animal and Machine." Wiley, New York.

Williams, P. L. (Ed.) (1995). "Gray's Anatomy," 38th (British) ed. Churchill Livingstone, New York.

Suggested Readings

Bergquist, H. and Källén, B. (1954). Notes on the early histogenesis and morphogenesis of the central nervous system in vertebrates. J. Comp. Neurol. 100, 627–659.

Björklund, A. and Hökfelt, T. (1983-present). "Handbook of Chemical Neuroanatomy." Elsevier, Amsterdam.

Descartes, R. (1972). "Treatise on Man." French text with translation by T. S. Steele. Harvard University Press, Cambridge.

Herrick, C. J. (1948). "The Brain of the Tiger Salamander." University of Chicago Press, Chicago.

Kingsbury, B. F. (1922). The fundamental plan of the vertebrate brain. J. Comp. Neurol. 34, 461–491.

Lorenz, K. (1978). "Behind the Mirror." Harcourt Brace Jovanovich, Orlando.

Russell, E. S. (1916). "Form and Function: A Contribution to the History of Animal Morphology." John Murray, London.

Tinbergen, N. (1951). "The Study of Instinct." Oxford University Press, London.

Larry W. Swanson

SECTION II

CELLULAR AND MOLECULAR NEUROSCIENCE

Cellular Components of Nervous Tissue

Several types of cellular elements are integrated to constitute normally functioning brain tissue. The neuron is the communicating cell, and many neuronal subtypes are connected to one another via complex circuitries, usually involving multiple synaptic connections. Neuronal physiology is supported and maintained by neuroglial cells, which have highly diverse and incompletely understood functions. These include myelination, secretion of trophic factors, maintenance of the extracellular milieu, and scavenging of molecular and cellular debris from it. Neuroglial cells also participate in the formation and maintenance of the blood–brain barrier, a multicomponent structure that is interposed between the circulatory system and the brain substance and that serves as the molecular gateway to brain tissue.

NEURONS

The neuron is a highly specialized cell type and is the essential cellular element in the CNS. All neurological processes are dependent on complex cell–cell interactions among single neurons as well as groups of related neurons. Neurons can be categorized according to their size, shape, neurochemical characteristics, location, and connectivity, which are important determinants of that particular functional role of the neuron in the brain. More importantly, neurons form circuits, and these circuits constitute the structural basis for brain function. *Macrocircuits* involve a population of neurons projecting from one brain region to another region, and *microcircuits* reflect the local cell–cell interactions within a brain region. The detailed analysis of these macro- and microcircuits is an essential step in understanding the neuronal basis of a given cortical function in the healthy and the diseased brain. Thus, these cellular characteristics allow us to appreciate the special structural and biochemical qualities of a neuron in relation to its neighbors and to place it in the context of a specific neuronal subset, circuit, or function.

Broadly speaking, therefore, there are five general categories of neurons: inhibitory neurons that make local contacts (e.g., GABAergic interneurons in the cerebral and cerebellar cortex), inhibitory neurons that make distant contacts (e.g., medium spiny neurons of the basal ganglia or Purkinje cells of the cerebellar cortex), excitatory neurons that make local contacts (e.g., spiny stellate cells of the cerebral cortex), excitatory neurons that make distant contacts (e.g., pyramidal neurons in the cerebral cortex), and neuromodulatory neurons that influence neurotransmission, often at large distances. Within these general classes, the structural variation of neurons is systematic, and careful analyses of the anatomic features of neurons have led to various categorizations and to the development of the concept of cell type. The grouping of neurons into descriptive cell types (such as chandelier, double bouquet, or bipolar cells) allows the analysis of populations of neurons and the linking of specified cellular characteristics with certain functional roles.

General Features of Neuronal Morphology

Neurons are highly polarized cells, meaning that they develop distinct subcellular domains that subserve different functions. Morphologically, in a typical neuron, three major regions can be defined: (1) the cell body (*soma* or *perikaryon*), which contains the nucleus and the major cytoplasmic organelles; (2) a variable number of dendrites, which emanate from the perikaryon and ramify over a certain volume of gray matter and which differ in size and shape, depending on the neuronal type; and (3) a single axon, which extends, in most cases, much farther from the cell body than the dendritic arbor (Fig. 3.1). Dendrites may be spiny (as in pyramidal cells) or nonspiny (as in most interneurons), whereas the axon is generally smooth and emits a variable number of branches (collaterals). In vertebrates, many axons are surrounded by an insulating myelin sheath, which facilitates rapid impulse conduction. The axon terminal region, where contacts with other cells are made, displays a wide range of morphological specializations, depending on its target area in the central or peripheral nervous system.

The cell body and dendrites are the two major domains of the cell that receive inputs, and dendrites play a critically important role in providing a massive receptive area on the neuronal surface. In addition, there is a characteristic shape for each dendritic arbor, which can be used to classify neurons into morphological types. Both the structure of the dendritic arbor and the distribution of axonal terminal ramifications confer a high level of subcellular specificity in the localization of particular synaptic contacts on a given neuron. The three-dimensional distribution of dendritic arborization is also important with respect to the type of information transferred to the neuron. A neuron with a dendritic tree restricted to a particular cortical layer may receive a very limited pool of afferents, whereas the widely expanded dendritic arborizations of a large pyramidal neuron will receive highly diversified inputs within the different cortical layers in which segments of the dendritic tree are present (Fig. 3.2) (Mountcastle, 1978). The structure of the dendritic tree is maintained by surface interactions between adhesion molecules and, intracellularly, by an array of cytoskeletal components (microtubules, neurofilaments, and associated proteins), which also take part in the movement of organelles within the dendritic cytoplasm.

An important specialization of the dendritic arbor of certain neurons is the presence of large numbers of dendritic spines, which are membranous protrusions. They are abundant in large pyramidal neurons and are much sparser on the dendrites of interneurons (see later).

The perikaryon contains the nucleus and a variety of cytoplasmic organelles. Stacks of rough endoplasmic reticulum are conspicuous in large neurons and, when interposed with arrays of free polyribosomes, are referred to as *Nissl substance*. Another feature of the perikaryal cytoplasm is the presence of a rich cytoskeleton composed primarily of neurofilaments and

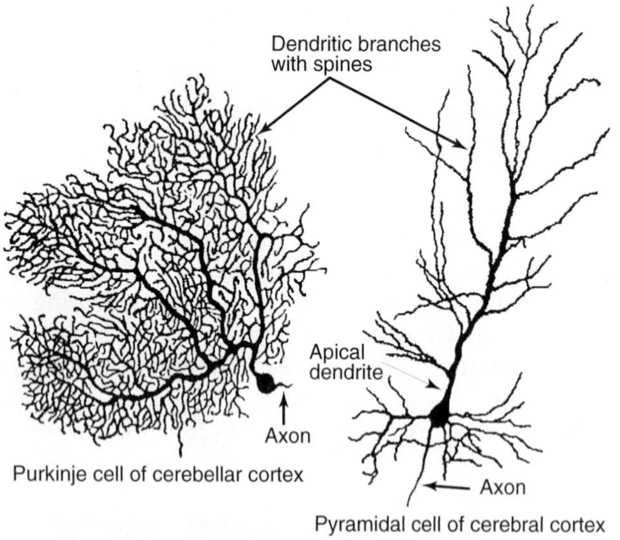

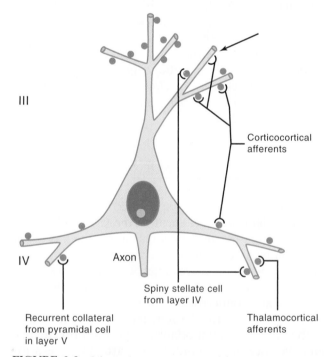

FIGURE 3.1 Typical morphology of projection neurons. (Left) A Purkinje cell of the cerebellar cortex and (right) a pyramidal neuron of the neocortex. These neurons are highly polarized. Each has an extensively branched, spiny apical dendrite, shorter basal dendrites, and a single axon emerging from the basal pole of the cell.

FIGURE 3.2 Schematic representation of four major excitatory inputs to pyramidal neurons. A pyramidal neuron in layer III is shown as an example. Note the preferential distribution of synaptic contacts on spines. Spines are labeled in red. Arrow shows a contact directly on the dendritic shaft.

microtubules, discussed in detail in Chapter 4. These cytoskeletal elements are dispersed in bundles that extend from the soma into the axon and dendrites.

Whereas dendrites and the cell body can be characterized as domains of the neuron that receive afferents, the axon, at the other pole of the neuron, is responsible for transmitting neural information. This information may be primary, in the case of a sensory receptor, or processed information that has already been modified through a series of integrative steps. The morphology of the axon and its course through the nervous system are correlated with the type of information processed by the particular neuron and by its connectivity patterns with other neurons. The axon leaves the cell body from a small swelling called the *axon hillock*. This structure is particularly apparent in large pyramidal neurons; in other cell types, the axon sometimes emerges from one of the main dendrites. At the axon hillock, microtubules are packed into bundles that enter the axon as parallel fascicles. The axon hillock is the part of the neuron where the action potential is generated. The axon is generally unmyelinated in local circuit neurons (such as inhibitory interneurons), but it is myelinated in neurons that furnish connections between different parts of the nervous system. Axons usually have higher numbers of neurofilaments than dendrites, although this distinction can be difficult to make in small elements that contain fewer neurofilaments. In addition, the axon may show extensive, spatially constrained ramified, as in certain local circuit neurons; it may give out a large number of recurrent collaterals, as in neurons connecting different cortical regions; or it may be relatively straight in the case of projections to subcortical centers, as in cortical motor neurons that send their very long axons to the ventral horn of the spinal cord. At the interface of axon terminals with target cells are the synapses, which represent specialized zones of contact consisting of a presynaptic (axonal) element, a narrow synaptic cleft, and a postsynaptic element on a dendrite or perikaryon.

Synapses and Spines

Synapses

Each synapse is a complex of several components: (1) a *presynaptic element*, (2) a *cleft*, and (3) a *postsynaptic element*. The presynaptic element is a specialized part of the presynaptic neuron's axon, the postsynaptic element is a specialized part of the postsynaptic somatodendritic membrane, and the space between these two closely apposed elements is the cleft. The portion of the axon that participates in the axon is the *bouton*, and it is identified by the presence of synaptic vesicles and a presynaptic thickening at the active zone (Fig. 3.3). The postsynaptic element is marked by a postsynaptic thickening opposite the presynaptic thickening. When both sides are equally thick, the synapse is referred to as *symmetric*. When the postsynaptic thickening is greater, the synapse is *asymmetric*. Edward George Gray noticed this difference, and divided synapses into two types: *Gray's type 1* synapses are symmetric, and have variably shaped, or pleomorphic, vesicles; *Gray's type 2* synapses are asymmetric, and have clear, round vesicles. The significance of this distinction is that research has shown that in general, Gray's type 1 synapses tend to be inhibitory, whereas Gray's type 2 synapses tend to be excitatory. This correlation greatly enhanced the usefulness of electron microscopy in neuroscience.

In cross-section on electron micrographs, a synapse looks like two parallel lines separated by a very narrow space (Fig. 3.3). Viewed from the inside of the axon or dendrite, it looks like a patch of variable shape. Some synapses are a simple patch, or *macule*. Macular synapses can grow fairly large, reaching diameters over $1\,\mu m$. The largest synapses have discontinuities or holes within the macule, and are called *perforated synapses* (Fig. 3.3). In cross-section, a perforated synapse may resemble a simple macular synapse, or several closely spaced smaller macules.

The portion of the presynaptic element that is apposed to the postsynaptic element is the *active zone*. This is the region where the synaptic vesicles are concentrated, and where at any time, a small number of vesicles are docked, and presumably ready for fusion. The active zone is also enriched with voltage gated calcium channels, which are necessary to permit activity-dependent fusion and neurotrans-mitter release.

The synaptic cleft is truly a space, but its properties are essential. The width of the cleft ($\sim 20\,\mu m$) is critical because it defines the volume in which each vesicle releases its contents, and therefore, the peak concentration of neurotransmitter upon release. On the flanks of the synapse, the cleft is spanned by adhesion molecules, which are believed to stabilize the cleft.

The postsynaptic element may be a portion of a soma or a dendrite, or rarely, part of an axon. In the cerebral cortex, most Gray's type 1 synapses are located on somata or dendritic shafts, and most Gray's type 2 synapses are located on dendritic spines, which are specialized protrusions of the dendrite. A similar segregation is seen in cerebellar cortex. In nonspiny neurons, symmetric and asymmetric synapses are often less well separated. Irrespective of location, a postsynaptic thickening marks the postsynaptic element. In Gray's type 2 synapses, the postsynaptic thickening (or postsynaptic density, PSD) is greatly enhanced. Among the molecules that are associated

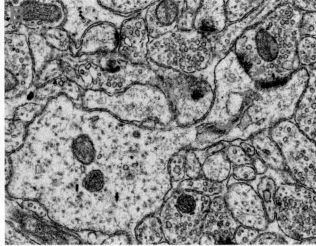

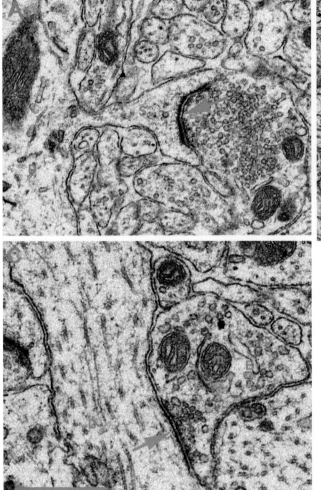

FIGURE 3.3 Ultrastructure of dendritic spines and synapses in the human brain. A and B: Narrow spine necks (asterisks) emanate from the main dendritic shaft (D). The spine heads (S) contain filamentous material (A, B). Some large spines contain cisterns of a spine apparatus (sa, B). Asymmetric excitatory synapses are characterized by thickened postsynaptic densities (arrows A, B). A perforated synapse has an electron-lucent region amidst the postsynaptic density (small arrow, B). The presynaptic axonal boutons (B) of excitatory synapses usually contain round synaptic vesicles. Symmetric inhibitory synapses (arrow, C) typically occur on the dendritic shaft (D) and their presynaptic boutons contain smaller round or ovoid vesicles. Dendrites and axons contain numerous mitochondria (m). Scale bar = 1 μm (A, B) and 0.6 μm (C). Electron micrographs courtesy of Drs S.A. Kirov and M. Witcher (Medical College of Georgia), and K.M. Harris (University of Texas – Austin).

with the PSD are neurotransmitter receptors (e.g., NMDA receptors) and molecules with less obvious function, such as PSD-95.

Spines

Spines are protrusions on the dendritic shafts of some types of neurons and are the sites of synaptic contacts, usually excitatory. Use of the silver impregnation techniques of Golgi or of the methylene blue used by Ehrlich in the late nineteenth century led to the discovery of spiny appendages on dendrites of a variety of neurons. The best known are those on pyramidal neurons and Purkinje cells, although spines occur on neuron types at all levels of the central nervous system. In 1896, Berkley observed that terminal axonal boutons were closely apposed to spines and suggested that spines may be involved in conducting impulses from neuron to neuron. In 1904, Santiago

Ramón y Cajal suggested that spines could collect the electrical charge resulting from neuronal activity. He also noted that spines substantially increase the receptive surface of the dendritic arbor, which may represent an important factor in receiving the contacts made by the axonal terminals of other neurons. It has been calculated that the approximately 20,000 spines of a pyramidal neuron account for more than 40% of its total surface area (Peters et al., 1991).

More recent analyses of spine electrical properties have demonstrated that spines are dynamic structures that can regulate many neurochemical events related to synaptic transmission and modulate synaptic efficacy. Spines are also known to undergo pathologic alterations and have a reduced density in a number of experimental manipulations (such as deprivation of a sensory input) and in many developmental, neurologic, and psychiatric conditions (such as dementing illnesses, chronic alcoholism, schizophrenia, trisomy 21). Morphologically, spines are characterized by a narrower portion emanating from the dendritic shaft, the neck, and an ovoid bulb or head, although spine morphology may vary from large mushroom-shaped bulbs to small bulges barely discernable on the surface of the dendrite. Spines have an average length of ~2 μm, but there is considerable variability in their dimensions. At the ultrastructural level (Fig. 3.3), spines are characterized by the presence of asymmetric synapses and contain fine and quite indistinct filaments. These filaments most likely consist of actin and α- and β-tubulins. Microtubules and neurofilaments present in dendritic shafts do not enter spines. Mitochondria and free ribosomes are infrequent, although many spines contain polyribosomes in their neck. Interestingly, most polyribosomes in dendrites are located at the bases of spines, where they are associated with endoplasmic reticulum, indicating that spines possess the machinery necessary for the local synthesis of proteins. Another feature of the spine is the presence of confluent tubular cisterns in the spine head that represent an extension of the dendritic smooth endoplasmic reticulum. Those cisterns are referred to as the *spine apparatus*. The function of the spine apparatus is not fully understood but may be related to the storage of calcium ions during synaptic transmission.

SPECIFIC EXAMPLES OF DIFFERENT NEURONAL TYPES

Inhibitory Local Circuit Neurons

Inhibitory Interneurons of the Cerebral Cortex

A large variety of inhibitory interneuron types is present in the cerebral cortex and in subcortical structures. These neurons contain the inhibitory neurotransmitter γ-aminobutyric acid (GABA) and exert strong local inhibitory effects. Their dendritic and axonal arborizations offer important clues as to their role in the regulation of pyramidal cell function. In addition, for several GABAergic interneurons, a subtype of a given morphologic class can be defined further by a particular set of neurochemical characteristics. Interneurons have been extensively characterized in the neocortex and hippocampus of rodents and primates, but they are present throughout the cerebral gray matter and exhibit a rich variety of morphologies, depending on the brain region, as well as on the species studied.

In the neocortex and hippocampus, the targets and morphologies of interneuron axons is most useful to classify them into morphological and functional groups. For example, *basket cells* have axonal endings surrounding pyramidal cell somata (Somogyi *et al.*, 1983) and provide most of the inhibitory GABAergic synapses to the somas and proximal dendrites of pyramidal cells. These cells are also characterized by certain biochemical features in that the majority of them contain the calcium-binding protein parvalbumin, and cholecystokinin appears to be the most likely neuropeptide in large basket cells.

Chandelier cells have spatially restricted axon terminals that look like vertically oriented "cartridges," each consisting of a series of axonal boutons, or swellings, linked together by thin connecting pieces. These neurons synapse exclusively on the axon initial segment of pyramidal cells (this cell is also known as *axoaxonic cell*), and because the strength of the synaptic input is correlated directly with its proximity to the axon initial segment, there can be no more powerful inhibitory input to a pyramidal cell than that of the chandelier cell (Freund *et al.*, 1983; DeFelipe *et al.*, 1989).

The double bouquet cells are characterized by a vertical bitufted dendritic tree and a tight bundle of vertically oriented varicose axon collaterals (Somogyi and Cowey, 1981). There are several subclasses of double bouquet cells based on the complement of calcium-binding protein and neuropeptide they contain. Their axons contact spines and dendritic shafts of pyramidal cells, as well as dendrites from nonpyramidal neurons.

Inhibitory Projection Neurons

Medium-sized Spiny Cells

These neurons are unique to the striatum, a part of the basal ganglia that comprises the caudate nucleus and putamen (see Chapter 31). Medium-sized spiny cells are scattered throughout the caudate nucleus and putamen and are recognized by their relatively large size, compared with other cellular elements of the basal ganglia, and by the fact that they are generally isolated neurons. They differ from all others in the striatum in that they have a highly ramified dendritic arborization radiating in all directions and densely covered with spines. They furnish a major output from the caudate nucleus and putamen and receive a highly diverse input from, among other sources, the cerebral cortex, thalamus, and certain dopaminergic neurons of the substantia nigra. These neurons are neurochemically quite heterogeneous, contain GABA, and may contain several neuropeptides and the calcium-binding protein calbindin. In Huntington disease, a neurodegenerative disorder of the striatum characterized by involuntary movements and progressive dementia, an early and dramatic loss of medium-sized spiny cells occurs.

Purkinje Cells

Purkinje cells are the most salient cellular elements of the cerebellar cortex. They are arranged in a single row throughout the entire cerebellar cortex between the molecular (outer) layer and the granular (inner) layer. They are among the largest neurons and have a round perikaryon, classically described as shaped "like a chianti bottle," with a highly branched dendritic tree shaped like a candelabrum and extending into the molecular layer where they are contacted by incoming systems of afferent fibers from granule neurons and the brainstem (see Chapter 32). The apical dendrites of Purkinje cells have an enormous number of spines (more than 80,000 per cell). A particular feature of the dendritic tree of the Purkinje cell is that it is distributed in one plane, perpendicular to the longitudinal axes of the cerebellar folds, and each dendritic arbor determines a separate domain of cerebellar cortex (Fig. 3.1). The axons of Purkinje neurons course through the cerebellar white matter and contact deep cerebellar nuclei or vestibular nuclei. These neurons contain the inhibitory neurotransmitter GABA and the calcium-binding protein calbindin. Spinocerebellar ataxia, a severe disorder combining ataxic gait and impairment of fine hand movements, accompanied by dysarthria and tremor, has been documented in some families and is related directly to Purkinje cell degeneration.

Excitatory Local Circuit Neurons

Spiny Stellate Cells

Spiny stellate cells are small multipolar neurons with local dendritic and axonal arborizations. These neurons resemble pyramidal cells in that they are the only other cortical neurons with large numbers of dendritic spines, but they differ from pyramidal neurons in that they lack an elaborate apical dendrite. The relatively restricted dendritic arbor of these neurons is presumably a manifestation of the fact that they are high-resolution neurons that gather afferents to a very restricted region of the cortex. Dendrites rarely leave the layer in which the cell body resides. The spiny stellate cell also resembles the pyramidal cell in that it provides asymmetric synapses that are presumed to be excitatory, and is thought to use glutamate as its neurotransmitter (Peters and Jones, 1984).

The axons of spiny stellate neurons have primarily intracortical targets and a radial orientation, and appear to play an important role in forming links among layer IV, the major thalamorecipient layer, and layers III, V, and VI, the major projection layers. The spiny stellate neuron appears to function as a high-fidelity relay of thalamic inputs, maintaining strict topographic organization and setting up initial vertical links of information transfer within a given cortical area (Peters and Jones, 1984).

Excitatory Projection Neurons

Pyramidal Cells

All cortical output is carried by pyramidal neurons, and the intrinsic activity of the neocortex can be viewed simply as a means of finely tuning their output. A pyramidal cell is a highly polarized neuron, with a major orientation axis perpendicular (or orthogonal) to the pial surface of the cerebral cortex. In cross-section, the cell body is roughly triangular (Fig. 3.1), although a large variety of morphologic types exist with elongate, horizontal, or vertical fusiform, or inverted perikaryal shapes. Pyramidal cells are the major excitatory type of neurons and use glutamate as their neurotransmitter. A pyramidal neuron typically has a large number of dendrites that emanate from the apex and form the base of the cell body. The span of the dendritic tree depends on the laminar localization of the cell body, but it may, as in giant pyramidal neurons, spread over several millimeters. The cell body and dendritic arborization may be restricted to a few layers or, in some cases, may span the entire cortical thickness (Jones, 1984).

In most cases, the axon of a large pyramidal cell extends from the base of the perikaryon and courses toward the subcortical white matter, giving off several collateral branches that are directed to cortical domains generally located within the vicinity of the cell of origin (as explained later). Typically, a pyramidal cell has a large nucleus, and a cytoplasmic rim that contains, particularly in large pyramidal cells, a collection of granular material chiefly composed of *lipofuscin*. Although all pyramidal cells possess these general features, they can also be subdivided into numerous classes based on their morphology, laminar location, and connectivity with cortical and subcortical regions (Fig. 3.4) (Jones, 1975).

Spinal Motor Neurons

Motor cells of the ventral horns of the spinal cord, also called α motoneurons, have their cell bodies within the spinal cord and send their axons outside the central nervous system to innervate the muscles. Different types of motor neurons are distinguished by their targets. The α motoneurons innervate skeletal muscles, but smaller motor neurons (the γ motoneurons, forming about 30% of the motor neurons) innervate the spindle organs of the muscles (see Chapter 28). The α motor neurons are some of the largest neurons in the entire central nervous system and are characterized by a multipolar perikaryon and a very

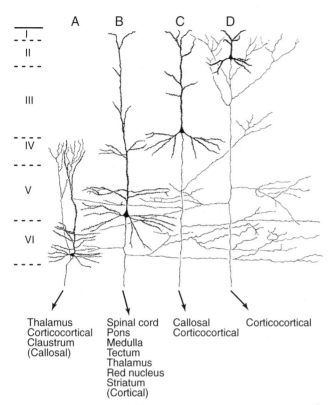

FIGURE 3.4 Morphology and distribution of neocortical pyramidal neurons. Note the variability in cell size and dendritic arborization, as well as the presence of axon collaterals, depending on the laminar localization (I–VI) of the neuron. Also, different types of pyramidal neurons with a precise laminar distribution project to different regions of the brain. Adapted from Jones (1984).

rich cytoplasm that renders them very conspicuous on histological preparations. They have a large number of spiny dendrites that arborize locally within the ventral horn. The α motoneuron axon leaves the central nervous system through the ventral root of the peripheral nerves. Their distribution in the ventral horn is not random and corresponds to a somatotopic representation of the muscle groups of the limbs and axial musculature (Brodal, 1981). Spinal motor neurons use acetylcholine as their neurotransmitter. Large motor neurons are severely affected in lower motor neuron disease, a neurodegenerative disorder characterized by progressive muscular weakness that affects, at first, one or two limbs but involves more and more of the body musculature, which shows signs of wasting as a result of denervation.

Neuromodulatory Neurons

Dopaminergic Neurons of the Substantia Nigra

Dopaminergic neurons are large neurons that reside mostly within the pars compacta of the substantia nigra and in the ventral tegmental area (van Domburg and ten Donkelaar, 1991). A distinctive feature of these cells is the presence of a pigment, *neuromelanin*, in compact granules in the cytoplasm. These neurons are medium-sized to large, fusiform, and frequently elongated. They have several large radiating dendrites. The axon emerges from the cell body or from one of the dendrites and projects to large expanses of cerebral cortex and to the basal ganglia. These neurons contain the catecholamine-synthesizing enzyme *tyrosine hydroxylase*, as well as the monoamine dopamine as their neurotransmitter. Some of them contain both calbindin and calretinin. These neurons are affected severely and selectively in Parkinson disease—a movement disorder different from Huntington disease and characterized by resting tremor and rigidity—and their specific loss is the neuropathologic hallmark of this disorder.

NEUROGLIA

The term neuroglia, or "nerve glue," was coined in 1859 by Rudolph Virchow, who conceived of the neuroglia as an inactive "connective tissue" holding neurons together in the central nervous system. The metallic staining techniques developed by Ramón y Cajal and del Rio-Hortega allowed these two great pioneers to distinguish, in addition to the ependyma lining the ventricles and central canal, three types of supporting cells in the CNS: oligodendrocytes, astrocytes, and microglia. In the peripheral nervous system (PNS), the Schwann cell is the major neuroglial component.

Oligodendrocytes and Schwann Cells Synthesize Myelin

Most brain functions depend on rapid communication between circuits of neurons. As shown in depth later, there is a practical limit to how fast an individual bare axon can conduct an action potential. Organisms developed two solutions for enhancing rapid communication between neurons and their effector organs. In invertebrates, the diameters of axons are enlarged. In vertebrates, the myelin sheath (Fig. 3.5) evolved to permit rapid nerve conduction.

Axon enlargement accelerates action potential propagation in proportion to the square root of axonal diameter. Thus larger axons conduct faster than small ones, but substantial increases in conduction velocity require huge axons. The largest axon in the invertebrate kingdom is the squid giant axon, which is about the thickness of a mechanical pencil lead. This axon

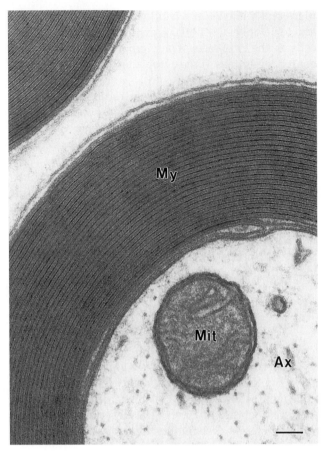

FIGURE 3.5 An electron micrograph of a transverse section through part of a myelinated axon from the sciatic nerve of a rat. The tightly compacted multilayer myelin sheath (My) surrounds and insulates the axon (Ax). Mit, mitochondria. Scale bar: 75 nm.

conducts the action potential at speeds of 10 to 20 m/s. As the axon mediates an escape reflex, firing must be rapid if the animal is to survive. Bare axons and continuous conduction obviously provide sufficient rates of signal propagation for even very large invertebrates, and many human axons also remain bare. However, in the human brain with 10 billion neurons, axons cannot be as thick as pencil lead, otherwise heads would weigh one hundred pounds or more.

Thus, along the invertebrate evolutionary line, the use of bare axons imposes a natural, insurmountable limit—a constraint of axonal size—to increasing the processing capacity of the nervous system. Vertebrates, however, get around this problem through evolution of the myelin sheath, which allows 10- to 100-fold increases in conduction of the nerve impulse along axons with fairly minute diameters.

In the central nervous system, myelin sheaths (Fig. 3.6) are elaborated by oligodendrocytes. During brain development, these glial cells send out a few cytoplasmic processes that engage adjacent axons and form myelin around them (Bunge, 1968). Myelin consists of a long sheet of oligodendrocyte plasma membrane, which is spirally wrapped around an axonal segment. At the end of each myelin segment, there is a bare portion of the axon, the node of Ranvier. Myelin segments are thus called *internodes*. Physiologically, myelin has insulating properties such that the action potential can "leap" from node to node and therefore does not have to be regenerated continually along the axonal segment that is covered by the myelin membrane sheath. This leaping of the action potential from node to node allows axons with fairly small diameters to conduct extremely rapidly (Ritchie, 1984), and is called *saltatory conduction*.

Because the brain and spinal cord are encased in the bony skull and vertebrae, CNS evolution has promoted compactness among the supporting cells of the CNS. Each oligodendrocyte cell body is responsible for the construction and maintenance of several myelin sheaths (Fig. 3.6), thus reducing the number of glial cells required. In both PNS and CNS myelin, cytoplasm is removed between each turn of the myelin, leaving only the thinnest layer of plasma membrane. Due to protein composition differences, CNS lamellae are approximately 30% thinner than in PNS myelin. In addition, there is little or no extracellular space or extracellular matrix between the myelinated axons passing through CNS white matter. Brain volume is thus reserved for further expansion of neuronal populations.

Peripheral nerves pass between moving muscles and around major joints, and are routinely exposed to physical trauma. A hard tackle, slipping on an icy sidewalk, or even just occupying the same uncomfortable seating posture for too long, can painfully compress peripheral nerves and potentially damage them. Thus, evolutionary pressures shaping the PNS favor robustness and regeneration rather than conservation of space. Myelin in the PNS is generated by Schwann cells (Fig. 3.7), which are different to oligodendrocytes in several ways. Individual myelinating Schwann cells form a single internode. The biochemical composition of PNS and CNS myelin differs, as discussed later. Unlike oligodendrocytes, Schwann cells secrete copious extracellular matrix components and produce a basal lamina "sleeve" that runs the entire length of myelinated axons. Schwann cell and fibroblast-derived collagens prevent normal wear-and-tear compression damage. Schwann cells also respond vigorously to injury, in common with astrocytes but unlike oligodendrocytes. Schwann cell growth factor secretion, debris removal by Schwann cells after injury, and the axonal guidance function of the basal lamina are responsible for the exceptional regenerative capacity of the PNS compared with the CNS.

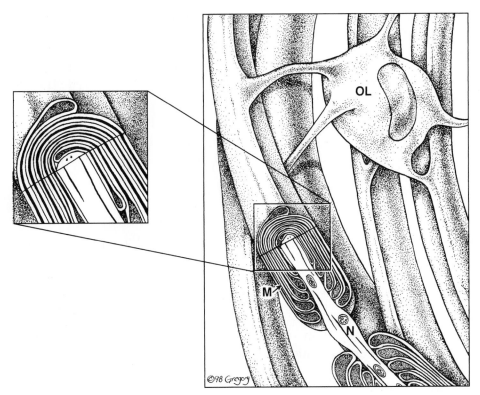

FIGURE 3.6 An oligodendrocyte (OL) in the central nervous system is depicted myelinating several axon segments. A cutaway view of the myelin sheath is shown (M). Note that the internode of myelin terminates in paranodal loops that flank the node of Ranvier (N). (Inset) An enlargement of compact myelin with alternating dark and light electron-dense lines that represent intracellular (major dense lines) and extracellular (intraperiod line) plasma membrane appositions, respectively.

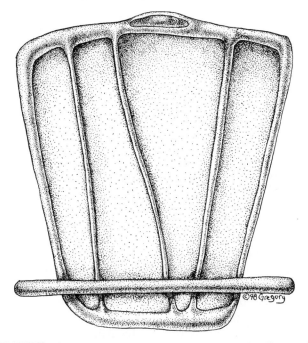

FIGURE 3.7 An "unrolled" Schwann cell in the PNS is illustrated in relation to the single axon segment that it myelinates. The broad stippled region is compact myelin surrounded by cytoplasmic channels that remain open even after compact myelin has formed, allowing an exchange of materials among the myelin sheath, the Schwann cell cytoplasm, and perhaps the axon as well.

The major integral membrane protein of peripheral nerve myelin is protein zero (P0), a member of a very large family of proteins termed the immunoglobulin gene superfamily. This protein makes up about 80% of the protein complement of PNS myelin. Interactions between the extracellular domains of P0 molecules expressed on one layer of the myelin sheath with those of the apposing layer yield a characteristic regular periodicity that can be seen by thin section electron microscopy (Fig. 3.5). This zone, called the intraperiod line, represents the extracellular apposition of the myelin bilayer as it wraps around itself. On the other side of the bilayer, the cytoplasmic side, the highly charged P0 cytoplasmic domain probably functions to neutralize the negative charges on the polar head groups of the phospholipids that make up the plasma membrane itself, allowing the membranes of the myelin sheath to come into close apposition with one another. In electron microscopy, this cytoplasmic apposition appears darker than the intraperiod line and is termed the major dense line. In peripheral nerves, although other molecules are present in small quantities in compact myelin and may have important functions, compaction (i.e., the close apposition of membrane surfaces without intervening cytoplasm) is accomplished solely by P0–P0 interactions at both extracellular and intracellular (cytoplasmic) surfaces.

Curiously, P0 is present in the CNS of lower vertebrates such as sharks and bony fish, but in terrestrial vertebrates (reptiles, birds, and mammals), P0 is limited to the PNS. CNS myelin compaction in these higher organisms is subserved by proteolipid protein (PLP) and its alternate splice form, DM-20. These two proteins are generated from the same gene, both span the plasma membrane four times, and differ only in that PLP has a small, positively charged segment exposed on the cytoplasmic surface. Why did PLP/DM-20 replace P0 in CNS myelin? Manipulation of PLP and P0 in CNS myelin established an axonotrophic function for PLP in CNS myelin. Removal of PLP from rodent CNS myelin altered the periodicity of compact myelin and produced a late onset axonal degeneration (Griffiths *et al.*, 1998). Replacing PLP with P0 in rodent CNS myelin stabilized compact myelin but enhanced the axonal degeneration (Yin *et al.*, 2006). These and other observations in primary demyelination and inherited myelin diseases have established axonal degeneration as the major cause of permanent disability in diseases such as multiple sclerosis.

Myelin membranes also contain a number of other proteins such as the myelin basic protein, which is a major CNS myelin component, and PMP-22, a protein that is involved in a form of peripheral nerve disease. A large number of naturally occurring gene mutations can affect the proteins specific to the myelin sheath and cause neurological disease. In animals, these mutations have been named according to the phenotype that is produced: the shiverer mouse, the shaking pup, the rumpshaker mouse, the jumpy mouse, the myelin-deficient rat, the quaking mouse, and so forth. Many of these mutations are well characterized, and have provided valuable insights into the role of individual proteins in myelin formation and axonal survival.

Astrocytes Play Important Roles in CNS Homeostasis

As the name suggests, astrocytes were first described as star-shaped, process-bearing cells distributed throughout the central nervous system. They constitute from 20 to 50% of the volume of most brain areas. Astrocytes appear stellate when stained using reagents that highlight their intermediate filaments, but have complex morphologies when their entire cytoplasm is visualized. The two main forms, protoplasmic and fibrous astrocytes, predominate in gray and white matter, respectively (Fig. 3.8). Embryonically, astrocytes develop from radial glial cells, which transversely

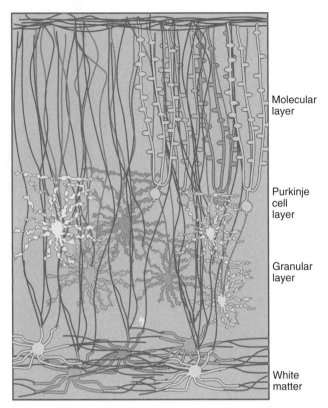

Molecular layer

Purkinje cell layer

Granular layer

White matter

FIGURE 3.8 The arrangement of astrocytes in human cerebellar cortex. Bergmann glial cells are in red, protoplasmic astrocytes are in green, and fibrous astrocytes are in blue.

compartmentalize the neural tube. Radial glial cells serve as scaffolding for the migration of neurons and play a critical role in defining the cytoarchitecture of the CNS (Fig. 3.9). As the CNS matures, radial glia retract their processes and serve as progenitors of astrocytes. However, some specialized astrocytes of a radial nature are still found in the adult cerebellum and the retina and are known as Bergmann glial cells and Müller cells, respectively.

Astrocytes "fence in" neurons and oligodendrocytes. Astrocytes achieve this isolation of the brain parenchyma by extending long processes projecting to the pia mater and the ependyma to form the glia limitans, by covering the surface of capillaries, and by making a cuff around the nodes of Ranvier. They also ensheath synapses and dendrites and project processes to cell somas (Fig. 3.10). Astrocytes are connected to each other by gap junctions, forming a syncytium that allows ions and small molecules to diffuse across the brain parenchyma. Astrocytes have in common unique cytological and immunological properties that make them easy to identify, including their star shape, the glial end feet on capillaries, and a unique population of large bundles of intermediate filaments. These filaments are composed of an astroglial-specific pro-

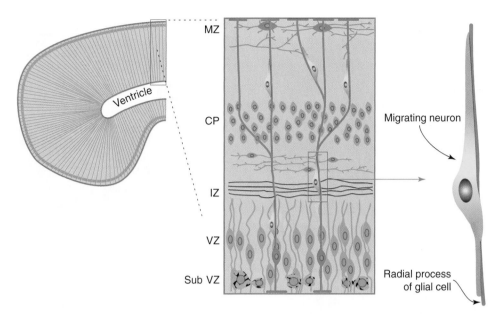

FIGURE 3.9 Radial glia perform support and guidance functions for migrating neurons. In early development, radial glia span the thickness of the expanding brain parenchyma. (Inset) Defined layers of the neural tube from the ventricular to the outer surface: VZ, ventricular zone; IZ, intermediate zone; CP, cortical plate; MZ, marginal zone. The radial process of the glial cell is indicated in blue, and a single attached migrating neuron is depicted at the right.

tein commonly referred to as glial fibrillary acidic protein (GFAP). S-100, a calcium-binding protein, and glutamine synthetase are also astrocyte markers. Ultrastructurally, gap junctions (connexins), desmosomes, glycogen granules, and membrane orthogonal arrays are distinct features used by morphologists to identify astrocytic cellular processes in the complex cytoarchitecture of the nervous system.

For a long time, astrocytes were thought to physically form the blood–brain barrier (considered later in this chapter), which prevents the entry of cells and diffusion of molecules into the CNS. In fact, astrocytes are indeed the blood–brain barrier in lower species. However, in higher species, astrocytes are responsible for inducing and maintaining the tight junctions in endothelial cells that effectively form the barrier. Astrocytes also take part in angiogenesis, which may be important in the development and repair of the CNS. Their role in this important process is still poorly understood.

Astrocytes Have a Wide Range of Functions

There is strong evidence for the role of radial glia and astrocytes in the migration and guidance of neurons in early development. Astrocytes are a major source of extracellular matrix proteins and adhesion molecules in the CNS; examples are nerve cell–nerve cell adhesion molecule (N-CAM), laminin, fibronectin, cytotactin, and the J-1 family members janusin and tenascin. These molecules participate not only in the migration of neurons, but also in the formation of neuronal aggregates, so-called nuclei, as well as networks.

Astrocytes produce, *in vivo* and *in vitro*, a very large number of growth factors. These factors act singly or in combination to selectively regulate the morphology, proliferation, differentiation, or survival, or all four, of distinct neuronal subpopulations. Most of the growth factors also act in a specific manner on the development and functions of astrocytes and oligodendrocytes. The production of growth factors and cytokines by astrocytes and their responsiveness to these factors is a major mechanism underlying the developmental function and regenerative capacity of the CNS. During neurotransmission, neurotransmitters and ions are released at high concentration in the synaptic cleft. The rapid removal of these substances is important so that they do not interfere with future synaptic activity. The presence of astrocyte processes around synapses positions them well to regulate neurotransmitter uptake and inactivation (Kettenman and Ransom, 1995). These possibilities are consistent with the presence in astrocytes of transport systems for many neurotransmitters. For instance, glutamate reuptake is performed mostly by astrocytes, which convert glutamate into glutamine and then release it into the extracellular space. Glutamine is taken up by neurons, which use it to generate glutamate and γ-aminobutyric acid, potent excitatory and inhibitory neurotransmitters, respectively (Fig. 3.11). Astrocytes contain ion

Pia mater

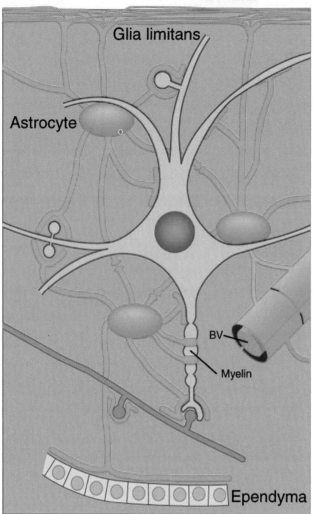

FIGURE 3.10 Astrocytes (in orange) are depicted *in situ* in schematic relationship with other cell types with which they are known to interact. Astrocytes send processes that surround neurons and synapses, blood vessels, and the region of the node of Ranvier and extend to the ependyma, as well as to the pia mater, where they form the glial limitans.

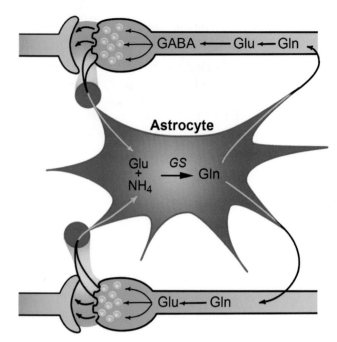

FIGURE 3.11 The glutamate–glutamine cycle is an example of a complex mechanism that involves an active coupling of neurotransmitter metabolism between neurons and astrocytes. The systems of exchange of glutamine, glutamate, GABA, and ammonia between neurons and astrocytes are highly integrated. The postulated detoxification of ammonia and the inactivation of glutamate and GABA by astrocytes are consistent with the exclusive localization of glutamine synthetase in the astroglial compartment.

channels for K⁺, Na⁺, Cl⁻, HCO₃, and Ca²⁺, as well as displaying a wide range of neurotransmitter receptors. K⁺ ions released from neurons during neurotransmission are soaked up by astrocytes and moved away from the area through astrocyte gap junctions. This is known as *spatial buffering*. Astrocytes play a major role in detoxification of the CNS by sequestering metals and a variety of neuroactive substances of endogenous and xenobiotic origin.

In response to stimuli, intracellular calcium waves are generated in astrocytes. Propagation of the Ca²⁺ wave can be visually observed as it moves across the cell soma and from astrocyte to astrocyte. The generation of Ca²⁺ waves from cell to cell is thought to be mediated by second messengers, diffusing through gap junctions (see Chapter 11). In the adult brain, gap junctions are present in all astrocytes. Some gap junctions also have been detected between astrocytes and neurons. Thus, they may participate, along with astroglial neurotransmitter receptors, in the coupling of astrocyte and neuron physiology.

In a variety of CNS disorders—neurotoxicity, viral infections, neurodegenerative disorders, HIV, AIDS, dementia, multiple sclerosis, inflammation, and trauma—astrocytes react by becoming hypertrophic and, in a few cases, hyperplastic. A rapid and huge upregulation of GFAP expression and filament formation is associated with astrogliosis. The formation of reactive astrocytes can spread very far from the site of origin. For instance, a localized trauma can recruit astrocytes from as far as the contralateral side, suggesting the existence of soluble factors in the mediation process. Tumor necrosis factor (TNF) and ciliary neurotrophic factors (CNTF) have been identified as key factors in astrogliosis.

Microglia Are Mediators of Immune Responses in Nervous Tissue

The brain traditionally has been considered an "immunologically privileged site," mainly because the blood–brain barrier normally restricts the access of immune cells from the blood. However, it is now known that immunological reactions do take place in the central nervous system, particularly during cerebral inflammation. Microglial cells have been termed the tissue macrophages of the CNS, and they function as the resident representatives of the immune system in the brain. A rapidly expanding literature describes microglia as major players in CNS development and in the pathogenesis of CNS disease.

The first description of microglial cells can be traced to Franz Nissl (1899), who used the term "rod cell" to describe a population of glial cells that reacted to brain pathology. He postulated that rod-cell function was similar to that of leukocytes in other organs. Cajal described microglia as part of his "third element" of the CNS—cells that he considered to be of mesodermal origin and distinct from neurons and astrocytes (Ramón y Cajal, 1913).

Del Rio-Hortega (1932) distinguished this third element into microglia and oligodendrocytes. He used silver impregnation methods to visualize the ramified appearance of microglia in the adult brain, and he concluded that ramified microglia could transform into cells that were migratory, ameboid, and phagocytic. Indeed, a hallmark of microglial cells is their ability to become reactive and to respond to pathological challenges in a variety of ways. A fundamental question raised by del Rio-Hortega's studies was the origin of microglial cells. Some questions about this remain even today.

Microglia Have Diverse Functions in Developing and Mature Nervous Tissue

On the basis of current knowledge, it appears that most ramified microglial cells are derived from bone marrow-derived monocytes, which enter the brain parenchyma during early stages of brain development. These cells help phagocytose degenerating cells that undergo programmed cell death as part of normal development. They retain the ability to divide and have the immunophenotypic properties of monocytes and macrophages. In addition to their role in remodeling the CNS during early development, microglia secrete cytokines and growth factors that are important in fiber tract development, gliogenesis, and angiogenesis. They are also the major

CNS cells involved in presenting antigens to T lymphocytes. After the early stages of development, ameboid microglia transform into the ramified microglia that persist throughout adulthood (Altman, 1994).

Little is known about microglial function in the healthy adult vertebrate CNS. Microglia constitute a formidable percentage (5–20%) of the total cells in the mouse brain. Microglia are found in all regions of the brain, and there are more in gray than in white matter. The neocortex and hippocampus have more microglia than regions like the brainstem or cerebellum. Species variations also have been noted, as human white matter has three times more microglia than rodent white matter.

Microglia usually have small rod-shaped somas from which numerous processes extend in a rather symmetrical fashion. Processes from different microglia rarely overlap or touch, and specialized contacts between microglia and other cells have not been described in the normal brain. Although each microglial cell occupies its own territory, microglia collectively form a network that covers much of the CNS parenchyma. Because of the numerous processes, microglia present extensive surface membrane to the CNS environment. Regional variation in the number and shape of microglia in the adult brain suggests that local environmental cues can affect microglial distribution and morphology. On the basis of these morphological observations, it is likely that microglia play a role in tissue homeostasis. The nature of this homeostasis remains to be elucidated. It is clear, however, that microglia can respond quickly and dramatically to alterations in the CNS microenvironment.

Microglia Become Activated in Pathological States

"Reactive" microglia can be distinguished from resting microglia by two criteria: (1) change in morphology and (2) upregulation of monocyte–macrophage molecules (Fig. 3.12). Although the two phenomena generally occur together, reactive responses of microglia can be diverse and restricted to subpopulations of cells within a microenvironment. Microglia not only respond to pathological conditions involving immune activation, but also become activated in neurodegenerative conditions that are not considered immune mediated. This latter response is indicative of the phagocytic role of microglia. Microglia change their morphology and antigen expression in response to almost any form of CNS injury.

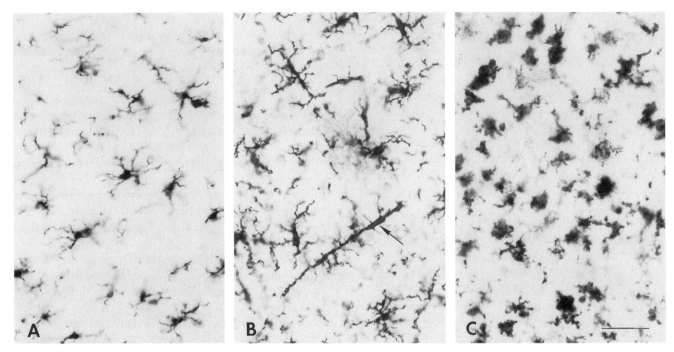

FIGURE 3.12 Activation of microglial cells in a tissue section from human brain. Resting microglia in normal brain (A). Activated microglia in diseased cerebral cortex (B) have thicker processes and larger cell bodies. In regions of frank pathology (C) microglia transform into phagocytic macrophages, which can also develop from circulating monocytes that enter the brain. Arrow in B indicates rod cell. Sections stained with antibody to ferritin. Scale bar = 40 μm.

CEREBRAL VASCULATURE

Blood vessels form an extremely rich network in the central nervous system, particularly in the cerebral cortex and subcortical gray masses, whereas the white matter is less densely vascularized (Fig. 3.13) (Duvernoy *et al.*, 1981). There are distinct regional patterns of microvessel distribution in the brain. These patterns are particularly clear in certain subcortical structures that constitute discrete vascular territories and in the cerebral cortex, where regional and laminar patterns are striking. For example, layer IV of the primary visual cortex possesses an extremely rich capillary network in comparison with other layers and adjacent regions (Fig. 3.13). Interestingly, most of the inputs from the visual thalamus terminate in this particular layer. Capillary densities are higher in regions containing large numbers of neurons and where synaptic density is high. Progressive occlusion of a large arterial trunk, as seen in stroke, induces an ischemic injury that may eventually lead to necrosis of the brain tissue. The size of the resulting infarction is determined in part by the worsening of the blood circulation through the cerebral microvessels. Occlusion of a large arterial trunk results in rapid swelling of the capillary endothelium and surrounding astrocytes, which may reduce the capillary lumen to about one-third of its normal diameter, pre-venting red blood cell circulation and oxygen delivery to the tissue. The severity of these changes subsequently determines the time course of neuronal necrosis, as well as the possible recovery of the surrounding tissue and the neurological outcome of the patient. In addition, the presence of multiple microinfarcts caused by occlusive lesions of small cerebral arterioles may lead to a progressively dementing illness, referred to as vascular dementia, affecting elderly humans.

The Blood–Brain Barrier Maintains the Intracerebral Milieu

Capillaries of the central nervous system form a protective barrier that restricts the exchange of solutes between blood and brain. This distinct function of brain capillaries is called the blood–brain barrier (Fig. 3.14) (Bradbury, 1979). Capillaries of the retina have similar properties and are termed the blood–retina barrier. It is thought that the blood–brain and blood–retina barriers function to maintain a constant intracerebral milieu, so that neuronal signaling can occur without interference from substances leaking in from the blood stream. This function is important because of the nature of intercellular communication in the CNS, which includes chemical signals across intercellular spaces. Without a blood–brain barrier, circulating factors in the blood, such as certain hormones,

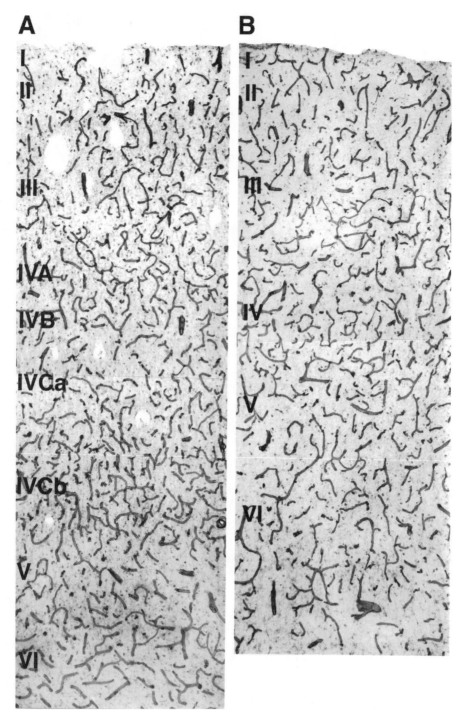

FIGURE 3.13 Microvasculature of the human neocortex. (A) The primary visual cortex (area 17). Note the presence of segments of deep penetrating arteries that have a larger diameter than the microvessels and run from the pial surface to the deep cortical layers, as well as the high density of microvessels in the middle layer (layers IVCa and IVCb). (B) The prefrontal cortex (area 9). Cortical layers are indicated by Roman numerals. Microvessels are stained using an antibody against heparan sulfate proteoglycan core protein, a component of the extracellular matrix.

which can also act as neurotransmitters, would interfere with synaptic communication. When the blood–brain barrier is disrupted, edema fluid accumulates in the brain. Increased permeability of the blood–brain barrier plays a central role in many neuropathological conditions, including multiple sclerosis, AIDS, and childhood lead poisoning, and may also play a role in Alzheimer's disease. The cerebral capillary wall is composed of an endothelial cell surrounded by a very thin (about 30 nm) basement membrane or basal

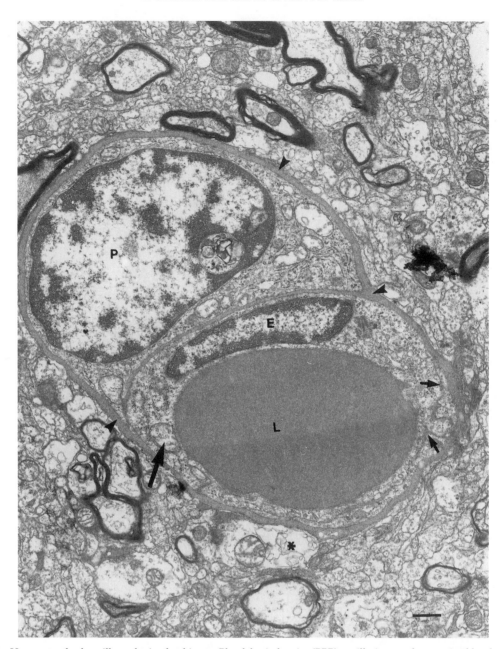

FIGURE 3.14   Human cerebral capillary obtained at biopsy. Blood–brain barrier (BBB) capillaries are characterized by the paucity of transcytotic vesicles in endothelial cells (E), a high mitochondrial content (large arrow), and the formation of tight junctions (small arrows) between endothelial cells that restrict the transport of solutes through the interendothelial space. The capillary endothelium is encased within a basement membrane (arrowheads), which also houses pericytes (P). Outside the basement membrane are astrocyte foot processes (asterisk), which may be responsible for the induction of BBB characteristics on the endothelial cells. L, lumen of the capillary. Scale bar = 1 μm. From Claudio *et al.* (1995).

lamina. End feet of perivascular astrocytes are apposed against this continuous basal lamina. Around the capillary lies a virtual perivascular space occupied by another cell type, the pericyte, which surrounds the capillary walls. The endothelial cell forms a thin monolayer around the capillary lumen, and a single endothelial cell can completely surround the lumen of the capillary (Fig. 3.14).

A fundamental difference between brain endothelial cells and those of the systemic circulation is the presence in brain of interendothelial tight junctions, also known as zonula occludens. In the systemic circulation, the interendothelial space serves as a diffusion pathway that offers little resistance to most blood solutes entering the surrounding tissues. In contrast, blood–brain barrier tight junctions effectively restrict the intercellular route of solute transfer. The blood–brain barrier interendothelial junctions are not static seals; rather they are a series of active gates that can allow certain small molecules to penetrate. One such molecule is the lithium ion, used in the control of manic depression.

Another characteristic of endothelial cells of the brain is their low transcytotic activity. Brain endothelium, therefore, is by this index not very permeable. It is of interest that certain regions of the brain, such as the area postrema and periventricular organs, lack a blood–brain barrier. In these regions, the perivascular space is in direct contact with the nervous tissue, and endothelial cells are fenestrated and show many pinocytotic vesicles. In these brain regions, neurons are known to secrete hormones and other factors that require rapid and uninhibited access to the systemic circulation.

Because of the high metabolic requirements of the brain, blood–brain barrier endothelial cells must have transport mechanisms for the specific nutrients needed for proper brain function. One such mechanism is glucose transporter isoform 1 (GLUT-1), which is expressed asymmetrically on the surface of blood–brain barrier endothelial cells. In Alzheimer's disease, the expression of GLUT-1 on brain endothelial cells is reduced. This reduction may be due to a lower metabolic requirement of the brain after extensive neuronal loss. Other specific transport mechanisms on the cerebral endothelium include the large neutral amino acid carrier-mediated system that transports, among other amino acids, L-3,4-dihydroxyphenylalanine (L-dopa), used as a therapeutic agent in Parkinson disease. Also on the surface of blood–brain barrier endothelial cells are transferrin receptors that allow the transport of iron into specific areas of the brain. The amount of iron that is transported into the various areas of the brain appears to depend on the concentration of transferrin receptors on the surface of endothelial cells of that region. Thus, the transport of specific nutrients into the brain is regulated during physiological and pathological conditions by blood–brain barrier transport proteins distributed according to the regional and metabolic requirements of brain tissue.

In general, disruption of the blood–brain barrier causes perivascular or vasogenic edema, which is the accumulation of fluids from the blood around the blood vessels of the brain. This is one of the main features of multiple sclerosis. In multiple sclerosis, inflammatory cells, primarily T cells and macrophages, invade the brain by migrating through the blood–brain barrier and attack cerebral elements as if these elements were foreign antigens. It has been observed by many investigators that the degree of edema accumulation causes the neurological symptoms experienced by people suffering from multiple sclerosis.

Studying the regulation of blood–brain barrier permeability is important for several reasons. Therapeutic treatments for neurological disease need to be able to cross the barrier. Attempts to design drug delivery systems that take therapeutic drugs directly into the brain have been made by using chemically engineered carrier molecules that take advantage of receptors such as that for transferrin, which normally transports iron into the brain. Development of an *in vitro* test system of the blood–brain barrier is of importance in the creation of new neurotropic drugs that are targeted to the brain.

References

Altman, J. (1994). Microglia emerge from the fog. *Trends Neurosci.* **17**, 47–49.

Bradbury, M. W. B. (1979). "The Concept of a Blood-Brain Barrier," pp. 381–407. Wiley, Chichester.

Brodal, A. (1981). "Neurological Anatomy in Relation to Clinical Medicine," 3rd Ed. Oxford Univ. Press, New York.

Bunge, R. P. (1968). Glial cells and the central myelin sheath. *Physiol. Rev.* **48**, 197–251.

Carpenter, M. B. and Sutin, J. (1983). "Human Neuroanatomy." Williams & Wilkins, Baltimore, MD.

Claudio, L., Raine, C. S., and Brosnan, C. F. (1995). Evidence of persistent blood-brain barrier abnormalities in chronic-progressive multiple sclerosis. *Acta Neuropathol.* **90**, 228–238.

del Rio-Hortega, P. (1932). Microglia. *In* "Cytology and Cellular Pathology of the Nervous System" (W. Penfield, ed.), Vol. 2, pp. 481–534. Harper (Hoeber), New York.

DeFelipe, J., Hendry, S. H. C., and Jones, E. G. (1989). Visualization of chandelier cell axons by parvalbumin immunoreactivity in monkey cerebral cortex. *Proc. Natl. Acad. Sci. USA* **86**, 2093–2097.

Dolman, C. L. (1991). Microglia. *In* "Textbook of Neuropathology" (R. L. Davis and D. M. Robertson, eds.), pp. 141–163. Williams & Wilkins, Baltimore, MD.

Duvernoy, H. M., Delon, S., and Vannson, J. L. (1981). Cortical blood vessels of the human brain. *Brain Res. Bull.* **7**, 519–579.

Freund, T. F., Martin, K. A. C., Smith, A. D., and Somogyi, P. (1983). Glutamate decarboxylase-immunoreactive terminals of Golgi-impregnated axoaxonic cells and of presumed basket cells in synaptic contact with pyramidal neurons of the cat's visual cortex. *J. Comp. Neurol.* **221**, 263–278.

Hudspeth, A. J. (1983). Transduction and tuning by vertebrate hair cells. *Trends Neurosci.* **6**, 366–369.

Jones, E. G. (1984). Laminar distribution of cortical efferent cells. *In* "Cellular Components of the Cerebral Cortex" (A. Peters and E. G. Jones, eds.), Vol. 1, pp. 521–553. Plenum, New York.

Jones, E. G. (1975). Varieties and distribution of non-pyramidal cells in the somatic sensory cortex of the squirrel monkey. *J. Comp. Neurol.* **160**, 205–267.

Kettenman, H. and Ransom, B. R., eds. (1995). "Neuroglia." Oxford University Press, Oxford.

Krebs, W. and Krebs, I. (1991). "Primate Retina and Choroid: Atlas of Fine Structure in Man and Monkey." Springer-Verlag, New York.

Mountcastle, V. B. (1978). An organizing principle for cerebral function: The unit module and the distributed system. *In* "The

Mindful Brain: Cortical Organization and the Group-Selective Theory of Higher Brain Function" (V. B. Mountcastle and G. Eddman, eds.), pp. 7–50. MIT Press, Cambridge, MA.

Nissl, F. (1899). Über einige Beziehungen zwischen Nervenzellenerkränkungen und gliösen Erscheinungen bei verschiedenen Psychosen. *Arch. Psychol.* **32**, 1–21.

Peters, A. and Jones, E. G., eds. (1984). "Cellular Components of the Cerebral Cortex," Vol. 1. Plenum, New York.

Peters, A., Palay, S. L., and Webster, H. deF. (1991). "The Fine Structure of the Nervous System: Neurons and Their Supporting Cells," 3rd ed. Oxford University Press, New York.

Ramón y Cajal, S. (1913). Contribucion al conocimiento de la neuroglia del cerebro humano. *Trab. Lab. Invest. Biol.* **11**, 255–315.

Ritchie, J. M. (1984). Physiological basis of conduction in myelinated nerve fibers. *In* "Myelin" (P. Morell, ed.), pp. 117–146. Plenum, New York.

Somogyi, P. and Cowey, A. (1981). Combined Golgi and electron microscopic study on the synapses formed by double bouquet cells in the visual cortex of the cat and monkey. *J. Comp. Neurol.* **195**, 547–566.

Somogyi, P., Kisvárday, Z. F., Martin, K. A. C., and Whitteridge, D. (1983). Synaptic connections of morphologically identified and physiologically characterized baket cells in the striate cortex of cat. *Neuroscience* **10**, 261–294.

van Domburg, P. H. M. F. and ten Donkelaar, H. J. (1991). The human substantia nigra and ventral tegmental area. *Adv. Anat. Embryol. Cell Biol.* **121**, 1–132.

Yin, X., Baek, R. C., Kirschner, D. A., Peterson, A., Fujii, Y., Nave, K. A., Macklin, W. B., and Trapp, B. D. (2006). Evolution of a neuroprotective function of central nervous system myelin. *J. Cell Biol.* **172**, 469–478.

Suggested Readings

Brightman, M. W. and Reese, T. S. (1969). Junctions between intimately apposed cell membranes in the vertebrate brain. *J. Cell Biol.* **40**, 648–677.

Broadwell, R. D. and Salcman, M. (1981). Expanding the definition of the BBB to protein. *Proc. Natl. Acad. Sci. USA* **78**, 7820–7824.

Fernandez-Moran, H. (1950). EM observations on the structure of the myelinated nerve sheath. *Exp. Cell Res.* **1**, 143–162.

Gehrmann, J., Matsumoto, Y., and Kreutzberg, G. W. (1995). Microglia: Intrinsic immune effector cell of the brain. *Brain Res. Rev.* **20**, 269–287.

Kimbelberg, H. and Norenberg, M. D. (1989). Astrocytes. *Sci. Am.* **26**, 66–76.

Kirschner, D. A., Ganser, A. L., and Caspar, D. W. (1984). Diffraction studies of molecular organization and membrane interactions in myelin. *In* "Myelin" (P. Morell, ed.), pp. 51–96. Plenum, New York.

Lum, H. and Malik, A. B. (1994). Regulation of vascular endothelial barrier function. *Am. J. Physiol.* **267**, L223–L241.

Rosenbluth, J. (1980). Central myelin in the mouse mutant shiverer. *J. Comp. Neurol.* **194**, 639–728.

Rosenbluth, J. (1980). Peripheral myelin in the mouse mutant shiverer. *J. Comp. Neurol.* **194**, 729–753.

Patrick R. Hof, Jean de Vellis,
Esther A. Nimchinsky, Grahame Kidd,
Luz Claudio, and Bruce D. Trapp

Subcellular Organization of the Nervous System: Organelles and Their Functions

Cells have many features in common, but each cell type also possesses a functional architecture related to its unique physiology. In fact, cells may become so specialized in fulfilling a particular function that virtually all cellular components may be devoted to it. For example, the machinery inside mammalian erythrocytes is completely dedicated to the delivery of oxygen to the tissues and the removal of carbon dioxide. Toward this end, this cell has evolved a specialized plasma membrane, an underlying cytoskeletal matrix that molds the cell into a biconcave disk, and a cytoplasm rich in hemoglobin. Modification of the cell machinery extends even to the discarding of structures such as the nucleus and the protein synthetic apparatus, which are not needed after the red blood cell matures. In many respects, the terminally differentiated, highly specialized cells of the nervous system exhibit comparable commitment—the extensive development of subcellular components reflects the roles that each plays.

The neuron serves as the cellular correlate of information processing and, in aggregate, all neurons act together to integrate responses of the entire organism to the external world. It is therefore not surprising that the specializations found in neurons are more diverse and complex than those found in any other cell type. Single neurons commonly interact in specific ways with hundreds of other cells—other neurons, astrocytes, oligodendrocytes, immune cells, muscle, and glandular cells. This chapter defines the major functional domains of the neuron, describes the subcellular elements that compose the building blocks of these domains, and examines the processes that create and maintain neuronal functional architecture.

AXONS AND DENDRITES: UNIQUE STRUCTURAL COMPONENTS OF NEURONS

Neural cells are remarkably complex (Peters *et al.*, 1991). As discussed in Chapter 3, the perikaryon, or cell body, contains the nucleus and the protein synthetic machinery. In neurons, nuclei are large and contain a preponderance of euchromatin. Because protein synthesis must be kept at a high level just to maintain the neuronal extensions, transcription levels in neurons are generally high. In turn, the variety of different polypeptides associated with cellular domains in a neuron requires that many different genes be transcribed constantly.

As mRNAs are synthesized, they move from the nucleus into a protein-synthesizing region termed the "translational cytoplasm," comprising cytoplasmic ("free") and membrane-associated polysomes, the intermediate compartment of the smooth endoplasmic reticulum, and the Golgi complex. Neurons have relatively large amounts of translational cytoplasm to accommodate high levels of protein synthesis. This protein synthetic machinery is arranged in discrete intracellular "granules," termed Nissl substance after the histologist who first discovered these structures in the nineteenth century. The Nissl substance is actually a combination of stacks of rough endoplasmic reticulum (RER), interposed with rosettes of free polysomes. This arrangement is unique to neurons, and its functional significance remains unknown. Most, but not all proteins used by the neuron are synthesized in the perikaryon. During or after synthesis and processing, proteins are packaged into membrane-limited

59

organelles, incorporated into cytoskeletal elements, or remain as soluble constituents of the cytoplasm. After packaging, membrane proteins are transported to their sites of function.

In general, neurons have two discrete functional domains, the axonal and somatodendritic compartments, each of which encompasses a number of microdomains (Fig. 4.1). The axon classically is defined as the cellular process by which a neuron makes contact with a target cell to transmit information. It provides a conduit for transmitting the action potential to a synapse, and acts as a specialized subdomain for transmission of a signal from neuron to target cell (neuron, muscle, etc.), usually by release of neurotransmitters. Consequently, most axons end in a presynaptic terminal, although a single axon may have hundreds or thousands of presynaptic specializations known as "en passant" synapses along its length. Characteristics of presynaptic terminals are presented in greater detail later.

The axon is the first neuronal process to differentiate during development. A typical neuron has only a single axon that proceeds some distance from the cell body before branching extensively. Usually the longest process of a neuron, axons come in many sizes. In a human adult, axons range in length from a few micrometers for small interneurons to a meter or more for large motor neurons, and they may be even longer in large animals (such as giraffes, elephants, and whales). In mammals and other vertebrates, the longest axons generally extend approximately half the body length.

Axonal diameters also are quite variable, ranging from 0.1 to 20 μm for large myelinated fibers in vertebrates. Invertebrate axons grow to even larger diameters, with the giant axons of some squid species achieving diameters in the millimeter range. Invertebrate axons reach such large diameters because they lack the myelinating glia that speed conduction of the action potential. As a result, axonal caliber must be large to sustain the high rate of conduction needed for the reflexes that permit escape from predators and capture of prey. Although axonal caliber is closely regulated in both myelinated and nonmyelinated fibers, this parameter is critical for those organisms that are unable to produce myelin.

The region of the neuronal cell body where the axon originates has several specialized features. This domain, called the axon hillock, is distinguished most readily by a deficiency of Nissl substance. Therefore, protein synthesis cannot take place to any appreciable degree in this region. Cytoplasm in the vicinity of the axon hillock may have a few polysomes but is dominated by the cytoskeletal and membranous organelles

that are being delivered to the axon. Microtubules and neurofilaments begin to align roughly parallel to each other, helping to organize membrane-limited organelles destined for the axon. The hillock is a region where materials either are committed to the axon (cytoskeletal elements, synaptic vesicle precursors, mitochondria, etc.) or are excluded from the axon (RER and free polysomes, dendritic microtubule-associated proteins). The molecular basis for this sorting is not understood. Cytoplasm in the axon hillock does not appear to contain a physical "sizing" barrier (like a filter) because large organelles such as mitochondria enter the axon readily, whereas only a small number of essentially excluded structures such as polysomes are occasionally seen only in the initial segment of the axon and not in the axon proper. An exception to this general rule is during development when local protein synthesis does take place at the axon terminus or growth cone. In the mature neuron, the physiological significance of this barrier must be considerable because axonal structures are found to accumulate in this region in many neuropathologies, including those due to degenerative diseases (such as amyotrophic lateral sclerosis) and to exposure to neurotoxic compounds (such as acrylamide).

The initial segment of the axon is the region of the axon adjacent to the axon hillock. Microtubules generally form characteristic fascicles, or bundles, in the initial segment of the axon. These fascicles are not seen elsewhere. The initial segment and, to some extent, the axon hillock also have a distinctive specialized plasma membrane. Initially, the plasmalemma was thought to have a thick electrondense coating actually attached to the inner surface of the membrane, but this dense undercoating is in reality separated by 5–10 nm from the plasma membrane inner surface and has a complex ultrastructure. Neither the composition nor the function of this undercoating is known. Curiously, the undercoating is present in the same regions of the initial segment as the distinctive fasciculation of microtubules, although the relationship is not understood.

FIGURE 4.1   Basic elements of neuronal subcellular organization. The neuron consists of a soma, or cell body, in which the nucleus, multiple cytoplasm-filled processes termed dendrites, and the (usually single) axon are placed. The neuron is highly extended in space; one with a cell body of the size shown here might maintain an axon several miles in length! The unique shape of each neuron is the result of a cooperative interplay between plasma membrane components and cytoskeletal elements. Most large neurons in vertebrates are myelinated by oligodendrocytes in the CNS and by Schwann cells in the PNS. The compact wraps of myelin encasing the axon distal to the initial segment permit rapid conduction of the action potential by a process termed "saltatory conduction" (see Chapter 3).

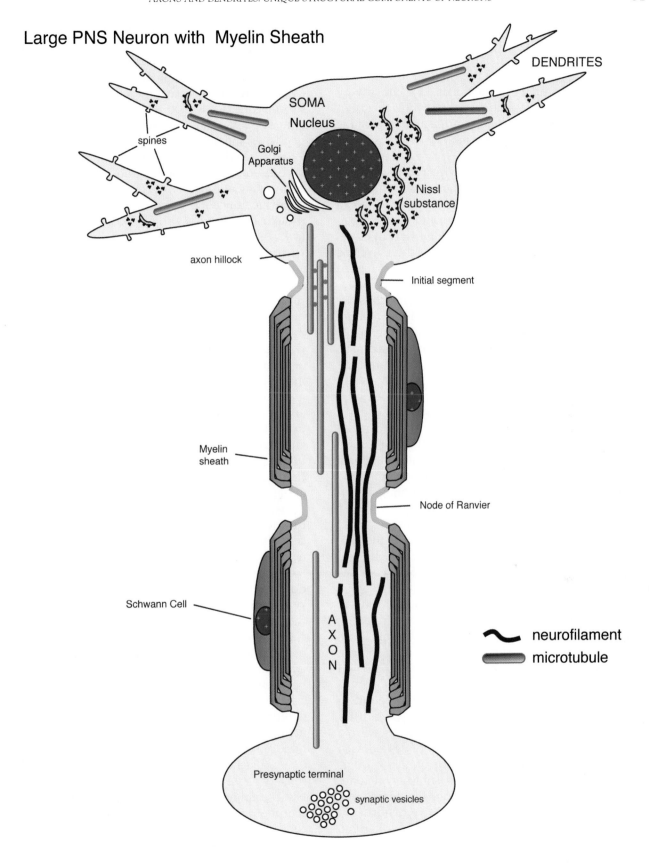

Large PNS Neuron with Myelin Sheath

DENDRITES

SOMA

Nucleus

spines

Golgi Apparatus

Nissl substance

axon hillock

Initial segment

Myelin sheath

Node of Ranvier

Schwann Cell

AXON

neurofilament

microtubule

Presynaptic terminal

synaptic vesicles

The plasma membrane is specialized in the initial segment and axon hillock in that it contains voltage-sensitive ion channels in large numbers, and most action potentials originate in this domain. The molecular composition of the axon initial segment is very similar to that of the node of Ranvier; however, evidence is growing that the mechanisms that govern the assembly of these components in the two locations are distinct.

Ultimately, axonal structure is geared toward the efficient conduction of action potentials at a rate appropriate to the function of that neuron. This can be seen from both the ultrastructure and the composition of axons. Axons are roughly cylindrical in cross-section with little or no taper. As discussed later, this diameter is maintained by regulation of the cytoskeleton. Even at branch points, daughter axons are comparable in diameter to the parent axon. This constant caliber helps ensure a consistent rate of conduction. Similarly, the organization of membrane components is regulated to this end. Voltage-gated ion channels are distributed to maximize conduction. Sodium channels are distributed more or less uniformly in small nonmyelinated axons, but are concentrated at high density in the regularly spaced unmyelinated gaps, known as nodes of Ranvier. An axon so organized will conduct an action potential or train of spikes long distances with high fidelity at a defined speed. These characteristics are essential for maintaining the precise timing and coordination seen in neuronal circuits.

Nodes of Ranvier in myelinated fibers are flanked by paranodal axoglial junctions comprised of the axolemmal proteins Caspr/Paranodin and Contactin and the glial isoform of neurofascin, Nfasc155. There has been considerable debate about the role of axoglial junctions in assembling the node of Ranvier, but, at least in the PNS, the nodal isoform of Neurofascin, Nfasc186, seems to be the crucial molecule that allows NrCAM, beta-IV spectrin, ankyrin-G and sodium channels to form a nodal complex (Sherman and Brophy, 2005).

Most vertebrate neurons have multiple dendrites arising from their perikarya. Unlike axons, dendrites branch continuously and taper extensively with a reduction in caliber in daughter processes at each branching. In addition, the surface of dendrites is covered with small protrusions, or spines, which are postsynaptic specializations. Although the surface area of a dendritic arbor may be quite extensive, dendrites in general remain in the relative vicinity of the perikaryon. A dendritic arbor may be contacted by the axons of many different and distant neurons or innervated by a single axon making multiple synaptic contacts.

The base of a dendrite is continuous with the cytoplasm of the cell body. In contrast to the axon, Nissl substance extends into dendrites, and certain proteins are synthesized predominantly in dendrites. There is evidence for the selective placement of some mRNAs in dendrites as well (Steward, 1995). For example, whereas RER and polysomes extend well into the dendrites, the mRNAs that are transported and translated in dendrites are a subset of the total neuronal mRNA, deficient in some mRNA species (such as neurofilament mRNAs) and enriched in mRNAs with dendritic functions (such as microtubule-associated protein, MAP2, mRNAs). Also, certain proteins appear to be targeted, postsynthesis, to the dendritic compartment as well.

The shapes and complexity of dendritic arborizations may be remarkably plastic. Dendrites appear relatively late in development and initially have only limited numbers of branches and spines. As development and maturation of the nervous system proceed, the size and number of branches increase. The number of spines increases dramatically, and their distribution may change. This remodeling of synaptic connectivity may continue into adulthood, and environmental effects can alter this pattern significantly. Eventually, in the aging brain, there is a reduction in complexity and size of dendritic arbors, with fewer spines and thinner dendritic shafts. These changes correlate with changes in neuronal function during development and aging.

As defined by classical physiology, axons are structural correlates for neuronal output, and dendrites constitute the domain for receiving information. A neuron without an axon or one without dendrites therefore might seem paradoxical, but such neurons do exist. Certain amacrine and horizontal cells in the vertebrate retina have no identifiable axons, although they do have dendritic processes that are morphologically distinct from axons. Such processes may have both pre-and postsynaptic specializations or may have gap junctions that act as direct electrical connections between two cells. Similarly, the pseudounipolar sensory neurons of dorsal root ganglia (DRG) have no dendrites. In their mature form, these DRG sensory neurons give rise to a single axon that extends a few hundred micrometers before branching. One long branch extends to the periphery, where it may form a sensory nerve ending in muscle spindles or skin. Large DRG peripheral branches are myelinated and have the morphological characteristics of an axon, but they contain neither pre- nor postsynaptic specializations. The other branch extends into the central nervous system, where it forms synaptic contacts. In DRG neurons, the action potential is generated at distal sensory nerve endings and then is transmitted along the peripheral branch to the central branch and the appropriate central nervous system (CNS) targets, bypassing the cell body. The functional and morphological hallmarks of axons and dendrites are listed in Table 4.1.

Summary

Neurons are polarized cells that are specialized for membrane and protein synthesis, as well as for conduction of the nerve impulse. In general, neurons have a cell body, a dendritic arborization that usually is located near the cell body, and an extended axon that may branch considerably before terminating to form synapses with other neurons.

PROTEIN SYNTHESIS IN NERVOUS TISSUE

Both neurons and glial cells have strikingly extended morphologies. Protein and lipid components are synthesized and assembled into the membranes of these cell extensions through pathways of membrane biogenesis that have been elucidated primarily in other cell types. However, some adaptations of these general mechanisms have been necessary, due to the specific requirements of cells in the nervous system. Neurons, for example, have devised mechanisms for ensuring that the specific components of the axonal and dendritic plasma membranes are selectively delivered (targeted) to each plasma membrane subdomain.

TABLE 4.1   Functional and Morphological Hallmarks of Axons and Dendrites[a]

Axons	Dendrites
With rare exceptions, each neuron has a single axon.	Most neurons have multiple dendrites arising from their cell bodies.
Axons appear first during neuronal differentiation.	Dendrites begin to differentiate only after the axon has formed.
Axon initial segments are distinguished by a specialized plasma membrane containing a high density of ion channels and distinctive cytoskeletal organization.	Dendrites are continuous with the perikaryal cytoplasm, and the transition point cannot be distinguished readily.
Axons typically are cylindrical in form with a round or elliptical cross-section.	Dendrites usually have a significant taper and small spinous processes that give them an irregular cross-section.
Large axons are myelinated in vertebrates, and the thickness of the myelin sheath is proportional to the axonal caliber.	Dendrites are not myelinated, although a few wraps of myelin may occur rarely.
Axon caliber is a function of neurofilament and microtubule numbers with neurofilaments predominating in large axons.	The dendritic cytoskeleton may appear less organized, and microtubules dominate even in large dendrites.
Microtubules in axons have a uniform polarity with plus ends distal from the cell body.	Microtubules in proximal dendrites have mixed polarity, with both plus and minus ends oriented distal to the cell body.
Axonal microtubules are enriched in tau protein with a characteristic phosphorylation pattern.	Dendritic microtubules may contain some tau protein, but MAP2 is not present in axonal compartments and is highly enriched in dendrites.
Ribosomes are excluded from mature axons, although a few may be detectable in initial segments.	Both rough endoplasmic reticulum and cytoplasmic polysomes are present in dendrites, with specific mRNAs being enriched in dendrites.
Axonal branches tend to be distal from the cell body.	Dendrites begin to branch extensively near the perikaryon and form extensive arbors in the vicinity of the perikaryon.
Axonal branches form obtuse angles and have diameters similar to the parent stem.	Dendritic branches form acute angles and are smaller than the parent stem.
Most axons have presynaptic specializations that may be *en passant* or at the ends of axonal branches.	Dendrites are rich in postsynaptic specializations, particularly on the spines that project from the dendritic shaft.
Action potentials usually are generated at the axon hillock and conducted away from the cell body.	Dendrites may generate action potentials, but more commonly they modulate the electrical state of perikaryon and initial segment.
Traditionally, axons are specialized for conduction and synaptic transmission, i.e., neuronal output.	Dendritic architecture is most suitable for integrating synaptic responses from a variety of inputs, i.e., neuronal input.

[a]Neurons typically have two classes of cytoplasmic extensions that may be distinguished using electrophysiological, morphological, and biochemical criteria. Although some neuronal processes may lack one or more of these features, enough parameters can generally be defined to allow unambiguous identification.

The distribution to specific loci of organelles, receptors, and ion channels is critical to normal neuronal function. In turn, these loci must be "matched" appropriately to the local microenvironment and specific cell–cell interactions. Similarly, in myelinating glial cells during the narrow developmental window when the myelin sheath is being formed, these cells synthesize sheets of insulating plasma membrane at an unbelievably high rate. To understand how the plasma membrane of neurons and glia might be modeled to fit individual functional requirements, it is necessary to review the progress that has been made so far in our understanding of how membrane components and organelles are generated in eukaryotic cells.

There are two major categories of membrane proteins: integral and peripheral. Integral membrane proteins, which include the receptors for neurotransmitters (e.g., the acetylcholine receptor subunits) and polypeptide growth factors (e.g., the dimeric insulin receptor), have segments that either are embedded in the lipid bilayer or are bound covalently to molecules that insert into the membrane, such as those proteins linked to glycosyl phosphatidylinositol at their C termini (e.g., Thy–1). A protein with a single membrane-embedded segment and an N terminus exposed at the extracellular surface is said to be of type I, whereas type II proteins retain their N termini on the cytoplasmic side of the plasma membrane. Peripheral membrane proteins are localized on the cytoplasmic surface of the membrane and do not traverse any membrane during their biogenesis. They interact with membranes either by means of their associations with membrane lipids or the cytoplasmic tails of integral proteins, or by means of their affinity for other peripheral proteins (e.g., platelet-derived growth factor receptor-Grb2-Sos-Ras complex). In some cases, they may bind electrostatically to the polar head groups of the lipid bilayer (e.g., myelin basic protein).

Integral Membrane and Secretory Polypeptides Are Synthesized *de Novo* in the Rough Endoplasmic Reticulum

The subcellular destinations of integral and peripheral membrane proteins are determined by their sites of synthesis. In the secretory pathway, integral membrane proteins and secretory proteins are synthesized in the rough endoplasmic reticulum, whereas the mRNAs encoding peripheral proteins are translated on cytoplasmic "free" polysomes, which are not membrane associated but which may interact with cytoskeletal structures.

The pathway by which secretory proteins are synthesized and exported was first postulated through the

elegant ultrastructural studies on the pancreas by George Palade and colleagues (Palade, 1975). Pancreatic acinar cells were an excellent choice for this work because they are extremely active in secretion, as revealed by the abundance of their RER network, a property they share with neurons. Nissl deduced, in the nineteenth century, that pancreatic cells and neurons would be found to have common secretory properties because of similarities in the distribution of the Nissl substance (Fig. 4.2).

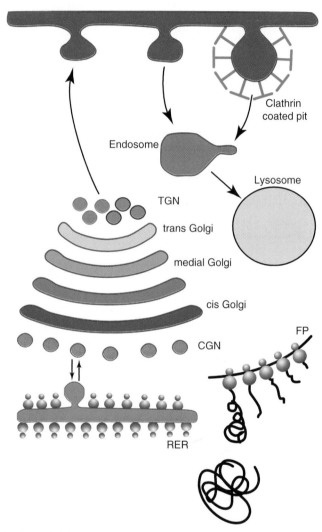

FIGURE 4.2 The secretory pathway. Transport and sorting of proteins in the secretory pathway occur as they pass through the Golgi before reaching the plasma membrane. Sorting occurs in the *cis*-Golgi network (CGN), also known as the intermediate compartment, and in the *trans*-Golgi network (TGN). Proteins exit from the Golgi at the TGN. The default pathway is the direct route to the plasma membrane. Proteins bound for regulated secretion or transport to endosomes are diverted from the default path by means of specific signals. In endocytosis, one population of vesicles is surrounded by a clathrin cage and is destined for late endosomes.

Pulse–chase radioautography has revealed that in eukaryotic cells newly synthesized secretory proteins move from the RER to the Golgi apparatus, where the proteins are packaged into secretory granules and transported to the plasma membrane across which they are released by exocytosis. Pulse–chase studies in neurons reveal a similar sequence of events for proteins transported into the axon. Unraveling of the detailed molecular mechanisms of the pathway began with the successful reconstitution of secretory protein biosynthesis *in vitro* and the direct demonstration that, very early during synthesis, secretory proteins are translocated into the lumen of RER vesicles, prepared by cell fractionation, termed microsomes. A key observation here was that the fate of the protein was sealed as a result of encapsulation in the lumen of the RER at the site of synthesis. This cotranslational insertion model provided a logical framework for understanding the synthesis of integral membrane proteins with a transmembrane orientation.

The process by which integral membrane proteins are synthesized closely follows the secretory pathway, except that integral proteins are of course not released from the cell, but instead remain bound to cellular membranes. Synthesis of integral proteins begins with synthesis of the nascent chain on a polysome that is not yet bound to the RER membrane (Fig. 4.3). Emergence of the N terminus of the nascent protein from the protein synthesizing machinery allows a ribonucleoprotein, a signal recognition particle (SRP), to bind to an emerging hydrophobic signal sequence and prevent further translation (Walter and Johnson, 1994). Translation arrest is relieved when SRP docks with its cognate receptor in the RER and dissociates from the signal sequence in a process that requires GTP. Synthesis of transmembrane proteins on RER is an extremely energy-efficient process. The passage of a fully formed and folded protein through a membrane is thermodynamically formidably expensive; it is infinitely "cheaper" for cells to thread amino acids, in tandem, through a membrane during initial protein synthesis. Protein synthesis then resumes, and the emerging polypeptide chain is translocated into the RER membrane through a conceptualized "aqueous pore" termed the "translocon."

A few polypeptides deviate from the common pathway for secretion. For example, certain peptide growth factors, such as basic fibroblast growth factor and ciliary neurotrophic factor, are synthesized without signal peptide sequences but are potent biological modulators of cell survival and differentiation. These growth factors appear to be released under certain conditions, although the mechanisms for such release are still controversial. One possibility is that release of

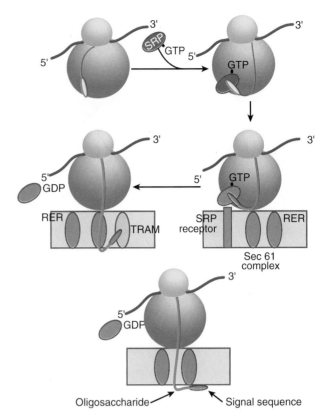

FIGURE 4.3 Translocation of proteins across the rough endoplasmic reticulum (RER). Integral membrane and secretory protein synthesis begins with partial synthesis on a free polysome not yet bound to the RER. The N-terminus of the nascent protein emerges and allows a ribonucleoprotein, signal recognition particle (SRP), to bind to the hydrophobic signal sequence and prevent further translation. Translation arrest is relieved once the SRP docks with its receptor at the RER and dissociates from the signal sequence in a GTP-dependent process. Once protein synthesis resumes, translocation occurs through an aqueous pore termed the translocon, which includes the translocating chain associating membrane protein (TRAM). The signal sequence is removed by a signal peptidase in the RER lumen.

these factors may be associated primarily with cellular injury.

Two cotranslational modifications commonly are associated with the emergence of the polypeptide on the luminal face of the RER. First, an N-terminal hydrophobic signal sequence that is used for insertion into the RER usually is removed by a signal peptidase. Second, oligosaccharides rich in mannose sugars are transferred from a lipid carrier, dolichol phosphate, to the side chains of asparagine residues (Kornfeld and Kornfeld, 1985). The asparagines must be in the sequence N X T (or S), and they are linked to mannose sugars by two molecules of *N*-acetylglucosamine. The significance of glycosylation is not well understood, and furthermore, it is not a universal feature of integral membrane proteins: some proteins, such as

A

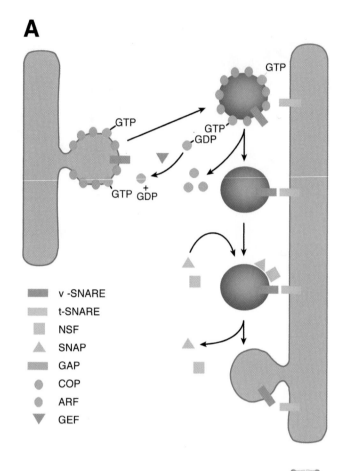

- v -SNARE
- t-SNARE
- NSF
- SNAP
- GAP
- COP
- ARF
- GEF

B

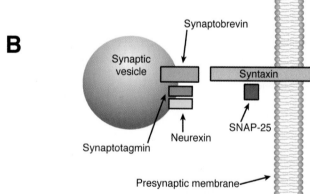

FIGURE 4.4 (A) General mechanisms of vesicle targeting and docking in the ER and Golgi. Assembly of coat proteins (COPs) around budding vesicles is driven by ADP-ribosylation factors (ARFs) in a GTP-dependent fashion. Dissociation of the coat is triggered by hydrolysis of GTP bound to ARF is stimulated by a GTPase-activating protein (GAP) in the Golgi membrane. The cycle of coat assembly and disassembly can continue when replacement of GDP on ARF by GTP is catalyzed by a guanine nucleotide exchange factor (GEF). Fusion of vesicles with target membrane in the Golgi is regulated by a series of proteins, N-ethyl-maleimide-sensitive factor (NSF), soluble NSF attachment proteins (SNAPs), and SNAP receptors (SNAREs), which together assist vesicle docking with target membrane. SNAREs on vesicles (v-SNAREs) are believed to associate with corresponding t-SNAREs on target membrane. (B) Mechanisms of vesicle targeting and docking in the synaptic terminal. The synaptic counterpart of v-SNARE is synaptobrevin (also known as VAMP), and syntaxin corresponds to t-SNARE. SNAP-25 is an accessory protein that binds to syntaxin. Synaptotagmin is believed to be the Ca^{2+} sensitive regulatory protein in the complex that binds to syntaxin. Neurexins appear to have a role in conferring Ca^{2+} sensitivity to these interactions.

ticularly important and diverse category of plasma membrane proteins in neurons and myelinating glial cells), however, many variations on this basic theme have been found. Simply stated:

1. Signal sequences for membrane insertions need not be only N-terminal; those that lie within a polypeptide sequence are not cleaved.
2. A second type of signal, a "halt" or "stop" transfer signal, functions to arrest translocation through the membrane bilayer. The halt transfer signal is also hydrophobic and usually is flanked by positive charges. This arrangement effectively stabilizes a polypeptide segment in the RER membrane bilayer.
3. The sequential display in tandem of insertion and halt transfer signals in a polypeptide as it is being synthesized ultimately determines its disposition with respect to the phospholipid bilayer, and thus its final topology in its target membrane. By synthesizing transmembrane polypeptides in this way, virtually any topology may be generated.

Newly Synthesized Polypeptides Exit from the RER and Are Moved Through the Golgi Apparatus

When the newly synthesized protein has established its correct transmembrane orientation in the RER, it is incorporated into vesicles and must pass through the Golgi complex before reaching the plasma membrane (Fig. 4.2). For membrane proteins, the Golgi serves two major functions: (1) it sorts and targets proteins and, (2) it performs further posttranslational modifications,

the proteolipid proteins of CNS myelin, neither lose their signal sequence nor become glycosylated.

In general, however, for the vast majority of polypeptides destined for release from the cell (secretory polypeptides), an N-terminal "signal sequence" first mediates the passage of the protein into the RER and is cleaved immediately from the polypeptide by a signal peptidase residing on the luminal side of the RER. For proteins destined to remain as permanent residents of cellular membranes (and these form a par-

particularly on the oligosaccharide chains that were added in the RER. Sorting takes place in the *cis*-Golgi network (CGN), and in the *trans*-Golgi network (TGN), whereas sculpting of oligosaccharides is primarily the responsibility of the *cis*-, *medial*-, and *trans*-Golgi stacks. The TGN is a tubulovesicular network wherein proteins are targeted to the plasma membrane or to organelles.

The CGN serves an important sorting function for proteins entering the Golgi from the RER. Because most proteins that move from the RER through the secretory pathway do so by default, any resident endoplasmic reticulum proteins must be restrained from exiting or returned promptly to the RER from the CGN should they escape. Although no retention signal has been demonstrated for the endoplasmic reticulum, two retrieval signals have been identified: a Lys-Asp-Glu-Leu or KDEL sequence in type I proteins and the Arg-Arg or RR motif in the first five amino acids of proteins with a type II orientation in the membrane. The KDEL tetrapeptide binds to a receptor called Erd 2 in the CGN, and the receptor–ligand complex is returned to the RER. There may also be a receptor for the N arginine dipeptide; alternatively, this sequence may interact with other components of the retrograde transport machinery, such as microtubules.

Movement of proteins between Golgi stacks proceeds by means of vesicular budding and fusion (Rothman and Wieland, 1996). The essential mechanisms for budding and fusion have been shown to require coat proteins (COPs) in a manner that is analogous to the role of clathrin in endocytosis. Currently, two main types of COP complex, COPI and COPII, have been distinguished. Although both have been shown to coat vesicles that bud from the endoplasmic reticulum, they may have different roles in membrane trafficking. Coat proteins provide the external framework into which a region of a flattened Golgi cisternae can bud and vesiculate. A complex of these COPs forms the coatamer (coat protomer) together with a p200 protein, AP-1 adaptins, and a family of GTP-binding proteins called ADP-ribosylation factors (ARFs). Immunolocalization of one of the coatamer proteins, β-COP, predominantly to the CGN and *cis*-Golgi indicates that these proteins may also take part in vesicle transport into the Golgi (Fig. 4.4). The function of ARF is to drive the assembly of the coatamer and therefore vesicle budding in a GTP-dependent fashion. Dissociation of the coat is triggered when hydrolysis of the GTP bound to ARF is stimulated by a GTPase-activating protein (GAP) in the Golgi membrane. The cycle of coat assembly and disassembly can continue when the replacement of GDP on ARF by GTP is catalyzed by a guanine nucleotide exchange factor (GEF).

Fusion of vesicles with their target membrane in the Golgi apparatus is believed to be regulated by a series of proteins, *N*-ethylmaleimide-sensitive factor (NSF), soluble NSF attachment proteins (SNAPs), and SNAP receptors (SNAREs), which together assist the vesicle in docking with its target membrane. The emerging view is that complementary SNARES on membranes destined to fuse (e.g., synaptic vesicles and the presynaptic membrane) are fundamentally responsible for driving membrane fusion. In addition, Rabs, a family of membrane-bound GTPases, act in concert with their own GAPs, GEFs, and a cytosolic protein that dissociates Rab–GDP from membranes after fusion called guanine-nucleotide dissociation inhibitor. Rabs are believed to regulate the action of SNAREs, the proteins directly engaged in membrane–membrane contact prior to fusion. The tight control necessary for this process and the importance of ensuring that vesicle fusion takes place only at the appropriate target membrane may explain why eukaryotic cells contain so many Rabs, some of which are known to take part specifically in the internalization of endocytic vesicles at the plasma membrane (Fig. 4.2).

Exocytosis of the neurotransmitter at the synapse must occur in an even more finely regulated manner than endocytosis. The proteins first identified in vesicular fusion events in the secretory pathway (namely NSF, SNAPs, and SNAREs or closely related homologues) appear to play a part in the fusion of synaptic vesicles with the active zones of the presynaptic neuronal membrane (Fig. 4.4) (Jahn and Scheller, 2006).

Originally a distinction was made between so-called v-SNARES and t-SNARES reflecting their different locations in the donor and acceptor compartments. An example of specifity is the fact that the synaptic counterpart of vSNARE is synaptobrevin (also known as vesicle-associated membrane protein (VAMP)), and syntaxin corresponds to t-SNARE. VAMP does not facilitate fusion with endocytotic vesicle compartments. SNAP-25 is an accessory protein that binds to syntaxin. In the constitutive pathway, such as between the RER and Golgi apparatus, assembly of the complex at the target membrane promotes fusion. However, at the presynaptic membrane, Ca^{2+} influx is required to stimulate membrane fusion. Synaptotagmin is believed to be the Ca^{2+}-sensitive regulatory protein in the complex that binds syntaxin. Neurexins appear to have a role in regulation as well, because, in addition to interacting with synaptotagmin, they are the targets of black widow spider venom (α)-latrotoxin, which deregulates the Ca^{2+}-dependent exocytosis of the neurotransmitter. However, a superficially disturbing lack of specificity in the ability of other membrane-bound SNARES to complex indicates that much remains to be learned about the regulation of SRAE-mediated membrane fusion.

When comparing secretion in slow-releasing cells, such as the pancreatic (β)-cell, and neurotransmitter release at the neuromuscular junction, two differences stand out. First, the speed of neurotransmitter release is much greater both in release from a single vesicle and in total release in response to a specific signal. Releasing the contents of a single synaptic vesicle at a mouse neuromuscular junction takes from 1 to 2 ms, and the response to an action potential involving the release of many synaptic vesicles is over in approximately 5 ms. In contrast, releasing the insulin in a single secretory granule by a pancreatic (β)-cell takes from 1 to 5 s, and the full release response may take from 1 to 5 min. A 103- to 105-fold difference in rate is an extraordinary range, making neurotransmitter release one of the fastest biological events routinely encountered, but this speed is critical for a properly functioning nervous system.

A second major difference between slow secretion and fast secretion is seen in the recycling of vesicles. In the pancreas, secretory vesicles carrying insulin are used only once, and so new secretory vesicles must be assembled *de novo* and released from the TGN to meet future requirements. In the neuron, the problem is that the synapse may be at a distance of 1 m or more from the protein synthetic machinery of the perikaryon, and so newly assembled vesicles even traveling at rapid axonal transport rates (see later) may take more than a day to arrive. Now, the number of synaptic vesicles released in 15 min of constant stimulation at a single frog neuromuscular junction has been calculated to be on the order of 105 vesicles, but a single terminal may have only a few hundred vesicles at any one time. These measurements would make no sense if synaptic vesicles had to be replaced constantly through new synthesis in the perikaryon, as is the case with insulin-carrying vesicles. The reason that these numbers are possible is that synaptic vesicles are taken up locally by endocytosis, refilled with neurotransmitter, and reutilized at a rate fast enough to keep up with normal physiological stimulation levels. This takes place within the presynaptic terminal, and evidence shows that these recycled synaptic vesicles are used preferentially. Such recycling does not require protein synthesis because the classical neurotransmitters are small molecules, such as acetylcholine, or amino acids, such as glutamate, that can be synthesized or obtained locally.

Significantly, neurons have fast and slow secretory pathways operating in parallel in the presynaptic terminal (Sudhof, 2004). Synapses that release classical neurotransmitters (acetylcholine, glutamate, etc.) with these fast kinetics also contain dense core granules containing neuropeptides (calcitonin generelated peptide, substance P, etc.) that are comparable to the secretory granules of the pancreatic (β)-cell. These are used only once because neuropeptides are produced from large polypeptide precursors that must be made by protein synthesis in the cell body. The release of neuropeptides is relatively slow; as is the case in endocrine release, neuropeptides serve primarily as modulators of synaptic function. The small clear synaptic vesicles containing the classic neurotransmitters can in fact be depleted pharmacologically from the presynaptic terminal, whereas the dense core granules remain. These observations indicate that even though fast and slow secretory mechanisms have many similarities and may even have common components, in neurons they can operate independent of one another.

Proteins Exit the Golgi Complex at the *trans*-Golgi Network

Most of the N-linked oligosaccharide chains acquired at the RER are remodeled in the Golgi cisternae, and while the proteins are in transit, another type of glycosyl linkage to serine or threonine residues through *N*-acetylgalactosamine can also be made. Modification of existing sugar chains by a series of glycosidases and the addition of further sugars by glycosyl transferases occur from the *cis* to the *trans* stacks. Some of these enzymes have been localized to particular cisternae. For example, the enzymes (β)-1,4-galactosyltransferase and (α)-2,6-sialyltransferase are concentrated in the *trans*-Golgi. How they are retained there is a matter of some debate. One idea is that these proteins are anchored by oligomerization. Another view is that the progressively rising concentration of cholesterol in membranes more distal to the ER in the secretory pathway increases membrane thickness, which in turn anchors certain proteins and causes an arrest in their flow along the default route.

The default or constitutive pathway seems to be the direct route to the plasma membrane taken by vesicles that bud from the TGN (Fig. 4.2). This is how, in general, integral plasma membrane proteins reach the cell surface. Proteins bound for regulated secretion or for transport to endosomes and from there to lysosomes are diverted from the default path by means of specific signals. It has been assumed that the sorting of proteins for their eventual destination takes place at the TGN itself. However, recent analyses of the three-dimensional structure of the TGN have provoked a revision of this view. These studies have shown that the TGN is tubular, with two major types of vesicles that bud from distinct populations of tubules. The

implication is that sorting may already have occurred in the *trans*-Golgi prior to the protein's arrival at the TGN. One population of vesicles consists of those surrounded by the familiar clathrin cage, which are destined for late endosomes. The other population appears to be coated in a lace-like structure, which may prove to be made from the elusive coat protein required for vesicular transport to the plasma membrane. The β-COP protein and related coatomer proteins active in more proximal regions of the secretory pathway are absent from the TGN.

Endocytosis and Membrane Cycling Occurs in the *trans*-Golgi Network

Two types of membrane invagination occur at the surface of mammalian cells and are clearly distinguishable by electron microscopy. The first type is a caveola, which has a thread-like structure on its surface made of the protein caveolin. Caveolae mediate the uptake of small molecules and may also have a role in concentrating proteins linked to the plasma membrane by the glycosylphosphatidylinositol anchor. Demonstration of the targeting of protein tyrosine kinases to caveolae by the tripeptide signal MGC (Met-Gly-Cys) also suggests that caveolae may function in signal transduction cascades.

The other type of endocytic vesicle at the cell surface is that coated with the distinctive meshwork of clathrin triskelions. The triskelion comprises three copies of a clathrin heavy chain and three copies of a clathrin light chain (Maxfield and McGraw, 2004). The ease with which these triskelions can assemble into a cage structure demonstrates how they promote the budding of a vesicle from a membrane invagination. Clathrin binds selectively to regions of the cytoplasmic surface of membranes that are selected by adaptins. The AP-2 complex, which is primarily active at the plasma membrane, consists of 100-kDa α and β subunits and two subunits of 50 and 17 kDa each. AP-1 complexes localize to the TGN and have γ and subunits of 100 kDa together with smaller polypeptides of 46 and 19 kDa. Adaptins bind to the cytoplasmic tails of membrane proteins, thus recruiting clathrin for budding at these sites.

A further component of the endocytic complex at the plasma membrane is the GTPase dynamin, which seems to be required for the normal budding of coated vesicles during endocytosis. Dynamins are a family of 100-kDa GTPases found in both neuronal and non-neuronal cells that may interact with the AP-2 component of a clathrin-coated pit (Murthy and De Camilli, 2003). Oligomers of dynamin form a ring at the neck of a budding clathrin-coated vesicle, and GTP hydrolysis appears to be necessary for the coated vesicle to pinch off from the plasma membrane. The existence of a specific neuronal form of dynamin (dynamin I) may be a manifestation of the unusually rapid rate of synaptic vesicle recycling.

The primary function of clathrin-coated vesicles at the plasma membrane is to deliver membrane proteins together with any ligands bound to them to the early endosomal apparatus. Regulation of membrane cycling in the endosomal compartment is likely to include the Rab family of small GTP-binding proteins. Indeed, each stage of the endocytic pathway may have its own Rab protein to ensure efficient targeting of the vesicle to the appropriate membrane. Rab6 is believed to have a role in transport from the TGN to endosomes, whereas Rab9 may regulate vesicular flow in the reverse direction. In neurons, Rab5a has a role in regulating the fusion of endocytic vesicles and early endosomes and appears to function in endocytosis from both somatodendritic domains and the axon. The association of the protein with synaptic vesicles in nerve terminals, attached presumably by means of its isoprenoid tail, also suggests that early endosomal compartments may have a role in the packaging and recycling of synaptic vesicles.

How Are Peripheral Membrane Proteins Targeted to Their Appropriate Destinations?

Peripheral membrane proteins are synthesized in the same type of free polysome in which the bulk of the cytosolic proteins are made. However, the cell must ensure that these membrane proteins are sent to the plasma membrane rather than allowed to attach in a haphazard way to other intracellular organelles. The fact that a complex machinery has evolved to ensure the correct delivery of integral membrane proteins suggests that some equivalent targeting mechanism must exist for proteins that attach to the cytoplasmic surface of the plasma membrane. Such proteins are translated on "free" polysomes, but these polysomes are associated with cytoskeletal structures and are not distributed uniformly throughout the cell body. In a number of cases, mRNAs that encode soluble cytosolic proteins are concentrated in discrete regions of the cell, resulting in a local accumulation of the translated protein close to the site of action. For some peripheral membrane proteins, this is the plasma membrane.

Evidence that this mechanism might operate in peripheral membrane protein synthesis came from studies showing biochemically and by *in situ* hybridization that mRNAs encoding the myelin basic proteins are concentrated in the myelinating processes that extend from the cell body of oligodendrocytes

(Colman *et al.*, 1982). As in oligodendrocytes, Schwann cells also transport MBP mRNA by microtubule-based transport, which also appears to require specialized cytoplasmic channels called Cajal bands (Court *et al.*, 2004).

Myelin basic protein may be a special case because of its very strong positive charge and consequent propensity for binding promiscuously to the negatively charged polar head groups of membrane lipids. Nevertheless, the fact that actin mRNAs are localized to the leading edge of cultured myocytes and mRNA for the microtubule-associated protein MAP2b is concentrated in the dendrites of neurons suggest that targeting by local synthesis is more common than originally thought. This mechanism is probably less important for peripheral membrane proteins that associate with the cytoplasmic surface of the plasma membrane by means of strong specific associations with proteins already located at the membrane because such proteins would act as specific receptors. Because only selected cytoplasmic mRNAs are localized to the periphery, the process is specific. However, no mRNAs are localized exclusively to the periphery, and a significant fraction typically is localized proximal to the nucleus in a region rich with the translational and protein-processing machinery of the cell (the Nissl substance or translational cytoplasm).

Summary

Membrane biogenesis and protein synthesis in neurons and glial cells are accomplished by the same mechanisms that have been worked out in great detail in other cell types. Integral membrane proteins are synthesized in the rough endoplasmic reticulum, and peripheral membrane proteins are products of cytoplasmic-free ribosomes that are found in the cell sap. For transmembrane proteins and secretory polypeptides, synthesis in the RER is followed by transport to the Golgi apparatus, where membranes and proteins are sorted and targeted for delivery to precise intracellular locations. It is likely that the neuron and glial cell have evolved additional highly specialized mechanisms for membrane and protein sorting and targeting because these cells are so greatly extended in space, although these additional mechanisms have yet to be fully described. The basic features of the process of secretion, which includes neurotransmitter delivery to presynaptic terminals, are beginning to be understood as well. The key features of this process are apparently common to all cells, including yeast, although the neuron has developed certain specializations and modifications of the secretory pathway that reflect its unique properties as an excitable cell.

CYTOSKELETONS OF NEURONS AND GLIAL CELLS

The cytoskeleton of eukaryotic cells is an aggregate structure formed by three classes of cytoplasmic structural proteins: microtubules (tubulins), microfilaments (actins), and intermediate filaments. Each of these elements exists concurrently and independently in overlapping cellular domains. Most cell types contain one or more examples of each class of cytoskeletal structure, but there are exceptions. For example, mature mammalian erythrocytes contain no microtubules or intermediate filaments, but they do have highly specialized actin cytoskeletons. Among cells of the nervous system, the oligodendrocyte is unusual in that it contains no cytoplasmic intermediate filaments. Typically, each cell type in the nervous system has a unique complement of cytoskeletal proteins that are important for the differentiated function of that cell type.

Although the three classes of cytoskeletal elements interact with each other and with other cellular structures, all three are dynamic structures rather than passive structural elements. Their aggregate properties form the basis of cell morphologies and plasticity in the nervous tissue. In many cases, the cytoskeleton is biochemically specialized for a particular cell type, function, and developmental stage. Each type of cytoskeletal element has unique functions essential for a functional nervous system.

Microtubules Are an Important Determinant of Cell Architecture

Microtubules are near ubiquitous cytoskeletal components in eukaryotes (Hyams and Lloyd, 1994). They play key roles in intracellular transport, are a primary determinant of cell morphology, form the structural correlate of the mitotic spindle, and are the functional core of cilia. Microtubules are very abundant in the nervous system, and tubulin subunits of microtubules may constitute more than 10% of total brain protein. As a result, many fundamental properties of microtubules were defined with microtubule protein from brain extracts. However, neuronal microtubules have biochemical specializations to meet the unique demands imposed by neuronal size and shape.

Intracellular transport and generation of cell morphologies are the most important roles played by microtubules in the nervous system. In part, this comes from their ability to organize cytoplasmic polarity. Microtubules *in vitro* are dynamic, polar structures with plus and minus ends that correspond to the fast- and slow-growing ends, respectively. In contrast, both

stable and labile microtubules can be identified *in vivo*, where they help define both microscopic and macroscopic aspects of intracellular organization in cells. Microtubule organization, stability, and composition are all highly regulated in the nervous system.

By electron microscopy, microtubules appear as hollow tubes 25 nm in diameter and can be hundreds of micrometers in length in axons. Microtubule walls typically comprise 13 protofilaments formed by a linear arrangement of globular subunits. Globular subunits in microtubule walls are heterodimers of α- and β-tubulin, with a variety of microtubule-associated proteins (MAPs) binding to microtubule surfaces.

Neuronal microtubules are remarkable for their genetic and biochemical diversity. Multiple genes exist for both α- and β-tubulins. These genes are expressed differentially according to cell type and developmental stage. Some genetic isotypes are expressed ubiquitously, whereas others are expressed only at specific times in development, in specific cell types, or both. Most tubulin genes are expressed in nervous tissue, and some are enriched or specific to neurons. Specific tubulin isotypes prepared in a pure form, vary in assembly kinetics and ability to bind ligands. However, when more than one isotype is expressed in a single cell, such as a neuron, they coassemble into microtubules with mixed composition.

The most common posttranslational modifications of tubulins are tyrosination–detyrosination, acetylation–deacetylation, and phosphorylation. The first two are intimately linked to assembled microtubules, but little is known about physiological functions for any tubulin modification. Most α-tubulin isotypes are synthesized with a Glu-Tyr dipeptide at the C terminus (Tyr-tubulin), but the tyrosine is removed by tubulin carboxypeptidase after incorporation into a microtubule, leaving a terminal glutamate (Glu-tubulin). Microtubules assembled for a longer time are enriched in Glu-tubulin, but when Glu-tubulin enriched microtubules are disassembled, liberated α-tubulins are rapidly retyrosinated by tubulin tyrosine ligase. The tyrosination state of α-tubulin does not affect assembly–disassembly kinetics *in vitro*, but detyrosination may affect interactions of microtubules with other cellular structures. Concurrent with detyrosination, α-tubulins can be subject to a specific acetylation. Tubulin acetylation was first described in flagellar tubulins, but this modification is widespread in neurons and many other cell types. Acetylase acts preferentially on α-tubulin in assembled microtubules, so long-lived or stable microtubules tend to be acetylated, but the distribution of microtubules rich in acetylated tubulin may not be identical to that of Glu-tubulin. Acetylated α-tubulin is rapidly deacetylated upon microtubule disassembly, but acetylation does not alter microtubule stability *in vitro*.

Tubulin phosphorylation involves β-tubulin and may be restricted to an isotype expressed preferentially in neurons and neuron-like cells. Various kinases can phosphorylate tubulin *in vitro*, but the endogenous kinase is unknown. Effects of phosphorylation on assembly are unknown, but phosphorylation is upregulated during neurite outgrowth. As with α-tubulin modifications, the physiological role of phosphorylation on neuronal β-tubulin has yet to be determined. Other posttranslational modifications have been reported, but their significance and distribution in the nervous system are not well documented.

The biochemical diversity of microtubules is increased through association of different MAPs with different populations of microtubules (Table 4.2). The significance of microtubule diversity is incompletely understood, but may include functional differences as well as variations in assembly and stability. In particular, MAP composition may define specific neuronal domains. For example, MAP-2 is restricted to dendritic regions of the neuron, whereas tau proteins are modified differentially in axons. Similarly, oligodendrocyte progenitors transiently express a novel MAP-2 isoform with an additional microtubule-binding repeat; that is, 4-repeat MAP-2c or MAP-2d. This MAP is in cell bodies but not in processes, suggesting that MAP-2d might have a role distinct from its capacity to bundle microtubules (Vouyiouklis and Brophy, 1995).

MAPs in nervous tissue fall into two heterogeneous groups: tau proteins and high molecular weight MAPs. Tau proteins have been of intense interest because posttranslationally modified tau proteins are the primary constituents of neurofibrillary tangles in the brains of Alzheimer patients. Tau proteins are primarily neuronal MAPs, although tau may be found outside neurons as well. Tau binds to microtubules during assembly–disassembly cycles with a constant stoichiometry and promotes microtubule assembly and stabilization. Tau exists in a number of molecular weight isoforms expressed differenctially in different regions of the nervous system and developmental stage. For example, tau proteins in the adult CNS are typically 60–75 kDa, whereas PNS axons contain a higher molecular mass tau of 100 kDa. Different isoforms of tau protein are generated from a single mRNA by alternative splicing, and additional heterogeneity is produced by phosphorylation.

High molecular weight MAPs are a diverse group of largely unrelated proteins found in various tissues, some of which are brain specific. All have molecular masses greater than 1300 kDa and form side arms protruding from microtubule surfaces. Many MAPs may

TABLE 4.2   Major Microtubule Proteins and Microtubule Motors in Mammalian Brain

	Location and Function
Tubulins	
α-and β-tubulins	Neurons, glia, and nonneuronal cells except mature mammalian erythrocytes. Multigene family with some genes expressed preferentially in brain, whereas others are ubiquitous. Primary structural polypeptides of microtubules.
γ-Tubulin	Present near microtubule-organizing center in all microtubule-containing cells. Needed for nucleation of microtubules.
Microtubule-Associated Proteins (MAPs)	
MAP-1a/1b	Widely expressed in neurons and glia, including both axons and dendrites; developmentally regulated phosphoproteins.
MAP-2a/2b MAP-2c	Dendrite-specific MAPs. The smaller MAP-2c is regulated developmentally, becoming restricted to spines in adults, whereas 2a and 2b are major phosphoproteins in adult brain.
LMW tau HMW tau	Tau proteins are enriched in axons with a distinctive phosphorylation pattern. A single tau gene is alternatively spliced to give multiple isoforms.
Microtubule Severing Proteins	
Katanin	Enriched at the microtubule organizing center and thought to be important in the release of microtubules for transport into axons and dendrites.
Motor Proteins	
Kinesins (kinesin-1s, kinesin-2s, kinesin-3s, and others)	Kinesin-1s are plus-end directed motors associated with membrane-bound organelles and moving them in fast axonal transport. The other members of the kinesin family are a diverse set of motor proteins with a kinesin-related motor domain and varied tails. Many are regulated developmentally and some are mitotic motors, restricted to dividing cells.
Axonemal dynein	A set of minus-end-directed microtubule motors associated with cilia and flagella, such as ependymal cells.
Cytoplasmic dynein	Cytoplasmic forms may be involved in the axonal transport of either organelles or cytoskeletal elements.

participate in microtubule assembly and cytoskeletal organization. Traditionally, high molecular weight MAPs comprise five polypeptides: MAPs 1a, 1b, 1c, 2a, and 2b. MAP-2 proteins are closely related and located primarily in dendrites. In contrast, the polypeptides known as MAP-1 are unique polypeptides with little sequence homology. MAPs 1a and 1b are expressed widely and regulated developmentally. MAPs 1a, 1b, and 2 are all thought to play important roles in stabilizing and organizing the microtubule cytoskeleton.

In most cell types, cytoplasmic microtubules are dynamic, although stable microtubule segments are found in all cells. In nonneuronal cells, such as astrocytes and other glia, microtubules typically are anchored in centrosomal regions that serve as microtubule-organizing centers. As a result, their cytoplasmic microtubules are oriented with plus ends at the cell periphery. The biochemistry of microtubule-organizing centers is not fully understood, but they contain a novel tubulin subunit, γ-tubulin, which functions as a microtubule nucleating protein. In contrast, dendritic and axonal microtubules of neurons are not continuous with a microtubule-organizing center, so alternate mechanisms must exist for their stabilization and organization. The situation is complicated further

because dendritic and axonal microtubules differ in both composition and organization. Both axonal and dendritic microtubules are nucleated at the microtubule-organizing center but are subsequently released for delivery to the appropriate compartment. The release of microtubules from the microtubule-organizing center appears to involve the microtubule severing protein, katanin (Baas, 2002). Surprisingly, axonal and dendritic compartments are not equivalent. First, dendritic and axonal MAPs differ in both identity and phosphorylation state. Second, microtubule orientation in axons has the plus end distal similar to other cell types, but microtubules in dendrites may exhibit both polarities. Finally, dendritic microtubules are less likely to be aligned with one another and are less regular in their spacing. As a result, dendritic diameters taper, whereas axons have a constant diameter as one proceeds away from the cell body.

Stabilization of axonal and dendritic microtubules is essential because of the volume of cytoplasm and the distance from sites of protein synthesis for tubulin. A common side effect of one class of antineoplastic drugs, the vinca alkaloids, underscores the importance of microtubule stability in axons. Vincristine and other vinca alkaloids act by destabilizing spindle microtubules, but dosage must be monitored carefully to prevent development of peripheral neuropathies due to loss of axonal microtubules. Microtubules play critical roles in both dendritic and axonal function, so mechanisms to ensure their proper extent and organization exist.

Axonal microtubules contain a particularly stable subset of microtubule segments resistant to depolymerization by antimitotic drugs, cold, and calcium. Stable microtubule segments are biochemically distinct and may constitute more than half of the axonal tubulin. Stable domains in microtubules may serve to regulate the axonal cytoskeleton by nucleating and organizing microtubules as well as stabilizing them. The biochemical basis of microtubule stability is not well understood but may include posttranslational modification of tubulins, presence of stabilizing proteins, or both. Relatively little is known about regulation of dendritic microtubules, but local synthesis of MAP-2 in dendrites may play a role.

Microfilaments and the Actin-Based Cytoskeleton Are Involved in Intracellular Transport and Cell Movement

The actin cytoskeleton is universal in eukaryotes, although microfilaments are most familiar as thin filaments in skeletal muscle. Microfilaments (Table 4.3) play critical roles in contractility for both muscle

TABLE 4.3 Selected Proteins of the Microfilament Cytoskeleton in Brain

Actins

 α-actin (smooth muscle)

 β-actin and γ-actin (neuronal and nonneuronal cells)

Actin monomer-binding proteins

 Profilin

 Thymosin 4 and 10

Capping proteins

 Ezrin/radixin/moesin

 Schwannomin/merlin

Gelsolin and other microfilament severing proteins

 Gelsolin

 Villin

Cross-linking and bundling proteins

 Spectrin (fodrin)

 Dystrophin, utrophin, and related proteins

 α-Actinin

Tropomyosin

Myosins I, II, II, V, VI, VII

and nonmuscle cells. Actin and its contractile partner myosin are particularly abundant in nervous tissue relative to other nonmuscle tissues. In fact, one of the earliest descriptions of nonmuscle actin and myosin was in brain. In neurons, microfilaments are most abundant in presynaptic terminals, dendritic spines, growth cones, and subplasmalemmal cortex. Although concentrated in these regions, microfilaments are present throughout the cytoplasm of neurons and glia as short filaments (4–6 nm in diameter and 400–800 nm long).

Multiple actin genes exist in both vertebrates and invertebrates. Four α-actin human genes have been cloned, each expressed specifically in a different muscle cell type (skeletal, cardiac, vascular smooth, and enteric smooth muscle). In addition, two nonmuscle actin genes (β- and γ-actin) are present in humans. β-actin and γ-actin genes are expressed ubiquitously and are abundant in nervous tissue. The functional significance of different genetic isotypes is not clear because actins are highly conserved. Across the range of known actin sequences, amino acids are identical at approximately two of three positions. Even the positions of introns within different actin genes are highly conserved across species and genes. Despite the high degree of conservation, differences in distribution of specific isotypes within a single neuron are seen. For example, β-actin may be enriched in growth cones. The prominent actin bundles seen in some nonneuronal cells in culture are not characteristic of neurons and most neuronal microfilaments are less than 1 μm in length.

Many microfilament-associated proteins are found in nervous tissue (myosin, tropomyosin, spectrin, α-actinin, etc.), but less is known about their distribution and normal function in neurons and glia. Myosins and myosin-associated proteins are considered in the section on molecular motors, but multiple categories of actin-binding proteins exist (Table 4.3). Monomer actin-binding proteins such as profilin and thymosins are abundant in the developing brain and are thought to help regulate assembly of microfilaments by sequestering actin monomers, which may be mobilized rapidly in response to appropriate signals. For example, phosphatidylinositol 4,5-bisphosphate causes the actin–profilin complex to dissociate, freeing monomer for explosive microfilament assembly. This may play a role in growth cone motility, where actin assembly is critical for filopodial extension.

Several proteins have been identified that cap microfilaments, serving to anchor them to other structures or regulate microfilament length. The ezrin–radixin–moesin gene family encodes barbed-end capping proteins that are concentrated at sites where the microfilaments meet the plasma membrane, suggesting a role in anchoring microfilaments or linking them to extracellular components through membrane proteins. They are prominent components of nodal and paranodal structures in nodes of Ranvier. A mutation in a member of this family expressed in Schwann cells, merlin or schwannomin, is responsible for the human disease neurofibromatosis type 2. Development of numerous tumors with a Schwann cell lineage in neurofibromatosis type 2 suggests that this microfilament-binding protein acts as a tumor suppressor.

Whereas some membrane proteins interact directly with microfilaments in the membrane cytoskeleton, others interact with the actin cytoskeleton through intermediaries. Proteins such as spectrin (fodrin), α-actinin, and dystrophins cross-link, or bundle, microfilaments, giving rise to higher order complexes. Spectrin is enriched in the cortical membrane cytoskeleton and is thought to have a role in localization of integral membrane proteins such as ion channels and receptors. Dystrophin is the best known member of a family of proteins that appear to be essential for clustering of receptors in muscle and nervous tissue. A mutation in dystrophin is responsible for Duchenne muscular dystrophy. Positioning of integral membrane proteins on the cell surface is an essential function of the actin-rich membrane cytoskeleton, acting in concert with a class of proteins that contain the protein-binding module, the PDZ domain.

Members of the gelsolin family have multiple activities. They not only cap the barbed end of a microfilament, but also sever microfilaments and can nucleate microfilament assembly. Severing-capping proteins may be critical for reorganizing the actin cytoskeleton. The Ca^{2+} dependence of gelsolin severing activity may provide a mechanism for altering the membrane cytoskeleton in response to Ca^{2+} transients. Other second messengers, such as phosphatidylinositol 4,5-bisphosphate, may also regulate gelsolin function, suggesting interplay between different classes of actin-binding proteins such as gelsolin and profilin. Oligodendrocytes are the only nonneural cells in the CNS that express significant amounts of the actin-binding and microfilament-severing protein gelsolin.

Proteins with other functions may interact directly with actin or actin microfilaments. For example, some membrane proteins, such as epidermal growth factor receptor, bind actin microfilaments directly, which may be important in anchoring these components at a particular location on the cell surface. Other cytoskeletal structures also interact with microfilaments. Both MAP-2 and tau microtubule-associated proteins can interact with microfilaments *in vitro* and may mediate interactions between microtubules and microfilaments. Finally, the synaptic vesicle-associated phosphoprotein, synapsin I, has a phosphorylation-sensitive inter-action with microfilaments that may be important for targeting and storage of synaptic vesicles in the presynaptic terminal (Murthy and De Camilli, 2003). Many of these interactions were defined by *in vitro* binding studies, and their physiological significance is not always established. The presence of actin as a major component of both pre- and postsynaptic specializations, as well as in growth cones, gives the actin cytoskeleton special significance in the nervous system (Murthy and De Camilli, 2003). The enrichment of the microfilament cytoskeleton at the plasma membrane makes them the cytoskeletal components most responsive to changes in the local external environment of the neuron. Microfilaments also play a critical role in positioning receptors and ion channels at specific locations on neuronal surfaces. Although we emphasize enrichment of the microfilament cytoskeleton at the plasma membrane, microfilaments are also abundant in the deep cytoplasm. The microfilaments are best regarded as a uniquely plastic component of the neuronal cytoskeleton that plays a critical role in local trafficking of cytoskeletal and membrane components.

Intermediate Filaments Are Prominent Constituents of Nervous Tissue

Intermediate filaments appear as solid, ropelike fibrils from 8 to 12 nm in diameter that may be many micrometers long (Lee and Cleveland, 1996). Intermediate filament proteins constitute a superfamily of five classes with expression patterns specific to cell type and developmental stage (Table 4.4). Type I and type II intermediate filament proteins are keratins, hallmarks of epithelial cells. Keratins are not associated with nervous tissue and will not be considered further. In contrast, all nucleated cells contain type V intermediate filament proteins, nuclear lamins. Lamins are the most evolutionarily divergent of intermediate filament genes, with regard to both intron/exon distribution and polypeptide domain structure. Cytoplasmic intermediate filaments in the nervous system are all either type III or type IV.

Type III intermediate filaments are a diverse family that includes vimentin (characteristic of fibroblasts and embryonic cells including embryonic neurons) and glial fibrillary acidic protein (GFAP, a marker for astrocytes and Schwann cells). Type III intermediate filament subunits are typically 45 to 60 kDa with a conserved rod domain and relatively small gene-specific amino- and carboxy-terminal sequences. As a result, type III intermediate filament subunits form smooth filaments without side arms. Type III polypeptides can form homopolymers or coassemble with other type III intermediate filament subunits.

TABLE 4.4   Intermediate Filament Proteins of the
Nervous System

Class and Name	Cell Type
Types I and II	
Acidic and basic keratins	Epithelial and endothelial cells
Type III	
Glial fibrillary acidic protein	Astrocytes and nonmyelinating Schwann cells
Vimentin	Neuroblasts, glioblasts, fibroblasts, etc.
Desmin	Smooth muscle
Peripherin	A subset of peripheral and central neurons
Type IV	
NF triplet (NFH, NFM, NFL)	Most neurons, expressed at highest level in large myelinated fibers
α-Internexin	Developing neurons, parallel fibers of cerebellum
Nestin	Early neuroectodermal cells
	The most divergent member of this class; some have classified it as a sixth type
Type V	
Nuclear lamins	Nuclear membranes

Type III intermediate filament proteins in the nervous system typically are restricted to glia or embryonic neurons. Vimentin is abundant in a many cells during early development, including both glioblasts and neuroblasts. Some Schwann cells and astrocytes also contain vimentin. Curiously, mature oligodendrocytes do not have intermediate filaments; an exception to the general rule that metazoan cells contain all three classes of cytoskeletal structures. Oligodendrocyte precursors, however, do express vimentin and may express GFAP transiently.

Peripherin is one type III intermediate filament protein unique to neurons. Peripherin has a characteristic expression during development and regeneration in specific neuronal populations and may be coexpressed with type IV neurofilament proteins. It can coassemble with type IV neurofilament subunits both *in vitro* and *in vivo*, where it can substitute for the low molecular weight neurofilament subunit (NFL). However, whether coassembly is generally the case is not known. Unlike type IV intermediate filaments, intermediate filaments made from type III subunits tend to disassemble more readily under physiological conditions. Thus, the presence of type III intermediate filament subunit proteins may produce more dynamic structures, which could be important during development or regeneration.

Neuronal intermediate filaments typically have side arms that limit packing density, whereas glial intermediate filaments lack side arms and may be very tightly packed. Neuronal intermediate filaments have an unusual degree of metabolic stability, which makes them well suited to the role of stabilizing and maintaining neuronal morphology. Due to this stability, the existence of neurofilaments was recognized long before much was known about their biochemistry or function. Neurofilaments were seen in early electron micrographs, and many traditional histological procedures to visualize neurons were based on a specific reaction of silver and other metals with neurofilaments.

Most neuronal intermediate filament have three distinct subunits present in varying stochiometries, all type IV polypeptides. Apparent molecular mass for neurofilament subunits vary widely across species, but mammalian forms are typically a triplet ranging from 180 to 200 kDa for the high molecular weight subunit (NFH), from 130 to 170 kDa for the medium subunit (NFM), and from 60 to 70 kDa for NFL. Neurofilament triplet proteins are each encoded by a separate type IV intermediate filament gene, which have a characteristic domain structure that can be recognized in both primary sequence and gene structure. Type IV genes typically are expressed only in neurons, although Schwann cells in damaged peripheral nerves may also transiently express NFM and NFL. Neurofilament polypeptides initially were identified from axonal transport studies. Neurofilament subunits are highly phosphorylated in axons, particularly NFM and NFH. In humans and some other species, NFH has more than 50 repeats of a consensus phosphorylation site at its carboxy terminus, and levels of NFH phosphorylation indicate that most are phosphorylated *in vivo*. This high level of phosphorylation in neurofilament tail domains is a distinctive characteristic of neurofilaments.

A second motif characteristic of neurofilaments is the presence of a glutamate-rich region in the tail adjacent to the core rod domain. This glutamate region has particular significance for neuroscientists because it appears to be the basis for reaction of the classic neurofibrillary silver stains for neurons. These stains were introduced in the late nineteenth century and used extensively by histologists and neuroanatomists from Ramon y Cajal's time to the present. The molecular basis of neurofibrillary stains was unknown until 1968, when F. O. Schmitt showed that neurofibrils were formed by neurofilaments. Remarkably, the ability of neurofilament subunits to react with silver histological stains is retained even after separation in gel electrophoresis for neurofilaments from organisms as diverse as human, squid, and the marine fanworm,

Myxicola. Conservation of this glutamate-rich domain suggests both an important functional role and early divergence of neurofilaments from the other intermediate filament families.

Neurofilaments and neurofilament triplet proteins play a critical role in determining axonal caliber. As noted earlier, neurofilaments have characteristic side arms, unique among intermediate filaments. Although all three subunits contribute to the neurofilament central core, side arms are formed only by carboxy-terminal regions of NFM and NFH. Phosphorylation of NFH and NFM side arms alters charge density on the neurofilament surface, repelling adjacent similarly charged neurofilaments. Although cross bridges between neurofilaments often are noted, direct studies of interactions between neurofilaments provide little evidence of stable crosslinks between neurofilaments or between neurofilaments and other cytoskeletal structures. The high density of surface charge due to phosphorylation of neurofilaments makes it difficult to imagine a stable interaction between neurofilaments and other structures of like charge. However, dynamic interactions between neurofilaments and cellular structures or proteins may be critical for neurofilament function and metabolism.

Altered expression levels of neurofilament subunits or mutations in neurofilament genes are associated with some neuropathologies. Disruption of neurofilament organization is a hallmark of pathology for many degenerative diseases of the nervous system, particularly those affecting large myelinated axons such as those of spinal motor neurons, such as amyotrophic lateral sclerosis. Overexpression of normal NFH or expression of some mutant NFL genes in transgenic mouse models leads to the accumulation of neurofilaments in the cell body and proximal axon of spinal motor neurons, similar to those seen in amyotrophic lateral sclerosis and related motor neuron diseases. Similarly, an early indicator of neuropathies due to neurotoxins such as acrylamide and hexanedione is accumulation of neurofilaments in either proximal or distal regions of axons. However, the question of whether neurofilament defects are a primary event in pathogenesis or reflect an underlying metabolic pathology remains unclear.

Another type IV intermediate filament gene expressed only in neurons is α-internexin. Unlike the triplet, α-internexin is expressed preferentially early in development and disappears from most neurons during maturation. Intermediate filament with α-internexin do persist in some adult neurons, such as the branched axons of granule cells in the cerebellar cortex. Although α-internexin can coassemble with neurofilament triplet subunits, it also forms homopol-

ymeric filaments. The primary sequence of α-internexin has features in common with NFL and NFM that are thought to confer assembly properties distinct from other type IV intermediate filaments.

The final intermediate protein expressed in the nervous system is nestin, which is seen transiently during early development. Nestin is expressed in neurons, Schwann cells and oligodendrocyte progenitors, which appear late in the development of the embryonic nervous system. Remarkably, nestin is expressed almost exclusively in ectodermal cells after commitment to the neuroglial lineage, but prior to terminal differentiation. At 1250 kDa, nestin is the largest intermediate filament subunit and the most divergent in sequence. Several distinctive features lead some to classify nestin as a sixth type of intermediate filament, whereas others group it with type IV genes. Relatively little is known about assembly properties of nestin *in vivo* or physiological functions of nestin filaments in neuroectodermal cells.

How Do the Various Cytoskeletal Systems Interact?

Each class of cytoskeletal structures may be found without the others in some cellular domains, but all three classes—microtubules, microfilaments, and intermediate filaments—coexist in many domains and inevitably interact. These interactions are typically dynamic, rather than through stable cross-links to one another. As mentioned earlier, microtubules and neurofilaments have highly phosphorylated side arms projecting from their surfaces. The high density of negative surface charge tends to repel structures with a like charge and rigidify microtubules and neurofilaments, affecting axon diameter.

The growth cone is a unique neuronal domain with distinctive cytoskeletal organization, such as longer microfilaments in filopodia and feurofilaments are excluded from growth cones, typically extending no further than the growth cone neck. In contrast, microtubules and microfilaments play complementary roles in growth cones. Microfilaments are critical in sprouting but less critical for elongation. In contrast, disrupting microtubules in distal neurites inhibits neurite elongation, but does not affect sprouting.

Summary

The intracellular framework giving shape to neurons and glia is the cytoskeleton, a complicated set of structures and their associated proteins. These organelles are also responsible for intracellular movement of materials and, during development, for cell migration and plasma membrane extension within nervous tissue.

MOLECULAR MOTORS IN THE NERVOUS SYSTEM

Until 1985, our knowledge of molecular motors in vertebrate cells of any type was restricted to myosins and flagellar dyneins. Myosins were identified in nervous tissue, but functions were uncertain. The preponderance of evidence indicated that fast axonal transport was microtubule based, so there was considerable interest in cytoplasmic dyneins, but initial studies failed to find a functional cytoplasmic dynein. However, a better understanding of molecular motors in the nervous system has now emerged, largely through studies on axonal transport (Brady and Sperry, 1995; Hirokawa and Takemura, 2005).

Myosins and dyneins can be distinguished pharmacologically by their differential susceptibility to inhibitors of ATPase activity, but the spectrum of inhibitors active against fast axonal transport fails to match properties of either myosin or dynein. The most striking difference between inhibitor effects on axonal transport and on myosin or dynein motors was seen with a nonhydrolyzable analog of ATP. Adenylylimidodiphosphate (AMP-PNP) is a weak competitive inhibitor of both myosin and dynein, requiring a 10- to 100-fold excess of analog. In contrast, both anterograde and retrograde axonal transport stop within minutes of AMP-PNP perfusion into isolated axoplasm, even in the presence of stoichiometric concentrations of ATP. Organelles moving in both directions freeze in place and remain attached to microtubules. AMP-PNP weakens interactions of myosin with microfilaments and of dynein with microtubules, but stabilizes binding of membrane-bound organelles to microtubules. Thus, effects of AMP-PNP indicated that axonal transport of membrane-bound organelles involved another type of motor, distinct from both myosins and dyneins.

The effects of AMP-PNP both demonstrated the existence of a new type of motor protein and provided a basis for identifying its constituent polypeptides. Binding of this ATPase to microtubules should be increased by AMP-PNP and decreased by ATP. Polypeptides meeting this criterion soon were identified and the new mechanochemical ATPase was named kinesin, based initially on an ability to move microtubules across glass coverslips as plus-end directed motor. Studies soon established that kinesin was a microtubule-activated ATPase with minimal basal activity. This combination of ATPase activity and motility *in vitro* confirmed it as the first member of a new class of microtubule-based motor, the kinesins (Brady and Sperry, 1995; Hirokawa and Takemura, 2005).

Kinesins are now known to comprise over 40 different genes in at least 14 subfamilies all with a highly conserved motor domain that includes ATP- and microtubule-binding domains, (Miki *et al.*, 2005). Multiple members of the kinesin superfamily are expressed in both adult and developing brains. Many kinesin family members are associated with mitosis, although some of these are also in postmitotic neurons. Members of kinesin families 1–6 are implicated in various neuronal functions ranging from transport of membrane-bounded organelles to translocation of microtubules in dendrites, and others may also have functions in the nervous system. This proliferation of motor proteins dramatically has altered the questions being asked about motor function in the brain. Most kinesins move toward the plus end of microtubules, but some move toward the minus end increasing the number of potential functions that kinesin family members might serve, including a role in transport of cytoskeletal structures. Studies continue to identify new functions for members of the kinesin superfamily expressed in neurons or glial cells.

Kinesin-1, the founding member of the kinesin superfamily, remains the most abundant class of kinesin expressed in brain and other tissues, leading to an extensive characterization of its biochemical, pharmacological, immunochemical, and molecular properties. Electron microscopic and biophysical analyses reveal kinesin as a long, rod-shaped protein, approximately 80 nm in length. Kinesin-1 is a heterotetramer with two heavy chains (115–130 kDa) and two light chains (62–70 kDa). Localization of antibodies specific for kinesin subunits by high resolution electron microscopy of brain kinesin indicates that two heavy chains arranged in parallel, forming the heads and much of the shaft, whereas light chains are localized to the fan-shaped tail region (Fig. 4.5).

ATP-binding and microtubule-binding domains of kinesin are in the heavy chain head regions. Axonal microtubules are oriented with plus ends distal from the cell body, so anterograde transport would require a motor that moves organelles toward the plus end direction. Three different genes for kinesin-1 heavy chain are expressed in neurons, one of which is neuron-specific, along with two light chain genes. Kinesin-1 appears associated with a variety of membrane-bound organelles, including synaptic vesicles or their precursors, mitochondria, and endosomes. Mechanisms for associating specific kinesin-1 isoforms with specific neuronal cargoes are incompletely understood. In the case of kinesin, the interaction is thought to involve both kinesin light chains and the carboxy termini of the heavy chains. Remarkably, mutations in the neuron-specific isoform of kinesin-1 can lead to a form

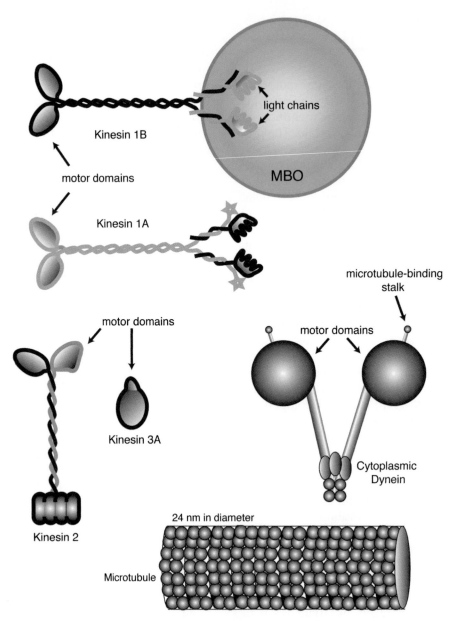

FIGURE 4.5 Examples of microtubule motor proteins in the mammalian nervous system. The first microtubule motor identified in nervous tissue was a kinesin 1, but showed 3 kinesin 1 genes including a neuron-specific form (kinesin 1A). Motor domains are well conserved by tail domains and appear to be specialized for interaction with various targets, such as different membrane-bound organelles. After the sequence of the kinesin heavy chain was established, the presence of additional genes that contained sequences homologous to the motor domain of kinesin was soon recognized. The molecular organization of these various motor proteins is diverse, including monomers (Kinesin 3A), trimers (Kinesin 2), and tetramers (ubiquitous and neuron-specific kinesins). Cytoplasmic dynein may interact with membrane-bound organelles and cytoskeletal structures. Genetic methods have established that there may be > 40 kinesin-related genes and 16 dynein heavy chains genes in a single organism.

of hereditary spastic paraplegia, an adult onset neurodegenerative disease of motor neurons.

An indirect result of the discovery of kinesin was the long sought cytoplasmic form of dynein, previously identified as a high molecular weight MAP in brain called MAP-1c,. Both cytoplasmic dynein and kinesin-1 can be isolated from brain by incubation of microtubules with nucleotide-free soluble extracts. Both are bound to microtubules under these conditions and released by ATP. MAP-1c dynein moved microtubules *in vitro* with a polarity opposite that seen with kinesin and was identified as a two-headed cytoplasmic dynein using both structural and biochemical criteria. Dynein heavy chains are also a gene family with 14 flagellar dyneins and two cytoplasmic dyneins (Pfister *et al.*, 2006). Some, but not all, functions of cytoplasmic dynein also involve another complex of polypeptides known as dynactin (Schroer, 2004).

Cytoplasmic dyneins are a 40-nm-long complex of molecular mass 1.6×10^6 Da, that include two heavy chains as well as multiple intermediate and light chains (Fig. 4.5). Nonneuronal cells show immunoreactivity for dynein on mitotic spindles and a punctate pattern of immunoreactivity present in interphase cells is thought to be dynein bound to membrane-bounded organelles. Dyneins are widely thought to be the motor for retrograde fast axonal transport, but are also implicated as motors for slow axonal transport (Baas and Buster, 2004). As with kinesin-1, partial loss of cytoplasmic dynein function leads to degeneration of motor neurons, reflecting the importance of dyneins in neuronal function (Levy and Holzbaur, 2006).

Myosins from muscle were the first molecular motors identified, but research in nonmuscle myosins has increased the number of myosins expressed in humans to some 40 different genes grouped in 18 different subfamilies (Berg *et al.*, 2001). As with kinesins, all myosins share considerable homology in their motor domains but diverge widely in other domains. Many nonmuscle myosins are expressed in neurons and glia (Brown and Bridgman, 2004). Myosins play critical roles in neuronal growth and development, as well as in specialized cells such as sensory hair cells of the cochlea and vestibular organs (Gillespie and Cyr, 2004).

The most familiar myosins are myosin II (Fig. 4.6), forming the thick filaments of smooth and skeletal muscle, but also present in nonmuscle cells. Myosin II heavy chains form a dimer that may interact with other myosin II dimers to form bipolar filaments. In tissue culture, many cells contain bundled actin microfilament stress fibers with a characteristic sarcomeric distribution of myosin II, but stress fibers are not apparent in neurons and glia *in situ*. However, bipolar thick filaments assembled from myosin II dimers can be isolated from nervous tissues. Although brain myosin II was one of the first nonmuscle myosins to be described, relatively little is known about myosin II function in neurons. Many cellular contractile events in nonneuronal cells, such as the contractile ring in mitosis, involve myosin II.

Myosin I proteins have a single, smaller heavy chain that does not form filaments but possesses a homologous actin-activated ATPase domain and has been purified from neural and neuroendocrine tissues (Fig. 4.6). Some myosin I motors have the ability to interact directly with membrane surfaces, which may generate

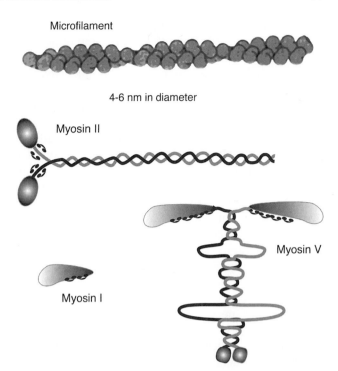

FIGURE 4.6 Examples of myosin motor proteins found in mammalian brain. Myosin heavy chains contain the motor domain, whereas light chains regulate motor function. Myosin II was the first molecular motor characterized biochemically from skeletal muscle and brain. Genetic approaches have now defined more than 15 classes of myosin, many of which are found in brain. Myosin II is a classic two-headed myosin forming thick filaments in nonmuscle cells. Myosin I motors have single motor domains, but may interact with actin microfilaments or membranes. Myosin V has multiple binding sites for calmodulin that act as light chains. Mutations in other classes of myosin have been linked to deafness. Myosins I, II, and V have been detected in growth cones as well as in mature neurons.

movements of plasma membrane components or intracellular organelles. Mammals have at least three myosin 1 genes and multiple forms are in brain. For example, myosin Ic is in cochlear and vestibular hair cell stereocilia and plays a key role in mechanotransduction (Gillespie and Cyr, 2004).

The mouse mutation *dilute*, which affects coat color, results from mutations in a myosin V gene, which is distinct from both myosins I and II (Fig. 4.6). Coat color changes in *dilute* mouse are due to ineffective pigment delivery to developing hairs by dendritic pigment cells, but *dilute* mutants also have complex neurological deficits, including seizures in early adulthood that may lead to death. The specific cellular localization and function of myosin V motors in neurons remain unclear. Genes for myosin VI and VIIA are implicated in some forms of congenital deafness, but are expressed in both brain and other tissues. Myosin VI is the gene responsible for Snell's Walzer deafness, and myosin

VIIA is associated with Usher syndrome type 1B, a human disease involving both deafness and blindness. Both of these myosins are expressed in cochlear and vestibular hair cells, but exhibit a different localization from each other and from myosin Ic.

The diversity of brain myosins and their distinctive localization suggests that the various myosins may have narrowly defined functions. However, relatively little is known about specific neuronal functions for most myosins despite intensive study of myosins in the nervous system. Myosins likely play roles in growth cone motility, synaptic plasticity, and even neurotransmitter release. The axonal transport of myosin II-like proteins was described, but relatively little progress has been made in defining functions of myosin II in the mature nervous system. Even less is known about myosin I in the nervous system beyond their role in hair cell function. There are few instances in our knowledge of neuronal function in which we fully understand the role played by specific molecular motors, but members of all three classes are abundant in nervous tissue. Proliferation of different motor molecules and isoforms suggests that some physiological activities may require multiple classes of motor molecules.

Summary

The concept is now firmly in place that neurons and glial cells, like other cells, contain multiple molecular motors responsible for moving discrete populations of molecules, particles, and organelles through intracellular compartments. The complex morphologies and diverse functional interactions of neurons mean that motor proteins and their regulation play a critical role in the nervous system.

BUILDING AND MAINTAINING NERVOUS SYSTEM CELLS

The functional architecture of neurons comprises many specializations in cytoskeletal and membranous components. Each of these specializations is dynamic, constantly changing, and being renewed at a rate determined by the local environment and cellular metabolism. Axonal transport processes represent a key to understanding neuronal dynamics and provide a basis for exploring neuronal development, regeneration, and neuropathology. Recent advances provide insight into the molecular mechanisms underlying axonal transport and its role in both normal neuronal function and pathology.

Slow Axonal Transport Moves Soluble Proteins and Cytoskeletal Structures

Slow axonal transport has two major components, both representing movement of cytoplasmic constituents (Fig. 4.7). Cytoplasmic elements in axonal transport move at rates comparable to the rate of neurite elongation. Slow component a (SCa) is movement of cytoskeletal elements, primarily neurofilaments and microtubules, SCa rates typically range from 0.1 to 1 mm/day and newly synthesized cytoskeletal proteins may take more than 1000 days to reach the end of a meter-long axon. Slow component b (SCb) is a complex and heterogeneous rate component, including hundreds of distinct polypeptides from cytoskeletal proteins such as actin (and sometimes tubulin) to soluble enzymes of intermediary metabolism (i.e., glycolytic enzymes). SCb moves at 2 to 4 mm/day and is the rate-limiting component for nerve growth or regeneration.

The coordinated movement of neurofilament and microtubule proteins provided strong evidence for the "structural hypothesis." For example, in pulse-labeling experiments labeled neurofilament proteins move as a bell-shaped wave with little or no trailing of neurofilament protein (Baas and Buster, 2004). Neurofilament stability under physiological conditions indicates that soluble neurofilament subunit pools are negligible, so coherent transport of neurofilament triplet proteins implied a transport complex—neurofilaments. Similarly, coordinate transport of tubulin and MAPs made sense only if microtubules move, because MAPs do not interact with unpolymerized tubulin. The simplest explanation is that neurofilaments and microtubules move as discrete cytological structures, but this idea was controversial for many years.

Development of fluorescently tagged neurofilament or microtubule subunits and methods for visualizing these structures in living cells resolved this issue by documenting movements of individual microtubules and neurofilaments neurites of cultured neurons (Brown, 2003). Direct observations of individual microtubule or neurofilament segments indicated that they move down axons as assembled polymers. Video images of fluorescently tagged microtubules or neurofilaments reveal discontinuous movements, with long pauses punctuated by brief, rapid translocations at 1 to 2 µm/sec. Due to long pauses, average rates are two to three orders of magnitude slower than instantaneous velocities (Baas and Buster, 2004; Brown, 2003). Remarkably, dynein plays a major role in slow axonal transport of microtubules and neurofilaments (He *et al.*, 2005).

SLOW AXONAL TRANSPORT

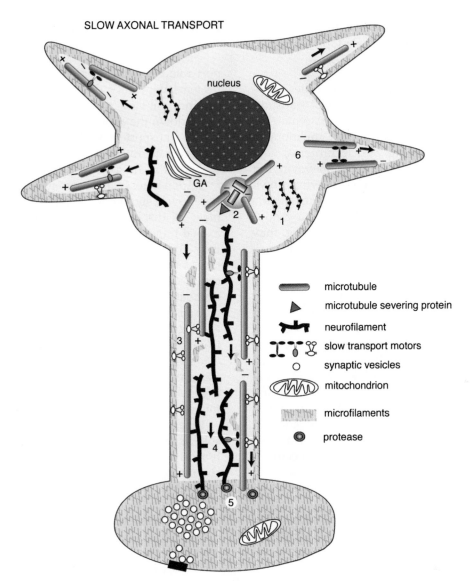

FIGURE 4.7 Slow axonal transport represents the delivery of cytoskeletal and cytoplasmic constituents to the periphery. Cytoplasmic proteins are synthesized on free polysomes and organized for transport as cytoskeletal elements or macromolecular complexes (1). Microtubules are formed by nucleation at the microtubule-organizing center near the centriolar complex (2) and then released for migration into axons or dendrites. Slow transport appears to be unidirectional with no net retrograde component. Studies suggest that cytoplasmic dynein may move microtubules with their plus ends leading (3). Neurofilaments may move on their own or may hitchhike on microtubules (4). Once cytoplasmic structures reach their destinations, they are degraded by local proteases (5) at a rate that allows either growth (in the case of growth cones) or maintenance of steady-state levels. The different composition and organization of cytoplasmic elements in dendrites suggest that different pathways may be involved in delivery of cytoskeletal and cytoplasmic materials to dendrites (6). In addition, some mRNAs are transported into dendrites, but not into axons.

Studies on transport of neurofilament proteins indicated that little or no degradation occurs until neurofilaments reach nerve terminals, where they are degraded rapidly. Comparable results were obtained with microtubule proteins. Differential metabolism appears to be a key to targeting of cytoplasmic and cytoskeletal proteins. Proteins with slow degradative rates accumulate, reaching higher steady-state levels. Altering degradation rates changes that steady-state concentration, so enrichment of actin in presynaptic terminals is due to slower turnover of actin than neurofilaments and tubulin. As a result, inhibiting calpain causes neurofilament accumulation in terminals. Differential turnover may involve specific proteases or posttranslational modifications that affect susceptibility to degradation. Regardless, cytoplasmic proteins are degraded in the distal axon and do not return in retrograde axonal transport.

Fast Axonal Transport Is the Rapid Movement of Membrane Vesicles and Their Contents Over Long Distances within a Neuron

Early biochemical and morphological studies established that material moving in fast axonal transport was associated with membrane-bound organelles (Fig. 4.8) (Brady, 1995). Mitochondria, membrane-associated receptors, synaptic vesicle proteins, neurotransmitters, and neuropeptides all move in fast anterograde transport. Many cargoes moving down axons in anterograde transport return by retrograde transport (Kristensson, 1987). In addition, exogenous materials taken up in distal regions of axons may be moved back to the cell body by retrograde transport (Fig. 4.8). Exogenous materials in retrograde transport include neurotrophins, such as nerve growth factor, and viral particles invading the nervous system.

Electron microscopic analysis of materials accumulated at a ligation or crush demonstrated that organelles moving in the anterograde direction were morphologically distinct from those moving in the retrograde direction (Tsukita and Ishikawa, 1980). Consistent with ultrastructural differences, radiolabel and immunocytochemical studies indicate quantitative and qualitative differences between anterograde and retrograde moving material. These differences indicate that processing or repackaging events must occur for turnaround in axonal transport. Both proteases and kinases may play a role in turnaround processing for retrograde transport.

Biochemical and morphological approaches resulted in a detailed description of materials moved in fast axonal transport but were not suitable for identifying molecular motors for axonal transport. Methods that permitted direct observation of organelle movements and precise control of experimental conditions were required. Development of video-enhanced contrast (VEC) microscopy allowed characterization of bidirectional movement of membrane-bounded organelles in giant axons from the squid *Loligo pealeii* (Brady, 1995). Years before, studies showed that axoplasm could be extruded from the giant axon as an intact cylinder. VEC microscopic analysis of axoplasm revealed that fast axonal transport continued unabated in isolated axoplasm for hours despite lacking a plasma membrane or other permeability barriers. Combining VEC microscopy with isolated axoplasm, complemented by biochemical and pharmacological approaches, permitted rigorous dissection of mechanisms for fast axonal transport and led to the discovery of kinesin molecular motors (Brady and Sperry, 1995) as well as allowing characterization of regulatory mechanisms associated with fast axonal transport.

How Is Axonal Transport Regulated?

The diversity of polypeptides in each axonal transport rate component and the coherent movement of proteins with very different molecular weights is a conundrum: How can so many different polypeptides move down the axon as a group? Rate components of axonal transport move as discrete waves, each with a characteristic rate and a distinctive composition (Figs. 4.7 and 4.8). The structural hypothesis was formulated in response to such observations. The hypothesis is deceptively simple: Axonal transport represents movement of discrete cytological structures. Proteins in axonal transport do not move as individual polypeptides. Instead, they move as part of a cytological structure or in association with a cytological structure. The only assumption made is that a limited number of elements can interact directly with transport motors so transported material must be packaged appropriately to be moved. Different rate components result from packaging of transported material into distinct cytological structures. In other words, membrane-associated proteins move as membrane-bounded organelles (vesicles, etc.), whereas tubulins move as microtubules.

Kinesin-1 isoforms appear to be the major (but not sole) motors for fast anterograde movement of membrane-bounded organelles such as vesicles and mitochondria. Similarly, cytoplasmic dynein appears to be the motor for fast retrograde transport of membrane-bounded structures. However, cytoplasmic dynein is also the involved in the anterograde transport of microtubules in slow axonal transport. Regulation of motor proteins is needed to assure that appropriate levels of axonal and synaptic components are delivered where needed in the neuron.

Because synthesis of proteins occurs at some distance from many functional domains of a neuron, transport to distal regions of a neuron is necessary, but not sufficient, for proper function. Specific materials must also be delivered to proper sites of utilization and not left in inappropriate locations. For example, synaptophysin has no known function in axons or cell body, so it must be delivered to a presynaptic terminal along with other components necessary for regulated neurotransmitter release. The traditional picture places the presynaptic terminal at the axon end. Such images imply that synaptic vesicles need only move along axonal microtubules until reaching their ends in the presynaptic terminal. However, many CNS synapses are not at axon ends. Many terminals may be located sequentially along a single axon, making en passant contacts with multiple target cells. Targeting of synaptic vesicles then becomes a more complex problem and

FAST AXONAL TRANSPORT

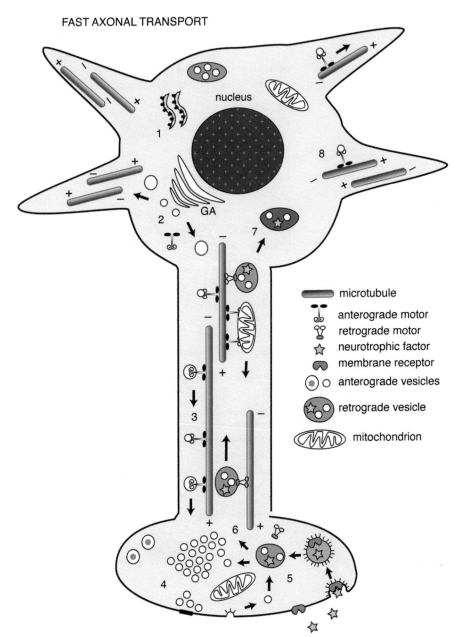

FIGURE 4.8 Fast axonal transport represents transport of membrane-associated materials, having both anterograde and retrograde components. For anterograde transport, most polypeptides are synthesized on membrane-bound polysomes, also known as rough endoplasmic reticulum (1), and then transferred to the Golgi for processing and packaging into specific classes of membrane-bound organelles (2). Proteins following this pathway include both integral membrane proteins and secretory polypeptides in the vesicle lumen. Cytoplasmic peripheral membrane proteins such as kinesins are synthesized on free polysomes. Once vesicles are assembled and appropriate motors associate with them, they move down the axon at a rate of 100–400 mm per day (3). Different membrane structures are delivered to different compartments and may be regulated independently. For example, dense core vesicles and synaptic vesicles are both targeted for presynaptic terminals (4), but release of vesicle contents involves distinct pathways. After vesicles merge with the plasma membrane, their protein constituents are taken up in coated vesicles via the receptor-mediated endocytic pathway and delivered to a sorting compartment (5). After proper sorting into appropriate compartments, membrane proteins either are committed to retrograde axonal transport or recycled (6). Retrograde moving organelles are morphologically and biochemically distinct from anterograde vesicles. These larger vesicles have an average velocity about half that of anterograde transport. The retrograde pathway is an important mechanism for delivery of neurotrophic factors to the cell body. Material delivered by retrograde transport typically fuses with cell body compartments to form mature lysosomes (7), where constituents are recycled or degraded. However, neurotrophic factors and neurotrophic viruses act at the level of the cell body and escape this pathway. Vesicle transport also occurs into dendrites (8); less is known about this process.

targeting ion channels to nodes of Ranvier or other appropriate sites on the neuronal surface is equally challenging.

Although specific details of targeting are not well understood, a simple model for targeting of synaptic vesicle precursors or ion channels serves to illustrate how such targeting may occur (Fig. 4.9). A local change in the balance between kinase and phosphatase activity in a subdomain like a node of Ranvier or presynaptic terminal can lead to phosphorylation of the motor

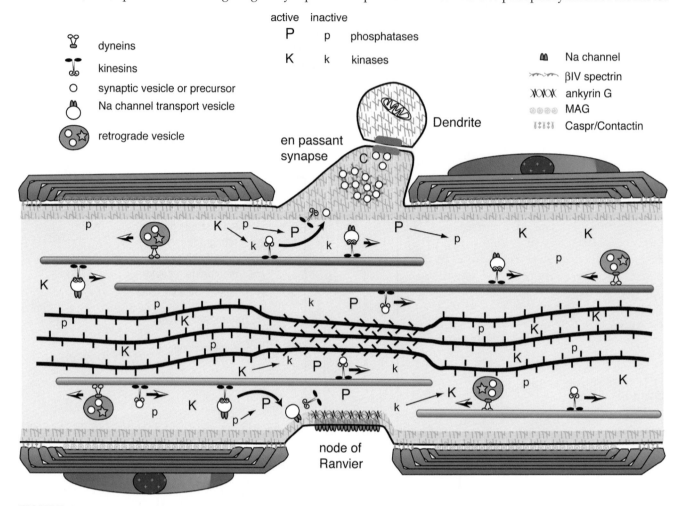

FIGURE 4.9 Axonal dynamics in a myelinated axon from the peripheral nervous system (PNS). Axons are in a constant flux with many concurrent dynamic processes. This diagram illustrates a few of the many dynamic events occurring at a node of Ranvier in a myelinated axon from the PNS. Axonal transport moves cytoskeletal structures, cytoplasmic proteins, and membrane-bound organelles from the cell body toward the periphery (from right to left). At the same time, other vesicles return to the cell body by retrograde transport (retrograde vesicle). Membrane-bound organelles are moved along microtubules by motor proteins such as the kinesins and cytoplasmic dyneins. Each class of organelles must be directed to the correct functional domain of the neuron. Synaptic vesicles must be delivered to a presynaptic terminal to maintain synaptic transmission. In contrast, organelles containing sodium channels must be targeted specifically to nodes of Ranvier for salta-tory conduction to occur. Cytoskeletal transport is illustrated by microtubules (rods in the upper half of the axon) and neurofilaments (bundle of rope-like rods in the lower half of the axon) representing the cytoskeleton. They move in the anterograde direction as discrete elements and are degraded in the distal regions. Microtubules and neurofilaments interact with each other transiently during transport, but their distribution in axonal cross-sections suggests that they are not stably cross-linked. In axonal segments without compact myelin, such as the node of Ranvier or following focal demyelination, a net dephosphorylation of neurofilament side arms allows the neurofilaments to pack more densely. Myelina-tion is thought to alter the balance between kinase (K indicates an active kinase; k is an inactive kinase) and phosphatase (P indicates an active phophatase; p is an inactive phosphatase) activity in the axon. Most kinases and phosphatases have multiple substrates, suggesting a mecha-nism for targeting vesicle proteins to specific axonal domains. Local changes in the phosphoryation of axonal proteins may alter the binding properties of proteins. The action of synapsin I in squid axoplasm suggests that dephosphorylated synapsin cross-links synaptic vesicles to microfilaments. When a synaptic vesicle encounters the dephosphorylated synapsin and actin-rich matrix of a presynaptic terminal, the vesicle is trapped at the terminal by inhibition of further axonal transport, effectively targeting the synaptic vesicle to a presynaptic terminal. Similarly, a sodium channel-binding protein may be present at nodes of Ranvier in a high-affinity state (i.e., dephosphorylated). Transport vesicles for nodal sodium channels (Na channel vesicle) would be captured upon encountering this domain, effectively targeting sodium channels to the nodal membrane. Interactions between cells could in this manner establish the functional architecture of the neuron.

protein on a vesicle carrying a cargo targeted to that domain. Thus, phosphorylation of kinesin-1 carrying Na channels at the node of Ranvier would allow delivery of Na channels to the nodal membrane. Evidence for such a mechanism exists for delivery of membrane proteins to growth cones (Morfini *et al.*, 2002) and other domains. A number of kinases have been identified that can regulate kinesin and/or dynein function. Significantly, many of these kinases are misregulated in neurodegenerative diseases such as Alzheimer's, Parkinson's, and Huntington's disease, raising the possibility that axonal transport is disrupted in these diseases.

Although such models are speculative, they satisfy criteria that any mechanism for targeting to specific neuronal subdomains must address. Specifically, mechanisms must act locally because distances to cell body can be great and the number of targets is large. There must be some means to connect a targeting signal to the external microenvironment, such as a glial or muscle cell. Finally, there must be a way of distinguishing subdomains. Thus, synaptic vesicles will not be delivered to nodes of Ranvier and voltage-gated sodium channels for nodes are not targeted to presynaptic terminals. Careful segregation of different organelles and proteins to different domain of a neuron suggests that highly efficient targeting mechanisms do exist.

Summary

A well-studied feature of the neuron is the phenomenon of axonal transport, which moves in both anterograde and retrograde directions. Axonal transport is responsible for delivery of both membrane-associated and cytoplasmic materials from the cell body to distant parts of the neuron, membrane retrieval and circulation, and uptake of materials from presynaptic terminals and dendrites as well as their delivery to the cell soma. Precise molecular mechanisms by which anterograde and retrograde transport are targeted within an individual dendrite or axon are still being defined.

Neurons and glial cells have unusually large cell volumes enclosed within extensive plasma membrane surfaces. Nature has evolved a number of "universal" mechanisms in other systems and adapted them for the special needs of nervous tissue cells. The synthesis and packaging of components, and in particular proteins, destined for cytoplasmic organelles and cell surface subdomains engage general and evolutionarily conserved molecular mechanisms and pathways that are employed in single-cell yeasts as well as in cells in complex nervous tissue.

Once synthesized and sorted, most intracellular organelles (vesicles destined for axonal or dendritic domains, mitochondria, cytoskeletal components) must be distributed, and targeted, to precise intracellular locations. Because they are so extended in space and exhibit exceptionally complex functional architecture, neurons and glial cells have adapted and developed to a high degree mechanisms that operate to distribute components within all cells. In neurons, movement of materials within the axon has been the central focus of most studies. The motors, cargoes, and regulation of axonal transport is now understood in some measure at the molecular level.

References

Baas, P. W. (2002). Microtubule transport in the axon. *Int Rev Cytol* **212**, 41–62.

Baas, P. W. and Buster, D. W. (2004). Slow axonal transport and the genesis of neuronal morphology. *J Neurobiol* **58**, 3–17.

Berg, J. S., Powell, B. C., and Cheney, R. E. (2001). A millennial myosin census. *Mol Biol Cell* **12**, 780–794.

Brady, S. T. (1995). A kinesin medley: Biochemical and functional heterogeneity. *Trends in Cell Biol* **5**, 159–164.

Brady, S. T. and Sperry, A. O. (1995). Biochemical and functional diversity of microtubule motors in the nervous system. *Curr Op Neurobiol* **5**, 551–558.

Brown, A. (2003). Live-cell imaging of slow axonal transport in cultured neurons. *Methods Cell Biol* **71**, 305–323.

Brown, M. E. and Bridgman, P. C. (2004). Myosin function in nervous and sensory systems. *J Neurobiol* **58**, 118–130.

Colman, D. R., Kreibich, G., Frey, A. B., and Sabatini, D. D. (1982). Synthesis and incorporation of myelin polypeptides into CNS myelin. *J Cell Biol* **95**, 598–608.

Court, F. A., Sherman, D. L., Pratt, T., Garry, E. M., Ribchester, R. R., Cottrell, D. F., Fleetwood-Walker, S. M., and Brophy, P. J. (2004). Restricted growth of Schwann cells lacking Cajal bands slows conduction in myelinated nerves. *Nature* **431**, 191–195.

Gillespie, P. G. and Cyr, J. L. (2004). Myosin-1c, the hair cell's adaptation motor. *Annu Rev Physiol* **66**, 521–545.

He, Y., Francis, F., Myers, K. A., Yu, W., Black, M. M., and Baas, P. W. (2005). Role of cytoplasmic dynein in the axonal transport of microtubules and neurofilaments. *J Cell Biol* **168**, 697–703.

Hirokawa, N. and Takemura, R. (2005). Molecular motors and mechanisms of directional transport in neurons. *Nat Rev Neurosci* **6**, 201–214.

Jahn, R. and Scheller, R. H. (2006). SNAREs—Engines for membrane fusion. *Nat Rev Mol Cell Biol* **7**, 631–643.

Kornfeld, R. and Kornfeld, S. (1985). Assembly of asparagine-linked oligosaccharides. *Annu Rev Biochem* **54**, 631–664.

Kristensson, K. (1987). Retrograde transport of macromolecules in axons. *Annu Rev Pharmacol Toxicol* **18**, 97–110.

Levy, J. R. and Holzbaur, E. L. (2006). Cytoplasmic dynein/dynactin function and dysfunction in motor neurons. *Int J Dev Neurosci* **24**, 103–111.

Maxfield, F. R. and McGraw, T. E. (2004). Endocytic recycling. *Nat Rev Mol Cell Biol* **5**, 121–132.

Miki, H., Okada, Y., and Hirokawa, N. (2005). Analysis of the kinesin superfamily: Insights into structure and function. *Trends Cell Biol* **15**, 467–476.

Morfini, G., Szebenyi, G., Elluru, R., Ratner, N., and Brady, S. T. (2002). Glycogen Synthase Kinase 3 Phosphorylates Kinesin Light Chains and Negatively Regulates Kinesin-based Motility. *EMBO Journal* **23**, 281–293.

Murthy, V. N. and De Camilli, P. (2003). Cell biology of the presynaptic terminal. *Annu Rev Neurosci* **26**, 701–728.

Palade, G. (1975). Intracellular aspects of the process of protein synthesis. *Science* **189**, 347–358.

Peters, A., Palay, S. L., and Webster, H. D. (1991). The Fine Structure of the Nervous System: Neurons and their supporting cells, 3rd ed. New York, NY, Oxford University Press.

Pfister, K. K., Shah, P. R., Hummerich, H., Russ, A., Cotton, J., Annuar, A. A., King, S. M., and Fisher, E. M. (2006). Genetic Analysis of the Cytoplasmic Dynein Subunit Families. *PLoS Genet* **2**, e1.

Rothman, J. E. and Wieland, F. T. (1996). Protein sorting by transport vesicles. *Science* **272**, 227–234.

Schroer, T. A. (2004). Dynactin. *Annu Rev Cell Dev Biol* **20**, 759–779.

Sherman, D. L. and Brophy, P. J. (2005). Mechanisms of axon ensheathment and myelin growth. *Nat Rev Neurosci* **6**, 683–690.

Sudhof, T. C. (2004). The synaptic vesicle cycle. Annu Rev Neurosci **27**, 509–547.

Tsukita, S. and Ishikawa, H. (1980). The movement of membranous organelles in axons. Electron microscopic identification of anterogradely and retrogradely transported organelles. *J Cell Biol* **84**, 513–530.

Walter, P. and Johnson, A. E. (1994). Signal sequence recognition and protein targeting to the endoplasmic reticulum membrane. *Annu Rev Cell Biol* **10**, 87–119.

*Scott T. Brady, David R. Colman,
and Peter J. Brophy*

Electrotonic Properties of Axons and Dendrites

The functional operations of neurons are the neural basis of behavior. In order to understand those operations, we need to understand how the different parts of the neuron interact. In this chapter we begin by considering how electrical current spreads.

Neurons characteristically have elaborate dendritic trees arising from their cell bodies and single axons with their own terminal branching patterns (see Chapters 3 and 4). With this structural apparatus, neurons carry out five basic functions (Fig. 5.1):

1. Generate intrinsic activity (at any given site in the neuron through voltage-dependent membrane properties and internal second-messenger mechanisms).
2. Receive synaptic inputs (mostly in dendrites, to some extent in cell bodies, and in some cases in axon hillocks, initial axon segments, and axon terminals).
3. Integrate signals by combining synaptic responses with intrinsic membrane activity (in dendrites, cell bodies, axon hillocks, and initial axon segments).
4. Encode output patterns in graded potentials or action potentials (at any given site in the neuron).
5. Distribute synaptic outputs (from axon terminals and, in some cases, from cell bodies and dendrites).

In addition to synaptic inputs and outputs, neurons may receive and send nonsynaptic signals in the form of electric fields, volume conduction through the extracellular environment of neurotransmitters and gases, and release of hormones into the bloodstream.

TOWARD A THEORY OF NEURONAL INFORMATION PROCESSING

A fundamental goal of neuroscience is to develop quantitative descriptions of these functional operations and their coordination within the neuron that enable the neuron to function as an integrated information processing system. This is the necessary basis for testing experiment-driven hypotheses that can lead to realistic empirical computational models of neurons, neural systems, and networks and their roles in information processing and behavior.

Toward these ends, the first task is to understand how activity spreads. To do this for a single process, such as the axon, is difficult enough; for the branching dendrites it becomes extremely challenging; and for the interactions between the two even more so. It is no exaggeration to say that the task of understanding how intrinsic activity, synaptic potentials, and active potentials spread through and are integrated within the complex geometry of dendritic trees to produce the input–output operations of the neuron is one of the main frontiers of molecular and cellular neuroscience.

This chapter begins with the passive properties of the membrane underlying the spread of most types of neuronal activity. Chapter 12 then considers the active membrane properties that contribute to more complex types of information processing, particularly the types that take place in dendrites. Together, the two chapters provide an integrated theoretical framework for understanding the neuron as a complex information processing system. Both draw on other chapters for the specific properties—membrane receptors (Chapter 9), internal

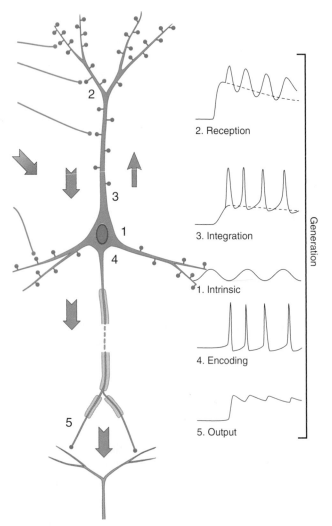

2. Reception

3. Integration

1. Intrinsic

4. Encoding

5. Output

Generation

FIGURE 5.1 Nerve cells have four main regions and five main functions. Electrotonic potential spread is fundamental for coordinating the regions and their functions.

of current spread in nerve cells and muscle with the development of cable theory for long distance transmission of electric current through cables on the ocean floor. The electrotonic properties of neurons therefore often are referred to as *cable properties*. Electrotonic theory was first applied mathematically to the nervous system in the late nineteenth century for spread of electric current through nerve fibers. By the 1930s and 1940s, it was applied to simple invertebrate (crab and squid) axons—the first steps toward the development of the Hodgkin–Huxley equations (Chapter 6) for the action potential in the axon.

Mathematically it is impractical to apply cable theory to complex branching dendrites, but in the 1960s, Wilfrid Rall showed how this problem could be solved by the development of computational compartmental models (Rall, 1964, 1967, 1977; Rall and Shepherd, 1968). These models have provided the basis for a theory of dendritic function (Segev *et al.*, 1995). Combined with mathematical models for the generation of synaptic potentials and action potentials, they provide the basis for a complete theoretical description of neuronal activity.

A variety of software packages now makes it possible for even a beginning student to explore functional properties and construct realistic neuron models. These tools are all freely accessible on the Web (see NEURON; GENESIS; ModelDB; Ziv *et al.*, 1994). We therefore present modern electrotonic theory within the context of constructing these compartmental models. Exploration of these models will aid the student greatly in understanding the complexities that are present in even the simplest types of passive spread of current in axons and dendrites.

receptors (Chapter 10), synaptically gated membrane channels (Chapter 9), intrinsic voltage-gated channels (Chapter 6), and second-messenger systems (Chapter 10)—that mediate the operations of the neuron.

BASIC TOOLS: CABLE THEORY AND COMPARTMENTAL MODELS

Slow spread of neuronal activity is by ionic or chemical diffusion or active transport. Our main interest in this chapter is in rapid spread by electric current. What are the factors that determine this spread? The most basic are *electrotonic properties*.

Our understanding of electrotonic properties arose in the nineteenth century from a merging of the study

SPREAD OF STEADY-STATE SIGNALS

Modern Electrotonic Theory Depends on Simplifying Assumptions

The successful application of cable theory to nerve cells requires that it be based as closely as possible on the structural and functional properties of neuronal processes. The problem confronting the neuroscientist is that processes are complicated. As discussed in Chapters 3 and 4, a segment of axon or dendrite contains a variety of molecular species and organelles, is bounded by a plasma membrane with its own complex structure and irregular outline, and is surrounded by myriad of neighboring processes (see Fig. 5.2A).

Describing the spread of electric current through such a segment therefore requires some carefully

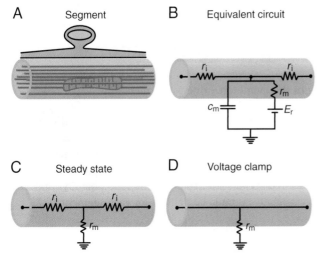

A Segment

B Equivalent circuit

C Steady state

D Voltage clamp

FIGURE 5.2 Steps in construction of a compartmental model of the passive electrical properties of a nerve cell process. (A) Identification of a segment of the process and its organelles. (B) Abstraction of an equivalent electrical circuit based on the membrane capacitance (c_m), membrane resistance (r_m), resting membrane potential (E_r), and internal resistance (r_i). (C) Abstraction of the circuit for steady-state electrotonus, in which c_m and E_r can be ignored. (D) The space clamp used in voltage-clamp analysis reduces the equivalent circuit even further to only the membrane resistance (r_m), usually depicted as membrane conductances (g) for different ions. In a compartmental modeling program, the equivalent circuit parameters are scaled to the size of each segment.

chosen simplifying assumptions, which allow the construction of an *equivalent circuit* of the electrical properties of such a segment. These are summarized in Box 5.1. Understanding them is essential for describing electrotonic spread under the different conditions that the nervous system presents.

Electrotonic Spread Depends on the Characteristic Length

We begin by using the assumptions in Box 5.1 to represent a segment of a process by electrical resistances: an internal resistance r_i connected to the r_i of the neighboring segments and through the membrane resistance r_m to ground (see Fig. 5.2B). Let us first consider the spread of electrotonic potential under steady-state conditions (Fig. 5.2C). In standard cable theory, this is described by

$$V = \frac{r_m}{r_1} \cdot \frac{d^2V}{dx^2}. \tag{5.1}$$

This equation states that if there is a steady-state current input at point $x = 0$, the electrotonic potential (V) spreading along the cable is proportional to the second derivative of the potential (d^2V) with respect to

distance and the ratio of the membrane resistance (r_m) to the internal resistance (r_i) over that distance. The steady-state solution of this equation for a cable of infinite extension for positive values of x gives

$$V = V_0 e^{-x/\lambda}, \tag{5.2}$$

where lambda is defined as the square root of r_m/r_i (in centimeters) and V_0 is the value of V at $x = 0$.

Inspection of this equation shows that when $x = \lambda$, the ratio of V to V_0 is $e - 1 = 1/e = 0.37$. Thus, lambda is a critical parameter defining the length over which the electrotonic potential spreading along an infinite cable decays (is attenuated) to a value of 0.37 of the value at the site of the input. It is referred to as the *characteristic length* (space constant, length constant) of the cable. The higher the value of the specific membrane resistance (R_m), the higher the value of r_m for that segment, the larger the value for λ, and the greater the spread of electrotonic potential through that segment (Fig. 5.3). Specific membrane resistance (R_m) is thus an important variable in determining the spread of activity in a neuron.

Most of the passive electrotonic current may be carried by K^+ "leak" channels, which are open at "rest" and are largely responsible for holding the cell at its resting potential. However, as mentioned earlier, many cells or regions within a cell are seldom at "rest" but are constantly active, in which case electrotonic current is carried by a variety of open channels. Thus, the effective R_m can vary from values of less than $1000\,\Omega\,cm^2$ to more than $100,000\,\Omega\,cm^2$ in different neurons and in different parts of a neuron. Note that lambda varies with the square root of R_m, so a 100-fold difference in R_m translates into only a 10-fold difference in lambda.

Conversely, the higher the value of the specific internal resistance (R_i), the higher the value of r_i for that segment, the smaller the value of λ, and the less the spread of electrotonic potential through that segment (see Fig. 5.3). Traditionally, the value of R_i has been believed to be in the range of approximately 50–$100\,\Omega\,cm$ based on muscle cells and the squid axon. In mammalian neurons, estimates now tend toward a value of $200\,\Omega\,cm$. This limited range may suggest that R_i is less important than R_m in controlling passive current spread in a neuron. The square-root relation further reduces the sensitivity of λ to R_i. However, as noted in assumption 7 in Box 5.1, the membranous and filamentous organelles in the cytoplasm may alter the effective R_i. The presence of these organelles in very thin processes, such as distal dendritic branches, spine stems, and axon preterminals, may thus have potentially significant effects on the spread of electrotonic current through them. Furthermore, the relative

BOX 5.1

BASIC ASSUMPTIONS UNDERLYING CABLE THEORY

1. **Segments are cylinders.** A segment is assumed to be a cylinder with constant radius.

This is the simplest assumption; however, compartmental simulations can readily incorporate different geometrical shapes with differing radii if needed (Fig. 5.2B).

2. **The electrotonic potential is due to a change in the membrane potential.** At any instant of time, the "resting" membrane potential (E_r) at any point on the neuron can be changed by several means: injection of current into the cell, extracellular currents that cross the membrane, and changes in membrane conductance (caused by a driving force different from that responsible for the membrane potential). Electric current then begins to spread between that point and the rest of the neuron, in accord with

$$V = V_m - E_{r'}$$

where V is the electrotonic potential and V_m is the changed membrane potential.

Modern neurobiologists recognize that the membrane potential is rarely at rest. In practice, "resting" potential means the membrane potential at any given instant of time other than during an action potential or rapid synaptic potential.

3. **Electrotonic current is ohmic.** Passive electrotonic current flow is usually assumed to be ohmic, i.e., in accord with the simple linear equation

$$E = IR,$$

where E is the potential, I is the current, and R is the resistance.

This relation is largely inferred from macroscopic measurements of the conductance of solutions having the composition of the intracellular medium, but rarely is measured directly for a given nerve process. Also largely untested is the likelihood that at the smallest dimensions ($0.1\,\mu m$ diameter or less), the processes and their internal organelles may acquire submicroscopic electrochemical properties that deviate significantly from macroscopic fluid conductance values; compartmental models permit the incorporation of estimates of these properties.

4. **In the steady state, membrane capacitance is ignored.** The simplest case of electrotonic spread occurs from the point on the membrane of a steady-state change (e.g., due to injected current, a change in synaptic conductance, or a change in voltage-gated conductance) so that time-varying properties (transient charging or discharging of the membrane) due to the membrane capacitance can be ignored (Fig. 5.2C).

5. **The resting membrane potential can usually be ignored.** In the simplest case, we consider the spread of electrotonic potential (V) relative to a uniform resting potential (E_r) so that the value of the resting potential can be ignored. Where the resting membrane potential may vary spatially, V must be defined for each segment as

$$V = E_m - V_r.$$

6. **Electrotonic current divides between internal and membrane resistances.** In the steady state, at any point on a process, current divides into two local resistance paths: further within the process through an internal (axial) resistance (r_i) or across the membrane through a membrane resistance (r_m) (see Fig. 5.2C).

7. **Axial current is inversely proportional to diameter.** Within the volume of the process, current is assumed to be distributed equally (in other words, the resistance across the process, in the Y and Z axes, is essentially zero). Because resistances in parallel sum to decrease the overall resistance, axial current (I) is inversely proportional to the cross-sectional area ($I \propto \dfrac{1}{A} \propto \dfrac{1}{m^2}$); thus, a thicker process has a lower overall axial resistance than a thinner process. Because the axial resistance (r_i) is assumed to be uniform throughout the process, the total cross-sectional axial resistance of a segment is represented by a single resistance,

$$r_i = R_i/A,$$

where r_i is the internal resistance per unit length of r_i cylinder (in ohms per centimeter of axial length), R_i is the specific internal resistance (in ohms centimeter, or ohm cm), and A ($= \pi r^2$) is the cross-sectional area.

The internal structure of a process may contain membranous or filamentous organelles that can raise the effective internal resistance or provide high-conductance submicroscopic pathways that can lower it. In voltage-clamp experiments, the space clamp eliminates current through r_i, so that the only current remaining is through r_m, thereby permitting isolation and analysis of different ionic membrane conductances, as in the original experiments of Hodgkin and Huxley (Fig. 5.2D; see also Chapter 6).

8. **Membrane current is inversely proportional to membrane surface area.** For a unit length of cylinder, the

---

BOX 5.1 (cont'd)

membrane current (i_m) and the membrane resistance (r_m) are assumed to be uniform over the entire surface. Thus, by the same rule of the summing of parallel resistances, the membrane current is inversely proportional to the membrane area of the segment so that a thicker process has a lower overall membrane resistance. Thus,

$$r_m = R_m/c$$

where r_m is the membrane resistance for unit length of cylinder (in ohm cm of axial length), R_m is the specific membrane resistance (in ohm cm), and $c(= 2\pi r)$ is the circumference. For a segment, the entire membrane resistance is regarded as concentrated at one point; that is, there is no axial current flow within a segment but only between segments (Fig. 5.2C).

Membrane current passes through ion channels in the membrane. The density and types of these channels vary in different processes and indeed may vary locally in different segments and branches. These differences are incorporated readily into compartmental representations of the processes.

9. **The external medium along the process is assumed to have zero resistivity.** In contrast with the internal axial resistivity (r_i), which is relatively high because of the small dimensions of most nerve processes, the external medium has a relatively low resistivity for current because of its relatively large volume. For this reason, the resistivity of the paths either along a process or to ground generally is regarded as negligible, and the potential outside the membrane is assumed to be everywhere equivalent to ground (see Fig. 5.2C). This greatly simplifies the equations that describe the spread of electrotonic potentials inside and along the membrane.

Compartmental models can simulate any arbitrary distribution of properties, including significant values for extracellular resistance where relevant. Particular cases in which external resistivity may be large, such as the special membrane caps around synapses on the cell body or axon hillock of a neuron, can be addressed by suitable representation in the simulations. However, for most simulations, the assumption of negligible external resistance is a useful simplifying first approximation.

10. **Driving forces on membrane conductances are assumed to be constant.** It usually is assumed that ion concentrations across the membrane are constant during activity.

Changes in ion concentrations with activity may occur, particularly in constricted extracellular or intracellular compartments; these changes may cause deviations from the assumptions of constant driving forces for the membrane currents, as well as the assumption of uniform E_r. For example, accumulations of extracellular K^+ may change local E, and intracellular accumulations of ions within the tiny volumes of spine heads may change the driving force on synaptic currents. These special properties are easily included in most compartmental models.

11. **Cables have different boundary conditions.** In classical electrotonic theory, a cable such as one used for long-distance telecommunication is very long and can be considered of infinite length (one customarily assumes a semi-infinite cable with $V = 0$ at $x = 0$ and only positive values of length x). This assumption carries over to the application of cable theory to long axons, but most dendrites are relatively short. This imposes boundary conditions on the solutions of the cable equations, which have very important effects on electrotonic spread.

In highly branched dendritic trees, boundary conditions are difficult to deal with analytically but are readily represented in compartmental models.

Gordon M. Shepherd

---

significance of R_i and R_m depends greatly on the length of a given process, as will be seen shortly.

Electrotonic Spread Depends on the Diameter of a Process

The space constant (λ) depends not only on the internal and membrane resistance, but also on the diameter of a process (Fig. 5.4). Thus, from the relations between r_m and R_m, and r_i and R_i, discussed in the preceding section,

$$\lambda = \sqrt{\frac{r_m}{r_i}} = \sqrt{\frac{R_m}{R_i} \cdot \frac{d}{4}}. \tag{5.3}$$

Neuronal processes vary widely in diameter. In the mammalian nervous system, the thinnest processes are the distal branches of dendrites, the necks of some dendritic spines, and the cilia of some sensory cells; these processes may have diameters of only $0.1\,\mu m$ or less (the thinnest processes in the nervous system are approximately $0.02\,\mu m$). In contrast, the thickest processes in the mammal are the largest myelinated axons

A

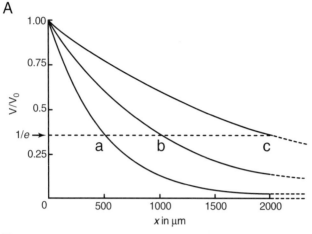

B

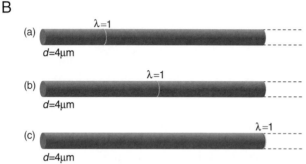

FIGURE 5.3 Dependence of the space constant governing the spread of electrotonic potential through a nerve cell process on the square root of the ratio between the specific membrane resistance (R_m) and the specific internal resistance (R_i). (A) Potential profiles for processes with three different values of λ. (B) Dotted lines represent the location of l on each of the three processes.

and the largest dendritic trunks, which may have diameters as large as 20 to 25 μm. This means that the range of diameters is approximately three orders of magnitude (1000-fold). Note, again, that the relation to λ is the square root; thus, over a 10-fold difference in diameter, the difference in λ is only about three-fold (Fig. 5.4).

Electrotonic Properties Must Be Assessed in Relation to the Lengths of Neuronal Processes

Application of classical cable theory to neuronal processes assumes that the processes are infinitely long (assumption 11 in Box 5.1). However, because neuronal processes have finite lengths, the length of a given process must be compared with λ to assess the extent to which λ accurately describes the actual electrotonic spread in that process. One of the largest processes in any nervous system, the squid giant axon, has

a diameter of approximately 1 mm. R_m for this axon has been estimated as $600\,\Omega\,cm^2$ (a very low value compared to most values of R_m in mammals), and R_i as approximately $80\,\Omega\,cm$, the value of Ringer solution (note that the very large diameter is counterbalanced by the very low R_m). Putting these values into Eq. (5.3) gives a λ of approximately 5.5 mm. The real length of the giant axon is several centimeters; to relate real length to characteristic length, we define *electrotonic length* (L) as

$$L = x/\lambda \qquad (5.4)$$

Thus, if $x = 30$ mm, then $L = 30\,mm/4.5\,mm = 7$. The electrotonic potential decays to a small percentage of the original value by only three characteristic lengths (see Fig. 5.4), so for this case the assumption of an infinite length is justified. In contrast to axons, dendritic branches have lengths that are usually much shorter than three characteristic lengths. In dendrites, therefore, the branching patterns come to dominate the extent of potential spread. We discuss the methods for dealing with these branching patterns later in this chapter.

A reason often given for why the nervous system needs action potentials is that they overcome the severe attenuation of passively spreading potentials that occurs over the considerable lengths required for transmission of signals by axons. This applies to the long axons of projection neurons, but not necessarily to shorter axons and their collaterals. Recent studies in fact have revealed that excitatory synaptic potentials in the soma may spread through the axon to reach terminal boutons onto nearby cells; the variable amount of synaptic depolarization thus acts as an analog signal to modify the digital signaling carried by the axonal action potentials. This mechanism has been shown in the mossy fiber terminals of dentate granule cells onto CA3 pyramidal cells in the hippocampus (Alle and Geiger, 2006), and in the axon terminals of layer 5 pyramidal neurons onto neighboring cells in the cerebral cortex (Shu *et al.*, 2006). The combined analog and digital signaling is computationally more powerful than digital signaling alone.

A reverse situation is seen in the retina, where a particular type of horizontal cell has elaborate branches of both its dendrites and its terminal axon, interconnected by a long thin axon. Physiological studies have shown that each branching system processes different properties of the visual signal, but they do not interact, because the axon has passive properties that give it a short length constant. This enables one cell to provide two distinct input–output processing systems (Nelson

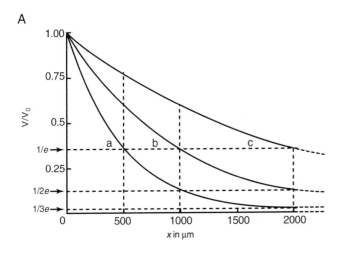

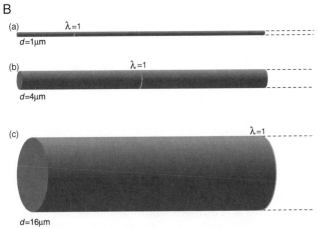

FIGURE 5.4 Dependence of the space constant on the square root of the diameter of the process. (A) Potential profiles for processes with three different diameters but fixed values of R_i and R_m. (B) The three axon profiles in A. Note that to double λ, the diameter must be quadrupled.

et al., 1975). Never underestimate the ingenuity of the nervous system!

Summary

Passive spread of electrical potential along the cell membrane underlies all types of electrical signaling in the neuron. It is thus the foundation for understanding the interactive substrate whereby the neuron can generate, receive, integrate, encode, and send signals.

Electrotonic spread shares properties with electrical transmission through electrical cables; the mathematical study of cable transmission has put these properties on a quantitative basis. The theoretical basis for

extension of cable theory to complex dendritic trees has been developed in parallel with compartmental modeling methods for simulating dendritic signal processing.

Cable theory depends on a number of reasonable simplifying assumptions about the geometry of neuronal processes and current flow within them. Steady-state electrotonus in dendrites depends on passive resistance of the membrane and of the internal cytoplasm and on the diameter and length of a nerve process.

SPREAD OF TRANSIENT SIGNALS

Electrotonic Spread of Transient Signals Depends on Membrane Capacitance

Until now, we have considered only the passive spread of steady-state inputs. However, the essence of many neural signals is that they change rapidly. In mammals, fast action potentials characteristically last from 1 to 5 ms, and fast synaptic potentials last from 5 to 30 ms. How do the electrotonic properties affect spread of these rapid signals?

Rapid signal spread depends not only on all the factors discussed thus far, but also on the membrane capacitance (c_m), which is due to the lipid moiety of the plasma membrane. Classically, the value of the specific membrane capacitance (C_m) has been considered to be $1\,\mu F\,cm^{-2}$. However, a value of $0.6–0.75\,\mu F\,cm^{-2}$ is now preferred for the lipid moiety itself, with the remainder being due to gating charges on membrane proteins (Jack *et al.*, 1975).

The simplest case demonstrating the effect of membrane capacitance on transient signals is that of a single segment or a cell body with no processes. This is a very unrealistic assumption, equivalent to the single node of neural network models, but a simple starting point. In the equivalent electrical circuit for a neural process, the membrane capacitance is placed in parallel with ohmic components of the membrane conductance and the driving potentials for ion flows through those conductances (see Fig. 5.2B). Again neglecting the resting membrane potential, we take as an example the injection of a current step into a soma; in this case, the time course of the current spread to ground is described by the sum of the capacitative and resistive current (plus the input current, I_{pulse}):

$$C\frac{dV_m}{dt} + \frac{V_m}{R} = I_{pulse}. \tag{5.5}$$

Rearranging,

$$RC\frac{dV_m}{dt} + V_m = I_{pulse} \cdot R \qquad (5.6)$$

where $RC = \tau$ (τ is the time constant of the membrane).

The solution of this equation for the response to a step change in current (I) is

$$V_m(T) = I_{pulse} R(1 - e^{-T}). \qquad (5.7)$$

where $T = t/\tau$.

When the pulse is terminated, the decay of the initial potential (V_0) to rest is given by

$$V_m(T) = V_0 e^{-T}. \qquad (5.8)$$

These "on" and "off" transients are shown in Figure 5.5. The significance of tau is shown in the diagram; it is the time required for the voltage change across the membrane to reach $1/e = 0.37$ of its final value. This time constant of the membrane defines the transient voltage response of a segment of membrane to a current step in terms of the electrotonic properties of the segment. It is analogous to the way that the length constant defines the spread of voltage change over distance.

A Two-Compartment Model Defines the Basic Properties of Signal Spread

These spatial and temporal cable properties can be combined in a two-compartment model (Shepherd, 1994) that can be applied to the generation and spread of any arbitrary transient signal (Fig. 5.6).

In the simplest case, current is injected into one of the compartments, as in an electrophysiological experiment. Positive charge injected into compartment A attempts to flow outward across the membrane, partially opposing the negative charge on the inside of the lipid membrane (the charge responsible for the negative resting potential), thereby depolarizing the membrane capacitance (C_m) at that site. At the same time,

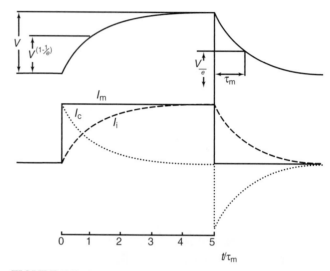

FIGURE 5.5 The equivalent circuit of a single isolated compartment responds to an injected current step by charging and discharging along a time course determined by the time constant, τ. In actuality, because nerve cell segments are parts of longer processes (axonal or dendritic) or larger branching trees, the actual time courses of charging or discharging are modified. V steady-state voltage; I_m, injected current applied to membrane; I_c, current through the capacitance; I_i, current through the ionic leak conductance; m, membrane time constant. From Jack et al. (1975).

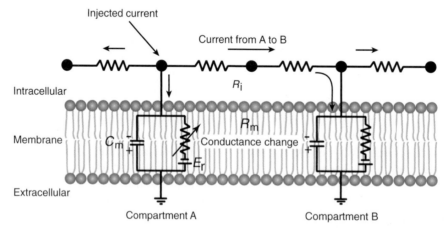

FIGURE 5.6 The equivalent circuit of two neighboring compartments or segments (A and B) of an axon or dendrite shows the pathways for current spread in response to an input (injected current or increase in membrane conductance) at segment A. See text for full explanation.

the charge begins to flow as current across the membrane through the resistance of the ionic membrane channels (R_m) that are open at that site. The proportion of charge divided between C_m and R_m determines the rate of charge of the membrane; that is, the membrane time constant, τ. However, charge also starts to flow through the internal resistance (R_i) into compartment B, where the current again divides between capacitance and resistance. The charging (and discharging) transient in compartment A departs from the time constant of a single isolated compartment, being faster because of the impedance load (e.g., current sink) of the rest of the cable (represented by compartment B). Thus, the time constant of the system no longer describes the charging transient in the system because of the conductance load of one compartment on another. The system is entirely passive; the response to a second current pulse sums linearly with that of the first.

This case is a useful starting point because an experimenter often injects electrical currents in a cell to analyze nerve function. However, a neuron normally generates current spread by means of localized conductance changes across the membrane. In Figure 5.6, consider such a change in the ionic conductance for Na^+, as in the initiation of an action potential or an excitatory postsynaptic potential, producing an inward positive current in compartment A. The charge transferred to the interior surface of the membrane attempts to follow the same paths followed by the injected current just described by opposing the negativity inside the membrane capacitance, crossing the membrane through the open membrane channels to ground, and spreading through the internal resistance to the next compartment, where the charge flows are similar.

Thus, the two cases start with different means of transferring positive charge within the cell, but from that point the current paths and the associated spread of the electrotonic potential are similar. The electrotonic current that spreads between the two segments is referred to as the *local current*. The charging transient in compartment A is faster than the time constant of the resting membrane; this difference is due both to the conductance load of compartment B (as in the injected current case) and to the fact that the imposed conductance increase in compartment A reduces the time constant of compartment A (by reducing effective R_m). This illustrates a critical point first emphasized by Wilfrid Rall (1964): changes in membrane conductance alter the system so that it is no longer a linear system, even though it is a passive system. Thus, passive electrotonic spread is not so simple as most people think! Nonlinear summation of

synaptic responses is discussed further later in this chapter.

Summary

In addition to the properties underlying steady-state electrotonus, passive spread of *transient* potentials depends on the membrane capacitance. Initiation of electrotonic spread by intracellular injection of a transient electrical current pulse produces an electrotonic potential that spreads by passive local currents from point to point. It is more attenuated in amplitude than the steady-state case as it spreads along an axon or dendrite due to the low-pass filtering action of the membrane capacitance. Simultaneous current pulses at that site or other sites produce potentials that add linearly because the passive properties are invariant. However, transient conductance changes, as in synaptic responses, generate electrotonic potentials that do not sum linearly because of the nonlinear interactions of the conductances.

ELECTROTONIC PROPERTIES UNDERLYING PROPAGATION IN AXONS

Impulses Propagate in Unmyelinated Axons by Means of Local Electrotonic Currents

We next apply our knowledge of electrotonic current properties to propagation of an action potential in an unmyelinated axon, that is, one that is not surrounded by myelin or other membranes that restrict the spread of extracellular current. Details on the ionic mechanisms of the nerve impulse can be found in Chapter 6. The local current spreading through the internal resistance to the neighboring compartment enables the action potential to propagate along the membrane of the axon. The rate of propagation is determined by both the passive cable properties and the kinetics of the action potential mechanism.

Each of the cable properties is relevant in specific ways. For brief signals such as the action potential, C_m is critical in controlling the rate of change of the membrane potential. For long processes such as axons, R_i increasingly opposes electrotonic current flow as the value of r_i increases beyond the characteristic length λ, whereas the effect of r_m decreases, due to the increased membrane area for parallel current paths (see earlier). This effect is greater in thinner axons, which have shorter characteristic lengths. Finally, R_m is a parameter that can vary widely. Thus, each of these parameters must be assessed in order to understand the exquisite effects of passive variables on the rates of impulse propagation in axons.

A high value of R_m, for example, forces current further along the membrane, increasing the characteristic length and consequently the spread of electrotonic potential, as we have seen; however, at the same time, it increases the membrane time constant, thus slowing the response of a neighboring compartment to a rapid change. Increasing the diameter of the axon lowers the effective internal resistance of a compartment, thereby also increasing the characteristic length, but without a concomitant effect on the time constant. Thus, changing the diameter is a direct way of affecting the rate of impulse propagation through changes in passive electrotonic properties. The conduction rate of any given axon depends on the particular combination of these properties (Rushton, 1951; Ritchie, 1995). For example, in the squid giant axon, the very large diameter (as large as 1 mm) promotes rapid impulse propagation; the very low value of R_m (600 gV cm) lowers the time constant (promoting rapid current spread) but also decreases the length constant (limiting the spatial extent of current spread).

The effects of these passive properties on impulse velocity also depend on other factors. For example, on the basis of the cable equations, we can show that the conduction velocity should be related to the square root of the diameter (Rushton, 1951). However, the density of Na^+ channels in fibers of different diameters is not constant; thus, the binding of saxitoxin molecules, for example, to Na^+ channels varies greatly with diameter, from almost $300 m^{-2}$ in the squid axon to only $35 \mu m^{-2}$ in the garfish olfactory nerve (Ritchie, 1995). Thus, both active and passive properties must be assessed in order to understand a particular functional property.

Myelinated Axons Have Membrane Wrappings and Booster Sites for Faster Conduction

The evolution of larger brains to control larger bodies and more complex behavior required communication over longer distances within the brain and body. This requirement placed a premium on the ability of axons to conduct impulses as rapidly as possible. As noted in the preceding section, a direct way of increasing the rate of conduction is by increasing the diameter, but larger diameters mean fewer axons within a given space, and complex behavior must be mediated by many axons. Another way of increasing the rate of conduction is to make the kinetics of the impulse mechanism faster; that is, to make the rate of

increase in Na^+ conductance with increasing membrane depolarization faster. The Hodgkin–Huxley equations (Chapter 6) for the action potential in mammalian nerves in fact have this faster rate.

As we have seen, the rapid spread of local currents is promoted by an increase in R_m but is opposed by an associated increase in the time constant. What is needed is an increase in R_m with a concomitant decrease in C_m. This is brought about by putting more resistances in series with the membrane resistance (because resistances in series add) while putting more capacitances in series with the membrane capacitance (capacitances in series add as the reciprocals, much like resistances in parallel, as noted earlier). The way the nervous system does this is through a special satellite cell called a Schwann cell, a type of glial cell. As described in Chapters 3 and 4, Schwann cells wrap many layers of their plasma membranes around an axon. The membranes contain special constituents and together are called myelin. Myelinated nerves contain the fastest conducting axons in the nervous system. A general empirical finding known as the Hursh factor (Hursh, 1939) states that the rate of propagation of an impulse along a myelinated axon in meters per second is six times the diameter of the axon in micrometers. Thus, the largest axons in the mammalian nervous system are approximately 20 m in diameter, and their conduction rate is approximately $120 ms^{-1}$, whereas the thin myelinated axons of about $1 \mu m$ in diameter have conduction rates of approximately 5 to $10 ms^{-1}$.

As discussed in Chapter 4, myelinated axons are not myelinated along their entire length; at regular intervals (approximately 1 mm in peripheral nerves), the myelin covering is interrupted by a node of Ranvier. The node has a complex structure. The density of voltage sensitive Na^+ channels at the node is high ($10,000 \mu m^{-2}$), whereas it is very low ($20 \mu m^{-2}$) in the internodal membrane. This difference in density means that the impulse actively is generated only at the node; the impulse jumps, so to speak, from node to node, and the process therefore is called *saltatory conduction*. A myelinated axon therefore resembles a passive cable with active booster stations.

In rapidly conducting axons the impulse may extend over considerable lengths; for example, in a 20-μm-diameter axon conducting at $120 ms^{-1}$, at any instant of time an impulse of 1-ms duration extends over a 120-mm length of axon, which includes more than 100 nodes of Ranvier. It is therefore more appropriate to conceive that the impulse is generated simultaneously by many nodes, with their summed local currents spreading to the next adjacent nodes to activate them.

The specific membrane resistance (R_m) at the node is estimated to be only $50\,\Omega\,cm^2$, due to a large number of open ionic channels at rest. This value of R_m reduces the time constant of the nodal membrane to approximately 50 ms, which enables the nodal membrane to charge and discharge quickly, aiding rapid impulse generation greatly. For axons of equal cross-sectional area, myelination is estimated to increase the impulse conduction rate 100-fold.

In all axons, a critical relation exists between the amount of local current spreading down an adjacent axon and the threshold for opening Na^+ channels in the membrane of the adjacent axon so that propagation of the impulse can continue. This introduces the notion of a *safety factor*—the amount by which the electrotonic potential exceeds the threshold for activating the impulse. The safety factor must protect against a wide range of operating conditions, including adaptation (during high frequency firing), fatigue, injury, infection, degeneration, and aging.

Normally, an excess of local current ensures an adequate margin of safety against these factors. In the squid axon, the safety factor ranges from 4 to 5. In myelinated axons, an exquisite matching between internodal electrotonic properties and nodal active properties ensures that the electrotonic potential reaching a node has an adequate amplitude and the node has sufficient Na^+ channels to generate an action potential that will spread to the next node. The safety factors for myelinated axons range from 5 to 10. Thus, the interaction of passive and active properties underlies the safety factors for impulse propagation in axons. Similar considerations apply to the orthodromic spread of signals in dendritic branches and the back-propagation of action potentials from the axon hillock into the soma and dendrites.

Theoretically, the conduction velocity, space constant, and impulse wavelength of myelinated fibers scale linearly with fiber diameter (Rushton, 1951; Ritchie, 1995), as indeed is indicated in the aforementioned Hursh factor. This difference between myelinated and unmyelinated fibers in their dependence on diameter thus is related to the scaling of the internodal length. At approximately $1\,\mu m$ in diameter, the Hursh factor breaks down; at less than $1\,\mu m$ in diameter, there is an advantage, all other factors being equal, for an axon to be unmyelinated. However, myelinated axons are found down to a diameter of only $0.2\,\mu m$, which has been correlated with shorter internodal distances (Waxman and Bennett, 1972). Thus, conduction velocity in myelinated nerve depends on a complex interplay between passive and active properties.

Summary: Passive Spread and Active Propagation

Impulses propagate continuously through unmyelinated fibers because the local currents spread directly to neighboring sites on the membrane. The rate of propagation is determined directly by the electrotonic properties of the fiber. In myelinated axons, the impulse propagates discontinuously from node to node. The electrotonic properties of both the nodal and internodal regions determine not only the rate of impulse propagation, but also the safety factor for impulse transmission.

Here and in Chapter 12, it will contribute to clarity to distinguish between passive spread and active propagation (see Box 5.2).

BOX 5.2

ELECTROTONIC POTENTIALS SPREAD, ACTION POTENTIALS PROPAGATE

It is important to distinguish between passive and active spread of potentials, which is helped by using different terms. Based on common dictionary definitions, "spreading" has a more general meaning of distributing something (in this case a current or potential) over an area or along an object. It applies specifically to passive electronic "spread" and to the local circuit currents that spread before an action potential, and can also be used in a general way to refer to spread of the action potential itself. "Conduction" also has a general meaning in the electrical sense. In contrast, "propagating" refers specifically to the action potential, because it carries the dictionary meaning of spreading by sequential active processes of reproducing oneself, which is what an action potential does along an axon or dendrite.

These distinctions of meaning as applied to nervous conduction date from the work of Wilfrid Rall in the 1960s, and continue to be useful.

ELECTROTONIC SPREAD IN DENDRITES

Dendrites are the main neuronal compartment for the reception of synaptic inputs. The spread of synaptic responses through the dendritic tree depends critically on the electrotonic properties of the dendrites. Because dendrites are branching structures, understanding the rules governing dendritic electrotonus and the resulting integration of synaptic responses in dendrites is much more difficult than understanding the rules of simple spread in a single axon.

Dendritic Electrotonic Spread Depends on Boundary Conditions of Dendritic Termination and Branching

As noted earlier, compared with axons, dendrites are relatively short, and their length becomes an important factor in assessing their electrotonic properties. Consider, in the mammalian nervous system, a moderately thin dendrite of $1\,\mu m$ (three orders of magnitude smaller than the squid axon) that has a typical R_m of $60,000\,\Omega\,cm^{-2}$ (two orders of magnitude larger than that of the squid axon) and an R_i of $240\,\Omega\,cm$ (three times the squid value). Inserting these values into the equation for characteristic length (Eq. 5.3) gives a λ of approximately $790\,\mu m$. This illustrates that lambda tends to be relatively long in comparison with the actual lengths of the dendrites; in other words, because of the relatively high membrane resistance, the electrotonic spread of potentials is relatively effective within a dendritic branching tree.

This essential property underlies the integration of signals in dendrites. The effective spread immediately leads to a second property. The assumption of infinite length no longer holds; dendritic branches are bounded by their terminations on the one hand, and the nature of their branching on the other. These are termed *boundary conditions*. The spread of electrotonic potentials is therefore exquisitely sensitive to the boundary conditions of the dendrites.

This problem is approached most easily by considering two extreme types of termination of a dendritic branch. First, consider that at $x = \lambda$; the branch ends in a sealed end with infinite resistance. In this case, the axial component of the current can spread no further and must therefore seek the only path to ground, which is across the membrane of the cylinder. This current is added to the current already crossing the membrane; in the equation for Ohm's law ($E = IR$), I is increased, giving a larger E. The membrane will thus be more depolarized up to the terminal point a; in fact, near point a, axial current is negligible and almost all the current is across the membrane, which amounts to a virtual space clamp (Fig. 5.7). If at point λ the infinite resistance is replaced by the more realistic assumption of an end that is sealed with surface membrane, only a small amount of current crosses this membrane and attenuation of electrotonic potential is only slightly greater. Infinite resistance is therefore a useful approximation for assessing the effects of a sealed end on electrotonic spread in a terminal dendritic branch.

At the other extreme, consider that at point λ, a small dendritic branch opens out into a very large conductance. Examples are, in the extreme, a hole in the membrane; less extreme are a very small dendritic branch on a large soma and a small twig or spine on a large dendritic branch. Recall that large processes sum their resistances in parallel, which gives low current density and small voltage changes. Therefore, a current spreading through the high resistance of a small branch into a large branch encounters a very low resistance. For steady-state current spread, this situation is referred to as a large conductance load; for a transient current, we refer to it as a low impedance (which includes the effect of the membrane capacitance).

This introduces the key principle of *impedance matching* between interacting compartments, an important principle generally in biological systems. In our example, an impedance mismatch exists between the high impedance thin branch and the lower impedance thick branch. This mismatch reduces any voltage change due to the current and, in the extreme, effectively clamps the membrane to the resting potential (E_r) at that point. The electrotonic potential thus is attenuated through the branch much more rapidly than would be predicted by the characteristic length (see Fig. 5.7). This does not invalidate λ as a measure of electrotonic properties; rather, it means that, as with the time constant, each cable property must be assessed within the context of the size and branching of the dendrites.

All the different types of branching found in neuronal dendrites lie between these two extremes, with a corresponding range of boundary conditions at $x = \lambda$. Consider a segment of dendrite that divides into two branches at $x = \lambda$. We can appreciate intuitively that the amount of spread of electrotonic potential into the two branches will be governed by the factors just considered. One possibility is that the two branches have very small diameters, so their input impedance is higher than that of the segment; in this case, the situation will tend toward the sealed end case (Fig. 5.7, top trace). In contrast, the segment may

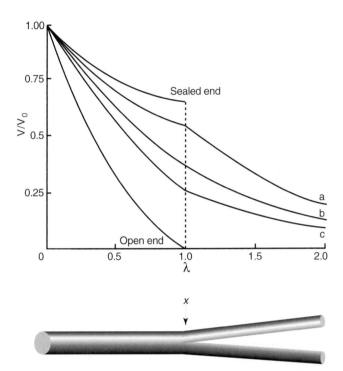

FIGURE 5.7 The spread of electrotonic potential through a short nerve cell process such as a dendritic branch is governed by the space constant and by the size of the branches; the latter imposes a boundary condition at the branch point. Curves a–c represent a range of realistic assumptions about the sizes of the branches relative to the size of the stem, together with the limiting conditions of an open circuit (corresponding to an infinite conductance load) and a closed circuit (corresponding to a sealed tip).

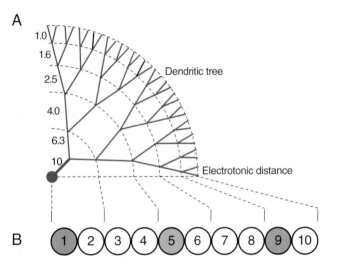

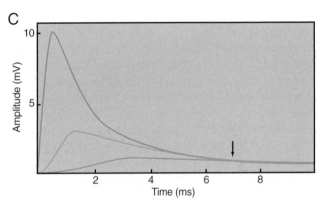

FIGURE 5.8 The spread of electrotonic potentials is accompanied by a delay and an attenuation of amplitude. (A) Dendritic diameters (left) satisfy the 3/2d rule so that the tree can be portrayed by an equivalent cylinder. An excitatory postsynaptic potential (EPSP) is generated in compartment 1, 5, or 9 (B) and recordings are made from compartment 1. (C) Short latency, large amplitude, and rapid transient response in compartment 1 at the site of input, as well as the later, smaller, and slower responses recorded in compartment 1 for the same input to compartments 5 and 9. Despite the initial differences in time course, the responses converge at the arrow to decay together. Based on Rall (1967).

give rise to two very fat branches, so the situation will tend toward the large conductance load case (Fig. 5.7, bottom trace).

For many cases of dendritic branching, the input impedance of the branches is between the two extremes (see Fig. 5.7, traces a–c), providing for a reasonable degree of impedance matching between the stem branch and its two daughter branches. This situation thus approximates the infinite cylinder case, in which by definition the input impedance at one site matches that at its neighboring site along the cylinder. The general rules for impedance matching at branch points were worked out by Rall (1959, 1964, 1967), who showed that the input conductance of a dendritic segment varies with the diameter raised to the 3/2 power. There is electrotonic continuity at a branch point equivalent to the infinitely extended cylinder if the diameter of the segment raised to the 3/2 power equals the sum of the diameters raised to the 3/2 power of all the daughter branches. An idealized branching pattern that satisfies this rule is shown in

Figure 5.8. When the branching tree reduces to a single chain of compartments, as in this case, it is called an "equivalent cylinder." When the branching pattern departs from the 3/2 rule, the compartment chain is referred to as an "equivalent dendrite" (Rall and Shepherd, 1968).

Dendritic Synaptic Potentials Are Delayed and Attenuated by Electrotonic Spread

We are now in a position to assess the effects of cable properties on the time course of the spread of synaptic potentials through dendritic branches and trees. Consider in Figure 5.8 the case of recording from a soma while delivering a brief excitatory synaptic conductance change to different locations in the dendritic tree. The response to the nearest site is a rapidly rising synaptic potential that peaks near the end of the conductance change and then decays rapidly toward baseline. When the input is delivered to the middle of the chain of compartments, the response in the soma begins only after a delay, rises more slowly, reaches a much lower peak (which is reached after the end of the conductance change in the soma), and decays slowly toward baseline. For input to the terminal compartment, the voltage delay at the soma is so long that the response has scarcely started by the end of the conductance change in the distal dendrite; the response rises slowly to a delayed (several milliseconds) and prolonged plateau that subsides very slowly (see Fig. 5.8).

Although the synaptic potentials thus decrease in amplitude as they spread, the rate of electrotonic spread can be calculated in terms of the half-amplitude at any point. If distance is expressed in units of λ and time in units of τ, then for spread through a semi-infinite cable, we have the simple equation (Jack *et al.*, 1975)

$$\text{Velocity} = 2\frac{\lambda}{\tau}. \tag{5.9}$$

Thus, if we ignore boundary effects, for the 10-mm process mentioned earlier in which $\lambda = 1500$ and $\tau = 10\,\text{ms}$, the velocity of spread would be $0.3\,\text{ms}^{-1}$, or $300\,\mu\text{m ms}^{-1}$. It can be seen that electrotonic spread can be relatively fast over short distances within a dendritic tree but is very slow in comparison with impulse transmission for an axon of this diameter ($60\,\text{ms}^{-1}$). Thus, both the severe decrement and the slow velocity make passive spread by itself ineffective for transmission over long distances.

These general rules of delay and attenuation govern the passive spread of all transient potentials in dendritic branches and trees. As a rule of thumb, spread within one space constant (see the decrement between compartments 1 and 5 in Fig. 5.8) mediates relatively effective linkage for rapid signal integration, whereas spread over one or two space constants (see the decrement between compartments 1 and 9 in Fig. 5.8) is limited to slower background modulation. In real dendrites, these limitations often are overcome through boosting the signals at intermediate sites by voltage-gated properties (see Chapter 12).

The spread of electrotonic potential from a point of input involves the *equalization of charge* on the membrane throughout the system. After cessation of the input, a time is reached when charge has become equalized and the entire system is equipotential; from this time on, the remaining electrotonic potential decays equally at every point in the system. This time is indicated by the vertical arrow in Figure 5.8C. Before this time, the decaying transients are governed by equalizing time constants, indicating electrotonic spread, which can be identified by "peeling" on semilogarithmic plots of the potentials (Rall, 1977). After this time, the decay of electrotonic potential is governed solely by the membrane time constant. In experimental recordings of synaptic potentials, the overall electrotonic length of the dendritic system, considered as an "equivalent cylinder" or "equivalent dendrite" (see earlier discussion), can be estimated from measurements of the membrane time constant and the equalizing time constants. The electrotonic lengths of the dendritic trees of many neuron types lie between 0.3 and 1.5.

What is the spread of the postsynaptic potential throughout the system when a synaptic input is delivered to only a single terminal dendritic branch (Fig. 5.9) (Rall and Rinzel, 1973; Rinzel and Rall, 1974). Let us begin by considering a steady-state potential. Two main factors are involved. First, in the terminal branch, both the effective membrane resistance and the internal resistance are very large; hence, the branch has a very high input resistance, which produces a very large voltage change for any given synaptic conductance change. Balanced against this high input resistance is a second factor: the small branch has a very large conductance load on it because of the rest of the dendritic tree. As a result, there is a steep decrement in the electrotonic potential spreading from the branch through the tree to the cell body (Fig. 5.9A). For comparison, a direct input to the soma produces only a small potential change there because of the relatively very low input resistance at that site.

For a transient synaptic input, a third factor—membrane capacitance—must be taken into account. The small surface area of a terminal branch has little capacitance, so the amplitude of a transient response differs little from a steady-state response in the branch. However, in spreading out from a small process (such as a distal dendritic twig or spine), the transient synaptic potential is attenuated by the impedance mismatch between the process and the rest of the dendritic tree. Spread of the transient through the dendritic tree is attenuated further by

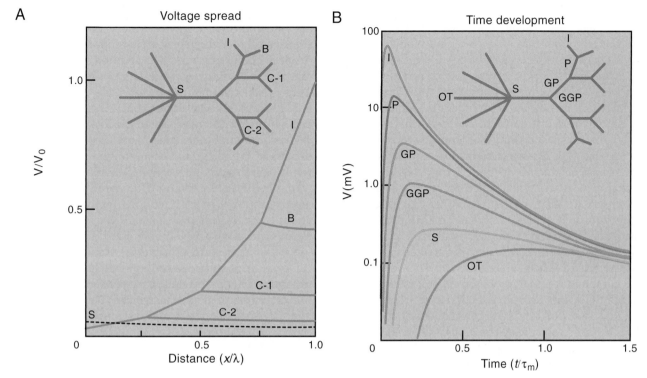

FIGURE 5.9 Electrotonic spread from a single small dendritic branch. (A) For steady-state input (I), the electrotonic potential (V), relative to the initial potential (V_0) at the site of input, spreads from the distal branch through the dendritic tree, with a large decrement into the parent branch (due to the large conductance load) but a small decrement into neighboring branches B, C–1, and C–2 (due to the small conductance loads). The resulting potential in the soma (S) is much reduced, as is the response to the same input delivered directly to the soma (because of the low input resistance at the soma and the large conductance load of the dendritic tree). The dashed line indicates the response when the same amount of current is injected into the soma. (B) For transient input (I) to a distal branch, transient electrotonic potentials decrease sharply in amplitude and are delayed and slower as they spread toward the soma through the parent (P), grandparent (GP), and great-grandparent (GGP) branches, eventually reaching the soma (S) and output trunk (OT). Modified from Segev (1995) based on Ralland Rinzel (1973, 1974).

the need to charge the capacitance of the dendritic membrane and is slowed by the time taken for the charging.

The amount of slowing is so precise that the relative distance of a synapse in the dendritic tree from the soma can be calculated from experimental measurements in the soma of the time to peak of the recorded synaptic potential (Rall, 1977; Johnston and Wu, 1995). For these reasons, the peak of a synaptic potential transient spreading from distal dendrites toward the soma may be severely attenuated, several-fold more than for the case of steady-state attenuation. This often is referred to as the *filtering effect* of the cable properties. However, the integrated response (the area under the transient voltage) is approximately equivalent to the steady-state amplitude, indicating that there is only a small loss of total charge (see Fig. 5.9B).

DYNAMIC PROPERTIES OF PASSIVE ELECTROTONIC STRUCTURE

Electrotonic Structure of the Neuron Changes Dynamically

These considerations show that, compared with the anatomical structure of a dendritic system, which is relatively fixed over short periods of time, the electrotonic structure continually shifts over time, producing complex effects on signal integration. The effects reflect different relations between the electrotonic and signaling properties, such as the direction of signal spread, inhomogeneities in passive properties, rates of signal transfer, and interactions between synaptic or active conductances, to name a few. The effects can be illustrated in graphic fashion for the entire soma–dendritic system by taking a stained neuron and modifying its size according to its electrotonic properties. This is termed a *morphoelectrotonic transform* (MET) or *neuromorphic transform*.

We illustrate three types of neuromorphic transforms, beginning with the direction of signal spread. Figure 5.10 illustrates a CA1 hippocampal pyramidal cell in which a comparison is made between spread of a signal from the soma to the dendrites (voltage out, V_{out}) with spread from the dendrites to the soma (voltage in, V_{in}). On the left is the stained neuron, with its long many-branched apical dendrite and shorter basal dendrites and their branches. In the right lower diagram is an electrotonic representation of the neuron for signals spreading from the distal dendrites toward the soma. There is severe decrement from each distal branch (cf. Fig. 5.9) so that apical and basal dendritic trees have electrotonic lengths of approximately 3 and 2, respectively. By comparison, in the right upper diagram is an electrotonic representation of this neuron for a signal spreading from the soma to the dendrites. The basal dendrites have shrunk to almost nothing, indicating that they are nearly isopotential. This is because they are relatively

short compared with their electrotonic lengths and because the sealed end boundary condition greatly reduces the decrement of electrotonic potential through them (cf. Fig. 5.7). The apical dendrite has shrunk to an electrotonic length of approximately 1. Thus, distal synaptic responses decay considerably in spreading all the way to the soma, which active properties help to overcome, as we shall see in Chapter 12, whereas signals at the soma "see" a relatively compact dendritic tree. This, for example, would be the case for a back-propagating action potential.

The analysis in Figure 5.10 applies to spread of steady-state or very slowly changing signals. What about spread of rapid signals? We have seen that membrane capacitance makes the dendrites act as a low-pass filter, further reducing rapid signals. The electrotonic transforms can assess this effect, as shown in Figure 5.11. On the left, the electrotonic representation of a pyramidal neuron is shown for a slow (100 Hz) current injected in the soma. The form is similar to that of the cell in Figure 5.10, with tiny, virtually isopotential basal dendrites and a longer apical dendritic tree of electrotonic length of approximately 1.5. By comparison, a rapid (500 Hz) signal is severely attenuated in spreading into the dendrites, as shown by the basal dendrites with L of approximately 1 and the apical dendritic tree electrotonic lengths of 4–5. Thus, a somatic action potential could back-propagate into the basal dendrites rather effectively, but would require active properties to invade very far into the apical dendrites. There is direct evidence for these properties underlying back-propagating action potentials in apical dendrites (Chapter 12).

The electrotonic structure of a neuron is not necessarily fixed, but may vary under synaptic control. Our final example is shown in Figure 5.12 for the case of a medium spiny cell in the basal ganglia. During low levels of resting excitatory synaptic input, the electrotonic transform of this cell type is relatively large (left) because of the action of a specific K$^+$ current (known as I_h) in the dendrites that holds them relatively hyperpolarized (see arrow at −90 mV). When synaptic excitation increases, the K$^+$ current is deactivated, reducing the membrane conductance and thereby increasing the input resistance of the cell; the dendritic tree becomes more compact electrotonically (middle) so that synaptic inputs are more effective in activating the cell. As the cell responds to the synaptic excitation, the resulting depolarization activates other K$^+$ currents, which expand the electrotonic structure again (right). This example illustrates how cable properties and voltage-gated properties interact to control the integrative actions of the neuron.

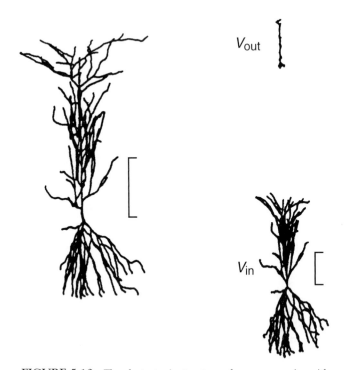

V_{out}

V_{in}

FIGURE 5.10 The electrotonic structure of a neuron varies with the direction of spread of signals. (Left) Stained CA1 pyramidal neuron. (Right) Electrotonic transform of the stained morphology for the case of a voltage spreading toward the cell body (bottom, V_{in}) and away from the cell body (top, V_{out}). Calibration, 1 electrotonic length. See text. From Carnevale *et al.* (1997).

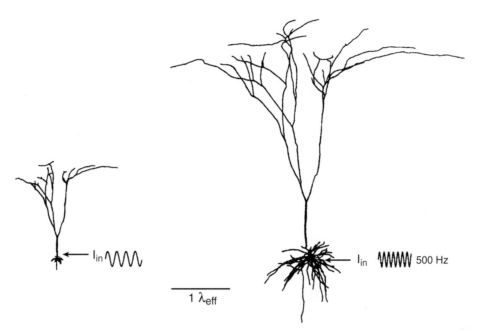

FIGURE 5.11 The electrotonic structure of a neuron varies with the rapidity of signals. (Left) Electrotonic transform of a pyramidal neuron in response to a sinusoidal current of 100 Hz injected into the soma (i.e., this is an example of V_{out}). (Right) Electrotonic transform of same cell in response to 500 Hz. Calibration, 1 electrotonic length. See text. From Zador *et al.* (1995).

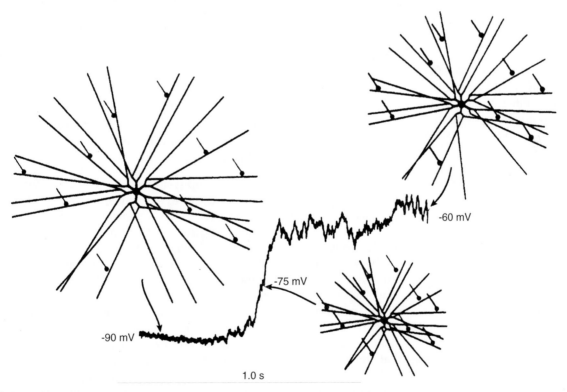

FIGURE 5.12 The electrotonic structure of a neuron can vary with shifts in the resting membrane potential. In this medium spiny cell, the electrotonic transform varies with the resting membrane potential, which in turn reflects the combination of resting voltage-gated K^+ currents and excitatory synaptic currents. See text. From Wilson (1998).

Synaptic Conductances in Dendrites Tend to Interact Nonlinearly

Dynamic interactions also occur between synaptic conductances. It often is assumed that synaptic responses sum linearly, but we have already noted that this is not generally true. In an electrical cable, responses to simultaneous current inputs sum linearly (they show "superposition") because the cable properties remain invariant at all times. However, as noted in relation to Figure 5.6, synaptic responses in real neurons generate current by means of changes in the membrane conductance at the synapse, which alters the overall membrane resistance of that segment and with it the input resistance, thereby changing the electrotonic properties of the whole system. As pointed out by Rall (1964), excitatory and inhibitory conductance changes involve "a change in a conductance which is an element of the system; the system itself is perturbed; the value of a constant coefficient in the linear differential equation is changed; hence the simple superposition rules do not hold."

This effect is illustrated by the two-compartment model of Figure 5.6. Consider a synaptic input to compartment A, which decreases the membrane resistance of that compartment. Now consider a simultaneous synaptic input to compartment B, which has the same effect on the membrane resistance of that compartment. The internal current flowing between the two compartments encounters a much lower impedance and hence has much less effect on the membrane potential than would have been the case for current injection. The integration of these two responses therefore gives a smaller summed potential than the summation of the two responses taken individually. This effect is referred to as *occlusion*. In essence, each compartment partially short-circuits the other through a larger conductance load, thus reducing the combined response.

These properties mean that, as noted earlier, synaptic integration in dendrites in general is not linear even for purely passive electrotonic properties. The further apart the synaptic sites, the fewer the interactions between the conductances, and the more linear the summation becomes (Fig. 5.13). These nonlinear properties of passive dendrites, combined with the nonlinear properties of voltage-gated channels at local sites on the membrane, contribute to the complexity of signal processing that takes place in dendrites, as will be discussed in Chapter 12. As we shall see, dendritic spines affect these nonlinear properties.

A

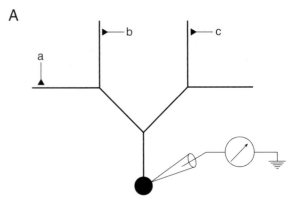

B

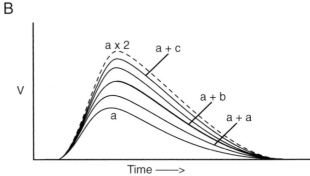

FIGURE 5.13 Schematic diagram of a dendritic tree to illustrate graded effects of nonlinear interactions between synaptic conductances. (A) Three sites of synaptic input (a–c) are shown, with a recording site in the soma. (B) The voltage response (*V*) is shown for the response to a single input at a, the theoretical linear summation for two inputs at a (a × 2), and the gradual reduction in summation from c to a due to increasing shunting between the conductances. See text. From Shepherd and Koch (1990).

Significance of Active Conductances in Dendrites Depends on Their Relation to Cable Properties

In electrophysiological recordings from the cell body, dendritic synaptic responses often appear small and slow (cf. Fig. 5.8). However, at their sites of origin in the dendrites, the responses tend to have a large amplitude (because of the high input resistances of the thin distal dendrites) and a rapid time course (because of the small membrane capacitance) (cf. Fig. 5.9). These properties have important implications for the signal processing that takes place in dendrites. In particular, the fact that distal dendrites contain sites of voltage-gated channels means that local integration, local boosting, and local threshold operations can take place. These most distal responses need spread no further than to neighboring local active sites to be boosted by these sites; thus, a rapid integrative sequence of these actions ultimately produces significant effects on signal integration at the cell body. These properties will be considered further in Chapter 12.

In addition to their role in local signal processing, the cable properties of the neuron are also important for (1) controlling the spread of synaptic potentials from the dendrites through the soma to the site of action potential initiation in the axon hillock initial segment and (2) back-propagation of an action potential into the soma–dendritic compartments, where it can activate dendritic outputs and interact with the active properties involved in signal processing. These properties are discussed further in Chapter 12.

Dendritic Spines Form Electrotonic and Biochemical Compartments

The rules governing electrotonic interactions within a dendritic tree also apply at the level of a spine, the smallest process of a nerve cell. A spine may vary from a bump on a dendritic branch to a twig to a lollipop-shaped process several micrometers long (Fig. 5.14). A dendritic spine usually receives a single excitatory synapse; an axonal initial segment spine characteristically receives an inhibitory synapse.

Dendritic spines receive most of the excitatory inputs to pyramidal neurons in the cerebral cortex and to Purkinje cells in the cerebellum, as well as to a variety of other neuron types, so an understanding of their properties is critical for understanding brain function (Shepherd, 1996; Araya *et al.*, 2006; Alvarez and Sabatini, 2007). As with the whole dendritic tree, one begins with their electrotonic properties. Given the rules we have built earlier in this chapter, by simple inspection of spine morphology as shown in Figure 5.14, we can postulate several distinctive features that may have important functional implications (see Box 5.3).

In addition to its electrotonic properties, the spine may have interesting biochemical properties. The same cable equations that govern electrotonic properties also have their counterparts in describing the diffusion of substances (as well as the flow of heat). Thus, as already noted, accumulations of only small numbers of ions are needed within the tiny volumes of spine heads to change the driving force on an ion species or to affect significant changes in the concentrations of subsequent second messengers. This interest is intensifying, as the ability to image ion fluxes, such as for Ca^{2+}, and to measure other molecular properties of individual spines increases with new technology such as two-photon microscopy. The interpretation of those results for the integrative properties of the neuron will

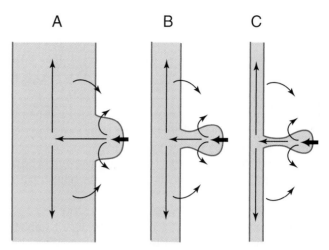

FIGURE 5.14 Diagrams illustrating different types of spines and current flows generated by a synaptic input. (A) Stubby spine arising from a thick process. (B) Moderately elongated spine from a medium diameter branch. (C) Spine with a long stem originating from a thin branch. Parallel considerations apply to diffusion between the spine head and dendritic branch. Modified from Shepherd (1974).

require considerations in the biochemical domain that parallel those discussed in the electrotonic domain. The range of properties and possible functions of spines are discussed further in Chapter 12.

Summary

In addition to membrane properties, the spread of electrotonic potentials in branching dendritic trees is dependent on the boundary conditions set by the modes of branching and termination within the tree. In general, other parts of the dendritic tree constitute a conductance load on activity at a given site; the spread of activity from that site is determined by the impedance match or mismatch between that site and the neighboring sites. Rules governing these impedance relations have been worked out relative to the case in which the sum of the daughter branch diameters raised to the 3/2 power is equal to that of the parent branch, in which case the system of branches is an "equivalent cylinder," resembling a single continuous cable. This provides a starting point in analyzing synaptic integration, which can be adapted for different types of branching patterns in terms of "equivalent dendrites."

Synchronous synaptic potentials in several branches spread relatively effectively through most dendritic trees. Responses in individual branches may be relatively isolated because of the decrement of passive spread and require local active boosting for effective communication with the rest of the tree. Passive spread

BOX 5.3

SOME BASIC ELECTROTONIC PROPERTIES OF DENDRITIC SPINES

a. High input resistance. The smaller the size and the narrower the stem, the higher the input resistance; this gives a large amplitude synaptic potential for a given synaptic conductance. Such a large depolarizing EPSP can have powerful effects on the local environment within the spine.

b. Low total membrane capacitance. The small size also means a small total membrane capacitance, implying that synaptic (and any active) potentials may be rapid; this means that spines on dendrites can potentially be involved in rapid information transmission.

c. Increases in total dendritic membrane capacitance. Although the membrane capacitance of an individual spine is small, the combined spine population increases the total capacitance of its parent dendrite. This increases the filtering effect of the dendrite on transmission of signals through it.

d. Decrement of potentials spreading from the spine. There is an impedance mismatch between the spine head and its parent dendrite; this means that potentials spreading from the spine to the dendrite will suffer considerable decrement unless there are active properties of the dendrite or of neighboring spines to boost the signal.

e. Ease of potential spread into the spine. The other side of the impedance mismatch is that membrane potential changes within the dendrite spread into the spine with little decrement; thus, the spine tends to follow the potential of its dendrite, except for the transient large-amplitude responses to its own synaptic input. This means that a spine can serve as a *coincidence detector* for nearby synaptic responses or for an action potential back-propagating into the dendritic tree.

f. Linearization of synaptic integration. The spine necks increase the anatomical and electrotonic distance between the spine synapses, thereby decreasing the interactions between their conductances, producing more linear superposition of the postsynaptic responses.

Gordon M. Shepherd

can be characterized in terms of several measures, including characteristic length of the equivalent cylinder. There is scaling within individual branches, such that electrotonic in finer branches spread is relatively effective over their shorter lengths. Integration of synaptic potentials in passive dendrites is fundamentally nonlinear because of interactions between the synaptic conductances. The rules for electrotonic spread in dendrites are the basis for understanding the contributions of active properties of dendrites (see Chapter 12).

RELATING PASSIVE TO ACTIVE POTENTIALS

We can now begin to gain insight into the relation between passive and active potentials in a neuron. We consider a model, the olfactory mitral cell, in which we apply the principles of this chapter and look forward to the principles underlying active properties in Chapter 12.

A basic problem is to understand the factors that decide where the action potential will be initiated with different levels of excitatory or inhibitory inputs. The possible sites are anywhere from the axon through the soma to the most distal dendrites. The mitral cell is advantageous for this analysis (1) because all the excitatory synaptic input is through olfactory nerve terminals that make their synapses on the distal dendritic tuft and (2) because the primary dendrite that connects the tuft to the cell body is an unbranched cylinder. Applying depolarizing current to distal dendrite or soma, the experimental findings were counterintuitive: with weak distal inputs the action potential initiation site is far away, in the axon, but with increasing excitation it shifts to the distal dendrite, as illustrated in Figure 5.15A (Chen *et al.*, 1997). How can the weak response spread so far passively, and why does it not excite the active dendrites along the way? Electrotonic spread is the key to the answer.

FIGURE 5.15 Interactions of passive and active potentials in the olfactory mitral cell. (A) Insets show diagrams of a mitral cell with recording sites at soma and distal dendrite. Curves show fitting of experimental and computed responses to weak and strong depolarizing currents injected into the distal primary dendrite. Note the nearly exact superposition of experimental (solid lines) and computed (dashed lines) responses. (B) Longitudinal distribution of membrane potential changes during responses to weak distal dendritic excitation. (C) Same to strong distal dendritic excitation. Blue lines, predominantly passively generated potentials; red lines, predominantly actively generated potentials. d, dendrite; s, soma; c, passive charging; o, onset of action potential; sp, spike peak; r, recovery. See text. Adapted from Shen *et al.* (1999).

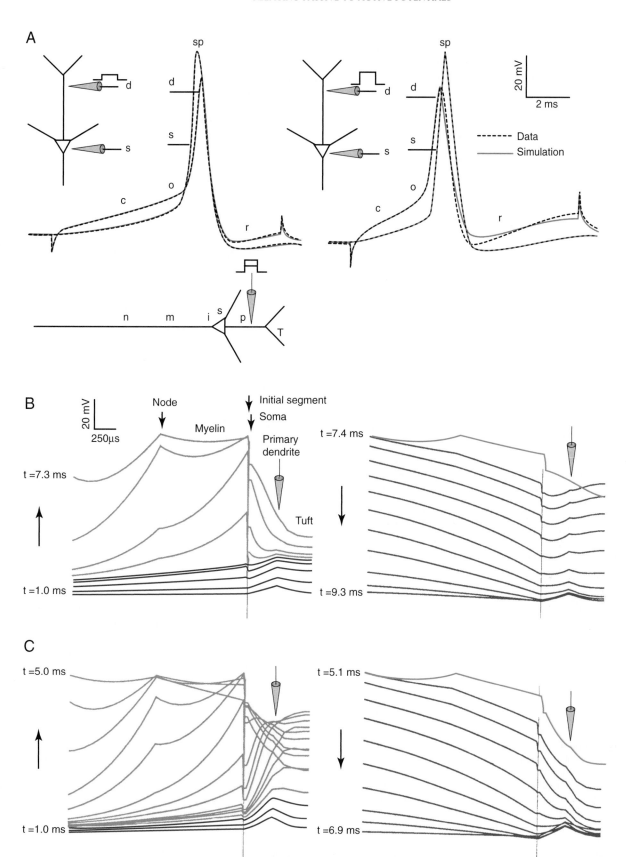

This is much too complex a problem to solve in your head or with "back of the envelope" calculations. The only effective method is a realistic computational simulation. A compartmental model of the mitral cell was therefore constructed, with Na^+ and K^+ conductances scaled to the structure of the mitral cell. Fitting of computed with experimental responses was carried out under stringent constraints, with minimization of eight simultaneous simulations (distal and soma recording sites, distal and soma sites of excitatory current input; strong and weak levels of excitation) (Shen *et al.*, 1999).

We will analyze the active properties in Chapter 12; here we focus on fitting the passive properties. Two steps were essential. First, each experimental recording began with a period of passive charging of the mitral cell membrane (c in Fig. 5.15A). Figure 5.15A shows that the model gave a very accurate simulation, even when the charging was long lasting (left, weak stimulation). This was a critical fit for giving the correct latency of action potential initiation. Second, the longitudinal spread of passive current between the axon and the distal dendrite was calculated. This showed that with weak distal excitation (Fig. 5.15B), the electrotonic current spread with a shallow gradient from the site of injection along the dendrite to the axon (bottom traces); the action potential arose first in the axon because of the much higher density of Na^+ channels there compared with the dendrite. However, with strong distal excitation (Fig. 5.15C), the direct depolarization of the less excitable distal dendrite led the weaker electrotonic depolarization of the more excitable axon, and dendritic action potential initiation occurred first. This can be explored online at senselab. med.yale.edu/modeldb.

The computational simulations thus show precisely how the interactions of passive and active potentials control the sites of action potential initiation in the neuron. This is a model for the complex integrative properties of the neuron, which are explored further in Chapter 12.

References

Alle, H. and Geiger, J. R. (2006). Combined analog and action potential coding in hippocampal mossy fibers. *Science* **311**, 1290–1293.

Alvarez, V. A. and Sabbatini, B. L. (2007). Anatomical and physiological plasticity of dendritic spines. *Annu. Rev. Neurosci.*

Araya, R., Eisenthal, K. B., and Yuste, R. (2006). Dendritic spines linearize the summation of excitatory potentials. *Proc. Natl. Acad. Sci. USA* **103**, 18799–18804.

Bower, J. and Beeman, D. (eds.) (1995). "The Book of Genesis." Springer-Verlag (Telos), New York.

Carnevale, N. T., Tsai, K. Y., Claiborne, B. J., and Brown, T. H. (1997). Comparative electrotonic analysis of 3 classes of rat hippocampal neurons. *J. Neurophysiol.*

Chen, W. R., Midtgaard, J., and Shepherd, G. M. (1997). Forward and backward propagation of dendritic impulses and their synaptic control in mitral cells. *Science* **278**, 463–467.

Hines, M. (1984). Efficient computation of branched nerve equations. *Int. J. Bio-Med. Comput.* **15**, 69–76.

Hursh, J. B. (1939). Conduction velocity and diameter of nerve fibers. *Am. J. Physiol.* **127**, 131–139.

Jack, J. J. B., Noble, D., and Tsien, R. W. (1975). "Electrical Current Flow in Excitable Cells." Oxford Univ. Press (Clarendon), London.

Johnston, D. and Wu, S. M. S. (1995). "Foundations of Cellular Neurophysiology." MIT Press, Cambridge.

Nelson, R., Lutzow, A. V., Kolb, H., and Gouras, P. (1975). Horizontal cells in cat retina with independent dendritic systems. *Science* **189**, 137–139.

Rall, W. (1959). Branching dendritic trees and motoneuron membrane resistivity. *Exp. Neurol.* **1**, 491–527.

Rall, W. (1964). Theoretical significance of dendritic trees for neuronal input-output relations. *In* "Neural Theory and Modeling" (R. F. Reiss, eds.), pp. 73–97. Stanford Univ. Press, Stanford, CA.

Rall, W. (1967). Distinguishing theoretical synaptic potentials computed for different soma-dendritic distributions of synaptic input. *J. Neurophysiol.* **30**, 1138–1168.

Rall, W. (1977). Core conductor theory and cable properties of neurons. *In* "The Nervous System, Cellular Biology of Neurons" (E. R. Kandel, ed.), Vol. 1; pp. 39–97. Am. Physiol. Soc., Bethesda, MD.

Rall, W. and Rinzel, J. (1973). Branch input resistance and steady attenuation for input to one branch of a dendritic neuron model. *Biophys. J.* **13**, 648–688.

Rall, W. and Shepherd, G. M. (1968). Theoretical reconstruction of field potentials and dendrodendritic synaptic interactions in olfactory bulb. *J. Neurophysiol.* **3**(6), 884–915.

Rinzel, J. and Rall, W. (1974). Transient response in a dendritic neuron model for current injected at one branch. *Biophys. J.* **14**, 759–790.

Ritchie, J. M. (1995). Physiology of axons. *In* "The Axon, Structure, Function, and Pathophysiology" (S. G. Waxman, J. D. Kocsis, and P. K. Stys, eds.), pp. 68–69. Oxford Univ. Press, New York.

Rushton, W. A. H. (1951). A theory of the effects of fibre size in medullated nerve. *J. Physiol. (Lond.)* **115**, 101–122.

Segev, I. (1995). Cable and compartmental models of dendritic trees. *In* "The Book of Genesis" (J. M. Bower and D. Beeman, eds.), pp. 53–82. Springer-Verlag (Telos), New York.

Segev, I., Rinzel, J., and Shepherd, G. M. (eds.) (1995). "The Theoretical Foundation of Dendritic Function." MIT Press, Cambridge, MA.

Shen, G., Chen, W. R., Midtgaard, J., Shepherd, G. M., and Hines, M. L. (1999). Computational analysis of action potential initiation in mitral cell soma and dendrites based on dual patch recordings. *J. Neurophysiol.* **82**, 3006–3020.

Shepherd, G. M. (1974). "The Synaptic Organization of the Brain." Oxford Univ. Press, New York.

Shepherd, G. M. (1996). The dendritic spine, A multifunctional integrative unit. *J. Neurophysiol.* **75**, 2197–2210.

Shepherd, G. M. and Brayton, R. K. (1979). Computer simulation of a dendrodendritic synaptic circuit for self- and lateral-inhibition in the olfactory bulb. *Brain Res.* **175**, 377–382.

Shepherd, G. M. and Koch, C. (1990). Dendritic electrotonus and synaptic integration. *In* "The Synaptic Organization of the Brain" (G. M. Shepherd, ed.), 3rd ed., pp. 439–574. Oxford Univ. Press, New York.

Shu, Y., Hasenstab, A., Duque, A., Yu, Y., and McCormick, D. A. (2006). Modulation of intracortical synaptic potentials by presynaptic somatic membrane potential. *Nature* **444**, 761–765.

Waxman, S. G. and Bennett, M. V. L. (1972). Relative conduction velocities of small myelinated and nonmyelinated fibres in the central nervous system. *Nature, New Biol.* **238**, 217.

Wilson, C. J. (1998). Basal ganglia. *In* "The Synaptic Organization of the Brain" (G. M. Shepherd, ed.), 5th ed., pp. 361–414. Oxford Univ. Press, New York.

Zador, A. and Koch, C. (1994). Linearized models of calcium dynamics, Formal equivalence to the cable equation. *J. Neurosci.* **14**, 4705–4715.

Zador, A. M., Agmon-Snir, H., and Segev, I. (1995). The morphoelectrotonic transform, A graphical approach to dendritic function. *J. Neurosci.* **15**, 1169–1682.

Ziv, I., Baxter, D. A., and Byrne, J. H. (1994). Simulator for neural networks and action potentials, Description and application. *J. Neurophysiol.* **71**, 294–308.

Gordon M. Shepherd

Membrane Potential
and Action Potential

The communication of information between neurons and between neurons and muscles or peripheral organs requires that signals travel over considerable distances. A number of notable scientists have contemplated the nature of this communication through the ages. In the second century AD, the great Greek physician Claudius Galen proposed that "humors" flowed from the brain to the muscles along hollow nerves. A true electrophysiological understanding of nerve and muscle, however, depended on the discovery and understanding of electricity itself. The precise nature of nerve and muscle action became clearer with the advent of new experimental techniques by a number of European scientists, including Luigi Galvini, Emil Du Bois-Reymond, Carlo Matteucci, and Hermann von Helmholtz, to name a few (Brazier, 1988). Through the application of electrical stimulation to nerves and muscles, these early electrophysiologists demonstrated that the conduction of commands from the brain to the muscle for the generation of movement was mediated by the flow of electricity along nerve fibers.

With the advancement of electrophysiological techniques, electrical activity recorded from nerves revealed that the conduction of information along the axon was mediated by the active generation of an electrical potential, called the action potential. But what precisely was the nature of these action potentials? To know this in detail required not only a preparation from which to obtain intracellular recordings but also one that could survive *in vitro*. The squid giant axon provided precisely such a preparation. Many invertebrates contain unusually large axons for the generation of escape reflexes; large axons conduct more quickly than small ones and so the response time for escape is

reduced (see Chapter 5). The squid possesses an axon approximately 0.5 mm in diameter, large enough to be impaled by even a course micropipette (Fig. 6.1). By inserting a glass micropipette filled with a salt solution into the squid giant axon, Alan Hodgkin and Andrew Huxley demonstrated in 1939 that axons at rest are electrically polarized, exhibiting a resting membrane potential of approximately −60 mV inside versus outside. In the generation of an action potential, the polarization of the membrane is removed (referred to as depolarization) and exhibits a rapid swing toward, and even past, 0 mV (Fig. 6.1). This depolarization is followed by a rapid swing in the membrane potential to more negative values, a process referred to as hyperpolarization. The membrane potential following an action potential typically becomes even more negative than the original value of approximately −60 mV. This period of increased polarization is referred to as the after-hyperpolarization or the undershoot.

The development of electrophysiological techniques to the point that intracellular recordings could be obtained from the small cells of the mammalian nervous system revealed that action potentials in these neurons are generated through mechanisms similar to that of the squid giant axon.

It is now known that action potential generation in nearly all types of neurons and muscle cells is accomplished through mechanisms similar to those first detailed in the squid giant axon by Hodgkin and Huxley. This chapter considers the cellular mechanisms by which neurons and axons generate a resting membrane potential and how this membrane potential briefly is disrupted for the purpose of propagation of an electrical signal, the action potential.

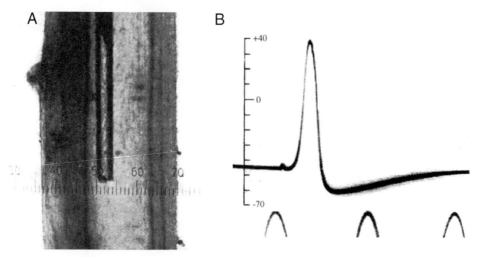

FIGURE 6.1   Intracellular recording of the membrane potential and action potential generation in the squid giant axon. (A) A glass micro-pipette, about 100 μm in diameter, was filled with seawater and lowered into the giant axon of the squid after it had been dissected free. The axon is about 1 mm in diameter and is transilluminated from behind. (B) One action potential recorded between the inside and the outside of the axon. Peaks of a sine wave at the bottom provided a scale for timing, with 2 ms between peaks. From Hodgkin and Huxley (1939).

MEMBRANE POTENTIAL

Membrane Potential Is Generated by the Differential Distribution of Ions

Through the operation of ionic pumps and special ionic buffering mechanisms, neurons actively maintain precise internal concentrations of several important ions, including Na^+, K^+, Cl^-, and Ca^{2+}. The mechanisms by which they do so are illustrated in Figures 6.2 and 6.3. The intracellular and extracellular concentrations of Na^+, K^+, Cl^-, and Ca^{2+} differ markedly (Fig. 6.2); K^+ is actively concentrated inside the cell, and Na^+, Cl^-, and Ca^{2+} are actively extruded to the extracellular space. However, this does not mean that the cell is filled only with positive charge; anions to which the plasma membrane is impermeant are also present inside the cell and almost balance the high concentration of K^+. The osmolarity inside the cell is approximately equal to that outside the cell.

Electrical and Thermodynamic Forces Determine the Passive Distribution of Ions

Ions tend to move down their concentration gradients through specialized ionic pores, known as ionic channels, in the plasma membrane. Through simple laws of thermodynamics, the high concentration of K^+ inside glial cells, neurons, and axons results in a tendency for K^+ ions to diffuse down their concentration gradient and leave the cell or cell process (Fig. 6.3). However, the movement of ions across the membrane also results in a redistribution of electrical charge. As

K^+ ions move down their concentration gradient, the intracellular voltage becomes more negative, and this increased negativity results in an electrical attraction between the negative potential inside the cell and the positively charged, K^+ ions, thus offsetting the outward flow of these ions. The membrane is selectively permeable; that is, it is impermeable to the large anions inside the cell, which cannot follow the potassium ions across the membrane. At some membrane potential, the "force" of the electrostatic attraction between the negative membrane potential inside the cell and the positively charged K^+ ions will exactly balance the thermal "forces" by which K^+ ions tend to flow down their concentration gradient (Fig. 6.3). In this circumstance, it is equally likely that a K^+ ion exits the cell by movement down the concentration gradient as it is that a K^+ ion enters the cell due to the attraction between the negative membrane potential and the positive charge of this ion. At this membrane potential, there is no net flow of K^+ (the same number of K^+ ions enter the cell as leave the cell per unit time) and these ions are said to be in equilibrium. The membrane potential at which this occurs is known as the equilibrium potential. (See Box 6.1 for calculation of the equilibrium potential.)

To illustrate, let us consider the passive distribution of K^+ ions in the squid giant axon as studied by Hodgkin and Huxley. The K^+ concentration $[K^+]$ inside the squid giant axon is about 400 mM, whereas the $[K^+]$ outside the axon is about 20 mM. Because $[K^+]_i$ is greater than $[K^+]_o$, potassium ions will tend to flow down their concentration gradient, taking positive charge with them. The equilibrium potential (at which the tendency for

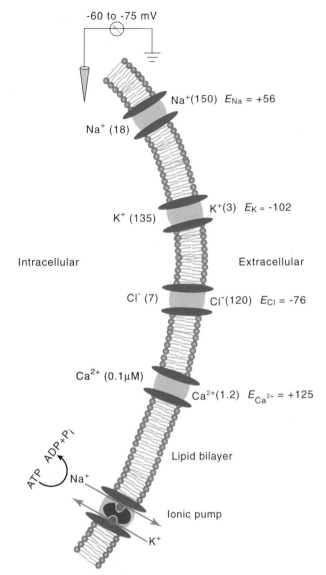

FIGURE 6.2 Differential distribution of ions inside and outside plasma membrane of neurons and neuronal processes, showing ionic channels for Na^+, K^+, Cl^-, and Ca^{2+}, as well as an electrogenic Na^+–K^+ ionic pump (also known as Na^+, K^+-ATPase). Concentrations (in millimoles except that for intracellular Ca^{2+}) of the ions are given in parentheses; their equilibrium potentials (E) for a typical mammalian neuron are indicated.

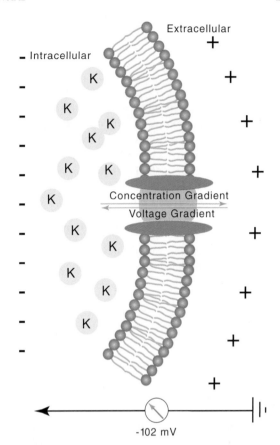

FIGURE 6.3 The equilibrium potential is influenced by the concentration gradient and the voltage difference across the membrane. Neurons actively concentrate K^+ inside the cell. These K^+ ions tend to flow down their concentration gradient from inside to outside the cell. However, the negative membrane potential inside the cell provides an attraction for K^+ ions to enter or remain within the cell. These two factors balance one another at the equilibrium potential, which in a typical mammalian neuron is $-102\,mV$ for K^+.

K^+ ions to flow down their concentration gradient will be exactly offset by the attraction for K^+ ions to enter the cell because of the negative charge inside the cell) at a room temperature of 20°C can be calculated by the Nernst equation as such:

$$E_K = 58.2 \log_{10}(20/400) = -76\,mV$$

Therefore, at a membrane potential of $-76\,mV$, K^+ ions have an equal tendency to flow either into or out of the axon. The concentrations of K^+ in mammalian neurons and glial cells differ considerably from that in the squid giant axon, which is adapted to live in sea water. By substituting 3.1 mM for $[K^+]_o$ and 140 mM for $[K^+]_i$ in the Nernst equation, with mammalian body temperature, T = 37°C, we obtain

$$E_K = 61.5 \log_{10}(3.1/140) = -102\,mV$$

Movements of Ions Can Cause Either Hyperpolarization or Depolarization

In mammalian cells, at membrane potentials positive to $-102\,mV$, K^+ ions tend to flow out of the cell. Increasing the ability of K^+ ions to flow across the membrane (i.e., increasing the conductance of the membrane to K^+ (gK)) causes the membrane potential to become more negative, or hyperpolarized, due to the exiting of positively charged ions from inside the cell (Fig. 6.4).

BOX 6.1

NERNST EQUATION

The equilibrium potential is determined by (1) the concentration of the ion inside and outside the cell, (2) the temperature of the solution, (3) the valence of the ion, and (4) the amount of work required to separate a given quantity of charge. The equation that describes the equilibrium potential was formulated by a German physical chemist named Walter Nernst in 1888:

$$E_{ion} = RT/zF \cdot \ln([ion]_o/[ion]_i)$$

Here, E_{ion} is the membrane potential at which the ionic species is at equilibrium, R is the gas constant [8.315 J per Kelvin per mole (J K^{-1} mol^{-1})], T is the temperature in Kelvins ($T_{Kelvin} = 273.16 + T_{Celcius}$), F is Faraday's constant [96,485 coulombs per mole (C mol^{-1})], z is the valence of the ion, and [ion]$_o$ and [ion]$_i$ are the concentrations of the ion outside and inside the cell, respectively. For a monovalent, positively charged ion (cation) at room temperature (20°C), substituting the appropriate numbers and converting natural log (ln) into log base 10 (log$_{10}$) results in

$$E_{ion} = 58.2 \log_{10}([ion]_o/[ion]_i);$$

at a body temperature of 37°C, the Nernst equation is

$$E_{ion} = 61.5 \log_{10}([ion]_o/[ion]_i).$$

David A. McCormick

At membrane potentials negative to −102 mV, K$^+$ ions tend to flow into the cell; increasing the membrane conductance to K$^+$ causes the membrane potential to become more positive, or depolarized, due to the flow of positive charge into the cell. The membrane potential at which the net current "flips" direction is referred to as the reversal potential. If the channels conduct only one type of ion (e.g., K$^+$ ions), then the reversal potential and the Nernst equilibrium potential for that ion coincide (Fig. 6.4A). Increasing the membrane conductance to K$^+$ ions while the membrane potential is at the equilibrium potential for K$^+$ (E_K) does not change the membrane potential because no net driving force causes K$^+$ ions to either exit or enter the cell. However, this increase in membrane conductance to K$^+$ decreases the ability of other species of ions to change the membrane potential because any deviation of the potential from E_K increases the drive for K$^+$ ions to either exit or enter the cell, thereby drawing the membrane potential back toward E_K (Fig. 6.4B). This effect is known as a "shunt" and is important for some effects of inhibitory synaptic transmission.

The exiting and entering of the cell by K$^+$ ions during generation of the membrane potential gives rise to a curious problem. When K$^+$ ions leave the cell to generate a membrane potential, the concentration of K$^+$ changes both inside and outside the cell. Why does this change in concentration not alter the equilibrium potential, thus changing the tendency for K$^+$ ions to flow down their concentration gradient? The reason is that the number of K$^+$ ions required to leave the cell to achieve the equilibrium potential is quite small. For example, if a cell were at 0 mV and the membrane suddenly became permeable to K$^+$ ions, only about 10^{-12} mol of K$^+$ ions per square centimeter of membrane would move from inside to outside the cell in bringing the membrane potential to the equilibrium potential for K$^+$. In a spherical cell of 25 μm diameter, this would amount to an average decrease in intracellular K$^+$ of only about 4 μM (e.g., from 140 to 139.996 mM). However, there are instances when significant changes in the concentrations of K$^+$ may occur, particularly during the generation of pronounced activity, such as an epileptic seizure. During the occurrence of a tonic–clonic generalized (grand mal) seizure, large numbers of neurons discharge throughout the cerebral cortex in a synchronized manner. This synchronous discharge of large numbers of neurons significantly increases the extracellular K$^+$ concentration, by as much as a couple of millimoles, resulting in a commensurate positive shift in the equilibrium potential for K$^+$. This shift in the equilibrium potential can increase the excitability of affected neurons and neuronal processes and thus promote the spread of the seizure activity. Fortunately, the extracellular concentration of K$^+$ is tightly regulated and is kept at normal levels through uptake by glial cells, as well as by diffusion through the fluid of the extracellular space.

As is true for K$^+$ ions, each of the membrane-permeable species of ions possesses an equilibrium potential that depends on the concentration of that ion inside and outside the cell. Thus, equilibrium potentials may

A

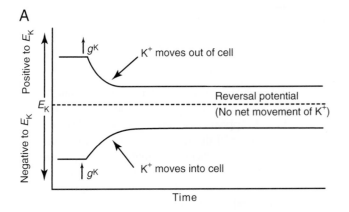

B

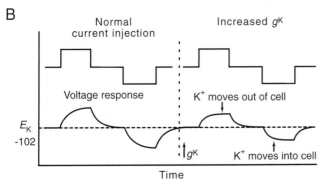

FIGURE 6.4 Increases in K$^+$ conductance can result in hyperpolarization, depolarization, or no change in membrane potential. (A) Opening K$^+$ channels increases the conductance of the membrane to K$^+$, denoted gK. *If* the membrane potential is positive to the equilibrium potential (also known as the reversal potential) for K$^+$, then increasing gK will cause some K$^+$ ions to leave the cell, and the cell will become hyperpolarized. If the membrane potential is negative to E_K when gK is increased, then K$^+$ ions will enter the cell, therefore making the inside more positive (more depolarized). If the membrane potential is exactly E_K when gK is increased, then there will be no net movement of K$^+$ ions. (B) Opening K$^+$ channels when the membrane potential is at E_K does not change the membrane potential; however, it reduces the ability of other ionic currents to move the membrane potential away from E_K. For example, a comparison of the ability of the injection of two pulses of current, one depolarizing and one hyperpolarizing, to change the membrane potential before and after opening K$^+$ channels reveals that increases in gK decrease the responses of the cell noticeably.

vary between different cell types, such as those found in animals adapted to live in salt water versus mammalian neurons. In mammalian neurons, the equilibrium potential is approximately 56 mV for Na$^+$, approximately −76 mV for Cl$^-$, and about 125 mV for Ca^{2+} (Fig. 6.2). Thus, increasing the membrane conductance to Na$^+$ (gNa) through the opening of Na$^+$ channels depolarizes the membrane potential toward 56 mV; increasing the membrane conductance to Cl$^-$ brings the membrane potential closer to −76 mV; and finally increasing the membrane conductance to Ca^{2+} depolarizes the cell toward 125 mV.

Na$^+$, K$^+$, and Cl$^-$ Contribute to the Determination of the Resting Membrane Potential

If a membrane is permeable to only one ion and no electrogenic ionic pumps are operating (see next section), then the membrane potential is necessarily at the equilibrium potential for that ion. At rest, the plasma membrane of most cell types is not at the equilibrium potential for K$^+$ ions, indicating that the membrane is also permeable to other types of ions. For example, the resting membrane of the squid giant axon is permeable to Cl$^-$ and Na$^+$, as well as K$^+$, due to the presence of ionic channels that not only allow these ions to pass but also are open at the resting membrane potential. Because the membrane is permeable to K$^+$, Cl$^-$, and Na$^+$, the resting potential of the squid giant axon is not equal to E_K, E_{Na}, or E_{Cl}, but is somewhere in between these three. A membrane permeable to more than one ion has a steady-state membrane potential whose value is between those of the equilibrium potentials for each of the permeant ions (Box 6.2).

Different Types of Neurons Have Different Resting Potentials

Intracellular recordings from neurons in the mammalian central nervous system (CNS) reveal that different types of neurons exhibit different resting membrane potentials. Indeed, some types of neurons do not even exhibit a true "resting" membrane potential; they spontaneously and continuously generate action potentials even in the total lack of synaptic input. In the visual system, intracellular recordings have shown that photoreceptor cells of the retina—the rods and cones—have a membrane potential of approximately −40 mV at rest and are hyperpolarized when activated by light. Cells in the dorsal lateral geniculate nucleus, which receive axonal input from the retina and project to the visual cortex, have a resting membrane potential of approximately −70 mV during sleep and −55 mV during waking, whereas pyramidal neurons of the visual cortex have a resting membrane potential of about −75 mV. Presumably, the resting membrane potentials of different cell types in the central and peripheral nervous system are highly regulated and are functionally important. For example, the depolarized membrane potential of photoreceptors presumably allows the membrane potential to move in both negative and positive directions in response to changes in light intensity. The hyperpolarized membrane potential of thalamic neurons during sleep (−70 mV) dramatically decreases the flow of information from the sensory periphery to the cerebral cortex,

---

BOX 6.2

GOLDMAN-HODGKIN-KATZ EQUATION

An equation developed by Goldman and later used by Alan Hodgkin and Bernard Katz describes the steady-state membrane potential for a given set of ionic concentrations inside and outside the cell and the relative permeabilities of the membrane to each of those ions:

$$V_m \frac{RT}{F} \cdot \ln\left(\frac{(p_K[K^+]_o + p_{Na}[Na^+]_o + p_{Cl}[Cl^-]_i)}{(p_K[K^+]_i + p_{Na}[Na^+]_i + p_{Cl}[Cl^-]_o)} \right).$$

The relative contribution of each ion is determined by its concentration differences across the membrane and the relative permeability (p_K, p_{Na}, p_{Cl}) of the membrane to each type of ion. If a membrane is permeable to only one ion, then the Goldman–Hodgkin–Katz equation reduces to the Nernst equation. In the squid giant axon, at resting membrane potential, the permeability ratios are

$$p_K : p_{Na} : p_{Cl} = 1.00 : 0.04 : 0.45.$$

The membrane of the squid giant axon, at rest, is most permeable to K^+ ions, less so to Cl^-, and least permeable to Na^+. (Chloride appears to contribute considerably less to the determination of the resting potential of mammalian neurons.) These results indicate that the resting membrane potential is determined by the resting permeability of the membrane to K^+, Na^+, and Cl^-. In theory, this resting membrane potential may be anywhere between E_K (e.g., $-76\,mV$) and E_{Na} ($55\,mV$). For the three ions at 20°C, the equation is

$$V_m = \frac{58.2 \log_{10}\{(1 \cdot 20 + 0.04 \cdot 440 + 0.45 \cdot 40)}{(1 \cdot 400 + 0.04 \cdot 50 + 0.45 \cdot 560)\}} = -62\,mV.$$

This suggests that the squid giant axon should have a resting membrane potential of $-62\,mV$. In fact, the resting membrane potential may be a few millivolts hyperpolarized to this value through the operation of the electrogenic Na^+–K^+ pump.

David A. McCormick

---

presumably to allow the cortex to be relatively undisturbed during sleep, and the 20-mV membrane potential between the resting potential and the action potential threshold in cortical pyramidal cells allows these cells to be strongly influenced by subthreshold barrages of synaptic potentials from other cortical neurons (see Chapters 5 and 12).

Ionic Pumps Actively Maintain Ionic Gradients

Because the resting membrane potential of a neuron is not at the equilibrium potential for any particular ion, ions constantly flow down their concentration gradients. This flux becomes considerably larger with the generation of electrical and synaptic potentials because ionic channels are opened by these events. Although the absolute number of ions traversing the plasma membrane during each action potential or synaptic potential may be small in individual cells, the collective influence of a large neural network of cells, such as in the brain, and the presence of ion fluxes even at rest can substantially change the distribution of ions inside and outside neurons. Cells have solved this problem with the use of active transport of ions against their concentration gradients. The proteins that actively transport ions are referred to as ionic pumps, of which

the Na^+–K^+ pump is perhaps the most thoroughly understood (Lauger, 1991). The Na^+–K^+ pump is stimulated by increases in the intracellular concentration of Na^+ and moves Na^+ out of the cell while moving K^+ into it, achieving this task through the hydrolysis of ATP (Fig. 6.2). Three Na^+ ions are extruded for every two K^+ ions transported into the cell. Because of the unequal transport of ions, the operation of this pump generates a hyperpolarizing electrical potential and is said to be electrogenic. The Na^+–K^+ pump typically results in the membrane potential of the cell being a few millivolts more negative than it would be otherwise.

The Na^+–K^+ pump consists of two subunits, α and β, arranged in a tetramer $(\alpha\beta)_2$. The Na^+–K^+ pump is believed to operate through conformational changes that alternatively expose a Na^+-binding site to the interior of the cell (followed by the release of Na^+) and a K^+ binding site to the extracellular fluid (Fig. 6.2). Such a conformation change may be due to the phosphorylation and dephosphorylation of the protein.

The membranes of neurons and glia contain multiple types of ionic pumps, used to maintain the proper distribution of each ionic species important for cellular signaling. Many of these pumps are operated by the Na^+ gradient across the cell, whereas others operate through a mechanism similar to that of the Na^+–K^+

pump (i.e., the hydrolysis of ATP). For example, the calcium concentration inside neurons is kept to very low levels (typically 50–100 nM) through the operation of both types of ionic pumps, as well as special intracellular Ca^{2+} buffering mechanisms. Ca^{2+} is extruded from neurons through both a Ca^{2+}, Mg^{2+}-ATPase, and a Na^{+}–Ca^{2+} exchanger. The Na^{+}–Ca^{2+} exchanger is driven by the Na^{+} gradient across the membrane and extrudes one Ca^{2+} ion for each Na^{+} ion allowed to enter the cell.

The Cl^{-} concentration in neurons is actively maintained at a low level through operation of a chloride-bicarbonate exchanger, which brings in one ion of Na^{+} and one ion of HCO_3^{-} for each ion of Cl^{-} extruded. Intracellular pH can also markedly affect neuronal excitability and is therefore tightly regulated, in part by a Na^{+}–H^{+} exchanger that extrudes one proton for each Na^{+} allowed to enter the cell.

Summary

The membrane potential is generated by the unequal distribution of ions, particularly K^{+}, Na^{+}, and Cl^{-}, across the plasma membrane. This unequal distribution of ions is maintained by ionic pumps and exchangers. K^{+} ions are concentrated inside the neuron and tend to flow down their concentration gradient, leading to a hyperpolarization of the cell. At the equilibrium potential, the tendency of K^{+} ions to flow out of the cell will be exactly offset by the tendency of K^{+} ions to enter the cell due to the attraction of the negative potential inside the cell. The resting membrane is also permeable to Na^{+} and Cl^{-} and therefore the resting membrane potential is approximately −75 to −40 mV, in other words, substantially positive to E_K.

ACTION POTENTIAL

An Increase in Na^{+} and K^{+} Conductance Generates Action Potentials

Hodgkin and Huxley not only recorded the action potential with an intracellular microelectrode (Fig. 6.1), but also went on to perform a remarkable series of experiments that explained qualitatively and quantitatively the ionic mechanisms by which the action potential is generated (Hodgkin and Huxley, 1952a, b). As mentioned earlier, these investigators found that during the action potential, the membrane potential of the cell rapidly overshoots 0 mV and approaches the equilibrium potential for Na^{+}. After generation of the action potential, the membrane potential repolar-

izes and becomes more negative than before, generating an after-hyperpolarization. These changes in membrane potential during generation of the action potential were associated with a large increase in conductance of the plasma membrane, but to what does the membrane become conductive in order to generate the action potential? The prevailing hypothesis was that there was a nonselective increase in conductance causing the negative resting potential to increase toward 0 mV. Since publication of the experiments of E. Overton in 1902, the action potential had been known to depend on the presence of extracellular Na^{+}. Reducing the concentration of Na^{+} in the artificial seawater bathing the axon resulted in a marked reduction in the amplitude of the action potential. On the basis of these and other data, Hodgkin and Katz proposed that the action potential is generated through a rapid increase in the conductance of the membrane to Na^{+} ions. A quantitative proof of this theory was lacking, however, because ionic currents could not be observed directly. Development of the voltage-clamp technique by Kenneth Cole at the Marine Biological Laboratory in Massachusetts resolved this problem and allowed quantitative measurement of the Na^{+} and K^{+} currents underlying the action potential (Cole, 1949; Box 6.3).

Hodgkin and Huxley (1952) used the voltage-clamp technique to investigate the mechanisms of generation of the action potential in the squid giant axon. Axons and neurons have a threshold for the initialization of an action potential of about −45 to −55 mV. Increasing the voltage from −60 to 0 mV produces a large, but transient, flow of positive charge into the cell (known as inward current). This transient inward current is followed by a sustained flow of positive charge out of the cell (the outward current). By voltage clamping the cell and substituting different ions inside or outside the axon or both, Hodgkin, Huxley, and colleagues demonstrated that the transient inward current is carried by Na^{+} ions flowing into the cell and the sustained outward current is mediated by a sustained flux of K^{+} ions moving out of the cell (Fig. 6.6) (Hodgkin and Huxley, 1952a, 1952b; Hille, 1977).

Na^{+} and K^{+} currents (I_{Na} and I_K, respectively) can be blocked, allowing each current to be examined in isolation (Fig. 6.6B). Tetrodotoxin (TTX), a powerful poison found in the puffer fish *Spheroides rubripes*, selectively blocks voltage-dependent Na^{+} currents (the puffer fish remains a delicacy in Japan and must be prepared with the utmost care by the chef). Using TTX, one can selectively isolate I_K and examine its voltage dependence and time course (Fig. 6.6B).

Another compound, tetraethylammonium (TEA), is a useful pharmacological tool for selectively blocking

---

BOX 6.3

VOLTAGE-CLAMP TECHNIQUE

In the voltage-clamp technique, two independent electrodes are inserted into the squid giant axon: one for recording the voltage difference across the membrane and the other for intracellularly injecting the current (Fig. 6.5). These electrodes are then connected to a feedback circuit that compares the measured voltage across the membrane with the voltage desired by the experimenter. If these two values differ, then current is injected into the axon to compensate for this difference. This continuous feedback cycle, in which the voltage is measured and current is injected, effectively "clamps" the membrane at a particular voltage. If ionic channels were to open, then the resultant flow of ions into or out of the axon would be compensated for by the injection of positive or negative current into the axon through the current-injection electrode. The current injected through this electrode is necessarily equal to the current flowing through the ionic channels. It is this injected current that is measured by the experimenter. The benefits of the voltage-clamp technique are twofold. First, the current injected into the axon to keep the membrane potential "clamped" is necessarily equal to the current flowing through the ionic channels in the membrane, thereby giving a direct measurement of this current. Second, ionic currents are both voltage and time dependent; they become active at certain membrane potentials and do so at a particular rate. Keeping the voltage constant in the voltage clamp allows these two variables to be separated; the voltage dependence and the kinetics of the ionic currents flowing through the plasma membrane can be measured directly.

David A. McCormick

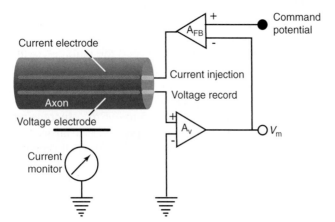

FIGURE 6.5 The voltage-clamp technique keeps the voltage across the membrane constant so that the amplitude and time course of ionic currents can be measured. In the two-electrode voltage-clamp technique, one electrode measures the voltage across the membrane while the other injects current into the cell to keep the voltage constant. The experimenter sets a voltage to which the axon or neuron is to be stepped (the command potential). Current is then injected into the cell in proportion to the difference between the present membrane potential and the command potential. This feedback cycle occurs continuously, thereby clamping the membrane potential to the command potential. By measuring the amount of current injected, the experimenter can determine the amplitude and time course of the ionic currents flowing across the membrane.

I_K (Fig. 6.6B). The use of TEA to examine the voltage dependence and time course of the Na^+ current underlying action-potential generation (Fig. 6.6B) reveals some fundamental differences between Na^+ and K^+ currents. First, the inward Na^+ current activates, or "turns on," much more rapidly than the K^+ current (giving rise to the name "delayed rectifier" for this K^+ current). Second, the Na^+ current is transient; it inactivates, even if the membrane potential is maintained at 0 mV (Fig. 6.6A). In contrast, the outward K^+ current, once activated, remains "on" as long as the membrane potential is clamped to positive levels; that is, the K^+ current does not inactivate, it is sustained. Remarkably, from one experiment, we see that the Na^+ current both activates and inactivates rapidly, whereas the K^+ current only activates slowly. These fundamental properties of the underlying Na^+ and K^+ channels allow the generation of action potentials.

Hodgkin and Huxley proposed that K^+ channels possess a voltage-sensitive "gate" that opens by depolarization and closes by the subsequent repolarization of the membrane potential. This process of "turning on" and "turning off" the K^+ current came to be known as activation and deactivation. The Na^+ current also exhibits voltage-dependent activation and deactivation (Fig. 6.6), but the Na^+ channels also become inactive despite maintained depolarization. Thus, the Na^+ current not only activates and deactivates, but also exhibits a separate process known as inactivation, whereby the channels become blocked even though

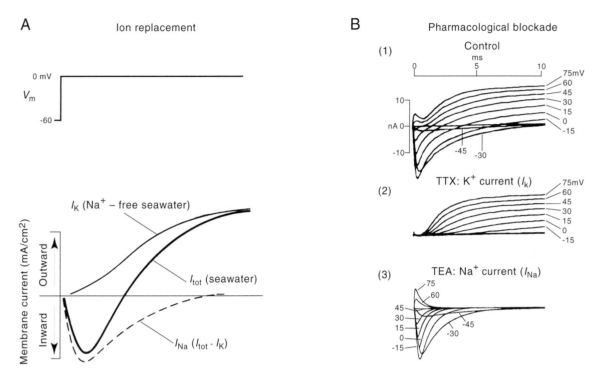

FIGURE 6.6 Voltage-clamp analysis reveals ionic currents underlying action potential generation. (A) Increasing the potential from −60 to 0 mV across the membrane of the squid giant axon activates an inward current followed by an outward current. If the Na^+ in seawater is replaced by choline (which does not pass through Na^+ channels), then increasing the membrane potential from −60 to 0 mV results in only the outward current, which corresponds to I_K. Subtracting I_K from the recording in normal seawater illustrates the amplitude–time course of the inward Na^+ current, I_{Na}. Note that I_K activates more slowly than I_{Na} and that I_{Na} inactivates with time. From Hodgkin and Huxley (1952). (B) These two ionic currents can also be isolated from one another through the use of pharmacological blockers. (1) Increasing the membrane potential from −45 to 75 mV in 15-mV steps reveals the amplitude–time course of inward Na^+ and outward K^+ currents. (2) After the block of I_{Na} with the poison tetrodotoxin (TTX), increasing the membrane potential to positive levels activates I_K only. (3) After the block of I_K with tetraethylammonium (TEA), increasing the membrane potential to positive levels activates I_{Na} only. From Hille (1977).

they are activated. Removal of this inactivation is achieved by removal of depolarization and is a process known as deinactivation. Thus, Na^+ channels possess two voltage–sensitive processes: activation–deactivation and inactivation–deinactivation. The kinetics of these two properties of Na^+ channels are different: inactivation takes place at a slower rate than activation. The functional consequence of the two mechanisms is that Na^+ ions are allowed to flow across the membrane only when the current is activated but not inactivated. Accordingly, Na^+ ions do not flow at resting membrane potentials because the activation gate is closed (even though the inactivation gate is not). Upon depolarization, the activation gate opens, allowing Na^+ ions to flow into the cell. However, this depolarization also results in closure (at a slower rate) of the inactivation gate, which then blocks the flow of Na^+ ions. Upon repolarization of the membrane potential, the activation gate once again closes and the inactivation gate once again opens, preparing the axon for generation of the next action potential (Fig. 6.7). Depolarization allows ionic current to flow by virtue of activation of the channel. The rush of Na^+ ions into

the cell further depolarizes the membrane potential and more Na^+ channels become activated, forming a positive feedback loop that rapidly (within 100 μs or so) brings the membrane potential toward E_{Na}. However, the depolarization associated with generation of the action potential also inactivates Na^+ channels, and, as a larger and larger percentage of Na^+ channels become inactivated, the rush of Na^+ into the cell diminishes. This inactivation of Na^+ channels and the activation of K^+ channels result in the repolarization of the action potential. This repolarization deactivates the Na^+ channels. Then, the inactivation of the channel is slowly removed, and the channels are ready, once again, for the generation of another action potential (Fig. 6.7).

By measuring the voltage sensitivity and kinetics of these two processes, activation–deactivation and inactivation–deinactivation of the Na^+ current, as well as the activation–deactivation of the delayed rectifier K^+ current, Hodgkin and Huxley generated a series of mathematical equations that quantitatively described the generation of the action potential (calculation of the propagation of a single action potential required

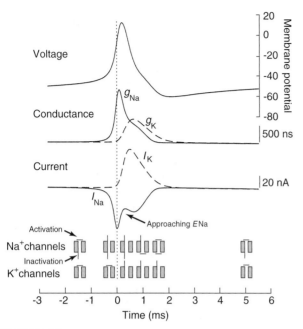

FIGURE 6.7 Generation of the action potential is associated with an increase in membrane Na$^+$ conductance and Na$^+$ current followed by an increase in K$^+$ conductance and K$^+$ current. Before action potential generation, Na$^+$ channels are neither activated nor inactivated (illustrated at the bottom of the figure). Activation of Na$^+$ channels allows Na$^+$ ions to enter the cell, depolarizing the membrane potential. This depolarization also activates K$^+$ channels. After activation and depolarization, the inactivation particle on the Na$^+$ channels closes and the membrane potential repolarizes. The persistence of the activation of K$^+$ channels (and other membrane properties) generates an after-hyperpolarization. During this period, the inactivation particle of the Na$^+$ channel is removed and the K$^+$ channels close.

an entire week of cranking a mechanical calculator). According to these early experimental and computational neuroscientists, the action potential is generated as follows. Depolarization of the membrane potential increases the probability of Na$^+$ channels being in the activated, but not yet inactivated, state. At a particular membrane potential, the resulting inflow of Na$^+$ ions tips the balance of the net ionic current from outward to inward (remember that depolarization will also increase K$^+$ and Cl$^-$ currents by moving the membrane potential away from E_K and E_{Cl}). At this membrane potential, known as the action potential threshold (typically about −55 mV), the movement of Na$^+$ ions into the cell depolarizes the axon and opens more Na$^+$ channels, causing yet more depolarization of the membrane; repetition of this process yields a rapid, positive feedback loop that brings the axon close to E_{Na}.

However, even as more and more Na$^+$ channels are becoming activated, some of these channels are also inactivating and therefore no longer conducting Na$^+$

ions. In addition, the delayed rectifier K$^+$ channels are also opening, due to the depolarization of the membrane potential, and allowing positive charge to exit the cell. At some point, close to the peak of the action potential, the inward movement of Na$^+$ ions into the cell is exactly offset by the outward movement of K$^+$ ions out of the cell. After this point, the outward movement of K$^+$ ions dominates, and the membrane potential is repolarized, corresponding to the fall of the action potential. The persistence of the K$^+$ current for a few milliseconds following the action potential generates the after-hyperpolarization. During this after-hyperpolarization, which is lengthened by the membrane time constant, inactivation of the Na$^+$ channels is removed, preparing the axon for generation of the next action potential (Fig. 6.7).

The occurrence of an action potential is not associated with substantial changes in the intracellular or extracellular concentrations of Na$^+$ or K$^+$, as shown earlier for the generation of the resting membrane potential. For example, generation of a single action potential in a 25-μm-diameter hypothetical spherical cell should increase the intracellular concentration of Na$^+$ by only approximately 6 μM (from about 18 to 18.006 mM). Thus, the action potential is an electrical event generated by a change in the distribution of charge across the membrane and not by a marked change in the intracellular or extracellular concentration of Na$^+$ or K$^+$.

Action Potentials Typically Initiate in the Axon Initial Segment and Propagate Down the Axon and Backward through the Dendrites

Neurons have complex morphologies including dendritic arbors, a cell body, and typically one axonal output, which branches extensively. In many cells, all these parts of the neuron are capable of independently generating action potentials. The activity of most neurons is dictated by barrages of synaptic potentials generated at each moment by a variable subset of the thousands of synapses impinging upon the cell's dendrites and soma. Where then is the action potential initiated? In most cells, each action potential is initiated in the initial portion of the axon, known as the axon initial segment (Coombs *et al.*, 1957; Stuart *et al.*, 1997; Shu *et al.*, 2007). The initial segment of the axon has the lowest threshold for action potential generation because it typically contains a moderately high density of Na$^+$ channels and it is a small compartment that is easily depolarized by the in-rush of Na$^+$ ions. Once a spike is initiated (e.g., about 30–50 microns down the axon from the cell body in cortical pyramidal cells), this action potential then propagates orthodromically down the axon to the synaptic terminals, where it causes release of transmitter, as well as antidromically back through the cell body and into the cells dendrites, where it can modulate intracellular processes.

Refractory Periods Prevent "Reverberation"

The ability of depolarization to activate an action potential varies as a function of the time since the last generation of an action potential, due to the inactivation of Na^+ channels and the activation of K^+ channels. Immediately after the generation of an action potential, another action potential usually cannot be generated regardless of the amount of current injected into the axon. This period corresponds to the absolute refractory period and largely is mediated by the inactivation of Na^+ channels. The relative refractory period occurs during the action potential after-hyperpolarization and follows the absolute refractory period. The relative refractory period is characterized by a requirement for the increased injection of ionic current into the cell to generate another action potential and results from persistence of the outward K^+ current. The practical implication of refractory periods is that action potentials are not allowed to "reverberate" between the axon initial segment and axon terminals.

The Speed of Action Potential Propagation Is Affected by Myelination

Axons may be either myelinated or unmyelinated. Invertebrate axons or small vertebrate axons are typically unmyelinated, whereas larger vertebrate axons are often myelinated. As described in Chapter 4, sensory and motor axons of the peripheral nervous system are myelinated by specialized cells (Schwann cells) that form a spiral wrapping of multiple layers of myelin around the axon (Fig. 6.8). Several Schwann cells wrap around an axon along its length; between the ends of successive Schwann cells are small gaps (nodes of Ranvier). In the central nervous system, a single oligodendrocyte, a special type of glial cell, typically ensheaths several axonal processes.

In unmyelinated axons, the Na^+ and K^+ channels taking part in action potential generation are distributed along the axon, and the action potential propagates along the length of the axon through local depolarization of each neighboring patch of membrane, causing that patch of membrane also to generate an action potential (Fig. 6.8). In myelinated axons, however, the Na^+ channels are concentrated at the nodes of Ranvier. The generation of an action potential at each node results in depolarization of the next node and subsequently generation of an action potential

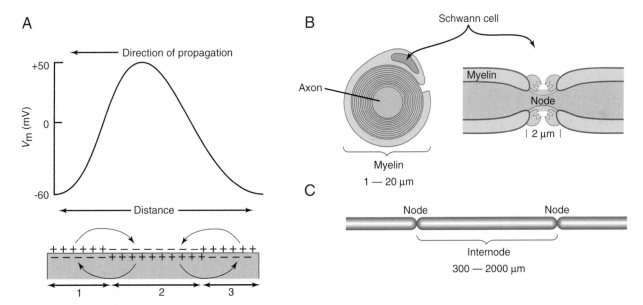

FIGURE 6.8 Propagation of the action potential in unmyelinated and myelinated axons. (A) Action potentials propagate in unmyelinated axons through the depolarization of adjacent regions of membrane. In the illustrated axon, region 2 is undergoing depolarization during the generation of the action potential, whereas region 3 already has generated the action potential and is now hyperpolarized. The action potential will propagate further by depolarizing region 1. (B) Vertebrate myelinated axons have a specialized Schwann cell that wraps around them in many spiral turns. The axon is exposed to the external medium at the nodes of Ranvier (Node). (C) Action potentials in myelinated fibers are regenerated at the nodes of Ranvier, where there is a high density of Na^+ channels. Action potentials are induced at each node through the depolarizing influence of the generation of an action potential at an adjacent node, thereby increasing conduction velocity.

with an internode delay of only about $20\,\mu s$ (see Chapter 5), referred to as saltatory conduction (from the Latin *saltare*, "to leap"). Growing evidence indicates that, near the nodes of Ranvier and underneath the myelin covering, K^+ channels may play a role in determining the resting membrane potential and repolarization of the action potential. A cause of some neurological disorders, such as multiple sclerosis and Guillain–Barre syndrome, is the demyelination of axons, resulting in a block of conduction of the action potentials.

Ion Channels Are Membrane-Spanning Proteins with Water-Filled Pores

The generation of ionic currents useful for the propagation of action potential requires the movement of significant numbers of ions across the membrane in a relatively short time. The rate of ionic flow during the generation of an action potential is far too high to be achieved by an active transport mechanism and results instead from the opening of ion channels. Although the existence of ionic channels in the membrane has been postulated for decades, their properties and structure only recently have become known in detail. The powerful combination of electrophysiological and molecular techniques has enhanced our knowledge of the structure–function relations of ionic channels greatly (Box 6.4).

Various neural toxins were particularly useful in the initial isolation of ionic channels. For example, three subunits (α, $\beta 1$, $\beta 2$) of the voltage-dependent Na^+ channel were isolated with the use of a derivative of a scorpion toxin. The α subunit of the Na^+ channel is a

---

BOX 6.4

ION CHANNELS AND DISEASE

Cells cannot survive without functional ion channels. It is therefore not surprising that an ever-increasing number of diseases have been found to be associated with defective ion channel function. There are a number of different mechanisms by which this may occur.

1. Mutations in the coding region of ion channel genes may lead to gain or loss of channel function, either of which may have deleterious consequences. For example, mutations producing enhanced activity of the epithelial Na^+ channel are responsible for Liddle's syndrome, an inherited form of hypertension, whereas other mutations in the same protein that cause reduced channel activity give rise to hypotension. The most common inherited disease in Caucasians is also an ion channel mutation. This disease is cystic fibrosis (CF), which results from mutations in the epithelial chloride channel, known as CFTR. The most common mutation, deletion of a phenylalanine at position 508, results in defective processing of the protein and prevents it from reaching the surface membrane. CFTR regulates chloride fluxes across epithelial cell membranes, and this loss of CFTR activity leads to reduced fluid secretion in the lung, resulting in potentially fatal lung infections.

2. Mutations in the promoter region of the gene may cause under- or overexpression of a given ion channel.

3. Other diseases result from defective regulation of channel activity by cellular constituents or extracellular ligands. This defective regulation may be caused by mutations in the genes encoding the regulatory molecules themselves or defects in the pathways leading to their production. Some forms of maturity-onset diabetes of the young (MODY) may be attributed to such a mechanism. ATP-sensitive potassium (K-ATP) channels play a key role in the glucose-induced insulin secretion from pancreatic β cells, and their defective regulation is responsible for one form of MODY.

4. Autoantibodies to channel proteins may cause disease by downregulating channel function—often by causing internalization of the channel protein itself. Well-known examples are myasthenia gravis, which results from antibodies to skeletal muscle acetylcholine channels, and Lambert–Eaton myasthenic syndrome, in which patients produce antibodies against presynaptic Ca^{2+} channels.

5. Finally, a number of ion channels are secreted by cells as toxic agents. They insert into the membrane of the target cell and form large nonselective pores, leading to cell lysis and death. The hemolytic toxin produced by the bacterium *Staphylococcus aureus* and the toxin secreted by the protozoan *Entamoeba histolytica*, which causes amebic dysentery, are examples.

Natural mutations in ion channels have been invaluable for studying the relationship between channel structure and function. In many cases, genetic analysis of a

BOX 6.4 (cont'd)

disease has led to the cloning of the relevant ion channel. The first K⁺ channel to be identified (Shaker), for example, came from the cloning of the gene that caused Drosophila to shake when exposed to ether. Likewise, the gene encoding the primary subunit of a cardiac potassium channel (KCNQ1) was identified by positional cloning in families carrying mutations that caused a cardiac disorder known as long QT syndrome (see later). Conversely, the large number of studies on the relationship between Na⁺ channel structure and function has greatly assisted our understanding of how mutations in Na⁺ channels produce their clinical phenotypes.

Many diseases are genetically heterogeneous, and the same clinical phenotype may be caused by mutations in different genes. Long QT syndrome is a relatively rare inherited cardiac disorder that causes abrupt loss of consciousness, seizures, and sudden death from ventricular arrhythmia in young people. Mutations in five different genes, two types of cardiac muscle K⁺ channels (HERG, KCNQ1, KCNE1, KCNE2) and the cardiac muscle sodium channel (SCN1A), give rise to long QT syndrome. The disorder is characterized by a long QT interval in the electrocardiogram, which reflects the delayed repolarization of the cardiac action potential. As therefore might be expected, mutations in the cardiac Na⁺ channel gene that cause long QT syndrome enhance the Na⁺ current (by reducing Na⁺ channel inactivation), whereas those in

potassium channel genes cause loss of function and reduce the K⁺ current.

Mutations in many different types of ion channels have been shown to cause human diseases. In addition to the examples listed earlier, mutations in water channels cause nephrogenic diabetes insipidus; mutations in gap junction channels cause Charcot–Marie–Tooth disease (a form of peripheral neuropathy) and hereditary deafness; mutations in the skeletal muscle Na⁺ channel cause a range of disorders known as periodic paralyses; mutations in intracellular Ca^{2+}-release channels cause malignant hyperthermia (a disease in which inhalation anesthetics trigger a potentially fatal rise in body temperature); and mutations in neuronal voltage-gated Ca^{2+} channels cause migraine and episodic ataxia. The list increases daily. As is the case with all single gene disorders, the frequency of these diseases in the general population is very low. However, the insight they have provided into the relationship between ion channel structure and function, and into the physiological role of the different ion channels, has been invaluable. As William Harvey said in 1657 "nor is there any better way to advance the proper practice of medicine than to give our minds to the discovery of the usual form of nature, by careful investigation of the rarer forms of disease."

Frances M. Ashcroft

large glycoprotein with a molecular mass of 270 kDa, whereas the $\beta 1$ and $\beta 2$ subunits are smaller polypeptides of molecular masses 39 and 37 kDa, respectively (Fig. 6.9). The α subunit, of which there are at least nine different isoforms, is the building block of the water-filled pore of the ionic channel, whereas the β subunits have some other role, such as in the regulation or structure of the native channel (Catterall, 2000a).

The α subunit of the Na⁺ channel contains four internal repetitions (Fig. 6.9B). Hydrophobicity analysis of these four components reveals that each contains six hydrophobic domains that may span the membrane as an α-helix. Of these six membrane–spanning components, the fourth (S4) has been proposed to be critical to the voltage sensitivity of the Na⁺ channels. Voltage-sensitive gating of Na⁺ channels is accomplished by the redistribution of ionic charge ("gating charge") in the channel. Positive charges in the S4 region may act as voltage sensors such that an increase

in the positivity of the inside of the cell results in a conformational change of the ionic channel. In support of this hypothesis, site-directed mutagenesis of the S4 region of the Na⁺ channel to reduce the positive charge of this portion of the pore also reduces the voltage sensitivity of activation of the ionic channel.

The mechanisms of inactivation of ionic channels have been analyzed with a combination of molecular and electrophysiological techniques. The most convincing hypothesis is that inactivation is achieved by a block of the inner mouth of the aqueous pore. Ionic channels are inactivated without detectable movement of ionic current through the membrane; thus inactivation is probably not directly gated by changes in the membrane potential alone. Rather, inactivation is triggered or facilitated as a secondary consequence of activation. Site-directed mutagenesis or the use of antibodies has shown that the part of the molecule between regions III and IV may be allowed to move to

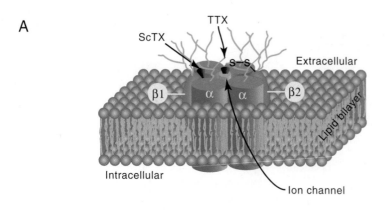

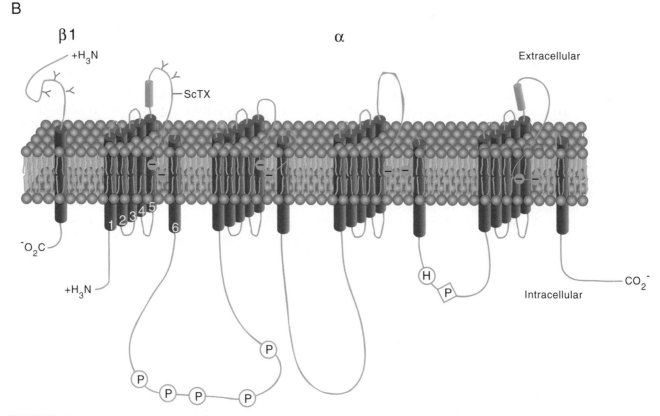

FIGURE 6.9 Structure of the sodium channel. (A) Cross-section of a hypothetical sodium channel consisting of a single transmembrane α subunit in association with a $\beta 1$ subunit and a $\beta 2$ subunit. The α subunit has receptor sites for α-scorpion toxins (ScTX) and tetrodotoxin (TTX). (B) Primary structures of α and $\beta 1$ subunits of sodium channel illustrated as transmembrane-folding diagrams. Cylinders represent probable transmembrane α-helices.

block the cytoplasmic side of the ionic pore after the conformational change associated with activation.

Neurons of the Central Nervous System Exhibit a Wide Variety of Electrophysiological Properties

The first intracellular recordings of action potentials in mammalian neurons by Sir John Eccles and col-

leagues revealed a remarkable similarity to those of the squid giant axon and gave rise to the assumption that the electrophysiology of neurons in the CNS was really rather simple: when synaptic potentials brought the membrane potential positive to action potential threshold, action potentials were produced through an increase in Na^+ conductance followed by an increase in K^+ conductance, as in the squid giant axon. The assumption, therefore, was that the complicated patterns of activity generated by the brain during the resting, sleeping, or active states were brought about as an interaction of the very large numbers of neurons present in the mammalian CNS. However, intracellular recordings of invertebrate neurons revealed that different cell types exhibit a wide variety of different electrophysiological behaviors, indicating that neurons may be significantly more complicated than the squid giant axon.

Elucidation of the basic electrophysiology and synaptic physiology of different types of neurons and neuronal pathways within the mammalian CNS was facilitated by the *in vitro* slice technique, in which thin (~0.5 mm) slices of brain can be maintained for several hours. Intracellular recordings from identified cells revealed that neurons of the mammalian nervous system, such as those of invertebrate networks, can generate complex patterns of action potentials entirely through intrinsic ionic mechanisms and without synaptic interaction with other cell types. For example, Rodolfo Llinás and colleagues discovered that Purkinje cells of the cerebellum can generate high-frequency trains (>200 Hz) of Na$^+$- and K$^+$-mediated action potentials interrupted by Ca^{2+} spikes in the dendrites, whereas a major afferent to these neurons, the inferior olivary cell, can generate rhythmic sequences of broad action potentials only at low frequencies (<15 Hz) through an interaction between various Ca^{2+}, Na$^+$, and K$^+$ conductances (Fig. 6.10). These *in vitro* recordings confirmed a major finding obtained with earlier intracellular recordings *in vivo*: each morphologically distinct class of neuron in the brain exhibits distinct electrophysiological features. Just as cortical pyramidal cells are morphologically distinct from cerebellar Purkinje cells, which are distinct from thalamic relay cells, the electrophysiological properties of each of these different cell types are also markedly distinct (Llinás, 1988).

Although no uniform classification scheme has been formulated in which all the different types of neurons of the brain can be classified, a few characteristic patterns of activity seem to recur. The first general class of action potential generation is characterized by those cells that generate trains of action potentials one spike at a time. The more prolonged the depolarization of these cells, the more prolonged their discharge. The more intensely these cells are depolarized, the higher the frequency of action potential generation. This type of relatively linear behavior is typical for brain stem and spinal cord motor neurons functioning in muscle contraction. A modification of this basic pattern of "regular firing" is characterized by the generation of trains of action potentials that exhibit a marked tendency to slow down in frequency with time, a process

known as spike frequency adaptation. Examples of cells that discharge in this manner are cortical and hippocampal pyramidal cells.

In addition to these regular firing cells, many neurons in the central nervous system exhibit the intrinsic propensity to generate rhythmic bursts of action potentials (Fig. 6.10). Examples of such neurons are thalamic relay neurons, inferior olivary neurons, and some types of cortical and hippocampal pyramidal cells. In these cells, clusters of action potentials can occur together when the membrane is brought above the firing threshold. These clusters of action potentials typically are generated through the activation of specialized Ca^{2+} currents that, through their slower kinetics, allow the membrane potential to be depolarized for a sufficient period to result in the generation of a burst of regular, Na$^+$- and K$^+$-dependent action potentials (discussed in the next section).

Yet another general category of neurons in the brain comprises cells that generate relatively short duration (<1 ms) action potentials and can discharge at relatively high frequencies (>300 Hz). Such electrophysiological properties often are found in neurons that release the inhibitory amino acid γ-aminobutyric acid (Fig. 6.10) including some types of interneurons in the cerebral cortex, thalamus, and hippocampus. Finally, the last general category of neurons consists of those that spontaneously generate action potentials at relatively slow frequencies (e.g., 1–10 Hz). This type of electrophysiological behavior often is associated with neurons that release neuromodulatory transmitters, such as acetylcholine, norepinephrine, serotonin, and histamine. Neurons that release these neuromodulatory substances often innervate wide regions of the brain and appear to set the "state" of the different neural networks of the CNS in a manner similar to the modulation of the different organs of the body by the sympathetic and parasympathetic nervous systems.

Each of these unique intrinsic patterns of activity in the nervous system is due to the presence of a distinct mixture and distribution of different ionic currents in the cells. As in classical studies of the squid giant axon, these different ionic currents have been characterized, at least in part, with voltage-clamp and pharmacological techniques, and the basic electrophysiological properties have been replicated with computational simulations (Figs. 6.7 and 6.12).

Neurons Have Multiple Active Conductances

The search for the electrophysiological basis of the varying intrinsic properties of different types of neurons of vertebrates and invertebrates revealed a wide variety of ionic currents. Each type of ionic

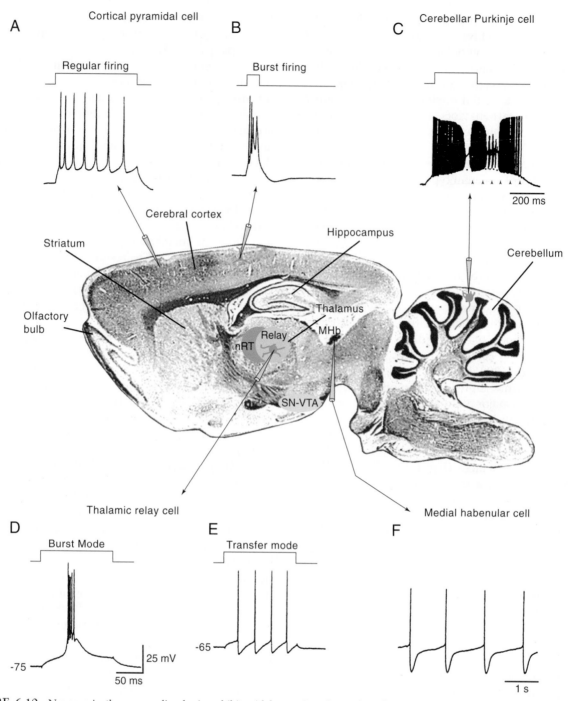

FIGURE 6.10 Neurons in the mammalian brain exhibit widely varying electrophysiological properties. (A) Intracellular injection of a depolarizing current pulse in a cortical pyramidal cell results in a train of action potentials that slow down in frequency. This pattern of activity is known as "regular firing." (B) Some cortical cells generated bursts of three or more action potentials, even when depolarized only for a short period of time. (C) Cerebellar Purkinje cells generate high-frequency trains of action potentials in their cell bodies that are disrupted by the generation of Ca^{2+} spikes in their dendrites. These cells can also generate "plateau potentials" from the persistent activation of Na^+ conductances (arrowheads). Thalamic relay cells may generate action potentials either as bursts (D) or as tonic trains of action potentials (E) due to the presence of a large low-threshold Ca^{2+} current. (F) Medial habenular cells generate action potentials at a steady and slow rate in a "pacemaker" fashion.

current is characterized by several features: (1) the type of ions conducted by the underlying ionic channels (e.g., Na⁺, K⁺, Ca²⁺, Cl⁻, or mixed cations), (2) their voltage and time dependence, and (3) their sensitivity to second messengers. In vertebrate neurons, two distinct Na⁺ currents have been identified and six distinct Ca²⁺ currents and more than seven distinct K⁺ currents are known (Fig. 6.11). This is a minimal number, as these currents are formed from a much greater pool of channel subunits. The following sections briefly review these classes of ionic currents and their ionic channels, relating them to the different patterns of behavior mentioned earlier for neurons in the mammalian CNS.

Na⁺ Currents Are Both Transient and Persistent

Depolarization of many different types of vertebrate neurons results not only in the activation of the rapidly activating and inactivating Na⁺ current (I_{Nat}) underlying action potential generation, but also in the rapid activation of a Na⁺ current that does not inactivate and is therefore known as the "persistent" Na⁺ current (I_{Nap}). The threshold for activation of the persistent Na⁺ current is typically about −65 mV; that is, below the threshold for the generation of action potentials. This property gives this current the interesting ability to enhance or facilitate the response of the neuron to depolarizing, yet subthreshold, inputs. For example, synaptic events that depolarize the cell will activate I_{Nap}, resulting in an extra influx of positive charge and

therefore a larger depolarization than otherwise would occur. Likewise, hyperpolarizations may result in deactivation of I_{Nap}, again resulting in larger hyperpolarizations than otherwise would occur. In this manner, the persistent Na⁺ current may play an important regulatory function in the control of the functional responsiveness of the neuron to synaptic inputs and may contribute to the dynamic coupling of the dendrites to the soma.

Persistent activation of I_{Nap} may also contribute to another electrophysiological feature of neurons: the generation of plateau potentials. A plateau potential refers to the ability of many different types of neurons to generate, through intrinsic ionic mechanisms, a prolonged (from tens of milliseconds to seconds) depolarization and action potential discharge in response to a short-lasting depolarization (Fig. 6.10C). One can wonder whether such plateau potentials contribute to persistent firing in neurons during the performance of working memory tasks, where groups of neurons have to retain information for brief (seconds) periods of time.

K⁺ Currents Vary in Their Voltage Sensitivity and Kinetics

Potassium currents that contribute to the electrophysiological properties of neurons are numerous and exhibit a wide range of voltage-dependent and kinetic properties (reviewed in Coetzee *et al.*, 1999). Perhaps the simplest K⁺ current is that characterized by Hodgkin and Huxley: this K⁺ current, I_K, activates rapidly on depolarization and does not inactivate (Fig. 6.11). Other K⁺ currents activate with depolarization but also inactivate with time. For example, the rapid activation and inactivation of I_A give this current a transient appearance (Fig. 6.11), and I_A is believed to be important in controlling the rate of action potential generation, particularly at low frequencies (Fig. 6.12). Like the Na⁺ channel, I_A channels are inactivated by the plugging of the inner mouth of the pore through movement of an inactivation particle.

Another broad class of K⁺ channels consists of those that are sensitive to changes in the intracellular concentration of Ca²⁺. These K⁺ currents collectively are referred to as I_{KCa} (Fig. 6.11). Still other K⁺ channels are not only activated by voltage, but are also modulated by the activation of various modulatory neurotransmitter receptors (e.g., I_M; Fig. 6.11). Between these classic examples of K⁺ currents are a variety of other types that have not been fully characterized, including K⁺ currents that vary from one another in their voltage sensitivity, kinetics, and response to various second messengers.

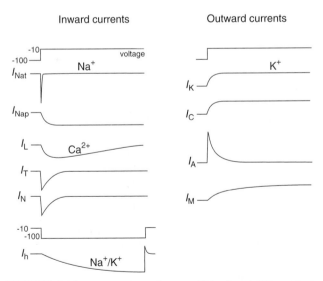

FIGURE 6.11 Voltage dependence and kinetics of different ionic currents in the mammalian brain. Depolarization of the membrane potential from −100 to −10 mV results in the activation of currents entering or leaving neurons.

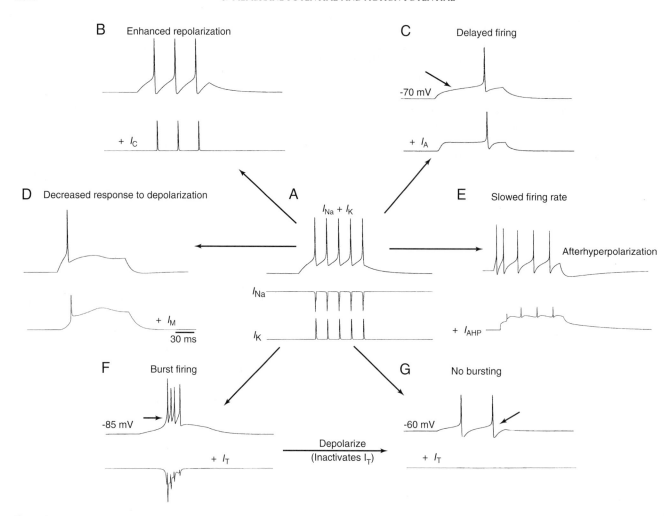

FIGURE 6.12 Simulation of the effects of the addition of various ionic currents to the pattern of activity generated by neurons in the mammalian CNS. (A) The repetitive impulse response of the classical Hodgkin–Huxley model (voltage recordings above, current traces below). With only I_{Na} and I_K, the neuron generates a train of five action potentials in response to depolarization. Addition of I_C (B) enhances action potential repolarization. Addition of I_A (C) delays the onset of action potential generation. Addition of I_M (D) decreases the ability of the cell to generate a train of action potentials. Addition of I_{AHP} (E) slows the firing rate and generates a slow after-hyperpolarization. Finally, addition of the transient Ca^{2+} current I_T results in two states of action potential firing: (F) burst firing at −85 mV and (G) tonic firing at −60 mV. From Huguenard and McCormick (1994).

Molecular biological studies of voltage-sensitive K⁺ channels, first done in *Drosophila* and later in mammals, have revealed the presence of a large number of genes that generate K⁺ channels. The voltage-gated K⁺ channel subunits are contained within nine distinct subfamilies: Kv1-9 (reviewed in Coetzee *et al.*, 1999). These genes generate a wide variety of different K⁺ channels, not only due to the large number of genes involved, but also due to alternative RNA splicing, gene duplication, and other posttranslational mechanisms. Functional expression of different K⁺ channels reveals remarkable variation in the rate of inactivation, such that some are rapidly inactivating (A current like), whereas others inactivate more slowly and, finally, some K⁺ channels do not inactivate, such as I_K.

One of the largest subfamilies of K⁺ channels are those that give rise to the resting membrane potential, so-called "leak channels." Interestingly, these channels appear to be opened by gaseous anesthetics, indicating that hyperpolarization of central neurons is a major component of general anesthesia. It is now clear that each type of neuron in the nervous system contains a unique set of functional voltage-sensitive K⁺ channels, selected, modified, and placed in particular spatial locations in the cell in a manner that facilitates the unique role of that cell type in neuronal processing.

An additional current that also regulates the responsiveness of neurons to depolarizing inputs is the voltage-sensitive K⁺ current known as the M current (Figs. 6.11 and 6.12D). By investigating the ionic mechanisms by which the release of acetylcholine from pre-ganglionic neurons in the brain results in prolonged changes in the excitability of neurons of the sympathetic ganglia, Brown and Adams (1980) discovered a unique K^+ current that slowly (over tens of milliseconds) turns on with depolarization of the neuron (Fig. 6.12D). The slow activation of this K^+ current results in a decrease in the responsiveness of the cell to depolarization, and therefore regulates how the cell responds to excitation. This K^+ current, like I_{AHP}, is reduced by the activation of a wide variety of receptors, including muscarinic receptors, for which it is named. Reduction of I_M results in a marked increase in responsiveness of the affected cell to depolarizing inputs and again may contribute to the mechanisms by which neuromodulatory systems control the state of activity in cortical and hippocampal networks (reviewed in McCormick, 1992).

Ca^{2+} Currents Control Electrophysiological Properties and Ca^{2+}-Dependent Second-Messenger Systems

Ionic channels that conduct Ca^{2+} are present in all neurons. These channels are special in that they serve two important functions. First, Ca^{2+} channels are present throughout the different parts of the neuron (dendrites, soma, synaptic terminals) and contribute greatly to the electrophysiological properties of these processes. Second, Ca^{2+} channels are unique in that Ca^{2+} is an important second messenger in neurons, and entry of Ca^{2+} into the cell can affect numerous physiological functions, including neurotransmitter release, synaptic plasticity, neurite outgrowth during development, and even gene expression. On the bases of their voltage sensitivity, their kinetics of activation and inactivation, and their ability to be blocked by various pharmacological agents, Ca^{2+} currents can be separated into at least six separate categories, three of which are I_T ("transient"), I_L ("long lasting"), and I_N ("neither"), illustrated in Fig. 6.11A. A fourth, I_P, is found in Purkinje cells of the cerebellum, as well as in many different cell types of the CNS. These Ca^{2+} channels are formed from at least 10 different α subunits as well as a variety of β and γ subunits, indicating that there are an even greater number of Ca^{2+} currents yet to be characterized.

Neurons Possess Multiple Subtypes of High-Threshold Ca^{2+} Currents

High voltage-activated Ca^{2+} channels are activated at membrane potentials positive to approximately $-40\,mV$ and include the currents I_L, I_N, and I_P. L-type calcium currents exhibit a high threshold for activation (about $-10\,mV$) and give rise to rather persistent, or long-lasting, ionic currents (Fig. 6.11A). Dihydropyridines, Ca^{2+} channel antagonists, are clinically useful for their effects on the heart and vascular smooth muscle (e.g., for the treatment of arrhythmias, angina, and migraine headaches) and selectively block L-type Ca^{2+} channels. In contrast with I_L, I_N is not blocked by dihydropyridines: it is blocked selectively by a toxin found in Pacific cone shells (ω-conotoxin-GVIA). N-type Ca^{2+} channels have a threshold for activation of about $-20\,mV$, inactivate with maintained depolarization, and are modulated by a variety of neurotransmitters. In some cell types, I_N has a role in the Ca^{2+}-dependent release of neurotransmitters at presynaptic terminals. The P-type calcium channel is distinct from N and L types in that it is not blocked by either dihydropyridines or ω-conotoxin-GVIA but is blocked by a toxin (ω-agatoxin-IVA) present in the venom of the Funnel web spider. This type of calcium channel activates at relatively high thresholds and does not inactivate. Prevalent in Purkinje cells as well as other cell types, as mentioned earlier, the P-type Ca^{2+} channel participates in the generation of dendritic Ca^{2+} spikes, which can strongly modulate the firing pattern of the neuron in which it resides (Fig. 6.10C; Llinás, 1988).

Collectively, high threshold-activated Ca^{2+} channels contribute to the generation of action potentials in mammalian neurons. The activation of Ca^{2+} currents adds somewhat to the depolarizing part of the action potential, but, more importantly, these channels allow Ca^{2+} to enter the cell, which has the secondary consequence of activation of various Ca^{2+}-activated K^+ currents and protein kinases (see Chapter 10). As mentioned earlier, activation of these K^+ currents modifies the pattern of action potentials generated in the cell (Figs. 6.10 and 6.12).

High-threshold Ca^{2+} channels are similar to the Na^+ channel in that they are composed of a central $\alpha 1$ subunit that forms the aqueous pore and several regulatory or auxiliary subunits. As in the Na^+ channel, the primary structure of the $\alpha 1$ subunit of the Ca^{2+} channel consists of four homologous domains (I–IV), each containing six regions (S1–S6) that may generate transmembrane α-helices. Genes for at least 10 different Ca^{2+} channel α subunits have been cloned and are separated into three subfamilies (Cav1, Cav2, and Cav3). The properties of the products of these genes indicate that I_L is likely to correspond to the Cav1 subfamily, whereas I_N corresponds to Cav2.2 and I_T is formed from the Cav3 subfamily (see Catterall, 2000b).

Low-Threshold Ca²⁺ Currents Generate Bursts of Action Potentials

Low-threshold Ca²⁺ currents (Fig. 6.11A) often take part in the generation of rhythmic bursts of action potentials (Figs. 6.10 and 6.12). The low-threshold Ca²⁺ current is characterized by a threshold for activation of about −65 mV, which is below the threshold for generation of typical Na⁺-K⁺-dependent action potentials (−55 mV). This current inactivates with maintained depolarization. Because of these properties, the role of low-threshold Ca²⁺ currents differs markedly from that of the high-threshold Ca²⁺ currents. Through activation and inactivation of the low-threshold Ca²⁺ current, neurons can generate slow (about 100 ms) Ca²⁺ spikes, which can result, due to their prolonged duration, in generation of a high-frequency "burst" of short-duration Na⁺-K⁺ action potentials (Fig. 6.10).

In the mammalian brain, this pattern is especially well exemplified by the activity of thalamic relay neurons; in the visual system, these neurons receive direct input from the retina and transmit this information to the visual cortex. During periods of slow wave sleep, the membrane potential of these relay neurons is relatively hyperpolarized, resulting in the removal of inactivation (deinactivation) of the low-threshold Ca²⁺ current. This deinactivation allows these cells to spontaneously generate low-threshold Ca²⁺ spikes and bursts of from two to five action potentials (Fig. 6.13). The large number of thalamic relay cells bursting during sleep in part gives rise to the spontaneous synchronized activity that early investigators were so surprised to find during recordings from the brains of sleeping animals. It has even proved possible to maintain one of the sleep-related brain rhythms (spindle waves) intact in slices of thalamic tissue maintained *in vitro*, due the generation of this rhythm by the interaction of a local network of thalamic cells and their electrophysiological properties.

The transition to waking or the period of sleep when dreams are prevalent (rapid eye movement sleep) is associated with a maintained depolarization of thalamic relay cells to membrane potentials ranging from about −60 to −55 mV. The low-threshold Ca²⁺ current is inactivated and therefore the burst discharges are abolished. In this way, the properties of a single ionic current (I_T) help explain in part the remarkable changes

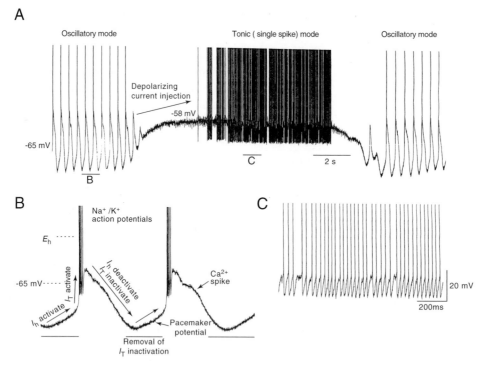

FIGURE 6.13 Two different patterns of activity generated in the same neuron, depending on membrane potential. (A) The thalamic neuron spontaneously generates rhythmic bursts of action potentials due to the interaction of the Ca²⁺ current I_T and the inward "pacemaker" current I_h. Depolarization of the neuron changes the firing mode from rhythmic burst firing to tonic action potential generation in which spikes are generated one at a time. Removal of this depolarization reinstates rhythmic burst firing. This transition from rhythmic burst firing to tonic activity is similar to that which occurs in the transition from sleep to waking. (B) Expansion of detail of rhythmic burst firing. (C) Expansion of detail of tonic firing. From McCormick and Pape (1990).

in brain activity taking place in the transition from sleep to waking (Fig. 6.13).

Hyperpolarization-Activated Ionic Currents Are Involved in Rhythmic Activity

In most types of neurons, hyperpolarization negative to approximately −60 mV activates an ionic current, known as I_h, that conducts both Na$^+$ and K$^+$ ions (Fig. 6.11). This current typically has very slow kinetics, turning on with a time constant on the order of tens of milliseconds to seconds. Because the channels underlying this current allow the passage of both Na$^+$ and K$^+$ ions, the reversal potential of I_h is typically about −35 mV—between E_{Na} and E_K. Because this current is activated by hyperpolarization below approximately −60 mV, it is typically dominated by the inward movement of Na$^+$ ions and is therefore depolarizing. For what purpose could neurons use a depolarizing current that activates when the cell is hyperpolarized? A clue comes from cardiac cells in which this current, known as I_f for "funny," is important for determining heart rate. Activation of I_f results in a slow depolarization of the membrane potential between adjacent cardiac action potentials. The more that I_f is activated, the faster the membrane depolarizes between beats and therefore the sooner the threshold for the next action potential is reached and the next beat is generated. In this manner, the amplitude, or sensitivity to voltage, of I_f can modify the heart rate. Interestingly, the sensitivity of I_f to voltage is adjusted by the release of noradrenaline and acetylcholine; the activation of adrenoceptors by noradrenaline increases I_f and therefore increases the heart rate, whereas the activation of muscarinic receptors decreases I_f, thereby decreasing the heart rate (see DiFrancesco, 2005). This continual adjustment of I_f results from a push–pull arrangement between β-adrenergic and muscarinic cholinergic receptors and is mediated by the adjustment of intracellular levels of cyclic AMP. Indeed, the recent cloning of h-channels reveals that their structure is similar to that of cyclic nucleotide-gated channels.

Could I_h play a role in neurons similar to that of I_f in the heart? Possibly. Synchronized rhythmic oscillations in the membrane potential of large numbers of neurons, in some respects similar to those of the heart, are characteristic of the mammalian brain. Oscillations of this type are particularly prevalent in thalamic relay neurons during some periods of sleep, as mentioned earlier. Intracellular recordings from these thalamic neurons reveal that they often generate rhythmic "bursts" of action potentials mediated by the activation of a slow spike that is generated through the activation of the low-threshold, or transient, Ca^{2+} current,

I_T (Fig. 6.13). Between the occurrence of each low-threshold Ca^{2+} spike is a slowly depolarizing membrane potential generated by activation of the mixed Na$^+$-K$^+$ current I_h, as with I_f in the heart. The amplitude, or voltage sensitivity, of I_h adjusts the rate at which the thalamic cells oscillate, and, as with the heart, this sensitivity is adjusted by the release of modulatory neurotransmitters. In a sense, the thalamic neurons are "beating" in a manner similar to that of the heart.

Summary

An action potential is generated by the rapid influx of Na$^+$ ions followed by a slightly slower efflux of K$^+$ ions. Although the generation of an action potential does not disrupt the concentration gradients of these ions across the membrane, the movement of charge is sufficient to generate a large and brief deviation in the membrane potential. Action potentials typically are initiated in the axon initial segment and the propagation of the action potential along the axon allows communication of the output of the cell to its distant synapses. Neurons possess many different types of ionic channels in their membranes, allowing complex patterns of action potentials to be generated and complex synaptic computations to occur within single neurons.

References

Brazier, M. A. B. (1988). "A History of Neurophysiology in the 19th Century." Raven Press, New York.

Brown, D. A. and Adams, P. R. (1980). Muscarinic suppression of a novel voltage sensitive K$^+$ current in a vertebrate neurone. *Nature (London)* **283**, 673–676.

Carbonne, E. and Lux, H. D. (1984). A low voltage-activated, fully inactivating Ca channel in vertebrate sensory neurones. *Nature (London)* **310**, 501–502.

Catterall, W. A. (2000a). From ionic currents to molecular mechanisms: The structure and function of voltage-gated sodium currents. *Neuron* **26**, 13–25.

Catterall, W. A. (2000b). Structure and regulation of voltage-gated Ca^{2+} channels. *Annu. Rev. Cell Dev. Biol.* **16**, 521–555.

Coetzee, W. A., Amarillo, Y., Chui, J., Chow, A., Lau, D., McCormack, T., Moreno, H., Nadal, M. S., Ozaita, A., Pountney, D., Saganich, M., Vega-Saenz de Miera, E., and Rudy, B. (1999). Molecular diversity of K$^+$ channels. *Ann. N.Y. Acad. Sci.* **868**, 233–285.

Cole, K. S. (1949). Dynamic electrical characteristics of the squid axon membrane. *Arch. Sci. Physiol.* **3**, 253–258.

Coombs, J. S., Curtis, D. R., and Eccles, J. C. (1957). The interpretation of spike potentials of motoneurons. *J. Physiol.* **139**, 198–231.

DiFrancesco, D. (2005). Physiology and pharmacology of the cardiac pacemaker ("funny") current. *Pharmacol. Ther.* **107**, 59–79.

Hille, B. (1977). Ionic basis of resting potentials and action potentials. *In* "Handbook of Physiology" (E. R. Kandel, ed.), Sect. 1, Vol. 1, pp. 99–136. Am. Physiol. Soc., Bethesda, MD.

Hodgkin, A. L. (1976). Chance and design in electrophysiology: An informal account of certain experiments on nerve carried out between 1934 and 1952. *J. Physiol. (Lond.)* **263**, 1–21.

Hodgkin, A. L. and Huxley, A. F. (1939). Action potentials recorded from inside a nerve fiber. *Nature (Lnd.)* **144**, 710–711.

Hodgkin, A. L. and Huxley, A. F. (1952). Currents carried by sodium and potassium ions through the membrane of the giant axon of *Loligo. J. Physiol. (Lond.)* **116**, 449–472.

Hodgkin, A. L. and Huxley, A. F. (1952). A quantitative description of membrane current and its application to conduction and excitation in nerve. *J. Physiol. (Lond.)* **117**, 500–544.

Lauger, P. (1991). "Electrogenic Ion Pumps." Sinauer, Sunderland, MA.

Llinás, R. R. (1988). The intrinsic electrophysiological properties of mammalian neurons: Insights into central nervous system function. *Science* **242**, 1654–1664.

Ludwig, A., Zong, X., Jeglitsch, M., Hofmann, F., and Biel, M. (1998). A family of hyperpolarization-activated mammalian cation channels. *Nature* **393**, 587–591.

McCormick, D. A. (1992). Neurotransmitter actions in the thalamus and cerebral cortex and their role in neuromodulation of thalamocortical activity. *Prog. Neurobiol.* **39**, 337–388.

McCormick, D. A. and Pape, H.-C. (1990). Properties of a hyperpolarization-activated cation current and its role in rhythmic oscillation in thalamic relay neurones. *J. Physiol. (Lond.)* **431**, 291–318.

Nicoll, R. A. (1988). The coupling of neurotransmitter receptors to ion channels in the brain. *Science* **241**, 545–551.

Nowycky, M. C., Fox, A. P., and Tsien, R. W. (1985). Three types of neuronal calcium channel with different calcium agonist sensitivity. *Nature (Lond.)* **316**, 440–443.

Reithmeier, R. A. F. (1994). Mammalian exchangers and co-transporters. *Curr. Opin. Cell Biol.* **6**, 583–594.

Shu, Y., Duque, A., Yu, Y., Haider, B., and McCormick, D. A. (2007). Properties of action potential initiation in neocortical pyramidal cells: evidence from whole cell axon recordings. *J. Neurophysiol.* **97**, 746–760.

Stuart, G., Spruston, N., Sakmann, B., and Hausser, M. (1997). Action potential initiation and backpropagation in neurons of the mammalian CNS. *Trends Neurosci.* **10**, 125–131.

Suggested Readings

Hille, B. (2001) "Ionic Channels of Excitable Membranes," 3rd Ed. Sinauer, Sunderland, MA.

Hodgkin, A. L. (1992). "Chance and Design." Cambridge Univ. Press, Cambridge.

Huguenard, J. and McCormick, D. A. (1994). "Electrophysiology of the Neuron." Oxford Univ. Press, New York.

Johnston, D. and Wu, S. M-S. (1995). "Foundations of Cellular Neurophysiology." MIT Press.

Koch, C. (1999). "Biophysics of Computation." Oxford Univ. Press, New York.

David A. McCormick

Neurotransmitters

A century ago the as yet unnamed field of neuroscience was having growth pains. It was a time marked by claim and counterclaim, confusion, and recrimination—not unlike politics today, or for that matter, science. One major reason for the tumult was the heretical idea that the brain is not one continuous network (a syncytium), with each cell in physical contact with its neighbors. The pioneering studies of Santiago Ramon y Cajal revealed a very different picture from the dogma, with each cell (neuron) of the brain being an independent structure (see Shepherd, 1991). Although final confirmation of this view awaited electron microscopic verification a half century later, a general acceptance of neurons as the independent building blocks of the nervous system emerged much more quickly. In turn, this resulted in a new debate: how do these neurons communicate? The answer to this question has evolved continuously, and become increasing complex, with different modes of communication identified. In this chapter we will explore how neurons communicate by chemical messengers.

Neurons vary widely in function but share certain structural characteristics. They have a cell body (soma) from which processes emerge. Axons can be short or long and can be local or project to distant areas. In contrast, dendrites are exclusively local structures. The identification of long (axon) and short (dendrite) processes led to the idea that the long axonal projections served to transmit information between neurons, often over long distances, and dendrites were the receptive elements of neurons, responding to the long information sent via axons. The gap over which information between neurons must be conveyed was termed the synapse by Charles Sherrington (see Shepherd, 1991), and came to identify pre- and postsynaptic cells. This general conceptual framework remains in place today, although there are many exceptions, including dendrites that release neuroactive substances and axons that receive inputs from other neurons. One other characteristic proposed by Sherrington that is central to the concept of chemical communication between neurons is that synaptic transmission does not follow all-or-none rules, but is graded in strength and is flexible.

SEVERAL MODES OF NEURONAL COMMUNICATION EXIST

Over the first half of the twentieth century there was a vigorous debate on the nature of neuronal communication: as one review of this debate described, it was a war of sparks (electrical communication) and soups (chemical messengers) (see Valenstein, 2005). This debate raged despite the fact some evidence was marshaled in support of chemical transmission in the mid-nineteenth century. In 1849, Claude Bernard noted that curare, the active constituent of a poison that was applied to arrows in South America, blocked nerve-to-muscle signaling. This effect subsequently was shown to be due to the binding of curare to postsynaptic (muscle) receptors for acetylcholine (ACh), thus blocking neuromuscular transmission. About 50 years later Thomas Elliott found that epinephrine caused the contraction of smooth muscle that had been deprived of its nerve inputs, suggesting that muscle contraction depended on the action of chemical molecules liberated from nerves. In a key series of studies using isolated frog hearts, Otto Loewi provided firm evidence in support of chemical neurotransmission by showing that ACh was released upon nerve stimulation and activated a target muscle.

These and other data suggested that the major means of interneuronal communication is chemical in nature, but that neurons also use other processes for intercellular communication. Among these are electrical synaptic transmission, ephaptic interactions, and autocrine, paracrine, and long-range signaling, to which molecules produced by both neural and non-neural cells contribute. The nonsynaptic mode of intercellular communication with the longest range (distance) is hormonal signaling. For example, some hormones made outside of the brain can enter the central nervous system (CNS) to exert effects on neurons that express receptors for the hormones. These hormone actions may occur over the short term (changes in neuronal activity) or, more often, over a longer period (long-lasting changes in gene expression).

Molecules released from neurons can also be used in intercellular signaling that does not require synaptic specializations. There are a host of molecules that are secreted by neurons or diffuse passively from cells, ranging from conventional neurotransmitters to gases such as nitric oxide. These factors may act through autocrine mechanisms (activating receptors on the same cell that releases them) or paracrine pathways to influence nearby cells. Some of these compounds can be retrograde signaling molecules, providing chemically-coded information to the presynaptic neuron, thus reversing the more common forward (anterograde) direction of information flow. The role of such molecules is thought to be primarily in modulating neural activity, although they may also provide guidance cues for neurons that are growing toward their final targets in the brain as well as for the establishment and maintenance of synaptic connections (see Chapter 19).

CHEMICAL TRANSMISSION

Chemically-mediated transmission is the major mode of neuronal communication. The general acceptance of chemical neurotransmission as the means of conveying information between neurons resulted in the establishment of specific criteria for designation of a compound as a neurotransmitter. A relatively small number of compounds have been designated as neurotransmitters based on these "classical" criteria. However, over the past generation it has become apparent that there is a large number of chemical messengers that broadly qualify as intercellular transmitters, although these compounds often do not meet—and even stand in sharp contrast to—the classical criteria.

Several Criteria Have Been Established for a Neurotransmitter

Neurotransmitters usually are considered to be endogenous substances that are released from neurons, act on receptor sites that are typically present on membranes of postsynaptic cells, and produce a functional change in the properties of the target cell. Over the years general agreement evolved that several criteria should be met for a substance to be designated a neurotransmitter:

1. A neurotransmitter must be synthesized by and released from neurons. This means that the presynaptic neuron should contain a transmitter and the appropriate enzymes need to synthesize that neurotransmitter. Synthesis in the *axon terminal* is not an absolute requirement. For example, peptide transmitters are synthesized in the *cell body* and transported to distant sites, where they are released.
2. The substance should be released from nerve terminals in a chemically- or pharmacologically-identifiable form. Thus, one should be able to isolate the transmitter and characterize its structure using biochemical or other techniques.
3. A neurotransmitter should reproduce at the postsynaptic cell the specific events (such as changes in membrane properties) that are seen after stimulation of the presynaptic neuron.
4. The effects of a putative neurotransmitter should be blocked by competitive antagonists of the receptor for the transmitter in a dose-dependent manner. In addition, treatments that inhibit synthesis of the transmitter candidate should block the effects of presynaptic stimulation.
5. There should be active mechanisms to terminate the action of the putative neurotransmitter. Among such mechanisms are uptake of the transmitter by the presynaptic neuron or glial cells through specific transporter molecules, or alternatively enzymatic inactivation of the chemical messenger.

The Process of Chemical Neurotransmission Can Be Divided into Five Steps

Synaptic transmission consists of a number of steps. The general mechanisms of chemical synaptic transmission are depicted in Figure 7.1.

1. *Synthesis of the neurotransmitter in the presynaptic neuron.* In order for the transmitter to be synthesized, precursors should be present in appropriate places within neurons. Enzymes

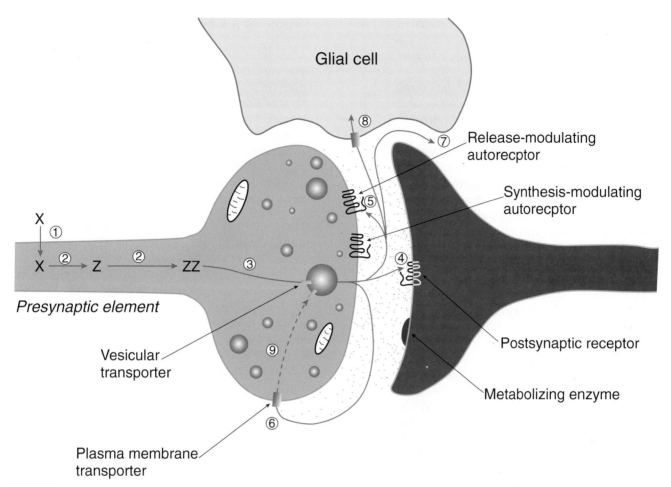

FIGURE 7.1 Schematic representation of the life cycle of a classical neurotransmitter. After accumulation of a precursor amino acid into the neuron (1), the amino acid precursor is metabolized sequentially (2) to yield the mature transmitter. The transmitter is then accumulated into vesicles by the vesicular transporter (3), where it is poised for release and protected from degradation. Once released, the transmitter can interact with postsynaptic receptors (4) or autoreceptors (5) that regulate transmitter release, synthesis, or firing rate. Transmitter actions are terminated by means of a high-affinity membrane transporter (6) that usually is associated with the neuron that released the transmitter. Alternatively, transmitter actions may be terminated by diffusion from the active sites (7) or accumulation into glia through a membrane transporter (8). When the transmitter is taken up by the neuron, it is subject to metabolic inactivation (9).

taking part in the conversion of the precursor(s) into the transmitter should be present in an active form and localized to the appropriate compartment in the neuron. For example, one would not expect synthetic enzymes for the transmitter to be found in the nucleus or the mitochondrion. In addition, any cofactors that are required for enzyme activity should be present. Drugs affecting the synthesis of neurotransmitters have long been of value in medicine. An example is α-methyl-p-tyrosine, a drug used to treat an adrenal gland tumor that causes very high blood pressure by releasing massive amounts of norepinephrine (NE). α-Methyl-p-tyrosine prevents the synthesis of NE, thereby lowering blood pressure.

2. *Storage of the neurotransmitter and/or its precursor in the presynaptic nerve terminal.* Classical and peptide transmitters are stored in synaptic vesicles, where they are sequestered and protected from enzymatic degradation and are ready for quick release. In the case of classical neurotransmitters such as acetylcholine, the synaptic vesicles are small (~50 nm in diameter), in contrast to the large dense-core vesicles (~100 nm in diameter) in which neuropeptide transmitters are stored. Vesicles are often found adjacent to the presynaptic membrane, where they are poised to be released in response to stimulation of the neuron. Because most neurotransmitters are synthesized in the cytosol of neurons, there must be some mechanism through which the

transmitter enters the vesicle. This mechanism is the vesicular transporter protein, which actively accumulates the transmitter though an energy-dependent process.

3. *Release of the neurotransmitter into the synaptic cleft.* The vesicle in which the transmitter is stored fuses with the cell membrane and releases the transmitter. Neurons use two pathways to secrete proteins. The release of most neurotransmitters occurs by a regulated pathway that is controlled by extracellular signals. The neurotransmitter release process is discussed more fully in Chapter 8. A second (constitutive) pathway of release is not triggered by extracellular stimulation and is used to secrete membrane components, viral proteins, and extracellular matrix molecules; some unconventional transmitters (such as growth factors) may be synthesized and released by both constitutive and regulated pathways.

4. *Binding and recognition of the neurotransmitter by target receptors.* Neurotransmitters that are released interact with receptors located on the target cell. Most transmitter receptors fall into two broad classes. The first are membrane proteins called metabotropic receptors, which are coupled to intracellular G proteins as effectors (see Chapter 9). Ionotropic receptors form channels through which ions such as Na^+ and Ca^{2+} flow. Receptors can be found on neurons that are postsynaptic to the cell releasing the transmitter. Receptors can also be located on the presynaptic neuron, where they can respond to the transmitter released from the same cell in the case of classical transmitters, or respond to a retrograde signal elaborated from the postsynaptic neuron; this latter process has not been described for classical or peptide transmitters. Presynaptic receptors that respond to the transmitter released by the same neuron are termed autoreceptors, and regulate transmitter release, synthesis, or impulse flow; these autoreceptors can be thought of as homeostatic feedback mechanisms.

5. *Termination of the action of the released transmitter.* If a cell cannot terminate the actions of neurotransmitters, dysfunction occurs: sustained activation of postsynaptic targets can result in tetanus (in muscles) or seizures (in the brain). If one thinks of chemical transmission as being the flow of information, continuous unregulated transmitter release is not conveying to the target cell temporal data, such as the rate of firing of the presynaptic neuron or the pattern of firing of the presynaptic neuron. Neurotransmitter actions may be terminated actively or passively. Among the active termination processes is reuptake of the neurotransmitter through specific transporter proteins on the presynaptic neuron or on glial cells. Another common means of the termination of action of a transmitter is by enzymatic degradation to an inactive substance. Finally, inactivation can occur by diffusion of the transmitter from the synaptic region. All three processes probably work cooperatively to terminate the action of a neurotransmitter.

These five steps form a logical scaffold for understanding classical and peptide chemical neurotransmission. There are, however, differences across the various transmitters, and our knowledge of the different transmitters varies widely. We discuss in detail one class of neurotransmitters, the catecholamines, to illustrate the process of chemical neurotransmission, and then examine more briefly other classical neurotransmitters. We also discuss differences between classical and nonclassical transmitters or chemical messengers, including peptide transmitters, and unconventional transmitters such as nitric oxide, growth factors, and endocannabinoids.

CLASSICAL NEUROTRANSMITTERS

The term classical is used to differentiate acetylcholine, the biogenic amines, and the amino acid transmitters from other transmitters. The designation is somewhat arbitrary, although these compounds were all accepted as neurotransmitters by the late 1950s. Several criteria can be used to distinguish classical from other transmitters, such as the peptide transmitters. First, storage vesicles for classical transmitters are smaller. Second, classical transmitters are subject to active reuptake by the presynaptic cell and thus can be viewed as homeostatically conserved; in contrast, there is no energy-dependent, high-affinity reuptake process for nonclassical transmitters. Third, most classical transmitters are synthesized in the nerve terminal by enzymatic action, and peptides are synthesized in the soma from a precursor protein and are then transported to the nerve terminal. The remainder of this chapter deals with the biochemistry of several different transmitter groups, discussing their synthesis, storage, and release. Other aspects of chemical transmission, including detailed discussions of transmitter release and the receptors for various transmitters is presented in Chapters 8 and 9.

Catecholamine Neurotransmitters

Catecholamines are organic compounds that contain a catechol nucleus (a benzene ring with two adjacent hydroxyl substitutions) and an amine group. In general practice the term normally is used to describe the three transmitters: dopamine (DA), NE, and epinephrine (Epi). These three neurotransmitters are formed by successive enzymatic steps requiring distinct enzymes (see Fig. 7.2). The presence of specific synthesizing enzymes in different cells results in separate groups of DA-, NE-, and Epi-containing neurons in the brain.

Catecholamines also have transmitter roles in the peripheral nervous system and certain hormonal functions. In the peripheral nervous system, DA has significant biological activity in the kidney but is mainly present as a NE precursor. NE is the transmitter of the sympathetic nervous system in mammals, whereas Epi is the sympathetic transmitter in frogs. Despite this species difference, biochemical aspects of neurotransmission are remarkably constant across different vertebrate species and even invertebrates.

Biosynthesis of Catecholamines

The amino acids phenylalanine and tyrosine are precursors for catecholamines. Both amino acids are found in high concentrations in the plasma and brain. In mammals, tyrosine can be formed from dietary phenylalanine by the enzyme phenylalanine hydroxylase, found in large amounts in the liver. Insufficient amounts of phenylalanine hydroxylase result in phenylketonuria, a metabolic disorder that leads to intellectual deficits unless treated by dietary manipulation.

Catecholamine synthesis usually is considered to begin with tyrosine. The enzyme tyrosine hydroxylase (TH) converts the amino acid L-tyrosine into 3,4-dihydroxyphenylalanine (L-DOPA). The hydroxylation of L-tyrosine by TH results in the formation of the DA precursor L-DOPA, which is metabolized by L-aromatic amino acid decarboxylase (AADC; see Cooper et al., 2002) to the transmitter dopamine. This step occurs so rapidly that it is difficult to measure L-DOPA in the brain without first inhibiting AADC. In neurons that use DA as the transmitter, the decarboxylation of L-DOPA to DA is the final step in transmitter synthesis. However, in those neurons using norepinephrine (also known as noradrenaline) or epinephrine (adrenaline) as transmitters, the enzyme dopamine β-hydroxylase (DBH), which converts DA to yield NE, is also present. In still other neurons in which epinephrine is the transmitter, a third enzyme (phenylethanolamine N-methyltransferase, PNMT) converts NE into Epi. Thus, a cell that uses Epi as its transmitter contains four enzymes (TH, AADC, DBH, and PNMT), whereas NE neurons contain only three enzymes (lacking PNMT) and DA cells only two (TH and AADC).

Tyrosine Hydroxylase

In human beings, a single TH gene is alternatively spliced to yield four TH mRNAs and four distinct TH protein isoforms. However, in most primates, only two TH isoforms are present, and the rat possesses but a single form of TH. It has been speculated that different TH forms in primates are associated with differences in activity of the enzyme, but conclusive data addressing this point are lacking.

The function of TH and other enzymes is determined by two factors: changes in enzyme activity (the rate at which the enzyme converts precursor into its product) and changes in the amount of enzyme protein. The major determinant of TH activity is phosphorylation of the enzyme. Another means of regulating enzyme activity is through end product inhibition: catecholamines can inhibit the activity of TH by competing for a required cofactor for the enzyme (Cooper et al., 2002).

Increased neuronal demand for catecholamines to be released can be accomplished by synthesizing new TH protein or increasing the activity of the TH activity. The degree to which increases in catecholamine synthesis depend on de novo synthesis of new enzyme protein or changes in the activity of existing enzymes differs across brain regions. For example, in most NE neurons, increases in TH gene expression and synthesis of the enzyme are readily seen when conditions require more catecholamine synthesis, but in midbrain DA neurons phosphorylation of TH is much more important.

The synthesis of catecholamines starts with the entry of tyrosine (or phenylalanine) into the brain. This process is an energy-dependent one in which tyrosine competes with large neutral amino acids as a substrate for a transporter. Because brain tyrosine levels are high enough to saturate TH, catecholamine synthesis cannot usually be increased by administration of tyrosine. Exceptions to this rule are catecholamine synthesis in cells that are very active, such as DA neurons that project to the prefrontal cortex, and under certain pathological conditions. Because TH is saturated by tyrosine, tyrosine hydroxylation is the rate-limiting step in catecholamine synthesis under basal conditions. However, when NE- or Epi-containing cells are activated, dopamine-β-hydroxylase (the enzyme responsible for NE and Epi synthesis) becomes rate-limiting, and thus tyrosine availability can regulate catecholamine synthesis under certain conditions.

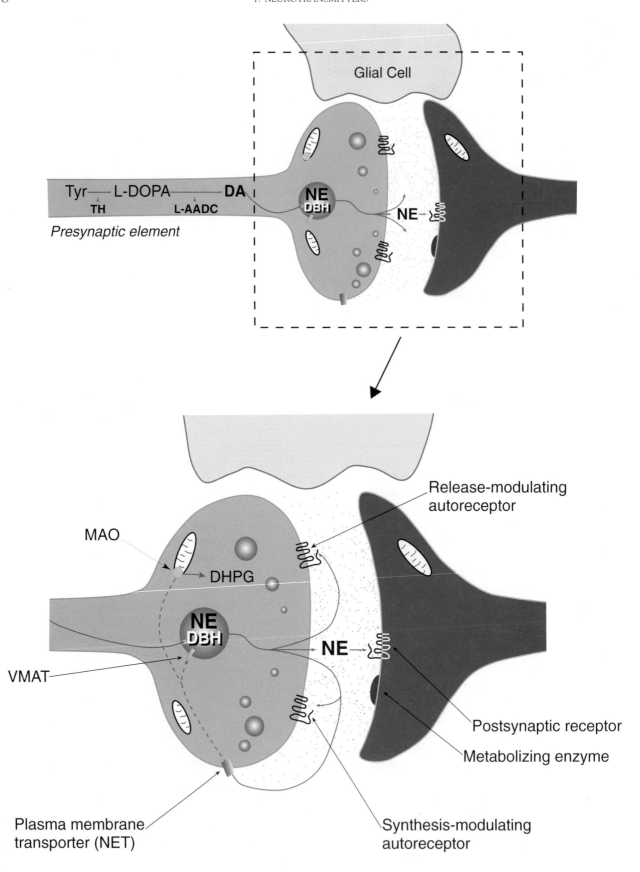

FIGURE 7.2 Characteristics of a norepinephrine (NE)-containing catecholamine neuron. Tyrosine (Tyr) is accumulated by the neuron and then is metabolized sequentially by tyrosine hydroxylase (TH) and L-aromatic amino acid decarboxylase (L-AADC) to dopamine (DA). The DA is then taken up through the vesicular monoamine transporter into vesicles. In DA neurons, this is the final step. However, in this NE-containing cell, DA is metabolized to NE by dopamine-β-hydroxylase (DBH), which is found in the vesicle. Once NE is released, it can interact with postsynaptic noradrenergic receptors or presynaptic noradrenergic autoreceptors. The accumulation of NE by the high-affinity membrane NE transporter (NET) terminates the actions of NE. Once taken back up by the neuron, NE can be metabolized to inactive compounds (DHPG) by degradative enzymes such as monoamine oxidase (MAO) or taken back up by the vesicle.

TH is a mixed function oxidase with moderate substrate specificity, hydroxylating phenylalanine as well as tyrosine. The actions of TH require the cofactor tetrahydrobiopterin (BH4) and iron (Fe^{2+}); BH4 is also an essential cofactor for tryptophan hydroxylase, the enzyme that synthesizes another transmitter, serotonin. Because BH4 is not present in saturating concentrations under basal conditions, it is crucial in regulating TH activity. Intracellular levels of BH4 are determined by its own synthesizing enzyme, GTP cyclohydrolase. Mutations in the gene encoding GTP cyclohydrolase result in DOPA-responsive dystonia, a movement disorder that is treated by administering L-DOPA (bypassing the disruption at the level of TH) to increase central DA levels.

L-Aromatic Amino Acid Decarboxylase

The hydroxylation of tyrosine by TH generates L-DOPA, which is then decarboxylated to dopamine (DA) by L-aromatic amino acid decarboxylase (AADC; also sometimes referred to as DOPA decarboxylase). AADC has low substrate specificity and decarboxylates tryptophan as well as tyrosine. Because AADC is found in both catecholamine and serotonin neurons, it plays an important role in the synthesis of both transmitters. In dopaminergic neurons, AADC is the final enzyme of the synthetic pathway.

DA does not cross the blood-brain barrier, but its precursor L-DOPA enters the brain readily. L-DOPA has achieved fame as a means of treating Parkinson's disease, which is due to loss of DA in the striatum (see Chapter 31). Although DOPA freely enters the brain, it is readily metabolized in the liver and bloodstream. It is therefore necessary to administer L-DOPA together with an inhibitor of decarboxylases that does not enter the brain but blocks metabolism in the liver and blood.

Dopamine β-Hydroxylase

Noradrenergic and adrenergic neurons contain the enzyme dopamine-β-hydroxylase (DBH), which converts DA into NE. In noradrenergic neurons, this is the final step in catecholamine synthesis. Humans have two different DBH mRNAs that are generated from a single gene. DBH lacks a high degree of substrate specificity and *in vitro* it oxidizes almost any phenylethylamine to its corresponding phenylethanolamine. For example, in addition to forming NE from DA, DBH converts tyramine into octopamine. Interestingly, receptors that have a high affinity for the trace amines tyramine and octopamine recently have been discovered (see Zucchi *et al.*, 2006) and may be involved in certain psychiatric disorders.

Phenylethanolamine N-Methyltransferase

Phenylethanolamine N-methyltransferase (PNMT) is present at high levels in a restricted number of brainstem neurons that use epinephrine as a transmitter, and in the inner portion (medulla) of the adrenal gland, where it methylates NE to form Epi, the major adrenal catecholamine. There is a single PNMT gene. The regulation of PNMT activity in the brain has not been as extensively studied as in the adrenal gland, where glucocorticoids and nerve growth factor regulate the enzyme.

Storage of Catecholamines and Their Enzymes

Catecholamines are stored in specialized subcellular organelles called vesicles. Accordingly, a specialized vesicular transporter protein is required in order for dopamine that is formed in the cytosol to enter the vesicle.

Vesicular Storage

The vesicle serves as a depot for the transmitter until it is released by appropriate physiological stimuli (see Chapter 8). In addition, vesicles isolate transmitters and protect them from metabolic inactivation by intraneuronal enzymes found in the cytosol, and from attack by toxins that have gained entry to the cell vesicle.

The NE-synthesizing enzyme DBH differs from the other enzymes in catecholamine synthesis by being localized to the vesicle rather than the cytosol. This means that only *after* DA is taken up by vesicles through the vesicular monoamine transporter (VMAT) is it metabolized to NE. This vesicular storage of DBH has an interesting consequence: DBH and dopamine can both be released into the synapse along with NE.

Vesicular Monoamine Transporters

The ability of vesicles to accumulate DA and other compounds depends on VMAT (Weihe and Eiden, 2000). The vesicular transporter differs from the neuronal membrane transporter that terminates the action of catecholamines in terms of substrate affinity and localization. Two VMAT genes have been cloned: VMAT1 is found in adrenal gland cells that synthesize and release catecholamines, and VMAT2 is found in the catecholamine and serotonin-containing neurons of the brain. VMAT2 is not very specific and transports catecholamines, indoleamines such as serotonin, and histamine into vesicles.

The VMATs are inhibited by reserpine. This drug has been used for centuries in India as a folk medicine to treat high blood pressure and psychoses. These uses of reserpine were reported in international journals in the early 1930s, but these therapeutic actions were not appreciated in Western medicine until a generation later; the unpleasant side effects of reserpine initially garnered more attention. In the 1950s Bernard Brodie and coworkers discovered that reserpine depleted brain serotonin. The contemporaneous discovery that the hallucinogen LSD is structurally similar to serotonin led to the proposal that the antipsychotic actions of reserpine were due to its ability to deplete serotonin in the brain. However, it was soon realized that reserpine depletes both serotonin and catecholamines in the brain; thus, antipsychotic effects of reserpine might be due to serotonin or catecholamine depletion, or both.

To determine if serotonin or catecholamines were more important for the antipsychotic effect of reserpine, Arvid more important, Arvid Carlsson and colleagues administered catecholamine and serotonin precursors to rats treated with reserpine in an attempt to replenish the levels of the transmitters. They then examined locomotor activity, which is severely depressed by reserpine. Motor function was restored by the DA precursor L-DOPA but not by the serotonin precursor 5-hydroxytryptophan. This work led to the characterization of DA as a neurotransmitter. The demonstration that L-DOPA increases brain DA concentrations in reserpinized animals suggested that the primary mechanism through which reserpine exerts antipsychotic effects was through its ability to disrupt DA transmission. This idea led directly to the hypothesis that the dysfunction of central DA systems underlies schizophrenia. Interestingly, studies have suggested that drugs that are antagonists at both certain DA and serotonin receptors may be better than DA antagonists in treating schizophrenia.

VMATs have significant homology with a group of bacterial antibiotic drug resistance transporters, suggesting a role of VMAT in detoxification. This is indeed the case: VMAT sequesters the transmitter in vesicles where it is protected from various toxins. This is perhaps best illustrated by studies of heterozygous mice with one copy of VMAT2. In these mice, the parkinsonian toxin MPTP (see Chapter 31) causes a greater loss of DA neurons than in wildtype control mice because there is less VMAT to package the toxin into vesicles and thus permits the toxin to disrupt mitochondrial function.

Release of Catecholamines

Catecholamine release usually operates through the same calcium-dependent exocytotic process used by other transmitters (see Chapter 8), but can also occur through at least two other mechanisms. First, catecholamines can be released by a reversal of the direction of transport through the membrane transporters DAT and NET. For example, this occurs in response to certain drugs (such as amphetamine). Second, DA and perhaps other catecholamines can be released from dendrites through a process that does not appear to involve Ca^{2+}.

Regulation of Catecholamine Synthesis and Release by Autoreceptors

Enzymes that control catecholamine synthesis can be regulated at both the transcriptional level and by post-translational modifications that alter enzymatic activity. In addition, DA and NE can be regulated by the interaction of released catecholamine with specific DA or NE "autoreceptors" located on the nerve terminal.

Autoreceptors are found on most parts of the neuron and are defined functionally by the events that they regulate. Thus, synthesis-, release-, and impulse-modulating DA autoreceptors have been described (see Cooper *et al.*, 2002). All three types of DA autoreceptors belong to the D2 family of DA receptors, which includes three different receptors (D_2, D_3, and D_4). It is clear that D_2 autoreceptors exist, but considerable controversy surrounds the presence of D_3 autoreceptors and there is no evidence for D_4 autoreceptors. The presence of three functionally different DA autoreceptors suggests that a single D_2 protein may have different autoreceptor roles that occur through distinct transduction mechanisms.

The autoreceptor can be thought of as a key part of a feedback mechanism. Thus, DA that is released from a neuron stimulates an autoreceptor to inhibit further DA release. *Release-modulating* autoreceptors are a common regulatory feature of all neurons that use classical transmitters. Because *intra*cellular DA levels regulate tyrosine hydroxylase activity by interfering with the binding of TH to its cofactor, changes in the release of DA may also alter transmitter synthesis. *Synthesis-modulating* autoreceptors directly regulate DA synthesis: DA release decreases DA synthesis whereas DA antagonists that block the autoreceptor increase synthesis. Interestingly, synthesis-modulating autoreceptors are not found on all DA neurons: some midbrain and hypothalamic DA neurons lack synthesis-modulating autoreceptors. Finally, *impulse-modulating* autoreceptors found on the soma and dendrites of DA neurons and regulate the firing rates of these cells. Because the release of DA can alter synthesis of the transmitter by changing feedback inhibition on TH, it follows that impulse-modulating autoreceptors also change DA synthesis. Thus, all three types of DA autoreceptors may ultimately regulate synthesis. The interdependence of regulatory processes governing DA cells is characteristic of monoaminergic neurons.

Constant levels of a neurotransmitter in the synapse will not accurately convey information about the dynamic state of the presynaptic neuron to its follower cell. In addition, continuous stimulation of certain receptors is pathological and can damage postsynaptic cells. For example, when certain receptors for the transmitter glutamate are activated continuous, the result is "excitotoxic" cell death.

In order to prevent continuous presynaptic signaling, different mechanisms for terminating the actions of a transmitter have evolved. The simplest is diffusion of the transmitter and its subsequent dilution in extracellular fluid to subthreshold concentrations. Believed to be more important are active modes of halting transmitter action, including enzymatic inactivation of the transmitter and uptake of the transmitter by membrane-associated transporter proteins.

Enzymatic Inactivation of Catecholamines

Enzymatic inactivation originally was thought to be the major means of terminating catecholamine actions in the CNS, although it now appears that reuptake of the transmitter is the primary mode of inactivation. Two enzymes contribute to catecholamine catabolism: monoamine oxidase (MAO) and catechol-0-methyltransferase (COMT). These enzymes can act independently or can act on the products generated by the other enzyme, leading to catecholamine metabolites that are deaminated, O-methylated, or both. COMT is a relatively nonspecific enzyme that transfers methyl groups from the donor S-adenosylmethionine to the *m*-hydroxy group of catechols. COMT is found in both peripheral tissues and central nervous system and is the major means of inactivating catecholamines that are released from the adrenal gland.

MAOs oxidatively deaminates catecholamines and their O-methylated derivatives to form inactive and unstable derivatives that can be further degraded by other enzymes. Two forms of MAO have been identified. MAO_A has high affinities for NE and serotonin and is selectively inhibited by drugs such as clorgyline. In contrast, MAO_B has a higher affinity for o-phenylethylamines and is selectively inhibited by different compounds, such as deprenyl. The MAOs are important targets of drugs used to treat several neuropsychiatric disorders (Box 7.1).

Neuronal Catecholamine Transporters

The reuptake of a transmitter released by a neuron is the major mode of transmitter inactivation in the brain. Accumulation of the transmitter also allows intracellular enzymes that degrade the transmitter to act, thus bolstering the actions of extracellular enzymes.

Several characteristics define the high-affinity neuronal reuptake of transmitters (Clark and Amara, 1993). The process is energy dependent and saturable, depends on Na^+ cotransport, and requires extracellular Cl^-. Because reuptake depends on coupling to the Na^+ gradient across the neuronal membrane, toxins that inhibit Na^+, K^+-ATPase inhibit reuptake. Under certain conditions, the coupling of transporter function to Na^+ flow may cause local changes in membrane Na^+ gradients and thereby paradoxically cause the transporter to operate in "reverse" to extrude ("release") the transmitter from the cell to the extracellular space.

Membrane catecholamine transporters are not Mg^{2+} dependent and are not inhibited by reserpine, which distinguishes neuronal and vesicular membrane transporters. Catecholamine transporters are localized to neurons: although there appears to be a reuptake process that accumulates catecholamines into glia, this is not a high affinity process and its functional significance is unclear.

Two distinct mammalian catecholamine transporters, one for dopamine (dopamine transporter, DAT) and one for norepinephrine (NET), have been identified. The two are closely related members of a class of transporter proteins (including serotonin and amino acid transmitter transporters) with 12 transmembrane domains. Neither transporter is specific, with each accumulating both DA and NE. In fact, NET has a higher affinity for dopamine than for NE. An epinephrine transporter has been identified in the frog but not in mammals.

The regional distribution of DAT and NET largely follows the expected localization to DA and NE neurons, respectively. However, DAT does not appear to be expressed in all DA cells. Certain hypothalamic cells that release DA into the blood system of the pituitary lack detectable DAT mRNA and protein. Because

BOX 7.1

MAO AND COMT INHIBITORS IN THE TREATMENT OF NEUROPSYCHIATRIC DISORDERS

One hypothesis of the pathophysiology of depression posits a decrease in noradrenergic levels in the brain. MAO_A inhibitors, such as tranylcypromine, effectively increase NE levels (as well as DA and 5-HT concentrations) and were once a mainstay in the treatment of depression. However, the use of MAO inhibitors in depression largely has been supplanted by the introduction of drugs that increase extracellular NE levels by blocking the NE transporter (tricyclic antidepressants) and other agents that increase 5-HT or DA levels by blocking SERT or DAT (such as Prozac and Welbutrin). The treatment of depression with MAO_A inhibitors, although still useful for certain patients who do not respond to other antidepressants, is marred by a large number of side effects. Among the most serious is hypertensive crisis. Patients treated with MAO_A inhibitors cannot metabolize tyramine efficiently, which is present in large amounts in certain foods, such as aged cheeses and red wines. Because tyramine releases catecholamines peripherally, small amounts of tyramine increase blood pressure significantly and may lead to a high risk for stroke.

Deprenyl, a specific inhibitor of MAO_B, has been used as an initial treatment for Parkinson's disease (PD; see Chapter 31). The use of deprenyl in the treatment of PD and the rationale for its use were based on data from studies of a neurotoxin, 1-methyl-4-pheny 1-1,2,3,6-tetrahydropyridine (MPTP). MPTP results in the degeneration of midbrain DA neurons and a parkinsonian syndrome. MPTP-induced parkinsonism was first noted in a group of opiate addicts. In an attempt to synthesize a designer drug, the structurally related MPTP was inadvertently produced; addicts who injected this drug developed a severe parkinsonian syndrome. Subsequent animal studies showed that MPTP itself is not toxic, but that its active metabolite, MPP^+, is highly toxic. The formation of MPP^+ from MPTP is catalyzed by MAO_B, and treatment

with MAO inhibitors such as deprenyl can prevent MPTP toxicity. The realization that MPTP administration rather faithfully reproduces the cardinal signs and symptoms of PD reawakened interest in environmental toxins as a cause of PD. The MPTP saga also led to the idea that deprenyl treatment might slow the progression of PD by preventing metabolism of an environmental compound to an active toxin such as MPP^+. Although clinical studies initially were interpreted to suggest that there was a slowing of clinical progression of Parkinson's disease in response to deprenyl, later studies showed that deprenyl increases DA levels slightly and thus gives some symptomatic relief.

Catechol O-methyl transferase, which together with MAO degrades catecholamines, also plays a role in the treatment of PD. Two COMT inhibitors are used to prevent the enzymatic inactivation of DOPA. By inhibiting COMT, these drugs prolong the therapeutic action of DOPA and may smooth out fluctuations in the therapeutic response to DOPA.

Changes in catecholamine function also have been the object of intense scrutiny in schizophrenia, with recent attention focusing on possible changes in dopamine. One allelic variant of the COMT gene results in a much reduced activity of the enzyme. Data have examined COMT alleles for full vs. low COMT activity in normal subjects and schizophrenics. Individuals bearing the allele that confers lower COMT activity display improved performance on cognitive tasks that involve DA actions in the prefrontal cortex; the performance of schizophrenic persons on these tasks is impaired. It therefore has been proposed that high COMT activity may confer an increased risk to schizophrenia.

Ariel Y. Deutch and Robert H. Roth

DA released from these neurons is carried away in the blood, the existence of a transporter protein on these DA cells would be superfluous.

Immunohistochemical studies of the subcellular localization of DAT led to an unexpected finding. DAT is not found at the synaptic junction, but just outside of this region. Thus, the transporter may be used to accumulate and therefore inactivate DA that has

escaped from the synaptic cleft. This suggests that diffusion is the initial process by which DA is removed from the synapse. The extrasynaptic localization of the catecholamine transporters, coupled with a similar extrasynaptic localization of dopamine receptors, suggests that extrasynaptic ("volume") neurotransmission may be of major importance in catecholamine signaling.

Mice in which the gene encoding DAT has been deleted have been particularly useful in clarifying the function of transporters. Transgenic mice that lack DAT have a remarkable number of changes in DA function, ranging from increased extracellular DA levels and delayed clearance of released DA to a striking decrease in tissue concentrations of DA in the face of increased DA synthesis (Gainetdinov *et al.*, 1998). There is also a complete loss of autoreceptor-mediated tone, including deficits in release-, synthesis-, and impulse-modulating autoreceptor function. These deficits in DAT knockout mice have been suggested to reflect a disinhibition of tyrosine hydroxylase due to a lack of intraneuronal DA (removing feedback inhibition of the enzyme), resulting in a marked increased in DA synthesis and release.

The catechoamine transporters are important targets of many drugs. Cocaine and amphetamine both increase extracellular levels of catecholamines by blocking transporters. In particular, cocaine shows a very high affinity for DAT; amphetamine is a less potent inhibitor but also induces "release" (via transporter reversal) of catecholamines. Drugs used in the treatment of attention deficit disorder, including amphetamine, methylphenidate, and atomoxetine also act by blocking catecholamine transporters. NET is the molecular target of tricyclic antidepressant drugs, which potently inhibit NE reuptake.

Serotonin

Well over a century ago scientists were aware of a substance in the blood that induced powerful contractions of smooth muscle organs. In the mid-twentieth century, Page and collaborators succeeded in isolating the compound, which they suggested to be a possible cause of high blood pressure, from platelets. At the same time, Italian researchers were studying a substance in intestinal mucosa that caused contractions of intestinal smooth muscle. The material isolated from platelets was called "serotonin," and the substance isolated from the intestinal tract was named "enteramine." Studies soon revealed that the two substances were the same compound, 5-hydroxytryptamine (5-HT), which now is commonly referred to as serotonin.

Serotonin is found in neurons and in several other types of cells in the body. In fact, the brain accounts for only about 1% of total body stores of serotonin. Although the isolation and identification of serotonin were from peripheral tissues, much of the subsequent interest in serotonin was based on its potential involvement in psychiatric disorders. The finding that serotonin's chemical structure is similar to that of LSD led to theories that associated abnormalities in serotonin function to schizophrenia and depression.

The basic process of serotonin biosynthesis is very similar to that of catecholamine transmitters: a peripheral amino acid (tryptophan) gains entry into the brain and is metabolized in serotonergic neurons via a series of enzymatic steps. Once tryptophan enters the serotonergic neuron, it is hydroxylated by tryptophan hydroxylase, the rate-limiting step in serotonin synthesis, giving rise to 5-hydroxytryptophan (5-HTP), which is then decarboxylated by L-aromatic amino acid decarboxylase to the transmitter serotonin. Thus, only two critical enzymes (tryptophan hydroxylase and AADC) are involved in the synthesis of 5-HT (Cooper *et al.*, 2002).

Alternative Tryptophan Metabolic Pathways

Although serotonin is usually the endpoint of tryptophan metabolism in brain, serotonin can be metabolized further to yield active products. In the pineal gland, 5-HT is metabolized to form 5-methoxy-N-acetyltryptamine (melatonin), a hormone thought to play an important role in sleep. In peripheral tissues, most tryptophan is not metabolized to 5-HT but instead is metabolized by the kynurenine pathway. This kynurenine shunt is also present in the brain and leads to the accumulation of several interesting active substances (see Schwarcz, 2004). The two major tryptophan metabolites generated by the kynurenine shunt are quinolinic and kynurenic acids. Quinolinic acid is a potent agonist at certain glutamate receptors, and acts through glutamate receptors to cause cell loss and convulsions; in contrast, kynurenine is an antagonist at these receptors. These compounds are the focus of considerable clinical interest in various neuropsychiatric disorders.

Inactivation of Released Serotonin

Serotonin is inactivated primarily by reuptake through SERT, the serotonin transporter that belongs to the same family of transporters as the catecholamine transporters. The most commonly used antidepressants today are a class of compounds known as serotonin selective reuptake inhibitors, or SSRIs; the prototypic SSRI is fluoxetine (Prozac). Because antidepressant drugs increase serotonin or norepinephrine levels, or both, current theories of depression posit critical modulatory roles for NE and 5-HT.

The enzymatic degradation of 5-HT is catalyzed by monoamine oxidase (MAO). The product of this reaction, 5-hydroxyindole acid aldehyde, is oxidized further to 5-hydroxyindoleacetic acid. MAO inhibitors increase 5-HT levels and have been widely used as antidepressants (Box 7.1).

Acetylcholine

Our basic ideas about chemical synaptic transmission are based on early studies of acetylcholine (ACh). Both pharmacological and electrophysiological studies at the neuromuscular junction led to key findings on ACh that remain the cornerstone for today's understanding of chemical neurotransmission. For example, electrophysiological studies revealed fast excitatory responses of muscle fibers to the stimulation of nerves innervating the muscle. The presence of miniature end plate potentials (mEPPs) in muscle fiber was noted, and in the early 1950s, Fatt and Katz demonstrated that these mEPPs resulted from the slow "leakage" of ACh, with each mEPP representing the release of transmitter in one vesicle (termed a quantum). Overt depolarization generated an increase in the number of quanta released over a given period of time (see Chapter 8). Over the past half century many of the rules that govern ACh neurotransmission have been found to be general principles that apply to many transmitters.

In large part the general principles that govern catecholamine and serotonin synthesis also apply to acetylcholine synthesis. The synthesis of ACh is simple, with only a single step: the acetyl group from acetyl-coenzyme A is transferred to choline by the enzyme choline acetyltransferase (ChAT). There are correspondingly few requirements for ACh synthesis: the presence of the substrate choline, the donor acetyl-coenzyme A, and ChAT. The acetyl-CoA that serves as the donor is derived from pyruvate generated by glucose metabolism. There is one interesting twist on ACh synthesis: acetyl-CoA is localized to mitochondria, but the synthetic enzyme ChAT is cytoplasmic, dictating that acetyl-CoA must exit the mitochondria to gain access to ChAT. Despite the fact that ChAT is the only enzyme involved in ACh synthesis, it is not the rate-limiting step in synthesis of the transmitter. The full enzymatic activity of ChAT is not expressed *in vivo*: ChAT activity measured *in vitro* is much greater than would be expected on the basis of ACh synthesis *in vivo*. The reason for this discrepancy has been suggested to be related to the requirement that acetyl-CoA be transported from the mitochondria to the cytoplasm. Alternatively, intracellular choline concentrations may determine the rate of ACh synthesis. This latter idea has led to the use of choline precursors in attempts to enhance ACh synthesis in Alzheimer disease, in which there is a marked decrease of cortical ACh levels. Unfortunately, these attempts have not proven very successful.

ACh is taken up into storage vesicles by the vesicular cholinergic transporter (VAChT). Cloning of human VAChT revealed that the gene is localized to chromosome 10, near the ChAT gene for ChAT. Additional studies then demonstrated that the entire VAChT coding **region** is contained in the first intron of the ChAT gene. This suggested that both genes are coordinately regulated, a suspicion that subsequently was confirmed.

The primary mode of inactivation of ACh appears to be enzymatic, which is simply the hydrolysis of ACh to choline. Two groups of cholinesterases have been defined on the basis of substrate specificity: acetylcholinesterases (AChEs) and butyrylcholinesterases (Taylor and Brown, 2006). The former are relatively specific for ACh and are found in high concentration in the brain, whereas butyrylcholinesterases are enriched in the liver and are present in lower levels in the brain.

AChE is found in high concentrations in cholinergic neurons. However, AChE also is found in moderately high concentrations in some noncholinergic neurons that receive cholinergic inputs. This observation is consistent with the fact that AChE is a secreted enzyme that is associated with the cell membrane. Thus, ACh hydrolysis takes place extracellularly, and the choline generated is conserved by the high-affinity reuptake process. In addition to its role in inactivating acetylcholine, AChE has been proposed to be a chemical messenger in the CNS (Greenfield, 1991).

γ-Aminobutyric Acid: The Major Inhibitory Neurotransmitter

Several amino acids fulfill most of the criteria for consideration as neurotransmitters. The best studied of these are γ-aminobutyric acid (GABA), the major inhibitory transmitter in brain, and glutamate, which is the major excitatory transmitter in brain. Many of the principles concerning transmitter synthesis and inactivation that were previously discussed apply to amino acid transmitters, however there are some key differences between the amino acid and other classical transmitters. The most obvious difference is that GABA is derived from glucose metabolism (Fig. 7.3). Accordingly, mechanisms must exist to segregate the transmitter and general metabolic pools of the amino acid transmitters (Olsen and Betz, 2006). A second difference between amino acid and catecholamine transmitters is that the former are taken up readily by glia as well as neurons.

GABA Synthesis

The GABA Shunt and GABA Transaminase

GABA ultimately is derived from glucose metabolism. α-Ketoglutarate formed by the Krebs cycle is

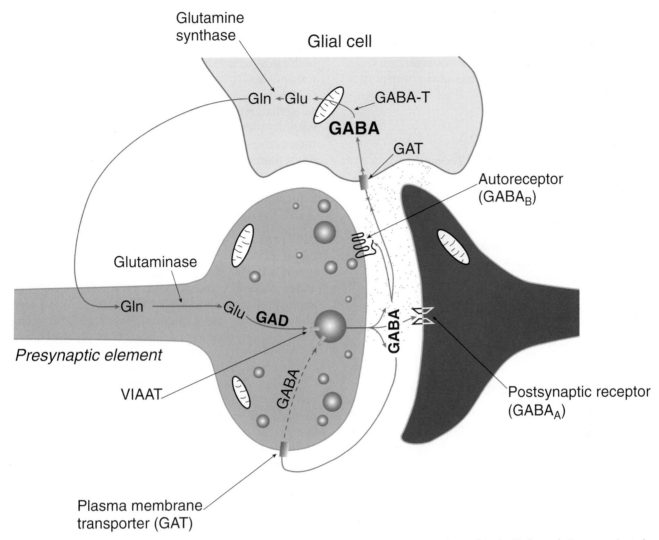

FIGURE 7.3 Schematic depiction of the life cycle of a GABAergic neuron. α-Ketoglutarate formed in the Krebs cycle is transaminated to glutamate (Glu) by GABA transaminase (GABA-T). The transmitter GABA is formed from the Glu by glutamic acid decarboxylase (GAD). GABA that is released is taken by high-affinity GABA transporters (GAT) present on neurons and glia.

transaminated to the amino acid glutamate by the enzyme GABA-transaminase (GABA-T). In those cells in which GABA is used as a transmitter, the presence of another enzyme, glutamic acid decarboxylase (GAD), generates GABA from glutamate. The presence of GAD therefore serves as an unambiguous marker of a neuron that uses GABA as a transmitter.

An unusual feature of GABA synthesis is that intraneuronal GABA is inactivated by the actions of GABA-T, which appears to be associated with mitochondria (Fig. 7.3). Thus, GABA-T is both a key synthetic enzyme and a degradative enzyme! GABA-T metabolizes GABA to succinic semialdehyde, but only if α-ketoglutarate is present to receive the amino group that is removed from GABA. This unusual GABA shunt serves to maintain supplies of GABA.

How Are Transmitter and Metabolic Pools Separate?

A major question concerning amino acid transmitters is how the transmitter pool is separately maintained from the general metabolic roles of amino acids. The GABA synthetic enzyme GAD is a cytosolic enzyme, but GABA-T, which converts α-ketoglutarate into the GAD substrate glutamate, is present in mitochondria. Thus, the metabolic pool is present in mitochondria, and glutamate destined for the transmitter pool must be exported from mitochondria to the cytosol. This process is poorly understood.

Glutamate is not only a precursor to GABA, but is also the major excitatory neurotransmitter. Why GABA neurons fail to use the precursor glutamate as a transmitter may involve different biosynthetic enzymes for the transmitter and metabolic pools of glutamate, as well as different vesicular transporters that accumulate glutamate and GABA into vesicles. A neuronal enzyme that is localized to vesicles has been proposed to be responsible for the synthesis of the transmitter pool of glutamate.

Storage and Release of GABA

Vesicular Inhibitory Amino Acid Transporter

A vesicular GABA transporter has been cloned on the basis of homology to a protein in the worm *Caenorhabditis* elegans. The strategy of moving from invertebrate to mammalian species has been very useful in identifying a variety of mammalian transmitter-related genes. The vesicular GABA transporter shares with the vesicular monoamine transporters a lack of substrate specificity, and will transport the inhibitory transmitter glycine as well as GABA. Some rare GABA cells may lack the transporter, raising the specter of a related transporter or some unique functional attribute of these cells.

Autoreceptor Regulation of GABA Release

As is the case with other classical transmitters, GABAergic cells have autoreceptors that regulate release of the transmitter. There are two major classes of GABA receptors. One type of GABA receptor forms ion channels, which are found mainly on cells postsynaptic to GABA terminals; the second class are G protein-coupled receptors. Pharmacological studies indicate that the autoreceptor-mediated regulation of GABA neurons takes place mainly through G protein-coupled GABA$_B$ receptors located on GABAergic axon terminals. Anatomical studies have revealed that GABA$_B$ receptors sometimes also are found on postsynaptic non-GABAergic neurons. It is possible that these postsynaptic GABA$_B$ sites respond to GABA released from a neuron that is presynaptic to another GABA neuron. Interestingly, when one GABA neuron terminates on another GABA cell, the inhibition of the second (postsynaptic cell) will have the same functional consequence as an autoreceptor (decreasing subsequent transmitter release) on the third cell in the chain.

Inactivation of GABA

Reuptake is the primary mode of inactivation of the transmitter GABA. There are four GABA transporters (GATs), providing a diverse means of regulating GABA neurons. Early studies defined the different GABA transporters as neuronal or glial based on pharmacological criteria. However, anatomical studies found that one GAT that had been defined on pharmacological grounds as a glial transporter is present in both neurons and glia. The reason for multiple GABA transporters is not clear. GATs are expressed in both GABAergic and non-GABAergic cells (presumably cells that receive a GABA innervation). Consistent with the promiscuous uptake of transmitters by other transporters, GATs also appear to take up amino acids other than GABA.

Glutamate and Aspartate: Excitatory Amino Acid Transmitters

Excitatory amino acid transmitters account for most of the fast synaptic transmission that occurs in the mammalian brain. Glutamate and aspartate are the major excitatory amino acid neurotransmitters, and several related amino acids, such as N-acetylaspartylglutamate, are also thought to have neurotransmitter roles. Excitatory amino acids such as glutamate participate in both intermediary metabolism and neuronal communication.

Neither glutamate nor aspartate crosses the blood–brain barrier, and in the brain these transmitters are derived only by local synthesis from glucose. Two processes lead to glutamate synthesis in the nerve terminal. As discussed earlier, glutamate is formed from glucose through the Krebs cycle and transamination of α-ketoglutarate. In addition, glutamate can be formed directly from glutamine. Because glutamine is synthesized in glial cells, both neurons and glia are important in determining the transmitter pool of glutamate. Glutamine is exported from glia and is transported into nerve terminals before being converted into glutamate by a glutaminase enzyme (Hassel and Dingledine, 2006).

Because of the intermingling of glial and neuronal contributions to glutamate synthesis, and the lack of specific enzymes or other proteins to distinguish the metabolic pool of glutamate from the transmitter glutamate, it was difficult to clearly identify a neuron as using glutamate as a transmitter. The identification of three vesicular glutamate transporters, VGluT1, 2, and 3, has resulted in a simple means of marking glutamatergic neurons. However, anatomical studies over the past several years have revealed that many glutamate neurons also contain one or more other transmitters (Fremeau *et al.*, 2004), with some even containing the inhibitory amino acid transmitter GABA. Because GABA hyperpolarizes cells in adult neurons, while glutamate depolarizes neurons, one would expect that the transmitter pools of GABA and glutamate are not found in the same neurons. The functional significance of such an arrangement is not clear.

Just as there are several vesicular glutamate receptors, there are several glutamate transporters that terminate the action of glutamate. An unusual aspect is that one of these transporters is expressed by astrocytes and is the major accumulator of glutamate from the extracellular space. Moreover, relatively new data point to glia as releasing glutamate in a signaling capacity. These findings have led to an increasing awareness of glia as being far more than structural support cells and argue for a much broader contribution of glia to neuronal communication (Volterra *et al.*, 2002).

NONCLASSICAL NEUROTRANSMITTERS

We have discussed classical transmitters, of which there are relatively few. Many more transmitters are peptides. There are some clear differences between peptide and classical transmitters, but the two groups have more in common. For example, both classical and peptide transmitters usually are very well conserved across species; many of the peptide transmitters initially were isolated from amphibians. In addition, both classical and peptide transmitters are synthesized in neurons, where they are stored in vesicles and released in a Ca^{2+}-dependent manner. However, the biosynthetic mechanisms and the modes of inactivation of peptide and classical transmitters differ. We will first consider the question of the significance of multiple neurotransmitters, and then turn to the general principles of peptide transmitter biosynthesis and inactivation.

Why Do Neurons Have So Many Transmitters?

About a dozen classical transmitters and dozens of neuropeptides function as transmitters. There are still more "unconventional" transmitters, which we will address later in this chapter. If transmitters simply serve as a chemical bridge that conveys information between two spatially distinct cells, why have so many chemical messengers?

Convergence of Different Transmitter-containing Axons on a Common Neuron

Perhaps the simplest explanation for multiple transmitters is that many nerve terminals synapse onto a single neuron. How can a neuron distinguish between multiple inputs that carry different information? One way is to segregate the place on the neuron at which an input arrives, such as the soma, axon, or dendrite. However, because many afferents terminate in close proximity, another means of distinguishing inputs and their information is necessary: chemical coding of the inputs by neurotransmitters.

Colocalization of Neurotransmitters

The idea that a neuron is limited to one transmitter can be traced to Henry Dale, or more accurately, to an informal restatement of what is termed Dale's principle. In the 1940s, Dale posited that a metabolic process that takes place in the cell body can reach or influence events in all processes of the neuron. Sir John Eccles restated Dale's view to suggest that a neuron releases the same transmitter at all its processes. Illustrating the dangers of scientific sound bites, this principle was soon misinterpreted to indicate that only a single transmitter can be present in a given neuron. This view held sway for about 40 years, when it was no longer possible to dismiss as artifact the data in support of peptide transmitters. We now know that neurons can contain multiple transmitters, including both a classical transmitter (such as DA) and a peptide transmitter (such as neurotensin). Indeed, it appears that few, if any, neurons contain only one transmitter, and in many cases three or more transmitters are found in a single neuron.

The presence of multiple transmitters in single neurons suggests that different transmitters are used by a neuron to signal different functional states to its target cell. For example, the firing rates of the neurons differ considerably, and it may be useful to encode fast firing by one transmitter and slower firing by a different transmitter. The firing *pattern* of neurons also conveys information. For example, a neuron may discharge five times every second on average. This may mean that the cell regularly discharges every 200 ms, but can also represent a cell that has a burst of five discharges during an initial 100-ms period followed by 900 ms of silence. Peptide transmitters often are released at higher firing rates and particularly under burst-firing patterns.

The different synthetic steps in peptides and classical transmitters lead to differential release. Classical transmitters can be replaced rapidly because their synthesis occurs in nerve terminals. In contrast, peptide transmitters must be synthesized in the cell body and transported to the terminal. Thus, it is useful to conserve peptide transmitters for situations of high demand because they would otherwise be depleted rapidly.

Transmitter Release from Different Processes

The restatement of Dale's principle by Eccles held that a transmitter is found in all processes of a neuron. In invertebrate species a transmitter can be localized to different parts of a neuron, and it is clear that many proteins in mammalian neurons are distributed in a highly specialized manner. Although differential targeting of transmitters to different processes in mammalian neurons remains to be demonstrated conclusively, if a transmitter were restricted to a particular part of a neuron, the neuron would need multiple transmitters to account for different release sites. Receptors are very specific in their locations on neurons, and certain receptors are recruited to "hot spots" on neuronal processes by the activity patterns of presynaptic inputs.

Synaptic Specializations versus Nonjunctional Appositions between Neurons

The anatomical relations between one cell and its follower may contribute to the need for different transmitters. We typically think of synaptic specializations as the physical substrate of communication between two neurons. However, there may also be nonsynaptic communication between two neurons. This could occur across distances that are larger than conventional synaptic arrangements. In such a situation the requirements for transmitter action would differ from those discussed previously because the distance traversed by the transmitter molecule would be farther than at a synaptic apposition. Thus, transmitters that lack an efficient reuptake system, such as peptide transmitters, might be favored at nonsynaptic sites. Because a single neuron can form both synaptic and nonsynaptic specializations, a single neuron may require more than one neurotransmitter.

Fast versus Slow Responses of Target Neurons to Neurotransmitters

Different firing rates or patterns may be accompanied by changes in the type or relative amounts of a transmitter being released from a neuron. For example, stimulation of receptors that form ion channels leads to very rapid changes, whereas actions at metabotropic receptors that are coupled to intracellular events through specific transduction molecules have slower response characteristics. Differences in temporal response characteristics allow the receptive neuron to respond differently to a stimulus depending on the antecedent activity in the cell. A transmitter can change the response characteristics of a particular cell to subsequent stimuli by seconds or even minutes, and thus short-term changes can occur independent of changes in gene expression.

PEPTIDE TRANSMITTERS

Synthesis and Storage of Peptide Transmitters

Classical transmitters typically are synthesized in the axon from which they are released. In contrast, genes encoding peptide transmitters give rise to a prohormone, which is incorporated into secretory granules, after which the prohormone is acted on by peptidases to form the peptide transmitter (Fig. 7.4). This process typically occurs in the cell body, and the peptide-containing vesicles then are transported to the axon; a small number of peptide transmitters are synthesized enzymatically.

In neurons that use classical transmitters, demands for increased amounts of transmitter are met by increasing local transmitter synthesis. However, increasing the amount of a peptide transmitter requires an increase in gene expression to yield a prohormone, with a subsequent delay in delivery to the axon terminal of the peptide. Thus, classical transmitters respond to increased demand rapidly, but peptide transmitters cannot.

The storage of peptides and classical transmitters also differs. Classical neurotransmitters generally are stored in small (approximately 50 nm) synaptic vesicles. In contrast, neuropeptide transmitters are stored in large (approximately 100 nm) dense core vesicles. Because peptide transmitters typically are released at a high neuronal firing frequency or in a burst-firing pattern, it is reasonable to assume that there are different mechanisms for the exocytosis and subsequent release of peptide and classical transmitter vesicles. Although the release of peptide transmitters, like that of classical transmitters, is calcium-dependent, distinct but related molecular mechanisms subserve the release of small and large dense core vesicles (Sieburth et al., 2007).

Inactivation of Peptide Transmitters

There are also differences in the inactivation of peptide and classical transmitters. Classical transmitters have high-affinity reuptake processes to remove the transmitter from the extracellular space. Peptide transmitters are inactivated enzymatically or by diffusion, but lack a high-affinity active reuptake process. Enzymatic inactivation of peptide transmitters also differs from that of classical transmitters. Peptide transmitters are short chains of amino acids, but the inactivating enzymes are specific for certain dipeptide sequences and are not specific to any single peptide. For example, an enzyme that inactivates opioid-like peptide transmitters called enkephalins usually is referred to as enkephalinase, but also cleaves other peptide transmitters.

A final difference between the inactivation of peptide and classical transmitters is in the final product. When classical transmitters are broken down enzymatically, the metabolites are inactive at the transmitter's receptor. However, certain peptide fragments derived from the enzymatic "inactivation" of peptide transmitters are biologically active. An example is angiotensin, in which the angiotensin I is metabolized to yield angiotensin II and III, each successively more active than the parent angiotensin I. It can therefore sometimes be difficult to distinguish between transmitter synthesis and transmitter inactivation. The peptide that is stored in vesicles and then released is therefore considered the transmitter, although the actions of certain peptidases may lead to other biologically active fragments.

Neurotensin Synthesis

Neurotensin (NT) is widely expressed in the central nervous system and in certain peripheral tissues, such as the small intestine. A related peptide, neuromedin N (NMN), is also transcribed from the mammalian gene that encodes NT (Fig. 7.4). A 170 amino acid prohormone precursor of NT is the product of a single gene that is transcribed to yield two mRNAs. The smaller transcript is the major form in the intestine; the two mRNA species are equally abundant in most brain areas. The precursor contains one copy each of NT and NMN. The molar ratios of NT/NMN differ across different tissues, suggesting different enzymatic processing of the prohormone or, alternatively, the generation of different transcripts. Because NT and NMN are contained in the same exon of the NT-NMN gene, differences in relative abundance of the two are due to differential processing of the precursor.

NT and other peptide transmitters are stored in large dense core vesicles, but the general characteristics of NT release are similar to those of classical transmitters. Thus, depolarization evokes the Ca^{2+}-dependent release of both NT and NMN (Kitabgi et al., 1992). The impulse-dependent release of NT varies as a function of frequency and pattern of impulses, with higher frequencies of firing or burst firing patterns eliciting greater peptide release (Bean and Roth, 1992).

Inactivation of Neurotensin

Peptides are inactivated by enzymatic actions or diffusion. NT is degraded by three endopeptidases, known with great flair as 24.11, 24.15, and 24.16 (Kitabgi et al., 1992). Endopeptidase 24.11 cleaves NT at two specific sites to yield a decapeptide, and endopeptidase 24.16 acts at the same two sites (Fig. 7.4); the other endopeptidase acts at a different set of amino acid residues. Because these enzymes act at dipeptides that are found in many peptides and proteins, and two of the three act at the same site, it is obvious that enzymes that inactivate peptides are not very specific.

Although there are no known membrane transporters for peptide transmitters, peptides can be accumulated by neurons. This occurs by internalization of the peptide bound to its receptor. G protein-coupled receptors undergo internalization via an endocytotic mechanism, where they are either recycled to the membrane after various steps or shipped to lysosomes for degradation (Tsao et al., 2001). Receptor-bound peptides can also be internalized through this process and once inside the cell dissociate from the receptor.

Coexistence of Neurotensin and Classical Transmitters

Neurotensin is colocalized with dopamine in certain hypothalamic and midbrain neurons. The colocalization of NT and DA has provided a useful system in which to explore the interrelationships between two colocalized transmitters. In the prefrontal cortex of the rat, NT is found only in DA axons. Neurotensin release in the prefrontal cortex is increased when neuronal firing is increased or when DA neurons enter into a burst-like firing pattern (Bean and Roth, 1992). In addition, the release-modulating dopamine autoreceptor on DA axons regulates NT release, but in a very different manner than it regulates DA release. Dopamine agonists (which decrease DA release) enhance NT release from colocalized DA-NT axons; conversely, antagonists at the DA autoreceptor decrease NT release but enhance DA release. Thus, the release of NT and DA are regulated reciprocally by actions at a DA autoreceptor.

UNCONVENTIONAL TRANSMITTERS

The differences between peptide and classical transmitters originally were viewed as inconsistent with a transmitter role of the peptides, and the common acceptance of peptides as transmitters met with considerable resistance. However, once this battle was won, the gate was opened for the consideration of radically different molecules as transmitters.

Remarkable technical advances have allowed us to measure substances in the brain that are present in minute quantities or are very unstable. This has forced

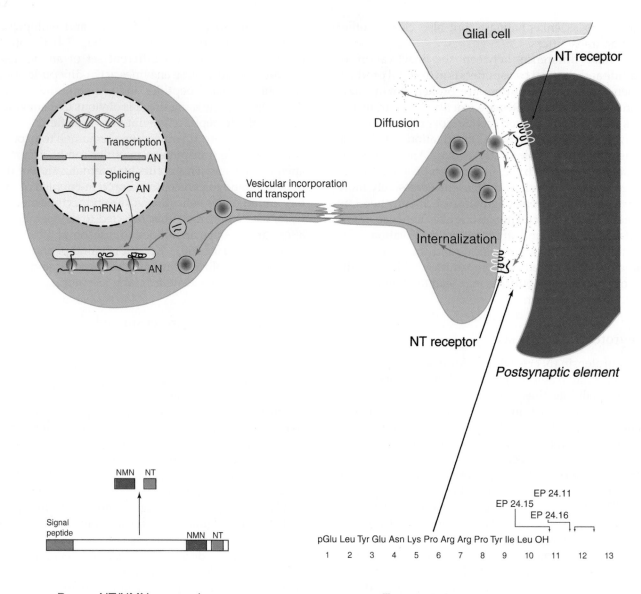

Prepro NT/NMN processing   **Enzymatic inactivation of NT**

FIGURE 7.4 Schematic illustration of the synthesis, release, and termination of action of the peptide transmitter neurotensin (NT). The illustrative panels of the bottom show processing of NT from the prohormone (left) and enzymatic inactivation of NT (right).

us to consider the possibility that substances that do not meet the established requirements for neurotransmitters may indeed be transmitters. What then should be the key attribute of a neurotransmitter? One simple approach would be to designate as a neurotransmitter any compound that permits information flow from one neuron to another. This definition circumvents the matter of glial contribution to the ionic milieu of the neuron, which certainly imparts information concerning the function of the glia. However, this definition is similar to that of a hormone and does not address the temporal characteristics of transmitter action and the distance of the target cell from the transmitter release

site. Finally, this definition does not accommodate unconventional roles for transmitters, such as the regulation of neuronal development or intracellular trafficking of proteins. We will now discuss what we call, for lack of a better term, "unconventional" transmitters.

Gases as Unconventional Transmitters

Nitrates such as nitroglycerine have been long been used to treat angina pectoris, in which chest pain arises because of insufficient blood delivery to the heart muscle. Nitrates dilate cardiac blood vessels, thereby increasing blood flow and relieving the pain. However, the mechanism by which nitrates dilate coronary arteries was not known until relatively recently. In 1980, an endothelial-derived relaxing factor contained in the cells lining blood vessels was found to potently and rapidly dilate blood vessels. This factor was soon shown to be the gas nitric oxide (NO). In addition, glutamate acting at one type of glutamate receptor was noted to release a factor that caused blood vessels in the brain to dilate. It became apparent that the endothelial-derived relaxing factor and the glutamate-induced factor that caused cerebrovascular vasodilation were the same compound, NO. Thus, NO was present in neurons as well as vasculature. These data led to the proposal that NO may take part in intercellular communication (see Box 7.2).

Nitric oxide is a well-known air pollutant. The idea that an unstable toxic gas could serve as a transmitter led to several incredulous questions, including "how can a gas be stored for release in an impulse-dependent manner?" The answer is simple: it is not stored. If one accepts the argument that NO is a transmitter, the classical definition of a neurotransmitter becomes untenable. Many theories can accommodate one exception. However, NO is not the only gas neurotransmitter; carbon monoxide and hydrogen sulfide play similar transmitter-like roles (Baranano *et al.*, 2001).

BOX 7.2

GOING FOR GASES AS NEUROTRANSMITTER

One of the most rewarding experiences for a scientist is to find that long-held prejudices are altogether wrong and that a new, correct insight reveals a novel scientific principle. In the late 1950s, only acetylcholine and the biogenic amines were known to be neurotransmitters. The next decade saw amino acids acknowledged as neurotransmitters. The discovery of enkephalins and endorphins in the mid-1970s reinforced gradually accumulating evidence that peptides are transmitters, and now we find that there are over 100 different bioactive brain peptides.

In the 1970s and early 1980s, nitric oxide (NO) was found to mediate the ability of macrophages to kill tumor cells and bacteria and to regulate blood vessel relaxation. A short report suggested that NO can be formed in brain tissue. Progress in the NO field at that time was slow because assaying NO synthase (NOS), the enzyme that oxidizes the amino acid arginine to NO, was quite tedious, based on the accumulation of nitrite formed from the NO. A much simpler approach was to monitor the conversion of [³H]arginine into [³H]citrulline, which is formed simultaneously with NO; this assay could process 100 or more samples in an hour. Research on blood vessels revealed that NO acts by stimulating cyclic GMP formation. In the brain, the excitatory neurotransmitter glutamate was known to augment cyclic GMP levels. Glutamate, acting through its NMDA receptor, triples NO synthase activity in a matter of seconds, and arginine derivatives that inhibit NOS activity block the elevation of cyclic GMP. This finding causally linked the actions of so prominent a neurotransmitter as glutamate to NO.

To determine if NO was a neurotransmitter, it was necessary to ascertain whether NOS was localized in neurons. The most straightforward approach would be to generate an antibody to use in anatomical studies. However, purifying NOS protein to generate an antibody proved very difficult because the enzyme lost its activity in attempts to purify it. The addition of calmodulin was found to stabilize the enzyme. Because calmodulin is a calcium-binding protein, this finding immediately explained how NO formation can be triggered rapidly by synaptic activation through glutamate. When glutamate activates its NMDA receptor, Ca^{2+} rushes into the cell, binds to calmodulin, and activates NOS.

The ability of purified NOS led to antibodies being developed and NOS being localized by immunohistochemistry. The neuronal form of NOS (nNOS) is present in only about 1% of the neurons in the brain. However, these cells give rise to processes that ramify so extensively that probably every neuron in the brain is exposed to NO. The purified NOS protein also allowed an amino acid sequence to be obtained and the gene for the enzyme cloned. The structure of NOS revealed that it is regulated by many more factors than virtually any other enzyme in biology, including at least five oxidative-reductive cofactors, four phosphorylating enzymes, and three binding proteins. This makes sense because of the unique properties of NO as a gaseous neurotransmitter. Most neurotransmitters are stored in vesicles with large storage pools so that only a small amount of the transmitter is released with each nerve impulse. In contrast, every time a neuron wishes to release a molecule of NO, it must

(Continues)

BOX 7.2   (cont'd)

activate NOS—hence, a requirement for exquisitely subtle regulation of the enzyme.

Neurotransmitters come in chemical classes such as biogenic amines, amino acids, and peptides. Might there be at least one other gaseous neurotransmitter? Carbon monoxide (CO) is normally formed in the body by the enzyme heme oxigenase, which is primarily responsible for degrading heme in aging red blood cells. It cleaves the heme ring to form biliverdin, which is reduced rapidly to bilirubin, the pigment that accounts for jaundice in patients with a degradation of red blood cells. When the enzyme cleaves the heme ring, CO is released as a single carbon fragment.

The biosynthetic enzyme for a transmitter should be localized to selected neuronal populations. Heme oxigenise-2, the neuronal form of the enzyme, was shown to be localized to discrete neuronal populations throughout the brain. To seek a neurotransmitter function, the peripheral nervous system (in which synaptic transmission is characterized more readily than in the brain) was used. The myenteric plexus of nerves regulates intestinal peristalsis. A previously unidentified neurotransmitter of myenteric

plexus neurons accounts for the relaxation phase of peristalsis. nNOS had already been localized to neurons of the myenteric plexus, and some functional evidence showed that NO might be a neurotransmitter of this pathway. Heme oxigenise-2 was found to be localized to the same myenteric plexus neurons as NOS. Mice with targeted deletions of the genes for nNOS or heme oxigenise-2 were used to elucidate function. In both types of gene knockout mice, intestinal relaxation evoked by neuronal depolarization was reduced about 50%, implying that NO and CO each contribute half of the relaxation. This finding, along with other evidence, established transmitter functions for both NO and CO and suggested that they are functioning as cotransmitters, although exactly how they interact remains a mystery. Such a cotransmitter role reminds us of the fact that most neurons in the brain contain at least two and sometimes more neurotransmitters. Thus, in addition to overturning a number of dogmas about neurotransmission, NO and CO may help resolve the riddle of cotransmission.

Solomon H. Snyder

The list of exceptions posed by NO to the dogma of traditional neurotransmitters is long. NO is not stored in cells, is not released in an exocytotic manner, lacks an active process that terminates its action, does not interact with specific membrane receptors on target cells, and often acts as a retrograde transmitter to regulate axon terminals presynaptic to the neuron in which NO is synthesized. It is therefore not difficult to understand the skepticism that first met the hypothesis that NO is a neurotransmitter, or to have some sympathy for those who expressed the view that NO is not a transmitter but an alien event, benign or otherwise, intent on making neuroscientists question their most cherished beliefs (Fig. 7.5).

Endocannabinoids

The psychoactive properties of marijuana have been known for thousands of years. However, how marijuana exerts these actions was not uncovered until 1990, when the receptor at which Δ-9-tetrahydrocannabinol, the major active component of marijuana, was cloned. Two years later an endogenous ligand that binds to the cannabinoid receptor was identified. Over

the past decade the identification of multiple endocannabinoids (ECs), the enzymes involved in synthesizing and degrading endocannabinoids, and the characterization of cannabinoid receptors have led to the realization that ECs serve as chemical messengers between neurons.

Endocannabinoids are highly hydrophobic, which allows them to pass easily through plasma membranes, but prevents them from being stored in vesicles. Instead, they are synthesized and released when needed from lipid precursors present in neuronal membranes. There are two major ECs in brain, anadamide and 2-arachidonoylglycerol (2-AG). Anandamide is a fatty acid amide and is the best characterized of the ECs. It is "stored" in the cell membrane in the form of a lipid precursor, from which it is cleaved by the enzyme phospholipase D. 2-AG is more abundant in brain than anadamide, and also is produced from membrane lipids, but via the action of two other enzymes.

Three proteins are involved in EC inactivation. Two are enzymes (fatty acid amide hydrolyase [FAAH] and monoacylglycerol lipase [MGL]), with the third protein an anadamide transporter. FAAH is found mainly in

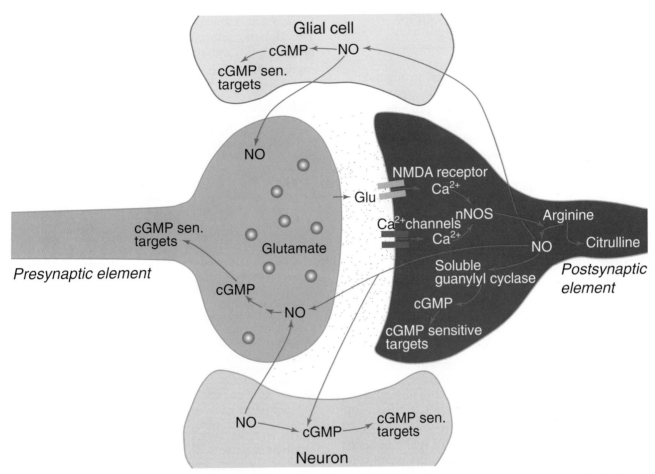

FIGURE 7.5 Schematic representation of a nitric oxide (NO)-containing neuron. NO is formed from arginine by the actions of different nitric oxide synthases (NOS). NO diffuses freely across cell membranes and can thereby influence both presynaptic neurons (such as the glutamatergic presynaptic neuron in the figure) or other cells that are not apposed directly to the NOS-containing neuron; these other cells can be neurons or glia.

the soma and dendrites of neurons, whereas MGL is more prominent in synaptic terminals. The ananamide transporter is a sodium- and energy-dependent carrier protein, acting by a process with the characteristics of facilitated diffusion.

Recent studies have demonstrated that ECs act as retrograde messengers at synapses. Thus, ECs are produced on demand in an activity-dependent manner by a neuron through the cleavage of membrane lipids. The resultant ECs travel backward (retrogradely) across the synapse to activate presynaptic cannabinoid receptors on axon terminals, resulting in a suppression of transmitter release. Because the presynaptic CB1 cannabinoid receptor receptor is found on both GABAergic and glutamatergic nerve endings, EC signaling can modulate both excitatory and inhibitory transmission. EC-mediated retrograde signaling appears to be crucial for certain types of short- and long-term synaptic plasticity that underlie learning and memory.

Although still early, studies of ECs already have inspired the development of novel treatment strategies for neuropsychiatric disorders. Consistent with subjective reports that marijuana increases eating (the "munchies"), early clinical trials have reported some success of CB1 receptor antagonists in the treatment of obesity. Similarly, CB1 antagonists have been advanced as an aide to smoking cessation.

Summary

We have discussed but two "unconventional" transmitters, one a gas, the other a lipid. This growing category of transmitters contains many other similarly unconventional molecules, including growth factors, neuroactive steroids, and others still to be uncovered. All offer a significant challenge to our definitions of what constitutes a neurotransmitter, but with the advantage of expanding our knowledge of how the nervous system works in health and disease.

SYNAPTIC TRANSMISSION IN PERSPECTIVE

Several of the key proteins involved in regulating chemical neurotransmission in mammals have been identified on the basis of homologies to proteins found in invertebrates such as the worm *C. elegans* and the fly *Drosophila melanogaster*. It now appears that some of the molecules used as neurotransmitters even are found in plants, as are their receptors! As nervous systems have evolved, many transmitter-related proteins have maintained roles that are not directly related to transmitter function or, alternatively, are involved in less discrete and more spatially elaborate signaling.

An example is acetylcholine, the synthesis of which depends on the cytosolic enzyme choline acetyltransferase. ChAT mRNA is also found in the testes, leading to the appearance of ACh in spermatozoa. There is even one form of ChAT that is targeted to the nucleus of cells. ChAT mRNA has also been reported to be present in lymphocytes, as have certain muscarinic cholinergic receptors. Also found in these white blood cells are AChE mRNAs, and decreases in both acetylcholinesterase and butyrylcholinesterase enzyme activity have been reported in Alzheimer's disease. Still another example is the presence of AChE in bone marrow cells and peripheral blood cells in certain leukemias; recent data indicate that inhibition of AChE gene expression in bone marrow suppresses apoptosis, or programmed cell death (see Chapter 19). These examples clearly indicate that although a neurotransmitter may have one role in the brain, a totally different role may emerge in other tissues—thus encouraging us to attend to developments in diverse fields.

We have discussed chemically-coded synaptic transmission. However, one cannot discuss the biochemistry and pharmacology of synaptic transmission without referring to and appreciating critical information about the structure (anatomy) and function (physiology and behavior) of neurons. Neuroscience is multidisciplinary, requiring an understanding of different aspects of cellular function to come to grips with the basic principles of synaptic communication.

Our concepts of synaptic transmission are in flux, requiring frequent reevaluation and revision. This can be seen most clearly in the evolving definition of a neurotransmitter. Classical transmitters are but one part of the family of transmitters, with other relatives being peptides, gases, growth factors, and lipid-derived transmitters such as endocannabinoids. The use of the terms "conventional" and "unconventional" in discussing transmitters is indicative of our current unease with the expanding definition of transmitters. This dynamic state of affairs is seen in all areas of neuroscience and helps make neuroscience such an exciting discipline.

References

Baranano, D. E., Ferris, C. D., and Snyder, S. H. (2001). Atypical neural messengers. *Trends Neurosci.* **24**, 99–106.

Bean, A. J. and Roth, R. H. (1992). Dopamine-neurotensin interactions in mesocortical neurons: Evidence from microdialysis studies. *Ann. N.Y. Acad. Sci.* **668**, 43–53.

Cartmell, J. and Schoepp, D. D. (2000). Regulation of neurotransmitter release by metabotropic glutamate receptors. *J. Neurochem.* **75**, 889–907.

Chaudhry, F. A., Lehre, K. P., van Lookeren Campagne, M., Otterson, O. P., Danbolt, N. C., and Storm-Mathisen, J. (1996). Glutamate transporters in glial plasma membranes: Highly differentiated localizations revealed by quantitative ultrastructural immunocytochemistry. *Neuron* **15**, 711–720.

Cooper, J. R., Bloom, F. E., and Roth, R. H. (2002). "The Biochemical Basis of Neuropharmacology," 8th ed. Oxford Univ. Press, New York.

DeLorcy, T. N. and Olsen, R. W. (1994). GABA and glycine. *In* "Basic Neurochemistry" (G. J. Siegel, B. W. Agranoff, R. W. Albers, and P. B. Molinoff, eds.), 5th ed., pp. 389–400. Raven Press, New York.

Erlander, M. G. and Tobin, A. J. (1991). The structural and functional heterogeneity of glutamic acid decarboxylase, A review. *Neurochem. Res.* **16**, 215–226.

Farhadi, H. E., Mowla, S. J., Petrecca, K., Morris, S. J., Seidah, N. G., and Murphy, R. A. (2000). Neurotrophin-3 sorts to the constitutive secretory pathway of hippocampal neurons and is diverted to the regulated secretory pathway by coexpression with brain-derived neurotrophic factor. *J. Neurosci.* **20**(11), 4059–4068.

Fremeau, R. T., Voglmaier S., Seal R. P., and Edwards, R. H. (2004) VGLUTs define subsets of excitatory neurns and suggest novel roles for glutamate. *TINS* **27**, 98–102.

Gainetdinov, R. R., Jones, S. R., Fumagalli, E., Wightman, R. M., and Caron, M. G. (1998). Re-evaluation of the role of the dopamine transporter in dopamine system homeostasis. *Brain Res. Rev.* **26**, 148–153.

Greenfield, S. A. (1991). A non-cholinergic role of AChE in the substantia nigra: From neuronal secretion to the generation of movement. *Mol. Cell. Neurobiol.* **11**, 55–77.

Hassel, B. and Dingledine, R. (2006). Glutamate. *In* "Basic Neurochemistry" (G. J. Siegel, R. W. Albers, S. Brady, and D. L. Price, eds.), 7th ed. Elsevier–Academic Press, San Diego, CA, pp. 267–290.

Kitabgi, P., De Nadal, F., Rovere, C., and Bidard, J.-N. (1992). Biosynthesis, maturation, release, and degradation of neurotensin and neuromedin N. *Ann. N.Y. Acad. Sci.* **668**, 30–42.

Kohara, K., Kitamura, A., Morishima, M., and Tsumoto, T. (2001). Activity-dependent transfer of brain-derived neurotrophic factor to postsynaptic neurons. *Science* **291**, 2419–2423.

Olsen, R. W. and Betz, H. (2006). GABA and Glycine. *In* "Basic Neurochemistry" (G. J. Siegel, R. W. Albers, S. Brady, and D. L. Price, eds.), 7th ed. Elsevier–Academic Press, San Diego, CA, pp. 291–302.

Paul, S. P. (1995). GABA and glycine. *In* "Neuropyschopharmacology: The Fourth Generation of Progress" (F. E. Bloom and D. J. Kupfer, eds.), pp. 87–94. Raven Press, New York.

Poo, M. M. (2001). Neurotrophins as synaptic modulators. *Nature Rev. Neurosci.* **2**, 24–32.

Schwarcz, R. (2004). The kynurenine pathway of tryptophan degradation as a drug target. *Curr. Opin. Pharmacol.* **4**, 12–17.

Shepherd, G. M. (1991). "Foundations of the Neuron Doctrine." Oxford Univ. Press, New York.

Sieburth, D., Madison, J. M., and Kaplan, J. M. (2007). PKC-1 regulates secretion of neuropeptides. *Nat. Neurosci.* **10**, 49–57.

Taylor, P., and Brown, J.H. (2006). Acetylcholine. *In* "Basic Neurochemistry," 7th ed. (Siegel, G. J., Albers, R. W., Brady, S., and Price, D. L., eds.). Elsevier-Academic Press, San Diego.

Tsao, P., Cao, T., and von Zastrow, M. (2001). Role of endocytosis in mediating downregulation of G-protein-coupled receptors. *Trends Pharmacol. Sci.* **22**, 91–96.

Valenstein, E. S. (2005). "The War of the Soups and the Sparks." Columbia University Press, New York.

Volterra, A., Magistretti, P., and Haydon, P. (2002). "The Tripartite Synapse: Glia in Synaptic Transmission." Oxford University Press.

Weihe, E. and Eiden, L. E. (2000). Chemical neuroanatomy of the vesicular amine transporters. *FASEB J.* **14**, 2435–2449.

Xu, E., Gainetdinov, R. R., Wetsel, W. C., Jones, S. R., Bohn, L. M., Miller, G. W., Wang, Y. M., and Caron, M. G. (2000). Mice lacking the norepinephrine transporter are supersensitive to psychostimulants. *Nature Neurosci.* **3**, 465–471.

Zucchi, R., Chiellini, G., Scanlan, T. S., and Grandy, D. K. (2006). Trace amine-associated receptors and their ligands. *Br. J. Pharmacol.* **149**, 967–978.

Ariel Y. Deutch and Robert H. Roth

Release of Neurotransmitters

The synapse is the primary place at which information is transmitted from neuron to neuron or from neuron to peripheral target, be it a gland or a muscle. Most synapses rely on a chemical intermediary, or transmitter, secreted in response to an action potential in the presynaptic cell in order to influence the activity of the postsynaptic cell. In chemical transmission, a single action potential in a small presynaptic terminal can generate a large *postsynaptic potential* (PSP) (as large as tens of millivolts). This is accomplished by the release of thousands to hundreds of thousands of molecules of transmitter that can bind to postsynaptic receptor molecules and open (or close) thousands of ion channels in about 1 ms. The effect can be either excitatory or inhibitory, depending on the ions that permeate the channels operated by the receptor. The resulting responses are either *excitatory postsynaptic potentials* (EPSPs) or *inhibitory postsynaptic potentials* (IPSPs), depending on whether they drive the cell toward a point above or below its firing threshold, as discussed in Chapter 11.

Why are most synapses chemical? The simpler alternative might appear to be the electrical synapse (see Chapter 11) in which the electrical signal from one cell can cross directly into the next through the electrically conducting pathway of the gap junction. Even when pre- and postsynaptic elements are coupled in this manner, however, a presynaptic spike of 100 mV would likely cause only a 1-mV change in the postsynaptic cell because relatively little charge can flow through these junctions to charge the large membrane capacitance of the postsynaptic cell. More efficient electrical transfer requires that the presynaptic terminal be as large as or larger than the postsynaptic element and therefore only a few presynaptic cells could converge on a given postsynaptic cell. Such synapses would

further limit the computational capacity of the brain because they could only induce excitation, not inhibition, in response to a presynaptic action potential. Chemical synapses are characterized by great flexibility. Different afferents can have different effects, with different strengths and time courses, on each other as well as on postsynaptic cells. These differences depend on the identity of the transmitter(s) released and the receptors present (see Chapters 7 and 9). Chemical synapses often are modified by prior activity in the presynaptic neuron. Chemical synapses are also particularly subject to the modulation of presynaptic ion channels by substances released by the postsynaptic or neighboring neurons. This flexibility is essential for the complex processing of information that neural circuits must accomplish, and it provides an important locus for modifying neural circuits in adaptive processes such as learning (Chapters 49 and 50).

TRANSMITTER RELEASE IS QUANTAL

In order to secrete thousands of transmitter molecules rapidly and simultaneously, nerve terminals, like other secretory cells, package the transmitter into membrane-enclosed organelles. When one of these *synaptic vesicles* fuses with the plasma membrane, the contents of the vesicle (approximately 5000 molecules) can diffuse into the extracellular space and encounter receptors on the postsynaptic cell within hundreds of microseconds. A consequence of releasing transmitter by the fusion of vesicles with the plasma membrane (called *exocytosis*) is that synaptic transmission is *quantal*—responses are built from signals with a discrete amplitude corresponding to a single vesicle. In

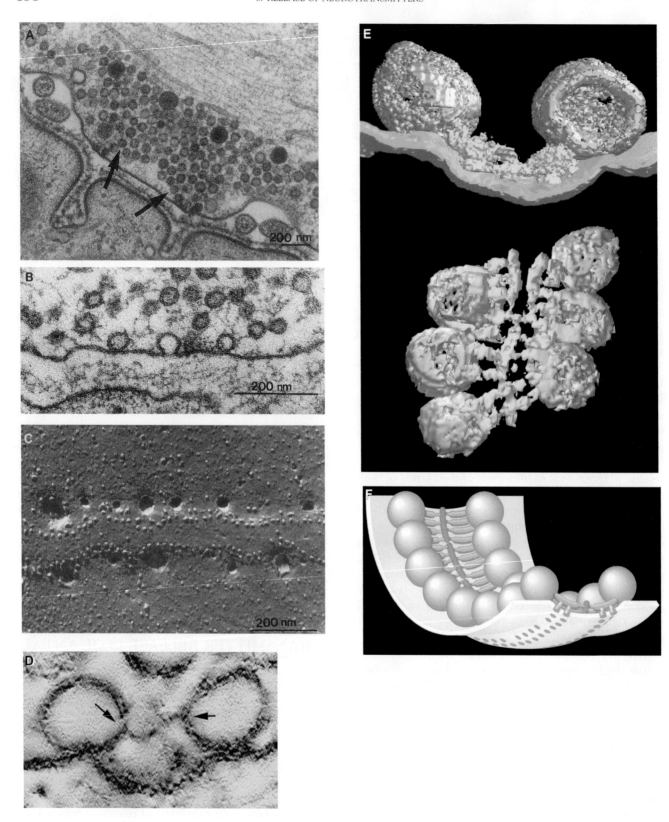

FIGURE 8.1 Ultrastructural images of exocytosis and active zones. (A–C) Synapses from frog sartorius neuromuscular junctions were quick-frozen milliseconds after stimulation in conditions that enhance transmission. (A) A thin section showing vesicles clustered in the active zone (arrows) some docked at the membrane. (B) Shortly (5 ms) after stimulation, vesicles were seen to fuse with the plasma membrane. (C) After freezing, presynaptic membranes were freeze-fractured and a platinum replica was made of the external face of the cytoplasmic membrane leaflet. Vesicles fuse about 50 nm from rows of intramembranous particles thought to include Ca^{2+} channels. (D–F) The fine structure of the active zone at a frog neuromuscular junction as seen with electron tomography. (D) In a cross-sectional image from tomographic data, two vesicles are docked at the plasma membrane and additional electron-dense elements are seen. When these structures are traced and reconstructed through the volume of the EM section (E), proteins of the active zone (gold) appear to form a regular structure adjacent to the membrane that connects the synaptic vesicles (silver) and plasma membrane (white). Viewed from the cytoplasmic side (E, lower image), proteins are seen to extend from the vesicles and connect in the center. (F) Schematic rendering of an active zone based on tomographic analysis. An ordered structure aligns the vesicles and connects them to the plasma membrane and to one another. Parts A and B from Heuser (1977); part C from Heuser *et al.* (1979). Part B reproduced from the *Journal of Cell Biology* **88**, 564–580 (1981). (D–F) After Harlow *et al.* (2001).

the absence of presynaptic electrical activity, transmitter is released spontaneously as individual quanta. Each packet generates a small postsynaptic signal—either a *miniature excitatory* or a *miniature inhibitory postsynaptic potential* (MEPSP or MIPSP, respectively, or just "mini")—that can be detected by microelectrode recording (Katz, 1969). An action potential accelerates tremendously, but very briefly, the rate of secretion of quanta and synchronizes them to evoke a PSP. At a synapse between two neurons, this might represent the release of 1 to 10 vesicles. At the vertebrate neuromuscular junction, a remarkably large and specialized synapse, hundreds of vesicles can be released, and the response is, to a first approximation, the sum of the individual quanta.

The anatomical specializations of the synapse and the properties of the presynaptic ion channels and postsynaptic receptors unite to achieve fast, quantal transmission (Figs. 8.1 and 8.2). The chief anatomical feature of the terminal is the profusion of synaptic

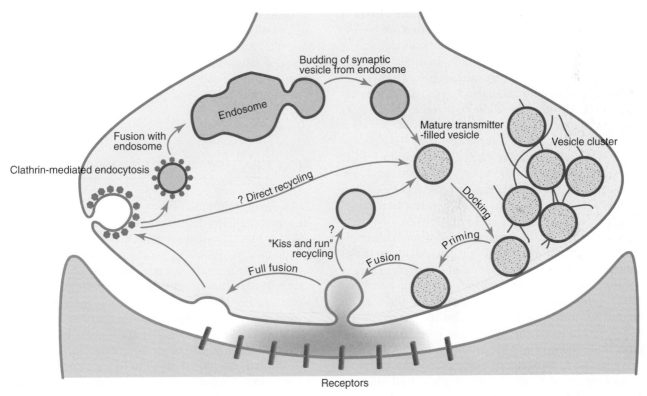

FIGURE 8.2 The life cycle of synaptic vesicles. Transmitter-filled vesicles can be observed in clusters in the vicinity of the active zone. Some vesicles are recruited to sites within the active zone in a process called docking. These vesicles subsequently are primed for release. The rise in cytosolic Ca^{2+} that occurs during an action potential triggers the opening of a fusion pore between some of the primed, docked vesicles and the plasma membrane. Transmitter exits through this fusion pore. Three pathways are proposed by which the now empty vesicle can be recovered and returned to the releasable pool: (1) by a direct reclosing of the fusion pore and reformation of the vesicle, often called "kiss and run"; (2) by complete fusion (i.e., the flattening of the vesicle onto the membrane surface) followed by clathrin-mediated endocytosis, removal of the clathrin coat, and return of the vesicle to the releasable pool; and (3) by complete fusion and recycling as in the second pathway, but the endocytosed vesicle fuses first with an endosome and mature vesicles are subsequently formed by budding from the endosome. After or during this recycling process, the vesicle must be refilled with transmitter.

vesicles, typically 50 nm in diameter, that cluster near the synapse and dock at specialized sites called *active zones* along the presynaptic membrane (Fig. 8.1). The vesicles may differ in their appearance depending on the transmitter they enclose: thus glutamate and acetylcholine are stored in small clear vesicles, whereas peptide neurotransmitters occupy large dense-cored vesicles. Although it may appear as indistinct fuzz on electron micrographs, the active zone is likely to be a highly structured specialization of the membrane and cytoskeleton (Fig. 8.1; also, see discussion later). Action potentials release transmitter by triggering the exocytosis of vesicles with the plasma membrane (Figs. 8.1B and 8.1C) and the release of their contents into the narrow *synaptic cleft* (about 100 nm wide) separating the presynaptic terminal from high concentrations of postsynaptic receptors. At neuromuscular junctions, one of the best studied synapses, transmitter from one vesicle diffuses across the synaptic cleft in 2 ms and reaches a concentration of about 1 mM at the postsynaptic receptors. A large number of these receptors (up to 2000) will bind transmitter rapidly and thereby open an ion channel (see Chapter 9). Each channel has a 25 pS conductance and remains open for about 1.5 ms, admitting a net inflow of 35,000 positive ions. Thus, a single vesicle can cause 70 million ions to cross the membrane and give rise to a mini of a few millivolts. The neuromuscular junction also is specialized for efficiency; a single nerve ending will have about 1000 active zones and a single action potential will cause a vesicle to fuse at about a third of those. The consequence is the release of about 300 quanta within 1.5 ms. The resulting postsynaptic depolarization begins after a synaptic delay of about 0.5 ms, reaches a peak of tens of millivolts, and is typically sufficient to generate an action potential in the muscle fiber.

The anatomy and physiology are somewhat different at fast central synapses. Postsynaptic cells make contact with presynaptic axon swellings that are called *varicosities* when they occur along fine axons and are called *boutons* when they are located at the tips of axons. The postsynaptic cell can be contacted on the cell body, but often the synapse is made onto a fine dendritic branch or tiny spine with a length of a few micrometers. A typical varicosity or bouton contains one to four active zones. At any single active zone, an action potential may release zero, one, or perhaps two vesicles. However, with multiple active zones between two cells, the action of each zone will be additive in determining the response of the postsynaptic cell. At a representative excitatory glutamatergic synapse, each action potential releases from 5 to 10 quanta. Each quantum that is released elevates transmitter concentration in the cleft to about 1 mM, as at the neuromus-

cular junction, but because there are fewer receptors clustered beneath these synapses, each quantum activates only 30 or so ion channels as compared with the 1000–2000 channels activated at the neuromuscular junction. At excitatory synapses, the release of a quantum may be sufficient to generate EPSPs of 1 mV or less in amplitude, clearly subthreshold for generating action potentials. However, central neurons often receive thousands of inputs, each of which has a "vote" on how the cell should respond.

A comparison of these excitatory synapses at small neurons and the vertebrate neuromuscular junction at the much larger skeletal muscle fiber illustrates two important points: (1) the fundamental quantal nature of the chemical synapse is universal and (2) the different physiological requirements of particular synapses, such as synapses onto small or large cells, can be met by differences in any of several parameters, including the number of receptors activated per quantum, the number of active zones connecting two cells, the probability of a given active zone releasing a vesicle, and the conductance of the postsynaptic receptor channel. In describing synaptic transmission, two parameters are particularly important. *Quantal size* or quantal amplitude describes the unitary response to release of a single quantum (vesicle), whereas *quantal content* refers to the average number of quanta released by a single impulse. These parameters are discussed further later.

EXCITATION–SECRETION COUPLING

The action potential is an electrical event—a change in the voltage gradient across the plasma membrane. How is this electrical change converted to the fusion of synaptic vesicles and the release of neurotransmitter? The classic studies of Bernard Katz and colleagues (Katz, 1969) established that this coupling is achieved by the use of Ca^{2+} as an intracellular messenger. Ca^{2+} inside the cell normally is buffered to very low levels (~100 nM), and both concentration and electrical gradients provide a strong driving force for Ca^{2+} entry. Thus, when a voltage-dependent Ca^{2+} channel opens in response to the depolarization of the membrane during an action potential, there is the potential for the intracellular Ca^{2+} concentration to increase 1000-fold. In this manner, the electrical signal is converted to a large chemical signal, and Ca^{2+} sensors in the fusion machinery can trigger vesicle fusion. The mechanism of fusion itself is discussed later, but the nature of the Ca^{2+} signal and the evidence for its centrality must be understood first.

It has been known for over 100 years that Ca^{2+} must be present in the extracellular saline for transmission to occur. Subsequently (Katz, 1969), it was shown that Ca^{2+} need be present only at the moment of invasion of the nerve terminal by the action potential. Changing the timing of the influx of Ca^{2+} or the amount of influx (by changing the extracellular Ca^{2+} concentration or blocking Ca^{2+} channels with divalent cations such as Co^{2+} and Mn^{2+}) will change the timing and amplitude of the synaptic response. Similarly, an ionophore that lets Ca^{2+} ions flow across the membrane will cause a sustained increase in intracellular Ca^{2+} and the sustained outpouring of quanta of transmitter. It has also become possible to fill a terminal with a caged form of the Ca^{2+} ion; that is, with Ca^{2+} ions bound up within a chemical carrier. A flash of light can rearrange this carrier, uncaging the Ca^{2+}, and thereby cause an abrupt increase in cytosolic Ca^{2+} and a sudden increase in secretion as well. Finally, loading the terminal with Ca^{2+} chelators to prevent increases in cytosolic Ca^{2+} can prevent release. Thus a rise in intracellular Ca^{2+} is the essential trigger for vesicle fusion.

Vesicles Are Released by Calcium Microdomains

The opening of Ca^{2+} channels during a single action potential allows in enough Ca^{2+} to raise the concentration in the bouton from approximately 100 to 110 nM. How can a mere 10% change in the concentration cause a 1000-fold shift in the activity of the fusion machinery? Furthermore, because Ca^{2+} pumps and exchangers on the plasma membrane work relatively slowly, this small rise in Ca^{2+}, which can be detected with fluorescent Ca^{2+} indicator dyes, persists for hundreds of milliseconds after the action potential. Why then does transmitter secretion occur only for a millisecond or so? The answer to both of these questions lies in the fact that the release mechanism is *not* responding to the general concentration of Ca^{2+} in the bouton, but rather to the concentration in a *microdomain* in the immediate vicinity of the Ca^{2+} channels (Fig. 8.3). In these microdomains, very high concentrations of Ca^{2+} are reached very quickly and drop rapidly to near resting levels within microseconds of the closing of the channels. Presently, there are no Ca^{2+} dyes that are sufficiently local, sensitive, and fast that one can measure the concentration changes in these microdomains directly. Instead, our understanding of these signals comes from diffusion modeling, from modifying the buffering capacity of the cytosol, and from using indirect indicators of the local Ca^{2+} concentration, such as the gating of Ca^{2+}-activated K^+

channels or transmitter release itself (Roberts, 1994; Schneggenburger and Neher, 2000).

In the brief period for which the Ca^{2+} channels are open, the cytosol that lies within 100 nm of the mouth of each channel is flooded with Ca^{2+}, and the local Ca^{2+} concentration is likely to reach 100 μM or higher; the closer to the channel, the higher the concentration (Figs. 8.3A and 8.3B). The further from the channel, the more dilute the Ca^{2+} becomes and the greater the chance that it has been bound up by the high Ca^{2+}-buffering capacity of the cytosol. After the channels close, diffusion and buffering bring the Ca^{2+} concentration of the microdomains to near resting levels within a few milliseconds: the concentration gradient that existed around the mouth of the channel completely dissipates, and only the small net rise of total Ca^{2+} in the terminal remains (Fig. 8.3C). This signal, sometimes called residual Ca^{2+}, is what is detected by fluorescent indicator dyes. The effect may be compared to dumping a bucket of water into a swimming pool—a dramatic rise in water level occurs at one spot but it is very transient and gives rise to just a small net rise in the level of the entire pool.

A single active zone may have more than 100 channels in its membrane and a single vesicle may therefore be within 50-nm of as many as 10 Ca^{2+} channels. Though not all these channels will open during each action potential, more than one channel is likely to open, and therefore Ca^{2+} entering through several nearby channels can sum in overlapping microdomains to influence a vesicle (Fig. 8.3). Of the subtypes of Ca^{2+} channels that have been described, N and P/Q types of Ca^{2+} channels appear to be the most prevalent at CNS synapses, although R and L types are also reported at certain synapses.

The Exocytosis Trigger Must Have Fast, Low-Affinity, Cooperative Ca^{2+} Binding

The sensor that detects the Ca^{2+} so as to trigger release must also have special properties to achieve fast and transient exocytosis. Estimates at several synapses suggest that the Ca^{++}-sensor has an affinity in the range of 10–100 μM, as appropriate for the relatively high concentrations of Ca^{2+} in the microdomains near channels. The on rate must be particularly fast. This is confirmed by the finding that presynaptic injection of relatively slow Ca^{2+} chelators such as ethylene glycol bis(β-aminoethyl ether)-N,N'-tetraacetic acid (EGTA) have almost no effect on transmitter release to single action potentials. Only fast Ca^{2+} chelators, such as 1,2-bis(2-aminophenoxy)ethane-N,N,N',N'-tetraacetic acid (BAPTA), with on rates of about $5 \times 10^8 M^{-1} s^{-1}$, can capture Ca^{2+} ions before they bind to the secretory

A

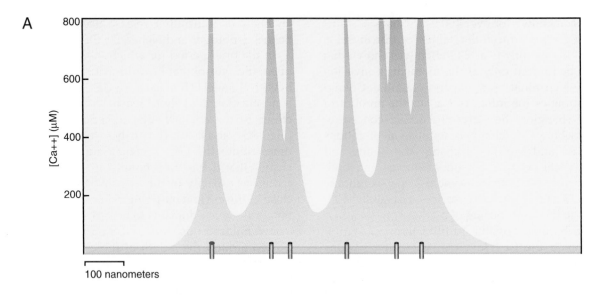

B

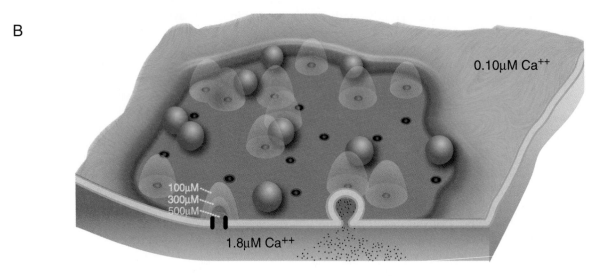

C

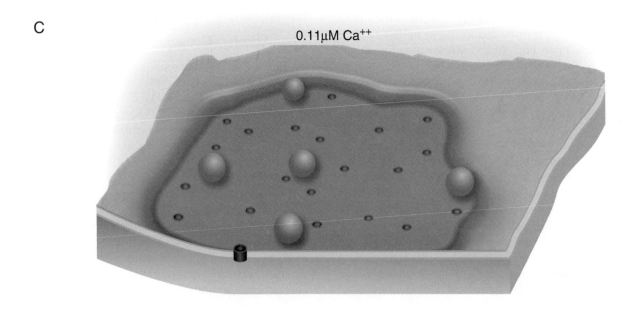

FIGURE 8.3 Microdomains with high Ca^{2+} concentrations form in the cytosol near open Ca^{2+} channels and trigger the exocytosis of synaptic vesicles. (A) In this adaptation of a model of Ca^{2+} dynamics in the terminal, a set of Ca^{2+} channels is spaced along the x axis, as if in a cross-section of a terminal. The channels have opened and, while they are open, the cytosolic Ca^{2+} concentration (y axis) is spatially inhomogeneous. Near the mouth of the channel, the influx of Ca^{2+} drives the local concentration to as high as $800\,\mu M$, but within just 50nm of the channel, the concentration drops off to $100\,\mu M$. Channels are spaced irregularly but are often sufficiently close to one another that their clouds of Ca^{2+} can overlap and sum. (B) In the active zone (gray), an action potential has opened a fraction of the Ca^{2+} channels, and microdomains of high cytosolic Ca^{2+} (pink) arise around these open channels as Ca^{2+} flows into the cell. In the rest of the cytoplasm, the Ca^{2+} concentration is at resting levels ($0.10\,\mu M$), but within these microdomains, particularly near the channel mouth, Ca^{2+} concentrations are much higher, as in (A). Synaptic vesicles docked and primed at the active zone may come under the influence of one or more of these microdomains and thereby be triggered to fuse with the membrane. (C) A few milliseconds after the action potential, the channels have closed and the microdomains have dispersed. The overall Ca^{2+} concentration in the terminal is now slightly higher ($0.11\,\mu M$) than before the action potential. If no other action potentials occur, the cell will pump the extra Ca^{2+} out across the plasma membrane and restore the initial condition after several 100ms.

Summary

Chemical synapses permit one neuron to rapidly and effectively excite or inhibit the activity of another cell. A diversity of transmitters and receptors allows varied postsynaptic responses. The packaging of transmitter into vesicles and its release in quanta enable a single action potential to secrete hundreds of thousands of molecules of transmitter almost instantaneously onto another cell. Ca^{2+} acts as an intracellular messenger tying the electrical signal of presynaptic depolarization to neurosecretion. At fast synapses, Ca^{2+} enters through clusters of channels near docked synaptic vesicles in active zones. This Ca^{2+} acts at extremely short distances (tens of nanometers) in remarkably little time ($200\,\mu s$) and at very high local concentrations ($\sim100\,\mu M$), in calcium microdomains, by binding cooperatively to a low-affinity receptor with fast kinetics to trigger exocytosis. When Ca^{2+} channels close, these microdomains of high Ca^{2+} return to near resting concentrations quickly and the evoked response is terminated. Speed, efficiency, and flexibility are the hallmarks of this process.

MOLECULAR MECHANISMS OF THE NERVE TERMINAL

To release neurotransmitter in response to an action potential, a synaptic vesicle must fuse with the plasma membrane with great rapidity and fidelity, and thus the synapse requires an effective and well-regulated molecular machine. This machine must include the means to load the vesicle with transmitter, to dock the vesicle near the membrane so that it can fuse with a short latency, to define a release site on the plasma membrane, and to cause the release itself. Additionally, a reserve of synaptic vesicles must be held near the active zone (e.g., Fig. 8.1A) and those vesicles must be recruited to the plasma membrane as needed. The number of vesicles that are ready to fuse must be strictly determined, and the protein and lipid components of the vesicle must be recycled to form a new vesicle after fusion has occurred. For each of these processes, a molecular understanding remains incomplete, but rapid progress in this field has resulted in considerable headway.

trigger, indicating that the on-rate of Ca^{2+} binding to this trigger is similarly fast. At a rate of $5 \times 10^{8}\,M^{-1}s^{-1}$, $100\,\mu M$ Ca^{2+} reaches equilibrium with its target in about $50\,\mu s$. The off rate of Ca^{2+} dissociation from these sites must also be fast, at least $10^{3}\,s^{-1}$, to account for the rapid termination of transmitter release (0.25ms time constant) after Ca^{2+} channels close and Ca^{2+} microdomains collapse.

Importantly, the relationship of Ca^{2+} influx or cytosolic Ca^{2+} in microdomains to the amount of release is not linear at most synapses. By plotting this relationship on a log-log plot, a Hill coefficient can be determined, and values as high as 3 or 4 commonly are obtained. This value implies a high cooperativity in the action of Ca^{2+} inside the terminal, perhaps because multiple Ca^{2+}-binding sites must be occupied in order to trigger release efficiently (Schneggenburger and Neher, 2000). The interesting physiological consequence of this steep relationship is that small modulations of Ca^{2+} channels that increase or decrease Ca^{2+} influx can have a large effect on the strength of a synapse. For example, at a synapse from a parallel fiber onto a Purkinje cell in the cerebellum, when the activation of modulatory presynaptic GABA receptors inhibits Ca^{2+} channels sufficiently to reduce Ca^{2+} influx by 25%, the amplitude of transmission decreases by 70%.

Most Neurons Require a Cycle of Membrane Trafficking

Active neurons are in constant need of transmitter-filled vesicles ready to release their contents. A bouton in the CNS, for example, may contain a store of 200 vesicles, but if it releases even one of these with each action potential and if the cell is firing at an unexceptional rate such as 5 Hz, the store of vesicles would be consumed within less than a minute. Transport of newly synthesized vesicles from the cell body would be far too slow to support such a demand and the axonal traffic needed to supply an entire arbor of nerve terminals would be staggering.

In short, active neurons need an efficient mechanism to recycle and reload vesicles within the terminal. The exception is peptidergic neurons because peptides must be synthesized in the endoplastic reticulum (ER), chiefly in the cell body. Not surprisingly, therefore, peptide vesicles are released at very low rates. For most neurotransmitters, however, the exocytosis of a synaptic vesicle is followed rapidly by endocytosis and within approximately 30 s the vesicle is again available for release. This pathway is sometimes referred to as the exo-endocytic cycle (Fig. 8.2). If we start with a transmitter-filled vesicle in the cytosol, we can outline its progression through this cycle. The vesicle will be mobilized from the reserve pool in the cytosol to the readily releasable pool by translocating to the active zone and becoming "docked" at the plasma membrane. Biochemical priming steps may occur at this point that will enable the vesicle to fuse within microseconds once the Ca^{2+} signal is given. Exocytosis often involves a complete merging of the vesicle membrane with the plasma membrane, though some research supports the coexistence of a second pathway involving only a transient connection of the two. In the latter case (see later), the vesicle, having emptied its contents, may return directly to the vesicle pool and be refilled with transmitter. If fusion is complete, however, the vesicle needs to be reformed with a clathrin cage and pinched off the plasma membrane. In some cases, these endocytosed vesicles seem to fuse with endosomes or large membranous sacs called cisternae, and mature synaptic vesicles then bud from this compartment. In other cases, the mature vesicle may be formed directly from the plasma membrane. In either case, the vesicle must be reloaded with transmitter prior to its reuse. The pathways for neurotransmitters are discussed in Chapter 7.

Histological tracers have provided a means to follow the vesicle cycle. One such tracer is the enzyme horseradish peroxidase (HRP), which could be placed in the extracellular solution and which would be internalized into intracellular compartments, including synaptic vesicles, when the synapse was stimulated. Another set of useful tracers are lipophilic dyes, including FM1-43, which partition into the surface membrane of the cell and are similarly internalized as membrane vesicles endocytosis. When the dye is washed from the extracellular space, the surface destains, but dye remains trapped in internal compartments. Loading of this dye upon stimulation and subsequent release of the dye with further stimulation has been an important method for examining the kinetics of the cycle.

A defect in any single step in the exo-endocytic cycle will halt transmitter release. Moreover, each step in this cycle represents a potential control point for modulating the efficacy of the synapse. For example, regulation by Ca^{2+} and other second messenger systems is known to affect the docking, fusing, and recycling of vesicles at some synapses and is also likely to regulate the balance between reserve stores and those vesicles actively engaged in the exo-endocytic cycle. Understanding the mechanisms of this modulation is an important goal and will certainly require the detailed understanding of the fundamental machinery itself.

The Fusion Mechanism Is Rapid

As discussed earlier in this chapter, the delay between the arrival of an action potential at a terminal and the secretion of the transmitter can be less than $200\,\mu s$. This places severe constraints on the fusion mechanism. Vesicles must already be present at the release sites, as there is no time to mobilize them from a distance. A catalytic cascade during fusion, such as that involved in phototransduction or in excitation-contraction coupling in smooth muscle, would also be far too slow for excitation-secretion coupling at the nerve terminal. Models therefore favor the idea that a fusion-ready complex of the vesicle and plasma membrane is preassembled at release sites and that Ca^{2+} binding need only trigger a simple conformation change in this complex to open a pathway for the transmitter to exit the vesicle.

Because the volume of the synaptic vesicle is small, the diffusion of transmitter from the vesicle proceeds almost instantaneously as soon as a pore has opened up between the vesicle lumen and the extracellular space. This structure is referred to as the fusion pore, but its biochemical nature is unknown. Thus, the time-critical steps come between the influx of Ca^{2+} and the formation of the fusion pore, and these steps establish the latency between arrival of the action potential and transmitter diffusion into the cleft. Any further step, such as the complete merging of the vesicle and plasma membrane, can occur on a slower time course, after the transmitter has left the vesicle. The tethering of the vesicle at the release site (often called docking) and any biochemical events that need to occur in order for the vesicle to reach the fusion-ready state (often called priming) can also be slower. Docking and priming a vesicle cannot be *too* slow, however; a central nervous system (CNS) synapse is estimated to have 2 to 20 vesicles in this fusion-ready state; therefore, if a synapse is to respond faithfully to a sustained train of action potentials, it must be able to replace the fusion-ready vesicles with a time course of seconds. The rate at which this occurs may determine some of the dynamic properties of the synapse.

The short latency of transmission would seem to preclude the involvement of ATP hydrolysis at the fusion step because such a reaction would be too slow. Indeed, fusion can proceed without ATP present and thus, if energy is needed to fuse the membranes, it must be stored in the fusion-ready state of the vesicle-membrane complex and released upon addition of Ca^{2+}.

Transmitter Release Is a General Cell Biological Question

Exocytosis and endocytosis are not unique to neurons; these processes go on in every eukaryotic cell. Moreover, exocytosis itself is only one representative of a general class of membrane-trafficking steps in which one membrane-bound compartment fuses with another. Other examples would include transport from the ER to the Golgi, or transport from endosomal compartments to lysosomes. In each case, the same biophysical problem must be overcome to bring the membranes close together and drive fusion.

At present, it appears that all the membrane-trafficking steps within the cell use a similar set of proteins to accomplish this task (Fig. 8.6). Indeed, it appears that representatives of this core set of proteins are present on every trafficking vesicle or target membrane and are required for every fusion step, whether it be in yeast, an epithelial cell, or a neuron. This discovery, which grew from the conjunction of independent studies of different model systems (Bennett and Scheller, 1993), has led to the exciting hypothesis that all intracellular membrane fusions will be united by a single and universal mechanism. To the extent that this proves true, it will be a great boon to cell biology: experiments in one model system, for example, the highly developed genetic analysis of membrane fusion in yeast, can be absorbed into neuroscience. Similarly, the abundance of synaptic vesicles for biochemical analysis and the unparalleled precision of electrophysiological assays of single vesicle fusions can deepen the understanding of other cellular events. Are these disparate membrane fusion events truly identical in their mechanisms? Most likely, different membrane fusions will present variations on a common theme;

the fundamental processes will be adapted to the specific requirements of each physiological step. The need for extremely tight control of fusion in the nerve terminal—for rapid rates of fusion with very short latencies to occur in brief bursts at precise points on the plasma membrane—may require some important differences in how the synaptic isoforms of these proteins function compared to the isoforms involved in general cellular traffic.

Synaptic Proteins Have Been Identified from Purified Vesicles

Synaptic vesicles are abundant in nervous tissue and, due to their unique physical properties (uniform small diameter and low buoyant density), they can be purified to homogeneity by subcellular fractionation techniques. As a result, at a biochemical level, synaptic vesicles are among the most thoroughly characterized organelles. The protein compositions of synaptic vesicles from different sources are remarkably similar, demonstrating that many of the proteins present on the synaptic vesicle membrane perform general functions that are not restricted to a single class of transmitter.

Our knowledge of many of the important proteins in vesicle fusion commenced with the characterization of purified synaptic vesicles from which individual proteins could be isolated and subsequently cloned. The major constituents of the synaptic vesicle are shown in Figure 8.4. For some of these proteins there are well-established functions, but for others, the functions remain uncertain (Table 8.1). The proton transporter, for example, is an ATPase that acidifies the lumen of the vesicle. The resulting proton gradient provides the energy by which transmitter is moved into the vesicle. This task is carried out by another vesicular protein, a vesicular transporter that allows protons to move down their electrochemical gradient and out of the vesicle in exchange for transmitter that moves into the lumen. Several vesicular transporters are known, all structurally related, and by their substrate selectivity they help specialize a terminal for release of the appropriate transmitter (see later). Two other large proteins with multiple transmembrane domains are abundant in vesicles: synaptic vesicle protein 2 (SV2) and synaptophysin. The mechanism of action of these proteins remains elusive despite extensive biochemical and genetic characterization. Other vesicular proteins are discussed later in this chapter.

Transmitter release depends on more than just vesicular proteins, however; proteins of the plasma membrane and cytoplasm are also important. In many cases, the identification of these additional

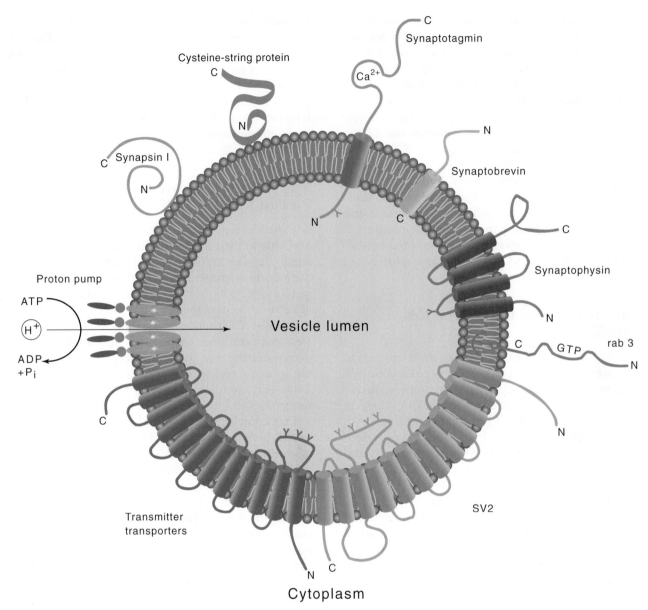

FIGURE 8.4 Schematic representation of the structure and topology of major synaptic vesicle membrane proteins (see also Table 8.1).

components or an appreciation of their importance to the synapse derived from investigations of vesicular proteins. Two examples will serve to illustrate this point. The first example begins with synaptotagmin, an integral membrane protein of the synaptic vesicle that was purified from synaptic vesicles and cloned. The portion of synaptotagmin that extends into the cytoplasm (the majority of the protein, see Fig. 8.4) was subsequently used for affinity column chromatography. In this manner, a protein called syntaxin was identified as a synaptotagmin-binding protein. This protein resides in the plasma membrane and is now appreciated as one of the critical players in vesicle fusion (see later). More recently, yeast two-hybrid

screens have been conducted and used to identify proteins that bind to syntaxin. One such protein, syntaphilin, may serve as a regulator or modulator of syntaxin function. Thus, subsequent to the isolation of an abundant vesicular protein, biochemical assays have led to a fuller picture.

A second and similar example (Gonzalez and Scheller, 1999) began with the realization that an abundant small GTP-binding protein, rab3, was present on the vesicle surface. This protein, discussed further later, may regulate vesicle availability or docking to release sites. Subsequently, rabphilin was identified on the basis of its affinity for the GTP-bound form of rab3. Rabphilin lacks a transmembrane domain but is

TABLE 8.1 Function of Synaptic Vesicle Proteins

Protein	Function
Proton pump	Generation of electrochemical gradient of protons
Vesicular transmitter transporter	Transmitter uptake into vesicle
VAMP/synaptobrevin	Component of SNARE complex; acts in a late, essential step in vesicle fusion
Synaptotagmin	Ca^{2+} binding; possible trigger for fusion and component of vesicle docking at release sites via interactions with SNARE complex and lipid; promotes clathrin-mediated endocytosis by binding AP-2 complex
Rab3	Possible role in regulating vesicle targeting and availability
Synapsin	Likely to tether vesicle to actin cytoskeleton
Cysteine string protein	Promotes reliable coupling of action potential to exocytosis
SV2	Unknown
Synaptophysin	Unknown

recruited to the surface of the synaptic vesicle by binding to rab3. Rabphilin may have a role in modulating transmission, particularly in mossy fiber terminals of the hippocampus. Still later, many additional rab3-binding proteins were identified, including RIM, a component of the active zone. Rab3 can exist in either GTP- or GDP-bound states, and additional factors that regulate these states were also identified: a GDP dissociation inhibitor (GDI), a GDP/GTP exchange protein (GEP), and a GTPase activating protein (GAP). Thus, from the identification of a synaptic vesicle protein, an array of additional factors has come to light, all of which are likely to figure in the exo-endocytic cycle.

Genetic Screens Have Led to the Identification of Synaptic Proteins

Genetic screens have provided an independent method for identifying the machinery of transmitter release. One of the most fertile screens was carried out not in the nervous system, but in yeast. Because membrane trafficking in yeast is closely parallel to vesicle fusion at the terminal, mutations that alter the secretion of enzymes from yeast can be a springboard for the identification of synaptic proteins. In the early 1980s, a series of such screens was carried out, and a collection of over 50 mutants was obtained. Many of these were shown to accumulate in post-Golgi vesicles

in the cytoplasm and thus appeared to block a late stage of transport, such as the targeting or fusion of these vesicles at the plasma membrane. Screens for suppressors and enhancers of these secretion mutations uncovered further components. Subsequently, excellent *in vitro* assays have been established in which to study the fusion of vesicles derived from yeast with their target organelles. Among the secretion mutants and their interacting genes were homologues of some of the proteins discussed earlier: sec4 encodes a small GTP-binding protein like rab3, and Sso1 and Sso2 encode plasma membrane proteins that are homologues of syntaxin. The sec1 gene encodes a soluble protein with a very high affinity for Sso1, and the mammalian homologue of this protein, n-sec1, is bound tightly to syntaxin in nerve terminals and has an essential function in transmission. The convergence of neuronal biochemistry with yeast genetics has built a strong case for a common mechanism of vesicle fusion.

Yeast is not the only organism in which a genetic screen uncovered an important protein for the synapse: the unc-13 mutation of *Caenorhabditis elegans*, e.g., identified a component of the active zone membrane that is important in priming vesicles for fusion and in modulation of the synapse.

Genetics has further contributed to our understanding of synaptic proteins by allowing tests of the significance of an identified protein for synaptic transmission. Such studies have been carried out in *C. elegans*, *Drosophila*, and mice, and can reveal either an absolute requirement for the protein (as in the case of syntaxin mutants in *Drosophila*) or relatively subtle effects (as in the case of rab3 mutations in mice).

From biochemical purifications, *in vitro* assays, genetic screens, and fortuitous discoveries, a list of nerve terminal proteins has been assembled (Tables 8.1 and 8.2). The manner in which these proteins coordinate the release of transmitter, as well as all the other cell biological functions of the exo-endocytic cycle, remains uncertain, but a consensus has emerged in recent years that puts one set of proteins at the core of the vesicle fusion.

SNAREs and the Core Complex Are Key to Membrane Fusions

Three synaptic proteins, vesicle-associated membrane protein (VAMP)/synaptobrevin, syntaxin, and the synaptosomal associated protein of 25 kDa (SNAP-25), are capable of forming an exceptionally tight complex with one another that generally is referred to as either the *core complex* or the *SNAP receptor (SNARE) complex* (Sollner *et al.*, 1993; Sutton *et al.*, 1998). The

TABLE 8.2 Additional Proteins Implicated in Transmitter Release

Protein	Function
Syntaxin	SNARE protein present on plasma membrane (and on synaptic vesicles to a lesser extent); forms core complex with SNAP-25 and VAMP/synaptobrevin; essential for late step in fusion
SNAP-25	SNARE protein present on plasma membrane (and on synaptic vesicles to a lesser extent); forms core complex with syntaxin and VAMP/synaptobrevin; essential for late step in fusion
Nsec-l/munc-18	Syntaxin-binding protein required for all membrane traffic to the cell surface; likely bound to syntaxin when syntaxin is not in a SNARE complex
Synaphin/complexin	Syntaxin-binding protein; may oligomerize core complexes
Syntaphilin	Binds syntaxin; prevents formation of SNARE complex (?)
Snapin	Binds SNAP-25; associated with synaptic vesicles; unknown function
NSF	ATPase that can disassemble SNARE complex; likely to disrupt complexes after exocytosis
α-SNAP	Cofactor for NSF in SNARE complex disassembly
unc-13/munc-13	Active zone protein; vesicle priming for release; modulation of transmission by diacyl glycerol
Rabphilin	C2 domain protein; Ca^{2+}-binding protein; binds rab3 and associates with synaptic vesicle; modulation of transmission (?)
DOC2	C2 domain protein; Ca^{2+}-binding protein; binds Munc-18; unknown function
RIM1 and related proteins	Active zone proteins; bind rab3; modulation of transmission (?)
Piccolo	Likely scaffolding protein to tether vesicles near active zone
Bassoon	Likely scaffolding protein to tether vesicles near active zone
Exocyst (sec6/8 complex)	Marks plasma membrane sites of vesicle fusion in yeast; synaptic role uncertain
Complexin	Binds to SNARE complex and arrests membrane fusion

interaction of these three proteins is essential for synaptic transmission and is likely to lie very close to or indeed at the final fusion step of exocytosis (Figs. 8.5 and 8.6). What are these proteins?

VAMP (also called synaptobrevin) was among the first synaptic vesicle proteins to be cloned. It is anchored to the synaptic vesicle by a single transmembrane domain and has a cytoplasmic domain that contributes a coiled-coil strand to the core complex. Syntaxin has a very similar structure but is located primarily in the plasma membrane (although some is present on vesicles as well). SNAP-25 is also a protein primarily of the plasma membrane but, unlike the others, lacks a transmembrane domain and instead is anchored in its central region by acylations. SNAP-25 contributes two strands to the SNARE coiled coil. The interactions of these proteins can be envisioned as closely juxtaposing the two membranes meant to fuse. The proteins of this complex are archetypes of a class of membrane-trafficking protein collectively called *SNAREs*. Vesicle-associated proteins, such as VAMP, are referred to as *v-SNAREs*, and those of the target membrane, such as SNAP-25 and syntaxin, are referred to as *t-SNAREs*.

A SNARE complex is found at each membrane-trafficking step within a eukaryotic cell (Fig. 8.6). Through a combination of SNARE proteins on the opposing membranes, a four-stranded coiled-coil is formed. Alongside the example of the synapse in Figure 8.6A, three analogous cases are shown from exocytosis in yeast (Fig. 8.6B), the fusion of late endosomes with one another (Fig. 8.6C), and the fusion of vesicles that form the yeast vacuole (Fig. 8.6D).

Abundant genetic and biochemical data argue for an essential role of SNAREs, and the combination of yeast genetics, *in vitro* assays, biochemical analyses, and synaptic physiology has had a synergistic effect in advancing the field. For example, the discovery that secretory mutations in yeast were homologues of synaptic SNAREs provided some of the first functional data on these proteins (Bennett and Scheller, 1993), whereas data from synapses first placed the proteins on vesicles and the plasma membrane. In addition, the crystal structure was solved first for synaptic proteins (Sutton *et al.*, 1998). Because transmembrane domains and of course the membranes themselves, were absent from the crystallized samples, it was not possible to conclude from the structure alone that SNAREs form a bridge between the membranes rather than forming complexes within the same membrane. Instead, an *in vitro* assay with purified yeast vesicles established this point.

Some of the strongest evidence for an essential role of SNAREs at the synapse has come from the study of a potent set of eight neurotoxins produced by clostridial bacteria. These toxins (tetanus toxin and the family of related botulinum toxins) have long been known to block the release of neurotransmitter from the terminal. The discovery that they do so by proteolytically cleaving individual members of the SNARE

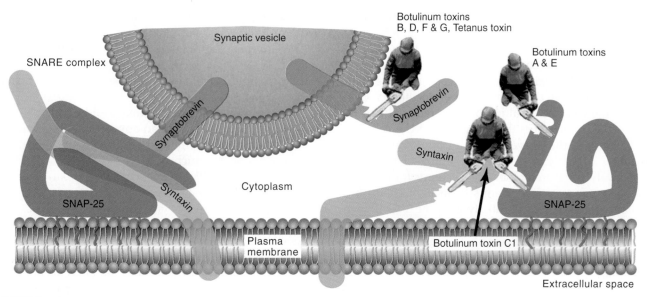

FIGURE 8.5 SNARE proteins and the action of clostridial neurotoxins. The SNARE complex shown at the left brings the vesicle and plasma membranes into close proximity and likely represents one of the last steps in vesicle fusion. Vesicular VAMP, also called synaptobrevin, binds with syntaxin and SNAP-25 that are anchored to the plasma membrane. Tetanus toxin and the botulinum toxins, proteases that cleave specific SNARE proteins as shown, can block transmitter release.

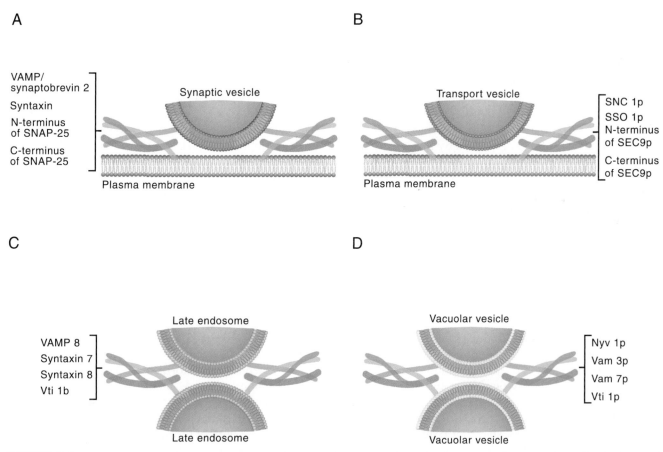

FIGURE 8.6 Neurotransmitter release shares a core mechanism with many membrane fusion events within eukaryotic cells. The fusion of synaptic vesicles (A) is driven by a particular complex of four coiled-coil domains contributed by three different proteins. Exocytosis in yeast (B), the fusion of late endosomes in mammalian cells (C), and the fusion of vacuolar vesicles in yeast (D) exemplify the closely related four-stranded coiled-coil complexes required to drive fusion in other membrane-trafficking steps.

complex provided neurobiologists with a set of tools with which to probe SNARE function (Fig. 8.5).

Each of the toxins comprises a heavy and a light chain that are linked by disulfide bonds. The heavy chain binds the toxin to surface receptors on neurons, thereby enabling the toxin to be endocytosed. Once inside the cell, the disulfide bond is reduced and the free light chain enters the cytoplasm of the cell. This light chain is the active portion of the toxin and is a member of the Zn^{2+}-dependent family of proteases. The catalytic nature of the toxin accounts for its astonishing potency; a few tetanus toxin light chains at a synapse can suffice to proteolyse all the VAMP/synaptobrevin, thereby shutting down transmitter release. The toxins are highly specific, recognizing unique sequences within an individual SNARE protein, as summarized in Fig. 8.5. VAMP/synaptobrevin can be cleaved not only by tetanus toxin but by the botulinum toxins of types B, D, F, and G, and each toxin cleaves at a different peptide bond within the structure. SNAP-25 is cleaved by botulinum toxins A and E, but again at different sites from one another. Botulinum toxin C1 cleaves both syntaxin and, less efficiently, SNAP-25.

What precisely is the function of SNARE proteins in promoting transmitter release? How does the assembly of this complex relate to membrane fusion? Studies with botulinum toxins, yeast mutants, mutants of *Drosophila* and *C. elegans*, and permeabilized mammalian cells all place SNAREs late in the process. Synapses that lack an individual SNARE, e.g., synapses whose VAMP has been mutated or cleaved by toxin, have the expected population of synaptic vesicles, and these vesicles accumulate at active zones in the expected manner. Indeed, as judged by electron microscopy, there may even be an excess of vesicles in close proximity to the plasma membrane at the active zone. Yet these synapses are incapable of secreting transmitter. Thus, SNAREs appear to be essential for a step that comes after the arrival of the vesicle to the release site, but before the fusion pore opens and transmitter can diffuse into the cleft. The details of how the complex functions, however, are less certain.

One attractive model is that the energy released by the formation of this very high-affinity complex is used to drive together the two membranes. A loose complex of the SNARES would form and then "zipper up" and pull the membranes together. This may correspond to the actual fusion of the membranes or, alternatively, to a priming step that requires a subsequent rearrangement of the lipids in order to open the pore that will connect the vesicle lumen to the extracellular space. Evidence from the fusion of yeast vacuolar vesicles implicates a distinct downstream step, regu-

lated by Ca^{2+}/calmodulin and involving subunits of the proton pump, but whether this is true of neurons as well remains unknown. Vesicles consisting of only lipid bilayers and SNAREs can fuse *in vitro*, suggesting that no other proteins will be essential for the fusion step.

One additional function may reside with SNARE proteins: identification of an appropriate target membrane. Within a cell, there are myriad membrane compartments with which a transport vesicle can fuse: how then is specificity achieved? The great diversity of SNAREs (Fig. 8.6) may account for some of this specificity because not all combinations of vesicle-associated and target-associated SNAREs (v- and t-SNAREs) will form functional complexes. This potential mechanism, however, is likely to be only a part of the story. Particularly in the nerve terminal, it appears that synaptic vesicles can find the active zone even in the absence of the relevant SNAREs; for example, when a SNARE has been cleaved by a clostridial toxin or in a mutant lacking a SNARE. In addition, the t-SNAREs syntaxin and SNAP-25 can be present along the entire axon and thus are inadequate to explain the selective release of transmitter at synapses and active zones.

NSF—An ATPase for Membrane Trafficking

At some point as vesicles move through their exo-endocytic cycle, energy must be added to the system. *N*-Ethylmaleimide sensitive factor (NSF), an ATPase involved in membrane trafficking, is one likely source. NSF was first identified as a required cytosolic factor in an *in vitro* trafficking assay (Block *et al.*, 1988). The importance of NSF was confirmed when it was found to correspond to the yeast sec18 gene, an essential gene for secretion. NSF hexamers bind a cofactor called α-SNAP (Soluble NSF Attachment Protein), or sec17, and this complex in turn can bind to the SNARE complex. When Mg-ATP is hydrolyzed, the SNARE complex is disrupted into its component proteins (Sollner *et al.*, 1993). If SNARE complexes form between VAMP, syntaxin, and SNAP-25 all in the same membrane, these futile complexes can be split apart by NSF so that productive complexes bridging the membrane compartments can be formed. After fusion, the tight SNARE complex needs to be disrupted so that the VAMP can be recycled to synaptic vesicles while the other SNAREs remain on the plasma membrane. If, indeed, the energy of forming a tight SNARE complex is part of the energy that drives fusion, NSF, by restoring the SNAREs to their dissociated, high-energy state, will be an important part of the energetics of membrane fusion.

Docking and Priming Vesicles Prepares Them for Fusion

As discussed earlier and outlined in Figure 8.2, many preparatory and regulatory steps may precede the action of SNAREs and the membrane fusion step at which SNAREs appear to act. These preparatory steps must tether the vesicle at an appropriate release site in the active zone and hold the vesicle in a fusion-ready state. These mechanisms remain among the most obscure of the processes involved in exocytosis, but there is a growing list of proteins that may participate (Table 8.2). One example is the syntaxin-binding protein n-sec1 (also called munc18), the neuronal homologue of the product of the yeast sec1 gene. Mutations of this protein prevent trafficking in both yeast and higher organisms. It appears likely that n-sec1 serves two functions: it may promote membrane fusion by priming syntaxin so that, once it has dissociated, it can participate correctly in fusion, but it may also be a negative regulator, keeping syntaxin inert until an appropriate vesicle or signal displaces n-sec1 and allows a SNARE complex to form.

Rab3, a small GTP-binding protein mentioned earlier, is another protein for which a priming or regulatory role often is invoked at the synapse. In yeast, and at other membrane-trafficking steps within mammalian cells, rab proteins have essential roles. They appear to help a vesicle to recognize its appropriate target and begin the process of SNARE complex formation. At the synapse, however, the significance of rab3 is still uncertain. Although it is clearly associated with synaptic vesicles, genetic disruption of rab3 has a surprisingly slight phenotype and causes only subtle alterations in synaptic properties. Whether this is due to additional, redundant rab proteins or whether the rab family has been relegated to a more minor role at the synapse remains to be determined.

Because synaptic vesicles dock and fuse specifically at the active zone, this region of the nerve terminal membrane must have unique properties that promote docking and priming. The special nature of this domain is easily discernible in electron micrographs: an area of electron-dense material can be observed opposite the postsynaptic density (Fig. 8.1). In some synapses, such as photoreceptors, hair cells, and many insect synapses, the structures are more elaborate and include ribbons, dense bodies, and T bars that extend into the cytoplasm and appear to have a special relationship with the nearby pool of vesicles.

Advances in electron microscopy have allowed a more detailed look at the association of vesicles and plasma membrane at the active zone (Harlow *et al.*, 2001). At the neuromuscular junction of the frog (Fig.

8.1), the electron-dense material adjacent to the presynaptic plasma membrane is actually a highly ordered structure—a lattice of proteins that connect the vesicles to a cytoskeleton, to one another, and to the plasma membrane. The molecules that correspond to these structures are not yet known. A few proteins, however, are known to be concentrated in the active zone or in the cloud of vesicles near the active zone. These proteins, piccolo, bassoon, RIM1, and unc-13, may be a part of the machinery that defines the active zone as the appropriate target for synaptic vesicle fusion.

In addition to the specializations of the active zone, additional machinery must be present to preserve a dense cluster of synaptic vesicles extending approximately 200 nm back from the active zone (Fig. 8.1). The vesicles in this domain are not likely to be releasable within microseconds of the arrival of an action potential but are instead likely to represent a reserve pool from which vesicles can be mobilized to release sites on the plasma membrane. The equilibrium between this pool and vesicles actually at the membrane may be an important determinant of the number of vesicles released per impulse, but remains poorly understood.

One protein family likely to play a role in the maintenance of the reserve pool is the synapsins, a set of peripheral membrane proteins on synaptic vesicles. Synapsins can also bind actin filaments and thus may provide a linker that tethers the vesicles in the cluster to the synaptic cytoskeleton. Disruption of this link can cause the vesicle cluster to be diminished and reduce the number of vesicles in the releasable pool of the terminal. Synapsin has attracted considerable interest because it is the substrate for phosphorylation by both cAMP and Ca^{2+}-dependent protein kinases. These phosphorylations may influence the availability of reserve vesicles for recruitment to release sites.

Ca^{2+}-Binding Proteins Are Candidates for Coupling the Action Potential to Exocytosis

The most striking difference between synaptic transmission and traffic between other cellular compartments is the rapid triggering of fusion by action potentials. As discussed previously, the opening of Ca^{2+} channels and the focal rise of intracellular Ca^{2+} activate the fusion machinery. How does the terminal sense the rise in Ca^{2+}? What is the Ca^{2+} trigger and how does it open the fusion pore? Is it a single Ca^{2+}-binding protein or do several components respond to the altered Ca^{2+} concentration? Does Ca^{2+} remove a brake that normally prevents a docked, primed vesicle from fusing or does Ca^{2+} induce a conformational change that is actively required to promote fusion? How does the steep, exponential relationship of release to intracellular Ca^{2+} arise? These questions are an active area of investigation and debate.

The leading candidate for being the synaptic Ca^{2+} sensor is synaptotagmin, an integral membrane protein of the synaptic vesicle (Fig. 8.4). Synaptotagmin has a large cytoplasmic portion that comprises two Ca^{2+}-binding C2 domains, called C2A and C2B. These domains can also interact with the SNARE complex proteins and with phospholipids in a Ca^{2+}-dependent manner. It has been hypothesized that one or more of these interactions is the molecular correlate of the triggering event for fusion. Consistent with this hypothesis, mutations that remove synaptotagmin profoundly reduce synaptic transmission in flies, worms, and mice, while having little effect on or enhancing the rate of spontaneous release of transmitter. Whether this reduction in evoked release is due specifically to the loss of the Ca^{2+} trigger, however, is harder to demonstrate. Synaptotagmin is likely to be involved in endocytosis and potentially in vesicle docking as well, which has complicated the analysis. These processes, as mentioned earlier, are also likely to be regulated by Ca^{2+}, and the Ca^{2+}-binding sites on synaptotagmin may be relevant for this regulation as well.

Recently, an intriguing model has emerged in which the fusion of membranes is arrested at an intermediate state by the binding of a small protein called complexin to the assembled SNARE complex. Synaptotagmin, in the presence of Ca^{++}, can bind to the SNARE complex and displace complexin, thereby allowing the fusion reaction to proceed. Evidence, both from in vitro studies of the biochemical interaction of complexin and synapatotagmin with the SNARE complex and from assays of membrane fusion suggests that complexin serves as a brake and that the primary function of synaptotagmin in exocytosis is to release that brake.

Transmitter Must Be Packaged into the Vesicle

To achieve the synchronous release of thousands of molecules of transmitter from the presynaptic nerve terminal, synaptic vesicles accumulate and store high concentrations of transmitter. In cholinergic neurons, for example, the concentration of acetylcholine within the synaptic vesicle can reach $0.6M$, more than 1000-fold greater than that in the cytoplasm. Two synaptic vesicle proteins mediate the uptake of transmitter: the vacuolar proton pump and a family of transmitter transporters. The vacuolar proton pump is a multisubunit ATPase that catalyzes the translocation of protons from the cytoplasm into the lumen of a variety of intracellular organelles, including synaptic vesicles. The resulting transmembrane electrochemical proton gradient is utilized as the energy source for the active uptake of transmitter by transmitter transporters.

Transmitter uptake has been characterized in isolated synaptic vesicle preparations in which at least four types of distinct transporters have been identified: one for acetylcholine, another for catecholamines and serotonin, a third for the excitatory amino acid glutamate, and the fourth for the inhibitory amino acids GABA and glycine. At least one gene for each of these classes of transmitter transporter has now been cloned. As expected, these distinct transporters are expressed differentially by neurons. The type of transporter in a cell dictates the type of transmitter stored in the synaptic vesicles of a particular neuron, and when investigators drive the expression of a glutamate transporter, e.g., in a GABA-releasing neuron, they can trick the cell into releasing glutamate.

Vesicular transporters are integral membrane proteins with 12 membrane-spanning domains that display sequence similarity with bacterial drug-resistance transporters. Synaptic vesicle transporters are clearly distinct from the plasma membrane transmitter transporters that remove transmitter from the synaptic cleft and thereby contribute to the termination of synaptic signaling. The distinguishing characteristics include their transport topology, energy source, pharmacology, and structure. The sequestration of transmitter into vesicles, in addition to its obviously essential role for transmitter exocytosis, has some further benefits to the cell. Sequestration can prevent high cytosolic concentrations of the transmitter from inhibiting the biosynthetic enzymes for transmitter synthesis and can protect the transmitter from catabolic enzymes. In addition, in some cell types, sequestration may protect the cell from damage caused by the oxidation of labile transmitters, particularly dopamine.

In contrast to small chemical transmitters, proteinaceous signaling molecules, including neuropeptides and hormones, typically are stored in granules that are larger and have a higher electron density than synaptic vesicles. The contents of these granules are not recycled at the release sites; as a result, their replenishment requires new protein synthesis followed by packaging into secretory vesicles in the cell body. Because of the slow kinetics of their release, the slow responsiveness of their postsynaptic receptors, and their inability to be recycled locally, proteinaceous signaling molecules typically mediate regulatory functions.

Endocytosis Recovers Synaptic Vesicle Components

After exocytosis, the components of the synaptic vesicle membrane must be recovered from the presynaptic plasma membrane, as discussed previously. Vesicle recycling theoretically could be accomplished by either of two mechanisms (Fig. 8.2). The first is simply a reversal of the fusion process. In this case, a fusion pore opens to allow transmitter release and then closes rapidly to reform a vesicle. Often nicknamed "kiss and run," it has the theoretical advantage that it would allow all the vesicular components to remain together on a single vesicle that would be available immediately for reloading with transmitter. The mechanism is potentially quick and energetically efficient. This mechanism is employed in some nonneuronal cells, but its relevance for the synapse is an unresolved, and highly controversial, issue.

The predominant pathway for synaptic vesicle recycling is more likely to be a second mechanism, endocytosis. Endocytosis of synaptic vesicle components, like receptor-mediated endocytosis in other cell types, is mediated by vesicles coated with the protein clathrin. Accessory proteins select the cargo incorporated into these vesicles as they assemble. The final pinching off of clathrin-coated vesicle requires the protein dynamin, which can form a ring-like collar around the neck of an endocytosing vesicle. The importance of dynamin is demonstrated most clearly in *Drosophila* via a temperature-sensitive dynamin mutant known as *shibire*: at the nonpermissive temperature, *shibire* flies become paralyzed rapidly due to a nearly complete depletion of synaptic vesicles from their nerve terminals. Once the components of the synaptic vesicle membrane have been recovered in clathrin-coated vesicles, recycling is completed by vesicle uncoating and, perhaps, passage through an endosomal compartment in the nerve terminal (Fig. 8.2).

Summary

The life of the synaptic vesicle involves much more than just the Ca^{2+}-dependent fusion of a vesicle with the plasma membrane. It is a cyclical progression that must include endocytosis, transmitter loading, docking, and priming steps as well. In many regards, this cell biological process shares mechanistic similarities with membrane trafficking in other parts of the cell and with simpler organisms such as yeast. Interaction of vesicular and plasma membrane proteins of the SNARE complex—VAMP/synaptobrevin, syntaxin, and SNAP-25—is an essential late step in fusion. Many other proteins have been identified that are likely to precede the action of the SNAREs, regulate the SNAREs, and recycle them. Components of the synaptic vesicle membrane, the presynaptic plasma membrane, and the cytoplasm all contribute to the regulation of synaptic vesicle function. Together, these proteins build on the fundamental core apparatus to create an astonishingly accurate, fast, and reliable means of delivering transmitter to the synaptic cleft.

QUANTAL ANALYSIS: PROBING SYNAPTIC PHYSIOLOGY

A quantitative description of the signal passing across a synapse can be a source of insight into the biophysics of transmission and its regulation. As discussed earlier, the response to an action potential in the presynaptic cell reflects the release of discrete packets of transmitter, called quanta, that are a direct consequence of the vesicular release of transmitter. A given synapse, however, will not secrete exactly the same number of vesicles in response to each stimulus. Instead, the probabilistic nature of the presynaptic apparatus typically causes the amount released to fluctuate from trial to trial in a quantal manner; that is, the postsynaptic response adopts preferred levels, which arise from the summation of various numbers of vesicle fusions (Katz, 1969). Thus, although a particular synapse may release an average of two vesicles per action potential, it may release one vesicle in response to one stimulation and three vesicles in response to the next. To the third stimulus, the synapse may fail to release any transmitter at all. From a full description of the properties of the synapse, the size of the individual quantum response can be estimated, as well as the average number of quanta released for a given presynaptic action potential. This approach, called quantal analysis, is also a source of insight into the probabilistic processes underlying transmitter release from the presynaptic terminal and into the mechanisms by which transmission can be modified by physiological, pharmacological, and pathological phenomena.

A Standard Model of Quantal Transmission Can Describe Many Synapses

The quantal nature of transmission was first demonstrated in the early 1950s by Bernard Katz and colleagues, who recorded from the frog neuromuscular junction. Spontaneous signals that were approximately 1/100 of the signal evoked by stimulating the presynaptic axon were observed to occur at random intervals.

These spontaneous minis were of roughly constant amplitude, with a coefficient of variation of 30%. As discussed previously, a neuromuscular junction normally releases hundreds of quanta per impulse, but when extracellular Ca^{2+} was lowered to reduce transmitter release to very low levels, the evoked response fluctuated from trial to trial between preferred amplitudes, which coincided with integral multiples of the mini amplitude (Fig. 8.7).

We now understand this to arise from the fact that minis are the fusion of a single vesicle and that evoked responses can represent one or more of these same vesicles. Before the vesicle hypothesis was established, however, Katz and colleagues inferred that the mini is the quantal building block, variable numbers of which are released to make up the evoked signal. The relative numbers of trials resulting in 0, 1, 2, . . . quanta were well described by a Poisson distribution—a statistical distribution that arises in many instances where a random process operates. On the basis of this evidence, Katz and colleagues proposed the following model of

transmission, which has gained wide acceptance and will be referred to as the standard Katz model.

- Arrival of an action potential at the presynaptic terminal briefly raises the probability of release of quanta of transmitter (i.e., the fusion of synaptic vesicles).
- Several quanta (synaptic vesicles) are available to be released, and every quantum gives roughly the same electrical signal in the postsynaptic cell. This is the quantal size or amplitude, Q, which sums linearly with all other quanta released.
- The average number of quanta released, m, is given by the product of n, the number of available quanta, and p, the average release probability: $m = np$. The average response to a stimulus is the product of the quantal size and the average number of quanta per stimulus, Qm or Qnp.
- The relative probability of observing $0, 1, 2, \ldots, n$ quanta released is then given by a binomial distribution, with parameters n and p.

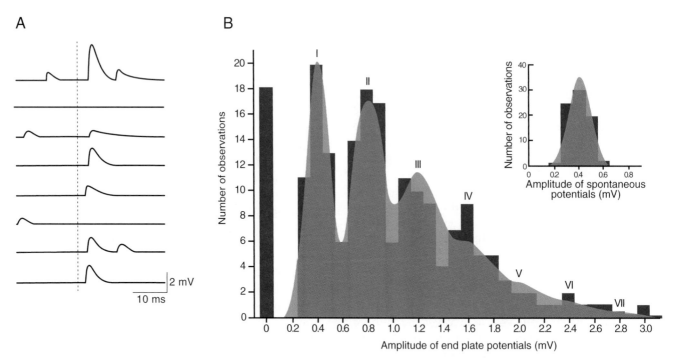

FIGURE 8.7 Quantal transmission at the neuromuscular junction. (A) Intracellular recordings from a rat muscle fiber in response to repeated presynaptic stimulation of the motor axon. Low extracellular [Ca^{2+}] and high [Mg^{2+}] restricted Ca^{2+} entry and so kept transmission to a very low level. The stimulus was given at the time marked by the dotted line. The size of the postsynaptic response fluctuated from trial to trial, with some trials giving failures of transmission. Spontaneous minis occurring in the background (e.g., those events that occur before the dotted line) had approximately the same amplitude as the smallest evoked responses, implying that they arose from the release of single quanta of acetylcholine. From Liley (1956). (B) Peak amplitudes of 200 evoked responses [end plate potentials (EPPs)] from a similar experiment, plotted as an amplitude histogram. Eighteen trials resulted in failures of transmission (indicated by the bar at 0 mV), and the rest gave EPPs whose amplitude tended to cluster at integral multiples of 0.4 mV. This coincides with the mean amplitude of the spontaneous minis, whose amplitude distribution is shown in the insert together with a Gaussian fit. Shading through the EPP histogram is a fit obtained by assuming a Poisson model of quantal release. Roman numerals indicate the number of quanta corresponding to each component in the distribution. From Boyd and Martin (1956).

The value of Q has a clear underlying basis: the efficacy of an individual synaptic vesicle fusion in altering the conductance of the postsynaptic membrane. The binomial parameters, n and p, however, are formalisms whose underlying meanings are harder to pin down. n may correspond to the number of vesicles in a primed, fusion-ready state at the active zone or to discrete release sites competent for release, or even to the number of active zones. Correspondingly, p may represent the probability of fusion of any individual vesicle or the probability of fusion at a given release site or active zone. For some mechanisms, therefore, it is difficult to say whether n or p is altered. For example, if additional vesicles are recruited to release sites, that change may be manifest as a change in n because there are more releasable quanta or as a change in p because the probability of fusion at a given release site has increased because it is more likely to be charged with a vesicle.

These quantal parameters remain a useful description of the synapse, however, despite their obscurity on a mechanistic level. For example, it is often useful to classify synapses as having a high or low probability of release (p) to explain their behavior during trains of stimuli (see later). In a synapse with few releasable vesicles and a high p, a train is predicted to deplete the releasable pool rapidly and the strength of transmission declines, a phenomenon known as synaptic depression. Another synapse might have the same average quantal content, but accomplish this with a large releasable pool and a low probability of fusion for any individual vesicle in the pool. Such a synapse would be less prone to depression and indeed both types of synapse are encountered.

A major impetus for the accurate measurement of quantal parameters is that it may help determine the locus of modulatory effects on synaptic transmission, or the site of action of a drug. A blocker of postsynaptic receptors, for example, will not alter the mini frequency or the number of quanta released, but will decrease the amplitudes of both spontaneous minis and the quantal components of evoked responses. A change in the loading of transmitter into a vesicle could cause a similar change. In contrast, a synaptic modulator that decreased Ca^{2+} channel opening or a blocker of those channels would decrease the probability of presynaptic release and be recognized as a decrease in quantal content, m. Many modulators of presynaptic transmission cause a parallel change in quantal content and in the frequency of spontaneous minis. This sort of analysis has been instrumental in understanding synaptic plasticity and in determining whether a given alteration occurs at the presynaptic terminal or the postsynaptic membrane.

The Standard Katz Model Does Not Always Apply

Incorrectly applying the Katz model and quantal analysis to synapses at which its assumptions do not hold has been a frequent source of error. Although a superb description of transmission at the neuromuscular junction, it is not uniformly applicable to central nervous system synapses, and the technical difficulties of isolating and accurately sampling the physiology of an individual synapse in the CNS can make the determination of quantal parameters exceptionally difficult.

In particular, five issues arise at CNS synapses with which the standard Katz model cannot deal adequately. These are *nonuniformity of quanta*, *nonuniformity of release sites*, *postsynaptic membranes that distort responses*, *saturation of receptors*, and *silent synapses*.

For quantal analysis to work, individual quanta should be roughly equivalent in size, that is, with the same amount of transmitter in each vesicle and with similar populations of receptors facing their release sites, as is true at the frog neuromuscular junction. But in the CNS, both vesicle content and receptor populations may differ widely, particularly where more than one synapse connects two cells. The resulting *nonuniformity of quanta* means that mini amplitudes will not have a normal distribution and will not sum to tidy peaks like those shown in Figure 8.7. Mini amplitude in the CNS can also be very hard to measure: the smallest events may be lost in the noise and the large variety of inputs converging on a postsynaptic cell makes it difficult to determine which responses derive from an individual presynaptic input. *Nonuniformity of release sites* can similarly prevent responses from falling in a simple binomial distribution. For example, the Ca channels at a nerve terminal may be of different isoforms or differentially phosphorylated and vesicles will therefore vary in their release probability depending on the channels closest to them.

Postsynaptic membranes that distort responses introduce another problem: the voltage-dependent properties of the postsynaptic cell and the particular electrical properties of dendritic spines can prevent vesicles from summing in a simple linear fashion. Equally problematic is the *saturation of receptors* on the postsynaptic membranes of many CNS synapses. A glutamatergic CNS synapse may have fewer than 100 receptors present, and if a single quantum activates most of these, it is impossible for quanta to sum linearly. Indeed, a major difference between some vertebrate CNS synapses and the neuromuscular junction is that the quantal amplitude often is determined not by the vesicle contents, but by the number of available receptors. In the extreme case of a developing synapse, the variation from impulse to impulse might depend as much on whether one, two, or three receptors were activated per vesicle fusion as whether one, two, or three vesicles fused.

Silent synapses represent an extreme form of the variability of postsynaptic responses and may be an important consideration for AMPA receptors at glutamatergic synapses (see Chapter 9). At these synapses, active zones may be secreting transmitter without producing a physiological response because no receptors are present in the postsynaptic membrane. These so-called silent synapses can spring to life when a set of receptors is inserted into the postsynaptic membrane in a concerted fashion. Because more active zones now are contributing to the detected physiological response, mini frequency will be higher and evoked release increased, although the quantal size will not have changed. Whereas the Katz model from the neuromuscular junction would consider these changes to be the hallmarks of a presynaptic change in release, the "unsilencing" of these CNS synapses is a postsynaptic event.

Thus, many of the requirements for the standard Katz model cannot realistically be expected to hold in all cases and the evidence that these simple probabilistic models are correct is far from compelling. However, despite the technical difficulties and theoretical caveats, the analysis of mini frequency, quantal size, quantal content, and synaptic probability has been the foundation of our description of synaptic transmission and the source of great insight into synaptic modulation.

It has been possible to augment this understanding with independent methods, including the ability to monitor vesicle fusion optically from the unloading of the lipophilic dyes discussed earlier, and thereby to describe presynaptic activity without depending on the postsynaptic membrane to report the event. Another method of indirectly estimating release probability at glutamatergic synapses makes use of a blocker of NMDA receptors, MK-801, that will act only in the open state. From the rate of blockade during a train of stimuli, the probability of the channel having been opened by a given impulse can be inferred.

Summary

Quantal analysis has improved our understanding of biophysical and pharmacological mechanisms of transmission by describing normal transmission and by offering a tool to distinguish and characterize presynaptic and postsynaptic actions of drugs or endogenous agents that modify the strength of a synapse.

Although these methods must be applied with caution, they yield a unique insight into the mechanisms of synaptic plasticity. The nature of transmission at CNS synapses can be quite different from what is encountered at the classical synapse of the neuromuscular junction.

SHORT-TERM SYNAPTIC PLASTICITY

Chemical synapses are not static transmitters of information. Their effectiveness waxes and wanes, depending on the frequency of stimulation and the history of prior activity. Because information processing in the nervous system in not conducted by individual action potentials at widely spaced intervals, this plasticity is of the utmost importance to the function of the synapse within its physiological circuit. At most synapses, repetitive high-frequency stimulation (called a *tetanus*) initially is dominated by a growth in successive PSP amplitudes, called synaptic *facilitation* (Fig. 8.8). This process builds to a steady state within about 1s and decays equally rapidly when stimulation stops. Decay is measured by single test stimuli given at various intervals after a conditioning train. Facilitation does not require a train to be observed; at many synapses it can be seen after a single action potential. Thus, when two stimuli are given with very close spacing, the second one can be as much as twice the amplitude of the first. This is called *paired-pulse facilitation* (Fig. 8.8). At most synapses, a slower phase of increase in efficacy, which has a characteristic time constant of several seconds (called *augmentation*), succeeds facilitation. Finally, with prolonged stimulation, some synapses display a third phase of growth in PSP amplitude that lasts minutes (called *potentiation*). Potentiation often is visible in isolation only long after a tetanus and is thus called *posttetanic potentiation* (PTP).

FIGURE 8.8　Short-term synaptic plasticity. (A) An idealized synapse in which facilitation predominates. Two stimuli (paired pulses) are given to the presynaptic nerve in each line and postsynaptic voltage responses are shown. Closely spaced stimuli (top line) facilitate strongly but as the interval between stimuli increases, the degree of facilitation diminishes. The fifth line illustrates how facilitation can accumulate during a train of stimuli. (B) An idealized synapse in which synaptic depression predominates. (C) At most synapses, both depression and facilitation occur and, depending on the nature of the individual synapse and the time course of facilitation, depression, augmentation, and potentiation, any of these phenomena may predominate at a given moment. Thus, in a realistic pattern of activity, with unevenly spaced action potentials arriving at a terminal, the strength of transmission can depend greatly on the previous history of the synapse.

A Facilitation

B Depression

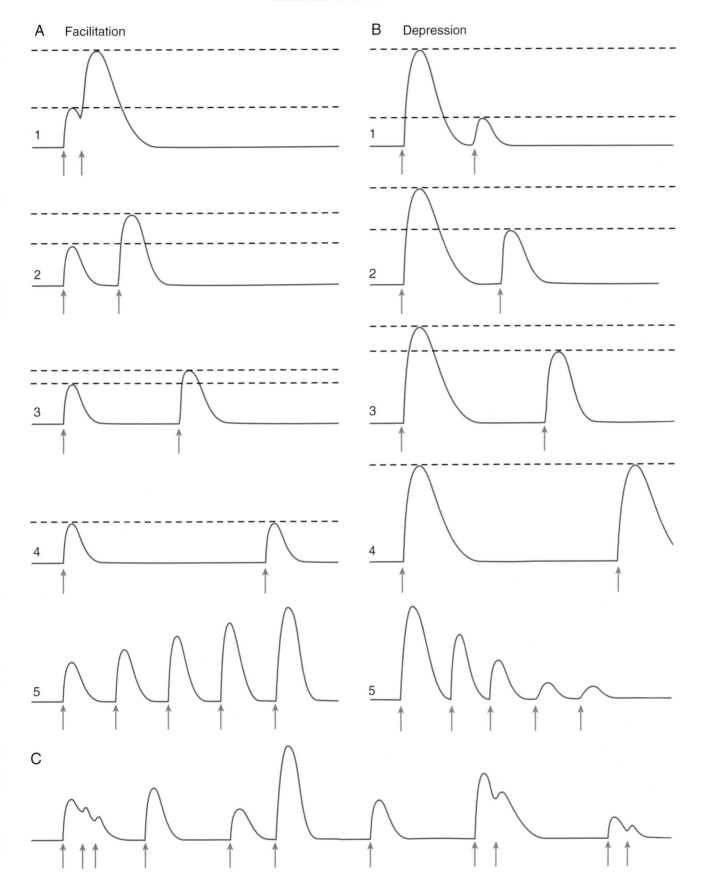

Often, a phase of decreasing transmission, called synaptic *depression*, is superimposed on these processes (Fig. 8.8). Synaptic depression leads to a dip in transmission during repetitive stimulation, which often tends to overlap and obscure the augmentation and potentiation phases. At some synapses, it is manifest after only a single action potential and gives rise to *paired-pulse depression*.

Additional forms of use-dependent plasticity, called long-term potentiation and long-term depression (LTP and LTD), have received considerable attention as likely correlates of some forms of learning and are discussed elsewhere (Chapter 49). Most frequently, at synapses in which a quantal analysis has been done, all these forms of synaptic plasticity (except some forms of cortical LTP) are due to changes in the number of quanta released by action potentials.

The propensity to facilitation or depression can be quite different at different types of synapse and thus contributes to the diversity of synaptic communication and the flexibility of information processing by synapses. At a given synapse, the frequency and duration of a train can alter whether facilitation or depression is most in evidence. In a physiologically realistic setting of irregularly spaced stimuli, all these phenomena may be simultaneously at play and one or another may predominate for any given response, depending on the preceding history of stimulation. Thus, responses in the postsynaptic cell will be anything but a constant response to each action potential in the train.

Synaptic Depression May Arise from Depletion of Readily Releasable Vesicles, Autoinhibition, or Receptor Desensitization

The rate at which synaptic depression develops usually depends on stimulation frequency. At many synapses, depression is relieved when transmission is reduced by lowering Ca^{2+} or raising Mg^{2+} in the medium—when little or no release occurs, there is little depression of the subsequent response. These characteristics are consistent with depression being due to the depletion of a readily releasable store of docked or nearly docked vesicles and recovery being due to their replenishment from a nearby supply (Fig. 8.9) (Zucker, 1989). The parameters of such a depletion model can be estimated from the rate of recovery from depression, which gives the rate of refilling the releasable store, and the fractional drop in PSPs given at short intervals, which gives the fraction of the releasable store liberated by each action potential.

As mentioned previously, synapses with a high probability of release from a relatively small pool of vesicles will be the most prone to this form of depres-

sion. Recovery from this form of depression appears to be influenced by cytosolic Ca^{2+} as well, consistent with a hypothetical ability of elevated Ca^{2+} in the cytosol to enhance the recruitment of reserve vesicles to release sites.

Vesicle depletion cannot account for all forms of synaptic depression. At some synapses, depression is due to an inhibitory action of released transmitter on presynaptic receptors called autoreceptors. For example, in rat hippocampal cortex, the depression of GABA responses is blocked by antagonists of presynaptic GABAB receptors. Presumably, GABA acts to hyperpolarize nerve terminals and block transmitter release. At other synapses, desensitization of postsynaptic receptors contributes heavily to the observed depression.

Facilitation, Augmentation, and Potentiation Are Due to Effects of Residual Ca^{2+}

With few exceptions, all the phases of increased short-term plasticity are Ca^{2+}-dependent in the sense that little or no facilitation, augmentation, or potentiation is generated by stimulation in Ca^{2+}-free medium. Originally, these phases of increased transmission were thought to be due to the effect of residual Ca^{2+} remaining in active zones after presynaptic activity and summating with Ca^{2+} influx during subsequent action potentials to generate slightly higher peaks of Ca^{2+} (Katz and Miledi, 1968; Zucker, 1989). Due to the highly nonlinear dependence of transmitter release on Ca^{2+}, a small residual amount could activate a substantial increase in phasic transmitter release during an action potential, while having only a small effect on mini frequency itself. However, as this relationship was examined more closely and with Ca^{2+} indicators to actually measure the residual intracellular Ca^{2+}, this hypothesis became unlikely. These findings led to the proposal that, in addition to summating with peak Ca^{2+} transients to drive fusion, Ca^{2+} acts to increase transmission at one or more targets distinct from the sites triggering exocytosis. Unlike the triggering site, the facilitation site for Ca^{2+} binding should be a high-affinity site that can be influenced by the relatively low levels of residual Ca^{2+}. Other second messenger systems and kinases may also be involved.

Potentiation lasts for a long period after a strong tetanus because residual Ca^{2+} is present for the duration of PTP, presumably due to overloading of the processes responsible for removing excess Ca^{2+} from neurons. These processes include Ca^{2+} extrusion pumps, such as plasma membrane ATPase and Na^+-Ca^{2+} exchange, and Ca^{2+} uptake into organelles such as endoplasmic reticulum and mitochondria.

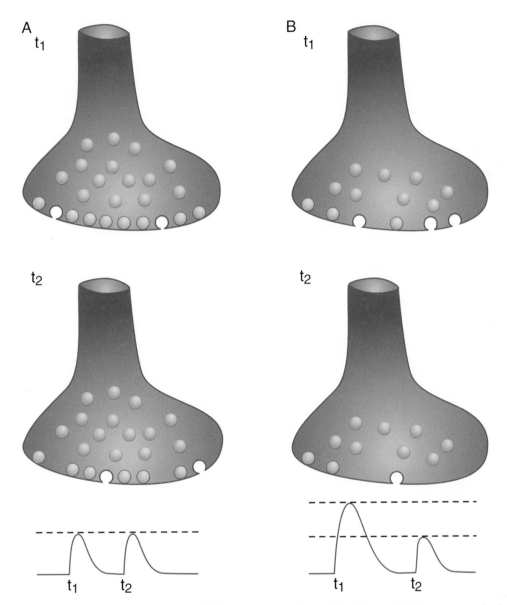

FIGURE 8.9 Depletion model of synaptic depression depicting the action of two stimuli (t_1 and t_2) that arrive so closely spaced that no vesicles have been recruited to the releasable pool to replace those that fused in response to the first stimulus. (A) A synapse in which the releasable pool (n) is large and the probability of fusion for a given vesicle (p) is low shows little or no depression. Because only a small fraction of the docked, releasable vesicles fused in response to the first stimulus, the second response is comparable in amplitude. (B) A synapse in which the releasable pool (n) is small but the probability of fusion (p) is high is likely to be strongly depressing. Vesicles released by the first impulse deplete half the fusion-competent pool, leaving n at t_2 equal to half its value at t_1. If p has not changed at t_2, the second stimulus will release half as many quanta.

Summary

Short-term synaptic plasticity allows synaptic strength to be modulated as a function of prior activity and can cause large changes in responses during physiologically relevant patterns of stimulation. Synapses may show a decline in transmission (depression) or an increase in synaptic efficacy with time constants ranging from seconds (facilitation and augmentation) to minutes (potentiation or PTP) to hours (LTP); many synapses show a mixture of several of these phases. Synaptic depression may be due to depletion of a readily releasable supply of vesicles, to the inhibitory action of transmitter on presynaptic autoreceptors, or to the desensitization of postsynaptic receptors. Depression makes synapses selectively responsive to brief stimuli or to changes in level of activity. Frequency-dependent increases in synaptic efficacy are due to the effects of residual presynaptic Ca^{2+} acting to modulate the release process, probably through a high-affinity binding site. These frequency-dependent increases in synaptic efficacy allow synapses to distinguish significant signals from noise and to respond to selected patterns of activity.

References

Bennett, M. K. and Scheller, R. H. (1993). The molecular machinery for secretion is conserved from yeast to neurons. *Proc. Natl. Acad. Sci. USA* **90**(7), 2559–2563.

Boyd, I. A. and Martin, A. R. (1956). Spontaneous subthreshold activity at mammalian neuromuscular junctions. *J. Physiol. (Lond.)* **132**, 74–91.

Davies, C. H., Davies, S. N., and Collingridge, G. L. (1990). Paired-pulse depression of monosynaptic GABA-mediated inhibitory postsynaptic responses in rat hippocampus. *J. Physiol.* **424**, 513–531.

Gonzalez, L. and Scheller, R. H. (1999). Regulation of membrane trafficking: Structural insights from a Rab/effector complex. *Cell* **96**(6), 755–758.

Harlow, M. L., Ress, D., Stoschek, A., Marshall, R. M., and McMahan, U. J. (2001). The architecture of active zone material at the frog's neuromuscular junction. *Nature* **409**(6819), 479–484.

Heuser, J. E. (1977). Synaptic vesicle exocytosis revealed in quick-frozen frog neuromuscular junctions treated with 4-aminopyridine and given a single electrical shock. *Soc. Neurosci. Symp.* **2**, 215–239.

Heuser, J. E., Reese, T. S., Dennis, M. J., Jan, Y., Jan, L., and Evans, L. (1979). Synaptic vesicle exocytosis captured by quick freezing and correlated with quantal transmitter release. *J. Cell Biol.* **81**(2), 275–300.

Katz, B. (1969). "The Release of Neural Transmitter Substances." Thomas, Springfield, IL.

Katz, B. and Miledi, R. (1968). The role of calcium in neuromuscular facilitation. *J. Physiol.* **195**(2), 481–492.

Liley, A. W. (1956). The quantal components of the mammalian endplate potential. *J. Physiol.* **133**, 560–573.

Roberts, W. M. (1994). Localization of calcium signals by a mobile calcium buffer in frog saccular hair cells. *J. Neurosci.* **14**(5 Pt 2), 3246–3262.

Schneggenburger, R. and Neher, E. (2000). Intracellular calcium dependence of transmitter release rates at a fast central synapse. *Nature* **406**(6798), 889–893.

Sollner, T., Bennett, M. K., Whiteheart, S. W., Scheller, R. H., and Rothman, J. E. (1993a). A protein assembly-disassembly pathway in vitro that may correspond to sequential steps of synaptic vesicle docking, activation, and fusion. *Cell* **75**(3), 409–418.

Sollner, T., Whiteheart, S. W., Brunner, M., Erdjument-Bromage, H., Geromanos, S., Tempst, P., and Rothman, J. E. (1993b). SNAP receptor implicated in vesicle targeting and fusion. *Nature* **362**(6418), 318–324.

Sutton, R. B., Fasshauer, Jahn, R., and Brunger, A. T. (1998). Crystal structure of a SNARE complex involved in synaptic exocytosis at 2.4 A resolution. *Nature* **395**(6700), 347–353.

Zucker, R. S. (1989) Short-term synaptic plasticity. *Annu. Rev. Neurosci.* **12**, 13–31.

Suggested Readings

Chen, Y. A. and Scheller, R. H. (2001). SNARE-mediated membrane fusion. *Nature Rev. Mol. Cell Biol.* **2**(2), 98–106.

De Camilli, P., Takei, K., and McPherson, P. S. (1995). The function of dynamin in endocytosis. *Curr. Opin. Neurobiol.* **5**(5), 559–565.

Dittman, J. S. and Regehr, W. G. (1998). Calcium dependence and recovery kinetics of presynaptic depression at the climbing fiber to Purkinje cell synapse. *J. Neurosci.* **18**(16), 6147–6162.

Liao, D., Hessler, N. A., and Malinow, R. (1995). Activation of postsynaptically silent synapses during pairing-induced LTP in CA1 region of hippocampal slice. *Nature* **375**(6530), 400–404.

Novick, P., Field, C., and Schekman, R. (1980). Identification of 23 complementation groups required for posttranslational events in the yeast secretory pathway. *Cell* **21**(1), 205–215.

Peters, C., Bayer, M. J., Buhler, S., Andersen, J. S., Mann, M., and Mayer, A. (2001). Trans-complex formation by proteolipid channels in the terminal phase of membrane fusion. *Nature* **409**(6820), 581–588.

Rothman, J. E. and Warren, G. (1994). Implications of the SNARE hypothesis for intracellular membrane topology and dynamics. *Curr. Biol.* **4**(3), 220–233.

Ryan, T. A., Li, L., Chin, L. S., Greengard, P., and Smith, S. J. (1996). Synaptic vesicle recycling in synapsin I knock-out mice. *J. Cell Biol.* **134**(5), 1219–1227.

Ryan, T. A., Reuter, H., Wendland, B., Schweizer, F. E., Tsien, R. W., and Smith, S. J. (1993). The kinetics of synaptic vesicle recycling measured at single presynaptic boutons. *Neuron* **11**(4), 713–724.

Schiavo, G., Rossetto, O., Tonello, F., and Montecucco, C. (1995). Intracellular targets and metalloprotease activity of tetanus and botulism neurotoxins. *Curr. Top. Microbiol. Immunol.* **195**, 257–274.

Thomas L. Schwarz

Neurotransmitter Receptors

Chemical synaptic transmission plays a fundamental role in neuron-to-neuron and neuron-to-muscle communication and is mediated by receptors embedded in the plasma membrane. These receptors can either mediate the direct opening of an ion channel (ionotropic receptors) or alter the concentration of intracellular metabolites (metabotropic receptor). The sign of a given response can be inhibitory or excitatory, and the response magnitude is determined by receptor number, the "state" of the receptors, and the amount of transmitter released. Finally, the temporal and spatial summation of information conveyed by the activation of multiple receptors ultimately determines whether that neuron will fire an action potential or the muscle will contract. As one can see, there is remarkable flexibility and diversity in molding the response to neurotransmitter by constructing a synapse with the desired receptor types.

Two broad classifications exist for receptors. An *ionotropic receptor* is a relatively large, multisubunit complex typically composed of four or five individual proteins that combine to form an ion channel through the membrane (Fig. 9.1A). In the absence of neurotransmitter, these ion channels exist in a closed state and are largely impermeable to ions. Neurotransmitter binding induces rapid conformational changes that open the channel, permitting ions to flow down their electrochemical gradients. Changes in membrane current resulting from ligand binding to ionotropic receptors generally are measured on a millisecond time scale. The ion flow ceases when transmitter dissociates from the receptor or when the receptor becomes desensitized, a process discussed in more detail later in this chapter.

In contrast, a *metabotropic receptor* is composed of a single polypeptide (Fig. 9.1B) and exerts its effects not through the direct opening of an ion channel, but through binding to and activating GTP-binding proteins (often referred to as G-proteins). Transmitters that activate metabotropic receptors typically produce responses of slower onset and longer duration (from tenths of seconds to potentially hours) due to the series of enzymatic steps necessary to produce a response. The metabotropic receptors have more recently been named G-protein-coupled receptors, or GPCRs for short, to more accurately capture their properties; the latter nomenclature is adopted in this chapter.

IONOTROPIC RECEPTORS

All ionotropic receptors are membrane-bound protein complexes that form an ion-permeable pore in the membrane. By comparing the amino acid sequence of cloned ionotropic receptors, one can deduce that they are similar in overall structure, although two independent ancestral genes have given rise to two distinct families. One family includes one of the receptors for acetylcholine (ACh), the nicotinic ACh receptor (nAChR), a receptor for γ-aminobutyric acid (GABA), the GABA$_A$ receptor, the glycine receptor, and one subclass of serotonin (5-HT) receptors, the 5HT3 receptor. The other family comprises the set of ionotropic glutamate receptors (Hollmann and Heinemann, 1994).

The nAChR Is a Heteromeric Protein Complex with a Distinct Architecture

The nAChR is so named because the plant alkaloid nicotine can bind to the ACh-binding site and activate

181

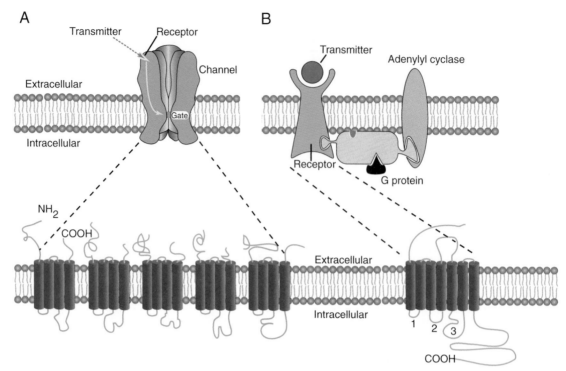

FIGURE 9.1 Structural comparison of ionotropic and metabotropic receptors. (A) *Ionotropic receptors* bind transmitter, and this binding translates directly into the opening of the ion channel through a series of conformational changes. Ionotropic receptors are composed of multiple subunits. The five subunits that together form the functional nAChR are shown. Note that each of the nAChR subunits wraps back and forth through the membrane four times and that the mature receptor is composed of five subunits. (B) *Metabotropic receptors* bind transmitter and, through a series of conformational changes, bind to G-proteins and activate them. G-proteins then activate enzymes such as adenylyl cyclase to produce cAMP. Through the activation of cAMP-dependent protein kinase, ion channels become phosphorylated, which affects their gating properties. Metabotropic receptors are single subunits. They contain seven transmembrane-spanning segments, with the cytoplasmic loops formed between the segments providing the points of interactions for coupling to G-proteins. Adapted from Kandel (1991).

the receptor. The nAChR purified as described from the electric organ of *Torpedo* ray is composed of five subunits (see Fig. 9.1A) and has a native molecular mass of approximately 290 kDa. The subunits are designated α, β, γ, and δ, and each receptor complex contains two copies of the α subunit. The subunits are homologous membrane-bound proteins that assemble in the bilayer to form a ring enclosing a central pore. The extracellular domain of each subunit together forms a funnel-shaped opening that extends approximately 100 Å outward from the outer leaflet of the plasma membrane. The funnel at the outer portion of the receptor has an inside diameter of 20–25 Å. The funnel shape is thought to concentrate and force ions and transmitter to interact with amino acids in the limited space of the pore without producing a major barrier to diffusion. This funnel narrows near the center of the lipid bilayer to form the domain of the receptor that determines the opened or closed state of the ion pore (Fig. 9.1). The intracellular domain of the receptor forms short exits for ions traveling into the

cell and an entrance for ions traveling out of the cell. The intracellular domain also establishes the association of the receptor with other intracellular proteins that determine the subcellular localization of the nAChR.

Each nAChR Subunit Has Multiple Membrane-Spanning Segments

The primary structure of each nAChR subunit was obtained by the efforts of Shosaka Numa and colleagues. The deduced amino acid sequence from cloned mRNAs indicates that nAChR subunits range in size from 40 to 65 kDa. Each subunit consists of four transmembrane (TM)-spanning segments referred to as TM1–TM4 (Fig. 9.2A). Each segment is composed mainly of hydrophobic amino acids that stabilize the domain within the hydrophobic environment of the lipid membrane. The four transmembrane domains are arranged in an antiparallel fashion, wrapping back and forth through the membrane. The N terminus of

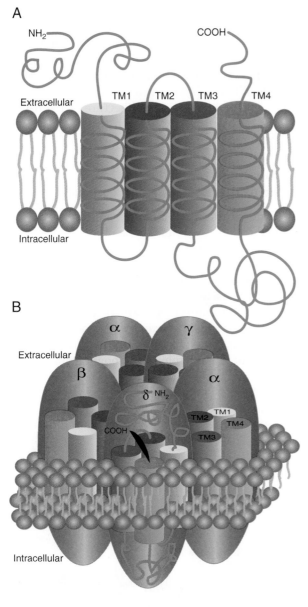

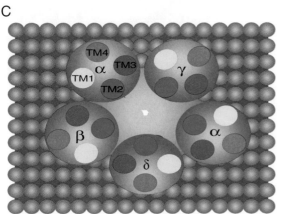

FIGURE 9.2 (A) Diagram highlighting the orientation of membrane-spanning segments of one subunit of the nAChR. The amino and carboxy termini extend in the extracellular space. The four membrane-spanning segments are designated TM1–TM4. Each forms an *a* helix as it traverses the membrane. (B) Side view of the five subunits in their approximate positions within the receptor complex. There are two *a* subunits present in each nAChR. (C) Top view of all five subunits highlighting the relative positions of their membrane-spanning segments, TM1–TM4, and the position of TM2 that lines the channel pore. Adapted from Kandel (1991).

each subunit extends into the extracellular space, as does the loop connecting TM2 and TM3, as well as the C terminus. Amino acids linking TM1 and TM2 and those linking TM3 and TM4 form short loops that extend into the cytoplasm.

Structure of the Channel Pore Determines Ion Selectivity and Current Flow

In the model shown in Figure 9.2C, each subunit of the nAChR can be seen to contribute one cylindrical component (representing a membrane-spanning segment) that presents itself to a central cavity that forms the ion channel through the center of the complex. The membrane-spanning segments that line the pore are the five TM2 regions, one contributed by each subunit. The amino acids that compose the TM2 segment are arranged in such a way that three rings of negatively charged amino acids are oriented toward the central pore of the channel (Fig. 9.3A). These rings of negative charge appear to provide much of the selectivity filter of the channel, ensuring that only cations can pass through the pore. The nAChR is permeable to most cations, such as Na^+, K^+, and Ca^{2+}, although monovalent cations are preferred. The restricted physical dimensions of the pore—9–10 Å in the open state—contribute greatly to the selectivity for particular ions (Unwin, 1995). A coarse filtering that also influences selectivity appears to be a shielding effect produced by other negatively charged amino acids surrounding the outer channel region of the receptor. Collectively, these physical characteristics of the pore—together with the electrochemical gradient across the plasma membrane—determine the possibility of ion movements. Thus, when the pore of the nAChR opens, anions remain restricted from movement across the membrane while positively charged cations move down their respective electrochemical gradients, resulting in an influx of Na^+ and Ca^{2+} and a small efflux of K^+.

Opening of the nAChR Occurs through Concerted Conformational Changes Induced by Binding of Two ACh Molecules

Each receptor complex has two ACh-binding sites that reside in the extracellular domain and are formed for the most part by six amino acids in α subunits; however, amino acids in both γ and δ subunits also contribute to binding. The two binding sites are not equivalent because of the asymmetry of the receptor due to the different neighboring subunits (either γ or δ) adjacent to the two α subunits. Significant cooperativity also exists within the receptor molecule, and so binding of the first molecule of ACh enhances binding of the second. When nAChR binds two molecules of ACh, the channel opens almost instantaneously (time constants for opening are approximately 20 μs), thus permitting the passage of ions. A model developed from electron micrographic reconstructions of the nicotine-bound form of the nAChR indicates that the closed-to-open transition is associated with a rotation of the TM2 segments (Fig. 9.3; Unwin, 1995). The TM2 segments are helical and exhibit a kink in their structure that forces a Leu residue from each segment into a tight ring that effectively blocks the flow of ions through the central pore of the receptor. When the TM2 segments rotate because of ACh binding, the

kinks also rotate; relaxing the constriction formed by the Leu ring, and ions can then permeate through the pore. The rotation also orients a series of Ser and Thr residues (amino acids with a polar character) into the central area of the pore (compare Figs. 9.3B to 9.3C), which facilitates the permeation of water-solvated cations. The well-establish architecture of the nAChR provides a structural framework to which all other ionotropic receptors can be compared.

The Muscle Form of the nAChR Is Very Similar to the nAChR from *Torpedo*

nAChRs at the neuromuscular junction are a concentrated collection of homogeneous receptors having a structure similar to that of the *Torpedo* electric organ. This similarity is not surprising because the electric organ is a specialized form of muscle tissue. The adult form of the muscle receptor has the pentameric structure $\alpha_2\beta\epsilon\delta$. An embryonic form of the receptor has an analogous structure, except that the ϵ subunit is replaced by a unique γ subunit. The embryonic and adult subunits of both mouse and bovine muscle receptors have been cloned and expressed and the receptors differ in both channel kinetics and channel conductance. These differences in channel properties appear to be necessary for the proper function of the nAChRs

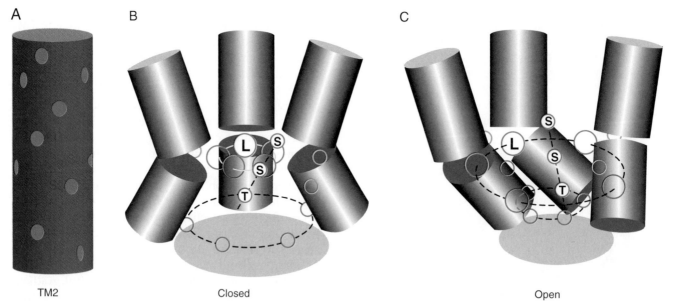

TM2   Closed   Open

FIGURE 9.3   (A) Relative positions of amino acids in the TM2 segment of one of the nAChR *a* subunits modeled as an *a* helix. Glutamate residues (E) that form parts of the negatively charged rings for ion selectivity are shown at the top and bottom of the helix. (B) Arrangement of three of the TM2 segments of the nAChR modeled with the receptor in the closed (ACh-free) configuration. In the closed configuration, leucine (L) residues form a right ring in the center of the pore that blocks ion permeation. (C) Arrangement of the three TM2 segments after ACh binds to the receptor. In the open configuration, construction formed by the ring of leucine (L) residues opens as the helices twist about their axes. Note that polar serine (S) and threonine (T) residues align when ACh binds, which apparently helps the water-solvated ions travel though the pore. Adapted from Unwin (1995).

as they undergo the transition from developing to mature neuromuscular junction synapses.

The nAChR Has Well-Ordered Assembly and Is Posttranslationally Modified

The pathway of nAChR assembly in muscle is a tightly regulated process. For example, the five subunits of nAChR have the potential to assemble randomly into 208 different combinations. Nevertheless, in vertebrate muscle, only one of these configurations ($\alpha_2\beta\epsilon\delta$) typically is found in mature tissue, indicating a very high degree of coordinated assembly and, ultimately, little structural variability. The well-ordered assembly takes place within the endoplasmic reticulum and during intracellular maturation, each subunit is glycosylated. Two highly conserved disulfide bonds in the N-terminal extracellular domain are essential for efficient assembly of the mature receptor. The first is between two adjacent Cys residues and resides very close to the ACh-binding site on the receptor. The second bond is between two Cys residues 15 amino acids apart, forming a loop in the extracellular domain.

Many ionotropic receptors, such as the nAChR, are phosphorylated, although the functional significance of the phosphorylation is not always evident. The nAChR is phosphorylated by at least three protein kinases. cAMP-dependent protein kinase (PKA) phosphorylates the γ and δ subunits, Ca^{2+}/phospholipid-dependent protein kinase (PKC) phosphorylates the δ subunit, and an unidentified tyrosine kinase phosphorylates the β, γ, and δ subunits. The phosphorylation sites are all found in the intracellular loop between TM3 and TM4 membrane-spanning segments. Phosphorylation by these three protein kinases increases the rate of desensitization of the receptor. Desensitization of receptors is a common observation, and this process limits the amount of ion flux through a receptor by producing transitions into a closed state (one that does not permit ion flow) in the continued presence of neurotransmitter. For the nAChR, the rate of desensitization has a time constant of approximately 50–100 ms.

Neuronal nAChRs Contain Two Types of Subunits and Their Assembly Leads to Functional Diversity

Neuronal nAChRs are similar, yet distinct in structure to the *Torpedo* isoform of the receptor. For example, the neuronal nAChR is composed of only two types of subunits, α and β, which combine to produce the functional receptor, and the majority of these receptors do not bind to α-bungarotoxin. At least nine different α subtypes ($\alpha1$ being the muscle α subunit) have been identified, and some are species-specific ($\alpha8$ is found only in chicken and $\alpha9$ is found only in rat). Four different β subtypes ($\beta1$ being the muscle β subunit) have been identified. The neuronal β subunits are distantly related to the muscle $\beta1$ subunit and sometimes are referred to simply as non-α subunits. All the α and β genes encode proteins with four transmembrane-spanning segments and, although the physical structure of this receptor family has not been well characterized, it appears that each functional receptor is a pentameric assembly.

Neuronal nAChRs have diverse functions and are the receptors presumed to be responsible for the psychophysical effects of nicotine addiction. One major function of nAChRs in the brain is to modulate excitatory synaptic transmission through a presynaptic action. The diversity in function of nAChRs can be related to the heterogeneous structure contributed by the thousands of possible combinations between the different α and β subunits. Functional neuronal nAChRs can be assembled from a single subunit (e.g., $\alpha7$, $\alpha8$, or $\alpha9$), and a single type of α subunit can also be assembled with multiple types of β subunits (e.g., $\alpha3$ with $\beta2$ or $\beta4$ or both) and vice versa. These additional possibilities produce a staggering array of potential receptor molecules, each with distinct properties, including differences in single-channel kinetics and rates of desensitization. It is also now established that subunit composition plays important roles in targeting the receptors to different intracellular locations.

Neuronal nAChRs exhibit a range of single-channel conductances between 5 and 50 pS, depending on the tissue or the specific subunits expressed. All the neuronal nAChRs are cation-permeable channels that, in addition to permitting the influx of Na^+ and the efflux of K^+, permit an influx of Ca^{2+}. The Ca^{2+} permeability for neuronal nAChR is greater than that for the muscle nAChR and is variable among the different neuronal receptor subtypes. Indeed, some receptors have very high Ca^{2+}/Na^+ permeability ratios; for example, $\alpha7$ nAChRs exhibit a Ca^{2+}/Na^+ permeability ratio of nearly 20, whereas other neuronal isoforms exhibit Ca^{2+}/Na^+ permeability ratios of about 1.0–1.5. The Ca^{2+} permeability of the $\alpha7$ nAChR can be eliminated by the mutation of a single amino acid residue in TM2 (Glu-237 for A1a) without significantly affecting other aspects of the receptor. This key Glu residue presumably lies within the pore of the receptor and enhances the passage of Ca^{2+} ions through an interaction with its negatively charged side chain. Activation of $\alpha7$ receptors through the binding of ACh therefore could produce a significant increase in the level of intracellular Ca^{2+} without the opening of voltage-gated Ca^{2+} channels. Subunits $\alpha 7$, $\alpha 8$, and $\alpha 9$ are also the α-bungarotoxin-binding subtypes of neuronal nAChRs.

For the nAChR from muscle, desensitization is minor and probably is not of physiological significance in determining the shape of the synaptic response at the neuromuscular junction. However, for some neuronal nAChRs, desensitization likely plays a major role in determining the effects of the actions of ACh. Receptors composed of $\alpha 7$, $\alpha 8$, and certain α/β combinations exhibit desensitization time constants of between 100 and 500 ms, whereas others exhibit desensitization constants between 2 and 20 s. Given the diverse functions of neuronal nAChRs, the variable rates of desensitization likely play important roles whereby this inherent property of the receptor shapes the physiological response generated from binding ACh.

One Serotonin Receptor Subtype, 5-HT3, Is Ionotropic and Is a Close Relative of the nAChR

5-HT historically is thought of as a metabotropic transmitter that binds to and activates only GPCRs (described in more detail later). The 5-HT3 subclass is an exception forming an ionotropic receptor activated by binding 5-HT. The 5-HT3 receptor is permeable to Na^+ and K^+ ions and is similar in many ways to nAChR in that both desensitize rapidly and are blocked by tubocurarine. From expression studies of the cloned cDNA, it appears that the 5-HT3 receptor is a homomeric complex composed of five copies of the same subunit. The deduced amino acid sequence of the cDNA indicates that the protein is 487 amino acids long (56 kDa) and has a structure most analogous to the $\alpha 7$ subtype of neuronal nAChRs, which also forms a homooligomeric receptor.

The 5-HT3 receptor is mostly impermeable to divalent cations. For example, Ca^{2+} is largely excluded from permeation and in fact effectively blocks current flow through the pore, even though the predicted pore size of the channel (7.6 Å) is approximately the same as that for the nAChR (8.4 Å). Apparently, other physical or electrochemical barriers limit the capacity of divalent ions to permeate the 5-HT3 pore. Dose–response studies indicate that at least two ligand-binding sites must be occupied for the channel to open; however, the binding of agonist and/or opening of the channel appears to be approximately 10 times slower than for most other ligand-gated ion channels. The functional significance or physical explanation of this slow opening is not known. The native 5-HT3 receptor also exhibits desensitization (time constant 1–5 s), although the reported desensitization rate varies widely,

depending on the methodology used for analysis and the source of receptor. Interestingly, this desensitization can be significantly slowed or enhanced by single amino acid substitutions at a Leu residue in the TM2 segment of the subunit.

5-HT3 receptors are distributed sparsely on primary sensory nerve endings in the periphery and are distributed widely at low concentrations in the mammalian CNS. The 5-HT3 receptor is clinically significant because antagonists of 5-HT3 receptors have important applications as antiemetics, anxiolytics, and antipsychotics.

GABA$_A$ Receptors Are Related in Structure to nAChRs, but Exhibit an Inhibitory Function

Synaptic inhibition in the mammalian brain is mediated principally by GABA receptors. The most widespread ionotropic receptor activated by GABA is designated GABA$_A$. The subunits composing the GABA$_A$ receptor have sequence homology with the nAChR subunit family, and the two families have presumably diverged from a common ancestral gene. The GABA$_A$ receptor is composed of multiple subunits, probably forming a heteropentameric complex of approximately 275 kDa. Five different types of subunits are associated with GABA$_A$ receptors and are designated α, β, γ, δ, and ε. An additional subunit, ρ, is found predominantly in the retina, whereas the other subunits are distributed widely in the brain. Each subunit group also has different subtypes; for example, six different α, four β, four γ, and two ρ subunits have been identified. The predicted amino acid sequences indicate that each of these subunits has a molecular mass ranging between 48 and 64 kDa. Like neuronal nAChR, these subunits mix in a heterogeneous fashion to produce a wide array of GABA$_A$ receptors with different pharmacological and electrophysiological properties. The predominant GABA$_A$ receptor in brain and spinal cord is $\alpha 1$, $\beta 2$, and $\gamma 2$ with a likely stoichiometry of two $\alpha 1$s, two $\beta 2$s, and one $\gamma 2$. Expression of subunit cDNAs in oocytes indicates that the α subunit is essential for producing a functional channel. The α subunit also appears to contain the high-affinity binding site for GABA.

The ion channel associated with the GABA$_A$ receptor is selective for anions (in particular, Cl^-), and the selectivity is provided by strategically placed positively charged amino acids near the ends of the ion channel. When GABA binds to and activates this receptor, Cl^- flows into the cell, producing a hyperpolarization by moving the membrane potential away from the threshold for firing an action potential. The neuronal

$GABA_A$ receptor exhibits multiple conductance levels, with the predominant conductance being 27–30 pS. Measurements and modeling of single-channel kinetics suggest that two sequential binding sites exist for anions within the pore.

The $GABA_A$ Receptor Binds Several Compounds That Affect Its Properties

The $GABA_A$ receptor is an allosteric protein, its properties are modulated by the binding of a number of compounds. Two well-studied examples are barbiturates and benzodiazepines, both of which bind to the $GABA_A$ receptor and potentiate GABA binding. The net result is that in the presence of barbiturates, benzodiazepines, or both, the same concentration of GABA will cause increased inhibition. Benzodiazepine binding is conferred on the receptor by the γ subunit, but the presence of α and β subunits is necessary for the qualitative and quantitative aspects of benzodiazepine binding. The benzodiazepine-binding site appears to lie along the interface between α and γ subunits and only certain subtypes are sensitive to benzodiazepines. Benzodiazepine binding to $GABA_A$ receptors requires α, 2, or 5 and $\gamma2$ or 3; other subunit combinations are insensitive to benzodiazepines.

Picrotoxin, a potent convulsant compound, appears to bind within the channel pore of the $GABA_A$ receptor and prevent ion flow. Single-channel experiments indicate that picrotoxin either slowly blocks an open channel or prevents the GABA receptor from undergoing a transition into a long-duration open state. Apparently, barbiturates produce similar changes in channel properties, but they potentiate rather than inhibit $GABA_A$ receptor function. Bicuculline, another potent convulsant, appears to inhibit $GABA_A$ receptor channel activity by decreasing the binding of GABA to the receptor. Steroid metabolites of progesterone, corticosterone, and testosterone also appear to have potentiating effects on GABA currents that are similar in many ways to the action of barbiturates; however, the binding sites for these steroids and the barbiturates are distinct. Finally, penicillin directly inhibits GABA receptor function, apparently by binding within the pore and thus is designated an open channel blocker.

The physiological effects of compounds such as picrotoxin, bicuculline, and penicillin are striking. Each of these compounds at a sufficiently high concentration can produce widespread and sustained seizure activity. Conversely, many, but not all, of the sedative properties associated with barbiturates and benzodiazepines can be attributed to their ability to augment inhibition in the brain through enhancing the inhibitory potency of GABA.

Interestingly, ρ-subunit-containing GABA receptors, found in abundance in the retina, are pharmacologically unique. They are resistant to the inhibitory action of bicuculline, although they remain sensitive to blockage by picrotoxin. In addition, these retinal receptors are not sensitive to modulation by barbiturates or benzodiazepines. Thus, ρ-containing receptors are distinct from $GABA_A$ receptors and are similar to receptors earlier designated $GABA_C$.

The Glycine Receptor Is Closely Related to the $GABA_A$ Receptor

Glycine receptors are the major inhibitory receptors in the spinal cord and the brain stem. Glycine receptors are similar to $GABA_A$ receptors in that both are ion channels selectively permeable to the anion Cl^-. The structure of the glycine receptor is indicative of this similarity in properties. The native complex is approximately 250 kDa and is composed of two main subunits: α (48 kDa) and β (58 kDa). The receptor appears to be pentameric, most likely composed of three α and two β subunits. The glycine receptor has an open channel conductance of approximately 35–50 pS, similar to that of the $GABA_A$ receptor. Strychnine is a potent antagonist of the glycine.

Four distinct α subunits and one β subunit of the glycine receptor have been cloned. Each exhibits the typical predicted four transmembrane segments and are approximately 50% identical with one another at the amino acid level. Expression of a single α subunit in oocytes is sufficient to produce functional glycine receptors, indicating that the α subunit is the pore-forming unit of the native receptor. β subunits play exclusively modulatory roles, affecting, for example, sensitivity to the inhibitory actions of picrotoxin, and they are widespread in the brain. Because their distribution does not specifically colocalize with glycine receptor α subunit mRNA, β subunits may serve other functions independent of their association with glycine receptor.

Certain Purinergic Receptors Are Also Ionotropic

Purinergic chemical transmission is distributed throughout the body and is considered in greater detail in a later section on GPCRs. Purinergic receptors bind to ATP (or other nucleotide analogs) or its breakdown product adenosine. ATP is released from certain synaptic terminals in a quantal manner and often is packaged within synaptic vesicles containing another neurotransmitter, the best described being ACh and catecholamines.

Two subtypes of ATP-binding purinergic receptors (P2x and P2z) have been discovered to be ionotropic receptors, but data on their functions and properties are sparse. P2x receptors appear to mediate a fast depolarizing response in neurons and muscle cells to ATP by the direct opening of a nonselective cation channel. cDNAs encoding the P2x receptor indicate that its structure comprises only two transmembrane domains, with some homology in its pore forming region with K⁺ channels. The P2z receptor is also a ligand-gated channel that permits permeation of either anions or cations and even molecules as large as 900 Da.

Glutamate Receptors Are Derived from a Different Ancestral Gene and Are Structurally Distinct from Other Ionotropic Receptors

Glutamate receptors are widespread in the nervous system where they are responsible for mediating the vast majority of excitatory synaptic transmission in the brain and spinal cord. In the 1970s, Jeffrey Watkins and colleagues advanced this field significantly by developing agonists that could pharmacologically distinguish between different glutamate receptor subtypes. Four of these agonists—N-methyl-D-aspartate (NMDA), amino-3-hydroxy-5-methylisoxazolepropri-onic acid (AMPA), kainate, and quisqualate—are distinct in the type of receptors to which they bind and have been used extensively to characterize the glutamate receptor family (Watkins *et al.*, 1990; Hollmann and Heinemann, 1994). A convenient distinction for describing ionotropic glutamate receptors has been to classify them as either NMDA or non-NMDA subtypes, depending on whether they bind the agonist NMDA. Non-NMDA receptors also bind the agonist kainite or AMPA. Both NMDA and non-NMDA receptors are ionotropic. Quisqualate is unique within this group of agonists in having the capacity to activate both ionotropic and GPCR glutamate receptor subtypes (Hollmann and Heinemann, 1994). A family tree highlighting the evolutionary relationship of the glutamate receptors is shown in Figure 9.4.

Non-NMDA Receptors Are a Diverse Family

In 1989, the isolation of a cDNA that produced a functional glutamate-activated channel when expressed in *Xenopus* oocytes was reported (Hollmann *et al.*, 1989). The initial glutamate receptor was termed GluR-K1, and the cDNA encoded a protein with an estimated molecular mass of 99.8 kDa. Not long after this original report, several groups (Boulter *et al.*, 1990; Keinanen *et al.*, 1990; Nakanishi *et al.*, 1990) indepen-

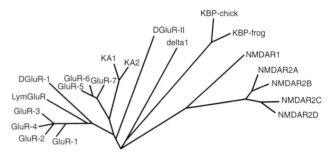

FIGURE 9.4 Evolutionary relationships of the ionotropic glutamate receptor family. Adapted from Hollmann and Heinemann (1994).

dently reported the isolation of families of glutamate receptor subunits, termed either GluR1–GluR4 or GluRA-GluRD. Each GluR subunit consists of approximately 900 amino acids and has four predicted membrane-spanning segments (TM1–TM4); however, there is an important distinction in the TM2 domain making the GluRs distinct from the nAChR family (Figs. 9.5A and 9.5B). The native form of GluR subunits appears to be a tetrameric complex with an approximate molecular mass of 600 kDa. Thus, the size of the glutamate receptor is almost twice that of the nAChR, mostly because of the large extracellular domain where glutamate binds to the receptor.

Unique Properties of Non-NMDA Receptors Are Determined by Assembly of Different Subunits

When cDNAs encoding these receptors were expressed in either oocytes or HeK-293 cells, application of the non-NMDA receptor agonist AMPA produced substantial inward currents. A striking observation from these expression studies was that when the GluR2 subunit alone was expressed in the oocytes, little current was obtained when the preparation was exposed to agonist, unlike the large currents found when either GluR1 or GluR3 was expressed (Verdoorn *et al.*, 1991; Nakanishi *et al.*, 1990). GluR2 subunits by themselves appear to form poorly conducting receptors. However, when GluR2 is expressed with either GluR1 or GluR3, the behavior of the heteromeric receptor is distinctly different.

Examination of *I/V* plots indicates that when GluR1 and GluR3 are expressed alone or together, they produce channels with strong inward rectification. Coexpression of GluR2 with either GluR1 or GluR3 produces a channel with little rectification and a near linear *I/V* plot. Further analyses indicated that GluR1 and GluR3, either independently or when coexpressed, exhibited channels permeable to Ca²⁺ (Hollmann and

A

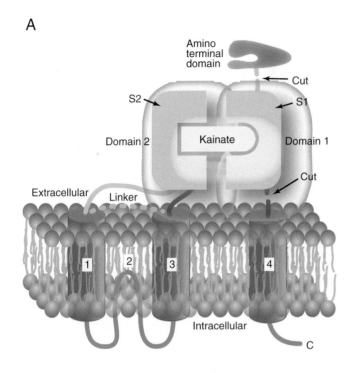

B

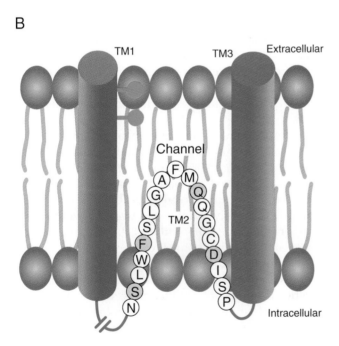

FIGURE 9.5 (A) Model of one of the subunits of the ionotropic glutamate receptor. Ionotropic glutamate receptors have four membrane-associated segments; however, unlike nAChR, only three of them completely traverse the lipid bilayer. TM2 forms a loop and reexits into the cytoplasm. Thus, the large N-terminal region extends into the extracellular space, whereas the C terminus extends into the cytoplasm. Two domains in the extracellular segments associate with each other to form the binding site for transmitter, in this example kainate, a naturally occurring agonist of glutamate. (B) Enlarged area of the predicted structure and amino acid sequence of the TM2 region of the glutamate receptor, GluR3. TM1 and TM3 are drawn as cylinders in the membrane flanking TM2. The residue that determines Ca^{2+} permeability of the non-NMDA receptor is the glutamine residue (Q) highlighted in gray. In NMDA receptors, an asparagine residue at this same position is the proposed site of interaction with Mg^{2+} ions that produce the voltage-dependent channel block. Serine (S) and phenylalanine (F), also shaded in gray, are highly conserved in the non-NMDA receptor family. The aspartate (D) residue is also conserved and is thought to form part of the internal cation-binding site. The break in the loop between TM1 and TM2 indicates a domain that varies in length among ionotropic glutamate receptors. Adapted from Wo and Oswald (1995).

non-Ca^{2+}-permeable to a Ca^{2+}-permeable channel (Burnashev *et al.*, 1992). Apparently, an Arg at this position blocks Ca^{2+} from traversing the pore formed in the center of the GluR channel.

Functional Diversity in GluRs Is Produced by mRNA Splicing and RNA Editing

Analysis of mRNAs encoding GluR subunits indicated that each could be expressed in one of two splice variants, termed flip and flop. These flip and flop modules are small segments (38 amino acids) just preceding the TM4 domain in all four GluR subunits. The receptor channel expressed from these splice variants has distinct properties, depending on which of the two modules is present. Specifically, flop-containing receptors desensitize more during glutamate application. Therefore, GluRs with flop modules express smaller steady-state currents than GluRs with flip modules. Both flip- and flop-containing GluRs are widely expressed in the brain with a few exceptions. One unique cell type appears to be pyramidal CA3 cells in the rat hippocampus, where the GluRs are deficient in flop modules. In neighboring CA1 pyramidal cells and dentate granule cells, flop-containing GluRs appear to dominate. The significance of these splice variations for information processing in the brain is not known, but the physiological prediction would be that CA3 neurons exhibit larger steady-state glutamate-activated currents due to decreased desensitization from the absence of flop modules.

Heinemann, 1994). In contrast, any combination of receptor that included the GluR2 subunit produced channels impermeable to Ca^{2+}. The replacement of a single amino acid (Arg for Gln) in TM2 of the GluR2 subunit (see Fig. 9.5 for identification of this amino acid) was shown to switch its behavior from a

Typically, one believes that there is absolute fidelity in the process of transcribing DNA into mRNA and then into protein; that is, that nucleotides present in the DNA are accurate predictors of the ultimate amino acid sequence of the protein. However, a novel mechanism in the neuronal nucleus was discovered that edits mRNAs post transcriptionally, and at least three of the four GluR subunits are subjected to this process (Sommer et al., 1991). In fact, one of the sites edited is the critical Arg residue regulating Ca^{2+} permeability in the GluR2 subunit. At another edited site, Gly replaces Arg-764 in the GluR2 subunit, and this editing also takes place in GluR3 and GluR4. The Arg-to-Gly conversion at amino acid 764 produces receptors that exhibit significantly faster rates of recovery from the desensitized state. The extent to which other receptors or other protein molecules undergo this form of editing is an area rich for investigation. At a minimum, this editing mechanism produces dramatic differences in the function of GluRs.

Glutamate Receptors Do Not Conform to the Typical Four Transmembrane-Spanning Segment Structures Described for nAChR

Although the field of glutamate receptors is advancing at a rapid pace, we have limited structural data on the native molecule or on the topology of any single GluR subunit as it exists in the membrane. One exception is the recent report of the 3D structure of the native AMPA receptor determined by electron microscopy reconstructions (Nakagawa et al., 2005). The study reveals the overall architecture of the tetrameric AMPA receptor complex and provides some global insights into how glutamate binding leads to channel opening. This low resolution structure compliments a significant amount of work analyzing crystal structures of the isolated glutamate receptor ligand binding domain with and without the presence of agonists and antagonists.

The glutamate receptor has a large extracellular domain that serves as the binding site for kainite and glutamate (Fig. 9.5A). Superficially, the remainder of the receptor originally was thought to resemble the nAChR in having four transmembrane segments that wrap back and forth through the membrane in an antiparallel fashion. However, that model now has been proven incorrect by a number of elegant molecular and biochemical studies. The most recent information indicates that the TM2 segment does not traverse the membrane completely (Fig. 9.5). Instead, it forms a kink within the membrane and enters back into the cytoplasm, similar in some ways to the pore-forming domain (P segment) of voltage-activated K^+ channels. An enlargement of this P segment (Fig. 9.5B) highlights the amino acids conserved in all the GluRs and further identifies the critical Gln residues responsible for Ca^{2+} permeability of the receptor. It also appears that glutamate receptors do not conform to the five-subunit structure of the nAChR. Both biochemical (Armstrong and Gouaux, 2000) and electrophysiological (Rosenmund et al., 1998) evidence indicates that functional glutamate receptors are composed of four, not five subunits. Thus, it appears that glutamate receptors are a rather highly divergent form of the nAChR receptor family. In fact, their general structural features conform more closely to the family of K^+ channels in that both appear tetrameric and both have a unique P segment that forms the selectivity filter.

Other Non-NMDA GluRs Have Poorly Characterized Functions

Three other members, GluR5–7, now form a second non-NMDA receptor subfamily (Fig. 9.4), whose contribution to producing functionally distinct receptors is less well understood. Their overall structure is similar to that of GluR1–4, and they exhibit about 40% sequence homology; however, their agonist-binding profile and their electrophysiological properties are distinct. They are expressed at lower levels in the brain than the GluR1–4 family (Hollmann and Heinemann, 1994).

Two members of the glutamate receptor family, KA-1 and KA-2, are the high-affinity kainate-binding receptors found in brain. Clearly distinct from the glutamate receptors discussed so far, KA-1 and KA-2 are more similar to the GluR5–7 subfamily than to the GluR1–4 subfamily as indicated by their evolutionary relatedness (Fig. 9.4). Neither KA-1 nor KA-2 produces a functional channel when expressed in cells or oocytes, even though high-affinity kainate-binding sites were detected. KA-1 does not appear to form functional receptors or channels with any of the other GluR subunits, and its physiological relevance remains obscure. It is expressed at high concentrations in only two cell types, hippocampal CA3 and dentate granule cells. KA-2 exhibits interesting properties when combined with other GluR subunits. For example, coexpression of GluR6 and KA-2 produces functional receptors that respond to AMPA, although neither subunit itself responds to this agonist. This information indicates that agonist-binding sites are at least partly formed at the interfaces between subunits.

NMDA Receptors Are a Family of Ligand-Gated Ion Channels That Are Also Voltage Dependent

NMDA receptors appear to be at least partly responsible for aspects of development, learning and memory, and neuronal damage due to brain injury. The particular significance of this receptor to neuronal function comes from two of its unique properties. First, the receptor exhibits associativity. For the channel to be open the receptor must bind glutamate and the membrane must be depolarized. This behavior is due to a Mg^{2+}-dependent block of the receptor at normal membrane resting potentials. Second, the receptor permits a significant influx of Ca^{2+}, and increases in intracellular Ca^{2+} activate a variety of processes that alter the properties of the neuron. Excess Ca^{2+} is also toxic to neurons, and the hyperactivation of NMDA receptors is thought to contribute to a variety of neurodegenerative disorders.

Many pharmacological compounds produce their effects through interactions with the NMDA receptor. For example, certain hallucinogenic compounds, such as phencyclidine (PCP) and dizocilpine (MK-801), are effective blockers of the ion channel associated with the NMDA receptor (Fig. 9.6). These potent antagonists require the receptor channel to be open to gain access to their binding sites and therefore are referred to as open-channel blockers. They also become trapped when the channel closes and therefore are difficult to wash out of the channel of the NMDA receptor. Antagonists for the glutamate-binding site also have been developed, and some of the most well known are AP-5 and AP-7. These and other antagonists specific for the glutamate-binding site also produce hallucinogenic effects in both animal models and humans. NMDA remains a specific agonist for this receptor; however, it is about one order of magnitude less potent than L-Glutamate for receptor activation. L-Glutamate is the predominant neurotransmitter that activates the NMDA receptor; however, L-aspartate can also activate the receptor, as can an endogenous dipeptide in the brain, N-acetylaspartylglutamate (Hollmann and Heinemann, 1994).

NMDA Receptor Subunits Show Similarity to Non-NMDA Receptor Subunits

The primary structure of the NMDA receptor was revealed in 1990 when the first cDNA encoding

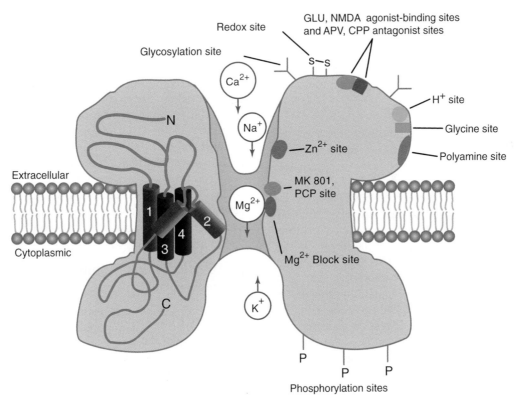

FIGURE 9.6 Diagram of a NMDA receptor highlighting binding sites for numerous agonists, antagonists, and other regulatory molecules. The location of these sites is a crude approximation for the purpose of discussion. Adapted from Hollmann and Heinemann (1994).

a subunit of the NMDA receptor was isolated (Moriyoshi *et al.*, 1991). The first cloned subunit was aptly named NMDAR1, and the deduced amino acid sequence indicated a protein of approximately 97 kDa, similar to other members of the GluR family. Four potential transmembrane domains were identified, and the current assumption is that four individual subunits compose the macromolecular NMDA receptor complex. However, recall that the transmembrane organization of GluR subunits indicates that TM2 does not fully transverse the membrane. It seems likely that the NMDA receptor subunits will also follow this recent modification of the model. The TM2 segment of each subunit clearly lines the pore of the NMDA receptor channel, as does the TM2 segment of the GluR subunits. In fact, a single Asn residue, analogous to that in the GluR2 subunit, regulates the Ca^{2+} permeability of the NMDA receptor. Mutation of this Asn residue reduces Ca^{2+} permeability markedly.

Three of the best-characterized facets of the NMDA receptor were found when the NMDAR1 subunit initially was expressed by itself in oocytes, although currents were relatively small. These characteristics are (1) a Mg^{2+}-dependent, voltage-sensitive ion channel block, (2) a glycine requirement for effective channel opening, and (3) Ca^{2+} permeability (Moriyoshi *et al.*, 1991). As described later, other NMDAR subunits contribute to assembly of the receptors thought to exist in the nervous system.

Functional Diversity of NMDA Receptors Occurs through RNA Splicing

At least eight splice variants now have been identified for the NMDAR1 subunit and these variants produce differences, ranging from subtle to significant, in the properties of the expressed receptor (Hollmann and Heinemann, 1994). For example, NMDAR1 receptors lacking a particular N-terminal insert due to alternative splicing exhibit enhanced blockade by protons and exhibit responses that are potentiated by Zn^{2+} in micromolar concentrations. Zn^{2+} classically has been described as an NMDA receptor antagonist that significantly blocks its activation. Clearly, the particular splice variant incorporated into the receptor complex affects the types of physiological response generated. Spermine, a polyamine found in neurons and in the extracellular space, also slightly increases the amplitude of NMDA responses, and this modulatory effect also appears to be associated with a particular splice variant. The physiologic role of spermine in regulating NMDA receptors remains unclear.

Multiple NMDA Receptor Subunit Genes Also Contribute to Functional Diversity

Four other members of the NMDA receptor family have been cloned (NMDAR2A–2D), and their deduced primary structures are highly related. These four NMDA receptor subunits do not form channels when expressed singly or in combination unless they are coexpressed with NMDAR1. Apparently, NMDAR1 serves an essential function for the formation of a functional pore by which activation of NMDA receptors permits the flow of ions. NMDA receptors 2A–2D play important roles in modulating the receptor activity when mixed as heteromeric forms with NMDAR1. Coexpression of NMDAR1 with any of the other subunits produces much larger currents (from 5- to 60-fold greater) than when NMDAR1 is expressed in isolation, and NMDA receptors expressed in neurons are likely to be heterooligomers of NMDAR1 and NMDAR2 subunits. The C-terminal domains of NMDAR2A–2D are quite large relative to the NMDAR1 C-terminus and appear to play roles in altering channel properties and in affecting the subcellular localization of the receptors. All the NMDAR subunits have an Asn residue at the critical point in the TM2 domain essential for producing Ca^{2+} permeability. This Asn residue also appears to form at least part of the binding site for Mg^{2+}, which suggests that the sites for Mg^{2+} binding and Ca^{2+} permeation overlap.

The distribution of NMDAR2 subunits generally is more restricted than the homogeneous distribution of NMDAR1, with the exception of NMDAR2A, which is expressed throughout the nervous system. NMDAR2C is restricted mostly to cerebellar granule cells, whereas 2B and 2D exhibit broader distributions. As noted, the large size of the C terminus of the NMDAR2 subunit suggests a potential role in association with other proteins, possibly to target or restrict specific NMDA receptor types to areas of the neuron. Mechanisms related to receptor targeting now are becoming understood and will clearly play major roles in determining the efficacy of synaptic transmission.

NMDA Receptors Exhibit Complex Channel Properties

The biophysical properties of the NMDA receptor are complex. Single-channel conductance has a main level of 50 pS; however, subconductances are evident, and different subunit combinations produce channels with distinct single-channel properties. A binding site for the Ca^{2+}-binding protein calmodulin also has been identified on the NMDAR1 subunit. Binding of Ca^{2+}–calmodulin to NMDA receptors produces a four-fold decrease in open-channel probability. Ca^{2+} influx through the NMDA receptor could induce calmodulin binding and lead to an immediate short-term feedback inhibition, decreasing ion flow through the receptor.

Summary

A general model for ionotropic receptors has emerged mainly from analyses of nAChR. Ionotropic receptors are large membrane-bound complexes generally composed of five subunits. The subunits each have four transmembrane domains, and the amino acids in TM2 form the lining of the pore. Transmitter binding induces rapid conformational changes that are translated into an increase in the diameter of the pore, permitting ion influx. Cation or anion selectivity is obtained through the coordination of specific negatively or positively charged amino acids at strategic locations in the receptor pore. How well the details of structural information obtained for the nAChR will generalize to other ionotropic receptors awaits structural analyses of these other members. However, it is already clear that this model does not adequately describe the orientation of the transmembrane domains or the subunit number of the glutamate receptor family. The TM2 domain of glutamate receptors forms a hairpin instead of traversing the membrane completely, causing the remainder of the receptor to adopt a different architecture than that described for the nAChR family. It also appears that glutamate receptors are composed of four, not five, subunits. These differences are perhaps not surprising given that the nAChR family and the glutamate receptor family appear to have arisen from two different ancestral genes.

G-PROTEIN COUPLED RECEPTORS

The number of members in the G-protein coupled receptor family is enormous, with over 1000 already identified. Historically, the term *metabotropic* was used to describe the fact that intracellular metabolites are produced when these receptors bind ligand. However, there are now clearly documented cases where the activation of "metabotropic" receptors does not produce alterations in metabolites but instead produces their effects by interacting with G-proteins that alter the behavior of ion channels. Thus, these receptors are now referred to as G-protein-coupled receptors (GPCRs).

When a GPCR is activated, it couples to a G-protein initiating the exchange of GDP for GTP, activating the G-protein (Fig. 9.1B). Activated G-proteins then couple to many downstream effectors and most alter the activity of other intracellular enzymes or ion channels. Many of the G-protein target enzymes produce diffusible second messengers (metabolites) that stimulate further downstream biochemical processes, including the activation of protein kinases (see Chapter 10). Time is required for each of these coupling events, and the effects of GPCR activation are typically slower in onset than those observed following the activation of ionotropic receptors. Because there is a lifetime associated with each intermediate, the effects produced by GPCR activation are also typically longer in duration than those produced by the activation of ionotropic receptors. Most small neurotransmitters, such as ACh, glutamate, 5-HT, and GABA, can bind to and activate both ionotropic and GPCRs. Thus, each of these transmitters can induce both fast responses (milliseconds), such as typical excitatory or inhibitory postsynaptic potentials, and slow-onset and longer duration responses (from tenths of seconds to, potentially, hours). Other transmitters, like neuropeptides, produce their effects largely by binding only to GPCRs. These effects across multiple time domains provide the nervous system with a rich source for temporal information processing that is subject to constant modification. Currently, the GPCR family can be divided into three subfamilies on the basis of their structures: (1) the rhodopsin-adrenergic receptor subfamily, (2) the secretin-vasoactive intestinal peptide receptor subfamily, and (3) the metabotropic glutamate receptor subfamily.

GPCR Structure Conforms to a General Model

A GPCR consists of a single polypeptide with a generally conserved structure. The receptor contains seven helical segments that wrap back and forth through the membrane (Fig. 9.7). G-protein-coupled receptors are homologous to rhodopsin from both mammalian and bacterial sources, and detailed structural information on rhodopsin has been used to provide a framework for developing a general model for GPCR structures. Aside from rhodopsin, two of the best structurally characterized GPCRs are the β-adrenergic receptor (βAR) and the muscarinic acetylcholine receptor (mAChR), and biochemical analyses to date support the use of rhodopsin as a structural framework for the family of GPCRs.

The most conserved feature of GPCRs is the seven membrane-spanning segments; however, other generalities can be made about their structure. The N terminus of the receptor extends into the extracellular space, whereas the C terminus resides within the cytoplasm (Fig. 9.7A). Each of the seven transmembrane domains between N and C termini consists of approximately 24

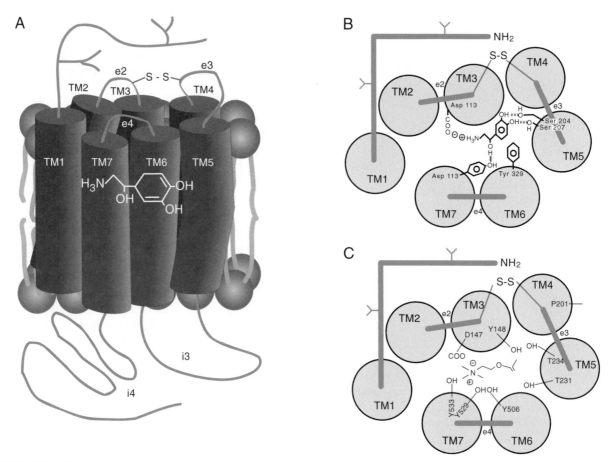

FIGURE 9.7 (A) Diagram showing the approximate position of the catecholamine-binding site in the βAR. The transmitter-binding site is formed by amino acids whose side chains extend into the center of the ring produced by the seven transmembrane domains (TM1–TM7). Note that the binding site exists at a position that places it within the plane of the lipid bilayer. (B) A view looking down on a model of the βAR identifying residues important for ligand binding. The seven transmembrane domains are represented as gray circles labeled TM1 though TM7. Amino acids composing the extracellular domains are represented as green bars labeled e1 through e4. The disulfide bond (–S–S–) that links e2 to e3 is also shown. Each of the specific residues indicated makes stabilizing contact with the transmitter. (C) A view looking down on a model of the mAChR identifying residues important for ligand binding. Stabilizing contacts, mainly through hydroxyl groups (-OH), are made with the transmitter on four of the seven transmembrane domains. The chemical nature of the transmitter (i.e., epinephrine versus Ach) determines the type of amino acids necessary to produce stable interactions in the receptor-binding site (compare B and C). Adapted from Strosberg (1990).

mostly hydrophobic amino acids. These seven domains associate together to form an oblong ring within the plasma membrane (Fig. 9.7B). Between each transmembrane domain is a loop of amino acids of various sizes. The loops connecting TM1 and TM2, TM3 and TM4, and TM5 and TM6 are intracellular and are labeled i1, i2, and i3, respectively, whereas those between TM2 and TM3, TM4 and TM5, and TM6 and TM7 are extracellular and are labeled e1, e2, and e3, respectively (see Fig. 9.7A).

The Neurotransmitter-Binding Site Is Buried in the Core of the Receptor

The neurotransmitter-binding site for many GPCRs (excluding the metabotropic glutamate, GABA_B, and

neuropeptide receptors) resides within a pocket formed in the center of the seven membrane-spanning segments (Fig. 9.7). In the βAR, this pocket resides at least 10.9 Å into the hydrophobic core of the receptor, placing the ligand-binding site within the plasma membrane lipid bilayer (Kobilka, 1992). Strategically positioned charged and polar residues in the membrane-spanning segments point inward into a central pocket that forms the binding site for the ligand. For example, Asn residues in the second and third segments, two Ser residues in the fifth segment, and a Phe residue in the sixth segment provide major contact points in the βAR-binding site for the transmitter (Fig. 9.7B; Kobilka, 1992). Replacing the Asp in TM3 with a Glu reduced transmitter binding by more than 100-fold, and replacement with a less conserved amino acid, such as Ser, reduces binding by more than 10,000-fold. Two Ser residues in TM5 are also essential for efficient transmitter binding and receptor activation, as is an Asp residue in TM2 and a Phe residue in TM6. In total, the two Asp, the two Ser, and the Phe residues are highly conserved in all receptors that bind catecholamines. Variations in the amino acids at these five positions appear to provide the specificity between binding of different transmitters to the individual GPCRs.

The neurotransmitter-binding site of mAChRs, like that of the β2AR, has been investigated in great detail (Fig. 9.7C). The Asp residue in TM3 is also critical for ACh binding to mAChRs. Mutagenesis studies indicate important roles for Tyr and Thr residues in TM3, TM5, TM6, and TM7 in contributing to the ligand-binding site for ACh. Interestingly, many of these mutations do not affect antagonist binding, indicating that distinct sets of amino acids participate in binding agonists and antagonists. When the transmembrane domains are examined from a side view (Fig. 9.7A), all the key amino acids implicated in agonist binding lie at about the same level within the core of the receptor structure, buried approximately 10–15 Å from the surface of the plasma membrane. An additional amino acid identified as essential for agonist binding of the mAChR is a Pro residue in TM4 (P201 in Fig. 9.7C). This residue is also highly conserved among GPCRs, and structural predictions suggest that it affects ligand binding not by interacting with agonist directly but by stabilizing a conformation essential for high-affinity binding. Structural predictions also place this Pro residue in the same plane as the Asp, Tyr, and Thr residues that form the ligand-binding site of the mAChR.

Transmitter Binding Causes a Conformational Change in the Receptor and Activation of G-Proteins

Proposed models for GPCR activation assume that the receptor can isomerize spontaneously between inactive and active states. Only the active state interacts with G-proteins in a productive fashion. This isomerization is analogous to the spontaneous isomerization proposed for ion channels as they oscillate between open and closed states. In the absence of agonist, the inactive state of GPCRs is favored, and little G-protein activation occurs. Agonist binding stabilizes the active conformation and shifts the equilibrium toward the active form, and G-protein activation ensues. This kinetic model indicates that agonist binding is not necessary for the receptor to undergo transition into the active state; instead, it stabilizes the

activated state of the receptor. This proposed model is supported by observations of both spontaneously arising and engineered mutants of βAR and αAR receptors. Specific amino acid replacements produced receptors that exhibit constitutive activity in the absence of agonists (Premont et al., 1995). The amino acid changes apparently stabilize the active conformation of the molecule in a state more similar to the agonist-bound form of the receptor, leading to productive interactions with G proteins in the agonist-free state.

The Third Intracellular Loop Forms a Major Determinant for G-Protein Coupling

Extensive studies using site-directed mutagenesis and the production of chimeric molecules have revealed the domains and amino acids essential for G-protein coupling to GPCRs. Receptor domains within the second (i2) and third (i3) intracellular loops (Fig. 9.7) appear largely responsible for determining the specificity and efficiency of coupling for adrenergic and muscarinic cholinergic receptors and are the likely sites for G-protein coupling of the entire GPCR family. In particular, the 12 amino acids of the N-terminal region of the third intracellular loop significantly affect the specificity of G-protein coupling. Other regions in the C terminus of the third intracellular loop and the N-terminal region of the C-terminal tail appear to be more important for determining the efficiency of G-protein coupling than for determining its specificity (Kobilka, 1992). The third intracellular loop varies enormously in size among the different G-protein-coupled receptors, ranging from 29 amino acids in the substance P receptor to 242 amino acids in the mAChR (Strader et al., 1994). The intracellular loop connecting TM5 and TM6 is the main point of receptor coupling to G-proteins, and ligand binding to amino acids in TM5 and TM6 may be responsible for triggering the G-protein–receptor interaction by transmitting a conformational change to the third intracellular loop.

Specific Amino Acids Are Involved in Transducing Transmitter Binding into G-Protein Coupling

Residues associated with transmitting the conformational change induced by ligand binding to the activation of G-proteins have been investigated with the use of mAChRs. These studies revealed that an Asp residue in TM2 is important for the receptor activation of G-proteins, and altering the Asp by site-directed mutagenesis has a major negative effect on G-protein–receptor activation. A Thr residue in TM5 (T231 in

Fig. 9.7C) and a Tyr residue in TM6 (Y506 in Fig. 9.7C) are also essential. Because these residues are connected by i3, they are assumed to play fundamental roles in transmitting the conformational change induced by ligand binding to the area of i3 essential for G-protein coupling and activation. When mutated, a Pro residue on TM7 produces a major impairment in the ability of the TM3 segment to induce the activation of phospholipase C through a G-protein and presumably is another key element in propagating the conformational changes necessary for efficient coupling to G-proteins.

As mentioned earlier, GPCRs are single polypeptides; however, they are clearly separable into distinct functional domains. For example, β2AR can be physically split, with the use of molecular techniques, into two fragments: one fragment containing TM1–TM5 and the other containing TM6 and TM7. In isolation, neither of these fragments can produce a functional receptor; however, when coexpressed in the same cell, functional β2ARs that can bind ligand and activate G-proteins are produced. This remarkable experiment indicates that physical contiguity in the primary sequence is not essential for producing functional β2ARs and also emphasizes the contribution of domains in the separate fragments (TM1–TM5 and TM6 and TM7) to both ligand binding and G-protein coupling.

GPCRs Also Exist as Homo- or Heterooligomers

The observation that GPCRs can be physically split through genetic engineering and produced functional channels when recombined provided the first hint that full-length GPCRs might also oligomerize with each other into functional molecules. A test of this hypothesis was accomplished by making chimeric receptors composed of the transmembrane domains 1–5 of the α2-AR and the transmembrane domains 6 and 7 of the m3 muscarinic receptors and vice versa. When either of these chimeric molecules was expressed in isolation, neither formed a functional receptor. However, when coexpressed, receptors were formed that bind both muscarinic and adrenergic ligands and ligand binding led to functional activation of downstream effectors. Through domain swapping, the ligand-binding sites for both receptor ligands were reconstituted by oligomerization of the two chimeric receptors into one bifunctional chimeric dimer. Although some debate remains, evidence now seems overwhelming that oligomerization of GPCRs is adding a new layer of complexity and diversity to the study of these receptors. Important functional consequences could relate to alterations in (1) ligand binding, (2) efficiency and specificity of coupling to downstream effectors, (3) subcellular localization, and (4) receptor desensitization. The evolving and apparently widespread nature of direct receptor–receptor interactions leads one to believe that our current understanding of neurotransmitter receptors and their biological impact will be undergoing continual modifications for many years to come.

G-Protein Coupling Increases Affinity of the Receptor for Neurotransmitter

The affinity of a GPCR for agonist increases when the receptor is coupled to the G-protein. This positive feedback effectively increases the lifetime of the agonist-bound form of the receptor by decreasing the dissociation rate of the agonist. An excellent demonstration of this effect comes from studies using engineered βAR receptors that are constitutively active in their ability to couple to G-proteins. These mutant receptors show a significantly increased affinity for agonists. When the G-protein dissociates, the agonist-binding affinity of the receptor returns to its original state. Changes induced by ligand binding apparently stabilize the receptor in a conformation with both higher affinity for ligand and higher affinity for coupling to G-proteins.

Specificity and Potency of G-Protein Activation Are Determined by Several Factors

GPCRs associate with G-proteins to transduce ligand binding into intracellular effects. This coupling step can lead to diverse responses, depending on the type of G-protein and the type of effector enzyme present. Ligand binding to a single subtype of GPCR can activate multiple G-protein-coupled pathways. For example, activated α2ARs have been shown to couple to as many as four different G-proteins in the same cell (Strader et al., 1994). Some of the specificity for G-protein activation can be determined by the specific conformations assumed by the receptor, and a single receptor can assume multiple conformations. For example, α2ARs can isomerize into at least two states. One state interacts with a G-protein that couples to phospholipase C, and a second state interacts with G-proteins that couple to both phospholipase C and phospholipase A_2. Thus, a single GPCR can produce a diversity of responses, making it difficult to assign specific biological effects to individual receptor subtypes in all settings.

Activated GPCRs are free to couple to many G-protein molecules, permitting a significant amplification of the initial transmitter-binding event. This catalytic mechanism is referred to as "collision coupling," whereby a transient association between the activated receptor and the G-protein is sufficient to produce the exchange of GDP for GTP, activating the G-protein. For example, because adenylate cyclase appears to be tightly coupled to the G-protein, the rate-limiting step in the production of cAMP is the number of successful collisions between the receptor and the G-protein. A constant GTPase activity hydrolyzes GTP, bringing the G-protein and therefore the adenylate cyclase back to the basal state. The apparent maximal rate is achieved when all the G-protein–cyclase complexes have become activated or, more accurately, when the rate of formation is maximal with respect to the rate of GTP hydrolysis. Additionally, the important process of receptor desensitization can also regulate the number of receptors capable of productive G-protein interactions.

Receptor Desensitization Is a Built-in Mechanism for Decreasing the Cellular Response to Transmitter

Desensitization is an important process whereby cells can decrease their sensitivity to a particular stimulus to prevent saturation of the system. Desensitization involves a complex series of events (Kobilka, 1992). For GPCRs, desensitization is defined as an increase in the concentration of agonist required to produce half-maximal stimulation of, for example, adenylate cyclase. In practical terms, desensitization of receptors produces less response for a constant amount of transmitter.

There are two known mechanisms for desensitization. One mechanism is a decrease in response brought about by the covalent modifications produced by receptor phosphorylation and is quite rapid (seconds to minutes). The other mechanism is the physical removal of receptors from the plasma membrane (likely through a mechanism of receptor-mediated endocytosis) and tends to require greater periods of time (minutes to hours). The latter process can be either reversible, sequestration, or irreversible, down-regulation (Fig. 9.8).

The Rapid Phase of GPCR Desensitization Is Mediated by Receptor Phosphorylation

Desensitization of the βAR appears to involve at least three protein kinases: PKA, PKC, and β-adrenergic receptor kinase (βARK; also referred to as a G-protein receptor kinase (GRK)). Phosphorylation of ARs by PKA does not require that the agonist be bound to the receptor and appears to be a general mechanism

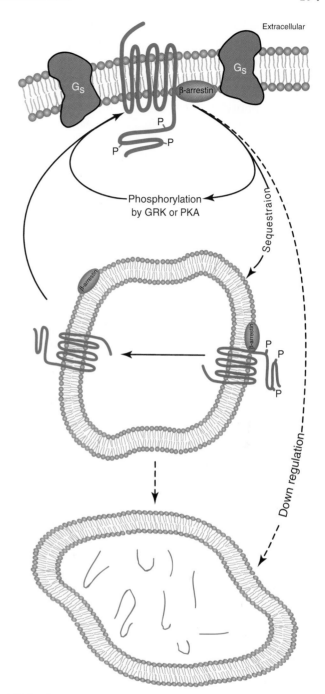

FIGURE 9.8 Intracellular pathways associated with desensitization of GPCRs. GPCRs are phosphorylated (noted with P) on their intracellular domains by PKA, GRK, and other protein kinases. The phosphorylated form of the receptor can be removed from the cell surface by a process called sequestration with the help of the adapter protein β-arrestin; thus fewer binding sites remain on the cell surface for transmitter interactions. In intracellular compartments, the receptor can be dephosphorylated and returned to the plasma membrane in its basal state. Alternatively, phosphorylated receptors can be degraded (downregulated) by targeting to a lysosomal organelle. Degradation requires replenishment of the receptor pool through new protein synthesis. Adapted from Kobilka (1992).

by which the cell can reduce the effectiveness of all receptors, independent of whether they are in the agonist-bound or unbound state. This process is also referred to as heterologous desensitization because the receptor does not require bound-agonist. PKA and PKC phosphorylate sites on the third intracellular loop and possibly the C-terminal cytoplasmic domain. Phosphorylation of these sites interferes with the ability of the receptor to couple to G-proteins, thus producing the desensitization.

GRKs can also phosphorylate GPCRs and lead to receptor desensitization. Six members of the GRK family of kinases have been identified: rhodopsin kinase (GRK1), βARK (GRK2), and GRK3 through GRK6 (Premont et al., 1995). GRK2 (originally called β-AR receptor kinase or βARK) is a Ser- and Thr-specific protein kinase initially identified by its capacity to phosphorylate βAR. GRK2 phosphorylates only the agonist-bound form of the receptor, usually when agonist concentrations reach the micromolar level, as typically found in the synaptic cleft. This process is referred to as homologous desensitization because the regulation is specific for those receptor molecules that are in the agonist-bound state. Phosphorylation of βAR by GRK2 does not interfere substantially with coupling to G-proteins. Instead, an additional protein, arrestin, binds the GRK2-phosphorylated form of the receptor, thus blocking receptor–G-protein coupling (Fig. 9.8). This process is analogous to the desensitization of the light-sensitive receptor molecule rhodopsin produced by GRK1 phosphorylation and the binding of arrestin. Phosphorylation sites on βAR for GRK2 reside on the C-terminal cytoplasmic domain and are distinct from those phosphorylated by PKA.

The cycle of homologous desensitization starts with the activation of a GPCR, which induces activation of G-proteins and dissociation of the βγ subunit complex from α subunits. At least one role for the βγ complex appears to be to bind to GRKs, which leads to their recruitment to the membrane in the area of the locally activated G-protein–receptor complex. The recruited GRK is then activated, leading to phosphorylation of the agonist-bound receptor and subsequent binding of arrestin. Arrestin binds to the same domains on the receptor necessary for coupling to G-proteins, thus terminating the actions of the activated receptor (Fig. 9.8). The ensuing process of sequestration follows GPCR phosphorylation and arrestin binding.

Desensitization Can Also Be Produced by Loss of Receptors from the Cell Surface

Desensitization of GPCRs is also produced by removal of the receptor from the cell surface. This process can be either reversible (sequestration, or internalization) or irreversible (downregulation). Sequestration is the term used to describe the rapid (within minutes) but reversible endocytosis of receptors from the cell surface after agonist application (Fig. 9.8). Neither G-protein coupling nor receptor phosphorylation appears to be essential for this process, but phosphorylation by GRKs clearly enhances the rate of sequestration. The binding of arrestins to the phosphorylated receptor also enhances sequestration (Ferguson et al., 1996). Thus, arrestin binding appears to promote not only rapid desensitization by disrupting the receptor–G-protein interaction, but also receptor sequestration. Receptor cycling through intracellular organelles is a trafficking mechanism that leads to an enhanced rate of dephosphorylation of the phosphorylated receptor, returning it to the cell surface in its basal state (Fig. 9.8).

Downregulation occurs more slowly than sequestration and is irreversible (Fig. 9.8). The early phase (within 4h) may involve both a PKA-dependent and a PKA-independent process. This early phase of downregulation is apparently due to receptor degradation after endocytotic removal from the plasma membrane. The later phases (>14h) of downregulation appear to be further mediated by a reduction in receptor biosynthesis through a decrease in the stability of the receptor mRNA and a decreased transcription rate.

Other Posttranslational Modifications Are Required for Efficient Metabotropic Receptor Function

Like many proteins expressed on the cell surface, GPCRs are glycosylated, and the N-terminal extracellular domain is the site of carbohydrate attachment. Relatively little is known about the effect of glycosylation on the function of GPCRs. Glycosylation does not appear to be essential to the production of a functional ligand-binding pocket (Strader et al., 1994), although prevention of glycosylation may decrease membrane insertion and alter intracellular trafficking of the β2AR.

Another important structural feature of most GPCRs is the disulfide bond formed between two Cys residues present on the extracellular loops (e2 and e3; Fig. 9.7). Apparently, the disulfide bond stabilizes a restricted conformation of the mature receptor by covalently linking the two extracellular domains, and this conformation favors ligand binding. Disruption of this disulfide bond significantly decreases agonist binding (Kobilka, 1992). A third Cys residue, in the C-terminal domain of GPCRs (in i4 Fig 9.7A), appears to serve as a point for covalent attachment of a fatty acid (often palmitate). Presumably, fatty acid attachment stabilizes an interaction between the C-terminal domain of a GPCR and the membrane.

GPCRs Can Physically Associate with Ionotropic Receptors

There is now good evidence that metabotropic and ionotropic receptors can interact directly with each other (Liu *et al.*, 2000). GABA$_A$ receptors (ionotropic) were shown to couple to DA (D5) receptors (metabotropic) through the second intracellular loop of the γ subunit of the GABA$_A$ receptor and the C-terminal domain of the D5, but not the D1, receptor. DA binding to D5 receptors produced downregulation of GABA$_A$ currents, and pharmacologically blocking the GABA$_A$ receptor produced decreases in cAMP production when cells were stimulated with DA. It further appeared that ligand binding to both receptors was necessary for their stable interaction. Whether this form of receptor regulation is unique to this pair of partners or is a widespread phenomenon remains an open question ripe for further investigation.

GPCRs All Exhibit Similar Structures

The family of GPCRs exhibits structural similarities that permit the construction of "trees" describing the degree to which they are related evolutionarily (Fig. 9.9). Some remarkable relations become evident in such an analysis. For example, the D1 and D5 subtypes of DA receptors are related more closely to the α2AR than to the D2, D3, and D4 DA receptors. The similarities and differences among GPCR families are highlighted in the remainder of this chapter.

Muscarinic ACh Receptors

Muscarine is a naturally occurring plant alkaloid that binds to muscarinic subtypes of the AChRs and activates them. mAChRs play a dominant role in mediating the actions of ACh in the brain, indirectly producing both excitation and inhibition through binding to a family of unique receptor subtypes. mAChRs are found both presynaptically and postsynaptically and, ultimately, their main neuronal effects appear to be mediated through alterations in the properties of ion channels. Presynaptic mAChRs take part in important feedback loops that regulate ACh release. ACh released from the presynaptic terminal can bind to mAChRs on the same nerve ending, thus activating enzymatic processes that modulate subsequent neurotransmitter release. This modulation is typically an inhibition; however, activation of m5 AChR produces an enhance-

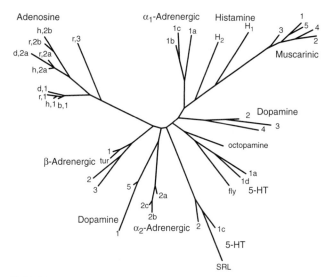

FIGURE 9.9 Evolutionary relationship of the GPCR family. To assemble this tree, sequence homologies in the transmembrane domains were compared for each receptor. Distance determines the degree of relatedness. r, rat; d, dog; h, human; tur, turkey; SRL, a putative serotonin receptor; and 5-HT, 5-hydroxytryptamine (serotonin). Adapted from Linden (1994). Original tree construction was by William Pearson and Kevin Lynch, University of Virginia.

ment in subsequent release. These *autoreceptors* are an important regulatory mechanism for the short-term (milliseconds to seconds) modulation of neurotransmitter release (see Chapter 7).

The family of mAChRs now includes five members (m1–m5), ranging from 55 to 70 kDa, and each of the five subtypes exhibits the typical architecture of seven transmembrane domains. Much of the diversity in this family of receptors resides in the third intracellular loop (i3) responsible for the specificity of coupling to G-proteins. The m1, m3, and m5 mAChRs couple predominantly to G-proteins that activate the enzyme phospholipase C. m2 and m4 receptors couple to G-proteins that inhibit adenylate cyclase, as well as to G-proteins that regulate K$^+$ and Ca^{2+} channels directly. As is the case for other GPCRs, the domain near the N terminus of i3 is important for the specificity of G-protein coupling. This domain is conserved in m1, m3, and m5 AChRs, but is unique in m2 and m4. Several other important residues also have been identified for G-protein coupling. A particular Asp residue near the N-terminus of the second intracellular loop (i2) is important for G-protein coupling, as are residues residing in the C-terminal region of the i3 loop.

Major mAChRs found in the brain are m1, m3, and m4, and each is distributed diffusely. The m2 subtype is the heart isoform and is not highly expressed in other organs. Genes for m4 and m5 lack introns, whereas those encoding m1, m2, and m3 contain introns, although little is known concerning alternatively spliced products of these receptors. Atropine is the most widely utilized antagonist for mAChR and binds to most subtypes, as does N-methylscopolamine. The antagonist pirenzipine appears to be relatively specific for the m1 mAChR, and other antagonists, such as AF-DX116 and hexahydrosiladifenidol, appear to be more selective for m2 and m3 subtypes.

Adrenergic Receptors

The catecholamines epinephrine (adrenaline) and norepinephrine (noradrenaline) produce their effects by binding to and activating adrenergic receptors. Interestingly, epinephrine and norepinephrine can both bind to the same adrenergic receptor. Adrenergic receptors are currently separated into three families: $\alpha 1$, $\alpha 2$, and β (Fig. 9.9). Each of the $\alpha 1$ and $\alpha 2$ families is further subdivided into three subclasses (Fig. 9.9). Similarly, the β family also contains three subclasses ($\beta 1$, $\beta 2$, and $\beta 3$; Fig. 9.9). The main adrenergic receptors in the brain are the $\alpha 1$ and $\beta 1$ subtypes. $\alpha 2$ARs have diverse roles, but the function that is best characterized (in both central and peripheral nervous tissue) is their role as autoreceptors. Different AR subtypes bind to G-proteins that can alter the activity of phospholipase C, Ca^{2+} channels, and, probably the best studied, adenylate cyclase. For example, activation of $\alpha 2$ARs produces inhibition of adenylate cyclase, whereas all βARs activate the cyclase.

Only a few agonists or antagonists cleanly distinguish the AR subtypes. One of them, isoproterenol, is an agonist that appears to be highly specific for βARs. Propranolol is the best-known antagonist for β receptors, and phentolamine is a good antagonist for α receptors but binds weakly at β receptors. The genomic organization of the different AR subtypes is unusual. Like many G-protein-coupled GPCRs, $\beta 1$ and $\beta 2$ARs are encoded by genes lacking introns. $\beta 3$ARs, which apparently have a role in lipolysis and are poorly characterized, are encoded by an intron-containing gene, as are αARs, providing an opportunity for alternative splicing as a means of introducing functional heterogeneity into the receptor.

Dopamine Receptors

Some 80% of the DA in the brain is localized to the corpus striatum, which receives major input from the substantia nigra and takes part in coordinating motor movements. DA is also found diffusely throughout the cortex, where its specific functions remain largely undefined. However, many neuroleptic drugs appear to exert their effects by blocking DA binding, and imbalances in the dopaminergic system have long been associated with neuropsychiatric disorders.

DA receptors are found both pre- and postsynaptically, and their structure is homologous to that of the receptors for other catecholamines (Civelli et al., 1993). Five subtypes of DA receptors can be grouped into two main classes: D1-like and D2-like receptors. D1-like receptors include D1 and D5, whereas D2-like receptors include D2, D3, and D4 (see Fig. 9.9). The main distinction between these two classes is that D1-like receptors activate adenylate cyclase through interactions with G_s, whereas D2-like receptors inhibit adenylate cyclase and other effector molecules by interacting with G_i/G_o. D1-like receptors are also slightly larger in molecular mass than D2-like receptors. An additional point of interest, as noted earlier, is that D5 receptors selectively associate with $GABA_A$ receptors, impacting their function and vice versa (Liu et al., 2000). The deduced amino acid sequence for the entire family ranges from 387 amino acids (D4) to 477 amino acids (D5). Main structural differences between D1-like and D2-like receptors are that the intracellular loop between the sixth and the seventh transmembrane segments is larger in D2-like receptors, and D2-like receptors have smaller C-terminal intracellular segments.

D1-like receptors, like βARs, are transcribed from intronless genes. Conversely, all D2-like receptors contain introns, thus providing for possibilities of alternatively spliced products. Posttranslational modifications include glycosylation at one or more sites, disulfide bonding of the two Cys residues in e2 and e3, and acylation of the Cys residue in the C-terminal tail (analogous to the $\beta 2$AR). The DA-binding site includes two Ser residues in TM5 and an Asp residue in TM3, analogous to the βAR.

Because of the presumed role of DA in neuropsychiatric disorders, enormous effort has been put into developing pharmacological tools for manipulating this system. DA receptors bind bromocriptine, lisuride, clozapine, melperone, fluperlapine, and haloperidol. Because these drugs do not show great specificity for receptor subtypes, their usefulness for dissecting effects specifically related to binding to one or another DA receptor subtype is limited. However, their role in the treatment of human neuropsychiatric disorders is enormous (see Chapter 43).

Purinergic Receptors

Purinergic receptors bind to ATP or other nucleotide analogs and to its breakdown product adenosine. Although ATP is a common constituent found within synaptic vesicles, adenosine is not and is therefore not considered a "classic" neurotransmitter. However, the multitude of receptors that bind and are activated by adenosine indicates that this molecule has important modulatory effects on the nervous system. Situations of high metabolic activity that consume ATP and situations of insufficient ATP-regenerating capacity can lead to the accumulation of adenosine. Because adenosine is permeable to membranes and can diffuse into and out of cells, a feedback loop is established in which adenosine can serve as a local diffusible signal that communicates the metabolic status of the neuron to surrounding cells and vice versa (Linden, 1994).

The original nomenclature describing purinergic receptors defined adenosine as binding to P1 receptors and ATP as binding to P2 receptors. Families of both P1 and P2 receptors have since been described, and adenosine receptors are now identified as A-type purinergic receptors, consisting of A1, A2a, A2b, and A3. ATP receptors are designated as P type and consist of P2x, P2y, P2z, P2t, and P2u. Recall that P2x and P2z subtypes are ionotropic receptors (see earlier discussion).

A-type receptors exhibit the classic arrangement of seven transmembrane-spanning segments but are typically shorter than most GPCRs, ranging in size between 35 and 46 kDa. The ligand-binding site of A-type receptors is unique in that the ligand, adenosine, has no inherent charged moieties at physiological pH. A-type receptors appear to utilize His residues as their points of contact with adenosine, and, in particular, a His residue in TM7 is essential because its mutation eliminates agonist binding. Other His residues in TM6 and TM7 are conserved in all A-type receptors and may serve as other points of contact with agonists. A1 receptors are highly expressed in the brain, and their activation downregulates adenylate cyclase and increases phospholipase C activity. The A2a and A2b receptors are not as highly expressed in nervous tissue and are associated with the stimulation of adenylate cyclase and phospholipase C, respectively. The A3 subtype exhibits a unique pharmacological profile in that binding of xanthine derivatives, which blocks the action of adenosine competitively, is absent. Very low levels of the A3 receptor are found in brain and peripheral nervous tissue. The A3 receptor appears to be coupled to the activation of phospholipase C.

The P-type receptors, P2y, P2t, and P2u, are typical G-protein-linked GPCRs, mostly localized to the periphery. However, direct effects of ATP have been detected in neurons, and often the response is biphasic; an early excitatory effect followed, with its break-down to adenosine, by a secondary inhibitory effect. Interestingly, P-type receptors exhibit a higher degree of homology to peptide-binding receptors than they do to A-type purinergic receptors. As in A-type receptors, P-type receptors have a His residue in the third transmembrane domain; however, other sites for ligand binding have not been specifically identified.

Serotonin Receptors

Cell bodies containing serotonin (5-HT) are found in the raphe nucleus in the brain stem and in nerve endings distributed diffusely throughout the brain. 5-HT has been implicated in sleep, modulation of circadian rhythms, eating, and arousal. 5-HT also has hormone-like effects when released in the bloodstream, regulating smooth muscle contraction and affecting platelet-aggregating and immune systems.

5-HT receptors are classified into four subtypes, 5-HT1 to 5-HT4, with a further subdivision of 5-HT1 subtypes. Recall that the 5-HT3 receptor is ionotropic (see earlier discussion). The other 5-HT receptors exhibit the typical seven transmembrane-spanning segments and all couple to G-proteins to exert their effects. For example, 5-HT1a, 1b, 1d, and 4 either activate or inhibit adenylate cyclase. 5-HT1c and 5-HT2 receptors preferentially stimulate activation of phospholipase C to produce increased intracellular levels of diacylglycerol and inositol 1,4,5-trisphosphate.

5-HT receptors can also be grossly distributed into two groups on the basis of their gene structures. Both 5-HT1c and 5-HT2 are derived from genes that contain multiple introns. In contrast, similar to the βAR family, 5-HT1 is coded by a gene lacking introns. Interestingly, 5-HT1a is more closely related ancestrally to the βAR family than it is to other membranes of the 5-HT receptor family and originally was isolated by utilizing cDNA for the β2AR as a molecular probe. This observation helps explain some pharmacological data suggesting that both 5-HT1a and 5-HT1b can bind certain adrenergic antagonists.

Glutamate GPCRs

GPCRs that bind glutamate (metabotropic glutamate receptors (mGluRs)) are similar in general structure in having seven transmembrane-spanning segments to other GPCRs; however, they are divergent enough to be considered to have originated from a separate evolutionary-derived receptor family (Hollmann and Heinemann, 1994; Nakanishi, 1994). In fact, sequence homology between the mGluR family and other GPCRs is minimal except for the GABA$_B$ receptor. The mGluR family is heterogeneous in size, ranging from 854 to 1179 amino acids. Both the N-terminal and the C-terminal domains are unusually large for G-protein-coupled receptors. One great difference in the structures of mGluRs is that the binding site for glutamate resides in the large N-terminal extracellular domain and is homologous to a bacterial amino acid-binding protein (Armstrong and Gouaux, 2000). In most of the other families of GPCRs, the ligand-binding pocket is formed by transmembrane segments partly buried in the membrane. Additionally, mGluRs exist as functional dimers in the membrane in contrast to the single subunit forms of most GPCRs (Kunishima *et al.*, 2000). These significant structural distinctions support the idea that mGluRs evolved separately from other GPCRs. The third intracellular loop, thought to be the major determinant responsible for G-protein coupling, of mGluRs is relatively small, whereas the C-terminal domain is quite large. The coupling between mGluRs and their respective G-proteins may be through unique determinants that exist in the large C-terminal domain.

Currently, eight different mGluRs can be subdivided into three groups on the basis of sequence homologies and their capacity to couple to specific enzyme systems. Both mGluR1 and mGluR5 activate a G-protein coupled to phospholipase C. mGluR1 activation can also lead to the production of cAMP and of arachidonic acid by coupling to G-proteins that activate adenylate cyclase and phospholipase A_2. mGluR5 seems more specific, activating predominantly the G-protein-activated phospholipase C.

The other six mGluR subtypes are distinct from one another in favoring either *trans*-1-aminocyclopentane-1,3-dicarboxylate (mGluR2, 3, and 8) or 1-2-amino-4-phosphonobutyrate (mGluR4, 6, and 7) as agonists for activation. mGluR2 and mGluR4 can be further distinguished pharmacologically by using the agonist 2-(carboxycyclopropyl)glycine, which is more potent at activating mGluR2 receptors. Less is known about the mechanisms by which these receptors produce intracellular responses; however, one effect is to inhibit the production of cAMP by activating an inhibitory G-protein.

mGluRs are widespread in the nervous system and are found both pre- and postsynaptically. Presynaptically, they serve as autoreceptors and appear to participate in the inhibition of neurotransmitter release. Their postsynaptic roles appear to be quite varied and depend on the specific G-protein to which they are coupled. mGluR1 activation has been implicated in long-term synaptic plasticity at many sites in the brain, including long-term potentiation in the hippocampus and long-term depression in the cerebellum (see Chapter 49).

GABA$_B$ Receptor

GABA$_B$ receptors are found throughout the nervous system, where they are sometimes colocalized with ionotropic GABA$_A$ receptors. GABA$_B$ receptors are present both pre- and postsynaptically. Presynaptically, they appear to mediate inhibition of neurotransmitter release through an autoreceptor-like mechanism by activating K^+ conductances and diminishing Ca^{2+} conductances. In addition, GABA$_B$ receptors may affect K^+ channels through a direct physical coupling to the K^+ channel, not mediated through a G-protein intermediate. Postsynaptically, GABA$_B$ receptor activation produces a characteristic slow hyperpolarization (termed the slow inhibitory postsynaptic potential) through the activation of a K^+ conductance. This effect appears to be through a pertussis toxin-sensitive G-protein that inhibits adenylate cyclase.

Cloning of the GABA$_B$ receptor (GABA$_B$R1) revealed that it has high sequence homology to the family of glutamate GPCRs, but shows little similarity to other G-protein-coupled receptors. The large N-terminal extracellular domain of the GABA$_B$ receptor is the presumed site of GABA binding. With the exception of this large extracellular domain, the GABA$_B$ receptor structure is typical of the GPCR family, exhibiting seven transmembrane domains. The initial cloning of the GABA$_B$ receptor was made possible by the development of the high-affinity, high-specificity antagonist CGP64213. This antagonist is several orders of magnitude more potent at inhibiting GABA$_B$ receptor function than the more widely known antagonist saclofen. Baclofen, an analog of saclofen, remains the best agonist for activating GABA$_B$ receptors.

Functional GABA$_B$ receptors appear to exist primarily as dimers in the membrane. Expression of the cloned GABA$_B$R1 isoform does not produce significant functional receptors. However, when coexpressed with the GABA$_B$R2 isoform, receptors that are indistinguishable functionally and pharmacologically from those in brain were produced. In addition, GABA$_B$ dimers exist in neuronal membranes, and all data point to the conclusion that GABA$_B$ receptors dimerize and that the dimer is the functionally important form of the receptor. As noted earlier, GPCRs can interact with themselves and other receptors. It is well to keep in mind that these types of direct receptor interactions may be more widespread than currently appreciated.

Peptide Receptors

Neuropeptide receptors form an immense family. Because of their diversity, they cannot be covered in detail in this chapter. Despite this diversity, however, none of the receptors that bind peptides appears to be coupled directly to the opening of ion channels. Neuropeptide receptors exert their effects either through the typical pathway of activation of G-proteins or through a more recently described pathway related to activation of an associated tyrosine kinase activity.

Summary

GPCRs are single polypeptides composed of seven transmembrane-spanning segments. In general, the binding site for neurotransmitter is located within the core of the circular structure formed by these segments. Transmitter binding produces conformational changes in the receptor that expose parts of the i3 region, among others, for binding to G-proteins. G-protein binding increases the affinity of the receptor for transmitter. Desensitization is common among GPCRs and leads to a decreased response of the receptor to neurotransmitter by several distinct mechanisms. mGluRs are structurally distinct from other GPCRs; mGluRs have large N-terminal extracellular domains that form the binding site for glutamate. Otherwise, the basic structure of mGluRs appears to be similar to that of the rest of the GPCR family.

References

Armstrong, N. and Gouaux, E. (2000). Mechanisms for activation and antagonism of an AMPA-sensitive glutamate receptor: Crystal structures of the GluR2 ligand binding core. *Neuron* **28**, 165–181.

Boulter, J., Hollmann, M., O'Shea-Greenfield, A., Hartley, M., Deneris, E., Maron, C., and Heinemann, S. (1990). Molecular cloning and functional expression of glutamate receptor subunit genes. *Science* **249**, 1033–1037.

Civelli, O., Bunzow, J. R., and Grandy, D. K. (1993). Molecular diversity of the dopamine receptors. *Annu. Rev. Pharmacol. Toxicol.* **33**, 281–307.

Ferguson, S. S. G., Downey, W. E., Colapietro, A.-M., Barak, L. S., Menard, L., and Caron, M. G. (1996). Role of arrestin in mediating agonist-promoted G-protein coupled receptor internalization. *Science* **271**, 363–366.

Hollmann, M. and Heinemann, S. (1994). Cloned glutamate receptors. *Annu. Rev. Neurosci.* **17**, 31–108.

Hollmann, M., O'Shea-Greenfield, A., Rogers, S. W., and Heinemann, S. (1989). Cloning by functional expression of a member of the glutamate receptor family. *Nature (Lond.)* **342**, 643–648.

Kandel, E. R., Schwartz, J. H., and Jessell, T. M. (1991). "Principles of Neurol Science," 3rd Ed. Elsevier, New York.

Keinanen, K., Wisden, W., Sommer, B., Werner, P., Herb, A., Verdoorn, T. A., Sakmann, B., and Seeburg, P. H. (1990). A family of AMPA-selective glutamate receptors. *Science* **249**, 556–560.

Kobilka, B. (1992). Adrenergic receptors as models for G-protein-coupled receptors. *Annu. Rev. Neurosci.* **15**, 87–114.

Kunishima, N., Shimada, Y., Tsuji, Y. *et al.* (2000). Structural basis of glutamate recognition by a dimeric metrabotropic glutamate receptor. *Nature* **407**, 971–977.

Linden, J. (1994). *In* "Basic Neurochemistry" (G. J. Siegel, B. W. Agranoff, R. W. Albers, and P. B. Molinoff, eds.), pp. 401–416. Raven Press, New York.

Liu, F., Wan, Q., Pristupa, Z. B., Yu, X.-M., Want, Y. T., and Niznik, H. B. (2000). Direct protein-protein coupling enables cross-talk between dopamine D5 and γ-aminobutyric acid A receptors. *Nature* **403**, 274–278.

Moriyoshi, K., Masu, M., Ishii, T., Shigemoto, R., Mizuno, N., and Nakanishi, S. (1991). Molecular cloning and characterization of the rat NMDA receptor. *Nature (Lond.)* **354**, 31–37.

Nakagawa, T., Cheng, Y., Ramm, E., Sheng, M., and Walz, T. (2005). Structure and different conformational sates of native AMPA receptor complexes. *Nature* **433**, 545–549.

Nakanishi, S. (1994). Metabotropic glutamate receptors: Synaptic transmission, modulation, and plasticity. *Neuron* **13**, 1031–1037.

Nakanishi, N., Shneider, N. A., and Axel, R. (1990). A family of glutamate receptor genes: Evidence for the formation of heteromultimeric receptors with distinct channel properties. *Neuron* **5**, 569–581.

Premont, R. T., Inglese, J., and Lefkowitz, R. J. (1995). Protein kinases that phosphorylate activated G-protein-coupled receptors. *FASEB J.* **9**, 175–182.

Rosenmund, C., Stern-Bach, Y., and Stevens, C. F. (1998). The tetrameric structure of a glutamate receptor channel. *Science* **280**, 1596–1599.

Sommer, B., Kohler, M., Sprengel, R., and Seeburg, P. H. (1991). RNA editing in brain controls a determinant of ion flow in glutamate-gated channels. *Cell (Cambridge, Mass.)* **67**, 11–19.

Strader, C. D., Fong, T. M., Tota, M. R., Underwood, D., and Dixon, R. A. (1994). Structure and function of G-protein-coupled receptors. *Annu. Rev. Biochem.* **63**, 101–132.

Strosberg, A. D. (1990). Biotechnology of β-adrenergic receptors. *Mol. Neurobiol.* **4**, 211–250.

Unwin, N. (1995). Acetylcholine receptor channel imaged in the open state. *Nature (Lond.)* **373**, 37–43.

Watkins, J. C., Krogsgaard-Larsen, P., and Honore, T. (1990). Structure activity relationships in the development of excitatory amino acid receptor agonists and competitive antagonists. *Trends Pharmacol. Sci.* **11**, 25–33.

Wo, Z. G. and Oswald, R. E. (1995). Unraveling the modulor design of glutamate-gated ion channels. *Trends Neurosci.* **18**, 161–168.

M. Neal Waxham

10

Intracellular Signaling

Almost all aspects of neuronal function, from its maturation during development, to its growth and survival, cytoskeletal organization, gene expression, neurotransmission, and use-dependent modulation, are dependent on intracellular signaling initiated at the cell surface. The response of neurons and glia to neurotransmitters, growth factors, and other signaling molecules is determined by their complement of expressed receptors and pathways that transduce and transmit these signals to intracellular compartments and the enzymes, ion channels, and cytoskeletal proteins that ultimately mediate the effects of the neurotransmitters. Cellular responses are determined further by the concentration and localization of signal transduction components and are modified by the prior history of neuronal activity. Several primary classes of signaling systems, operating at different time courses, provide great flexibility for intercellular communication.

One class comprises ligand gated ion channels, such as the nicotinic receptor considered in Chapter 9. This class of signaling system provides fast transmission that is activated and deactivated within 10 ms. It forms the underlying "hard wiring" of the nervous system that makes rapid multisynaptic computations possible. A second class consists of receptor tyrosine kinases, which typically respond to growth factors and to trophic factors and produce major changes in the growth, differentiation, or survival of neurons (Chapter 19). A third and largest class utilizes G-protein-linked signals in a multistep process that slows the response from 100 to 300 ms to many minutes. The relatively slow speed is offset, however, by a richness in the diversity of its modulation and capacity for amplification. The initial steps in this signaling system typically generate a second messenger inside the cell, and this

second messenger then activates a number of proteins, including protein kinases that modify cellular processes. Signal transduction also modulates the level of transcription of genes, which determine the differentiated and functional state of cells.

SIGNALING THROUGH G-PROTEIN-LINKED RECEPTORS

Signal transduction through G-protein-linked receptors requires three membrane-bound components:

1. A cell surface receptor that determines to which signal the cell can respond
2. A G protein on the intracellular side of the membrane that is stimulated by the activated receptor
3. Either an effector enzyme that changes the level of a second messenger or an effector channel that changes ionic fluxes in the cell in response to the activated G protein

The human genome encodes for more than 800 receptors for catecholamines, odorants, neuropeptides, and light that couple to one or more of the 16 identified G proteins. These, in turn, regulate one or more of more than two dozen different effector channels and enzymes. The key feature of this information flow is the ability of G proteins to detect the presence of activated receptors and to amplify the signal by altering the activity of appropriate effector enzymes and channels. A nervous system with information flow by fast transmission alone would be capable of stereotyped or reflex responses. Modulation of this transmission and changes in other cellular functions by G-protein-linked systems and by receptor-tyrosine kinase-linked systems enables an orchestrated response. The large diversity of signaling molecules and their intracellular targets offer nearly unlimited flexibility of response over a broad time scale and with high amplification.

G proteins are GTP-binding proteins that couple the activation of seven-helix receptors by neurotransmitters at the cell surface to changes in the activity of effector enzymes and effector channels. A common effector enzyme is adenylate cyclase, which synthesizes cyclic AMP (cAMP)—an intracellular surrogate, or second messenger, for the neurotransmitter, the first messenger. Phospholipase C (PLC), another effector enzyme, generates diacylglycerol (DAG) and inositol 1, 4, 5-trisphosphate (IP$_3$), the latter of which releases intracellular stores of Ca^{2+}. Information from an activated receptor flows to the second messengers that typically activate protein kinases, which modify a host of cellular functions. Ca^{2+}, cAMP, and DAG have in common the ability to activate protein kinases with broad substrate specificities. They phosphorylate key intracellular proteins, ion channels, enzymes, and transcription factors taking part in diverse cellular biological processes. The activities of protein kinases and phosphatases are in balance, constituting a highly regulated process, as revealed by the phosphorylation state of these targets of the signal transduction process.

In addition to regulating protein kinases, second messengers such as cAMP, cyclic GMP (cGMP), Ca^{2+}, and arachidonic acid can directly gate, or modulate, ion channels. G proteins can also couple directly to ion channels without the interception of second messengers or protein kinases. In these diverse ways, a neurotransmitter outside the cell can modulate essentially every aspect of cell physiology and encode the history of cell stimuli in the form of altered activity and expression of its cellular constituents. An overview of G-protein signaling to protein kinases is presented in Figure 10.1.

Receptors Catalyze the Conversion of G Proteins into the Active GTP-Bound State

G proteins undergo a molecular switch between two interconvertible states that are used to "turn on" or "turn off" downstream signaling. G proteins taking part in signal transduction utilize a regulatory motif that is seen in other GTPases engaged in protein synthesis and in intracellular vesicular traffic. G proteins are switched on by stimulated receptors, and they switch themselves off after a time delay. G proteins are inactive when GDP is bound and are active when GTP is bound. The sole function of seven-helix receptors in activating G proteins is to catalyze an exchange of GTP

for GDP. This is a temporary switch because G proteins are designed with a GTPase activity that hydrolyzes the bound GTP and converts the G protein back into the GDP-bound, or inactive, state. Thus, a G protein must continuously sample the state of activation of the receptor, and it transmits downstream information only while the neuron is exposed to neurotransmitter. The GTPase activity of G proteins thus serves both as a regulatable timer and as an amplifier (Fig. 10.2).

The G-Protein Cycle

G proteins are trimeric structures composed of two functional units: (1) an α subunit (39–52 kDa) that catalyzes GTPase activity and (2) a $\beta\gamma$ dimer (35 and 8 kDa,

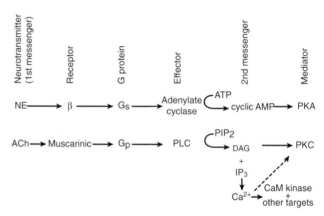

FIGURE 10.1 Overview of G-protein signaling to protein kinases. Norepinephrine (NE) and acetylcholine (ACh) can stimulate certain receptors that couple through distinct G proteins to different effectors, which results in increased synthesis of second messengers and activation of protein kinases (PKA and PKC). PLC, phospholipase C; P1P$_2$, phosphatidylinositol bisphosphate; DAG, diacylglycerol; CaM, Ca^{2+}-calmodulin dependent; IP$_3$, inositol 1, 4, 5-triphosphate.

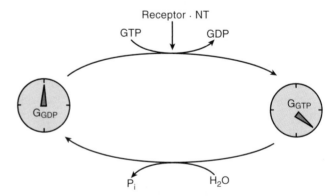

FIGURE 10.2 GTPase activity of G proteins serves as a timer and amplifier. Receptors activated by neurotransmitters (NT) initiate the GTPase timing mechanism of G proteins by displacement of GDP by GTP. Neurotransmitters thus convert G—GDP ("turned-off state") to GGTP (time-limited "turned-on" state).

respectively) that interacts tightly with the α subunit when bound to GDP (Stryer and Bourne, 1986; Birnbaumer, 2007). The role of the three subunits in the G-protein cycle is depicted in Figures 10.3 and 10.4. In the basal state, GDP is bound tightly to the α subunit, which is associated with the $\beta\gamma$ pair to form an inactive G protein. In addition to blocking interaction of the α subunit with its effector, the $\beta\gamma$ pair increases the affinity of the α subunit for activated receptors. Binding of the neurotransmitter to the receptor produces a conformational change that positions previously buried residues that promote increased affinity of the receptors for the inactive G protein. A given receptor can interact with only one or a limited number of G proteins, and the α subunit produces most of this specificity. Coupling with the activated receptor reduces the affinity of the α subunit for GDP, facilitating its dissociation and replacement with GTP. Thus, the receptor effectively catalyzes an exchange of GTP for GDP. GTP-GDP exchange is inherently very slow and ensures that very little of the G protein is in the on state under basal conditions. The level of G protein in the on state can increase from being 1% to being more than 50% of all G protein (Stryer and Bourne, 1986).

Information Flow through G-Protein Subunits

One of the more tense and public debates in signal transduction has been the question whether the α subunit alone conveys information that specifies which effector is activated or whether the $\beta\gamma$ pair can also interact with effectors. One of the contestants even paid for a vanity license plate proclaiming "α not $\beta\gamma$." This notion was eventually changed because of the finding that $\beta\gamma$ can directly activate certain K$^+$ channels. It is now apparent that α and $\beta\gamma$ subunits can

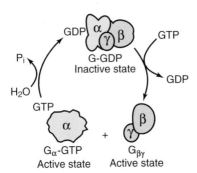

FIGURE 10.3 Interconversion, catalyzed by excited receptors, of G-protein subunits between inactive and active states. Displacement of GDP with GTP dissociates the inactive heterotrimeric G protein, generating α-GTP and $\beta\gamma$, both of which can interact with their respective effectors and activate them. The system converts into the inactive state after GTP has been hydrolyzed and the subunits have reassociated. From Stryer (1995). Used with permission of W. H. Freeman and Company.

both modify effector enzymes, but the historic association of G-protein function with α has persisted for the purpose of nomenclature, with G$_s$ and α_s referring to the G protein and its corresponding α subunit, which stimulates adenylate cyclase. The α subunits may act either independently or in concert with $\beta\gamma$ (Clapham and Neer, 1993). Furthermore, β and γ subunits in a $\beta\gamma$ pair can combine in many different ways. Other legacy terms include G$_i$, G$_p$, and G$_o$ used for G-protein activities that inhibited adenylate cyclase, stimulated phospholipase, or were presumed to have other effects, respectively.

Effector Enzymes, Channels, and Transporters Decode Receptor-Mediated Cell Stimulation in the Cell Interior

The function of the trimeric G proteins is to decode information about the concentration of neurotransmitters bound to appropriate receptors on the cell surface and convert this information into a change in the activity of enzymes and channels that mediate the effects of the neurotransmitter. The known effector functions of α include both stimulation and inhibition of adenylate cyclases that is sensitive to cholera toxin and pertussis toxin, respectively. In addition, it modulates activation of cGMP phosphodiesterase, PLC, and regulation of Na$^+$/K$^+$ exchange, PI3K, RhoGEF, and rasGAP. The effector functions of β/γ dimers include inhibition of many adenylate cyclase and stimulation of adenylate cyclase types II and IV (with α). In addition, they regulate stimulation of phospholipase Cβ, K$^+$, and Ca^{2+} channels, phospholipase A$_2$, phosphatidylinositol-3-kinase, PKD, and dynamin in vesicle budding.

Response Specificity in G-Protein Signaling

Signals originating from activated receptors can either converge or diverge, depending on the receptor and on the complement of G proteins and effectors in a given neuron (Fig. 10.5). How can a neurotransmitter produce a specific response if G-protein coupling has the potential for such a diversity of effectors? A given neuron has only a subset of receptors, G proteins, and effectors, thereby limiting possible signaling pathways. Transducin, for example, is confined to the visual system, where the predominant effector is the cGMP phosphodiesterase and not adenylate cyclase. Signal specificity is further refined by selective affinities between cognate sets of receptors, G proteins, and effector(s); and by spatial compartmentalization (e.g., at nerve terminals).

Furthermore, the intrinsic GTPase activity can be modulated by GTPase-activating proteins (GAPs),

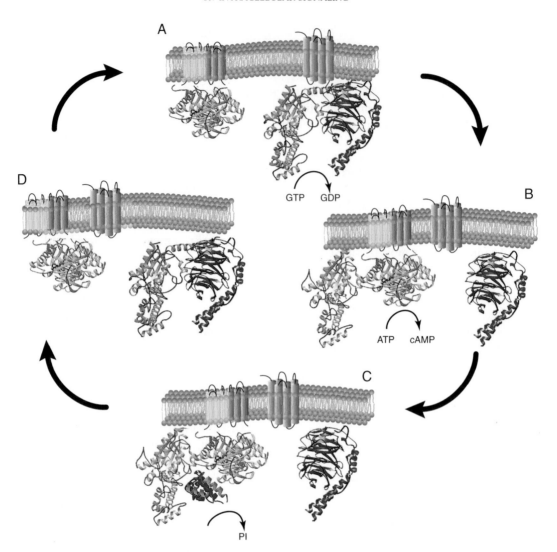

FIGURE 10.4 (A) G proteins are held in an inactive state because of very high affinity binding of GDP to their α subunits. When activated by agonist, membrane-bound seven helical receptors (right, glowing magenta) interact with heterotrimeric G proteins (α, amber; β, teal; γ, burgundy) and stimulate dissociation of GDP. This permits GTP to bind to and activate α, which then dissociates from the high-affinity dimer of β and γ subunits. (B) Both activated (GTP-bound) α (lime) and $\beta\gamma$ are capable of interacting with downstream effectors. This figure shows the interaction of GTP-α_s with adenylate cyclase (catalytic domains are mustard and ash). Adenylate cyclase then catalyzes the synthesis of the second messenger cyclic AMP (cAMP) from ATP. (C) Signaling is terminated when α hydrolyzes its bound GTP to GDP. In some signaling systems, GTP hydrolysis is stimulated by GTPase-activating proteins or GAPs (cranberry) that bind to α and stabilize the transition state for GTP hydrolysis. (D) Hydrolysis of GTP permits GDP-α to dissociate from its effector and associate again with $\beta\gamma$. The heterotrimeric G protein is then ready for another signaling cycle if an activated receptor is present. This figure is based on the original work of Mark Wall and John Tesmer.

which terminate its active state more quickly, and selectively affect signal output.

Fine-Tuning of cAMP by Adenylate Cyclases

The level of cAMP is highly regulated due to a balance between synthesis by adenylate cyclases and degradation by cAMP phosphodiesterases (PDEs). Each of these enzymes can be regulated and manipulated independently. Adenylate cyclase was the first G-protein effector to be identified, and now a group of related adenylate cyclases are known to be regulated differentially by both α and $\beta\gamma$ subunits (Taussig and Gilman, 1995). G proteins can both activate and inhibit adenylate cyclases either synergistically or antagonistically.

Adenylate cyclases are large proteins of approximately 120 kDa. All the known classes of adenylate cyclase consist of a tandem repeat of the same structural motif—a short cytoplasmic region followed by six putative transmembrane segments and then a

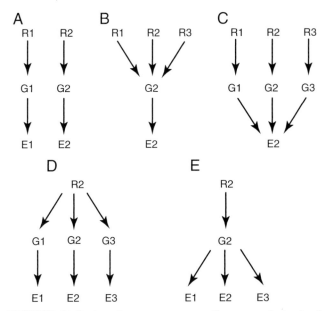

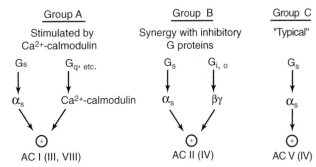

FIGURE 10.6 Isoforms of adenylate cyclase (AC). All isoforms are stimulated by α_s but differ in the degree of interaction with Ca^{2+}-calmodulin and with $\beta\gamma$ derived from inhibitory G proteins. Not shown is the ability of excess $\beta\gamma$ to complex with α_s and inhibit group A and group C adenylate cyclases. Adapted from Taussig and Gilman (1995).

FIGURE 10.5 Signals can converge or diverge on the basis of interactions between receptors (R) and G proteins (G) and between G proteins and effectors (E). The complement of receptors, G proteins, and effectors in a given neuron determines the degree of integration of signals, as well as whether cell stimulation will produce a focused response to a neurotransmitter or a coordination of divergent responses. Adapted from Ross (1989).

highly conserved catalytic domain of approximately 35 kDa on the cytoplasmic side that bind ATP and catalyze its conversion into cAMP. Some isoforms are activated by calmodulin.

Differential Regulation of Adenylate Cyclase Isoforms

All adenylate cyclase isoforms are stimulated by G_s through its α_s subunit. Known isoforms can be divided minimally into at least three groups on the basis of additional regulatory properties (Fig. 10.6). Group A (types I, III, and VIII) possesses a calmodulin-binding domain and is activated by Ca^{2+}-calmodulin. Group B (types II and IV) is weakly responsive to direct interaction with α_s or $\beta\gamma$ but is highly activated when both are present. As described later, this synergistic effect enables this cyclase to function as a coincidence detector. Group C is typified by types V and VI (and IX), which differ from group A cyclases in their inhibitory regulation.

Inhibition of Adenylate Cyclases Adenylate cyclases are also subject to several forms of inhibitory control. First, activation of all adenylate cyclases can be antagonized by $\beta\gamma$ released from abundant G proteins, such as G_i, G_o, and G_z, which complex with α_s-GTP and shift the equilibrium toward an inactive trimer by mass

action. Second, either α or $\beta\gamma$ subunits derived from G_i, G_o, or G_z can directly inhibit group A cyclases, and the α subunit from G_i or G_z can inhibit group C cyclases. The level of G_s in particular is low; thus, α_s derived from G_s is sufficient to activate adenylate cyclases, but the $\beta\gamma$ derived from it is insufficient to directly inhibit or activate adenylate cyclases. This explains the apparent paradox that receptors that couple to G_s produce effects only through α_s, whereas receptors that couple to G_i produce effects through both α_i and $\beta\gamma$ even though they can share the same $\beta\gamma$.

Receptors Coupling to Adenylate Cyclase Dozens of neurotransmitters and neuropeptides work through cAMP as a second messenger and by G-protein-linked activation or inhibition of adenylate cyclase. Among the neurotransmitters that increase cAMP are the amines norepinephrine, epinephrine, dopamine, serotonin, and histamine, and the neuropeptides vasointestinal peptide (VIP) and somatostatin. In the olfactory system, a special form of G-protein α subunit, termed α_{olf}, serves the same function as α_s and couples several hundred seven-helix receptors to type III adenylate cyclase in the neuroepithelium.

Adenylate Cyclases as Coincidence Detectors Type I and type II adenylate cyclases can integrate concurrent stimulation of neurons by two or more neurotransmitters (Bourne and Nicoll, 1993). Type I adenylate cyclase is stimulated by neurotransmitters that couple to G_s and by neurotransmitters that elevate intracellular Ca^{2+}. This adenylate cyclase can convert the depolarization of neurons into an increase in cAMP. Its role in associative forms of learning may be related to its ability to link cAMP-based and Ca^{2+}-based signals. Stimulation of type II adenylate cyclase by α_s is conditional on the presence of $\beta\gamma$ derived from an abundant G protein (i.e., other than G_s), thus enabling the cyclase to serve as a coincidence detector. Thus, activation of a second receptor, presumably coupled to the abundant G_i and G_o, is needed to provide the $\beta\gamma$.

Sources of Second Messengers: Phospholipids

Two phospholipids, phosphatidylinositol 4,5-bisphosphate (PIP_2) and phosphatidylcholine (PC), are primary precursors for a G-protein-based second-messenger system. Three second messengers, diacylglycerol, arachidonic acid and its metabolites, and elevated Ca^{2+}, ultimately are produced. A single step converts inert phospholipid precursors into lipid messengers. DAG action is mediated by protein kinase C (PKC) (Tanaka and Nishizuka, 1994). The elevation of Ca^{2+} levels is accomplished by the regulated entry of Ca^{2+} from a concentrated pool sequestered in the endoplasmic reticulum or from outside the cell. Ca^{2+} has many direct cellular targets but mediates most of its effects through calmodulin, a Ca^{2+}-binding protein that activates many enzymes after it binds Ca^{2+}. One class of calmodulin-dependent enzymes is a family of protein kinases that enable Ca^{2+} signals to modulate a large number of cellular processes by phosphorylation (Hudmon and Schulman, 2002).

Generation of DAG and IP_3 from G_q and G_i Coupled to $PLC\beta$

The phosphatidylinositide-signaling pathway is just as prominent in neuronal signaling as the cAMP pathway and is similar to it in overall design. Stimulation of a large number of neurotransmitters and hormones (including acetylcholine (Ml, M3), serotonin ($5HT_2$, $5HT_{1C}$), norepinephrine (α_{1A}, α_{1B}), glutamate (metabotropic), neurotensin, neuropeptide Y, and substance P) is coupled to the activation of a phosphatidylinositide-specific PLC.

Phosphatidylinositol (PI) is composed of a diacylglycerol backbone with myoinositol attached to the sn-3 hydroxyl by a phosphodiester bond (Fig. 10.7). The six positions of the inositol are not equivalent: the 1 position is attached by a phosphate to the DAG moiety. PI is phosphorylated by PI kinases at the 4 position and then at the 5 position to form PIP_2. In response to the appropriate G-protein coupling, PLC hydrolyzes the bond between the sn-3 hydroxyl of the DAG backbone and the phosphoinositol to produce two second messengers—DAG, a hydrophobic molecule, and IP_3, which is water soluble (Fig. 10.8). Three classes of PLC that hydrolyze PIP_2, $PLC\beta$, $PLC\gamma$, and $PLC\delta$ are soluble enzymes that have in common a catalytic domain structure but differ in their regulatory properties. G proteins couple to several variants of $PLC\beta$. $PLC\gamma$ is regulated by growth factor tyrosine

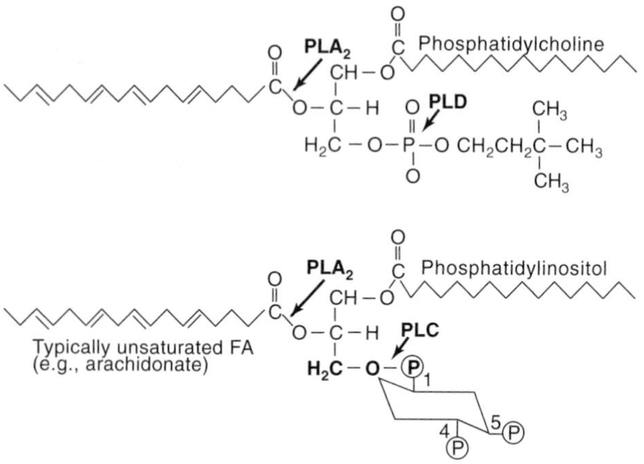

FIGURE 10.7 Structures of phosphatidylinositol and phosphatidylcholine. The sites of hydrolytic cleavage by PLC, PLD, and PLA2 are indicated by arrows. FA, fatty acid.

kinases. In contrast, $PLC\delta$ in brain is primarily glial, and its mode of regulation is not well understood.

$PLC\beta$ is coupled to neurotransmitters by G_i and G_q. A pertussis toxin-sensitive pathway is mediated by a number of isoforms referred to as G_q and mediated by their α_q. G_i is coupled to $PLC\beta$ via its $\beta\gamma$ rather than α subunit in a pathway that is insensitive to pertussis toxin. Receptor tyrosine kinases can regulate $PLC\gamma$ by a G-protein-independent pathway involving their recruitment to the receptor and activation via phosphorylation.

DAG Derived from Activation of Phospholipase D

A slower but larger increase in DAG can be generated by activation phospholipase D (PLD), which cleaves phosphatidylcholine to produce phosphatidic acid and choline. Dephosphorylation of phosphatidic acid produces DAG. The PLD pathway may be used by some mitogens and growth factors and likely contains a variety of activation schemes that may include G proteins.

Additional Lipid Messengers

DAG is itself a source of another lipid messenger, by the action of phospholipase A_2 (PLA_2), which releases the fatty acid, typically arachidonic acid, from the sn-2 position of the DAG backbone (Fig. 10.7). Arachidonic acid has biological activity of its own in addition to serving as a precursor for prostaglandins and leukotrienes. Arachidonic acid and other cis-unsaturated fatty acids can modulate K^+ channels, $PLC\gamma$, and some forms of PKC.

A subfamily of lipid kinases that are specific for addition of a phosphate moiety on the 3 position, phosphoinositide 3-kinases (PI-3 kinases), also play a regulatory role. Depending on their preferred lipid

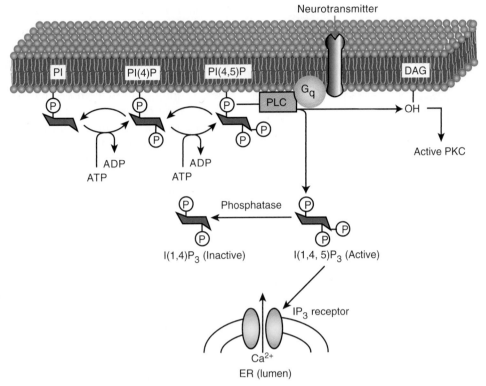

FIGURE 10.8 Schematic pathway of IP$_3$ and DAG synthesis and action. Stimulation of receptors coupled to G$_q$ activates PLCβ, which leads to the release of DAG and IP$_3$. DAG activates PKC, whereas IP3 stimulates the IP$_3$ receptor in the endoplasmic reticulum (ER), leading to mobilization of intracellular Ca^{2+} stores. Adapted from Berridge (1993).

substrate, they can produce PI-3-P, PI-3,4-P$_2$, PI-3,5-P$_2$, and PI-3,4,5-P$_3$. A number of signals, including growth factors, activate PI-3 kinases to generate these lipid messengers. In turn, these lipids then bind directly to a number of proteins and enzymes to modify vesicular traffic, protein kinases involved in survival and cell death. There is also evidence that another lipid, sphingomyelin, is a precursor for intracellular signals as well.

IP$_3$, a Potent Second Messenger That Produces Its Effects by Mobilizing Intracellular Ca^{2+} The main function of IP$_3$ is to stimulate the release of Ca^{2+} from intracellular stores. Ca^{2+} levels are kept low in the cytosol by its sequestration in the ER where it is complexed with low-affinity-binding proteins. The ER is the major IP$_3$-sensitive Ca^{2+} store in cells (Fig. 10.8) and Ca^{2+} readily flows down its concentration gradient into the cell lumen upon opening of Ca^{2+} channels in the ER.

The IP$_3$ receptor is a macromolecular complex that functions as an IP$_3$ sensor and a Ca^{2+} release channel. It has a broad tissue distribution but is highly concentrated in the cerebellum. The IP$_3$ receptor is a tetramer of 313-kDa subunits with a single IP$_3$-binding site at its

N-terminal of each subunit, facing the cytoplasm. Ca^{2+} release by IP$_3$ is highly cooperative so that a small change in IP$_3$ has a large effect on Ca^{2+} release from the ER. The mouse mutants *pcd* and *nervous* have deficient levels of the IP$_3$ receptor and exhibit defective Ca^{2+} signaling, and a genetic knockout of the IP$_3$ receptor leads to motor and other deficits.

Termination of the IP$_3$ Signal IP$_3$ is a transient signal terminated by dephosphorylation to inositol. Inactivation is initiated either by dephosphorylation to inositol 1,4-bisphosphate (Fig. 10.9) or by an initial phosphorylation to a tetrakisphosphate form that is dephosphorylated by a different pathway. Both pathways have in common an enzyme that cleaves the phosphate on the 1 position. Complete dephosphorylation yields inositol, which is recycled in the biosynthetic pathway. Recycling is important because most tissues do not contain *de novo* biosynthetic pathways for making inositol. Salvaging inositol may be particularly important when cells are actively undergoing PI turnover. It is intriguing that Li$^+$, the simple salt used to treat bipolar disorder, selectively inhibits the salvage of inositol by inhibiting the enzyme that dephosphorylates the 1 position and is common to the two

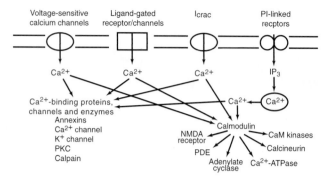

FIGURE 10.9 Multiple sources of Ca²⁺converge on calmodulin and other Ca²⁺-binding proteins. Cellular levels of Ca²⁺ can rise either by influx (e.g., through voltage-sensitive channels or ligand-gated channels) or by redistribution from intracellular stores triggered by IP₃. Calcium modulates dozens of cellular processes by the action of the Ca²⁺-calmodulin complex on many enzymes, and calcium has some direct effects on enzymes such as PKC and calpain. CaM kinase, Ca²⁺-calmodulin-dependent kinase.

pathways. At therapeutic doses of Li⁺, the reduced salvage of inositol in cells with high phosphoinositide signaling may lead to a depletion of PIP₂ and a selective inhibition of this signaling pathway in active cells.

Calcium Ion

Calcium has a dual role as a carrier of electrical current and as a second messenger. Its effects are more diverse than those of other second messengers such as cAMP and DAG because its actions are mediated by a much larger array of proteins, including protein kinases (Carafoli and Klee, 1999). Furthermore, many signaling pathways directly or indirectly increase cytosolic Ca²⁺ concentration from 100 nM to 0.5–1.0 mM. The source of elevated Ca²⁺ can be either the ER or the extracellular space (Fig. 10.9). In addition to IP₃-mediated release, Ca²⁺ can activate its own mobilization through the ryanodine receptor on the ER. Mechanisms for Ca²⁺ influx from outside the cell include several voltage-sensitive Ca²⁺ channels and ligand-gated cation channels that are permeable to Ca²⁺ (e.g., nicotinic receptor and N-methyl-D-aspartate (NMDA) receptor).

Dynamics of Ca²⁺ Signaling Revealed by Fluorescent Ca²⁺ Indicators We know a great deal about the spatial and temporal regulation of Ca²⁺ signals because of the development of fluorescent Ca²⁺ indicators. A variety of fluorescent compounds selectively bind Ca²⁺ at physiological concentration ranges and rapidly change their fluorescent properties upon binding Ca²⁺ to a fairly accurate measurement of ionized Ca²⁺.

Calmodulin-Mediated Effects of Ca²⁺

Ca²⁺ acts as a second messenger to modulate the activity of many mediators. The predominant mediator of Ca²⁺ action is calmodulin, a ubiquitous 17-kDa calcium-binding protein. Ca²⁺ binds to calmodulin in the physiological range and converts it into an activator of many cellular targets (Cohen and Klee, 1988). Binding of Ca²⁺ to calmodulin produces a conformational change that greatly increases its affinity for more than two dozen eukaryotic enzymes that it activates, including cyclic nucleotide PDEs, adenylate cyclase, nitric oxide synthase, Ca²⁺-ATPase, calcineurin (a phophoprotein phosphatase), and several protein kinases (Fig. 10.9). This activation of calmodulin allows neurotransmitters that change Ca²⁺ to affect dozens of cellular proteins, presumably in an orchestrated fashion.

Regulation of Guanylate Cyclase by Nitric Oxide An important target of Ca²⁺-calmodulin is the enzyme nitric oxide synthase (NOS) (Chapter 8). This enzyme synthesizes one of the simplest known messengers, the gas NO (Baranano *et al.*, 2001). In the pathway that led to its discovery, acetylcholine stimulates the PI signaling pathway in the endothelium to increase intracellular Ca²⁺, which activates NOS so that more NO is made. NO then diffuses radially from the endothelial cells across two cell membranes to the smooth muscle cell, where it activates guanylate cyclase to make cGMP. This in turn activates a cGMP-dependent protein kinase that phosphorylates proteins, leading to a relaxation of muscle. In 1998, Robert F. Furchgott, Louis J. Ignarro, and Ferid Murad received the Nobel Prize for their discoveries concerning nitric oxide as a signaling molecule and therapeutic mediator in the cardiovascular system.

Let us now turn to the details of the NO pathway. Nitric oxide is derived from L-arginine in a reaction catalyzed by NOS, a complex enzyme that converts L-arginine and O₂ into NO and L-citrulline. NOS likely produces the neutral free radical NO as the active agent. NO lasts only a few seconds in biological fluids and thus, no specialized processes are needed to inactivate this particular signaling molecule. As a gas, NO is soluble in both aqueous and lipid media and can diffuse readily from its site of synthesis across the cytosol or cell membrane and affect targets in the same cell or in nearby neurons, glia, and vasculature (Baranano *et al.*, 2001). NO produces a variety of effects, including relaxation of smooth muscle of the vasculature, relaxation of smooth muscle of the gut in peristalsis, and killing of foreign cells by macrophages. It was first recognized as a neuronal messenger that couples glutamate receptor stimulation to increases in cGMP. NO produced by Ca²⁺-calmodulin-dependent activation of NO synthase concentrated in cerebellar granule cells activates guanylate cyclase in nearby Purkinje cells during the induction of long-term depression in the cerebellum (Chapter 32).

Activation of Guanylate Cyclases Two types of guanylate cyclase, a soluble one regulated by NO and a membrane-bound enzyme regulated directly by neuropeptides, synthesize cGMP from GTP in a reaction similar to the synthesis of cAMP from ATP. NO activates the soluble enzyme by binding to the iron atom of the heme moiety. This is the basic mechanism for the regulation of soluble guanylate cyclases. A number of therapeutic muscle relaxants, such as nitroglycerin and nitroprusside, are NO donors that produce their effects by stimulating cGMP synthesis. Membrane-bound guanylate cyclases are transmembrane proteins with a binding site for neuroendocrine peptides on the extracellular side of the plasma membrane and a catalytic domain on the cytosolic side.

Cyclic GMP Phosphodiesterase, an Effector Enzyme in Vertebrate Vision

The versatility of G-protein signaling is illustrated in vertebrate phototransduction, in which a specialized G protein called transducin (G_t) is activated by light rather than by a hormone or neurotransmitter. Transducin stimulates cGMP phosphodiesterase, an effector enzyme that hydrolyzes cGMP and ultimately turns off the dark current (Chapter 27). Nature has evolved an elegant mechanism for using photons of light to modify a hormone-like molecule, retinal, that activates a seven-helix receptor called rhodopsin. Activated rhodopsin dissociates α_t from transducin, which then activates a soluble cGMP phosphodiesterase.

Rods can detect a single photon of light because the signal-to-noise ratio of the system is very low and the amplification factor in phototransductin is quite high; one rhodopsin molecule stimulated by a single photon can activate 500 transducins. Transducin remains in the "on" state long enough to activate 500 PDEs. PDE can hydrolyze about 100 cGMP molecules in the second before it is deactivated. cGMP in rods regulates a cGMP-gated cation channel, leading to additional amplification of the signal.

Modulation of Ion Channels by G Protein

Each type of neuron has a repertoire of ion channels that give it a distinct response signature, and it is not surprising that several types of mechanisms regulate these channels. Channel modulation occurs via G proteins, second messengers and their cognate protein kinases that phosphorylate ion channels as well as by direct effects of G proteins.

The first ion channel demonstrated to undergo regulation by G proteins was the cardiac K^+ channel that mediates slowing of the heart by acetylcholine released from the vagus nerve. When this I_{KACh} channel is examined in a membrane patch delimited by the seal of a cell-attached electrode, the addition of acetylcholine within the electrode increases the frequency of channel opening dramatically, whereas the addition of acetylcholine to the cell surface outside the seal does not. The process is therefore described as membrane delimited, with a direct interaction between the G protein (either α_i or $\beta\gamma$) and the channel.

There is also compelling evidence for the stimulation or inhibition of Ca^{2+} channel subtypes by G proteins. The central role played by Ca^{2+} in muscle contraction, in synaptic release, and in gene expression makes Ca^{2+} influx a common target for regulation by neurotransmitters. In the heart, where L-type Ca^{2+} channels are critical for the regulation of contractile strength, the Ca^{2+} current is enhanced by α_s formed by β-adrenergic stimulation of G_s. In contrast, N-type Ca^{2+} channels, which modulate synaptic release in nerve terminals, often are inhibited by muscarinic and α-adrenergic agents and by opiates acting at receptors coupled to G_i and G_o.

G-Protein Signaling Gives Special Advantages in Neural Transmission

The G-protein-based signaling system provides several advantages over fast transmission (Hille, 1992; Birnbaumer, 2007). These advantages include amplification of the signal, modulation of cell function over a broad temporal range, diffusion of the signal to a large cellular volume, cross talk, and coordination of diverse cell functions. The sacrifice in speed relative to signaling by ligand gated ion channels is compensated by a broad range of signaling that facilitates integration of signals by the G-protein system. A slower time frame means that cellular processes that are quite distant from the receptor can be modulated. Diffusion of second messengers such as IP_3, Ca^{2+}, and DAG can extend neurotransmission through the cell body and to the nucleus to alter gene expression and via NO to other cells. Neurotransmitters acting through G proteins can elicit a coordinated response of the cell that can modulate synaptic release, resynthesis of neurotransmitter, membrane excitability, the cytoskeleton, metabolism, and gene expression.

Summary

A major class of signaling utilizing G-protein-linked signals affords the nervous system a rich diversity of modulation, amplification, and plasticity. Signals are mediated through second messengers activating proteins that modify cellular processes and gene transcription. A key feature is the ability of G proteins to detect the presence of activated receptors and to amplify the signal through effector enzymes and channels. Phosphorylation of key intracellular proteins, ion channels, and enzymes activates diverse, highly regulated cellular processes. The specificity of response is ensured through receptors reacting only with a limited number of G proteins. Coupling between receptor, G protein, and its effector(s), and spatial compartmentalization of the system enables specificity and localized control of signaling. Phospholipids and phosphoinositols provide substrates for second-messenger signaling for G proteins. Stimulation of release of intracellular calcium is often the mediator of the signal. Calcium itself has a dual role as a carrier of electrical current and as a second messenger. Calmodulin is a key regulator that provides complexity and enhances specificity of the signaling system. Sensitivity of the system is imparted by an extremely robust amplification system, as seen in the visual system, which can detect single photons of light.

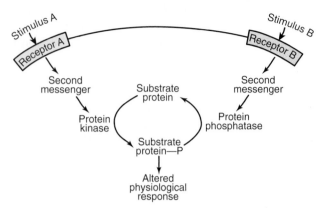

FIGURE 10.10 Regulation by protein kinases and protein phosphatases. Enzymes and other proteins serve as substrates for protein kinases and phosphoprotein phosphatases, which modify their activity and control them in a dynamic fashion. Multiple signals can be integrated at this level of protein modification. Adapted from Svenningsson *et al.* (2004).

MODULATION OF NEURONAL FUNCTION BY PROTEIN KINASES AND PHOSPHATASES

Protein phosphorylation and dephosphorylation are key processes that regulate cellular function. They play a fundamental role in mediating signal transduction initiated by neurotransmitters, neuropeptides, growth factors, hormones, and other signaling molecules (Fig. 10.10). The functional state of many proteins is modified by phosphorylation-dephosphorylation, the most ubiquitous post-translational modification in eukaryotes. A fifth of all proteins may serve as targets for kinases and phosphatases. Phosphorylation or dephosphorylation can rapidly modify the function of enzymes, structural and regulatory proteins, receptors, and ion channels taking part in diverse processes, without a need to change the level of their expression. It can also produce long-term alterations in cellular properties by modulating transcription and translation and changing the complement of proteins expressed by cells.

Protein kinases catalyze the transfer of the terminal, or γ, phosphate of ATP to the hydroxyl moieties of Ser, Thr, or Tyr residues at specific sites on target proteins. Most protein kinases are either Ser/Thr kinases or Tyr kinases, with only a few designed to phosphorylate both categories of acceptor amino acids. Protein phos-

phatases catalyze the hydrolysis of the phosphoryl groups from phosphoserine, phosphothreonine, phosphotyrosine, or both types of phosphorylated amino acids on phosphoproteins.

The activity of protein kinases and protein phosphatases often is regulated either by a second messenger (e.g., cAMP or Ca^{2+}) or by an extracellular ligand (e.g., nerve growth factor). In general, second-messenger-regulated kinases modify Ser and Thr, whereas receptor-linked kinases modify Tyr. Among the many protein kinases and protein phosphatases in neurons, a relatively small number serve as master regulators to orchestrate neuronal function. The cAMP-dependent protein kinase (PKA) is a prototype for the known regulated Ser/Thr kinases; they are similar in overall structure and regulatory design. PKA is the predominant mediator for signaling through cAMP, the only others being a cAMP-liganded ion channel in olfaction and exchange protein directly activate by cAMP (Epoc) which modulate GDP-GTP exchange in cell adhesion and exocytosis. In a similar fashion, the related cGMP-dependent protein kinase (PKG) mediates most of the actions of cGMP. Ca^{2+}-calmodulin-dependent protein kinase II (CaM kinase II) and several other kinases mediate many of the actions of stimuli that elevate intracellular Ca^{2+}. Finally, the PI signaling system increases both DAG and Ca^{2+}, which activate any of a family of protein kinases collectively called protein kinase C.

The activities of these second messenger-regulated protein kinases are countered by a relatively small number of phosphatases, exemplified by protein phosphatase 1 (PP–1), protein phosphatase 2A (PP–2A), and protein phosphatase 2B (PP–2B, or calcineurin). Phosphorylation and dephosphorylation are reversible processes, and the net activity of the two processes determines the phosphorylation state of each substrate. The Nobel Prize for Physiology and Medicine was awarded to Edwin G. Krebs and Edmund H. Fischer in 1992 for their pioneering work on the regulation of cell function by protein kinases and phosphatases.

Certain Principles Are Common in Protein Phosphorylation and Dephosphorylation

Protein kinases and protein phosphatases are described either as multifunctional if they have a broad specificity and therefore modify many protein targets, or as dedicated if they have a very narrow substrate specificity. Spatial positioning of kinases and their substrates in the cell either increases or decreases the likelihood of phosphorylation-dephosphorylation of a given substrate.

The amplification of signal transduction described earlier is continued during transmission of the signal by protein kinases and protein phosphatases. In some cases, the kinases are themselves subject to activation by phosphorylation in a cascade in which one activated kinase phosphorylates and activates a second, and so on, to provide amplification and a switch-like response termed ultrasensitivity.

Kinases and phosphatases integrate cellular stimuli and encode the stimuli as the steady-state level of phosphorylation of a large complement of proteins in the cell (Hunter, 1995). Distinct signal transduction pathways can converge on the same or different target substrates (Fig. 10.10). In some cases, these substrates can be phosphorylated by several kinases at distinct sites. Phosphorylation can alter cellular processes over broad time scales, from milliseconds to hours and much longer by altering gene expression.

Phosphorylation produces specific changes in the function of a target protein, such as increasing or decreasing the catalytic activity of an enzyme, conductance of an ion channel, or desensitization of a receptor. Kinases and phosphatases modulate proteins by regulating the presence of a highly charged and bulky phosphoryl moiety on Ser, Thr, or Tyr at a precise location on the substrate protein. The phosphate may elicit a conformational change or alter interaction with other proteins.

Finally, each of the three kinases described here is capable of functioning as a cognitive kinase—that is, a kinase capable of a molecular memory. Although each is activated by its respective second messenger, it can undergo additional modification that reduces its requirement for the second messenger. This molecular memory potentiates the activity of these kinases and may enable them to participate in aspects of neuronal plasticity.

cAMP-Dependent Protein Kinase Was the First Well-Characterized Kinase

Neurotransmitters that stimulate the synthesis of cAMP exert their intracellular effects primarily by activating PKA. The functions (and substrates) regulated by PKA include gene expression (cAMP response element-binding protein (CREB)), catecholamine synthesis (tyrosine hydroxylase), carbohydrate metabolism (phosphorylase kinase), cell morphology (microtubule-associated protein 2 (MAP–2)), postsynaptic sensitivity (AMPA receptor), and membrane conductance (K channel). Paul Greengard and Eric Kandel received the Nobel Prize for Medicine in 2000 (along with Arvid Carlsson) for their discoveries concerning signal transduction via PKA and phosphoprotein phosphatases in the nervous system. PKA is a tetrameric protein composed of two types of subunits: (1) a dimer of regulatory (R) subunits (either two RI subunits for type I PKA or two RII subunits for type II PKA) and (2) two catalytic subunits (C subunit). Two or more isoforms of the RI, RII, and C subunits have distinct tissue and developmental patterns of expression but appear to function similarly. The C subunits are 40-kDa proteins that contain the binding sites for protein substrates and ATP. The R subunits are 49- to 51-kDa proteins that contain two cAMP-binding sites. In addition, the R subunit dimer contains a region that interacts with cellular anchoring proteins that serve to localize PKA appropriately within the cell.

The binding of second messengers by PKA and the other second-messenger-regulated kinases relieves an inhibitory constraint and thus activates the enzymes (Fig. 10.11). cAMP binding leads to subunit dissociation thereby relieving the C subunit of its inhibitory R subunits, thereby activating the kinase. The steady-

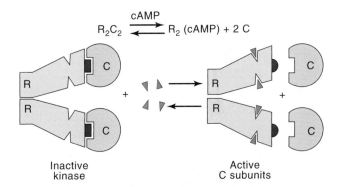

FIGURE 10.11 Activation of PKA by cAMP. An autoinhibitory segment (blue) of the regulatory subunit (R) dimer interacts with the substrate-binding domain of the catalytic (C) subunits of PKA, blocking access of substrates to their binding site. Binding of four molecules of cAMP reduces the affinity of R for C, resulting in dissociation of constitutively active C subunits.

state level of cAMP determines the fraction of PKA that is in the dissociated or active form. In this way PKA decodes cAMP signals into the phosphorylation of proteins and the resultant change in various cellular processes.

PKA is a member of a large family of protein kinases that have in common a significant degree of homology in their catalytic domains and are likely derived from an ancestral gene (Fig. 10.12). This homology extends to the three-dimensional crystal structure based on X-ray crystallography of PKA and other kinases. The catalytic domain may be in a subunit distinct from the regulatory domain, as in PKA, or in the same subunit, as in PKC and the CaM kinases. The crystal structure of the C subunit complexed to a segment of protein kinase inhibitor (PKI), a selective high-affinity inhibitor of PKA, reveals that the C subunit is composed of two lobes. The N-terminal lobe contains a highly conserved region that binds Mg^{2+}-ATP in a cleft between the two lobes. A larger C-terminal lobe contains the protein-substrate recognition sites and the appropriate

amino acids for catalyzing transfer of the phosphoryl moiety from ATP to the substrate. Inhibition by PKI is diagnostic of PKA involvement; PKI contains an auto-inhibitory sequence resembling PKA substrates and is positioned in the catalytic site like a substrate, thus blocking access for substrates.

PKA phosphorylates Ser or Thr at specific sites in dozens of proteins. The sequences of amino acids at the phosphorylation sites are not identical. Each kinase has a characteristic consensus sequence that forms the basis for distinct substrate specificities.

A regulatory theme common to PKA, CaM kinase II, and PKC is that their second messengers activate them by displacing an autoinhibitory domain from the active site (Kemp et al., 1994). The R subunit blocks access of substrates by positioning a pseudosubstrate or autoinhibitory domain in the catalytic site. Binding of cAMP to the R subunit near this autoinhibitory domain must disrupt its binding to the C subunit, thus leading to dissociation of an active C subunit. CaM kinase II and PKC likewise have autoinhibitory segments that are near the second-messenger-binding sites and may be activated similarly (Fig. 10.12).

Multifunctional CaM Kinase II Decodes Diverse Signals That Elevate Intracellular Ca^{2+}

Most of the effects of Ca^{2+} in neurons and other cell types are mediated by calmodulin, and many of the effects of Ca^{2+}-calmodulin are mediated by protein phosphorylation-dephosphorylation (Schulman and Braun, 1999). The Ca^{2+}-signaling system contains a family of Ca^{2+}-calmodulin-dependent protein kinases with broad substrate specificity, including CaM kinases I, II, and IV; of these, CaM kinase II is the best characterized. CaM kinase II phosphorylates tyrosine hydroxylase, MAP-2, synapsin I, calcium channels, Ca^{2+}-ATPase, transcription factors, and glutamate receptors and thereby regulates synthesis of catecholamines, cytoskeletal function, synaptic release in response to high-frequency stimuli, calcium currents, calcium homeostasis, gene expression, and synaptic plasticity, respectively. This kinase is found in every tissue but is particularly enriched in neurons. It is found in the cytosol, in the nucleus, in association with cytoskeletal elements, and in postsynaptic thickening termed the postsynaptic density found in asymmetric synapses. It is a large multimeric enzyme, consisting of 12 subunits derived from four homologous genes (α, β, γ, and δ) that encode different isoforms of the kinase that range from 54 to 65 kDa per subunit.

The catalytic, regulatory, and targeting domains of CaM kinase II are all contained within a single polypeptide (Fig. 10.12). Following the catalytic domain on

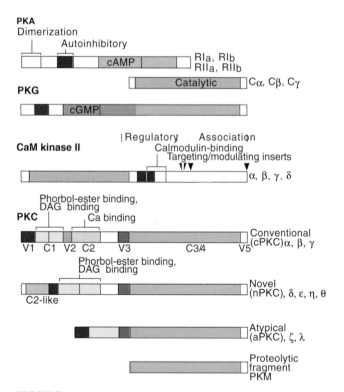

FIGURE 10.12 Domain structure of protein kinases. Protein kinases are encoded by proteins with recognizable structural sequences that encode specialized functional domains. Each of the kinases (PKA, PKG, CaM kinase II, and PKC) has homologous catalytic domains that are kept inactive by the presence of an autoinhibitory segment (blue lines). Regulatory domains contain sites for binding second messengers such as cAMP, cGMP, Ca^{2+}-calmodulin, DAG, and Ca^{2+}-phosphatidylserine. Alternative splicing creates additional diversity.

the N-terminal half of each isoform is the regulatory domain, which contains an autoinhibitory domain with an overlapping calmodulin-binding sequence. The C-terminal end contains an association domain that allows 12 subunits (two rings of six catalytic domains each) to assemble into a multimer, as well as targeting sequences that direct the kinase to distinct intracellular sites.

Regulation of the kinase by autophosphorylation is a critical feature of CaM kinase II. The kinase is inactive in the basal state because an autoinhibitory segment distorts the active site and sterically blocks access to its substrates. Binding of Ca^{2+}-calmodulin to the calmodulin-binding domain of the kinase displaces the autoinhibitory domain from the catalytic site and thus activating the kinase by enabling ATP and protein substrates to bind. Displacement of this domain also exposes a binding site for anchoring proteins that the activated kinase can bind. If the kinase is activated, it can autophosphorylate Thr-286 (in α-CaM kinase II). Phosphorylation disables the autoinhibitory segment by preventing it from reblocking the active site after calmodulin dissociates and thereby locks the kinase in a partially active state that is independent, or autonomous, of Ca^{2+}-calmodulin and can anchor to additional targets. Autophosphorylation prolongs the active state of the kinase, a potentiation that led to its description as a cognitive kinase.

CaM kinase II is targeted to distinct cellular compartments. Differences between the four genes encoding CaM kinase II and between the two or more isoforms that are encoded by each gene by apparent alternative splicing reside primarily in a variable region at the start of the association domain (Fig. 10.12). In some isoforms, this region contains an additional sequence that targets those isoforms to the nucleus. The major neuronal isoform, α-CaM kinase II, is largely cytosolic but is also found associated with postsynaptic densities synaptic vesicles and may therefore have several targeting sequences. Targeting to the NMDA type glutamate receptor occurs only after calmodulin activates the kinase and exposes a binding site.

Protein Kinase C Is the Principal Target of the PI Signaling System

Protein kinase C is a collective name for members of a relatively diverse family of protein kinases most closely associated with the PI-signaling system. PKC is a multifunctional Ser/Thr kinase capable of modulating many cellular processes, including exocytosis and endocytosis of neurotransmitter vesicles, neuronal plasticity, gene expression, regulation of cell growth and cell cycle, ion channels, and receptors. The role of

DAG generated during PI signaling was unclear until its link to PKC was established. Many PKC isoforms also require an acidic phospholipid such as phosphatidylserine for appropriate activation. The kinase is also of interest because it is the target of a class of tumor promoters called phorbol esters. They activate PKC by simulating the action of DAG, bypassing the normal receptor-based pathway, and inappropriately stimulating cell growth.

We now understand that the PKC family of kinases is diverse in structure and regulatory properties. PKC is monomeric (78–90 kDa) with catalytic, regulatory, and targeting domains all on one polypeptide (Fig. 10.12). The conventional isoforms (or cPKC), have all the following domains:

- VI, which contains the autoinhibitory or pseudosubstrate sequence
- C1, a cysteine rich domain that binds DAG and phorbol esters
- C2, a region necessary for Ca^{2+} sensitivity and for binding to phosphatidylserine and to anchoring proteins
- V3, a protease-sensitive hinge
- C3/4, the catalytic domain
- V5, which may also mediate anchoring

Another class of isoforms, termed novel PKCs (nPKC), lacks a true C2 domain and is therefore not Ca^{2+} sensitive. Another class is considered atypical (aPKC) because it lacks C2 and the first of two cysteine-rich domains that are necessary for DAG (or phorbol ester) sensitivity. This class is neither Ca^{2+} nor DAG sensitive. Not included is a DAG-interacting kinase originally designated as PKC-μ and now termed PKD because its catalytic domain is different from the other PKC isoforms.

Activation of PKC is best understood for the conventional isoforms. Generation of DAG resulting from stimulation of the PI-signaling pathway increases the affinity of cPKC isoforms for Ca^{2+} and phosphatidylserine. DAG, or specifically its sn-1,2-diacylglycerol isomer, is derived only from PI turnover, and it is the only isomer effective in activating PKC. Cell stimulation results in the translocation of cPKC from a variety of sites to the membrane or cytoskeletal elements where it interacts with PS-Ca^{2+}-DAG at the membrane. Binding of the second messengers to the regulatory domain disrupts the nearby autoinhibitory domain, leading to a reversible activation of PKC by deinhibition, as is found for PKA and CaM kinase II. Translocation is not restricted to the plasma membrane. Upon activation some PKC isoforms reversibly translocate to intracellular sites enriched with anchoring proteins, termed receptors for activated C kinase (RACK).

Prolonged activation of PKC can be produced by the addition of phorbol esters, which simulate activation by DAG but remain in the cell until they are washed out. In a matter of hours to days, such persistent activation by phorbol esters leads to a degradation of PKC. This phenomenon is sometimes used experimentally to produce a PKC-depleted cell (at least for phorbol esterbinding isoforms) and thereafter to test for a loss of putative PKC functions.

Spatial Localization Regulates Protein Kinases and Phosphatases

Protein kinases and protein phosphatases often are positioned spatially near their substrates or they translocate to their substrates upon activation to improve speed and specificity in response to neurotransmitter stimulation. For example, A Kinase Anchoring Protein 79 (AKAP79), although first identified with PKA binding, also has binding site for PKC and calcineurin (Klauck et al., 1996). Another example of a signaling complex is the protein termed yotiao, which binds to the NMDA type glutamate receptor and serves as an anchor for both PKA and a phosphatase (PP–1).

The use of anchoring proteins has several consequences. First, rate of phosphorylation and specificity are enhanced when kinases or phosphatases are concentrated near intended substrates. Second, it increases the signal-to-noise ratio for substrates that are not near anchoring proteins by reducing basal state phosphorylation. For example, PKA is anchored on the Golgi away from the nucleus so that phosphorylation in the basal state or even after a brief stimulus produces little phosphorylation of nuclear proteins. Prolonged stimuli, however, enable some C subunits to diffuse passively through nuclear pores and regulate gene expression. Termination of the nuclear action of C subunits is aided by PKI, which acts to inhibit and export it back out of the nucleus. Third, anchoring can enable significant phosphorylation of nearby substrates at basal cAMP, such as a Ca^{2+} channel phosphorylated when its phosphorylation site is exposed during depolarization.

The Cognitive Kinases

The ability of three major Ser/Thr kinases (PKA, CaM kinase II, and PKC) in brain to initiate or maintain synaptic changes that underlie learning and memory may require that they themselves undergo some form of persistent change in activity. Both their functional and molecular properties led to their description as cognitive kinases.

cAMP-Dependent Protein Kinase

A role for PKA as a cognitive kinase can be seen in long-term facilitation of the gill-withdrawal reflex in *Aplysia* and in long-term potentiation in the rodent hippocampus. In motor neuron cultures, repeated or prolonged exposure to serotonin or cAMP leads to long-term facilitation because PKA becomes persistently active despite the fact that cAMP is no longer elevated (Chain et al., 1999). During such activation there is a preferential degradation and decrease in the inhibitory RII subunits and thus a slight excess of C subunits that remain persistently active because of insufficient RII subunits. The C subunit then enters the nucleus and induces expression of one protein that facilitates further proteolysis of RII. In this interesting process, a molecular memory of appropriate stimulation by serotonin is encoded by a persistence of PKA activity that is regenerative.

Ca^{2+}-Calmodulin-Dependent Protein Kinase

CaM kinase II has features of a cognitive kinase because it has a molecular memory based on autophosphorylation and it phosphorylates proteins that modulate synaptic plasticity (Lisman et al., 2002). The biochemical properties of CaM kinase II suggest mechanisms by which appropriate stimulus frequencies can generate an autonomous enzyme (Fig. 10.13). At low stimulus frequency, the time between stimuli is sufficient for calmodulin to dissociate and the kinase to

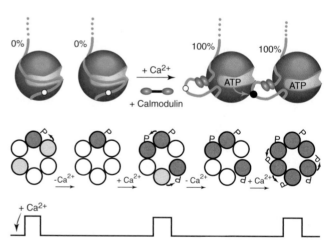

FIGURE 10.13 Frequency-dependent activation of CaM kinase II. Autophosphorylation occurs when both neighboring subunits in a holoenzyme are bound to calmodulin. At a high frequency of stimulation (rapid Ca^{2+} spikes), the interspike interval is too short to allow significant dephosphorylation or dissociation of calmodulin, thereby increasing the probability of autophosphorylation with each successive spike. In a simplified CaM kinase with only six subunits, calmodulin-bound subunits are shown in pink and autophosphorylated subunits with trapped calmodulin are shown in red. Adapted from Hudmon and Schulman (2002).

be dephosphorylated, and the same submaximal activation will occur with each stimulus. At higher frequencies, however, some subunits will remain autophosphorylated and bound to calmodulin so successive stimuli will result in more calmodulin bound per holoenzyme, which will make autophosphorylation more probable because it requires two active proximate neighboring subunits. The enzyme is therefore able to decode the frequency of cellular stimulation and translate this into a prolonged activated state.

CaM kinase II phosphorylates a number of substrates that affect synaptic strength. Inhibition of CaM kinase II in hippocampal slices or just elimination of its autophosphorylation by an α-CaM kinase II mouse knock-in in which the critical Thr was replaced by Ala blocks autonomy and the induction of long-term potentiation. These mice are deficient in learning spatial navigational cues, one of the functions of the rodent hippocampus. The basis for its role is uncertain but may be the phosphorylation of AMPA receptors and their recruitment to the membrane, leading to a greater postsynaptic response (Lisman *et al.*, 2002).

Protein Kinase C

PKC can also be converted into a form that is independent, or autonomous, of its second messenger and can be described as a cognitive kinase. Physiological activation of PKC can lead to proteolyic removal of its inhibitory domain, thus converting it to a constitutively active kinase termed protein kinase M (PKM). However, during the persistent phase of long-term potentiation, some of the PKC is converted to PKM (Serrano *et al.*, 2005). PKC (and PKM) substrates associ-

ated with long-term potentiation include NMDA and AMPA receptors.

Protein Tyrosine Kinases Take Part in Cell Growth and Differentiation

Protein kinases that phosphorylate tyrosine residues usually are associated with the regulation of cell growth and differentiation. Signal transduction by protein tyrosine kinases often includes a cascade of kinases phosphorylating other kinases, eventually activating Ser/Thr kinases, which carry out the intended modification of a cellular process. There are both receptor tyrosine kinases, activated by the binding of extracellular growth factors such as nerve growth factor and epidermal growth factor and soluble ones, activated indirectly by extracellular ligands such as c-Src.

Protein Phosphatases Undo What Kinases Create

Protein phosphatases in neuronal signaling are categorized as either phosphoserine-phosphothreonine phosphatases (PSPs) or phosphotyrosine phosphatases (PTPs) (Hunter, 1995; Mansuy and Shenolikar, 2006). The enzymes catalyze the hydrolysis of the ester bond of the phosphorylated amino acids to release inorganic phosphate and the unphosphorylated protein. A limited number of multifunctional PSPs account for most of such phosphatase activity in cells (Hunter, 1995). They are categorized into six groups (1, 4, 5, 2A, 2B, and 2C) on the basis of their substrates, inhibitors, and divalent cation requirements (Table 10.1). Of these

TABLE 10.1 Categories of Protein Phosphatases[a]

Phosphatase	Characteristic	Other Inhibitors
PP-1	Sensitive to phospho-inhibitor-1, phospho-DARPP-32, and inhibitor-2; has targeting subunits	Weakly sensitive to okadaic acid
PP-4	Nuclear	Highly sensitive to okadaic acid
PP-5	Nuclear	Mildly sensitive to okadaic acid
PP-2A	Regulatory subunits Does not require divalent cation	Highly sensitive to okadaic acid
PP-2B (calcineurin)	Ca^{2+}/calmodulin-dependent CnB regulatory subunit	FK506, cyclosporin
PP-2C	Requires Mg^{2+}	EDTA
Receptor PTPs[b]	Plasma membrane	Vanadate, tyrphosphtin, erbstatin
Nonreceptor PTPs	Various cellular compartments	Vanadate, tyrphosphtin
Dual specificity PTPs	Nuclear (e.g., cdc25A/B/C and VH family)	Vanadate

[a]From Hunter (1995).
[b]Protein phosphotyrosine phosphatases.

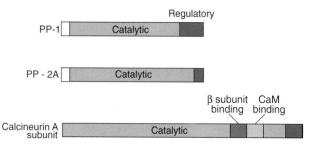

FIGURE 10.14 Domain structure of the catalytic subunits of some Ser/Thr phosphatases. The three major phosphoprotein phosphatases, PP-1, PP-2A, and calcineurin, have homologous catalytic domains but differ in their regulatory properties.

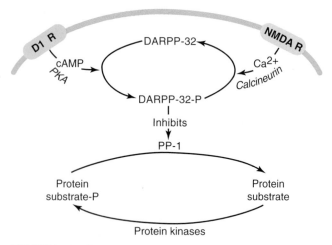

FIGURE 10.15 Cross talk between kinases and phosphatases. The state of phosphorylation of protein substrates is regulated dynamically by protein kinases and phosphatases. In the striatum, for example, dopamine stimulates PKA, which converts DARPP-32 into an effective inhibitor of PP-1. This increases the steady-state level of phosphorylation of a hypothetical substrate subject to phosphorylation by a variety of protein kinases. This action can be countered by NMDA receptor stimulation by another stimulus that increases intracellular Ca^{2+} and activates calcineurin. PP-1 is deinhibited and dephosphorylates the phosphorylated substrate when calcineurin deactives DARPP-32-P. Adapted from Svenningsson *et al.* (2004).

PSPs, only protein phosphatase 2B (PP-2B, or calcineurin) responds directly to a second messenger, Ca^{2+}. The specificity of PP-1 and PP-2A is particularly broad, and each can remove phosphates that were transferred by any of the protein kinases discussed herein as well as many other kinases. Phosphotyrosine phosphatases constitute a distinct and larger class of phosphatases, including PTPs with dual specificity for both phosphotyrosines and phosphoserine-phosphothreonines. PTPs are either soluble enzymes or membrane proteins with variable extra cellular domains that enable regulation by extracellular binding of either soluble or membrane-bound signals.

Structure and Regulation of PP-1 and Calcineurin

PP-1 and calcineurin are the best characterized phosphatases with regard to both structure and regulation. The domain structures of the catalytic subunits of PP-1 and calcineurin are depicted in Figure 10.14. PP-1 is a protein of 35–38 kDa; most of the sequence forms the catalytic domain; its C-terminal is the site of regulatory phosphorylation. The catalytic domains of PP-1, PP-2A, and calcineurin are highly homologous.

PP-1 and PP-2A normally are complexed in cells with specific anchoring or targeting subunits. Targeting of PP-1 can be modulated by phosphorylation of its targets. As PP-1 dissociates from targeting subunits, it becomes susceptible to inhibition by inhibitor-2.

Inhibition of PP-1 by two other inhibitors, inhibitor-1 and its homologue DARPP-32 (dopamine and cAMP-regulated phosphoprotein; M_r 32,000), is conditional on their phosphorylation by either PKA or PKG (Fig. 10.15). Because the substrates for PKA and PP-1 overlap to a great extent, the rate and extent of phosphorylation of such substrates are enhanced by the ability of PKA to catalyze their phosphorylation while blocking their dephosphorylation via PP-1.

Inhibitor-1, DARPP-32, and inhibitor-2 are all selective for PP-1. Highly selective inhibitors capable of penetrating the cell membrane are available for these phosphatases. Okadaic acid, a natural product of marine dinoflagellates, is a tumor promoter but, unlike phorbol esters, it acts on PP-2A and PP-1 rather than on PKC.

Protein Phosphatase 1 The X-ray structure of the catalytic subunit of PP-1 bound to the toxin microcystin, a cyclic peptide inhibitor, reveals PP-1 to be a compact ellipsoid with hydrophobic and acidic surfaces forming a cleft for binding substrates. PP-1 is a metalloenzyme requiring two metals in the active site that likely take part in electrostatic interactions with the phosphate on substrates that aid in catalyzing the hydrolytic reaction. Substrate binding is blocked when phospho-inhibitor-1 or microcrystin LR binds to this surface.

Calcineurin (PP-2B) Calcineurin is a Ca^{2+}-calmodulin-dependent phosphatase that is highly enriched in the brain. It is a heterodimer with a 60-kDa subunit (CnA) that contains an N-terminal catalytic domain similar to PP-1 and a C-terminal regulatory domain that includes an autoinhibitory segment, a calmodulin-binding domain, and a binding site for the 19-kDa regulatory B subunit (CnB). CnB is a calmodulin-like Ca^{2+}-binding protein that binds to a hinge region of CnA. Some activation of calcineurin is attained by binding of Ca^{2+} to CnB. Stronger activation is obtained by the binding of Ca^{2+}-camodulin. Additional regulation may be accorded by interaction of its hinge region with cyclophilin and FKBP (FK506-binding protein), proteins that bind the immunosuppressive agents cyclosporin and FK506, respectively.

The Ca^{2+}-calmodulin sensitivity of calcineurin and CaM kinase II are quite different. Weak or low-frequency stimuli may selectively activate calcineurin whereas strong or high-frequency stimuli activate CaM kinase II and calcineurin. This difference may play a role in the bidirectional control of synaptic strength (depression vs. potentiation) by low-and high-frequency stimulation.

Protein Kinases, Protein Phosphatases, and Their Substrates Are Integrated Networks

Cross talk between protein kinases and protein phosphatases is critical to their ability to integrate inputs into neurons (Cohen, 1992). Such cross talk is exemplified by the interaction of cAMP and Ca^{2+} signals through PKA and calcineurin, respectively. The medium spiny neurons in the neostriatum receive cortical inputs from glutamatergic neurons that are excitatory and nigral inputs by dopaminergic neurons that inhibit them. A possible signal transduction scheme for this regulation is shown in Figure 10.15. The key to regulation is the bidirectional control of DARPP-32 phosphorylation (Svenningsson et al., 2004). Glutamate activates calcineurin by increasing intracellular Ca^{2+}, leading to the dephosphorylation and inactivation of phospho-DARPP-32. This releases inhibition of PP-1, which can then dephosphorylate a variety of substrates, including Na^+, K^+-ATPase, and lead to membrane depolarization. This is countered by dopamine, which stimulates cAMP formation and activation of PKA, which then converts DARPP-32 into its phosphorylated (i.e., PP-1 inhibitory) state. There are many other receptors and signaling integrated by these pathways. For example, adenosine, serotonin, and VIP act through their cognate receptors to elevate cAMP, similarly to dopamine at D1 receptors, whereas opiates can signal to inhibit the action of dopamine at D1 and adenosine at A2A receptors. Although PKA and calcineurin are acting in an antagonistic manner, they are not doing it by phosphorylating and dephosphorylating the ATPase. By their actions upstream, at the level of DARPP-32, the regulation of numerous target enzymes (e.g., Ca^{2+} channels

and Na^+ channels) in addition to the ATPase can be coordinated.

Studying Cellular Processes Controlled by Phosphorylation-Dephosphorylation

Major goals of signal transduction research are to delineate pathways by which signals such as neurotransmitters transduce their signals to modify cellular processes. This is often the start of a process to identify targets for therapeutic intervention in disease. Cellular and biochemical assays can often identify the entire signaling pathway, from stimulation of receptor, to generation of a second-messenger activation of a kinase or phosphatase, change in the phosphorylation state of the substrate, and an ultimate change in its functional state. Such investigations utilize a variety of pharmacological inhibitors or activators of the signaling molecules complemented by genetic approaches that utilize transfection of activated forms of the kinases or phosphatases in question, siRNAs, transgenic animals, and mice with individual signaling components knocked out.

Summary

The morphology of a cell is determined by protein constituents. Its function is regulated by the phosphorylation or dephosphorylation of the proteins. Phosphorylation modifies the function of regulatory proteins subsequent to their genetic expression. The activities of the protein kinases and protein phosphatases typically are regulated by second messengers and extracellular ligands. Kinases and phosphatases integrate and encode stimulation of a large group of cellular receptors. The number of possible effects is almost limitless and enables the tuning of cellular processes over a broad time scale. Most of the effects of Ca^{2+} in cells are mediated by calmodulin, which in turn mediates changes in protein phosphorylation-dephosphorylation. The phosphoinositol signaling system is mediated through PKC, which modulates many cellular processes from exocytosis to gene expression. All three classes of enzymes discussed have been described as cognitive kinases because they are capable of sustaining their activated states after their second-messenger stimuli have returned to basal levels. PKA has been implicated in learning and memory in *Aplysia* and in hippocampus, where it is involved in long-term potentiation. Protein phosphatases play an equally important role in neuronal signaling by dephosphorylating proteins. Cross talk between protein kinases and protein phosphatases is key to their ability to integrate inputs into neurons.

INTRACELLULAR SIGNALING AFFECTS NUCLEAR GENE EXPRESSION

The first part of this chapter describes how signaling systems regulate the function of cellular proteins already expressed; another critical level of control exerted by these systems is their ability to regulate the synthesis of cellular proteins by regulating the expression of specific genes. For all living cells, regulation of gene expression by intracellular signals is a fundamental mechanism of development, homeostasis, and adaptation to the environment. Protein phosphorylation and regulation of gene expression by intracellular signals are the most important mechanisms underlying the remarkable degree of plasticity exhibited by neurons. Alterations in gene expression underlie many forms of long-term changes in neural functioning, with a time course that ranges from hours to many years.

Interactions of Specific DNA Sequences with Regulatory Proteins Control Both Basal and Signal-Regulated Transcription

Information contained within DNA must be expressed through other molecules: RNA and proteins. The human genome contains approximately 25,000 genes that encode structural RNAs or protein-coding messenger RNAs (mRNAs). Regulated gene expression conferred by the nucleotide sequence of the DNA itself is called *cis* regulation because the control regions are linked physically on the DNA to regions that can potentially be transcribed. The *cis* regulatory sequences function by serving as high-affinity binding sites for regulatory proteins called transcription factors.

The transcription of specific genes into mRNA is carried out by a complex enzyme called RNA polymerase II. Roger Kornberg won the Nobel Prize in Chemistry in 2006 for his studies on the molecular basis for eukaryotic transcription. Transcription often is divided into three steps: initiation of RNA synthesis, RNA chain elongation, and chain termination. Extracellular signals, such as neurotransmitters, hormones, drugs, and growth factors generally control the transcription initiation step.

Transcription initiation requires two critical processes: (1) positioning of RNA polymerase II at the correct start site of the gene to be transcribed and (2) controlling the efficiency of initiations to produce the appropriate transcriptional rate for the circumstances of the cell (Tjian and Maniatis, 1994). The *cis*-regulatory elements that set the transcription start sites of genes are called the basal promoter. Other *cis*-

regulatory elements tether additional activator and represser proteins to the DNA to regulate the overall transcriptional rate (Fig. 10.16).

Sequence-Specific Transcription Factors

The promoters for RNA polymerase II transcribed genes contain a distinct basal promoter element on which a basal transcription complex is assembled. The basal promoter of most of these genes contains a sequence called a TATA box that is rich in the nucleotides adenine (A) and thymine (T) located between 25 and 30 bases upstream of the transcription start site.

To achieve significant levels of transcription, this multiprotein assembly requires help from sequence-specific transcriptional activators that recognize and bind distinct *cis*-regulatory elements. Functional *cis*-regulatory elements are generally 7–12 in length and structured as a palindrome, each of which is a specific binding site for one or more transcription factors. Each gene has a particular combination of *cis*-regulatory elements, the nature, number, and spatial arrangement of which determine the gene's unique pattern of expression, including the cell types in which it is expressed, the times during development in which it is expressed, and the level at which it is expressed in adults both basally and in response to physiological signals (Tjian and Maniatis, 1994).

Many transcription factors are active only as dimers or higher order complexes formed via a multimerization domain. Both partners in a dimer commonly contribute jointly to both the DNA-binding domain and the activation domain. Dimerization can be a mechanism of either positive or negative control of transcription.

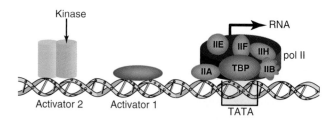

FIGURE 10.16 Schematic of a generalized RNA polymerase II promoter showing three separate *cis*-regulatory elements along a stretch of DNA. These elements are two hypothetical activator protein-binding sites and the TATA element. The TATA element is shown binding the TATA-binding protein (TBP). Multiple general transcription factors (IIA, IIB, etc.) and RNA polymerase II (pol II) associate with TBP. Each transcription factor comprises multiple individual proteins complexed together. This basal transcription apparatus recruits RNA polymerase II into the complex and also forms the substrate for interactions with the activator proteins binding to the activator elements shown. Activator 2 is shown to be a substrate for a protein kinase.

The effects of sequence-specific transcriptional activator and represser factors frequently are mediated by adapter proteins (Fig. 10.17). In many cases, these adapter proteins are enzymes, such as histone acyl transferases and deacetylases, with the ability to modify the structure of the proteins associated with the DNA and modulate transcription (Fig. 10.17).

A Significant Consequence of Intracellular Signaling Is the Regulation of Transcription

Intracellular signals play a major role in the regulation of gene expression, for example via nuclear translocation and/or phosphorylation of activator proteins. Signal-directed change in location or conformation of these proteins permit information obtained by the cell from its different signaling systems to regulate gene expression appropriate to the status of the cell.

Transcriptional Regulation by Intracellular Signals

Extracellular control of transcription requires a translocation step by which the signal is transmitted through the cytoplasm to the nucleus. Some transcription factors are themselves translocated to the nucleus. For example, the transcription factor NF-κB is retained in the cytoplasm by its binding protein IκB, which masks the NF-κB nuclear localization signal. Signal-regulated phosphorylation of IκB by PKC and other protein kinases leads to dissociation of NF-κB, permitting it to enter the nucleus. Other transcription factors must be directly phosphorylated or dephosphorylated to bind DNA. In many cytokine-signaling

pathways, plasma membrane receptor tyrosine phosphorylation of transcription factors known as signal transducers and activators of transcription (STATs) permits their multimerization, which in turn permits both nuclear translocation and construction of an effective DNA-binding site within the multimer. Yet other transcription factors, such as CREB (Fig. 10.17), already are bound to their cognate cis-regulatory elements and become able to activate transcription after phosphorylation by a kinase that translocates to the nucleus.

Role of cAMP and Ca²⁺ in the Activation Pathways of Transcription

The cAMP second-messenger pathway regulates expression of a large number of genes via cAMP response elements (CREs). Phosphorylation of CRE-bound CREB on its Ser-133 by activated PKA that translocates to the nucleus recruits the adapter protein CBP. This, in turn, interacts with the basal transcription complex and modifies histones to enhance the efficiency of transcription. CREB also serves to illustrate the convergence of signaling pathways (Fig. 10.18). CREB Ser-133 phosphorylation not only is mediated by PKA but also by CaM kinase types II and IV and by RSK2, a kinase activated in growth factor pathways, including Ras and MAP kinase.

CREB illustrates yet another important principle of transcriptional regulation: CREB is a member of a family of related proteins. Many transcription factors are members of families; this permits complex forms of positive and negative regulation. CREB is closely related to other proteins called activating transcription factors (ATFs) and CRE modulators (CREMs). The dimerization domain used by CREB-ATF proteins and several other families of transcription factors is called a leucine zipper. The dimerization motif is an α helix in which every seventh residue is a leucine; based on the periodicity of α helices, the leucines line up along one face of the helix two turns apart. Dimerization juxtaposes the adjacent basic, DNA binding, regions of each of the partners. This combination of motifs is why this superfamily of proteins is referred to as basic leucine zipper proteins (bZIPs).

AP-1 Transcription Factors

Activator protein 1 (AP-1) is another family of bZIP transcription factors that play a central role in the regulation of neural gene expression by extracellular signals. The AP-1 family comprises multiple proteins that bind as heterodimers (and a few as homodimers) to the DNA sequence TGACTCA. Although the AP-1 sequence differs from the CRE sequence (TGACGTCA) by only a single base, this one-base difference strongly

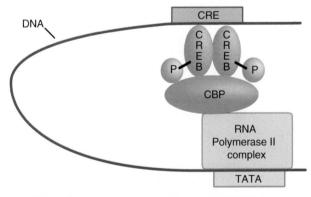

FIGURE 10.17 Looping of DNA permits activator (or represser) proteins binding at a distance to interact with the basal transcription apparatus. The basal transcription apparatus is shown as a single box (pol II complex) bound at the TATA element. The activator protein (CREB) is shown as having been phosphorylated. On phosphorylation, many activators, such as CREB, are able to recruit adaptor proteins that mediate between the activator and the basal transcription apparatus. An adaptor protein that binds phosphorylated CREB is called a CREB-binding protein (CBP).

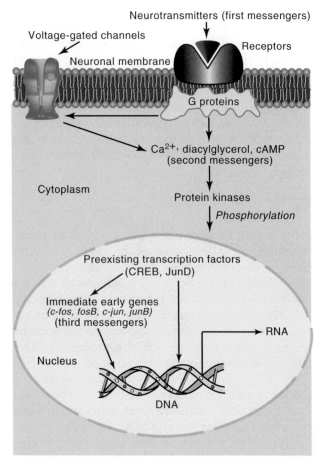

FIGURE 10.18 Signal transduction to the nucleus. In this schematic, activation of a neurotransmitter receptor activates cellular signals (G proteins and second-messenger systems). These signals, in turn, regulate the activation of protein kinases, which translocate to the nucleus. Within the nucleus, protein kinases can activate genes regulated by constitutively synthesized transcription factors. A subset of these genes encodes additional transcription factors (third messengers), which can then activate multiple downstream genes.

(within minutes), transiently, and without requiring new protein synthesis often are described as cellular immediate-early genes (IEGs). Genes that are induced or repressed more slowly (within hours) and are dependent on new protein synthesis have been described as late-response genes. Several IEGs have been used as cellular markers of neural activation because they are markedly induced by depolarization (the critical signal being Ca^{2+} entry) and second-messenger and growth factor pathways, permitting novel approaches to functional neuroanatomy (Chapter 39, Box 39.4).

The protein products of those cellular IEGs that function as transcription factors bind to *cis*-regulatory elements contained within a subset of late-response genes to activate or repress them (Fig. 10.18). In sum, neural genes that are regulated by extracellular signals are activated or repressed with varying time courses by reversible phosphorylation of constitutively synthesized transcription factors and by newly synthesized transcription factors, some of which are regulated as IEGs.

Activation of the c-*fos* Gene

The c-*fos* gene is activated rapidly by neurotransmitters or drugs that stimulate the cAMP pathway or Ca^{2+} elevation. Both pathways produce phosphorylation of transcription factor CREB. The c-*fos* gene contains three binding sites for CREB. The c-*fos* gene can also be induced by the Ras/MAP kinase pathway, which is activated by a number of growth factors. For example, neurotrophins, such as nerve growth factor (NGF), bind a family of receptor tyrosine kinases (Trks); NGF interacts with Trk A, which activates Ras. Ras then acts through a cascade of protein kinases (Chapter 21). Cross talk between neurotransmitter and growth factor-signaling pathways has been documented with increasing frequency and likely plays an important role in the precise tuning of neural plasticity to diverse environmental stimuli.

Regulation of c-Jun

Expression of most of the proteins of the Fos and Jun families that constitute transcription factor AP-1 and the binding of AP-1 proteins to DNA is regulated by extracellular signals. Phosphorylation and activation of c-Jun can result from the action of Jun N-terminal kinase (JNK). JNK is a member of the mitogen activated protein kinase (MAPK) family of protein kinases. JNK also has been shown to be activated by neurotransmitters, including glutamate. Thus, AP-1-mediated transcription within the nervous system requires multiple steps, beginning with the activation of genes encoding AP-1 proteins.

biases protein binding away from the CREB family of proteins. AP-1 sequences confer responsiveness to the PKC pathway. AP-1 proteins generally bind DNA as heterodimers composed of one member each of two different families of related bZIP proteins, the Fos family (c-Fos, Fra-1, Fra-2, and FosB) and the Jun family (c-Jun, JunB, and JunD), providing for a multiplicity of regulatory control.

Cellular Immediate-Early Genes

Genes that are activated transcriptionally by synaptic activity, drugs, and growth factors often have been classified roughly into two groups. Genes, such as the c-*fos* gene itself, that are activated rapidly

Cytokines as Inducers of Gene Expression in the Nervous System

With regard to function, the boundary between trophic, or growth, factors and cytokines in the nervous system has become increasingly arbitrary. However, cell-signaling mechanisms offer a useful means of distinction. Growth factors, such as neurotrophins (e.g., nerve growth factor, brain-derived neurotrophic factor, and neurotrophin 3), epidermal growth factor (EGF), and fibroblast growth factor (FGF), act through receptor protein tyrosine kinases, whereas cytokines, such as leukemia inhibitory factor (LIF), ciliary neurotrophic factor (CNTF), and interleukin-6 (IL-6), act through nonreceptor protein tyrosine kinases.

LIF, CNTF, and IL-6 subserve a wide array of overlapping functions inside and outside the nervous system, including hematopoietic and immunologic functions outside the nervous system and regulation of neuronal survival, differentiation, and, in certain circumstances, plasticity within the nervous system. Their receptors contain a common signal-transducing subunit, gp130. Receptors for these cytokines consist of a signal-transducing β component, which includes gp130, and interacts with nonreceptor protein tyrosine kinases (PTKs) of the Janus kinase (Jak) family (e.g., Jakl, Jak2, and Tyk2). Some cytokine receptors interact only with a single Jak PTK whereas others interact with multiple Jak PTKs.

Signal transduction to the nucleus includes tyrosine phosphorylation by the Jak PTKs of one or more of the STAT proteins mentioned earlier. Upon phosphorylation, STAT proteins form dimers through the association of SH2 domains, an important type of protein interaction domain. Dimerization is thought to trigger translocation to the nucleus, where STATs bind their cognate cytokine response elements. Different STATs become activated by different cytokine receptors, not because of differential use of Jak PTKs, but because of specific coupling of certain STATs to certain receptors. Thus, for example, the IL-6 receptor preferentially activates STAT1 and STAT3; the CNTF receptor preferentially activates STAT3. Cytokine response elements, to which STATs bind, have now been identified within many neural genes, including vasoactive intestinal polypeptide and several other neuropeptide genes.

Steroid Hormone Receptors

The differentiation of many cell types in the brain is established by exposure to steroids. Steroid hormones, including glucocorticoids, sex steroids, mineralocorticoids, retinoids, thyroid hormone, and vitamin D, are small lipid-soluble ligands that can diffuse across cell membranes. They act on their receptors within the cell cytoplasm in marked distinction to the other types of intercellular signals described herein. Another unique feature of steroid hormones is that their receptors are themselves transcription factors. Each has a transcriptional-activation domain at its amino terminus, a DNA-binding domain, and a hormone-binding domain at its carboxy terminus. DNA-binding domains recognize specific palindromic DNA sequences, steroid hormone response elements, within the regulatory regions of specific genes.

After having been bound by hormone, activated steroid hormone receptors translocate into the nucleus, where they bind to their cognate response elements. Such binding then increases or decreases the rate at which these target genes are transcribed, depending on the precise nature and DNA sequence context of the element.

Summary

The formation of long-term memories requires changes in gene expression and new protein synthesis. It is at transcription initiation that extracellular signals such as neurotransmitters, hormones, drugs, and growth factors exert their most significant control. The transcription is modulated by transcription factors that recruit the RNA polymerases to the DNA. For example, the critical nuclear translocation step in the activation of transcription factor CREB involves the catalytic subunit of PKA, which can phosphorylate CREB on entering the nucleus. In addition, increasing evidence indicates that at least some forms of long-term memory require new gene expression.

Genes that encode the transcription factors themselves may respond quickly or slowly. These genes have been coined third messengers in signal transduction cascades. Cross talk between neurotransmitter and growth factor-signaling pathways is likely to play an important role in the precise tuning of neuronal plasticity to diverse environmental stimuli.

The active, mature transcription complex is a remarkable architectural assembly of RNA polymerase II, transcription factors, and adaptors assembled at the basal promoter. Cells can exert exquisite control of the genes being transcribed in a variety of situations; for example, to govern appropriate entry or exit from the cell cycle, to maintain appropriate cellular identity, and to respond appropriately to extracellular signals.

Transcription can be regulated by many different extracellular signals modulated by a large array of signaling pathways (many including reversible phosphorylation) and a complex array of typically dimeric transcription factors. In this chapter, regulation has been illustrated by only a few of the families of transcription factors. Those chosen appear to play important roles in the nervous system and illustrate many of the basic principles of gene regulation.

References

Berridge, M. J. (1993). Inositol trisphosphate and calcium signalling. *Nature* **361**, 315–325.

Birnbaumer, L. (2007). Expansion of signal transduction by G proteins. The second 15 years or so: From 3 to 16 α subunits plus $\beta\gamma$ dimers. *Biochem. Biophys. Acta* **1768**, 772–793.

Bourne, H. R. and Nicoll, R. (1993). Molecular machines integrate coincident synaptic signals. *Cell* **72**, 65–75.

Chain, D. G., Casadio, A., Schacher, S., Hegde, A. N., Valbrun, M., Yamamoto, N., Goldberg, A. L., Bartsch, D., Kandel, E. R., and Schwartz, J. H. (1999). Mechanisms for generating the autonomous cAMP-dependent protein kinase required for long-term facilitation in *Aplysia*. *Neuron* **22**, 147–156.

Clapham, D. E. and Neer, E. J. (1993). New roles for G-protein $\beta\gamma$ dimers in transmembrane signalling. *Nature* **365**, 403–406.

Cohen, P. (1992). Signal integration at the level of protein kinases, protein phosphatases and their substrates. *TIBS* **17**, 408–413.

Greengard, P. (2001). The neurobiology of slow synaptic transmission. *Science* **294**, 1024–1030.

Hille, B. (1992). G protein-coupled mechanisms and nervous signaling. *Neuron* **9**, 187–195.

Hudmon, A. and Schulman, H. (2002) Neuronal Ca^{2+}/calmodulin-dependent protein kinase II: The role of structure and autoregulation in cellular function. *Annu. Rev. Biochem.* **71**, 473–510.

Hunter, T. (1995). Protein kinases and phosphatases: The yin and yang of protein phosphorylation and signaling. *Cell* **80**, 225–236.

Kemp, B. E., Faux, M. C., Means, A. R., House, C., Tiganis, T., Hu, S.-H., and Mitchelhill, K. I. (1994). Structural aspects: Pseudo-substrate and substrate interactions. *In* "Protein Kinases" (J. R. Woodgett, ed.), pp. 30–67.

Klauck, T. M., Faux, M. C., Labudda, K., Langeberg, L. K., Jaken, S., and Scott, J. D. (1996). Coordination of three signaling enzymes by AKAP79, a mammalian scaffold protein. *Science* **271**, 1589–1592.

Lisman, J., Schulman, H., and Cline, H. (2002). The molecular basis of CaMKII function in synaptic plasticity and behavioural memory. *Nat. Rev. Neurosci.* **3**, 175–190.

Mansuy, I. M. and Shenolikar, S. (2006). Protein serine/threonine phosphatases in neuronal plasticity and disorders of learning and memory. *Trends Neurosci.* **29**, 679–686.

Ross, E. M. (1989). Signal sorting and amplification through G protein-coupled receptors. *Neuron* **3**, 141–152.

Schulman, H. and Braun, A. (1999). Ca^{2+} calmodulin-dependent protein kinases. *In* "Calcium as a Cellular Regular" (E. Carafoli and C. Klee, eds.), pp. 311–343. Oxford Univ. Press, New York.

Serrano, P., Yao, Y., and Sacktor, T. C. (2005). Persistent phosphorylation by protein kinase Mζ maintains late-phase long-term potentiation. *J. Neurosci.* **25**,1979–1984.

Stryer, L. and Bourne, H. R. (1986). G proteins: A family of signal transducers. *Ann. Rev. Cell Biol.* **2**, 391–419.

Svenningsson, P., Nishi, A., Fisone, G., Girault, J.–A., Nairn, A. C., and Greengard, P. (2004). DARPP-32: An Integrator of Neurotransmission. *Annu. Rev. Pharmacol. Toxicol.* **44**, 269–296.

Tanaka, C. and Nishizuka, Y. (1994). The protein kinase C family for neuronal signaling. *Annu. Rev. Neurosd.* **17**, 551–567.

Taussig, R. and Gilman, A. G. (1995). Mammalian membrane-bound adenylyl cyclases. *J. Bio. Chem.* **270**, 1–4.

Tjian, R. and Maniatis, T. (1994). Transcription activation: A complex puzzle with few easy pieces. *Cell* **77**, 5–8.

Stryer, L. (1995). "Biochemistry," 4th ed. Freeman, New York.

Suggested Readings

Boehning, D. and Snyder, S. H. (2003). Novel neural modulators. *Annu. Rev. Neurosci.* **26**, 105–131.

Carafoli, E. and Klee, C. (1999). "Calcium as a Cellular Regulator," Oxford Univ. Press, New York.

Cohen, P. and Klee, C. B. (1988). Calmodulin. In "Molecular Aspects of Cellular Regulation," Vol. 5. Elsevier, Amsterdam.

Nairn, A. C., Hemmings, H. C., Jr., and Greengard, P. (1985). Protein kinases in the brain. *Annu. Rev. Biochem.* **54**, 931–976.

Ubersax, J. A. and Ferrell, J. E. Jr. (2007). Mechanisms of specificity in protein phosphorylation. *Nature Rev. Cell Biol.* **8**, 530–541.

Howard Schulman and James L. Roberts

11

Postsynaptic Potentials and Synaptic Integration

The study of synaptic transmission in the central nervous system (CNS) provides an opportunity to learn more about the diversity and richness of mechanisms underlying this process and to learn how some of the fundamental signaling properties of the nervous system, such as action potentials and synaptic potentials, work together to process information and generate behavior.

Postsynaptic potentials (PSPs) in the CNS can be divided into two broad classes on the basis of mechanisms and, generally, duration of these potentials. One class is based on the direct binding of a transmitter molecule(s) with a receptor-channel complex; these receptors are ionotropic. The structure of these receptors is discussed in detail in Chapter 9. The resulting PSPs are generally short-lasting and hence sometimes are called fast PSPs; they have also been referred to as "classical" because they were the first synaptic potentials to be recorded in the CNS (Eccles, 1964; Spencer, 1977). The duration of a typical fast PSP is about 20 ms.

The other class of PSPs is based on the indirect effect of a transmitter molecule(s) binding with a receptor. The receptors that produce these PSPs are metabotropic. As discussed in Chapter 9, the receptors activate G proteins (G-protein-coupled receptors; GPCRs) that affect the channel either directly or through additional steps in which the level of a second messenger is altered. The changes in membrane potential produced by metabotropic receptors can be long-lasting and therefore are called slow PSPs. The mechanisms for fast PSPs mediated by ionotropic receptors are considered first.

IONOTROPIC RECEPTORS: MEDIATORS OF FAST EXCITATORY AND INHIBITORY SYNAPTIC POTENTIALS

The Stretch Reflex Is Useful to Examine the Properties and Functional Consequences of Ionotropic PSPs

The stretch reflex, one of the simpler behaviors mediated by the central nervous system, is a useful example with which to examine the properties and functional consequences of ionotropic PSPs. The tap of a neurologist's hammer to a ligament elicits a reflex extension of the leg, as illustrated in Figure 11.1. The brief stretch of the ligament is transmitted to the extensor muscle and is detected by specific receptors in the muscle and ligament (Chapter 29). Action potentials initiated in the stretch receptors are propagated to the spinal cord by afferent fibers (Chapter 29). The receptors are specialized regions of sensory neurons with somata located in the dorsal root ganglia just outside the spinal column. Axons of the afferents enter the spinal cord and make excitatory synaptic connections with at least two types of postsynaptic neurons. First, a synaptic connection is made to the extensor motor neuron. As the result of its synaptic activation, the motor neuron fires action potentials that propagate out of the spinal cord and ultimately invade the terminal regions of the motor axon at neuromuscular junctions. There, acetylcholine (ACh) is released, nicotinic ACh receptors are activated, an end plate potential (EPP) is produced, an action potential is initiated in the muscle cell, and the muscle cell is contracted, producing the

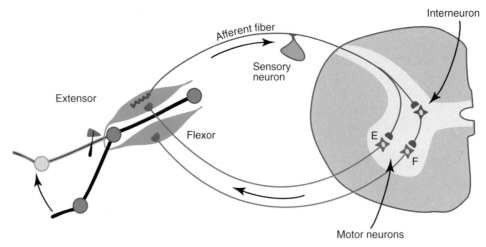

FIGURE 11.1   Features of the vertebrate stretch reflex. Stretch of an extensor muscle leads to the initiation of action potentials in the afferent terminals of specialized stretch receptors. The action potentials propagate to the spinal cord through afferent fibers (sensory neurons). The afferents make excitatory connections with extensor motor neurons (E). Action potentials initiated in the extensor motor neuron propagate to the periphery and lead to the activation and subsequent contraction of the extensor muscle. The afferent fibers also activate interneurons that inhibit the flexor motor neurons (F).

reflex extension of the leg. Second, a synaptic connection is made to another group of neurons called interneurons (nerve cells interposed between one type of neuron and another). The particular interneurons activated by the afferents are inhibitory interneurons because activation of these interneurons leads to the release of a chemical transmitter substance that inhibits the flexion motor neuron. This inhibition tends to prevent an uncoordinated (improper) movement (i.e., flexion) from occurring. The reflex system illustrated in Figure 11.1 also is known as the monosynaptic stretch reflex because this reflex is mediated by a single ("mono") excitatory synapse in the central nervous system. Spinal reflexes are described in greater detail in Chapter 29.

Figure 11.2 illustrates procedures that can be used to experimentally examine some of the components of synaptic transmission in the reflex pathway for the stretch reflex. Intracellular recordings are made from one of the sensory neurons, the extensor and flexor motor neurons, and an inhibitory interneuron. Normally, the sensory neuron is activated by stretch to the muscle, but this step can be bypassed by simply injecting a pulse of depolarizing current of sufficient magnitude into the sensory neuron to elicit an action potential. The action potential in the sensory neuron leads to a potential change in the motor neuron known as an excitatory postsynaptic potential (EPSP; Fig. 11.2).

Mechanisms responsible for fast EPSPs mediated by ionotropic receptors in the CNS are fairly well known. Moreover, the ionic mechanisms for EPSPs in the CNS are essentially identical with the ionic mechanisms at the skeletal neuromuscular junction. Specifically, the transmitter substance released from the presynaptic terminal (Chapters 7 and 8) diffuses across the synaptic cleft, binds to specific receptor sites on the postsynaptic membrane (Chapter 9), and leads to a simultaneous increase in permeability to Na^+ and K^+, which makes the membrane potential move toward a value of about 0 mV. However, the processes of synaptic transmission at the sensory neuron–motor neuron synapse and the motor neuron–skeletal muscle synapse differ in two fundamental ways: (1) in the transmitter used and (2) in the amplitude of the PSP. The transmitter substance at the neuromuscular junction is ACh, whereas that released by the sensory neurons is an amino acid, probably glutamate. Indeed, glutamate is the most common transmitter that mediates excitatory actions in the CNS. The amplitude of the postsynaptic potential at the neuromuscular junction is about 50 mV; consequently, each PSP depolarizes the postsynaptic cell beyond threshold so there is a one-to-one relation between an action potential in the spinal motor neuron and an action potential in the skeletal muscle cell. Indeed, the EPP must depolarize the muscle cell by only about 30 mV to initiate an action potential, allowing a safety factor of about 20 mV. In contrast, the EPSP in a spinal motor neuron produced by an action potential in an afferent fiber has an amplitude of only about 1 mV. The mechanisms by which these small PSPs can trigger an action potential in the postsynaptic neuron are discussed in a later section of this chapter and in Chapter 12.

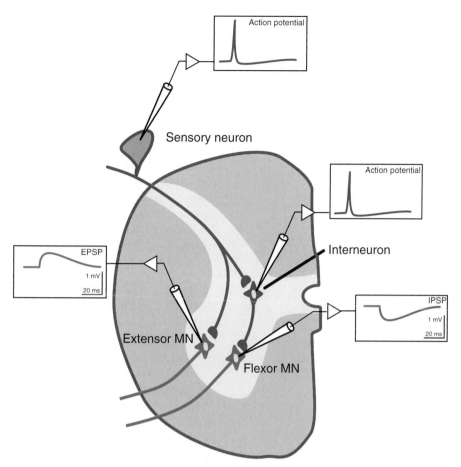

FIGURE 11.2 Excitatory (EPSP) and inhibitory (IPSP) postsynaptic potentials in spinal motor neurons. Idealized intracellular recordings from a sensory neuron, interneuron, and extensor and flexor motor neurons (MNs). An action potential in the sensory neuron produces a depolarizing response (an EPSP) in the extensor motor neuron. An action potential in the interneuron produces a hyperpolarizing response (an IPSP) in the flexor motor neuron.

Macroscopic Properties of PSPs Are Determined by the Nature of Gating and Ion-Permeation Properties of Single Channels

Patch-Clamp Techniques

Patch-clamp techniques (Hamill *et al.*, 1981), with which current flowing through single isolated receptors can be measured directly, can be sources of insight into both the ionic mechanisms and the molecular properties of PSPs mediated by ionotropic receptors. This approach was pioneered by Erwin Neher and Bert Sakman in the 1970s and led to their being awarded the Nobel Prize in Physiology or Medicine in 1991.

Figure 11.3A illustrates an idealized experimental arrangement of an "outside-out" patch recording of a single ionotropic receptor. The patch pipette contains a solution with an ionic composition similar to that of the cytoplasm, whereas the solution exposed to the outer surface of the membrane has a composition similar to that of normal extracellular fluid. The electrical potential across the patch, and hence the transmembrane potential (V_m), is controlled by the patch-clamp amplifier. The extracellular (outside) fluid is considered "ground." Transmitter can be delivered by applying pressure to a miniature pipette filled with an agonist (in this case, ACh), and the current (I_m) flowing across the patch of membrane is measured by the patch-clamp amplifier (Fig. 11.3). Pressure in the pipette that contains ACh can be continuous, allowing a constant stream of ACh to contact the membrane, or can be applied as a short pulse to allow a precisely timed and discrete amount of ACh to contact the membrane. The types of recordings obtained from such an experiment are illustrated in the traces in Figure 11.3. In the absence of ACh, no current flows through the channel (Fig. 11.3A). When ACh is applied continuously, current flows across the membrane (through the channel), but the current does not flow continuously; instead, small step-like changes in current are observed (Fig. 11.3B). These changes represent the probabilistic (random) opening and closing of the channel.

A

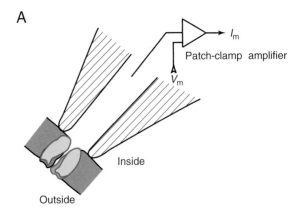

Patch-clamp amplifier

I_m

V_m

Inside

Outside

Single-channel current

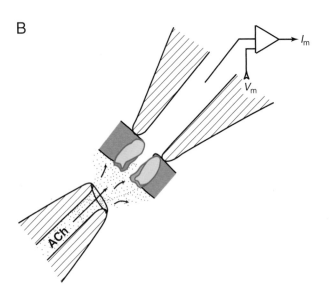

B

I_m

V_m

ACh

Closed   Closed   Closed   Closed

Open   Open   Open   Open | 4 pA

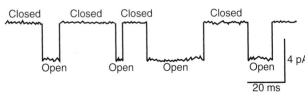

20 ms

FIGURE 11.3 Single-channel recording of ionotropic receptors and their properties. (A) Experimental arrangement for studying properties of ionotropic receptors. (B) Idealized single-channel currents in response to application of ACh.

Channel Openings and Closings

As a result of the type of patch-recording techniques heretofore described, three general conclusions about the properties of ligand-gated channels can be drawn. First, ACh, as well as other transmitters that activate ionotropic receptors, causes the opening of individual ionic channels (for a channel to open, usually two mol-

ecules of transmitter must bind to the receptor). Second, when a ligand-gated channel opens, it does so in an all-or-none fashion. Increasing the concentration of transmitter in the ejection microelectrode does not increase the permeability (conductance) of the channel; it increases its probability (P) of being open. Third, the ionic current flowing through a single channel in its open state is extremely small (e.g., 10^{-12} A); as a result, current flowing through any single channel makes only a small contribution to the normal postsynaptic potential. Physiologically, when a larger region of the postsynaptic membrane, and thus more than one channel, is exposed to a released transmitter, the net conductance of the membrane increases due to the increased probability that a larger population of channels will be open at the same time. The normal PSP, measured with standard intracellular recording techniques (e.g., Fig. 11.2), is then proportional to the sum of the currents that flow through these many individual open channels. The properties of voltage-sensitive channels (see Chapter 6) are similar in that they, too, open in all-or-none fashion, and, as a result, the net effect on the cell is due to the summation of currents flowing through many individual open ion channels. The two types of channels differ, however, in that one is opened by a chemical agent, whereas the other is opened by changes in membrane potential.

Statistical Analysis of Channel Gating and Kinetics of the PSP

The experiment illustrated in Figure 11.3B was performed with continuous exposure to ACh. Under such conditions, the channels open and close repeatedly. When ACh is applied by a brief pressure pulse to more accurately mimic the transient release from the presynaptic terminal, the transmitter diffuses away before it can cause a second opening of the channel. A set of data similar to that shown in Figure 11.4A would be obtained if an ensemble of these openings were collected and aligned with the start of each opening. Each individual trace represents the response to each successive "puff" of ACh. Note that, among the responses, the duration of the opening of the channel varies considerably—from very short (less than 1 ms) to more than 5 ms. Moreover, channel openings are independent events. The duration of any one channel opening does not have any relation to the duration of a previous opening. Figure 11.4B illustrates a plot that is obtained by adding 1000 of these individual responses. Such an addition roughly simulates the conditions under which a transmitter released from a presynaptic terminal leads to the near simultaneous activation of many single channels in the postsynaptic membrane. (Note that the addition of 1000 channels would produce

A

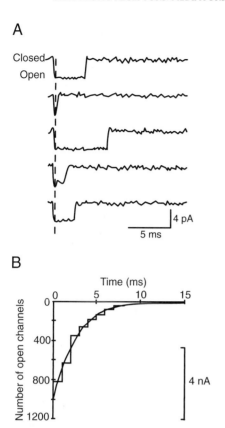

B

FIGURE 11.4 Determination of the shape of the postsynaptic response from single-channel currents. (A) Each trace represents the response of a single channel to a repetitively applied puff of transmitter. Traces are aligned with the beginning of the channel opening (dashed line). (B) The addition of 1000 of the individual responses. If a current equal to 4 pA were generated by the opening of a single channel, then a 4-nA current would be generated by 1000 channels opening at the same time. Data are fitted with an exponential function having a time constant equal to $1/\alpha$ (see text). Reprinted with permission from Sakmann (1992). American Association for the Advancement of Science, © 1992 The Nobel Foundation.

a synaptic current equal to about 4nA.) This simulation is valid given the assumption that the statistical properties of a single channel over time are the same as the statistical properties of the ensemble at one instant of time (i.e., an ergotic process). The ensemble average can be fit with an exponential function with a decay time constant of 2.7ms. An additional observation (discussed later) is that the value of the time constant is equal to the mean duration of the channel openings. The curve in Figure 11.4B is an indication of the probability that a channel will remain open for various times, with a high probability for short times and a low probability for long times.

The ensemble average of single-channel currents (Fig. 11.4B) roughly accounts for the time course of the EPSP. However, note that the time course of the aggregate synaptic current can be somewhat faster than that

of the excitatory postsynaptic potential in Figure 11.2. This difference is due to charging of the membrane capacitance by a rapidly changing synaptic current. Because the single-channel currents were recorded with the membrane voltage clamped, the capacitive current $[I_c = C_m * (dV/dt)]$ is zero. In contrast, for the recording of the postsynaptic potential in Figure 11.2, the membrane was not voltage clamped, and therefore as the voltage changes (i.e., dV/dt), some of the synaptic current charges the membrane capacitance (see Eq. 11.7).

Analytical expressions that describe the shape of the ensemble average of the open lifetimes and the mean open lifetime can be derived by considering that single-channel opening and closing is a stochastic process (Johnston and Wu, 1995; Sakmann, 1992). Relations are formalized to describe the likelihood (probability) of a channel being in a certain state. Consider the following two-state reaction scheme:

$$C \underset{\alpha}{\overset{\beta}{\Leftrightarrow}} O$$

In this scheme, α represents the rate constant for channel closing and β the rate constant for channel opening. The scheme can be simplified further if we consider a case in which the channel has been opened by the agonist and the agonist is removed instantaneously. A channel so opened (at time 0) will then close after a certain random time (Fig. 11.4). It can be shown that the mean open time = $1/\alpha$ (Johnston and Wu, 1995; Sakmann, 1992).

Gating Properties of Ligand-Gated Channels

Although statistical analysis can be a valuable source of insight into the statistical nature of the gating process and the molecular determinants of the macroscopic postsynaptic potential, the description in the preceding section is a simplification of the actual processes. Specifically, a more complete description must include the kinetics of receptor binding and unbinding and the determinants of the channel opening, as well as the fact that channels display rapid transitions between open and closed states during a single agonist receptor occupancy. Thus, the open states illustrated in Figures 11.3B and 11.4A represent the period of a burst of extremely rapid openings and closings. If the bursts of rapid channel openings and closings are thought of, and behave functionally, as a single continuous channel closure, the formalism developed in the preceding section is a reasonable approximation for many ligand-gated channels. Nevertheless, a more complex reaction scheme is necessary to quantitatively explain available data. Such a scheme would include the following states,

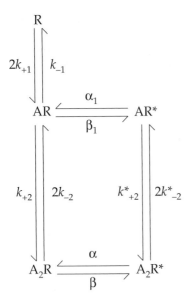

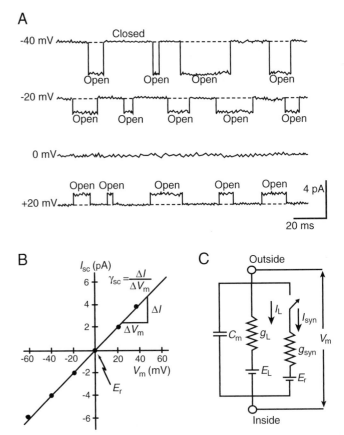

where R represents the receptor, A the agonist, and the α, β, and k values the forward and reverse rate constants for the various reactions. A_2R^* represents a channel opened as a result of the binding of two agonist molecules. The asterisk indicates an open channel. Note that the lower part of the reaction scheme is equivalent to the one developed earlier; that is,

$$C \underset{\alpha}{\overset{\beta}{\Leftrightarrow}} O$$

With the use of probability theory, equations describing transitions between the states can be determined. The approach is identical to that used in the simplified two-state scheme. However, the mathematics and analytical expressions are more complex. For some receptors, additional states must be represented. For example, as described in Chapter 9, some ligand-gated channels exhibit a process of desensitization in which continued exposure to a ligand results in channel closure.

Null (Reversal) Potential and Slope of I-V Relations

What ions are responsible for the synaptic current that produces the EPSP? Early studies of the ionic mechanisms underlying the EPSP at the skeletal neuromuscular junction yielded important information. Specifically, voltage-clamp and ion-substitution experiments indicated that the binding of transmitter to receptors on the postsynaptic membrane led to a simultaneous increase in Na^+ and K^+ permeability that depolarized the cell toward a value of about 0 mV (Fatt and Katz, 1951; Takeuchi and Takeuchi, 1960). These findings are applicable to the EPSP in a spinal motor

FIGURE 11.5   Voltage dependence of the current flowing through single channels. (A) Idealized recording of an ionotropic receptor in the continuous presence of agonist. (B) I–V relation of the channel in A. (C) Equivalent electrical circuit of a membrane containing that channel. γ_{sc}, single-channel conductance; I_L, leakage current; I_{sc}, single-channel current; g_L, leakage conductance; g_{syn}, macroscopic synaptic conductance; E_L, leakage battery; E_r, reversal potential.

neuron produced by an action potential in an afferent fiber and have been confirmed and extended at the single-channel level.

Figure 11.5 illustrates the type of experiment in which the analysis of single-channel currents can be a source of insight into the ionic mechanisms of EPSPs. A transmitter is delivered to the patch while the membrane potential is varied systematically (Fig. 11.5A). In the upper trace, the patch potential is –40 mV. The ejection of transmitter produces a sequence of channel openings and closings, the amplitudes of which are constant for each opening (i.e., about 4 pA). Now consider the case in which the transmitter is applied when the potential across the patch is –20 mV. The frequency of the responses, as well as the mean open lifetimes, is about the same as when the potential was at –40 mV, but now the amplitude of the single-channel currents is decreased uniformly. Even more interesting, when the patch is depolarized artificially to a value of about 0 mV, an identical puff of transmitter produces no current in the patch. If the patch potential is depolarized to a value of about 20 mV and the puff is delivered again, openings are again observed, but the flow of current through the channel is reversed in sign; a series of upward deflections indicate outward single-channel currents. In summary, there are downward deflections (inward currents) when the membrane potential is at −40 mV, no deflections (currents) when the membrane is at 0 mV, and upward deflections (outward currents) when the membrane potential is moved to 20 mV.

The simple explanation for these results is that no matter what the membrane potential, the effect of the transmitter binding with receptors is to produce a permeability change that tends to move the membrane potential toward 0 mV. If the membrane potential is more negative than 0 mV, an inward current is recorded. If the membrane potential is more positive than 0 mV, an outward current is recorded. If the membrane potential is at 0 mV, there is no deflection because the membrane potential is already at 0 mV. At 0 mV, the channels are opening and closing as they always do in response to the agonist, but there is no net movement of ions through them. This 0-mV level is known as the synaptic null potential or reversal potential because it is the potential at which the sign of the synaptic current reverses. The fact that the experimentally determined reversal potential equals the calculated value obtained by using the Goldman-Hodgkin-Katz (GHK) equation (Chapter 6) provides strong support for the theory that the EPSP is due to the opening of channels that have equal permeabilities to Na^+ and K^+. Ion-substitution experiments also confirm this theory. Thus, when the concentration of Na^+ or K^+ in the extracellular fluid is altered, the value of the reversal potential shifts in a way predicted by the GHK equation. (Some other cations, such as Ca^{2+}, also permeate these channels, but their permeability is low compared with that of Na^+ and K^+.)

Different families of ionotropic receptors have different reversal potentials because each has unique ion selectivity. In addition, it should now be clear that the sign of the synaptic action (excitatory or inhibitory) depends on the value of the reversal potential relative to the resting potential. If the reversal potential of an ionotropic receptor channel is more positive than the resting potential, opening of that channel will lead to depolarization (i.e., an EPSP). In contrast, if the reversal potential of an ionotropic receptor channel is more negative than the resting potential, opening of that channel will lead to hyperpolarization; that is, an inhibitory postsynaptic potential (IPSP), which is the topic of a later section in this chapter.

Plotting the average peak value of single-channel currents (I_{sc}) versus the membrane potential (trans-patch potential) at which they are recorded (Fig. 11.5B) can be a source of quantitative insight into the properties of the ionotropic receptor channel. Note that the current-voltage (I-V) relation is linear; it has a slope, the value of which is the single-channel conductance, and an intercept at 0 mV. This linear relation can be put in the form of Ohm's law ($I = G * \Delta V$). Thus,

$$I_{sc} = \gamma_{sc} * (V_m - E_r), \qquad (11.1)$$

where γ_{sc} is the single-channel conductance and E_r is the reversal potential (here, 0 mV).

Summation of Single-Channel Currents

We now know that the sign of a synaptic action can be predicted by knowledge of the relation between the resting potential (V_m) and the reversal potential (E_r), but how can the precise amplitude be determined? The answer to this question lies in understanding the relation between synaptic conductance and extra synaptic conductances. These interactions can be rather complex (see Chapter 12), but some initial understanding can be obtained by analyzing an electrical equivalent circuit for these two major conductance branches. We first need to move from a consideration of single-channel conductances and currents to that of macroscopic conductances and currents. The postsynaptic membrane contains thousands of any one type of ionotropic receptor, and each of these receptors could be activated by transmitter released by a single action potential in a presynaptic neuron. Because conductances in parallel add, the total conductance change produced by their simultaneous activation would be

$$g_{syn} = \gamma_{sc} P * N* \qquad (11.2)$$

where γ_{sc}, as before, is the single-channel conductance, P is the probability of opening of a single channel (controlled by the ligand), and N is the total number of ligand-gated channels in the postsynaptic membrane. The macroscopic postsynaptic current produced by the transmitter released by a single presynaptic action potential can then be described by

$$I_{syn} = g_{syn} * (V_m - E_r). \qquad (11.3)$$

Equation 11.3 can be represented physically by a voltage (V_m) measured across a circuit consisting of a resistor (g_{syn}) in series with a battery (E_r). An equivalent circuit of a membrane containing such a conductance is illustrated in Figure 11.5C. Also included in this circuit is a membrane capacitance (C_m), a resistor representing the leakage conductance (g_L), and a battery (E_L) representing the leakage potential. (Voltage-dependent Na^+, Ca^{2+}, and K^+ channels that contribute to the generation of the action potential have been omitted for simplification.)

The simple circuit allows the simulation and further analysis of the genesis of the PSP. Closure of the switch simulates the opening of the channels by transmitter released from some presynaptic neuron (i.e., a change in P of Eq. (2) from 0 to 1). When the switch is open (i.e., no agonist is present and the ligand-gated channels are closed), the membrane potential (V_m) is equal to the value of the leakage battery (E_L). Closure of the switch (i.e., the agonist opens the channels) tends to polarize the membrane potential toward the value of the battery (E_r) in series with the synaptic conductance. Although the effect of the channel openings is to depolarize the postsynaptic cell *toward* E_r (0 mV), this value is never achieved because ligand-gated receptors are only a small fraction of the ion channels in the membrane. Other channels (such as the leakage channels, which are not affected by the transmitters) tend to hold the membrane potential at E_L and prevent the membrane potential from reaching the 0-mV level. In terms of the equivalent electrical circuit (Fig. 11.5C), g_L is much greater than g_{syn}.

An analytical expression that can be a source of insight into the production of an EPSP by the engagement of a synaptic conductance can be derived by examining the current flowing in each of the two conductance branches of the circuit in Figure 11.5C. As shown previously (Eq. 11.3), current flowing in the branch representing the synaptic conductance is equal to

$$I_{syn} = g_{syn} * (V_m - E_r).$$

Similarly, the current flowing through the leakage conductance is equal to

$$I_L = g_L * (V_m - E_L) \qquad (11.4)$$

By conservation of current, the two currents must be equal and opposite. Therefore,

$$g_{syn} * (V_m - E_r) = -g_L * (V_m - E_L)$$

Rearranging and solving for V_m, we obtain

$$V_m = \frac{g_{syn} E_r + g_L E_L}{g_{syn} + g_L} \qquad (11.5)$$

Note that when the synaptic channels are closed (i.e., switch open), g_{syn} is 0 and

$$V_m = E_L$$

Now consider the case of ligand-gated channels being opened by the release of transmitter from a presynaptic neuron (i.e., switch closed) and a neuron with $g_L = 10$ nS, $E_L = -60$ mV, $g_{syn} = 0.2$ nS, and $E_r = 0$ mV.

Then

$$V_m = \frac{(0.2 \times 10^{-9} * 0) + (10 \times 10^{-9} * -60)}{10.2 \times 10^{-9}}$$
$$= -59 \text{ mV}$$

Thus, as a result of the closure of the switch, the membrane potential has changed from its initial value of −60 mV to a new value of −59 mV; that is, an EPSP of 1 mV has been generated.

The preceding analysis ignored membrane capacitance (C_m), the charging of which makes the synaptic potential slower than the synaptic current. Thus, a more complete analytical description of the postsynaptic factors underlying the generation of a PSP must account for the fact that some of the synaptic current will flow into the capacitive branch of the circuit. Again, by conservation of current, the sum of the currents in the three branches must equal 0. Therefore,

$$0 = C_m \frac{dV_m}{dt} + I_L + I_{syn}, \qquad (11.6)$$

$$0 = C_m \frac{dV_m}{dt} + g_L * (V_m - E_L) + g_{syn}(t) * (V_m - E_r), \qquad (11.7)$$

where $C_m (dV_m/dt)$ is the capacitive current.

By solving for V_m and integrating the differential equation, we can determine the magnitude and time course of a PSP. An accurate description of the kinetics of the PSP requires that the simple switch closure (all-or-none engagement of the synaptic conductance) be replaced with an expression [$g_{syn}(t)$] that describes the dynamics of the change in synaptic conductance with time.

Nonlinear I-V Relations of Some Ionotropic Receptors

For many PSPs mediated by ionotropic receptors, the current-voltage relation of the synaptic current is linear or approximately linear (Fig. 11.5B). Such ohmic relations are typical of nicotinic ACh channels and AMPA (alpha-amino-3-hydroxyl-5-methyl-4-isoxazolepropionate) glutamate channels (as well as many receptors mediating IPSPs). The linear I–V relation is indicative of a channel whose conductance is not affected by the potential across the membrane. Such linearity should be contrasted with the steep voltage dependency of the conductance of channels underlying the initiation and repolarization of action potentials (Chapter 6).

NMDA (*N*-methyl-D-aspartate) glutamate channels are a class of ionotropic receptors that have nonlinear current-voltage relations. At negative potentials, the channel conductance is low even when glutamate is bound to the receptor. As the membrane is depolarized, conductance increases and current flowing through the channel increases, resulting in the type of I–V relation illustrated in Figure 11.6A. This nonlinearity is represented by an arrow through the resistor representing this synaptic conductance in the equivalent circuit of Figure 11.6B. The nonlinear I–V relation of the NMDA receptor can be explained by a voltage-dependent block of the channel by Mg^{2+} (Fig. 11.7). At normal values of the resting potential, the pore of the channel is blocked by Mg^{2+}. Thus, even when glutamate binds to the receptor (Fig. 11.7B), the blocked channel prevents ionic flow (and an EPSP). The block can be relieved by depolarization, which presumably displaces Mg^{2+} from the pore (Fig. 11.7B). When the pore is unblocked, cations (i.e., Na^+, K^+, and Ca^{2+}) can flow readily through the channel, and this flux is manifested in the linear part of the I–V relation (Fig. 11.6A). AMPA channels (Fig. 11.7A) are not blocked by Mg^{2+} and have linear I–V relations (Fig. 11.5B).

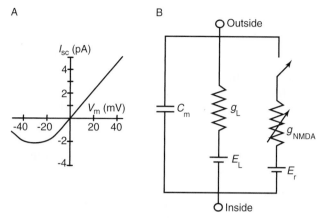

FIGURE 11.6 (A) I–V relation of the NMDA receptor. (B) Equivalent electrical circuit of a membrane containing NMDA receptors.

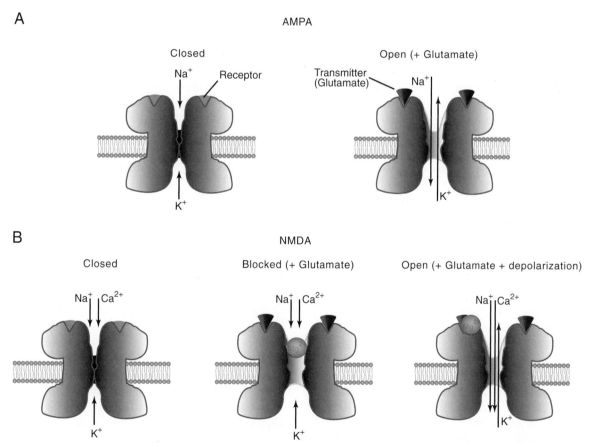

FIGURE 11.7 Features of AMPA and NMDA glutamate receptors. (A) AMPA receptors: (left) in the absence of agonist, the channel is closed, and (right) glutamate binding leads to channel opening and an increase in Na^+ and K^+ permeability. AMPA receptors that contain the GluR2 subunit are impermeable to Ca^{2+}. (B) NMDA receptors: (left) in the absence of agonist, the channel is closed; (middle) the presence of agonist leads to a conformational change and channel opening, but no ionic flux occurs because the pore of the channel is blocked by Mg^{2+}; and (right) in the presence of depolarization, the Mg^{2+} block is removed and the agonist-induced opening of the channel leads to changes in ion flux (including Ca^{2+} influx into the cell).

Inhibitory Postsynaptic Potentials Decrease the Probability of Cell Firing

Some synaptic events decrease the probability of generating action potentials in the postsynaptic cell. Potentials associated with these actions are called inhibitory postsynaptic potentials. Consider the inhibitory interneuron illustrated in Figure 11.2. Normally, this interneuron is activated by summating EPSPs from converging afferent fibers. These EPSPs summate in space and time such that the membrane potential of the interneuron reaches threshold and fires an action potential. This step can be bypassed by artificially depolarizing the interneuron to initiate an action potential. The consequences of that action potential from the point of view of the flexor motor neuron are illustrated in Figure 11.2. The action potential in the interneuron produces a transient increase in the membrane potential of the motor neuron. This transient hyperpolarization (the IPSP) looks very much like the EPSP, but it is reversed in sign.

What are the ionic mechanisms for these fast IPSPs and what is the transmitter substance? Because the membrane potential of the flexor motor neuron is about −65 mV, one might expect an increase in the conductance to some ion (or ions) with an equilibrium potential (reversal potential) more negative than −65 mV. One possibility is K⁺. Indeed, the K⁺ equilibrium potential in spinal motor neurons is about −80 mV; thus, a transmitter substance that produced a selective increase in K⁺ conductance would lead to an IPSP. The K⁺-conductance increase would move the membrane potential from −65 mV toward the K⁺ equilibrium potential of −80 mV. Although an increase in K⁺ conductance mediates IPSPs at some inhibitory synapses (see later), it does not at the synapse between the inhibitory interneuron and the spinal motor neuron. At this particular synapse, the IPSP seems to be due to a selective increase in Cl⁻ conductance. The equilibrium potential for Cl⁻ in spinal motor neurons is about −70 mV. Thus, the transmitter substance released by the inhibitory neuron diffuses across the cleft and interacts with receptor sites on the postsynaptic membrane. These receptors are normally closed, but when opened they become selectively permeable to Cl⁻. As a result of the increase in Cl⁻ conductance, the membrane potential moves from a resting value of −65 mV toward the Cl⁻ equilibrium potential of −70 mV.

As in the sensory neuron–spinal motor neuron synapse, the transmitter substance released by the inhibitory interneuron in the spinal cord is an amino acid, but in this case the transmitter is glycine. The toxin strychnine is a potent antagonist of glycine receptors. Although glycine originally was thought to be localized to the spinal cord, it is also found in other regions of the nervous system. The most common transmitter associated with inhibitory actions in many areas of the brain is γ-aminobutyric acid (GABA; see Chapter 7).

GABA receptors are divided into three major classes: GABA$_A$, GABA$_B$, and GABA$_C$ (Bormann and Fiegenspan, 1995; Billinton *et al.*, 2001; Bowery, 1993; Cherubini and Conti, 2001; Gage, 1992; Moss and Smart, 2001). As discussed in Chapter 9, GABA$_A$ receptors are ionotropic receptors, and, like glycine receptors, binding of transmitter leads to an increased conductance to Cl⁻, which produces an IPSP. GABA$_A$ receptors are blocked by bicuculline and picrotoxin. A particularly striking aspect of GABA$_A$ receptors is their modulation by anxiolytic benzodiazepines. Figure 11.8 illustrates the response of a neuron to GABA before and after treatment with diazepam (Bormann, 1988). In the presence of diazepam, the response is potentiated greatly. In contrast to GABA$_A$ receptors that are pore-forming channels, GABA$_B$ receptors are G-protein coupled (see also Chapter 9). GABA$_B$ receptors can be coupled to a variety of different effector mechanisms in different neurons. These mechanisms include decreases in Ca^{2+} conductance, increases in K⁺ conductance, and modulation of voltage-dependent A-type K⁺ current. In hippocampal pyramidal neurons, the GABA$_B$-mediated IPSP is due an increased in K⁺ conductance. Baclofen is a potent agonist of GABA$_B$ receptors, whereas phaclofen is a selective antagonist. GABA$_C$ receptors are pharmacologically distinct from GABA$_A$ and GABA$_B$ receptors and are found predominantly in the vertebrate retina. GABA$_C$ receptors, like GABA$_A$ receptors, are Cl⁻ selective pores.

Ionotropic receptors that lead to the generation of IPSPs and ionotropic receptors that lead to the generation of EPSPs have biophysical features in common. Indeed, analyses of the preceding section are generally applicable. A quantitative understanding of the effects of the opening of glycine or GABA$_A$ receptors can be obtained by using the electrical equivalent circuit of Figure 11.5C and Eq. 11.5, with the values of g_{syn} and E_r appropriate for the respective ionotropic receptor. Interactions between excitatory and inhibitory conductances can be modeled by adding additional branches to the equivalent circuit (see Fig. 11.15D and Chapter 12).

Some PSPs Have More Than One Component

The transmitter released from a presynaptic terminal diffuses across the synaptic cleft, where it binds to ionotropic receptors. In many cases, the postsynaptic

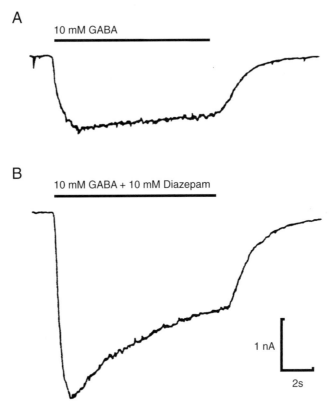

FIGURE 11.8 Potentiation of GABA responses by benzodiazepine ligands. (A) Brief application (bar) of GABA leads to an inward Cl⁻ current in a voltage-clamped spinal neuron. (B) In the presence of diazepam, the response is enhanced significantly. From Bormann (1988).

receptors are homogeneous. In other cases, the same transmitter activates more than one type of receptor. A major example of this type of heterogeneous post-synaptic action is the simultaneous activation by glutamate of NMDA and AMPA receptors on the same postsynaptic cell. Figure 11.9 illustrates such a dual-component glutamatergic EPSP in the CA1 region of the hippocampus. The cell is voltage clamped at various fixed holding potentials, and the macroscopic synaptic currents produced by activation of the presynaptic neurons are recorded. The experiment is performed in the presence and absence of the agent 2-amino-5-phosphonovalerate (APV), which is a specific blocker of NMDA receptors. When the cell is held at a potential of 20 or −40 mV, APV leads to a dramatic reduction of the late, but not the early, phase of the excitatory postsynaptic current (EPSC). In contrast, when the potential is held at −80 mV, the EPSC is unaffected by APV. These results indicate that PSP consists of two components: (1) an early AMPA-mediated

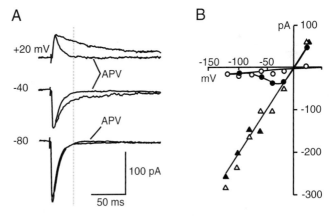

FIGURE 11.9 Dual-component glutamatergic EPSP. (A) The excitatory postsynaptic current was recorded before and during the application of APV at the indicated membrane potentials. (B) Peak current-voltage relations are shown before (▲) and during (△) the application of APV. Current-voltage relations measured 25 ms after the peak of the EPSC (dotted line in (A)); before (●) and during (○) application of APV are also shown. Reprinted with permission from Hestrin *et al.* (1990).

component and (2) a late NMDA-mediated component. In addition, results indicate that conductance of the non-NMDA component is linear, whereas conductance of the NMDA component is nonlinear. The I–V relations of the early (peak) and late (at approximately 25 ms) components of the EPSC are plotted in Figure 11.9 (Hestrin *et al.*, 1990).

Dual-component IPSPs also are observed in the CNS, but here the transmitter (GABA) that mediates the inhibitory actions may be released from different neurons that converge on a common postsynaptic neuron. Stimulation of afferent pathways to the hippocampus results in an IPSP in a pyramidal neuron, which has a fast initial inhibitory phase followed by a slower inhibitory phase (Fig. 11.10). Application of GABA$_A$ antagonists blocks the early inhibitory phase, whereas the GABA$_B$ receptor antagonist phaclofen blocks the late inhibitory phase (not shown). Early and late IPSPs can also be distinguished based on their ionic mechanisms. Hyperpolarizing the membrane potential to −78 mV nulls the early response, but at this value of membrane potential the late response is still hyperpolarizing (Figs. 11.10A and 11.10B). Hyperpolarizing the membrane potential to values more negative than −78 mV reverses the sign of the early response, but the slow response does not reverse until the membrane is made more negative than about −100 mV (Thalmann, 1988). The reversal potentials are consistent with a fast Cl⁻-mediated IPSP, mediated by fast opening of GABA$_A$ receptors and a slower K⁺-mediated IPSP mediated by G-protein GABA$_B$ receptors.

Dual-component PSPs need not be strictly inhibitory or excitatory. For example, a presynaptic cholinergic neuron in the mollusk *Aplysia* produces a diphasic excitatory-inhibitory (E-I) response in its postsynaptic follower cell. The response can be simulated by local discrete application of ACh to the postsynaptic cell (Fig. 11.11) (Blankenship *et al.*, 1971). The ionic mechanisms underlying this synaptic action were investigated in ion-substitution experiments, which revealed that the dual response is due to an early Na⁺-dependent component followed by a slower Cl⁻-dependent component. Molecular mechanisms underlying such slow synaptic potentials are discussed next.

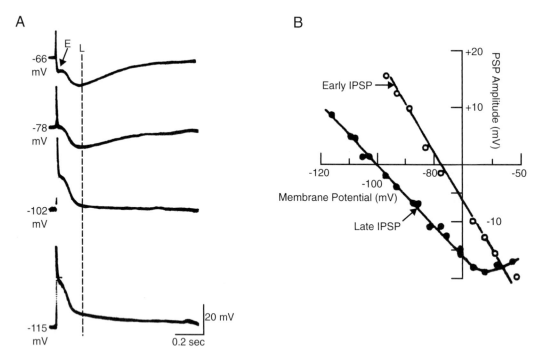

FIGURE 11.10 Dual-component IPSP. (A) Intracellular recordings from a pyramidal cell in the CA3 region of the rat hippocampus in response to activation of mossy fiber afferents. With the membrane potential of the cell at the resting potential, afferent stimulation produces an early (E) and late (L) IPSP. With increased hyperpolarizing produced by injecting constant current into the cell, the early component reverses first. At more negative levels of the membrane potential, the late component also reverses. This result indicates that the ionic conductance underlying the two phases is distinct. (B) Plots of the change in amplitude of the early (measured at 25 ms) and the late (measured at 200 ms, dashed line in A) response as a function of membrane potential. Reversal potentials of the early and late components are consistent with a GABA$_A$-mediated chloride conductance and a GABA$_B$-mediated potassium conductance, respectively. From Thalmann (1988).

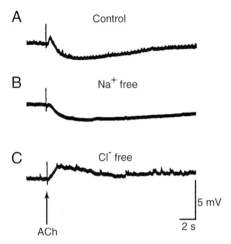

FIGURE 11.11 Dual-component cholinergic excitatory-inhibitory response. (A) Control in normal saline. Ejection of ACh produces a rapid depolarization followed by a slower hyperpolarization. (B) In Na$^+$-free saline, ACh produces a purely hyperpolarizing response, indicating that the depolarizing component in normal saline includes an increase in g_{Na}. (C) In Cl-free saline, ACh produces a purely depolarizing response, indicating that the hyperpolarizing component in normal saline includes an increase in g_{Cl}. Reprinted with permission from Blankenship *et al.* (1971).

Summary

Synaptic potentials mediated by ionotropic receptors are the fundamental means by which information is transmitted rapidly between neurons. Transmitters cause channels to open in an all-or-none fashion, and the currents through these individual channels summate to produce the macrosynaptic postsynaptic potential. The sign of the postsynaptic potential is determined by the relationship between the membrane potential of the postsynaptic neuron and the ion selectivity of the ionotropic receptor.

METABOTROPIC RECEPTORS: MEDIATORS OF SLOW SYNAPTIC POTENTIALS

A common feature of the types of synaptic actions heretofore described is the direct binding of the transmitter with the receptor-channel complex. An entirely separate class of synaptic actions has as its basis the indirect coupling of the receptor with the channel. Two major types of coupling mechanisms have been identified: (1) coupling of the receptor and channel through an intermediate regulatory protein, such as a G-protein; and (2) coupling through a diffusible

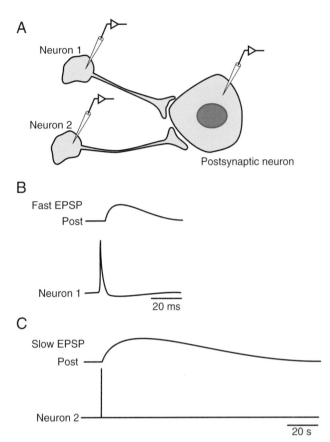

FIGURE 11.12 Fast and slow synaptic potentials. (A) Idealized experiment in which two neurons (1 and 2) make synaptic connections with a common postsynaptic follower cell (Post). (B) An action potential in neuron 1 leads to a conventional fast EPSP with a duration of about 30 ms. (C) An action potential in neuron 2 also produces an EPSP in the postsynaptic cell, but the duration of this slow EPSP is more than three orders of magnitude greater than that of the EPSP produced by neuron 1. Note the change in the calibration bar.

second-messenger system. Because coupling through a diffusible second-messenger system is the most common mechanism, it is the focus of this section.

A comparison of the features of direct, fast ionotropic-mediated and indirect, slow metabotropic-mediated synaptic potentials is shown in Figure 11.12. Slow synaptic potentials are not observed at every postsynaptic neuron, but Figure 11.12A illustrates an idealized case in which a postsynaptic neuron receives two inputs, one of which produces a conventional fast EPSP and the other of which produces a slow EPSP. An action potential in neuron 1 leads to an EPSP in the postsynaptic cell with a duration of about 30 ms (Fig. 11.12B). This type of potential might be produced in a spinal motor neuron by an action potential in an afferent fiber. Neuron 2 also produces a postsynaptic potential (Fig. 11.12C), but its duration (note the calibration bar) is more than three orders of magnitude greater than that of the EPSP produced by neuron 1.

How can a change in the postsynaptic potential of a neuron persist for many minutes as a result of a single action potential in the presynaptic neuron? Possibilities include a prolonged presence of the transmitter due to continuous release, to slow degradation, or to slow reuptake of the transmitter, but the mechanism here involves a transmitter-induced change in the metabolism of the postsynaptic cell. Figure 11.13 compares the general mechanisms for fast and slow synaptic potentials. Fast synaptic potentials are produced when a transmitter substance binds to a channel and produces a conformational change in the channel, causing it to become permeable to one or more ions (both Na^+ and K^+ in Fig. 11.13A). The increase in permeability leads to a depolarization associated with the EPSP. The duration of the synaptic event critically depends on the amount of time during which the transmitter substance remains bound to the receptors. Acetylcholine, glutamate, and glycine remain bound only for a very short period. These transmitters are removed by diffusion, enzymatic breakdown, or reuptake into the presynaptic cell. Therefore, the duration of the synaptic potential is directly related to the lifetimes of the opened channels, and these lifetimes are relatively short (see Fig. 11.4B).

One mechanism for a slow synaptic potential is shown in Figure 11.13B. In contrast with the fast PSP for which the receptors are actually part of the ion channel complex, channels that produce slow synaptic potentials are not coupled directly to the transmitter receptors. Rather, the receptors are separated physically and exert their actions indirectly through changes in metabolism of specific second-messenger systems. Figure 11.13B illustrates one type of response in *Aplysia* for which the cAMP-protein kinase A (PKA) system is the mediator, but other slow PSPs use other second-messenger kinase systems (e.g., the protein kinase C system). In the cAMP-dependent slow synaptic responses in *Aplysia*, transmitter binding to membrane receptors activates G-proteins and stimulates an increase in the synthesis of cAMP. Cyclic AMP then leads to the activation of cAMP-dependent protein kinase (PKA), which phosphorylates a channel protein or protein associated with the channel (Siegelbaum *et al.*, 1982). A conformational change in the channel is produced, leading to a change in ionic conductance. Thus, in contrast with a direct conformational change produced by the binding of a transmitter to the receptor-channel complex, in this case, a conformational change is produced by protein phosphorylation. Indeed, phosphorylation-dependent channel regulation is a fairly general feature of slow PSPs. However, channel regulation by second messengers is not produced exclusively by phosphorylation. In one family of ion channels, the channels are gated or regulated directly by cyclic nucleotides. These cyclic nucleotide-gated channels require cAMP or cGMP to open but have other features in common with members of the superfamily of voltage-gated ion channels (Kaupp, 1995; Zimmermann, 1995).

Another interesting feature of slow synaptic responses is that they are sometimes associated with decreases rather than increases in membrane conductance. For example, the particular channel illustrated in Figure 11.13B is selectively permeable to K^+ and is normally open. As a result of the activation of the second messenger, the channel closes and becomes less permeable to K^+. The resultant depolarization may seem paradoxical, but recall that the membrane potential is due to a balance between resting K^+ and Na^+ permeability. K^+ permeability tends to move the membrane potential toward the K^+ equilibrium potential ($-80\,mV$), whereas Na^+ permeability tends to move the membrane potential toward the Na^+ equilibrium potential ($55\,mV$). Normally, K^+ permeability predominates, and the resting membrane potential is close to, but not equal to, the K^+ equilibrium potential. If K^+ permeability is decreased because some of the channels close, the membrane potential will be biased toward the Na^+ equilibrium potential and the cell will depolarize.

At least one reason for the long duration of slow PSPs is that second-messenger systems are slow (from seconds to minutes). Take the cAMP cascade as an example. Cyclic AMP takes some time to be synthesized, but, more importantly, after synthesis, cAMP levels can remain elevated for a relatively long period (minutes). The duration of the elevation of cAMP depends on the actions of cAMP-phosphodiesterase, which breaks down cAMP. However, duration of an effect could outlast the duration of the change in the second messenger because of persistent phosphorylation of the substrate protein(s). Phosphate groups are removed from substrate proteins by protein phosphatases. Thus, the net duration of a response initiated by a metabotropic receptor depends on the actions of not only the synthetic and phosphorylation processes, but also the degradative and dephosphorylation processes.

Activation of a second messenger by a transmitter can have a localized effect on the membrane potential through phosphorylation of membrane channels near the site of a metabotropic receptor. The effects can be more widespread and even longer lasting than depicted in Figure 11.13B. For example, second messengers and

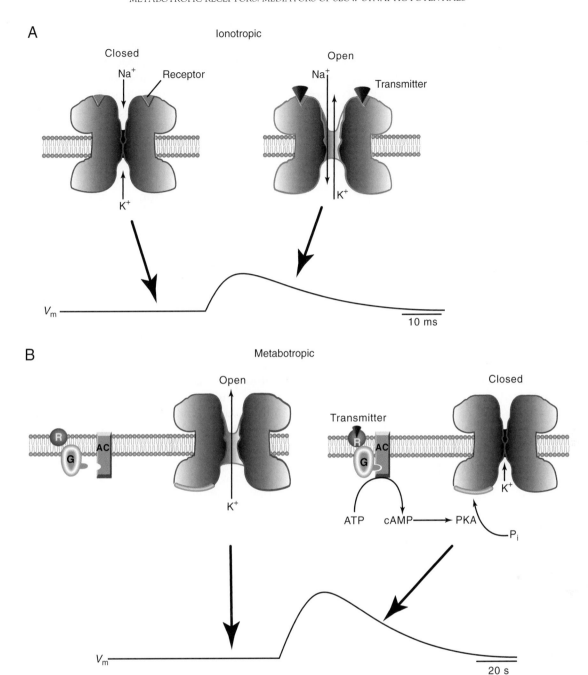

FIGURE 11.13 Ionotropic and metabotropic receptors and mechanisms of fast and slow EPSPs. (A, left) Fast EPSPs are produced by binding of the transmitter to specialized receptors that are directly associated with an ion channel (i.e., a ligand-gated channel). When the receptors are unbound, the channel is closed. (A, right) Binding of the transmitter to the receptor produces a conformational change in the channel protein such that the channel opens. In this example, the channel opening is associated with a selective increase in the permeability to Na^+ and K^+. The increase in permeability results in the EPSP shown in the trace. (B, left) Unlike fast EPSPs, which are due to the binding of a transmitter with a receptor-channel complex, slow EPSPs are due to the activation of receptors (metabotropic) that are not coupled directly to the channel. Rather, coupling takes place through the activation of one of several second-messenger cascades, in this example, the cAMP cascade. A channel that has a selective permeability to K^+ is normally open. (B, right) Binding of the transmitter to the receptor (R) leads to the activation of a G-protein (G) and adenylyl cyclase (AC). The synthesis of cAMP is increased, cAMP-dependent protein kinase (protein kinase A, PKA) is activated, and a channel protein is phosphorylated. The phosphorylation leads to closing of the channel and the subsequent depolarization associated with the slow EPSP shown in the trace. The response decays due to both the breakdown of cAMP by cAMP-dependent phosphodiesterase and the removal of phosphate from channel proteins by protein phosphatases (not shown).

protein kinases can diffuse and affect more distant membrane channels. Moreover, a long-term effect can be induced in the cell by altering gene expression. For example, protein kinase A can diffuse to the nucleus, where it can activate proteins that regulate gene expression. Detailed descriptions of second messengers and their actions are given in Chapter 10.

Summary

In contrast to the rapid responses mediated by ionotropic receptors, responses mediated by metabotropic receptors are generally relatively slow to develop and persistent. These properties arise because metabotropic responses can involve the activation of second-messenger systems. By producing slow changes in the resting potential, metabotropic receptors provide long-term modulation of the effectiveness of responses generated by ionotropic receptors. Moreover, these receptors, through the engagement of second-messenger systems, provide a vehicle by which a presynaptic cell cannot only alter the membrane potential, but also produce widespread changes in the biochemical state of a postsynaptic cell.

INTEGRATION OF SYNAPTIC POTENTIALS

The small amplitude of the EPSP in spinal motor neurons (and other cells in the CNS) poses an interesting question. Specifically, how can an EPSP with an amplitude of only 1 mV drive the membrane potential of the motor neuron (i.e., the postsynaptic neuron) to threshold and fire the spike in the motor neuron that is necessary to produce the contraction of the muscle? The answer to this question lies in the principles of temporal and spatial summation.

When the ligament is stretched (Fig. 11.1), many stretch receptors are activated. Indeed, the greater the stretch, the greater the probability of activating a larger number of the stretch receptors; this process is referred to as recruitment. However, recruitment is not the complete story. The principle of frequency coding in the nervous system specifies that the greater the intensity of a stimulus, the greater the number of action potentials per unit time (frequency) elicited in a sensory neuron. This principle applies to stretch receptors as well. Thus, the greater the stretch, the greater the number of action potentials elicited in the stretch receptor in a given interval and therefore the greater the number of EPSPs produced in the motor neuron

from that train of action potentials in the sensory cell. Consequently, the effects of activating multiple stretch receptors add together (spatial summation), as do the effects of multiple EPSPs elicited by activation of a single stretch receptor (temporal summation). Both of these processes act in concert to depolarize the motor neuron sufficiently to elicit one or more action potentials, which then propagate to the periphery and produce the reflex.

Temporal Summation Allows Integration of Successive PSPs

Temporal summation can be illustrated by firing action potentials in a presynaptic neuron and monitoring the resultant EPSPs. For example, in Figures 11.14A and 11.14B, a single action potential in sensory neuron 1 produces a 1-mV EPSP in the motor neuron. Two action potentials in quick succession produce two EPSPs, but note that the second EPSP occurs during the falling phase of the first, and the depolarization associated with the second EPSP adds to the depolarization produced by the first. Thus, two action potentials produce a summated potential that is about 2 mV in amplitude. Three action potentials in quick succession would produce a summated potential of about 3 mV. In principle, 30 action potentials in quick succession would produce a potential of about 30 mV and easily drive the cell to threshold. This summation is strictly a passive property of the cell. No special ionic conductance mechanisms are necessary. Specifically, the postsynaptic conductance change (g_{syn} in Eq. 11.3) produced by the second of two successive action potentials adds to that produced by the first. In addition, the postsynaptic membrane has a capacitance and can store charge. Thus, the membrane temporarily stores the charge of the first EPSP, and the charge from the second EPSP is added to that of the first.

However, the "time window" for this process of temporal summation very much depends on the duration of the postsynaptic potential, and temporal summation is possible only if the presynaptic action potentials (and hence postsynaptic potentials) are close in time to each other. The time frame depends on the duration of changes in the synaptic conductance and the time constant (Chapter 5). Temporal summation, however, rarely is observed to be linear as in the preceding examples, even when the postsynaptic conductance change (g_{syn} in Eq. 11.3) produced by the second of two successive action potentials is identical with that produced by the first (i.e., no presynaptic facilitation or depression) and the synaptic current is slightly less because the first PSP reduces the driving force

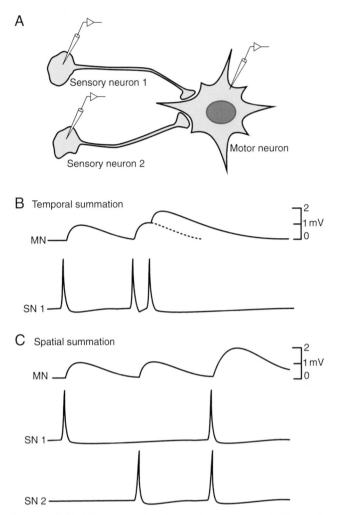

FIGURE 11.14 Temporal and spatial summation. (A) Intracellular recordings are made from two idealized sensory neurons (SN1 and SN2) and a motor neuron (MM). (B) Temporal summation. A single action potential in SN1 produces a 1-mV EPSP in the MN. Two action potentials in quick succession produce a dual-component EPSP, the amplitude of which is approximately 2mV. (C) Spatial summation. Alternative firing of single action potentials in SN1 and SN2 produce 1-mV EPSPs in the MN. Simultaneous action potentials in SN1 and SN2 produce a summated EPSP, the amplitude of which is about 2mV.

$(V_m\text{-}E_r)$ for the second. Interested readers should try some numerical examples.

Spatial Summation Allows Integration of PSPs from Different Parts of a Neuron

Spatial summation (Fig. 11.14C) requires a consideration of more than one input to a postsynaptic neuron. An action potential in sensory neuron 1 produces a 1-mV EPSP, just as it did in Figure 11.14B.

Similarly, an action potential in a second sensory neuron by itself also produces a 1-mV EPSP. Now, consider the consequences of action potentials elicited simultaneously in sensory neurons 1 and 2. The net EPSP is equal to the summation of the amplitudes of the individual EPSPs. Here, the EPSP from sensory neuron 1 is 1mV, the EPSP from sensory neuron 2 is 1mV, and the summated EPSP is approximately 2mV (Fig. 11.14C). Thus, spatial summation is a mechanism by which synaptic potentials generated at different sites can summate. Spatial summation in nerve cells is influenced by the space constant—the ability of a potential change produced in one region of a cell to spread passively to other regions of a cell (see Chapter 5).

Summary

Whether a neuron fires in response to synaptic input depends, at least in part, on how many action potentials are produced in any one presynaptic excitatory pathway and on how many individual convergent excitatory input pathways are activated. The summation of EPSPs in time and space is only part of the process, however. The final behavior of the cell is also due to the summation of inhibitory synaptic inputs in time and space, as well as to the properties of the voltage-dependent currents (Fig. 11.15) in the soma and along the dendrites (Koch and Segev, 1989; Ziv et al., 1994). For example, voltage-dependent conductances such as A-type K^+ conductance have a low threshold for activation and can thus oppose the effectiveness of an EPSP to trigger a spike. Low-threshold Na^+ and Ca^{2+} channels can boost an EPSP. Finally, we need to consider that spatial distribution of the various voltage-dependent channels, ligand-gated receptors, and metabotropic receptors is not uniform. Thus, each segment of the neuronal membrane can perform selective integrative functions. Clearly, this system has an enormous capacity for the local processing of information and for performing logical operations.

The flow of information in dendrites and the local processing of neuronal signals are discussed in Chapter 12. Several software packages are available for the development and simulation of realistic models of single neurons and neural networks. One, Simulator for Neural Networks and Action Potentials (SNNAP) (http://snnap.uth.tmc.edu/), provides mathematical descriptions of ion currents, intracellular second messengers, and ion pools, and allows simulation of current flow in multicompartment models of neurons.

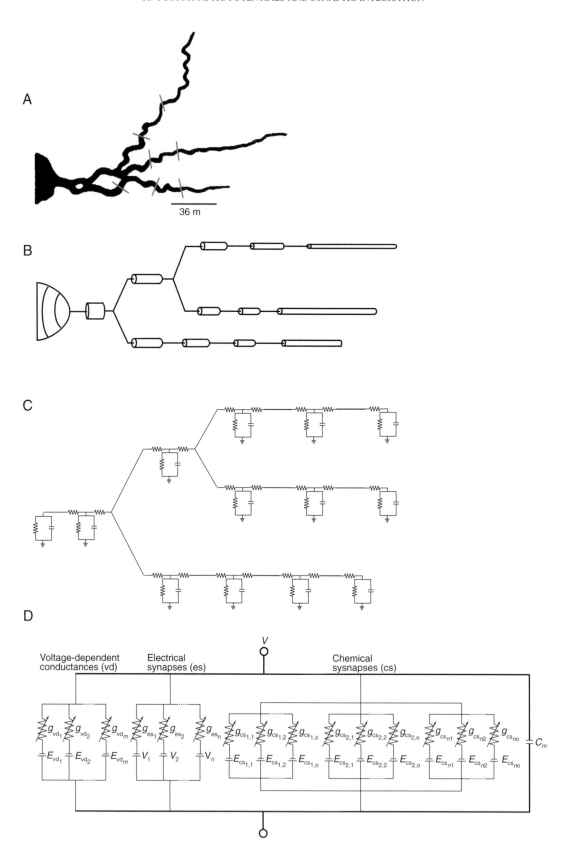

◀ **FIGURE 11.15** Modeling the integrative properties of a neuron. (A) Partial geometry of a neuron in the CNS revealing the cell body and pattern of dendritic branching. (B) The neuron modeled as a sphere connected to a series of cylinders, each of which represents the specific electrical properties of a dendritic segment. (C) Segments linked with resistors representing the intracellular resistance between segments, with each segment represented by the parallel combination of the membrane capacitance and the total membrane conductance. Reprinted with permission from Koch and Segev (1989). Copyright 1989 MIT Press. (D) Electrical circuit equivalent of the membrane of a segment of a neuron. The segment has a membrane potential V and a membrane capacitance C_m. Currents arise from three sources: (1) m voltage-dependent (vd) conductances $(g_{vd_1}-g_{vd_m})/(2)$ n conductances due to electrical synapses (es) $(g_{es_1}-g_{es_n})$, and (3) n times o time-dependent conductances due to chemical synapses (cs) with each of the n presynaptic neurons $(g_{cs_{1,1}}-g_{cs_{n,o}})$. E_{vd} and E_{cs} are constants and represent the values of the equilibrium potential for currents due to voltage-dependent conductances and chemical synapses, respectively. V_1-V_n represent the value of the membrane potential of the coupled cells. Reprinted with permission from Ziv *et al.* (1994).

References

Billinton, A., Ige, A. O., Bolam, J. P., White, J. H., Marshall, F. H., and Emson, P.C. (2001). Advances in the molecular understanding of GABA_B receptors. *Trends Neurosci.* **24**, 277–282.

Blankenship, J. E., Wachtel, H., and Kandel, E. R. (1971). Ionic mechanisms of excitatory, inhibitory and dual synaptic actions mediated by an identified interneuron in abdominal ganglion of *Aplysia. J. Neurophysiol.* **34**, 76–92.

Bormann, J. (1988). Electrophysiology of GABA_A and GABA_B receptor subtypes. *Trends Neurosci.* **11**, 112–116.

Bormann, J. and Feigenspan, A. (1995). GABA_C receptors. *Trends Neurosci.* **18**, 515–519.

Bowery, N. G. (1993). GABA_B receptor pharmacology. *Annu. Rev. Pharmacol. Toxicol.* **33**, 109–147.

Cherubini, E. and Conti, F. (2001). Generating diversity at GABAergic synapses. *Trends Neurosci.* **24**, 155–162.

Eccles, J. C. (1964). "The Physiology of Synapses." Springer-Verlag, New York.

Fatt, P. and Katz, B. (1951). An analysis of the end-plate potential recorded with an intra cellular electrode. *J. Physiol. (Lond.)* **115**, 320–370.

Gage, P. W. (2001). Activation and modulation of neuronal K+ channels by GABA. *Trends Neurosci.* **15**, 46–51.

Hamill, O. P., Marty, A., Neher, E., Sakmann, B., and Sigworth, J. (1981). Improved patch-clamp techniques for high-resolution current recording from cells and cell-free membrane patches. *Pflüg Arch.* **391**, 85–100.

Hestrin, S., Nicoll, R. A., Perkel, D. J., and Sah, P. (1990). Analysis of excitatory synaptic action in pyramidal cells using whole-cell recording from rat hippocampal slices. *J. Physiol. (Lond.)* **422**, 203–225.

Johnston, D. and Wu, S. M.-S. (1995). "Foundations of Cellular Neurophysiology." MIT Press, Cambridge, MA.

Kaupp, U. B. (1995). Family of cyclic nucleotide gated ion channels. *Curr. Opin. Neurobiol.* **5**, 434–442.

Koch, C. and Segev, I. (1989). "Methods in Neuronal Modeling." MIT Press, Cambridge, MA.

Moss, S. J. and Smart, T. G. (2001). Constructing inhibitory synapses. *Nature Rev. Neurosci.* **2**, 240–250.

Sakmann, B. (1992). Elementary steps in synaptic transmission revealed by currents through single ion channels. *Science* **256**, 503–512.

Siegelbaum, S. A., Camardo, J. S., and Kandel, E. R. (1982). Serotonin and cyclic AMP close single K+ channels in *Aplysia* sensory neurones. *Nature (Lond.)* **299**, 413–417.

Spencer, W. A. (1977). The physiology of supraspinal neurons in mammals. *In* "Handbook of Physiology" (E. R. Kandel, ed.), Vol. 1, Part 2, Sect. 1, pp. 969–1022. American Physiological Society, Bethesda, MD.

Takeuchi, A. and Takeuchi, N. (1960). On the permeability of end-plate membrane during the action of transmitter. *J. Physiol. (Lond.)* **154**, 52–67.

Thalmann, R. H. (1988). Evidence that guanosine triphosphate (GTP)-binding proteins control a synaptic response in brain: Effect of pertussis toxin and GTPγS on the late inhibitory postsynaptic potential of hippocampal CA3 neurons. *J. Neurosci.* **8**, 4589–4602.

Zimmermann, A. L. (1995). Cyclic nucleotide gated channels. *Curr. Opin. Neurobiol.* **5**, 296–303.

Ziv, I., Baxter, D. A., and Byrne, J. H. (1994). Simulator for neural networks and action potentials: Description and application. *J. Neurophysiol.* **71**, 294–308.

Suggested Readings

Burke, R. E. and Rudomin, P. (1977). Spatial neurons and synapses. *In* "Handbook of Physiology" (E. R. Kandel, ed.), Sect. 1, Vol. 1, Part 2, pp. 877–944. American Physiological Society, Bethesda, MD.

Byrne, J. H. and Schultz, S. G. (1994). "An Introduction to Membrane Transport and Bioelectricity," 2nd ed. Raven Press, New York.

Cowan, W. M., Sudhof, T. C., and Stevens, C. F., eds. (2001). "Synapses." Johns Hopkins Univ. Press.

Hille, B., ed. (2001). "Ion Channels of Excitable Membranes," 3rd ed. Sinauer, Sunderland, MA.

Shepherd, G. M., ed. (2004). "The Synaptic Organization of the Brain," 5th ed. Oxford Univ. Press, New York.

John H. Byrne

12

Complex Information Processing
in Dendrites

A hallmark of neurons is the variety of their dendrites. The branching patterns are dazzling and the size range astounding, from the large trees of cortical pyramidal neurons to the tiny size of a retinal bipolar cell, which would fit comfortably within the cell body of a pyramidal neuron (see Fig. 12.1). A main challenge in modern neuroscience is understanding the molecular and functional properties of these structures, and their significance for information processing by neurons.

Information processing by spread of electrical current through passive branching structures has already been discussed in Chapter 5. It is being increasingly recognized that the methods introduced by Wilfrid Rall for passive properties provide the essential basis for understanding the much more complex types of information processing that can occur through the distribution of active properties within the branching tree.

In this chapter we apply the Rall approach to ask the fundamental questions: (1) what are the principles of information processing in these dendritic trees with their elaborate branching patterns, distributed connectivity, and nonlinear properties, and (2) how are these dendrites with these properties adapted for the operational tasks of a specific neuron type within the microcircuits characteristic of that region?

STRATEGIES FOR STUDYING
COMPLEX DENDRITES

Strategies for answering these two questions may be illustrated by the synaptic organization of two cell types, the mitral and granule cells of the olfactory bulb,

shown in the lower left-hand corner of Figure 12.1. A first step in the modern approach is to break down a complex dendritic tree into functional compartments so that the integrative actions within the tree can be identified at successive levels of functional organization. As illustrated in Figure 12.2, these levels start with the individual synapse, which may be on a dendritic branch, as in the mitral cell, or on a dendritic spine, as in the granule cell. The next level is in terms of local patterns of synaptic connections. Successive levels involve larger extents of dendritic branches, until one reaches the level of distinct dendritic compartments, as in the case of the mitral cell, of distal tuft, primary dendrite, and secondary dendrites. At each level a dendritic compartment includes as well the cells interacting synaptically with that compartment. At the highest level is the global summation at the axon hillock and the global output through the axon. A similar analysis applies to the granule cell, except that it lacks an axon.

This approach allows one to identify the synaptic interactions within a dendritic tree as constituting a hierarchy, within which the specific pattern of interactions at a given level forms the fundamental integrative unit for the next level in the hierarchy. These integrative units are sometimes referred to as *microcircuits*, defined as a specific pattern of interactions performing a specific functional operation (Shepherd, 1978). In a computer microcircuit, a particular circuit configuration can be useful in many different contexts; similarly, in the brain, this gives the hypothesis that a particular microcircuit may be useful in different cells in different contexts at the equivalent level of organization. By this means the principles of information processing across different cell types in different regions and phyla can be identified. We will use this

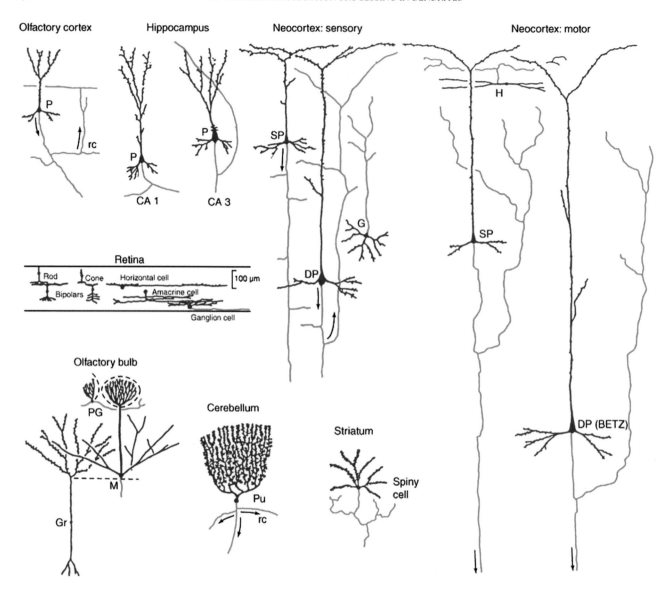

FIGURE 12.1 Varieties of neurons and dendritic trees. P, pyramidal neuron; re, recurrent collateral; SP, small pyramidal neuron; DP, deep pyramidal neuron; G, granule cell; Gr, granule cell (olfactory); M, mitral cell; PG, periglomerular cell; Pn, Purkinje cell. Modified from Shepherd (1992).

approach to parse the organization of the dendrites of representative cell types, including those in Figure 12.1.

BUILDING PRINCIPLES STEP BY STEP

As discussed in Chapter 5, the neuron processes information through five basic types of activity: intrinsic, reception, integration, encoding, and output. We saw that understanding how these activities are integrated within the neuron starts with the rules of passive current spread. Many of the principles were worked out first in the dendrites of neurons that lack axons or the ability to generate action potentials. There are many examples in invertebrate ganglia. In vertebrates, they include the retinal amacrine cell and the olfactory granule cell. These studies have shown that a dendritic tree by itself is capable of performing many basic functions required for information processing, such as the generation of intrinsic activity, input-output functions for feature extraction, parallel processing, signal-to-noise enhancement, and oscillatory activity. These cells demonstrate that there is no one thing that dendrites do; they do whatever is required to process

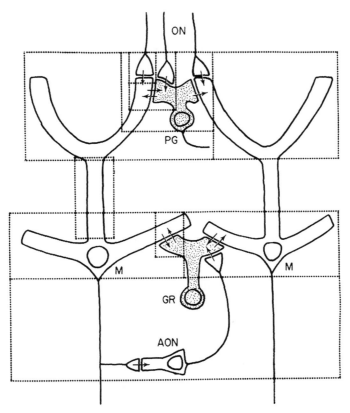

FIGURE 12.2 Compartmentalization (dashed lines) of olfactory bulb neurons to identify the functional subunits of dendrites and their relation to different levels of synaptic organization. ON, olfactory nerve; PG, periglomerular cell; M, mitral cell; GR, granule cell; AON, anterior olfactory nucleus. From Shepherd (1977).

information within their particular neuron or neuronal circuit, with or without an axon.

We also need to recognize that information in dendrites can take many forms. There are actions of neuropeptides on membrane receptors and internal cytoplasmic or nuclear receptors; actions of second and third messengers within the neuron; movement of substances within the dendrites by diffusion or by active transport; and changes occurring during development. All these types of cellular traffic and information flow in dendrites are coming under direct study (Matus and Shepherd, 2000; Stuart *et al.*, 2008). The student should review these subjects in earlier chapters. This chapter focuses on information processing in dendrites involving electrical signaling mechanisms by synapses and voltage-gated channels.

We will focus on how this takes place in neurons with axons. Neurons with axons may be classified into two groups, as suggested originally by Camillo Golgi in 1873: those with long axons and those with short axons. Long axon (output) cells tend to be larger than short axon (local) cells, and therefore have been more accessible to experimental analysis. Indeed, virtually everything known about the functional relations

between dendrites and axons has been obtained from studies of long axon cells. Consequently, much of what we *think* we understand about those relations in short axon cells is only by inference.

As noted in the analysis of the passive properties of neurons in Chapter 5, there are a number of sites on the Web that support the analysis of complex neurons and their active dendrites. For orientation to the molecular properties of dendritic compartments of different neurons discussed in this chapter, consult senselab.med.yale.edu/neurondb; for computational models based on those properties, consult senselab.med.yale.edu/modeldb. For the structures of dendrites, see synapse-web.org; cell centered database, and neuromorpho.org.

AN AXON PLACES CONSTRAINTS ON DENDRITIC PROCESSING

As we saw in Chapter 5 (Fig. 5.1), the neuron has five essential functions related to signal processing: generation, reception, integration, encoding, and output. The presence of an axon places critical constraints on the dendritic processing that leads to axonal output.

The first principle is: If a neuron has an axon, it has only one. This near universal "single axon rule" is remarkable and still little understood. It results from developmental mechanisms that provide for differentiation of a single axon from among early undifferentiated processes. These mechanisms are being analyzed especially in neuronal cultures (Craig and Banker, 1994). The principle means that for dendritic integration to lead to output from the neuron to distant targets, all the activity within the dendrites eventually must be funneled into the origin of the axon in the single axon hillock. Therefore, in these cells the flow of information in dendrites has an overall orientation. Ramon y Cajal and the classical anatomists called this the Law of Dynamic Polarization of the neuron (Cajal, 1911; summarized in Shepherd, 1991). We thus have a **principle of global output**:

> In order to transfer information between regions, the information distributed at different sites within a dendritic tree of an output neuron must be encoded, for global output at a single site at the origin of the axon.

A related principle is that the main function of the axon in long axon cells is to support the generation of action potentials in the axon hillock-initial segment region. By definition, action potentials there have thresholds for generation; thus, the **principle of frequency encoding of global output** in an axonal neuron is:

> The results of dendritic integration affect the output through the axon by initiating or modulating action potential generation in the axon hillock-initial segment. Global output from dendritic integration is therefore encoded in impulse frequency in a single axon.

Classically it has been known that the axons of most output neurons are so long that the only significant signals reaching their axon terminals are the digital all-or-nothing action potentials carrying a frequency code. However, biology always produces exceptions. Recent research has shown that the synaptic potentials within the soma-dendrites may spread sufficiently in some axons to modulate the membrane potentials of the axon terminals (Shu *et al.*, 2006). This effect would presumably be most significant in short axon cells, where, as noted earlier, our understanding of signal processing is most limited. In these cells, it appears that the axon may carry the outcome of soma-dendritic integration mainly in a digital (impulse frequency) form, but with a contribution from analog (synaptic potential amplitude) signals.

A further consequence of the spatial separation of dendrites and axon is that some of the activity within a dendritic tree will be below threshold for activating an axonal action potential; we thus have the **principle of subthreshold dendritic activity**:

> A considerable amount of subthreshold activity, including local active potentials, can affect the integrative states of the dendrites and any local outputs, but not necessarily directly or immediately affect the global output of the neuron.

We turn now to the functional properties that allow dendritic trees to process information within these constraints.

DENDRODENDRITIC INTERACTIONS BETWEEN AXONAL CELLS

We first recognize, from the example in Figure 12.2, that axonal cells as well as anaxonal cells can have outputs through their dendrites. This is against the common wisdom, which assumes that if a neuron has an axon, all the output goes through the axon. There are many examples in invertebrates.

Neurite-Neurite Synapses in Lobster Stomatogastric Ganglion

One of the first examples in invertebrates was in the stomatogastric ganglion of the lobster (Selverston *et al.*, 1976). Neurons were recorded intracellularly and stained with Procion yellow. Serial electron micrographic reconstructions showed the synaptic relations between stained varicosities in the processes and their neighbors (the processes are equivalent to dendrites, but often are referred to as neurites in the invertebrate literature). In many cases, a varicosity could be seen to be not only presynaptic to a neighboring varicosity, but also postsynaptic to that same process. It was concluded that synaptic inputs and outputs are distributed over the entire neuritic arborization. Polarization was not from one part of the tree to another. Bifunctional varicosities appeared to act as local input-output units, similar to the manner in which granule cell spines appear to operate (see later). Similar organization has been found in other types of stomatogastric neurons (Fig. 12.3A).

Sets of these local input-output units, distributed throughout the neuritic tree, participate in the generation and coordination of oscillatory activity involved in controlling the rhythmic movements of the stomach. In a current model of this oscillatory circuit, these interactions are mutually inhibitory (Fig. 12.3B).

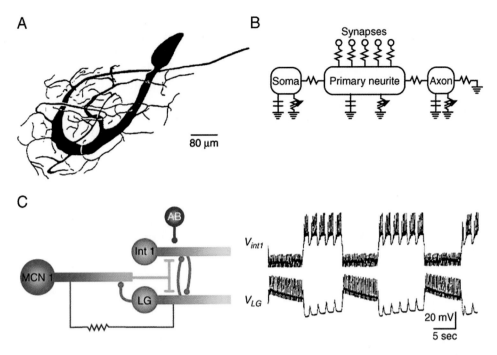

FIGURE 12.3   Local synaptic input-output sites are widely found within the neuropil of invertebrate ganglia. (A) Output neuron with many neurite branches in the gastric mill ganglion of the lobster, (B) compartmental representation of stomatogastric neuron, (C) model of rhythm generating circuit of the gastric mill of the lobster, involving neurite-neurite interactions. A and B from Golowasch and Marder (1992); C from Manor *et al.* (1999).

In summary, a cell with an axon can have local outputs through its dendrites, as well as distant outputs through its axon, which may be involved in specific local functions such as generating oscillatory circuits.

PASSIVE DENDRITIC TREES CAN PERFORM COMPLEX COMPUTATIONS

Another principle that carries over from axonless cells is the ability of the dendrites of axonal cells to carry out complex computations with mostly passive properties. This is exemplified by neurons that are motion detectors.

Motion detection is a fundamental operation carried out by the nervous systems of most species; it is essential for detecting prey and predator alike. In invertebrates, motion detection has been studied especially in the brain of the blowfly. In the lobula plate of the third optic neuropil are tangential cells (LPTCs) that respond to preferential direction (PD) of motion with increased depolarization due to sequential responses across their dendritic fields. This response has been modeled by Reichardt and colleagues by a series of elementary

motion detectors (HMDs) in the dendrites. A compartmental model (Single and Borst, 1998) reproduces the experimental results and theoretical predictions by showing how local modulations at each HMD are smoothed by integration in the dendritic tree to give a smoothed high-fidelity global output at the axon (Fig. 12.4A). In the model, spatial integration is largely independent of specific electrotonic properties but depends critically on the geometry and orientation of the dendritic tree.

In vertebrates, motion detection is built into the visual pathway at various stages in different species: the retina, midbrain (optic tectum), and cerebral cortex. Studies in the optic tectum have revealed cells with splayed uniplanar dendritic trees and specialized distal appendages that appear highly homologous across reptiles, birds, and mammals (Fig. 12.4B) (Luksch *et al.*, 1998). Physiological studies are needed to test the hypothesis that these cells perform operations through their dendritic fields similar to those of LPTC cells in the insect. To the extent that this is borne out, it will support a principle of motion detection through spatially distributed dendritic computations that is conserved across vertebrates and invertebrates.

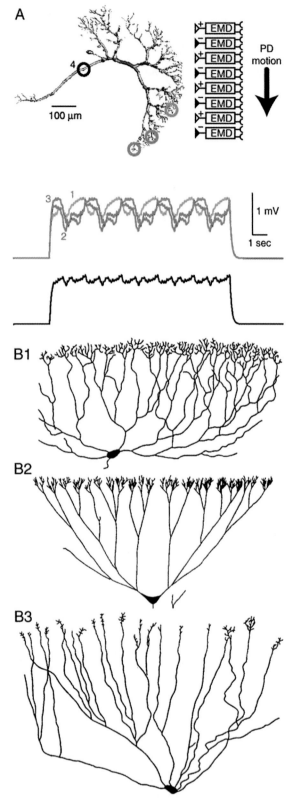

Directional selectivity of dendritic processing was predicted by Rall (1964) from his studies of dendritic electrotonus (Chapter 5). In an electrotonic cable, summation of EPSPs moving away from a recording site produces a plateau of ever-decreasing potentials, whereas summation of EPSPs moving toward a recording site produces an accumulating peak of potential. This is one of several possible mechanisms that are under current investigation in different cell types.

SEPARATION OF DENDRITIC FIELDS ENHANCES COMPLEX INFORMATION PROCESSING

An important feature of many types of neuron is a separation of their dendritic fields, which has important functional consequences. We saw this in the compartmental organization of the mitral cell (Figs. 12.1 and 12.2), where the primary dendritic tuft receives the olfactory nerve input whereas the secondary dendritic branches are specialized for a completely different function, self and lateral inhibition, as we explain later.

Pyramidal cells in the cerebral cortex also show a clear separation into apical and basal dendrites (Figs. 12.1 and 12.2). The apical dendrite extends across different layers, allowing fibers within those layers from different cells to modulate the transfer of activity from the distal tuft toward the cell body. This kind of modulation is absent in the mitral cell but key in the pyramidal neuron.

Within the basal dendrites, the placement of inputs is critical. An example has been shown in experiments in which excitatory and inhibitory inputs can be independently targeted to the same or different dendritic branches. As illustrated in Figure 12.5, synaptic inhibition has little effect on synaptic excitation when the two are targeted to different branches, but a profound effect when on the same branches. This is a clear example of the interaction of synaptic conductances illustrated in Figure 5.13.

FIGURE 12.4 Dendritic systems as motion detectors. (A) A computational model of a motion detector neuron in the visual system of the fly, consisting of elementary motion detector (EMD) units in its dendritic tree activated by the preferential direction (PD) of motion. Local modulations of the individual EMDs are integrated in the dendritic tree to give smooth global output in the axon (Single and Borst, 1998). (B) Dendritic trees of neurons in the optic tectum of lizard (Bl), chick (B2), and gray squirrel (B3). The architecture of the dendritic branching patterns and distal specialization for the reception of retinal inputs is highly homologous (references in Luksch *et al.*, 1998).

Location, location, location

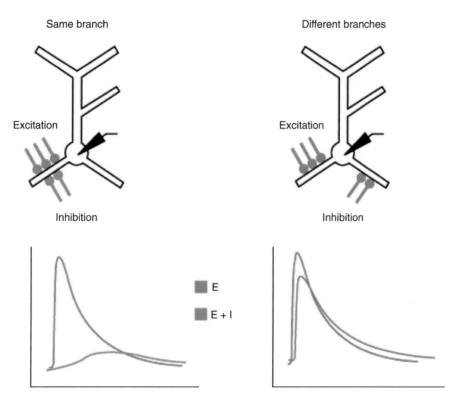

FIGURE 12.5 Importance of the locations of interacting synaptic responses within a dendritic field. Left, Excitation and inhibition converging on the same dendritic branches produces sharp reduction of the excitatory response recorded at the soma (see electrode). Right, inhibition on a different set of branches has little effect in reducing the excitatory response recorded at the soma. This illustrates a practical application of the principle illustrated in Figure 5.13 for electrotonic relations between synaptic conductances in individual branches. From Mel and Schiller (2004).

As a final example, cells with separate dendritic fields are critical for selectively summing their synaptic inputs to mediate directional selectivity in the auditory system (Overholt *et al.*, 1992).

DISTAL DENDRITES CAN BE CLOSELY LINKED TO AXONAL OUTPUT

An obvious problem for a neuron with an axon is that the distal branches of dendritic trees are a long distance from the site of axon origin at or near the cell body. The common perception is that these distal dendrites are too distant from the site of axonal origin and impulse generation to have more than a slow and weak background modulation of impulse output, and that the only synapses that can bring about rapid signal processing by the neuron are those located on the soma or proximal dendrites.

This perception is so ingrained in our visual impression of what is near and far that it is difficult to accept that it is wrong. It is disproved, however, by many kinds of neurons in which specific inputs are located preferentially on their distal dendrites. An example is the mitral (and tufted) cells in the olfactory bulb, which we have met in Figures 12.1 and 12.2. In this cell the input from the olfactory nerves ends on the most distal dendritic branches in the glomeruli; in rat mitral cells, this may be 400–500 μm or more from the cell body, in turtle, 600–700 μm. The same applies to their targets, the pyramidal neurons of the olfactory cortex, where the input terminates on the spines of the most distal dendrites in layer I. In many other neurons, a given type of input terminates over much or all of the dendritic tree; such is the case, for example, for climbing fiber and parallel fiber inputs to the cerebellar Purkinje cells. All these neuron types are shown in Figure 12.1.

How do distal dendrites in these neurons effectively control axonal output? Some of the important properties underlying this ability are summarized in Table 12.1. We consider several examples next, and in later sections.

Large Diameter Dendrites

The simplest way to enhance spread of a signal through dendrites is by a large diameter. It already has been illustrated for the spread of current in a branching cable in Chapter 5 (Fig. 5.4). However, space is at a premium in the central nervous system, so there is a tradeoff between diameter and length. This is why other membrane properties become important in overcoming distance.

High Specific Membrane Resistance

A key property is the specific membrane resistance (R_m) of the dendritic membrane. The functional significance of R_m is discussed in Chapter 5 (see Fig. 5.3). Traditionally, the argument was that if R_m is relatively low, the characteristic length of the dendrites will be relatively short, the electrotonic length will be correspondingly long, and synaptic potentials will therefore decrement sharply in spreading toward the axon hillock. However, as discussed in Chapter 5, intracellular recordings indicated that R_m is sufficiently high that the electrotonic lengths of most dendrites are relatively short, in the range of 1–2 (Johnston and Wu, 1995), and patch recordings suggest much higher R_m values, indicating electrotonic lengths less than 1. Thus, a relatively high R_m seems adequate for close electrotonic linkage between distal dendrites and somas, at least in the steady state.

Low K Conductances

An important factor controlling effective membrane resistance is K conductances. Chapter 5 discusses how a K channel, I_h, can affect the summation of EPSPs in striatal spiny cells. There is increasing evidence that dendritic input conductance is controlled by different types of K currents (Midtgaard *et al.*, 1993; Magee,

1999). When dendritic K conductances are turned off, R_m increases and dendritic coupling to the soma is enhanced. These conductances also control back-propagating action potentials, as discussed in Chapter 5 and earlier chapters.

Large Synaptic Conductances

A potentially important property is the amplitude of the conductance generated by the synapse itself. Early studies showed that in motor neurons, distal excitatory synaptic potentials were many times the amplitude of proximal synapses (Redman and Walmsley, 1983). This would account for the fact that the unitary synaptic response recorded at the soma slows with increasing distance in the dendrites, but maintains a constant amplitude of approximately $100 \mu V$. A corresponding increase in synaptic conductance has been shown in the distal dendrites of cortical pyramidal neurons (Magee, 2000). Patch recordings show that, whereas inhibitory postsynaptic currents (IPSCs) are similar in amplitude whether recorded from the distal dendrites or the soma, excitatory postsynaptic currents (EPSCs) are larger when recorded from distal dendrites than from the soma (Fig. 12.6). It is hypothesized that this reflects receptor channels composed of different subunits, for which there is increasing evidence. Research is needed to determine which synaptic protein subunits are involved to give these differences in conductance in specific cells.

Voltage-Gated Depolarizing Conductances

For transient responses, the electrotonic linkage becomes weaker because of the filtering effect of the capacitance of the membrane, and it is made worse by a higher R_m, which increases the membrane time constant, thereby slowing the spread of a passive potential (Chapter 5). This disadvantage can be overcome by depolarizing voltage-gated conductances: Na, Ca, or both. These add a wide variety of signal processing mechanisms to dendrites. As indicated in Table 12.1, they include boosting EPSP amplitudes, generating large-amplitude slow pacemaker potentials underlying the spontaneous activity of a neuron, supporting full back-propagating or forward propagating action potentials in the dendrites, forming "hot spots" that set up fast prepotentials at branch points in the distal dendrites, and functioning as coincidence detectors. Interactions between active sites, such as spines with voltage-gated conductances, can give rise to a sequence of activation of those sites, resulting in "pseudosaltatory conduction" through the dendritic tree between active sites or active clusters of sites.

TABLE 12.1 Properties That Increase the Effectiveness of Distal Synapses in Effecting Axonal Output

---

Higher membrane resistance

Larger distal synaptic conductances

Voltage-gated channels:

  increase EPSP amplitude

  generate large amplitude slow action potentials

  give rise to forward propagating full action potential

  are local "hot spots" that set up fast prepotentials

  function as coincidence detectors to summate responses

  mediate "pseudosaltatory conduction" toward the soma through individual active sites or clusters

---

Postsynaptic currents at different distances in the dendritic tree

Excitatory

Inhibitory

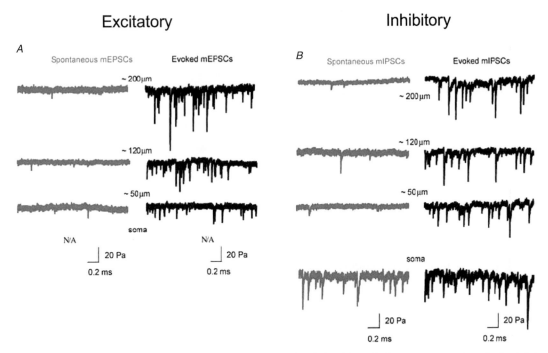

FIGURE 12.6 Larger excitatory synaptic currents may be present in distal dendrites. A. Spontaneous and evoked miniature excitatory postsynaptic currents (mEPSCs) in patch recordings at three distances from the soma. Note the increase in amplitude with increasing distance. B. Same for inhibitory mIPSCs. Note relatively constant amplitudes with distance. From Andrasfalvy and Mody (2006).

Some of these active properties also contribute to complex information processing capabilities, including logic operations, as discussed further later.

Summary

These examples illustrate an important principle of distal dendritic processing:

> Distal dendrites can mediate relatively rapid, specific information processing, even at the weakest levels of detection, in addition to slower modulation of overall neuronal activity. The spread of potentials to the site of global output from the axon is enhanced by multiple passive and active mechanisms.

DEPOLARIZING AND HYPERPOLARIZING DENDRITIC CONDUCTANCES INTERACT DYNAMICALLY

We see that depolarizing conductances increase the excitability of distal dendrites and the effectiveness of distal synapses, whereas K conductances reduce the excitability and control the temporal characteristics of

the dendritic activity. This balance is thus crucial to the functions of dendrites. Figure 12.7 summarizes data showing how these conductances vary along the extents of the dendrites of mitral cells, hippocampal and neocortical pyramidal neurons, and Purkinje cells.

The significance of a particular density of channel needs to be judged in relation to the electrotonic properties discussed in Chapter 5. For instance, a given conductance has more effect on membrane potential in smaller distal branches because of the higher input resistance (Fig. 5.14). Dendritic conductances are crucial in setting the intrinsic excitability state of the neuron. In the motor neuron, for example, the neuron can alternate between bistable states dependent on the activation of dendritic metabotropic glutamate receptors (Svirskie et al., 2001). The significance of these and other conductance interactions for the firing properties of different cell types is discussed later.

These combinations of ionic conductances occur within the larger framework of the morphological types of dendritic trees, particularly whether they arise from thick or thin trunks. This has given rise to a classification of dendritic types on integrative principles that cuts across the traditional classification of neuron types (Migliore and Shepherd, 2002, 2005). This is a

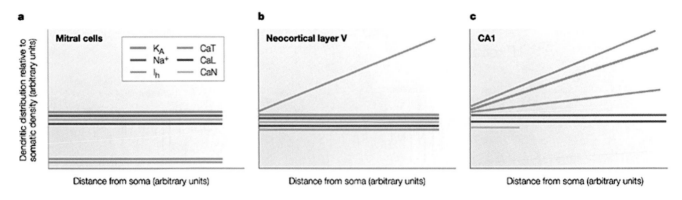

Nature Reviews | Neuroscience

FIGURE 12.7   Graphs of the distribution of different types of intrinsic membrane conductances along the dendritic trees in different types of neurons. From Migliore and Shepherd (2002).

first step toward a deeper insight into canonical types of input-output operations that are carried out by dendritic trees.

As noted earlier, the combinations of properties within different dendritic compartments of a given neuron type can be searched in an online database (senselab.med.yale.edu/neurondb) and models based on these compartmental representations of many types of neurons can be accessed and run at senselab.med.yale.edu/modeldb.

Summary

The combination of conductances at different levels of the dendritic tree involves a delicate balance between depolarizing and hyperpolarizing actions acting over different time periods. These combinations vary in different morphological types of neurons. They also contribute to a new classification of dendrites according to a principle of multiple criteria for dendritic classification:

> Dendritic trees can be categorized functionally on the basis of a combination of branch morphology, functional ionic current type, and genetic channel subunit type. These categories appear to define canonical integrative properties that extend across classical morphological categories.

THE AXON HILLOCK-INITIAL SEGMENT ENCODES GLOBAL OUTPUT

In cells with long axons, activity in the dendrites eventually leads to activation and modulation of action potential output in the axon. A key question is the precise site of origin of this action potential. This question was one of the first to be addressed in the rise of modern neuroscience; the historical background is summarized in Box 12.1.

These studies established the classical model: the lowest threshold site for action potential generation is in the *axonal initial segment*.

Definitive analysis was achieved by Stuart and Sakmann (1994) using dual patch recordings from cortical pyramidal neurons under differential contrast microscopy. This approach has provided the breakthrough for subsequent analyses of dendritic properties and their coupling to the axon (see later). As shown in Figure 12.10, with depolarization of the distal dendrites by injected current or excitatory synaptic inputs, a large amplitude depolarization is produced in the dendrites, which spreads to the soma. Despite its lower amplitude, soma depolarization is the first to initiate the action potential. Subsequent studies with triple patch electrodes have shown that the action potential actually arises first in the initial segment and first node.

MULTIPLE IMPULSE INITIATION SITES ARE UNDER DYNAMIC CONTROL

In addition to the evidence for action potential initiation in the axon hillock, another line of work has provided evidence for shifting of the site under dynamic conditions. This line began with extracellular recordings of a "population spike" that appears to propagate along the apical dendrites toward the cell body in hippocampal pyramidal cells (Andersen, 1960). This was supported by the recording in these

BOX 12.1

CLASSICAL STUDIES OF THE ACTION POTENTIAL INITIATION SITE

Fuortes and colleagues (1957) were the first to deduce that an EPSP spreads from the dendrites through the soma to initiate the action potential in the region of the axon hillock and the initial axon segment. They suggested that the action potential has two components: (1) an A component that normally is associated with the axon hillock and initial segment and (2) a B component that normally is associated with retrograde invasion of the cell body. Because the site of action potential initiation can shift under different membrane potentials, they preferred the noncommittal terms "A" and "B" for the two components as recorded from the cell body. In contrast, Eccles (1957) referred to the initial component as the initial-segment (IS) component and to the second component as the somadendritic (SD) component (Fig. 12.8).

Apart from the motor neuron, the best early model for intracellular analysis of neuronal mechanisms was the crayfish stretch receptor, described by Eyzaguirre and Kuffler (1955). Intracellular recordings from the cell body showed that stretch causes a depolarizing receptor potential equivalent to an EPSP, which spreads through the cell to initiate an action potential. It was first assumed that this action potential arose at or near the cell body. Edwards and Ottoson (1958), working in Kuffler's laboratory, tested this postulate by recording the local extracellular current in order to locate precisely the site of inward current associated with action potential initiation. Surprisingly, this site turned out to be far out on the axon, some $200\,\mu\text{m}$ from the cell body (Fig. 12.9). This result showed that potentials generated in the distal dendrites can spread all the way through the dendrites and soma well out into the initial segment of the axon to initiate impulses. It further showed that the action potential recorded at the cell body is the backward spreading impulse from the initiation site. Edwards and Ottoson's study was important in establishing the basic model of impulse initiation in the axonal initial segment.

Gordon M. Shepherd

References

Eccles, J. C. (1957). "The Physiology of Nerve Cells." Johns Hopkins Univ. Press, Baltimore.

Edwards, C. and Ottoson, D. (1958). The site of impulse initiation in a nerve cell of a crustacean stretch receptor. *J. Physiol. (Lond.)* **143**, 138–148.

Eyzaguirre, C. and Kuffler, S. W. (1955). Processes of excitation in the dendrites and in the soma of single isolated sensory nerve cells of the lobster and crayfish. *J. Gen. Physiol.* **39**, 87–119.

Fuortes, M. G. E., Frank, K., and Banker, M. C. (1957). Steps in the production of motor neuron spikes, *J. Gen. Physiol.* **40**, 735–752.

cells of "fast prepotentials" at dendritic "hot spots" (Spencer and Kandel, 1961), and by current source density calculations in cortical pyramidal neurons (Herreras, 1990). In recordings from dendrites in tissue slices in CA1 hippocampal pyramidal neurons, weak synaptic potentials elicited action potentials near the cell body (Richardson *et al.*, 1987), but this site shifted to proximal dendrites with stronger synaptic excitation (Turner *et al.*, 1991). This confirmed the suggestion of M.G.F. Fuortes and K. Frank that the site can shift under different stimulus conditions, and was consistent with the stretch receptor, where larger receptor potentials shift the initiation site closer to the cell body.

The olfactory mitral cell is a favorable model for studying this question, because it is unusual in that all its excitatory inputs are restricted to its distal dendritic tuft. As illustrated in Chapter 5 (Fig. 5.15), at weak levels of electrical shocks to the olfactory nerves, the site of action potential initiation is at or near the soma, as in the classical model. The action potential is due to Na channels distributed along the extent of the primary dendrite.

As the level of distal excitatory input is increased, dual-patch recordings show clearly that the action potential initiation site shifts gradually from the soma to the distal dendrite (Fig. 5.15). Thus the site of impulse initiation is not fixed in the mitral cell, but varies with the intensity of distal excitatory input balanced against the difference in density of Na channels between initial segment and the apical dendrite. This shift was discussed in Chapter 5 because it is governed by the longitudinal gradient of spread of the passive electrotonic potential along the dendrite. The site can also be shifted to distal dendrites by synaptic inhibition applied to the soma through dendrodendritic synapses.

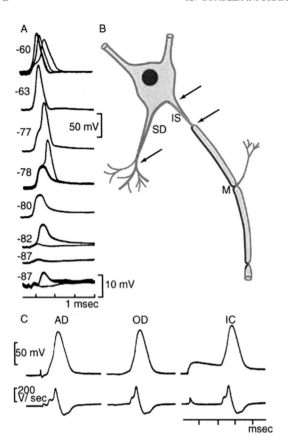

FIGURE 12.8 Classical evidence for the site of action potential initiation. Intracellular recordings were from the cell body of the motor neuron of an anesthetized cat. (A) Differential blockade of an antidromic impulse by adjusting the membrane potential by holding currents. Recordings reveal the sequence of impulse invasion in the myelinated axon (recordings at −87mV, two amplifications), the initial segment of the axon (first component of the impulse beginning at −82mV), and the soma–dendritic region (large component beginning at −78mV). (B) Sites of the three regions of impulse generation (M, myelinated axon; IS, initial segment; SD, soma and dendrites; arrows show probable sites of impulse blockade in A. (C) Comparison of intracellular recordings of impulses generated antidromically (AD), synaptically (orthodromically, OD), and by direct current injection (IC). Lower traces indicate electrical differentiation of these recordings showing the separation of the impulse into the same two components and indicating that the sequence of impulse generation from the initial segment into the soma–dendritic region is the same in all cases. From Eccles (1957).

Summary

The low threshold of the initial axonal segment favors it being the site of action potential output for a wide range of dendritic activity, but the site can shift with strongly depolarizing dendritic input. This introduces the **principle of the dynamic control of action potential initiation**:

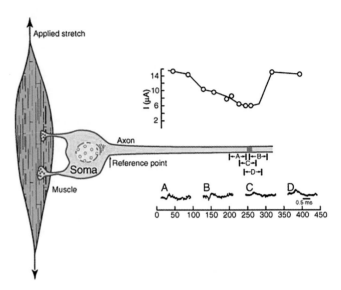

FIGURE 12.9 Classical demonstration of the site of impulse initiation in the stretch receptor cell of the crayfish. Moderate stretch of the receptor muscle generated a receptor potential that spread from the dendrites across the cell body into the axon. Paired electrodes recorded the longitudinal extracellular currents at positions A–D, showing the site of the trigger zone (green region). The excitability curve (shown at the top), obtained by passing current between the electrodes and finding the current (I) intensity needed to evoke an impulse response, also shows the trigger zone to be several hundred micrometers out on the axon. From Ringham (1971).

The site of global output through action potential initiation from a neuron can shift between first axon node, initial segment, axon hillock, soma, proximal dendrites, and distal dendrites, depending on the dynamic state of dendritic excitability.

RETROGRADE IMPULSE SPREAD INTO DENDRITES CAN HAVE MANY FUNCTIONS

In addition to identifying the preferential site for action potential initiation in the axonal initial segment, the experiments of Stuart and Sackmann (1994) showed clearly that the action potential does not merely spread passively back into the dendrites but actively back-propagates. Note that we distinguish between passive electrotonic "spread" and active "propagation" of the action potential (see Box 5.3 in Chapter 5).

What is the function of the dendritic action potential? Experimental evidence shows that it can have a variety of functions.

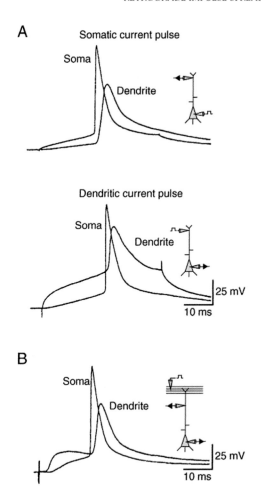

FIGURE 12.10 Direct demonstration of the impulse-initiation zone and back-propagation into dendrites using dual-patch recordings from soma and dendrites of a layer V pyramidal neuron in a slice preparation of the rat neocortex. (A) Depolarizing current injection in either the soma or the dendrite elicits an impulse first in the soma. (B) The same result is obtained with synaptic activation of layer I input to distal dendrites. Note the close similarity of these results to the earlier findings in the motor neuron (Fig. 12.8). From Stuart and Sakmann (1994).

Dendrodendritic Inhibition

A specific function for an action potential propagating from the soma into the dendrites was first suggested for the olfactory mitral cell, where mitral-to-granule dendrodendritic synapses are triggered by the action potential spreading from the soma into the secondary dendrites (Fig. 12.11A, B). Because of the delay in activating the reciprocal inhibitory synapses from the granule cells, self-inhibition of the mitral cell occurs in the wake of the passing impulse; the two do not collide. The mechanism operates similarly with both active back-propagation and passive electrotonic spread into the dendrites, as tested in computer simulations. Functions of dendrodendritic inhibition include center-surround antagonism mediating the abstraction of molecular determinants underlying the discrimination of different odor molecules, storing of olfactory memories at the reciprocal synapses, and generation of oscillating activity in mitral and granule cell populations (Shepherd et al., 2004; Egger and Urban, 2006).

Intercolumnar Connectivity

Recent research has given new insight into the function of the action potential in the mitral cell lateral dendrite. Because the action potential can propagate away from the cell body throughout the length of the dendrite (Fig. 12.11; Xiong and Chen, 2002), it enables activation of granule cells independent of distance. Connectivity of mitral cells to distant groups of granule cells, arranged in columns in relation to glomeruli, has been demonstrated by pseudorabies viral tracing (Willhite et al., 2006), and activation of distant granule cells by means of such connectivity has been shown in realistic computational studies (Migliore and Shepherd, 2007). This has led to the hypothesis that the lateral dendrite can function to activate ensembles of granule cell columns processing similar aspects of an odor map, with the added flexibility that the dendrite can be modulated by granule cell inhibition throughout its length (Fig. 12.11A). The diagram in B thus provides an updated representation of the functional subunits and microcircuits formed by the mitral and granule cells shown in Figures 12.1 and 12.2.

Boosting Synaptic Responses

In several types of pyramidal neurons, active dendritic properties appear to boost action potential invasion so that summation with EPSPs occurs that makes the EPSPs more effective in spreading to the soma.

Resetting Membrane Potential

A possible function of a back-propagating action potential is that the Na⁺ and K⁺ conductance increases associated with active propagation wipe out the existing membrane potential, resetting the membrane potential for new inputs.

Synaptic Plasticity

The action potential in the dendritic branches presumably depolarizes the spines (because of the favorable impedance matching, as discussed in Chapter 5), which means that the impulse depolarization would

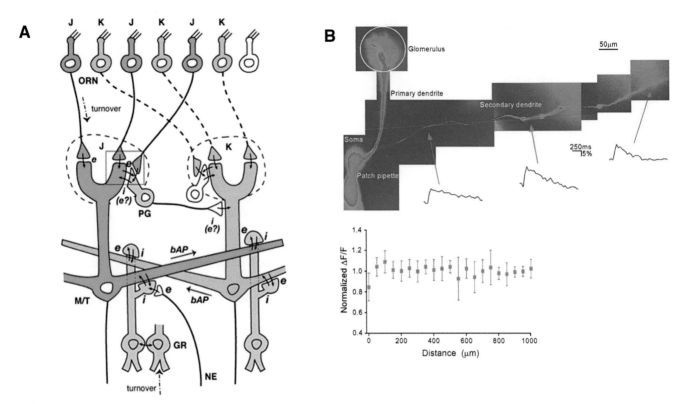

FIGURE 12.11 Dendrodendritic interactions in the olfactory bulb. (A) An action potential in the mitral cell body sets up a back-spreading/back-propagating impulse into the secondary dendrites, activating both feedback and lateral inhibition of the mitral cells by columns of granule cells acting through the dendrodendritic pathway. From Shepherd *et al.* (2007). (B) Ability of an action potential to invade the length of a secondary dendrite, as shown by Ca fluorescence. Flourescence measurements are plotted in the graph below, showing full propagation up to 1000 microns. From Xiong and Chen (2002).

summate with the synaptic depolarization of the spines. This process would enable the spines to function as coincidence detectors and implement changes in synaptic plasticity (see Hebbian synaptic mechanisms in Chapter 49). This postulate has been tested by electrophysiological recordings (Spruston *et al.*, 1995) and Ca^{2+} imaging (Yuste *et al.*, 1994). Activity-dependent changes of dendritic synaptic potency are not seen with passive retrograde depolarization but appear to require actively propagating retrograde impulses (Spruston *et al.*, 1995).

Frequency Dependence

Trains of action potentials generated at the soma-axon hillock can invade the dendrites to varying extents. Proximal dendrites appear to be invaded throughout a high-frequency burst, whereas distal dendrites appear to be invaded mainly by the early action potentials (Regehr *et al.*, 1989; Callaway and Ross, 1995; Yuste *et al.*, 1994; Spruston *et al.*, 1995).

Activation of Ca^{2+}-activated K$^+$ conductances by early impulses may effectively switch off the distal dendritic compartment.

Retrograde Actions at Synapses

The retrograde action potential can contribute to the activation of neurotransmitter release from the dendrites. The clearest example of this is the olfactory mitral cell as already described. Dynorphin released by synaptically stimulated dentate granule cells can affect the presynaptic terminals (Simmons *et al.*, 1995). In the cerebral cortex there is evidence that GABAergic interneuronal dendrites act back on axonal terminals of pyramidal cells and that glutamatergic pyramidal cell dendrites act back on axonal terminals of the interneurons (Zilberter, 2000). The combined effects of the axonal and dendritic compartments of both neuronal types regulate the normal excitability of pyramidal neurons and may be a factor in the development of cortical hyperexcitability and epilepsy.

Conditional Axonal Output

Because of the long distance between distal dendrites and initial axonal segment, we may hypothesize that the coupling between the two is not automatic. Indeed, conditional coupling dependent on synaptic inputs and intrinsic activity states at intervening dendritic sites appears to be fundamental to the relation between local dendritic inputs and global axonal output (Spruston, 2000).

Summary

The action potential arising at the initial axonal segment has two functions: propagating into the axon to carry the global output to the axon terminals, and propagating retrogradely through the soma into the dendrites. In the dendrites the action potential can carry out many distinct functions, as described above.

When the retrograde action potential has been activated by EPSPs spreading from the dendrites, we call the action potential back-propagating; that is, back toward the site of the initial input. When the retrograde action potential propagates through the soma into previously unactivated dendrites, we can still call it back-propagating, in the sense of backward with regard to the law of dynamic polarization, which, when applied to the overall flow of activity, is from distal dendrites to soma and axon; or we can consider it as propagating retrogradely, to distinguish it from back-propagating toward a distal input site.

EXAMPLES OF HOW VOLTAGE-GATED CHANNELS ENHANCE DENDRITIC INFORMATION PROCESSING

It is commonly believed that active dendrites are a modern concept, but in fact this idea is as old as Cajal; he assumed that dendrites conduct impulses like axons do. However, with the first intracellular recordings in the 1950s, it appeared that dendritic membranes were mostly passive. We have noted that studies since then increasingly have documented the widespread distribution and numerous functions of voltage-gated channels in dendritic membranes.

These channels are the principle means for enhancing the information processing capabilities of complex dendrites. Detailed analysis of active dendritic properties began with computational studies of olfactory mitral cells and experimental studies of cerebellar Purkinje cells. Since then, studies of active dendritic prop-

erties have proliferated, particularly since introduction of the patch recording method. Several types of neurons have provided important models for the possible functional roles of active dendritic properties.

Purkinje Cells

The cerebellar Purkinje cell has the most elaborate dendritic tree in the nervous system, with more than 100,000 dendritic spines receiving synaptic inputs from parallel fibers and mossy fibers. The basic distribution of active properties in the Purkinje cell was indicated by the pioneering experiments of Llinas and Sugimori (1980) in tissue slices (Fig. 12.12). The action potential in the cell body and axon hillock is due mainly to fast Na^+ and delayed K^+ channels; there is also a Ca^{2+} component. The action potential correspondingly has a large amplitude in the cell body and decreases by electrotonic decay in the dendrites. In contrast, recordings in the dendrites are dominated by slower "spike" potentials that are Ca^{2+} dependent due to a P-type Ca^{2+} conductance (Fig. 12.12). These spikes are generated from a plateau potential due to a persistent Na_p current.

There are two distinct operating modes of the Purkinje cell in relation to its distinctive inputs. Climbing fibers mediate strong depolarizing EPSPs throughout most of the dendrites that appear to give rise to

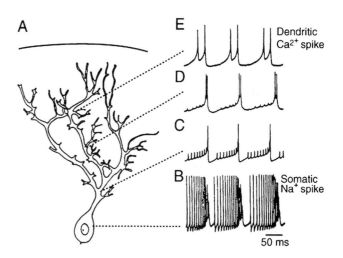

FIGURE 12.12 Classical demonstration of the difference between soma and dendritic action potentials. (A) Drawing of a Purkinje cell in the cerebellar slice. (B) Intracellular recordings from the soma showing fast Na^+ spikes. (C–E) Intracellular recordings from progressively more distant dendritic sites; fast soma spikes become small due to electrotonic decrement and are replaced by large-amplitude dendritic Ca^{2+} spikes. Spread of these spikes to the soma causes an inactivating burst that interrupts the soma discharge. Adapted from Llinas and Sugimori (1980).

synchronous Ca²⁺ dendritic action potentials through-out the dendritic tree, which then spread to the soma to elicit the bursting "complex spike" in the axon hillock. In contrast, parallel fibers are active in small groups, giving rise to smaller populations of individual EPSPs possibly targeted to particular dendritic regions (compartments).

The Purkinje cell thus illustrates several of the principles we have discussed. *Subthreshold amplification* through active dendritic properties may enhance the effect of a particular set of input fibers in controlling or modulating the frequency of Purkinje cell action potential output in the axon hillock. The Purkinje cell is subjected to *local inhibitory control* by stellate cell synapses targeted to specific dendritic compartments, and to *global inhibitory control* of axonal output by basket cell synapses on the axonal initial segment.

Medium Spiny Cell

A different instructive example of the role of active dendritic properties is found in the medium spiny cell of the neostriatum (Figs. 12.13A). The passive electrotonic properties of this cell are described in Chapter 5 (Fig. 5.12). Inputs to a given neuron from the cortex are widely distributed, meaning that a given neuron must summate a significant number of synaptic inputs before generating an impulse response. The responsiveness of the cell is controlled by its cable properties; individual responses in the spines are filtered out by the large capacitance of the many dendritic spines so that individual EPSPs recorded at the soma are small.

With synchronous specific inputs, larger summated EPSPs depolarize the dendritic membrane strongly. The dendritic membrane contains inwardly rectifying channels (I_h) (Fig. 12.13C), which reduce their conductance upon depolarization and thereby increase the effective membrane resistance and shorten the electrotonic length of the dendritic tree. Large depolarization also activates HT Ca²⁺ channels, which contribute to large-amplitude, slow depolarizations. These combined effects change the neuron from a state in which it is insensitive to small noisy inputs into a state in which it gives a large response to a specific input and is maximally sensitive to additional inputs. Through this voltage-gated mechanism, a neuron can enhance the effectiveness of distal dendritic inputs, not by boosting inward Na⁺ and K⁺ currents, but by reducing outward shunting K⁺ currents. This exemplifies the principle of dynamic control over dendritic properties through interactions involving K conductances mentioned earlier.

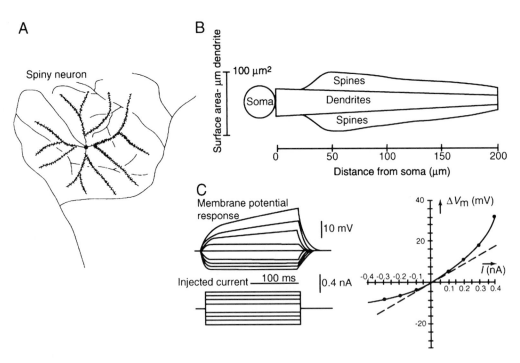

FIGURE 12.13 Dendritic spines and dendritic membrane properties interact to control neuronal excitability. (A) Diagram of a medium spiny neuron in the caudate nucleus; (B) plot of surface areas of different compartments showing a large increase in surface area due to spines; and (C) intracellular patch-clamp analysis of medium spiny neuron showing inward rectification of the membrane that controls the response of the dendrites to excitatory synaptic inputs (*cf.* Chapter 5, Fig. 5.12). From Wilson (1998).

Pyramidal Neurons

Active properties of the apical dendrite of hippocampal pyramidal neurons have been documented amply by patch recordings (Magee and Johnston, 1995). In contrast to the Purkinje cell, both fast Na^+ and Ca^{2+} conductances have been shown throughout the dendritic tree of the pyramidal neuron by electrophysiological and dye-imaging methods (Fig. 12.7). Activation of low-threshold Na^+ channels is believed to play an important role in triggering the higher-threshold Ca^{2+} channels. Similar results have been obtained in studies of pyramidal neurons of the cerebral cortex.

At the simplest level, the output pattern of a neuron depends on its dendritic properties and their interaction with the soma. This is exemplified by the generation of a burst response in a pyramidal neuron. EPSPs spread through the dendrite, activating fast Na^+ and then high-threshold (HT) Ca^{2+} channels that give a subthreshold boost to the EPSP. The enhanced EPSP spreads to the soma-axon hillock, triggering a Na^+ action potential. This propagates into the axon and also back-propagates into the dendrites, eliciting a slower all-or-nothing Ca^{2+} action potential. This large-amplitude, slow depolarization then spreads through the dendrites and back to the soma, triggering a train of action potentials that form a burst response.

This sequence of events is simulated most accurately by a realistic multicompartmental model of the dendritic tree. However, the essence can be contained in a two-compartment model representing the soma and dendritic compartments (Fig. 12.14). The model sequence emphasizes not only the importance of the interplay between the different types of channels, but also the critical role of the compartmentalization of the neuron into dendritic and somatic compartments so that they can interact in controlling the intensity and time course of the impulse output.

This simpler model would argue that the specific form of the input-output transformation does not depend on a specific distribution of active channels in the dendritic tree. Na^+ and Ca^{2+} channels in fact are distributed widely in pyramidal neuron dendrites. In computational simulations, grouping channels in different distributions may have little effect on the input-output functions of a neuron (Mainen and Sejnowski, 1995). However, there is evidence that subthreshold amplification by voltage-gated channels may tend to occur in the more proximal dendrites of some neurons (Yuste and Denk, 1995). In addition, the dendritic trees of some neurons clearly are divided into different anatomical and functional subdivisions, as discussed in the next section.

Summary

These are only a few examples of the range of operations carried out by complex dendrites. These dendritic operations are embedded in the circuits that control behavior. Thus, for each neuron, the dendritic tree constitutes an expanded unit essential to the circuits' underlying behavior.

DENDRITIC SPINES ARE MULTIFUNCTIONAL MICROINTEGRATIVE UNITS

Much of the complex processing that takes place in dendrites involves inputs through dendritic spines, the tiny outcroppings from the dendritic surface. Their electrotonic properties were described in relation to Fig. 5.14. The very small size of dendritic spines has made it difficult to study them directly. However, examples already have been given of spines with complex information processing capacities, such as granule cell spines in the olfactory bulb and spines of medium spiny neurons in the striatum. In cortical neurons, spines have been implicated in cognitive functions from observations of dramatic changes in spine morphology in relation to different types of mental retardation and different hormonal exposures.

One of the most fertile hypotheses, by Rall and Rinzel (1974), is that changes in the dimensions of the spine stem control the effectiveness of coupling of

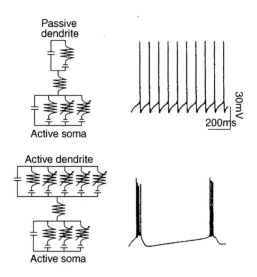

Passive dendrite

Active soma

30mV

200ms

Active dendrite

Active soma

FIGURE 12.14 Generation of a burst response by interactions between soma and dendrites. From Pinsky and Rinzel (1994).

the synaptic response in the spine head to the rest of the dendritic branch, and could therefore provide a mechanism for learning and memory (see also Harris and Krater, 1994; Shepherd, 1996; Yuste and Denk, 1995). For example, an activity-dependent decrease in stem diameter could increase the input resistance of the spine head, increasing an EPSP amplitude, which could have local effects on subsequent responses, and it can also decrease the coupling to the parent dendrite. In addition to these electrotonic effects, a decrease in stem diameter could also increase the biochemical compartmentalization of the spine head (Fig. 5.14).

Computational models have been very useful in testing these hypotheses, as well as suggesting other possible functions, such as the dynamic changes of electrotonic structure in medium spiny cells of the basal ganglia (see earlier discussion). With the development of more powerful light microscopic methods, such as two-photon laser confocal microscopy, it has become possible to test these hypotheses directly by imaging Ca^{2+} fluxes in individual spines in relation to synaptic inputs and neuronal activity (Fig. 12.15).

Evidence for active properties of dendrites has suggested that the spines may also have active properties. Thus, spines may be devices for nonlinear thresholding operations, either through voltage-gated ion channels (Fig. 12.7) or through voltage-dependent synaptic properties such as N-methyl-D-aspartate (NMDA) receptors. This could powerfully enhance the information processing capabilities of spiny dendrites. As an example, computational simulations have shown that logic operations are inherent in coincidence detection by active dendritic sites such as spines. The example is an AND operation performed by two dendritic spines

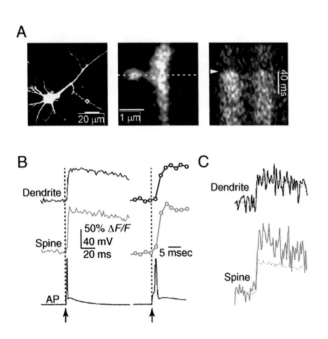

FIGURE 12.15 Calcium transients can be imaged in single dendritic spines in a rat hippocampal slice. (A) Fluo-4, a calcium-sensitive dye, injected into a neuron enables an individual spine to be imaged under two-photon microscopy. (B) An action potential (AP) induces an increase in Ca^{2+} in the dendrite and a larger increase in the spine (averaged responses). (C) Fluctuation analysis indicated that spines likely contain up to 20 voltage-sensitive Ca channels; single channel openings could be detected, which had a high (0.5) probability of opening following a single action potential. From Sabatini and Svoboda (2000).

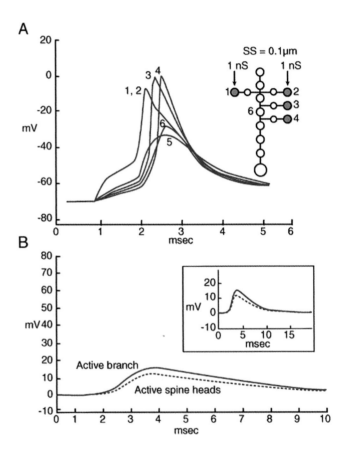

FIGURE 12.16 Logic operations are inherent in coincidence detection by active dendritic sites. The example is an AND operation performed by two dendritic spines with Hodgkin–Huxley-type active kinetics, with intervening passive dendritic membrane. (A) Simultaneous synaptic input of 1 nS conductance to spines 1 and 2 gives rise to action potentials within both spines, which spread passively to activate action potentials in spines 3 and 4. Sequential coincidence detection by active spines can thus bring boosted synaptic responses close to the soma. From Shepherd and Brayton (1987). (B) Recording of boosted spine responses at the soma shows their similarity to the slow time course of classical EPSPs due to the electrotonic properties of the intervening dendritic membrane (see text). SS, spine stem diameter. From Shepherd et al. (1989).

with Hodgkin–Huxley-type active kinetics, with intervening passive dendritic membrane. As illustrated in Figure 12.16, simultaneous synaptic input of 1 nS conductance to spines 1 and 2 gives rise to action potentials within both spines, which spread passively to activate action potentials in spines 3 and 4. Sequential coincidence detection by active spines can thus bring boosted synaptic responses close to the soma.

Further computational experiments have shown that spines can function as OR gates or as AND-NOT gates, which together with AND gates, provide the basic operations for a digital computer. This shows that simple logic operations are inherent in dendrites, a starting point for investigating the actual kinds of information processing that the brain uses.

In addition to these functions underlying normal functioning of dendrites, the morphological characteristics of a spine may be used to isolate functional properties that result from pathological processes. One such suggestion is that spines may function as compartments to isolate changes at the synapse, such as influx of excess Ca^{2+} that occurs in ischemia due to stroke, which lead to degenerative changes that are harmful to the rest of the neuron (Volfovsky et al., 1999).

The range of functions that have been hypothesized for spines is partly a reflection of how little direct evidence we have of specific properties of spines. It also indicates that the answer to the question "What is the function of the dendritic spine?" is unlikely to be only one function, but rather a range of functions that is tuned in a given neuron to the specific operations of that neuron. The spine is increasingly regarded as a microcompartment that integrates a range of functions (Harris and Kater, 1994; Shepherd, 1996; Yuste and Denk, 1995). A spiny dendritic tree thus is covered with a large population of microintegrative units. As discussed previously, the effect of any given one of these units on the action potential output of the neuron therefore should not be assessed with regard only to the far-off cell body and axon hillock, but rather with regard first to its effect on its neighboring microintegrative units.

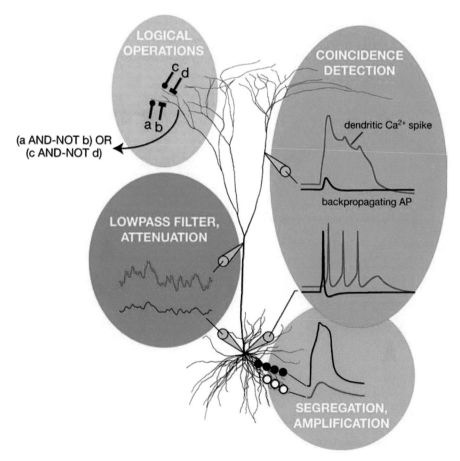

FIGURE 12.17 Summary of some of the functions of dendritic tree of cortical pyramidal neurons that have been demonstrated experimentally and computationally and discussed in this chapter. From Mel and Schiller (2003).

SUMMARY: THE DENDRITIC TREE AS A COMPLEX INFORMATION PROCESSING SYSTEM

Dendrites are the primary information processing substrate of the neuron. They allow the neuron wide flexibility in carrying out the operations needed for processing information in the spatial and temporal domains within nervous centers. The main constraints on these operations are the rules of passive electrotonic spread (Chapter 5), and the rules of nonlinear thresholding at multiple sites within the complex geometry of dendritic trees. Many specific types of information processing can be demonstrated in dendrites, such as logic operations, motion detection, oscillatory activity, lateral inhibition, and network control of sensory processing and motor control. These types are possible for both cells without axons and cells with axons, the latter operating in addition within constraints that govern

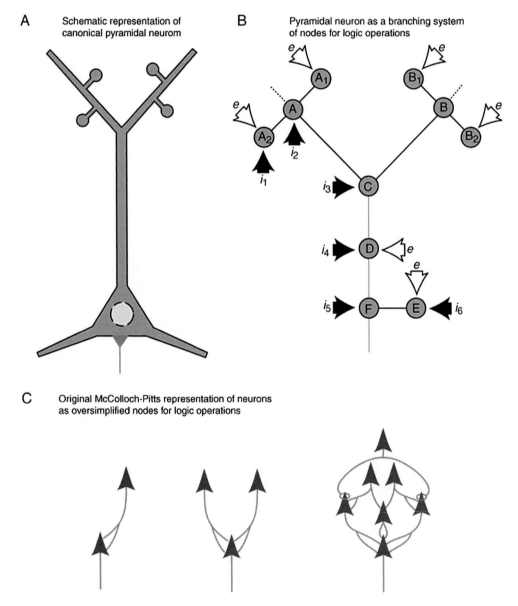

FIGURE 12.18 The dendritic tree as a complex system of logic nodes. (A) A simplified representation of a cortical pyramidal cell. (B) Conversion to a representation in terms of logic nodes and interconnections. (C) Comparison with the concept introduced by McCulloch and Pitts (1943) of the neuron as a functional node for carrying out logic operations, but in which the dendritic tree is ignored and the entire neuron is reduced to a single computational node, *e*, excitatory synapse; *i*, inhibitory synapse. From Shepherd (1994).

Mapping a multinode dendrite neuron Into a two-layer neural network

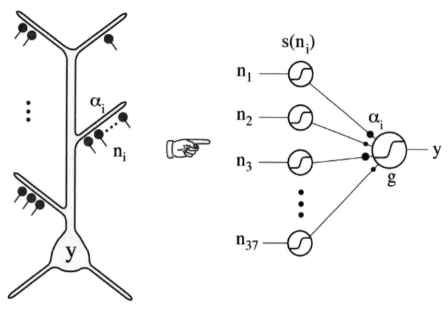

FIGURE 12.19 How can the kinds of complex operations illustrated in Figures 12.17 and 12.18 be incorporated into network models? One way is illustrated here, in which the input-output operations of the thin oblique dendritic branches of the apical dendrite of a CA1 pyramidal neuron are represented, together with the summing node at the soma, as a two-layer neural network. From Poirazi *et al.* (2003).

local vs. global outputs and sub- vs. suprathreshold activities. Several of these operations, as reviewed earlier in the chapter, are summarized in the diagram of Figure 12.17.

Spines add a dimension of local computation to dendritic function that is especially relevant to mechanisms for learning and memory. Although spines seem to distance synaptic responses from directly affecting axonal output, many cells demonstrate that distal spine inputs carry specific information.

The key to understanding how all parts of the dendritic tree, including its distal branches and spines, can participate in mediating specific types of information processing is to recognize the tree as a complex system of active nodes. From this perspective, if a spine can affect its neighbor, and that spine has its neighbor, a dendritic tree becomes a cascade of decision points, with multiple cascades operating over multiple overlapping time scales (Fig. 12.18A, B). Far from being a single node, as in the classical concept of McCulloch and Pitts (1943) (Fig. 12.18C) and classical neural network models, the complex neuron is a system of nodes in itself, within which the dendrites constitute a kind of neural microchip for complex computations.

The global output becomes the summation of all the logic operations taking place in the dendrites (Shepherd and Brayton, 1987). The neuron as a single node, so feeble in its information processing capacities, is replaced by the neuron as a powerful complex multinodal system.

The range of operations of which this complex system is capable continues to expand. A formal representation of the dendrite as a multinodal system has been applied to the ensemble of thin oblique dendrites that are emitted by the apical dendrite of a CA1 pyramidal neuron. As shown in Figure 12.19, these thin dendrites, as with spines in Figure 12.18, can be represented by individual summing nodes. The cell can then be mapped onto a two-layer "neural network," in which the first layer consists of the synaptic inputs to the oblique nodes, whose outputs are then summed at the cell body for final thresholding and global output (Poirazi *et al.*, 2003). Most of the input to the oblique dendrites is believed to be involved in the generation of long-term potentiation, a candidate model for learning and memory. The two-layer conceptual approach thus may be a bridge between realistic multicompartmental models and single node neural networks in the study of brain mechanisms in learning and memory. Exploring the information processing capacities of the brain at the level of real dendritic systems, by both experimental and theoretical methods, thus presents one of the most exciting challenges for neuroscientists at present and into the future.

References

Andersen, P. (1960). Interhippocampal impulses. II. Apical dendritic activation of CA1 neurons. *Acta Physiol. Scand.* **48**, 178–208.

Andrásfalvy, B. K. and Mody, I. (2006). Differences between the scaling of miniature IPSCs and EPSCs recorded in the dendrites of CA1 mouse pyramidal neurons. *J. Physiol.* **576**, 191–196.

Craig, A. M. and Banker, G. (1994). Neuronal polarity. *Annu. Rev. Neurosci.* **17**, 267–310.

Egger, V. and Urban, N. N. (2006). Dynamic connectivity in the mitral cell-granule cell microcircuit. *Semin. Cell Dev. Biol.* **17**, 424–432.

Golowasch, J. and Marder, E. (1992). Ionic currents of the lateral pyloric neuron of the stomatogastric ganglion of the crab. *J. Neurophysiol.* **67**, 2, 318–331.

Harris, K. M. and Kater, S. B. (1994). Dendritic spines: Cellular specializations imparting both stability and flexibility to synaptic function. *Annu. Rev. Neurosci.* **17**, 341–371.

Herreras, O. (1990). Propagating dendritic action potential mediates synaptic transmission in CA1 pyramidal cells in situ. *J. Neurophysiol.* **64**, 1429–1441.

Johnston, D. A. and Wu, S. M.-S. (1995). "Foundations of Cellular Neurophysiology." MIT Press, Cambridge, MA.

Kayadjanian, N., Lee, H. S., Pina-Crespo, J., and Heinemann, S. F. (2007). Localization of glutamate receptors to distal dendrites depends on subunit composition and the kinesin motor protein KIF17. *Mol. Cell. Neurosci.* **34**, 219–230.

Koch, C. (1999). "Biophysics of Computation: Information Processing in Single Neurons." Oxford Univ. Press, New York.

Liu, G. (2004). Local structural balance and functional interaction of excitatory and inhibitory synapses in hippocampal dendrites. *Nature Neurosci.* **7**, 373–379.

Llinas, R. and Sugimori, M. (1980). Electrophysiological properties of in vitro Purkinje cell dendrites in mammalian cerebellar slices. *J. Physiol. (Lond.)* **305**, 197–213.

Luksch, H., Cox, K., and Karten, H. J. (1998). Bottlebrush dendritic endings and large dendritic fields: motion-detecting neurons in the tectofugal pathway. *J. Comp. Neurol.* **396**, 399–414.

Magee, J. C. (1999). Voltage-gated ion channels in dendrites. *In* "Dendrites" (G. Stuart, N. Spruston, and M. Hausser, eds.), pp. 139–160. Oxford Univ. Press, New York.

Magee, J. C. (2000). Dendritic integration of excitatory synaptic input. *Nature Neurosci.* **1**, 181–190.

Magee, J. C. and Johnston, D. (1995). Characterization of single voltage-gated Na$^+$ and Ca^{2+} channels in apical dendrites of rat CA1 pyramidal neurons. *J. Physiol. (Lond.)* **487**, 67–90.

Mainen, Z. E. and Sejnowski, T. J. (1995). Influence of dendritic structure on firing pattern in model neocortical neurons. *Nature* **382**, 363–365.

Manor, Y., Nadim, E., Epstein, S., Ritt, J., Marder, E., and Kopell, N. (1999). Network oscillations generated by balancing graded asymmetric reciprocal inhibition in passive neurons. *J. Neurosci.* **19**, 2765–2779.

Matus, A. and Shepherd, G. M. (2000). The millennium of the dendrite? *Neuron* **27**, 431–434.

McCulloch, W. S. and Pitts, W. H. (1943). A logical calculus of the ideas immanent in nervous activity. *Bull. Math. Biophys.* **5**, 115–133.

Mel, B. W. and Schiller, J. (2004). On the fight between excitation and inhibition: Location is everything. *Sci. STKE.* Sept. 7 (250), PE44.

Midtgaard, J., Lasser-Ross, N., and Ross, W. N. (1993). Spatial distribution of Ca^{2+} influx in turtle Purkinje cell dendrites in vitro: Role of a transient outward current. *J. Neurophysiol.* **70**, 2455–2469.

Migliore, M. and Shepherd, G. M. (2002). Emerging rules for the distributions of active dendritic conductances. *Nature Neurosci. Revs.* **3**, 362–370.

Migliore, M. and Shepherd, G. M. (2007). Dendritic action potentials connect distributed dendrodendritic microcircuits. *J. Comput. Neurosci.* Aug 3; [Epub ahead of print].

Overholt, E. M., Rubel, E. W., and Hyson, R. L. (1992). A circuit for coding interaural time differences in the chick brainstem. *J. Neurosci.* **12**, 1698–1708.

Pinsky, P. E. and Rinzel, J. (1994). Intrinsic and network rhythmogenesis in a reduced Traub model for CAS neurons. *J. Comput. Neurosci.* **1**, 39–60.

Poirazi, P., Brannon, T., and Mel, B. W. (2003). Pyramidal neuron as two-layer neural network. *Neuron* **37**, 989–999.

Polsky, A., Mel, B. W., and Schiller, J. (2004). Computational subunits in thin dendrites of pyramidal cells. *Nature Neurosci.* **7**, 621–627.

Rall, W. (1964). Theoretical significance of dendritic trees for neuronal input-output relations. In "Neural Theory and Modelling" (R. E Reiss, ed.), pp. 73–97. Stanford University Press.

Rall, W. (1974). Dendritic spines and synaptic potency. *In:* "Studies in Neurophysiology" (Porter, R. ed.). Cambridge: Cambridge University Press, pp. 203–209.

Rail, W. and Shepherd, G. M. (1968). Theoretical reconstruction of field potentials and dendrodendritic synaptic interactions in olfactory bulb. *J. Neurophysiol.* **31**, 884–915.

Redman, S. J. and Walmsley, B. (1983). Amplitude fluctuations in synaptic potentials evoked in cat spinal motoneurons at identified group in synapses. *J. Physiol. (Lond.)* **343**, 135–145.

Regehr, W. G., Connor, J. A., and Tank, D. W. (1989). Optical imaging of calcium accumulation in hippocampal pyramidal cells during synaptic activation. *Nature (Lond.)* 533–536.

Richardson, T. L., Turner, R. W., and Miller, J. J. (1987). Action-potential discharge in hippocampal CA1 pyramidal neurons. *J. Neurophysiol.* **58**, 98–996.

Ringham, G. L. (1971). Origin of nerve impulse in slowly adapting stretch receptor of crayfish. *J. Neurophysiol.* **33**, 773–786.

Sabatini, B. L. and Svoboda, K. (2000). Analysis of calcium channels in single spines using optical fluctuation analysis. *Nature* **408**, 589–593.

Segev, L., Rinzel, J., and Shepherd, G. M. (eds.) (1995). "The Theoretical Foundation of Dendritic Function. Selected Papers of Wilfrid Rall." MIT Press, Cambridge.

Selverston, A. L., Russell, D. E., and Miller, J. P. (1976). The stomatogastric nervous system: Structure and function of a small neural network. *Prog. Neurobiol.* **37**, 215–289.

Shepherd, G. M. (1977). The olfactory bulb: A simple system in the mammalian brain. *In* "Handbook of Physiology, Sect. l, The Nervous System; Part l, Cellular Biology of Neurons" (Kandel, E. R., ed.). Bethesda, MD: American Physiological Society: Bethesda, pp. 945–968.

Shepherd, G. M. and Brayton, R. K. (1987). Logic operations are properties of computer-simulated interactions between excitable dendritic spines. *Neurosci.* **21**, 151–166.

Shepherd, G. M. (1991). "Foundations of the Neuron Doctrine." Oxford Univ. Press, New York.

Shepherd, G. M. (1992). Canonical neurons and their computational organization. In "Single Neuron Computation" (T. McKenna, J. Davis, and S. E. Zornetzer, eds.), pp. 27–59. MIT Press, Cambridge.

Shepherd, G. M. (1994). "Neurobiology," 3rd ed. Oxford Univ. Press, New York.

Shepherd, G. M. (1996). The dendritic spine: A multifunctional integrative unit. J. Neurophysiol. 75, 2197–2210.

Shepherd, G. M. (2004). "The Synaptic Organization of the Brain," 5th ed. New York: Oxford University Press.

Shepherd, G. M., Chen, W. R., Willhite, D., Migliore, M., and Greer, C. A. (2007). The olfactory granule cell: From classical enigma to central role in olfactory processing. Brain Res. Rev.

Shu, Y., Hasenstaub, A., Duque, A., Yu, Y., and McCormick, D. A. (2006). Modulation of intracortical synaptic potentials by presynaptic somatic membrane potential. Nature 441, 761–765.

Simmons, M. L., Terman, G. W., Gibbs, S. M., and Chavkin, C. (1995). L-type calcium channels mediate dynorphin neuro-peptide release from dendrites but not axons of hippocampal granule cells. Neuron 14, 1265–1272.

Single, S. and Borst, A. (1998). Dendritic integration and its role in computing image velocity. Science 281, 1848–1850.

Spruston, N., Schiller, Y., Stuart, G., and Sakmann, B. (1995). Activity-dependent action potential invasion and calcium influx into hippocampal CA1 dendrites. Science 268, 297–300.

Spruston, N. (2000). Distant synapses raise their voices. Nature Neurosci. 3, 849–851.

Stuart, G., Spruston, N., and Hausser, M. (2007). "Dendrites." Oxford Univ. Press, New York.

Stuart, G., Spruston, N., Sakmann, B., and Hausser, M. (1997). Action potential initiation and backpropagation in neurons of the mammalian central nervous system. Trends Neurosci. 20, 125–131.

Stuart, G. J. and Sakmann, B. (1994). Active propagation of somatic action potentials into neocortical pyramidal cell dendrites. Nature (Lond.) 367, 6–72.

Svirskie, G., Gutman, A., and Hounsgaard, J. (2001). Electrotonic structure of motoneurons in the spinal cord of the turtle: Inferences for the mechanisms of bistability. J. Neurophysiol. 85, 391–399.

Turner, R. W., Meyers, E. R., Richardson, D. L., and Barker, J. L. (1991). The site for initiation of action potential discharge over the somatosensory axis of rat hippocampal CA1 pyramidal neurons. J. Neurosci. 11, 2270–2280.

Volfovsky, N., Parnas, H., Segal M., and Korkotian, E. (1999). Geometry of dendritic spines affects calcium dynamics in hippocampal neurons: Theory and experiments. J. Neurophysiol. 82, 450–462.

Willhite, D. C., Nguyen, K. T., Masurkar, A. V., Greer, C. A., Shepherd, G. M., and Chen, W. R. (2006). Viral tracing identified distributed columnar organization in the olfactory bulb. Proc. Natl. Acad. Sci, U.S.A. 103, 12592–12597.

Wilson, C. (1998). Basal ganglia. In "The Synaptic Organization of the Brain" (G. Shepherd, ed.), 4th ed., pp. 329–375. Oxford Univ. Press, New York.

Xiong, W. and Chen, W. R. (2002). Dynamic gating of spike propagation in the mitral cell lateral dendrites. Neuron 34, 115–126.

Yuste, R. and Denk, W. (1995). Dendritic spines as basic functional units of neuronal integration in dendrites. Nature (Lond.) 375, 682–684.

Yuste, R., Gutnick, M. J., Saar, D., Delaney, K. D., and Tank, D. W. (1994). Calcium accumulations in dendrites from neocortical neurons: An apical band and evidence for functional compartments. Neuron 13, 23–43.

Zilberter, Y., Harkany, T., and Holmgren, C. D. (2005). Dendritic release of retrograde messengers controls synaptic transmission in local neocortical networks. Neuroscientis 11, 334–344. Review.

Gordon M. Shepherd

Brain Energy Metabolism

All the processes described in this textbook require energy. Ample clinical evidence indicates that the brain is exquisitely sensitive to perturbations of energy metabolism. This chapter covers the topics of energy delivery, production, and utilization by the brain. Careful consideration of the basic mechanisms of brain energy metabolism is an essential prerequisite to a full understanding of the physiology and pathophysiology of brain function. Abnormalities in brain energy metabolism are observed in a variety of pathological conditions such as neurodegenerative diseases, stroke, epilepsy, and migraine. The chapter reviews the features of brain energy metabolism at the global, regional, and cellular levels and extensively describes recent advances in the understanding of neuro-glial metabolic cooperation. A particular focus is the cellular and molecular mechanisms that tightly couple neuronal activity to energy consumption. This tight coupling is at the basis of functional brain-imaging techniques, such as positron emission tomography (PET) and functional magnetic resonance imaging.

ENERGY METABOLISM OF THE BRAIN AS A WHOLE ORGAN

Glucose Is the Main Energy Substrate for the Brain

The human brain constitutes only 2% of the body weight, yet the energy-consuming processes that ensure proper brain function account for approximately 25% of total body glucose utilization. With a few exceptions that will be reviewed later, glucose is the obligatory energy substrate of the brain. In any tissue, glucose can follow various metabolic pathways; in the brain, glucose is almost entirely oxidized to CO_2 and water through its sequential processing by glycolysis (Fig. 13.1), the tricarboxylic acid (TCA) cycle (Fig. 13.2), and the associated oxidative phosphorylation, which yield, on a molar basis, between 30 and 36 ATP per glucose, depending on the coupling efficiency of oxidative phosphorylation. Indeed, the oxygen consumption of the brain, which accounts for almost 20% of the oxygen consumption of the whole organism, is 160 mmol per 100 g of brain weight per minute and roughly corresponds to the value determined for CO_2 production. This O_2/CO_2 relation corresponds to what is known in metabolic physiology as a respiratory quotient of nearly 1 and demonstrates that carbohydrates, and glucose in particular, are the exclusive substrates for oxidative metabolism.

This rather detailed information of whole brain energy metabolism was obtained using an experimental approach in which the concentration of a given substrate in the arterial blood entering the brain through the carotid artery is compared with that present in the venous blood draining the brain through the jugular vein (Kety and Schmidt, 1948). If the substrate is utilized by the brain, the arteriovenous (A-V) difference is positive; in certain cases, the A-V difference may be negative, indicating that metabolic pathways resulting in the production of the substrate predominate. In addition, when the rate of cerebral blood flow (CBF) is known, the steady-state rate of utilization of the substrate can be determined per unit time and normalized per unit brain weight according to the following relation: CMR = CBF (A-V), where CMR is the cerebral metabolic rate of a given substrate.

This approach was pioneered by Seymour Kety and C. F. Schmidt in the late 1940s and was further developed in the 1950s and 1960s. In normal adults,

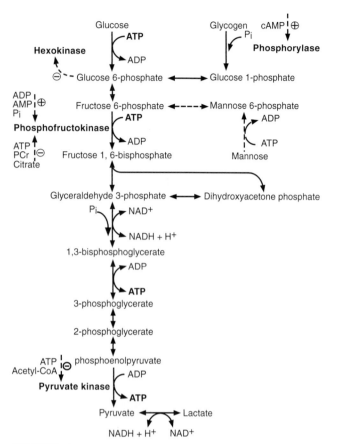

FIGURE 13.1 Glycolysis (Embden-Meyerhof pathway). Glucose phosphorylation is regulated by hexokinase, an enzyme inhibited by glucose 6-phosphate. Glucose must be phosphorylated to glucose 6-phosphate to enter glycolysis or to be stored as glycogen. Two other important steps in the regulation of glycolysis are catalyzed by phosphofructokinase and pyruvate kinase. Their activity is controlled by the levels of high-energy phosphates, as well as of citrate and acetyl-CoA. Pyruvate, through lactate dehydrogenase, is in dynamic equilibrium with lactate. This reaction is essential to regenerate NAD+ residues necessary to sustain glycolysis downstream of glyceraldehyde 3-phosphate. PCr, phosphocreatine.

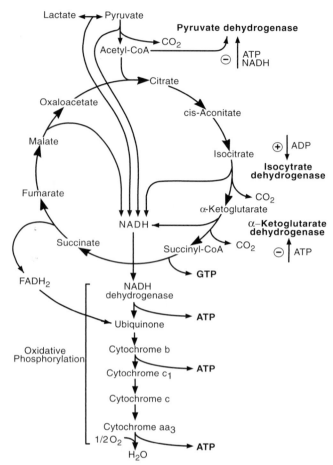

FIGURE 13.2 Tricarboxylic acid cycle (Krebs' cycle) and oxidative phosphorylation. Pyruvate entry into the cycle is controlled by pyruvate dehydrogenase activity that is inhibited by ATP and NADH. Two other regulatory steps in the cycle are controlled by isocitrate and α-ketoglutarate dehydrogenase, whose activity is controlled by the levels of high-energy phosphates.

CBF is approximately 57 ml per 100 g of brain weight per minute, and the calculated glucose utilization by the brain is 31 mmol per 100 g of brain weight per minute, as determined with the A-V difference method (Kety and Schmidt, 1948). This value is slightly higher than that predicted from the rate of oxygen consumption of the brain. Thus, in an organ such as the brain with a respiratory quotient of 1, the stoichiometry would predict that 6 mmol of oxygen are needed to fully oxidize 1 mmol of the six-carbon molecule of glucose; given an oxygen consumption rate of 160 mmol per 100 g of brain weight per minute, the predicted glucose utilization would be 26 mmol per 100 g of brain

weight per minute (160:6), yet the actual measured rate is 31 mmol. What then is the fate of the excess 4.4 mmol? First, glucose metabolism may proceed, to a very limited extent, only through glycolysis, resulting in the production of lactate without oxygen consumption (see Fig. 13.1); glucose can also be incorporated into glycogen (Fig. 13.1). Second, glucose is an essential constituent of macromolecules such as glycolipids and glycoproteins present in neural cells. Finally, glucose enters the metabolic pathways that result in the synthesis of three key neurotransmitters of the brain: glutamate, GABA, and acetylcholine (see Chapter 7).

Ketone Bodies Become Energy Substrates for the Brain in Particular Circumstances

In particular circumstances, substrates other than glucose can be utilized by the brain. For example, breastfed neonates have the capacity to utilize the ketone bodies acetoacetate (AcAc) and D-3-hydroxybutyrate (3-HB), in addition to glucose, as energy substrates for the brain. This capacity is an interesting example of a developmentally regulated adaptive mechanism because maternal milk is highly enriched in lipids, resulting in a lipid-to-carbohydrate ratio much higher than that present in postweaning nutrients. Indeed, lipids account for approximately 55% of the total calories contained in human milk, in contrast with 30 to 35% for a balanced postweaning diet. In addition to the ketone bodies AcAc and 3-HB, other products of lipid metabolism, relevant to brain metabolic processes, are free fatty acids. Acetoacetate, 3-HB, and free fatty acids can all be processed to acetyl-CoA, thus providing ATP through the TCA cycle (Fig. 13.3). We will see later that brain energy metabolism is highly compartmentalized, with certain metabolic pathways specifically localized in a given cell type. It is therefore not surprising that whereas ketone bodies can be oxidized by neurons, oligodendrocytes, and astrocytes, the β-oxidation of free fatty acids is localized exclusively in astrocytes.

Another consideration regarding the lipid-rich diet provided during the suckling period relates to its contribution to the process of myelination. The question is whether the polar lipids and cholesterol that make up myelin are derived from dietary sources or are synthesized within the brain. Evidence shows that brain lipids can be synthesized from blood-borne precursors such as ketone bodies. In addition, when suckling rats are fed a diet low in ketones, carbon atoms for lipogenesis can also be provided by glucose. To summarize, ketone bodies and AcAc are energy substrates, as well as precursors for lipogenesis during the suckling period; however, the developing brain appears to be metabolically quite flexible because glucose, in addition to its energetic function, can be metabolized to generate substrates for lipid synthesis.

Starvation and diabetes are two situations in which the availability of glucose to tissues is inadequate and in which plasma ketone bodies are elevated because of enhanced lipid catabolism. Under these conditions, the adaptive mechanisms described for breast-fed neonates become operative in the brain, allowing it to utilize AcAc or 3-HB as energy substrates.

Mannose, Lactate, and Pyruvate Serve as Instructive Cases

A number of metabolic intermediates have been tested as alternative substrates to glucose for brain energy metabolism. Among the numerous molecules tested, mannose is the only one that can sustain normal brain function in the absence of glucose. Mannose crosses the blood–brain barrier readily and, in two enzymatic steps, is converted into fructose 6-phosphate, an intermediate of the glycolytic pathway (Fig. 13.1). However, mannose is not normally present in the blood and therefore is not considered a physiological substrate for brain energy metabolism.

Lactate and pyruvate can be sources of insight into the intrinsic properties of isolated brain tissue versus those of the brain as an organ receiving substrates from the circulation. Lactate and pyruvate can sustain the synaptic activity of isolated brain preparations, usually thin slices, maintained *in vitro* in a physiological medium lacking glucose (Schurr, 2006). *In vivo*, until recently, it was thought that their permeability across the blood–brain barrier was limited, hence preventing circulating lactate or pyruvate to substitute for glucose to maintain brain function adequately. However, evidence from magnetic resonance spectroscopy (MRS) experiments indicates that the permeability of circulating lactate across the blood–brain barrier may actually be higher than previously thought (Hassel and Brathe, 2000); in addition, the presence of monocarboxylate transporters on intraparenchymal brain capillaries has been documented (Pierre and Pellerin, 2005). Thus there is a need for the reappraisal of the

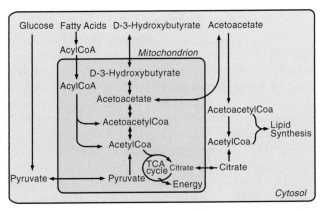

FIGURE 13.3 Relationship between lipid metabolism and the TCA cycle. Under particular dietary conditions, such as lactation in newborns or fasting in adults, the ketone bodies acetoacetate and D-3-hydroxybutyrate and circulating fatty acids can provide substrates to the TCA cycle after conversion into acetyl-CoA. Carbon atoms for lipid synthesis can be provided by glucose through citrate produced in the TCA cycle, a particularly relevant process for the developing brain.

use by the brain of monocarboxylates. For example, during vigorous exercise resulting in increases in blood lactate, the brain takes up lactate and glucose in equal amounts; lactate is then fully oxidized by the brain parenchyma (Dalsgaard, 2006). Furthermore, artificially raising lactate concentration from 0.6 (physiological value) to 4mM (as observed in moderate-to-high exercise) markedly decreases glucose utilization by the brain as determined in humans with 18F-2-deoxyglucose Positron Emission Tomography (Smith et al., 2003). Overall these data support the notion that plasma lactate can be an energy substrate for the human brain. In addition. if formed within the brain parenchyma from glucose that has crossed the blood–brain barrier, lactate and pyruvate may in fact become the preferential energy substrates for activated neurons (see later).

Summary

Glucose is the obligatory energy substrate for brain, and it is almost entirely oxidized to CO_2 and H_2O. This simple statement summarizes, with few exceptions, over four decades of careful studies of brain energy metabolism at organ and regional levels. Under ketogenic conditions, such as starvation and diabetes and during breastfeeding, ketone bodies may provide an energy source for the brain. Lactate and pyruvate, formed from glucose within the brain parenchyma, are adequate energy substrates as well.

TIGHT COUPLING OF NEURONAL ACTIVITY, BLOOD FLOW, AND ENERGY METABOLISM

A striking characteristic of the brain is its high degree of structural and functional specialization. Thus, when we move an arm, motor areas and their related pathways are activated selectively (see Chapter 28); intuitively, one can predict that as "brain work" increases locally (e.g., in motor areas), the energy requirements of the activated regions will increase in a temporally and spatially coordinated manner. Because energy substrates are provided through the circulation, blood flow should increase in the modality-specific activated area. More than a century ago, the British neurophysiologist Charles Sherrington showed, in experimental animals, increases in blood flow localized to the parietal cortex in response to sensory stimulation (Roy and Sherrington, 1890). He postulated that "the brain possesses intrinsic mechanisms by which its vascular supply can be varied

locally in correspondence with local variations of functional activity." With remarkable insight, he also proposed that "chemical products of cerebral metabolism" produced in the course of neuronal activation could provide the mechanism to couple activity with increased blood flow.

Which Mechanisms Couple Neuronal Activity to Blood Flow?

Since Sherrington's seminal work, the search for the identification of chemical mediators that can couple neuronal activity with local increases in blood flow has been intense. These signals can be broadly grouped into two categories: (1) molecules or ions that transiently accumulate in the extracellular space *after* neuronal activity and (2) specific neurotransmitters that mediate the coupling *in anticipation* or at least *in parallel* with local activation (neurogenic mechanisms). The increases in extracellular K^+, adenosine, and lactate and the related changes in pH are all a consequence of increased neuronal activity, and all have been considered mediators of neuro-vascular coupling because of their vasoactive effects (Villringer and Dirnagl, 1995). However, the spatial and temporal resolution achieved by these mediators may not be sufficient to entirely account for the activity-dependent coupling between neuronal activity and blood flow. Indeed, these vasoactive agents are formed with a certain delay (seconds) after the initiation of neuronal activity and can diffuse at considerable distance. In this respect, neurogenic mechanisms appear to be better fitted. Brain microvessels are richly innervated by neuronal fibers. These fibers may have an extrinsic origin (e.g., in the autonomic ganglia) or be part of neuronal circuits intrinsic to the brain, such as local interneurons or long projections that originate in the brainstem (e.g., those containing monoaminergic neurotransmitters). In addition, functional receptors coupled to signal transduction pathways have been identified for several neurotransmitters on intraparenchymal microvessels. Neurotransmitters with potential roles in coupling neuronal activity with blood flow include the amines noradrenaline, serotonin, and acetylcholine and the peptides vasoactive intestinal peptide, neuropeptide Y (NPY), calcitonin gene-related peptide (CGRP), and substance P (SP). The neurogenic mode of neurovascular coupling implies that vasoactive neurotransmitters are released from perivascular fibers as excitatory afferent volleys activate a discrete and functionally defined brain volume (Hamel, 2006).

An attractive addition to the list of potential mediators for coupling neuronal activity to blood flow is nitric oxide (NO). Indeed, NO is an ideal candidate; it is formed locally by neurons and glial cells under the action of a variety of neurotransmitters likely to be released by depolarized afferents to an activated brain area. Nitric oxide is a diffusible and potent vasodilator whose short half-life spatially and temporally restricts its domain of action. However, in several experimental models in which the activity of NO synthase, the enzyme responsible for NO synthesis, was inhibited, a certain degree of coupling was still observed, indicating that NO is probably only one of the regulators of local blood flow acting in synergy with others (Hamel, 2006).

Recent *in vitro* and *in vivo* experiments have provided evidence that astrocytes may play a key role in neurovascular coupling. Indeed as will be elaborated in greater detail in the section devoted to neurometabolic coupling (e.g., Figure 13.8), astrocytes occupy a strategic position between capillaries and the neuropil. Through receptors and reuptake sites for neurotransmitters, notably glutamate, they can sense synaptic activity and couple it to vascular response. The molecular mediators of the astrocyte-dependent hyperemia that accompanies activation include prostanoids and adenosine (Koheler *et al.*, 2006).

In summary, several products of activity-dependent neuronal and glial metabolism such as lactate, H+, adenosine, prostanoids, and K+ have vasoactive effects and are therefore putative mediators of coupling, although the kinetics and spatial resolution of this mode do not account for all the observed phenomena. As attractive as it is, an exclusively neurogenic mode of coupling neuronal activity to blood flow is unlikely and, moreover, still awaits firm functional confirmation *in vivo*. Nitric oxide is undoubtedly a key element in coupling, particularly in view of the fact that glutamate, the principal excitatory neurotransmitter, triggers a receptor-mediated NO formation in neurons and glia; this is consistent with the view that whenever a functionally defined brain area is activated and glutamate is released by the depolarized afferents, NO may be formed, thus providing a direct mechanism contributing to the coupling between activity and local increases in blood flow. Astrocytes appear to function as intermediary processor in neurovascular coupling (Koheler *et al.*, 2006).

Through the activity-linked increase in blood flow, more substrates—namely, glucose and oxygen—necessary to meet the additional energy demands are delivered to the activated area per unit time. The cellular and molecular mechanisms involved in oxygen consumption and glucose utilization are treated in a later section.

Blood Flow and Energy Metabolism Can Be Visualized in Humans

Modern functional brain-imaging techniques enable the *in vivo* monitoring of human blood flow and the two indices of energy metabolism: glucose utilization and oxygen consumption (Raichle and Mintun, 2006). For instance, with the use of PET and appropriate positron-emitting isotopes such as ^{18}F and ^{15}O, basal rates, as well as activity-related changes in local blood flow or oxygen consumption, can be studied using ^{15}O-labeled water or ^{15}O, respectively. Local rates of glucose utilization (also defined as local cerebral metabolic rates for glucose (LCMRglu)) can be determined with ^{18}F-labeled 2-deoxyglucose (2-DG) (Phelps *et al.*, 1979). The use of 2-DG as a marker of LCMRglu was pioneered by Louis Sokoloff and associates at the National Institutes of Health, first in laboratory animals (Sokoloff, 1981). The method is based on the fact that 2-DG crosses the blood–brain barrier, is taken up by brain cells, and is phosphorylated by hexokinase with kinetics similar to that for glucose; however, unlike glucose 6-phosphate, 2-deoxyglucose 6-phosphate cannot be metabolized further and therefore accumulates intracellularly (Fig. 13.4).

For studies in laboratory animals, tracer amounts of radioactive 2-DG are injected intravenously; the animal is subjected to the behavioral paradigms of interest and sacrificed at the end of the experiment. Serial thin sections of the brain are prepared and processed for autoradiography. This autoradiographic method provides, after appropriate corrections, an accurate measurement of LCMRglu with a spatial resolution of approximately 50–100 mm. Using this method, researchers have determined LCMRglu in virtually all structurally and functionally defined brain structures in various physiological and pathological states, including sleep, seizures, and dehydration, and after a variety of pharmacological treatments (Sokoloff, 1981). Furthermore, glucose utilization increases in the pertinent brain areas during motor tasks or activation of pathways subserving specific modalities, such as visual, auditory, olfactory, or somatosensory stimulation (Sokoloff, 1981). For example, in mice, sustained stimulation of the whiskers results in marked increases in LCMRglu in discrete areas of the primary sensory cortex called the barrel fields, where each whisker is represented with an extreme degree of topographical specificity (see Chapter 25). Basal glucose utilization of the gray matter as determined by 2-DG autoradiography varies, depending on the brain structure, between 50 and 150 mmol per 100 g of wet weight per minute in the rat.

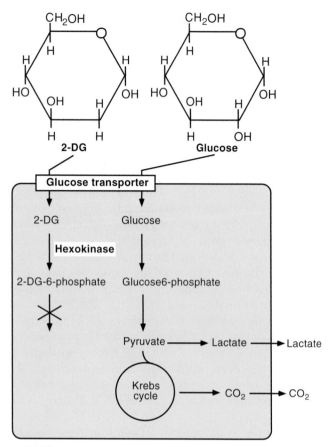

FIGURE 13.4 Structure and metabolism of glucose and 2-deoxyglucose (2-DG). 2-DG is transported into cells through glucose transporters and phosphorylated by hexokinase to glucose 6-phosphate without significant further processing or dephosphorylation back to glucose. Therefore, when labeled radioactively, 2-DG used in tracer concentrations is a valuable marker of glucose uptake and phosphorylation, which directly indicates glucose utilization.

In humans, LCMRglu determined by PET with the use of ^{18}F-2-DG is approximately 50% lower than that in rodents, and physiological activation of specific modalities increases LCMRglu in discrete areas of the brain that can be visualized with a spatial resolution of a few millimeters. For example, visual stimulations presented to subjects as checkerboard patterns reversing at frequencies ranging from 2 to 10 Hz selectively increase LCMRglu in the primary visual cortex and a few connected cortical areas. With the use of this stimulation paradigm, the combined PET analysis of local cerebral blood flow (LCBF) and local oxygen consumption (LCMRO$_2$), in addition to LCMRglu, has revealed a unique and unexpected feature of human brain energy metabolism regulation. The canonical view was that the three metabolic parameters were tightly coupled, implying that if, for example, CBF increased locally during physiological activation, LCMRglu and LCMRO$_2$ would increase in parallel. In what is now referred to as the phenomenon of "uncoupling," physiological stimulation of the visual system increases LCBF and LCMRglu (both by 30–40%) in the primary visual cortex without a commensurate increase in LCMRO$_2$ (which increases only 6%) (Raichle and Mintun, 2006), indicating that the additional glucose utilized during neuronal activation can be processed through glycolysis rather than through the tricarboxylic acid (TCA) cycle and oxidative phosphorylation. The phenomenon of uncoupling has been confirmed in other cortical areas, although its magnitude may differ depending on the modality, and may actually be absent in certain cases.

A glance at the metabolic pathways reveals that if glucose does not enter the TCA cycle to be oxidized, then lactate will be produced (see Figs. 13.1 and 13.2). Lactate, like several other metabolically relevant molecules, can be determined with the technique of magnetic resonance imaging (MRI) spectroscopy for ^{1}H, which provides a means of unequivocally identifying in living tissues the presence of molecules that bear the naturally occurring isotope ^{1}H. Consistent with the prediction that if during activation glucose is predominantly processed glycolytically, then lactate should be produced locally in the activated region, a transient increase in the lactate signal is detected with ^{1}H MRI spectroscopy in the human primary visual cortex during appropriate visual stimulation (Prichard et al., 1991). These observations support the view that to face the local increases in energy demands linked to neuronal activation, the brain transiently resorts to an integrated sequence of glycolysis and oxidative phosphorylation (Magistretti and Pellerin, 1999). This transient uncoupling may vary in amplitude depending on the modalities of activation (Frackowiak et al., 2001) and is likely to occur in different cellular compartments; that is, astrocytes vs. neurons (Pellerin and Magistretti, 1994; Kasischke et al., 2004).

Summary

Studies at the whole organ level, based on the A-V differences of metabolic substrates, have revealed a great deal about the global energy metabolism of the brain. They have indicated that, under normal conditions, glucose is virtually the sole energy substrate for the brain and that it is entirely oxidized. New techniques that allow imaging of the three fundamental parameters of brain energy metabolism—namely, blood flow, oxygen consumption, and glucose utilization—provide a more refined level of spatial resolution and demonstrate that brain energy metabolism is regionally heterogeneous and is coupled tightly to the functional activation of specific neuronal pathways (Magistretti *et al.*, 1999).

ENERGY-PRODUCING AND ENERGY-CONSUMING PROCESSES IN THE BRAIN

What are the cellular and molecular mechanisms that underlie the regulation of brain energy metabolism revealed by the foregoing studies at global and regional levels? In particular, what are the metabolic events taking place in the cell types that make up the brain parenchyma? How is it possible to reconcile whole organ studies indicating complete oxidation of glucose with transient activation-induced glycolysis at

the regional level? These and other related questions will be addressed here and in the next sections.

Glucose Metabolism Produces Energy

Before we move on to an analysis of the cell-specific mechanisms of brain energy metabolism, it seems appropriate to briefly review some basic aspects of the energy balance of the brain. Because glucose, in normal circumstances, is the main energy substrate of the brain, the overview will be restricted to its metabolic pathways. Glucose metabolism in the brain is similar to that in other tissues and includes three principal metabolic pathways: glycolysis, the tri-carboxylic acid cycle, and the pentose phosphate pathway. Because of the global similarities with other tissues, these pathways are simply summarized in Figures 13.1, 13.2, and 13.5, and only a few aspects specific to the nervous tissue will be discussed.

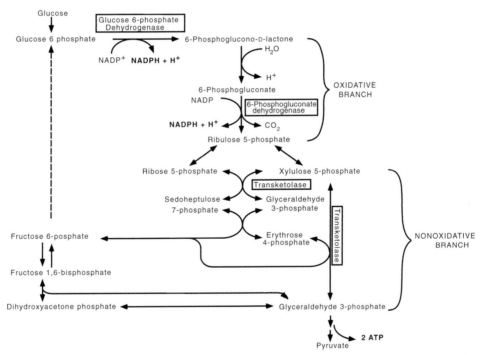

FIGURE 13.5 The pentose phosphate pathway. In the oxidative branch of the pentose phosphate pathway, two NADPH are generated per glucose 6-phosphate. The first rate-limiting reaction of the pathway is catalyzed by glucose-6-phosphate dehydrogenase; the second NADPH is generated through the oxidative decarboxylation of 6-phosphogluconate, a reaction catalyzed by glucose-6-phosphogluconate dehydrogenase. The nonoxidative branch of the pentose phosphate pathway provides a reversible link with glycolysis by regenerating the two glycolytic intermediates glyceraldehyde 3-phosphate and fructose 6-phosphate. This regeneration is achieved through three sequential reactions. In the first, catalyzed by transketolase, xylulose 5-phosphate and ribose 5-phosphate (which originate from ribulose 5-phosphate, the end product of the oxidative branch) yield glyceraldehyde 3-phosphate and sedoheptulose 7-phosphate. Under the action of transaldolase, these two intermediates yield fructose 6-phosphate and erythrose 4-phosphate. This latter intermediate combines with glyceraldehyde 3-phosphate, in a reaction catalyzed by transketolase, to yield fructose 6-phosphate and glyceraldehyde 3-phosphate. Thus, through the nonoxidative branch of the pentose phosphate pathway, two hexoses (fructose 6-phosphate) and one triose (glyceraldehyde 3-phosphate) of the glycolytic pathway are regenerated from three pentoses (ribulose 5-phosphate).

Glycolysis

Glycolysis (Embden-Meyerhof pathway) is the metabolism of glucose to pyruvate (Fig. 13.1). It results in the net production of only two molecules of ATP per glucose molecule; indeed, four ATPs are formed in the processing of glucose to pyruvate, whereas two ATPs are consumed to phosphorylate glucose to glucose 6-phosphate and fructose 6-phosphate to fructose 1,6-bisphosphate, respectively (Fig. 13.1). Under anaerobic conditions, pyruvate is converted into lactate, allowing the regeneration of nicotinamide adenine dinucleotide (NAD^+), which is essential to maintain a continued glycolytic flux. Indeed, if NAD^+ were not regenerated, glycolysis could not proceed beyond glyceraldehyde 3-phosphate (Fig. 13.1). Another situation in which the end product of glycolysis is lactate rather than pyruvate is when oxygen consumption does not match glucose utilization, implying that the rate of pyruvate production through glycolysis exceeds pyruvate oxidation by the TCA cycle (Fig. 13.2). This condition has been well described in skeletal muscle during intense exercise and appears to share similarities with the transient uncoupling observed between glucose utilization and oxygen consumption that has been described in the human cerebral cortex during activation with the use of PET (Raichle and Mintun, 2006).

Tricarboxylic Acid Cycle

Under aerobic conditions, pyruvate is oxidatively decarboxylated to yield acetyl-CoA in a reaction catalyzed by the enzyme pyruvate dehydrogenase (PDH). Acetyl-coenzyme A condenses with oxaloacetate to produce citrate (Fig. 13.2). This is the first step of the tricarboxylic acid cycle, in which three pairs of electrons are transferred from NAD^+ to NADH—and one pair from flavin adenine dinucleotide (FAD) to its reduced form ($FADH_2$)—through four oxidation-reduction steps (Fig. 13.2). NADH and $FADH_2$ transfer their electrons to molecular O_2 through the mitochondrial electron transfer chain to produce ATP in the process of oxidative phosphorylation. Thus, under aerobic conditions (i.e., when glucose is fully oxidized through the TCA cycle to CO_2 and H_2O), NAD^+ is regenerated, and glycolysis proceeds to pyruvate, not lactate. However, as soon as a mismatch, even a transient one, occurs between glucose utilization and oxygen consumption, lactate is produced. As discussed earlier, such a transient production of lactate appears to occur in the human brain during activation. Experiments performed in freely moving rats also have demonstrated a transient increase in lactate content in the extracellular space of discrete brain regions during physiological sensory stimulation (Hu and Wilson, 1997).

Pentose Phosphate Pathway

Although glycolysis, the TCA cycle, and oxidative phosphorylation are coordinated pathways that produce ATP, using glucose as a fuel, ATP is not the only form of metabolic energy. Indeed, for several biosynthetic reactions in which the precursors are in a more oxidized state than the products, metabolic energy in the form of reducing power is needed in addition to ATP. This is the case for the reductive synthesis of free fatty acids from acetyl-CoA, which are components of myelin and of other structural elements of neural cells, such as the plasma membrane. In cells of the brain, as in other organs, the reducing power is provided by the reduced form of nicotinamide adenine dinucleotide phosphate (NADPH). The processing of glucose through the pentose phosphate pathway produces NADPH. The first reaction in the pentose phosphate pathway is the conversion of glucose 6-phosphate into ribulose 5-phosphate (Fig. 13.5). This dehydrogenation, in which two molecules of NADPH are generated per molecule of glucose 6-phosphate, is the rate-limiting step of the pentose phosphate pathway. The NADP/NADPH ratio is the single most important factor regulating the entry of glucose 6-phosphate into the pentose phosphate pathway. Thus, if a high reducing power is needed, NADPH levels decrease and the pentose phosphate pathway is activated to generate new reducing equivalents. The pentose phosphate pathway is also tightly connected to glycolysis through two enzymes, transketolase and transaldolase, which recycle ribulose 5-phosphate to fructose 6-phosphate and glyceraldehyde 3-phosphate, two intermediates of glycolysis (Fig. 13.5).

Glucose Metabolism, Reactive Oxygen Species, and the Protective Role of Glutathione

In addition to reductive biosynthesis, NADPH is needed for the scavenging of reactive oxygen species (ROS). The superoxide radical anion (O_2^-), hydrogen peroxide (H_2O_2), and the hydroxy radical (HO) are three ROS, generated by the transfer of single electrons to molecular oxygen as by-products of several physiological cellular processes. A considerable contribution to the generation of ROS is the oxidative metabolism of glucose taking place in the mitochondrial electron transfer chain associated with oxidative phosphorylation. Other ROS-generating reactions include the activities of monoamine oxidase, tyrosine hydroxylase, nitric oxide synthase, and the eicosanoid-forming enzymes lipoxygenases and cyclooxygenases. Reactive oxygen species are highly damaging to cells because they can cause DNA disruption and mutations, as well as activation of enzymatic cascades, including proteases and lipases that can eventually lead to cell death.

Thus, oxidative metabolism, which is so essential to cell viability by generating large amounts of the cellular fuel ATP, implies as a by-product a potentially harmful activity such as ROS generation. The coordinated activity of two molecules is essential in protecting cells against ROS-mediated damage, or oxidative stress: NADPH and glutathione. As we have seen, NADPH is produced through a particular arm of glucose metabolism, the pentose phosphate pathway. Interestingly, therefore, glucose metabolism provides two forms of energy, high energy phosphates such as ATP and reducing power such as $NAD(P)H$, the latter contributing to the neutralization of ROS, the harmful by-products of the process (oxidative phosphorylation) which produces the former. It is, however, through its combined action with glutathione that NADPH contributes to ROS scavenging. Scavenging of ROS is ensured by the sequential action of superoxide dismutase (SOD) and glutathione peroxidase (Fig. 13.6). Thus, two superoxide anions formed by the aforementioned cellular processes are converted by SOD into H_2O_2, still a ROS. Glutathione peroxidase converts H_2O_2 into H_2O and O_2 at the expense of reduced glutathione, which is regenerated by glutathione reductase in the presence of NADPH.

The metabolism of glutathione is tightly regulated and implies yet another example of neuron-astrocyte cooperation. Glutathione is a tripeptide (GSH; gamma-L-glutamyl-L-cysteinylglycine) synthesized through the concerted action of two enzymes, gammaGluCys synthase, which combines glutamate and cysteine to yield the dipeptide gammaGlu Cys, and glutathione synthase, which adds a glycine to the dipeptide to yield GSH (Fig. 13.6).

The glutathione content and reducing potential are considerably higher in astrocytes compared to neurons; this fact, combined with the much higher oxidative activity of neurons vs. astrocytes, makes neurons more vulnerable to oxidative stress as well as highly dependent on astrocytes for their protection (Dringen, 2000). Indeed, a cooperativity between astrocytes and neurons appears to exist for glutathione metabolism; astrocytes release GSH, which is cleaved by the ectoenzyme gamma-Glutamyl Transferase (gamma-GT), which releases CysGly. The dipeptide is transported into neurons (note that neurons cannot take up GSH), pro-

viding two precursors for GSH synthesis glutamate, the third precursor of GSH, also is provided by astrocytes to neurons under the form of glutamine, from which glutamate is produced through the action of glutaminase (Fig. 13.6).

Several neurodegenerative disorders appear to involve a dysfunction in the ability of neural cells to control oxidative stress. For example, a familial form of amyotrophic lateral sclerosis is due to a SOD mutation; evidence for a decrease in GSH content in the substantia nigra has been described in Parkinson's disease (Beal, 2005).

The Wernicke-Korsakoff Syndrome: A Neuropsychiatric Disorder Due to a Dysfunction of Energy Metabolism

A well-characterized neuropsychiatric disorder, the Wernicke-Korsakoff syndrome, is caused by transketolase hypoactivity. The Wernicke-Korsakoff syndrome is characterized by a severe impairment of memory and of other cognitive processes accompanied by balance and gait dysfunction and by paralysis of oculomotor muscles. The syndrome is due to a lack of thiamine (vitamin B_1) in the diet; it affects only susceptible persons who are also alcoholics or chronically undernourished. Thiamine pyrophosphate is a thiamine-containing cofactor essential for the activity of transketolase. In patients with the Wernicke-Korsakoff syndrome, thiamine pyrophosphate binds 10 times less avidly to transketolase compared with the enzyme of normal persons. This enzymatic dysfunction renders patients with the Wernicke-Korsakoff syndrome much more vulnerable to thiamine deficiency. This syndrome illustrates how an anomaly in a discrete metabolic pathway of energy metabolism may result in severe alterations in behavior and motor function.

Processes Linked to Neuronal Function Consume Energy

The main energy-consuming process of the brain is the maintenance of ionic gradients across the plasma membrane, a condition that is crucial for excitability. Maintenance of these gradients is achieved predominantly through the activity of ionic pumps fueled by ATP, particularly Na^+, K^+-ATPase, localized in neurons as well as in other cell types such as glia. Activity of these pumps accounts for approximately 50% of basal glucose oxidation in the nervous system. Very recently theoretical calculations of the cost of synaptic transmission have been provided by Attwell and Laughlin (2001). These calculations are based on a number of assumptions and therefore should be taken with some

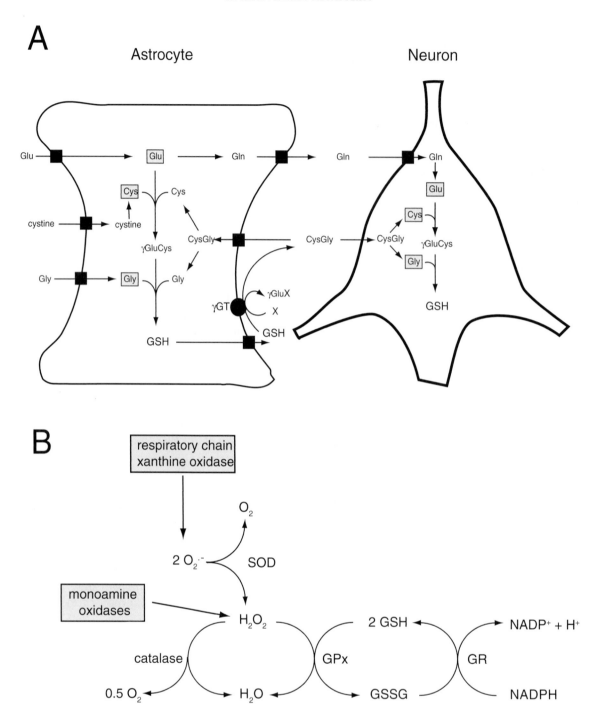

FIGURE 13.6 (A) Metabolic interaction between astrocytes and neurons in the synthesis of glutathione. In astrocytes, glutathione (GSH) is synthesized from cysteine (produced from cystine), glycine (Gly) and glutamate (Glu). GSH is released from astrocytes into the extracellular space; the membrane-bound astrocytic ectoenzyme, gamma-glutamyl transpeptidase (gamma-GT) releases the dipeptide CysGly, which along with glutamine (Gln, also released by astrocytes and taken up by neurons to yield glutamate) provides the precursors for neuronal glutathione synthesis. Neurons are highly dependent on astrocytes for GSH synthesis. (B) Enzymatic reactions for scavenging reactive oxygen species (ROS). The toxic superoxide anion (O_2^-) formed by a variety of physiological reactions, including respiratory chain and oxidase-mediated reactions (e.g., xanthine and monoamine oxydases), is scavenged by superoxide dismutase (SOD), which converts the superoxide anion into hydrogen peroxide (H_2O_2) and molecular oxygen. Glutathione peroxidase (GPx) converts the still toxic hydrogen peroxide into water; reduced glutathione (GSH) is required for this reaction, in which it is converted into its oxidized form (GSSG). GSH is regenerated through the action of glutathione reductase (GR), a reaction requiring NADPH.

caution; they nevertheless provide a valuable frame-work for further experimental studies on brain's energy budget. The energy budget of an average glutamatergic pyramidal neuron firing at 4 Hz was estimated, with the assumption that >80% of cortical neurons are pyramidal cells and that >90% of the synapses release glutamate. First, the cost of the recycling of released glutamate via reuptake and metabolism in astrocytes and the restoration of the postsynaptic ion gradient has been estimated. Glutamate recycling requires 2.67 ATP/glutamate molecule; since one vesicle contains 4×10^3 molecules of glutamate, the cost of transmitter recycling is $\sim 1.1 \times 10^4$ ATP/vesicle. The restoration of postsynaptic ionic gradients disrupted by the activity of NMDA and non-NMDA receptors is $\sim 1.4 \times 10^5$ ATP/vesicle, giving a total of 1.51×10^5 ATP/vesicle. By estimating the total number of synapses formed by a single pyramidal neuron at 8×10^3 and a firing rate of 4 Hz (implying a 1:4 chance that an active potential releases one vesicle), the figure of 3.2×10^8 ATP/action potential/neuron is obtained.

Contrary to previous estimates based on the measurement of heat production in peripheral *unmyelinated* nerves, the cost of action potential propagation is rather elevated. Thus by considering that an action potential actively depolarizes the cell body and axons by 100 mV and passively the dendrites by 50 mV, the calculation yields a value of 3.8×10^8 ATP/neuron. This calculation is based on the estimate of the minimal Na^+ influx required to depolarize the cell (Attwell and Laughlin, 2001). If calculations also include Ca^{2+}-mediated depolarization of dendrites, the cost is increased by 7%. Remember that these energetic costs are due to the activation of ATPases needed to restore ion gradients. Thus, the overall cost of synaptic transmission plus action potential propagation for a pyramidal neuron firing at 4 Hz would be 2.8×10^9 ATP/neuron/s. The basal energy consumption for maintenance of the resting potential based on the estimates of input resistance, reversal potential, and membrane conductance yields values of 3.4×10^8 ATP/cell/s for neurons and 1×10^8 ATP/cell/s for glia, thus a combined consumption of 3.4×10^9 ATP/cell/s assuming a 1:1 ratio between neurons and glia. On the basis of this calculation, one can conclude that approximately 87% of total energy consumed reflects the activity of glutamate-mediated neurotransmission and 13% reflects the energy requirements of resting potential maintenance (Fig. 13.7). This value is in remarkable agreement with estimates made *in vivo* using MRS. If the total energy consumption per neuron and the associated glia is compounded per gram of tissue per minute (the conventional form for expressing glucose utilisation), the figure obtained is 30 μM ATP/g/min,

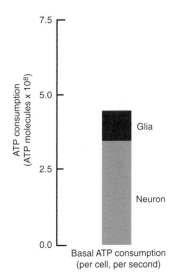

FIGURE 13.7 Energy budget for the rodent central cortex (Attwell and Laughlin, 2001). Relative rates of ATP consumption by resting neurons and glia (modified from Frackowiak *et al.*, 2001).

a value that is very close to that determined *in vivo* for brain glucose utilization—that is, 30 to 50 μM ATP/g/min (Sokoloff, 1981).

In addition to the maintenance of ionic gradients that are disrupted during activity, other energy-consuming processes exist in neurons. Thus, the permanent synthesis of molecules needed for communications, such as neurotransmitters, or for general cellular purposes consumes energy. Axonal transport of molecules synthesized in the nucleus to their final destination along the axon or at the axon terminal is yet another process fueled by cellular energy metabolism.

Summary

Exactly as in other tissues, the metabolism of glucose, the main energy substrate of the brain, produces two forms of energy: ATP and NADPH. Glycolysis and the TCA cycle produce ATP, whereas energy in the form of reducing equivalents stored in the NADPH molecule is produced predominantly through the pentose phosphate pathway. Reduced glutathione provides a major defense against oxidative stress. Maintenance of the electrochemical gradients, particularly for Na^+ and K^+, needed for electrical signaling via the action potential and for chemical signaling through synaptic transmission is the main energy-consuming process of neural cells.

BRAIN ENERGY METABOLISM AT THE CELLULAR LEVEL

Glia and Vascular Endothelial Cells, in Addition to Neurons, Contribute to Brain Energy Metabolism

Neurons exist in a variety of sizes and shapes and express a large spectrum of firing properties (Chapter 6). These differences are likely to imply specific energy demands; for example, large pyramidal cells in the primary motor cortex, which must maintain energy-consuming processes such as ion pumping over a large membrane surface or axonal transport along several centimeters, have considerably larger energy requirements than local interneurons. However, it is now clear that other cell types of the nervous system—glia and vascular endothelial cells—not only consume energy but also play a crucial role in the flux of energy substrates to neurons. Arguments for such an active role for nonneuronal cells—in particular, glia—are both quantitative and qualitative. Glial cells make up approximately half the brain volume. A conservative figure is a 1:1 ratio between the number of astrocytes, one of the predominant glial cell types (see Chapters 1 and 2), and neurons. Higher ratios have been described, depending on the regions, developmental ages, or species. Indeed, the astrocyte-to-neuron ratio increases with the size of the brain and is thus high in humans. It is therefore clear that glucose reaching the brain parenchyma provides energy substrates to a variety of cell types, only some of which are neurons.

Even more compelling for the realization of the key role that astrocytes play in providing energy substrates to active neurons are the cytological relations that exist among brain capillaries, astrocytes, and neurons. These relations, which are illustrated in Figure 13.8, are as follows. First, through specialized processes, called end feet, astrocytes surround brain capillaries (Kacem *et al.*, 1998). This implies that astrocytes form the first cellular barrier that glucose entering the brain parenchyma encounters and make them a likely site of prevalent glucose uptake and energy substrate distribution. More than a century ago, the Italian histologist Camillo Golgi and his pupil Luigi Sala sketched such a principle. In addition to perivascular end feet, astrocytes bear processes that ensheathe synaptic contacts. Astrocytes also express receptors and uptake sites with which neurotransmitters released during synaptic activity can interact (Chapters 6 and 7). These features endow astrocytes with an exquisite sensitivity to detect increases in synaptic activity. In summary, because of the foregoing structural and functional characteristics, astrocytes are ideally suited to couple local changes in

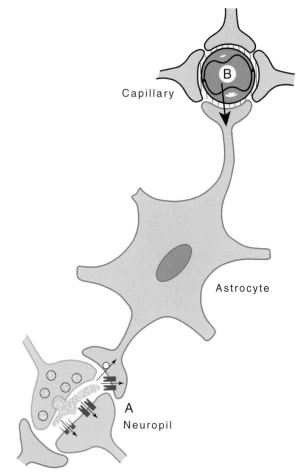

FIGURE 13.8 Schematic representation of cytological relations existing among intraparenchymal capillaries, astrocytes, and the neuropil. Astrocyte processes surround capillaries (end feet) and ensheathe synapses; in addition, receptors and uptake sites for neurotransmitters are present on astrocytes. These features make astrocytes ideally suited to sense synaptic activity (A) and to couple it with uptake and metabolism of energy substrates originating from the circulation (B).

neuronal activity with coordinated adaptations in energy metabolism (Fig. 13.8).

A Tightly Regulated Glucose Metabolism Occurs in All Cell Types of the Brain, Neuronal and Nonneuronal

Given the high degree of cellular heterogeneity of the brain, understanding the relative role played by each cell type in the flux of energy substrates has depended largely on the availability of purified preparations, such as primary cultures enriched in neurons, astrocytes, or vascular endothelial cells. Such preparations have some drawbacks because they may not necessarily express all the properties of the cells *in situ*. In addition, one of the parameters of energy metabolism *in vivo*—namely, blood flow—cannot be examined in cultures. Despite these limitations, *in vitro* studies in primary cultures have proved very useful in identifying the cellular sites of glucose uptake and its subsequent metabolic fate, particularly, glycolysis and oxidative phosphorylation, thus providing illuminating correlations of two parameters of brain energy metabolism that are monitored *in vivo*: (1) glucose utilization and (2) oxygen consumption.

Glucose Transporters in the Brain

Glucose is a highly hydrophilic molecule that enters cells through a facilitated transport mediated by specific transporters. Twelve genes, encoding glucose transporter proteins, have been identified and cloned so far; these are designated GLUT1 to GLUT12 (Uldry and Thorens, 2004). Glucose transporters belong to a family of rather homologous glycosylated membrane proteins with 12 transmembrane-spanning domains, and both amino and carboxyl terminals are exposed to the cytoplasmic surface of the membrane. In the brain, seven transporters are expressed predominantly in a cell-specific manner, GLUT1 (two isoforms), GLUT2 through GLUT5, and GLUT8 (McEwen and Reagan, 2004).

Two isoforms of GLUT1 with molecular masses of 55 and 45 kDa, respectively, are detected in the brain, depending on their degree of glycosylation. The 55-kDa form of GLUT1 essentially is localized in brain microvessels, choroid plexus, and ependymal cells. In microvessels, the distribution of GLUT1 is asymmetric, with a higher density on the ablumenal (parenchymal) side than on the vascular side. An intracellular pool of GLUT1 also has been identified in vascular endothelial cells. In the brain *in situ*, the 45-kDa form of GLUT1 is localized predominantly in astrocytes. Under culture conditions, all neural cells, including neurons and other glial cells, express GLUT1; however, this phenomenon appears to be due to the capacity of GLUT1 to be induced by cellular stress.

The glucose transporter specific to neurons is GLUT3. Its cellular distribution appears to predominate in the neuropil. This distrubution contrasts with that of GLUT 4 and 8, which appears to predominate on the cell body and proximal dendrites.

GLUT5 is localized to microglial cells, the resident macrophages of the brain, taking part in the immune and inflammatory responses of the nervous system. In peripheral tissues, particularly in the small intestine (from which it was cloned), GLUT5 functions as a transporter for fructose, whose concentrations are very low in the brain. In the nervous system, therefore, GLUT5 may have diverse transport functions.

Another glucose transporter, GLUT2, has been localized selectively in astrocytes of discrete brain areas, such as certain hypothalamic and brain stem nuclei, which participate in the regulation of feeding behavior and in the central control of insulin release.

It is clear that glucose uptake into the brain parenchyma is a highly specified process regulated in a cell-specific manner by glucose transporter subtypes. Figure 13.9 summarizes this process: Glucose enters the brain through 55-kDa GLUT1 transporters localized on endothelial cells of the blood–brain barrier. Uptake into astrocytes is mediated by 45-kDa GLUT1 transporters, whereas GLUT3 transporters mediate this process in neurons. GLUT2 transporters on astrocytes may "sense" glucose, a function of this glucose transporter subtype in pancreatic β cells. Finally, GLUT5 mediates the uptake of an unidentified substrate into microglial cells. Other glucose transporters identified in the brain with cellular and regional distributions not yet clearly defined are GLUT 6 and 10 (McEwen and Reagan, 2004).

Cell-Specific Glucose Uptake and Metabolism

As we have seen, glucose utilization can be assessed with radioactively labeled 2-DG. To determine the cellular site of basal and activity-related glucose utilization, this technique has been applied to homogeneous cultures of astrocytes or neurons. For quantitative purposes and to allow comparisons with *in vivo* studies, these *in vitro* experiments, in which radioactive 2-DG is used as a tracer, must be conducted in a medium containing a concentration of glucose near that measured *in vivo* in the extracellular space of the brain (0.5 to 2 mM). The basal rate of glucose utilization is higher in astrocytes than in neurons, with values of about 20 and 6 nmol per milligram of protein per minute, respectively (Magistretti and Pellerin, 1999). These values are of the same order as those determined *in vivo* for cortical gray matter (10–20 nmol mg^{-1} min^{-1}) with the 2-DG autoradiographic technique. In view of this difference and of the quantitative preponderance of astrocytes compared with neurons in the gray matter, these data reveal a significant contribution by astrocytes to basal glucose utilization as determined by 2-DG autoradiography or PET *in vivo*. Recent high-resolution microautoradiographic imaging *ex vivo* has indicated an approximately even distribution of 2-DG in neurons and astrocytes (Nehlig *et al.*, 2004).

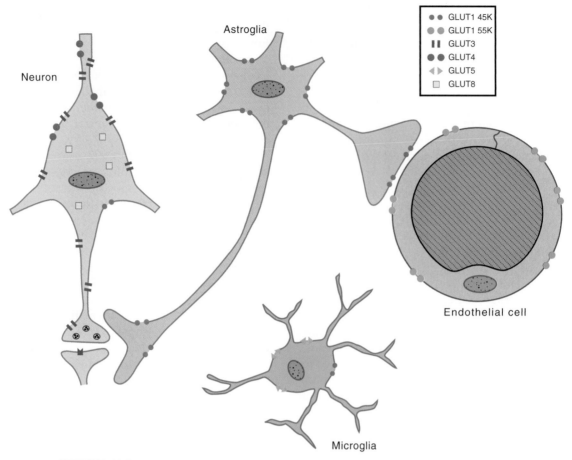

Astroglia

Neuron

Endothelial cell

Microglia

	GLUT1 45K
	GLUT1 55K
	GLUT3
	GLUT4
	GLUT5
	GLUT8

FIGURE 13.9 Cellular distribution of the principle glucose transporters in the nervous system.

The contribution of astrocytes to glucose utilization during activation appears to be even more striking. *In vitro*, activation can be mimicked by exposure of the cells to glutamate, the principal excitatory neurotransmitter (Chapter 7), because, during activation of a given cortical area, the concentration of glutamate in the extracellular space increases considerably due to its release from the axon terminals of activated pathways. As shown in Figure 13.10A, L-glutamate stimulates 2-DG uptake and phosphorylation by astrocytes in a concentration-dependent manner, with an EC_{50} of 60 to 80 mM (Pellerin and Magistretti, 1994; Takahashi *et al.*, 1995). Unlike other actions of glutamate, stimulation of glucose utilization in astrocytes is mediated not by specific glutamate receptors, but by glutamate transporters. Indeed, in addition to the maintenance of extracellular K^+ homeostasis, one of the well-established functions of astrocytes is to ensure the reuptake of certain neurotransmitters, particularly, that of glutamate at excitatory synapses. Five glutamate transporter subtypes have been cloned in various species,

including humans (Beart and O'Shea, 2007). The EAAT-1 and EAAT-2 subtypes are localized exclusively in astrocytes, whereas the EAAT 3 and 4 are localized predominantly in neurons, with a widespread distribution for EAAT 3 and a localized distribution to Purkinje neurons for EAAT 4. EAAT 5 is localized in rod photoreceptors and in bipolar cells of the retina. The density of EAAT 1 and 2 is particularly high on astrocytes that surround nerve terminals and dendritic spines, consistent with the prominent role of these transporters in the reuptake of synaptically released glutamate. The driving force for glutamate uptake through the specific transporters is the transmembrane Na^+ gradient; indeed, glutamate is cotransported with Na^+ in a ratio of one glutamate for every two or three Na^+ ions. The selective loss of EAAT 2, the astrocyte-selective glutamate transporter, has been demonstrated in the motor cortex and spinal cord of patients who died of amyotrophic lateral sclerosis, a neurodegenerative disease affecting motor neurons.

A

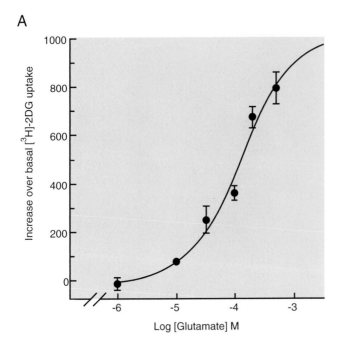

B

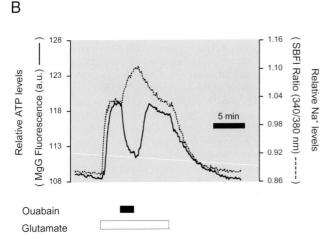

Ouabain

Glutamate

FIGURE 13.10 (A) Stimulation by glutamate of glucose uptake and phosphorylation in astrocytes. This effect is concentration-dependent with an EC$_{50}$ of ~60 mM. This process is dependent on sodium signaling associated with glutamate uptake and is energy consuming as the increase in sodium activates the sodium-potassium ATPase (see also Fig. 13.11). (B) Temporal coincidence in the increase in sodium concentration and ATP consumption triggered by glutamate in astrocytes. These processes, which are dependent on the activity of the sodium-potassium ATPase as they are ouabain-sensitive, are the effectors of the glutamate-stimulated glycolysis in astrocytes (see also Fig. 13.11). Modified from Magistretti and Chatton (2005).

Glutamate-Stimulated Uptake of Glucose by Astrocytes Is a Source of Insight into the Cellular Bases of ^{18}F-2-DG PET *in Vivo*

The glutamate-stimulated uptake of glucose by astrocytes is a source of insight into the cellular bases of the activation-induced local increase in glucose utilization visualized with ^{18}F-2-DG PET *in vivo*. As we have seen, focal physiological activation of specific brain areas is accompanied by increases in glucose utilization; because glutamate is released from excitatory synapses when neuronal pathways subserving specific modalities are activated, the stimulation by glutamate of glucose utilization in astrocytes provides a direct mechanism for coupling neuronal activity to glucose utilization in the brain (Fig. 13.11). The intracellular molecular mechanism of this coupling requires Na$^+$, K$^+$-ATPase because ouabain completely inhibits the glutamate-evoked 2-DG uptake by astrocytes (Pellerin and Magistretti, 1994). The astrocytic Na$^+$, K$^+$-ATPase responds predominantly to increases in intracellular Na$^+$ (Na^+_i) for which it shows a K_m of about 10 mM. In astrocytes, the Na^+_i concentration ranges between 10 and 20 mM, and so Na$^+$, K$^+$-ATPase is set to be activated readily when Na^+_i rises concomitantly with glutamate uptake (Magistretti and Chatton, 2005) (Figure 13.10B). These observations indicate that a major determinant of glucose utilization is the activity of Na$^+$, K$^+$-ATPase.

In this context, we should note that, *in vivo*, the main mechanism that accounts for activation-induced 2-DG uptake is the activity of Na$^+$, K$^+$-ATPase. It is important here to briefly consider the relative participation of the neuronal and astrocytic Na$^+$, K$^+$-ATPases in glucose utilization. When glutamate is released from depolarized neuronal terminals, it is taken up predominantly into astrocytes. The stoichiometry of glutamate reuptake being one molecule of glutamate cotransported with three Na$^+$ ions, the increase in intracellular astrocytic Na$^+$ concentration associated with glutamate reuptake massively activates the pump. Thus, although the tonic activity of the Na$^+$, K$^+$-ATPase is needed to maintain the transmembrane neuronal and glial ionic gradients and accounts for basal glucose utilization, on a short-term temporal scale (from milliseconds to seconds), when glutamate is released from depolarized axon terminals of modality-specific afferents, the astrocytic Na$^+$, K$^+$-ATPase is briskly activated, due to the massive increase (by at least 10 μM) in intracellular Na$^+$ associated with glutamate reuptake, providing the signal for the activation-dependent glucose utilization. Increases of glutamate as small as 10 μM are sufficient to double the activity of Na$^+$, K$^+$-ATPase.

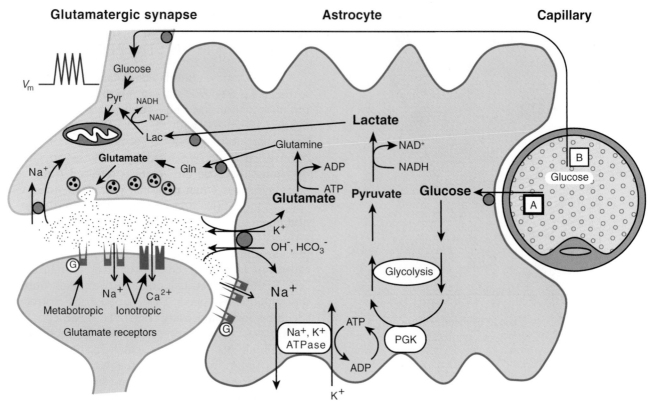

FIGURE 13.11　Schematic representation of the mechanism for glutamate-induced glycolysis in astrocytes during physiological activation. At glutamatergic synapses, presynaptically released glutamate depolarizes postsynaptic neurons by acting at specific receptor subtypes. The action of glutamate is terminated by an efficient glutamate uptake system located primarily in astrocytes. Glutamate is cotransported with Na^+, resulting in an increase in the intraastrocytic concentration of Na^+, leading to an activation of the astrocyte Na^+, K^+-ATPase. Activation of Na^+,K^+-ATPase stimulates glycolysis (i.e., glucose utilization and lactate production). The stoichiometry of this process is such that for one glutamate molecule taken up with three Na^+ ions, one glucose molecule enters astrocytes, two ATP molecules are produced through glycolysis, and two lactate molecules are released. Within the astrocyte, one ATP fuels one "turn of the pump," and the other provides the energy needed to convert glutamate to glutamine by glutamine synthase (Fig. 13.13). Once released by astrocytes, lactate can be taken up by neurons and serve as an energy substrate. (For graphic clarity only lactate uptake into presynaptic terminals is indicated. However, this process could also take place at the postsynaptic neuron.) This model, which summarizes *in vitro* experimental evidence indicating glutamate-induced glycolysis, is taken to show cellular and molecular events occurring during activation of a given cortical area (arrow labeled A, activation). Direct glucose uptake into neurons under basal conditions is also shown (arrow labeled B, basal conditions). Pyr, pyruvate; Lac, lactate; Gln, glutamine; G, G protein. Modified from Pellerin and Magistretti (1994).

How does activation of Na^+, K^+-ATPase cause increased glucose utilization? The mechanism was explained by pioneering studies on erythrocytes by Joseph Hoffmann and colleagues at Yale University, which have been confirmed in a number of other cell systems, including brain and vascular smooth muscle. The increase in pump activity consumes ATP (Fig. 13.10B), which is a negative modulator of phosphofructo-kinase, the principal rate-limiting enzyme of glycolysis (Fig. 13.1). Thus, when ATP concentration is low, phosphofructokinase activity is stimulated, resulting in increased glucose utilization. The activity of hexokinase, the enzyme responsible for glucose and 2-DG phosphorylation (Fig. 13.4), is also increased

under these conditions. This explains why the increase in glucose utilization, associated with the stimulation of Na^+, K^+-ATPase, can be monitored with 2-DG, which is not processed beyond the hexokinase step.

A compartmentalization of glucose uptake during activation also has been unequivocally found by Marco Tsacopoulos and colleagues in the honeybee drone retina (Tsacopoulos *et al.*, 1988). In this highly organized, crystal-like nervous tissue preparation, photoreceptor cells form rosette-like structures that are surrounded by glial cells. In addition, mitochondria are exclusively present in the photoreceptor neurons. Light activation reveals an increase in radioactive 2-DG uptake in the glial cells surrounding the rosettes but not in the photoreceptor neurons. An increase in O_2 consumption nevertheless is measured in photoreceptor neurons. After activation of photoreceptors by light, glucose probably is taken up predominantly by glial cells, which then release a metabolic substrate to be oxidized by photoreceptor neurons.

In summary, as indicated in the operational model described in Figure 13.11, upon activation of a particular brain area, glutamate released from excitatory terminals is taken up by a Na^+-dependent transporter located on astrocytes. The ensuing local increase in intracellular Na^+ concentration activates Na^+, K^+-ATPase, which in turn stimulates glucose uptake by astrocytes. The key role of glial glutamate transporters in activity-dependent glucose uptake by the brain has been demonstrated *in vivo* (Cholet *et al.*, 2001). This model delineates a simple mechanism for coupling synaptic activity to glucose utilization; in addition, it is consistent with the notion that the signals detected during physiological activation in humans with ^{18}F-2-DG PET and autoradiography in laboratory animals may predominantly reflect uptake of the tracer into astrocytes. This conclusion does not question the validity of the 2-DG-based techniques; rather, it provides a cellular and molecular basis for these functional brain-imaging techniques (Magistretti *et al.*, 1999).

Lactate Released by Astrocytes May Be a Metabolic Substrate for Neurons

The fact that the increase in glucose uptake during activation can be ascribed predominantly, if not exclusively, to astrocytes indicates that energy substrates must be released by astrocytes to meet the energy demands of neurons. As indicated earlier, lactate and pyruvate are adequate substrates for brain tissue *in vitro* (Schurr 2006). In fact, synaptic activity can be maintained *in vitro* in cerebral cortical slices with only lactate or pyruvate as a substrate. Lactate is quantitatively the main metabolic intermediate released by cultured astrocytes at a rate of 15 to 30 nmol per milligram of protein per minute. Other quantitatively less important intermediates released by astrocytes are pyruvate (approximately 10 times less than lactate) and α-ketoglutarate, citrate, and malate, which are released in marginal amounts. For lactate (or pyruvate) to be a metabolic substrate for neurons, particularly during activation, two additional conditions must be fulfilled: (1) that indeed during activation lactate release by astrocytes increases and (2) lactate uptake by neurons must be demonstrated. Both mechanisms have been demonstrated. Mimicking activation *in vitro* by exposing cultured astrocytes to glutamate results in a marked release of lactate and, to a lesser degree,

pyruvate (Pellerin and Magistretti, 1994). This glutamate-evoked lactate release shows the same pharmacology and time course as glutamate-evoked glucose utilization and indicates that glutamate stimulates the processing of glucose through glycolysis. *In vivo* [1]H MRI studies in humans that show a transient lactate peak in the primary visual cortex during physiological stimulation (Prichard *et al.*, 1991) are consistent with the notion of activation-induced glycolysis. In addition, lactate levels in the rat hippocampus transiently increase upon stimulation (Hu and Wilson 1997). Finally, monocarboxylate transporters have been demonstrated on neurons and astrocytes in addition to capillaries (Pierre and Pellerin, 2005).

Thus, a metabolic compartmentation whereby glucose taken up by astrocytes and metabolized glycolytically to lactate is then released in the extracellular space to be utilized by neurons is consistent with biochemical and electrophysiological observations (Tsacopoulos and Magistretti, 1996; Magistretti 2006; Hyder *et al.*, 2006). This array of *in vitro* and *in vivo* experimental evidence is summarized in the model of cell-specific metabolic regulation illustrated in Figure 13.11.

Studies of the well-compartmentalized honeybee drone retina and of isolated preparations of guinea pig retina containing photoreceptors attached to Mueller (glial) cells corroborate the existence of such metabolic fluxes between glia and neurons. In addition to the glial localization of glucose uptake during activation, glycolytic products have been shown to be released. In particular, during activation, glial cells in the honeybee drone retina release alanine produced from pyruvate by transamination; the released alanine is taken up by photoreceptor neurons and, after reconversion into pyruvate, can enter the TCA cycle to yield ATP through oxidative phosphorylation (Fig. 13.2). In the guinea pig retina, lactate, formed glycolytically from glucose, is released by Mueller cells to fuel photoreceptor neurons (Tsacopoulos and Magistretti, 1996).

Although plasma lactate, except under particular conditions such as vigorous exercise, contributes only marginally as a metabolic substrate for the brain, lactate formed *within* the brain parenchyma (e.g., through glutamate-activated glycolysis in astrocytes) can fulfill the energetic needs of neurons. Lactate, after conversion into pyruvate by a reaction catalyzed by lactate dehydrogenase (LDH), can provide, on a molar basis, 15–18 ATP through oxidative phosphorylation. Conversion of lactate into pyruvate does not require ATP, and, in this regard, lactate is energetically more favorable than the first obligatory step of glycolysis in which glucose is phosphorylated to glucose 6-phosphate at the expense of one molecule of ATP

(Fig. 13.1). In addition, lactate may contribute to the redox potential of neurons, since through its conversion to pyruvate it generates NADH (Cerdan *et al.*, 2006), hence providing reducing equivalents useful for ROS scavenging (Fig. 13.6). Another metabolic fate for lactate has been shown *in vitro* and *in vivo* by MRS. Thus, once converted to pyruvate in neurons, lactate may enzymatically yield glutamate and hence be a substrate for the replenishment of the neuronal pool of glutamate. Because this reaction is not associated with oxygen consumption, part of the uncoupling between glucose utilization and oxygen consumption described in certain paradigms of activation may be explained by the processing of glucose-derived lactate into the glutamate neuronal pool.

Glycogen, the Storage Form of Glucose, Is Localized in Astrocytes

Glycogen is the single largest energy reserve of the brain (Magistretti *et al.*, 1993); it is localized mainly in astrocytes, although ependymal and choroid plexus cells, as well as certain large neurons in the brain stem, contain glycogen. When compared to the contents in liver and muscle, the glycogen content of the brain is exceedingly small, about 100 and 10 times inferior, respectively. Thus, the brain can hardly be considered a glycogen storage organ, and here the function of glycogen should be viewed as that of providing a metabolic buffer during physiological activity.

Glycogen Metabolism Is Coupled to Neuronal Activity

Glycogen turnover in the brain is extremely rapid, and glycogen levels are finely coordinated with synaptic activity (Magistretti *et al.*, 1993). For example, during general anesthesia, a condition in which synaptic activity is markedly attenuated, glycogen levels rise sharply. Interestingly, however, the glycogen content of cultures containing exclusively astrocytes is not increased by general anesthetics; this observation indicates that the *in vivo* action of general anesthetics on astrocyte glycogen is due to the inhibition of neuronal activity, stressing the existence of a tight coupling between synaptic activity and astrocyte glycogen. Accordingly, reactive astrocytes, which develop in areas where neuronal activity is decreased or absent as a consequence of injury, contain high amounts of glycogen.

In addition to glycogen, glucose is incorporated into other macromolecules such as proteins (glycoproteins) and lipids (glycolipids) at rates specific for the turnover of each macromolecule, which can span from a few minutes to a few days.

Certain Neurotransmitters Regulate Glycogen Metabolism in Astrocytes

Glycogen levels in astrocytes are tightly regulated by various neurotransmitters. Several monoamine neurotransmitters—namely, noradrenaline, serotonin, and histamine—are glycogenolytic in the brain, in addition to certain peptides, such as vasoactive intestinal peptide (VIP) and pituitary adenylate cyclase activating peptide (PACAP), and adenosine and ATP (Magistretti *et al.*, 1993). The effects of all these neurotransmitters are mediated by their cogent specific receptors coupled to second messenger pathways that are under the control of adenylate cyclase or phospholipase C. The initial rate of glycogenolysis activated by VIP and noradrenaline is between 5 and 10 nmol per milligram of protein per minute, a value that is remarkably close to glucose utilization of the gray matter, as determined by the 2-DG autoradiographic method. This correlation indicates that glycosyl units mobilized in response to glycogenolytic neurotransmitters can provide quantitatively adequate substrates for the energy demands of the brain parenchyma. At present, whether the glycosyl units mobilized through glycogenolysis are used by astrocytes to meet their energy demands during activation or are metabolized to a substrate such as lactate, which is then released for the use of neurons, is not clear. It appears, however, that glucose is not released by astrocytes after glycogenolysis, supporting the view that the activity of glucose-6-phosphatase (Fig. 13.1) in astrocytes is very low. Lactate may be the metabolic intermediate produced through glycogenolysis and exported from astrocytes (Tekkok *et al.*, 2005).

These observations show that neuronal signals (e.g., certain neurotransmitters) can exert receptor-mediated metabolic effects on astrocytes in a manner similar to peripheral hormones on their target cells. However, the action of this type of neurotransmitter is temporally specified and spatially restricted to activated areas. Indeed, brain glycogenolysis visualized by autoradiography in laboratory animals also has been demonstrated *in vivo* after physiological activation of a modality-specific pathway (Swanson *et al.*, 1992). Repeated stimulation of whiskers resulted in a marked decrease in the density of glycogen-associated autoradiographic grains in the somatosensory cortex of rats (barrel fields), as well as in the relevant thalamic nuclei (Swanson *et al.*, 1992). These observations indicate that the physiological activation of specific neuronal circuits results in the mobilization of glial glycogen stores.

Summary

Under basal conditions, glucose uptake and metabolism occur in every brain cell type. Glucose uptake is mediated by specific transporters that are distributed in a cell-specific manner. Astrocytes play a critical role in the utilization of glucose coupled to excitatory synaptic transmission. The molecular mechanisms of this coupling are stoichiometrically directed: for each synaptically released glutamate molecule taken up with three Na^+ ions by an astrocyte, one glucose molecule enters the same astrocyte, two ATP molecules are produced through glycolysis, and two lactate molecules are released and consumed by neurons to yield 15–18 ATPs through oxidative phosphorylation. Neuronal signals (e.g., certain neurotransmitters) can exert receptor-mediated glycogenolysis in astrocytes in a manner similar to peripheral hormones on their target cells. However, this type of effect by neurotransmitters is temporally specified and spatially restricted within activated areas, possibly to provide additional energy substrates in register with local increases in neuronal activity.

Astrocyte

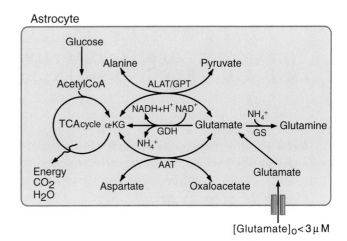

[Glutamate]$_O$ < 3 μM

FIGURE 13.12 Metabolic fate of glutamate taken up by astrocytes. ALAT, alanine aminotransferase; GDH, glutamate dehydrogenase; GS, glutamine synthase; AAT, aspartate aminotransferase; GPT, glutamate dehydrogenase; α-KG, α-ketoglutarate.

GLUTAMATE AND NITROGEN METABOLISM: A COORDINATED SHUTTLE BETWEEN ASTROCYTES AND NEURONS

As has been shown, synaptically released glutamate is removed rapidly from the extracellular space by a transporter-mediated reuptake system that is particularly efficient in astrocytes (Beart and O'Shea, 2007). This mechanism contributes in a crucial manner to the fidelity of glutamate-mediated neurotransmission. Indeed, glutamate levels in the extracellular space are low (<3 μM), allowing for optimal glutamate-mediated signaling after depolarization while preventing overactivation of glutamate receptors, which eventually could result in excitotoxic neuronal damage.

One may wonder how astrocytes dispose of the glutamate that they take up, because, unlike carbohydrates or lipids, amino acids cannot be stored. The predominant pathway in peripheral tissues for disposing of amino acids is the transfer of their α amino group to a corresponding α-keto acid; this reaction is catalyzed by aminotransferases (Fig. 13.12). In astrocytes, the α amino group of glutamate can be transferred to oxaloacetate to yield α-ketoglutarate (α-KG) and aspartate in a reaction catalyzed by aspartate amino transferase (AAT). The α-KG generated is an intermediate of the TCA cycle and is therefore oxi-

dized further. Another transamination reaction catalyzed by alanine amino transferase (ALAT) transfers the α amino group of glutamate to pyruvate, resulting in the formation of alanine and α-KG.

Two other pathways exist in astrocytes to metabolize glutamate. First, glutamate can be converted directly into α-KG through an NAD-requiring oxidative deamination catalyzed by glutamate dehydrogenase (GDH) (Fig. 13.12). Glutamate, by entering the TCA cycle indirectly (through AAT or ALAT) or directly (through GDH), is an energy substrate for astrocytes. Second, the quantitatively predominant metabolic pathway of glutamate in astrocytes is its amidation to glutamine, an ATP-requiring reaction in which an ammonium ion is fixed on glutamate (Fig. 13.12) (Van den Berg and Garfinkel, 1971). This reaction is catalyzed by glutamine synthase (GS), an enzyme almost exclusively localized in astrocytes, and provides an efficient means of disposing not only of glutamate but also of ammonium (Box 13.2). Glutamine is released by astrocytes and is taken up by neurons, where it is hydrolyzed back to glutamate by the phosphate-dependent mitochondrial enzyme glutaminase (Erecinska and Silver, 1990). This metabolic pathway, often referred to as the glutamate-glutamine shuttle, is a clear example of cooperation between astrocytes and neurons (Fig. 13.13). It allows the removal of potentially toxic excess glutamate from the extracellular space, while returning to the neuron a synaptically inert (glutamine does not affect neurotransmission) precursor with which to regenerate the neuronal pool of glutamate.

BOX 13.1

HEPATIC ENCEPHALOPATHY IS A DISORDER OF ASTROCYTE FUNCTION RESULTING IN A NEUROPSYCHIATRIC SYNDROME

Hepatic encephalopathy is observed in patients with severe liver failure. The disease can be in one of two forms: an acute form, called fulminant hepatic failure, and a chronic form, portosystemic encephalopathy. The neuropsychiatric symptoms of fulminant hepatic failure are delirium, coma, and seizures associated with acute toxic or viral hepatic failure. Patients having portosystemic encephalopathy may present personality changes, episodic confusion, or stupor and, in the most severe cases, coma. The current view on the pathophysiology of hepatic encephalopathy is that, due to liver failure, "toxic" substances that affect brain function accumulate in the circulation.

One of the substances thought to be responsible for neuropsychiatric "toxicity" is ammonia. The neuropathological findings are rather striking: astrocytes are the brain cells that appear principally affected. In the acute form, astrocyte swelling is prominent and likely to be the cause of the observed acute brain edema. In portosystemic encephalopathy, astrocytes adopt morphological features characteristic of what is defined as an Alzheimer type II astrocyte: in these cells, the nucleus is pale and enlarged, chromatin is marginated, and a prominent

nucleolus often is observed. Lipofuscin deposits may be present, and the amount of the astrocyte-specific protein glial fibrillary acidic protein (Chapter 4) is decreased. Neurons appear structurally normal. All the foregoing histopathological changes have been reproduced *in vitro* by acutely or chronically applying ammonium chloride to primary astrocyte cultures. As mentioned earlier, detoxification of ammonium is an ATP-requiring, astrocyte-specific reaction catalyzed by glutamine synthase (Fig. 13.12). It is therefore not surprising that excess ammonia perturbs energy metabolism; indeed, ammonia stimulates glycolysis whereas it inhibits TCA cycle activity. In addition, ammonia decreases the glycogen content of astrocytes markedly.

In summary, though the precise pathophysiological mechanisms of the neuropsychiatric syndrome in hepatic encephalopathy are still unknown, this clinical condition provides a striking illustration of the fundamental importance of neuron-astrocyte metabolic interactions because structural and functional alterations apparently restricted to astrocytes result in severe behavioral perturbations.

Pierre J. Magistretti

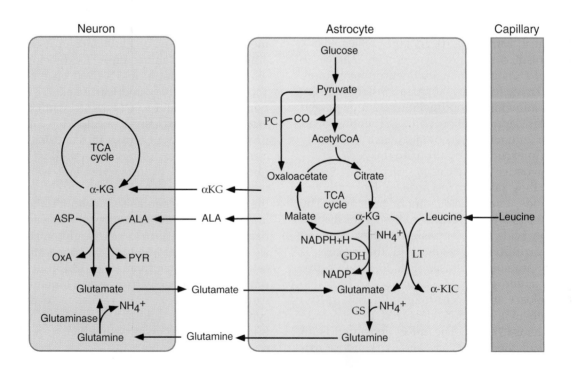

However, not all glutamate is regenerated through the glutamate-glutamine shuttle because some of the glutamate released by neurons enters at the α-KG level, the TCA cycle in astrocytes; therefore, *de novo* synthesis is required to maintain the neuronal glutamate pool. Glutamate can be synthesized through NADPH-dependent reductive amination of α-KG catalyzed by GDH (note that here the cofactor is NADPH, whereas, for the opposite reaction also catalyzed by GDH, the oxidant is NAD; see Figs. 13.12 and 13.13) (Erecinska and Silver, 1990). For the synthesis of glutamate, glucose provides the carbon backbone as α-KG through the TCA cycle, whereas an exogenous source of nitrogen is necessary (Fig. 13.13). Convincing evidence, obtained by using ^{15}N-labeled amino acids whose metabolic fate was determined by gas chromatography and mass spectrometry, indicates that plasma leucine provides the nitrogen required for net glutamate synthesis from α-KG. Thus, leucine taken up from the circulation at astrocytic end feet provides the amino group to α-KG in a reaction catalyzed by leucine transaminase (LT), resulting in the formation of glutamate and α-ketoisocaproate (α-KIC) (Fig. 13.13). Because this reaction takes place in astrocytes, to replenish the neuronal glutamate pool, the astrocytes export glutamate as glutamine. As noted earlier, the neuronal glutamate pool could also be replenished by lactate released by astrocytes.

Finally, another potential pathway described by Arne Schousboe and colleagues exists for the *de novo* synthesis of glutamate in neurons from substrates provided by astrocytes. With the use of uniformly labeled ^{13}C compounds in combination with magnetic resonance spectroscopy, astrocytes have been shown to release significant amounts of alanine and α-KG. Both metabolic intermediates are taken up by neurons and can be converted into glutamate and pyruvate in a transamination reaction catalyzed by ALAT (Fig. 13.12). In this case, as for the glutamate-glutamine shuttle (Fig. 13.13), astrocytes provide the substrate(s) necessary for glutamate synthesis in neurons.

Note that because α-KG is used for glutamate synthesis, metabolic intermediates downstream of α-KG

must be available to maintain a sustained flux through the TCA cycle in astrocytes (Fig. 13.13). This need is met by the activity of the enzyme pyruvate carboxylase (PC), which fixes CO_2 on pyruvate to generate oxaloacetate, which, by condensing with acetyl-CoA, maintains the flux through the TCA cycle. The carboxylation of pyruvate to oxaloacetate is referred to as an anaplerotic (Greek for "fill up") reaction. Interestingly, like glutamine synthase, PC is selectively localized in astrocytes. The fact that these two enzymes are localized in astrocytes in conjunction with the existence of a glutamate-glutamine shuttle stresses that astrocytes are essential for maintaining the neuronal glutamate pool used for neurotransmission (Fig. 13.13).

As noted earlier, the metabolic intermediate α-KG lies at the branching point of glucose and glutamate metabolism (Fig. 13.12). Any change in the activities of the enzymes that convert α-KG into glutamate or into succinyl-CoA, the next intermediate in the TCA cycle, may affect the efficacy of the TCA cycle or glutamate levels. Interestingly, a marked decrease in the activity of α-ketoglutarate dehydrogenase (α-KGDH), the enzyme catalyzing the conversion of α-KG into succinyl-CoA, was found in a very high proportion of postmortem brains from patients with Alzheimer disease; in addition, a similar decrease in α-KGDH activity has been demonstrated in the fibroblasts of patients affected by the familial form of Alzheimer disease.

Summary

A key function of astrocytes is to remove synaptically released glutamate. A large proportion of glutamate is transformed to glutamine through an energy-requiring process that also allows for the detoxification of ammonium. Glutamine released by astrocytes regenerates the neuronal glutamate pool. Some of the glutamate also is regenerated from lactate and through fixation of the amino group of leucine onto the TCA intermediate α-KG, providing another indication of the tight link existing between glutamate

FIGURE 13.13 Metabolic intermediates are released by astrocytes to regenerate the glutamate neurotransmitter pool in neurons. Glutamine, formed from glutamate in a reaction catalyzed by glutamine synthase (GS), is released by astrocytes and taken up by neurons, which convert it into glutamate under the action of glutaminase. GS is an enzyme selectively localized in astrocytes. This metabolic cycle is referred to as the glutamate-glutamine shuttle. Other quantitatively less important sources of neuronal glutamate are lactate, alanine, and α-ketoglutarate (α-KG). In astrocytes, glutamate is synthesized *de novo* from α-KG in a reaction catalyzed by glutamate dehydrogenase (GDH). The carbon backbone of glutamate is exported by astrocytes after conversion into glutamine under the action of GS; the conversion of leucine into α-ketoisocaproate (α-KIC), catalyzed by leucine transaminase (LT), provides the amino group for the synthesis of glutamine from glutamate. Carbons "lost" from the TCA cycle as α-KG is converted into glutamate are replenished by oxaloacetate (OxA) formed from pyruvate in a reaction catalyzed by pyruvate carboxylase, another astrocyte-specific enzyme.

and nitrogen metabolism and of the crucial function that astrocytes play in maintaining the neuronal glutamate pool at levels that ensure the maintenance of synaptic transmission.

THE ASTROCYTE-NEURON METABOLIC UNIT

From a strictly energetic viewpoint, the brain can be seen as an almost exclusive glucose-processing machine producing H_2O and CO_2. However, the metabolism of glucose in the brain is specified temporally, spatially, and functionally. Thus, glucose metabolism increases with exquisite spatiotemporal precision in register with neuronal activity. The site of this increase is not the neuronal cell body; rather, it is the neuropil, where presynaptic terminals, postsynaptic elements, and astrocytes ensheathing synaptic contacts are localized (Sokoloff, 1981). This cytological relation between astrocytes and neurons also is manifested by a functional metabolic partnership: in response to a neuronal signal (glutamate), astrocytes release a glucose-derived metabolic substrate for neurons (lactate). Glucose also provides the carbon backbone for regeneration of the neuronal pool of glutamate. This process results from a close astrocyte-neuron cooperation. Indeed, the selective localization of pyruvate carboxylase in astrocytes, indicating the need to replenish the TCA cycle with carbon backbones, strongly suggests that glucose-derived metabolic intermediates are used for glutamate (and other amino acid) synthesis. The newly synthesized glutamate is not provided as such by astrocytes to neurons; rather, it is converted into glutamine by glutamine synthase, another enzyme localized selectively in astrocytes. Glutamate, taken up by astrocytes during synaptic activity, undergoes the same metabolic process, also being released as glutamine (the glutamate-glutamine shuttle).

Summary

In conclusion, the axon terminal of glutamatergic neurons, which are the main communication lines in the nervous system, and the astrocytic processes that surround them should be viewed as a metabolic unit in which the neuron furnishes the activation signal (glutamate) to the astrocyte and the astrocyte provides not only the precursors needed to maintain the neurotransmitter pool (glutamine and, in part, lactate and alanine), but also the energy substrate (lactate) (Fig. 13.14). The efficacy of the predominant excitatory synapse in the brain, the glutamatergic synapse, cannot be maintained without a close astrocyte-neuron interaction.

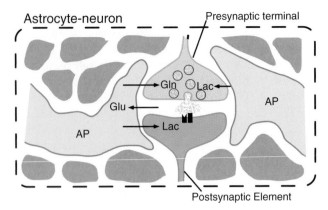

FIGURE 13.14 The astrocyte-neuron metabolic unit. Glutamatergic terminals and the astrocytic processes that surround them can be viewed as a highly specialized metabolic unit in which the activation signal (glutamate) is furnished by the neuron to the astrocyte, whereas the astrocyte provides the precursors needed to maintain the neurotransmitter pool (glutamine, lactate, alanine), as well as the energy substrate (lactate). AP, astrocyte process.

References

Attwell, D. and Laughlin, S. B. (2001). An energy budget for signaling in the grey matter of the brain. *J Cereb Blood Flow Metab.*

Beal, F. M. (2005). Mitochondria take center stage in aging and neurodegeneration. *Ann Neurol* **58**, 495–505.

Beart, P. M. and O'Shea, R. D. (2007). Transporters for L-glutamate: An update on their molecular pharmacology and pathological involvement. *Br J Pharmacol.* **150**, 5–17.

Cerdan, S., Rodrigues, T. B., Sierra, A., Benito, M., Fonseca, L. L., Fonseca, C. P., and Garcia-Martin, M. L. (2006). The redox switch/redox coupling hypothesis. *Neurochem Int.* **48**, 523–530.

Cholet, N., Pellerin, L., Welker, E., Lacombe, P., Seylaz, J., Magistretti, P., and Bonvento, G. (2001). Local injection of antisense oligonucleotides targeted to the glial glutamate transporter GLAST decreases the metabolic response to somatosensory activation. *J Cereb Blood Flow Metab.* **21**, 404–412.

Dalsgaard, M. K. (2006). Fuelling cerebral activity in exercising man. *J Cereb Blood Flow Metab.* **26**, 731–750.

Dringen, R. (2000). Metabolism and functions of glutathione in brain. *Prog Neurobiol.* **62**, 649–671.

Erecinska, M. and Silver, I. A. (1990). Metabolism and role of glutamate in mammalian brain. *Prog Neurobiol.* **35**, 245–296.

Frackowiak, R. S. J., Magistretti, P. J., Shulman, R. G., and Adams, M. (2001). "Neuroenergetics: Relevance for Functional Brain Imaging." HFSP, Strasbourg.

Hamel E. (2006). Perivascular nerves and the regulation of cerebrovascular tone. *J Appl Physiol.* **100**, 1059–1064.

Hassel, B. and Brathe, A. (2000). Cerebral metabolism of lactate in vivo: Evidence for neuronal pyruvate carboxylation. *J Cereb Blood Flow Metab.* **20**, 327–336.

Hu, Y. and Wilson, G. S. (1997). A temporary local energy pool coupled to neuronal activity: Fluctuations of extracellular lactate levels in rat brain monitored with rapid-response enzyme-based sensor. *J Neurochem.* **69**, 1484–1490.

Hyder, F., Patel, A. B., Gjedde, A., Rothman, D. L., Behar, K. L., and Shulman, R. G. (2006). Neuronal-glial glucose oxidation and glu-tamatergic-GABAergic function. *J Cereb Blood Flow Metab.* **26**, 865–877.

Kacem, K., Lacombe, P., Seylaz, J., and Bonvento, G. (1998). Struc-tural organization of the perivascular astrocyte endfeet and their relationship with the endothelial glucose transporter: A confocal microscopy study. *Glia* **23**, 1–10.

Kety, S. S. and Schmidt, C. F. (1948). The nitrous oxide method for the quantitative determination of cerebral blood flow in man: Theory, procedure, and normal values. *J Clin Invest.* **27**, 476–483.

Koehler, R. C., Gebremedhin, D., and Harder, D. R. (2006). Role of astrocytes in cerebrovascular regulation. *J Appl Physiol.* **100**, 307–317.

Magistretti, P. and Pellerin, L. (1999). Cellular mechanisms of brain energy metabolism and their relevance to functional brain imaging. *Phil Trans R Soc Lond. B* **354**, 1155–1163.

Magistretti, P. J., Pellerin, L., Rothman, D. L., and Shulman, R. G. (1999). Energy on demand. *Science* **283**, 496–497.

Magistretti, P. J., Sorg, O., and Martin, J. L. (1993). Regulation of glycogen metabolism in astrocytes: Physiological, pharmacologi-cal, and pathological aspects. *In* "Astrocytes: Pharmacology and Function" (S. Murphy, ed.), pp. 243–265. Academic Press, San Diego.

Magistretti, P. J. and Chatton, J. Y. (2005). Relationship between L-glutamate-regulated intracellular Na+ dynamics and ATP hydrolysis in astrocytes. *J Neural Transm.* **112**, 77–85.

Magistretti, P. J. (2006). Neuron-glia metabolic coupling and plastic-ity. *J Exp Biol.* **209**, 2304–2311.

McEwen, B. S. and Reagan, L. P. (2004). Glucose transporter expres-sion in the central nervous system: Relationship to synaptic function. *Eur J Pharmacol.* **490**, 13–24.

Nehlig, A., Wittendorp-Rechenmann, E., and Lam, C. D. (2004). Selective uptake of [14C]2-deoxyglucose by neurons and astro-cytes: High-resolution microautoradiographic imaging by cellu-lar 14C-trajectography combined with immunohistochemistry. *J Cereb Blood Flow Metab.* **24**, 1004–1014.

Pellerin, L. and Magistretti, P. J. (1994). Glutamate uptake into astro-cytes stimulates aerobic glycolysis: A mechanism coupling neu-ronal activity to glucose utilization. *Proc Natl Acad Sci USA* **91**, 10625–10629.

Phelps, M. E., Huang, S. C., Hoffman, E. J., Selin, C., Sokoloff, L., and Kuhl, D. E. (1979). Tomographic measurement of local cerebral glucose metabolic rate in humans with (F-18)2-fluoro-2-deoxy-D-glucose: Validation of method. *Ann Neurol.* **6**, 371–388.

Pierre, K. and Pellerin, L. (2005). Monocarboxylate transporters in the central nervous system: Distribution, regulation and func-tion. *J Neurochem.* **94**, 1–14.

Prichard, J., Rothman, D., Novotny, E., Petroff, O., Kuwabara, T., Avison, M., Howseman, A., Hanstock, C., and Shulman, R. (1991). Lactate rise detected by 1H NMR in human visual cortex during physiologic stimulation. *Proc Natl Acad Sci USA* **88**, 5829–5831.

Raichle, M. E. and Mintun, M. A. (2006). Brain work and brain imaging. *Annu Rev Neurosci.* **29**, 449–476.

Roy, C. S. and Sherrington, C. S. (1890). On the regulation of the blood supply of the brain. *J Physiol (Lond.)* **11**, 85–108.

Schurr, A. (2006). Lactate: The ultimate cerebral oxidative energy substrate? *J Cereb Blood Flow Metab.* **26**, 142–152.

Smith, D., Pernet, A., Hallett, W. A., Bingham, E., Marsden, P. K., Amiel, S. A. (2003). Lactate: A preferred fuel for human brain metabolism in vivo. *J Cereb Blood Flow Metab.* **23**, 658–664.

Sokoloff, L. (1981). Localization of functional activity in the central nervous system by measurement of glucose utilization with radioactive deoxyglucose. *J Cereb Blood Flow Metab.* **1**, 7–36.

Swanson, R. A., Morton, M. M., Sagar, S. M., and Sharp, F. R. (1992). Sensory stimulation induces local cerebral glycogenolysis: Dem-onstration by autoradiography. *Neuroscience* **51**, 451–461.

Takahashi, S., Driscoll, B. F., Law, M. J., and Sokoloff, L. (1995). Role of sodium and potassium ions in regulation of glucose metabolism in cultured astroglia. *Proc Natl Acad Sci USA* **92**, 4616–4620.

Tekkok, S. B., Brown, A. M., Westenbroek, R., Pellerin, L., and Ransom, B. R. (2005). Transfer of glycogen-derived lactate from astrocytes to axons via specific monocarboxylate transporters supports mouse optic nerve activity. *J Neurosci Res.* **81**, 644–652.

Tsacopoulos, M., Evequoz-Mercier, V., Perrottet, P., and Buchner, E. (1988). Honeybee retinal glial cells transform glucose and supply the neurons with metabolic substrates. *Proc Natl Acad Sci USA* **85**, 8727–8731.

Tsacopoulos, M. and Magistretti, P. J. (1996). Metabolic coupling between glia and neurons. *J Neurosci.* **16**, 877–885.

Uldry, M. and Thorens, B. (2004). The SLC2 family of facilitated hexose and polyol transporters. *Pflugers Arch.* **447**, 480–489.

Van den Berg, C. J. and Garfinkel, D. (1971). A simulation study of brain compartments. Metabolism of glutamate and related sub-stances in mouse brain. *Biochem J.* **123**, 211–218.

Pierre J. Magistretti

NERVOUS SYSTEM DEVELOPMENT

14

Neural Induction and Pattern Formation

This chapter covers some of the key events that take place in the early stages of development of the vertebrate nervous system, a period during which structures such as the neural tube, placodes, and neural crest are formed, setting in place the foundations on which a functioning nervous system is subsequently built. The first part of the chapter describes how these embryonic structures first are specified by inductive tissue interactions and how they form by the process of morphogenesis. The second part of the chapter describes the extensive early developmental events required for regionalizing the nervous system along its different axes. Regionalization requires the complex processes of neural patterning that endow neural precursor cells with the ability to give rise to correct types of neuron in appropriate locations in the adult nervous system. These processes are gradual, continuous, and begin when neural tissue first forms. We describe some of the processes underlying neural patterning, beginning with how polarity along each of the neuraxes is first established, and progressing to more fine levels of regional organization.

NEURAL INDUCTION

Embryonic Origins of the Nervous System

The progenitor cells that form the vertebrate nervous system can be traced back in development to an epithelial cell layer, called the ectoderm, that covers the outside of the embryo during gastrulation (Fig. 14.1). Ectodermal cells give rise to different tissue derivatives depending on axial position. The dorsal-most ectoderm thickens to form the *neural plate*, a structure in the shape of a key-hole with the broad end located

anteriorly. During a complex morphogenetic process called *neurulation*, cells in the neural plate give rise to the neural tube and, subsequently, the central nervous system (CNS). Ectodermal cells lying more ventrally at the edges of the neural plate, the neural folds, come to lie at the dorsal surface of the neural tube during neurulation, form the *neural crest* and emigrate, subsequently giving rise to most of the peripheral nervous system. The ectodermal cells lying even more ventral around the edge of the cranial neural plate constitute a domain where various sensory structures such as the ear, nose, and cranial sensory ganglia will arise from isolated ectodermal areas called placodes. Finally, ectodermal cells on the extreme ventral side of the embryo give rise to the skin or epidermis. The first step in forming the nervous system, therefore, is to establish these different regions of ectoderm along the dorsoventral (DV) axis of the embryo soon after gastrulation is complete. The key mechanism that establishes these different subregions of ectoderm involves inductive interactions that were discovered in the early part of the last century. More recently, the molecules that underlie embryonic induction have been defined and their action understood.

Neural Induction and the Organizer

In the 1930's, Mangold and Spemann discovered neural induction during experiments in which they transplanted small pieces of tissues from one amphibian embryo to another at pregastrulae stages. The key observation was made when they transplanted a small piece of tissue from a region called the dorsal blastopore lip (DBL), and the host embryo responded to the grafted tissue by forming a complete secondary dorsal axis (Fig. 14.2). Importantly, most of the tissues

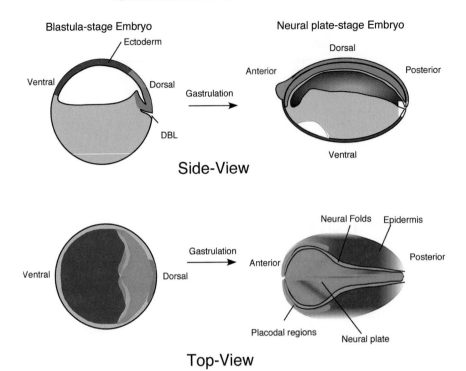

FIGURE 14.1 Ectoderm is subdivided into different fates during gastrulation. Shown is a cross-sectional side view, and top view of an amphibian embryo before (blastula-stage) and after gastrulation (neural plate-stage). During gastrulation, the grey and green (dorsal meso-derm) regions of the blastula involute inward, and the ectoderm (blue/red) at the top spreads over the outside of the embryo, and extends along anterior-posterior axis. As this process occurs, signals emanating from the dorsal and ventral sides of the embryo specify different fates, so that dorsal ectoderm (red) becomes neural tissue while ventral ectoderm (blue) becomes epidermis.

in the secondary dorsal axes were not derived from the transplanted tissue but rather from the tissue in the host embryo. In particular, the secondary dorsal axis contained a complete nervous system that was derived entirely from the ventral ectoderm of the host embryo, a tissue that would have differentiated into skin in the absence of a graft. The implication of this observation was that the transplanted tissue can act as a source of inducing signals that can cause ventral ectoderm to form neural tissue, and that this induc-tive interaction normally occurs on the dorsal side of the embryo. Tissue in the DBL was later termed the organizer because of its ability, when transplanted, to reprogram the ventral side of the embryo to form dorsal tissues, not only in the ectoderm but also in the internal mesodermal tissues. Following Mangold and Spemann's lead, it was subsequently found all verte-brate embryos appear to contain a region, called *Spe-mann's organizer*, which can induce ectoderm to form neural tissue.

Organizer transplantation experiments also gave the first indication that signals produced by Spemann's organizer were responsible for inducing different regions of the CNS. In these experiments, smaller regions of the DBL were used, and taken from embryos at different stages. The DBL of **younger** embryos con-tains the first involuting tissue and, when transplanted, induces head structures that contained neural tissue from the anterior portions of the neuraxis. Conversely, the DBL from **older** embryos involutes later and, when transplanted, induces tail structures that contained neural tissue from just the posterior portions of the neuraxis. The organizer, therefore, can be subdivided into two parts, a head and trunk/tail organizer, which specify different regions of the CNS along the AP axis. The role of the organizer tissue in the regionalization of the CNS is discussed further in a later section.

The Molecular Nature of Neural Inducers

The molecular nature of the inducers produced by organizer tissue remained elusive until the 1990's, when key biochemical pathways that mediate cell-cell signaling in animal development were identified using the tools of molecular genetics. One such pathway is a subfamily of TGFB-like factors, called the bone morphogenetic proteins (BMPs), which has proven to be remarkably conserved in action across the animal kingdom, from flies to humans (Bier, 1997). The core components of this pathway are the BMP

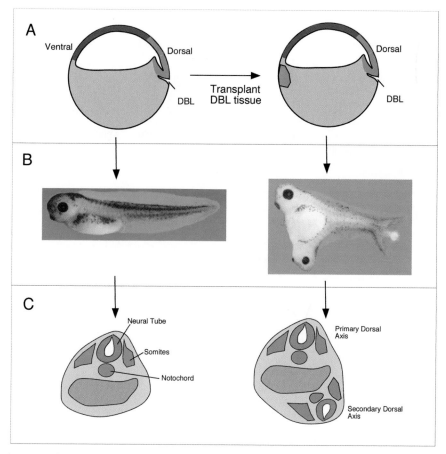

FIGURE 14.2 Organizer transplant experiment of Mangold and Spemann. (A) Tissue around the DBL was removed from one embryo and placed into the ventral side of another. (B) The transplanted DBL, if large enough, will cause a complete second dorsal axis to form on the host embryo, resulting in twinning. (C) Cross-section through the tadpoles shows that the second dorsal axis contains a complete nervous system. Importantly, by using tracers, one can show that the nervous system in this new dorsal axis is not derived from the transplanted tissue, but rather from host tissue, fated to give rise to ventral tissues in the absence of a graft.

ligands, secreted extracellular proteins that bind and activate a small family of heterodimeric cell surface receptors, which in turn transduce a signal by intracellular phosphorylation events that ultimately lead to changes in the activities of transcription factors, the SMAD proteins (Fig. 14.3). Importantly, the core activity of this signaling pathway is subjected to complex layers of regulation, both positive and negative (Fig. 14.3). This complex regulation is required to induce neural tissue by creating a gradient of BMP signaling activity across the DV axis of the ectoderm, although for a purpose not envisioned by the early embryologists (Weinstein and Hemmati-Brivanlou, 1999).

The Default Model

The current view of neural induction, the default model, stems from experiments using the amphibian *Xenopus laevis*, mainly because of an assay where the ectoderm can be explanted into culture before gastrulation into a simple salt solution. Ectoderm isolated before gastrulation differentiates into ventral epidermal tissue in culture. However, if BMP signaling is inhibited experimentally in these ectodermal explants, epidermal differentiation is suppressed, and neural tissue forms instead, presumably as a default pathway. The converse result can also be obtained using the fact that when isolated ectodermal cells are dissociated into single cells, they form nerve cells. However, if BMP4 is added to these cells, they revert back to epidermis. These early experiments were instrumental in showing that BMPs are potent epidermalizing agents and that blocking BMP signaling was sufficient to induce the ectoderm to form neural tissue as a default pathway. The ventral ectoderm of *Xenopus* and other vertebrate embryos was subsequently shown to express several BMPs ligands, consistent with their epidermalizing effects (Fig. 14.3).

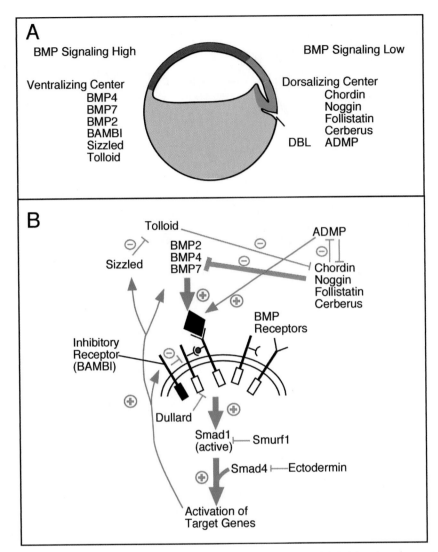

FIGURE 14.3   The BMP signalling pathway and its role in DV patterning of the ectoderm (A) At blastula-stages embryos, have a ventralizing center that promotes BMP signalling, and a dorsalizing center that induces neural tissue by inhibiting BMP signalling. (B) Schematic showing the various components of the BMP signalling pathway. The core components are the ligands (BMP2,4,7) that activate a signal transduction pathway (large green arrows), by binding transmembrane receptors. The BMP receptors signal by activating Smad1, allowing it to dimerize with Smad4 and activate gene expression. Transcriptional targets are genes involved in epidermal differentiation, but also components in the pathway as indicated by the green arrows. Also shown are various molecules that regulate the activity of the pathway, through complex interactions that allow self-regulation, as indicated by positive (green) and red (negative) arrows. The key negative regulators are BMP antagonists, such as chordin, noggin, follistatin and cereberus, which bind to ligand and inhibit their activity (large red arrow). Chordin activity on the ventral side is negatively regulated by a metalloprotease, Tolloid, which in turn is negatively regulated by Sizzled. ADMP is a BMP ligand that is expressed on the dorsal side, and show cross-inhibitory interactions with Chordin. BAMBI is an inhibitory receptor expressed on the ventral side to dampen BMP signalling. Finally Smurf1, Ectodermin, and Dullard target Smad1, Smad4, and the receptors, respectively, to degradation.

The default model gained further support from the discovery that Spemann's organizer is a rich source of BMP antagonists (Fig. 14.3A). At least four secreted polypeptide—follistatin, noggin, chordin, and Cerberus—are expressed in the organizer region of *Xenopus*. All four can bind tightly to the BMP ligands and prevent activation of the receptors, and all four act as potent neural inducers when expressed in the iso-lated ectoderm assay (Lamb *et al.*, 1993). At first, genetic tests of these putative neural inducers in other species were unimpressive since mutations that eliminate only one of these inhibitors tend to have relatively mild phenotypes on their own. For example, a loss-of-function mutation in Zebrafish chordin (the *chordino* mutant) causes only a reduction in the size of the neural plate. Similarly, mouse embryos that lack just one the BMP antagonists, chordin or noggin, by knockout mutations (Box 14.1) have a relatively normal nervous system. However, the full potential of these antagonists has only become apparent when multiple antagonists are removed at the same time, mainly by using an approach where embryos are injected with small, stable olignucleotides, called morpholinos, which bind to and inhibit the expression of target mRNAs. A complete loss of neural tissue is then observed when all three BMP antagonists, Chordin, Follastatin and Noggin, are simultaneously targeted using morpholinos, both in *Xenopus* and in Zebrafish. Thus, the production of multiple BMP antagonists within Spemann's organizer creates a dorsalizing center on one side of the embryo that inhibits the epidermalizing effects of BMPs produced in a ventralizing center on the other (Fig. 14.3A).

Why does Spemann's organizer express multiple BMP antagonists that apparently compensate for each other, when in principle one would do? Spemann coined the phrase "double-assurance" to emphasize the importance of redundancy in developmental mechanism as a way to reliably pattern the early embryo. Redundancy in the BMP pathway, moreover, is used as a mechanism to self-regulate, thus ensuring the appropriate gradient of BMP signaling is established across the DV axis of the ectoderm reliably, time after time (De Robertis, 2006). For example, targeting three BMPs, BMP2/4/7, with morpholinos in *Xenopus* embryos results in marked expansion of neural tissue, but some ventral, epidermal tissue remains. The surprising result is that ubiquitous neural tissue along the entire DV axis is only observed when a fourth ligand, ADMP, is also targeted, except that ADMP is not expressed in the ventralizing center with BMP2/4/7 but on the dorsal side where neural tissue forms. This paradox is resolved by the observation that BMP signaling promotes the expression of BMP ligands ventrally but represses the expression of ADMP dorsally. Thus, dorsal expression of ADMP is strongly upregulated when BMP2/4/7 are targeted by morpholinos, thereby compensating for the decrease in the source of BMPs ventrally, and reestablishing ventral fates. This is one of several such regulatory interactions that enables the BMP patterning system to self-regulate, thus ensuring that embryos consistently form proper amounts of neural and epidermal tissue.

Additional Pathways Involved in Neural Induction

While the basic tenets of the default model have stood up over time, the current picture of neural induction has increasingly become more complicated as additional neural inducers have been identified. One of the most important of these is fibroblast growth factor (FGF), a ligand that binds to its receptors and signals via the MAP kinase cascade. FGF has been proposed to have several potential roles during neural induction, as a "priming" signal and as an inducer of posterior neural tissue, which forms from a region of the blastula that lie at some distance from the organizer region. How FGF induces neural tissue remains somewhat controversial. One model is that FGF signaling acts by suppressing BMP signaling by targeting Smad1, either directly by phosphorylating a critical linker region, or indirectly by inducing the expression of a protein called SIP1, a zinc-finger, homedomain protein, that binds to and represses the transcriptional activity of the Smad proteins. Alternatively, FGF signaling is likely to induce neural tissue independent of its effects on BMP signaling, although the exact details of this mechanism remain unexplored. Currently it remains unclear whether there are other signaling pathways such as FGF that induce neural tissue by novel mechanisms or whether all inducers ultimately act by somehow inhibiting BMP signaling.

The default model correctly underscores the importance of inhibiting BMP activity in neural induction, but unfortunately de-emphasizes other changes that are required for neural tissue to form (Stern, 2006). Experiments in the chick embryo for example have shown that inhibiting BMP signaling is some regions of the ectoderm is not sufficient to induce a neural fate, suggesting that additional signals may act earlier in embryogenesis to create the potential for forming neural tissue. In addition, the ectodermal cells that form the neural plate activate the expression of genes that mark the formation of neural precursors, such as those encoding the Sox family of the HMG transcription factors. The transcriptional activation of these and other neural precursor genes requires a complex constellation of transcriptional regulators that need to be deployed when ectodermal cells take on a neural fate as a "default" pathway.

Summary

The nervous system first arises in the vertebrate embryo from a region of ectoderm that is induced to form the neuroepithelium of the neural plate and tube rather than differentiating into epidermis. Classical transplantation studies have shown that dorsal ectoderm forms neural tissue in response to signals from the organizer. More recent studies have shown that ventral ectoderm undergoes epidermal differentiation in response to multiple, endogenously produced BMP ligands. Spemann's organizer expresses multiple BMP antagonists that act redundantly to induce neural tissue. This redundancy is the bases for the complex mechanisms of self-regulation ensuring that the BMP signaling pathway is activated stably and correctly along the DV axis. Additional pathways are likely required during neural induction, including ones that establish gene expression required for neural precursor formation.

Early Neural Morphogenesis

Among the earliest changes that occur during the formation of the vertebrate nervous system are ones associated with morphological changes in cellular and tissue structure. At the cellular level, ectoderm cells lose their cuboidal shape and elongate into the pseudostratified, neuroepithelial cells that characterizes the morphology of neural precursors cells along the entire axis of the embryo. At a tissue level, the neuroepithelium of the neural plate undergoes the complex morphogenetic movements of *neurulation* to form the neural tube, which invaginates into the embryo, pinches off from the surrounding ectoderm and forms a separate tissue anlage. Neurulation is an exceedingly complex process and prone to failure for multiple reasons. Indeed, a failure to close the neural tube has devastating downstream consequences on neural development and the cause of neural tube defects (NTDs), a relatively common class of human birth defects.

In higher vertebrates, neurulation can be divided into two phases which differ in their morphological movements: a primary phase involving the brain and most of the spinal cord and a secondary phase involving more posterior regions of the spinal cord. While we are still quite far from understanding the complex tissue mechanics that underlie primary neurulation, some progress has been made in breaking down aspects of neurulation into cellular behaviors that occurs within subregions of the neuroepithelium. One of the earliest acting of these is convergent-extension (C-E), which occurs during gastrulation and neurulation and extends the neuroepithelium along the AP axis (Fig. 14.4). C-E is driven by organized cell intercalation much like the behavior of movie-goers when they exit a theater using only a central aisle (Keller, 2002). In a similar manner, neural plate cells intercalate pass each other to narrow the neural anlage medio-laterally by about ten-fold, while extending it at least as much along the AP axis of the embryo. C-E is known to require the planar cell polarity (PCP) pathway, a conserved group of proteins that were originally identified in the fly, Drosophila, as mutations that cause a defect in the planar organization of hairs on the wing (Wang and Nathans, 2007). These proteins localize asymmetrically to opposite sides of epithelial cells along the planar polar axis, where they influence the planar orientation of the cytoskeleton. In a similar manner, vertebrate homologs of these proteins are required during C-E to organize the cytoskeleton of migrating cells along the medial-lateral axis of the neural plate in a manner that orients polarized cell rearrangements. Consequentially, mutations of genes in the PCP pathway in *Xenopus*, Zebrafish or the Mouse disable C-E, resulting in an abnormally wide neural plate, a subsequent failure to elevate the neural folds at the edges, and an open neural tube.

As C-E extends the neural plate along the AP axis, other localized cell behaviors along the DV axis contribute to the buckling of the plate into a tube. A hinge point forms at the midline of the neural plate as well as at more dorsal positions (Fig. 14.4), by apical constriction of actin micofilaments in neuropithelial cells to produce a wedge-shaped morphology. Accordingly, mutations in various proteins that regulate microfilament dynamics are commonly associated with NTDs in the mouse. Shaping of the neural tube also requires localized regions of apoptosis, which if blocked, leads to NTDs. Finally, neural tube closure culminates with the fusion of the neural folds, which also give rise to a population of migrating neural progenitors, the neural crest. Mutations in the mouse that alter neural crest formation and emigration are also capable of producing NTDs. Even with this short summary one can appreciate that integrating these various aspects of cell behavior into a unifying mechanical model of neurulation is a formidable task although one required for a better understanding of this common source of human birth defects.

In the next section, we will discuss the various mechanisms that subdivide the neuroepithelium into different regions along the neuraxis during neural patterning. At this point, we emphasize that among the earliest readouts of this patterning are regional differences in morphogenetic behavior. For instance, the neural tube is wider at its amterior end where the brain forms than at the posterior end, which forms the spinal cord. This difference in shape arises in part because C-E is much more pronounced in posterior regions, presumably as a result of early patterning events along the AP axis. Another obvious example is the early steps in eye formation. The eye anlagen are first apparent during neurulation, when the neuroepithelium at the level of the prospective diencephalon evaginates to form the bilaterally paired eye vesicles. At the point where the neural tube and eye vesicle join, the neuroepithelium eventually pinches together to form the optic stalk while the eye vesicle forms a cup and

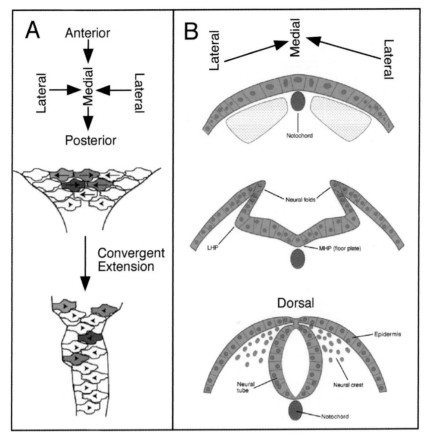

FIGURE 14.4 Morphogenesis of the neural plate and tube. (A) The formation of the neural plate and tube requires an organized intercalation of cells along the medial-lateral axis (convergence) which extends the neural anlage along the anterior-posterior axis (extension). The polarized cell behaviours that drive this process require the PCP pathway. (B) Neurulation of the neural plate into a tube requires several hinge points along the DV axis, including one at the forming ventral midline (MHP) which becomes the floor plate, (comma) as well as more dorsal ones (LHP). These hinge points allow the neural plate to buckle and the neural folds to elevate and fuse at the dorsal midline. At the same time, the neural crest forms (red cells) and emigrate.

divides further into an inner layer that gives rise to the retina and an outer layer that will form the pigmented epithelium. Again, these regionalized morphogenetic processes are downstream consequences of the early patterning of the neural plate into different regions although how patterning controls morphogenesis remains largely unexplored.

Summary

Formation of the vertebrate nervous system begins with a series of striking changes in cell and tissue morphology as the ectoderm forms the neural plate and tube. Neural tube formation involves the complex morphogenetic process of neurulation that requires coordinated changes in cell shape, cell division, cell migration, cell death, and cell–cell contacts. One prominent cell behavior during neural plate and tube formation is required for CE, which narrows the neural anlage and extends it along the AP axis.

CE requires the PCP pathway to orient cell movement and intercalation behavior. Much remains to be learned, however, about how changes at the molecular, cellular, and tissue level combine to produce the complex morphological changes that occur during neural plate and tube formation.

EARLY NEURAL PATTERNING

As discussed earlier, the neural plate is a morphologically homogeneous sheet of epithelial cells derived from dorsal ectoderm, which acquires its neural potential and fate as a result of inductive signaling. As the neural plate rolls up and closes into a tube, a series of constrictions appear in its wall, subdividing the anterior end of the tube into a series of vesicles representing the anlagen of fore-, mid-, and hindbrain

(Fig. 14.8). Further subdivision ensues, most conspicuously in the hindbrain region (rhombencephalon), where a series of segment-like swellings, rhombomeres, are formed. Caudal to the hindbrain, the neural tube forms a long narrow cylinder that is the precursor of the spinal cord. These early morphological features of the neural tube dictate the overall plan of the CNS and predict its later regional specializations. The neuroepithelium then commences with the production of a huge diversity of region-specific cell types, each having a distinct identity in terms of morphology, axonal trajectory, synaptic specificity, neurotransmitter content, and so on. Different neuronal cell types also carry distinctive surface labels that may ensure accuracy of axonal navigation and the formation of appropriate connections with other cells.

Perhaps most strikingly, individual neurons or groups of similar neurons originate at predictable times and at precise positions within the various regions of the neural tube. In some cases, neurons remain in their position of origin during and following differentiation; in other cases, young neurons or their precursors are directed to migrate along stereotypic paths to settle in locations distant from their position of origin. Correct specification of this intricate spatial ordering, or *pattern*, of cells is crucial to later events in CNS development when neurons establish complex arrays of specific interconnection that constitute functional networks. Activity-dependent processes and regressive events, such as the pruning of axons and cell death, later reinforce and refine initial patterns of connectivity, but a high degree of precision is achieved from the outset, dependent on, and as a direct result of, appropriate cell patterning. How the different regions of the CNS, and the individual cell types they each contain, are assigned their identity in early development remains an outstanding problem in neurobiology.

Until recently, studies of the earliest developmental stages have been hampered severely by our inability to detect nascent pattern. However, discovery of a multitude of molecular markers that reveal subregions of the neuroepithelium has made it possible to visualize emergent heterogeneity in what was previously seen only as a "white sheet" of cells.

From its inception, the central nervous system is organized along orthogonal axes: longitudinal (anteroposterior or AP) and transverse. Neural tube folding and closure deflect the transverse axis from its original lateromedial orientation so that it becomes dorsoventral (DV) with respect to the body. The third axis is radial (apical–basal) and has little involvement in early patterning, as the radial organization of the tube is largely uniform throughout (with the principal exception of the cerebral and cerebellar cortices). In contrast, both the AP and DV axes are nonuniform. Different neuronal types appear at different positions in these two dimensions as if reading their grid references on a map. Indeed, a Cartesian coordinate system of *positional information* is a useful framework with which to visualize neural pattern. As shown later, various morphogens act as *positional signals* on one or other of the two main axes, usually by establishing gradients of activity as their concentration falls with distance from a localized source. Uncommitted neural precursor cells respond to the local morphogen concentration by expressing specific transcriptional mediator genes (e. g., *homeobox* genes) that encode the *positional value* of the region. In effect, cells measure their position by reading the strength of signals and finally adopt a specific fate that is appropriate for their AP/DV grid reference in the neuroepithelium. An assigned positional value not only directs differentiation but also, for migrating neurons, translocation away from the site of origin to a new location. As we will see in the following sections, the patterning mechanisms used to regionalize the DV axis are remarkably uniform along the length of the neural tube, but those that operate on the AP axis vary considerably between telencephalon and spinal cord.

Establishment of the AP Axis

The initial establishment of AP polarity along the neuraxis is coupled intimately to the establishment of the main body axis during early embryonic development. Although AP axis formation in the vertebrate embryo remains poorly understood, it may have conceptual similarities with the strategies used for axis determination in the *Drosophila* embryo, where genetic studies have produced a detailed understanding. In *Drosophila*, AP polarity is first established by a gradient of positional information produced by the maternal morphogen Bicoid emanating from the anterior pole of the egg. The gradient of the Bicoid transcription factor initiates a cascade of transcription factor activation that progressively subdivides the body axes further into smaller segmental units. As these repeat units are established, genes of the *Antennapedia/Bithorax* homeotic complex (HOM-C) act in a well-ordered manner to define unique segment identities. Similarly, formation of the AP body axis in vertebrates is likely to involve the imposition of a crude polarity of transcription factor expression, which is then refined at later stages into smaller domains of gene expression. Despite these conceptual similarities, the vertebrate embryo differs markedly from Drosophila where the AP axis is laid down essentially at the same time along the entire cylindrical-shaped embryo. In higher vertebrates, those parts of the neuraxis lying posterior to the hindbrain are "added-on" by dividing cells that extend the axis in an anterior to posterior direction in a progressive manner during gastrulation. As a consequence, AP patterning of the vertebrate nervous system is temporally more asynchronous and protracted, and spatially more complex (Stern *et al.,* 2006).

Embryology of Early AP Patterning

As discussed in an earlier section, the transplant experiments of Mangold and Spemann also revealed the role of organizer tissues in imparting AP polarity on the developing CNS. Specifically, organizer tissue can be subdivided into two parts: a head organizer that induces anterior structures (i.e., brain) and a trunk/tail organizer that induces posterior structures (i.e., spinal cord). The two types of organizer tissue differ in their position within blastula stage embryos, their time of involution during gastrulation, where they come to lie after gastrulation is complete, and what tissue derivatives they form (Fig. 14.5). In amphibian and fish embryos, head organizer tissue lies adjacent to the prospective anterior neural plate on the dorsal most side of the blastula (Niehrs, 2004). During gastrulation, this tissue moves internally in an anterior direction to form the prechordal mesendoderm (PME), lying underneath the ectoderm that is induced to form the *prechordal* neural plate and give rise to the most anterior end of the brain (forebrain) (Fig. 14.5A). In contrast, trunk/tail organizer tissue lies more superficially and laterally on the blastula fate map, involutes later during gastrulation, forms derivatives of the chordal mesoderm, such as the notochord and somites, and underlies the posterior neural plate, referred to as the *epichordal* neural plate (Fig. 14.5A. In the mouse and chick embryo, the prospective prechordal neural plate receives anteriorizing signals before the onset of gastrulation from underlying anterior visceral endoderm (AVE), or the hypoblast, respectively (Fig. 14.5B). These early anteriorizing signals along with later signals from involuting PME cells and/or from an organizer region called Hensen's node are required to induce the anterior neural plate. Much of the trunk/tail organizer in mouse and chick embryos resides in Hensen's node when it extends posteriorly during gastrulation, inducing epichordal neural plate in an anterior to posterior progression (Fig. 14.5B). Finally, another important source of anteriorizing signals in Zebrafish at later stages of neural patterning is the anterior edge of the forming neural plate (Wilson and Houart, 2004). Thus, depending on the vertebrate species, head and trunk/tail organizer capabilities may reside in multiple tissues that act at different times to influence the initial division of the neural plate along the AP axis into pre- and epichordal domains.

Molecular Basis of Early AP Patterning

As described above for neural induction, the study of early neural patterning has been aided by molecular tools to manipulate various signaling pathways in embryos, as well as the ability to reveal the formation of a nascent pattern based on the expression of genes that mark different regions of neural tissue. One major insight from these recent studies is that members of the secreted Wnt family act in combination with the anti-BMP neuralizing agents to induce posterior neural (Niehrs, 2004). The added twist is that head organizer tissues are a rich source of antagonists that inhibit the activity of the Wnt pathway, and play an important role in protecting the anterior neural plate from posteriorization.

The identification of signaling molecules that could posteriorize neural tissue was aided by assays using explants of isolated ectoderm from Xenopus embryos neuralized by BMP inhibitors, or explants of the early, anterior chick neural plate dissected from embryos. These explants typically express anterior neural markers, such as the homeobox gene, *Otx2*, but not more posterior neural markers indicative of the hindbrain or spinal cord. Treating these explants with members of the Wnt family however, suppresses anterior neural marker expression and upregulates the expression of posterior markers. The WNTs are expressed by trunk/tail organizer tissue, making them strong candidates as signals required for posteriorizing the central nervous system when it first forms (Fig. 14.5C).

A role for Wnt signaling in posteriorizing the nervous system is also supported by mutations that deregulate Wnt activity in Zebrafish embryos, thereby causing a loss of prechordal neural plate and expansion of the epichordal neural plate. This phenotype is observed in Zebrafish *masterblind* and in *headless*, both of which encode molecules that act to keep the Wnt pathway off in the anterior neural plate (Wilson and Houart, 2004). In addition, a number of secreted proteins have been identified that act as molecular sinks for the Wnt signaling pathway and are produced by various inducing tissues involved in anteriorizing the neural plate (Niehrs, 2004). Dickkopf is a secreted protein produced by the head organizer tissue in the frog, blocks Wnt signaling at the level of the receptors,

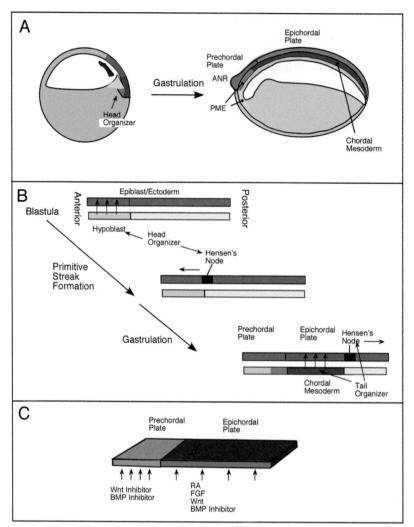

FIGURE 14.5 Early AP patterning of the vertebrate nervous system. (A) Early AP patterning of the amphibian nervous system involves signals emanating from different organizer tissues. The head organizer consists of anterior endoderm and head mesoderm that involutes underneath the prechordal neural plate during gastrulation. The epichordal neural plate is patterning by the chordal mesoderm that involutes later. (B) Similar mechanisms pattern the chick embryo. Early anteriorizing signals from the hypoblast begin to induce the prechordal neural plate, which are reinforced by signals from Hensen's node during primitive streak formation. During gastrulation, the regressing Hensen's node patterns the epichordal plate, along with the underlying chordal mesoderm. (C) Schematic representing the signals that are thought to divide the forming nervous system into a prechordal and epichordal neural plate.

and when mutant causes a loss of anterior neural tissue in the mouse. In Zebrafish, the anterior ridge of the neural plate has anteriorizing properties that can be attributed to the production of a secreted Wnt inhibitor (Wilson and Houart, 2004). The picture from these studies is that Wnt signaling needs to be suppressed during neural plate formation at the extreme anterior end in order for the prechordal neural plate to form (Fig. 14.5C).

In addition to the WNTs, the trunk/tail organizer produces other secreted molecules that may impart initial AP polarity on the neural plate as it first forms. One of these, FGF8, is required for extending the posterior axis, by maintaining a population of stem cells in Hensen's node that progressively emerge to form posterior neural tissue (Diez del Corral and Storey, 2004). One model is that exposure time to FGF8 defines AP axial position in the spinal cord, based on when stem cells leave the node and form the neural plate. Another molecule produced by the trunk/tail organizer is retinoic acid (RA) which signals through nuclear hormone receptors, has potent posteriorizing activity, and may have several roles in AP patterning of the hindbrain and spinal cord.

Summary

AP patterning of the CNS begins during the process of neural induction as dorsal ectoderm takes on a neural fate. This process divides nascent neural tissue into prechordal (anterior) and epichordal (posterior) neural plate regions based on signals that come from adjacent head and tail organizing tissues. Formation of the prechordal plate requires two inhibitory signals produced by the head organizer: one that inhibits BMP and the other WNT signaling. Tail organizer tissue produces potent posteriorizing agents, including WNTs, FGFs, and RA. The extent of signals required for generating AP polarity, however, is not fully known and the details of their action remain to be explored. Interestingly, both neural and anterior-neural are default states, requiring specific molecular activity to become nonneural (BMP) or posterior-neural (WNT).

REGIONALIZATION OF THE CENTRAL NERVOUS SYSTEM

Following the establishment of polarity along the AP axis of the embryo and the delineation of prechordal and epichordal regions of the neural plate, the AP axis becomes further regionalized into smaller and smaller domains, as revealed by the expression of developmental control genes. This process of progressive regional refinement involves two general classes of mechanism—the establishment of local organizers as sources of diffusible factors (morphogens) that inform neighboring cells about their position and fate, and the partitioning of the neuroepithelium into small modules or segments in which development can proceed with a degree of autonomy. In both cases, a conspicuous and important feature is the setting up of boundaries, which position a local organizer, contain cells within a compartment, or both. We will illustrate these patterning mechanisms by reference to selected examples. Our intention is to outline general principles rather than to provide a comprehensive review of nervous system patterning.

Definition of AP Pattern by Differential Homeobox Gene Expression

Central to the illumination of vertebrate CNS pattern formation has been the discovery that developmental control genes related to genes with a known patterning role in the Drosophila embryo are expressed in spatially restricted domains of the neural plate and tube. These regulatory genes, many of which encode homeodomain proteins, include the *Hox* genes, which specify positional value along the AP body axis of the fly embryo. In both flies and vertebrates the *Hox*

genes have a clustered chromosomal organization in which the relative position of a gene in the cluster reflects its boundary of expression along the AP axis. In vertebrates, the *Hox* genes are expressed in overlapping, or nested, domains along the AP axis of the early embryo, with those at the 3' ends of the cluster being expressed most anteriorly, in the hindbrain, where there is a precise correspondence between their anterior expression borders and the boundaries between the rhombomeres of the hindbrain neuroepithelium (Fig. 14.6).

Accumulating evidence supports the view that, by encoding positional value, *Hox* genes control the identity and phenotypic specializations of subregions of the epichordal neural tube in which they are expressed. Loss-of-function mutations of anteriorly expressed *Hox* genes result in malformations that represent a transformation of rhombomere identity. In the *Hoxb1* mutant mouse, for example, rhombomere (r) 4 (where the gene is normally expressed at a high level) loses its r4-specific character and takes on that of r2, where the gene is not normally expressed (Studer *et al.*, 1996). Similarly, overexpression of *Hoxb1* in r2 causes it to adopt phenotypic characters of r4 (Bell *et al.*, 1999).

Anterior to r2, the brain does not express *Hox* genes but displays spatially restricted expression of other transcriptional control genes whose homeoboxes are divergent from the *Hox* type. These genes are also highly conserved between flies and vertebrates. Two homologues of the Drosophila segmentation gene, *engrailed (en)*, are expressed in a broad region either side of the midbrain–hindbrain boundary (MHB), which later forms the cerebellum and the optic tectum (Fig. 14.7). Expression of the *En* genes is graded, being strongest at the MHB and declining both anteriorly and posteriorly. Morphological derivatives of the entire domain of *En* expression are deleted in *En1* knockout mice, showing that *En* function is crucially involved in the morphogenetic specification of the region (Wurst *et al.*, 1994).

The gap genes *orthodenticle* and *empty spiracles*, which function as homeotic selectors in the specification of particular head segments in Drosophila, have vertebrate homologues that are expressed in overlapping domains that encompass the entire rostral extremity of the AP axis with the exception of the ventral forebrain. The two *empty spiracles* homologues, *Emx*, are forebrain specific, whereas the two orthodenticle homologues, *Otx*, have wider expression, encompassing both forebrain and midbrain. The *Otx2* knockout mouse has an extreme phenotype that shows the importance of the gene but betrays little about its function—the entire head rostral to r3 is deleted. This is

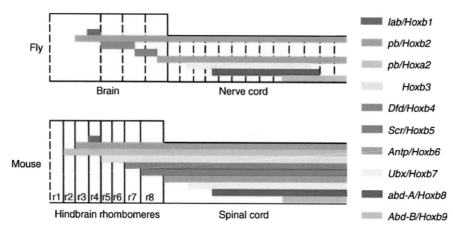

FIGURE 14.6 *Hox* gene expression domains in the CNS of fly and mouse. Nested domains of *Hox* genes along the AP axis of the Drosophila CNS closely parallel those of their homologues in mouse. Compare, for example, the fly gene *labial (lab)* with its mouse homologue *Hoxb1*. *Hox* genes specify a positional value along the AP axis, which is interpreted differently in fly and mouse in terms of downstream gene activation, resulting in neural structure; shared between the two organisms is the means of encoding the position of a cell along the AP axis. After Hirth *et al.*, (1998).

explained by the fact that *Otx2* is also expressed in the visceral endoderm that lies beneath the anterior ectoderm, where it is required for the induction of anterior neural structure during gastrulation (Simeone *et al.*, 2002). However, the gastrulation phenotype can be rescued by expressing *Otx* in the visceral endoderm while the ectoderm remains functionally null for *Otx2*. This reveals the brain-specific function of the gene, which is to maintain anterior neural identity; in the absence of neuroectodermal *Otx2*, the anterior brain becomes converted into hindbrain, with an enlarged cerebellum rather than forebrain forming the anterior end of the CNS.

An important later function of *Otx2* is to set the position of the midbrain–hindbrain boundary, whose local organizer functions are discussed later. The sharp posterior border of *Otx2* expression coincides with the anterior expression border of another homeobox gene, *Gbx2*, in the anterior hindbrain. Mutual or cross-repression between the two genes stabilizes the interface, forming a boundary where specialized signaling cells are generated. Experimentally extending the expression domain of *Otx2* into hindbrain territory causes the *Gbx2* border and MHB differentiation to retreat posteriorly, whereas extending the *Gbx2* domain into midbrain territory results in a corresponding anterior shift of the *Otx2* border (Rhinn and Brand, 2001).

Summary

Gene expression patterns have illuminated the time course and mechanisms underlying neural patterning. Functional studies show that homeobox genes direct AP pattern formation, establishing head-to-tail

regionalization. In the hindbrain, *Hox* genes encode subregional (rhombomere) identity. At the midbrain–hindbrain boundary, *Otx* and *Gbx* genes set up and position the MHB organizer, and *En* genes induced at the MHB confer AP polarity on the optic tectum, crucial to formation of the retinotopic map.

Local Organizers of AP Pattern

Homeobox genes and other classes of transcriptional mediator regulate position-specific development. How is the expression of these genes, or their upstream regulators, directed to specific domains of the AP axis? In many cases, it appears that long-range signals from local organizers are involved, directly or indirectly, in activating specific regional control elements of these genes at appropriate levels of the axis. In the case of *Hox* genes, a gradient of retinoic acid (RA) signaling has this role of positional signal (Glover *et al.*, 2006).

RA exerts its effects via multiple types of RA receptors (RARs and RXRs), members of the nuclear receptor superfamily. RA receptors are ligand-dependent transcription factors that bind as hetero-dimers (RAR+RXR) to RA response elements in the promoters of target genes, including *Hox* genes. In the early embryo, RA is produced by the somites that lie alongside the caudal hindbrain (r7, r8) and spinal cord, through the activity of a synthetic enzyme Raldh2. Expressed anterior to the hindbrain is another enzyme, Cyp26, which degrades RA. These appear to act as source and sink for a gradient of RA activity that traverses the AP length of the hindbrain. Thus, targeted mutation of *Raldh2* in mice and treatment of

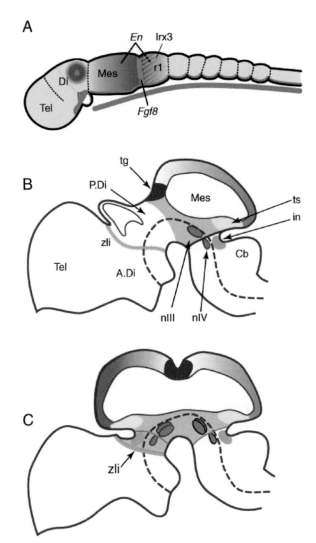

FIGURE 14.7 Role of FGF8 in mid/hindbrain patterning. (A) Implanting a bead that releases FGF8 protein (red) in the posterior diencephalon of an early chick embryo results in the induction of *En* (blue) and the transformation of normal posterior diencephalic territory (B) into midbrain (C). In the treated embryo (C), the posterior diencephalon forms a set of midbrain structures laid out in reverse AP polarity to the normal midbrain. This is thought to be due to the induced anterior-to-posterior gradient of EN protein that is the mirror image of the endogenous EN gradient. Also shown in (A) is the expression domain of the prepattern gene *Irx2* (green hatched), which establishes cerebellar competence in rhombomere 1. A.Di, anterior diencephalon; in, isthmic nuclei; Mes, midbrain; nIII, oculomotor nucleus; nIV, trochlear nucleus; P.Di, posterior diencephalon; r1, rhombomere 1; Tel, telencephalon; tg, tectal griseum; ts, torus semicircularis; zli, zona limitans intrathalamica. Data from Crossley *et al.,* (1996) and Matsumoto *et al.,* (2004).

chick embryos with an antagonist that specifically blocks all RARs both cause anteriorization of the hindbrain.

Immediately anterior to the expression domain of *Hox* genes is a territory, r1 and midbrain, that forms the cerebellum and optic tectum. A unitary process of specification for these adjacent regions is reflected both at the molecular level and by aspects of their developmental potential: the Drosophila *engrailed* homologues, *En1* and *En2* (as mentioned previously) and the *paired* homologues, *Pax2*, *Pax5*, and *Pax8*, are expressed in this domain. The development of midbrain and r1 is coordinated by the MHB: heterotopic grafts of MHB cells locally induce *En* expression in the host and change the fate of the host neuroepithelium such that it ultimately forms tectal structures (when grafted to the posterior diencephalon) or cerebellar structures (when grafted to the posterior hindbrain). These findings provide compelling evidence that a signal emanating from the MHB is involved in local AP patterning.

There are two prominent components of this organizing activity: WNT1, expressed in a transverse ring just anterior to the MHB, acts as a mitogen and to maintain the expression of *En* but is unable to mimic the activity of MHB grafts. FGF8, expressed immediately posterior to the *Wnt1* domain, has midbrain-inducing and polarizing abilities (Fig. 14.7). When a bead coated with recombinant FGF8 is implanted in the posterior diencephalon of chick embryos, *Fgf8*, *Wnt1*, and *En2* expression is induced in the surrounding cells. The posterior diencephalon then becomes completely transformed into the midbrain, whose AP polarity is reversed with respect to that of the normal host midbrain in accord with the gradient of induced *En* expression (Crossley *et al.*, 1996). Loss of FGF8 function in the zebrafish mutant *Ace* results in loss of the MHB, cerebellum, and part of the tectum. Thus, *Fgf8* expression at the MHB is necessary and sufficient to establish the polarized pattern of adjoining regions. The different responses of midbrain and hindbrain to the FGF8 signal is at least partially due to the prior expression of competence factors: expression of *Irx2* (a homologue of Drosophila *Iroquois*) in r1 is necessary and sufficient to direct the tissue to respond to FGF8 by making cerebellum. Ectopic expression of IRX2 in the midbrain transforms presumptive optic tectum into cerebellum on reception of the FGF8 signal, whereas expressing a constitutively repressive form of IRX2 in r1 transforms presumptive cerebellum into optic tectum. Interestingly, the role of FGF8 in this mechanism is to convert IRX2 protein from a repressor into an activator, which it does by phosphorylation via the MAP kinase pathway (Matsumoto *et al.*, 2004).

Diffusion of FGF8 away from the MHB (Scholpp and Brand, 2004) can be seen to act as an on-switch, setting the timing and extent of development of regions whose final identity has already been established by prepattern gene expression. An important feature of the FGF8-secreting cells is that they are positionally stabilized in the neuroepithelium by lineage restriction (Langenberg and Brand, 2005). Positional stabilization may be seen generally as a requirement for local signalling centers, ensuring that they can produce stereotypic and predictable molecular gradients in adjacent tissues (Irvine and Rauskolb, 2001).

Two further local organizers operate in more rostral regions of the neural tube where a number of early expressed developmental control genes are expressed in a patchwork quilt of small domains, subdividing the diencephalon and telencephalon (the main posterior and anterior subdivisions of the forebrain) into a number of transverse and longitudinal domains. Signaling at the anterior end of the axis involves FGF secretion from a region known as the anterior neural ridge (ANR), which is responsible for inducing and/or maintaining telencephalic (e.g., *Foxg1*) gene expression and for maintaining telencephalic identity (Rubenstein and Beachy, 1998). The ANR is probably a derivative of those early cells (identified as Row 1 in zebrafish, Houart *et al.*, 1998) that secrete antagonists to WNTs, whose posteriorizing activity must be reduced in order to establish anterior neural identity.

A second local organizer of forebrain pattern, the zona limitans intrathalamica (ZLI), lies at the middle of the prospective diencephalon, roughly midway between the ANR and MHB (Fig. 14.8). The ZLI secretes the signal molecule Sonic Hedgehog (SHH), which is responsible for inducing the expression of regional determinant genes both anteriorly (e.g., *Dlx2* in the

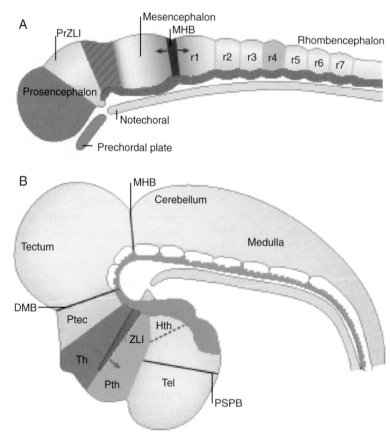

FIGURE 14.8 Boundaries and compartments in the avian embryo brain. (A) At an early stage cell lineage restriction exists on either side of the presumptive zona limitans intrathalamica (PrZLI), at the midbrain-hindbrain boundary (MHB) and between rhombomeres (r1-r7). Dark blue arrows represent FGF8 signaling to both the presumptive optic tectum (anterior) and presumptive cerebellum (r1, posterior). The presumptive ZLI is a wedge-shaped area devoid of Lunatic Fringe expression (blue) and lineage-restricted at its A and P borders. The pattern gene *Irx3* is expressed posterior to the presumptive ZLI (green hatched). (B) At a later stage, the wedge-shaped pr-ZLI has narrowed to form the definite ZLI, which now expresses Sonic hedgehog (red), inducing the expression of regional determinants in the prethalamus (Pth, anterior, light blue) and thalamus (Th, posterior, dark blue). DMB, diencephalic-mesencephalic boundary; Hth, hypothalamus; PSPB, pallial-subpallial boundary; Ptec, pretectum; Tel, telencephalon. After Kiecker and Lumsden (2005).

prethalmus) and posteriorly. (e.g., *Gbx2* in the thalamus). Like the MHB, the ZLI is positionally stabilized by cell lineage restriction and like the action of FGF8 at the MHB, the different inductive responses to SHH signaling from the ZLI are due, at least in part, to prepatterned differences in competence of the flanking regions: the *Iroquois* gene *Irx3* is expressed only posterior to the ZLI and directs thalamic differentiation via *Gbx2* on receipt of the Shh signal (Kiecker and Lumsden, 2005).

Summary

Local signaling mechanisms direct the spatial expression of homeobox genes, which appear to be crucial to the delineation and subsequent development of specific anteroposterior subregions of the CNS. Retinoic acid, diffusing from the cervical somites, is in part responsible for establishing the nested expression domains of the *Hox* homeobox genes in the hindbrain. FGF8, produced by cells at the midbrain/hindbrain boundary (MHB), is involved in the elaboration of midbrain and cerebellar structure. SHH, produced by cells at the Zona Limitans Intrathalamica (ZLI), is responsible for elaborating prethalamic and thalamic structure. For both the MHB and the ZLI, tissues lying anterior and posterior to the signaling center respond differently to the secreted morphogen because they have different competences, endowed by the expression of *Irx* prepattern genes.

Hindbrain Segmentation

Shortly after closure, a series of seven varicosities appear in the hindbrain neural tube. Although transient, these *rhombomeres* are true segments that play a crucial role in patterning (Lumsden and Keynes, 1989). Thus, the earliest formed neurons are laid out in stripes that match the morphological repeat pattern, with neurogenesis starting within the confines of alternate, even-numbered rhombomeres, and only appearing later in odd-numbered rhombomeres—a 'two-segment repeat' pattern (Fig. 14.9). Later in development, these segmental origins become obscured as the motor nuclei condense and migrate bodily to new positions. Segmentation is a developmental mechanism for specifying the pattern of developing structures, not necessarily for deploying those structures in the adult. Subdivision of a tissue or large region by axial segmentation (metamerism) involves the allocation of defined sets of precursor cells into an axially repeated set of similar modules. A developmental strategy adopted independently by many animal phyla, segmentation offers the advantages that organizational fields remain small and specializations of cell type and pattern can be generated as individual segmental variations on the repetitive theme. In a segmented system, precise boundaries can be set for both cellular assemblies and realms of gene action.

Segmentation of the neural tube cannot proceed in the same way as for the mesoderm, where physically separate somites are formed, because the tube has to retain epithelial continuity, not least as a conduit for extending axons. Rather, the process must involve some mechanism that restricts cell mingling. Vital dye-marking experiments have revealed such a lineage restriction mechanism in the hindbrain: clonal descendants of single marked cells disperse widely within the neuroepithelium, but the spreading cell clone always remains within a single rhombomere, confined at its boundaries (Fraser *et al.*, 1990). The vertebrate hindbrain thus shares with insects the phenomenon of modular construction using cell-tight developmental compartments, where lineage restriction prevents cells wandering from one rhombomere to another and blurring the resulting pattern. As we saw above, lineage restriction acts to stabilize the position of morphogen gradients emanating form local signaling centers; emerging evidence suggests that rhombomere boundaries also act as local signal centers, influencing neurogenesis within the rhombomere bodies (Amoyel *et al.* 2005).

What mechanism is responsible for segregating hindbrain cells into compartments? One possibility is that cells of adjacent rhombomeres have a different specification state, involving the expression of surface molecules that would favor affinity between the cells within a rhombomere but reduce affinity for cells in adjacent rhombomeres. Consistent with this, cultured cell aggregate experiments have shown that cells from even-numbered rhombomeres mix evenly with cells from other evens (as do odds with odds), whereas odd and even cells segregate from each other. Thus, rhombomeres may partition according to an adhesion differential, obeying a two-segment repeat rule. Candidate molecules for mediating reduced intercellular affinity or adhesion at rhombomere boundaries include the Eph family of receptor tyrosine kinases and their ephrin ligands (Xu *et al.*, 1999). The receptors are expressed in odd-numbered prerhombomeres, whereas the ligands are expressed in evens. This mutually exclusive pattern means that ligand–receptor interaction can only occur at forming boundaries, where it results in sharpening of the initially fuzzy interface between the adjacent domains (Fig. 14.10). Little is known about how the segmental pattern of the hindbrain is initially set out. Among the few known candidate regulators of segmental pattern is zinc finger protein Krox20, an upstream transcriptional regulator

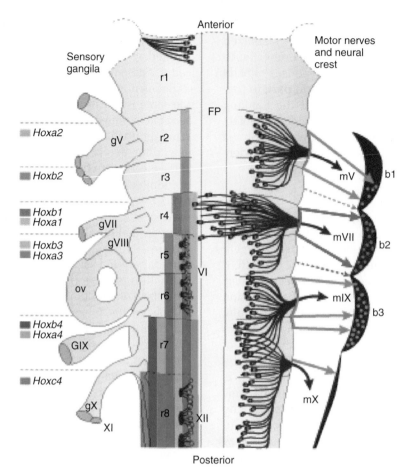

FIGURE 14.9 Distribution of motor neuronal types in the chick embryo hindbrain in relation to rhombomeres. Shown on the right side are branchiomotor neurons, forming in r2 + r3 (Vth cranial nerve, trigeminal), r4 + r5 (VIIth nerve, facial), and r6 + r7 (IXth nerve, glossopharyngeal), and contralaterally migrating efferent neurons of the VIIIth nerve (vestibuloacoustic), which are in the floor plate (FP) of r4 at the stage shown. Shown on the left side are somatic motor neurons, forming in r1 (IVth nerve, trochlear), r5 + r6 (VIth nerve, abducens), and r8 (XIIth nerve, hypoglossal). Cranial nerve entry/exit points and sensory ganglia associated with r2 (trigeminal), r4 (geniculate, vestibuloacoustic), r6 (superior), and r7 (jugular) are shown, as is the otic vesicle (ov). Colored bars represent the AP extent of *Hox* gene expression domains; note that one of these, *Hoxb1*, is expressed at a high level only in r4. Modified from Lumsden and Keynes (1989).

of Eph receptors, which is expressed in two stripes in the neural plate that later become r3 and r5.

Summary

Shortly after closure, the hindbrain neural tube becomes subdivided by transverse boundaries to form rhombomeres, a series of repetitive elements or metameres that have compartment properties. The existence of compartmental organization in the hindbrain suggests that cells become specified in segmental groups during their confinement in the ventricular zone and that they initially share a common identity, or ground state. First a row of similar boxes is formed and then the boxes are assigned individual identities according to their positional value, encoded by *Hox*

genes. The lineage-restricted boundaries between rhombomeres may also act as local signaling centers, influencing neurogenesis.

AP Pattern of Spinal Nerves

Spinal nerves form a ladder-like array on either side of the spinal cord, an obvious manifestation of segmentation, but here, in contrast to the intrinsic patterning mechanism that operates in the hindbrain, spinal segmentation is imposed extrinsically by somatic mesoderm that lies alongside. A serially reiterated asymmetry in the sclerotomal component of the somites allows axon to grow, and neural crest cells to migrate, only within the anterior half of each somite,

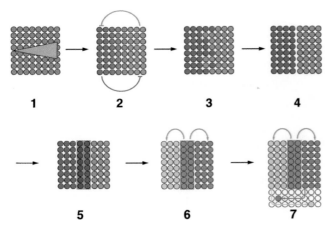

FIGURE 14.10 Model for neural boundary formation. 1. An initially uniform sheet of cells is polarized by a global signalling gradient (yellow), e.g., by WNT from the posterior pole of the neuraxis. 2. This results in a coarse prepattern of transcription factor expression, with blue induced at low signal level and red at a higher level. 3. Cross-repression between blue and red establishes two distinct populations of cells separated by a fuzzy interface. 4. Cell sorting processes depending on affinity molecules regulated by the red and blue transcription factors then sharpen the interface, generating a sharp interface and reducing adhesion red and blue cells. 5. Boundary cell phenotype involves expression of boundary-specific markers (shaded area). 6. The boundary cells then express signal molecules (green) that spread laterally and induce prepattern-dependent cell fates (yellow, violet). Later (7), postmitotic cells may migrate to new positions in the non-lineage restricted mantle layer. After Kiecker and Lumsden (2005).

where neural crest cells then condense to form dorsal root ganglia (DRG). The posterior half of each somite expresses glycoproteins that inhibit cell migration and cause growth cone repulsion or collapse. Thus, microsurgically substituting posterior halves for anterior halves results in the local absence of peripheral nerves, whereas the converse experiment results in enlarged DRG and motor nerves. Subdivision of the paraxial mesoderm into AP-polarized somites thus ensures a positional correspondence between the segmented dermomyotome on the one hand and its sensorimotor innervation on the other (Fig. 14.11).

Although it lacks intrinsic segmentation and has a superficial uniformity of organization, the developing spinal cord does manifest distinct AP variations in cellular subtype composition, particularly with respect to motor neurons arranged in discontinuous longitudinal columns. Posterior Hox genes have domains of expression in the spinal region that underlie this regional cellular diversity.

Local Organizers of DV Pattern

The entire neural tube has a characteristic dorsoventral zonation where different cell types differentiate stereotypically at different DV positions. The ventral midline of the spinal cord, hindbrain, midbrain and diencephalon, is formed by a palisade of specialized glia, the floor plate. Above the floor plate are five zones containing, in ventral to dorsal sequence, V3 interneurons, motor neurons, and three other types of interneuron, V2, V1, and V0 (Fig. 14.13A). Further subtypes of interneuron differentiate in the dorsal half, whereas the dorsal-most region, represented early on by neural folds that mark the transition between cells with neural and epidermal fates, forms the migratory neural crest cells that give rise to the glia and the majority of neurons in the peripheral nervous system. Later, after the neural crest has departed, the dorsal midline is populated by nonneurogenic roof plate cells.

The patterning of neuronal cell types has been studied intensively in the spinal neural tube, but the same molecules appear to be involved in regionalizing the DV axis of more anterior regions of the neuraxis. Unlike the AP axis, where local organizers signal position and fate to neighboring cells through single morphogen gradients, the DV axis has organizers at both dorsal and ventral poles and their respective morphogen gradients appear to counteract one another.

Ventral Organizers

Crucially involved in patterning the ventral neural tube is the notochord, a mesodermal skeletal structure that occupies the midline of the embryo directly beneath the neurectoderm. Grafting experiments in avian embryos have shown that both floor plate and motor neuron differentiation depend on notochord signals. Early removal of the notochord results in a normal-sized spinal cord in which both of these ventral cell types are absent, with dorsal cell types and dorsal-specific markers appearing in their place. Similarly, implanting a supernumerary notochord alongside and in contact with the lateral neural plate results in formation of an additional group of floor plate cells at the point of contact, with clusters of motor neurons on either side (Fig. 14.12A). These experiments show not only the power of the ventral midline signal to influence fate choice, but also the multipotent competence of responding neural tube cells at different DV positions. At a slightly later developmental stage, the floor plate itself acquires the same inductive capabilities—it can also induce motor neurons and will induce itself homeogenetically. The floor plate thus becomes an organizing center for a ventral pattern that is built into the neural tube itself.

Although it first appeared that motor neuron and floor plate induction would require different signals, one diffusible and the other contact dependent, a

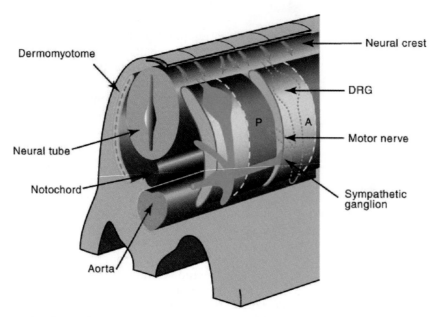

FIGURE 14.11 Segmented pathways for motor axon growth and neural crest migration in the trunk region. The AP positions at which motor axons collect to form a ventral root motor nerve and crest cell aggregate to form a dorsal root ganglion (DRG) are determined by somites. By the stage at which these constituents of the peripheral nervous system appear, the medial (sclerotomal) region of somites has dispersed and surrounds the notochord and ventral neural tube. The sclerotome is divided into anterior (light gray) and posterior (dark gray) halves, which are distinct from each other according to cell density and molecular markers. All components of the peripheral nervous system are confined to the anterior half sclerotome of each somite, which is permissive for the migration of neural crest cells (arrows), some of which condense to form DRG, and the ingrowth of motor neuron growth cones.

single molecule can account for both processes (for a review, see Briscoe and Ericson, 2001). *Sonic hedgehog (Shh)* is expressed first in the notochord and then in the floor plate and can elicit ectopic floor plate and motor neuron differentiation when misexpressed in the dorsal neural tube. Similarly, an activated form of Smoothened, the signal-transducing component of the SHH receptor, acts in cell autonomous fashion to produce ventral cell types ectopically. Explant studies have shown that the choice of cell fate by progenitor cells is influenced by the concentration of the diffusible Shh protein to which they are exposed: threefold incremental change in Shh concentration can result in the generation of five distinct classes of ventral neural tube cell types in vitro, including, at the higher end of the range, motor neurons, V3 interneurons, and floor plate cells (Fig. 14.13B). Midventral neural plate cells that are contacted by the notochord are thus likely to be exposed to a high local concentration of SHH, exceeding the threshold for floor plate induction, whereas the lower levels of SHH that diffuse from floor plate are sufficient to induce motor neuron and ventral interneuron differentiation, but insufficient to induce the floor plate. Antibodies that block SHH activity abolish the induction of ventral cell types showing that SHH is necessary as well as sufficient for establishing the ventral polarity of the neural tube.

The ability of midline signals to influence development of the ventral neural tube is not restricted to the spinal cord. In addition to inducing motor neurons and floor plate in the hindbrain and midbrain, midline signals also are involved in the development of AP region-specific neuronal subpopulations. Serotonergic neurons of the hindbrain raphé nucleus and dopaminergic neurons of the midbrain substantia nigra both develop close to the floor plate and can be induced to form in competent neuroepithelium by the notochord, floor plate, or SHH protein. The issue of how ventral midline signaling can elicit different responses at different AP positions is considered later.

Dorsal Organizers

In genetically or surgically notochordless animals, dorsal markers are expressed in the ventral spinal cord, showing that cell pattern in the dorsal half of the spinal cord does not require notochord signals. However, dorsal cell types such as neural crest and roof plate, do not develop by default, but are induced by an interaction that initially takes place between the neurectoderm and the flanking epidermal ectoderm when the two epithelia are still contiguous (Fig. 14.12B). This dorsalizing activity appears to be mediated by multiple types of BMP protein, which are expressed in

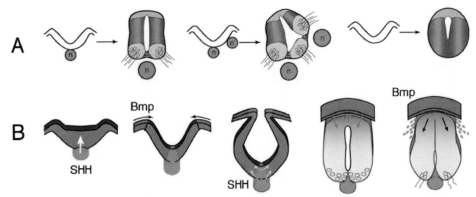

FIGURE 14.12 (A) Influence of ventral midline signals on spinal cord pattern. Cross sections through the developing chick spinal cord at the neural plate stage and resulting spinal cord, showing the effect of adding or removing notochord. (Left) Normal development: the floor plate (red) develops above the notochord (n) and motor neurons (yellow) differentiate in the adjacent ventrolateral region of the neural tube; *Pax6* (blue) is expressed in more dorsal regions. (Center) Grafting a donor notochord (n') alongside the folding neural plate results in formation of an additional floor plate and a third column of motor neurons. *Pax6* expression retreats from the transformed region. (Right) Removing the notochord from beneath the neural plate results in the permanent absence of both floor plate and motor neurons in the region of the extirpation. *Pax6* expression extends through the ventral region of the cord. (B) Sequence of stages in the formation of DV pattern in the spinal cord and hindbrain. Notochord-derived Shh protein (red) induces differentiation of the floor plate, which also expresses *Shh*. BMPs (dark blue) are expressed in epidermal ectoderm adjoining the neural plate. As the neural plate closes, neural crest cells individuate (light blue) at the junction between neural and epidermal ectoderm. At the early neural tube stage, *Isl1*–expressing motor neuron precursors (yellow) appear close to the floor plate and neural crest cells leave the dorsal tube and midline ectoderm through breaks in the basal lamina. *BMP* expression transfers to the dorsal neural tube Finally, motor neurons differentiate in the ventral cord. White and yellow arrows denote Shh signaling; black and purple arrows denote Bmp signaling.

the dorsal ectoderm and can mimic the ability of epidermal ectoderm to induce neural crest cells and roof plate differentiation. As for the ventral neural tube, where initial SHH activation is passed on to the floor plate by homeogenetic induction, so for the dorsal neural tube where BMPs are subsequently expressed in the roof plate and dorsal neural tube itself. BMP signaling is then responsible for inducing the several types of interneuron that differentiate at progressively more ventral positions in the dorsal half of the cord (Fig. 14.13A; Helms and Johnson, 2003; Liu and Niswander, 2005).

The DV axis is thus patterned by two counteracting gradients, emanating from the ventral and dorsal poles. At either pole, the transfer of signal molecule expression from notochord or epidermal ectoderm to the neuroepithelium attends the physical separation of the neural tube from the initial source, presumably ensuring a more precise control of concentration within the tube.

Definition of DV Pattern by Differential Homeobox Gene Expression

At the time of neuronal differentiation, each of the five zones of neuronal cell types in the ventral neural

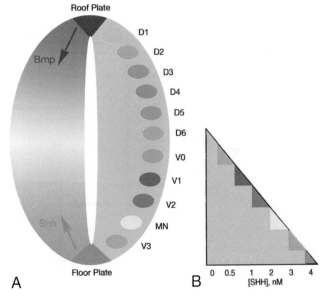

FIGURE 14.13 (A) Formation of different cell types at different DV positions in the spinal cord. Sonic hedgehog (SHH) is responsible for generating interneuron (V0 – V3) and motor neuron (MN) diversity in the ventral half of the spinal cord whereas BMPs are responsible for interneuron (D1-D6) diversity in the dorsal half. SHH protein (red), produced by the floor plate, and BMP protein (blue) by the roof plate act as counteracting graded signals, inducing different cell types at different DV positions *in vivo*. (B) Recombinant SHH protein has been shown to induce a similar range of responses and at different concentrations *in vitro*.

tube is defined by the combinatorial expression of homeobox genes. These genes are either induced (*Nkx2.2*, *Nkx6.1*) or repressed (*Pax6*, *Pax7*, *Dbx1*, *Dbx2*) at defined concentrations of SHH protein (Fig. 14.14). The differential responses to a graded SHH signal is thought to set up a pattern of fuzzy zones in the ventral half of the cord, which is later refined and sharpened by cross-repressive interactions between the homeobox genes themselves (see Briscoe and Ericson, 2001): the expression borders of complementary genes are seen to move following gain and loss of function of the opposing gene, as for *Otx2/Gbx2* at the MHB. Such shifts in the expression domains of these and other homeobox genes result in corresponding and predictable changes in neuronal identity, demonstrating the critical role of these homeobox genes as determinants of cell fate.

As for the AP axis, where positional determinants of identity are conserved with Drosophila, there are also striking similarities for the DV axis. For example, the homologue of *Nkx2.2*, *vnd*, is expressed close to the ventral midline of the fly embryo and is required for establishing the identity of its ventral-most neuroblasts (Cornell and von Ohlen, 2000).

Summary

Patterned cell differentiation along the DV axis of the neural tube involves the initial medial–lateral polarization of the neural plate and the later generation of distinct cell types at different DV positions under the spatial control and coordinate actions of an SHH-mediated ventralizing signal from the notochord (later from the floor plate) and BMP-mediated dorsalizing signals from the epidermal ectoderm (later from the roof plate). SHH is both necessary and sufficient for inducing a range of ventral cell types, including floor plate cells, motor neurons and interneurons, at different concentrations—it therefore has the prime characteristic of a morphogen. Patterning along the entire DV axis involves the intersection of opposing signals emanating from the two poles: BMPs act to limit the ventralizing activity of SHH, whereas SHH acts to limit the dorsalizing activity of BMPs.

Intersection of AP and DV Patterning Mechanisms

The early inductive signals that establish DV cell fate appear to be similar along the entire epichordal neuraxis. However, at a same DV position, there are marked differences in the identity of neurons at different AP positions, such as between oculomotor neurons and dopaminergic neurons of the midbrain and branchiomotor neurons of the hindbrain. That these distinct cell types arise in response to apparently similar levels of SHH signaling suggests that the choice of fate depends on the competence of the responding tissue—SHH must act within the context of previously established AP positional cues and previously specified AP positional values (Simon *et al.*, 1995).

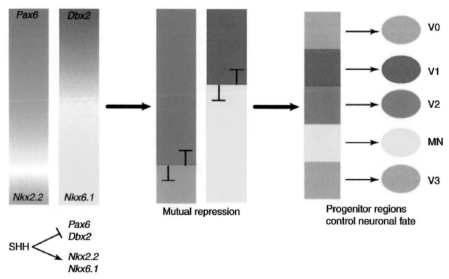

FIGURE 14.14   Model for ventral neural patterning. (Left) Graded SHH signaling from the ventral pole induces new expression of some homeobox genes (e.g., *Nkx2.2*, *Nkx6.1*) and represses the existing expression of others (e.g., *Pax6*, *Dbx2*). (Center) Cross-repressive interactions between pairs of transcription factors sharpen mutually exclusive expression domains. (Right) Profiles of homeobox gene expression define progenitor zones and control neuronal fate. After Briscoe and Ericson (2001).

DV Pattern in the Telencephalon

The DV axis of the telencephalon is conspicuously subdivided into a dorsal pallial region, which becomes the neocortex and archicortex (hippocampus) in mammals, and a ventral subpallial region that forms the basal ganglia. While the pallium remains a sheet-like roof over the lateral ventricles, proliferation in the subpallial region thickens the wall considerably, forming two swellings from which the basal ganglia arise: the lateral ganglionic eminence (LGE), which gives rise to the striatum, and the medial ganglionic eminence (MGE), which produces the globus pallidus. The pallial and subpallial moieties of the telencephalic vesicle are segregated by a longitudinal boundary, which marks the domain border of several dorsoventrally expressed developmental control genes and which may also act as a lineage restriction, at least during early stages of development (Fig. 14.15; Schuurmans and Guillemot, 2002; Campbell, 2003).

The telencephalon is devoid of motor neurons, has no floor plate, and is not underlain by notochord. The absence of these midline structures raises the question of how the bilateral organization of the forebrain and the differentiation of its ventral cell types are controlled. It appears, however, that even in this terminal expansion of the central nervous system a common mechanism is used for ventral patterning. SHH expression spreads rostrally in planar fashion from the diencephalic floor region into the ventral tel-

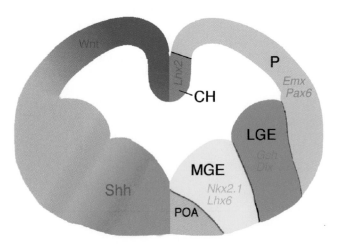

FIGURE 14.15 Cross section of the telencephalon of an early mouse embryo, with signalling on the left and DV regionalization on the right. Ventromedial patterning of medial and lateral ganglionic eminences (MGE, LGE) is influenced by SHH signalling from the ventral midline. BMPs and WNTs, produced by dorsomedial regions (roof plate, and medial pallium) and later from the cortical hem (CH), are implicated in pallium (P) specification. POA, preoptic area.

encephalon where it induces expression of the transcriptional mediator *Nkx2.1*, characteristic of the MGE (Fuccillo *et al.*, 2006). SHH from the MGE is subsequently required for generation of the more lateral region, the LGE, though some LGE markers, such as *Gsh* are still expressed in *Shh* null mutants. The different responses of medial versus lateral regions appear to reflect a change in the competence of telencephalic cells with time, with the resulting DV pattern reflecting the delay in reception of the SHH signal rather than a difference in SHH concentration. BMP and WNT signals emanating from the roof plate and medial pallium regulate tissue growth and the expression of dorsal telencephalic markers such as *Lhx2*, *Emx1* and *Emx2* (Campbell, 2003).

Patterning the Cortex

The adult mammalian neocortex, developing from the telencephalic pallium, comprises a very large number of anatomically and functionally discrete areas—more than 50 in human. One difficulty in addressing the question of how the elaborate areal pattern of the neocortex develops is that it is virtually uniform in structure and cytoarchitecture from one end to the other, at least in lower mammals such as mouse and rat. Furthermore, few known developmental control genes are expressed early enough to be candidate area determinants, and none of these is unique to a single prospective area. Thus, it has long been thought that the areal pattern is imposed on the neocortex by extrinsic elements, principally via the thalamic afferents that penetrate the cortex in an area-specific manner. However, it now appears that the cortical neuroepithelium acquires regional pattern prior to afferent innervation and in a manner not unlike other regions of the neural tube, with transcriptional mediators being expressed in a graded fashion in response to a local concentration of morphogens diffusing from focal sources. Putative signaling centers include the ANR, which secretes FGFs, and the posteromedial cortical hem region, which secretes WNTs and BMPs. In the absence of area-specific transcription factors it is likely that areal specification involves threshold levels of transcriptional control genes such as *Emx2* and *Pax6*, which are expressed in opposing gradients along the AP extent of the cortex (Fig. 14.16), and others such as *Lef1* (a mediator of Wnt signaling) that display graded expression from medial to lateral. In support of this, loss of *Emx2* function shrinks visual cortex (posterior) and expands sensory (mid) and motor (anterior) areas, whereas loss of *Pax6* function results in the opposite alteration. Thus, in contradiction to longstanding ideas, the cortical pattern is prob-

BOX 14.1

TRANSGENIC MICE AND ENGINEERED MUTATIONS

Rapid and complementary advances in the fields of molecular biology and experimental embryology have combined to offer neuroscience researchers unprecedented power to manipulate the mammalian genome. The technologies used for these manipulations have been worked out primarily in the laboratory mouse and fall into two basic classes: those used for transgenic mice and those used for embryonic stem cell chimeric mice.

Transgenic mice are created by the injection of a cloned DNA fragment into the male pronucleus of a recently fertilized mouse embryo. The fragment will integrate into the host genome and be passed through subsequent mitoses to all the cells of the adult, including the gametes. The integrated DNA fragment, now known as a "transgene," usually is engineered to contain a promoter and associated regulatory sequences, a structural gene, and a 3' polyadenylation signal. Transgenes add to the genome. As a genetic element, the chromosomal site of integration is random, and there is no wild-type allele on the sister chromosome. As an expressed locus, the transgene message is made over and above the endogenous gene expression pattern. This technique can be used as both an analytical and an experimental tool. Used as an analytical tool, the potency of a certain genetic element to direct cell- or tissue-specific gene expression can be determined by using the element to regulate marker genes such as β-galactosidase or green fluorescent protein. The genetic elements that regulate the temporal and spatial expression pattern of *Hox* genes in hindbrain, tyrosine hydroxylase in adrenergic neurons, and L7 in Purkinje cells have all been explored by this means. Used as an experimental tool, transgenes can exploit a genetic element with known specificity to deliver a gene product to an ectopic cell site or developmental time. Thus, the PDGF promoter has been used to drive the expression of human β-amyloid precursor protein, the L7 promoter has been used to deliver diphtheria toxin to differentiating Purkinje cells, and the β-actin promoter has been used to deliver *Hox-A1* to inappropriate sites in the developing embryo.

Embryonic stem cells (ES cells) are stable cell lines derived from the inner cell mass of the preimplantation embryo. They are totipotent, which means that if they are introduced into a host embryo, they can contribute to all cell types in the resulting chimera (including gametes). The use of homologous recombination in ES cells allows changes to be engineered in specific genetic loci in culture. By using modified cells to create chimeras, changes can be introduced into the mouse germline and propagated as new mutations. The mutations can be insertions, deletions, modifications, or any combination of the three. When the engineered insertion/deletion disables the normal allele, the resulting mutation often is referred to by the slang term "knockout." These techniques alter the genome. As a genetic element, the engineered locus replaces a specific gene locus, and there is a normal wild-type allele on the sister chromosome. Used in this way, a knockout mutation can be used to model an inherited disease. Lesch-Nyhan (HPRT-null), ataxia-telangectasia (ATM-null), and fragile-X mental retardation syndrome (FMR1-null) have all been modeled in this way. As an expressed locus, the knockout transcript is made instead of the wild-type gene product. This technique can thus be used to create a modified locus such that the targeted gene is mutated rather than destroyed, in the same way an endogenous transcript of one gene can be replaced with that of a different one. Thus, sequences encoding *Engrailed-2* have been inserted into the *Engrailed-1* locus in such a way that the *Engrailed-1* transcript is lost (a null mutation) and *Engrailed-2* is made in its place.

These are only some of the ways in which the powerful new technologies of transgenic and knockout mice are providing powerful genetic tools for use in neuroscience research.

Karl Herrup

ably set up early within the parent neuroepithelium and the role of thalamic innervation is secondary reinforcing, refining, or modifying this initial pattern (Ragsdale and Grove, 2001; O'Leary and Nakagawa, 2002).

CONCLUSIONS

Neural-inducing factors and modifiers produced during gastrulation have a basic role in establishing an initial crude AP regional identity in the neural plate

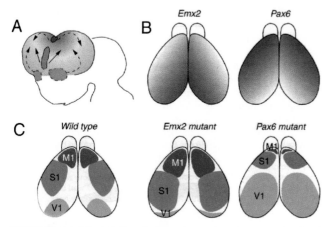

FIGURE 14.16 (A) Regulation of neocortical pattern by local signals: FGF8 (green) from the anterior neural ridge (ANR), WNTs and Bmps (blue) from the cortical hem; and SHH (red) from the medial ganglionic eminence. (B) These and other signals are thought to induce expression of the transcriptional factors *Emx2* and *Pax6* in opposed gradients. that control regional identity. (C) Loss of function of either of these genes results a specific shift of area identity from wild type (left), implying that visual cortex (V1) is specified by *Emx2* [high], *Pax6* [low] and that of motor cortex is specified by *Emx2* [low], *Pax6* [high]. After O'Leary and Nakagawa (2002).

that emerges from the dorsal surface of the embryo toward the end of this period. Abrogation of Wnt signaling, by depletion of ligand or blockade of signal transduction pathways, is important in conferring anterior neural identity. Wnts, Fgfs and RA cooperate to posteriorize the neuraxis. The resulting coarse-grained AP pattern is subsequently refined by the action of local organizers. Acting on both the AP and DV axes local organizers exert regional growth control and either induce or repress transcriptional mediators of regional fate. These mediators, in turn, may also interact among themselves, sharpening domains of action by cross-repression. The combined processes of segmentation and compartition provide an alternative or perhaps additional mechanism for sharpening gene expression borders and cellular domains. These events result in the precise delineation of regions and subregions. Although we tend to think of neuraxial patterning as a series of discrete steps leading to greater refinement, it is a continuous process that extends over a protracted period of development and involves a continuity of signaling systems (e.g., RA, FGFs, SHH, BMPs). We have also presented this subject as if pattern is acquired separately on AP and DV axes, whereas it is clear that cell specification on the DV axis must intersect with and act on already existing AP positional values. It is also clear that patterning mechanisms must be integrated with those that control neurogenesis, a topic that is considered in Chapter 15.

References

Amoyel, M., *et al.* (2005) Wnt regulates neurogenesis and mediates lateral inhibition of boundary cell specification in the zebrafish hindbrain. *Development* **132**, 775–785

Ashe, H.L., and Briscoe, J. (2006) The interpretation of morphogen gradients. *Development* **133**, 385–394.

Bier, E. (1997) Anti-neural-inhibition: a conserved mechanism for neural induction. *Cell* **89**, 681–684

Bell, E. *et al.* (1999) Homeotic transformation of rhombomere identity after localized *Hoxb1* misexpression. *Science* **284**, 2168–2171.

Briscoe, J., and Ericson, J. (2001) Specification of neuronal fates in the ventral neural tube. *Curr. Opin. Neurobiol.* **11**, 43–49.

Campbell, K. (2003) Dorsal-ventral patterning in the mammalian telencephalon. *Curr. Opin. Neurobiol.* **13**, 50–56.

Cornell, R. A., and von Ohlen, T. (2000) Vnd/nkx, ind/gsh, and msh/msx: Conserved regulators of dorsoventral neural patterning? *Curr. Opin. Neurobiol.* **10**, 63–71.

Crossley, P. H., *et al.* (1996) Midbrain development induced by Fgf8 in the chick embryo. *Nature* **380**, 66–68.

Diez del Corral, R, and Story, K. G. (2004) Opposing FGF and retinoid pathways: a signalling switch that controls differentiation and patterning onset in the extending vertebrate body axis. *Bioessay.* **26**, 857–869.

DeRobertis, E.M. (2006) Spemann's organizer and self-regulation in amphibian embryos. *Nat. Rev. Mol. Cell Bioo.* **7**, 265–267

Fraser, S. E., *et al.* (1990) Segmentation in the chick embryo hindbrain is defined by cell lineage restrictions. *Nature* **344**, 431–435.

Fucillo, M., *et al.* (2006) Morphogen to mitogen: the multiple roles of hedgehog signalling in vertebrate neural development. *Nature Rev. Neurosci.* **7**, 772–783.

Glover, J., *et al.* (2006) Retinoic acid and hindbrain patterning. *J. Neurobiol.* **66**, 705–725.

Helms, A.W., and Johnson, J. (2003) Specification of dorsal spinal cord interneurons. *Curr. Opin. Neurobiol.* **13**, 42–49.

Hirth, F., *et al.* (1998) Homeotic gene action in embryonic brain development of Drosophila. *Development* **125**, 1579–1589.

Houart, C., *et al.* (1998) A small population of anterior cells patterns the forebrain during zebrafish gastrulation. *Nature* **391**, 788–792.

Irvine, K. D., and Rauskolb, C. (2001) Boundaries in development: Formation and function. *Annu. Rev. Cell Dev. Biol.* **17**, 189–214.

Keller, R. (2002) Shaping the vertebrate body plan by polarized embryonic cell movements *Science* **298**, 1950–1954.

Kiecker, C., and Lumsden, A. (2005) Compartments and their boundaries in vertebrate brain development. *Nature Rev. Neurosci.* **6**, 553–564.

Khotz, J., *et al.* (1998) Regionalization within the mammalian telencephalon is mediated by changes in responsiveness to Sonic hedgehog. *Development* **125**, 5079–5089.

Langenberg, T., and Brand, M. (2005) Lineage restriction maintains a stable organizer cell population at the zebrafish midbrain-hindbrain boundary. *Development* **132**, 3209–3216.

Lamb, T.M., Knecht, A.K., Smith, W.C., Stachel, S.E., Economides, A.N., Stahl, N., Yancopolous, G.D., Harland, R.M. (1993) Neural induction by the secreted polypeptide Noggin. *Science* **262**: 713–718.

Lee, A., and Niswander, L.A. (2005) Bone morphogenetic protein signalling and vertebrate nervous system development. *Nature Rev. Neurosci.* **6**, 945–954.

Lumsden, A., and Keynes, R. (1989) Segmental patterns of neuronal development in the chick hindbrain. *Nature* **337**, 424–428.

Lumsden, A., and Krumlauf, R. (1996) Patterning the vertebrate neuraxis. *Science* **274**, 1109–1115.

Lupo, G., *et al.* (2006) Mechanisms of ventral patterning in the vertebrate nervous system. *Nature Rev. neurosci.* **7**, 103–114.

Matsumoto, K., *et al.* (2004) The prepattern transcription factor Irx2, a target of the FGF8/MAPK cascade, is involved in cerebellum formation. *Nature Neurosci.* **7**, 605–612.

O'Leary, D.D.M., and Nakagawa, Y. (2002) Patterning centers, regulatory genes and extrinsic mechanisms controlling arealization of the neocortex. *Curr. Opin. Neurobiol.* **12**, 14–25.

Niehrs, C. (2004) Regional specific induction by Spemann-Mangold organizer. *Nat. Rev. Genet.* **5**, 425–434.

Ragsdale, C. W., and Grove, E. A. (2001) Patterning the mammalian cerebral cortex. *Curr. Opin. Neurobiol.* **11**, 50–58.

Rhinn, M., and Brand, M. (2001) The midbrain-hindbrain boundary organizer. *Curr. Opin. Neurobiol.* **11**, 34–42.

Rhinn, M., *et al.* (2006) Global and local mechanisms of forebrain and midbrain patterning. *Curr. Opin. Neurobiol.* **16**, 5–12.

Rubenstein, J. L. R., and Beachy, P. A. (1998) Patterning the embryonic forebrain. *Curr. Opin. Neurobiol.* **8**, 18–26.

Scholpp, S.,and Brand, M. (2004) Endocytosis controls spreading and effective signalling range of Fgf8 protein. *Curr. Biol.* **14**, 1834–1841.

Schuurmans, C., and Guillemot, F. (2002) Molecular mechanisms underlying cell fate specification in the developing telencephalon. *Curr. Opin. Neurobiol.* **12**, 26–34.

Simeone, A., *et al.* (2002) The Otx family. *Curr. Opin. Genet. Dev.* **12**, 409–415.

Simon, H., *et al.* (1995) Independent assignment of anteroposterior and dorsoventral positional values in the developing chick hindbrain. *Curr. Biol.* **5**, 205–214.

Stern, C.D. (2006) Neural induction: 10 years after the default model. *Curr. Opin. Cell Biol.* **18**, 692–697

Stern CD, Charite J, Deschamps J, Duboule D, Durston AJ, Kmita M, Nicolas JF, Palmeirim I, Smith JC, Wolpert L. (2006) Head-tail patterning of the vertebrate embryo: one, two or many unresolved problems. *Int. J. Dev. Biol.* **50**, 3–15.

Studer, M., *et al.* (1996) Altered segmental identity and abnormal migration of motor neurons in mice lacking Hoxb1. *Nature* **384**, 630–634.

Wang, Y. and Nathans, J. (2007) Tissue/planar cell polarity in vertebrates: new insights and new questions. *Development* **134**, 647–658.

Weinstein, D.C., and Hemmati-Brivanlou, A. (1999) Neural Induction. *Ann. Rev. Cell Dev. Biol.* **15**: 411–433.

Wilson, S.W. and Houart, C. (2004) Early steps in the formation of the forebrain. *Dev. Cell* **6**, 167–181.

Wurst, W., *et al.* (1994) Multiple developmental defects in *Engrailed-1* mutant mice: An early midhindbrain deletion and patterning defects in forelimbs and sternum. *Development* **120**, 2065–2075.

Xu, Q., *et al.* (1999) In vivo cell sorting in complementary segmental domains mediated by Eph receptors and ephrins. *Nature* **400**, 267–271.

Andrew Lumsden and Chris Kintner

15

Cellular Determination

In order to make a functional nervous system, the progenitor cells that proliferate throughout the induction and patterning of the neural primordium must eventually exit the cell cycle and differentiate into mature neurons of different types. There are a huge variety of cells types in the nervous system. Differences can be dramatic, such as neurons and glia, or even between types of neurons. Compare, for example, a cone cell in the retina and a cerebellar Purkinje cell. They have completely different shapes, patterns of connectivity, neurochemistry, and physiological characteristics. Differences can also be subtle, such as between types of motor neurons, which may differ only with regard to their axonal projection to different muscles and the synaptic inputs they receive.

The distribution of different types of neuron follows a highly invariant pattern in a given species. Neurons located at a given position within the nervous system are always the same type. In vertebrates, this is true on the level of populations of neurons (e.g., cells of layer III of the cortical area 17 are glutaminergic pyramidal cells projecting to cortical areas 18 and 19); in many invertebrates, it is true even on the level of individual, uniquely identifiable cells. Experimental studies carried out over the last few decades have provided substantial progress in understanding how the right types of neuron develop in the right place in the brain. This chapter attempts to summarize some of the main insights gained into the mechanisms controlling neuronal fate.

ORIGINS AND GENERATION OF NEURONAL PROGENITORS

Neurons are specified gradually by a series of fate choices that begin with their distant progenitors who make the first decision between neural and ectodermal fates. Progenitors of the nervous system arise from specialized regions of the ectoderm called the neurectoderm (Fig. 15.1A). A highly conserved molecular network, which includes the localized expression of BMP signaling molecules and their antagonists, specifies the size and position of these regions (Stern, 2006). In vertebrates, BMP is expressed in the ectoderm of the gastrulating embryo (Fig. 15.1A). Target genes of BMP signaling promote the differentiation of epidermis, one of the two major organ systems that develop from the ectoderm. BMP signaling is inhibited by BMP-antagonists, notably chordin and noggin, which are secreted from the dorsal part of the mesoderm, the so-called organizer. Ectoderm cells within reach of BMP antagonists do not embark on the epidermal pathway of differentiation. They instead develop as neurectoderm. In vertebrates, the determined neurectoderm, called the neural plate, invaginates to become the elongated, dorsally located neural tube from which brain and spinal cord develop (Fig. 15.1A–D).

In many invertebrates, including the fruit fly, *Drosophila*, most of the neurectoderm is located on the ventral side of the ectoderm (Fig. 15.1E) and gives rise to a ventral chain of segmental ganglia, called the ventral nerve cord, which is comparable to the vertebrate spinal cord, and the posterior part of the brain, which is comparable to the vertebrate hindbrain. The anterior brain of the fly, which is comparable to the mid/forebrain of vertebrates, is derived from the procephalic neurectoderm (Fig. 15.1E).

The neural tube of a vertebrate embryo forms a continuous epithelium that lines an inner lumen (ventricle; Fig. 15.1C,D). Cells of the neural tube first divide symmetrically in the plane of the epithelium (Fig. 15.1C'). After variable periods of time, they spin off postmitotic daughters that differentiate into neurons and glia. The generation of postmitotic neurons in some parts of the vertebrate neuroepithelium, such as

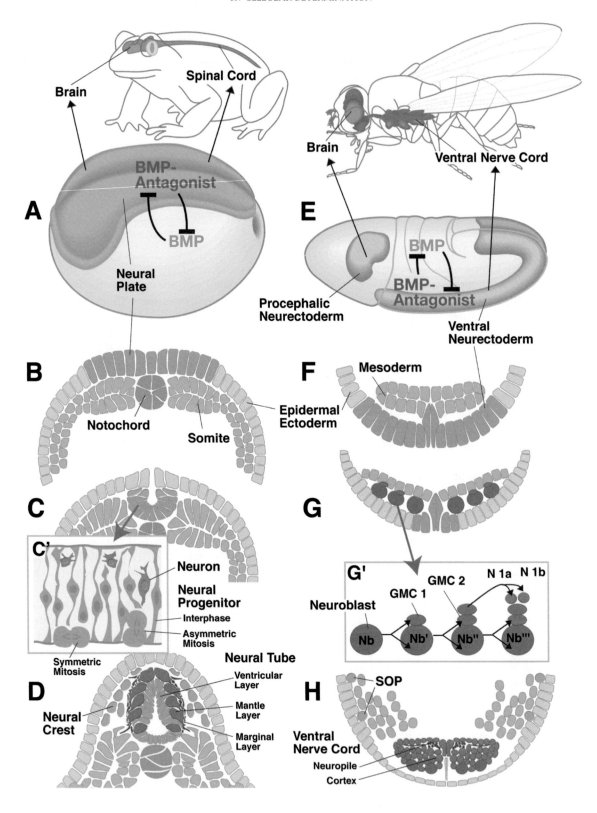

FIGURE 15.1 Synopsis of early neural development in vertebrates (example: the amphibian *Xenopus laevis*; A–D) and insects (example: the fruit fly *Drosophila melanogaster*; E–H). (A and E) Dorso-lateral views of embryos at the onset of neurulation. The neurectoderm is shaded light blue. The expression of the signal bone morphogenetic protein (BMP) and its antagonists is indicated. (B and F) Schematic cross-sections of embryos depicted in A and E. (C and G) Cross-sections at a later stage when neurulation is well under way. In vertebrates, the neural plate folds in to become the neural tube. Cells at the junction between the neural tube and the epidermal ectoderm (green) form the neural crest, which gives rise to the peripheral and autonomic nervous systems. The neural tube is a proliferating neuroepithelium (C'); symmetric mitoses increase the number of epithelial neural progenitor cells, asymmetric mitoses result in daughter cells that delaminate from the epithelium and become neurons (red). In insects, individual neural progenitors (neuroblasts; purple) delaminate from the neurectoderm. They divide in a stem cell mode (G'), producing stacks of daughter cells called ganglion mother cells (GMC). Each ganglion mother cell divides into two neurons (e.g., N1a and N1b). (D and H) Cross-sections of late embryos in which some neurons (red) have differentiated. In vertebrates, these neurons form the mantle layer that surrounds the neuroepithelium, now called ventricular layer. Neurites gather at the outside of the neural tube (marginal layer). In insects neuronal cell bodies form an outer layer (cortex); neurites gather in the center, forming the neuropile. Progenitors of the peripheral nervous system (sensory organ progenitors (SOPs); green) segregate from different locations in the epidermis.

the cortex, has been seen to be associated with asymmetric cleavages. The cell that remains next to the ventricle continues to divide as a neuroepithelial progenitor cell, while its sibling exits the cell cycle and migrates toward the outer surface of the neural tube (Fig. 15.1C'). Here young neurons form dendrites and axons. At this stage, the nervous system has reached a three-layered configuration of a ventricular layer, mantle layer, and marginal layer (Fig. 15.1D).

Progenitors of the central nervous system of *Drosophila*, called neuroblasts, separate from the neurectoderm and move inside the embryo as individual cells (Fig. 15.1G). Other ectodermal cells, which stay behind after the neuroblasts have migrated, are called dermoblasts and give rise to the epidermis. Mitotic divisions of *Drosophila* neuroblasts are asymmetric throughout development. Thus, each neuroblast divides unequally into one large and one small daughter cell. The large cell remains as a neuroblast and continues to divide asymmetrically for a number of rounds (Fig. 15.1G'). The small cell, called the ganglion mother cell (GMC), typically divides one more time and both of its daughter cells differentiate into mature neurons. Often, GMCs and immature neurons form a stack on top of the neuroblast from which they originated. Postmitotic neurons generally do not migrate, so the progeny of a neuroblast remain spatially close to one another, with the position of each neuron dependent on the position of the neuroblast and the time at which it was born. Thereby, a two-layered cortex-neuropil architecture typical of the mature ganglion is generated (Fig. 15.1H).

SPATIAL AND TEMPORAL COORDINATES OF NEURONAL SPECIFICATION

Neurogenesis proceeds quite differently in different parts of the neural primordium. For example, neuro-epithelial cells in the anterior part of the vertebrate neural tube (which includes the forebrain) undergo many more rounds of division than those of the posterior part (spinal cord). In addition, the maturing neurons have distinct phenotypes and establish different types of connections depending on where they are located. When pieces of neural tube are transplanted to new positions, features such as proliferative potential and cellular phenotype often are carried to the new position, implying that, in the neural primordium, intrinsic determinants profoundly influence the fate of neural progenitors. These determinants subdivide the neural primordium into a "geographic map" with many different domains, each one characterized by a unique identity. The mechanisms underlying the spatial regionalization, discussed in more detail in Chapter 14, are intricately involved in neural fate specification.

Along its anterior-posterior axis the neural tube is subdivided into transverse domains that give rise to morphologically defined partitions of the CNS, including the spinal cord, hindbrain, midbrain, and forebrain (Fig. 15.2B). Each of these major sections of the neural tube becomes further partitioned into neuromeres, each neuromere delineated by its own peripheral nerve. In the spinal cord of mammals, one distinguishes seven cervical (neck) neuromeres whose neurons innervate the upper extremity, 12 thoracic neuromeres, five lumbar neuromeres belonging to the lower extremity, and a variable number of sacral neuromeres. Seven or eight neuromeres are present in the hindbrain, where they are called rhombomeres. Similar to spinal neuromeres, rhombomeres are characterized by the pattern of peripheral nerves they emit. The midbrain is thought to represent a single neuromere. The forebrain possesses no peripheral nerves; however, based upon other morphological landmarks, as well as the expression of molecular markers, it is subdivided into six prosomeres. The posterior three prosomeres (P4, P5, and P6) give rise to the diencephalon and prosomeres 1, 2, and 3 form the telencephalon.

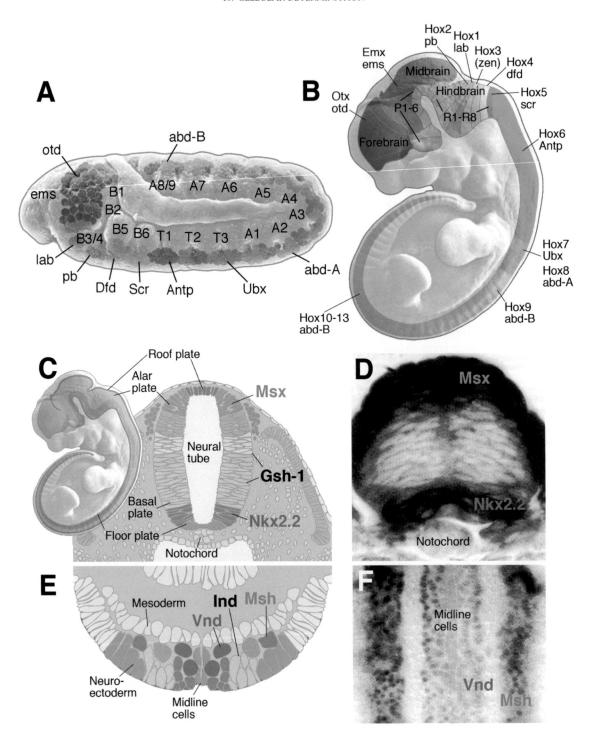

Along this anterior-posterior axis, Hox genes, which code for Homeobox transcription factors, and segment polarity genes (Chapter 14) are expressed in sharply demarcated domains in the neural primordia of both vertebrates and invertebrates. These genes provide neural progenitors with intrinsic "spatial information" that reflects their location along this axis (Fig. 15.2A,B).

Each neuromere is characterized by the expression of a distinct set of homeobox genes (a "Hox code") and thereby acquires its unique anterior posterior identity. Thus, knock out of a Hox gene in posterior segment of the nervous systems transforms the neurons there into more anterior types, whereas ectopic activation of these genes has the opposite effect. The Hox genes are

FIGURE 15.2 Specification of regional identity of the neural primordium. (A and B) Homeobox genes control regional identity along the antero-posterior axis. (A) *Drosophila*. The neurectoderm and neuroblast layer is subdivided into a series of neuromeres, each of which gives rise to a segmental ganglion. Transcription factors of the Hox family are expressed in blocks of neuromeres and provide neuroblasts with a segment-specific identity. In the segments of the abdomen (A1–A9), thorax (T1–T3), and posterior head (B3–B6) genes of the Hox complex are expressed (*lab, pb, Dfd, Scr, Antp, Ubx, AbdA, AbdB*); head gap genes (*otd, ems*) are found in the anterior head (B1–2). B: Mouse. Homologous Hox genes and Head gap genes are expressed in sets of vertebrate neuromeres in the same antero-posterior sequence (P1–P6: prosomeres of forebrain; R1–7: rhombomeres of hindbrain). Otx is expressed in the dorsal telencephalon (P1–3) that gives rise to cerebral cortex. Emx defines alar plate of prosomeres 4–6, that will form the dorsal thalamus and ventral thalamus, as well as the dorsal midbrain. Hox genes are expressed in hindbrain and spinal cord. Note that for the Hox genes, the level indicated represents the anterior boundary of a wider domain that extends posteriorly for an unspecified distance. (C–F) Regional identity along the dorso-ventral axis. (C and D): Vertebrate. The neural tube (shown in C for mouse) is subdivided into four domains, the floor plate, basal plate, alar plate, and roof plate. Floor plate and roof plate do not give rise to neurons. The basal plate produces motoneurons and interneurons. Interneurons formed by the alar plate are the target of sensory input to the spinal cord and brain. The Homeobox gene Nkx2.2 is expressed in the ventral part of the basal plate; Msx is expressed in the dorsal part of the alar plate. The expression of these two genes in a cross-section of the zebrafish spinal cord are shown in (D). A third homeobox gene, Gsh-1, is expressed at a lateral level in between Msx and Nkx2.2. (E and F) Homologs of these genes are expressed in the same dorso-ventral succession in *Drosophila*. The schematic cross-section of the fly neuroectoderm (E) shows that the Nkx2.2 homolog Vnd forms a medial stripe adjacent to the midline cells; Ind (homolog of Gsh-1) and Msh (homolog of Msx) are expressed at an intermediate and lateral level, respectively. Panel F shows a ventral view of a wholemount *in situ* preparation demonstrating the expression of Vnd and Msh.

arranged in evolutionarily conserved genetic complexes, such that the more anterior and more posterior parts of the nervous system are specified by homologous Hox transcription factors. This is dramatically evident in animals where the patterning of the nervous system has been studied in detail such as *Drosophila* and mice (Fig. 15.2).

The dorsal-ventral axis of the neurectoderm in both vertebrates and *Drosophila* is divided up by the expression of three conserved homeodomain transcription factors, called, *vnd*, *ind*, and *msh* in flies, which are expressed in longitudinal stripes along the extent of the neural primordium (Cornell and Ohlen, 2000). In *Drosophila*, *vnd* gene expression specifies the neuroblasts close to the ventral midline, *msh* specifies the neuroblasts in the most dorsal domain and *ind* neuroblasts in an intermediate stripe (Fig. 15.2E,F). These dorsal-ventral genes are, like the Hox genes, functionally conserved in evolution such that in the vertebrate neural tube, the more ventral regions are specified by the *vnd*, *msh*, and *ind* homologues *Nkx2.2*, *msx1/2*, and *Gsh-1*, respectively (Fig. 15.2C,D). The expression of dorsal-ventral genes subdivides the vertebrate neural tube into four longitudinal domains, called the floor plate, basal plate, alar plate, and roof plate. The floor plate is characterized by expression of the Sonic hedgehog (Shh). The basal plate next to the floor plate includes all motor neurons. In the lateral wall of the neural tube, a furrow called sulcus limitans separates the basal plate ventrally from the alar plate dorsally. The alar plate later gives rise to the dorsal sensory column.

Although the mechanisms that originally set up the polarity of the Hox gene expression in flies and vertebrates are quite distinct, it appears that high dorsal concentrations of the signaling molecule BMP in vertebrates, or its equivalent Dpp in *Drosophila*, is key to establishing the relative domains of these dorsal-ventral specifying transcription factors (Mizutani *et al.*, 2006), although in vertebrate the ventral gradient of Shh also has an important role in specifying particular fates (see section on motor neuron determination, later). In addition, in both vertebrates and invertebrates, these spatially arrayed transcription factors of the anterior-posterior and the dorsal-ventral axes collaborate to divide the developing nervous system into a grid of molecular coordinates that specify the positional identity of neural progenitors down to the level of small groups or, in some invertebrates, even single cells.

Neurogenesis is patterned in time as well as space. This is exemplified in *Drosophila* neuroblasts, which go through a program of proliferation where each neuroblast gives rise to a series of distinct GMCs. The first GMCs of a neuroblast lineage tend to lie deeper in the CNS and usually generate neurons with long axons, whereas the later arising GMCs stay closer to the edge of the CNS and produce neurons with short axons. In the stages when the first GMCs are generated, most neuroblasts express *hunchback* (*hb*), and GMCs that arise at this time inherit this *hb* expression. Later, the same neuroblasts turn off *hb* and express *Krueppel* (*Kr*) instead, and GMCs generated at this stage inherit *Kr* expression. Experiments in which *hb* or *Kr* is eliminated or expressed at the wrong time lead to predictable switches in the fates of early and later born descendants of these neuroblasts (Isshiki *et al.*, 2001). Another gene that contributes to temporal patterning is called *Chinmo*, named for *chronologically inappropriate morphogenesis*. The identity of neurons issuing from a particular neuroblast can be advanced or retarded by reducing or increasing the expression of *Chinmo* (Doe, 2006). It is likely that a temporal succession of transcription factors like those in *Drosophila* also pattern neurogenesis in vertebrates.

The expression of both spatially and temporally coordinated transcription factors in neuroblasts, preserved in their progeny, forms part of the rich intrinsic inheritance that influences the phenotype of each developing neuron.

THE PRONEURAL AND NEUROGENIC GENES

The neurectoderm of *Drosophila* is a mixed population of cells. Only some of these cells become neuroblasts; others develop as dermoblasts. Cell–cell interactions that take place within the neurectoderm define the number and pattern of neuroblasts. Experimental and genetic studies suggest a two-step mechanism for this process (Campos-Ortega, 1995). First, discrete groups of neurectodermal cells are made competent to become neuroblasts. These groups of cells, called proneural clusters, represent "equivalence groups" in which all cells initiate a neural rather than an epidermal fate (Fig. 15.3A,B). In *Drosophila*, a group of proneural genes expressed in the proneural clusters is responsible for making these cells competent to become neural progenitors. In a second step, cells of the proneural cluster interact to sort out which of them will become neurons and which will fall back to become dermoblasts (Fig. 15.3A). As the first cell in each cluster begins to differentiate as a neuroblast, it sends out inhibitory signals to its neighbors, inhibiting them from also becoming neuroblasts. If the delaminating neuroblasts are ablated by a laser microbeam, then neighboring cells of the equivalence group, which would otherwise have developed into dermoblasts, replace the lost neuroblast. Genetic studies in *Drosophila* led to the discovery of a group of genes, called neurogenic genes, which mediate this inhibitory cell–cell interaction.

Most proneural genes are transcription factors of the basic helix-loop-helix (bHLH) family. In *Drosophila*, four of these genes form a complex called the achaete-scute complex (AS-C). Loss of proneural gene function leads to a reduction of neuroblasts. The neurogenic genes encode a cell communication mechanism that is activated within the proneural clusters and that acts to restrict the number of neural progenitors developing from each cluster to one or a few cells. The neurogenic genes *Notch* and *Delta* encode membrane proteins that function as signal and receptor, respectively. Activation of Notch by Delta sets in motion a signal transduction cascade (Fig. 15.3C) that involves another neurogenic gene called *suppressor of hairless* (Su(H)). Su(H) forms a complex with the intracellular domain

FIGURE 15.3 Function of proneural and neurogenic genes in neural development. (A) The sequence of events that leads to the segregation of individual neural progenitors (in insects) or neurons (in vertebrates). First, proneural genes are turned on in discrete cell populations, called proneural clusters (purple). In *Drosophila*, a number of genes (called "regional identity genes" in the diagram) have been identified. They are expressed in distinct pattern of transverse and longitudinal stripes before proneural genes appear and are proven or likely factors to control the expression of proneural genes. Proneural genes in turn trigger different events in the proneural clusters. One outcome of their expression is a cell–cell interaction process, mediated by the neurogenic genes, that selects a single cell (or a smaller group of cells) from each proneural cluster. Only this cell continues along the neural pathway and segregates from the neurectoderm as a neuroblast (right side of panel; intense purple), which divides and produces a neural lineage (red); the other cells of the proneural cluster remain within the neurectoderm. In vertebrates (left side of panel), the entire neurectoderm (i.e., the neural plate/neural tube) is formed by neural progenitors. Here, the proneural/neurogenic gene cassette seems to select cells that become postmitotic and differentiate from other (neural progenitor) cells, which remain in the neuroepithelium and continue to proliferate. (B) Detail of proneural gene expression in the neurectoderm of the *Drosophila* embryo (from Skeath and Carroll, 1992). (C) Interaction between proneural and neurogenic genes in the *Drosophila* neurectoderm. The upper panel shows two cells of a proneural cluster in which proneural genes of the *achaete-scute* complex (AS-C) are expressed. Genes of the AS-C activate the expression of the signaling molecule Delta (Dl). Dl activates the receptor Notch (N) in neighboring cells; this leads to cleavage of the N molecule and the translocation of the cytoplasmic domain of N along with the transcription factor Suppressor of Hairless (Su(H)) to the nucleus, where they upregulate the expression of proteins of the Enhancer of split complex (E(spl)-C). E(spl) initiates or maintains the development of an undifferentiated neurectodermal cell; at the same time, it directly inhibits the expression of AS-C genes, which promote differentiation of a cell as a neural progenitor or neuron. Imbalances introduced by an unknown mechanism into the levels of AS-C or E(spl)-C expression, respectively, are rapidly amplified (lower half of C) and lead to the segregation of a single neural progenitor (high levels of AS-C) from other cells that stay epithelial (high levels of E(spl)-C). (D-G) The proneural gene turns epidermal cells into neurons. (D) A normal *Xenopus* embryo stained for the neural marker NCAM shows no staining in the epidermis. (E) A neuroD-injected embryo has NCA-stained cells with neuronal morphology in the entire epidermis (F) Wild-type *Drosophila* embryo labeled with the neuronal marker 22C10 contains discrete clusters of neurons forming the central nervous system (CNS) and peripheral nervous system (PNS). (G) Loss of the signaling molecule Dl causes a massive increase in neurons at the expense of epidermis.

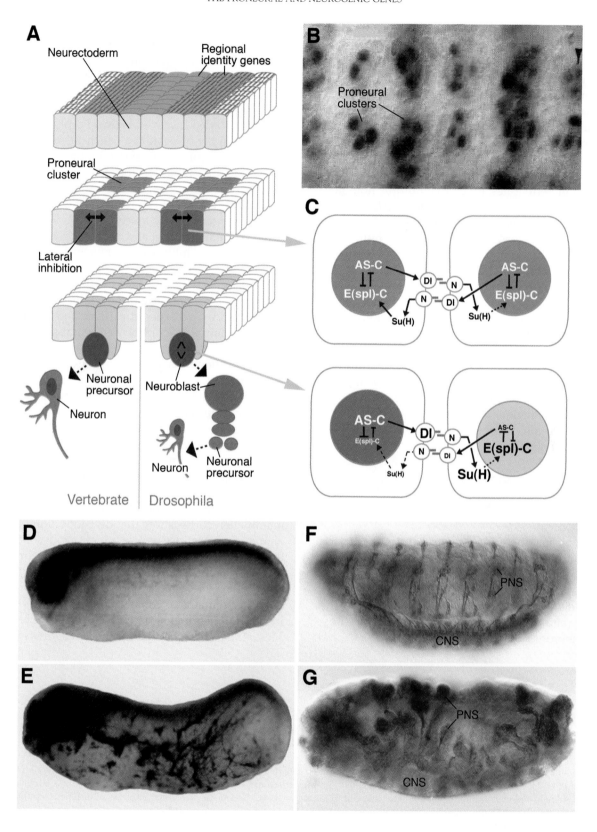

of Notch, which is released from the membrane when the pathway is activated. This complex moves into the nucleus and activates the *E(spl)-C* genes. E(spl)-C proteins are bHLH transcription factors that antagonize and repress the proneural bHLH genes, completing an inhibitory feedback loop. Loss of function of any of the neurogenic genes results in a higher number of neural progenitors, at the expense of dermoblasts (Fig. 15.3F,G).

Several AS-C homologues (*ASH* genes) have been identified in vertebrates. Likewise, multiple homologs of the *Drosophila* proneural gene *atonal* (*ato*) exist in vertebrates. In *Xenopus*, injection of *XASH-3* mRNA into blastomeres leads to an increased number of neural progenitor cells in the neural tube. In similar experiments, an *atonal* bHLH proneural homologue, called *neuroD*, powerfully transforms epidermal cells into neurons (Fig. 15.3D,E). Conversely, knockout of *ASH* genes in mouse or zebrafish leads to the absence of populations of neural cells. Homologues of the neurogenic genes *Notch*, *Delta*, *Su(H)*, and *E(spl)-C* also have been identified in all vertebrates. In *Xenopus*, injection of active forms of *Notch* or *Delta* mRNAs resulted in decreased formation of primary neurons. Lack of Delta function has the opposite effect, increasing the formation of primary neurons at the expense of neuroepithelial cells.

ASYMMETRIC CELL DIVISION AND CELL FATE

Neuroblasts divide asymmetrically and differentially distribute intrinsic determinants to their two daughter cells to produce distinct fates, a large neuroblast that remains in the neuroectoderm to divide again and again, and a small GMC that moves further interiorly and divides just once more to produce postmitotic neurons.

The initial positions of the two different daughters give a clue to the way in which asymmetric divisions happen. The machinery involved in asymmetric cell division (Betschinger and Knoblich, 2004) involves a number of proteins at the apical cortex of the neuroblast (Fig. 15.4A). One of the key proteins is the membrane-associated Inscuteable (Insc) protein, and so the whole complex of proteins, which includes a protein called Partner of Inscuteable (aka Pins), is called the Insc complex. Pins locally activates a G-protein-coupled pathway that attracts one of the centrioles toward this complex. The result of this is that the axis of the division plane is apicobasal. The different fates of the apical neuroblast and the basal GMC are due to the differential inheritance of cytoplasmic determinants, Numb and Prospero, which move to the basal pole of the parent neuroblast just before cytokinesis (Fig. 15.4B). As Numb is a negative regulator of the Notch pathway (discussed earlier), and Prospero is a homeodomain transcription factor that specifies GMC fate, the fact that these proteins end up in the GMC and not the neuroblast make the fates of these two cells very different.

In the vertebrate CNS, neuroepithelial cells first tend to divide symmetrically, giving rise to two dividing progenitor cells, but as neurogenesis proceeds, some neuroepithelial cells, like *Drosophila* neuroblasts, begin to divide asymmetrically, in this case giving rise to one neuroepithelial progenitor and one differentiated neuron. Time-lapse observations of neuroepithelial cells in organotypic slices of the developing mammalian cortex show that divisions along the horizontal plane produce two cells that stay in contact with the ventricular (apical) surface, whereas more apicobasal divisions produce a basal daughter that migrates away from the ventricular surface while the other daughter remains in contact, consistent with it retaining a neuroepithelial fate (Fig. 15.4C) (Chenn and McConnell, 1995). Homologues of the *Drosophila* genes

FIGURE 15.4 Control of asymmetric cell division in neural progenitors. (A) Schematic section of *Drosophila* neurectoderm at the stage of neuroblast delamination. The Inscuteable protein complex (green) is expressed apically in the neurectoderm and is carried interiorly by delaminating neuroblasts. (B) Schematic (left) and microphotograph (right) of neuroblast before (top), during (middle), and after (bottom) mitosis. The Inscuteable complex controls asymmetric distribution of intrinsic fate determinants, such as Numb (red), by orienting the mitotic spindle vertically and by localizing the Miranda protein (yellow) basally. Miranda traps Numb at the basal pole of the dividing neuroblast (middle panel) and thereby channels it into the ganglion mother cell (bottom panel). (C) Schematic cross-section of the mouse neuroepithelium. The mouse Numb homolog, mNumb, is expressed apically in neural progenitors (red). In asymmetric divisions, mNumb is maintained only in the epithelial daughter cell and not the delaminating neural precursor (right). D–F: Loss of mNumb (right panels) results in loss of progenitors compared to wild type control (left panels). All panels show part of cross-sections of the neural tube. (D) Histological section shows overall thinning of the ventricular layer (compare insets in the two photographs). (E) The differentiation marker Hu reveals that most neuroepithelial cells of the mutant have started to differentiate, as opposed to wild type where only cells of the mantle layer express this molecule. (F) BrdU incorporation visualizes dividing progenitors, which are strongly reduced in the mutant compared to wild type (from Petersen *et al.*, 2002).

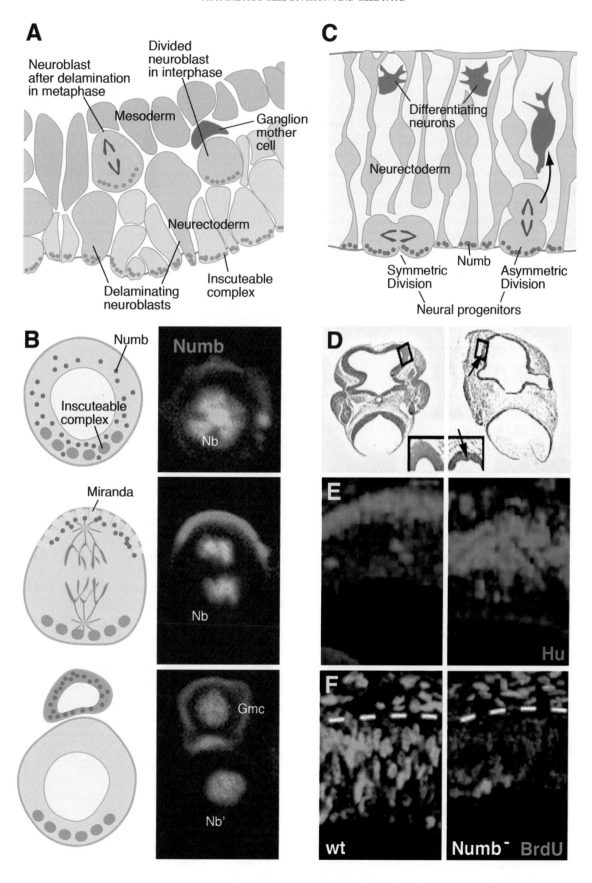

involved in asymmetric cell divisions work similarly in vertebrates. mInsc is the mammalian homologue of Insc and ASG3 is the mammalian homolog of Pins. Interfering with the expression or function of these homologues leads to the loss of apicobasal spindle orientation in the mammalian nervous system, resulting in too many cells dividing in the horizontal plane. In the mouse retina this causes the production of large clones arising from neuroepithelial cells, which can be explained if these cells divide symmetrically to produce two progenitors rather than one. The mammalian homologue of Numb may also play a key role in the distinct fates of daughter cells, as knockout mice suggest that mNumb is essential for cortical neuroepithelial cells to remain in a progenitor state (Fig. 15.4D,E, F) (Petersen et al., 2002).

The orientation of division can thus have an important role in determining cell fate, though which specific fates, of course, depend on where and when these cells arise. Interestingly the environment plays a role in the orientation of cell division lineages, and recent studies in both insects and vertebrates have shown that extrinsic signals, perhaps coming from recently postmitotic cells, can influence the plane of cleavage of a neighboring progenitor. This is one mechanism by which the specific types of neuron may work through a feedback control mechanism to control the appropriate number of neurons of a particular type.

CENTRAL NEURONS AND GLIA

The nervous system is composed of two main classes of cells: neurons and glia. In both vertebrates and invertebrates, the application of lineage tracers to individual progenitors often yields clones that contain both neurons and glia, suggesting that these two quite distinct cell types often are produced from a common progenitor, a "neuroglioblast."

In Drosophila, intrinsic determinants of glial fate are expressed at an early stage in neuroglioblasts (Jones, 2005). Two genes encoding transcriptional regulators, glial cells missing (gcm) and reversed polarity (repo), are expressed in most glial precursors in insects. Their crucial involvement in glial fate is attested to by the fact that ectopic expression of these genes converts neurons into glial cells and that loss-of-function mutants lack glial cells. Gcm is expressed first and activates Repo by direct binding to the repo promoter. The directed expression of gcm in glial precursors has been studied in a neuroglioblast called NB6-4 (Akiyama-Oda et al., 2000) (Fig. 15.5A). Following delamination from the neuroectoderm, this neuroglio-

blast cell divides asymmetrically using the Insc complex to localize the Gcm protein to the medial side of NB6-4. The medial daughter of this division "inherits" this glial determinant and develops as a pure glioblast, NB6-4G. The other daughter, NB6-4N, which does not receive Gcm protein at this division, acts as a "pure" neuroblast.

In the central nervous system of vertebrates, multipotent stem cells often give rise to glia such as oligodendrocytes, as well as neurons (Fig. 15.5B,C). It is usually the case that neurons arise first and glia latter. Time-lapse studies of these stem cells in culture show that they tend to go through several rounds of division that give rise to neuroblasts, and then they suddenly make a transition to divisions that generate glioblasts (Qian et al., 2000) (Fig. 15.5B). PDGF is a mitogen for the oligodendrocyte precursors in the spinal cord and the receptor for this mitogen is specifically expressed in a restricted ventral domain of the neural tube that gives rise to the oligodendrocytes. The oligodendrocyte specific transcription factors, Olig1 and Olig2, are expressed in this same ventral domain. Interestingly, however, motoneurons also arise from this domain suggesting that the same progenitors make both motoneurons and oligodendrocytes. Indeed, cells that express Olig1 initially also express Neurogenin2, a proneural bHLH transcription factor (Kessaris et al., 2001; Zhou et al., 2001). These progenitors that express both Neurogenin 2 and Olig1 generate motoneurons. Later in development, the expression domain of Neurogenin 2 moves dorsally and no longer overlaps with the Olig1/2 expression domain. When this happens, the neuroepithelial cells in this region express Olig1/2 and not Neurogenin 2 and thus start producing glia.

The Notch signaling pathway also plays an important role in gliogenesis. Transient activation of Notch promotes the differentiation of glia in the CNS. Activation of the Notch pathway suppresses the transcription of proneural bHLH transcription factors like Neurogenin 2, and unblocks the transcription of factors that are necessary for gliogenesis.

SENSORY NEURONS OF THE PERIPHERAL NERVOUS SYSTEM

Sensory neurons, which are by and large peripheral, do not arise from neuroectodermal regions that generate the CNS. With the exception of the eye, all sensory organs of the vertebrate are derived from sensory placodes and neural crest. A placode is a morphologically visible thickening of an epithelium that often subse-

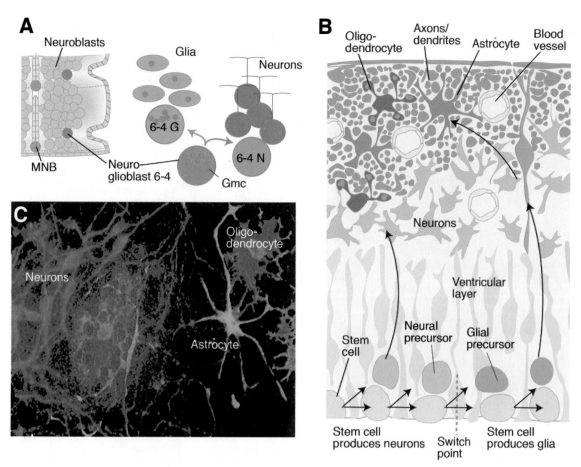

FIGURE 15.5 Specification of glial lineages. (A) Position of two identified neuroglioblasts, Nb6-4 and MNB, in the *Drosophila* neuroblast map (left). Nb6-4 expresses the glial regulatory protein, Gmc (green). When this cell divides into two daughter cells, 6-4 G and 6-4 N, the Inscuteable complex and Miranda segregate Gmc into 6-4 G, which thereby becomes specified as glioblast. 6-4 N generates neurons. (B) Glial cell development in vertebrates. Diagram shows cross-section of neural tube containing neurons and two main types of glial cells, astrocytes and oligodendrocytes. Oligodendrocytes form processes that wrap around axons and give rise to the myelin sheath. Astrocyte processes connect to capillaries and neurites. Glial progenitors and neural progenitors are derived from the same pool of stem cells that divide in the ventricular layer of the neural tube (bottom of B). At early stages, a stem cell generates neural precursors. Later it undergoes a specific asymmetric division (the "switch point") at which it changes from making neurons to making glia. (C) Photograph of cell lineage obtained by injecting an individual neuroepithelial cell from mouse cortical progenitor cultured in a dish. Neural and glial-specific antibodies reveal the presence of both neurons (purple) and glia (astrocytes: green; oligodendrocytes: blue) in the lineage (from Qian *et al.*, 2000).

quently invaginates into the interior and transforms into a distinct tissue or organ (Schlosser, 2006). Sensory placodes appear during the late stages of neurulation in the dorsal ectoderm that flanks the invaginating neural tube (Fig. 15.6A). The most anterior placode is the olfactory placode that gives rise to the olfactory sensory neurons of the nose. Further posterior is the otic placode that forms the inner ear. Slightly more laterally is the placode of the trigeminal system that produces the sensory neurons innervating the receptors of the skin and muscles of the face and the head. Developing later and at a more lateral level in contact with the pharyngeal clefts (the gill slits in fish and

amphibians) are the epibranchial placodes. They give rise to sensory neurons that innervate the taste receptors of the mouth, and receptors in the inner organs, such as the heart, large blood vessels, and digestive tract.

Sensory neurons of the trunk are derived from the neural crest (Fig. 15.6B). The neural crest is a transient population of cells that arises along the lateral edges of the neural plate (Chapter 14). Crest cells become localized to the dorsal part of the neural tube as it folds up, and then leave the neural tube and migrate along several well-defined pathways. Neural crest stem cells give rise to a large variety of cell types, among them

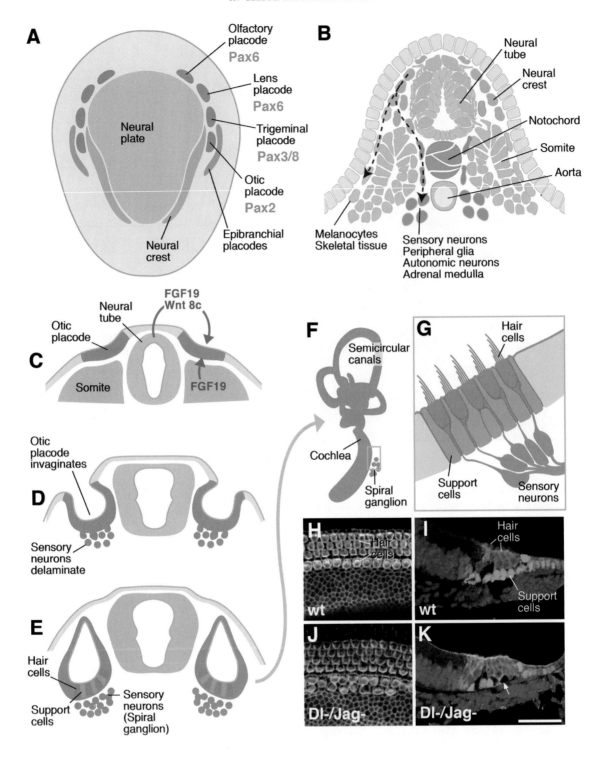

skeletal tissue, melanocytes, sensory neurons, and glial cells (Schwann cells) of the peripheral nervous system, adrenergic and cholinergic neurons of the autonomic nervous system, and the endocrine chromaffin cells of the adrenal medulla. Sensory neurons stay close to the neural tube where they aggregate in segmentally reit-

erated dorsal root ganglia. Long peripheral processes extend from the dorsal root ganglia toward the skin, muscles, and skeleton. Here, dendritic endings of sensory neurons interact with specialized, mesodermally derived support cells, forming small "corpuscles" that are tuned to a specific type of stimulus,

FIGURE 15.6 Sensory organ development in vertebrates. (A) Schematic view of neurula stage embryo, showing neural plate (light blue) flanked by sensory placodes (dark blue; anteriorly) and neural crest (green; posteriorly). Placodes express specific transcription factors of the Pax family. (B) Schematic cross-section of vertebrate embryo in which migrating neural crest cells (green) are indicated. These cells follow two different pathways: a dorsal one (light green), giving rise to melanocytes, connective tissue and skeletal tissue, and a ventral one (dark green), giving rise to sensory and autonomic ganglia, as well as the adrenal medulla. (C–F) Schematic cross-sections of vertebrate embryo at successive stages, showing development of the otic placode. (C) Otic placode is induced from dorsal ectoderm by inductive signals (e.g., FDG19, Wnt8c) from neural tube and somites. (D) Otic placode invaginates. Precursors of sensory neurons (red) delaminate from placodal epithelium. (E) Cells of the placodal epithelium differentiate. Precursors of hair cells (sensory receptors; magenta) and support cells (green) appear. (F) Otic placode grows and forms cochlea and semicircular canals. (G) Magnified view of cross-section of part of cochlea. Dendrites of sensory neurons (spiral ganglion) form synaptic contacts with sensory hair cells. These specialized, ciliated cells are surrounded by support cells. (H and I) Surface view (H) and section (I) of mouse cochlea, showing normal number and pattern of hair cells and support cells. (J and K) In Dl/Jag double mutants, hair cells are increased in number, and support cells are decreased (from Kiernan *et al.*, 2005).

including touch, vibration, stretch, temperature, and pain. Peripheral sensory neurons of pladodal origin also are associated with specialized nonneuronal support cells that mechanically anchor the sensory apparatus, provide the proper ionic environment for sensory transduction, and "tune" the receptor cell to a special stimulus.

Classical embryological experiments in which parts of donor embryos were transplanted to different sites in host embryos indicated that as they invaginate, placodes come in contact with the neural tube, the archenteron, the mesoderm, and the epidermal ectoderm. These structures all have inductive capacities and influence the fate of the pladodal cells. In some instances, specific signaling molecules have been identified. For example, BMP4 and BMP7, secreted by the pharyngeal endoderm, are involved in the induction the epibranchial placodes; and FGF3 and Wnt8c from the hindbrain and FGF19 from the axial mesoderm work together to induce the otic placode (Fig. 15.6C). A number of transcription factors are expressed under the control of these signaling proteins. These factors, like those in the CNS involved in regional identity, act in specifying sensory organ fate. The regional identity genes expressed in sensory placodes are members of several families: the *distalless* family (*dlx3/5*), the pax family (*pax2, 3, 5, 6, 8*), the *eyes absent* (*eya*) family (*eya1/2*), and the *six* family (*six1, 3, 6*). Initially, many of these genes are expressed in widespread domains surrounding the neural plate (a.k.a. the preplacodal ectoderm). Later, the genes become restricted to specific sensory placodes. Notably, *pax2* expression is strongly correlated with the otic placode; *pax6* appears in the olfactory and lens placode, and *pax3* and *pax8* in the trigeminal placode (Fig. 15.6A). Absence or downregulation of these genes results in loss of the corresponding sensory organ.

Let us follow the determination of sensory cells of the inner ear. Following invagination from the dorsal ectoderm, the otic placode forms a hollow vesicle, the otocyst (Fig. 15.6C–E). The lumen of the otocyst ultimately becomes the endolymph filled cavity of the inner ear. Initially, the wall of the otocyst is formed by a homogenous epithelial layer. Subsequently, the otocyst folds in a complex manner to give rise to the cochlea for sensing sound, the utricle for sensing gravity, and the semicircular canals for sensing rotational head movements in the three dimensions (Fig. 15.6F).

As the inner ear takes shape, the cells that form the octocyst wall diversify. In the same way that we learned about for the neural tube, different cell types delaminate from the epithelium and form additional layers basally adjacent to the epithelium (Fig. 15.6D). These cells become sensory neurons (forming the spiral ganglion), whereas the neural cells that remain epithelial form sensory hair cells and support cells (Fig. 15.6G). Hair cells, the receptor cells of the inner ear, have specialized cilia and microvilli ("stereocilia") that sense movement of the endolymph. Deflection of stereocilia generates a receptor potential that is transmitted to the sensory neuron connected to the hair cell.

The domain of the otocyst that produces sensory neurons expresses the proneural gene *Ath1*, as well as signals of the Notch pathway, which suggests that a lateral inhibition mechanism is active to control the number of neural precursors that form at any given time. Notch activity is also required in the epithelial layer of the otocyst to distinguish between hair cells and support cells. Thus, precursors of hair cells appear at regular intervals, separated by support cell precursors (Fig. 15.6G). A Notch dependent lateral inhibition mechanism operates to determine the spacing of hair cells. Thus, at an early stage, the Notch receptor and the Delta ligand are expressed in all cells of the otic epithelium. Subsequently, Delta becomes restricted to the developing hair cells, which thereby inhibit their neighbors from choosing the same fate; these neighbors become the support cells. Genetic studies, using mouse and zebrafish, show that loss of Delta in the developing inner ear results in an increased number of hair cells (Fig. 15.6H–K)

Homologues of most of the molecules involved in specification of peripheral sensory neurons also operate in the *Drosophila* sensory system. Indeed, molecular evidence shows that the proteins encoded by the *pax*, *eya*, and *six* genes form complexes that as a whole unit bind to DNA and regulates transcription. Genes that in this manner encode cooperating proteins are called gene networks. The fact that their function is dependent on all members of the complex explains why entire gene networks often are conserved throughout evolution. The *pax/six/eya* network, activated by BMP and/or Wnt signals, is one of the best-studied examples of such a conserved gene network.

In *Drosophila* and many other invertebrates, sensory organs are comprised of small clusters of sensory neurons and support cells, which together are called sensilla (Fig. 15.7). Sensilla are scattered over the entire body surface of the larval and adult fly. The cells that compose each sensillum are clonal descendents of a single Sensory Organ Progenitor cell (SOP) (Lai and Orgogozo, 2004). The special morphology and molecular composition of sensory neurons and accessory cells adapt sensilla toward the reception of a wide range of stimuli, including mechanical stimuli (touch, pressure, vibration, stretching), chemical stimuli (olfactory, gustatory), temperature, and humidity. Most sensilla have contact to the external surface; some sensillum types, in particular stretch receptors, are attached to the inner surface of the body wall or internal organs. Despite their great diversity, sensilla are built according to only two main schemes (Fig. 15.7A) (Hartenstein, 2005). The large majority of sensilla (type I) contain bipolar sensory neurons surrounded by a pair of inner and outer support cells (Fig. 15.7A, top). The inner support cells form a sheath around the dendrite, and sometimes around the soma and proximal axon of the sensory neuron. The outer support cells are arranged concentrically around the inner ones. They form processes that, in case of sensilla situated at the surface (external sensilla or ES organs), secrete cuticle in the shape of hairs or bristles. Outer accessory cells of subepidermal stretch receptors, called chordotonal organs, form ligaments that attach the sensory neurons to the body wall. The second major type of sensilla (type II) consists of subepidermally located multidendritic neurons that are either "naked" (Fig. 15.7A, bottom) or are associated with variable types of accessory cells.

In most cases, all cells of a sensillum are born in three consecutive divisions. The SOP, also called the primary precursor cell or pI, divides into two pII daughters called pIIa and pIIb (Fig. 15.7B). pIIa divides one more time and gives rise to the outer accessory cells, the tormogen and trichogen cells. pIIb gives rise to the sensory neuron and inner accessory cells in two consecutive divisions. The first division produces precursor pIIIb, which divides into sensory neuron and thecogen cell, and the glial cell. Given that SOPs produce the different sensillum cells by a fixed lineage mechanism, it seems likely that intrinsic determinants expressed in the SOP and "forwarded" in a specific

FIGURE 15.7 Sensory organ development in *Drosophila*. (A) Structure of mature sensilla. Schematic section of two main classes of sensilla. Top: external mechanoreceptor (microchaete). Bipolar sensory neuron (red) is surrounded by two inner support cells, the thecogen cell (purple) and glial cell (yellow). Two outer support cells, the tormogen cell (light blue) and trichogen cell (dark blue) form the socket and shaft of the bristle-shaped cuticular component of the sensillum. Bottom: multidendritic neuron. This type of sensillum does not possess external support cells; multidendritic neurons (stretch receptors) are attached to the basal surface of the epidermis and inner organs. (B) Tree diagram of canonical cell lineage of mechanoreceptive sensillum (microchaete). Progenitor of sensillum (SOP or pI; top) divides into two daughter cells, pIIa and pIIb. Division of pIIa gives rise to outer support cells (trichogen and tormogen cell). PIIb divides into progenitor pIIIb that then produces sensory neuron and thecogen cell; the other daughter of pIIb develops as glial cell (as shown here for the microchaete), or expresses a number of other different fates (not shown). (C and D) Sensillum patterning. (C) shows schematic dorsal view of *Drosophila* thorax (notum) indicating invariant pattern of mechanoreceptive bristles. Some of the large bristles (macrochaetes) are identified individually. Anp, anterior notopleural; apa, anterior postalar; pdc, posterior dorsocentral; pnp, posterior notopleural; psc, posterior scutellar. (D) Expression of the proneural gene *scute* (gray) in wing imaginal disc. In the proximal region of the disc, which gives rise to the notum, *scute* is expressed in proneural clusters that can be assigned to individual macrochaetes. (E–J) Distribution of fate determinants and Notch signaling in the sensillum lineage. (E) During the pI mitosis, Numb, an inhibitor of Notch signaling, is distributed to pIIb. This activates determinants of neurons like Prospero and Phyllopod. Loss of Numb in pI causes transformation of pIIb into pIIa, with a concomitant doubling of outer support cells and absence of neuron/inner support cells. (F, G) Examples of this transformation are the mechanoreceptive Keilin's organ, which, in wild type (F1) has three sets of shaft/socket and, in numb mutant (G1), 4–5 sets of these outer support cells. Likewise, duplication of shaft/socket occurs in a basiconical sensillum (F2: wild-type; G2: numb mutant). F3 shows molecular marker Cut expressed in two outer support cells of a mechanoreceptive sensory organ of the embryo (arrowhead). In the numb mutant (G3), these cells are doubled in number (arrow). (H) Following the pIIa mitosis, determination of the trichogen cell (tr) requires Numb, Hairless (H) and the Pax2 homolog, Sparkling. In the precursor of the tormogen cell (to), Su(H) and the N signaling pathway are active. (I) External structure of wild-type sensillum, consisting of shaft and socket. (J) In loss of Hairless mutation, or overexpression of Su(H), the shaft forming trichogen cell is transformed into a second tormogen cell, resulting in double-socket and lack of shaft.

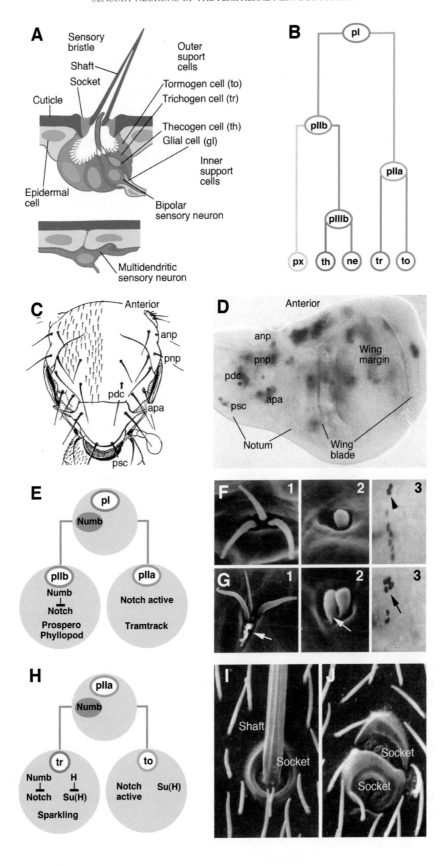

sequence of events to the individual daughter cells play an important role. Several determinants of sensillum cell fate have been identified in genetic screens. Most of them have to do with the Notch signaling pathway, indicating that this signaling pathway is involved not only in selecting SOPs, and thereby determining the sensillum pattern, but also in differentiating sensory neurons and accessory cells. We will look at two examples of how regulation of the Notch pathway acts in the *Drosophila* sensillum lineage.

The first step in sensillum development, the determination of SOPs, is controlled in a manner similar to that regulating neuroblast determination. Thus, the expression of proneural genes of the AS-C defines proneural clusters in the ectoderm that are competent to form SOPs (Fig. 15.7C,D). At the same time, proneural genes activate the Notch pathway, mediated by the neurogenic genes, which single out one SOP and prompt all other cells of the proneural cluster to abort neural development. The identities of individual SOPs are specified by different proneural genes. For example, *Cut* (*ct*) is a homeobox gene expressed in precursors of all external sensilla. If *ct* is absent, external sensilla do not develop and chordotonal organs appear instead. *Poxn* and *Amos* are expressed and required in precursors of multiply innervated sensilla, in particular chemosensory receptors. *Atonal* (*ato*) is expressed and required in the proneural clusters that give rise to chordotonal organs. Loss of *ato* function results in the absence of these cell types; likewise ectopic expression of *ato* in other SOPs induces the production of chordotonal organs.

Once determined, SOPs divide asymmetrically. A crucial factor in generating this asymmetry is the protein Numb (Fig. 15.7E). In pI, the Numb protein becomes localized to one side of the cytocortex, so that when this cell divides, only one daughter "inherits" Numb. The cell containing Numb becomes pIIb. Numb disrupts the Notch signaling pathway by preventing Notch from translocating to the nucleus. Thus, upon activation of the Notch receptor in both pIIa and pIIb, the signal can get through only in pIIa. If Numb is deleted, Notch can act in both pII cells, which changes the fate of pIIb to that of pIIa, and causes a duplication of outer support cells (the descendants of pIIa) and loss of neuron/thecogen cell (the descendants of pIIb; Fig. 15.7G).

Asymmetric activation of the Notch pathway also regulates cell fate at later divisions in the sensillum lineage. For example, pIIa starts out with an activated Notch signaling pathway. As the division into trichogen and tormogen cell occurs, Notch becomes inhibited in the trichogen cell. Again, asymmetrically inherited Numb is an essential player in inhibiting

Notch signaling. Numb appears in pIIa shortly before its mitosis and localizes to its anterior pole, whereby it ends up in the trichogen cell (Fig. 15.7H). Transcription factors that are downstream of Notch or collaborate with it are expressed differentially in the tormogen and trichogen cells. One protein, Suppressor of Hairless (Su(H)), becomes restricted to the tormogen cell. Loss of Su(H) or Notch causes a duplication of trichogen cell and loss of socket; overexpression of Su(H) in the trichogen cell causes its transformation into a second socket cell (Fig. 15.7J).

It is clear that recurring themes, such as spatially restricted homeobox containing transcription factors, proneural bHLH genes, asymmetric cell divisions, and Notch signaling, play a role in sensory neuron determination in vertebrates and invertebrates just as they do in the central nervous system of these animals.

THE RETINA: A COLLABORATION OF INTRINSIC AND EXTRINSIC CUES

The retina is a sensory organ with different origins. The *Drosophila* retina arises from an imaginal disc, whereas the vertebrate retina arises from an outpocketing of the neural plate. Studies of how specific neurons arise and diversify in the retinas of these distinct animals reveals surprising similarities despite their very different embryological origins. These similarities may be due to the shared molecular evolution of visual organs that predates the separation of vertebrates and invertebrates.

The insect compound eye is built of a large number of identical facets, called ommatidia. *Drosophila* possesses approximately 800 ommatidia in each eye. Each contains several different types of photoreceptors and accessory cells. There are eight identifiably unique photoreceptors in each ommatidium; six of them (R1–R6) form an outer trapezoidal array, whereas two (R7 and R8) are located in the center (Fig. 15.8A). Among the accessory cells are cone cells, which form the lens of each ommatidium, and pigment cells, which surround the photoreceptors and optically shield the ommatidia from one another.

Cells of the ommatidia are formed from clonally unrelated, uncommitted precursor cells that are generated from the proliferation of cells in the eye imaginal disc (Mollereau and Domingos, 2005). The ommatidial cells do not appear all at once, but follow a reproducible temporal sequence (Fig. 15.8B). During late larval life, a wave of differentiation passes over the eye disc in a posterior-to-anterior direction. In front of the wave, cells are still dividing and uncommitted, whereas

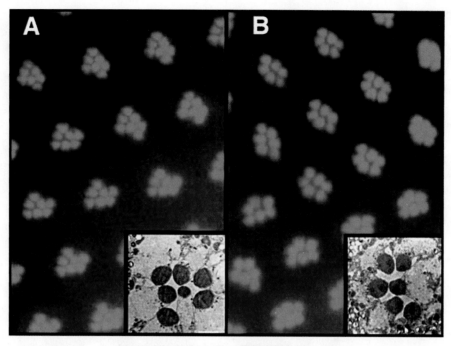

FIGURE 15.9 Photoreceptors in the eye of normal and *sevenless* mutants. (A) If a light is shone from the back of a fly's head and focused in the facets of the eye, individual photoreceptors can be seen because of their ability to pipe light. The wild-type animal has the normal pattern of seven photoreceptors visible in each facet. The small one in the center is photoreceptor R7. (B) The same technique used in a *sevenless* mutant shows only the six large photoreceptors R1–6 in each facet. R7 is missing. Insets show electron micrographs through single facets.

A genetic screen for R7-less flies yielded identification of a membrane bound signaling molecule, called Bride of sevenless (Boss), which is expressed specifically in R8 cells and serves as the ligand for Sev. These experiments provided an impressive demonstration of the awesome power of genetic screens combined with precise knowledge of the developing system in identifying new genes and proteins controlling development.

The vertebrate retina, like the *Drosophila* retina, develops from a population of pluripotent neuroepithelial progenitors, which produce a diversity of neurons and glia (Fig. 15.10A). As in *Drosophila*, vertebrate neurogenesis and determination in the retina follows a temporal order (Livesey and Cepko, 2001). Thus, at any given time during retinogenesis, only a few fates are available to differentiating cells. Cells born early in mammals generally adopt fates as retinal ganglion cells (Fig. 15.11A–C). Horizontal cells and cones are also born early. Amacrine cells are produced subsequently, followed by rods, bipolars, and Müller cells. This is also similar to *Drosophila*, in which first R8s and then sequentially other cell types differentiate. This progressive shift in cell type genesis in vertebrates

is supported by dissociation experiments in which cells are put into culture at low density at various stages of development. If progenitors are isolated at the time when retinal ganglion cells (RGCs) are normally born, they tend to turn into ganglion cells in culture. If such cells are isolated at later stages, they become rod photoreceptors in the same culture medium. These results are consistent with an intrinsic progression of cell fates, such as seen in *Drosophila* CNS neuroblasts.

There is an intrinsic element to the succession of cellular fates in the vertebrate retina (Cayouette *et al.*, 2006), which is revealed when late progenitors are isolated and allowed to divide on their own in culture. These isolated progenitors give rise to clones of cells of same general birth order, size, and composition as clones *in vivo*. A vertebrate homologue of *ato*, *ath5*, is certainly one intrinsic factor in the process of RGC fate specification in the vertebrate retina. *ath5* is turned on in a subset of retinal progenitors that give rise to RGCs and it is essential for the production of RGCs. Some *ath5* expressing progenitors were followed by time-lapse analysis in the zebrafish retina (Poggi *et al.*, 2005). They were found to divide just once to generate one

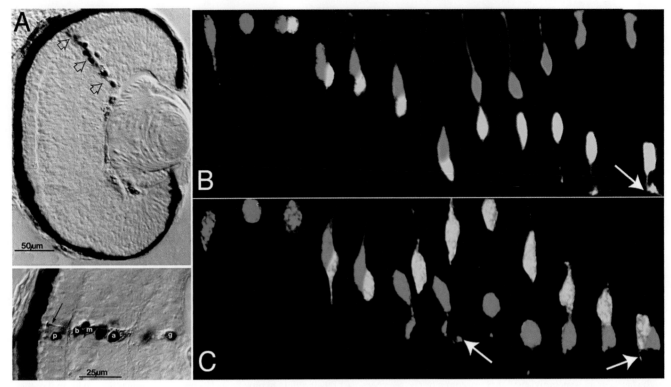

FIGURE 15.10 Clones of cells in the *Xenopus* and zebrafish retina. (A top) Daughters of a single retina progenitor in *Xenopus* injected with horseradish peroxidase are seen to form a column that spans the retinal layers and contributes many distinct cell types; (A bottom) p, photoreceptor; b, bipolar cell; m, Muller cell; a, amacrine cell; and g, ganglion cell. (B) Time-lapse images of an RGC progenitor in zebrafish dividing at the apical surface of the retina to produce two neurons, one of which (yellow) becomes an RGC. The other cell (red) is probably a photoreceptor. (C) A similar RGC progenitor in an *ath5* mutant retina in which there are no host RGCs. This cell divided to produce two daughters, both of which become RGCs (arrows point to emerging axons).

RGC and one other postmitotic retinal neuron by an asymmetric cell division (Fig. 15.10B). Seeing such a lineage pattern *in vivo*, however, does not mean that this pattern is immune from extrinsic influence. Indeed, when retinal cells are removed at the stage when RGCs are being born, mixed into aggregates, and cultured *in vitro*, they differentiate into RGCs as one would expect. However, if the same cells are mixed with an excess of retinal cells several days older (i.e., when photoreceptors are generated), they generally become photoreceptors (Fig. 15.11D). This work shows that individual cells have the capacity to differentiate into different cell types, and the fate they choose depends on the environment in which they are born. This makes good sense, for as retinal development proceeds, the environment changes simply as a result of various retinal cell types being generated, much as in the *Drosophila* retina. Indeed, there is strong evidence for feedback in influencing the fate of RGCs. Young embryonic chick progenitors are inhibited in their ability to produce RGCs when cultured adjacent to older retinas in which

there are lots of RGCs, and depletion of the RGCs from these older retinal cell populations abolishes this inhibition. Similarly, whereas *ath5*-positive cells in normal embryos generate only a single RGC daughter (see earlier), these same cells when transplanted into mutant retinas that lack RGCs often generate two RGCs daughters, providing evidence that such lineages are sensitive to feedback signals (Fig. 15.10C).

What could these feedback signals be? RGCs express Shh, and the Shh knockout mouse contains increased number of RGCs. These results suggest that Shh provides feedback inhibition signal in the developing mouse retina, echoing what happens in the *Drosophila* retina where R8 cells express Hh, which regulates the expression of *ato*. The growth differentiation factor 11 (GDF11) is another factor secreted by RGCs, and it acts by limiting the temporal window during which progenitors are competent to express ath5 and thus produce RGCs. Finally, as in the *Drosophila* retina, the Notch signaling pathway contributes to retinal diversity in the vertebrate retina (Fig. 15.11E,F). When Notch

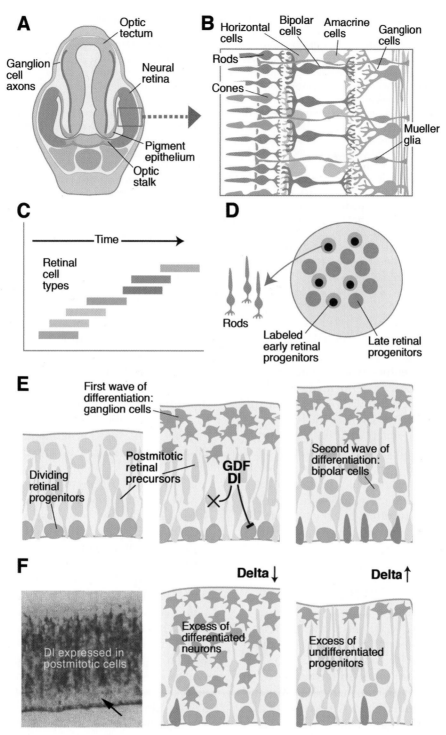

FIGURE 15.11 Vertebrate retina development. (A) The neural retina bulges out of the ventral neural tube at the level of the diencephalon. It is joined to the brain by the optic stalk along which the axons of retinal ganglion cells will course on their way to the tectum. (B) The seven major cell types in the retina coded by color. Their laminar arrangement by cell type is evident. (C) Birth dating studies in the retina show that different cell types are born in different but overlapping periods of development. (D) When early generated retinal precursor cells (labeled by a pulse of BrdU) are mixed with older cells in culture, they show an increased probability of turning into late cells, such as rods. (E) Development of the retina. Schematic section of retinal neuroepithelium at three different stages (early: left; late: right). Dividing retinal progenitors (purple) produce different cell types depending on the developmental stage. Differentiation inhibitors (Dl) released from postmitotic neural precursors inhibit retinal progenitors from producing more ganglion cells. Postmitotic cells also secrete differentiation factors such as GDF, which limit the time window during which postmitotic cells can become retinal ganglion cells. (F) Expression and function of Dl in the developing retina. Left panel: expression of Dl in postmitotic neural precursor; dividing progenitors of the ventricular layer (arrow) or differentiated neurons of the basal mantle layer (green) do not express Dl. Middle: Loss of Dl results in increased number of early born retinal cells (retinal ganglion cells), at the expense of undifferentiated retinal progenitors. Right: Overexpression of Dl causes the opposite phenotype.

FIGURE 15.12 Specification of motor neurons in the vertebrate spinal cord. (A) Schematic cross-section of the neural tube. The notochord, which is located underneath the floor plate, releases the signal Sonic hedgehog (Shh). Consecutively, Shh is released from the floor plate and forms a gradient with high concentrations ventrally and low concentrations dorsally. BMP molecules released from the dorsal epidermis and dorsal neural tube form an opposing gradient. (B) In a concentration-dependent manner, Shh directs expression domains of the class I and class II homeodomain genes (see text for details). (C) Polarized expression of FGF, among other signaling pathways, is responsible for the regionally specific expression of Hox genes along the antero-posterior axis of the neural tube. In addition, inhibitory interactions among the Hox genes themselves, sharpen the boundaries within which a given Hox gene is expressed. In such manner, the anterior Hox gene HoxC6 is expressed in the cervical spinal cord, and the posterior Hox gene HoxC9 is expressed in the thoracic cord. These genes direct the differentiation of the motor column into specific domains, such as the cervical lateral motor column (LMC) innervating forelimb muscles, or the column of Terni (CT) in the thoracic spinal cord that innervates sympathetic ganglia. (D) The motor column is subdivided into pools of motor neurons innervating specific muscles. These pools, including the Tri motor neuron pool and the CM motor neuron pool, occupy discrete locations in the ventral horn of the spinal cord and show distinct dendritic branching patterns. These properties, which depend on the expression of specific transcription factors such as the LIM gene PEA3, can be visualized by backfilling the motor neurons with fluorescent labels injected in the corresponding muscles. (E and F) Two different markers visualize the dendritic arborization (E) and cell body position (F) of the Tri motor neurons (green) and CM motor neurons (red). (G and H) Knockout of the PEA3 gene results in ventral shift of CM motor neuron location, and dorsal expansion of Tri motor neuron dendrites. (I) Reflex circuits between sensory afferents (from muscle spindles) and motor neurons innervating the corresponding muscle are formed because both spindle afferents and motor neurons targeted by them express a common transcription factor, such as ETS ER81 shown here. Among the downstream targets of ETS ER81 are homophilic adhesion molecules that allow the ingrowing sensory axon to "recognize" the proper motor neuron target. (E–H) (from Vrieseling and Arber, 2006).

signaling is compromised in early retinal progenitors, too many cells differentiate as early cell types such as RGCs and cone photoreceptors, whereas later interference with Notch signaling leads to an increase in later-born cell types such as rod cells. When cells are released from the inhibition mediated by the Notch pathway, proneural basic helix-loop-helix (bHLH) transcription factors, such as ath5, bias cells towards particular fates, such as RGCs. The Notch signaling pathway, by allowing only a certain number of cells to differentiate at any one time, is thus important in creating cellular diversity in the vertebrate retina (Dorsky *et al.*, 1997).

COMBINATORIAL CODING IN MOTOR NEURONS DETERMINATION

The most ventral part of the vertebrate spinal cord is the "floor plate" and just dorsal to this are the motor neurons (Fig. 15.12A). Transplantation experiments have shown that the floor plate is induced by a Shh signal emitted from the underlying notochord (Chapter 14). Once induced, the floor plate serves as a secondary source of Shh. As a result a gradient of the Shh signal percolates dorsally through the cord. At least five different neuronal types are generated in response to this single gradient. The most ventral neurons including motor neurons, require the high doses of Shh, and successively more dorsal neurons require correspondingly less (Jessell, 2000). When Shh is missing or antagonized with an antibody, there is no floor plate, nor indeed any ventral neuronal type, and in particular no motor neurons in the spinal cord.

How do cells at different dorso-ventral levels interpret their exposure to different levels Shh to acquire different fates? Threshold levels of Shh turn on some homeobox genes such as Nkx2.2 and Nkx6.1 (Class II) and turn off others such as Pax6 and Dbx2 (Class I). Thus neural tube neurons in different dorso-ventral regions express specific homeodomain proteins, which either are turned on or off at particular Shh thresholds (Fig. 15.12B). The boundaries between these domains are sharpened through reciprocal-repression of the two classes of genes. The ventral border of the class I Pax-6 gene initially overlaps the dorsal border of the class II Nkx2.2 gene, but then reciprocal repression sets in, so that only one of these genes is expressed in any particular cell. By this process, specific domains uniquely express particular combinations of Class I and Class II homeodomain transcription factors. Motor neurons arise from the domain that uniquely expresses *Nkx6.1* but not *Irx3* and *Nkx2.2*. *Nkx6.1* turns on OLIG2, a bHLH transcription factor that is required for motor neuron differentiation, and Nkx2.2 and Irx3 repress OLIG2 expression. OLIG2 in turn activates the expression of motor neuron specific transcription factors, such as Mnr2, HB9, Lim3, and Isl1/2, that are involved in motor neuron differentiation (Fig. 15.12B).

Motor neurons in the spinal cord are organized in columns that project to muscle groups along the anterior to posterior axis of the body (Dasen *et al.*, 2003). Thus, in the cervical region, the Lateral Motor Column (LMC) innervates forelimb muscles, and in the mid-thoracic region, the Column of Terni (CT) innervates the sympathetic chain (Fig. 15.12C). The patterning of motor neurons into motor columns is accomplished in response to a gradient of FGF8 (high posteriorly to low

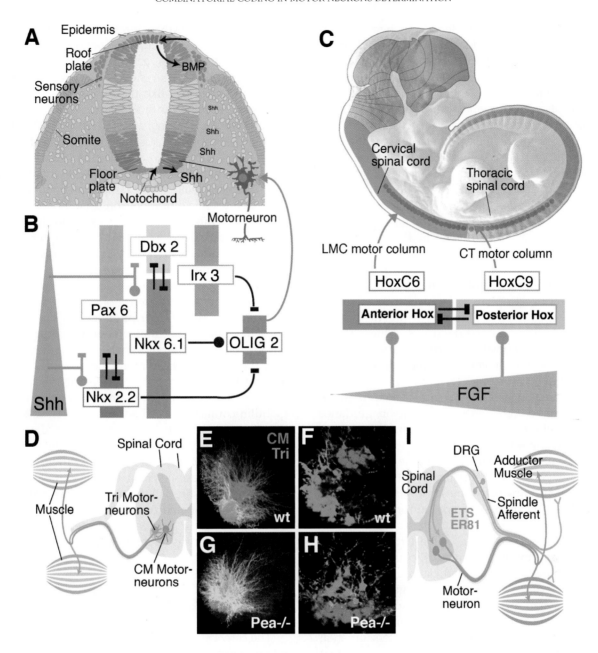

anteriorly) secreted by the paraxial mesoderm. This gradient establishes domains of different Hox genes at different levels, boundaries between which are sharpened by reciprocal inhibition. This anterior-posterior logic of motor neuron identity is thus strikingly similar to that governing the positioning of the neurons in the dorsal ventral axis of the spinal cord.

Motor columns are subdivided into classes that innervate different groups of muscles. Transcription factors of the LIM homeodomain (LIM-HD) family are involved in this further specification, as motor neurons expressing different combinations of LIM-HD factors such as Islet-1, Islet-2, and Lim-3 are expressed in dif-

ferent motor neuron classes that innervate different groups muscles. This is particularly well illustrated in the primary motor neurons (Rop, Mip, and Cap) of zebrafish, which can easily be distinguished by their relative rostral to caudal position in each spinal segment as well as their axon trajectories to distinct muscle regions. These three neurons express different combinations of LIM transcription factors. The identity of these neurons, as judged by their projection to different muscle regions, is influenced by the position in which they develop. These projection patterns can therefore be changed by transplantation in which the position of these neurons is changed prior to axonongenesis. Transplantations that lead to such switches in projection correlate with induced changes in LIM-HD code suggesting a causal link.

Motor neurons that innervate specific muscles are grouped into motor pools, which distinguished by the expression of distinct members of the ETS family of transcription factors. For example the ETS gene *ER81* is expressed in the motor neurons that innervate the limb adductor muscle in chicks, whereas the iliotrochanter motor neurons express the *PEA3* ETS gene (Fig. 15.12I). The initial expression of these ETS genes depends on peripherally derived signals such as GDNF and coincides with the arrival of motor neuron terminals in the vicinity of their specific targets. In PEA3 knockout mice, the motoneurons that are normally PEA3 positive fail to develop normally so muscles that are normally targets of these neurons are hypoinnervated and become severely atrophic. Not only do these motor neuron pools fail to branch normally within their target muscles, the cell bodies and dendrites of these motor neurons also are mispositioned within the spinal cord, which is critical for the control of their reflexive and descending innervation (Fig. 15.12D–H).

Recent work suggests that these ETS genes regulate the expression of cell adhesion and axon guidance factors that help motor neurons recognize their target. Interestingly, the sensory afferents that innervate the stretch receptors, in particular muscle, express the same ETS gene as the motor neurons that innervate that muscle, and this helps the sensory neuron axons find the dendrites of these motor neurons completing the monosynaptic stretch reflex.

The motor neuron system of the *Drosophila* larva, responsible for crawling, is relatively simple and has a reproducible segmental organization located in the ventral nerve cord of the larva (Landgraf and Thor, 2006). In each half segment, there are usually 36 motoneurons arising from about 10 neuroblasts (Fig. 15.13A,B). Some project through the intersegmental nerve (ISN) to innervate the more interior muscles that span the length of each segment and cause the animal to shorten, and others project through the segmental nerve (SN) to innervate the more exterior set of muscles that are more radial in orientation and cause the animal to lengthen. This is comparable to the division in vertebrate limbs, where the sets of (usually antagonistic) muscles derived from the ventral and dorsal muscle masses are innervated by motoneurons, whose axons project along divergent paths.

As in the vertebrate, combinatorial codes specify different motor neuronal types. In *Drosophila*, this subdivision works down to the level of single cells. The first subdivision in *Drosophila* is between ventral projecting motor neurons, vMNs, and dorsal projecting

motor neurons, dMNs. Interesting vMN identity is controlled by the combinatorial action of the *Nkx6* and *HB9* homeobox genes, a similar combination to that used to specify motor neurons in vertebrates. The vMNs in *Drosophila*, like the motor neurons in vertebrates, are specified into subclasses that project along specific nerves to innervate specific muscles by the further expression of combinatorial codes of LIM-HD factors, just as in vertebrates (Fig. 15.13C). In *Drosophila*, motor neurons often arise sequentially from a common neuroblast. For example, the neuroblast NB 7-1 generates five motor neurons that innervate nearby muscles. In this case, the temporal coordinate genes that were described earlier in this chapter (Kr, Hb, Pdm, Cas, and Gh) are involved in the specification of unique identities of these motor neurons.

What aspects of neuronal phenotype do these combinatorial codes of transcription factors control? Islet and Lim3 are two transcription factors expressed in a particular subclass of motor neuron that sends axons out the transverse nerve. When either of these genes is lost, the axons of these motor neurons defasciculate, leading to the failure of transverse nerve formation. These motor neurons all express a specific immunoglobulin-containing cell-adhesion molecule called Beat-1c, and the *beat-ic* gene is likely a direct target of these transcription factors. The defasciculation observed in islet and Lim3 mutants can be rescued by the forced expression of Beat-1c in these motor neurons, strongly suggesting that the combinatorial code in *Drosophila* motor neurons, like that in the vertebrate motor neurons, regulates cell adhesion factors that help motor neurons navigate to their targets.

CELLS OF THE CEREBRAL CORTEX

The cerebral cortex of mammals is the neural tissue where the conscious processing of information occurs. So, it is natural to want to know about the various neuronal types here and how they arise. Neuroepithelial cells in anterior-dorsal region of neural tube divide extensively to cause the ballooning out of two telencephalic vesicles. Further proliferation eventually produces an enormous folded sheet of tissue (about 4 mm thick and about 1.6 m^2 in area when unfolded in humans) that forms the cerebral cortex. The neocortex, which forms most of the cerebral cortex in humans, is a six-layered structure, with each layer having its own particular array of neuronal cell types. Differences in the thicknesses and cellular densities of particular layers have enabled histologists to distinguish as many as 100 cytoarchitectonically distinct cortical

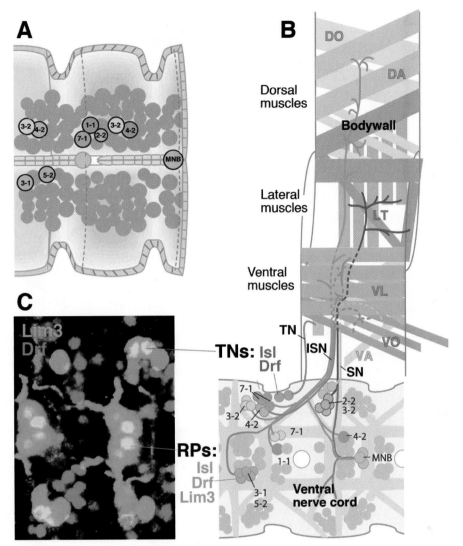

FIGURE 15.13 Specification of motor neurons in the *Drosophila* ventral cord. (A) Neuroblast map of two embryonic segments, showing in color the subset of neuroblasts that give rise to motor neurons. (B) Schematic flattened view of one body segment, showing in different colors the pattern of bodywall muscles and motor neurons innervating these muscles. The specific motor neuron(s) innervating each of the 30 muscle fibers comprising the musculature has been mapped. Motor neurons are color coded such that their color matches the one of the neuroblast from which it descends (see A), and the muscle it innervates. As in vertebrates, distinct groups of motor neurons are specified by expressing transcription factors of the LIM and POU family. Shown here and, photographically, in panel C are the RP neurons, expressing the combination Isl, Drf, and Lim3, and the transverse nerve (TN) motor neurons that express only Isl and Drf. Mutations in any of these genes result in motor neuronal abnormalities. (B) (From Certel and Thor, 2004).

areas in humans. Most neocortical cells and all glutamatergic ones originate from neuroepithelial cells at the apical surface or ventricular zone of the telencephalic vesicles. Asymmetric divisions of these cells generate one daughter that remains neuroepithelial and apical, and another daughter that migrates basally. Most of the time, the migrating daughter becomes a postmitotic cortical neuron, but in some of cases and increasing as development proceeds, the migrating daughter travels only a short distance to the subventricular zone, where it divides symmetrically, generating two neurons.

Cortical precursors, once generated, must then migrate basally away from the ventricular and subventricular zones, through an intermediate zone that will later be filled with axons, until they arrive at the cortical plate where they stop migrating and begin to differentiate into an enormous variety of specialized cortical cell types. This basal migration is coupled with an inside-out pattern of histogenesis in which the first-born cells take up the deep layers of the cortical plate and the later born cells migrate past the earlier generated neurons to occupy more superficial layers (Fig. 15.14A) (Angevine and Sidman, 1961). Cortical cells

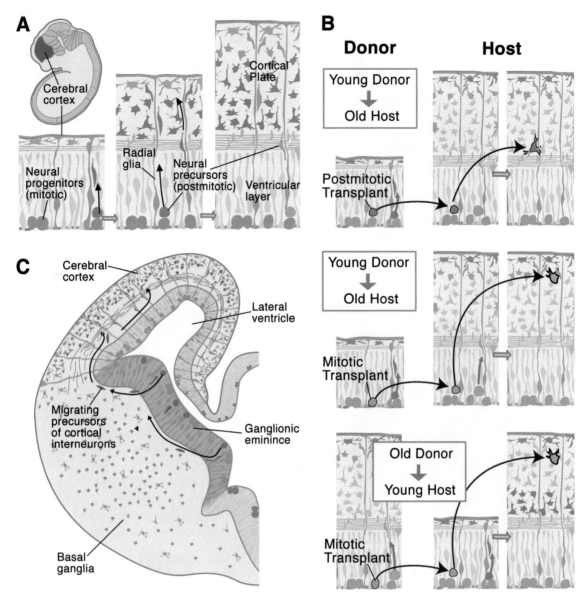

FIGURE 15.14 Laminar fate determination in the cerebral cortex. (A) Morphogenesis of the mammalian cerebral cortex. Neural precursors are born in the ventricular layer and migrate away from the ventricular surface, following tracks provided by radial glial cells. The first born cells are the Cajal-Retzius neurons (left in figure). Later born neurons accumulate in a dense matrix of cells, the cortical plate (middle of figure). In this plate, neurons are ordered by birth date in such a way that older neurons (magenta) remain in deep layers, and younger neurons (blue) migrate through the deep layers to attain a superficial position (right in figure). (B) Top: If postmitotic ventricular cells from young donors (which normally would become deep cells) are transplanted into an old host, they do not adapt to their new environment and develop as deep layer neurons (arrow). However, if ventricular cells that have not yet exited the cell cycle are transplanted from young donors when into an old host (middle panel), they adapt to their new environment and develop as superficial neurons (arrow). If, on the other hand, premitotic progenitors from old donors are transplanted into young brains, they can take later fates and become upper layer neurons. (C) Migration route of GABAergic interneurons from the ganglionic eminences in the ventral telencephalon into the cortical plate in a mammalian embryo.

are specified with respect to their fates before they migrate, for if postmitotic cells from the ventricular zone of a young embryo (destined for deep layers) are transplanted into the ventricular zone of an older one that is generating superficial cells, then the transplanted cells will still migrate to deep layers and have deep layer fates (Fig. 15.14B, top panel). Consistent with this, markers of specific fates, such as the transcription factor NEX, first appear before cell migration (Wu *et al.*, 2005). However, if the progenitors from a young ventricular zone are in their last mitotic cycle when they are transplanted to the older environment, they can change their laminar identity, suggesting that cortical cell fate is still undetermined until the cells exit

BOX 15.1

NEURAL STEM CELLS

The proliferating progenitor cells or neuroblasts that give rise to more differentiated progeny but themselves remain in the cell cycle are called neural stem cells (NSCs) (Gage, 2000). Some frogs and fish continue to grow bigger throughout their lifetimes, and their brains grow with their bodies, adding new cells from localized stem cell populations. However, in most animals, including humans, most proliferating neural cells use themselves up during development so the sources of new neurons in an adult animal are extremely limited. This is why damage to the central nervous system is medically much more serious than damage to some other organs, such as the skin or liver, where stem cells that persist into adulthood can replace injured tissue. Nevertheless, it has been found that even in adult mammals, NSCs persist in certain "niches" near ventricular layers. Interestingly, one pocket of rich stem cell activity is in the hippocampus, where learning takes place. There is excitement about the potential of using NSC cells in a replacement strategy for brain damage due to injury, stroke, or degenerative diseases such as Parkinson's disease, or retinal degeneration. In the future, one could imagine harvesting NSCs from a patient, proliferating these cells in culture under conditions where they begin to differentiate as dopaminergic neurons, and using these neurons to replace ones lost due to damage or disease.

Important questions about neural stem cells in the adult brain abound. Why do these cells remain undifferentiated and capable of division when their neighbors have exited the cell cycle and differentiated? What signals do these cells need in order to stimulate their proliferation and differentiation? How can their differentiation toward particular types of neurons or glia be controlled experimentally? Although there is much speculation on the role of various signaling and growth factors in regulating these processes, more work needs to be done to identify the molecules that prevent stem cells from differentiating and the signals that release their potential.

Interestingly, the environment of the organism may play a key role regulating NSCs. Rats raised in complex environments containing toys and exercise wheels show more proliferation and survival of hippocampal stem cells than rats raised in simple cages. Stress works in the other direction and seems to inhibit stem cell proliferation and survival in the hippocampus. The identification of the intrinsic regulatory factors that regulate these NSC activities might prove invaluable in treating neural or glial degenerative conditions without the need for transplants.

Another source of stem cells in mammals is the inner cell mass of the early conceptus. These are the cells that give rise to all the tissues of the embryo proper (Vallier and Pedersen, 2005). When these embryonic stem cells (ESCs) are grown in culture, some of them turn into neural precursors. It seems that the lessons learned about the mechanisms of neural induction are paying off in this context, because neural inducing signals increase the probability that ESCs will follow a neural pathway. ESCs may have an advantage over neural stem cells from adults because they come from a stage in development when their potential fates are less restricted by the inheritance of intrinsic determinants.

Various laboratories are trying to turn ESCs into various types of neurons. As a great deal is known about the inductive signals that are critical to the generation of motor neurons in the vertebrate spinal cord, it may be asked whether this knowledge can be used to direct stem cells to a motor neuron fate. Jessell and colleagues (Wichterle et al., 2002) combined inducers that first turn mouse ESCs into neural progenitors with factors such as retinoic acid that drives posterior regional identity such as spinal cord and Shh that drives ventral fates in the spinal cord to push these ESCs to differentiate into neural then spinal and finally into motor neurons that express the same combinatorial codes of transcription factors as motor neurons in vivo. These ESC-derived motor neurons can be labeled by Green Fluorescent Protein (GFP) under the control of the HB9 promoter (expressed only in motor neurons), and when these fluorescent cells are transplanted into the embryonic spinal cord of a chick embryo, they extend axons, and form synapses with target muscles. Moreover such ES-derived motor neurons develop appropriate transmitter enzymes and receptors, and electrical properties, such that they can make perfectly functional synapses with muscle fibers. Thus, by using signaling pathways discovered by studying normal cell determination in the developing nervous system, biologists may be able to direct stem cells to form specific types of neurons.

Clearly, the more we know about neuronal determination, the more likely we will be able to direct stem cells down appropriate developmental pathways that will be useful in treating the damaged nervous system.

William A. Harris and Volker Hartenstein

the cell cycle (Fig. 15.14B, middle panel). Interestingly, the reverse experiment of transplanting premitotic progenitors of old donors to younger ventricular zones does not result in these cells assuming earlier, deeper-layer fates (Fig. 15.14B, bottom panel), suggesting that progression of determination is a one-way street on which there is no going back (McConnell, 1995).

These findings demonstrate a relationship between the cell cycle exit and cell determination in the cortex reminiscent of what was described earlier in this chapter for the *Drosophila* CNS and the vertebrate retina, and provide a plausible mechanism for how cytoarchitectonic differences in cortical lamination can be achieved simply by altering cell cycle kinetics in a regionally specific way. Thalamic afferents that innervate the cortex may have a role in this, as the invading axons release trophic factors that promote proliferation. Thus, different regions of the cortex could have differences in the number of cells in specific layers, due to the timing and density of the innervation that they receive.

Not all cells in the neocortex originate from local ventricular zone. The oligodendrocytes and many classes of GABAergic interneurons migrate from more ventral part of the telencephalic vesicles, the ganglionic eminences (Fig. 15.14C) (Anderson *et al.*, 1997). We do not know very much about how these subtypes are generated, except that regional and temporal identity is also of significance here (Butt *et al.*, 2005). Once these cells exit the cell cycle, they begin a long tangential migration to the cortical plate where they join contemporaneously and more locally generated precursors on their radial migration to the developing cortical plate. Arriving together in the right layers, these cells from different location begin to differentiate and interconnect, building the local circuits that process conscious information.

CONCLUSIONS

The generation of neurons involves a variety of extrinsic signals and intrinsic transcription factors that confer individual identities on postmitotic neurons. Whether vertebrate or invertebrate, this process of determination involves a similar set of molecular and cellular mechanisms, such as regional and temporal identity genes, proneural genes, signaling molecules and receptors, the Notch pathway, and asymmetric cell division. All these conspire, over the course of neuronal generation, to create progenitors with distinct combinations of transcription factors that uniquely specify the fates of the differentiating neurons down by regulating the expression of genes that affect

specific features of the cell such as axon and dendrite growth and targeting, the topics of Chapters 17, 18 and 21.

References

Akiyama-Oda, Y., Hotta, Y., Tsukita, S., and Oda, H. (2000). Mechanism of glia-neuron cell-fate switch in the Drosophila thoracic neuroblast 6-4 lineage. *Development* **127**, 3513–3522.

Anderson, S. A., Eisenstat, D. D., Shi, L., and Rubenstein, J. L. (1997). Interneuron migration from basal forebrain to neocortex: Dependence on Dlx genes. *Science* **278**, 474–476.

Angevine, J. B. J. and Sidman, R. L. (1961). Autoradiographic study of cell migration during histogenesis of cerebral cortex in the mouse. *Nature* **192**, 766–768.

Betschinger, J. and Knoblich, J. A. (2004). Dare to be different: Asymmetric cell division in Drosophila, C. elegans and vertebrates. *Curr Biol* **14**, R674–685.

Butt, S. J., Fuccillo, M., Nery, S., Noctor, S., Kriegstein, A., Corbin, J. G., and Fishell, G. (2005). The temporal and spatial origins of cortical interneurons predict their physiological subtype. *Neuron* **48**, 591–604.

Campos-Ortega, J. A. (1995). Genetic mechanisms of early neurogenesis in Drosophila melanogaster. *Mol Neurobiol* **10**, 75–89.

Cayouette, M., Poggi, L., and Harris, W. A. (2006). Lineage in the vertebrate retina. *Trends Neurosci* **29**, 563–570.

Certel, S. J. and Thor, S. (2004). Specification of Drosophila motoneuron identity by the combinatorial action of POU and LIM-HD factors. *Development* **131**, 5429–5439.

Chenn, A. and McConnell, S. K. (1995). Cleavage orientation and the asymmetric inheritance of Notch1 immunoreactivity in mammalian neurogenesis. *Cell* **82**, 631–641.

Cornell, R. A. and Ohlen, T. V. (2000). Vnd/nkx, ind/gsh, and msh/msx: Conserved regulators of dorsoventral neural patterning? *Curr Opin Neurobiol* **10**, 63–71.

Dasen, J. S., Liu, J. P., and Jessell, T. M. (2003). Motor neuron columnar fate imposed by sequential phases of Hox-c activity. *Nature* **425**, 926–933.

Doe, C. Q. (2006). Chinmo and neuroblast temporal identity. *Cell* **127**, 254–256.

Dorsky, R. I., Chang, W. S., Rapaport, D. H., and Harris, W. A. (1997). Regulation of neuronal diversity in the Xenopus retina by Delta signaling. *Nature* **385**, 67–70.

Flores, G. V., Duan, H., Yan, H., Nagaraj, R., Fu, W., Zou, Y., Noll, M., and Banerjee, U. (2000). Combinatorial signaling in the specification of unique cell fates. *Cell* **103**, 75–85.

Gage, F. H. (2000). Mammalian neural stem cells. *Science* **287**, 1433–1438.

Hartenstein, V. (2005). Development of Sensilla. In "Comprehensive Insect Physiology, Biochemistry, Pharmacology and Molecular Biology" (Gilbert, L., Gill, S., Kostas, I., eds.).

Isshiki, T., Pearson, B., Holbrook, S., and Doe, C. Q. (2001). Drosophila neuroblasts sequentially express transcription factors which specify the temporal identity of their neuronal progeny. *Cell* **106**, 511–521.

Jessell, T. M. (2000). Neuronal specification in the spinal cord: Inductive signals and transcriptional codes. *Nat Rev Genet* **1**, 20–29.

Jones, B. W. (2005). Transcriptional control of glial cell development in Drosophila. *Dev Biol* **278**, 265–273.

Kessaris, N., Pringle, N., and Richardson, W. D. (2001). Ventral neurogenesis and the neuron-glial switch. *Neuron* **31**, 677–680.

Kiernan, A. E., Cordes, R., Kopan, R., Gossler, A., and Gridley, T. (2005). The Notch ligands DLL1 and JAG2 act synergistically to regulate hair cell development in the mammalian inner ear. *Development* **132**, 4353–4362.

Lai, E. C. and Orgogozo, V. (2004). A hidden program in Drosophila peripheral neurogenesis revealed: Fundamental principles underlying sensory organ diversity. *Dev Biol* **269**, 1–17.

Landgraf, M. and Thor, S. (2006). Development and structure of motoneurons. *Int Rev Neurobiol* **75**, 33–53.

Livesey, F. J. and Cepko, C. L. (2001). Vertebrate neural cell-fate determination: Lessons from the retina. *Nat Rev Neurosci* **2**, 109–118.

McConnell, S. K. (1995). Constructing the cerebral cortex: neurogenesis and fate determination. *Neuron* **15**, 761–768.

Mizutani, C. M., Meyer, N., Roelink, H., and Bier, E. (2006). Threshold-dependent BMP-mediated repression: a model for a conserved mechanism that patterns the neuroectoderm. *PLoS Biol* **4**, e313.

Mollereau, B. and Domingos, P. M. (2005). Photoreceptor differentiation in Drosophila: From immature neurons to functional photoreceptors. *Dev Dyn* **232**, 585–592.

Petersen, P. H., Zou, K., Hwang, J. K., Jan, Y. N., and Zhong, W. (2002). Progenitor cell maintenance requires numb and numblike during mouse neurogenesis. *Nature* **419**, 929–934.

Poggi, L., Vitorino, M., Masai, I., and Harris, W. A. (2005). Influences on neural lineage and mode of division in the zebrafish retina in vivo. *J Cell Biol* **171**, 991–999.

Qian, X., Shen, Q., Goderie, S. K., He, W., Capela, A., Davis, A. A., and Temple, S. (2000). Timing of CNS cell generation: A programmed sequence of neuron and glial cell production from isolated murine cortical stem cells. *Neuron* **28**, 69–80.

Schlosser, G. (2006). Induction and specification of cranial placodes. *Dev Biol* **294**, 303–351.

Skeath, J. B. and Carroll, S. B. (1992). Regulation of proneural gene expression and cell fate during neuroblast segregation in *Drosophila*. *Development* **114**, 939–946.

Stern, C. D. (2006). Neural induction: 10 years on since the "default model." *Curr Opin Cell Biol* **18**, 692–697.

Vallier, L. and Pedersen, R. A. (2005). Human embryonic stem cells: An in vitro model to study mechanisms controlling pluripotency in early mammalian development. *Stem Cell Rev* **1**, 119–130.

Vrieseling, E. and Arber, S. (2006). Target-induced transcriptional control of dendritic patterning and connectivity in motor neurons by the ETS gene Pea3. *Cell* **127**, 1439–1452.

Wichterle, H., Lieberam, I., Porter, J. A., and Jessell, T. M. (2002). Directed differentiation of embryonic stem cells into motor neurons. *Cell* **110**, 385–397.

Wu, S. X., Goebbels, S., Nakamura, K., Nakamura, K., Kometani, K., Minato, N., Kaneko, T., Nave, K. A., and Tamamaki, N. (2005). Pyramidal neurons of upper cortical layers generated by NEX-positive progenitor cells in the subventricular zone. *Proc Natl Acad Sci U S A* **102**, 17172–17177.

Zhou, Q., Choi, G., and Anderson, D. J. (2001). The bHLH transcription factor Olig2 promotes oligodendrocyte differentiation in collaboration with Nkx2.2. *Neuron* **31**, 791–807.

Zipursky, S. L. and Rubin, G. M. (1994). Determination of neuronal cell fate: Lessons from the R7 neuron of Drosophila. *Annu Rev Neurosci* **17**, 373–397.

William A. Harris and Volker Hartenstein

16

Neurogenesis and Migration

One of the remarkable features of the developing nervous system is the wide-ranging migration of precursor cells. Over the past decade, genetic studies in invertebrate and live imaging of vertebrates have given insights into the mechanisms responsible for neuronal motility. In particular, signal pathways have been identified that initiate cell movements and govern the dynamics of cell motility. A conserved family of polarity protein complexes, and their signaling pathways, orient the mitotic spindle in dividing precursor cells during neurogenesis, and induce postmitotic progenitor cell to the extend motile protrusions, driven by actin polymerization dynamics and stabilized by adhering to adjacent cells or to the extracellular matrix (ECM) via transmembrane receptors linked to the actin cytoskeleton. The molecular nature of the cues guiding cell movement are only beginning to be understood and appear to differ between different regions of the nervous system. This chapter compares the strategies used to generate final cell pattern in the peripheral and central nervous systems, considering classical views of brain development that resulted from anatomical studies, as well as emerging molecular programs of neurogenesis and cell migration that have resulted from genetic and molecular genetic studies.

INTRODUCTION

In the late nineteenth century, the Swiss histologist Wilhelm His (His, 1886) developed methods to section embryonic brain and resolve individual cells in the *neurospongium*. Subsequently the Spanish neuroanatomist Santiago Ramon y Cajal (Ramon y Cajal, 1995) developed protocols to fix nervous tissues of embryos

with barbiturates, and to modify methods discovered by the Italian histologist, Camillo Golgi, to stain cells with nitrates of silver, by immersing the tissue into photographic reagents. By examining embryos of different vertebrate species, Ramon y Cajal was able to discern detailed features of immature neurons during development and to chronicle the growth and connectivity of the major classes of nerve cells. In studies of CNS development, he imagined the extension of growth cones from nascent neurons, which he called "soft battering rams," and chronicled the arborization patterns of major classes of neurons in different brain regions.

The idea that CNS neurogenesis occurs in the thin epithelium of the embryonic brain vesicles became apparent when Fred and Mary Sauer showed that cells in the neuroepithelium contain different amounts of DNA, with diploid cells located on the ventricular surface (Sauer and Chittenden, 1959). Clear images of the pseudo-columnar character of the neuroepithelium of embryonic rat brain were achieved in ^{3}H-Thymidine labeling studies of the neocortex (Berry and Rogers, 1965), and in serial scanning and electron micrographs (Fig. 16.9A), which demonstrated movements of the nuclei during the principal phases of the cell cycle (Hinds and Ruffett, 1971). Cells in interphase extended processes across the thin wall of the epithelium, one cell deep at this stage. Cells entering mitosis retracted the processes and moved to the ventricular surface to divide. The changing shapes of the cells, and the apparent movement of the nucleus within the cells as they undergo morphological changes during the phases of the cell cycle are termed "interkinetic movements." A cardinal feature of the neuroepithelium is its "upside down" orientation relative to epithelia of other tissues. This relates to the fact that the neuroepithelium forms

351

a tube, which orients the apical surface of neuroblasts in the epithelial sheet toward the lumen of the neural tube. This "ventricular zone" (i.e., zone of the epithelium facing the brain vesicles) is the site of primary neurogenesis in the CNS.

The polarization of the epithelium of the neural tube depends on the formation of adhesion junctions between cells (Joberty *et al.*, 2000; Macara, 2004) (Fig. 16.9B). Two classes of trans-membrane adhesion proteins, cadherins and nectins, establish adhesions between epithelial cells. Aggregation of adhesion sites to an adherens or gap junction in the basal portion of the cells connects the cells, and initiates the movement of cytoskeletal elements to the junctions. A second adhesion site forms at the apical side of the cell, which typically becomes a tight junction. In polarized mammalian epithelial cells, the conserved polarity proteins PAR3/ASIP, PAR-6, and aPKC form a complex that is localized to the tight junction. Cdc42, a polarity regulator from yeast to humans, and Rac GTPases activate the Par-3/Par-6/aPKC-complex, which binds other polarity proteins in some cells (Gao and Macara, 2004; Macara, 2004; Chen and Macara, 2005). A key aspect of "tight" junctions is the requirement for Par-3.

During neuroblast division in ventricular zones, the microtubule-organizing center (MTOC) nucleates astral microtubules required for spindle orientation. Cdc42 is a master polarity regulator from yeast to humans. Cdc42, aPKC, and dynein/dynactin activity control spindle orientation and govern symmetric versus asymmetric patterns of neuroblast division (Yu *et al.*, 2006). Thus, Cdc42-induced MTOC orientation may contribute to whether a cell undergoes symmetric versus asymmetric divisions, and as discussed later, to the "to and fro" movements of dividing cells within the neuroepithelium. The mechanism of "checkpoint control" is critical for CNS neurogenesis as this step governs assembly of the mitotic spindle, and subsequent segregation of chromosomes into daughter cells. In *Drosophila*, Lis1/dynactin complexes physically associate and colocalize on centrosomes, spindle microtubules, and kinetochores to regulate spindle formation and cell cycle checkpoint release (Siller *et al.* 2005).

The neurons of both the peripheral and central nervous systems originate from the neural tube and the folds of the neural tube. The cells of the PNS are generated first, as they delaminate in a specialized region of the dorsal aspect of the neural tube called the neural crest. Neural crest neuroblasts migrate out into the developing nonneuronal tissues, where they will form the axon tracts of the periphery, as the progenitors of the CNS are beginning to establish the regions of the brain and spinal cord that will receive the sensory information from the PNS and establish brain circuitry. We will therefore discuss neurogenesis and migration in the peripheral nervous system first, followed by patterns of neurogenesis and migration that establish the CNS.

DEVELOPMENT OF THE PERIPHERAL NERVOUS SYSTEM

The Neural Crest Is a Migratory Cell Population That Forms Multiple Derivatives

The *neural crest* is a transient population of cells, so named because they arise on the "crest" of the closing neural tube. This cell population is unique to vertebrates and forms most of the peripheral nervous system (PNS). Neural crest cells migrate extensively along characteristic pathways and give rise to diverse and numerous derivatives. All the *dorsal root*, *sympathetic*, *parasympathetic*, and *enteric ganglia* are derived from neural crest cells. Furthermore, most *cranial sensory ganglia* receive a contribution from the neural crest, with the remaining cells derived from *ectodermal placodes*. In addition to forming neurons and glia of the peripheral nervous system, neural crest cells form melanocytes, cranial cartilage, and adrenal chromaffin cells. This wide variety of cell types arises from precursors in the neural folds and neural tube that are multipotent and have stem cell properties (Bronner-Fraser and Fraser, 1988; Stemple and Anderson, 1992) (see Chapter 15 for a discussion of stem cells).

The neural crest originates at the border between the neural plate and the nonneural ectoderm by an inductive interaction between these two tissues (Selleck and Bronner-Fraser, 1995) (Chapter 14). Precursors with the potential to form neural crest initially are contained within the dorsal portion of the neural tube, and these premigratory neural crest cells subsequently emerge from the neural tube. Cells in the neural tube are epithelial and look much like contiguous soda cans that have a defined top (apical) and bottom (basal) side. They are closely apposed to one another and are connected by various types of adhesive junctions. In contrast, migratory cells such as neural crest cells are mesenchymal, having a fibroblast-like morphology that facilitates their movement. Thus, precursor cells within the neural tube change from an epithelial to a mesenchymal morphology as they turn into migratory neural crest cells. Such an *epithelial-to-mesenchymal conversion* is a common event in development during the formation of tissues and organs. The transcription factor *Slug* (also called *Snail2*) is expressed in premigratory and early migrating cells and represents an early known neural crest marker. *Slug* has been associated with epithelial–mesenchymal conversions in a number of cell types and its function appears to be necessary for the emigration of neural crest cells. Initiation of neural crest cell migration proceeds in a rostral-to-caudal progression along most of the neural axis, following upon the heels of the head-to-tailward closure of the neural tube. After emigration, these cells move in a highly patterned fashion through neighboring tissues and localize in diverse sites.

Initiation of Migration

As they change from epithelial to migratory mesenchymal cells, neural crest cells undergo changes in adhesive properties. While within the neuroepithelium, neural tube cells express high levels of the *cell adhesion molecules* N-cadherin and cadherin. Migrating cells down-regulate these cadherins during migration and up-regulate cadherin-7. Upon coalescing into ganglia and ceasing migration, cadherin-7 is down-regulated and N-cadherin-6b are again up-regulated (Nakagawa and Takeichi, 1998). This suggests that a shift in cell surface and adhesive properties may accompany the onset and cessation of migratory behavior. Neural crest cells also turn on RhoB during the initiation of migration.

After leaving the neural tube, neural crest cells encounter extracellular spaces that are rich in *extracellular matrix* (ECM) *molecules*, such as fibronectin, laminin, collagens, and proteoglycans. These may serve as a good migratory substrate and, indeed, the neural crest cell surface has abundant integrin receptors that mediate adhesion to ECM molecules. Furthermore, antibodies to the β_1 *subunit of integrin* cause severe perturbations in neural crest development in the head. They appear to prevent the migration of some neural crest cells from the cranial neural tube. In the trunk, function-blocking antibodies that interfere with the a_4 *subunit of integrin* cause defects in neural crest cell movement, but fail to alter the segmental pattern of migration through the somites. These results suggest that perturbing integrin function alters the properties of migratory cells, but not their overall metameric pattern of migration. Thus, cell–matrix interactions may play a permissive role in the migration of neural crest cells through the somites, but cannot play an instructive role in directing the precise patterns or pathways for crest cell migration.

Techniques for Following Neural Crest Cells

When neural crest cells emerge from the neural tube, they first enter a cell-free space in which they are easily identifiable (Fig. 16.1). Subsequently, they invade

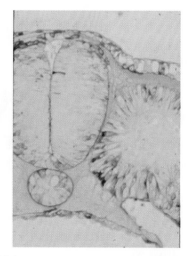

FIGURE 16.1 A transverse section through a chick embryo showing neural crest cells initiating migration from the dorsal neural tube and into an ECM-filled space. Courtesy of Jan Lofberg.

other tissues in which they are difficult to distinguish. Therefore, it is necessary to *mark* neural crest cells in order to study their migratory patterns and derivatives. A variety of techniques have emerged for this purpose, ranging from transplantation of tissue containing premigratory neural crest cells to lineage tracers and molecular markers. *Neural tube transplantations* have provided a wealth of information about neural crest migratory pathways and, in particular, derivatives arising from this population. Although this approach was widely used in amphibians for studying numerous embryonic processes, it has been applied most successfully to the analysis of neural crest migratory pathways and derivatives in avian embryos (LeDouarin and Kalcheim, 1999). Initial experiments in birds involved transplanting a neural tube from a donor labeled with the radioactive marker [³H] thymidine into an unlabeled host (Weston, 1963). Neural crest cells generated from the labeled neural tube also were labeled and could be identified readily in the periphery. This technique yielded important information about early stages of neural crest migration, but the label became diluted with further cell division and was not useful for looking at long-term differentiation of neural crest cells into diverse derivatives.

To circumvent this problem, LeDouarin took advantage of the ability to perform grafts between related species of birds, such that the grafted cells were indelibly marked (Fig. 16.2). This created an interspecific *chimera*, from the Greek meaning "fabulous monster," containing a donor quail portion of the neural tube and neural crest in an otherwise normal chick host embryo. Quail neural crest cells migrate away from the grafted neural tubes and can be recognized easily within the

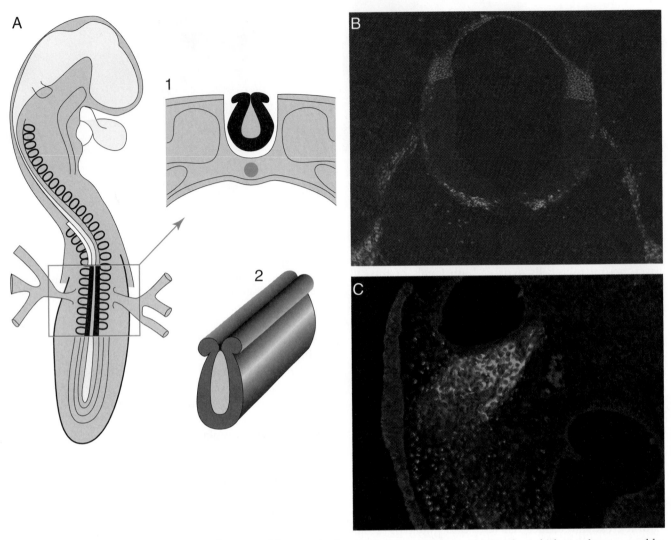

FIGURE 16.2 Procedures for grafting a fragment of the neural primordium from a donor quail into a host chicken embryo as used by LeDouarin and colleagues. (A) View of an avian embryo with anterior at the top. Neural folds are shown in black in the boxed region; this structure is removed and transplanted to a host embryo. (1) Cross-section through the embryo in regions shown in the box with the neural tube (2) shown in black. From LeDouarin (1982). (B) An example of a section through an embryo after grafting of a quail neural tube into a chick host. Quail cells are recognized by a quail-specific antibody (red), whereas neurons are marked in green with a neurofilament marker. (C) A higher magnification section showing quail cells (red nuclei) incorporated into a neural crest-derived ganglion, stained green with a neurofilament antibody. Courtesy of Anne Knecht and Clare Baker.

host chick embryo by staining for condensed heterochromatin, which characterizes the quail but not chick cells.

More recently, this technique has been facilitated greatly by the advent of quail-specific antibodies. Use of the chick–quail marking system made it possible to demonstrate that neural crest cell populations originating from different axial levels follow distinct migratory pathways and give rise to different progeny once they reach their destinations. In addition to grafting experiments, antibodies that recognize neural crest cells, such as HNK-1 and NC-1 antibodies, made it possible to identify early migrating neural crest cells without the necessity of performing microsurgery. The exclusive use of neural crest antibodies has several pitfalls because these antibodies are neither entirely specific nor stain the full complement of neural crest cells. However, they do provide important confirmatory information about the pathways followed by neural crest cells. More recently, a number of other molecular markers for early neural crest populations have become available. These include the cell adhesion molecule cadherin-6b and the transcription factor Slug, expressed in the neural folds and neural crest cells during early stages of migration and Sox10, expressed in early migrating neural crest cells and later in neural crest-derived glia.

One problem with using antibodies or molecular markers to follow cell migratory patterns is that these do not represent true lineage markers. Shifts in their expression patterns could just as easily reflect up- or down-regulation of these molecules as changes in cell position. Therefore, approaches for labeling small groups as well as large populations of neural crest cells have been employed to follow individual cell movements and interactions within the population. One successful approach has been to inject the *lipophilic vital dye DiI* or electroporating GPF constructs into the neural tube or neural folds. Because the dye is hydrophobic and lipophilic, it intercalates into all cell membranes that it contacts. Injection into the neural tube marks all neural tube cells, including presumptive neural crest cells, within its dorsal aspect. Because the time and location of injection can be controlled, this provides a direct approach for following migratory pathways. When dye injections are made focally into neural folds, this technique can be used to label very small numbers of cells. It has the further advantage of being applicable to almost all vertebrates. A disadvantage, however, is that the dye becomes diluted with each cell division and, therefore, is not useful for examining the long-term differentiation of neural crest derivatives.

Because most studies of cell migration *in vivo* look only at the beginning and end point, little is known about the dynamics of neural crest cell movement. Advances in imaging technologies have made it possible to acquire more refined images *in ovo*. Time-lapse movies of cranial neural crest migration can be generated by injecting DiI into the lumen of the chick neural tube, and high-resolution confocal images can be made of the labeled cells over time (Fig. 16.3). These movies make it possible to visualize the migratory behavior of neural crest cells as they emerge from the cranial neural tube and migrate toward the branchial arches. As the embryo develops, it is possible to follow individual cell movements within the branchial arches, which will give rise to the bone and cartilage of the jaw. Neural crest cells move in defined streams and appear to form distinct chains in which cells are oriented along the same trajectory and remain in contact via their processes. This suggests a large degree of intercommunication between migrating populations of cranial neural crest cells.

A number of different methods have proved useful for following the pathways of neural crest migration, including neural tube transplantations, vital dye labeling, and antibody staining. These have made it possible not only to establish the various routes of neural crest migration occurring at different axial levels, but also to follow the derivatives of neural crest cells arising from distinct locations.

Regionalization of the Neural Crest Along the Body Axis

Despite differences in the nature of the techniques involved, quail/chick chimeras, molecular markers, and cell labeling have provided similar pictures of the migratory pathways followed by neural crest cells. These migratory pathways and the derivatives formed by neural crest cells are regionalized according to their original position along the anterior/posterior axis, such that cells from a given axial level give rise to a characteristic array of progeny and follow distinct pathways from those arising at other axial levels (LeDouarin and Kalcheim, 1999; Noden, 1975). The different populations of neural crest cells arising along the neural axis have been designated as cranial, vagal, trunk, and lumbosacral (Fig. 16.4). Distinct cell types differentiate from these different populations.

At cranial levels, some neural crest cells contribute to the *cranial sensory ganglia* and the *parasympathetic ciliary ganglion* of the eye, whereas others migrate ventrally to form many of the *cartilaginous elements of the facial skeleton*. One of the interesting things about neural crest-derived bones of the head is that these are the only skeletal elements in the body that are derived from ectoderm. Precise quail/chick grafting experiments have determined the regions of neural tube from which neural crest cells arise to contribute to cartilaginous elements and cranial ganglia (Noden, 1975). Neural crest cells originating in the midbrain migrate primarily as a broad, unsegmented sheet under the ectoderm; they contribute to derivatives ranging from the skeleton around the eye, connective tissue, and membranous bones of the face, to the ciliary ganglion and trigeminal ganglia (LeDouarin and Kalcheim, 1999). Neural crest cells arising in the hindbrain migrate ventrally and enter the branchial arches to form the bones of the jaw.

Vagal neural crest cells migrate long distances to form the *enteric nervous system*, which also receives a contribution from the lumbosacral neural crest. Within the gut, the earliest generated crest cells move as a wave from anterior to posterior to populate the bowel, which they appear to populate in sequence such that the anterior portions are populated by neural crest cells first and the cells move to progressively more posterior sites. Mice bearing a "lethal spotted" mutation lack neural crest cells in a portion of their bowel. This leads to a lack of innervation in the region called aganglionic bowel. As a consequence, food waste fails

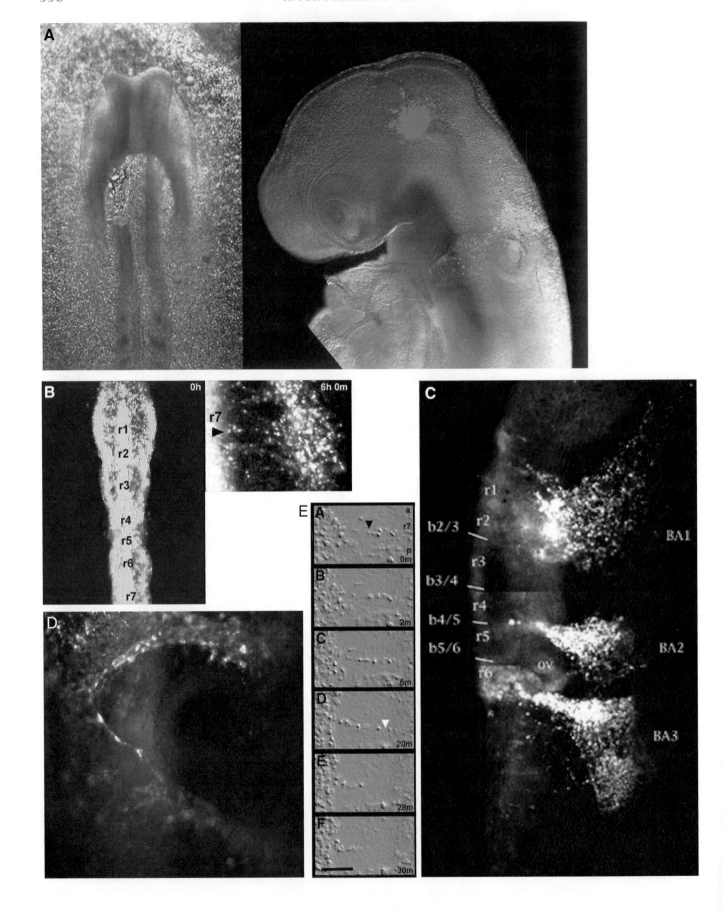

◀ **FIGURE 16.3** Labeling of neural crest cells with liphophilic tracer dyes. (A) On the left is a chick embryo viewed shortly after unilateral ablation of the neural folds of roughly half the neural tube at the level of the presumptive midbrain. A spot of DiO (green) was focally injected into the remaining neural tube, and DiI (red) was injected into the bordering intact neural folds. On the right, the same embryo after 48 h of further development. Cells that were originally in the neural tube had dispersed as migratory neural crest cells from both the green- and the red-labeled spots. (B–D) Different views of time-lapse movies of embryos in which the neural tube and premigratory neural crest cells were labeled with DiI. (B) Two views of the same embryo immediately after DiI labeling (left) and several hours after neural crest migration from the hindbrain. Rhombomeres (r1–r7) are indicated. (C) A similar embryo at a later time point by which time neural crest cells have migrated into the branchial arches (BA). (D) At higher magnification of neural crest streams in the branchial arches, cells seem to be in close contact as if following each other in narrow streams. (E) Following individual cells by time-lapse cinematography demonstrates close and maintained connections between individual cell pairs. Courtesy of Paul Kulesa and Scott Fraser.

to move through this portion of the bowel, leading to a disorder called "mega-colon." A similar defect is observed in humans leading to Hirschsprung's disease. One cause of this disorder is from a defect in molecules called endothelins and their receptors. The failure of neural crest migration in the aganglionic bowel of lethal spotted mutant mice is caused by a defect in mesenchymal components of the gut, whereas the neural crest cells themselves are normal.

Trunk neural crest cells follow two primary migratory pathways (Fig. 16.5): a *dorsolateral pathway* between the ectoderm and the somite and *a ventral pathway* through the rostral half of each sclerotome, the mesenchymal portion of the somite that will go on to form the vertebrae.

Cells following the dorsolateral stream give rise to *melanocytes*. Those cells following the ventral pathway give rise to the peripheral nervous system of the trunk, including the chain of *sympathetic ganglia* and *dorsal root ganglia*, as well as *chromaffin cells* of the adrenal medulla. In addition to these neurons, these cells generate *glia* of the peripheral ganglia and *Schwann cells* that ensheathe and myelinate peripheral axons.

The neural crest can be subdivided into specific subpopulations based upon their level of origin along the neural tube. Different populations follow different migratory pathways and form characteristic types of derivatives. For example, cranial bone and cartilage only arise from the cranial neural crest.

Grafting of Neural Crest Cells to New Locations— Intrinsic versus Extrinsic Cues?

Neural crest cells that arise and migrate at different axial levels assume different fates. Thus, it is possible that they are specified at early times to take on particu-

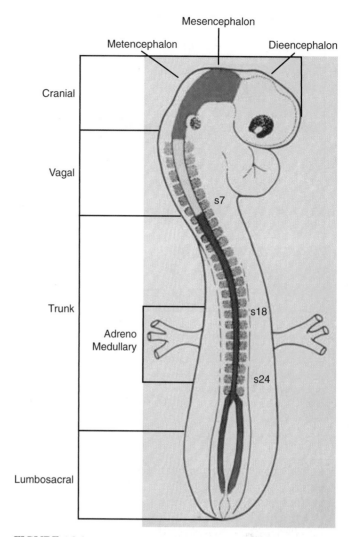

FIGURE 16.4 Schematic diagram showing different levels of the neural axis from which neural crest cells arise. From anterior to posterior, the neural axis can be divided into cranial, vagal, trunk, and lumbosacral levels. Each gives rise to distinct derivatives.

lar fates. For example, one possibility is that neural crest cells are preprogrammed by their axial level to populate specific derivatives; alternatively, neural crest cells might be multipotent and migrate naively into available locations, where local cues provide them with instructions about their fates. To test the role of migratory pathways in choice of fate, neural tubes from particular axial levels have been grafted to new locations. This is referred to as a "heterotopic" graft. These experiments have revealed, for example, that when vagal neural crest cells are grafted to trunk regions, they form normal trunk derivatives (dorsal root ganglia, sympathetic ganglia, etc.) and normal vagal derivatives (enteric ganglia of the gut) (Fig. 16.6). The gut is immediately ventral to the trunk region and is connected to it by a narrow piece of tissue called the dorsal mesentery. Despite their close proximity, trunk

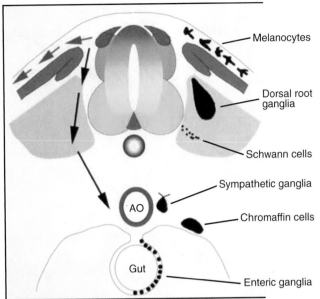

FIGURE 16.5 Schematic diagram of an idealized embryo in cross-section showing pathways of neural crest migration in trunk and derivatives formed. Neural crest cells migrate along two primary pathways: dorsally under the skin or ventrally through the sclerotome. Dorsal migrating cells form pigment cells, whereas ventrally migrating cells give rise to dorsal root and sympathetic ganglia, Schwann cells, and cells of the adrenal medulla. Drawn by Mark Selleck.

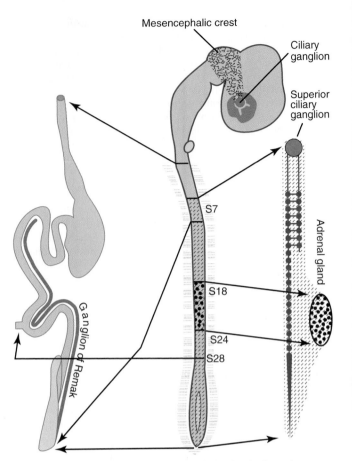

FIGURE 16.6 Neural crest cells at different levels along the anterior–posterior axis give rise to distinct autonomic and adrenomedullary derivatives in avian embryos. In the cephalic region (center), mesencephalic crest cells populate the ciliary ganglion. Ganglia of the sympathetic chain (right), including the superior ciliary ganglion, are formed from spinal neural crest cells originating caudal to somite 5. Cells of the adrenal medulla (right) originate exclusively from neural crest cells between somites 18 and 24. Vagal neural crest cells generated between somites 1 and 7 form enteric ganglia (left), whereas cells of the ganglion of Remak (left) are derived from the lumbosacral neural crest posterior to somite 28. Drawing provided by Dr. M. Bronner-Fraser.

neural crest cells normally fail to invade the gut where enteric ganglia form. When vagal neural tubes are grafted in place of trunk neural tubes, however, donor vagal neural crest cells do invade the gut and form enteric ganglia. Therefore, these cells can respond to normal trunk neural crest migratory pathways, but, in addition, some cells behave in a "uniquely vagal" fashion and migrate directionally to the gut. This indicates some intrinsic differences between the two populations.

Along similar lines, when cranial neural tubes are grafted in place of the trunk neural tube, donor cranial neural crest cells form some normal derivatives in the trunk-like dorsal root and sympathetic ganglia. Other cells, however, fail to migrate and instead differentiate into ectopic cartilage. In the reciprocal experiment, avian trunk neural crest cells grafted to the head region appear unable to generate cartilage at all, although they can participate in the formation of elements of the cranial ganglia (LeDouarin and Kalcheim, 1999) (Fig. 16.6).

These experiments suggest two things: (1) environmental factors can influence neural crest cells from different axial levels to express a broader range of fates than they would normally express when left *in situ* and (2) some neural crest cells undergo their intrinsic

program even when grafted to an ectopic site. Thus, some combination of *intrinsic* and *extrinsic information* is likely to govern neural crest cell fate decisions. The migratory pathways taken by neural crest cells are likely to play an important regulatory role in cell fate specification. Indeed, different migratory pathways contain different distribution patterns of important inducing factors. A number of factors, including BMPs, neuregulins, and glucocorticoids, have been demonstrated to influence the choice of neural crest cells into neural, glial, or chromaffin lineages (see Chapter 15). When added to multipotent neural crest stem cells in culture, these factors can drive these cells into neural, glial, or chromaffin lineages, respectively. Accordingly, BMP-7 is present in the dorsal aorta adjacent to the location where trunk neural crest cells differentiate into sympathetic neurons. Similarly, glucocorticoids are produced by the adrenal cortex, which surrounds the adrenal medullary cells.

In addition to specific growth factors, the timing of emigration may play an important role in eventual neural crest cell fate decisions. Neural crest cells exhibit an orderly pattern in the timing of migration, with cells initially following the ventral pathway and later contributing to progressively more dorsal derivatives. Although early and late migrating cranial neural crest cells appear to have a similar developmental potential, it is likely that the cell population does undergo a change in the range of fates that the cells can assume. For example, the last emigrating trunk neural crest cells in the bird give rise to pigment cells and appear to have a limited capacity to form sympathetic neurons. However, *stem cells* appear to persist in the neural tube and retain the potential to form neural crest. Well past the normal time of neural crest cell emigration, cells with the full range of neural crest potential can be isolated from various neural crest derivatives (Kruger *et al.*, 2002; Joseph *et al.*, 2004). In the embryo, these late-emigrating cells appear to contribute to a subpopulation of neural crest derived cells in the dorsal root ganglia.

Summary

Neural crest cells are somewhat plastic with respect to their prospective fates. When put into a new environment, they sometimes behave according to their new location. However, some cells act according to their original location and thus have some "intrinsic" information. With time, the developmental potential of neural crest populations becomes restricted, although a subpopulation of "stem cells" may remain until late stages.

Segmental Migration of Neural Crest Cells

A hallmark of the developing peripheral nervous system is its inherent *segmentation*. After neural crest cells migrate from the neural tube and through the somites, they condense to form segmentally arranged sensory and sympathetic ganglia. For each somite, a single sensory and sympathetic ganglion forms. This exquisite and reproducible pattern suggests the presence of some inherent segmental information in the embryo that is responsible for segmental migration and gangliogenesis of neural crest cells. In longitudinal sections through early embryos, it is clear that neural crest cells migrate in a metameric pattern, moving exclusively through the rostral half of each somite while failing to enter the caudal half (Fig. 16.7).

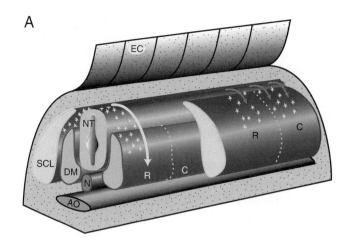

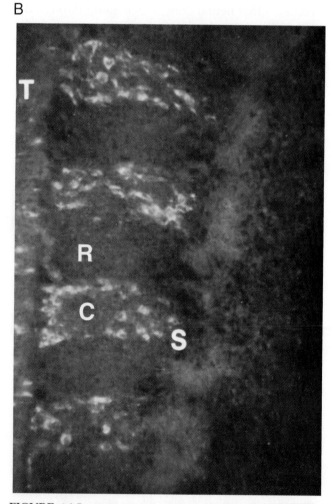

FIGURE 16.7 Trunk neural crest cells migrate in a segmental fashion. (A) Schematic diagram demonstrating that neural crest cells migrate through the sclerotomal portion of the somites, but only through the rostral half of the sclerotome. (B) In longitudinal section, neural crest cells (green) can be seen migrating selectively through the rostral half of each somitic sclerotome (S). From Bronner-Fraser (1986).

Such a segmental pattern of migration could be caused by inherent cues within the neural tube that direct neural crest cells to migrate segmentally. Alternatively, tissues through which neural crest cells migrate may contain patterning information that results in segmental migration. The relationship between neural crest cells and their surrounding tissues has been explored by manipulating the neural tube and/or somites in a series of grafting experiments. Removal of the somites results in the formation of huge, unsegmented neural crest-derived ganglia, suggesting that somites are necessary for the segmental migration of neural crest cells. However, inversion of the segmental plate in the rostrocaudal dimension reverses the pattern of neural crest migration, such that neural crest cells migrate through the half of the rotated sclerotomes that was originally rostral but is now caudal. Motor axons, which also traverse the rostral sclerotome in normal animals, exhibit similar behavior to neural crest cells after segmental plate rotation. This suggests that the information necessary to guide neural crest cells and motor axons is intrinsic to the somites. Furthermore, these experiments show that the rostrocaudal polarity of the somites is already established at the segmental plate stage.

Other experimental manipulations demonstrate that the caudal half of each somite is inhibitory whereas the rostral half somite is permissive for neural crest migration and motor axon guidance. If somites are constructed to contain only caudal sclerotome tissue, neural crest cells and motor axons fail to migrate altogether. Conversely, an all-rostral somite results in the absence of segmentation for both neural crest cells and motor axons. Taken together, these findings demonstrate that the segmental migration of both neural crest cells and motor axons is due to cues inherent in the somite. These cues may be caused by attractive cues in the rostral-half sclerotome, inhibitory cues in the caudal-half sclerotome, or a combination of both. The finding that somites containing all caudal-half sclerotomes cannot support neural crest migration strongly suggests that some of the guidance cues are inhibitory. This segmentally arranged streams of migrating neural crest cells, in turn results in the segmental organization of the neural crest-derived ganglia of the peripheral nervous system.

Although it was suggested that Eph/ephrin signaling might direct the pattern of trunk neural crest migration (Krull *et al.*, 1997; Wang and Anderson, 1997), the Eph and ephrin mutant mice that have been examined fail to exhibit trunk neural crest migration defects (Oriolli *et al.*, 1996; Wang *et al.*, 1998). Another receptor expressed by neural crest cells is Neuropilin-1 (Npn-2), which is expressed in premigratory and migratory neural crest in both chick and mouse. Strikingly, trunk neural crest cells no longer migrate segmentally through the somites of *Npn-2* mutants and are distributed equally in rostral and caudal sclerotome. These defects are the result of Npn-2/Sema3F signaling, as *Sema3F* mutants have an identical phenotype. These results demonstrate that Npn-2/Sema3F signaling is necessary to restrict neural crest migration to the rostral somite (Gammill *et al.*, 2006). Interestingly, metameric dorsal root ganglia still form in Npn-2 mutants in the absence of segmental neural crest migration, suggesting that segmental migration may be separable from ganglion formation.

As in the trunk, the migration of neural crest cells in the hindbrain is also segmented. Three broad streams of migrating cells are found adjacent to rhombomere 2 (r2), r4, and r6, whereas no neural crest cells are apparent adjacent to r3 and r5. Focal injections of DiI at the levels of r3 and r5 have demonstrated that both of these rhombomeres generate neural crest cells and that the apparent segmental pattern results from the DiI-labeled cells that originated in r3 and r5 deviating

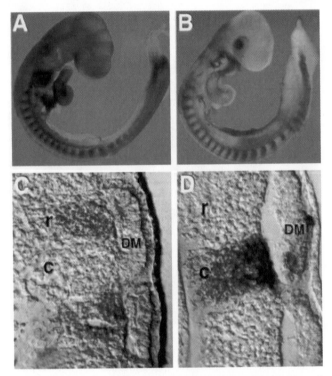

FIGURE 16.8 Distribution of Eph receptors (A and C) on neural crest cells in the rostral sclerotome and ephrin ligands in the caudal sclerotome (B and D) of chick embryos. Eph receptors are on neural crest cells in the rostral half of each somite, whereas inhibitory ephrin ligands are expressed in the caudal halves of each sclerotome. From Krull *et al.* (1997).

rostrally or caudally and failing to enter the adjacent preotic mesoderm or otic vesicle region.

Trunk neural crest cells move in a segmental pattern, which is controlled by inhibitory molecules present in the caudal-half of each somite. Neural crest cells have Neuropilin-2 receptors on their surface. When they encounter Sempahorin 3F ligands in the caudal somite, they are diverted from this location, leading to selective migration through only the rostral-half somite.

Cessation of Neural Crest Migration

Neural crest cells exhibit an orderly pattern in their migration. Whereas the first cells to migrate tend to move most ventrally, later migratory cells contribute to progressively more dorsal derivatives. The different destinations of neural crest derivatives are therefore populated in a sequential order during development. One possibility is that the early migrating cells "fill" the more ventral sites, effectively clogging up the pathway. Alternatively, the sites themselves may change with time so that they can no longer support migration.

Surprisingly little is known about how neural crest cells know that they have reached the appropriate destination and stop their migration. This is in marked contrast to the developing CNS, where a number of mutations affect the ability of neuroblasts to cease migrating. It is clear that cell adhesion molecules such as N-cadherin and N-CAM are up-regulated after cells reach their final sites and condense to form peripheral ganglia. However, there is no evidence for a causal role for cell adhesion molecules in this process.

Neurogenesis in the PNS

Some of the factors and signaling cascades that result in neurogenesis in the developing PNS are beginning to be understood (see Chapter 15). Many neural crest cells appear to be multipotent and not yet committed to a neural fate. However, particular external factors can influence their fate decisions. For example, clonal populations of neural crest cells become neural in the presence of BMPs and glial in the presence of glial growth factor. Similarly, activation of Notch signaling promotes gliogenesis at the expense of neurogenesis in neural crest cells and their derivatives. Certain transcription factors may bias cells toward certain fates; for example, whereas mash-1 is essential for sympathetic neuron formation, neurogenins are essential for sensory fates (Christiansen et al., 2000).

An interesting contrast between the developing PNS and the CNS is that migrating neural crest cells proliferate rapidly as they move. In fact, even after exhibiting defined neuronal characteristics, some neural crest derivatives continue to divide. For example, in the developing sympathetic ganglia, neural crest-derived cells express neurotransmitters and other proteins characteristic of sympathetic neurons but remain actively mitotic. However, other neural crest cells such as sensory neurons appear to withdraw from the cell cycle well before they express neuronal traits.

Summary

Neural crest cells migrate over long distances throughout the body to form diverse cell types, including neurons, glia, melanocytes, and cells of the adrenal gland. The migratory pathways of neural crest cells vary along the rostrocaudal body axis, and the local environments through which cells migrate and eventually differentiate play an important role in phenotypic specification. The migration of neural crest cells is guided by both positive (attractive or permissive) and inhibitory (repulsive) cues that are found in the environment.

CELL MIGRATION IN THE CNS

In developing CNS, a series of complex cell movements organizes young neurons into a series of layers, or neuronal laminae. The relationship of neurogenesis to the formation of neural layers in developing cortex emerged from the pioneering work of Richard Sidman who discovered the technique of thymidine "birthdating" (Sidman, 1970). The broad application of this approach to different brain regions and times of development demonstrated the spatiotemporal pattern of neurogenesis in the vertebrate CNS.

The discovery of directed neuronal migration along glial fibers by Pasko Rakic in the 1970s (Rakic, 1972; Sidman and Rakic, 1973) set the stage for a period of intense focus on the process of neuronal migration in cortical histogenesis. Their review of neuronal migration in the human neocortex demonstrated that directed migrations establish the principal neuronal layers by the end of the second trimester (Sidman and Rakic, 1973). The studies on the architectonics of human cortical malformations (Dobyns, 1995; Barkovich, 1996) revealed populations of heterotopic cells and suggested that pathogenic processes disrupted normal glial-guided neuronal migration (Rakic, 1988). The hypothesis that cortical malformations resulted from migration defects became even more attractive with the analysis of spontaneously occurring neurological mutations. Cell and molecular studies of identified neurons from the cerebellar cortex and hippocampal formation provided real time imaging of neuronal migration along glia (Edmondson, 1987; Gasser and Hatten, 1990a). All these studies underscore the critical role of neuronal migration in brain development.

Cortical Histogenesis

The histogenesis of the cortex requires both directed migrations and the formation of two intermediate scaffolding zones, the preplate and the cortical plate. The preplate forms in the initial cell cycles of cortical neurogenesis (between embryonic days 11 and 12 in the mouse) and gradually transforms into the cortical plate. During the preplate stage, pioneer axons begin to extend from neurons within the preplate zone, and somatosensory and motor areas begin to emerge within the neocortex. The preplate contains cells that express specific markers and project axons along defined pathways (Valverde *et al.*, 1995; Super and Uylings, 2001; Jimenez *et al.*, 2003; Garcia-Moreno *et al.*, 2007; Meyer, 2007). As the preplate matures, extensive cell movements occur across the *dorsoventral* (DV) and radial planes of the preplate. The latter apparently orients the cells into a colonnade, which splits into a superficial marginal zone of Cajal-Retzius cells and small neurons, and the cortical plate (Rakic and Zecevic, 2003). The latter contains the immature Pyramidal neurons destined for layers V and VI as well as a smaller population of interneurons.

As first proposed by Shatz, in studies that demonstrated the critical role of the subplate in establishing cortical architectonics (McConnell *et al.*, 1989; Ghosh and Shatz, 1993; Allendoerfer and Shatz, 1994; Kanold and Shatz, 2006), and more recently as a site of action of the *reeler* gene, the cortical plate is an essential transient structure in cortical development. When the preplate fails to form the cortical plate, as in the *reeler* mutant mouse, the normal laminar architecture of the neocortex is disrupted. As development proceeds, directed migrations of Pyramidal neuron precursors along the radial glial fibers establish layers IV, I, and II with vast populations of interneurons (80%) entering the cortex tangentially from their site of origin in basal forebrain, as discussed in the next section.

Detailed studies of the cell cycle by Takahashi and Caviness (Caviness *et al.*, 2003) demonstrate that the murine neocortex is generated over an epoch of 11 cell cycles. Recent studies show that the polarity protein Numb is expressed in early phases of cortical histogenesis (Petersen *et al.*, 2002) that colocalizes with EGFR by a process that is actin-dependent (Sun *et al.*, 2005). To determine whether the roles of Numb and Numb-like (*Numbl*) proteins change as neurogenesis progresses, Zhong and colleagues (Petersen *et al.*, 2004) conditionally ablated both genes in the neocortex at later phases of neurogenesis. The loss of *Numb* and *Numbl* caused premature progenitor cell depletion and malformations of the neocortex. This finding led to the idea that *Numb*-mediated asymmetric cell divisions provide a general mechanism for cell cycle control during cell fate allocation events in the developing mammalian brain (Petersen *et al.*, 2004). An especially interesting aspect of cortical polarity in CNS neurogenesis is the fact that proteins related to the *Drosophila Notch* and *Numb* function both as polarity proteins and as mitogens (Solecki *et al.*, 2001).

Within the neocortical VZ, one of the first cells to express markers of differentiation is a specialized form of glial cell, called the radial glial cell. This cell, recognized by its expression of the RC2 antigen, extends long processes perpendicular to the ventricular surface toward the overlying cerebral wall, and continues to grow as the laminae develop such that it spans from the ventricular surface to the pial surface throughout development (Misson *et al.*, 1991). As discussed later, these radial processes presage the basic columnar plan of development, providing a scaffold "in plane" with the layer of dividing cells for a subpopulation of neurons to migrate away from the primary germinal matrix.

Formation of the Basic Embryonic Zones

As development proceeds, rapid cell division thickens the proliferative zone, and several layers of cells develop above the neuroepithelium. The first step in this process is the creation of the marginal zone (M), a cell-sparse zone from which nuclei are apparently excluded during interkinetic movements in the epithelium. Second, a zone of growing, afferent axons appears, forming an intermediate zone (I) between the VZ and the marginal zone (Fig. 16.11). In the cortex, these axons are pioneering the connections in both directions between the cortex and the thalamus. The neuroepithelium thus progresses from a single layer of dividing cells to three and then four zones, termed the ventricular zone (proliferating cells), intermediate zone (axons), preplate zone (postmitotic neuronal precursors), and the marginal zone (Committee, 1970).

Cell Movements within the Primary Proliferative Matrix

Imaging of dye-labeled cells in slices of embryonic cortex reveals extensive movements of the neuroblasts within the neuroepithelium. Labeled cells move intermittently within the plane of the epithelium at speeds between 10 and $100 \mu m/h$ (Fishell *et al.*, 1993). Since cells can move within the plane of the neuroepithelium, the spatial organization of the neuroblasts in the VZ does not provide a "protomap" of the adult neocortex. Experiments using retroviral markers to follow the dispersion of clonally related cells in the developing brain confirmed this result, demonstrating widespread dispersion of cells within the developing cortex (Walsh and Cepko, 1992).

Molecular Mechanisms of Cell Motility: General Features

Neuronal motility in the developing CNS, like that of all metazoan cells, occurs in response to a migration-promoting signal(s), which polarizes the cell and induce the extension of protrusions in the direction of migration (Ridley *et al.*, 2003) (Fig. 16.13). These protrusions can be large, broad lamellipodia or spike-like filopodia, driven by actin polymerization and microtubule dynamics, and stabilized by adhering to the extracellular matrix (ECM) in the case of migrating neural crest cells or adjacent to radial glial cells or to bundles of axons in the case of radial, glial-guided migrations and tangential migrations. Adhesions serve as traction sites for migration as the cell moves forward.

A number of labs have provided evidence that the small GTPase Cdc42, and homologs of the *C. elegans* polarity signaling complex Par6, orient

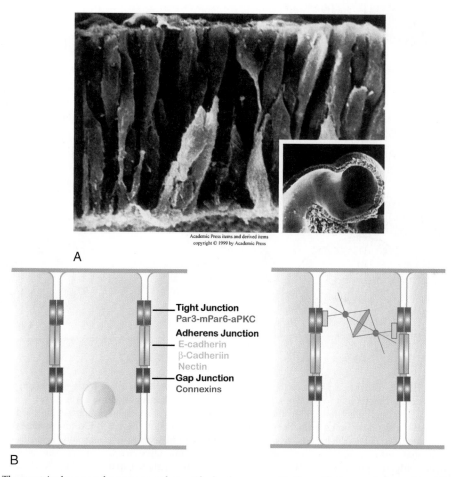

FIGURE 16.9 (A) The ventricular zone forms a pseudostratified columnar epithelium. Here neural progenitor cells have been visualized in the cerebral vesicle of a hamster embryo using scanning electron microscopy. Neuroepithelial cells are elongated bipolar cells that, at this early stage of development (E9.25), span the entire wall of the cerebrum. Some of the cells at the ventricular surface (bottom) appear spherical; these cells have retracted their cytoplasmic processes and are presumably rounding up in preparation for mitosis. Other rounded cells at the external surface (top) may be young neurons beginning to differentiate. (Inset) A low-power view of the hamster cerebral vesicle, corresponding roughly to that of a human embryo at the end of the first month of gestation. From Sidman and Rakic (1973). (B) Polarity proteins and vertebrate epithelia. Three classes of cell–cell junctions form between cells in a vertebrate epithelium. Conserved polarity proteins function in the formation of tight junctions (red). The cadherin and nectin cell adhesion proteins are required to form adherens junctions (green) and connexins form gap junctions (purple). During cell division in the neuroepithelium, the mPar6a polarity complex positions the spindle of the dividing cell (see text for details).

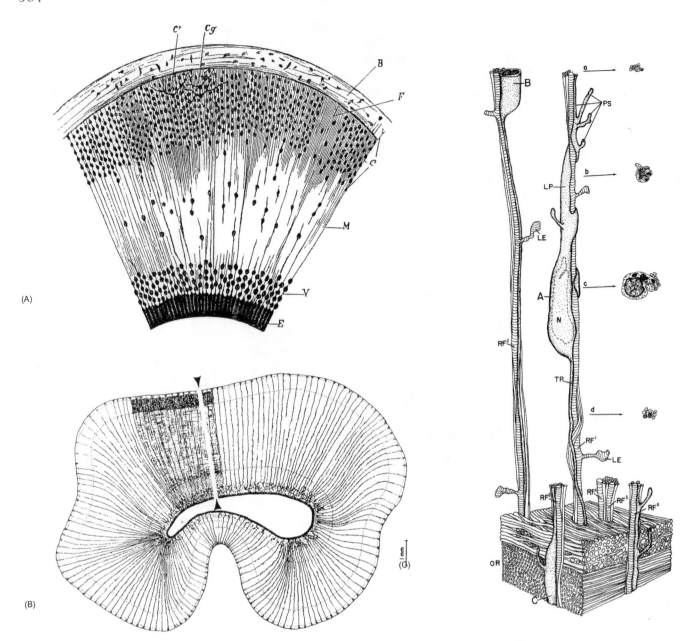

FIGURE 16.10 Radial glial cells provide a pathway for neuronal migration. (A) In the late nineteenth century, the Italian histologist G. Magini used Golgi impregnation methods to visualize a system of glial fibers that spanned the cortical wall. He proposed that the glial fibers provided a scaffold for neuronal migration. (B) In the late twentieth century, Rakic and Sidman again used the Golgi impregnation method to map radial glial cells in the developing primate neocortex. As described later, they used EM to visualize the relationship of young neurons to the glial fiber system. (C) Three-dimensional reconstruction of serial EM sections of a migrating neurons in the intermediate zone of the primate neocortex illustrate the cytology and neuron-glia relationships of migrating neurons *in vivo*. The cell soma of the migrating cell apposes to radial glial fibers (striped vertical shafts, RF1-6), which extend short lamellate expansions (LE). Nuclei (N) of migrating neurons are elongated, and their leading processes (LP) are thicker and richer in organelles than their railing processes (TP). The lower part of the diagram depicts the numerous parallel axons of the optic radiations (OR). These axons have been deleted from the upper portion of the figure to reveal the radial glial fibers. The leading process extends several pseudopodial endings (PS), which appear to explore the territory through which the neuron is migrating. In cross-section (a–d), a migrating neuron partially encircles the shaft of the radial glial fiber. From Sidman and Rakic (1973).

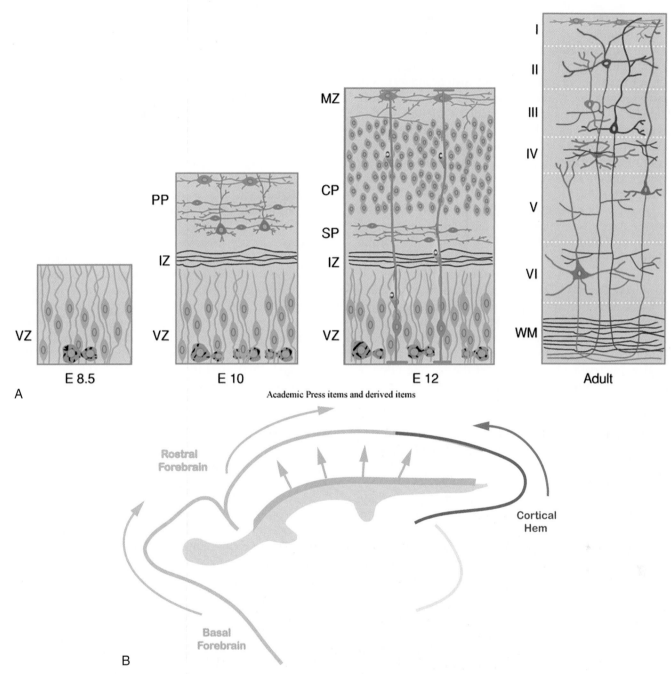

FIGURE 16.11 Development of the cerebral cortex. (A) *Formation of the Fundamental Layers:* The ventricular zone (VZ) contains the progenitors of neurons and glia. The first neurons to be generated establish the preplate (PP); their axons, as well as in growing axons from the thalamus, establish the intermediate zone (IZ). The subsequently generated neurons of cortical layers II–VI establish the cortical plate (CP), which splits the preplate into the marginal zone (MZ), or future layers I, and the subplate (SP), a transient population of neurons. After the completion of neuronal migration and differentiation, six cortical layers are visible overlying the white matter (WM) and the subplate has largely disappeared. Neural precursors in the subventricular zone (SVZ) continue to generate neurons that migrate rostrally into the olfactory bulb, even during postnatal life. (B) *Emerging Complexity on the Origin of Layer 1 Neurons.* Recent experiments in rodents and primates reveal multiple origins and migratory pathways of the precursor neurons in layer 1 of the developing neocortex. As discussed earlier, the neocortical VZ is a primary source of preplate neurons, which contains layer 1 cells. However, subsets of layer 1 neurons also arise in the rostral aspect of the neocortex, the basal forebrain, and the cortical hem. The latter progenitor cells migrate along the surface of the emerging brain from their sites of origin into the superficial layer of the neocortex. (This population of cells migrates several days before progenitors destined to become cortical interneurons). See text for details.

the nucleus, and centrosome/Golgi apparatus in both nonneuronal and neuronal motility (Etienne-Manneville and Hall, 2001; Burakov *et al.*, 2003; Etienne-Manneville and Hall, 2003). The spatial localization of protrusive activity in the extending growth cone occurs by a mechanism that involves Cdc42 signaling that localizes Rac activity to the leading edge, and recruits ζPIX Rac-GEF to the front of the cell. Thus, Cdc42 signaling orchestrates the polarization of actin and microtubule dynamics in migrating cells through separate signaling pathways (Cau and Hall, 2005).

Actin polymerization in lamellipodia is mediated by the Arp2/3 complex, which binds to the sides or tip of a preexisting actin filament and induces the formation of a new daughter filament that branches off the mother filament. Activation of the Arp2/3 complex is localized by WASP/WAVE family members, which are also activated at the cell surface. During migration, an "elastic Brownian ratchet" mechanism, in which thermal energy bends nascent short actin filaments, and pushes the membrane forward. Thus, unbending a growing axon filament against the leading edge of the cell provides a driving force for the motility of neural crest cells and extending axons. The local activation of the Arp2/3 complex, through downstream signaling events, induces lamellopodia to extend in a particular direction (Krause *et al.*, 2003), providing the basis for directional migration (reviewed in Ridley *et al.*, 2003).

The binding of myosin II to the actin cytoskeleton regulates the contraction of the actin cytoskeleton, which transmits tensile force and moves the cell toward adhesion sites (Kim and Chang, 2004; Gomes *et al.*, 2005). Myosin II activity therefore regulates the generation of force necessary to move the cell along a glial fiber, in the case of glial-guided migration (Fig. 16.13A), or along axons or other substrates discussed later. Myosin II activation involves phosphorylation of the myosin light-chain (MLC) by the MLC kinase (MLCK) and the Rho kinase (ROCK1). MLC phosphatase dephosphorylates MLC, which decreases Myosin II activity and slows cell motility. Recent studies by Clace Waterman-Storer reveal that microtubule-based contractions also regulate the tensile force of the cell, by a complex feedback system that uses Myosin II concentrations to measure the density of adhesion sites and regulate the speed of migration (Gupton, 2006). Thus, Myosin II-actin contractile systems and microtubule contractile systems move cells toward adhesions on cellular or extracellular pathways. In addition to generating force at sites of adhesion, Myosin II also localizes the nucleus in migrating cells, a feature that will be discussed again later.

Migration through the Intermediate Zone and Glial-Guided Migrations

As neurons in the VZ of cortical regions of brain exit the cell cycle, polarity proteins and their signaling cascades establish the polarity of the young neuron and control the extension of a short leading process. As the post-mitotic cell delaminates from the epithelium, movement of the cell soma occurs by interkinetic nuclear movement (Nadarajah *et al.*, 2001). Subsequently, postmitotic neurons bind to the processes of radial glial cells and use the glial cell as a substrate to migrate through the thickening cortical wall. During glial-guided migrations, immature CNS neurons express a highly polarized morphology and mode of movement. During migration, the Golgi, centrosome, and associated microtubule networks polarize the neuron in the direction of movement (Fig. 16.3; Edmondson and Hatten, 1987; Gregory *et al.*, 1988; Hatten, 1999). Whereas the extending growth cone forms focal adhesions at the leading edge, migrating cells establish a broad interstitial junction beneath the cell soma (Gregory *et al.*, 1988). In contrast to extending growth cones, the movement of the tip of the leading process does not correlate with the movement of the cell soma (Edmondson and Hatten, 1987; Gregory *et al.*, 1988), arguing that there is no "leading edge" at the tip of the leading process. Instead, the "motors" for glial-guided migration appear to be localized in the cell soma.

To examine the role of polarity proteins in neuronal migration, Solecki *et al.* (2004) imaged the conserved polarity complex mPar6α in migrating and stationary cerebellar granule neurons. As the neuron migrates, Par6α/PKCζ localizes to the centrosome (Fig. 16.12A) and signals forward movement of the centrosome prior to the translocation of the cell soma. Thus, migration occurs by a two-stroke mechanism controlled by mPar6a signaling (Fig. 16.3; Solecki *et al.*, 2004). Rather than localizing actin assembly and regulatory elements to the front of the cell, neurons migrating on glial fibers assemble a specialized motility apparatus in the cell soma and proximal region of the leading process. This apparatus includes a perinuclear tubulin cage that holds the nucleus in the rear of the cell and a specialized *adherens* junction beneath the cell soma (rather than focal adhesions at the front of the cell). As the neuron moves, bundles of F-actin form in the proximal portion of the leading process. Activation of Myosin IIB moves the cell forward, suggesting that Myosin IIB binding to the actin cytoskeleton in this zone generates tensile force that releases the adhesion site and controls the forward movement of the nucleus within the perinuclear tubulin cage (Fig. 16.13A).

The specific roles of dynein and microtubule contractions are being studied by a number of labs, as are the mechanisms that control signaling events that initiate the dissolution of the adhesion site via endocytosis of the neuron-glial adhesion ligands, including ASTN1/2 (Stitt et al., 1991; Zheng, 1996). The ability to combine the tools of molecular biology, using retroviruses or electroporation methods to express cytoskeletal proteins and their downstream regulators, and state of the art imaging methods have been key steps in deciphering the mechanism of glial-guided neuronal migration. Understanding this critical step in cortical histogenesis in detail is an important tool in the discovery of genes that cause human cortical malformations (Hatten, 1999; Ross and Walsh, 2001; Bielas et al., 2004; Hatten, 2005; Tsai et al., 2005).

Radial Glial System of the Neocortex

Radial glial cells are a transient population of cells that provide scaffolding for neuronal migration. The histologists Albert Kölliker and Wilhelm His described the development of glial cells in the late nineteenth century. The silver impregnation method, described earlier, played a crucial role in the detection of radial glial processes. The Italian anatomist Giuseppe Magini used Golgi's silver stain to visualize radial fibers that extended from the ventricular epithelium to the surface of the neocortex or mammalian embryos (Fig. 16.10A). Moreover, Magini observed cells along these processes and proposed that the neurons migrate on radial glia from the ventricular zone out through the thickening cortical wall.

Ramon y Cajal rejected Magini's theory of neuronal migration along glial fibers, which discouraged further studies on glial guidance of migration, until the 1970s, when Richard Sidman and Pasko Rakic confirmed the idea that young neurons migrate along radially arranged glial processes. In a series of studies, Rakic and Sidman used EM studies to document the proposal that the neurons were closely apposed to the radial glial fibers (Rakic, 1972; Sidman, 1973) (Fig. 16.10B,C). Rakic and Sidman further proposed that the radial glia guide the migrations of young neurons from the VZ to the neural layers, in a spatiotemporal order that positioned the youngest born neurons in the most superficial layers. This "inside-out" model relied upon coordinating 3H-thymidine labeling of neuroblast proliferation in the VZ with light and EM analysis (Sidman, 1970). Cells born at successively later times after administration of labeled thymidine would be lighter than those born the day of the injection (their "birthday"), due to dilution of the label with continued proliferation.

Experiments by Nowakowsi and Rakic (Nowakowski and Rakic, 1979) on the developing hippocampus provided the most compelling argument for glial guidance of migration, as young neurons in the hippocampal formation followed the undulating paths of the glial fibers, rather than a straight radial line, during their migrations. Real time imaging of the migration of identified neurons—purified from cerebellum, hippocampus, and neocortex—on glial processes provided critical support for this model (Hatten and Edmondson, 1987) and revealed the dynamics of cell movement along the glial guide described earlier (Fig. 16.12B). Cell-based assays further showed that neurons from one brain region could migrate freely on glia from other brain regions (Fig. 16.12C), suggesting that the glial fibers were a generic "monorail" system, rather than a track that specified the duration and direction of migration (Gasser and Hatten, 1990a, 1990b; Hatten, 1999). In addition, migration assays with purified cell populations provided cell-based assays for neuron-glial adhesion systems, such as the astrotactin (Stitt et al., 1991; Zheng, 1996), for growth factor systems that regulate glial differentiation (Gierdalski et al., 2005) and thereby influence migration such as Neuregulin (Rio et al., 1997), and for soluble factors that act as chemoattractants or chemorepellants.

As methods became available to culture acute slices of developing brain, it became feasible to acquire real time images of neurons migrating in the complex setting of developing brain tissue. Lipophilic dyes (Rivas and Hatten, 1995), retroviral constructs that tagged the cells with fluorescent markers such as EGFP (Bhatt et al., 2000) and in vivo electroporation of retroviral constructs into embryonic neocortex followed by the generation of acute slices several days later made it possible to alter gene expression (Zhang et al., 2002) of particular cohorts of developing neurons and to follow their fate in the emerging neocortex. Moreover, this approach facilitated real time imaging of cells with altered levels of gene expression, by the introduction of shRNAs to knockdown gene expression, to test the function of proteins associated with human cortical malformations (Tsai et al., 2005), identify transcription factors (Hand et al., 2005), polarity proteins (Solecki et al., 2004), cytoskeletal proteins (Tsai et al., 2005), and components of downstream signaling pathways.

Methods for real time imaging, such as spinning disc confocal imaging, and for imaging cells deep in living tissue with multiphoton systems are evolving rapidly, providing neurobiologists with the tools to study the molecular regulation of CNS migrations. These experiments have occurred in parallel with cell biological studies of the fundamental mechanisms of metazoan motility, described briefly earlier. The availability of a host of new methods for marking proteins with fluorescent tags by genetic methods, combined with the discovery of more sensitive fluorescent tags to measure protein-protein interactions (FRET) and protein turnover rates (FRAP) in organelles within migrating neurons, are rapidly accelerating the range of experimental approaches to elucidate the mechanisms of neurogenesis and motility of particular classes of neural progenitors in specific regions of the developing nervous system.

Lineage Analysis of Radial Glial Cells: Glia Generate Neurons in Late Phases of Cortical Development

Ultrastructural, molecular, and physiological characteristics of radial glial cells distinguish them from neuronal progenitors in the ventricular zone and migrating neurons in the intermediate zone. During the past five years, real time imaging of labeled radial glial cells and genetic experiments on the lineal descent of radial glia have shown that a large number of neurons are descended from glial progenitor cells. In the developing neocortex, imaging experiments by Noctor et al. (2002) and Kriegstein and Noctor (2004) revealed the surprising finding that radial glial cells can also generate neuronal precursors, which immediately, or after several rounds of cell division, migrate along the radial processes of the glial mother cells that stretch across the expanding cerebral wall (Fig. 16.14).

Genetic studies by Anthony and Heintz (Anthony et al., 2004) used Cre/loxP fate mapping and clonal analysis to demonstrate that radial glia throughout the CNS serve as neuronal progenitors and that radial glia within different regions of the CNS pass through their neurogenic stage of development at distinct time points. They went on to show that the promoter region of the radial glial cell marker gene Blbp (Feng et al., 1994) contains a binding site for the Notch effector CBF1 that is essential to Blbp transcription in radial glia (Anthony et al., 2005). These results identified Blbp as the first predominantly CNS-specific Notch target gene. Their results also confirmed earlier studies by Gaiano and Fishell (2002), who showed that Notch signaling regulates radial glial differentiation in the neocortex.

Tangential Migration Pathways

Cortical Interneurons Although radial migration was once thought to be the primary mode of CNS migrations in cortical regions of vertebrate brain, we now realize that a small subset of the total population of neurons in the murine neocortex migrate from the neocortical VZ to form the layers of the mature cerebral cortex (Tan et al., 1995, 1998). These neurons are the large output neurons, the pyramidal cells (Parnavelas et al., 1991), and a subpopulation of cortical interneurons. The majority of cortical interneurons originate in the basal forebrain and migrate into the neocortex (Fig. 16.15). The importance of this ventral to dorsal mode of migration is evidenced by the rather remarkable fact that the GABAergic neurons are about 20% of the total neuronal population in the cortex. A detailed view of the migratory routes of cells migrating from the basal forebrain into the neocortex emerged from genetic and transplantation studies. Initial experiments with mice lacking the transcription factors Dlx1/2 suggested that cortical interneurons originate in the medice ganglionic eminence (MGE) (Anderson et al., 1997; Tamaki et al., 1997). More detailed transplantation and fate mapping studies refined this original view, providing support for the general model that majority of cortical interneurons originate in the MGE and migrate into the neocortex (Polleux et al., 2002; Xu et al., 2004). The LGE likely contributes a small population of interneurons to the cortical interneuron population, but current experiments show that the many LGE cells migrate into the olfactory bulb intermingling with cells that migrate from the cortical SVZ.

The Rostral Migratory Stream Between E12.5 and E14 in the mouse (E14–16 in the rat), a secondary proliferative zone is formed along the third ventricle of the forebrain (Pencea and Luskin, 2003). This progenitor zone, known as the subventricular zone (SVZ), persists in the adult, where it continues to generate neurons. The subventricular zone has been the subject of intense interest, because it continues to generate new neurons into adulthood, providing GABAergic neurons for the olfactory bulb (Luskin, 1998). As a renewable population of partially committed, CNS precursors, the cells from the RMS are an ideal system for studying adult neural "stem" cells (as indicated, they are not "stem cells" in the strict definition of the term). RMS cells differentiate into interneurons when transplanted into the wide variety of target areas, including the septum, thalamus, hypothalamus, and midbrain of embryonic brain (Alvarez-Buylla et al., 2000). Thus, many labs are studying the molecular mechanisms that control the differentiation of these cells when placed in ectopic positions, especially in the adult brain.

RMS neurons migrate from the SVZ to the olfactory bulb by a unique mechanism. As the neurons move, they use other migrating neurons in the stream as a substrate (Lois and Alvarez-Buylla, 1994; Lois et al.,

1996). This mode of neuronal migration is termed "chain migration" as the neurons form chains within glial sheaths. Recent experiments by Muller and colleagues show that neuroblasts express integrins containing the β1 and β5 subunits (Belvindrah *et al.*, 2007). Using genetic and cell biological approaches, they demonstrated that β5 integrins are dispensable for chain migration, whereas β1 integrins promote cell–cell interactions that link neuroblasts into chains. The β1 integrin ligand laminin is recruited to the cell surface of migrating neuroblasts and induces chain formation in SVZ explants and the aggregation of purified neuroblasts. Thus, β1 integrins and their laminin ligands promote the formation of cell chains in the adult RMS (Belvindrah *et al.*, 2007).

Several secreted signaling molecules are thought to guide the migration of RMS neuroblasts to the olfactory bulb, including the diffusible chemoattractants netrin-1 (Murase and Horwitz, 2002), the glial cell line-derived neurotrophic factor (Paratcha *et al.*, 2006), and the chemorepellent slit1/2 (Wu *et al.*, 1999; Nguyen-Ba-Charvet *et al.*, 2004). In addition, receptor tyrosine kinases of the Eph families and their ephrin ligands (Conover *et al.*, 2000), and Neuregulin and Erb-B ligands (Ghashghaei *et al.*, 2006) apparently guide neuronal migration in the RMS.

Layer 1 Cells Follow Diverse Migratory Pathways

The neurons of the marginal zone, later the preplate and layer 1 of the neocortex, have long been assumed to be the earliest generated neurons in the cortical VZ. Classical studies held that these cells exited the cell cycle beginning on about E10 in the mouse and then migrated away from the VZ to form the overlying preplate between days 11 and 12 (Valverde *et al.*, 1995; Jimenez *et al.*, 2003) (Fig. 16.11B). Recent evidence from genetic and molecular genetic studies supports a long discounted view that some preplate cells originate in the rostral forebrain (or even the basal forebrain) (Meyer *et al.*, 1998), whereas others emerge from the cortical hem, and migrate across the surface of the developing cortex (Garcia-Moreno *et al.*, 2007; Meyer, 2007). Support for this view came from studies on gene expression patterns in Egfp-BAC transgenic mice of the GENSAT project (Gong *et al.*, 2003), which showed that cells marked by the gene Pde1c migrate from the rostral forebrain back over the surface of the emerging neocortex.

Similarly studies on *Wnt3a+* cells showed that cells of the cortical hem also contribute to the preplate via a superficial migratory pathway across the surface of the neocortex. Recently cloning of genes related to migration genes in *C. elegans* led to the discovery that preplate neurons express Mig-13A, a gene related to

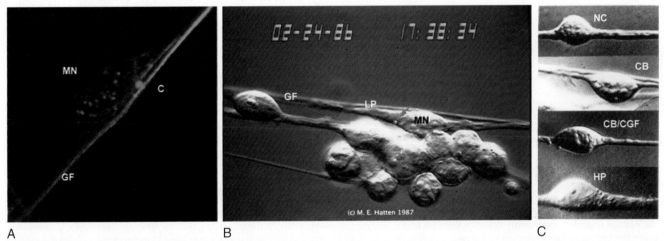

FIGURE 16.12 Live imaging of glial-guided neuronal migration. Live imaging of cells with vide-enhanced differential contrast (VEC-DIC) optics. The cytology and neuron-glial relationship of a cerebellar granule neuron migrating on a glial fiber. With live imaging of neuronal migration along glial fibers, the granule neuron closely apposes the cell glial fiber (GF) along the length of the neuronal soma (MN) and extends a leading process (LP) in the direction of migration along the glial guide. The nucleus is in the posterior aspect of the cell during periods of movement, the centrosome is just forward of the neuronal nucleus. (A) In migrating neurons, the polarity signaling complex mPar6a localizes to the centrosome, and apparently orchestrates the movement of the neuron along the glial guide. Labeling with antibodies against p50 dynactin labels the centrosome and dynein on the surface of the nucleus. Antibodies against tubulin reveal a perinuclear "cage" of tubulin in migrating neurons. (B) Live imaging of a cerebellar granule neuron migrating on a glial fiber. (C) Cytology of cerebellar and hippocampal neurons migrating along homotypic and heterotypic astroglial fibers: Cerebellar granule neuron migrating along a hippocampal glial fiber (CB/HP), a cerebellar granule neuron migrating on a cerebellar glial fiber (CB/CB), a hippocampal neuron migrating along a cerebellar glial fiber (HP/CB), or a hippocampal neuron migrating on a hippocampal glial fiber (HP/HP). In heterotypic cocultures, migrating neurons have a stereotyped cytology, extending their cell soma along the glial guide and extending a leading process in the direction of migration.

the *C. elegans* Mig-13 gene. The dynamics of cell migrations and patterns of axon outgrowth of apparent precursors of neurons within sensory areas of the neocortex are revealed by imaging the movements of labeled cells in *Egfp-Lrp12*-BAC transgenic mice (Schneider *et al.*, 2007).

Cerebellar System During embryogenesis, the cerebellar territory arises from rhombomere 1, delineated by the expression of Hoxa2 (posterior) and Otx2 (anterior). In the mouse, between E9.0 and E9.5, the pontine flexure creates an area along the border of the mesencephalon and metencephalon, where the neural tube fails to close, the rhombic lip (Harkmark, 1954). The embryonic rhombic lip is a specialized germinative epithelium that arises relatively late in development at the interface between the neural tube and the roofplate of the fourth ventricle. The rostral (cerebellar) and

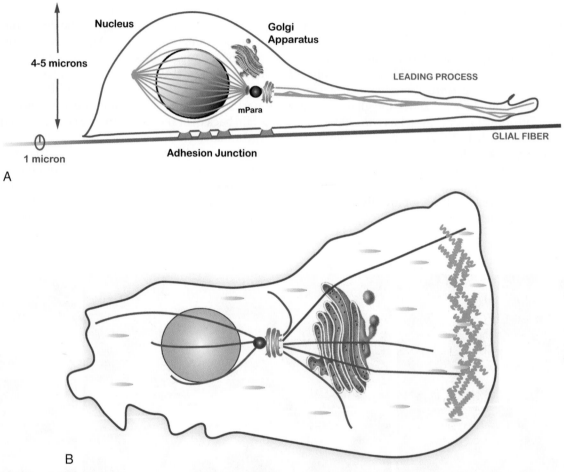

FIGURE 16.13 Comparison of the cellular organization of neurons migrating on glial fibers (A) and fibroblasts migrating *in vivo* (B). As the neuron migrates, the polarity complex mPar6α and signaling pathways, polarize the migrating cell. mPar6α localizes to the centrosome, paces the timing of the movement of the neuron along the glial fiber via Myosin IIB-mediated contraction of F-actin in the proximal aspect of the leading process. A perinuclear cage of tubulin positions the nucleus in the posterior aspect of the cell. Microtubules grow out of the centrosome into the leading process and support vesicle transport during movement. As the neurons moves, it forms attachments beneath the cell soma (blue), which provides traction for neuronal locomotion along the fiber. (B) The model shown represents a polarized cell that has distinct leading and trailing edges. This is a common feature of fibroblastic motility. The leading edge points in the direction of movement and is driven by actin-polymerization-mediated protrusion. Red spots represent points of interaction of the cell with the substrate. The larger spots represent stable adhesions (a classic feature of fibroblastic motility that is absent in faster-moving cells), and smaller spots at the periphery represent nascent adhesion complexes. Color gradients within the spot represent the dynamics of adhesion turnover (at the front) and disassembly (at the back). Other structures depicted include the nucleus (light brown), the Golgi apparatus (dark brown), and the microtubule-organizing center (MTOC), from which the microtubule network (gray) radiates, as well as an actin-rich lamellipodium at the front. Insets show specific features within the migrating cell, such as the regulation of actin polymerization at the protrusion sites, adhesion dynamics, MTOC- and nucleus-based cell polarity and tail retraction, as well as a node map depicting some of the key molecules involved in regulation of the process Reproduced from Vicente-Manzanares *et al.* (2005).

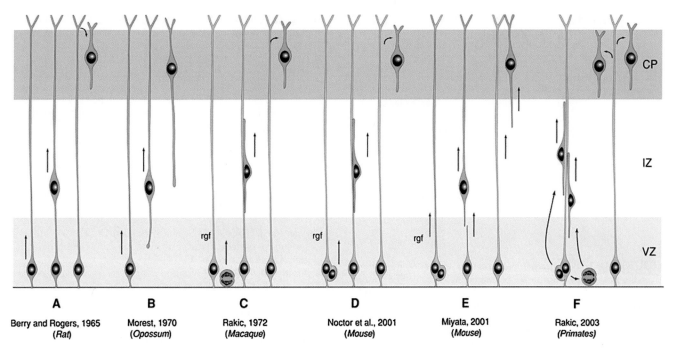

FIGURE 16.14 (A–F) Schema of the evolving concepts of the relationship between radial glial cells and migrating neurons in the developing mammalian cerebral. As discussed in the text, initial theories of neuronal migration ignored the glial cells. In the early 1970s Sidman and Rakic proposed that neurons migrate along glial fibers. Live imaging of labeled radial glia and migrating neurons revealed that during the epoch of migration, glial cells produce neuronal progeny, which immediately or after several rounds of division use the mother glial cell as a guide for their migration. As indicated in the diagram, there are some differences in this process among vertebrate species. Neurons, red; radial glial cells, green. Reprinted from Rakic (2003).

hindbrain domains of the rhombic lip form a continuous proliferative epithelium, which contains overlapping pools of progenitor cells that express combinatorial patterns of transcription factors. Recent lineage tracing experiments demonstrate that progenitor cell populations in the rhombic lip generate both cerebellar granule cells and neurons of the cerebellar nuclei, the lateral pontine nucleus and the cochlear nucleus (Machold and Fishell, 2005; Wang *et al.*, 2005) Moreover, lineage tracing experiments define the neurons of the precerebellar nuclei, which project afferent axons to cerebellar neurons (Rio *et al.*, 1997).

Thus, experiments on the allocation of cell fate among RL precursors destined for the cerebellar cortex, the cerebellar nuclei, and the precerebellar nuclei of the chick and mouse demonstrate that combinatorial patterns of transcription factor expression mark progenitors of the cerebellar nuclei (Irx3/Meis2/Lhx2/9), granule neurons (Meis1) of the cerebellar cortex, and the lateral pontine nucleus (*Pcsk9*, *Pde1c*, *Math1*). In addition to marking classes of cerebellar progenitors, patterns of transcription factor expression illustrate the complex migrations that generate the nuclear and cortical regions of the cerebellum (Morales and Hatten, 2006). These patterns include both radial and tangential migratory pathways.

Summary

During CNS development, a remarkable range of neuronal migrations set forth the architectonics of the brain. In general, these migrations can be seen as DV or AP migrations, pathways thought to be prominent in lower organisms but not in vertebrates. Indeed, genes discovered in *C. elegans* and *Drosophila* provide molecular cues for guiding migrations in higher vertebrates, and conserved polarity proteins and their signaling pathways play a key role sensing directional cues, polarizing cellular extrusions, and generating forces required to propel cells along pathways defined by specific adhesions and/or chemoattractants. Insights from studies on fibroblast motility have been especially informative to general aspects of neuronal motility, and will likely continue to provide novel methods for studying the molecular controls of neuronal migrations. The relationship of cell migrations, the establishment of the neuronal layers and brain circuitry will ultimately require novel insights into the role of the transient scaffolding systems of cortical development, the preplate, subplate, and cortical plate, which appear to pattern the connections between laminar, cortical areas of the brain, subcortical nuclei, and the formation of sensorimotor pathways.

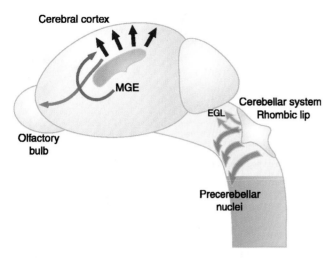

FIGURE 16.15 Nonradial pathways of neuronal migration. Over the past decade, a number of tangential migratory pathways have been described, in addition to radial pathways (black), including the migration of interneurons from the basal forebrain into the developing neocortex (blue), the migration of neurons from the SVZ to the olfactory bulb along the rostral migratory stream (red), and migrations of progenitors from the rhombic lip to cerebellar cortex (green), and the precerebellar nuclei of the brainstem (purple). Reprinted from Hatten (2002).

References

Allendoerfer, K. L. and Shatz, C. J. (1994). The subplate, a transient neocortical structure: Its role in the development of connections between thalamus and cortex. *Annu Rev Neurosci* **17**, 185–218.

Alvarez-Buylla, A., Herrera, D. G., and Wichterle, H. (2000). The subventricular zone: Source of neuronal precursors for brain repair. *Prog Brain Res* **127**, 1–11.

Anderson, S. A., Eisenstat, D. D., Shi, L., and Rubenstein, J. L. (1997). Interneuron migration from basal forebrain to neocortex: Dependence on Dlx genes. *Science* **278**, 474–476.

Anthony, T. E., Klein, C., Fishell, G., and Heintz, N. (2004). Radial glia serve as neuronal progenitors in all regions of the central nervous system. *Neuron* **41**, 881–890.

Anthony, T. E., Mason, H. A., Gridley, T., Fishell, G., and Heintz, N. (2005). Brain lipid-binding protein is a direct target of Notch signaling in radial glial cells. *Genes Dev* **19**, 1028–1033.

Arndt, K. and Redies, C. (1996). Restricted expression of R-cadherin by brain nuclei and neural circuits of the developing chicken brain. *J Comp Neurol* **373**(3), 373–399.

Barkovich, A. J., Kuzniecky, R., Dobyns, W. B., Jackson, G., Becker, L. E., and Evrard, P. (1996). Malformations of the cortical development. *Neuropediatrics* **27**, 59–63.

Belvindrah, R., Hankel, S., Walker, J., Patton, B. L., and Muller, U. (2007). Beta1 integrins control the formation of cell chains in the adult rostral migratory stream. *J Neurosci* **27**, 2704–2717.

Berry, M. and Rogers, A. W. (1965). The migration of neuroblasts in the developing cerebral cortex. *J Anat* **99**, 691–709.

Bhatt, R. S., Tomoda, T., Fang, Y., and Hatten, M. E. (2000). Discoidin domain receptor 1 functions in axon extension of cerebellar granule neurons. *Genes Dev* **14**, 2216–2228.

Bielas, S., Higginbotham, H., Koizumi, H., Tanaka, T., and Gleeson, J. G. (2004). Cortical neuronal migration mutants suggest separate but intersecting pathways. *Annu Rev Cell Dev Biol* **20**, 593–618.

Bronner-Fraser, M. and Fraser, S. (1988). Cell lineage analysis shows multipotentiality of some avian neural crest cells. *Nature* **335**(8), 161–164.

Burakov, A., Nadezhdina, E., Slepchenko, B., and Rodionov, V. (2003). Centrosome positioning in interphase cells. *J Cell Biol* **162**, 963–969.

Cau, J. and Hall, A. (2005). Cdc42 controls the polarity of the actin and microtubule cytoskeletons through two distinct signal transduction pathways. *J Cell Sci* **118**, 2579–2587.

Caviness, V. S., Jr., Goto, T., Tarui, T., Takahashi, T., Bhide, P. G., and Nowakowski, R. S. (2003). Cell output, cell cycle duration and neuronal specification: A model of integrated mechanisms of the neocortical proliferative process. *Cereb Cortex* **13**, 592–598.

Chen, X. and Macara, I. G. (2005). Par-3 controls tight junction assembly through the Rac exchange factor Tiam1. *Nat Cell Biol* **7**, 262–269.

Christensen, J. H., Coles, E. G., and Wilkinson, D. G. (2000). Molecular control of neural crest formation, migration and differentiation. *Curr Opin Cell Biol* **12**, 719–724.

Committee, B. (1970). Embryonic vertebrate central nervous system: Revised terminology. *Anat Rec* **166**, 257–262.

Conover, J. C., Doetsch, F., Garcia-Verdugo, J. M., Gale, N. W., Yancopoulos, G. D., and Alvarez-Buylla, A. (2000). Disruption of Eph/ephrin signaling affects migration and proliferation in the adult subventricular zone. *Nat Neurosci* **3**, 1091–1097.

Dobyns, W. B. and Truwit, C. L. (1995). Lissencephaly and other malformations of cortical development. update. 1995 update; *Neuropediatrics* **26**, 132–147.

Edmondson, J. C. and Hatten, M. E. (1987). Glial-guided granule neuron migration in vitro: A high-resolution time-lapse video microscopic study. *J Neurosci* **7**, 1928–1934.

Etienne-Manneville, S. and Hall, A. (2001). Integrin-mediated activation of Cdc42 controls cell polarity in migrating astrocytes through PKCzeta. *Cell* **106**, 489–498.

Etienne-Manneville, S. and Hall, A. (2003). Cell polarity: Par6, aPKC and cytoskeletal crosstalk. *Curr Opin Cell Biol* **15**, 67–72.

Fekete, D. M., Perez-Miguelsanz, J., *et al.* (1994). Clonal analysis in the chicken retina reveals tangential dispersion of clonally related cells. *Dev Biol* **166**(2), 666–682.

Feng, L., Hatten, M. E., and Heintz, N. (1994). Brain lipid-binding protein (BLBP): A novel signaling system in the developing mammalian CNS. *Neuron* **12**, 895–908.

Fishell, G., Mason, C. A., and Hatten, M. E. (1993). Dispersion of neural progenitors within the germinal zones of the forebrain. *Nature* **362**, 636–638.

Gaiano, N. and Fishell, G. (2002). The role of notch in promoting glial and neural stem cell fates. *Annu Rev Neurosci* **25**, 471–490.

Gao, L. and Macara, I. G. (2004). Isoforms of the polarity protein par6 have distinct functions. *J Biol Chem* **279**, 41557–41562.

Garcia-Moreno, F., Lopez-Mascaraque, L., and De Carlos, J. A. (2007). Origins and migratory routes of murine Cajal-Retzius cells. *J Comp Neurol* **500**, 419–432.

Gasser, U. and Hatten, M. (1990a). Neuron-glia interactions of rat hippocampal cells in vitro: Glial-guided neuronal migration and neuronal regulation of glial differentiations. *J Neurosci* **10**, 1276–1285.

Gasser, U. E. and Hatten, M. E. (1990b). Central nervous system neurons migrate on astroglial fibers from heterotypic brain regions in vitro. *Proc Natl Acad Sci USA* **87**, 4543–4547.

Ghashghaei, H. T., Weber, J., Pevny, L., Schmid, R., Schwab, M. H., Lloyd, K. C., Eisenstat, D. D., Lai, C., and Anton, E. S. (2006). The role of neuregulin-ErbB4 interactions on the proliferation and organization of cells in the subventricular zone. *Proc Natl Acad Sci USA* **103**, 1930–1935.

Ghosh, A. and Shatz, C. J. (1993). A role for subplate neurons in the patterning of connections from thalamus to neocortex. *Development* **117**, 1031–1047.

Gierdalski, M., Sardi, S. P., Corfas, G., and Juliano, S. L. (2005). Endogenous neuregulin restores radial glia in a (ferret) model of cortical dysplasia. *J Neurosci* **25**, 8498–8504.

Gomes, E. R., Jani, S., and Gundersen, G. G. (2005). Nuclear movement regulated by Cdc42, MRCK, myosin, and actin flow establishes MTOC polarization in migrating cells. *Cell* **121**, 451–463.

Gong, S., Zheng, C., Doughty, M. L., Losos, K., Didkovsky, N., Schambra, U. B., Nowak, N. J., Joyner, A., Leblanc, G., Hatten, M. E., and Heintz, N. (2003). A gene expression atlas of the central nervous system based on bacterial artificial chromosomes. *Nature* **425**, 917–925.

Hand, R., Bortone, D., Mattar, P., Nguyen, L., Heng, J. I., Guerrier, S., Boutt, E., Peters, E., Barnes, A. P., Parras, C., Schuurmans, C., Guillemot, F., and Polleux, F. (2005). Phosphorylation of Neurogenin2 specifies the migration properties and the dendritic morphology of pyramidal neurons in the neocortex. *Neuron* **48**, 45–62.

Harkmark, W. (1954). Cell migrations from the rhombic lip to the inferior olive the nucleus raphe and the pons: A morphological and experimental investigation on chick embryos. *J Comp Neurol* **100**, 115–209.

Hatten, M. E. (1999). Central nervous system neuronal migration. *Annu Rev Neurosci* **22**, 511–539.

Hatten, M. E. (2002). New directions in neuronal migration. *Science* **297**, 1660–1663.

Hatten, M. E. (2005). LIS-less neurons don't even make it to the starting gate. *J Cell Biol* **170**, 867–871.

Hinds, J. W. and Ruffett, T. L. (1971). Cell proliferation in the neural tube: An electron microscopic and Golgi analysis in the mouse cerebral vesicle. *Z Zellforsch Mikrosk Anat* **115**, 226–264.

His, W. (1886). *Zur Geschichte des menschlichen Ruckenmarks und der Nervenwurzeln Ges Wissensch. BD* **13** S, 477.

Hosoda, K., Hammer, R. E., Richardson, J. A., Baynash, A. G., Cheung, J. C., Giaid, A., and Yanagisawa, M. (1994). Targeted and natural (piebald-lethal) mutations of endothelin-B receptor gene produce megacolon associated with spotted colon color in mice. *Cell* **79**, 1267–1276.

Jimenez, D., Rivera, R., Lopez-Mascaraque, L., and De Carlos, J. A. (2003). Origin of the cortical layer I in rodents. *Dev Neurosci* **25**, 105–115.

Joberty, G., Petersen, C., Gao, L., and Macara, I. G. (2000). The cell-polarity protein Par6 links Par3 and atypical protein kinase C to Cdc42. *Nat Cell Biol* **2**, 531–539.

Joseph, N. M., Mukouyama, Y. S., Mosher, J. T., Jaegle, M., and Crone, S. A. (2004). Neural crest stem cells undergo multilineage differentiation in developing peripheral nerves to generate endoneurial fibroblasts in addition to Schwann cells. *Development* **131**, 5599–5612.

Kanold, P. O. and Shatz, C. J. (2006). Subplate neurons regulate maturation of cortical inhibition and outcome of ocular dominance plasticity. *Neuron* **51**, 627–638.

Kim, Y. and Chang, S. (2004). Modulation of actomyosin contractility by myosin light chain phosphorylation/dephosphorylation through Rho GTPases signaling specifies axon formation in neurons. *Biochem Biophys Res Commun* **318**, 579–587.

Krause, M., Dent, E. W., Bear, J. E., Loureiro, J. J., and Gertler, F. B. (2003). Ena/VASP proteins: regulators of the actin cytoskeleton and cell migration. *Annu Rev Cell Dev Biol* **19**, 541–564.

Kriegstein, A. R. and Noctor, S. C. (2004). Patterns of neuronal migration in the embryonic cortex. *Trends Neurosci* **27**, 392–399.

Kruger, G. M., Mosher, J. T., Bixby, S., Joseph, N., Iwashita, T., Morrison, S. J., Dormand, E. L., Lee, K. F., Meijer D., Anderson, D. J., and Morrison, S. J. (2002). Neural crest stem cells persist in the adult gut but undergo changes in self-renewal, neuronal subtype potential, and factor responsiveness. *Neuron* **35**, 657–659.

Krull, C. E., Lansford, R., Gale, N. W., Marcelle, C., Collazo, A., Yancopoulos, G., Fraser, S. E., and Bronner-Fraser, M. (1997). Interactions between Eph-related receptors and ligands confer rostrocaudal polarity to trunk neural crest migration. *Curr Biol* **7**, 571–580.

Larsen, C. W., Zeltser, L. M. *et al.* (2001). Boundary formation and compartition in the avian diencephalon. *J Neurosci* **21**(13), 4699–4711.

Leber, S. M. and Sanes, J. R. (1995). Migratory paths of neurons and glia in the embryonic chick spinal cord. *J Neurosci* **15**(2), 1236–1248.

LeDouarin, N. M. (1982). "The Neural Crest." Cambridge Univ. Press, New York.

LeDouarin, N. M. and Kalcheim, C. (1999). "The Neural Crest." Cambridge Univ. Press, New York.

Lois, C. and Alvarez-Buylla, A. (1994). Long-distance neuronal migration in the adult mammalian brain. *Science* **264**, 1145–1148.

Lois, C., Garcia-Verdugo, J-M., and Alvarez-Buylla, A. (1996). Chain migration of neuronal precursors. *Science* **271**, 978–981.

Luskin, M. B. (1998). Neuroblasts of the postnatal mammalian forebrain: Their phenotype and fate. *J Neurobiol* **36**, 221–233.

Macara, I. G. (2004). Par proteins: partners in polarization. *Curr Biol* **14**, R160–162.

Machold, R. and Fishell, G. (2005). Math1 is expressed in temporally discrete pools of cerebellar rhombic-lip neural progenitors. *Neuron* **48**, 17–24.

McConnell, S. K., Ghosh, A., and Shatz, C. J. (1989). Subplate neurons pioneer the first axon pathway from the cerebral cortex. *Science* **245**, 978–982.

Meyer, G. (2007). Genetic control of neuronal migrations in human cortical development. *Adv Anat Embryol Cell Biol* **189**, 1 p preceding 1, 1–111.

Meyer, G., Soria, J. M., Martinez-Galan, J. R., Martin-Clemente, B., and Fairen, A. (1998). Different origins and developmental histories of transient neurons in the marginal zone of the fetal and neonatal rat cortex. *J Comp Neurol* **397**, 493–518.

Misson, J. P., Austin, C. P., Takahashi, T., Cepko, C. L., and Caviness, V. S., Jr. (1991). The alignment of migrating neural cells in relation to the murine neopallial radial glial fiber system. *Cereb Cortex* **1**, 221–229.

Morales, D. and Hatten, M. E. (2006). Molecular markers of neuronal progenitors in the embryonic cerebellar anlage. *J Neurosci* **26**, 12226–12236.

Murase, S. and Horwitz, A. F. (2002). Deleted in colorectal carcinoma and differentially expressed integrins mediate the directional migration of neural precursors in the rostral migratory stream. *J Neurosci* **22**, 3568–3579.

Nadarajah, B., Brunstrom, J. E., Grutzendler, J., Wong, R. O., and Pearlman, A. L. (2001). Two modes of radial migration in early development of the cerebral cortex. *Nat Neurosci* **4**, 143–150.

Nakagawa, S. and Takeichi, M. (1998). Neural crest emigration from the neural tube depends on regulated cadherin expression. *Development* **125**, 2963–2971.

Nguyen-Ba-Charvet, K. T., Picard-Riera, N., Tessier-Lavigne, M., Baron-Van Evercooren, A., Sotelo, C., and Chedotal, A. (2004). Multiple roles for slits in the control of cell migration in the rostral migratory stream. *J Neurosci* **24**, 1497–1506.

Noctor, S. C., Flint, A. C., Weissman, T. A., Wong, W. S., Clinton, B. K., and Kriegstein, A. R. (2002). Dividing precursor cells of the embryonic cortical ventricular zone have morphological and molecular characteristics of radial glia. *J Neurosci* **22**, 3161–3173.

Noden, D. M. (1975). An analysis of the migratory behavior of avian cephalic neural crest cells. *Devl. Biol.* **42**, 106–130.

Nowakowski, R. S. and Rakic, P. (1979). The mode of migration of neurons to the hippocampus: A Golgi and electron microscopic analysis in foetal rhesus monkey. *J Neurocytol* **8**, 697–718.

Paratcha, G., Ibanez, C. F., and Ledda, F. (2006). GDNF is a chemoattractant factor for neuronal precursor cells in the rostral migratory stream. *Mol Cell Neurosci* **31**, 505–514.

Parnavelas, J. G., Barfield, J. A., Franke, E., and Luskin, M. B. (1991). Separate progenitor cells give rise to pyramidal and nonpyramidal neurons in the rat telencephalon. *Cereb Cortex* **1**, 463–468.

Pencea, V. and Luskin, M. B. (2003). Prenatal development of the rodent rostral migratory stream. *J Comp Neurol* **463**, 402–418.

Petersen, P. H., Zou, K., Krauss, S., and Zhong, W. (2004). Continuing role for mouse Numb and Numbl in maintaining progenitor cells during cortical neurogenesis. *Nat Neurosci* **7**, 803–811.

Petersen, P. H., Zou, K., Hwang, J. K., Jan, Y. N., and Zhong, W. (2002). Progenitor cell maintenance requires numb and numblike during mouse neurogenesis. *Nature* **419**, 929–934.

Polleux, F., Whitford, K. L., Dijkhuizen, P. A., Vitalis, T., and Ghosh, A. (2002). Control of cortical interneuron migration by neurotrophins and PI3-kinase signaling. *Development* **129**, 3147–3160.

Ponti, A., Machacek, M., Gupton, S. L., Waterman-Storer, C. M., and Danuser, G. (2004). Two distinct actin networks drive the protrusion of migrating cells. *Science* **305**, 1782–1786.

Rakic, P. (1971). Neuronglia relationship during granule cell migration in developing cerebellar cortex. A Golgi and electronmicroscopic study in macacus rhesus. *J Comp Neurol* **141**(3), 283–312.

Rakic, P. (1972). Mode of cell migration to the superficial layers of fetal monkey cortex. *J Comp Neurol* **145**, 61–84.

Rakic, P. (1988). Defects of neuronal migration and the pathogenesis of cortical malformations. *Progress in Brain Res* **73**, 15–37.

Rakic, P. (2003). Elusive radial glial cells: Historical and evolutionary perspective. *Glia* **43**, 19–32.

Rakic, S. and Zecevic, N. (2003). Emerging complexity of layer I in human cerebral cortex. *Cereb Cortex* **13**, 1072–1083.

Ramon y Cajal, S. (1995). "Histology of the nervous system of man and vertebrates."

Reese, B. E. and Tan, S. S. (1998). Clonal boundary analysis in the developing retina using X-inactivation transgenic mosaic mice. *Semin Cell Dev Biol* **9**(3), 285–292.

Ridley, A. J., Schwartz, M. A., Burridge, K., Firtel, R. A., Ginsberg, M. H., Borisy, G., Parsons, J. T., and Horwitz, A. R. (2003). Cell migration: Integrating signals from front to back. *Science* **302**, 1704–1709.

Rio, C., Rieff, H., Pelmin, Q., and Corfas, G. (1997). Neuregulin and erbB receptors play a critical role in neuronal migration. *Neuron* **19**, 39–50.

Rivas, R. and Hatten, M. (1995). Motility and cytoskeletal organization of migrating cerebellar granule neurons. *Journal of Neuroscience* **15**, 981–989.

Rodriguez, C. I. and Dymecki, S. M. (2000). Origin of the precerebellar system. *Neuron* **27**, 475–486.

Ross, M. and Walsh, C. (2001). Human brain malformations and their lessons for neuronal migration. *Annu Rev Neurosci* **24**, 1041–1070.

Sauer, M. E. and Chittenden, A. C. (1959). Deoxyribonucleic acid content of cell nuclei in the neural tube of the chick embryo: Evidence for intermitotic migration of nuclei. *Exp Cell Res* **16**, 1–6.

Sechrist, J., Serbedzija, G., Scherson, T., Fraser, S., and Bronner-Fraser, M. (1993). Segmental migration of the hindbrain neural crest does not arise from segmental generation. *Development* **118**, 691–703.

Selleck, M. A. J. and Bronner-Fraser, M. (1995). Origins of the avian neural crest: The role of neural plate-epidermal interactions. *Development* **121**, 526–538.

Serbedzija, G., Bronner-Fraser, M., and Fraser, S. E. (1989). Vital dye analysis of the timing and pathways of avian trunk neural crest cell migration. *Development* **106**, 806–816.

Sharma, K., Korade, Z., and Frank, E. (1995). Late-migrating neuroepithelial cells from the spinal cord differentiate into sensory ganglion cells. *Neuron* **14**, 143–152.

Sidman, R. L. (1970). Autoradiographic methods and principles for study of the nervous system with thymidine-H3. *In* "Contemporary research techniques of Neuroanatomy" (Nauta, W. J., Ebbesson, S. O. E., eds.), pp. 252–274. New York, Springer-Verlag.

Sidman, R. L. and Rakic, P. (1973). Neuronal migration with special reference to developing human brain: A review. *Brain Res* **62**, 1–35.

Solecki, D. J., Govek, E. E., Tomoda, T., and Hatten, M. E. (2006). New insights on neuronal polarity in neurogenesis, neuronal migration, and the establishment of brain circuity. *Genes & Develop* **24**, 883–886.

Solecki, D. J., Liu, X. L., Tomoda, T., Fang, Y., and Hatten, M. E. (2001). Activated Notch2 signaling inhibits differentiation of cerebellar granule neuron precursors by maintaining proliferation. *Neuron* **31**, 557–568.

Solecki, D. J., Model, L., Gaetz, J., Kapoor, T. M., and Hatten, M. E. (2004). Par6alpha signaling controls glial-guided neuronal migration. *Nat Neurosci* **7**, 1195–1203.

Stemple, D. L. and Anderson, D. J. (1993). Lineage diversification of the neural crest: *In vitro* investigations. *Dev Biol* **159**, 12–23.

Stitt, T. N., Gasser, U. E., and Hatten, M. E. (1991). Molecular mechanisms of glial-guided neuronal migration. *Ann N Y Acad Sci* **633**, 113–121.

Sun, Y., Goderie, S. K., and Temple, S. (2005). Asymmetric distribution of EGFR receptor during mitosis generates diverse CNS progenitor cells. *Neuron* **45**, 873–886.

Super, H. and Uylings, H. B. (2001). The early differentiation of the neocortex: A hypothesis on neocortical evolution. *Cereb Cortex* **11**, 1101–1109.

Tamaki, K., Fujimori, E., and Takauji, R. (1997). Origin and route of tangentially migrating neurons in the developing neocortical intermediate zone. *J Neurosci* **17**, 8313–8323.

Tan, S-S., Faulkner-Jones, B., Breen, S. J., Walsh, M., Bertram, J. F., and Reese, B. E. (1995). Cell dispersion patterns in different cortical regions studied with an X-inactivated transgenic marker. *Devel* **121**, 1029–1039.

Tan, S. S., Kalloniatis, M., Sturm, K., Tam, P., Reese, B., and Faulkner-Jones, B. (1998). Separate progenitors for radial and tangential cell dispersion during development of the cerebral neocortex. *Neuron* **21**, 295–304.

Tsai, J. W., Chen, Y., Kriegstein, A. R., and Vallee, R. B. (2005). LIS1 RNA interference blocks neural stem cell division, morphogenesis, and motility at multiple stages. *J Cell Biol* **170**, 935–945.

Vallee, R. B. and Tsai, J.-W. (2006). The cellular roles of the lissencephaly gene LIS1, and what they tell us about brain development. *Genes & Dev* **20**, 1384–1393.

Valverde, F., De Carlos, J. A., and Lopez-Mascaraque, L. (1995). Time of origin and early fate of preplate cells in the cerebral cortex of the rat. *Cereb Cortex* **5**, 483–493.

Vicente-Manzanares, M., Webb, D. J., and Horwitz, A. R. (2005). Cell migration at a glance. *J Cell Sci* **118**, 4917–4919.

Walsh, C. and Cepko, C. L. (1992). Widespread dispersion of neuronal clones across functional regions of the cerebral cortex [see comments]. *Science* **255**, 434–440.

Wang, H. U. and Anderson, D. J. (1997). Roles of Eph family transmembrane ligands in repulsive guidance of trunk neural crest migration and motor axon outgrowth. *Neuron* **18**, 383–396.

Wang, V. Y., Rose, M. F., and Zoghbi, H. Y. (2005). Math1 expression redefines the rhombic lip derivatives and reveals novel lineages within the brainstem and cerebellum. *Neuron* **48**, 31–43.

Weston, J. A. (1963). A radiographic analysis of the migration and localization of trunk neural crest cells in the chick. *Dev Biol* **6**, 279–310.

Wilkinson, D. G. (2001). Multiple roles of EPH receptors and ephrins in neural development. *Nature Rev Neurosci* **2**, 155–164.

Wu, W., Wong, K., Chen, J., Jiang, Z., Dupuis, S., Wu, J. Y., and Rao, Y. (1999). Directional guidance of neuronal migration in the olfactory system by the protein Slit. *Nature* **400**, 331–336.

Xu, Q., Cobos, I., De La Cruz, E., Rubenstein, J. L., and Anderson, S. A. (2004). Origins of cortical interneuron subtypes. *J Neurosci* **24**, 2612–2622.

Yu, F., Kuo, C. T., and Jan, Y. N. (2006). Drosophila neuroblast asymmetric cell division: Recent advances and implications for stem cell biology. *Neuron* **51**, 13–20.

Zhang, J., Campbell, R. E., Ting, A. Y., and Tsien, R. Y. (2002). Creating new fluorescent probes for cell biology. *Nat Rev Mol Cell Biol* **3**, 906–918.

Zheng, C., Heintz, N., and Hatten, M. E. (1996). CNS gene encoding Astrotactin, which supports neuronal migration along glial fibers. *Science* **272**, 417–419.

Suggested Readings

Anderson, S. A., Marin, O. *et al.* (2001). Distinct cortical migrations from the medial and lateral ganglionic eminences. *Development* **128**, 353–363.

Wang, V. and Zoghbi, H. (2001). Genetic regulation of cerebellar development. *Nature Rev Neurosci* **2**, 484–491.

Marianne Bronner-Fraser
and Mary E. Hatten

Growth Cones and Axon Pathfinding

Approximately 100 years ago, Ramón y Cajal described the trajectories taken by nerve cell processes as they extend toward their intermediate and final targets. From these observations he inferred that multiple influences likely guide growing axons. Cajal also observed for the first time the growing tips of axons and named them growth cones. In an unparalleled feat of scientific conjecture based on morphological observations of fixed material, he described their behavior:

> From the functional point of view, one might say that the growth cone is like a club or battering ram endowed with exquisite chemical sensitivity, rapid ameboid movements, and a certain motive force allowing it to circumvent obstacles in its path, thus coursing between various cells until reaching its destination. (Ramon y Cajal, 1890)

Decades later, Harrison developed the technique of growing living tissue in culture and demonstrated the truth of Cajal's description of a highly motile, ameboid specialization at the tips of growing axons. Shortly after, Speidel took advantage of the thinness and transparency of tadpole fins to examine living growth cones extending *in situ*. Viewed in real time, their shape changes very slowly, at a rate just detectable by an observer. Viewed with modern time-lapse techniques, however, the dynamism of their ever-changing morphology as they crawl forward is striking. The pioneering studies of Cajal, Harrison, and Speidel identified the growth cone as the key decision-making component in the elaboration of axonal pathways and inspired subsequent studies of the cell biology and behavior of growth cones *in vivo* and *in vitro*.

The growth cone at the distal tip of an axon extending in tissue culture is flattened into a thin fan-shaped sheet with many long, very thin spikes radiating forward (Fig. 17.1). The fan-shaped sheets are called lamellipodia, and the spikes are called filopodia or microspikes. Many growth cones growing *in situ* have a similar appearance, however it is not unusual to observe spindle-shaped growth cones with tufts of forward-directed filopodia within axon bundles. More complex growth cone shapes are often characteristic of slowly extending growth cones choosing between possible routes of extension, whereas simpler morphologies are characteristic of rapidly extending growth cones coursing along permissive tracts.

GROWTH CONES ARE ACTIVELY GUIDED

Growth cones crawl forward as they elaborate the axons trailing behind them, and their extension is controlled by cues in their outside environment that ultimately direct them toward their appropriate targets. What is the nature of these guidance cues and how do they affect growth cone behavior? One view to account for the ability of the nervous system to wire itself, already implicit in Ramon y Cajal's writings and reinforced by Speidel's observations, held that axonal growth is highly directed, with each class of axons navigating along a distinctive prescribed pathway to reach its target. For example, during neural development spinal commissural axons extend seemingly unerringly toward the ventral-most region of the spinal cord, cross to the contralateral side, and then extend rostrally to targets in the brain (Fig. 17.2A). These neurons could be influenced by a variety of repulsive and attractive cues at different points along their trajectory, with each portion of the journey directed

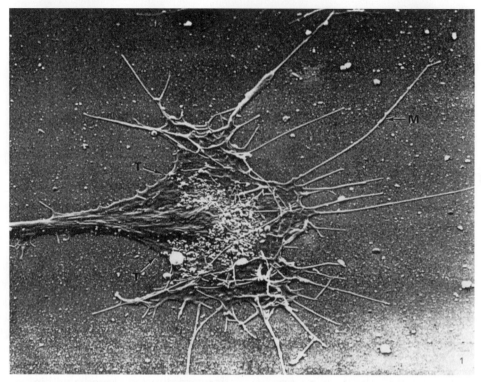

FIGURE 17.1   Scanning electron micrograph of a growth cone in culture. Growth cones extending on a flat surface are typically very thin, with broad lamellae and numerous filopodia. From Wessells and Nuttall (1973).

anew by distinct guidance mechanisms (Fig. 17.2B). An alternative view for wiring the nervous system postulated that axonal growth involves a random wandering of axons throughout the embryonic environment with connections appropriate for proper functioning of the nervous system somehow being maintained and reinforced at the expense of inappropriate connections.

A number of experiments provided evidence supporting directed axon growth, including the classic experiments of Sperry, started in the 1940s, demonstrating highly specific targeting of regenerating retinal ganglion cell axons to the optic tectum (Sperry, 1963). The application of axonal tracing techniques has made it possible over the past 30 years to observe in multiple species a variety of different classes of axons en route to their targets. These studies show that the establishment of complex neural wiring patterns occurs through a combination of initial neuronal activity-independent guidance events and subsequent refinement of connections requiring electrical signaling among neurons (Goodman and Shatz, 1993). When projecting to the vicinity of their target field, axons grow along very stereotyped trajectories, making few projection errors. This growth is highly directed and selection of the target cell is equally precise. A good example is the highly stereotyped and reproducible extension of indi-

vidual axons arising from defined insect neurons as they grow within the central and peripheral nervous systems (Raper *et al.*, 1984; Bastiani *et al.*, 1986; Caudy and Bentley, 1986). Neuronal ablation experiments demonstrate that individual neurons show selective affinity for certain axon bundles and in the absence of these targets fail to extend upon alternative neuronal, or nonneuronal, substrates (Fig. 17.3). Studies in vertebrates have also supported the idea that neurons are programmed to connect with specific targets. One of the first was the demonstration that axons of vertebrate motoneurons that have been surgically moved to abnormal nearby locations can reach their normal and appropriate targets even when they must traverse unusual routes to do so (Lance-Jones and Landmesser, 1980). Thus guidance to target fields occurs prior to and is independent of the functioning of neuronal circuits. However, once many similar axons arrive at a target field containing numerous similar target cells, as occurs when groups of motor axons all innervate many muscle fibers in a single muscle or when thalamic neurons receiving input from both eyes establish overlapping axonal arborizations in the visual cortex, neuronal function does come into play. In these circumstances, individual axons initially arborize widely within the target field and contact many target cells, only later refining their pattern of connections in a

A

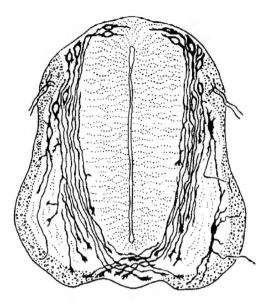

B

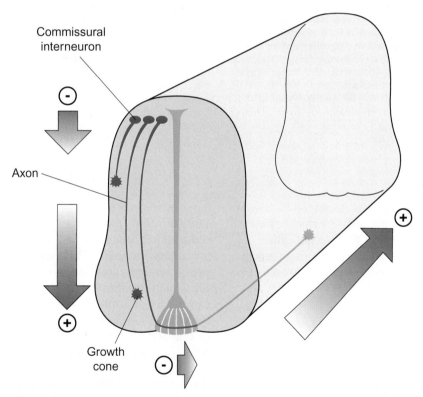

Commissural
interneuron

Axon

Growth
cone

FIGURE 17.2 Multiple guidance cues direct spinal cord commissural axons during neural development. (A) This original drawing by Cajal illustrates neuronal pathways in the developing chick spinal cord, showing several commissural axons extending to the ventral spinal cord and crossing the floor plate. Note the distinct growth cone morphologies observed at different portions of the pathway, with growth cones at or crossing the midline exhibiting complex and expanded morphologies. From Cajal, (1890). (B) At least four classes of guidance cues are now known to instruct commissural axons during development: repellents from the dorsal spinal cord, initially directing commissural axons on a ventral course; attractive cues emanating from the ventral spinal cord guiding commissural axons toward the ventral spinal cord; repulsive cues in the ventral spinal cord, facilitating midline crossing and preventing re-crossing; and attractants that guide post-crossing commissural axons anteriorly toward targets in the brain.

A

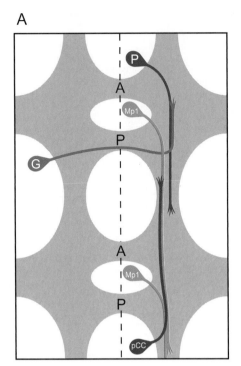

B

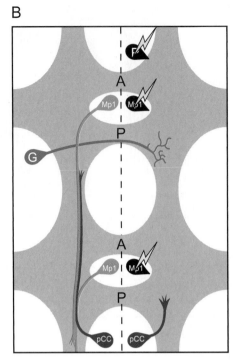

C

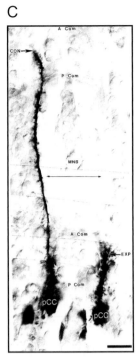

FIGURE 17.3   **Extending axons demonstrate exquisite selectivity in choosing their pathways during neural development.** (A) A simplified diagrammatic representation of the embryonic grasshopper central nervous system shows the establishment of two distinct axon pathways (or fascicles): the joining of the G neuron with the P trajectory, and the joining of the pCC neuron with the MP1 pathway. The G neuron extends across the CNS midline (dotted line) within the posterior commissure ("P") and selectively joins (or fasciculates) with axons from the P neurons (one of which is shown here), extending anteriorly along these descending P axons. The pCC neuron does not cross the midline and instead extends anteriorly along descending MP1 axons. (B) Ablation of neurons that contribute the P or MP1 pathways results in the inability of the G or pCC neurons, respectively, to extend anteriorly—these neurons do not fasciculate with numerous other axon pathways present very close by at this stage of development (not shown), and they instead stall at the point where they would have selectively fasciculated with their axonal targets. (C) Dye filling of pCC neurons on the experimental (right) and control (left) sides of a grasshopper embryo shows that when pCC does not encounter descending MP1 neurons, which are absent on the right side owing to ablation of these neurons, its growth cone stalls near the point where normally this contact would take place and no directed extension on alternative substrates occurs. From Bastiani *et al.* (1986) and Raper *et al.* (1984).

process that depends on the precise patterns of electrical activity in these neurons and their target cells.

The rest of this chapter focuses on the first step in the formation of connections: the directed growth of axons to target fields. The selection of target cells and the role of electrical activity in this process are discussed in Chapters 18 and 20.

GUIDANCE CUES FOR DEVELOPING AXONS

How do axons succeed in navigating through the embryonic environment to targets that in some cases can be many centimeters away? The trajectories of many axons appear to be broken up into short segments, each perhaps a few hundred micrometers long.

The daunting task of reaching a distant target is then reduced to the simpler task of navigating each of these successive segments. This task is well illustrated by the projections of commissural neurons in the spinal cord, illustrated in Figure 17.2. The pathway of these axons can be divided into discrete segments, each bounded by a specific cell or group of cells that marks the end of one segment and the beginning of the next. Such cells, forming an intermediate target for the axons, are sometimes termed "guidepost" cells. For example, commissural axons initially grow from the dorsal spinal cord to a set of guidepost cells at the ventral midline termed "floor plate cells," which serves as an important intermediate target. Commissural axons cross the midline at the floor plate, then turn and extend alongside the midline, before eventually leaving the midline to seek out their ultimate targets. In mice, genetic ablation of floor plate cells results in profound misrouting of the axons when they reach the midline (Matise *et al.*, 1999). Evidence for the existence of such guidepost cells harboring important guidance information in fact has been obtained throughout the animal kingdom.

The appreciation that axonal trajectories are formed in small segments pushes the question back one step: How do axons navigate each small segment of their trajectory? Axons appear to be guided along their appropriate trajectories by their responses to selectively distributed molecular signals within the developing embryo. Studies over the past three decades have led to the view that axon guidance involves the coordinate action of four types of cues: short-range (or local) cues and long-range cues, each of which can be either positive (attractive) or negative (repellent) (Fig. 17.4). The operations of short- and long-range guidance mechanisms and of attraction and repulsion are not mutually exclusive. Rather, axons may generally be guided over each individual segment of their trajectories by several different types of mechanisms acting in concert to ensure reproducible and high-fidelity guidance. Many experiments have shown that axons may navigate short segments by using several and, in some cases, even all four types of cues (Fig. 17.4): a repellent from behind the axons to "push," a corridor marked by a permissive local cue and bounded by an inhibitory local cue to "hem in" the growth, and an attractant at the end of the corridor to "pull" (Tessier-Lavigne and Goodman, 1996). Push, pull, and hem: these forces working together can ensure accurate guidance.

The identification of signaling molecules that function as guidance cues has depended primarily on three experimental approaches: (1) pairing biochemistry and *in vitro* tissue culture assays to detect proteins with either attractive or repellent properties, (2) using forward genetics to identify mutations that affect axon trajectories *in vivo*, or (3) using genetic and tissue culture approaches to characterize the functions of molecules with distributions or molecular structures that make them attractive candidate guidance cues.

Studies over the past decade and a half, in this way, have led to the identification of several prominent families of signaling molecules that function as instructive guidance cues. A first wave of studies in the 1990s identified four major families that have well-established and widespread roles in axon guidance: semaphorins, netrins, slits, and ephrins (Fig. 17.5A). These are sometimes referred to as the "canonical" axon guidance molecules. More recently, two additional classes of molecules initially identified in other contexts were shown to function as guidance cues as well: proteins that are usually classified as morphogens (including members of the Bone Morphogenetic Protein, Hedgehog, and Wnt families), as well as some proteins initially identified as growth factors (Fig. 17.5B). The identification and characterization of these cues and their receptors have led to several important generalizations about guidance mechanisms (Tessier-Lavigne and Goodman, 1996; Chisholm and Tessier-Lavigne, 1999; Dixon 2002). First, guidance cues come in families, which may, in some cases, comprise both diffusible members that can function in long-range axon guidance, as well as nondiffusible members functioning at short range. Second, many guidance cues are multifunctional, attracting some axons, repelling other axons, and sometimes controlling other aspects of axonal morphogenesis such as axonal branching or arborization. Different axons may respond to the same cue differently because they express a distinct complement of surface receptors or contain particular components of the relevant signal transduction pathways. Third, many (although not all) cues are evolutionarily conserved between vertebrates and simpler invertebrate organisms, with species homologues often performing similar roles in axon guidance. Higher vertebrates typically have many more members within a given family of guidance cues, and these cues are likely to have overlapping functions.

These different families of guidance cues are discussed briefly to illustrate these features of wiring mechanisms. We start by reviewing each of the four families of "canonical" guidance cues, before briefly describing the roles of morphogens and growth factors in guidance. We also focus specifically on members of the immunoglobulin gene superfamily, which have diverse roles as ligands and receptors in axons guidance.

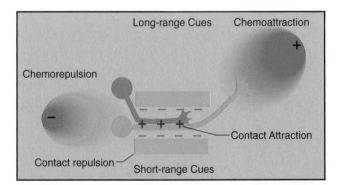

FIGURE 17.4 **Axons are guided by the simultaneous and coordinate actions of four types of guidance mechanisms:** contact attraction, chemoattraction, contact repulsion, and chemorepulsion. Individual growth cones might be "pushed" from behind by a chemorepellent, "pulled" from in front by a chemoattractant, and "hemmed in" by attractive and repulsive local cues (cell surface or extracellular matrix molecules). Adapted from Tessier-Lavigne and Goodman (1996).

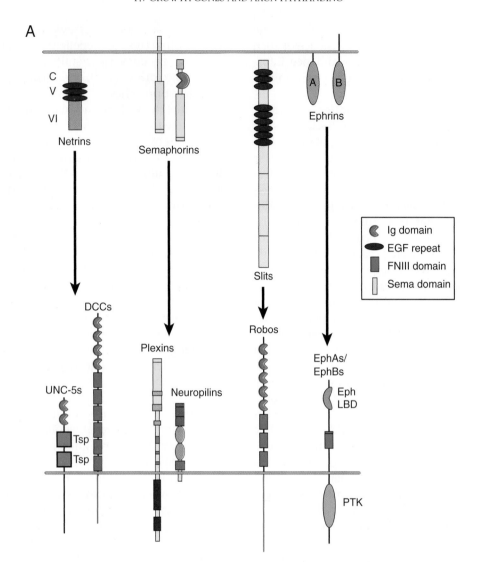

FIGURE 17.5 Multiple families of molecules serve as guidance cues. (A) Structures of the "canonical" axon guidance cues: netrins, semaphorins, slits, ephrins, and their receptors. Netrin receptors include members of the DCC and UNC5 family. There are seven classes of semaphorins (two invertebrate and five vertebrate). Of the ~20 mammalian semaphorins, 5 are secreted (class 3) and the others are divided into four families of transmembrane or GPI-linked proteins (represented here by a single diagram). Semaphorin receptors include plexins (of which nine are known in mammals, divided into classes A–D) and neuropilins (two in mammals). Class 3 semaphorin signaling is mediated by complexes of neuropilins (which serve as binding moieties) and plexins (signaling moieties). Slit receptors are members of the Roundabout (Robo) family. The receptors for Ephrins are members of the Eph family. Ephrin-As are GPI-anchored and predominantly bind EphAs, whereas Ephrin-Bs are transmembrane proteins and preferentially bind EphBs. Adapted from Winberg *et al.* (1998) and Chisholm and Tessier-Lavigne (1999). (B) Morphogens (members of the Wnt, Hedgehog and BMP families) and various growth factors listed here also serve to guide neuronal processes. Known receptors functioning to mediate their effects in guidance are indicated.

Netrins

The netrin family (Fig. 17.5) comprises about half a dozen members in vertebrates. The first vertebrate members were identified in studies on the guidance of commissural axons in the spinal cord. As described earlier, these axons project from the dorsal spinal cord to floor plate cells at the ventral midline (Fig. 17.6A). Experiments in tissue culture using microdissected pieces of rat spinal cord showed that floor plate cells secrete a diffusible outgrowth-promoting and chemoattractive activity for commissural axons, and netrins were identified as proteins that can mimic these activities (Tessier-Lavigne *et al.*, 1988; Serafini *et al.*, 1994). In the spinal cord, a dorsal-to-ventral increasing gradient of netrin-1 protein has been detected, as diagrammed in Figure 17.6Ai and visualized by immunohistochemistry in Figure 17.6Aii. This gradient appears to provide guidance information that is interpreted by commissural axons and that guides them to the floor plate.

Two types of experiments have supported this view. First, netrin-1 can clearly attract the axons in various assays. For example, when pieces of dorsal spinal cord are placed in tissue culture and viewed from the side, commissural axons grow within the tissue along a vertical dorso-ventral trajectory, but when cells secreting netrin-1 are placed alongside the dorsal spinal cord pieces, the axons are deflected from this trajectory and turn toward the source (Fig. 17.6B). Second, removal of netrin-1 function, which occurs in a netrin-1 knockout mouse, results in significant misrouting of commissural axons, with many failing to reach the floor plate (Fig. 17.6C), supporting a key role for netrin-1 in guiding the axons. Interestingly, a few axons do reach the floor plate even in the absence of netrin-1 (Fig. 17.6C), showing that one or more additional cues must collaborate to guide the axons. As we will see later, the morphogen sonic hedgehog is one of these additional cues. The collaboration of multiple guidance cues even over short segments of an axon's trajectory helps reduce the risk of guidance errors and contributes to the high degree of accuracy of wiring of the nervous system.

In addition to attracting commissural axons, netrins have been shown to function in attracting other subsets of axons to various targets in the retina and brain. Remarkably, netrin-1 also appears to function as a long-range repellent, providing a push from behind for a group of axons in the hindbrain that grow away from the midline, thus illustrating the bifunctionality of guidance cues.

Strikingly, netrins are vertebrate homologues of the UNC-6 protein of the nematode Caenorhabditis elegans, a protein similarly involved in both attract-ing some axons toward the nervous system midline and in repelling others away from it (Fig. 17.6Aiii) (Wadsworth *et al.*, 1996). Netrin homologues are also expressed at the midline of the nervous system of Drosophila melanogaster, where they contribute to attracting axons to the midline. In all of these organisms, the attractive effects of netrins on axons are mediated by receptors of the DCC family, whereas repulsive actions require receptors of the UNC5 family; both sets of receptors are members of the immunoglobulin gene superfamily (Fig. 17.5) (Chisholm and Tessier-Lavigne, 1999). In fact, as first shown in C. elegans, and illustrated with a vertebrate neuron in Fig. 17.7, it is possible to convert netrin attraction, mediated by DCC, to repulsion by introducing an UNC5 family member into the neuron (Hamelin *et al.*, 1993; Hong *et al.*, 1999). These findings on netrins and their receptors vividly illustrate both the bifunctionality of guidance cues and the remarkable conservation of axon guidance mechanisms (ligands and receptors) during evolution.

The signaling mechanisms mediating attraction and repulsion are currently the subject of intense investigation. One interesting finding suggests that the signaling machinery within growth cones is organized to allow rapid switching from attractive to repulsive responses. Specifically, the responses of growth cones that are attracted to netrin-1 can be converted to repulsive responses through pharmacological manipulations that result in a reduction in cyclic AMP (cAMP) signaling (for instance, incubation with a membrane permeable inhibitor of the cAMP effector protein kinase A (PKA)) (Fig. 17.7) (Ming *et al.*, 1997). Although it is not known how cAMP signaling impinges on the netrin response, the fact that the response can be converted provides evidence that the growth cone turning machinery is organized to facilitate such switching.

Identification of netrins as long-range guidance cues also illustrates the fact that long-range and short-range guidance mechanisms can be closely related. Although netrins are capable of long-range attraction, they are closely related in structure to one region of the archetypal nondiffusible extracellular matrix (ECM) molecule laminin-1. In fact, the extent to which netrins can diffuse in the embryo appears to be regulated so that in some circumstances they function as local rather than long-range cues, as shown in some regions in the nematode C. elegans, in the vertebrate retina, and at the Drosophila midline (Wadsworth *et al.*, 1996; Deiner *et al.*, 1997; Brankatschk and Dickson, 2006). Thus, as in the distinction between attractive and repulsive cues, there is not a hard and fast distinction between local and long-range cues.

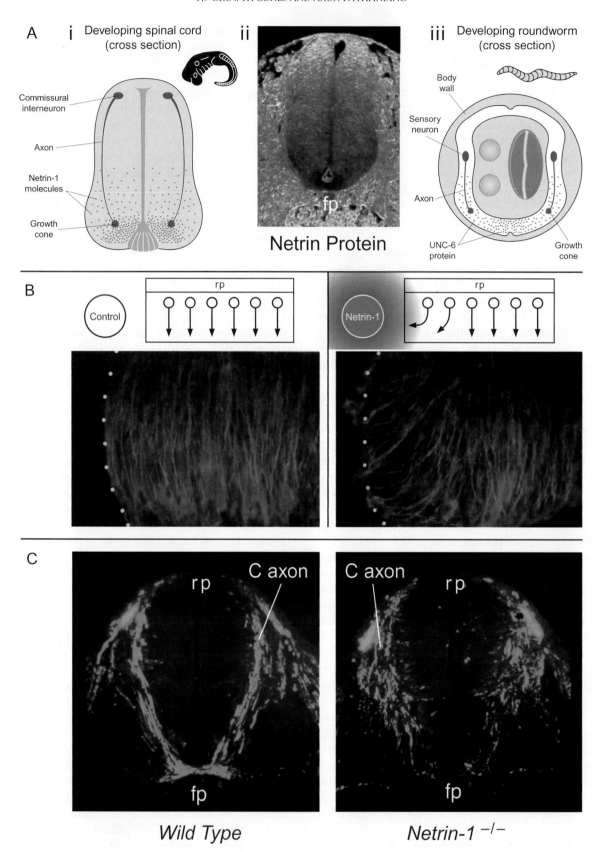

A i Developing spinal cord (cross section)

Commissural interneuron

Axon

Netrin-1 molecules

Growth cone

ii Netrin Protein

fp

iii Developing roundworm (cross section)

Body wall

Sensory neuron

Axon

UNC-6 protein

Growth cone

B Control

rp

Netrin-1

rp

C

rp   C axon

fp

C axon   rp

fp

Wild Type

Netrin-1 −/−

◀ FIGURE 17.6 **Netrins are phylogenetically conserved axon attractants.** (A) Distribution and function of netrins in axon guidance. (i) Diagram of a cross section through the embryonic vertebrate spinal cord depicting the role of netrin-1 (here and in all subsequent schematic figures shown in green) in guiding commissural axons from the dorsal spinal cord to the ventral midline. (ii) A ventral to dorsal gradient of netrin protein in the developing mouse spinal cord visualized using an immunohistochemical stain (from Kennedy *et al.* (2006)). (iii) Diagram of a cross section through the developing nematode *C. elegans* illustrating the role of the netrin/UNC-6 in guiding axons along a dorso-ventral trajectory. (B) Attractive effect of netrin-1 on rodent commissural axons in an in vitro assay for chemoattraction. Pieces of the dorsal half of the spinal cord, viewed from the side, were placed in culture for 40 hours with either control COS cells or COS cells secreting Netrin-1. In control conditions (left), all the commissural axons grow from their cell bodies at the top (dorsal) side of the explants along a vertical (dorso-ventral) trajectory (shown in diagrammatic form at the top, and visualized within the spinal cord explants using an immunohistochemical axonal marker (red) at bottom). In the presence of cells secreting netrin-1, however, the axons within ~150 μm of the cells (arrowheads) turn towards the netrin source. Dots mark the boundary between COS cells and spinal cord tissue. In the presence of netrin-1 axons can be seen invading the netrin-secreting COS cell clump. (From Kennedy *et al.*, 1994.) (C) Netrin-1 is required for guidance of commissural axons to the ventral midline. Shown are cross-sections through the spinal cord of embryonic day 11.5 mice. Left: wild-type (control); right: *netrin-1* knock-out mouse. Axons are visualized with an axonal marker. Normally (left), commissural axons (c) extend along a smooth dorsal-to ventral trajectory to the floor plate (fp) at the ventral midline. In the absence of netrin-1 (right), they extend normally in the dorsal half of the spinal cord, but when their trajectory through the ventral half is profoundly perturbed, with many projecting medially or laterally, and only a very few axons reaching the floor plate. (From Serafini *et al.*, 1996.)

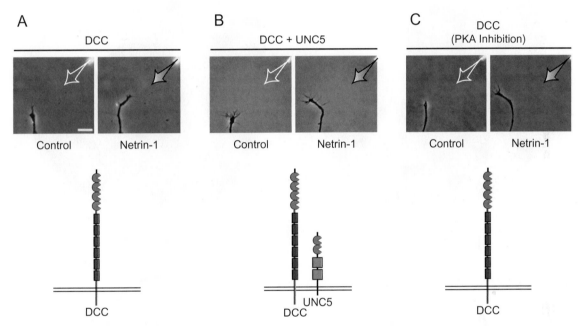

FIGURE 17.7 **Netrin bifunctionality depends upon both receptor composition and intracellular signaling.** (A) Single Xenopus spinal neurons in culture are attracted and steer along an increasing gradient of netrin-1 originating from a micropipette (hollow arrow, upper right: green arrow, netrin; white outlined arrow, control). These neurons express DCC on their surface. (B) If a Xenopus neuron expresses both DCC and UNC5 receptors, these proteins form a signaling complex which now interprets netrin-1 as a repellent, and they steer down the gradient. (C) Inhibition of cAMP-mediated signaling using a protein kinase A (PKA) inhibitor also converts netrin-1 attraction to repulsion, even though these neurons only express DCC. (From Ming *et al.*, 1997 and Hong *et al.*, 1999.)

Semaphorins

The semaphorin family of proteins includes many secreted and membrane-associated proteins, suggesting that these cues also act at both long- and short-range (Raper, 2000). The first known semaphorin was grasshopper semaphorin 1a, a transmembrane protein required for correct pathfinding of pioneer sensory axons in the developing limb (Kolodkin *et al.*, 1992). The first vertebrate semaphorin, semaphorin 3A

(Sema3A, originally called collapsin-1), was identified utilizing an *in vitro* axon repulsion assay to facilitate the biochemical purification of axonal repellents from brain extracts (Fig. 17.8A) (Luo *et al.*, 1993). There are approximately 20 distinct semaphorin family members in higher vertebrates and all contain the family's signature semaphorin domain, an approximately 500 amino acid domain that is the key extracellular signaling domain of these proteins (Fig. 17.5). Semaphorin biological specificity is in part determined by a

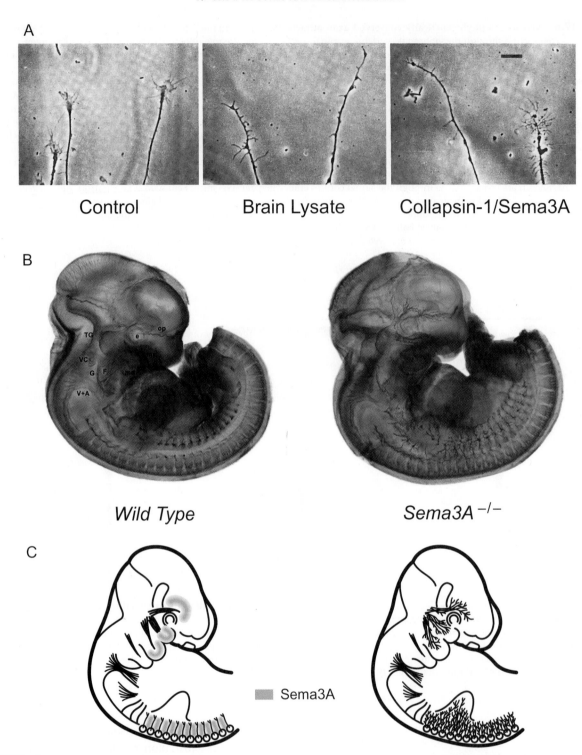

FIGURE 17.8 **Secreted semaphorins are potent repellents.** (A) Dorsal root ganglion neurons in culture exhibit a dramatic collapse when exposed to biochemically enriched fractions derived from chick brain lysates or to conditioned medium from tissue culture cells engineered to express recombinant Sema3A/collapsin-1. Medium from control non-transfected cells has no effect on DRG growth cone morphology. From Luo *et al.* (1993). (B) Mouse whole mount, embryonic day 11.5, wild-type and *Sema3A* mutant embryos immunostained for neurofilaments to illuminate spinal and cranial peripheral nerve projections. In *Sema3A* mutants cranial nerves including the trigeminal nerve (TG), are dramatically defasciculated, extend branches into ectopic locations, and overshoot their normal target regions. Spinal nerves, too, are defasciculated and extend into intersomitic regions they normally avoid (cranial nerves: ophthalmic, op; maxillary, mx; mandibular, md; facial, F, glossopharyngeal, G; vestibulocochlear, VC; vagus and accessory, V + A. Eye, e. Spinal nerves, arrow). Observations first reported by Taniguchi *et al.* (1997); whole-mount photographs courtesy of J. Merte and D. Ginty. (C) Schematic of whole mount embryos in B showing cranial and spinal nerve projections, and also Sema3A localization (gray).

relatively short stretch of amino acids within the semaphorin domain.

Most semaphorins have the capacity to act as potent inhibitory cues for neuronal or nonneuronal cells, and in a variety of assays semaphorins have been shown to repel very specific subsets of cultured axons. *In vivo* genetic analyses in invertebrates and vertebrates demonstrate conclusively that semaphorins serve as key axon repellents during neural development (Tran *et al.*, 2007). For example, mice in which the *Sema3A* gene has been disrupted by homologous recombination exhibit dramatic axon guidance defects (Fig. 17.8B) (Taniguchi *et al.*, 1997). How does Sema3A-mediated repulsion contribute to normal axon guidance? Once axons have begun to grow, often they are observed to extend upon the surfaces of axons that have preceded them on the same route. Thick bundles of axons, called fascicles, are built up over time, and many closely associated fascicles generally make up a nerve. Growth cones grow within fascicles in part because axons can be a preferred substratum compared to surrounding tissues. Though various permissive cues expressed on axons promote axon outgrowth on axonal surfaces, the presence of repellents in the surrounding tissue also restrains axon outgrowth into these regions, thereby driving axon–axon fasciculation—a process termed "surround repulsion." Sema3A is expressed in tissues surrounding many peripheral nerves and acts as a repellent for sensory and motor axons. In *Sema3A* mutants many, but not all, peripheral nerves are defasciculated, extending and branching excessively into regions where normally Sema3A would inhibit their growth (Fig. 17.8B and C). Interestingly, many of these axons in *Sema3A* mutants with abnormal and incorrect trajectories are eliminated later in development, very likely through the mechanisms described in Chapter 20. Transmembrane semaphorins expressed on axons within fascicles also serve as repellents, however rather than providing "surround repulsion" like secreted Sema3A, they serve to counter attractive axon–axon interactions, allowing axons to respond to other cues and defasciculate from main bundles in order to extend toward their specific targets (see later, and Fig. 17.9B).

The major receptors for semaphorins are members of the plexin family, transmembrane proteins that are distant relatives of the semaphorins themselves and include nine different proteins in higher vertebrates (Fig. 17.5A) (Tamagnone and Comoglio, 2000). Many transmembrane and some secreted semaphorins bind plexins directly. However, several secreted vertebrate semaphorins, including Sema3A, do not. Instead, these secreted semaphorins bind to the obligate coreceptors neuropilin-1 or neuropilin-2, which, together with a plexin family member, form an active receptor complex (Fig. 17.5A). Different secreted semaphorins appear to require specific combinations of neuropilin-1 or neuropilin-2 and select plexins in order to evoke guidance responses in distinct neuronal subtypes. This was shown through analysis of *plexin* mutants in invertebrates and vertebrates, and *neuropilin* mutants in mice, which display axon guidance phenotypes quite similar to those observed in the absence of the corresponding semaphorin ligand (Tran *et al.*, 2007).

Plexin receptor activation can initiate a series of intracellular signaling events that results in the local disassembly of growth cone cytoskeletal components and substrate attachments, ultimately producing inhibition of axon outgrowth or repulsive steering (Kruger *et al.*, 2005). However, as has been observed for many guidance cues, semaphorins include both attractants and repellents, and the same semaphorin under certain circumstances may serve both functions. For example, the secreted semaphorin Sema3B in the mouse embryo functions as an attractant for one subset of axons in the developing anterior commissure tract in the brain, and as a repellent for a different subset of axons in this same tract (Falk *et al.*, 2005). Remarkably, Sema3A can act as both an attractant and a repellent for distinct neural processes extending from mouse cortical pyramidal neurons (Polleux *et al.*, 1998, 2000). These neurons normally extend axons and dendrites in opposite directions, and in a neuronal culture paradigm Sema3A repels cortical pyramidal neuron axons but attracts dendrites from these same neurons. *Sema3A* mutant mice display misoriented cortical pyramidal neuron axons and dendrites, demonstrating how guidance cue bifunctionality participates in organization of cortical connectivity. Like netrin-1, Sema3A repulsive steering of isolated growth cones *in vitro* also is modulated by cyclic nucleotide signaling, however it is modulation of cGMP signaling, not cAMP signaling, that switches Sema3A repulsion to attraction (Song *et al.*, 1998). Stimulation of cGMP-dependent signaling switches Sema3A repulsion to attraction, which is intriguing since this same signaling pathway is required for Sema3A-mediated attraction of cortical neuron dendrites (Polleux *et al.*, 2000).

Semaphorins facilitate the formation of central and peripheral axon pathways by regulating axon pathfinding and fasciculation, however semaphorins also have been implicated in targeting axons to the specific locations of their synaptic partners, in directing the pruning of exuberant projections in the hippocampus, and in regulating the morphology of cortical pyramidal neuron dendrites (Tran *et al.*, 2007). The multiple roles played by semaphorin-plexin interactions in the establishment of neuronal connectivity showcases the versatility of plexin signaling.

Slits

Slits are large secreted proteins (Fig. 17.5A) that are also multifunctional. They were identified as axonal repellents through studies of chemorepellent factors in *Drosophila* and in vertebrates (Kidd *et al.*, 1999; Brose *et al.*, 1999; Li *et al.*, 1999), and as stimulators of axon branching through studies of sensory axon arborization in vertebrates (Wang *et al.*, 1999). A single slit gene is found in *Drosophila* and in *C. elegans*, whereas there are three in mammals. Extensive studies in all three organisms have demonstrated key roles for slit-mediated repulsion in guidance of many axonal classes, including a variety of axons in the mammalian forebrain, as well as guidance at the midline of the spinal cord (discussed later). The branching activity of slits, so far documented only in vertebrates (Ma and Tessier-Lavigne, 2007), has been implicated in regulating dendritic branching as well (Whitford *et al.*, 2002).

In all species, the repulsive actions of slit proteins are mediated by receptors of the Roundabout or Robo family (Kidd *et al.*, 1998) (Fig. 17.5A), which, like DCC and UNC5 family netrin receptors, are also members of the immunoglobulin superfamily. Interestingly, whereas in the case of netrins where different receptors mediate attraction and repulsion, in the case of slits it appears that both repulsion and branching are mediated by Robo family receptors, presumably activating distinct downstream effectors within the axons.

Ephrins

The fourth family of "canonical" guidance cues are the ephrins, cell surface signaling molecules that play important roles in a large number of developmental events including axon guidance (Klein, 2004). There are two subfamilies (Fig. 17.5A): eight class A ephrins are tethered to the cell surface via GPI linkages, and six class B ephrins are transmembrane molecules. Ephrins must be clustered together to activate their receptors and do not appear to be active if released from the cell surface, so they are thought to function exclusively as short-range guidance cues. These ligands bind receptor tyrosine kinases of the Eph family. Class A ephrins interact with various degrees of selectivity with five class A Eph receptors, whereas class B ephrins interact with three class B Eph receptors (Fig. 17.5A).

Ephrins have been shown to play an essential role in organizing topographic projections that connect, for example, retinal ganglion cells in the eye with their target cells in the appropriate portion of the optic tectum in lower vertebrates, or the lateral geniculate nucleus of the thalamus in higher vertebrates. These

mapping functions demonstrate the versatility of ephrins, which can function as attractants for some axons and repellents for other, as well as either positive or negative regulators of axonal branching. These functions in topographic mapping are discussed in detail in Chapter 18. In addition to topographic mapping, ephrins are implicated as short-range attractants and repellents in the guidance of a variety of central and peripheral axons.

Morphogens and Growth Factors

As important as the "canonical" guidance cue families are, they do not account for all of axon guidance. Studies over the past decade have revealed that various morphogens and growth factors have also been coopted to guide axons (Fig. 17.5B). We are still learning how widely these molecules are used in this role.

Morphogens are proteins that function to specify cell fates at earlier stages of embryonic development; the term refers specifically to factors that can induce different fates in a dose-dependent fashion. Three key families of developmental morphogens have been implicated in axon guidance as well: members of the hedgehog (Hh), bone morphogenetic protein (BMP), and Wnt families. These families were introduced in Chapter 15 in the discussion of cell fate determination. The evidence implicating these factors in axon guidance is growing, and is particularly well established in the vertebrate spinal cord. Here, members of all three families having been shown to collaborate with classic guidance cues of the netrin and slit families, as we will discuss in detail in a later section of this chapter. Outside the spinal cord, evidence for guidance functions of members of the BMP and hedgehog families is still limited. An example has been described in the vertebrate retina (involving sonic hedgehog), but just how widely these factors function as guidance cues remains to be determined. The evidence implicating Wnt family members in guidance is, however, already quite extensive, as they appear to play key roles in axon guidance along the longitudinal (anterior-posterior) axis in *C. elegans*, *Drosophila*, and vertebrates, and they are also implicated in topographic mapping in the retinotectal system (Zou, 2006). A list of these factors, and the receptors implicated in mediating their guidance effects, is provided in Figure 17.5B.

Whereas morphogens that function at early stages of development have been coopted to guide axons at later stages, various growth factors initially known for their functions at later stages of nervous system development—particularly in regulating the survival of neurons (Chapter 19)—have been found also to play an earlier role in axon guidance. So far, these factors are all implicated as attractants: hepatocyte growth factor (HGF), fibroblast growth factors (FGFs) and glial cell-derived neurotrophic factor (GDNF) for subsets of motor axons, neurotrophins (BDNF and NT3) for sensory axons, and neuregulins for thalamocortical axons (Fig. 17.5B). In each case, the factors have been implicated in guidance of just one subset of axons, and just in vertebrates. Thus, it remains an open question how widely growth factors are used to guide axons, and whether they can also function as repellents *in vivo*. There is, however, already considerable evidence that many of these factors function to stimulate branching of diverse classes of axons at their targets (for example, the neurotrophin NGF regulates branching of sensory axons at their targets, whereas BDNF regulates branching of thalamocortical axons in the cortex).

The following sections of this chapter explain how the guidance cues just discussed are deployed *in vivo* to affect accurate guidance. Before turning to this, however, we describe one family of proteins, already discussed to a limited extent, that has diverse roles in regulating nervous system wiring: the Ig superfamily proteins.

Ig Superfamily Proteins: Cell Adhesion Molecules, Guidance Cue Receptors, and Repellents

Initial work directed toward understanding nervous system connectivity in part focused on identifying factors that promote neuronal adhesion to substrates, with the hope this might explain complex wiring patterns. In the 1980s several cell adhesion molecules (CAMs) were identified, including a protein called neural cell adhesion molecule (NCAM) that was shown to mediate *in vitro* adhesive interactions in a homophilic (like-binds-like) fashion. NCAM contains a large extracellular domain structure that includes multiple repeats of an immunoglobulin (Ig)-C2 motif, and also several fibronectin type II repeats (Edelman, 1986). In several cell culture settings NCAM can regulate axon–axon associations, but mice harboring mutations in *NCAM*, or in other vertebrate Ig superfamily adhesion molecules such as L1, do not exhibit guidance defects indicative of these proteins playing major instructive roles in nervous system wiring. However, the Drosophila cell adhesion molecule fasciclin II (FasII), the fly orthologue of NCAM, provides insight into how selective CAM expression can contribute to more subtle aspects of nervous system wiring, namely fasciculation (or bundling) of axons.

FasII is expressed on a defined subset of fascicles in the developing insect CNS and PNS. Axons that contribute to FasII-expressing pathways still extend in approximately correct directions in *FasII* mutants. However, ultrastructural analysis reveals that in *FasII* mutants, axons that normally express FasII on their surface and maintain very close contact now fail to bundle together, and instead proceed independently (Fig. 17.9A) (Lin *et al.*, 1994). In insects, FasII also is expressed on motor axons as they extend into the periphery, and overexpression of FasII on motor axons *in vivo* causes them to remain bundled where normally they would defasciculate from the main fascicle and extend to their individual muscle targets (Fig. 17.9B). Interestingly, removal of the axonal repellent Sema-1a through genetic means results in a similar failure of motor neuron defasciculation, however simultaneous removal of both FasII *and* Sema-1a significantly restores normal patterns of axon–axon association and motor neuron pathfinding (Yu *et al.*, 2000) (Fig. 17.9B). These results highlight the role played by adhesive and repulsive guidance molecules in maintaining the balance of attraction and repulsion critical for allowing axons to extend as part of the same bundle toward target regions but then separate so as to contact unique synaptic targets.

A myriad of different Ig superfamily proteins is expressed in developing neurons, including CAMs and also receptors for canonical guidance cues (such as DCC, Robo, and Unc5; see earlier). Recent work also demonstrates a novel repulsive function for homophilic interactions involving the CAM Dscam (Down Syndrome CAM) in Drosophila neural development (Schmucker *et al.*, 2000; Zinn, 2007). Dscam is distinguished from all other Ig superfamily members by its genomic structure; alternative splicing of numerous exons has the capacity, at least in Drosophilia, to result in the generation of close to 40,000 different Dscam isoforms (Fig. 17.9C). Homophilic Dscam interactions occur only between identical Dscam isoforms; a difference of even seven amino acids in otherwise identical isoforms precludes homophilic association. This astounding molecular diversity, coupled with expression of only a few isoforms in any one neuron, has the potential to allow processes from individual neurons, either axon branches or dendrites, to distinguish between self and nonself. This is because it is extremely unlikely that any two neurons will express the same complement of Dscam isoforms. Indeed, in the mushroom body of the *Drosophila* brain, isoform-specific Dscam interactions are required for axon branches from the same neuron to separate from one another and bifurcate at a single choice point (Zhan *et al.*, 2004) (Fig. 17.9C). Further, the elaboration of individual

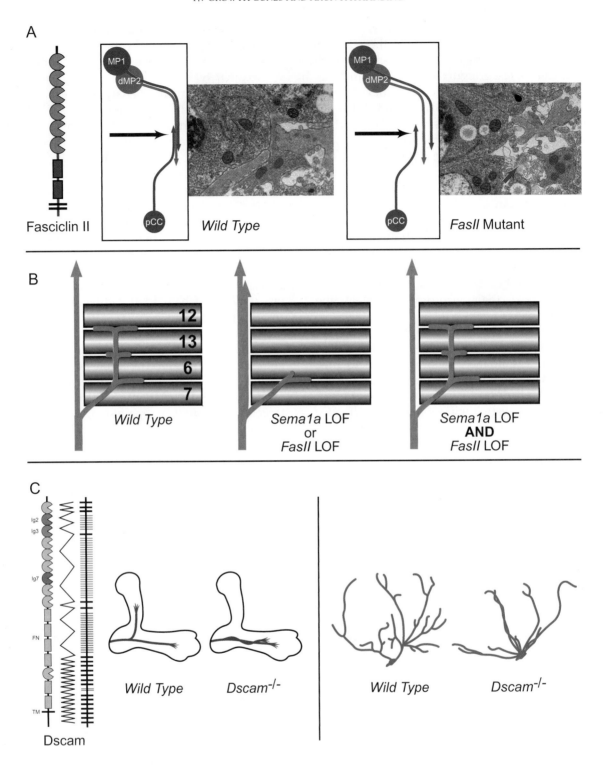

sensory neuron dendritic arborizations with nonoverlapping branches (allowing for an even distribution of sensory endings), also requires homophilic interactions between identical Dscam isoforms. Both of these examples demonstrate that Dscam isoform diversity allows neuronal processes emanating from a single neuron to be repelled from one another while simultaneously maintaining contact with processes from other neurons. Though in vertebrates Dscam is not extensively alternatively spliced, several other verte-

FIGURE 17.9 **Ig superfamily members regulate axon fasciculation and can also serve as repellents.** (A) The fasciclin II (FasII) protein, an Ig superfamily member, is required for axons that make up the FasII-expressing MP1 fascicle to remain in close apposition to one another. Wild-type (left) and *FasII* mutant (right) embryonic CNS regions that include the MP1 pathway are depicted schematically as viewed from above, and are viewed in cross section in an electron microscopy section (one of a complete serial reconstruction) taken at the location of the arrow, and with MP1, dMP2, and pCC axonal projections colored to highlight their position. In wild-type embryos, the dMP2, MP1, and pCC neurons are closely associated in the same fascicle and share extensive cell membrane contacts. However, in *FasII* mutants these same neurons do not share extensive membrane contact and extend separately (blue arrow indicates tip of pCC axon). Modified from Lin *et al.* (1994). (B) Schematic representation of embryonic Drosophila motor neurons extending from the CNS onto muscles (rectangles) in the periphery. In wild-type embryos (left) motor axons exit the CNS as part of a single bundle but then, at specific choice points, subsets of these axons defasciculate from the main bundle and enter regions that contain the different muscles upon which they will elaborate synaptic arborizations. When FasII is over-expressed in all motor neurons (*FasII* gain-of function (GOF)), motor axons often fail to defasciculate and instead bypass their normal choice points (middle). A similar result is observed in the absence of the axonal repellent Sema 1a: *Sema-1a* loss of function (LOF) (middle). However, the combination of *Sema-1a* LOF and *FasII* LOF restores the balance of attraction and repulsion along these axons and results in significant resuce of normal motor axon pathfinding (Lin and Goodman, 1994; Yu *et al.*, 2000). (C) Dscam is required for homophilic repulsive interactions that allow neuronal processes extending from individual neurons to distinguish self from non-self. Dscam is an Ig superfamily member whose extracellular domain contains Ig and fibronectin repeats, however extensive alternative splicing from a large number of exons (color-coded to illustrate Ig domains that contain one from among multiple possible exons) generates ~40,000 different Dscam isoforms (left). In the developing Drosophila brain, axons that project into the mushroom body bifurcate in order to extend branches into different lobes of this structure. Individual neurons lacking Dscam, however, generate branches that fail to separate from one another and instead extend in close association within the same lobe (middle) (Zhan *et al.*, 2004). Dendrites of sensory neurons in the body wall of Drosophila larvae elaborate arborizations in which processes do not overlap. If Dscam is not present in a sensory neuron, however, dendritic arborizations are less elaborate and processes overlap (right). For both axons and dendrites, ectopic expression of a single Dscam isoform in *Dscam* mutant neurons (not shown) rescues these Dscam LOF phenotypes (Hughes *et al.*, 2007; Mathews *et al.*, 2007; and Soba *et al.*, 2007).

brate neuronal proteins, and also families of proteins, have the potential to be represented by comparable diversity. Therefore employing extreme protein isoform diversity in the service of distinguishing neuronal uniqueness may be generally applicable across species.

Summary

Several prominent classes of guidance cues have been identified: the four families of canonical guidance cues (netrins, semaphorins, slits, and ephrins), whose functions in guidance are widespread and well established; three families of morphogens (Hedgehogs, BMPs, and Wnts), whose functions are just being defined; a growing set of growth factors implicated in select attractive guidance events; and members of the Ig superfamily, which, beyond their roles as receptors for other guidance cues, are implicated in axon fasciculation and, recently, in axonal repulsion. These guidance cues can function in long-range or short-range axon guidance. Many guidance cues are multifunctional and can function in attraction, in repulsion, or in regulating branch formation. Furthermore, many cues are conserved evolutionarily, both in structure and in function, between vertebrates and invertebrates. In addition to these molecules, yet other molecules appear to contribute to guidance in ways that are just being defined; these include other extracellular matrix components, transmembrane phosphatases, and cadherins (Dickson, 2002).

GUIDANCE CUES AND THE CONTROL OF CYTOSKELETAL DYNAMICS

Guidance cues are signaling molecules that influence the cell biological mechanisms by which growth cones extend, turn, and retract. The forward-crawling motion of a growth cone depends on its own intrinsic motile mechanism interacting with a permissive outside environment. The extension and withdrawal of the leading edge and filopodia are an intrinsic property of a healthy growth cone. Once the leading edge has extended, an appropriate substratum on which it can attach and become stabilized must be present. Once attached, an inherent traction-generating mechanism within the growth cone causes tension to develop. Unattached or poorly attached processes are thereby withdrawn, whereas tension exerted against attached processes helps draw the body of the growth cone forward. The continuous repeated cycling of extension, attachment, retraction of poorly attached processes, and tension generation produces net forward extension. In principle, guidance cues could act at any one of these steps to affect the direction in which growth cones extend.

Actin Cycle, Microtubules, and Axon Growth

Within growth cones the distribution of actin and microtubules is dramatically segregated, with micro-

FIGURE 17.10   Distributions of microtubules and fibrillar actin in a neuronal growth cone. Microtubules and fibrillary actin are highly compartmentalized within growth cones. Microtubules (green) are a key structural component of the axon and though tightly bundled within the axon shaft, a dynamic sub-population of microtubules actively explores the peripheral domain of the growth cone by extending along filopodia. All of the growing ends of microtubules are pointed toward the leading edge. In contrast, actin (red) is highly concentrated in the filopodia and in the leading edges of lamellae. Within filopodia, actin fibrils are oriented with their growing tips pointed distally. The same is true of many fibrils within lamellae, although many additional fibrils are oriented randomly and form a dense meshwork. Modified from Lin *et al.* (1994). (B) The organization of cytoplasmic domains and cytoskeletal components in an *Aplysia* bag cell growth cone. Fluorescent labeling of microtubules (green) and fibrillar actin (red) shows their segregation within the growth cone central and peripheral domains, respectively. Note the extension of a small number of microtubules into the actin-rich peripheral domain. From Suter and Forscher (2000).

tubules primarily occupying the axon shaft and the central domain of the growth cone, and actin residing in the peripheral domain (Fig. 17.10). One way in which guidance cues can affect the direction of growth cone advance is by controlling the actin polymerization that helps drive protrusion of the leading edge of the growth cone. A dense mesh-work of fibrillar actin (F-actin) is concentrated at the leading edge of the growth cone. New actin polymerization just behind the leading edge effectively helps push it forward, while, on average, F-actin is simultaneously depolymerized at an equal rate more centrally. A second important component of growth cone motility is a continuous rearward flow of polymerized actin away from the leading edge toward the central domain. This actin flow can be visualized in cultured growth cones with the aid of a drug that blocks actin polymerization (Forscher and Smith, 1988). Actin monomers generated by depolymerization in the body of the growth cone are recycled to the front, polymerized again at the leading edge, and swept rearward, where they are depolymerized once again. This continuous rearward flow of polymerized actin is called *treadmilling*, and it results from the action of myosin-based motors driving F-actin toward the growth cone central domain (Lin *et al.*, 1996) (Fig. 17.11A). In the absence of adhesive contacts with the substratum, a steady state arises whereby the actin polymerization–depolymerization cycle and tread-milling do not contribute to growth cone advance. If polymerized F-actin is linked to a permissive substratum through cell surface receptors, however, growth cone actin dynamics advance the leading edge and withdraw the trailing edge.

Particular representatives of several families of proteins have been shown to provide permissive substrata for growth cone extension. Among these are extracellular matrix molecules such as laminin-1 and fibronectin, and also specialized cell surface molecules from either the immunoglobulin superfamily or the cadherin family. These molecules are bound by specific cell surface molecules on the growth cone, including members of the integrin, Ig, and cadherin superfamilies. Although these proteins provide potent permissive substrates for cultured neurons, it is still unknown whether they or other classes of molecules are the major permissive factors for axon growth *in vivo*. At least two of the sets of canonical guidance cues, the netrins and semaphorins, are known to include members that have a permissive role for axon growth *in vivo*.

Microtubules also play a key role in regulating growth cone advance. In the axon and proximal region of the growth cone's central domain, microtubule polymers are bundled and relatively stable. However, toward the peripheral region a highly dynamic population of microtubule polymers extends individually toward the leading edge along filopodia (Fig. 17.10). Subsequent stabilization and bundling of these dynamic microtubules within the advancing growth cone leads to growth cone advance and defines the location of the newly generated trailing axon shaft.

Linking Cytoskeleton and Permissive Substratum Molecules

Cell surface receptors that bind permissive substratum molecules must in turn be linked to the cellular cytoskeleton. Indirect linkages between the cytoskeleton and members of the integrin, cadherin, and immunoglobulin families can give the cytoskeleton traction on the substratum and facilitate growth cone advance (Fig. 17.11B). This traction can be modulated by several signaling events, including the phosphorylation of specific cytoplasmic proteins. For example, activity of the nonreceptor tyrosine kinase Src reduces the strength of cadherin and integrin mediated cell adhesion. Localized activation of these kinds of signaling pathways could control the direction of growth cone advance by modulating adhesion to the substratum.

The linkage of F-actin to the substratum generates tension within the growth cone and causes processes that are poorly attached to the substratum to shrink or withdraw. When the retrograde movement of actin is impeded by a relatively strong attachment between the actin cytoskeleton and the surface substratum, not only is the leading edge of the growth cone advanced,

A

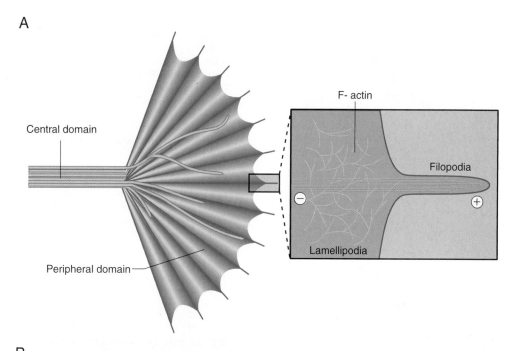

B

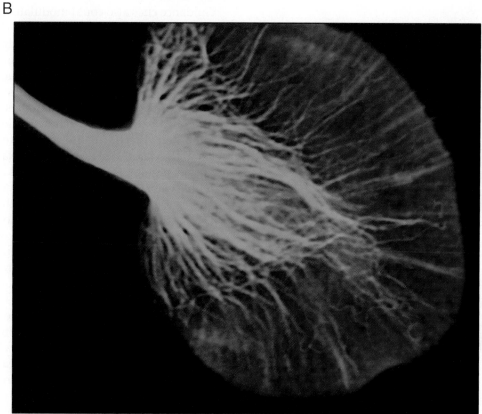

but microtubules tend to be drawn forward. The strong attachment of individual filopodia to appropriate cellular targets induces additional localized actin accumulation in the contacting process, and these processes are preferentially invaded by dynamic microtubules. In this way, filopodia can act as scouts for less advanced portions of the growth cone, either seizing hold of permissive substrata or avoiding nonpermissive or inhibitory substrates, and thereby reorienting subsequent process extension.

A

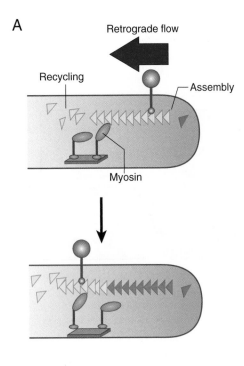

B

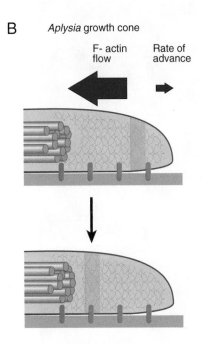

FIGURE 17.11   Actin dynamics regulate growth cone advance. (A) A schematic representation of a growth cone filopodium, depicting polarized assembly and disassembly of actin monomers at the distal (red triangles) and proximal (yellow triangles) ends of a fibrillar actin filament. Retrograde flow (large arrow) of fibrillar actin polymers occurs through the action of non-muscle myosin. This can be visualized by the proximal translocation of a particle that is linked to the actin cytoskeleton (sphere), which fails to occur if myosin function in is inhibited. Modified from Lin *et al.* (1996). (B) Linkage of the actin cytoskeleton to a permissive surface is required for forward advance. Actin polymerized at the leading edge of the growth cone (right) is swept toward the rear. If the actin meshwork is not linked to cell surface receptors that bind permissive molecules on adjacent cell surfaces, the actin cycles from front to rear but does not advance the growth cone. However, when the actin meshwork becomes attached to these receptors and thereby linked to the substrate, the meshwork remains in place and newly polymerized actin helps advance the leading edge. Modified from Lin *et al.* (1994).

leading edge by increasing the local rate of actin polymerization in the preferred direction or by decreasing its rate of depolymerization in other directions. Guidance cues also could modulate the rate of myosin-mediated F-actin translocation, thereby stimulating or inhibiting filopodia and leading edge retraction. It is therefore reasonable to expect that many signaling pathways activated by guidance cues ultimately converge on actin dynamics. Studies on the receptor mechanisms for the several families of guidance cues described earlier show that activation of these receptors often directly or indirectly alters the activity of small Rho family GTPases such as Rac and Rho, which are key regulators of actin dynamics and impact on cell motility in a large variety of cell types (Dickson, 2001; Huber *et al.*, 2002). For example, the netrin receptor DCC activates Rac, whereas activation of certain semaphorin receptors results in activation of Rho through the action of an adaptor protein that associates with the cytoplasmic domain of Plexin B receptors and stimulates the exchange of GDP for GTP on the Rho GTPase. Activation of both Eph and Robo receptors can also modulate Rho GTPases through the action of distinct adaptor proteins that associate with these receptors and either stimulate or inactivate Rho GTPases. In addition to modulating signaling cascades that regulate actin dynamics, guidance cue receptor activation can also directly regulate microtubule dynamics through activation of microtubule binding proteins that either promote or inhibit microtubule extension (Fig. 17.12).

Interactions between the Cytoskeleton and Guidance Receptors

How do cues such as semaphorins, netrins, slits, and ephrins tap into these mechanisms for growth cone extension to affect guidance? Attractive and repulsive signals could promote the extension of the

Summary

Growing axons utilize an intrinsic, actin-based treadmilling mechanism to extend filopodia and advance the leading edge of the growth cone. Growth cone extension requires an appropriate substratum for attachment and stabilization, which allows an inherent traction-generating mechanism within the growth cone to generate tension. This tension results in the withdrawal of poorly attached processes and the growth of processes that form attachments with the substratum. Net forward extension results from repeated cycles of extension, attachment, and the retraction of poorly attached processes. Cell surface receptors that bind substratum molecules are linked indirectly to the cellular cytoskeleton, and these linkages can be modulated by the activation of signaling pathways. The binding of extracellular guidance cues to receptors on the growth cone surface modulates axon growth through several mechanisms, including regulation of actin polymerization, F-actin treadmilling, linkage between the growth cone cytoskeleton and the substrate, and modulation of microtubule dynamics. Many guidance receptors directly or indirectly modulate the activity of small Rho family GTPases, which are key regulators of actin polymerization.

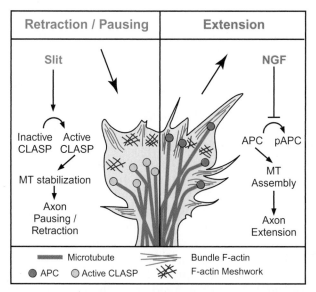

FIGURE 17.12 **Microtubule dynamics can be directly regulated by extracellular cues.** Two different microtubule binding proteins, CLASP and APC, bind to the distal (+) end of microtubules and are intracellular targets for guidance cue signaling. Slit-mediated repulsion results in activation of CLASP, which in turn stabilizes microtubules and prevents their extension into the growth cone periphery (Lee *et al.* 2004). In contrast, NGF-mediated neurite outgrowth results in promotion of microtubule assembly and extension into the peripheral region of the growth cone (Zhou *et al.* 2004). Modified from Kalil and Dent (2004).

GUIDANCE AT THE MIDLINE: CHANGING RESPONSES TO MULTIPLE CUES

How are guidance cues interpreted by growth cones *in vivo* and how do they direct the appropriate wiring together of the nervous system? The guidance of an axon from its inception through its acquisition of a target is a complex process that can be broken down into a series of discrete decisions. This section illustrates how multiple guidance cues and receptors cooperate closely to direct a complex series of guidance events at the nervous system midline.

Single guidance cues probably rarely or never act by themselves, and the integration of several simultaneously active cues is required for most, if not all, guidance decisions. This can be illustrated by interactions between the activities of many different guidance cues at the midline of the developing nervous systems of both invertebrates and vertebrates, which have been elucidated through genetic, biochemical, and embryological studies in rodents, Drosophila and C. elegans. Focusing on the mammalian spinal cord, we describe four important regulatory mechanisms that are required for reaching, crossing, and leaving the midline: initial attraction, switching on of repulsion, switching off of attraction, and prevention of premature repulsion. These mechanisms involve both canonical guidance cues, illustrated in Figure 17.13, and morphogens, illustrated in Figure 17.14.

Initial Attraction

As mentioned earlier, in both vertebrates and invertebrates the so-called commissural axons are attracted to the nervous system midline by members of the netrin family, which activate receptors of the DCC family (Tessier-Lavigne and Goodman, 1996; Chisholm and Tessier-Lavigne, 1999). We also mentioned that netrins do not function alone in guiding commissural axons to the midline. In the mammalian spinal cord, the morphogen sonic hedgehog is also made by floor plate cells at the ventral midline and collaborates with netrin-1 to attract the axons to the midline. In addition, morphogens of the BMP family are made at the dorsal midline (by so-called "roof plate cells") and function as repellents for commissural axons, providing a "push from behind" to help set them out on their initial ventral trajectory. Thus, even over a relatively simply trajectory, multiple cues collaborate to ensure accurate guidance of axons.

FIGURE 17.13 A switch from attraction to repulsion allows commissural axons to enter then leave the CNS midline. (A) Schematic of a cross-section through the mammalian spinal cord (as in Figure 1), illustrating the trajectory of commissural axons from the dorsal spinal cord to, and across, the ventral midline. The axons cross the midline at the floor plate; upon exiting the floor plate, they make a sharp right angle turn and project in a rostral (anterior) direction. A long-range gradient of netrin-1 protein (green dots) and a shorter-range gradient of slit protein (red dots) are shown. Right: Illustration of the "open-book" configuration, in which the spinal cord is opened at the dorsal midline and flattened, which helps to visualize the crossing and turning behavior of the axons at the midline. (B) As the axons project to the midline ("precrossing"), they are attracted by netrin-1 acting on the attractive netrin receptor DCC. They are insensitive to the repulsive action of slit proteins despite expression of the repulsive slit receptors (Robo1 and Robo2) because they also express Robo3, a slit receptor that suppresses the activity of Robo-1 and Robo-2. This enables these axons to enter the midline upon reaching it. However, upon crossing the midline, commissural axons down-regulate expression of Robo3 and increase expression of Robo1 and Robo2 (not shown), which allows them to sense the repulsive action of slits and results in their expulsion from the midline. (C) In *Robo3* mutant mice, the precrossing commissural axons are prematurely sensitive to slits, which prevents them from entering the midline. Shown are cross sections through a normal E11.5 mouse embryo (left) and a stage-matched *Robo3* mutant embryo (right), illustrating the inability of commissural axons in the mutant to cross the midline. (From Sabatier *et al.*, 2004). (D) Human patients with Horizontal Gaze Palsy and Progressive Scoliosis (HGPPS) carry mutations in the human *ROBO3* gene. They are capable of coordinated eye movements along the vertical axis (center), but not along the horizontal axis (left and right panels are attempts by this HGPPS patient to look left and right). The deficits in horizontal gaze result from a defect in neuronal connections across the midline in the hindbrain, which is presumed to arise from aberrant axon guidance during development since a similar defect is observed in *Robo3* mutant mice. (From Jen *et al.*, 2004.)

Switching on Repulsion

The midline is not, however, the final destination for commissural axons: upon reaching it, axons cross the midline, then they turn at right angles and project alongside the midline to other levels of the embryo, and finally they leave the midline area altogether to reach their eventual targets (Fig. 17.14). This behavior immediately raises a paradox: if the midline is such an attractive environment for the axons, how can they leave it? The answer is that midline cells, in addition to making attractive netrin proteins, also make repellents of the slit family (and, in vertebrates, of the semaphorin family) (Fig. 17.14A) (Kidd *et al.*, 1999; Zou *et al.*, 2000). Axons can approach and then cross the midline a first time because they are initially insensitive to these repellents and thus respond only to the attractive effects of the netrins. However, and quite remarkably, upon crossing the midline the axons become responsive to the repellents by increasing the expression or function of receptors for the repellents, including upregulating expression of the slit receptors of the Robo family on their surfaces (Fig. 17.14A) (Kidd *et al.*, 1998; Zou *et al.*, 2000). What causes this dramatic change in receptor expression and function is not yet known, but its net effect is to make the axons interpret the midline as a repulsive environment, which helps expel the axons from the midline and move on to the next leg of their trajectory.

Switching off Attraction

The upregulation of responsiveness to midline repellents is only half of the equation, however; to be efficiently expelled from the midline, it would be desirable for the axons to stop being attracted. Indeed, the attractive response of these axons to netrins is switched off (silenced) as the axons cross the midline. There is evidence that the silencing of netrin attraction is caused by activation of the slit receptor Robo: activation of Robo by slit causes a specific region of its cytoplasmic domain to bind a specific region of the cytoplasmic domain of the netrin receptor DCC and to prevent it from transducing an attractive response to netrin (Stein and Tessier-Lavigne, 2001). How silencing of attraction by sonic hedgehog is achieved is not known. Together, these two mechanisms (upregulation of a response to midline repellents and silencing of the attractive receptor) ensure that the growth cones perceive a once attractive environment, the midline, as unambiguously repulsive.

Preventing Premature Repulsion

For guidance at the midline to occur accurately, it is essential that the switch from attraction to repulsion be carefully choreographed. In particular, it is essential that growth cones upregulate responses to repellents only after crossing the midline, not before, otherwise the axons would not be able to enter the midline. It is perhaps not surprising, then, that there are mechanisms that ensure these axons do not become responsive to repellents too soon. One mechanism that regulates the switch, already mentioned, is that the Robo receptors are expressed only at low levels before crossing, being upregulated only after crossing. How surface expression of these receptors is controlled is poorly understood, although in *Drosophila* this appears to occur at the level of regulated trafficking of the receptor to the neuronal cell surface under the control of a protein called Commissureless. However, the low level of Robo receptor that is present on the axons before crossing appears to be capable nonetheless of transducing a repulsive slit signal that can prevent

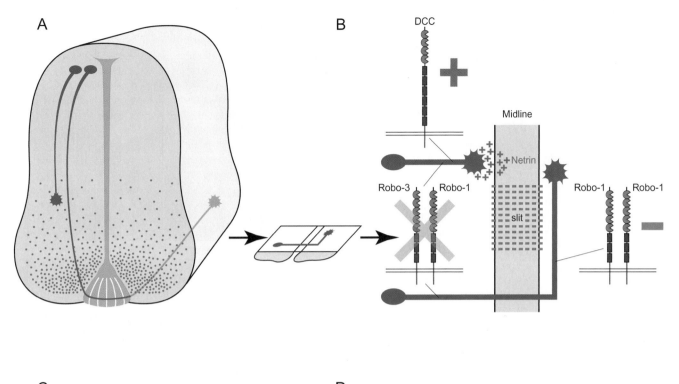

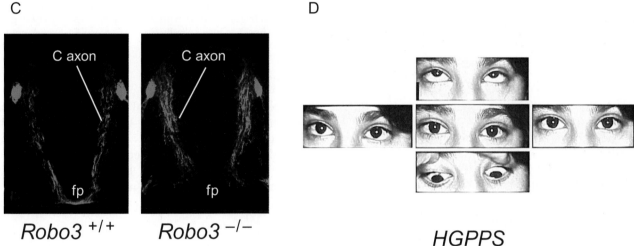

Robo3 +/+   Robo3 −/−

HGPPS

midline crossing, and which must be silenced. In vertebrates, this is achieved by a more divergent member of the Robo family called Robo3 (also known as Rig1). Robo3 is expressed on commissural axons before and during crossing, and it silences the repulsive slit signal mediated by other Robo receptors, allowing the axons to cross (Sabatier *et al.*, 2004). Thus, in *Robo3* knock-out mice, which lack Robo3 function, the axons can sense slit repulsion before midline crossing. As a result, commissural axons fail to cross the midline in these mutant mice, resulting in a dramatic "commissureless" phenotype (Figure 17.13B).

Remarkably, a similar midline crossing defect is caused by mutations in the human *ROBO3* gene, found in patients with a syndrome known as Horizontal Gaze Palsy with Progressive Scoliosis (HGPPS) (Jen *et al.*, 2004). These patients are unable to move their eyes in a coordinated fashion along the horizontal axis (horizontal conjugate gaze), a behavior that requires the function of commissural interneurons linking the two sides of the hindbrain (Fig. 17.13C). Imaging and evoked potential studies have shown that, as in *Robo3* knock-out mice, there is a failure of midline crossing by axons in the hindbrains of these patients. Thus, the

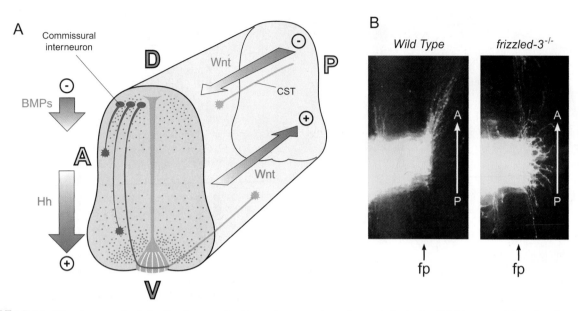

FIGURE 17.14 **Morphogens also help direct pre- and postcrossing commissural axon trajectories.** (A) Diagram of the roles of morphogens in commissural axon guidance to the midline and beyond. BMP proteins made at the dorsal midline repel the axons from the dorsal spinal cord, providing a "push from behind," whereas Sonic Hedgehog (Shh) made by floor plate cells attracts the axons, providing a "pull from afar" in collaboration with Netrin-1 (see Figure 17.13). After midline crossing, the axons turn in a rostral (anterior) direction in response to an attractive rostral-to-caudal decreasing gradient of Wnt protein. In contrast, cortical spinal tract (CST) axons extending caudally in the dorsal spinal cord are repelled by different Wnt family members. (B) The rostral Wnt signal is mediated by the Frizzled3 (Fz3) receptor. The trajectory of commissural axons at the midline is examined in the "open-book" configuration (Figure 17.13A), with the axons visualized by labeling with the fluorescent dye DiI. In wild-type mice (left), commissural axons cross the midline then turn rostrally (up), whereas in *Frz3* knock-out mice (right) the axons are disorganized after crossing (From Lyuksyutova *et al.*, 2003).

role of Robo3 proteins in allowing midline crossing is conserved in humans.

The ability illustrated here of growth cones to change their responses to guidance cues makes it possible for them to move from one intermediate target (such as the midline) to the next along their often complex and lengthy trajectories. Guidance at the midline also has shed light on the mechanisms involved in longitudinal guidance (i.e., along the anterior-posterior axis). Upon crossing the midline, commissural axons turn either anteriorly or posteriorly; in particular, the earliest born commissural axons in the mammalian spinal cord turn in an anterior direction. This anterior turn appears to be directed by an anterior-to-posterior gradient of Wnt protein in the spinal cord, which the growth cones sense as an attractive cue (Fig. 17.14A). Transduction of the attractive Wnt signal involves a receptor called Frizzled 3 (Fz3), so that loss of Fz3 function in a *Fz3* knock-out mouse causes the growth cones to become misrouted after midline crossing, unable as they are to sense the Wnt gradient (Lyuksyutova *et al.*, 2003). Anterior-posterior gradients of Wnt proteins provide a mechanism conserved between vertebrates and invertebrates to guide a variety of classes of axons along the longitudinal axis.

Some axons, like the spinal commissural axons just described, respond to the Wnts with an attractive response, and extend up the gradient. Others have repulsive responses and therefore grow down gradient. This has been shown, for example, for corticospinal tract axons, which navigate in the opposite direction to commissural axons (Figure 17.14) (Schmitt *et al.*, 2006).

Summary

The study of the growth of commissural axons to and beyond the midline exemplifies how guidance of axons *in vivo* involves the simultaneous interpretation of multiple cues, including canonical guidance cues and morphogens, and how dynamic changes in axon responsiveness enables axons to navigate successive intermediate targets. Before reaching the midline, commissural axons are guided through the combined actions of repellent BMPs, providing a "push from behind," and the attractive effects of netrin-1 and sonic hedgehog from the ventral midline. Just as important, these axons are insensitive to midline repellents, which allows them to enter and cross the midline; this is achieved in part by Robo3, expressed on the axons before crossing, which prevents them from sensing repellent midline slits. Upon reaching the midline, the axons acquire responsiveness to the repellent molecules, including slits, through upregulation of the function of repellent receptors, including Robo receptors for midline slits. They concomitantly *lose* responsiveness to the midline attractants. The coordinated switch from attraction to repulsion ensures that the axons can move on from the midline. Finally, the precise positioning of the axons in axon tracts after midline crossing involves a code of Robo proteins, and the anterior turn is directed by a Wnt gradient.

References

Bastiani, M. J., du Lac, S., and Goodman, C. S. (1886). Guidance of neuronal growth cones in the grasshopper embryo. I. Recognition of a specific axonal pathway by the pCC neuron. *J. Neurosci.* **6**, 3518–3531.

Brankatschk, M. and Dickson, B. J. (2006). Netrins guide Drosophila commissural axons at short range. *Nat. Neurosci.* **9**, 188–194.

Brose, K., Bland, K. S., Wang, K. H., Arnott, D., Henzel, W., Goodman, C. S., Tessier-Lavigne, M., and Kidd, T. (1999). Slit proteins bind Robo receptors and have an evolutionarily conserved role in repulsive axon guidance. *Cell* **96**, 795–806.

Caudy, M. and Bentley, D. (1986). Pioneer growth cone steering along a series of neuronal and non-neuronal cues of different affinities. *J. Neurosci.* **6**, 1781–1795.

Charron, F., Stein, E., Jeong, J., McMahon, A. P., and Tessier-Lavigne, M. (2003). The morphogen sonic hedgehog is an axonal chemoattractant that collaborates with netrin-1 in midline axon guidance. *Cell* **113**, 11–23.

Chisholm, A. and Tessier-Lavigne, M. (1999). Conservation and divergence of axon guidance mechanisms. *Curr. Opin. Neurobiol.* **9**, 603–615.

Deiner, M. S., Kennedy, T. E., Fazeli, A., Serafini, T., Tessier-Lavigne, M., and Sretavan, D. W. (1997). Netrin-1 and DCC mediate axon guidance locally at the optic disc: Loss of function leads to optic nerve hypoplasia. *Neuron* **19**, 575–589.

Dickson, B. J. (2001). Rho GTPases in growth cone guidance. *Curr. Opin. Neurobiol.* **11**, 103–110.

Dickson, B. J. (2002). Molecular mechanisms of axon guidance. *Science* **298**, 1959–1964.

Edelman, G. M. (1986). Cell adhesion molecules in the regulation of animal form and tissue pattern. *Annu. Rev. Cell Biol.* **2**, 81–116.

Falk, J., Bechara, A., Fiore, R., Nawabi, H., Zhou, H., Hoyo-Becerra, C., Bozon, M., Rougon, G., Grumet, M., Puschel, A. W. *et al.* (2005). Dual functional activity of semaphorin 3B is required for positioning the anterior commissure. *Neuron* **48**, 63–75.

Forscher, P. and Smith, S. J. (1988). Actions of cytochalasins on the organization of actin filaments and microtubules in a neuronal growth cone. *J. Cell Biol.* **107**, 1505–1516.

Goodman, C. S. and Shatz, C. J. (1993). Developmental mechanisms that generate precise patterns of neuronal connectivity. *Cell* **72**, 77–98.

Hamelin, M., Zhou, M., Su, M. W., Scott, I. M., and Culotti, J. G. (1993). Expression of the UNC-5 guidance receptor in the touch neurons of C. elegans steers their axons dorsally. *Nature* **364**, 327–330.

Hong, K., Hinck, L., Nishiyama, M., Poo, M. M., Tessier-Lavigne, M., and Stein, E. (1999). A ligand-gated association between cyto-plasmic domains of UNC5 and DCC family receptors converts netrin-induced growth cone attraction to repulsion. *Cell* **97**, 927–941.

Huber, A. B., Kolodkin, A. L., Ginty, D. D., and Cloutier, J. F. (2003). Signaling at the growth cone: Ligand-receptor complexes and the control of axon growth and guidance. *Annu. Rev. Neurosci.* **26**, 509–563.

Hughes, M. E., Bortnick, R., Tsubouchi, A., Baumer, P., Kondo, M., Uemura, T., and Schmucker, D. (2007). Homophilic Dscam interactions control complex dendrite morphogenesis. *Neuron* **54**, 417–427.

Jen, J. C., Chan, W. M., Bosley, T. M., Wan, J., Carr, J. R., Rub, U., Shattuck, D., Salamon, G., Kudo, L. C., Ou, J. *et al.* (2004). Mutations in a human ROBO gene disrupt hindbrain axon pathway crossing and morphogenesis. *Science* **304**, 1509–1513.

Kalil, K. and Dent, E. W. (2004). Hot +TIPS: Guidance cues signal directly to microtubules. *Neuron* **42**, 877–879.

Kennedy, T. E., Serafini, T., de la Torre, J. R., and Tessier-Lavigne, M. (1994). Netrins are diffusible chemotropic factors for commissural axons in the embryonic spinal cord. *Cell* **78**, 425–435.

Kennedy, T. E., Wang, H., Marshall, W., and Tessier-Lavigne, M. (2006). Axon guidance by diffusible chemoattractants: A gradient of netrin protein in the developing spinal cord. *J. Neurosci.* **26**, 8866–8874.

Kidd, T., Bland, K. S., and Goodman, C. S. (1999). Slit is the midline repellent for the robo receptor in Drosophila. *Cell* **96**(6), 785–794.

Kidd, T., Brose, K., Mitchell, K. J., Fetter, R. D., Tessier-Lavigne, M., Goodman, C. S., and Tear, G. (1998). Roundabout controls axon crossing of the CNS midline and defines a novel subfamily of evolutionarily conserved guidance receptors. *Cell* **92**, 205–215.

Klein, R. (2004). Eph/ephrin signaling in morphogenesis, neural development and plasticity. *Curr. Opin. Cell. Biol.* **16**, 580–589.

Kolodkin, A. L., Matthes, D. J., O'Connor, T. P., Patel, N. H., Admon, A., Bentley, D., and Goodman, C. S. (1992). Fasciclin IV: Sequence, expression, and function during growth cone guidance in the grasshopper embryo. *Neuron* **9**, 831–845.

Kruger, R. P., Aurandt, J., and Guan, K. L. (2005). Semaphorins command cells to move. *Nat. Rev. Mol. Cell Biol.* **6**, 789–800.

Lance-Jones, C. and Landmesser, L. (1980). Motoneuron projection patterns in the chick hind limb following early partial reversals of the spinal cord. *J. Physiol. (Lond.)* **302**, 581–602.

Lee, H., Engel, U., Rusch, J., Scherrer, S., Sheard, K., and Van Vactor, D. (2004). The microtubule plus end tracking protein Orbit/MAST/CLASP acts downstream of the tyrosine kinase Abl in mediating axon guidance. *Neuron* **42**, 913–926.

Li, H. S., Chen, J. H., Wu, W., Fagaly, T., Zhou, L., Yuan, W., Dupuis, S., Jiang, Z. H., Nash, W., Gick, C. *et al.* (1999). Vertebrate slit, a secreted ligand for the transmembrane protein roundabout, is a repellent for olfactory bulb axons. *Cell* **96**, 807–818.

Lin, C. H., Espreafico, E. M., Mooseker, M. S., and Forscher, P. (1996). Myosin drives retrograde F-actin flow in neuronal growth cones. *Neuron* **16**, 769–782.

Lin, C.-H., Thompson, C. A., and Forscher, P. (1994). Cytoskeletal reorganization underlying growth cone motility. *Curr. Opin. Neurobiol.* **4**, 640–647.

Lin, D. M., Fetter, R. D., Kopczynski, C., Grenningloh, G., and Goodman, C. S. (1994). Genetic analysis of fasciclin II in Drosophila: Defasciculation, refasciculation, and altered fasciculation. *Neuron* **13**, 1055–1069.

Lin, D. M. and Goodman, C. S. (1994). Ectopic and increased expression of fasciclin II alters motoneuron growth cone guidance. *Neuron* **13**, 507–523.

Luo, Y., Raible, D., and Raper, J. A. (1993). Collapsin: A protein in brain that induces the collapse and paralysis of neuronal growth cones. *Cell* **75**, 217–227.

Lyuksyutova, A. I., Lu, C. C., Milanesio, N., King, L. A., Guo, N., Wang, Y., Nathans, J., Tessier-Lavigne, M., and Zou, Y. (2003). Anterior-posterior guidance of commissural axons by Wnt-frizzled signaling. *Science* **302**, 1984–1988.

Ma, L. and Tessier-Lavigne, M. (2007). Dual branch-promoting and branch-repelling actions of Slit/Robo signaling on peripheral and central branches of developing sensory axons. *J. Neurosci.* **27**, 6843–6851.

Matise, M. P., Lustig, M., Sakurai, T., Grumet, M., and Joyner, A. L. (1999). Ventral midline cells are required for the local control of commissural axon guidance in the mouse spinal cord. *Development* **126**, 3649–3659.

Matthews, B. J., Kim, M. E., Flanagan, J. J., Hattori, D., Clemens, J. C., Zipursky, S. L., and Grueber, W. B. (2007). Dendrite self-avoidance is controlled by Dscam. *Cell* **129**, 593–604.

Ming, G.-L., Song, H.-J., Berninger, S., Holt, C. E., Tessier-Lavigne, M., and Poo, M.-M. (1997). cAMP-dependent growth cone guidance by netrin-1. *Neuron* **19**, 1225–1235.

Polleux, F., Giger, R. J., Ginty, D. D., Kolodkin, A. L., and Ghosh, A. (1998). Patterning of cortical efferent projections by semaphorin-neuropilin interactions. *Science* **282**, 1904–1906.

Polleux, F., Morrow, T., and Ghosh, A. (2000). Semaphorin 3A is a chemoattractant for cortical apical dendrites. *Nature* **404**(6778), 567–573.

Ramon y Cajal, S. (1890). Sur l'origine et les ramifications des fibres nerveuses de la moelle embryonaire. *Anat. Anz.* **5**, 609–613. Extract from Ramon y Cajal, S. (1909). "Histology of the Nervous System" (N. Swanson and L. W. Swanson, transl.). Oxford Univ. Press, Oxford, 1995.

Raper, J. A., Bastiani, M. J., and Goodman, C. S. (1984). Pathfinding by neuronal growth cones in grasshopper embryos. IV. The effects of ablating the A and P axons upon the behavior of the G growth cone. *J. Neurosci.* **4**, 2329–2345.

Raper, J. A. (2000). Semaphorins and their receptors in vertebrates and invertebrates. *Opin. Neurobiol.* **10**, 88–94.

Sabatier, C., Plump, A. S., Le, M., Brose, K., Tamada, A., Murakami, F., Lee, E. Y., and Tessier-Lavigne, M. (2004). The divergent Robo family protein rig-1/Robo3 is a negative regulator of slit responsiveness required for midline crossing by commissural axons. *Cell* **117**, 157–169.

Schmitt, A. M., Shi, J., Wolf, A. M., Lu, C. C., King, L. A., and Zou, Y. (2006). Wnt-Ryk signalling mediates medial-lateral retinotectal topographic mapping. *Nature* **439**, 31–37.

Schmucker, D., Clemens, J. C., Shu, H., Worby, C. A., Xiao, J., Muda, M., Dixon, J. E., and Zipursky, S. L. (2000). Drosophila Dscam is an axon guidance receptor exhibiting extraordinary molecular diversity. *Cell* **101**, 671–684.

Serafini, T., Kennedy, T. E., Galko, M. J., Mirzayan, C., Jessell, T. M., and Tessier-Lavigne, M. (1994). The netrins define a family of axon outgrowth-promoting proteins homologous to C. elegans UNC-6. *Cell* **78**, 409–424.

Serafini, T., Colamarino, S. A., Leonardo, E. D., Wang, H., Beddington, R., Skarnes, W., and Tessier-Lavigne, M. (1996). Netrin-1 is required for commissural axon guidance in the developing vertebrate nervous system. *Cell* **87**, 1001–1014.

Song, H., Ming, G., He, Z., Lehmann, M., McKerracher, L., Tessier-Lavigne, M., and Poo, M. (1998). Conversion of neuronal growth cone responses from repulsion to attraction by cyclic nucleotides. *Science* **281**, 1515–1518.

Sperry, R. W. (1963). Chemoaffinity in the orderly growth of nerve fiber patterns and connections. *Proc. Natl. Acad. Sci. USA* **50**, 703–710.

Stein, E. and Tessier-Lavigne, M. (2001). Hierarchical organization of guidance receptors: Slit silences netrin attraction through a Robo/DCC receptor complex. *Science* **291**, 1847–2034.

Suter, D. M. and Forscher, P. (2000). Substrate-cytoskeletal coupling as a mechanism for the regulation of growth cone motility and guidance. *J. Neurobiol.* **44**, 97–113.

Tamagnone, L. and Comoglio, P. M. (2000). Signaling by semaphorin receptors: Cell guidance and beyond. *Trends Cell Biol* **10**, 377–383.

Taniguchi, M., Yuasa, S., Fujisawa, H., Naruse, I., Saga, S., Mishina, M., and Yagi, T. (1997). Disruption of semaphorin III/D gene causes severe abnormality in peripheral nerve projection. *Neuron* **19**, 519–530.

Tran, T. S., Kolodkin, A. L., and Bharadwaj, R. (2007). Semaphorin regulation of cellular morphology. *Annu. Rev. Cell Dev. Biol.*

Tessier-Lavigne, M. and Goodman, C. S. (1996). The molecular biology of axon guidance. *Science* **274**, 1123–1133.

Tessier-Lavigne, M., Placzek, M., Lumsden, A. G., Dodd, J., and Jessell, T. M. (1988). Chemotropic guidance of developing axons in the mammalian central nervous system. *Nature (Lond.)* **336**, 775–778.

Wadsworth, W. G., Bhatt, H., and Hedgecock, E. M. (1996). Neuroglia and pioneer neurons express UNC-6 to provide global and local netrin cues for guiding migrations in C. elegans. *Neuron* **16**, 35–46.

Wang, K. H., Brose, K., Arnott, D., Kidd, T., Goodman, C. S., Henzel, W., and Tessier-Lavigne, M. (1999). Biochemical purification of a mammalian slit protein as a positive regulator of sensory axon elongation and branching. *Cell* **96**, 771–784.

Wessells, N. K. and Nuttall, R. P. (1978). Normal branching, induced branching, and steering of cultured parasympathetic motor neurons. *Exp. Cell Res.* **115**, 111–122.

Whitford, K. L., Dijkhuizen, P., Polleux, F., and Ghosh, A. (2002). Molecular control of cortical dendrite development. *Annu. Rev. Neurosci.* **25**, 127–149.

Winberg, M. L., Noordermeer, J. N., Tamagnone, L., Comoglio, P. M., Spriggs, M. K., Tessier-Lavigne, M., and Goodman, C. S. (1998). Plexin A is a neuronal semaphorin receptor that controls axon guidance. *Cell* **95**, 903–916.

Yu, H. H., Huang, A. S., and Kolodkin, A. L. (2000). Semaphorin-1a acts in concert with the cell adhesion molecules fasciclin II and connectin to regulate axon fasciculation in Drosophila. *Genetics* **156**, 723–731.

Zhan, X. L., Clemens, J. C., Neves, G., Hattori, D., Flanagan, J. J., Hummel, T., Vasconcelos, M. L., Chess, A., and Zipursky, S. L. (2004). Analysis of Dscam diversity in regulating axon guidance in Drosophila mushroom bodies. *Neuron* **43**, 673–686.

Zhou, F. Q., Zhou, J., Dedhar, S., Wu, Y. H., and Snider, W. D. (2004). NGF-induced axon growth is mediated by localized inactivation of GSK-3beta and functions of the microtubule plus end binding protein APC. *Neuron* **42**, 897–912.

Zinn, K. (2007). Dscam and neuronal uniqueness. *Cell* **129**, 455–456.

Zou, Y., Stoeckli, E., Chen, H., and Tessier-Lavigne, M. (2000). Squeezing axons out of the gray matter: A role for Slit and Semaphorin proteins from midline and ventral spinal cord. *Cell* **102**, 363–365.

Alex L. Kolodkin and Marc Tessier-Lavigne

18

Target Selection, Topographic Maps, and Synapse Formation

A fundamental issue in neurobiology is defining the mechanisms by which neurons recognize and innervate their targets. The formation of a proper functioning nervous system depends on the development of precise connectivity between appropriate sets of neurons or neurons with peripheral targets such as muscles, tendons, skin, and various organs. The development of appropriate synaptic connections requires a series of steps including the specification and generation of neurons and their target cells (Chapters 14, 15, and 16), the guidance of axons to their targets (Chapter 17), the selection of appropriate targets, the formation of orderly specific projections within the target, and ultimately the induction of a specialized presynaptic terminal and postsynaptic membrane (this chapter). The first section of this chapter will focus on the mechanisms and molecules that govern target selection by cortical and retinal axons, and the formation of topographically ordered connections. The second section will focus on the signaling molecules and mechanisms that induce presynaptic and postsynaptic differentiation at the neuromuscular synapse. The third section will summarize current views of how presynaptic and postsynaptic differentiation is initiated in the central nervous system (CNS).

TARGET SELECTION

Target Selection by Delayed Interstitial Axon Branching

As discussed in Chapter 17, the growth cone makes navigational decisions in response to axon guidance molecules. This process of pathfinding by the growth cone is crucial to bring the axon within the vicinity of its targets. The growth cone has also long been considered to be responsible for the process of target selection itself. In some vertebrate systems, and especially in many of the projections studied in Drosophila and other invertebrates, the growth cone tipping the primary axon does ultimately select its target. Many neurons, though, in the vertebrate brain innervate multiple, widely separated targets by axon collaterals, and therefore face a unique problem of target selection during development. It has now become clear that in these situations, the process of target selection is accomplished by a distinct mechanism, often referred to as *interstitial axon branching*, a phenomenon first described as a targeting mechanism by Dennis O'Leary and colleagues in mid-late 1980s.

A prominent example of neurons that employ delayed interstitial branching as a mechanism of target selection is layer 5 neurons in the mammalian neocortex (Fig. 18.1). Layer 5 neurons form the major output projection of the cortex and establish connections with several targets in the midbrain, hindbrain, and spinal cord. Studies by O'Leary and colleagues show that during development, layer 5 axons extend out of the cortex along a spinally-directed pathway; their growth cones ignore several potential targets as they grow past them and continue to extend caudally through the corticospinal tract. Axon collaterals extended by layer 5 axons later innervate the brainstem and spinal targets. One of the major brainstem targets is the basilar pons, a prominent nucleus that lies at the ventral surface of the anterior hindbrain, and is particularly well suited for studying this mechanism of target selection. Evidence obtained from the examination of fixed tissue sections suggests that the collaterals to the basilar pons, as well as to the other subcortical targets of layer

A   Primary axon extension

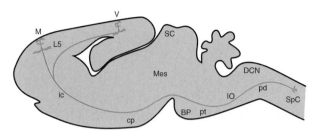

B   Delayed collateral branch formation

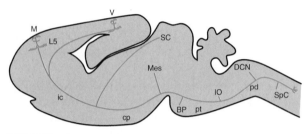

C   Selective axon elimination

Motor cortex

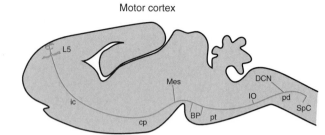

Visual cortex

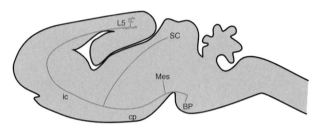

FIGURE 18.1   Area-specific subcortical projections of layer 5 neurons of the neocortex develop by delayed interstitial branching and selective axon elimination. The three main phases of this mechanism is illustrated in schematics of a sagittal view of the developing rat brain. (A) Primary axon extension: Layer 5 neurons (L5) extend a primary axon out of the cortex along a pathway that directs them toward the spinal cord (SpC) passing by their subcortical targets. (B) Delayed collateral branch formation: Subcortical targets are later contacted exclusively by axon collaterals that develop by a delayed extension of collateral branches interstitially along the spinally-directed primary axon. As a population, layer 5 neurons in all areas of rat neocortex develop branches to a common set of targets. (C) Selective axon elimination: As illustrated for visual and motor cortex, specific collateral branches or segments of the primary axon are selectively eliminated to generate the mature projections functionally appropriate for the area of neocortex in which the layer 5 neuron is located. Abbreviations: BP, basilar pons; cp, cerebral peduncle; DCN, dorsal column nuclei; ic, internal capsule; M, motor cortex; Mes, mesencephalon; pd, pyramidal decussation; pt, pyramidal tract; SC, superior colliculus; V, visual cortex. Adapted from O'Leary and Koester (1993).

5 axons, develop by a delayed interstitial branching from the axon shaft millimeters behind the growth cone, often days after the parent axons have grown past the target. Time-lapse imaging in hemi-brain slice preparations from neonatal mice has provided further evidence that the branches form *de novo* along the portion of the axon shaft overlying the target. Evidence from collagen gel assays suggests that the target releases a diffusible activity with chemoattractant properties that induces branching along layer 5 axons, and directs the branches into the target. Thus, the axon shaft millimeters behind the growth cone is actively involved in target selection.

Examples of interstitial branching as the primary mechanism of target selection in vertebrates are accumulating. These examples include the extensive analysis by Katherine Kalil and coworkers on the development of cortical callosal projections, as well as development of axonal projections from the hippocampal formation (i.e., the subiculum) to the mammillary bodies, dorsal root ganglion neurons to the spinal gray matter, the innervation of the cortical plate by thalamic axons and

of the dorsal lateral geniculate nucleus by retinal axons. In addition, as discussed later, the formation of topographic connections by retinal axons in the optic tectum of chicks and superior colliculus of rodents also occurs by interstitial branching along the axon shaft.

Exuberant Axonal Connections and Collateral Elimination

The development of many axonal projections in the brain is characterized by an initially exuberant, or widespread, growth of axons, followed by the elimination of functionally inappropriate axon segments and branches. For example, this mechanism is used to generate the adult patterns of callosal, intracortical, and subcortical projections of the mammalian neocortex (O'Leary and Koester, 1993). In the adult cortex, neurons that send an axon through the midline corpus callosum to the opposite cortical hemisphere, termed callosal neurons, have a limited, discontinuous distribution. Giorgio Innocenti and colleagues were the first to show that the limited adult distribution of callosal neurons emerges from an early widespread, continuous distribution of cortical neurons that send an axon through the corpus callosum. Subsequently, Innocenti, O'Leary, and others used fluorescent retrograde tracers as fate markers to show that the developmental restriction in the distribution of callosal neurons is due to the loss of callosal axons rather than to the death of the parent neurons, which maintain an ipsilateral cortical connection. Collateral elimination also has been implicated in establishing the mature connections between cortical areas in the same hemisphere and in the refinement of horizontal connections within an area. For example, in adult visual cortex, layer 2/3 neurons have discrete horizontal projections to groups of other layer 2/3 cells with similar receptive field properties. However, Larry Katz and others have shown that the initial axonal outgrowth from layer 2/3 cells is very widespread in the horizontal plane, and the mature pattern of connections emerges in part through collateral elimination.

The organization of the adult neocortex into functionally specialized areas requires that each area establishes projections to specific subsets of targets in the brainstem and spinal cord. During development, though, layer 5 neurons project more broadly, and form collateral projections to a larger set of layer 5 targets than they will retain in the adult (Fig. 18.1). The functionally appropriate patterns of layer 5 projections characteristic of the adult later are pruned from this initial widespread pattern through selective axon elimination. Depending upon the cortical area in which the layer 5 neuron is located, it will eliminate different subsets of the initial complement of branched projections, and retain only those that are functionally appropriate. For example, layer 5 neurons in motor cortex lose their collateral branch to the superior colliculus, but retain branches to other targets including the basilar pons, inferior olive, dorsal column nuclei, and spinal gray matter. In contrast, layer 5 neurons in visual cortex lose the entire segment of their primary axon and its branches caudal to the basilar pons, and retain branches to the pons and superior colliculus. Although functionally inappropriate for the proper operation of the adult brain, the eliminated collateral branches or axon segments should not be viewed as projection errors, since they seem to be elaborated according to a specific axonal growth program characteristic of that general class of neuron.

The cellular mechanisms that control axon elimination and the final patterning of cortical projections are not well understood. In general, axon elimination occurs through one of two distinct phenomena that relate to the scale of the event: small-scale elimination typically occurs by retraction, whereas large-scale elimination, as discussed earlier for layer 5 axon patterning, relies primarily on degeneration, which is similar in appearance to Wallerian degeneration but is mechanistically distinct (Hoopfer et al., 2006). The triggers and selection process are also poorly understood, but the available evidence for large-scale cortical pruning indicates a role for neural activity, specifically the sensory information being relayed by thalamocortical input. In contrast, the elimination of callosal axons is perturbed by a variety of peripheral manipulations of either visual or somatosensory input, that alters either patterns of neural activity (e.g., strabismus) or absolute levels of activity (e.g., dark-rearing, eyelid suture, or silencing of retinal activity with the sodium channel blocker, tetrodotoxin). In these instances, callosal axon elimination is abnormal, resulting in the retention of callosal connections in parts of cortex that would normally lose them. Similar findings have been obtained for the development of layer 2/3 horizontal connections. Thus sensory input plays an important role in developing the adult pattern of callosal and intracortical connections by influencing the pattern of axon elimination.

Heterotopic transplant experiments by O'Leary and Brent Stanfield show that collateral elimination by layer 5 neurons is also plastic during development. Developing layer 5 neurons transplanted from visual cortex to motor cortex permanently retain their normally transient spinal axon, whereas layer 5 neurons transplanted from motor cortex to visual cortex lose their normally permanent spinal axon and retain their transient axon collateral to the superior colliculus. Thus, the projections retained by the transplanted layer 5 neurons are appropriate for the cortical area in which the transplanted neurons develop, not where they were born.

Summary

Most projections in the vertebrate brain are formed by the mechanism of delayed interstitial branching along the length of the axon shaft rather than by direct axon targeting. During development, neurons often project to more targets than in the adult, and generate their adult pattern of connections through a process of selective axon or collateral elimination. This developmental phenomenon of transiently "exuberant" axonal projections may provide a substrate for developmental plasticity. For example, alterations in axon elimination may be a source of functional sparing or recovery following neural insults during development. In addition, this mechanism may contribute to differences between species in axonal connections as suggested by studies of the projection from the subiculum to the

FIGURE 18.2   Development of topographic order in the retinotectal projection of model species. (A) Photographs, at critical ages during topographic map formation, of the primary model systems currently used for the examination of retinocollicular/retinotectal map development. All photographs are at the same scale. The mouse superior colliculus (SC), on the dorsal surface of the midbrain (bracket denotes size), is approximately as large as an entire developing zebrafish and the chick optic tectum (ot) is approximately the size of an entire Xenopus tadpole at these ages. The Xenopus and zebrafish optic tectum sizes are denoted by the small white bars adjacent to the brain. The mouse and zebrafish are used commonly for the potential genetic manipulations available in these organisms. Chick and Xenopus are especially useful for overexpression, misexpression, and knock-down studies. Retinal ganglion cells from these organisms can be cultured and manipulated *in vitro*. (B and C) Mechanisms and molecules controlling retinotopic mapping in (B) mouse and chick and (C) frogs and zebrafish. The names and distributions of molecules and activities demonstrated *in vivo* to control, in part, the mapping mechanisms used at each stage are listed. The gradients may represent the consensus distribution for a combination of related molecules (i.e., EphAs), which are not listed individually due to distinctions in the individual members expressed and the precise distributions between species. Molecules and activities other than those listed are likely to participate in topographic mapping. (B) In mouse and chick RGC axons originating from a small focal retinal site enter the OT/SC over a broad L-M extent and significantly overshoot their future termination zone (TZ, circle). Interstitial branches form *de novo* from the axon shaft well behind the primary growth cone; their formation is topographically biased for the anterior-posterior (A-P) position of the future TZ. Branches are directed along the lateral-medial (L-M) axis of the OT/SC toward the position of the future TZ, where they arborize in a domain encompassing the forming TZ. The broad, loose array of arbors is refined to a dense TZ in the topographically appropriate location. (C) In frogs and fish the RGC growth cone extends to the TZ where it forms a terminal arbor through a process of backbranching from the base of the primary growth cone. The tectum expands as terminal arborizations elaborate and refine into a mature TZ. Abbreviations: cb, cerebellum; ctx, cortex; D, dorsal; fb, forebrain; hb, hindbrain; N, nasal; T, temporal; V, ventral. Adapted and updated from McLaughlin, T., Hindges, R, and O'Leary, D. D. M. (2003). *Current Opinion in Neurobiology*.

mammillary bodies in mammals ranging from rodents to rabbits to elephants (O'Leary, 1992).

Topographic Map Development

Once axons reach their targets, they must select appropriate target cells with which to form synaptic connections. Many axonal projections within the brain establish an orderly arrangement of connections within their target field, termed a topographic map. These maps are arranged such that the spatial order of the cells of origin is reflected in the order of their axon terminations, thereby neighboring cells project to neighboring parts of the target to form a smooth and continuous map. Topographic projections are especially evident in sensory systems, such as the somatosensory and visual. In the somatosensory system, a map of the sensory receptors distributed on the body is reiterated multiple times at various levels of the neuraxis. In the visual system, the main objective is to represent the visual world in the brain; that is, to reconstruct a topographic representation of the visual world that projects onto the retina and is remapped multiple times in the brain, initially through direct retinal projections to the dorsal thalamus and midbrain, and subsequently by higher order projections. This precise mapping requires the maintenance of the spatial ordering of the axons of retinal ganglion cells (RGCs) within their central targets in a pattern that reflects their origins in the retina. The projection from the retina to its major midbrain target, the superior colliculus (SC) of mammals, or its nonmammalian homologue, the optic tectum (OT), has been the predominant model system for studying the mechanisms that account for the development of topographic axonal connections. The relative size of the OT and SC

in the species predominantly used as models to study the development of topographic maps is shown in Figure 18.2A.

A large body of evidence developed over the past 50 years has shown that the target presents guidance information to incoming axons that controls their development of topographic connections. This work has culminated in the discovery of topographic guidance molecules over the past decade. Further, the development of a fully refined map requires activity-dependent mechanisms that are influenced by retinal waves that correlate the patterns of spiking activity between neighboring RGCs. These molecules and mechanisms will be discussed in detail in the following sections and placed into a historical context.

The Chemoaffinity Hypothesis

The mechanisms that control the establishment of topographic maps have been intensively studied for many decades, but only in recent years has the molecular control of this process begun to be defined. The chemoaffinity hypothesis, formally proposed by Roger Sperry over 40 years ago (Sperry, 1963), has been a driving force in the field. Sperry designed clever experiments to address earlier theories such as that of Weiss, which posited that the specificity of neuronal connections in an adult resulted from the "functional molding" of circuits formed more or less at random; connections that are functionally appropriate are retained and the others are eliminated. Sperry studied the regeneration of the retinotectal projection in newts and frogs; in these amphibians, unlike in birds or mammals, cut RGC axons are capable of regenerating to reestablish functional connections with target

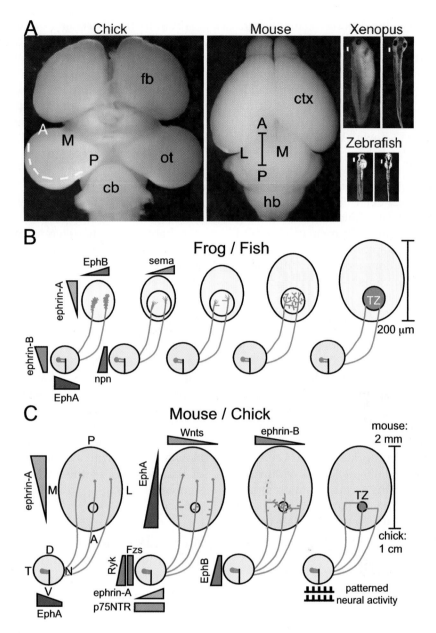

neurons in the tectum. In one particularly conclusive experiment, Sperry cut the optic nerve, rotated the eye in its orbit by 180°, and allowed the RGC axons to regrow to the tectum. According to earlier theories, one would predict that the axons eventually should form a novel pattern of connections that could generate appropriate behavioral responses to visual stimuli. Instead, Sperry showed that the frogs behaved as if their visual world had been rotated 180°. For instance, when a fly was presented in the upper left-hand quadrant of the visual field of the rotated eye, the frog responded by diving down to the right. This inappropriate response was retained throughout the frog's life, even after attempts to train the animals to compensate

for the rotation. These findings suggested that the regenerated RGC axons had reestablished their original pattern of connections, which was subsequently confirmed by Sperry and others using anatomical and electrophysiological analyses. These findings provided experimental evidence inconsistent with the prevailing theories of the time that the adult pattern of axonal connections is established by the selective retention of functionally appropriate connections, and eventually led to Sperry's chemoaffinity hypothesis.

Sperry proposed that molecular tags on projecting axons and their target cells determine the specificity of axonal connections within a neural map. Further, he suggested that these molecular tags might establish topography through their distribution in complementary gradients that mark corresponding points in both sensory and target structures. The representation of the retina onto the tectum (or SC) typically is simplified to the mapping of two sets of orthogonally-oriented axes: the temporal (T)-nasal (N) axis of the retina along the anterior (A)-posterior (P) axis of the tectum, and the ventral (V)-dorsal (D) axis of the retina along the medial (M)-lateral (L) axis (also referred to as the D-V axis) of the tectum. Based on the chemoaffinity hypothesis, each point in the tectum would have a unique molecular address determined by the graded distribution of topographic guidance molecules along the two tectal axes, and similarly each RGC would have a unique profile of receptors for those molecules that would result in a position-dependent, differential response to them by RGC axons.

Over the next half century, the specificity of the projections of RGC axons to tectal cells was investigated further by the tracing of axonal projections following experimental manipulations, first in the regenerating retinotectal system and later during the development of the projection. The manipulations included rotations or transplantations of the retina, tectum, and even the optic pathway, using either the whole structure or parts of it. These experiments showed that regenerating and developing RGC axons formed topographically appropriate connections even when they are experimentally deflected within the tectum, or forced to enter the tectum from abnormal positions or with a reversal in the relative time of arrival of populations of RGC axons. Studies of this sort done by Bill Harris and Christine Holt were particularly compelling, and provided a body of evidence supporting the basic tenet of the chemoaffinity hypothesis that the establishment of topographic projections involves the recognition of positional information on the tectum.

Prior to the discovery of the ephrins (described later), the most compelling evidence for topographic guidance molecules came from the work of Friedrich Bonhoeffer and coworkers using several elegant *in vitro* assays, for example the membrane stripe assay, as well as the growth cone collapse assay codeveloped by Jonathon Raper. Using the membrane stripe assay, Bonhoeffer showed that chick temporal RGC axons given a choice between growing on alternating lanes of anterior and posterior tectal membranes, show a strong preference to grow on their topographically appropriate anterior membranes, whereas nasal RGC axons exhibit no preference (Fig. 18.3). A critical finding was that the growth preference of temporal axons was not due to an attractant or growth-promoting activity associated with anterior tectal membranes, but instead to a repellent activity associated with posterior tectal membranes. This was the first demonstration of a role for repellent activities in axon guidance. By taking advantage of the finding that posterior tectal membranes also preferentially collapse the growth cones of temporal axons, the repellent activity was biochemically isolated to a 33 kDa, GPI-anchored protein referred to as RGM (Repulsive Guidance Molecule). An interesting twist is that the search for RGM led to the cloning of ephrin-A5 (described later), and only several years after that was RGM itself cloned. However, mice with a targeted deletion of RGM1, one of the members of the RGM family, do not exhibit significant mapping defects in the retinocollicular projection.

Axonal Behaviors during Development of Topographic Maps and Their Implications for Actions of Topographic Guidance Molecules

Following the pioneering work of Sperry and Bonhoeffer, two major sets of events directly led to the

FIGURE 18.3 **Repellent effects of ephrin-As on retinal axons *in vitro*.** (A, B) Summary of use of the membrane stripe assay to analyze the effects of membrane-associated molecules in the optic tectum, such as the ephrin-A ligands, on the guidance of retinal ganglion cell (RGC) axons. This elegant *in vitro* assay, developed by Friedrich Bonhoeffer, has been used extensively by many investigators to characterize repellent versus attractant effects of membrane-associated molecules on axon guidance. A retinal strip oriented along the nasal-temporal axis is explanted on carpets consisting of alternating 90 μm wide lanes of membranes derived from the anterior (A) or posterior (P) third of the tectum, or from heterologous cell lines (e.g., Cos or 293T cells) mock-transfected or transfected with *ephrin-A2* or *ephrin-A5* cDNA. RGC axons growing out of the temporal half of the retinal strip show a strong preference to grow on anterior tectal membranes, whereas those growing out of the nasal half show no preference. In contrast, temporal axons do not show a preference for anterior membranes when posterior membranes are pretreated with heat or proteases, indicating that the preference is due to a repellent in posterior membranes. ephrin-A2 and ephrin-A5 are candidate mediators of this repellent activity, since they are enriched in posterior membranes and repel retinal axons in the stripe assay. Adapted from O'Leary *et al.* (1999). (C, D) Examples of RGC growth preferences in the stripe assay. The lanes containing *ephrin-A2* or *ephrin-A5* transfected cell membranes are labeled with rhodamine isothiocyanate (RITC) fluorescent beads, visualized as red lanes in the lower part of each panel. Temporal RGC axons grow predominantly on membranes from mock transfected cells, while nasal RGCs axons grow equally well on membranes from ephrin-A and mock transfected cells. These growth preferences exhibit an abrupt transition at mid-retina on *ephrin-A2* transfected membranes, similar to the preferences observed on anterior and posterior tectal membranes, and in contrast to the more gradual transition in growth preferences observed on *ephrin-A5* transfected membranes. Adapted from Monschau *et al.* (1997).

A

Tectum
P

OR

A

EphrinA
transfected
cells

Mock
transfected
cells

Membrane
preparation

B

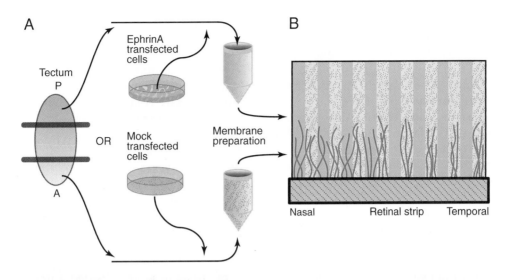

Nasal   Retinal strip   Temporal

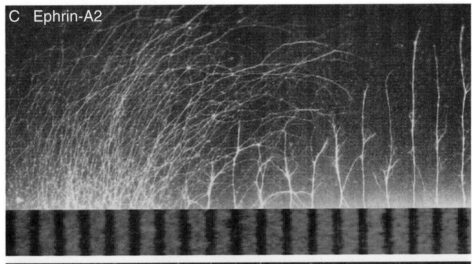

C Ephrin-A2

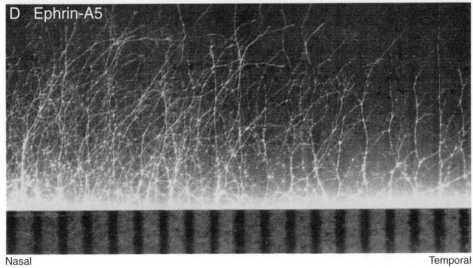

D Ephrin-A5

Nasal   Temporal

current understanding of the development of retino-topic maps. One was the major breakthrough toward understanding the molecular control of topographic mapping that came in the mid-1990s with the cloning of two closely related genes, ephrin-A2 (originally called Eph Ligand Family-1, or ELF-1) by John Flanagan and colleagues, and ephrin-A5 (originally called Repulsive Axon Guidance Signal, or RAGS) by Bonhoeffer, Uwe Drescher, and colleagues. These proteins, like all members of the ephrin-A family, are anchored to the cell membrane by a GPI-linkage and bind with similar affinities and activate the same receptors, members of the EphA subfamily of receptor tyrosine kinases (Fig. 18.4). As discussed in the following sections, these proteins are prominent players in topographic mapping.

The other major set of events was the result of the introduction of high resolution axon tracers, the lipophilic fluorescent dyes DiI and DiO (and subsequently related molecules such as DiA and DiD), in the late 1980s that allowed for the first time detailed analyses of the morphologies and behaviors exhibited by RGC axons in chicks and rodents as they form topographically ordered projections. This was important for several reasons, including that it allowed for the analysis of chicks and mice, which was critical to take advantage of genetic manipulations, and showed that interstitial axon branching was a key mechanism of map development in warm-blooded vertebrates. Prior to this work, the behavior of RGC axons during map development was based largely upon the studies of amphibians and fish done by many investigators, including Hollis Cline, Martha Constantine-Paton, Scott Fraser, Haijme Fujisawa, Bill Harris, Christine Holt, and Claudia Stuermer. Among the reasons for the initial focus on fish and amphibians is that they were already widely used for studies of retinotopic mapping because of their ability to regenerate retinal connections, the relative ease with which their visual systems could be experimentally studied, but most importantly, preexisting axon tracers (e.g., HRP, cobalt chloride) provided high resolution labeling of RGC

axons in these cold-blooded animals although these and other available tracers did not in developing chicks and rodents. Numerous studies in frogs and fish concluded that topographic retinotectal connections develop by the direct topographic targeting of RGC axon growth cones (Fig. 18.2B); therefore, similar mechanisms were assumed to account for mapping in other vertebrates. These findings and assumptions led to models of mapping and roles for topographic guidance molecules, such as the ephrin-As, to account for a mechanism of direct topographic targeting of the primary growth cone.

Studies by O'Leary and colleagues using the new age high resolution lipophilic tracers, DiI and related compounds, revealed an unexpected and unique picture of the development of topographic retinotectal projections in warm-blooded animals, which in turn required a substantial reassessment of the mechanisms of mapping and roles for topographic guidance molecules in controlling them. They showed that development of retinotectal topography in chicks, rats, and mice is a multistep process that involves axon overshoot of the topographically correct position of the termination zone along the A-P axis of the target followed by topographically appropriate interstitial branching along the axon shaft.

As schematized in Figure 18.2C, in chicks and mice, RGC axons originating from a focal site in the retina enter the chick OT and mouse SC at its anterior edge with an aberrantly broad distribution along its L-M axis, with a peak in their distribution centered on the location of the future termination zone. The axons extend posteriorly across the OT/SC well past the topographically appropriate site for their termination; during this process they form branches *de novo* interstitially from the axon shaft hundreds of microns or even millimeters behind the primary growth cone. Interstitial branching exhibits a significant degree of topographic specificity along the A-P axis, with the highest percentage of branches found near the A-P position of the future termination zone (Yates *et al.*, 2001). The interstitial branches form roughly perpen-

FIGURE 18.4 Eph and ephrin families. Eph receptors comprise the largest family of receptor tyrosine kinases, currently numbering 15 ▶ members, and are divided into two subfamilies, EphA (A1 to A9; EphA9 is not depicted) and EphB (B1 to B6) with distinct binding specificities that correlate with structural similarities. The nine known ephrins are divided into the ephrin-A (A1–A6) and ephrin-B (B1–B3) subfamilies on the basis of sequence homology and their membrane anchoring: ephrin-As are anchored by a glycosyl phosphatidylinositol (GPI) linkage, and the ephrin-Bs by a transmembrane domain. Within each receptor-ligand subfamily, the ligands bind, albeit with different affinities, and activate, with few exceptions (EphB5 and EphB6), all the receptors, but only very limited interaction occurs between subfamilies (for example, EphA4 interacts with ephrin-As and two ephrin-B proteins; dashed lines indicate relatively weak interactions). Although membrane-anchored ephrins activate Eph receptors, soluble forms of an ephrin can do so only when artificially clustered. Clustered ligand binding results in the formation of receptor multimers and activates the catalytic kinase domain on the cytoplasmic portion of the Eph receptor. In addition, the ephrin-B ligands themselves can also transduce signals; thus receptor-ligand binding can result in bidirectional signaling into the receptor-expressing and ligand-expressing cells. Recent evidence suggests that the ephrin-A ligands are also capable of bidirectional signaling.

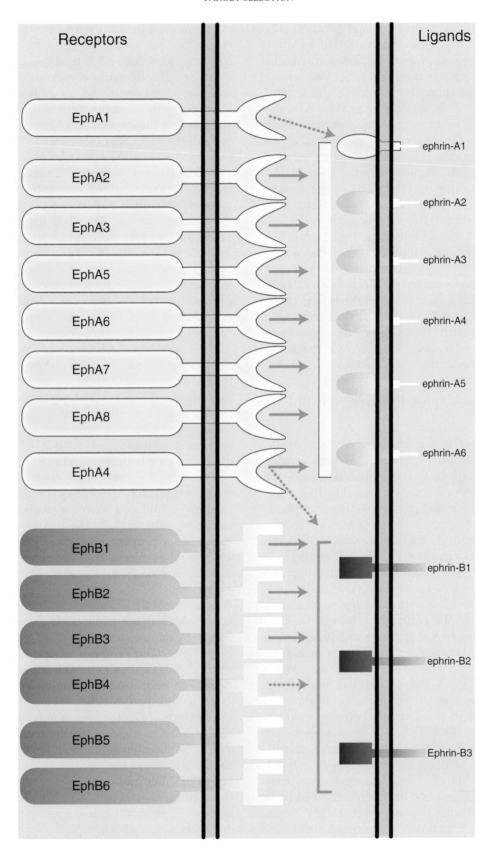

dicular to the primary axon shaft and preferentially extend along the L-M axis to their future termination zone where they go on to arborize. These interstitial branches are the exclusive means by which RGCs form permanent, topographically ordered connections (Yates *et al.*, 2001). This mechanism of map development results in an initial map that is very coarse. In mice, for example, these features result in RGC axons originating from a focal source in the retina covering virtually the entire SC at perinatal ages (Fig. 18.8A). The refined topography characteristic of the mature projection finally is achieved through the preferential arborization of appropriately positioned branches and the elimination of the overshooting segments of the primary axons and any ectopic branches and arbors (Fig. 18.8A–C). These analyses of topographic mapping of RGC axons in chicks and mice indicate that a primary role for topographic guidance molecules is to regulate topographic branching along RGC axons, and guide the branches to the correct site in the target: processes that impose unique requirements on the molecular control of map development (Yates *et al.*, 2001).

Role for Ephrin-A/EphA Forward Signaling in A-P Mapping

Since the discovery of ephrin-As in the mid-1990s, six ephrin-As have been identified, as well as nine EphA receptors (Fig. 18.4). Figure 18.5 illustrates the ephrin-As and EphAs that are expressed by RGCs and in the OT/SC in chicks and mice, as well as an approximation of their expression patterns. In both chicks and mice, multiple ephrin-As are expressed in the OT/SC, but together they form a low-to-high A-P gradient. In addition, RGCs express multiple EphAs that together form a high-to-low T-N gradient of expression.

In chick, ephrin-A2 is expressed in an increasing A-P gradient across the entire OT, and ephrin-A5 is expressed in a steeper A-P gradient largely limited to posterior OT; together they combine to form an increasing gradient of ephrin-As across the A-P axis. RGCs express EphA3, EphA4, and EphA5, but only EphA3 is expressed in a gradient, which is highest in temporal retina and lowest in nasal retina. Because temporal retina with high levels of EphA3 maps to anterior OT with low levels of ephrin-As, and vice versa, ephrin-A2 and ephrin-A5 were predicted to act as axon repellents that more strongly affect temporal axons than nasal axons. This suggestion has been confirmed by *in vitro* and *in vivo* studies (Nakamoto *et al.*, 1996; Monschau *et al.*, 1997). Experiments employing membrane stripe and growth cone collapse assays using membranes from transfected cell lines show that both ephrin-A2 and ephrin-A5 preferentially repel tempo-

ral axons, and that at the appropriate concentrations, retinal axons exhibit a graded temporal-to-nasal response to ephrin-A5 (Fig. 18.3). *In vivo*, temporal axons specifically avoid ectopic patches of ephrin-A2 that were overexpressed in anterior OT following infection with recombinant retrovirus. These studies suggest that the interaction of EphA3 with ephrin-A2 and ephrin-A5 could determine the specificity of RGC projections along the A-P tectal axis.

Although species-specific differences are apparent in the particular EphAs and ephrin-As that are expressed, or in their patterns of expression (Fig. 18.5), the basic theme just described for chicks is constant across all vertebrate species examined. For example, as in chick tectum, ephrin-A2 and ephrin-A5 combine to form a smooth increasing A-P gradient in the mouse SC. However, their pattern of expression in the SC differs substantially from chick tectum: ephrin-A5 is expressed in an increasing A-P gradient across the SC, resembling ephrin-A2 in chick tectum, whereas ephrin-A2 is expressed at high levels in a broad domain centered on mid-posterior SC and shows a graded decline to low or no expression in the anterior third and far-posterior SC (Fig. 18.6). The expression of EphA receptors by RGCs also differs between chick and mouse. In mouse, EphA3 is not expressed by RGCs, EphA4 is expressed uniformly, and both EphA5 and EphA6 are expressed in a high temporal to low nasal gradient.

The graded expression of ephrin-As and their differential repulsion of temporal versus nasal RGC axons strongly implicated them as topographic guidance molecules. The first genetic test of whether they are required for proper topographic mapping came from an analysis of mice with a targeted deletion of ephrin-A5 (Frisen *et al.*, 1998) (Fig. 18.6). The mapping of RGC axons in the SC of ephrin-A5 null mice is topographically aberrant in a manner consistent with the loss of ephrin-A5 and the maintained expression of ephrin-A2. For example, temporal RGC axons form ectopic projections to far-posterior SC and within the anterior third of the SC. As expected, topographic mapping defects in ephrin-A2/A5 double knockout mice are more severe than in either ephrin-A single knockout (Feldheim *et al.*, 2000). Surprisingly, though, in ephrin-A2/A5 double knockout mice, temporal RGC axons form a termination zone, albeit smaller, in anterior SC at approximately the correct position, suggesting the action of other molecules or mechanisms. This led to the finding that ephrin-A3 also is expressed in the mouse SC, although at low levels uniformly across its A-P axis. Mice with a triple knockout of ephrin-A2, -A3, and -A5 exhibit a near complete loss of topography. When this ephrin-A triple knockout is combined with mice that lack patterned RGC activity (see follow-

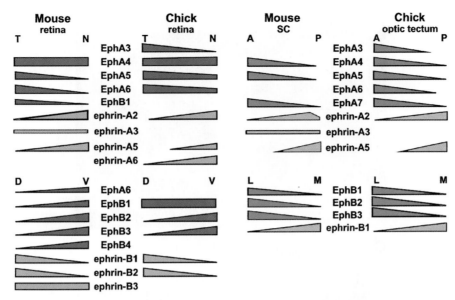

FIGURE 18.5 Distribution of Ephs and ephrins in the retina and optic tectum/superior colliculus (OT/SC) of chick and mouse. The table represents generalized patterns for individual Ephs and ephrins in the retina and OT/SC during critical stages in the development of the retinotopic map (E15-P7 in mouse and E6-E14 in chick). The chart is likely incomplete and the distributions for several molecules is dynamic during development. The sizes and shapes of the gradients are generalized and relative expression levels should not be inferred. A, anterior; D, dorsal; L, lateral; M, medial; N, nasal; P, posterior; T, temporal; V, ventral. Adapted and updated from McLaughlin, T. and O'Leary, D.D.M. (2005). *Annual Review of Neuroscience*.

ing section), topography is essentially completely lost (Pfeiffenberger *et al.*, 2006).

Although it is well established that ephrin-As mediate the repulsion of RGC axons by forward signaling through EphA receptors, genetic studies showing roles for EphAs have been few. Mice with a targeted deletion of EphA5 have mapping defects consistent with it mediating a repellent activity through forward signaling. A particularly intriguing study of EphA function was a gain-of-function strategy that took advantage of the features that ephrin-As bind and activate with similar efficacy most EphA receptors, and that EphA3 is not expressed by RGCs in mice. Mice were generated in which EphA3 was ectopically expressed in about half of the RGCs uniformly distributed across the retina, thus producing two subpopulations of RGCs, one which has the wild-type gradient of EphA receptors (EphA5 and EphA6), and one with an elevated gradient of overall EphA expression (Brown *et al.*, 2000). In these mice, the projection of the EphA3 RGCs is compressed to the anterior half of the SC, indicating that the level of EphA receptor dictates the degree to which an RGC axon is repelled by ephrin-As. Surprisingly, though, the projection of the wild-type RGCs is compressed to the posterior half of the SC, and is likely excluded from the anterior SC by competitive interactions with the EphA3 RGCs. Thus, the mapping of the two RGC subpopulations is not determined solely by the absolute level of EphA receptor signaling, but instead is plastic and can be strongly influenced by the relative difference in EphA signaling. Importantly, these findings also reveal a hierarchy in the mechanisms for generating topographic maps, showing that molecular axon guidance information, such as that signaled by ephrin-As and EphAs, dominates over activity-dependent patterning mechanisms that are based on near neighbor relationships and correlated activity.

Countergradients of Repellent Activities in A-P Mapping and Roles for Ephrin-A Reverse Signaling

Historically, models of topographic mapping have been based upon the action of molecular activities that promote axon growth. However, with Bonhoeffer's identification in the late 1980s of repellent activities concentrated in posterior chick OT that preferentially affect temporal RGC axons, it became clear that topographic mapping was controlled in a different fashion than previously recognized. These findings eventually resulted in the incorporation of repellent activities into models of topographic mapping, but until recently these models focused on guiding RGC axons to their correct TZ in the target. These models were based upon a gradient determined repulsion, in which an RGC axon would stop its growth when it reaches a threshold level of repulsion found at the AP position

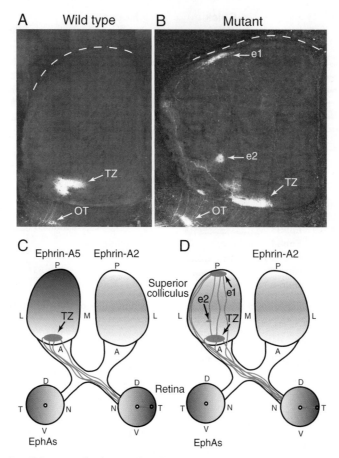

FIGURE 18.6 **RGC projections in wild-type and *ephrin-A5* knockout mice related to the expression patterns of *ephrin-A5* and *ephrin-A2*.**
(A, B) Anterograde DiI labeling of RGC axons from peripheral temporal retina in wild-type and *ephrin-A5* knockout mice. Dorsal views of whole mounts of the superior colliculus (SC) are shown; midline is to the right, dashed lines indicate the posterior SC border. (A) In wild-type mice, temporal RGC axons end and arborize in a densely labeled termination zone (TZ) in anterior SC. The RGC projection to optic tract nuclei is also evident (OT). (B) In *ephrin-A5* null mutant mice, temporal RGC axons also end and arborize in a densely labeled TZ at the topographically appropriate site in anterior SC, but in addition temporal axons project to and arborize at topographically inappropriate sites in posterior-most SC (e1) and anterior SC (e2). (C, D) Schematic representations summarizing temporal RGC axon mapping in wild-type and *ephrin-A5* mutant mice in relationship to the expression of *ephrin-A5* and *ephrin-A2*. (C) In wild-type mice, a focal DiI injection in peripheral temporal retina labels axons (red lines) that form a dense TZ (red oval) in the topographically correct anterior SC. *Ephrin-A5* (blue shading) is expressed in a low anterior (A) to high posterior (P) gradient across the SC. *Ephrin-A2* (orange shading) is expressed highest in mid-posterior parts of the SC, and declines to low levels in more anterior and far-posterior SC. Together, ephrin-A5 and ephrin-A2 form a smooth gradient of repellent activity across the SC. EphA receptors are expressed in a high temporal (T) to low nasal (N) gradient by RGCs. (D) In *ephrin-A5* mutant mice, *ephrin-A2* (orange shading) is expressed in the same pattern as in wild-type. A focal DiI injection in temporal retina labels axons that form a dense TZ in topographically correct anterior SC, but in addition, aberrant terminations (e1, e2) form at topographically incorrect locations. The pattern of ectopic arbors relates to the maintained expression pattern of *ephrin-A2*: ectopic arbors are typically present in far-posterior and anterior SC where *ephrin-A2* expression is low, but are rare in mid-SC where *ephrin-A2* expression is highest. This distribution suggests that in the absence of ephrin-A5, ectopic arbors are present where the levels of repellent activity due to ephrin-A2 are too low to prevent their formation and stabilization. Abbreviations: D, dorsal; L, lateral; M, medial; V, ventral. Adapted from Frisén *et al.* (1998). *Neuron*.

of the future TZ (Fig. 18.7). Although such models could explain retinotectal map development in amphibians and fish, they cannot account for aspects of map development along the A-P axis in chick OT and mouse SC, including the initial A-P overshoot of RGC axons and the subsequent topographic interstitial branching along RGC axons.

Models for A-P mapping in mouse SC and chick OT must explain the topographic bias in the formation of interstitial branches, which arguably is the key feature of A-P map development in warm-blooded vertebrates. Such a model must be able to account for the paucity of branching both posterior and anterior to the appropriate A-P location of the termination zone along each axon. Models that in principle can account for topographic branching have been recently proposed; for example models based upon parallel gradients of a branch repellent (or inhibitory) activity and a branch promoting activity (Fig. 18.7). However, the field has moved toward favoring a model based upon opposing gradients along the A-P axis, each of which inhibits branching. As described in the preceding section, the low-to-high A-P repellent gradient is due to EphA forward signaling, which acts to inhibit interstitial branching posterior to the correct termination zone along the A-P axis. Indeed, use of a modified membrane stripe assay shows that temporal axons preferentially branch on their topographically-appropriate anterior tectal membranes, and that this branching specificity is due to the inhibition of branching on posterior tectal membranes by ephrin-As.

Emerging evidence indicates that the opposing high-to-low A-P repellent gradient, which would serve to prevent branching anterior to the correct termination zone (Fig. 18.7), is due to ephrin-A reverse signaling. A key feature of the countergradients of ephrin-A/EphA mediated repellent activities is that EphAs and ephrin-As both are expressed by RGCs and in the OT/SC, and exhibit opposing gradients of expression along both the T-N retinal axis and the A-P OT/SC axis (Figs. 18.2, 18.5). Recent experimental data has shown that ephrin-As can reverse signal, and that this reverse signaling has a repellent effect on RGC axons. The critical findings, which come from the work of Uwe Drescher, show that RGC axons are repelled by EphA7 in the "Bonhoeffer" stripe assay, a repulsion blocked by ephrin-A5-fc that blocks EphA function, and that EphA7 knockout mice have retinocollicular mapping defects consistent with the loss of the high-to-low AP graded expression of EphA7 in the SC (EphA7 is not expressed by RGCs; Fig. 18.5). Thus, the low-to-high T-N gradient of ephrin-As on RGC axons and a high-to-low A-P gradient of EphAs expressed in the SC generates through reverse signaling a high-to-low A-P gradient of repellent activity.

Because ephrin-As are anchored to the cell membrane by a GPI linkage and lack an intracellular domain, to reverse signal they must associate with transmembrane proteins capable of activating intracellular signaling pathways. Such "coreceptors" that mediate the repellent effect of ephrin-A reverse signaling on RGC axons upon binding EphAs are being studied.

Molecular Control of Mapping along D-V (M-L) Axis of the Target

As described previously, RGC axons enter the OT/SC with a aberrantly broad distribution along the L-M axis; topographic specificity along the L-M axis emerges through the bidirectional guidance of interstitial branches that form along RGC axons with an A-P topographic bias (Fig. 18.2C). Branches that extend from RGC axons located lateral to their future termination zone grow medially whereas branches that extend from RGC axons located medial to their future termination zone grow laterally. Branches that reach the vicinity of the nascent termination zone selectively form complex arbors. Therefore, not only is the directed guidance of interstitial branches a critical feature for D-V retinotopic mapping, but the molecular mechanisms that control it must account for the bidirectional guidance of branches along the L-M axis of branches formed by RGC axons that have the same retinal origin, and presumably express the same set of receptors at similar levels.

Studies of the molecular control of D-V mapping in multiple species (chick, mouse, and frog) independently by the O'Leary and Holt labs have shown roles for EphBs and ephrin-Bs (Fig. 18.3), and that they exhibit bidirectional signaling (forward and reverse) as well as bifunctional action (acting as repellents and attractants). In addition, Wnts and their receptors also have been implicated in D-V mapping based on studies in chick. In retina, EphB receptors are expressed by RGCs during map development in an overall low-to-high D-V gradient, complemented by an overall high-to-low D-V gradient of ephrin-B1. In both chick OT and mouse SC, ephrin-B1 is expressed in a low-to-high L-M gradient, complemented by an overall high-to-low L-M EphB gradient (Figs. 18.2, 18.5). Analyses of mice with targeted deletions of EphB2 and EphB3, with and without reverse signaling intact, show aberrant L-M mapping due to defects in the guidance of interstitial branches (Hindges et al., 2002). Other analyses show that for a given population of RGC axons, ephrin-B1 acts via EphB2/B3 forward signaling as a branch attractant at lower levels of ephrin-B1 expression and as a branch repellent at higher levels of expression (McLaughlin et al., 2003a). Thus, an interstitial branch that forms lateral to its appropriate termination zone is attracted up the gradient of ephrin-B1 to its termination zone, whereas a branch located medial to its nascent termination zone is repelled down the ephrin-B1 gradient to its termination zone. The primary RGC axons are not influenced by ephrin-B1, but instead the effects of ephrin-B1 are specific to interstitial branches and is context dependent—the location of the branch on the ephrin-B1 gradient in relation to the location of its future termination zone and its EphB level determine its response (McLaughlin et al., 2003a). In frogs, ephrin-B reverse signaling also has been implicated in retinotopic mapping (Mann et al., 2002), but such a role has yet to be shown in mice and chicks.

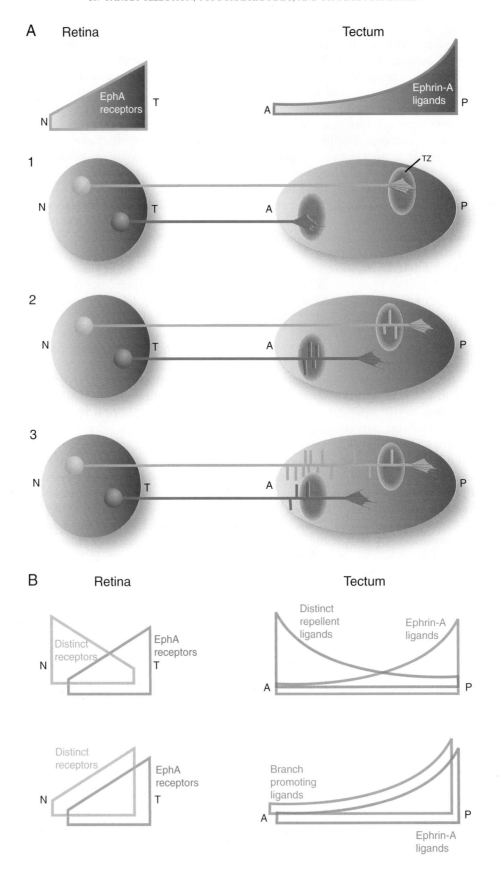

◀ **FIGURE 18.7   Actions and limitations of ephrin-As in retinotectal map development.** (A) The top panel schematizes the approximate gradient profiles for EphA receptors and ephrin-A ligands in the chick retina and optic tectum, respectively. RGCs show a high temporal (T) to low nasal (N) gradient of EphA receptor expression, due mainly to EphA3. Ephrin-A2 and ephrin-A5 combine to form a low anterior (A) to high posterior (P) gradient of expression across the tectum. Panels A1–A3 schematize observed and theoretical behaviors of RGC axons. Growth cones stop at A-P positions in the tectum where they reach a threshold level of repellent activation following a mass action law of receptor-ligand interactions. Temporal growth cones, which have high levels of EphA receptors, reach threshold levels of repellent at anterior (A) positions in the tectum, with low levels of ephrin-A ligand. Growth cones from nasal retina (N), which have low levels of EphA receptors, will reach threshold levels of activation at more posterior (P) positions in the tectum. This mechanism can account for the guidance of RGC growth cones to their topographically correct TZ, as is observed in amphibians and fish (A1), as well as the position-dependent overshoot of RGC axons, as observed in chicks (A2). However, a single repellent gradient, such as that formed by ephrin-A ligands in the tectum, alone is insufficient to generate topographic branching along RGC axons observed in chick (A2). The ephrin-As can inhibit branching along the segment of the overshooting axons posterior to their correct TZ, but anterior to the correct TZ, the level of ephrin-A repellent signal would be below the threshold required to inhibit branching. Thus, if only the tectal ephrin-As regulated branching, all RGC axons should exhibit increased branching at more anterior positions in the tectum, which have the lowest levels of ephrin-A repellent signal (A3). (B) Illustrated are two potential models, among many, that can account for topographic branching along RGC axons. Both models incorporate the graded ephrin-A repellent, and a distinct graded activity that cooperates with it to generate topographic branching. In each case, the ephrin-A repellent prevents branching along the axon shaft posterior to the TZ and the distinct graded activity regulates branching along axons anterior to their TZ. One model includes a distinct repellent in a gradient that opposes the ephrin-A gradient, and acts by inhibiting branching along the axon shaft anterior to the TZ. Thus, branching along the axon shaft occurs at an A-P tectal position below threshold for branch inhibition for both of the repellent signals. Recent findings have indicated that ephrin-A reverse signaling may account for the distinct opposing repellent gradient: ephrin-As expressed by RGCs are the "Distinct receptors" and EphAs expressed in the chick tectum, or mouse superior colliculus, are the "Distinct repellent ligands" depicted in panel B (see text for details). The other model includes a branch promoting activity in a gradient that roughly parallels the ephrin-A gradient. In this model, branching along the axon shaft occurs at an A-P tectal position above threshold for the branch promoting signal, but below threshold for branch inhibition by the ephrin-A repellent signal. Adapted from O'Leary, D. D. M., Yates, P., McLaughlin, T. *Cell* **96**, 255–269.

In principle, EphBs and ephrin-B1 could account for D-V mapping through their actions as bifunctional and bidirectional topographic guidance molecules. However, Yimin Zhou has reported that Wnt signaling also is involved in D-V mapping (Schmitt *et al.*, 2006). Wnt3 is expressed in a high-to-low M-L gradient in chick OT and mouse SC, and Wnt receptors, Ryk and Frizzled family members, are expressed in an overall high-to-low V-D gradient by RGCs (Fig. 18.2). Functional studies show that Ryk mediates RGC axon repulsion by higher levels of Wnt3, whereas Frizzled receptors mediate an attractant effect of lower levels of Wnt3 on dorsal RGC axons. Thus, Wnt3, which is classically known as a secreted, diffusible signaling molecule, appears to act as a topographic guidance molecule in the OT and cooperate with EphB-ephrin-B1 to regulate D-V mapping.

Role for Patterned Activity in Map Refinement

It is well established that neural activity plays an important role in the development of axonal connections in many neural systems including the segregation of RGC axons into eye-specific laminae in the lateral geniculate nucleus or eye-specific stripes in experimentally created three-eyed frogs (see Chapters 20 and 22). In addition, correlated activity has long been postulated to refine topographic connections by strengthening coordinated inputs and weakening

uncorrelated inputs. However, for the development of retinotopic projections, roles for neural activity appear to vary considerably depending upon the species. Among the first studies to address this issue were those done by Bill Harris, who homotopically transplanted axolotl eyes into the California newt, a species that produces endogenous tetrodotoxin (TTX), a neurotoxin that blocks Na+ channels. The transplanted axolotl RGCs were silenced by the TTX, but their projection to the tectum developed an appropriately ordered topographic map.

Similarly, in other species in which the map develops by the direct topographic targeting of RGC axons, such as fish and frogs, activity blockade has little affect on topography. For example, RGC axonal arbors are reportedly unaffected in zebrafish bathed in either TTX or AP5, and blocking the NMDA class of glutamate receptors in Xenopus results only in a slowing of map development. Even in chicks and rats, in which the early retinotectal projection is topographically very diffuse, a considerable degree of order emerges under pharmacological activity blockade. In chick, TTX and grayanotoxin, which interfere with Na+ currents, prevent the elimination of only a small proportion of overshooting axon segments and aberrant branches and arbors. Similarly, in rats, chronic application to the SC of AP5, an antagonist of the NMDA receptor, at levels that block the activation of SC neurons by RGCs (which use glutamate as their neurotransmitter), also results in the abnormal retention of only a small proportion of topographically aberrant axons and arbors. These studies, though, have caveats associated with them, including questions about the effectiveness of the block and potential side effects of activity block on gene expression. However, recent analyses of mice lacking cholinergic-mediated retinal waves has circumvented these problems and show that correlated patterns of RGC activity are required for the large-scale remodeling of the retinocollicular projection into a refined map.

Interestingly, the large-scale remodeling of the retinocollicular projection required to generate refined topography in the mouse SC occurs well before the onset of visually evoked activity and eye opening. However, this remodeling is coincident with a period of correlated waves of spontaneous neural activity that propagate across the retina, a phenomenon first described by Carla Shatz. These waves are mediated by a network of cholinergic amacrine cells and correlate the activity of neighboring RGCs, thereby relating an RGC's position to its pattern of activity. Mice lacking the $\beta2$ subunit of the nicotinic acetylcholine receptor maintain normal levels of spontaneous activity, but the correlation in activity patterns among neighboring RGCs is lost. The topographic projection in $\beta2$ mutant mice fails to properly refine and RGC axons from a given retinal location do not form a dense termination zone but rather maintain a loose collection of diffuse arborizations around the appropriate location of their termination zone (Fig. 18.8) (McLaughlin *et al.*, 2003b). However, some remodeling does occur and may be driven by the same EphA-ephrin-A mediated repellent activites that establish the initial coarse map. In $\beta2$ mutant mice, correlated RGC activity does resume during the second postnatal week through a glutamatergic process, and patterned visually evoked activity begins soon thereafter, but neither process leads to proper map refinement, indicating a brief early critical period for activity-dependent map remodeling in the mouse retinocollicular projection (McLaughlin *et al.*, 2003b).

Summary

The formation of topographic maps involves the establishment of an initial, coarse map that subsequently is refined. Analysis of the retinotectal system in amphibians, fish, birds, and mammals shows that the initial map is formed based on positional information present in the tectum. In birds and mammals, a critical mechanism in map development is the topographic specific branching of RGC axons that overshoot their correct termination zone. Topographic guidance information is encoded in the form of gradients of signaling molecules, Ephs and ephrins, along both the A-P and D-V axes of the tectum. Other receptor-ligands systems , as well as patterned RGC activity, are also involved in map development.

DEVELOPMENT OF THE NEUROMUSCULAR SYNAPSE

Once axons have arrived at their appropriate target destination, synapse formation ensues. Much of our understanding about the mechanisms of synapse formation arises from studies of the neuromuscular synapse. These studies have benefited from (1) the relative ease of experimentally manipulating developing and regenerating neuromuscular synapses *in vivo*, (2) cell culture systems for both motor neurons and skeletal muscle cells, (3) the *Torpedo* electric organ, an abundant and homogeneous source of neuromuscular-like synapses (see Box 18.1), and (4) transgenic and mutant mice for studying and altering gene expression. Consequently, we have a good, although incomplete, understanding of the mechanisms that lead to the formation of the neuromuscular synapse.

Substructural Organization of the Neuromuscular Synapse

An adult myofiber, a syncitial cell containing several hundred to several thousand nuclei, is innervated by a single motor axon that terminates and arborizes over approximately 0.1% of the muscle fiber's cell surface. The neurotransmitter receptor, acetylcholine receptors (AChRs), is localized to this small patch of the muscle fiber membrane, and its precise localization to synaptic sites during development is a hallmark of the inductive events of synapse formation. Although other proteins are likewise concentrated at synaptic sites, much of our knowledge about synaptic differentiation has come from studies aimed at understanding how AChRs accumulate at synaptic sites.

Nerve terminals are situated in shallow depressions of the muscle cell membrane, which is invaginated further into deep and regular folds, termed postjunctional folds (Fig. 18.9). AChRs and additional proteins (see later) are localized to the crests of these postjunctional folds, whereas other proteins, including sodium channels, are enriched in the troughs of the postjunctional folds. The nerve terminal likewise is organized spatially, and its substructural organization reflects that of the postsynaptic membrane. Synaptic vesicles are sparse in the region of the nerve terminal underlying Schwann cells and are abundant in the region of the nerve terminal facing the muscle fiber. Moreover,

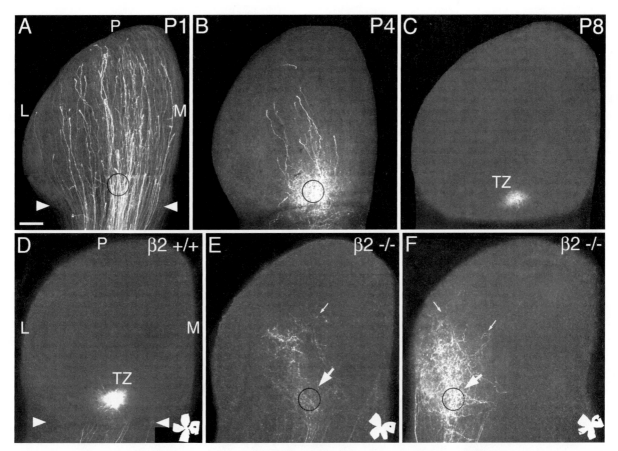

FIGURE 18.8 **Normal development of the topographic retinocollicular map requires large-scale remodeling mediated by retinal waves that correlate RGC activity patterns.** (A–C) Normal development of the topographic retinocollicular map requires large-scale remodeling. Focal injection of the fluorescent lipophilic tracer DiI into temporal retina of WT mice labels RGC axons from a single retinal location. Shown are fluorescence images of DiI labeled RGC axons in WT SC at the ages indicated. (A) At P1 RGC axons are dispersed across the entire lateral-medial (L-M) and anterior (arrowheads)–posterior (P) axes of the SC. Essentially all RGC axons extend well posterior to the location of their future TZ (circle). Branches form *de novo* from the axon shaft in a distribution biased for the A-P location of the future TZ. In addition, branches are directed toward their future TZ along the L-M axis. (B) At P4 RGC axons with branches near the future TZ preferentially elaborate arbors. Note that many RGC axons have eliminated their initial posterior overshoot. (C) At P8, a dense focal TZ is labeled at the topographically appropriate location in the SC. The initial axon overshoot is eliminated and no arbors persist outside the TZ. At P8 the retinocollicular map resembles its mature form. (D–F) β2 –/– mice that lack retinal waves have defective topographic remodeling of the retinocollicular projection. Fluorescence images of DiI labeled RGC axons in SC of P8 β2 +/+ (D) and β2 –/– (E and F) mice. (D) Focal injection of DiI into temporal retina of a P7 β2 +/+ mouse reveals a single, densely labeled TZ in anterior SC at P8 (arrowheads mark anterior border of SC) characteristic of refined topographic organization of the mature retinocollicular projection in WT. (E and F) Focal DiI injections, similar in size to that in panels A–D, made into temporal retina of P7 β2 –/– mice reveal at P8, TZs characterized by aberrantly large domains of loosely organized arborizations in the vicinity of the appropriate topographic position but have not refined into a single dense TZ characteristic of WT mice at this age. Large arrows point to black circles that indicate the approximate size and position of the appropriate TZ predicted by the injection site. Many RGC axons persist in mid-SC with elaborate arbors (small arrows). The inset in each panel is a tracing of the flat-mounted retina and injection site (black spot) for each case. The cases shown in panels A–C have an injection similar to that in panel D. Temporal is to the right and dorsal is to the top for each retinal tracing. L, lateral; M, medial; P, posterior. Scale bar = 250 μm. Adapted from McLaughlin, T., Torborg, C. L., Feller, M. B., and O'Leary, D. D. M. (2003). *Neuron* **40**, 1147–1146.

synaptic vesicles are clustered adjacent to a poorly characterized specialization of the presynaptic membrane, termed active zones, which are the sites of synaptic vesicle fusion. Active zones are organized at regular intervals and are aligned precisely with the mouths of the postjunctional folds. This precise registration of active zones and postjunctional folds ensures that acetylcholine encounters a high concentration of

AChRs within microseconds after release, thereby facilitating synaptic transmission. The alignment of structural specializations in pre- and postsynaptic membranes, separated by a 500-Å synaptic cleft, suggests that spatially restricted signaling between pre- and postsynaptic cells is important to coordinate pre- and postsynaptic differentiation. The mechanisms that align active zones and postjunctional folds are not understood, but $\alpha 3/\beta 1$ integrin is concentrated at active zones and several laminins are found in the synaptic basal lamina, raising the possibility that laminin/integrin interactions have a role in positioning and organizing active zones (see later).

The precise organization of molecules in pre- and postsynaptic membranes belies the concept that the neuromuscular synapse is a simple synapse. Rather, the substructure of pre- and postsynaptic membranes, together with the faithful registration of pre- and postsynaptic specializations, suggests that complex mecha-

nisms are required to assemble the synapse and to coordinate pre- and postsynaptic differentiation.

Muscle-Autonomous Patterning of AChR Expression

As motor axons extend toward developing muscle, muscle cells are themselves undergoing differentiation. Myotubes, formed by fusion of precursor myoblasts, develop prior to innervation and continue to grow after innervation by further myoblast fusion.

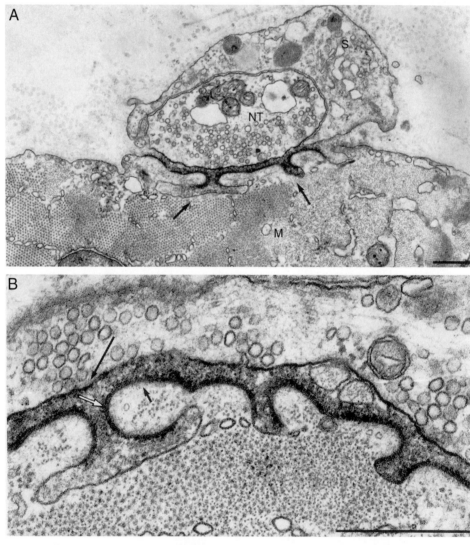

FIGURE 18.9 **Pre- and postsynaptic membranes at the neuromuscular synapse are highly specialized.** (A) An electron micrograph of a neuromuscular synapse shows that the nerve terminal is capped by a Schwann cell and is situated in a shallow depression of the muscle cell membrane, which is invaginated further into deep and regular folds, termed postjunctional folds (arrows). AChRs, labeled with α-bungarotoxin coupled to horseradish peroxidase, are concentrated at the synaptic site. (B) A higher magnification view shows that AChRs are concentrated at the crests and along the sides of the postjunctional folds (white arrow). Rapsyn, NRG receptors, and MuSK are also concentrated in the postsynaptic membrane, whereas Agrin, NRG-1, acetylcholinesterase, S-laminin, and certain isoforms of collagen are localized to the synaptic basal lamina. The postjunctional folds of the myofiber are spaced at regular intervals and are situated directly across from active zones and clusters of synaptic vesicles in the nerve terminal (arrows).

Because myoblasts fuse to developing myotubes at their growing ends, the central region of the muscle is more mature than the distal ends of the muscle. Recent studies have shown that the central region of muscle is regionally specialized prior to innervation. For example, AChRs are already clustered in the central region of the muscle prior to and independent of innervation (Yang *et al.*, 2000, 2001; Lin *et al.*, 2001). As such, motor axons approach muscles that are regionally specialized, or prepatterned, prior to innervation, in a manner that preconfigures the prospective zone of innervation (Fig. 18.10).

How the prospective synaptic region of the muscle becomes prepatterned is not well understood, but muscle prepatterning restricts motor axon growth and promotes synapse formation in the central region of mammalian muscle (Kim and Burden, 2008).

The Synaptic Basal Lamina Contains Signals for Synaptic Differentiation

Following contact with the growth cone of a developing motor neuron, developing muscle fibers undergo a still further complex differentiation program in the synaptic region, which is dependent upon motor neuron-derived signals. In addition, signals from the muscle further regulate differentiation of presynaptic nerve terminals. The idea that signals for regulating both presynaptic and postsynaptic differentiation are contained in the synaptic basal lamina arose from studies of regenerating neuromuscular synapses. Following damage to a motor axon, the distal portion of the motor axon degenerates, and the proximal end regenerates to muscle. The regenerated axon precisely reinnervates the original synaptic site and forms a synapse that is indistinguishable from the original synapse. Original synaptic sites are not thought to provide guidance cues to motor axons; rather, it is believed that the vacated perineurial tubes, containing Schwann cells and their basal lamina, provide a favorable substrate for motor axons and have a role in directing regenerating motor axons to original synaptic sites (Son and Thompson, 1995). Although motor axons precisely reinnervate original synaptic sites, there appears to be little, if any, selectivity among motor neurons or among Schwann cells in assuring accurate regeneration; indeed, both original and foreign motor neurons can accurately and functionally reinnervate original synaptic sites in denervated muscle.

Following damage to motor axons and muscle, nerve terminals and muscle fibers degenerate and are phagocytized, but the basal lamina of the muscle fiber remains intact. Even in the absence of nerve terminals

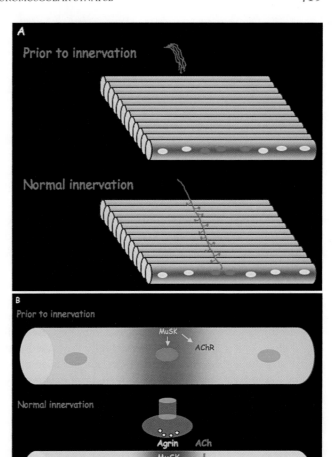

FIGURE 18.10 Muscle pre-patterning. (A) Motor axons approach muscles that already express *MuSK and acetylcholine receptor (AChR)* genes (green) in the central, prospective synaptic region of the muscle. (B) This aspect of muscle pre-patterning requires MuSK. Following innervation, this zone of *AChR* expression is refined and sharpened, so that *AChR* expression and AChR clustering is restricted to nascent synaptic sites. Two neural signals refine AChR expression: Agrin/MuSK signaling stabilizes and stimulates AChR clustering, whereas acetylcholine (ACh) disperses AChR clusters. As a consequence of these two signals, AChR clusters are maintained only at nascent synaptic sites.

and muscle fibers, several structures, including the terminal Schwann cells, the basal lamina of the postjunctional folds, and AChE, remain at the original synaptic site and allow for its identification.

Axons eventually regenerate into the muscle, and new myofibers regenerate within the basal lamina of the original myofiber. The regenerated motor axons form synapses with the regenerated myofibers precisely at the original synaptic sites. If axons regenerate into muscle, but muscle regeneration is prevented,

BOX 18.1

TORPEDO ELECTRIC ORGAN

The majority of proteins known to be localized to neuromuscular synapses were first identified in postsynaptic membranes isolated from the electric organ of the marine ray *Torpedo*. Indeed, this specialized tissue has been essential for the identification and purification of the AChR, AChE, Rapsyn, Syntrophin, Agrin, MuSK and several synaptic vesicle proteins. The electric organ is a particularly homogeneous and abundant source of pre-and post-synaptic membranes that are similar in structure and function to those at neuromuscular synapses. The biochemical advantages of the electric organ are evident from its anatomy. The postsynaptic cell, which differentiates initially as a syncitial skeletal muscle fiber but subsequently loses its contractile machinery, is termed an electroplaque. Each electroplaque is a thin, elongated cell (1 cm × 1 cm × 10 mm) that is so densely innervated that nearly one-half of the electrocyte membrane is studded by nerve terminals. In contrast, nerve terminals occupy less than 0.1% of the cell surface of a skeletal myofiber.

Because of this dense innervation and because the electric organ from a moderate size ray weighs several kilograms, it is possible to obtain several mg of purified postsynaptic proteins from a single electric organ.

Innervation is restricted to the ventral surface of the electroplaque, whereas the dorsal surface is enriched for the sodium/potassium ATPase which maintains the resting potential. Because the electroplaque lacks action potentials, activation of AChRs on the innervated ventral surface results in a voltage drop across the innervated but not the non-innervated membrane of the electrocyte. Since thousands of electroplaques are stacked closely one upon another, the potential difference across a single electrocyte is summed by the stack of electroplaques, resulting in a several thousand volt potential difference across the entire electric organ, a voltage that is sufficient to stun prey.

Steven J. Burden, Dennis D. M. O'Leary and Peter Scheiffele

axons still unerringly reinnervate the original synaptic site on the basal lamina, and active zones form in register with the basal lamina of the original postjunctional folds (Fig. 18.11) (Sanes *et al.*, 1978). Thus, the presence of the myofiber is necessary neither for precise reinnervation nor for the morphological differentiation of regenerated nerve terminals. The vacated perineurial tubes, containing Schwann cells and their basal lamina, may direct regenerating motor axons to original synaptic sites, but the induction of active zones suggests that cues in the synaptic basal lamina have a role in organizing the presynaptic terminal. Notably, laminin β2, which is expressed by muscle and found in the synaptic basal lamina, binds presynaptic calcium channels, and this interaction has a role in anchoring active zones and coordinating pre- and postsynaptic differentiation (Nisimune *et al.*, 2004).

If regeneration of motor axons is prevented, but myofibers are allowed to regenerate, AChRs accumulate and membrane folds form in the regenerated myofiber precisely at the original synaptic site on the basal lamina (Burden *et al.*, 1979). Thus, in the absence of nerve terminals, myofibers, and terminal Schwann cells, information that remains at the original synaptic site instructs both presynaptic and postsynaptic differentiation (Fig. 18.11). Because the synaptic basal lamina is the most prominent extracellular structure

remaining at neuromuscular synapses following removal of all cells, these results indicate that the synaptic basal lamina contains signals that can induce differentiation of both nerve terminals and myofibers.

Agrin Is a Signal for Postsynaptic Differentiation

Because clustering of AChRs, unlike the formation of active zones, can be studied readily in cell culture, it has been far simpler to identify the basal lamina signals that induce postsynaptic rather than presynaptic differentiation. Extracellular matrix from the *Torpedo* electric organ, a tissue that is homologous to muscle but more densely innervated, contains an activity that stimulates AChR clustering in cultured myotubes (see Box 18.1). McMahan and colleagues purified the electric organ activity, which they termed Agrin, and showed that Agrin is synthesized by motor neurons, transported in motor axons to synaptic sites, and deposited in the synaptic basal lamina. Agrin also stimulates the clustering of other synaptic proteins (see later), including AChE, Rapsyn and Utrophin (see later), indicating that Agrin has a central role in synaptic differentiation.

Agrin is a ~200-kDa protein containing multiple epidermal growth factor (EGF)-like signaling domains,

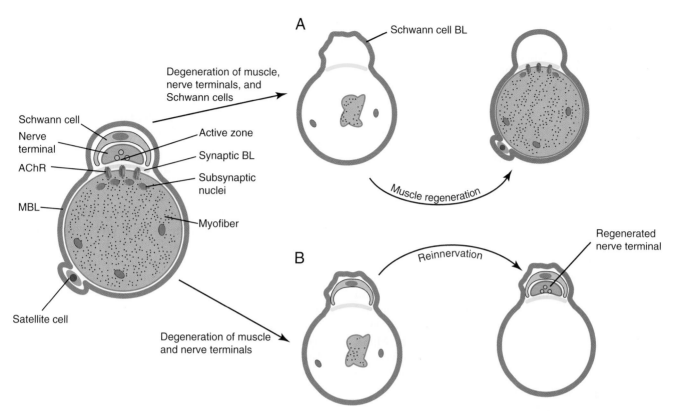

FIGURE 18.11 **The synaptic basal lamina contains signals for presynaptic and postsynaptic differentiation.** A cross-section of normal muscle containing a nerve terminal, Schwann cell, and myofiber is shown at the far left. *AChR* genes are induced in synaptic nuclei (red) and AChRs (green) are concentrated at synaptic sites. Active zones and clusters of synaptic vesicles in the nerve terminal are aligned with postjunctional folds in the myofiber. Each myofiber is ensheathed by a basal lamina (MBL), which is specialized at the synaptic site (synaptic BL) and which remains intact following degeneration of the original myofiber. The MBL serves as a scaffold for regenerating myofibers, which form from the fusion of satellite cells that proliferate following damage to the original myofiber. (A) Experiments showing that the synaptic basal lamina contains signals for clustering AChRs and activating *AChR* genes in synaptic nuclei. Following damage and degeneration of axons, Schwann cells and myofibers, new myofibers regenerate within the basal lamina of the original myofiber in the absence of the nerve. AChRs cluster and *AChR* gene expression is reinduced at original synaptic sites in myofibers that regenerate in the absence of the nerve and other original presynaptic cells. (B) Experiments showing that the synaptic basal lamina contains signals for inducing presynaptic differentiation. Following damage and degeneration of axons and myofibers, motor axons regenerate to original synaptic sites in the absence of the myofiber and accumulate synaptic vesicles precisely across from the sites of the original postjunctional folds.

two different laminin-like domains, and multiple follistatin-like repeats. The four EGF-like domains and three laminin G domains are contained in the carboxyl-terminal region, which is sufficient for inducing AChR clusters in cultured myotubes; sequences in the amino-terminal region are responsible for the association of Agrin with the extracellular matrix.

The *agrin* gene is expressed in a variety of cell types. Alternative splicing results in multiple Agrin isoforms that differ in their AChR clustering efficiency. The isoform that is most active in clustering AChRs is expressed in neurons, including motor neurons, whereas other Agrin isoforms are expressed in additional cell types, including skeletal muscle cells. The active, neuronal-specific isoforms of Agrin contain 8, 11, or 19 amino acids at a splice site, referred to as the Z site in rat Agrin and the B site in chick Agrin.

The *agrin* gene contains multiple promoters. Although the promoter used in motor neurons leads to expression of isoforms that are associated with the extracellular matrix, the promoter used in the CNS leads to expression of membrane-bound isoforms (Burgess *et al.*, 2000).

In mice lacking Agrin, AChRs are initially prepatterned, and synapses appear to form, but only transiently (Lin *et al.*, 2001). Over the course of several days, the AChR prepattern is dispersed, synapses are lost, and motor axons begin to grow throughout the muscle. Thus, Agrin is necessary to stabilize the AChR prepattern and nascent synapses. Because the AChR prepattern is stable in the absence of innervation but dispersed by motor axons lacking Agrin, neural signals, other than Agrin, extinguish the AChR prepattern Yang *et al.*, 2001; Lin *et al.*, 2001). Genetic evidence indicates that acetylcholine is a key neural signal that disperses AChR clusters (Lin *et al.*, 2005; Misgeld *et al.*, 2005). Thus, during normal development, Agrin counteracts acetylcholine-mediated dispersion, thereby stabilizing AChR clusters selectively at nascent synapses.

MuSK Is Required for Agrin-Mediated Signaling and Synapse Formation

The mechanisms of Agrin-mediated AChR clustering are not known, but a receptor tyrosine kinase, termed MuSK, is a critical component of an Agrin receptor complex. MuSK is expressed in *Torpedo* electric organ and in skeletal muscle, where it is concentrated in the postsynaptic membrane. Mice deficient in MuSK (DeChiara *et al.*, 1996), like *agrin* mutant mice, lack normal neuromuscular synapses. Both *agrin* and *MuSK* mutant mice are immobile, cannot breathe, and die at birth. Muscle differentiation is normal in *agrin* and *MuSK* mutant mice, but muscle fibers in *MuSK* mutant mice lack all known features of postsynaptic differentiation. Muscle-derived proteins, including AChRs and AChE, which are concentrated at synapses in normal mice, are distributed uniformly in *MuSK* mutant myofibers (Fig. 18.12). In addition, *AChR* genes, which normally are transcribed selectively in synaptic nuclei of normal muscle fibers (see later), are transcribed at similar rates in synaptic and nonsynaptic nuclei of muscle fibers from *MuSK* mutant mice. Unlike *agrin* mutant mice, the AChR prepattern is absent from *MuSK* mutant mice at all stages, indicating that MuSK is essential not only for synapse formation but also to establish muscle prepatterning.

Five lines of evidence indicate that MuSK is required for Agrin-mediated signaling and is a component of the Agrin receptor complex: (1) Agrin can be chemically cross-linked to MuSK in cultured myotubes; (2) Agrin induces rapid tyrosine phosphorylation of MuSK in cultured myotubes; (3) a recombinant, soluble extracellular fragment of MuSK inhibits Agrin-induced AChR clustering in cultured muscle cells; (4) cultured *MuSK* mutant muscle cells, unlike normal muscle cells, do not cluster AChRs in response to Agrin; and (5) dominant-negative forms of MuSK inhibit Agrin-induced AChR clustering in cultured myotubes. MuSK itself, however, does not bind Agrin, indicating that other activities or additional proteins are required for Agrin to activate MuSK. The additional activities in muscle that are required for Agrin to activate MuSK are not known, but Lrp4, a member of the low density lipoprotein receptor family, is essential for synapse formation, and the synaptic defects in *lrp4* mutant mice appear identical to those in *MuSK* mutant mice

(Weatherbee *et al.*, 2006). Thus, Lrp4 may be a component of a MuSK protein complex.

Interestingly, a pair of proteins with homology to MuSK and Lrp4 are required for clustering AChRs at *C. elegans* neuromuscular synapses: CAM-1, a receptor tyrosine kinase that shares homology with MuSK, and LEV-10, a transmembrane protein that contains a LDLa domain (Gally *et al.*, 2004; Francis *et al.*, 2005). However, the kinase activity of CAM-1 is dispensable for synaptic differentiation, pointing to a different mechanism of action than MuSK at vertebrate neuromuscular synapses.

The signaling mechanisms that regulate neuromuscular synapse formation during embryogenesis in Drosophila are poorly understood, but two ligands, Wingless and TGF-β, unrelated to Agrin, are involved in the growth of Drosophila neuromuscular synapses during larval development. Wingless is secreted by motor neurons and thought to activate Frizzled-2, a Wingless receptor expressed in muscle (Packard *et al.*, 2002); Glass bottom boat, a ligand from the TGF-β family, is synthesized by muscle and thought to activate Wishful thinking, a type II receptor, in motor neurons (Aberle *et al.*, 2002).

How does MuSK activation lead to postsynaptic differentiation? Agrin stimulates the rapid phosphorylation of MuSK, and the kinase activity of MuSK is essential for Agrin to stimulate clustering and tyrosine phosphorylation of AChRs. Signaling downstream from MuSK depends on phosphorylation of a tyrosine residue (Y553) in the juxtamembrane region of MuSK (Herbst *et al.*, 2002). Phosphorylation of this tyrosine leads to recruitment of Dok-7 (Okada *et al.*, 2006), which engages additional, but unknown downstream signaling pathways essential for synapse formation. Mice lacking Dok-7 have defects identical to *MuSK* mutant mice (Okada *et al.*, 2006), and mutations in human Dok-7 are a cause of congenital myasthenic syndromes with defects in neuromuscular synapses (Beeson *et al.*, 2006). Rac and Rho, small GTP-binding proteins that regulate actin organization, are activated by Agrin and required for Agrin to stimulate AChR clustering (Weston *et al.*, 2000, 2003). Rac acts early and is important for the formation of AChR microclusters, whereas Rho is activated more slowly and is important for the consolidation of microclusters into larger AChR clusters. Moreover, at least one kinase acts downstream from MuSK and is recruited and/or activated by Agrin-activated MuSK, as staurosporine, a protein kinase inhibitor, inhibits Agrin-induced AChR tyrosine phosphorylation and clustering without blocking tyrosine phosphorylation of MuSK.

Following the recruitment of Dok-7 to phosphorylated MuSK, the signaling pathways downstream from

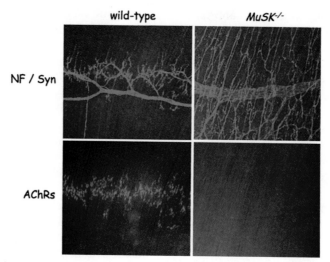

FIGURE 18.12 **Presynaptic and postsynaptic differentiation are defective in mice lacking MuSK.** Whole mounts of muscle from wild-type and *MuSK* mutant mice were stained with antibodies to neurofilament (NF) and synaptophysin (Syn) to label motor axons and nerve terminals (green), respectively, and with α-bungarotoxin to label acetylcholine receptors (AChRs) (red). In *MuSK* mutant mice, motor axons fail to stop and differentiate adjacent to the main intramuscular nerve and instead wander aimlessly over the muscle; AChRs, as well as other postsynaptic proteins, are expressed at normal levels but they fail to cluster.

MuSK are poorly understood. Abl 1/2, Src-family kinases and Pak have been proposed to act downstream from MuSK, but functional evidence for Abl and Pak in forming synapses *in vivo* is currently lacking. Genetics studies point to a role for Src-family kinases, as well as dystrobrevin, in stabilizing rather than forming neuromuscular synapses, but how these kinases stabilize synapses is not known (Smith *et al.*, 2001).

Muscle-Autonomous and Neural-Dependent Patterning of AChR Expression

The preceding sections have emphasized the role of muscle prepatterning and motor neuron-derived signals in regulating the distribution of AChR expression. Taken together, these findings suggest a model for the sequential steps involved in establishing the pattern of AChR expression on developing skeletal muscle fibers (Arber *et al.*, 2002; Kummer *et al.*, 2006). An initial, spatially restricted pattern of AChR expression is generated in muscle independent of neurally derived Agrin, or indeed of any other neural signal (Fig. 18.10). Because this prepattern is not observed in mice lacking MuSK, the emergence of this muscle AChR prepattern, nonetheless, coopts at least some of the same molecules used during normal synaptogenesis. Arrival of the nerve, and its attendant signals, converts the AChR prepattern into the more refined pattern of *AChR* transcription and AChR clustering characteristic of mature synapses. This conversion appears to be dependent on two separable nerve-

dependent programs: one program appears to utilize neurally derived Agrin to maintain AChR expression at nascent synaptic sites and a second program, which appears to be triggered by acetylcholine, extinguishes AChR clustering at sites that are not stabilized by Agrin/MuSK signaling (Fig. 18.10). These two programs thus ensure the stable expression of AChR clusters selectively at nascent synapses. The mechanisms that are intrinsic to muscle and responsible for establishing regional differences in muscle in the absence of innervation are not well understood but appear to depend upon the pattern of muscle growth and positive feedback loops that sustain MuSK activation (Kim and Burden, 2008).

Rapsyn Is Required for Postsynaptic Differentiation and Is Downstream of Agrin and MuSK

A 43-kDa protein, termed Rapsyn, has an important role in Agrin-mediated signaling. Rapsyn is a myristolated, peripheral membrane protein that is present at 1:1 stoichiometry with AChRs at synaptic sites and interacts directly with AChRs and potentially other synaptic proteins, including Dystroglycan and Dystrophin, Utrophin, Syntrophin, and Dystrobrevin, proteins that are components of a subsynaptic, cytoskeletal complex.

Agrin stimulates the clustering of Rapsyn in myotubes grown in cell culture, and clustering of Rapsyn and AChRs occurs coincidentally at developing synapses. Rapsyn is critical for synapse formation, as mice lacking Rapsyn die within hours after birth and have difficulty moving and breathing (Gautam *et al.*, 1995). Mutations in human Rapsyn are a cause of congenital myasthenic syndromes with defects in AChR clustering that lead to muscle weakness (Engel *et al.*, 2003). Although the mechanisms by which Rapsyn regulates the clustering of synaptic proteins is poorly understood, the clustering of most synaptic proteins, including AChRs, Utrophin and Dystroglycan, depends upon Rapsyn. Nonetheless, MuSK is clustered at synapses in rapsyn mutant mice, raising the possibility that engagement with Agrin is sufficient to stabilize MuSK at rapsyn mutant synapses. Synapse-specific gene expression (see below) is also normal in rapsyn mutant mice, suggesting that MuSK may trigger this pathway independent of Rapsyn.

Agrin and MuSK Are Required for Retrograde Signaling and Presynaptic Differentiation

Although pathfinding of motor axons to muscle is normal in mice lacking Agrin or MuSK, *agrin* and *MuSK* mutant mice lack normal nerve terminals. In the mutant mice, branches of the main intramuscular nerve fail to stop and differentiate and instead wander aimlessly across the muscle (Fig. 18.12). Because MuSK is expressed in skeletal muscle and not in motor neurons, it seems likely that the aberrant behavior of presynaptic terminals in *MuSK* mutant mice is due to indirect actions of the Agrin/MuSK signaling system. Indeed, restoring MuSK expression selectively in muscle of *MuSK* mutant mice rescues synapse formation and neonatal lethality (Herbst *et al.*, 2002). These results indicate that Agrin, released from nerve terminals, causes the muscle cell, via MuSK activation, to reciprocally release a recognition signal back to the nerve to indicate that a functional contact has occurred. In response to this putative muscle-derived recognition or adhesion signal, the nerve undergoes presynaptic differentiation and stops growing.

Certain Genes Are Expressed Selectively in Synaptic Nuclei of Myofibers

Like AChR protein, mRNAs encoding the different AChR subunits (α, β, γ, or ε and δ) are concentrated at synaptic sites. Studies with transgenic mice that harbor gene fusions between regulatory regions of *AChR* subunit genes and reporter genes have shown that *AChR* genes are transcribed selectively in myofiber nuclei near the synaptic site (Fig. 18.13). Thus, localized transcription of *AChR* genes in synaptic nuclei is responsible, at least in part, for the accumulation of

AChR mRNA at synaptic sites. This pathway is important for ensuring that AChRs are expressed at the required density in the postsynaptic membrane, as defects in synapse-specific gene expression of the *AChR* ε subunit gene are the cause of a congenital myaesthenia (see later).

Like *AChR* subunit genes, the *utrophin* gene is transcribed selectively in synaptic nuclei, resulting in an accumulation of *utrophin* mRNA and protein at synaptic sites. mRNAs encoding *Rapsyn*, NCAM, MuSK, sodium channels, the catalytic subunit of AChE, LL5beta and CD24 are also concentrated in the synaptic region of skeletal myofibers, raising the possibility that these genes are likewise transcribed preferentially in synaptic nuclei. Thus, synapse-specific transcription may be a common and important mechanism for expressing and localizing a variety of gene products at high levels at the neuromuscular synapse.

The neural signals that regulate synapse-specific transcription remain elusive. Until recently, Neuregulin-1, which stimulates tyrosine phosphorylation of ErbB2, a member of the EGF receptor family, was considered the best candidate. However, neuromuscular synapses appear normal in mice lacking ErbBs in skeletal mice as well as in mice lacking Neuregulin-1 in motor neurons and skeletal muscle (Escher *et al.*, 2005; Jaworski and Burden, 2006). Synapse-specific transcription, however, requires Agrin and MuSK, indicating that this signaling pathway has a critical role in regulating synaptic gene expression.

Ets Domain Transcription Factors Regulate Synapse-Specific Transcription

A binding site for Ets domain proteins in the *AChR* δ subunit gene is critical for synapse-specific gene expression in mice. Importantly, mutation of an Ets-binding site in the human *AChR* ε subunit gene leads to a myopathy, termed congenital myasthenic syndrome, due to decreased AChR expression. This Ets site can bind GABP, a complex containing GABPα, an Ets protein, and GABPβ, a protein that lacks an Ets domain but dimerizes with GABPα, suggesting that GABP may be a transcriptional regulator that stimulates transcription of *AChR* genes in synaptic nuclei. In addition, the gene encoding another Ets domain protein, Erm, is expressed selectively by myofiber synaptic nuclei (Hippenmeyer *et al.*, 2002), suggesting that Erm may also have a role in synapse-specific expression. Genetic analysis of *GABP* and *erm* mutant mice indicates that Erm has a critical role, whereas GABP has a more subsidiary role in synapse-specific transcription.

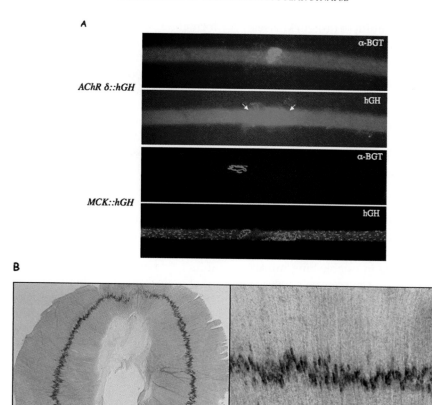

FIGURE 18.13 *Acetylcholine receptor (AChR) genes are expressed selectively by myofiber synaptic nuclei, leading to enrichment of AChR mRNA in the central, synaptic region of the muscle.* (A) Muscle from transgenic mice harboring a gene fusion between the regulatory region from the *AChR δ* subunit gene and *human growth hormone (hGH)* gene were stained with α-bungarotoxin (α-BGT) to label AChRs and with antibodies to hGH. Intracellular hGH is concentrated in the ER/Golgi apparatus near myofiber synaptic nuclei of *AChR δ* transgenic mice but expressed uniformly in muscle from control transgenic mice carrying a gene fusion between the regulatory region of the *muscle creatine kinase (MCK)* gene and *hGH*. (B) A whole mount of a mouse diaphragm muscle was processed for *in situ* hybridization to reveal the pattern of *AChR δ* mRNA expression. *AChR δ* mRNA expression is concentrated in a narrow band in the central, synaptic region of the muscle.

Electrical Activity Regulates Gene Expression

Changes in the pattern of muscle electrical activity have an important role in regulating the electrophysiological and structural properties of muscle, as well as the ability of motor axons to innervate muscle. The expression of several genes, including *AChR* genes, is repressed by electrical activity, and this repression, together with focal activation in synaptic nuclei, as described earlier, contributes to the disparate levels of AChR expression in synaptic and nonsynaptic regions of the muscle. A binding site for myogenic bHLH transcription factors, or E box, in the proximal promoter of *AChR* subunit genes is essential for electrical activity-dependent transcription, as transgenes containing a mutation in this E box, unlike wild-type transgenes, are not induced following denervation. These results

suggest that electrical activity decreases the level and/or activity of E box binding proteins, notably Myogenin, leading to decreased *AChR* expression. The transcriptional mechanisms that control expression of *myogenin* in an activity-dependent manner are just beginning to be explored. Electrical activity induces expression of two transcriptional repressors, HDAC-9 (aka MITR) and Dach2, which act to repress *myogenin* expression (Schaeffer *et al.*, 2004; Tang and Goldman, 2006). A reduction in myofiber electrical activity leads to a reduction in *HDAC-9* and *Dach2* expression, causing an elevation in *myogenin* expression. Thus, the mechanisms that couple changes in electrical activity to changes in *HDAC-9* and *Dach2* expression are critical to regulating genes, such as *AChR* genes, which are dependent upon Myogenin. The mechanisms that regulate *HDAC-9* and *Dach2* expression are not understood, but the involvement of multiple protein kinases, including PKA, CAM kinase II, and PKC, has been suggested.

Summary

Signals exchanged at nascent synaptic sites ensure that differentiated presynaptic terminals are aligned precisely with a highly specialized postsynaptic membrane. Analysis of mice lacking innervation demonstrates that a coarse pattern of postsynaptic differentiation is present prior to innervation and that neuronal signals, including Agrin, selectively maintain, rather than induce, postsynaptic differentiation at sites of nerve–muscle contact (Fig. 18.10). MuSK, a receptor tyrosine kinase expressed by muscle and activated by Agrin, has a key role in establishing muscle prepatterning and in stabilizing nascent synapses in response to Agrin.

SYNAPSE FORMATION IN THE CENTRAL NERVOUS SYSTEM

Compared to synapse formation between neurons and muscle cells, we know only relatively little about how neuron-neuron synapses in the central nervous system are put together. The reason for this lack of knowledge is the complexity of the problem. CNS synapses are much smaller (about one hundredths of a neuromuscular junction), a single neuron typically receives hundreds to thousands of synapses and the synaptic inputs are derived from multiple different synaptic partners. Finally, these synapses employ not just one but a variety of neurotransmitters, some excitatory that depolarize the postsynaptic partner, and some that are inhibitory and lead to hyper-polarization. The placement of excitatory and inhibitory synapses on the postsynaptic cell is a critical parameter for integration of these counteracting inputs. Another complicating factor is that the synaptic connectivity in the central nervous system is not static but undergoes extensive plastic changes that modify the function of neuronal circuits in response to experience. All of these properties pose intriguing challenges regarding the synthesis of synaptic components, their targeting and assembly at functionally distinct synaptic sites, and the recognition of the appropriate synaptic partners in a sea of candidate partners within the target area.

While the assembly of the vertebrate neuromuscular junction represents a much simpler problem there are several principles discovered in work on the NMJ that appear to apply to CNS synapse formation: (1) Cells express most synaptic components before synapse formation, (2) bi-directional signaling organizes synaptic components at emerging synaptic contacts, (3) synapses undergo extensive functional maturation, (4) during development an exuberant number of synaptic contacts is formed, some of which later undergo elimination.

Building on these parallels to NMJ formation significant progress in understanding CNS synapse formation has been made. Other invaluable information for these studies has come from the previous elucidation of the molecular composition of pre- and postsynaptic structures (Chapters 8 and 9) and the recent advances in understanding the mechanisms of synaptic plasticity of mature synapses (Chapters 50 and 51), some of which are likely to be also important for their initial formation. This section focuses on the formation of chemical synapses between neurons in the central nervous system of vertebrates. It examines emerging evidence on the mechanisms controlling recruitment and assembly of pre- and postsynaptic components and discusses some candidate molecules that may contribute to the selectivity of synaptic wiring in development.

The First Contact between Axons and Their Neuronal Targets

Neurons express both pre- and postsynaptic components prior to synapse formation. For example, growing axons and growth cones contain synaptic vesicles and release neurotransmitter before interacting with a postsynaptic target cell. Similarly, target cells express functional neurotransmitter receptors on their surface before synapse formation. Cell–cell contact, however, is essential for the concentration and juxtaposition of pre- and postsynaptic components in opposing membranes and for achieving the mature complement of both pre- and postsynaptic proteins.

Many initial contacts between synaptic partners are established through filopodia, tiny membrane extensions that rapidly extend and retract from a cell (Chapter 17). Importantly, filopodia extend from both, axons and dendrites and increase the extracellular space sampled by a cell during development (Jontes and Smith, 2000). Imaging studies, both in culture and *in vivo* indicate that when filopodia contact potential synaptic partners, they can lose their motility and become stabilized. These contacts are then thought to transform into synaptic structures (McAllister, 2007). Indeed, synaptic specializations have been seen on filopodia themselves and neurotransmitter release from axons is likely to play an important role for the regulation of filopodial dynamics (Tashiro *et al.*, 2003).

A second important realization from imaging studies has been that the assembly of synapses occurs very rapidly. The time required from initial contact to establishment of a functional synapse is only in the range of 1–2 hours, with the formation of a presynaptic terminal occurring in only 10–20 minutes and the recruitment of postsynaptic neurotransmitter receptors lagging somewhat behind (Friedman et al., 2000) (Fig. 18.14). This rapid timecourse is plausible if one considers that most synaptic components are "ready-to-go" before contact and primarily need to be redistributed in response to a cell surface signal. The key molecules mediating stabilization of filopodial contacts between pre- and postsynaptic partners and rapid assembly of synaptic structures are not known but recent studies identified several candidate proteins for these events.

Target-derived Signals for Presynaptic Development

What are the trans-synaptic interactions that mediate the stabilization of filopodial contacts and rapid recruitment of presynaptic components? A striking realization emerging during the past years has been that single target-derived signals are sufficient to induce a substantial degree of presynaptic organization. The first proteins with such an activity to be identified were the neuroligins. Neuroligins are a small family of postsynaptic adhesion molecules that can bind to neurexins, a second family of neuronal receptors, that is present on axons (Brose, 1999). Neuroligins and neurexins are expressed in the nervous system at the right time and place to influence synaptogenesis. Work in cultured cerebellar and hippocampal neurons demonstrated that the heterologous expression of neuroligin in nonneuronal cells induces focal accumulations of neurexins and synaptic vesicle proteins such as synapsin, synaptotagmin, and synaptophysin in axons (Dean et al. 2003; Scheiffele et al., 2000) (Fig. 18.15). These presynaptic structures were capable of vesicle exocytosis upon depolarization indicating that the machinery for regulated exocytosis had been recruited to the neuroligin-induced presynaptic terminals. Subsequent studies identified additional synaptic adhesion molecules with similar "synaptogenic" or "synapse-organizing" functions (Biederer et al., 2002; Kim et al., 2006). These include the Ig-domain protein SynCAM and Netrin-G-ligand, both of which have similar activitites as neuroligins with respect to the induction of presynaptic structures. While these studies provided convincing evidence that each of these molecules is sufficient to induce presynaptic specializations, it is not clear how these multiple trans-synaptic systems cooperate during development in

vivo and whether any one of these systems is absolutely essential for the assembly of synapses in vivo.

How might adhesion molecules on the postsynaptic cell induce presynaptic development? For neuroligins, presynaptic induction is mediated by recruitment and local aggregation of neurexins that act as a neuroligin receptor. Neurexins bind intracellularly to several cytoplasmic scaffolding molecules including Mint/Lin-10 and CASK/Lin-2, which may couple neurexins to voltage-gated calcium channels at the active zone (Maximov et al., 1999; Missler et al., 2003). Consistent with such a model, voltage-gated calcium channel function is perturbed in knock-out mice lacking a subset of neurexin isoforms. Despite these advances, it remains a major question how the enormous number of presynaptic active zone components required for regulated vesicular release (discussed in Chapter 8) can be recruited by an adhesion complex within the 10–20 minute timeframe that has been observed in imaging studies. One intriguing model for this rapid assembly of functional presynaptic terminals is that presynaptic components are recruited in pre-assembled units. Morphological and imaging experiments revealed that synaptic vesicles move in axons before synapse formation in grape-like clusters (Kraszewski et al., 1995). Also active zone components can be found pre-complexed in so-called "active zone precursor vesicles" (Garner et al., 2002). These vesicles could be rapidly captured and inserted in response to a target-derived signal, thereby greatly simplifying the process of presynaptic assembly (Fig. 18.14). However, whether any of the identified "synaptogenic" signals such as the neuroligins, SynCAM, or netrin-G-ligands specifically defines the insertion site of these vesicles remains to be shown.

Additional studies identified secreted growth factors that might act as target-derived signals upstream of synaptogenic cell-adhesion molecules. Using the cerebellar mossy fiber-granule cell synapse as a model system, Salinas and co-workers showed that WNT-7a derived from granule cells can induce the characteristic spreading of pontine mossy fiber axon terminals in vitro (Hall et al., 2000). Normal presynaptic development is delayed in WNT-7a knockout mice, resulting in reduced synapsin clustering and simplified presynaptic membrane structures. Sanes and colleagues discovered that fibroblast growth factor 22 (FGF22) is a second granule cell-derived growth factor that contributes to mossy fiber development (Umemori et al., 2004). Suppression of FGF signaling by injection of blocking proteins into the cerebellar cortex or ablation of FGF signaling receptors results in a profound reduction of presynaptic vesicles clustering in mossy fiber axons. Similar to WNT-7a, also FGF22 alters cytoskeletal

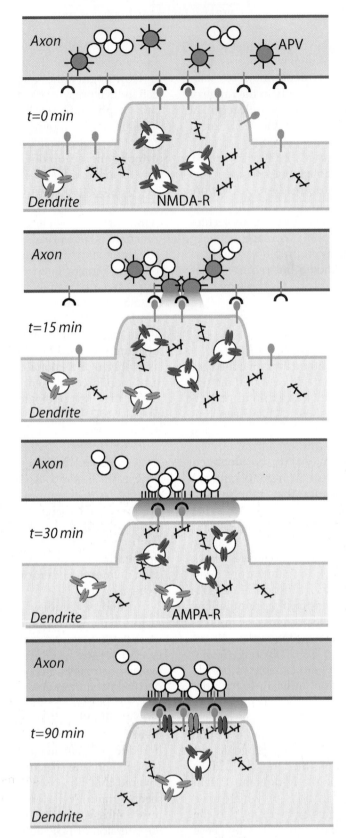

FIGURE 18.14 Model for the assembly of central synapses. Before synapses formation, axons and dendrites contain synaptic vesicles, active zone components, and neurotransmitter receptors, respectively. Cell–cell contact is initiated via cell surface adhesion and/or signaling molecules (red, $t = 0$ min). Contact triggers fusion of active-zone precursor vesicles (APV, green) with the plasma membrane ($t = 15$ min), which leads to the deposition of active zone material (black spikes). Subsequently, synaptic vesicles (clear circles) are recruited to the cell–cell contact sites ($t = 30$ min), and postsynaptic scaffolding molecules (black), NMDA-type (blue), and AMPA-type neurotransmitter receptors (red) are delivered to the maturing postsynaptic membrane. The surface molecules that mark the insertion sites for APVs are not known and further work will be needed to confirm whether APV fusion represents the main mechanism for active zone assembly. Adapted from Garner *et al.* (2002).

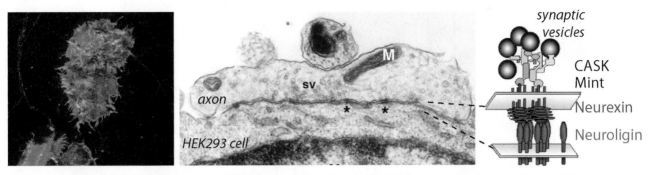

FIGURE 18.15 **Synapse-organizing function of the postsynaptic adhesion molecule neuroligin.** When HEK293 cell fibroblasts overexpressing neuroligin (green) are brought into contact with pontine axons, synaptic vesicles (vesicle protein synapsin, red) accumulate at the contact sites in punctate structures in the axon. Electron microscopy of the axon-HEK293 cell contacts shows clusters of synaptic vesicles (sv) and electron-dense material (*). The right panel shows a cartoon of the transcellular arrangement of the neuroligin-neurexin complex, with cytoplasmic scaffolding molecules CASK and Mint attached to the neurexin cytoplasmic tail. In this cell culture assay, the proteins SynCAM and Netrin-G-ligand have activities that are similar to neuroligin. How the synapse-organizing function of all three proteins is used in the intact brain *in vivo* is still an area of active studies. Adapted from Scheiffele *et al.* (2000).

properties and promotes axonal branching. These morphological rearrangements could shape the presynaptic partner and facilitate the recruitment of appropriate organelles in response to trans-synaptic adhesion molecules.

Control of Postsynaptic Development

NMJs and central synapses differ most dramatically in their postsynaptic structures. This is not surprising considering the difference in cell type of the postsynaptic cell, neurons in the CNS *versus* muscle cells at the NMJ. While the cleft of neuron-neuron synapses is filled with proteinaceous and carbohydrate material it does not contain a thick basal lamina like the one seen at the NMJ. With 20 nm, the width of the synaptic cleft is much narrower and the postsynaptic membrane is closely apposed to the presynaptic active zone, lacking the junctional infoldings of the NMJ. Despite these substantial differences, both synapses share the accumulation of postsynaptic neurotransmitter receptors as the hallmark of postsynaptic differentiation.

Excitatory neuron-neuron synapses using the transmitter glutamate contain multiple types of neurotransmitter receptors. So-called metabotropic glutamate receptors (mGluRs) that couple to G-proteins and modulate synaptic function and the ionotropic NMDA- and AMPA-type receptors that carry ion currents for synaptic transmission which depolarize and excite the postsynaptic cell. During development NMDA- and AMPA-receptors are thought to be recruited to synapses in a stepwise manner, with an initial accumulation of NMDA receptors at postsynaptic sites and the subsequent addition of AMPA receptors. This view is supported both by electron microscopy using immu-

nogold labeling and by electrophysiology in which NDMA receptor-rich "silent synapses" are identified initially in development and the AMPA-receptor content increases at later stages.

How can postsynaptic neurotransmitter receptors be recruited to a newly forming neuron-neuron contact? One model for neurotransmitter recruitment in analogy to the NMJ is that trans-synaptic signals could recruit a rapsyn-like cytoplasmic protein that provides a scaffold underneath the postsynaptic membrane. This scaffold would then also interact with postsynaptic neurotransmitter receptors and selectively retain receptor molecules at the appropriate sites. Biochemical studies have identified a vast array of such scaffolding proteins, many of which contain multiple protein-protein interaction domains and are ideally suited to cross-link multiple postsynaptic membrane proteins into a larger macro-molecular complex (Kim and Sheng, 2004) (Fig. 18.16). Interestingly, neuroligins, SynCAM and Netrin-G-ligands all interact with such scaffolding proteins via their cytoplasmic tail. For example neuroligins and Netrin-G-ligand bind via PDZ-domain interactions to PSD-95, a major component of the glutamatergic postsynaptic scaffold. Overexpression or aggregation of neuroligins with recombinant neurexins results in the recruitment of the scaffolding proteins PSD-95, Homer, GKAP and NMDA-type glutamate receptors (Chih *et al.*, 2005; Graf *et al.*, 2004). An important role for neuroligin in neurotransmitter receptor recruitment is also supported by knock-out mice. Mice lacking most neuroligin proteins show a 20% reduction in synapse density and a significant loss of postsynaptic neurotransmitter receptor proteins from synapses (Varoqueaux *et al.*, 2006). Mouse knockouts lacking the scaffolding protein

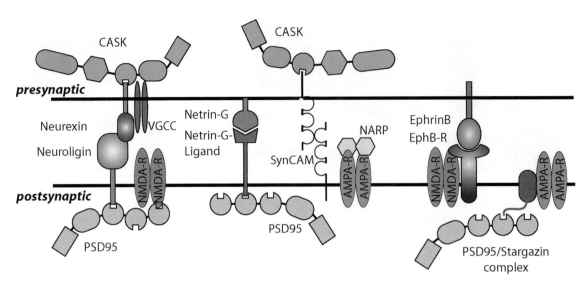

FIGURE 18.16 **A simplified model for transsynaptic interactions that may coordinate assembly of presynaptic and postsynaptic elements of glutamatergic synapses and recruitment of NMDA- and AMPA-type glutamate receptors.** Neurexin-neuroligin interactions can organize pre- and postsynaptic structures. In this process presynaptic neurexin might couple to voltage-gated calcium channels (VGCC) through the scaffolding molecule CASK. Postsynaptic neuroligin binds via PDZ-domain interactions to PSD95, which itself can directly interact with NMDA-receptors. Heterophilic interactions through Netrin-G–Netrin-G-ligand and homophilic interactions via SynCAM-SynCAM pairs provide two additional transsynaptic links that organize pre- and postsynaptic sites (note that Netrin-G is a glycosyl-phosphatidylinositol-anchored protein that does not have an intracellular domain). Presynaptic EphrinB binding to postsynaptic EphB-receptors can regulate the clustering of NMDA receptors through interactions via the extracellular domains of the proteins. The secreted protein Narp provides a second example for an extracellular mechanism for clustering of certain populations of AMPA receptors at synapses. Another critical mechanism for synaptic trafficking of AMPA-receptors is via the Stargazin (blue)/PSD95 complex. This cartoon provides only a working model for how transsynaptic adhesion molecules might be coupling to neurotransmitter receptors. The importance of some of these interactions for synapse formation *in vivo* remains to be determined.

PSD95, however, indicate that PSD-95 is not essential for NMDA receptor localization at synapses. This means, either there are other scaffolding molecules that are more important with respect to NMDA-receptor recruitment or perhaps multiple PDZ-containing proteins have overlapping functions in this process and can substitute for the loss of PSD-95 in the mutant mice.

In a complex with another postsynaptic protein called stargazin, PSD-95 has also been implicated in synaptic recruitment of AMPA-type glutamate receptors. Mice lacking the stargazin protein have severely reduced AMPA-receptor currents in cerebellar granule cells (Chen *et al.*, 2000). Stargazin itself is a membrane protein that co-assembles with AMPA-receptors in the endoplasmic reticulum. Upon transport to the cell surface, PSD-95 might then capture these AMPA-receptor-stargazin complexes at glutamatergic synapses.

The concept that scaffolding molecules contribute to the clustering of postsynaptic neurotransmitter receptors applies to both, excitatory and inhibitory synapses. Importantly, these synapses differ not only in their neurotransmitter receptors but also in their scaffolding proteins. In fact, the scaffolding molecules

may ensure that the appropriate neurotransmitter receptors are selectively recruited to excitatory and inhibitory synapses, respectively. The scaffolding protein gephyrin represents one of the best-understood examples for the scaffold-mediated postsynaptic receptor clustering at inhibitory synapses. A host of studies, including mouse knockouts, demonstrates that gephyrin, a cytoplasmic protein associated with the glycine receptor, is essential for receptor clustering at synaptic sites (Kneussel and Betz, 2000b). Gephyrin is also concentrated at GABAergic postsynaptic sites but there it might not be absolutely essential for the recruitment of GABA receptors. Notably, gephyrin is well equipped to serve as a scaffolding protein that links receptors to the cytoskeleton; it binds with high affinity to polymerized tubulin, the actin-binding protein profilin, and the lipid-binding protein collybistin. Thereby, gephyrin may link membrane lipids, neurotransmitter receptors and cytoskeletal elements to stabilize postsynaptic specializations.

Work on glycine receptor transport also uncovered another key mechanism for regulating neurotransmitter receptor recruitment to synapses, namely that the function of the receptor itself can regulate its concentration at synapses. Pharmacological blockade of glycine receptors with strychnine or the blockade of sodium channel dependent action potentials with tetrodotoxin (TTX) prevents postsynaptic accumulation of the receptors. Further studies on the activity-dependent glycine receptor clustering at synapses indicate that calcium influx is essential for clustering, possibly by initiating a calcium-activated signaling cascade in the cytoplasm (Kneussel and Betz, 2000a). These mechanisms may ensure the accumulation of glycine receptors only at active synapses or at synapses that release the appropriate neurotransmitter.

A second mechanism for the recruitment of postsynaptic neurotransmitter receptors that is independent of scaffolding molecules is the direct interaction of membrane and secreted proteins with the extracellular domain of neurotransmitter receptors. Receptor tyrosine kinases of the EphB-family and their cognate ligands, called ephrinB proteins, were shown to form a tripartite complex with NMDA receptors (Dalva et al., 2000) (see Fig. 18.4 for more detail on Eph receptors). EphrinB-EphB interactions trigger the clustering of NMDA receptors and stimulation of EphB-receptors increases density of glutamatergic synapses in cultured neurons. The coclustering depends only on the extracellular portions of the NMDA and EphB receptors and may reflect a direct interaction between them. In addition, EphB-receptor kinase activation initiates a cytoplasmic signaling cascade which leads to phosphorylation of NMDA-receptors and increases calcium flux through these receptors (Takasu et al., 2002). Studies on EphB-receptor knock-out mice support an important role in the synaptic recruitment and function of NMDA-receptors as well for the structural assembly of synapses (Pasquale, 2005). This synaptic role for the Eph-receptor tyrosine kinases provides an interesting mechanistic parallel to the role of the receptor tyrosine kinase MuSK in AChR recruitment at the NMJ.

Notably, EphB-receptors had been initially recognized for their critical roles in axon guidance and topographic mapping and in the past years novel synaptic functions have also emerged for other axon guidance receptors. This suggests that many of the proteins that function in axon guidance also contribute to synapse formation. Notably, in many cases the signaling readout downstream of these receptors differs at the different steps of neural development: many receptors that mediate repulsive signaling during axon guidance have adhesive and stabilizing functions at synapses. How this functional switch is achieved is an important focus of ongoing work.

Besides Eph-receptors there are additional proteins that contribute to the synaptic clustering neurotransmitter receptors through extracellular interactions. One of these factors is the protein Narp (O'Brien et al., 2002). Narp is a member of the pentraxin family, extracellular components capable of head-to-head aggregation. O'Brien and colleagues discovered that Narp can interact directly with the extracellular domain of AMPA-type glutamate receptors resulting in their clustering at synapses. Because patterned electrical activity upregulates Narp expression, it may contribute to activity-dependent synapse maturation.

Role of Neuronal Activity in Synapse Formation

Understanding how synaptic activity influences the formation and stabilization of synapses is an area of intense study. During development, each axon comes into contact with a large number of potential synaptic partners but the final connectivity pattern is remarkably selective. In the section on topographic mapping above we already discussed two principle mechanisms that might underlie selective interactions during development: Sperry's chemo-affinity hypothesis which suggested that molecular recognition labels specify the positioning and connectivity of neurons and Paul Weiss' hypothesis that neuronal activity would provide instructive information on which neurons are wired into a functional circuit. The essence of both of these hypotheses, put forward more than 50 years ago, is still very much alive and ongoing studies continue to identify wiring decisions that require synaptic activity or cell surface recognition molecules that might act as chemoaffinity tags.

It should be noted that there is evidence that synaptic activity is not absolutely required for the formation of synaptic structures during development. Ablation of Munc18–1, a key component of the presynaptic release machinery results in a complete loss of synaptic transmission but Munc18-1 knock-out mice form synaptic structures. Neurons of the null mutants fail to display evoked and spontaneous transmitter release, thereby resulting in a complete loss of synaptic transmission. Munc18-1 null mutants display apparently normal brain development during embryonic stages (Verhage et al., 2000). Due to the complete loss of synaptic transmission, the mutant animals can not breathe and die at birth. This precludes an analysis of more mature synaptic networks. Nonetheless, morphological development of synaptic connections appears normal, though at later stages of embryonic development many neurons undergo apoptosis. It will be informative to examine more mature synaptic networks in conditional Munc18-1 mutants where the protein is only ablated in small populations of cells. This should clarify how far synaptic development can proceed in the absence of synaptic transmission.

Another approach to understanding activity-dependent mechanisms of synapse formation has been to search for proteins that are specifically up- or down-regulated by synaptic activity. In fact, NARP, the secreted protein discussed above that clusters AMPA receptors was originally identified in such a screen based on its upregulation with strong synaptic stimulation. Another particularly, intriguing example of an activity-induced gene is CGP15, a small cell surface protein that is upregulated by activity in adult hippocampus (Nedivi et al., 1998). In vivo expression of CPG15 increases axonal arborization by stabilizing branches and promotes the maturation of synapses, as evidenced by a precocious increase in the ratio of AMPA/NMDA-mediated currents. Moreover, CPG15 expression decreases the number of silent synapses, consistent with an early acquisition of AMPA receptors at such sites (Cantallops et al., 2000).

Many additional activity-induced genes have been identified and the analysis of their regulation and function is likely to yield important new insights into the mechanisms of synapse formation.

Surface Recognition Molecules in the Synaptic Specificity

Understanding the molecular mechanisms of synaptic specificity is arguably the most exciting and important unresolved question in the field of CNS synaptogenesis. A leading model is that selective trans-synaptic interactions are achieved by an adhesive code. Such a code could be generated by matching expression of adhesion molecules in the pre- and postsynaptic partners that underlie a key-lock mechanism for selective interactions. The first proteins for which such a function was proposed are the cadherins, a family of homophilic adhesion molecules (Fannon and Colman, 1996). Cadherins are concentrated at CNS synapses and pre- and postsynaptic partners often contain the same cadherin family member, suggesting that the homophilic cadherin interactions may specifically link them together. Since there are 20 different ("classical") cadherin isoforms expressed in different neuronal populations these proteins could encode a number of different synaptic interactions. Recent work in the Drosophila visual system provided evidence for this hypothesis. Zipursky and colleagues demonstrated an important role for N-cadherin in selection of the appropriate target layer of photoreceptor axons in vivo (Clandinin and Zipursky, 2002). However, direct evidence for a role for cadherins in synaptic specificity in vertebrates is still lacking.

The laminar organization of the visual system has also lent itself for studies on synaptic specificity in vertebrates. Sanes and colleagues discovered a family of proteins called Sidekicks which mediate homophilic binding (Yamagata et al., 2002). As seen for the cadherins, Sidekick proteins show matching expression in pre- and postsynaptic partners and concentrate at synapses (Fig. 18.17). Misexpression of a Sidekick protein in a cell that normally does not express it is sufficient to stabilize ectopic axonal arborization on the misexpressing cells strongly suggesting that Sidekick-mediated adhesion can alter the outcome of innervation.

Another recent exciting discovery with potentially important implications for synaptic specificity was the identification of several neuronal cell surface protein families that encode hundreds or even thousands of different isoforms. The best characterized of these polymorphic surface molecules in vertebrates are related to cadherins and were initially named cadherin-related neuronal receptors (CNRs) but are now primarily referred to as protocadherins (Kohmura et al., 1998). The human genome contains 52 protocadherins genes. These genes are arranged into gene clusters that are reminiscent of the immunoglobulin family (Wu and Maniatis, 1999). By alternative splicing variable axons which encode unique extracellular domains are joined to axons encoding a common intracellular domain. Thereby each protocadherin gene can give rise to numerous different recognition molecules and this molecular diversity encouraged speculations that protocadherins may encode distinct recognition events that specify synaptic connections. Ablation of a large number of protocadherin isoforms results in apoptotic cell death of neurons and a loss of synapses (Wang et al., 2002; Weiner et al., 2005). While it is clear that protocadherin isoforms are selectively expressed in neuronal populations, cell- and isoform-specific functions have not been identified yet.

Two families with even higher molecular diversity than protocadherins are DSCAM and the neurexins. Like for the protocadherins, diversity is generated through a combination of expression form multiple genes and alternative splicing. In Drosophila, alternative splicing generates 38,016 DSCAM variants. Remarkably, all variants tested so far engage in exclusively homophilic binding, that means the DSCAM variants theoretically have the ability to encode thousands of specific cell recognition events (Schmucker et al., 2000; Wojtowicz et al., 2004). The functional relevance of DSCAM molecules is beginning to emerge and recent experiments suggest that DSCAM-mediated interactions in the Drosophila nervous system allow a cell to distinguish its own axonal and dendritic processes from those of a neighboring cell (Matthews et al., 2007). While there are vertebrate homologues of

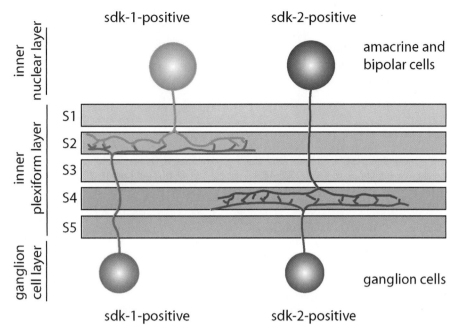

FIGURE 18.17 Model for lamina-specific synaptic connectivity through homophilic adhesion molecules. Pre- and postsynaptic partners express the same homophilic adhesion molecules and thereby preferentially form synapses with each other. In this example, subsets of amacrine and bipolar cells in the inner nuclear layer express different isoforms of the protein Sidekick (sdk-1 or sdk-2) and form lamina-specific synaptic connections on the dendrites of sdk-1 or sdk-2-positive ganglion cells, respectively. The layered organization of the inner plexiform layer (sublaminae S1-S5) is an excellent system for the analysis of this type of synaptic specificity. In principle, other homophilic adhesion molecules such as cadherins with corresponding expression in pre- and postsynaptic partners may function in a similar manner. Adapted from Yamagata *et al.* (2002).

DSCAM, these homologues do not show a comparable diversity and it remains to be shown what aspects of DSCAM function are conserved in higher organisms.

Vertebrate neurexins encode nearly 4,000 isoforms and as observed for protocadherins and DSCAM the sequence diversity of neurexins is primarily found in their extracellular domains (Ushkaryov *et al.*, 1992). The previously characterized functions of neurexins in the organization and function of synapses (see above) make these proteins excellent candidates for synapse-specific recognition events. However, understanding the selective biochemical interactions of these proteins and their function in recognition events at synapsis remains a challenge for the future.

An important conclusion derived from the characterization of the protocadherin, DSCAM, and neurexin gene families is that anatomically apparently identical cells often express different cell surface recognition molecules. This implies that within one anatomically defined cell population there might be subpopulations containing different cell surface recognition markers. While we are only beginning to understand the functional relevance of these highly diverse gene families their analysis should provide important new insights into the organization and function of neuronal networks.

Summary

The diversity of synapses found in the nervous system is likely to be matched by an equally diverse array of mechanisms producing the synapses. These mechanisms are only now beginning to be elucidated. Prime candidates at present for synaptogenic roles include transmembrane proteins interacting with cognate receptors on the apposing cell surface (e.g., neuroligin/β-neurexin; SynCAM; ephrinB/EphB receptors), diffusible molecules that instruct synaptic development in target cells (e.g., FGF22, WNT-7a) or interact directly with synaptic components (e.g., Narp), membrane-attached components that act to promote maturation both pre- and postsynaptically (e.g., CPG15), and intracellular components that act locally to organize synaptic constituents (e.g., gephyrin). Testing the effects of such components *in vivo* will be a high priority in the near term. Neuronal activity is also important for certain aspects of synapse formation, but exactly how it participates and whether it employs mechanisms shared with those governing synaptic plasticity remain intriguing questions. Finally, we are beginning to identify molecular families (e.g., protocadherins and neurexins) that may contribute to achieving the remarkable synaptic specificity found in the nervous system. Determining how specific components participate in and direct the continuing "handshake" of trans-synaptic signals governing synapse formation is one of the many challenges for the future.

References

Biederer, T., Sara, Y., Mozhayeva, M., Atasoy, D., Liu, X., Kavalali, E. T., and Sudhof, T. C. (2002). SynCAM, a synaptic adhesion molecule that drives synapse assembly. *Science* **297**, 1525–1531.

Brose, N. (1999). Synaptic cell adhesion proteins and synaptogenesis in the mammalian central nervous system. *Naturwissenschaften* **86**, 516–524.

Brown, A., Yates, P. A., Burrola, P., Ortuño, D., Vaidya, A., Jessell, T. M., Pfaff, S. L., O'Leary, D. D. M., and Lemke, G. (2000). Topographic mapping from the retina to the midbrain is controlled by relative but not absolute levels of EphA receptor signaling. *Cell* **102**, 77–88.

Cantallops, I., Haas, K., and Cline, H. T. (2000). Postsynaptic CPG15 promotes synaptic maturation and presynaptic axon arbor elaboration in vivo. *Nature Neuroscience* **3**, 1004–1011.

Chen, L., Chetkovich, D. M., Petralia, R. S., Sweeney, N. T., Kawasaki, Y., Wenthold, R. J., Bredt, D. S., and Nicoll, R. A. (2000). Stargazin regulates synaptic targeting of AMPA receptors by two distinct mechanisms. *Nature* **408**, 936–943.

Chih, B., Engelman, H., and Scheiffele, P. (2005). Control of excitatory and inhibitory synapse formation by neuroligins. *Science* **307**, 1324–1328.

Clandinin, T. R. and Zipursky, S. L. (2002). Making connections in the fly visual system. *Neuron* **35**, 827–841.

Dalva, M. B., Takasu, M. A., Lin, M. Z., Shamah, S. M., Hu, L., Gale, N. W., and Greenberg, M. E. (2000). EphB receptors interact with NMDA receptors and regulate excitatory synapse formation. *Cell* **103**, 945–956.

Dean, C., Scholl, F. G., Choih, J., DeMaria, S., Berger, J., Isacoff, E., and Scheiffele, P. (2003). Neurexin mediates the assembly of presynaptic teminals. *Nature Neurosclence* **6**, 708–716.

DeChiara, T. M., Bowen, D. C., Valenzuela, D. M., Simmons, M. V., Poueymirou, W. T., Thomas, S., Kinetz, E., Compton, D. L., Park, J. S., Smith, C., DiStefano, P. S., Glass, D. J., Burden, S. J., and Yancopoulos, G. D. (1996). The receptor tyrosine kinase, MuSK, is required for neuromuscular junction formation in vivo. *Cell* **85**, 501–512.

Fannon, A. M. and Colman, D. R. (1996). A model for central synaptic junctional complex formation based on the differential adhesive specificities of the cadherins. *Neuron* **17**, 423–434.

Feldheim, D. A., Kim, Y. I., Bergemann, A. D., Frisen, J., Barbacid, M., and Flanagan, J. G. (2000). Genetic analysis of ephrin-A2 and ephrin-A5 shows their requirement in multiple aspects of retinocollicular mapping. *Neuron* **25**, 563–574.

Friedman, H. V., Bresler, T., Garner, C. C., and Ziv, N. E. (2000). Assembly of new individual excitatory synapses: Time course and temporal order of synaptic molecule recruitment. *Neuron* **27**, 57–69.

Frisen, J., Yates, P. A., McLaughlin, T., Friedman, G. C., O'Leary, D. D. M., and Barbacid, M. (1998). Ephrin-A5 (AL-1/RAGS) is essential for proper retinal axon guidance and topographic mapping in the mammalian visual system. *Neuron* **20**, 235–243.

Garner, C. C., Zhai, R. G., Gundelfinger, E. D., and Ziv, N. E. (2002). Molecular mechanisms of CNS synaptogenesis. *Trends in Neurosciences* **25**, 243–251.

Graf, E. R., Zhang, X., Jin, S. X., Linhoff, M. W., and Craig, A. M. (2004). Neurexins induce differentiation of GABA and glutamate postsynaptic specializations via neuroligins. *Cell* **119**, 1013–1026.

Hall, A. C., Lucas, F. R., and Salinas, P. C. (2000). Axonal remodeling and synaptic differentiation in the cerebellum is regulated by WNT-7a signaling. *Cell* **100**, 525–535.

Hindges, R., McLaughlin, T., Genoud, N., Henkemeyer, M., and O'Leary, D. D. M. (2002). EphB forward signaling controls directional branch extension and arborization required for dorsal ventral retinotopic mapping. *Neuron* **35**, 475–487.

Hoopfer, E. D., McLaughlin, T., Watts, R. J., Schuldiner, O., O'Leary, D. D. M., and Luo, L. (2006). Wld^s protection distinguishes axon degeneration following injury from naturally-occurring developmental pruning. *Neuron* **50**, 883–895.

Jontes, J. D. and Smith, S. J. (2000). Filopodia, spines, and the generation of synaptic diversity. *Neuron* **27**, 11–14.

Kim, E. and Sheng, M. (2004). PDZ domain proteins of synapses. *Nat Rev Neurosci* **5**, 771–781.

Kim, S., Burette, A., Chung, H. S., Kwon, S. K., Woo, J., Lee, H. W., Kim, K., Kim, H., Weinberg, R. J., and Kim, E. (2006). NGL family PSD-95-interacting adhesion molecules regulate excitatory synapse formation. *Nature Neuroscience* **9**, 1294–1301.

Kim, N. and Burden, S. J. (2008). MuSK controls where motor axons grow and form synapses. *Nature neuroscience* **11**, 19–27.

Kneussel, M. and Betz, H. (2000a). Clustering of inhibitory neurotransmitter receptors at developing postsynaptic sites: the membrane activation model. *Trends in Neurosciences* **23**, 429–435.

Kneussel, M. and Betz, H. (2000b). Receptors, gephyrin and gephyrin-associated proteins: Novel insights into the assembly of inhibitory postsynaptic membrane specializations. *Journal of Physiology* **525** Pt 1, 1–9.

Kohmura, N., Senzaki, K., Hamada, S., Kai, N., Yasuda, R., Watanabe, M., Ishii, H., Yasuda, M., Mishina, M., and Yagi, T. (1998). Diversity revealed by a novel family of cadherins expressed in neurons at a synaptic complex. *Neuron* **20**, 1137–1151.

Kraszewski, K., Mundigl, O., Daniell, L., Verderio, C., Matteoli, M., and De Camilli, P. (1995). Synaptic vesicle dynamics in living cultured hippocampal neurons visualized with CY3-conjugated antibodies directed against the lumenal domain of synaptotagmin. *J Neurosci* **15**, 4328–4342.

Mann, F., Ray, S., Harris, W. A., and Holt, C. E. (2002). Topographic mapping in dorsoventral axis of the *Xenopus* retinotectal system depends on signaling through ephrin-B ligands. *Neuron* **35**, 461–473.

Matthews, B. J., Kim, M. E., Flanagan, J. J., Hattori, D., Clemens, J. C., Zipursky, S. L., and Grueber, W. B. (2007). Dendrite self-avoidance is controlled by Dscam. *Cell* **129**, 593–604.

Maximov, A., Sudhof, T. C., and Bezprozvanny, I. (1999). Association of neuronal calcium channels with modular adaptor proteins. *Journal of Biological Chemistry* **274**, 24453–24456.

McAllister, A. K. (2007). Dynamic aspects of CNS synapse formation. *Annu Rev Neurosci*.

McLaughlin, T., Hindges, R., Yates, P. A., and O'Leary, D. D. M. (2003a). Bifunctional action of ephrin-B1 as a repellent and attractant to control bidirectional branch extension in dorsal ventral retinotopic mapping. *Development* **130**, 2407–2418.

McLaughlin, T., Torborg, C. L., Feller, M. B., and O'Leary, D. D. M. (2003b). Retinotopic map refinement requires spontaneous retinal waves during a brief critical period of development. *Neuron* **40**, 1147–1146.

Missler, M., Zhang, W., Rohlmann, A., Kattenstroth, G., Hammer, R. E., Gottmann, K., and Sudhof, T. C. (2003). Alpha-neurexins couple Ca2+ channels to synaptic vesicle exocytosis. *Nature* **423**, 939–948.

Monschau, B., Kremoser, C., Ohta, K., Tanaka, H., Kaneko, T., Yamada, T., Handwerker, C., Hornberger, M. R. *et al.* (1997). Shared and distinct functions of RAGS and ELF-1 in guiding retinal axons. *Embo J* **16**, 1258–1267.

Nakamoto, M., Cheng, H. J., Friedman, G. C., McLaughlin, T., Hansen, M. J., Yoon, C. H., O'Leary, D. D. M., and Flanagan, J. G. (1996). Topographically specific effects of ELF-1 on retinal axon guidance in vitro and retinal axon mapping in vivo. *Cell* **86**, 755–766.

Nedivi, E., Wu, G. Y., and Cline, H. T. (1998). Promotion of dendritic growth by CPG15, an activity-induced signaling molecule. *Science* **281**, 1863–1866.

O'Brien, R., Xu, D., Mi, R., Tang, X., Hopf, C., and Worley, P. (2002). Synaptically targeted narp plays an essential role in the aggregation of AMPA receptors at excitatory synapses in cultured spinal neurons. *J Neurosci* **22**, 4487–4498.

O'Leary, D. D. M. (1992). Development of connectional diversity and specificity in the mammalian brain by the pruning of collateral projections. *Current Opinion in Neurobiology* **2**, 70–77.

O'Leary, D. D. M. and Koester, S. E. (1993). Development of projection neuron types, axonal pathways and patterned connections of the mammalian cortex. *Neuron* **10**, 991–1006.

Pasquale, E. B. (2005). Eph receptor signalling casts a wide net on cell behaviour. *Nature Reviews* **6**, 462–475.

Pfeiffenberger, C., Yamada, J., and Feldheim, D. A. (2006). Ephrin-As and patterned retinal activity act together in the development of topographic maps in the primary visual system. J. Neurosci 26, 12873-12884.

Rashid, T., Upton, A. L., Blentic, A. *et al.* (2005). Opposing gradients of ephrin-As and EphA7 in the superior colliculus are essential for topographic mapping in the mammalian visual system. *Neuron* **47**, 57–69.

Scheiffele P, Fan J, Choih J, Fetter R, Serafini T. (2000). Neuroligin expressed in nonneuronal cells triggers presynaptic development in contacting axons. *Cell* **101**:657–669.

Schmitt, A. M., Shi, J., Wolf, A. M. *et al.* (2006). Wnt-Ryk signalling mediates medial-lateral retinotectal topographic mapping. *Nature* **439**, 31–37.

Schmucker, D., Clemens, J. C., Shu, H., Worby, C. A., Xiao, J., Muda, M., Dixon, J. E., and Zipursky, S. L. (2000). Drosophila Dscam is an axon guidance receptor exhibiting extraordinary molecular diversity. *Cell* **101**, 671–684.

Sperry, R. (1963). Chemoaffinity in the orderly growth of nerve fiber patterns and connections. *Proceedings of the National Academy of Sciences USA* **50**, 703–710.

Takasu, M. A., Dalva, M. B., Zigmond, R. E., and Greenberg, M. E. (2002). Modulation of NMDA receptor-dependent calcium influx and gene expression through EphB receptors. *Science* **295**, 491–495.

Tashiro, A., Dunaevsky, A., Blazeski, R., Mason, C. A., and Yuste, R. (2003). Bidirectional regulation of hippocampal mossy fiber filopodial motility by kainate receptors: A two-step model of synaptogenesis. *Neuron* **38**, 773–784.

Umemori, H., Linhoff, M. W., Ornitz, D. M., and Sanes, J. R. (2004). FGF22 and its close relatives are presynaptic organizing molecules in the mammalian brain. *Cell* **118**, 257–270.

Ushkaryov, Y. A., Petrenko, A. G., Geppert, M., and Sudhof, T. C. (1992). Neurexins: Synaptic cell surface proteins related to the alpha- latrotoxin receptor and laminin. *Science* **257**, 50–56.

Varoqueaux, F., Aramuni, G., Rawson, R. L., Mohrmann, R., Missler, M., Gottmann, K., Zhang, W., Sudhof, T. C., and Brose, N. (2006). Neuroligins determine synapse maturation and function. *Neuron* **51**, 741–754.

Verhage, M., Maia, A. S., Plomp, J. J., Brussaard, A. B., Heeroma, J. H., Vermeer, H., Toonen, R. F., Hammer, R. E., van den Berg, T. K., Missler, M. *et al.* (2000). Synaptic assembly of the brain in the absence of neurotransmitter secretion. *Science* **287**, 864–869.

Wang, X., Weiner, J. A., Levi, S., Craig, A. M., Bradley, A., and Sanes, J. R. (2002). Gamma protocadherins are required for survival of spinal interneurons. *Neuron* **36**, 843–854.

Weiner, J. A., Wang, X., Tapia, J. C., and Sanes, J. R. (2005). Gamma protocadherins are required for synaptic development in the spinal cord. *Proc Natl Acad Sci USA* **102**, 8–14.

Wojtowicz, W. M., Flanagan, J. J., Millard, S. S., Zipursky, S. L., and Clemens, J. C. (2004). Alternative splicing of Drosophila Dscam generates axon guidance receptors that exhibit isoform-specific homophilic binding. *Cell* **118**, 619–633.

Wu, Q., and Maniatis, T. (1999). A striking organization of a large family of human neural cadherin-like cell adhesion genes. *Cell* **97**, 779–790.

Yamagata, M., Weiner, J., and Sanes, J. (2002). Sidekicks. Synaptic adhesion molecules that promote lamina-specific connectivity in the retina. *Cell* **110**, 649.

Yates, P. A., Roskies, A. R., McLaughlin, T., O'Leary, D. D. M. (2001). Topographic specific axon branching controlled by ephrin-As is the critical event in retinotectal map development. *Journal of Neuroscience* **21**, 8548–8563.

Suggested Readings

Brown, M., Keynes, R., Lumsden, A. (2001). "The Developing Brain." Chapter 10, pp. 261–280. Oxford Univ Press.

Flanagan, J. G. (2006). Neural map specification by gradients. *Current Opinion in Neurobiology* **16**, 59–66.

McLaughlin, T. and O'Leary, D. D. M. (2005). Molecular gradients and development of retinotopic maps. *Annual Review of Neuroscience* **28**, 327–355.

Steven J. Burden, Dennis D.M. O'Leary, and Peter Scheiffele

Programmed Cell Death and Neurotrophic Factors

One of the hallmarks of embryonic development is the enormous production of new cells and the acquisition of new cellular and morphological properties (phenotypes). Accordingly, in the past, a major focus of developmental neurobiologists has been the study of these progressive events, including proliferation and migration (Chapter 15), determination and differentiation (Chapter 16), pathway formation (Chapter 17), and synaptogenesis (Chapter 18). In this context, the concept of significant regressive events, such as programmed cell death (PCD), occurring during development, initially was considered counterintuitive (Oppenheim, 1981, 1991). Subsequently, however, it has been demonstrated that cell loss and other regressive events, including synapse elimination (Chapter 20) are the rule rather than the exception. In most developing tissues that have been examined in multicellular organisms, substantial cell loss occurs (Buss et al., 2006). As we discuss later, even a proportion of adult-generated neurons in the hippocampus and olfactory system undergo normal PCD (Kempermann, 2006). Genetic programs that result in cell death have been suggested to be a default pathway for all cells (Raff, 1992). In this view neurons would escape this default fate only by receiving the appropriate survival signals (e.g., neurotrophic factors, NTFs) that induce changes in the molecular and biochemical events required for cell survival and neuronal differentiation. If the notion that cells are "born to die" is correct, then understanding the cellular and molecular mechanisms regulating cell death programs is critical. Aberrations in the mechanisms regulating the balance between cell proliferation and cell elimination can lead to pathologies of tissue growth and differentiation (i.e., cancer) or pathological neuronal death (e.g., neurodegenerative disease). Mechanisms that regulate the balance

between cell production and death also serve to determine the ultimate number and maintenance of neurons in the nervous system. Genes regulating these mechanisms provide an underlying substrate for evolutionary changes in brain size and structure and ultimately in adaptive behavior (Buss et al., 2006; von Bartheld and Fritzsch, 2006).

Both progressive and regressive events during development are regulated by intercellular signals. Somewhat surprisingly, many of the same intercellular signals (e.g., NTFs) that contribute to the regulation of progressive events during nervous system development (cell proliferation, migration, differentiation, axonal and dendritic growth, synaptogenesis, synaptic plasticity) also contribute to the control of regressive events (cell death/survival, axon collateral and synapse elimination, dendritic pruning). Neurotrophic factors are now appreciated for their role in survival- as well as nonsurvival-related activities in both developing and mature nervous systems where they play important roles in the regulation of activity-dependent anatomical and functional plasticity of the nervous system. Accordingly, we include here a review of both survival and nonsurvival functions of NTFs.

Because neurons may die for a variety of reasons and in many different situations, it is important to describe the type of cell death observed most often in the developing nervous system. Although the loss of cells during normal development has been called by many different names (normal cell death, spontaneous cell death, naturally occurring cell death, and developmental cell death), this chapter uses the term programmed cell death (PCD). PCD is defined as the spatially and temporally reproducible and species-specific loss of large numbers of individual cells during development. Accidental, injury-induced, pathological, and disease-related forms of cell death are not included even though we recognize that the biochemical and molecular mechanisms used to kill cells in these situations may overlap with those involved in developmental PCD. This definition of PCD also makes no *a priori* assumptions about either the morphological or the biochemical pathways by which cells die or the stimuli that trigger cell death. The use of the word "programmed" refers to the reproducible, spatiotemporally specific occurrence of cell loss and is not meant to imply that the cell loss is genetically predetermined, inherited from precursor cells, or inevitable. In fact, PCD is clearly not predetermined in most cases, but instead is critically dependent on extrinsic signals arising from diverse kinds of cellular interactions. Finally, and as we discuss in more detail later, the term PCD is not synonymous with the term apoptosis, which refers only to one specific, albeit common, mode of cell death.

CELL DEATH AND THE NEUROTROPHIC HYPOTHESIS

Early embryological studies of the interactions between developing neurons and their peripheral targets laid the foundation for the discovery of cell death and NTFs (Cowan, 2001). These studies formed the conceptual framework for the neurotrophic ("nerve feeding") hypothesis (Fig. 19.1). There is general consensus that the neurotrophic hypothesis and the modern history of cell death research began together in the mid-1930s (Table 19.1 and Box 19.1).

By 1949, Viktor Hamburger and Rita Levi-Montalcini had provided compelling evidence to support the significance of normal embryonic neuronal death and postulated that target-derived signals act to regulate the number of neurons that survive embryonic development. In subsequent studies, they and their colleagues identified a specific protein, nerve growth factor (NGF), that influenced development of the same populations of neurons (sensory and sympathetic) that their earlier studies had suggested were regulated by target-derived signals (Table 19.1).

By 1960, the preparation of specific antibodies that block NGF activity allowed Stanley Cohen, Levi-Montalcini, and coinvestigators to demonstrate the almost total degeneration of sympathetic ganglia *in vivo* following the specific deprivation of NGF activity (Fig. 19.2; Table 19.1). Only with the evidence from this "immunosympathectomy" (the deletion of the sympathetic system by antibody treatment) was NGF considered to be an endogenous survival or maintenance factor for these neurons. Three decades after the original discovery that an unknown chemical substance produced by tumor cells affects the development of sensory and sympathetic neurons, it was finally recognized that NGF was the hypothetical target-derived trophic signal first postulated by Hamburger and Levi-Montalcini in 1949 to be involved in regulating the number of surviving neurons in these populations during normal development (Oppenheim, 1996). Historically then, the appreciation of the role of PCD in the developing nervous system and the discovery of the first target-derived NTF (NGF) were inextricably linked.

THE ORIGINS OF PROGRAMMED CELL DEATH AND ITS WIDESPREAD OCCURRENCE IN THE DEVELOPING NERVOUS SYSTEM

PCD has been identified in several species of unicellular eukaryotes, including yeast, as well as in prokaryotes such as bacteria, one of the oldest forms of life on

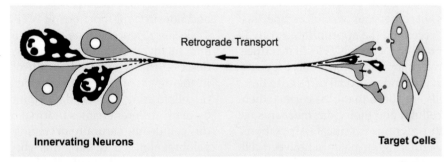

FIGURE 19.1 The neurotrophic hypothesis postulates that developing neurons survive (green neurons) only when they successfully compete for target-derived trophic molecules (blue) that are internalized at the nerve terminal and are transported retrogradely along the axon (arrow) to the cell body. Neurons that fail to obtain a sufficient flow of trophic molecules from the target die by programmed cell death (black neurons with condensed chromatin). Modified from Barde, 1989.

TABLE 19.1 Major Events in the Discovery and Characterization of Neurotrophins*

1934	*Hamburger* discovered that removal of the limb bud in the chick embryo resulted in reduced numbers of sensory and motor neurons in the spinal cord and suggested that targets in the limb are the source of signals that control neuronal development and that travel retrogradely in axons to their respective centers.
1942	*Levi-Montalcini* and *Levi* showed that early limb bud removal reduced the number of sensory and motor neurons and proposed that the hypothetical target-derived signals act to maintain the survival of differentiating neurons.
1953	*Levi-Montalcini* and *Hamburger* carried out transplantation experiments with a tumor, sarcoma 180, and discovered that sympathetic and sensory ganglia remote from the tumor and not connected with it by nerve fibers were also enlarged greatly. This suggested involvement of a diffusible factor.
1954	*Levi-Montalcini*, *Meyer*, and *Hamburger* developed an *in vitro* assay that used explanted sympathetic ganglia. Ganglia cocultured with fragments of tumor cells lacking any physical contacts between the two exhibited massive outgrowth of nerve fibers (Fig. 19.2). The extent and density of outgrowth provided a rapid quantitative bioassay for subsequent attempts to isolate and purify the tumor factor.
1960	*Cohen* discovered that the salivary gland of male mice was an extremely rich source of the same growth-promoting activity found in the sarcoma 180. When an antiserum to the mouse factor was injected into newborn mice, all sympathetic neurons were lost.
1969	*Bocchini* and *P. Angeletti* described a method for the purification of biologically active NGF from male mouse submaxillary glands. This activity is the β subunit of NGF, also known as 2.5S NGF. It has been estimated that to purify NGF from relevant target organs would have required a purification factor of 100 million, whereas a purification factor of only 100–200 was sufficient to purify NGF from the mouse salivary gland. Salivary glands from female mice or other mammals do not contain this extremely high concentration of NGF.
1971	*P. Angeletti* and *Bradshaw* identified the amino acid sequence of 2.5S NGF purified from the mouse submaxillary gland.
1982	*Barde* and colleagues isolated a novel neurotrophic factor from the mammalian brain (brain-derived neurotrophic factor) (BDNF). Unlike the original purification of NGF from a fortuitous rich source—the salivary gland—this factor was isolated from many brains by an amazing purification factor of several millionfold. BDNF was identified and isolated using an *in vitro* survival assay with sensory neurons.
1983	*Korsching* and *Thoenen* developed a sensitive two-site immunoassay, allowing for the first time the detection of NGF in target organs. With this method it was possible to demonstrate a strong correlation between the density of sympathetic innervation and target levels of NGF, a finding consistent with the neurotrophic theory.
1986	*Chao* and his colleagues identified the first neurotrophin receptor, at the time believed to be a receptor specific for NGF. The Nobel Prize in medicine was awarded to *Levi-Montalcini* and *Cohen* for the discovery of NGF and EGF.
1989	The sequencing and molecular cloning of BDNF by *Barde* and colleagues revealed a high degree of sequence homology between the new factor and NGF. This finding enabled the rapid molecular cloning and sequencing of other related *neurotrophins* containing the conserved regions without prior protein purification. These and related studies on the characterization of neurotrophins were carried out by several prominent researchers trained in Eric Shooter's lab or in Hans Thoenen's lab during the 1980s.
1990/1991	*Several groups* discovered additional members of the "neurotrophin" family by molecular cloning, including NT-3 (in several species), NT-4 (in Xenopus), and NT-5 (in mammals). Research by *Kaplan*, *Parada*, and colleagues identified the trk receptor (an "orphan" receptor) as a member of a family of tyrosine kinase receptors that are specific for neurotrophins, rapidly leading to the identification of the three members trkA (for NGF), trkB (for BDNF and NT-4), and trkC (for NT-3).
1993	*Poo* and his colleagues discovered that the neurotrophins BDNF and NT-3 have rapid and substantial effects on the strength of synapses, showing for the first time that neurotrophins have important effects in regulating synaptic plasticity.
1996	Work by *Barde* and his colleagues revealed that neurotrophins can induce cell death (apoptosis) by binding to the "common" neurotrophin receptor, p75NTR.
2001	*Hempstead* and her colleagues discovered that proneurotrophins, the larger precursor forms believed to have no significant biological function, have important physiological effects.

*For references cited in this table, see Barde, 1989; Cowan, 2001; Huang and Reichardt, 2001.

earth (Ameisen, 2004). One likely explanation for the evolutionary origins of PCD in eukaryotes was the appropriation of cell death associated genes from bacterial mitochondria by a process of endosymbiosis. After being incorporated into the eukaryotic genome this core cell death machinery expanded in complexity over time to include the diversity found in extant multicellular organisms (Buss *et al.*, 2006). Once the pro- and anti-apoptotic machinery was in place in the genome of multicellular animals, it could then be coopted to mediate the many diverse roles now served by PCD, including the PCD of developing neurons.

BOX 19.1

LEVI-MONTALCINI, HAMBURGER, AND THE NEUROTROPHIC HYPOTHESIS[1]

In the early 1930s Rita Levi-Montalcini was a research associate working in the laboratory of the eminent neuroanatomist Giuseppe Levi in Turin, Italy. Viktor Hamburger sent a personal copy of his 1934 paper on wing removal to his acquaintance Levi, who shared it with Levi-Montalcini. She was inspired by the results but was dubious about the validity of the recruitment hypothesis. Together with Levi, she reexamined the effects of limb bud removal on spinal ganglia in the chick embryo. By examining embryos at regular intervals after limb removal, she showed that neuron numbers develop normally up to a certain stage, after which a gradual neuron loss ensues—a finding incompatible with Hamburger's recruitment hypothesis, which predicted a deficit of neurons immediately after limb removal. Levi-Montalcini and Levi postulated that after the loss of peripheral targets, neurons develop normally but then later regress or degenerate. In other words, peripheral targets were apparently regulating the *survival* or *maintenance*, rather than the recruitment, proliferation, or differentiation of innervating populations of neurons. That targets might act to keep neurons alive was a novel perspective that ushered in a fundamentally new line of investigation of cell death and neurotrophic factors.

Although this important paper by Levi-Montalcini and Levi was published in 1942, Hamburger first saw it after World War II. Intrigued by the novel interpretation, in 1946 he invited Levi-Montalcini to join him in St. Louis to resolve their differences. She accepted and by 1947 they had begun experiments that resulted in their first joint publication in 1949. This paper, which soon became a landmark in the history of this field, fully vindicated the original conclusion of Levi-Montalcini and Levi that regression and cell death, not the failure of recruitment, correctly explained the results of the limb removal experiments. More importantly, this same paper showed that many neurons normally die during development and that a major effect of altering target size is the perturbation of this normal regressive process. They proposed that both normal cell death and death following limb removal result from a lack of target-derived substances necessary for growth and survival. A year later, they used the term trophic or neurotrophic to designate these hypothetical signals or substances. In this context, "trophic" and "neurotrophic" refer to signals that mediate long-term dependencies between neurons and the cells they innervate (e.g., survival) and are distinct from "tropic" and "neurotropic," which refer to diffusible chemoattractant signals derived from target cells that guide or orient the migration of other cells or cell processes (e.g., axons) toward the target. By 1950, therefore, the basic tenets of what was subsequently called the *trophic theory* were established: developing neurons are overproduced and compete for limiting amounts of target-derived molecules that provide retrograde signal(s) for their survival. Despite this early conceptual breakthrough it would be more than 30 years before convincing evidence in support of all the tenets of the trophic theory was available.

Historically, the prospect of large losses of cells during normal development was conceptually unpalatable to most biologists. Early in the 20th century, German anatomists M. Ernst and A. Glücksmann and the French anatomist R. Collin described widespread cell death in the nervous system, but their observations either went unrecognized or were ignored by other investigators of this period. For this reason, the description of normal neuronal death and its proposed role in neuron-target interactions by Hamburger and Levi-Montalcini marked a watershed in the history of this field. Although it would be many more years before cell death was fully accepted as a fundamental process in nervous system development, after 1949 it was no longer possible to exclude it completely from consideration in discussions of the development of neuron-target interactions. Additionally, for the first time it also became possible to conceive of chemical substances (trophic factors) as the source of signals that regulate developmental events in the nervous system such as neuronal survival. Therefore, although Hamburger and Levi-Montalcini were not the first to observe normal cell death in the nervous system, they were the first to draw attention to its significance; they were the first to provide a plausible explanation for its occurrence; and they were the first to postulate a possible mechanism for its regulation. For these reasons, they deserve equal credit along with Collin, Ernst, and Glücksmann as pioneers in the history of this field.

[1]For additional details and references see W. M. Cowan (2001). *Annu. Rev. Neurosci.* **24**, 551–600.

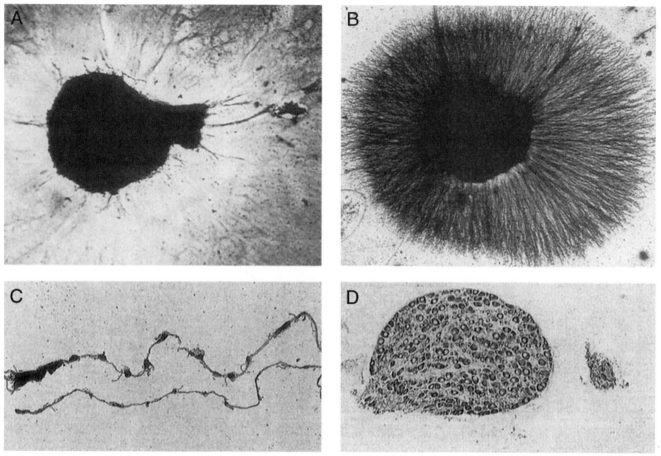

FIGURE 19.2 Biological activity of nerve growth factor (NGF). Explanted sensory and sympathetic ganglia or dissociated neurons were used in bioassays detecting neurotrophic activity. (A) Control ganglion 24 h in culture without NGF and (B) experimental ganglion 24 h after NGF treatment. Treatment with NGF (100 ng/ml) causes the formation of a "halo" of axonal growth from sensory neurons in the ganglion. Why the factor was named NGF is obvious. Reprinted with permission from Levi-Montalcini. (C and D) Experimental immunosympathectomy. Antibodies that selectively block NGF activity were administered to newborn mice to deprive the developing animals of endogenous factor. Sympathetic ganglia were examined several weeks after the treatment. Note the marked atrophy of the entire sympathetic chain ganglia (lower in C) and the almost complete loss of sympathetic neurons in a histological section of a single ganglion (right side in D) after antibody treatment. Controls are in the upper part of C and on the left in D. Photographs kindly provided by Viktor Hamburger and Rita Levi-Montalcini.

Although a systematic taxonomic survey of all representative species with a nervous system has not been done, nonetheless, the available evidence is consistent with the idea that some PCD of developing neurons occurs in all such organisms (Buss *et al.*, 2006). For many invertebrate and vertebrate species, the PCD of neurons involves virtually all regions and cell types in the central and peripheral nervous system. Motoneurons, sensory neurons, autonomic neurons, and both long projection and local circuit interneurons in the brain and spinal cord all undergo restricted periods of PCD (Buss *et al.*, 2006). Although the magnitude of neuronal cell death varies from population to population, as many as one-half or more of all cells in a population will die during development (Fig. 19.3). In some special cases, such as the loss of transient neuronal structures during insect and amphibian metamorpho-

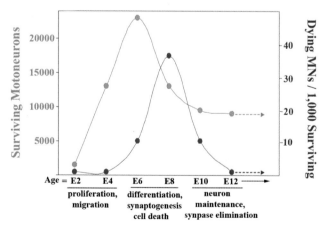

FIGURE 19.3 Changes in the number of developing lumbar spinal motoneurons in the chick embryo (in red) compared to the number of dying motoneurons (in blue) during the first half of the 21 day incubation period. Approximately 50–60% of these motoneurons undergo PCD between Embryonic day (E)6 and E12.

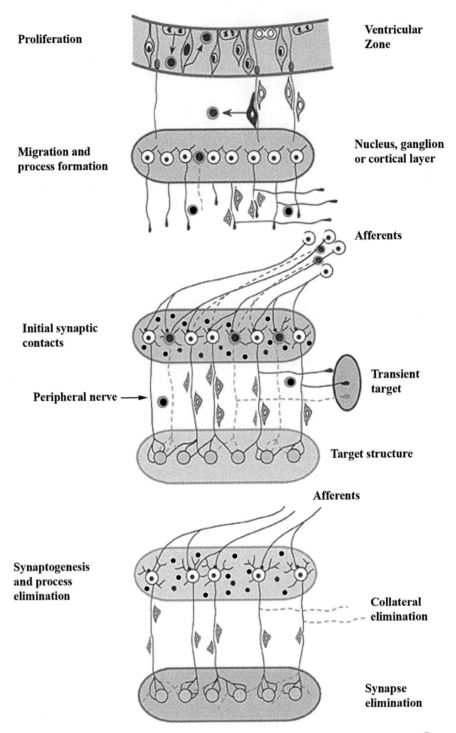

FIGURE 19.4 Schematic illustration of some key steps in neuronal development. Neurons undergoing PCD (⬤) are observed during neurogenesis in the ventricular zone, during migration and while establishing synaptic contacts. Schwann cells in developing nerves also undergo PCD. ⬦ Represents peripheral glial (Schwann) cells; ● represents CNS glia (astrocytes, oligodendrocytes); ⊙ represents surviving, differentiating neurons (e.g., motoneurons whose targets are skeletal muscle and other neurons in the CNS, with neuronal targets).

sis (e.g., the Rohon-Beard sensory neurons in fish and frogs), most or all cells die. Cell death in the nervous system thus clearly occurs on a very large scale, indicating that it plays a fundamental and essential role in normal development.

Studies of cell death in the nervous system have focused on the loss of developing postmitotic neurons as they form synaptic connections with targets and afferents. However, extensive PCD also occurs during neurulation and in mitotically active progenitor cells, as well as in postmitotic but undifferentiated neurons in the early neural tube (Fig. 19.4). Accordingly, PCD in the nervous system is not limited to any particular stage of development. The loss of cells at different

developmental stages probably serves distinct functions and may be mediated by different mechanisms (Oppenheim *et al.*, 2001). PCD also occurs in central and peripheral glial cells. For example, many myelin-forming oligodendrocytes in the optic nerve and Schwann cells in peripheral nerves die by PCD. Their loss is thought to reflect a competition for axon-derived trophic signals, the end result of which is the survival of an appropriate number of glial cells for optimum myelination of the available axons (Fig. 19.5). Because glial PCD has been studied much less extensively, it is not known whether the death of other nonneuronal cells in the nervous system (e.g., astrocytes or microglia) is as common as the death of neurons.

Because PCD is the normal differentiated or terminal fate of many developing cells, commitment to this fate occurs in much the same way as the phenotypic fate of cells destined to survive in the embryo. Developmental biologists have identified two major ways in which the commitment of a cell to a particular differentiated phenotype occurs (Gilbert, 2006). The first mechanism, intrinsic or *autonomous specification*, involves the segregation of critical cytoplasmic molecules by asymmetric cell division (Fig. 19.6); cell death thus is programmed into the lineages that generate somatic cells. The second mechanism of commitment involves extrinsic signals from other cells and is called *conditional specification*. Initially, the cells have the potential to follow more than one path of differentiation. As development proceeds, however, signals from

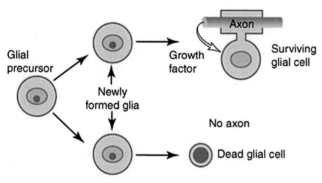

FIGURE 19.5 The PCD of glial cells is regulated by axonally derived signals. Glial cells that fail to compete successfully for these signals undergo PCD.

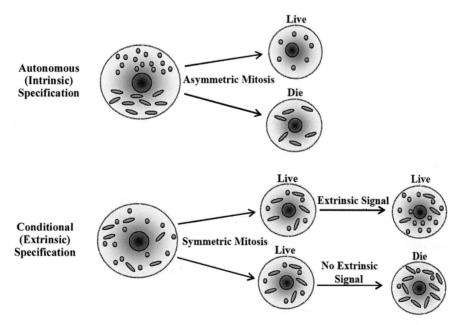

FIGURE 19.6 Schematic illustration of the phenotypic determination of the progeny of dividing precursor cells by autonomous vs. conditional specification. The red and green symbols represent hypothetical cytoplasmic molecules (e.g., mRNAs) whose relative expression in this example specifies survival or death.

other cells act to gradually limit and specify cell fate. Although all organisms use a combination of autonomous and conditional developmental strategies, as a general rule invertebrates are more likely to utilize autonomous specification, whereas most vertebrates exhibit conditional specification. Despite this distinction, as described later, many of the genetic and molecular pathways for PCD are remarkably similar in invertebrates and vertebrates.

Summary

The PCD of developing neurons appears to occur in virtually all vertebrate and invertebrate species. More generally, PCD also occurs in many different cells and tissues of unicellular and multicellular organisms and thus may have arisen very early during evolution. With few exceptions, PCD occurs in virtually all types of developing neurons and can take place at stages of development from the time of proliferation until the establishment of synaptic connections. Developing glial cells also exhibit PCD. Cell death is the terminal phenotypic fate of subpopulations of developing neuronal and glial cells and, like other cell fate decisions, is controlled by both intrinsic and extrinsic signals.

FUNCTIONS OF NEURONAL PROGRAMMED CELL DEATH

Why does PCD occur? This is a reasonable question to ask because the loss of large numbers of developing neurons is counterintuitive. Why should embryos invest precious resources in generating cells and tissues only to later cast many of these aside? A satisfactory answer to this apparent paradox requires an evolutionary perspective that addresses two central aspects of the problem. First, how did the biochemical machinery (the cell death program) needed to actively kill cells arise? Second, why, in many developing tissues, are more cells generated than are apparently needed? Because PCD acts to delete these excess cells, an understanding of the overproduction is critical if one is ever to understand cell death from an evolutionary perspective. The first question was addressed in the previous section; however, understanding the evolution of the biochemical cell death program does not help answer the second question of why there is often a massive overproduction of neurons during development that are later eliminated by PCD. Two explanations have been offered (Oppenheim, 1991). First, each case of PCD may have evolved to serve a distinct biological function. For example, according to this view,

in the case of spinal motoneurons, natural selection is thought to be directly responsible for both the overproduction and the subsequent death of neurons as a means for creating an optimal (adaptive) level of functional muscle innervation (e.g., systems-matching).

The second view is that the overproduction of neurons (or other cells) is an inevitable outcome of the imprecise kinetics of proliferation of precursor cells. Once excess cells are available and the pro- and anti-apoptotic machinery is in place, however, natural selection then acts via regulated cell survival and death to mediate a variety of different adaptive needs. For example, following the loss or absence of essential survival signals, the death of excess cells could be accomplished easily by coopting the cell death machinery that evolved to kill cells in early eukaryotes. In reality, both of the proposed mechanisms may occur. For example, it seems highly likely that the creation of transient structures that function at one stage of development but later regress and are discarded (e.g., the tail of tadpoles, larval muscles of insects, and transient neuronal structures such as sensory Rohon-Beard cells in frogs and fish) reflects the direct selection of PCD as a means for mediating adaptive life history transitions (e.g., metamorphosis). In contrast, the presence of increased numbers of neurons in limb versus nonlimb spinal segments of vertebrates may result from the unselected outcome of an overproduction (proliferation) of neurons at all spinal levels, followed by increased survival in limb compared to nonlimb regions.

Many of the circumstances in which the PCD of neurons occurs are thought to mediate distinct adaptive functions, as summarized in Table 19.2. In many of these examples, the production of excess neurons provides a substrate on which differential survival or PCD can then act to meet a variety of adaptive needs. Although the biological functions attributed to PCD in these situations are quite plausible, some of them have not been directly demonstrated experimentally to serve a specific adaptive role. As new genes are identified that regulate PCD, and as we gain a better understanding of how cellular and molecular signals control cell death and survival, there will be increased opportunities for preventing PCD in select neuronal populations *in vivo* and directly assessing whether its occurrence is, in fact, adaptive (Buss *et al.*, 2006).

Summary

The biochemical and molecular pathways that regulate cell death and survival arose early in the evolution of animal life. Once this cellular capacity arose, however, it is likely that it was coopted to serve a variety of biological functions. In the nervous system,

TABLE 19.2   Some Possible Functions of Developmental PCD in the Nervous System*

1. Differential removal of cells in males and females (sexually dimorphic spinal motor nucleus in many mammals).

2. Deletion of some of the progeny of a specific sublineage that are not needed (loss of specific progeny of the AB blastomere that is involved in generating ring ganglia in *C. elegans*).

3. Negative selection of cells of an inappropriate phenotype (ligand/receptor-induced cell death in the early chick embryo retina?).

4. Pattern formation and morphogenesis (neurulation/neural tube closure; differential thickness of cortical layers).

5. Deletion of cells that act as transient targets or that provide transient guidance cues for axon projections (death of pioneer neurons/glia in insects).

6. Removal of cells and tissues that serve a transient physiological or behavioral function (loss of Roh-Beard sensory neurons during metamorphosis in frogs).

7. "Systems"-matching by creating optimal quantitative innervation between interconnected groups of neurons and between neurons and their targets (see text and Fig. 19.4).

8. Systems-matching between neurons and their glial partners by regulated glial PCD (Schwann cells and peripheral axons; see Figs. 19.5 and 19.15).

9. Error correction by the removal of ectopically positioned neurons or of neurons with misguided axons or inappropriate synaptic connections (loss of ipsilaterally projecting retinal ganglion cells).

10. Removal of damaged or harmful cells (death of cells with DNA damage).

11. Regulation of the size of mitotically active neural progenitor populations (see text and Fig. 19.4).

12. The production of excess neurons may serve as an ontogenetic buffer for accommodating mutations that require changes in neuronal numbers in order to be evolutionary adaptive (evolutionary increases in limb size may require increased sensory and motoneuron survival).

13. Regulated survival of subpopulations of adult-generated neurons as a means of experience-dependent plasticity (see text and Fig. 19.18).

*For references see Buss *et al.*, 2006; Oppenheim *et al.*, 2001.

some major functions include establishing optimal levels of connectivity between neuronal populations, eliminating aberrant cells or connections, regulating the size of progenitor populations, and serving transient functional or other needs of immature animals.

MODES OF CELL DEATH IN DEVELOPING NEURONS

The specific morphological appearance exhibited by degenerating neurons can provide insight into the cellular and molecular mechanisms by which the cells are destroyed. Historically, pathologists were the first to be interested in this issue, and they focused on distinguishing different kinds of cell and tissue degeneration following disease, injury, and trauma (Clarke and Clarke, 1996). Over 125 years ago, the term necrosis was coined to describe what today comprises the major form of accidental or pathological degeneration. The pathological necrotic death of neurons following injury usually involves the degeneration of groups of contiguous cells in a region that initiates an inflammatory response that can be discerned easily in tissue sections (Fig. 19.7). At about the same time that necrosis was first described, however, another form of cell degeneration, called spontaneous cell death, was observed

in normal adult mammals. Spontaneous cell death (now called PCD) was thought to provide a means for counterbalancing mitosis in adult tissues in which the turnover of cells normally occurs. PCD typically involves the sporadic loss of individual cells in a population that often degenerate by a different mode from necrosis, one that does not involve inflammation. Early steps in the spontaneous PCD cascade leading up to when histological signs of frank degeneration first occur may take many hours or days. Once that point is reached, however, the degenerative process is rapid, with individual cells dying and being removed in minutes or a few hours. The loss of thousands of neurons over several days is the consequence of many rapid individual cell deaths that at any moment in time represent only a small minority (~1%) of all the cells in the population. For this reason, historically, the occurrence and magnitude of even massive PCD can (and often did) go unnoticed.

Two different strategies can be used to quantify neuron death: One can determine the total number of neurons at different stages and quantify the decrease (Fig. 19.3), or one can determine an increase in the number of dying (pyknotic) cells during the period of PCD and determine the timing and (with less precision) the extent of cell death (Fig. 19.7). The counting of neurons originally was done by counting profiles in thin sections and application of correction factors to

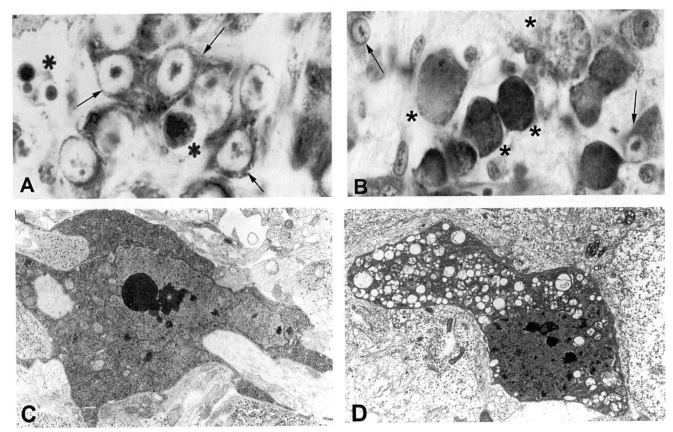

FIGURE 19.7 Spinal motoneurons in the chick embryo. (A) Ventral horn from a control embryo. Note that only two cells (asterisks) are undergoing apoptotic PCD; the others appear normal (arrows). (B) Ventral horn from an embryo following treatment with an excitotoxin. Most neurons are undergoing a necrotic type of cell death (asterisks) whereas some appear normal (arrows). Apoptotic (C) and necrotic motoneurons (D) in the chick embryo spinal cord as seen with an electron microscope.

account for multiple profiles per particle (Clarke and Oppenheim, 1995); such counting is now increasingly replaced by stereological methods that count particles in thicker sections, a method that is less affected by confounding changes in particle size but that still presents challenges due to other sources of potential errors if not executed properly (Schmitz and Hof, 2005).

Until quite recently, cell degeneration in adult and developing tissues generally has been dichotomized into death by either apoptosis or necrosis, a distinction based initially on morphological differences and later on other apparent differences between the two (Fig. 19.8). *Apoptosis* is a Greek word indicating the seasonal piecemeal dropping of leaves from a tree and originally was coined to describe all forms of PCD that share certain morphological characteristics. Cells dying by apoptosis shrink in size and the nuclear chromatin condenses and becomes pyknotic, whereas the cell membrane and cytoplasmic organelles tend to remain relatively intact. Eventually the cytoplasm and nucleus break up into membrane-bound apoptotic bodies that

are phagocytized either by professional phagocytes (macrophages) or by healthy adjacent cells (e.g., glia that serve as transient phagocytes). In contrast, necrosis involves an initial swelling of the cell, only modest condensation of chromatin, cytoplasmic vacuolization, breakdown of organelles, and rupture of the cell membrane allowing the release of cellular contents (causing inflammation), followed by shrinkage and loss of nuclear chromatin. Because necrotic cell death elicits an inflammatory response in adjacent tissue, macrophages derived from the immune system attack and phagocytize cellular debris. In contrast, cell death by apoptosis usually involves individual cells that are engulfed and phagocytized before they can release their cellular contents or induce an inflammatory response in adjacent tissue. Phagocytes recognize dying cells by their expression of death-related cell surface signals.

Another feature that has been used to distinguish between apoptotic and necrotic cell death is the occurrence during apoptosis of a specific form of chromosomal DNA fragmentation and degradation that is

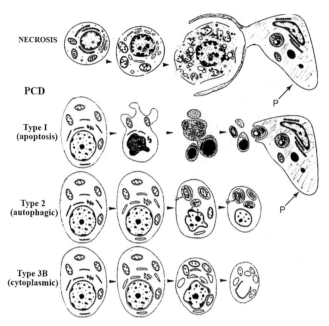

FIGURE 19.8 Schematic representation of cellular changes during necrotic cell death and during three of the most common types of PCD observed at the ultrastructural level. Only type 1 PCD meets most of the criteria for defining apoptosis. The cells on the right marked P represent phagocytic cells engulfing necrotic cell corpses and apoptotic bodies. Phagocytosis also occurs in the other types of PCD but is not shown.

mediated by DNA-specific proteases. DNA digestion occurs at internucleosomal sites, producing small, double-stranded fragments of DNA that migrate in a ladder pattern in multiples of 180–200 bp after electrophoresis in agarose gels. This form of DNA fragmentation can also be visualized in tissue sections by a technique that labels the double-stranded DNA breaks associated with apoptosis.

It is widely believed that apoptosis and necrosis reflect mechanistically distinct cell death pathways that are triggered by different stimuli. However, our basic understanding of these pathways is still limited, and caution should be exercised in drawing too fine of a distinction between them. For example, a variety of toxic and traumatic stimuli, previously thought to induce necrosis, can also induce morphological signs of apoptosis and may be associated with changes in PCD- or apoptosis-associated genes (Bredesen et al., 2006). Necrotic cell death may also occur by a distinct molecular pathway that involves endogenous release of calcium. The PCD of developing vertebrate neurons provides a striking example of the problems encountered in attempting to rigidly classify the pathway of degeneration as being either necrotic or apoptotic (Clarke, 1990; Bredesen et al., 2006). Developing neurons may adopt one of at least three different

morphological modes during PCD: (1) apoptotic, (2) autophagic, and (3) cytoplasmic (Fig. 19.8). Only the first fits the classic morphological definition of apoptosis. Inhibition of the apoptotic mode of PCD often results in death of cells by alternative pathways (e.g., autophagic). Despite the occurrence of these different morphological types of death, it is clear that they all reflect normal PCD, as they involve the stereotypic loss of individual cells at specific times during development in the absence of injury or inflammation. Within the context of the developing nervous system, the dichotomy of apoptosis versus necrosis is an oversimplification that should be abandoned. Instead, developing cells undergoing normal PCD as well as neurons dying following injury should be categorized by operational definitions that use morphological, genetic, and biochemical criteria (Bredesen et al., 2006; Blomgren et al., 2007). As described in the following section, one of the major success stories begun in the early 1990s has been the remarkable progress made in understanding the biochemical pathways of PCD and identifying the specific genes involved.

Summary

Historically, degenerating cells have been categorized into two classes: death by apoptosis or death by necrosis. Although the situation is more complex than is reflected in this simple dichotomy, apoptosis in general is more characteristic of PCD, whereas necrosis is more characteristic of cells that die following injury or trauma. A variety of morphological, biochemical, and molecular features have been used to distinguish between these two types of cell death. However, the occurrence of certain features of apoptosis in neurons following injury and the occurrence of several types of PCD other than apoptosis indicates that a more accurate means of identifying and defining distinct forms of cell death should now be employed.

THE MODE OF NEURONAL CELL DEATH REFLECTS THE ACTIVATION OF DISTINCT BIOCHEMICAL AND MOLECULAR MECHANISMS

As described previously, the normal death of cells in the developing nervous system has long been thought to be regulated by competition for NTFs (the Neurotrophic Hypothesis). Until quite recently, investigators agreed that the doomed neurons, lacking sufficient amounts of a NTF to sustain normal metabolic events, passively degenerated by a process analogous to starvation. However, PCD of some nonneuronal cells was known previously to be a metabolically active, ATP-dependent process. For example, it had been shown that RNA or protein synthesis inhibitors prevent the programmed death of muscle cells in metamorphic insects and amphibians and also block the hormone-induced death of thymocytes in the mammalian immune system. Additionally, in the early 1980s, genetic mutations in the nematode worm *Caenorhabditis elegans* that prevent PCD had been described, thereby providing further evidence that neuronal death is a genetically regulated, metabolically active process. Beginning in the late 1980s, these various lines of evidence forced a reappraisal of the view that cell death in the nervous system is a passive process and led to the demonstration that for many types of neurons, PCD is regulated by the interaction of specific genetic programs that either inhibit or induce degeneration. Neurotrophic survival molecules are thought to act as extracellular signals that when present in sufficient amounts result in intracellular signaling that either inhibits the expression or activity of pro-apoptotic genes or induces the expression or activity of anti-apoptotic gene products. Considerable progress has been made since the early 1990s in identifying cell death-associated genes and their pathways of action. Due to historical precedent, these have been classified as pro- and anti-apoptotic genes. Although we retain this terminology, in some instances the genes involved may actually induce both apoptotic and non-apoptotic modes of degeneration (see the previous section).

Although early genetic studies of cell death in C. *elegans* demonstrated that PCD is regulated by specific genes, the first indication that the PCD of developing vertebrate neurons may also be controlled by similar so-called "killer" or "death" genes appeared in 1988 (Johnson and Deckwerth, 1993). Cultured neonatal rat sympathetic neurons, which normally require NGF for survival, remain viable following NGF removal if mRNA or protein synthesis inhibitors were added to the cultures. Subsequently, other types of developing neurons also were shown to be rescued by these drugs both *in vitro* and *in vivo* following trophic factor deprivation. These findings were interpreted as evidence that one or more steps in the neuronal PCD pathway requires the *de novo* transcription of genes and the regulated expression of proteins that actively destroy the cell. Because the genetics, morphology, and cell lineages in C. *elegans* have been so well defined, this organism provides a particularly informative and powerful model for analyzing the molecular genetics of PCD (Horvitz, 2003). Of the approximately 1000 somatic cells generated (of which 302 are neurons and

56 are glial cells), 131 undergo embryonic PCD and many of these are neuron precursors. In each individual, the same cells die at specific times in development and these corpses are then engulfed and degraded by neighboring cells. Despite the enormous evolutionary gap that separates the appearance of worms, flies, and vertebrates, significant homology exists in the structure and function of specific cell death pathways between these taxonomic groups.

As summarized in Figure 19.9, there is a sequential cascade of genetically regulated steps involved in PCD. Upstream of the actual execution of the death process, transcription factors specify certain cell types for death while sparing others. In C. *elegans*, manifestation of the death fate requires expression of two pro-apoptotic genes, ced–3 and ced–4 (ced, cell death), which together mediate the actual breakdown of cellular constituents. Prevention of cell death induced by ced–3 and ced–4 can occur by activation of the anti-apoptotic gene ced–9. Separate genes, such as nuclease–1 (nuc-1), are required for the degradation of DNA in dying cells, and several additional genes are involved in the recognition and engulfment of dead cellular corpses by phagocytes. Evidence for the involvement of this genetic cascade in cell death and survival in C. *elegans* comes from several different approaches, most notably from genetic studies of loss-of-function mutants. For example, in the absence of ced–9, many cells that normally survive die, whereas in the absence of ced–3 or ced–4, all PCD is prevented.

Identification of the DNA sequences of the major cell death genes ced–3, ced–4, and ced–9 in the worm in the 1980s and early 1990s resulted in the subsequent discovery of vertebrate and insect homologues that serve similar functions (Kornbluth and White, 2005; Johnson and Deckwerth, 1993) (Fig. 19.9). In mammals, ced–3 is represented by a large family of related cysteine proteases called caspases, whereas ced–4 is represented by a single vertebrate homologue, Apaf–1 (apoptosis protease activating factor), that is required for caspase activation. The survival-promoting function of ced–9 originally was represented by a single vertebrate homologue, bcl–2 (B-cell lymphoma-related gene), but subsequently other bcl–2 family members have been identified with similar anti-apoptotic functions (e.g., bcl–x) as well as other bcl–2 family members that are pro-apoptotic (e.g., bax, bak, bim). Loss-of-function mutations in mice by targeted gene deletion (gene knockout) of many of these vertebrate homologues have confirmed their role as important regulators of PCD.

This increased genetic complexity in vertebrates in general and in neurons in particular probably reflects a need for multiple levels of control of death and sur-

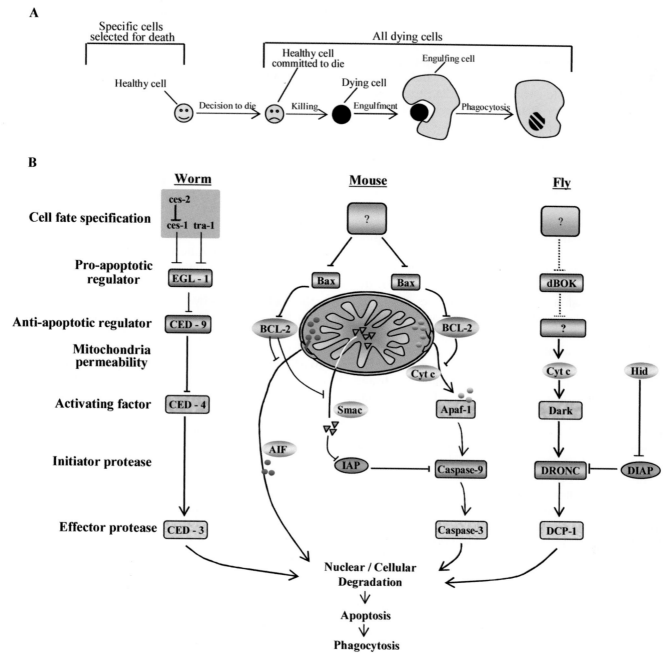

FIGURE 19.9 (A) Schematic representation of the major steps in the developmental PCD pathway of neurons in the nematode worm *C. elegans*. (B) The evolutionary conserved core cell death pathway in three diverse species. The role of structurally or functionally homologous genes in the three species are indicated by color coding. Transcription factors such as ces-1, ces-2, and tra-1 that specify cell fate in worms have not yet been identified in developing fly or vertebrate neurons. In the worm, these act upstream of Egl-1 but only on specific, not all, cells that die. Ces-1 directly represses ces-2 and thereby blocks ces-1 from preventing Egl-1 activation. Tra-1 can also prevent Egl-1 activation. Egl-1 is pro-apoptotic by its role in preventing ced-9 from preventing ced-4 from inducing the killing activity of ced-3. In the mouse, once neurons are selected to undergo PCD (e.g., by failure to obtain NTF support), the pro-apoptotic gene Bax interacts with and inhibits the anti-apoptotic gene Bcl–2 in mitochondria. This results in the release from mitochondria of cytochrome-c, which forms a complex with Apaf-1 and caspase-9 that in turn activates downstream caspases such as caspase-3 that ultimately directly or indirectly degrade both the nucleus and diverse cytoplasmic targets. These degradative changes are what define apoptosis and that result in eventual engulfment and phagocytosis of the apoptotic cell. In some situations two additional molecules released from mitochondria (along with cytochrome-c) are the pro-apoptotic proteins AIF, that can degrade the nucleus independent of caspases, and Smac, which can inhibit IAP and promote the apoptotic pathway via caspase-9 and caspase-3. Although not shown here, in some situations developing neurons undergoing PCD activate cell cycle proteins that also serve a signaling function required for apoptosis (Copani *et al.*, 2001). Part A modified and redrawn from Horvitz (2003). Part B modified and redrawn from Sanes *et al.* (2006). For additional details including definitions of abbreviations see Bredesen *et al.* (2006), Horvitz (2003), Kornbluth and White (2005), and Johnson and Deckworth (1993).

vival at the cellular level. It also provides for diverse pathways of PCD in different cells and tissues, at different stages of development, and in response to different death and survival signals. In some neurons, NTFs and other extracellular signals may induce death rather than promote survival by binding to so-called death receptors. For example, immature cells in the retina may be induced to die following activation of the common neurotrophin receptor (p75NTR) by NGF, and PCD in this situation involves intracellular pathways partly distinct from those used when neurons die following the loss of neurotrophic support. Additionally, although apparently not a common occurrence, neighboring nonneuronal cells (e.g., microglia) may actively participate in the PCD of developing neurons.

A central integrator of cell death in vertebrates is the mitochondrion (Fig. 19.9). By monitoring the relative expression and activation of pro- and anti-apoptotic family members (e.g., bcl–2, bax), the mitochondrion modulates cell survival by the regulated release of molecules (e.g., cytochrome c) that activates the downstream cell death machinery (e.g., caspases). Although the evidence for the role of mitochondria in regulating PCD in flies and worms is less clear than for vertebrates, it seems likely that mitochondria or other intracellular organelles (e.g., the endoplasmic reticulum) are involved as key control points for death or survival. Most developing neurons appear to share a core cell death program, but there is also increasing evidence that alternative, mitochondrial-independent and caspase-independent pathways may exist for mediating the normal PCD of some neurons. In addition to the increased complexity in vertebrates of the core pro- and anti-apoptotic machinery, a variety of additional biochemical and molecular mechanisms have been described that are involved in pathological, injury-induced forms of PCD/necrosis in the vertebrate nervous system (Bredesen *et al.*, 2006; Blomgren *et al.*, 2007; Vila and Przedborski, 2003).

Summary

PCD is a metabolically active process that involves specific genetic pathways necessary for the cascade of events leading to degeneration. Several genes in the PCD pathway were first identified in *C. elegans* and homologues have been found in insects and vertebrates. The mitochondrion is a central integrator of cell death signaling. Although most developing neurons appear to share a common core PCD pathway, morphological and biochemical evidence also indicates the presence of alternatives to the core PCD machinery. One goal of the increased evolutionary

complexity in the regulation of neuronal PCD is the need for additional checks and balances before a cell passes the irreversible point of commitment to die. Because virtually all neurons are postmitotic and neuron numbers finite, multiple, relatively fail-safe, mechanisms are required so that accidental death occurs only as a last resort.

NERVE GROWTH FACTOR: THE PROTOTYPE TARGET-DERIVED NEURONAL SURVIVAL FACTOR

Following the discovery of NGF by Levi-Montalcini, Cohen and Hamburger in the 1950s and 1960s, and its possible role in the survival of developing sympathetic neurons (Table 19.1, Box 19.1), the characterization of the functional role of NGF became the archetypical model for the investigation of other putative NTFs—but other NTF families are also important as discussed later. Analysis of the function(s) of a putative NTF involved several steps. Typically, the biological activities and target cell specificity of putative factors were first investigated *in vitro*. Primary cultures of neurons dissociated from peripheral ganglia were ideal for these early assays. Analysis of the developmental expression of specific NTFs and their corresponding receptors has been used to determine if both the ligand and the receptor are normally present at the appropriate time and place for the putative factor to serve in the regulation of a specific subpopulation of neurons. In gain-of-function approaches, treatment of embryos with excess exogenous NTF has been used to determine if the survival or differentiation of responsive neurons is restricted by either the production or the access to a limited quantity of endogenous factor. Finally, in loss-of-function approaches, methods that inhibit or prevent the function of specific NTFs or their receptors have allowed investigators to determine whether the perturbation of endogenous NTF signaling alters normal development. These factor/receptor deprivation experiments have included treatment with activity-blocking antibodies that prevent trophic signaling, treatment with soluble receptor-derived antagonists that compete for and adsorb endogenous ligands, and the generation of transgenic mice with null mutations of either the factors or their receptors.

Sympathetic and sensory ganglia removed from developing animals have been shown to produce a dense halo of axonal outgrowth when treated with NGF (Fig. 19.2). In fact, this "axonal halo" assay—not a neuronal survival assay—was the original biological activity first used to purify and characterize NGF as a

NTF. NGF was shown to be required for the survival of dissociated sympathetic and some sensory neurons when they were grown in the absence of nonneuronal cells. This demonstrated that NGF could prevent cell death by the direct activation of receptors on isolated neurons. When developing embryos were treated with excess exogenous NGF, sympathetic and sensory ganglia were enlarged significantly, and axonal growth from these neurons increased markedly. In addition, ganglia in NGF-treated embryos contain many more neurons than normal, because naturally occurring cell death had been prevented. The soma was enlarged significantly by NGF treatment, and the dendritic arbors of sympathetic neurons were more complex. These studies indicated that the supply or access to endogenous NGF in sympathetic and sensory targets was likely to be rate limiting for the survival and growth of these dependent populations. The most convincing evidence that NGF is required for neuron survival has been gained from NGF deprivation experiments. Embryos treated with antibodies that selectively block NGF activity as well as the null mutation of either NGF or its trkA receptor in transgenic mice have both confirmed that sympathetic as well as some sensory neurons require NGF for survival (Fig. 19.10). NGF was localized to the peripheral targets of these neurons at the time of their normal innervation, con-

sistent with its role as a target-derived survival factor. Further, the level of NGF synthesis was correlated with the density of target innervation, and the NGF receptor, trkA, was localized to dependent afferent neurons at the times and places appropriate for regulating normal PCD. The localization of NGF synthesis in sympathetic targets and the loss of these afferent neurons with NGF deprivation firmly established NGF as a prototype target-derived NTF required for the survival of sympathetic neurons and a subset of sensory neurons.

One prediction of the neurotrophic hypothesis is that access to NGF in only the distant target region is adequate to support the survival of the remote cell body. This idea has been tested by Robert Campenot and colleagues *in vitro*. NGF has been applied selectively only to local axon terminals in a three-compartment tissue culture chamber to determine the long- and short-range effects of NGF treatment. NGF-dependent neuronal cell bodies in the central chamber survived when only their terminals were treated with the factor, indicating that target-derived NGF available only to neurites can generate and retrogradely transport the signaling required for cell body survival. Peripheral processes were lost rapidly and selectively in outer chambers where NGF was withdrawn, but were maintained and grew in the outer chambers where NGF was

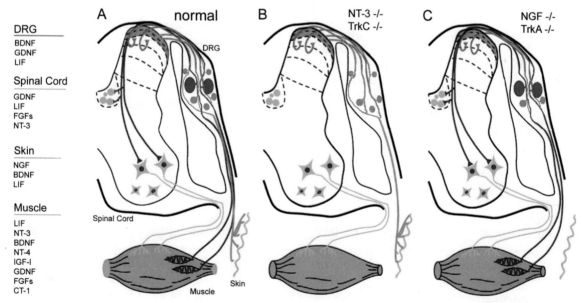

FIGURE 19.10 Phenotypic alterations in sensory/motor pathways caused by null mutations in NGF/trkA and NT-3/trkC. In the dorsal root ganglia (DRG) of normal mice (A), small-diameter (red), medium-diameter (green), and large-diameter (blue) neurons are present. Large-diameter neurons innervate muscle spindles and other proprioceptive end organs and have axon terminations in the lowest laminae of the dorsal horn and in the ventral horn. These neurons are lost when NT-3 or trkC is absent (compare A and B). Many of the small-diameter neurons innervate skin, respond to temperature and pain, and have terminations in the dorsal-most laminae of the dorsal horn. These neurons are lost when NGF or trkA is absent (compare A and C). DRG neurons indicated in green are neurotrophin-independent mechanoreceptive neurons (peripheral projections not shown). Modified and redrawn from Snider (1994).

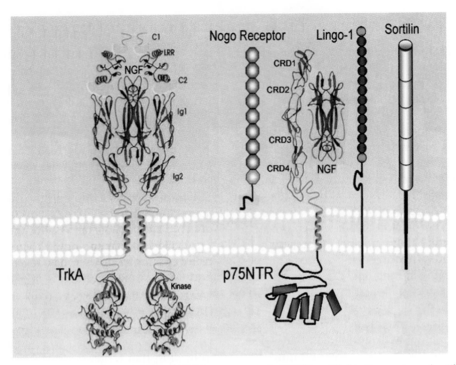

FIGURE 19.11 Models of NGF, the catalytic (full-length) TrkA receptor, and p75NTR and its binding partners (sortilin, Nogo receptor and Lingo 1). Note the ability of one ligand (NGF) to bring together two trk receptor molecules to initiate signaling. P75NTR lacks a cytosolic kinase domain, but associates with partners (including trk receptors, interaction not shown) to assemble signaling platforms. Note that the NGF dimer binds to a p75NTR monomer in an opposite orientation compared to TrkA. C1, C2, cysteine clusters 1, 2; Ig1, Ig2, IgG-like domains 1, 2; LRR, leucine-rich repeat; CRD 1–4, cysteine-rich domains 1–4. Modified from McDonald and Rust (1995) and Barker (2004).

added. This important demonstration illustrates the capacity of target-derived NGF to have both direct long distance effects on the survival of neurons and direct local effects on the growth, maintenance, and sprouting of axonal branches. NGF-responsive neurons possess both a "ligand-specific" receptor, trkA, and a "common" receptor, p75 (Fig. 19.11) that binds all neurotrophins with a similar affinity. Many of the biological activities of NGF have been attributed to the ligand-induced transduction of trkA.

Summary

The modern study of neuronal cell death began with investigations of how synaptic targets of sensory and motor neurons regulate their development. Viktor Hamburger and Rita Levi-Montalcini, beginning in the 1930s, ultimately showed that targets promote the survival and maintenance of innervating neurons. This notion, in turn, provided a conceptual framework for the discovery of the first target-derived NTF, nerve growth factor. From these beginnings, the neurotrophic hypothesis was formulated: neurons compete for limiting amounts of target-derived survival promoting (trophic) agents during development. NGF was established as the prototypical target-derived NTF.

THE NEUROTROPHIN FAMILY

Only a few subpopulations of peripheral neurons, including sympathetic and some sensory neurons, are exclusively dependent on NGF for survival during development. While cholinergic neurons in the mammalian basal forebrain and noradrenergic neurons in the avian locus coeruleus are NGF responsive, the survival of CNS neurons is largely unchanged following the null mutation of NGF. Other survival factors are now known to regulate neuron survival elsewhere in the nervous system. Extracts made from a number of tissues, as well as media containing proteins secreted by a variety of cultured neuronal and nonneuronal cells, all have been shown to support the survival of many different classes of neurons that are not NGF dependent and that do not express the NGF receptor trkA, suggesting that additional NTFs exist. Because NTFs are made in extremely low quantities, the biochemical isolation of NGF-related molecules using conventional protein purification methods proved to be difficult.

A significant breakthrough occurred with the purification of a second NGF-related NTF by Yves Barde and colleagues in 1982. Unlike NGF, which was purified several hundredfold from an extraordinarily rich biological source unrelated to the nervous system (Table 19.1), endogenous brain-derived neurotrophic factor (BDNF) was purified several millionfold from adult pig brains. Each kilogram of starting material yielded only a microgram of factor which over time was eventually sequenced and cloned to produce recombinant factor. The molecular cloning and expression of BDNF opened the door for an accelerated period of research on NGF-related factors (Table 19.1). When the protein structure of BDNF was compared with NGF, they were both found to encode homodimers of small, very basic secreted peptide ligands with an amino acid homology of approximately 50%. Using polymerase chain reaction primers prepared from homologous domains to search for other related proteins, investigators rapidly identified additional neurotrophin family members in multiple species (Lewin and Barde, 1996). Described as neurotrophins or nerve feeding factors (i.e., NT-3 and NT-4/5, NT-6), these additional proteins were cloned and sequenced without the requirement of exhaustive protein purification.

Each neurotrophin family member is synthesized as an approximately 250 amino acid precursor (pro-neurotrophin) that is processed into a roughly 120 amino acid protomer. Homologous regions of the several different family members are concentrated in six hydrophobic domains containing cysteine residues. The linkage formed by each homodimer ligand utilizes these regions to form a "cysteine knot" that maintains the twin protomers in juxtaposition. The secreted dimer appears as a symmetrical twin with variable regions containing basic amino acid residues exposed on the surface (Fig. 19.11) (McDonald and Rust, 1995). Because all family members share this core structure, they are remarkably similar, with three-dimensional symmetry around two axes. The symmetry of this twin structure allows the neurotrophin ligand to activate trk receptors by binding separate receptor molecules together in the membrane, permitting docking of additional, intracellular signaling molecules and initiation of signal transduction cascades (Figs. 19.11, 19.14). The exposed outer regions that vary between neurotrophin family members are responsible for receptor-binding specificity.

Summary

The purification, molecular cloning, and expression of BDNF opened a floodgate of research on an NGF-related family of NTFs called neurotrophins or nerve feeding factors. Each neurotrophin family member is released as a homodimer with a conserved region containing a cysteine knot in the core of the molecule. The secreted factor is a symmetrical twin with duplicate sites used for bivalent receptor binding and formation of signaling platforms.

NEUROTROPHIN RECEPTORS

NGF binds to a relatively small number of very high-affinity-binding sites and a second set of about 10-fold more abundant, but lower affinity, binding sites at higher concentrations (Roux and Barker, 2002). A 75-kDa protein (p75) was purified and cloned first. It is a transmembrane glycoprotein with extracellular cysteine repeat motifs that share structural homology with the tumor necrosis factor receptor family. The cytoplasmic domain of p75 lacks the kinase domain present in most growth factor receptors for intracellular signal transduction, but it can associate with several other signaling proteins (discussed later) (Fig. 19.11). When expressed in fibroblasts, this receptor has low-affinity NGF-binding properties (ligand binding is rapidly on and off) and therefore also has been called the low-affinity NGF receptor. This name has proven to be a misnomer, as other neurotrophin family members also bind p75 with a similar affinity. It is therefore more appropriately named the "common" neurotrophin receptor (p75NTR).

A major breakthrough in the characterization of the NGF receptors came with the fortuitous discovery and cloning of an oncogene identified in a human colon cancer. The sequence of this 140-kDa transmembrane protein contained a cytoplasmic kinase common to many growth factor receptors. The corresponding proto-oncogene was named trk (pronounced "track"—for tropomyosin-related kinase). It rapidly was appreciated as a member of the tyrosine kinase-containing receptor superfamily (Huang and Reichardt, 2001) with trk mRNA expression localized to NGF responsive neurons. Low-stringency screening of cDNA libraries with the original trk proto-oncogene probes led to the discovery of other related neurotrophin receptors. The NGF binding receptor was called trkA, whereas two additional 145-kDa members of this protein family were named trkB and trkC (Fig. 19.12). Expression of trkA in a mouse fibroblast cell line or in frog oocytes conferred specific high-affinity NGF binding and NGF-induced receptor phosphorylation. NGF-signaling properties have been examined most extensively in the NGF responsive pheochromocytoma (PC 12) cell line, derived from adrenal medullary cells. Mutant PC12 cell lines that have lost their capacity to respond to NGF contain many p75NTR receptors but lack trkA. Transfection of these mutant cells with trkA

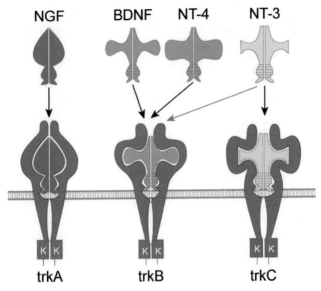

FIGURE 19.12 Ligand binding preferences of neurotrophins for each member of the trk receptor family. Not shown are the truncated (kinase deleted) isoforms of trkB and trkC. Other isoforms containing inserts and deletions also exist, providing a wide variety of receptors. NT-4 is also named NT-4/5. K = tyrosine kinase.

restores their biological responses to NGF treatment. The most convincing evidence for the necessity of trkA comes from the analysis of transgenic mice lacking functional trkA receptors (Huang and Reichardt, 2001). As expected, these mice have a phenotype that is similar to that of transgenic animals that have a null mutation for NGF (Table 19.3).

The trkB receptor is activated specifically by low concentrations of BDNF or NT-4/5 and, to a lesser extent, by higher concentrations of NT-3. NT-3 activates the trkC receptor most effectively. All trk receptors contain three leucine-rich motifs, two cysteine clusters, and two immunoglobulin-like motifs in the extracellular region, a transmembrane domain and a tyrosine kinase domain in the cytosolic region (Fig. 19.11). The unusual combination of extracellular motifs makes up the ligand-binding region and places this family in a novel class of tyrosine kinase receptors. The region of highest sequence homology among family members and other growth factor receptors is in the kinase domain.

TABLE 19.3   Percentage of Neurons Lost in Neurotrophic Factor or Receptor Deficient Mice*

Neurotrophic factor/receptor null mutation	Viability	Losses in PNS ganglia						Losses in CNS nuclei	
		Dorsal root ganglia	Trigeminal ganglia	Nodose petrosal ganglia	Vestibular ganglia	Cochlear ganglia	Superior cervical ganglia	Spinal moto neurons	Facial moto neurons
TrkA	Poor	−70–90%	−70–82%	?	−0%?	−0%?	−95%	−0%?	?
NGF	Poor	−70%	−75%	−0–15%	?	−0%	−82–95%	−0%?	?
TrkB	Very poor	−0–41%	−39–60%	−90–94%	−56–85%	−15–20%	−0%	−0%**	−0%**
BDNF	Moderate	−0–44%	−0–45%	−39–66%	−82–87%	−7%	−0%	−0%	−0–3%
NT-4	Good	−0–14%	−0–5%	−40–59%	−0–21%	?	−0%	−0%	−0–8%
BDNF/NT-4	Good	−0%	−9–34%	−79–90%	−82–90%	?	−0%	−0%	−0–11%
TrkC	Moderate	−17–38%	−21–48%	−14–18%	−15–29%	−50–85%	−0%	−0%	−0%
NT-3	Very poor	−36–79%	−61–68%	−30–47%	−15–34%	−85%	−48–53%	?	−0%
TrkB/TrkC	Very poor	−41%	?	−95%	−58–100%	−61–100%	−0–4%	−0–5%	−0–5%
BDNF/NT-3	Very poor	−84%	−74%	−62%	−99%	−100%	?	?	?
BDNF/NT-4/NT-3	Very poor	−92%	−88%	−96%	−100%	?	−47%	−20%	−22%
gp130 (IL receptor)	Very poor	−21%						−41%	−31%
CNTFRα	Very poor	−0%	−0%	?	?	?	−0%	?	−40%
LIFR	Very poor	?	?	?	?	?	?	−40%	−35%
c-ret	Very poor	−0%	?	?	?	?	−100%	−50%	−30%
GFRα1	Very poor	−0%	−0%	−15%	−0%	−0%	−0%	−24%	−0%
GFRα2	Poor	−0%	−0%	−0%	−0%	−0%	−0%	−24%	?
GFRα3	Good	−0%	−0%	?	?	?	−50–95%	?	?
GDNF	Very Poor	−0–23%	?	−0–40%	−0%	?	−35%	−22–37%	−18%
Neurturin	Good	−0%	?	−0%	?	?	−0%	−0%	?

*For references to original studies, see Huang and Reichardt, 2001; Airaksinen and Saarma, 2002; von Bartheld and Fritzsch, 2006.

Losses reported by Klein *et al.*, 1993 (*Cell* **75, 113–122) could not be replicated.

Receptor isoforms resulting from splice variants of trk mRNA transcripts exist for each family member. A variety of both full-length and kinase-deleted or truncated receptors are widely expressed on neurons throughout the nervous system. Truncated receptors, which also are expressed on glial cells, can bind and internalize their cognate ligand and are capable of limited signaling, but they cannot initiate the phosphorylation events required for mainstream receptor signal transduction (Fig. 19.14). As a result, the distribution and membrane concentration of truncated receptors could potentially modulate neurotrophin activity.

Although trk receptors account for many of the biological responses of neurons to neurotrophins, p75NTR can modify trk ligand binding and neurotrophin specificity, associates with additional signaling partners, and can initiate pathways for intracellular signaling independent of trk receptors (Huang and Reichardt, 2001; Barker, 2004). Sensory neurons from transgenic mice that lack p75NTR require higher NGF concentrations for survival than sensory neurons from normal animals. Antibodies that block NGF binding to p75NTR but not trkA reduce high-affinity NGF-binding sites, and p75NTR has been demonstrated to enhance trkA receptor phosphorylation. Several mechanisms have been proposed to account for an accessory role of p75NTR. The fast on and fast off kinetics of p75NTR could maintain and increase the local concentration of neurotrophins at the membrane surface and thereby increase the access of trk receptors to ligand. Alternatively, p75NTR could form a transient heterodimer with trk receptors and thereby "hand off" the factor for trk binding (Roux and Barker, 2002). In addition to enhancing NGF binding to and activation of trkA, p75NTR initiates NGF responses in cells that lack trkA. NGF binding to p75NTR causes activation and nuclear translocation of a transcription factor, nuclear factor kappa B (NFκB), that promotes cell survival. However, p75NTR is related structurally to members of the tumor necrosis factor receptor (TNFR) family, many of which regulate the onset of cell death programs in the immune system and, for that reason, are termed death receptors. The cytoplasmic domain of p75NTR contains a "death domain" sequence similar to active sequences found in the TNFR family, and the p75NTR mechanisms of inducing PCD appear to be shared with other death receptors. In some neuronal cells that express p75NTR but not trkA, NGF or proNGF induce cell death via p75NTR binding (Roux and Barker, 2002). Once bound, p75NTR may activate a PCD signaling pathway via the Jun kinase cascade. The Jun kinase cascade activates the transcription factor p53 with gene targets that include the proapoptotic gene BAX. It should be noted that NGF and other trophic factors can be utilized in different cells or in different epochs of time to induce many diverse biological activities. It is noteworthy that under specific circumstances (such as the absence of trk activity), a factor originally identified for the capacity to prevent PCD instead is utilized to execute programs promoting neuronal death.

Summary

Many neurons have both specific and common binding sites for neurotrophins. Biological responses primarily are associated with high-affinity binding and rapid phosphorylation signaling events. All neurotrophins bind p75NTR, the common neurotrophin receptor. p75NTR lacks a cytoplasmic kinase domain but can independently initiate signaling, can associate with additional signaling partners, and enhances signaling through trkA. There are three tyrosine receptor kinase (trk) family members: trkA, trkB, and trkC. Each receptor binds one or more members of the neurotrophin family. Splice variants of trks result in isoforms that include truncated receptors with reduced signaling capabilities. p75NTR is related to the TNFR family of death receptors and contains a cytosolic "death domain." Activation of p75NTR, especially by pro-neurotrophins, may serve to kill neurons or other cell types through well-established signaling pathways used to promote PCD.

SECRETION AND AXONAL TRANSPORT OF NEUROTROPHINS AND PRO-NEUROTROPHINS

Processing and Secretion of Neurotrophins

Neurotrophins are expressed as larger precursors that are cleaved by furins and prohormone convertases. The extent of cleavage differs between cell types. The precursor forms were believed for a long time to be of little functional significance, and it was thought that the mature neurotrophins were the only important moiety that was secreted. Research by Barbara Hempstead and her colleagues in the early 2000s demonstrated that many cells secrete pro-neurotrophins, and that the release of both pro-neurotrophins and mature neurotrophins contributes to the spectrum of physiological functions. Pro-neurotrophins bind with higher affinity to p75NTR than mature neurotrophins and appear to be the "preferred" ligand for p75NTR. Whether endogenous pro-neurotrophins bind with lower affinity to trk receptors is still controversial. To activate different signaling pathways, the neurotrophin-bound p75NTR can associate with several signaling partners, including Nogo receptor (a GPI-linked protein), sortilin (a member of the VPS10 family), and LINGO-1 (involved in myelin-based growth inhibition, Fig. 19.11). This complex is involved in regulating neurite outgrowth in response to myelin proteins such as Nogo, MAG, and OMgp. Formation of these different platforms may explain the multiple effects of p75NTR in different cell types and contexts. The local balance and concentrations of pro- and mature neurotrophins as well as the availability and activity of binding partners determines physiological functions (Barker, 2004).

Neurotrophic Factors and Synaptic Plasticity

The classic neurotrophic hypothesis postulates that NTFs, including neurotrophins, are released by postsynaptic target cells and bind to presynaptic receptors for internalization, retrograde axonal transport and activation of signal transduction cascades after arrival at the neuronal soma (Fig. 19.1). Yet NT receptors are present on both post- and presynaptic sites of the synaptic cleft. *In vitro* studies have shown that NTs, and in particular BDNF, have rapid effects on synaptic transmission (Lohof *et al.*, 1993). Numerous studies have confirmed that NTFs have important functions in synaptic plasticity, meaning that these "synaptotrophins" regulate the functional strength of synaptic transmission including long-term potentiation (LTP), an experience-dependent, long-lasting increase in chemical strength of neurotransmission that is a crucial mechanism in learning and memory (see also Table 19.5).

The regulation of secretion of neurotrophins is particularly important for mechanisms of synaptic plasticity (Lessmann *et al.*, 2003). Neurotrophins can be released by constitutive modes (from cell bodies and processes) and by activity-dependent modes of secretion (from dendrites and axons). The pattern of activity, such as high frequency stimulation, is crucial for the release. LTP is relatively synapse-specific—only active and successfully transmitting synapses are strengthened, so the regulated release of neurotrophins and subsequent signaling should be restricted to active synapses. Depolarization and cAMP-dependent pathways are thought to regulate the locally restricted release of neurotrophins such as BDNF. This may contribute to the local "tagging" or marking of the active synapses destined for strengthening. Trafficking of BDNF mRNA through the neuron might also contribute to synapse-specific plasticity, if the mRNA is "captured" for local translation at active synapses.

Interestingly, a naturally occurring polymorphism of the BDNF gene (a Val-Met substitution in the 5′pro-region of BDNF) gives rise to a reduced capacity for activity-dependent (but not constitutive) secretion of BDNF, causing abnormal hippocampal function and some learning and memory deficits in humans (Egan *et al.*, 2003).

Endogenous neurotrophins may be derived from either pre- or postsynaptic sites and they may act on either side of the synaptic cleft in paracrine or autocrine modes to implement functional and structural changes in synaptic plasticity (Poo, 2001). Trophic responses can also be modified by the recruitment of receptors to surface membranes from internal storage sites. Furthermore, after release and binding of newly synthesized neurotrophins, recycling of neurotrophins occurs, and the internalized neurotrophin is not necessarily degraded immediately after the initial binding and internalization steps. Recycled neurotrophins have been shown to play functionally significant roles in LTP.

The Endosome Signaling Hypothesis of Retrograde Axonal Signaling

The neurotrophic hypothesis postulates retrograde axonal movement of trophic signals. Evidence for retrograde transport of exogenous neurotrophins (NGF) was established in the 1970s (Oppenheim, 1996). Yet the precise mechanisms of axonal transport of trophic signals from the nerve terminal to the neuronal soma remained largely elusive. The prevailing hypothesis is that neurotrophins bind presynaptic surface membrane receptors on nerve terminals, are internalized, and form a ligand-receptor complex with the neurotrophin inside transport vesicles, leaving the trk receptor with its kinase domain on the outside of the vesicle or endosome (Fig. 19.13). The trk receptors may remain "active" (phosphorylated) during axonal transport along microtubules, because the ligand remains bound to the receptor within the vesicle. When the endosome, with the ligand-receptor complex attached, arrives at the neuronal soma, the phosphorylated trk receptor initiates a signal transduction cascade that results in changes of gene expression. Experimental evidence from the Mobley and Ginty labs supports this "signaling endosome" hypothesis (Zweifel *et al.*, 2005). Signaling endosomes can be isolated from axons; trk receptors and other signaling proteins interact at the level of the axon; trks bind to the dynein light chain subunit (part of the dynein motor complex) directly, and thus partake in the axonal transport machinery (Fig. 19.13). Furthermore, endosomes that reach the soma may generate a different signal (activate a differ-

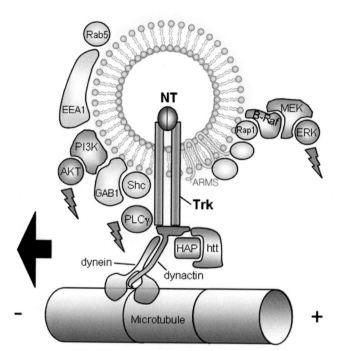

FIGURE 19.13 The signaling endosome. The internalized neurotrophin (NT)–receptor (Trk) complex is transported retrogradely from the nerve terminal to the cell body. The internalized vesicle serves as a platform for docking of adaptor proteins (light blue) that elicit signal transduction cascades via kinases (red); molecular motors and regulators (green) attach for transport along microtubules (brown). The arrow indicates the direction of transport toward the minus end of microtubules. The main signal transduction cascades are via the phospholipase C-γ (PLCγ), Raf-MAPK-ERK, and PI3K signaling pathways (red lightning symbol). For details of signal transduction, see legend to Fig. 19.14. Rap1 and Rab5 are small G-proteins. Other abbreviations: Akt, v-akt murine thymoma viral oncogene homologue (protein kinase B); ARMS, ankyrin-rch membrane spanning protein; B-Raf, v-raf murine sarcoma viral oncogene homologue B1; EEA1, early endosome antigen 1; ERK, extracellular signal-regulated kinase; GAB1, GRB2 (growth factor receptor bound protein 2)-associated binding protein1; HAP, huntingtin-associated protein 1; htt, huntingtin; MEK, MAPK (mitogen-activated protein kinase); NT, neurotrophin; PI3K, phosphatidylinositol 3-kinase; Shc, Src homology 2 domain-containing transforming protein C; Trk, trk tyrosine kinase receptor. For details and references, see Zweifel *et al.* (2005) and Gauthier *et al.* (2004).

ent ERK) than those generated by binding to cell surface receptors on the soma.

However, not all trophic signals from the nerve terminal that arrive at the soma may require the neurotrophin to be physically transported in a ligand-receptor complex. In addition to retrograde axonal transport from nerve terminals to the soma, trophic signals also are transported anterogradely along axons (possibly even a quantitatively more important route) from soma to nerve terminals, mediated by kinesin motors, followed by activity-dependent release of neurotrophins from axon terminals. Indeed, some data

suggest that BDNF and NT-3 may rapidly activate tetrodotoxin-insensitive sodium channels, which would blur the distinction between neurotransmitters, neuromodulators, and NTFs.

Summary

Expression and secretion of pro-neurotrophins has significant physiological functions in the regulation of cell death. Pro-neurotrophins bind preferentially to the p75NTR. The p75NTR associates with several different signaling partners to form multiple signaling platforms that include sortilin and Nogo receptors. Secretion of neurotrophins requires activity-dependent stimulation patterns that may specifically strengthen active synapses. Neurotrophins are transported retrogradely along axons in signaling endosomes, but they can also be transported anterogradely for release from axon terminals. Trk receptors are expressed both pre- and postsynaptically, forming paracrine as well as autocrine loops of synaptic signaling. Synaptically released BDNF may have rapid effects on sodium channels, acting in a fashion similar to classical peptide neurotransmitters.

SIGNAL TRANSDUCTION THROUGH TRK RECEPTORS

Neurotrophin binding to trk receptors at the cell surface causes the formation of receptor dimers and coactivation of their tyrosine kinase activity. The homodimeric structure of the factors allows each bivalent ligand to bring two separate receptor molecules into close proximity. Aggregated receptors phosphorylate each other on specific tyrosine substrates within intracellular domains. The generation of phosphotyrosine residues in turn activates the receptor kinase and further catalyzes the formation of large signaling complexes through the recruitment of adaptor proteins. These proteins link the activated receptor kinase with intracellular signaling pathways shared by other growth factors (Ip and Yancopoulos, 1996). Once activated, the receptor initiates intracellular signals both locally in the cytoplasm and by a series of enzymatic cascades that eventually produce changes in gene transcription within the nucleus (Segal and Greenberg, 1996).

Receptor signal transduction involves multiple signaling pathways that can differ between individual neurons or in the same neurons at different periods of time (depending on recent events). Thus, the response of neurons to trophic factor stimuli is dependent on the intracellular status of the cell in a dynamic fashion. The molecular components of these pathways are so well conserved that many of the signaling proteins are interchangeable among invertebrate and vertebrate species. Three of these pathways have been identified to implement trk signal transduction events that mediate survival as well as many other cellular responses to neurotrophins (Fig. 19.14). They are (1) the *phospholipase C (PLC-γ) pathway*, (2) *Ras-MAP kinase pathway*, and (3) *phosphatidylinositol-3 kinase (PI-3K) pathway*. The latter two pathways begin with adaptor proteins that contain a structural motif, the src homology domain 2 (SH2), which specifically recognizes the phosphotyrosine residue and flanking sequences. PLC-γ activity generates two distinct second messenger signals: inositol triphosphate (IP$_3$) and diacylglycerol (DAG). IP$_3$ rapidly releases intracellular Ca^{2+} sequestered in local membrane compartments. This signal initiates the activity of local Ca^{2+} dependent enzymes (protein kinases and phosphatases). Similarly, DAG regulates the activity of DAG-dependent enzymes. Both Ras and PI-3 kinase pathways are engaged via adaptor and associated linker or extender proteins.

One of the best known Ras-dependent signaling pathways involves the ERK family of MAP kinases. This cascade is composed of serine/threonine kinases that are serially phosphorylated and activated. The initial member of this cascade is Raf, which phosphorylates and activates MEK, which phosphorylates the MAP kinases (ERK1 and ERK2). Once activated, these MAP kinases in turn phosphorylate a number of cytoplasmic and nuclear effectors. Importantly, activation of MAP kinases and their substrate, the protein kinase RSK, results in the phosphorylation of a number of transcription factors, including CREB (cAMP response element-binding protein), which controls expression of immediate-early genes (i.e., *c-fos* and *c-jun*) and delayed response genes. Once activated, transcription factors cause rapid and long-lasting changes in gene expression regulating cell survival, axonal and dendritic growth, neuronal differentiation, synaptic potentiation, and plasticity. The ultimate biological response differs with the duration of Ras activity in response to receptor phosphorylation. TrkA activation of Ras persists much longer compared to only a rapid and transient Ras activation following receptor transduction by epidermal growth factor, suggesting that the duration of Ras activation is a critical determinant of transcriptional activity and biological responses (Huang and Reichardt, 2001). In addition, the requirement on Ras-dependent pathways for biological responses varies among neuron types.

Neurotrophic factor receptor activation of the PI-3 kinase pathway is essential for the normal survival of many neurons. The phosphatidyl inositides made by PI-3 kinase regulate in part the activity of AKT (protein kinase B). This important protein kinase plays a critical role in controlling the biological activity of several regulatory proteins that govern normal PCD (see Fig. 19.14 and previous sections). BAD is phosphorylated by AKT, but when dephosphorylated by apoptotic stimuli, BAD associates with Bcl–2 proteins, which leads to the release of cytochrome C from mitochondria, resulting in caspase activation and DNA degradation (Fig. 19.14). Trk receptor activation of both the PI-3 kinase and the Ras-ERK pathways suppresses the capacity of activated p75NTR to induce cell death programs via Jun kinase. In some cases, p75NTR may therefore induce neuronal death where appropriate neurotrophin binding to trk receptors is absent. Interactions between the multiple trk receptor signaling pathways in different cell phenotypes and under a variety of dynamic receptor signaling conditions provide a rich assortment of interrelated mechanisms to govern the PCD machinery.

Summary

Three tyrosine receptor kinases and trk family members—trkA, trkB, and trkC—transduce neurotrophin signals. Signaling pathways used by the trk family members are shared with those activated by many other growth factor receptors. Neurotrophin binding to trk causes receptor dimerization and phosphorylation of cytoplasmic tyrosine residues. Phosphotyrosines recruit cytosolic adaptor proteins that couple the activated receptor with intracellular signaling pathways. The biological response of the cell to a neurotrophic factor is dependent on the dynamic status of the pathway that can vary with recent cell history. Three of the best investigated signaling pathways are PLC-γ, Ras-ERK kinase, and PI-3K. Normal PCD is governed in many neurons by PI-3 kinase activation of the protein kinase Akt. Trophic factor deprivation can lead to the release of pro-apoptotic mechanisms that promote cell death.

CYTOKINES AND GROWTH FACTORS HAVE MULTIPLE ACTIVITIES

Several Cytokines Mediate Cell Interactions within the Nervous System

We have emphasized neurotrophins as an example of the activities of NTFs in the developing and mature nervous system. Yet there are several other important families of NTFs. Often NTF families utilize receptors

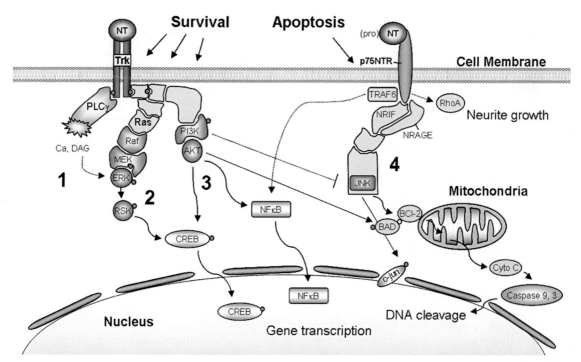

FIGURE 19.14 Trk and p75NTR signaling pathways. Neurotrophin (NT) dimers bind two trk receptor monomers or one p75NTR to activate survival (1–3) or apoptotic (4) signaling pathways. Ligand binding initiates trk receptor transduction by the phosphorylation (indicated by small red dot) in the cytoplasmic domains. The activated Trk kinase docks adaptor and linker proteins (light blue), which engage signaling cascades often containing multiple kinases (red). The three principal signaling pathways illustrated are the (1) phospholipase C pathway, (2) Ras-MAP kinase pathway; and (3) PI-3 kinase pathway. These pathways lead to nuclear translocation of transcription factors (green), such as CREB and NFκB, and ultimately regulation of gene expression. The binding of neurotrophins or proneurotrophins to p75NTR can activate BAD via the JNK cascade and eventually engages, via release of effectors from mitochondria, caspases involved in apoptosis (see also Fig. 19.9). There are several venues for crosstalk between the survival- and pro-apoptotic signaling cascades as indicated. Abbreviations not explained in Legend 19.13: BAD, bcl2-antagonist of cell death; Bcl2, B-cell lymphoma 2; Ca, calcium; CREB, cyclic AMP response element binding protein; Cyto C, cytochrome C; DAG, diacylglycerol; JNK, C-Jun N-terminal kinases; NFκB, nuclear factor-kappa B; NRAGE, neurotrophin receptor interacting MAGE (melanoma antigen gene expression) homolog; NRIF, neurotrophin receptor interacting factor; p75NTR, p75 neurotrophin receptor; (pro)NT, pro-neurotrophin; Raf, Ras, and RhoA, small GTPases; RSK, ribosomal S6 kinase. For details, see text, and Huang and Reichardt (2001), Chao (2003), and Barker (2004).

with coreceptor complexes (e.g., neurotrophins—trk, p75NTR; and GDNF family ligands—ret and GFRs). In many tissues, intercellular induction factors are important for governing the proliferation and differentiation of both embryonic and adult stem cells. In addition, these factors also play an important role in the response of tissues to trauma, inflammation, infection, or tumor growth. In 1974, Stanley Cohen proposed that both lymphocyte-derived and nonlymphocyte-derived chemotactic and migration inhibitory factors be grouped into families of cytokines ("cell movement factors"). More recently, this term has been adopted as a general umbrella for many families of secreted proteins that mediate diverse biological responses, including changes in the immune system (interleukins), tumor cytotoxicity (tumor necrosis factors), and inhibition of viral replication or cell growth (interferons) (Table 19.4).

Many cytokines originally were named according to the particular biological activity that was utilized for their isolation, only to be later rediscovered or renamed as important mediators of other physiological processes. For example, some factors were isolated on the basis of their ability to enhance the survival of specific populations of neurons isolated *in vitro*. These include ciliary neurotrophic factor (CNTF); glial cell line-derived neurotrophic factor (GDNF); and the other GDNF family members, neurturin, persephin, and artemin. CNTF is a member of a broader family of neuropoietic cytokines, including leukemia inhibitory factor (LIF), oncostatin M, cardiotrophin-like cytokine/cytokine-like factor (CLC/CLF) and cardiotrophin-1 (CT-1), that share a common three-dimensional structure and receptor subunits. CNTF originally was isolated and named because it supports the survival of neurons cultured from the parasympathetic ciliary ganglion. The GDNF family primarily supports enteric neurons, dopaminergic neurons, and some motoneurons (Airaksinen and

TABLE 19.4   Cytokine and Growth Factor Families

Family	Representative members	Original biological activities
Neurotrophins	NGF, BDNF, NT-3, NT-4/5, NT-6	Neuronal survival and differentiation
Neuropoietic cytokines	CNTF, LIF, CT-1, CLC/CLF, oncostatin M	Survival of ciliary neurons and moto neurons, leukemia inhibitory activity, increased cholinergic properties
Tissue growth factors	TGF-α, TGF-β, FGFs, IGF-Iα, IGF-Iβ, IGF-II, EGF, PDGF	Cell proliferation and differentiation in diverse tissues and organs
GDNF family	GDNF, neurturin, artemin, persephin	Neuronal survival and dopaminergic cell differentiation, and morphogens in kidney and sperm development
Interleukins	IL-1α, IL-1β, IL-2 through IL-15	Immunoregulation, diverse activities in the immune system
Tumor necrosis factors	TNF-α, TNF-β	Tumor cytotoxicity
Chemokines	MCAF, MGSA, RANTES, NAP-1, NAP-2, MIP-1	Leukocyte chemotaxis and cell activation
Colony-stimulating factors	G-CSF, M-CSF, GM-CSF	Hematopoietic cell proliferation and differentiation
Interferons	IFN-α, IFN-β, IFN-γ	Inhibition of viral replication, cell growth, or immunoregulation

Saarma, 2002). Other proteins that originally were identified as mitogens or chemotactic factors in nonneuronal tissues also have been shown to affect either the survival or the differentiation of neurons, including the fibroblast growth factors (FGFs), insulin-like growth factors (IGFs), and hepatocyte growth factor (HGF). The first neuronal function identified for LIF was the induction of cholinergic properties in cultured sympathetic neurons, but it also supports the survival of several classes of neurons and induces neural precursors to become astrocytes.

LIF, which has a number of actions in the immune system and other nonneuronal tissues, shares receptor subunits with the CNTF family and therefore can mimic both the survival and the cholinergic differentiation activities of CNTF observed in culture. Many cytokines important for the development or maintenance of other organs and tissues are also widely expressed within the nervous system. Their roles in the nervous system, however, remain to be defined. Likewise, factors first recognized as neuronal survival factors have mitogenic properties for either nonneuronal cells or neuronal precursors. As a result, these pleiotropic factors are grouped into families based on their protein sequences and receptor usage rather than on their biological properties. Table 19.4 summarizes major known ligands, receptors, and biological functions for the CNTF-related family of cytokines, the GDNF/TGF-related superfamily, and FGFs.

Neurotrophic Factors Have Multiple Activities

Neurotrophic factors that prevent neuronal death during development appear to have many other important biological activities, including effects on cell proliferation, migration, differentiation, axonal growth and sprouting, alterations in dendritic arbors, and synaptic plasticity of the nervous system (Table 19.5). Frequently, different populations of neurons respond to the same factor in distinct ways, and the same neuron may respond differently to the same factor at different developmental stages. Variations in neuronal responses to the same factor appear to depend not only on modifications of trophic receptors and their binding partners, but also on potential differences in the intracellular context of downstream signaling pathways. The distinctive activities of a NTF on different cells or during different epochs of development reflect the intrinsic properties of differentiating neurons or a dynamic change in a trophic response due to alterations in neural activity or other signaling events.

As described earlier, neurotrophins have been shown repeatedly to play fundamental roles in the survival of many peripheral neuronal populations. Null mutations of neurotrophins or their cognate trk receptors manifest substantial deficits in dependent populations (Table 19.3; Fig. 19.10) (Huang and Reichardt, 2001). However, relatively few changes in the number of neurons have been observed in the CNS of neurotrophin or trk receptor null mutant mice. This observation was surprising because significant cell death occurs in the CNS, and NTFs and their receptors are widely expressed in the CNS during this period. The complexity and number of synaptic relationships established by CNS compared to those established by PNS neurons may partially explain why CNS neurons are not as sensitive to the loss of a single neurotrophin or neurotrophin receptor (Fig. 19.15). The trophic support to CNS neurons is likely to arise from multiple families of NTFs with possible synergistic and/or compensatory effects.

The traditional view of the neurotrophic hypothesis has been that trophic support is derived from target tissues, but other sources of trophic support are now recognized (Figs. 19.15, 19.16). Interpretation of mRNA localization studies is complicated by the fact that NTFs may not simply be released locally in the region of the cell body, but may also (or instead) be delivered to distant regions of the nervous system by anterograde or retrograde axonal transport. Thus, although the original neurotrophic hypothesis that neurons depend on target-derived trophic factors still holds true for many neurons during critical periods of development (including sympathetic neuron dependence for target-derived NGF), other sources of trophic support appear to play an important functional role in development, as briefly described earlier.

Finally, although beyond the scope of the present chapter, neurotrophins and other NTFs and their receptors also are distributed throughout the mature brain, and NTFs can alter neural activity by rapid and long-term changes in synaptic transmission. Neurotrophins and other NTFs also play a role in modulating long-term changes in functional and anatomical plasticity in the developing and mature brain by altering long-term potentiation, synaptic connectivity, and responses to stress, inflammation, and trauma (Chao, 2003) (Table 19.5).

Summary

There are multiple families of NTFs. CNTF and LIF belong to a neuropoietic cytokine family. The expression and biological properties of these cytokines distinguish them from neurotrophins. They possess widespread neurotrophic activity for many different neuronal and nonneuronal populations *in vitro* and facilitate the cholinergic differentiation of sympathetic and motor neurons. The GDNF family (GDNF, neurturin, artemin, and persephin) is related to the TGF-β gene family and has effects on enteric, dopaminergic

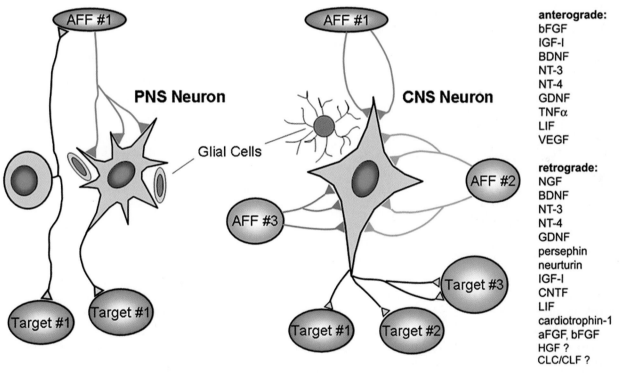

FIGURE 19.15 Possible sources of trophic support for peripheral (PNS) and central (CNS) neurons. Many peripheral neurons such as sympathetic and bipolar sensory neurons (left side) have only two sources of support: one in the periphery (Target #1) for retrograde support and one from afferents (AFF #1) or the central target (for bipolar sensory neurons). Glial cells (gray) may also provide trophic factors. In contrast, central neurons (right side) receive synaptic input from many different types of neurons (AFF #1, 2, and 3), which may serve as a source of anterograde trophic support. Central neurons may also project to several different targets (Targets #1, 2, and 3), which each may provide retrograde trophic support. Trophic factors are listed that have been demonstrated to be either anterogradely or retrogradely transported, with question marks when transport is suspected, but not proven. In addition, glial cells in the CNS may also provide trophic factors to CNS neurons. Adapted from Snider (1994) and von Bartheld *et al.* (2001).

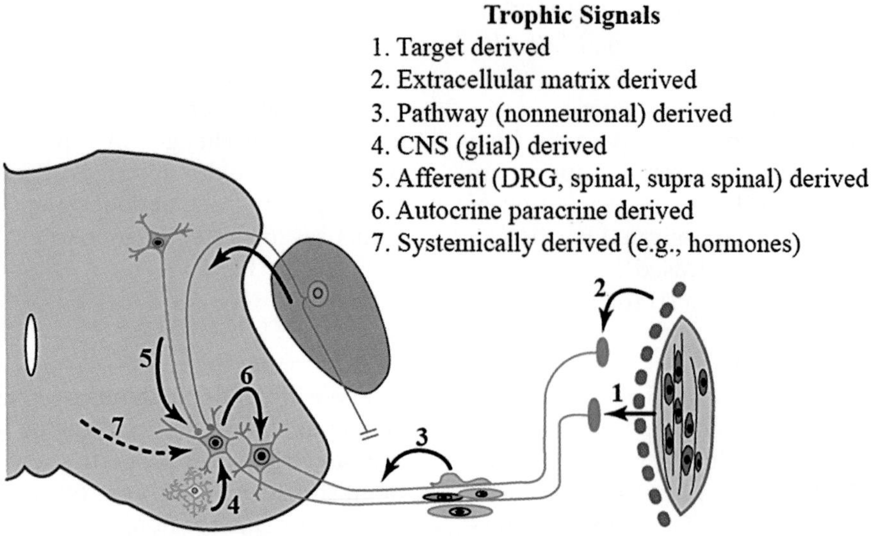

Trophic Signals
1. Target derived
2. Extracellular matrix derived
3. Pathway (nonneuronal) derived
4. CNS (glial) derived
5. Afferent (DRG, spinal, supra spinal) derived
6. Autocrine paracrine derived
7. Systemically derived (e.g., hormones)

FIGURE 19.16 Schematic illustration of different sources of potential trophic signals acting on motor neurons in the spinal cord. Motor neurons (green cell bodies) can receive trophic support from a number of different sources. Axon terminals of the motor neurons have access to diffusible muscle-derived (1) or extracellular matrix-associated (2) trophic factors. Schwann cells (3) in the peripheral nerve or ventral root are another source of trophic support. Glial cells (4) in the spinal cord (astrocytes and/or oligodendrocytes) may also influence motor neuron survival. Motor neurons receive afferent input from several sources, including descending fibers from the brain, spinal interneurons, and dorsal root axons (5), which could supply trophic support. Finally, motor neurons could influence their own growth and survival via autocrine trophic support (6), as well as respond to trophic support provided by circulating hormones (7).

TABLE 19.5 Nonsurvival Functions of Neurotrophic Factor*

Function	Examples of relevant trophic factors and cytokines
Cell fate decisions (stem cell/neuronal precursor derivatives)	BDNF, NGF, EGF, NT-3, CNTF, LIF, oncostatin M, CT-1
Axon guidance	NGF, BDNF, HGF, GDNF, neurturin
Neurite growth, branching	BDNF, NT-3, NT-4, CNTF, CT-1, HGF, GDNF, LIF, IGF-I
Dendrite development	BDNF, NT-3, NGF, GDNF
Synapse development	BDNF, NT-3, GDNF, neurturin, NT-4, CNTF
Synapse stabilization and plasticity	BDNF, NT-3, GDNF, FGFs, HGF, VEGF, CNTF, LIF, NGF, IGF-I
Long-term potentiation, learning, and memory	BDNF, NT-3, NGF, NT-4, IGF-I, interleukins
Pain physiology and pathophysiology	NGF, BDNF, GDNF, artemin, NT-3
Neurotransmitter phenotype	NGF, LIF, FGFs, GDNF, CNTF, BDNF, NT-3, IGF-I
Ion channel regulation/function	BDNF, interleukins
Neurological disease and neurodegeneration	
Seizure, epilepsy	BDNF, NGF, NT-3, VEGF, GDNF, LIF, CT-1, oncostatin M, CNTF
Stress, ischemia	NGF, NT-3, BDNF, IGF-I, VEGF, GDNF, HGF, BMPs
Traumatic injury, axonal regeneration, inflammation, stress	GDNF, IGF-I, LIF, CNTF, oncostatin M, CT-1, NGF
Psychiatric disease: depression	BDNF, NGF, FGFs, IGF-I, GDNF

*For references, see Snider, 1994; McAllister *et al.*, 1999; Huang and Reichardt, 2001; Poo, 2001; Chao, 2003; Zweifel *et al.*, 2005.

and motor neurons. FGFs may play important roles during development or after injury. NTFs have different functions during different developmental stages: After early effects on proliferation and phenotype determination (stem cells), NTFs influence neuronal migration, survival, differentiation, and dendritic as well as axonal growth. At later stages, NTFs regulate synaptic competition and plasticity, LTP, responses to injury, regeneration, and adult neurogenesis. The activity of multiple trophic factors from multiple sources is integrated to achieve a functional circuitry.

PROGRAMMED CELL DEATH IS REGULATED BY INTERACTIONS WITH TARGETS, AFFERENTS, AND NONNEURONAL CELLS

The PCD of vertebrate neurons and their precursors can occur at any stage of neuronal development from neurogenesis and neurulation to the time of establishment of synaptic connections with targets and afferents, and can involve mitotically active cells and migrating neurons, as well as undifferentiated and immature postmitotic cells (Fig. 19.4). However, the most common and historically the best studied type of PCD of neurons involves postmitotic, differentiating cells that die while establishing synaptic connections with other neurons and target cells (Oppenheim, 1991; Pettmann and Henderson, 1998). Because massive neuronal death and its regulation were first clearly recognized in early studies of neuron-target interactions (Table 19.1, Box 19.1), the role of targets in controlling PCD historically has received the most attention. However, as discussed in the previous section, signals derived from afferent inputs, as well as from nonneuronal cells such as central and peripheral glia and endocrine glands (e.g., steroid hormones), have now been shown to be possible sources of trophic regulation of cell death and survival (Fig. 19.16).

Studies of the regulation of vertebrate neuronal PCD have shown that targets are critically involved in regulating how many postmitotic cells in the innervating population survive or die. Complete or partial deletion of targets reduces survival, whereas increasing the size or number of available targets results in increased survival (Oppenheim, 1991; Buss *et al.*, 2006). Although it is thought that the relationship between neuronal survival and the availability of synaptic targets is proportional and linear (this has been called numerical-, quantitative-, size-, or systems-matching), there may also be other ways in which the number of neurons that innervate a target can be regulated. For example, signals from glial cells along axon pathways may modulate neuron numbers prior to target innervation. Spinal motoneurons provide one of the best examples of how systems-matching is regulated by a competition for target-derived signals (NTFs) that promote survival (the Neurotrophic Hypothesis).

In the case of avian spinal motoneurons that innervate limb skeletal muscle, the number of neurons that survive the period of PCD bears a 1:1 relationship with the number of primary myotubes present in individual muscle precursors during the period of cell death, rather than being correlated with the final number of myotubes or myofibers (i.e., muscle size)

present after the cessation of cell death. Accordingly, in this situation, motoneuron numbers are controlled, in part, by signals that are limited by the number of primary myotubes available. Although for many populations of neurons the essential factors provided by targets that mediate neuronal survival are not known, extrapolation from what is known for sensory, sympathetic, and motor neurons suggests that specific target-derived NTFs are involved. Target-dependent motoneuron survival is mediated by muscle-derived proteins. Several candidate motoneuron NTFs have been identified and include BDNF, NT-4/5, IGF, HGF, and CNTF and GDNF family members. The PCD of CNS neurons in the avian isthmo-optic nucleus (ION) and neurons in the mammalian thalamus and substantia nigra are controlled by many of the same mechanisms involved in the PCD of peripheral neurons, including a need for target-derived NTFs such as BDNF. As discussed previously, the survival of myelin-forming glial cells also involves competition, in this case, competition for trophic signals derived from axons. In both neurons and glia, the final outcome of this competitive process (Figs. 19.4 and 19.5) is the survival of optimal numbers of cells for innervation (neurons) or myelination (glia) (Oppenheim *et al.*, 2001).

Motoneurons (and some other neuronal populations as well, including retinal ganglion cells, isthmo-optic neurons (ION) cells, and ciliary ganglion neurons) have another interesting property in that their target dependency appears to be regulated by physiological synaptic interactions with their targets (Oppenheim, 1991). For example, following the formation of initial synaptic contacts between motoneurons and target muscles, the initiation of synaptic transmission activates the muscle and results in overt embryonic movements. Chronic blockade of this activity during the cell death period with specific drugs or toxins that cause paralysis prevents the death of all motoneurons (Fig. 19.17). Although the cellular and molecular mechanisms that mediate this effect are not entirely clear, two major hypotheses have been proposed: the *production hypothesis*, which predicts that the production (expression) of NTFs by the target is regulated inversely by target muscle activity; and the *access hypothesis*, which argues that sufficient NTF initially is produced by targets to maintain all motoneurons, but that activity regulates access to NTFs by modulating axonal branching and synaptogenesis, thus restricting the uptake of the NTFs to axon terminals or synapses. At present, the available evidence favors the access hypothesis (Terrado *et al.*, 2001).

Because PCD is enhanced after the pharmacological blockade of afferent synaptic activity, the functional

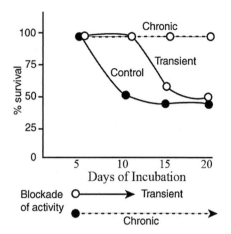

FIGURE 19.17 Daily treatment of the chick embryo with neuromuscular blocking agents during the major period of spinal motoneuron cell death (days 5–10) results in the inhibition of muscle activity and the rescue of motoneurons from PCD (hatching occurs on day 21). Following the cessation of treatment on E10 (transient blockade of activity), the rescued motoneurons undergo delayed PCD as muscle activity gradually recovers, whereas treatment from days 5–20 (chronic blockade of activity) maintains the rescued motoneurons after day 10.

input provided by afferents also appears to be of fundamental importance in this situation (Linden, 1994; Oppenheim *et al.*, 2001). Functional afferent input may act to regulate the survival of both developing and adult-generated postsynaptic neurons by several different mechanisms:

1. Neurotransmitter release results in depolarization by afferents that can alter intracellular calcium levels in postsynaptic cells, which in turn can independently modulate survival.
2. Afferent activity can regulate the expression of NTFs and their receptors in postsynaptic cells.
3. The activity-dependent release of NTFs from terminals of afferent axons or from adjacent glial cells may also provide survival signals to postsynaptic cells.

Because the PCD of many developing neurons may be coordinately regulated by activity-dependent and independent signals derived from targets, afferents, and nonneuronal cells, an important unresolved issue is how these different sources interact to control survival. In the final analysis, the various target- and afferent-derived extrinsic signals influence survival by modulating intracellular signaling pathways that regulate the expression of pro- and anti-apoptotic gene products (Figs. 19.9, 19.14). Although targets and afferents appear to be the major source of signals that regulate the survival of differentiating neurons, they are not likely to be the only source (Fig. 19.16).

Although the cellular mechanisms are not yet well established, the survival and death of newly generated neurons in the adult hippocampus and olfactory system provide another example of how afferent activity can regulate these events. For example, learning, stress, motor activity, sensory stimuli, and other environmental influences appear to control granule cell numbers in the dentate gyrus of the hippocampus by activity-dependent modulation of neuronal survival (Fig. 19.18). Similar to developing neurons during embryonic and postnatal stages, the default state of developing adult-generated neurons is PCD and (as also occurs during embryonic and postnatal stages) their survival and incorporation into functional circuits requires synaptic input and NTF signaling (Kempermann, 2006).

Summary

During development, the fate of a cell, including the decision to live or die, can be determined by intrinsic cell-autonomous mechanisms or by extrinsic signals derived from cell–cell interactions. The survival of most developing neurons is likely to be dependent on a variety of signals, including multiple NTFs derived from diverse sources, that serve to maintain survival and regulate differentiation in complex ways that reflect the specific requirements of neurons at each step in their development. Although developing neurons can undergo PCD at any stage of differentiation, the best studied type of PCD occurs as neurons are establishing connections with targets and afferents. Targets and afferents provide critical survival-promoting signals that prevent PCD. One major class of such signals are NTFs. Neurons are thought to compete for limiting amounts of these NTFs. For many populations of neurons, synaptic transmission and physiological activity also plays an important role in regulating survival. Accordingly, both NTFs and electrical activity/synaptic transmission are important for the regulation of PCD and are closely linked functionally.

THE ROLE OF PROGRAMMED CELL DEATH IN NEUROPATHOLOGY

The widespread occurrence of PCD during normal development indicates that cell loss, together with cell production (proliferation), is a fundamental mechanism for controlling final cell numbers in many tissues. Accordingly, the dysregulation of normal PCD could be maladaptive and pathological. For example, devel-

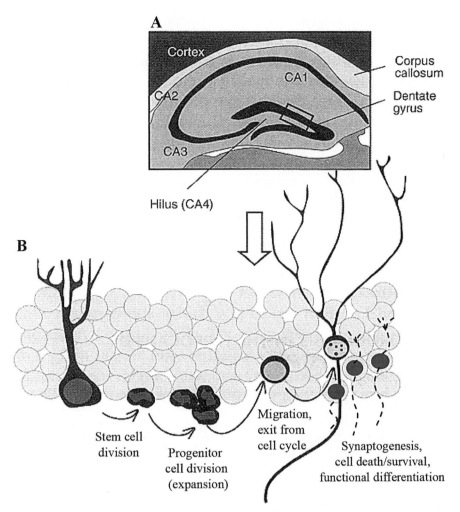

FIGURE 19.18 (A) Schematic illustration of the hippocampus including the dentate gyrus a portion of which is shown enlarged in B. (B) The dentate gyrus is comprised of several layers of differentiated granule cells (light gray circles and dark cell with dendrites on the left). Stem cells in the lower subgranule layer of the dentate gyrus undergo cell division and a subset of these migrate into the granule cell layer and differentiate (cell with dendrites on the right), whereas others migrate and begin differentiation but then undergo PCD (small dark cells with dashed axons and dendrites). Modified and redrawn from Kempermann (2006).

opmental, genetic, or congenital neurological defects may be caused by perturbations of PCD. However, there is relatively little solid evidence on this point. Although there are genetic mutations in animals and humans that involve significant alterations in neuron numbers, whether these always reflect the loss of control of normal PCD or reflect other abnormalities (e.g., altered neurogenesis) that could also influence final cell numbers often is not known. Nonetheless, the idea that aberrant cell death in the developing and adult nervous system may reflect a loss of normal control mechanisms for survival is appealing in that the increased understanding of the genetic regulation of PCD provides a potentially powerful and rational approach for the development of therapeutic treatment strategies.

A hopeful sign in this regard comes from reports that neuronal (or glial) death in patients with Alzheimer's, Parkinson's, Huntington's and Motor Neuron Disease, Down syndrome, multiple sclerosis, retinal degeneration, epilepsy, and in some forms of traumatic CNS injury (e.g., head trauma, ischemia, cerebral stroke, spinal cord injury, perinatal asphyxia) exhibit some characteristics of apoptosis, such as the stereotyped fragmentation of DNA and the expression of genes or gene products associated previously with developmental PCD (e.g., the bcl–2 gene family) (Vila and Przedborski, 2003; Blomgren et al., 2007). In some of these diseases, specific gene mutations have been identified that may directly or indirectly perturb inter- and intracellular mechanisms involved in the maintenance and survival of specific populations of neurons.

Other brain pathologies in which neuron loss by PCD may be involved include fetal alcohol syndrome, schizophrenia, major depression disorders, Shy-Drager syndrome, narcolepsy, and fetal exposure to excitotoxic drugs. Therefore, in addition to the important diverse roles of PCD during normal development and in adult tissue homeostasis, studies leading to a better understanding of the cellular and molecular mechanisms of PCD may ultimately shed new light on the causes and prevention of a variety of neurological disorders and brain pathologies that affect large numbers of the human population. As we learn more about the molecular components of the cell death pathways and their role in neurological disease and pathology, there will be increased opportunities for therapeutic intervention.

Despite this promising perspective for utilizing the increased understanding of the biology of PCD for developing therapeutic strategies to treat pathological neuron death, it is important to consider the validity of the widely held assumption that it is the death of neurons that is the primary cause of the pathophysiology in these diseases. In fact, there is growing evidence that neuronal and synaptic dysfunction that precedes any significant neuronal loss may be a major factor in the behavioral pathology that characterizes many neurodegenerative diseases (Gould and Oppenheim, 2007; Palop *et al.*, 2006).

Summary

Because PCD is primarily a developmental phenomenon, historically the major focus of investigation has been the normal biology of cell death in the embryo, fetus, and newborn. However, with the growing recognition that pathological neuronal cell death may share certain biochemical and molecular features with PCD, there is growing hope that a better understanding of PCD may reveal potential therapeutic strategies for the treatment of developmental neuropathologies, neurodegenerative disease, and neuronal loss following CNS trauma.

References

Airaksinen, M. S. and Saarma, M. (2002). The GDNF family: Signaling, biological functions and therapeutic value. *Nat. Rev. Neurosci.* 3, 383–394.

Ameisen, J. C. (2004). Looking at death at the core of life in the light of evolution. *Cell Death Differ.* 11, 4–10.

Barde, Y. A. (1989). Trophic factors and neuronal survival. *Neuron* 2, 1525–1534.

Barker, P. A. (2004). p75NTR is positively promiscuous: novel partners and new insights. *Neuron* 42, 529–533.

Blomgren, K., Leist, M., and Groc, L. (2007). Pathological apoptosis in the developing brain. *Apoptosis* 12, 993–1010.

Bredesen, D. E., Rao, R. V., and Mehlen, P. (2006). Cell death in the nervous system. *Nature* 443, 796–802.

Buss, R. R., Sun, W., and Oppenheim, R. W. (2006). Adaptive roles of programmed cell death during nervous system development. *Annu. Rev. Neurosci.* 29, 7–35.

Chao, M. V. (2003). Neurotrophins and their receptors: A convergence point for many signaling pathways. *Nat. Rev. Neurosci.* 4, 299–309.

Clarke, P. G. H. (1990). Developmental cell death: Morphological diversity and multiple mechanisms. *Anat. Embryol.* 181, 195–213.

Clarke, P. G. H. and Oppenheim, R. W. (1995). Neuron death in vertebrate development: In vivo methods. *Methods in Cell Biol.* 46, 277–323.

Clarke, P. G. H. and Clarke, S. (1996). Nineteenth century research on naturally occurring cell death, and related phenomena. *Anal. Embryol.* 193, 81–99.

Copani, A., Uberti, D. Sortino, M. A., Bruno, V., Nicoletti, F., and Memo, M. (2001). Activation of cell-cycle-associated proteins in neuronal death. *Trends Neurosci.* 24, 25–31.

Cowan, W. M. (2001). Viktor Hamburger, and Rita Levi-Montalcini: The path to the discovery of nerve growth factor. *Annu. Rev. Neurosci.* 24, 551–600.

Egan, M. F., Kojima, M., Callicott, J. H., Godlgerg, T. E., Kolachana, B. D., Bertolino, A., Zaitsev, E., Gold, B., Goldman, D., Dean, M., Lu, B., and Weinberger, D. R. (2003). The BDNF val66met polymorphism affects activity-dependent secretion of BDNF and human memory and hippocampal function. *Cell* 112, 257–269.

Gauthier, L. R., Charrin, B. C., Borrell-Pages, M., Dompierre, J. P., Rangone, H., Cordelieres, F. P., DeMey, J., MacDonald, M. E., Lessmann, V., Huymbert, S., and Saudou, F. (2004). Huntingtin controls neurotrophic support and survival of neurons by enhancing BDNF vesicular transport along microtubules. *Cell* 118, 127–138.

Gilbert, S. F. (2006). "Developmental Biology," 8th ed. Sinauer, Sunderland, MA.

Gould, T. W. and Oppenheim, R. W. (2007). Synaptic dysfunction vs cell death in the pathogenesis of neurological disease and following injury in the developing and adult nervous system: Caveats in the choice of therapeutic strategies. *Neuroscience and Biobehavioral Reviews* doi:10.1016/j.neubiorev.2007.04.015.

Horvitz, H. R. (2003). Worms, life and death (Nobel lecture). *Chembiochem.* 4, 697–711.

Huang, E. J. and Reichardt, L. F. (2001). Neurotrophins: Roles in neuronal development and function. *Annu. Rev. Neurosci.* 24, 677–736.

Ip, N. and Yancopoulos, G. (1996). The neurotrophins and CNTF: Two families of collaborative neurotrophic factors. *Ann. Rev. Neurosci.* 19, 491–515.

Johnson, E. M. and Deckwerth, T. L. (1993). Molecular mechanisms of developmental neuronal death. *Annu. Rev. Neurosci.* 16, 31–46.

Kempermann, G. (2006). "Adult Neurogenesis: Stem Cells and Neuronal Development in the Adult Brain." Oxford University Press, New York.

Kornbluth, S. and White, K. (2005). Apoptosis in *Drosophila*: Neither fish nor fowl (nor man, nor worm). *J. Cell Sci.* 118, 1779–1787.

Lessmann, V., Gottmann, K., and Malcangio, M. (2003). Neurotrophin secretion: Current facts and future prospects. *Prog. Neurobiol.* 69, 341–374.

Lewin, G. R. and Barde, Y. A. (1996). Physiology of the neurotrophins. *Annu. Rev. Neurosci.* 19, 289–317.

Linden, R. (1994). The survival of developing neurons: A review of afferent control. *Neuroscience* 58, 671–682.

Lohof, A. M., Ip, N. Y., and Poo, M. M. (1993). Potentiation of developing neuromuscular synapses by the neurotrophins NT-3 and BDNF. *Nature* **363**, 350–353.

McAllister, A. K., Katz, L. C., and Lo, D. C. (1999). Neurotrophins and synaptic plasticity. *Annu. Rev. Neurosci.* **22**, 295–318.

McDonald, N. Q. and Rust, J. M. (1995). Insights into neurotrophin function from structural analysis. *In* "Life, and Death in the Nervous System" (C. F. Ibanez, T. Hokfelt, L. Olson, K. Fuxe, H. M. Jornvall, and D. Ottoson, eds.), pp. 3–18. Pergamon, Oxford.

Oppenheim, R. W. (1981). Neuronal cell death and some related regressive phenomena during neurogenesis. *In* "Studies in Developmental Neurobiology: Essays in Honor of Viktor Hamburger" (W. M. Cowan, ed.), pp. 74–133, Oxford, New York.

Oppenheim, R. W. (1996). The concept of uptake and retrograde transport of neurotrophic molecules during development: History and present status. *Neurochem. Res.* **21**, 769–777.

Oppenheim, R. W. (1991). Cell death during development of the nervous system *Annu. Rev. Neurosci.* **14**, 453–501.

Oppenheim, R. W., Caldero, J., Esquerda, J., and Gould, T. (2001). Target-independent programmed cell death in the developing nervous system. *In* "Brain, and Behaviour in Human Development" (A. F. Kalverboer and A. Gramsbergen, eds.), pp. 343–407, Kluwer, Dordrecht, The Netherlands.

Palop, J. J., Chin, J., and Muckey, L. (2006). A network dysfunction perspective on neurodegenerative disease. *Nature* **443**, 768–773.

Pettmann, B. and Henderson, C. E. (1998). Neuronal cell death. *Neuron* **20**, 633–647.

Poo, M. M. (2001). Neurotrophins as synaptic modulators. *Nat. Rev. Neurosci.* **2**, 24–32.

Raff, M. C. (1992). Social controls on cell survival, and cell death. *Nature* **356**, 397–400.

Roux, P. P. and Barker, P. A. (2002). Neurotrophin signaling through the p75 neurotrophin receptor. *Prog. Neurobiol.* **67**, 203–233.

Sanes, D., Reh, T. A., and Harris, W. A. (2006). "Development of the Nervous System." *Academic Press*, New York.

Schmitz, C. and Hof, P. R. (2005). Design-based stereology in neuroscience. *Neuroscience* **130**, 813–831.

Segal, R. and Greenberg, M. (1996). Intracellular signaling pathways activated by neurotrophic factors. *Annu. Rev. Neurosci.* **19**, 463–489.

Snider, W. D. (1994). Functions of the neurotrophins during nervous system development: What the knockouts are teaching us. *Cell* **77**, 627–638.

Terrado, J., Burgess, R. W., DeChiara, T., Yancopoulos, G., Sanes, J. R., and Kato, A. C. (2001). Motoneuron survival is enhanced in the absence of neuromuscular junction formation in embryos. *J. Neurosci.* **21**, 3144–3150.

Vila, M. and Przedborski, S. (2003). Targeting programmed cell death in neurodegenerative diseases. *Nature Rev. Neurosci.* **4**, 365–375.

von Bartheld, C. S. and Fritzsch, B. (2006). Comparative analysis of neurotrophin receptors and ligands in vertebrate neurons: Tools for evolutionary stability or changes in neural circuits? *Brain Behav. Evol.* **68**, 157–172.

von Bartheld, C. S., Wang, X., and Butowt, R. (2001). Anterograde axonal transport, transcytosis, and recycling of neurotrophic factors: The concept of trophic currencies in neural networks. *Mol. Neurobiol.* **24**, 1–28.

Zweifel, L. S., Kuruvilla, R., and Ginty, D. D. (2005). Functions and mechanisms of retrograde neurotrophin signaling. *Nat. Rev. Neurosci.* **6**, 615–625.

Suggested Readings

Bothwell, M. (2006). Evolution of the neurotrophin signaling system in invertebrates. *Brain Behav. Evol.* **68,** 124–132.

Danial, N. N. and Korsmeyer, S. J. (2004). Cell death: Critical control points. *Cell* **116**, 205–219.

Ellis, R. E., Yuan, J. Y., and Horvitz, H. R. (1991). Mechanisms and functions of cell death. *Annu. Rev. Cell Biol.* **7**, 663–698.

Hamburger, V. (1992). History of the discovery of neuronal death in embryos. *J. Neurobiol.* **23**, 1116–1123.

Jacobson, M. (1991). "Developmental Neurobiology." *Plenum*, New York.

Lee, R., Kermani, P., Teng, K. K., and Hempstead, B. L. (2001). Regulation of cell survival by secreted proneurotrophins. *Science* **294**, 1945–1948.

Levi-Montalcini, R. (1987). The nerve growth factor 35 years later. *Science* **237**, 1154–1162.

Purves, D. (1988). "Body and Brain: A Trophic Theory of Neural Connections." Harvard, Cambridge, MA.

Yuan, J. and Yankner, B. A. (2000). Apoptosis in the nervous system. *Nature* **407**, 802–809.

Ronald W. Oppenheim and
Christopher S. von Bartheld

Synapse Elimination

OVERVIEW

In the development of both central and peripheral nervous systems, synapse formation generates some connections that exist only transiently. Proof that synapses are eliminated comes from the finding that in many parts of the mammalian nervous system, presynaptic axons are transiently connected to postsynaptic partners during development. One unambiguous example is the loss of synapses between motoraxons and muscle fibers. At birth, each neuromuscular junction receives convergent innervation from multiple motor axons (Fig. 20.1A,B). In adult muscles, however, only a single motor axon innervates each neuromuscular junction (Fig. 20.1C). In another part of the developing peripheral nervous system, autonomic ganglia, axons also transiently supply extra inputs to each ganglion cell (Fig. 20.2). In the developing central nervous system (CNS), several cases of synapse elimination are also well known (Table 20.1). For example, in the cerebellum multiple climbing fibers innervate each Purkinje cell at birth, whereas all but one is removed in early postnatal life. In the auditory system, developing neurons in the magnocellularis nucleus receive innervation from several inputs but in later life only one axonal input per neuron remains. Perhaps the best-known examples of synapse elimination occur in the visual system.

In all these cases, synapse elimination reduces the number of axons that innervate target cells, making the term *input elimination* perhaps a better way of describing the phenomenon. In most of the developing CNS, however, it remains unknown whether synapse elimination is occurring. The difficulty is that *total* number of synapses may not be decreasing at the stage when axonal inputs are being eliminated. Thus assaying for synapse elimination requires quantifying a change in the number of innervating axons, which in turn requires a means of counting them, or at least having a situation where distinct but overlapping pathways can be separately labeled or stimulated (such as the two eyes). It is therefore possible that there is far more synapse elimination during development than presently documented. For example, stimulation of the optic nerve with progressively larger voltages to count the number of retinal ganglion cells innervating individual lateral geniculate neurons has revealed a dramatic change from more than 20 innervating axons to only 1 or 2 during the first postnatal month (Chen and Regehr, 2000).

Importantly, synapse elimination does not lead to a net weakening of synaptic drive on target cells. In the example just mentioned, the retinogeniculate axons that persist undergo a greater than 50-fold increase in the efficacy, which more than compensates for the loss of other axonal inputs to the same target cells. Analogous increases in synaptic efficacy of inputs that persist have been described for the excitatory drive to neurons of the magnocellularis nucleus (Lu and Trussell, 2007) and for the climbing fiber input to Purkinje cells in the cerebellum (Hashimoto and Kano, 2003). In each case the inputs that remains increase their efficacy by 10-fold or more as other axons are eliminated. The same trend was seen many years ago in the parasympathetic submandibular ganglion, where the remaining single input was always far more powerful than any of the inputs at the earlier stage when five to six multiple axons converged to the same single target cell

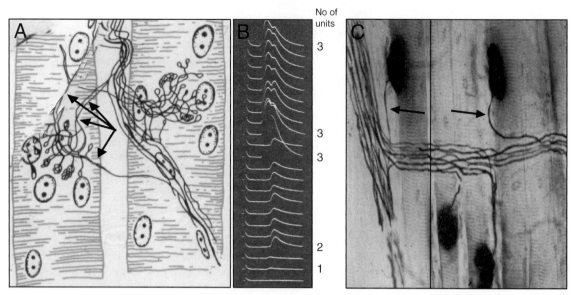

FIGURE 20.1   Evidence of synapse elimination at the neuromuscular junction. (A) Drawing of a silver stain made by J. Boeke in 1932 of the motor nerve terminals converging on two adjacent muscle fibers from the tongue muscle of a 4-day-old mouse. Each fiber is innervated by endings from several different axons (see arrows). (B) Intracellular recordings from a muscle fiber in the neonatal rat diaphragm showing multiple innervation of a skeletal muscle fiber. Gradually increasing the strength of motor nerve stimulation elicits three different postsynaptic responses in this fiber. Similar assays in adult animals give only a single postsynaptic response that cannot be fractionated by graded stimulation. The preparation was partially curarized to prevent the endplate potentials from reaching threshold. (C) Photomicrograph of several adult muscle fibers from the rat extensor digitorum longus muscle. Each adult fiber receives innervation from a single axon (arrows). Axons are impregnated with silver and stained for cholinesterase (dark oval smudge on each muscle fiber). Figures adapted from (A) Boeke (1932); (B) Redfern (1970); and Gorio *et al.*, (1983).

(Lichtman, 1977). Similarly, at the neuromuscular junction, the remaining motor axon increases its quantal content as other inputs are eliminated (Colman *et al.*, 1997). It thus appears that although some inputs are being removed, the survivors are potentiated. Such results raise the outstanding question of the degree to which the compensation is explained by changes in the efficacy of the maintained synapses or rather the elaboration of new synapses.

A partial answer comes from studies in the peripheral submandibular ganglion. Here, the simplicity of the neuropil (most synapses are axosomatic) allows for direct counts of synapse number during the period of synapse elimination. In this ganglion, input elimination was accompanied by concurrent synapse addition by the remaining input. In particular, over the first postnatal month, there was approximately a twofold increase in the total number of synapses per ganglion cell, whereas the number of innervating axons was reduced fivefold (Fig. 20.2). This compensation indicates that the remaining axon on each ganglion cell must be increasing the number of synaptic contacts by approximately a factor of 10 as other axons lose all

their connections to the same neuron. Such compensatory synaptogenesis means that assays of the total *number* of synapses at various developmental stages or following learning paradigms may belie a rather dramatic change in the *source* of the synapses.

It is also clear that in many situations the alterations occurring during this developmental reorganization are better described as a *redistribution of synapses* than as loss per se. The view is based on the fact that counts of the total number of neurons that innervate the target is not changing during the stage when the number of axons innervating each postsynaptic cell is decreasing (see for example, Brown *et al.*, 1976). Moreover, in several regions of the nervous system, it is known that synapse elimination occurs after the period of *naturally occurring cell death* is complete (Chapter 19). Thus, as the elimination process reduces the number of axons converging on postsynaptic cells, it does not entirely remove axons from the postsynaptic target region but rather reduces the number of target cells innervated by each axon—that is, it reduces the axonal divergence as individual axons restrict their synapses to a smaller number of postsynaptic target cells (Fig. 20.3). In

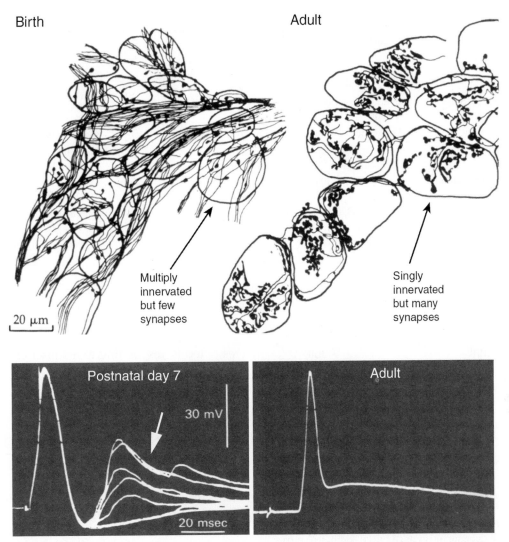

FIGURE 20.2 In the developing nervous system synapse number is increasing while axonal convergence is falling. The total number of preganglionic synapses on ganglion neurons in the rat submandibular ganglion increases in early postnatal life during the period of synapse elimination. Camera lucida drawings of clusters of ganglion neurons at birth and in adult animals that were treated with zinc iodide osmium. This reagent selectively fills terminal axons and synaptic boutons with a dense black precipitate. The number of synaptic boutons increases (compare the amount of synaptic boutons (arrows) at birth and on adult ganglion cells), and the number of preganglionic axons innervating each ganglion cell, identified by electrophysiological measurements, decreases (see the number of electrophysiological steps (arrow), each representing an input, at early postnatal life and in adult ganglion neurons, see lower panels). Thus, axons that are not removed create new synapses to more than compensate for the loss of synapses from other axons. From Lichtman (1977).

muscle, axonal branch trimming during development has been long appreciated by an electrophysiological assay: the proportion of the total twitch tension an individual axon exerts decreases gradually over early postnatal life (Brown *et al.*, 1976).

Newer techniques of visualizing all the branches of single axons by transgenic expression of GFP have allowed this *branch retraction* to be observed directly (Fig. 20.4). Time-lapse imaging shows that when two axons coinnervate the same neuromuscular junction in development one axon atrophies, then disconnects

from the postsynaptic specialization and temporarily appears as a bulb-tipped free axonal ending called a *retraction bulb*. The withdrawing axon sheds small spherical membrane-enclosed axonal fragments filled with synaptic organelles (axosomes) to be engulfed by surrounding Schwann cells (Fig. 20.5).

Summary

In the developing nervous system both the number of axons that converge on postsynaptic cells, and the

TABLE 20.1   Synapse Elimination in the Mammalian Nervous System

Visual cortex (layer IV)	Binocular to monocular	Postnatal monkey, cat, ferret	Hubel and Wiesel (1963); Hubel *et al.* (1977); Wiesel (1982)
Thalamus (lateral geniculate nucleus)	Binocular to monocular, >20 to ~1–2 retinal axons	Postnatal cat and rat	Shatz (1990); Chen and Regehr (2000)
Retina (retinal ganglion cells)	On/off to on- or off-center receptive fields	Postnatal ferret	Wang *et al.* (2001)
Cerebellum (Purkinje cell)	>3 to 1 climbing fibers	Postnatal rat	Mariani (1983)
Cochlea (Magnocellularis cells)	>4 to 1 cochlear axon	Chick	Jackson and Parks (1982); Lu and Trussell (2007)
Parasympathetic (submandibular ganglion)	~5 to ~1 preganglionic axons	Postnatal rat	Lichtman (1977)
Sympathetic (superior cervical ganglion)	~14 to ~7 preganglionic axons	Postnatal hamster	Lichtman and Purves (1980)
Neuromuscular junction	2–6 to 1 motor axons	Postnatal rat, mouse	Redfern (1970); Brown *et al.* (1976)

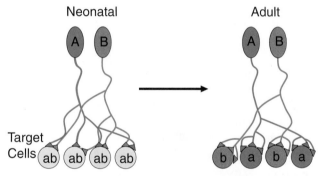

FIGURE 20.3   **Changes in fan-in and fan-out at developing circuits.** In neonatal vertebrates, neurons (A, green and B, red) project their axons to fan-out to many target cells (each labeled ab). These nascent connections are typically weak and mediated only by a small number of synaptic contacts (triangles from the green and red axons). Axons undergo several structural rearrangements before reaching adulthood. First, axons disconnect from many target cells reducing their fan-out or divergence. Second, this rearrangement leads to less fan-in or convergence. Third, at the same time as axons disconnect from some target cells they are strengthening their connection with other target cells by increasing the number of synapses at their remaining targets.

number of postsynaptic cells an axon contacts decreases. The removal of inputs is due to a process of branch trimming. Concurrently, the remaining inputs appear to compensate by adding synaptic strength at least in part by the formation of new synapses. This redistribution refines synaptic circuitry by allowing an axon to strongly focus its innervation on a subset of the cells it initially contacted while each postsynaptic cell is

restricted to responding to a subset of the axons that initially innervated it.

THE PURPOSE OF SYNAPSE ELIMINATION

Although at first sight it might seem reasonable that synapse elimination is some form of error correction to rid the developing nervous system of connectivity mistakes, this does not seem to be the case. In muscle for example, the lost connections are from exactly the motor neurons as those connections that are maintained because each neuron in the pool is losing some connections. In some situations the outcome of synapse elimination may *sharpen specificity* based on topographic maps (see for example, Laskowski *et al.*, 1998). It is thus possible that the outcome of synapse elimination is biased by the same kind of cues that promote selective synapse formation rather than synapse elimination being the mechanism that achieves the specificity. This view may explain why in some cases there seems to be little evidence of intrinsic positional or other qualitative differences between axons that are maintained and those that are lost from a particular postsynaptic cell.

An alternative hypothesis is that the loss of connections is a consequence of the extreme degree to which mammalian (and other vertebrate) nervous systems are composed of *duplicated neurons* (Lichtman and Colman, 2000). For example, pools of motor neurons that may number in the hundreds innervate individual

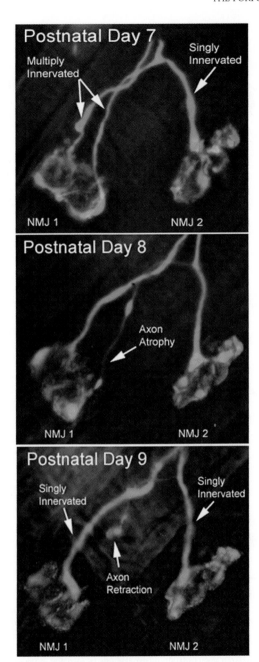

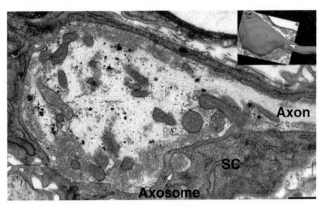

FIGURE 20.5   Retracting axons shed material as they disappear. The inset shows a surface rendering of a serial electron microscopy reconstruction of a retraction bulb (green, arrowhead) during synapse elimination at a developing neuromuscular junction. The electron micrograph shows that the retracting axon contains clusters of vesicles and mitochondria. In the vicinity of the bulb but not connected are small spherical axonal fragments (axosomes, arrow) that are engulfed by glial cells (i.e., Schwann cells (SC), darker cytoplasm). Scale bars, $1 \mu m$. From Bishop *et al.* (2004).

FIGURE 20.4   Time-lapse imaging shows how axonal branches are lost during synapse elimination at the neuromuscular junction. Two neuromuscular junctions (NMJ1 and NMJ2) were viewed *in vivo* on postnatal days 7, 8, and 9 in a transgenic mouse that expresses YFP in its motor axons (green). The acetylcholine receptors at the muscle fiber membrane are labeled with rhodamine tagged α-bungarotoxin (red). On day 7, NMJ1 is multiply innervated whereas NMJ2 is singly innervated by a branch of one of the axons that innervates NMJ1. One day later (postnatal day 8), one of the motor axons innervating NMJ1 becomes thinner all the way back to its branch point. When viewed on postnatal day 9, this thin branch is no longer connected to NMJ1 and now ends in a bulb-shaped swelling that is called a retraction bulb. Modified from Keller-Peck *et al.* (2001).

skeletal muscles containing thousands of muscle fibers. This differs from invertebrates that have single identified motor neurons innervating single muscle fibers (Fig 20.6A). Thus nearly identical motor neurons seem to have approximately equivalent roles as does the large population of comparable postsynaptic cells that constitute a muscle (Fig. 20.6B).

There are two potential consequences of such redundancy in pre- and postsynaptic populations. First, many presynaptic neurons could be appropriate matches for each postsynaptic cell causing substantial axonal convergence or fan-in. Second, many postsynaptic neurons are appropriate matches for each presynaptic axon causing substantial axonal divergence or fan-out. For example, in mammalian muscles, because the many nearly identical motor neurons projecting to one muscle may all be equally appropriate presynaptic partners for each muscle fiber, it is not unexpected that multiple axons can converge on the same target cell (Fig. 20.6C). At the same time, the multiple duplicated muscle fibers may all be equally appropriate targets for each motor neuron, causing each motor neuron to diverge to innervate many muscle fibers (Fig. 20.6D). Given the redundancy in both pre- and postsynaptic populations, it would not be surprising that in the mammalian motor system there are overlapping *converging and diverging pathways* (Fig. 20.6E). What is surprising, however, is that this overlap is short-lived. During neuromuscular development, axonal branches are pruned in a highly selective way, causing the projection of each motor

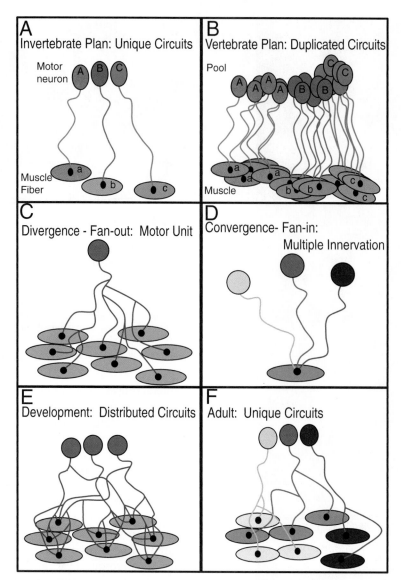

FIGURE 20.6 A diagram showing differences between the vertebrate and invertebrate synaptic circuits. (A) Invertebrates have small numbers (for example sometimes just one motor neuron) of identifiable neurons innervating small numbers of identifiable target cells (e.g., sometimes just one muscle fiber) whereas in vertebrates, pools of similar neurons innervating targets contain hundreds or thousands of similar postsynaptic cells (B). The redundancy in the vertebrate nervous system allows a neuron to diverge (fan-out) and innervate many equally appropriate target cells (in the neuromuscular system this is called a motor unit), (C). The homogeneity of innervating neurons allows multiple axons to converge (fan-in) on each target cell (D). As a result vertebrate circuits contain a substantial amount of fan-in and fan-out (E), at least initially. Synapse elimination's purpose may be to transform such a set of redundant circuits into multiple unique ones by trimming away the multiple innervation of target cells (F). The result of the widespread loss of synapses is the generation of thousands of nonredundant circuits from an initially much less specific innervation pattern. From Lichtman and Colman (2000).

axon to become completely nonoverlapping, and thus completely distinct from other axons' projections (Fig. 20.6F). From a functional standpoint, once synapse elimination is complete, the recruitment of each motor axon gives rise to activation of a different set of muscle fibers and consequently a substantive increase in tension of the muscle. This stepwise tension increase is a necessary aspect of the *size principle* (Chapter 19).

By parsing highly redundant circuitry into multiple, unique, functionally distinct circuits, synapse elimination may be an essential maturation step for synaptic circuits that begin with considerable redundancy, such as our own nervous systems and those of other terrestrial vertebrates. It is possible that this paring down strategy is less relevant in other kinds of animals.

Indeed a comparison between animals, such as insects, that show little evidence of neuronal redundancy and little evidence of input elimination with animals, such as mammals, with large redundant pools of neurons and extensive elimination during development suggests very different neurodevelopmental strategies. Humans and other mammals seem highly dependent on experience for the acquisition of their behavioral repertoire, whereas invertebrate behavior is, to a greater degree, intrinsic. Compare, for example, the ease with which a newly hatched dragonfly takes wing versus the protracted period necessary for a human child to learn to walk. It is possible that the neuronal redundancy found in higher vertebrates that give rise to overlapping convergent and divergent pathways are used in the acquisition of skills by the experience-mediated *selection* of connections.

This difference between mammals and invertebrates does not mean that synapses cannot be eliminated in invertebrates. Indeed, many studies have made the point that synaptic remodeling through addition and removal occurs in insect nervous systems (see for example, Eaton and Davies, 2003; Ding *et al.*, 2007). The molecular mechanisms involved in invertebrate synaptic remodeling may provide insights into the mechanisms of synaptic loss and maintenance in higher vertebrate nervous systems. In worms for example, the ubiquitin proteosome complex has been shown to regulate the elimination of motoneuron synapses onto vulval muscles (Ding *et al.*, 2007). Similar degradation mechanisms may also be involved in synaptic remodeling of vertebrate neurons. The important point here, however, is that these mechanisms are used in mammals to alter the connectivity of the nervous system (i.e., to alter the number and identity of an axon's postsynaptic targets). In most invertebrate systems and perhaps lower vertebrates such as fish these same mechanisms may play more of a role in modifying the strength of existing connections without altering the circuit's wiring diagram.

Summary

In contrast to invertebrates, the nervous systems of terrestrial vertebrates contain reduplicated populations of neurons that serve each function. Elimination of synaptic connections during development may be an adaptation that converts highly overlapping connections of redundant neurons into unique circuits. Because this conversion may be based on experiences that affect the development of the nervous system this process may tune the nervous systems of higher animals to the particular environment of each individual animal.

A STRUCTURAL ANALYSIS OF SYNAPSE ELIMINATION AT THE NEUROMUSCULAR JUNCTION

Thanks in large part to the power of fluorescence microscopy and the accessibility of the neuromuscular junction; it has been possible to describe the physical changes in axons and synapses that take place during synapse elimination. Interestingly this purely descriptive analysis has yielded important clues and some mechanistic insights into the process of synapse elimination. As Yogi Berra said, "You can observe a lot by just watching." Imaging neuromuscular junctions at various ages in early postnatal life makes clear the fact that the process of synapse elimination is not a sudden calamitous event but rather protracted process. At birth, multiple axons converging at a neuromuscular junction are highly intermingled and the extent of the receptor areas occupied by each is similar (Balice-Gordon *et al.*, 1993). These highly intermingled connections eventually become progressively segregated over several days (Fig. 20.7). The partitioning of synaptic areas associated with different axons suggests that there is a spatial component to the mechanism. One scenario for example is that each axon *locally destabilizes* other inputs in their immediate vicinity but cannot as efficiently affect slightly more distant branches of the same axons. If the destabilization of an input was followed in turn by the takeover of its synaptic sites by the remaining axon, then an axon with a larger consolidated area might begin to destabilize more distant inputs. The idea of progressive consolidation was put to the test by the use of transgenic mice expressing different color fluorescent proteins in individual axons. With these animals it is possible to obtain a precise map of the territories occupied by two axons at the same neuromuscular junction. *In vivo* time-lapse imaging of a multiple innervated neuromuscular junction over several days showed both gradual withdrawal of one axon from postsynaptic sites and a corresponding expansion of another axon to takeover those synaptic sites (Fig. 20.8). This result may mean that axons are vying to occupy the same sites. Interestingly, it was not inevitable that the withdrawal of one axon was followed by the takeover of its sites by another input. In some cases an axon that was already somewhat segregated from the other input would vacate its postsynaptic territory but rather than being replaced by the remaining axon, its acetylcholine receptor-rich postsynaptic site would disappear. Synaptic loss without reoccupation by another input suggested that the takeover process per se was not causing withdrawal of the other axon. These

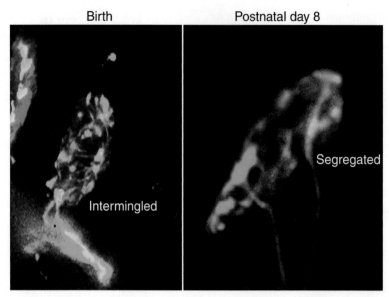

FIGURE 20.7 Segregation of synaptic territory as axons compete during synapse elimination. Double transgenic mice expressing YFP and CFP in different motor axons were used to observe axons nerve terminal interactions. At birth is shown two competing nerve terminals, one expressing CFP (blue) and the other YFP (green) at this neuromuscular junction, which are intermingled. By the second postnatal week, however, YFP and CFP axon terminals are generally completely segregated from each other in multiply innervated junctions. Red is fluorescently tagged alpha-bungarotoxin that binds to AChRs. Modified from Gan and Lichtman (1998).

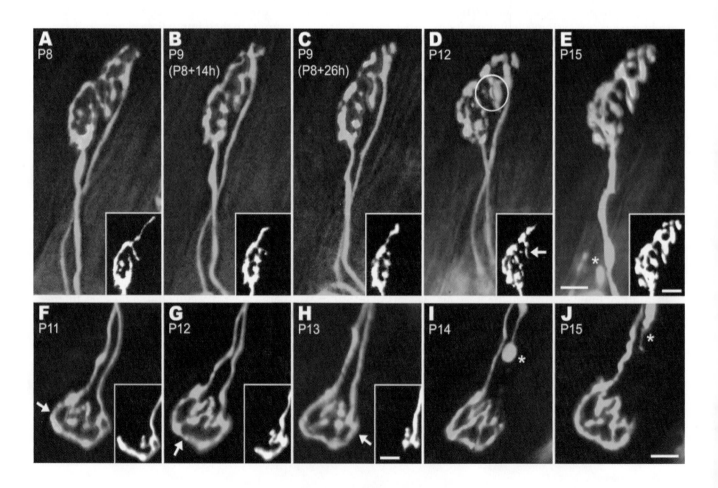

observations suggest that synaptic takeover might be a response to the recent synaptic vacancy.

These time-lapse studies also demonstrated that the shift in favor of one axon was not irreversible. In some junctions for example, it was hard to predict which axon might ultimately be maintained because the input with the majority of the territory shifted back and forth. Such flips in which axon seemed to be dominant suggest that the outcome is not preordained. Instead, axons at multiply innervated neuromuscular junctions appear may be in the midst of a highly dynamic interaction to determine which axon is most likely to stay.

Insights into the regulation of this dynamism come from analyses of each of the branches of individual motor axons. In lines of transgenic mice in which fluorescent proteins are expressed in a very small subset of motor neurons, it has been possible to examine all the branches of one axon during the developmental period when branches are being pruned. These studies show that axonal branch loss is occurring asynchronously among all the branches of one axon (Fig. 20.9). Thus while some branches are recently eliminated (i.e., retraction bulbs), other branches are still connected to neuromuscular junctions that are multiply innervated. This range suggests that the fate of each branch is controlled independently. If axonal branches of one neuron are interacting with different axons at each of its neuromuscular junctions, then perhaps the rate of synapse elimination is regulated by which particular axons that are coinnervating each of these junctions. These interactions might not only determine the rate but also the fates of the terminals; that is, which neuron's branches are maintained and which are eliminated. For example an axon might be eliminated quickly at neuromuscular junctions innervated by some other neurons, but fare better at junctions innervated by other axons.

To explore this possibility, transgenic mice with only two labeled axons (one yellow and one cyan) were generated to examine each of the neuromuscular junctions cooccupied by the same two axons within a developing muscle. The result was dramatic: at each of the shared neuromuscular junctions, the two axons

seemed to be in the same relative state (Fig. 20.10). Thus if the cyan axon was occupying only a small amount of territory at one junction that it shared with the yellow axon, then it occupied a small amount of territory at all the other junctions it shared with the same yellow axon. However, where the cyan axon was interacting with other axons, its fate could be quite

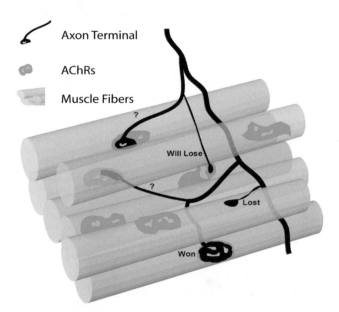

FIGURE 20.9 **Asynchronous synapse elimination among the branches of one axon.** Using transgenic animals that express YFP in a small subset of motor axons it was possible to monitor the behavior of multiple branches of the same axon. This diagram shows the typical result for an axon in the midst of the synapse elimination process. Represented in red are the AChRs on each muscle fiber (represented as gray tubes). This motor axon (black) has won the competition on the bottom muscle fiber and occupies the entire receptor plaque. The same axon has lost the competition for the adjacent muscle fiber, where only a retraction bulb remains. The three other neuromuscular junctions that this axon innervates are still undergoing competition. In one case a small portion of AChRs is being innervated by this axon. It is likely that this synapse will be eliminated. On the other two muscle fibers, each axon terminal occupies ~50% of the AChRs. This indicates that its fate is not yet determined in these junctions. Adapted from Lichtman and Colman (2000).

◄ FIGURE 20.8 *In vivo* **imaging shows takeover of synaptic territory by the remaining axon during the period of synapse elimination.** Neuromuscular junctions in transgenic mice that express YFP (yellow) and CFP (blue) were imaged multiple times during early postnatal life. (A–E) views of one neuromuscular junction between P8 and P15. The CFP axon takes over occupancy of the postsynaptic sites (labeled red) in the upper parts of the junction that were formerly innervated by the YFP axon. The YFP labeled axon withdrew until only a retraction bulb remained (E, asterisk). At P12, a process of the CFP axon had begun to invade the territory of the YFP axon (D, circle and arrow in inset). (F–J) Although the CFP axon (blue and insets) has greater terminal area (~70%) at the first view, it progressively withdraws from the junction (arrows). Its retraction bulb can be seen in (I) and (J) (asterisks). Scale bars = 10 μm. Insets show the blue axons. From Walsh and Lichtman (2003).

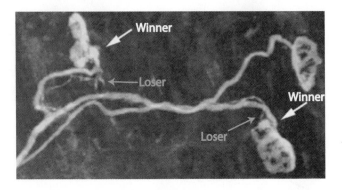

FIGURE 20.10 **Axonal fate at individual neuromuscular junctions is related to the identity of the competing axons.** Using transgenic animals that express YFP (yellow) and CFP (blue) in a small subset of motor axons was possible to reconstruct two motor units in which several neuromuscular junctions are innervated by the exact same two axons, one expressing YFP and the other CFP. At these coinnervated junctions it appears that the same outcome is occurring at each. In this case, the YFP axon always occupied more territory than the CFP axon. Note also that the axon calibers of the YFP axon branches are thicker than CFP branches to the same junctions. Thus synapse elimination seems to be biased such that when the same two axons compete the fate is the same at each coinnervated neuromuscular junction. Image from Kasthuri and Lichtman (2003).

different. This result argued that the fate of axonal branches and the rate at which they are eliminated is related strictly to the identity of the coinnervating axons. But why might the yellow axon be consistently "better" than the blue at all the coinnervated neuromuscular junctions they share? Counts of the total number of neuromuscular junctions each axon innervated (i.e., its motor unit size) provided a hint; the outcome of synapse elimination seemed to depend on the relative sizes of the two axons' motor units. In particular, neurons with larger motor units (such as the cyan axon in Fig. 20.10) were at a disadvantage when confronting neurons with smaller arborizations (such as the yellow axon in Fig. 20.10). One interpretation of this result is that axons with few branches in a muscle could dedicate more resources to a multiply innervated neuromuscular junction than neurons with a larger number of branches that are in some sense overextended. If this idea is correct then the consequence for an axon (such as the cyan one in Fig. 20.10) of losing branches is that it can now dedicate more resources to its remaining multiply innervated junctions shared with other axons.

These results raise the obvious question of what resources might be in limited supply in an axon that could affect synapse maintenance. One popular idea is that the synaptic activity of an axon terminal is a key

determinant in its ultimate fate. Is it possible that large motor units have fewer synaptic resources (e.g., synaptic vesicles per synapse) than small motor units? This idea was tested by generating mice in which a subset of axons had diminished choline acetyltransferase (ChAT, the enzyme that synthesizes the acetylcholine; Buffelli *et al.*, 2003). The results showed that when axons containing normal levels of ChAT coinnervated junctions with axons that had subnormal amounts, the subnormal axons typically occupied smaller territories. This result is consistent with the idea that axons that have the greatest number of synaptic branches may be at a disadvantage because they cannot maintain sufficient resources to drive postsynaptic cells as well as axons with fewer branches.

Summary

Structural studies of synapse elimination at the neuromuscular junction have provided a number of clues as to how and why branches are pruned. One important insight is that synapse elimination at individual neuromuscular junctions may be part of a larger scale circuit optimization. This optimization may be occurring to assure that all the connections that are maintained into adulthood are sufficiently strong that they can consistently drive the postsynaptic cell to threshold. Perhaps such an optimization requires axons to forfeit some of their synaptic connections to assure that the ones that remain have sufficient resources to be consistently efficacious.

A ROLE FOR INTERAXONAL COMPETITION AND ACTIVITY

A number of lines of evidence suggest that the loss of synapses is the consequence of *competition*. The word "competition," however, has many different meanings, and which of these definitions is most relevant to synaptic development is an area of some debate (Lichtman and Colman, 2000; Ribchester, 1992; Ribchester and Barry, 1994). In broad terms, competition occurs when more than one individual (in this case, more than one axon) is capable of having the same fate (e.g., sole occupation of a neuromuscular junction), and the probability of having that fate is related inversely to the number of such individuals. The point of defining competition so broadly is to emphasize that competition does not imply what kind of mechanism drives the outcome—even lotteries, where winners are picked by random, are competitions. Synapse elimination occurring on muscle fiber cells or cerebellar Purkinje cells is thought to be competitive because there is always only one axon remaining at the completion of the process and therefore more than one axon cannot share the same fate.

Importantly, which particular axon is maintained does not seem to be preordained. There is no evidence of an extensive molecular specificity that uniquely matches each motor axon to an exclusive subset of muscle fibers or each climbing fiber to a matching set of Purkinje cells. But if it is a competition, how is this competition being driven? Competitive mechanisms range from situations where the contestants interact with each other directly (e.g., a sumo wrestling match) to mechanisms where the contestants have little or anything to do with each other and a third party (a judge) decides the outcome (e.g., competition for a Pulitzer prize). Evidence in the neuromuscular system suggests that synaptic competition may be more akin to the latter alternative with the muscle fiber playing the role of judge.

Neuromuscular System

In 1970, Paul Redfern published the first physiological report of synapse elimination. He found that rat diaphragm muscle fibers were innervated by several motor neurons in the first postnatal week but this multiple innervation was short lived; by two weeks of age all muscle fibers were singly innervated (Fig. 20.1). At the time, he suggested that the extra innervation may be explained by the presence of motor neurons that project to a muscle in early life but subsequently undergo *cell death*. A few years later this idea was shown to be incorrect when the tension elicited by activating single motor axons was shown to drop precipitously over the same period of early postnatal life indicating that synapse elimination was accompanied by motor unit branch trimming rather than wholesale motor neuron loss (an event that occurs in the prenatal period) (Brown *et al.*, 1976). Brown *et al.* (1976) also made the important point that because all neuromuscular junctions ended up with exactly one innervating axon, the elimination process was likely explained by interaxonal competition—otherwise all axons should leave a neuromuscular junction or more than one converging axon should persist into adulthood. These two seminal reports stimulated many investigators to seek an understanding of the underlying mechanisms. Much of this effort has focused on the role of activity in synapse elimination. Many lines of evidence show that modifying neuromuscular activity has a large effect on synapse elimination (Thompson, 1985). However, it has remained unclear how activity might mediate interaxonal competition, as conflicting evidence suggested that either active (Ridge and Betz,

1984) or surprisingly, inactive axons (Callaway *et al.*, 1987) are at an advantage. At first glance, it may seem improbable that inactive axons could out-compete active ones (a result that is opposite to one discussed later, on the effects of monocular deprivation in visual system synapse elimination). Callaway *et al.* (1987) argued that the way muscles are used, the axons that are recruited most infrequently were the ones that in adults had the largest motor units. If their large sizes resulted from less branch trimming during development then inactive axons might indeed have the advantage in synaptic competitions. Further complicating matters is the fact that the total number of axons innervating the muscle does not appear to change (Brown *et al.*, 1976). This constancy implies that no individual axon's activity pattern could be considered *the* worst, because all axons maintain some neuromuscular junctions and thus all axons must out-compete other axons at some junctions.

Other experiments have questioned whether activity is even necessary for synapse elimination. For example, after reinnervation of adult muscle, evidence has been obtained showing that electrically silent inputs can displace other electrically silent inputs (Costanzo *et al.*, 2000). Moreover, in some cases the presence of activity does not invariably lead to neuromuscular synapse elimination (Costanzo *et al.*, 1999).

As these conflicting experimental results demonstrate, the role of activity in synapse elimination is not easily summarized. One way to obtain a clearer idea of the role of activity is to view "activity" not as one phenomenon but as several different influences. For example,

- Synaptic competition may be affected by the relative efficacy (i.e., amplitude of the postsynaptic potential) of different axons attempting to drive the postsynaptic cell to threshold, with the most powerful input being favored.
- Alternatively, the firing frequency of an axon (i.e., action potentials per second) may have a negative impact especially if the axon has a large arbor and thus insufficient resources to maintain efficacious synaptic transmission throughout its terminal branches.
- Finally, activity may affect the electrical properties of the postsynaptic cell or cell region either encouraging synapse elimination (excitable encouraging membrane) or preventing it (inexcitable postsynaptic membrane) (Fig. 20.11).

The most accepted hypothesis is that activity differences between axons is critical for synapse elimination to occur. Indeed, because an axon has many synaptic release sites impinging on one postsynaptic cell that

Tonic fiber: Multiply Innervated   Twitch fiber: Singly Innervated

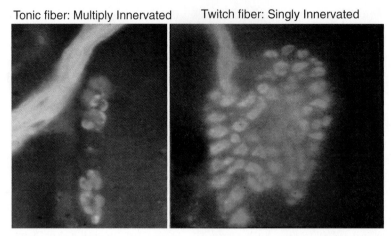

FIGURE 20.11 **Postsynaptic cells that don't fire action potentials don't undergo synapse elimination.** Photomicrographs of multiply and singly innervated snake muscle fibers. The singly innervated neuromuscular junctions are on twitch muscle fibers that have voltage-sensitive sodium channels. The multiply innervated neuromuscular junctions are found on tonic muscle fibers that do not have regenerative potentials. Labeling of different axons with different colors was accomplished by activity-dependent fluorescent labeling of axon terminals. From Lichtman *et al.* (1985).

BOX 20.1

α-BUNGAROTOXIN

A number of different dyes, stains, and markers are useful in revealing synaptic structure and function. Some of the more powerful have been borrowed from nature. Many toxins and poisons bind to specific proteins. One example is the snake toxin α-bungarotoxin (α-btx). This toxin is a constituent of the venom of a *Krait* snake of the species *bungarus*. The lethality of α-btx is the consequence of its ability to bind to the α subunits of nicotinic AChRs in skeletal muscle cells of vertebrates (with a few notable exceptions: snakes, mongooses, and hedgehogs). Because α-btx is an irreversible competitive antagonist of the AChR, its binding thus paralyzes and suffocates the prey. Researchers have taken advantage of this snake toxin to study many aspects of the AChR. For example, α-btx was used to purify AChRs from *Torpedo* membranes. Further-

more, the toxin can be conjugated to radioactive markers or to fluorescent molecules that can be seen in the microscope to stain receptors on muscle cells to determine their distribution, stability, and motility in the membrane. Because the toxin binds essentially irreversibly to the receptor in the muscle fiber membrane, receptors can be labeled once and then their behavior followed over time. This approach has provided substantial evidence about the stability of synaptic regions on muscle fibers, the lifetime of receptors in the membrane at the junctional sites, and how AChRs move within the plane of the membrane. For studies of synapse elimination in particular, the toxin also has been used to selectively inactivate some regions of a synapse by "puffing" it locally over a small region of a junction.

---

do not seem to compete with themselves, interaxonal activity differences are assumed to be a crucial element of the competition. An experiment in which one part of a neuromuscular junction was desynchronized from the rest by silencing its acetylcholine receptors with alpha bungarotoxin (Box 20.1) examined how activity differences might give rise to synapse elimination. Focal postsynaptic silencing at one site within an otherwise normally active adult neuromuscular junction

induced synapse withdrawal from the silenced site (Fig. 20.12). Postsynaptic silencing of an entire neuromuscular junction, however, did not cause synapse loss. These results suggested that synaptic activity at some synaptic sites can induce synaptic destabilization of other sites if they are not active at the same time. This view of the way activity modifies synaptic connections is related to the well-known theory of plasticity: *Hebb's postulate* (Hebb, 1949; see Chapter 50).

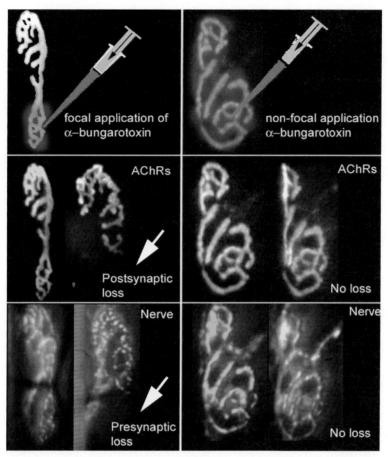

FIGURE 20.12 An experimental test of the role of postsynaptic activity in synapse elimination. If axons converging on the same postsynaptic cell were competing based on their efficacy in activating the postsynaptic cell, then locally blocking synaptic transmission at one axon's synaptic site should lead to its elimination and the maintenance of the other unblocked inputs. In this test, focal blockade of neuromuscular transmission was accomplished by applying saturating doses of α-bungarotoxin with a fine glass pipette (upper panels, red) to a neuromuscular junction in a living mouse. The AChR sites (middle panels) and the nerve terminals (lower panels) were viewed at the time of blockade and several times over the next few weeks. Following focal blockade (left panels) progressive loss of the nerve terminal staining and AChRs (arrows) was observed in regions previously saturated with α-bungarotoxin. However, when the entire neuromuscular junction was blocked (right panels) there was no loss. These results argue that active postsynaptic synaptic sites can cause the disassembly of inactive sites. Adapted from Balice-Gordon and Lichtman (1994).

Although Hebb argued that inputs that are consistently active when the postsynaptic cell is active, are strengthened, this idea is the logical obverse: synapses that consistently fail to excite the postsynaptic cell when the cell is being activated are eliminated (Lichtman and Balice-Gordon, 1990; Stent, 1973). How might active synaptic sites destabilize silent ones? One suggestion is that active synapses generate two kinds of postsynaptic signals: one that protects them from the destabilizing effects of activity and the other that punishes other inputs that are not active at the same time (Fig. 20.13). The physical basis of these protective and punishment signals remains unclear. Since neurotransmitter receptors are sometimes permeable to calcium, and the activity-induced depolarization can also raise

intracellular calcium levels, one idea is that calcium signaling serves one or both of these roles.

This view of the role of activity implies that if all axons were firing synchronously then synapse elimination would not occur, which is exactly the conclusion reached using a regeneration model for synapse elimination (Busetto *et al.*, 2000). Interestingly, during the period of naturally occurring synapse elimination there appears to be a switch from synchronous to asynchronous activity patterns (Personius and Balice-Gordon, 2001; Buffelli *et al.*, 2002). It has been suggested that a gradual loss of electrical coupling among motor neurons may be the reason synapse elimination begins. This hypothesis was recently tested in mice that lack a gap junction protein (connexin 40) in which motor

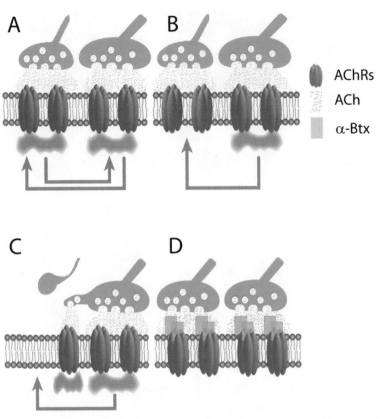

FIGURE 20.13　Putative mechanism for a postsynaptic role in synapse elimination. Postsynaptic receptor activation may elicit two opposing signals within a muscle fiber. One consequence of receptor activation is a "punishment" signal (here designated as red arrows) that causes destabilization of synaptic sites. Receptor activation may also generate a "protective" signal (here designated as blue clouds) that locally prohibits the punishment signals from destabilizing synapses in the vicinity of where receptor activation recently occurred. (A) When all the receptors are activated synchronously, as might occur before birth when motor neurons are electrically coupled (Personious *et al.*, 2001), there is no synaptic destabilization (due to protective blue clouds everywhere). (B) However, later in development, when two inputs are activated asynchronously, the active synaptic sites at any time point are protected (i.e., the synaptic sites beneath the active green axon are protected by a local blue cloud) whereas the asynchronously activated synapses (i.e., the synaptic sites under the pink axon) are not protected. (C) This asynchrony allows the more powerful input to destabilize the weaker one. This destabilization can lead to nerve and postsynaptic disappearance and/or nerve withdrawal followed quickly by takeover of its former synaptic sites by the remaining axon, which would restabilize the site. (D) If all synaptic sites are inactive, as might occur with α-bungarotoxin application that inactivates nicotinic AChRs, then there are no blue clouds to protect synaptic sites, but also no red arrows to destabilize them. Thus synaptically silent and synchronously active synaptic sites (see panel A) do not undergo synapse elimination. Idea adapted from Jennings (1994).

neuronal electrical coupling is reduced. In $Cx40^{-/-}$ muscles, synapse elimination was significantly accelerated, suggesting that asynchronous firing of neurons enhances synapse elimination (Personius *et al.*, 2007).

Such an activity-based mechanism for synapse elimination would tend to pit the synchronously active synaptic terminals of one axon on a given postsynaptic cell (a *synaptic "cartel"*) against the terminals of other axons contacting the same target cell. In this way an axon's terminals on one postsynaptic cell cannot compete against themselves but rather serve as a competitive unit vying against the cartels of other axons.

Visual Cortex

Classic studies on the visual system by Hubel and Wiesel were the first to suggest that competition in fact was driving synaptic reorganization in the developing brain (Hubel and Wiesel, 1963; Hubel *et al.*, 1977). In most species of young mammals, input neurons to layer IV in the visual cortex can be activated by inputs driven from both the left and the right eye; that is, they are driven binocularly. Subsequently, however, in many species, cortical input neurons become strongly dominated by either the right or left eye but not both (Fig. 20.14). In agreement with this physiological result, the terminal arbors of the geniculocortical axons from

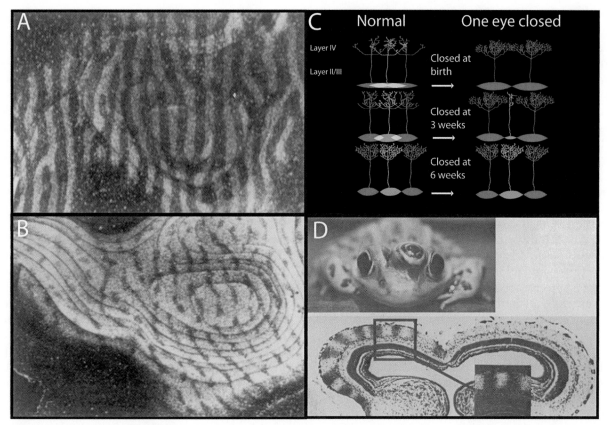

FIGURE 20.14 Synapse elimination in the visual system. (A) Ocular dominance columns of the neonatal monkey primary visual cortex in layer IVC, revealed by injecting [³H]-proline into the vitreous of one eye. Light stripes (columns) represent sites containing the anterograde transported ³H-amino acid from the injected eye. Dark regions are occupied by axons driven by the other eye. (B) Monocular deprivation by lid suture of one eye (2 weeks after birth for a period of 18 months) resulted in the shrinkage of the columns representing the deprived eye (dark stripes) and an expansion of the columns of the nondeprived eye (light stripes). (C) A schematic representation of ocular dominance column development represents the way in which a gradual segregation ocular dominance columns could lead to the end of the critical period and progressively more modest effects of monocular deprivation as development ensues. (Top) At birth the afferents from the two eyes (red and green ovals) overlap completely in layer IV and thus each eye is capable of maintaining inputs everywhere. At this young age, monocular deprivation would allow the nondeprived eye to remain in all parts of layer IV so that the entire cortex would be dominated by the red inputs. (Middle) In nondeprived animals, the two sets of afferents become progressively more segregated with age, meaning that by three weeks there would be regions of layer IV that are exclusively driven by the red or, as shown, green afferents. Once an eye's inputs are removed from a territory, it can no longer reoccupy that territory when the other eye is silenced. Hence monocular deprivation (that begins at three weeks) will spare a small strip of the inactive eyes territory (in this case the green regions). (Bottom) Once segregation is complete then monocular deprivation has no effect and the critical period is over. (D) A remarkable example showing how interactions between two eyes can cause segregation was found in frogs in which a third eye is implanted (at the tadpole stage) next to one of the eyes and projects with the native eye to the same optic tectum (ordinarily each frog tectum is monocular). After injection of H³-proline into the normal eye, one optic tectum of a three-eyed frog shows dark and light bands strikingly similar to the ocular dominance columns observed in monkeys (D). A–C adapted from Hubel *et al.* (1977); D from Constantine-Paton and Law (1978).

the two eyes overlap in early postnatal life (Hubel *et al.*, 1977; but see Horton and Hocking, 1996). However with time, their projections appear to resolve into a striking pattern of alternating stripes known as *ocular dominance columns* (Fig. 20.14). This anatomical result was based on anterograde transneuronal transport of radioactive amino acid that was injected into one eye and passed through the thalamus to label the eye specific axons in the optic radiation (Hubel *et al.*,

1977). Despite the remarkable clarity of these stripes, their functional significance is not well understood and in some mammals such as rodents and new world monkeys, ocular dominance columns are absent (Horton and Adams, 2005; Livingstone, 1996).

During development the ocular dominance columns' organization is less obvious because there is overlap in the thalamic inputs driven by the right and left eye to layer IV. Anatomically as development proceeds each eye's columns become narrower and eventually almost nonoverlapping (especially in primates). Even though the narrowing in the widths of right and left eye ocular dominance columns is likely due to the loss of axonal branches in overlapping regions (Antonini and Stryker, 1993), this retraction of branches should not be taken to mean that the total number of thalamo-cortical synapses in layer IV is decreasing during this period. Similar to the peripheral nervous system, the axons associated with each eye elaborate many new synapses that more than compensate numerically for the lost connections of the withdrawing axons (Crowley and Katz, 2000; Erisir and Dreusicke, 2005). In other words, the process of ocular dominance column formation is one in which individual arbors lose synaptic connections with some targets but gain connections with others. Thus elimination restricts the neuronal population that is directly driven by each eye, but the synaptic addition strengthens the influence of one eye on the regions of cortex it continues to drive.

The most interesting aspect of this segregation is that the gradual removal of overlap in the two eyes' input streams leading to equally sized ocular dominance columns is not inevitable. Hubel and Wiesel showed that during a developmental *critical period* in early postnatal life, the widths of these columns can be dramatically and permanently changed by alterations in the relative amounts of visual experience in the two eyes. In particular, the outcome of the segregation can be radically skewed in favor of one eye if the activity of the other eye is decreased (e.g., by patching one eye). This monocular deprivation results in larger columns for the open eye and smaller columns for the deprived eye (Fig. 20.14). Once the critical period of sensitivity is passed (approximately the seventh postnatal week in kittens, the tenth week in ferrets, and the twentieth week in rhesus monkeys), the widths of the ocular dominance columns are fixed and no longer subject to shifts based on visual experience. Remarkably, even long-term monocular deprivation (of decades or more) apparently has little effect on the width of ocular dominance columns if the visual deprivation is begun after the critical period is over (approximately six years in humans; Keech and Kutschke, 1995). For example, in human patients, in which one eye was removed for surgical reasons in adulthood or late childhood, postmortem analysis of the visual cortex indicated that eye removal has little or no effect on the width of the ocular dominance columns dominated by the removed eye (Horton and Hocking, 1998). What might account for this dramatic change in sensitivity?

One idea is that during the critical period, thalamic afferents driven by the two eyes *compete* for control of the cortical neurons that they share temporarily. If each eye has the same average amount of activity, each ended up with similar amounts of cortical territory. However, if there were imbalances between the eyes in terms of visual experience, the outcome tipped the segregation in favor of one eye over the other. The skewing that resulted from depriving one eye of vision was due both to additional losses in the connections driven by the inactive eye (shrinking its ocular dominance columns) and to additional maintenance of the connections from the normally active nondeprived eye (maintaining its columns at the wider width it had at an earlier age) (Fig. 20.14). Ordinarily, each eye's afferents would relinquish its connections with approximately half of its postsynaptic target cells in visual cortex. Furthermore, binocular eye closure during the critical period appears to have far less serious effects than monocular occlusion. These results support the idea that synapse elimination is due to an activity-mediated competitive interaction between the connections driven by the two eyes. Because binocular deprivation has less dramatic effects on ocular dominance columns than monocular deprivation, here, as at the neuromuscular junction, active synaptic inputs seem to play a role in destabilizing inactive inputs.

Though these conclusions appear straightforward, the roles of activity in cortical refinements may need reevaluation as new investigations show that the mechanisms may be more complicated than originally imagined. In studies of the rodent visual system, evidence suggests roles for both activity dependent and independent factors in developmental axonal refinements. Although mice lack ocular dominance columns, they do have a binocular cortical region that becomes progressively smaller as development proceeds. This shrinkage can be shifted with monocular deprivation during a critical period (Antonini et al., 1999). This system has been used to show an important role of inhibitory circuits in both the establishment and maintenance of the critical period (Hensch, 2005). For example, deletion of the gene for glutamic acid decarboxylase, the enzyme that is responsible for the synthesis of the inhibitory neurotransmitter, GABA, prevents visual cortical refinements (Fagiolini and Hensch, 2000). Interestingly, these refinements can be reinitiated at any age by injecting GABA receptor agonists such as benzodiazepines into visual cortex. Thus intracortical inhibitory circuitry may be sufficient to trigger the opening or closure of the critical period in mice (Hensch et al., 1998; Hensch, 2004).

In mouse visual cortex it has also been possible to study the sharpening of the retinotopic map in early postnatal life. The small receptive fields seen in adult visual cortex emerge from shrinkage of receptive field size in development (Issa *et al.*, 1999). Activity plays a complex role in this sharpening; the effects of deprivation of formed vision (by lid suture) are different than the effects of pharmacological blockade or enucleation (Smith and Trachtenburg, 2007). A contralateral eye that is sensing light through a sutured lid impedes the refinement of the open eye's central projection, whereas either an open, or completely silent, contralateral eye does not. This result suggests that when inputs are synchronous (i.e., both eyes open), the activity mediated refinements that shrink receptive fields occur more efficiently than when the same cortical neurons are receiving inputs with different activity patterns (i.e., in an animal with one open and one sutured eye). On the other hand when one eye is entirely silent (i.e., by enucleation or pharmacological blockade), then the refinement of the open eye's inputs can still occur because in this case, there is no competing activity pattern from another eye. Thus some kinds of refinements may be mediated by cooperative interactions between the two eyes rather than competitive ones. One recent trend is that experiments that might have previously been interpreted strictly in terms of activity mediated competition between different axons (à la Hubel and Wiesel) are now recast in terms of *homosynaptic* mechanisms of potentiation and depression or mechanisms of *synaptic homeostasis* (Chapter 50). These newer frameworks for thinking about critical periods suggest that many different regulatory mechanisms working simultaneously may help to assure that the right numbers and kinds of synapses survive the period of developmental refinements.

Thalamus

Separation of the inputs from the two eyes occurs twice in the visual system. Prior to the emergence of cortical ocular dominance columns, eye input to the lateral geniculate nucleus of the thalamus segregates into layers rather than columns. In embryonic cats, axon terminals of ganglion cells from the two eyes overlap extensively within the lateral geniculate nucleus before gradually segregating to form the characteristic *eye-specific* layers by birth. As in the cortex, this refinement process involves both the retraction of axonal branches from inappropriate regions of the geniculate nucleus and the elaboration of processes within the correct eye layer (Shatz, 1990). Physiological studies support anatomical observations that geniculate neurons initially are driven binocularly but maintain the axonal input from only one eye at maturity (Shatz, 1990). There is also a dramatic change in the convergence of retinal ganglion cell input to thalamic neurons related to a shrinkage in receptive field size

(Tavazoie and Reid, 2000; Chen and Regehr, 2000) (Fig. 20.15).

It is likely that *spontaneous activity* as opposed to actual visual experience is important in the segregation of retinogeniculate connections in the thalamus. This view is based on the fact that in cats, ferrets, and monkeys the eye specific layers are established well before the retina is sensitive to light and is dependent on the spontaneous activity of these inputs. What information may be contained in the spontaneous activity patterns of immature retinas that could lead to eye-specific lamination? Electrophysiological recordings and Ca^{++} imaging studies demonstrate that each immature retina generates correlated propagating *waves* (Fig. 20.16), which have no preferred direction of propagation and which occur periodically, about once a minute. It is possible that retinal waves contain temporal and spatial cues that guide activity-dependent refinement of retinogeniculate connections. For example, because waves are generated independently in each retina, activities from the two eyes are unlikely to be coincident. Asynchrony between the inputs of the two eyes could account for the segregation of inputs into different eye-specific layers in the thalamus. Since ocular dominance column formation is initiated before birth, the spontaneous patterns of activity from the two eyes could be responsible for the eye specific pathways throughout the visual system. Moreover, because the waves ensure that nearby retinal ganglion cells are better synchronized than more distant cells, geniculate neurons are able to gauge neighbor relationships in the retina by their sequential activation. This feature could be useful for refinements of the retinotopic map in both thalamus and cortex. Pharmacological blockade of retinal waves during the period of eye-specific segregation prevents the emergence of these layers, suggesting that activity from the retinas is involved (Wong, 1999). However, these activity patterns must be only part of the story: the laminar organization of the lateral geniculate is stereotyped from animal to animal suggesting that other developmental mechanisms are also at play.

Cerebellum

A particularly clear example of synapse elimination in the developing CNS is the *climbing fiber input* onto cerebellar Purkinje cells. Climbing fibers are the terminals of the axons arriving from inferior olive neurons that form strong synaptic connections to Purkinje cells. In adults only one climbing fiber innervates each Purkinje cell. That input might contain 500 synaptic boutons that tightly invest the large ascending proximal dendrite. Immature climbing fibers on the other hand form fewer synapses, mostly on the

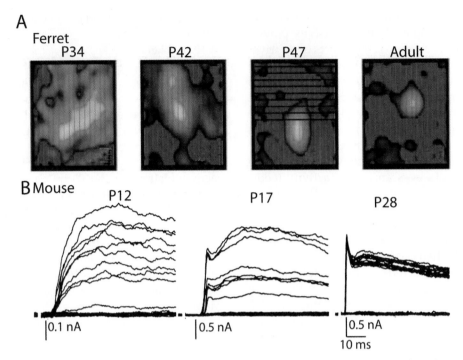

FIGURE 20.15   Synapse elimination in the lateral geniculate nucleus of the thalamus. (A) In the ferret lateral geniculate nucleus, receptive fields of geniculate neurons are larger and much more diffuse at one month of age than those receptive fields observed in adult neurons. Red regions represent the receptive field map of geniculate neurons that correspond to areas excited by bright stimuli. Note that red areas become smaller as development proceeds. The shrinkage in the receptive field is likely to result from the elimination of the convergence of multiple retinal afferents onto each geniculate neuron (B). At P12 in mouse, multiple retinogeniculate axons are recruited as stimulation intensities to the bundle of axons is increased (see also Figure 20.1C). At P17 there are fewer steps and after P28, only one or two inputs innervate each geniculate neuron (i.e., no steps in the evoked-synaptic currents are seen even though optic nerve stimulation is increased). (A) Adapted from Tavazoie and Reid (2000). (B) Adapted from Chen and Regehr (2000).

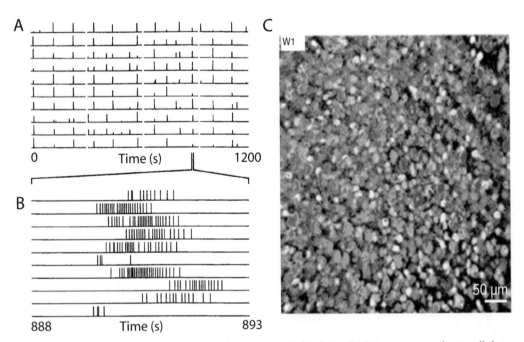

FIGURE 20.16   Immature retinal ganglion cells show correlated patterns of activity. (A) Using an array of extracellular recordings, rhythmic bursts of action potentials (indicated by vertical lines) are synchronized between neighboring retinal ganglion cells before eye opening in ferret. (B) Action potential bursts expanded in time scale shows that each burst corresponds to ten or more action potentials. (C) Using calcium indicators has been possible to observe waves of neuronal activity in immature retinal ganglion cells (pseudo-colored image). In this example, a wave of neuronal activity propagates through the retina. Pseudo-colored cells indicating the temporal firing pattern of retinal ganglion neurons, cells in green fire before yellow ones, and lastly red cells. (A, B) from Meister *et al.* (1991). (C) from Wong (1999).

Purkinje cell soma, and many climbing fibers project to each Purkinje cell in the first postnatal week (in rodents). The transition from multiple innervation to single innervation of individual Purkinje cells occurs at the same time there is a change in the number of Purkinje cells innervated by each climbing fiber. This "neural unit" shrinkage is remarkably analogous to the reduction in the size of motor units in the peripheral nervous system. Thus, one olivocerebellar axon may give rise to branches that innervate more than 100 Purkinje cells during the first postnatal week but over the next several weeks its projection is trimmed to only ~7 Purkinje cells (Fig. 20.17), albeit these connections are much more powerful.

It has long been appreciated that the loss of climbing fiber inputs depends on the presence of parallel fiber innervation from granule cells of the distal part of the Purkinje cell arbor. Elimination of granule cell inputs to Purkinje cells by X irradiation, viral infection, or in mutants such as *reeler*, *weaver*, and *staggerer* results

in a higher incidence of Purkinje cells that are multiply innervated by climbing fibers in adulthood. Some studies suggest that the activity of the parallel fiber input is the important parameter. Perturbation of activity along the parallel fiber—Purkinje cell pathway in mGluR1 and GluRδ2 knockouts mice or application of NMDA receptor antagonists all inhibit the elimination of climbing fibers (Hashimoto and Kano, 2005). In addition disruption of one calcium binding kinase (PKC gamma) appears to selectively prevent climbing fiber elimination (Hashimoto and Kano, 2005). Although the mechanism by which this kinase alters synapse elimination is not known, these animals recently have been shown to have a profound deficit in vestibulo-ocular reflex (VOR) motor learning but not other kinds of cerebellar learning (Kimpo and Raymond, 2007). These results imply that synapse elimination may be important in generating the circuitry for some kinds of adult learning. As we will mention later, however, synapse elimination itself may be a form of learning.

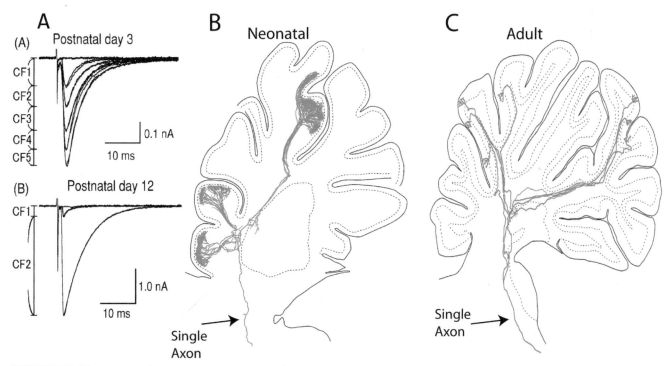

FIGURE 20.17 **Synapse elimination and axonal pruning of climbing fibers in the neonatal period.** (A) Recordings from Purkinje cells while olivocerebellar axons are stimulated in the inferior olive show functional evidence of synapse elimination. At birth, each Purkinje cell is innervated by several different climbing fibers, as the strength of stimulation is increased additional inputs are recruited (compare with Fig. 20.1C and Fig. 20.15B). The average number of climbing fibers innervating each cerebellar Purkinje cell, in the rat, decreases gradually as the animal matures until most are singly innervated. (B, C) Reconstructions of the trajectory of single neonatal (B) and adult (C) olivocerebellar climbing fiber axons. Both neonatal and adult axons terminate in several separate lobules in the hemisphere. However, neonatal olivocerebellar axons have much more branches than those in adult axons and presumably innervate many more Purkinje cells than those in adult animals. After pruning is complete, each axon gives rise to ~7 climbing fibers that each singly innervates a different Purkinje cell. (A) Adapted from Hashimoto and Kano (2003). (B, C) Modified from Sugihara (2005).

Summary

In many parts of the central and peripheral nervous system the divergence and convergence of synaptic circuits is decreased. A popular hypothesis, albeit still somewhat perplexing, is that neural activity differences between axons underlies synapse elimination. This hypothesis has driven researchers to try many kinds of experiments to test activity's role. At present, the precise way in which electrical activity exerts an influence on the synapse elimination process remains a central and unresolved question.

IS SYNAPSE ELIMINATION STRICTLY A DEVELOPMENTAL PHENOMENON?

In this chapter, we presented evidence to demonstrate that synapse elimination is a powerful force that can refine synaptic circuits in young animals, based on interneuronal competition. Is there any reason to think it has more than a strictly developmental phenomenon? The most important form of adult plasticity must certainly be memory. Might synapse elimination have something to do with memory? A number of neurobiologists, including Kandel (1967), Toulouse et al. (1986), and Edeleman (1988) have explicitly made arguments for selection (as opposed to instruction) as potentially playing an important role in learning. The idea is that in the brain synaptic circuitry exists *a priori* for many things that may ultimately be learned, so that learning might occur by the selection of synaptic pathways that already exist rather than construction of new circuits. Although such selection could occur by increasing the strength of one set of synaptic interconnections or weakening of others, it could also occur by completely eliminating some circuits. It is important to emphasize the distinction between plasticity that alters the strengths of existing connections and the more extreme kind of plasticity, analogous to the developmental synaptic processes' described here, that causes permanent eradication of an axon's input to particular postsynaptic cells.

Because postsynaptic cells appear to be the intermediary in synaptic competition leading to axonal removal (see earlier), once an axon's synaptic drive to a postsynaptic cell is removed, it can no longer have any influence on the synaptic connections of the other axons that remain connected. Complete loss of influence following synaptic disconnection is thus a plausible explanation for the finite length of critical periods. For example, once all the inputs driven by an eye deprived of vision are eliminated, return of visual experience in that eye can no longer cause a shift in ocular dominance columns if that shift is mediated via activation of postsynaptic cells. The same argument could also be made for memory. Memories have a kind of indelibility that prevents more recent memories from "overwriting" prior ones. Input elimination is an attractive means of assuring indelibility because by eliminating competing (i.e., asynchronously firing) inputs, a circuit becomes sheltered from disruption by different activity patterns.

A model of memory based on this kind of synapse elimination, however, would require that axonal inputs continue to be eliminated in the adult brain. That critical periods in the visual system are strictly developmental can be used as an argument against the idea that these kinds of changes may underlie adult memory. On the other hand, critical periods in the visual system tend to be prolonged in proportion to their distance from the input. For example, critical periods for higher visual processing areas occur later in development than in those areas that are more proximal in the visual pathway. The loss of overlapping connections is known to occur prenatally within thalamic circuitry well before segregation in layer IV primary visual cortex, and higher anatomical levels in the visual cortex that receive input from layer IV segregate out later than layer IV. A possible explanation for this sequential crystallization of brain regions may be that synapse elimination can occur only when a cohort of synchronous inputs work together to drive the elimination of competing inputs. Such a collection of synchronously active neurons requires that the presynaptic input to these cells has itself sorted out. It also is the case that the length of the critical periods for vision are vastly longer in humans than other mammals, as is the rest of our neotenic development. For example, whereas our closest animal relatives finish the critical period for monocular deprivation by 7 months of age, in humans monocular deprivation can affect visual acuity even in children 6 to 7 years old.

SUMMARY

We would not like to give the impression that naturally occurring synapse elimination at developing systems is the equivalent of learning and memory. But, as neurobiologists who have studied this phenomenon and mulled these ideas over for many years, we have come to the conclusion that permanent loss of axonal input is an attractive mechanism for information storage. Whether our bias is reasonable based on the data or rather due to structural elimination of competing hypotheses from our brains, we do not know.

References

Antonini, A., Fagiolini, M., and Stryker, M. P. (1999). Anatomical correlates of functional plasticity in mouse visual cortex. *J Neurosci* **19**, 4388–4406.

Antonini, A. and Stryker, M. P. (1993). Rapid remodeling of axonal arbors in the visual cortex. *Science* **260**, 1819–1821.

Balice-Gordon, R. J., Chua, C. K., Nelson, C. C., and Lichtman, J. W. (1993). Gradual loss of synaptic cartels precedes axon withdrawal at developing neuromuscular junctions. *Neuron* **11**, 801–815.

Balice-Gordon, R. J. and Lichtman, J. W. (1994). Long-term synapse loss induced by focal blockade of postsynaptic receptors. *Nature* **372**, 519–524.

Bishop, D. L., Misgeld, T., Walsh, M. K., Gan, W. B., and Lichtman, J. W. (2004). Axon branch removal at developing synapses by axosome shedding. *Neuron* **44**, 651–661.

Boeke, J. (1932). Nerve endings, motor and sensory., Vol. 1. New York, Hafner Press.

Brown, M. C., Jansen, J. K., and Van Essen, D. (1976). Polyneuronal innervation of skeletal muscle in new-born rats and its elimination during maturation. *J Physiol* **261**, 387–422.

Buffelli, M., Burgess, R. W., Feng, G., Lobe, C. G., Lichtman, J. W., and Sanes, J. R. (2003). Genetic evidence that relative synaptic efficacy biases the outcome of synaptic competition. *Nature* **424**, 430–434.

Buffelli, M., Busetto, G., Cangiano, L., and Cangiano, A. (2002). Perinatal switch from synchronous to asynchronous activity of motoneurons: Link with synapse elimination. *Proc Natl Acad Sci USA* **99**, 13200–13205.

Busetto, G., Buffelli, M., Tognana, E., Bellico, F., and Cangiano, A. (2000). Hebbian mechanisms revealed by electrical stimulation at developing rat neuromuscular junctions. *J Neurosci* **20**, 685–695.

Callaway, E. M., Soha, J. M., and Van Essen, D. C. (1987). Competition favouring inactive over active motor neurons during synapse elimination. *Nature* **328**, 422–426.

Chen, C. and Regehr, W. G. (2000). Developmental remodeling of the retinogeniculate synapse. *Neuron* **28**, 955–966.

Colman, H., Nabekura, J., and Lichtman, J. W. (1997). Alterations in synaptic strength preceding axon withdrawal. *Science* **275**, 356–361.

Constantine-Paton, M. and Law, M. I. (1978). Eye-specific termination bands in tecta of three-eyed frogs. *Science* **202**, 639–641.

Costanzo, E. M., Barry, J. A., and Ribchester, R. R. (1999). Co-regulation of synaptic efficacy at stable polyneuronally innervated neuromuscular junctions in reinnervated rat muscle. *J Physiol* **521 Pt 2**, 365–374.

Costanzo, E. M., Barry, J. A., and Ribchester, R. R. (2000). Competition at silent synapses in reinnervated skeletal muscle. *Nat Neurosci* **3**, 694–700.

Crowley, J. C. and Katz, L. C. (2000). Early development of ocular dominance columns. *Science* **290**, 1321–1324.

Ding, M., Chao, D., Wang, G., and Shen, K. (2007). Spatial regulation of an E3 ubiquitin ligase directs selective synapse elimination. *Science*.

Eaton, B. A. and Davis, G. W. (2003). Synapse disassembly. *Genes Dev* **17**, 2075–2082.

Edelman, G. M. (1988). Neural Darwinism: the theory of neuronal group selection. New York: Basic Books.

Erisir, A. and Dreusicke, M. (2005). Quantitative morphology and postsynaptic targets of thalamocortical axons in critical period and adult ferret visual cortex. *J Comp Neurol* **485**, 11–31.

Fagiolini, M. and Hensch, T. K. (2000). Inhibitory threshold for critical-period activation in primary visual cortex. *Nature* **404**, 183–186.

Gan, W. B. and Lichtman, J. W. (1998). Synaptic segregation at the developing neuromuscular junction. *Science* **282**, 1508–1511.

Gorio, A., Marini, P., and Zanoni, R. (1983). Muscle reinnervation—III. Motoneuron sprouting capacity, enhancement by exogenous gangliosides. *Neuroscience* **8**, 417–429.

Hashimoto, K. and Kano, M. (2003). Functional differentiation of multiple climbing fiber inputs during synapse elimination in the developing cerebellum. *Neuron* **38**, 785–796.

Hashimoto, K. and Kano, M. (2005). Postnatal development and synapse elimination of climbing fiber to Purkinje cell projection in the cerebellum. *Neurosci Res* **53**, 221–228.

Hebb, D. O. (1949). "The organization of behavior." New York, Wiley.

Hensch, T. K. (2004). Critical period regulation. *Annu Rev Neurosci* **27**, 549–579.

Hensch, T. K. (2005). Critical period plasticity in local cortical circuits. *Nat Rev Neurosci* **6**, 877–888.

Hensch, T. K., Fagiolini, M., Mataga, N., Stryker, M. P., Baekkeskov, S., and Kash, S. F. (1998). Local GABA circuit control of experience-dependent plasticity in developing visual cortex. *Science* **282**, 1504–1508.

Horton, J. C. and Adams, D. L. (2005). The cortical column: A structure without a function. *Philos Trans R Soc Lond B Biol Sci* **360**, 837–862.

Horton, J. C. and Hocking, D. R. (1996). An adult-like pattern of ocular dominance columns in striate cortex of newborn monkeys prior to visual experience. *J Neurosci* **16**, 1791–1807.

Horton, J. C. and Hocking, D. R. (1998). Effect of early monocular enucleation upon ocular dominance columns and cytochrome oxidase activity in monkey and human visual cortex. *Vis Neurosci* **15**, 289–303.

Hubel, D. H. and Wiesel, T. N. (1963). Receptive fields of cells in striate cortex of very young, visually inexperienced kittens. *J Neurophysiol* **26**, 994–1002.

Hubel, D. H., Wiesel, T. N., and LeVay, S. (1977). Plasticity of ocular dominance columns in monkey striate cortex. *Philos Trans R Soc Lond B Biol Sci* **278**, 377–409.

Issa, N. P., Trachtenberg, J. T., Chapman, B., Zahs, K. R., and Stryker, M. P. (1999). The critical period for ocular dominance plasticity in the Ferret's visual cortex. *J Neurosci* **19**, 6965–6978.

Jackson, H. and Parks, T. N. (1982). Functional synapse elimination in the developing avian cochlear nucleus with simultaneous reduction in cochlear nerve axon branching. *J Neurosci* **2**, 1736–1743.

Jennings, C. (1994). Developmental neurobiology. Death of a synapse. *Nature* **372**, 498–499.

Kandel, E. (1967). "Cellular Studies of Learning." New York, Rockefeller University Press.

Kasthuri, N. and Lichtman, J. W. (2003). The role of neuronal identity in synaptic competition. *Nature* **424**, 426–430.

Keech, R. V. and Kutschke, P. J. (1995). Upper age limit for the development of amblyopia. *J Pediatr Ophthalmol Strabismus* **32**, 89–93.

Keller-Peck, C. R., Walsh, M. K., Gan, W. B., Feng, G., Sanes, J. R., and Lichtman, J. W. (2001). Asynchronous synapse elimination in neonatal motor units: Studies using GFP transgenic mice. *Neuron* **31**, 381–394.

Kimpo, R. R. and Raymond, J. L. (2007). Impaired motor learning in the vestibulo-ocular reflex in mice with multiple climbing fiber input to cerebellar Purkinje cells. *J Neurosci* **27**, 5672–5682.

Laskowski, M. B., Colman, H., Nelson, C., and Lichtman, J. W. (1998). Synaptic competition during the reformation of a neuromuscular map. *J Neurosci* **18**, 7328–7335.

Lichtman, J. W. (1977). The reorganization of synaptic connexions in the rat submandibular ganglion during post-natal development. *J Physiol* **273**, 155–177.

Lichtman, J. W. and Balice-Gordon, R. J. (1990). Understanding synaptic competition in theory and in practice. *J Neurobiol* **21**, 99–106.

Lichtman, J. W. and Colman, H. (2000). Synapse elimination and indelible memory. *Neuron* **25**, 269–278.

Lichtman, J. W. and Purves, D. (1980). The elimination of redundant preganglionic innervation to hamster sympathetic ganglion cells in early post-natal life. *J Physiol* **301**, 213–228.

Lichtman, J. W., Wilkinson, R. S., and Rich, M. M. (1985). Multiple innervation of tonic endplates revealed by activity-dependent uptake of fluorescent probes. *Nature* **314**, 357–359.

Livingstone, M. S. (1996). Ocular dominance columns in New World monkeys. *J Neurosci* **16**, 2086–2096.

Lu, T. and Trussell, L. O. (2007). Development and elimination of end bulb synapses in the chick cochlear nucleus. *J Neurosci* **27**, 808–817.

Mariani, J. (1983). Elimination of synapses during the development of the central nervous system. *Prog Brain Res* **58**, 383–392.

Meister, M., Wong, R. O., Baylor, D. A., and Shatz, C. J. (1991). Synchronous bursts of action potentials in ganglion cells of the developing mammalian retina. *Science* **252**, 939–943.

Personius, K. E. and Balice-Gordon, R. J. (2001). Loss of correlated motor neuron activity during synaptic competition at developing neuromuscular synapses. *Neuron* **31**, 395–408.

Personius, K. E., Chang, Q., Mentis, G. Z., O'Donovan M, J., and Balice-Gordon, R. J. (2007). Reduced gap junctional coupling leads to uncorrelated motor neuron firing and precocious neuromuscular synapse elimination. *Proc Natl Acad Sci USA* **104**, 11808–11813.

Redfern, P. A. (1970). Neuromuscular transmission in new born rats. *J Physiol* **209**, 701–709.

Ribchester, R. R. (1992). Cartels, competition and activity-dependent synapse elimination. *Trends Neurosci* **15**, 389; author reply 390–381.

Ribchester, R. R. and Barry, J. A. (1994). Spatial versus consumptive competition at polyneuronally innervated neuromuscular junctions. *Exp Physiol* **79**, 465–494.

Ridge, R. M. and Betz, W. J. (1984). The effect of selective, chronic stimulation on motor unit size in developing rat muscle. *J Neurosci* **4**, 2614–2620.

Shatz, C. J. (1990). Impulse activity and the patterning of connections during CNS development. *Neuron* **5**, 745–756.

Smith, S. L. and Trachtenberg, J. T. (2007). Experience-dependent binocular competition in the visual cortex begins at eye opening. *Nat Neurosci* **10**, 370–375.

Stent, G. S. (1973). A physiological mechanism for Hebb's postulate of learning. *Proc Natl Acad Sci USA* **70**, 997–1001.

Sugihara, I. (2005). Microzonal projection and climbing fiber remodeling in single olivocerebellar axons of newborn rats at postnatal days 4–7. *J Comp Neurol* **487**, 93–106.

Tavazoie, S. F. and Reid, R. C. (2000). Diverse receptive fields in the lateral geniculate nucleus during thalamocortical development. *Nat Neurosci* **3**, 608–616.

Thompson, W. J. (1985). Activity and synapse elimination at the neuromuscular junction. *Cell Mol Neurobiol* **5**, 167–182.

Toulouse, G., Dehaene, S., and Changeux, J. P. (1986). Spin glass model of learning by selection. *Proc Natl Acad Sci USA* **83**, 1695–1698.

Walsh, M. K. and Lichtman, J. W. (2003). In vivo time-lapse imaging of synaptic takeover associated with naturally occurring synapse elimination. *Neuron* **37**, 67–73.

Wang, G. Y., Liets, L. C., and Chalupa, L. M. (2001). Unique functional properties of on and off pathways in the developing mammalian retina. *J Neurosci* **21**, 4310–4317.

Wiesel, T. N. (1982). Postnatal development of the visual cortex and the influence of environment. *Nature* **299**, 583–591.

Wong, R. O. (1999). Retinal waves and visual system development. *Annu Rev Neurosci* **22**, 29–47.

Juan C. Tapia and Jeff W. Lichtman

Dendritic Development

Dendrites play a critical role in information processing in the nervous system as substrates for synapse formation and signal integration. Neurons have highly branched, cell type-specific dendritic trees (or dendritic arbors), that determine the spatial extent and types of afferent input that the neurons receive (Cline, 2001; Wong and Ghosh, 2002). It is now widely recognized that dendrites do not develop in a void, but do so in constant interaction with other neurons and glia. The signals from these other cells affect dendritic arbor development in different spatial domains and time scales. For instance, synaptic inputs may increase calcium influx rapidly and locally to enhance rates of branch addition and stabilization (Lohmann et al., 2005). This dynamic morphological remodeling allows individual neurons to constantly adapt and respond to external stimuli (Cline, 2001; Ruthazer et al., 2003). In contrast, calcium signals with slower temporal dynamics may selectively signal to the nucleus to trigger gene transcription (Dolmetsch et al., 2001; Wu et al., 2001; Kornhauser et al., 2002). Such activity-induced genes can then have profound effects on dendritic arbor structure and function (Nedivi et al., 1998; Nedivi, 1999; Cantallops et al., 2000b; Redmond et al., 2002). Dendrite arbor structure and plasticity are altered under a variety of neurological disorders such as mental retardation (Benavides-Piccione et al., 2004; Govek et al., 2004; Bagni and Greenough, 2005; Newey et al., 2005) and can be affected by exposure to drugs including nicotine (Gonzalez et al., 2005) and cocaine (Kolb et al., 2003; Morrow et al., 2005). The study of dendritic arbor development can therefore provide important insight into the cellular basis of normal brain development, as well as neurological and psychiatric disorders.

This chapter is written with a special emphasis on the regulation of dendritic arbor structure in the context of the developing neuronal circuits. After a brief description of dendritic arbor development, we present representative molecular and cellular mechanisms governing dendrite arbor architecture.

DYNAMICS OF DENDRITIC ARBOR DEVELOPMENT

Dendritic arbor development requires global, arbor-wide architectural modifications (e.g., generalized arbor growth) as well as localized structural changes (e.g., sprouting and retraction of high order branches). Furthermore, some changes in dendritic arbor morphology may occur rapidly in response to synaptic inputs or growth factors, whereas others may occur with some time-delay, secondary to new gene transcription. Figure 21.1A illustrates the generalized growth of differentiating optic tectal neurons *in vivo* over three days, and Figure 21.1B shows an optic tectal neuron imaged in the intact *Xenopus* tadpole over about one day, but at shorter intervals. Apart from the widespread changes in dendritic architecture, short interval imaging unveils sites of local branch dynamics and growth. Although the initial development of the dendritic tree is under the control of genetic and molecular programs, the growth and refinement of the dendritic tree after synapse formation is strongly influenced by sensory input and calcium signaling. In the dendrites of *Xenopus* tectal neurons, there appears to be a developmental refinement in the spatial spread of Ca^{++} in response to retinal axon stimulation. At early

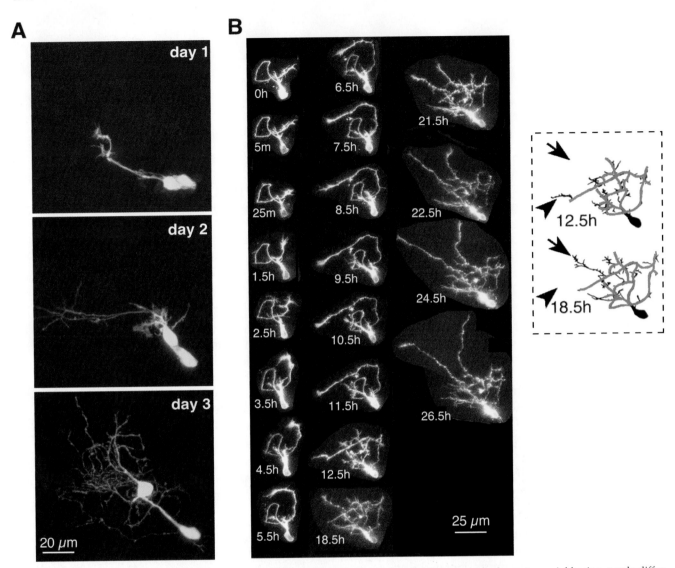

FIGURE 21.1 Time-lapse images of *Xenopus* optic tectal neurons, collected *in vivo*. (A) This example shows two neighboring, newly differentiated neurons, close to the proliferative zone, imaged over a period of 3 days. These cells initially present glial-like morphologies (day 1). By the next day, these cells have migrated elaborated complex dendritic arbors (day 2), which continue to grow to the end of imaging period (day 3). (B) Example showing an optic tectal interneuron imaged at short intervals over about 1 day. Dendritic trees develop as neurons differentiate within the tectum. Although portions of the dendritic arbor are stable over time (red outlined skeleton at 12.5 h and 18.5 h), other areas are very dynamic as dendritic branches are both added (arrows) and retracted (arrowheads) over time (in this case between 12.5 h and 18.5 h). From Bestman *et al.* (2008).

stages in neuronal development, when the dendritic arbor is still very simple, retinal axon stimulation results in Ca^{++} signals that spread throughout the cell, but Ca^{++} signals become more spatially restricted as neurons mature (Tao *et al.*, 2001). These changes in the spatial distribution of Ca^{++} signals in response to retinal stimulation could represent changes in dendritic integration as well as synapse-to-nucleus signaling by calcium during development.

GENETIC CONTROL OF DENDRITE DEVELOPMENT IN DROSOPHILA

Studies using *Drosophila* genetics have been instrumental in the identification of core programs that control dendrite development. The dendritic arborization (da) neurons, a group of *Drosophila* sensory neurons with a stereotyped dendritic branching pattern, have provided a useful assay system for the genetic dissection of dendrite development (Gao *et al.*, 1999; Grueber *et al.*, 2002).

Transcription Factors Regulate Cell Type Specific Dendritic Morphology

A striking feature of the nervous system is that there are many different types of neurons, each with a characteristic and recognizable dendritic arborization pattern. How do neurons acquire their type-specific dendritic morphology? Studies with *Drosophila* da neurons indicate that transcription factors are important regulators of the size and complexity of dendritic fields and the logic of their usage is beginning to emerge. Each hemi-segment of the abdomen of the *Drosophila* embryo or larva has 15 da neurons that can be subdivided into four classes based on their dendritic morphology (Grueber *et al.*, 2002). Each da neuron occupies an invariant position and has a highly stereotyped and unique dendritic branching pattern (Fig. 21.2) (Grueber *et al.*, 2002, 2003b). Class I and II have relatively simple dendritic branching patterns and small dendritic fields. In contrast, class III and IV neurons have more complex dendritic branching patterns and large dendritic fields.

In some cases, the "dendritic fate" of a particular neuron can be specified by a single transcription factor. For example, Hamlet functions as a binary switch between the elaborate multiple dendritic morphology of da neuron and the single, unbranched dendritic morphology of external sensory (es) neuron (Moore *et al.*, 2002). *Hamlet* encodes a multiple-domain, evolutionarily conserved, Zn finger containing nuclear protein that is transiently expressed in a subset of neurons at the time of dendrite outgrowth. In a loss-of-function *hamlet* mutant, the es neurons are transformed into neurons with an elaborate dendrite arbor. Conversely, ectopic expression of *hamlet* even in post-mitotic da neurons causes the opposite transformation.

In most cases, however, the dendritic fate is determined by the combined action of multiple transcription factors. Expression of the gene *cut* in the da neurons differs such that neurons with small and simple dendritic arbors either do not express Cut (class I neurons) or express low levels of Cut (class II), whereas neurons with more complex dendritic branching patterns and lxarger dendritic fields (class III and IV) express higher levels of Cut. Analysis of loss-of-function mutations and class-specific overexpression of Cut demonstrated that the level of Cut expression controls the distinct, class-specific patterns of dendritic branching (Grueber *et al.*, 2003a). Loss of Cut reduced dendrite growth and class-specific terminal branching and converted class III and IV neurons to class I and II morphologies such that they have relatively simple dendritic branching pattern and small dendritic fields. Conversely, overexpression of Cut in neurons that express lower levels of endogenous cut resulted in transformations toward the branch morphology of high-Cut expressing neurons. Furthermore, a human Cut homologue, CDP, can substitute for *Drosophila* Cut in promoting the dendritic morphology of high-Cut neurons (Fig. 21.2). Thus, Cut may function as an evolutionarily conserved regulator of neuronal-type specific dendrite morphologes.

In contrast to Cut, Spineless (ss), the *Drosophila* homologue of the mammalian dioxin receptor, is expressed at similar levels in all da neurons. In *ss* mutants, different classes of da neurons elaborate dendrites with similar branch numbers and complexities (Fig. 21.2), suggesting that da neurons might reside in a common "ground state" in the absence of *ss* function. Studies of the epistatic relationship between Cut and Spineless indicate that these transcription factors likely are acting in independent pathways to regulate morphogenesis of da neuron dendrites (Kim *et al.*, 2006). A comprehensive analysis of transcription factors with RNAi screens has revealed more than 70 transcription factors regulate dendritic arbor development of class I neurons in *Drosophila*. These findings suggest that complicated networks of transcriptional regulators likely regulate neuron-specific dendritic arborization patterns (Parrish *et al.*, 2007).

Dendro-Dendritic Interactions Regulate the Shape and Organization of Dendritic Fields

Dendro-dendritic interaction can have a profound influence on determining the size and shape of the dendritic field as well as the spatial relationship between different dendritic fields. In many areas of the nervous system, dendrites of different types of neurons are intermingled and packed into a tight space. This arrangement is not random but well organized. At least three mechanisms contribute to the orderly organization of dendritic fields: self-avoidance, tiling, and coexistence. Dendrites of a neuron rarely bundle together or crossover one another (self-avoidance). Presumably, self-avoidance contributes to maximal dispersion of a neuron's dendritic arbor for efficient and unambiguous signal processing. Certain types of neuron also exhibit a phenomenon known as tiling, which refers to the avoidance between the dendrites of adjacent neurons of the same type. This proper tie allows neurons to cover large areas of the nervous system like tiles covering a floor, completely and without redundancy. Tiling was first discovered in

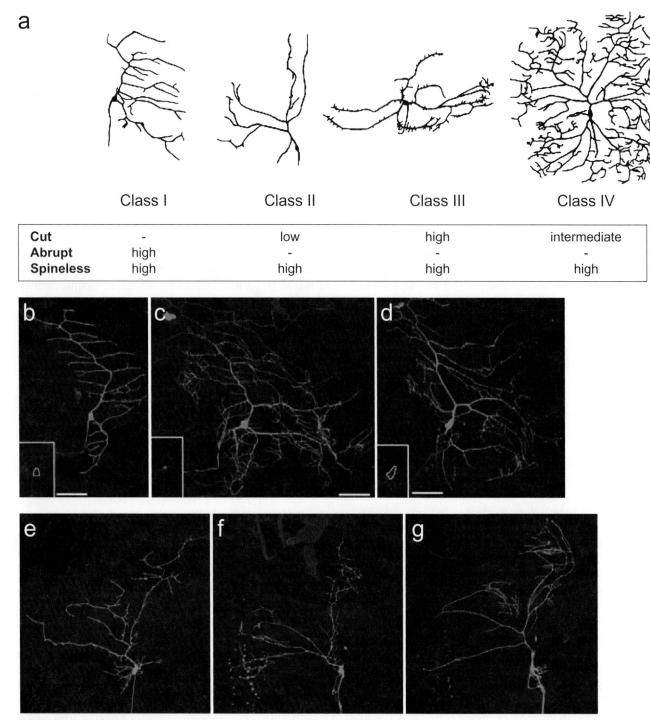

FIGURE 21.2 Transcription factors regulate the diversity and complexity of dendrites. (*a*) Dendrite morphologies of representative class I, II, III, and IV dendritic arborization (da) sensory neurons in the *Drosophila* PNS and a summary of the relative levels of expression of the transcription factors Cut, Abrupt, and Spineless in these neurons. (*b–d*). Ectopic expression of *cut* increases the dendritic complexity of class I da neurons. (*b*) Wild-type dendritic morphology of the ventral class I neuron vpda. Cut is normally not expressed in vpda (*inset*). (*c*) Ectopic expression of Cut in vpda leads to extensive dendritic outgrowth and branching. (*d*) Ectopic expression of CCAAT-displacement protein (CDP), a human homolog of *Drosophila cut*, also induces overbranching. (*e–g*) Loss of *spineless* function leads to a dramatic reduction in the dendritic diversity of different classes of da neurons. In loss-of-function *spineless* mutants, class I (*e*), class II (*f*), and class III (*g*) da neurons begin to resemble one another. From Parrish *et al.* (2007b).

mammalian retina (Wassle *et al.*, 1981). The dendrites of the same type of neurons may repel one another based on self-avoidance or tiling, however it is essential that dendritic fields of different types of neurons don't repel each other so different types of neurons can process different aspects of inputs. Thus, a neuron's dendritic branches needs to be able to recognize other branches of the same neuron, and in the case of tiling, branches of other neurons of its own kind. In addition, branches of different types of neurons need to be able to ignore each other to coexist. How do neurons manage such a plethora of dendritic interactions? What are the underlying molecular mechanisms?

These general organizational principles of dendritic fields can be studied in Drosophila da neurons. Their dendrites show self-avoidance and tend to spread out. Of the four classes of da neurons, class III and class IV neurons show tiling (Grueber *et al.*, 2003b; Sugimura *et al.*, 2003). Further, different classes of da neurons don't repel each other and they can coexist with their dendritic fields superimposed on each other. Recent studies of the organization of da dendritic fields have begun to reveal the molecular mechanisms including the roles of Dscam (Down syndrome cell adhesion molecule) (Hughes *et al.*, 2007; Matthews *et al.*, 2007; Soba *et al.*, 2007), Tricornered, and Furry (Emoto *et al.*, 2004).

Dscam, a member of the immunoglobulin superfamily, originally was identified as an axon guidance receptor. Alternative splicing can potentially generate over 38 thousand isoforms (Schmucker *et al.*, 2000). A neuron typically expresses only a small subset (a couple dozen) of those isoforms (Neves *et al.*, 2004). Dscam appears to be involved in self-recognition of certain Drosophila neurites (Hummel *et al.*, 2003; Wang *et al.*, 2002; Zhu *et al.*, 2006). Biochemical studies showed that Dscam exhibits isoform-specific homophilic binding. Strong homophilic interactions are observed only between the same isoforms, and differences of even a few amino acids greatly reduced the strength of the interactions (Wojtowicz *et al.*, 2004). These results suggest that only when neurite express the same set of Dscam isoforms, there is high level of signaling resulting in repulsion. If they express different isoforms, neurites don't repel each other (Wojtowicz *et al.*, 2004).

Dscam is necessary for da neuron dendrite self-avoidance. Mutant neurons devoid of Dscam exhibit dendrite bundling and a crossing-over phenotype. This self-avoidance phenotype can be rescued largely by expressing a randomly selected single isoform in the neuron, suggesting that it is necessary to have Dscam in the da neuron for their dendritic self-avoidance but the particular isoform is not important

(Hughes *et al.*, 2007; Matthews *et al.*, 2007; Soba *et al.*, 2007).

In contrast, tiling does not seem to be affected in Dscam mutants. Thus, tiling requires some cell surface recognition molecules other than Dscam to mediate the homotypic repulsion between neurons of the same type. Although the signal(s) that mediate tiling behavior remain to be identified, the evolutionarily conserved protein kinase Tricornered (Trc) and the putative adaptor protein Furry (Fry), have been identified as important components of the intracellular signaling cascade involved in tiling (Emoto *et al.*, 2004; Gallegos and Bargmann, 2004). In *trc* or *fry* mutants, dendrites no longer show their characteristic turning or retracting response when they encounter dendrites of the same type of neuron. In the mutants, unlike in the wild-type, there is extensive overlap of dendrites between adjacent neurons of the same kind. As a result, the mutant neurons have enlarged dendritic fields (Emoto *et al.*, 2004).

Given that a neuron's dendrites can self-avoid as long as it has at least one Dscam isoform and it doesn't matter what particular isoform is expressed, what might be the reason for having such large number of potential isoforms? One idea is that a given neuron would express a small number of Dscam isoforms (a dozen or so) more or less stochastically (Neves *et al.*, 2004). Since there are a large number of isoforms (over 38,000), the chance of two adjacent neurons expressing the same set of isoforms and therefore repelling others is very small. This notion predicts that overexpression of the same Dscam isoform in two different kinds of neurons that normally have overlapping dendritic fields would cause the dendrites to repel each other. Indeed, overexpression of the same Dscam isoform in different classes of da neurons whose dendritic fields normally overlap extensively leads to their mutual repulsion (Hughes *et al.*, 2007; Matthews *et al.*, 2007; Soba *et al.*, 2007). The idea that the diversity of Dscam is essential for overlapping dendritic fields is further supported by another experiment in which the Dscam diversity was reduced so that a single isoform is expressed in all da neurons and different classes of da neurons repel each other (Soba *et al.*, 2007). Although single Dscam isoform is sufficient for dendrite self-avoidance, different neurons need to express different isoforms so they can share the same space. Thus Dscam functions as a tag for neuron to recognize itself and the diversity is needed for coexistence.

The Maintenance of Dendritic Fields

Dendrite development is a dynamic process involving both growth and retraction. Thus, selective stabilization or destabilization of branches might be one important mechanism to shape dendritic arbors. Studies of Drosophila class IV da neurons revealed that dendritic fields are actively maintained and there is a genetic program used to maintain dendritic fields. The tumor suppressor Warts (Wts), as well as the Polycomb group of genes are required for the maintenance of the class IV da dendrites.

Drosophila has two NDR (nuclear Dbf2-related) families of kinase: Trc and Wts. Wts and its positive regulator Salvador originally were identified as tumor suppressor genes that function to coordinate cell proliferation and cell death. Loss-of-function mutants of either gene causes a progressive defect in the maintenance of the dendritic arbors, resulting in large gaps in the receptive fields (Fig. 21.3). Time-lapse studies suggest that the primary defect is in the maintenance of terminal dendrites, so Wts may normally function to stabilize these dendrites.

How are the establishment and maintenance of dendritic fields coordinated? In Drosophila class IV neurons, the Ste-20-related tumor suppressor kinase Hippo (Hpo) can directly phosphorylate and regulate both Trc, which functions in the establishment of dendritic tiling, and Wts, which functions in the maintenance of dendritic tiling (Emoto et al., 2006). Furthermore, hpo mutants have defects in both establishment and maintenance of dendritic fields. How Hpo regulates the transition from establishment to maintenance of dendritic fields remains to be determined.

What might be the downstream genes regulated by Wts? In the Drosophila retina, Wts regulates cell prolif-

eration and apoptosis by phosphorylating the transcriptional coactivator Yorkie (Huang et al., 2005). However, Yorkie does not appear to function in dendrite maintenance. Instead, the Polycomb genes are good candidates as targets for Wts/Sav for dendritic maintenance. The Polycomb genes are known to regulate gene expression by establishing and maintaining repression of developmentally regulated genes. PcG genes can be separated into two multiprotein complexes: Polycomb repressor complex 1 (PRC 1) and PRC2. PRC2 is thought to mark the genes to be silenced by methylating histone H3, and PRC1 then comes in and blocks transcription. Mutants of several members of PRC1 and PRC2 have dendrite maintenance phenotype very similar to that of Wts. Further, genetic and biochemical experiments suggest a functional link between Hpo/Wts signaling and the PcG and that PcG genes regulate the dendritic field in part through Ultrabithorax (Ubx), one of the Hox genes in Drosophila (Parrish et al., 2007).

EXTRACELLULAR REGULATION OF DENDRITIC DEVELOPMENT IN THE MAMMALIAN BRAIN

Regulation of Dendrite Orientation

Much of our understanding of the molecular mechanisms of dendritic growth control in vertebrates comes from investigations in the developing cerebral cortex. Most cortical neurons are generated from precursors proliferating in the germinal zones lining the ventricle (Fig. 21.4). Once the cells become postmitotic, they migrate from the ventricular zone to the cortical plate.

Dendritic differentiation, as determined by expression of dendrite-specific genes such as MAP-2, does not begin until the cells have completed their migration. Following migration, pyramidal neurons extend an axon toward the ventricle and an apical dendrite toward the pial surface. To test the role of the local cortical environment in directing the growth of nascent axons and dendrites, Polleux and Ghosh developed an in vitro assay in which dissociated neurons from a donor cortex were plated onto cortical slices in organotypic cultures. Strikingly, neurons plated on cortical slices behave just like the endogenous pyramidal neurons and extend an axon toward the ventricular zone and an apical dendrite toward the pial surface. Both the oriented growth of the axon and the apical dendrite are regulated by the chemotropic signal Sema 3A, which is present at high levels near the pial surface, and acts as a chemorepellant for axons and a chemoattractant for dendrites (Polleux et al., 1998, 2000).

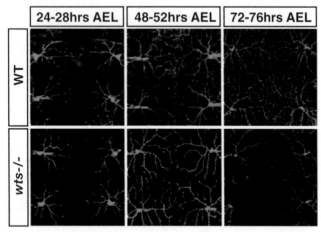

FIGURE 21.3 Dendritic fields are largely unchanged once established during development. Late-onset dendritic loss in Drosophila warts mutants (wts-/-) in late larval stages. Live images of wild-type (WT) and wts mutant (wts) dendrites of class IV da neurons at different times after egg laying (AEL). In wts mutants, dendrites initially tile the body wall normally but progressively lose branches at later larval stages. Adapted from Emoto et al. (2006).

Sema3A mRNA

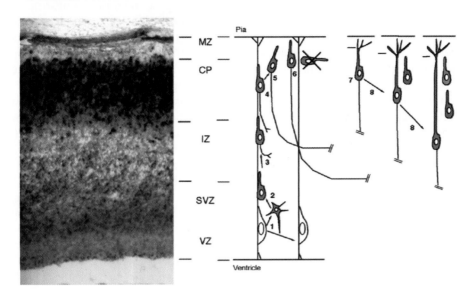

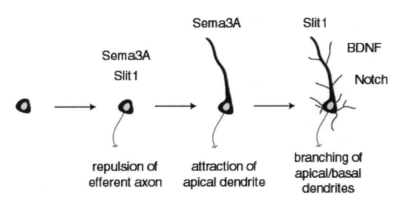

FIGURE 21.4 Upper Panel: Development of the dendritic morphology of cortical pyramidal neurons. Pyramidal neurons are generated from radial glial precursors in the dorsal telencephalon during embryonic development. Upon cell cycle exit from the ventricular zone (VZ), young post-mitotic neurons migrate along the radial glial scaffold and display a polarized morphology with a leading process directed toward the pial surface and sometimes a trailing process directed toward the ventricle. The leading process later becomes the apical dendrite. The trailing process of some neurons (but not all) develop into an axon that grows toward the intermediate zone (IZ; the future white matter) once cells reach the cortical plate (CP). Upon reaching the top of the cortical plate, postmitotic neurons detach from the radial glial processes and have to maintain their apical dendrite orientation toward the pial surface and axon outgrowth orientation toward the ventricle, which appears to be regulated by Sema3A, which acts as a chemoattractant for the apical dendrite and a chemorepellant for the axon. Adapted from Polleux and Ghosh (2008). Lower Panel: A model of how sequential action of extracellular factors might specify cortical neuron morphology. A newly post-mitotic neuron arrives at the cortical plate, where it encounters a gradient of Sema3A (Polleux *et al.* 1998), which directs the growth of the axon towards the white matter. The same gradient of Sema3A attracts the apical dendrite of the neuron toward the pial surface (Polleux *et al.*, 2000). Other factors, such as BDNF and Notch, control the subsequent growth and branching of dendrites. Adapted from Polleux and Ghosh (2008).

The differential response of axons and dendrites to Sema3A led Polleux *et al.* to explore the mechanisms that might lead to the generation of opposite responses in two compartments of the same neuron. They discovered that the enzyme that regulates cGMP production, soluble guanylate cyclase (sGC), was localized asymmetrically in immature cortical neurons and was preferentially targeted to the emerging apical dendrite (Polleux *et al.*, 2000). Pharmacological inhibition of sGC activity

or one of its downstream targets, cGMP-dependent protein kinase (PKG), abolishes the ability of Sema3A to attract apical dendrites, but does not affect the axons. Thus the basis of the differential response of axons and dendrites to Sema3A appears to be asymmetric targeting of sGC to the emerging dendrite.

The nonreceptor tyrosine kinases Fyn and Cdk5 also play important roles in mediating the effects of Sema3A on cortical dendrite orientation (Sasaki *et al.*, 2002). Fyn is a member of the Src family of nonreceptor tyrosine kinases. Cyclin-dependent kinase 5 (Cdk5), a member of the serine/threonine kinase Cdk family, has enzymatic activity only in postmitotic neurons due to a neuron-specific expression of the regulatory subunit p35 (Lew and Wang, 1995). Cdk5 and p35 play critical roles in the laminar organization of the cerebral cortex by regulating the migration of neurons (Chae *et al.*, 1997; Ohshima *et al.*, 1996). Sasaki *et al.* provided genetic evidence that Fyn acts downstream of Sema3A by showing that the apical dendrite orientation of layer 5 and layer 2/3 pyramidal neurons is not different from wild-type controls in Sema3A(+/−) and Fyn (+/−) single heterozygous mice but is significantly impaired in Sema3A(+/−)/Fyn(+/−) double heterozygous mice.

The generality of the concept of dendritic guidance is supported by studies in Drosophila on the role of Netrin–Frazzled signaling in axonal and dendritic development (Huber *et al.*, 2003; Yu and Bargmann, 2001). Netrin-A and Netrin-B, two netrin-family proteins in Drosophila, are diffusible glycoproteins produced by specialized midline cells. The activation of Frazzled, a cell-surface receptor for netrins (known in vertebrates as deleted in colorectal cancer (DCC)), causes chemoattraction and midline-crossing of axons from neurons located near the midline (Huber *et al.*, 2003; Yu and Bargmann, 2001). In single-cell analysis of bilaterally paired neurons, axons and dendrites show cell-autonomous use of Frazzled at the midline (Furrer *et al.*, 2003). For example, the RP3 motoneuron in wild-type Drosophila extends axons across the midline and extends dendrites on both sides the midline. However, in both frazzled-null and netrinA/netrinB double-null mutants, the RP3 neuron fails to direct its axon or dendrite toward the midline in three-quarters of the cases examined, suggesting that the midline-directed outgrowth of the RP3 axons and dendrite requires the Netrin–Frazzled signaling (Furrer *et al.*, 2003). In robo mutants RP3 dendrites converge at the midline, indicating that Robo signaling also regulates dendritic guidance.

Regulation of Dendritic Growth and Branching

The growth and branching of dendrites can be influenced by a large number of extracellular signals (Fig. 21.5). In this section we discuss how specific extracellular factors regulate the development of the dendritic tree.

Neurotrophins

Studies from the last few years provide compelling evidence that the growth and branching of dendritic arbors is regulated by extracellular signals, including neurotrophic factors. Neurotrophins (NGF, BDNF, NT-3, and NT-4) exert their effects through the Trk family of tyrosine kinase receptors. Experiments in which the effects of neurotrophins on dendritic growth control have been examined in slice cultures indicate that in general, neurotrophins increase the dendritic complexity of pyramidal neurons by increasing total dendritic length, the number of branchpoints, and/or the number of primary dendrites (Baker *et al.*, 1998; McAllister *et al.*, 1995; Niblock *et al.*, 2000). The response is rapid and an increase in dendritic complexity is readily apparent within 24 hours of neurotrophin exposure. There is a clear specificity in the short-term response of pyramidal neurons of different cortical layers to each of the neurotrophins. For instance, NT-3 strongly increases dendritic complexity in layer 4 neurons, but has no apparent effect on layer 5 neurons. In addition, basal dendrites in specific layers respond most strongly to single neurotrophins whereas apical dendritic growth is increased by a wider array of neurotrophins. Live imaging of layer 2/3 neurons expressing BDNF show a high level of dendrite dynamics. Both dendritic branches and spines are rapidly lost and gained in BDNF transfected neurons (Baker *et al.*, 1998; McAllister *et al.*, 1995; Niblock *et al.*, 2000). BDNF overexpression favors addition of primary dendrites and proximal branches at the expense of more distal segments. Similarly, overexpression of TrkB in layer 6 pyramidal neurons results in a predominance of short proximal basal dendrites (Yacoubian and Lo, 2000).

Recently, Osteogenic Protein-1 (OP-1), which is a member of the transforming growth-factor-beta superfamily, was shown to increase total dendritic growth and branching from dissociated embryonic cortical neurons (Le Roux *et al.*, 1999). Furthermore, insulin-like growth factor-1 (IGF-1) was shown to affect dendrite growth and branching of postnatal layer 2 cortical neurons (Niblock *et al.*, 2000). In contrast to neurotrophins, IGF affects both basal and apical dendritic growth and remodeling, illustrating that the final dendritic complexity of pyramidal neurons is likely to be influenced by the action of multiple neurotrophic factors.

How do neurotrophic factors mediate the morphological changes linked with dendritic remodeling? The observed short-term dynamics indicate a rapid modulation of cytoskeletal elements by neurotrophic factor signaling. Of the major signaling pathways activated by Trk receptors and most other tyrosine kinase receptors, the MAP kinase and PI-3Kinase pathways have been implicated in neurite formation in both neuronal cell lines and primary neurons (Posern *et al.*, 2000; Wu

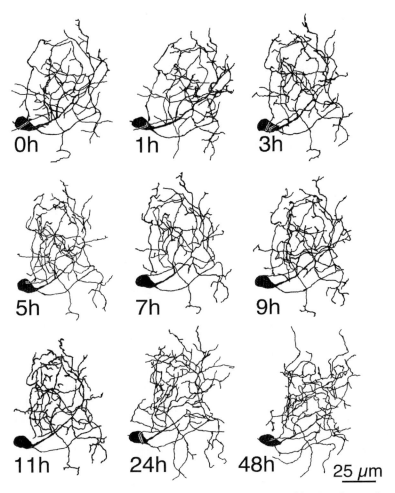

FIGURE 21.5 Reconstructions of the dendritic arbor of a xenopus tectal neuron imaged by time-lapse microscopy. (From Bestman *et al.* 2008).

et al., 2001; Dijkhuizen *et al.*, 2005). It is likely that these signaling pathways influence neuronal morphology by regulating the activity of the Rho family GTPases, which mediate actin cytoskeleton dynamics and are known to induce rapid dendritic remodeling (Box 21.1; Fig. 21.6). Experiments in neuronal cell lines show that NGF can activate the small GTPase Rac1 in a PI-3 Kinase dependent manner, and this activation is necessary for neurite elaboration (Kita *et al.*, 1998; Posern *et al.*, 2000; Yasui *et al.*, 2001).

Part of the neurotrophic factor effect on dendritic morphogenesis may also include their control of expression of structural proteins, since long-term exposure to neurotrophins leads to net dendritic growth. It was reported recently that BDNF can upregulate local protein synthesis in dendrites within hours (Aakalu *et al.*, 2001). In addition, specific mRNAs for several cytoskeletal proteins are present in dendrites (Kuhl and Skehel, 1998). This raises the interesting possibility that local synthesis of structural compo-

nents may be involved in neurotrophic factor control of dendritic growth.

Notch Signaling

The diversity of signals that can influence dendritic morphology is underscored by a series of studies on the role of mammalian Notch proteins in regulating dendritic growth and branching. Originally identified in Drosophila, Notch is a type I cell-surface protein, which functions as a receptor. Proteolytic processing of full-length Notch generates two fragments that associate at the plasma membrane to form a receptor complex. The mechanism of Notch receptor activation involves cleavage and nuclear translocation of the intracellular domain of the receptor (reviewed in Weinmaster, 2000). The intracellular domain of Notch enters the nucleus and binds the transcription factor Suppressor of Hairless (Su(H)) activating gene transcription. Mammalian homologs of Notch (Notch1–4), the Notch ligands Delta (Delta1–3) and Serrate (Jagged1, Jagged2), the

BOX 21.1

Rho GTPases CONTROL THE STRUCTURE OF THE DENDRITIC CYTOSKELETON

Since all the pathways necessarily converge on the regulation of the cytoskeleton, we start by outlining the role of the Rho GTPases in controlling the dendritic cytoskeleton and then examine some of the extracellular cues that regulate GTPases. It is noteworthy that many, if not all, of these pathways also affect gene expression and different aspects of dendritic function, such as synaptic transmission, calcium signaling, and neuronal excitability. Such divergence of signaling from extracellular cues assures that the development of dendritic structure and function are tightly coregulated.

The dendritic cytoskeleton is composed of bundles of microtubules extending within the center of the dendritic shaft, a cortex of actin sandwiched between the microtubular bundles and the plasma membrane, and an actin matrix at the tip of dendritic processes. Fine terminal dendritic branches, or filopodia, have actin filaments as their sole cytockeletal component (reviewed in Van Aelst and Cline, 2004). Considerable effort has been devoted to understand the interaction between extracellular signaling events and the cytoskeleton, since these interactions are likely to be essential for the highly stereotyped and yet plastic elaboration of the dendritic arbor structure. A general scenario is emerging in which an extracellular signal interacts with a cell surface receptor that activates a cascade controlling the RhoA GTPases that, in turn, affect both the actin and microtubule based cytoskeleton in dendrites (Newey et al., 2005). The significance of these molecules for dendritic morphogenesis is perhaps best illustrated by the fact that the abnormal development of dendritic trees, a hallmark of several different types of mental retardation, is at least in part caused by deficient signaling via Rho GTPases (Govek et al., 2005).

The Rho GTPases regulate the cytoskeleton in all cell types, however the elaborate and plastic structure of neurons poses particularly fascinating regulatory constraints on GTPase signaling (Luo, 2000; da Silva and Dotti, 2002; Van Aelst and Cline, 2004). The Rho GTPases function as bimodal switches, cycling between inactive, GDP-bound and active, GTP-bound conformations. RhoA, Rac1, and cdc42 are arguably the best studied of the small Rho GTPases. These molecules regulate both actin and microtubule dynamics (Gundersen et al., 2004; Zheng, 2004) and the manipulation of their individual activities has shown that each plays a particular role in dendritic structure development (reviewed in Newey

et al., 2005). The interplay of these effects on the cytoskeleton is key in shaping the intricacy of dendritic trees.

As more refined methods are used to probe the molecular and cellular basis of structural plasticity, our understanding of the intricate web of control becomes more complete. A striking example is that of the *in vivo* dendritic arbor development of optic tectal neurons in *Xenopus*. These neurons respond to stimulation of the tadpole visual system with an increased dendritic arbor growth rate, which requires glutamate receptor activity (Sin et al., 2002). Experimental paradigms in which dominant negative or constitutively active forms of RhoA, Rac, and Cdc42 were expressed, demonstrated the participation of the Rho GTPases in the activity-dependent enhanced dendritic arbor growth rate. Expression of dominant negative forms of Rac or Cdc42, or expression of active RhoA blocked dendritic arbor elaboration in response to visual stimulation (Li et al., 2000, 2002; Sin et al., 2002) (Fig. 21.4A). These data suggest that glutamatergic synaptic input regulates the development of dendritic arbor structure by controlling cytoskeletal dynamics, and that the Rho GTPases are an interface between glutamate receptor activity and the cytoskeleton. Furthermore, these data and reports from other systems support a model in which Rac and Cdc42 activity regulate rates of terminal branch dynamics (Ruchhoeft et al., 1999; Li et al., 2000, 2002; Wong et al., 2000; Hayashi et al., 2002; Ng et al., 2002; Sin et al., 2002; Scott et al., 2003) whereas RhoA regulates extension of branches (Ruchhoeft et al., 1999; Lee et al., 2000; Li et al., 2000, 2002; Nakayama et al., 2000; Wong et al., 2000; Sin et al., 2002; Ahnert-Hilger et al., 2004; Pilpel and Segal, 2004) in response to activity.

Branch formation requires the regulation of local cortical actin dynamics to create protrusive forces that allow filopodial sprouting (Luo, 2002; Nimchinsky et al., 2002). Time-lapse imaging indicates that filopodia are extremely dynamic, consistent with the rapid assembly and disassembly of actin filaments. The stabilization of filopodia and their extension as branches likely depends on their invasion by microtubules. Although the invasion of filopodia by microtubules is a key regulatory event in dendritic arbor development, the mechanisms regulating this process are unknown. Microtubules generate the mechanical forces necessary for branch elongation and can serve as tracks for the delivery of new membrane as new branches extend (Horton and Ehlers, 2003, 2004). A close

BOX 21.1 *(cont'd)*

interplay between actin and microtubules is then fundamental for the cellular events leading to changes in dendritic architecture. Importantly, in nonneuronal cells, the activity of Rho GTPases not only induces changes in both actin and microtubules but is also itself modified by alterations in the dynamics of the two cytoskeletons, thus serving as the regulator for the interplay between actin and microtubules (Wittmann and Waterman-Storer, 2001; Fukata *et al.*, 2003; Etienne-Manneville, 2004; Zheng, 2004).

It is tempting to hypothesize that in response to activity, the role of RhoA in regulating extension of branches relies on its capacity to control microtubule stabilization (e.g., via mDia; Palazzo *et al.*, 2001) and favor polymerization of cortical actin (Nobes and Hall, 1995; Da Silva *et al.*, 2003), whereas Rac and cdc42 act on branch dynamics by regulating actin (e.g., supporting filopodial formation; Luo, 2002) and by favoring microtubular dynamics (e.g., by regulating catastrophe rates; Daub *et al.*, 2001; Kuntziger *et al.*, 2001). The description of dendritic roles for other Rho GTPases, such as that of Rnd2 in regulating branching via its effector Rapostlin (Negishi and Katoh, 2005), will help in the detailed understanding of how Rho GTPases regulate dendritic arbor development.

Rho GTPases are distributed ubiquitously throughout the neuronal cytoplasm (Govek *et al.*, 2005) and, consequently, the activity of these proteins must be restrained in dendrites by resident upstream regulators. The link between incoming signals, for instance by activation of neurotransmitter receptors, and GTPase activity is mediated by GTPase regulatory proteins, which are particularly interesting because they are capable of integrating extracellular signaling with other signaling events relevant to neuronal structure. These include the guanine exchange factors (GEFs), which activate GTPases by favoring the substitution of GDP for GTP and GTPase activating proteins (GAPs), which inactivate GTPases by inducing GTP hydrolysis (Schmidt and Hall, 2002; Bernards and Settleman, 2004). A rush of recent papers has examined the potential participation of several GEFs and GAPs in activity-dependent dendritic structural plasticity. GAPs can regulate dendritic development, as is exemplified by

the observation that p190 RhoGAP is necessary for the dendritic remodeling that allows the shift from pyramidal to nonpyramidal morphologies in cortical cultures (Threadgill *et al.*, 1997). Importantly, p190 RhoGAP is likely to exert its effect in an activity-dependent manner as indicated by its importance in fear memory formation in the lateral amygdala (Lamprecht *et al.*, 2002).

GEFs also play important roles in regulating dendrite arbor structure. For instance, Tiam1, a Rac-GEF, is located in dendrites and in particular in spines in cortical and hippocampal neurons. Tiam is noteworthy because it associates with the NMDA receptor, is phosphorylated in a calcium- and NMDA receptor-dependent manner and is required for dendritic arbor development (Tolias *et al.*, 2005). This is particularly interesting if one considers that the closely related member of the Dbl family of GEFs (Rossman and Sondek, 2005), Trio, regulates the development of axons in a potentially calcium-dependent manner (Debant *et al.*, 1996), indicating the subcellular localization within different neuronal compartments is key to the specificity of GEF function. Kalirin, another example of a RhoGEF, in this case a dual RhoA- and Rac1-GEF, has been shown to regulate the development and maintenance of dendritic arbors by modulation of RhoA and Rac activities (Penzes *et al.*, 2001). Recruitment of this Rho GTPase regulator in dendrites depends on the ephrin-EphB transynaptic signaling pathway, another cell surface signaling system linked to the actin cytoskeleton regulatory machinery. EphrinB-EphB receptor signaling may coordinate pre- and postsynaptic structural and functional development (Palmer and Klein, 2003). Its activation results in the translocation of Kalirin to synaptic sites and the activation of a signaling pathway involving Rac1 and the specific downstream effector PAK (Penzes *et al.*, 2003). One intriguing possibility is that the Ephrin-EphR signaling could be coordinated with regulation of NMDA receptor distribution and calcium-permeability in postsynaptic sites (Dalva *et al.*, 2000; Takasu *et al.*, 2002). This kind of crosstalk between proteins involved in cell–cell contact and neurotransmitter receptors provides evidence for coregulation of development of dendritic arbor structure and synaptic communication.

---

transcription factors Su(H) (CBF1/RBP-Jk) and E(Spl) (Hes1–5) have been isolated (reviewed in Weinmaster, 2000). Several of these genes are expressed in the developing brain and spinal cord and are likely to control various aspects of neural development.

The possibility that Notch might play a role in regulating dendritic patterning was suggested by immunocytochemical localization studies that showed that mammalian Notch1 is expressed by both dividing cells in the ventricular zone (VZ) and postmitotic neurons

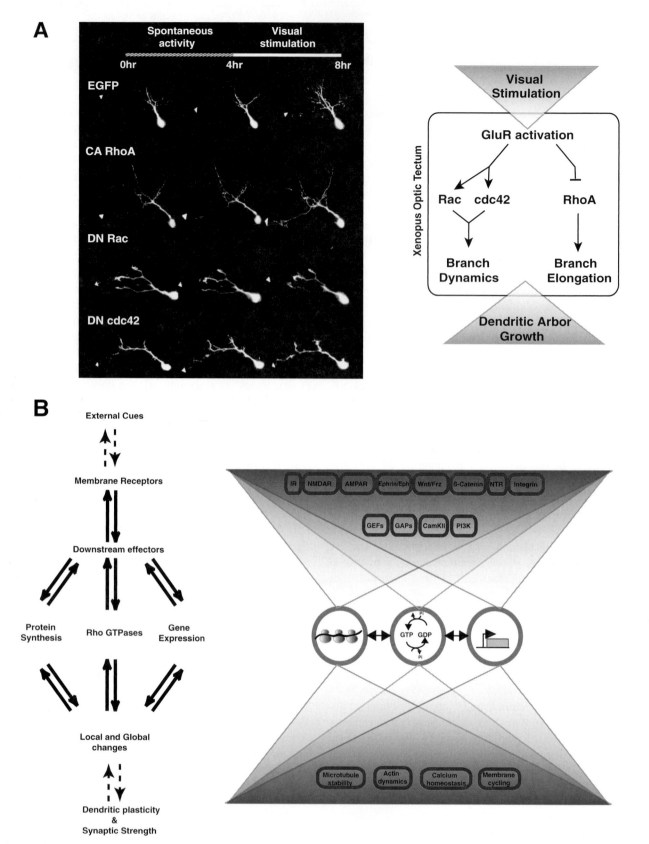

FIGURE 21.6 Effect of Rho, Rac, and Cdc42 on dendritic growth. (A) Visual stimulation over a 4-hour period increases the rate of dendrite growth in optic tectal neurons from Xenopus, Expression of constitutively active (CA) RhoA, dominant negative (DN) Rac, or cdc42 affect specific aspects of dynamic dendritic arbor growth, as shown in the images of GFP-expressing tectal neurons collected *in vivo*. As summarized in the diagram on the right, visual stimulation, acting through glutamate receptors, triggers enhanced dendrite growth by regulating the rho GTPases. (B) External cues acting through membrane receptors and downstream signaling pathways regulate functional and structural dendritic plasticity. Multiple mechanisms operate in parallel to accomplish and control neuronal plasticity. (From Bestman *et al.* 2008)

in the cortical plate (CP) (Redmond *et al.*, 2000; Sestan *et al.*, 1999). Several observations suggest that Notch signaling might mediate contact-dependent inhibition of neurite outgrowth. For example, in postmitotic neurons there is an inverse correlation between Notch1 expression and total neurite length, and overexpression of a constitutively active Notch1 construct leads to a reduction in the total neurite length (Sestan *et al.*, 1999). Cocultures of cortical neurons with Delta- or Jagged-expressing cell lines, or addition of soluble ligands leads to a decrease in total neurite length, suggesting that Delta or Jagged are the relevant Notch1 ligands (Sestan *et al.*, 1999). Also, overexpression of Numb and Numblike, intracellular modulators that inhibit Notch activation via Su(H)/CBF1, leads to an increase in total neurite length (Sestan *et al.*, 1999). Berezovska and coworkers (Berezovska *et al.*, 1999) also have found that expression of constitutively active Notch1 in hippocampal neurons leads to an inhibition of neurite outgrowth. A study examining Notch function in neuroblastoma cells came to a similar conclusion regarding the effects of Notch signaling on neurite length (Franklin *et al.*, 1999). Together these observations indicate that Notch signaling has an inhibitory effect on process outgrowth.

In addition to restricting length, Notch signaling in cortical neurons has a major influence on dendritic branching (Redmond *et al.*, 2000). Inhibition of Notch1 signaling by overexpression of a dominant negative Notch1 construct or with antisense oligonucleotide treatment leads to a decrease in dendritic branching in neurons, and overexpression of a constitutively active Notch1 construct decreases average dendrite length but increases the branching index, resulting in an overall increase in dendritic complexity. Taken together these experiments reveal a positive role for Notch in dendrite branching and a negative role in dendrite and total neurite length.

Slit/Robo Signaling

In a search for other extracellular cues that regulate dendrite development, Whitford *et al.* discovered that Slit proteins simultaneously repel pyramidal neuron axons and stimulate dendrite growth and branching (Whitford *et al.*, 2002). The Slits are a well-studied family of multifunctional guidance cues that have been shown to both repel axons and migrating cells, as well as promote elongation and branching of developing sensory axons (reviewed in Huber *et al.*, 2003). Generally, Slits exert their effects through binding to specific members of the Roundabout, or Robo, family of receptors (Huber *et al.*, 2003). Whitford *et al.* (2002) demonstrated that one of the three vertebrate Slits, Slit1, and two of the three Robo receptors, Robo1 and

Robo2, are expressed in the developing cortex during the time of initial axon and dendrite differentiation. Using the slice overlay assay developed by Polleux *et al.* (1998, 2000), Whitford *et al.* demonstrated that Slit1 is a chemorepellant for cortical axons. Interestingly, in addition to repelling cortical axons, Slit1 also potently increases dendritic growth and branching of both pyramidal and nonpyramidal cortical neurons, paralleling a similar role for Slit proteins in the regulation of axonal branching. These effects of Slit1 are mediated by the Robo1 and 2 receptors since transfection of neurons with dominant-negative forms of these receptors in dissociated cultures and slices decreases dendritic branching. Thus, in contrast to the guidance role that Sema3A plays in orienting apical dendrites, Slit1 acts as a more general dendrite growth and branching signal for cortical neurons (Whitford *et al.*, 2002).

WNT Signaling

Another illustration of the ability of individual signals to control diverse biological responses including the control of axonal and dendritic development is provided by the WNT family of secreted proteins. The WNTs represent a large family of extracellular cues initially identified as potent morphogens involved in patterning organ development in both invertebrates and vertebrates. WNTs also have been shown to regulate cell proliferation, migration, and survival (Ciani and Salinas, 2004, 2005). Recently, a role for WNT proteins in the regulation of the neuronal cytoskeleton has emerged (reviewed in Ciani and Salinas, 2005). WNT proteins can function as axon-guidance molecules and as target-derived signals that regulate axonal remodeling and synapse formation (Hall *et al.*, 2000; Krylova *et al.*, 2002). WNT proteins signal through at least three different pathways. The binding of WNT proteins to Frizzled receptors results in the activation of the scaffolding protein Dishevelled (Dvl). In the so-called canonical pathway, WNT proteins signal through Dvl to inhibit GSK3-β a serine/threonine kinase. Inhibition of GSK3-β, in turn, activates β-catenin-T-cell–specific transcription factor mediated transcription. WNT proteins can also signal through Dvl to regulate Rho GTPases during convergent extension movements and tissue polarity during early development. Finally, WNT proteins can activate a Ca^{2+}-dependent pathway, again through Dvl. WNT proteins induce axonal remodeling through the activation of Dvl and the subsequent inhibition of GSK3-β (Hall *et al.*, 2000). Dvl has been shown to act locally to regulate microtubule stability by inhibiting a pool of GSK3-β through a β-catenin- and transcriptional-independent pathway (Ciani *et al.*, 2004).

WNT proteins recently have been implicated in the control of dendritic arborization of hippocampal neurons during development (Rosso *et al.*, 2005). Salinas and colleagues have found that *Wnt7b* is expressed in the mouse hippocampus and induces dendritic arborization of hippocampal neurons during development. This effect is mimicked by the expression of DVL. Importantly, analyses of the *Dvl1* mutant mouse revealed that DVL1 is crucial for dendrite development, as hippocampal neurons developed shorter and less complex dendrites in the mutant than in the wild-type. Neither inhibition of GSK3-β nor expression of GSK3-β affects dendritogenesis. Moreover, a dominant-negative β-catenin does not block DVL function in dendrites. These results suggest that the WNT canonical pathway is not involved. In this case WNT7B and DVL signal through a noncanonical pathway in which the small GTPase Rac, but not Rho, is involved. First, endogenous DVL associates with Rac but not Rho in hippocampal neurons. Second, WNT7B or expression of DVL activates Rac in hippocampal neurons. Last, expression of a dominant-negative Rac blocks DVL function in dendrites.

Inhibition of the WNT pathway by SFRP1 (a secreted antagonist of WNT proteins) decreases Rac activation by WNT7B and blocks the effect of WNT7B in dendrite development (Rosso *et al.*, 2005). Rosso *et al.* also report that WNT7B and DVL activate JNK, a downstream effector of Rac. Inhibition of JNK blocks DVL function in dendrites, whereas pharmacological activation of JNK enhances dendrite development. It remains to be determined whether JNK acts downstream of Rac, or whether the WNT–DVL pathway regulates Rac and JNK independently. Although Rho GTPases are well-known modulators of dendrite development and maintenance (see following section), the mechanisms by which extracellular factors modulate these molecules during dendrite morphogenesis has remained poorly understood. The findings reported by Rosso *et al.* (2005) demonstrate that DVL functions as a link between WNT factors and Rho GTPases in dendrites, and they reveal a novel role for JNK in dendrite development.

Cadherins and β-catenin

One of the central challenges in the study of dendritic development is to understand how extracellular cues that regulate dendritic branching are integrated with Ca^{2+} activity-dependent signals. A potential clue comes from recent results exploring the role of β-catenin and cadherins in dendritic branching (Yu and Malenka, 2003). Yu and Malenka report that overexpression of β-catenin (and other members of the cadherin/catenin complex) enhances dendritic arborization, whereas sequestering endogenous β catenin causes a decrease in dendritic branching. Importantly, the authors show that blocking β catenin prevents the enhancement of dendritic morphogenesis caused by neuronal depolarization. Yu and Malenka (2003) also show that the release of secreted WNT, which occurs during normal neuronal development, is enhanced by manipulations that mimic increased activity, and that WNTs contributes to the effects of neural activity on dendritic arborization. These results demonstrate that β catenin is an important mediator of dendritic morphogenesis and that WNT β catenin signaling is likely to be important during critical stages of dendritic development (Yu and Malenka, 2003).

These observations are reinforced by the recent demonstration that another class of cadherins (*Celsr1–3*) regulates dendritic branching of cortical pyramidal neurons (Shima *et al.*, 2004). This class of cadherins has been identified in vertebrates as orthologs of *flamingo*, a gene previously implicated in the control of dendritic development in *Drosophila* (Gao *et al.*, 2000). Shima *et al.* (2004) combined loss-of-function techniques including RNAi-mediated gene silencing of single neurons using biolistic-delivery in P8 organotypic slice cultures to demonstrate that knocking-down *Celsr2* expression in both layer 5 pyramidal neurons and Purkinje cerebellar neurons significantly reduces dendritic branching (Shima *et al.*, 2004). Furthermore, using the same technique, Shima *et al.* performed a structure-function analysis demonstrating that these effects of *Celsr2* on dendritic branching require the integrity of a site on the extracellular portion outside the cadherin-domain as well as a portion of the intracellular domain called the EGF-HRM region. These results indicate that cadherins play an important role in the control of dendritic complexity.

EFFECT OF EXPERIENCE ON DENDRITIC DEVELOPMENT

One of the most striking manifestations of the effects of experience on brain development is the effect sensory input has on dendritic development. Dendrites tend to extend toward sources of afferent input, for instance in the barrel field of somatosensory cortex (Greenough and Chang, 1988), at borders of ocular dominance columns (Katz and Constantine-Paton, 1988; Katz *et al.*, 1989; Kossel *et al.*, 1995), in the olfactory system (Malun and Brunjes, 1996), in the auditory system (Schweitzer, 1991), and in the spinal cord (Inglis *et al.*, 2000). A particularly clear example of this phenomenon comes from study of the retina, where developing retinal ganglion cells transiently respond to both light on and light off events and their dendritic arbors terminate in both On and Off neuropil laminae. With normal visual experience, the bistratified dendritic arbors are pruned to On and Off laminae. Dark-rearing blocks the normal pruning of dendritic branches in retinal ganglion cells so that retinal ganglion cells remain responsive to On and Off visual stimuli (Tian and Copenhagen, 2003). Bodnarenko and Chalupa have shown that glutamate receptor activity is required for the development of stratification in retinal ganglion cell dendrites (Bodnarenko and Chalupa, 1993).

The advent of *in vivo* time-lapse imaging has allowed the assessment of cellular events underlying the development of dendritic arbors. *In vivo* imaging of dendritic arbor development was pioneered in the 1980s by Purves and colleagues, who discovered that peripheral neurons could be imaged repeatedly in developing mice (Hume and Purves, 1981). These studies revealed considerable heterogeneity in the dynamics of dendritic arbor structure and suggested that dendrite structure was affected significantly by differences in afferent inputs to different neurons. More than a decade passed before the methods were developed to reliably label single central neurons and image them over time without damage (Wu and Cline, 1998). *In vivo* time-lapse imaging of neurons is now routine in a few experimental systems, including some translucent fish and amphibian tadpoles, and *Drosophila*. More recent technical advances, including 2-photon microscopy, have permitted imaging of CNS neurons in rodent brain, which was previously refractory to *in vivo* imaging because of the light-scattering nature of cortical tissue and the difficulty of imaging through the skull (Brecht *et al.*, 2004).

In vivo time-lapse imaging studies of newly differentiated neurons in *Xenopus* and Zebrafish optic tectal neurons and peripheral sensory neurons in *Drosophila* demonstrate that the dendritic arbor develops as a result of a gradual process in which fine branches are added and retracted rapidly (Fig. 21.1, Fig. 21.5). Many more branches are added to the arbor than are ultimately maintained, so that the net elaboration of the dendritic arbor occurs as a result of the stabilization of a tiny fraction of the newly added branches. These branches then become the substrate for further branch additions. In this way the complex arbor develops gradually as a result of concurrent and iterative branch addition, stabilization, and extension (Cline, 2001; Hua and Smith, 2004) (Fig. 21.1, Fig. 21.5). A similar pattern of dendritic arbor growth has been observed in hippocampal neurons in slice culture (Dailey and Smith, 1996), suggesting that this is a plan that applies widely. Time-lapse imaging also reveals the changes

in arbor structure over time. In some cases (Figs. 21.1A and 21.1B), the neuron dramatically changes shape by adding or retracting large parts of the arbor (Wu and Cline, 2003). Although this has not been directly tested *in vivo*, it is expected that afferent inputs affect dendritic arbor development through a combination of cell adhesion and diffusible signals, such as neurotransmitters and growth factors. Constant conversation between presynaptic axons and dendrites and back again may result in the mutual stabilization of pre- and postsynaptic structures (Hua and Smith, 2004).

Synaptic inputs and neurotransmitter activity play a significant role in controlling dendritic arbor development. Whole cell recordings from newly differentiated optic tectal neurons in *Xenopus* or cortical neurons in turtle demonstrate that they have glutamatergic and GABA-ergic synaptic inputs as soon as they extend their first dendrites (Blanton and Kriegstein, 1991; Wu *et al.*, 1996). In addition, ambient neurotransmitters, including dopamine, serotonin, glutamate, and GABA may provide low levels of tonic extrasynaptic neurotransmitter receptor activation and signaling in the developing nervous system (Lauder, 1993). These data suggest that neurotransmitters, acting through nonsynaptic or synaptic receptors may affect dendritic arbor development. Synaptic communication allows pre- and postsynaptic neurons to assess potential partners, to establish and strengthen optimal connections, and to prune back or eliminate suboptimal contacts. Considerable evidence supports a model in which newly generated glutamatergic synapses are mediated by the NMDA type of glutamate receptor and that AMPA type glutamate receptors are trafficked into developing synapses as they mature. Addition of AMPA receptors to synapses renders them functional at resting potentials and may stabilize the synaptic structure (Cline, 2001) (Fig. 21.7). If this is true, then conditions that affect AMPA receptor trafficking to synaptic sites would be expected to affect synapse stability and the development of the dendritic arbor.

Previous work has established a link between NMDA and AMPA receptor function and dendritic branch stabilization, suggesting that synaptic strengthening and structural stability may be mechanistically linked. LTP-inducing stimuli increase both synaptic strength and spine formation (Engert and Bonhoeffer, 1999; Maletic-Savatic *et al.*, 1999). Similarly, LTD has been shown to lead to the internalization of AMPAR (Snyder *et al.*, 2001) and a reduction in dendritic spine number, which may lead to synapse elimination (Zhou and Poo, 2004). In a recent study LTP and LTD stimuli resulted in bidirectional changes in dendritic spine structure (Zhou *et al.*, 2004). Genetic

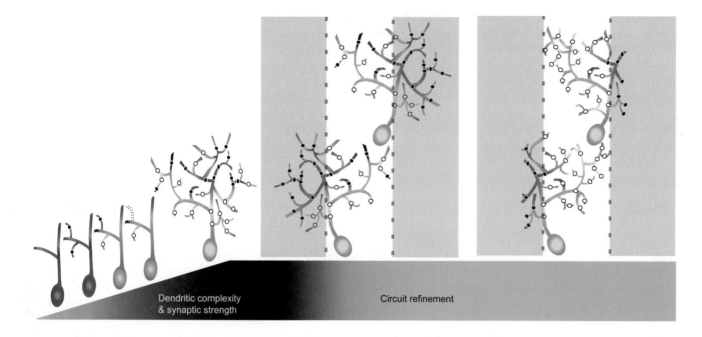

TIME

FIGURE 21.7 During dendritic development, newly generated glutamatergic synapses are mediated primarily by NMDA type glutamate receptors (black dots). With the addition of AMPA type glutamate receptors (white dots), synapses can be active at resting membrane conditions and overall synaptic strength is increased leading to dendritic branch stabilization. Dendrites with relatively weak synapses lacking AMPA receptors are retracted (dotted line). The resulting local increases in calcium influx through AMPA/NMDA containing synapses (indicated with "hot" colors) may enhance rates of branch addition and further stabilization. Newly stabilized branches become the substrate for further branch additions. It is the interplay between the dendrites and their synaptic partners that leads to selective stabilization and elaboration of branches toward appropriate target areas (gray and white bars) that is essential for circuit refinement. These processes are not just important during development, but underlie changes in map refinement in the mature nervous system. (From Bestman *et al.*, 2008)

manipulations of NMDA receptor subunits also demonstrate that NMDA receptor function is required for dendritic arbor development (Lee *et al.*, 2005).

The preceding suggests that the molecular mechanisms involved in dendritic morphogenesis concomitantly control glutamate receptor trafficking to synaptic sites. This suggests a close link between the relative strength of synapses, which depends on the relative contribution of AMPA receptors and the morphology of dendrites (Cline, 2001). Evidence for this idea would strengthen the case for intimate interaction between the development and plasticity of synaptic strength and neuronal structure. Indeed, recently it has been demonstrated that a dendrite-specific immunoglobulin family protein, called dendrite arborization and synapse maturation 1 (Dasm1) promotes AMPA receptor trafficking to excitatory synapses and affects dendrite arborization (Shi *et al.*, 2004a, 2004b).

These data are among a growing number of reports that support a model in which afferent input operating

through AMPA receptors increases dendritic arbor elaboration, potentially through the stabilization of new synapses and new dendritic branches on which they reside. This model is well supported. Glutamate receptors, though favoring arborization at early development stages, actually stimulate branch stabilization at later times, concomitantly with stabilization of local synapses (Rajan and Cline, 1998). This developmental shift reflects the progressive enrichment of synapses with AMPA receptors and an increase in synaptic strength (Wu *et al.*, 1996; Cantallops *et al.*, 2000a; Cline, 2001; Sin *et al.*, 2002). It is noteworthy that this interplay between synapses and the dendrites on which they form is a calcium-dependent event. In the *Xenopus* optic projection, CaMKII is expressed at times when synapses are maturing and dendritic arbors are stabilizing (Wu and Cline, 1998) and its activity was found to slow dendritic growth rates and increase synaptic strength (Wu *et al.*, 1996; Wu, 1998; Zou, 1999). Since these events require modifications in intracellular calcium and this can be achieved by the same molecular pathways discussed later, namely those targeting the Rho GTPases (Li *et al.*, 2002) (Box 21.1), the interplay between synaptic and structural plasticity in response to signals received by the dendrite seems a logical contraption. Indeed, synaptic strength and dendritic arbor development share the same molecular regulatory pathways (Luo, 2002) and examples of cellular events regulated by synaptic transmission and capable of concurrently governing dendritic arborization are at hand.

In addition to synaptic activity, neurons exhibit a variety of types of electrical and biochemical activity during the period of dendritic arbor growth. For instance, waves of calcium spread across large populations of neurons throughout the brain, including retina, cortex, and hippocampus (Firth *et al.*, 2005). Clusters of electrically coupled cortical neurons show pulses of calcium at early stages of development when cortical circuits are developing (Yuste *et al.*, 1992, 1995). These electrical events may be important for the initial formation of neural circuits that underlie normal brain development and plasticity.

MECHANISMS THAT MEDIATE ACTIVITY-DEPENDENT DENDRITIC GROWTH

The effects of neuronal activity on dendritic development are mediated by calcium signaling. Calcium levels in neurons are regulated by influx through calcium channels as well as by release of calcium from intracellular stores. Calcium influx is mediated mainly by voltage-sensitive calcium channels (VSCC), and NMDA receptors. Release from internal stores principally involves calcium-induced calcium release (CICR) or activation by ligands that lead to the production of IP3, which acts on internal stores. Two major signaling targets of calcium influx are calcium/calmodulin dependent protein kinases (CaMKs) and mitogen-activated kinase (MAPK). Upon calcium entry via VSCC or NMDA receptors, calmodulin binds multiple calcium ions and can activate various intracellular effectors, including CaMKs (reviewed in Ghosh and Greenberg, 1995).

Of the CaMKs, CaMKII has been most extensively studied in relation to a role in dendritic development and function. Two isoforms of CaMKII, CaMKIIα and CaMKIIβ, mediate contrasting outcomes on dendrites. CaMKIIα has been reported to stabilize or restrict dendritic growth of frog tectal neurons *in vivo* and mammalian cortical neurons *in vitro* (Wu and Cline, 1998; Redmond *et al.*, 2002). CaMKIIβ, however, has a posi-

tive effect on filopodia extension and fine dendrite development mediated by direct interaction with cytoskeletal actin (Fink *et al.*, 2003). CaMKI recently also has been shown to alter cerebellar granule cell dendrite growth and hippocampal neurons process formation (Wayman *et al.*, 2004).

Unlike CaMKI, CaMKIIα, and CaMKIIβ, CaMKIV is predominately localized in the nucleus. Mice lacking CaMKIV have a defect in dendritic development (Ribar *et al.*, 2000). Pharmacological blockade of CaMKs inhibits calcium-induced dendritic growth in cortical neurons, and expression of an activated form of CaMKIV mimics the dendritic growth effects induced by calcium influx (Redmond *et al.*, 2002). The nuclear localization of CaMKIV suggests that it mediates its effect on dendrite growth via transcriptional events. The best-characterized target of CaMKIV is the transcription factor CREB, which is phosphorylated by CaMKIV at Ser-133. The effects of calcium influx and constitutively active CaMKIV on cortical dendrites are suppressed by dominant negative mutants of CREB, suggesting that CREB-dependent transcription is required for activity-dependent dendritic growth (Redmond *et al.*, 2002). The small GTP-binding protein Rap1 also appears to be an important effector of calcium-dependent activation of CREB and activity-dependent dendrite development. Rap1 is rapidly activated by calcium influx and inhibition of Rap1 suppresses activity-induced CREB phosphorylation and dendritic growth (Chen *et al.*, 2005) (Fig. 21.8).

Although CREB is required for calcium-dependent dendritic growth, activation of CREB is not sufficient to induce dendritic growth. This suggests that other transcription factors are likely to be involved in mediating activity-dependent dendritic growth. A novel approach for identifying activity-induced transcription factors in cortical neurons led to the discovery of CREST, a calcium-activated transactivator required for dendritic growth (Aizawa *et al.*, 2004). CREST is a CBP-interacting protein that is expressed at high levels in the early postnatal cortex. Dendritic growth is severely compromised in the cortex and hippocampus in CREST knockout mice, indicating that CREST function is required for dendritic development *in vivo*. Importantly, depolarization-induced dendritic growth is abolished in cortical neurons from CREST-mutant animals, supporting a key role for CREST in calcium-dependent dendritic growth. These observations suggest that activity-induced dendritic development in early postnatal life requires activation of a transcriptional program, which is likely to be regulated by CREB, CREST, and CBP.

Mitogen-activated kinases also are activated by calcium influx via NMDA receptors and VSCCs.

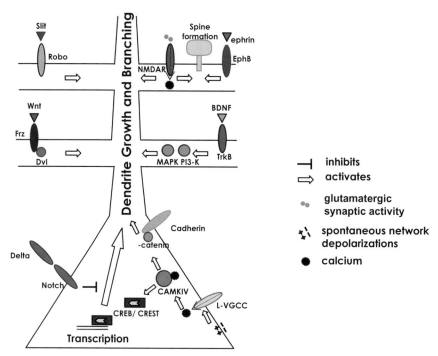

FIGURE 21.8 A summary of signaling pathways by which neuronal activity influences dendritic development. The effects of neuronal activity on dendritic development are mediated by calcium influx via voltage-gated calcium channels (VGCCs) and NMDA receptors, as well as release from internal stores. Local calcium signals act via Rho family proteins to regulate dendritic branch dynamics and stability, whereas global calcium signals recruit transcriptional mechanisms to regulate dendritic growth.

Repeated depolarization can lead to sustained MAPK activation and influences the formation and stability of dendritic filopodia (Wu *et al.*, 2001). In addition MAPK has been implicated in mediating growth of SCG dendrites in response to activity by phosphorylating MAP2 (Vaillant *et al.*, 2002). MAPK signaling also has been implicated in mediating the effects of calcium influx on dendrite growth in cortical neurons (Redmond *et al.*, 2002). Thus CaMKs and MAPKs appear to be key mediators of calcium-dependent dendritic growth.

CONVERGENCE AND DIVERGENCE

A wide variety of signaling pathways participate in dendritic arbor development by regulating different downstream events such as microtubule stability, actin dynamics, Ca^{++} homeostasis, and membrane cycling (Fig. 21.9). In addition, many signaling cascades diverge, for example to regulate somatic gene expression, local protein synthesis as well as the Rho GTPases. These diverging pathways ultimately result in modifications of dendritic plasticity. For example, the ephrin-EphB cascade mentioned earlier is one example in which ephrin-EphB receptor interaction results in rapid, local signaling events as well as longer-term

changes in neuronal structure and gene expression. EphB receptors cooperate with NMDA receptors to regulate not only calcium influx but also the subsequent changes in NMDA-receptor dependent gene expression (Takasu *et al.*, 2002). In fact, local changes in Ca^{++} are responsible for the long-term changes induced by activity-driven gene expression dependent on transcription regulators such as CREB or CREST (West *et al.*, 2001; Aizawa *et al.*, 2004).

Another example of divergence comes from the activation of the Ras superfamily GTPases that regulate membrane cycling downstream of neurotransmitter receptor activity. Synaptic NMDA receptors activate the Rab GTPase Rab5 (Pfeffer and Aivazian, 2004) that in turn regulates the internalization and dephosphorylation of AMPA receptors (Brown *et al.*, 2005), an event associated with changes in dendritic structure (Ikegaya *et al.*, 2001) and synaptic strength. Similarly, decreasing or increasing the activity of the Ras GTPase Rap1 decreases or increases, respectively, dendritic arbor elaboration *in vitro* (Chen *et al.*, 2005). One way Rap1 works is by regulating how calcium influx affects CREB-dependent transcription (Chen *et al.*, 2005). Rap1 also mediates the internalization of AMPA type glutamate receptors from local synaptic sites in response to NMDA receptor signaling (Zhu *et al.*, 2002). Since blocking glutamate receptor activity also decreases

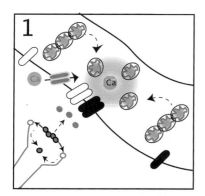

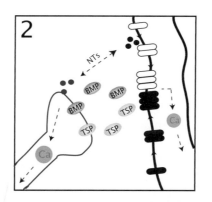

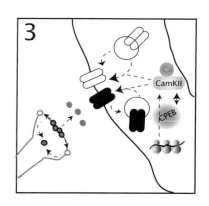

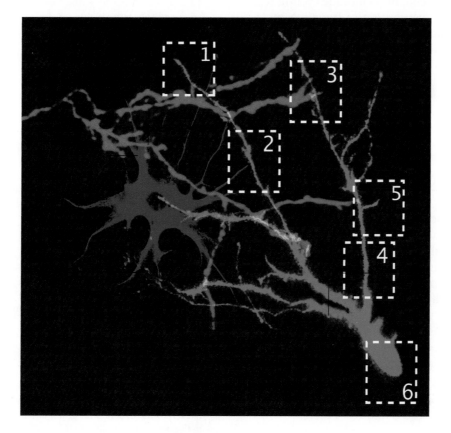

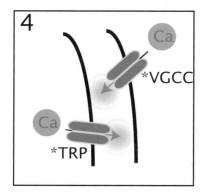

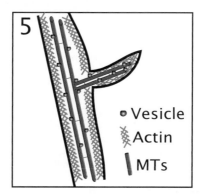

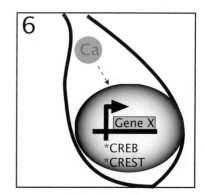

FIGURE 21.9 Summary of multiple mechanisms that influence dendritic growth and remodeling. Dendritic arbors develop within a complex environment in which they contact and receive signals from afferent axons, glial cells, and other dendrites. Several changes occur at sites of contact between axons and dendrites, marked by 1 and 3 in the image, including local changes in enzyme activity, such as CaM kinase and phosphatases, receptor trafficking, and local protein synthesis. Interactions between glia and neurons, marked by #2, include release of trophic substances that regulate synapse formation and maintenance. Process outgrowth, represented by #4 and #5, is mediated by cytoskeletal rearrangements. Finally activity-induced gene transcription, #6, can change the constellation of protein components in neurons in response to growth factors or synaptic inputs. (From Bestman *et al.*, 2008)

dendritic arbor development (Rajan and Cline, 1998), it is possible that Rap1 may act principally to control cell surface levels of glutamate receptors and this, in turn, affects dendritic arbor elaboration. As with the Rho GTPases, Rap1 and the other GTPases regulating membrane cycling are at nodal points in the cascades controlling cell surface distribution of membrane proteins. For instance, Rap1 is downstream of other receptors, such as the neurotrophin receptors (Huang *et al.*, 2003), plus GTPase-dependent membrane cycling regulates the cell surface distribution of signaling molecules, other than glutamate receptors. As illustrated in Figure 21.9, the pathways regulating dendritic arbor development are nonlinear and neither their regulation nor their targets are independent. Consequently, it is essential to investigate the function of these complex molecules in their normal *in vivo* context in order to dissect their endogenous functions.

CONCLUSION

Studies such as those summarized here, aided by computational approaches to map biochemical events in space and time will ultimately lead to a complete appreciation of the signaling pathways that regulate dendritic structural and functional plasticity. Elucidation of these pathways will reveal mechanisms by which activity can rapidly affect dendritic structure, and over more-prolonged time frames, exert profound changes in the so-called intrinsic states of the dendritic tree by modifying gene expression (Goldberg, 2004). Knowledge of these events, together with an understanding of the integrative capacity of dendrites, is essential to grasp the function of neuronal circuits and the brain.

References

Aakalu, G., Smith, W. B., Nguyen, N., Jiang, C., Schuman, E. M. (2001). Dynamic visualization of local protein synthesis in hippocampal neurons. *Neuron* **30**, 489–502.

Adams, J. C. (2001). Thrombospondins: Multifunctional regulators of cell interactions. *Annu Rev Cell Dev Biol* **17**, 25–51.

Ahnert-Hilger, G., Holtje, M., Grosse, G., Pickert, G., Mucke, C., Nixdorf-Bergweiler, B., Boquet, P., Hofmann, F., and Just, I. (2004). Differential effects of Rho GTPases on axonal and dendritic development in hippocampal neurones. *J Neurochem* **90**, 9–18.

Aizawa, H., Hu, S. C., Bobb, K., Balakrishnan, K., Ince, G., Gurevich, I., Cowan, M., and Ghosh, A. (2004). Dendrite development regulated by CREST, a calcium-regulated transcriptional activator. *Science* **303**, 197–202.

Aizenman, C. D., Munoz-Elias, G., and Cline, H. T. (2002). Visually driven modulation of glutamatergic synaptic transmission is mediated by the regulation of intracellular polyamines. *Neuron* **34**, 623–634.

Aizenman, C. D., Akerman, C. J., Jensen, K. R., and Cline, H. T. (2003). Visually driven regulation of intrinsic neuronal excitability improves stimulus detection in vivo. *Neuron* **39**, 831–842.

Araque, A. and Perea, G. (2004). Glial modulation of synaptic transmission in culture. *Glia* **47**, 241–248.

Atkins, C. M., Nozaki, N., Shigeri, Y., and Soderling, T. R. (2004). Cytoplasmic polyadenylation element binding protein-dependent protein synthesis is regulated by calcium/calmodulin-dependent protein kinase II. *J Neurosci* **24**, 5193–5201.

Bagni, C. and Greenough, W. T. (2005). From mRNP trafficking to spine dysmorphogenesis: the roots of fragile X syndrome. *Nat Rev Neurosci* **6**, 376–387.

Bagni, C., Mannucci, L., Dotti, C. G., and Amaldi, F. (2000). Chemical stimulation of synaptosomes modulates alpha -Ca2+/calmodulin-dependent protein kinase II mRNA association to polysomes. *J Neurosci* **20**, RC76.

Baker, R.E., Dijkhuizen, P. A., Van Pelt, J., and Verhaagen, J. (1998). Growth of pyramidal, but not non-pyramidal, dendrites in long-term organotypic explants of neonatal rat neocortex chronically exposed to neurotrophin-3. *Eur J Neurosci* **10**, 1037–1044.

Benavides-Piccione, R., Ballesteros-Yanez, I., de Lagran, M. M., Elston, G., Estivill, X., Fillat, C., Defelipe, J., and Dierssen, M. (2004). On dendrites in Down syndrome and DS murine models: A spiny way to learn. *Prog Neurobiol* **74**, 111–126.

Berezovska, O., McLean, P., Knowles, R., Frosh, M., Lu, F. M. *et al.* (1999). Notch1 inhibits neurite outgrowth in postmitotic primary neurons. *Neuroscience* **93**, 433–439.

Bernards, A. and Settleman, J. (2004). GAP control: Regulating the regulators of small GTPases. *Trends Cell Biol* **14**, 377–385.

Bestman, J., Santos da Silva, J., and Cline, H. T. (2008). Dendrite Development. in "Dendrites", Stuart, Spruston, Hauser (Eds.) (Oxford University Press).

Bezzerides, V. J., Ramsey, I. S., Kotecha, S., Greka, A., and Clapham, D. E. (2004). Rapid vesicular translocation and insertion of TRP channels. *Nat Cell Biol* **6**, 709–720.

Blanton, M. G. and Kriegstein, A. R. (1991). Spontanous action potential activity and synaptic currents in the embryonic turtle cerebral cortex. *J Neurosci* **11**, 3907–3923.

Bodian, D. (1965). A suggestive relationship of nerve cell RNA with specific synaptic sites. *Proc Natl Acad Sci USA* **53**, 418–425.

Bodnarenko, S. R. and Chalupa, L. M. (1993). Stratification of On and Off dendrites depends on glutamate-mediated afferent activity in the developing retina. *Nature* **364**, 144–146.

Brecht, M., Fee, M. S., Garaschuk, O., Helmchen, F., Margrie, T. W., Svoboda, K., and Osten, P. (2004). Novel approaches to monitor and manipulate single neurons in vivo. *J Neurosci* **24**, 9223–9227.

Brown, T. C., Tran, I. C., Backos, D. S., and Esteban, J. A. (2005). NMDA receptor-dependent activation of the small GTPase Rab5 drives the removal of synaptic AMPA receptors during hippocampal LTD. *Neuron* **45**, 81–94.

Bruckner, K., Pablo Labrador, J., Scheiffele, P., Herb, A., Seeburg, P. H., and Klein, R. (1999). EphrinB ligands recruit GRIP family PDZ adaptor proteins into raft membrane microdomains. *Neuron* **22**, 511–524.

Cantallops, I., Haas, K., and Cline, H. T. (2000a). Postsynaptic CPG15 promotes synaptic maturation and presynaptic axon arbor elaboration in vivo. *Nat Neurosci* **3**, 1004–1011.

Cantallops, I., Haas, H., and Cline, H. T. (2000b). Postsynaptic CPG15 expression enhances presynaptic axon growth and retinotectal synapse maturation. *Nat Neurosci* **3**, 498–503.

Chae, T., Kwon, Y. T., Bronson, R., Dikkes, P., Li, E., and Tsai, L. H. (1997). Mice lacking p35, a neuronal specific activator of Cdk5, display cortical lamination defects, seizures, and adult lethality. *Neuron* **18**, 29–42.

Chen, L., El-Husseini, A., Tomita, S., Bredt, D. S., and Nicoll, R. A. (2003). Stargazin differentially controls the trafficking of alpha-amino-3-hydroxyl-5-methyl-4-isoxazolepropionate and kainate receptors. *Mol Pharmacol* **64**, 703–706.

Chen, L., Chetkovich, D. M., Petralia, R. S., Sweeney, N. T., Kawasaki, Y., Wenthold, R. J., Bredt, D. S., and Nicoll, R. A. (2000). Stargazin regulates synaptic targeting of AMPA receptors by two distinct mechanisms. *Nature* **408**, 936–943.

Chen, Y., Wang, P. Y., and Ghosh, A. (2005). Regulation of cortical dendrite development by Rap1 signaling. *Mol Cell Neurosci* **28**, 215–228.

Christopherson, K. S., Ullian, E. M., Stokes, C. C., Mullowney, C. E., Hell, J. W., Agah, A., Lawler, J., Mosher, D. F., Bornstein, P., and Barres, B. A. (2005). Thrombospondins are astrocyte-secreted proteins that promote CNS synaptogenesis. *Cell* **120**, 421–433.

Ciani, L., Krylova, O., Smalley, M. J., Dale, T. C., and Salinas, P. C. (2004). A divergent canonical WNT-signaling pathway regulates microtubule dynamics: dishevelled signals locally to stabilize microtubules. *J Cell Biol* **164**, 243–253.

Ciani, L. and Salinas, P. C. (2005). Signaling in neural development: WNTS in the vertebrate nervous system: From patterning to neuronal connectivity. *Nat Rev Neurosci* **6**, 351–362.

Cline, H. T. (2001). Dendritic arbor development and synaptogenesis. *Curr Opin Neurobiol* **11**, 118–126.

Colledge, M., Snyder, E. M., Crozier, R. A., Soderling, J. A., Jin, Y., Langeberg, L. K., Lu, H., Bear, M. F., and Scott, J. D. (2003). Ubiquitination regulates PSD-95 degradation and AMPA receptor surface expression. *Neuron* **40**, 595–607.

Colomar, A. and Robitaille, R. (2004). Glial modulation of synaptic transmission at the neuromuscular junction. *Glia* **47**, 284–289.

Crino, P. B. and Eberwine, J. (1996). Molecular characterization of the dendritic growth cone: Regulated mRNA transport and local protein synthesis. *Neuron* **17**, 1173–1197.

da Silva, J. S. and Dotti, C. G. (2002). Breaking the neuronal sphere: Regulation of the actin cytoskeleton in neuritogenesis. *Nat Rev Neurosci* **3**, 694–704.

Da Silva, J. S., Medina, M., Zuliani, C., Di Nardo, A., Witke, W., and Dotti, C. G. (2003). RhoA/ROCK regulation of neuritogenesis via profilin IIa-mediated control of actin stability. *J Cell Biol* **162**, 1267–1279.

Dailey, M. E. and Smith, S. J. (1996). The dynamics of dendritic structure in developing hippocampal slices. *J Neurosci* **16**, 2983–2994.

Dalva, M. B., Takasu, M. A., Lin, M. Z., Shamah, S. M., Hu, L., Gale, N. W., and Greenberg, M. E. (2000). EphB receptors interact with NMDA receptors and regulate excitatory synapse formation. *Cell* **103**, 945–956.

Daub, H., Gevaert, K., Vandekerckhove, J., Sobel, A., and Hall, A. (2001). Rac/Cdc42 and p65PAK regulate the microtubule-destabilizing protein stathmin through phosphorylation at serine 16. *J Biol Chem* **276**, 1677–1680.

Davis, H. P. and Squire, L. R. (1984). Protein synthesis and memory: A review. *Psychol Bull* **96**, 518–559.

Debant, A., Serra-Pages, C., Seipel, K., O'Brien, S., Tang, M., Park, S. H., and Streuli, M. (1996). The multidomain protein Trio binds the LAR transmembrane tyrosine phosphatase, contains a protein kinase domain, and has separate rac-specific and rho-specific guanine nucleotide exchange factor domains. *Proc Natl Acad Sci USA* **93**, 5466–5471.

Dijkhuizen, P. A. and Ghosh, A. (2005). Regulation of dendritic growth by calcium and neurotrophin signaling. *Prog Brain Res* **147**, 17–27.

Dolmetsch, R. E., Pajvani, U., Fife, K., Spotts, J. M., and Greenberg, M. E. (2001). Signaling to the nucleus by an L-type calcium channel-calmodulin complex through the MAP kinase pathway. *Science* **294**, 333–339.

Dreier, L., Burbea, M., and Kaplan, J. M. (2005). LIN-23-mediated degradation of beta-catenin regulates the abundance of GLR-1 glutamate receptors in the ventral nerve cord of C. elegans. *Neuron* **46**, 51–64.

Ehrlich, I., and Malinow, R. (2004). Postsynaptic density 95 controls AMPA receptor incorporation during long-term potentiation and experience-driven synaptic plasticity. *J Neurosci* **24**, 916–927.

El-Husseini Ael, D., Schnell, E., Dakoji, S., Sweeney, N., Zhou, Q., Prange, O., Gauthier-Campbell, C., Aguilera-Moreno, A., Nicoll, R. A., and Bredt, D. S. (2002). Synaptic strength regulated by palmitate cycling on PSD-95. *Cell* **108**, 849–863.

Emoto, K., He, Y., Ye, B., Grueber, W. B., Adler, P. N., Jan, L. Y., and Jan, Y. N. (2004). Control of dendritic branching and tiling by the Tricornered-kinase/Furry signaling pathway in Drosophila sensory neurons. *Cell* **119**, 245–256.

Emoto, K., Parrish, J. Z., Jan, L. Y., and Jan, Y. N. (2006). The tumour suppressor Hippo acts with the NDR kinases in dendritic tiling and maintenance. *Nature* **443**, 210–213.

Engert, F. and Bonhoeffer, T. (1999). Dendritic spine changes associated with hippocampal long-term synaptic plasticity. *Nature* **399**, 66–70.

Etienne-Manneville, S. (2004). Actin and microtubules in cell motility: Which one is in control? *Traffic* **5**, 470–477.

Firth, S. I., Wang, C. T., Feller, M. B. (2005). Retinal waves: Mechanisms and function in visual system development. *Cell Calcium* **37**, 425–432.

Franklin, J. L., Berechid, B. E., Cutting, F. B., Presente, A., Chambers, C. B. *et al.* (1999). Autonomous and non-autonomous regulation of mammalian neurite development by Notch1 and Delta1. *Curr Biol* **9**, 1448–1457.

Frey, U., Krug, M., Reymann, K. G., and Matthies, H. (1988). Anisomycin, an inhibitor of protein synthesis, blocks late phases of LTP phenomena in the hippocampal CA1 region in vitro. *Brain Res* **452**, 57–65.

Frick, A. and Johnston, D. (2005). Plasticity of dendritic excitability. *J Neurobiol* **64**, 100–115.

Fukata, M., Nakagawa, M., and Kaibuchi, K. (2003). Roles of Rho-family GTPases in cell polarisation and directional migration. *Curr Opin Cell Biol* **15**, 590–597.

Furrer, M. P., Kim, S., Wolf, B., and Chiba, A. (2003). Robo and Frazzled/DCC mediate dendritic guidance at the CNS midline. *Nat Neurosci* **6**, 223–230.

Gallegos, M. E. and Bargmann, C. I. (2004). Mechanosensory neurite termination and tiling depend on SAX-2 and the SAX-1 kinase. *Neuron* **44**, 239–249.

Gao, F. B., Brenman, J. E., Jan, L. Y., and Jan, Y. N. (1999). Genes regulating dendritic outgrowth, branching, and routing in Drosophila. *Genes Dev* **13**, 2549–2561.

Gao, F. B., Kohwi, M., Brenman, J. E., Jan, L. Y., and Jan, Y. N. (2000). Control of dendritic field formation in Drosophila: The roles of flamingo and competition between homologous neurons. *Neuron* **28**, 91–101.

Gardiol, A., Racca, C., and Triller, A. (1999). Dendritic and postsynaptic protein synthetic machinery. *J Neurosci* **19**, 168–179.

Goldberg, J. L. (2004). Intrinsic neuronal regulation of axon and dendrite growth. *Curr Opin Neurobiol* **14**, 551–557.

Gonzalez, C. L., Gharbawie, O. A., Whishaw, I. Q., and Kolb, B. (2005). Nicotine stimulates dendritic arborization in motor cortex and improves concurrent motor skill but impairs subsequent motor learning. *Synapse* **55**, 183–191.

Govek, E. E., Newey, S. E., and Van Aelst, L. (2005). The role of the Rho GTPases in neuronal development. *Genes Dev* **19**, 1–49.

Govek, E. E., Newey, S. E., Akerman, C. J., Cross, J. R., Van der Veken, L., and Van Aelst, L. (2004). The X-linked mental retardation protein oligophrenin-1 is required for dendritic spine morphogenesis. *Nat Neurosci* **7**, 364–372.

Greenough, W. T. and Chang, F. L. (1988). Dendritic pattern formation involves both oriented regression and oriented growth in the barrels of mouse somatosensory cortex. *Brain Res* **471**, 148–152.

Grueber, W. B., Jan, L. Y., and Jan, Y. N. (2002). Tiling of the Drosophila epidermis by multidendritic sensory neurons. *Development* **129**, 2867–2878.

Grueber, W. B., Jan, L. Y., and Jan, Y. N. (2003a). Different levels of the homeodomain protein cut regulate distinct dendrite branching patterns of Drosophila multidendritic neurons. *Cell* **112**, 805–818.

Grueber, W. B., Ye, B., Moore, A. W., Jan, L. Y., and Jan, Y. N. (2003b). Dendrites of distinct classes of Drosophila sensory neurons show different capacities for homotypic repulsion. *Curr Biol* **13**, 618–626.

Gundersen, G. G., Gomes, E. R., and Wen, Y. (2004). Cortical control of microtubule stability and polarization. *Curr Opin Cell Biol* **16**, 106–112.

Guo, X., Lin, Y., Horbinski, C., Drahushuk, K. M., Kim, I. J., Kaplan, P. L., Lein, P., Wang, T., and Higgins, D. (2001). Dendritic growth induced by BMP-7 requires Smad1 and proteasome activity. *J Neurobiol* **48**, 120–130.

Hall, A. C., Lucas, F. R., and Salinas, P. C. (2000). Axonal remodeling and synaptic differentiation in the cerebellum is regulated by WNT-7a signaling. *Cell* **100**, 525–535.

Hayashi, K., Ohshima, T., and Mikoshiba, K. (2002). Pak1 is involved in dendrite initiation as a downstream effector of Rac1 in cortical neurons. *Mol Cell Neurosci* **20**, 579–594.

Hering, H., Lin, C. C., and Sheng, M. (2003). Lipid rafts in the maintenance of synapses, dendritic spines, and surface AMPA receptor stability. *J Neurosci* **23**, 3262–3271.

Hickmott, P. W. and Steen, P. A. (2005). Large-scale changes in dendritic structure during reorganization of adult somatosensory cortex. *Nat Neurosci* **8**, 140–142.

Holtmaat, A. J., Trachtenberg, J. T., Wilbrecht, L., Shepherd, G. M., Zhang, X., Knott, G. W., and Svoboda, K. (2005). Transient and persistent dendritic spines in the neocortex in vivo. *Neuron* **45**, 279–291.

Horton, A. C. and Ehlers, M. D. (2003). Dual modes of endoplasmic reticulum-to-Golgi transport in dendrites revealed by live-cell imaging. *J Neurosci* **23**, 6188–6199.

Horton, A. C. and Ehlers, M. D. (2004). Secretory trafficking in neuronal dendrites. *Nat Cell Biol* **6**, 585–591.

Hua, J. Y., and Smith, S. J. (2004). Neural activity and the dynamics of central nervous system development. *Nat Neurosci* **7**, 327–332.

Huang, C. S., Shi, S. H., Ule, J., Ruggiu, M., Barker, L. A., Darnell, R. B., Jan, Y. N., and Jan, L. Y. (2005). Common molecular pathways mediate long-term potentiation of synaptic excitation and slow synaptic inhibition. *Cell* **123**, 105–118.

Huang, Y. S., Carson, J. H., Barbarese, E., and Richter, J. D. (2003). Facilitation of dendritic mRNA transport by CPEB. *Genes Dev* **17**, 638–653.

Huber, A. B., Kolodkin, A. L., Ginty, D. D., and Cloutier, J. F. (2003). Signaling at the growth cone: Ligand-receptor complexes and the control of axon growth and guidance. *Annu Rev Neurosci* **26**, 509–563.

Hughes, M. E., Bortnick, R., Tsubouchi, A., Baumer, P., Kondo, M., Uemura, T., and Schmucker, D. (2007). Homophilic Dscam inter-

actions control complex dendrite morphogenesis. *Neuron* **54**, 417–427.

Hume, R. I. and Purves, D. (1981). Geometry of neonatal neurones and the regulation of synapse elimination. *Nature* **293**, 469–471.

Hummel, T., Vasconcelos, M. L., Clemens, J. C., Fishilevich, Y., Vosshall, L. B., and Zipursky, S. L. (2003). Axonal targeting of olfactory receptor neurons in Drosophila is controlled by Dscam. *Neuron* **37**, 221–231.

Ikegaya, Y., Kim, J. A., Baba, M., Iwatsubo, T., Nishiyama, N., and Matsuki, N. (2001). Rapid and reversible changes in dendrite morphology and synaptic efficacy following NMDA receptor activation: Implication for a cellular defense against excitotoxicity. *J Cell Sci* **114**, 4083–4093.

Inglis, F. M., Zuckerman, K. E., and Kalb, R. G. (2000). Experience-dependent development of spinal motor neurons. *Neuron* **26**, 299–305.

Juo, P. and Kaplan, J. M. (2004). The anaphase-promoting complex regulates the abundance of GLR-1 glutamate receptors in the ventral nerve cord of C. elegans. *Curr Biol* **14**, 2057–2062.

Katz, L., Gilbert, C., and Wiesel, T. (1989). Local circuits and ocular dominance columns in monkey striate cortex. *J Neurosci* **9**, 1389–1399.

Katz, L. C. and Constantine-Paton, M. (1988). Relationships between segregated afferents and postsynaptic neurons in the optic tectum of three-eyed frogs. *J Neurosci* **8**, 3160–3180.

Kim, M. D., Jan, L. Y., and Jan, Y. N. (2006). The bHLH-PAS protein Spineless is necessary for the diversification of dendrite morphology of Drosophila dendritic arborization neurons. *Genes Dev* **20**, 2806–2819.

Kita, Y., Kimura, K. D., Kobayashi, M., Ihara, S., Kaibuchi, K. *et al.* (1998). Microinjection of activated phosphatidylinositol-3 kinase induces process outgrowth in rat PC12 cells through the Rac-JNK signal transduction pathway. *J Cell Sci* **111**, 907–915.

Kolb, B., Gorny, G., Li, Y., Samaha, A. N., and Robinson, T. E. (2003). Amphetamine or cocaine limits the ability of later experience to promote structural plasticity in the neocortex and nucleus accumbens. *Proc Natl Acad Sci U S A* **100**, 10523–10528.

Kornhauser, J. M., Cowan, C. W., Shaywitz, A. J., Dolmetsch, R. E., Griffith, E. C., Hu, L. S., Haddad, C., Xia, Z., and Greenberg, M. E. (2002). CREB transcriptional activity in neurons is regulated by multiple, calcium-specific phosphorylation events. *Neuron* **34**, 221–233.

Kossel, A., Lowel, S., and Bolz, J. (1995). Relationships between dendritic fields and functional architecture in striate cortex of normal and visually deprived cats. *J Neurosci* **15**, 3913–3926.

Kossut, M. (1998). Experience-dependent changes in function and anatomy of adult barrel cortex. *Exp Brain Res* **123**, 110–116.

Krylova, O., Herreros, J., Cleverley, K. E., Ehler, E., Henriquez, J. P., Hughes, S. M., and Salinas, P. C. (2002). WNT-3, expressed by motoneurons, regulates terminal arborization of neurotrophin-3-responsive spinal sensory neurons. *Neuron* **35**, 1043–1056.

Kuhl, D. and Skehel, P. (1998). Dendritic localization of mRNAs. *Curr Opin Neurobiol* **8**, 600–606.

Lardelli, M., Dahlstrand, J., and Lendahl, U. (1994). The novel Notch homologue mouse Notch 3 lacks specific epidermal growth factor-repeats and is expressed in proliferating neuroepithelium. *Mech Dev* **46**, 123–136.

Kuntziger, T., Gavet, O., Manceau, V., Sobel, A., and Bornens, M. (2001). Stathmin/Op18 phosphorylation is regulated by microtubule assembly. *Mol Biol Cell* **12**, 437–448.

Lamprecht, R., Farb, C. R., and LeDoux, J. E. (2002). Fear memory formation involves p190 RhoGAP and ROCK proteins through a GRB2-mediated complex. *Neuron* **36**, 727–738.

Lauder, J. M. (1993). Neurotransmitters as growth regulatory signals: Role of receptors and second messengers. *TINS* **16**, 233–239.

Le Roux, P., Behar, S., Higgins, D., and Charette, M. (1999). OP-1 enhances dendritic growth from cerebral cortical neurons in vitro. *Exp Neurol* **160**, 151–163.

Ledesma, M. D. and Dotti, C. G. (2003). Membrane and cytoskeleton dynamics during axonal elongation and stabilization. *Int Rev Cytol* **227**, 183–219.

Lee, L. J., Lo, F. S., and Erzurumlu, R. S. (2005). NMDA receptor-dependent regulation of axonal and dendritic branching. *J Neurosci* **25**, 2304–2311.

Lee, T., Winter, C., Marticke, S. S., Lee, A., and Luo, L. (2000). Essential roles of Drosophila RhoA in the regulation of neuroblast proliferation and dendritic but not axonal morphogenesis. *Neuron* **25**, 307–316.

Lee-Hoeflich, S. T., Causing, C. G., Podkowa, M., Zhao, X., Wrana, J. L., and Attisano, L. (2004). Activation of LIMK1 by binding to the BMP receptor, BMPRII, regulates BMP-dependent dendritogenesis. *Embo J* **23**, 4792–4801.

Lein, P. J., Beck, H. N., Chandrasekaran, V., Gallagher, P. J., Chen, H. L., Lin, Y., Guo, X., Kaplan, P. L., Tiedge, H., and Higgins, D. (2002). Glia induce dendritic growth in cultured sympathetic neurons by modulating the balance between bone morphogenetic proteins (BMPs) and BMP antagonists. *J Neurosci* **22**, 10377–10387.

Lendvai, B., Stern, E. A., Chen, B., and Svoboda, K. (2000). Experience-dependent plasticity of dendritic spines in the developing rat barrel cortex in vivo. *Nature* **404**, 876–881.

Li, Z., Van Aelst, L., and Cline, H. T. (2000). Rho GTPases regulate distinct aspects of dendritic arbor growth in Xenopus central neurons in vivo. *Nat Neurosci* **3**, 217–225.

Li, Z., Aizenman, C. D., and Cline, H. T. (2002). Regulation of rho GTPases by crosstalk and neuronal activity in vivo. *Neuron* **33**, 741–750.

Li, Z., Okamoto, K., Hayashi, Y., and Sheng, M. (2004). The importance of dendritic mitochondria in the morphogenesis and plasticity of spines and synapses. *Cell* **119**, 873–887.

Ligon, L. A. and Steward, O. (2000). Role of microtubules and actin filaments in the movement of mitochondria in the axons and dendrites of cultured hippocampal neurons. *J Comp Neurol* **427**, 351–361.

Lin, H., Huganir, R., and Liao, D. (2004). Temporal dynamics of NMDA receptor-induced changes in spine morphology and AMPA receptor recruitment to spines. *Biochem Biophys Res Commun* **316**, 501–511.

Lisman, J., Schulman, H., and Cline, H. (2002). The molecular basis of CaMKII function in synaptic and behavioural memory. *Nat Rev Neurosci* **3**, 175–190.

Lohmann, C., Finski, A., and Bonhoeffer, T. (2005). Local calcium transients regulate the spontaneous motility of dendritic filopodia. *Nat Neurosci* **8**, 305–312.

Lordkipanidze, T. and Dunaevsky, A. (2005). Purkinje cell dendrites grow in alignment with Bergmann glia. *Glia*.

Luo, L. (2000). Rho GTPases in neuronal morphogenesis. *Nat Rev Neurosci* **1**, 173–180.

Luo, L. (2002). Actin cytoskeleton regulation in neuronal morphogenesis and structural plasticity. *Annu Rev Cell Dev Biol* **18**, 601–635.

Maier, D. L., Grieb, G. M., Stelzner, D. J., and McCasland, J. S. (2003). Large-scale plasticity in barrel cortex following repeated whisker trimming in young adult hamsters. *Exp Neurol* **184**, 737–745.

Maletic-Savatic, M., Malinow, R., and Svoboda, K. (1999). Rapid dendritic morphogenesis in CA1 hippocampal dendrites induced by synaptic activity. *Science* **283**, 1923–1927.

Malinow, R. and Malenka, R. C. (2002). AMPA receptor trafficking and synaptic plasticity. *Annu Rev Neurosci* **25**, 103–126.

Mallardo, M., Deitinghoff, A., Muller, J., Goetze, B., Macchi, P., Peters, C., and Kiebler, M. A. (2003). Isolation and characterization of Staufen-containing ribonucleoprotein particles from rat brain. *Proc Natl Acad Sci USA* **100**, 2100–2105.

Malun, D. and Brunjes, P. C. (1996). Development of olfactory glomeruli: temporal and spatial interactions between olfactory receptor axons and mitral cells in opossums and rats. *J Comp Neurol* **368**, 1–16.

Marrs, G. S., Green, S. H., and Dailey, M. E. (2001). Rapid formation and remodeling of postsynaptic densities in developing dendrites. *Nat Neurosci* **4**, 1006–1013.

Martin, K. C., Barad, M., and Kandel, E. R. (2000). Local protein synthesis and its role in synapse-specific plasticity. *Curr Opin Neurobiol* **10**, 587–592.

Matthews, B. J., Kim, M. E., Flanagan, J. J., Hattori, D., Clemens, J. C., Zipursky, S. L., and Grueber, W. B. (2007). Dendrite self-avoidance is controlled by Dscam. *Cell* **129**, 593–604.

Mattson, M. P. (1999). Establishment and plasticity of neuronal polarity. *J Neurosci Res* **57**, 577–589.

Mayford, M., Baranes, D., Podsypanina, K., and Kandel, E. R. (1996). The 3'-untranslated region of CaMKII alpha is a cis-acting signal for the localization and translation of mRNA in dendrites. *Proc Natl Acad Sci USA* **93**, 13250–13255.

McAllister, A. K., Lo, D. C., and Katz, L. C. (1995). Neurotrophins regulate dendritic growth in developing visual cortex. *Neuron* **15**, 791–803.

Mendez, R. and Richter, J. D. (2001). Translational control by CPEB: A means to the end. *Nat Rev Mol Cell Biol* **2**, 521–529.

Merzenich, M. M. and Jenkins, W. M. (1993). Reorganization of cortical representations of the hand following alterations of skin inputs induced by nerve injury, skin island transfers, and experience. *J Hand Ther* **6**, 89–104.

Miller, S., Yasuda, M., Coats, J. K., Jones, Y., Martone, M. E., and Mayford, M. (2002). Disruption of dendritic translation of CaMKIIalpha impairs stabilization of synaptic plasticity and memory consolidation. *Neuron* **36**, 507–519.

Miyashiro, K., Dichter, M., and Eberwine, J. (1994). On the nature and differential distribution of mRNAs in hippocampal neurites: Implications for neuronal functioning. *Proc Natl Acad Sci USA* **91**, 10800–10804.

Miyashiro, K. Y., Beckel-Mitchener, A., Purk, T. P., Becker, K. G., Barret, T., Liu, L., Carbonetto, S., Weiler, I. J., Greenough, W. T., and Eberwine, J. (2003). RNA cargoes associating with FMRP reveal deficits in cellular functioning in Fmr1 null mice. *Neuron* **37**, 417–431.

Mizrahi, A. and Katz, L. C. (2003). Dendritic stability in the adult olfactory bulb. *Nat Neurosci* **6**, 1201–1207.

Moore, A. W., Jan, L. Y., and Jan, Y. N. (2002). Hamlet, a binary genetic switch between single- and multiple-dendrite neuron morphology. *Science* **297**, 1355–1358.

Mori, Y., Imaizumi, K., Katayama, T., Yoneda, T., and Tohyama, M. (2000). Two cis-acting elements in the 3' untranslated region of alpha-CaMKII regulate its dendritic targeting. *Nat Neurosci* **3**, 1079–1084.

Morrow, B. A., Elsworth, J. D., and Roth, R. H. (2005). Prenatal exposure to cocaine selectively disrupts the development of parvalbumin containing local circuit neurons in the medial prefrontal cortex of the rat. *Synapse* **56**, 1–11.

Muller, M., Mironov, S. L., Ivannikov, M. V., Schmidt, J., and Richter, D. W. (2005). Mitochondrial organization and motility probed by two-photon microscopy in cultured mouse brainstem neurons. *Exp Cell Res* **303**, 114–127.

Murai, K. K. and Pasquale, E. B. (2005). New exchanges in eph-dependent growth cone dynamics. *Neuron* **46**, 161–163.

Nakayama, A. Y., Harms, M. B., and Luo, L. (2000). Small GTPases Rac and Rho in the maintenance of dendritic spines and branches in hippocampal pyramidal neurons. *J Neurosci* **20**, 5329–5338.

Nedivi, E. (1999). Molecular analysis of developmental plasticity in neocortex. *J Neurobiol* **41**, 135–147.

Nedivi, E., G. Y., W., and Cline, H. T. (1998). Promotion of dendritic growth by CPG15, an activity-induced signaling molecule. *Science* **281**, 1863–1866.

Negishi, M. and Katoh, H. (2005). Rho family GTPases and dendrite plasticity. *Neuroscientist* **11**, 187–191.

Neves, G., Zucker, J., Daly, M., and Chess, A. (2004). Stochastic yet biased expression of multiple Dscam splice variants by individual cells. *Nat Genet* **36**, 240–246.

Newey, S. E., Velamoor, V., Govek, E. E., and Van Aelst, L. (2005). Rho GTPases, dendritic structure, and mental retardation. *J Neurobiol* **64**, 58–74.

Newman, E. A. and Volterra, A. (2004). Glial control of synaptic function. *Glia* **47**, 207–208.

Ng, J., Nardine, T., Harms, M., Tzu, J., Goldstein, A., Sun, Y., Dietzl, G., Dickson, B. J., and Luo, L. (2002). Rac GTPases control axon growth, guidance and branching. *Nature* **416**, 442–447.

Niblock, M. M., Brunso-Bechtold, J. K., and Riddle, D. R. (2000). Insulin-like growth factor I stimulates dendritic growth in primary somatosensory cortex. *J Neurosci* **20**, 4165–4176.

Nimchinsky, E. A., Sabatini, B. L., and Svoboda, K. (2002). Structure and function of dendritic spines. *Annu Rev Physiol* **64**, 313–353.

Nobes, C. D. and Hall, A. (1995). Rho, rac, and cdc42 GTPases regulate the assembly of multimolecular focal complexes associated with actin stress fibers, lamellipodia, and filopodia. *Cell* **81**, 53–62.

Ohshima, T., Ward, J. M., Huh, C. G., Longenecker, G., Veeranna, Pant, H. C., Brady, R. O., Martin, L. J., and Kulkarni, A. B. (1996). Targeted disruption of the cyclin-dependent kinase 5 gene results in abnormal corticogenesis, neuronal pathology and perinatal death. *Proc Natl Acad Sci USA* **93**, 11173–11178.

Oliet, S. H., Piet, R., Poulain, D. A., and Theodosis, D. T. (2004). Glial modulation of synaptic transmission: Insights from the supraoptic nucleus of the hypothalamus. *Glia* **47**, 258–267.

Ostroff, L. E., Fiala, J. C., Allwardt, B., and Harris, K. M. (2002). Polyribosomes redistribute from dendritic shafts into spines with enlarged synapses during LTP in developing rat hippocampal slices. *Neuron* **35**, 535–545.

Ouyang, Y., Rosenstein, A., Kreiman, G., Schuman, E. M., and Kennedy, M. B. (1999). Tetanic stimulation leads to increased accumulation of Ca(2+)/calmodulin-dependent protein kinase II via dendritic protein synthesis in hippocampal neurons. *J Neurosci* **19**, 7823–7833.

Palazzo, A. F., Cook, T. A., Alberts, A. S., and Gundersen, G. G. (2001). mDia mediates Rho-regulated formation and orientation of stable microtubules. *Nat Cell Biol* **3**, 723–729.

Palmer, A. and Klein, R. (2003). Multiple roles of ephrins in morphogenesis, neuronal networking, and brain function. *Genes Dev* **17**, 1429–1450.

Parri, H. R., Gould, T. M., and Crunelli, V. (2001). Spontaneous astrocytic Ca2+ oscillations in situ drive NMDAR-mediated neuronal excitation. *Nat Neurosci* **4**, 803–812.

Parrish, J. Z., Emoto, K., Jan, L. Y., and Jan, Y. N. (2007a). Polycomb genes interact with the tumor suppressor genes hippo and warts in the maintenance of Drosophila sensory neuron dendrites. *Genes Dev* **21**, 956–972.

Parrish, J. Z., Emoto, K., Kim, M. D., and Jan, Y. N. (2007b). Mechanisms that regulate establishment, maintenance, and remodeling of dendritic fields. *Annu Rev Neurosci* **30**, 399–423.

Passafaro, M., Nakagawa, T., Sala, C., and Sheng, M. (2003). Induction of dendritic spines by an extracellular domain of AMPA receptor subunit GluR2. *Nature* **424**, 677–681.

Patrick, G. N., Bingol, B., Weld, H. A., and Schuman, E. M. (2003). Ubiquitin-mediated proteasome activity is required for agonist-induced endocytosis of GluRs. *Curr Biol* **13**, 2073–2081.

Penzes, P., Beeser, A., Chernoff, J., Schiller, M. R., Eipper, B. A., Mains, R. E., and Huganir, R. L. (2003). Rapid induction of dendritic spine morphogenesis by trans-synaptic ephrinB-EphB receptor activation of the Rho-GEF kalirin. *Neuron* **37**, 263–274.

Penzes, P., Johnson, R. C., Sattler, R., Zhang, X., Huganir, R. L., Kambampati, V., Mains, R. E., and Eipper, B. A. (2001). The neuronal Rho-GEF Kalirin-7 interacts with PDZ domain-containing proteins and regulates dendritic morphogenesis. *Neuron* **29**, 229–242.

Pfeffer, S. and Aivazian, D. (2004). Targeting Rab GTPases to distinct membrane compartments. *Nat Rev Mol Cell Biol* **5**, 886–896.

Pierce, J. P., van Leyen, K., and McCarthy, J. B. (2000). Translocation machinery for synthesis of integral membrane and secretory proteins in dendritic spines. *Nat Neurosci* **3**, 311–313.

Pierce, J. P., Mayer, T., and McCarthy, J. B. (2001). Evidence for a satellite secretory pathway in neuronal dendritic spines. *Curr Biol* **11**, 351–355.

Pilpel, Y. and Segal, M. (2004). Activation of PKC induces rapid morphological plasticity in dendrites of hippocampal neurons via Rac and Rho-dependent mechanisms. *Eur J Neurosci* **19**, 3151–3164.

Polleux, F., Giger, R. J., Ginty, D. D., Kolodkin, A. L., and Ghosh, A. (1998). Patterning of cortical efferent projections by semaphorin-neuropilin interactions. *Science* **282**, 1904–1906.

Polleux, F., Morrow, T., and Ghosh, A. (2000). Semaphorin 3A is a chemoattractant for cortical apical dendrites. *Nature* **404**, 567–573.

Polleux, F. and Ghosh, A. (2008). Molecular determinants of dendrite and spine development. In "Dendrites", Stuart, Spruston, Hauser (Eds.) (Oxford University Press).

Posern, G., Rapp, U. R., and Feller, S. M. (2000). The Crk signaling pathway contributes to the bombesin-induced activation of the small GTPase Rap1 in Swiss 3T3 cells. *Oncogene* **19**, 6361–6368.

Rajan, I. and Cline, H. T. (1998). Glutamate receptor activity is required for normal development of tectal cell dendrites in vivo. *J Neurosci* **18**, 7836–7846.

Redmond, L., Oh, S. R., Hicks, C., Weinmaster, G., and Ghosh, A. (2000). Nuclear Notch1 signaling and the regulation of dendritic development. *Nat Neurosci* **3**, 30–40.

Richter, J. D. and Lorenz, L. J. (2002). Selective translation of mRNAs at synapses. *Curr Opin Neurobiol* **12**, 300–304.

Rintoul, G. L., Filiano, A. J., Brocard, J. B., Kress, G. J., and Reynolds, I. J. (2003). Glutamate decreases mitochondrial size and movement in primary forebrain neurons. *J Neurosci* **23**, 7881–7888.

Rossman, K. L. and Sondek, J. (2005). Larger than Dbl: New structural insights into RhoA activation. *Trends Biochem Sci* **30**, 163–165.

Rosso, S. B., Sussman, D., Wynshaw-Boris, A., and Salinas, P. C. (2005). Wnt signaling through Dishevelled, Rac and JNK regulates dendritic development. *Nat Neurosci* **8**, 34–42.

Ruchhoeft, M. L., Ohnuma, S., McNeill, L., Holt, C. E., and Harris, W. A. (1999). The neuronal architecture of Xenopus retinal ganglion cells is sculpted by rho-family GTPases in vivo. *J Neurosci* **19**, 8454–8463.

Ruthazer, E. S., Akerman, C. J., and Cline, H. T. (2003). Control of axon branch dynamics by correlated activity in vivo. *Science* **301**, 66–70.

Sasaki, Y., Cheng, C., Uchida, Y., Nakajima, O., Ohshima, T., Yagi, T., Taniguchi, M., Nakayama, T., Kishida, R., Kudo, Y. *et al.* (2002). Fyn and Cdk5 mediate semaphorin-3A signaling, which is involved in regulation of dendrite orientation in cerebral cortex. *Neuron* **35**, 907–920.

Scheetz, A. J., Nairn, A. C., and Constantine-Paton, M. (2000). NMDA receptor-mediated control of protein synthesis at developing synapses. *Nat Neurosci* **3**, 211–216.

Schmidt, A. and Hall, A. (2002). Guanine nucleotide exchange factors for Rho GTPases: Turning on the switch. *Genes Dev* **16**, 1587–1609.

Schmucker, D., Clemens, J. C., Shu, H., Worby, C. A., Xiao, J., Muda, M., Dixon, J. E., and Zipursky, S. L. (2000). Drosophila Dscam is an axon guidance receptor exhibiting extraordinary molecular diversity. *Cell* **101**, 671–684.

Schratt, G. M., Nigh, E. A., Chen, W. G., Hu, L., and Greenberg, M. E. (2004). BDNF regulates the translation of a select group of mRNAs by a mammalian target of rapamycin-phosphatidylinositol 3-kinase-dependent pathway during neuronal development. *J Neurosci* **24**, 7366–7377.

Schuman, E. and Chan, D. (2004). Fueling synapses. *Cell* **119**, 738–740.

Schweitzer, L. (1991). Morphometric analysis of developing neuronal geometry in the dorsal cochlear nucleus of the hamster. *Developmental Brain Research* **59**, 39–47.

Scott, E. K., Reuter, J. E., and Luo, L. (2003). Small GTPase Cdc42 is required for multiple aspects of dendritic morphogenesis. *J Neurosci* **23**, 3118–3123.

Sestan, N., Artavanis-Tsakonas, S., and Rakic, P. (1999). Contact-dependent inhibition of cortical neurite growth mediated by notch signaling. *Science* **286**, 741–746.

Shen, W., Finnegan, S., Lein, P., Sullivan, S., Slaughter, M., and Higgins, D. (2004). Bone morphogenetic proteins regulate ionotropic glutamate receptors in human retina. *Eur J Neurosci* **20**, 2031–2037.

Shi, S., Hayashi, Y., Esteban, J. A., and Malinow, R. (2001). Subunit-specific rules governing AMPA receptor trafficking to synapses in hippocampal pyramidal neurons. *Cell* **105**, 331–343.

Shi, S. H., Cheng, T., Jan, L. Y., and Jan, Y. N. (2004a). The immunoglobulin family member dendrite arborization and synapse maturation 1 (Dasm1). controls excitatory synapse maturation. *Proc Natl Acad Sci USA* **101**, 13346–13351.

Shi, S. H., Cox, D. N., Wang, D., Jan, L. Y., and Jan, Y. N. (2004b). Control of dendrite arborization by an Ig family member, dendrite arborization and synapse maturation 1 (Dasm1). *Proc Natl Acad Sci USA* **101**, 13341–13345.

Shima, Y., Kengaku, M., Hirano, T., Takeichi, M., and Uemura, T. (2004). Regulation of dendritic maintenance and growth by a mammalian 7-pass transmembrane cadherin. *Dev Cell* **7**, 205–216.

Sin, W. C., Haas, K., Ruthazer, E. S., and Cline, H. T. (2002). Dendrite growth increased by visual activity requires NMDA receptor and Rho GTPases. *Nature* **419**, 475–480.

Snyder, E. M., Philpot, B. D., Huber, K. M., Dong, X., Fallon, J. R., and Bear, M. F. (2001). Internalization of ionotropic glutamate receptors in response to mGluR activation. *Nat Neurosci* **4**, 1079–1085.

Soba, P., Zhu, S., Emoto, K., Younger, S., Yang, S. J., Yu, H. H., Lee, T., Jan, L. Y., and Jan, Y. N. (2007). Drosophila sensory neurons require Dscam for dendritic self-avoidance and proper dendritic field organization. *Neuron* **54**, 403–416.

Soderling, T. R. (2000). CaM-kinases: modulators of synaptic plasticity. *Curr Opin Neurobiol* **10**, 375–380.

Stefani, G., Fraser, C. E., Darnell, J. C., and Darnell, R. B. (2004). Fragile X mental retardation protein is associated with translating polyribosomes in neuronal cells. *J Neurosci* **24**, 7272–7276.

Steward, O. and Levy, W. B. (1982). Preferential localization of polyribosomes under the base of dendritic spines in granule cells of the dentate gyrus. *J Neurosci* **2**, 284–291.

Steward, O. and Schuman, E. M. (2001). Protein synthesis at synaptic sites on dendrites. *Annu Rev Neurosci* **24**, 299–325.

Sugimura, K., Yamamoto, M., Niwa, R., Satoh, D., Goto, S., Taniguchi, M., Hayashi, S., and Uemura, T. (2003). Distinct developmental modes and lesion-induced reactions of dendrites of two classes of Drosophila sensory neurons. *J Neurosci* **23**, 3752–3760.

Sutton, M. A. and Schuman, E. M. (2005). Local translational control in dendrites and its role in long-term synaptic plasticity. *J Neurobiol* **64**, 116–131.

Suzuki, T., Ito, J., Takagi, H., Saitoh, F., Nawa, H., and Shimizu, H. (2001). Biochemical evidence for localization of AMPA-type glutamate receptor subunits in the dendritic raft. *Brain Res Mol Brain Res* **89**, 20–28.

Tailby, C., Wright, L. L., Metha, A. B., and Calford, M. B. (2005). Activity-dependent maintenance and growth of dendrites in adult cortex. *Proc Natl Acad Sci USA* **102**, 4631–4636.

Takasu, M. A., Dalva, M. B., Zigmond, R. E., and Greenberg, M. E. (2002). Modulation of NMDA receptor-dependent calcium influx and gene expression through EphB receptors. *Science* **295**, 491–495.

Tao, H. W., Zhang, L. I., Engert, F., and Poo, M. (2001). Emergence of input specificity of ltp during development of retinotectal connections in vivo. *Neuron* **31**, 569–580.

Threadgill, R., Bobb, K., and Ghosh, A. (1997). Regulation of dendritic growth and remodeling by Rho, Rac, and Cdc42. *Neuron* **19**, 625–634.

Tian, N., and Copenhagen, D. R. (2003). Visual stimulation is required for refinement of ON and OFF pathways in postnatal retina. *Neuron* **39**, 85–96.

Tiedge, H. and Brosius, J. (1996). Translational machinery in dendrites of hippocampal neurons in culture. *J Neurosci* **16**, 7171–7181.

Tolias, K. F., Bikoff, J. B., Burette, A., Paradis, S., Harrar, D., Tavazoie, S., Weinberg, R. J., and Greenberg, M. E. (2005). The Rac1-GEF Tiam1 couples the NMDA receptor to the activity-dependent development of dendritic arbors and spines. *Neuron* **45**, 525–538.

Ullian, E. M., Christopherson, K. S., Barres, B. A. (2004). Role for glia in synaptogenesis. *Glia* **47**, 209–216.

Van Aelst, L., Cline, H. T. (2004). Rho GTPases and activity-dependent dendrite development. *Curr Opin Neurobiol* **14**, 297–304.

Volterra, A. and Steinhauser, C. (2004). Glial modulation of synaptic transmission in the hippocampus. *Glia* **47**, 249–257.

Wall, J. T., Kaas, J. H., Sur, M., Nelson, R. J., Felleman, D. J., and Merzenich, M. M. (1986). Functional reorganization in somatosensory cortical areas 3b and 1 of adult monkeys after median nerve repair: Possible relationships to sensory recovery in humans. *J Neurosci* **6**, 218–233.

Wang, J., Zugates, C. T., Liang, I. H., Lee, C. H., and Lee, T. (2002). Drosophila Dscam is required for divergent segregation of sister branches and suppresses ectopic bifurcation of axons. *Neuron* **33**, 559–571.

Wassle, H., Peichl, L., and Boycott, B. B. (1981). Dendritic territories of cat retinal ganglion cells. *Nature* **292**, 344–345.

Weinmaster, G. (2000). Notch signal transduction: a real rip and more. *Curr Opin Genet Dev* **10**, 363–369.

Wells, D. G., Dong, X., Quinlan, E. M., Huang, Y. S., Bear, M. F., Richter, J. D., and Fallon, J. R. (2001). A role for the cytoplasmic polyadenylation element in NMDA receptor-regulated mRNA translation in neurons. *J Neurosci* **21**, 9541–9548.

West, A. E., Chen, W. G., Dalva, M. B., Dolmetsch, R. E., Kornhauser, J. M., Shaywitz, A. J., Takasu, M. A., Tao, X., and Greenberg, M. E. (2001). Calcium regulation of neuronal gene expression. *Proc Natl Acad Sci USA* **98**, 11024–11031.

Whitford, K. L., Marillat, V., Stein, E., Goodman, C. S., Tessier-Lavigne, M., Chedotal, A., and Ghosh, A. (2002). Regulation of cortical dendrite development by Slit-Robo interactions. *Neuron* **33**, 47–61.

Withers, G. S., Higgins, D., Charette, M., and Banker, G. (2000). Bone morphogenetic protein-7 enhances dendritic growth and receptivity to innervation in cultured hippocampal neurons. *Eur J Neurosci* **12**, 106–116.

Wittmann, T. and Waterman-Storer, C. M. (2001). Cell motility: Can Rho GTPases and microtubules point the way? *J Cell Sci* **114**, 3795–3803.

Wojtowicz, W. M., Flanagan, J. J., Millard, S. S., Zipursky, S. L., and Clemens, J. C. (2004). Alternative splicing of Drosophila Dscam generates axon guidance receptors that exhibit isoform-specific homophilic binding. *Cell* **118**, 619–633.

Wong, R. O. and Ghosh, A. (2002). Activity-dependent regulation of dendritic growth and patterning. *Nat Rev Neurosci* **3**, 803–812.

Wong, W. T., Faulkner-Jones, B. E., Sanes, J. R., and Wong, R. O. (2000). Rapid dendritic remodeling in the developing retina: Dependence on neurotransmission and reciprocal regulation by Rac and Rho. *J Neurosci* **20**, 5024–5036.

Wu, G., Malinow, R., and Cline, H. T. (1996). Maturation of a central glutamatergic synapse. *Science* **274**, 972–976.

Wu, G. Y. and Cline, H. T. (1998). Stabilization of dendritic arbor structure in vivo by CaMKII. *Science* **279**, 222–226.

Wu, G. Y. and Cline, H. T. (2003). Time-lapse in vivo imaging of the morphological development of Xenopus optic tectal interneurons. *J Comp Neurol* **459**, 392–406.

Wu, G. Y., Deisseroth, K., and Tsien, R. W. (2001). Spaced stimuli stabilize MAPK pathway activation and its effects on dendritic morphology. *Nat Neurosci* **4**, 151–158.

Yacoubian, T. A. and Lo, D. C. (2000). Truncated and full-length TrkB receptors regulate distinct modes of dendritic growth. *Nat Neurosci* **3**, 342–349.

Yasuda, R., Sabatini, B. L., and Svoboda, K. (2003). Plasticity of calcium channels in dendritic spines. *Nat Neurosci* **6**, 948–955.

Yasui, H., Katoh, H., Yamaguchi, Y., Aoki, J., Fujita, H. *et al.* (2001). Differential responses to nerve growth factor and epidermal growth factor in neurite outgrowth of PC12 cells are determined by Rac1 activation systems. *J Biol Chem* **276**, 15298–15305.

Yu, T. W. and Bargmann, C. I. (2001). Dynamic regulation of axon guidance. *Nat Neurosci* **4** Suppl, 1169–1176.

Yu, X. and Malenka, R. C. (2003). Beta-catenin is critical for dendritic morphogenesis. *Nat Neurosci* **6**, 1169–1177.

Yuste, R., Peinado, A., and Katz, L. C. (1992). Neuronal domains in developing neocortex. *Science* **257**, 665–669.

Yuste, R., Nelson, D. A., Rubin, W. W., and Katz, L. C. (1995). Neuronal domains in developing neocortex: Mechanisms of coactivation. *Neuron* **14**, 7–17.

Zheng, Y. (2004). G protein control of microtubule assembly. *Annu Rev Cell Dev Biol* **20**, 867–894.

Zhou, Q. and Poo, M. M. (2004). Reversal and consolidation of activity-induced synaptic modifications. *Trends Neurosci* **27**, 378–383.

Zhou, Q., Homma, K. J., and Poo, M. M. (2004). Shrinkage of dendritic spines associated with long-term depression of hippocampal synapses. *Neuron* **44**, 749–757.

Zhu, H., Hummel, T., Clemens, J. C., Berdnik, D., Zipursky, S. L., and Luo, L. (2006). Dendritic patterning by Dscam and synaptic partner matching in the Drosophila antennal lobe. *Nat Neurosci* **9**, 349–355.

Zhu, J. J., Qin, Y., Zhao, M., Van Aelst, L., and Malinow, R. (2002). Ras and Rap control AMPA receptor trafficking during synaptic plasticity. *Cell* **110**, 443–455.

Zuo, Y., Lin, A., Chang, P., and Gan, W. B. (2005). Development of long-term dendritic spine stability in diverse regions of cerebral cortex. *Neuron* **46**, 181–189.

Hollis Cline, Anirvan Ghosh, and Yuh-Nung Jan

Early Experience and Sensitive Periods

The nervous system has evolved to cope with an environment that is in many ways largely predictable. Therefore, much of the architecture and functional properties of the brain can be specified by genetic determinants that reflect the common experience of previous generations. Most of this circuitry is established prenatally, guided by genetically determined molecular mechanisms and shaped by patterns of spontaneous impulse activity that propagate through the central nervous system (CNS) in unborn animals, as discussed in Chapters 17–20.

Not all aspects of an animal's world are certain, however. Details of an animal's physical characteristics vary, as do habitats and social conditions. To deal with such uncertainties, the CNS maintains the capacity to modify its connections based on the interactions of an animal with its environment. Through adaptive adjustments based on use or quality of performance, the developing nervous system customizes its functional properties to the needs and environment of the individual animal with a precision that does not need to be, and sometimes cannot be, encoded in the genome.

Although the nervous system is capable of making adaptive adjustments throughout the lifetime of an animal, many neural circuits pass through a period during their development when the capacity for adjustment in response to experience is substantially greater than it is after the circuit has matured. This period is referred to as a *sensitive period* (Knudsen, 2004). During a sensitive period, information derived from experience selects particular functional properties from a range of possible properties that a circuit could adopt. If appropriate experience is not gained during a sensitive period, many circuits never attain the ability to process information in a typical fashion and,

as a result, perception or behavior may be impaired permanently.

This chapter introduces four examples of circuits that have been relatively well studied with respect to their dependence on instruction by early experience for normal development. These examples are the circuits involved in (1) song learning in songbirds, (2) sound localization in owls, (3) binocular representation in the visual cortex, and (4) temperament in rats. Sensitive periods for language learning in humans and filial imprinting in birds are also discussed. These examples are used to illustrate principles that govern sensitive periods. Finally, factors that contribute to the extraordinary capacity of the nervous system for adaptive change during sensitive periods are discussed.

BIRDSONG: LEARNED BY EXPERIENCE

The song of most songbirds depends on learning that occurs early in life during a sensitive period (Konishi, 1985; Doupe and Kuhl, 1999). Birdsong is a special form of vocal communication used by certain species of birds to identify neighbors, defend territories, and attract mates. Songs are distinguished from other communication sounds by their length, spectral complexity, and periodic structure—properties that give birdsong its melodic quality. The songs sung by birds are characteristic of the species (conspecific song); dialects of the species' song often reflect the geographical area in which the bird was raised (Fig. 22.1A).

Songs are passed on from one generation to the next by a combination of genetic instruction and learning. In a few species of songbirds, the influence of genetic instruction is strong and learning plays a relatively

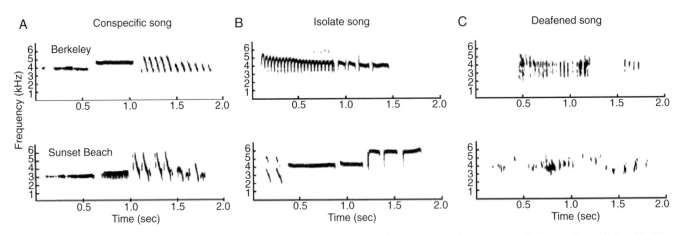

FIGURE 22.1 Songs of white-crowned sparrows. These are sonograms (time-frequency sound spectrograms) of songs from birds with different kinds of early experience. Sound energy in each frequency band is indicated by the darkness of the trace. (A) Song dialects. Birds raised in different areas sing slightly different songs. These dialects are stable for many years and are transmitted by learning. (B) Isolate songs. These simpler songs develop in birds raised in acoustic isolation or in birds that fail to copy a tutor song. (C) Songs of deafened birds. These kinds of songs develop in birds that are deafened after the sensitive period for song memorization but before the period of vocal learning. The birds need to hear their own voice to develop normal song. From Konishi (1985).

minor role. In most, however, the role of learning is paramount. Some of these species learn new songs each year (seasonal learners), whereas others learn their songs only once early in life. In the latter case, the birds memorize the song they will sing as adults during a sensitive period.

White-crowned sparrows and zebra finches are species that learn their songs during a sensitive period (Immelmann, 1972; Konishi, 1985). The extent of the sensitive period in each species has been determined by raising birds in acoustic isolation and then exposing them to conspecific song for brief periods in development. The effect of this experience on song learning is assessed by observing the song that the male eventually sings (in these species, only the male sings). Birds raised in acoustic isolation throughout the sensitive period sing an "isolate" song (Fig. 22.1B) that lacks the spectral and temporal complexity typical of normal song. When baby birds are allowed to hear conspecific song even for a few days during the sensitive period, however, they memorize that particular song and reproduce it accurately when they later learn to sing.

Song learning in these species illustrates an important principle that pertains to most sensitive period learning: The nervous system is genetically predisposed to accept only a limited range of potential stimuli as appropriate for learning (Doupe and Kuhl, 1999). For example, baby birds that are allowed to hear only alien songs that differ substantially from their conspecific song develop isolate song, indicating that they reject these distinctly alien songs as models for learning. Even within the range of songs that a bird

will learn, it strongly prefers a conspecific song when given a choice of several similar song types. Moreover, babies learn a conspecific song rapidly, whereas they learn slightly different alien songs only after much longer periods of experience. Thus, the circuits responsible for song memorization contain genetically determined filters that require certain spectral and temporal features before the stimulus is accepted as appropriate, and within the range of stimuli that is deemed acceptable, some song patterns are preferred over others.

Song learning involves two components: song memorization and vocal learning. In white-crowned sparrows, these components are separated by many months (Fig. 22.2A), whereas in zebra finches, which develop much more rapidly, they overlap. During the sensitive period for song memorization, according to a current hypothesis, high-order sensory neurons become tuned to respond selectively to the acoustic patterns of the songs that the bird memorizes. During the period of vocal learning, these high-order neurons act as templates for evaluating the bird's own song, guiding the development of song so that it eventually matches the previously memorized song pattern.

A Sensitive Period Exists for Song Memorization

The sensitive period for song memorization begins at about 2 weeks of age and lasts for about 8 weeks in both zebra finches and white-crowned sparrows (Immelmann, 1972; Doupe and Kuhl, 1999). Baby birds that are exposed to a conspecific song before the sensi-

A

White-crowned Sparrow

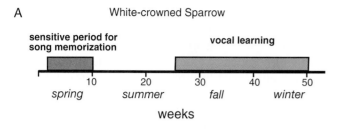

B

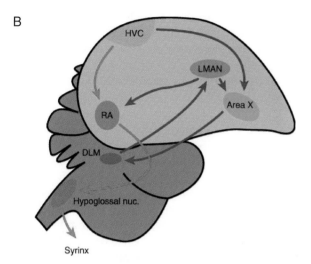

FIGURE 22.2 The sensitive period for song memorization and the set of nuclei that comprise the song system in songbirds. (A) Time-line for song learning for the white-crowned sparrow. For this species, the sensitive period for song memorization does not overlap with the period when the bird learns to sing. (B) A schematic diagram of a side view of the brain of a songbird. The vocal motor pathway, shown in red, consists of the HVC, the robust nucleus of the archopallium (RA), and the hypoglossal nucleus. The anterior pathway, shown in blue, consists of Area X, the medial portion of the dorsolateral nucleus of the thalamus (DLM), and the lateral portion of the magnocellular nucleus of the anterior nidopallium (LMAN).

tive period opens do not learn the song, even though they can hear at this early age. This suggests that the neuronal substrate for song memorization is not yet ready to be shaped by experience. Similarly, babies that do not hear a normal song until after 3 to 4 months of age do not learn to sing a normal song. Apparently, by this age the influence of experience on the neuronal substrate for song memorization has become reduced greatly.

The time at which learning occurs during the sensitive period for song memorization is determined by the individual's own experience. Once the sensitive period has opened, exposure of a baby bird to normal song for 1 week is sufficient for the bird to learn the song, although subsequent exposure to other songs can still modify the song that the bird comes to sing. If a baby bird is kept in acoustic isolation (or hears

only songs that are suboptimal as models for learning) for many weeks past the opening of the sensitive period and then hears normal song, it learns the normal song. Thus, during the sensitive period the nervous system waits in a receptive state for appropriate experience-dependent instruction. Once this instruction is received, a particular pattern of connectivity becomes established and the sensitive period closes (Knudsen, 2004).

When a baby bird is deprived continuously of appropriate auditory experience, its capacity to memorize song eventually diminishes with age. Under these conditions, the sensitive period closes gradually because of additional, age-dependent factors (discussed later) that reduce the plasticity of the relevant circuits. As a bird approaches this age, experience with appropriate stimuli must be richer in order to have an effect. For example, white-crowned sparrows raised in acoustic isolation until 50 days of age no longer memorize songs presented from loudspeakers, but do memorize songs presented by live tutors. Thus, enrichment of the sensory experience provided by social interactions with the tutor overcomes the decline in the facility of the pathway for song memorization.

Vocal Learning

Learning to sing requires a combination of vocal practice and auditory feedback. A young bird that is deafened after song memorization but before the onset of vocal learning, and is thereby prevented from hearing its own voice, will not develop a normal song (Fig. 22.1C). Clearly, auditory feedback is essential for shaping the patterns of connectivity in the vocal motor pathway while the bird is learning to sing. Presumably, auditory feedback is necessary for the bird both to learn how motor system commands correspond with the sounds that it produces and to compare the sounds it produces with its memorized song template.

A Neural Pathway Exists for Song Learning

The neural mechanisms that underlie song memorization are being explored in many laboratories but, as yet, we know little about this aspect of song learning. In contrast, a great deal is known about the neural mechanisms that underlie song production. The pathway for song production was identified by its sexual dimorphism in species in which only males sing. In these species, a distinct set of nuclei, referred to as the song system, is conspicuously hypertrophied in males (Fig. 22.2B). The song system consists of two distinct groups of nuclei: one group in the posterior forebrain that is responsible for song production and another in the anterior forebrain that is critical for song learning and maintenance.

The posterior, vocal motor pathway consists of three serially connected nuclei: the HVC and the robust nucleus of the archopallium (RA) in the forebrain and the hypoglossal nucleus in the brain stem (Fig. 22.2B). As a bird prepares to sing, a wave of neural activity spreads from the HVC to the RA, and finally to the hypoglossal nucleus, which contains the motor neurons that control the vocal musculature. Bilateral lesions of the HVC or the RA leave birds permanently incapable of producing song, although they still can make other kinds of unlearned vocalizations.

The anterior pathway (Fig. 22.2B) consists of Area X, a thalamic nucleus (DLM), and the lateral portion of the magnocellular nucleus of the anterior nidopallium (LMAN), and is essential for experience-dependent adjustments of song (Brainard and Doupe, 2000). Lesions made in the LMAN of a young bird that is just learning to sing cause a dramatic cessation of song development, freezing the bird's song in an immature state. This freezing of song resembles song crystallization. In adult birds, LMAN lesions result in birds that sing normal songs, but can no longer adjust the quality of their songs based on experience.

In young birds, the anterior pathway provides information to the vocal motor pathway for the purpose of vocal learning. According to one hypothesis, the anterior pathway compares auditory feedback about the song that a bird produces with a stored template of the memorized song. The result of this comparison is then used to instruct the development and maintenance of connections in the vocal motor pathway. Consistent with this hypothesis, Area X and the LMAN in the anterior pathway contain neurons that respond maximally to the sound of the bird's own song. During development, axons from the LMAN are the first to innervate the RA in the vocal motor pathway. Information transmitted by the LMAN-RA pathway is mediated predominantly by NMDA receptors, a class of glutamate receptors that, when activated, is known in other systems to induce synaptic plasticity. As birds begin learning to sing, a second set of axons enters the RA from the HVC in the vocal motor pathway and begins making glutamatergic synapses. These later connections may be guided by the activity of the pre-existing, LMAN-RA synapses.

Hormonal Regulation of Learning

The period of vocal learning closes as birds reach sexual maturity and circulating levels of steroid hormones rise. An abrupt increase in androgen hormones triggers the termination of the period of vocal learning. Exposure of a juvenile bird to high levels of testosterone causes its song to "crystallize" (become stable) prematurely in an abnormal state (Konishi, 1985). Conversely, juvenile birds that are castrated before they learn to sing produce inconsistent song patterns throughout life.

A cellular link between sex hormones and song plasticity has been found in the LMAN. Neurons in the LMAN, as well as in the HVC, RA, and hypoglossal nuclei, bind and accumulate androgens. As the period of vocal learning closes, the density of dendritic spines on LMAN neurons decreases dramatically, suggesting that synaptic selection has taken place. In addition, the total volume of the LMAN regresses precipitously and the influence of LMAN activity on the song motor nuclei declines (Wallhausser-Franke et al., 1995). Thus, the close of the period for vocal learning may be due to synaptic selection and stabilization in the LMAN, triggered by a rise in steroid hormone levels.

Seasonal song learners, such as canaries, appear to recapitulate the process of song memorization and vocal learning each year. This relearning is linked with, and could result from, the waxing and waning of steroid hormone levels. This raises the intriguing possibility that these sensitive periods can be opened and closed by hormonal or environmental factors.

Summary

Song learning in birds shares many characteristics with language learning in humans (Box 22.1). Song is a learned form of vocal communication used by certain species of birds to attract mates, defend their territories, and identify their neighbors. Songs are learned in two phases. The first is song memorization, which for many species takes place during a sensitive period. The second is vocal learning, during which vocal practice and auditory feedback shape the bird's song to match the memorized song. The neural pathway for song production is sexually dimorphic in many species in which only males sing. A separate pathway, in the anterior forebrain, plays a special role in vocal learning, guiding adjustments in the song production pathway as the bird learns to sing.

SOUND LOCALIZATION: CALIBRATED BY EARLY EXPERIENCE IN THE OWL

A pathway that is highly modifiable during a sensitive period is the auditory pathway that creates a map of space in the midbrain of the barn owl (Fig. 22.3).

BOX 22.1

SENSITIVE PERIODS IN HUMANS

Many human capabilities depend critically on experience gained during early life. These capabilities range from fundamental capacities, such as stereoscopic vision, visual acuity, and binocular coordination, to high-level capacities, such as social behavior, language, and the ability to perceive forms and faces. In each case, normal experience during a restricted period in early life is essential for the normal development of the capacity. The rules that govern these sensitive periods appear to be the same as those that govern sensitive periods in other animals, as described in the text.

The best known and most thoroughly studied sensitive period in humans is for language (Newport *et al.*, 2001). A clear relationship exists between the age of exposure to a language and the level of proficiency achieved in that language. This relationship holds for the learning of both first and second languages. Acquisition of a first language has been assessed in children who have been raised in the absence of any language (feral or abused children) or, more frequently, in congenitally deaf children who have been raised without the aid of sign language. Much more data are available for people who began learning a second language at different ages. For both first and second languages, a thorough command of the language is attained by those who learn the language before 7 years of age. The degree of language proficiency that is eventually achieved decreases progressively with age of exposure and reaches adult levels by the end of adolescence.

Only certain aspects of language are affected by learning during sensitive periods (Newport *et al.*, 2001; Kuhl, 2000). Full proficiency with grammar (the classes of words, their functions, and relations in a sentence), syntax (the way in which words are put together in a sentence), and the production and comprehension of phonetics (the speech sounds of a language) are each dependent on early exposure to language. In contrast, semantics (word meaning) and size of vocabulary are not affected by the age of exposure. Thus, sensitive periods seem to affect the formal and subtle aspects of language, whereas the capacity to learn new words and their meanings continues unabated throughout life.

Physiological measures reveal an age dependence in the way in which language is processed and represented in the brain (Weber-Fox and Neville, 1996; Dehaena *et al.*, 1997). Various techniques have been used to assess brain activity while human subjects make perceptual judgments in language tasks. These techniques include functional magnetic resonance imaging, position emission tomography, and event-related potentials. In normal adults, language is processed in specific areas, primarily in the left or "dominant" hemisphere. In people who have learned a second language prior to the age of 7 years, the brain areas that are involved in processing the first and second languages overlap extensively. In contrast, in people who have learned a second language later in life, the areas of the brain that are activated by the second language do not overlap, or they overlap little, with those that are activated by the first language. The brain areas activated by the second language are less lateralized to the left hemisphere and are more variable across subjects. The effect of age at the time of learning on the brain areas activated by language is far more conspicuous for tasks requiring grammatical and phonic judgments than for tasks requiring semantic judgments. Thus, consistent with the behavioral observations of the age dependence for learning grammar and phonetics, the regions of the brain that contribute to the processing of grammar and phonetics are shaped in a unique way during sensitive periods.

Detailed knowledge of the mechanisms that control sensitive periods and of the plasticity that occurs during sensitive periods will provide a basis for formulating optimal therapeutic procedures to help minimize long-term harmful effects of early abnormal experience, associated with neonatal and childhood disabilities, for example, and maximize the acquisition of normal function once normal conditions are restored. Such knowledge may also lead to improved methods of rearing and teaching normal children that take advantage of the full capacity of the central nervous system to learn from experience.

Eric I. Knudsen

References

Dehaena, S., Doupoux, E., Mehler, J., Cohen, L., Perani, D., van de Moortele, P.-F., Leherici, S., and Le Bihan, D. (1997). Anatomical variability in the cortical representation of first and second languages. *Neuroreport* 17, 3809–3815.

Kuhl, P. K. (2000). A new view of language acquisition. *Proc. Natl. Acad. Sci. USA* 97, 11850–11857.

Newport, E. L., Bavelier, D., and Neville, H. J. (2001). Critical thinking about critical periods: Perspectives on a critical period for language acquisition. *In* "Language, Brain and Cognitive Development: Essays in Honor of Jacques Mehler" (E. Doupoux, ed.), pp. 481–502. MIT Press, Cambridge, MA.

Weber-Fox and Neville, H. J. (1996). Maturational constraints on functional specializations for luanguage processing: ERP and behavioral evidence in bilingual speakers. *J. Cognit. Neurosci.* 8, 231–256.

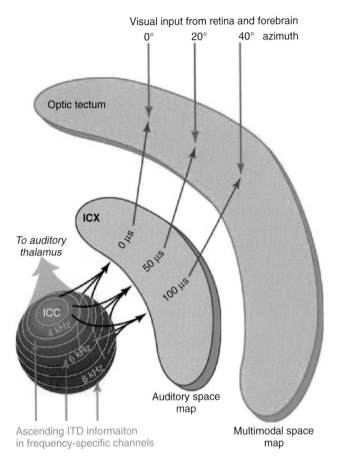

FIGURE 22.3 The ascending auditory pathway to the optic tectum in the barn owl. Auditory inputs enter the optic tectum from the brain stem (bottom arrows); these inputs already encode frequency-specific information about interaural time difference (ITD). These inputs project into the central nucleus of the inferior colliculus (ICC), where they are organized topographically by frequency. Bands in the ICC represent these frequencies (e.g., 4, 6, 8 kHz). Neurons in the ICC convey information both to the auditory thalamus (the primary pathway) and to the external nucleus of the inferior colliculus (ICX). In the ICX, ITD information is combined across frequency channels to synthesize a map of auditory space. For example, an ITD of 0 μs is generated by sound stimuli directly in front of the animal, such that sound reaches the two ears at exactly the same time. At the position marked 0 μs in the ICX are neurons that respond maximally to sounds with an ITD value of 0 μs and thus respond selectively to sounds originating in front of the animal. Sounds originating, for example, from positions that are further to the left-hand side will reach the ears with progressively greater left ear-leading ITDs and thus stimulate neurons with progressively larger best ITDs. From the ICX, the auditory map of space is conveyed via a topographic projection to the optic tectum. Here the auditory map is aligned and merged with a visual map of space (top arrows, representing inputs from the retina and the forebrain) to produce a multimodal space map.

This pathway transforms a representation of auditory spatial cues that exists in the central nucleus of the inferior colliculus (ICC; Chapter 26) into a topographic representation of space in the external nucleus of the inferior colliculus (ICX). The auditory map of space is

then sent on to the optic tectum, the avian analog of the mammalian superior colliculus, where it aligns with and is integrated with a visual map of space. The function of this pathway is to extract spatial information from sound that can be used to direct orienting movements of the eyes and head toward auditory stimuli (Chapter 33).

The pathway derives the location of a sound source by evaluating spatial cues that are present in the auditory signals at the two ears. The most reliable cues for sound localization are interaural timing differences (ITDs) and interaural level differences (ILDs). ITDs are due to the difference in the path length that sound must travel to reach the near versus the far ear. Because the ears are on the sides of the head, ITD varies systematically with the horizontal (azimuthal) location of a sound source (Fig. 22.4A; contour lines). ILDs result from the fact that each ear is most sensitive to sound coming from certain directions. Consequently, a sound from a particular direction will usually produce a higher sound level in one ear than the other. ILDs vary both with the azimuthal and with the elevational location of a sound source in spatial patterns that depend on sound frequency.

In creating the map of auditory space, the nervous system can only roughly anticipate the relationship between encoded values of ITD and ILD and the locations of sound sources that produce them. The correspondence of ITDs and ILDs with locations in space changes with the size and shape of the head and ears, features that vary across individuals, as well as for a given individual during growth. Moreover, the encoded values of sound timing and level that are transmitted to the CNS depend on the sensitivity and transduction properties of each ear, and these properties can change over time. Therefore, to establish and maintain an accurate map of space in the optic tectum, this midbrain pathway must learn the exact relationship between the encoded cue values and the locations of sound sources that produce them.

The influence of early experience on the owl's auditory space map has been demonstrated using a variety of techniques that change the relationship between cue values and locations in space: For example, the external ears have been altered drastically or the auditory canal of one ear has been plugged chronically (Knudsen, 1999). The midbrain pathway responds adaptively to such manipulations by adjusting the tuning of neurons in the ICX and optic tectum to ITDs and ILDs that restore an accurate map of space. In young animals, this plasticity enables the recovery of a substantially normal auditory map even after severe disruptions of hearing. In adult animals, plasticity is far more limited in extent.

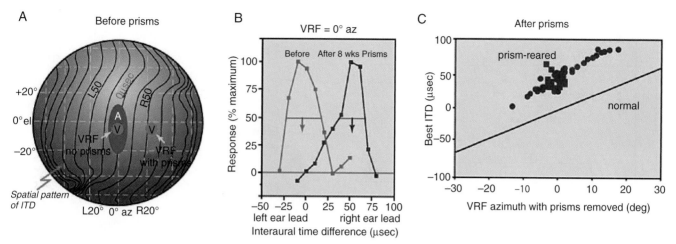

FIGURE 22.4 Rearing owls with laterally displacing, optical prisms causes an adaptive shift in the tuning of neurons in the optic tectum for ITD. (A) This map represents the space in front of the owl, showing both the elevation and the azimuth of a stimulus in space. Contour lines indicate the correspondence of ITD values (in microseconds) with particular locations in space. The point at which the 0° axes intersect represents the point in space directly in front of the owl's head. The auditory (A) and visual (V) receptive fields of one tectal neuron are shown in the center of the map. This neuron responds optimally when the stimulus is directly in front of the animal. Normally, the auditory and visual receptive fields are aligned. Optical prisms induce a horizontal displacement of the neuron's visual receptive field (VRF), resulting in a misalignment between A and V. (B) Tuning for ITD is shifted by prism experience. These ITD tuning curves were recorded from similar sites in the optic tectum before (blue) and after (purple) 8 weeks of prism experience. Both sites had a VRF at 0° azimuth. After 8 weeks of experience, the neuron is tuned for the ITD produced by an acoustic stimulus at the location of the optically displaced VRF, as shown in A. Arrows indicate the best ITD for each site; the best ITD is defined as the center of the range of ITDs to which the neuron responded with more than 50% of its maximum response. (C) The relationship between best ITD and VRF azimuth is shifted systematically from normal in prism-reared owls. The black line indicates the regression of best ITD on VRF azimuth that is observed in normal owls. Dots represent individual sites in a prism-reared owl. The map of ITD is shifted systematically relative to the visual map of space.

An instructive signal that adjusts the tuning of ICX and tectal neurons is provided by the visual system. The instructive role of vision in guiding the tuning of ICX and tectal neurons has been demonstrated in experiments in which owls wear optical displacing prisms that chronically shift the visual field (Knudsen, 2002). The effect of experience with displacing prisms on auditory spatial tuning is most apparent in the optic tectum, where the visual receptive field of each neuron indicates the location in auditory space to which that neuron should normally be tuned. When prisms that displace the visual field horizontally are placed in front of the eyes, the visual receptive fields of tectal neurons are shifted horizontally and out of alignment with the auditory receptive fields of these neurons (Fig. 22.4A). In juvenile birds, continuous experience with such prisms over a period of 6 to 8 weeks causes the auditory receptive fields of tectal neurons to realign with their visual receptive fields: The tuning of tectal neurons to ITDs and ILDs changes so that they respond to auditory cue values representing the locations of their optically displaced visual receptive fields (Figs. 22.4B, C). As a result, the auditory map of space shifts to match the optically shifted visual map. This adjustment is adaptive because, by making it, the animal alters its orientation toward sounds so that it sees the source of the sound through the prisms.

A Sensitive Period for Neuronal Adjustments

This adaptive auditory plasticity is regulated developmentally. The magnitude of the shift in neuronal ITD tuning that is induced under standard conditions of prism experience depends greatly on the age of the animal (Fig. 22.5). Large shifts in ITD tuning, of up to 70 μs, occur only in juvenile owls. In adult owls, equivalent conditions rarely shift ITD tuning by more than 10 μs, even after many months of prism experience. Larger adaptive shifts in ITD and ILD tuning have been induced in adult owls that are required to hunt live prey to survive, but even under these conditions, the adaptive shifts that occur in adults are smaller than those that occur in juvenile owls that do not hunt. The period during which prism experience induces large changes in ITD tuning, the sensitive period, ends as the owls approach sexual maturity, at about 200 to 250 days old.

Although an experience-dependent shift in ITD tuning is observed most easily in the optic tectum (due to the physiological reference provided by the visual

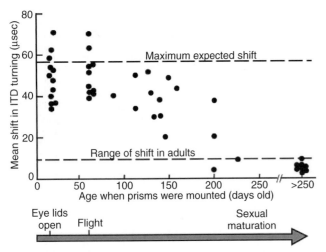

FIGURE 22.5   The sensitive period for visual calibration of neuronal ITD tuning in the optic tectum. Each dot represents data from a single owl. The large arrow below is a timeline, indicating important developmental stages in an owl's life. Each owl experienced a 23° displacement of the visual field for at least 60 days. ITD tuning was then measured at 15 to 23 sites in the superficial layers of the optic tectum. The difference between the best ITD measured and the best ITD expected normally, based on the location of the site's VRF (see Fig. 22.4C), was taken as the "shift in ITD tuning." The mean shift in ITD tuning for the population of sampled sites as a function of the age of the owl when prisms were first mounted is plotted.

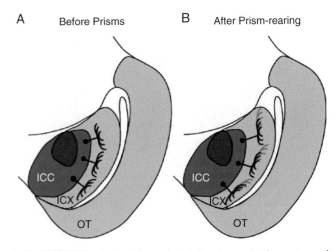

FIGURE 22.6   Schematic model of the change in the pattern of axonal projections from the ICC to the ICX that accompanies the shift in the map of ITD in the ICX. Based on DeBello and Knudsen (2001). (A) The initial state of the projection before prism experience. (B) After prism experience, axonal projections from the ICC to the ICX are shifted systematically as indicated by the red arbors.

receptive field of the neuron), the site in the pathway where the plasticity actually takes place is in the ICX: The maps of ITD in the ICX and in the optic tectum are shifted by equivalent amounts in prism-reared owls, whereas the representation of ITD in the ICC remains unaltered.

Mechanisms of Plasticity

A shift of the auditory space map that occurs during the sensitive period is associated with a change in the architecture of the neurons that project from the ICC to the ICX (DeBello *et al.*, 2001). A topographic projection from the ICC to the ICX brings ITD information to the appropriate site in the ICX, where the information is integrated across frequency channels to create spatial receptive fields (Fig. 22.3). A topographic projection is established early in development, before prism experience exerts its effects (Fig. 22.6A). Prism experience causes neurons in the portion of the ICC that represents the shifted values of ITD to project axons to neurons in novel regions of the ICX (Fig. 22.6B; red lines) that then become tuned to those abnormal values of ITD. Thus, experience induces the elaboration of axons at sites in the ICX where they support appropriate responses. In owls that have

acquired shifted space maps, the learned anatomical circuit coexists with the normal circuit.

The newly learned responses that result from prism experience are mediated differentially by a special class of glutamate receptor, the *n*-methyl-D-aspartate (NMDA) receptor. Drugs that specifically block this receptor, such as AP5, eliminate or severely reduce the responses of ICX neurons to newly learned values of ITD while having substantially less effect on their responses to the normal value. Thus, the expression of newly learned responses in this pathway depends heavily on the activation of NMDA receptors. The action of NMDA receptors plays a critical role in many other examples of experience-dependent plasticity as well.

As mentioned earlier, large shifts in ITD tuning in response to visual field displacement occur only in juvenile owls during a sensitive period. In contrast, removal of prisms from adult owls that have been raised from the day of eye opening wearing prisms results in a shift of the map of ITD back to normal. The genetically programmed, normal circuitry persists into adulthood even without validation by experience. In owls that have been raised with prisms, the adult circuit is able to switch back and forth, over a period of weeks, from an abnormal representation of ITD to a normal representation of ITD and vice versa, depending on the visual world the animal experiences. In this case, acquired alterations in circuit architecture resulting from experience during the sensitive period, together with the genetically programmed circuit architecture, influence the range of connectional states that the circuit can assume later in adult life.

Summary

The brain derives the location of a sound source by evaluating spatial cues that are present in the auditory signals at the two ears. Interaural timing differences and interaural level differences are used to determine the position of a sound in space. In owls, this information is used to create a map of auditory space, which is aligned closely with a visual map of space in the optic tectum. Establishing and maintaining the alignment of the auditory space map with the visual space map in the optic tectum is an active process, guided by information provided by experience. The capacity of this pathway to change adaptively in response to experience is particularly great during a sensitive period. Plasticity during the sensitive period involves changes in circuit architecture and depends on the action of the NMDA subtype of glutamate receptor.

Experience Shapes Functional Organization in the Visual Cortex

In the mammalian nervous system, visual information from the two eyes first comes together at the level of the primary visual cortex. As described in detail in Chapter 27, visual experience during a limited period early in life has an enormous impact on how much of the visual cortex is devoted to processing input from each eye and the degree to which binocular inputs are combined (Hubel and Wiesel, 1970; Hubel *et al.*, 1977). Early in development (before birth in many species), afferents that provide inputs from the left and right eyes, respectively, begin to cluster in separate, interleaved areas in layer 4 of the primary visual cortex. This early clustering of eye-specific inputs from the lateral geniculate nucleus (LGN) is driven by a combination of molecular mechanisms and patterns of spontaneous neuronal activity (Chapter 20).

Soon after birth, a developmental period opens during which visual experience influences the competition among LGN afferents for territory in layer 4. As long as the eyes are coordinated and used equally, the typical final state, consisting of equally wide ocular dominance columns, is achieved (Fig. 22.7). If, however, vision is impaired in one eye, due to monocular eyelid closure in an experimental animal or to a cataract in a human, for example, the balance between LGN afferents in their competition for layer 4 territory is disrupted: LGN afferents that convey input from the impaired eye lose the ability to drive layer 4 neurons in an abnormally large region of the cortex, whereas LGN afferents that convey input from the normal eye gain the ability to drive layer 4 neurons in an

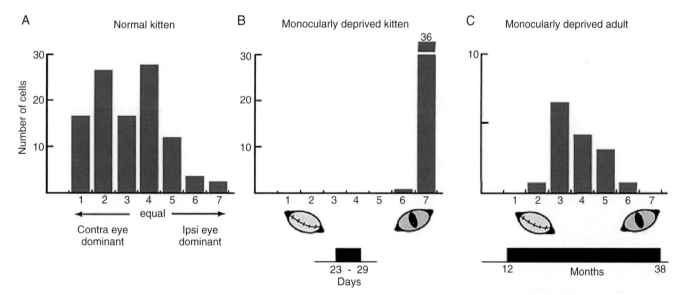

FIGURE 22.7 Effect of chronic closure of one eye on the responsiveness of visual cortical neurons to input from each eye. (A) Ocular dominance distribution in the primary visual cortex of two normal kittens, 3 to 4 weeks old. Cells in group 1 were driven only by the contralateral eye; for group 2, the contralateral eye was markedly dominant; for group 3, the contralateral eye was slightly dominant; for group 4, there was no apparent difference in the drive from the two eyes; for group 5, the ipsilateral eye dominated slightly; for group 6, it dominated markedly; and for group 7, cells were driven only by the ipsilateral eye. (B) Ocular dominance distribution was altered dramatically in a kitten exposed to contralateral eye closure for 1 week (from 23 to 29 days of age). (C) Ocular dominance distribution was essentially normal in an adult cat exposed to contralateral eye closure for 26 months. From Hubel and Wiesel (1970).

abnormally large portion of the cortex. As a consequence, activity throughout most of the visual cortex becomes driven by LGN afferents from the normal eye (Fig. 22.7).

The opening of the critical period requires the maturation of inhibitory circuitry in the visual cortex (Hensch, 2005). The processing of visual information depends on a precise balance of excitatory and inhibitory influences. The balance of excitation and inhibition is regulated carefully and dynamically in mature circuits. In the visual cortex, the critical period for ocular representation does not open until inhibitory connections, which develop after excitatory connections, become effective. The critical period opens prematurely in mice in which inhibition has been enhanced pharmacologically, and it is delayed in mice in which GABA levels have been reduced by genetic manipulations. Thus, the critical period in the visual cortex depends on the ability of this circuit to process information.

Mechanisms of Plasticity

Cellular mechanisms by which binocular experience during this sensitive period shapes the architecture and functional properties of neurons in the visual cortex are described in Chapter 20. In brief, the dramatic change in functional properties in layer 4 is accompanied by an equally dramatic change in the axonal architecture of the LGN neurons that project to layer 4: axonal arbors conveying input from the normal eye expand while those conveying input from the impaired eye shrink. The anatomical remodeling of LGN axons depends on the availability of neurotrophins—BDNF and NT-4—as well as on the expression of their cognate receptor, TrkB (Lein and Shatz, 2001). In addition, the adjustments of synaptic drive by the left and right eyes result from long-term potentiation and long-term depression of synapses in the cortex, and requires the activation of NMDA receptors.

A Critical Period for Ocular Representation Exists in the Visual Cortex

Because the effects of disruptions of binocular vision on ocular representation in the visual cortex are apparently irreversible when they occur during this limited period in early life, this period has been referred to in the literature as a *critical period*. The critical period for ocular representation in the visual cortex has been studied particularly carefully in cats by measuring the age dependence of the effects of monocular eyelid closure (Fig. 22.8). The onset of the critical period is rapid, beginning at about 3 weeks in cats. By 4 to 6

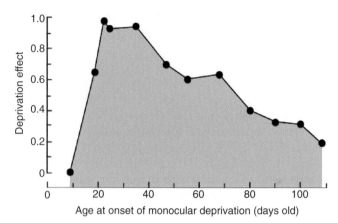

FIGURE 22.8   The critical period for ocular representation in the primary visual cortex of the cat. The degree of functional disconnection of cortical neurons from the deprived eye is quantified and plotted as a function of the kitten's age at the time of monocular closure. Chronic monocular closure lasted 10 to 12 days. Each point represents data from a single animal. Functional disconnection was based on the ocular dominance distribution (see Fig. 22.7) and indicated the degree to which the influence of the closed eye was weakened or lost. The index was defined such that the mean value for normal cats was 0, whereas total disconnection resulted in a value of 1. From Olson and Freeman (1980).

weeks of age, the cortex is maximally sensitive to monocular deprivation: A few days of monocular deprivation causes a complete shift in ocular dominance, leaving the cortex almost entirely driven by input from the nondeprived eye (Fig. 22.7B). Beyond 6 weeks of age, the critical period gradually closes: Over the next 10 months, the rate at which ocular dominance can be shifted by monocular deprivation and the degree to which it can be shifted both decrease. Once a cat is about 1 year old, monocular deprivation even for months no longer affects ocular dominance in the cortex (Fig. 22.7C). Similar critical periods, but extending later in life, exist for the visual cortex in monkeys and humans.

The Critical Period Can Be Prolonged

The close of the critical period in cats can be delayed substantially by raising animals in complete darkness. At the beginning of the critical period, most cortical neurons respond to inputs from either eye, but the responses tend to be weak. In cats that are raised in the dark until well past the end of the critical period (as defined by monocular occlusion), cortical neurons continue to be driven binocularly and their responses remain weak. When these animals are finally allowed visual experience, the responses of these neurons gradually increase in strength and the relative representations of the left and right eyes is shaped by the animal's experience: In monocularly deprived cats, nonde-prived eye inputs become predominant in the cortex, and in cats that experience binocular vision, normal ocular dominance columns develop. This indicates that the critical period has remained open. Thus, without experience-driven input, the mechanisms that control ocular representation in the cortex remain in an uncommitted state, waiting for instruction for a prolonged, if not indefinite, period of time.

The critical period for ocular representation in the cortex is atypical in one respect: There is no predisposi-tion to establish a normal pattern of connectivity based on normal experience. Even after baby cats or monkeys have experienced normal binocular vision for several weeks, monocular deprivation still causes the responses in the cortex to become dominated by the nondeprived eye (Fig. 22.7). Moreover, once the responses in the cortex become dominated by one eye, reinstating normal visual input to the previously deprived eye does not readily restore normal binocular responses in the cortex. Instead, the nondeprived eye continues to dominate the responses of cortical neurons for as long as animals have been studied (up to 5 years after res-toration of binocular input). Thus, unlike in the case of song learning in birds, the pattern of neural connectiv-ity that supports normal function in the visual cortex is not stabilized immediately by exposure to normal binocular input. Instead, normal binocular vision must persist throughout the entire critical period to prevent the cortex from becoming dominated by monocular responses. An adaptive advantage of this characteristic of the binocular pathway has yet to be recognized.

Summary

The representation of the two eyes in layer 4 of the primary visual cortex is shaped by binocular experi-ence during a sensitive period. During prenatal devel-opment and before the onset of vision, thalamic inputs representing the left and right eyes segregate from each other to form ocular dominance columns of roughly equal width in the visual cortex. Soon after birth, a period opens during which the visual experi-ence of the animal shapes the representations of the two eyes in the cortex. Normal binocular vision con-solidates and refines the established patterns of ocular representation. Impaired vision in one eye, however, causes the LGN axons in layer 4 carrying information from the impaired eye to shrink and their synapses to loose efficacy, whereas axons carrying information from the normal eye expand and their synapses increase in efficacy. In contrast, equivalent, monocular deprivation in adult animals has no apparent effect. The critical period for ocular representation in the

visual cortex can be extended by rearing animals in complete darkness.

A Sensitive Period for Shaping the Temperament of Rats

The way in which a rat pup is cared for by its mother during a sensitive period has an enormous impact on a rat's temperament as an adult (Weaver *et al.*, 2004). In this example, the primary influence of early experi-ence is to set the expression level of a gene, the gene for the glucocorticoid receptor (GR), in a particular circuit in the hippocampus. The protein product of this gene strongly influences a rat's emotional res-ponses to stressful conditions. The effect of mothering on the expression level of this gene occurs only during the first week of a rat's life, and the behavioral consequences of this experience last throughout adulthood.

Different rats respond differently to threatening situations. For example, when introduced into a new environment, some rats are calm and adventurous, whereas others are anxious and fearful. This funda-mental difference in emotional responses to stressors reflects, in part, a rat's interactions with its mother during the first week after birth (Fig. 22.9). Rats raised by a mother who groomed them extensively (high-grooming) and nursed them in a way that facilitated their access to milk (arched-back nursing) are less fearful and less reactive to stressors as adults than are rats raised by a mother who did not treat them in this way.

Cross-fostering experiments demonstrate that the transmission of these traits is dominated by

FIGURE 22.9 A rat mother grooms her pups during the first week after their birth. This interaction permanently affects the tem-perament of these rat pups.

experience, and not by genetics. Rats born to low-grooming mothers (nonattentive and no arched-back nursing), but raised by high-grooming mothers, become themselves calm, adventurous, high-grooming mothers. Conversely, rats born to high-grooming mothers, but raised by low-grooming mothers, become anxious, low-grooming mothers. Thus, the transmission of these traits depends on mother-infant interactions during the first week of a rat's life. The experiments demonstrate that, although genetics constrains the ranges of these traits, early experience can modify them dramatically.

The emotional responses of a rat to stressors reflect the reactivity of the hypothalamic-pituitary-adrenal (HPA) system. Activation of the HPA system leads to the release of a special class of stress hormones, glucocorticoid hormones, from the adrenal glands. Animals with high circulating levels of glucocorticoid hormones are anxious and fearful, those with low circulating levels of this hormone are calm.

The hippocampus exerts a powerful negative feedback on the release of glucocorticoid hormones from the adrenal glands. High levels of GRs in the hippocampus result in low basal levels of glucocorticoids and tight regulation of glucocorticoid release when an animal is stressed. Low levels of GRs in the hippocampus have the opposite effects.

Mechanisms of Plasticity

The expression level of GRs in the hippocampus is adjusted by interactions of a rat pup with its mother. Rats raised by high-grooming mothers express high levels of GRs in the hippocampus, whereas rats raised by low-grooming mothers express low levels of GRs. The same effect is observed in experiments in which mothering behavior is manipulated. For example, removing rat pups from a mother for 15 minutes and then returning them to the mother, causes the mother to increase her grooming of those pups. Rats that experience this increase in grooming express increased levels of GRs in the hippocampus compared with control animals; correspondingly, they are less reactive to stressors, less fearful, and more adventurous. This manipulation has these effects only when applied during the first two weeks after birth.

One mechanism that mediates the effect of mothering on GR levels in the hippocampus, and therefore on an animal's reactivity to stressors, is the methylation of the GR gene. Methylation silences the GR gene by blocking access of transcription factors to a promoter region of the gene. Gene promoter methylation is adjusted during development and is stable thereafter. Experience during the first week after birth with a high-grooming mother (Fig. 22.9) causes the GR gene promoter to become demethylated, resulting in high levels of GR gene expression and high levels of GRs in the hippocampus. Conversely, experience with low-grooming mothers causes the GR gene promoter to become methylated, leading to low GR levels in the hippocampus. The sensitive period for the effect of mothering on the temperament of rats is accounted for by the narrow window in development when the methylation state of the GR gene is adjusted by experience. The persistence of the effects of this experience is due to the subsequent stability of the gene's methylation state.

A second mechanism that contributes to the effect of mothering on GR levels in the hippocampus is the acetylation of histones. Acetylation of histones in the chromatin of a gene increases the access of transcription factors to promoters. Histones within the nucleosome core that includes the GR gene are more acetylated in rats reared by high-grooming mothers than in rats reared by low-grooming mothers, an effect that increases the transcription of the GR gene and raises GR levels in the hippocampus. The difference in histone acetylation is thought to be driven by the difference in the methylation state of the GR gene, discussed earlier.

The contribution of histone acetylation to controlling GR levels in the hippocampus has been tested by administering an inhibitor of histone deacetylation to adult rats that had been reared by low-grooming mothers. The inhibitor increased acetylation of the histones surrounding the GR gene, increasing access of demethylating enzymes to the gene. The inhibitor thereby caused a decrease in the methylation state of the gene, which increased GRs in the hippocampus and changed the temperament of the adult rats into the calm, adventurous temperament typical of rats raised by high-grooming mothers.

Summary

Interactions of a rat with its mother during the first week of life shape the way in which a rat responds to stressful situations as an adult. The primary mechanism for this effect is an experience-dependent adjustment of the expression level of the GR gene in the hippocampus. High maternal care during the first postnatal week causes a demethylation of a GR gene promoter, resulting in a stable increase in GR gene expression in the hippocampus and a consequent decrease in basal levels of glucocorticoid hormones and a tighter regulation glucocorticoid hormone release in response to stressors. As a result, the animal is calm, adventurous, and reacts in a measured fashion to stressors. Maternal behavior has these effects only during a sensitive period in a rat's development when the methylation state of the GR gene can be altered. Thereafter, the methylation state of the GR gene remains stable throughout adulthood.

PRINCIPLES OF DEVELOPMENTAL LEARNING

During the later stages in the maturation of many neural circuits, patterns of neuronal activity, driven by stimuli or the animal's behavior, shape the circuit's functional properties, architecture, and/or biochemistry. This shaping process selects circuit properties that are appropriate for the individual's experience.

Circuit changes that occur during sensitive periods differ from those that occur in adulthood in terms of their magnitude and persistence and the behavioral conditions under which the changes can occur. As illustrated by the preceding examples in this chapter, sensitive period experience can cause changes in the nervous system that are beyond the range of changes that occur in adults, and the changes that result typically persist for the lifetime of the animal.

Changes in circuit properties are far more readily induced during sensitive periods than in mature circuits. During sensitive periods, circuits may be altered simply by exposing animals to unusual conditions such as monocular deprivation, for example. In mature circuits, equivalent conditions either have no effect or require the attention of the animal to the conditions in order for plasticity to occur.

Sensitive period learning is influenced heavily by genetic predispositions. Only a limited range of stimuli is allowed to operate as an instructive influence for a particular circuit. Within this acceptable range, some stimuli are preferred over others. This property is well illustrated by song learning in birds and by imprinting (Box 22.2). The predisposition of the nervous system to be instructed by "normal" experience probably originates in the selectivity of the response properties, genetically determined as well as shaped by experience, of the neurons that provide input to the sites in the pathway where the learning take place. As learning progresses, the selectivity of the pathway for acceptable input becomes progressively higher.

Whether a particular pathway passes through a sensitive period can vary across species. For example, some songbirds, such as canaries and mockingbirds, learn new songs seasonally throughout life, whereas others, such as white-crowned sparrows and zebra finches, learn their songs only during a sensitive period. Such species differences may provide a useful tool for uncovering the mechanisms that are responsible for sensitive periods.

The magnitude of changes that may result from experience-driven adjustments varies greatly across circuits and across species. The magnitude of changes depends on the degree of genetic specification of the inputs to the site of change. When the selection of appropriate inputs is from a large potential range of inputs, the effect of experience can have a profound influence on a circuit. Conversely, when the range of potential inputs is highly restricted by genetic specification, the effect of experience is correspondingly small.

The duration of different sensitive periods also varies greatly. At one end of the spectrum are the sensitive periods for imprinting, which may open and close within hours (Box 22.2). At the other end are sensitive periods for acquiring complex cognitive capabilities, such as language (described in Chapter 51), which involve sensitive periods that last for many years.

The Opening of Sensitive Periods Depends on Pathway Maturation

Sensitive periods cannot open until the brain has matured to the point where three conditions are met. First, the information provided to the circuit from lower level circuits must be sufficiently reliable and precise to allow the circuit to carry out its function. Second, the circuit's connectivity must have matured adequately for it to process the information. For example, the critical period in the visual cortex does not open until and unless inhibitory as well as excitatory connections are effective. Third, the mechanisms that enable plasticity must be active.

Because complex behaviors depend on information that is processed through hierarchies of circuits, the first condition implies that sensitive periods for circuits at higher levels in these hierarchies cannot open until the information from circuits at lower levels has become reliable. Sensitive periods for low-level circuits, such as the LGN, occur earlier than sensitive periods for higher level circuits, such as those in the visual cortex. The same principle holds for circuits at different levels in the owl's sound localization pathway. Thus, pathways that support complex behaviors, such as human language (Box 22.1) or object recognition, may rely on circuits that pass through sensitive periods that end at very different stages in an animal's life.

Experience cannot shape a circuit until the mechanisms that enable plasticity are active. Experience-driven changes in functional properties can involve a

BOX 22.2

FILIAL IMPRINTING: BABIES LEARN TO RECOGNIZE THEIR PARENTS

For many species of birds and mammals, including ducks, geese, mice, and monkeys, parental care is essential for the survival of the young. The young of these species learn rapidly to distinguish their parents from all other individuals and form a unique and close relationship with their parents from that point on—a process referred to as filial imprinting (Horn, 2004; Hess, 1973). Filial imprinting can involve the learning of visual, auditory, olfactory, and gustatory cues that identify a parent. The learning of these cues takes place during short, well-defined sensitive periods early in postnatal life.

The visual component of the learning process usually is preceded by auditory, olfactory, and/or gustatory components. In many species, the babies learn to recognize the vocalizations of the mother based on experience that begins before or soon after the animal is born. In addition, babies may learn the odor and/or taste of the mother from the odors and tastes experienced immediately after birth. The recognition of the parent based on acoustic and/or chemical cues helps the young select the correct individual for visual imprinting, once the eyes and nervous system are capable of adequate form vision.

Sensitive periods for filial imprinting are relatively discrete and can be as short as a few hours in some species (Ramsay and Hess, 1954). For example, the sensitive period for filial imprinting in ducklings occurs during the first day of life. When a duckling is exposed once, for 10 min, to a model of a male duck on the first day after hatching, the duckling imprints on this particular model. When the duckling is tested 5 to 70 h later, by being offered a choice between the previously presented model and a model of a female duck, the duckling prefers to follow the model of the male duck over that of the female duck, even when the female model is much closer to the duckling and makes louder calls. Measured by the following response, imprinting in ducks is most effective between 10 and 20 h after hatching.

As in other examples of learning during sensitive periods, young animals exhibit an innate preference to imprint on normal stimuli. When given a choice, babies in the process of imprinting attend preferentially to images that more closely resemble members of their own species. Thus, when baby ducks are given the choice of imprinting on geese or on people, they imprint on the (duck-like) geese. This predisposition is based on genetically programmed preferences for simple, conspicuous features, referred to as sign stimuli by ethologists, that tend to distinguish the species from all others. This implies that the neural circuitry involved in filial imprinting, like that involved in song learning in songbirds, contains genetically determined neuronal filters that help identify stimuli that are appropriate models for learning. As imprinting proceeds, learning causes these filters to become more selective until ultimately the young are capable of discriminating one individual from all others.

Imprinting results in rapid functional and structural changes in a specific part of the forebrain of birds (Horn, 2004). In chicks that have been imprinted on an artificial object, neurons in the intermediate and medial mesopallium (IMM) become responsive to the imprinted stimulus, and some become highly selective for the stimulus. Associated with the acquisition of selective neuronal responses in the IMM are local increases in the number of NMDA receptors, increases in the size of certain synapses, and a temporary surge in inhibitory activity. The increased inhibitory activity, which begins soon after exposure to the imprinting stimulus and lasts for many hours, is thought to contribute critically to the shaping of the specificity of IMM neurons for the imprinted stimulus.

Eric I. Knudsen

References

Hess, E. H. (1973). "Imprinting: Early Experience and the Developmental Psychobiology of Attachment." Van Nostrand-Reinhold, New York.

Horn, G. (2004). Pathways of the past: the imprint of memory. *Nature Rev. Neurosci.* **5**, 108–120.

Ramsay, A. O. and Hess, E. H. (1954). A laboratory approach to the study of imprinting. *Wilson Bull.* **66**, 196–206.

remodeling of axons and dendrites, an elaboration of new synapses, an adjustment of synaptic efficacies, and/or a regulation of gene expression. Anatomical remodeling and the establishment of new synapses may well be guided by the same mechanisms that control synaptogenesis during circuit development (Chapter 18). Neurotrophins (described in Chapter 19), particularly BDNF for example, are essential for anatomical remodeling during the critical period in the visual cortex (Lein and Shatz, 2001). Mechanisms that regulate synaptic efficacy during a sensitive period include those that underlie synaptic plasticity in the adult nervous system (Chapter 50). Long-term potentiation (LTP) and long-term depression (LTD), for example, have been shown to operate in many models of developmental learning. Moreover, the ubiquitous presence of NMDA receptors at sites of change in the various models of developmental learning indicates that this class of glutamate receptor plays a key role in adjusting patterns of connectivity during sensitive periods.

The Closing of Sensitive Periods May Involve Several Mechanisms

Sensitive periods end once an animal has received adequate experience and the relevant circuit is irreversibly committed to a pattern of connectivity. The factors that render the commitment irreversible are not known. Many factors may play a role, and the factors that are most important may differ across different pathways.

One factor that may contribute to the closing of sensitive periods is the age- or experience-dependent alteration of molecular mechanisms that support changes in synaptic efficacy such as LTP and LTD, described in Chapter 49. In addition, in those circuits in which axonal elaboration is a necessary component of experience-dependent changes in connectivity, loss of the mechanisms that support axonal growth would end the sensitive period. In the visual cortex of cats, for example, levels of the growth-associated protein GAP-43, which is thought to be necessary for axonal growth, decrease precipitously during the critical period. A host of other molecular mechanisms could contribute to a decline in plasticity, including the loss of responsiveness of presynaptic axonal arbors to neurotrophins secreted by postsynaptic neurons, the stabilization of synapses by the extracellular matrix or by proteoglycans, the myelination of axons (preventing them from growing or retracting), the appearance of molecules that prevent growth, and the disappearance of molecules that enable growth.

Another major factor that decreases the plasticity of circuits is the experience-driven sharpening of functional tuning. Initially in development, neuronal responses are relatively weak and broadly tuned. Experience causes selective changes in anatomical connections and synaptic efficacy that refine the patterns of both excitatory and inhibitory connections. These changes are self-reinforcing due to the action of self-organizational mechanisms that operate by Hebbian principles (Chapter 49). As a result, once a circuit has been shaped to process information in a certain way,

it becomes far more difficult for altered experience to induce new patterns of connectivity.

Deprivation Prolongs Sensitive Periods

Sensitive periods typically close once an animal receives adequate experience. Therefore, when an animal is deprived of appropriate experience, the sensitive period is prolonged. For example, raising songbirds in acoustic isolation prolongs the sensitive period for song memorization, and raising cats in complete darkness prolongs the critical period for ocular representation in the visual cortex. This characteristic of sensitive periods indicates that the event that triggers sensitive period adjustments is the powerful and repeated activation of neurons at the site where changes take place. Without the vigorous activation of these neurons, the pathway remains in an uncommitted state and capable of adjusting in response to experience when it becomes available. This characteristic suggests that for young animals (including humans) suffering from a peripheral or central abnormality, a total absence of relevant input is far better than abnormal input. This implies that the optimal therapeutic strategy for such individuals is to deprive them of relevant sensory input until the abnormality is corrected. Otherwise, an abnormal sensory experience may close the sensitive period, resulting in a commitment to an abnormal pattern of connectivity that cannot later be reversed.

Summary

Early experience shapes the functional properties, architecture, and biochemistry of many circuits so that the properties of these circuits are appropriate for the needs and the environment of the individual animal. The changes in circuit properties that occur during a sensitive period differ quantitatively, if not qualitatively, from those that occur in mature circuits. First, they occur readily only during a restricted period in the lifetime of the animal. Second, they involve the selection of particular circuit properties from a wide range of possible properties. Third, the changes in circuit properties that occur during sensitive periods typically do not require the attention of the animal. Fourth, the changes persist throughout life.

Sensitive periods vary in timing and duration across pathways and across species; some last only a few hours, whereas others last until the individual reaches sexual maturity. Sensitive periods open once the information conveyed to a circuit is sufficiently precise and the circuit is competent to process the information and to undergo plastic change. The signal that induces change is probably the repeated, vigorous activation of postsynaptic neurons by presynaptic activity representing the occurrence of an appropriate stimulus or the execution of adaptive behavior. The range of stimuli that are effective in driving plastic change is specified by genetic preprogramming, with most pathways biased heavily to prefer normal patterns of stimulation.

After a sensitive period closes, equivalent conditions have much less effect on the architecture, biochemistry, and functional properties of a circuit. The closure of a sensitive period may be triggered by experience itself or as a consequence of circuit maturation. The mechanisms that close a sensitive period probably vary for different circuits. When experience-induced changes require anatomical remodeling, the end of the sensitive period may be controlled by factors that regulate cell growth; when the induced changes require adjustments in synaptic efficacy, the sensitive period will be controlled by factors that influence the capacity of synapses to modify their efficacy; and when the induced changes involve adjustments in the levels of specific receptors, the sensitive period will be controlled mechanisms that regulate the expression of receptors.

References

Brainard, M. S. and Doupe, A. J. (2000). Auditory feedback in learning and maintenance of vocal behaviour. *Nature Rev. Neurosci.* **1**, 31–40.

DeBello, W. M., Feldman, D. E., and Knudsen, E. I. (2001). Adaptive axonal remodeling in the midbrain auditory space map. *J. Neurosci.* **21**, 3161–3174.

Doupe, A. J. (1997). Song- and order-selective neurons in the songbird anterior forebrain and their emergence during vocal development. *J. Neurosci.* **17**, 1147–1167.

Doupe, A. J. and Kuhl, P. K. (1999). Birdsong and human speech: Common themes and mechanisms. *Annu. Rev. Neurosci.* **22**, 567–631.

Hensch, T. K. (2005). Critical period plasticity in local cortical circuits. *Nature Rev. Neurosci.* **6**, 877–888.

Hubel, D. H. and Wiesel, T. N. (1970). The period of susceptibility to the physiological effects of unilateral eye closure in kittens. *J. Physiol. (Lond.)* **206**, 419–436.

Hubel, D., Wiesel, T., and LeVay, S. (1977). Plasticity of ocular dominance columns in the monkey striate cortex. *Philos. Trans. R. Soc. Land. Ser. B* **278**, 377–409.

Immelmann, K. (1972). Sexual imprinting in birds. *Adv. Study Behav.* **4**, 147–174.

Knudsen, E. I. (1999). Mechanisms of experience-dependent plasticity in the auditory localization pathway of the barn owl. *J. Comp. Physiol. A* **185**, 305–321.

Knudsen, E. I. (2002). Instructed learning in the auditory localization pathway of the barn owl. *Nature* **417**, 322–328.

Knudsen, E. I. (2004). Sensitive periods in the development of the brain and behavior. *J. Cogn. Neurosci.* **16**, 1412–1425.

Knudsen, E. I., Esterly, S. D., and Olsen, J. F. (1994). Adaptive plasticity of the auditory space map in the optic tectum of adult and baby barn owls in response to external ear modification. *J. Neurophysiol.* **71**, 79–94.

Konishi, M. (1985). Birdsong: From behavior to neuron. *Annu. Rev. Neurosci.* **8**, 125–170.

Leiderman, P. (1981). Human mother-infant social bonding: Is there a sensitive phase? *In* "Behavioral Development" (K. Immelmann, G. W. Barlow, L. Petrinovich, and M. Main, eds.), pp. 454–468. Cambridge Univ. Press, Cambridge.

Lein, E. S. and Shatz, C. J. (2001). Neurotrophins and refinement of visual circuitry. *In* "Synapses" (W. M. Cowan, T. C. Sudhog, and C. F. Stevens, eds.), pp. 613–649. Johns Hopkins Univ. Press, Baltimore, MD.

Newport, E. L., Bavelier, D., and Neville, H. J. (2001). Critical thinking about critical periods: Perspectives on a critical period for language acquisition. *In* "Language, Brain and Cognitive Development: Essays in Honor of Jacques Mehler" (E. Doupoux, ed.), pp. 481–502. MIT Press, Cambridge, MA.

Olson, C. R. and Freeman, R. D. (1980). Profile of the sensitive period for monocular deprivation in kittens. *Exp. Brain Res.* **39**, 17–21.

Wallhausser-Franke, E., Nixdorf-Bergweiler, B. E., and DeVoogd, T. J. (1995). Song isolation is associated with maintaining high spine frequencies on zebra finch LMAN neurons. *Neurobiol. Learn. Mem.* **64**, 25–35.

Weaver, I. C. G., Cervoni, N., Champagne, F. A., D'Alessio, A. C., Sharma, S., Seckl, J. R., Dymov, S., Szyf, M., and Meaney, M. J. (2004). Epigenetic programming by maternal behavior. *Nature Neurosci.* **7**, 847–854.

Eric I. Knudsen

SECTION IV

SENSORY SYSTEMS

23

Fundamentals of Sensory Systems

In bringing information about the world to an individual, sensory systems perform a series of common functions. At its most basic, each system responds with some specificity to a stimulus and each employs specialized cells—the peripheral receptors—to translate the stimulus into a signal that all neurons can use. Because of their physical or chemical specialization, the many types of receptors transduce the energy in light, heat, mechanical, and chemical stimulation into a change in membrane potential. That initial electrical event begins the process by which the central nervous system (CNS) constructs an orderly representation of the body and of things visible, audible, or chemical. To bridge the distance between peripheral transduction and central representation, messages are carried along lines dedicated to telling the CNS what has taken place in the external world and where it has happened. Such precision requires that labor be divided among neurons so that not only different stimulus energies (light vs. mechanical deformation) but also different stimulus qualities (steady indentation vs. high-frequency vibration of the skin) are analyzed by separate groups of neurons.

In addition to their organization along labeled lines, sensory systems perform common types of operations. Foremost among these is the ability of each system to compare events that occur simultaneously at different receptors, a process that serves to bring out the greatest response where the difference in stimulus strength (contrast) is greatest. At late stages in sensory processing, systems make comparisons with past events and with sensations received by other sensory systems. These comparisons are the fundamental bases of perception, recognition, and comprehension.

This chapter gives an overview of the functional attributes and patterns of organization displayed by the auditory, olfactory, somatosensory, gustatory, and visual systems; and it outlines the physiological and anatomical principles common to all sensory systems. When variations on a common theme exist, they are discussed with the goal of bringing the general pattern into sharper focus.

SENSATION AND PERCEPTION

The Function of Each Sensory System Is to Provide the CNS with a Representation of the External World

Because of the changes that occur around an individual, each sensory system has the task of providing a constantly updated representation of the external world. Accomplishing this task is no simple feat because it requires a close interaction between ascending or stimulus-driven mechanisms and descending or goal-directed mechanisms. Together these two mechanisms evoke sensations, give rise to perceptions, and activate stored memories to form the basis of conscious experience. Ascending mechanisms begin with the activity of peripheral receptors, which together form an initial neural representation of the external world. Descending mechanisms work to sort out from the large amount of sensory input those events that require immediate attention. In doing so, the descending mechanisms alter ascending inputs in ways that optimize perception.

Perception of a sensory experience can change even though the input remains the same. A classic example is seen in the image of a vase that can also be perceived as two faces, pointed nose to nose (Fig. 23.1). In this

FIGURE 23.1 An example of a figure that can elicit different perceptions (faces or vase) even though stimulus and sensation remain constant. The mind can "see" purple figures against a blue background or a blue figure against a purple background.

case the image remains the same—the sensory input remains constant—but the perception of what is being viewed changes as the goal of the viewer changes or as his or her attention wanders. Using this example, it is apparent that detection of a stimulus and recognition that an event has occurred usually are called sensation; interpretation and appreciation of that event constitute perception.

Psychophysics Is the Quantitative Study of Sensory Performance

A psychophysical experiment determines the quantitative relationship between a stimulus and a sensation in order to establish the limits of sensory performance (Stevens, 1957). Such an experiment relies on reports from a subject who is asked to judge quantitatively the presence or magnitude of a stimulus as careful adjustments in the physical attributes are made. One example of threshold detection is the *two-point limen*, in which two blunt probes, separated by a distance that is progressively enlarged or reduced over a series of trials, are applied to the skin surface. The minimum separation distance at which a subject reports two stimuli half the time and one stimulus the other half is taken as the detection threshold. That distance can be measured accurately and is found to vary markedly across the body surface; the two-point limen is smallest for the fingertips and largest for the

skin of the back. Other studies, such as those exploring the detection of relative magnitudes of stimuli, can include assessments of object heaviness, loudness of sound, or brightness of light. Studies of this sort have been combined with neurophysiological experiments to compare reports from subjects (sensory behavior) with the responses of single cells (neuronal physiology). Through this procedure the neural mechanisms underlying sensory perception can be examined.

Some general principles hold for all sensation measured in psychophysical experiments. As pointed out in the preceding paragraph, one principle is that of threshold for detecting a difference between stimuli. Studies look to determine a difference threshold by asking what a just noticeable difference (JND) between two stimuli (two lights of different brightness) is that an observer can detect. E. Weber was the first to formally recognize that small differences between two minimal stimuli are easier to detect than small differences between two robust stimuli. One example is the ability to detect a difference between two light objects that weigh 0.1 and 0.2 kilograms versus a difference between two heavy objects that weigh 10.1 and 10.2 kilograms. The former is much easier than the latter. Weber's law states that the difference threshold for a stimulus is a constant fraction of intensity (the Weber fraction). The formula, $\Delta I/I = k$, describes that law, where I is the intensity of a baseline stimulus, ΔI is the JND between baseline and a second stimulus, and k is the Weber fraction. It is important to recognize that k may be a constant of a particular value for one feature, such as the frequency of sound, but the value changes markedly for another feature of the same sense (sound pressure level). The Weber fraction for sound frequency is exceedingly low (roughly .003) whereas that for sound pressure level is relatively high (0.15). For purposes of comparison, the Weber fraction for luminance is 0.02 and that for concentration of an odorant molecule is 0.10.

G. Fechner proposed that every JND between one stimulus and the next is an equal increment in the magnitude of sensation. That would mean a JND is proportional to a physical variable. His law is formalized in the equation, $S = k/\log I$, where S is the sensory experience in terms of magnitude, I is the physically measured intensity of a stimulus and k is a constant. There is strong intuitive value to this equation as it states the magnitude of a sensory experience is related logarithmically to the physical intensity. Lifting 1 kg and 2 kg produces very different sensory experiences, but lifting 10 kg and 11 kg produces almost the same experience, even though the added weight was equal in the two cases. S. Stevens recognized a century later that rather than a logarithmic relation, perceived sensation and physical intensity were related by a power function, described by the equation $S=kI^P$. Yet the exponent of I could be infinite, depending on the relationship of neural response to stimulus intensity. V. Mountcastle and his colleagues proposed that for mechanosensation the relationship between the physical properties of a stimulus and the response of individual neurons is linear. And most recently, K. Johnson and his colleagues have shown that for the complex percept of roughness perception, a linear relationship exists between subjective experience and neural activity. Thus a basic law of psychophysics emerges: how an observer perceives a stimulus is a linear function of the intensity of that stimulus.

RECEPTORS

Receptors Are Specific for a Narrow Range of Input

Neurons of the brain and spinal cord do not respond when they are touched or when they are exposed to sound or light or odors. Each form of energy must be transduced by a population of specialized cells, which converts the stimulus into a signal that all neurons understand. In every sensory system, cells that perform this transduction step are called *receptors* (Fig. 23.2). For each of the fundamental types of stimuli (mechanical, chemical, or thermal energy or light) there is a separate population of receptors selective for the particular form of energy. Even within a single sensory system, there are classes of receptors that are particularly sensitive to one stimulus (e.g., heat or cold) and not another (muscle stretch). This specificity in the receptor response is a direct function of differences in receptor structure and chemistry.

Receptor Types Vary Across Sensory Systems

Systems differ in the number of distinct receptor types they incorporate, and a correlation exists between the number of receptor types displayed by a system and the types of stimuli that system is able to detect. In the somatosensory system, a large number of receptor types exist to detect many types of stimuli. Separate receptors exist to transduce a variety of mechanical stimuli, including steady indentation of hairless skin, deformation of hair, vibration, increased or decreased skin temperature, tissue destruction, and stretch of muscles or tendons (Fig. 23.2). In the auditory system, two classes of receptor—the inner and outer hair cells of the cochlea—transduce mechanical energy of the

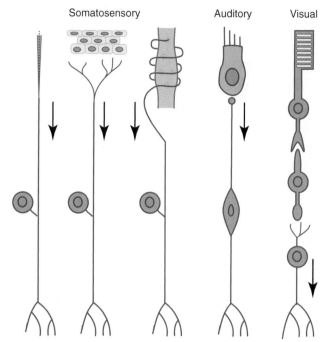

FIGURE 23.2 Receptor morphology and relationship to ganglion cells in the somatosensory, auditory, and visual systems. Receptors are specialized structures that adopt different shapes depending on their function. In the somatosensory system the receptor is a specialized peripheral element that is associated with the peripheral process of a sensory neuron. In the auditory and visual systems, a distinct type of receptor cell is present. In the auditory system, the receptor (hair cell) synapses directly on the ganglion cell, whereas in the visual system, an interneuron receives synapses from the photoreceptor and in turn synapses on the retinal ganglion cell. Adapted from Bodian (1967).

basilar membrane, which is set in motion by sound waves (Dallos, 1996). Here, the motility of outer hair cells provides an additional amplification of the basilar membrane motion to increase sensitivity and allow sharp tuning to sound frequency. The inner hair cells respond to the amplified vibrations and excite the large population of neurons upon which they synapse. Thus the two types of receptors act in concert to transduce a single type of stimulus. In the visual system, transduction is performed by two broad classes of receptor in the retina: rods and cones. The number of cone types varies from one in some species to two in many species to three in a few species; in general cones are tuned to ranges of wavelengths of light. Rods are more sensitive to light and enable vision when light levels are dim. Olfactory receptors also vary in number from a few hundred in primates to more than a thousand in rodents. Here the difference between one olfactory receptor neuron (ORN) and the next is a subtle variation on a common theme, as each ORN differs from its neighbor in the primary sequence of a single receptor protein. As that protein varies so does the ORN's sensitivity to odorant molecules.

Receptors Perform a Common Function in Unique Fashion

All receptors transduce the energy to which they are sensitive into a change in membrane voltage. The task of the receptor is to transmit that voltage change by one route or another to a class of neurons—usually referred to as ganglion cells—that send their axons into the brain or spinal cord (Fig. 23.2). Systems vary in the mechanism whereby receptors and ganglion cells interact. Most receptors in the somatosensory system are part of multicellular organs, the neural components of which are the terminal specializations of dorsal root ganglion cell axons. An appropriate stimulus applied to a somatosensory receptor produces a generator potential—a graded change in membrane voltage (Katz, 1950)—that, when large enough, leads to action potentials that can be carried over a considerable distance into the central nervous system (CNS). The same approach is used by the olfactory system, as ORNs not only transduce the stimulus of an odorant molecule into a change in membrane potential but also conduct those action potentials into the CNS.

Receptors of the auditory, visual, and gustatory systems are separate, specialized cells that transduce a stimulus and then transmit the resulting signal to the nearby process of a neuron. Because the distances between receptor and target neuron are short, auditory hair cells, photoreceptors, and taste receptors do not generate action potentials but signal their response by a passive flow of current. These systems differ, however, in the path between receptor and ganglion cell. In the cochlea, auditory receptors form chemical synapses directly with the processes of ganglion cells so that the response properties of inner hair cells are conveyed directly to the ganglion cells on which they synapse. A similar arrangement is seen for taste receptors (which are epithelial cells and not neurons) and the axons of ganglion cells from cranial nerves VII, IX, and X. Taste receptors synapse directly onto ganglion cell axons. For these systems, the synapse between a receptor and a ganglion cell is little more than the conversion of an analog signal (graded changes in membrane potential) into a digital signal (action potentials). That is not the case in the retina, where photoreceptors relay their response through populations of interneurons interposed between them and retinal ganglion cells (Dowling, 1987) (Fig. 23.2). Because of this additional synapse and the opportunity it affords for summation and comparison of receptor signals, the retinal ganglion cell response differs appreciably from that of photoreceptors.

The mechanisms whereby receptors transduce and transmit signals are known in greater or lesser detail for each system. Visual transduction is a well-understood, rapid process in which a weak signal (a single photon) can be amplified greatly through a biochemical cascade, leading to the closure of thousands of Na^+ channels and a hyperpolarizing response (Yau and Baylor, 1989). For auditory hair cells and somatosensory mechanoreceptors, the mechanical deformation of a part of the cell is transduced into a change in membrane voltage (Hudspeth, 1985). The response of mechanoreceptors is similar to that of photoreceptors in being of one sign only, but it is a sign opposite to that of photoreceptors, as an appropriate tactile stimulus leads to the opening of Na^+ channels and a depolarizing response. This requirement for depolarization may result from the demands placed on the somatosensory ganglion cell to generate action potentials and transmit information over long distances. In contrast, auditory receptors and those of the vestibular system can generate a biphasic response. When protruding villi of a hair cell, called stereocilia, are deflected in one direction, transducer channels open and the cell is depolarized. Yet with deflection of stereocilia in the opposite direction, the same channels close and the cell is hyperpolarized, although to a lesser extent (Hudspeth, 1985). Because sound usually produces a back-and-forth deflection of stereocilia, the result is a back-and-forth movement of the receptor potential—at least for low and moderate frequencies of sound. Thus, the receptor output contains temporal information about the waveform of an acoustic stimulus.

Receptors Have Characteristic Patterns of Position and Density

Receptors are not scattered randomly across the sensory surface. An orderly arrangement of receptors exists along the skin, basilar membrane, retina, olfactory epithelium, and the lining of the tongue and throat. In the retina, for example, photoreceptors adopt a hexagonal packing array (Wassle and Boycott, 1991) in the region of highest density, called the fovea. Moreover, only cones are found in the fovea and for that of humans and other Old World primates, only red and green cones are found in very center of the fovea. Hair cells of the vestibular system are even more tightly sequestered, as they occupy very small regions in the semicircular canals and the otolith organs. In the skin, the arrangement of receptors is not nearly so orderly but the density of cutaneous receptors varies markedly across the skin surface. By far the greatest density of receptor terminals is found at the fingertips and the mouth, whereas receptors along the surface of the back are at least an order of magnitude less frequent. In each system, the differences in peripheral innervation density are tightly correlated with spatial acuity. Regions of highest receptor density are also the regions of highest acuity in vision (fovea) and somatic sensation (fingertip).

A perfect test case for innervation density and acuity is seen in the auditory system of microbats. Because these animals use echolocation to navigate and find prey the auditory system greatly overrepresents the frequencies of sound a bat emits as a probing signal and the surrounding frequencies of Doppler-shifted sounds that echo from objects. Throughout most of the cochlea, the physical properties of the basilar membrane change in a steady fashion so that cochlear hair cells display a progressive shift in the frequency that excites them best. But at the frequencies represented in the Doppler-shifted echo, both the amount of basilar membrane and the density of inner hair cells along that region increase markedly. The result is a much greater acuity for those information-rich frequencies than for all other frequencies.

Receptors Are the Sites of Convergence and Divergence

The relationship between receptor and ganglion cell is seldom exclusive. Most commonly, a single ganglion cell receives input from several receptors and, in many cases, a single receptor sends information to two or more ganglion cells. *Convergence* and *divergence* go hand in hand for the somatosensory system as an individual receptor often is innervated by axons of several ganglion cells while the axon of a single ganglion cell can branch to end as part of several receptor organs. In the somatosensory system, however, the amount of divergence and convergence varies with the class of receptor involved (e.g., thermal receptor vs. mechanoreceptor) and the location of the receptor on the body surface (e.g., shoulder vs. fingertip). Similar features are seen in the visual system, as divergence and convergence dominate different parts of the retina populated by different receptor types. In the cone-rich central retina, each cone provides as many as five ganglion cells with their main visual drive, whereas in the rod-rich periphery, a few dozen rods supply each ganglion cell with its visual input. In its precision and in its implications for sensory processing, nothing approaches the divergence seen in the cochlea, where a single inner hair cell can be the source of all input received by at least 20 ganglion cells (humans) or as

many as 35 ganglion cells (gerbils). Thus, what emerges from a comparison across systems is that convergence and divergence from receptor to ganglion cell vary directly with the demands placed on the system at the specific location. When spatial resolution is a requirement, the convergence of receptor inputs onto individual ganglion cells is low. When detection of weak signals is necessary, convergence is high. When receptor input is used for a complex function or for multiple functions, divergence of input from a single receptor onto many ganglion cells occurs.

Receptors Vary in Their Embryonic Origin

For auditory, vestibular, somatosensory, and olfactory systems, the various classes of receptors and ganglion cells are part of the peripheral nervous system, generated as progeny of neuroblasts located in neural crests and sensory placodes. That is not the case for photoreceptors and retinal ganglion cells. The retina is generated as a protrusion of the embryonic diencephalon and thus all its neurons and supporting cells are CNS derivatives of neural tube origin. As a result of their origin, receptors and ganglion cells of the auditory, vestibular, and somatosensory systems and the ORNs of the nasal epithelium are supported by classes of nonneuronal cells that include modified epithelial supporting cells and Schwann cells. Photoreceptors and retinal ganglion cells, by contrast, are supported by CNS neuroglial cells. Most dramatic of all the consequences resulting from this difference in origin is the ability of axons in somatosensory peripheral nerves to regenerate and reinnervate targets after they are damaged, as opposed to the complete and permanent loss of visual function when optic nerves are cut or crushed.

For all systems except the olfactory, the receptor neurons you were born with are the ones you will live it. Nothing new is added. ORNs, however, have short lives, as they die off and are replaced every six weeks or so. What is seen for all systems, however, is the progressive decline in the number of receptors and in sensory acuity with normal aging. Somatosensory mechanoreceptors in humans are reduced by more than half from the age of 25 years to 65 years. Degeneration of photoreceptors is common as is a progressive reduction in hair cells of both the cochlea and the vestibular organs. Even the ORNs fail to keep up with the ravages of age, as the rate of generation does not match the rate of degeneration. The result in every case is a marked reduction in acuity that can lessen the hedonic value of foods and flowering plants, render a person deaf or blind, and leave him unsteady while standing or walking.

PERIPHERAL ORGANIZATION AND PROCESSING

Sensory Information Is Transmitted Along Labeled Lines

A long-appreciated principle that unites structure and function in a sensory system is the doctrine of specific energy, or the *labeled line* principle. This principle states that when a particular population of neurons is active, the conscious perception is of a specific stimulus (Fig. 23.3). For example, in one particular population of somatosensory neurons (colored orange on Fig. 23.3), activity is always interpreted by the CNS as a painful stimulus, no matter whether the stimulus

is natural (a sharp instrument jabbed into the skin) or artificial (electrical stimulation of the appropriate axons). An entirely separate population of neurons (colored blue on Fig. 23.3) would signal light pressure. Why this is so can be seen from the fact that receptors are selective not only in what drives them, but also in the postsynaptic targets with which they communicate. Each ganglion cell transmits its activity into a well-defined region of the CNS, after which a strictly organized series of synaptic connections relays information in a sequence that eventually leads to the thalamus and then to the cerebral cortex (Darian-Smith *et al.*, 1996). It is this orderly relay from receptor to ganglion cell to central neurons at each of several stations that makes up a labeled line. All sensory informa-

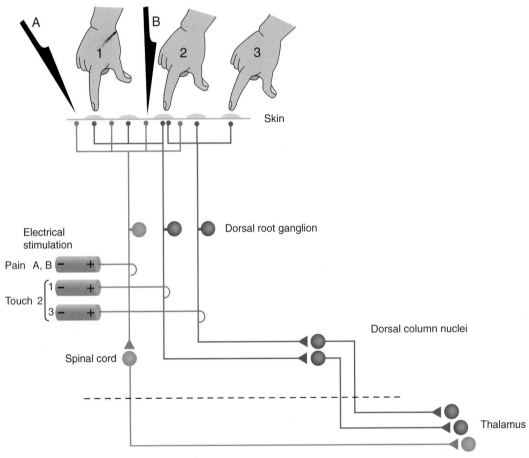

FIGURE 23.3   Example of labeled lines in the somatosensory system. Two dorsal root ganglion (DRG) cells (blue) send peripheral axons to be part of a touch receptor, whereas a third cell (red) is a pain receptor. By activating the neurons of touch receptors, direct touching of the skin or electrical stimulation of an appropriate axon produces the sensation of light touch at a defined location. The small receptive fields of touch receptors in body areas such as the fingertips permit distinguishing the point at which the body is touched (e.g., position 1 vs. position 2). In addition, convergence of two DRG axons onto a single touch receptor on the skin permits touch stimulus 2 to be localized precisely. Electrical stimulation of both axons produces the same sensation, although localized to somewhat different places in the skin. Sharp stimuli (A, B) applied to nearby skin regions selectively activate the third ganglion cell, eliciting the sensation of pain. Electrical stimulation of that ganglion cell or of any cell along that pathway also produces a sensation of pain along that region of skin. Stimulus A and B, however, cannot be localized separately with the pain receptor circuit that is drawn. As the labeled lines project centrally, they cross the midline (decussate) and project to separate centers in the thalamus.

tion arising from a single class of receptors is referred to as a *modality* (e.g., the sensations of pain and light pressure involve distinct modalities). Thus, the existence of labeled lines means that neurons in sensory systems carry specific modalities.

Topographic Projections Dominate the Anatomy and Physiology of Sensory Systems

Receptors in the retina and body surface are organized as two-dimensional sheets, and those of the cochlea form a one-dimensional line along the basilar membrane. Receptors in these organs communicate with ganglion cells and those ganglion cells with central neurons in a strictly ordered fashion, such that relationships with neighbors are maintained throughout. This type of pattern, in which neurons positioned side by side in one region communicate with neurons so positioned in the next region, is called a *topographic pattern*. As an example, the two touch-sensitive neurons in Figure 23.3 innervate somewhat different positions in the skin. Thus, light touch at position 3 will activate the right-most ganglion cell in the dorsal root ganglion, whereas touch at position 1 will activate the neighboring ganglion cell. The central projections of these cells are kept separate and activate different targets in the thalamus and above. The end result is a map of the sensory surface of the skin. A different topography is seen in the olfactory system. ORNs divide the nasal epithelium into zones in which sensitivity to a particular range of odorants is maximally represented. A large-scale map has the ORNs is neighboring zones of the epithelium communicate with neighboring regions in the olfactory bulb. Yet the small-scale map from epithelium to bulb displays a fundamental rearrangement in which all ORNs expressing a particular receptor protein send convergent inputs to a single synaptic region (a glomerulus) in the bulb. In general topographies are a place code for sensory information in which the location of a particular neuron tells you what that neuron responds to, both in place along the sensory surface and in modality.

Neural Signaling Is by a Combination of Rate and Temporal Codes

A great deal of research has been aimed at determining the codes by which neurons signal the presence and the intensity of a stimulus. In addition to place codes, neurons can signal information in the rate at which they respond and in the temporal pattern of their response. For a given receptor, the firing rate or frequency of action potentials signals the strength of

the sensory input. The perceived intensity arises from an interaction between this firing rate and the number of neurons activated by a stimulus. Together the number of neurons active with any sensory stimulus and the level of their activity gives rise to an intensity code. This is the kind of code used by retinal ganglion cells to signal the intensity (luminance) of light and by spiral ganglion cells to signal the intensity (sound pressure level) of sound. Temporal codes are also used in some systems. For instance, the phase-locking ability of auditory neurons extends to sound frequencies up to several thousand cycles per second (kHz), and this code is used (largely by accident) for the perception of the pitch of sounds. In addition, all sensory systems must deal with the fact that stimuli can move, as with vibratory stimuli on the skin. This temporal information in a stimulus is carried by the time-varying pattern of activity in small groups of receptors and central neurons.

Lateral Mechanisms Enhance Sensitivity to Contrast

A hallmark of all sensory systems is the ability of neurons at even the earliest stages of central processing to integrate the activity of more than one receptor. The most common and easily understood of these mechanisms is *lateral* (or surround) *inhibition* (Fig. 23.4). By this mechanism, a sensory neuron displays a receptive field with an excitatory center and an inhibitory surround (Kuffler, 1953). Such a mechanism serves to enhance contrast: each neuron responds optimally to a stimulus that occupies most of its center but little of its surround. In some cases the comparisons involve receptors of different types so that the center and surround differ not only in sign (excitation vs. inhibition, ON vs. OFF), but also in the stimulus quality to which they respond. One such example occurs in visual neurons that possess centers and surrounds responsive to stimulation of different types of cones. These cells display a combination of spatial contrast (the difference in the location of cones that produce center and surround) and chromatic contrast (the difference in the visible wavelengths to which these cones respond best).

Similar types of responses are evident in the somatosensory system, where the difference between center and surround is the location on the skin from which each is activated. In this case, skin mechanics produce receptive fields with a central hot spot of activity and a surrounding inactive zone. In the auditory system, lateral suppressive areas (Fig. 23.4) also are produced by mechanics, in this case the mechanics of the basilar membrane of the cochlea. These two-tone suppression

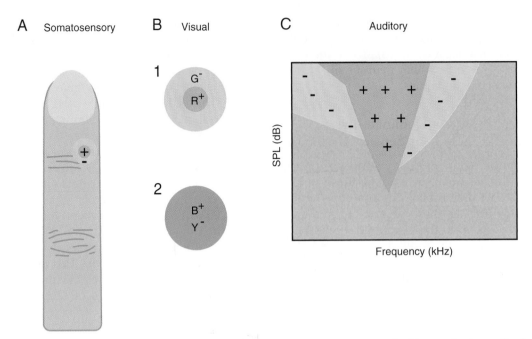

FIGURE 23.4 Center/surround organization of receptive fields is common in sensory systems. In this organization, a stimulus in the center of the receptive field produces one effect, usually excitation, whereas a stimulus in the surround area has the opposite effect, usually inhibition. (A) In the somatosensory system, receptive fields display antagonistic centers and surrounds because of skin mechanics. (B) In the retina and visual thalamus, a common type of receptive field is antagonistic for location and for wavelength. Receptive field 1 is excited by turning on red light (R) at its center and is inhibited by turning on green light (G) in its surround. Receptive field 2 is less common and is antagonistic for wavelength (blue vs. yellow) without being antagonistic for the location of the stimuli. Both are generated by neural processing in the retina. (C) In the auditory system, primary neurons are excited by single tones. The outline of this excitatory area is known as the tuning curve. When the neuron is excited by a tone in this area, the introduction of a second tone in flanking areas usually diminishes the response. This "two-tone suppression" is also generated mechanically, as is seen in motion of the basilar membrane of the cochlea. All these center/surround organizations serve to sharpen responses over that which would be achieved by excitation alone.

areas are demonstrated by exciting the auditory nerve fiber with a tone in the central excitatory area and observing the response decrease caused by a second tone in flanking areas (Sachs and Kiang, 1968). In all these sensory systems, center/surround organization serves to sharpen the selectivity of a neuron either for the position of the stimulus or for its exact quality by subtracting responses to stimuli of a general or diffuse nature.

CENTRAL PATHWAYS AND PROCESSING

Axons in Each System Cross the Midline on Their Way to the Thalamus

Axons of ganglion cells entering the CNS form the initial stage in a pathway through the thalamus to the cerebral cortex (Fig. 23.5). Axons in visual, somatosensory, and auditory systems cross the midline—they *decussate*—prior to reaching the thalamus, but those of the olfactory and gustatory systems do not. A single, incomplete decussation occurs in the visual system of primates and carnivores, where slightly more than half the axons of the optic nerve cross the midline at the optic chiasm. Near complete decussations at the optic chiasm are seen in animals with laterally placed eyes. Decussation in the somatosensory system is nearly total, as all but a small group of axons cross the midline in the spinal cord or brain stem. These decussations serve two broad functions. They bring together into one hemisphere all axons carrying information from half the visual world or they bring somatosensory information into alignment with visual input and motor output. In contrast, multiple decussations occur in the auditory system prior to the thalamus, as comparison of input from the two ears is the dominant requirement of sound source localization. Nevertheless, at high levels in the auditory system, one side of the brain is concerned mainly with processing information about sound sources located toward the opposite side of the body, as demonstrated by lesion/behavioral studies of sound localization.

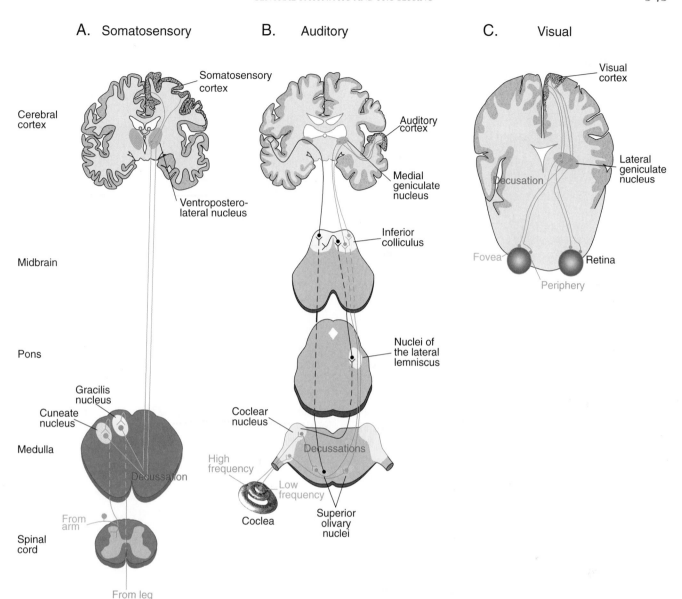

A. Somatosensory

Cerebral cortex

Somatosensory cortex

Ventroposterolateral nucleus

Midbrain

Pons

Gracilis nucleus

Cuneate nucleus

Medulla

Decussation

From arm

Spinal cord

From leg

B. Auditory

Auditory cortex

Medial geniculate nucleus

Inferior colliculus

Nuclei of the lateral lemniscus

Coclear nucleus

Decussations

High frequency

Low frequency

Coclea

Superior olivary nuclei

C. Visual

Visual cortex

Lateral geniculate nucleus

Decusation

Fovea

Periphery

Retina

FIGURE 23.5 Comparison of central pathways of sensory systems. In every case, soon after peripheral input arrives in the brain, decussations result in one hemifield being represented primarily by the brain on the opposite side. Each pathway has a unique nucleus in the thalamus and several unique fields in the cerebral cortex. Within each of these areas, the organized mapping that is established by receptors in the periphery is preserved.

Specific Thalamic Nuclei Exist for Each Sensory System

Information from all sensory systems except the olfactory are relayed through the thalamus on its way to the cerebral cortex (Fig. 23.5). Olfactory information reaches primary olfactory cortex without a relay in thalamus. Yet even in this sense, perception of odorants and discrimination of one odorant from another occurs only after a thalamic relay. This relay of sensory input through the thalamus involves either a single large nucleus or, in the case of the somatosensory

systems, two nuclei: one for the body and one for the face. In each nucleus, synaptic circuits are said to be secure because activity in presynaptic axons usually leads to a postsynaptic response. Within the thalamic nuclei, neurons performing one function (e.g., relay of discriminative touch) are segregated from those performing another (e.g., relay of pain and temperature). Even within one function, mappings of neurons are preserved so that there is separation of neurons providing for touch information from the arm vs. from the leg and of neurons responding to low vs. high sound frequencies (Fig. 23.5). Usually, for each thalamic nucleus, there is a population of large neurons and one or two populations of small neurons (Jones, 1981). In each case the larger neurons carry the most rapidly transmitted signals from the periphery to the cortex.

Multiple Maps and Parallel Pathways

Nuclei in the central pathways often contain multiple maps. For instance, in the auditory system, axons of spiral ganglion cells divide into branches as they enter the CNS and terminate in three subdivisions of the cochlear nucleus. Each division contains its own map of sound frequency (tonotopic map) that was originally established by the cochlea. The greatest number of maps generated from ganglion cell input is found in the visual system of primates, in which as many as six separate retinotopic maps are stacked on top of one another in the lateral geniculate nucleus (LGN, in the thalamus), which receives direct input from the retina (Kaas *et al.*, 1972). Such a large number of distinct maps in the LGN is indicative of inputs from ganglion cells that vary in location, structure, and function. For vision, one idea is that surface features such as color and form are carried along a path separate from the one that handles three-dimensional features of motion and stereopsis. The functional significance of multiple maps in general, however,

remains to be clarified. Perhaps the need for multiple parallel paths exists because of the relatively slow speed and the limited capacity of single neurons. So rather than have the same group of neurons perform different functions in serial order, each of several parallel groups performs a separate function. This leads eventually to the problem of binding together all features of a stimulus into a coherent percept, the neural basis for which may be the synchronized activity of neurons across several areas of the cerebral cortex (Singer, 1995).

SENSORY CORTEX

Sensory Cortex Includes Primary and Association Areas

Axons of sensory relay nuclei of the thalamus project to a single area or a collection of neighboring areas of the cerebral cortex, thereby providing them with a precise topographic map of the sensory periphery. These parts of cortex are frequently referred to as *primary sensory areas* (Fig. 23.6). Neighboring areas with which the primary areas communicate directly or by a single intervening relay area are sensory *association areas*.

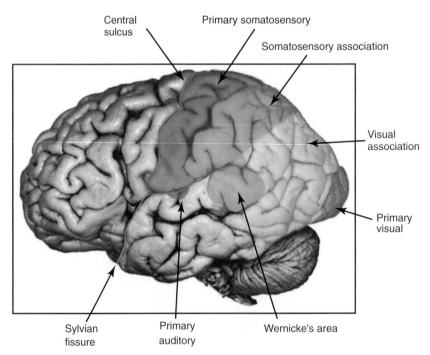

FIGURE 23.6   The location of primary sensory and association areas of the human cerebral cortex. The primary auditory cortex is mostly hidden from view within the Sylvian fissure. From Guyton (1987).

Response Mappings and Plasticity

Each area of sensory cortex shares with its subcortical components a map of at least part of the sensory periphery. Thus, retinotopic, somatotopic, and tonotopic maps are evident in the relevant areas of cortex. The retina and skin are two-dimensional sheets so the map of the sensory periphery on the surface of the cortex is a simple transformation of the peripheral representation onto the cortex. In the auditory periphery, there is a one-dimensional mapping of frequency. This tonotopic mapping is represented faithfully along one dimension of cortical distance, and the orthogonal direction may map a second, as yet undiscovered, property.

As previously indicated, the distribution of receptors is uneven for most systems. Thus, the fovea of the primate retina and the fingertips of the primate hand are regions that possess a high density of receptors with small receptive fields. Such an uneven distribution of neurons devoted to a structure is further amplified in the CNS. A much greater percentage of neural machinery subserves the representation of the retinal fovea or the fingertips than deals with the representation of other regions of the retina or body surface (Fig. 23.7). This expansion of a representation in the CNS, referred to as a *magnification factor*, appears particularly impressive in humans, in whom a very large part of primary visual cortex is devoted to the couple of millimeters of retina in and around the fovea. In the auditory cortex, such magnification of frequency representation is again seen in echolating bats, as the region devoted to the Doppler-shifted signaling frequency is much enlarged over that for any other set of frequencies.

It is now clear that mappings of sensory cortex are not fixed and immutable but rather plastic. In the somatosensory system, if input from a restricted area of the body surface is removed by severing a nerve or by amputation of a digit, that portion of the cortex that was previously responsive to that region of the body surface becomes responsive to neighboring regions (Merzenich *et al.*, 1984). In the auditory system, following high-frequency hearing loss, the portion of cortex previously responsive to high frequencies becomes responsive to middle frequencies. Frequency tuning of neurons or auditory cortex can also be shifted with classical conditioning methods (Weinberger, 2007). Such plasticity of cortical maps requires some time to be established and could result from strengthening of already established lateral connections or from growth of new connections. It is likely but not firmly established that the same types of mechanisms cause cortical changes during the processes of learning and memory.

A Common Structure Exists for Sensory Cortex

Neurons in areas of the sensory cortex (and most other areas of the cerebral cortex) are organized into six layers. The middle layers (III and IV) are the main site of termination of axons from the thalamus (Fig. 23.8). In the primary sensory cortices, these middle layers are enlarged and contain many small neurons. Because the small cells resemble grains of sand in standard histological preparations, the sensory areas are themselves referred to as granular areas of cortex.

Columnar Organization

Properties other than place in the periphery are mapped in primary sensory areas of the cortex. The third spatial dimension of the cortex, that of depth, arranges neurons in adjacent 0.5- to 1-mm-wide regions, referred to as *columns* (Mountcastle, 1997). In these columns, neurons stacked above and below one another are fundamentally similar but differ significantly from neurons on either side of them. One example of columns with a clear anatomical correlate is the division of the primary visual cortex of most primates and some carnivores into a series of alternating regions dominated by the right and left retinas. Each ocular dominance column contains cells driven exclusively or predominantly by one eye; adjacent columns are dominated by the other eye. Other properties, such as selectivity for the orientation of a visual

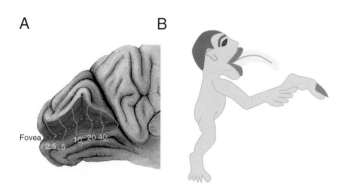

A   **B**

Fovea

FIGURE 23.7 Examples of sensory magnification in the visual and somatosensory systems. (A) Determination of a visual field map in the human primary visual cortex shows that more than half this area is devoted to the central 10° of the visual field. Very little is devoted to the visual periphery beyond 40°. From Horton and Hoyt (1991). (B) Figure of how the human body would appear if the body surface were a perfect reflection of the map in the first somatosensory cortex. The mouth and tongue and the tip of the index finger enjoy a greatly enlarged representation in the thalamus and cortex.

Cortical
layer

I

II

III

IV

V

VI

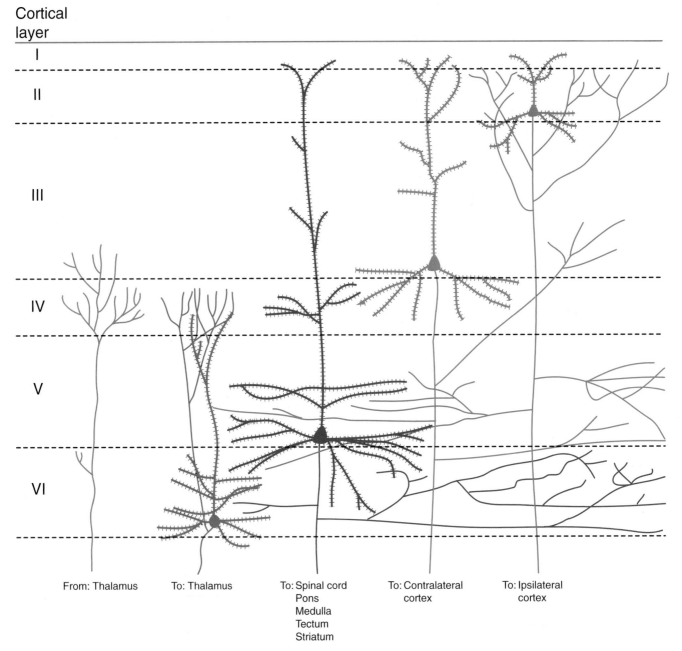

From: Thalamus   To: Thalamus   To: Spinal cord   To: Contralateral   To: Ipsilateral
                                                    Pons   cortex   cortex
                                                    Medulla
                                                    Tectum
                                                    Striatum

FIGURE 23.8   Cellular organization of the sensory cortex into six layers (I–VI). Inputs from the thalamus terminate mainly in layers III and IV. The main output neurons of the cortex are pyramidal cells, which are distributed in different layers according to their projections. Descending projections are to the thalamus (neurons in layer VI) or to the spinal cord, pons and medulla, tectum, or striatum (neurons at various levels in layer V). Ascending projections to other "higher" cortical centers are often from neurons located above layer IV; some of these projections are to the same hemisphere (layer II) or to the opposite hemisphere (layer III). Adapted from Jones (1985).

stimulus and the contrast between it and the surround, are also arranged in columns of primary visual cortex (Hubel, 1988). Similar types of columns are evident in the somatosensory system, as regions of modality and place specificity. They are most clearly seen in the representation of mystacial vibrissae in rodent somato-

sensory cortex, where a one-to-one matching of vibrissa with cortical territory gives rise to structures called barrels.

In the auditory system, neurons within a column generally share the same best frequency and the same type of binaural interaction characteristic: either one ear excites the neurons and the other inhibits or suppresses the response to the first ear (suppression column) or one ear excites and the other ear also excites or facilitates the response to the first ear (summation column). Moreover, in areas of nonprimary cortex, the feature displayed most commonly by neurons of a particular area is one that often comes to occupy columns. A good example is found in the middle temporal area (MT) of visual association cortex, where neurons are tuned for the direction of a moving visual stimulus. Neurons selective for one particular direction of visual stimulus movement are organized into columns through the depth of MT; these are flanked by columns of neurons tuned for other directions of movement (Albright *et al.*, 1984). So consistent are these findings among sensory, motor, and association areas that columnar organization is viewed as a principal organizing feature for all of the cerebral cortex (Mountcastle, 1997).

Stereotyped Connections Exist for Areas of Sensory Cortex

Neurons of the cerebral cortex send axons to subcortical regions throughout the neuraxis and to other areas of the cortex (Fig. 23.8). Subcortical projections are to those nuclei in the thalamus and brain stem that provide ascending sensory information. By far the most prominent of these is to the thalamus: the neurons of a primary sensory cortex project back to the same thalamic nucleus that provides input to the cortex. This system of descending connections is truly impressive, as the number of descending corticothalamic axons greatly exceeds the number of ascending thalamocortical axons. These connections permit a particular sensory cortex to control the activity of the very neurons that relay information to it. One role for descending control of thalamic and brain stem centers is likely to be the focusing of activity so that relay neurons most activated by a sensory stimulus are more strongly driven and those in surrounding less well activated regions are further suppressed.

The overwhelming majority of cortical neurons project to other areas of cortex (Fig. 23.8). Corticocortical projections link primary and association areas of the sensory cortex and establish parallel paths so that different aspects of vision, audition, and somatic sensation come to be handled by different areas of cortex. These connections establish a hierarchy within a system, such that "ascending" or "forward" connections begin with neurons from superficial cortical layers (I–III) and end with axonal terminations mainly in layers III and IV of higher cortical regions. Similarly, the ascending projection from the thalamus terminates mainly in these layers in primary sensory cortices. Corresponding descending projections from higher to lower cortical regions begin in the deep or superficial cortical layers and project to layers outside of III and IV. In addition to projections to the ipsilateral hemisphere of cortex, there are also projections to the contralateral hemisphere via the corpus callosum and other commissures. In visual and somatosensory systems, these commissural connections are restricted in origin and termination; they exist to unite the representation of midline structures into a coherent percept (Hubel, 1988).

Response Complexity of Cortical Neurons

Responses of cortical neurons in primary sensory cortices are more complex than those seen for neurons in the periphery. One example is seen in the primary visual area of the cerebral cortex, where neurons are responsive to stimuli that are not concentric circles (center and surround) but elongated lines possessing a specific orientation. Comparable synthesis of simpler inputs to reconstruct more complex features of stimulus is apparent at higher levels in the visual system (Logothetis and Sheinberg, 1996) and in the somatosensory and auditory areas of the cerebral cortex.

Physiologically, processing, and selectivity for stimulus features become progressively more complex within the hierarchically organized pathways that connect primary with association areas of the cortex (Gallant and Van Essen, 1994). In the visual system, separate "streams" involved in visuosensory and eventually visuomotor functions have been described; one is responsible for using visual cues to drive appropriate eye movements and the other for dealing with the tasks of visual perception (Gallant and Van Essen, 1994). In the somatosensory system, separate motor and limbic paths exist to perform much the same functions for the entire body, supplying sensory input to coordinate and adjust motor output and using complex input from many receptor types to match the shape of a tactual stimulus with one already stored in memory (Johnson and Hsiao, 1992). In the association pathways of the human auditory system, a specialized area of cortex, Wernicke's area, plays a fundamental role in processing speech and language information and in communicating with Broca's area to form a speech motor response. These streams are not separate, as traditionally viewed "motor" areas such as Broca's are now known to become activated in comprehension tasks. Apparent from this pattern in association areas of the cortex is the continued pressure for a division of labor within each sensory system; not one that produces separate paths for analyzing elemental features of a stimulus but one that combines those features either to elicit appropriate movements or to match a stimulus with an internal representation of the world.

SUMMARY

The functional organization of sensory systems shares common themes of transduction, relay, organized mappings, parallel processing, and central modification. It is no surprise that a case has been made for a common phylogenetic origin of sensory systems. Differences among the systems, however, demonstrate that each has existed and operated independently for as long as there have been vertebrates. What remains in overview is a well-ordered basic plan from periphery to perception that has been modified in its details as variations in niche have led to specializations in function.

References

Albright, T. D., Desimone, R., and Gross, C. G. (1984). Columnar organization of directionally selective cells in visual area MT of the macaque. *J. Neurophysiol.* **51**, 16–31.

Gallant, J. L. and Van Essen, D. C. (1994). Neural mechanisms of form and motion processing in the primate visual system. *Neuron* **13**, 1–10.

Hudspeth, A. J. (1985). The cellular basis of hearing: The biophysics of hair cells. *Science* **230**, 745–752.

Johnson, K. O., Hsiao, S. S., and Yoshioka, T. (2002). Neural coding and the basic law of psychophysics. *Neuroscientist* **6**, 111–121.

Johnson, K. O. and Hsiao, S. S. (1992). Tactual form and texture perception. *Annu. Rev. Neurosci.* **15**, 227–250.

Jones, E. G. (1981). Functional subdivision and synaptic organization of the mammalian thalamus. *Int. Rev. Physiol.* **25**, 173–245.

Kaas, J. H., Guillery, R. W., and Allman, J. M. (1972). Some principles of organization in the dorsal lateral geniculate nucleus. *Brain Behav. Evol.* **6**, 253–299.

Katz, B. (1950). Depolarization of sensory terminals and the initiation of impulses in the muscle spindle. *J. Physiol. (Lond.)* **111**, 261–282.

Kuffler, S. W. (1953). Discharge patterns and functional organization of mammalian retina. *J. Neurophysiol.* **16**, 37–68.

Logothetis, N. K. and Sheinberg, D. L. (1996). Visual object recognition. *Annu. Rev. Neurosci.* **19**, 577–621.

Merzenich, M. M., Nelson, R. J., Stryker, M. P., Cynader, M. S., Schoppmann, A., and Zook, M. (1984). Somatosensory cortical map changes following digit amputation in adult monkeys. *J. Comp. Neurol.* **224**, 591–605.

Sachs, M. B. and Kiang, N. Y. S. (1968). Two-tone inhibition in auditory-nerve fibers. *J. Acoust. Soc. Am.* **43**, 1120–1128.

Singer, W. (1995). Time as coding space in neocortical processing: A hypothesis. *In* "The Cognitive Neurosciences" (M. S. Gazzaniga, ed.), pp. 91–104. MIT Press, Cambridge, MA.

Stevens, S. S. (1957). On the psychophysical law. *Psychol. Rev.* **64**, 153–181.

Wassle, H. and Boycott, B. B. (1991). Functional architecture of the mammalian retina. *Physiol. Rev.* **71**, 447–480.

Weinberger, N. M. (2007). Auditory associative memory and representational plasticity in primary auditory cortex. *Hear. Res.* **70**, 226–251.

Yau, K.-W. and Baylor, D. A. (1989). Cyclic GMP-activated conductance of retinal photoreceptor cells. *Annu. Rev. Neurosci.* **12**, 289–328.

Suggested Readings

Dallos, P. (1996). Overview: Cochlear neurobiology. *In* "The Cochlea" (P. Dallos, A. N. Popper, and R. R. Fay, eds.), pp. 1–43. Springer-Verlag, New York.

Darian-Smith, I., Galea, M. P., Darian-Smith, C., Sugitani, M., Tan, A., and Burman, K. (1996). The anatomy of manual dexterity: The new connectivity of the primate sensorimotor thalamus and cerebral cortex. *Adv. Anat. Cell Biol.* **133**, 1–142.

Dowling, J. E. (1987). "The Retina: An Approachable Part of the Brain." Belknap Press, Cambridge, MA.

Hubel, D. H. (1988). "Eye, Brain and Vision." Freeman, New York.

Mountcastle, V. B. (1997). The columnar organization of the neocortex. *Brain* **120**, 701–722.

Stewart H. Hendry, Steven S. Hsiao, and M. Christian Brown

Chemical Senses: Taste and Olfaction

Some of the most remarkable feats in the animal kingdom are accomplished using chemical detection. For example, the male silkmoth can follow the scent of a female for several miles to reach its mate. A great white shark can detect one part blood in 1 million parts water, or one drop of blood in an Olympic-size swimming pool, and swims toward its injured prey from a quarter mile away. Strychnine, a plant alkaloid from the Strychnos nux vomic plant in Southern Asia and Australia, tastes bitter to humans in minute quantities (10^{-6} M), allowing us to detect and avoid consumption of toxic quantities that produce a convulsive violent death approximately 20 minutes after swallowing.

The number of chemical compounds in our environment is vast, with more than 30 million compounds catalogued. Different combinations of these chemicals are emitted by foods, predators, and mates and act as signatures that animals use to distinguish among them. The chemical senses of olfaction and gustation allow for the detection of an enormous number of chemical cues and translate this information into meaningful behaviors. The identification of large families of chemoreceptors, the ability to monitor cell activity in the periphery and central nervous system and the ability to genetically manipulate chemosensory systems in model organisms are rapidly revolutionizing our understanding of chemosensory systems and beginning to elucidate the brain's mechanisms for encoding flavors and fragrances.

TASTE

The sense of taste is involved primarily in feeding, allowing animals to identify food that is nutrient-rich and avoid toxic substances. Like the other primary senses, the best definition of the gustatory system is that it has specialized sensory cells in the periphery and unique regions in the brain dedicated to sensory processing. The sensory cues detected by the gustatory system are soluble chemicals, limiting detection to a short range by direct contact with a chemical source. The concentration range for taste detection is broad and depends on the nature of the chemical stimulus. At one extreme, taste cells detect sugars and amino acids at very high concentrations (100 millimolar), allowing animals to detect only the most caloric foodstuffs instead of food with little nutritional value. At the other extreme, taste cells can also detect minute amounts of noxious substances or toxins, compounds that are harmful at very low concentrations. Mammals are thought to perceive only five taste modalities: sweet, bitter, sour, salty and umami (the taste of the amino acid glutamate), with several chemicals in most categories.

How the brain translates chemical detection into the perception of different taste modalities and taste behaviors is a basic problem in neural coding. Of course, the gustatory system does not work in isolation and the smell, texture, and sight of food also contributes to flavor perception. For example, without our olfactory neurons, we are unable to recognize coffee, chocolate, or wine! However, it is the gustatory system that acts as the final checkpoint controlling food acceptance or rejection and is essential for the basic recognition of taste modalities. Understanding the neural coding of taste information begins with knowledge about how taste receptor cells detect chemical cues, an area of rapid advances. How taste quality is encoded in higher brain areas is an area of active investigation that will ultimately require unraveling neural circuits and neural activity underlying taste behavior.

Taste Cells Are Situated within Taste Buds Located in Several Distinct Subpopulations

In mammals, taste recognition begins on the tongue, where specialized cells detect chemical cues. Mammalian taste cells are derived from epithelial tissue rather than neural tissue, although they display neural properties such as depolarization and release of neurotransmitters. The apical tip of each receptor cell contains microvilli that project into the mucus of the oral environment. The microvilli are the site of taste detection, where taste ligands interact with membrane-bound receptors or ion channels. The basolateral membrane of taste cells forms chemical synapses with primary gustatory nerve fibers. The nerve fibers enter the base of the taste bud and transmit taste detection to the brain.

Groups of 50 to 100 taste cells are clustered together into onion-shaped organs called taste buds (Fig. 24.1). The apical center of the taste buds is called the taste pore, where the microvilli of taste cells interact with the environment. Groups of taste buds are located pri-

marily within papillae on the tongue although they are distributed on other parts of the oral cavity as well (Fig. 24.2). Three classes of papillae are defined based on their location on the lingual epithelium: fungiform papillae are found on the anterior part of the tongue, foliate papillae are on the posterior sides of the tongue and circumvallate papillae are located at the midline on the posterior tongue. In addition, taste cells are found on the palate and the epiglottis.

Taste buds are innervated by dendrites of neurons that travel in the facial (VIIth), glossopharyngeal (IXth), or vagus (X) nerves to gustatory nuclei in the brain stem. Different regions of the tongue are innervated by different nerves. Fungiform papillae and the foliate papillae on the anterior part of the tongue are innervated by the chorda tympani branch of the facial nerve. Taste buds on the soft palate receive fibers from the greater superficial petrosal (GSP) branch of the facial nerve. Circumvallate and foliate papillae on the posterior tongue contain taste buds innervated by the lingual-tonsillar branch of the glossopharyngeal nerve. Taste buds of the epiglottis and the esophagus are

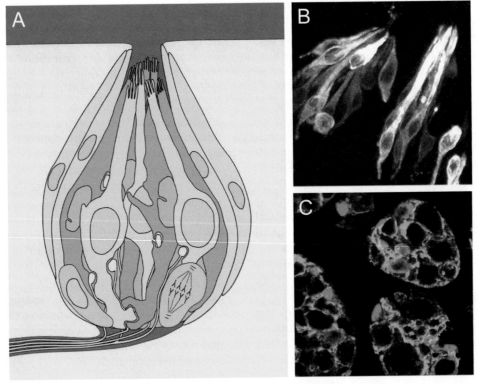

FIGURE 24.1 Cell types in mammalian taste buds. (A) The taste bud is a barrel-shaped structure containing approximately 50–100 taste cells. These epithelial receptor cells make synaptic contact with distal processes of cranial nerves VII, IX, or X, whose cell bodies lie within the cranial nerve ganglia. Microvilli of the taste receptor cells project into an opening in the epithelium, the taste pore, where they make contact with gustatory stimuli. (B) The characteristic spindle shape of taste receptor cells is revealed when a subset of taste cells is immunoreacted to an antibody against α-gustducin, a gustatory G protein. (C) When sectioned transversely, gustducin-positive cells appear round in cross-section, as shown by α-gustducin immunoreactivity (red), whereas other cells are revealed with an antibody against the H blood group antigen (green).

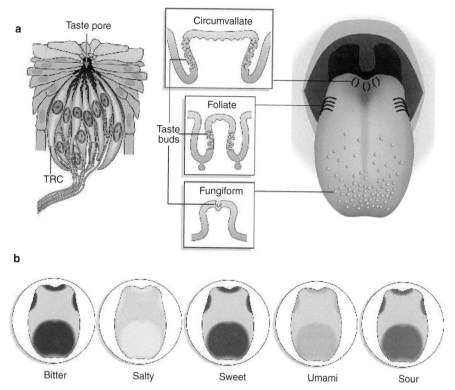

FIGURE 24.2 Diagram of a tongue showing the distribution of various taste bud populations, which are found in the fungiform (F) papillae on the anterior tongue, the vallate (V) and foliate (FO) papillae on the posterior tongue. These taste buds are innervated by branches of the VIIth, IXth, and Xth cranial nerves (see text). From Chandrashekar *et al.*, 2006.

innervated by the internal branch of the superior laryngeal nerve (SLN), which is a branch of the vagus (Xth) nerve. A single nerve fiber may innervate cells in more than one taste bud, and each taste bud is innervated by several different afferent fibers.

Turnover and Replacement of Taste Bud Cells Are Continuous Processes

Taste receptor cells arise continually from an underlying population of basal epithelial cells. In rats, the life span of a taste cell in a fungiform papilla is approximately 10 days. Although the mechanisms of renewal are unknown, processes underlying taste cell survival are becoming clear. In particular, it is well established that the nerves that synapse with taste cells are required for taste cell survival. Severing the chordate tympani nerve (CT) that innervates fungiform papillae results in rapid degeneration of taste papillae and taste buds. Thus, gustatory nerves maintain a trophic influence over taste buds, which degenerate when their nerve supply is removed. Although taste nerves are required for survival, they do not instruct the differentiation of taste cells. The electrophysiological properties of taste cells are not altered when the nerves that innervate

them are cross-wired (for example, by redirecting the chordate tympani to invade circumvallate papillae instead of fungiform papillae). Together, these experiments demonstrate that synaptic connections are necessary for cell survival but not for taste cell development.

Many interesting questions remain to be explored regarding taste cell turnover. For example, how is taste cell turnover regulated? Are all taste cells renewed with equal probability or do representations of different taste cell populations change over time? In other words, does a tongue maintain or change its taste-responsiveness over time? How does a dendrite know which taste cells to contact in the changing milieu of living and dying cells? How is synapse formation and retraction controlled? The renewal of taste cells may provide an ideal system to study tissue regeneration and stem cell differentiation as well as synapse plasticity.

Mechanisms of Taste Detection by Receptor Cells

Recent fundamental discoveries have begun to identify the receptor molecules that detect taste ligands.

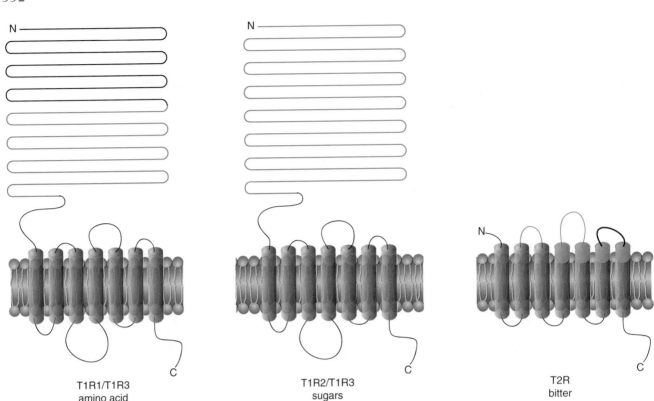

FIGURE 24.3 Two families of taste receptors mediate detection of amino acids, sugars, and bitter compounds. The T1R family consists of three members. T1R1 and T1R2 mediate amino acid detection and T1R2 and T1R3 mediate the recognition of sugars. The T2R family of receptors consists of approximately 30 different receptors that respond to bitter stimuli.

There are two different families of G protein-coupled receptors that mediate the detection of sugars, amino acids and bitter compounds (Fig. 24.3). In addition, candidate ion channels for the recognition of salts and sour compounds have been identified. The diversity of receptor-types used by the gustatory system (different GPCRs, ion channels) suggests that mechanisms of detecting different taste modalities have evolved independently, unlike the olfactory system where one GPCR family mediates the detection of all odors. The total number of receptors used by the gustatory system is far fewer than for the olfactory system: under 50 taste receptors versus approximately 1000 odor receptors. This suggests that the gustatory system recognizes fewer chemical cues, perhaps with a greater diversity of molecular structures, than the olfactory system. The molecular mechanisms of taste detection are described next.

The T1R Receptor Family Mediates the Taste of Sugars and Amino Acids

In mammals, a small family containing three genes mediates the detection of sugars and amino acids. This receptor family was identified by isolation of genes

expressed in taste tissue and by bioinformatics searches for related genes in the genome. The Taste Receptor 1 family is comprised of the T1R1, T1R2, and T1R3 receptors. These receptors are all G protein coupled receptors and contain large extracellular domains similar to the metabotropic glutamate receptors.

What constitutes proof that a protein is a taste receptor? First, the protein should be in the right place at the right time to do the job. Second, loss of the protein should lead to loss of taste detection. Third, addition of the protein into novel cells should cause these cells to respond to taste cues. These criteria have been firmly established for the T1R family of taste receptors and have demonstrated that the combination of T1R2 plus T1R3 detects sugars and the combination of T1R1 plus T1R3 recognizes amino acids.

The following evidence demonstrates that the combination of T1R2 plus T1R3 recognizes sugars. First, as expected for taste receptors, the T1Rs are expressed in taste cells on the tongue. Second, mice engineered to lack either T1R2 or T1R3 do not detect sugars, as determined by electrophysiological recordings of taste nerves (Fig. 24.4) and by behavioral taste tests. Whereas normal mice prefer sugar water to water, mice lacking these receptors do not. Given a choice between water

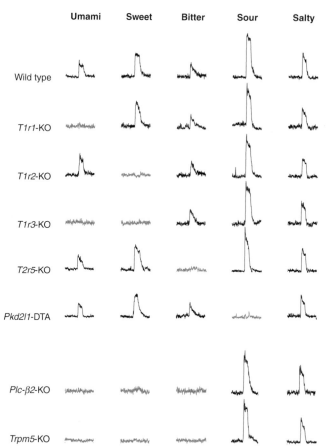

FIGURE 24.4 Mice lacking specific taste receptors have specific taste defects. Shown are nerve recordings showing taste-induced activity to amino acids, sugars, bitter compounds, sour compounds, and salts. Wild-type mice respond to all compounds. Mice lacking T1R1 do not detect amino acids, mice lacking T1R2 do not detect sugars, and mice lacking T2R5 do not detect specific bitter compounds. From Chandrashekar *et al.*, 2006.

and sugar water in a short-term taste preference test, they drink equal amounts of water and sugar water. Third, heterologous cells that coexpress T1R2 and T1R3 respond to sugars, but those that express T1R2 or T1R3 alone do not. This argues that T1R2 and T1R3 act together to detect sugars. Equivalent experiments demonstrate that the combination of T1R1 and T1R3 recognizes amino acids. Thus, T1R1 is selective for amino acids, T1R2 sugars, and T1R3 is an obligate partner for both amino acid and sugar detection.

Mammals detect a wide array of sugars and amino acids only at high concentrations (10–100 mM), allowing them to recognize compounds that provide nutritional value. The three T1R receptors apparently have solved the problem of detection by having loose ligand-binding sites that recognize several ligands with low affinity. Differences in the binding sites of the T1R receptors are found in different species. For example, the taste of umami in humans is the taste of the amino

acid glutamate. The human T1R1 plus T1R3 amino acid receptor binds to glutamate with a 10-fold higher affinity than other amino acids. In contrast, the mouse T1R1 plus T1R3 receptor recognizes all amino acids with similar affinity. Thus, mice have an amino acid taste whereas humans have an umami taste. Although this may sound like humans are short-changed, it may not be the case. Because foods that have one amino acid generally contain many, the enhanced ability to detect glutamate in humans may actually enhance detection of all amino acids. The observation that T1R sequence variation underlies differences in taste detection also occurs for sugars. For example, we find aspartame (Nutrasweet) sweet, but mice and rats do not detect it. Mice have been genetically engineered to express the human T1R2 and T1R3 in taste cells and these mice are now able to detect aspartame. Interestingly, cats lack a functional T1R2 receptor, providing a plausible explanation for why they do not taste sugars. These studies highlight the role of T1Rs in sugar and amino acid taste detection in mammals and demonstrate that sequence differences in T1Rs can account for species-specific taste preferences.

Studies of the ligand-binding properties of the T1Rs argue that all T1Rs contribute to binding. Taking advantage of the observation that rat T1R2/T1R3 and human T1R2/T1R3 recognize some different sweeteners, experiments exchanging extracellular domains of the human and rodent T1R2/T1R3 receptors were used to dissect receptor domains that recognize sweeteners. For example, as mentioned earlier, human T1R2/T1R3 recognizes aspartame and rat T1R2/T1R3 does not. Importantly, a rat receptor in which the extracellular domain of T1R2 has been replaced by the human T1R2 extracellular domain does detect aspartame. This argues that the extracellular domain of human T1R2 detects aspartame. A series of these domain-swapping studies demonstrated that the T1R2 extracellular amino-terminal domain binds several sugars, the T1R2 transmembrane carboxyl-terminal domain binds the G-protein and the T1R3 transmembrane carboxyl-terminal domain binds some artificial sweeteners. Thus, different regions of T1Rs mediate taste detection and both T1R2 and T1R3 participate in ligand recognition. This strongly argues that T1R2 and T1R3 are coreceptors for sugars.

The T2R Receptor Family Mediates Bitter Taste Detection

A common variation in human behavior is the ability to detect the bitter compound phenylthiocarbamide (PTC). In 1931, a scientist at the Dupont chemical company synthesized this compound and quickly

BOX 24.1

IDENTIFYING A GENETIC LOCUS LINKED TO BITTER TASTE PERCEPTION

The ability to taste the compound phenylthiocarbamide (PTC) varies drastically among humans. Approximately 75% of people find PTC intensely bitter whereas 25% do not detect it. The ability to detect PTC was linked to a small region of chromosome 7 in human genetic studies. The taste receptor gene T2R38 is located within this interval and variations in its sequence perfectly correlate with the ability to taste PTC. In the human population, there are two major sequence variations in T2R38, which differ by three amino acids. One variant (containing the amino acid substitutions AVI) accounts for the

nontasters and the other variant (containing the amino acid substitutions PAV) is the major taster allele (see Kim *et al.*, 2003). Thus, people with two AVI alleles are nontasters, those with one PAV and one AVI are tasters, and those with two PAV alleles are super-tasters. In addition, there are five other forms of T2R38 in the human population, but these are found in low frequency in the population. These studies demonstrate that humans vary in their ability to detect taste compounds because of variations in the sequence of taste receptors.

Kristin Scott

came to the realization that although it was flavorless to him, his colleague found it unbearably bitter. Since then, numerous human taste tests (often done in high school classes) have revealed that the ability to detect PTC and a related compound n-propyl uracil (PROP) behaves as a classic inherited recessive trait with taster and nontaster alleles (see Box 24.1). The identification of the genomic location linked to PTC detection enabled the identification of genes associated with this chromosomal interval and the seminal discovery of a large family of related mammalian bitter taste receptors.

The Taste Receptor 2 family (T2R) of bitter receptors is a family of G protein-coupled receptors whose members, unlike T1Rs, contain a small extracellular domain and most resemble the opsin family of GPCRs. There are 25 functional T2R receptors in human, 35 in mouse and rat, and only three in chicken. (Perhaps this explains why chickens will eat anything!) Most T2R genes are linked in large chromosomal arrays, suggesting rapid expansion of this gene family.

A number of lines of evidence conclusively demonstrate the T2Rs are bitter receptors. First, T2Rs are specifically expressed in taste cells on the tongue. Second, the receptors are localized to taste cell dendrites decorating the taste pore, the site of taste detection. Third, heterologous cells expressing T2Rs respond to bitter compounds; for example, cells expressing mT2R5 respond specifically to cyclohexamide, whereas cells containing mT2R8 respond strongly to denatonium and weakly to PROP. Fourth, mice lacking specific T2Rs show specific bitter taste defects (Fig. 24.4). As predicted from the expression studies of mT2R5 in heterologous cells, mice lacking mT2R5 do not detect

cyclohexamide but do detect a myriad of other bitter compounds. Even more remarkably, mice engineered to contain a human T2R that recognizes phenyl-β-D-glucopyranoside in place of the mouse T2R now avoid phenyl-β-D-glucopyranoside whereas wild-type mice do not detect it.

Studies of the T2R receptor family provide important insights into how animals recognize a large number of bitter compounds and are able to detect them at very low concentrations. Each T2R receptor generally recognizes only a few bitter compounds with very high affinity, allowing for very specific detection. In addition, the large number of T2R receptors ensures that animals can recognize several bitter compounds.

Mechanisms of Taste Transduction for T1Rs and T2Rs

Activation of the T1R and T2R receptors ultimately leads to cell depolarization and neurotransmitter release. In general, signal transduction mediated by G-protein coupled receptors involves a conformational change in the receptor leading to activation of a heterotrimeric G protein, in turn activating enzymes whose byproducts regulate the activity of ion channels. Many signal transduction cascades have been proposed to mediate sweet and bitter taste detection. However, key experiments pinpoint a few critical transducers for sweet and bitter signaling (Fig. 24.5).

Gustducin is the $G\alpha$ subunit of a heterotrimeric protein that participates in taste transduction. Gustducin is found in taste cells on the tongue that contain

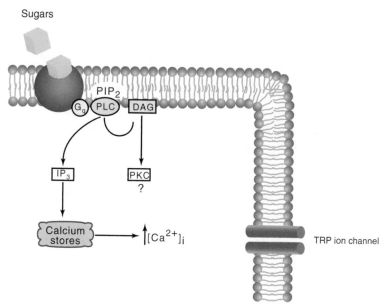

FIGURE 24.5 Taste transduction for sugars, amino acids, and bitter compounds uses a common signaling pathway. Activation of the receptor activates a heterotrimeric G protein. This activates phospholipase C-β2 (PLC-β2), producing the second messengers IP$_3$ and diacylglycerol (DAG). Activation of PLC-β2 leads to opening of the TRPM5 ion channel and cell depolarization.

T1R and T2R receptors. Confusingly, it is not found in all T1R-containing cells, suggesting that either a subset of T1R-containing cells is nonfunctional or other Gα subunits contribute to taste transduction. Animals lacking gustducin show behavioral and electrophysiological defects in the detection of bitter and sugar compounds, arguing that this Gα participates in taste transduction.

Downstream of G protein activation, the enzyme phospholipase C-β2 and the ion channel TRPM5 mediate taste transduction. PLC-β2 and TRPM5 are coexpressed in all T1R and T2R-containing cells. Transgenic mice that lack the PLC-β2 enzyme do not taste sugar or bitter compounds. Similarly, loss of the ion channel TRPM5 results in mice that do not taste sugars or bitter compounds. The simplest model for taste transduction is that activation of T1Rs or T2Rs leads to activation of gustducin, leading to activation of PLC-β2 and depolarization by TRPM5. PLC-β2 catalyzes the breakdown of phosphoinositol 1,4 bisphosphonate into inositol-1,4,5,-triphosphate (IP$_3$) and diacylglycerol (DAG). IP$_3$ can release calcium from the endoplasmic reticulum. Whether IP$_3$, DAG, calcium, or another second messenger activates TRPM5 is unknown. Indeed, it has not been formally demonstrated that activation of PLC-β2 directly activates TRPM5 or that these signaling molecules are in a linear pathway. Nevertheless, the most parsimonious explanation is that gustducin, PLC-β2, and TRPM5 are signaling molecules that form the taste transduction cascade.

Acids Depolarize Taste Cells by Modulating Ion Channels

Unlike sweet and bitter taste detection, which relies on activation of G-protein coupled receptors and downstream signaling molecules to regulate channel activity, the tastes of sour and salty are likely to be directly mediated by ion channels. Sour taste is produced by acids, and the degree of sourness depends primarily on proton concentration. Patch-clamp studies suggest that several different ion channels may participate in sour transduction, which is not surprising because protons are capable of modulating most ion channels. A recent study has identified a candidate sour receptor, comprised of PKD1L3 and PKD2L1, that is expressed in taste cell subsets necessary for sour detection.

PKD1L3 and PKD2L1 are cationic ion channels of the TRP family (transient receptor potential) of ion channels. TRP channels are necessary signaling molecules in *Drosophila* phototransduction, mammalian pheromone detection, *C. elegans* chemosensation, as well as thermosensation and hearing. They can be directly activated by ligands or downstream transducers of signaling. PKD1L3 and PKD2L1 are coexpressed in taste cells on the tongue that do not contain T1Rs or T2Rs. These channels are localized to the taste pore, the site of ligand binding. In heterologous cells, they act together to detect sour compounds such as citric acid, hydrochloric acid, and malic acid. Mice lacking these channels have not yet been generated. However, transgenic mice have been produced such that the PKD2L1 promoter drives expression of a toxin gene to kill all cells that express the channel. Mice lacking these cells do not show electrophysiological responses to sour compounds, but still respond to sugars, salts, and bitter compounds. These experiments argue that cells containing PKD2L1 are necessary for sour detection and suggest that PKD2L1 and PKD1L3 are the mammalian sour detectors.

Apically Located Na⁺ Channels May Mediate Salt Taste Detection

Molecules that mediate salt taste are not clearly defined. The best candidate ion channel that may mediate sodium salt taste is the epithelial sodium channel ENaC. This channel, which is responsible for sodium transport in a variety of epithelial tissues, is expressed on the apical membrane of salt-sensitive taste cells, where it mediates the passive influx of sodium. Sodium simply diffuses through the open channels to depolarize taste cells, and is presumably pumped out by a NaK-ATPase on the basolateral membrane. The first evidence for a role of ENaC in taste came from experiments showing that the gustatory nerve response to NaCl was inhibited by amiloride, a diuretic drug known to block these channels in other transporting epithelial tissues. More recently, patch-clamp recordings have directly demonstrated the presence of amiloride-sensitive sodium channels in taste cell membranes. However, the taste cells that express ENaC channels have not been defined, animals lacking ENaCs in taste cells have not been generated to determine their function, and ENaCs have not been misexpressed in bitter or sugar cells and shown to confer salt detection to these cells. Thus, although ENaCs do depolarize in response to salts and are expressed in taste cells, it has not yet been determined whether they are necessary and sufficient for salt taste detection.

The Organization of Taste Receptor Genes on the Tongue

Taste receptors are segregated into different taste cells on the tongue. *In situ* hybridization experiments as well as immunohistochemistry have been used to determine the expression patterns of receptor mRNAs and proteins. These experiments demonstrated that T1Rs and T2Rs are not found in the same taste cells (Fig. 24.6). Among the T1Rs, T1R1 plus T1R3 and T1R2 plus T1R3 are segregated into different cells. In addition, the PKD1L3 and PKD2L1 candidate sour ion channel subunits are found in different cells than the T1Rs or T2Rs. This demonstrates that there are modality-specific taste cells on the tongue, with one taste cell population recognizing sugars, another amino acids (or glutamate for humans), a third bitter compounds, and a fourth sour cues. An important conclusion is that different taste modalities are encoded by the activation of different taste cells. Our perception of only a few different tastes may stem from the fact that there are only a few different types of taste cells on the tongue. Sugar-sensing and bitter-sensing cells also are found in *Drosophila melanogaster*, suggesting that modality-specific cells are a common principle of taste systems (Box 24.2).

Most taste receptors are expressed in subsets of taste buds on the tongue and on the palate. T1R1 plus T1R3 (amino acid receptor) are coexpressed in fungiform papillae and palate; T1R2 and T1R3 (sugar receptor) are coexpressed in the circumvallate, foliate, and palate papillae. T2Rs (bitter) are expressed on the circumvallate, foliate, and palate papillae; and PKD2L1 (sour) on the circumvallate, foliate, fungiform, and

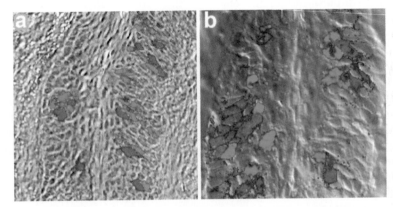

FIGURE 24.6 Different taste cells express different taste receptor genes. *In situ* hybridization experiments showing expression of T1Rs (green) and T2Rs (red) in mammalian tongue sections. Data from Nelson *et al.* (2001).

BOX 24.2

TASTE CELLS IN FRUITFLIES ARE ALSO TUNED BY TASTE MODALITY

Drosophila melanogaster, like mammals, recognize sugars, salts, and bitter compounds, making them a model system for comparative studies of taste recognition. *Drosophila* taste with sensory neurons on their proboscis (mouth), legs, and wings. A family of 68 candidate Gustatory Receptor genes, unrelated to mammalian T1R or T2R receptors, may mediate taste detection as some members are expressed in taste tissue and recognize taste ligands in heterologous systems. Similar to mammalian taste, multiple receptors are expressed in each *Drosophila* taste cell. Two different classes of taste cells have been identified based on the receptor subsets they contain. One population detects bitter compounds and a second population detects sugars. Additional subpopulations of taste neurons likely recognize other taste categories. Thus, although flies and mammals are separated by more than 500 million years of evolution, they both have taste cells that selectively recognize bitter compounds or sugars. This suggests that modality-specific taste cells may be a general strategy that organisms use to distinguish nutrients from toxins.

Kristin Scott

palate papillae. This argues that different parts of the tongue are not dedicated to detecting a single taste modality. Although many textbooks include a tongue "taste map" suggesting that the front of the tongue detects sugars, the side salt and sour, and the back bitter, this is not true. Instead, the topographic distribution of taste receptors on the tongue suggests there are more subtle differences in the relative sensitivities of different tongue regions.

Activation of Taste Neurons Is Sufficient to Generate Taste Behavior

The finding that different taste compounds activate different populations of taste cells suggests that activation of different cells in the periphery leads to different taste percepts and different taste behaviors. To test this, transgenic mice were engineered to express novel receptors in taste cells. In one derivation of this experiment, mice were engineered to contain a modified G protein coupled receptor activated solely by a synthetic ligand (RASSL) in all T1R2 containing taste cells. This exogenous receptor is activated only by a compound called spiradoline, which animals do not detect. These mice then were given a choice between drinking water or water with the synthetic ligand, spiradoline. The mice with the RASSL in T1R2 cells showed a strong preference for spiradoline, in contrast to normal mice. In another experiment, the RASSL was put in T2R taste cells; these mice now avoid spiradoline. In a third extremely elegant experiment, transgenic mice were engineered to contain the human bitter receptor (hT2R16) for phenyl-β-D-glucopyranoside (a com-

pound that humans perceive as bitter) in either T2R (bitter) or T1R2 (sugar) cells (Fig. 24.7). If hT2R16 is expressed in T2R cells, mice avoid phenyl-β-D-glucopyranoside. If this human bitter receptor is expressed in T1R2 cells, however, mice prefer phenyl-β-D-glucopyranoside. Thus, expression of a bitter receptor in sugar cells triggers attraction to the bitter compound. These experiments demonstrate that activation of sugar cells leads to taste acceptance behavior and activation of bitter cells leads to avoidance. This argues that the activity of different taste cells is hard-wired to different behaviors.

Transmission of Taste Information from the Tongue to the Brain

The three nerves that contact taste cells have cell bodies in ganglia and send axons to the solitary tract nucleus (NST) of the medulla in a topographic order (Fig. 24.8). Dendrites from the chorda tympani (a branch of the facial nerve VII) contact the anterior tongue, cell bodies reside in the geniculate ganglion and axons terminate in the rostral pole of the NST. Dendrites from the lingual-tonsillar branch (glossopharyngeal nerve IX) contact the posterior tongue, have cell bodies in the petrosal ganglion, and axons invade the NST at intermediate regions. Dendrites from the superior laryngeal branch (vagus nerve X) innervate the epiglottis and esophagus, have cell bodies in the nodose ganglion, and send axons to the caudal regions of the NST. Thus information from the anterior-posterior tongue is represented from the rostral to caudal NST.

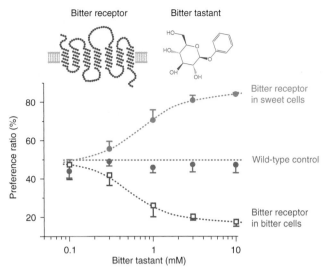

FIGURE 24.7 Activation of different taste cells is tethered to specific taste behaviors. Mice were engineered to contain a human bitter receptor in either T1R2 sugar-sensing or T2R bitter-sensing cells. Mice avoid the specific bitter compound when the receptor is expressed in T2R cells and prefer the same compound when the receptor is expressed in T1R2 cells. This elegant experiment demonstrates that the same compound can elicit different behaviors depending on which cells detect it. Figure adapted from *Nature* (2006); **444(7117)**, 283–284.

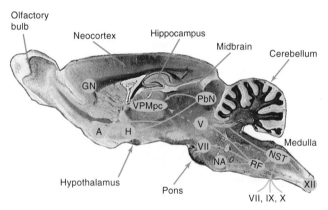

FIGURE 24.8 Schematic diagram of the ascending gustatory pathway; descending projections are not shown. Connections of the rodent gustatory system within the CNS are shown by solid lines; the projection from NST to VPMpc in primates is indicated by a dashed line. NST, nucleus of the solitary tract; PbN, parabrachial nuclei; VPMpc, venteroposteromedial nucleus (parvi cellularis) of the thalamus; GN, gustatory neocortex; A, amygdala; H, hypothalamus; NA, nucleus ambiguous; RF, reticular formation; V, VII, and XII, trigeminal, facial, and hypoglossal motor nuclei; VII, IX, and X, axons of peripheral gustatory fibers in the facial, glossopharyngeal, and vagal cranial nerves.

This segregation continues throughout the gustatory pathway to the cortex, where there are separate terminal fields for VIIth and IXth nerve inputs. From the NST, ascending fibers project in most species to third-order cells within the parabrachial nuclei (PbN)

of the pons. A thalamocortical projection arises from the PbN to carry taste information to the parvicellular portion of the ventroposteromedial nucleus of the thalamus (VPMpc) and on to the gustatory neocortex (GN), located in rodents within the agranular insular cortex. In primates, taste fibers bypass the pontine relay and project directly to the VPMpc.

Arising in parallel with the thalamocortical projection is a second projection that carries gustatory afferent information into limbic forebrain areas involved in feeding and autonomic regulation, including the lateral hypothalamus, the central nucleus of the amygdala, and the bed nucleus of the stria terminalis. Descending axons within the gustatory system arise from the insular cortex and several ventral forebrain areas and project to the PbN and NST. There are also numerous local connections among neurons within the NST and with cells of the oral, facial, and pharyngeal motor nuclei (V, VII, ambiguous, and XII), either directly (as with XII) or via interneurons in the reticular formation. These hindbrain systems form the substrate for many taste-mediated somatic and visceral responses related to ingestion and rejection of tastants.

Gustatory Afferent Neurons Extract Several Types of Sensory Information

Three features of taste ligands are thought to be encoded by the activation of gustatory neurons: intensity, quality, and hedonic value. Intensity refers to the perceived strength of the stimulus; for example, low concentrations of sugar taste less sweet than high concentrations. Quality refers to the nature of the sensory stimulus; mammals are thought to perceive five different taste qualities—sweet, bitter, sour, salty, and amino acids. Hedonic value is the perceived pleasantness or unpleasantness of a taste ligand. Thus, gustatory afferent input provides at least three types of information that are interrelated in complex ways. How taste intensity, quality, and hedonic value are represented in the nervous system is the problem of gustatory neural coding.

Gustatory stimulus intensity, the magnitude of the evoked sensation, generally is assumed to be encoded by neural impulse frequency and numbers of responding neurons. All neurons responsive to taste stimuli show some modulation by stimulus concentration, with increased firing rates as concentration increases. In addition, at high taste ligand concentrations, more cells are responsive than at low concentrations. These studies suggest that a simple rate code, and cell recruitment at high concentrations, allow the discrimination of different concentrations of a taste stimulus.

The encoding of taste quality and hedonic value are more complicated. Taste quality and hedonic value are strongly related. Most animals find sugars pleasant-tasting and bitter compounds unpleasant. However, species-specific predispositions toward the hedonic value of a stimulus can be overcome in response to metabolic or pharmacologic manipulations. Thus, the hedonic value of taste qualities can be influenced by experience and physiologic state. Unlike gustatory intensity and quality, the issue of hedonic coding has not been addressed systematically in neurobiological studies of the taste system, probably because hedonic value is not independent of either quality or intensity and can be modified by both experience and physiologic state.

Mechanisms underlying taste quality encoding are controversial. Different models are described next.

Models of Encoding Taste Quality

The nature of the neural coding of taste quality has been debated vigorously for many years, with considerable disagreement about whether taste quality is represented by activity in specific neural channels (a labeled line code) or by the relative activity across the responsive neurons (a population code) (Fig. 24.9).

Labeled Lines

The labeled line model of taste coding proposes that different taste qualities are encoded by the activation of different cells. In this model, cells respond to selective cues in the periphery and this information remains

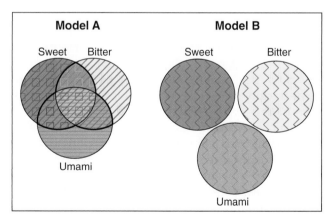

FIGURE 24.9 Two models of taste coding have been proposed. The left panel depicts the labeled line model in which different taste cells recognize different taste qualities, such that one cell population is activated by sugars and different cells are activated by bitter compounds. The right panel depicts the population coding model in which cells respond to multiple taste modalities. The pattern of activity across the ensemble of taste neurons encodes for specific tastes. Adapted from Amrein and Bray, 2003.

segregated in the brain. This hypothesis suggests that "sweetness" is coded by activity of neurons that respond selectively to sugars, "saltiness" by activity of neurons that respond selectively to salt, and so forth. Activity in a given cell type provides complete information about the quality of the stimulus. In this scheme, the animal then discerns different tastes by the activation of different neurons.

Population Coding of Taste Quality (Across-Fiber Patterning)

In this model of taste coding, each cell responds to multiple taste modalities with different activity levels for different taste ligands, such that taste quality is coded by the pattern of activity across taste fibers. In this coding hypothesis, the pattern of activity generated across the entire array of taste neurons encodes taste quality, whereas activity in any one cell cannot unambiguously represent both stimulus quality and intensity. This population approach to quality coding makes the multiple sensitivity of gustatory neurons an essential part of the neural code for taste quality; it stresses that the code for quality is given in the response of the entire population of cells, placing little or no emphasis on the role of an individual neuron. In this scenario, the animal then discerns different tastes by differences in the ensemble activity of taste neurons.

Taste Coding in the Periphery

The studies of taste receptor genes strongly argue for the labeled line encoding of taste information in the periphery. Different taste cells contain different taste receptor genes and respond to different taste ligands. T1R2 plus T1R3 cells detect sugars, T1R1 plus T1R3 cells detect amino acids, T2R cells detect bitter compounds, and PKD2L1 cells detect acids. There is no overlap in the expression of different receptor-types and thus no cell that detects both sugar and bitter compounds. Thus, "sweetness" is encoded by the activation of sugar cells; "bitter" by the activation of bitter-sensing cells.

Not only do different cells identify different tastes, but these different cell-types are necessary and sufficient for taste detection and behavior. Mice lacking T1R2 do not detect sugars, but still detect bitter, salts, sour, and amino acids. Similarly, mice lacking cells containing PKD2L1 do not detect sour, but do detect sugars, bitter, salt, and amino acids. Moreover, simply activating the sugar cells artificially generates taste acceptance behavior and activating the bitter cells generates avoidance behavior. These experiments are difficult to reconcile with population coding models. In population coding, the relative activity of nonselective neurons dictates taste quality and loss of any subpopulation of taste cells would affect detection of all compounds. In addition, in population coding, activation of a subset of taste cells would be meaningful only in the context of the ensemble activity. Instead, different taste cells recognize different taste modalities and mediate specific taste behaviors, arguing that there are labeled lines of taste information from peripheral activation to behavior.

Taste Coding in the Central Nervous System

Although the selectivity of taste cells in the periphery strongly argues for labeled lines of taste information, neurophysiological recordings of both primary gustatory nerves and central gustatory neurons typically show that individual fibers or cells respond to more than one of the stimuli representing the salty, sweet, sour, or bitter taste qualities, often to as many as three or four. Accumulating evidence suggests that there are functional classes of neurons that correspond in some way to primary taste qualities, for example, "sucrose-best" cells, "sodium-best" cells. However, analyses show that no single class of neurons in isolation can discriminate well between different taste qualities. As a result, an "across neuron pattern" theory suggesting that taste quality is coded by the relative activity across a population of neurons was proposed.

The current debate in the mammalian taste field is how to reconcile the evidence of labeled lines for different tastes in the periphery with the evidence for mixed lines in the central nervous system. One possibility is that a single nerve fiber synapses onto taste cells of many different taste qualities on the tongue, such that the labeled lines in the periphery become mixed at the first synapse. Alternatively, electrophysiological studies of single neurons with defined connections to taste cells, rather than random recordings of neurons in the solitary tract nucleus or gustatory cortex, may reveal populations of higher-order taste neurons that show selective responses to single taste qualities, supporting labeled line encoding.

Summary

The sense of taste allows animals to recognize different chemical compounds and gives rise to a limited number of taste sensations (saltiness, sweetness, sourness, bitterness, and umami/amino acids). Two families of taste receptors, the T1R and T2R families, mediate the detections of sugars, amino acids, and bitter compounds. These receptors are G protein-coupled receptors that activate a heterotrimeric G protein, phospholipase C, and TRPM5, leading to cell depolarization and neurotransmitter release. Candidate ion channels have been implicated in the detections of salts and sour compounds. Compared to mammalian olfaction, the gustatory system utilizes a relatively small number of receptors with very diverse structures to detect chemical compounds.

The transduction mechanisms for taste stimuli are located on receptor cells within taste buds distributed in several subpopulations, innervated by different peripheral nerves. These nerves project into the nucleus of the solitary tract in the medulla. From there, projections arise to the parabrachial nuclei in the pons and then to the thalamus and gustatory neocortex. A parallel pathway carries taste information into the ventral forebrain to areas involved in autonomic regulation. The gustatory system extracts information about taste intensity, quality and the hedonic value of the stimuli. How gustatory information is encoded in the brain is an area of active investigation. In the periphery, an individual taste cell is tuned to a single taste quality, arguing that there are labeled lines of taste information. More centrally, gustatory neurons appear broadly tuned to stimuli of different taste qualities, leading to a debate whether the labeled lines in the periphery become mixed centrally. The ability to determine the connectivity of higher order taste neurons and their taste response profiles will enhance our understanding of taste neural circuits and the coding mechanisms underlying taste perception.

OLFACTION

The number of volatile cues in the environment that animals can detect is colossal. Humans, although not known to have particularly keen noses, can sense over 10,000 different odors, with detection thresholds in the parts per million and even parts per billion. Scent-tracking dogs, especially bloodhounds, have even more sensitive noses, approximately 10 million times more sensitive than humans, allowing them to follow scent trails a few days old. The ability to detect chemicals a distance from the source allows animals to direct their movement toward food and potential mates and away from harmful or dangerous environments. Although many animals rely on their olfactory system for survival, the sense of smell in humans is primarily an aesthetic sense, enhancing our enjoyment of foods and evoking memories of past experiences.

How does the olfactory system allow for the recognition and discrimination of thousands of different odors? A large family of 1000 olfactory receptors provides the molecular diversity necessary for recogniz-

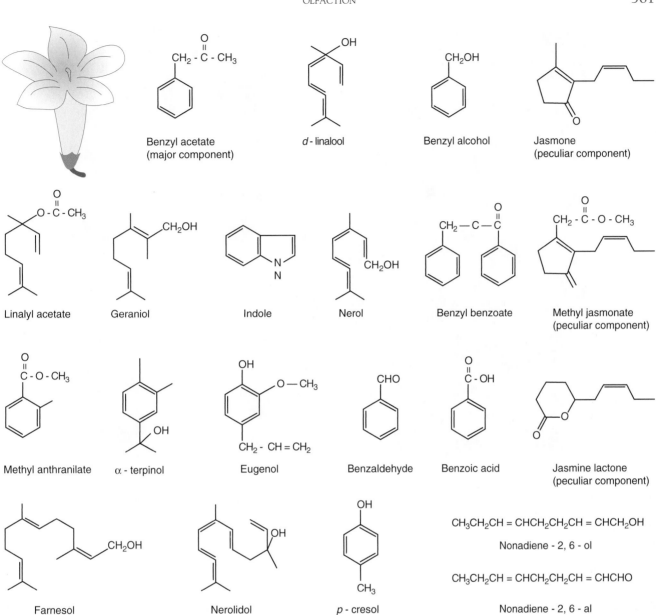

FIGURE 24.10 Odor molecules given off by the jasmine flower that constitute the smell of jasmine as an odor object (Mori and Yoshihara, 1995).

ing numerous ligands. The discovery of the olfactory receptor genes has provided molecular tools to probe the molecular underpinnings of olfactory detection. The logic of olfactory detection in the periphery is discussed next.

Odor Stimuli Consist of a Wide Range of Small Signal Molecules

Odor signals are low molecular weight molecules that fall into several broad classes. In terrestrial animals, the molecules tend to be small (under 200 Da) and volatile so that they can vaporize readily and be carried in the air; they also tend to be lipid soluble. Some of these are acids, alcohols, and esters found in various plant and animal foods. Some are essential oils. Others are aromatic compounds given off by flowering plants (Fig. 24.10). They may function for long distance signaling of the presence and palatability of food in the environment.

An important category consists of molecules used in reproductive activities, including signals used in attracting mates, identifying them, copulating, blocking pregnancy, facilitating nipple attachment by infants, and infant identification. Some of these activities are mediated by complex mixtures of acids, esters,

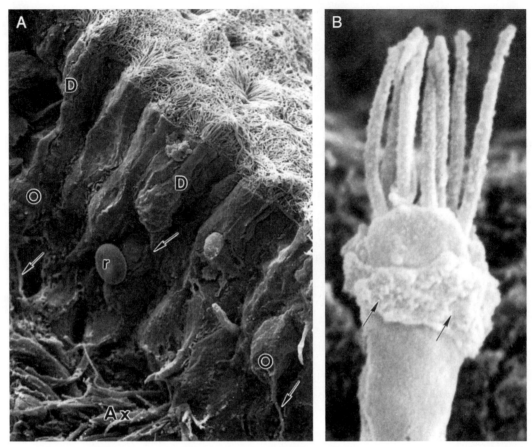

FIGURE 24.11 Olfactory receptor neurons are bipolar cells within the pseudostratified olfactory epithelium. (A) Scanning electron micrograph of the human olfactory epithelium, showing cell bodies of the olfactory receptor neurons (O) with their dendrites (D) ending in cilia that form a mat within the mucus layer overlying the epithelium. From the deeper aspect an axon (arrows) arises, forming bundles (Ax) in the submucosa. Red blood cells (r). (B) High magnification view of the distal dendritic knob giving rise to olfactory cilia. The terminal web is visible encircling the know (arrows). From Morrison and Costanzo (1990).

and other types of common molecules; others by individual larger and more complex molecules such as musks; and still others by specific types of molecules for conspecific signaling known as *pheromones*.

Odor Molecules Are Transduced by Olfactory Receptor Neurons

Most vertebrate animals sense odor molecules by means of olfactory receptor neurons (ORNs) located in a pseudostratified epithelium within the nasal cavity. These are bipolar neurons; a thin dendrite arises from one pole, ending in a knob with 6 to 12 cilia (Fig. 24.11). These contain the 9 + 2 pairs of microtubules characteristic of true cilia in other cells of the body. The cilia are thin (0.2 μm in diameter near the knob, tapering to 0.1 μm near their tips) and vary in length in different species, from 5 to 10 μm in humans to 200 μm in frogs. They contain no other organelles. The knobs and cilia are embedded in the mucus overlying the epithelium.

The cilia greatly increase the surface area containing the olfactory receptors. The cilia form a dense mat within the mucus layer that provides an effective device for capturing odor molecules that are absorbed from the air into the mucus. The mucus is viscous (secreted by the supporting cells) except for a watery surface layer (secreted by Bowman's glands). From the other pole of the neuron, a thin unmyelinated axon arises and joins other axons in the submucosa to form bundles that connect to the olfactory bulb.

G-Protein-Coupled Receptors Are the Initial Site of Odor Transduction in Mammals

The 2004 Nobel Prize in Medicine was awarded to Dr. Richard Axel and Dr. Linda Buck for their seminal discovery of a large family of mammalian olfactory receptors. Their search for olfactory receptors was based on three assumptions: first, that olfactory receptors would be G protein-coupled receptors (GPCRs) similar to visual receptors; second, that they would be exclusively expressed in nasal epithelium; and third, that there would be a large number of receptors, given the large number of odors that mammals smell. This led Axel and Buck to search for novel GPCRs in nasal tissue and to the discovery of the odorant receptor gene (OR) family.

There are approximately 1300 olfactory genes in mouse (5% of their genome), 500 in humans (2% of the genome), and 100 in zebrafish and catfish (Box 24.4). They are members of the class A family of GPCRs, which includes opsin and β-adrenergic receptors. These receptors contain a short extracellular amino terminus, seven membrane spanning (TM) domains, and an intracellular carboxyl terminus. Hypervariable regions in TM 3, 4, and 5 likely form the ligand-binding pockets, based on the ligand binding domains of other GPCRs such as the β-adrenergic receptor. Individual human olfactory receptor genes share between approximately 50 and 95% sequence identity. Large gene families comprising olfactory receptors also have been identified in nematodes (Box 24.3), fish (Box 24.4), and insects (see later).

Olfactory receptor genes are organized in the genome in large linked arrays on several chromosomes. The size of the arrays ranges from 6 to 138 in humans and up to 15 in *C. elegans*. The *Drosophila* genome, with fewer olfactory receptors, contains small arrays of two to three genes as well as many singly distributed genes. The large arrays in mammals and nematodes suggest that odor receptor gene families are evolving rapidly by duplication and expansion.

The identification of odor receptors provided fundamental insight into the molecular basis of odor perception. The large number of receptors dedicated to binding odor molecules provides an elegant solution to the problem of how animals detect a large number of smells. In addition, identifying receptors opened the door to examine interactions of different receptors with different odors, the organization of olfactory receptors in olfactory neurons, and the representations of olfactory receptors in the olfactory bulb.

Evidence That ORs Detect Odors

The large number of ORs and the large number of odor molecules, as well as technical difficulties in expressing ORs in heterologous cells, has limited the ability to determine the ligands that different ORs recognize. In the first demonstration that ORs detect volatile cues, the "I7" odor receptor was overexpressed in most olfactory neurons of the rat. The electro-olfactogram (EOG) response of the epithelium to a battery of

---

BOX 24.3

SENSING THOUSANDS OF CHEMICALS WITH A HANDFUL OF NEURONS

The nemode worm *Caenorhabditis elegans* lives in the soil, lacks a visual system, and relies largely on chemical detection to survive in the environment. This small animal has only 302 neurons, but displays complex olfactory behaviors, making it a model system to study chemosensory processing. Thirty-two sensory neurons mediate chemical detection and contain members of a large chemoreceptor gene family. These receptors are G protein-coupled receptors, but are not related to mammalian olfactory receptors. There are approximately 1500 chemoreceptor genes in *C. elegans*, comprising 7% of all its genes. Thus, of all animals whose genomes have been sequenced, *C. elegans* dedicates the most genes, and the highest fraction of its genes, to chemical sensing.

Many receptors are found in the same chemosensory cell, as might be expected given the large number of receptors and the small number of sensory cells. This organization differs from the olfactory systems of mammals and flies, but resemble mammalian and fly gustatory systems. Parceling many receptors into a single cell allows these animals to detect many chemical compounds despite the limited number of chemosensory neurons.

C. elegans shows sophisticated behavioral responses to chemical cues, including attraction and avoidance of different volatiles cues, adaptation to specific odors as well as learned associations. For example, worms can learn to avoid pathogenic bacteria based on its smell after one exposure to the bacteria. This is similar to conditioned taste aversion in mammals where one encounter with a nausea-producing food leads to aversion of the smell and taste of the food.

Kristin Scott

BOX 24.4

CHEMOSENSATION IN FISH

The distinction between taste and smell may seem blurry in the fish. All chemicals that fish encounter are water-soluble, eliminating the distinction between volatile odors and soluble taste cues. Nevertheless, fish have distinct senses for taste and smell. Chemicals that aquatic animals detect include various amino acids that are important for food recognition as well as bile salts that may function as alarm signals to warn of the presence of predators.

Similar to mammals, the sensory organs for taste and smell are separate in the fish. Olfactory neurons are found in structures called olfactory rosettes in facial pits. Openings called nares lead to the external environment for smell detection. Taste cells are found on lips, gill rakers, pharynx, oral cavity, and distributed on the body surface.

Fish also contain receptors that are similar to mammalian olfactory and gustatory receptors. Zebrafish contain approximately 143 genes with sequence similarity to mammalian olfactory receptors, type A receptors with short amino-termini. This is about 10-fold fewer receptors than in mice. In addition, zebrafish have 54 type C receptors, similar to mammalian V2R pheromone receptors. These are expressed in olfactory neurons, and one receptor has been shown to detect an amino acid in heterologous expression systems. There are 4 T1R genes in the zebrafish expressed in gustatory neurons. In addition, only 2 T2R-like genes have been identified in the zebrafish genome.

The molecular similarity of taste receptors and olfactory receptors from fish to man, and the anatomical segregation into different sensory structures, argues that taste and smell are different senses for fish.

Kristin Scott

---

odors was examined. Among some 80 compounds tested, only longer chain aldehydes gave increased responses over controls, with octyl aldehyde giving the peak response, and lesser responses for flanking long-chain aldehydes. This experiment demonstrated that a single OR detects a small subset of odors.

More recently, advances in the ability to express receptors in heterologous cells has resulted in the identification of receptor-ligand pairs for approximately 20 different ORs. Some ORs bind very few ligands (specialists) whereas others are broadly tuned (generalists). These receptor-ligand studies have shown that one OR will bind more than one odor and one odor will activate more than one OR. These experiments provide direct evidence that members of the large gene family of GPCR receptors are specifically sensitive to odors.

One Olfactory Receptor Gene Is Expressed in Each Olfactory Neuron

Studies of olfactory receptor expression in mammals revealed that each olfactory neuron contains only one member of the OR gene family. OR gene expression was determined by *in situ* hybridization experiments, showing that each OR is found in approximately 0.1% of olfactory neurons. In addition, isolating OR receptor sequences from single olfactory neurons demonstrated that each neuron expresses mRNA for only one OR.

Finally, several pairs of ORs were examined to determine if two receptors were found in the same cell, but none of the ORs examined were coexpressed. Instead, receptors label nonoverlapping cell populations.

Not only is one receptor expressed per cell, but only one allele of a receptor is expressed per cell. All diploid organisms contain paired homologous chromosomes and two copies of each gene, the maternal allele and the paternal allele. For most genes, a cell that expresses the maternal allele also expresses the paternal allele. However, olfactory receptor genes show allelic exclusion. One demonstration of this came from engineering two sets of mice: one transgenic mouse contained the olfactory "P2" receptor linked to a green fluorescent protein (P2 green) and the other transgenic mouse contained the "P2" receptor linked to a second protein visualized with red fluorescence (P2 red). Offspring of these mice contained one allele of P2 green and one allele of P2 red. These mice had red olfactory neurons and green olfactory neurons, but no neurons that were both red and green. This experiment directly showed that only one allele of an olfactory receptor is expressed per cell. The consequence of allelic exclusion is that each olfactory neuron contains only genetically identical receptors. One could imagine that small changes in the nucleotide sequence of an allele might alter its ligand-binding properties, such that the maternal and paternal alleles recognize different odors. Segregating these alleles into different cells might expand the repertoire of odors that different olfactory neurons detect.

The general rule that each olfactory neuron expresses only one olfactory receptor gene has important implications for olfactory coding in the periphery. The organization allows the activation of different olfactory receptors to be equivalent to the activity of different cells. Thus, the brain can know which odors are present by which neurons are activated.

How is the choice of olfactory receptor controlled to ensure that one and only one receptor is expressed per cell? Although the determinants of receptor choice are not yet clear, a feedback mechanism has been uncovered that ensures that once one odorant receptor is expressed in an olfactory neuron, no other receptors are expressed. Simply put, the expression of a functional OR turns off the expression of all other ORs. The main experimental support for this is the observation that neurons expressing a receptor that does not function (either by deletion of the coding region or by missense mutation) turn on the expression of a second receptor. In contrast, olfactory neurons expressing a functional receptor do not express other receptors. This argues that expression of a functional odorant receptor elicits a feedback signal that eliminates expression of other receptors.

Odor-Binding Proteins May Link Odor Molecules to Odor Receptors

A class of secreted, soluble proteins called odor binding proteins (OBP) is abundantly expressed in olfactory sensory epithelium in mammals and flies. OBPs are small, globular proteins that are produced by support cells and released into the extracellular space. Insect OBPs are predominantly helical proteins whereas the vertebrate OBPs have an eight-stranded barrel structure that is unrelated to the sequence and structure of the insect OBPs. Both vertebrate and invertebrate OBPs have been shown to bind odors *in vitro*. In general, vertebrate OBPs have broad odor–ligand affinities, and insect OBPs have narrower binding affinities.

Because OBP molecules have been shown to bind odorants, a number of models regarding their function have been proposed. One model suggests that OBPs function to partition hydrophobic ligands from the air to the aqueous phase. Alternatively, OBPs may directly present odors to the olfactory receptors and act as co-receptors. Third, OBPs may sequester the odor away from the site of odor recognition, mediating odor clearance. Recent studies on an odor binding protein in Drosophila argue that it is involved in presenting the odor to the olfactory receptor (Box 24.5). Whether different OBPs serve different functions remains to be determined.

Olfactory Signal Transduction Utilizes Adenylate Cyclase and a Cyclic Nucleotide Gated Ion Channel

A combination of molecular biology and biochemistry has uncovered the olfactory signal transduction cascade (Fig. 24.12). Similar to other G protein-coupled signaling cascades, ligand-binding to an odorant receptor induces a conformational change that results in activation of a heterotrimeric G protein. The $G\alpha$ subunit is an olfactory-specific G_s protein (G_{olf}) that activates an adenylate cyclase type III (AC3). AC3 is an enzyme that catalyzes the conversion of ATP to 3′,5′-cyclic AMP (cAMP). A cyclic nucleotide-gated (CNG) channel is activated in response to increases in cAMP, leading

---

BOX 24.5

A PHEROMONE-BINDING PROTEIN IN DROSOPHILA

There are 35 odor-binding proteins in the fruit fly *Drosophila melanogaster*. Each OBP is expressed in a spatially restricted region of the olfactory epithelium, suggesting that different OBPs serve different functions. One of the best studied odorant binding proteins is a Drosophila OBP called lush. Fly mutants lacking lush do not detect the compound 11-cis-vaccenyl acetate (cVA). cVA acts as a pheromone in Drosophila mediating aggregation and aspects of courtship behavior. Olfactory neurons of lush mutants do not respond to cVA, demonstrating the lush is essential for detection of this pheromone. Remarkably, application of lush protein to the mutant olfactory neurons rescues the ability of these neurons to respond to cVA. This experiment demonstrates that lush is necessary for olfactory neurons to detect cVA. It argues that lush is involved in presenting the odor to the odor receptor and is not involved in clearance of the odor away from the receptor. For more information, please refer to Xu *et al.*, 2005.

Kristin Scott

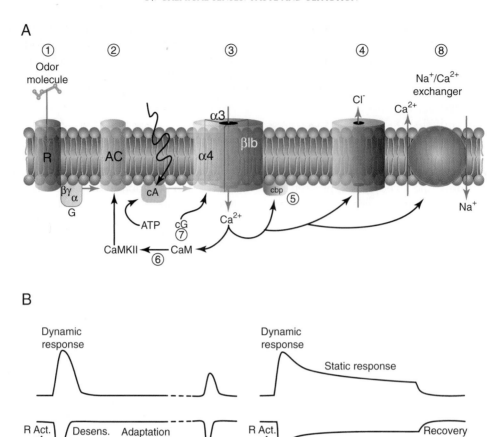

FIGURE 24.12 Sensory transduction of odor molecules involves a cyclic AMP second messenger pathway. Odor molecules initially are absorbed into the olfactory mucus, where they may bind to olfactory-binding protein (OBP), which carries them to the olfactory cilia. Activation of a receptor by odor molecules activates a GTP-binding protein (G); an adenylate cyclase (AC3), which produces cyclic AMP (cA); and a cationic cyclic nucleotide-gated (CNG) channel. Steps 1–3 generate the initial sensory response. Olfactory adaptation occurs in several steps. Step 4: Calcium activates a chloride conductance, which amplifies the sensory response. Step 5: Ca^{2+} activates a calcium binding protein (cbp), which produces immediate adaptation from the initial dynamic response peak. Step 6: Ca^{2+} activates a calcium/calmodulin-dependent protein kinase II, which produces short-term adaptation (LTA). Step 7: Ca^{2+} activates a cyclic GMP second messenger pathway that activates CO, which acts on the CNG channel to produce long-term adaptation (LTA). Step 8: A Ca/Na exchanger restores ion balance. Abbreviations, see text. Based on Menini (1999); Zufall and Leinders-Zufall (2000), and others.

to cell depolarization, the generation of action potentials, and neurotransmitter release.

Mice lacking the G_{olf}, the AC3, and the CNG channel have been generated and the phenotypes of these mutants highlight the critical roles of olfactory signaling. Mice lacking G_{olf}, AC3, or CNG channel generally die within one to two days of birth without milk in their stomach, unable to nurse. This reduced survival rate is consistent with the idea that olfactory cues are necessary for newborn pups to suckle. The activity of olfactory neurons in these mutant mice was monitored by extracellular recording from the nasal epithelium in

response to a panel of odors. Olfactory neurons from the AC3 mutants and the CNG channel mutants do not respond to odors. Olfactory neurons from mice lacking G_{olf} show significantly reduced odor responses, suggesting that a second $G\alpha$ subunit found in olfactory neurons, G_s, may mediate the residual odor response in the G_{olf} mutants. These studies demonstrate that the G_{olf}, the AC3, and the CNG channel are essential components of the olfactory signal transduction cascade.

The olfactory CNG channel is structurally related to the mammalian photoreceptor transduction channel. Similar to the photoreceptor channel, the olfactory channel is nonselective for cations. Unlike the photoreceptor, the olfactory channel is closed in the absence of sensory stimulation. When odor molecules bind to the receptor and lead to the production of cAMP, the cAMP activates the channel, causing a net inflow of cations that depolarizes the membrane toward an equilibrium potential around zero. Calcium entering through the channel acts as a second messenger activating a chloride channel to increase the outflow of chloride down its gradient from a high internal concentration; this amplifies the depolarization, thereby maximizing the amount of depolarizing sensory current while minimizing calcium inflow that would be harmful to the cilia. In addition, the chloride channel has a very small unitary conductance (less than 1 pS), and hence amplification is achieved with minimal noise.

The combined sensory current causes a depolarization that spreads through the dendrite to the cell body and axon hillock, activating voltage-gated channels that generate action potentials. In this way, the amplitude and time course of the graded sensory potentials generated by odor stimuli are transduced into a frequency code of impulses that propagate through the axon to its terminals in the olfactory bulb.

Olfactory Adaptation Occurs in Several Stages

The decline of a sensory response during sustained stimulation is referred to as adaptation. In olfactory sensory neurons, adaptation proceeds in several stages. Experimental studies have quickly identified calcium as a key player. The first stage (step 5 in Fig. 24.12), of decline from the initial peak in the sensory response, is associated with the action of calcium; in the absence of calcium there is no decline, and patch recordings show continued channel activity with no desensitization. Single channel analysis has shown that an increase in internal calcium reduces the channel open probability. This effect appears to be mediated by an intermediate calcium-binding protein, possibly calcium–calmodulin, which binds to the channel protein to reduce its affinity for cyclic nucleotides.

The second stage of adaptation is called short-term adaptation (STA). This occurs in response to cAMP alone and suggests a feedback pathway from calcium involving calcium/calmodulin-dependent protein kinase II acting on adenylate cyclase (step 6 in Fig. 24.12). This is followed by long-term adaptation (LTA), which involves the activation of guanylate cyclase and the production of cyclic GMP (step 7 in Fig. 24.12). Another longer term contributor is a Na/Ca exchanger, which restores the ion balance (step 8 in Fig. 24.12).

Calcium ions also act externally to reduce the conductance of the channel. In the absence of calcium, the channel has a unitary conductance of some 45 pS; in normal concentrations of extracellular calcium, calcium entering the channel induces a "flicker block," reducing the conductance to less than 1 pS. This block is removed by depolarizing the membrane, suggesting that in normal calcium concentrations, olfactory sensory neurons act as coincidence detectors for cyclic nucleotide production plus membrane depolarization. In normal calcium and at normal resting potential, most of the channels appear to be in the blocked state. This mechanism may enhance the signal-to-noise ratio of the sensory response.

The Spatial Organization of ORNs in the Olfactory Epithelium

The olfactory epithelium of mammals is distributed over the medial septal wall and the lateral turbinates toward the back of the nasal cavity. Each olfactory neuron expresses one of a thousand olfactory receptors. Neurons that express the same receptor are not tightly clustered together in the epithelium nor are they stochastically distributed throughout the epithelium. Instead, the ORNs that express a given receptor gene are located within one of roughly four zones that run anterior to posterior in the epithelium. Within a zone, the organization of ORNs with the same receptor appears random. The pattern of receptor expression suggests that there is positional information in the nasal epithelium that limits receptor expression to only one of four broad zones, with receptors randomly selected within a zone. How receptor expression is determined is an area of active investigation.

Axons of Olfactory Neurons Project to the Olfactory Bulb

Olfactory neurons in the nasal epithelium send axons directly to the mammalian brain, to paired structures in the most rostral region called the olfactory bulbs (Fig. 24.13). The olfactory bulb is the primary relay station for smell. It is comprised of multiple cellular layers. Most peripheral is the glomerular layer where the axons of olfactory neurons reside, with deeper layers containing cells that synapse with olfactory neurons.

Axons of olfactory neurons terminate in the rounded regions of neuropil termed glomeruli. Anatomical studies from the time of Ramon y Cajal have shown that the glomerulus is an anatomical unit for the convergence of axons from many ORNs. In rats, the 15 million olfactory neurons converge onto 1500 glomeruli, giving an average overall convergence of some

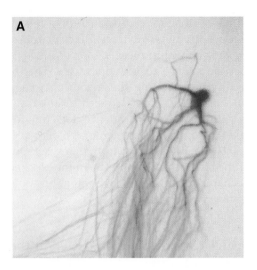

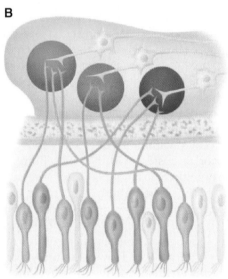

FIGURE 24.13 (A) Whole mount of the nose of a rat shows olfactory receptor axons converging on a single glomerulus. This subset of cells is stained for a lacZ reporter linked to mRNA for one type of olfactory receptor and the microtubule-associated protein tau. (B) Diagram of projections from the olfactory epithelium to the olfactory bulb. Neurons with the same receptor are scattered in the epithelium, but the send axons that converge onto 1–2 glomeruli in the olfactory bulb. From Mombaerts, 1999.

10,000:1. *In situ* hybridization and gene-targeting methods showed that ORNs expressing the same odorant receptor protein converge onto one or a few glomeruli in the olfactory bulb (Fig. 24.13). Neurons with different ORs send axons to different glomeruli. These experiments thus provide strong experimental evidence for the concept that subsets of ORNs expressing a single receptor gene project their axons as a labeled line onto one or a few target glomeruli. Activation of different olfactory receptors in the periphery leads to the activation of different glomeruli in the olfactory bulb. This suggests that the animal knows what it is smelling by the combination of glomeruli that are activated.

Different Odors Activate Different Combinations of Glomeruli

The functional significance of glomeruli is seen when an animal is exposed to an odor and the activity patterns in the olfactory bulb are observed (Box 24.6). The key observation is that different odors evoke different patterns of active glomeruli located in distinct domains within the olfactory bulb. Thus, the focal glomerular activity elicited by the odor of amyl acetate is localized in two broad zones, one medial and one lateral, within the olfactory bulb glomerular sheet (Fig. 24.14). In contrast, the activity elicited by camphor is distributed in a curving line of smaller glomerular patches. Although the two domains overlap, their overall patterns are distinct and different. These results suggest that each type of odor elicits a characteristic pattern of glomerular activation in the olfactory bulb. The patterns may be considered to constitute odor images in neural space.

These maps were obtained with the original 2-deoxyglucose (2DG) mapping method, which has been confirmed and extended by other techniques (Table 24.1). Each method has its advantages and disadvantages. Two main categories are optical methods and activity mapping methods. Optical methods give high resolution at the level of single glomeruli. Many of these studies have focused on the local architecture of relations between individual glomeruli activated by odors with related molecular structures (Fig. 24.15A). This is key for understanding the relationships of odor molecules and the glomeruli they activate. However, these methods image only a small dorsal portion (10–15%) of the total olfactory bulb. Global activity mapping methods such as 2DG are complementary in that they image the entire glomerular sheet, including areas in the 85% of the bulb not accessible to optical methods (Fig. 24.15B).

A different methodological approach has been to focus on the time course of the electrophysiological responses of olfactory cells. These studies (Laurent *et al.*, 1996) have suggested that the temporal spiking patterns contain information that itself could encode the identity of the stimulating odor molecule. This approach is still in its early stages and more information is needed, such as where in the odor maps the responses are recorded; how the temporal patterns relate to the spatial maps; and how long it takes in the temporal response for the response pattern to transmit decipherable information. This approach has thus

BOX 24.6

PRINCIPLES OF ODOR MAPS

Odor stimulation gives rise to spatial patterns of activity in the glomerular layer of the olfactory bulb due to the differential activation of glomeruli.

The glomerulus is the basic molecular, anatomical, developmental, and functional unit for odor mapping and odor processing.

The pattern of activated glomeruli for a given odor is relatively constant across animals for equivalent odor stimulation conditions. However, the pattern can change under different experimental conditions (e.g., anesthesia, adaptation).

The pattern for a given odor characteristically includes sites in the medial and lateral bulb, reflecting the pattern of projections of olfactory sensory neuron subsets. The patterns are characteristically bilaterally symmetrical, dependent on equivalent stimulating conditions on the two sides of the nose.

Identified glomeruli can be correlated with specific odors. This is true of the modified glomerular complex in mammals, and particularly true of identifiable glomeruli in the insect.

Different odors elicit activity in different, often overlapping, patterns. It is hypothesized that processing of the differing patterns by olfactory bulb circuits provides the basis for olfactory discrimination.

A given homologous chemical series (such as alcohols, acids, aldehydes) activates overlapping shifted patterns, reflecting similarities of chemical structure in the series.

At a weak concentration, an odor elicits activity in the single or small group of glomeruli receiving input from the olfactory receptor neuron subset whose receptor type is most sensitive to that odor. Higher odor concentrations activate increasing numbers of glomeruli. Changes in odor pattern with increasing concentration may be correlated with perceptual changes. It is therefore hypothesized that the odor patterns also are involved in the encoding of odor concentration. For references, see Xu *et al.* (2000).

Gordon M. Shepherd

TABLE 24.1 Methods for Odor Mapping in the Olfactory Bulb

Single unit recordings

Focal field potential recordings

2-Deoxyglucose

c-fos RNA or protein

Voltage sensitive dyes

Intrinsic imaging

Calcium imaging

fMRI (functional magnetic resonance imaging)

stimulated the field with new questions requiring further study.

How Do Olfactory Neurons Target to the Appropriate Glomerulus?

Neurons with the same olfactory receptor are scattered in the sensory epithelium, yet all target to only one of a thousand possible glomeruli in the olfactory bulb. Axon targeting of olfactory neurons is a very complicated wiring problem. One model for targeting is that concentration gradients of axon guidance receptors in the nasal epithelium and gradients of guidance cues in the olfactory bulb direct targeting. This approach is used in the formation of the topographic map of retinal projections in the visual system, where neurons that occupy nearby positions in the periphery (and express similar levels of guidance molecules) synapse at nearby locations. Varying the concentration of a guidance cue or guidance receptor along a sheet of tissue is unlikely to be sufficient to direct targeting the olfactory system, as neurons containing the same receptor are found at very different positions yet project to the same glomerulus. An alternative model is that neurons with the same olfactory receptor express the same axon guidance receptor, and the matching of a thousand different guidance cues and a thousand different guidance receptors is used to direct each neuron to its appropriate glomerulus. This begs the question: how are the expression of an odorant receptor and a guidance receptor coordinately controlled? An elegant solution that is emerging is that the olfactory receptors themselves may participate in axon guidance.

A number of lines of evidence argue that the olfactory receptors participate in axon guidance. First, olfactory receptor protein is localized to the axon

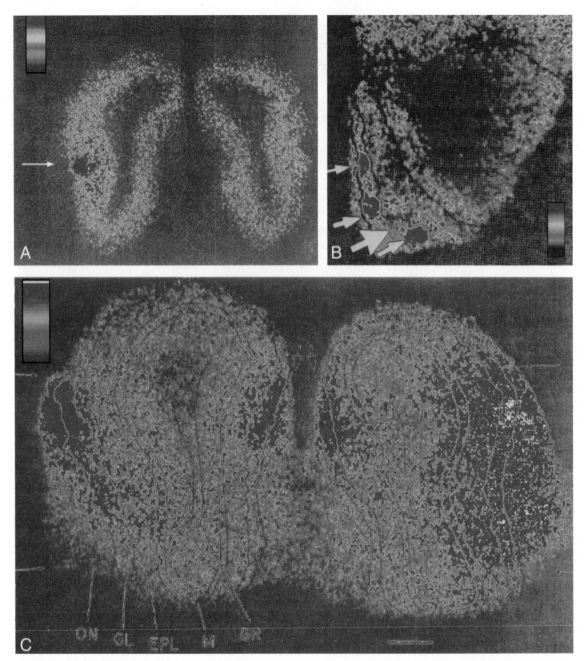

FIGURE 24.14 The 2-deoxyglucose (2DG) method reveals the functional organization of the olfactory glomerular sheet. (A) Patterns of 2-deoxyglucose utilization in an X-ray film autoradiograph of a frontal section through the olfactory bulb of a rat exposed to a low concentration of the odor of amyl acetate. A single focus associated with one glomerulus or a small group of neighboring glomeruli is seen in one of the olfactory bulbs. (B) With a moderate concentration of amyl acetate, the activity induced in the glomerular layer is characterized by activation of several glomeruli or groups of glomeruli (white lines outline the histological layers). (C) With a high concentration of odor of amyl acetate, the induced activity consists of broad regions of increased 2DG uptake, centered on the glomerular layer in medial and lateral regions of the olfactory bulb, that are roughly bilaterally symmetrical.

termini as well as to the dendrites, in the right place to detect guidance cues. Second, olfactory neurons in which the OR has been deleted do not go to a single glomerulus. Third, transgenic mice have been generated that misexpress olfactory receptors and they show specific olfactory targeting defects. As an illustration, mice can be genetically modified to express OR B in neurons that usually express OR A. Remarkably, these neurons no longer go to the A glomerulus. Instead, they go to a novel glomerulus or the B glomerulus. Because the neurons do not go to the A glomerulus, this argues that olfactory receptors are involved in

A

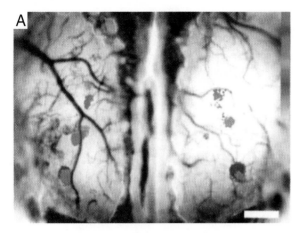

B

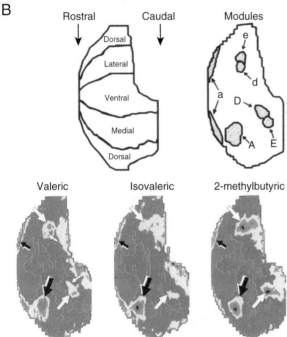

FIGURE 24.15 Odor maps. (A) Visualization of odor-elicited activity in glomeruli of the dorsal olfactory bulb by intrinsic imaging. From Belluscio and Katz (2001). (B) Global maps of odor-elicited activity patterns in the glomerular layer by 2-deoxyglucose. From Johnson and Leon (2000).

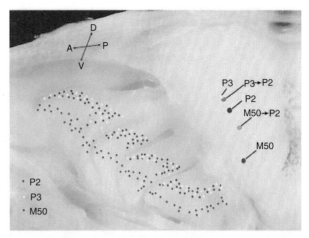

FIGURE 24.16 Diagram showing the olfactory epithelium with neurons expressing the M50, P2, or P3 receptors in the periphery and the glomeruli they project to in the olfactory bulb. Replacing the P2 receptor with the P3 receptor (P3→P2) results in a novel glomerulus very near the original P3 glomerulus. This argues that the P3 receptor can direct targeting to the P3 glomerulus, and that the receptor is instructive for guidance. Replacing the P2 receptor with the M50 receptor (M50→P2) results in a novel glomerulus in between the P2 and M50 glomeruli. This argues that factors other than the receptor also influence guidance. From Wang *et al.*, 1998.

guidance. Because the neurons do not necessarily go to the B glomerulus, other factors besides the olfactory receptor must also be involved in axon guidance (Fig. 24.16).

Recent experiments suggest that activation of signaling molecules directs axon guidance. Like many G protein-coupled receptors, the odor receptors contain a three amino acid motif (Asp-Arg-Tyr (DRY)) at the cytoplasmic loop between transmembranes 3 and 4, implicated in coupling to G proteins. Olfactory neurons containing a receptor that lacks this domain do not target to a single glomerulus. Expression of a constitu-

tively-activated G protein (always ON) rescues this defect, arguing that coupling of the receptor to the G protein is necessary for axon targeting. Interestingly, the G protein is not G_{olf}, the G protein that is involved in olfactory transduction, but rather G_s, another G protein expressed in olfactory neurons. The use of different G proteins may allow activation of the same odor receptor to produce different responses for signal transduction and axon guidance. Increasing or decreasing the levels of cAMP production also changes the position of the glomerulus, such that neurons with the same receptors but different cAMP levels project to different glomeruli. The model that emerges from these studies is that activation of olfactory receptors on axon termini activates Gs and increases cAMP levels, leading to glomerular targeting. Although this model still requires rigorous testing, the implication is that activity of each odor receptor produces a unique level of cAMP, and that different cAMP levels ensure the proper targeting of neurons containing 1000 different receptors to 1000 glomeruli. Whether and how receptor activity generates different cAMP levels remains to be determined.

The Olfactory Bulb Is the Primary Relay for Smell

The olfactory bulb provides the first stage of synaptic processing of the sensory information in the olfac-

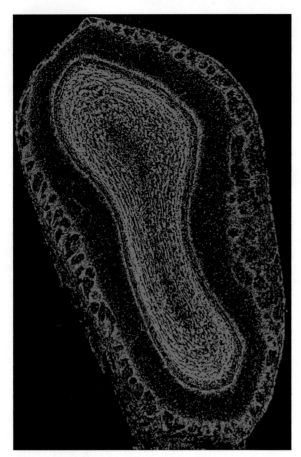

FIGURE 24.17 Layers of the mouse olfactory bulb. The most superficial layer is the glomerular layer (blue) containing olfactory axons and periglomerular cells. Next (in red) is the external plexiform layer containing fibers and the mitral/tufted cell layer, containing neurons that synapse with olfactory sensory neurons. In the deep layers of the bulb (green) are the fiber-filled internal plexiform layer and the cell bodies of the granules cells, which make inhibitory connections with the mitral/tufted cells.

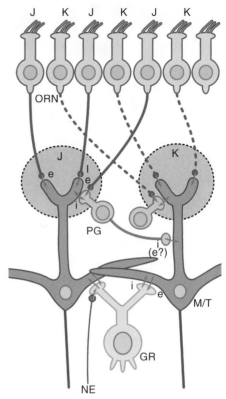

FIGURE 24.18 Diagram summarizing the synaptic organization of the glomerular layer. Olfactory neurons with the same receptor project to the same glomerulus. Mitral/tufted cells synapse onto a single glomerulus. Periglomerulur cells are inhibitory interneurons that synapse within and between glomeruli. Granule cells are inhibitory interneurons that synapse between mitral/tufted cells.

tory pathway. The olfactory bulb is comprised of several cell layers (Fig. 24.17). The most superficial layer is the glomerular layer, where the axons of olfactory neurons with the same odorant receptor converge onto one or two glomeruli. This layer also contains periglomerular cells, inhibitory interneurons that make connections within and between glomeruli. Next is the external plexiform layer, which contains axons and dendrites. Third is the mitral/tufted cell layer, containing the cell bodies of mitral/tufted cells, which synapse onto the olfactory axons within the glomerulus and relay information to the olfactory cortex. Fourth is the internal plexiform layer, which contains axons and dendrites. The final cell layer is the granule cell layer. Granule cells are interneurons that synapse onto mitral/tufted cells. The synaptic connections of different cell types in the olfactory bulb serve as the initial stage for olfactory information processing (Fig. 24.18).

Within a glomerulus, olfactory axons make glutamatergic excitatory synapses onto the dendrites of the mitral/tufted neurons. Each mitral/tufted cell sends dendrites to one glomerulus, and thus relays the activity of one olfactory receptor to the brain. The dendrites of mitral/tufted cells also form reciprocal synapses with the periglomerular cell. The mitral/tufted dendrites make glutamatergic connections onto periglomerular cell dendrites, and dendrites of periglomerular cells form both GABAergic and dopaminergic synapses onto mitral/tufted cell dendrites. These connections are believed to mediate numerous types of interactions, including serial excitatory synapses (which spread excitation widely within a glomerulus), as well as recurrent and lateral inhibitory synaptic circuits.

In addition to intraglomerular processing, there is interglomerular processing. The axons of the periglomerular cells make synapses onto periglomerular cells in neighboring glomeruli and onto the dendritic shafts of neighboring mitral/tufted cells. One action of these synapses may be to inhibit periglomerular cells and mitral/tufted dendrites, thus providing contrast enhancement between neighboring glomeruli.

The second level of synaptic processing in the olfactory bulb occurs through inhibitory connections between granule cells and mitral cells. Reciprocal dendrodendritic synapses that mediate mitral/tufted-to-granule excitation and granule-to-mitral/tufted inhibition cause activated mitral/tufted cells to mediate feedback inhibition on themselves and lateral inhibition on their neighbors. Granule cells may participate in lateral inhibition onto a large population of neighboring cells that belong to neighboring glomerular units. It has been postulated that this inhibition contributes a more complex, more contextual type of contrast enhancement between cells belonging to different glomerular modules.

The current model is that the main function of the dendrodendritic synapses present at both the glomerular and granule cell levels is to mediate lateral inhibition and the role of this lateral inhibition is to enhance contrast between different odors.

Olfactory Bulb Output Goes Directly to Olfactory Cortex

The output of the olfactory bulb is carried by the axons of mitral cells and their smaller counterparts, the tufted cells. These axons project directly to olfactory cortex in the forebrain, the only sensory system to have this immediate access to the forebrain. The olfactory cortex has a three-layer structure that represents the primitive anlage of forebrain cortex found in fish, amphibia, and reptiles (Fig. 24.19). The principal neuron is the pyramidal cell, with apical and basal dendrites bearing spines and recurrent collaterals connecting both to inhibitory interneurons and directly to other pyramidal cells. Mitral/tufted axons make excitatory connections to spines on the distal apical dendrites. Processing in the cortex takes place by means of the intrinsic excitatory and inhibitory synaptic circuits. Together these neural elements and their connections constitute a basic, canonical circuit. This type of circuit is present in other types of cortex, such as the hippocampus and neocortex, suggesting that it represents the simplest type of cortical circuit, which is adapted and elaborated in the other types to carry out different or more complex types of functional operations.

Like cortical regions in other systems, the olfactory cortex is differentiated into several different areas. The main area is the pyriform (sometimes called the prepyriform) cortex. This area receives input from mitral/tufted cells. It projects to the mediodorsal thalamus, which in turn projects to medial and lateral orbitofrontal areas of the neocortex. It is at this level that conscious perception of odors presumably takes place. Second is the olfactory tubercle, which receives input mainly from tufted cells. Third is the cortico-medial group of amygdalar nuclei, which receive specific input from the accessory olfactory bulb. Fourth is the lateral entorhinal area, which projects to the hippocampus. Last is the anterior olfactory nucleus, a sheet of cells just posterior to the olfactory bulb in subprimates.

Stem Cells and Olfactory Function

There is much current interest in stem cells and the possibilities they raise for maintaining or repairing brain function. The olfactory system is unique in the adult brain in being supplied by two sources of stem cells.

First is the olfactory epithelium, where new ORNs arise from basal stem cells during development and throughout the adult life of the animal. During development, new ORNs expressing specific ORs differentiate from basal cells in the olfactory placode. During early life the process of neurogenesis is activated by dying ORNs, leading to a constant turnover of ORNs.

The second example is the anterior migratory stream. Stem cells in the ventricular epithelium of the basal forebrain give rise to neural progenitor cells, which migrate in an anterior direction to the olfactory bulb, where they differentiate and become incorporated into the populations of granule cells and periglomerular cells. This process also occurs throughout adult life. Migration involves homotypic interactions between migrating cells, rather than migration along radial glia as in the cerebral cortex. Current studies are aimed at understanding how the new cells are incorporated into the processing circuits of the bulb.

Almost nothing is known about the mechanisms controlling these processes. Further work is thus needed to understand the relevance of these mechanisms to olfactory processing, as well as gaining insights into the fundamental problems of stem cell functions in the brain.

The Drosophila Olfactory System

The olfactory system of Drosophila shares many common principles with the mammalian olfactory system, although scaled down to size. The smell organs of the fly are its antennae and maxillary palp (Fig. 24.20). Olfactory neurons have dendrites that are exposed to the external environment for the detection of odors and axons that travel to the antennal lobe, the

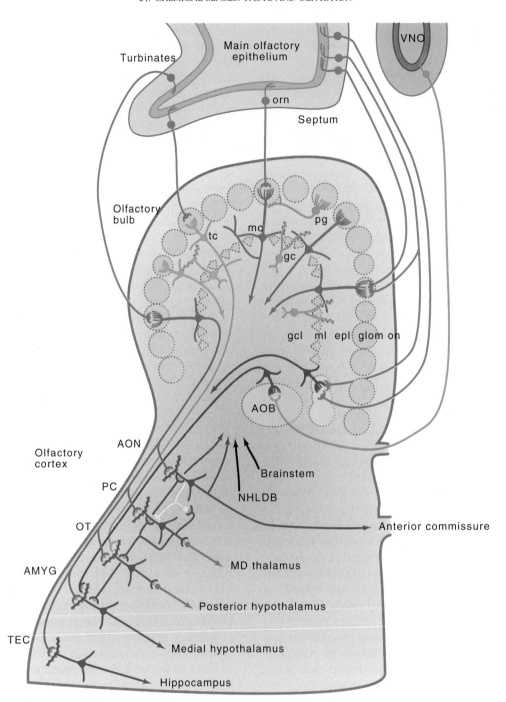

FIGURE 24.19 Summary of main projection pathways in the olfactory system. AON, anterior olfactory nucleus; PC, pyriform cortex; OT, olfactory tubercle; AMYG, amygdala; TEC, transitional entorhinal cortex; NHLDB, nucleus of horizontal limb of diagonal band; MD, mediodorsal.

insect equivalent of the olfactory bulb. A family of approximately 60 Drosophila odorant receptors (dOR) detects volatile cues. Because the number of odor receptors in Drosophila is relatively small, it has been possible to unambiguously determine the expression pattern of each OR in the periphery. About 44 dORs

are expressed in the adult and 20 are expressed in larvae. Each receptor is expressed in a small subset of olfactory neurons, except for one receptor that is found in nearly all olfactory neurons. This receptor, dOR83b, is necessary to localize other odor receptors to the dendritic membrane and is essential for olfactory

| Antenna | Antennal lobe | Lateral horn |

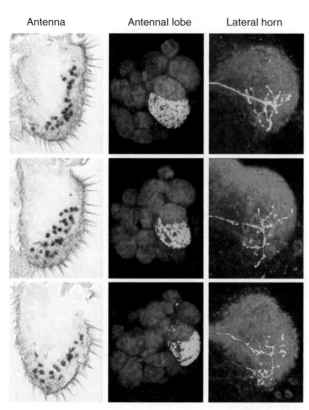

Current Opinion in Neurobiology

FIGURE 24.20 The fly olfactory system. The left panel shows a diagram of the fly head, highlighting olfactory regions. The antenna containing olfactory neurons is in light blue, and the first relay, the antenna lobes, in dark blue. Neurons with the same receptor project to a single glomerulus. A single projection neuron innervates one glomerulus and sends axons to the protocerebrum. Olfactory neurons containing Or47a (blue) in the antenna. The projections of Or47a neurons (green) in the antennal lobe. The projection neuron that synapses onto Or47a neurons arborizes in a stereotyped fashion in the protocerebrum of the fly brain. From Keller and Vosshell, 2003.

function. Each olfactory neuron contains dOR83b as well as one other specific olfactory receptor.

Neurons with the same specific receptor project to a single glomerulus in the antennal lobe. Consistent with the number of receptors, there are 49 glomeruli in the adult and 21 glomeruli in larvae. A complete map of the larval and adult antennal lobes has been made, identifying each receptor by the glomerulus that innervates it. Thus the organization of the olfactory system in Drosophila is remarkably similar to the mammalian system where neurons with different receptors project to different glomeruli.

Because the numbers of receptors are much smaller, it has been possible to examine the selectivity of receptors to different odors. To determine the selectivity of different receptors, the responses of 24 different odor receptors to 110 different chemicals were examined. These studies revealed that approximately 70% of the

odor-receptor combinations did not show increased firing rates, 30% showed increased firing rate, and 10% showed inhibitory responses. The number of odors that a receptor recognizes (out of the 100 tested) varied from none to thirty, with receptor response profiles ranging continuously from very narrow to broadly tuned. In general, receptors that are broadly tuned respond to structurally similar compounds. A single odor activated from one to 16 different receptors. Overall, these studies demonstrate that a given odor is sparsely encoded by the activation of a few receptors.

Is there a chemotopic map in the Drosophila antennal lobe? Knowing the odors that different receptors recognize as well as the glomeruli associated with all receptors has made it possible to map the distribution of odor responses in the antennal lobe. These studies argue that broadly tuned receptors occupy medial regions of the antennal lobe, and narrowly tuned receptors occupy more lateral regions. However, there is no obvious chemotopic organization: neurons responding to the same stimuli are not clustered together.

Olfactory receptor neurons send axons to the antennal lobe. The second order neurons are projection neurons and local inhibitory interneurons. Each projection neuron sends dendrites to a single glomerulus and axons that branch to terminate in the protocerebrum and mushroom bodies (involved in associative learning) of the fly brain. The pattern of projections is invariant: each projection neuron sends dendrites to a specific glomerulus and has a characteristic, defined axon branching pattern in the protocerebrum and mushroom bodies that is invariant from individual to individual. These arborizations occur in the absence of olfactory neurons. Studies of the development of projection neurons argue that projection neuron identity is defined by the birth-order of the neuron. The picture that has emerged from these studies is that the connectivity of projection neurons is hard-wired.

In general, although flies and mammals are separated by approximately 400 million years in evolution, the principles of olfaction are conserved in these organisms. In fly olfaction, as in mammalian olfaction, one receptor is expressed per cell and neurons with the same receptor project to the same glomerulus. Moreover, the simplicity of the Drosophila system allows a complete description of the ligand-binding properties of ORs and their targeting in the olfactory bulb. Comparative studies of the olfactory systems of different organisms highlight the general strategies that animals use for chemical recognition. Interestingly, even organisms without a nervous system may have a sense of smell (see Box 24.7).

PHEROMONE DETECTION

The Accessory Olfactory System May Mediate Pheromone Detection

In addition to the main olfactory pathway, an accessory olfactory pathway exists for the detection of pheromones in many species. Pheromones are species-specific and gender-specific chemical cues emitted by an individual and detected by other members of the species that provide information about the individual's social, sexual, and reproductive status. In general, pheromones cause stereotyped and innate changes in conspecifics' behavior, either by direct, short-term effects on animal behavior or longer-term changes that alter the endocrine state. For example, one of the few vertebrate pheromones isolated is the sex pheromone of the red-sided garter snake. Female snakes produce this substance that is sufficient to induce courting behavior in males. Another well-described pheromone is made by slave-maker ants. They emit a blend of chemicals that serve as an alarm signal, causing the slave-makers to attack other ants' nests, and the resident ants to flee. An uncharacterized pheromone is thought be responsible for the Bruce effect in mice: a behavior in which the scent of a novel male will terminate the pregnancy of a recently inseminated female, providing the novel male the opportunity to produce offspring.

The sensory structure for pheromone detection is the vomeronasal organ (VNO), also referred to as Jacobson's organ (Fig. 24.21). The VNO is a cigar-shaped cavity in the anterior nasal septum surrounded by bone. Openings in the anterior cavity contact either the oral cavity or the nasal cavity, depending on the species, for the detection of pheromonal cues. The sensory epithelium of the VNO is divided into anterior and posterior layers. Each layer contains sensory neurons that detect soluble and volatile cues. Different receptors and signaling molecules are found in the two layers, suggesting that they serve different functions.

Axons of sensory neurons travel from the VNO to the accessory olfactory bulb (AOB), which is located at the dorsal posterior surface of the main olfactory bulb (MOB) in rodents. Projection neurons then send axons to the bed nuclei of the accessory olfactory tract and stria terminalis, and the posteromedial cortical and medial nuclei of the amygdala. These projections do not overlap with projections from the main olfactory bulb, demonstrating that there is segregation of processing from the two systems.

Information Flow from the VNO to the AOB

The sensory epithelium of the VNO is divided into an apical and a basal layer. Neurons in the apical layer express V1R receptors and those in the basal layer express V2R receptors (see next). In general, each sensory neuron expresses one receptor, similar to the olfactory system of mammals. This suggests that each sensory neuron detects a small subset of chemical cues.

Neurons from the apical layer project to the anterior region of the AOB and neurons from the basal layer project to the posterior AOB. Unlike the olfactory system, neurons with the same receptor do not project to one or a few glomeruli (Fig. 24.21). Instead, neurons with the same receptor converge onto multiple glomeruli (from 6–30) in spatially restricted domains forming a complex pattern that is loosely conserved from animal to animal. Dendrites of mitral cells synapse onto several glomeruli in a complex fashion. Mitral cell dendrites have been reported to contact glomeruli from neurons with the same receptor as well as from neurons with different receptors. Thus unlike the olfactory system, a single mitral cell does not synapse onto a single glomerulus. The complex connectivity suggests that integration of different sensory inputs occurs in the AOB. One hypothesis is that this organization allows the animal to respond only to the relevant mixture of pheromone cues. For example, the simultaneous activity of different sensory receptors may be necessary to activate mitral cells, such that mitral cells respond to specific blends of pheromones. In this model, only the right mix of pheromones will trigger stereotyped innate behavior.

Signal Transduction in the VNO

Different signaling molecules reside in the apical and basal compartments of the VNO. In the apical layer, the V1R family of candidate pheromone receptors is expressed. There are approximately 165 full-

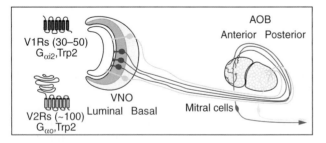

FIGURE 24.21　The accessory olfactory system of the mouse. The vomeronasal organ in the nasal cavity contains two layers. The apical layer expresses the V1R family of pheromone receptors and the basal layer contains the V2R family. These neurons project to the accessory olfactory bulb (AOB) where they form several glomeruli. From Dulac, 2000.

length V1R genes in mice, 106 in rat, and only two in humans. These G-protein coupled receptors all contain a short amino-terminus extracellular domain and are distantly related to the T2R family of mammalian bitter receptors. Only one receptor-ligand pair has been identified for this gene family: the receptor V1Rb2 recognizes hepatanone in isolated VNO neurons. Mice lacking 12 V1Rs do not detect a subset of candidate pheromones by electrophysiology studies of VNO field responses. These mice also show behavioral defects in stereotyped behaviors thought to be mediated by pheromones: for example, the female mice do not show maternal aggression (normally, nursing mothers fend off intruders). These findings argue that the V1R family detects pheromones and mediates behavior. Neurons that express V1Rs also contain the Gαi2 alpha subunit of heterotrimeric G proteins and the ion channel TRPC2 as shown by *in situ* hybridization and immunocytochemical staining. This suggests that these molecules transduce information from the activation of V1Rs to cell depolarization.

The V2R family of chemoreceptors is expressed in the basal layer of the VNO. This receptor family has a long amino-terminal extracellular domain and shares sequence similarity with the mammalian T1R family of taste receptors. There are approximately 60 intact V2R receptor genes in mice and rats. All genes are expressed exclusively in the VNO suggesting that they mediate pheromone detection. However, no ligands have been identified for these receptors and mice lacking V2Rs have not yet been engineered. The Gαo subunit of heterotrimeric G proteins is expressed in the basal layer along with V2Rs. In addition, the TRPC2 ion channel that is expressed in the apical layer also is found in the basal layer.

Mice Lacking TRPC2 Show Behavioral Defects

Because the TRPC2 ion channel is expressed in both the apical and basal layer of the VNO, it may act as an ion channel mediating detection of pheromones. To test this, mice were generated that lack the TRPC2 channel. The VNO is not activated by candidate pheromones in these mice by electrophysiological recordings, demonstrating a deficit in VNO function. If the VNO is not functional, one prediction would be that these mice would lack pheromone-driven social behaviors such as mating and aggression. Consistent with this, mice lacking TRPC2 do not show aggressive behavior. Surprisingly, male mice without TRPC2 indiscriminately mate both females and males. The simplest interpretation of these studies is that mating does not require the VNO; instead, the VNO is required

to inhibit male–male courtship behavior and promote discrimination between the sexes. One caveat with these studies is that it is not clear that loss of TRPC2 eliminates all VNO function; therefore it will be important to generate mice specifically lacking the VNO to determine its role in social and sexual behavior.

Pheromone Detection in Humans?

Although evidence is accumulating that the VNO participates in pheromone-driven behavior in mice, the role of the VNO in human behavior is far less certain. Anatomical studies of the vomeronasal organ argue that this structure exists but lacks sensory neurons and nerve bundles in the adult. The accessory olfactory bulb exists in the fetus but regresses and is not present in the adult. Genomic studies demonstrate the TRPC2 gene is a nonfunctional pseudogene in humans. Moreover, almost all V1R pheromone receptors are pseudogenes (115/117). There are only two intact V1R sequences and no functional V2R sequences in the human genome. Taken together, these studies strongly argue that the accessory olfactory system does not function in humans.

The loss of VNO in humans can be interpreted in two ways. One possibility is that humans do not detect pheromones and have evolved other strategies to ensure appropriate detection of mates and attackers. Alternatively, it is possible that humans detect odors through the main olfactory system that act as pheromones to elicit innate, stereotypical behaviors involved in reproduction and aggression.

Comparison of the Main Olfactory System and the Accessory Olfactory System

Although the accessory olfactory system has been classically thought to detect pheromones and the main olfactory system to detect a vast array of volatile cues, the distinction between the two noses is blurred both in terms of the compounds they detect and the behaviors they mediate. For example, electrophysiological studies show that sensory neurons in the VNO respond to volatile and nonvolatile compounds. Similarly, olfactory neurons depolarize in response to compounds classified as pheromones. Mice that have a defective VNO (lacking TRPC2) lack many pheromone mediated behaviors, but not all. For example, pup suckling is a stereotyped behavior that is normal in TRPC2 mutants. Mice with a defective MOE (lacking the olfactory cyclic nucleotide-gated channel) do not mate, fight, or suckle, showing that the olfactory system also is required for innate stereotyped behavior. Thus, the best ways to distinguish the main and

---

BOX 24.7

DO PLANTS HAVE A SENSE OF SMELL?

Plants have devised a number of strategies to detect information about the outside world, and have mechanisms for detecting light, nutrients, and mechanical stimulation. Recent studies suggest that they might have a sense of smell as well (Runyon et al., 2006). The dodder (also called stranglethread or witches shoelaces) is a small parasitic plant that does not contain machinery for photosynthesis. Instead, this major agricultural pest must attach onto host plants such as tomatoes for nutrients. In order to examine how the dodder finds its way to its host, experiments were done to test the sensory cues necessary to direct growth. A dodder plant directs its growth toward a tomato plant, and the scent, but not the sight, of a tomato is sufficient to direct dodder growth. Several individual volatile compounds were also able to direct growth, suggesting that dodders can sense multiple odors. Whether it is a general principle that plants recognize odors remains to be seen. The diversity of odors recognized by plants and the mechanism are also unknown. However, it is interesting to consider that plants also have primary senses and translate information about the outside world into behavior, even without a brain.

Kristin Scott

---

accessory olfactory systems are not by the ligands they recognize or the behaviors they mediate, but by their anatomical segregation in the periphery and central nervous system.

Summary

The olfactory system enables animals to detect and discriminate between thousands of different odors. The enormous number of olfactory receptors allows for the ability to detect a vast array of odors. Each olfactory receptor recognizes a subset of chemical cues, with one odor activating more than one receptor and one receptor recognizing more than one odor. In the periphery, each olfactory neuron expresses only one olfactory receptor. Neurons with the same OR are distributed randomly in the nasal epithelium but they all project to the same glomerulus in the olfactory bulb. Thus, the activation of different olfactory receptors leads to the activation of different glomeruli in the central nervous system. Different combinations of activated glomeruli represent different smells. In *Drosophila* as well as mammals, one receptor is expressed per cell and neurons with the same receptor project to the same glomerulus, demonstrating that the logic of olfaction has been maintained through evolutionary time. In addition to the main olfactory pathway in vertebrates, a parallel accessory pathway exists for transmitting signals from less volatile odorous compounds called pheromones.

Higher order processing begins in the olfactory bulb, where mitral/tufted cells synapse onto olfactory neurons and transmit this information to the olfactory cortex. Periglomerular cells and granule cells provide inhibitory connections in the bulb that shape olfactory responses. Information is then relayed to five different brain regions where it is ultimately translated into different odor percepts and behavior.

References

Adler, E., Hoon, M. A., Mueller, K. L., Chandrashekar, J., Ryba, N. J. P., and Zuker, C. S. (2000). A novel family of mammalian taste receptors. *Cell* **100**, 693–702.

Adrian, E. D. (1950). The electrical activity of the mammalian olfactory bulb. *Electroencephalogr. Clin. Neurophysiol.* **2**, 377–388.

Amrein, H. and Bray, S. (2003). Bitter-sweet solution in taste transduction. *Cell* **112**, 283–284.

Belluscio, L., Gold, G. H., Nemes, A., and Axel, R. (1998). Mice deficient in G(olf) are anosmic. *Neuron* **20**, 69–81.

Belluscio, L. and Katz, L. C. (2001). Symmetry, stereotypy, and topography of odorant representations in mouse olfactory bulbs. *J. Neurosci.* **21**, 2113–2122.

Brunet, L. J., Gold, G. H., and Ngai, J. (1996). General anosmia caused by a targeted disruption of the mouse olfactory cyclic nucleotide-gated cation channel. *Neuron* **17**, 681–693.

Buck, L. and Axel, R. (1991). A novel multigene family may encode odorant receptors: a molecular basis for odor recognition. *Cell* **65**, 175–187.

Duchamp-Viret, P., Chaput, M. A., and Duchamp, A. (1999). Odor response properties of rat olfactory receptor neurons. *Science* **284**, 2171–2174.

Dulac, C. (2000). Sensory coding of pheromone signals in mammals. *Curr Opin Neurobiol* **10**, 511–518.

Dulac, C. and Axel, R. (1995). A novel family of genes encoding putative pheromone receptors in mammals. *Cell* **83**, 195–206.

Gogos, J. A., Osborne, J., Nemes, A., Mendelsohn, M., and Axel, R. (2000). Genetic ablation and restoration of the olfactory topographic map. *Cell* **103**, 609–620.

Hallem, E. A., Ho, M. G., and Carlson, J. R. (2004). The molecular basis of odor coding in the Drosophila antenna. *Cell* **117**, 965–979.

Hamilton, K. and Kauer, J. S. (1989). Patterns of intracellular potentials in salamander mitral tufted cells in response to odor stimulation. *J. Neurophysiol.* **62**, 609–625.

Heck, G. L., Mierson, S., and DeSimone, J. A. (1984). Salt taste transduction occurs through an amiloride-sensitive sodium transport pathway. *Science* **223**, 403–405.

Hildebrand, J. G. and Shepherd, G. M. (1997). Molecular mechanisms of olfactory discrimination: Converging evidence for common principles across phyla. *Annu. Rev. Neurosci.* **20**, 595–631.

Hoon, M. A., Adler, E., Lindemeier, J., Battey, J. F., Ryba, N. J., and Zuker, C. S. (1999). Putative mammalian taste receptors: a class of taste-specific GPCRs with distinct topographic selectivity. *Cell* **96**, 541–551.

Huang, A. L., Chen, X., Hoon, M. A., Chandrashekar, J., Guo. W., Trankner, D., Ryba, N. J., and Zuker, C. S. (2006). The cells and logic for mammalian sour taste detection. *Nature* **442**, 934–938.

Imai, T., Suzuki, M., and Sakano, H. (2006). Odorant receptor-derived cAMP signals direct axonal targeting. *Science* **314**, 657–661.

Johnson, B. A. and Leon, M. (2000). Modular representations of odorants in the glomerular layer of the rat olfactory bulb and the effects of stimulus concentration. *J. Comp. Neurol.* **422**, 496–509.

Keller, A. and Vosshall, L. B. (2003). Decoding olfaction in Drosophila. *Current Opinion in Neurobiol.* **13**, 103–110.

Kim, U. K., Jorgenson, E., Coon, H., Leppart, M., Risch, N., and Drayna, D. (2003). Positional cloning of the human quantitative trait locus underlying taste sensitivity to phenylthiocarbamide. *Science* **229**, 1221–1225.

Laurent, G., Wehr, M., and Davidowitz, H. (1996). Temporal representations of odors in an olfactory network. *J. Neurosci.* **16**, 3837–3847.

Li, X., Staszewski, L., Xu, H., Durick, K., Zoller, M., and Adler, E. (2002). Human receptors for sweet and umami taste. *Proc Natl Acad Sci USA* **99**, 4692–4696.

Malnic, B., Hirono, J., Sato, T., and Buck, L. B. (1999). Combinatorial receptor codes for odors. *Cell* **96**, 713–723.

Meredith, M. (2001). Human vomeronasal organ function: A critical review of best and worst cases. *Chem. Senses* **26**, 433–445.

Mombaerts, P. (1999). Seven-transmembrane proteins as odorant and chemosensory receptors. *Science* **286**, 707–711.

Mombaerts, P., Wang, F., Dulac, C., Chao, S. K., Nemes, A., Mendelsohn, M., Edmondson, J., and Axel, R. (1996). Visualizing an olfactory sensory map. *Cell* **87**, 675–686.

Mori, K. and Shepherd, G. M. (1994). Emerging principles of molecular signal processing by mitral/tufted cells in the olfactory bulb. *Semin Cell Biol* **5**, 65–74.

Mori, K. and Yoshihara, Y. (1995). Molecular recognition and olfactory processing in the mammalian olfactory system. *Prog. Neurobiol.* **45**, 585–619.

Mueller, K. L., Hoon, M. A., Erlenbach, I., Chandrashekar, J., Zuker, C. S., and Ryba, N. J. (2005). The receptors and coding logic for bitter taste. *Nature* **434**, 225–229.

Nelson, G., Chandrashekar, J., Hoon, M. A., Feng, L., Zhao, G., Ryba, N. J., and Zuker, C. S. (2002). An amino-acid taste receptor. *Nature* **416**, 199–202.

Nelson, G., Hoon, M. A., Chandrashekar, J., Zhang, Y., Ryba, N. J., and Zuker, C. S. (2001). Mammalian sweet taste receptors. *Cell* **106**, 381–390.

Oakley, B. (1967). Altered taste responses from cross-regenerated taste nerves in the rat. *In* "Olfaction and Taste II" (T. Hayashi, ed.), pp. 535–547. Pergamon, London.

Pelosi, P. (1998). Odorant-binding proteins: structural aspects. *Ann NY Acad. Sci.* **855**, 281–293.

Pfaffmann, C. (1955). Gustatory nerve impulses in rat, cat and rabbit. *J. Neurophysiol.* **18**, 429–440.

Rall, W. and Shepherd, G. M. (1968). Theoretical reconstruction of field potentials and dendrodendritic synaptic interactions in olfactory bulb. *J. Neurophysiol.* **31**, 884–915.

Runyon, J. B., Mescher, M. C., and DeMorales, C. M. (2006). Volatile chemical cues guide host location and host selection by parasitic plants. *Science* **313**, 1964–1967.

Sengupta, P., Chou, J. H., Bargmann, C. I. (1996). odr-10 encodes a seven transmembrane domain olfactory receptor required for responses to the odorant diacetyl. *Cell* **84**, 899–909.

Serizawa, S., Miyamichi, K., Nakatani, H., Suzuki, M., Saito, M., Yoshihara, Y., and Sakano, H. (2003). Negative feedback regulation ensures the one receptor-one olfactory neuron rule in mouse. *Science* **302**, 2088–2094.

Shepherd, G. M. and Greer, C. A. (1998). Olfactory bulb. *In* "The Synaptic Organization of the Brain" (G. M. Shepherd, ed.), 4th ed., pp. 159–203. Oxford Univ. Press, New York.

Shykind, B. M., Rohani, S. C., O'Donnell, S., Nemes, A., Mendelsohn, M., Sun, Y., Axel, R., and Barnea, G. (2004). Gene switching and the stability of odorant receptor gene choice. *Cell* **117**, 801–815.

Smith, D. V., Van Buskirk, R. L., Travers, J. B., and Bieber, S. L. (1983b). Coding of taste stimuli by hamster brainstem neurons. *J. Neurophysiol.* **50**, 541–558.

Stowers, L., Holy, T. E., Meister, M., Dulac, C., and Koentges, G. (2002). Loss of sex discrimination and male-male aggression in mice deficient for TRP2. *Science* **295**, 1493–1500.

Vosshall, L. B. (2000). Olfaction in *Drosophila. Current Opin. Neurobiol.* **10**, 498–503.

Wang, F., Nemes, A., Mendelsohn, M., and Axel, R. (1998). Odorant receptors govern the formation of a precise topographic map. *Cell* **93**, 47–60.

Wong, G. T., Gannon, K. S., and Margolskee, R. F. (1996) Transduction of bitter and sweet taste by gustducin. *Nature* **381**, 796–800.

Wong, S. T., Trinh, K., Hacker, B., Chan, G. C., Lowe, G., Gaggar, A., Xia, Z., Gold, G. H., and Storm, D. R. (2000). Disruption of the type III adenylyl cyclase gene leads to peripheral and behavioral anosmia in transgenic mice. *Neuron* **27**, 487–497.

Xu, F. Q., Greer, C. A., and Shepherd, G. M. (2000). Odor maps in the olfactory bulb. *J. Comp. Neurol.* **422**, 489–495.

Xu, P., Atkinson, R., Jones, D. N., and Smith, D. P. (2005). Drosophila OBP LUSH is required for activity of pheromone-sensitive neurons. *Neuron* **45**, 193–200.

Yokoi, M., Mori, K., and Nakanishi, S. (1995). Refinement of odor molecule tuning by dendrodendritic synaptic inhibition in the olfactory bulb. *Proc. Natl. Acad. Sci. U.S.A.* **92**, 3371–3375.

Zhang, Y., Hoon, M. A., Chandrashekar, J., Mueller, K. L., Cook, B., Wu, D., Zuker, C. S., and Ryba, N. J. (2003). Coding of sweet, bitter, and umami tastes: Different receptor cells sharing similar signaling pathways. *Cell* **112**, 293–301.

Zhao, G. Q., Zhang, Y., Hoon, M. A., Chandrashekar, J., Erlenbach, I., Ryba, N. J., and Zuker, C. S. (2003). The receptors for mammalian sweet and umami taste. *Cell* **115**, 255–266.

Zhao, H., Ivic, L., Otaki, J. M., Hashimoto, M., Mikoshiba, K., and Firestein, S. (1998). Functional expression of a mammalian odorant receptor. *Science* **279**, 237–242.

Suggested Readings

Axel, R. (1995). The molecular logic of smell. *Sci. Am.* 154–159.

Bargmann, C. I. (2006). Comparative chemosensation from receptors to ecology. *Nature* **444**, 295–301.

Buck, L. B. (2000). The molecular architecture of odor and phero-mone sensing in mammals. *Cell* **100**, 611–618.

Chandrashekar, J., Hoon, M. A., Ryba, N. J., and Zuker C. S. (2006). The receptors and cells for mammalian taste. *Nature* **444**, 288–294.

Dulac, C. and Torello, A. T. (2003). Molecular detection of phero-mone signals in mammals: From genes to behaviour. *Nat Rev Neuroscience* **4**, 551–562.

Scott, K. (2005). Taste recognition: Food for thought. *Neuron* **48**, 455–464.

Shepherd, G. M. (1994). Discrimination of molecular signals by the olfactory receptor neuron. *Neuron* **13**, 771–790.

Kristin Scott

Somatosensory System

The somatic sensory system is burdened with many responsibilities. Beyond the obvious need to bring to consciousness events that occur along the skin surface, this system provides an organism with knowledge of where it is in space and it provides the motor system with feedback to control and coordinate action. Those are functions so diverse that minute variations on a common theme cannot accomplish them all. That is, the many demands on the somatosensory system cannot be met by subtle changes in gene expression or pattern of synapses or location of cell bodies. As a result, single neurons, large ensembles, and groups of interconnected regions use different tactics and strategies to achieve a final goal.

PERIPHERAL MECHANISMS OF SOMATIC SENSATION

All Somatic Sensation Begins with Receptors and Ganglion Cells

Neurons of the dorsal root and trigeminal ganglia are the only routes by which mammals receive information from the periphery of the body and of the face. These are pseudo-unipolar cells, with a single process—an axon—that divides close to the cell body and sends separate branches out to the periphery and into the central nervous system (Fig. 25.1). Under normal circumstances neurons in the 31 pairs of dorsal root ganglia and the single pair of trigeminal ganglia receive no synapses. That situation can change, however, when the peripheral branches of ganglion cells are cut and axons of neurons in nearby sympathetic ganglia are attracted to enter and innervate. In the absence of

this pathology, the cell body of a somatic sensory ganglion cell is properly considered a factory for all that must be made to keep its two branches functioning properly. Its major function is to keep a very long axon healthy enough to avoid conduction failures, particularly at the T junction near the cell body. This is a function large ganglion cell bodies do very well, as the incidence for conduction failures in their axons is reported to be zero. Even where failures at the T-junction are reported, as in the small ganglion cell bodies of some vertebrate species, the effect appears deliberate and is tied to the difference in action potential shape seen in large and small cells.

Neuronal cell bodies in the dorsal root and trigeminal ganglia vary in size and in gene expression (Fig. 25.2). Although the details of size vary among species according to differences in body size, all species examined to date include two populations, one at least half again larger than the other. Variations in the density of Nissl staining combined with the size difference lead to a simple division of ganglion cells into large light and small dark neurons. The group with the larger somata is outnumbered by at least 3-to-1 by those with the smaller somata.

Peripheral branches of dorsal root ganglion cells enter peripheral nerves. Those of the trigeminal ganglion enter the trigeminal nerve. In a mixed nerve, with both sensory and motor axons, the ganglion cells provide more than a fair share of the large, myelinated axons and all of the thinly myelinated and unmyelinated axons. Conduction velocities vary because of these differences in axon diameter and degree of myelination. Classic studies divide the sensory axons in human peripheral nerve into the following groups with conduction velocities in parentheses: Group I or Aα (70–120 meters/second), Group II or Aβ (40–70

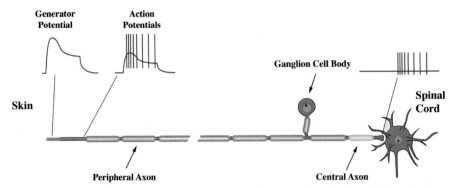

FIGURE 25.1 A dorsal root ganglion cell is a pseudo-unipolar neuron with an axon that divides at T-junction into a peripheral branch and a central branch. At the tip of the peripheral branch are receptor proteins that, through opening of cation channels, produce a depolarization called a generator potential. With sufficient depolarization, voltage-gated Na+ channels open to initiate action potentials. These action potentials are conducted down the axon and into the central branch that innervates second-order neurons in the spinal cord (in this case) or in the medulla.

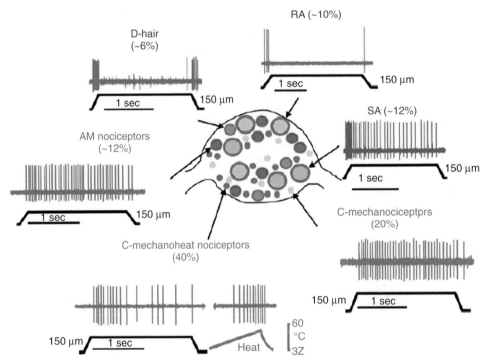

FIGURE 25.2 A somatosensory ganglion (center) is populated by a broad range of neurons that differ in size, gene expression, and receptive field properties. Mechanoreceptors (in blue) differ in how they respond to a sustained stimulus. Rapidly adapting (RA) afferents of the glaborous skin and D-hair receptors or hair follicle receptors generate short bursts of action potentials at stimulus onset and offset. Slowly adapting (SA) afferents respond to a sustained indentation of skin with a prolonged series of action potentials. Nociceptors (in red) also vary in size and conduction velocity of their axons (Aδ vs. C) and in their response to noxious mechanical stimulation, such as a hard pinch. Some receptors respond specifically to this stimulus (AM nociceptors and C-mechanonociceptors) whereas others respond to a broad range of noxious stimuli (C-mechanoheat nociceptors). From Lewin, G. R. and Moshourab, R. (2004).

meters/second), Group III or Aδ (12–36 meters/second), and Group IV or C (0.5–2.0 meters/second). These groups correspond to functional classes that carry proprioceptive, mechanosensory, thermoreceptive, or nociceptive information from muscles, tendons, and skin to the spinal cord (Table 25.1).

What sets apart ganglion cells of the somatosensory system from all other neurons carrying the name ganglion cell is the requirements for double duty. As ganglion cells of the somatosensory system they carry into the brain and spinal cord all the information about skin and deep tissues the CNS will ever have. Unlike

TABLE 25.1 Summary of Primary Afferent Fibers and Their Roles

Modality	Submodality	Receptor	Fiber type	Conduction velocity (m s⁻¹)	Role in perception
Mechanoreception	SAI	Merkel cell	$A\beta$	42–72	Pressure, form, texture
	RA	Meissner corpuscle	$A\beta$	42–72	Flutter, motion
	SAII	Ruffini corpuscle	$A\beta$	42–72	Unknown, possibly skin stretch
	PC	Pacinian corpuscle	$A\beta$	42–72	Vibration
Thermoreception	Warm	Bare nerve endings	C	0.5–1.2	Warmth
	Cold	Bare nerve endings	$A\delta$	12–36	Cold
Nociception	Small, myelinated	Bare nerve endings	$A\delta$	12–36	Sharp pain
	Unmyelinated	Bare nerve endings	C	0.5–1.2	Burning pain
Propioception	Joint afferents	Ruffini-like and paciniform-like endings, bare nerve	$A\beta$	42–72	Protective function against hyperextension
	Golgi tendon organs	Golgi endings	$A\alpha$	72–120	Muscle tension
	Muscle spindles	Type I	$A\alpha$	72–120	Muscle length and velocity
		Type II	$A\beta$	42–72	Muscle length
	SAII	Ruffini corpuscle	$A\beta$	42–72	Joint angle?

the ganglion cells of auditory, vestibular, visual, and gustatory systems (and much like the olfactory sensory neurons that carry information about odors into the brain), ganglion cells of the trigeminal and dorsal root ganglia also transduce all forms of somatosensory stimulation. Somatosensory ganglion cells are also receptors. Inserted at the peripheral tips of the ganglion cells' axons are two types of cation channels. One is either physically tethered to peripheral tissue (and is therefore a mechanoreceptor protein) or responsive to changes in temperature, pH, or the concentration of circulating factors that signal tissue damage (thermoreceptors and nociceptors). By way of these channels the tips of ganglion cells are depolarized to produce generator potentials. In addition, near the peripheral terminals of all sensory axons are voltage-gated Na⁺ channels, responsible for initiation and conduction of action potentials. The best evidence suggests these voltage-gated channels are present at the first node of Ranvier in myelinated axons or at a similar distance from axon tip in unmyelinated axons (Fig. 25.1).

Many Types of Somatosensory Receptors Innervate Skin and Deep Tissues

The modern tradition breaks up the axons of sensory ganglia into 13 varieties of receptors. These include four types of proprioceptors, three types of nociceptors, two groups of thermoreceptors (one each for cooling and warming), and four types of mechanoreceptors. That is a decidedly conservative point of view. At a minimum it leaves unmentioned the class of D-hair receptor or hair follicle afferent and it recognizes only the major type of nociceptors. Molecular and functional distinctions among the nociceptors alone add enough additional types to make the list approach 20. The point here is not to split peripheral receptors into the greatest number of types, but to emphasize the unusual nature of somatosensory transduction.

Mechanosensory Axons of Glaborous Skin Are Split into Four Types

Four types of mechanosensory axons are found in the hairless or glaborous skin and immediately adjacent deeper tissues of the human body (Fig. 25.3). Whereas a decade ago it would have seemed a radical thought, we now recognize the four receptors are responsible for specific functions. These include peripheral events that lead to perception of form and texture, detection of object slip leading to adjustment of grip, sensory feedback necessary for the use of tools, perception of vibration, and perception of hand shape and limb position (Table 25.1).

All mechanoreceptors conduct action potentials in the $A\beta$ (type II) range, so no distinction can be made among them in that property. Rather it is in the response of the axon to a ramp-and-hold stimulus that

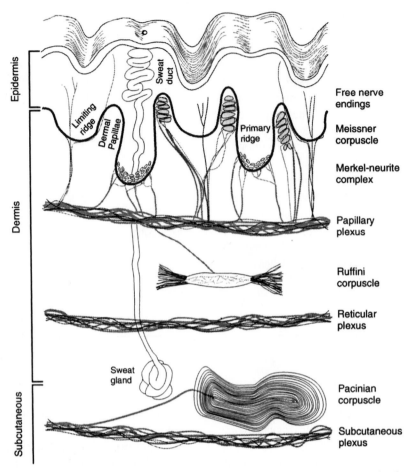

FIGURE 25.3 Peripheral receptors of the hairless (glaborous) skin are present in dermis, epidermis, and subcutaneous tissue. Superficial receptors at the dermis-epidermis border include the free nerve endings of nociceptors and thermoreceptors, the rapidly adapting afferents associated with Meissner's corpuscle, and the type I slowly adapting afferents that end as Merkel's disks. Deep receptors include a rapidly adapting receptor enclosed by a Pacinian corpuscle and a type II slowly adapting afferent in some species. Those SAII afferents are associated with Ruffini corpuscles in domestic cats but appear to have some other arrangement in most of the human hand. SAII afferents are missing from the skin of macaques and mice. From Johnson, K. (2002).

the four types become clear. Two types of afferents respond at elevated rates for as long as a blunt mechanical stimulus is applied to its receptive field. These are slowly adapting (SA) afferents, and they can be further divided into SAI and SAII types based on differences in their response to the applied stimuli (Fig. 25.1). Two other types of mechanoreceptor afferents are frequently called rapidly adapting (RA) or sometimes fast adapting or quickly adapting because their response to a prolonged indentation of skin is a spike or two at onset of the stimulus, perhaps a single spike at offset and nothing in between. They, too, are split into two varieties, RA and PC, differing principally in their sensitivity and response to vibrating stimuli (see later). We should also point out some investigators object to the use of the term, adaptation, to describe this change in a mechanoreceptor axon's response because, as we

will see, it is not the axon that adapts but the tissue around it. Nevertheless, the terms rapidly adapting and slowly adapting are illustrative and are in common usage, so will we use them here.

Pacinian Corpuscles

The work of Iggo and Muir in the late 1960s established a consistent relationship between the four types of peripheral mechanoreceptor axons and the morphology of axon tips, or more properly of the axons' relationship with nonneural cells. For three of the axons this structure-function relationship has held up well. The clearest example is that of the PC afferent, the unmyelinated tip of which ends as part of a Pacinian corpuscle. Unmistakable in its onion skin-like structure, the Pacinian corpuscle serves as a connective tissue filter of low frequency mechanical stimuli such as sustained pressure applied to the skin (Fig. 25.3).

PC afferents are exquisitely sensitive to vibration. They display a peak response near 200 Hz with skin indentations of no more than 10 nm. Yet as pointed out earlier, a single indentation of the skin surface produces only a couple of spikes from these axons. Careful dissection of the connective tissue that surrounds the PC axon shows the axon itself is capable of generating a steady burst of action potentials with continued application of a blunt probe. The axon does not adapt. Rather, a change in structure of the fluid-filled capsule carries the energy of a continually applied probe away from the axon tip and closes the cation channels responsible for mechanical transduction. By contrast, repeated application of a mechanical stimulus, such as occurs with a tuning fork vibrating at 200 Hz, produces a series of discrete transduction events and a series of action potentials. We can say with great confidence, then, that PCs are responsive to high frequency vibration at even the smallest magnitude. This extreme sensitivity to vibration turns the PC afferent into a detector of remote events. These are the receptors, for example, that respond as hands gripping a steering wheel vibrate when a car travels over a rough road. As a more common and practical matter the minute vibrations transduced by PC afferents provide information about the texture of surfaces during the manipulation of tools.

Meissner's Corpuscles

Lower frequency vibration, sometimes called flutter, produces a maximal response in RA afferents. As is the case of PCs, the correlation between this type of response and the structure of the afferent axon and its surrounding tissue is consistent. Each RA afferent ends as a stack of broad terminal disks within a Meissner's corpuscles. Both divergence and convergence is seen in the relationship between corpuscle and axon. Two RA afferents end in a Meissner's corpuscle whereas each afferent innervates anywhere between 20 and 50 separate corpuscles. In addition to the $A\beta$ axons, C fibers are also present in Meissner's corpuscles of monkey glaborous skin. Whether these axons play a role in mechanosensation or provide the Meissner's corpuscle with nociceptive and thermoreceptive properties is not yet known.

The anatomy of the RA afferent says a great deal about what this mechanoreceptor does. Meissner's corpuscles are found in dermal pockets of the adhesive ridges, as close to the epidermis as any dermal structure can be (Fig. 25.3). And their density is extraordinary, approaching 50 mm² in the index fingertip of a young adult. The result is an afferent very sensitive to even the slightest stretch of skin, as happens when a

slippery object moves in the hand. Yet the levels of divergence and convergence from a single RA afferent lead to large receptive fields (5 mm²). That feature and the filtering properties of the connective tissue capsule make them inappropriate for form and texture perception. RA afferents are responsible, instead, for the detection of objects slipping across the hand and fingers. They provide the sensory information that leads to the adjustment of grip force.

Merkel's Disks

Of the two slowly adapting afferents, SAIs end in a manner that prompts no arguments. A single SAI axon breaks into several branches that end in several closely packed dermal ridges. Each branch ends in a series of axon terminals and each terminal is enfolded by an epidermal Merkel cell. This confluence of dozens of axon terminals and their Merkel cells produces a surface elevation called a touch dome in the skin of cats and in the hairy skin of humans.

Years of study have brought us little closer to answering the question of what the Merkel cell does. Merkel cells contact the tips of SAI afferents by what look to be chemical synapses, and vesicles in the Merkel cell are filled with the amino acid neurotransmitter, glutamate. There are as yet, however, no clear indications that Merkel cells are capable of transducing a mechanical stimulus or that SAI afferents respond to chemical agents Merkel cells might release. Most data point to the axon tips of the SAI afferent as the site of mechanical transduction and so the prevailing wisdom at this time suggests Merkel cells work as supporting elements only.

SAI Afferents Are Responsible for Form, Texture, and Curvature

Studies by K. Johnson and S. Hsiao and their colleagues established the SAI afferent as the source of peripheral information used by the CNS to perceive both form and texture. Parallel studies by R. LaMotte and T. Goodwin and their colleagues documented a role of SAIs in the perception of object curvature. Form perception is approachable by a combination of psychophysics and neurophysiology because its dimensions are readily quantifiable. When the performance of individual receptors on the surface of a monkey's finger pad is compared with the performance of the monkey itself, only one type of afferent is sensitive enough to account for the animal's ability to discriminate the form of a mechanical stimulus. That receptor is at the tip of SAI afferents ending in Merkel's disk. One example is seen in the response to the type of dots used in Braille (Fig. 25.4). Only the SAI afferents respond to the minute edges present in these embossed

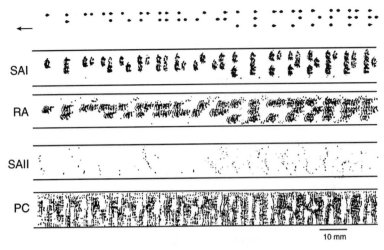

FIGURE 25.4 Response of peripheral axons to a Braille pattern of dots scanned over the surface of a human fingertip at a rate of 60 mm/s, with 200-μm shifts in position after each pass. Dots represent individual action potentials. Only the response of the SAI afferents (Merkel disk receptors) follows the Braille pattern faithfully, whereas RA afferents and Pacinians (PC) produce a response that distorts the input. SAIIs display little response to this stimulus. Adapted from Phillips *et al.* (1990).

dots (6.0 mm high or greater) so accurately as to account for the ability a monkey or a human has in telling one pattern from the next.

Peripheral factors contribute significantly to the SAI's contribution to form and texture perception. One of these factors is the unique sensitivity of SAI afferents to strain energy density, a term that Johnson describes as "the energy required to produce local deformation" per unit volume of skin. Thus rather than responding to the total energy generated by application of a uniform stimulus, SAI afferents respond to even the slight imperfections in that surface and the local deformation of skin they produce. Movement of the skin across a surface, as in scanning Braille, greatly increases the energy produced by any imperfection and leads to a greatly enhanced response by SAI afferents. This finding has a strong intuitive feel to it, given the much greater tactile acuity each of us has when we move a fingertip along a surface with even slight imperfections.

A second peripheral factor is that of surround suppression. Skin mechanics lead not only to local peaks in strain energy density but also to broader troughs. The result for any single SAI afferent are hot spots in its receptive field surrounded by tissue that, when simultaneously probed, leads to a reduction in the response of that afferent. An average terminal domain for an SAI afferent on the finger tip may be as much as 5 mm², but the presence of hot spots and surround

suppression permits these afferents to signal the presence of a stimulus (such a gap between two elevations) of as little as 0.5 mm.

In contrast to tactile form, texture has relatively few dimensions (rough-smooth, hard-soft, and sticky-slippery are the most prominent) but they are difficult to quantify. Nevertheless, a series of careful studies has documented that for variations in roughness, only the response of SAI afferents matches human perceptual ability. That is, the variation in firing rates among SAI follows precisely the perception human subjects have of surface roughness. Much the same can be said for the detection of surface hardness. Only the response of SAI afferents can account for human perception of how hard or soft is the surface of an object scanned by the fingertips. In summary, a combination of physiological and psychophysical studies leads one to conclude that SAI's provide the central somatosensory system with all the information it needs to detect the shape, hardness, and roughness of objects pressed or scanned across the skin.

SAII Afferents

A second slowly adapting afferent in human skin, the SAII afferent, differs from an SAI in the greater size of its receptive field, the reduced sensitivity to simple indentation of the skin, and the greater sensitivity to skin stretch. One surprising feature of these afferents is the less than universal presence of SAIIs across the short range of well-studied mammals. Direct recordings from the peripheral nerves of humans and of domestic cats show these afferents to be a commonly encountered feature of both species, but they do not exist in monkeys or mice. Just as perplexing is the poor correlation between structure and function with this receptor. The original correlative work in cats found SAII responses to arise from axons terminating in skin as Ruffini endings. These structures were first described in human skin at the turn of the twentieth century, but a twenty-first century study by M. Pare and colleagues found true Ruffini endings in human skin very rarely and only in the bed of fingernails. These findings indicate SAII responses over most of the human hand arise from some other arrangement of mechanoreceptor axon and connective tissue sheath. Recent recordings from human peripheral nerves have added to the conundrum by documenting the presence of a third SA variety—labeled SA3—that has properties intermediate between the other two slowly adapting types. A conservative conclusion from these findings is that some arrangement of nonneural cells, perhaps a classic Ruffini in cats but another configuration in humans, leads to the cardinal feature of SAII afferents, namely a robust response to skin stretch. That configuration could vary from one species to another and from one area of skin surface to the next, but in the end the non-neural tissue serves as a mechanical filter. Given the task of subtracting from the activity of a SAII afferent a response to simple deformation of the skin, the non-neural tissue leaves behind an unambiguous response to anything that stretches the skin.

Hair Follicle Afferents

In addition to receptor types found in glaborous skin, hairy skin is innervated by a separate receptor, called the D-hair receptor or hair follicle afferent (HFA). It is the most sensitive receptor in hairy skin. The HFA threshold is said to be one-tenth that of any other afferent in mouse skin, and displacement of a hair follicle by as little as $1\mu m$ produces robust responses in this population of receptors. Single afferents innervate more than one hair follicle and as a result, the receptive field of an HFA is large ($>10\,mm^2$ in mice). Unlike other mechanoreceptors, HFA's conduct action potentials in the $A\delta$ range, which translates into a velocity of 20–25 m/s for humans but a much lower velocity in mice.

Receptor Density

A plot of SAI and SAII receptors along the human skin surface shows receptive field size and density varies by a factor of four in the short distance from the fingertip to the wrist. As is the case for other sensory systems, wherever receptor density declines receptive field size ascends. This inverse variation in receptive size and density is reflected in the ability of a human subject to discriminate the two-dimensional shape of an object. Human performance in this area has been measured classically as two-point discrimination (Fig. 25.5). Two blunt probes applied to a skin surface can be moved together to produce a perception of a single probe. How close they can be and still produce the percept of two separate probes says something about the receptive field size and density of receptors that underlie the function referred to as fine touch. Normal human subjects are able to detect two probes separated by as little as one millimeter on the surface of the distal pad of the index finger and face. Acuity declines in other regions of the body and is poorest on the back where two points cannot be distinguished from one until they are about 70 mm apart. Regions of high spatial acuity (the hand and face) are where form perception is greatest and are analogous to the fovea of the retina.

Muscle Spindles and Golgi Tendon Organs Are Proprioceptors

The most prominent receptors fulfilling the function of sensing position and movement are muscle

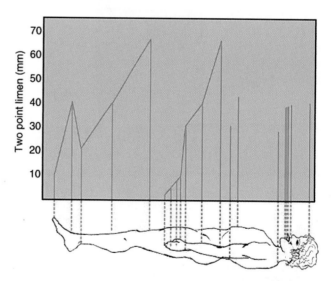

FIGURE 25.5 Variation in two-point limen (threshold) across the body surface. The graph plots the distance necessary for a human subject to detect two blunt probes as separate stimuli. That distance is lowest for the fingertips and mouth (approximately 10 mm) and highest for the legs, shoulders, and back (as much as 70 mm). From Patton, H. D., Sundsten, J. W., Crill, W. E., and Swanson, P. D. (eds.) (1976). "Introduction to Basic Neurology," p. 160. Saunders, Philadelphia.

receptors and tendon receptors. These two have in common their sensitivity to stretch and the large diameter of the axons that carry the receptors' activity into the CNS. The manner in which they are arranged, however, makes all the difference in the world. Muscle receptors are arranged in parallel with the muscle fibers and as a result, these afferents respond when the muscle is stretched. As a common occurrence, muscle fibers stretch when load is added to them in the form of weight or resistance. The resulting stretch of the extrafusal or work muscle fibers produces a simultaneous stretch of the much smaller intrafusal muscle fibers. The intrafusal fibers have their own motor

innervation and a sensory innervation from the largest diameter axons in a sensory nerve (Fig. 25.6). These sensory axons are called Ia afferents and they end in one of two configurations around the noncontractile portion of an intrafusal muscle fiber, where they signal the static or dynamic aspects of muscle stretch. All of these—sensory axons, intrafusal muscle fibers, and motor axons—are surrounded by a connective tissue capsule to form a muscle spindle. The contribution of the motor axons to all this is rather simple to envision—by adjusting the contractile state of the intrafusal muscle fiber, they adjust the sensitivity of the muscle afferent. When the intrafusal fiber is contracted

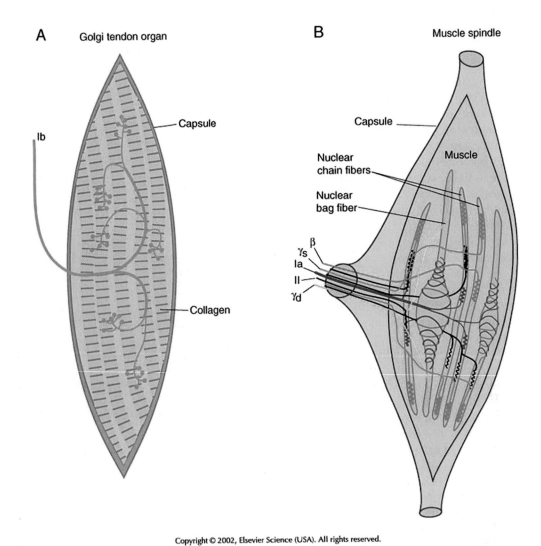

FIGURE 25.6 Proprioceptive afferents. (A) Golgi tendon organs and their termination along collagen fibers of the tendon capsule. These afferents respond when the entire capsule is stretched, usually by overvigorous contraction of the muscle. (B) Muscle spindle afferents (Ia and II) terminate on the noncontractile portions of intrafusal muscle fibers. They are arranged in parallel with work muscle fibers and respond to stretch of the entire muscle. Specialized motoneurons (γ) provide the motor innervation of the intrafusal muscle fibers and control the overall sensitivity of the muscle spindle.

the spindle is at its most sensitive and when the intrafusal fiber is relaxed, the spindle is least sensitive.

Much as muscle receptors are sensitive to muscle stretch, tendon receptors (referred to as Golgi tendon organs) are sensitive to tendon stretch and provide information about muscle force. Because tendons are arranged in series with the muscles, so too are tendon receptors. What stretches the tendon is muscle contraction. In this regard, muscle spindles and Golgi tendon organs signal the opposite trends. Spindles fire when a muscle is relatively inactive and stretched whereas tendon organs fire when the muscle is most active and contracted.

Proprioception Involves More Than Proprioceptors

By a definition of proprioception as an awareness of the position of the body and limbs, the two most prominent proprioceptors are necessary but not sufficient. More than just muscle and tendon receptors are at work. The related sense of kinesthesia, or awareness of position of moving body parts, is equally complex. Often used interchangeably for position sense, these two are complex percepts that require the contribution of receptors in muscle, skin, and joints as well as a sense of muscle exertion. Studies of the 1950s and 1960s focused on the contribution of joint afferents to a sense of limb position. The connective tissue and bones of joints are richly innervated and would seem to be in an ideal location to signal limb position. Yet a series of findings in the 1970s made clear that a sense of position survived joint removal; and the joint afferents, themselves, responded only at the extreme limits of joint flexion. With no response in the usual, midrange of joint movement, these afferents could not signal position under most circumstances.

Studies of more recent vintage have dealt principally with muscle spindle afferents as a major source of the position signal. Since all muscles are organized as antagonistic pairs, contraction of one delivers a robust stimulus to afferents of the antagonistic, stretched muscle. From these sorts of inputs, limb position and movement would appear to require a simple neural computation. That position sense is significantly affected by stimulation, anesthesia and disengagement of spindle afferents adds weight to the argument that the burden of signaling limb position and movement falls on these receptors. Perhaps the most convincing data are from studies of illusory movements produced by tendon vibration. This type of stimulus selectively activates spindle afferents and leads to the perception of limb movement when none has occurred. Illusory movement is muscle specific, so that activation of arm flexors gives rise to the percept of an arm that has extended (a movement that normally produces activation of flexor muscle afferents). The illusion is so strong it gives rise to the Pinocchio effect: if arm flexor afferents are activated when a subject is touching his or her nose with an index finger, the nose itself, is perceived to grow.

Only in the past several years has proper attention been paid to the role of cutaneous receptors in position sense. In recording from peripheral nerve axons of human subjects, B. Edin has made a strong case for a unique response by an ensemble of SAII afferents to any position a limb or digit might adopt. Particularly compelling is the role these skin stretch receptors play in signaling the position of fingers. For the hand and fingers, therefore, a conscious sense of position appears to arise from the cooperative activity of SAII afferents, muscle afferents, and (at the extremes of movement) joint afferents.

NOCICEPTION, THERMORECEPTION, AND ITCH

Nociceptors Respond to Noxious Stimuli

For most purposes anything that has produced tissue damage or that threatens do so in the immediate future can be defined as noxious and the type of axon that responds selectively to the noxious quality of a stimulus is, by definition then, a nociceptor. These are not pain receptors because nociception is not pain just as sensitivity to the wavelength of light is not color perception. Both are CNS constructs of peripheral events. Unlike color, pain has not only a perceptual component that involves comparison across receptors but also rich psychological and cognitive components. We will deal with those elements of pain perception but for now the most important principle to grasp is also intuitively obvious: the range of stimuli that are perceived as painful is very broad, from heat above 42°C to acids below pH6, from a sharp pinch on the fingertip to a swollen ankle. Each of these is tied to the activity in a variety of nociceptors.

True for all nociceptors is the simple morphology of their axon terminations. These are usually described as free nerve endings because unlike mechanoreceptive afferents, nociceptors end in no specialized capsule of nonneural cells. Another way of looking at this relationship is to notice that nothing extraneuronal serves as a filter or buffer between the nociceptive axon tip and its immediate environment. The only thing that determines the response of a nociceptor, then, is the type of protein receptor it inserts into its membrane.

Nociceptors also differ from mechanoreceptors in how broadly the terminals of the peripheral axon branch as they reach target. Take the tip of the human index finger as an example. An individual SAI and SAII axon ends in a well-confined cluster of terminals over a distance as small as a few millimeters. Terminals of a single C fiber, by contrast, end over an area of more than a dozen millimeters. This is the first of several anatomical features along the nociceptive pathway that, together, produce a much coarser spatial sense for pain than exists for mechanosensation. Only the simultaneous activation of mechanoreceptors as occurs with puncture wounds or damaging compression permits a person to accurately detect the location of a nociceptive stimulus.

The afferents that make up the nociceptive population can be subdivided into groups that are named by their axon conduction velocity (Aδ vs. C) and the response to noxious mechanical stimuli and noxious heat. Thus, an AM receptor conducts in the Aδ range and responds to intense mechanical stimuli whereas a CMH receptor conducts in the C range and responds to both noxious mechanical energy and noxious heat (Fig. 25.1). Other permutations of conduction velocity and response type are evident in the peripheral nerves of humans and other mammals, but these two are most common. They are frequently referred to as specific mechanical nociceptors and polymodal nociceptors.

First and Second Pain

Many afferents responding specifically to a mechanical stimulus (AM receptors) carry the more rapid signals into the CNS, whereas those responding to the broad range of noxious stimuli (CMH receptors) conduct action potentials more slowly. These are the peripheral components to the two very different qualities of pain perceived by humans (Fig. 25.7). First pain or epicritic pain is rapidly perceived and carries with it much that is discriminative. A person can quickly and with some ease figure out what has happened and where it has happened when he or she drops a heavy object onto a toe or touches a hot stove surface with a couple of fingers on the right hand. First pain is informative and the peripheral component of it is the population of Aδ nociceptors. What follows later is second pain or protopathic pain. This is agonizing pain that carries much less information about location or source of energy. Second pain is punishing pain that serves to change the behavior of a person. Its peripheral component is the population of C nociceptors.

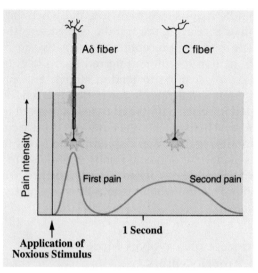

FIGURE 25.7 The two classes of nociceptors that conduct action potentials in the C and Aδ ranges are peripheral components for two types of pain. First pain carried by Aδ axons reaches consciousness rapidly and is discriminative. Both the location and the subjective intensity of the stimulus can be judged with relatively good precision in first pain. Second pain, in contrast, is much slower and is agonizing pain, with greatly reduced discriminative value.

Two varieties of C nociceptors are found in mammalian skin. One of these is characterized by the presence of fluoride-resistant acid phosphatase (FRAP) in its cytoplasm and cell-surface glycoproteins recognized by the isolectin I-B4 and the monoclonal antibody LA4. As none of these proteins is known to contribute to the physiological features of this C fiber type their presence is currently a convenient feature that allows anyone studying them to recognize and target them. The situation is very different in the second type of C nociceptor. Present in its cytoplasm are two neuroactive peptides, calcitonin gene-related peptide (CGRP) and substance P. These play major roles in the function of the peptide-containing type of C nociceptor.

The Axon Reflex

Release of neuropeptides from the second type of C nociceptor is responsible for the axon reflex (Fig. 25.8). Injury to the skin surface is often well confined, as happens with a paper cut, for example. Yet within a short time of that injury, tissue surrounding the cut becomes reddened in what is referred to as flair and edema or swelling sets in as the tissue fills with fluid. Most importantly, the region surrounding a punctate wound becomes painful to touch even though it is outside the zone of direct damage.

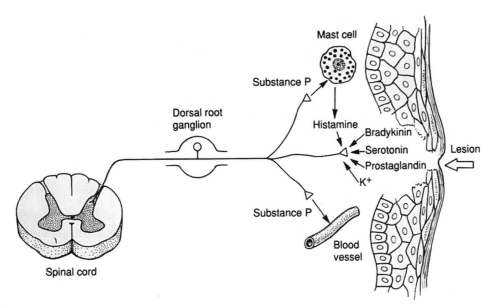

FIGURE 25.8 The axon reflex is a mechanism by which a class of C fibers communicates with both the spinal cord and peripheral cells. Action potentials are generated at one branch of a peptide-containing C fiber through tissue damage and the release of chemical signaling factors, including bradykinin and prostaglandin. Those action potentials invade not only the central branch of the axon but also the other peripheral branches. Substance P and CGRP released from the terminals of the C fiber induce mast cells to release histamine and promote swelling by widening arterioles and shrinking the diameter of venules.

Recall that an individual C fiber terminates over a wide area in skin, and so it is very likely that a punctate wound directly affects only a fraction of all that fiber's many branches. Action potentials generated at those directly affected branches invade all the peripheral branches as well as the parent axon that conducts the signal to the CNS. At all the peripheral terminals, substance P and CGRP are released onto two principal targets, the smooth muscle surrounding peripheral blood vessels and histamine-rich mast cells. By causing the arterial smooth muscles to relax, the peptides increase the flow of blood into the neighborhood of damaged tissue and produce a flow of water and electrolytes out of capillaries and into extracellular space. This process is referred to as extravasation. Histamine released from mast cells leads to a pronounced inflammatory response. All of this is important for the infiltration of damaged tissue with cellular elements that will protect against infection and promote repair. Yet what is most salient about the axon reflex is the much greater sensitivity of the tissue surrounding a wound to anything that might be noxious. This primary hyperalgesia is a direct result of the axon reflex, produced because the protein receptors inserted into nociceptive axons are sensitive to chemical changes produced by that reflex.

Both the histamine released by mast cells and the edema resulting from extravasation affect the response of nociceptors. Histamine's effects are more selective,

as only a subclass of the most slowly conducting C fibers insert histamine receptors into the membranes of their axon terminals. A considerable body of evidence suggests these are peripheral receptors whose activity leads to the unpleasant percept of itch. A parallel system for itch that operates independently of histamine release has been discovered recently but the population of afferents involved is not yet known.

Edema causes a general reduction in pH of extracellular fluid from 7.4 to below 6.0. As outlined later, a general feature of protein receptors inserted into nociceptive axons is their sensitivity to the concentration of H^+. By this path, activation of one branch of a C fiber leads to increased sensitivity of all its branches and all neighboring nociceptors to noxious stimulation. At least some of those neighboring nociceptors are likely to be silent nociceptors, so named because they are unresponsive to intense mechanical stimulation under normal circumstances. When activated by changes in pH or the presence of factors that signal the presence of tissue damage (see later) the silent nociceptors become responsive to noxious stimuli.

Chemical Sensitivity of Nociceptors

Two of the most powerful pain-inducing chemicals that can be delivered to a human observer are natural products of tissue damage. They are lipids of the prostaglandin family and the nonapeptide bradykinin

(Fig. 25.8). Prostaglandins are derivatives of the membrane fatty acid, arachidonic acid, which is itself a major component of the lipid bilayer. Damage to tissue and the resulting disruption of cell membranes releases arachidonic acid into extracellular fluid, where it is broken down by the enzyme, cyclo-oxygenase (COX), to form prostaglandin. Inhibitors that target both forms of COX or more specifically target the COX-2 enzyme are taken by the hundreds of thousands every day to relieve pain. These are referred to as nonsteroidal anti-inflammatory drugs (NSAIDS) and they include over-the-counter drugs such as aspirin and ibuprofen. Their popularity is a powerful indication of effectiveness in suppressing pain by interfering with the production of prostaglandins.

Nociceptor Proteins

Polymodal nociceptor activation occurs by way of a widespread family of receptor channels referred to as transient receptor potential (TRP). Relevant to the function of nociceptors is the subfamily of TRP-V channels, particularly of the channel protein TRP-V1 (Fig. 25.9). The V in the channel's name refers to its sensitivity to vanilloids, the active ingredients of chili peppers. TRP-V1 is a nonspecific cation channel with

approximately 10 times more permeability to Na^+ than to Ca^{2+}. Like other members of the TRP-V subfamily, TRP-V1 is a molecular thermometer that opens in response to increased temperature. What makes TRPV-1 a receptor responsive to other noxious stimuli is the effect these stimuli have on the threshold for channel opening. When exposed to high concentrations of H^+ or to agents such as prostaglandins or bradykinin the threshold temperature at which the channel opens descends to normal body temperature. By this mechanism, each of several distinct noxious stimuli elicits the same neural response by opening the same depolarizing channels in the population of polymodal C nociceptors.

CNS COMPONENTS OF SOMATIC SENSATION

The central paths taken by large diameter afferents and small diameter afferents tell two stories worth paying attention to. The first is that the path for mechanosensation and proprioception are separate and largely distinct from the path for nociception and thermoreception. These separate paths continue through the CNS to cerebral cortex and represent one of the clearest divisions of labor seen in any sensory system. The second story is seen in the multiple targets contacted by each population of axons. Here the lesson repeats a statement made at the beginning of this chapter. The somatosensory system has many tasks to perform, from the input side of motor reflexes to the higher order functions of perception, comprehension, and emotion. To accomplish them, mechanosensory/proprioceptive and nociceptive/thermoreceptive axons send information along several pathways performing separate functions.

An Outline of Ascending Paths to Perception
(Fig. 25.10)

Primary afferents in the somatosensory system terminate on second order neurons in either the spinal cord (nociceptors and thermoreceptors) or medulla (mechanoreceptors and proprioceptors). Second order neurons in the spinal cord and medulla send their axons across the midline to terminate in thalamus. Convergence is kept to a minimum so that second order mechanosensory and nociceptive neurons end in separate nuclei and subnuclei of the thalamus. In addition, axons of particular submodalities (e.g., muscle spindle afferents vs. cutaneous afferents) end in different subnuclei of the thalamus. The major thalamic

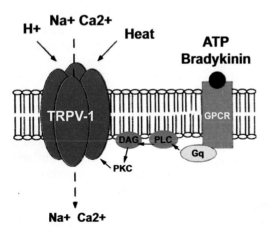

FIGURE 25.9 The receptor protein, TRPV-1, provides a nociceptor with the ability to respond to many noxious stimuli. In addition to heat, TRPV-1 is directly gated by a reduction in pH (the presence of H^+) produced in response to tissue swelling. Either or both opens a nonspecific cation channel that, through an influx of Na^+, depolarizes the nociceptor axon. Circulating agents that signal the presence of tissue damage (ATP and bradykinin) bind to a G-protein coupled receptor (GPCR). Through a series of steps PKCε is activated and TRPV-1 subunits are phosphorylated, leading to a sensitization of the receptor. One result of that cascade is an opening of the cation channel at body temperature. Adapted from Caterina, M. J. and Julius, D. (2001).

A

Postcentral gyrus (first somatosensory cortex)

Ventral posterolateral nucleus of thalamus

Medial lemniscus

Gracile nucleus

Cuneate nucleus

Internal arcuate fibers

Decussation of the medial lemnisci

Cuneate fasciculus

Dorsal part of lateral funiculus

Gracile fasciculus

B

Second somatosensory cortex

Intralaminar and posterior groups of thalamic nuclei

Periaqueductal gray matter

Spinal lemniscus

Pontine reticular formation

Medullary reticular formation

Spinothalamic tract

Ventral white commissure

Cervical level

Thoracic level

Lumbosaccral level

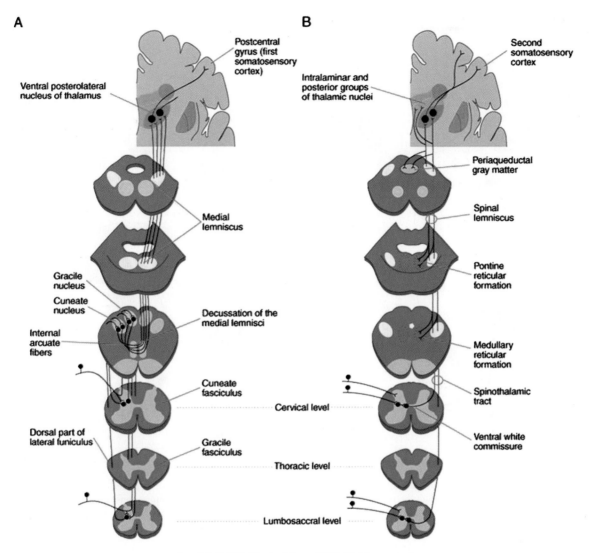

FIGURE 25.10 Anatomy of ascending somatosensory paths. (A) Organization of the dorsal column-medial lemniscal system from entry of large-diameter afferents into the spinal cord to the termination of thalamocortical axons in the first somatosensory area of the cerebral cortex. An obligatory synapse occurs in the gracile and cuneate nuclei, from which second-order axons cross the midline and ascend to the ventral posterolateral nucleus of the thalamus (VPL) by way of the medial lemniscus. (B) Organization of the spinothalamic tract and the remainder of the anterolateral system. Primary axons terminate the spinal cord itself. Second-order axons cross the midline and ascend through the spinal cord and brain stem to terminate in VPL and other nuclei of the thalamus. Collaterals of these axons terminate in the reticular formation of the pons and medulla.

nuclei involved are the various parts of the ventral posterior complex (referred to as ventral caudal in human studies). The lateral and medial subnuclei of VP, logically named VPL and VPM, are the recipients of discriminative inputs from the body and face, respectively. Both innervate the first somatosensory cortex (SI) and a third subnucleus, VPI, sends its axons to the second somatosensory area (SII).

SI and SII are recognized in the cerebral cortex of all mammals that have been examined (Fig. 25.11). Yet the organization of each appears to vary considerably

across that class. If we take human cerebral cortex as a starting point, SI is traditionally divided into four structurally distinct areas named (from rostral to caudal) areas 3a, 3b, 1, and 2. The same is seen for the best-studied genus of monkeys, the macaques. What stands out in the organization of these four is the combination of serial and parallel processing. Parallel processing is seen in the types of thalamic neurons that innervate each area and the receptive field properties displayed by those cortical neurons. Areas 3b and 1 most often are characterized by the response of neurons to cutaneous

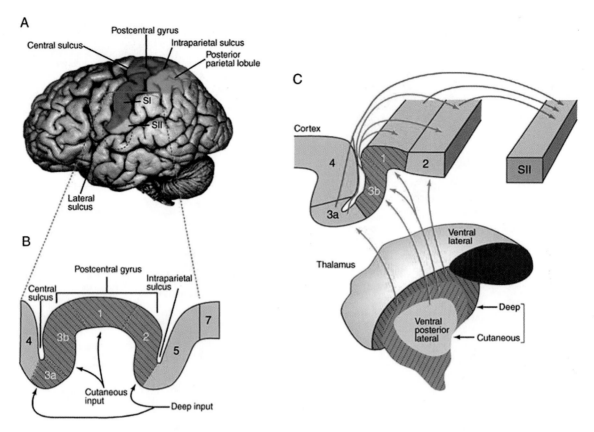

FIGURE 25.11 Functional organization of the ventrobasal complex and first somatosensory cortex (SI). (A) Location of SI in the postcentral gyrus and its relationship to SII and the somatosensory association cortex in the posterior parietal lobe. (B) Cross-section through the postcentral gyrus, cut orthogonal to the central sulcus. SI is divided into four anatomically and functionally distinct areas. They are bordered by area 4 of the precentral motor cortex and by area 5 of the parietal association cortex. (C) Relationship between regions of cutaneous and deep input to VPL and the termination of thalamocortical axons in SI. The serial processing of somatosensory inputs also is indicated by the projections from one area of SI to others and from all areas in SI to the second somatosensory area (SII). Adapted from Jones and Friedman (1982).

inputs, SAs and RAs, whereas neurons in area 3a are found to respond exclusively to stimulation of deep receptors and neurons in area 2 respond to both.

Serial processing of information is evident in both the physiology and connectivity of these areas. Area 3b is much more richly innervated by VPL and VPM than any of the other three areas of SI and, in terms of its structure and connectivity appears much more of a primary sensory area. Comparative studies suggest that area 3b of rhesus monkeys and humans is in most ways equivalent to all of SI in other mammals and so this area often is referred to as SI Proper. The principal outputs of area 3b are directed caudally to areas 1 and 2; and in a similar fashion, area 1 sends its own rich set of axons to area 2. From these considerations, a hierarchy emerges, with area 2 at something of a pinnacle, in that it receives a direct deep receptor input from the thalamus, indirect deep input from 3a and an indirect cutaneous input from areas 3b and 1.

The output of SI as a whole is in two directions (Fig. 25.12):

- A ventral path to SII and from there to caudal insula, areas of the temporal lobe and to premotor and prefrontal cortical areas. Eventual convergence of somatosensory, auditory, and visual information in medial temporal lobe is viewed as the path toward shape and form processing. This ventral path is vital for inserting new information into declarative memory and accessing established memories for comparison with ongoing events.
- A dorsal path to superior parietal lobule, providing areas in that lobule with somesthetic information for control of voluntary movements, selective attention and information about how to perform different tasks, sometimes known as the "how" pathway.

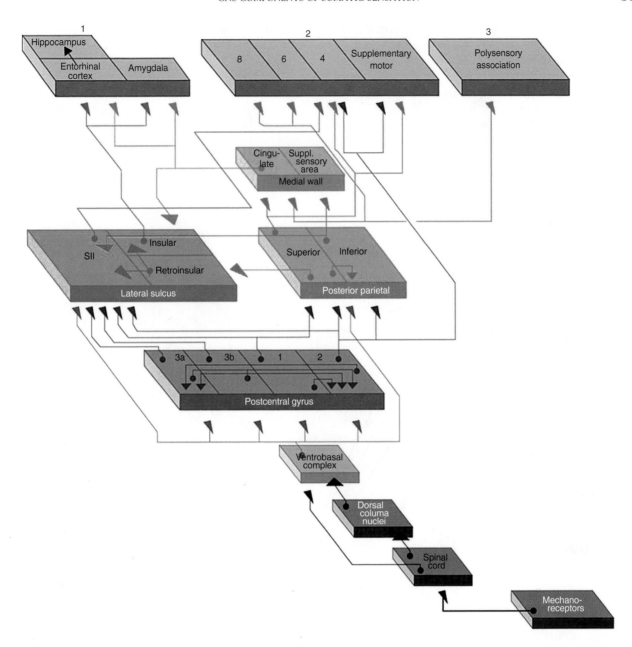

FIGURE 25.12 Schematic representation of the path taken by mechanoreceptor input to eventually reach three cortical targets. All relevant information reaches the ventrobasal complex and most is relayed to the areas of SI. From there, by steps through SII and the posterior parietal areas, somatosensory information reaches (1) the limbic system (entorhinal cortex and hippocampus), as a means for becoming part of or gaining access to stored memories; (2) the motor system (primary and supplementary motor cortex), where the continuous sensory feedback onto motor system occurs; and (3) the polysensory cortex in the superior temporal gyrus, in which creation of a complete and abstract sensory map of the external world is thought to occur. Adapted from Wall (1988), with permission.

With these essential elements in mind we will consider separately the paths for mechanosensation and nociception from the entry of primary afferents into the spinal cord to the areas of cortex in which the elements of sensation are brought together into a perceptual whole.

The Path for Mechanosensation for the Body

Dorsal columns are the principal routes for spinal mechanoreceptor axons. All axons of sensory ganglion cells enter the CNS by way of dorsal roots and the trigeminal nerves. At a gross level the spinal cord is segmented by the existence of separate dorsal root ganglia. Because the peripheral tissue innervated by any one dorsal root ganglion is restricted, the entire innervation of the body surface can be seen as a series of overlapping bands. These are the dermatomes (Fig. 25.13). As a practical matter, then, a mechanical stimulus applied to a restricted region of the body surface leads to action potentials conducted in one or at most two dorsal roots.

Mechanosensory and Nociceptive Axons Enter the Cord Along Different Paths

The dorsal roots divide after they have penetrated the dura mater but before they have entered the spinal cord itself. A medial division of large-diameter, heavily myelinated axons enters at the dorsal funiculus of the cord and the majority of those axons turn at right angles after they have entered the cord and ascend

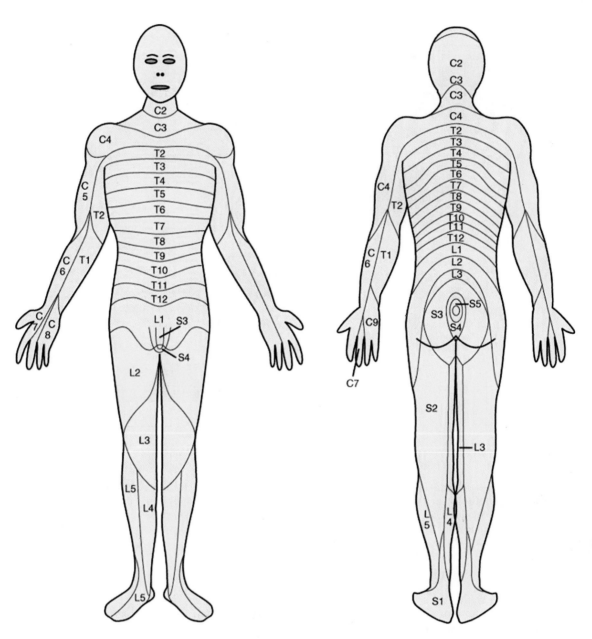

FIGURE 25.13 Classic dermatomal map showing the distribution of spinal nerves and the segments from which they arise. Despite extensive overlap between nerves arising from adjacent segments, this map permits localization of injuries and other conditions that give rise to restricted sensory deficits.

toward the brain. From level T7 of the cord to the terminal coccygeal segment, the large diameter axons enter one fiber tract called the gracile fasciculus. And as they do so they stack up as thin sheets with the earliest entering axons (coccygeal) occupying the most medial part of the funiculus and the last entering (7th thoracic) occupying the most lateral part. That same medial-to-lateral pattern continues with large axons that enter from T6 up to the first cervical segment but at those levels the axons form a separate fasciculus, called the cuneate fasciculus. Together the gracile and cuneate fasciculi are called the dorsal columns. The orderly pattern established by the axons of the dorsal columns as they ascend means there is a body map or somatotopy to the ascending mechanosensory system. Large diameter axons ascend on the same side of the spinal cord as the one they entered to reach the lowest levels of the medulla.

Dorsal Column Nuclei

In a path toward sensory perception, the majority of axons in the dorsal columns ascend to reach the dorsal column nuclei at the junction of the spinal cord and medulla. The dorsal column nuclei are neither simple nor homogeneous. At the grossest level, each nucleus is divided into at least three regions in which group I afferents (mainly muscle spindle afferents), PC axons, and SA axons terminate separately. In some species the neurons receiving spindle inputs are sufficiently distinct to be given their own name (nucleus Z). All this serves to accent the role played by subcortical somatosensory nuclei in keeping segregated the input from different functional classes of receptors. At a finer level the dorsal column nuclei are broken up into clusters of large relay neurons separate from one another by bundles of axons and groups of small cells. These also serve to keep afferents segregated since, as seen in one example, mechanosensory axons from adjacent toes terminate in adjacent but separate clusters.

Despite an anatomical convergence of primary afferents onto single neurons of the dorsal column nuclei, convergence and comparison of receptor input appears minimal. Studies in the laboratory of M. Rowe have shown that the coupling between a single neuron in the gracile or cuneate nucleus and a single afferent axon is extremely tight. As a result of this coupling, a single action potential in the afferent axon reliably produces an action potential in the postsynaptic cell. That phenomenon is seen for RA, PC and SAI afferents of the glaborous skin and hair follicle afferents of the hairy skin. These findings indicate that for a neuron in the gracile and cuneate nuclei receptive field locations, sizes and properties are a matter of which primary afferent innervates that neuron.

Role of Inhibition in the Dorsal Column Nuclei

The dominant input from a single afferent leaves room for synaptic processing to play a role in the function of the dorsal column nuclei. An anatomical basis for this inhibition is the significant population of inhibitory interneurons scattered among the clusters of relay neurons. Several studies suggest inhibition suppresses a response to any input other than the most powerful afferent ending on a relay neuron. Thus, the convergence of multiple afferents on a single neuron is whittled down to the contribution of one dominant afferent by a powerful synaptic inhibition, mediated by GABAergic synaptic transmission. The peripheral source of this suppressive inhibition appears to be nociceptors, so that when C fiber activity is eliminated, receptive fields in neurons of the dorsal column nuclei grow. The route by which nociceptive input reaches the dorsal column nuclei is indirect, most likely through a population of spinal cord neurons that receive convergent mechanosensory and nociceptive innervation. Because axons of these spinal cord neurons enter the dorsal columns they are called postsynaptic dorsal column axons.

The Lateral Cervical Nucleus

A cluster of mechanosensitive relay neurons is found in the lateral neck of the dorsal horn at cervical segments 1 and 2 and extends into the caudal one-third of the medulla. This is called the lateral cervical nucleus. Innervated by mechanosensory neurons in the dorsal horn throughout all segments of the spinal cord, the neurons in the lateral cervical nucleus display predominantly cutaneous receptive fields that cover large patches of hairy skin and usually some part of glaborous skin. In addition, responses to intense mechanical stimulation such as that carried into the cord by Aδ mechano-nociceptors, are common in the lateral half of the nucleus. Most axons of cells in the nucleus decussate immediately and reach the contralateral VPL by way of the medial lemniscus. In VPL, the cervicothalamic axons responding to low-threshold mechanical stimulation of skin or hair appear to converge with axons that arise in dorsal column nuclei.

Nociceptive inputs to the lateral cervical nucleus are relayed not only to thalamus (VPL and POm) but also the periaqueductal gray (PAG) region in cats. The PAG is a complex region of the midbrain, populated by neurons that control functions as diverse as defense responses and sexual behavior. As outlined next, some neurons in this region are an essential part of a descending system for control of pain. The innervation of PAG by lateral cervical neurons that respond to intense mechanical stimulation is one route by which this descending system is informed of damaging events along the surface of the skin. Moreover, this input appears to end on PAG neurons that drive a flight response and can be seen, then, as the afferent side of a reflexive response to danger.

Unlike other nuclei of the somatosensory system, the lateral cervical nucleus is an inconstant feature of the mammalian nervous system. Whereas it is a robust part of the spinal cord and medulla in rodents and carnivores and has been found consistently in nonhuman primates, it is present in only half the humans studied and is well defined in very few of them. Comparison across species also shows a marked difference in the density of cervicothalamic terminations in VPL, with the density in cats far exceeding that in rhesus monkeys. These findings lead to the assumption that functions parceled out to the lateral cervical nucleus in rodents and carnivores are sequestered in the dorsal column-medial lemniscal system of primates, and particularly of humans.

THALAMIC MECHANISMS OF SOMATIC SENSATION

Segregation of Place and Modality Continues in Thalamus

Medial lemniscal axons terminate on large neurons of VPL and VPM. As in the dorsal column nuclei, the synaptic means to generate lateral inhibitory influences exists in the thalamus but evidence for robust lateral inhibition is seldom found in these nuclei. Where inhibition has been detected, its strength across the neuron's receptive field matches that of the excitation so that the region of greatest inhibition is also the region of greatest excitation. Missing from this stage of somatosensory processing, then, is the synaptic means to produce contrast, by which activity produced by stimulation of one spot on the skin surface is compared with activity produced by stimulation of surrounding regions.

Neurons in VPL and VPM display a cluster of features labeled as lemniscal properties. Static lemniscal properties are place and modality specificity. Thalamic circuits maintain these features much as they exist in the lemniscal afferents, themselves. As a result, mixing of inputs is avoided and the specific properties of where on the body surface something has occurred and what exactly has occurred are kept anatomically distinct. Dynamic properties characteristic of these

cells include the ability to follow a rapid train of stimuli as well as the incoming lemniscal afferent. Here the synaptic organization of the thalamus favors the ability to behave much like the lemniscal afferent. Axons of the medial lemniscus terminate as a series of very large axon terminals that form synaptic contacts with broad active zones on the primary dendrites of thalamocortical neurons. These synaptic arrangements in VPL and VPM insure a faithful transfer of information.

Anatomical and physiological studies demonstrate that as lemniscal afferents terminate in VPL they do so as clusters of axon terminals elongated in the rostrocaudal dimension. They are referred to as rods. This appears to be the major organizing principle to somatosensory thalamus as lemniscal axons carrying information of the same type (e.g., SAI input) from the same part of the body terminate in a rod-like formation. On the output side, thalamocortical neurons that send axons to a 1 mm-wide patch in SI occupy the same sort of formation. Thus, thalamic rods are the anatomical basis for the most salient feature of somatosensory thalamus, the strict segregation of place- and modality-specific responses.

THE PATH FROM NOCICEPTION TO PAIN

Spinal Cord Pathways

In stark contrast to the pattern of input by mechanosensory axons, nociceptive axons terminate in the dorsal horn of the spinal cord. These lightly myelinated and unmyelinated axons enter the cord by way of a lateral division of the dorsal root and divide to send branches up and down the cord for a segment or two. The tract they form is a cap on the surface of the dorsal horn called Lissauer's tract. Nociceptive axons terminate, therefore, on dorsal horn neurons across four or five segments and a single dorsal horn neuron is innervated by nociceptors that cover a broad swath of the body or limb surface. A majority of second-order neurons innervated by the nociceptive axons then decussate and ascend in the anterolateral quadrant of white matter to reach the brainstem and thalamus.

Spinal cord circuits for nociception form two ascending routes with distinct functions (Fig. 25.14). Recall the peripheral basis for first and second pain is the division of nociceptors into relatively fast-conducting Aδ axons and slow-conducting C fibers. Yet the discriminative versus agonizing components of pain is a CNS construct, produced by the difference in central circuits driven by Aδ and C fibers. To appreciate that

Lateral Path

Medial Path

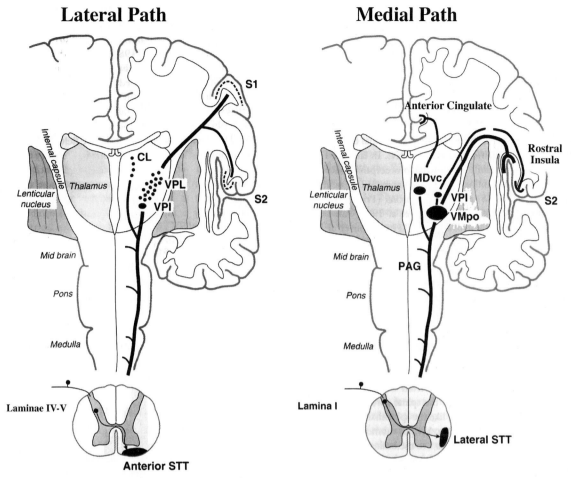

FIGURE 25.14 The path taken by spinothalamic neurons in lamina I (driven by C fibers) differs from that taken by neurons in laminae IV and V (driven by C and Aδ nociceptors and Ab mechanoreceptors). Anterior spinothalamic tract axons are given off by the deeper neurons and terminate in lateral thalamus (VPL and VPI and the centrolateral nucleus). Lateral spinothalamic tract axons given off by lamina I neurons innervate medial thalamus, including the ventral caudal division of the mediodorsal nucleus (MDvc) and a region in posterolateral thalamus. In this figure, the region is labeled a separate, nociceptive/thermoreceptive-specific nucleus called VMpo, but several lines of evidence indicate lamina I neurons also innervate VPL and VPM, as well as VPI. From this varied thalamic innervation, nociceptive information reaches SI and SII for discriminative aspects of pain and temperature and the anterior cingulate and rostral insula for the affective, punishing aspects of pain. Modified from Craig, A. D. and Dostrovsky, J. O. (1999).

difference it is best to start with the population of spinothalamic neurons.

Spinothalamic Neurons

Spinal cord neurons that directly innervate thalamus (therefore, spinothalamic neurons) occupy several laminae in rhesus monkey spinal cord. The major populations, however, are found in the most superficial layer of dorsal horn (lamina I) and a second region deeper in the dorsal horn (laminae IV and V). Nociceptive innervation of these two populations differs fundamentally. Lamina I neurons are innervated pre-dominantly by C fibers, both directly and indirectly by way of excitatory interneurons in the immediately adjacent superficial half of lamina II. They also receive an indirect innervation from Aδ fibers by those same interneurons and can be thought of as nociceptive-specific cells. Laminae IV and V neurons, on the other hand, are innervated predominantly by Aδ fibers that end deep in lamina II on a population of excitatory interneurons. Many of these deeper spinothalamic neurons also receive a convergent input from mechanosensory axons (Aβ axons). Because they respond to both nociceptive and mechanosensory stimulation these neurons often are referred to as wide dynamic range cells.

Ascending Paths to Thalamus

Axons from both lamina I neurons and laminae IV/V neurons decussate and enter fiber tracts in the anterolateral quadrant of spinal cord, but the tracts they enter differ from one another in location and termination (Fig. 25.14). Neurons of laminae IV and V enter the anterior spinothalamic tract and terminate in lateral parts of the thalamus, including VPL and VPI, and an intralaminar nucleus, the central lateral nucleus. The innervation of ventral posterior thalamus does not mean nociceptive and mechanosensory information converge in the thalamus. The two remain segregated as medial lemniscal axons terminate on groups of large neurons that express the calcium-binding protein parvalbumin whereas the spinothalamic axons innervate clusters of smaller neurons that express a second calcium binding protein, calbindin. Response properties of the calbindin-rich neurons in VPL are much like those of the wide dynamic range neurons that innervate them, both in the type of stimuli to which they respond and the size of their receptive fields.

Nociceptive specific neurons of lamina I enter the lateral spinothalamic tract and terminate in several nuclei of the thalamus, including two that receive few deep lamina inputs (Fig. 25.14). Targets in which convergence of lamina I and lamina IV–VI inputs are strongly suspected or known to occur are the nucleus of the ventral posterior complex. Although the most compelling data in support of convergence has been reported for VPI, several careful studies have documented termination of lamina I inputs in the calbindin-rich clusters of VPL and VPM. This represents what has become a traditional view of the nociceptive pathway, that discriminative pain is driven by nociceptive inputs to both lamina I and laminae IV–VI and through a relay in the ventral posterior nuclei, that nociceptive information reaches SI and SII.

Outside the ventral posterior complex are two sites of spinothalamic terminations predominantly from lamina I (Fig. 25.14). One of these is a subnucleus of the mediodorsal nucleus (Mdvc) in which spinothalamic terminations are surprisingly rich. The other is posterior to the classically drawn borders of VP and includes the medial nucleus of the posterior group (POm). Neurons in this region are innervated by spinothalamic axons and display the same chemical signature (calbinding immunoreactivity) as spinothalamic-recipient neurons of VPL and VPM.

Considerable disagreement exists with the source and meaning of the spinothalamic innervation in posterolateral thalamus of monkeys and humans. One part of the disagreement deals with the proportion of lamina I afferents that end inside the bounds of VPL and VPM and those that end caudal and medial to it. A. D. Craig and his colleagues indicate a majority of lamina I output targets a neurochemically distinct region outside VPL and VPM. They have given this region the name VMpo in apparent recognition of the data suggesting it is a nucleus separate and distinct from VPL and VPM. E. G. Jones and W. D. Willis and their colleagues hold to the traditional view that termination of axons from lamina I neurons is widespread, and includes VPL, VPM, and other thalamic nuclei, including POm. Jones, in addition, argues that what has been called VMpo is simply the most medial tip of VPM. That disagreement is not minor in its implications. An exclusive lamina I innervation of the region labeled VMpo turns it into a nociceptive- and thermoreceptive-specific relay nucleus, whereas the traditional view sees a major role for VPL and VPM in nociceptive and thermoreceptive relay. Resolution of this question has not been a simple matter, as studies favoring and refuting the central premise of a nociceptive- and thermoreceptive-specific VMpo are common in the experimental and clinical literature. If the most conservative conclusions are adopted, we are left to recognize spinothalamic innervation of four thalamic regions, each with its own cortical target. Spinothalamic terminations in VPL and VPM are relayed to SI and those in VPI to SII. Together, they can be seen as a lateral path to first or discriminative pain. A second medial path to anterior cingulate by way of MDvc and to the insula by way of POm and the neurons of the region designated VMpo appears to be the central route for second or punishing pain.

SI and Pain

Nociceptive responses in SI have been difficult to record and where neurons responding to noxious stimuli have been encountered, their location (e.g., area 3a) has not been easy to fit into a general scheme of what these areas do. Nevertheless, ablation studies in monkeys show SIs involvement in pain perception and studies of human cerebral cortex (see later) regularly show SI responds to stimuli judged to be painful. Perhaps the most compelling story can be told for area 1 of SI in monkeys, where evidence exists of clustered organization for nociceptive neurons. Yet these neurons do not form columns. They are largely confined to layer IV of that area and are intermixed with mechanosensory neurons. These data suggest that nociceptive and mechanosensory inputs converge in area 1, perhaps as a means to accurately locate the source of pain.

The Human Axis of Pain

Functional imaging studies of the human brain indicate four areas of cerebral cortex are active during (and often just prior to) the application of a painful stimulus. Activation of SI and of SII occurs as part of a discriminative component to painful stimuli. A subject's ability to report the location and grade the intensity of a stimulus is correlated with activity in these areas. Two other areas appear tied to the cognitive and emotional content of pain. These are the rostral half of the anterior cingulate gyrus and the rostral insula, both of which display elevated activity during the anticipation of painful stimuli and the infliction of pain on a loved one (empathetic pain). And it is these areas that show a reduction in activity when the administration of a placebo produces reports of lessened pain. As a group, then, studies of the cortical representation of pain point to a distributed network of four main areas (SI, SII, rostral anterior cingulate, and rostral insula) in which the noxious stimulus applied to some peripheral region becomes a painful sensation with both discriminative and punishing components.

Nonperceptual Elements of Nociception

Several paths are taken by first- and second-order nociceptor axons that reach neither spinothalamic neurons nor regions of the thalamus. Targets of these axons include the following: (1) spinal interneurons that mediate the withdrawal and crossed-extensor reflexes; (2) the pontine reticular formation, where the startle reflex is generated; and (3) the midbrain periaqueductal gray (PAG) as a means to control pain. The last of these is a well-studied mechanism (Fig. 25.15). Neurons of the PAG are innervated by second-order nociceptive neurons in laminae IV and V. They provide PAG neurons with an indication of the source and intensity of nociceptive input. By way of a relay from serotonin neurons in nucleus raphe magnus and norepinephrine neurons in the locus coeruleus, activity in PAG drives a population of interneurons in the lamina II of the spinal cord. Two characteristics of those interneurons are worth some consideration. First, they end not only on the populations of spinothalamic neurons, but also on the axon terminals of C and Aδ axons. Through postsynaptic inhibition these inhibitory interneurons are able to suppress the response of spinothalamic cells and through presynaptic inhibition they are able to suppress primary nociceptive afferents. The combination of pre- and postsynaptic inhibition reduces activity in the population of spinal neurons that carries nociceptive information to the brain. Second, the inhibitory neurons release the pentapep-

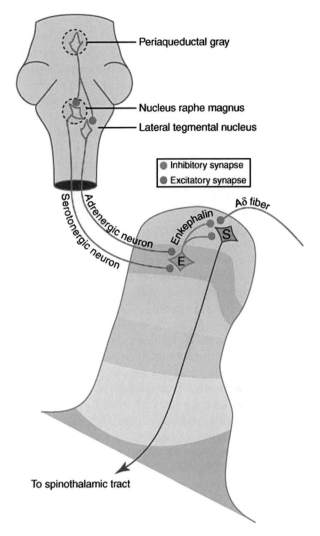

FIGURE 25.15 Descending control of pain. Serotoninergic axons arise from neurons in the nucleus raphe magnus and adrenergic axons from neurons in the lateral tegmental nucleus. Neurons in each nucleus are innervated by neurons of the periaqueductal gray area and both form excitatory synapses onto spinal interneurons (E). Those interneurons use opiate-like peptides (enkephalins) as neurotransmitters; release of enkephalins inhibits both the incoming nociceptive axons and the spinothalamic neurons (S) on which they synapse.

tide, met-enkephalin as a neurotransmitter. Met-enkephalin is an opioid peptide that binds to a member of the opiate receptor family. Its actions are mimicked by the administration of morphine and synthetic opiates accounting for at least part of the analgesic properties those compounds possess.

Pain is such a rich experience with an obvious impact on the well-being of a person that unusual phenomena associated with pain have been studied in considerable depth. Among the best studied are referred pain, secondary hyperalgesia/allodynia and phantom limb pain. Each has been explained by referring to the circuitry and chemistry of normal nociceptive processing. Referred pain is the experience in which noxious stimuli in the viscera (e.g., ischemia of cardiac muscle) is felt as pain in a peripheral location such as the shoulder. A classic explanation for referred pain notes the widely branching nature of C and Aδ fibers as they enter the cord, so that lamina I neurons innervated by nociceptors of the shoulder are also innervated by nociceptors of the pericardium. When the latter are driven to fire a series of action potentials the percept is one of a more common occurrence, shoulder pain.

Central or secondary hyperalgesia and allodynia are related phenomena that occur in response to synaptic plasticity at the level of the spinal cord. Secondary hyperalgesia is a phenomenon in which a greater degree of pain is felt upon application of a noxious stimulus. Allodynia, by contrast, is the perception of pain when the stimulus itself is not noxious. Each occurs when the application of an intense or prolonged noxious stimulus leads in spinal circuits to a rearrangement in synaptic strength, much like long-term potentiation. With secondary hyperalgesia, the potentiation is homosynaptic, leading to a greater postsynaptic response in spinal neurons to a noxious stimulus. Plasticity with allodynia is thought to be heterosynaptic. Under normal circumstances, activity in Aβ fibers to a benign mechanical stimulus (e.g., the movement of a cotton swab across the skin) modulates a response to noxious stimuli but fails to drive spinothalamic cells. Yet when the synapses formed by the Aβ fibers are potentiated in response to an intense or prolonged barrage of C fiber activity spinothalamic neurons are driven by that mechanical stimulus. The result is perception of pain where only nonnoxious mechanical stimulation has occurred.

Unlike allodynia, phantom limb pain appears to involve synaptic plasticity at each of several levels in the somatosensory system. Removal of a digit or limb in experimental animals and in humans leads to a robust functional plasticity. In cases first documented by J. Kaas and M. Merzenich and their colleagues, surgical removal of digits produces a short-lived silent zone in the somatotopically appropriate part of SI. Quickly thereafter, the previously silent region becomes responsive to tactile stimulation. Yet the newly acquired response is to adjacent body parts, as the activity evoked in neighboring receptors fills in the zone where the cortex had been silent. Subsequent work has shown that following amputation of a digit or limb, this process of filling in takes place through synaptic changes in the spinal cord, medulla, and thalamus, as well as in cerebral cortex. It is assumed that phantom limb pain works much the same way. Thus, stimulation of the shoulder and chest in amputees drives regions of cerebral cortex that had been responsive to stimulation of an arm prior to its amputation. The result is the percept of arm pain even in the absence of an arm.

THE TRIGEMINAL SYSTEM (Fig. 25.16)

Mechanoreceptive, nociceptive, and thermoreceptive afferents for the face have their cell bodies in the pair of trigeminal ganglia. The central processes of trigeminal ganglion cells enter the mid-pons as the trigeminal nerve. In many ways, the central trigeminal system is organized along parallel lines with the spinal somatosensory system.

Three nuclei make up the somatosensory part of the trigeminal system. The largest of these is the principal or main sensory nucleus, the cell bodies of which are in mid-pons, at the level of entry for the trigeminal nerve. Acting much like the dorsal column nuclei, the principal sensory nucleus is innervated by large diameter afferents of the ipsilateral half of the face. Its neurons respond to skin indentation and to vibrotactile stimuli. Most of these neurons send their axons across the midline in the pons, where they join the fibers of the medial lemniscus. The trigemino-thalamic axons ascend in the most medial part of the lemniscus (next to the axons that carry information about the hand) and terminate in VPM.

The spinal trigeminal nucleus is an elongated nucleus, split into three subnuclei distinguished geographically by reference to the obex. Pars oralis occupies the lower pons and upper medulla to the level of the obex. It gives way to the pars interpolaris at the obex and then to pars caudalis through the lower medulla and into the first two cervical segments of the spinal cord. Borders between these subdivisions are hardly distinct as afferents innervating a restricted part of the face may end across two or all three parts of the spinal trigeminal nucleus. The traditional subdivision of the spinal nucleus into functional units accents the mechanosensory functions of pars oralis, the deep receptor functions of pars interpolaris, and the nociceptive and thermoreceptive functions of pars caudalis. In this context, the relay of input from pars caudalis to VPM is viewed as the equivalent of the spinothalamic system for the face. Yet studies of C fibers that innervate tooth pulp find these afferents terminate in a long, continuous sheet from the caudal

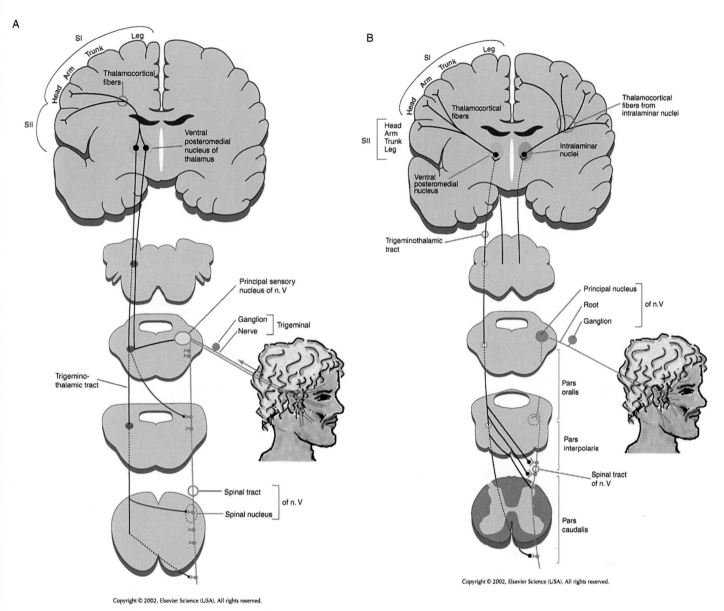

FIGURE 25.16 Sensory components of the trigeminal system. (A) Path for discriminative touch. Large-diameter afferents from the face innervate second-order neurons in the spinal trigeminal nucleus (pars oralis) and the principal sensory nucleus. Neurons in these nuclei give rise to axons that cross the midline, ascend in the trigeminothalamic tract, and terminate in the ventral posteromedial (VPM) nucleus of the thalamus. (B) Path for pain and temperature in the trigeminal system. Small-diameter afferent axons descend in the spinal trigeminal tract and terminate in the pars caudalis of the spinal nucleus. Second-order axons cross the midline and ascend to the thalamus.

half of the principal nucleus to the caudal-most aspect of pars caudalis. These data indicate a great deal more intermixing of mechanosensory neurons and nociceptive neurons takes place in the trigeminal system than is seen for the spinal somatosensory system.

The third nucleus of the sensory trigeminal system is the most unusual. Incorporated into the CNS as the mesencephalic nucleus of the trigeminal system is a collection of ganglion cells that give rise to muscle spindle afferents. This nucleus is one of only two in the CNS to contain neurons of neural crest origin. The central processes of the ganglion cells innervate several targets, the most prominent of which is the motor trigeminal nucleus.

CORTICAL REPRESENTATION OF TOUCH

Neurons of SI Are the First in the Somatosensory System to Show Clear Signs of Lateral Inhibition

Between the level of the peripheral afferent, where skin mechanics produces surround suppression and the level of the cerebral cortex, the somatosensory system is unusual in the sparseness and weakness of lateral inhibition. As discussed earlier, the synaptic means to generate lateral inhibition exist in the dorsal column nuclei and thalamus, but the response of a single neuron in those regions appears very much like the response of the one axon that drives it best. Neurons in SI, however, display true inhibitory surrounds (Fig. 25.17). For area 3b in monkeys such surrounds are one-third larger than the central excitatory region and can occupy one or more (but rarely all four) sides of the excitatory center. Many neurons in area 3b also exhibit a strong temporal component to their response. Application of a peripheral stimulus produces an initial excitatory response that evolves over a 25 ms period into inhibitory response, referred to as replacing inhibition. Receptive fields in area 3b neurons possess three distinct components: (1) a central region of excitation that comes on rapidly following application of a stimulus; (2) a rapidly occurring lateral inhibition that diminishes the response to a stimulus that is applied to it and the excitatory center; and (3) a delayed inhibition that overlaps the central excitatory region in part or in whole. These properties of spatial and temporal inhibition in area 3b are emergent properties—ones found in cerebral cortex but not earlier in the somatosensory system—that produce responses tied to the spatial details of a tactile stimulus.

Spatial and Temporal Inhibition (Fig. 25.14)

Spatial inhibition maintains the response selectivity of neurons when objects scanned across the skin surface at different speeds (as in reading Braille). In short, the receptive fields of neurons in area 3b are velocity insensitive. An increase in the velocity of scanning strengthens the response of a neuron but that increased response occurs in both the excitatory and the inhibitory subfields and it occurs without a change in the geometry of either field. The result is a population of neurons in SI more responsive to an ideal stimulus but no less tightly tuned to the spatial features of the stimulus. Temporal inhibition performs a parallel function by producing a progressive increase in response rate to surface elements that are scanned

A

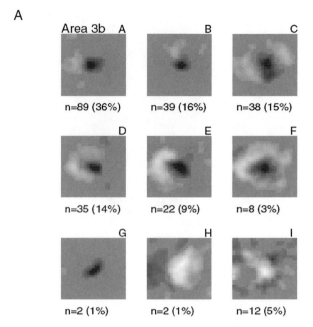

B

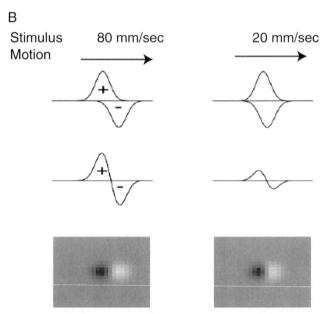

FIGURE 25.17 Receptive fields of neurons in area 3b. Examples of receptive fields displayed by neurons in area 3b of an alert rhesus monkey show clear signs of lateral inhibition (A). In this figure, black represents regions of excitation and white regions of inhibition. An inhibitory part of the receptive field is rarely missing in neurons of this area, as most neurons have inhibitory regions on one or more sides of the excitatory center. Below each subplot is the percentage of times each neuron type was observed in the population. In (B) a hypothetical example is shown illustrating the velocity invariance of a neuron's spatial receptive field. In the top of this figure, overlapping regions of excitation and inhibition are shown at two velocities. The degree of overlap decreases as scan velocity increases, but the spatial profile of the RF is unchanged. Part A is adapted from DiCarlo, Johnson and Hsiao, 1998 and DiCarlo and Johnson 1999.

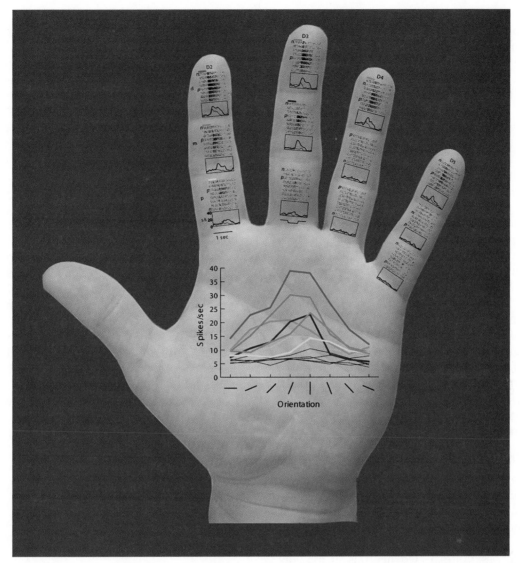

FIGURE 25.18 The receptive field of a neuron in area SII from a macaque monkey is overlayed on the glaborous surface of a human hand. On each finger pad are rasters showing the response of the SII neuron to a bar indented at 8 orientations. Histograms are also shown for the most responsive and least responsive orientation. Shown on the palm are tuning curves for each finger pad. These results demonstrate that neurons in SII have large receptive fields so that tuning to orientation of a bar on one finger pad is the same as the tuning across other pads. Adapted from Cover illustration of J. Neurosci. vol 26 2006.

more quickly. The lag between the initial excitatory response and the delayed inhibitory one liberates more activity when the velocity of scanning increases. This more than makes up for the reduction in contact time between peripheral receptor and surface feature that accompanies any increase in scanning velocity.

Orientation Selectivity

As a result of circuits in cerebral cortex, the roughly circular receptive fields that enter SI by way of thalamocortical axons become orientation-selective responses in 70% of cortical neurons. The synaptic

mechanism for generating a receptive field tuned to the orientation of a stimulus could be convergence of excitatory inputs, much like the mechanism proposed originally for orientation tuning in visual cortex. Yet examination of spatial–temporal response of neurons in SI indicates orientation selectivity is best explained by the presence and spatial distribution of inhibitory regions.

Serial Processing in SI

Area 3b is considered an early step in the cortical processing of tactile information. Much of the data supporting that conclusion come from studies in the lab of Randolph and Semmes, showing that ablation of area 3b produces a general decline in tactile discrimination behavior whereas ablation in other areas of SI leads to losses that are more selective. These include difficulties in discriminating the texture of a surface following ablation of area 1 and deficits in three-dimensional form perception with ablation of area 2. From these data come the conclusion that areas of SI are organized as a hierarchy. A similar conclusion is seen in physiological properties of SI neurons as receptive fields become larger and their properties become more complex in going from area 3b to one of the more caudal somatosensory areas. One example is selectivity for the direction of object movement across the skin. Neurons in area 3b show no sign of direction selectivity but in areas 1 and 2 this property emerges from the cortical circuitry. For example, a neuron in areas 1 and 2 with a receptive field on the palmar surface of a monkey's hand show tuned responses to movements over a 90° range but respond poorly and often not at all to movements over the remaining 270°. The ability of subjects to detect movements with much finer detail than that of single neurons occurs through a population code much like the directions of arm movement are encoded by a group of broadly tuned neurons in the precentral motor area.

Columnar Organization

Responses of like type are organized as columns in SI. The original description of cortical columns was made by V. Mountcastle in his 1953 study of cat somatosensory cortex. For SI in macaques one of obvious place for columnar processing is in the RA and SA responses. As expected neurons in monkey area 3b directly innervated by thalamocortical axons show strong specificity for either RA or SA inputs. These neurons are in layer IV and deep layer III. Yet superficial to this layer, where neurons of layer IV send their axons, the response is predominantly slowly adapting. In other words, the response of neurons outside layer IV makes it appear as though rapidly adapting inputs are incorporated into a slowly adapting output.

Nevertheless neurons in areas 3b and 1 are clearly involved in the discrimination of vibrotactile stimuli at low frequencies, such as those carried into the CNS by RA afferents (Meissner's corpuscles). The work of R. Romo and his colleagues documents a response of SI neurons in macaques that is tied to the frequency of a stimulus applied to the fingertips. A monkey's ability to discriminate between two stimuli of different frequencies is only as good as an SI's neuron's ability to signal that difference. And when electrical stimula-

tion of SI is used as a replacement for one of those vibratory stimuli (e.g., electrical stimulation at 30 Hz is used instead of a 30 Hz vibrotactile stimulus), a monkey will indicate that he perceived a stimulus of that frequency. This perceptual ability is not tied to the frequency of the SI neuron's response but to the number of spikes generated in the first 250 milliseconds. Thus, not only does a population of SI neurons contribute to the perception of flutter but also the neural code used by those neurons is an intensity code.

The Role of SII in Somatic Sensation

Several Body Representations in SII

At first glance the division of labor that is a hallmark of SI contrasts strongly with the presence of a single area called SII. Recent functional and anatomical studies show, however, that across the 10 mm in the upper bank of the lateral fissure that normally defines SII in macaques are multiple functionally distinct areas. In monkeys three separate regions have been described, referred to as SIIa, SIIc, and SIIp (some researchers prefer the terms SII, PV and PR). Anatomical studies in humans suggest there are also four separate regions called OP1-OP4, yet how the human studies relate to the monkey studies is not known.

Ablation studies in monkeys reveal three features of SII's role in somesthesis:

- Removal of all SII leaves a macaque incapable of discriminating the shape and texture of a tactile stimuli. These data show any or all subdivisions of SII play a pivotal role in tactile behavior.
- Removal of SII has no effect on the response of neurons in SI. So, even though SII provides feedback projections to SI, the dominant input to SI comes from neurons of VPL and VPM.
- Removal or cooling of SI eliminates the tactile responses in SII.

These results extend to modality-specific loss in SII following ablations restricted to one area in SI. A modern view of SII suggests it is a complex of at least three functionally distinct areas. The anterior and posterior portions (SIIa and SIIp) are involved in integrating proprioceptive inputs with cutaneous input as a means for representing the size and shape of objects held in the hand. In addition, a central field (SIIc) responds to cutaneous inputs and is important for processing information about 2D shape and texture discrimination.

SII Receptive Fields

Receptive field properties of neurons in macaque SII support the conclusions that SII is part of a serial processing scheme (Fig. 25.18). Unlike the partially iso-morphic responses to raised patterns that are seen in many neurons of SI, neurons in SII appear selective for more complex features of a stimulus. They are any-thing but isomorphic. Moreover, orientation selectivity is seen in fewer than one out of three neurons in SII as receptive fields of single neurons grow to encompass over two or more fingers. Neurons such as these appear to represent a sparse code for objects such as the edge of a keyboard or large curved shapes that span several digits.

SII and Attention

Whereas only a small percentage of neurons in SI are affected by attentional focus of animals, practically all the neurons in SII show attention modulated responses. Psychophysical studies show attention is a spotlight with finite boundaries that can improve per-formance on each of many tactile discrimination tasks. These studies raise questions as to the targets and mechanism of an attentional effect. Single unit studies show SII to be one of the major targets as the response of individual neurons changes with shifts in attention. Mechanisms by which attention appears to exert its effects include a change in the response rate (about 90% of SII neurons in macaques increase their response to a tactile stimulus) and a shift to synchronous activ-ity. Pairs of neurons in macaque SII recorded at the same time are found to become more synchronous in their response to a tactile stimulus when an animal has been cued to pay attention to that stimulus. As the task becomes more difficult, when the difference between two stimuli is deliberately reduced, the degree of syn-chrony increases by a factor of four. These data show that not only is SII a nodal point for the discrimination of form and texture but it is also a site at which cogni-tive mechanism like selective attention are likely to exert their greatest influence.

References

Braz, J. M., Nassar, M. A., Wood, J. N., and Basbaum, A. I. (2005). Parallel "pain" pathways arise from subpopulations of primary afferent nociceptor. *Neuron* **47**, 787–793.

Craig, A. D. and Dostrovsky, J. O. (1999). Medulla to thalamus. *In* "Textbook of Pain," 4th edition, pp. 183–214. Churchill Livingstone, Philadelphia.

Craig, A. D., Bushnell, M. C., Zhang, E. T., and Blomqvist, A. (1994). A thalamic nucleus specific for pain and temperature sensation. *Nature* **372**, 770–773.

Edin, B. B. (2004). Quantitative analyses of dynamic strain sensitiv-ity in human skin mechanoreceptors. *J. Neurophysiol.* **92**, 3233–3243.

Fitzgerald, P. J., Lane, J. W., Thakur, P. H., and Hsiao, S. S. (2006). Receptive field (RF) properties of the macaque second somato-sensory cortex: RF shape, size and somatotopic organization. *J. Neurosci.* **26**, 6485–6495.

Friedman, R. M., Khalsa, P. S., Greenquist, K. W., and LaMotte, R. H. (2002). Neural coding of the location and direction of moving object by a spatially distributed population of mechano-receptors. *J. Neurosci.* **22**, 9556–9666.

Goodwin, A. W., Browning, A. S., and Wheat, H. E. (1995). Repre-sentation of curved surfaces in responses of mechanoreceptive afferent fibers innervating the monkey's fingerpad. *J. Neurosci.* **15**, 798–810.

Grazziano, A. and Jones, E. G. (2004). Widespread thalamic termina-tions of fibers arising in the superficial medullary dorsal horn of monkeys and their relation to calbindin immunoreactivity. *J. Neurosci.* **24**, 248–256.

Hsiao, S. S. and Vega-Bermudez, F. (2002). Attention in the somato-sensory system in "The somatosensory System:Deciphering the Brain's Own body Image" Nelson R.J. Ed CRC Press Boca Raton.

Iggo, A. and Muir, A. R. (1969). The structure and function of a slowly adapting touch corpuscle in hairy skin. *J. Physiol. Lond.* **200**, 763–796.

Johnson, K. O. (2001). Neural basis for haptic perception. *In* "Stevens Handbook of Experimental Psychology," 3rd edition, Volume 1: Sensation and Perception, 537–583. Pashler, H. and Yantis, S. (eds.). Wiley, New York.

Jones, E. G. (1983). Organization of the thalamocortical complex and its relation to sensory processes. *In* "Handbook of of Physiol-ogy: Neurophysiology/Sensory Processes," 149–212, Darian-Smith, I. (ed.). American Physiological Society, Washington, D. C.

Kaas, J. H., Florence S. L., and Jain, N. (1999). Subcortical con-tributions to massive cortical reoganization. *Neuron* **22**, 657–660.

Kenshalo, D. R., Iwata, K., Sholas, M., and Thomas D. A. (2000). Response properties and organization of nociceptive neurons in area 1 of monkey primary somatosensory cortex. *J. Neurophysiol.* **84**, 719–729.

Kew, J. J., Mulligan, P. W., Marshall, J. C., Passingham, R. E., Rothwell, J. C., Ridding, M. C., Marsden, C. D., and Brooks, D. J. (1997). Abnormal access of axial vibrotactile input to deaffer-ented cortex in human upper limb amputees. *J. Neurophysiol.* **77**, 2753–2764.

Krubitzer, L. A., Clarey, J., Tweedale, R., Elston, G., and Calford, M. B. (1995). A redefinition of somatosensory areas in the lateral sulcus of macaque monkeys. *J. Neurosci.* **15**, 3821–3839.

LaMotte, R. H. and Srinivasan, M. A. (1987). Tactile discrimination of shape: Responses of rapidly adapting mechanoreceptive affer-ents to a step stroked across the monkey fingerpad. *J. Neurosci.* **7**, 1672–1681.

Lewin, G. R. and Moshourab, R. (2004). Mechanosensation and pain. *J. Neurobiol.* **61**, 30–44.

Luna, R., Hernandez, A., Brody, C. D., and Romo, R. (2005). Neural codes for perceptual discrimination in primary somatosensory cortex. *Nature Neuroscience* **8**, 1210–1219.

Mountcastle, V. B. (1957). Modality and topographic properties of single neurons of cat's somatic sensory cortex. *J. Neurophysiol.* **20**, 408–434.

Oouchida, Y., Okada, T., Nakashima, T., Matsumura, M., Sadato, N., and Naito, E. (2004). Your hand movements in my somato-sensory cortex: A visuo-kinesthetic function in human area 2. *NeuroReport* **15**, 2019–2023.

Pare, M., Behets, C., and Cornu, O. (2003). Paucity of presumptive ruffini corpuscles in the index finger pad of humans. *J. Comp. Neurol.* **456**, 260–266.

Perl, E. R. and Kruger, L. (1996). Nociception and pain: Evolution of concepts and observations. *In* "Touch and Pain," 180–212, Kruger, L. (ed.). Academic Press, New York.

Pons, T. P., Garraghty, P. E., and Mishkin, M. (1992). Serial and parallel processing of tactual information in somatosensory cortex of rhesus monkeys. *J. Neurophysiol.* **68**, 518–527.

Prud'homme, M. J. L. and Kalaska, J. F. (1994). Proprioceptive activity in primate somatosensory cortex during arm reaching movements. *J. Neurosci.* **72**, 2280–2301.

Rowe, M. J. (2002). Synaptic transmission between single tactile and kinesthetic sensory nerve fibers and their central target neurons. *Behav. Brain Res.* **135**, 192–212.

Sahai, V., Mahns, A., Perkins, M., Robinson, L., Perkins, N. M., Coleman, G. T., and Rowe, M. J. (2006). Processing of vibrotactile inputs from hairy skin by neurons of the dorsal column nuclei. *J. Neurophysiol.* **95**, 1451–1464.

Sur, M., Wall, D. T., and Kaas, J. H. (1984). Modular distribution of neurons with slowly adapting and rapidly adapting responses in area 3b of somatosensory cortex in monkeys. *J. Neurophysiol.* **51**, 724–744.

Tominaga, M., Caterina, M. J., Malmberg, A. B., Rosen, T. A., Gilbert, H., Skinner, K., Raumanns, B. E., Basbaum, A. I., and Julius, D. (1998). The cloned capsaicin receptor integrates multiple pain-producing stimuli. *Neuron* **21**, 531–543.

Willis, W. D., Zhang, X., Honda, C. N., and Giesler, G. J. (2001). Projections from the marginal zone and deep dorsal horn to the ventrobasal nuclei of the primate thalamus. *Pain* **92**, 267–276.

Suggested Readings

Apkarian, A. V., Bushnell, M. C., Treede, R-D., and Zubieta, J-K. (2005). Human brain mechanisms of pain perception and regulation in health and disease. *Euro. J. Pain* **9**, 463–484.

Clark, F. J. and Horch, K. W. (1986). Kinesthesia. *In* "Handbook of perception and human performance," Volume 1: Sensory processes and perception, 1–62, Boff, K. R., Kaufman, L., and Thomas, J. P. (eds.). Wiley and Sons, New York.

Kaas, J. H. (2000). Organizing principles of sensory representations. *Novartis Found. Symp.* **228**, 188–198.

Mountcastle, V. B. (2005). "The Sensory Hand." Harvard University Press, Cambridge, MA.

Stewart Hendry and Steven Hsiao

Audition

The auditory system detects sound and uses acoustic cues to identify and locate sounds in the environment. The auditory system shares functional and evolutionary similarities with other mechanoreceptive systems, such as the vestibular system and the lateral line system of lower vertebrates. All these systems use the same type of receptor cell—the hair cell—and all are specialized to detect an external stimulus that eventually causes the stereocilia of the hair cells to be displaced. Unlike other mechanoreceptive systems, the auditory system is sensitive to sound. This chapter explores the characteristics of the auditory system that make it sensitive to sound and that make it capable of localizing a sound source. The generalized mammalian auditory system is emphasized, but examples also are taken from studies of animals with specialized auditory systems, such as those of bats and owls, which have greatly advanced our knowledge of audition.

EXTERNAL AND MIDDLE EAR

The peripheral auditory system is divided into the external, middle, and inner ears (Fig. 26.1). The external ear is composed of the pinna and external auditory canal. These structures convey sound to the middle ear, but not without influencing it. This influence emphasizes and deemphasizes certain sound frequencies, resulting in peaks and notches in the sound spectrum. Because positions of the peaks and notches depend on the location of the sound source, they provide information about location, even when using only one ear (monaural sound localization). Such information is especially important for determining the elevation of a sound source because spectral peaks and notches are one of the few cues that depend strongly on elevation.

Mechanosensitive organs in lower vertebrates, such as the lateral line of fish, are sensitive to vibrations in aquatic environments. When vertebrates evolved onto land, they faced the problem of converting sound in air to sound in fluid, as the inner ear of vertebrates remains fluid filled. Sound at an air–fluid interface is almost completely reflected back into the air, rather than being transmitted into the fluid. The function of the middle ear is to ensure efficient transmission of sound from air into the fluid of the inner ear (reviewed by Geisler, 1998).

The middle ear begins at the tympanic membrane (eardrum), continues with the three middle ear ossicles (malleus, incus, and stapes), and ends at the footplate of the stapes, which contacts the inner ear fluids at the oval window of the cochlea (Fig. 26.1). The middle ear also has air spaces and two middle ear muscles. There are several ways in which the middle ear increases the transmission of sound into the inner ear. Most importantly, because the area of the eardrum is larger (by about 35 times) than the area of the stapes footplate, there is a corresponding increase in pressure from the eardrum to the stapes footplate. Additionally, there may be small contributions from a lever action of the ossicles and a buckling motion of the tympanic membrane, both increasing the force applied to the footplate. Together, these mechanisms provide a pressure gain of about 25 to 30 dB in the middle frequencies over what would be achieved by sound striking the oval window directly. This amount of gain means that for middle frequencies, most of the acoustic energy that strikes the eardrum is transmitted through the middle ear. For lower and higher frequencies, however,

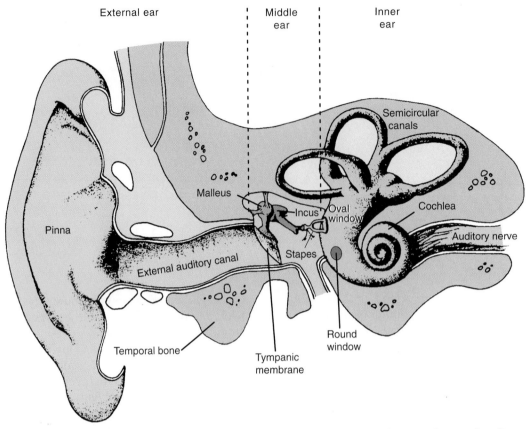

External ear | Middle ear | Inner ear

Semicircular canals

Malleus

Incus | Oval window | Cochlea

Pinna

Auditory nerve

Stapes

External auditory canal

Round window

Temporal bone

Tympanic membrane

FIGURE 26.1 Drawing of the auditory periphery within the human head. The external ear (pinna and external auditory canal) and the middle ear (tympanic membrane or eardrum, and the three middle ear ossicles: malleus, incus, and stapes) are indicated. Also shown is the inner ear, which includes the cochlea of the auditory system and the semicircular canals of the vestibular system. There are two cochlear windows: oval and round. The oval window is the window through which the stapes conveys sound vibrations to the inner ear fluids. From Lindsey and Norman (1972).

energy is lost. When sound conduction through the middle ear is compromised, a person has a conductive hearing loss.

One disease that causes a conductive hearing loss is otosclerosis, in which bony growths around the stapes cause it to adhere to surrounding bone and lessen its transmission of vibration. A surgical procedure called a stapedectomy usually restores hearing by replacing the native stapes with a prosthetic one.

THE COCHLEA

The inner ear is located deep within the head (Fig. 26.1). The inner ear contains the cochlea, which is the sensory endorgan for the auditory system. It also contains the utricle, saccule, and cristae of the three semicircular canals, which are the sensory end-organs for the vestibular system (see Chapter 33). The word cochlea comes from the Greek word *kokhlias*, meaning

"snail," as the cochlea is coiled like the shell of a snail. In most species, the long spiraled tube of the cochlea has two to four turns (depending on species), which are visible in cross-section (Fig. 26.2A). The sensory organ, the organ of Corti, contains the receptor cells (hair cells) and supporting cells. The organ of Corti rests on the basilar membrane and is covered by the tectorial membrane (Fig. 26.2B). Hair cells are of two types: inner hair cells and outer hair cells. Their names are derived from the position of the hair cells along the cochlear spiral: inner hair cells are located innermost along the spiral and outer hair cells are located outermost. There is one row of inner hair cells and usually three rows of outer hair cells. In the human cochlea, there are about 3500 inner hair cells and about 14,000 outer hair cells.

FIGURE 26.3 Illustration of hair cell stimulation, receptor potential response, and afferent fiber discharge. From A. Flock.

A

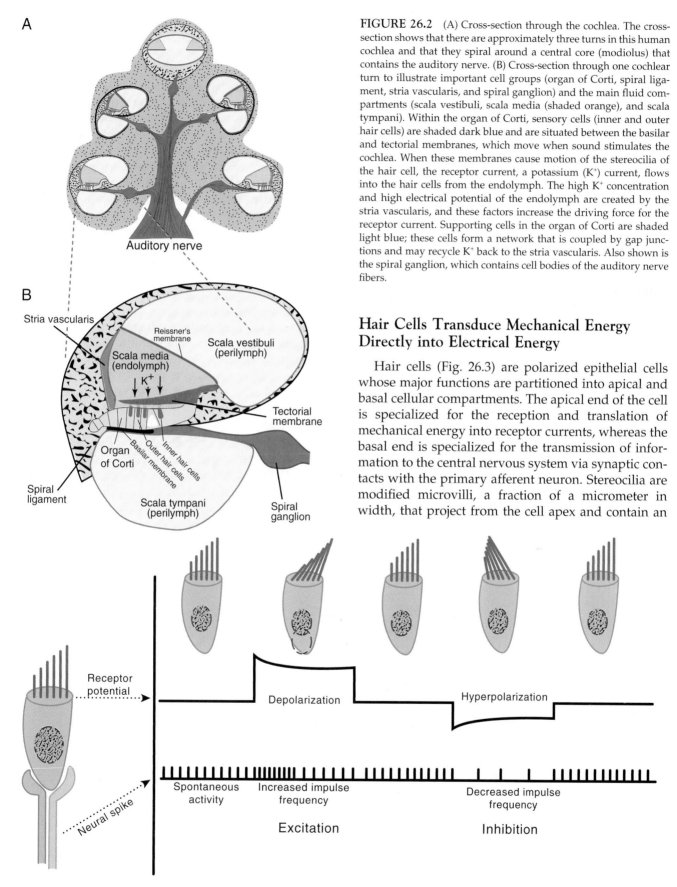

Auditory nerve

B

Stria vascularis

Reissner's
membrane

Scala vestibuli
(perilymph)

Scala media
(endolymph)

K⁺

Tectorial
membrane

Organ
of Corti

Inner hair cells

Outer hair cells

Basilar membrane

Spiral
ligament

Scala tympani
(perilymph)

Spiral
ganglion

Receptor
potential

Depolarization

Hyperpolarization

Neural spike

Spontaneous
activity

Increased impulse
frequency

Decreased impulse
frequency

Excitation

Inhibition

FIGURE 26.2 (A) Cross-section through the cochlea. The cross-section shows that there are approximately three turns in this human cochlea and that they spiral around a central core (modiolus) that contains the auditory nerve. (B) Cross-section through one cochlear turn to illustrate important cell groups (organ of Corti, spiral ligament, stria vascularis, and spiral ganglion) and the main fluid compartments (scala vestibuli, scala media (shaded orange), and scala tympani). Within the organ of Corti, sensory cells (inner and outer hair cells) are shaded dark blue and are situated between the basilar and tectorial membranes, which move when sound stimulates the cochlea. When these membranes cause motion of the stereocilia of the hair cell, the receptor current, a potassium (K⁺) current, flows into the hair cells from the endolymph. The high K⁺ concentration and high electrical potential of the endolymph are created by the stria vascularis, and these factors increase the driving force for the receptor current. Supporting cells in the organ of Corti are shaded light blue; these cells form a network that is coupled by gap junctions and may recycle K⁺ back to the stria vascularis. Also shown is the spiral ganglion, which contains cell bodies of the auditory nerve fibers.

Hair Cells Transduce Mechanical Energy Directly into Electrical Energy

Hair cells (Fig. 26.3) are polarized epithelial cells whose major functions are partitioned into apical and basal cellular compartments. The apical end of the cell is specialized for the reception and translation of mechanical energy into receptor currents, whereas the basal end is specialized for the transmission of information to the central nervous system via synaptic contacts with the primary afferent neuron. Stereocilia are modified microvilli, a fraction of a micrometer in width, that project from the cell apex and contain an abundant supply of tightly packed actin filaments, bound by fimbrin, coursing along their length. Depending on which specific organ the hair cells reside in, their lengths range from a fraction (in the bat cochlea) to tens of micrometers (in vestibular organs). The number of stereocilia per hair cell ranges from about 10 to 300. Together the stereocilia from one cell form a "hair bundle" and are bundled together by extracellular filamentous linkages. They taper rapidly as they insert into the hair cell's cuticular plate, an actin-rich apical cytoplasmic structure. Because the stereocilia are stiff, rod-like structures, they pivot at their insertion when deflected.

Receptor Potentials Are Evoked by Mechanically Gated Ion Channels within Stereocilia

With the hair bundle in its resting, unperturbed position (Fig. 26.3), a standing inward current exists through a small proportion (10–25%) of mechanically activated channels. Because these channels are nonselective for cations, this inward, positively charged flux tends to depolarize the hair cells. In many hair cell sensory systems, including the organ of Corti, the ionic milieu surrounding the hair bundle is richest in potassium (Fig. 26.2B); thus, the major charge carrier for the transduction current is potassium. However, small amounts of calcium are also required to sustain stereociliar channel activity. Displacement of the hair bundle toward the tallest stereocilium increases the proportion of open channels, thereby producing an increase of the inward resting current. The subsequent flow of current across the basolateral membrane produces a depolarizing voltage change, a receptor potential, capable of activating a variety of voltage-dependent conductances in that membrane. Conversely, closure of transduction channels during bundle movement away from the tallest stereociliar row reduces the inward current, effectively hyperpolarizing the basolateral membrane. Several lines of evidence indicate that the flow of transduction current into hair cells occurs near the top of the hair bundle, i.e., channels are located near the tips of the stereocilia. Maximal transducer conductance changes up to about 10 nS have been observed. Single channel conductances on the order of 20–200 pS either have been measured or estimated and may increase along the tonotopic axis of the cochlea. From such conductance measures, the number of channels per stereocilium has been computed to be one to two.

The relationship between degree of bundle deflection and receptor potential magnitude is neither linear nor symmetric (Fig. 26.4). Bundle displacements in the depolarizing direction produce larger responses than equal displacements in the opposite direction. The displacement–response function is sigmoidal and shifted from its midpoint. Thus, symmetrical sinusoidal deflections of the bundle (as might occur with acoustic stimuli) will produce both sinusoidal (ac) and superimposed depolarizing steady-state (dc) changes in membrane potential. Saturating electrical responses are evoked by deflections as small as 300 nm (Hudspeth and Corey, 1977).

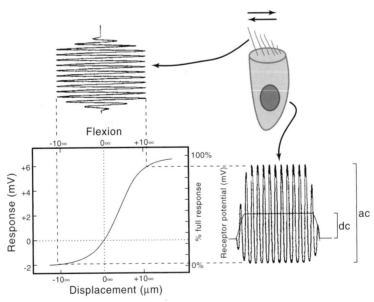

FIGURE 26.4 Sigmoidal input-output function of hair cell. Symmetrical sinusoidal displacement of stereocilia produces ac and dc receptor potential components. From Hudspeth and Corey (1977).

Stereocilia Tip Links May Underlie Transducer Gating

The molecular basis of the hair bundle's response polarity appears to reside in specialized structural attachments at the tips of stereocilia (Pickles *et al.*, 1984) (Fig. 26.5). These "tip links" are elastic filaments ("springs") that link the top of each stereocilium with a dense membranous plaque on the upper side of the adjacent taller stereocilium, and occur in line with the axis of maximal bundle sensitivity. They are believed to provide the tension required to open transduction channels during bundle deflection, with one end of the filament being anchored and the other pulling on the channel gate. As the hair bundle tilts during deflection in the excitatory direction, adjacent stereocilia shear against one another, stretching and increasing the tension of tip links, thereby increasing the probability that transducer channels will open. When the deflection is in the hyperpolarizing direction, tip link tension slackens and the channels tend to close. In line with tip link or gating spring hypothesis, destruction of the tip links by enzymatic (elastase) or chemical (calcium chelators) treatments can abolish mechanical transduction. In addition, if tip link tension accounts for much of the bundle's stiffness, as expected, bundle compliance varies with the extent of deflection. The deflection versus bundle compliance function is bell-shaped, with maximum compliance occurring when half the transduction channels are open. Furthermore, compliance changes are abolished by blocking stereociliar transduction channels with aminoglycoside antibiotics.

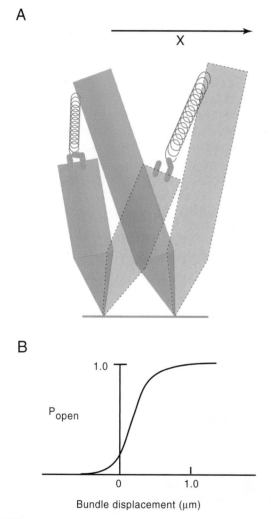

FIGURE 26.5 Gating spring model of hair cell transduction. (A) Tip links connecting channel gate to adjacent taller stereocilium tense during displacement toward the taller stereocilium, thus increasing the probability that the channel will open (B). From Pickles and Corey (1992).

Electrical Properties of the Basolateral Membrane Shape Receptor Potentials

The speed of the hair cell transduction process is incredibly fast compared to transduction in other sensory systems, such as vision, olfaction, and the taste modalities of sweet and bitter. The delay between a bundle deflection and the onset of receptor current is estimated to be about $10\,\mu s$ at 37°C (Corey and Hudspeth, 1983). This rapid response is a consequence of direct gating of transduction channels, and such speed is essential for auditory hair cells to detect sound frequencies in the kilohertz (thousand per second) range. Humans can detect sound frequencies up to about 20 kHz, and some mammals, such as bats, can hear above 100 kHz. Although receptor currents may be generated without attenuation across frequency, receptor potentials, which ultimately are responsible for the release of neurotransmitter at the hair cell synapse, are susceptible to the RC filter characteristics of the basolateral membrane. The RC time constant of hair cell membranes ranges from a fraction of a millisecond to a few milliseconds at the resting potential. This translates in the frequency domain to a low-pass filter whose cutoff frequency (fc, the frequency at which the response energy is halved) ranges from tens of hertz to about 1 kHz. This filtering reduces the ac receptor potential in half for every octave increase in frequency above the cutoff. Ultimately, at very high frequencies, ac responses will be negligible, but the dc component will remain unperturbed—a process termed rectification (Fig. 26.6). The actual cutoff frequency may be considered dynamic because receptor potentials themselves may activate voltage-dependent ionic conductances in the basolateral membrane that will modify the resistive component of the RC product. In outer hair cells from the organ of Corti, the capacitance of the basolateral membrane is also highly voltage dependent (Santos-Sacchi, 1992), and therefore in this cell type variations in both resistive and capacitive components may influence the membrane filter.

The mammalian organ of Corti rests upon an acellular basilar membrane (Fig. 26.2B) that extends along the coiled cochlea. Cells at the basal end of the organ of Corti have high characteristic frequencies, whereas those at the apex have low ones. Von Bekesy, who won the Nobel prize in 1968, discovered that different regions of the basilar membrane are tuned to particular frequencies; low-frequency tones cause maximal vibrations of the basilar membrane near the apex, whereas higher-frequency tones cause vibrations more basally. This tuning results from the mechanical characteristics of the basilar membrane, which becomes stiffer and narrower with distance from the apex to the base. As demonstrated in early experiments, the intrinsic frequency selectivity afforded by basilar membrane tuning is not great enough to account for the very selective responses observed in hair cells and auditory nerve fibers. Later experiments demonstrated that active mechanisms are necessary to achieve this high frequency selectivity.

Active Mechanical Properties of the Outer Hair Cells Drive the Mammalian Cochlea Amplifier

Outer hair cells (OHCs) are cylindrically shaped and decrease in resting length by a factor of about four from the apical to the basal ends of the cochlea. When an isolated OHC is stimulated electrically, it responds by altering its length (Brownell *et al.*, 1985). No other auditory cell type responds in this manner. The mechanical response is voltage dependent; depolarizing stimuli induce contractions and hyperpolarizing stimuli induce elongations. The length change versus voltage (dL vs dV) function (Fig. 26.7) is sigmoidal, and like the stereocilia transducer function, it is operatively offset from its midpoint; the midpoint voltage is near –30 mV (OHC resting potential is near –70 mV). The slope of the function (dL/dV; Fig. 26.7) indicates the sensitivity of the mechanical response to voltage change, and responses as large as 30 nm/mV have been found. *In vitro*, at least, this maximum sensitivity or gain resides at a voltage that is depolarized relative to the resting potential of the cell. The mechanical activity of the cell is not akin to any other known form of cellular motility and is governed directly by voltage-dependent, integral membrane protein motors, recently identified as the gene product of Prestin (Zheng *et al.*, 2000), one of a family of sulfate transporter genes (SLC26). *In vivo*, the motor action of OHCs probably is responsible for otoacoustic emissions (Box 26.1).

Through a variety of experimental approaches, motor activity has been shown to be restricted to the lateral membrane of the OHC. As might be expected for a voltage-dependent process that resides within the membrane, a charged voltage sensor must exist, just as voltage-dependent ion channels have voltage sensors. The existence of an OHC motility voltage

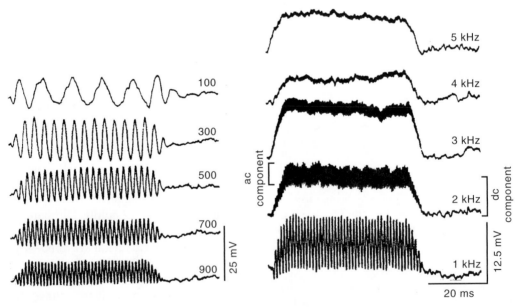

FIGURE 26.6 Receptor potentials from the mammalian inner hair cell in response to acoustic bursts of increasing frequency. Because of the cell's *RC* time constant, the ac component diminishes as frequency increases. However, the dc component remains intact. From Russell and Sellick (1983).

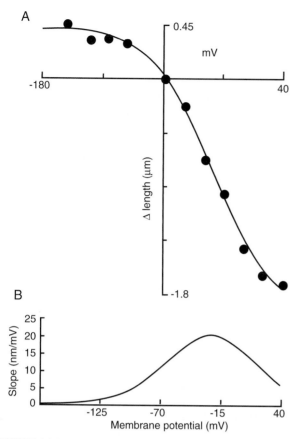

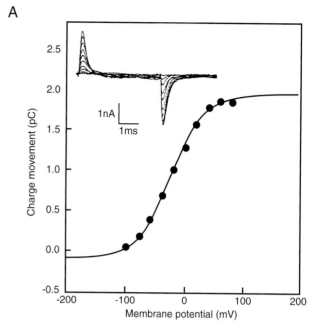

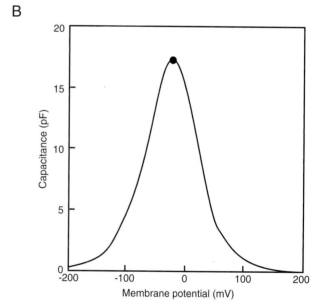

FIGURE 26.7 Mechanical response of mammalian outer hair cell (OHC) under voltage clamp. The OHC changes its length when the cell is held at different membrane potentials (A). The slope of the sigmoidal input-output function defines the cell's sensitivity to membrane potential change (B). From Santos-Sacchi (1992). Used with permission.

FIGURE 26.8 Gating charge associated with OHC motility voltage sensor. When the membrane potential is stepped with increasingly larger depolarizing voltages from a negative holding potential, nonlinear capacitive currents are generated and are obvious after linear capacitive currents are subtracted (inset). Integrating the onset currents, a measure of the amount of charge moved within the membrane can be obtained (A). The function is sigmoidal and has characteristics similar to the mechanical response. The first derivative of the charge with respect to membrane voltage defines the cell's nonlinear capacitance (B). From Santos-Sacchi (1991).

sensor is confirmed by measuring gating charge movements (capacitive-like currents) under voltage clamp while blocking ionic conductances. These gating currents represent the restricted movement of the charged voltage sensor within the plane of the lateral membrane; increasing voltage will move more charge (thus activating more motors) according to Boltzmann statistics. Maximum charge moved is about $7500e/\mu^2$. This value also is believed to characterize the density of motor molecules in the lateral plasma membrane. The plot of charge versus voltage (Q/V) is sigmoidal (Fig. 26.8A) and has the same shape and characteristics as the dL versus dV function. The slope of the Q/V function is defined as capacitance, and thus the capacitance of OHC is a bell-shaped function of voltage (Fig. 26.8B). This nonlinear capacitance rides atop the cell's intrinsic linear membrane capacitance of $1\,\mu F/cm^2$. Thus, in an OHC of about $70\,\mu m$ in length, at the point where motile gain is maximum (i.e., where half the motors are activated), the capacitance peaks at about double the cell's linear capacitance. As mentioned

earlier, such nonlinear capacitance may have significant effects on membrane-filtering characteristics. Although the mechanical response is voltage dependent and is not evoked by activation of any particular

BOX 26.1

OTOACOUSTIC EMISSIONS

In 1978, David Kemp made a surprising discovery: the ear can sometimes *emit* sounds. These "otoacoustic emissions" are now the focus of much interest because basic researchers can use them to study the function of the ear and because clinicians can use them as tests of hearing (reviewed by Shera, 2004). Emissions are almost always so low in level that they are inaudible unless the individual is in an exceptionally quiet environment. Thus, measurement of emissions requires a sensitive, low-noise microphone that is placed in the external ear canal. Many studies indicate that the cochlea (inner ear) is the source of emissions. Within the cochlea, the outer hair cells are likely to be the generators of emissions. After these movements are generated, they travel in reverse of the normal pathway for sound into the inner ear: the emissions propagate from their point of generation along the basilar membrane to the oval window, then through the middle ear via the ossicles, and finally they move the tympanic membrane to result in airborne sound.

There are two main types of emissions—spontaneous and evoked—the latter are evoked in response to an externally presented sound. Spontaneous emissions are detected from the ears of about one-third of normal hearing humans, but these emissions are rare in laboratory animals. They are almost always pure tones. Transient-evoked emissions are evoked by a short sound such as a click and appear several to tens of milliseconds later. These emissions were first called the "cochlear echo," but they are not a real echo because more energy can appear in the emission than was present in the evoking sound. Distortion product-evoked emissions (Fig. 26.9) are evoked by two tones ("primaries" of frequencies f_1 and f_2) and occur at combinations of the primary frequencies (such as at frequency $2f_1-f_2$). The two primary tones each produce traveling waves along the basilar membrane, and a likely point of generation of the emission is where these waves overlap maximally.

Otoacoustic emissions can be used clinically as a screening tool in hearing tests, as subjects with sensory hearing losses greater that 30 dB typically lack emissions. Transient-evoked or distortion product-evoked emissions are used in such tests. Emission-based tests are especially valuable for individuals such as infants who are otherwise hard to test by conventional audiometry. Otoacoustic emissions only rarely are the cause of tinnitus, the sensation of ringing in one's ears. Most individuals who have tinnitus do not have emissions corresponding to the tinnitus. Thus, tinnitus arises by some other mechanism, almost certainly within the nervous system.

M. Christian Brown

Reference

Shera, C. A. (2004). Mechanisms of mammalian otoacoustic emission and their implications for the clinical utility of otoacoustic emissions. *Ear and Hearing* **25**, 86–97.

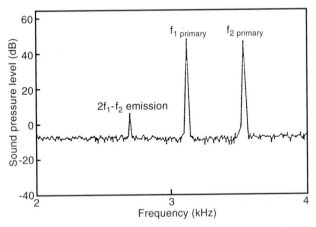

FIGURE 26.9 Distortion product otoacoustic emission from a human subject measured with a microphone placed in the ear canal. The microphone records the sound pressure of the two primary tones (frequencies $f_1 = 3.164$ kHz and $f_2 = 3.828$ kHz, each 50 dB SPL) that were used to evoke the emission at $2f_1-f_2$ (2.5 kHz at 12 dB SPL). From Lonsbury-Martin and Martin (1990).

voltage-dependent ionic conductance, the activity of the voltage sensor requires intracellular chloride, and in intact OHCs a stretch-activated chloride conductance appears crucial for maintaining and modulating motor activity (Rybalchenko and Santos-Sacchi, 2003).

Because hearing sensitivity and frequency selectivity in mammals span the kiloHertz range, any mechanism designed to augment hearing must function at these rates. Indeed, OHC motility has been demonstrated, *in vitro*, to extend well into the tens of kiloHertz range. Nevertheless, because the mechanical response is voltage dependent, it will be affected by the RC time constant of the OHC. In this scenario, at high frequencies, transmembrane ac receptor potentials, which presumably drive the mechanical response *in vivo*, will be attenuated greatly. Consequently, a current debate focuses on the relative contributions of lateral membrane activity and stereocilia activity, each of which is capable of force production. Active mechanical responses of the stereocilia bundle may be driven by calcium influx through calcium-sensitive, mechanically activated stereocilia channels, and thus may not be limited by the membrane filter. Of course, identification of the important role of chloride in the function of prestin *in vivo* and its mechanically activated flux through the lateral plasma membrane may underlie a similar voltage independence. In this way the OHCs of the mammalian inner ear may have overcome the limiting effects of the membrane filter at high-frequency acoustic stimulation.

To summarize, OHCs, probably through a mechanical feedback scheme, are required to boost BM motion and enhance frequency selectivity, a process termed the "cochlear amplifier." In the absence of OHCs or with a deletion of the gene for Prestin (Liberman *et al.*, 2002), hearing sensitivity and frequency selectivity are likewise impaired. OHCs are unique because they function as both receptors and effectors, transducing mechanical stimuli via hair bundle displacement and changing length and other mechanical properties in response to the generated receptor potentials. These mechanical events, via an energetic boost to BM motion, provide an enhanced stimulus to the inner hair cells, which are the cells that provide synapses to the overwhelming majority of auditory nerve afferents.

Cochlear Endolymph Increases the Sensitivity of Hearing

An increase in the sensitivity of hair cells is made possible by the unique composition of the inner ear fluids. Within the inner ear, fluids are contained in three compartments known as scalae: scala tympani, scala media, and scala vestibuli (Fig. 26.2B). These scalae extend in parallel along the length of the cochlea from the base to the apex. Scala tympani and scala vestibuli contain perilymph, which is high in Na^+ and low in K^+, similar to other extracellular fluids. Scala media contains endolymph, a specialized fluid with a low concentration of Na^+ and a high concentration of K^+ (about 160 mM). The endolymph is also unusual because it has a positive electrical potential (about 90 mV), observable even in the absence of sound. The endolymph is generated by a highly vascularized tissue in the lateral edge of the cochlea, the stria vascularis (Fig. 26.2B).

The organ of Corti, which contains the hair cells, is located at the junction between endolymph and perilymph (Fig. 26.2B). Stereocilia are surrounded by endolymph, whereas the remainder of the hair cell (basolateral membrane) is surrounded by perilymph. The endolymph increases the sensitivity of the hair cells because it increases the transduction current, a K^+ current that flows from the endolymph into the hair cells (Fig. 26.2B). The endolymph increases this current because (1) the high concentration of K^+ in endolymph forms a concentration gradient that favors K^+ flow into the cells and (2) the positive potential of the endolymph forms a large electrical gradient that also favors K^+ flow into the hair cells. Once the chemical and electrical properties of the endolymph are established and sound stimulates the cochlea, K^+ flows into the hair cells with little energy expenditure by the hair cells, as K^+ is flowing down its electrochemical gradient. Thus, they do not require a high blood flow that might bring noise to the organ of Corti, interfering with sound reception. In fact, there are only a few blood vessels near the organ of Corti.

Sensorineural Hearing Loss Often Results from Damage to Hair Cells

Sensorineural hearing loss results from damage to hair cells or, less commonly, from damage to afferent nerve fibers (reviewed by Schuknecht, 1993). Hair cells can be destroyed or their hair bundles damaged by intense sound such as those generated by guns, jet engines, or even earphones operated at high sound levels. Hair cell loss is permanent in the mammalian cochlea, but in the bird cochlea, hair cells are regenerated from nearby supporting cells (Corwin and Cotanche, 1988). Damage from intense sounds can be minimized by (1) decreasing the duration of the sound exposure and (2) decreasing the level of the sound at the tympanic membrane (by wearing protectors such as earguards and earplugs). Hair cells can also be destroyed by chemical agents like aminoglycoside antibiotics (e.g., streptomycin, kanamycin). These agents block the transduction channels directly, but their mechanism of hair cell destruction is independent. In individuals with sensorineural hearing loss, some hearing can be restored with a cochlear implant (Box 26.2).

Summary

Current research on the cochlea is centered on the mechanisms of outer hair cell motility and how this motility contributes to basilar membrane motion. The mechanisms of how sound causes active stereocilia motion remain to be worked out. Substantial interest focuses on the processes and factors involved in hair cell regeneration with the eventual hope that approaches like stem cell implants might encourage restoration of

---

BOX 26.2

COCHLEAR IMPLANTS

The cochlear implant is one of the most successful prostheses used to stimulate the nervous system (Niparko et al., 2000). The cochlear implant can provide partial restoration of hearing for individuals with sensorineural hearing loss. In sensorineural hearing loss, there is usually a partial or complete loss of the sensory cells (hair cells) in the inner ear (cochlea). Hair cells normally transduce the mechanical energy of sound into the electrical energy of receptor potentials. They synapse on primary auditory nerve fibers, which send information to the brain. Hair cells can be damaged irreversibly by intense sound, ototoxic drugs, or the aging process. Once lost, hair cells are not regenerated in mammals. Often, however, individuals with hair cell loss retain a significant complement of auditory nerve fibers. It is these fibers that are stimulated by the cochlear implant.

The cochlear implant consists of a microphone to detect sound, an electronic "processor" that transforms the sound waveform into a code of electrical stimuli, and an array of stimulating electrodes in the cochlea. The electrode array (Fig. 26.10) is inserted through or near the round window into scala tympani where it lies close to the peripheral axons of primary auditory neurons. Usually, the implant consists of about a dozen electrodes that begin in the base and are spaced apically along the cochlear spiral. Those electrodes in the most basal regions are positioned to stimulate nerve fibers that originally responded to high frequencies and those in more apical regions are positioned to stimulate fibers that originally responded to lower frequencies.

In some individuals that have received implants, comprehension of speech is possible, even when there are no other cues such as lip reading. These individuals can carry on a normal conversation, even via telephone. In other individuals, however, full speech comprehension is not restored but the implant, together with lip-reading ability, assists with spoken conversation. It also provides for the detection of important sounds such as the ring of a telephone and the approach of a vehicle. Variability in the success of the implant from patient to patient probably depends on the number of surviving primary auditory neurons, the exact orientation of the electrodes with respect to the neurons, and the coding scheme of the processor that is used. The individual's motivation and the assistance received from clinicians are also likely to be important factors. Finally, individuals who have become deaf after the acquisition of spoken language reacquire language ability more easily than prelingually deafened individuals, almost certainly because of differences in the central nervous system. Cochlear implant research is focused on making improved designs for the processor and electrodes so that in the future implant users may be able to more fully comprehend speech.

M. Christian Brown

Reference

Niparko, J. K., Kirk, K. I., Mellon, N. K., Robbins, A. M., Tucci, D. L., and Wilson, B. S., Eds. (2000). "Cochlear Implants. Principles and Practices." Philadelphia, Lippincott Williams & Wilkins.

---

function in the mammalian cochlea. The improvement of cochlear implants is also a very active area of clinical research.

THE AUDITORY NERVE

Hair cells receive their primary afferent innervation from neurons of the spiral ganglion, located in the central core, or modiolus, of the cochlea (Figs. 26.2, 26.10). These bipolar auditory neurons send peripheral axons to the hair cells and central axons into the brain by way of the auditory nerve, a subdivision of the eighth cranial nerve. Two types of afferent neurons separately innervate the inner and outer hair cells (Fig. 26.11). The first type (the type I neuron) sends processes to contact inner hair cells, almost always contacting a single hair cell, whereas the second type (the type II neuron) sends processes to contact from 5 to 100 outer hair cells. Fibers from type I neurons are relatively large in diameter and myelinated; thus their information reaches the brain quickly, within a few tenths of a millisecond. Fibers from type II neurons are thin and unmyelinated and transmit information much more slowly. Both types of afferent fibers project centrally into the cochlear nucleus in the brain stem. Interestingly, type I neurons total about 95% of the afferent population (about 30,000 in humans), whereas type II neurons total only about 5%. Thus, outer hair cells,

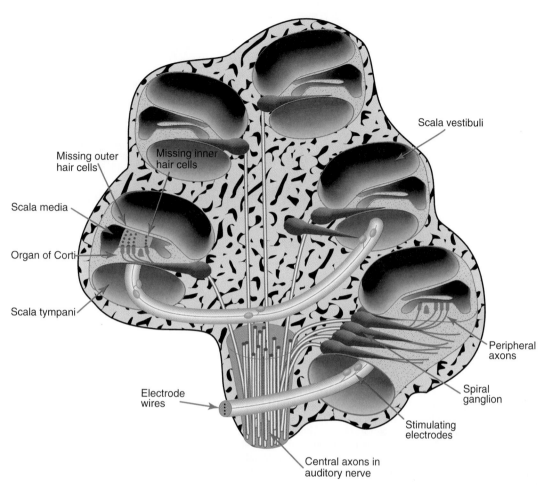

Scala vestibuli

Missing outer
hair cells

Missing inner
hair cells

Scala media

Organ of Corti

Scala tympani

Peripheral
axons

Electrode
wires

Spiral
ganglion

Stimulating
electrodes

Central axons in
auditory nerve

FIGURE 26.10 Drawing of the human cochlea showing cochlear implant electrodes. The electrode array is inserted through the round window of the cochlea into the fluid-filled space called scala tympani. It stimulates peripheral axons of the primary auditory neurons, which send messages via the auditory nerve into the brain. In the normal cochlea, frequency is mapped along the cochlear spiral, with the lowest frequencies at the apex of the cochlea (at the top of the figure). The different electrodes of the cochlear implant are designed to stimulate different groups of nerve fibers that originally responded to different frequencies, although because of spatial constraints the implant is not inserted all the way to the cochlear apex. From Loeb (1985).

which number over three-quarters of the receptor cell population, are innervated by only a small minority of the afferent neurons. This innervation plan strongly suggests that the functional role for inner hair cells and type I neurons is to serve as the main channel for sound-evoked information flow into the brain.

Responses Are Sharply Tuned to Frequency

Type I auditory nerve fibers respond to sound and transmit these responses to the brain via discrete action potentials (reviewed by Ruggero, 1992; Geisler, 1998). The brain must extract information from these spikes and process this information to eventually form a percept of the stimulus. The information available to the brain via the auditory nerve consists of which nerve fibers are responding and the rate and time

pattern of the spikes in each fiber. The response area for a nerve fiber usually is plotted as a graph of sound pressure level vs. frequency with the line showing the threshold contour; such a graph is known as a tuning curve (Fig. 26.12). The lowest point on the tuning curve is at the characteristic frequency (CF). This is the frequency that evokes a response at the lowest sound pressure level; at CF, auditory nerve fibers can respond to sound levels as low as 0 dB in the most sensitive range of hearing. At low sound levels, the tuning curve is impressively narrow, indicating that the fiber responds only to a narrow band of frequencies near CF. This sharply tuned "tip" region (Fig. 26.12) is likely generated by the active motility of the outer hair cells. At high sound levels, the tuning curve becomes much wider, especially for frequencies below CF. This response to a broad range of frequencies likely reflects

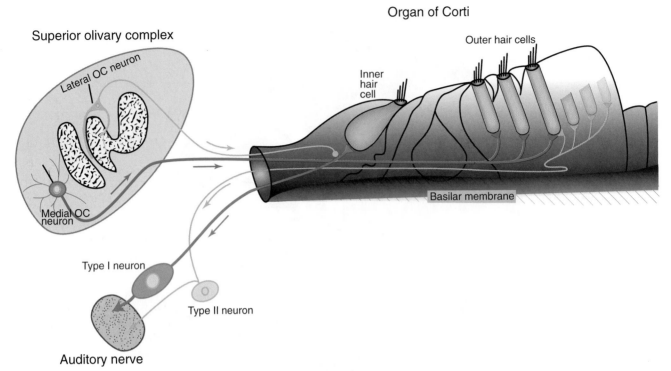

FIGURE 26.11 Innervation patterns of afferent and efferent neurons in the organ of Corti. Afferent innervation is provided by ganglion cells of the spiral ganglion in the cochlea, which have central axons that form the auditory nerve. There are two types of afferent neurons: (1) type I neurons, which synapse with inner hair cells, and (2) type II neurons, which synapse with outer hair cells. Efferent innervation is provided by a subgroup of neurons in the superior olivary complex that send axons to the cochlea and are hence called olivocochlear (OC) neurons. There are two types of OC neurons: (1) lateral OC neurons, which innervate type I dendrites near inner hair cells, and (2) medial OC neurons, which innervate outer hair cells. Lateral OC neurons are distributed mainly ipsilateral to the innervated cochlea, whereas medial OC neurons are distributed bilaterally to the innervated cochlea, with approximately two-thirds from the contralateral side (not illustrated) and one-third from the ipsilateral side of the brain. From Warr *et al.* (1986).

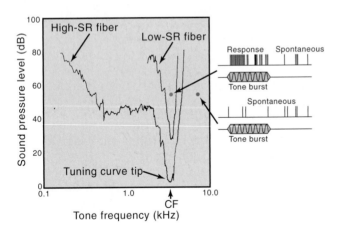

Figure 26.12 Tuning curves of two type I auditory nerve fibers. These curves plot the sound pressure level necessary to cause a response as a function of sound frequency. Within the tuning curve, the fiber responds to sound, whereas outside the tuning curve, there is only spontaneous firing (insets at right). The lowest point on the tuning curve corresponds to the characteristic frequency (CF); it is the point of maximal sensitivity. The sharply tuned region near the CF is called the tuning curve tip. The sharp tuning and high sensitivity of this region are generated by the mechanical properties of the outer hair cells. Tuning curves from two fibers are shown, one from a high SR (spontaneous rate) fiber and another from a low SR fiber. As is typical, the high SR fiber has the highest sensitivity. From Kiang (1984).

the passive mechanical characteristics of basilar membrane motion with little contribution from outer hair cells.

The relationship between CF and its point of innervation in a type I fiber has been studied by single unit labeling (Liberman, 1982). The distance to the point of innervation along the length of the cochlea is logarith-

mically related to fiber CF. Fibers with the lowest CFs innervate the apex of the cochlea, and fibers with progressively higher CFs innervate progressively more basal positions, as expected from the pattern of basilar membrane vibration. This precise mapping of frequency to position is known as tonotopic mapping. It is preserved as the auditory nerve projects centrally into the cochlear nucleus and for much of the central pathway. This observation strongly suggests that frequency is coded via a place code, with neurons at different places coding for different frequencies.

Phase Locking of Responses and Codes for Sound Frequency

Responses of auditory nerve fibers can show time-locked discharges at particular phases within the cycle of the sound waveform, a property known as phase locking (Fig. 26.13). Although locked to a particular phase, there is generally not a spike for every waveform peak. Phase locking in nerve fibers results from the phasic release of neurotransmitter as dictated by the ac receptor potential (Fig. 26.6). Phase locking and the ac receptor potential decrease for frequencies above 1 kHz. At low sound frequencies, phase locking is an important temporal code for sound frequency and the sensation of pitch. At high frequencies where phase locking is diminished, the only code for sound frequency is the place code.

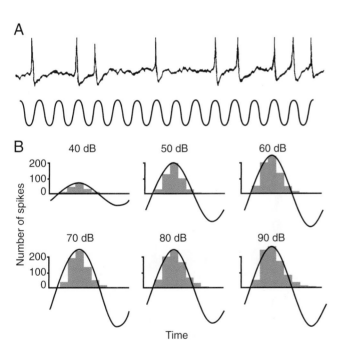

Figure 26.13 (A) Preferential firing of an auditory nerve fiber at a certain phase of the sound waveform. This pattern is called "phase locked," although the firing is not on every cycle of the waveform. Stimulus frequency was 0.3 kHz. (B) Histograms that quantify the time of firing plotted within one cycle of the sound waveform for many repeated cycles of the stimulus. Response of the fiber is phase locked at the moderate and high SPLs shown, even on this very fast time scale (time for one period was about 0.9 ms, given the stimulus frequency of 1.1 kHz). (A) From Evans (1975). (B) From Rose *et al.* (1971).

Neural Response Is a Function of Sound Level and Spontaneous Activity

The response of a single auditory nerve fiber increases with sound level until a point at which the rate of the fiber no longer increases and is saturated (Fig. 26.14A, solid line). The dynamic range over which the rate of most fibers increases is generally between 20 and 30 dB, with some fibers showing somewhat greater dynamic ranges. How then can the auditory nerve signal the large range in level of audible sound from 0 to 100 dB? First, it is likely that as the level of a tone increases, more and more fibers that are tuned to other CFs begin to respond, because tuning curves become broader at higher sound levels (Fig. 26.12). Second, auditory nerve fibers vary in their sensitivity to sound, and as sound level is increased, the less sensitive fibers begin to respond. Sensitivity of fibers at a given CF varies by as much as 70 dB. The sensitivity of response is correlated with the rate of spontaneous firing, which is the rate of firing when there is no stimulus or when the stimulus is outside the tuning curve (Fig. 26.12). Spontaneous rates (SRs) vary from one fiber to another over the range of 0 to 100 spikes/s.

Although there may be a continuum of SR, three main groups of fibers have been defined (low SR: <0.5 spikes/s; medium SR: 0.5 to 17.5 spikes/s; high SR: >17.5 spikes/s), and these groups predict many physiological and anatomical characteristics of auditory nerve fibers. The high SR fibers have higher sensitivities than medium and low SR fibers (Fig. 26.12). Low and medium SR fibers give off the largest number of terminals in the cochlear nucleus of the brain stem and preferentially innervate certain regions, such as the peripheral cap of small cells, suggesting that information carried by the different SR groups may be kept somewhat separate in the brain stem. Low SR fibers may be less sensitive, but they likely play important roles in detecting changes in sounds at high sound levels. Low SR fibers can signal changes at high sound levels because their low sensitivity causes them to respond mostly at higher sound levels and because they have less tendency to saturate, as their response often grows more slowly with sound level.

Masking Involves Adaptation and Suppression

The auditory nerve response to one stimulus may be masked, or hidden, by the presence of other stimuli. Masking involves properties of auditory nerve response such as adaptation and two-tone suppression, and possibly refractory properties. In adaptation, firing to a tone burst is high initially and then lessens, or adapts,

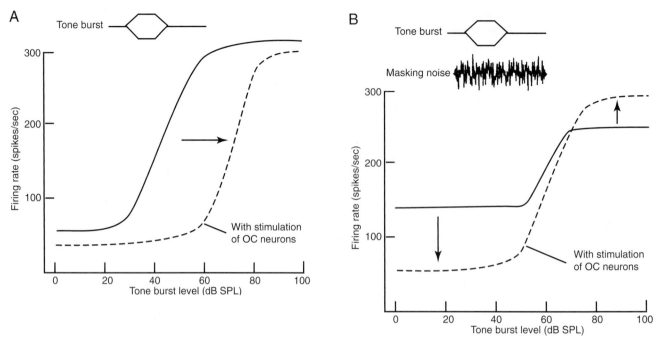

Figure 26.14 Rate level function for an auditory nerve fiber in response to tone bursts without (solid lines) and with (dashed lines) electrical stimulation of olivocochlear (OC) neurons. (A) With tone bursts alone, the discharge rate rises with sound level until it reaches a maximum and no longer increases (saturation). Stimulation of the OC neurons shifts the function to the right toward higher tone burst levels (arrow). This shift adjusts the dynamic range of the fiber so that it can signal changes in the tone burst level even for high sound levels; this is likely to be an important function of OC neurons. (B) When the tone bursts are accompanied by continuous masking noise (insets at top), the function in response to tone bursts is changed (solid curve). At low levels of tone bursts, the fiber has a significant firing rate because it is responding mainly to the noise. Even at high levels of tone bursts, the fiber is still responding to the noise in between tone bursts, which causes adaptation, and the fiber responds less to the tone burst than with tone bursts alone. In this case, stimulation of OC neurons (dashed line) decreases the response to the noise, thus decreasing the rate of the fiber at low levels of tone bursts (left arrow). Because the fiber is responding less to the noise, it is less adapted and has a greater response at high levels of the tone burst (right arrow). The fiber now has a greater ability to signal changes in level of the tone burst; this effect has been called "antimasking" and is likely to be another important function of the OC system. Adapted from Winslow and Sachs (1987).

to a steady state over time (Fig. 26.15). Adaptation probably takes place at the hair cell/nerve fiber synapse, as there is little adaptation in receptor potentials of hair cells for these short times (Fig. 26.6). Because of adaptation to one stimulus, the fiber is less likely to respond to a second stimulus; that is, the first stimulus masks the second stimulus. Masking by continuous noise changes the responses to tone bursts greatly. With tone bursts alone (Fig. 26.14A, solid line), there is a moderate dynamic range and a large difference in firing rate from low to high tone burst levels. With tone bursts in masking noise (Fig. 26.14B, solid line), there is a substantial rate at low tone burst levels because the fiber is now responding to the masking noise. At high tone burst levels, the rate is decreased because the fiber is adapted by the noise and is less likely to respond to the tone burst. Thus, with masking noise, there is much less difference in the rate of the fiber as a function of tone burst level as well as a lower dynamic range: these effects decrease the ability of the fiber to signal changes in the tone burst level.

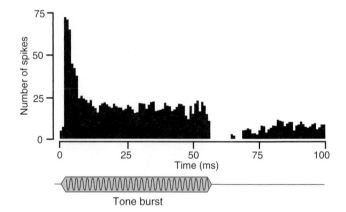

Figure 26.15 Poststimulus time (PST) histogram from an auditory nerve fiber in response to a tone burst (outline indicated below). A PST histogram is constructed by repeatedly presenting a stimulus while counting the number of action potentials that fall into bins of time during and after the stimulus. The histogram can be thought of as the probability of firing as a function of time. This function has an initial peak and then a decrease in firing (adaptation) during the burst. After the burst ends, spontaneous firing is lessened but then returns gradually.

An additional type of masking, referred to as suppressive masking, involves a phenomenon called two-tone suppression. In two-tone suppression, one tone lowers the response to a second tone even though the first tone does not excite the auditory nerve fiber. Two-tone suppression is present in the motion of the basilar membrane. It may help control the amount of gain provided by the cochlear amplifier, because as stimuli composed of several frequencies increase in sound level, two-tone suppression decreases the response of the nerve fiber so that it does not saturate. Whether there is adaptive masking or suppressive masking depends on the frequencies and levels of signals relative to the response area of the fiber. Masking is important in the auditory system because many auditory processes, such as speech comprehension, are affected greatly by maskers and because background or masking signals are common in many everyday situations.

Descending Systems to the Hair Cells and Nerve Fibers

The auditory periphery and other hair cell organs receive an efferent innervation from the brainstem (Fig. 26.11). The cochlear efferent neurons have cell bodies in the superior olivary complex and project to the cochlea and hence are called olivocochlear (OC) neurons (reviewed by Warr, 1992; Guinan, 1996). There are two groups of OC neurons—medial and lateral (Fig. 26.11)—named according to the positions of their cell bodies in the superior olive. These olivocochlear fibers separately innervate the two types of hair cells. Medial OC neurons innervate outer hair cells, whereas lateral OC neurons innervate the inner hair cell region by synapsing on dendrites of type I auditory nerve fibers. Little is known about the lateral OC neurons, as their very thin axons are difficult to study. Medial OC neurons can be activated experimentally by electrical stimulation, which causes them to release the neurotransmitter acetylcholine. At the outer hair cell, acetylcholine acts on a nicotinic receptor that allows Ca^{2+} influx, which then opens Ca^{2+}-activated K^+ channels, allowing K^+ efflux that hyperpolarizes the cell. The hyperpolarization and accompanying decrease in input resistance of the cell probably reduces the electromotility of the outer hair cell, decreases basilar membrane motion, and reduces the responses of inner hair cells and auditory nerve fibers. These decreases shift responses of auditory nerve fibers to higher sound levels (Fig. 26.14A, dashed line). This shift means that sound levels that previously saturated the discharge rates are now within the increasing portion of the rate level curve of the fiber, so the fiber can now signal

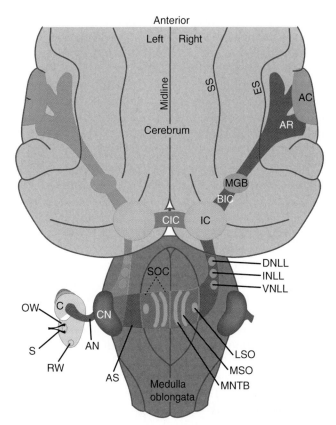

Figure 26.16 Simplified schematic of the pathways of the ascending auditory system of a generalized mammal. The pathway begins at the lower left with the cochlea and ends at top right with the auditory cortex. The large band crossing the midline indicates that the bulk of the pathway is crossed by the level of the inferior colliculus. Some ipsilateral pathways and frequent commissural pathways explain the responses of some units at higher centers to ipsilateral sound. AC, auditory cortex; AN, auditory nerve; AR, auditory radiations; AS, acoustic striae; BIC, brachium of the inferior colliculus; C, cochlea; CIC, commissure of the inferior colliculus; CN, cochlear nucleus; DNLL, dorsal nucleus of the lateral lemniscus; ES, ectosylvian sulcus; IC, inferior colliculus; INLL; inferior nucleus of the lateral lemniscus; LSO, lateral superior olive; MGB, medial geniculate body; MNTB, medial nucleus of the trapezoid body; MSO, medial superior olive; OW, oval window; RW, round window; S, stapes; SOC, superior olivary complex; SS, suprasylvian sulcus; VNLL, ventral nucleus of the lateral lemniscus. From Kiang and Peake (1988).

changes in sound level even at higher sound levels. Thus, one function of medial OC neurons may be to control the gain of the cochlear amplifier to prevent saturation of responses.

Medial OC neurons may also reduce the effects of masking noise on the responses of auditory nerve fibers. Decreases in responses by maskers can occur by adaptation and suppression. The effect of OC stimulation on masked responses to tone bursts can be to *enhance* the response (Fig. 26.14B, dashed line). Responses at low tone burst levels are decreased because the response of the fiber to the noise is decreased (left arrow on Fig. 26.14B); responses at high tone burst levels are increased because there is less noise-induced adaptation and a greater response to the tone bursts (right arrow on Fig. 26.14B). The fiber now has a greater ability to signal changes in level of the tone burst; this effect has been called antimasking, and may be an important function of the OC system. An additional function of medial OC neurons may be to protect hair cells in the cochlea from damage due to intense sounds via their large synaptic endings on outer hair cells.

Summary

Although the function of type I auditory fibers is fairly well understood, we do not yet know how afferent fibers of the outer hair cells, the type II auditory nerve fibers, function in the hearing process. Future experiments will provide insight into the roles played by different spontaneous rate groups of type I auditory nerve fibers. More work is needed to identify the mechanisms used to prevent the degradation of signals by masking noise. These mechanisms could potentially be applied clinically in the design of hearing aids and cochlear implants.

CENTRAL NERVOUS SYSTEM

Auditory Pathways Are Tonotopically Organized

The auditory nerve terminates centrally in the cochlear nucleus, which in turn projects to the other auditory nuclei of the brain stem: the superior olivary complex, nuclei of the lateral lemniscus, and inferior colliculus (Fig. 26.16). These multiple brain stem nuclei are important for determination of the location of a sound source (see discussion later). The external location of a sound source is not represented directly along the receptor organ of the auditory system in contrast to other systems such as the somatosensory system, where position of a stimulus is mapped by sensory endings along the body surface. Instead, the cochlea maps frequency. Thus, directional information must then be determined by neural processing that compares interaural differences in responses; this processing is accomplished mainly in the brain stem. Above the brain stem, auditory information proceeds to the thalamus and cortex, analogous to other sensory systems. At these highest stages of the auditory pathway are the medial geniculate body of the thala-

mus and the auditory fields of cerebral cortex (Fig. 26.16). In addition to the ascending pathways shown in Figure 26.16, descending systems link "higher" to "lower" centers in the auditory pathway. The functions of these descending systems have not been well explored, except for the olivocochlear perhaps system (see earlier discussion).

An important characteristic of most central auditory nuclei is tonotopic organization, the mapping of neural CF onto position (reviewed by Webster *et al.*, 1992; Popper and Fay, 1992; Ehret and Romand, 1997). This tonotopy is established by the basilar membrane and is relayed into the central nervous system by the auditory nerve. Such inputs result in isofrequency laminae, sheets of tissue in which neurons have the same CF. These observations support a place code for sound frequency in much of the auditory central nervous system. This is especially true in higher nuclei because these neurons have a decreased ability to phase lock to the sound waveform, and thus a temporal code does not seem to be as robust. In most auditory nuclei there are not large overrepresentations of certain frequencies, although, as noted later, a given nucleus may be more devoted to low frequencies (e.g., medial superior olive) or to high frequencies (e.g., lateral superior olive). An important exception is the case of echolocating bats that use a constant frequency in their echolocating pulse; in these species, there is an overrepresentation of one of the constant frequencies. This overrepresentation begins in the cochlea and is preserved in the cortex (e.g., Fig. 26.23C). In these bats much cortical area is devoted to the constant frequency, analogous to the large somatosensory cortical areas devoted to important areas such as the hands and face.

Under certain conditions, tonotopic mappings have been shown to be capable of plastic changes (Robertson and Irvine, 1989; reviewed by Buonomano and Merzenich, 1998). After peripheral hearing loss, tonotopic mapping of the auditory cortex is altered such that a region that originally processed frequencies within the hearing loss begins to respond to adjacent frequencies. For instance, when the cochlea is damaged so that it no longer responds to high frequencies, the high frequency portion of the cortex does not stay unresponsive, but over time becomes responsive to middle frequencies where hearing is still normal. Plasticity is of interest to humans because of a common condition called presbycusis, which is hearing loss with advanced age. Presbycusis generally begins with a loss at high frequencies and spreads to lower frequencies with advancing age. Although presbycusis is likely caused by peripheral changes, it may result in plastic changes in the central pathways that change the way sound is processed in the brain.

Units Are Classified by Their Responses to Sound

Classification by PST Histograms

The cochlear nucleus is the auditory center that is understood best at the cellular level (reviewed by Rhode and Greenberg, 1992). Cochlear nucleus neurons have been well classified both anatomically and physiologically, and structure–function correlations can be made between these classifications. Physiologically, excitation of cochlear nucleus neurons arises from their auditory nerve inputs. Single-unit recordings from cochlear nucleus neurons reveal that the auditory nerve spike pattern is changed to many new patterns. The patterns may be used to classify units on the basis of the shape of the poststimulus time (PST) histogram, which plots the spike pattern of the unit as a function of time for short-duration tone bursts. Unit types are called "pauser," "onset," "primary-like with notch," "chopper," and "primary-like" (Fig. 26.17). A correspondence has been established between these unit types and anatomical cell types of the cochlear nucleus. The correspondence was first suggested by the regional distributions of unit and cell types; for example, there is a region of the cochlear nucleus in which one finds mostly "octopus" cells and from which mostly onset units are recorded. Direct correspondence has been made by single-unit labeling, in which a single unit is first classified physiologically according to PST type and subsequently injected with a neural tracer (e.g., horseradish peroxidase) that fills the neuron and its processes. The anatomical cell type can then be determined from postexperiment histology. These types of experiments are difficult and low yielding, but they firmly establish the structure–function correspondences, at least for neurons large enough to record and

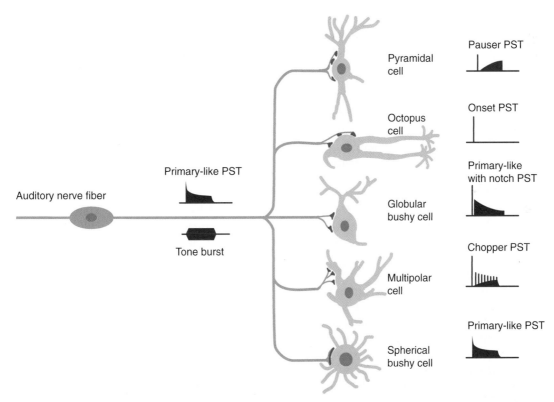

Figure 26.17 Schematic of the main anatomical cell types of the cochlear nucleus and their corresponding poststimulus time (PST) histograms. (Left) An auditory nerve fiber is shown with its typical response, a primary-like PST histogram (shown in Fig. 26.15). (Center) The auditory nerve fiber divides to innervate the main cochlear nucleus cell types. (Right) PST histograms corresponding to these cell types are shown. In their PST histograms, pauser units fire an initial spike and then have a distinct pause before a slow resumption of activity. Onset units fire mainly at the tone burst onset. Primary-like units get their name from the similarity of their PSTS to those of primary auditory nerve fibers, but the primary-like with notch type additionally has a brief notch following the initial spike. Chopper units have regular interspike intervals that result in regular peaks in the PST. Most of these patterns are very different from the primary-like PST and irregular interspike intervals of the auditory nerve fibers. For histograms, the sound stimulus is typically a 25-ms tone burst with frequency at the CF of the neuron and sound level at 30 dB above threshold. From Kiang (1975).

label *in vivo*. For instance, a major cell type in the ventral subdivision of the cochlear nucleus, "spherical bushy" cells, corresponds to a major unit type, the "primary-like" units (Fig. 26.17).

Use of PST histograms is helpful not only in classifying units, but also in revealing functional properties of the neurons. Let us consider the example of primary-like units. The primary-like pattern reveals that the spherical bushy cell faithfully has preserved the temporal properties of the auditory nerve fiber response and implies that these units receive end-bulbs from nerve fibers (see Box 26.3). In contrast, other types of cochlear nucleus neurons alter the spike patterns of the auditory nerve fibers and presumably have distinct functions. For instance, onset units respond mainly at the onset of a short tone burst (Fig. 26.17). One subclass of onset units can increase its response over a large range of SPL and might be involved in signaling sound level. Relative to mammals, tests of such hypotheses are easier in birds, where there is much more regional segregation of neurons by anatomical and physiological type in the cochlear nucleus. In birds, bushy cells are segregated in one region (nucleus magnocellularis), whereas other cell types, such as multipolar cells, are confined to a second region (nucleus angularis). The bushy cells in nucleus magnocellularis have excellent phase locking, but a poor dynamic range of response when sound level is increased. Multipolar neurons in nucleus angularis have complementary properties: poor phase locking, but excellent dynamic ranges. Overall, these results suggest that in the auditory system of diverse animal species there are separate neural pathways for time coding vs. sound level coding as well as for other types of coding. Thus, the cochlear nucleus is the level where parallel pathways in the auditory system begin.

Classification by Response Map

Another important way to classify neurons in the cochlear nucleus, as well as throughout the auditory nervous system, is by the response map of a neuron (Fig. 26.19). Response maps are plotted on graphs of sound level versus frequency; like tuning curves, these maps show areas of excitation but they additionally show areas of inhibition. They are especially valuable where inhibitory influences play a role in shaping responses, such as in the dorsal subdivision of the cochlear nucleus, the inferior colliculus, and at higher stages of the auditory system. In the dorsal cochlear nucleus, five response types have been defined (Fig. 26.19). Type I neurons have excitatory tuning curves similar to those of the auditory nerve (Fig. 26.12) and have no inhibitory areas.

Other response types have progressively larger inhibitory areas; for example, type IV neurons have only a small excitatory area near CF and a narrow, knife-sharp excitatory area above CF, with the rest of their area dominated by inhibition. This inhibition is generated by inhibitory circuits within the dorsal cochlear nucleus, in part from type II neurons. Type IV neurons correspond to the main projection neurons of the dorsal cochlear nucleus, the pyramidal neurons, whereas type II neurons are likely to be inhibitory interneurons. One hypothesis for the functional role of type IV neurons is that their sharp borders between excitatory and inhibitory areas serve to detect the spectral notches that result from the acoustic characteristics of the external ear, especially the pinna. These notches could be used for sound localization, as their frequencies depend on sound source location. The notches also depend on position of the pinna; in fact, type IV neurons receive input from brain stem somatosensory nuclei that may inform the type IV neurons about the position of the animal's moveable pinna. Animals that lack a moveable pinna, such as humans and cetaceans (whales and porpoises), have a dorsal cochlear nucleus that differs greatly from that of other mammals by being unlayered and possibly lacking in several cell types (granule and cartwheel cells).

Classification by Laterality of Response: The Response to the Contralateral versus Ipsilateral Ear

A final important way to classify neurons is by their laterality of response, which is defined as whether the neuron responds to the contralateral or ipsilateral ear and whether the response is excitatory or inhibitory (reviewed by Irvine, 1986). Many neurons in central auditory nuclei above the cochlear nucleus are binaural and can be influenced by sound presented to either ear. A predominant pattern, however, is for the neuron to be excited by sound in the contralateral ear (the side opposite to where the neuron is located). The influence of the ipsilateral ear can be excitatory, inhibitory, or mixed. The contralateral response results from the fact that many central auditory pathways cross to the opposite side of the brain (Fig. 26.16). There are also uncrossed pathways; these pathways generate the response to the ipsilateral ear. Despite an influence of the ipsilateral ear, lesion studies indicate the functional importance of excitation from the contralateral ear. For instance, damage to the inferior colliculus or auditory cortex on one side decreases the ability to localize sounds on the opposite side. Thus, as in other sensory and in motor systems, one side of the brain is concerned primarily with function on the opposite side of the body.

BOX 26.3

GIANT SYNAPTIC TERMINALS: ENDBULBS AND CALYCES

The largest synaptic terminals in the brain are contained in the central auditory pathway. There are two types of these giant synaptic terminals: (1) endbulbs of Held, which are found in the ventral cochlear nucleus (Fig. 26.18A); and (2) calyceal endings, which are found in the medial nucleus of the trapezoid body. Calyces are so large that it is possible to use patch electrodes to record and clamp the presynaptic terminal while simultaneously doing the same with their postsynaptic target in *in vitro* preparations (Forsythe, 1994). This type of study has given insight into presynaptic and postsynaptic regulation of transmitter release at this glutamatergic synapse.

Endbulbs and calyces enable secure transmission of information to their postsynaptic neurons. Endbulbs of Held are formed by primary auditory nerve fibers; each nerve fiber forms one or sometimes two endbulbs. The endbulbs contact and completely encircle their postsynaptic target, the spherical bushy cells of the cochlear nucleus (Fig. 26.18A). The endbulb probably forms hundreds of synapses directly onto the soma of the spherical bushy cell. In single-unit recordings near bushy cells, a metal microelectrode records a complex waveform that differs from recordings in other regions of the brain (Fig. 26.18B). The waveform consists of a prepotential followed about 0.5ms later by a spike. The prepotential is likely from the presynaptic endbulb and the spike is from the postsynaptic bushy cell. The delay between the two events is the synaptic delay. This unusual synapse has several important properties. First, the prepotential is almost always followed by a spike, indicating that the discharge in the endbulb is securely followed by a discharge in the bushy cell. Second, the delay between the prepotential and the spike is almost always the same, indicating that the synapse has a low jitter, or variability in time (Fig. 26.18C).

The large influence of the endbulb would, for most neurons, last many milliseconds so that two closely spaced presynaptic spikes would produce only one postsynaptic spike or a second spike that was delayed, thus smearing in time the pattern of input spikes. However, the bushy cell membrane potential recovers quickly because its membrane contains specialized K^+ channels that allow the cell to repolarize after firing an impulse so it is "reset" and ready to fire again (Manis and Marx, 1991). The overall effect of endbulb–bushy cell specializations are to replicate the spike pattern of the auditory nerve fiber, producing a PST histogram and phase-locking pattern like that of an auditory nerve fiber. These characteristics are important because the end bulb–bushy cell synapse is the only central synapse in the pathway to the medial superior olivary nucleus, a nucleus where timing information from the two ears is compared in order to localize sound sources (Fig. 26.20A).

M. Christian Brown

References

Forsythe, I. D. (1994). Direct patch readings from identified presynaptic terminals mediating glutamatergic EPSCs in the rat CNS, *in vitro*. *J. Physiol.* **479**, 381–387.

Manis, P. B. and Marx, S. D. (1991). Outward currents in isolated ventral cochlear nucleus neurons. *J. Neurosci.* **11**, 2865–2880.

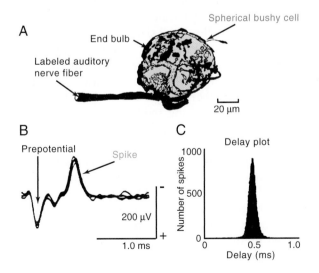

Figure 26.18 (A) Drawing of a labeled auditory nerve fiber forming an endbulb of Held that completely envelops a spherical bushy cell in the anteroventral cochlear nucleus. From Rouiller *et al.* (1986). (B) Superimposed waveforms recorded by an extracellular metal electrode in the anteroventral cochlear nucleus. The endbulb is such a large synaptic ending that it generates a prepotential, which is followed after a synaptic delay by the spike from the bushy cell. At this synapse, prepotentials are usually accompanied by spikes, demonstrating the reliability of the synapse. From Pfeiffer (1966). (C) Histogram of the delay time between prepotential and spike. This represents the synaptic delay between endbulb and bushy cell. The delay of this synapse has an exceptionally low jitter in time. From Molnar and Pfeiffer (1968).

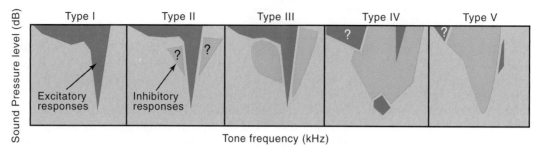

Figure 26.19 Unit classification by "response map." Shown are response areas for the five response types (types I–V) typically found in the cochlear nucleus. Response types are distinguished by the positions of their response areas: excitatory (green shading) and inhibitory (pink shading). Question marks show variable or uncertain areas. Type I neurons have excitatory response areas similar to the tuning curves of auditory nerve fibers. Type II neurons have similar excitatory areas and are inferred to have inhibitory flanking areas because they have little response to broadband signals like noise. Type III neurons have similar excitatory areas and definite inhibitory areas on either side. Type IV neurons have a small excitatory area at low levels (near the characteristic frequency), as well as a knife-sharp excitatory area at higher frequencies. Inhibition dominates much of the remainder of their response area. Type V neurons are similar to type IV neurons but lack a low-level excitatory area. From Young (1984).

The Auditory Brain Stem Uses Binaural Cues to Determine Sound Location

An important function of the auditory system is to determine the location of sound sources in space. Binaural sound localization occurs when cues from both ears are used to locate sounds (reviewed by Wightman and Kistler, 1993). This is the predominant type of localization for determining the azimuthal position of a sound source. Binaural localization uses two cues: interaural time differences (ITDs) and interaural level differences (ILDs) (Fig. 26.20). ITDs result because sound reaches the ear closest to the source sooner than the ear farther from the source (Fig. 26.20A). ITDs are a good cue for localization because they depend greatly on the azimuth of the source. For ongoing sounds, such as pure tones, they can be translated into phase differences in the sound waveforms at the two ears. These phase differences are useful at low frequencies, but become ambiguous for frequencies above about 1.5 kHz (for a human-sized head), because by the time sound reaches the ear away from the source, the waveform has repeated by a cycle or more. Auditory neurons can phase lock to the sound waveform, as noted earlier. The decline in phase locking for frequencies above 1 to 3 kHz is a second reason that phase differences are less important for localizing sounds at high frequencies.

Interaural level differences (ILDs) result when the head forms a "sound shadow," reducing the level of sound at the ear away from the source (Fig. 26.20B). ILDs vary greatly with sound-source azimuth, but due to the directional characteristics of sound, ILDs are significantly large only at high frequencies and are much smaller at low frequencies. Thus, for sound localization at low frequencies (<1 kHz), ITDs are the major cues, but for localization at high frequencies

(>3 kHz), ILDs are the major cues. For human performance using pure tones, the accuracy of azimuthal localization is good at low frequencies and at high frequencies, but somewhat less accurate at middle frequencies, perhaps because the cues are more ambiguous in this range. In psychophysical experiments in humans under optimal conditions, the minimum discriminable angle for localization of a sound source approaches one degree of azimuth. The physical cues corresponding to this angle are about 10 μs in ITD or 1 dB in ILD. When experiments are conducted with headphones to manipulate the ITDs and ILDs, such interaural differences are indeed discriminable in human subjects.

Mechanisms of Interaural Time Sensitivity

Two neural circuits that provide sensitivity to ITD or ILD are within the superior olivary complex (Fig. 26.21; Irvine, 1986; Yin and Chan, 1990). The neural mechanisms that generate sensitivity to ITDs are impressive given the small sizes of the differences. For instance, the central nervous system must be able to detect ITDs of 10 μs by comparing spikes coming in from the neural channels of the two sides that differ in time by 10 μs. However, 10 μs is less than the rise time of neural spikes, making the incoming spikes almost identical. The neural circuit that is sensitive to ITDs in the mammal is the medial superior olive (MSO) and its inputs (Fig. 26.21A); a similar circuit is present in birds in nucleus laminaris. The MSO inputs are from primary-like units (spherical bushy cells) of left and right cochlear nuclei. These inputs preserve the phase-locking and timing characteristics of the auditory nerve because of the low jitter in the endbulb/bushy cell synapse (see Box 26.3). For a low frequency sound with

A Interaural time differences

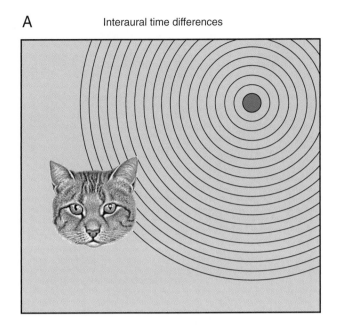

B Interaural level differences

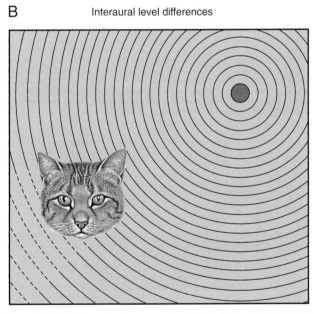

Figure 26.20 The two cues for binaural localization of sound. A sound source is shown as a solid dot to the right of the animal's head and sound waves are shown as concentric lines. (A) Interaural time differences result from the longer time it takes sound to travel from the source to the ear away from the source. (B) Interaural level differences result from the head forming a "sound shadow," reducing the level of sound at the ear away from the source.

an ITD, phase-locked spikes from one side will have a time difference relative to the other side. Phase-locked spikes will repeat this time difference many times during the many waveforms of a continuous sound.

Neural delay lines that "make up" for this time difference are formed by axons for both contralateral and ipsilateral inputs, in the model originally proposed by Lloyd Jeffress (Fig. 26.21A). An axon forms a delay line simply because it takes time for an impulse to travel along the axon. Within the MSO, neurons respond best when they receive coincident input from the two sides; thus a neuron in the middle of the drawing of the MSO in Figure 26.21A would respond best if the delays were about equal. A neuron at the bottom of the drawing has a long contralateral axonal delay that makes up for stimuli coming in later from the ipsilateral side. Thus, this neuron would respond best if the sound to the contralateral ear were leading, which occurs for a sound source located on the contralateral side (Fig. 26.21B). Medial superior olive neurons thus are tuned to a particular ITD and respond less to other ITDs.

The Jeffress model recently has been challenged: new data in small mammals demonstrates that the best delays are outside the physiological range for the size of the animal's head and the neural inhibition in the MSO is not explained in the Jeffress model (Brand *et al.*, 2002). As mentioned earlier, ITDs are most important for sound localization at low frequencies and the MSO has a tonotopic organization that is composed predominantly of neurons with low characteristic frequencies.

Mechanisms of Interaural Level Sensitivity

A second neural circuit in the brain stem, within the lateral superior olive (LSO), generates responses that are sensitive to ILDs (Fig. 26.21C). Input to the LSO from the ipsilateral side is excitatory by way of spherical bushy cells of the cochlear nucleus. Input from the contralateral side is inhibitory. This input originates from globular bushy cells of the cochlear nucleus and synapses by way of giant calyceal endings on neurons in the medial nucleus of the trapezoid body (MNTB; Fig. 26.21C). Many of the neurons in the medial nucleus of the trapezoid body are inhibitory and use the neurotransmitter glycine, which is a common inhibitory transmitter throughout the auditory brain stem. These inhibitory neurons then project to the LSO. Because of these inputs, LSO neurons compare the difference in levels of the sound at the two ears: they are excited when sound in the ipsilateral ear is of higher level, but are inhibited when sound in the contralateral ear is of higher level (Fig. 26.21D).

When the sound is of equal level in the two ears, the strong contralateral inhibition usually dominates and there is little response. These neurons are thus excited by sound sources located on the ipsilateral side of the head. As the lateral superior olive projects centrally, this ipsilateral-side response is transformed to a contralateral-side response by crossing to the inferior colliculus on the opposite side. (There is also an uncrossed

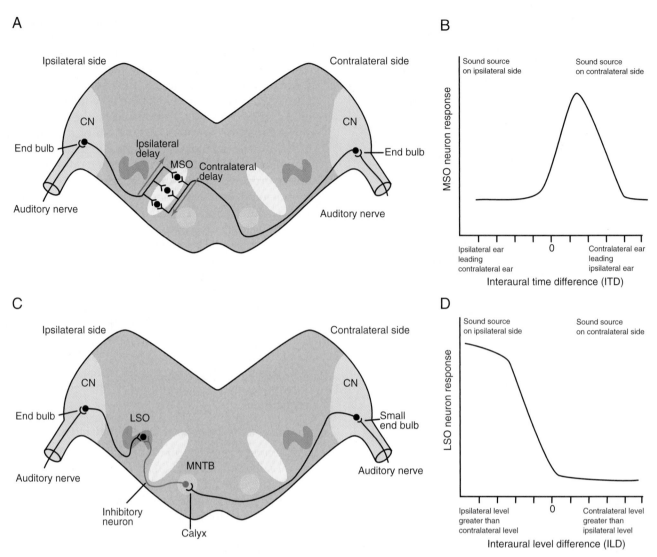

Figure 26.21 Innervation schematics and responses of two circuits in the lower brain stem that are important in binaural sound localization. Neuronal cell bodies are shown as dots, and fiber pathways are shown as lines; positions of large synaptic terminals (endbulbs and calyces) are indicated. (A) Circuit of the medial superior olive (MSO), which is sensitive to interaural time differences (ITD). Input to the cochlear nucleus (CN) from the auditory nerve terminates at the large endbulbs of Held that synapse onto spherical bushy cells (see Fig. 26.18). Bushy cells project bilaterally such that a single MSO receives input from both sides. Bushy cell inputs form delay lines such that ITD is mapped along the MSO. Data suggest that the delay line is oriented rostrocaudally and that only contralateral inputs are delayed. (B) Response of an MSO neuron as a function of ITD. Neurons within the MSO respond when spikes from their two inputs arrive at the same time. The response plotted is of a neuron in the lower part of the MSO drawn in part A; there is a large response when the ipsilateral input lags so that early contralateral input has time to proceed down the axonal delay line to reach the neuron at the same time as the lagging ipsilateral input. This type of lagging ipsilateral input would be produced by a sound source located on the contralateral side, as would be the case for a MSO on the left side of the animal shown in Fig. 26.20. (C) Circuit of the lateral superior olive (LSO), which is sensitive to interaural level differences (ILD). Excitatory input arises from the ipsilateral CN. Inhibitory input (red line) from the contralateral side is through the medial nucleus of the trapezoid body (MNTB), a nucleus of inhibitory neurons. The large synaptic ending in the MNTB is called a calyx (similar to the endbulb pictured in Fig. 26.18). (D) Response of an LSO neuron as a function of ILD. There is a large response when sound is of higher level on the ipsilateral side and no response when sound is of higher level on the contralateral side. Thus, a response is produced by a sound source located on the ipsilateral side.

projection, but it is inhibitory.) As mentioned earlier, ILDs are most important for sound localization at high frequencies and the LSO has a tonotopic organization that is composed predominantly of neurons with high characteristic frequencies.

Inferior Colliculus

Ascending input from lower brain stem centers converges at the inferior colliculus, which is an obligatory synaptic station for almost all ascending neurons. The inferior colliculus consists of several subdivisions, the best studied of which is the large, laminated, central nucleus. In the central nucleus, direct input from the cochlear nucleus interacts with binaurally responsive input from the MSO and LSO. Terminals from the MSO and LSO may have limited spatial overlap, however, as the central nucleus is organized tonotopically. Low CF input (including that from the MSO) projects to the dorsolateral part of the colliculus and high CF input (including that from the LSO) projects to the ventromedial part. Many low CF collicular neurons are sensitive to ITDs, like their MSO inputs, whereas many high CF collicular neurons are sensitive to ILDs, like their LSO inputs. Interestingly, large lesions of the superior olivary complex do not completely disrupt ILD sensitivity in the colliculus; this is evidence that ILD sensitivity is created anew at levels above the LSO. ILD sensitivity may be created in part by the dorsal nucleus of the lateral lemniscus, a nucleus within the lateral lemniscus that sends a large inhibitory projection to the colliculus. ILD sensitivity may also be created anew by inhibitory mechanisms within the colliculus. Additional circuits for the generation of ITD sensitivity have not been identified, thus the colliculus appears to be sensitive to ITD because of its inputs from the MSO.

An important question is whether the colliculus uses its inputs to form a "space map"—a mapping of sound source location to position within the brain. Such a mapping has not been reported in the mammalian inferior colliculus, but a mapping exists in the deep layers of the superior colliculus, where it is in register with a mapping of the visual field. These deep layers of the superior colliculus have sensorimotor functions concerning orientation movements of the head, eyes, and pinnae. Another auditory space map has been observed in the barn owl, which hunts for prey in darkness using acoustic cues (reviewed by Cohen and Knudsen, 1999, see Chapter 22). In the owl, the mapping has been observed within a nucleus homologous to the external nucleus of the inferior colliculus (nucleus mesencephalicus lateralis dorsalis). Here, spatial receptive fields of neurons are narrow in both azimuth and elevation. These receptive fields are arranged such that there is a mapping of sound source location within the brain. This mapping demonstrates that space maps do exist, but they do not appear to be a common feature of the auditory system of nonspecialized mammals.

The Medial Geniculate and Auditory Cortex Are the Highest Stages of the Auditory Pathway

The medial geniculate and auditory cortex contain subdivisions that have clear tonotopic organization,

as well as subdivisions with less obvious tonotopy (reviewed by Clarey *et al.*, 1992; de Ribaupierre, 1996). For instance, in the cortex of the cat, there are four fields that have clear tonotopic mappings (fields AI, A, P, VP), but other fields that do not (Fig. 26.22). The tonotopic axis of field AI runs from high frequencies rostrally to low frequencies caudally; other adjacent fields (such as A) have mirror-image tonotopy (Fig. 26.22B). The major ascending pathway that connects tonotopic areas at the highest levels of the pathway begins in the central nucleus of the inferior colliculus,

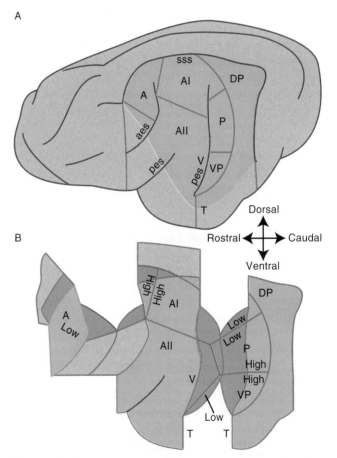

Figure 26.22 Auditory cortical fields in the temporal cortex of the cat. (A) Lateral view. (B) Lateral view that is "unfolded" to show the part of the fields that normally are hidden within the sulci (orange shading), as well as the high- and low-frequency limits of the tonotopic fields. The four tonotopic fields are the anterior (A), primary (AI), posterior (P), and ventroposterior (VP). Positions of the lowest and highest CFs in these fields are indicated in B. Note that at the boundaries of the tonotopic fields, the direction of tonotopy is reversed so that adjacent fields have "mirror-image" tonotopy. Other cortical fields have less rigidly organized tonotopy or little tonotopy. These fields are secondary (AII), ventral (V), temporal (T), and dorsoposterior (DP). Also indicated are suprasylvian sulcus (sss) and anterior and posterior ectosylvian sulci (aes, pes). From Imig and Reale (1980).

forms synapses in the laminated ventral division of the geniculate, and then continues to most of the tonotopic cortical fields. Other parallel pathways exist. A parallel pathway connecting less tonotopic areas begins in the dorsal cortex of the colliculus, forms synapses in the dorsal division of the medial geniculate, and projects mainly to cortical field AII. Finally, a polysensory pathway begins in the external and dorsal nuclei of the colliculus and projects via the medial division of the geniculate to almost all the auditory cortical fields.

In general, the physiology of tonotopic areas has been explored better than that of other areas. Units from the tonotopic areas, such as the ventral division of the geniculate and cortical field AI, tend to have short latencies and sharply tuned tuning curves. Neurons in areas with less obvious tonotopy tend to have longer latencies, broader tuning curves and responses that can habituate or stop responding after multiple presentations of the stimulus, especially in anesthetized preparations. In humans, the primary auditory cortex is located on Heschl's gyrus, which is on the superior surface of the temporal lobe (Fig. 26.16). Sound-evoked activation is seen in this region with positron emission tomography (PET) and functional magnetic resonance imaging (fMRI).

The medial geniculate exerts its influence on ascending information, but it does so with extensive influence from the cortex. The medial geniculate receives a large number of projections from the auditory cortex: it probably receives more input from auditory cortex than from lower centers. One large projection is from the tonotopic cortical fields AI and A to the tonotopic ventral division of the medial geniculate. Apparently, however, the geniculate mediates some functions that do not require the cortex. For example, fear conditioning is a behavior that can be established by pairing a sound with a painful electric shock in rats. After conditioning is established, conditioned responses such as an increase in blood pressure can be elicited by sound alone. Work by Joseph LeDoux has shown that lesions of the auditory pathway up to and including the geniculate have a large effect on such conditioning; however, lesions of the auditory cortex do not produce large alterations. Pathways directly from the geniculate to the amygdala mediate such conditioned responses. Cortical neuron responses are also changed by conditioning; for example, the pairing of an acoustic stimulus with a noxious stimulus greatly alters the response of cortical neurons. Such observations indicate that responses at these high levels of the auditory pathway are context dependent.

Cortical Columns

A fundamental feature of cortical organization is the cortical column, which is oriented normal to the cortical surface and runs across all six of the cortical layers. In the cortex, neurons within a column tend to have similar response characteristics. For instance, in the auditory cortex, neurons within a given column generally have similar CFs. Neurons also tend to have similar types of responses to binaural sounds; these binaural interaction characteristics are tested in preparations in which each ear is stimulated with a separate sound source. Usually one ear, the main ear, excites a cortical neuron; most often this ear is the contralateral ear. The effect of the opposite ear can be either to excite by itself or to facilitate the main ear response of the neuron (summation interaction), or to inhibit by itself or to suppress the main ear response (suppression interaction). Typically, either summation interactions or suppression interactions are found for neurons within a column. For the high-CF part of AI, summation and suppression columns show some organization, although not nearly as ideal as a checkerboard pattern with CF on one axis and summation/suppression interaction on the perpendicular axis.

The binaural interaction classification is useful, but oversimplified, as some neurons can show both summation and suppression, depending on factors such as sound level. However, one finding that supports summation/suppression columns as an organizational plan is the pattern of projections from a column in one hemisphere to the opposite cortical hemisphere. Neurons within a summation column tend to have large projections to the opposite hemisphere. Suppression columns tend to have few projections (with the exception of less common columns in which the contralateral ear is inhibitory and the ipsilateral ear is excitatory). These findings support binaural interaction characteristics as an important property of cortical columns.

Cortical Field AI and Sound Localization

Physiological studies suggest a role for cortical field AI in sound localization. In AI, many neurons are sensitive to interaural time and level differences, much as for neurons at lower stages of the pathway. When tested with a sound source in space, the response of many neurons depends on the azimuth of the source. As expected from the fact that auditory pathways are predominantly crossed, a large number of cortical neurons respond to sound sources centered on the contralateral side. The receptive fields for many neurons occupy much of the azimuth on the contralateral side and some have even wider fields that encompass both sides (omnidirectional fields). Other neurons, however, have receptive fields that are very narrow. Units within a given column tend to have receptive fields that are similar, probably because of the similar binaural interaction characteristics within a column. It is easy to imagine that neurons with summation characteristics would tend to have receptive fields that would be large and encompass both contralateral and ipsilateral sides because of their bilateral excitatory input. Units with suppression characteristics, however, would tend to have narrower fields located mainly on one side because they receive inhibitory input from the other side. Finally, the auditory cortex has not yet been shown to have an organized space map of a sound's position in external space onto a dimension within the cortex.

Behavioral studies also indicate a strong role of AI in sound localization (Jenkins and Merzenich, 1984). Cats with lesions of AI on one side have a deficit for localizing sounds on the contralateral side. Lesions of the posterior auditory field (a tonotopic field just caudal to the primary auditory field) and an area near the anterior ectosylvian sulcus, but not other fields, also produce deficits (Malhotra and Lomber, 2006). Furthermore, the deficit is frequency specific: if the lesion is in areas of AI tuned to certain frequencies, the deficit is observed only for those frequencies. The deficit is most pronounced when the task is for a subject to localize and move toward the sound source and is less obvious when the task is simply to lateralize the sound; that is, to press a bar on the right or left side corresponding to the side of the sound source. This finding suggests that the simpler lateralization task is processed at a subcortical location, but the more difficult task of forming an image of a sound source's position in space and moving toward that position is processed at the cortex.

Coding of Complex Signals in the Cortex

Animals emit a variety of vocal signals. These signals make possible a wide range of behaviors, including communication and even echolocation in bats (Box 26.4). Are there specific "call detectors" in the auditory cortices of these animals? This area is relatively unexplored; however, available evidence suggests that this type of detector may exist only rarely and that the response to vocalizations is probably represented by spatially dispersed, synchronized assemblies of cortical neurons rather than by individual neurons. However, cortical neurons do show selectivity to complex sounds such as noise bands and species-specific calls over what would be predicted from their

pure tone frequency selectivity (Rauschecker et al., 1995). For instance, a neuron that responds well to a particular type of call does not usually respond as well to the call played backward in time even though it has an identical frequency content. Similar results in response to speech stimuli have been found in recordings of units from auditory areas in the superior temporal gyrus of human patients undergoing surgery for epilepsy.

A number of studies using lesions of the auditory cortex indicate a strong role of the cortex in processing complex acoustic signals. After bilateral lesions of the auditory cortex, experimental animals cannot discriminate between different temporal patterns of sounds, such as a sequence in which the sound frequency pattern is a repeated "low–high–low" and a sequence in which the pattern is a repeated "high–low–high." In primates, lesions of the cortex impair discrimination of species-specific vocalizations. This impairment is more pronounced after lesion of the left cortex, indicating lateralization of processing of these vocalizations at the cortical level. Human speech is an especially complex acoustical signal; its intelligibility and production can be decreased greatly by a stroke that creates a lesion of the cortex in humans. Although there is great variability in the extent and effect of such lesions, language processing is most interrupted by lesions of perisylvian cortex regions, especially in Wernicke's area and Broca's area (Chapter 51). Wernicke's area is especially close to auditory cortex and can be considered an auditory association area. Lesions lateralized in the left hemisphere in right-handed individuals are most disruptive of speech comprehension and production. The lateralization of language processing in one hemisphere is a unique finding of asymmetry in brain function, very different from the apparent symmetry of subcortical nuclei and primary auditory cortical fields.

Imaging studies (PET, fMRI) in normal subjects suggest acoustic stimuli such as noise and modulated tones activate primary auditory cortical fields, but do not activate surrounding areas. These surrounding areas, including Wernicke's and Broca's areas, can be activated by speech stimuli. Furthermore, the activation by speech stimuli usually is lateralized to the left hemisphere of right-handed individuals. Taken together, the lesion and imaging studies suggest a hierarchical pattern of cortical activation with simple stimuli being processed in the primary cortical fields and more complex stimuli, such as speech, processed in association areas that are lateralized in one hemisphere. Further imaging studies will be very useful in showing more specifically the functions of the auditory areas in the cortex.

BOX 26.4

BAT ECHOLOCATION

Bats offer unique opportunities for researchers studying the auditory system. Echolocating bats emit high-frequency pulses of sound and use the return echoes to locate and capture their flying insect prey. These pulses are above the upper frequency limit of human hearing. They were discovered in the 1940s by Donald Griffin, who was then an undergraduate student at Harvard University. The echolocation performance of the bat is impressive: bats are adept enough to find a mosquito above a golf course at night. Indeed, Griffin demonstrated that bats can avoid wires as thin as 0.3 mm diameter while flying in a darkened room.

There are more than 800 species of echolocating bats, all within the suborder Microchiroptera. One of the best-studied bats is the mustached bat, *Pteronotus parnellii*. It emits echolocating pulses that have an initial part of constant frequency, followed by a part of decreasing frequency (frequency modulated). This bat is thus called a CF-FM bat (Fig. 26.23B). Some other species of bats emit only the FM portion and are called FM bats. Bats use differences between the emitted pulse and the returned echo to locate targets. Relative to the pulse, the returned echo is delayed because of the sound's round-trip travel time from the bat to the target. The delay between the pulse and echo can thus be used as a measure of target range.

The echo can also be changed in frequency, or Doppler-shifted, because the bat is moving relative to the reflecting surface. Bats that are moving toward a reflecting surface will have an echo that is Doppler-shifted toward higher frequency. Most background surfaces will be Doppler shifted about the same. However, targets that are moving differently from the background will generate different Doppler shifts. For instance, a moth with moving wings will reflect an echo with a moving Doppler shift that is quite different from the stationary background. Such differences are presumably used by the CF-FM bat to locate moving targets. Insects are not always passive in the face of such predatory behavior. For instance, some moths have good ultrasonic hearing and take evasive flight maneuvers when exposed to a bat's pulse.

Research on the auditory system of bats has been greatly aided by knowledge of the "relevant" stimulus, since much of the time the bat hears an emitted pulse followed a short time later by an echo. Pioneering work by Suga (1990) has demonstrated that the auditory cortex of the mustached bat contains many specialized areas (Fig. 26.23C). One large area (DSCF area) is devoted to processing the Doppler-shifted echo for the strongest harmonic of the pulse, near 60 kHz. These frequencies also have large representations throughout the bat's auditory system beginning at the cochlea. Within the DSCF area, there is a mapping of neuronal best level as well as characteristic frequency. In other cortical regions, there are neurons whose response properties cause them to detect various features of the pulse and echo. For instance, neurons in the FM-FM region respond preferentially to two FM pulses separated by a specific delay. The best delays for these neurons range from 0.4 to 18 ms, which would correspond to target ranges of 7 to 310 cm. Furthermore, these best delays are mapped along cortical distance. Work on the bat cortex has resulted in specific hypotheses about the function of most of their auditory cortical fields. Our state of knowledge of the cortical fields in other animals is by contrast much more primitive.

M. Christian Brown

Reference

Suga, N. (1990). Biosonar and neural computation in bats. *Sci. Am.* June: 60–68.

Summary

Much about the central auditory pathways remains to be explored. Although these pathways are mapped tonotopically, whether there are other mappings in orthogonal dimensions is unknown. Much current research is focused on the plasticity of central auditory responses with auditory experience and after hearing loss and whether this plasticity is generally present at the many subcortical auditory centers. The fact that there are descending systems at many levels of the pathways is known, but research is necessary to show how these systems alter the processing of auditory information. Finally, the medial geniculate and auditory cortex are likely to play a role in sound localization, but the specifics of this role and their functions in many of the other hearing processes remain to be discovered.

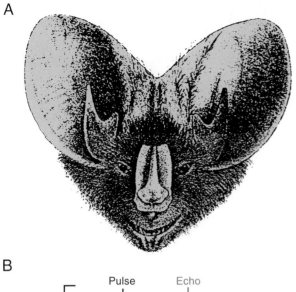

A

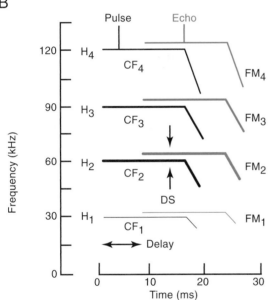

B

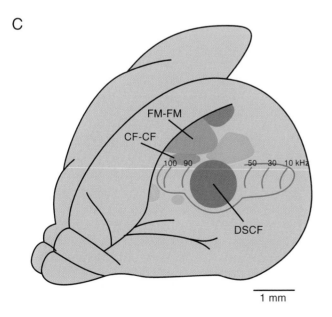

C

1 mm

Figure 26.23 (A) Close-up view of an echolocating bat, *Megaderma lyra*. Like many echolocating bats, this bat has an enlarged nose leaf that probably focuses the echolocating pulse ahead. Also like many echolocating bats, this bat has enlarged external ears that make the bat more sensitive to echoes coming from ahead. From Griffin (1958). (B) Schematic of emitted pulse and returned echo of the mustached bat, *Pteronotus parnellii*. The emitted pulse consists of four harmonics (H_1–H_4), the strongest of which is H_2 at about 60 kHz. Each harmonic has an initial part of constant frequency (CF) and a later part of changing frequency (frequency modulation, FM). The echoes are returned after a travel time that causes a delay relative to the pulse. Additionally, if the target is moving relative to the bat, the echo is returned with a Doppler shift (DS) in frequency. (C) Dorsolateral view of the auditory cortex of the mustached bat showing several of the areas specialized for processing the echolocation signals. The primary auditory cortex, AI, is delineated by a red line and its isofrequency contours are indicated in kHz. It contains a region called the Doppler-shifted CF region (DSCF), which is a greatly expanded region devoted to the most prominent component of the Doppler-shifted echo (60 to 63 kHz). Other shaded regions contain neurons that are combination sensitive and respond to combinations of the pulse and echo, often at certain delays. Neurons that respond to pulse/echo CF portions are in the CF/CF region. Those that respond to pulse/echo FM portions are in the FM/FM region. From Fitzpatrick *et al.* (1993).

References

Brownell, W. E., Bader, C. R., Bertrand, D., and de Ribaupierre, Y. (1985). Evoked mechanical response of isolated hair cells. *Science* **227**, 194–196.

Brand, A., Behrend, O., Marquardt, T., McAlpine, D., and Grothe, B. (2002). Precise inhibition is essential for microsecond interaural time difference coding. *Nature* **417**, 543–547.

Buonomano, D. V. and Merzenich, M. M. (1998). Cortical plasticity: From synapses to maps. *Annu. Rev. Neurosci.* **21**, 149–186.

Clarey, J. C., Barone, P., and Imig, T. J. (1992). Physiology of thalamus and cortex. *In* "The Mammalian Auditory Pathway: Neurophysiology" (A. N. Popper and R. R. Fay, eds.), pp. 232–334. Springer-Verlag, New York.

Cohen, Y. E. and Knudsen, E. I. (1999). Maps versus clusters: Different representations of auditory space in the midbrain and forebrain. *Trends Neurosci.* **12**, 128–135.

Corey, D. P. and Hudspeth, A. J. (1983). Kinetics of the receptor current in bullfrog saccular hair cells. *Neurosci.* **3**, 962–976.

Corwin, J. T. and Cotanche, D. A. (1988). Regeneration of sensory hair cells after acoustic trauma. *Science* **240**, 1772–1774.

de Ribaupierre, F. (1997). Acoustical information processing in the auditory thalamus and cerebral cortex. *In* "The Central Auditory System" (G. Ehret and R. Romand, eds.), pp. 317–397. Oxford Univ. Press, New York.

Guinan, J. J., Jr. (1996). The physiology of olivocochlear efferents. *In* "The Cochlea" (P. Dallos, A. N. Popper, and R. R. Fay, eds.), pp. 435–502. Springer-Verlag, New York.

Hudspeth, A. J. and Corey, D. P. (1977). Sensitivity, polarity, and conductance change in the response of vertebrate hair cells to controlled mechanical stimuli. *Proc. Natl. Acad. Sci. USA* **74**, 2407–2411.

Liberman, M. C. (1982). The cochlear frequency map for the cat; Labeling auditory-nerve fibers of known characteristic frequency. *J. Acoust. Soc. Am.* **72**, 1441–1449.

Liberman, M. C., Gao, J., He, D. Z. Z., Wu, X., Jia, S., and Zuo, J. (2002). Prestin is required for electromotility of the outer hair cell and for the cochlear amplifier. *Nature* **419**, 300–304.

Malhotra, S. and Lomber, S. G. (2006). Sound localization during homotopic and heterotopic bilateral cooling deactivation of primary and non-primary auditory cortical areas in the cat. *J. Neurophysiol.* **97**, 26–43.

Pickles, J. O., Comis, S. D., and Osborne, M. P. (1984). Cross-links between stereocilia in the guinea-pig organ of Corti, and their possible relation to sensory transduction. *Hearing Res.* **15**, 103–112.

Rauschecker, J. P., Tian, B., and Hauser, M. (1995). Processing of complex sounds in the macaque nonprimary auditory cortex. *Science* **268**, 111–114.

Rhode, W. S. and Greenberg, S. (1992). Physiology of the cochlear nuclei. *In* "The Mammalian Auditory Pathway, Neurophysiology" (A. N. Popper and R. R. Fay, eds.), pp. 94–152. Springer-Verlag, New York.

Robertson, D. and Irvine, D. R. F. (1989). Plasticity of frequency organization in auditory cortex of guinea pigs with partial unilateral deafness. *J. Comp. Neurol.* **282**, 456–471.

Ruggero, M. A. (1992). Physiology and coding of sound in the auditory nerve. *In* "The Mammalian Auditory Pathway, Neurophysiology" (A. N. Popper and R. R. Fay, eds.), pp. 34–93. Springer-Verlag, New York.

Rybalchenko, V. and Santos-Sacchi, J. (2003). Cl-flux through a nonselective, stretch sensitive conductance influences the outer hair cell motor of the guinea pig. *J. Physiol.* **547.3**, 873–891.

Santos-Sacchi, J. (1992). On the frequency limit and phase of outer hair cell motility: Effects of the membrane filter. *J. Neurosci.* **12**, 1906–1916.

Warr, W. B. (1992). Organization of olivocochlear efferent systems in mammals. *In* "The Mammalian Auditory Pathway, Neuroanatomy" (D. B. Webster, A. N. Popper, and R. R. Fay, eds.), pp. 410–448. Springer-Verlag, New York.

Wightman, F. L. and Kistler, D. J. (1993). Sound localization. *In* "Human Psychophysics" (W. A. Yost, A. N. Popper, and R. R. Fay, eds.), pp. 155–192. Springer-Verlag, New York.

Yin, T. C. T. and Chan, J. C. K. (1990). Interaural time sensitivity in medial superior olive of cat. *J. Neurophysiol.* **64**, 465–488.

Zheng, J., Shen, W., He, D. Z., Long, K. B., Madison, L. D., and Dallos, P. (2000). Prestin is the motor protein of cochlear outer hair cells. *Nature* **405**, 149–155.

Suggested Readings

Dallos, P., Popper, A. N., and Fay, R. R. (eds.) (1996). "The Cochlea." Springer-Verlag, New York.

Ehret, G. and Romand, R. (eds.). (1997). "The Central Auditory System." Oxford Univ. Press, New York.

Geisler, C. D. (1998). "From Sound to Synapse." Oxford Univ. Press, Oxford.

Griffin, D. R. (1958). "Listening in the Dark." Yale University Press, New Haven.

Irvine, D. R. F. (1986). "The Auditory Brainstem." Berlin, Springer-Verlag.

Jahn, A. F. and Santos-Sacchi, J. (eds.) (2001). "Physiology of the Ear," 2nd ed. Raven Press, New York.

Pickles, J. O. (1988). "An Introduction to the Physiology of Hearing," 2nd ed. Academic Press, London.

Popper, A. N. and Fay, R. R. (eds.) (1992). "The Mammalian Auditory Pathway: Neurophysiology." Springer-Verlag, New York.

Popper, A. N. and Fay, R. R. (eds.) (1995). "Hearing by Bats." Springer-Verlag, New York.

Schuknecht, H. F. (1993). "Pathology of the Ear," 2nd ed. Lea & Febiger, Philadelphia.

Von Bekesy, G. (1960). "Experiments in Hearing." McGraw-Hill, New York.

Webster, D. B., Popper, A. N., and Fay, R. R. (eds.) (1992). "The Mammalian Auditory Pathway: Neuroanatomy." Springer-Verlag, New York.

M. Christian Brown and Joseph Santos-Sacchi

Vision

OVERVIEW

Vision is the most studied and perhaps the best understood topic in sensory neuroscience. This chapter is concerned primarily with vision in mammals and focuses on the pathway in the visual system that has to do with perception: from the retina to the lateral geniculate nucleus of the thalamus (LGN) and on to the multiple areas of visual cortex. Other regions of the brain that receive visual input (such as the superior colliculus) are dealt with briefly in this chapter and more extensively in the chapter on eye movements (Chapter 33).

Studying vision provides the opportunity to explore the brain at many different levels, from the physical and biochemical mechanisms of phototransduction to the boundary between psychology and physiology (Chapters 51 and 52). At each of these levels, the visual system has evolved to solve a number of difficult problems. In terms of the physical stimulus, vision operates over extremely wide ranges of illumination. The visual system detects single photons in the dark but can also see clearly in bright sunlight, when the retina is bombarded with over 10^{14} photons per second. At a much higher level of complexity, ensembles of neurons in the cerebral cortex are able to solve extremely difficult problems, such as extracting the three-dimensional motion of an object from two-dimensional retinal images.

At every moment, the visual system is confronted with the vast amount of information present in visual scenes. The complex circuitry of the retina has evolved so that much of this information is extensively processed and relayed to the rest of the central nervous system, both efficiently and with great fidelity. Vision, however, has not evolved to treat all of this information equally; instead, it appears to be best suited to extract the sort of information that may be useful to animals, including humans, in a natural environment. Vision allows animals to navigate in the world; to judge the speed and distance of objects; and to identify food, members of other species, and familiar or unfamiliar members of the same species.

In many animals, primates in particular, more of the brain is devoted to vision than to any other sensory function. This is perhaps because of the extreme complexity of the task required of vision: to classify and to interpret the wide range of visual stimuli in the physical world. At the highest levels of processing, the cerebral cortex extracts from the world the diverse qualities experienced as visual perception: from motion, color, texture, and depth to the grouping of objects, defined by the combination of simple features.

The Receptive Field Is the Fundamental Concept in Visual Physiology

The strategies that the brain uses to solve the problems of vision can be understood at a very intuitive level. The most useful concept to aid this intuition is that of the *receptive field*, which is the cornerstone of visual physiology. As defined by H. K. Hartline in 1938 (Ratliff, 1974, p. 167), a visual receptive field is the "region of the retina which must be illuminated in order to obtain a response in any given fiber." In this case, "fiber" refers to the axon of a retinal neuron, but any visual neuron, from a photoreceptor to a visual cortical neuron, has a receptive field. The definition was later extended to include not only the region of the retina that excited a neuron, but also the specific properties of the stimulus that evoked the strongest response. Visual neurons can respond preferentially to the turning on or turning off of a light stimulus—termed *on-and-off* responses—or to more complex features, such as color or the direction of motion. Any of these preferences can be expressed as attributes of the receptive field.

Sensory Systems Detect Contrast or Change

In the 1930s and 1940s, Hartline developed the concept of the receptive field with studies of the axons of individual neurons that project from the lateral eye of the horseshoe crab (*Limulus*) and from the frog's eye (Ratliff, 1974). The lateral eye of the *Limulus* is a compound eye made up of about 300 *ommatidia* arranged in a roughly hexagonal array. Each ommatidium contains optical elements, photoreceptors, and a single neuron whose axon joins the optic nerve. Hartline found that when an isolated ommatidium was illuminated, the firing rate of its axon increased. More surprisingly, the firing of the same axon was decreased by a light stimulus in any adjacent ommatidium. This form of antagonistic behavior, known as *lateral inhibition*, serves to enhance responses to edges while reducing responses to constant surfaces. Without it, visual neurons would be just as sensitive to a featureless stimulus, such as a clean white wall, as to stimuli defined by edges, such as a white square on a black wall. Similar spatially antagonistic visual responses were found in mammals, as first demonstrated by Kuffler (1953) in the retina of the cat (Box 27.1).

Lateral inhibition represents the classic example of a general principle: most neurons in sensory systems are best adapted for detecting changes in the external environment. This principle can be explained in behavioral terms. As a rule, it is change that has the greatest significance for an animal, for example, the edge of an extended object or a static object beginning to move. This principle can also be explained in terms of information processing. Given a world that is filled with constants—with uniform objects, with objects that move only rarely—it is most efficient to respond only to changes.

Several types of visual responses can be discussed in terms of the detection of change, or of *contrast*—defined as the fractional difference in luminance between two stimuli. There are several forms of contrast. The first is spatial contrast, the detection of which is enhanced by neurons in the retina that have lateral inhibition (center-surround organization; Box 27.1). Next, there is temporal contrast, or change over time. Starting in the retina, visual neurons are affected very little by slow changes in illumination, but are extremely sensitive to more rapid changes. Finally, there is motion, which is distinguished by characteristic changes in a stimulus over both space and time. Many neurons in the visual system are excited selectively by objects that have a certain rate or direction of motion. In summary, contrast sensitivity can take on at least three forms: sensitivity to spatial variations in a stimulus (spatial contrast), sensitivity to changes over time (temporal contrast), and sensitivity to changes in both space and time (motion; Box 27.4).

Receptive Fields Encode Increasingly High-Order Features of the Visual World

From the photoreceptors to the multiple visual cortical areas, the visual system is hierarchical. One level provides input to the next in a feedforward progression, although lateral interactions and feedback are almost always present as well. As a general rule, receptive fields at successive stages of processing (from photoreceptors, bipolar cells, ganglion cells, and geniculate neurons, through neurons in multiple visual cortical areas) encode increasingly high-level features of the visual stimulus. The outer segment of a photoreceptor, which contains the visual pigment, is influenced only by a small point in visual space. It is therefore almost entirely insensitive to the spatial structure of a stimulus. At the opposite extreme, neurons in the inferior parietal region of visual cortex seem to respond best when the animal is viewing a specific face (Chapter 46).

These two extremes of visual responses illustrate the visual system's dual task: to maintain generality—the ability to respond to any stimulus—while being able to represent specific, environmentally important classes of stimuli. High-level neurons classify visual stimuli by integrating information that is present in the earlier stages of processing, but also by ignoring information that is independent of that classification. For instance, motion-sensitive neurons in area MT of the cerebral cortex (see later) are exquisitely sensitive to the direction and rate of motion of an object, but very poor at distinguishing the object's color or its position. This lack of localization is quite common in high-level neurons: receptive fields become larger as the features they represent become increasingly complex. Thus, for instance, neurons that respond to faces typically have receptive fields that cover most of visual space. For these cells, large receptive fields have a distinct advantage: the preferred stimulus can be identified no matter where it is located on the retina.

Summary

The mammalian visual system is a complex, hierarchical system that can be studied from a number of

BOX 27.1

KUFFLER'S STUDY OF CENTER-SURROUND RETINAL GANGLION CELLS

The classic experiments that Kuffler (1953) performed on retinal ganglion cells have formed the foundations for much of the subsequent physiological analysis of the mammalian visual system. Even beyond the study of vision, they represent a model for understanding the neurobiology of sensory systems. These experiments were performed *in vivo* in anesthetized cats. The first step in this sort of experiment is the careful placement of a fine micro-electrode close to a single neuron so that action potentials can be recorded extracellularly. An oscilloscope trace of the firing pattern of this neuron is important, but the sound of action potentials on an audio monitor is even more critical. This immediate feedback allows the researcher to search for visual stimuli that excite or inhibit the neuron. Kuffler's first finding was that there were two categories of ganglion cells, as Hartline had seen in the retina of the frog. The cells were either *on*, i.e., excited by light increment, or *off*, i.e., excited by light decrement.

One of Kuffler's most important contributions was the careful mapping of lateral interactions in the retina, or what he termed the *center-surround* structure of the receptive field. When an *on* ganglion cell was being studied, a small light spot placed in the center of its receptive field would cause an immediate increase in firing rate of the cell (Fig. 27.3B, top). The center of the receptive field of an *on* ganglion cell was defined as all positions where the small spot evoked an *on* excitatory response. When the same spot was placed just beyond the center, in the region termed the *surround*, the neuron decreased its firing rate

(Fig. 27.3B, bottom). Kuffler mapped the spatial extent of the regions that evoked excitation or inhibition simply by listening to the responses to spots flashed at many different locations. Alternatively, Kuffler studied receptive fields by searching for an optimal stimulus—one that increased the firing of a ganglion cell most effectively. The strongest stimulus for an *on* center cell was a spot of light that filled the receptive field center entirely (Fig. 27.3C). Similarly, the most effective inhibitory stimulus was a bright annulus shown to the surround alone. Following such strong inhibition, the cell had an excitatory response when the stimulus was turned off (Fig. 27.3D). Finally, Kuffler studied interactions between receptive field subregions. A large bright stimulus that covered both center and surround was found to evoke a much weaker response than a smaller spot confined to the center; the surround inhibition weakened or altogether eliminated the central excitation (Fig. 27.3E).

Kuffler's early experiments, along with those of Hartline, established the technical and conceptual foundations for the field of visual physiology. Almost all subsequent work in this field can be seen as falling into the three broad categories of experiments, exemplified by Kuffler's 1953 study: (1) the mapping of responses with isolated, suboptimal stimuli, (2) the search for an optimal stimulus, and (3) the study of interactions between responses evoked by two or more stimuli.

R. Clay Reid

different viewpoints. This chapter concentrates on the physiology of visual neurons, which is based on the study of receptive fields. Originally, receptive fields were defined as the area of the retina that could evoke responses in a visual neuron. The concept has evolved to include the stimulus attributes that lead to a neural response, such as color, motion, or even the complex features of a specific physical object. Within the hierarchy of the visual system—from retina to the multiple areas of the visual cortex—neurons are selective for increasingly complex or high-order features.

THE EYE AND THE RETINA

The Optics of the Eye Project an Inverted Visual Image on the Retina

The study of vision begins with the eye (Fig. 27.1), whose refractive properties are determined by the curvature of the cornea and the lens behind it. These optical elements act to focus an inverted image on the retina, where the first stages of neural visual processing take place. The curvature of the cornea is fixed, but the curvature of the lens is adjusted by smooth muscles that

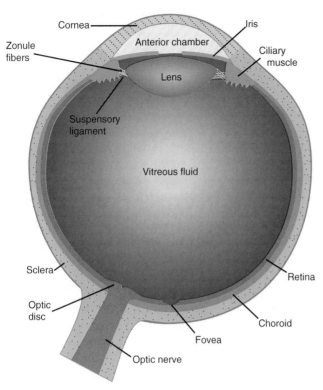

FIGURE 27.1 Schematic diagram of the human eye.

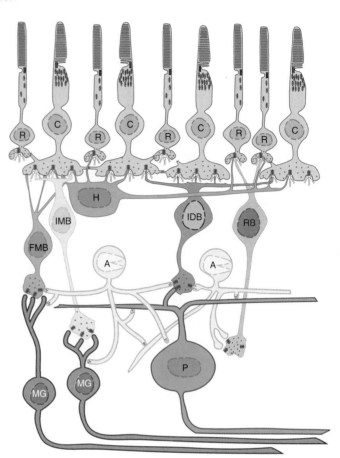

FIGURE 27.2 Summary diagram of the cell types and connections in the primate retina. R, rod; C, cone; H, horizontal cell; FMB, flat midget bipolar; IMB, invaginating midget bipolar; IDB, invaginating diffuse bipolar; RB, rod bipolar; A, amacrine cell; P, parasol cell (also confusingly called an M cell because of its thalamic targets; see text for details); MG, midget ganglion cell (also confusingly called a P cell). Adapted from Dowling (1997).

flatten the lens when they relax, thus bringing more distant objects into focus. The amount of light that reaches the retina is controlled by the iris, whose aperture is the pupil. The iris, which is situated between the cornea and the lens in the *anterior chamber* of the eye, contracts at high light levels and expands in the dark.

The Retina Is a Three-layered Structure with Five Types of Neurons

The anatomy of the retina has an almost crystalline beauty, a beauty that is enhanced by the clear relationships between form and function (Dowling, 1997). It is composed of five principal layers: three layers of cell bodies separated by two layers of neural processes, dendrites and axons (Fig. 27.2). The vertebrate retina is oriented within the eye so that light must travel through the entire thickness of the neuropil to reach the photoreceptors. Of the three cell layers, the first is farthest from the center of the eye and thus is called the *outer nuclear layer*. It contains the cell bodies of the photoreceptors, the rods and cones. The next cell layer is the *inner nuclear layer*, which contains the cell bodies of the interneurons of the retina, both excitatory and inhibitory. These include horizontal cells, bipolar cells, and amacrine cells. Finally, the *ganglion cell layer* is home to the retinal neurons whose axons form the

optic nerve, the sole pathway from the retina to the rest of the central nervous system. Interposed between the cell body layers are two layers of cell processes: *inner* and *outer plexiform layers*. The two plexiform layers are the sites of all interactions between the neurons of the retina.

The retina is one of the few circuits in the nervous system simple enough that cell types and connections can be learned without great effort. This is made even easier since the role of each of the five cell types can be placed within a simple functional scheme. The two main attributes of the output of the retina—the point-to-point representation of the visual image and the spatially antagonistic center-surround interactions in the receptive field (Box 27.1)—can be understood in terms of the anatomy. The direct pathway, photo-receptor → bipolar cell → ganglion cell, is the substrate for the center of the receptive field of the ganglion cell

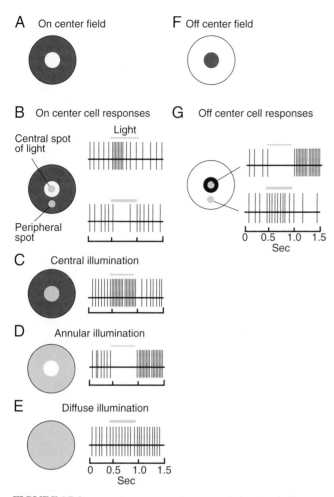

FIGURE 27.3 Visual responses of *on*-center (white) and *off*-center (dark gray) retinal ganglion cells. Visual stimuli are indicated in yellow, and the responses to these stimuli are shown to the right. See text for details.

and thus for its spatial resolution. Lateral interactions in the retina, most notably the center-surround antagonism, or lateral inhibition, are mediated by horizontal cells and amacrine cells.

Photoreceptors Are Hyperpolarized by Light

The two major types of photoreceptors in the vertebrate eye are the rods and cones (Fig. 27.2). Both types of photoreceptor have an outer segment that contains the molecular machinery for phototransduction, an inner segment that contains densely packed mitochondria, a cell body that contains the nucleus and other important organelles, and a terminal process that releases neurotransmitter. Although rods and cones were once viewed as the only cells in the vertebrate eye to transduce light energy into an electrical response, recent work shows that phototransduction also takes place in melanopsin-containing ganglion cells that

likely convey nonspatial information about light levels to regions of the brain involved with circadian rhythms.

Rods and cones work together to allow the visual system to operate over a wide range of luminance conditions. Under scotopic conditions when luminance levels are very low (e.g., starlight), only the rods are active. As luminance levels increase to mesopic conditions (e.g., moonlight), both the rods and cones contribute to vision. As luminance levels increase further yet, to scotopic condition (e.g., sunlight), rod responses saturate and only the cones contribute to vision. Because the center of our retina (the fovea) contains very few, if any rods, we have a foveal blind spot under very dim conditions. Consequently, a dim star will disappear when we move our eyes to look at it directly.

Phototransduction occurs when photons of light are absorbed by photopigments located in the outer segments of photoreceptors. Photopigments have an organic, light-absorbing component (11-*cis* retinal) and a protein component (opsin). The opsin determines the wavelength specificity of the photopigment and has the ability to interact with G proteins. Rods contain a single photopigment, rhodopsin, whereas cones contain one of three cone opsins (the relationship between these opsins and color vision is discussed later). Once a photon is absorbed, a cascade of events occurs that ultimately affects the membrane potential of the photoreceptor. This cascade begins with the activation of the G protein (transducin) that then activates a phosphodiesterase that hydrolyzes cGMP and reduces the intracellular concentration of cGMP. Because the outer membrane of a photoreceptor contains many cGMP-gated cation channels, a decrease in the intracellular concentration of cGMP will cause the photoreceptor to hyperpolarize. In this way, light increments lead to hyperpolarization and a reduction in neurotransmitter release, whereas light decrements lead to depolarization and an increase in transmitter release. Importantly, photoreceptors do not produce action potentials, but rather have graded potentials that are modulated around a mean level.

Bipolar Cells Can Be Hyperpolarized or Depolarized by Light

Similar to photoreceptors, bipolar cells generate graded potentials rather than action potentials. Bipolar cells are divided into two broad classes, termed *on* and *off* bipolars, that respond to light stimuli with depolarization and hyperpolarization, respectively. *On* bipolars are also known as *invaginating bipolars*, and *off* bipolars are known as *flat bipolars* because of the shape of the synaptic contacts they receive from photoreceptors (Fig. 27.2). Because all photoreceptors use glutamate as their neurotransmitter, these opposite responses require that *on*-and-*off* bipolar cells respond to the same neurotransmitter in opposite ways (Wu, 1994). Glutamate, typically an excitatory transmitter, is inhibitory for *on* bipolars. It acts by closing a cGMP-gated sodium channel, similar to that seen in photoreceptors. The depolarization of *on* bipolars by light results not from excitation, but from the removal of inhibition, which occurs when photoreceptors are hyperpolarized by light. *Off* bipolars are excited by glutamate via more typical kainate receptors (named for the pharmacological agent that selectively activates them; Chapter 9). When these cells are inhibited by light, the inhibition is in fact due to the removal of tonic excitation, which again occurs when the photoreceptors are hyperpolarized.

The functional importance of the *on* and the *off* pathways can best be understood in terms of contrast. From the bipolar cells onward (i.e., once on-and-*off* pathways have been established), visual neurons respond best to spatial and temporal contrast rather than to absolute light levels. Objects in the world are visible by the light they reflect; their borders are discerned usually because of different degrees of reflectance, which leads to contrast (changes in illumination, except for shadows, tend to be much more gradual). Because the visual system is adapted for seeing objects that can be either brighter or darker than their backgrounds, it is not too surprising that it is equally sensitive to positive and negative contrast steps.

Most ganglion cells, the output cells of the retina, receive their main excitatory input from bipolar cells. Thus these ganglion cells also have either *on* or *off* responses in the center of their receptive fields, according to which class of bipolar cells provide their input (Box 27.1). Other ganglion cells, known as *on-off* cells, respond to both light onset and to light offset. These latter cells receive input from both classes of bipolar cells. Many of these cells respond preferentially to stimuli moving in a particular direction of motion (Box 27.4).

Horizontal and Amacrine Cells Mediate Lateral Interactions in the Retina

There is a complex three-way synaptic relationship among the terminals of the photoreceptors, the processes of horizontal cells, and the dendrites of bipolar cells (Fig. 27.2). Horizontal cells are inhibitory (GABAergic) and, as the name implies, they contact photoreceptors over a much larger horizontal extent of retina than bipolar cells. Thus, because they are the only neurons that sample over a sufficiently large area in the outer plexiform layer, horizontal cells are thought to be responsible for the antagonistic surround seen in the receptive fields of bipolar cells.

The second family of lateral interneurons in the retina, the amacrine cells (literally, cells with no axons), is a diverse class of cells that exhibits a wide range of morphologies. As is true of horizontal cells, all processing takes place on the dendrites of amacrine cells (in the inner plexiform layer), which contain both pre- and postsynaptic elements (Fig. 27.2). Most amacrine cells use GABA, glycine, or both acetylcholine and GABA as neurotransmitters, but it is thought that virtually every transmitter found in the brain is used by some amacrine cell. The functions of most of these diverse subclasses remain poorly understood. In addition to a possible role in the creation of the antagonistic surround, it has been proposed that some amacrine cells are responsible for the nonlinear responses in Y cells of the cat (Box 27.2) or for the direction selectivity seen in the rabbit retina (Box 27.4).

Retinal Ganglion Cells Provide the Output of the Retina

Visual information leaves the eye via the optic nerve, which is composed of the axons of all the different classes of ganglion cells. The optic nerve begins at the *optic disc* (Fig. 27.1). Because there are no photoreceptors at the optic disc, this circular region constitutes a blind spot in the retina. The routing of ganglion cell axons to different parts of the brain provides a number of interesting problems for developmental neurobiology (Chapters 19 and 23). The routing of axons is both macroscopic—different types of ganglion cells project to different regions of the brain—and microscopic: within any given target region, axons are sorted out in a precise manner. In each of two principal targets of retinal axons, the superior colliculus and lateral geniculate nucleus of the thalamus (LGN), a topographical map of visual space is created in which spatial relations are maintained between neighboring neurons.

This chapter is concerned primarily with the classes of retinal ganglion cells that project to LGN neurons, which in turn project to the primary visual (striate) cortex. Although cells in the LGN are more than mere relays of information from the retina, their receptive fields are quite similar to those of their ganglion cell inputs. The following discussion of retinal receptive fields, therefore, can serve equally well to describe receptive fields in the LGN.

BOX 27.2

QUANTITATIVE METHODS IN THE STUDY OF VISUAL NEURONS CLASSIFICATION OF RETINAL GANGLION CELLS

Although neuroscientists have learned a tremendous amount about the visual system using the tools developed by Kuffler, researchers have been following a parallel line of studies of the visual system by asking different sorts of questions with the aid of more quantitative methods. In Kuffler's experiments and, most notably, many of Hubel and Wiesel's (see later), easily produced stimuli (such as spots, edges, and bars) were used to stimulate visual neurons. Action potentials were detected primarily with an oscilloscope and audio monitor. In the more quantitative studies of visual physiology, stimuli are shown primarily on video monitors and data are recorded with computers. The quantitative study of visual physiology constitutes a large field, which stems from the work of Hartline, Ratliff, Campbell, Barlow, and others (see Ratliff, 1974). Analysis of the receptive fields of retinal ganglion cells in the cat performed by Enroth-Cugell and Robson (1966) represents an early and influential example of this line of research.

Stimuli used by Enroth-Cugell and Robson, and in many studies that followed theirs, consisted of *gratings*: light and dark bands whose luminance varied in a sinusoidal fashion across a screen. Gratings were not designed to be the most effective stimuli for visual neurons. The broad goal of this sort of experiment is to study not just what makes neurons respond best, but to study how they would respond to *any* stimulus and to probe what mechanisms they use to produce these responses. The analytical framework that goes along with these experiments is called *systems theory*, the study of input-output (or stimulus-response) systems. *Linear systems,* those that simply add up all their inputs to produce an output, constitute an important part of this theory.

Enroth-Cugell and Robson used a simple test of linearity in their study of retinal ganglion cells in the cat.

A grating was presented at different positions (called phases), and the responses to its introduction and removal were recorded. Two questions were addressed for each cell studied: (1) If a certain phase (whose bars were arranged white, black, and white; as in Fig. 27.4, 0°) excited the cell, did the opposite grating (black, white, and black; Fig. 27.4, 180°) inhibit it? (2) More importantly, were there null positions for the grating that evoked no response, as excitation and inhibition were perfectly balanced (Fig. 27.4, 90° and 270°)?

One class of ganglion cells, called X *cells*, passed these tests for linear spatial summation (Fig. 27.4, left) A second class of cells, Y *cells*, behaved differently. Like X cells, these cells also had a center-surround receptive field when mapped with spots and annuli. When studied with two opposite gratings, however, the responses evoked were not equal and opposite. Instead, the introduction and removal of the gratings at all positions resulted in two peaks of excitation, and no null position could be found (Fig. 27.4, right).

The use of systems theory and quantitative techniques in the study of the visual system has had a number of notable successes. It has helped in the classification of different families of visual neurons, such as X and Y cells. Further, it has been used to elucidate some of the mechanisms responsible for the responses of visual neurons (such as directionally selective cells, see Box 27.4). Perhaps most importantly, along with a large body psychophysical experiments that employ the same stimuli, it has helped create a common framework in which the responses of neurons can be related to perception.

R. Clay Reid

Parallel Pathways Are Composed of Distinct Classes of Retinal Ganglion Cells in the Cat

Kuffler found that retinal ganglion cells in the cat have a stereotyped center-surround organization. Similar cells (both *on* and *off*) were found at every position across the entire retina. Researchers later discovered, however, that there are in fact several different functional classes of ganglion cells, each with distinct response properties. Although different cell classes have been found in different species, in none is the retina composed of a single mosaic of identical cells. The fact that there are multiple types of output from the retina has had a profound effect on the study of visual processing in the brain. The degree to which these *parallel pathways* (Livingstone and Hubel, 1984;

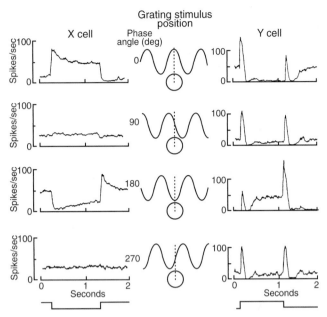

FIGURE 27.4 Visual responses of X and Y cells to contrast-reversing sine-wave gratings at different spatial phases (different positions, indicated schematically in the middle column). Visual responses, in spikes per second, are shown as poststimulus time histograms synchronized to the repetitive stimulus. Upward deflection of the lowest trace indicates introduction of the grating pattern (contrast on), and downward deflection indicates removal of the pattern (contrast off, but no change in mean luminance). Adapted from Enroth-Cugell and Robson (1966).

Merigan and Maunsell, 1994; Sincich and Horton, 2005) are combined or kept separate has been a major theme in the study of central visual pathways, particularly the visual cortex.

The idea that there are parallel sensory pathways into the brain is not new. At the beginning of the nineteenth century, both Bell and Müller argued for what Müller termed the "specific energy of nerves." This term is a precursor of the idea that axon sensory neurons are "labeled lines," each of which conveys distinct signals to the brain. A surprisingly modern version of this idea was given in 1860 by Helmholtz in the *Handbook of Physiological Optics* (2000), in which he expressed Thomas Young's theory of color vision (1807) in the following manner: "The eye is provided with three distinct sets of nervous fibers. Stimulation of the first excites the sensation of red, stimulation of the second the sensation of green, and stimulation of the third the sensation of violet."

In one of the earlier studies using a quantitative approach to receptive field mapping (Box 27.2), Enroth-Cugell and Robson (1966) found that cat retinal ganglion cells could be classified into two categories: X and Y (more categories have been found subsequently).

This classification was based on a simple criterion: did ganglion cells simply add together all their excitatory and inhibitory inputs (linear summation) or were the interactions between inputs more complicated (nonlinear summation)? The linear/nonlinear distinction between X and Y cells (Box 27.2) may seem fairly abstract, but these cells turned out to be different in a number of other ways. Most notably, Y cell receptive fields are, on average, three times larger than those of neighboring X cells. X cells are therefore far more numerous, because many more X cell receptive fields are needed to effectively tile, or cover, the retina. One view is that the large, nonlinear receptive fields of Y cells make them well suited to detect change, at the cost of their ability to signal the exact location or nature of the stimulus; similarly, X cells are better at localizing more specific stimulus features, but less sensitive in detecting change.

Primate Ganglion Cells Project to Three Subdivisions of the Lateral Geniculate Nucleus: Parvocellular, Magnocellular, and Koniocellular

Primates have a number of different classes of retinal ganglion cells, each with distinct cellular morphology, projection patterns, and visual response properties. This discussion focuses on ganglion cells that project to the lateral geniculate nucleus (LGN). There are many differences among primate species (nocturnal versus diurnal, Old World versus New World); this section concentrates on the macaque; a diurnal, Old World primate. The macaque visual system has been studied extensively with anatomical, physiological, and behavioral methods. Most importantly, macaque vision is very similar to human vision.

The most numerous class of ganglion cells in the macaque retina are sometimes referred to as P *cells*, so-called because they project to the dorsal-most layers of the LGN, the *parvocellular layers* (small cells; Fig. 27.9). M *cells*, which project to the two ventral layers of the LGN, the *magnocellular layers* (large cells), constitute a second class of retinal neurons. P and M cells have distinct morphologies (Fig. 27.2). P cells (also known as P/3 or, in the central retina, midget ganglion cells) have very small dendritic fields. M cells (also known as *Pa* or parasol cells) have much larger dendritic fields. Finally, members of a third class of retinal cells project to the *intercalated layers* between principal parvocellular and magnocellular layers. These intercalated layers are also termed *koniocellular* (dust-like, or tiny cells). There are roughly 1,000,000 P cells in the

BOX 27.3

INHERITED AND ACQUIRED DEFECTS OF COLOR VISION: RETINAL AND CORTICAL MECHANISMS

Inherited defects of color vision have been studied for over 200 years, originally with psychophysical methods but more recently with the tools of molecular genetics. Normal color vision is *trichromatic* (literally, three-colored) because there are three different classes of cone photoreceptors. In the most common form of color vision defect, there are three pigments, but one of them is abnormal, or anomalous. More severe defects are found in *dichromats*, who lack a single cone type entirely. The current terminology for the three types of dichromacy was proposed by von Kries in 1897: *protanopia* (literally, the first type of blindness, or "red-blindness"), *deuteranopia* (second type, or "green-blindness"), and *tritanopia* (third type, or "blue-blindness"). Whereas dichromats are unable to make certain color distinctions, only the very rare *monochromats*—people with one cone type or only rods—are truly color blind.

The genetics of red-green defects have long been known to be X-linked: they are inherited from the mother and are fairly common in men, but they are uncommon in women. Roughly 2% of European white males are protanopes or deuteranopes and, depending on the population, between 2 and 6% are trichromats with a single anomalous pigment (either protanomalous or, more commonly, deuteranomalous). In contrast, only 0.4% of women in similar populations have red-green defects. Tritanopia, found equally in men and women, is significantly rarer. It is inherited as an autosomal-dominant trait thought to occur at low frequencies, estimated variously between 1 in 500 and 1 in over 10,000.

Because of the overlap in cone pigments, inherited color vision defects are not as simple as terms such as "red blindness" might suggest. In broad terms, protanopes and deuteranopes are unable to make specific discriminations along the red-green axis. Depending on the defect, they are unable to distinguish certain reds and greens from gray or from each other. Tritanopes are unable to discriminate between colors with and without a short wavelength component, such as between gray and certain shades of violet or yellow.

In the 1980s, Nathans, Hogness, and colleagues cloned and sequenced the three cone pigment genes and related

them to color defects in humans. The cloning strategy was based on the pigments predicted close homology to rhodopsin (Fig. 27.7B). As expected from the genetics of protanopia and deuteranopia, two very similar genes for L and M pigments were found on the X chromosome. The third gene, which codes for the S pigment, was found on chromosome 7. In subsequent work by this group and others, molecular genetics of the inherited color defects, including anomalous trichromacy, have been studied in great detail.

Although the inherited color deficiencies are fairly simple and well understood, acquired defects in color vision—found in a number of ocular, neurological, and systemic diseases—are far more varied. One such syndrome, *cerebral achromatopsia*, was described in the neurological literature in the 1880s. In rare patients with certain cortical lesions, the discrimination of colors and the naming of colors were severely impaired. In the majority of cases, there were other associated deficits—either an area of complete blindness (a scotoma) or an inability to recognize faces (prosopagnosia)—but in some cases, vision was otherwise quite normal. From the location of the lesions, some neurologists early on inferred the existence of a region of cortex required for color, vision, near the inferior border of the primary visual cortex.

The literature of cerebral achromatopsia, quite developed in the late nineteenth century, fell into eclipse for the first half of the twentieth century when highly specific functional divisions of the cortex were questioned. With the discovery in the 1970s of multiple visual areas in primate cerebral cortex (Felleman and Van Essen, 1993), however, the existence of a cortical region devoted to color in humans seemed less far-fetched. With recent advances in noninvasive brain imaging there is little doubt that such regions exist in humans, although their homology to brain regions in nonhuman primates remains controversial.

R. Clay Reid

BOX 27.4

THREE TYPES OF SELECTIVITY FOR MOTION

Motion is one of the most important aspects of the visual world. It is a powerful cue for navigating in the world, for segregating figures from their background, and for predicting the trajectory of objects. It is not surprising, therefore, that sensitivity to motion is a highly developed feature of the mammalian visual system. At the lowest level, a class of neurons in the retina respond best to stimuli moving in a specific direction (Barlow *et al.*, 1964). These neurons give *on-off* responses to flashed lights (denoted ± in Fig. 27.16A), but are best excited by the motion of an object anywhere in their receptive fields. They can signal that motion in a particular range of directions is present, but because their receptive fields are relatively large and have both *on-and-off* responses, they cannot resolve the details of an object. Directionally selective retinal neurons do not project to the LGN so directional neurons found in the visual cortex (Hubel and Wiesel, 1962) create their selectivity independently and with a different mechanism.

Unlike directionally selective ganglion cells, which signal that motion is present somewhere within a large region, directionally selective simple cells in the visual cortex maintain detailed spatial information as well. A moving stimulus, such as a light bar moving to the left (Fig. 27.16B, left), is described by its trajectory in space over time. If the trajectory is plotted in space versus time, the slope represents the direction and velocity of the object. For instance, a bar moving to the right traces an oblique trajectory (up and to the left) in such a space-time plot (Fig. 27.16B, middle). When represented in a similar plot, the receptive fields of directionally selective simple cells can show a similar orientation (Fig. 27.16B, right). This plot is exactly analogous to Kuffler's maps of ganglion cell receptive fields. It represents the responses of the cell to a bright or dark bar at different positions, but it includes the time course of the responses as well. The *on* responses of this receptive field (the responses to a bright bar) are indicated with solid contour lines and are shown in white for added emphasis; *off* flanks (the responses to dark bars) are indicated with dotted contour lines.

For directionally selective simple cells, the timing of the response is directly related to the position of the stimulus. In the illustration, the cell is sensitive to a bright bar at a range of positions, but the timing of the peak response changes with position. These responses to flashed bars explain the responses to moving stimuli. Just as *on*-center/*off*-surround ganglion cells respond best to a bright spot on a dark background (Fig. 27.3C), a directionally selective simple cell responds best to a bar moving in a specific direction. A moving bar is the stimulus that best matches the template formed by its receptive field. In this manner, simple cell receptive fields can be strongly direction selective, but their responses preserve precise information about the position and structure of the stimulus.

A third kind of directionally selective cell is found in cortical area MT of the macaque. Receptive fields of individual neurons in MT integrate motion information over large regions of visual space. By comparison, receptive fields in the retina, LGN, and V1 can be thought of as viewing the world through much smaller apertures. If the goal of vision is to extract the properties of objects, rather than isolated features, then the existence of small receptive fields can lead to what is known as the *aperture problem*. As illustrated in Figure 27.16C, two objects moving in different directions can appear to have the same direction of motion when viewed through an aperture. In a dual study of perception in humans and of MT neurons in macaques, Movshon and colleagues (1985) analyzed the responses to complex stimuli whose components moved in different directions (such as the edges of the square in the bottom half of Fig. 27.16C), but which were perceived as coherent patterns moving in an intermediate "pattern direction." Unlike neurons in V1, many neurons in MT responded to the pattern direction rather than to the components. This would imply that these MT neurons combine their inputs in a complex manner to achieve a selectivity for the motion of extended objects rather than primitive features.

R. Clay Reid

---

retina and parvocellular neurons in the LGN; there are 100,000 M and magnocellular neurons. Although the intercalated layers appear quite sparse and thin in standard histological sections (Fig. 27.9), there are as many intercalated cells as there are magnocellular cells. Due to their size and location, intercalated cells have been difficult to study and little is known about them in the macaque. Therefore the following section discusses only the functional properties of P and M pathways.

P and M Pathways Have Different Response Properties

As noted earlier, the receptive field properties of the LGN relay cells match closely those of their retinal afferents, but the two parallel pathways, P → parvocellular and M → magnocellular, are quite different. Five main characteristics distinguish the responses of P cells and M cells (Figs. 27.5 and 27.6):

1. P cell receptive fields are smaller than M cell receptive fields at the same retinal position.
2. M cell axons conduct impulses faster than P cell axons.
3. The responses of P cells to a prolonged visual stimulus, particularly a color stimulus, can be very sustained, whereas M cells tend to respond more transiently.
4. Most P cells are sensitive to the color of a stimulus; M cells are not (Figs. 27.5A–27.5C).
5. M cells are much more sensitive than P cells to low-contrast, black-and-white stimuli (Fig. 27.5D).

These last two differences, which have strong implications for the functional role of these pathways, are discussed next.

Color-Selective Responses in the P Pathway Are Derived from Antagonistic Inputs from L and M Cones

Color vision depends on the distinct sensitivities of the three classes of cone photoreceptors to different wavelengths of light (the rods are active only under low light levels and are not involved in color vision). Although fine distinctions can be made between different wavelengths, it is important to emphasize that each of the light-sensitive pigments in the three cone classes is sensitive to a broad range of wavelengths. The spectral sensitivities of these pigments overlap considerably (Fig. 27.6A). Consequently, the names sometimes used for these pigments—red, green, and blue sensitive—are inaccurate. For instance, the "red" pigment is most sensitive to light whose wavelength is 564 nm, or yellow, although it is more sensitive than the other pigments to red light. A more accurate terminology, one that identifies the relative sensitivities of the three cone absorption spectra, usually is employed: long, middle, and short wavelength sensitive (or L, M, and S).

Because green light will excite both long and middle wavelength cones, the color green can be distinguished only by the fact that it excites middle wavelength cones more strongly than it excites long wavelength cones. This distinction can be made by neurons that are sensitive to the difference between the signals from two cone classes, or neurons that are *color opponent*. In the retina and LGN, there are two categories of color-opponent cells: red-green and blue-yellow (yellow is made of the sum of long and middle wavelength cone signals; Dacey, 2000). These are sometimes also termed red-minus-green cells and blue-minus-yellow cells. The most numerous color-opponent neurons in the primate retina are red-green opponent P cells. These cells receive antagonistic input from long and middle wavelength-sensitive (L and M) cones.

Near the fovea, the center of the retina, a P cell receives input from only one bipolar cell. This *midget bipolar* in turn receives input from a single cone (Fig. 27.2). This arrangement ensures not only that the center of the ganglion cell's receptive field is as small as possible, but also that the center receives input from only one cone type: L or M. Classically, the antagonistic surround of these P cells (or their counterparts in the LGN) has been thought to be dominated by the other main cone class, M or L, respectively (Wiesel and Hubel, 1966). Although the exact nature of the surround in red-green P cells has been somewhat controversial in recent years (Dacey, 2000), this classical view is shown in Fig. 27.6B.

There are four different types of red-green opponent P cells in the retina (Fig. 27.6B), and hence parvocellular neurons in the LGN. L-*on* center and M-*off* center cells are excited by red and inhibited by green (and some blues; see Fig. 27.5A). M-*on* and L-*off* cells are excited by green and inhibited by red. As Wiesel and Hubel showed (1966), parvocellular neurons in the LGN are sensitive to small bright spots presented in their receptive field centers, but they are most selective for the color of the stimulus when larger stimuli are used (see Figs. 27.5A and 27.5B). This is because only larger stimuli are effective in stimulating simultaneously the color-opponent center and surround. Because they absorb a broad range of wavelengths (Fig. 27.6), single cones are not very color selective. Parvocellular neurons are most color selective when the stimulus evokes antagonistic influences from two cone classes: one found in the center and the other found only in the surround.

M Cells Are Highly Sensitive to Contrast

Because of their marked color sensitivity and small receptive fields, P cells have long been thought to be involved with the ability to make color discriminations and to see the finest details. When tested with black-and-white stimuli, however, P cells responded to low contrast very poorly compared to M cells. In a study that used the quantitative methodology similar

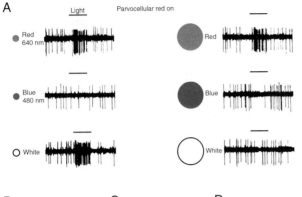

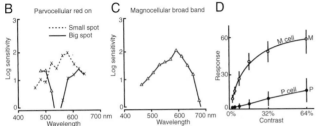

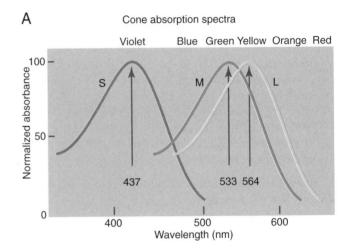

to Enroth-Cugell and Robson's (Box 27.2), Kaplan and Shapley (1986) measured the responses of P and M cells to sine-wave gratings of various contrasts (Fig. 27.5D). Perceptually, stimuli can be detected that have less than 1% contrast (i.e., 1% luminance deviation from the mean). Kaplan and Shapley (1986) found that M cells often responded well to contrasts of under 5% and that they gave their strongest responses at contrasts as low as 20%. P cells tended to respond poorly to contrasts below 10%, a stimulus that is quite salient,

FIGURE 27.5 Visual responses of magnocellular and parvocellular neurons in the macaque. (A) Responses of a color-opponent (red-*on*/green-*off*) parvocellular neuron to small and large stimuli. For small spots, roughly the size of the receptive field center, the neuron was excited by both red and white and was weakly inhibited by blue. The responses to large spots were more selective. Red was excitatory, blue was strongly inhibitory, and white stimuli were ineffective. (B) Responses of the same red-*on*/green-*off* parvocellular neuron to different wavelengths of light. For small spots, the neuron was excited (X) for a broad range of wavelengths. For big spots, it was excited above 550 nm, from yellow to red, and inhibited (D) below 550 nm, from blue to green. (C) Responses of a magnocellular neuron to different wavelengths of light presented in the receptive field center. The neuron had *off* (D) responses at all wavelengths, i.e., no wavelength specificity was found. (D) Average contrast response functions of P (•) and M (o) cells measured in spikes per second. P cells respond poorly to low contrasts and do not saturate at high contrasts. M cells respond better to low contrasts, but saturate by 20–30%. A–C: From Wiesel and Hubel (1966). D: From Kaplan and Shapley (1986).

FIGURE 27.6 Aspects of color vision. (A) Absorption spectra of the three cone photoreceptors in humans: long, middle, and short wavelength sensitive (L, M, and S). L and M cones in particular are sensitive to overlapping ranges of wavelengths. (B) Receptive field of the four types of red-green parvocellular cells in the macaque (labeled L+ and M+ or L– and M– according to their on-or-off centers, respectively). Classically, these small receptive fields have been described as receiving antagonistic input from L cones in their centers and M cones in their surrounds, or vice versa. When probed with color stimuli, L-*on* center and M-*off* center cells are excited by red and inhibited by green (and some blues: see Fig. 27.5A). M-*on* and L-*off* cells are excited by green and inhibited by red. When probed with small white spots, L-*on* and M-*on* cells are excited when the stimulus is turned on; L-*off* and M-*off* cells are excited when the stimulus is turned off. (C) Receptive fields of the two main types of magnocellular cells. These larger receptive fields receive mixed L and M input to both center and surround. There are two types: *on* center (M+ L+) and *off* center (L–M–).

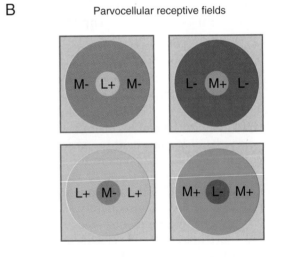

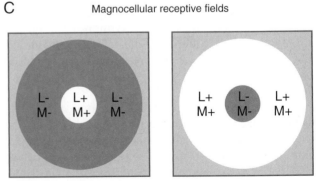

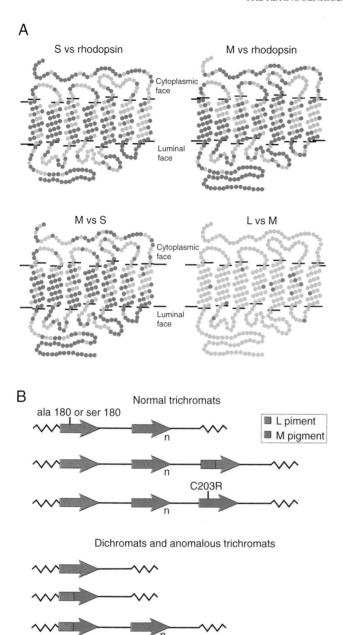

FIGURE 27.7 Molecular biology of photopigment genes. (A) Transmembrane model of cone photopigments. Homologies between the different photopigments are shown, with differences between proteins highlighted. The amino acid sequences of M and S cone pigments and rhodopsin are all approximately 40% identical (and 75% homologous), whereas L and M pigments are 96% identical (99% homologous). (B) Structure of the portion of the X chromosome coding for L and M pigments. Normally, there is one L pigment gene and several M pigment genes. Because not all these genes are expressed, different variants can result in normal color vision (top). Many different patterns of recombination and deletion can be associated with a defect in red-green vision: either dichromacy or anomalous trichromacy (bottom). A: From Nathans *et al.* (1986). B: From Nathans (1994).

and rarely reached their strongest responses below 64%. From this, Kaplan and Shapley concluded that the ability to see low contrasts is due primarily to the M cell system.

The relatively poor sensitivity of P cells to low-contrast stimuli is not well understood, but it may be partially the result of two factors: receptive-field centers receive input from a very small region of the retina, and the inputs from different cone classes are subtracted from each other, rather than added together. Whatever the reason, the poor performance of the P pathway in detecting low contrast implies an important role for M cells in the detection of the form of objects (which is robust at low contrasts), in addition to their role in the motion pathway (discussed later).

Summary

The retina, part of the central nervous system, is a self-contained neuronal circuit whose anatomy and physiology have been studied in great detail. There are five types of cells in the retina. Photoreceptors, bipolar cells, and ganglion cells constitute the direct, feed-forward pathway. Horizontal cells and amacrine cells subserve lateral interactions within the retina. Photoreceptors are all hyperpolarized by light, but bipolar cells can be either hyperpolarized (*off* cells) or depolarized by light (*on* cells). There is a great variety of retinal ganglion cells, but most have an antagonistic center-surround organization. In the primate, the two best studied classes of ganglion cells are P cells and M cells, which project to the parvocellular and magnocellular layers of the LGN, respectively. P cells have small receptive fields, are sensitive to the color of a stimulus, and are relatively insensitive to low contrasts. M cells have larger receptive fields, are insensitive to color, and are very sensitive to low contrasts.

THE RETINOGENICULOCORTICAL PATHWAY

Visual Information Is Relayed to the Cortex via the Lateral Geniculate Nucleus

In mammals with forward-facing eyes, such as most carnivores and the primates, retinal axons are routed so that visual information from the same points in space coming from the two eyes can be combined. From both eyes, ganglion cells whose receptive fields are in one half of the visual field project to the opposite cerebral hemisphere (Fig. 27.8). This cross-routing occurs at the optic chiasm (from the Greek letter *chi*,

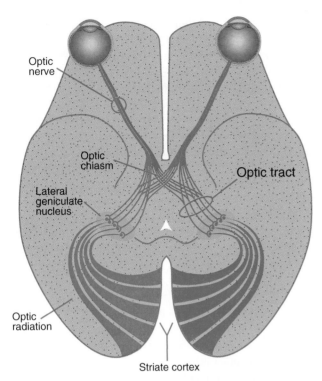

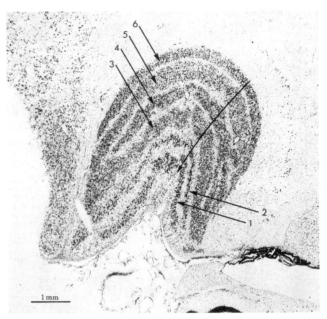

FIGURE 27.8 The retino-geniculo-cortical pathway in the human. Optic nerve axons from the nasal retina cross at the optic chiasm and join axons from the temporal retina of the other eye. Together, these contralateral and ipsilateral axons make up the optic tract, which projects to the LGN. Each of the six layers of the LGN receives input from only one eye. Axons from the LGN make up the optic radiations, which project to the striate cortex. Adapted from Polyak (1941).

FIGURE 27.9 The six-layered LGN of the macaque monkey. The top four parvocellular layers (6, 5, 4, and 3) receive input from the ipsilateral and contralateral eye in the order of contra, ipsi, contra, and ipsi. The bottom two magnocellular layers (2 and 1) receive ipsi and contra input, respectively. In between these principal layers are intercalated or koniocellular layers. The arrow from layer 6 to 1 indicates organization of the precisely aligned retinotopic maps of the six layers. The receptive fields of neurons found along this line are located at the same position in visual space. From Hubel and Wiesel (1977).

χ). Here, axons from the medial (nasal, or nearer the nose) half of one retina cross over to join the axons from the lateral (temporal, or nearer the temples) half of the other retina. In other words, the temporal retina projects to the *ipsilateral* (same-side) hemisphere and the nasal retina projects to the *contralateral* (opposite-side) hemisphere. Unlike somatic sensation, which is entirely crossed (Chapter 25), only half of the retinal axons cross. Like somatic sensation, however, the crossing of retinal axons results in each half of the external visual field being represented in the opposite cerebral hemisphere.

The LGN Is a Layered Structure That Receives Segregated Input from the Two Eyes

As they leave the optic chiasm, retinal axons from the two eyes travel together in the optic tract, which terminates in the LGN (Fig. 27.8). The LGN is composed of layers, each of which receives input from only one eye, although the number varies between species. In the macaque, the six principal layers contain most of the relay cells to primary visual cortex (Fig. 27.9).

The four parvocellular layers are found dorsally. They receive P cell input from the contralateral and ipsilateral eye in the order of contra, ipsi, contra, and ipsi. The magnocellular layers are found more ventrally. They receive M cell input in the order of ipsi and contra. Interposed between these principal layers are the intercalated layers, populated by konio-cellular neurons.

In the cat, the LGN has only two principal layers, called A and A1, which receive input from the contralateral and ipsilateral retinas, respectively. The neurons are again quite similar to either the retinal X cells or the Y cells that innervate them.

The Primary Visual Cortex in the Cat Is an Example of a Functional Hierarchy

Cat primary visual cortex (area 17) is perhaps the best understood neocortical area in any species. The first studies of Hubel and Wiesel (1962) form the foundation of this understanding. Hubel and Wiesel found cells whose responses differed dramatically from those in the retina and thalamus. Many of these cells could be excited by stimuli presented to either eye. The primary visual cortex thus represents the first level of the visual system at which binocular interactions could form a substrate for depth perception. Most strikingly, Hubel and Wiesel found that the vast majority of cortical cells respond best to elongated stimuli at a specific orientation and give no response at the orthogonal orientation. This is in sharp contrast to cells in the retina and LGN, which are not selective for orientation.

Hubel and Wiesel described two broad classes of neurons in area 17: *simple cells* and *complex cells*. The receptive fields of simple cells were segmented into oriented *on-and-off* subregions (Figs. 27.10C and 27.10D) and were therefore most sensitive to similarly oriented stimuli. In these receptive field subregions, responses were evoked by turning a light stimulus *on* or *off*, but not by both. Given these well-defined *on-and-off* subregions in simple cells, Hubel and Wiesel suggested a straightforward hierarchical model to explain orientation selectivity: neurons in the lateral geniculate nucleus whose receptive fields are arranged in a row could all converge to excite a specific simple cell (Fig. 27.10E). Although this model has generated much controversy over the ensuing years, many of its elements have been demonstrated directly (Reid and Alonso, 1995; Ferster *et al.*, 1996).

The second class of cortical neurons, complex cells, also responded best to oriented stimuli, but their receptive fields were not elongated along a preferred orientation, nor were they divided into distinct *on-and-off* subregions. It would therefore be difficult to imagine how the responses of these neurons could be constructed directly from the center/surround receptive fields in the thalamus. Given the properties of simple cells, Hubel and Wiesel proposed a hierarchical scheme. Complex cells receive their orientation selectivity from convergent input from simple cells whose receptive fields have the same orientation preference, but have slightly different spatial locations (Fig. 27.10F).

This original model is hierarchical or serial; information flows from one level to the next in a well-defined series. In contrast to this hierarchical model of cortical processing, it also has been proposed that each region of the cerebral cortex is made up of several different streams of information that are all processed in parallel. As with most dichotomies, elements of both hierarchical and parallel processing have been found in the visual cortex of all mammals studied. For simplicity, an outline of the organization of visual cortex in the cat is discussed here in terms of a hierarchical model. Parallel processing will be illustrated with the example of a primate visual cortex.

The hierarchical scheme has received support from several correlations between anatomy and physiology

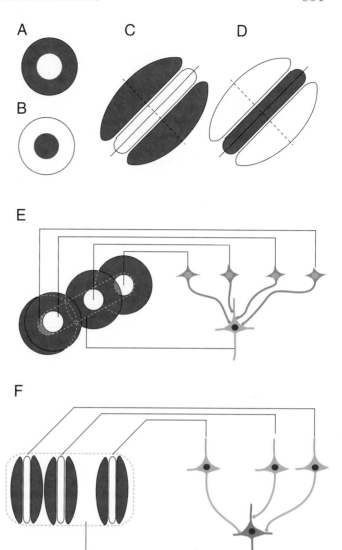

FIGURE 27.10 Hubel and Wiesel's original models of visual cortical hierarchy. (A and B) Receptive field maps of center-surround receptive fields in LGN. White: *on* responses; dark gray: *off* responses. (C and D) Receptive field maps of simple cells. (E) Model of convergent input from LGN neurons onto the cortical simple cell. (F) Model of convergent input from simple cells onto complex cells. Adapted from Hubel and Wiesel (1962).

of area 17. In broad terms, simple cells are more common in the layers of cortex that receive direct input from the thalamus (layer 4 and, to a lesser extent, layer 6). Complex cells are found more frequently in layers that are more distant from the thalamic input in layers 2 + 3, which receive input primarily from layer 4, and in layer 5, which receives most of its input from layers 2 + 3 (Gilbert and Wiesel, 1985; Fig. 27.11).

In terms of their probable function in perception, the receptive fields of complex cells furnish an early example of what was meant in the introduction by a higher order representation. Complex cells convey

A

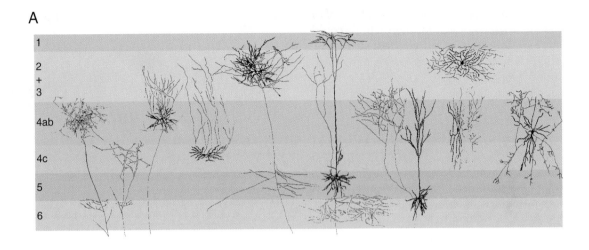

B

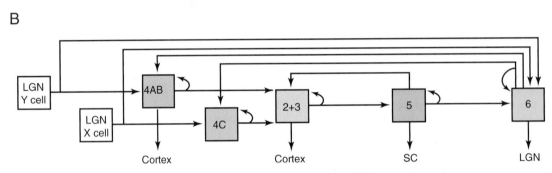

FIGURE 27.11 Circuitry of cat cortical area 17. (A) Morphology of individual cells stained with horseradish peroxidase. Thick lines: dendrites; thin lines: axons. (B) Schematic diagram of connections. Most thalamic (LGN) input is concentrated in layer 4 and, to a lesser degree, in layer 6. Y cells tend to project more superficially in layer 4 than X cells (note that sublayers 4ab and 4c are not strictly analogous to 4A, 4B, and 4C in the macaque). There is a fairly strict hierarchical pathway from layers 4 Æ 2 + 3 Æ 5 Æ 6 with feedback connections from 5 Æ 2 + 3 and 6 JE 4. There are also many lateral connections between neurons within the same layer. Adapted from Gilbert and Wiesel (1985).

information about orientation to later stages of processing, but that information has been combined and generalized. A simple cell responds to an oriented stimulus of a specific configuration and a specific location; in particular, it has separate *on-and-off* subregions. A complex cell also responds to stimuli at one orientation, but the receptive field is not segregated into *on-and-off* subregions. Instead, it can respond to a light or dark stimulus of the correct orientation, independent of the exact location of the stimulus within the receptive field. This sort of generalization is a recurrent theme in the cortical processing of visual information.

Several Parallel Streams Are Found in the Macaque Primary Visual Cortex

In the macaque primary visual cortex (V1, or striate cortex), a hierarchical organization is certainly present, but it is also clearly composed of several parallel streams. As discussed earlier, at least three types of inputs to visual cortex (parvocellular, magnocellular, and koniocellular) are first segregated within the LGN. Each class of neurons projects to a specific subdivision of primary visual cortex. Most thalamic afferents terminate in layer 4C, which is split into two divisions. 4Cα receives its input from magnocellular neurons and 4C/3 from parvocellular neurons. Intercalated, or koniocellular, geniculate neurons project to layers 2 + 3, specifically to regions known as "blobs" (discussed later) that stain densely for the enzyme cytochrome oxidase (Livingstone and Hubel, 1984)

Thus visual information from functionally and anatomically distinct M and P retinal neurons are kept separate through the LGN and at least up to the cortical neurons that receive direct thalamic input. The degree to which these pathways—as well as the more poorly understood koniocellular pathway—are kept separate within visual cortex remains an area of active research (Fig. 27.15).

Functional Architecture Can Be Seen in the Columnar Structure of the Visual Cortex

So far, the physiology of the visual cortex has been considered in terms of individual neurons and their response properties. Another major contribution of Hubel and Wiesel was their demonstration that cells with similar receptive field properties tend to be found near each other in the cortex. Further, they found that physiological response properties, such as orientation selectivity, are organized in an orderly fashion across the cortical surface. They termed the relationship between anatomy and physiology the *functional architecture* of visual cortex.

Hubel and Wiesel's key observation in this arena was that when an electrode is advanced through the cortex perpendicular to its surface, all neurons encountered have similar response properties. If a cell near the surface has a specific orientation and is dominated by input from one eye, then cells below it share these preferences. This finding is consistent with the idea, proposed by Lorente de Nó, that the fundamental unit in cortical architecture is a vertically oriented *column* of neurons. Hubel and Wiesel proposed that a cortical column is both a physiological and an anatomical unit, as had Mountcastle in his study of somatosensory cortex (Chapter 26). A second observation was that the property of orientation preference varies smoothly over the cortical surface. If an electrode takes a tangential path through the cortex through many different columns, the orientation preference generally changes in a steady clockwise or counterclockwise progression, although occasionally there are discontinuous jumps (Fig. 27.12A).

In addition to orientation, at least two other parameters are mapped smoothly across the cortical surface. Cells in the visual cortex receive inputs from the two eyes in varying proportion, a property that Hubel and Wiesel termed *ocular dominance*. In tangential microelectrode penetrations, cells are found that are dominated first by input from one eye and then the other. This provides evidence for *ocular dominance columns*, which have subsequently been demonstrated anatomically as well as physiologically. Second, prior to Hubel and Wiesel's work, a precise map of visual space across the surface of the cortex had been demonstrated. For any cortical column, receptive fields are all located at roughly the same position on the retina. Nearby columns represent nearby points in visual space in a precise and orderly arrangement. The position of a stimulus on the retina is termed its *retinotopy*; thus a region of the brain (such as the superior colliculus, LGN, or the visual cortex) that maintains the relations between adjacent retinal regions is said to have a *retinotopic map*.

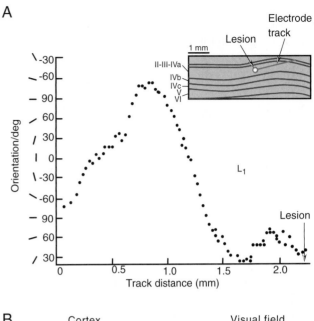

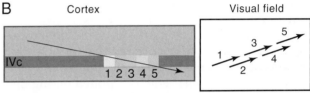

FIGURE 27.12 Two aspects of the functional architecture of the macaque primary visual cortex. (A) Graph of the preferred orientation of neurons encountered in a long microelectrode penetration through layers 2 + 3 (inset). There was a steady, slow progression of preferred orientations, although there were a few positions where the orientations changed more abruptly. (B) Schematic diagram of an electrode penetration through layer 4C (left) and the retinotopic positions of receptive fields (right). Numbers (1–5) indicate regions dominated by input from alternating eyes. At the border between ocular dominance columns (e.g., instance between 1 and 2), the location of receptive fields jumps back to a point represented near the middle of the previous ocular dominance column. There is a complete representation of visual space in columns dominated by each eye (1,3,5 or 2,4) and these representations are spatially interleaved. Adapted from Hubel and Wiesel (1977).

Given the existence of multiple functional maps, the obvious question is: How do these maps relate to each other? This question has been answered most definitively for the relationship between retinotopy and ocular dominance. Layer 4 in the primate is ideal for studying this question, as the receptive fields are quite small and the borders between ocular dominance columns are well defined. By making long, tangential penetrations through layer 4 of the striate cortex, Hubel and Wiesel found a precise interdigitating map from each eye. When eye dominance shifts, the receptive field location shifts to a point corresponding to the middle of the previous ocular dominance column (Fig. 27.12B). Thus there is a 50% overlap in the spatial locations represented by adjacent ocular dominance columns. In this manner, a complete representation of space is attained for both eyes while ensuring that cells that respond to overlapping points in visual space are always nearby within the cortex.

Ocular Dominance and Orientation Columns Can Be Revealed with Optical Imaging

The technique of optical imaging has proven extremely useful in the study of the functional architecture of the visual cortex. Optical imaging allows direct visualization of the relative activity of small cortical ensembles rather than relying on inferences made from single-unit studies (Blasdel and Salama, 1986; Grinvald et al., 1986). The technique uses dyes that change their optical properties with neural activity. Even if no dyes are used, brain activity can be mapped with an intrinsic signal, caused primarily by changes in blood flow and blood oxygenation.

A typical optical imaging experiment works as follows (Fig. 27.13A). First, a series of digitized images is taken of a region of visual cortex (Fig. 27.13B) while the animal is presented with visual stimuli through the left eye. Next, a similar series is captured during right eye stimulation. When the right eye images are subtracted digitally from the left eye images, a striking picture is created that reveals the functional architecture of ocular dominance (Fig. 27.13C). Previously, anatomical methods had been used to produce maps of ocular dominance, and these maps appear similar to those found with optical imaging, but an important feature of optical imaging is that multiple images can be obtained from the same region of the cortex. For instance, in addition to ocular dominance, maps of the preferred orientation can also be made. A useful way to present these data is in terms of a color code, in which each color represents a different preferred orientation (Fig. 27.13D). The orientation columns revealed by optical imaging were consistent with earlier microelectrode studies (compare Fig. 27.13E with Fig. 27.12A), but the images for the first time gave a detailed picture of the layout of orientation across the cortical surface. Details of these orientation maps—and their relationships with ocular dominance—were entirely new and had not been demonstrated with previous techniques.

Cytochrome Oxidase Staining Reveals Blobs and Stripes in Cortical Areas V1 and V2

In the primate, there is a fourth feature of the functional architecture of visual cortex in addition to retinotopy, orientation, and ocular dominance columns.

When stained for the enzyme cytochrome oxidase (a metabolic enzyme whose presence indicates high activity), histological sections of V1 revealed a regular pattern of patches, or blobs (Fig. 27.14). In layers 2 + 3, these blobs were reported to contain neurons whose receptive fields are color selective, poorly oriented, and monocular (Livingstone and Hubel, 1984).

In V2, the second visual area in the cerebral cortex (see later), cytochrome oxidase staining reveals regions of high and low activity arranged in parallel stripes (thick, thin, and pale stripes; Fig. 27.14, along the top). The anatomical connections between V1 and V2 are strongly constrained by the subdivisions revealed by cytochrome oxidase staining. In an early view, V1 neurons in the blobs projected to the thin stripes of V2, neurons in the interblobs projected to the pale stripes, and neurons in layer 4B projected to the thick stripes (Livingstone and Hubel, 1984). More recently, this organizational scheme has been revised and simplified such that V1 neurons in blob (or patch) columns, including underlying regions of layer 4B, project to the thin stripes of V2, whereas neurons in the interblob (interpatch) columns, again including underlying regions of layer 4B, project to both the pale stripes and thick stripes (Sincich and Horton, 2005).

One of the reasons it is important to understand the anatomical relationship between V1 and V2 is that neurons in the thick, thin and pale stripes of V2 give rise to projections that appear selective for either the dorsal or ventral streams of visual processing in extrastriate cortex (described later), whereby neurons in the thick stripes provide input to the dorsal stream and neurons in the thin and pale stripes provide input to the ventral stream.

Many Extrastriate Visual Areas Perform Different Functions

In early studies of the cytoarchitecture of cerebral cortex, the visual cortex was divided into three areas according to Brodmann's classification: areas 17, 18, and 19. These divisions were based on differences in cell cytoarchitecture, or differences in the size, morphology, and distributions of cells within the six cortical laminae. Area 17 is primary visual cortex, or striate cortex, so named because of a heavily myelinated sublamina within layer 4 ($4C\alpha$), prominently visible as a stripe in transverse section. Areas 18 and 19 were known simply as the visual association cortex. An assumption behind these designations was that areas defined by anatomical criteria would ultimately prove to be functionally specialized.

The demonstration of multiple visual cortical areas has been one of the important discoveries of the past

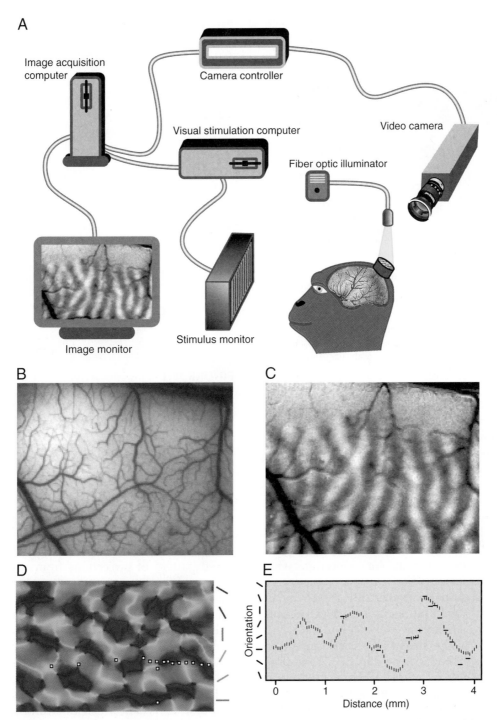

FIGURE 27.13 Optical imaging of functional architecture in the primate visual cortex. (A) Schematic diagram of the experimental setup for optical imaging. Digitized images of a region of visual cortex (as in B) are taken with a CCD camera while the anesthetized, paralyzed animal is viewing a visual stimulus. These images are stored on a second computer for further analysis. (B) Individual image (9 by 6mm) of a region of V1 and a portion of V2 taken with a special filter so that blood vessels stand out. (C) Ocular dominance map. Images of the brain during right-eye stimulation were subtracted digitally from images taken during left-eye stimulation. (D) Orientation map. Images of the brain were taken during stimulation at 12 different angles. The orientation of stimuli that produced the strongest signal at each pixel is color coded, as indicated to the right. The key at the right gives the correspondence between color and the optimal orientation. (E) Comparison of the preferred orientation of single neurons with the optical image. At each of the locations indicated by squares in D, single neurons were recorded with microelectrodes. The preferred orientations of the neurons (dashes) were compared with the preferred orientations measured in the optical image, sampled along the line connecting the recording sites (dots). A: Adapted From Grinvald *et al.* (1988). B and C: Adapted from Ts'o *et al.* (1990). D and E: Adapted from Blasdel and Salama (1986).

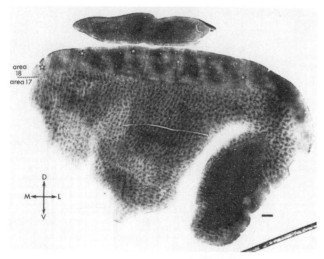

FIGURE 27.14 Macaque visual cortex stained for the metabolic enzyme, cytochrome oxidase. A tangential section through layers 2 + 3 of V1 (below) and V2 (above) is shown. In V1, blobs are seen in a regular array with a spacing of 400 mm. In V2, there are cytochrome oxidase-rich stripes with much coarser spacing. The distinction between thick and thin stripes (see text) is seen less readily in the macaque than in the owl monkey. Scale bar: 1 mm. From Livingstone and Hubel (1984).

quarter century in the field of sensory neurobiology. A vast expanse of cerebral cortex—greater than 50% of the total in many primate species—is involved primarily or exclusively in the processing of visual information. The extrastriate cortex now includes areas 18 and 19, as well as large regions of the temporal and parietal lobes (see Fig. 27.15 for a diagram of the four main regions of cerebral cortex: occipital, parietal, temporal, and frontal). It is composed of some 30 subdivisions that can be distinguished by their physiology, cytoarchitecture, histochemistry, and/or connections with other areas (Felleman and Van Essen, 1991). Each of these extrastriate visual areas is thought to make unique functional contributions to visual perception and visually guided behavior.

As a testament to the increasing complexity of the field, the naming of visual areas (Fig. 27.15) has progressed from the use of simple labels—V1 through V4—to the use of more complex terms specifying anatomical location of each new area, such as MT (medial temporal, or V5), MST (medial superior temporal), or IT (inferotemporal, itself made up of several distinct areas).

The dual themes of parallel and hierarchical processing are central to the understanding of the extrastriate cortex. Figure 27.15 is a vastly simplified version of a wiring diagram between visual areas (Felleman and Van Essen, 1991). Several criteria have been used to define new visual cortical areas. First, extrastriate regions have retinotopic maps of the visual world that

A

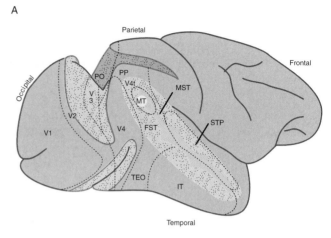

B

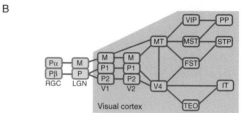

FIGURE 27.15 Extrastriate cortical regions. (A) Lateral view of the macaque brain, with the sulci partially opened to expose the areas within them. Shown are the rough outlines of the main visual areas, which take up all of the occipital cortex and much of the parietal and temporal cortex. (B) Partial diagram of the connections between visual areas. Emphasis is placed on the hierarchical organization of the connections and on the partially segregated P parvocellular and M magnocellular pathways. Adapted from Albright (1993).

can be demonstrated by physiological recordings. Second, a clear hierarchy between many cortical areas can be demonstrated anatomically. There is a stereotyped pattern of projections from one visual area to the next. Where strong connections between two areas exist, these connections tend to be bidirectional. The feedforward and feedback connections are distinguished by the layers that send and receive the connections. In such a manner, a clear hierarchy can be traced, for instance, along the pathway V1 → V2 → V3 → MT → MST (with several shortcuts, such as V1 → MT).

Physiological studies of cortical areas revealed profound differences between them, and a series of studies by Ungerleider and Mishkin (1982) uncovered a higher order dichotomy between two types of processing in the extrastriate cortex. Using behavioral analyses of animals with anatomically defined cortical lesions, Ungerleider and Mishkin found a strong dissociation between the types of deficits exhibited by animals with lesions in either their parietal cortex or their temporal cortex (Fig. 27.15). Animals with temporal lesions were often much worse at recognizing objects visually, although lower level visual function, such as acuity,

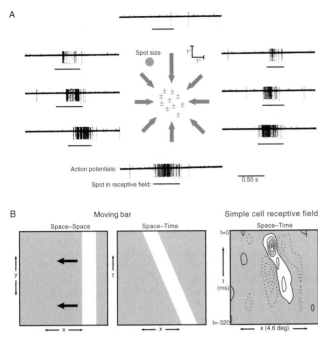

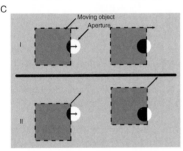

FIGURE 27.16 Three types of direction selectivity. (A) Receptive field map of a directionally selective ganglion cell in the rabbit (±indicates where neuron responded in an *on-off* manner). Surrounding it are individual responses to stimuli moving in eight different directions. The smooth traces below each train of action potentials correspond to the position of the stimulus as it moves in the indicated direction. Horizontal bars indicate when the spot was in the receptive field. (B) Direction selectivity in cortical simple cells of the cat. A bar moving to the left (shown in a space-time plot in the middle panel) matches the template formed by a directionally selective simple receptive field (shown in a space-time plot at right). The vertical time axis in the right-hand panel corresponds to the stimulus location 0 to 320 ms *before* the neuron fired. Bright regions in the receptive field correspond to the best location for a bright stimulus at each delay between stimulus and response. (C) The ambiguous motion of an extended object when viewed through a small aperture—known as the aperture problem—is partially resolved by some neurons in macaque cortical area MT. The squares in I and II (each shown at two successive time points) are moving in different directions, but they appear identical when viewed through a small aperture. A: Adapted from Barlow *et al.* (1964). B: Far right adapted from McLean *et al.* (1994).

was not appreciably lessened. Parietal lesions led to little or no deficit in object recognition, but visuospatial tasks, such as visually guided behavior, were impaired profoundly. From these studies and the growing body of physiological evidence, two distinct streams were postulated: a *temporal* or *ventral stream* devoted to object recognition and a *parietal* or *dorsal stream* devoted to action or spatial tasks. Ungerleider and Mishkin termed these the "what" and "where" pathways.

The ventral stream, V1 → V2 → V3 → V4 → IT . . . , is discussed in Chapter 55. This chapter considers only the dorsal stream, V1 → V2 → MT → MST. . . . The dorsal stream is dominated by magnocellular cells and the ventral stream by parvocellular cells, although the segregation is far from strict (Merigan and Maunsell, 1994).

In the Parietal Cortex, Neurons Are Selective for Higher Order Motion Cues

MT, or V5, is perhaps the best understood of the extrastriate visual areas. It has held particular interest ever since its was discovered in 1971 (Allman and Kaas, 1971; Dubner and Zeki, 1971), primarily because it was the first area found that was strongly dominated by one visual function. Fully 95% of the neurons in MT are highly selective for the direction of motion of a stimulus (Dubner and Zeki, 1971). In V1, a significant fraction of neurons are selective for the direction of motion, but the optimal speed may vary depending on the spatial structure of the object that is moving. In MT, speed tuning is less dependent on other stimulus attributes. Receptive fields of individual neurons in MT integrate motion information over large regions of visual space and are qualitatively less selective than neurons in the primary visual cortex. This generalization of motion signals can be achieved in a simple manner, such as by adding together inputs over space or, in a complex manner, by combining two component motions in different directions into a single coherent motion (Box 27.4). Of even greater interest, neurons in MT (and, to a greater extent, those in MST) appear to be sensitive to more complex aspects of visual motion, such as the motion of extended objects rather than isolated features (Albright, 1993).

The range of stimuli that a given MT neuron can respond to is impressively broad—the attributes, or form cues, that define a figure can be luminance, texture, or relative motion—but the preferred direction and speed are always the same for that neuron (Albright, 1993). This is an extreme example of what were termed high-order responses in the introduction to this chapter. Neurons at lower levels in the visual system are sensitive to isolated and specific features in visual scenes. Higher visual areas respond to very specific attributes, but these attributes are increasingly remote from the physical stimulus. Instead, they represent increasingly complex concepts, such the motion of an extended object or the identity of a face.

Summary

The lateral geniculate nucleus of the thalamus is a layered structure that receives segregated input from the two eyes and projects to the primary visual cortex (V1). Geniculate neurons have center-surround receptive fields that are similar to those of their retinal inputs. In contrast, most neurons in the primary visual cortex are sensitive to the orientation of a stimulus. In mammalian species, there are elements of both hierarchical and parallel processing in the primary visual cortex. This section has emphasized that the visual cortical circuit in the cat can be seen as a functional hierarchy that transforms geniculate input into simple and then complex receptive fields. In the macaque monkey the focus has been on the parallel pathways that are kept relatively separate in the striate and extrastriate cortex.

The visual cortex has an orderly functional architecture. Neurons within a cortical column have similar receptive field attributes, such as orientation selectivity, ocular dominance, and receptive field location. Each of these receptive field attributes varies smoothly over the cortical surface. The organization of ocular dominance columns and orientation columns across the cortical surface can be visualized with optical imaging.

Finally, in the macaque monkey, there are more than 30 extrastriate visual cortical regions, each of which performs different functions. These extrastriate regions can be divided into two pathways: the ventral stream, devoted to form recognition, and the dorsal stream, devoted to action or to spatial tasks. Areas in each stream form a functional and anatomical hierarchy. Neurons in successive visual areas respond to increasingly high-order or abstract features of the visual world.

References

Albright, T. D. (1993). Cortical processing of visual motion. In "Visual Motion and Its Role in the Stabilization of Gaze" (F. A. Miles and J. Wallman, eds.), pp. 177–201. Elsevier Science, Amsterdam.

Allman, J. M. and Kaas, J. H. (1971). A representation of the visual field in the caudal third of the middle temporal gyrus of the owl monkey (Aotus trivirgatus). Brain Res. 31, 85–105.

Barlow, H. B., Hill, R. M., and Levick, W. R. (1964). Retinal ganglion cells responding selectively to direction and speed of image motion in the rabbit. J. Physiol. 173, 377–407.

Berson, D. M. (2003). Strange vision: ganglion cells as circadian photoreceptors. Trends Neurosci. 26, 314–320.

Blasdel, G. G. and Salama, G. (1986). Voltage-sensitive dyes reveal a modular organization in monkey striate cortex. Nature 321, 579–585.

Dacey, D. M. (2000). Parallel pathways for spectral coding in primate retina. Annu. Rev. Neurosci. 23, 743–775.

Dowling, J. E. (1997). Retina. In "Encyclopedia of Human Biology," 2nd ed., Vol. 7, pp. 571–587. Academic Press, New York.

Dubner, R. and Zeki, S. M. (1971). Response properties and receptive fields of cells in an anatomically defined region of the superior temporal sulcus in the monkey. Brain Res. 35, 528–532.

Enroth-Cugell, C. and Robson, J. G. (1966). The contrast sensitivity of retinal ganglion cells of the cat. J. Physiol. 187, 517–552.

Felleman, D. J. and van Essen, D. C. (1991). Distributed hierarchical processing in the primate cerebral cortex. Cerebral Cortex 1, 1–47.

Ferster, D., Chung, S., and Wheat, H. (1996). Orientation selectivity of thalamic input to simple cells of cat visual cortex. Nature 380, 249–252.

Gilbert, C. D. and Wiesel, T. N. (1985). Intrinsic connectivity and receptive field properties in visual cortex. Vision Res. 25, 365–374.

Grinvald, A., Lieke, E., Frostig, R. D., Gilbert, C. D., and Wiesel, T. N. (1986). Functional architecture of cortex revealed by optical imaging of intrinsic signals. Nature 324, 361–364.

Helmholtz, H. (2000). "Handbook of Physiological Optics" (J. P. C. Southall, ed.), Vol. 2, p. 143. Thoemmes Press, Bristol, UK.

Hubel, D. H. and Wiesel, T. N. (1962). Receptive fields, binocular interaction and functional architecture in the cat's visual cortex. J. Physiol. 160, 106–154.

Kaplan, E. and Shapley, R. M. (1986). The primate retina contains two types of ganglion cells, with high and low contrast sensitivity. Proc. Natl. Acad. Sci. USA 83, 2755–2757.

Kuffler, S. W. (1953). Discharge patterns and functional organization of the mammalian retina. J. Neurophysiol. 16, 37–68.

Livingstone, M. S. and Hubel, D. H. (1984). Anatomy and physiology of a color system in the primate visual cortex. J. Neurosci. 4, 309–356.

McLean, J., Raab, S., and Palmer, L. A. (1994). Contribution of linear mechanisms to the specification of local motion by simple cells in areas 17 and 18 of the cat. Visual Neurosci. 11, 271–294.

Merigan, W. H. and Maunsell, J. H. R. (1994). How parallel are the primate visual pathways? Annu. Rev. Neurosci. 16, 369–402.

Movshon, J. A., Adelson, E. H., Gizzi, M., and Newsome, W. T. (1985). The analysis of moving visual patterns. In "Pattern Recognition Mechanisms" (C. Chagas, R. Gattass, and C. G. Gross, eds.), pp. 117–151. Vatican Press, Rome.

Nathans, J., Thomas, D., and Hogness, D. S. (1986). Molecular genetics of human color vision: The genes encoding blue, green, and red pigments. Science 232, 193–202.

Ratliff, F. (1974). "Studies on Excitation and Inhibition in the Retina. A Collection of Papers from the Laboratories of H. Keffer Hartline." Rockefeller Univ. Press, New York.

Reid, R. C. and Alonso, J. M. (1995). Specificity of monosynaptic connections from thalamus to visual cortex. Nature 378, 281–284.

Sincich, L. C. and Horton, J. C. (2005). The circuitry of V1 and V2: Integration of color, form, and motion. Annu. Rev. Neurosci. 28, 303–326.

Ts'o, D. Y., Frostig, R. D., Lieke, E. E., and Grinvald, A. (1990). Functional organization of primate visual cortex revealed by high resolution optical imaging. Science 249, 417–420.

Ungerleider, L. G. and Mishkin, M. (1982). Two cortical visual systems. *In* "Analysis of Visual Behavior" (D. J. Ingle, M. A. Goodale, and R. J. W. Mansfield, eds.), pp. 549–586. MIT Press, Cambridge, MA.

Wiesel, T. N. and Hubel, D. H. (1966). Spatial and chromatic interactions in the lateral geniculate body of the rhesus monkey. *J. Neurophysiol.* **29**, 1115–1156.

Wu, S. M. (1994). Synaptic transmission in the outer retina. *Annu. Rev. Physiol.* **56**, 141–168.

Suggested Readings

Boynton, R. M. and Kaiser, P. K. (1996). "Human Color Vision," 2nd ed. Optical Society of America, Washington, DC.

Chalupa, L. M. and Warner, J. S. (2004). "The Visual Neurosciences." MIT Press, Cambridge, Massachusetts.

Hubel, D. H. (1988). "Eye, Brain, and Vision." Freeman, New York.

Hubel, D. H. and Wiesel, T. N. (2005). "Brain and visual perception." Oxford University Press, New York.

Marr, D. (1982). "Vision: A Computational Investigation into the Human Representation and Processing of Visual Information." Freeman, New York.

Rodieck, R. W. (1998). "The First Steps in Seeing." Sinauer Associates, Sunderland, Massachusetts.

Wandell, B. A. (1993). "Foundations of Vision." Sinauer Associates, Sunderland, MA.

R. Clay Reid and W. Martin Usrey

SECTION V

MOTOR SYSTEMS

28

Fundamentals of Motor Systems

"All mankind can do is to move things . . . whether whispering a syllable or felling a forest."—Sherrington

This brief quote is a reminder of the basic fact that all interactions with the surrounding world are through the actions of the motor system. When a human baby is born it is a sweet but very immature survival machine, with a limited behavioral repertoire. It is able to breathe and has searching and sucking reflexes so that it can be fed from the mother's breast. It can swallow, vomit and process food, and cry to call for attention if something is wrong. A baby also has a variety of protective reflexes that mediate coughing, sneezing, and touch avoidance. These different patterns of motor behavior are thus available at birth and are due to innate motor programs (Fig. 28.1).

During roughly the first 15 years of life the motor system continues to develop through maturation of neuronal circuitry and by learning through different motor activities. Playing represents an important element both in children and in young mammals such as kittens and pups. During the first year of life the human infant matures progressively. It can balance its head at 2–3 months, is able to sit at around 6–7 months, and stand with support at approximately 9–12 months. The coordination of different types of posture, such as standing, is a complex motor task to master, with hundreds of different muscles taking part in a coordinated fashion. Sensory information contributes importantly, in particular from the vestibular apparatus, eyes, muscle, and skin receptors located at the soles of the feet. This development represents to a large degree a maturation process following a given sequence, but with individual variability among different children. When the postural system has evolved to a sufficient

degree, a child is able to start walking, which requires that the body posture be maintained while the points of support are changing by the alternating movements of the two legs. In common language the child is said to "learn" to walk, but in reality a progressive maturation of the nervous system is taking place. Identical twins start to walk essentially at the same time, even if one has been subjected to training and the other has not. At this point the motor pattern is still very immature. Proper walking coordination followed by running appears later, and the basic motor pattern actually continues to develop until puberty. The fine details of the motor pattern are adapted to the surrounding world, but also to modification by will. The basic motor coordination underlying reaching and the fine control of hands and fingers undergo a similar characteristic maturation process over many years.

The newborn human infant is comparatively immature, but other mammals, such as horses and deer, represent another extreme. The gnu, an African buffalo-like antelope, needs to run away in order to survive attacks of predators such as lions. The young calf of the gnu can stand and run directly after birth and has been reported to be able to gallop ahead 10 minutes after delivery, tracking the running mother (Fig. 28.2). Clearly the neural networks underlying locomotion, equilibrium control, and steering must be sufficiently mature and available at birth, needing minimal calibration. This is astounding. A similar range of maturity is present in birds. To get out of the egg, a chick makes coordinated hatching movements to open up the eggshell to subsequently lift off the top of the egg and to stand up to walk away on two legs following the mother hen and start picking at food grains. Most birds are more immature when hatching, but after a

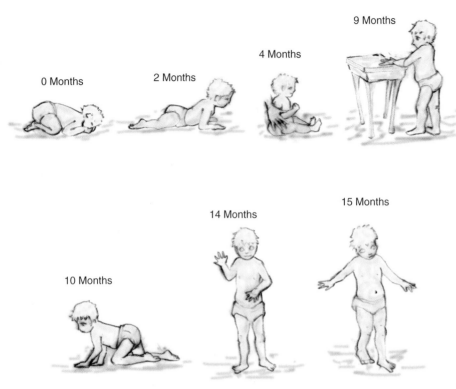

FIGURE 28.1 Motor development of the infant and young child. The pattern of maturation of the motor system follows a characteristic evolution. Two months after birth a child can lift its head, at 4 months it sits with support, and subsequently it is able to stand with support; later it crawls, stands without support, and finally walks independently. The approximate time at which a child is able to perform these different motor tasks is indicated above each figure. The variability in the maturation process is substantial. Modified from M. M. Shirley.

FIGURE 28.2 Some animals are comparatively mature when they are born. Ten minutes after the calf of the gnu, a buffalo-like antelope, is born it is able to track its mother in a gallop. This means that the postural and locomotor systems are sufficiently mature to allow the young calf to generate these complex patterns of motor coordination at birth. There is thus little time to calibrate the motor system after birth and obviously no time for learning. Courtesy of Erik Tallmark.

few weeks they leave the nest flying rather successfully, for the first time in their life, and thus without any previous experience.

In addition to the basic motor skills such as standing, walking, and chewing, humans also develop skilled motor coordination, allowing delicate hand and finger movements to be used in handwriting or playing an instrument or utilizing the air flow and shape of the oral cavity to produce sound as in speech or singing. The neural substrates allowing learning and execution of these complex motor sequences are expressed genetically and characteristic of our species. What is learned, however, such as which language one speaks or the type of letters one writes, is obviously a function of the cultural environment.

BASIC COMPONENTS OF THE MOTOR SYSTEM

Motoneurons and Motor Units

The motoneurons that control different muscles are located in different motor nuclei along the spinal cord and in the brain stem. Each motoneuron sends its axon to one muscle and innervates a limited number of muscle fibers. A motoneuron with its muscle fibers is referred to as a motor unit. The muscle fibers of each motor unit have similar contractile properties and metabolic profile. The muscle fibers in different muscles are composed of three main types specialized for different demands, such as a continuous effort as in long-distance running (slow motor units) or fast explosive movements, such as lifting a heavy object (fast motor units (two subtypes)).

Motoneurons are activated by interneurons of different motor programs or reflex centers and by descending tracts from the forebrain and the brain stem. Thus motoneurons supplying different muscles can be activated with great precision by these different sources that together determine the degree of activation, as well as the exact timing of the motoneurons of a given muscle.

Sensory Receptors Are Important in Movement Control

Signals from sensory receptors are used by the motor system in a number of ways.

1. Sensory signals can trigger behaviorally meaningful motor acts such as withdrawal, coughing, or swallowing reflexes.
2. Sensory receptors may contribute to the control of an ongoing motor pattern and influence the switch from one phase of movement to another. For instance, in the case of breathing, sensory signals from lung volume receptors control when inspiration is terminated. In walking, sensory signals related to hip position and load on the limb help regulate the duration of the support and swing phases of the step cycle and correct for perturbations.
3. Sensory signals may also be more specific and influence only the level of activity of one muscle or a group of close synergists. Muscle receptors play a particularly important role in this context. They are of two types: (1) Golgi tendon organs that sense the degree of contraction in a muscle (located in series with the muscle fibers at their insertion into the tendon) and (2) muscle spindles that signal the length of a muscle (located in parallel to the muscle fibers), as well as the speed of changes in length. Moreover, the sensitivity of muscle spindles can be regulated actively through a separate set of small motoneurons, referred to as γ-motoneurons (in contrast to the larger α-motoneurons that control muscle contraction). These two muscle receptors have fast conducting afferent axons that provide rapid feedback to the spinal cord and take part in the autoregulation of the motor output to a given muscle. Fast muscle spindle afferents provide direct monosynaptic excitation to the α-motoneurons that control the muscle in which the muscle spindle is located. Thus, lengthening a muscle will lead to excitation of its motoneurons, counteracting the lengthening. This stretch reflex thus involves negative feedback. This reflex arc also provides the basis of the muscle contraction elicited by a brief tendon tap, which is used as a clinical test to investigate whether responsiveness of the motoneuronal pool is normal (the tendon reflex) (Fig. 28.3).
4. Sensory signals from a number of different receptor systems help detect and counteract any disturbance of body posture during standing or maintenance of another body position. Skin and muscle receptors and receptors signaling joint position, as well as vestibular receptors and vision, contribute to different aspects of the dynamic and static control of body position.
5. When an object is held by the fingers, skin receptors at the contact points at the fingertips play an important role. If the object tends to slip, the skin receptors become activated and signal rapidly to the nervous system that there is a need for extra muscle force.
6. A great variety of sensory signals provide information about the position of different parts

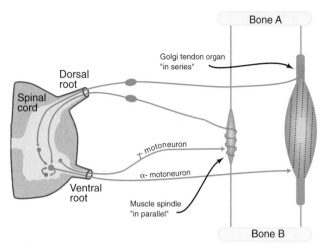

FIGURE 28.3 Feedback from muscle receptors to motoneurons: the stretch reflex. In a muscle connected between two bones, the Golgi tendon organ is located at the transition between muscle and tendon. It senses any active tension produced by the muscle fibers, being in series between muscle and tendon. The muscle spindle is located in parallel with the muscle fibers. It signals the muscle length and dynamic changes in muscle length. Both the muscle spindle and the Golgi tendon organ have fast conducting afferent nerve fibers in the range of ~100 m/s. The muscle spindle activates α-motoneurons directly, which causes the muscle fibers to contract. The sensitivity of the muscle spindle can be actively regulated by γ-motoneurons, which are more slowly conducting. The muscle spindle provides negative feedback. If the muscle with its muscle spindle is lengthened, the afferent activity from the muscle spindle increases, exciting the α-motoneurons and leading to an increased muscle contraction, which in turn counteracts the lengthening. This is called the stretch reflex. The Golgi tendon organ provides force feedback. The more the muscle contracts, the more the Golgi tendon organ and its afferent are activated. In the diagram the intercalated interneuron between afferent nerve fiber and motoneuron is inhibitory. Thus increased muscle force leads to inhibition of the α-motoneuron, which results in a decrease of muscle force. The efficacy of length and force feedback can be regulated independently in the spinal cord and via γ-motoneurons. Thus their respective contributions can vary considerably between different patterns of motor behavior.

of the body in relation to each other and to the external world. Such information is critical when initiating a voluntary movement. For example, to move a hand toward a given object, it is important to know the initial position of the arm and the hand in space. Depending on whether the hand is located to the left or the right of the object, different types of motor commands must be given to bring the hand to the target. Sensory information about the initial position of the hand (or any other body part) in relation to the rest of the body, as well as the target for the movement,

is thus important for eliciting an effective neural command signal.

To summarize, the sensory contribution to motor control is very important in many different contexts. If sensory control is incapacitated, motor performance will in most cases be degraded. Without sensory information in different forms, movements can usually still be executed, but as a rule with much less perfection.

Feedback and Feed Forward Control

If a perturbation occurs when an object is held in the hand, or when standing in a moving bus, it will be detected by different sensory receptors. This sensory signal will be fed back to the nervous system and be used to counteract the perturbation rapidly. This represents feedback control that is a correction of the actual perturbation after it has occurred. A limiting factor for the efficacy of feedback control in biological systems is the delay involved. A sensory afferent signal must first be elicited in the receptors concerned. It then has to be conducted to the central nervous system and be processed there to determine the proper response. The correction signal must subsequently be sent back to the appropriate muscle(s) and make the muscle fibers build up the contractile force required. In large animals, including humans, the delays involved can be substantial, and during fast movement sequences there may be little or no time for feedback corrections. In less demanding, more static situations, such as standing, sensory feedback is of critical importance.

In many cases a perturbation is anticipated before it is initiated, and correction begins before it actually has occurred. This type of control often is called feed forward control, in contrast to feedback. Such anticipatory control mechanisms are often automatic and involuntary, and are part of an inherent control strategy. For example, when standing on two legs and planning to lift up one leg to stand on the other, the body position begins to shift over to the supporting leg before the other leg is lifted. The projection of the center of gravity will fall between the two feet initially and will have to shift over to a projection through the supporting limb. In most instances, movement commands are designed so that the corrections of body posture required for a stable movement occur before the particular movement has started, as when lifting a heavy object. If the converse situation occurred, and body position was corrected only after a perturbation had taken place, movements would become much less precise than with feed forward control.

MOTOR PROGRAMS COORDINATE BASIC MOTOR PATTERNS

Both vertebrates and invertebrates have preformed microcircuits—neuronal networks that contain the necessary information to coordinate a specific motor pattern such as swallowing, walking, or breathing. When a given neuronal network is activated, the particular behavior it controls will be expressed. A typical network consists of a group of interneurons that activate a specific group of motoneurons in a certain sequence and inhibit other motoneurons that may counteract the intended movement. Such a group of interneurons often is referred to as a central pattern generator (CPG) or motor program. This CPG can be activated by will, as at the start of walking, or be triggered by sensory stimuli, as in a protective reflex or swallowing. In most types of motor behavior, sensory feedback may also form an integral part of the motor control circuit and may determine the duration of the motor activity such as the inspiratory phase of breathing (Fig. 28.4).

There are several different types of motor coordination, as described next.

Motor Patterns Triggered as a Reflex Response

Protective skin reflexes lead to withdrawal of the stimulated part of the body from a stimulus that may cause pain or tissue damage. The central program consists in this case of a group of interneurons that activate motoneurons and muscles, giving rise to an appropriate withdrawal movement. Motoneurons to antagonistic muscles are inhibited (reciprocal inhibition). The withdrawal reflexes of the limbs often are referred to as flexion reflexes, but in reality they represent a family of specific reflexes, each of which is designed to remove the specific skin region involved from the painful stimulus.

Coughing and sneezing reflexes remove an irritant from the nasal or tracheal mucosa by inducing a brief pulse of air flow at very high velocity (at storm or hurricane speeds). This is caused by an almost synchronized activation of abdominal and respiratory muscles, coordinated by a CPG, the activity of which is triggered by afferents activated by the irritant.

Swallowing reflexes are activated when food is brought in contact with mucosal receptors near the pharynx. This leads to a coordinated motor act with sequential activation of different muscles that propel the food bolus through the pharynx down the esophagus to the stomach. In this case the CPG is able to

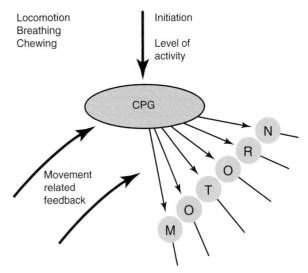

FIGURE 28.4 Motor coordination through interneuronal networks: central pattern generators. The brain stem and spinal cord contain a number of networks that are designed to control different basic patterns of the motor repertoire, such as breathing, walking, chewing, or swallowing. These networks often are referred to as central pattern generator networks (CPGs). CPGs contain the necessary information to activate different motoneurons and muscles in the appropriate sequence. Some CPGs are active under resting conditions, such as that for breathing, but most are actively turned on from the brain stem or the frontal lobes. For instance, the CPG coordinating locomotor movements is turned on from specific areas in the brain stem, referred to as locomotor centers. These descending control signals not only turn on the locomotor CPG, but also determine the level of activity in the CPG and whether slow or fast locomotor activity will occur. In order to have the motor pattern well adjusted to external conditions and different perturbations, sensory feedback acts on the CPG and can modify the duration of different phases of the activity cycle, providing feedback onto motoneurons. Although vertebrates as well as invertebrates have CPGs for a great variety of motor functions, the intrinsic operation of these networks of interneurons constituting the CPG is unclear in most cases. In invertebrates, a few CPG networks have been studied in great detail, such as the stomatogastric system of the lobster and networks coordinating the activity of the heart and locomotion in the leech. In vertebrates detailed information concerning locomotor networks is only available in lower forms, such as the frog embryo and the lamprey.

coordinate the motor act over several seconds, in contrast to the previous motor acts in which the activity occurs almost simultaneously.

Rhythmic Movements

Walking movements (or locomotor behaviors in general) are produced by CPGs in the spinal cord (or ganglia of invertebrates). These spinal CPGs are turned on by descending control signals from particular areas in the brain stem (locomotor areas), which also determine the level of locomotor activity (e.g., slow walking versus trot or gallop). The CPG then sequentially activates motoneurons/muscles that support the limb during the support phase and then move it forward during the swing phase. Sensory stimuli from the moving limb also contribute importantly to regulate the duration of the support phase and the degree of activation of different muscles that are sequentially activated in each step cycle. The relative importance of the sensory component varies between species and with the speed of locomotion. During very fast movements, there is actually no time for sensory feedback to act, and adjustments need instead to be performed in a predictive mode.

Chewing movements are controlled by brain stem circuits that in principle are designed in a similar manner to those of locomotion, and generate alternating activity between jaw-opener and -closer muscles.

Breathing movements are continuously active from the instant of birth to the time of death, except for short voluntary interventions when speaking, singing, or choosing not to breathe for some other reason, like diving. The level of respiration (depth and frequency) is driven by metabolic demands (e.g., pCO_2) detected by chemosensors in the brain stem where the respiratory CPG is also located. Sensory stimuli from lung volume receptors contribute by setting the level at which the inspiration is terminated and expiration takes over.

Eye Movements

Fast saccadic eye movements are used when looking rapidly from one object to another and are represented with a much more complex organization in the superior colliculus (mesencephalon) than the motor patterns discussed earlier. Saccadic eye movements with different directions and amplitudes can be generated by the stimulation of different microregions in the superior colliculus, according to a well-defined topographical map. Each microregion is responsible for activating a subset of brain stem interneurons, which in turn activates the appropriate combination of eye motoneurons in order to move the two eyes in a coordinated fashion from one position to another. The purpose is to bring the visual object of interest into the foveal region so that it can be scrutinized in the greatest detail. To recruit a saccadic eye movement to a specific site, cortical or subcortical areas have to access the appropriate microregions within the superior colliculus to trigger the specific eye movement desired.

Other types of eye movement control are used when tracking an object that is moving, as when watching a game of tennis. These tracking movements are represented in the cerebral cortex and the brain stem.

Posture

Animals, including humans, have the ability to position the body in a variety of postures. The most stable position is lying horizontally in bed, which requires little activity from the nervous system. The situation is very different when standing and even more so when supporting oneself on one leg. This requires a very structured command to handle the activation of different muscles in the legs, trunk, shoulders, and neck. The nervous system issues this complex motor command to achieve a certain position. The ability to maintain a stable position standing on one leg is then critically dependent on feedback from a variety of sensory receptors that detect falling to one side or the other. These receptors activate reflex circuits that try to counteract the impending loss of balance. Of great importance are the vestibular receptors located in the head, which sense head movements in the different directions. Vestibular reflexes act on the neck muscles to correct the position of the head, and vestibulospinal reflexes act on the trunk and the extensor muscles of the limbs. In addition, vestibular effects also are mediated indirectly by the reticulospinal system. Muscle receptors in neck and limb muscles are of equal importance. They also play a critical role in detecting and compensating for postural perturbations, in particular around the ankle, where skin receptors on the bottoms of the feet are activated. Finally, visual feedback may contribute as when one is standing on a moving surface such as the deck of a boat.

The body can be arranged in a number of ways, sitting in a relaxed way on a sofa, stiffly on an uncomfortable chair, or standing with the body and the legs held in a range of postures. In one sense the postural system is essentially automatic in that there is no conscious thought about how to achieve a given position. However, one can decide at will what posture to assume. In that sense it is a voluntary motor act organized to a large extent on the brain stem level, but is also affected by areas in the frontal lobe located just in front of the motor cortex.

ROLES OF DIFFERENT PARTS OF THE NERVOUS SYSTEM IN THE CONTROL OF MOVEMENT

The different basic motor patterns are present in different forms in most vertebrates and invertebrates. In the former, the underlying neural networks are located either in the brain stem or in the spinal cord, and in the latter, in the chain of ganglia. In vertebrates, from fish to mammals the basic organization of the nervous system is similar with regard to the spinal cord, brain stem (medulla oblongata and mesencephalon), and diencephalon. Each species and larger group of vertebrates may have specializations related to particular types of functions, and complexity increases during evolution. The cerebral cortex, with its distinct lamination, is present in mammals and most highly developed in primates, including humans (Fig. 28.5).

Basic Motor Programs Are Located at the Brainstem–Spinal Cord Level

The brainstem–spinal cord is, to a large extent, responsible for the coordination of different basic motor patterns. The spinal cord contains the motor programs (CPGs) for protective reflexes and locomotion, whereas those for swallowing, chewing, breathing, and fast saccadic eye movements are located in the brain stem (mesencephalon and medulla oblongata). In most cases, however, both the brain stem and the spinal cord are involved to some degree. In mammals as well as lower vertebrates, the brain stem–spinal cord, isolated from the forebrain (di- and telencephalon), is able to produce breathing movements and swallowing, as well as walking and standing. These brain stem–spinal cord animals (referred to as decerebrate models) can thus be made to walk, trot, and

gallop with respiration adapted to the intensity of the movements and to swallow when food is put in their mouths. However, they perform these maneuvers in a stereotyped fashion, such as a robot or a reflex machine. The movements are thus not goal directed nor adapted to the surrounding environment, but nevertheless are coordinated in an appropriate way.

Diencephalon and Subcortical Areas of Telencephalon Are Involved in Goal-Directed Behavior—Hypothalamus and Basal Ganglia

Mammals that are devoid of the cerebral cortex but have the remaining parts of the nervous system intact have been used to investigate the importance of these areas. Such animals display a more advanced behavioral repertoire when compared to the brain stem animals described earlier. At first glance, these decorticate animals look surprisingly normal. They move around spontaneously in a big room and can avoid obstacles to some extent. They eat and drink spontaneously and can learn where to obtain food and search for food when supposedly hungry. They may also display emotions such as rage and attack other animals; however, they appear unable to interact in a normal way with other individuals of the same species. Even without the cerebral cortex, a surprisingly large part of the normal motor repertoire can be performed,

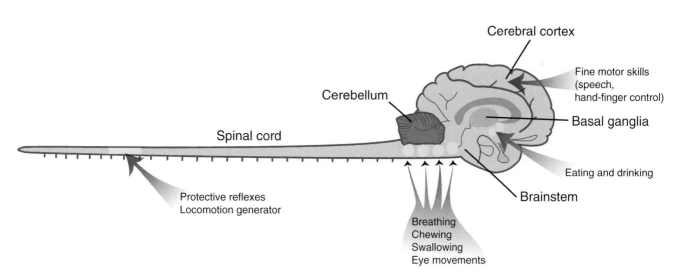

FIGURE 28.5 Location of different networks (CPGs) that coordinate different motor patterns in vertebrates. The spinal cord contains CPGs for locomotion and protective reflexes, whereas the brain stem contains CPGs for breathing, chewing, swallowing, and saccadic eye movements. The hypothalamus in the forebrain contains centers that regulate eating and drinking. These areas can coordinate the sequence of activation of different CPGs. For instance, if the fluid intake area is activated, the animal starts looking around for water, walks toward the water, positions itself to be able to drink, and finally starts drinking. The animal will continue to drink as long as the stimulation of the hypothalamic area is sustained. This is an example of recruitment of different CPGs in a behaviorally relevant order. The cerebral cortex is important, particularly for fine motor coordination involving hands and fingers and for speech.

including some aspects of goal-directed behavior. Thus the cerebral cortex is not required to achieve this level of complex motor behavior (however, see later discussion).

The diencephalon and subcortical areas of the telencephalon of the forebrain contain two major structures that are important in this context: the hypothalamus and the basal ganglia. The hypothalamus is composed of a number of nuclei that control different autonomic functions, including temperature regulation and intake of fluid and food. The latter nuclei become activated when the osmolality is increased (fluid is needed) or the glucose levels become low (food is required). Stimulation of the paraventricular nucleus involved in the control of fluid intake results in a sequence of motor acts involving a number of different motor programs. Continuous activation of this nucleus by electrical stimulation or local ejection of a hyper-osmolalic physiological solution leads to an alerting reaction. The animal first starts looking for water (perhaps experiencing a feeling of thirst). It then starts walking toward the water, positions itself at the water basin, bends forwards, and starts drinking. The animal will continue drinking as long as the nucleus is stimulated. These hypothalamic structures thus are able to recruit a sequence of motor acts that appear in a logical order. Other parts of this region can elicit rage and attack behavior, as was first described by the Nobel laureate Walter Hess.

Basal ganglia constitute a second major structure. They are present in all vertebrates and are of critical importance for the normal initiation of motor behavior. They are subdivided into an input region (the dorsal and ventral striatum) activated by the cerebral cortex, thalamus, and the brain stem and an output region (the pallidum). The output neurons are inhibitory and have a very high level of activity at rest. They inhibit a number of different motor centers in the di- and mesencephalon and also influence the motor areas of the cerebral cortex via the thalamus. The input stage of the striatum determines the level of activity in the different pallidal output neurons. When a motor pattern, such as a saccadic eye movement, is going to be initiated, the pallidal output neurons that are involved in eye motor control become inhibited by the striatum. This means that the tonic inhibition produced by these neurons at rest is removed, and the saccadic motor centers in mesencephalon (superior colliculus) are relieved from tonic inhibition and become free to operate and induce an eye saccade to a new visual target.

In contrast to the tonically active output neurons of the basal ganglia (pallidum), the striatal neurons are silent at rest and have membrane properties that make it difficult to activate them from cortex or thalamus unless the dopamine system is active. The level of activity in the striatum is strongly modulated by dopaminergic neurons in the substantia nigra and associated nuclei. If the level of dopamine activity is reduced, as in Parkinson's disease, it becomes very difficult to activate the striatal neurons and thereby to initiate and carry out most types of movements. This is a severe handicap characterized by paucity of movement (hypokinesia). The converse condition with enhanced levels of dopamine in the basal ganglia (which can occur as a side effect of medication) results instead in a richness of movements, and even in unintended movements (hyperkinesias). These movements may be well coordinated, but are without a purpose, and occur without the conscious involvement of the patient. Dopamine neurons are thus of critical importance for the operation of the striatum and they regulate the responsiveness of the striatal circuitry to input from the cerebral cortex and thalamus. The striatum in itself clearly has a key role for the operation of the entire motor system.

The Cerebral Cortex and Descending Motor Control

In the frontal lobe there are several major regions that are involved directly in the execution of different complex motor tasks, such as the skilled movements used to control hands and fingers when writing, drawing, or playing an instrument. Another skilled type of motor coordination is that underlying speech, which is served by Broca's area. These different regions are organized in a somatotopic fashion. In the largest area, referred to as the primary motor cortex or M1 in the precentral gyrus, the legs and feet are represented most medially and the trunk, arm, neck, and head are represented progressively more laterally. Areas taken up by the hands and the oral cavity are very large in humans, and are much larger than that for the trunk. This is explained by the fact that speech and hand motor control require a greater precision and thus a larger cortical processing area than the trunk. The latter is important for postural control but is less involved in the type of skilled movement controlled by the motor cortex (Fig. 28.6).

The large pyramidal cells in the motor cortex send their axons to the contralateral side of the spinal cord and are able to activate their target motoneurons directly, but a number of interneurons in the brain stem or spinal cord also are influenced. These long projection neurons are called corticospinal neurons. Corticospinal neurons of the arm region in M1 thus project to arm motoneurons in the spinal cord, as well

Gyrus precentralis (M)

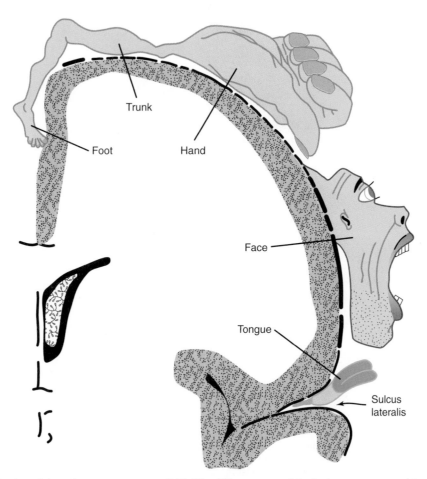

FIGURE 28.6 Organization of the primary motor cortex (M1). The different parts of the body are represented in a somatotopic fashion in M1, with the legs represented most medially, arms and hand more laterally, and the oral cavity and face even more so. Note that the two areas that are represented with disproportionally large regions are the hand with fingers and the oral cavity. They are represented in this way due to the fine control required in speech and fine manipulation of objects with the fingers. Adapted from Penfield and Rasmussen (1950).

as interneurons involved in the control of arm, hand, and fingers. In addition to M1, other areas in the frontal lobe, such as the supplementary motor area and prefrontal areas, are involved in other aspects of motor coordination (Fig. 28.7).

The cortical control of movement is executed in part by the direct corticospinal neurons but also by cortical fibers that project to brain stem nuclei, which in turn contain neurons that project to spinal motor centers such as the fast rubrospinal, vestibulospinal, and reticulospinal pathways. The brain stem contains a large number of descending pathways that can initiate movements and correct motor performance, or provide more subtle modulation of the spinal circuitry. The former act rapidly in the millisecond time frame that is required in motor control. Examples of the latter are

the slow-conducting and slow-acting noradrenergic and serotonergic pathways. These pathways set the responsiveness of different types of neurons, synapses, and spinal circuits. In addition, there are direct projections from the cerebral cortex to the input area of the basal ganglia, the striatum. The cortical control of motor coordination thus is achieved through both direct action on spinal and brain stem motor centers but also to a significant degree by parallel action on a variety of brain stem nuclei. The execution of movements that are initiated from the cortex is, to a large degree, a collaborative effort of many parts of the nervous system. The reticulospinal and vestibulospinal pathways mediate the control of posture, and the former take part in the initiation of locomotion, via the CPG located in the spinal cord (Fig. 28.8).

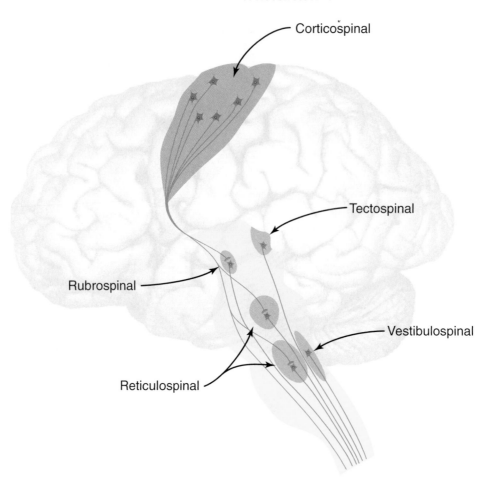

FIGURE 28.7 Descending pathways projecting from the brain that mediate motor actions to the spinal cord. From the cerebral cortex, including M1, a number of neurons project directly to the spinal cord to both motoneurons and interneurons. In addition, they also project to different motor centers in the brain stem, which in turn send direct projections to the spinal cord. One of these centers is the red (rubrospinal) nucleus, which influences the spinal cord via the rubrospinal pathway. Corticospinal and rubrospinal pathways act on the contralateral side of the spinal cord and have somewhat overlapping functions. They are sometimes referred to as the lateral system, as most of the fibers descend in the lateral funiculus of the spinal cord. Vestibulospinal and reticulospinal pathways are important both for regulating posture and for correcting perturbations, and mediate commands initiating locomotion. The different reticulo- and vestibulospinal pathways are sometimes lumped together as the medial system, as most of the axons project in the medial funiculus of the spinal cord.

Judging from experiments on mammals and primates, and patients that have suffered focal lesions of the frontal lobe, the cortical control of movement is of particular importance for dexterous and flexible motor coordination, such as the fine manipulatory skills of fingers and hands and also for speech. It is very difficult for a casual observer to see the difference between a normal monkey moving around in a natural habitat and a monkey that is lacking the corticospinal direct projections to the spinal cord but has all other cortical circuits intact. They are able to move around, climb, and locomote as normal monkeys. It is only when special tests are performed that one can see that delicate independent movements of the individual fingers

are incapacitated. This is a type of motor coordination that is needed in monkeys and other primates for skilled manipulations of the environment, such as picking fine food objects from small holes.

The most flexible types of motor coordination, such as the skilled movements of the fingers and hands or in speech, involve the frontal lobes. These types of movements often are referred to as voluntary because they are performed at will. This terminology is commonly used, although inappropriate in that more basic types of motor coordination, such as the ones used when walking, chewing, or positioning ourselves in a relaxed posture, are also controlled by will, although motor coordination is handled to a larger degree by

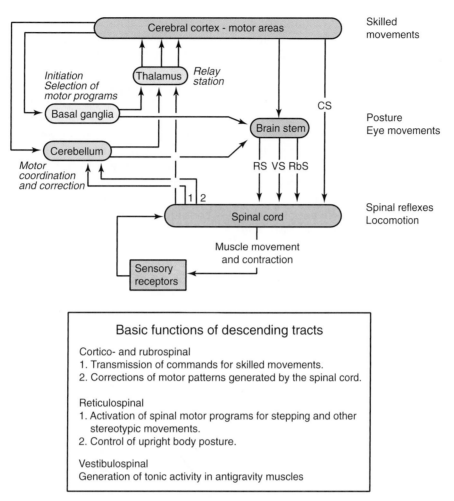

FIGURE 28.8 Summarizing scheme of the interaction between different motor centers. The different major compartments of the motor system and their main pathways for interaction are indicated. The basic functions of the different compartments and descending tracks are summarized on and below the scheme. CS, corticospinal; RbS, rubrospinal; VS, vestibulospinal; RS, reticulospinal.

motor programs located at the brain stem–spinal cord level. Thus it appears important not to draw a clear-cut dividing line between the different types of movements, at least with respect to the voluntary aspect of their control.

It is also useful to realize that, to some degree, simple reflexes can also be modulated by will. For instance, the simple skin avoidance reflex (flexion reflex) that causes rapid withdrawal of a finger after touching something painful is subject to such control. Knowledge that the object will be sufficiently hot to be somewhat painful enables this modulation if it is important. If, however, the object is touched without knowing that it is hot, the hand will be withdrawn with the shortest possible latency. Thus, in the entire motor apparatus from the simplest reflex to a skilled movement, there is the possibility for modulation and flexibility. The human nervous system allows a remark-

able combination of movements, adapted to different situations. This flexibility is perhaps greater for primates and humans than for any other species. However, other species may perform much better in particular types of specialized movements. A cheetah runs faster, a hawk catches prey at higher speeds, and a fly walks upside down on the ceiling.

Visuomotor Coordination—Reaction Time

Visuomotor coordination is very demanding and requires complex processing. When visualizing an approaching object like a ball, a rapid calculation provides information to bring the arms to the appropriate spot to grasp the ball at just the right time. This is an amazing achievement considering the necessity of predicting where and when the ball will arrive, the details about joint angles from hand to head that need to be considered to elicit the appropriate commands to the different muscles involved, and the time constraints. There are extensive projections to the frontal lobes from the areas in the parietal lobe processing visual information about movement.

It is clear that in order to arrive at an accurate and precise motor command, a large amount of information of different types needs to be processed about the dynamic and static conditions of the different parts of the body in relation to each other and the surrounding world. The processing time under different conditions has been investigated experimentally by having a subject respond to a signal and perform a movement, such as pressing a button when a light goes on. The delay is around 0.1 to 0.2 s and is called the reaction time. It represents a minimal delay for a given test situation that cannot be shortened by training (Fig. 28.9).

The more complex the situation is, the longer the reaction time. If a choice is involved, such as responding by pressing different buttons to light stimuli of different colors, the time delays become much longer than in the simple reaction time task. The choice reaction task time increases in proportion to the level of complexity and the number of choices (Fig. 28.9). When young children are tested, their reaction times are much longer than in adults. The simple reaction time for a 6-year-old can be three times that of a 14-year-old, and the difference with choice reaction times may be even larger. This condition may have severe consequences in everyday life. In an unexpected traffic situation, a 6-year-old requires at least three times longer time to interpret what he or she sees, such as an approaching car. Whereas an adult may have sufficient time to respond, a child may have no chance.

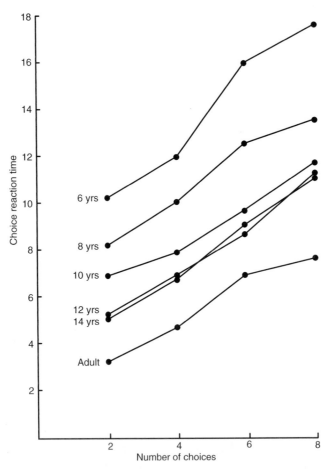

FIGURE 28.9 Reaction time tasks of different complexity and in different age groups. The choice reaction time is plotted versus the number of choices the subject is exposed to for six different age groups. A simple experimental situation with two choices takes an adult 0.3 s to initiate, but a 6-year-old needs around 1.0 s. A 10-year-old falls in between at around 0.7 s. An increase in the number of choices causes a marked increase in reaction time. From a practical standpoint, this provides important information when one considers the possibility for a child to cope with different demanding situations, such as moving around in a modern city. Data replotted from Conolly.

This is the basis for recommending that children below 11 years of age not bicycle in open traffic. It is noteworthy that this is a gradual maturation process and that training cannot shorten the reaction time. During aging the choice reaction time tends to increase again, a factor that can be important when driving a car.

Cerebellum

The cerebellum is a structure that is also involved in the coordination of movement. Although the cerebellum is smaller than the cerebral cortex in volume, it contains as many nerve cells as the latter. It is characterized by a very stereotyped neuronal organization with two types of inputs from mossy and from climbing fibers. These inputs not only carry information from the spinal cord about ongoing movements in all different parts of the body, but also information from the different motor centers about planned movements even before a movement has been executed. The cerebellum also interacts with practically all parts of the cerebral cortex. This means that it is updated continuously about what goes on in all parts of the body with regard to movement and also about the movements that are planned in the immediate future. Lesions of the cerebellum lead initially to great problems with postural stability and with the accuracy of different types of movements. It is noteworthy that movements in general can be carried out, but their quality is reduced drastically. Therefore, the cerebellum originally was thought to be involved exclusively in the coordination of movement. In addition to motor control, evidence has accumulated that the lateral parts of cerebellum may also be involved in different cognitive tasks.

The cerebellum is subdivided into a great number of different regions that process different kinds of information. These different regions respond via their output neurons, called Purkinje cells. The cerebellar cortex contains only inhibitory interneurons, and the two input systems, climbing fibers and mossy fibers (via granule cells), provide the excitatory components. All Purkinje cells are inhibitory and project to the cerebellar nuclei, which in turn are excitatory and project to different motor centers in the brain stem and to the cortex via the thalamus.

Each granule cell receives very specific input from only a few mossy fibers, which are able to make them fire. The granule cell axons form parallel fibers in the cerebellar cortex, and thousands of parallel fibers impinge on each Purkinje cell that has a very extensive dendritic tree. It is believed that only a small portion of these parallel synapses are functional in any given moment, but that they are available to become recruited in learning situations. In contrast, there is only one climbing fiber for each Purkinje cell. The climbing fibers can serve as "error detectors," at least under some conditions. Through the interaction of climbing fibers with the parallel input, the synaptic efficacy of the input of the latter synapse onto the Purkinje cell can be regulated (depressed). This change in efficacy can last over many hours and possibly much longer and is therefore referred to as long-term depression (LTD). It is generally thought that this LTD can contribute to motor learning (see later). With regard to basic movements such as posture, walking, and the much-studied eye-blink reflex, the cerebellum clearly is involved through each movement phase. It is likely that the fine-tuning and modification of movements that are required in different behavioral contexts involve the cerebellum. With regard to the eye-blink reflex, motor learning has been demonstrated in terms of associating an unconditioned stimulus with a conditioned one. It requires that the cerebellar cortex is intact, which strongly suggests that the cerebellar cortex contributes importantly to this type of motor learning. The cerebellum is clearly of importance for the long-term adaptation of different patterns of basic motor coordination.

There is still, however, much to learn about the actual processing that goes on in the cerebellum, a structure that regulates the quality of motor performance and appears to be involved in one form of motor learning. A variety of complex motor tasks, including speech and handwriting, can still be performed, but with less accuracy. Thus, the motor programs for these learned tasks cannot be stored exclusively in the cerebellum. When new types of movements are learned, such as riding a bicycle, playing an instrument, or typing at a keyboard, the new stored programs for sequences of motor acts must also involve other structures in the brain, most likely the basal ganglia and the cerebral cortex.

Motor Learning

All vertebrate and invertebrate species with a nervous system have a motor infrastructure that is determined evolutionarily. In the case of primates including humans, there are preformed circuits (e.g., CPGs) that allow performance of a basic movement repertoire (e.g., locomotion, posture, breathing, eye movements), as well as basic circuits that underlie reaching, hand and finger movements, and sound production as in speech. This constitutes the motor infrastructure that is available to a given individual, after maturation of the nervous system has occurred.

The motor system is in addition characterized by its ability to learn, in what is referred to as motor or procedural learning. The cortex and basal ganglia play an important role for learning of new motor sequences and skills, and cerebellum is indispensable for the quality of the different motor patterns and the ability to associate two stimuli as during conditioned reflexes. For instance, one can learn how to play an instrument like the flute. The particular finger settings that produce a given tone can be retained in memory, along with the sequence of tones that produce a certain melody. Thus particular muscle combinations that produce a sequence of well-timed motor patterns are stored. Similarly, when learning to pronounce different sounds (phonemes) and combine them in words, muscle activation patterns are stored. A simpler case occurs when learning to recombine basic movement patterns to be able to maintain equilibrium while bicycling. There are a large number of examples involving different types of motor learning and parts of the body.

It is characteristic of a given learned motor program that one can "call" upon it to perform a given motor act over and over again. Despite this, there is generally little awareness of the way in which the movement actually is performed in terms of the muscles involved and their timing. Information thus has been stored in the nervous system with regard to which parts of the infrastructure of the motor system should be used to produce a given learned motor pattern, as well as the exact timing between the different components. However, the details of the complex storage process are not yet fully understood.

CONCLUSION

After this overview of the motor control system, it should be apparent that biological evolution has succeeded in refining remarkable sensorimotor machinery capable of executing the motor tasks required for the full behavior of a given species. The degree of difficulty involved is apparent when watching the clumsiness and lack of flexibility in robots, characteristic of even the most advanced models that have been developed. Details of the different neural subsystems involved in motor control are dealt with in the following chapters.

References

Deliagina, T. G., Orlovsky, G. N., Zelenin, P. V., and Beloozerova, I. N. (2006). Neural bases of postural control. *Physiology* (Bethesda) **21**, 216–225.

Georgopoulos, A. P. (1999). News in motor cortical physiology. *News Physiol. Sci.* **14**, 64–68.

Grillner, S. (2006). Biological pattern generation: The cellular and computational logic of networks in motion. *Neuron* **52**, 751–766.

Grillner, S., Hellgren, J., Menard, A., Saitoh, K., and Wikstrom, M. A. (2005). Mechanisms for selection of basic motor programs—Roles for the striatum and pallidum. *Trends Neurosci.* **28**, 364–370.

Hikosaka, O., Takikawa, Y., and Kawagoe, R. (2000). Role of the basal ganglia in the control of purposive saccadic eye movements. *Physiol. Rev.* **80**, 953–978.

Ito, M. (2002). The molecular organization of cerebellar long-term depression. *Nat. Rev. Neurosci.* **3**, 896–902.

Johansson, R. S. (2002). Dynamic use of tactile afferent signals in control of dexterous manipulation. *Adv. Exp. Med. Biol.* **508**, 397–410.

Jorntell, H., and Ekerot, C. F. (2002). Reciprocal bidirectional plasticity of parallel fiber receptive fields in cerebellar Purkinje cells and their afferent interneurons. *Neuron* **34**, 797–806.

Lawrence, D. G. and Kyupers, H. G. J. M. (1968). The functional organization of the motor system in the monkey. I. The effects of lesions of the descending brainstem pathways. *Brain* **91**, 15–36.

Marder, E. (1998). From biophysics to models of network function. *Annu. Rev. Neurosci.* **21**, 25–45.

Orlovsky, G. N., Deliagina, T. G., and Grillner, S. (eds.) (1999). "Neuronal Control of Locomotion: From Mollusc to Man." Oxford Univ. Press, Oxford.

Pearson, K. G. (2000). Neural adaptation in the generation of rhythmic behavior. *Annu. Rev. Physiol.* **62**, 723–753.

Rossignol, S., Dubuc, R., and Gossard, J. P. (2006). Dynamic sensorimotor interactions in locomotion. *Physiol. Rev.* **86**, 89–154.

Sten Grillner

29

The Spinal and Peripheral Motor System

Most coordinated movements involve the use of several muscles, each with a precisely orchestrated timing and pattern of activation and relaxation. Contraction of an individual muscle is controlled by a pool of motor neurons that innervate that muscle. Patterned contraction of several muscles requires coordination between pools of motor neurons. For many movements, interneurons in the spinal cord provide the means for coordinating activity of different motor neuron pools. A key feature of spinal interneurons is their convergent and divergent connections that allow them to integrate signals from different sources. Interneurons that receive inputs from descending motor systems and sensory afferents, for example, can update a voluntary command for limb movement to incorporate feedback from the moving limb. Spinal interneurons also form connections with other spinal interneurons to create networks that generate rhythmic patterns of excitatory and inhibitory activity. These networks play a key role for organizing rhythmic, repetitive movements such as scratching or locomotion. In this chapter, we will use locomotion as an example of a motor behavior to illustrate how the neural elements of the spinal cord and peripheral neuromuscular system operate together to carry out a movement. As details of the motor neurons, spinal interneurons, muscles, and sensory afferents are presented, the primary focus will be on mammals, but variations that occur in other vertebrates and in invertebrates are highlighted in Box 29.1.

LOCOMOTION IS A CYCLE

Locomotion is the act of moving from place to place. Walking, swimming, and flying are modes of locomotion. All modes of locomotion consist of a cycle of patterned and rhythmic activity of different muscle groups. Locomotion can be described by the phase relationships between limbs and between muscle groups during the cycle. Nonlimbed vertebrates such as lamprey swim by alternately contracting and relaxing trunk muscles on the two sides of the body—the sides are in opposite phases of the cycle. Along the length of the lamprey's body, there is a slight phase lag between segments, creating the propulsive wave that moves it forward. In contrast, birds activate their wings synchronously in flight—the limbs are in phase, with no lag.

Stepping, the major form of mammalian locomotion, is a cycle with two phases: stance (the extension phase) and swing (the flexion phase). In people, the stance phase begins when the heel strikes the ground and the leg begins to bear weight. As stance progresses, the weight is gradually shifted forward, rolling from the heel to the ball of the foot. The swing phase begins when the toe leaves the ground and the leg moves forward. To accomplish this motor sequence within one leg, muscles acting across the hip, knee, or ankle are activated in a characteristic pattern (Fig. 29.1). In general, muscles that flex the joint are more active

677

BOX 29.1

SPECIES DIVERSITY IN NEUROMUSCULAR SYSTEMS

Not all animals control muscle contraction in the same way as mammals. Differences are found in the proteins and structure of the muscle fiber, the number of neuromuscular junctions on each muscle fiber, the neurotransmitters used at the neuromuscular junction, and the number of motor neurons contacting each muscle fiber. Major differences are found in invertebrates, which often have evolved adaptations that suit their movement patterns or constraints imposed by their environments.

Some invertebrates require muscles that generate great forces or provide mechanical stiffness. One solution used by mussels, the nematode *C. elegans*, and the fruit fly *Drosophila* is the incorporation of molecules of paramyosin into thick filaments, so that additional cross-bridges can be formed. Mussel retractor muscles also have a specialized catch property, in which cross-bridges are tightly bound with low energy expenditure until the catch state is released by neurotransmitter action that leads to phosphorylation of twitchin, a titin-like molecule. Very slow movements can be achieved by long sarcomeres, since Ca^{2+} diffusion throughout the sarcomere takes longer. In mammalian muscle fibers, sarcomeres are generally 2–3 microns long, although longer sarcomeres occasionally are found in muscles that stiffen but do not twitch, such as the tensor tympani muscle of the cat eardrum. In contrast, claw muscles of crabs may have sarcomeres from 6–18 microns, and one marine worm was found to have sarcomere lengths of 30 microns.

Most mammalian muscles usually contain a mixture of muscle fiber types, but some animals have distinct muscles for fast and slow movements. Trout, for example, use red muscles composed of slow twitch fibers for sustained swimming and white, fast twitch muscles for leaping and rapid bursts of speed.

Mammals and birds use acetylcholine as the transmitter at the neuromuscular junction. So does *C. elegans*, but many other invertebrates use glutamate, including arthropods such as *Drosophila*, locust, and crayfish. Corelease of glutamate and peptides occurs at some neuromuscular junctions, such as the peptide FMRF-amide in the sea snail Aplysia. Crustacean muscles have separate excitatory and inhibitory neuromuscular junctions that use glutamate and GABA.

In mammals, most muscle fibers are innervated by a single motor neuron, with a few exceptions, such as laryngeal muscles and spindles. Innervation by more than one motor neuron is seen in some avian muscles, and many invertebrate muscles. The motor neurons may merely have an additive function, as in *Drosophila*, where recruitment of motor neurons produces graded contractions of the muscle. In other animals, the different motor neurons may play distinct roles. The locust jumping muscle has a dual arrangement, with one motor neuron that produces a rapid and powerful excitation and another that produces graded excitation and contraction.

Lastly, the central control of motor neurons differs. In mammals, a motor neuron action potential leads inevitably to a muscle fiber action potential. In crayfish, presynaptic inhibition of motor terminals allows modulation at a more distant site. Structural specializations, such as electrical synapses onto motor neurons and gap junctions between motor neurons, often occur in invertebrates, but generally are found only during development in mammals. The distributed structure of invertebrate systems and rich palette of ion channels expressed by motor neurons is favorable to the formation of motor pattern generating circuits.

Mary Kay Floeter

during the swing phase and extensor muscles during the stance phase. However, within each phase, the individual muscles have a more elaborate and finely graded timing and pattern of activation. Stepping can also be performed with different gaits. Gaits can be characterized by the cycle rate, the relative proportion of the cycle spent in swing and stance phases, and the phase relationships between limbs. At slow or moderate walking speeds (i.e., slow cycle rates), the extensor phase is longer than flexion and the extension phases of opposite legs overlap. In a walking gait both feet are in simultaneous contact with the ground for some period during the cycle. As the speed of stepping increases and the cycle is shorter, the duration of the extension phase shortens, with briefer periods of double foot contact. At high speeds where there is no overlap of extension phases between the legs, the gait switches recognizably from walking to running.

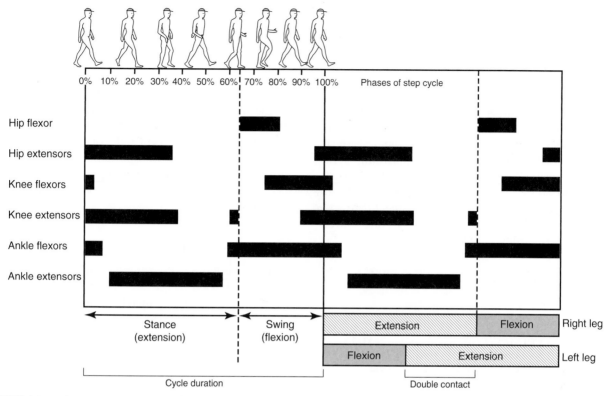

FIGURE 29.1 Schematic diagram of two step cycles in human locomotion. Drawings at the top illustrate the position of the limbs during one cycle. Dark horizontal bars below the drawings show the timing of activity in flexor and extensor muscles of the right leg. The division of each step cycle into stance (extension) and swing (flexion) phases is indicated at the bottom of the figure. During walking, the stance phases of the two legs overlap, with both feet in contact with the ground.

Central Pattern Generating Circuits

Although we are able to exert voluntary control over locomotion, the basic patterns of movement for locomotion can be generated by the spinal cord itself, without descending or sensory inputs. The first demonstration came from Graham Brown (1911), who observed that cats in which the spinal cord and dorsal roots were cut were still able to walk on a treadmill as long as their weight was supported. Because there was no sensory feedback or descending input he suggested that the rhythmic activity in the hindlimb muscles must be generated by a motor program within the spinal cord. The motor program, also known as a central pattern generator (CPG), needed to explain three important features of locomotion: (1) its rhythmic and cyclical occurrence, (2) coordinated alternation between flexor and extensor activity within the same limb, and (3) alternation between opposite limbs. Brown proposed that the CPG consisted of two mutually inhibitory "half-centers," both under tonic excitation: one half-center drove muscles that flexed the legs

and the other drove muscles that extended the legs. The mutual inhibition between the two half-centers would cause activity to oscillate between them, producing alternating limb flexion and extension.

CPGs are now thought to consist of networks of neurons that are synaptically or electrically connected, with cellular properties that generate rhythmic firing. CPGs are found in many areas of the central nervous system and are responsible for different types of patterned rhythmic activities such as breathing, swallowing, chewing, and walking. The identification and characterization of the neurons that comprise the locomotor CPG in the spinal cord is an active area of research that will be discussed later in this chapter.

Sensory Contributions to Movement

During a movement, the sensory afferents from muscles, joints, and skin will each be activated in a distinct way, reflecting how that movement displaces joints, produces pressure on regions of skin, and stretches individual muscles. For rhythmic movements such as walking, these recurring "multisensory" feedback signals often reinforce activity of CPG circuits. However, when a perturbation occurs during the movement, the resulting sensory signal may evoke a reflex motor response that alters the movement. Many of these reflex responses adapt the movement in an appropriate way to proceed under the changed environmental conditions. Most sensory afferents produce these reflex effects through interneuronal circuits. (The stretch reflex is an exception.) The excitability of many of these interneuronal circuits usually is modulated during movement, allowing reflexes to be suppressed or enhanced at different phases. CPGs are one source of this phase modulation. Many reflexes can also be elicited when the organism is not moving, and in these situations the reflex tends to exhibit a more stereotyped motor pattern.

CONNECTING THE SPINAL CORD TO THE PERIPHERY

Ultimately movement is carried out by muscles, and the motor commands from the central nervous system must be transmitted to motor neurons that innervate the muscles. The organization of motor neurons into pools, and the manner in which motor neurons properties are matched to contractile and metabolic properties of the muscle fibers simplify many of the variables needed to be controlled to produce smooth movements with precise amounts of force. Feedback from sensory receptors of the muscle play a key role in monitoring the force and stretch of the muscle. This section will cover some of the basic properties of motor neurons, muscle, and muscle receptors.

Motor Neurons

Motor neurons are located in the ventral horn of the spinal cord. The pool of motor neurons that innervates an individual muscle forms an elongated column that typically extends over two or three spinal cord segments. Intermingled among the large α-motor neurons that innervate skeletal muscle fibers are smaller γ-motor neurons, which innervate muscle spindles, as well as β-motor neurons, which dually innervate skeletal muscle and spindles. The α-motor neurons are the largest neurons in the spinal cord, with myelinated axons that exit the spinal cord through the ventral roots and travel in peripheral nerves to innervate muscles. Their axons also give off small branches within the spinal cord, called recurrent collaterals, that synapse on inhibitory interneurons, called Renshaw cells, and motor neurons. The extensive dendritic tree of motor neurons extends into the dorsal and ventral horns, receiving thousands of synaptic inputs from excitatory and inhibitory neurons (Fig. 29.2). At the neuromuscular junction, the synapse between the motor axon and the muscle fiber, acetylcholine is the transmitter used by mammals, but other transmitters, such as the peptide CGRP or glutamate (Mentis et al., 2005; Nishimaru et al., 2005), coexist in many motor neurons, and may play a particular role during development and at the central synapses formed by recurrent collaterals.

Muscle Contraction

Movement is achieved by the contraction of muscles that are attached to the skeleton or soft tissues, such as muscles producing facial expressions. A skeletal muscle is made up of thousands of individual muscle fibers. Each muscle fiber is a multinucleated cell, a few centimeters long, that contains the contractile proteins, actin and myosin, arranged into thick and thin filaments in repeating units called sarcomeres (Fig. 29.3). The myosin molecule head region contains an ATP-ase enzymatic activity that is sensitive to levels of intracellular calcium. Myosin heads are loosely bound to actin at rest, but when intracellular calcium rises, the cross-bridge strengthens, causing myosin to break down ATP. When this occurs the cross-bridge loosens, and the myosin head swivels along the actin filament to another binding site. The repetition of this swiveling movement in the presence of elevated intracellular calcium and ATP causes thick and thin filaments to slide along each other and shorten the sarcomere length. Shortening of sarcomeres shortens the muscle, or if the muscle is carrying a heavy load, the muscle may only stiffen in an isometric contraction.

The process of muscle contraction begins with an action potential in the motor axon that causes release of acetylcholine at the neuromuscular junction. In mammals, normally every spike in the motor axon produces an action potential in the muscle fiber because an excess amount of acetylcholine is released (Box 29.2). In adult mammals, each skeletal muscle fiber has one neuromuscular junction innervated by one motor neuron. Muscle fiber action potentials lead to the activation of voltage-gated calcium channels in the membrane that interact with calcium release proteins in the sarcoplasmic reticulum, an organelle in the muscle fiber that stores and releases calcium near the sarcomeres. This interaction couples excitation to muscle contraction. Other proteins in the sarcoplasmic reticulum reuptake calcium, leading to relaxation.

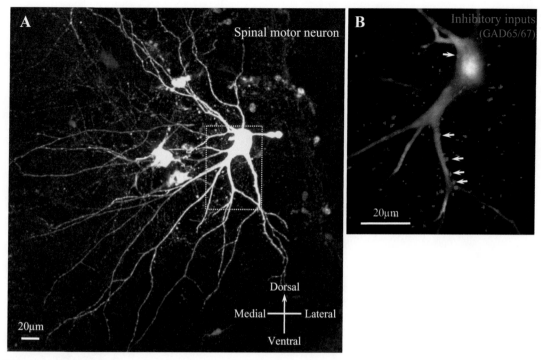

FIGURE 29.2 Motor neuron from a neonatal mouse spinal cord that has been filled with biocytin through an intracellular electrode and visualized by immunocytochemistry. (A) The extensive dendritic arbor is directed into the dorsal, medial, and ventral grey matter of the spinal cord, where dendrites receive thousands of synapses from interneurons, sensory afferents, and descending pathways. (B) Double staining for a marker of GABA-ergic inhibitory terminals (GAD65/.67) shows some of the inhibitory synapses along the proximal dendrites and cell body.

Three Basic Muscle Fiber Types

Muscle fibers are classified into types according to their contractile and metabolic properties. Contractile properties refer to the muscle fiber's speed of contraction and relaxation, and the specific force it can produce. Contractile properties are mainly determined by the isoform of myosin that is expressed by the muscle fiber, particularly the myosin heavy chain gene, although the genes for other proteins associated with the sarcomere and the sarcoplasmic reticulum are expressed in fiber-type specific patterns that match the properties of the myosin isoform. The contractile properties of a muscle fiber are classified as slow twitch or fast twitch. Metabolic properties refer to whether aerobic or anaerobic pathways are used to produce energy (ATP) from fuels. Muscle fibers with abundant mitochondria are able to use oxidative (aerobic) metabolism, which provides relative resistance to fatigue as long as there is a good blood supply. Muscle fibers with few mitochondria rely on stored fuel, such as glycogen, which can more easily be depleted and also produces lactic acid as a byproduct of glycolysis.

There are three basic muscle fibers types in adult skeletal muscle. Type I fibers have a slow twitch but have a good capacity for oxidative metabolism. Type IIA fibers have a fast twitch and can use both aerobic and anaerobic metabolism, giving them an intermediate resistance to fatigue. Type IIB fibers produce a fast twitch, but rely mainly on anaerobic glycolysis for their energy. Most muscles contain a mixture of all three fiber types, in proportions that reflect the typical usage of the muscle.

Motor Unit Types

A "motor unit" is defined as one motor neuron and all the muscle fibers it innervates. In mammals, each muscle fiber is innervated by only one motor neuron, but one motor neuron innervates many muscle fibers as its axon branches in the muscle. However, all muscle fibers of one motor unit are of the same fiber type. Motor units cluster into three basic types based on the three physiological properties of twitch speed, the amount of force produced, and fatigability (Fig. 29.4). Type S motor units are defined by the properties of

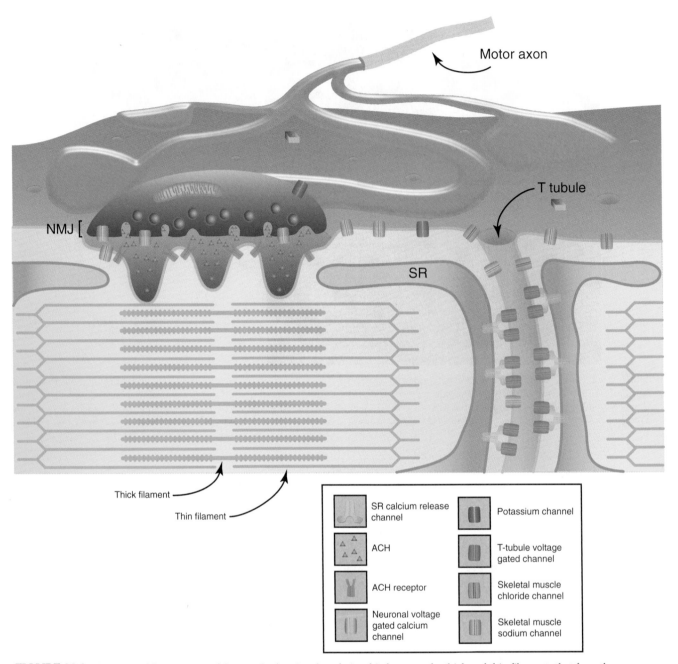

FIGURE 29.3 Diagram of the structure of the muscle showing the relationship between the thick and thin filaments that form the sarcomere, the sarcoplasmic reticulum (SR), and the location of the ion channels of the motor axon, neuromuscular junction, and muscle membrane. With permission, modified from Cooper and Jan (1999).

slow twitch, small amounts of force, and high resistance to fatigue. Type FR motor units are defined by fast twitch, moderately large force, and relative resistance to fatigue. Type FF motor units have fast twitch, produce the highest forces, and fatigue quickly.

The physiological properties of the motor neurons and the muscle fibers they innervate are closely matched. Type S motor neurons innervate Type I fibers. Type S motor neurons have small cell bodies and dendritic arbors, allowing them to be depolarized easily. Afterhyperpolarization following each motor neuron spike promotes a steady rate of firing that allows forces produced by successive twitches of the muscle fibers to summate (Fig. 29.4B). With their

BOX 29.2

MYATHENIA AND MYASTHENIC DISORDERS

Normally every action potential of the motor axon releases more than enough transmitter than is needed to produces a muscle fiber action potential. This occurs because both the amount of transmitter released from the presynaptic motor axon terminal and the density of AChRs on the postsynaptic muscle fiber exceed that needed to produce an end plate potential (EPP) large enough to initiate a muscle fiber action potential. Genetic or acquired diseases that reduce the safety factor of neuromuscular transmission are called myasthenic disorders (Fig. 29.12). Myasthenic disorders cause weakness that fluctuates with use of the muscle. The most common is the acquired auto-immune disorder myasthenia gravis, caused by antibodies that bind to AChRs at the neuromuscular junction. The antibodies interfere with ACh binding and lead to internalization of AChRs and, ultimately, the loss of junctional folds. The reduced numbers of functioning AChRs produce a smaller EPP that fails intermittently to trigger a muscle fiber action potential. However, in myasthenia gravis the presynaptic terminal remains normal, such that a short burst of stimulation will produce facilitation of transmitter release, increasing the concentration of ACh in the cleft and transiently improving neurotransmission. Acetylcholinesterase inhibitors provide symptomatic improvement by blocking the breakdown of ACh, allowing a longer effect on residual receptors. Another autoimmune myasthenic disorder, Lambert–Eaton myasthenic syndrome, is caused by antibodies that bind to presynaptic voltage-gated calcium channels. Less ACh is released at the neuromuscular junction, producing small EPPs that fail to trigger action potentials.

Congenital myasthenic syndromes (CMS) are extremely rare, but provide important insights into the function of components of the neuromuscular junction. Postsynaptic CMS stem from a deficiency and/or altered kinetic properties of the AChR. More than 40 different mutations in subunits of the AChR have been described. Most mutations decrease subunit expression or prevent subunit assembly or glycosylation, leading to reduced numbers of functioning AChRs and thereby smaller EPPs. In fast channel CMS, mutations produce decreased binding affinity for ACh, leading to a smaller quantal response and miniature EPP amplitude, without reducing the numbers of AChRs. In contrast, the slow channel CMS stem from mutations that prolong the channel opening episodes. Slow channel mutations cause depolarization block due to temporal summation of the prolonged EPPs. In the long term, cationic overloading of the postsynaptic region leads to destruction of the junctional folds and loss of AChRs. Presynaptic CMS are less common. They are caused by mutations that lead to a paucity of synaptic vesicles or defects in ACh resynthesis or vesicular packaging. As a result, fewer vesicles are released or fewer ACh molecules are contained in each vesicle, and EPPs are smaller. Identifying CMS as involving pre- or postsynaptic components and understanding how these alterations affect neuromuscular transmission are necessary for developing rational therapies for these disorders.

Mary Kay Floeter

steady firing rates and resistance to fatigue, Type S motor units are optimal for the sustained firing needed for postural and tonic movements. Each Type S unit individually contributes only a small amount of force to a muscle's contraction, but typically Type S units make up a large proportion of the motor units innervating a muscle. In contrast, Type F motor neurons have large cell bodies and dendritic arbors, shorter afterhyperpolarizations and somewhat faster firing rates. Type FR motor neurons innervate Type IIA muscle fibers, and Type FF motor neurons innervate

Type IIB muscle fibers. Type FR motor units are optimal for fast, powerful movements. Type FF units are used mainly in movements that require brief bursts of muscle strength. Most muscles contain a mixture of motor unit types, intermingled in a mosaic pattern.

The Size Principle and Orderly Recruitment

Most movements don't require a muscle to contract at its maximal strength, and only a fraction of the hundred or more motor neurons in the pool that

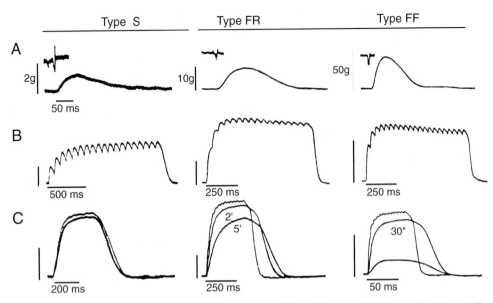

FIGURE 29.4 The three motor unit types (S, FR, and FF) can be defined experimentally by measuring their contractile properties and fatigability. Panels A–C show recordings of muscle force, with insets in A showing recordings of motor neuron action potentials. Note different time and amplitude calibration scales for each of the motor units. (A) Single twitches produced by one action potential of the motor neuron. (B) Maximal force produced by repetitive stimulation of the motor neuron to produce an unfused tetanus. In addition to differences in maximal force, the "sag" property, a dropping off of tension during maintained stimulation, is seen in FR and FF units. (C) Fatigability is demonstrated by a drop in the tension produced by a single twitch after short periods of activation, as noted. Note that S units show little fatigue, whereas FF units fatigue within 30 s. Reproduced with permission from Burke *et al.* (1973).

innervates that muscle will need to fire. For most movements, motor units are recruited in an orderly sequence, first described by Henneman (1957), as based on the "size" of the motor neuron. According to the size principle, motor units that produce the smallest amounts of force are the first to begin firing, and motor units that produce larger forces will be progressively recruited as the muscle makes progressively stronger contractions. Many of the anatomical and physiological properties of the motor unit types correlate with measures of size (Table 29.1). These size-related properties favor activation of motor unit types in the following order: Type S → Type FR → Type FF. The pool of motor neuron innervating each muscle contains varying proportions of Type S and Type F motor neurons, although generally Type S motor neurons are more abundant.

Many of the properties that produce orderly activation by size are stable and intrinsic to the motor neuron—cell body diameter, for example. Most extrinsic synaptic inputs also reinforce recruitment according to the size principle. This occurs because a synaptic current will produce a larger postsynaptic potential in a smaller, compact neuron with a higher input resistance. Because most sources of synaptic input are distributed uniformly to a motor pool, EPSP or IPSP

TABLE 29.1   Size-Related Properties of Motor Neurons

Properties that increase with size	Properties that decrease with size
Diameter of soma and axon	Resistance to fatigue
Conduction velocity	Ia EPSP amplitude
Complexity of axonal collaterals	Input resistance
Membrane area, dendritic extent	Membrane resistance
Rheobase	Time constant
Muscle fiber diameter	Duration of afterhyperpolarization
Maximum force output	Twitch contraction time
	Twitch relaxation time

amplitudes correlate with the motor neuron size—larger in Type S motor neurons than Type F motor neurons. This means, for example, that stretch of a muscle that activates the primary muscle spindle (Ia) afferent will produce the largest Ia EPSPs in Type S motor neurons, and these will be the first units recruited to fire in the stretch reflex. In the less common situation in which synaptic inputs are not uniformly distributed to every motor neuron in the pool, alternative recruitment sequences are possible.

Stronger muscle contractions are produced not only by recruitment of additional motor units, but also by increasing the firing rates of motor units that are already actively firing. The relative use of recruitment and rate modulation to maintain a desired profile of force varies according to the muscle and the level of force. In some muscles, such as the small hand muscles, several motor units begin firing at low levels of force, and fine gradations of force are controlled by modulating their firing rates. In other muscles, recruitment and rate coding occur together in a more balanced fashion throughout the working force range of the muscle.

Muscle Afferents

Muscles contain two kinds of specialized sensory receptors that provide feedback to the spinal cord and brain. Muscle spindles sense muscle stretch or length, and Golgi tendon organs (GTOs) sense the load upon the muscle (Fig. 29.5). The spindle consists of an encapsulated bundle of modified muscle fibers, called intrafusal fibers, that lie within the muscle in parallel to the skeletal muscle fibers. The spindle receives both sensory and motor innervation. Sensory innervation is supplied by one primary afferent, the Ia fiber, and several secondary, group II afferents. Ia afferents are more sensitive to quick stretches, such as a tendon tap or vibration, and group II afferents maintain a firing rate proportional to the level of sustained stretch. Intrafusal muscle fibers are innervated by γ-motor neurons and β-motor neurons. The motor innervation of intrafusal fibers adjusts the tension of the spindle to compensate for changes in muscle length when a muscle contracts and shortens. When the muscle shortens the spindle can become momentarily slack, causing a pause in spindle afferent firing. However, during most voluntary movements, α and γ-motor neurons are coactivated such that the sensitivity of the spindle is maintained during the muscle shortening. Nevertheless, because γ-motor neurons have a separate innervation of intrafusal muscle fibers, the nervous system can separately adjust the sensitivity of the spindle for different types of movement.

Golgi tendon organs lie at the junction of the muscle and tendon, in series with the muscle. GTOs are sensitive to the forces generated by muscle contraction against a load. Each GTO senses the force generated by a small number of motor units that insert onto a local region of the tendon. The axons of the GTOs are called Ib afferents. Like Ia afferents, they are large myelinated axons with fast conduction veloc-

ities. Unlike spindles, GTOs do not have motor innervation.

Muscles are innervated by smaller diameter and unmyelinated afferents, called group III and IV afferents, that respond to extracellular substances, such as the acidic substances released during fatiguing exercise. These slowly conducting afferents relay the sense of muscle pain and soreness and produce reflex actions with a relatively slow time course.

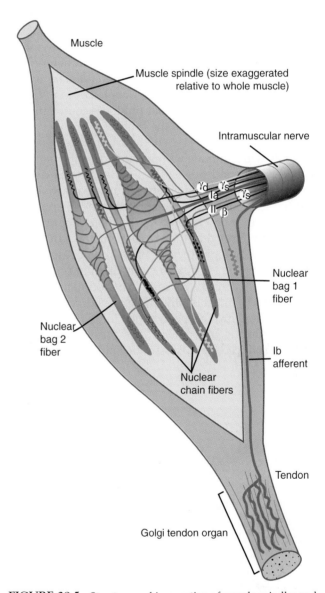

FIGURE 29.5 Structure and innervation of muscle spindles and Golgi tendon organs. The sensory innervation of the spindle is through primary (Ia) spindle afferents that innervate both nuclear bag and nuclear chain intrafusal muscle fibers and secondary (II) afferents that innervate nuclear chain fibers. Motor innervation of the spindle is supplied by static and dynamic γ-motor neurons and by β-motor neurons.

Stretch Reflexes

The muscle stretch reflex plays an important role in providing feedback during movements. The stretch reflex consists of a monosynaptic response from the direct connection between Ia afferents and motor neurons, which may be followed by polysynaptic reflex activity. At rest, the stretch reflex can be obtained by a quick stretch, such as by tapping on a muscle tendon, which produces a burst of action potentials in the Ia spindle afferents (Fig. 29.6). In the spinal cord, the Ia afferents synapse on the motor neurons of the muscle containing the spindle (the homonymous motor neurons) and, to a lesser extent, on motor neurons that innervate other muscles (heteronymous motor neurons). The rapid volley of glutamatergic Ia EPSPs caused by a quick stretch can summate in the motor neuron, and if sufficient depolarization occurs, an action potential occurs that produces contraction of the muscle, with a jerk of the limb.

If the stretch reflex is elicited during movement, the Ia EPSPs will summate with inputs from other sources such as CPGs and descending inputs. Stretch during the phase of movement when the muscle is active will add to other sources of motor neuron depolarization, and the reflex motor response will enhance the ongoing contraction. Stretch of the muscle when it is out of its active phase will produce a smaller motor response, if any. In this way, the spatial summation of postsynaptic potentials contributes to phase modulation of the stretch reflex. Additionally, there is modulation of the stretch reflex through presynaptic inhibition of the Ia afferents during movement. This presynaptic inhibition is produced by GABA-ergic spinal interneurons

that synapse on the Ia afferent terminals, forming axo-axonic synapses. Presynaptic inhibition reduces the release of glutamate from Ia afferents, leading to smaller Ia EPSPs in the motor neuron, but does not affect the size of other inputs to the motor neuron. The GABA-ergic interneurons mediating presynaptic inhibition are strongly activated by antagonist Ia afferents; that is, spindles from muscles that produce the opposite movement, as well as by CPGs and descending inputs. Thus, presynaptic inhibition of a muscle's stretch reflex is strongest when the muscle contraction is out of phase with the movement cycle, when its antagonist muscle is contracting.

Summary

Muscles are innervated by motor neurons in the ventral horn of the spinal cord. The motor unit consists of one motor neuron and all the muscle fibers it innervates. Physiological properties of each motor neuron match the contractile and metabolic properties of its muscle fibers. Type S motor units have slow twitch, high aerobic capabilities, and fatigue resistance. These smaller motor units are the first recruited for contraction, and are heavily used for slow, sustained postural movements. Type F motor units innervate muscle fibers with a fast twitch and are recruited for more forceful or ballistic movements. Type FR motor units use aerobic and anaerobic metabolism and are relatively resistant to fatigue. Type FF motor units use anaerobic metabolism and fatigue quickly. In the pool of motor neurons innervating a muscle, motor neurons are recruited in an orderly fashion, beginning with Type S units that generate small amounts of force, followed by units that generate greater forces. As the muscle contracts, changes in its length are detected by spindles, specialized receptors that lie within the muscle. The load on the muscle is detected by a second specialized receptor, the Golgi tendon organ. These sensory receptors of the muscle provide feedback to the spinal cord allowing reflex changes in drive to the muscles.

SPINAL INTERNEURON NETWORKS

So far, this chapter has discussed how the output of the spinal cord produces muscle actions though the motor neurons. This section will focus on spinal interneurons —neurons with a cell body in the spinal cord and an axon that synapses on another spinal neuron. Spinal interneurons have three important functions for motor control:

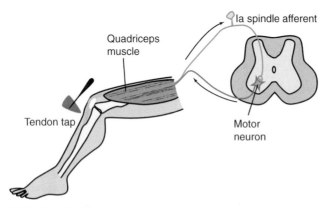

FIGURE 29.6 Basic circuitry underlying the knee-jerk stretch reflex. Ia spindle afferents from the quadriceps muscle make monosynaptic, excitatory connections on α-motor neurons that innervate the quadriceps.

- They relay sensory inputs from the periphery that may modulate the motor output.
- They relay as well as modulate signals from the brain via descending pathways.
- They form networks that produce patterned rhythmic activity.

It is important to note that some interneurons may participate in all three functions, switching from one role to another during specific movements or carrying on two roles simultaneously.

Historically, interneurons were identified and named by their axonal projections, the location of their soma, their neurotransmitter, or by a physiological effect or reflex that they mediated. Later, populations of neurons were identified by their pattern of activity during movements, visualized using activity-dependent dyes. More recently, molecular biology tools have enabled scientists to identify interneurons based on their gene expression pattern and cell lineage. Molecular markers have been used to identify several classes of interneurons and motor neurons. Four distinct populations of interneurons, designated as V0, V1, V2, and V3, have been identified in the ventral spinal cord by their developmental pattern of expression of transcription factors, proteins that regulate gene expression (Jessell, 2000; Goulding and Pfaff, 2005). The "V" denotes the *ventral* location, and each subclass appears at a distinct position along the dorso-ventral axis (0 most dorsal, 3 most ventral) of the ventral spinal cord (Fig. 29.7).

Models for Studying the Locomotor CPG Network

At the beginning of the chapter we noted that the CPG network needed to produce three fundamental features of locomotion: generation of a cyclical rhythm, coordinated alternation between antagonist flexor and extensor muscles, and coordination between limbs. Figuring out what makes up a CPG—the neurons involved, the connections between these neurons, the synaptic interactions, and the contributions from cellular ion channels—is challenging. To tackle this problem, researchers construct models that can be tested and refined by experimental observations. As noted earlier, Brown (1911) first proposed a model

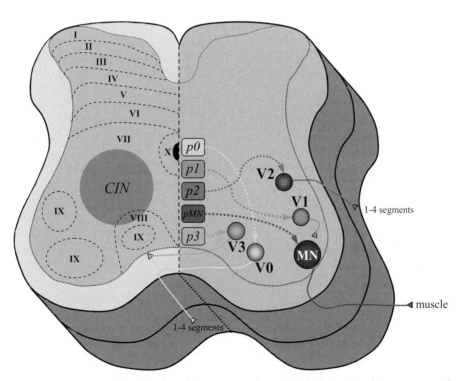

FIGURE 29.7 Cross-sectional schematic of the spinal cord depicting on the right side the origin of interneuron and motor neuron populations defined in developmental studies using molecular markers for cell identification. The ventral interneurons (V) are marked as V0–V3 based on the distinct position along the dorso-ventral axis (0 most dorsal, 3 most ventral) of the ventral horn. The left side of the spinal cord depicts the general location of commissural interneurons (CINs) and the Rexed's laminae. CINs are characterized based on the projection of their axons, as ascending (a), descending (d), or both (ad).

BOX 29.3

MOTOR NEURON DISEASES

There are several neurological disorders that selectively affect motor neurons. Spinal muscular atrophy (SMA) is a hereditary disorder in which motor neurons develop normally but begin to degenerate not long after birth. Autosomal recessive SMA was described by neurologists long before its cause was known, and several different clinical phenotypes were described. SMA causing weakness in infancy was called Werdnig-Hoffman disease; SMA causing weakness later in childhood was called Kugelberg-Welander disease. Both forms of SMA are caused by a mutation in a gene called "survival of motor neuron" or SMN gene. The SMN protein is widely expressed in neurons and has a role in processing mRNAs. It is not known why motor neurons are more vulnerable to mutations of SMN than other cells. In an X-linked SMA, called Kennedy's disease or spinal and bulbar muscular atrophy (SBMA), weakness begins in adult life in affected males. Motor neurons innervating facial and jaw muscles often are affected early on. The mutation in SBMA is a triplet repeat expansion in the androgen receptor gene that leads to expression of an abnormal androgen receptor. The androgen receptor is expressed in motor neurons, and over time accumulation of the abnormal receptor leads to degeneration of the motor neuron.

Amyotrophic lateral sclerosis (ALS), also known as Lou Gehrig's disease, is a disease with degeneration not only of the motor neurons, but also of corticospinal neurons of the motor cortex. In most patients with ALS, the disease is not hereditary and its cause is unknown. There are a few hereditary forms of ALS that make up a small percentage of all patients with ALS. Nevertheless, the identification of familial ALS (FALS) genes has provided ideas about pathways that may be responsible for causing sporadic ALS. The first gene mutation identified in FALS encoded the enzyme superoxide dismutase, SOD1. Studies using transgenic mice expressing the mutant SOD1 gene showed that FALS was not caused by loss of enzyme activity, but by abnormal properties of the mutant SOD1 protein. Studies expressing mutant SOD1 in different cell types are revealing that glial and inflammatory cells may also contribute to motor neuron degeneration in FALS (Boillee *et al.*, 2006). The ability to manipulate the expression of mutant genes in animal models promises to provide an invaluable contribution to finding causes and treatment of ALS.

Mary Kay Floeter

Reference

Monani, U. R. (2005). Spinal muscular atrophy: a deficiency in a ubiquitous protein, a motor neuron-specific disease. *Neuron.* **48**, 885–896.

with mutually inhibitory half-centers driving flexors and extensors. Grillner (1985) proposed three important modifications: (1) that at each joint, movement in one direction was produced by a "burst generator" unit, a half-center capable of independent rhythmic activity; (2) as in Brown's model, that flexion and extension at each joint were controlled by reciprocally connected burst generators to form a unit CPG; and (3) that multijoint coordination was achieved by loose or adjustable coupling between the mosaic of unit CPGs controlling each of the joints (Fig. 29.8).

Experimental findings have led to further refinements of the proposed organization of the CPG network. For example, the rhythm and timing of the network may be controlled separately from the patterning of the network output. In studies of locomotor rhythms in the neonatal rodent spinal cord, the CPG appears to be distributed among several spinal segments. Studies in the isolated neonatal rodent spinal cord preparation maintained *in vitro* (Fig. 29.9) show a difference in the rhythmogenic capacity between the rostral (upper) and the caudal (lower) segments of the lumbar spinal cord. The upper lumbar (L1–L3 in rodents) segments seem to be most important for rhythm generation compared to caudal lumbar segments. In experiments using calcium dyes to visualize motor neuron activity, rhythmic signals from different parts of the lumbar (L1, L2) and sacral (S1–S3) segments rose, peaked, and decayed in a rostrocaudal sequence, giving rise to a rostrocaudal "wave" of motor neuron activation during each locomotor cycle (Bonnot *et al.*, 2002) (see video*). The importance of the upper and mid-lumbar cord for rhythm generation has also been observed in the spinalized cat and is consistent with case reports in humans noting that irritative lesions of the upper lumbar cord produce stepping movements.

* Videos and other content available via the website information found in the front of this book.

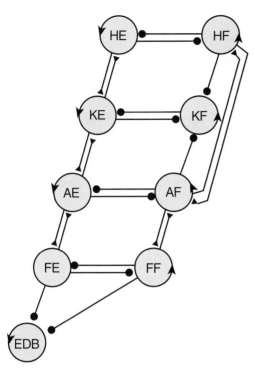

FIGURE 29.8 Scheme for multijoint coordination within the leg using a mosaic of "unit CPGs." Burst generators produce movement in one direction, extension or flexion, at the hip (HE, HF), knee (KE, KF), ankle (AE, AF), or foot (FE, FF), with a separate generator for the muscle that lifts the toes (EDB). Burst generators for flexion and extension at each joint are mutually inhibitory, forming a unit CPG that produces alternation. The strength of excitatory (triangles) and inhibitory (circles) connections between unit CPGs can vary to create a variety of different gaits. Reproduced with permission from Grillner *et al.* (1985).

Excitatory Interneurons Responsible for the Rhythm

Maintaining rhythmic bursting in an interneuronal network requires a source of excitatory drive. Although this drive could be provided by tonic excitation, such as from descending pathways, a more robust rhythm is possible if there are excitatory cells within the network that have intrinsic properties that allow rhythmic and repetitive firing (Box 29.4). At least three classes of rhythmically active excitatory interneurons have been identified in the spinal cord and proposed as candidates for CPG rhythm generators. The first is a population of glutamatergic neurons located throughout the rostrocaudal extent of the spinal cord that are mutually excitatory and project to neurons that inhibit their counterparts on the opposite segment of the cord. In lamprey and other nonlimbed vertebrates, these neurons play a role in generating the swimming rhythms (Grillner, 2003). Another candidate popula-

tion of excitatory interneurons were first identified in mice with deletions in the genes for two axon guidance molecules, *EphA4* and *ephrinB3* (Kullander *et al.*, 2003). These mice had an abnormal hopping pattern of gait. Subsequent study of interneurons expressing *EphA4* receptors showed that these were glutamatergic and rhythmically active during locomotor behavior. Most of the interneurons were "last order" interneurons, interneurons that synapse on motor neurons. The third candidate thought to participate in the CPG is a recently defined excitatory interneuron in the lumbar spinal cord that is rhythmically active in locomotion and expresses the transcription factor HB9, which is also expressed in the motor neuron lineage during development (Fig. 29.10).

Although these interneurons have been proposed as candidates forming the rhythm generating circuitry of the CPG, it is important to note that more work needs to be done to prove their essential role in rhythmogenesis. Not all rhythmically active neurons are components of the CPG network. Some may receive rhythmic output from CPG neurons. The critical test is to show that temporarily removing a population of interneurons affects the locomotor rhythm. Recent strategies using genetic markers that identify specific populations of interneurons to drive selective expression of reversible inhibitors of activity provide a potential tool to test the contribution of distinct interneuron populations to CPG networks (Lechner *et al.*, 2002).

Reciprocal Inhibition between Flexor and Extensor Muscles

Alternation between flexor and extensor muscles that act across the same joint is another fundamental feature of locomotion, as well as of many other limb movements. There are several spinal interneuron circuits that act in different ways to produce reciprocal inhibition between antagonist muscles. Presynaptic inhibition of antagonist Ia afferents by GABA-ergic interneurons, described earlier, prevents spindle afferents from activating a muscle when it is stretched by the contraction of its antagonist. Another interneuron, the glycinergic Ia inhibitory interneuron (IaIN), produces postsynaptic reciprocal inhibition. The IaIN was so named because it is activated by the muscle's Ia afferents and inhibits motor neurons innervating the antagonist muscle. The IaIN synapses on the soma and main dendrites of motor neurons, producing a fast, potent IPSP that lasts only for several milliseconds. The IaIN develops from the V1 cell lineage, and can be recognized by its expression of the transcription factor *En1*. Almost every input that excites a motor neuron will also excite the IaINs that project to its antagonist's

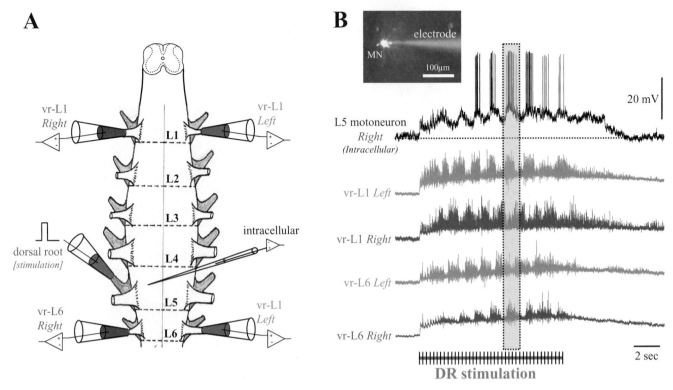

FIGURE 29.9 (A) Drawing of an isolated neonatal mouse spinal cord used in *in vitro* experiments. Suction electrodes are placed in the ventral lumbar (L) roots to record motor neuron activity (blue and red) or in the dorsal root (gray) to stimulate sensory afferents. Glass electrodes are used to record intracellularly from individual neurons in the spinal cord. (B) The image shows a motor neuron (MN) filled with a fluorescent dye and visualized with a microscope during intracellular recording. Sustained electrical stimulation of sensory afferents can evoke rhythmic activity in motor neurons (intracellular record) characterized by rhythmic membrane depolarization, which reaches the threshold for action potentials. This stimulation also results in alternating rhythmic activity between the left and right side of the spinal cord (compare the activity in L1-left (red) and L1-right (blue)). It also produces alternating activity between the upper (L1) and lower (L6) segments of the spinal cord (compare the 2 red traces or the 2 blue traces), as highlighted in the dotted box. The duration of afferent stimulation is indicated by the graded bar.

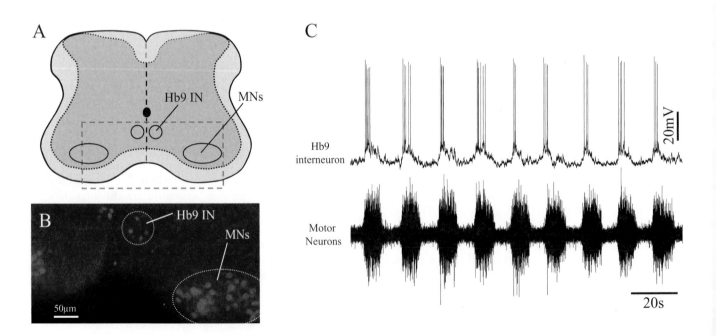

BOX 29.4

NONLINEAR PROPERTIES OF MOTOR NEURONS AND CPG INTERNEURONS

Experimental evidence from animal work indicates that motor neurons can participate actively in generating the rhythmic pattern observed during locomotion. Some motor neuron properties can be turned ON during locomotion due to synaptic inputs, which in turn produce nonlinear input-output relationships.

For example, there is a locomotor-dependent reduction in the threshold for spike generation, resulting in higher motor neuronal excitability. Two other motor neuronal properties susceptible to interact significantly with afferent inputs are the membrane potential oscillations known as *locomotor-drive-potentials* (LDP) and *plateau potentials*. LDPs are membrane potentials in motor neurons, in which a wave of depolarization during the active phase alternates with a wave of hyperpolarization during the antagonist phase of locomotion. Experiments in the cat have shown that the depolarizing phase is due to an excitatory synaptic drive and the hyperpolarizing phase is due to an inhibitory drive.

Plateau potentials are a behavior of the membrane potential in which the potential jumps between two states (bistable behavior). This motor neuron behavior has been observed in animals such as the turtle, mouse, and cat. This behavior is mediated mostly by a persistent inward current which at least in part is mediated by calcium channels. Ionic currents responsible for the generation of plateau potentials in motor neurons are predominantly occurring in dendrites because they posses voltage-sensitive L-type calcium conductances. These currents are turned ON by some neuromodulators such as serotonin or noradrenalin. Data from animal experiments have shown that synaptic inputs on dendrites can recruit plateau potentials and evoke sustained discharge. It is possible therefore that dendritic activation can amplify the effect of synaptic inputs from either sensory or descending afferents. Although the role of plateau potential is still unclear during normal development, one hypothesis is that motor tasks like posture or stepping recruit fatigue-resistant motor units readily with sustained discharge produced by persistent inward currents without continuous network inputs.

From animal experiments there are some preliminary data that indicate that plateau potentials and sustained discharge are also found in some spinal interneurons. As in motor neurons, these plateaus are facilitated by serotonin and noradrenalin. In a similar fashion as in motor neurons, plateau potentials activated in interneurons by neuromodulators released during locomotion may act as an amplification step of synaptic inputs in particular dendrites. It is possible therefore, because interneurons integrate sensory and supraspinal information, for this mechanism to be very effective in the transmission of inputs relevant for the generation of motor neuronal activities.

George Z. Mentis

motor neurons. For this reason, voluntary contraction of a muscle will produce inhibition of its antagonist muscle, and, likewise, the stretch reflex of a muscle will produce inhibition of its antagonist. The excitability of IaINs is determined by their many inputs, and the strength of reciprocal inhibition can be varied to allow movements that use cocontraction. IaINs are rhythmically active during locomotion, even without feedback from muscle stretch (Pratt and Jordan, 1987). This observation led to the proposal that IaINs were involved in producing the flexor-extensor alternation of the locomotor CPG. However, this proposal is not supported by more recent studies using neonatal mice in which the V1 lineage neurons were reversibly

FIGURE 29.10 (A) Drawing of a transverse section of the upper lumbar spinal cord from a neonatal mouse showing the location of the Hb9 interneuron nuclei bilaterally (annotated as Hb9 IN) as well as the motor neuron nuclei (annotated as MNs). The dotted box is shown in (B) as a fluorescence image revealing the Hb9 protein visualized by immunohistochemistry. The protein is present in motor neurons (dotted oval) and the small but distinct population of Hb9 interneurons in the ventro-medial part of the spinal cord (dotted circle). (C) Rhythmic firing in an Hb9 interneuron recorded intracellularly, and correlated rhythmic activity from a ventral root. The Hb9 interneuron activity was characterized by rhythmic membrane depolarizations underlying action potentials. This activity was in phase with the activity recorded from motor neurons (ventral root recording). Modified from Hinckley *et al.* (2005).

inactivated. In this study, the locomotor rhythms became slower, but still exhibited alternation of flexors and extensors (Gosgnach *et al.*, 2006).

Interneurons Involved in Left–Right Coordination

Interlimb coordination is the third fundamental feature of locomotion. To date, the leading candidate interneurons for coordinating activity between the limbs during locomotion are several populations of commissural interneurons (Fig. 29.11). Commissural interneurons have axons that cross the midline of the spinal cord at the level of the ventral commissure. They can be subdivided into three classes based on the direction of their axon: descending, ascending, or both (bifurcated axon). Descending commissural interneurons have been extensively studied in the rodent spinal cord during early development. Their axons project at least two spinal segments away from their soma to make contacts with other ventral horn interneurons and motor neurons. Some ascending commissural interneurons are cholinergic and are thought to transmit information to the upper parts of the spinal cord. Many of the commissural interneurons arise from the V0 and V3 cell lineage. The V0 lineage gives rise to ascending commissural interneurons, whereas the V3 lineage comprises a mixed population of commissural as well as ipsilaterally projecting neurons. In transgenic mice lacking the V0 commissural interneurons, the coordination between left and right legs was disrupted during locomotor rhythms, with episodic synchronous activation between sides (Lanuza *et al.*, 2004). These results suggest that this class of ascending commissural interneurons contributes, in part, to side-to-side interlimb coordination.

Recurrent (Renshaw) Inhibition

The recurrent collaterals of motor neurons synapse on a population of inhibitory interneurons in the ventral horn, called Renshaw cells. Renshaw cells use glycine or GABA as their transmitter and during development may contain both transmitters. Renshaw cells project back to homonymous motor neurons and synergist motor neurons in adjacent segments. They provide a negative feedback reflex called recurrent inhibition that limits the firing of motor neurons. During locomotor rhythms, Renshaw cells are rhythmically active, as expected since they receive rhythmic synaptic inputs from motor neurons. Their role in voluntary movements is uncertain, but one possibility is that they enhance the contrast between the active motor neurons and quiescent synergist motor neurons in adjacent segments. Renshaw cells arise from the V1 cell lineage, as do IaINs. During development, Renshaw cells, like IaINs, receive input from Ia afferents (Mentis *et al.*, 2006). In the studies in neonatal mice, mentioned earlier, reversible inactivation of the V1 lineage neurons that produced slowing of locomotor rhythms included Renshaw cells and IaINs (Gosgnach *et al.*, 2006).

Summary

In addition to serving as relays for sensory and descending pathways, interneurons of the spine interconnect to form CPG networks that can produce rhythmic firing. Several models of the CPG network have been proposed to guide experimenters. There has been significant progress in unraveling the interneuronal network that forms the spinal central pattern generator for locomotion. New experimental techniques have

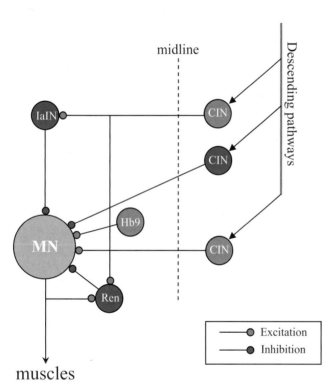

FIGURE 29.11 Schematic model with some of the known candidate interneurons that are involved in the modulation of the mammalian locomotor CPG. Excitatory interneurons (red) with rhythmic activity include classes of commissural interneurons (CIN) and the interneurons identified by expression of the HB9 transcription factor. Inhibitory interneurons (blue) that participate in termination of excitatory bursts include the Renshaw (Ren), the Ia interneuron (IaIN), and some commissural interneurons (CIN). Modified from Grillner (2006).

allowed the identification in neonatal rodents of candidate interneurons for mediating some of the fundamental properties of locomotion: rhythmicity, alternation of antagonist muscles, and interlimb coordination.

DESCENDING CONTROL OF SPINAL CIRCUITS

Neural circuits in the brainstem, cerebellum, and cerebral cortex are activated with locomotion and other movements. For locomotion, the cerebral cortex not only formulates conscious decisions about initiating and maintaining walking, but plays a role in adaptations of gait, particularly visually guided walking. Recordings from corticospinal neurons in cats as they stepped over obstacles placed in their field of vision showed that many corticospinal neurons changed their firing rates, often with a timing appropriate for driving the altered contraction patterns of specific muscles (Drew et al., 2004). Corticospinal input, affecting motor neurons directly or through spinal CPG interneurons, can alter the limb's trajectory with a high degree of precision, in a similar manner to a goal-directed reaching movement. In people, clinical observations and experimental studies suggest that supraspinal regions play a more active role in locomotor control than in lower mammals (Box 29.5).

Brain Stem Control of Spinal Interneuron Networks

The brain transmits signals for movement to the spinal cord through several descending fiber tracts: corticospinal, vestibulospinal, and reticulospinal, as described in earlier chapters. For locomotion, supraspinal regions are thought to activate spinal cord CPGs through brainstem relays. In decerebrate cats, electrical stimulation of several areas of the brain can elicit stepping. These include the diencephalon and cerebellum, but the most critical region is an area in the brain stem near the pedunculopontine nucleus known as the mesencephalic locomotor region (MLR). The MLR receives inputs from several forebrain regions, including the basal ganglia. The MLR does not connect directly to spinal CPG neurons, but synapses on reticulospinal neurons in the pons and medulla that receive a wide variety of sensory and descending inputs. Some of the reticulospinal axons in the pontomedullary formation travel in the ventral and ventrolateral tracts (funiculi) of the spinal cord. Lesions of these funiculi disrupt the ability to initiate and maintain locomotion.

Sparing of at least some fibers in the ventrolateral funiculus markedly improves locomotor recovery after spinal lesions in cats (Rossignol et al., 1999) and could be critical for the recovery of locomotor movements after spinal injury in humans.

In addition to their role in triggering rapid changes in locomotion and regulating postural balance, descending fiber systems also modulate physiological properties of spinal neurons. For instance, the plateau potentials and bistability in motor neurons (Box 29.4) depend on tonic activity in descending serotonergic pathways. Stimulation of the dorsolateral funiculus of the spinal cord in turtles, an area which is thought to contain descending pathways from the raphe nucleus, increased the excitability of motor neurons and promoted plateau potentials.

SENSORY MODULATION

Although spinal CPGs can generate the basic pattern of neural activity that organizes muscle contractions into a recognizable movement such as locomotion, the final pattern is fine-tuned by sensory inputs as well as descending systems. As noted earlier, most sensory inputs affect motor neurons through polysynaptic, interneuronal circuits. The interneurons that process sensory information typically integrate sensory signals with inputs from a variety of sources, including from CPG networks and descending pathways. As a result the reflex motor response can vary depending on the phase of the movement cycle, or it may occur only when the animal is in the "state" of moving, or it may even be reversed between resting and active states. Earlier we discussed the muscle sensory receptors and some of the reflex changes they mediate. This last section will present a sampling of other sensory reflex pathways.

Flexor Reflexes

A wide variety of sensory stimuli can evoke reflex flexion of the limb. Flexor reflexes may have several roles: they serve a protective function, reflexively withdrawing the limb from potentially harmful stimuli, and they may be used in movements that require coordinated multijoint flexor movements. Usually the reflex consists of contraction of muscles that flex several joints coupled with relaxation of muscles that extend those joints. Extension of the contralateral limb may also occur. Flexor reflexes can be elicited by group II and III muscle afferents, joint afferents, and high- and low-threshold cutaneous afferents. Because these

BOX 29.5

NEURAL CONTROL OF HUMAN WALKING

Human walking is a complex behavior that is based on integration of activity from descending supraspinal motor commands, spinal neuronal circuitries, and sensory feedback. Accumulating evidence suggests that humans, as well as other species, have a network in the spinal cord that is capable of generating basic rhythmic walking activity. Rhythmic leg movements can be induced by epidural electrical stimulation after a clinically complete spinal cord injury (SCI). Also, spontaneous involuntary rhythmic leg movements have been reported after a clinically complete and incomplete SCI. However, even though this evidence points to the existence of a CPG network in humans, in all these cases sensory input likely contributed to the rhythm generation. Furthermore, clinically complete injuries are not always anatomically complete. Therefore, the evidence for existence of a CPG in humans remains indirect.

Noninvasive electrophysiological techniques have provided information about how certain brain structures contribute to human walking. For example, the involvement of the primary motor cortex, where the corticospinal tract originates, has been demonstrated, in part, with transcranial magnetic stimulation (TMS). Several groups have found changes in the size of motor evoked potentials (MEP) in lower limb muscles elicited by TMS during walking, showing that transmission in the corticospinal tract is modulated during the gait cycle. However, TMS of the motor cortex activates cells with monosynaptic or polysynaptic connections to the spinal motor neurons. The size of the MEPs elicited by TMS reflects not only cortical excitability changes, but also changes at a subcortical level. To more precisely identify the various sites activated by the magnetic stimulus, Petersen and collaborators (1998) tested the effect of TMS on the H-reflex, an electrical homologue of the monosynaptic stretch reflex, during walking. They found that at least part of the MEP modulation was caused by changes in the excitability of corticospinal cells with direct monosynaptic projections to the spinal motor neurons. They also found that TMS over the motor cortex at intensities below the threshold for activating spinal motor neurons depressed ongoing muscle activity in the tibialis anterior muscle of the leg during walking. This effect was thought to be caused by activation of cortical inhibitory interneurons targeting cortical pyramidal tract neurons (Petersen et al., 2001).

During human walking, sensory feedback plays an important role in driving the active motor neurons and providing error signals to the brain that can be used to adapt and update the gait pattern. In healthy humans, a study in which a plantarflexion perturbation was induced during the stance phase of the gait cycle provides clear evidence for this. This experiment showed a drop in electromyographic (EMG) activity in the plantarflexor muscles, even when the common peroneal nerve that innervates the ankle dorsiflexor muscles was blocked by local anesthesia (Sinkjaer et al., 2000). These results demonstrated that sensory feedback from plantarflexor muscle contributed to the active drive of motor neurons during walking. Additional information about the contribution of sensory feedback for the control of locomotion has been gathered from studies that investigate changes in EMG responses in lower limb muscles after stimulation of cutaneous and proprioceptive afferents during walking. The amplitudes of the EMG responses to electrical stimulation are dependent on the phase (or time) of stimulation during the gait cycle. For example, electrical stimulation of the human sural nerve results in facilitation of the ankle flexor muscle tibialis anterior during early swing, but leads to suppression when delivered during late swing (reflex reversal). It has been proposed that the phase-dependent modulation of EMG responses may have functional relevance at various times in the step cycle. For example, a flexor reflex is appropriate at the onset of the swing phase when the leg is flexed, but the same reflex is not convenient at the end of the swing phase when the foot is ready to take up body weight. To some extent this modulation is mediated by supraspinal centers. To examine this possibility, a combination of cutaneous input and activation of supraspinal pathways by TMS has been used during the gait cycle. Available data suggest that long-latency reflexes elicited by cutaneous electrical stimulation are in part transcortically meditated.

In summary, in recent years a combination of neurophysiological techniques has provided important insights that have improved our understanding about the neural control of human walking. Human walking is based on the integration of activities from spinal circuitry, sensory feedback, and descending motor commands.

Monica A. Perez

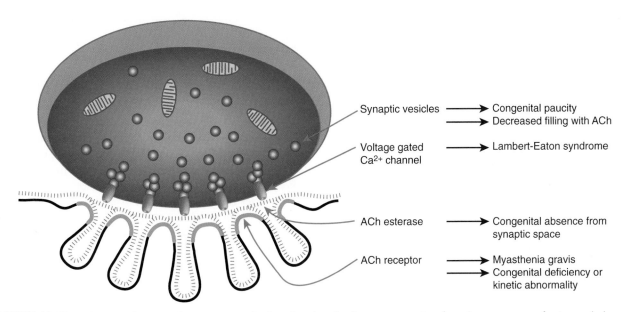

FIGURE 29.12 Schematic diagram of a neuromuscular junction showing key components subserving neuromuscular transmission, and the myasthenic disorders associated with these components.

afferents elicit a common action, they are sometimes called *flexor reflex afferents* (FRAs). The muscles activated by the sensory stimulus respond with short and long latency bursts of contraction. These contraction components are thought to be generated by distinct interneuron circuits. The late components can be differentially enhanced by treatment with L-dopa in spinalized animals. It has been suggested that interneurons mediating the late flexor reflex could function as a half-center for flexion, as proposed in models of the locomotor CPG.

Cutaneous Reflexes

In contrast to the widely distributed flexor reflex, there are a number of reflexes in which stimulation of small areas of skin evokes a reflex limited to a few muscles. One example of a local cutaneous reflex is the stumbling corrective reaction. When the skin on the top of the foot is stimulated during walking, the foot is lifted higher, as if to step over an obstacle. The stumbling corrective reaction is phase dependent, occurring only when the limb is in the swing phase of the step cycle. This and several other local cutaneous reflexes are oligosynaptic, involving only a few interneurons. The last-order interneurons in these cutaneous reflexes connect with relatively limited sets of motor neurons. It has been proposed that such interneurons offer a means to control or alter activity of specific muscles during movement without involving the interneurons of pattern generating circuitry.

Summary

Several areas of the brain give rise to descending pathways that connect to spinal interneurons and, to varying extents in different species, motor neurons. Areas of the brainstem are critical for relaying descending commands for locomotion. Volitional movements also seem to employ spinal interneuron circuits for organizing routine patterns of muscle contraction, within the context of a planned goal. Sensory feedback from skin, joints, and most muscle afferents also provide fine-tuning of basic movement patterns. The reflex motor responses evoked by most sensory afferent are relayed through interneurons.

BOX 29.6

PLASTICITY IN SPINAL CORD CIRCUITS

Over the last decade many studies have reported adaptive changes in spinal cord reflex circuits in animal and humans. Short- and long-term adaptations have been associated with the acquisition and learning of new motor skills. One of the pathways that has been investigated is the spinal stretch reflex (SSR) and its electrical analog, the H-reflex (Hoffman reflex). Operant conditioning has been used to show that descending activity from the brain can induce adaptations of the SSR and the H-reflex. In an operant conditioning protocol, monkeys, humans, and rats can gradually decrease (down-conditioning) or increase (up-conditioning) the size of the SSR or the H-reflex (Wolpaw and Tennissen, 2001). The acquisition of these simple motor skills can occur over days and weeks of practice.

The size of the SSR and the H-reflex is also affected by the nature, intensity, and duration of past physical activity and by the specificity of training. For instance, a cross-sectional study demonstrated smaller H-reflex sizes in ballet dancers than in nonathletic persons. This may be related to the frequent use of cocontraction of antagonistic ankle muscles to ensure stability when dancing (Nielsen et al., 1993). A short-term depression of the H-reflex size has been shown to occur during learning a novel motor task that requires a high degree of precision and attention. Individuals that trained to perform a task that required complex association between a visual and a motor response showed smaller reflexes than individuals training a task of similar duration and intensity but without any learning requirements (Perez et al., 2005). Other complex training paradigms, such as walking backward, stepping over obstacles, and cycling against changing levels of resistance have been shown to induce temporary changes in the H-reflex that are partly associated with the skill acquisition process.

The mechanisms underlying short- and long-term adaptations in the H-reflex during motor skill learning are incompletely understood. Since changes in descending drive and in the spinal cord itself occur during the learning process, it is likely that the mechanisms involve changes at multiple sites. On one hand, modulation in descending drive may contribute to maintain a tight control of spinal circuits during acquisition of novel motor skill tasks as part of the sensorimotor integration process. Also changes in presynaptic inhibition of the synapses between sensory afferents and motoneurones are important in the adaptation of the reflex circuitry during motor learning. Habituation of the monosynaptic gill-withdrawal reflex in Aplysia is caused by a depression of synaptic transmission between the sensory afferents and motoneurones through changes in presynaptic inhibition (Kandel et al., 2000). In rats and monkeys, short-term down-regulation of the H-reflex is likely to be related to changes in presynaptic inhibition, whereas a change in motor neuron firing threshold seems to be a mechanism associated with long-term down-regulation of the H-reflex during classical operant conditioning (Wolpaw and Tennissen, 2001). In humans, recent evidence suggests that acquisition of a visuo-motor skill also is associated with changes in presynaptic inhibition at the terminals of Ia afferent fibers (Perez et al., 2005).

Even though the role of spinal cord plasticity in motor learning remains largely unknown, accumulating evidence has demonstrated that spinal cord plasticity occurs in the course of motor learning and it contributes to skill acquisition. Plasticity in spinal cord circuits may be an important mechanism for establishing motor recovery after motor disorders including spinal cord injury.

Monica A. Perez

References

Kandel, E. R., Schwartz, J. H., and Jessel, T. M. Principles of neural science. 4. USA: The Mcgraw-Hill Companies, Inc; 2000. pp. 1247–1279. Chapter 63.

Nielsen, J., Crone, C., and Hultborn, H. (1993). H-reflexes are smaller in dancers from The Royal Danish Ballet than in well-trained athletes. Eur J Appl Physiol Occup Physiol. **66**, 116–121.

Perez, M. A., Lungholt, B. K., and Nielsen, J. B. (2005). Presynaptic control of Ia afferents in relation to acquisition of a novel visuo-motor skill in healthy humans. J Physiology. **568**, 343–354.

Wolpaw, J. R., and Tennissen, A. M. (2001). Activity-dependent spinal cord plasticity in health and disease. Annu Rev Neurosci. 24, 807–843.

References

Brown, T. G. (1911). The intrinsic factors in the act of progression in the mammal. *Proc. R. Soc. London* **84**, 308–319.

Boillee, S., Vande Velde, C., and Cleveland, D. W. (2006). ALS: A disease of motor neurons and their nonneuronal neighbors. *Neuron* **52**, 39–59.

Bonnot, A., Whelan, P. J., Mentis, G. Z., and O'Donovan, M. J. (2002). Locomotor-like activity generated by the neonatal mouse spinal cord. *Brain Res. Rev.* **40**, 141–151.

Burke, R. E., Levine, D. N., Tsairis, P., and Zajac, F. E. (1973). Physiological types and histochemical profiles in motor units of the cat gastrocnemius. *J. Physiol. (Lond.)* **234**, 723–748.

Cooper, E. C. and Jan, L. Y. (1999). Ion channel genes and human neurological diseases: Recent progress, prospects, and challenges, *Proc. Natl. Acad. Sci. USA* **96**, 4759–4766.

Drew, T., Prentice, S., and Schepens, B. (2004). Cortical and brainstem control of locomotion. *Prog. Brain Res.* **143**, 251–261.

Gosgnach, S., Lanuza, G. M., Butt, S. J., Saueressig, H., Zhang, Y., Velasquez, T., Riethmacher, D., Callaway, E. M., Kiehn, O., and Goulding, M. (2006). V1 spinal neurons regulate the speed of vertebrate locomotor outputs. *Nature* **440**, 215–219.

Goulding, M. and Pfaff, S. L. (2005). Development of circuits that generate simple rhythmic behaviors in vertebrates. *Current Opinion Neurobiology* **15**, 4–20.

Grillner, S. (2003). The motor infrastructure: from ion channels to neuronal networks. *Nature Review Neuroscience* **4**, 573–586.

Grillner, S. (1985). Neurobiological bases of rhythmic motor acts in vertebrates. *Science* **228**, 143–149.

Henneman, E. (1957). Relation between size of neurons and their susceptibility to discharge. *Science* **126**, 1345–1347.

Hinckley, C. A., Hartley, R., Wu, L., Todd, A., and Ziskind-Conhaim, L. (2005). Locomotor-like rhythms in a genetically distinct cluster of interneurons in the mammalian spinal cord. *J. Neurophysiol.* **93**, 1439–1449.

Jessell, T. M. (2000). Neuronal specification in the spinal cord: Inductive signals and transcriptional codes. *Nature Review Genetics* **1**, 20–29.

Kullander, K., Butt, S. J., Lebret, J. M., Lundfald, L., Restrepo, C. E., Rydstrom, A., Klein, R., and Kiehn, O. (2003). Role of EphA4 and EphrinB3 in local neuronal circuits that control walking. *Science* **299**, 1889–1892.

Lanuza, G. M., Gosgnach, S., Pierani, A., Jessell, T. M., and Goulding, M. (2004). Genetic identification of spinal interneurons that coordinate left-right locomotor activity necessary for walking movements. *Neuron* **42**, 375–386.

Lechner, H. A., Lein, E. S., and Callaway, E. M. (2002). A genetic method for selective and quickly reversible silencing of Mammalian neurons. *J. Neurosci.* **22**, 5287–5290.

Mentis, G. Z., Alvarez, F. J., Bonnot, A., Richards, D. S., Gonzalez-Forero, D., Zerda, R., and O'Donovan, M. J. (2005). Noncholinergic excitatory actions of motoneurons in the neonatal mammalian spinal cord. *Proc. Natl. Acad. Sci. USA* **102**, 7344–7349.

Mentis, G. Z., Siembab, V. C., Zerda, R., O'Donovan, M. J., and Alvarez, F. J. (2006). Primary afferent synapses on developing and adult Renshaw cells. *J. Neurosci.* **26**, 13297–13310.

Nishimaru, H., Restrepo, C. E., Ryge, J., Yanagawa, Y., and Kiehn, O. (2005). Mammalian motor neurons corelease glutamate and acetylcholine at central synapses. *Proc. Natl. Acad. Sci. USA* **102**, 5245–5249.

Nielsen, J., Crone, C., and Hultborn, H. (1993). H-reflexes are smaller in dancers from The Royal Danish Ballet than in well-trained athletes. *Eur. J. Appl. Physiol. Occup. Physiol.* **66**, 116–121.

Petersen, N., Christensen, L. O., and Nielsen, J. (1998). The effect of transcranial magnetic stimulation on the soleus H reflex during human walking. *J. Physiol. (Lond.)* **513**, 599–610.

Petersen, N. T., Butler, J. E., Marchand-Pauvert, V., Fisher, R., Ledebt, A., Pyndt, H. S., Hansen, N. L., and Nielsen, J. B. (2001). Suppression of EMG activity by transcranial magnetic stimulation in human subjects during walking. *J. Physiol. (Lond.)* **537**, 651–656.

Pratt, C. A. and Jordan, L. M. (1987). Ia inhibitory interneurons and Renshaw cells as contributors to the spinal mechanisms of fictive locomotion. *J. Neurophysiol.* **57**, 56–71.

Rossignol, S., Drew, T., Brustein, E., and Jaing, W. (1999). Locomotor performance and adaptation after partial or complete spinal cord lesions in the cat. *Prog. Brain Res.* **123**, 349–365.

Sinkjaer, T., Andersen, J. B., Ladouceur, M., Christensen, L. O., and Nielsen, J. B. (2000). Major role for sensory feedback in soleus EMG activity in the stance phase of walking in man. *J. Physiol. (Lond.)* **523**, 817–827.

Wolpaw, J. R. and Tennissen, A. M. (2001). Activity-dependent spinal cord plasticity in health and disease. *Ann. Rev. Neurosci.* **24**, 807–843.

Suggested Readings

Engel, A. G. and Sine, S. M. (2005). Current understanding of congenital myasthenic syndromes. *Curr. Opin. Pharmacol.* **5**, 308–321.

Grillner, S. (2006). Biological pattern generation: the cellular and computational logic of networks in motion. *Neuron* **52**, 751–766.

Kiehn, O. (2006). Locomotor circuits in the mammalian spinal cord. *Annu. Rev. Neurosci.* **29**, 279–306.

O'Donovan, M. J., Bonnot, A., Wenner, P., and Mentis, G. Z. (2005). Calcium imaging of network function in the developing spinal cord. *Cell Calcium* **37**, 443–450.

Schwab, J. M., Brechtel, K., Mueller, C. A., Failli, V., Kaps, H. P., Tuli, S. K., and Schluesener, H. J. (2006). Experimental strategies to promote spinal cord regeneration—An integrative perspective. *Prog. Neurobiol.* **78**, 91–116.

Mary Kay Floeter and George Z. Mentis

30

Descending Control of Movement

Four major pathways descend from the brain to the spinal cord to control muscles that move the skeleton. These pathways arise in the vestibular nuclei (vestibulospinal), in the brain stem reticular formation (reticulospinal), in the red nucleus (rubrospinal), and in the cerebral cortex (corticospinal). Although these four descending pathways work together to provide seamless control of movements from postural reflexes to delicate manipulation, they can be separated broadly into two main systems. Vestibulospinal and reticulospinal pathways can be grouped together as a medial system: their axons descend through the brain stem and spinal cord close to the midline, and chiefly innervate axial and proximal limb musculature whose motoneurons lie medially in the ventral horn. The corticospinal and rubrospinal pathways together can be considered a lateral system: their axons descend in the lateral column of the spinal cord and chiefly innervate limb musculature, particularly distal musculature, whose motoneurons lie laterally in the ventral horn.

More that just an anatomical distinction, the medial and lateral systems have different principal functions. The medial system provides postural control (Box 30.1). Monkeys with medial system lesions fall over when they attempt to ambulate and climb, but when supported, they can use their hands and fingers adeptly in retrieving food pieces from narrow holes (Lawrence and Kuypers, 1968b). The lateral system provides fine control of voluntary movement. Monkeys with lateral system lesions rapidly recover the use of all four extremities in activities such as ambulation and climbing, but they are unable to make the fine finger movements needed to extract small pieces of food from narrow holes (Lawrence and Kuypers, 1968a). This chapter examines the contributions of first the medial

and then the lateral descending systems to control of movement.

THE MEDIAL POSTURAL SYSTEM

Vestibular and Reticular Nuclei Control Posture and Reflex Behaviors

Early experimenters found that some degree of postural and reflex control remains after higher brain centers are completely cut off from the brain stem and spinal cord, and that the extent of preserved motor function is determined by the level of the transection that disconnects the upper portions of the brain (Fig. 30.1). There is no functional regeneration of central neuronal pathways after damage, and when the spinal cord is separated from all the brain above it, as can occur in motor vehicle or equestrian accidents, the result is quadriplegia, a loss of all voluntary movement and postural control. Only the spinal reflexes discussed in the preceding chapter return, and those return only when spinal cord neurons recover excitability after a prolonged period of limp (flaccid) paralysis called spinal shock.

Transections at higher levels of the neuraxis leave part of the brain stem connected with the spinal cord. Chronic bulbospinal cats, those with medulla and spinal cord below the transection, show primitive attempts at righting themselves when placed on one side. Effective righting may be observed in chronically surviving animals subjected to transection that leaves the mesencephalon intact below the level of the cut. As the level of transection is raised, posture and balance become progressively closer to normal. The decorticate

---

BOX 30.1

NEUROTRANSMITTERS AND POSTURE

Although most of the work on postural reflexes has concerned mammalian responses and neurophysiological mechanisms, invertebrate studies offer simpler systems that may be more directly manipulated, especially by neuropharmacological agents. The work of Kravitz and colleagues exemplifies this approach. They found that systemic administration of the aminergic neurotransmitter serotonin (Chapter 7) causes a lobster to assume a flexed posture characteristic of defensive responses. Octopamine, in contrast, causes the animal to lie flat with its legs extended. These responses are not due to the direct action of serotonin and octopamine on muscles; both

agents stimulate muscular contraction. However, serotonin and octopamine have opposite effects on extensor and flexor motor neurons: serotonin excites flexor motor neurons and is selectively localized in a subset of interneurons. The conclusion is that serotonin acts on the lobster central nervous system to trigger the flexion posture, whereas octopamine triggers extension. In mammals, axons descending from the reticular formation release neurotransmitters, including serotonin, that have analogous modulatory functions.

Marc H. Schieber and James F. Baker

---

animal produced by the ablation of cerebral cortical tissue has many apparently normal postural reactions, lacking only certain placing and stepping reactions. Progressively higher brain transections enable more of the vestibular (balance) and proprioceptive (muscle and position sense) circuitry that participates in postural control, as well as improving the overall balance between excitatory and inhibitory influences on the spinal cord.

Nuclei of the brain stem reticular formation with intact spinal projections after high brain stem transection include the locus ceruleus and raphe nuclei, the origins of descending noradrenergic and serotonergic neuromodulatory axonal projections that alter excitability of neurons in the spinal cord, as discussed in the preceding chapter. Parts of the raphe nuclei also have a role in circadian rhythms (Chapters 41 and 42), and projections of the locus ceruleus are involved in attention (Chapter 48) and internal reward (Chapter 43). Other areas of the reticular formation involved in motor control have specific functions in generation of the locomotor rhythm (MLR, mesencephalic locomotor region, Chapter 29) and rapid eye movements (PPRF and MRF, Chapter 33). Nuclei of the reticular formation are studied most often from the perspective of their contributions to specific functional systems described in other chapters and through the actions of their monoamine neuromodulators noradrenalin (norepinephrine) and serotonin. Vestibular nuclei are studied for their contributions to postural and oculomotor reflexes (Chapter 33).

The Vestibular Apparatus Senses Head Rotation and Tilt

The sensory inputs most important for postural responses are vision, proprioception, and vestibular sense. Vestibular sensory signals from the labyrinth of the inner ear provide direct information about rotations of the head and its orientation with respect to gravity. Two types of vestibular transducer organs— semicircular canals and otolithic maculae—are located within the labyrinth of the ear and respond to accelerations of the head in space. The three semicircular canals and the two otolithic maculae are stimulated by rotational and linear acceleration, respectively. The mechanical structure of the sense organ determines the nature of the effective stimulus (Wilson and Melvill-Jones, 1979).

The labyrinth (Fig. 30.2) (Box 30.2) is outlined by the bony labyrinth, a set of passages in the skull. The bony labyrinth is lined with membranes that contain endolymph (Chapter 26) and that make up the membranous labyrinth. The membranous labyrinth has a large central vessel called the utricle, another vessel called the saccule, and three narrow passages that emerge from the utricle and loop around to rejoin it. These three loops form circular passages for endolymph and are known as semicircular canals. The canals are oriented orthogonal to one another: there is a lateral or horizontal canal, an anterior or superior canal, and a posterior canal. The signals from the semicircular canals are discussed along with the eye movement

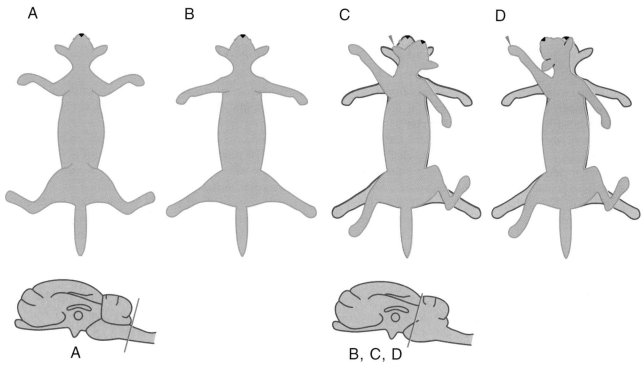

FIGURE 30.1 Posture following transection of the neuraxis. (A) The limp posture of a cat immediately after separation of the spinal cord from the brain. Limbs are neither flexed nor extended. The level of transection is shown below in a sagittal view. (B) The rigid posture of the decerebrate quadruped. All four limbs are extended. The transection, shown below, has been made through the pons. (C) Adjustment of the posture of the decerebrate preparation in response to stimulation of the pinna. All four limbs and the head alter their posture. (D) Adjustment of decerebrate posture in response to stimulation of the forepaw. Again, limbs and neck adjust. In this case the head turns toward the stimulus. Based on reports by Sherrington.

reflexes they generate (Chapter 33), and are introduced here only briefly here. A semicircular canal is excited by rotational accelerations, and the mechanical structure of the canal converts acceleration into a rotational velocity signal. The three canals in each labyrinth are oriented for excitation by rotations toward that side of the body, horizontally, diagonally forward, and diagonally backward for the horizontal, anterior, and posterior canals, respectively. Thus, each labyrinth is excited by head rotation toward its side, called ipsilateral rotation. Canal afferents fire action potentials at a rate approximately proportional to the velocity of head rotation in the direction of the canal orientation. Both labyrinths are excited by purely forward or backward rotation, the anterior canals excited during the forward phase and the posterior canals excited during the backward phase.

There are two otolith organs in mammals, the utricle (utriculus) and saccule (sacculus). The names apply to the fluid chambers in which they lie, and the specific locations of the receptors are called the maculae. The two maculae are small regions that contain hair cells innervated by vestibular afferents. Like auditory and semicircular canal hair cells, bending of their cilia is

the excitatory stimulus for macular hair cells, and the cilia are imbedded in a gelatinous mass above the cell bodies. Unlike canal hair cells, the otolithic hair cell maculae rest in large chambers, in which fluid motion is thought to exert little or no force on the cilia. Instead, the gelatinous mass holding the cilia contains dense crystals, the otoconia or statoconia, and the entire mass above the hair cells of a macula is acted on by gravity or linear acceleration because of its higher density than the endolymph. The otoconial mass sags in the direction of a head tilt. This sagging, or the equivalent lagging of the mass behind a linear acceleration, provides the natural stimulus for otoliths.

The utricular macula lies horizontally on the floor of its cavity. When the head is still and held approximately erect with respect to gravity, there is no stimulus to this organ, but tilts from the erect posture will excite utricular afferents. Microscopic examination of the utricular macula shows that the hair cells are not all oriented with their cilia arranged in the same direction, and different directions of tilt excite different populations of utricular afferents. The direction of the hair cell indicated by its structural orientation is called its morphological polarization vector and corresponds

BOX 30.2

POSITIONAL NYSTAGMUS

As many as one person in ten suffers from repeated or persistent dizziness at some point in life, typically accompanied by nystagmus (see Chapter 33 for a description of these eye rotations). Often, the dizziness and nystagmus are evoked by a sudden change in head orientation, identifying the disorder as positional nystagmus, also known as benign positional paroxysmal nystagmus (BPPN) or benign positional vertigo (BPV). The usual cause of BPV is debris in the posterior semicircular canal. This debris stimulates the canal receptors when the head is rotated, eliciting a sensation of rotation and a vestibulo-ocular reflex nystagmus. The debris may be otoconia dislodged from the otoliths, and it can be either free-floating (canalithiasis), or stuck to the canal receptor surface (cupulolithiasis). When the head is repositioned, the debris either stirs the endolymph as it falls to a new location in the canal semicircle, or its weight shifts on the receptors, in either case causing abnormal canal stimulation. BPV may be self-correcting after days to months, or its treatment may require a sequence of head rotations aimed at emptying the debris back into the utricle where it will not stimulate the canal.

A related positional nystagmus experienced by many party-goers is positional alcohol nystagmus (PAN), which is similar to the condition where debris is stuck to the canal receptor surface. Rapid consumption of alcohol leads to decreased blood density, which makes the canal receptors less dense than the more slowly absorptive endolymph. When the intoxicated person lies down and turns the head to one side, the canal receptors float up in the endolymph, as if pushed by a head rotation, and a vigorous nystagmus can occur. PAN is peripheral in origin, not among the central effects of the drug. This has been shown by having subjects drink deuterium water ("heavy water"), which makes the canal receptors sink in the endolymph and causes nystagmus in the opposite direction to PAN.

Marc H. Schieber and James F. Baker

---

to the best direction of tilt for exciting the receptor. Each utricular macula has hair cell orientations capable of responding to any direction of head tilt, but ipsilateral tilt excites a predominant number of afferents. Thus, as for the semicircular canals, we say that ipsilaterally directed head motion is excitatory. The saccular macula is placed vertically on the side of the saccule, approximately in a parasagittal plane. There is a strong acceleration stimulus to the saccular macula when the head is erect and little or no stimulus when the head is lying on either side. The sacculus responds vigorously when the head bobs up and down, as during locomotion, but the exact functions of the sacculus are debated and are beyond the scope of this chapter. In summary, two types of vestibular signals are available to assist balance: ipsilaterally directed rotational head velocity is signaled by the semicircular canals and ipsilaterally directed head tilt or an equivalent linear acceleration is signaled by the utricular otoliths.

Vestibulocervical and Vestibulospinal Reflexes Stabilize Head and Body Posture

Vestibulocervical (neck or collic reflexes) and vestibulospinal reflexes make use of canal and otolith signals to stabilize the posture of the head and body. These reflexes are thought to serve two functions, both as stabilizing, or "negative feedback" systems (Schor et al., 1988). First, when the head and body rotate or tilt in any direction, the vestibular stimulus excites pathways that contract neck and limb muscles that oppose the motion so that the undesired movement is reduced or corrected. Second, the biomechanical components of the head have a characteristic resonant frequency of around 2 to 3 Hz at which oscillation is especially likely to occur, and the vestibulocervical reflex effectively dampens this tendency for oscillatory motion. The importance of vestibulocervical and vestibulospinal reflexes is demonstrated by damage to the labyrinth, section of the VIIIth nerve, or mechanical blockage of the semicircular canal passages. Unilateral labyrinthectomy causes an initial postural disability, with leaning or falling toward the side of the lesion. Bilaterally symmetric semicircular canal plugging, which removes head velocity signals with little loss or imbalance in tonic vestibular nerve activity, produces head instability with oscillations that may persist for several days. Note that the vestibulocervical reflex, although discussed separately, is as much a vestibulospinal reflex as vestibular limb reflexes, the difference being primary action at the cervical versus lumbar spinal levels.

The neuroanatomical pathways that contribute to the vestibulocervical and other vestibulospinal reflex responses (Fig. 30.3) have been elucidated primarily through the use of electrical stimulation and other physiological methods. Of the four brain stem vestibular nuclei that receive the VIIIth nerve, the medial, lateral (Deiters' nucleus), and descending (inferior nucleus) contribute to the medial and lateral vestibulospinal tracts that provide direct vestibular influence on the spinal cord. The superior vestibular nucleus is concerned primarily with vestibulo-ocular reflexes (Chapter 33). Electrical activation of the vestibular pathways shows that vestibulocervical reflexes are mediated primarily by excitatory and inhibitory connections of the bilaterally projecting medial vestibulospinal tract and that ipsilateral excitation of limb extensors is carried via the lateral vestibulospinal tract. The bulk of the lateral vestibulospinal tract projects

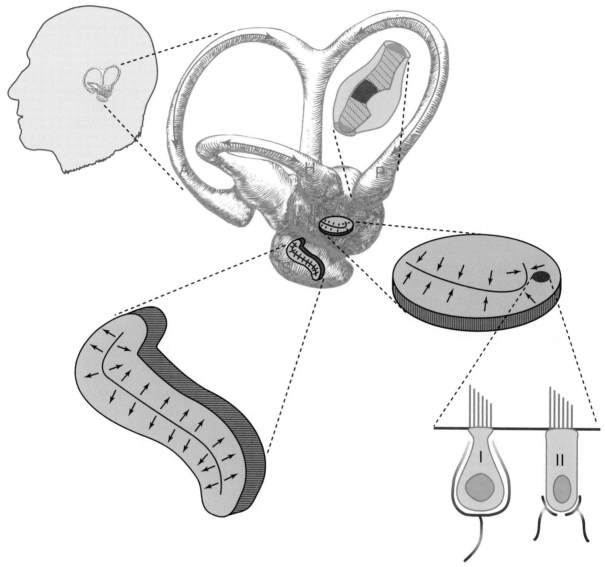

FIGURE 30.2 Vestibular canals and otoliths. The position and orientation of the labyrinth (not to scale) in the head are shown at the upper left of the figure. An enlarged view of the labyrinth shows the directions of head rotation and endolymph flow (red arrows) that excite each of the three semicircular canals. The horizontal orientation of the utricular macula and vertical orientation of the saccular macula are shown schematically. More highly magnified views of the receptor regions of a canal and of the otolith organs are shown with the cupula of the canal colored dark gray and the best directions for excitation of otolith hair cells marked by black arrows. At the lower right are two anatomical types of hair cell: the calyx or type I and bouton-ending or type II receptor. The tallest cilial extension on each cell is the kinocilium. A, anterior semicircular canal; H, horizontal semicircular canal; P, posterior semicircular canal; S, saccule containing saccular macula; U, utricle containing utricular macula; I, type I receptor; II, type II receptor. Based on studies reviewed by Wilson and Melvill-Jones.

from the lateral vestibular nucleus as far as the lumbar spinal cord for vestibulospinal reflex control of upright body posture. Vestibular nuclei also project heavily to reticular nuclei, and only lesions that impinge on both vestibulospinal and reticulospinal pathways substantially reduce the strength of vestibular reflex responses.

The bilateral, mixed excitatory and inhibitory connections of the vestibulocervical reflex circuits indicate that its actions are complex, and the considerable effort devoted to explicating the exact nature of these actions will be summarized here. The excitatory lateral vestibulospinal tract projection to ipsilateral extensor motor neurons offers a simple example of negative feedback: When the head and body tilt to one side, canals and otoliths of the labyrinth on that side are excited, and the lateral vestibulospinal tract carries this excitation to the ipsilateral leg extensor motor neurons. The ipsilateral leg is thus extended to oppose the tilt and maintain upright posture. An upcoming section will discuss studies of this response in the context of coordinated postural reactions.

Initial quantitative studies of the vestibulocervical reflex used sinusoidally oscillating stimuli and concentrated on the timing of the reflex head torques or the electromyographic activity that accompanied neck muscle contractions, just as response timing had been the focus in studies of vestibular primary afferents. The conclusion from these studies was that vestibulocervical reflex circuitry in the vestibular nuclei introduces central neural processing with two important features. First, at low frequencies a portion of the velocity signal from the semicircular canals is converted to a head position signal, appropriate for repositioning the head in response to a slow change in head angle. Second, at high frequencies, vestibulocervical reflex circuitry appears to introduce an anticipatory or "phase lead" response, which works in conjunction with the canal afferent signal to make very rapid responses so that the behavioral output of the reflex opposes head angular acceleration, as would be appropriate if the inertia of the head were an important factor to be overcome for compensation during rapid head motion. The mechanisms by which neurons in the brain stem accomplish these functions are unknown.

Timing of vestibulocervical responses is reasonable given the function of the reflex and the mechanical properties of the load presented by the head. The mechanical properties of the multijoint system of the limbs and trunk are far more complex, and it is correspondingly more difficult to predict what the timing of the controlling neural signals should be. Electromyographic recordings during vestibulospinal reflexes in decerebrated animals have shown that limb extensors at low frequencies of oscillation are excited in phase with the position of the head. For example, when the head rolls slowly to the left, the left forelimb extends as if to brace against further displacement. The likely source of signals exciting these responses is the otolith organs. As the frequency of head oscillation increases, the timing of vestibulospinal responses corresponds more closely to head rotation velocity. The value of this timing is debatable, but the implication is that semicircular canal signals begin to predominate over otolith signals in vestibulospinal responses at higher frequencies.

A major focus of work on both vestibulocervical and vestibulospinal reflexes is the spatial organization of the responses and the possible interaction of spatial properties and timing (Baker et al., 1985). Early studies had concentrated on left-to-right motion (yaw) for vestibulocervical reflex studies and on left ear down to right ear down rolling motion for vestibulospinal limb reflex studies. Of course, the vestibular reflexes work to compensate for motion in any direction. The problem of producing the correct direction of reflex response to any direction of disturbance has been considered thoroughly for the case of the vestibulo-ocular reflex that stabilizes the eyes (Chapter 33), but vestibulospinal reflexes present a more difficult situation.

In the case of the vestibulocervical reflex, there are many ways that the more than 30 muscles of the neck could be used to compensate for a particular direction of head rotation. What synergies or coordination of muscle groups does the brain use to generate compensatory vestibulocervical responses? The principles underlying the brain's pattern of coordinated activation of neck muscles to compensate for head rotation are unknown, although some features of spatial organization have been elucidated. Each of the major neck muscles exhibits a characteristic directionality of excitation in response to vestibular stimulation by rotation in many different directions, but that direction does not match the directional sensitivity of any single semicircular canal. Instead, the responses to rapid motions reflect a weighted sum of canal inputs, and various proposals attempt to account for the particular weightings observed.

A complicating feature of vestibulocervical spatial organization is the dependence in some cases of the timing of vestibulocervical neuron or neck muscle activation upon the direction of the vestibular stimulus. This dependence has been termed spatial-temporal convergence and may reflect canal and otolith signal addition to match a varying mechanical load presented by the head for different directions of rotation. Another difficulty in the analysis of vestibulocervical responses is the presence of signals related to the velocity of eye movement or the direction in which the eyes are pointed within their orbits. Add to this the visual and proprioceptive signals present in vestibulospinal neural circuitry, discussed in the following sections, and the overall picture is one of a rich and complex control system that we are only beginning to understand. Comparable information on the spatial organization of vestibulospinal control of the limbs has begun to reveal a similar complexity.

The Cervicocervical Reflex Stabilizes the Head by Opposing Lengthening of Neck Muscles

The proprioceptive contribution to reflex stabilization of the head and body has been examined in many of the same ways as vestibular contributions (Peterson *et al.*, 1985). The sensory situation for proprioceptive signals is more complicated than for vestibular signals, as any of the muscles of the neck, trunk, or limbs could provide an input to influence any other muscle or synergistic group of muscles. Only the signals from neck proprioception that activate neck muscles (i.e., the neck stretch reflex or cervicocervical reflex, also called the cervicocollic reflex) have been studied extensively. Like the stretch reflex for limb extensors, the cervicocervical reflex opposes lengthening of the muscles concerned and so is a negative feedback compensatory system. Other proprioceptive systems often are considered postural reactions rather than simple reflexes and are described in turn.

The cervicocervical reflex in alert animals is not consistent, so the reflex has been analyzed by rotating the trunk of decerebrated cats while holding the head fixed in space. This stimulates neck proprioceptors without any vestibular input. The results are reminiscent of the vestibulocervical reflex. At low frequencies of neck rotation cervicocervical reflex muscle contractions correspond to the extent of muscle stretch, but at higher frequencies the responses occur more rapidly, corresponding to the velocity or acceleration of muscle stretch. The directionality of neck muscle responses to stretch matches that of vestibulocervical responses fairly well. Common directionality implies that signals for vestibular and stretch reflexes to neck muscles share some central circuitry. Not only does this directionality not match that of any single semicircular canal, it also does not appear to match the direction of pulling actions of the neck muscles under study. We are left to speculate on the principles of central organization of these reflexes. The central pathways mediating the cervicocervical reflex are more varied still, and less well mapped, than those for vestibulocervical reflex responses. There are homonymous monosynap-

tic stretch reflex connections, heteronymous connections to a muscle from other muscles, and longer pathways that may involve connections through medial, lateral, and descending vestibular nuclei as well as reticulospinal neurons.

The Brain Stem Controls Coordinated Postural Reactions

A variety of reflexes contribute to the maintenance of overall body posture. All benefit from the integrity of connections with the brain stem, and in some cases higher centers, but not all are characterized as readily by objective measures as the vestibulocervical, vestibulospinal, and cervicocervical reflexes already discussed. The tonic neck or cervicospinal reflexes are the best known of these, but the supporting reactions, placing reactions, righting reactions, hopping or stepping reactions, and others are useful parts of the neural control of posture. The tonic neck reflex adjusts the extension of the limbs in response to the angle of the head on the trunk. Roberts (1967) showed that stretching the neck by rolling the spine to one side elicited tonic flexion and extension of the forelimbs, as if the support surface had been tilted to produce the rotation (Fig. 30.4). The placement of a limb on the ground initiates a set of reflex reactions that stiffen the limb into a supporting pillar. This response is called the "positive supporting reaction" and depends on the integrity of the brain stem. As discussed earlier, righting reactions also depend on the brain stem, and "optical righting reflexes" mediated by vision require an intact cerebral cortex.

Stable posture requires that the feet be not only rigid but placed correctly on a supporting surface, and the placing reactions of quadrupeds contribute to this by moving the feet toward a visible surface (visual placing reaction) or onto a surface that has tactile contact with the top of the foot, chin, or whiskers (tactile placing reactions). Although a tactile placing reaction may be evoked in primitive form in a spinalized animal, it is generally held that at least the lower brain stem must be intact for tactile placing to occur in an effective form. Finally, if balance reactions are insufficient to maintain stable posture, as when the support surface is moved beneath the foot, the limb may hop to a new position where stable posture is possible (hopping reaction).

The action of the righting reflex of cats during falling is familiar to all. Humans show a similar reaction to an unexpected drop, measurable as a short latency electromyographic response in the gastrocnemius muscle. In cats, this response survives blockage of the semicircular canals, but not total labyrinthectomy, and so appears to be otolith mediated. The action of these many postural reactions in the context of maintenance of posture and balance in intact humans has been the focus of several studies utilizing movable posture platforms.

Balance Depends on Context-Dependent Postural Strategies

How then do these descending pathways from the brain stem coordinate to produce the postural behaviors of intact animals? The basic problem of standing posture is keeping the center of mass of the body, also called center of gravity, positioned over the base of support provided by the feet so that the body does not topple. The ranges of body positions that rest over the base of support define the limits of stability. When the extremes of body sway extend beyond those limits, the body must reduce them, take a step, or fall. An important tool for exploring body stability is the posture platform, a base on which subjects can stand and be subjected to displacements of the underlying supporting surface. The posture platform is used with body position sensors and electromyography to evaluate standing posture (Nashner, 1982; Horak and Nashner, 1986; Horak and Shupert, 1994). Even on a perfectly stable support surface with vision, somesthesis, and vestibular sense all active, a standing person will sway slightly. This sway will increase as the context is altered by distorting or removing sensory input, showing the importance of the multiple avenues of balance information but also showing the adequacy of limited sensory input in maintaining upright posture. When the eyes are closed (the Romberg test used in clinical evaluation), sway is greater. When the visual surround is linked to the subject's head position in the condition called "sway referencing of vision," visual information is misleading and sway is increased. When the support surface is compliant, proprioceptive feedback is much less useful, and sway is greater. When the support surface tilts along with the body in the condition called "sway referencing of the support surface," proprioceptive information is misleading and sway is increased. When the preceding conditions are combined, sway is greater still but normal subjects can maintain upright posture through the use of vestibular information.

Sudden displacement of the platform on which a subject stands allows measurement of the latencies and patterns of electromyographic activity in the postural muscles of the legs. When the subject is standing on a large platform that is displaced forward or backward, the leg muscles contract in sequence from ankle to thigh to hip, and motion occurs primarily about the

ankle joint. When the platform displacement is forward, the ankle flexor tibialis anterior on the front of the calf is first to be excited (flexion is the toe-up direction), at about 80–100 ms after the displacement. This is followed about 20 ms later by contraction of the quadriceps thigh muscles and then later still by contraction of trunk musculature. Backward displacement first elicits excitation of the gactrocnemius muscle at the back of the calf to extend the ankle, followed by excitation of thigh and trunk muscles antagonistic to those excited by forward displacement (Fig. 30.5). The overall postural reaction of distal-to-proximal excita-

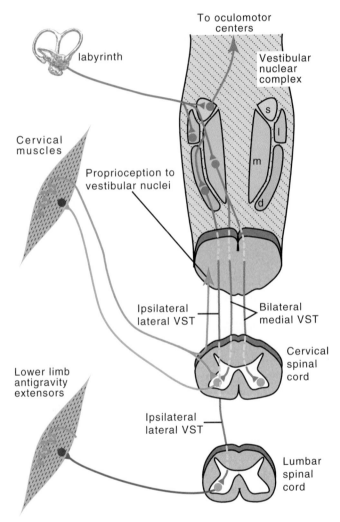

FIGURE 30.3 Vestibular and proprioceptive reflex signal inputs and major pathways from the brain stem vestibular nuclei. The medial vestibulospinal tract projects bilaterally to the cervical spinal cord to mediate the vestibulocollic reflex. The lateral vestibulospinal tract descends to lumbar levels of the spinal cord to influence limb extensors involved in balance. Neck muscle proprioceptors send signals to vestibular nuclei to participate in cervicocollic reflexes and interactions among reflexes, d, descending vestibular nucleus; l, lateral vestibular nucleus; m, medial vestibular nucleus; s, superior vestibular nucleus; VST, vestibulospinal tract.

tion has been termed an "ankle strategy" for maintaining balance.

Although the ankle strategy was confirmed in all subjects tested with simple forward–backward displacement, other strategies are available, and postural reactions adapt to alterations in the support surface or sensory inputs. Displacement of a support surface that is short compared to the foot, or rotation of the support instead of displacement, demands a different postural response. A second strategy that can be adopted when ankle strategy is contraindicated is the hip strategy, in which the body is bent at the hips so that the lower half of the body moves in the same direction as it does during ankle strategy reactions while the upper body moves in the opposite direction. Thus, during a forward, front downward tilt of a supporting platform, the hips move backward and the head forward. This shifts the center of gravity, which has moved forward due to the displacement, backward so that it is placed over the support again. Other work has questioned the prevalence of ankle strategy and documented a stiffening strategy of muscle cocontraction in response to base rotation and a multilink strategy that includes ankle, hip, and neck joint motions in response to translation. Stiffening, ankle, and hip or multilink strategies are applicable to the sway that occurs during normal standing on commonly encountered kinds of surfaces and may represent basic units of postural reaction.

Like simple reflexes, these postural reactions represent neurally mediated negative feedback responses. However, postural reactions differ from ordinary reflex responses in the degree of their dependence on context and recent experience. It also appears that the postural strategies are not mutually exclusive, but are more like additive units or building blocks. When Horak and Nashner placed subjects on support surfaces intermediate in length between the long surfaces associated with ankle strategy and the short surfaces associated with hip strategy, they found that subjects adopted complex strategies that they believed represented combinations of ankle and hip strategies with different magnitudes and temporal relations. In addition, they found that during the first few trials after switching from one length of support surface to another, the subjects' responses in part reflected the previous surface, only gradually shifting from one strategy to another over several trials.

The range of postural reactions recorded from standing human subjects argues for a complexity beyond that observed in the vestibulocervical or vestibulospinal reflexes studied during rotation or tilting of animal subjects, but it is not yet clear whether this complexity reflects the rich variety of strategies that can achieve the single goal of balance or whether there are strong constraints on postural responses that arise from the complicated biomechanics of the multijoint system of the body. Balance is subject not only to the static requirement for location of the center of mass over the support provided by the feet, but also to dynamic factors that accompany the rapid movement of massive body parts. Inertia, viscosity, and elasticity act at each of the major joints, and the task of measuring or modeling these factors and their interactions has barely begun.

Experimental animals can also be placed on posture platforms, which in the case of quadrupeds can consist of four small pads for the limbs (Macpherson and Ingliss, 1993). Among other findings, these experiments confirm the disability of animals with spinal transection in maintaining erect posture. A surprising finding from these animal studies is that even though vestibular information from head motion accompanying body displacement is vigorous and early enough to guide reflex responses, cats are able to maintain apparently good quadrupedal standing posture and responses to unexpected motion after total labyrinthectomy. Clearly, we have a great deal to learn about the management by the brain of the many systems that contribute to posture and balance (Box 30.2).

Vestibular Damage Results in Disorders of Postural Control

Control of posture involves many levels within the nervous system, and it is not surprising that disorders of posture can result from damage to the sensory periphery or to telencephalic, cerebellar, brain stem, or spinal centers. The striking motor consequences of lesions of the basal ganglia may include profound effects on posture, such as the rigidity and general poverty of movement associated with Parkinson's disease (Chapter 31). Damage to the anterior vermis of the cerebellum can exaggerate decerebrate rigidity, and cerebellar patients show poorer performance in posture platform situations, with less adaptive modification of responses, as might be expected from our knowledge of cerebellar function (Chapter 32).

Many of the more plainly visible postural disorders stem from damage to the vestibular system. Diseases of the vestibular system that affect posture (and generally also cause dizziness) include vestibular neuritis, peripheral or central tumors or infarction, and Meniere's syndrome (Baloh and Honrubia, 1990) (Box 30.4). Bilateral involvement and acute, episodic, or chronic time courses can occur in many vestibular disorders. The most obvious form of vestibular system damage is unilateral labyrinthectomy, performed

BOX 30.2

VESTIBULAR PLASTICITY

The experience of sailors getting their "sea legs" suggests that, in addition to the role of context and choice of strategies we have discussed, postural reflexes and balance are adapted gradually over time for better performance in new circumstances. Clear evidence for this general phenomenon comes from the study of postural control in astronauts. After a space flight, astronauts initially rely more heavily than before on visual cues for postural orientation, and they show degraded performance in their responses to disturbances generated by a posture platform. Sway during standing is increased dramatically when visual cues are removed, and the body segments move in a less coordinated manner than before space flight. Still, the performance of astronauts after the experience of space flight is quite remarkable; they are able to balance adequately within hours of landing and regain their preflight postural performance within a few days.

Marc H. Schieber and James F. Baker

BOX 30.3

MENIERE SYNDROME

The typical patient with Meniere syndrome develops a sensation of fullness and pressure along with decreased hearing and tinnitus in one ear. Vertigo follows rapidly, reaching a maximum intensity within minutes and then slowly subsiding over several hours. Often the patient is left with a sense of unsteadiness and nonspecific dizziness that can go on for days after the acute vertiginous spell. In the early stages, hearing loss is completely reversible, but as the disease progresses, residual hearing loss becomes a prominent feature. The tinnitus typically is described as a roaring sound similar to the sound of the ocean. These episodes occur at irregular intervals over years, with periods of remission unpredictably intermixed. Eventually, most patients reach the so-called "burnt out phase," where the episodic vertigo disappears and severe permanent hearing loss remains.

The clinical syndrome was first described by Prosper Meniere in 1861, but Hallpike and Cairns made the initial clinical-pathological correlation with hydrops of the labyrinth in 1938. Patients with Meniere syndrome invariably show an increase in volume of endolymph associated with distention of the entire endolymphatic system. Herniations and ruptures in the membranous labyrinth commonly occur, which may explain the episodes of hearing loss and vertigo.

Delayed endolymphatic hydrops occurs in an ear that has been damaged years before, usually by infection. With this disorder, the patient reports a long history of hearing loss, typically since childhood, followed many years later by episodic vertigo but without the typical auditory symptoms. The pathologic findings are remarkably similar to idiopathic Meniere syndrome, suggesting a common etiology. A subclinical viral infection could damage the resorptive mechanism of the inner ear, leading to an eventual decompensation in the balance between secretion and resorption of endolymph.

The key to the diagnosis of Meniere syndrome is to document fluctuating hearing levels in a patient with the characteristic clinical history. In the early stages, the sensorineural hearing loss is usually greater in the low frequencies. Some patients with Meniere syndrome develop abrupt episodes of falling to the ground without loss of consciousness or associated neurologic symptoms. These episodes have been called "otolithic catastrophes" because they are thought to result from a sudden mechanical deformation of the otolith receptor organ. Patients often report feeling as though they were pushed to the ground by some external force.

Because the cause of Meniere syndrome is usually unknown, treatment is empiric. Medical management consists of symptomatic treatment with antivertiginous drugs and long-term prophylaxis with salt restriction and diuretics. Many different surgical procedures have been tried but none has been consistently effective. Shunt operations to decrease the endolymph pressure have not been successful because the implanted drain devices are encapsulated rapidly by fibrous tissue. Destructive surgeries (removing the labyrinth or cutting the vestibular nerve) can stop the episodes of vertigo but do not change the tinnitus and progressive hearing loss.

Robert W. Baloh

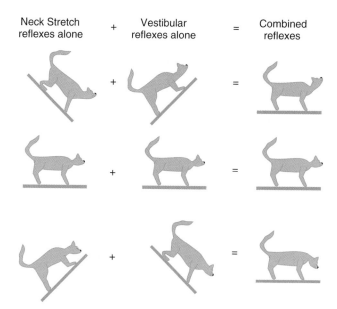

Neck Stretch reflexes alone + Vestibular reflexes alone = Combined reflexes

FIGURE 30.4 Operation of the tonic neck reflex in quadrupeds. The left column shows the operation of neck-to-limb proprioceptive reflexes (cervicospinal reflexes), the center column the static vestibulospinal reflexes, and the right column the antagonistic nature of their summed responses. A straight limb orthogonal to the trunk indicates a neutral posture, as in the middle row and right column of stick figures. A limb extended away from the center of the stick figure represents reflex extension (e.g., forelimb in upper left stick figure), and a reflexively flexed limb is shown in two segments (e.g., hindlimb in upper left stick figure). Based on work by Roberts.

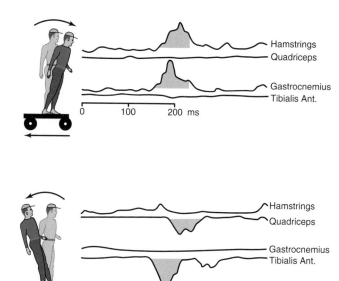

Hamstrings
Quadriceps
Gastrocnemius
Tibialis Ant.

Hamstrings
Quadriceps
Gastrocnemius
Tibialis Ant.

FIGURE 30.5 Sequencing of muscle activation in response to displacement of a supporting platform. When the supporting surface is displaced backward (at 0 ms), flexor muscles are excited first in the distal lower limb segments (gastrocnemius, about 80 ms latency) and then in the proximal segment (hamstrings, about 100 ms latency). Forward displacement of the platform activates lower limb extensors, again in a distal (tibialis anterior) to proximal (quadriceps) sequence. Black arrows mark the first detected electromyographic response to displacement. Based on studies by Horak and Nashner.

either experimentally in animals or as the result of disease or surgical intervention against disease in humans. The postural symptoms immediately following loss of vestibular input to one side of the brain vary across species, with nonmammalian and "lower" mammalian species typically showing a more prominent tilting of the head and body toward the side of damage. Vestibular afferents have a high level of resting activity, and leaning toward the side of lost vestibular signals makes sense in that the spontaneous or resting activity of the remaining vestibular apparatus is signaling, in effect, motion toward the intact side because it is not balanced by equal resting activity from the damaged side. To counter this neural stimulus, the affected individual tilts away from the directions of apparent motion. This static postural effect of unilateral labyrinthectomy is accompanied by dynamic postural deficits, seen in experimental animals as weakened ipsilateral limb extensor responses and delayed responses to sudden drops. Compensation for unilateral labyrinthectomy occurs over a period of a few weeks, during which there is considerable recovery.

Patients with long-standing bilateral loss may perform well on posture platforms when visual and somatosensory cues are present, but fail completely to maintain upright stance when the support surface and visual surround both sway with the patient so that only vestibular information is accurate. Patients with recent bilateral vestibular loss or who have not yet compensated for vestibular loss also do poorly when vision and support surface are unreliable, but these patients perform poorly if only one sense, be it vision or somatic sense, is made an unreliable indicator of balance.

Summary

Axons from vestibular and reticular nuclei in the brain stem descend in pathways located ventromedially in the spinal cord to provide postural tone and balance. Balance is maintained by negative feedback reflexes that are stimulated by the vestibular canals and otoliths. The vestibulocervical reflex excites neck muscles via the medial vestibulospinal tract to oppose head tilt, and vestibulospinal reflexes carried by the lateral vestibulospinal tract excite ipsilateral limb extensors to oppose body tilt. Vestibular, muscle stretch, and other reflexes work together to coordinate posture and balance, and the medial system adopts different strategies for balance control depending on the body support surface and information from sensory inputs. The medial postural system can adapt to different postural situations, but its actions can be compromised by central neural damage or unbalanced by peripheral vestibular loss. The reflexive control of body stability and posture by the medial descending system allows the lateral descending system to specialize in execution of precise voluntary movements of the extremities.

THE LATERAL VOLUNTARY SYSTEM

The medial system thus provides underlying postural control for stance, ambulation, and orientation of the head; the lateral system superimposes the ability to make more sophisticated, voluntary movements in response to complex features of the external environment (perceived through the senses) and internal state (stored memories, knowledge, and emotion). Much of this control is mediated by specialized regions of the cerebral cortex. Outflow from this motor cortex dominates the lateral descending system, particularly in primates and humans.

Components of the Lateral Voluntary System

The Corticospinal Projection Is the Most Direct Pathway from the Cerebral Cortex to Spinal Motor Neurons

For more than a century, electrical stimulation of a limited portion of the frontal lobe of the cerebral cortex has been known to evoke movements. Electrical stimulation of this "excitable cortex" evokes movements readily because this region has relatively direct connections to spinal motor neurons. In primates with a central sulcus (Rolandic fissure) in the neocortex, cortical neurons with axons projecting to the spinal cord are found most densely in the anterior bank of the central sulcus. The density of such neurons decreases from there rostrally to the precentral sulcus (posterior bank of the arcuate sulcus in macaque monkeys) and medially to the cingulate sulcus (Fig. 30.6). This territory corresponds to cytoarchitectonic areas 4 and 6 of Brodmann. Additional corticospinal neurons in areas 1, 2, 3, 5, and 7 of the parietal lobe project to the dorsal horn of the spinal cord to regulate sensory inflow.

The axons of neurons that project from the cerebral cortex to the spinal cord constitute the corticospinal tract. Corticospinal neurons have large pyramid-shaped somata in cortical layer Vb. Their axons leave the cortex, pass through the centrum semiovale, and

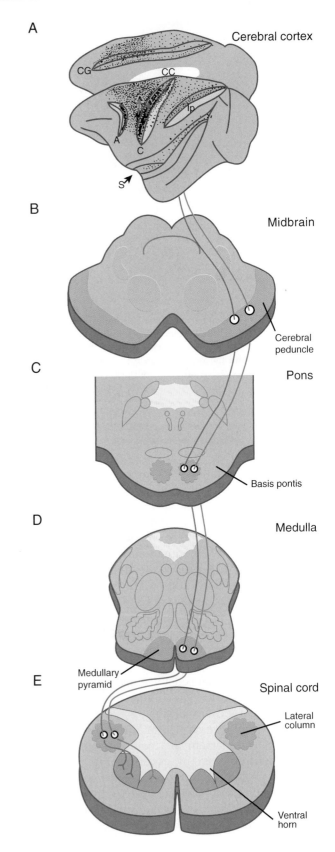

FIGURE 30.6 The corticospinal projection in the macaque monkey. (A) The density of corticospinal neuronal somata is shown by stippling in this lateral view of the left cerebral hemisphere; the superior medial surface of the hemisphere is also shown (above) as if reflected in a mirror. The central sulcus (C), arcuate sulcus (A), cingulate sulcus (Cg), intraparietal sulcus (Ip), and Sylvian fissure (S) are drawn as if pulled open to reveal the neurons in their banks. Two schematic corticospinal neurons, one relatively posterior and the other relatively anterior in area 4, send their axons down through the midbrain (B), pons (C), medulla (D), and spinal cord (E), which are drawn in cross-section. In the spinal cord, the former corticospinal axon leaves the lateral column to terminate in the dorsolateral ventral horn, whereas the latter axon terminates in the ventromedial ventral horn.

enter the internal capsule, along with many other axons from other cortical areas. The corticospinal axons are concentrated most heavily in the middle third of the posterior limb of the internal capsule. As axons descend from the internal capsule below the thalamus, they come to lie on the ventral surface of the brain stem. Here, they form the cerebral peduncle of the midbrain, where the corticospinal axons are concentrated in the middle third. The axons of the peduncle become intermixed with pontine nuclear neurons in the base of the pons (basis pontis). Most of these axons synapse on pontine neurons and end here, providing input from the cerebral cortex to the cerebellum. The continuing corticospinal axons collect to form the medullary pyramid (Box 30.4). As the medulla blends into the spinal cord, the vast majority of corticospinal axons cross the midline and enter the lateral column of white matter on the opposite side of the spinal cord. Because of this decussation of the pyramidal tract, the left

motor cortex controls movements on the right side of the body and vice versa. A small minority of corticospinal axons remain uncrossed and continue caudally as the ventral (or anterior) corticospinal tract near the ventral midline of the spinal cord.

As descending corticospinal axons reach their target levels in the spinal cord, they enter the spinal gray matter, where they ramify and synapse. Whereas the majority of corticospinal axons synapse on premotor interneurons in the intermediate zone (Rexed's laminae VII and VIII), a minority of these axons synapse directly on motor neurons in Rexed's lamina IX. The somata of most corticospinal neurons that make such monosynaptic connections to motor neurons lie posteriorly in area 4, and their monosynaptic connections are made on the motor neurons of distal limb muscles, whose somata are clustered in the dorsolateral ventral horn. Corticospinal axons from neurons located more anteriorly typically synapse in the ventromedial portion of the ventral horn, where the motor neurons of proximal limb muscles and axial muscles are located. Some corticospinal axons cross the midline spinal gray matter to reach the ventromedial ventral horn ipsilateral to their origin; such doubly decussating corticospinal axons, along with the uncrossed ventral corticospinal tract, may be partly responsible for the relative preservation of trunk and proximal limb movements after unilateral damage to the cortex.

Indirect Pathways to the Spinal Cord Involve Centers in the Brain Stem

The corticospinal tract is not the only output pathway through which the cerebral cortex contributes to motor control (Kuypers, 1987). The presence of

---

BOX 30.4

PYRAMIDS IN THE BRAIN

Inspection of cross-sections through the ventral surface of the medulla reveals a structure that looks like a small pyramid. This medullary pyramid turns out to be the collected bundle of corticospinal axons. The corticospinal tract therefore often is referred to as the pyramidal tract.

Axons of noncorticospinal descending pathways do not pass through the medullary pyramid and therefore can be described collectively as extrapyramidal pathways. At one time these pathways were thought to receive their major controlling inputs from the basal ganglia. Human neurologic disorders that result from dysfunction of the basal ganglia thus came to be known as "extrapyramidal

syndromes." This term is misleading, however. More recent work has shown that many of the brain stem nuclei giving rise to extrapyramidal pathways receive cortical input and that much of the output of the basal ganglia is directed via the thalamus back to cortical motor areas. Many manifestations of extrapyramidal syndromes may therefore be played out through cortical motor areas, even via the pyramidal tract. Indeed, surgical lesions of the pyramidal tract once were used to ameliorate some extrapyramidal syndromes.

Marc H. Schieber and James F. Baker

other, indirect pathways in the macaque has been demonstrated by cutting the medullary pyramid on one side. Stimulation of the cortex on that side still evoked contralateral movements. The somatotopic organization of the cortex in the operated animals was similar to that in normal macaques, although the thresholds for stimulation were raised and distal movements were evoked less often.

Intermixed with corticospinal neurons in layer V of areas 4 and 6 are corticorubral neurons, whose axons project to the red nucleus (RN). Some corticospinal axons also send collaterals to the RN. In addition to cortical inputs, the RN also receives considerable input from the cerebellum (Chapter 32). Many RN neurons, particularly those in the caudal, magno-cellular portion, in turn send their axons across the midline and into the lateral column of the spinal cord, terminating most heavily in the dorsolateral region of the ventral horn. These descending axons from the RN constitute the rubrospinal tract. An indirect pathway thus exists from the cortex to the RN to the spinal cord. Although the rubrospinal tract is less prominent in humans than in nonhuman primates and carnivores, the activity of rubrospinal neurons indicates that they also play a significant role during voluntary movements of the arm, hand, and fingers. The rubrospinal pathway thus works synergistically with the corticospinal pathway in controlling limb movements.

A second indirect pathway involves neurons scattered in the medial reticular formation of the pons and medulla. The medial reticular formation receives input from cortical motor areas and projects via the ventral column of the spinal cord to the ventral horn, chiefly its ventromedial portion. The axons that make this projection constitute the reticulospinal tract. Although the reticulospinal tract is also part of the medial descending system, the activity of some reticular formation neurons indicates that they also participate in controlling reaching movements.

Organization of the Motor Cortex

The original "motor cortex" is now appreciated to be comprised of several cortical areas. These areas are considered "motor" because (1) they project to other motor structures, (2) their ablation causes deficits in movement, and (3) their stimulation evokes or alters movements.

The Motor Cortex Is Subdivided into Multiple Cortical Motor Areas

In addition to its descending projections to spinal motor neurons, the motor cortex has cytoarchitectonic features that distinguish it from other regions of the cerebral cortex. The somata of many pyramidal neurons in layer V, the output layer, are exceptionally large (Betz cells). In addition, the neurons of layer IV, the granular layer that receives thalamic input, are very sparse compared to other areas of the neocortex, and hence the motor cortex has been described as dysgranular (area 6) or agranular (area 4) cortex. Based on cytoarchitectonics, myeloarchitectonics, and histochemical features, the motor cortex of macaque monkeys can be subdivided into the cortical motor areas shown in Figure 30.7. In general, these subdivisions show three mediolaterally oriented strips of cortex, with the primary motor cortex (M1) most posterior and two premotor strips progressively more anterior. Each of these premotor strips has a subdivision on the medial wall of the hemisphere, a second subdivision on the dorsal convexity, and a third on the ventral convexity. In addition to these subdivisions of Brodmann's areas 4 and 6, areas 23 and 24 in the banks of the cingulate sulcus and on the medial surface of the hemisphere in the cingulate gyrus contain at least two additional cortical motor areas. Cortical motor areas differ, not only in their intrinsic composition, but also in their interconnections with other parts of the central nervous system.

Cortical motor areas differ in their connections with the thalamus. M1, PMv, and SMA, for example, connect primarily with VPLo/VLc, area X, and VLo in the thalamus, respectively. These differences are significant because thalamic nuclei VPLo/VLc and area X receive major inputs from the cerebellum, whereas VLo receives major input from the basal ganglia. Thus, information processed by the cerebellum is directed largely to M1 and PMv, whereas information from the basal ganglia is sent largely to SMA.

Cortical motor areas also differ in their connections with one another and with other regions of the cortex (Fig. 30.7). For example, when horseradish peroxidase (HRP) is injected into the hand region of M1, separate pockets of retrogradely labeled neuronal somata are found in PMv, PMd, SMA, CGc, and CGr, indicating that each of these areas sends a separate projection to the hand region of M1. In contrast, PMvr, PMdr, and pre-SMA do not project directly to M1. Instead, PMvr projects to PMvc, PMdr projects to PMdc, and pre-SMA projects to PMvr.

Different cortical motor areas also receive input from different cortical sensory and association areas: M1 receives input from the primary somatosensory area (S1), PMv receives input from visual association area 7b, and PMd receives input from somatosensory association area 5. Because of these inputs, many motor cortex neurons have somatosensory receptive fields. Neurons located caudally in M1 tend to respond to

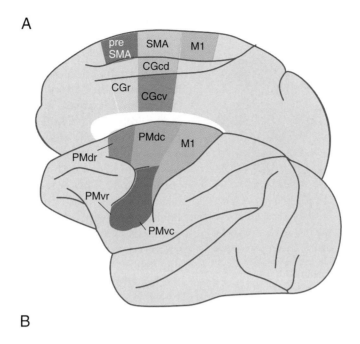

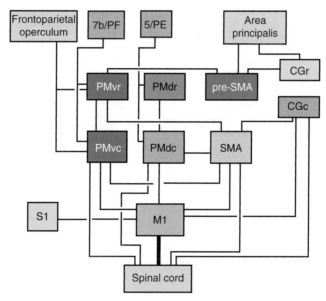

FIGURE 30.7 Cortical motor areas. (A) Diagram of a macaque brain showing current parcellation of cortical motor areas in the frontal lobe. Modified from Matelli *et al.* (1991). (B) Connections of the cortical motor areas. Most corticocortical connections are reciprocal. CGc,-cingulate motor area, caudal; CGr,-cingulate motor area, rostral; M1, primary motor cortex; PMdc, premotor cortex, dorsal caudal; PMdr, premotor cortex, dorsal, rostral; PMvc, premotor cortex, ventral, caudal; PMvr, premotor cortex, ventral, rostral; Pre-SMA, presupplementary motor area; SMA supplementary motor area proper. Thin lines to the spinal cord from PMvc, PMdc, SMA, and CGc indicate that corticospinal projections from these areas are not as strong as that from M1.

cutaneous modalities, whereas neurons located rostrally in M1 tend to respond to deep modalities. The somatosensory input of an M1 neuron is generally related to the output function of that neuron. For example, caudally located M1 neurons that control fingers may have cutaneous receptive fields on the fingers (Fig. 30.8A), whereas more rostrally located M1

neurons that control the biceps or triceps may have deep receptive fields in those muscles.

Somatosensory inputs to M1 also help mediate long loop responses; that is, muscular responses that occur too slowly to be mediated by the monosynaptic stretch reflex (Chapter 29) but too quickly to be considered voluntary reactions. For example, when a subject holds

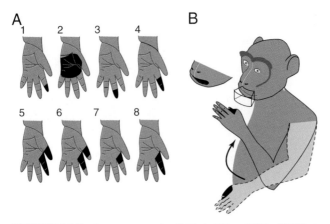

FIGURE 30.8 Sensory receptive fields in M1 and PMv. (A) Black regions show the tactile receptive fields of eight M1 neurons recorded at loci where intracortical microstimulation evoked flexion of the monkey's index and middle fingers. Other neurons at the same loci responded to passive extension of those fingers. From Rosen and Asanuma (1972). (B) A single PMv neuron responded to visual stimuli moving near the mouth, to tactile stimulation of the lips and of the skin between the thumb and the index finger, and to flexion of the elbow. From Rizzolatti *et al.* (1981).

a handle that suddenly jerks away, extending the elbow and stretching the biceps, three distinguishable waves of contraction may be recorded in the biceps. The first wave is the monosynaptic reflex produced by stretching the muscle; the third wave is the subject voluntarily pulling back on the handle. The middle wave occurs at a latency consistent with the conduction of impulses from muscle afferents to the cortex and then almost directly back from M1 to spinal motor neurons. Many biceps related M1 neurons discharge a burst of impulses at a time appropriate to contribute to this long loop response. Interestingly, the strength of the long loop response may be increased if the subject plans to pull on the handle by contracting the biceps and decreased if the subject plans to push on the handle by relaxing the biceps; parallel changes may occur in the bursts of the M1 neurons. Thus, long loop responses are affected by the motor plans of the subject.

PMv receives both somatosensory and visual input via area 7b. Both the somatosensory and the visual receptive fields of PMv neurons tend to be large, and when single PMv neurons receive both types of sensory information, the fields tend to be related (Fig. 30.8B). A neuron with a somatosensory receptive field covering the forearm, for instance, may respond to visual stimuli moving near the forearm. Interestingly, if the forearm is moved to a different position, the visual receptive field of the PMv neuron moves with the forearm.

Somatotopic Organization in the Motor Cortex Is Not a One-to-One Map of Body Parts, Muscles, or Movements

The organization of the primary motor cortex, M1, has been studied more extensively than that of other cortical motor areas. Within the M1 of primates, the face is represented laterally, the lower extremity (or hindlimb and tail) medially, and the upper extremity (or forelimb) in between (Fig. 30.9). This overall organization of M1 according to major body parts is termed somatotopic. The somatotopic organization of M1 becomes evident in clinical neurology. Lesions on the lateral convexity of human M1 cause weakness or paralysis of the contralateral face, more medial lesions on the convexity affect the contralateral hand and arm, and lesions on the medial wall of the hemisphere affect the leg and foot.

Moreover, distal parts of the extremities and acral parts of the face (lips and tongue) are most heavily represented caudally, whereas proximal parts of the extremities and axial movements are most heavily represented rostrally. Those parts of the body that are used for fine manipulative movements (such as lips, tongue, and fingers in primates) are generally represented over a wider cortical territory than body parts used in gross movements such as ambulation. Penfield, in his cartoon of the motor homunculus (little man), and Woolsey, in his cartoon of the motor simiusculus (little monkey), conveyed this apparent magnification of certain body parts with respect to cortical territory by distorting the size of body parts relative to their normal proportions.

The somatotopic organization of M1 is not, however, a one-to-one mapping of body parts, muscles, or movements. Within the arm representation, for example, the

FIGURE 30.9 Somatotopic maps in M1. (A) Map by Woolsey and coworkers (1952) in which each figurine represents in black and gray the body parts that moved a lot or a little, respectively, when the cortical surface at that site was stimulated. In addition to the primary representation on the convexity, their map shows a secondary representation on the medial surface of the hemisphere, called the supplementary motor area (SMA). As defined in this study, M1 and SMA each included several of the currently defined cortical motor areas. (B) Intracortical microstimulation of M1 in an owl monkey produced this map, consisting of a complex mosaic of different body parts. In this species, the central sulcus is only a shallow dimple, and M1 is entirely on the surface of the hemisphere. Each dot represents a stimulated locus, and lines surround adjacent loci from which movement of the same body part was evoked. Note that the forepaw digits (purple) and the hindpaw digits (green) are represented in multiple areas separated by areas representing nearby body parts. Modified from Gould *et al.* (1986). In both A and B, the inset at the top indicates the region of the frontal lobe enlarged below.

A

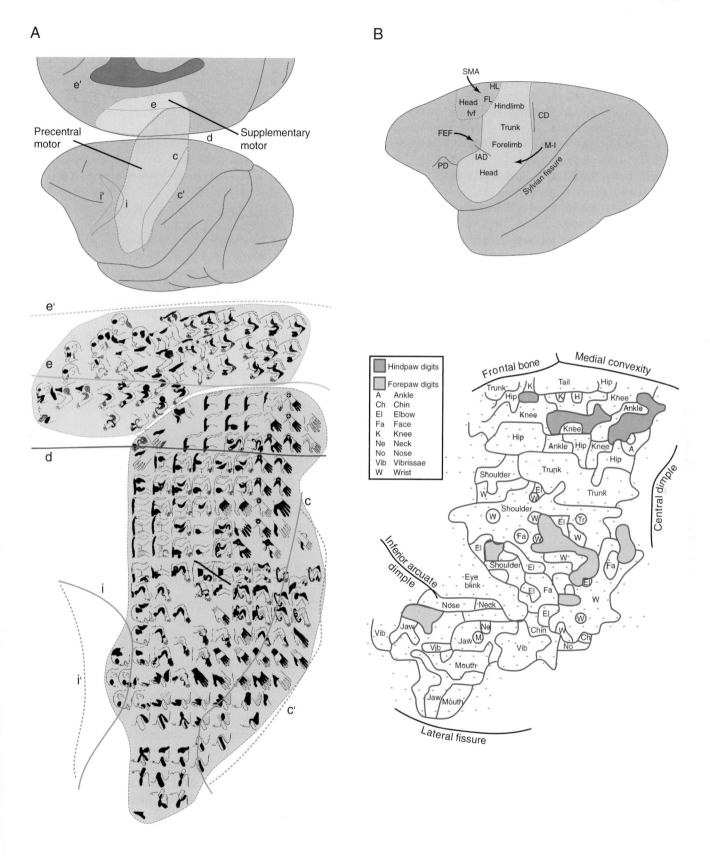

Precentral motor

Supplementary motor

B

SMA

HL
Head   FL
fvf   Hindlimb   CD

                    Trunk

FEF   Forelimb   M-I

PD   IAD

        Head   Sylvian fissure

	Hindpaw digits
	Forepaw digits
A	Ankle
Ch	Chin
El	Elbow
Fa	Face
K	Knee
Ne	Neck
No	Nose
Vib	Vibrissae
W	Wrist

Frontal bone   Medial convexity

Trunk   K   Tail   Hip

Central dimple

Inferior arcuate dimple

Lateral fissure

cortical territory representing any particular part of the arm overlaps considerably with the territory representing nearby parts. This overlap results from three features of M1's organization: convergence, divergence, and horizontal interconnection.

M1 outputs arising from a wide cortical territory converge on the spinal motor neuron pool of single muscles. Convergence of M1 outputs has been shown by delivering intracortical microstimulation while observing the body for evoked movements and/or recording electromyographic (EMG) activity from multiple muscles. Maps of which body parts move or which muscles contract upon threshold stimulation at different cortical sites look like a complex mosaic (Fig. 30.9B). As the stimulation current is increased, the mosaic pieces representing a given muscle coalesce into a larger and larger territory that overlaps more and more with the territories of other muscles (Fig. 30.10A). These findings indicate that any given muscle is controlled by a large territory in M1 and that the territories for different muscles overlap (Fig. 30.10B). Using retrograde transneuronal tracing with rabies virus, these large and overlapping cortical territories for different muscles recently have been demonstrated anatomically as well.

A second factor contributing to the overlap of territories in M1 is the divergence of output from single cortical neurons to multiple motor neuron pools. Intracellular staining has shown that a single corticospinal axon may have terminal ramifications within the motor neuron pools of multiple muscles over several segmental levels of the spinal cord (Fig. 30.11A). Moreover, spike-triggered averaging of EMG activity has demonstrated that single M1 neurons can have relatively direct effects on the motoneuron pools of multiple muscles (Fig. 30.11B). These findings indicate that single M1 neurons have output connections to the spinal motoneuron pools of multiple muscles (Fig. 30.11C).

A third factor contributing to overlapping representation is the horizontal interconnectivity intrinsic to M1. Although most horizontal axon collaterals of neurons in layer V extend only 1–2 mm, some extend across the upper extremity representation, interconnecting even the territories of proximal and distal muscles.

Because of convergence, divergence, and horizontal interconnections, neuronal activity is widely distributed in M1 during natural movements, even discrete movements of single body parts. In monkeys trained to perform individuated movements of each finger, single M1 neurons are active during movements of multiple fingers, and neurons throughout the M1 hand area are active during movements of any given finger.

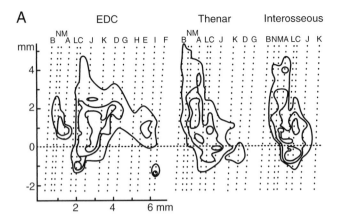

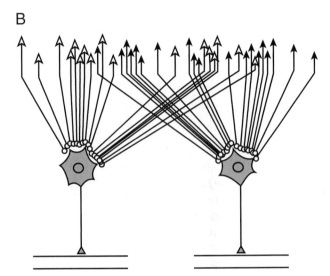

FIGURE 30.10 Convergence of M1 outputs to single muscles. (A) Isothreshold contours show the points at which EMG responses were evoked in three different muscles—extensor digitorum communis (EDC), thenar, and first dorsal interosseus—by intracortical microstimulation in the anterior bank of the central sulcus. (B) Data shown in A indicate that the cortical input to the motor neurons of any muscle originates from a wide territory in M1 and that the cortical territory providing input to a given muscle overlaps extensively with the cortical territory providing input to other muscles in the same part of the body. Modified from Andersen et al. (1975).

Likewise, in humans performing movements of different fingers, activity is distributed over the entire M1 hand representation whether the subject is moving a single finger or the whole hand. Although the entire hand representation may be activated for movement of any given finger, monkey and human studies have shown a tendency for the center of activation during movements of the thumb to be located laterally to that for movements of other fingers, consistent with the somatotopic orientation of the hand in the classic simiusculus and homunculus.

A

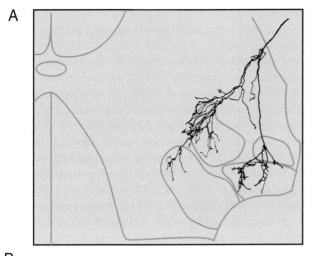

FIGURE 30.11 Divergence of M1 outputs to multiple muscles. (A) Tracing of a single corticospinal axon ramifying in the ventral horn of the spinal cord shows terminal fields in the motor neuron pools of four forearm muscles. From Shinoda *et al.* (1981). (B) Action potentials in a cortical neuron (top trace) are followed at a fixed latency by peaks of postspike facilitation in EMGs recorded from four of six forearm muscles (lower traces), consistent with monosynaptic excitation of all four motor neuron pools by that cortical neuron. The EMGs are rectified and averaged responses to 7051 action potentials in the cortical neuron. From Fetz and Cheney (1980). (C) These anatomic and physiologic findings indicate that the output of single corticospinal neurons often diverges to influence multiple muscles. From Cheney *et al.* (1985).

B

Unit

ED 2, 3

ECU

ED 4, 5

EDC

ECR-L

ECR-B

25 ms

The somatotopic representation in M1, like that in the primary somatosensory cortex, has a certain degree of plasticity. The cortical territory representing a given muscle can enlarge when nearby body parts are denervated experimentally. Threshold stimulation in the cortical territory that had represented a denervated body part then comes to evoke movements in nearby body parts. Enlargement of a given muscle's cortical territory occurs as well when that muscle is stretched passively or used intensely for a prolonged period. Furthermore, in normal humans who are actively practicing a complex sequence of finger movements, the cortical territory representing a given finger muscle enlarges as the subjects become skilled at performing the sequence. Because such changes can occur within several minutes, they probably are mediated by long-term potentiation and/or depression at existing synapses. This ability of the M1 cortex to reorganize may in part underlie motor recovery seen in humans after damage to M1 or the corticospinal tract.

Control of Voluntary Movements by the Motor Cortex

Several methods have enabled investigators to explore how cortical motor areas contribute to the production and control of voluntary movement in awake, behaving subjects. Whereas microelectrodes record the action potentials of single neurons, measurements made by functional imaging or by recording surface potentials reflect the net activity of thousands of cortical neurons and millions of synapses.

Multiple Cortical Areas Are Active When the Brain Generates a Voluntary Movement

These different methodologies now make it clear that many cortical areas in addition to the primary motor cortex are active during the planning and execution of voluntary movements. During performance of

C

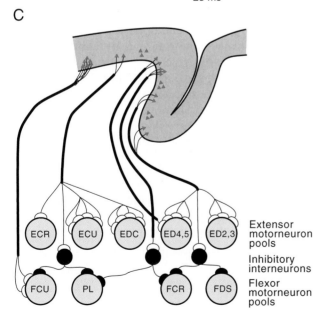

Extensor motorneuron pools

Inhibitory interneurons

Flexor motorneuron pools

either a simple key press or a complex movement sequence, for example, functional imaging studies in humans have shown bilateral activation of the primary sensorimotor hand representation, the supplementary motor area, and the ventrolateral premotor cortex, plus contralateral activation of the dorsolateral premotor cortex and the medial cortex rostral to the SMA. Whereas such techniques provide information on the parts of the cortex that are active in a given situation, studies of electrical potentials provide information on the time course of activation. For simple self-paced movements, cortical electrical potentials over the SMA and M1 begin to change as early as 1 s prior to the movement. As the time of movement onset approaches, the amplitude of these bilateral electrical potential shifts increases (the Bereitschaft potential). At the time of the movement, a further increase in amplitude occurs over the somatotopically appropriate region of M1 contralateral to the moving body part. Given that many cortical areas are active in controlling movement raises the question: What does each area contribute to control?

M1 Neurons Control Movement Kinematics and Dynamics

In one of the earliest studies of single neuron activity in awake behaving animals, Evarts recorded the activity of M1 pyramidal tract neurons in monkeys trained to raise and lower weights using flexion and extension wrist movements (Fig. 30.12). The discharge frequency of many M1 neurons changed systematically in temporal relation to either flexion or extension. A typical flexion-related neuron, for example, began to discharge several hundred milliseconds before wrist flexion began. As the onset of a flexion movement approached, the discharge frequency of the neuron increased, accelerating still more as the flexion movement was made. The neuron was silent during extension. The time course of these movement-related changes in single neuron activity parallels the time course of cortical electrical potential shifts.

Evarts went on to demonstrate that the discharge frequency of M1 pyramidal tract neurons (PTNs) varied in relation to a number of mechanical parameters of the movement the monkey was making. By changing the weights the monkey had to lift, Evarts showed that PTN discharge frequency varied with the force the monkey exerted. Comparison of PTN discharge with simultaneous force recordings revealed that bursts of PTN firing were correlated with sudden increases in the exerted force, indicating a relationship between firing frequency and the rate of change of force. Subsequent studies have confirmed these findings and have also shown that the discharge of M1 neurons can be related to the direction of movement, the position of a particular joint, or the velocity of movement. Cortical neurons whose firing is related to kinematic and dynamic parameters of movement are not found only in M1. Many neurons in the SMA, PMd, and parietal area 5 (which projects to SMA and PMd) fire in relation to movement direction, and some PMv neurons fire in relation to movement force. Therefore, neurons in several nonprimary cortical motor areas participate with those in M1 to control movement parameters such as direction, force, position, and velocity.

Although very strong correlations can be found between the discharge frequency of a given neuron

---

BOX 30.5

BRAIN-MACHINE INTERFACES: MAKING USE OF THE MOTOR CORTEX

Technological advances now enable our knowledge of neuronal activity during voluntary movements to be applied in creating useful brain-machine interfaces. The activity of dozens of neurons can be recorded simultaneously from electrode arrays implanted in the primary motor cortex, premotor cortex, and/or parietal cortex. The spike discharges of these neuron populations, passed into computers, can be decoded in real time to obtain specific information about the direction and speed of voluntary movements the subject intends to make. This decoded information then can be used online to control a cursor on a computer monitor, or even a robotic arm. For patients with an amputated extremity or a spinal cord lesion, such brain-machine interfaces eventually may drive neuroprosthetic devices that can be used independently by the patient, replacing the function of missing or paralyzed arms and hands for many activities of daily living.

Marc H. Schieber and James F. Baker

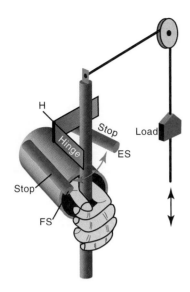

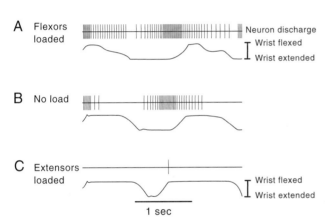

FIGURE 30.12 Discharge of a single M1 neuron in a monkey making flexion and extensor muscles wrist movements with wrist flexor muscles loaded (A) and unloaded (B) and with wrist extensors loaded (C). The discharge rate of this neuron was greatest when the monkey used its wrist flexor muscles against a load. Modified from Evarts (1968).

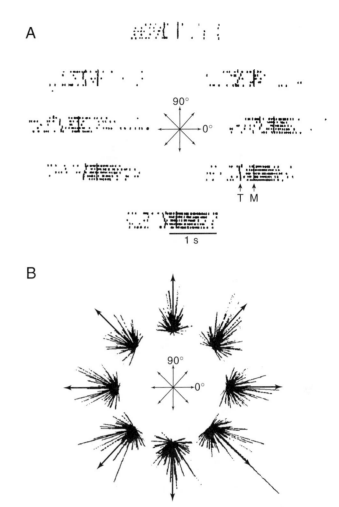

FIGURE 30.13 (A) Discharge of a single M1 neuron before and during arm movements in a monkey. Movements (represented by arrows) started from the same central point and ended at eight different points on a circle. The eight rasters show that the activity of this neuron was related to movements in four of the eight directions. The neuron discharged most intensely for movements down and to the right and was inhibited during movements up and to the left. (B) For each of the eight movements, the discharge of each M1 neuron is shown as a line pointing in the preferred direction of the neuron. Each line starts at the movement end point, and its length is proportional to the intensity of the discharge of that neuron during movement in that direction. Although the discharge of single neurons rarely identified any single movement direction with accuracy, the population vectors (arrows) summing the discharge of an ensemble of M1 neurons adequately specify each of the eight movement directions. From Georgopoulos (1988).

and a particular movement parameter, the discharge of a M1 neuron does not simply encode a single parameter. Rather, single M1 neurons show various degrees of correlation with direction, force, position, and velocity. Thus, the discharge of a single neuron in the motor cortex may influence several movement parameters.

Conversely, any given parameter or other feature of a movement is represented not by the discharge of a single M1 neuron, but by the ensemble activity of a large population of cortical neurons. Although a single M1 neuron typically discharges during reaching movements in several directions, for example, when the activity of many M1 neurons is combined appropri-

ately, the population of M1 neurons can be shown to represent movement direction precisely (Fig. 30.13). Researchers have correlated the force, rate of change of force, position, and velocity of movements more accurately with the summed, weighted activity of a number of simultaneously recorded M1 neurons than with the discharge of any single neuron.

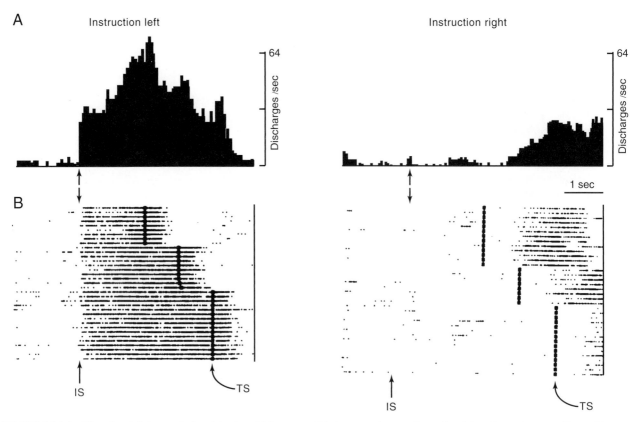

FIGURE 30.14 Directional delay period activity in a PM neuron. (A) As a monkey performed a delayed-reaction paradigm, this neuron began to discharge shortly after receiving instructions (IS) to perform a leftward movement. Discharge continued until after the monkey had subsequently received a separate triggering signal (TS, which occurred at three different time intervals after the IS) and performed the movement. During the delay between IS and TS, while the monkey did not move, the discharge of the neuron encoded the direction of the instruction, the direction of the impending movement, or both. (B) When the instruction was for rightward movement, this neuron did not discharge until after the movement had been made, presumably as the monkey was then preparing to move back to its original position. From Wise and Strick (1996).

Cortical Motor Areas Prepare Voluntary Movements Based on a Variety of Cues

Cortical areas outside the primary motor cortex seem to be especially concerned with using a wide variety of sensory and other information as "cues" to select and guide movements. Insight into these cortical processes has been obtained by separating in time a cue instructing which movement to make from a second cue to execute the instructed movement. This creates a period between the two cues during which the subject knows what movement to make, but is not making the movement per se. During such an instructed delay period, many neurons in M1, SMA, and PMd discharge at their highest rate while the subject waits to move in a particular direction (Fig. 30.14). Such directional delay period activity is more common in PMd and SMA than in M1, where activity during movement execution predominates. During the delay between instruction and trigger, PMd, SMA, and other areas appear to store information on the direction of the impending movement. Indeed, neurons sometimes discharge in error during the delay, as if the monkey has seen a cue other than the one actually given and is preparing to move in the wrong direction. When such error discharges occur, the monkey often does make the wrong movement. The delay period discharge of such neurons thus represents stored information, not about which cue the monkey has seen, but rather about what movement the monkey will make.

The direction instructed usually is the same as the direction of the actual movement. To pick up a pencil, for example, you look at the pencil and then move your hand to the same place. Somehow your brain transforms the visual location of the pencil into the location to which your hand is moved. Insight into how cue direction is transformed into movement direction has come from tasks in which these two features were experimentally dissociated, as if you looked at a pencil in a mirror and then reached to pick up the real

BOX 30.6

MIRROR NEURONS: FUNDAMENTAL NEUROSCIENCE

Mirror neurons are a distinct class of neurons, originally discovered in a sector of the ventral premotor cortex (area F5) of the monkey, that discharge both when the monkey *executes* a given motor act (e.g., grasping an object) and when it *observes* another individual (a human being or another monkey) performing the same or a similar motor act. Mirror neurons do not discharge in response to the simple presentation of food or other interesting objects. They also do not discharge, or discharge much less, when the monkey observes hand actions mimed without the target object. Thus, the effective visual stimulus for monkey mirror neurons is the observation of an effector (hand or mouth) interacting with an object.

All mirror neurons have motor properties and, as other neurons located in area F5, discharge in relation to the goal of the motor act, such as grasping an object, rather than to the movements that form it (e.g., individual finger movements). On the basis of their motor properties mirror neurons can be subdivided into various classes. Among them the most common are grasping, manipulating, tearing, and holding neurons.

Mirror neurons are present not only in the premotor area F5, but have also been found in the inferior parietal lobule (IPL). This region receives input from the cortex of the superior temporal sulcus (STS) and sends output to ventral premotor cortex, including area F5. STS neurons are known to respond to the observation of actions done by others, but they are not endowed with motor properties. Thus, the cortical mirror neuron system is formed by two main sectors: the ventral premotor cortex and the inferior parietal lobule.

A large amount of evidence indicates that a mirror neuron system also exists in humans. Evidence in this sense comes from neurophysiological (EEG, MEG and TMS) and brain-imaging experiments (PET, fMRI). The neurophysiological experiments showed that the mere observation of others' actions induces an increase of excitability of the motor cortex. This excitability increase manifests itself in a desynchronization of EEG rhythms recorded from the motor cortex (the "mu" rhythm in particular) similar, although of less intensity, to that observed during active movements. The TMS experiments confirmed these findings. They showed that, during the observation of actions done by others, there is a specific increase of motor evoked potentials in the muscles of the observer that corresponds to those used by the action's agent.

The neurophysiological techniques have been instrumental in demonstrating that a mirror neuron system exists in humans. They could not show, however, its extent and precise localization. This gap has been filled by brain imaging experiments (PET and fMRI). These experiments showed that the observation of object-directed actions activate two cortical sectors that are also active during action execution: the inferior parietal lobule and a caudal sector of the frontal lobe. There is therefore a good correspondence between the localization of mirror neuron system in humans and monkeys.

With brain imaging techniques it has been possible to demonstrate that the human mirror neuron system have a somatotopic, albeit rough, organization. In the frontal lobe, mouth movements are located ventrally in the posterior part of the inferior frontal gyrus (IFG), the hand movement are located more dorsally, partly in IFG and partly in the ventral premotor cortex, and leg movements even more dorsally, in dorsalmost part of the ventral premotor cortex. In the parietal lobe the observation of mouth actions determines activation in the rostral part of the inferior parietal lobule, that of hand/arm actions activates a more caudal region corresponding to area PFG, while that of foot actions produces a signal increase more caudally and dorsally.

Another important results of brain imaging studies was the demonstration that the presentation of emotional stimuli such as disgust or pain elicits, in the observer, the activations, of the same cortical regions that become active when the individual feels the same emotions. The cortical regions forming this "emotional mirror neuron system" are the anterior insula and the rostral cingulate cortex.

What is the functional role of the mirror neurons? A series of hypotheses have been formulated. Among them are action understanding, imitation, intention understanding, and empathy. The question, however, of which is *the* functional role of mirror neurons is an ill posed question. Mirror neurons do not have a unique function. Their properties indicate that they represent a *mechanism* that maps the pictorial description of actions carried out by others onto their motor counterpart. Their functional role depends, first of all, on where this mapping occurs.

This becomes very clear if one compares the activity of the traditional parieto-frontal mirror neuron circuit with that of the insula and anterior cingulate cortex. The

(continues)

BOX 30.6   (cont'd)

activation of parieto-premotor circuit elicits the motor representation of the observed action. In this way it gives a first person representation of what the others are doing. In contrast, the activation of the insula and cingulate, elicits the viscero-motor representation of the observed emotion. In this way, it gives the first person representation of what the others are feeling.

The functional role of mirror neurons is not limited to the recognition of motor acts and emotions. Recent experiments showed that the parieto-premotor circuit, besides the recognition of the "what" of a motor act (e.g., grasping an object), is also involved in understanding the "why" this motor act is done (e.g., grasping for placing or for eating, etc.), that the agent's intentions. The mechanism underlying this capacity is related to action organization of the parietal and premotor cortices, where successive motor acts are specifically linked in goal-directed action chains. These chains are activated not only during action execution, but also during motor act observation, thus allowing the individual, if relevant clues are available, to "play" internally the not-yet performed series of motor acts of the agent and to understand in this way his or her intention.

Finally, there is evidence that the mirror neuron system plays a crucial role both in imitation and imitation learning. As far as imitation is concerned, the capacity to repeat a simple motor act (e.g., finger lifting) relies on a direct translation of the observed act in the relative motor representation. For imitation learning, the situation is more complex and requires both the mirror neuron system, which codes the observed motor acts, and additional structures that reorganize these acts in new patterns or sequences.

Giacomo Rizzolatti and Maddalena Fabbri Destro

References

Buccino, G., Vogt, S., Ritzl, A., Fink, G., Zilles, K., Freund, H., Rizzolatti G. Neural circuits underlying imitation learning of hand actions: an event-related fMRI study. *Neuron* 42 (2004) 323–34.

Gallese, V., Keysers, C., Rizzolatti, G. A unifying view of the basis of social cognition. *Trends in Cognitive Sciences* 8 (2004) 396–403.

Rizzolatti, G., Craighero, L. The mirror-neuron system. *Annual Rev. Neurosci.* 27 (2004) 169–92.

Rizzolatti, G., Fogassi, L., Gallese, V. Mirrors of the mind. *Scientific American* 295 (2006) 54–61.

---

pencil instead of the mirror image. To dissociate cue direction and movement direction experimentally, researchers have trained monkeys to perform mental rotation tasks in which a cue at one location instructs a movement to a different location.

Investigators have used such tasks to study neurons in the area principalis of the dorsolateral frontal lobe, in PMd, in SMA, and in M1. In each of these cortical areas, the activity of some neurons correlates best with the direction of the instructional cue, whereas the activity of other neurons correlates best with the direction of the actual movement. In general, from the area principalis, through the PMd and SMA to M1, the percentage of cue-related neurons decreases and the percentage of movement-related neurons increases. Moreover, as time progresses from the appearance of the instructional cue to the execution of the movement, the discharge of the neuronal populations encode cue direction initially and movement direction subsequently. The transformation of cue direction into movement direction thus involves many cortical motor areas and progresses throughout the reaction time of the subject.

Some Cortical Motor Areas Are Involved in Movements in Response to Either Internal or External Cues

Another aspect of movement preparation that differentially involves certain cortical motor areas has to do with whether the instructions about what to do come from internally remembered or externally delivered cues. In one experiment, for example, monkeys were trained to touch three of four pads in randomly selected sequences that were cued when the pads were lit in sequence. Once a monkey was accustomed to a given sequence, the lights gradually were dimmed until the monkey was performing the correct sequence based only on internally remembered cues. After several of these remembered trials, the pads were lit in a different sequence for several trials. Whereas neurons in M1 showed similar discharge rates during the internally remembered and externally cued trials for a given sequence, SMA neurons were generally more active during the internally remembered trials, and PMv neurons were generally more active during the externally cued trials (Fig. 30.15).

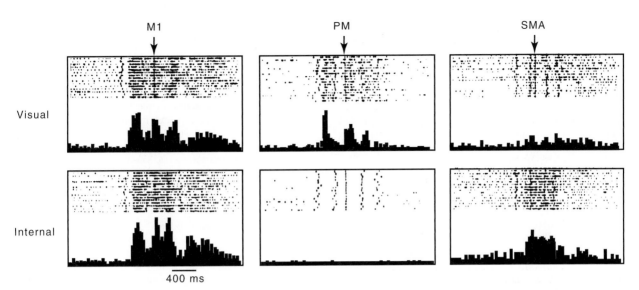

FIGURE 30.15 Activity of three neurons—one in M1, one in PM, and one in SMA—recorded as a monkey pressed three buttons in sequence. The sequence was first cued visually by lighting the buttons and was then cued internally. The M1 neuron showed similar activity whether the monkey performed from visual or internal cues. The PM neuron, however, was much more active in response to visual than internal cues, whereas the opposite was true for the SMA neuron. Modified from Mushiake *et al.* (1991).

Summary

Axons from the cerebral cortex and from the red nucleus in the brain stem descend in pathways located dorsolaterally in the spinal cord to control voluntary movements. In addition to corticospinal and rubrospinal pathways, the motor cortex also sends projections to the red nucleus and to the pontomedullary reticular formation, providing indirect pathways for voluntary control. Multiple cortical motor areas—distinguished by their inputs, their cytoarchitecture, and their interconnections with other cortical areas and with the thalamus—make different contributions to the control of voluntary movements. The most detailed motor map is found in the primary motor cortex, M1, but the somatotopic organization is limited by convergence and divergence in the corticospinal projection. Populations of M1 neurons are also in most direct control of movement kinematics and dynamics. Other cortical motor areas participate differentially in the selection of, and preparation for, voluntary movements based on a variety of internal and external cues.

SUMMARY

The motor cortex, the red nucleus, the pontomedullary reticular formation, and the vestibular nuclei each send major axonal projections descending from the brain to the spinal cord to control bodily movements.

Although these pathways normally provide seamless movement control, different contributions of the medial and lateral descending pathways have been identified experimentally.

The medial system—vestibulospinal and reticulospinal—receives sensory input from the vestibular apparatus and mediates postural responses that keep the head and body stabilized for stance and gait. Additional visual and proprioceptive inputs that reach the vestibular and reticular nuclei via brain stem pathways or after processing in the cerebellum (Chapter 32) also influence postural control. Context-dependent strategies established via centers including the cerebellum and motor cortex further adapt postural responses to the needs of particular complex situations.

The lateral system—corticospinal and rubrospinal—receives, in addition to somatosensory inputs, information processed by the cerebellum (Chapter 32), the basal ganglia (Chapter 31), and associated cortical areas to mediate voluntary movements of the face and extremities. Different cortical motor areas receive different portions of these inputs and therefore make different contributions to controlling voluntary movements. The most elaborate somatotopic representation is found in the primary motor cortex, M1. While the convergence, divergence, and horizontal interconnections of the intrinsic organization of M1 result in considerably distributed activation during voluntary movements, the same substrates enable substantial plasticity. Whereas M1 neurons most directly control the kinematics and dynamics of movement execution, other cortical motor areas participate in selecting which movement(s) to make on the basis of external cues and internal states.

References

Andersen, P., Hagan, P. J., Phillips, C. G., and Powell, T. P. (1975). Mapping by microstimulation of overlapping projections from area 4 to motor units of the baboon's hand. *Proc. R. Soci. Lond. Seri. B Biol. Sci.* **188**, 31–36.

Baker, J., Goldberg, J., and Peterson, B. (1985). Spatial and temporal response properties of the vestibulocollic reflex in decerebrate cats. *J. Neurophysiol.* **54**, 735–756.

Baloh, R. and Honrubia, V. (1990). "Clinical Neurophysiology of the Vestibular System," 2nd ed. F. A. Davis, Philadelphia.

Cheney, P. D., Fetz, E. E., and Palmer, S. S. (1985). Patterns of facilitation and suppression of antagonist forelimb muscles from motor cortex sites in the awake monkey. *J. Neurophysiol.* **53**, 805–820.

Clarke, E. and O'Malley, C. D. (1996). "The Human Brain and Spinal Cord," 2nd ed. Norman Publishing, San Francisco.

Evarts, E. V. (1968). Relation of pyramidal tract activity to force exerted during voluntary movement. *J. Neurophysiol.* **31**, 14–27.

Fetz, E. E. and Cheney, P. D. (1980). Postspike facilitation of forelimb muscle activity by primate corticomotoneuronal cells. *J. Neurophysiol.* **44**, 751–772.

Georgopoulos, A. P. (1988). Neural integration of movement: Role of motor cortex in reaching. *FASEB J.* **2**, 2849–2857.

Gould, H. J., Cusick, C. G., Pons, T. P., and Kaas, J. H. (1986). The relationship of corpus callosum connections to electrical stimulation maps of motor, supplementary motor, and the frontal eye fields in owl monkeys. *J. Comp. Neurol.* **247**, 297–325.

Horak, F. and Nashner, L. (1986). Central programming of postural movements: Adaptation to altered support-surface configurations. *J. Neurophysiol.* **55**, 1369–1381.

Horak, F. and Shupert, C. (1994). Role of the vestibular system in postural control. *In* "Vestibular Rehabilitation" (S. Herdman, eds.). F. A. Davis, Philadelphia.

Kuypers, H. G. (1987). Some aspects of the organization of the output of the motor cortex. *Ciba Found. Symp.* **132**, 63–82.

Lawrence, D. G. and Kuypers, H. G. (1968a). The functional organization of the motor system in the monkey. I. The effects of bilateral pyramidal lesions. *Brain* **91**, 1–14.

Lawrence, D. G. and Kuypers, H. G. (1968b). The functional organization of the motor system in the monkey. II. The effects of lesions of the descending brain-stem pathways. *Brain* **91**, 15–36.

Macpherson, J. and Inglis, J. (1993). Stance and balance following bilateral labyrinthectomy. *Prog. Brain Res.* **97**, 219–228.

Matelli, M., Luppino, G., and Rizzolatti, G. (1991). Architecture of superior and mesial area 6 and the adjacent cingulate cortex in the macaque monkey. *J. Comp. Neurol.* **311**, 445–462.

Mushiake, H., Inase, M., and Tanji, J. (1991). Neuronal activity in the primate premotor, supplementary, and precentral motor cortex during visually guided and internally determined sequential movements. *J. Neurophysiol.* **66**, 705–718.

Nashner, L. (1982). Adaptation of human movement to altered environments. *Trends Neurosci.* **5**, 358–361.

Paloski, W., Black, F., Reschke, M., Calkins, D., and Shupert, C. (1993). Vestibular ataxia following shuttle flights: Effects of microgravity on otolith-mediated sensorimotor control of posture. *Am. J. Otol.* **14**, 9–17.

Peterson, B., Goldberg, J., Bilotto, G., and Fuller, J. (1985). Cervicocollic reflex: Its dynamic properties and interaction with vestibular reflexes. *J. Neurophysiol.* **54**, 90–109.

Rizzolatti, G., Scandolara, C., Matelli, M., and Gentilucci, M. (1981). Afferent properties of periarcuate neurons in macaque monkeys. II. Visual responses. *Behav. Brain Res.* **2**, 147–163.

Roberts, T. (1967). "Neurophysiology of Postural Mechanisms." Plenum Press, New York.

Rosen, I. and Asanuma, H. (1972). Peripheral afferent inputs to the forelimb area of the monkey motor cortex: Input-output relations. *Exp. Brain Res.* **14**, 257–273.

Schor, R., Kearney, R., and Dieringer, N. (1988). Reflex stabilization of the head. *In* "Control of Head Movement" (B. Peterson and F. Richmond, eds.). Oxford Univ. Press, New York.

Shinoda, Y., Yokota, J., and Futami, T. (1981). Divergent projection of individual corticospinal axons to motoneurons of multiple muscles in the monkey. *Neurosci. Lett.* **23**, 7–12.

Wilson, V. and Melvill-Jones, G. (1979). "Mammalian Vestibular Physiology." Plenum Press, New York.

Wise, S. P. and Strick P. L. (1996). Anatomical and physiological organization of the non-primary motor cortex. *TINS* **7**, 442–446.

Woolsey, C. N., Settlage, P. H., Meyer, D. R., Sencer, W., Hamuy, T. P., and Travis, A. M. (1952). Patterns of localization in precentral and "supplementary" motor areas and their relation to the concept of a premotor area. *Res. Pub. Assoc. Res. Nerv. Ment. Dis.* **30**, 238–264.

Marc H. Schieber and James F. Baker

31

The Basal Ganglia

The basal ganglia are large subcortical structures comprising several interconnected nuclei in the forebrain, midbrain, and diencephalon (Fig. 31.1). Because of certain features, it is generally agreed that basal ganglia participate in the control of movement. The largest portion of basal ganglia inputs and outputs are connected with motor areas. Second, the discharge of many basal ganglia neurons correlates with movement. Third, basal ganglia lesions cause severe movement abnormalities. In addition to their role in motor control, basal ganglia are involved in cognitive and affective functions. This chapter concentrates on the motor functions of the basal ganglia, but participation of the basal ganglia in cognition is discussed after the basic framework of basal ganglia function is presented for motor control. The anatomy of the basal ganglia is discussed first, as it is the anatomy that provides the infrastructure for the physiology. Next we consider the activity of the individual components of the basal ganglia during movement. Then we discuss the effect of placing a lesion in one component while leaving the rest of the nervous system intact. The combined strategy of measuring the activity of a structure during movement and then determining the effect a selective lesion on that movement has been fruitful in studying the function of the motor system. After presentation of current models of basal ganglia motor function, the chapter concludes with a discussion of basal ganglia participation in cognitive function.

BASAL GANGLIA ANATOMY

The basal ganglia receive a broad spectrum of cortical inputs. The information conveyed to the basal ganglia then is processed by the basal ganglia circuitry to produce a more focused output to areas of the frontal lobes and brain stem that are involved in the planning and production of movement. The basal ganglia output is entirely inhibitory. Thus, an *increase* in basal ganglia output leads to a *reduction* in the activity of its targets. The fact that the basal ganglia output is inhibitory to thalmocortical and brainstem targets is important to understanding its normal motor function.

At first glance, the anatomy of the basal ganglia may seem confusing. There are several component nuclei at different levels in the brain, and two of the nuclei (substantia nigra (SN) and globus pallidus (GP)) are divided into functionally different components. The names of some structures are different in primates than in other mammals. Furthermore, the terminology has changed over the years so that an individual structure may be known by more than one name. However, with consistent use of modern terminology and a functional context in which to place the anatomy, the organization of basal ganglia can be learned readily. Moreover, once the circuitry is learned, many aspects of its normal function and of the response to injury become much easier to understand.

The basal ganglia include the striatum (caudate, putamen, nucleus accumbens), the subthalamic nucleus (STN), the globus pallidus (internal segment or GPi; external segment or GPe; and ventral pallidum), and the substantia nigra (pars compacta or SNpc and pars reticulata or SNpr) (Fig. 31.2). The striatum and STN are the primary recipients of information from outside of the basal ganglia. Most of those inputs come from the cerebral cortex, but thalamic nuclei also provide strong inputs to striatum. There are no direct inputs to the basal ganglia from spinal or brainstem sensory or

A

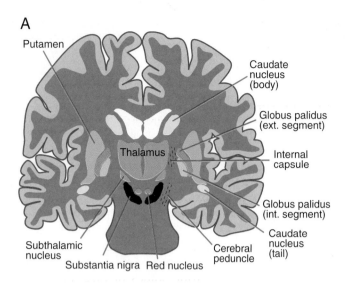

B

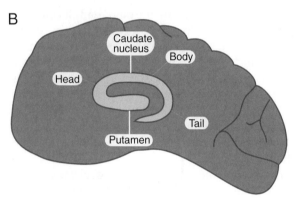

FIGURE 31.1 Location of basal ganglia in the human brain. (A) Coronal section. (B) Parasagittal section.

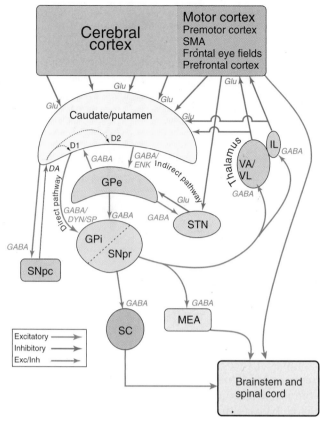

FIGURE 31.2 Simplified wiring diagram of basal ganglia circuits. Simplified schematic diagram of basal ganglia circuitry. Excitatory connections are indicated by green arrows, inhibitory connections by red arrows, and the modulatory dopamine projection is indicated by a red and green arrow. GPe, globus pallidus, external segment; GP$_i$, globus pallidus, internal segment; IL, intralaminar thalamic nuclei; MD, mediodorsal thalamic nucleus; MEA, midbrain extrapyramidal area; SC, superior colliculus; SNpc, substantia nigra pars compacta; SNpr, substantia nigra, pars reticulata; STN, subthalamic nucleus; VA, ventral anterior thalamic nucleus; VL, ventral lateral thalamic nucleus; DA, dopamine (with D1 and D2 receptor subtypes); Dyn, dynorphin; Enk, enkephalin; GABA, γ-aminobutyric acid; Glu, glutamate; SP, substance P.

motor systems. The major outputs from basal ganglia arise from GPi and SNpr and are inhibitory to thalamic nuclei and to brain stem. There are no direct outputs from basal ganglia to spinal motor circuitry.

The Striatum Receives Most of the Inputs to the Basal Ganglia

The striatum is located in the forebrain and comprises the caudate nucleus and putamen (neostriatum) and nucleus accumbens (ventral striatum). This section focuses on the neostriatum (see Box 31.3 and Chapter 43 for a discussion of the ventral system of the basal ganglia). Embryologically, the striatum develops from a basal region of the lateral telencephalic vesicle. It is named the striatum because axon fibers passing through the striatum give it a striped appearance. In rodents, the caudate and putamen are a single struc-

ture with fibers of the internal capsule coursing through, but in carnivores and primates, the caudate and putamen are separated by the internal capsule. The caudate and putamen receive input from the neocortex, and the size of these nuclei increases in parallel with the neocortex throughout phylogeny.

Four major types of neurons have been described in the striatum. The typing is based on the size of the cell body, presence or absence of dendritic spines in individual neurons that have been stained with the Golgi technique or filled with horseradish peroxidase, and other staining properties. The first and by far the most numerous neuron type is the *medium spiny neuron* (Fig.

31.3). Medium spiny neurons make up 80 to 95% of the total number of striatal neurons, depending on the species. They utilize γ-aminobutyric acid (GABA) as a transmitter and constitute the output of the striatum, sending axonal projections to the GP and SN. Medium spiny neurons have large dendritic trees that span 200 to 500 μm (Wilson and Groves, 1980). They also have extensive local axon collaterals that may inhibit neighboring striatal neurons. The medium spiny neurons are morphologically homogeneous, but chemically heterogeneous, as discussed later. Second, *large aspiny neurons* make up 1 to 2% of the striatal population. They are interneurons and use acetylcholine (ACh) as a neurotransmitter. They have extensive axon collaterals in the striatum that terminate on medium spiny neurons. The third type of neuron is the *medium aspiny cell*. These are also interneurons and are thought to use somatostatin as a neurotransmitter. The fourth type of neuron is a *small aspiny cell* that uses GABA as a neurotransmitter. These inhibitory interneurons are more prevalent in primates than in rodents and play an important role in regulating the activity of the medium spiny output neurons.

The striatum receives excitatory input from nearly all the cerebral cortex. The cortical input uses glutamate as its neurotransmitter and terminates largely on the heads of the dendritic spines of medium spiny neurons (Fig. 31.4). The projection from the cerebral cortex to the striatum has a roughly topographical organization. For example, the somatosensory and motor cortex project to the posterior putamen and the prefrontal cortex projects to the anterior caudate. Within the somatosensory and motor projection to the

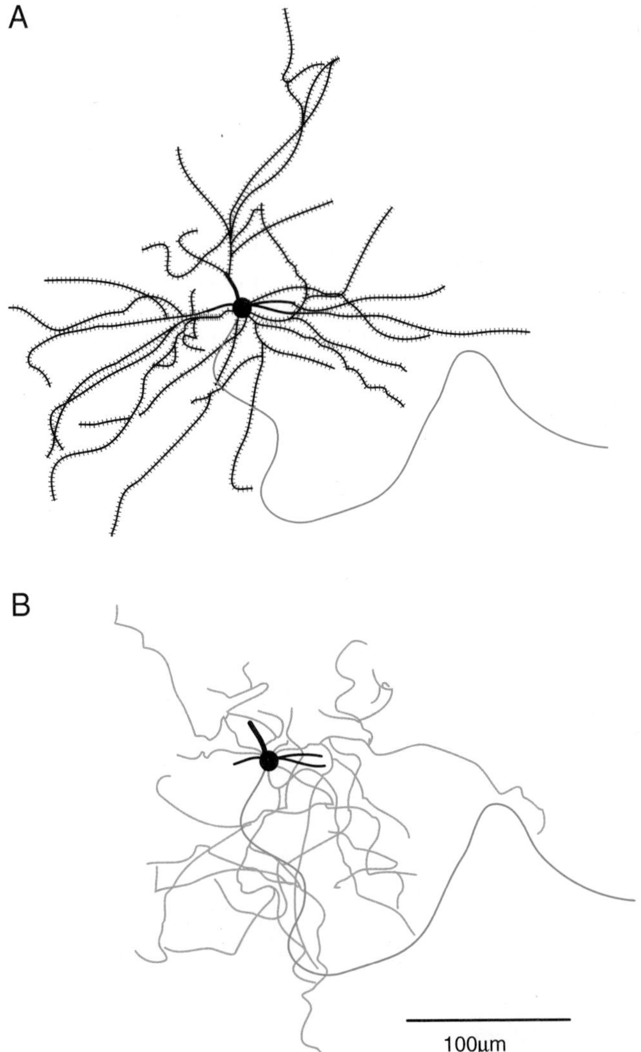

A

B

100μm

FIGURE 31.3 Two representations of a striatal medium spiny neuron that has been filled with HRP. (A) The soma and dendritic tree with numerous dendritic spines. The thin process is the axon, which has been drawn without its collaterals. (B) The same neuron drawn to show the axonal collaterals, which branch extensively within the same field as the dendritic tree. From Wilson and Groves (1980).

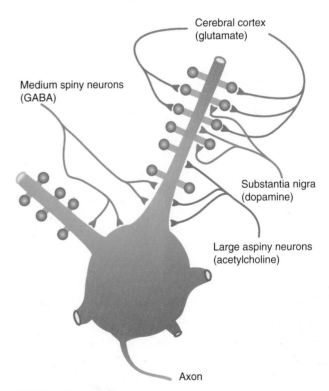

FIGURE 31.4 Pattern of termination of afferents on a medium spiny neuron. The soma and the proximal dendrites with their spines are shown.

striatum, there is a preservation of somatotopy. It has been suggested that the topographic relationship between the cerebral cortex and the striatum provides a basis for the segregation of functionally different circuits in the basal ganglia (Alexander *et al.*, 1986) (Fig. 31.5). These circuits include somatomotor, oculomotor, cognitive, and limbic connections. Within each circuit there appear to be subcircuits such that the primary motor cortex and premotor cortex have non-identical connections with basal ganglia structures. Likewise, dorsolateral and orbitofrontal circuits have distinct connectivity patterns.

Although the topography and somatotopy imply a certain degree of parallel organization, there is also convergence and divergence in the corticostriatal projection. The large dendritic fields of medium spiny neurons allow them to receive input from adjacent projections, which arise from different areas of cortex. Inputs from more than one cortical area overlap (Flaherty and Graybiel, 1991), and input from a single cortical area projects divergently to multiple striatal zones (Fig. 31.6). This convergent and divergent organization provides an anatomical framework for the integration and transformation of information from several areas of the cerebral cortex.

In addition to cortical input, medium spiny striatal neurons receive a number of other inputs: (1) excitatory and presumed glutamatergic inputs from intralaminar and ventrolateral nuclei of the thalamus; (2) cholinergic input from large aspiny neurons; (3) GABA, substance P, and enkephalin input from adjacent medium spiny striatal neurons; (4) GABA input from small interneurons; and (5) a large input from dopamine (DA)-containing neurons in the SNpc. The DA input is of particular interest because of its role in Parkinson's disease (PD, see later).

The DA input to the striatum terminates largely on the shafts of the dendritic spines of medium spiny neurons (Fig. 31.4). The location of dopaminergic terminals puts them in a position to modulate transmission from the cerebral cortex to the striatum. The action of DA on striatal neurons depends on the type of DA receptor involved. Five types of G protein-coupled DA receptors have been described (D1–D5). These have been grouped into two families based on their response to agonists. The D1 family includes D1 and D5 receptors, and the D2 family includes D2, D3, and D4 receptors. D1 receptors potentiate the effect of cortical input to striatal neurons. D2 receptors decrease the effect of cortical input to striatal neurons (Chapters 8 and 9).

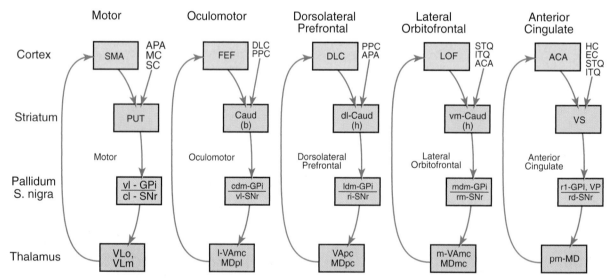

FIGURE 31.5 Parallel circuits connecting the basal ganglia, thalamus, and cerebral cortex. The five circuits are named according to the primary cortical target of the output from the basal ganglia: motor, oculomotor, dorsolateral prefrontal, lateral orbitofrontal, and anterior cingulate. ACA, anterior cingulate area; APA, arcuate premotor area; CAUD, caudate; b, body; h, head; DLC, dorsolateral prefrontal cortex; EC, entorhinal cortex; FEF, frontal eye fields; GPi, internal segment of globus pallidus; HC, hippocampal cortex; ITG, inferior temporal gyrus; LOF, lateral orbitofrontal cortex; MC, motor cortex; MDpl, medialis dorsalis pars paralamellaris; MDme, medialis dorsalis pars magnocellularis; MDpc, medialis dorsalis pars parvocellularis; PPC, posterior parietal cortex; PUT, putamen; SC, somatosensory cortex; SMA, supplementary motor area; SNr, substantia nigra pars reticulate; STG, superior temporal gyrus; VAmc, ventralis anterior pars magnocellularis; Vapc, ventralis anterior pars parvocellularis; VLm, ventralis lateralis pars medialis; VLo, ventralis lateralis pars oralis; VP, ventral pallidum; VS, ventral striatum; cl, caudolateral; cdm, caudal dorsomedial; d1, dorsolateral; 1, lateral; 1dm, lateral dorsomedial; m, medial; mdm, medial dorsomedial; pm, posteromedial; rd, rostrodorsal; rl, rostrolateral; rm, rostromedial; vm, ventromedial; vl, ventrolateral.

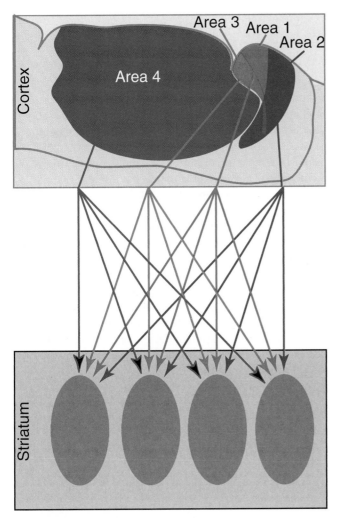

FIGURE 31.6 Schematic representation of the sensorimotor cortical projection to the striatum from arm areas in the somatosensory cortex (areas 1, 2, and 3) and motor cortex (area 4). Note that each cortical area projects to several striatal zones and several functionally related cortical areas project to a single striatal zone. After Flaherty and Graybiel (1991).

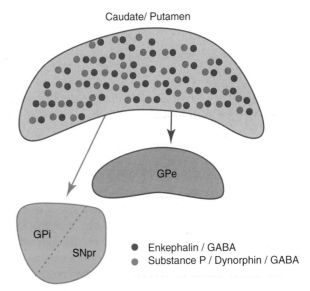

FIGURE 31.7 The two chemically different populations of striatal medium spiny neurons are intermixed. One population (blue) projects to the globus pallidus external segment (GPe) and contains GABA and enkephalin. The other population (red) projects to the globus pallidus internal segment (GPi) and substantia nigra pars reticulata (SNpr) and contains GABA, dynorphin, and substance P (red).

Medium spiny neurons contain the inhibitory neurotransmitter GABA. In addition, medium spiny neurons have peptide neurotransmitters that are colocalized with GABA. Based on both the type of neurotransmitters and the type of DA receptor they contain, medium spiny neurons can be divided into two populations. One population contains GABA, dynorphin, and substance P and primarily expresses D1 receptors. These neurons send axons to GPi and to SNpr. Although substance P generally is thought to be an excitatory neurotransmitter, the predominant effect of these neurons is inhibition of their targets. The second population contains GABA and enkephalin and primarily expresses D2 receptors. These neurons

project to GPe and are inhibitory. The two populations of medium spiny neurons are morphologically indistinguishable and are not segregated topographically within the striatum (Fig. 31.7). This commingling of neurons suggests that they receive similar types of input and thus they may convey similar information to their respective targets. However, there is emerging evidence that the two classes of medium spiny neurons may receive input from different classes of cortical neurons within the same cortical area. The neurotransmitter type, dopamine receptor type, and possible differences in cortical input provide one level of functional segregation within the striatum.

Although there are no apparent regional differences in the striatum based on cell morphology, an intricate internal organization has been revealed with special stains. When the striatum is stained for acetyl-cholinesterase (AChE), the enzyme that inactivates acetylcholine (Chapter 8), there is a patchy distribution of lightly staining regions within more heavily stained regions. The AChE-poor patches have been called *striosomes* and the AChE-rich areas have been called the extrastriosomal *matrix* (Graybiel *et al.*, 1981). The matrix forms the bulk of the striatal volume and receives input from most areas of the cerebral cortex. Within the matrix are clusters of neurons with similar

BOX 31.1

CONCEPT OF THE EXTRAPYRAMIDAL MOTOR SYSTEM

Participation of the basal ganglia in normal motor control is an old idea dating back to the early part of this century. The British neurologist S. A. Kinnier Wilson described a disease that included muscular rigidity, tremor, and weakness that was associated with pathological changes in the liver and in the putamen (hepatolenticular degeneration). Wilson noted that several features seen with damage to the corticospinal tracts were not present in this disease and postulated that the motor abnormalities were due to disease of an "extrapyramidal" motor system. Wilson postulated that basal ganglia (striatum and globus pallidus) were the major constituents of this "extrapyramidal" motor system that he viewed to be independent of the "pyramidal" (corticospinal) motor system. In a later writing, Wilson developed the view of two motor systems, the phylogenetically "old" extrapyramidal system and the phylogenetically "new" pyramidal system (Wilson, 1928). He thought that the extrapyramidal system had an automatic, postural, and static function, which was minimally modifiable, and that the pyramidal system had a voluntary, phasic function that could be modified. At the time, it was thought that the output of the basal ganglia went directly to the spinal cord.

In the 1960s, more modern anatomical techniques showed that the bulk of basal ganglia output went via the thalamus to motor cortical areas. In this sense, basal ganglia output was "prepyramidal," not extrapyramidal. With the developing model of initiation of movement by the basal ganglia, the prepyramidal location was emphasized as the most important basal ganglia output. More recently, it has been possible to inject a different colored fluorescent dye into each of two downstream targets to find out if individual upstream neurons project to both targets. It was found that the vast majority of neurons in GPi branch and project both to the thalamus and to the brain stem. Hence, the basal ganglia are both prepyramidal and extrapyramidal. Although we now know that the basal ganglia can act through the pyramidal system and that there are many extrapyramidal motor systems that project to the spinal cord independent of the pyramidal tract, the phrase "extrapyramidal motor system" is still used to refer to the basal ganglia and to the disorders that result from basal ganglia damage.

Jonathan W. Mink

Reference

Wilson, S. A. K. (1928). "Modern Problems in Neurology." Arnold, London.

---

inputs that have been termed *matrisomes*. The bulk of the output from cells in the matrix is to both segments of the GP and to SNpr. The striosomes receive input from the prefrontal cortex and send output to SNpc. Immunohistochemical techniques have demonstrated that many substances, such as substance P, dynorphin, and enkephalin, have a patchy distribution that may be partly or wholly in register with the striosomes (Graybiel *et al.*, 1981). The striosome-matrix organization provides the substrate for another level of functional segregation within the striatum.

The STN Receives Inputs from the Frontal Lobe

The STN is located at the junction of the diencephalon and midbrain, ventral to the thalamus and rostral and lateral to the red nucleus. Embryologically, it develops from the lateral hypothalamic cell column. Phylogenetically, it increases in size in proportion to the neocortex. The STN receives an excitatory, glutamatergic input from the frontal cortex with large contributions from motor areas, including the primary motor cortex (area 4), premotor and supplementary motor cortex (area 6), and frontal eye fields (area 8). The STN also receives an inhibitory GABA input from GPe. The output from the STN is excitatory and glutamatergic. STN projects to GPi and SNpr as well, projecting back to GPe.

Although the STN receives input from the neocortex and projects to both segments of GP and to SNpr, it is different from striatum in several ways. Unlike striatum, the cortical input to the STN is from the frontal lobe only. The output from STN is excitatory, whereas the output from striatum is inhibitory. Of the two routes from the cortex to GPi, GPe, and SNpr, the excitatory route through STN (5–8 ms) is faster than the inhibitory route through striatum (15–20 ms). The pathway from cerebral cortex via STN to GPi and SNpr has been called the "hyperdirect pathway" because it is the fastest route for information flow from cerebral cortex to the basal ganglia output (Fig. 31.2) (Nambu et al., 2002).

GPi Is the Primary Basal Ganglia Output for Limb Movements

GP lies medial to the putamen and rostral to the hypothalamus. Embryologically, it arises from the lateral hypothalamic cell column. In primates, GP is separated into an internal segment and an external segment by a fiber tract called the internal medullary lamina. In rodents and carnivores, the internal segment lies within the internal capsule and is called the entopeduncular nucleus. GPe is not a principal source of output from the basal ganglia and it is considered later.

The GPi is composed primarily of large neurons that project outside of the basal ganglia. The dendritic trees of GPi neurons radiate from the cell body in a disc-like distribution and are oriented so that the faces of the disc are perpendicular to incoming axons from the striatum. The dendrites of an individual GPi cell can span up to 1 mm in diameter and therefore GPi neurons have the potential to receive a large number of converging inputs.

The principal inputs to GPi are from striatum and the STN. These two types of inputs differ in the neurotransmitters used, in their physiological action, and in their pattern of termination. As noted earlier, neurons projecting from striatum to GPi contain GABA, substance P, and dynorphin, and are inhibitory. Each axon from the striatum enters GPi and sparsely contacts several neurons in passing before ensheathing a single neuron with a dense termination. This pathway has been called the "direct" striatopallidal pathway (Fig. 31.2). The excitatory, glutamate projection from STN to GPi is highly divergent such that each axon from the STN ensheathes many GPi neurons (Fig. 31.8). The striatal projection to GPi is about 10 to 15 msec slower than the STN projection to GPi. Thus, the striatal and STN inputs to GPi form a pattern of fast, widespread, divergent excitation from STN and slower, focused, convergent inhibition from striatum.

The output from GPi is inhibitory and, like the input to this structure from striatum, uses GABA as its neurotransmitter. The majority (70%) of the GPi output is sent via collaterals to both the thalamus and the brain stem. In the thalamus, axons from GPi terminate in the oral part of the ventrolateral nucleus (VLo) and in the

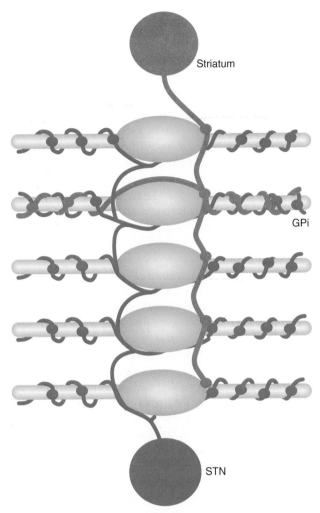

FIGURE 31.8 The two primary inputs to globus pallidus internal segment (GPi) have different patterns of termination. Axons from the subthalamic nucleus (green) are excitatory and terminate extensively on multiple GPi neurons. Axons from the striatum (red) contact several GPi neurons weakly in passing before terminating densely on a single neuron.

principal part of the ventral anterior nucleus (VApc). In turn, these thalamic nuclei project to the motor, premotor, supplementary motor, and prefrontal cortex. They also project back to striatum in a potential feedback loop. Some evidence suggests that an individual GPi neuron sends output via the thalamus to just one area of the cortex (Hoover and Strick, 1993). In other words, GPi neurons that project to the motor cortex are adjacent to, but separate from, those that project to the premotor cortex. Thus, just as there are parallel inputs from the cortex to striatum, there appear to be potential functionally parallel outputs from GPi.

Collaterals of the axons projecting from GPi to the thalamus project to an area at the junction of the midbrain and pons near the pedunculopontine nucleus,

BOX 31.2

PARKINSON'S DISEASE

Parkinson's disease (PD) is the second most common progressive neurodegenerative disorder. Patients with PD experience slowness of movement, rigidity, a low-frequency rest tremor, and difficulty with balance. These major motoric features of PD are due to the degeneration of dopamine (DA) containing neurons in the substantia nigra pars compacta (SNC) with an accompanying loss of DA and its metabolites in the striatum. PD patients may exhibit other symptoms, including cognitive dysfunction, depression, anxiety, autonomic problems, and disturbances of sleep, which may be due to the degeneration of non-DA-containing neurons. PD is characterized pathologically by the cytoplasmic accumulation of aggregated proteins with a halo of radiating fibrils and a less defined core known as Lewy bodies. Measures of increased oxidative stress are also seen, including glutathione depletion, iron deposition, increased markers of lipid peroxidation, oxidative DNA damage and protein oxidation, and decreased expression and activity of mitochondrial complex 1 in the SNC. This mitochondrial complex 1 defect is specific to the SNC, as it has not been observed in other neurodegenerative diseases. Thus, in sporadic PD, oxidative stress and mitochondrial dysfunction appear to play prominent roles in the death of DA neurons. The majority of PD appears to be sporadic in nature, however, an estimated 10% of cases are familial, with specific genetic defects. The identification of these rare genes and their functions has provided tremendous insight into the pathogenesis of PD and opened up new areas of investigation.

Two genes have been clearly linked to PD: α-synuclein and parkin. α-Synuclein is a presynaptic protein and a primary component of Lewy bodies. Very rare missense mutations at A53T and A30P cause autosomal-dominant PD. How derangements in α-synuclein lead to dysfunction and death of neurons is not known. However, aggregation and fibrillization are thought to play a central role, perhaps leading to a dysfunction in protein handling by inhibiting the proteasome. The second "PD gene" parkin is an E3 ligase for the ubiquitination of proteins. Mutations in parkin include exonic deletions, insertions, and several missense mutations and result in autosomal recessive PD. Parkin mutations are now considered to be one of the major causes of familial PD. Substrates for parkin include the synaptic vesicle-associated protein CDCrel-1, which may regulate synaptic vesicle release in the nervous system. Whether CDCrel-1 is involved in the release of dopamine is not yet known. Synphilin-1 is a protein of unknown function that was identified as an α-synuclein

interacting protein. It is a synaptic vesicle-enriched protein and is present in Lewy bodies. Synphilin-1 is a target of parkin, and expression of synphilin-1 with α-synuclein results in the formation of Lewy body-like aggregates that are ubiquitinated in the presence of parkin. Parkin is upregulated by unfolded protein stress, and expression of parkin suppresses unfolded protein stress-induced toxicity. The unfolded putative G-protein-coupled transmembrane receptor, the parkin-associated, endothelial-like receptor (Pael-R), is a parkin substrate. When overexpressed, Pael-R becomes unfolded, insoluble, and causes unfolded protein-induced cell death. Parkin ubiquitinates Pael-R, and coexpression of parkin results in protection against Pael-R-induced cell toxicity. Pael-R accumulates in the brains of autosomal-recessive Parkinson's disease patients and thus may be an important parkin substrate.

Ultimately, it would be of interest to link all the multiple genes and the sporadic causes of PD into a common pathogenic biochemical pathway. Derangements in protein handling seem to be central in the pathogenesis of familial PD. Oxidative stress seems to play a prominent role in sporadic PD, and oxidative stress leads to synuclein aggregation and/or proteasomal dysfunction. Data suggest that there may be selective derangements in the proteasomal system in the substantia nigra of sporadic PD. Thus, protein mishandling may be central to the pathogenesis of PD.

In contrast to most other neurodegenerative disorders, there is effective temporary symptomatic treatment for PD consisting of DA replacement with levodopa or DA agonists and adjunctive medications or surgical approaches. Because the neurodegeneration in PD is progressive and there is no proven preventative, restorative, or regenerative therapy, patients eventually become quite disabled. Ultimately, to affect a true cure, the underlying mechanisms of neuronal cell death need to be understood, strategies for enhancing neuronal survival and regrowth need to be developed, and consideration needs to be made toward the replacement of cells lost during the degenerative process. If these goals can be obtained, a full recovery could be achieved.

Valina L. Dawson and Ted M. Dawson

Adapted from Dawson T. M. (2000). "New Animal Models for Parkinson's Disease." *Cell* **101**(2), 115–118.
Adapted from Zhang Y., Dawson V. L., Dawson T. M. "Oxidative Stress and Genetics in the Pathogenesis of Parkinson's Disease." *Neurobiol. Dis.* **7**(4), 240–250.

which has been termed the "midbrain extrapyramidal area." The midbrain extrapyramidal area projects in turn to the reticulospinal motor system. Other GPi neurons (20%) project to intralaminar nuclei of the thalamus or to the lateral habenula.

Substantia Nigra Pars Reticulata Is the Basal Ganglia Output for Eye Movements

Like the GP, the SN is divided into two segments. One is a densely cellular region called the pars compacta (SNpc) and the other is more sparsely cellular and is called the pars reticulata (SNpr). The pars compacta contains DA cells and is not a principal output nucleus of the basal ganglia. It is considered later. SNpr has features similar to GPi, including cell size, histochemistry, and connectional anatomy.

Like GPi, the SNpr contains large neurons that project outside of the basal ganglia. The dendritic trees are less disc-like than those of GPi neurons, but like GPi neurons they extend up to 1 mm and thus receive a wide field of inputs. As in the case of GPi, the SNpr receives inhibitory inputs from striatum direct pathways that contain GABA, SP, and dynorphin and excitatory inputs from STN that contain glutamate and provides inhibitory, GABAergic outputs. In the case of SNpr, these outputs are to the medial part of the ventro-lateral thalamus (VLm) and the magnocellular part of the ventral anterior thalamus (VAmc). These thalamic areas in turn project to the premotor and prefrontal cortex. Like the GPi output, SNpr sends collaterals to the midbrain extrapyramidal area and also has a projection to intralaminar nuclei of the thalamus. A primary difference in the outputs of GPi and SNpr is that the lateral portion of SNpr sends an inhibitory projection to the superior colliculus and to the paralaminar part of the medial dorsal thalamus (MDpl). MDpl projects in turn to the frontal eye fields. Thus, the lateral portion of SNpr is connected with cortical and brain stem areas that control eye movements.

GPe Is Connected to Several Other Basal Ganglia Nuclei

The GPe is one of two nuclei that may be viewed as intrinsic nuclei of the basal ganglia. The other is the SNpc, which is considered in the next section. Both GPe and SNpc receive the bulk of their input from and send the bulk of their output to other basal ganglia nuclei.

GPe receives an inhibitory projection from the striatum and an excitatory one from STN. The patterns of termination of the striatal and STN afferents are similar in that the striatal input is focused and convergent while the STN input is divergent (Parent and Hazrati, 1993). Unlike GPi, the striatal projection to GPe contains GABA and enkephalin but not substance P. The majority of the output of GPe projects to STN. The connections from striatum to GPe, from GPe to STN, and from STN to GPi and SNpr thus are referred to as the "indirect" pathway to GPi and Snpr to distinguish it from the "direct" pathway that runs from striatum to GPi and SNpr (Fig. 31.2) (Alexander and Crutcher, 1990). In addition, there is a monosynaptic GABAergic inhibitory output from GPe directly to GPi and to SNpr and a GABAergic projection back to striatum. Thus, GPe neurons are in a position to provide feedback inhibition to neurons in striatum and STN and feed-forward inhibition to neurons in GPi and SNpr. This circuitry suggests that GPe may act to oppose, limit, or focus the effect of the striatal and STN projections to GPi and SNpr, as well as focus activity in these output nuclei.

SNpc Provides DA Input to the Striatum

The SNpc is perhaps the most studied structure in the basal ganglia. The SNpc is made up of large DA-containing cells and it is these neurons that degenerate in PD, which is characterized by abnormal movement. DA neurons contain a substance called neuromelanin, which is a dark pigment that gives the SNpc a blackish appearance and appears to represent an oxidation product of DA. This is the basis for its name (substantia nigra = black substance). SNpc receives input from the striatum, specifically from the striosomes and a more sparse input from prefrontal frontal cortex. The SNpc DA neurons project to all of caudate and putamen in a topographic manner. However, nigral DA neurons receive inputs from one striatal circuit and project back to the same and to adjacent circuits (Haber et al., 2000). Thus, they are in a position to modulate activity across circuits. The action of DA depends on the receptors located on target neurons, as was discussed earlier.

Summary

To review:

1. The striatum receives input from nearly all the cerebral cortex such that several functionally related cortical areas project to overlapping striatal zones and that an individual cortical area projects to several striatal zones. Cortical areas that are not functionally related project to separate zones of the striatum, although there may be some interaction between adjacent zones.

2. The striatum sends a focused and convergent inhibitory projection to the basal ganglia output nuclei, GPi and SNpr.

3. The STN receives input from the frontal lobe, especially from the motor, premotor, and supplementary motor cortex and from the frontal eye fields.

4. The STN sends a fast divergent excitatory projection to GPi and SNpr.

5. There are reciprocal and loop-like connections among basal ganglia nuclei that may play a negative or positive feedback role or may result in focusing of signals.

6. The output from the basal ganglia (GPi and SNpr) is inhibitory and projects to motor areas in the brain stem and thalamus.

7. There are no direct connections between the basal ganglia and the spinal sensory or motor apparatus.

There is some disagreement among basal ganglia experts whether to view the overall anatomic organization of the basal ganglia as convergent or as multiple parallel segregated loops through the basal ganglia, each with a separate output. In support of convergence are (1) the reduction of the number of neurons at each level from the cortex to striatum to GPi and SNpr; (2) the large number of synapses on each striatal neuron; (3) the large dendritic trees of striatal, GPi, and SNpr neurons; and (4) interaction across circuits mediated by SNpc and GPe neurons. In support of parallel segregated loops are (1) the preserved somatotopy with separate representations of the face, arm, and leg in the cortex, striatum, and GPi/SNpr; (2) the relative preservation of topography through the basal ganglia (e.g., the prefrontal cortex to caudate to SNpr to VA thalamus to the prefrontal cortex and motor cortex to putamen to GPi to VLo thalamus to motor cortex); and (3) the finding that separate groups of GPi neurons project via the thalamus to separate motor areas of the cortex. It appears that there is local convergence and global parallelism, but it remains unknown whether the parallel pathways are functionally segregated or to what degree there is interaction at the interfaces between functionally different pathways.

SIGNALING IN BASAL GANGLIA

Although the anatomic organization may provide some clues as to what might be the function of basal ganglia circuits, inference of function from anatomy is speculative. One approach to studying the function of an area of the central nervous system is to record the electrical activity of individual neurons with an extracellular electrode in awake, behaving animals. Other approaches involve inferences of neuronal signaling from imaging studies of blood flow and metabolism, or of changes in gene expression. By sampling the signal of a part of the brain during behavior, one can gain some insight into what role that part might play in behavior. Neurons within different basal ganglia nuclei have characteristic baseline discharge patterns that change with movement. If an animal is trained to perform a task consistently, the activity of single neurons can be correlated with individual aspects of the task performance. Furthermore, the timing of neural activity in one part can be compared to the timing of another and to the timing of the movement. This section emphasizes signals in the basal ganglia that are correlated with movement or the preparation for movement.

The Striatum Has Low Spontaneous Activity That Increases during Movement

In awake animals, the majority of striatal neurons (80–90%) have low baseline discharge rates of 0.1 to 1 Hz (Fig. 31.9). These neurons project outside of the striatum and therefore are probably medium spiny neurons. In general, the activity of these neurons reflects the activity of areas of cerebral cortex from which they receive inputs, but striatal neurons are less modality specific. Thus, neurons in areas of the putamen that receive input from the somatosensory

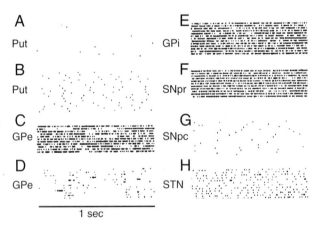

FIGURE 31.9 Representative neural discharge patterns from several basal ganglia nuclei. In the raster displays, each dot indicates the occurrence of an action potential. Each horizontal raster line represents a 1-s period of discharge. Several such periods are arranged vertically for each nucleus.

and motor cortex have activity correlated with active and passive movement, but not with specific tactile modalities such as light touch, vibration, or joint position. The activity of movement-related putamen neurons is different from corticospinal neurons in the motor cortex, but is similar to the activity of cortico-striatal neurons.

Striatal neurons related to movement have a distinct somatotopy with the face represented ventromedially and the leg dorsolaterally. They tend to occur in clusters of neurons with similar discharge characteristics. These clusters of physiologically similar neurons may correspond to the anatomically defined matrisomes described earlier. Movement-related activity changes typically are observed as increases above the very low baseline. On average, movement-related striatal neurons fire 20 msec prior to movement (Fig. 31.10). About half of these neurons fire in relation to the direction of movement. Some neurons fire in relation to the start of movement and others in relation to the stop. Some neurons in the putamen fire in relation to self-initiated movements, some in relation to stimulus-triggered movements, and some in relation to both (Romo *et al.*, 1992). Some neurons in the anterior part of the putamen and in the caudate nucleus that receive input from the premotor or prefrontal cortex have

activity during the preparation for movement. This activity is time locked to instructional cues but not to the movement itself. Some signals correlate with the instruction of whether to move ("set"). Other signals correlate with the signal to move ("go"). Although some individual neurons have signals that differentially relate to specific tasks or task parameters, as a whole the striatum is not exclusively active in relation to any of these tasks or parameters.

Striatal neurons active in relation to eye movements are located in a longitudinal zone in the central part of the caudate nucleus. This region receives input from the frontal and supplementary eye fields in the cortex. Similar to striatal neurons related to limb movements, striatal neurons related to eye movements may discharge during preparation for the movement or during the movement. Caudate neurons involved in saccades are strongly modulated by reward expectation (Hikosaka, 2007). This context-dependent modulation of activity may be a fundamental property of medium spiny neuronal activity in all circuits, but has been demonstrated best in the oculomotor circuit.

The timing of neuronal activity related to limb or eye movements is important to understanding the function of those neurons in the production of movement. In the putamen, most movement-related activity is late, occurring on average after the onset of muscle activity responsible for producing the movement. When compared to activity in the cerebral cortex, movement-related putamen neurons fire after those in the motor cortex and supplementary motor area. Putamen neurons related to movement preparation also fire substantially later than cortical neurons involved in movement preparation. Caudate neurons related to eye movements fire late relative to oculomotor areas of cortex. The late timing of striatal discharge during movement and movement preparation relative to discharge in cortical areas suggests that the striatum is not involved primarily in the initiation of movement. Instead, it receives information from cortical areas that are responsible for movement generation and may use that information for facilitation, gating, or scaling of cortically initiated movement.

A subset of striatal neurons (10%) is tonically active with discharge rates of 2 to 10 Hz. These neurons are distributed throughout the striatum and their discharge bears no specific relation to movement. These tonically active neurons (TANs) fire in relation to certain sensory stimuli that are associated with reward. For example, a TAN will fire in relation to a clicking sound if the click precedes a fruit juice reward but will not fire to the click alone or to the reward alone (Aosaki *et al.*, 1994). It has been suggested that these neurons signal aspects of tasks that are related to learning and

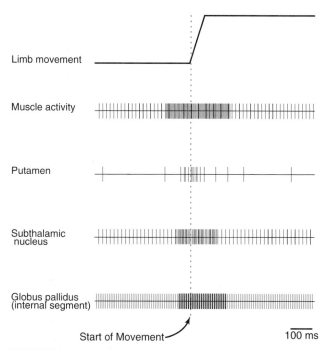

FIGURE 31.10 Schematic representation of the timing of neuronal activity in three nuclei of the basal ganglia in relation to limb movement and the muscle activity used to make the movement.

reinforcement. TANs are not activated by electrical stimulation of the GP and thus are probably not projection neurons. It is thought that TANs are cholinergic large aspiny interneurons that innervate the medium spiny neurons (see earlier discussions), which would place them in a position to modify the sensitivity of the medium spiny neurons to cortical input in relation to specific behavioral contexts.

Another subset of striatal neurons are the fast-spiking GABAergic inhibitory interneurons (Tepper *et al.*, 2004). These neurons correspond to the parvalbumin-positive neurons described previously. Their role in motor control has not yet been elucidated, in part because their small size makes recording in awake animals difficult. Nonetheless, these neurons are likely to play an important role in regulating the activity of medium spiny striatal neurons.

STN Has Moderate Spontaneous Activity That Increases during Movement

Neurons in the STN are tonically active with average baseline discharge rates of about 20 Hz (Fig. 31.9). They are organized somatotopically and change activity in relation to eye or limb movement. For 90% of these neurons the change is an increase, occurring on average 50 msec prior to the movement (Fig. 31.10). The majority of movement-related neurons in STN have signals related to movement direction, but little is known about whether they discharge in relation to movement parameters such as amplitude or velocity.

GPi and SNpr Have High Spontaneous Activity That Increases or Decreases after Movement Initiation

As described earlier, GPi and SNpr form the output of the basal ganglia. Just as they are similar anatomically, they are also similar physiologically. Neurons in these output structures are tonically active with average firing rates of 60 to 80 Hz (Fig. 31.9). They are organized somatotopically with the leg and arm in GPi and the face and eyes in SNpr. Due to the somatotopy, studies of limb movements have focused on GPi and studies of eye movements have focused on SNpr. Because limb and eye movements are controlled differently, these studies are discussed separately.

The discharge of many GPi neurons correlates with the direction of limb movement, but most GPi activity is not correlated with other physical parameters of movement, including joint position, force production, movement amplitude, or movement velocity. Few GPi neurons have activity correlated with the pattern of muscle activity. Like striatum, some GPi neurons have

activity related to movement preparation. During a reaching movement, 25% of GPi neurons are related to movement preparation and 50% are related to movement, but many neurons have more than one type of response. Approximately 70% of arm movement-related GPi neurons increase activity and 30% decrease activity from this tonic baseline during movement. Because the output from GPi is inhibitory, the large proportion of activity increases during movement translates to broad inhibition of thalamic and brain stem targets with more restricted facilitation during movement.

The timing of movement-related GPi activity is late compared to the activation of agonist muscles. The average onset of GPi activity is after the onset of EMG but before the onset of movement (Fig. 31.10). There is a tendency for GPi movement-related increases to occur earlier than decreases, which probably reflects the faster speed of the excitatory pathway from the cortex to GPi via STN compared with the slower inhibitory pathway from the cortex to GPi via striatum. The late timing of GPi movement-related activity suggests that the output of the basal ganglia is unlikely to initiate movement.

Like GPi neurons, SNpr neurons are tonically active. SNpr is the principal basal ganglia output for the control of eye movements, and most studies of SNpr activity have been in eye movement tasks (Hikosaka *et al.*, 2000). Most saccade-related SNpr neurons fire in relation to the direction of the eye movements. Unlike GPi neurons during limb movements, virtually all saccade-related SNpr neurons decrease activity during the saccade. It is not known whether the fact that most GPi limb movement neurons increase and most SNpr saccade neurons decrease indicates a fundamental difference between eye and limb movement control or if it reflects task differences. Other differences between GPi limb movement neurons and SNpr saccade neurons have been noted. GPi limb movement neurons have weak sensory responses, whereas some SNpr neurons have strong responses to visual stimuli. For the visually responsive SNpr neurons, the spatial location of the stimulus seems to be the most salient feature.

Some SNpr neurons have their strongest relation to saccades made to remembered targets. Thus, if a visual target is presented and then removed and a monkey is trained to look to where the target had been, some SNpr cells will fire more in this condition as compared to when a saccade is made to a visible target. Some SNpr neurons have activity related to the preparation for eye movement. One-third of SNpr cells have activity related to saccades to a visual target. Another third of SNpr cells have activity related to saccades to a remembered target location. During saccades, the time of SNpr activity change is just after that in the superior colliculus, a structure that is known to be involved in the initiation of saccades. Thus, for eye movements, as well as limb movements, the basal ganglia output acts after structures that initiate movement.

GPe Has Irregular Activity That Increases or Decreases after Movement Initiation

Two types of neurons have been described in GPe based on their baseline activity patterns. Most fire at a high frequency (70 Hz) that is interrupted by long pauses. A smaller number fire at a low frequency (10 Hz on average) and have frequent spontaneous bursts of activity. Both types of neurons change activity in relation to limb movement and, for the majority, these changes are increases in activity. As has been described for GPi, GPe neurons weakly and inconsistently code for movement amplitude, velocity, muscle length, or force. Like the other structures of the basal ganglia, the activity of movement-related activity is late.

SNpc Has Low Spontaneous Activity That Does Not Change with Movement, but Changes with Significant Environmental Stimuli

The activity of single neurons in the SNpc of trained animals is different from the activity of single neurons in the other basal ganglia structures. At baseline, SNpc neurons fire at a low rate (2 Hz on average). The activity of these neurons is not related to movement itself and there is no apparent somatotopy in SNpc. The neurons carry little specific information regarding sensory modality or spatial properties. The activity of SNpc neurons does change in relation to behaviorally significant events such as reward or the presentation of instructional cues. The responses to stimuli occur only if the stimulus is presented in the context of a movement task. Furthermore, the activity changes with conditioning. As an example, consider an SNpc neuron that fires in relation to an unexpected reward. If one records from this neuron over several trials while the reward is paired with a tone that precedes the reward, the SNpc neuron will gradually stop firing in relation to the reward and instead will begin to fire in relation to the tone that predicts the reward.

Thus, it appears that SNpc DA neurons can predict the occurrence of a significant behavioral event. In this way, SNpc neurons are similar to the TANs neurons described earlier. Remember that SNpc neurons also

synapse extensively on medium spiny striatal neurons and that DA terminals are on the shafts of dendritic spines. It has been suggested that the DA input changes the sensitivity of striatal neurons to cortical inputs that terminate on the heads of dendritic spines. If so, the activity of SNpc neurons could modify the response of striatal neurons to cortical input that occurs in a specific behavioral context. Such influences may be related to the apparent ability of DA to mediate both long-term potentiation and long-term depression in striatal neurons, which in turn is likely to be an important mediator of learning and plasticity in the striatum (see Chapters 49 and 50).

Summary

Several general statements about basal ganglia movement-related neuronal discharge can be made.

1. Movement-related neurons in the striatum, STN, GP, and SNpr are arranged somatotopically.
2. Neurons in the striatum are quiet at rest and increase during movement. Neurons in STN are tonically active and increase during movement. Recalling the anatomy, this means that GPi neurons receive a widespread, tonically active excitatory input and a focused, intermittent inhibitory input.
3. Neurons in both GPe and GPi are tonically active. Most increase activity, but up to one-third decrease activity during limb movement.
4. Neurons in SNpr are also tonically active. Those that are related to saccadic eye movements decrease activity during the movement.
5. Neurons in SNpc discharge in relation to rewards and behaviorally relevant stimuli, but not to movement. They are likely to play a critical role in some types of motor learning.
6. Changes in the activity of basal ganglia occur at the onset of movement but after the muscles are already active. Thus, they are unlikely to initiate movement.

THE EFFECT OF BASAL GANGLIA DAMAGE ON MOVEMENT

Valuable clues to the function of the basal ganglia have come from recording the activity of single neurons during behavior. However, correlation of neural activity with an aspect of behavior does not necessarily mean that neural activity causes that aspect of behavior. To better determine the role of the basal ganglia in behavior, one would want to selectively remove a specific component from an otherwise intact system. Certain human neurological diseases involve the degeneration of neurons in the basal ganglia. Historically, these diseases fueled great interest and have provided some insight into basal ganglia function. The movement disorders that result from basal ganglia damage are often dramatic. Depending on the site of the pathology, some basal ganglia diseases cause extreme slowness of movement and rigidity and others cause uncontrollable involuntary movements. Although human diseases are of great interest, they often affect more than one structure. This necessarily limits the power of functional models derived from the study of human basal ganglia diseases. In experimental animals, more selective lesions can be made, but until recently it was difficult to reproduce the movement disorders associated with human basal ganglia disease.

Experimental lesions can be made in a number of ways. Permanent lesions can be made by passing electric current through an electrode to destroy both cell bodies and axons in the area surrounding the electrode. This has the disadvantage of destroying axon fibers that are passing through the area but that are not necessarily part of the structure of interest. Injection of a small amount of certain toxic chemicals into the brain (e.g., kainic acid or ibotenic acid) results in the focal death of neurons with cell bodies in the area of the injection but axons passing through the area are spared. Agonists and antagonists of the GABA have been injected into to the brain to cause temporary, focal inactivation or disinhibition of neurons. The effect of these chemicals lasts for several hours followed by complete return to the normal state. Later we discuss the use of toxins to produce a third type of lesion, one that can remove a particular type of neurons, while leaving others intact. This section reviews the results of selective basal ganglia lesions in animals produced by a variety of techniques. We discuss human basal ganglia diseases in the context of these experiments.

Damage to the Striatum Causes Slow Voluntary Movements or Involuntary Postures and Movements

Lesions in the striatum produce variable results that depend on the location of the lesion, the lesion method, and what is measured. Many studies of unilateral striatal lesions in monkeys have described only minimal deficits. In human patients, strokes localized to the putamen often cause dystonia, a syndrome characterized by abnormal sustained muscle contraction causing twisting postures. If the putamen is inactivated phar-

macologically with the GABA agonist muscimol unilaterally, the result is slow movement of the contralateral limb. The slowed movement is associated with an increased activity of antagonist muscles. Although movement is slow after putamen lesions, reaction time is generally normal, indicating movement initiation is intact. Large bilateral electrolytic lesions result in the paucity of movement, severe slowness of movement bilaterally, and postural abnormalities. In other studies, there appears to be little effect of experimental bilateral electrolytic putamen lesions. The reason for this discrepancy is not clear, but it may be due to differences in lesion size.

Huntington's disease (HD) is a genetically based, degenerative disease in humans that results in disabling involuntary movements. These movements are called chorea (Greek for "dance"). They are frequent, brief, sudden, random twitch-like movements that involve all parts of the body and resemble fragments of normal voluntary movement. As the disease progresses, muscular rigidity appears. Despite the excessive involuntary movements, the voluntary movements of patients with HD are slower than normal. The pathologic hallmark of HD is a marked loss of neurons in the striatum. Studies have shown that not all striatal neurons are equally affected in HD (Albin et al., 1989). In early adult-onset HD, the enkephalin-containing neurons that project to GPe are lost first. Clinically, this is associated with the presence of chorea. In late adult-onset and in juvenile-onset HD, both the enkephalin-containing and the substance P-containing striatal neurons that project to GPi and SNpr are lost. Clinically, this loss is associated with the presence of rigidity and dystonia (sustained, abnormal postures). Interestingly, experimental destructive lesions of the striatum in monkeys do not result in chorea. This is probably due to the nonselective destruction of both GPe-and GPi projecting striatal neurons. More selective disruption of the striatal–GPe pathway with a GABA antagonist administered into the GPe does produce chorea. The suggested mechanism for chorea is that disinhibition of GPe neurons (pharmacologically, or as a result of selective striatal cell death) causes inhibition of STN and GPi. This results in abnormal overactivity of motor cortical and brain stem mechanisms, resulting in chorea.

Damage to STN Causes Large-Scale Involuntary Movements

In human patients, stroke involving the STN and in monkeys, electrolytic or pharmacological lesions in the STN cause dramatic involuntary flinging movements of the contralateral arm and leg. These movements have been called hemiballismus. They resemble chorea in their brief, random, and sudden nature, but they tend to be much larger in amplitude. After a lesion in STN, the hemiballism lasts for days to weeks before gradually resolving. Animals and humans with hemiballism can still make voluntary movements and often the voluntary movements appear to be quite normal. It has been suggested that the mechanism underlying the hemiballism is a loss of excitatory input to GPi, resulting in decreased GPi activity and ultimately in the disinhibition of cortical and brain stem motor mechanisms (Albin *et al.*, 1989). Several lines of evidence support this. First, blocking the excitatory transmission from STN to GPi with a glutamate antagonist administered into GPi causes involuntary movements. Second, metabolic activity patterns as shown with 2–deoxyglucose uptake indicate decreased GPi output in hemiballism. Third, neuronal recording during stereotaxic surgical treatment of movement disorders has shown decreased GPi activity in patients with chorea or hemiballism. However, it is likely that chorea or hemiballism is due to abnormal patterns of basal ganglia output signals and not just tonically decreased activity. Thus, if hemiballism or chorea is produced by an STN lesion in monkeys, a second lesion placed in GPi abolishes the involuntary movements. Furthermore, in monkeys with STN lesions, the level of GPi activity is reduced even after the involuntary movements resolve. Finally, destruction of GPi in humans with hemiballism or chorea abolishes the involuntary movements and experimental lesions of GPi in monkeys but do not cause chorea. Therefore, it seems likely that chorea and hemiballism reflect an abnormal fluctuating or bursting output from the basal ganglia rather than a simple reduction of output.

Damage to GP Causes Slow Voluntary Movements and Involuntary Postures and Does Not Delay Movement Initiation

Lesions of the GPi result in slowing of movement but normal movement initiation. Because of their close proximity to each other, it is difficult to lesion GPi without including part of GPe or vice versa. Therefore, caution must be used in interpreting results of GP lesions. However, in most cases, combined GPe and GPi lesions have a similar effect to the lesion of GPi alone. Unilateral lesions of GPe and GPi result in slowness of movement, with abnormal cocontraction of agonist and antagonist muscles, but normal initiation of movement. Bilateral lesions of GPe and GPi result in even more severely abnormal flexed postures with an apparent inability to move out of them. Electrolytic lesions producing this severe abnormality have tended

to be large, involving both GPe and GPi and part of the internal capsule. Similar abnormalities are seen with carbon monoxide or carbon disulfide poisoning, which results in neuron death in GPi and SNpr.

Lesions restricted to GPi result in slowness of movement of the contralateral limbs. After GPi lesions, reaction time is normal, which indicates that mechanisms involved in the initiation of movement are intact. In some studies, the slowness of movement is accompanied by a cocontraction of agonist and antagonist muscles, which results in abnormal postures (dystonia). In most cases, there is relatively more activity of flexor muscles that is reflected in a tendency of the limbs to assume an abnormally flexed posture. Monkeys with GPi lesions are more impaired when movements are made by turning off active muscles than when movements are made by further turning on active muscles (Mink and Thach, 1991) (Fig. 31.11). Thus, lesions that remove the inhibitory output of the basal ganglia appear to interfere with the ability to turn off unwanted muscle activity.

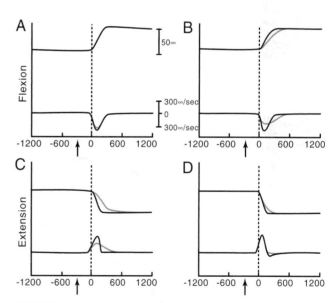

FIGURE 31.11 Wrist position and velocity in visually guided wrist movements before (black traces) and after (blue traces) a lesion of the globus pallidus internal segment. In each graph, the top traces represent wrist position, the lower traces represent wrist velocity. (A) Flexion with the flexor muscles loaded (movement made by further activating the loaded muscles). (B) Flexion with extensor muscles loaded (movement made by turning off the loaded muscles). (C) Extension with flexor muscles loaded (movement made by turning off the loaded muscles). (D) Extension with extensor muscles loaded (movement made by further activating the loaded muscles). After the lesion, the peak velocity was slower when the movements were made by decreasing the activity of the loaded muscle than when made by increasing the activity of the loaded muscle. From Mink and Thach (1991).

Damage to SNpr Causes Involuntary Eye Movements

Due to its proximity to SNpc, it is difficult to produce electrolytic lesions exclusively in SNpr. However, with the use of the GABA agonist muscimol, it has been possible to inactivate neurons focally in SNpr. Injection of muscimol into the lateral SNpr inactivates neurons that are normally involved in saccadic eye movements. Inactivation in this area results in an inability to maintain visual fixation because of involuntary saccades (Fig. 31.12). The inability to suppress involuntary saccades appears to result from disinhibition of the superior colliculus. Injection of the GABA antagonist bicuculline into the superior colliculus mimics the effects of muscimol in SNpr. Thus, just as GPi inactivation results in abnormal excess limb and trunk muscle activity, SNpr inactivation results in abnormal excess eye movements.

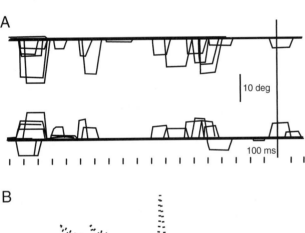

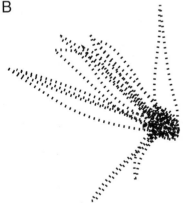

FIGURE 31.12 After inactivation of substantia nigra pars reticulata, monkeys are unable to maintain fixation of gaze because of involuntary contraction of the eye muscles. (A) The top line represents vertical eye position, and the bottom line represents horizontal eye position during attempted visual fixation. (B) Lines represent the trajectory of the involuntary eye movement when the monkey was instructed to maintain its gaze in the center dot. From Hikosaka and Wurtz (1985).

Damage to SNpc Causes Symptoms of Parkinson's Disease

Lesions of SNpc are of particular interest because the DA neurons in this nucleus die in Parkinson's disease (PD). PD is a neurological disease that affects older adults. The main symptoms of PD are (1) tremor at rest that decreases during movement, (2) slowness of movement (bradykinesia), (3) paucity of movement (akinesia), (4) muscular rigidity, and (5) unstable posture. The primary pathology in PD is a progressive degeneration of neurons in the SNpc. Additionally, there is some degree of loss of DA neurons in the ventral tegmental area and of norepinephrine neurons in the locus coeruleus. Although PD has been recognized for over a century, it only has been since the early 1980s that a complete animal model has been available. As noted earlier, the interdigitation of SNpc and SNpr has made selective electrolytic lesions of SNpc or SNpr difficult to produce. In the 1960s, it was found that 6-hydroxydopamine (6-OHDA) produces specific lesions of catecholamine neurons. Therefore, injection of 6-OHDA into the rat SN bilaterally produces a selective loss of SNpc DA neurons. Lesions made in this way result in some of the abnormalities of PD, namely the slowness of movement and rigidity, but they did not produce tremor. In the 1980s, a street drug contaminant called MPTP (Box 31.3) was found to produce all the symptoms of PD in humans and in certain species of monkeys. The MPTP monkey has proven an extremely valuable model with which to explore this disease.

Despite extensive investigation, the fundamental mechanism of the tremor in PD is not known. Some evidence suggests that it results from abnormal bursting of neurons in the thalamus. Other evidence suggests it is due to abnormal synchronous bursting of GPi neurons. The slowness of movement in PD has been associated with a reduced magnitude and duration of muscle activity during movement. In monkeys that have been given 6-OHDA or MPTP, the activity of motor cortex neurons during arm movements is also reduced compared with normal monkeys. In the MPTP monkey model of PD, some animals have abnormally increased activity of STN and GPi and decreased activity of GPi. GPi neurons in the MPTP monkey have abnormal bursting activity and abnormally increased responses to somatosensory stimuli. It has been suggested that the increased activity of GPi causes an excess inhibition of motor mechanisms in the cortex and brain stem and that this excess inhibition causes movements to be slow (Albin *et al.*, 1989). However, recent studies have shown that GPi firing rate can be normal in PD or in MPTP monkeys, but that abnormal

BOX 31.3

THE MPTP STORY

Up until the early 1980s, the quest for a complete animal model of Parkinson's disease (PD) was largely unsuccessful. Although some of the abnormalities of PD could be caused by electrolytic lesion of SNpc or by injecting the neurotoxin 6-hydroxydopamine into SNpc, neither of these methods produced the full syndrome of PD. Then, in 1982 an unfortunate, but fortuitous, accident happened. Four young drug users in northern California developed symptoms of PD. Because PD is highly unusual in young adults, the neurologists caring for these patients began a search for the cause (Langston *el al.*, 1983). They discovered that each of the patients had recently obtained a "new" synthetic heroin. This "new heroin" contained an analog of the arcotic meperidine, 1-methyl-4-phenyl-proprionoxy-piperidine (MPPP), and a contaminant, 1-methyl-4-phenyl-1,2,5,6-tetrahydropyridine (MPTP). It turned out that MPTP was the agent responsible for the parkinsonism. MPTP is oxidized in the brain to MPP^+ by monoamine oxidase. MPP^+ is taken up by DA neurons where it inhibits oxidative metabolism in the mitochondria and ultimately leads to cell death.

Subsequently, it has been found that MPTP produces a Parkinson's-like syndrome in several species of monkeys. Unlike previous models, monkeys given MPTP had not only slowness of movement, paucity of movement, and rigidity, but also tremor. The Parkinson's-like syndrome in monkeys given MPTP is associated with a nearly complete degeneration of DA neurons in SNpc and the ventral tegmental area and variable degeneration in the locus coeruleus. MPTP monkeys improve when given the DA precursor L-DOPA and have side effects of chorea when they are given too much L-DOPA, similar to what happens in people with PD. Hence, by behavioral, pharmacological, and pathological measures, the MPTP monkey is an excellent model of human PD. It is currently used to study the pathophysiology and pharmacology of PD and it has lead to new ideas about possible causes of PD. One such idea is that environmental toxins may play a role in the etiology of PD. In this regard, it is important to note that paraquat and rotenone, two compounds that are used commonly as pesticides and/or herbicides, both have MPTP-like structures and have also been shown to damage DA neurons.

Jonathan W. Mink

Reference

Langston, J. W., Ballard, P. *et al.* (1983). Chronic parkinsonism in humans due to a product of meperidine-analog synthesis. *Science* **219**, 979–980.

---

bursting and oscillatory firing is seen more consistently. Thus, it appears that the abnormal patterns of GPi activity may be more important than the more modest increase in average firing rate.

The rigidity of PD has been attributed in part to hyperactivity of the transcortical stretch reflex. Normally, the transcortical stretch reflex is active to resist displacement from an actively held posture. It is inhibited when subjects are instructed not to resist the displacement. People with PD have abnormally increased transcortical stretch reflexes and are unable to suppress them in response to instruction. The inability to inhibit long-loop reflexes may also account for the postural instability of PD. Patients with PD have an inappropriate cocontraction of leg and back muscles in response to perturbation from an upright stance. When the same subjects perturbed from a sitting position, they are not able to inhibit the postural reflexes that were active during stance. This suggests that the mechanism of rigidity and postural instability may be similar and that they reflect an inability to suppress unwanted reflex activity.

Paradoxically, it is known that lesions of GPi or STN can ameliorate many or all of the signs and symptoms of PD without producing the abnormalities associated with lesions in these areas in normal animals or humans. This is particularly paradoxical for GPi lesions that can produce the signs of PD (flexed posture, slow movement, rigidity) in normal monkeys and in some rare human diseases, but can reduce or eliminate those signs in monkeys and humans with preexisting lesions of the DA system. Why this is the case is not known, but it may relate to long-term changes in basal ganglia physiology that result from chronic DA depletion.

Chronic, high-frequency stimulation of the STN or GPi has been found to be effective for the treatment of symptoms in advanced PD. This has been called deep brain stimulation (DBS). There is active investigation into mechanisms by which DBS works and the exact mechanism is not fully understood. However, it appears that DBS activates efferents from the site of stimulation and disrupts abnormal patterns of neural transmission (Perlmutter and Mink, 2006). This disruption of abnormal patterns mimics the effects of a lesion in the stimulated area. DBS is now the most commonly used neurosurgical treatment for PD. The growth of neurosurgical treatment of PD is a direct result of knowledge about basal ganglia circuitry.

Summary

To review the preceding section:

1. Damage to any basal ganglia structure may cause slowness of voluntary movement, involuntary movements, involuntary postures, or a combination of these.
2. Damage to the striatum causes voluntary movements to be slow and may produce involuntary movements or postures depending on the mechanism of damage.
3. Damage to the STN causes large amplitude involuntary limb movements.
4. Damage to the GP causes slowness of movement, abnormal postures, and difficulty relaxing muscles, but does not delay movement initiation.
5. Damage to SNpr causes abnormal eye movements, but does not delay the initiation of eye movements.
6. Damage to SNpc causes tremor at rest, slowness of movement, rigidity, and postural instability, which are the main features of PD.

FUNDAMENTAL PRINCIPLES OF BASAL GANGLIA OPERATION FOR MOTOR CONTROL

How do the basal ganglia participate in motor control? Several hypotheses have been advanced over the past century, many of which have been mutually contradictory. This has been due in part to models that infer normal function from the abnormalities resulting from basal ganglia disease. There is growing consensus that models based on human disease states may not be sufficient to explain normal basal ganglia function. Regardless, the models based on human disease have been useful in developing new treatments for movement disorders and for the development of testable hypotheses relating to basal ganglia function.

An old model of basal ganglia function is that the basal ganglia initiate movement. This model was based in large part on the manifestation of basal ganglia dis-

eases. The paucity and slowness of movement in PD were attributed to an inability to initiate movements, whereas the involuntary movements of chorea and hemiballism were attributed to a release of normal motor systems from basal ganglia control. This model gained support from the fact that much of the output from the basal ganglia goes to parts of the thalamus that project to the premotor and motor cortex. It was argued that motor programs are stored in the basal ganglia and are called up and sent to the motor cortex for execution. This model is no longer widely accepted because it is now apparent that basal ganglia are active relatively late in relation to movement and to the activation of those brain mechanisms that are known to be involved in initiation. Furthermore, lesions of basal ganglia output nuclei do not delay the initiation of movement.

If basal ganglia do not initiate movement, what do they do? We consider three current hypotheses. One hypothesis states that although basal ganglia do not initiate movement, they contribute to the automatic execution of movement sequences. This hypothesis suggests that other mechanisms initiate the first component in a sequence, but that basal ganglia contain the programs for completion of the sequence. The second hypothesis states that the basal ganglia circuitry is made up of opposing parallel pathways that adjust the magnitude of the inhibitory GPi output in order to increase or decrease movement. According to this hypothesis, increased GPi output slows movements and decreased GPi output increases movement. The third hypothesis states that basal ganglia act to permit desired movements and to inhibit unwanted competing movements.

Do Basal Ganglia Automatically Generate Learned Movement Sequences?

A popular hypothesis states that basal ganglia are responsible for the automatic execution of learned movement sequences (Marsden, 1987). It has been pointed out that patients with PD have difficulty moving several body parts simultaneously or sequentially, and that this difficulty is more than one would expect from a simple addition of the deficits of each component of the movement. An example of an apparent problem with sequential movements in PD is the phenomenon of micrographia (or small writing). A patient begins to write a sentence with nearly normal-sized writing, but within several letters, the writing begins to get smaller so that by the end of the sentence, it may be illegible. It has been emphasized that the early components of the sequence are larger and faster than are the subsequent components. One experiment compared the performance of elbow flexion and hand grip individually or in sequence (Benecke *et al.*, 1986). Patients with PD performed each movement more slowly than normal subjects. However, when the movement was part of a sequence, it was slowed to an even greater degree than when it was performed separately.

Another experiment involved recording the activity of GPi neurons in monkeys trained to perform two successive prompt wrist movements. It was found that some GPi neurons fired after the first component of the movement but before the second component (Brotchie *et al.*, 1991). Proponents of the sequencing hypothesis speculate that the loss of this GPi output signal in PD is responsible for the relatively greater difficulty in producing sequential movements than in producing individual movements.

Additional support for this hypothesis comes from experimental work showing correlation of striatal neuron activity patterns with sequence learning (Fig. 31.13) or with learning a new skill and from a human functional imaging studies showing correlation of blood flow patterns in striatum with procedural learning (Graybiel, 2005).

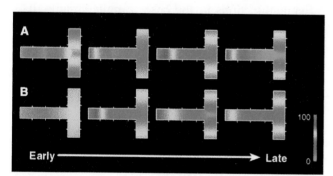

FIGURE 31.13 Reorganization of neuronal activity in the sensorimotor striatum during habit learning in a T-maze task. (A) Schematic activity maps representing the average proportion of task-related units responsive to different parts of the task from early to late in training. The maps are based on neuronal firing data and proportion of neurons active according to the color scale shown to the right (red is high and blue is low). From left to right, the maps activity patterns in relation to the different parts of the T-maze for training stage 1 (day 1), stage 3 (first criterial performance day), stage 5 (average stage of stable learning), and stage 9 (seventh day with above-criterion performance). (B) Schematic activity maps representing the population spike discharge as the percentage of total spike discharges of striatal neurons simultaneously recorded during the same four training stages illustrated in (A). The activity maps were made as those in (A), and the color scales used for (A) and (B) were identical. Note the progressive change in activity such that activity increases at the beginning and end of the behavior as learning progresses. This is thought to represent transition from representation of individual components to representation of the entire sequence (or chunk) of behavior. From Jog *et al.* (1999).

Do Basal Ganglia Produce or Prevent Movement by Using Opposing Direct and Indirect Parallel Pathways?

This hypothesis emphasizes the two major paths of information flow from the striatum to GPi and SNpr that were described earlier (Fig. 31.14) (Alexander and Crutcher, 1990). To recapitulate, one is an inhibitory "direct" pathway from striatum to GPi/SNpr and the other is a net excitatory "indirect" pathway from striatum to GPe (inhibitory), from GPe to STN (inhibitory),

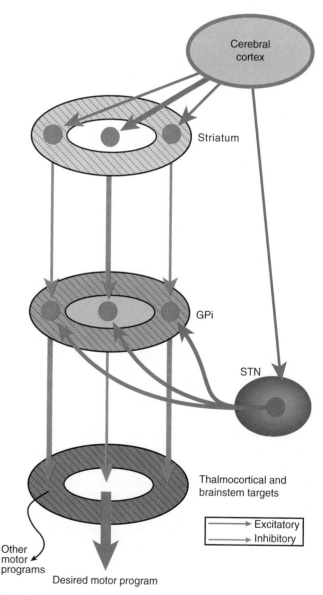

FIGURE 31.14 Schematic of proposed "direct" and "indirect" pathways from putamen to GPi. See the text for a description. Red symbols represent inhibitory pathways, and green symbols represent excitatory pathways. From Alexander and Crutcher (1990).

and from STN to GPi/SNpr (excitatory). In this hypothesis, the two pathways are in balance in such a way that increased activity in the "direct" pathway causes decreased GPi/SNpr output and increased activity in the "indirect" pathway causes increased GPi/SNpr output. By adjusting the balance, cortical targets of the basal ganglia can be facilitated or inhibited. The hypothesis predicts that abnormally decreased output results in excessive movements (chorea) and abnormally increased output results in a decreased movement (PD). The "direct/indirect pathway" model has been useful for understanding and developing certain treatments of movement disorders, but it has some shortcomings when it comes to explaining normal basal ganglia function and the mechanism of treatments for involuntary movements. Despite the shortcomings, this has been a central hypothesis in the field of basal ganglia research for over 20 years. If the success of a model can be measured by the amount of research it stimulates, this one has been highly successful.

Do Basal Ganglia Select and Inhibit Competing Motor Patterns?

The output of the basal ganglia is inhibitory to posture and movement pattern generators in the cerebral cortex (via thalamus) and in the brain stem. The inhibitory output neurons fire tonically at high frequencies. In this hypothesis, the motor output of the basal ganglia is analogous to a brake (Mink, 1996). The hypothesis states that when a movement is initiated by a particular motor pattern generator, GPi neurons projecting to that generator decrease their discharge, thereby removing tonic inhibition and "releasing the brake" on that generator. GPi neurons projecting to other movement pattern generators increase their firing rate, thereby increasing inhibition and applying a "brake" on those generators. Thus, other postures and movements are prevented from interfering with the one selected.

How might this mechanism work? When one makes a voluntary movement, that movement is initiated by the prefrontal, premotor, and motor cortex and by the cerebellum. The premotor and motor cortex send a corollary signal to STN, exciting it. STN projects to GPi in a widespread pattern and excites GPi. In parallel, signals are sent from the cortex to the striatum, which inhibits GPi focally via a direct pathway. Striatum can also disinhibit GPi via two indirect pathways (striatum → GPe → GPi and striatum → GPe → STN → GPi). The indirect pathways further focus the effects of the fast excitatory cortico-STN pathway and the slower inhibitory cortico-striatal pathway to GPi. The net result is to release the "brake" from the selected

voluntary movement pattern generator and to apply the "brake" to potentially competing posture-holding pattern generators (transcortical, vestibular, tonic neck, and other postural reflexes). The result is the focused selection of desired motor patterns and surround inhibition of competing patterns. Disruption of this organization at different nodes can lead to specific movement disorders.

BASAL GANGLIA PARTICIPATION IN NONMOTOR FUNCTIONS

Although the focus of this chapter has been on motor control, it has become increasingly clear that the basal ganglia participate in a variety of nonmotor functions. These include functions of the limbic system (see Box 31.4 and Chapters 43 and 44) and cognitive functions. The anatomy of basal ganglia circuits has revealed outputs going to all areas of frontal cortex, placing the basal ganglia in a position to influence a wide variety of behaviors. As discussed earlier, specific areas of the basal ganglia are connected preferentially with specific areas of the cerebral cortex. This provides the anatomic substrate for the localization of different functions in the basal ganglia circuits. It is known that focal lesions of the striatum tend to produce deficits similar to those seen after lesions of afferent areas of the cortex. Thus, lesions of posterior putamen cause movement deficits, lesions of inferior caudate produce deficits similar to lesions of the orbitofrontal cortex, and lesions of the dorsolateral caudate produce deficits similar to lesions of dorsolateral prefrontal cortex. The basal ganglia have been implicated in a variety of nonmotor disorders, including depression, obsessive-compulsive disorder (Box 31.6), attention deficit hyperactivity disorder, and schizophrenia. Regardless of the type of function, it is becoming clear that the basic underlying principles of basal ganglia function are similar. Thus, the intrinsic circuitry is the same for cognitive and motor parts of the basal ganglia, the basic neurophysiology appears to be the same, and the effects of lesions are analogous. For example, chorea is characterized by excessive involuntary movements, and obsessive-compulsive disorder is characterized by excessive involuntary thoughts and complex behaviors. The hypothesis of focused selection and surround inhibition has been applied to nonmotor basal ganglia functions (Redgrave *et al.*, 1999) and data support each hypothesis from cognitive science and psychiatric research.

There is increasing evidence for a role of the basal ganglia in procedural learning. Much of the research

BOX 31.4

VENTRAL SYSTEM OF THE BASAL GANGLIA

There are nuclei in the brain that historically have not been included with the basal ganglia, but that are analogous to the traditional components. These lie ventral to the neostriatum and GP and are called the ventral striatum (nucleus accumbens and olfactory tubercle) and the ventral pallidum. The ventral striatum receives input from limbic and olfactory areas of the cortex, including the amygdala and hippocampus. Like the neostriatum, the ventral striatum receives a *DA* input, but it is from the ventral tegmental area (VTA), which lies medial to the SN. The ventral striatum sends a projection back to VTA and to the adjacent SNpr. The ventral pallidum is analogous to both GPe and GPi. It receives input from the ventral striatum and possibly from STN. Unlike the GP, the ventral pallidum receives direct input from the amygdala. The output of the ventral pallidum projects to the dorsomedial nucleus of the thalamus (DM) and from there to limbic areas of the cortex. By virtue of its inputs and outputs, the ventral system is closely linked to the limbic system. This linkage has led to the suggestion that the ventral system is involved in motivation and emotion (see Chapter 43). The exact nature of this role is not known, but it may be analogous to the motor role of the basal ganglia, with the inhibitory output of the ventral pallidum acting to suppress or select potentially competing limbic mechanisms. In addition, Haber and colleagues have described circuitry by which activity in the ventral system ultimately influences the other circuits of the basal ganglia (Haber *et al.*, 2000). Such circuitry may underlie a more integrative function across basal ganglia circuitry.

Jonathan W. Mink

has focused on the learning of tasks or of sequential behavior. There is evidence for a role of the basal ganglia in procedural learning that leads to the formation of habits and the performance of behavioral routines once they are learned (Jog *et al.*, 1999). As described earlier, TANs and SNpc DA neurons fire in relation to behaviorally significant events. The activity patterns of these neurons change as the task becomes learned and when novel stimuli or events are introduced. There is also growing evidence for DA-mediated long-term potentiation and long-term depression in striatal neurons that is likely to play an important role in learning. It has been shown that non-TAN striatal neurons also change activity in relation to learning. In rats performing a T-maze task, striatal neurons changed activity patterns as the task became learned and more automatic (Jog *et al.*, 1999). In monkeys performing a discrimination learning task, striatal neurons changed activity as the animal learned new associations between stimuli and reward (Tremblay *et al.*, 1998). Functional imaging studies have also shown basal ganglia activity correlated with the learning of new tasks. Striatal lesions or focal striatal DA depletion impairs the learning of new movement sequences, and procedural learning has been shown to be impaired in people with PD. These findings together support a role for the basal ganglia in certain types of procedural learning. However, other brain structures have also been implicated in procedural learning, including the cerebellum (see Chapter 50). Just as the cerebellum and basal ganglia play different roles in motor control, they are likely to play different, but complementary, roles in procedural learning. The exact nature of those different roles is the subject of current research.

Summary

It is appropriate to consider the basal ganglia as part of the motor system because the largest portion of the basal ganglia is devoted to motor control and the most prominent deficits resulting from basal ganglia damage are motor. However, it has become clear that the basal ganglia play substantial roles in nonmotor function and in more cognitive aspects of movement. These roles are the subject of active research. Ultimately, a unifying theory of basal ganglia function that will encompass all the subdivisions is likely to emerge. Based on the circuitry and known physiology, it is likely that different cortico-basal ganglia-thalamo-cortical circuits employ similar mechanisms.

BOX 31.5

HUNTINGTON'S DISEASE

Huntington's disease (HD) is an inherited progressive neurodegenerative disorder that affects about 1 in 10,000 people. Symptoms include abnormal movements (comprising both involuntary movements termed chorea or dystonia and motor incoordination), cognitive difficulties, and emotional difficulties, including depression, apathy, and irritability. Onset is usually in midlife, but can range between childhood and old age. HD is generally fatal within 15 to 20 years after onset. The disease is caused by an expanded CAG repeat coding for polyglutamine in the HD gene product, termed the "huntingtin" protein. CAG repeat lengths between about 10 and 25 are normal, whereas those above 36 cause HD. Within the expanded range, the longer the repeat, the earlier the age of onset. The normal function of huntingtin is poorly understood, but it may be involved with cytoskeletal function, or cellular transport.

Pathologically, HD is characterized by selective neuronal vulnerability. The caudate and putamen of the corpus striatum are most affected in early stages, but as the disease progresses, other areas of the brain also become affected. Within the striatum, medium spiny GABA neurons are severely affected, with up to 95% loss in advanced cases, whereas large interneurons are relatively spared. In addition, there are intranuclear inclusion bodies and perinuclear and neuritic aggregates of huntingtin in HD neurons.

The pathogenesis of HD is still incompletely understood and represents a challenge for neuroscience. The huntingtin protein is widely expressed throughout the brain and is expressed at lower levels in other regions of the body as well. HD, like other neurodegenerative diseases, involves abnormal protein folding and aggregation. The inclusions themselves appear not to be directly toxic, but they likely represent a late stage of an abnormal process, whose earlier steps are pathogenic. The study of the disease has been facilitated by the generation of cell models, invertebrate models, and mouse models. Chaperone proteins can ameliorate the pathology, perhaps by refolding abnormally folded huntingtin. Huntingtin may normally be degraded by intracellular proteolytic machinery, termed the proteasome, and inhibition of the proteasome may be one of the toxic effects of the abnormally expanded polyglutamine.

A number of lines of evidence have pointed to a possible role for abnormal gene transcription in HD. Several studies have suggested that targeting of huntingtin to the nucleus enhances toxicity. Huntingtin may normally have a role in regulation of gene transcription, although this is poorly understood. Expression array studies show that mouse and cell models have an alteration of normal gene expression patterns. Abnormal interactions with coactivator and corepressor complexes may contribute.

Whatever the initial mechanisms, symptoms appear to be caused by both cell dysfunction and cell death, with cell death correlating best with functional disability. Thus, preventing cell death is a major goal of experimental therapeutics in HD and other neurodegenerative disorders. Cell death may partly involve apoptotic mechanisms with activation of caspases and may also involve mechanisms such as excitotoxicity, metabolic, or free radical stress. Therapeutic strategies under investigation include protection of neurons with neurotrophins, altering abnormal gene transcription patterns, and reducing cell death with inhibitors of caspase activation or blockers of excitotoxicity.

Christopher A. Ross

BOX 31.6

OBSESSIVE-COMPULSIVE DISORDER

Obsessive-compulsive disorder (OCD) is a chronic disorder characterized by recurrent intrusive thoughts and ritualistic behaviors that consume much of the afflicted individual's attentional and goal-directed processes. Among the more agonizing illnesses in clinical medicine, OCD is classified as an anxiety disorder because of the marked tension and distress produced by resisting the obsessions and compulsions. Common obsessions involve thoughts of harming oneself or others, of being afflicted by various illnesses, or of being contaminated by germs. Compulsions may be a response to obsessive thoughts (e.g., repetitive hand-washing following skin contact with any object due to obsessive worries of having contacted germs) or may instead reflect cognitive-behavioral rituals performed according to stereotyped rules (e.g., counting one's footsteps to avoid ending on certain numbers). Although individuals with OCD recognize that such thoughts and behaviors are irrational, they feel irresistibly compelled to engage in them.

The onset of OCD usually occurs between late childhood and early adulthood. Without treatment, OCD is often disabling. However, chronic treatment with drugs that potently inhibit serotonin reuptake can reduce the amount of time engaged in obsessions and compulsions and the magnitude of the associated anxiety. In addition, behavioral therapy involving repeated *in vivo* exposure and response prevention (e.g., having patients touch a toilet seat and subsequently preventing them from hand washing) may facilitate extinction of the anxiety responses generated by resisting obsessive thoughts and compulsive behaviors.

The etiology and pathophysiology of OCD are unknown. Converging evidence from analysis of the lesions that result in obsessive-compulsive symptoms, functional neuroimaging studies of OCD, and observations regarding the neurosurgical interventions that can ameliorate OCD implicate prefrontal cortical-striatal circuits in the pathogenesis of obsessions and compulsions. The neurological conditions associated with the development of secondary obsessions and compulsions include lesions of the globus pallidus and putamen, Sydenham chorea (a poststreptococcal autoimmune disorder associated with neuronal atrophy in the caudate and putamen), Tourette disorder (an idiopathic syndrome characterized by motoric and phonic tics that may have a genetic relationship with OCD, see Box 31.7), chronic motor tic disorder, and lesions of the ventromedial prefrontal cortex. Several of these conditions are associated with complex motor tics (repetitive, coordinated, involuntary movements occurring in patterned sequences in a spontaneous, unpredictable, and transient manner). Complex tics and obsessive thoughts may thus both reflect aberrant neural processes originating in distinct portions of the cortical-striatal-pallidal-thalamic circuitry that are manifested

within the motor and cognitive-behavioral domains, respectively. Compatible with this hypothesis, neurosurgical procedures that interrupt the white matter tracts carrying projections between the frontal lobe, basal ganglia, and thalamus are effective at reducing obsessive-compulsive symptoms in OCD cases that prove intractable to other treatments.

Functional neuroimaging studies have also implicated prefrontal cortical-striatal circuits in the pathophysiology of OCD. In primary OCD, "resting" cerebral blood flow and metabolism are increased abnormally in the orbitofrontal cortex and the caudate nucleus, and increase further during symptom provocation in these areas, as well as in the putamen, thalamus, and anterior cingulate cortex. During effective pharmacotherapy, orbital metabolism decreases toward normal, and both drug treatment and behavioral therapy are associated with a reduction of caudate metabolism. In contrast, imaging studies of obsessive-compulsive syndromes arising in the setting of Tourette syndrome or basal ganglia lesions have found reduced metabolism in the orbital cortex in such subjects relative to controls. Differences in the functional anatomical correlates of primary versus secondary OCD suggest that dysfunction arising at multiple points within the ventral prefrontal cortical-striatal-pallidal-thalamic circuitry may result in pathological obsessions and compulsions.

This circuitry in general appears to be involved in the organization of internally guided behavior toward a reward, switching of response strategies, habit formation, and stereotypic behavior. Electrophysiological and lesion analysis studies indicating that parts of the orbitofrontal cortex specifically are involved in the correction of behavioral responses that become inappropriate as reinforcement contingencies change. Such functions could potentially become disturbed at the level of this cortex itself or within the circuits conveying projections from the ventral PFC through the basal ganglia, conceivably giving rise to the perseverative patterns of nonreinforced thought and behavior that characterize the cognitive-behavioral state of OCD.

Wayne C. Drevets

Suggested Readings

Baxter, L. R. (1995). Neuroimaging studies of human anxiety disorders. *In* "Psychopharmacology: The Fourth Generation of Progress" (F. E. Bloom and D. J. Kupfer, eds.), Chap. 80, pp. 921–932. Raven Press, New York.

LaPlane, D. *et al.* (1989). Obsessive-compulsive and other behavioural changes with bilateral basal ganglia lesions. *Brain* **112**, 69–725.

McDougle, C. J. (1999). The neurobiology and treatment of obsessive-compulsive disorder. *In* "Neurobiology of Mental Illness" (D. S. Charney, E. J. Nestler, B. J. Bunney, eds.), pp. 518–533. Oxford Univ. Press, New York.

BOX 31.7

TOURETTE SYNDROME (TS)

A "tic" is a sudden, rapid, recurrent, nonrhythmic, stereotyped movement or sound. Individuals with TS exhibit prominent and persistent motor tics and at least one phonic tic, beginning in childhood. Current estimates of TS prevalence range between 0.1 and 1.0%, with a 3.51 male:female ratio. Symptoms wax and wane; roughly half of all TS patients experience a significant symptom decline in their early 20s, whereas others experience lifelong symptoms.

Tics differ in complexity (simple, complex) and domain of expression (motor, phonic). In addition to their outward manifestations, a variety of sensory and mental states are associated with tics. "Simple" sensory tics are rapid, recurrent, and stereotyped, and are sensations at or near the skin. "Premonitory urges" are more complex phenomena, which often include both sensory and psychic discomfort that may be momentarily relieved by a tic. The full elaboration of tics, therefore, can include a sequence: (1) a sensory event or premonitory urge, (2) a complex state of inner conflict over if and when to yield to the urge, (3) the motor or phonic production, and (4) a transient sensation of relief.

Functional impairment in TS often results from comorbid conditions. More than 40% of individuals with TS experience symptoms of obsessive-compulsive disorder (OCD), and a comparable number have symptoms of attentional deficits and hyperactivity (ADHD).

Limited neuropathological evidence suggests four locations of potential pathology within cortico-striato-pallido-thalamic (CSPT) circuitry in TS: (1) intrinsic striatal neurons (increased neuronal packing density); (2) diminished striato-pallidal "direct" output (reduced dynorphin-like immunoreactivity in the lenticular nuclei); (3) increased dopaminergic innervation of the striatum (increased density of dopamine transporter sites); and (4) reduced glutamatergic output from the STN (reduced lenticular glutamate content).

Volumetric neuroimaging studies report an enlarged corpus callosum, reduced caudate volume, or diminished striatal and lenticular asymmetry; these changes are subtle and are not replicated uniformly. Metabolic neuroimaging studies in TS report a reduced glucose uptake in the orbitofrontal cortex, caudate, parahippocampus, and midbrain regions and reduced blood flow in the caudate nucleus, anterior cingulate cortex, and temporal lobes. Regional glucose uptake patterns suggest distributed CSPT dysfunction, based on covariate relationships between reduced glucose uptake in striatal, pallidal, thalamic, and hippocampal regions. Tic suppression is associated with increased right caudate activation, measured by functional MRI, and by bilaterally diminished activity in the putamen, globus pallidus, and thalamus. Neurochemical imaging studies have reported relatively subtle abnormalities in the levels of dopamine receptors, dopamine release, DOPA decarboxylase, and dopamine transporter in the striatum of some individuals with TS. A report of greater caudate D2 receptor binding among more symptomatic TS identical twins may be relevant to factors contributing to heterogeneity of the TS phenotype.

There are a number of reported subtle abnormalities in the levels of major neurotransmitters, precursors, metabolites, biogenic amines, and hormones in blood, cerebrospinal fluid (CSF), and urine of TS subjects compared to controls. One qualitatively different finding is that of approximately 40% elevations of serum antiputamen antibodies in TS children. This finding may have particular importance, based on growing evidence for autoimmune contributions to at least some forms of TS.

Converging evidence for CSPT pathology in TS comes from neuropsychological and psychophysiological studies. Some forms of TS appear to be accompanied by abnormalities in sensorimotor gating, oculomotor functions, and visuospatial priming consistent with mild cortico-striatal dysfunction.

The mode of TS inheritance remains elusive. First-degree relatives of TS probands are 20–150 times more likely to develop TS compared to unrelated individuals. Concordance rates for TS among monozygotic twins approach 90%, if the phenotypic boundaries include chronic motor or vocal tics, versus 10–25% concordance for dizygotic twins across the same boundaries. One affected sibpair study with a total of 110 sibpairs yielded a multipoint maximum-likelihood scores (MLS) for two regions (4q and 8p) suggestive of high sharing (MLS>2.0). Four additional regions also gave multipoint MLS scores between 1.0 and 2.0.

Treatments for TS focus on education, comorbid conditions, and direct tic suppression. Education is a critically important intervention for both family members and affected individuals. Treatments for comorbid conditions, particularly OCD and ADHD, are highly effective and can provide significant relief. Dopamine antagonists, particularly high potency, D2 preferential blockers, are the

---

BOX 31.7 (cont'd)

most potent and rapid acting tic-suppressing agents, but have important, undesirable side effects. Newer, "atypical" antipsychotics are better tolerated, and some are effective in suppressing tics. The DA depleter tetrabenazine also offers significant tic suppression, but is not yet approved for use in the United States. α_2-Adrenergic agonists are often used as antitic agents; compared to dopamine antagonists, these drugs have relatively weaker antitic abilities and their benefit generally evolves more gradually. New therapeutic avenues for TS are being explored in controlled studies, including dopamine agonists such as pergolide, nicotinic manipulations, and Δ^9-tetrahydro-cannabinol. Intractable, localized tics also have been treated successfully with injections of botulinum toxic. An effective nonpharmacologic therapy, habit reversal therapy, involves the application of cognitive and behavioral therapy principles to TS, analogous to the

successful use of these therapies in the treatment of OCD.

Neal R. Swendlow

Suggested Readings

Cohen, D. J., Leckman, J. F., and Pauls, D. (1997). Neuro-psychiatric disorders of childhood: Tourette's syndrome as a model. *Acta Paediatr.* **422**, 106–111.

Kurlan, R. (1997). Treatment of tics. *Neuro. Clin. North Am.* **15**, 403–409.

Leckman, J. F., Walker, D. E., and Cohen, D. J. (1993). Premonitory urges in Tourette's syndrome. *Am. J. Psychiatry* **150**, 98–102.

Swerdlow, N. R. and Young, A. B. (1999). Neuropathology in Tourette syndrome. *CNS Spectrums* **4**, 65–74.

The Tourette Syndrome Association International Consortium for Genetics. (1999). A complete genome screen in sib pairs affected by Gilles de la Tourette syndrome. *Am. J. Hum. Genet.* **65**, 1428–1436.

References

Albin, R. L., Young, A. B. *et al.* (1989). The functional anatomy of basal ganglia disorders. *Trends Neurosci.* **12**, 366–375.

Alexander, G. E. and Crutcher, M. D. (1990). Functional architecture of basal ganglia circuits: Neural substrates of parallel processing. *Trends Neurosci.* **13**, 266–271.

Alexander, G. E., DeLong, M. R. *et al.* (1986). Parallel organization of functionally segregated circuits linking basal ganglia and cortex. *Annu. Rev. Neurosci.* **9**, 357–381.

Aosaki, T., Tsubokawa, H. *et al.* (1994). Responses of tonically active neurons in the primate's striatum undergo systematic changes during behavioral sensorimotor conditioning. *J. Neurosci.* **14**, 3969–3984.

Benecke, R., Rothwell, J. C. *et al.* (1986). Performance of simultaneous movements in patients with Parkinson's disease. *Brain* **109**, 739–757.

Brotchie, P., Iansek, R. *et al.* (1991). Motor function of the monkey globus pallidus. 2. Cognitive aspects of movement and phasic neuronal activity. *Brain* **114**, 1685–1702.

Flaherty, A. W. and Graybiel, A. M. (1991). Corticostriatal transformations in the primate somatosensory system: Projections from physiologically mapped body-part representations. *J. Neurophysiol.* **66**, 1249–1263.

Graybiel, A. M. (2005). The basal ganglia: Learning new tricks and loving it. *Curr. Opin. Neurobiol.* **15**, 638–644.

Graybiel, A. M., Ragsdale, C. W. *et al.* (1981). An immunohistochemical study of enkephalins and other neuropeptides in the striatum of the cat with evidence that the opiate peptides are arranged to form mosaic pattes in register with striosomal compartments visible with acetylcholinesterase staining. *Neuroscience* **6**, 377–397.

Haber, S. N., Fudge, J. L. *et al.* (2000). Striatonigrostriatal pathways in primates form an ascending spiral from the shell to the dorsolateral striatum. *J. Neurosci.* **20**, 2369–2382.

Hikosaka, O. (2007). Basal ganglia mechanisms of reward-oriented eye movement. *Ann. N.Y. Acad. Sci.* **1104**, 229–249.

Hoover, J. E. and Strick, P. L. (1993). Multiple output channels in the basal ganglia. *Science* **259**, 819–821.

Jog, M., Kubota, Y. *et al.* (1999). Building neural representations of habits. *Science* **286**, 1745–1749.

Langston, J. W., Ballard, P. *et al.* (1983). Chronic parkinsonism in humans due to a product of meperidine-analog synthesis. *Science* **219**, 979–980.

Marsden, C. D. (1987). What do the basal ganglia tell premotor cortical areas? *Ciba Found. Symp.* **132**, 282–300.

Mink, J. W. and Thach, W. T. (1991). Basal ganglia motor control. III. Pallidal ablation: Normal reaction time, muscle cocontraction, and slow movement. *J. Neurophysiol.* **65**, 330–351.

Nambu, A., Tokuno, H., and Takada, M. (2002). Functional significance of the cortico-subthalamo-pallidal "hyperdirect" pathway. *Neurosci. Res.* **43**, 111–117.

Parent, A. and Hazrati, L.-N. (1993). Anatomical aspects of information processing in primate basal ganglia. *Trends Neurosci.* **16**, 111–116.

Perlmutter, J. S. and Mink, J. W. (2006). Deep brain stimulation. *Annu. Rev. Neurosci.* **29**, 229–257.

Redgrave, P., Prescott, T. *et al.* (1999). The basal ganglia: A vertebrate solution to the selection problem? *Neuroscience* **89**, 1009–1023.

Romo, R., Scarnati, E. *et al.* (1992). Role of primate basal ganglia and frontal cortex in the internal generation of movements. II. Movement-related activity in the anterior striatum. *Exp. Brain Res.* **91**, 385–395.

Tepper, J. M., Koos, T., and Wilson, C. J. (2004). GABAergic micro-circuits in the neostriatum. *Trends Neurosci.* **27**, 662–669.

Tremblay, L., Hollerman, J. *et al.* (1998). Modifications of reward expectation-related neuronal activity during learning in primate striatum. *J. Neurophysiol.* **80**, 964–977.

Wilson, C. J. and Groves, P. M. (1980). Fine structure and synaptic connections of the common spiny neuron of the rat neostriatum: A study employing intracellular injection of horseradish peroxidase. *J. Comp. Neurol.* **194**, 599–614.

Wilson, S. A. K. (1928). "Modern Problems in Neurology." Arnold, London.

Suggested Readings

Bolam, J. P., Hanley, J. J. *et al.* (2000). Synaptic organisation of the basal ganglia. *J. Anat.* **196**, 527–542.

Hikosaka, O., Takikawa, Y. *et al.* (2000). Role of the basal ganglia in the control of purposive saccadic eye movements. *Physiol. Rev.* **80**, 953–978.

Kelly, R. M. and Strick, P. L. (2004). Macro-architecture of basal ganglia loops with cerebral cortex: Use of rabies virus to reveal multisynaptic circuits. *Prog. Brain Res.* **143**, 449–459.

Mink, J. W. (2003). The Basal Ganglia and involuntary movements: Impaired inhibition of competing motor patterns. *Arch. Neurol.* **60**, 1365–1368.

Tepper, J. M., Abercrombie, E. D., and Bolam, J. P. (2007). "GABA and the Basal Ganglia," *Prog. Brain Res.*, Vol. 160. Elsevier, New York.

Watts, R. L. and Roller, W. C. (2004). "Movement Disorders: Neurologic Principles and Practice." 2nd ed. McGraw-Hill, New York.

Jonathan W. Mink

Cerebellum

The cerebellum is a softball-sized structure located at the base of the skull. Grab the back of your head just above where it meets your neck: your hand is now cupped around your cerebellum. As with most brain systems, much of what we know about the cerebellum stems from symptoms of damage or pathology and from its connectivity with the rest of the brain. From such evidence it has long been clear that the cerebellum is an important component of the motor system. Severe abnormalities of movement are produced by pathologies of the cerebellum. Cerebellar outputs influence systems that are unambiguously motor, such as the rubro-spinal tract, and inputs to the cerebellum convey information known to be essential for movement such as joint angles and loads on muscles. More recently it has become equally clear that the role of the cerebellum is not limited to movement. This is indicated by its interconnections with nonmotor structures, by the more subtle, nonmotor deficits seen with cerebellar lesions, and by functional imaging studies where regions of the cerebellum show activation during nonmotor tasks. In moving toward a stronger understanding of the cerebellum one obvious task will be to identify the common aspects or computational demands that these motor and nonmotor functions share.

Research on the anatomy, physiology, and function of the cerebellum has been complemented and enhanced by computational approaches that emphasize the rules for input/output transformation and how the cerebellar neurons and synapses implement this transformation. This interdisciplinary approach was made possible by the seminal work of Eccles, Ito, and Szentagothai, who in 1967 published "The cerebellum as a neuronal machine," which described most of the essential aspects of the cellular and synaptic organization of the cerebellum. Soon after, a remarkable paper was published in 1969 by David Marr who inferred some basic computational properties of the cerebellum solely on the basis of the wiring diagram that Eccles and associates had worked out. Since then, the numerous conceptual and practical advantages of working on the cerebellum have enabled a more detailed understanding of what the cerebellum computes and how its neurons and synapses produce this computation.

ANATOMY AND PHYLOGENETIC DEVELOPMENT OF THE CEREBELLUM

The cerebellum is present in all vertebrates and in the most primitive prevertebrates (myxinoids) up through primates (Jansen and Brodal, 1954; Larsell, 1967, 1970, 1972). In agnathans (lampreys and hagfish), it is a rudimentary structure that assists the functions of the well-developed vestibulo-ocular, vestibulospinal, and reticulospinal systems. The cerebellum is somewhat larger in fishes where, on the input side, it processes sensory information from the vestibular, lateral line, and to a lesser extent proprioceptive and somatosensory systems. On the output side it is connected to the vestibular and reticular nuclei (Box 32.1). In amphibians the region of the cerebellum that receives proprioceptive and other sensory information is expanded. This region, called the corpus cerebella, increases further in reptiles, birds, and mammals. In these vertebrate classes, it constitutes the largest portion of the cerebellum, receiving proprioceptive, somatosensory, visual, and auditory information and projecting to the tectum, the red nucleus, and the cerebral cortex via the thalamus. In primates, the hugely expanded lateral hemispheres of the corpus cerebella are connected with the enlarged cerebral cortex (Fig. 32.1). The hemispheres receive information from the frontal parietal, and visual cortices by way of pontine nuclei and project to the motor and premotor cortices as well as to more anterior portions of the frontal lobe (Fig. 32.2). (Asanuma *et al.*, 1983a, 1983b, 1983c, 1983d; Middleton and Strick, 1994; Orioli and Strick, 1989; Sasaki *et al.*, 1976; Schell and Strick, 1984)

The Cerebellum Can Be Subdivided on the Basis of Phylogeny, Anatomy, and the Effect of Lesions

Superficially, the cerebellum consists of a three-layered cortex folded in thin, parallel strips called folia (leaves), which in most species run roughly transverse to the long axis of the body (Fig. 32.2). The cerebellar cortex surrounds three pairs of deep cerebellar nuclei. From medial to lateral, the deep cerebellar nuclei are the fastigius, the interpositus (which is further divided into the globose and emboliform nuclei in humans), and the dentate. The deep cerebellar nuclei and the vestibular nuclei constitute the output structures of the cerebellum. An inner mass of white matter contains axons that run between the cerebellar cortex and the deep cerebellar nuclei.

The cerebellum is divided into three lobes (Jansen and Brodal, 1954). The flocculonodular lobe (vestibulocerebellum) is located on the inferior surface and is separated from the posterior lobe via the posterolateral fissure. Superior to the posterior lobe is the anterior lobe: these two lobes are separated by the primary fissure. The lobes are divided further into lobules (Larsell, 1967, 1970, 1972) (Fig. 32.2), which are numbered I–X beginning at the dorsal anterior vermis and ending at the inferior posterior vermis. Each lobule contains a number of folia. Lobulation is fairly consis-

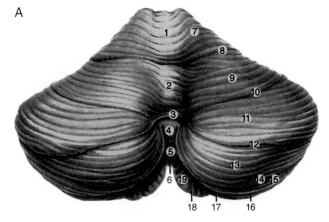

A

1. culmen	7. quadrangular lobule	13. inferior semilunar lobule
2. declive	8. primary fissure	14. ansoparamedian fissure
3. vermal folium	9. simplex lobule	15. gracile lobule
4. vermal tuber	10. superior posterior fissure	16. prebiventral fissure
5. vermal pyramis	11. superior semilunar lobule	17. biventral lobule
6. vermal uvula	12. horizontal fissure	18. secondary fissure
		19. cerebellar tonsil

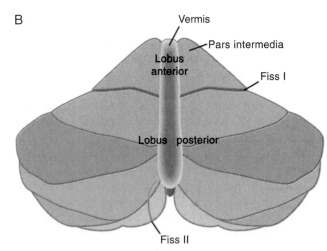

FIGURE 32.1 Cerebellar structure and functional subdivisions. (A) Dorsal view of the human cerebellum. (B) Functional subdivisions. Adapted from Nieuwenhuys, Voogd, and van Huijzen (1979).

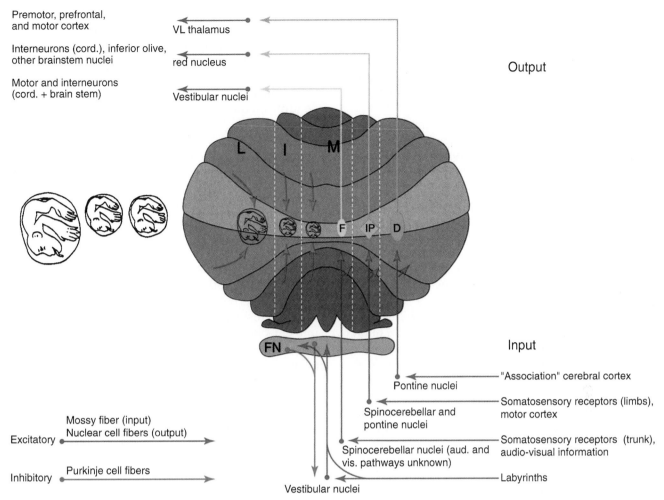

FIGURE 32.2 Cerebellar inputs and outputs. The cerebellum is shown schematically with the flocculonodular (FN) lobe unfolded and represented at the inferior surface. Green arrows indicate excitatory projections into and out of the cerebellum; red arrows indicate the inhibitory projections from cerebellar cortex to the deep cerebellar nuclei and the vestibular nuclei. Adapted from Thach, Fig. 31-3. In: Mountcastle, J. B. (1980). *Medical Physiology*, Volume I. 14 Ed. C. V. Mosby Co., St. Louis.

tent across individuals of the same species and extends with few exceptions across species, despite great variation in hemispheric development.

The corpus cerebella can be divided into three longitudinal zones (medial, intermediate, and lateral) based on the projection of the cortex onto the three deep nuclei (Jansen and Brodal, 1954). These zones differ in the type of information that comes into them over mossy fibers. The medial zone is dominated by information from vestibular, somatosensory, visual, and auditory regions. The intermediate zone receives proprioceptive and somatosensory information from the spinal cord, as well as information from the motor cortex via the pontocerebellar nuclei (Allen *et al.*, 1977). The lateral zone receives information via the pontine nuclei from the motor cortex, the premotor cortex (area 6), and perhaps all the rest of the cerebral cortex, with the possible exception of portions of the temporal lobes

(Allen *et al.*, 1978). It was once thought that the outputs of the zones (and nuclei) were also segregated—flocculonodular lobe to vestibular nuclei, fastigius to the vestibular and reticulospinal systems, interpositus to red nucleus, and dentate (via thalamus) to motor cortex—but we now know that this picture is greatly oversimplified. Although there may be a predominance to projection along these lines, there is also a great deal of overlap. Thus, vestibular, fastigial, interposed, and dentate nuclei all project via the thalamus to the motor cortex, and vestibular, fastigial, and interposed nuclei all project directly to the spinal cord (Asanuma *et al.*, 1983a, 1983b, 1983c, 1983d).

Staining for acetylcholinesterase, zebrin, and other markers reveals strips running sagittally within the longitudinal zone (Bloedel and Courville, 1981; Groenewegen and Voogd, 1977; Groenewegen *et al.*, 1979; Voogd and Bigare, 1980). The strips, which have been given names (cl, c2, etc.), are constant in their presence and location within a species and to some extent across species. They have been used chiefly as anatomical markers for physiological experiments, although whether they represent afferent mossy and climbing fibers, efferent Purkinje fibers, or boundaries between them is not clear.

Various attempts have been made to divide the cerebellum according to its functional organization. Fulton and Dow (Botterell and Fulton, 1938a, 1938b; Dow, 1938) described three divisions in the monkey based on different behavioral abnormalities that resulted when each was damaged. Lesion of the vestibulocerebellum (flocculonodular lobe) caused ipsilateral jerk nystagmus, head tilt (occiput to the side of the lesion), and circling gait (Dow, 1938). These deficits were interpreted as consistent with disturbed vestibular function. Lesion of the spinocerebellum (anterior lobe and lobulus simplex) impaired gait without impairing reaching, deficits interpreted as consistent with abnormal spinal control of walking (Botterell and Fulton, 1938b). Lesions of the cerebrocerebellum (pontocerebellum) caused inaccuracy in reaching and clumsiness of hand movements, suggesting impaired voluntary control of body parts (Botterell and Fulton, 1938a). These results were confirmed and extended with lesions of the deep nuclei in the cat (Chambers and Sprague, 1955a, 1955b; Sprague and Chambers, 1953).

Another scheme based on functional organization arose from cerebellar input mapping studies that revealed a somatotopic map in the anterior lobe of the cerebellum, with trunk in the midline, limbs lateral, tail anterior, and head posterior (Adrian, 1943; Snider and Eldred, 1952; Snider and Stowell, 1944). Each paramedian lobule also has a small ipsilateral map whose head region is contiguous with that of the anterior lobe map. It has been argued from these maps that the midline cerebellum controls the trunk, the lateral cerebellum controls distal limb parts, and the intermediate zone controls proximal limb parts. This scheme differs from that of Fulton and Dow, which holds that each major subdivision controls all parts of the body for a specific range of behaviors (e.g., behaviors under vestibular control, reflex control, and voluntary control).

The Cerebellar Cortex Contains Seven Types of Neurons

The three-layered cerebellar cortex contains a highly regular arrangement of seven distinct cell types (Eccles *et al.*, 1967; Llinas, 1981; Ramon y Cajal, 1911) (Fig. 32.3). The innermost layer, the granule cell layer, is home to an enormous number of small granule cells,

as well as other inhibitory neurons called Golgi and Lugaro cells. Granule cells have compact dendritic arbors consisting of three to five arms, each of which terminates in a glomerulus, a specialized synaptic cluster containing synaptic terminals from mossy fibers (the major input to the cerebellum), granule cell dendrites, and Golgi cell axons and dendrites. Each mossy fiber excites up to 30 granule cells, and each granule cell receives input from about four mossy fibers. Despite the compact electrotonic structure of granule cells, their excitability is lowered by tonic GABAergic currents such that a granule cell must receive three to four near-simultaneous inputs to produce an action potential. Golgi neurons, also located in the granule layer, receive excitatory input from both mossy fibers and granule cell axons, so that they provide both feed-forward and feedback inhibition to granule cells. Nearby Lugaro cells form inhibitory synaptic contacts onto Golgi, stellate, and basket neurons; a distinguishing characteristic is that Lugaro cells, unlike Golgi neurons, are sensitive to serotonin. Both Golgi and Lugaro cells can be found at the outermost edge of the granule cell layer.

Granule cell axons ascend radially through the next layer of the cerebellum, the Purkinje cell layer, where the large Purkinje cell somata are located. The ascending granule cell axons make a few synaptic contacts onto Purkinje cells before bifurcating in the outermost molecular layer to form the parallel fibers (so named because they run parallel to the folia), each of which extends 5 to 10 mm though the cortex and releases glutamate onto the dendritic spines of 2000 to 3000 Purkinje cells (Mugnaini, 1983). The parallel fibers in any one region of the cortex may come from granule cells in the medial, intermediate, and lateral zones of the cerebellum. Thus, each Purkinje cell may receive information about sensory conditions, internal states, external states, and the plans of the organism.

The outer molecular layer contains two additional types of inhibitory interneurons, called stellate and basket cells. Both interneuron types are excited by parallel fibers and form a feed-forward inhibitory circuit by synapsing onto Purkinje cells, whose dendritic arbors extend into the outer molecular layer. Relatively little is known about how the recruitment of these interneurons influences Purkinje cell firing.

The unipolar brush cell is a more recently characterized cerebellar neuron that is found mostly in vestibular regions of the cerebellum (Nunzi *et al.*, 2001). It interacts with the mossy fiber/granule cell system with an unusual connectivity. Unipolar brush cells receive excitatory input from mossy fibers, as granule cells do, but their axons branch like mossy fibers and form synapses onto granule cells. *In vitro* studies have

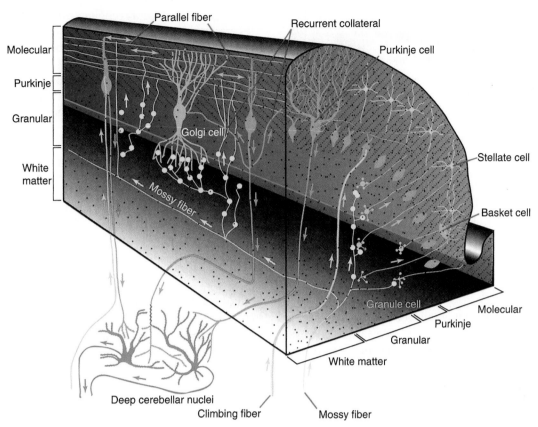

FIGURE 32.3 Schematic view of cerebellar cortex. This section shows the three-dimensional relationship between the different cell types—specifically, the medullary layer or white matter, granular layer, basket cells, stellate cells, Purkinje cells, Golgi cells, granular cells, parallel fibers, climbing fibers, mossy fibers, deep cerebellar nuclei, and recurrent collaterals. Adapted from Fox, C. A. (1962). The structure of the cerebellar cortex. In *Correlative Anatomy of the Nervous System*. New York, MacMillan [Fig. 16.10A].

shown that unipolar brush cells respond to brief bursts of excitatory input with prolonged bursts of output that far outlast the input (Diño *et al.*, 2000). The influences of these properties on coding in the mossy fiber/granule cell system are still being investigated.

Most Cerebellar Input Arrives over Mossy Fibers and Climbing Fibers

The mossy fiber is the main operational input to the cerebellum, carrying afferent information both from the periphery and from other brain centers (Ramon y Cajal, 1911). Mossy fibers that project to the vestibulocerebellum originate predominantly in neurons in the vestibular nuclei, whereas those that project to vermal cerebellar cortical regions and the fastigius nucleus come from the spinocerebellar pathway, the vestibular nuclei, and the reticular nuclei. The intermediate zone of the cerebellar cortex and the interpositus nuclei receive mossy fibers from the periphery via the spinocerebellar tracts and from the motor cortex via the pons (Eccles *et al.*, 1967; Llinas,

1981). Finally, mossy fibers that project to the lateral zone of the cerebellar cortex and the dentate nucleus arise from the basal pontine nuclei. The basal pontine nuclei receive information from virtually the entire cerebral cortex but primarily from the premotor, supplementary motor area, sensorimotor, parietal association, and visual cortices (Allen and Tsukahara, 1974; Brodal, 1978; Brooks and Thach, 1981; Glickstein *et al.*, 1985).

Because mossy fibers originate from second-order sensory neurons, some processing of even the most direct afferent information occurs before that information is sent to the cerebellum. All mossy fiber input into the cerebellum is excitatory (glutamatergic) (Eccles *et al.*, 1967; Llinas, 1981), and many mossy fibers send a branch to excite the deep nucleus cells before the main trunk ascends to the cortex. Mossy fiber excitation of granule cells causes them to fire, leading to transmitter release from parallel fiber synapses onto Purkinje cells (Figs. 32.3 and 32.4). Each parallel fiber synapse has a small unitary conductance (~1 nS), which alone is insufficient to influence Purkinje cell firing.

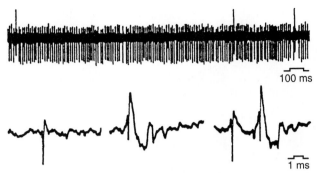

100 ms

1 ms

FIGURE 32.4 Maintained discharge of a Purkinje cell recorded extracellularly from an awake, behaving monkey. The "simple" and "complex" spikes are shown. The top trace shows the relationship between simple and complex spikes over a few seconds. The bottom traces show simple and complex spikes individually and together over smaller periods of time to demonstrate the different waveform patterns. Adapted from Thach.

However, because Purkinje cells receive approximately 150,000 parallel fiber synapses on their large dendritic arbors, the concurrent arrival of many such inputs will affect Purkinje activity. Purkinje cells fire "simple spikes" spontaneously at rates of about 30 to 50 Hz, even in the absence of excitatory input; therefore, parallel fiber synaptic currents likely affect the precise timing of these simple spikes.

The second major type of input to the cerebellum is the climbing fiber, which arises from the inferior olivary complex located in the brainstem (Eccles *et al.*, 1967). Climbing fibers branch in the sagittal plane and make contact with up to 10 Purkinje cells in the molecular layer of the cortex. Each Purkinje cell receives input from only one climbing fiber, which makes about 300 synaptic contacts on the soma and proximal dendrites. The simultaneous release of multiple vesicles of glutamate from each of these synapses results in a powerful postsynaptic excitatory current that drives an unusual action potential termed a "complex spike" because of its multiple components, including several spikelets riding on the back of a larger depolarization. Olivary neurons fire at about 1 Hz (Keating and Thach, 1995; Thach, 1968) and increase firing as an animal encounters novel conditions or tries to adapt its movement (Gellman *et al.*, 1985; Gilbert and Thach, 1977; Ojakangas and Ebner, 1992; Simpson and Alley, 1974).

Neuromodulatory Brain Stem Afferents Are a Third Type of Cerebellar Inputs

A class of cerebellar afferents with nonlaminar distribution originates in the locus ceruleus, raphe nuclei, and other undefined brainstem nuclei. These afferents innervate the deep cerebellar nuclei and all cortical layers, with some preference for the molecular layer. Afferent fibers from the locus ceruleus release norepinephrine (Hoffer *et al.*, 1972, 1973; Mugnaini, 1972), fibers from the raphe nuclei release serotonin, and fibers from undefined brain stem nuclei release acetylcholine. All three nonlaminar fiber systems synapse on Purkinje cells, although they also target cortical neurons in all layers.

Synapses of nonlaminar afferents have been difficult to define at the electron microscopic level for several reasons (Mugnaini, 1972; Palay and Chan-Palay, 1974). The afferents are rather sparse and form terminal branches that mimic those of other cerebellar fibers. They may ascend radially from the granular layer and divide into two branches that run longitudinally in the folia, where they resemble parallel fibers; they may ascend into the molecular layer and branch along the dendritic arbors of Purkinje cells, resembling stunted climbing fibers; or they may form ascending and descending collaterals in the molecular layer, resembling stellate cell axons. In the granular layer, the branches of nonlaminar afferents intermingle with the axonal plexus of Golgi cells. With standard preparation methods the synaptic vesicles contained in the varicosities of nonlaminar afferents are difficult to distinguish from the vesicles of granule cell axons. Some confusion also arises because certain mossy fiber endings in the caudal cerebellum have serotonin uptake properties similar to those of nonlaminar afferents from the raphe nuclei (Mugnaini, 1972; Palay and Chan-Palay, 1974). Distinct active zones and postsynaptic densities have been identified for all three nonlaminar afferent systems, although one cannot exclude that these afferents may also release their transmitters at nonspecialized regions of their terminal branches.

Purkinje Cells Inhibit Deep Nuclear and Vestibular Neurons and Exhibit Plasticity

Purkinje cell output exclusively targets deep cerebellar and vestibular nucleus neurons. Their GABAergic synapses are specialized for faithful high-frequency response, with large boutons that each contain multiple vesicle release sites, ensuring successful transmission (Telgkamp *et al.*, 2004). Deep nuclear and vestibular neurons, like Purkinje cells, also fire spontaneous action potentials even in the absence of glutamatergic excitation. Therefore, modulations of Purkinje cell activity sculpt the firing rates of their targets via powerful inhibition. Many deep nuclear neurons recorded respond to prolonged intracellular hyperpolarization by firing at exceptionally high rates afterward, providing a potential mechanism for the coordination of target neuronal firing by synchronous activity of multiple Purkinje cells.

When a Purkinje cell fires a complex spike in close temporal register with parallel fiber input, the strength of the parallel fiber contact onto the Purkinje cell is reduced (Ekerot and Kano, 1985; Ito *et al.*, 1982; Kano and Kato, 1987; Strata, 1985). This phenomenon, which is an example of long-term depression (Chapter 49), is thought to be an important mechanism in cerebellar learning. Conversely, when parallel fibers are stimulated without matched complex spikes, their strength is increased in a form of long-term potentiation (LTP). In contrast to hippocampus and cortex, where high postsynaptic calcium levels can induce LTP and low levels LTD, Purkinje neurons appear to have an inverted relationship, such that low calcium levels (from concurrent parallel fiber activation alone) cause LTP, and high calcium levels (from climbing fiber activation) yield LTD. NMDA receptors, a primary requirement for cerebral cortical synaptic plasticity, are absent from Purkinje cells, which instead rely on metabotropic glutamate receptors to induce LTD. Interestingly, not all parallel fiber synapses are equally plastic. In recordings *in vitro*, zebrin II positive Purkinje cells appear less susceptible to LTD induction than those in zebrin negative areas (Wadiche and Jahr, 2005). Several studies have shown that abolishing parallel fiber LTD leads to impairments in short-term motor learning, but the role of LTD in long-term memory storage, as well as the implications of disinhibition of deep nuclear neurons for cerebellar function and plasticity, remain to be elucidated (DeZeeuw and Yeo 2005).

Cerebellar Development Proceeds from Output Structures to Input Structures

Cerebellar cell types develop at different times and at different places (Jacobson, 1985; Rakic, 1982; Verbitskaya, 1969). The first cells to be formed are the neurons of the deep cerebellar nuclei. They are followed soon after by Purkinje cells, which originate in the ventricular epithelium and migrate to their ultimate location in the cortex. Golgi cells, basket cells, stellate cells, astrocytes, and Bergmann glia also originate at the ventricular epithelium and migrate to their final positions after the migration of the Purkinje cells. The Bergmann glia later guide the descent of the granule cells from the external granule cell layer.

Once Purkinje, Golgi, basket, and stellate cells have formed, climbing fibers enter the cerebellum from the inferior olive and begin to innervate the Purkinje cells. Much later, after the Purkinje cells have begun to receive synapses from parallel fibers, most of the climbing fiber contacts with Purkinje cells are eliminated. Mossy fibers also enter the cerebellum and grow to the level just below the Purkinje cell layer. They will ultimately synapse on granule cells, which have yet to arrive.

The granule cells first develop at a very distant site, posterior and lateral in the anlagen at the rhombic lip. They then crawl over the Purkinje cells to form an external granule cell layer. Some granule cells may cross the midline to the other side of the cerebellum as they make this migration. The granule cells extend their axons, which branch to form parallel fibers that run as coronal beams through the dendrites of the Purkinje cells. Only then do the granule cells descend, guided by the Bergmann glia, to form the internal granule cell layer. Each granule cell is connected by the proximal portion of its axon to the branch point of its parallel fibers, which remain in the external granule cell layer. In the internal granule cell layer, the cell's two to seven dendrites meet the terminals of the mossy fibers and gradually make connections with them. Connections also are made between granule cells and the intrinsic cortical inhibitory neurons—the Golgi, stellate, and basket cells. It is interesting that the output side of the cerebellar circuit (the deep nuclear I cells and Purkinje cells) forms first, the input side (mossy fibers) then arrives, and the "matrix" that connects the two (the granule cells and intrinsic inhibitory neurons) is the last to develop.

Human Cerebellar Development Is Not Complete at Birth

In humans, the first cerebellar structures develop at approximately 32 days after fertilization, and development is not completed until after birth. The cerebellar cortex of the early embryo has six distinct layers, but this number ultimately is reduced to three. The cortex begins to differentiate slightly earlier in the vermis, flocculus, and median sections of the hemispheres than in the lateral hemispheres. By seven months after fertilization, the deep cerebellar nuclei have attained the shape and location they will have in the adult. At birth, the cerebellar cortex consists of four uneven layers, and the Purkinje cells and basket cells are weakly developed. The fourth layer (the external granule cell layer) disappears within the first postnatal year. In humans, full myelination of cerebellar connections is not complete until the second year of life (Brody *et al.*, 1987).

Summary

The cerebellum is a prominent structure in the nervous systems of all vertebrates. The cerebellum receives input from many parts of the nervous system but projects mainly to motor and frontal lobe cognitive areas. This suggests that many different kinds of information are brought together to help control movement and certain cognitive functions. The intrinsic structure is relatively simple and stereotyped throughout the cerebellum, very different from that of the cerebrum. The ontogenetic development is conspicuously late: cell migrations and fiber connections continue to occur after birth and the development of the rest of the motor nervous system.

ASSESSING CEREBELLAR FUNCTION

The Deep Nuclei Generate the Output of the Cerebellum

All the cerebellum's output is produced by the deep cerebellar nuclei and vestibular nuclei. Each deep nucleus has a separate somatotopic representation of the body in which the head, tail, trunk, and extremities are represented in the caudal, rostral, lateral, and medial regions of the nucleus, respectively (Allen *et al.*, 1977, 1978; Asanuma *et al.*, 1983a, 1983b, 1983c, 1983d; Ramon y Cajal, 1911; Rispal-Padel *et al.*, 1982; Thach *et al.*, 1992b, 1993). The deep nuclei exert control over movement of ipsilateral parts of the body. Nearly every motor center of the CNS receives input from the deep nuclei, including the spinal cord, the vestibular, reticular, and red nuclei, the superior colliculus, and (via the thalamus) the primary motor and premotor cortices, the primary and secondary frontal eye fields, and even the prefronial cortex. The projections from the deep nuclei to these centers are thought to be glutamatergic and excitatory. However, a set of small GABAergic neurons in the deep nuclei makes inhibitory projections to the inferior olivary complex (Mugnaini, 1972). These inhibitory projections may be important for maintaining balanced induction of plasticity in the cerebellar cortex and thus for preventing spontaneous drift of synaptic strengths (Kenyon *et al.*, 1998; Medina *et al.*, 2002).

All cerebellar target structures receive other excitatory inputs in addition to those from the cerebellum. There are no direct projections from the deep nuclei to the basal ganglia (Table 32.1). The deep nuclei differentially control the medial and lateral motor systems and their respective functions. The vestibular nucleus and the fastigius control eye and head movements, equilibrium, upright stance, and gait; the interpositus controls the stretch, contact, placing, and other reflexes; and the dentate controls voluntary movements of the

extremities, such as those involved in reaching and grasping for objects (Fig. 32.5).

In the absence of movement, deep nuclear cells fire at maintained rates of approximately 40 to 50 Hz (Fortier *et al.*, 1989; Thach, 1968, 1970a, 1970b, 1978). During movement, their firing rates increase and decrease above and below their baseline. Through these variations in firing rate, the cerebellum modulates the activity of motor pattern generators (MPGs). In addition, increases in cerebellar nuclear firing rate precede and help increase the discharge frequency in MPGs, thus facilitating the initiation of movement (Lamarre *et al.*, 1983; Meyer-Lohmann *et al.*, 1975; Spidalieri *et al.*, 1983; Strick, 1983; Thach, 1978). Deep cerebellar nuclear cells may also combine the functions resident in MPGs into new or novel combinations (Babinski, 1899, 1906; Flourens, 1824; Thach *et al.*, 1992a). Thus, an MPG by itself could initiate movements that are within its repertoire, but only the cerebellum could initiate movements that are combinations of components within and across movement patterns.

Activity in Individual Deep Cerebellar Nuclei Correlates with Behavior

The fastigius controls all musculature involved in stance and gait. It receives input from the vermal cortex, vestibular complex, lateral reticular nucleus, and (indirectly) spinocerebellar pathways. Single-unit recordings in the fastigius and vermal cortex of decerebrate cats have shown neural discharge that is correlated with both walking and scratching movements (Andersson and Armstrong, 1987; Antziferova *et al.*, 1980). Discharge in the interpositus and dentate is not correlated with these behaviors (Arshavsky *et al.*, 1980). Fastigial neurons also are involved in the generation of saccadic eye movements (Chapter 33).

Interpositus neurons fire when the holding position of a limb is perturbed. In doing so, these neurons appear to control antagonist muscles that check the reflex movement of the limb to its prior hold position (Flament and Hore, 1986; Strick, 1983; Thach, 1978; Vilis and Hore, 1977, 1980). Interpositus neurons also modulate their activity in relation to sensory feedback, including that from tremor accompanying movement (Schieber and Thach, 1985; Soechting *et al.*, 1978; Thach, 1978). This finding is consistent with the idea that the interpositus is involved in somesthetic reflex behaviors, controlling the antagonist muscle to dampen the tremor (Elble *et al.*, 1984; Flament and Hore, 1986; Mauritz *et al.*, 1981; Vilis and Hore, 1977, 1980). Other evidence suggests that the interpositus is important in deter-

F   I   D

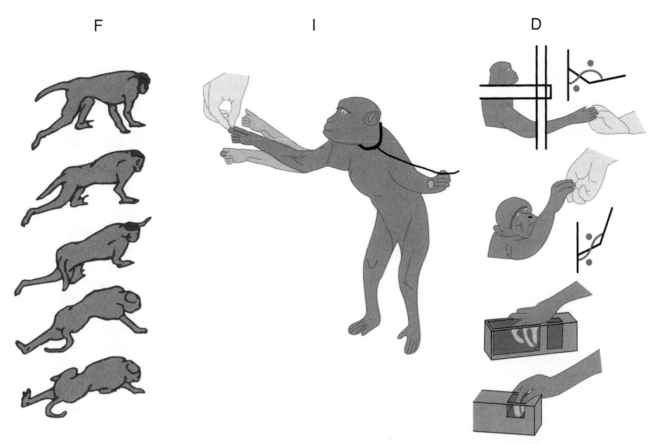

FIGURE 32.5 Major deficits produced by microinjection of muscimol and kainic acid into different regions of the cerebellum. F indicates the standing and walking deficit seen after muscimol injections into the fastigius. I indicates arm positon (tremor) during reaching after muscimol injection into the interpositus. D indicates the deficits in reaching and pinching after muscimol injection into the dentate. Adapted from Thach *et al.*

mining whether the pattern of activity in muscles acting at a joint represents reciprocal activation or cocontraction (Frysinger *et al.*, 1984; Smith and Bourbonnais, 1981; Wetts *et al.*, 1985). During behaviors that involve cocontraction, interpositus cells fire as if activating both agonist and antagonist muscles, and Purkinje cells are silent. In behaviors where agonists and antagonists are reciprocally active, both interpositus cells and Purkinje cells fire in similar patterns.

One interpretation of these results is that alternating firing in Purkinje cells creates (through inhibition) a similar pattern of activity the interpositus cells, which in turn produces the alternation between agonist and antagonist muscles. Other work suggests that the interpositus contributes to stretch reflex excitability by controlling the discharge of gamma motor neurons (Soechting *et al.*, 1978) (Fig. 32.6).

Neuronal activity in the dentate precedes the onset of movement and may also precede firing in the motor cortex (Lamarre *et al.*, 1983; Thach, 1978). Dentate cells fire preferentially at the onset of movements that are triggered by mental associations, with either visual or auditory stimuli. Single-unit recordings in the motor cortex, dentate, and interpositus were correlated with electromyograms as monkeys made wrist movements in response to stimuli. In tasks where movements were triggered by light, the order of activity is dentate, motor cortex, interpositus, and muscles. When a transient force perturbs the wrist, however, the firing order is muscles, interpositus, motor cortex, and dentate. These results suggest that the dentate helps initiate movements triggered by stimuli that are mentally associated with the movement, whereas the interpositus is more involved in compensatory or corrective movements initiated via feedback from the movement itself. In other studies, the dentate responded strongly when movements were triggered by either visual or auditory signals but not somesthetic signals (Lamarre *et al.*, 1983). Activity in both the dentate and the interpositus is thought to relate more to movements involving multiple joints than to those involving single joints. Neither nucleus codes for any specific parameter (e.g., velocity, amplitude, or duration) during single jointed movements (Elble *et al.*, 1984;

A

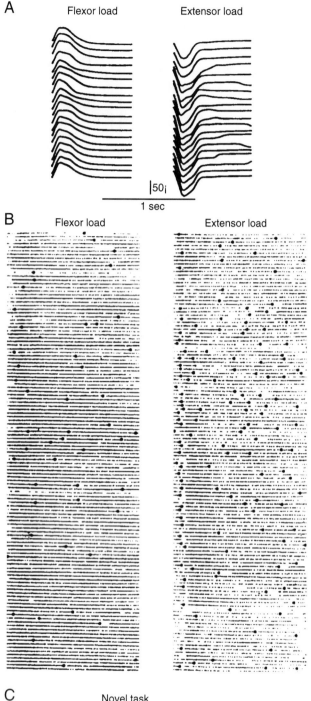

Flexor load   Extensor load

|50ᵢ

1 sec

B

Flexor load   Extensor load

C

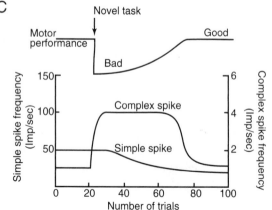

Mauritz *et al.*, 1981; Schieber and Thach, 1985; Thach *et al.*, 1992a; van Kan *et al.*, 1993). In sum, the dentate plays a role in initiating movements that require a mental interpretation of the visual or auditory signal, and both the dentate and the interpositus are increasingly active during multijointed movements.

Damage to the Cerebellar Nuclei Causes Unique Behavioral Deficits

Ablations of the fastigius in cat and monkey dramatically impair movements requiring control of equilibrium, such as unsupported sitting, stance, and gait (Botterell and Fulton, 1938b; Sprague and Chambers, 1953; Thach *et al.*, 1992b). Longitudinal splitting of the cerebellum along the midline also produces very significant and long lasting disturbances of equilibrium. In humans, lesions in the vermal and intermediate zones of the anterior cerebellar lobe preferentially impair movements requiring equilibrium control (Horak and Diener, 1994; Mauritz *et al.*, 1979). These data suggest that the fastigius may be preferentially involved in movements like gait and stance.

Ablations of the interpositus in monkeys primarily cause tremor (Dow, 1938). Temporary inactivation of both the interpositus and the dentate with cooling probes elicits tremor that is dependent on proprioceptive feedback but is influenced by vision (Vilis and Hore, 1977). In monkeys, interpositus inactivation disturbs gait minimally but causes a large-amplitude, 3- to 5-Hz action tremor as the animals reach for food (Dow, 1938). These studies support the idea that the interpositus in monkeys is most concerned with balancing the agonist-antagonist muscle activity of a limb as it moves. The interpositus may use the abundant afferent input it receives from the periphery to generate predictive signals that decrease alternating stretch reflexes capable of causing limb oscillation.

Ablations of the dentate nucleus in monkeys produce slight reaction time delays, poor endpoint control, and impaired multijointed movements far beyond any deficits found in single-jointed movements. In several studies, lesions of the dentate produced very slight delays in the onset of motor cortex cell activity (Beaubaton and Trouche, 1982; Meyer-Lohmann *et al.*, 1975; Spidalieri *et al.*, 1983; Thach *et al.*, 1992b). These delays were due to the loss of phasic dentate activity rather than tonic support, because

FIGURE 32.6 Simple and complex spikes recorded during motor learning task. Adapted from Ito; original reference Gibert and Thach.

there was no change in the resting firing rate of motor cortical cells after dentate cooling (Meyer-Lohmann et al., 1975). Lesions of the dentate also produce a slight delay in the reaction time of movements triggered by light or sound (Beaubaton and Trouche, 1982; Spidalieri et al., 1983). In single-jointed movements, dentate ablation causes monkeys to overshoot very slightly (Thach et al., 1992b) or moderately (Flament and Hore, 1986), whereas in multijointed movements, it results in profoundly impaired reaching patterns with abnormally increased angulation of the shoulder and elbow and excessive overshoot of the target (Thach et al., 1992b). Dentate inactivation also causes monkeys to have difficulty pinching small bits of food out of a narrow well; instead, the animals use one finger as a scoop to retrieve the food (Fig. 32.5). In sum, dentate ablation profoundly impairs movements requiring coordination of multiple joints but affects single-jointed movements only slightly. Dentate inactivation also slightly impairs the initiation of movements that are triggered by vision or mental percepts.

Damage to the cerebellar cortex causes disability similar to but less severe than that resulting from damage to the deep nuclei. Damage to the cortex and the inferior olive also prevents many kinds of motor adaptation, including the acquisition of new and novel muscle synergies.

Parallel Fibers and Purkinje Cell Beams Act in the Coordination of Linked Nuclear Cells

The relationship between cerebellar somatotopic maps and parallel fiber morphology is consistent with a cerebellar role in movement coordination. In Figure 32.2, in each body representation in the deep nuclei, the rostrocaudal axis of the body is mapped onto the sagittal axis of the nucleus. The hindlimbs are represented anteriorly, the head posteriorly, distal parts medially, and proximal parts laterally. This orientation would suggest that the myotomes, which are arranged orthogonal to the rostrocaudal axis of the body, are represented primarily in the coronal dimension of the cerebellum and thus roughly parallel to the trajectory of the parallel fibers. Because the parallel fibers are connected to the deep nuclear cells by Purkinje cells, a coronal "beam" of parallel fibers would control the nuclear cells that influence the synergistic muscles in a myotome. In this way, the parallel fiber would be a single neural element spanning and coordinating the activities of multiple synergistic muscles and joints.

Anatomical studies indicate that parallel fibers in the monkey are about 6 mm long (Mugnaini, 1983), spanning stretch of cerebellar cortex that projects across the width of one deep nucleus. Thus, a strip of Purkinje cells under the influence of a set of parallel fibers of the same origin and length will control a strip of nuclear cells across an entire nucleus. Depending on which portion of the somatotopic map is involved, that nuclear strip may influence synergistic muscles across several joints in a limb or the muscles of the eyes, head, neck, and arm, for example.

Parallel fiber beams can also serve to link activity in different deep cerebellar nuclei. A link occurring across the two fastigial nuclei would effectively couple the two sides of the body in stance and gait. A link across the fastigius and interpositus would couple locomotion and reflex sensitivity. A link across the interpositus and dentate would couple reach and reflex sensitivity. In a recent study in trained reaching rats (Heck et al., 2007), simple spikes of pairs of Purkinje cells were recorded that, with respect to each other, either were aligned on a beam of shared parallel fibers or instead were located off beam. Firing rates of simple spikes firing in both on-beam and off-beam Purkinje cell pairs commonly showed large variation during reaching behavior. But with respect to timing, on-beam Purkinje cell pairs had simple spikes that were tightly time-locked to each other and to movement, despite the variability in rate. By contrast, off-beam Purkinje cell pairs had simple spikes that were not time-locked to each other.

What is the consequence of the on-beam synchronicity? Parallel fibers in the cerebellar cortex run roughly parallel to the myotomes of the body maps in the deep nuclei. On-beam parallel fiber activity may thus recruit Purkinje cell inhibition to modulate coordination and timing of the natural myomeric synergists mapped within the deep nuclei. In addition, parallel fibers would be long enough to span (via Purkinje cell projections) two or more adjacent nuclei, fastigius controlling stance and gait, interpositus agonist/antagonist coordination at a single joint, and dentate eye-limb coordination in reaching and of digits in multidigit movements (Heck et al., 2007). Bastian et al. reported how this might explain the inability to coordinate bilateral body movements in gait and balance after vermal section in humans (Bastian et al., 1998). In five children who had surgical midline division of the posterior cerebellar vermis to remove fourth ventrical tumors, individuals could hop on one leg as well as normal controls. Yet these same individuals could not do tandem heel-to-toe walking along a straight line without falling to one side or the other. Fortunately, regular gait was little impaired These observations suggested that vestibular information could be used to control one leg, but that the two legs could not be so coordinated. The division of parallel fibers crossing the midline seemed best able to explain this dissociation.

Several Models Attempt to Explain the Operation of the Cerebellum

Tonic Reinforcer Model

Given its tonic excitatory effect on motor pattern generators in the vestibular nuclei, reticular nuclei, and the motor cortex, the cerebellum may be responsible for tuning these structures so that they respond optimally to noncerebellar inputs. Luciani championed the idea of a reinforcing tone with his interpretation of the behavioral deficits produced by cerebellar ablation in animals, a principal component of which was atonia. Holmes endorsed this interpretation based on his own analysis of behavioral deficits in human subjects with cerebellar lesions (Dow, 1987; Dow and Moruzzi, 1958; Homes, 1939). We now know that the predominant output of the cerebellum is excitatory and that the cerebellum is tonically active even in the absence of movement (Eccles et al., 1967; Thach, 1968). Granit et al., (1955) and Gilman (1969) emphasized how tone could be controlled by the cerebellum's excitation of gamma motor neurons, which modulate stretch reflexes. However, because the cerebellum also projects to alpha motor neurons (Dow, 1938) and because cerebellar nuclear discharge changes tonically in relation to different held postures and phasically in relation to different movements, basic cerebellar function must be more complex than this model supposes.

Timer Model

Numerous investigators have proposed that the cerebellum performs a pure timing function. Braitenberg and colleagues (Braitenberg, 1967) proposed that a wave of activity propagated along a parallel fiber could be "tapped off" by successive Purkinje cells, each tap occurring at an incremental delay after the onset of the wave. This delay could be used to time movements for up to 50 ms or so. Llinas and Lamarre (Lamarre, 1984; Lamarre and Mercier, 1971; Llinas and Yarom, 1986) independently proposed motor clock functions for the cerebellum on the basis of presumed periodic discharge the inferior olive. This proposal was founded on the effects of the drug harmaline, which induced a whole-body, 10-Hz tremor in experimental animals (Lamarre and Mercier, 1971) and a correlated synchronous discharge in inferior olive cells in slice preparations (Llinas and Yarom, 1986). It was also supported by the tendency for olivary neurons to fire periodically and in synchrony, and on the electronic coupling between olivary neurons, which might synchronize their discharge. However, this interpretation is contradicted by the finding that olivary discharge in the awake, performing monkey is not only

nonperiodic but indeed random (Keating and Thach, 1995).

Ivry et al. (1988) proposed, based on different evidence, that the cerebellum functions as a general purpose timer. Patients with lateral cerebellar injury are impaired in their ability to perceive differences in intervals between tone pairs of the order of half a second. This observation has been interpreted as indicative of a general clock not only for movement but also for perception. These timer/clock models maintain that the cerebellum (or the inferior olive) controls only the timing of muscle activity. Arguing against these ideas is the absence of any demonstrated clocklike periodicity of cerebellar nuclear cell discharge. Instead, these cells produce a graded, tonic discharge that correlates with muscle pattern and force, limb position, and movement direction (Keating and Thach, 1995, 1997).

Command-Feedback Comparator Model

Several models of cerebellar function incorporate linear systems principles (Allen and Tsukahara, 1974; Brooks and Thach, 1981; Evarts and Thach, 1969). According to these models the lateral cerebellum receives commands from the association cortex and feedback from the motor cortex, and it projects back to the motor cortex to help initiate and correct the commands. The intermediate cerebellum receives commands from the motor cortex and feedback from the spinal cord, and it projects to the red nucleus to help execute and correct errors in ongoing movements. These models are consistent with the early discharge of dentate neurons in the initiation of movement and the later discharge of interpositus neurons. However, the models do not specify how relatively slow neural feedback signals could make the same kind of error corrections that electronic feedback can.

Combiner-Coordinator Model

The combiner-coordinator model proposes that the role of the cerebellum is to coordinate movements (Flourens, 1824). This model is supported by the finding in cerebellar patients of a loss of coordination of compound movements without any impairment of force in simpler movements (Babinski, 1899, 1906). Mathematical descriptions of the model have been developed for translating one reference frame (such as body musculature) to another (such as movements in space), for both static and dynamic aspects of motor control (Fujita, 1982; Pellionisz and Llinas, 1980). The model has been modified to explain how parallel fibers might function in combining the activities of lower MPGs, motor neurons, and muscles to provide many unique patterns of coordinated movement (Thach *et al.*, 1992a). A learning capability, as developed in the Marr-Albus-Ito hypothesis (discussed next), is another feature of the model.

Motor Learning Model

According to the Marr-Ito motor learning hypothesis (Albus, 1971; Ito, 1972, 1984, 1989; Marr, 1969; Thach *et al.*, 1992a; Thompson, 1990), the climbing fiber input to the cerebellum controls the induction of synaptic plasticity in the cerebellar cortex. This plasticity mediates the unconscious correction or expression of movements through trial-and-error linking of a certain behavioral context to the movement response. The cerebellar circuitry seems well suited to interpret the various aspects of the context, to implement the many components in a response, to formulate the context-response linkages that constitute our learned movement repertoire, and to carry out the process of learning them. Damage to the cerebellum abolishes the ability to make automatic, rapid, smoothly coordinated movements and to learn new complex movements. Patients with cerebellar damage can still move, by virtue of the activity of the frontal cerebral cortex and the actions of sensory inputs on each of the MPGs, but each component of a movement must be thought out; movements are slow and irregular, because the agonist, antagonist, synergist, and fixator muscles cannot be linked together in time, amplitude, and combination.

The Cerebellum Contributes to Cognitive Processes

For many years the cerebellum was thought to be involved only or mostly in the control and adaptability of movement. This notion was based on the profound motor deficits seen in cerebellar patients and on the fact that cerebellar projections had been traced only to motor areas. Recently, however, this view of the cerebellum has changed (Leiner *et al.*, 1993; Thach, 1996). Sophisticated anatomical and brain imaging techniques, including positron emission tomography (PET) and magnetic resonance imaging (MRI), have provided evidence for cerebellar contributions to cognition (Dum *et al.*, 2002; Dum and Strick, 2003; Kelly and Strick, 2003; Middleton and Strick, 2001).

Initial suggestions that the cerebellum participates in cognition arose from anatomic connections that were thought to exist due to the parallel expansion of the frontal lobe, lateral cerebellum, and dentate (Leiner *et al.*, 1993). Transneuronal transport of herpes simplex virus type 1 (HSVl) has revealed these connections: injection of HSVl into the principle sulcus (area 46) of the cerebral cortex in cebus monkeys retro-

gradely labels neurons in the dentate (Middleton and Strick, 1994). Area 46 lies in the prefrontal cortex, which is not thought to be directly involved in motor performance.

Human brain imaging studies indicate that the cerebellum participates in some types of cognitive functions. One study revealed cerebellar activation when subjects imagined or passively observed movement without moving themselves (Decety *et al.*, 1990). In another study, the lateral cerebellum was activated in a language task in which subjects were asked to generate appropriate verbs for visually presented nouns (Petersen *et al.*, 1989). The cerebellum was not activated when the nouns were simply read, indicating that the cerebellum somehow contributed to the generation of the appropriate verb. Interestingly, once the verb generation task had been well practiced, cerebellar activation diminished (Raichle *et al.*, 1994). A third study showed that the dentate was activated bilaterally when subjects worked to solve a pegboard puzzle, and this activation was three to four times greater in magnitude than during simple peg movements (Kim *et al.*, 1994). Overall, these findings indicate that the lateral cerebellum and dentate are activated in ways that would not be expected to account for movement alone.

Behavioral observations of humans with cerebellar damage have also provided evidence that the cerebellum participates in nonmotor aspects of cognition. In one case study, a cerebellar patient was unable to detect errors in a verb-generating task and did not exhibit normal practice-related learning, although the patient performed normally on standard intelligence and memory tests (Fiez *et al.*, 1992). In another study, cerebellar patients were deficient in both the production and the perception of a timing task. Lateral cerebellar lesions interfered with the perception of time intervals, whereas medial cerebellar lesions affected the timing of implemented responses.

Clinical investigations of adults and children with diseases confined to the cerebellum have defined a cerebellar cognitive affective syndrome characterized by impairments of executive processing, working memory, visual spatial reasoning, language disturbances (ranging from autism to agrammatism), and a flattened or inappropriate affect. The net effect of these deficits is a lowering of overall intellectual ability. The posterior lobe of the cerebellum appears particularly important in generating the deficits, and the vermis is consistently involved when the affective component is pronounced. Elements of this clinical syndrome have also been noted in patients with developmental cerebellar anomalies, cerebellar degeneration, autism, and fragile X syndrome. In addition, cerebellar abnormalities, particularly in the vermis, have been observed in patients with schizophrenia.

Curmudgeonly Caveats

There is now little question that the cerebellum projects to many areas of cerebral cortex outside of primary motor cortex, and that many of these areas are known to have cognitive as well as motor functions, however it remains to be seen whether the cerebellum targets the "cognitive" functions of these areas and not the "motor." The studies of Schmahman and Sherman (1998), for example, include a wide variety of types of mental disturbance and of cerebellar location. In proposing that there is indeed a "cognitive affective cerebellar syndrome," one would like to be assured that there is indeed a "syndrome"—that is, a constellation of behavioral and anatomic features that consistently and reliably "run together." The studies of Courchesne and associates, which document the attention deficits and the focal cerebellar atrophy in autism, do seem to constitute a syndrome (Courchesne et al., 1975). Nevertheless, subjects with autism have abnormalities of movement in addition to abnormalities of mentation.

Since there is no consensus that focal cerebellar lesions in previously normal individuals consistently produce such a syndrome, one may be skeptical that the atrophy in autism produces the disorder of mentation as well as that of movement. Similarly for work on schizophrenia, which has documented a common association of focal cerebellar atrophy (Okugawa et al., (2002). Yet schizophrenic patients commonly also demonstrate disorders of movement. Thus, it remains to be seen whether a focal cerebellar lesion in a previously normal individual consistently produces the mental defects as well as the motor. Finally, there have been many studies as discussed earlier that have implicated a fundamental role of the cerebellum in governing abstract timing. The work of Ivry documents that patients with lateral cerebellar lesions do indeed have deficits in estimating differing time intervals and the speed of moving objects. Yet other studies of Decety and Ingvar and others show that the cerebellum is active in imagined movement as well as in overt movement. It also has been pointed out that one of the ways of estimating and comparing time intervals is to create a mental motor replica, as in the silent speech mechanism of estimating seconds by silently counting "one thousand one, one thousand two, one thousand three...." It remains to be seen therefore whether this is the timing mechanism within the cerebellum rather than or in addition to something more abstract and fundamental.

The Cerebellum Exerts Oculomotor Control

Eye movements are also differentially controlled by specific parts of the cerebellum (Chapter 33). The vestibular nuclei and the vestibulocerebellum control the vestibulo-ocular and optokinetic reflexes (VOR and OKR, respectively) and participate in smooth pursuit eye movements. The cerebellar cortex is necessary for visual smooth-pursuit tracking, which requires VOR cancellation. The interpositus and intermediate cerebellar cortex participate in conditioned eye-blink. The dentate and lateral cerebellar cortex are involved in voluntary gaze, saccades, and eye-hand coordination (see Fig. 32.7 for a test).

Clinical Testing Can Reveal Cerebellar Damage

In the early 1900s, Gordon Holmes carefully described the movement deficits associated with discrete cerebellar lesions caused by gunshot wounds (Dow, 1987; Dow and Moruzzi, 1958; Homes, 1939). His descriptions provide the fundamental basis for our understanding of clinical cerebellar syndromes and a rationale for classifying the manifestations of cerebellar disease according to seven basic deficits. These deficits may be attributed to cerebellar disease only in patients who have normal strength and somesthesis, because similar symptoms may arise from damage of peripheral nerves, the corticospinal tract, and lemniscalthalamocortical sensory systems (Fig. 32.8).

Ataxia is a condition that involves lack of coordination between movements of body parts. The term often is used in reference to gait or movement of a specific body part, as in "ataxic arm movements."

Dysmetria is an inability to make a movement of the appropriate distance. Hypometria is undershooting a target, and hypermetria is overshooting a target. Patients with cerebellar damage tend to make hypermetric movements when they move rapidly and hypometric movements when they move more slowly and wish to be accurate.

Dysdiadochokinesia is an inability to make rapidly alternating movements of a limb. It appears to reflect abnormal agonist-antagonist control.

Asynergia is an inability to combine the movements of individual limb segments into a coordinated, multisegmental movement.

Hypotonia is an abnormally decreased muscle tone. It is manifest as a decreased resistance to passive movement, so that a limb swings freely upon external perturbation. Hypotonia often is limited to the acute phase of cerebellar disease.

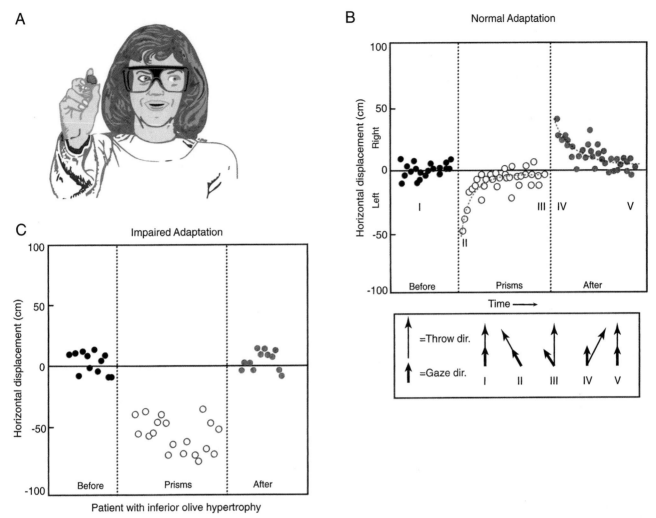

FIGURE 32.7 Prism adaptation test, control subject. (A) Eye-hand positions after adaptation to base right prisms. The optic path is bent to the subject's right, giving a larger view of right side of her face. Her gaze is shifted left along the bent light path to foveate the target in front of her. Her hand position is ready for a throw at the target in front of her. (B) Horizontal locations of throw hits displayed sequentially by trial number. Deviations to the left are negative values; deviations to the right, positive. While the subject is wearing the prisms (gaze shifted to the left), the first hit is displaced 60 cm left of center. Thereafter, hits trend toward 0. After the prisms are removed, the first hit is 50 cm right of center. Thereafter, hits trend toward 0. Data during and after prism use have been fitted with exponential curves. The decay constant is a measure of the rate of adaptation. The standard deviation of the last eight preprism throws is a measure of performance. Gaze and throw directions are schematized with arrows. Inferred gaze (eye and head) direction assumes the subject is foveating the target. Roman numerals beneath the arrows indicate times during the prism adaptation experiment (see B). (C) Failure of adaptation in a patient with bilateral infarctions in the territory of the posterior inferior cerebellar artery. Adapted from Martin *et al*.

Nystagmus is an involuntary and rhythmic eye movement that usually consists of a slow and a fast phase. In an unilateral cerebellar lesion, the fast phase of nystagmus is toward the side of the lesion.

Action tremor, or **intention tremor**, is an involuntary oscillation that occurs during limb movement and disappears when the limb is at rest. Cerebellar action tremor is generally of high amplitude and low frequency (3–5 Hz). **Titubation** is a tremor of the entire trunk during stance and gait. Lesions of

cerebellar target structures (e.g., the red nucleus and the thalamus) often result in cerebellar outflow tremor, or postural tremor. Most prominent when a limb is actively held in a static posture, postural tremor attenuates during limb movement and disappears when the limb is at rest.

Ataxia of stance and gait can result from a loss of equilibrium and vestibular reflexes produced by lesions of the vestibular nuclei and vestibulocerebellum. Abnormal synergies during standing and walking

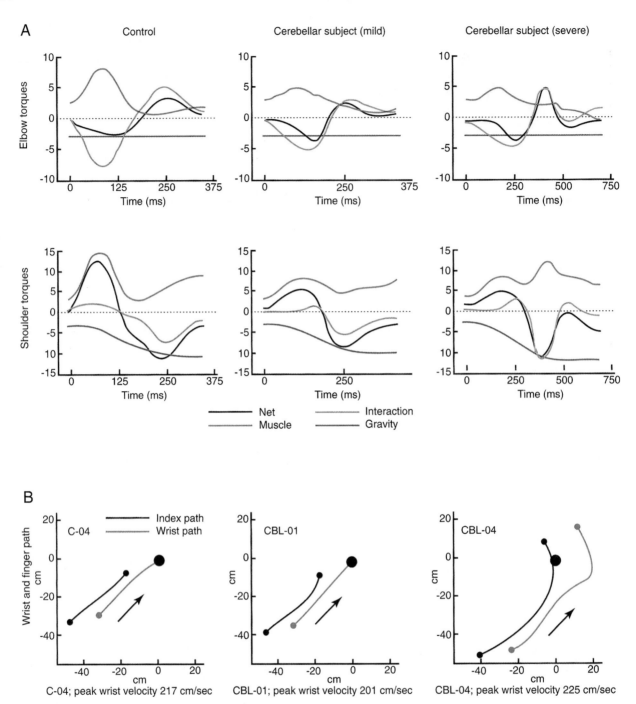

FIGURE 32.8 (A) Single trials of the torques produced by one control subject and two cerebellar subjects, all moving in the fast-accurate condition. One cerebellar subject (mild) produced mildly abnormal reaches by kinematic measures. The other cerebellar subject (severe) produced reaches that were profoundly abnormal by kinematic measures. The torques produced by the control subject and the mildly impaired cerebellar subject were similar, although the mildly impaired cerebellar subject's elbow and shoulder net torques (bold lines) more closely followed the dynamic interaction torque (small dotted lines). The severely impaired cerebellar subject produced torques that were abnormal, with the net elbow and shoulder torques (bold line) closely following the dynamic interaction torque at the respective joints (small dotted line). (B) The wrist and index finger paths are shown moving toward the target (large dot) for the same trials as in A. The control subject makes straight line wrist and index paths and stops right on the target (bold circle). One cerebellar subject (mild) produced slightly curved wrist and index paths and stopped on target. The other cerebellar subject (severe) produced curved wrist and index paths and overshot the target by a great extent. Adapted from Bastian *et al.*

can also be produced by lesions of the fastigius and vermal cerebellar cortex. The fastigius ultimately influences muscles associated with posture and gait through the vestibulospinal and reticulospinal pathways. Finally, degeneration of the medial anterior lobe of the cerebellar cortex, which often results from thiamin deficiency in alcoholism (Victor *et al.*, 1959), can cause ataxic gait patterns.

Action tremor probably is caused by instability of stretch reflexes, due to an inability to adjust the gain and sensitivity of the reflexes. Dysdiadochokinesis may be due to a failure to coordinate the activation of a muscle with the inhibition of its antagonists and antagonistic stretch reflexes. Damage to the interpositus, alone or in combination with damage to the dentate, is speculated to cause this deficit.

Asynergy of voluntary movements is thought to result mostly from damage to the dentate and the lateral cerebellar cortex, because loss of the dentate significantly impairs the coordination of multijointed movements but has much less effect on single-jointed movements (Gilman *et al.*, 1976; Goodkin *et al.*, 1993). The increased number of degrees of freedom in multijointed movements adds greatly to the mechanical complexity of these movements. When multiple joints are moved together, the motor system must compensate for interaction torques, forces that are generated at one joint when a segment that is directly or indirectly linked to it moves. Cerebellar patients are unable to correct for interaction torques generated during multijointed movements (Bastian *et al.*, 1996) (Fig. 32.8), such that their movements can become dominated by interaction torques, appearing ataxic and inaccurate.

It has been proposed that a major function of the lateral cerebellum is to generate predictive and feedback corrections that are used to compensate for extraneous forces generated by one's own body movements. Lesions of the lateral cerebellum also produce modest increases in voluntary reaction time. This defect may reflect the loss of the early cerebellar signals that generate movements. Cerebellar damage can also result in the loss of adjustability of movement. Studies of prism adaptation have shown that some patients with cerebellar damage are unable to adjust their hand-eye coordination appropriately (Martin *et al.*, 1996; Weiner *et al.*, 1983) (Fig. 32.7).

A Cerebellar Role in Rehabilitation? Functional Recovery through Relearning after Damage of Other Parts of the Nervous System

A series of patients sustained infarct of the left cerebral cortical frontal speech area with immediate onset of aphasia. After time and speech training, they recovered the ability to speak. FMRI study showed that upon speaking, symmetrically opposite portions of the opposite right cerebral hemisphere were active. Further, the regions of the cerebellum now active in speech had shifted to the left cerebellar hemisphere. This suggests that (1) the primary property of the cerebellar hemispheres is not speech (right) or spatial operations (left), but rather (2) their cooperation/training of the opposite cerebral hemisphere cortex, whatever activities it is engaged in (Connor *et al.*, 2006).

An ongoing series of studies is exploring this potential role in neurorehabilitation of gait after hemiplegia due to cerebral hemispheric cortical infarcts or hemispherectomy for seizure control. Subjects repeatedly walk on a split treadbelt. The belt on the defective side can be made to move faster to bring gait in the impaired leg up to match that on the normal side. This training appears to carry over to over ground gait for a period of time. Further studies will determine whether additional intensive training can induce a permanent learning effect, as in the prism gaze-arm throwing experiments.

Computation in the Cerebellum

The preceding sections illustrate that the contributions of the cerebellum are diverse and that there are a variety of ideas about the function of the cerebellum that at the surface appear somewhat mutually exclusive. Presumably however, what these contributions share is the computation that the cerebellum accomplishes—that is, the rules that describe the transformation of inputs to outputs. Like all brain systems, the cerebellum receives inputs from other parts of the brain, acts on these inputs with rules determined by its connectivity and synaptic/cellular properties, and produces output to other brain systems. Understanding the cerebellum, or any brain system, thus involves accomplishing three interrelated goals:

1. **What the cerebellum computes**, which involves a thorough characterization of the rules for input/output transformation. In essence, this goal states that we understand the cerebellum when given an input to it (and information about past history) we can reliably predict the output.
2. **How the cerebellum computes**, which means to understand the interactions between synaptic organization and synaptic physiology that mediates the input/output transformations.
3. **Why the cerebellum is involved in various aspects of behavior** should then be derivable from their common need for the computation that the cerebellum accomplishes.

In this section we outline progress toward these goals. We will limit discussion to research related to the motor function of the cerebellum, mainly because movement is inherently more experimentally tractable than cognition. Thus, much more is known about the cerebellar computations that are applied to motor coordination. Given that the synaptic organization of the cerebellum is relatively uniform it is likely that the same computational principles apply qualitatively for cerebellar regions involved in cognitive functions.

Motor Coordination, Motor Learning or Timing?

For cerebellar contributions to the motor system, three functions generally dominate discussion: (1) motor coordination, which is an obvious inference given the severe motor deficits described earlier; (2) motor learning, which is inferred from cerebellar involvement in several experimentally tractable forms of motor learning (Pavlovian eyelid conditioning, adaption of the VOR, adaptation of saccades and learning of smooth pursuit eye movements to name four); and (3) timing, which is suggested by timing-related deficits in cerebellar patients, functional imaging in humans, and by certain rhythmic properties of cerebellar neurons. This list is not complete, but it can serve as an illustration of how a focus on cerebellar computation can tie together many, if not all of the functions that have been proposed for the cerebellum. From analysis of the input/output transformations of the

cerebellum enabled largely by analysis of eyelid conditioning, this computation appears to be feed-forward prediction. Thus, motor coordination is enhanced by cerebellar feed-forward prediction, which requires associative learning with timing (Ohyama et al., 2003).

In principle there are two ways to make use of sensory input to make decisions about output: feed-back and feed-forward. Feedback is familiar and simple—a thermostat is an example of feedback. The measured temperature (sensory input) is compared to a target and differences are translated into turning on the heater or cooler. Feedback systems of this sort can be accurate and reliable, but they cannot be fast. Attempts to make a feedback system fast result in oscillations as, in the thermostat example, small deviations from the target produce alternating activations of the heater and cooler as each output causes overshoot of the target temperature (Fig. 32.9). Feed-forward control can be faster because it involves anticipation of errors from previous experience. A hypothetical feed-forward thermostat would be outfitted with a variety of sensory for temperature, humidity, number of people in the room, position of the sun, number of windows open, and so on. When a rapid change in room temperature is required, this hypothetical feed-forward thermostat would predict, from previous experience, the burst of hot or cold air required to produce the rapid change.

If, after producing its response, the wrong outcome is detected, then the feed-forward thermostat should learn from this mistake so that subsequent predictions

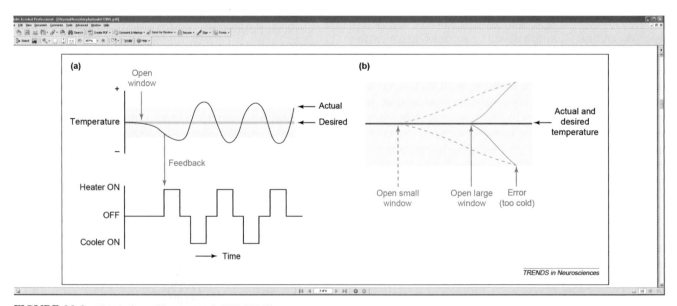

FIGURE 32.9　This is from Ohyama et al. 2003 (TINS).

in similar circumstances are better. The properties required of this learning are fairly specific. Since different mossy fiber inputs may have preceded the error signal by different temporal intervals, learning should not produce changes in output time locked to any particular mossy fiber input, but should instead produce changes that are timed to occur just before the error occurred (since that was the errant output). This hypothetical example reveals the need for specific learning features:

1. The learning should be associative: it should modify only the future output of the system for inputs that are similar to the ones that were present just before the error.
2. The learning should show temporal flexibility: the learned changes in output should not be time locked to the onset of any given mossy fiber input, but should be delayed after their onset so that the changed responses fix the errant prediction—namely, the one that occurred just before the arrival of the error signal. In other words, the error signal should fix the output that just occurred, not the output present at the onset of the mossy fiber inputs that preceded and predicted the error.

Eyelid conditioning studies reveal precisely these properties in cerebellar learning. It is associative in that it takes mossy fiber inputs predicting climbing fiber inputs to occur. Moreover, the timing of the responses is precisely as described earlier. Pairing a mossy fiber input with a climbing fiber input produces a change in cerebellar output that is not time locked to the onset of the mossy fiber input, but is instead timed to peak just before the time when the climbing fiber input arrives. This shows that cerebellar learning can develop predictions, through trial and error learning with climbing fiber error signals, that improve the prediction about the proper cerebellar output for a given mossy fiber input.

This may tie together many key ideas about cerebellar function. Motor coordination is improved by feed-forward learning feed-forward predictions. This learning requires associative learning with quite specific temporal properties. It requires motor learning with timing. An interesting challenge for the future will be to determine how feed-forward prediction can contribute to the various cognitive processes that appear to engage the cerebellum. The most important information that currently is lacking is what activates the climbing fiber (error) inputs for regions of the cerebellum involved in cognition and how do learned changes in output contribute to or improve cognitive processes?

Summary

Ideas about cerebellar function have been driven largely by the effects of lesion or dysfunction of the cerebellum and by the anatomical and physiological properties of cerebellar inputs, outputs, and intrinsic circuitry. This has led to a wide variety of models of cerebellar function and to debates about whether it is a motor structure, a sensory structure, or a cognitive structure. A view of cerebellar function that incorporates cerebellar computations may be more informative and may also help us understand how the various theories of cerebellum function are interrelated.

BOX 32.2

ANIMAL MODELS OF CEREBELLAR DEVELOPMENT

Several genetic mutations in mice perturb the normal development of the cerebellum. In the *weaver* mutation, granule cells (and thus parallel fibers) are absent, causing Purkinje cells to develop abnormally organized dendrites with an excessive number of spines. The *reeler* mutation causes Purkinje cells to migrate abnormally, resulting in a large displacement of Purkinje cells within the central cerebellar mass. In the *staggerer* mutation, most Purkinje cells are absent before granule cell migration, while in the *nervous* mutation, Purkinje cells begin to develop normally but then degenerate after most synaptic connections with parallel fibers have been formed. Studies of these and other mutations that cause specific cellular components to be deleted at different developmental stages are furthering our understanding of the complex cellular interactions that take place during normal synaptogenesis.

BOX 32.3

CEREBELLUM—THE TRUE THINKING MACHINE

It no longer seems reasonable to consider the function of the cerebellum as being confined to the control of voluntary movement, speech, and equilibrium. Considerable evidence suggests that the cerebellum is critical also for thought, behavior, and emotion. Anatomical studies demonstrate that the cerebellum is an important part of the distributed neural circuitry that subserves cognitive processing. The association and paralimbic cerebral cortices known to subserve higher order functions are linked with the cerebellum in a precisely organized system of feedforward and feedback loops, and physiological studies indicate that these pathways are functionally relevant. Cerebellar ablation and stimulation experiments in animals have demonstrated cerebellar influences on many nonmotor functions, including classical conditioning, navigational skills, cognitive flexibility, sham range, predatory attack, and aggression. Functional neuroimaging investigations of the morphologic correlates of cognitive processing using PET and fMRI in humans have revealed sites of activation in the cerebellum in a number of cognitive tasks. These include linguistic processing, verbal working memory, shifting attention, mental imagery, classical conditioning, motor learning, sensory processing, and modulation of emotion. Furthermore, there appears to be a topographic organization of the sites within the cerebellum activated by these different cognitive processes.

Clinical investigations of adults and children with diseases confined to the cerebellum have defined a cerebellar cognitive affective syndrome characterized by impairments of executive processing, working memory, visual spatial reasoning, language disturbances (ranging from mutism to agrammatism), and a flattened or inappropriate affect. The net effect of these deficits is a lowering of overall intellectual ability. The posterior lobe of the cerebellum appears particularly important in the generation of this syndrome, and the vermis is consistently involved when the affective component is pronounced. Elements of this clinical syndrome have also been noted in patients with developmental cerebellar anomalies, cerebellar degeneration, autism, and fragile X syndrome. In addition, cerebellar abnormalities, particularly in the vermis, have been observed in patients with schizophrenia.

The mechanisms whereby the cerebellum influences higher function are still debated. The organization of the cerebellar corticonuclear microcomplex and the interactions between the mossy and climbing fiber systems prompted the hypothesis that the cerebellum provides an error detection mechanism for the motor system. This mechanism may also be relevant for mental operations.

The relationship between the cerebellum and nonmotor function has been conceptualized as follows.

1. The cerebellum is able to subserve cognitive functions because it is anatomically interconnected with the associative and paralimbic cortices.
2. Cognitive and behavioral functions are organized topographically within the cerebellum.
3. The convergence of inputs from multiple associative cerebral regions to adjacent areas within the cerebellum facilitates cerebellar regulation of supramodal functions.
4. The cerebellar contribution to cognition is one of modulation rather than generation.
5. The cerebellum performs computations for cognitive functions similar to those for the sensorimotor system —but the information being modulated is different.
6. The disruption of the cerebellar influences on higher functions leads to dysmetria of thought, impairment of mental agility that manifests, at least in part, as the cerebellar cognitive affective syndrome.

The potential for further discovery in this field places the cerebellum, previously thought of as a motor control device, in the forefront of current behavioral neuroscience research.

Jeremy D. Schmahmann

References

Adrian, E. D. (1943). Afferent areas in the cerebellum connected with the limbs. *Brain* **66**, 289–315.

Albus, J. S. (1971). A theory of cerebellar function. *Math Biosci* **10**, 25–61.

Allen, G. I., Gilbert, P. F., Marini, R., Schultz, W., and Yin, T. C. (1977). Integration of cerebral and peripheral inputs by interpositus neurons in monkey. *Exp Brain Res* **27**, 81–99.

Allen, G. I., Gilbert, P. F., and Yin, T. C. (1978). Convergence of cerebral inputs onto dentate neurons in monkey. *Exp Brain Res* **32**, 151–170.

Allen, G. I. and Tsukahara, N. (1974). Cerebrocerebellar communication systems. *Physiol Rev* **54**, 957–1006.

Andersson, G. and Armstrong, D. M. (1987). Complex spikes in Purkinje cells in the lateral vermis (b zone) of the cat cerebellum during locomotion. *J Physiol* **385**, 107–134.

Antziferova, L. I., Arshavsky, Y. I., Orlovsky, G. N., and Pavlova, G. A. (1980). Activity of neurons of cerebellar nuclei during fictitious scratch reflex in cat. I. Fastigial nucleus. *Brain Res* **200**, 239–248.

Arshavsky, Y. I., Orlovsky, G. N., Pavlova, G. A., and Perret, C. (1980). Activity of neurons of cerebellar nuclei during fictitious scratch reflex in the cat. II. Interpositus and lateral nuclei. *Brain Res* **200**, 249–258.

Asanuma, C., Thach, W. T., and Jones, E. G. (1983a). Anatomical evidence for segregated focal groupings of efferent cells and their terminal ramifications in the cerebellothalamic pathway of the monkey. *Brain Res* **286**, 267–297.

Asanuma, C., Thach, W. T., and Jones, E. G. (1983b). Brainstem and spinal projections of the deep cerebellar nuclei in the monkey, with observations on the brainstem projections of the dorsal column nuclei. *Brain Res* **286**, 299–322.

Asanuma, C., Thach, W. T., and Jones, E. G. (1983c). Cytoarchitectonic delineation of the ventral lateral thalamic region in the monkey. *Brain Res* **286**, 219–235.

Asanuma, C., Thach, W. T., and Jones, E. G. (1983d). Distribution of cerebellar terminations and their relation to other afferent terminations in the ventral lateral thalamic region of the monkey. *Brain Res* **286**, 237–265.

Babinski, J. (1899). *De l'asynergie cerebelleuse. Rev Neurol (Paris)* **7**, 806–816.

Babinski, J. (1906). *Asynergie et inertie cerebelleuses. Rev Neurol (Paris)* **14**, 685–686.

Bastian, A. J., Martin, T. A., Keating, J. G., and Thach, W. T. (1996). Cerebellar ataxia: Abnormal control of interaction torques across multiple joints. *J Neurophysiol* **76**, 492–509.

Bastian, A. J., Mink, J. W., Kaufman, B. A., and Thach, W. T. (1998). Posterior vermal split syndrome. *Ann Neurol* **44**, 601–610.

Beaubaton, D. and Trouche, E. (1982). Participation of the cerebellar dentate nucleus in the control of a goal-directed movement in monkeys. Effects of reversible or permanent dentate lesion on the duration and accuracy of a pointing response. *Exp Brain Res* **46**, 127–138.

Bell, C. C. and Szabo, T. (1986). Electroreception in mormyrid fish: Central anatomy. *In* "Electroreception" (T. H. Bullock and W. Heilingenberg, eds.), pp. 375–421. Wiley, New York.

Bloedel, J. R. and Courville, J. (1981). Cerebellar afferent systems. *In* "Handbook of Physiology, The Nervous System, Sect. 1" (V. B. Brooks, ed.), Vol. 2, pp. 735–831. American Physiological Society, Bethesda.

Botterell, E. H. and Fulton, J. F. (1938a). Functional localization in the cerebellum of primates. III. Lesions of the hemispheres (neo-cerebellum). *J Comp Neurol* **69**, 47–62.

Botterell, E. H. and Fulton, J. F. (1938b). Functional localizaton in the cerebellum of primates. II. Lesions of midline structures (vermis) and deep nuclei. *J Comp Neurol* **69**, 47–62.

Braitenberg, V. (1967). Is the cerebellar cortex a biological clock in the millisecond range? *Prog Brain Res* **25**, 334–346.

Brodal, P. (1978). The corticopontine projection in the rhesus monkey. Origin and principles of organization. *Brain* **101**, 251–283.

Brody, A. B., Kinney, H. C., Kloman, A. S., and Gilles, F. H. (1987). Sequence of central nervous system myelination in human infancy. I. An autopsy study of myelination. *J Neuropath Exp Neurol* **46**, 283–301.

Brooks, V. B. and Thach, W. T. (1981). Cerebellar control of posture and movement. *In* "Handbook of Physiology, The Nervous System, Sect. 1, Vol. II" (V. B. Brooks, ed.), pp. 877–946.

Chambers, W. W. and Sprague, J. M. (1955a). Functional localization in the cerebellum. I. Organization in longitudinal cortico-nuclear zones and their contribution to the control of posture, both extrapyramidal and pyramidal. *J Comp Neurol* **103**, 105–129.

Chambers, W. W. and Sprague, J. M. (1955b). Functional localization in the cerebellum. II. Somatotopic organization in cortex and nuclei. *AMA Arch Neurol Psychiatry* **74**, 653–680.

Connor, L. T., DeShazo, B. T., Snyder, A. Z., Lewis, C., Blasi, V., and Corbetta, M. (2006). Cerebellar activity switches hemispheres with cerebral recovery in aphasia. *Neuropsychologia* **44(2)**, 171–177.

Decety, J., Sjoholm, H., Ryding, E., Stenberg, G., and Ingvar, D. H. (1990). The cerebellum participates in mental activity: tomographic measurements of regional cerebral blood flow. *Brain Res* **535**, 313–317.

Diño, M. R., Schuerger, R. J., Liu, Y., Slater, N. T., and Mugnaini, E. (2000). Unipolar brush cell: A potential feedforward excitatory interneuron of the cerebellum. *Neuroscience* **98(4)**, 625–636.

Dow, R. S. (1938). Effect of lesions in the vestibular part of the cerebellum in primates. *Arch Neurol Psychiatry* **40**, 500–520.

Dow, R. S. (1987). Cerebellum, pathology: Symptoms and signs. *In* "The Encyclopedia of Neuroscience" (G. Adelman, ed.), Vol. 1, pp. 203–206. Birkhauser Boston, Inc., Boston.

Dow, R. S. and Moruzzi, G. (1958). "The Physiology and Pathology of the Cerebellum." Univ. of Minnesota Press, Minneapolis.

Dum, R. P., Li, C., and Strick, P. L. (2002). Motor and nonmotor domains in the monkey dentate. *Ann NY Acad Sci* **978**, 289–301.

Dum, R. P. and Strick, P. L. (2003). An unfolded map of the cerebellar dentate nucleus and its projections to the cerebral cortex. *J Neurophysiol* **89(1)**, 634–639.

Eccles, J. C., Ito, M., and Szentagothai, J. (1967). "The cerebellum as a neuronal machine." Springer-Verlag, Inc., New York.

Ekerot, C. F. and Kano, M. (1985). Long-term depression of parallel fibre synapses following stimulation of climbing fibres. *Brain Res* **342**, 357–360.

Elble, R. J., Schieber, M. H., and Thach, W. T., Jr. (1984). Activity of muscle spindles, motor cortex and cerebellar nuclei during action tremor. *Brain Res* **323**, 330–334.

Fiez, J. A., Petersen, S. E., Cheney, M. K., and Raichle, M. E. (1992). Impaired non-motor learning and error detection associated with cerebellar damage. *Brain* **115**, 155–178.

Flament, D. and Hore, J. (1986). Movement and electromyographic disorders associated with cerebellar dysmetria. *J Neurophysiol* **55**, 1221–1233.

Flourens, P. (1824). "*Recherches experimentales sur les proprietes et les fonctions du systeme nerveux, dans les animaux vertebres.*" Crevot, Paris.

Fortier, P. A., Kalaska, J. F., and Smith, A. M. (1989). Cerebellar neuronal activity related to whole-arm reaching movements in the monkey. *J Neurophysiol* **62**, 198–211.

Frysinger, R. C., Bourbonnais, D., Kalaska, J. F., and Smith, A. M. (1984). Cerebellar cortical activity during antagonist cocontraction and reciprocal inhibition of forearm muscles. *J Neurophysiol* **51**, 32–49.

Fujita, M. (1982). Adaptive filter model of the cerebellum. *Biol Cybern* **45**, 195–206.

Gellman, R., Gibson, A. R., and Houk, J. C. (1985). Inferior olivary neurons in the awake cat: Detection of contact and passive body displacement. *J Neurophysiol* **54**, 40–60.

Gilbert, P. F. and Thach, W. T. (1977). Purkinje cell activity during motor learning. *Brain Res* **128**, 309–328.

Gilman, S. (1969). The mechanism of cerebellar hypotonia. An experimental study in the monkey. *Brain* **92**, 621–638.

Gilman, S., Carr, D., and Hollenberg, J. (1976). Kinematic effects of deafferentation and cerebellar ablation. *Brain* **99**, 311–330.

Glickstein, M., May, J. G., 3rd, and Mercier, B. E. (1985). Corticopontine projection in the macaque: The distribution of labeled cortical cells after large injections of horseradish peroxidase in the pontine nuclei. *J Comp Neurol* **235**, 343–359.

Goodkin, H. P., Keating, J. G., Martin, T. A., and Thach, W. T. (1993). Preserved simple and impaired compound movement after infarction in the territory of the superior cerebellar artery. *Can J Neurol Sci* **20 Suppl 3**, S93–104.

Granit, R., Holmgren, B., and Merton, P. A. (1955). The two routes for excitation of muscle and their subservience to the cerebellum. *J Physiol* **130**, 213–224.

Groenewegen, H. J. and Voogd, J. (1977). The parasagittal zonation within the olivocerebellar projection. I. Climbing fiber distribution in the vermis of cat cerebellum. *J Comp Neurol* **174**, 417–488.

Groenewegen, H. J., Voogd, J., and Freedman, S. L. (1979). The parasagittal zonation within the olivocerebellar projection. II. Climbing fiber distribution in the intermediate and hemispheric parts of cat cerebellum. *J Comp Neurol* **183**, 551–601.

Heck, D. H., Thach, W. T., and Keating, J. G. (2007). On-beam synchrony in the cerebellum as the mechanism for the timing and coordination of movement. *Proc Natl Acad Sci USA* **104(18)**, 7658–7663.

Herrup, K. and Mullen, R. J. (1981). Role of the Staggerer gene in determining Purkinje cell number in the cerebellar cortex of mouse chimeras. *Brain Res* **227**, 475–485.

Hoffer, B. J., Siggins, G. R., Oliver, A. P., and Bloom, F. E. (1972). Cyclic AMP-mediated adrenergic synapses to cerebellar Purkinje cells. *Adv Cyclic Nucleotide Res* **1**, 411–423.

Hoffer, B. J., Siggins, G. R., Oliver, A. P., and Bloom, F. E. (1973). Activation of the pathway from locus coeruleus to rat cerebellar Purkinje neurons: pharmacological evidence of noradrenergic central inhibition. *J Pharmacol Exp Ther* **184**, 553–569.

Homes, G. (1939). The cerebellum of man. The Hughlings Jackson memorial lecture. *Brain* **62**, 1–30.

Horak, F. B. and Diener, H. C. (1994). Cerebellar control of postural scaling and central set in stance. *J Neurophysiol* **72**, 479–493.

Ito, M. (1972). Neural design of the cerebellar control system. *Brain Res* **40**, 80–82.

Ito, M. (1984). "The cerebellum and neural control." Appleton-Century-Crofts, New York.

Ito, M. (1989). Long-term depression. *Ann Rev Neurosci* **12**, 85–102.

Ito, M., Sakurai, M., and Tongroach, P. (1982). Climbing fibre induced depression of both mossy fibre responsiveness and glutamate sensitivity of cerebellar Purkinje cells. *J Physiol* **324**, 113–134.

Ivry, R. B., Keele, S. W., and Diener, H. C. (1988). Dissociation of the lateral and medial cerebellum in movement timing and movement execution. *Exp Brain Res* **73**, 167–180.

Jacobson, M. (1985). Clonal analysis and cell lineages of the vertebrate central nervous system. *Ann Rev Neurosci* **8**, 71–102.

Jansen, J. and Brodal, A. (1954). "Aspects of cerebellar anatomy." Johan Grundt Tanum Forlag, Oslo.

Kano, M. and Kato, M. (1987). Quisqualate receptors are specifically involved in cerebellar synaptic plasticity. *Nature* **325**, 276–279.

Keating, J. G. and Thach, W. T. (1995). Nonclock behavior of inferior olive neurons: Interspike interval of Purkinje cell complex spike discharge in the awake behaving monkey is random. *J Neurophysiol* **73**, 1329–1340.

Keating, J. G. and Thach, W. T. (1997). No clock signal in the discharge of neurons in the deep cerebellar nuclei. *J Neurophysiol* **77**, 2232–2234.

Kelly, R. M. and Strick, P. L. (2003). Cerebellar loops with motor cortex and prefrontal cortex of a nonhuman primate. *J Neurosci* **23(23)**, 8432–8444.

Kenyon, G. T., Medina, J. F., and Mauk, M. D. (1998). A mathematical model of the cerebellar-olivary system I: Self-regulating equilibrium of climbing fiber activity. *J Comput Neurosci* **5(1)**, 17–33.

Kim, S. G., Ugurbil, K., and Strick, P. L. (1994). Activation of a cerebellar output nucleus during cognitive processing. *Science* **265(5174)**, 949–951.

Lamarre, Y. (1984). Animal models of physiological, essential, and parkinsonian-like tremors. *In* "Movement Disorders: Tremor" (L. J. Findlay and R. Capildeo, eds.), pp. 183–194. Oxford University Press, New York.

Lamarre, Y. and Mercier, L. A. (1971). Neurophysiological studies of harmaline-induced tremor in the cat. *Can J Physiol Pharmacol* **49**, 1049–1058.

Lamarre, Y., Spidalieri, G., and Chapman, C. E. (1983). A comparison of neuronal discharge recorded in the sensori-motor cortex, parietal cortex, and dentate nucleus of the monkey during arm movements triggered by light, sound or somesthetic stimuli. *Exp Brain Res Suppl* **7**, 140–156.

Landis, S. C. (1973). Ultrastructural changes in the mitochondria of cerebellar Purkinje cells of nervous mutant mice. *J Cell Biol* **57**, 782–797.

Larsell, O. (1967). "The comparative anatomy and histology of the cerebellum from myxinoids through birds." University of Minnesota Press, Minneapolis.

Larsell, O. (1970). "The comparative anatomy and histology of the cerebellum from monotremes through apes." University of Minnesota Press, Minneapolis.

Larsell, O. (1972). "The comparative anatomy and histology of the cerebellum: The human cerebellum, cerebellar connections, and cerebellar cortex." University of Minnesota Press, Minneapolis.

Leiner, H. C., Leiner, A. L., and Dow, R. S. (1993). Cognitive and language functions of the human cerebellum. *Trends Neurosci* **16**, 444–447.

Llinas, R. (1981). Electrophysiology of cerebellar networks. *In* "Handbook of Physiology, Section 1, Volume II, Part 2" (V. B. Brooks, ed.), pp. 831–876.

Llinas, R. and Yarom, Y. (1986). Oscillatory properties of guinea-pig inferior olivary neurones and their pharmacological modulation: An in vitro study. *J Physiol* **376**, 163–182.

Mariani, J., Crepel, F., Mikoshiba, K., and Changeux, J. P. (1977). Anatomical, physiological, and biochemical studies of the cerebellum from reeler mutant mouse. *Phil Trans R Soc Lond Ser B* **281**, 1–28.

Marr, D. (1969). A theory of cerebellar cortex. *J Physiol* **202**, 437–470.

Martin, T. A., Keating, J. G., Goodkin, H. P., Bastian, A. J., and Thach, W. T. (1996). Throwing while looking through prisms. I. Focal olivocerebellar lesions impair adaptation. *Brain* **119(Pt 4)**, 1183–1198.

Mauritz, K. H., Dichgans, J., and Hufschmidt, A. (1979). Quantitative analysis of stance in late cortical cerebellar atrophy of the anterior lobe and other forms of cerebellar ataxia. *Brain* **102**, 461–482.

Mauritz, K. H., Schmitt, C., and Dichgans, J. (1981). Delayed and enhanced long latency reflexes as the possible cause of postural tremor in late cerebellar atrophy. *Brain* **104**, 97–116.

Medina, J. F., Nores, W. L., and Mauk, M. D. (2002). Inhibition of climbing fibres is a signal for the extinction of conditioned eyelid responses. *Nature* **416(6878)**, 330–333.

Meyer-Lohmann, J., Conrad, B., Matsunami, K., and Brooks, V. B. (1975). Effects of dentate cooling on precentral unit activity following torque pulse injections into elbow movements. *Brain Res* **94**, 237–251.

Middleton, F. A. and Strick, P. L. (1994). Anatomical evidence for cerebellar and basal ganglia involvement in higher cognitive function. *Science* **266**, 458–461.

Middleton, F. A. and Strick, P. L. (2001). Cerebellar projections to the prefrontal cortex of the primate. *J Neurosci* **21(2)**, 700–712.

Mugnaini, E. (1972). The comparative anatomy and histology of the cerebellum: The human cerebellum, cerebellar connections, and cerebellar cortex. *In* "The Cerebellar Cortex, Part II" (J. Jansen, ed.), pp. 201–264. University of Minnesota Press, Minneapolis.

Mugnaini, E. (1983). The length of cerebellar parallel fibers in chicken and rhesus monkey. *J Comp Neurol* **220**, 7–15.

Nunzi, M. G., Birnstiel, S., Bhattacharyya, B. J., Slater, N. T., and Mugnaini, E. (2001). Unipolar brush cells form a glutamatergic projection system within the mouse cerebellar cortex. *J Comp Neurol* **434(3)**, 329–341.

Ohyama, T., Nores, W. L., Murphy, M., and Mauk, M. D. (2003). What the cerebellum computes. *Trends in Neuroscience* **26(4)**, 222–227.

Ojakangas, C. L. and Ebner, T. J. (1992). Purkinje cell complex and simple spike changes during a voluntary arm movement learning task in the monkey. *J Neurophysiol* **68**, 2222–2236.

Orioli, P. J. and Strick, P. L. (1989). Cerebellar connections with the motor cortex and the arcuate premotor area: An analysis employing retrograde transneuronal transport of WGA-HRP. *J Comp Neurol* **288**, 612–626.

Palay, S. L. and Chan-Palay, V. (1974). "Cerebellar Cortex. Cytology and Organization." Springer-Verlag, New York.

Paulin, M. G. (1993). The role of the cerebellum in motor control and perception. *Brain Behav Evol* **41**, 39–50.

Pellionisz, A. and Llinas, R. (1980). Tensorial approach to the geometry of brain function: cerebellar coordination via a metric tensor. *Neuroscience* **5**, 1125–1138.

Petersen, S. E., Fox, P. T., Posner, M. I., Mintun, M., and Raichle, M. E. (1989). Positron emission tomographic studies of the processing of single words. *J Cognitive Neurosci* **1**, 153–170.

Raichle, M. E., Fiez, J. A., Videen, T. O., MacLeod, A. M., Pardo, J. V., Fox, P. T., and Petersen, S. E. (1994). Practice-related changes in human brain functional anatomy during nonmotor learning. *Cereb Cortex* **4**, 8–26.

Rakic, P. (1982). Early developmental events: Cell lineages, acquisition of neuronal positions, and areal and laminar development. *Neurosci Res Program Bull* **20**, 439–451.

Rakic, P. and Sidman, R. L. (1973). Organization of cerebellar cortex secondary to deficit of granule cells in weaver mutant mice. *J Comp Neurol* **152**, 133–161.

Ramon y Cajal, S. (1911). "*Histologie du Systeme Nerveux.*" Maloine, Paris.

Rispal-Padel, L., Cicirata, F., and Pons, C. (1982). Cerebellar nuclear topography of simple and synergistic movements in the alert baboon (Papio papio). *Exp Brain Res* **47**, 365–380.

Sasaki, K., Kawaguchi, S., Oka, H., Sakai, M., and Mizuno, N. (1976). Electrophysiological studies on the cerebellocerebral projections in monkeys. *Exp Brain Res* **24**, 495–507.

Schell, G. R. and Strick, P. L. (1984). The origin of thalamic inputs to the arcuate premotor and supplementary motor areas. *J Neurosci* **4**, 539–560.

Schieber, M. H. and Thach, W. T., Jr. (1985). Trained slow tracking. II. Bidirectional discharge patterns of cerebellar nuclear, motor cortex, and spindle afferent neurons. *J Neurophysiol* **54**, 1228–1270.

Simpson, J. I. and Alley, K. E. (1974). Visual climbing fiber input to rabbit vestibulo-cerebellum: a source of direction-specific information. *Brain Res* **82**, 302–308.

Smith, A. M. and Bourbonnais, D. (1981). Neuronal activity in cerebellar cortex related to control of prehensile force. *J Neurophysiol* **45**, 286–303.

Snider, R. S. and Eldred, E. (1952). Cerebrocerebellar relationships in the monkey. *J Neurophysiol* **15**, 27–40.

Snider, R. S. and Stowell, A. (1944). Receiving areas of the tactile, auditory, and visual systems in the cerebellum. *J Neurophysiol* **7**, 331–357.

Soechting, J. F., Burton, J. E., and Onoda, N. (1978). Relationships between sensory input, motor output and unit activity in interpositus and red nuclei during intentional movement. *Brain Res* **152**, 65–79.

Spidalieri, G., Busby, L., and Lamarre, Y. (1983). Fast ballistic arm movements triggered by visual, auditory, and somesthetic stimuli in the monkey. II. Effects of unilateral dentate lesion on discharge of precentral cortical neurons and reaction time. *J Neurophysiol* **50**, 1359–1379.

Sprague, J. M. and Chambers, W. W. (1953). Regulation of posture in intact and decerebrate cat. I. Cerebellum, reticular formation, vestibular nuclei. *J Neurophysiol* **16**, 451–463.

Strata, P. (1985). Inferior olive: Functional aspects. *In* "Cerebellar Functions" (J. R. Bloedel, J. Dichgans, and W. Precht, eds.), pp. 231–246. Springer-Verlag, Berlin.

Strick, P. L. (1983). The influence of motor preparation on the response of cerebellar neurons to limb displacements. *J Neurosci* **3**, 2007–2020.

Thach, W. T. (1968). Discharge of Purkinje and cerebellar nuclear neurons during rapidly alternating arm movements in the monkey. *J Neurophysiol* **31**, 785–797.

Thach, W. T. (1970a). Discharge of cerebellar neurons related to two maintained postures and two prompt movements. I. Nuclear cell output. *J Neurophysiol* **33**, 527–536.

Thach, W. T. (1970b). Discharge of cerebellar neurons related to two maintained postures and two prompt movements. II. Purkinje cell output and input. *J Neurophysiol* **33**, 537–547.

Thach, W. T. (1978). Correlation of neural discharge with pattern and force of muscular activity, joint position, and direction of intended next movement in motor cortex and cerebellum. *J Neurophysiol* **41**, 654–676.

Thach, W. T. (1996). On the specific role of the cerebellum in motor learning and congnition: Clues from pet activation and lesion studies in humans. *Behav Brain Res* **19**, 411–431.

Thach, W. T., Goodkin, H. P., and Keating, J. G. (1992a). The cerebellum and the adaptive coordination of movement. *Annu Rev Neurosci* **15**, 403–442.

Thach, W. T., Kane, S. A., Mink, J. W., and Goodkin, H. P. (1992b). Cerebellar output: Multiple maps and motor modes in movement coordination. *In* "The cerebellum revisted" (R. Llinas and C. Sotelo, eds.), pp. 283–300. Springer-Verlag, New York.

Thach, W. T., Perry, J. G., Kane, S. A., and Goodkin, H. P. (1993). Cerebellar nuclei: Rapid alternating movement, motor somatotopy, and a mechanism for the control of muscle synergy. *Rev Neurol (Paris)* **149**, 607–628.

Thompson, R. F. (1990). Neural mechanisms of classical conditioning in mammals. *Phil Trans R Soc Lond Ser B* **329**, 161–170.

van Kan, P. L., Houk, J. C., and Gibson, A. R. (1993). Output organization of intermediate cerebellum of the monkey. *J Neurophysiol* **69**, 57–73.

Verbitskaya, L. B. (1969). Some aspects of the ontophylogenesis of the cerebellum. *In* "Neurobiology of Cerebellar Evolution and Development" (R. Llinas, ed.), pp. 859–878. American Medical Association, Chicago.

Victor, M., Adams, R. D., and Mancall, E. L. (1959). A restricted form of cerebellar cortical degeneration occurring in alcoholic patients. *Arch Neurol* **1**, 579–688.

Vilis, T. and Hore, J. (1977). Effects of changes in mechanical state of limb on cerebellar intention tremor. *J Neurophysiol* **40**, 1214–1224.

Vilis, T. and Hore, J. (1980). Central neural mechanisms contributing to cerebellar tremor produced by limb perturbations. *J Neurophysiol* **43**, 279–291.

Voogd, J. and Bigare, F. (1980). Topographical distribution of olivary and cortico nuclear fibers in the cerebellum. *In* "The inferior olivary nucleus: Anatomy and physiology" (J. Courville, C. D. Montigny, and Y. Lamarre, eds.), pp. 207–234. Raven Press, New York.

Weiner, M. J., Hallett, M., and Funkenstein, H. H. (1983). Adaptation to lateral displacement of vision in patients with lesions of the central nervous system. *Neurology* **33**, 766–772.

Wetts, R., Kalaska, J. F., and Smith, A. M. (1985). Cerebellar nuclear cell activity during antagonist cocontraction and reciprocal inhibition of forearm muscles. *J Neurophysiol* **54**, 231–244.

Michael D. Mauk and W. Thomas Thach

33

Eye Movements

As we learned in Chapter 27, the photoreceptor mosaic of the vertebrate retina transduces light energy in the form of photons into neural activity, ultimately in the form of action potentials. The spatial resolution of this transduction system is limited by the mosaic of photoreceptors and the optics of the eye, but only if the eye is held stationary with respect to the objects of interest in the external world. Thus, stabilizing the eyes with regard to the outside world and aiming the eyes toward moving or stationary targets are critical challenges to effective vision. Evolutionary pressures have shaped the eye movement systems of animals to meet this challenge in ways that are tailored to the visual structures and behavioral needs of each species. In this chapter we examine the behavioral and neural systems used by vertebrates to control eye movements in the support of vision.

As we shall see, the control of eye movements provides a window into several fundamental aspects of neuroscience. Because the eyes are a relatively simple mechanical system, eye movements provide an excellent system for investigating the neural mechanisms of motor control, and also offer some of the clearest views of other complex processes, such as neural plasticity, sensory-motor interactions, and the higher-order control of behavior.

EYE MOVEMENTS ARE USED TO STABILIZE GAZE OR TO SHIFT GAZE

Similar to the handling of a camera, the images that fall on the retina depend on how the eyes are held and moved. The orientation of the eyes in the head and the orientation and position of the head in space together determine the gaze direction (i.e., gaze = eye + head) and, consequently, control the retinal image. These

spatial arrangements underlie the two general classes of eye movements in vertebrates. One class is responsible for *gaze stabilization*. When animals move with respect to their surroundings, they risk degrading their visual acuity, because moving the head necessarily moves the eyes, and can cause images to streak across the retina. Gaze stabilization mechanisms have evolved to solve this problem. They maintain visual acuity during self-motion by stabilizing the retinal image of the world with rotations of the eyes that exactly compensate for head and body movements. The neural mechanisms for gaze stabilization are highly conserved across vertebrates, reflecting the widespread need to stabilize visual inputs despite other sensory and motor differences between species.

The second class is responsible for *gaze shifting*. When animals visually inspect their surroundings, they may use these eye movements to direct the line of sight to objects or features of particular interest, even when those items move. In contrast to stabilization mechanisms, gaze shifting mechanisms involve selectively sampling the animal's visual environment, because tracking one visual object often causes the retinal images of other objects to become blurred. Accordingly, the mechanisms for gaze shifting are found only in animals that have retinal specializations, such as the primate fovea, that can be used to examine a limited region of visual space at a higher spatial resolution.

Gaze Stabilization

There are two types of mechanisms for stabilizing gaze (Fig. 33.1): the vestibulo-ocular reflex (VOR) and the optokinetic response (OKR). The VOR uses signals from the vestibular labyrinth to counter-rotate the eyeballs to keep retinal images stable during head move-

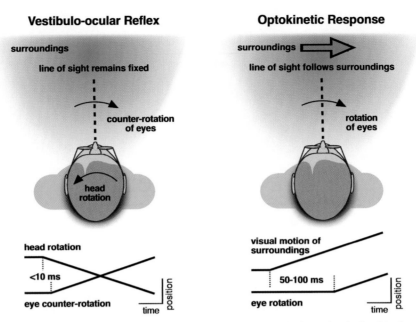

FIGURE 33.1 Eye movements that stabilize gaze. The vestibulo-ocular reflex keeps the line of sight fixed in the world by counter-rotating the eyes during movements of the head. Here, the eyes rotate rightward at a short latency after the beginning of the leftward head movement. The optokinetic response stabilizes the line of sight with respect to the moving visual surround, but does so after a longer latency.

ments. The *rotational VOR* is driven by head-rotation signals from the semicircular canals and represents a phylogenetically old reflex present in all vertebrate species. The fast transduction properties of the vestibular periphery, combined with a direct 3-neuron pathway, give the rotational VOR a very short latency (<10 ms). However, because the canals do not accurately transduce constant velocity head rotations, the output of the VOR decays over several seconds. Some vertebrates also possess a *translational VOR*, which is driven by head-movement signals from the otoliths. Unlike the image motion caused by head rotations, which is uniform across the visual field and can be fully compensated for by counter-rotating the eyes, the image motion caused by translations are complex and can only partly be eliminated by eye movements. This explains why the translational VOR is found only in animals with central retinal specializations (e.g., foveal vision).

The optokinetic response (OKR) is driven by visual signals about the en masse movement of images across the retina. Because of the slow transduction of visual signals, the OKR has a longer latency (50–100 ms) than the VOR, but the use of visual motion signals makes it able to compensate for the constant motions and low speeds that the VOR cannot. The properties of the OKR therefore complement those of the VOR in maintaining stable gaze during movements, and the visual signals used by the OKR play a major role in guiding adaptive changes in the VOR. Animals with foveal vision also possess an ocular following response that is driven at very short latencies (~50 ms) by shifts in

the retinal image, and that serves as the visual complement to the translational VOR (Miles, 1997).

Gaze Shifting

Animals with foveal vision also possess several mechanisms for shifting gaze (Fig. 33.2). Saccadic eye movements act to quickly move the image of a visual target from an eccentric retinal location to the center of the retina where it can be seen best. Saccades are very fast movements, reaching peak speeds greater than 500 deg/s and lasting only tens of milliseconds. Consequently, although saccades often are triggered by visual stimuli (at typical latencies of 150–250 ms), the trajectory of saccades is not guided by visual feedback, but instead follows a stereotyped profile that is largely determined by the size of the movement. Saccades do not require a visual stimulus, but can be guided by other modalities, and can even be directed toward the location of imagined or remembered targets.

Smooth pursuit eye movements slowly rotate the eyes to compensate for any motion of the visual target, and thus act to minimize the blurring of the target's retinal image that would otherwise occur. Smooth pursuit is a relatively slow movement whose trajectory is determined by the moving stimulus. Pursuit is triggered by visual stimuli at latencies similar to, but somewhat shorter than, the latencies of saccades (typically 100–200 ms), and generates eye velocities up to ~50 deg/s, continuously adjusted by visual feedback about the target's retinal image. Unlike saccades,

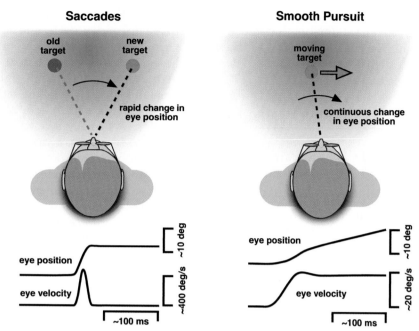

FIGURE 33.2 Eye movements that shift gaze. Saccades shift the line of sight to place the retinal image of visual targets onto the fovea. They are characterized by rapid changes in eye position (upward deflection in eye position trace) involving very high eye velocities (brief pulse in eye velocity trace). Smooth pursuit continuously changes the line of sight to minimize blurring of the target's retinal image. These movements are characterized by smooth and continuous changes in eye position (ramp in eye position trace) involving lower eye velocities (smooth step in eye velocity trace).

pursuit cannot be generated in the absence of a stimulus, although moving targets that are sensed through modalities other than vision can also guide pursuit.

Saccades and smooth pursuit produce conjugate, or yoked, movements of the two eyes. These are also referred to as *versional* eye movements. In contrast, a third type of mechanism for shifting gaze, called *vergence* eye movements, produces disconjugate movements—the two eyes move in opposite directions or by different amounts. Vergence eye movements act to change the depth at which the eyes' lines of sight meet. When we look at a near object our eyes converge (rotate inward), and when we look at a far object our eyes diverge (rotate outward). Before shifting our gaze to an object at a different distance, the pair of images produced by that object will be spatially mismatched between the two eyes, causing binocular disparity. Also, because the curvature of the two eyes' lenses are set to focus the currently viewed object, the image of the new object will also be blurred. These two signals, binocular disparity and blur, drive the combination of vergence eye movements and *accommodation* (which changes the dioptric power of the lenses by contracting or relaxing the ciliary muscle) that bring the new object into focus on the fovea of both eyes. Vergence and accommodation are tightly linked and often are referred together as the *near response*. When evoked by themselves (e.g., by binocular disparity), vergence movements are very slow, taking up to 1 second to

converge or diverge the eyes. However, when they occur in combination with faster versional eye movements such as saccades, vergence movements speed up, and interestingly, so does accommodation.

Fixation and Fixational Eye Movements

In between gaze shifts, the eyes are actively held steady by a fixation mechanism. By preventing the eyes from shifting gaze to another visual target, fixation makes it possible to visually inspect a particular object at length. Periods of fixation typically last about 200 ms, but vary depending on what the animal is doing. Fixation does not necessarily act to keep the eyes from moving, but to hold the eyes steady with respect to the environment. For example, gaze stabilization mechanisms such as the VOR are active during periods of fixation.

Even during periods of steady fixation of a stationary object, the eyes are not completely still. The eyes typically make tiny movements during fixation in seemingly random directions that cause very slight changes in how images fall on the retinal mosaic but without displacing images from the fovea. *Microsaccades* are very small saccades with amplitudes typically less than 0.1 degree of visual angle, and often are interspersed with slow drifts (speeds less than ¼ degree per second) of similar amplitude. The function of these movements remains unclear. They may prevent the adaptation of retinal signals that would occur with perfectly stabilized images, but the frequency of these small movements is in fact reduced during tasks that place high demands on foveal vision, such as threading a needle.

THE MECHANICS OF MOVING THE EYES

Each Eye Is Controlled by Three Pairs of Muscles

Moving the eye involves rotating the globe of the eye in the orbital socket of the skull. Because the eye undergoes minimal translation during movement, it can be viewed as a spherical joint with an orientation defined by three axes of rotation (horizontal, vertical, and torsional). Correspondingly, the orientation and movement of each eye is accomplished by three complementary pairs of extraocular muscles (Fig. 33.3). The lateral and medial rectus muscles form an antagonistic pair that controls the horizontal position of the eye. Contraction of the lateral rectus (and commensurate relaxation of the medial rectus) causes the eye to abduct, or rotate outward. Conversely, contraction of the medial rectus (and relaxation of the lateral rectus) causes the eye to adduct, or rotate inward. The actions of the other two pairs of muscles are more complex. When the eye is centered in the orbit, the primary

effect of the superior and inferior recti is to elevate (rotate up) or depress (rotate down) the eye. However, when the eye is deviated horizontally in the orbit, these muscles also contribute to torsion, the rotation of the eye around the line of sight that determines the orientation of images on the retina. The primary effect of the superior and inferior obliques is intortion (the top of the eye rotates toward the nose) or extortion (the top of the eye rotates away from the nose), but depending on the horizontal deviation of the eye, these muscles may also help determine the vertical orientation of the eye. Eye movements therefore involve redefining the three-dimensional orientation of the eye by changing the balance of forces applied to the globe by all three pairs of extraocular muscles.

The mechanics of moving the eye may seem simple, but the details are quite complicated. Because the eye is free to rotate in three dimensions, the final orientation of the eye depends on the order in which the rotations are carried out; for example, "turn-then-spin" gives a different outcome than "spin-then-turn." This property is referred to as *noncommutativity*. In fact, when the eye moves, the final three-dimensional orientation of the eye and the axis of rotation used to get there are not arbitrary, but form an orderly set. *Donder's law* states that the orientation of the eye is always the same when the eye is aimed in a particular direction. *Listing's law* specifies these orientations by stating that the axes used to rotate the eye are confined to a single plane, referred to as Listing's plane. These orderly relationships are achieved, at least in part, by a mechanism that involves the muscles themselves. The extraocular muscles consist of two distinct layers: global fibers, which attach directly to the globe of the eye, and orbital fibers, which terminate in connective tissue that ensheathes the body of the muscle and acts like a moveable pulley for the global fibers. Because the pulleys themselves move when the muscle is contracted or relaxed, the pulling direction of the extraocular muscles changes with eye position. These gaze-dependent mechanical changes in muscle action largely account for Listing's law (Demer, 2006).

Interestingly, Listing's law is obeyed during saccades and smooth pursuit, but is violated during the VOR and the OKR. This difference presumably stems from the different inputs for these movements. Saccades and smooth pursuit may follow Listing's law because they involve local two-dimensional retinal inputs that do not uniquely specify the orientation of the eye. In contrast, the VOR is driven by three-dimensional inputs from the semicircular canals for which the extra degree of freedom is not redundant. Thus, the mechanics of the eye muscles impose kinematic constraints that are helpful during visually guided eye movements, but

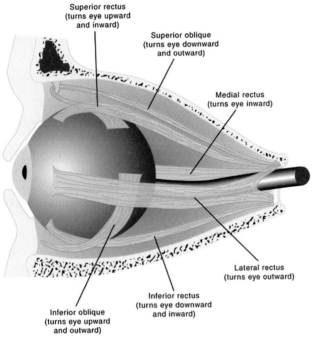

Superior rectus
(turns eye upward and inward)

Superior oblique
(turns eye downward and outward)

Medial rectus
(turns eye inward)

Lateral rectus
(turns eye outward)

Inferior rectus
(turns eye downward and inward)

Inferior oblique
(turns eye upward and outward)

FIGURE 33.3 Muscles of the eye. Six muscles, arranged in three pairs, control the movements of the eye as shown here in a cutaway view of the eye in its socket, or orbit.

these constraints can be altered or lifted by neural control signals during gaze stabilization.

Central Control of the Extraocular Muscles

The six extraocular muscles are innervated by three of the bilaterally paired cranial nerves (Fig. 33.4). The oculomotor nerve (cranial nerve III) innervates the medial rectus, the superior and inferior rectus, and the inferior oblique on one side of the head. The cell bodies for these motor neurons thus lie in the third cranial nerve nucleus. The oculomotor nerve also contains the fibers that control accommodation of the lens. The trochlear nerve (cranial nerve IV) innervates the superior oblique muscle, and the abducens nerve (cranial nerve VI) innervates the lateral rectus. These three pairs of nuclei, distributed through the brain stem, contain all the oculomotor motor neurons and are heavily interconnected by a pathway called the medial longitudinal fasciculus. This interconnection is crucial for the precise coordination of the extraocular muscles that is necessary for the control of eye movements.

The firing rates of the oculomotor motor neurons determine the forces applied by the extraocular muscles and, consequently, whether the eye moves or is held in place (Fuchs and Luschei, 1970). When the eye is stationary, the firing rate of oculomotor motor neurons is proportional to eye position, because under static conditions the balance of forces exerted on the globe determines the orientation of the eye (Fig. 33.5). Each motor neuron has a recruitment point (an eye position at which the motor neuron begins to fire) and a characteristic slope (the change in firing rate as the eye rotates in the pulling direction of the relevant muscle). Thus, as in the skeletal system, oculomotor motor neurons have a fixed recruitment order. When the eye moves in the pulling direction of the motor unit, oculomotor motor neurons exhibit a high-frequency pulse of activity that drives the high-velocity phase of the eye movement. The high-frequency pulse is followed by a sustained step-like increase in firing rate appropriate for the new static position of the eye. When the eye moves in the opposite direction, the motor neurons pause their firing during the movement and then resume their activity at a lower sustained level.

The neural commands from these motor neurons therefore take the form of a pulse-step during rapid eye movements such as saccades. The amplitude and duration of the pulse determine the speed and duration of the movement, and the amplitude of the step determines the final static eye position. The need for the pulse derives from the sluggish dynamics of the

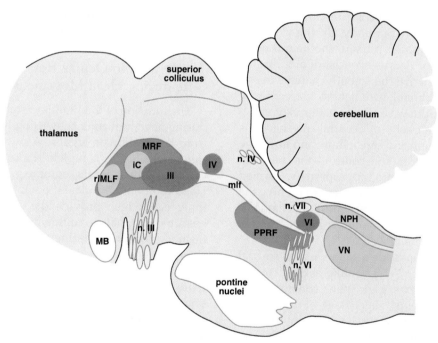

FIGURE 33.4 Oculomotor nuclei in the brainstem. Parasagittal section through the brainstem, cerebellum, and thalamus of a rhesus monkey, showing the location of the major brainstem nuclei involved in the control of eye movements. Motor neurons for the eye muscles are located in the oculomotor nucleus (III), trochlear nucleus (IV), and abducens nucleus (VI), and reach the extraocular muscles via the corresponding nerves (n. III, n. IV, n. VI). Premotor neurons for controlling eye movements are located in the paramedian pontine reticular formation (PPRF), the mesencephalic reticular formation (MRF), rostral interstitial nucleus of the medial longitudinal fasciculus (riMLF), the interstitial nucleus of Cajal (IC), the vestibular nuclei (VN), and the nucleus prepositus hypoglossi (NPH). The medial longitudinal fasciculus (mlf) is a major fiber tract containing the axons of these neurons. Other abbreviation: MB, mammillary body of the hypothalamus.

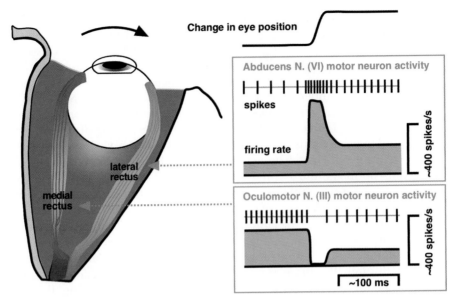

FIGURE 33.5 The firing rate patterns of eye movement motor neurons during horizontal eye movements. The rapid rightward change in eye position, indicted by the upward deflection in the eye position trace, involves contraction of the right lateral rectus and relaxation of the right medial rectus. This is accomplished by a pulse-step increase in activity of the agonist motor neurons in the abducens nucleus, and a pause-step decrease in activity of the antagonist motor neurons in the oculomotor nucleus.

eye and orbital tissues, which is dominated by the viscosity of the extraocular muscles. If the eye were moved by only a step change in force, it would take nearly a second for the eye to settle at its new orientation, which would obviously be incompatible with the goal of supporting vision. Similarly, during slow eye movements such as smooth pursuit, the neural commands from the motor neurons consist of a ramp-like change in activity to drive the smooth changes in eye position, plus an additional increment to overcome the sluggish dynamics of the eye. These patterns of activity are adjusted precisely by central mechanisms involving the brain stem and cerebellum, and illustrate a general problem in motor control—the need to convert desired actions into appropriately formed motor commands. The solution appears to involve the use of internal models that allow central mechanisms to take into account the dynamic properties of the body part or object to be moved.

Oculomotor motor units are also unusual in several ways (Buttner-Ennever, 2007). Unlike skeletal motor units, eye motor neurons participate equally in all types of eye movements, regardless of the speed involved. The extraocular muscles contain muscles spindles, but no Golgi tendon organs, and there are no ocular stretch reflexes. Instead, eye proprioception appears to come mainly from palisade endings on the eye muscle tendons. The function of this input is unclear, but it is distributed via the trigeminal nucleus to many key structures in the oculomotor system (including the vestibular nuclei, cerebellum, superior colliculus, and frontal eye fields) and may play a role in the long-term adaptive control of eye movements.

THE FUNDAMENTAL CIRCUITS FOR STABILIZING GAZE

The Vestibulo-Ocular Reflex (VOR) Stabilizes Gaze during Head Movements

The function of the VOR is to stabilize retinal images during head movements by using head-motion signals from the vestibular organs. A purely rotational VOR can be elicited in complete darkness by rotating the head roughly about its center (e.g., the interaural axis). Head rotations stimulate the semicircular canals of the inner ear, which provide the sensory signals about head motion that drive this highly conserved mechanisms for counter-rotating the eyes. During sustained rotations of the head, the eyes cannot continue to counter-rotate endlessly in the orbit, but are intermittently and rapidly reset to a central position during the ongoing compensatory eye rotation. The profile of eye position over time during the VOR therefore exhibits a characteristic saw-tooth pattern consisting of the slow-phase counter-rotations and quick-phase resetting eye movements. This pattern is called *nystagmus*, and is described by the direction of the quick phases (Fig. 33.6 is an example of right-beat nystagmus).

The speed of the slow phases during nystagmus indicates how effectively the VOR is compensating for

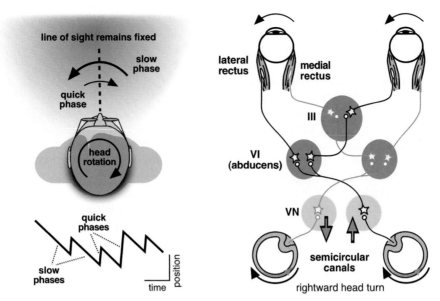

FIGURE 33.6 The eye movements and neural circuits for the VOR. During a sustained rotation of the head (here, rightward), the eyes rotate slowly to the left (downward deflections in eye position trace labeled "slow phases"), and then abruptly shift rightward to reset the eyes toward a central position (upward deflections in eye position trace labeled "quick phases"). This is an example of right-beat nystagmus. The diagram of the basic VOR circuit shows the push-pull organization of the horizontal VOR. When the head rotates rightward, activity increases in the right semicircular canal, right vestibular nucleus (VN), and left abducens nucleus (VI). This leads to increased activation of the left eye lateral rectus and, through a crossed projection to the oculomotor nucleus (III), increased activation of the right eye medial rectus. The complementary set of pathways (colored in gray) show decreases in activity.

the head rotation. The degree of compensation often is characterized by the VOR gain, which is defined as the ratio of slow-phase eye speed over the speed of head rotation. Perfect compensation occurs when the eye rotates in the opposite direction at exactly the same speed as the head rotation, which corresponds to a gain of 1. The ability of the VOR to compensate for head motion is limited by the dynamics of the signals from the semicircular canals, which act as "high-pass" sensors of head velocity. Thus, the VOR is very effective (its gain is near 1) for fast back-and-forth head rotations that take a second or less to complete (i.e., frequencies over 1 Hz), but it is much less effective for slow head rotations that continue over many seconds (e.g., 0.1 Hz). Under natural conditions, this limitation of the VOR does not pose a problem for vision, because at these lower speeds visual signals can be used to stabilize gaze very effectively through the optokinetic response.

The fundamental neural mechanism for the VOR consists of brain stem circuits linking the vestibular afferents to the extraocular muscles, including a direct 3-neuron arc pathway (Fig. 33.6). For example, during a rightward head rotation, vestibular afferents from the horizontal semicircular canal on the right side increase their firing rates, and this increase is passed on to neurons in the vestibular nucleus on the same side. These vestibular nucleus neurons, in turn, increase the activity of motor neurons and interneurons in the abducens nucleus on the opposite (left) side. Motor neurons

in the abducens innervate the lateral rectus of the left eye, completing the three-neuron arc. Interneurons in the abducens provide a crossed projection back to the right oculomotor nucleus, where motor neurons innervate the medial rectus of the right eye. By contracting both the left lateral rectus and the right medial rectus, the two eyes rotate leftward together in response to the rightward head rotation. In addition, because the vestibular afferents on the left side decrease their activity during a rightward rotation, a complementary circuit causes relaxation of the right lateral rectus and the left medial rectus. The VOR therefore is controlled in a push-pull fashion that yokes the movements of the two eyes through the bilateral symmetry of these basic circuits.

These direct pathways are key components of the VOR mechanism, but the VOR also requires indirect pathways to function correctly. As discussed in the previous section, the firing rate of oculomotor motor neurons is related primarily to eye position, but the afferent signals from the semicircular canals are related primarily to the head velocity. In order to construct the motor commands needed to hold, as well as move, the eyes during the VOR, the brain needs to transform the dynamic signals provided by the vestibular afferents into the combination of dynamic and static commands found on motor neurons. Similar to the operation of mathematical integration in calculus, this process involves the temporal integration of velocity inputs into position signals.

The neural integrator for eye movements is accomplished by neurons in the vestibular nuclei and two other brain stem nuclei—the nucleus prepositus hypoglossi (for horizontal components) and the interstitial nucleus of Cajal (for vertical)—and is fine-tuned by connections between these brain stem nuclei and the vestibulocerebellum (Cannon and Robinson, 1987). The output from the neural integrator pathway sums with the signals from the direct pathway to provide the motor neurons with the complete set of velocity and position commands needed to accomplish the VOR. Importantly, all conjugate eye movements also use this same indirect integrator pathway. Thus, although the source of the velocity signals for the VOR, OKR, saccades, and pursuit are quite different, they appear to share a common brain stem mechanism for transforming these signals into the final motor commands.

The Optokinetic Response (OKR) Uses Visual Inputs to Stabilize Gaze

The OKR also functions to stabilize retinal images during head movements, but uses visual inputs to infer the direction and speed of head motion rather than responding directly to head-velocity signals. The optimal stimulus for the OKR is full-field motion of the visual environment that, like head rotations, causes a saw-tooth pattern of eye movements referred to as *optokinetic nystagmus*. Like the VOR, the slow phase movements during nystagmus compensate for the head rotation that presumably caused the full-field motion. However, because the OKR is driven by visual signals, which are processed much more slowly than vestibular signals, the range of motions that are effectively compensated by the OKR is quite different from the VOR. The OKR is very effective (its gain is near 1) for slow rotations that continue over many seconds, but it is much less effective for very fast back-and-forth rotations. These properties of the OKR therefore complement those of the VOR and together, the two types of mechanisms can effectively stabilize gaze over a much wider range of speeds and behavioral conditions than either could achieve alone.

The interlocking nature of the two systems is also evident in the neural pathways for the OKR, despite the difference in their input signals. In vertebrate species without foveal vision, such as the rabbit, the visual processing for the OKR is accomplished entirely within the brain stem. Direction-selective retinal ganglion cells project directly to a set of visual nuclei in the midbrain called the pretectum and accessory optic nuclei. In species with foveal vision, the visual properties of these nuclei are determined not only by inputs from the retina, but also are shaped by inputs from visual areas of the cerebral cortex. Neurons in the accessory optic nuclei have very large receptive fields and respond selectively to global visual motion that would result from rotating the head about a particular spatial axis. Remarkably, the selectivity of individual neurons is restricted to one of the three axes of rotation that are also detected by the semicircular canals. This common rotational coordinate system makes it easier to combine visual *reafferent* signals about head motion from the accessory optic nuclei with the vestibular *afferent* signals provided by the labyrinths. Accordingly, neurons in the pretectum and accessory optic nuclei project to the vestibular nuclei and nucleus prepositus hypoglossi, where these visual signals join the same final pathways used by the VOR. These nuclei also project to the vestibulocerebellum, where the visual signals contribute to the velocity commands for the OKR and other smooth eye movements, and also act as feedback signals that play a critical role in the long-term adjustment of the VOR.

THE COMMANDS FOR SHIFTING GAZE ARE FORMED IN THE BRAIN STEM

The commands for shifting gaze use the same final motor pathways that are used for stabilizing gaze, but the signals and mechanisms giving rise to the gaze-shifting commands are quite different. For smooth pursuit, they are based on motion signals about the visual target that are first extracted in specialized areas of the cerebral cortex and then used to commandeer the brain stem and cerebellar pathways for the OKR. For saccades, which are not guided by visual feedback, a specialized circuit in the brain stem reticular formation produces the stereotyped velocity command that determines the trajectory of the saccade.

The Motor Commands for Saccades Are Constructed in the Brain Stem Reticular Formation

The control signals for saccades descending from higher centers specify the location of the target, which is then transformed by circuits in the brain stem reticular formation into the motor commands for saccades (Fig. 33.7). The horizontal and vertical components of saccades are formed by separate but interconnected circuits—the horizontal commands are formed in the paramedian pontine reticular formation (PPRF), and the vertical commands are formed in a part of the mesencephalic reticular formation called the rostral interstitial nucleus of the medial longitudinal fasciculus (riMLF).

Within these brain stem circuits, often referred to as the *saccadic burst generator*, several different classes of neurons work together to construct the saccade motor

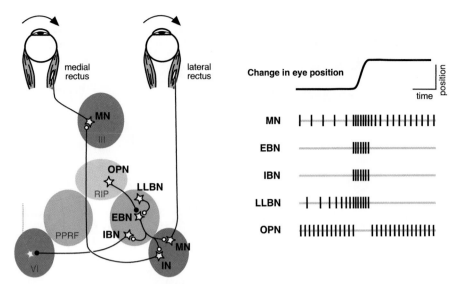

FIGURE 33.7 The brainstem circuits for controlling saccadic eye movements. This diagram of the saccade burst generator outlines the major classes of neurons involved in constructing the motor command for horizontal saccades. Omnipause neurons (OPN) located near the midline in the nucleus raphe interpositus (RIP) tonically inhibit excitatory burst neurons (EBN) located in the paramedian pontine reticular formation (PPRF). When OPNs pause, the EBNs emit a burst of spikes, which activate motor neurons (MN) in the abducens nucleus (VI) innervating the lateral rectus muscle. The burst also activates interneurons (IN) which, in turn, activate motor neurons on the oculomotor nucleus (III) on the opposite side, innervating the medial rectus. Inhibitory burst neurons (IBN) show a pattern of activity similar to EBNs, but provide inhibitory inputs to decrease activation in the complementary circuits and antagonist muscles. Long-lead burst neurons (LLBN) show activity long before movement onset, and provide an excitatory input to EBNs.

command (Keller, 1974). *Burst neurons* fire a high-frequency volley of action potentials that begins ~10 ms before the onset of the saccade and ends slightly before the saccade lands on the target. The firing rate during the burst is related to the instantaneous velocity of the saccade, and the number of spikes in the burst is related to the amplitude of the saccade. There are several types of burst neurons. Excitatory burst neurons (EBNs), sometimes called "medium-lead burst neurons," provide the velocity signal for the eye muscles that will contract during the saccade, and specify the "pulse" portion of the saccade motor command. The EBNs provide a copy of this velocity signal to the brain stem nuclei of the neural integrator, which then generate the "step" portion of the motor command, just like they do for other eye movements. EBNs also contact inhibitory burst neurons (IBNs), which provide a crossed projection that inhibits motor neurons for the antagonist eye muscles that will relax during the saccade. Finally, long-lead burst neurons show changes in their activity well in advance of the saccade, presumably reflecting inputs received from higher centers, and provide an excitatory input to the EBNs.

The other major class of neurons in the saccadic burst generator are the *omnipause neurons* (OPNs). These neurons fire at a constant high rate (~100 spikes/s) during fixation, but completely pause their activity just before and during saccades in all directions. Because the OPNs directly inhibit both EBNs and IBNs, their tonic activity prevents unwanted saccades during fixation, and the pause in their activity unleashes the synchronized bursts of action potentials that initiate and drive the saccade. The onset of the pause in OPN activity appears to be triggered by inhibitory signals originating from higher brain centers, but the duration of the pause does not determine when the saccade ends. Instead, stopping saccades involves another inhibitory influence on burst neurons. One likely candidate is the oculomotor cerebellum (lobules VI and VII of the vermis and the caudal fastigial nucleus). The oculomotor vermis is interconnected with the saccadic burst generator and plays a crucial role in adjusting the metrics of saccades so that the eyes land accurately on the target at the end of the gaze shift.

The Motor Commands for Pursuit Are Formed in the Brain Stem and Cerebellum

The control signals for pursuit descending from higher centers continuously specify the motion of the target to be followed with the eyes. Thus, in contrast to saccades, pursuit does not require motor circuitry to construct the velocity command from scratch. Instead, the pursuit pathways in the brain stem and cerebellum sculpt the descending control signals to form an appropriately shaped velocity command, and provide a copy of this velocity signal to the neural integrator to generate the position component of the motor command.

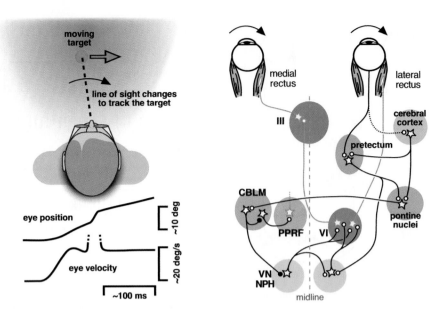

FIGURE 33.8 The brainstem and cerebellar circuits for controlling pursuit eye movements. During pursuit, the line of sight changes smoothly to follow the motion of the target. The smooth changes in eye velocity are often accompanied by small "corrective" saccades, indicated by the upward deflection in the eye velocity trace (dotted lines). The diagram outlines the circuits involved in constructing the motor command for horizontal pursuit. The major pathways involve projections from areas of the cerebral cortex via the pontine nuclei to the cerebellum (CBLM), which then modulates activity in nuclei associated with the VOR – namely, the vestibular nuclei (VN) and nucleus prepositus hypoglossi (NPH) – and also modulates activity in nuclei associated with saccades, such as the paramedian pontine reticular formation (PPRF). A second pathway involves projections from the pretectum to the pontine nuclei, as well as a direct projection to the vestibular nuclei and prepositus hypoglossi.

The velocity commands for pursuit arise in areas of the cerebral cortex and reach the final motor circuits through several routes (Fig. 33.8). One pair of pathways involves projections from the cerebral cortex to the cerebellum via a large set of relay neurons on the floor of the brain stem called the basal pontine nuclei. Areas in extrastriate visual cortex provide signals related to visual motion through the dorsolateral pontine nuclei to a portion of the vestibulocerebellum called the ventral paraflocculus. The frontal cortex provides signals related to target and eye velocity through the dorsomedial pontine nuclei, and the adjacent nucleus reticularis tegmenti pontis, to the oculomotor cerebellum (vermis), the same portion of the cerebellum involved in adjusting the metrics of saccades. Both of these cerebellar regions provide inputs to parts of the brain stem motor pathways that we already have discussed. The output of the ventral paraflocculus is closely related to eye velocity and terminates in the vestibular nuclei, where it contributes to the velocity command for pursuit, and also provides an input to the neural integrator. The output of the oculomotor vermis is less well defined, but appears to be related to eye acceleration as well as eye velocity. The outputs of the vermis and the associated deep cerebellar nucleus (the caudal fastigial nucleus) are sent to neurons in the brain stem reticular formation, and may play a role in shaping the velocity command appropriately when the target motion changes.

Another route for conveying motion signals for pursuit involves the pretectum. As described earlier, the pretectum plays a key role in the OKR, but in primates the pretectum receives inputs from areas of the cerebral cortex that are important for pursuit. Accordingly, some neurons in the primate pretectum have smaller receptive fields and respond well to the small moving stimuli typically used to elicit pursuit. The pretectum provides an input to the cerebellum through the basal pontine nuclei, but also projects directly to the vestibular nuclei and nucleus prepositus hypoglossi, offering a relatively direct route to the final motor pathways.

The brain stem control of pursuit also appears to involve some of the same neurons that are part of the saccadic burst generator. The EBNs and IBNs are not involved in forming the pursuit motor command, but some of the long-lead burst neurons in the brain stem reticular formation increase their tonic firing during pursuit, and show a build-up of activity before the onset of pursuit and saccades. Even more surprisingly, OPNs—the "gatekeepers" for saccades—are also modulated during pursuit. They do not completely pause their activity as they do during saccades, but decrease their activity by about one-third. Moreover, electrical stimulation applied to the OPNs slows down pursuit eye movements, in addition to stopping saccades (Missal and Keller, 2002). The function of this overlap between saccades and pursuit in these premotor circuits is not yet fully understood, but one possibility is that descending information about the target does not directly trigger saccades and pursuit, but instead a motor decision process at the level of the brain stem and cerebellum determines which type or combination of movements is most appropriate under the particular circumstances.

GAZE SHIFTS ARE CONTROLLED BY THE MIDBRAIN AND FOREBRAIN

The motor commands for gaze shifts formed in the brain stem depend on control signals descending from higher centers, in particular, from the superior colliculus (SC) and several eye-movement related areas of the cerebral cortex (Fig. 33.9). These brain regions make it possible to bring the well-honed motor circuits in the brain stem under the control of higher-order processes such as attention, perception, and cognition.

The Superior Colliculus Contains a Retinotopic Map for Controlling Gaze

The superior colliculus (SC) is a multilayered structure lying on the roof of the midbrain. In nonmammals, the homologous structure is called the optic tectum, which serves as the primary center for processing retinal inputs and for transforming sensory inputs into

commands for orienting movements. In mammals, the SC has a similar functional role, and consists of superficial layers that receive direct inputs from the retina and striate cortex, and intermediate and deep layers that receive inputs from extrastriate, parietal, and frontal cortex, and the basal ganglia. Neurons in the superficial layers have responses to visual stimuli that are enhanced when the stimulus is the target of a saccade. Neurons in the intermediate and deep layers may also respond to visual stimuli, and also to auditory stimuli from corresponding locations, so that the visual and auditory maps of the animal's surroundings are in register. It is these intermediate and deep layers that also possess the movement-related activity that plays a crucial role in the control of gaze shifts (Wurtz and Goldberg, 1972). These movement-related layers provide a major descending projection to the oculomotor nuclei in the brain stem reticular formation, and also a major ascending projection to the frontal cortex through the medial dorsal nucleus of the thalamus.

The most distinctive feature of the superior colliculus is that it contains a retinotopic map of contralateral visual space and this topography applies to the movement-related activity in the deeper layers as well as to the visual activity in the superficial layers (Fig. 33.10). The significance of this retinotopic map becomes evident when neural activity is artificially manipulated by passing small amounts of electrical current through the tip of an electrode placed in the intermediate layers (Robinson, 1972). Such stimulation elicits a saccadic eye movement with an amplitude and direction that depends on the placement of the electrode in the map. If the stimulation is applied as a long train, rather than as a brief pulse, a series of saccades is elicited, forming a "staircase" of eye movements, each with the same characteristic amplitude and direction. Thus, activity in the superior colliculus map is related to changes in eye position and not with absolute eye position, consistent with the organization of the superior colliculus in retinotopic (or oculocentric) coordinates.

During naturally occurring eye movements, activity in the superior colliculus is not restricted to just the few neurons that best match the endpoint of the saccade, but is distributed across a large number of neurons across the retinotopic map. This point was elegantly demonstrated by testing what happens to saccades when the injection of chemical agents into the superior colliculus temporarily inactivates neurons at a particular location in the map (Lee *et al.*, 1988). Remarkably, it is still possible to make saccades directed to the center of the inactivated region, because the broad distribution of activity across the map leaves a halo of neurons that can still accurately guide the saccade. Moreover, saccades directed to locations slightly beyond the inactivated site

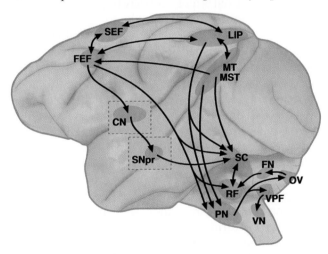

FIGURE 33.9 The descending pathways for controlling gaze shifts, depicted schematically on a lateral view of the monkey brain. The superior colliculus (SC), located on the roof of the midbrain, is a major source of descending control signals. In the cerebral cortex, the major areas involved are the frontal eye fields (FEF), supplementary eye fields (SEF), lateral intraparietal area (LIP), middle temporal area (MT), and medial superior temporal area (MST). In the basal ganglia, a cascade through the caudate nucleus (CN) and substantia nigra pars reticulata (SNpr) provides inhibitory control over activity in the superior colliculus. These descending control signals are then converted into motor commands by circuits involving regions such as the reticular formation (RF), pontine nuclei (PN), vestibular nuclei (VN), and parts of the cerebellum such as the ventral paraflocculus (VPF), oculomotor vermis (OV) and fastigial nucleus (FN).

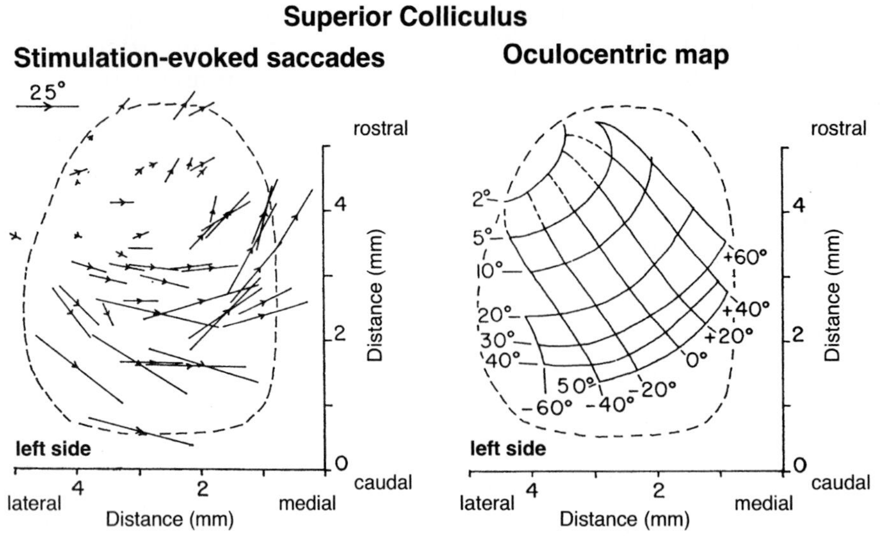

Superior Colliculus

Stimulation-evoked saccades   **Oculocentric map**

FIGURE 33.10   The superior colliculus and its retinotopic (or oculocentric) map of saccades and target positions. The left plot shows that electrical stimulation in the superior colliculus causes eye movements whose direction and amplitude depends on the location. Each arrow corresponds to a stimulation site in the left superior colliculus, and the length and direction of the arrow indicate the metrics of the evoked saccade. The evoked saccades are all directed rightward, but rostral sites are associated with smaller movements than caudal sites, and medial sites are associated with upward movements whereas lateral sites are associated with downward movements. The right plot summarizes these results in the form of a continuous map of saccades and target positions. As in the visual cortex, the representation of the center of the visual field is amplified. Adapted from Robinson (1972).

become hypermetric (bigger than normal), because the inactivation removes the contribution from neurons that code for more proximal locations and smaller saccades. These results show that saccades are guided by the vector average of population activity in the superior colliculus, and were the first demonstration of vector averaging in a mammalian motor system.

The intermediate layers of the superior colliculus contain two main classes of neurons related to the control of gaze (Munoz and Wurtz, 1995). Burst neurons in the superior colliculus fire a volley of action potentials that begins ~20 ms before the onset of saccades, somewhat earlier than the burst neurons located downstream in the reticular formation, and often also have a visual response to the onset of visual targets. Unlike the burst neurons in the reticular formation, those in the superior colliculus fire for saccades directed to a portion of the visual field referred to as the neuron's movement field, and the size of the burst depends on where the target falls with respect to the neuron's movement field. Buildup neurons also have visual responses and saccade-related activity, but they are distinguished by the presence of sustained activity when the subject is required to delay eye movement response after the presentation of a visual stimulus. This sustained activity typically increases over time as the time of the saccade draws near, and is believed to be important for the preparation of saccades, perhaps by facilitating the occurrence of the saccade-related burst. Recent results show that the activity of neurons in these layers of the

superior colliculus not only is involved in motor preparation, but also is involved in the antecedent step of selecting the visual target, and may be part of the neural circuits that control visual attention.

In the part of the superior colliculus map that represents the central visual field, the neuronal activity is somewhat different. Like the rest of the map, this part also contains burst neurons that fire during saccades, although the preferred saccades are smaller, as would be expected. In place of buildup activity, however, there are neurons that show tonic activity during fixation and decrease their activity during most saccades (Munoz and Wurtz, 1993). Because of this fixation-related activity, and a prominent projection to the omnipause neurons, this region has been referred to as a fixation zone. However, these same neurons increase their activity during small contralateral saccades and are also modulated during pursuit eye movements, which, like fixation, also involve foveal target positions (Krauzlis et al., 1997). Although the issue is not yet settled, this argues that these neurons do not provide a separate motor command for fixation, but instead represent the foveal region of a single target position map extending across the superior colliculus.

The Frontal Eye Field Is the Primary Cortical Area for Controlling Gaze Shifts

The frontal eye field (FEF) exerts its control on gaze shifts through several descending pathways, including a direct projection to the intermediate layers of the superior colliculus, an indirect projection to the superior colliculus via the basal ganglia, and also projections to the cerebellum and the brain stem reticular formation. The FEF is best known for its role in the generation of saccades, but it also contains zones that are involved in the control of pursuit and vergence eye movements. Like the superior colliculus, electrical stimulation of the FEF causes saccades to a particular location in the contrateral visual hemifield, although the topography in the FEF is less orderly than in the superior colliculus. Electrical stimulation of the pursuit zone of the FEF causes smooth eye movements toward the ipsilateral side, and is the only cortical area where stimulation can directly evoke smooth eye movements (Gottlieb *et al.*, 1993). Damage to the FEF produces deficits in saccades, and the impairments are most noticeable when the target stimulus is accompanied by irrelevant distracter stimuli. For pursuit, disruption of the FEF reduces pursuit eye velocity and, in particular, affects the ability of pursuit to follow targets using prediction. The impairments after FEF damage are usually temporary, but combined damage to the FEF and SC results in a permanent deficit in the ability to generate voluntary gaze shifts (Schiller *et al.*, 1980).

The FEF contains neurons with different combinations of motor and visual activity. Movement-related neurons have movement fields like those found in the superior colliculus—they are active for saccades within a particular range of directions and amplitudes. Movement-related neurons project to the superior colliculus and show changes in activity that are related with when and if voluntary gaze shifts will be initiated. In particular, the variability in the timing of saccade initiation is closely correlated with when the firing rates of these neurons reach a fairly constant "threshold" value (Hanes and Schall, 1996). Fixation neurons show the opposite pattern, decreasing their activity for saccades in all directions. In the pursuit zone of the FEF, neurons exhibit directionally selective responses to moving visual stimuli that are appropriate for determining the velocity command for pursuit eye movements.

Most neurons in the FEF also have visual responses, including a large class of visuo-movement neurons that show activity for both visual stimuli and the eye movements that follow. These visually responsive neurons often show activity related to the selection of the target. Prior to the gaze shift, the activity evoked by the non-selected stimulus is rapidly suppressed, whereas the activity for the selected stimulus remains conspicuous. Some FEF neurons discriminate visual targets even in the complete absence of an eye movement response, indicating that their activity is related to the allocation of visual attention rather than to the preparation of eye movements. Indeed, stimulation within the FEF with currents too weak to directly cause eye movements can nonetheless enhance the visual responses of neurons located in other visual areas of cortex, and can improve the behavioral performance on visual discrimination tasks (Moore and Fallah, 2004).

Another region of frontal cortex, the supplementary eye field (SEF), plays a less direct role in the control of saccades but is important for movements that are guided by cognitive factors. Neurons in the SEF are preferentially modulated in tasks that involve more abstract instructions, such as when saccades are directed to a particular side of an object rather than to a spatial location, or when saccades are directed to the location opposite the visual stimulus (the "anti-saccade" task). During pursuit eye movements, SEF neurons are especially active when the target moves along a predictable trajectory.

The Parietal Cortex Contributes to Eye Movements by Controlling Visual Attention

The lateral intraparietal area (LIP) plays a major role in the process of visual selection that precedes the generation of gaze shifts. Reversible inactivation of neurons in area LIP does not cause large deficits in the timing or accuracy of gaze shifts, but it dramatically reduces the frequency of movements toward the affected portions of the visual field, especially when competing stimuli are present elsewhere (Wardak *et al.*, 2002). These deficits suggest a visual neglect of stimuli in the affected region, rather than a motor deficit in the generation of eye movements. Damage to the parietal cortex in humans, especially to the right hemisphere, also produces a neglect syndrome with similar deficits in attention.

Neurons in LIP show activity related to visual attention and to the intention to make gaze shifts to particular locations in visual space. When an attention-grabbing, but irrelevant, distracter stimulus appears while the animal is preparing a saccade directed elsewhere, the activity of LIP neurons closely tracks the short-lived shift of attention to the distracter and its subsequent return to the intended location of the saccade (Bisley and Goldberg, 2003). Activity in LIP therefore often is described as a salience map that helps to determine the location of spatial attention, as well as the endpoint of gaze shifts.

Motion-Processing Areas of Cortex Provide Crucial Signals for Pursuit Eye Movements

The middle temporal area (MT) and the medial superior temporal area (MST) are the primary sources of visual motion information that are used to construct the motor commands for pursuit. This motion information also is used to adjust the size of saccades made to moving visual targets. These cortical areas provide these inputs through several pathways, including projections to other cortical areas involved in controlling gaze shifts (in particular, the FEF), descending projections to the vestibulocerebellum via the basal pontine nuclei, and projections to the pretectum, which contains neurons with activity related to the OKR and pursuit.

Neurons in MT and MST have activity related to the velocity of visual stimuli. In MT, neurons have smaller receptive fields and activity that depends on the presence of a visual stimulus. Lesions in MT produce blind spots for seeing motion that affect the motor commands for pursuit and saccades, and the perception of motion as well. In MST, neurons have larger receptive fields and prefer more complex patterns of visual motion. Lesions in MST also produce blind spots for motion, but in addition, cause a directional deficit in pursuit for movements directed to the ipsilateral side (Dürsteler and Wurtz, 1988). Stimulation of MT and MST can affect pursuit eye movements, but only if pursuit is already ongoing, unlike the effects of stimulation in the FEF. Thus, MT and MST are crucial sources of the motion signals needed to drive pursuit eye movements, but other areas appear necessary to determine when and if the movement should be started.

The Basal Ganglia Regulate Saccades by Inhibiting the Superior Colliculus

The caudate nucleus (CN) and the substantia nigra pars reticulata (SNpr) are parts of the basal ganglia that form an important circuit for controlling when gaze shifts are initiated. The SNpr applies tonic inhibition to the superior colliculus using GABA-ergic synapses, suppressing the response to the many excitatory inputs to the superior colliculus that might otherwise evoke gaze shifts. When SNpr neurons reduce or pause their activity, there is a disinhibition of activity in the superior colliculus, which then permits gaze shifts to be initiated. The SNpr itself is under inhibitory control from the CN, which contains neurons that increase their activity with gaze shifts. Together, this series of inhibitory connections through the basal ganglia act to gate the generation of gaze shifts by the superior colliculus.

The basal ganglia appear to be especially important for gaze shifts involving reward and motivation. For example, the visual responses of CN neurons is larger for stimuli that are associated with larger rewards (Lauwereyns et al., 2002). This modulation in excitability may be caused by dopaminergic inputs from neurons in the substantia nigra pars compacta (SNpc), the same neurons whose degeneration is implicated in Parkinson's disease.

The Thalamus Provides Feedback about the Control of Gaze Shifts

In addition to these descending pathways, there are also ascending pathways to eye movement-related areas of cortex from nuclei in the central thalamus. These nuclei receive inputs from a wide range of oculomotor structures, including the superior colliculus, substantia nigra pars reticulata, cerebellum, the nucleus prepositus hypoglossi, and the brain stem reticular formation. Feedback through the thalamus appears to convey corollary discharge signals about the gaze shifts that are executed by downstream motor structures, and is used to update the control signals formed in the cortex. Thus, when the central thalamus is inactivated by chemical agents, single saccades can still be made accurately, even in the dark, but subsequent saccades are spatially inaccurate, as though the intervening movement had not been taken into account (Sommer and Wurtz, 2002).

THE CONTROL OF GAZE SHIFTS INVOLVES HIGHER-ORDER PROCESSES

Saccades and Pursuit Are Coordinated during Gaze Shifts

The functions of saccades and pursuit are distinct, and yet during normal gaze shifts the two types of movements are tightly coordinated and almost always select the same visual target. How is this coordination achieved? One possibility is that target selection for pursuit is determined by the generation of saccades. When subjects are presented with two moving stimuli simultaneously, they almost always begin smooth pursuit in a direction that is the average of the two motions. After a brief period of this "vector average" pursuit, subjects make a saccade toward one stimulus or the other and then pursue that one selectively. These observations suggest that pursuit target selection is serially linked to saccade execution (Gardner and Lisberger, 2002). On the other hand, pursuit can selectively follow one stimulus in the presence of distracters without a targeting saccade, showing that saccades are not strictly necessary for pursuit target selection.

Another possibility is that targets are selected in common for pursuit and saccades. Pursuit typically starts earlier than saccades, and the stimulus selected by pursuit usually predicts the target of the saccade choice by ~50 ms. This suggests that the two choices are guided by the same decision signals, but saccades might use a somewhat more stringent decision criterion. Consistent with this view, pursuit and saccades exhibit nearly the same tradeoff between speed and accuracy, but saccades tend to be delayed and some-what more accurate (Liston and Krauzlis, 2005). The exact mechanism for this putative shared decision process is not yet known, but one possibility is that it is related to visual attention.

Voluntary Gaze Shifts Are Tightly Linked to Shifts of Visual Attention

A crucial aspect of voluntary gaze shifts is that they are selectively guided by objects of interest to the observer, despite the fact that our surroundings usually contain many other distracting signals. This selectivity is achieved by a tight functional link between the control of gaze shifts and the mecha-nisms of visual attention. When gaze shifts, the move-ment of the eyes always is preceded by a shift of visual attention to the new location. For example, when subjects are asked to detect visual targets that appear just before a saccade is made, perform-ance is much better when the target and saccade locations coincide than when they do not (Deubel and Schneider, 1996). Similarly during pursuit, sub-jects make more accurate perceptual judgments about the stimulus they are smoothly tracking with their eyes than about other, nontracked stimuli in the visual display. However, judgments about non-tracked stimuli are still better than chance, showing that the selection of targets for gaze shifts can occur in parallel with the perceptual processing of other visual stimuli. Also, although gaze shifts typically require visual attention, the converse is not true—visual attention can shift to new locations in the absence of gaze shifts; these are referred to as covert shifts of attention.

The tight linkage between visual attention and the planning of eye movements provides support for the premotor theory of attention, which posits that the mechanisms for spatial attention and eye motor pro-gramming are essentially the same. The theory is controversial, because the degree of overlap between attention and eye movements depends on which level of brain organization is considered relevant. As described in the previous section, there is substantial overlap between the areas of the brain involved in the control of attention and eye movements, consistent with the theory. On the other hand, the activity of single neurons in these areas can often be attributed to either visual attention or motor planning, but not both. Resolving these issues will require understanding how the properties of individual neurons are brought together into the larger functional circuits that span brain areas.

Visual Perception and Cognition Contribute to the Control of Gaze Shifts

Much of what we know about the neural control of gaze shifts was learned using very simple visual, often small bright spots in an otherwise dark room. These studies made it possible to test and identify the basic feedback mechanisms and motor circuits that control gaze shifts. However, it is now generally recognized that under more natural conditions, eye movements show properties that are more complex and flexible. One striking set of examples is provided by the work of Yarbus (Yarbus, 1967), who showed that the pattern of saccades made while looking at a visual scenes depends heavily on the instructions given to the observer (Fig. 33.11). These results illustrate that gaze shifts are normally not just reactions to retinal events, but instead are guided by longer-term goals and serve to collect relevant visual information from the environment.

Gaze shifts also reveal properties of the visual processing steps required to segment and interpret the visual scene. In one demonstration, a few lights are attached to the rim of a wagon wheel, which is then rolled along in an otherwise dark display. Although the retinal stimulus consists of several spots, each undergoing cycloidal motion, subjects perceive the horizontal motion of the wheel and are able to easily track this motion with their eyes (Steinbach, 1976). More recent studies have examined how these proc-esses evolve over time. For many moving objects, the true trajectory of the object cannot be calculated locally in the retinal image, but requires integrating informa-tion over regions of visual space. This integration evi-dently takes some time (for example, ~100 ms), and the transition from local-motion signals to object-motion signals can be found in both the eye motor output and perceptual judgments.

Visual perception also is changed by eye move-ments. In particular, the world is not perceived as moving during saccadic eye movements, despite the rapid motion of the retinal image. You can demon-strate this to yourself by watching yourself make sac-cades in a mirror—you can see that your eyes change position, but you cannot see them actually move. This phenomenon is referred to as saccadic suppression, and it refers to the observation that the ability to detect visual stimuli is reduced dramatically during saccadic eye movements, especially for visual stimuli with low spatial frequencies (Burr et al., 1994). The mechanism of saccadic suppression is not fully understood, but its effect is to help the visual system reduce its sensitivity to the unreliable visual signals that occur during gaze shifts.

"An Unexpected Visitor"

"Give the ages of the people."

"Surmise what the family had
been doing before the arrival
of the unexpected visitor."

FIGURE 33.11 The scan path depends on the goals of the observer. Among other experiments, Yarbus (1967) monitored the eye movements of subjects as they viewed Repin's painting "An Unexpected Visitor" and found that the pattern of eye movements depended on the task. When subjects were instructed to give the ages of the people, the scan path mostly settled on the faces, whereas when subjects were instructed to surmise what the family had been doing, the scan path included many more objects in the room. Adapted from Figure 109 of Yarbus (1967) with kind permission of Springer Science and Business Media.

THE CONTROL OF EYE MOVEMENTS CHANGES OVER TIME

Different Eye Movements Develop over Different Timescales

Studying the development of eye movements is especially challenging, because infants are not always cooperative subjects, and the observed changes in eye movements are due to developmental changes that occur at sensory as well as motor stages of processing. Nonetheless, in humans, all eye movements appear to be different in infants than they are in adults, and the development of eye movements that stabilize gaze (e.g., VOR and OKR) follow a faster time course than those that act to shift gaze (e.g., saccades and pursuit).

Human infants are born with an excellent VOR. In fact, the gain of the VOR (the ratio of eye velocity to head velocity) in infants is approximately 1, which is better than that typically found in adults. The temporal dynamics of the VOR are somewhat different at birth, possibly reflecting an immature neural integrator, but these achieve near-adult values within the first six months of life. The strong VOR at birth compensates for the fact that the OKR is not fully developed until age 5–6 months. The OKR in infants is symmetric for rightward and leftward motion, but only when viewed binocularly (i.e., with both eyes). With monocular viewing, the amplitude of the OKR is markedly larger for horizontal motion directed toward the nose than for motion directed away from the nose. This directional asymmetry is like that found in the adult OKR of "lower" mammals such as the rabbit, and in cats that have been reared with disrupted visual inputs from one eye. The asymmetry in infants is believed to be due to the delayed maturation of the cortical inputs to the visual nuclei in the brain stem, which confer binocularity and other visual properties to the neurons in the pretectum important for generating the OKR.

As might be expected given the relative immaturity of the visual system at birth, saccades and pursuit take many months to approach adult-like characteristics. Newborns can generate saccades in response to a visual target, but the latencies are long and the movements fall well short of the target. In fact, 1-month old infants typically acquire the target by making a series of three to five saccades, each of which is approximately the same size and which scales with the initial eccentricity of the visual target. Over time during the first year, the number of sequential saccades decreases and their amplitude increases. Similarly, newborns have a limited ability to generate pursuit eye movements, and the responses are better for larger stimuli that likely also activate the OKR. The gain of pursuit to small target spots is very low (0.5 or less) in the first few postnatal months, and saccades often are used to keep up with the target. Interestingly, the pursuit velocity generated at these young ages is not only low, but does not scale with the velocity of the visual target, presumably due to the delayed maturation of the cortical circuits involved in motion processing. The ability to match target speed improves markedly by the end of the first year, as does the ability to use prediction to guided pursuit. However, for both saccades and pursuit, the exact age at which the movements become adult-like is difficult to determine.

Eye Movements Continuously Undergo Plastic Changes

Even in adulthood, the circuits for controlling gaze movements are continually adjusted to compensate for changes in the motor output and the sensory inputs. As many of us have experienced after getting a new pair of eye glasses, which changes the VOR output needed to stabilize the visual world, the inaccuracies in the programming of eye movements usually are corrected automatically and fairly rapidly. This type of adaptive control is especially important for rapid movements like the VOR and saccades, because the movements occur too rapidly to be corrected mid-flight by visual feedback.

For saccades, the adaptation is guided by visual signals about whether the eyes have landed on the target. This feedback is used by structures such as the oculomotor vermis in the cerebellum to adjust the "pulse" component of the command for saccades. Saccadic adaptation also depends on visual signals about whether the eye stays on target after the saccade or drifts off. This visual motion feedback is used by the vestibulocerebellum to adjust the size of the "step" component of the saccade command. Adaptive changes in saccades also tend to be spatially specific rather than parametric—the changes affect a limited range of saccade amplitudes and directions near the adapted saccade, rather than simply scaling all saccades. This property illustrates the great flexibility and precision of the output motor pathways in fine-tuning the control of saccades.

Adaptive changes to the VOR have been extensively studied and provide a general model for understanding the neural basis of motor learning. It is well established that the vestibulocerebellum plays a fundamental role in motor learning for the VOR, but the exact role of the cerebellum remains controversial. Early studies emphasized the role of the cerebellar cortex as the principal site of learning. Theories put forward by Marr and Albus, and elaborated by Ito, drew attention to the crystalline structure of the cerebellar cortex and highlighted the synaptic interactions that might take place on the output neurons from the cerebellar cortex, the Purkinje cells. In particular, they proposed that climbing fiber inputs to the cerebellar cortex provide visual "error signals" that modify the strength of parallel fiber synapses providing vestibular information to Purkinje cells. Subsequent studies have supported this proposal by showing that paired stimulation of the climbing fibers and parallel fibers induces a long-term depression (LTD) of the parallel fiber synapse. However, other studies have shown that although the cerebellum is critical for VOR adaptation, it could be removed after the adaptation was complete without changing the state of the VOR. This led later researchers to propose that the cerebellum does not store the motor memory needed for VOR plasticity but instead computes an instructive signal that induces changes elsewhere, such as in the vestibular nuclei that receive the cerebellar output. These careful studies of the VOR circuit have made it possible now to begin probing the cellular and molecular mechanisms that underlie this basic form of learning.

CONCLUSIONS

Eye movements serve two different functions. First, eye movements stabilize gaze when animals move about in their environment. Unpredictable high-speed rotations of the head, which are produced whenever an animal moves, shift the line of gaze and, by smearing the optical image, reduce the resolution of the visual system. These high-velocity movements are compensated for by counter-rotations of the eyes generated by the VOR. Working in tandem with the VOR is the optokinetic system. The optokinetic system compensates for the slower movements of the head, which happen so gradually that the vestibular system cannot accurately detect them. Second, in animals with retinas that have a small region of high resolution, eye movements shift gaze to redirect the line of sight to objects or features in the visual scene that are of particular interest. Saccades are fast, discrete movements that quickly move the image of a visual target to the center of the retina where acuity is best. Pursuit is a slow, continuous movement that compensates for the motion of the visual target, minimizing the blurring that would otherwise limit acuity. Although the properties of saccades and pursuit are distinct, the two types of movements are closely coordinated during visual search of the environment and both are tightly linked to visual attention. Together, the different types of eye movements make it possible to efficiently gather the visual information that is important for achieving other, longer-term behavioral goals.

Eye movements are controlled by circuits that span the brain. The eyes themselves are a relatively simple motor apparatus, and they are controlled by a final motor pathway in the brain stem that is shared by all classes of eye movements. The construction of the motor commands involves generating an eye velocity command to move the eyes, and an integrated version of this command to hold the eyes at their new position. These motor commands are constructed by circuits in the brain stem reticular formation, and fine-tuned by additional circuits involving several parts of the cerebellum. The descending control signals that drive these motor commands are provided by multiple brain regions, but the superior colliculus in the midbrain and the frontal eye fields in the cortex are especially important. These brain regions help integrate sensory inputs with a variety of higher-order influences so that eye movements are directed to the appropriate visual stimuli.

Overall, eye movements provide a window into the brain mechanisms that accomplish motor control, but also the mechanisms of neural plasticity, sensory-motor interactions, and higher-order process like attention and perception.

References

Bisley, J. W. and Goldberg, M. E. (2003). Neuronal activity in the lateral intraparietal area and spatial attention. *Science* **299**, 81–86.

Burr, D. C. *et al.* (1994). Selective suppression of the magnocellular visual pathway during saccadic eye movements. *Nature* **371**, 511–513.

Buttner-Ennever, J. A. (2007). Anatomy of the oculomotor system. *Dev. Ophthalmol.* **40**, 1–14.

Cannon, S. C. and Robinson, D. A. (1987). Loss of the neural integrator of the oculomotor system from brain stem lesions in monkey. *J. Neurophysiol.* **57**, 1383–1409.

Demer, J. L. (2006). Current concepts of mechanical and neural factors in ocular motility. *Curr. Opin. Neurol.* **19**, 4–13.

Deubel, H. and Schneider, W. X. (1996). Saccade target selection and object recognition: evidence for a common attentional mechanism. *Vision Res.* **36**, 1827–1837.

Dürsteler, M. R. and Wurtz, R. H. (1988). Pursuit and optokinetic deficits following chemical lesions of cortical areas MT and MST. *J. Neurophysiol.* **60**, 940–965.

Fuchs, A. F. and Luschei, E. S. (1970). Firing patterns of abducens neurons of alert monkeys in relationship to horizontal eye movement. *J. Neurophysiol.* **33**, 382–392.

Gardner, J. L. and Lisberger, S. G. (2002). Serial linkage of target selection for orienting and tracking eye movements. *Nat. Neurosci.* **29**, 29.

Gottlieb, J. P. *et al.* (1993). Smooth eye movements elicited by microstimulation in the primate frontal eye field. *J. Neurophysiol.* **69**, 786–799.

Hanes, D. P. and Schall, J. D. (1996). Neural control of voluntary movement initiation. *Science* **274**, 427–430.

Keller, E. L. (1974). Participation of medial pontine reticular formation in eye movement generation in monkey. *J. Neurophysiol.* **37**, 316–332.

Krauzlis, R. J. *et al.* (1997). Shared motor error for multiple eye movements. *Science* **276**, 1693–1695.

Lauwereyns, J. *et al.* (2002). A neural correlate of response bias in monkey caudate nucleus. *Nature* **418**, 413–417.

Lee, C. *et al.* (1988). Population coding of saccadic eye movements by neurons in the superior colliculus. *Nature* **332**, 357–360.

Liston, D. and Krauzlis, R. J. (2005). Shared decision signal explains performance and timing of pursuit and saccadic eye movements. *J. Vis.* **5**, 678–689.

Miles, F. A. (1997). Visual stabilization of the eyes in primates. *Curr. Opin. Neurobiol.* **7**, 867–871.

Missal, M. and Keller, E. L. (2002). Common inhibitory mechanism for saccades and smooth-pursuit eye movements. *J. Neurophysiol.* **88**, 1880–1892.

Moore, T. and Fallah, M. (2004). Microstimulation of the frontal eye field and its effects on covert spatial attention. *J. Neurophysiol.* **91**, 152–162.

Munoz, D. P. and Wurtz, R. H. (1993). Fixation cells in monkey superior colliculus I. Characteristics of cell discharge. *J. Neurophysiol.* **70**, 559–575.

Munoz, D. P. and Wurtz, R. H. (1995). Saccade-related activity in monkey superior colliculus. I. Characteristics of burst and buildup cells. *J. Neurophysiol.* **73**, 2313–2333.

Robinson, D. A. (1972). Eye movements evoked by collicular stimulation in the alert monkey. *Vis. Res.* **12**, 1795–1808.

Schiller, P. H. *et al.* (1980). Deficits in eye movements following frontal eye field and superior colliculus ablations. *J. Neurophysiol.* **44**, 1175–1189.

Sommer, M. A. and Wurtz, R. H. (2002). A pathway in primate brain for internal monitoring of movements. *Science* **296**, 1480–1482.

Steinbach, M. (1976). Pursuing the perceptual rather than the retinal stimulus. *Vision Res.* **16**, 1371–1376.

Wardak, C. *et al.* (2002). Saccadic target selection deficits after lateral intraparietal area inactivation in monkeys. *J. Neurosci.* **22**, 9877–9884.

Wurtz, R. H. and Goldberg, M. E. (1972). Activity of superior colliculus in behaving monkey: III. Cells discharging before eye movements. *J. Neurophysiol.* **35**, 575–586.

Yarbus, A. L. (1967). "Eye Movements and Vision." Plenum, New York, pp. 1–222.

Suggested Readings

Carpenter, R. H. S. (1988). "Movements of the Eyes," 2nd ed. Pion Limited, London.

Fuchs, A. F., Kaneko, C. R. S., and Scudder, C. A. (1985). Brainstem control of saccadic eye movements. *Annu. Rev. Neurosci.* **8**, 307–337.

Krauzlis, R. J. and Stone, L. S. (1999). Tracking with the mind's eye. *Trends Neurosci.* **22**, 544–550.

Krauzlis, R. J. (2005). The control of voluntary eye movements: New perspectives. *Neuroscientist* **11**, 124–137.

Leigh, R. J. and Zee, D. S. (2006). "The Neurology of Eye Movements," 4th ed. F. A. Davis, Philadelphia.

Lisberger, S. G., Morris, E. J., and Tychsen, L. (1987). Visual motion processing and sensory-motor integration for smooth pursuit eye movements. *Annu. Rev. Neurosci.* **10**, 97–129.

Mays, L. E. and Gamlin, P. D. R. (1995). Neuronal circuitry controlling the near response. *Curr. Opin. Neurobiol.* **5**, 763–768.

Raphan, T. and Cohen, B. (1978). Brainstem mechanisms for rapid and slow eye movements. *Annu. Rev. Physiol.* **40**, 527–552.

Schall, J. D. and Thompson, K. G. (1999). Neural selection and control of visually guided eye movements. *Annu. Rev. Neurosci.* **22**, 241–259.

Simpson, J. I. (1984). The accessory optic system. *Annu. Rev. Neurosci.* **7**, 13–41.

Sommer, M. A. (2003). The role of the thalamus in motor control. *Curr. Opin. Neurobiol.* **13**, 663–670.

Sparks, D. L. (2002). The brainstem control of saccadic eye movements. *Nat. Rev. Neurosci.* **3**, 952–964.

Richard J. Krauzlis

SECTION VI

REGULATORY SYSTEMS

The Hypothalamus: An Overview of Regulatory Systems

The hypothalamus is an integrative center essential for survival of an organism and reproduction of its species. Regulatory systems emerged as each organism adapted to its environment and have evolved to control the complex interactions of physiology and behavior.

The diversity in regulatory function of the hypothalamus is reflected in its structure, neurochemistry, and connections. Almost every major subdivision of the central nervous system (CNS) communicates with the hypothalamus and is subject to its influence. In addition, the hypothalamus communicates with peripheral organ systems by converting synaptic information to blood-borne hormonal signals. In turn, the hypothalamus responds to feedback from the peripheral systems that it regulates. The chapters in this section discuss the processes by which the hypothalamus influences physiology and behavior, and the purpose of this chapter is to provide an introductory context for those presentations. We begin by providing a historical perspective on research that has defined the hypothalamus as an important regulatory center, continue with an overview of its structural and functional organization, and conclude with a discussion of how this region of the diencephalon functions within brain circuitry devoted to the control of motivated behaviors.

HISTORICAL PERSPECTIVE

Research on the hypothalamus has a rich and storied history. Although depicted in early brain atlases, the hypothalamus was not recognized as a distinct division of the diencephalon until the work of Wilhelm His (1893). Our understanding of the structure and function of the hypothalamus developed as basic and clinical neurosciences were integrated. Harvey Cushing, a neurosurgeon who studied endocrine disorders from a neuroscience perspective provided important and influential early insights into hypothalamic function. He described diabetes insipidus, a neuroendocrine dysfunction of uncontrolled urinary water excretion and thirst, and identified the importance of injury to the pituitary stalk in the condition. Cushing also described the syndrome that bears his name, Cushing's syndrome, a disorder characterized by excessive secretion of adrenal corticoids, and he was among the first to recognize that the hypothalamus and pituitary form a functional unit. He was a pioneer in the development of pituitary surgery and is viewed as a founder of the fields of neurosurgery and endocrinology. In 1929, summing up his work in this area shortly before his retirement, Cushing wrote of the hypothalamus, "*Here in this well concealed spot, almost to be covered with a thumb nail, lies the very mainspring of primitive existence—vegetative, emotional, reproductive—on which, with more or less success, man has come to superimpose a cortex of inhibitions.*"

Gurdijian (1927), Kreig (1932), and Le Gros Clark *et al.* (1938) build upon the clinical observations of Cushing and others, using histological procedures to describe the cytoarchitecture of hypothalamic nuclei. Their observations proved to be remarkably predictive of functional specializations revealed by the work of Ranson, Hess, and others during the same period. These investigators demonstrated that lesions and localized stimulation of the hypothalamus could selectively affect appetite, weight control, water balance, autonomic control, reproductive function, and emotional behavior. This work progressed so well that in 1940 it could be summarized only in a large volume of the annual series of the "Association for Research in Nervous and Mental Disease." In that volume, Ernst and Berta Scharrer (1940) presented evidence for "neurosecretory neurons," providing the foundation for a fundamental tenet of hypothalamic function. That is, there are neurons—in the brain—that secrete hormones directly into the bloodstream. In 1949, Wolfgang Bargmann published his studies documenting neurosecretory neurons with cell bodies in the hypothalamus and axons projecting to the posterior lobe of the pituitary gland. Bargmann's work provided a framework for our current understanding of neuroendocrine regulation, that is, that release of peptides into the peripheral circulation allows the hypothalamus to exert a profound influence on the physiology of peripheral organs.

In the 1950s Harris and Green defined the means through which the hypothalamus regulates the anterior lobe of the pituitary. Their identification of a neurovascular link, the portal plexus, through which the hypothalamus regulates the anterior pituitary, fundamentally altered concepts of hypothalamic function and neuroendocrine regulation. The neurochemical basis of anterior pituitary regulation was revealed in the 1970s when Guillemin and Schally independently isolated and characterized peptide hormones that acted on the anterior pituitary. In subsequent years the field of neuroendocrinology built enormously on these observations, and the resulting literature documents a diverse family of hypothalamic peptides that are transported through the portal plexus to stimulate or inhibit the release of other hormones from the anterior lobe of the pituitary gland.

We have made great progress in improving our understanding of Cushing's "mainspring of primitive existence." Tremendous insights into CNS function have emerged from the analysis of hypothalamic structure and function, underscoring the importance of this small, complex region of the diencephalon in the regulation of homeostasis and behavioral functions essential for survival.

HYPOTHALAMIC CYTOARCHITECTURE

Early studies of hypothalamic organization defined subdivisions that were largely based upon landmarks visible on the ventral surface of the brain. On this basis, three major subdivisions were defined based upon their associations with the optic chiasm (chiasmatic), the tuberal eminence containing the pituitary stalk (tuberal), and the mammillary bodies (mammillary) (Fig. 34.1). Contemporary research indicates that the hypothalamus is organized into three longitudinal columns on either side of the third ventricle (Fig. 34.2). These columns are largely consistent with the initial descriptions of Gurdjian (1927) and Krieg (1932) and consist of *periventricular*, *medial*, and *lateral* zones that extend through the entire rostrocaudal extent of the hypothalamus. The periventricular zone, as the name implies, lies adjacent to the ependymal lining of the third ventricle. Its constituent neurons are densely packed within a narrow zone adjacent to the ependyma and occasionally expand into cytoarchitecturally distinct nuclear groups such as the suprachiasmatic

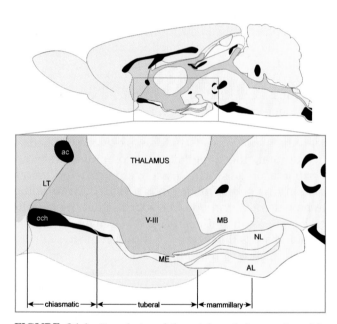

FIGURE 34.1 Boundaries of the rat hypothalamus viewed in midsagittal section. The ventricular system is shaded gray. The boxed area shows the approximate location of the hypothalamus in the ventral quadrant of the diencephalon, below the thalamus. The floor of the third ventricle (V–III), formed by the optic chiasm (och), median eminence (ME), and rostral portion of the mammillary body (MB), defines a large portion of the rostrocaudal extent of the hypothalamus. These landmarks also define the major subdivisions (chiasmatic, tuberal, and mammillary) classically used to subdivide the hypothalamus. The tuberal hypothalamus is distinguished by the infundibular stalk, which connects the ventral diencephalon to the pituitary. The rostral limit of the hypothalamus is the lamina terminalis (LT), a thin strip of tissue extending between the optic chasm and anterior commissure (ac) that also forms the rostral wall of the third ventricle. NL, neural lobe of posterior pituitary; AL, anterior lobe of pituitary.

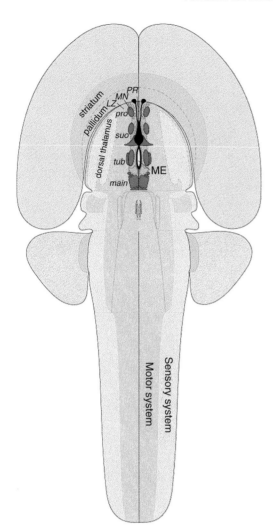

FIGURE 34.2 The columnar organization of hypothalamic cell groups is illustrated on a flat map of the rat central nervous system. Three longitudinal zones form columns that extend through the rostrocaudal extent of the hypothalamus. The periventricular region (PR; light red) lies immediately adjacent to the third ventricle and contains cell groups responsible for neuroendocrine control of the pituitary gland (black). The medial nuclei consist of cytoarchitecturally distinct cell groups (dark red) between the periventricular and lateral zones. The diffusely organized but phenotypically distinct neurons of the lateral zone (LZ) are shown in yellow. Contemporary research also recognizes four transverse subdivisions of hypothalamus (pro, preoptic; supraoptic or anterior, SUO; tuberal, tub; mammillary, mam). Reprinted with permission from Swanson (2000).

and arcuate nuclei. The medial zone lies adjacent to the periventricular zone and contains cell groups that vary in size and cell morphology, but are organized into cytoarchitecturally distinct nuclei. This contrasts with the diffusely distributed neurons of the lateral zone, which is the least distinct of the three columns. Neurons in all three regions have been further subdivided on

the basis of cellular phenotype, connectivity, and cellular markers of cellular activation that allow investigators to impose functional definitions.

FUNCTIONAL ORGANIZATION OF THE HYPOTHALAMUS

To effectively integrate diverse physiological processes and behaviors, the hypothalamus must communicate with systems in both the brain and periphery. More than any other region of the nervous system, the hypothalamus depends on conversion of synaptic information to hormonal signals. In addition, the hypothalamus is subject to feedback regulation by the physiological processes that it controls. The following sections describe basic organization principles that govern hypothalamic function.

Integrative Processing in the Hypothalamus

Projections from all major subdivisions of the nervous system terminate within the hypothalamus. In many instances, such as vision, monosynaptic projections bring first-order sensory information from the periphery directly to hypothalamic nuclei. In other cases, such as limbic projections, multisynaptic cerebral pathways relay processed information to the hypothalamus. Integration of this sensory information is an essential function of the hypothalamus and has profound influence on the regulatory outputs of hypothalamic nuclei. These diverse systems are best understood by looking at the functions that they modulate.

Olfaction

Multisynaptic pathways allow the olfactory system to influence neurons in many parts of the hypothalamus. Olfactory information passes directly to the olfactory tubercle, piriform cortex, and amygdala, which in turn relay it to the hypothalamus through the medial forebrain bundle (corticohypothalamic projections), stria terminalis, ventral amygdalofugal pathway, and fornix. In many vertebrates, olfaction is integral to behaviors involved in reproduction, defense, and feeding, and this is reflected in the large amount of forebrain dedicated to the processing of olfactory stimuli. Insight into the complexity of these circuits is apparent in recent rodent studies employing transneuronal tracing technology to define polysynaptic olfactory circuits that innervate hypothalamic neurons involved in the regulation of reproduction and fertility (Boehm *et al.*, 2005; Yoon *et al.*, 2005). These and other findings illustrate the importance of hypothalamic integration of olfactory information in functions essential to reproduction and survival.

Visual Projections

A principal function of the hypothalamus is to impose temporal organization on hormonal and behavioral processes through the timekeeping properties of the biological clock in the rostral hypothalamus. The neurons that constitute this clock are found within the suprachiasmatic nuclei (SCN). The genetically determined circadian rhythmicity of SCN neurons provides a timekeeping capacity that is essential to successful adaptation to the environment and, in many instances, reproduction of organisms. Importantly, the rhythmic activity of the biological clock is responsive to environmental cues, which is accomplished through central visual projections of the retina. A unique population of ganglion cells that contain light sensitive melanopsin transduces light in the retina. The central projections of these ganglion cells through the retinohypothalamic projection synchronize the circadian activity of SCN neurons with the daily cycle of light and dark. Polysynaptic connections from the clock that influence melatonin secretion from the pineal gland also provide a feedback mechanism for measuring day length that is important for the timing of reproduction in photoperiodic species. The responsiveness of this regulatory system to variations in photoperiod is also responsible for the "jet lag" that humans suffer when traveling across time zones.

Surprisingly, the neural projections of the SCN are largely confined to the hypothalamus. Therefore, the SCN is dependent upon relays that integrate a variety of sensory information important for the temporal organization of diverse functions regulated by the hypothalamus. A more detailed description of this "circadian timing system" is presented in Chapter 41.

Visceral Sensation

Sensory information also reaches the hypothalamus through ascending projections arising from neurons in the nucleus of the solitary tract (NTS) of the caudal brain stem. The NTS is the principal visceral sensory nucleus that receives topographically organized input from the major organ systems of the body by way of cranial nerves X and IX, the vagus and glossopharyngeal nerves. As such, it is the first region in the CNS to process information about visceral, cardiovascular, and respiratory functions, as well as taste. The NTS coordinates reflex modulation of peripheral organ function and sends processed sensory information to forebrain nuclei, where it is integrated in the control of more complex physiological processes and behaviors.

Neurons in the paraventricular hypothalamic nucleus (PVH) and the lateral hypothalamic area (LHA) are prominent among a subset of hypothalamic nuclei that receive direct projections from the NTS and indirect projections reach the hypothalamus through the ventrolateral medulla and parabrachial nucleus in the pons. Many of these projections are bidirectional and are part of a larger system that includes the central nucleus of the amygdala, bed nuclei of the stria terminalis, and insular cortex. In addition, the NTS and parabrachial nuclei relay splanchnic visceral and nociceptive sensory information from the spinal cord. This "interoceptive" sensory feedback has an important influence upon autonomic and endocrine outputs of the hypothalamus essential for regulation of homeostasis, and recent evidence suggests that it also influences mood. Evidence for the latter is particularly prevalent in literature examining sensory integration related to stress induced activation of the endocrine and autonomic axes.

Multimodal Brainstem Afferents

The medial forebrain bundle (mfb) is a group of thin fibers that traverse the lateral zone of the hypothalamus. This bidirectional fiber tract provides extrinsic inputs to a number of hypothalamic nuclei and is used as a conduit for hypothalamic neurons to communicate with other regions of the neuraxis. Early nonspecific stimulation and lesion experiments incorrectly attributed certain functions to the lateral hypothalamus that actually are associated with axons passing through the mfb to other regions of the brain.

Monoaminergic axons arising from neurons in the midbrain and medulla are a major constituent of this fiber tract. They include axons from noradrenergic neurons in the locus coeruleus and lateral tegmental cell groups, as well as serotoninergic neurons of the brain stem raphe nuclei. Many of these axons generate collaterals that terminate throughout the hypothalamus as they course rostrally to innervate other forebrain regions. Serotoninergic neurons in the midbrain raphe nuclei also generate a dense plexus of axons that enter the cerebral ventricles and arborize on the luminal surface of the ependyma. The precise function of this intraventricular plexus is not known.

A number of systems not functionally associated with the hypothalamus also pass through the mfb. Prominent among these are dopaminergic axons arising from the substantia nigra and ventral tegmental area (VTA). The nigral component of this projection plays an important role in the regulation of movement whereas the VTA exerts a widespread influence on the cortex as well as ventral striatal regions involved in motivated behaviors and addiction. These projection systems are distinct from the dopaminergic neurons in the hypothalamus (e.g., arcuate nuclei), which play an important role in neuroendocrine regulation.

Projections from Limbic Regions

In the 1930s, James Papez (1937) suggested that components of "limbic circuitry" form part of a multi-synaptic pathway, classically known as Papez's circuit, responsible for the expression of emotion. Although this function has not been confirmed, contemporary studies have demonstrated that both the hippocampus and amygdala give rise to extensive topographically organized projections to the hypothalamus. Prominent among the observations that have forged this vision are studies demonstrating that the fornix influences the activity of the hypothalamus through both direct and indirect pathways. The direct path is a prominent component of Papez's original circuit in that it is composed of axons that arise from neurons in the subicular complex of the hippocampal formation and terminate in the mammillary bodies. The indirect, or "precommissural," fornix influences the hypothalamus through a synaptic relay in the septum.

Risold and Swanson (1996) demonstrated that the precommissural projection arises from the hippocampus proper, is organized topographically, and terminates densely within the hypothalamus (Fig. 34.3). The

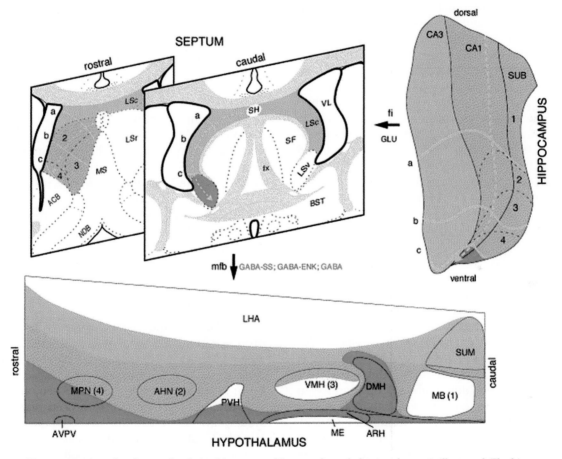

FIGURE 34.3 The organization of pathways that bring hippocampal input to hypothalamic columns is illustrated. The hippocampal formation projects topographically upon the lateral subdivision of the septum which, in turn, projects topographically upon the periventricular (red), medial (blue), and lateral (green) zones of the hypothalamus. Subfields of the hippocampal formation (CA1, CA3, subiculum) are represented on the flat map of the unfolded hippocampus in the upper right. The topography of the projection systems in coded in the colors, numbers (1, 2, 3, 4), and small letters (a, b, c) included in the maps of hippocampus, septum, and hypothalamus. ACB, nucleus accumbens; AHN, anterior hypothalamic nucleus; ARH, arcuate nucleus; AVPV, anteroventral periventricular nucleus; BST, bed nucleus of stria terminalis; DMH, dorsomedial nucleus; ENK, enkephalin; fi, fimbria of fornix; fx, fornix; GLU, glutamate; LS, lateral septum (small letters designate rostral, caudal, and ventral subdivisions); MB, mammillary body; ME, median eminence; mfb; medial forebrain bundle; MPN, medial preoptic nucleus; MS, medial septum; NDB, nucleus of the diagonal band; PVH, paraventricular nucleus; SH, septohippocampal nucleus; SS, somatostatin; SUM, supramammillary nucleus; VL, lateral ventricle; VMH, ventromedial nucleus. Reprinted with permission from Risold and Swanson (1996).

topography of this projection is apparent at its origin in the hippocampus and is preserved in the lateral septal nuclei and in the hypothalamus where second order projections innervate the three longitudinally organized columns of the hypothalamus, although not equally. Thus, this projection system is in a position to regulate neuroendocrine and autonomic function and ingestive behavior (periventricular zone), modulate motivated reproductive and agonistic behaviors (rostral medial zone), and also provide polysynaptic feedback projections to the hippocampus via the mammillary complex. The demonstration of "place" and "head-direction" cells in the hippocampal formation suggests that this projection is the source of adaptive multimodal sensory information relevant to a variety of hypothalamic functions (e.g., foraging).

Limbic projections arising from the amygdala also have a prominent influence on hypothalamic function through two projection pathways. The stria terminalis arises from neurons in the amygdala and follows a looping trajectory similar to the course of the fornix. Axons coursing through this pathway ramify within the bed nucleus and then enter the hypothalamus rostrally. Amygdala neurons also innervate the hypothalamus via the ventral amygdalofugal bundle, which passes directly over the optic tract to the hypothalamus. The demonstrated role of the amygdaloid complex in the expression of fear suggests that these projections play an important role in orchestrating the endocrine and autonomic components of the fear response.

Circumventricular Organs

Not all sensory information reaching the hypothalamus is of synaptic origin. Some regions of the CNS are chemosensitive to blood-borne molecules. Such information about body fluid composition has important implications for the maintenance of homeostasis. Circumventricular organs (CVOs) are integral to this regulation. Seven CVOs surround the ventricular system in the diencephalon, midbrain, and hindbrain (Fig. 34.4). These include the subfornical organ (SFO), the vascular organ of the lamina terminalis (OVLT), the median eminence (ME), the pituitary gland (PG), the pineal gland (P), the subcommissural organ (SCO), and the area postrema (AP) (Fig. 34.3). Three of these regions, the OVLT, ME, and PG, are components of the hypothalamus. Two others, the SFO and AP, have extensive connections with hypothalamic nuclei involved in neuroendocrine and homeostatic function. Paul Ehrlich identified these regions in the late 1800s when he noted that they selectively accumulated vital dyes injected into the peripheral vasculature. Later

studies demonstrated that the dye accumulated due to fenestrations in the capillaries in these regions that permitted relatively large molecules to leave the vascular lumen and enter the extracellular milieu. As a result, CVO neurons are accessible to blood-borne molecules that cannot enter other regions of the brain. The absence of a blood–brain barrier also allows CVOs to influence peripheral function through the release of bioactive molecules. Thus, CVOs represent important "windows" through which the brain controls peripheral function and, importantly, that render CNS neurons responsive to feedback regulation by the secretory products of the target organs that they modulate. This type of architecture is fundamental to the way in which the hypothalamus achieves tightly regulated control over the pituitary and, by extension, its peripheral targets.

EFFECTOR SYSTEMS OF THE HYPOTHALAMUS ARE HORMONAL AND SYNAPTIC

The extensive influence of the hypothalamus on behavior and physiology is reflected in its effector systems. The hypothalamus has conventional synaptic connections with essentially every major subdivision of the CNS and plays a major role in the control of behavioral state. As noted earlier, it also has regulatory control over the pituitary gland and thereby exerts a profound influence over peripheral physiology. Thus, the output of hypothalamic systems is integral to the integrated expression of physiology and behavior.

Behavioral State Control

Regulation of the sleep–wake cycle is a distributed process that involves the integrated action of the reticular activating system and the diencephalon. Early studies employing lesions and knife cuts conducted by Moruzzi, Magoun, and Bremer demonstrated the importance of the brain stem reticular activating system in the generation of consciousness. Subsequently, Nauta postulated the existence of sleep centers in the hypothalamus based upon the consequences of lesions to the rostral or caudal hypothalamus. In those rodent studies, he demonstrated that lesions of the preoptic region produced insomnia and that somnolence resulted from lesions of the caudal hypothalamus.

Until recently, little was known regarding the specific hypothalamic cell groups involved in sleep regulation. However, identification of the circadian

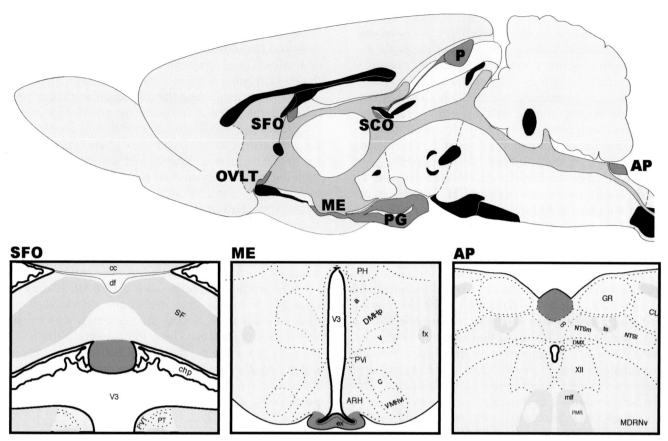

FIGURE 34.4 The location of the seven circumventricular organs (shown in red) in the rat brain (midsaggital section) are illustrated. Three CVOs that have an intimate functional association with the hypothalamus are illustrated in transverse sections in the lower figures. AP, area postrema; ARH, arcuate nucleus; cc, corpus callosum; CU, cuneate nucleus; df, dorsal fornix; DMHa, anterior portion of the dorsomedial nucleus; DMHp, posterior portion of dorsomedial nucleus; DMHv, ventral portion of dorsomedial nucleus; DMX, dorsal motor vagal nucleus; GR, gracile nucleus; ME, median eminence; mlf, medial longitudinal fasiculus; co, commissural portion of the nucleus of the solitary tract; NTSl, lateral portion of the nucleus of the solitary tract; NTSm, medial portion of the nucleus of the solitary tract; OVLT, vascular organ of the lamina terminalis; PH, posterior hypothalamus; P, pineal gland; PMR, paramedian reticular nucleus; PVi, intermediate part of periventricular nucleus; SCO, subcommissural organ; SF, septofimbrial nucleus; SFO, subfornical organ; ts, tractus solitarius; V–III, third ventricle; VMHc, central part of ventromedial nucleus; VMHvl, ventrolateral part of ventromedial nucleus; XII, hypoglossal nucleus.

pacemaker in the early 1970s marked the beginning of an era in which the hypothalamic contributions to behavioral state regulation have become increasingly clear. This literature has confirmed the early hypotheses of Nauta and identified cell groups in the rostral and caudal hypothalamus whose activity correlates with sleep or arousal.

The rostral hypothalamus is resident to two cell groups that are influential in the regulation of the sleep–wake cycle. Of these, the suprachiasmatic nuclei, or "Mind's Clock," are well known for their genetically determined oscillatory activity. A large literature has established that SCN activity opposes the homeostatic drive for sleep (Chapter 42). In contrast, Saper and colleagues (Sherin *et al.*, 1996) identified a group of hypothalamic neurons that become active at the onset

of sleep in rats. These cells, located in the ventrolateral preoptic region, contain the inhibitory neurotransmitter GABA and project to the tuberomammillary nuclei in the caudal hypothalamus. The tuberomammillary nuclei, a histaminergic cell group, diffusely innervate the cerebral cortex and have been implicated in arousal. These observations reveal a circuitry whose function coincides with the opposing effects of the rostral and caudal hypothalamus in sleep regulation documented by Nauta.

Compelling evidence also supports the conclusion that the caudal hypothalamus contains another population of neurons that, like the tuberomammillary nuclei, is diffusely projecting and implicated in the control of arousal. These neurons are found within the posterior and lateral hypothalamus and contain a recently discovered peptide known as hypocretin (or orexin). Studies of the causal mechanisms underlying the sleep disorder known as narcolepsy implicate this population of neurons in the regulation of behavioral state. Narcolepsy is a disorder characterized by excessive daytime sleepiness and sudden inappropriate intrusion of sleep during waking. A defect in one of the hypocretin receptors has been demonstrated in a canine model widely employed to study the disease, and transgenic mice lacking the hypocretin gene display sleep–wake disturbances similar to narcolepsy. Additionally, cerebrospinal fluid (CSF) levels of hypocretin are reduced in humans suffering from this disorder, and postmortem analysis of the hypothalamus of narcoleptics has demonstrated substantial reductions in the number of hypocretin neurons.

These findings, along with the aforementioned literature and the documented influence of the biological clock on the temporal organization of the sleep–wake cycle, are consistent with an important role for the hypothalamus in the regulation of behavioral state. Nevertheless, it is important to emphasize that the hypothalamus functions within a larger ensemble of neurons devoted to behavioral state regulation. Particularly important in this regard are monoaminergic and cholinergic brain stem neurons, as well as thalamic and basal forebrain cell groups (Chapter 42). Understanding how these distributed systems function in an integrated fashion to control behavioral state remains an important goal.

Neuroendocrine Regulation through the Pituitary Gland

The median eminence is the gateway through which the hypothalamus exerts regulatory control over peripheral organ systems. This CVO is continuous with the pituitary gland through a thin tissue bridge known as the infundibular stalk. Hypothalamic neurons exercise this regulatory control via the release of peptides and dopamine (DA) into the vasculature and the infundibular stalk is essential to this regulation.

Magnocellular neurons in the supraoptic and paraventricular nuclei give rise to axons that project through the median eminence and infundibular stalk to terminate in the posterior lobe of the pituitary (Fig. 34.5). This projection system, the tuberohypophyseal tract, is the only direct neural connection between the hypothalamus and the pituitary and permits magnocellular neurons to release vasopressin or oxytocin into the peripheral circulation in the posterior pituitary. These peptides are well known for their influences on fluid homeostasis (vasopressin) and milk letdown in lactating females (oxytocin).

Extensive hypothalamic control over peripheral organ systems is achieved through the release of peptides and DA into fenestrated capillaries in the median eminence. These fenestrated capillaries loop through the median eminence and coalesce to form long portal vessels that travel along the infundibular stalk to terminate in the vascular sinuses in the anterior pituitary. A variety of parvicellular (small) neurosecretory neurons distinguished by peptide phenotype contribute to this regulation either by stimulating or by inhibiting the release of hormones from the anterior lobe of the pituitary. The PVH and arcuate nuclei contain a large proportion of these neurons. PVH neurons are found within the parvicellular subdivisions of the nucleus and produce a diverse group of peptides, such as corticotropin-releasing hormone and thyrotropin-releasing hormone. Similarly, neurons in the arcuate nucleus produce peptides, such as growth hormone-releasing hormone, and neurotransmitters, such as dopamine, that influence the secretory activity of the anterior pituitary. Thus, the synaptic activity of neurons projecting on these hypothalamic neurons is integrated, and the information is converted to a hormonal signal that ultimately influences peripheral physiology.

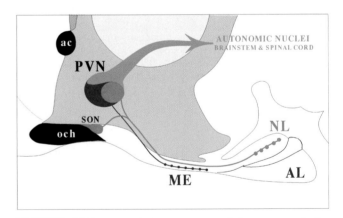

FIGURE 34.5 Functional associations of the paraventricular nucleus (PVH) with endocrine and autonomic systems are illustrated. Axons of magnocellular neurons in the PVH and supraoptic nuclei (SON) (shown in red) traverse the internal layer of the median eminence (ME) and the infundibular stalk to terminate on a vascular plexus in the neural (posterior) lobe of the pituitary gland (NL). Parvicellular PVH neurons influence the functional activity of the anterior lobe of the pituitary gland (AL) through projections (shown in blue) that terminate on a fenestrated capillary plexus, the portal plexus, in the external zone of the median eminence. The terminals of these axons release peptides and neurotransmitters into the portal vasculature, which carries them to the anterior lobe of the pituitary. Parvicellular PVH neurons also give rise to descending projections (shown in turquoise) to autonomic nuclei in the brain stem and spinal cord. ac, anterior commissure; och, optic chiasm.

Reproductive Function

The ability of mammals to reproduce depends on the function of the hypothalamus and its ability to communicate with the pituitary gland. This is especially true for the production of gametes, whose maturation and release in both males and females require a functionally intact hypothalamic–pituitary axis. The hypothalamus also plays an essential role in the organization and expression of the complex behaviors necessary for copulation (see Chapter 40).

Experimental studies in rodents have demonstrated that the rostral medial zone of the hypothalamus plays an important role in orchestrating reproductive behaviors. The medial preoptic region participates in the control of masculine sexual behavior, such as erection, mounting, and ejaculation, whereas ventral regions of the hypothalamus play a major role in the control of feminine sexual behaviors, such as lordosis.

A number of hypothalamic cell groups implicated in reproductive function are sexually dimorphic. For example, ultrastructural studies of a preoptic region in male and female rats have demonstrated differences in synaptology that develop in response to perinatal exposure to androgens. Also, differences in the size and serotonergic innervation of the medial preoptic nuclei of male and female rats have been demonstrated. The possibility that similar dimorphisms are present in the human hypothalamus is supported by the identification of a region of the human preoptic region that is approximately twice as large in males as in females. Although the functional significance of sexual dimorphism in the structure and connectivity of the hypothalamus remains to be established, the gender of the brain is clearly determined by the hormonal milieu present during a critical period of development.

Much is known about the hypothalamic cell groups that influence the production of gametes. The production of ova and sperm is controlled by the release of gonadotropins from the anterior pituitary gland, which is in turn is regulated by the hypothalamus in a gender specific manner. For example, spermatogenesis occurs in response to constant release of gonadotropins in males, whereas ovulation is a cyclic event initiated by a rhythmic surge in gonadotropin release in females. In both cases, hypothalamic neurons produce gonadotropin-releasing hormone (GnRH) to stimulate the release of gonadotropin from the anterior lobe of the pituitary. The gender-specific patterns of gonadotropin release imply that there are sexual differences in the functional organization of circuitry that controls the activity of GnRH neurons. One postulated sexually dimorphic difference is a central pattern generator, thought to be responsible for the cyclic release of GnRH in females. The identity, organization, and regulation of the cell groups that contribute to this pattern generator remain an active and important focus of investigation.

Central Integration of Autonomic Function

The hypothalamus is a prominent component of the central network of neurons that regulates autonomic function. Descending projections from the hypothalamus to autonomic cell groups in the brain stem and spinal cord exert a strong influence upon the sympathetic (thoracic) and parasympathetic (cranial-lumbosacral) divisions of the autonomic nervous system (ANS). Several hypothalamic cell groups also indirectly influence the ANS through relays that integrate processed sensory information and project to the preganglionic neurons of the spinal cord that are the final common pathway for autonomic outflow. Brain stem neurons involved in cardiovascular regulation are among the best-characterized examples of such relays. For example, brain stem neurons in the rostroventrolateral medulla (RVLM) exert control over mean arterial pressure through direct projections to sympathetic preganglionic neurons in the intermediolateral cell column of thoracic spinal cord. The activity of these neurons is controlled in a reflex manner through a caudal brain stem circuit that integrates baroreceptor sensory input terminating in the NTS (Chapter 36). However, descending projections from the hypothalamus (e.g., paraventricular nucleus) and other CNS cell groups important for coordinating autonomic function and behavioral responses to environmental challenges also influence the activity of RVLM neurons. This is one of many relays through which the central network devoted to autonomic control provides a dynamic regulatory capacity ideally suited for maintaining homeostasis and ensuring adaptive changes in physiology to environmental challenges.

Immune Function and Thermoregulation

Like homeostasis, the hypothalamus exerts an influence over the immune system through neuroendocrine output and through the ANS. Szentivanyi and colleagues examined the effects of hypothalamic lesions and stimulation on anaphylactic responses and found that the diencephalon may influence immune function. Felten and colleagues (1987) have shown that the activity of immune cells of the spleen is influenced directly by "synaptic-like" contacts of noradrenergic neurons of the sympathetic ANS. Neuroendocrine influences also have been demonstrated in work pioneered by Hugo Besedovksy. Nevertheless, the mechanisms through which this regulation is achieved remain unclear, and the extent to which the nervous system controls immune function remains to be established. This remains an active and important area of investigation (for a review, use Ader, 1996).

Fever is one consequence of an activated immune system. Thus, the neural circuits that control temperature regulation provide important insights into neural-immune interactions. In this regard, it is well known that the hypothalamus contains populations of neurons that either stimulate or inhibit heat production. These include heat sensitive neurons in the medial preoptic area that act to reduce core body temperature by inhibiting other neurons in the hypothalamus and caudal brainstem that induce thermogenesis when activated (Morrison, 2005). In rodents and other small mammals, nonshivering thermogenesis is achieved through the activation of sympathetic outflow to brown adipose tissue (BAT). The mitochondria of BAT adipocytes contain an uncoupling protein that converts energy from fatty acid metabolism to heat, a process that is increased by sympathetic activity. Understanding the identity and organization of the central pathways involved in the control of core body temperature has benefited considerably from analysis of this regulatory circuit. Interestingly, there is considerable overlap of cell groups that regulate energy balance, stress, and fever. Although there is much to be learned regarding the neural systems involved in temperature regulation, it is clear that the hypothalamus is at the center of the sensory integration necessary to respond to challenges generated within (e.g., inflammatory responses to infection) as well as those encountered in the environment.

Fluid Homeostasis and Thirst

Insight into the regions of the brain involved in fluid homeostasis has come from observing the response of neurons to dehydration and other experimental changes in the internal milieu. The hypothalamus is a prominent component of a set of CNS nuclei in which neurons respond to dehydration (an increase in fluid osmolarity) and changes in blood volume by increasing their metabolic activity. Activated neurons have been demonstrated in the subfornical organ, the paraventricular and supraoptic nuclei, and an area of the rostral hypothalamus known as the AV3V region, which includes the median preoptic nucleus and the OVLT. These regions interact with neurons in the area postrema, the nucleus of the solitary tract, the noradrenergic cell groups of the brain stem, and the parabrachial nucleus to integrate peripheral stimuli and elicit adaptive changes in the fluid environment of the

organism. CVOs subserve prominent regulatory functions within this circuitry.

The SFO has dense projections to the supraoptic and paraventricular nuclei, to the region adjacent to the OVLT, and to the perifornical region of the lateral hypothalamic area. Such projections are consistent with postulated roles of the SFO in autonomic and homeostatic functions. The peptide hormone angiotensin II (AII) is integral to the way in which the SFO transduces peripheral signals and then produces compensatory neural responses. When blood pressure falls, the kidneys release renin into the bloodstream. Renin triggers a biochemical cascade that produces AII. Because of the absence of a blood–brain barrier in the SFO, circulating AII can enter the SFO and bind AII receptors on SFO neurons. Interestingly, SFO neurons appear to use AII as a neurotransmitter in projections to hypothalamic cells that are involved in the control of fluid dynamics. In this example, activation of hypothalamic systems that elicit drinking and reduce fluid excretion at the kidneys would produce the adaptive compensatory homeostatic increase in blood pressure. Further details on the components of the regulatory system are provided in Chapter 36.

Different populations of hypothalamic neurons elicit adaptive responses to changes in fluid osmolarity. Osmosensitive neurons are located primarily in the AV3V region, which includes the OVLT. The response of these neurons to alterations in fluid osmolarity elicits an adaptive compensatory thirst. The neural circuitry that modulates this response is not fully understood but is known to involve input from the OVLT, SFO, and NTS. Further, it takes advantage of the same output regulatory pathway, hypothalamic magnocellular neurons, utilized by the SFO. Chapter 39 considers the hypothalamic role in fluid homeostasis in greater detail.

Food Intake

The consumption of food is another complex behavior that involves several regions of the brain. The hypothalamus plays an essential role in this circuitry as an integrative center for a variety of sensory stimuli, both synaptic and hormonal. It also serves to modulate autonomic outflow to the viscera. This modern view of neural regulation of food intake is a substantial modification of the simple concept of opposing centers for hunger and satiety that was popular in the older literature. Chapter 38 contains a detailed discussion of literature that has contributed to the modern view of the neural control of food intake. The following description summarizes the functional organization of hypothalamic nuclei that regulate food intake.

When food is consumed, hormones are released from the viscera, and sensory pathways innervating the alimentary tract are stimulated. These sensory pathways regulate neurons in the area postrema and NTS that are connected bidirectionally with hypothalamic nuclei. Hypothalamic nuclei are also subject to feedback regulation from hormonal signals arising in the gut. At least four hypothalamic nuclei are involved in the control of feeding, and a prominent part of their influence is exerted via descending projections to autonomic nuclei in the brain stem and spinal cord. Paraventricular nuclei provide a prominent descending hypothalamic influence. In particular, parvicellular oxytocinergic neurons in the PVH provide an important descending projection to the dorsal vagal complex (including the NTS) and the sympathetic preganglionic neurons of the thoracic spinal cord. At least some of this descending influence appears to be under the control of neuropeptide Y (NPY)-containing neurons that project from the arcuate nucleus to the PVH. In experiments, infusion of NPY into the PVH stimulates food intake and NPY-containing neurons in the arcuate nucleus are inhibited by insulin released into the peripheral circulation after a meal and by leptin, a peptide released from adipose tissue. Thus, hypothalamic nuclei regulate food intake through the autonomic nervous system, and a subset of these neurons is influenced by peripheral signals from the periphery (see Chapter 38).

Summary

It is clear that the hypothalamus plays a pivotal regulatory role central to the survival and propagation of vertebrate species. This regulation is achieved by output pathways that utilize both synaptic and hormonal mechanisms and is subject to feedback regulation by sensory information derived from the systems that they regulate. The integrative capacities of the hypothalamus are fundamental to its ability to orchestrate the complex and diverse adaptive behaviors that underlie homeostasis, reproduction of the species, and motivated behaviors. Within this context it is important to view the hypothalamus as an integral component of ensembles of neurons devoted to specific functions rather than an individual unit. Swanson has advanced a conceptual framework consistent with this philosophy. In this formulation, the hypothalamus is part of a "behavioral control column" that functions as a topographically organized integrative center subject to the control of cerebral hemispheres. The functional organization of this column is consistent with the dis-

tributed localization of functions classically associated with the hypothalamus. For example, the rostral portion of the column is devoted to the control of ingestive (eating and drinking) and social (defensive and reproductive) behaviors, whereas the caudal portion, which is continuous with the substantia nigra, functions in support of exploration and foraging. Modulation of these functions is under the control of the cerebral hemispheres through three descending projections involving the cortex, striatum, and pallidum. The role of the hypothalamus in imparting temporal organization on the behavioral state and physiology is an important component of this organizational framework. The explosion of organizational and functional data that has accumulated over recent decades has expanded our understanding of hypothalamic function and established this small but complex subdivision of the diencephalon as an important integrative unit for the expression of behavior and maintenance of homeostasis.

References

Ader, R. (1996). Historical perspectives on psychoneuroimmunology. In "Psychoneuroimmunology, Stress, and Infection" (H. Friedman, T. W. Klein, and A. L. Friedman, eds.), pp. 1–24. CRC Press, Boca Raton, FL.

Boehm, U., Zou, Z., and Buck, L. B. (2005). Feedback loops link odor and pheromone signaling with reproduction. *Cell* **123**, 683–695.

Ehrlich, P. (1956). *Uber die methylenblaureaction der lebenden Nervensubstanz. In* "The Collected Papers of Paul Ehrlich" (F. Himmelweit, ed.), Vol. 1, pp. 500–508. Pergamon Press, London.

Felten, D. L., Felten, S. Y., Bellinger, D. L., Carlson, S. L., Akerman, K. D., Madden, K. S., Olschowka, J. A., and Livnat, S. (1987). Noradrenergic sympathetic neural interactions with the immune system: Structure and function. *Immunol. Rev.* **100**, 225–260.

Fulton, J. F., Ranson, S. W., and Frantz, A. M. (eds.) (1940). "The Hypothalamus." Williams & Wilkins, Baltimore.

Gurdijian, E. S. (1927). The diencephalon of the albino rat. *J. Comp. Neurol.* **43**, 1–114.

Harris, G. W. (1948). Neural control of the pituitary gland. *Physiol. Rev.* **28**, 139–179.

Hess, W. R. (1957). "The Functional Organization of the Diencephalon." Grune & Stratton, New York.

His, W. (1893). *Vorschlage zur Einteilung des Gehirns. Arch. Anat. Entwicklungs Gesch.* (Leipzig) **17**, 157–171.

Johnson, A. K. and Gross, P. M. (1993). Sensory circumventricular organs and brain homeostatic pathways. *FASEB J.* **7**, 678–686.

Krieg, W. J. S. (1932). The hypothalamus of the albino rat. *J. Comp. Neurol.* **55**, 19–89.

Le Gros Clark, W. E., Beattie, W. E., Riddoch, W. E., and Dott, N. M. (1938). "The Hypothalamus." Oliver & Boyd, Edinburgh.

Morrison, S. F. (2004). Central pathways controlling brown adipose tissue thermogenesis. *News in Physiological Sciences* **19**, 67–74.

Nauta, W. J. H. (1946). Hypothalamic regulation of sleep in rats: An experimental study. *J. Neurophysiol.* **9**, 285.

Papez, J. W. (1937). A proposed mechanism of emotion. *Arch. Neurol. Psychiatry* **38**, 725–743.

Risold, P. Y. and Swanson, L. W. (1996). Structural evidence for functional domains in the rat hippocampus. *Science* **272**, 1484–1486.

Scharrer, B. and Scharrer, E. (1940). Sensory cells within the hypothalamus. *In* "The Hypothalamus" (J. F. Fulton, S. W. Ranson, and A. M. Frantz, eds.), pp. 170–194. Williams & Wilkins, Baltimore.

Sherin, J. E., Shiromani, P. J., McCarley, R. W., and Saper, C. B. (1996). Activation of ventrolateral preoptic neurons during sleep. *Science* **217**, 216–219.

Yoon, H., Enquist, L. W., and Dulac, C. (2005) Olfactory inputs to hypothalamic neurons controlling reproduction and fertility. *Cell* **123**, 669–682.

Suggested Readings

Rinaman, L. (2007). Visceral sensory inputs to the endocrine hypothalamus. *Front. Neuroendocrinol.* **28**, 50–60.

Saper, C. B. (1990). Hypothalamus. *In* "The Human Nervous System" (G. Paxinos, ed.), pp. 389–414. Academic Press, San Diego.

Swanson, L. W. (2000). Cerebral hemisphere regulation of motivated behavior. *Brain Res.* **886**, 113–164.

Swanson, L. W. and Cowan, W. M. (1975). Hippocampo-hypothalamic connections—Origin in subicular cortex, not Ammons Horn. *Science* **189**, 303–304.

J. Patrick Card, Larry W. Swanson, and
Robert Y. Moore

Central Control of Autonomic Functions: Organization of the Autonomic Nervous System

The autonomic nervous system (ANS) consists of the neural circuitry that controls the body's physiology. This network is a morphologically, embryologically, functionally, and pharmacologically distinct division of the nervous system. Working in concert with neuroendocrine mechanisms, the ANS is responsible for homeostasis.

Unlike the skeletal motor system, which innervates striated muscles, the autonomic nervous system innervates the body's smooth muscle organs and tissues. The ANS projects to, and forms plexuses within, the hollow organs, or viscera, such as the heart and lungs in the thorax, and the gastrointestinal, genital, and urinary tracts in the abdomen. It also projects to the blood vessels, glands, and other target tissues found within trunk muscles, limb muscles, and skin.

The ANS coordinates what Walter B. Cannon (1939) referred to as the "wisdom of the body"; specifically it provides the tissue-, organ-, and system-level integration of the body's physiology to achieve homeostasis and support behavior. In tandem with endocrine controls, the ANS orchestrates the continuous adjustments in blood chemistry, respiration, circulation, digestion, reproductive status, and immune responses that protect the integrity of the internal milieu and enable the functioning and coordination of skeletal muscles and exteroceptive senses into what we know as behavior. Autonomic adjustments are often phasic responses that occur with latencies typical of neural reflexes; in contrast, endocrine responses usually occur more slowly, taking anywhere from minutes to hours to seasons for full expression.

Autonomic impairments lead to the loss of ability to mobilize physiological responses and adaptations to challenges to homeostasis. For example, Cannon (1939), in his landmark physiological experiments on the ANS, found that destruction of different autonomic pathways devastated the ability of an animal to regulate its body temperature, to respond to perturbations of its fluid and electrolyte balance, to control its blood sugar homeostasis, or to mobilize in response to threats.

A familiar experience illustrates autonomic operations: Occasionally, when you have been lying down resting and then jump up suddenly, you may feel dizzy, your vision may become blurred, and you may momentarily have trouble maintaining your balance—let alone doing something active and coordinated. This phenomenon is postural, or orthostatic, hypotension and is caused by insufficient oxygenated blood reaching your brain under the altered hemodynamic conditions produced by the rapid change in posture. Most people experience postural hypotension only occasionally, though for some it is a chronic medical problem. Indeed, it is a debilitating condition in people with various forms of dysautonomia (Bannister, 1989).

The principle underscored by postural hypotension is that even something as mundane as a change in posture requires a near-instantaneous redistribution of blood flow throughout the body to protect the privileged flow of blood, and thus oxygen and glucose, to the brain. The demands of postural adjustments require unceasing monitoring and rebalancing of circulatory loads. The fact that these adjustments happen continuously and automatically is testimony to the efficiency of the ANS in producing homeostatic adjustments. This example also illustrates two other characteristics of many autonomic responses: (1) Often the responses either are initiated in anticipation of the perturbation or implemented so rapidly that the individual does not experience a deficit and (2) the activity of the ANS is linked to and coordinated with activity of the somatic nervous system.

The range of autonomic functions is hard to overemphasize. They are by no means limited to cardiovascular and metabolic operations, to "fight or flight" situations, or to short-term adjustments. The ANS has equally essential roles, for example, in all reproductive behavior and physiology, including the choreographies of copulation, gestation, parturition, and lactation. Autonomic reflexes control, on a continuous basis, all bodily interfaces with the environment (e.g., the dermal, respiratory, and digestive epithelia). And the ANS plays critical roles in immune system responses; it also coordinates the elimination of metabolic waste products.

The autonomous nature of ANS function is another characteristic of autonomic control. We are rarely aware of ongoing reflex adjustments made to maintain cardiovascular and regulatory dynamics, fluid and electrolyte homeostasis, energy balance, immune system operations, and our reproductive functions. Like the batch files of a computer, which execute without normally intruding on the screen or requiring attention, the autonomic reflex adjustments run unceasingly in the "background," whereas other sensory events and functional decisions occupy the "foreground." In one of his delightful and trenchant essays, Lewis Thomas (1974) speculates on how disastrous and overwhelmingly confusing it would be if he were put in charge of his liver—if he had to consider all visceral inputs and make a conscious decision for all adjustments in liver function. The ANS takes care of such hepatic, as well as all other visceral decisions automatically, allowing the individual to focus on the behavioral and cognitive functions that typically require awareness.

The hub of the autonomic nervous system is the visceral motor outflow, which is divided into *sympathetic* (SNS) and *parasympathetic* (PSNS) divisions. Each division is organized hierarchically into pre- and postganglionic levels (Fig. 35.1). The cell bodies of the preganglionic neurons lie within the central nervous

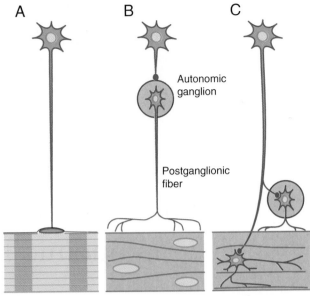

FIGURE 35.1   Somatic and autonomic styles of motor innervation are different. In the somatic motor pattern (A), motor neurons of the spinal cord and cranial nerve nuclei project directly to striated muscles to form neuromuscular junctions. In the autonomic or visceral pattern (B and C), in contrast, motor neurons in the central nervous system project to peripheral postganglionic neurons that in turn innervate smooth muscles. (B) The SNS has preganglionic neurons in the thoracic and lumbar spinal cord (intermediolateral cell column) that project to postganglionic motor neurons in para- and prevertebral autonomic ganglia. (C) The PSNS consists of preganglionic neurons in brain stem cranial nerve nuclei and in the sacral spinal cord (intermediomedial cell column) that project to postganglionic motor neurons in ganglia located near or inside the viscera. From Nauta and Feirtag (1986).

system, specifically in the brain stem and spinal cord. Descending projections from more rostral levels of the neuroaxis converge on these ANS preganglionic motor neurons in a pattern similar to that of the centrifugal projections to somatic motor neurons of the brain stem and spinal cord. However, unlike the outflow pattern of the skeletal motor system, in which neurons of the ventral horn or the cranial nerve nuclei project directly to the effector or muscle target, autonomic preganglionic neurons project to peripheral ganglia located between the last CNS station and the target tissues. These ganglia of the ANS contain the postganglionic motor neurons that project to the effector tissues. The extra synapse interrupting the autonomic outflows at peripheral ganglia allows for more divergence, as well as the possibility of more local integrative circuitry to impinge on the outflow. Different locations both of the preganglionic motor neurons in the CNS and of the

TABLE 35.1 Classical Comparisons of Sympathetic and Parasympathetic Divisions of the ANS[a]

	SNS	PSNS
Location of preganglionic somata	Thoracolumbar cord	Cranial neuroaxis; sacral cord
Location of ganglia (and postganglionic somata)	Distant from target organ	Near or in target organ
Postganglionic transmitter	Norepinephrine	Acetylcholine
Length of preganglionic axon	Relatively short	Relatively long
Length of postganglionic axon	Relatively long	Relatively short
Divergence of preganglionic axonal projection	One to many	One to few
Functions	Catabolic	Anabolic
Innervates trunk and limbs in addition to viscera	Yes	No

[a]Langley (1921) stressed the contrasting and distinguishing traits of the two divisions of the ANS. Since his time, most discussions have identified several contrasting features. Most of these distinctions apply generally but not universally. Although many of these distinctions have blurred (e.g., transmitter) or may be completely unfounded (e.g., divergence), they often are discussed.

ganglia containing the somata of the postganglionic motor neurons in the periphery are distinguishing features of the two principal divisions of the autonomic nervous system (Langley, 1921; see Table 35.1).

SYMPATHETIC DIVISION: ORGANIZED TO MOBILIZE THE BODY FOR ACTIVITY

Sympathetic Responses Predominately Produce Energy Expenditure, Catabolic Functions, and Adjustments for Intense Activity

"Fight or flight," Cannon's prototypical example, epitomizes the operation of the sympathetic division of the ANS: You are walking alone at night in an unfamiliar part of town. Although the area is somewhat unsafe and it is too dark to see much of your surroundings, the night is quiet, and you manage to relax. After walking for several minutes, however, there is a sudden, loud, and unfamiliar noise close by and behind you. In literally the time of a heart beat or two, your physiology moves into high gear. Your heart races; your blood pressure rises. Blood vessels in your skeletal muscles dilate, increasing the flow of oxygen and energy. At the same time, blood vessels in your gastrointestinal tract and skin constrict, making more blood available to be shunted to skeletal muscle. Digestion is inhibited; release of glucose from the liver is facilitated. Your pupils dilate, improving vision. You begin to sweat, a response serving several functions, including promoting additional dissipation of heat so

muscles can work efficiently. Multiple other smooth and cardiac muscle adjustments occur automatically and almost all of them are affected by the sympathetic division of the ANS.

This fight-or-flight example illustrates important features of the ANS generally and its SNS division more particularly. First, the situation points out the utility of the housekeeping functions of the ANS: If the alarming noise turns out to be a threat and it is necessary to fight or flee, skeletal muscles must be optimally tuned and provisioned. Extra quantities of oxygen and energy are essential. Second, the synergy of adjustments indicates a coordinated and adaptive program of responses, which in this case happen immediately, almost before cognitive evaluation. Third, the short latency of the response is characteristic (and in such a case possibly critical). Some of the physiological adjustments could be accomplished by hormonal mechanisms, but such adjustments would be too slow to be of much immediate help if the noise should signal a bona fide threat.

Though the flight-or-fight scenario highlights short-latency responses, not all sympathetic (or parasympathetic) responses happen on such short time scales. ANS operations also involve continuous, ongoing adjustments. Textbook accounts of the emergency functions of the ANS and of single, isolated reflex adjustments sometimes leave the incorrect impression that the ANS is quiescent in the absence of a threat or specific stimulus. Instead, the autonomic nervous system, like the somatic nervous system, always maintains an operating tone. Even during rest, the system maintains appropriate homeostatic balances and autonomic programs.

Preganglionic Neurons of the Sympathetic Nervous System Are Located in the Thoracic and Lumbar Spinal Cord

Sympathetic preganglionic neurons occupy the *intermediolateral nucleus*, a columnar grouping of cells running longitudinally through much of the spinal cord, in the lateral horn of the spinal gray. In humans, these cells are found between the first thoracic spinal segment (T1) and the third lumbar segment (L3) (Fig. 35.2A). The location of these preganglionic neurons and their axonal outflow gives the SNS another of its names, the *thoracolumbar division*. Cells within the intermediolateral column show a tendency to segmental aggregation. There is a rostral-to-caudal viscerotopic organization to the distribution (Fig. 35.2A). The column is found bilaterally, and preganglionics on one side send their axons out the ipsilateral ventral root. The axon of a preganglionic neuron typically exits from the segment in which its soma is located. Many of the preganglionic axons are lightly myelinated, and, as they separate from the ventral root to project to the appropriate peripheral ganglion, they form a *white ramus*, a connective that takes its descriptive name from myelinated fibers.

Sympathetic postganglionic neurons are found in two distinct types of peripheral ganglia: paravertebral and prevertebral. *Paravertebral ganglia*, as the name implies, are adjacent to the spinal cord bilaterally, in a position slightly ventral and lateral to the vertebral column (Figs. 35.3 and 35.4). Each ganglion receives a white ramus from the appropriate ventral root. These ganglia also are interconnected longitudinally into a sympathetic chain linked by axons from the preganglionic neurons that run rostrally or caudally to neighboring ganglia of the chain. Axons of the postganglionic neurons of the paravertebral ganglia course centrifugally out of the ganglion to join the peripheral nerves that will carry the individual fibers to their targets. Because most postganglionic axons in the sympathetic nervous system are unmyelinated, these connectives joining the ventral roots are called *gray rami*.

Paravertebral ganglia of the sympathetic chain supply the postganglionic innervation of the head, thorax, trunk, and limbs. For the trunk and limbs, postganglionic axons course both alongside and within the somatic peripheral nerve innervating the region. These autonomic fibers innervate the blood vessels in the muscles (vasoconstrictor fibers), as well as different targets within the skin, including the sweat glands (sudomotor fibers) and erector pili muscles of erectile hairs (pilomotor fibers).

At the rostral end of the sympathetic chain, individual segmental ganglia are fused into aggregate

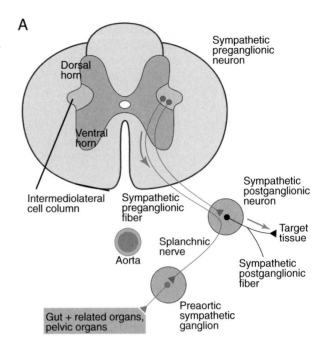

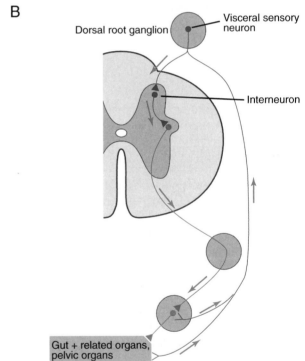

FIGURE 35.2 Details of the organization of the SNS and its sensory inputs. In thoracic and lumbar levels of the spinal cord, preganglionic neuronal somata located in the intermediolateral cell column project through ventral roots to either paravertebral chain ganglia or prevertebral ganglia, as illustrated for the splanchnic nerve (A). Visceral sensory neurons located in the dorsal root ganglia transmit information from innervated visceral organs to interneurons in the spinal cord to complete autonomic reflex arcs at the spinal level (B). From Loewy and Spyer (1990).

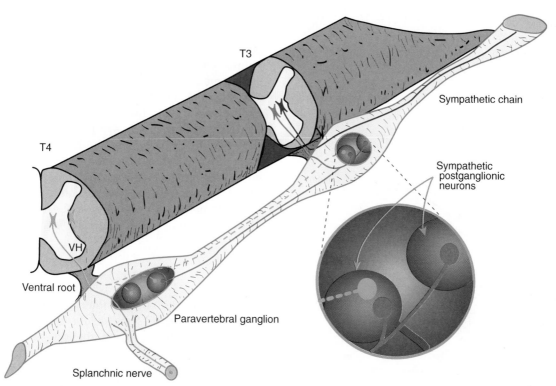

FIGURE 35.3 The SNS is organized segmentally. As illustrated here for segments T3 and T4 of the thoracic spinal cord, preganglionic axons exit through the ventral root of the segment in which the preganglionic cell body is located. These preganglionic axons can be myelinated (solid red lines) or unmyelinated (dashed red lines) and can project to the paravertebral ganglion associated with the segment, to neighboring paravertebral ganglia through the sympathetic chain, to postganglionic neurons located distally in the prevertebral ganglia, or through one of the autonomic nerves (splanchnic nerve is illustrated). T3 and T4, segments of the thoracic spinal cord; VH, ventral horn of spinal cord. From Loewy and Spyer (1990).

ganglionic groupings. The sympathetic chain, in a modification of the paravertebral chain pattern (Figs. 35.4 and 35.6), innervates the head and the thoracic viscera. The most rostral group, the *superior cervical ganglion* (supplied from the first and second thoracic segments), supplies the head and neck. The targets of its projections are extensive, including the eyes, salivary glands, lacrimal glands, blood vessels of the cranial muscles, and even the blood vessels of the brain. The paravertebral ganglia just caudal to the superior cervical ganglion form the *middle cervical ganglion* and, moving caudally, the *stellate ganglion*. These latter ganglia supply the postganglionic sympathetic outflows that innervate the heart, lungs, and bronchi.

Unlike the trunk, limbs, head, heart, and lungs, organs of the abdominal cavity are innervated by post-ganglionic neurons in ganglia situated distal to both the spinal cord and the paravertebral sympathetic chain in locations closer to the target tissue. These sites give the ganglia their distinguishing name, *prevertebral ganglia*. The preganglionic neurons innervating these

sets of postganglionic neurons are located in the inter-mediolateral column of the spinal cord, like those innervating the paravertebral chain, but their axons pass through the white rami, the chain ganglia, and the gray rami without synapsing. They then make contact with the peripherally situated prevertebral ganglion cells (Fig. 35.3).

There are four major prevertebral postganglionic stations. The more rostral three are located at points where major abdominal arteries separate from the descending aorta, and the ganglia take their names from their associated arteries (Fig. 35.5). Moving from more rostral to caudal, the *celiac, superior mesenteric,* and *inferior mesenteric ganglia* innervate the gastrointestinal tract. The celiac ganglion (also commonly called the solar plexus because its connectives radiate from it) projects predominantly to the stomach and rostral-most parts of the foregut and its embryological derivatives, such as the liver and pancreas. Consistent with their progressively more caudal locations, the middle and inferior mesenteric ganglia innervate the mid- and

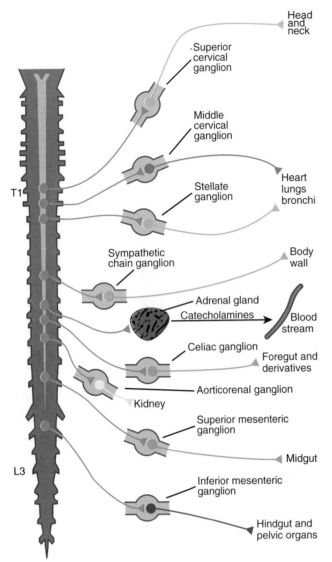

FIGURE 35.4 Summary of the major SNS ganglia and their target organs or tissues. Spinal cord is illustrated on the left. From Loewy and Spyer (1990).

The adrenal medulla is a unique variant of the sympathetic postganglionic pattern. By several criteria, the adrenal medulla is a sympathetic prevertebral ganglion containing postganglionics. However, neurons of this medullary ganglion, rather than issuing axons to innervate target organs, function as a neuro-endocrine organ. The adrenal medulla secretes the classical postganglionic sympathetic transmitter nor-epinephrine, as well as epinephrine, directly into the bloodstream. These catecholamines then circulate as hormones. Release of catecholamines from the adrenals during sympathetic activation provides powerful reinforcement and modulation of the more focal release of catecholamines at traditional—and local—effector sites.

Summary

Overall, the two major motor outflows, or divisions, of the ANS are responsible for the maintenance of homeostasis, but the sympathetic nervous system is the branch specialized for the mobilization of energy. The SNS provides the vascular, glandular, metabolic, and other physiological adjustments that optimize behavioral responses, particularly in emergency situations and conditions requiring activity. The SNS efferent outflow consists of preganglionic neurons, which are located in the intermediolateral column of the thoracic and lumbar spinal cord. SNS efferent axons course to paravertebral and prevertebral ganglia containing the postganglionic neurons, the axons of these postganglionic neurons project to the target tissues.

hindgut tissues, respectively. The inferior mesenteric ganglion also supplies the postganglionic innervation to the pelvic organs. The fourth, and most caudal, prevertebral ganglion is the *pelvic–hypogastric plexus*, which innervates urinary and genital tissues. Perhaps because these prevertebral ganglia are located farther from the spinal cord and closer to the target organs (than the paravertebral ganglia) and because they coordinate responses of visceral organs containing separate nerve networks (the enteric nervous system is described later), these prevertebral ganglia may have more complex neural organization than the paravertebral chain ganglia.

PARASYMPATHETIC DIVISION: ORGANIZED FOR ENERGY CONSERVATION

Parasympathetic Nervous System Functions Reduce Energy Expenditure and Increase Energy Stores

Functionally, the parasympathetic nervous system mobilizes homeostatic adjustments that are often opposite and reciprocal to those activated by the SNS (see Table 35.1). In particular, PSNS adjustments usually are considered "rest and digest" responses, in contrast to the "fight or flight" sympathetic activation. The PSNS promotes anabolic processing, whereas the sympathetic nervous system augments catabolic activity. The PSNS promotes the gastrointestinal processes required to digest and absorb nutrients effectively. This branch of the ANS also augments the efficient use

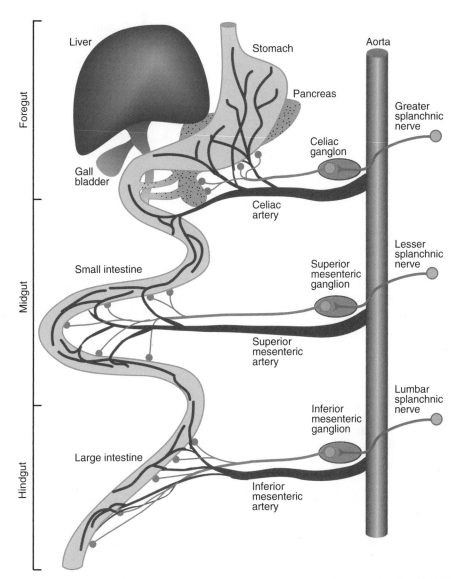

FIGURE 35.5 SNS innervation of the gastrointestinal tract. This more detailed schematic (compared to Fig. 35.4) illustrates how separate segmental levels of the spinal cord and the associated prevertebral ganglia innervate targets in a topographic, specifically viscerotopic, pattern. The association of autonomic pathways and vasculature is also shown. From Loewy and Spyer (1990).

of energy and the storage of extra calories as fat and glycogen. The PSNS, particularly the vagus nerve, plays a central role in controlling the flow of body reserves or energy between storage depots and the tissues that use the energy. The body's energy reserves are fat in adipose tissue and glycogen in liver and muscle. After a meal, the gastrointestinal tract serves as an additional reservoir of potential energy for the body. By mobilizing triglycerides from adipose tissue or glycogen from liver and muscle, the ANS can make reserves available for metabolism; similarly, by moving nutrient stores (e.g., from the stomach) and promoting

their digestion and absorption from the intestines into the bloodstream, the ANS can supply energy to fuel metabolism or to restock triglyceride and glycogen stores, as needed.

Preganglionic Neurons of the Parasympathetic Nervous System Are Located in the Brain Stem and the Sacral Spinal Cord

One of the features distinguishing the PSNS from the SNS is the location of their respective preganglionic neurons. PSNS preganglionic neurons are located in two longitudinal columns of neurons: one in the brain stem and the other in the sacral spinal cord. These locations of PSNS preganglionic neurons are responsible for the alternative name of this division of the ANS, the *craniosacral* (or *bulbosacral*) *division*. The longitudinal column of motor neurons in the brain stem is the general visceral cell column, which condenses into a number of nuclei ventrolateral to the cerebral aqueduct system within the brain. These nuclei then project through cranial nerves to targets in the head (cranial nerves III, VII, and IX) or thorax and abdomen (cranial nerve X) (Fig. 35.6). Cranial nerve III, the oculomotor nerve, carries the axons of the accessory, or autonomic, nucleus of the Edinger–Westphal complex, which participate in the control of the pupillary sphincter and ciliary muscles. Cranial nerve VII, the facial nerve, carries preganglionic axons of the superior salivatory nucleus and controls the submaxillary and sublingual salivary glands, as well as the lacrimal glands. Cranial nerve IX, the glossopharyngeal nerve, carries axons of the inferior salivatory nucleus, which control the parotid salivary glands and mucus secretion.

Preganglionic motor neurons of the Xth cranial nerve, the vagus nerve, control smooth muscles and glands throughout the digestive tract, from the pharynx to the distal colon, including viscera such as the liver and pancreas. These Xth nerve neurons occupy the dorsal motor nucleus of the vagus, a long fusiform nucleus in the medulla oblongata (Fig. 35.7). They control numerous motor and secretomotor responses that participate in the ingestion and digestion of food, as well as in the assimilation of the nutrients. The vagus nerve also carries motor fibers of a special visceral nucleus, the nucleus ambiguus or ventral vagal nucleus, which controls the striated muscles of the pharynx, larynx, and esophagus and the cardiac muscle of the heart. Much like the case described for sympa-

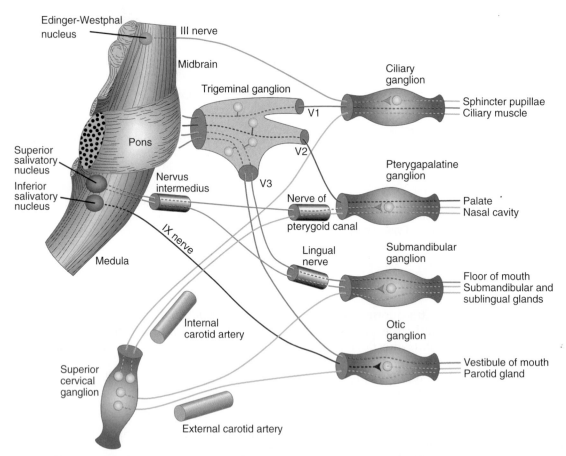

FIGURE 35.6 Autonomic innervation of the head by both sympathetic (superior cervical ganglion) and parasympathetic (cranial nerve nuclei in the medulla and midbrain projecting to the several prevertebral ganglia) pathways. Visceral afferents associated with the autonomic nervous system are found in the trigeminal nerve (cranial nerve V) ganglion. From Loewy and Spyer (1990).

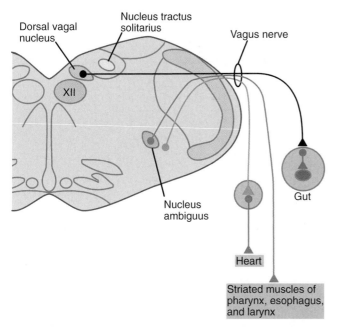

FIGURE 35.7 Parasympathetic projections of the vagal motor pathways, the most extensive of the PSNS circuits, to the viscera of the thorax and abdomen. From Loewy and Spyer (1990).

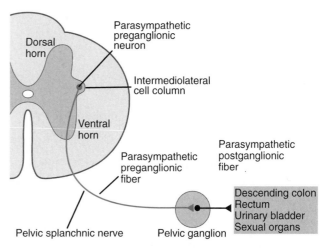

FIGURE 35.8 Spinal cord organization of the sacral division of the craniosacral, or parasympathetic, nervous system. From Loewy and Spyer (1990).

thetic neurons, motor neurons of the dorsal motor nucleus of the vagus and the nucleus ambiguus are organized into viscerotopic subnuclei (e.g., Bieger and Hopkins, 1987; Fox and Powley, 1992).

The central representation of the more caudal part of the PSNS is found in the sacral spinal cord, in humans the second (S2) through fourth (S4) sacral segments (Fig. 35.8). The parasympathetic preganglionic somata occupy two longitudinal groupings: one column immediately dorsolateral to the central canal and the other in roughly the same transverse location as that of the intermediolateral column of cells, which sympathetic preganglionics occupy in the thoracic and lumbar cord. Axons of the sacral preganglionics exit the cord in the ventral roots, run in a sacral plexus, and project to the target organs. The sacral preganglionic cell columns control parasympathetic motor, vasomotor, and secretomotor functions of the kidneys, bladder, transverse and distal colon, and reproductive organs.

The locations of their respective postganglionic neurons constitute another feature that distinguishes the PSNS from the SNS. Ganglia containing the PSNS postganglionic neurons are juxtaposed to, on the surface of, or even in, the target organ, in contrast to the para- and prevertebral locations of the SNS (Fig. 35.1c). These juxta- and intramural stations often are called *plexuses*, rather than ganglia. Like SNS prever-

tebral ganglia, the PSNS postganglionic plexuses are complex integrative sites that typically receive inputs not only from preganglionic neurons, but also from afferents within the plexuses or target tissues. Because the PSNS postganglionic neurons are situated so near their targets, the parasympathetic division of the ANS has, perforce, longer preganglionic axons and shorter postganglionic fibers than the sympathetic division. The relative lengths of the postganglionic fibers in the two divisions have been taken as evidence that individual axons of the SNS may diverge more extensively and innervate larger projection fields, whereas the PSNS may exhibit less divergence and more localized projections. (For counterarguments, see Wang *et al.*, 1995.)

Summary

The parasympathetic nervous system or craniosacral division (i.e., the second major outflow of the ANS) is organized to digest, assimilate, and conserve energy as well as promote rest. Its anabolic functions include not only those associated with the metabolism of nutrients, but also numerous protective reflexes, such as those that limit heat loss, reduce energy expenditure, and slow the heart. The PSNS outflow consists of preganglionic neurons located in brain stem cranial nerve nuclei or the autonomic columns of the sacral spinal cord. Their axons project to the postganglionic neurons located in ganglia situated near or in the target organs. Postganglionic neurons, in turn, project to the smooth muscle of the target viscera.

THE ENTERIC DIVISION OF THE ANS: THE NERVE NET FOUND IN THE WALLS OF VISCERAL ORGANS

Discussions of the ANS sometimes focus on the SNS and PSNS, but there is a third division as well. When Langley articulated the classic definition of the ANS, he identified the *enteric nervous system* (ENS) as the third division. The alimentary canal, or "entrum," and the tissues derived from it, such as those in the pancreas and liver, contain extensive and well-formed neural networks. In particular, the gastrointestinal tract contains two well-organized major plexuses that contain as many neurons as the entire spinal cord (Fig. 35.9). Each plexus consists of an extensive sheet of small nodes of cells linked by connectives. One plexus, the myenteric plexus, is situated between outer longitudinal and inner circular muscle layers of the viscus. The other plexus, the submucosal or submucous plexus, is located between the circular muscle layer and the lumenal mucosal layer of the viscera. As illustrated in Figure 35.9, there are additional, less extensive, enteric networks as well.

The number of neurons in the enteric plexuses, their extensive interconnections, and their capacity to support motility have led to proposals that the enteric nervous system serves as a more or less independent "brain" in the gut. Early descriptions emphasized the autonomy of the ENS by observing that the gastrointestinal tract could exhibit movement, peristalsis, even after all extrinsic connections to it were cut. However, fully integrated or coordinated responses of the gastrointestinal tract, including motor responses other than peristalsis, as well as absorptive and secretory responses, involve extrinsic projections from the central nervous system to the enteric nervous system.

Because the preganglionic neurons of the vagus nerve project to the enteric ganglia rather than to independent postganglionic stations, the enteric nervous system serves as an extensive postganglionic relay station for the vagus.

Summary

The ENS or enteric nervous system, the third division of the ANS, consists of ganglia and plexuses of intrinsic neurons in the walls of the viscera such as the gastrointestinal tract. These nerve networks contain intrinsic afferents, interneurons, and efferents that control local functions; they also receive extrinsic inputs from autonomic preganglionic efferents, particularly those PSNS projections coursing in the vagus nerve.

ANS PHARMACOLOGY: TRANSMITTER AND RECEPTOR CODING

Early in the twentieth century, pharmacological studies performed with naturally occurring agonists and antagonists provided the basis for the original differentiation of two major ANS divisions—the SNS and PSNS. Prior to the pharmacological analyses, the two often had been lumped together. Dale (the contributor of "Dale's law" of neurotransmitters), Langley, and a number of other early investigators applied natural extracts such as muscarine, nicotine, *d*-tubocurarine, and others to autonomic ganglia while measuring responses in order to infer the chemical taxonomy of the ANS. Subsequent experiments have replicated the basic distinctions made by these investigators, but this newer research, with access to a pharmacopoeia of synthetic agonists and antagonists, as well as to immunocytochemistry for the characterization of transmitter substances, has described a much richer and more complicated multiplicity of neurotransmitters, neuromodulators, and receptor subtypes in autonomic nerves (Lundberg, 1996).

Preganglionic Neurons Use Acetylcholine as a Transmitter

Most preganglionic neurons of both the SNS and PSNS have a cholinergic transmitter phenotype. Only

FIGURE 35.9   (A) The enteric nervous system plexuses in the wall of the intestine. Enteric neurons form two conspicuous and extensive networks, or plexuses, of ganglia and connectives located between the longitudinal and circular muscle layers of the wall (the myenteric plexus) and the circular muscle layer and the inner mucosal layer (the submucous plexus). The deep muscular plexus, periglandular plexus, and villous plexus are additional networks of enteric neurons and their processes. From Costa *et al.* (1987). (B) Autonomic preganglionic terminals innervating postganglionic neurons. This example of vagal preganglionic projections (labeled with the tracer Phaseolus vulgaris) innervating myenteric ganglion neurons (stained with Cuprolinic blue) in the stomach wall illustrates that autonomic preganglionic axons can form extensive networks controlling postganglionics. From Holst *et al.* (1997).

A

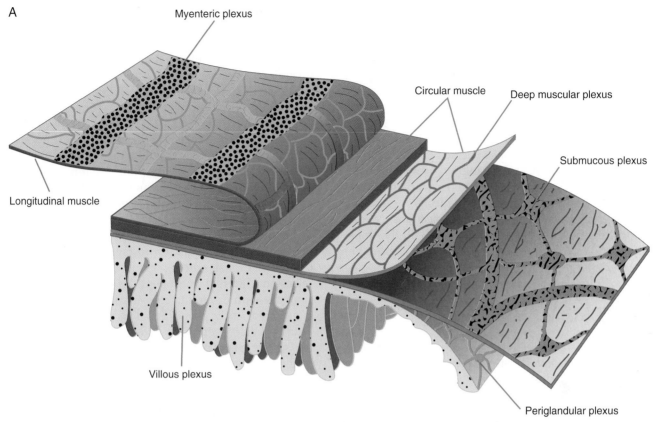

Myenteric plexus

Circular muscle

Deep muscular plexus

Submucous plexus

Longitudinal muscle

Villous plexus

Periglandular plexus

B

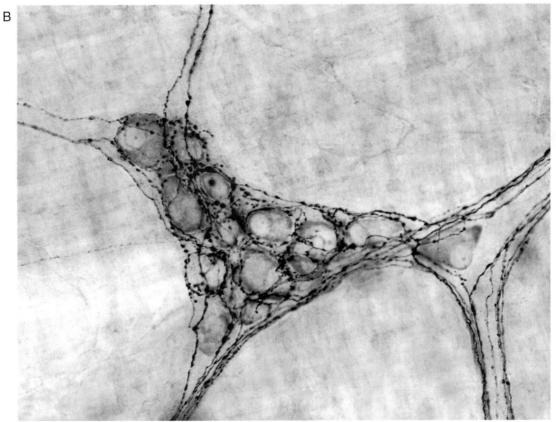

relatively small subpopulations of preganglionic neurons express other (e.g., dopaminergic or adrenergic) phenotypes. In the peripheral ganglia of these two divisions of the ANS, most postganglionic neurons, which receive inputs from the preganglionics, express a predominance of a nicotinic form of the cholinergic receptor on their somata. As discovered by the early autonomic pharmacologists, nicotine can serve as a general "ganglionic blocker" to thwart transmission at these synapses. Additional specificity and response selectivity, or coding, at the ganglionic level normally are maintained by the segregation of sympathetic and parasympathetic postganglionic neurons in separate ganglia, by point-to-point axonal projections, by distinguishing neuropeptides colocalized with the transmitter, and by the spatial separation of synapses within ganglia.

SNS and PSNS Postganglionic Neurons Use Different Transmitters

The respective transmitter phenotypes of the postganglionic neurons constitute one of the cardinal defining differences between the sympathetic and parasympathetic nervous system. The majority of SNS postganglionic neurons release the transmitter norepinephrine, whereas PSNS postganglionic neurons release primarily acetylcholine. Such a two-transmitter chemical code produces push–pull or positive bidirectional control of individual targets.

The biomechanics of smooth muscle are quite different from the biomechanics of striated muscle. In contrast to skeletal muscle, smooth muscle is not always organized into antagonistic pairs, with each member of the pair being innervated by a different motor neuron. Also, the smooth (and cardiac) muscles forming the walls of the viscera are not differentially attached to a hard frame to provide mechanical leverage through opposal activity. Reciprocal motor programs of the viscera, for the most part, involve active, phased contraction and relaxation of the same muscle to achieve coordination. To produce these patterns of excitation and inhibition, the target organs and tissues express both adrenergic and cholinergic receptors and have different, often antagonistic or complementary, responses to selective activation of the different receptors.

Response patterns of smooth muscle are further differentiated by specializations of the receptors. Adrenergic receptors, influenced by the catecholaminergic transmitter released by sympathetic fibers, are differentiated into at least four different metabotropic subtypes linked to different intracellular pathways. Similarly, the muscarinic acetylcholine receptor, influenced by the cholinergic transmitter released by parasympathetic postganglionic neurons, is differentiated into at least three ionotropic subtypes with differential influences on intracellular transduction. These heterogeneities of transmitter species and receptor types make it practical to mobilize multiple, different, and potentially highly differentiated responses from one effector tissue (Box 35.1).

Response selection may also be facilitated by neuropeptides that the postganglionic neurons synthesize and corelease during activity. Somatostatin, neuropeptide Y, or both are found in many sympathetic neurons; vasoactive intestinal polypeptide and calcitonin gene-related peptide are often coexpressed in parasympathetic neurons. These peptides may serve as neuromodulators that vary postsynaptic responses to the conventional autonomic neurotransmitters. The release of different neuropeptides and cotransmitters can vary with different rates or patterns of firing and is not invariantly proportional to transmitter release. Thus, a heterogeneity of different cotransmitter and transmitter release patterns can be produced by a single fiber or fiber type, depending on firing pattern and other local factors at the synapse (e.g., Lundberg, 1996).

Although dual innervation in an oppositional or push–pull pattern is presumed to yield faster, more responsive adjustments and to provide a mechanism that can adjust response gain, some tissues are innervated by only one arm of the ANS. The SNS has a wider distribution than the PSNS. In particular, white and brown adipose tissue, peripheral blood vessels, and sweat glands are innervated by the SNS without reciprocal PSNS projections. One possible explanation for this arrangement is that the functions of these particular effectors do not require the speed associated with push–pull projection patterns, particularly in terminating responses once activated. Furthermore, in the evolution of the autonomic nervous system, the SNS is thought to be the older, and correspondingly more extensive, network, whereas the PSNS occurred more recently in phylogeny and therefore has less widespread projections.

Finally, the enteric nervous system also has an extensively varied set of transmitters, cotransmitters, and neuromodulators, providing highly differentiated chemical coding of local functions (Furness, 2006; Lundberg, 1996).

Summary

Both the SNS and the PSNS use acetylcholine as a preganglionic transmitter, but they are distinguished by different postganglionic phenotypes: most SNS

BOX 35.1

AUTONOMIC POSTGANGLIONIC NEURONS CAN CHANGE THEIR TRANSMITTER PHENOTYPES

Transmitter phenotypes of autonomic neurons generally seem fixed, dictated by both intrinsic and extrinsic developmental cues. At least some ANS neurons, however, can alter their phenotypes under appropriate environmental conditions. This conclusion first was reached when populations of postmitotic adrenergic and cholinergic neurons were cocultured *in vitro* in different media or in the presence of different tissues. Without changes in overall cell number, the percentages of adrenergic and cholinergic neurons in a given culture changed over time. Even more definitively, when single postganglionic cells were grown in microculture wells, the presence of a medium conditioned by heart cells caused some of these mature neurons to change from an adrenergic to a cholinergic phenotype (Potter *et al.*, 1980).

This phenotypic switching does not appear to be an artifact of cell culture situations; the change appears to occur normally *in vivo* in development to generate the specialized minority of sympathetic postganglionic neurons that are cholinergic (e.g., for innervation of sweat glands). Landis and Keefe (1983) have verified this hypothesis by using rat foot pad sweat gland innervation *in vivo* and characterizing the transmitter phenotype of the autonomic projections at different stages of development. The developing sympathetic postganglionic innervation of the sweat glands initially expresses a typical adrenergic phenotype as the axon grows toward its target. Once the sweat gland target is innervated, however, the

tissue induces the change in neurotransmitter phenotype to a cholinergic one.

Such changes in transmitter are not unique to these examples. Perhaps reflecting pluripotential patterns seen earlier in phylogeny, at least some cholinergic neurons (e.g., avian ciliary ganglion cells) appear to remodel into an adrenergic phenotype, whereas other cells (e.g., maturing neural crest cells in the developing gut wall) exhibit a transient catecholaminergic stage that is lost during the ingrowth of extramural innervation.

Finally, in addition to such switching, autonomic postganglionic neurons also have been shown to exhibit plasticity in altering their levels of transmitters and cotransmitters as a function of the activity of their preganglionic inputs. For example, after preganglionic inputs are blocked or transected pharmacologically, postganglionic neurons of the superior cervical ganglion decrease their levels of tyrosine hydroxylase while increasing their level of substance P.

Terry L. Powley

References

Landis, S. C. and Keefe, D. (1983). Evidence for transmitter plasticity *in vivo*: Developmental changes in properties of cholinergic sympathetic neurons. *Dev. Biol.* **98**, 349–372.

Potter, D. D., Landis, S. C., and Funshpan, E. J. (1980). Dual function during development of not sympathetic neurones in culture. *J. Exp. Biol.* **89**, 57–71.

postganglionic neurons are catecholaminergic; most PSNS postganglionics are cholinergic. The different neurotransmitter and cotransmitter elements expressed in the postganglionic neurons, as well as numerous postsynaptic receptor subtypes, provide the chemical coding for responses of the PSNS and SNS. This functional organization ensures a varied, powerful, and dynamic push–pull operation of homeostatic systems.

AUTONOMIC COORDINATION OF HOMEOSTASIS

The "self-governing" or autonomic characteristic that gives its name to the ANS is based largely on

reflexes that involve more than peripheral efferent pathways. Though the early history of autonomic neuroscience emphasized this motor outflow, homeostasis is maintained through the cross-coupling and coordination of multiple reflex arcs. These reflexes involve afferent inputs as well as efferent outputs.

Two major inflows of visceral, or autonomic, afferents (Box 35.2) establish the input arms of the basic reflex arcs. These visceral afferents also provide critical cross-linked feedback between the sympathetic and the parasympathetic outflows, keeping them in balance. For example, sympathetic activity can increase heart rate (tachycardia) and produce constriction of vascular beds, thus leading to an increase in blood pressure. This increase in blood pressure, in turn, is detected by vagus nerve baroreceptors in the aortic arch and other sites. Visceral afferents in the vagus can

---

BOX 35.2

AUTONOMIC REFLEXES: ACTIVATED BY VISCERAL AFFERENTS, AUTONOMIC AFFERENTS, OR JUST AFFERENTS?

A long-standing dispute about terminology still influences discussions of autonomic reflexes. Some authors (and textbooks) adhere to the original autonomic classification of Langley, who considered the ANS a motor system without a corresponding sensory inflow. This classic view acknowledges that autonomic effectors are influenced by afferents but considers visceral afferents, as well as somatic afferents, as independent of the ANS. Conversely, other scientists (and textbooks) consider afferents arising in and associated with the viscera as the necessary counterpart to autonomic efferents and label them autonomic afferents.

The controversy goes back to Langley's concentration on a pharmacological definition of the ANS. In his initial studies, Langley speculated about the existence of autonomic afferents, but was unable to find a neurochemical marker to distinguish visceral from somatic afferents. He sidelined the issue pending more information (his seminal book is subtitled Part I), focused his studies on efferents, and never returned to the search for an afferent marker

(Part II never appeared). Adherents to his original motor-only classification stress that somatic and visceral afferents can elicit autonomic responses and that both classes of afferents elicit somatic responses, as well.

However, advocates for including visceral afferents within the autonomic schema argue that these afferents innervate the target tissues of the ANS efferents, share similar embryological histories with autonomic efferents, course in the same peripheral nerves as the autonomic efferents (e.g., the vagus), and form mono- and oligo-synaptic reflex arcs influencing autonomic preganglionic neurons. These proponents for revision of the autonomic terminology also point out that more recent analyses have identified neuropeptide markers shared by visceral afferents (but not somatic afferents) and autonomic motor neurons. This more inclusive view of the autonomic nervous system seems to be gaining ground. It works more naturally for the discussion of reflexes, and we have adopted it in this text.

Terry L. Powley

---

then reflexively stimulate vagal parasympathetic efferents that slow heart rate (bradycardia). The visceral afferents monitoring the effects of autonomic activity produce positive, reciprocal, and dynamic regulation of physiological responses. When a more sustained adjustment in blood pressure is required (e.g., in the general arousal of the fight-or-flight response discussed earlier), these outflows are adjusted centrally in a process akin to the α–γ linkage so that sympathetic activation is not nullified by parasympathetic responses damping the needed activation.

Visceral Afferents Link Target Tissues and the CNS

In the case of the sympathetic nervous system, motor neurons receive direct afferent input from most of the autonomic targets. The cell bodies of these primary afferents are located in the dorsal root ganglia of the spinal cord segment(s) in which the corresponding preganglionic neurons are located (Figs. 35.2B and 35.10). Afferent somata are similar to other dorsal root

ganglion cells, although they frequently are distinguishable as small, dark neurons or B afferents. From endings in the innervated viscera or tissue, the centripetally directed peripheral processes of the afferents typically course in mixed nerves that also contain the motor outflow. These processes then reach the dorsal roots through the major peripheral nerves. The central processes of these visceral afferents enter the spinal cord in association with Lissauer's tract, ending in laminae I and V of the dorsal horn (Cervero and Foreman, 1990). For many of these visceral afferents, their endings in the periphery and in the spinal cord contain substance P (SP) and other neuropeptides of the tachykinin family, such as neurokinin A (NKA) and neurokinin B (NKB) (see Lundberg, 1996).

Visceral afferent nerves relay sensory information about visceral volume, pressure, contents, or nociceptive stimuli to spinal centers, where automatic responses are interpreted and functional responses are generated.

In the parasympathetic limb of the ANS, there are two major afferent inflows with different organization.

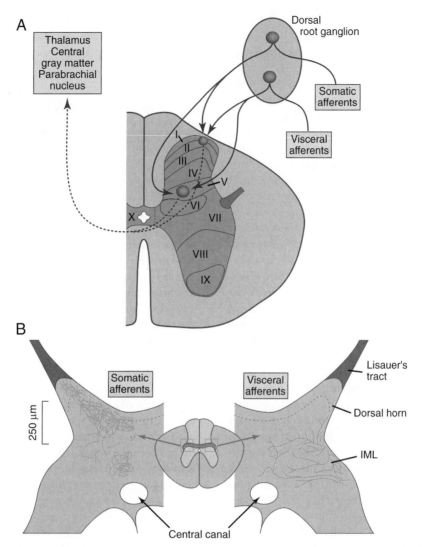

FIGURE 35.10 Visceral and somatic afferents follow parallel but distinct paths in the central nervous system. Visceral and somatic afferents consist of different populations of dorsal root ganglion neurons that project to laminae I and V of the dorsal horn of the spinal cord. These relay sites provide local spinal reflexes and also project to higher autonomic and somatic sites, respectively, in the brain (A). Although visceral and somatic afferents follow similar trajectories, more detailed analyses indicate the two types of afferents end in distinctly different distributions and densities within the spinal cord (B). IML, intermediolateral cell column. From Cervero and Foreman (1990).

In the cranial division, visceral afferents most often are found in the same cranial nerve as their corresponding motor counterparts. These mixed cranial nerves have sensory ganglia located outside the cranium, and the cell bodies of the afferents are found in these ganglia. Like other visceral afferents, these neurons are typically pseudounipolar neurons with a peripheral neurite extending to the target tissue and a central process projecting to the CNS. The central terminals of these afferents end in a cranial nerve sensory nucleus. Visceral afferents associated with the outflow of cranial nerve III are complex and enter through different channels (e.g., cranial nerve V, the trigeminal nerve).

Most of the visceral afferents associated with cranial nerve autonomic reflexes terminate in the extensive nucleus of the solitary tract, which is located immediately dorsal to the general visceral column of the brain stem. Specifically, afferents of VII, IX, and X all end in the nucleus of the solitary tract. These inputs form a visual viscerotopy in the nucleus, with facial gustatory information projecting most rostrally and medially, the glossopharyngeal information somewhat more caudally, and the afferents of the different branches of the vagus nerve terminating most caudally in different subnuclei of the nucleus of the solitary tract. As the largest and most complex of the visceral afferent relays in the brain, the nucleus of the solitary tract also receives afferent inputs from the spinal division of the trigeminal nerve and second-order inputs from the dorsal column nuclei.

Like visceral afferents associated with the SNS, visceral afferents associated with the PSNS contain SP and other tachykinins. In addition, several neuropeptides found in the gut are also found in some of these afferents. For example, the gut hormone and neuropeptide cholecystokinin (CCK) is found in vagal afferents relaying information from the gastrointestinal tract to the brain. CCK is also found in higher order ascending relays of vagal projections in the neuraxis, leading some to argue that CCK provides chemical coding for visceral afferents associated with the gastrointestinal tract.

The second major inflow of afferents to the parasympathetic limb of the ANS is associated with the sacral division. Visceral afferents in this division are organized much like the spinal afferents associated with the thoracolumbar or sympathetic division of the ANS (e.g., Morgan *et al.*, 1991).

At the central relays of autonomic reflex arcs, the thoracolumbar circuitry of sympathetic reflexes and the craniosacral circuitry of parasympathetic responses have similar morphological elements. The SNS motor neurons preferentially distribute their dendrites within the long longitudinally oriented column that constitutes the intermediolateral nucleus of the spinal cord. The neurons also distribute a subset of their dendrites dorsolaterally in an arch that brings the dendrites into contact with the incoming visceral afferents (e.g., Dembowsky *et al.*, 1985). Parasympathetic preganglionics exhibit similar dendroarchitecture (e.g., Fox and Powley, 1992). This characteristic dendroarchitecture

is consistent with the ideas that local mono- or oligosynaptic reflexes are organized segmentally and at the preganglionic outflows.

Visceral Afferents Also Organize Axon Reflexes and Signal Visceral Pain

In addition to forming conventional reflex circuits throughout the ANS, at least some visceral afferents support effector responses known as an *axon reflexes*. Unlike a conventional reflex, which typically involves an afferent neuron, at least one central nervous system synapse, and an efferent neuron, axon reflexes involve only the afferent neuron and they occur in the target tissue—without a CNS relay. The phenomenon can be seen clearly in the type of experiment that was instrumental in establishing the existence of axon reflexes. If all motor axons projecting to a target tissue are eliminated by surgery or other appropriate means while sparing the afferent innervation of the site, the afferent axons can be stimulated selectively. When the afferent axons are stimulated such that their peripheral processes innervating the target are invaded antidromically, one can measure an effector response or assay the release of a neuropeptide from the peripheral afferent neurite. Such experiments performed on a variety of different visceral and cutaneous afferent systems have demonstrated that axon reflexes produce a number of inflammatory and vascular effector responses (Lundberg, 1996).

When a peripheral ending of a visceral afferent transduces a stimulus, the resulting action potential can release the neuropeptide contained in vesicles in that ending (Fig. 35.11). The released compound then acts as a neuromodulator or neurotransmitter to

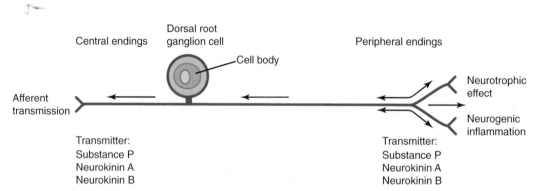

FIGURE 35.11 The architecture of visceral afferents that produce axon reflexes. Visceral and cutaneous afferents release transmitters from the tachykinin family. When action potentials (indicated by arrows) are generated peripherally, they are propagated centrally. Peripherally generated action potentials also produce a local release of tachykinins in the affected terminals and in terminals of the peripheral collaterals. From Cervero and Foreman (1990).

produce local changes on smooth muscle—a local, or peripheral, axon reflex. The response is also relayed to neighboring tissues. When an action potential is transmitted centrally in an afferent axon, it can also be transmitted centrifugally by collaterals of the same fiber, and release of the neuromodulator from these collateral endings can propagate the reflex in the immediate area. The inflammatory responses and extravasation associated with the classic "wheal and flare" reaction to skin damage were the first axon reflexes analyzed and are a classic illustration of the phenomenon. Such local responses have been widely documented not only in the skin, but also in viscera such as the lungs.

Finally, visceral afferents, particularly those associated with SNS pathways, are responsible for visceral pain (Craig, 2002; Jänig, 1996).

Summary

Visceral afferents innervate the target tissues of the ANS, and their axons frequently course in the same peripheral nerves as autonomic efferents. Also commonly called autonomic afferents, these sensory elements associated with the ANS constitute the afferent limbs of autonomic reflexes and carry the inputs recognized as visceral pain. Visceral afferents also mediate axon reflexes and are cross-linked with somatic, as well as autonomic, nervous system activity.

HIERARCHICALLY ORGANIZED ANS CIRCUITS IN THE CNS

Homeostasis no more occurs through isolated autonomic reflexes than posture is maintained, or movement is executed, through isolated somatic reflexes. Nevertheless, methodological and historical considerations have yielded a tendency to focus on individual autonomic reflexes. Methodologically, the need for experimental control typically dictates that experiments isolate a reflex (or small set of reflexes) for examination. In addition, historically, the ANS was conceptualized primarily as a motor branch of the peripheral nervous system.

Though the separate-reflex perspective shapes a view of ANS functions as a collection of individual motor responses, autonomic activity typically involves finely coordinated, fully integrated adjustments of multiple outflows. Just as somatic posture and movement are coordinated by supraspinal CNS controls, autonomic integration is achieved by supraspinal controls. The preganglionic motor pools of the cranio-

sacral and thoracolumbar outflows are final common paths for descending projections from the CNS. Similarly, the first- and second-order visceral afferent relays (e.g., the nucleus of the solitary tract and the spinal lamina V), which form short, mono-, or oligosynaptic reflex arcs with these preganglionic motor pools, are targets of rostral CNS stations that coordinate and modulate the autonomic outflows.

The CNS coordination of autonomic activity achieves a number of integrative functions:

- Much like the suprasegmental organization of skeletal motor responses, central ANS stations provide coordination and sequencing of different local autonomic reflexes. These stations provide, for example, the efferent choreography necessary to coordinate the sequential autonomic responses in the mouth, stomach, intestines, and pancreas during and after a meal. Such programs coordinate brain stem and spinal cord efferents, as well as reflexes within the SNS and PSNS divisions of the autonomic outflow.
- Central ANS circuitry also cross-links autonomic activity and somatic motor activity. The example used earlier of cardiovascular adjustments occurring in concert with postural adjustments illustrates such linkage between autonomic and somatic nervous system outflows.
- The CNS stations of the autonomic nervous system also provide the information, as well as the organization and planning, that enables an organism to mobilize autonomic adjustments or responses in anticipation of environmental contingencies and developing needs. For example, both fluid and energy homeostasis involve physiological adjustments in anticipation of major imbalances or deficits, responses that cannot be explained solely by reactive reflexes.

The importance and the nature of central autonomic controls are underscored by clinical disorders that result from interruptions in the connections between the central autonomic circuitry and the preganglionic motor neurons (Box 35.3). Quadriplegia and paraplegia, which result from injuries that divide the separate thoracic and lumbar sympathetic loops and sacral parasympathetic pathways from higher central regions, illustrate how essential these longitudinal connections are. Interruption of the longitudinal connections causes loss of voluntary bladder and bowel control. Disruption of the spinal cord also causes men to become impotent because penile tumescence is a predominantly parasympathetic response and ejaculation is a sympathetic response. Although erections can occasionally be elicited as local reflexes caused by direct

BOX 35.3

DISEASES AND AGING IMPAIR THE AUTONOMIC WISDOM OF THE BODY

The importance of the autonomic nervous system to health often goes unnoticed. When ANS functions operate normally, they typically occur automatically, without the individual being aware of any adjustments. Nevertheless, loss of an autonomic response can be disruptive, and autonomic disorders can be debilitating (Bannister, 1989). In a classic example of ANS degeneration known as multiple system atrophy and autonomic failure, or the Shy–Drager syndrome, individuals exhibit postural hypotension, urinary and fecal incontinence, sexual impotence, loss of sweating, cranial nerve palsies, and a movement disorder similar to Parkinson's disease. Disconnection of the suprasegmental autonomic stations from the spinal cord can occur in neurological disorders, such as multiple sclerosis, and traumatic spinal cord injury. As a result, bowel and bladder control are lost, and impotence is caused by a loss of autonomic genital reflexes.

Some autonomic disorders are less conspicuous. In part, the apparently more subtle deficits may be ones that are less important in the carefully controlled environments found in modern societies. For example, impaired thermoregulatory responses may be tolerated well by a person living in a climate-controlled environment. Also, metabolic emergencies are unlikely to occur and need autonomic compensation if a person regularly eatsenough food. Nevertheless, autonomic disorders occasion a number of serious, even life-threatening, health problems.

Autonomic imbalances can take many different forms. Excess activation of the ANS is implicated in various stress-related disorders, including ulcers, colitis, high blood pressure, and heart attacks. As discussed earlier,

chronic failure of autonomic cardiovascular homeostasis can cause debilitating postural hypotension. Developmental autonomic disorders, such as immature or anomalous medullary respiratory circuitry, are thought to contribute to sudden infant death syndrome. Some sleep disorders, such as sleep apnea, are also thought to have autonomic elements. Furthermore, widespread autonomic dysfunction resulting from autonomic neuropathies of diabetes, alcoholism, and Parkinson's disease complicates the primary diseases. Autonomic disturbances can also underlie metabolic disorders, such as stress-induced diabetes and reactive hypoglycemia. Autonomic disturbances have also been hypothesized to be a cause of obesity.

Autonomic function is not only impaired by diseases, it is also compromised by aging processes—even those of healthy aging (Amenta, 1993). With aging, both cell losses and neuropathies of surviving neurons occur in ANS circuits. Such age-related changes in the ANS reduce an individual's ability to optimize and sustain homeostatic responses. These consequences of aging produce both primary autonomic disorders and secondary complications and vulnerabilities when the individual is challenged by disease or injury.

Terry L. Powley

References

Amenta, F. (ed.) (1993). "Aging of the Autonomic Nervous System." CRC Press, Boca Raton, Florida.

Bannister, R. (1989). "Autonomic Failing: A Textbook of Clinical Disorders of the Autonomic Nervous System," 2nd ed. Oxford Medical Publications, Oxford, UK.

---

mechanical stimulation, ejaculations, which require suprasegmental coordination, almost universally disappear. In addition, numerous vascular and glandular reflexes are also disordered by the interruption of long autonomic pathways connecting the brain and spinal cord (Bannister, 1989).

Early Stimulation and Lesion Experiments Identified a Central Component of the ANS

Like the concept of the ANS itself, the corollary that the ANS includes a hierarchy of central mechanisms

was first demonstrated by physiological experiments. Nineteenth- and early twentieth-century experiments manipulating the brain caused autonomic disturbances. In one classical experiment, for example, Claude Bernard demonstrated that localized mechanical lesions, or stab wounds, in the floor of the fourth ventricle near the dorsal vagal complex produced a "piqure glycosurique," a disturbance of blood glucose homeostasis characterized by a diabetes-like condition in which glucose spills into the urine. Subsequent lesion and stimulation studies identified a number of "centers" for micturition, respiration, and other autonomic functions in the medulla oblongata and pons.

Other experiments implicated the hypothalamus in even more extensively coordinated and cross-linked autonomic patterns. This work led to the concept of the hypothalamus as the "head ganglion" of the autonomic nervous system. Following an earlier application of electrical stimulation to the hypothalamus by Karplus and Kreidl, Walter Hess found that focal electrical stimulation of regions of the hypothalamus elicited coordinated patterns of sympathetic and parasympathetic adjustments (e.g., changes in pupil size, piloerection, and respiration) and affective responses that were appropriate to behavioral responses elicited from the same loci. In fact, central manipulations that affect autonomic function seem practically invariably to produce adjustments of both SNS and PSNS outputs, an observation that underscores the integrative role the brain plays in ANS function.

Brain lesions in virtually all limbic system regions, and notably in the septal area, amygdala, hippocampus, frontal cortex, cingulate cortex, and insular cortex, have been shown to exaggerate, dissociate, blunt, or in other ways distort autonomic responses to environmental situations.

Anatomical Mapping Experiments Have Identified Multiple Hierarchically Organized, Reciprocally Interconnected Stations of the Visceral Neuroaxis That Control Autonomic Activity

As neuroscience tracing tools have become more powerful, they have revealed an extensive CNS autonomic control hierarchy coupling visceral afferent inputs with autonomic efferent outflows, linking the SNS and PSNS divisions of the ANS, and interconnecting the central autonomic stations with somatic and endocrine pathways in the brain (Box 35.4). This central autonomic circuitry has been considered a "central visceromotor system" (Nauta, 1972) or a "central autonomic network" (Loewy, 1981). These studies revealed a network of multisynaptic relays descending from the hypothalamus and midbrain to preganglionic neurons in the brain stem and spinal cord. Similar projections, both direct and relayed through the hypothalamus, were also found for a number of limbic system nuclei, including most prominently the amygdala (Nauta, 1972). Neural tracers, immunocytochemical techniques, and electron microscopy subsequently have increased the resolution of the maps of this circuitry and its transmitters (e.g., Loewy, 1981).

The paraventricular nucleus of the hypothalamus is a prototype of the extensive and profound central autonomic coordination. Parvocellular neurons in the paraventricular nucleus project monosynaptically to preganglionic neurons in the dorsal motor nucleus of the vagus and preganglionics in the intermediolateral column of the spinal cord. The paraventricular nucleus also projects to the visceral afferent relay nuclei associated with each of these efferent outflows (the nucleus of the solitary tract and spinal lamina V, respectively). These descending projections influence several cardiovascular and gastrointestinal responses. Illustrating the extent of the central integration of responses, additional neurons of the paraventricular nucleus control corticotropin-releasing factor and oxytocin neuron responses, which often are activated or modulated in association with autonomic functions.

Most mesencephalic, diencephalic, and telencephalic nuclei considered part of the limbic system affect visceral motor outflows and are now included in the concept of the central visceromotor system. In addition to the central gray and paramedian regions of the mesencephalon and the entire hypothalamus, including the preoptic hypothalamus, these limbic sites include the amygdala, bed nucleus of the stria terminalis, septal region, hippocampus, cingulate cortex, orbital frontal cortex, and insular and rhinal cortexes.

Summary

Multiple subcortical and diencephalic structures impose special controls over the ANS. This network of hierarchically organized central stations, many of them part of the limbic system, forms a central visceral neuroaxis that receives inputs from visceral afferents and issues descending projections to the ANS preganglionic neurons in the brain stem and spinal cord. This central autonomic circuitry organizes and sequences sets of separate autonomic reflexes, coordinates SNS and PSNS responses so that they are synergistic, cross-links autonomic and skeletal responses, and integrates autonomic activity with ongoing and anticipated behavior of the individual.

PERSPECTIVE: FUTURE OF THE AUTONOMIC NERVOUS SYSTEM

Our understanding of the autonomic nervous system is incomplete and evolving, driven by the technological advances in neuroscience. Molecular biology, immunocytochemistry, electron microscopy, and other

---

BOX 35.4

CENTRAL ANS CIRCUITS INTEGRATE AUTONOMIC REFLEXES WITH AFFECTIVE OR EMOTIONAL RESPONSES

The autonomic nervous system participates in emotion and motivation. Consider the fight-or-flight example that Cannon used in his characterizations of autonomic adjustments and that was discussed earlier in this chapter: The somatic and autonomic adjustments in this emergency situation contain elements we associate with anger or fear. The central relays, including the limbic circuitry just discussed, organize the autonomic components of these responses to external environmental stimuli and coordinate them with the appropriate somatic responses. Thus, the central ANS circuitry is involved in the *expression* of emotional reactions. This central circuitry has been called the "emotional motor system."

Some have hypothesized that in addition to being involved in the expression of emotions, the ANS is the substrate for the *experience* of emotions. The classic theory of this type was suggested over a century ago independently by James and Lange, who speculated that afferent experience resulting from autonomic adjustments might be the basis of emotional experiences. For the fight-or-flight example, the James–Lange idea could be considered the proposition that "you do not run because you are afraid; you are afraid because you run (and mobilize the requisite autonomic adjustments)." Cannon, in the early twentieth century, offered an influential refutation of the James–Lange theory and delayed general acceptance of these ideas, but most of his argument was based on now-outmoded ideas about the ANS and on a lack of distinction between expression and experience.

In an experiment designed to examine the James–Lange idea, Hohmann (1962) surveyed army veterans who had sustained spinal cord injuries at different levels. Consistent with a prediction of the theory, Hohmann found that, for the veterans' self-reported experiences of both fear and anger, the higher their spinal cord lesions (and thus the more of their visceral afferent inflow that was disconnected from the brain), the greater the reduction in their affective experiences.

More recently, extensive observations with neuropsychological analyses of patients with brain damage and spinal injury, with fMRI and other scanning techniques, and with modern electrophysiological techniques combined with tracing, have converged to suggest that emotions or "feelings" are central—specifically limbic system—representations of visceral afferent inputs. Recently, in views compatible with the James-Lange account of emotion, Damasio (1999), Craig (2002), and others have suggested that emotional experience is to the autonomic adjustments of the body what proprioception and somatosensation are to skeletal nervous system operations.

Terry L. Powley

References

Craig, A. D. (2002). How do you feel? Introception: The sense of the physiological condition of the body. *Nature Neuroscience Reviews* **3**, 655–666.

Damasio, A. (1999). "The Feeling of What Happens: Body and Emotion in the Making of Consciousness." Harcourt Brace, New York, New York.

Hohmann, G. W. (1962). Some effects of spinal cord lesions on experienced emotional feelings. *Psychophysiology* **3**, 143–156.

---

modern techniques are changing the autonomic nervous system definition that was derived with the techniques available to Langley and contemporaries at the beginning of the twentieth century. The distinction that sympathetic postganglionic neurons are noradrenergic whereas parasympathetic postganglionics are cholinergic has been blurred by the recognition of nonadrenergic noncholinergic (NANC) neurons, nitric oxide synthetase (NOS)-containing neurons, puriner-gic synapses, a large number of colocalized and co-released neuropeptides, postganglionic neurons changing their neurotransmitter phenotypes, and other exceptions that broaden the view developed by Langley's pharmacological studies. The canon that the ANS is strictly a motor system without an afferent counterpart is challenged by many who argue on functional and molecular biological grounds that "visceral afferents" belong to the ANS (Box 35.2).

The proposal that functional differences between the SNS and PSNS divisions of the ANS can be attributed to their contrasting patterns of projection (the sympathetic system preganglionics projecting to postganglionics in a one-to-many pattern; the parasympathetic system projecting in a one-to-few pattern) has not been substantiated by modern analyses of divergence (Wang *et al.*, 1995). Continued application of modern technologies and the prospect of continuing developments promise to define an autonomic nervous system quite different from that envisioned by Langley.

SUMMARY AND GENERAL CONCLUSIONS

The autonomic nervous system is the neural circuitry that maintains homeostasis and health. It is responsible for the physiological adjustments that achieve what Cannon recognized as the "wisdom of the body." With its sophisticated motor repertoire and complexes of reflexes, including local axon reflexes, segmental or oligosynaptic reflexes, and polysynaptic suprasegmental cascades, the ANS continuously adjusts and defends the body's physiology. The importance of these processes was summarized succinctly by Nauta and Feirtag (1986): "Life depends on the innervation of the viscera; in a way all the rest is biological luxury." The processes go on, for the most part, without awareness or cognitive representations.

The hub of the ANS consists of two separate motor outflows, each hierarchically organized with preganglionic neurons stationed in the CNS and postganglionic neurons located in peripheral ganglia. The sympathetic, or thoracolumbar, division facilitates the mobilization of energy, increases catabolism, and promotes physiological responses that support activity, including emergency responses such as "fight" or "flight." The parasympathetic, or craniosacral, division facilitates conservation of energy, increases anabolism, and supports physiological responses that typically promote rest, digestion, and restoration of body reserves. The adrenergic phenotype of sympathetic postganglionic neurons, the cholinergic phenotype of parasympathetic neurons, and distinguishing complements of neuropeptides and receptor specializations provide neurochemical coding for the two divisions of the ANS. A key peripheral pole—and third distinct division of the ANS—is the enteric nervous system found in the wall of the digestive tract.

The motor outflows of the ANS are efferent limbs of reflexes triggered by visceral and, in some cases, somatic afferents. These reflex circuits have second-order visceral afferent nuclei located adjacent to spinal and medullary nuclei of preganglionic motor neurons (laminae I and V and the nucleus of the solitary tract, respectively).

Higher order CNS circuitry, including the hypothalamus, limbic system, and a variety of cortical sites, provides hierarchical control and integration of autonomic reflexes. This hierarchy is responsible for coordinating different autonomic reflexes, integrating autonomic function with somatic activity, and executing response programs that anticipate needs and regulate physiology over more extended time scales than those represented by isolated reflexes.

References

Amenta, F. (ed.). (1993). "Aging of the Autonomic Nervous System." CRC Press, Boca Raton, Florida.

Bannister, R. (1989). "Autonomic Failure: A Textbook of Clinical Disorders of the Autonomic Nervous System," 2nd ed. Oxford Medical Publications, Oxford, UK.

Bieger, D. and Hopkins, D. A. (1987). Viscerotopic representation of the upper alimentary tract in the medulla oblongata in the rat: Nucleus ambiguus. *J. Comp. Neurol.* **262**, 546–562.

Cannon, W. B. (1939). "The Wisdom of the Body," 2nd ed. Norton, New York.

Cervero, F. and Foreman, R. D. (1990). Sensory innervation of the viscera. *In* "Central Regulation of Autonomic Functions" (A. D. Loewy and K. M. Spyer, eds.), pp. 104–125. Oxford Univ. Press, New York.

Costa, M., Furness, J. B., and Llewellyn-Smith, I. J. (1987). Histochemistry of the enteric nervous system. *In* "Physiology of the Gastrointestinal Tract" (L. R. Johnson, ed.), Vol. 1. Raven Press, New York.

Craig, A. D. (2002). How do you feel? Introception: The sense of the physiological condition of the body. *Nature Neuroscience Reviews* **3**, 655–666.

Damasio, A. (1999). "The Feeling of What Happens: Body and Emotion in the Making of Consciousness." Harcourt Brace, New York, New York.

Dembowsky, K., Czachurski, J., and Seller, H. (1985). Morphology of sympathetic preganglionic neurons in the thoracic spinal cord of the cat: An intracellular horseradish peroxidase study. *J. Comp. Neurol.* **238**, 453–465.

Fox, E. A. and Powley, T. L. (1992). Morphology of identified preganglionic neurons in the dorsal motor nucleus of the vagus. *J. Comp. Neurol.* **322**, 79–98.

Furness, J. B. (2006). "The Enteric Nervous System." Blackwell Publishing, Malden, Massachusetts.

Hohmann, G. W. (1962). Some affects of spinal cord lesions on experienced emotional feelings. *Psychophysiology* **3**, 143–156.

Holst, M.-C., Kelly, J. B., and Powley, T. L. (1997). Vagal preganglionic projections to the enteric nervous system characterized with PHA-L. *J. Comp. Neurol.* **381**, 81–100.

Jänig, W. (1996). Neurobiology of visceral afferent neurons: Neuroanatomy, functions, organ regulations and sensations. *Biol. Psychol.* **42**, 29–51.

Landis, S. C. and Keefe, D. (1983). Evidence for transmitter plasticity *in vivo*: Developmental changes in properties of cholinergic sympathetic neurons. *Dev. Biol.* **98**, 349–372.

Langley, J. N. (1921). "The Autonomic Nervous System," Part I. Heffer, Cambridge, UK.

Loewy, A. D. (1981). Descending pathways to sympathetic and parasympathetic preganglionic neurons. *J. Auton. Nerv. Syst.* **3**, 265–275.

Lundberg, J. (1996). Pharmacology of cotransmission in the autonomic nervous system: Integrative aspects on amines, neuropeptides, adenosine triphosphate, amino acids and nitric oxide. *Pharmacol. Rev.* **48**, 113–178.

Morgan, C., Nadelhaft, I., and de Groat, W. C. (1981). The distribution of visceral primary afferents from the pelvic nerve within Lissaure's tract and the spinal gray matter and its relationship to the sacral parasympathetic nucleus. *J. Comp. Neurol.* **201**, 415–440.

Nauta, W. J. H. (1972). The central visceromotor system: A general survey. *In* "Limbic System Mechanisms and Autonomic Function" (C. H. Hockman, ed.), pp. 21–40. Thomas, Springfield, IL.

Nauta, W. J. H. and Feirtag, M. (1986). "Fundamental Neuroanatomy." Freeman, New York.

Potter, D. D., Landis, S. C., and Furshpan, E. J. (1980). Dual function during development of rat sympathetic neurones in culture. *J. Exp. Biol.* **89**, 57–71.

Strack, A. M., Sawyer, W. B., Marubio, L. M., and Loewy, A. D. (1988). Spinal origin of sympathetic preganglionic neurons in the rat. *Brain Res.* **455**, 187–191.

Thomas, L. (1974). "Autonomy: The Lives of a Cell," pp. 64–68. Viking Press, New York.

Wang, F. B., Holst, M.-C., and Powley, T. L. (1995). The ratio of pre-to postganglionic neurons and related issues in the autonomic nervous system. *Brain Res. Rev.* **21**, 93–115.

Wood, J. D. (1987). Physiology of the enteric nervous system. *In* "Physiology of the Gastrointestinal Tract" (L. R. Johnson, J. Christensen, M. J. Jackson, E. D. Jacobson, and J. H. Walsh, eds.), 2nd ed., Vol. 1, pp. 67–110. Raven Press, New York.

Suggested Readings

Barraco, I. R. A. (ed.). (1994). "Nucleus of the Solitary Tract." CRC Press, Boca Raton, FL.

Björklund, A., Hökfelt, T., and Owman, C. (eds.). (1988). "Handbook of Chemical Neuroanatomy," Vol. 6. Elsevier Science, Amsterdam.

Brading, A. (1999). "The Autonomic Nervous System and Its Effectors." Blackwell Science, Oxford, UK.

Burnstock, G. (1992–2000). "The Autonomic Nervous System," Vols. 1–12. Harwood Academic, Chur, Switzerland.

Cameron, O. G. (2002). "Visceral Sensory Neuroscience Interoception." Oxford University Press, New York.

Cannon, W. B. (1939). "The Wisdom of the Body," 2nd ed. Norton, New York.

Gabella, G. (1976). "Structure of the Autonomic Nervous System." Chapman & Hall, London.

Hockman, C. H. (ed.). (1972). "Limbic System Mechanisms and Autonomic Function." Thomas, Springfield, IL.

Jänig, W. (2006). "The Integrative Action of the Autonomic Nervous System: Neurobiology of Homeostasis." Cambridge University Press, Cambridge, UK.

Kuntz, A. (1953). "The Autonomic Nervous System," 4th ed. Lea & Febiger, Philadelphia.

Loewy, A. D. and Spyer, K. M. (eds.). (1990). "Central Regulation of Autonomic Functions." Oxford Univ. Press, New York.

Pick, J. (1970). "The Autonomic Nervous System." Lippincott, Philadelphia.

Ritter, S., Ritter, R. C., and Barnes, C. D. (eds.). (1992). "Neuroanatomy and Physiology of Abdominal Vagal Afferents." CRC Press, Boca Raton, FL.

Terry L. Powley

Neural Regulation of the Cardiovascular System

The circulatory system, which includes the heart and the vascular system, distributes oxygen and nutrients and eliminates waste products of metabolically active cells in all organs. The heart's synchronized muscular pumping action provides kinetic energy to circulate blood that is distributed in accordance with the particular needs of each organ. Organ blood flow is critically dependent on driving pressure (blood pressure) and regional vascular resistance. The nervous system regulates vascular resistance, cardiac output, and hence arterial blood pressure.

The body must cope with diverse conditions varying from bed rest to maximal exercise, cold to hot environments, ocean depths to sea level to high altitude, including space flight, and fear to arousal. Clinical conditions like hypotension or hypertension, myocardial infarction and congestive heart failure, acute asthma (bronchoconstriction), and chronic renal failure represent a few of the varied conditions to which the cardiovascular system must respond to provide optimal organ function.

Short- and long-term regulatory mechanisms allow the cardiovascular system to meet these challenges. Two are critically important for short-term cardiovascular adjustments. These include neural reflexes activated by sensory systems that monitor the internal and external environment and rapidly alter cardiovascular function. Second, behavioral changes adjust cardiovascular function to cope with the environment, including, among others, changes in body position, activity level, and conscious changes in environmental exposure, such as avoidance. Other behavioral changes like the defense reaction or central command are involuntary and constitute primitive survival mechanisms, allowing higher organisms to respond quickly to adverse or stressful conditions like

exercise. This chapter focuses on neural regulation of the cardiovascular system, rather than on behavioral modifications.

AN ANATOMICAL FRAMEWORK

Neural regulation of the cardiovascular system includes two sets of interactive functional elements. One set of regulatory elements consists of regions in the central nervous system (CNS) that maintain basal or tonic output that, through the autonomic nervous system, continuously influences effector organ function, the heart and vascular system. The second system provides phasic regulation of cardiovascular function and consists of a large number of neural reflexes, each of which is composed of an afferent or sensory arm, regions in the CNS that integrate multiple sensory inputs, and an efferent or autonomic arm. *Cardiovascular reflexes* are differentiated by unique features of their sensory component and effector organ responses. Short and long-loop reflexes are initiated by stimulation of sensory nerve endings positioned in locations that provide the ability to sense the local environment. These afferent endings are depolarized by chemical, mechanical, or thermal stimuli or to a combination of events (i.e., bimodal or polymodal function). Afferent impulse activity is transmitted through finely myelinated or unmyelinated, group III/Aδ and group IV/C somatic and visceral nerve fibers. The sympathetic nervous system responds not globally but with individual patterns of activity depending on the reflex (e.g., baroreceptor, chemoreceptor, and cardiopulmonary) as a result of supraspinal organization. Several CNS regions are involved in the integrative aspect of the reflexes, as well as the autonomic component of the responses.

Anatomy of Somatic and Sympathetic Afferent Reflexes

Sympathetic or spinal afferent systems initiating somatic and visceral cardiovascular reflexes are stimulated by *nociceptive* and *non-nociceptive signals*. Primary afferents, with cell bodies in dorsal root ganglia, transmit nociceptive signals to the spinal cord through Lissauer's tract to laminae I and V of the dorsal horn gray matter. Fine afferents may contain substance P, calcitonin gene-related peptide, somatostatin, and vasoactive intestinal polypeptide. Spinal pathways ascend one or two segments before crossing to the contralateral side. Through *spinothalamic, spinoreticular, spinomesencephalic*, and *spinosolitary tracts*, these systems transmit information to the brain stem and thalamus, ultimately influencing autonomic outflow to the cardiovascular system, in addition to providing sensory perception. Many cells in these tracts respond to noxious stimuli and are classified as low-threshold, wide dynamic range (most common) and high-threshold (nociceptive). Neurons in the lateral spinothalamic tract ascend to the ventral and ventral posterior lateral thalamus where they mediate discriminative or localization of pain. Neurons in the medial spinothalamic tract projecting to the medial and intralaminar nuclear complex of the thalamus transmit information leading to affective responses, including autonomic adjustments. Cells in the spinoreticular tract respond to chemical (bradykinin) and mechanical (premature beats) stimulation of cardiac ventricular afferents. They project to the gigantocellular tegmental field (paramedian reticular formation) and, to a lesser extent, to the caudal raphe nuclei and magnocellular tegmental field (ventromedial reticular formation), which, in turn, project to the intralaminar region of the thalamus.

The medial reticular formation mediates motor responses to visceral (including cardiac) pain and altered sympathetic function through premotor collaterals to spinal intermediolateral column (IML) containing *sympathetic preganglionic neurons* or through interneurons projecting to other regions, such as the nucleus tractus solitarii (NTS) and rostral ventral lateral medulla (rVLM), concerned with integration of sensory signals that drive autonomic outflow. The spinomesencephalic tract sends information to the central periaqueductal gray (PAG, also called the central gray matter) and parabrachial nucleus (PBN) in the rostral pons, and possibly the ventral lateral nucleus of the thalamus. The spinosolitary tract projects to the caudal NTS (Loewy and Spyer, 1990).

Anatomy and Pharmacology of Baro- and Chemoreflex Pathways

The IX (glossopharyngeal) and X (vagus) cranial nerves transmit signals from arterial baroreceptors and chemoreceptors, as well as cardiac receptors, including both atrial and ventricular receptors to the CNS. Temporal and spatial integration of sensory information occurs in the CNS, in addition to modification resulting from convergent input from other afferent pathways. Important "cardiovascular" CNS regions that process sensory input include, among others, the NTS, caudal ventral lateral medulla (cVLM), rVLM, PBN, raphe nuclei, and lateral tegmental field (LTF). Other areas in the medulla, hypothalamus, and midbrain identified as important sites of integration include the area postrema, PAG, vestibular region, arcuate nucleus, and paraventricular nucleus. Each region receives input either directly from afferents or, more commonly, through second or higher order interconnections (interneurons) from other nuclei. Each participates in signal conditioning to an extent dictated by the overall input and underlying conditions. Thus, many central autonomic pathways form the anatomical substrate for reflexes involved in cardiovascular regulation. Although these can be separated broadly into CNS regions subserving spinal or sympathetic sensory pathways and regions that integrate visceral vagal afferent input, there is significant overlap of regions that process information from the two systems.

Nucleus Tractus Solitarii

The NTS receives direct input from vagal and glossopharyngeal sensory nerves originating in the heart and large arteries (mechano- or baro- and chemosensitive regions). It lies just ventral to the dorsal columns and is divided into the rostral, intermediate, and caudal NTS, relative to the obex. The intermediate NTS contains a central fiber bundle or solitary tract formed by sensory input from VII, IX, and X cranial nerves. Visceral afferents from the heart, lungs, and gastrointestinal regions project to neurons in the commissural medial NTS. Cardiovascular afferents from the heart and great vessels project to the dorsolateral, medial, and commissural regions of the NTS. Few solitary tract neurons have pulse synchronous activity, possibly due to convergent input from baroreceptor and nonbaroreceptor inputs. Ionotropic glutamatergic receptors, including *N*-methyl-D-aspartate (NMDA) and non-NMDA (e.g., α-amino-3-hydroxy-5 methyl-4-isoxazolepropionate, AMPA) subtypes, excite neurons in the NTS and throughout many other regions of the

CNS that process cardiovascular reflexes. Non-NMDA glutamate receptors mediate fast synaptic transmission, whereas NMDA receptors provide slower and longer signaling. Although glutamate appears to be the most important neurotransmitter for reflex transmission and processing in the NTS, other neurotransmitters, neuropeptides, and neuromodulators, including catecholamines, acetylcholine, γ-aminobutyric acid (GABA), substance P, angiotensin, nitric oxide (NO), and opioids, regulate activity of these interneurons.

There are important interactions between the NTS and other nearby nuclei like the area postrema, which functions as a *circumventricular organ*, since it is devoid of tight endothelial junctions that form the blood brain barrier. Neurons in the NTS project to forebrain nuclei including the limbic system, central nucleus of the amygdala, paraventricular, and median hypothalamic preoptic nuclei. There also are reciprocal connections between the NTS and A5 cell group, caudal raphe nuclei, rVLM, PAG, paraventricular, and lateral hypothalamic regions. Reciprocal connections are a major feature of this central autonomic network, which serves as a microprocessor to integrate signals from receptive regions important in cardiovascular control that change autonomic and neuroendocrine outflow and modify behavior ultimately to regulate cardiovascular function (Undem and Weinreich, 2005).

Caudal Ventrolateral Medulla

Second-order neurons from the NTS project to the cVLM where they provide tonic excitation mediated by an NMDA glutamatergic mechanism. Excitation from angiotensin II and inhibition from GABA, glycine (or a glycine-like compound), and opioids also tonically regulate activity of cVLM neurons. Third-order neurons from the cVLM, through a GABAergic mechanism, influence the activity of sympathetic premotor neurons in the rVLM and A5 cell group (Fig. 36.1). The cVLM contains noradrenergic (A1) and nonadrenergic (retro or caudal periambigual area) neurons. Lesions of the cVLM increase arterial blood pressure, whereas microinjections of glutamate into this region cause depressor responses resulting from sympathoinhibition and, to a lesser extent, activation of vagal motoneurons in the nearby nucleus ambiguus. Mapping studies suggest that cVLM depressor sites do not involve the A1 noradrenergic cell group.

Rostral Ventrolateral Medulla

The rVLM, also called the *subretrofacial nucleus*, located rostral to the lateral reticular nucleus, caudal to the facial nucleus, and lateral to the olivary nucleus, is an important site of integration. The medial region of this nucleus is the source of many bulbospinal sympathetic premotor cells that project to the IML located in the lateral horn of spinal gray matter. The rVLM is subdivided into a lateral subretrofacial nucleus and a medial nucleus paragigantocellularis (PGL), called pressor regions because chemical lesions elevate blood pressure. The rostral extension of the PGL is concerned with pain control rather than cardiovascular regulation. In addition to reciprocal connections between the rVLM and the medial NTS, the rVLM receives inputs from the area postrema, cVLM, and is connected reciprocally to midline (raphe) medullary nuclei, parabrachial complex, PAG, lateral hypothalamic area, zona incerta, and paraventricular hypothalamus.

A number of sympathetic premotor neurons receive convergent input from somatic and visceral afferents. Although many rVLM neurons are phenotypically adrenergic (C1 group), glutamate is the principal excitatory neurotransmitter. Blockade of excitatory amino acids does not alter resting blood pressure, indicating that tonic excitatory outflow does not exist. Conversely, blockade of GABA, which serves as the inhibitory neurotransmitter in barosensitive premotor sympathetic neurons projecting from cVLM to rVLM, increases their discharge activity and blood pressure. Other important neurotransmitters in the rVLM include acetylcholine and angiotensin, which also may be tonically active. Opioids (enkephalins, endorphins, endomorphin, and dynorphin), serotonin, neuropeptide Y, nociceptin, substance P, somatostatin, nitric oxide, and/or thyrotropin-releasing hormone are not tonically active and may act as neuromodulators.

Some vasomotor cells in the rVLM discharge with a cardiovascular rhythm, which can be measured directly or over time with pulse or spike triggered averaging, a time domain analysis relating intermittent unit activity to the cardiovascular cycle or sympathetic activity (Fig. 36.2). Many nonadrenergic rVLM cells display intrinsic pacemaker activity with spontaneous depolarization (see later for more detail). Most C1 cells containing phenylethanolamine-N-methyltransferase or tyrosine hydroxylase do not show pacemaker activity. Clusters of rVLM neurons provide selective input to regional vascular beds like skin and muscle (Fig. 36.3) (Loewy and Spyer, 1990).

A5 Cell Group

The A5 noradrenergic cell group located in the rostral-most portion of the ventrolateral medulla is distinct from the rVLM. These neurons receive input

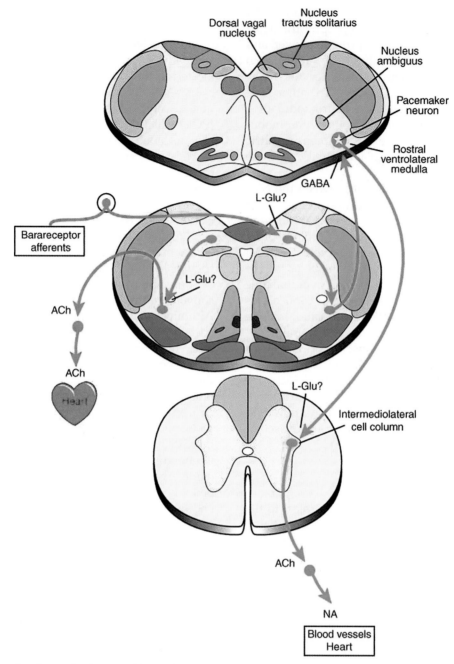

FIGURE 36.1   Neural pathways for the arterial baroreflex. Primary afferents in the IX and X cranial nerves project to the NTS. As shown on the right, interneurons forming sympathetic pathways project from the NTS to the caudal ventrolateral medulla, which in turn project to the rVLM, the source of reticulospinal projections to the intermediolateral columns in the spinal cord. Pathways from the NTS to the nucleus ambiguus form the major parasympathetic arm of the reflex, GABA, γ-aminobutyric; L-glu, L-glutamate; ACH, acetylcholine; NA, norepinephrine (Guyenet, 1990).

from the paraventricular hypothalamic nucleus, parafornical hypothalamic area, PBN, NTS, raphe obscurus, and cVLM. In turn, A5 neurons project to the NTS, dorsal vagal nucleus, PBN, lateral hypothalamic area, paraventricular thalamic nucleus, PAG, central nucleus of the amygdala, and the IML. The function of this cell

group is unclear, although it probably does not tonically regulate blood pressure (Loewy and Spyer, 1990). Stimulation of cells in this region decreases blood pressure and redistributes blood flow. A5 cells are inhibited by baroreceptor input and thus are sympathoexcitatory (Dampney, 1994a).

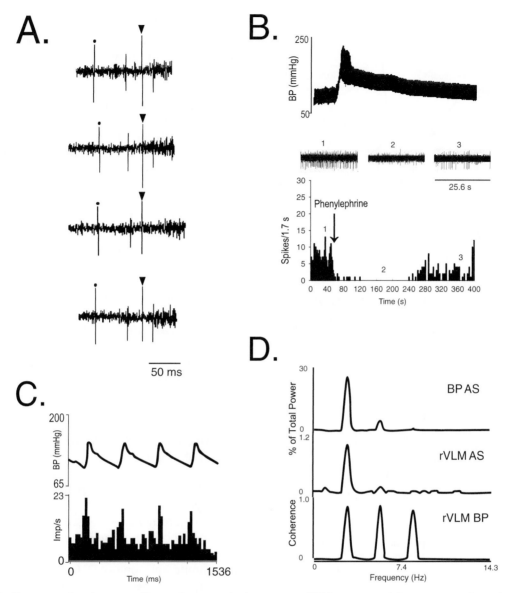

FIGURE 36.2 Responses of excitatory cardiovascular sympathetic premotor rVLM neuron receiving convergent input from ventrolateral PAG, median and splanchnic nerves. The bulbospinal projection of this neuron was demonstrated when an evoked orthodromic rVLM action potential (●) collided with an antidromic potential from the IML (▼), canceling the orthodromic spike (A, third sweep). Stimulation of baroreceptors with phenylephrine-evoked increase in blood pressure transiently reduced discharge of the sympathoexcitatory cardiovascular rVLM neuron (B) that discharged with a cardiovascular rhythmicity, shown by time domain pulse triggered averaging (C) and frequency domain coherence analyses (D). See text and Figure 36.6 for further explanation of mathematical methods used to assess relationship between neuronal and cardiovascular activity. (Modified from Tjen-A-Looi *et al.*, 2006.)

Intermediate Ventrolateral Medulla

The iVLM, located between the cVLM and the rVLM, is an essential component of the baroreflex in rabbits and possibly rats. Like the cVLM, the iVLM projects to the rVLM. Most neurons are not immunoreactive for tyrosine hydroxylase and therefore are distinguished from A1 cells located in this region (Dampney, 1994a).

A1 Cell Group

The noradrenergic A1 cell group in the cVLM is an important central component of the baroreflex. These neurons project directly to supraoptic and paraventricular hypothalamic regions and mediate vasopressin release during baroreflex unloading. Cells in the A1 region are under tonic GABA inhibition from the NTS.

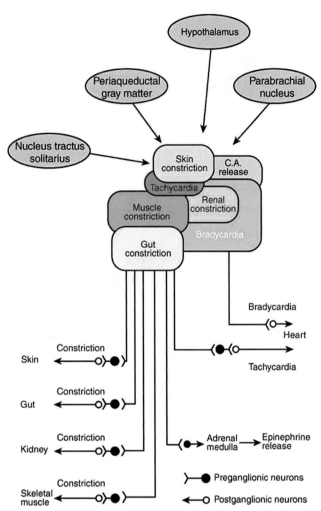

FIGURE 36.3 Inputs and outputs from the rVLM. A number of regions in the hypothalamus, midbrain, and medulla project either directly or indirectly to presympathetic neurons in the rVLM. Distinct but overlapping pools of neurons are present in this region that provide separate sympathetic inputs to different regional vascular beds, the heart, and the adrenal medulla. C.A., catecholamine (Lovick, 1987).

Periaqueductal Gray

The PAG, surrounding the cerebral aqueduct in the midbrain, receives inputs from NTS, hypothalamus, and PBN and projects to the medullary reticular formation. Activation of the dorsolateral PAG leads to a defense reaction, including an increase in blood pressure (Box 36.1). The ventrolateral PAG is a depressor region that receives input from the arcuate nucleus, medullary raphe nuclei, visceral and somatosensory systems. It regulates sympathetic outflow to muscle and kidney in a viscerotopic manner by influencing activity in the rVLM (Tjen-A-Looi et al., 2006).

Hypothalamus

Several specialized groups of hypothalamic neurons are important in cardiovascular regulation. The hypothalamus is divided into the anterior, posterior, and middle regions. The anterior region, overlying the optic chiasm, includes the circadian pacemaker (suprachiasmatic nucleus) as part of the preoptic nucleus, which controls blood pressure cycles, body temperature, and a number of hormones. The middle portion of the hypothalamus, overlying the pituitary stalk, is composed of dorsomedial, ventromedial, paraventricular, supraoptic, and arcuate nuclei. The paraventricular nucleus projects to sympathetic and parasympathetic premotor neurons in the medulla and spinal cord. The arcuate nucleus synthesizes opioid peptides, which modulate neuronal function in regions like the ventrolateral PAG and rVLM. The arcuate nucleus is part of a long-loop pathway that modulates visceral afferent related increases in sympathetic outflow during somatic afferent stimulation (see section on reflex interactions).

The hypothalamus directly and indirectly controls endocrine function. Large (magnocellular) neurons in paraventricular and supraoptic nuclei release vasopressin or oxytocin into the circulation from the posterior pituitary. *Vasopressin* neurons in magnocellular and paraventricular nuclei have a phasic pacemaker pattern of discharge activity. Vasopressin is produced as a prohormone in the cell bodies. Prohormones are cleaved as they are transported down axons in vesicles to the posterior pituitary where they are released as the active hormone. As suggested by its name, vasopressin vasoconstricts many regional circulations, particularly skin and muscle. Vasopressin also is called antidiuretic hormone (ADH) since it promotes water reabsorption in renal collecting tubules and expands plasma volume; both actions augment blood pressure. Hypothalamic input from atrial, baroreceptor, chemoreceptor, and somatic (muscle) afferents leads to vasopressin release.

ANATOMY AND CHEMICAL PROPERTIES OF AUTONOMIC PATHWAYS

Sympathetic

Outputs from the CNS to effector organs involved in cardiovascular reflex regulation pass through sympathetic and parasympathetic branches of the autonomic nervous system. *Cardiovascular sympathetic premotor neurons* in the brain stem and the diencepha-

BOX 36.1

DEFENSE REACTION

The central and autonomic nervous systems play crucial roles in modulating cardiovascular activity for appropriate responses to aversive stimuli. Behaviors studied include the *flight or fight* response, frequently called the defense reaction, and the *playing dead* response (Jordan, 1990). A number of stimuli, including visual, auditory, and somatosensory threats, cause the defense response in several species in the awake condition. This response can also be conditioned to occur. It consists of stereotypical behaviors and cardiovascular responses, including posturing in the crouched position, retraction of the head and ears, vocalization, piloerection, pupillary dilation, heightened alertness, increased blood pressure, heart rate, myocardial contractility, and differential activation of sympathetic output to regional circulatory beds, including arteriolar constriction in mesenteric, renal, and cutaneous circulations and, in the naive condition, vasodilation of vessels in skeletal muscle and widespread venoconstriction. This preparatory reflex sets the stage for maximal action. It is evoked by stimulating a number of regions in the brain, including the dorsolateral PAG and other hypothalamic and midbrain regions ranging from the limbic system amygdala to the medulla. The PAG likely plays a central role in organizing and integrating this response (Fig. 36.7). Outflow from the PAG projects to neurons in the NTS (inhibiting neurons that receive baroreceptor input) and to the rVLM. Lesions in substantia nigra dorsal to the cerebral peduncles abolish vasodilation in skeletal muscle. This reaction may conserve cardiac output by redirecting it to those regions that promote escape and ensure survival. In a few species, like the rabbit and opossum, the playing dead fear response is characterized by pronounced bradycardia and hypotension (along with increased muscle blood flow), locomotor paralysis, pupillary dilation, and rapid shallow breathing. Stimulation of the central nucleus of the amygdala largely replicates the response observed in the awake animal.

John C. Longhurst

Reference

Jordan, D. (1990). Autonomic changes in affective behavior. In: "Central Regulation of Autonomic Function" (A. D. Loewy and K. M. Spyer, eds.), pp. 349–366. Oxford Univ. Press, New York.

lon, including regions in the medulla, pons, and hypothalamus, project to the spinal cord IML. Retrograde transneuronal viral labeling of cell bodies has identified five principal areas in the brain that innervate all levels of sympathetic outflow. These include the paraventricular hypothalamic nucleus, A5 noradrenergic cell group, caudal raphe region, rVLM, and ventromedial medulla (Fig. 36.4). Hindbrain pontine projections to the IML originate from the Kölliker–Fuse and A5 cell groups. Paraventricular and lateral hypothalamic nuclei also provide input to sympathetic preganglionic neurons in the IML. To a lesser extent they also innervate the lateral funicular area (lateral and dorsal to IML in white matter of lateral funiculus), intercalated cell group (medial to IML), and central autonomic nucleus (also referred to as nucleus intercalatus pars paraependymalis, located dorsal to the spinal central canal). There is *viscerotopographic clustering* of spinal sympathetic preganglionic neurons in restricted regions of the spinal cord, including C8, entire thoracic, and upper lumbar (thoracolumbar) and caudal lumbar regions (L$_3$–L$_5$). Sympathetic neurons exit the spinal cord through the ventral roots and synapse with postganglionic neurons in para- or prevertebral ganglia. Phenotypically, many terminals surrounding sympathetic preganglionic cells are catecholaminergic (from rVLM) or serotinergic (from caudal raphe, including raphe pallidus and obscurus, and ventromedial medulla). However, the principal neurotransmitter in terminals synapsing with sympathetic preganglionic cells likely is glutamate (two-thirds of synaptic boutons) or an inhibitory neurotransmitter, GABA (one-third of axons). Serotonin, acetylcholine, enkephalins, substance P, neurotensin, leuteinizing hormone-releasing hormone, and somatostatin are colocated in many cells and presumably function as pre- or postsynaptic neuromodulators. Ultrastructural studies have demonstrated over 20 other transmitter-specific synapses on sympathetic preganglionic neurons.

Parasympathetic

Cardiac preganglionic parasympathetic neurons originate in the *nucleus ambiguus* in the ventrolateral medullary reticular formation and the *dorsal motor nucleus of the vagus* in the dorsomedial region of the

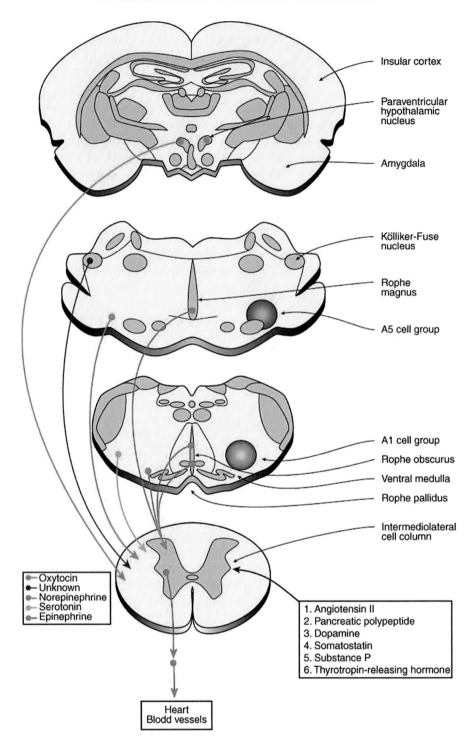

FIGURE 36.4 Direct preganglionic sympathetic inputs to the IML of the spinal cord. Projections from the raphe pallidus, obscurus, magnus, and the ventral medulla contain serotonin, whereas catecholaminergic A1 and A5 contain epinephrine and norepinephrine, respectively. The neurotransmitter for many of these neurons is glutamate or another excitatory amino acid (Loewy and Neil, 1981).

caudal medulla near the fourth ventricle (Fig. 36.5). Myelinated B fiber neurons from ventrolateral nucleus ambiguus and unmyelinated C fibers from dorsal vagal nucleus control heart rate and, to a lesser extent, contractility and coronary flow. Many preganglionic

neurons discharge in synchrony with the cardiac cycle because they receive input from arterial baroreceptors. They also are excited by input from peripheral chemoreceptors, cardiac receptors, and trigeminal receptors (diving reflex). The NTS provides the major afferent

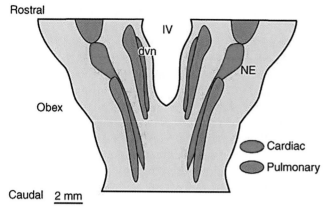

FIGURE 36.5 Horizontal view of preganglionic parasympathetic cardiac and pulmonary motoneurons in the nucleus ambiguus (nA) and dorsal motor nucleus of the vagus (dvn). The fourth (IV) ventricle and the obex are shown for reference (Jordan, 1995).

input into these regions. Fewer afferents originate from the medial and lateral PBN, paraventricular hypothalamic nucleus, Kölliker–Fuse cell group, ventrolateral nucleus of the solitary tract, caudal portion of the nucleus ambiguus, and contralateral rostral nucleus ambiguus. Although neurons in the dorsal motor nucleus and the nucleus ambiguus are phenotypically cholinergic, excitation from second order projections in the baroreflex from the NTS to the nucleus ambiguus is mediated by an excitatory amino acid. Some cardioinhibitory neurons in the nucleus ambiguus are under tonic GABA inhibitory control. Calcitonin gene related peptide, galanin, serotonin, and opioid peptides are colocated in some cholinergic neurons and likely function as neuromodulators. Laterality of cardiac motor neurons exists in the nucleus ambiguus. Electrical stimulation on the right inhibits the sinoatrial node, whereas stimulation of the left side inhibits atrioventricular (A-V) node conduction (Dampney, 1994a; Loewy and Spyer, 1990; Izzo, Deuchars and Spyer, 1993).

Summary

Substantial information is available about the principal components of the anatomical system responsible for neural reflex control of the cardiovascular system, particularly the better studied visceral and somatic cardiovascular reflexes. Concepts of central integration have undergone substantial refinement over the last several decades. However, despite the substantial new knowledge that has been gained, we are just beginning to understand the interactive nature of the various central nuclei, pathways, and processes responsible for controlling the cardiovascular system in many conditions associated with afferent activation.

NETWORK GENERATORS

Blood pressure is governed by circulating blood and plasma volume and the caliber of blood vessels. Neural mechanisms regulate vascular resistance/conductance through tonic sympathetic activity called *sympathetic tone* that actively regulates basal constriction (*vascular smooth muscle tone*) of resistance vessels. These small arterioles of 250–100 μm diameter are heavily innervated by a surrounding sympathetic neural meshwork. Neurally mediated changes in arteriolar lumen diameter occur in virtually all regional circulatory systems and hence are important in the regulation of arterial blood pressure. Sympathetic tone originates from higher brain centers in the medulla and medial diencephalon and exists even in the absence of input into these regions from external sensors (Dampney, 1994a).

Chemosensitivity and pacemaker activity contribute to tonic activity of premotor sympathetic neurons that generate basal sympathetic tone (Dampney, 1994a). Networks of cells in the rVLM, caudal medullary raphe (CMR), caudal ventrolateral (cVLP) and rostral dorsolateral pons (rDLP), LTF, and possibly the cVLM function as *network oscillators* that maintain tonic sympathetic activity (Barman and Gebber, 2000). Time and frequency domain analysis of neural activity in each of these regions in relation to cardiovascular activity, including coherence analysis and pulse or spike-triggered averaging, has demonstrated cardiovascular and/or 10-Hz rhythms that likely underlie tonic sympathetic activity (Fig. 36.6). The *cardiovascular rhythm* of 2–6 Hz largely originates from the LTF, rVLM, cVLM, cVLP, and rDLP. These rhythms are modified by baroreceptor input, but are apparent in its absence. The *10-Hz rhythm* originating mainly in the cVLM, cVLP, and rDLP and, to a lesser extent, the rVLM and CMR, is less well understood. The change from a cardiovascular to a 10-Hz rhythm is associated with an increase in blood pressure, whereas elimination of this rhythm leads to a fall in blood pressure. This rhythm may underlie the differential regional vascular responses during certain alerting conditions, like the defense reaction (Loewy and Spyer, 1990).

Summary

The cardiovascular system is controlled by networks of oscillators that help maintain blood pressure and the cardiovascular responses to a complex environment. Central generators responsible for cardiovascular and 10-Hz rhythms are distributed over a wide region in the brain stem. Additional investigation is required to understand how the rhythms are created and the mechanisms underlying the contribution of

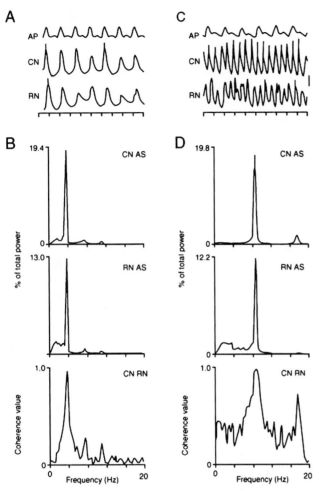

FIGURE 36.6 Spontaneous arterial pulse pressure (AP) and activity in the inferior cardiac (CN) and renal (RN) sympathetic nerves in an anesthetized baroreceptor-intact cat. Wide band-pass recordings of the time domain demonstrate activity phase locked with the cardiac rhythm (A) when mean blood pressure was 149 mm Hg or with a peak at 10 Hz when blood pressure was 103 mm Hg. Frequency domain autospectra (AS) of the cardiac and renal nerves and coherence between the two nerves (CN-RN) are shown in B (2–6 Hz) and D (10 Hz). Coherence analysis allows linear correlation of two signals, as a function of frequency, with 1.0 representing perfect correlation and 0 the absence of any relationship. Calibration in C is 50 μV (Barman and Gebber, 2000).

these rhythms to altered sympathetic outflow in diverse states requiring immediate or long-term cardiovascular responses.

SHORT-TERM CONTROL MECHANISMS

The cardiovascular system must respond quickly to cope with demands originating from changes in the internal and external environments. Rapid changes in autonomic tone occur principally by one of two mechanisms, alterations in activity originating from the

brain or through a reflex event. Alterations in autonomic activity originating as the result of voluntary or involuntary actions like exercise (Box 36.2) or the defense reaction, in part, originate from the brain and are termed feed-forward mechanisms, to be distinguished from the feedback component involving true reflexes. Feed-forward events lack an afferent limb characteristic of reflex events, although visual, auditory, or other forms of sensory input provide the input leading to initiation of the response. Reflex events involve afferent and efferent neural limbs and a central neural integration component.

REFLEX CONTROL OF THE CARDIOVASCULAR SYSTEM

The primary function of various reflex mechanisms concerned with cardiovascular control is to maintain homeostasis. Blood pressure, blood flow, and the local metabolic environment, especially blood pH and oxygen tension, are some of the most important variables controlled. Sensory endings of afferent fibers initiating these reflexes function as mechano- or chemoreceptors. This section describes several of the more important reflex pathways, including the sensory transduction mechanism, afferent limb, central neural integration, autonomic outflow, and effector organ responses.

ARTERIAL BARORECEPTORS

Anatomy

The heart and blood vessels distribute oxygen and nutrients to every cell in the body and, in turn, eliminate metabolic waste products through the lungs and kidneys. These actions are accomplished by appropriate function of the heart, which pumps blood, and optimal distribution of cardiac output to organs that need a variable supply of blood, depending on the underlying conditions. Maintenance of an adequate blood pressure as the driving force for blood flow allows appropriate perfusion of organs. Mechanosensitive nerve endings in baroreceptor regions are present predominately in two locations: bifurcation of common and internal carotid arteries (carotid sinus) and aortic arch. Although the vessel walls are elastic in nature, they also contain smooth muscle. Baroreceptor nerve endings branch to form loops and rings existing as either large compact endings or terminal fibrillar expansions in adventitia or at the border of adventitia and media between muscle cells, elastic, and collagen

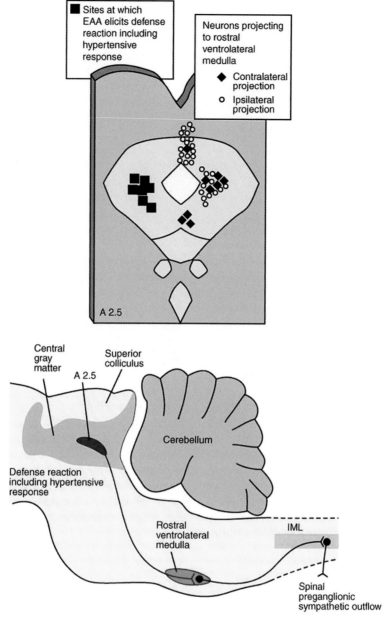

FIGURE 36.7 Coronal section displaying sites in PAG (also called the central gray matter) that elicited a defense reaction after microinjection of excitatory amino acid (EAA) (top). Sagittal section of the area in the PAG that causes the hypertensive response, overlapping with the area projecting to sympathoexcitatory cells in the rVLM that, in turn, project to the IML (bottom) (Carrive *et al.*, 1988).

fibers in vessel walls; a few endings terminate in the media. End organ anatomy of baroreceptors is speculative, as there are no recordings of intracellular activity in stretch sensitive endings whose anatomy has been described. Impulse activity in both A and C fiber afferent pathways is generated by changes in pressure or, more correctly, the resulting stretch and tension in vessel walls. Unmyelinated baroreceptor afferents greatly outnumber myelinated fibers. Carotid baroreceptor afferents course in the carotid sinus nerve to the glossopharyngeal or IX cranial nerve in most species;

cell bodies are in the *petrosal ganglion*. Afferents from the aortic arch are located in either a separate aortic depressor pathway or as part of the common vagus (cranial nerve X) bundle of nerves; cell bodies of aortic baroreceptor afferents reside in the *nodose ganglion*. Primary baroreceptor afferent fibers terminate mainly in the ipsilateral NTS and, to a lesser extent, in the paramedian reticular nucleus and area postrema. Secondary projections extend to the pontoreticular region (cVLM and rVLM), nucleus ambiguus, and dorsal motor nucleus of the vagus (Loewy and Spyer, 1990).

BOX 36.2

EXERCISE

Voluntary exercise leads to a number of hemodynamic changes, including increases in blood pressure, heart rate, myocardial contractility, stroke volume, cardiac output, and vasoconstriction of inactive regional circulations. Vasoconstrictor responses compete with vasodilation caused by increased metabolic activity and flow-mediated vasodilation in active regions of the body. These changes are the result of increased neurohumoral activity. In particular, changes in autonomic neural tone include increased sympathetic and withdrawal of parasympathetic activity to the heart and vasculature. Alterations in autonomic tone, particularly decreased activity in vagal motor nerves to the heart, begin immediately after exercise starts. Two processes drive these autonomic changes, including the *feed forward activity* or *central command* and the *exercise pressor reflex*. Central command or "cortical irradiation" originates in a number of supraspinal regions in the CNS, including motor cortex, premotor cortex, and supplementary motor cortex, subthalamic (hypothalamic) locomotor region or H1 and H2 fields of Forel in the diencephalon, midbrain mesencephalic locomotor region, and the pontomedullary locomotor strip. The hypothalamic locomotor region likely is part of the defense area (Box 36.1). Ultimately, one or more of these sites acts through medullary centers to regulate autonomic neural tone. Stimulation of the mesencephalic locomotor region in paralyzed animals either electrically or with a GABA antagonist evokes locomotion, hemodynamic, and respiratory changes quite similar to those observed during voluntary exercise. Stimulation of the subthalamic locomotor region also inhibits the baroreceptor reflex, consistent with observed suppression of this reflex during high-intensity exercise. In humans, central command is responsible for early cardiorespiratory changes that occur during dynamic and static exercise. In addition to central command, stimulation of both mechano- and chemosensitive receptors in active muscle, mediated largely by group III and IV afferents, also controls cardiovascular and respiratory changes associated with static and dynamic exercise in experimental models and humans. Changes in muscle tension and/or stretch predominately activate finely myelinated afferents to evoke early changes in heart rate and respiration. Production of a number of chemical metabolic factors produced after a few seconds or minutes leads to slightly later activation of unmyelinated muscle afferents and helps maintain cardiorespiratory responses. The nature of metabolic products that stimulate muscle afferents during exercise likely is linked to an imbalance between oxygen supply and demand—that is, a mismatch or error signal. Potential candidates include lactic acid, changes in pH, potassium, and perhaps kinins. Prostaglandins sensitize these sensory endings to the action of other metabolic products. Other metabolites, including adenosine, vasopressin, nicotine, angiotensin, catecholamines, and angiotensin, either do not cause cardiovascular responses when injected into muscle or are associated with pain rather than nonpainful reflex responses.

John C. Longhurst

Reflex Cardiovascular Responses

Stretching a vessel wall containing baroreceptor afferent endings, most commonly resulting from an increase in blood pressure, causes negative chronotropic and inotropic events, systemic arteriolar and venular vasodilation, as well as changes in the rate of secretion of vasopressin. The resulting decrease in blood pressure opposes the original change in transmural pressure and reduces baroreceptor stimulation, leading to the term *closed loop negative feedback reflex* (Sagawa, 1983).

Chronotropic Responses

Decreased heart rate, in response to increased blood pressure, is a function of an immediate increase in vagal motor tone and a slower decrease in sympathetic activity, depending on the initial heart rate. At low basal or resting heart rates, with significant parasympathetic and little sympathetic tone, decreased chronotropy chiefly results from a further increase in vagal activity. At high heart rates, increased parasympathetic activity and withdrawal of sympathetic activity contribute to the negative chronic response.

Baroreflex sensitivity in human subjects is increased at slow heart rates and during expiration, providing a partial explanation for sinus arrhythmia (Mark and Mancia, 1983). The influence of respiration on baroreflex sensitivity continuously oscillates throughout the respiratory cycle, either from interaction of the central respiratory generators with the reflex or from the influence of changes in arterial Pco_2 on the reflex–heart rate response. Altered vagal activity in humans mediates heart rate responses to increased carotid sinus pressure (after 200–600 ms latency), whereas alterations in both vagal and sympathetic activity cause chronotropic changes during carotid sinus hypotension. Slowing of the heart rhythm depends on the timing of carotid sinus stimulation relative to the cardiac cycle and the duration of stimulation, with maximal changes occurring 750 ms before the P wave and with stimuli lasting 1.25 s. Phasic enhancement and rapid adaptation provide a substantial advantage of the reflex to regulate beat-to-beat changes in heart rate. Temporal summation occurs during maintained stimulation of the reflex, providing prolonged chronotropic suppression. However, even a single brief stimulation can elicit a chronotropic response that outlasts the stimulus due to central or effector organ prolongation of the reflex.

Dromotropic Responses

Arterial baroreceptors, through a vagal mechanism, tonically influence the A-V cardiac conduction system and influence conduction through the His bundle during increases or decreases in pressure. This mechanism coordinates conduction between the upper and the lower cardiac chambers as heart rate varies, to optimize preload and stroke volume. Conversely, baroreceptors provide little regulation of atrial and ventricular conduction (bundle branches and Purkinje system).

Inotropic Responses

Baroreceptor-mediated changes in sympathetic tone lead to large changes in myocardial ventricular contractility, whereas small reciprocal changes in myocardial function occur during changes in vagal motor activity. Contrasting influences of sympathetic and vagal efferent activity on inotropic performance are largely a function of differences in ventricular innervation by the two systems.

Regional Vascular Responses

Studies in animals and humans have examined the influence of high-pressure baroreceptors on regional vascular resistance (Sagawa, 1983). Arterial vascular resistance in skeletal muscle and splanchnic regions is strongly regulated by baroreflexes. Baroreceptors control splanchnic vascular resistance in conscious but not anesthetized human subjects. Anesthesia, presumably through its action on central vasomotor centers, variably enhances or blunts the influence of baroreceptors on regional vascular resistance. Arterial baroreceptors do not regulate cutaneous sympathetic activity or arterial resistance, although more investigation is warranted. Such studies are difficult because multiunit recordings include cutaneous sympathetic activity monitor sudomotor, thermoregulatory, and vasomotor fibers.

Integrated Cardiovascular Responses

Arterial hypertension reflexly decreases heart rate and myocardial contractility, resulting in reduced cardiac output. Along with cardiac pump function changes, increased baroreceptor activity leads to withdrawal of sympathetic tone to arteries and veins, causing corresponding increases in caliber of resistance and capacitance vessels. Thus, increased arterial baroreceptor afferent activity in response to hypertension causes reflex arterial vasodilation, which, along with the decreased cardiac output, helps return blood pressure back toward normal. Concurrent venodilation, in most animal species studied, reduces venous return and cardiac preload. In contrast to observations in animals, human studies of baroreflex function have not documented significant changes in central venous pressure nor sustained changes in cutaneous venomotor tone (Mark and Mancia, 1983). During hypotension, reduced carotid baroreceptor stimulation causes reflex arterial vasoconstriction, indicating tonic regulation of vascular caliber by these receptors. High-pressure baroreceptors maintain systemic arterial pressure around a set point. The influence of carotid baroreflex deactivation on arterial blood pressure in humans is almost 50% greater than baroreflex activation. Conversely, the chronotropic influence of carotid baroreceptors is greater with reflex stimulation compared to inhibition. In contrast to the immediate influence on heart rate, the latency of the blood pressure response to baroreceptor stimulation is long (3 s) and the response is prolonged (10–30 s), reflecting the slower onset-offset of sympathetic-effector organ versus vagal neuroeffector responses; that is, vascular smooth muscle versus cardiac sinoatrial node. The adrenal medulla does not appear to contribute to reflex vascular responses to carotid baroreceptor stimulation or inactivation in humans.

Baroreflex Interactions

Significant interactions exist between the right and left baroreceptor regions and between carotid and aortic regions (Sagawa, 1983). For example, summation of reflex responses occurs during bilateral carotid sinus input. Summation is nonlinear and central neural occlusion causes inhibition at high intrasinus pressures. Interaction tends to be facilitatory at low pressures although, depending on the underlying experimental conditions, there can be mild inhibitory, additive or facilitatory interaction.

Effective Stimuli

Carotid sinus and aortic arch baroreceptors respond to changes in mean pressure, pulse pressure, and the rate of rise of blood pressure (Fig. 36.8). Hence, they demonstrate dynamic sensitivity with greater responsiveness on the rising phase and decreased frequency on the falling phase of the sinusoidal pressure–response curve (Sagawa, 1983). Discharge activity adapts rapidly to maintained arterial pressure due to viscoelastic properties of the vascular wall and remains constant above 70 mm Hg. Compared to the carotid sinus, aortic

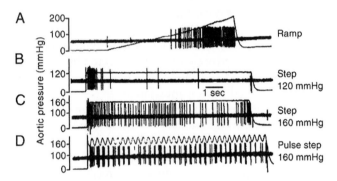

FIGURE 36.8 Responses of unmyelinated aortic baroreceptor afferent fiber to ramp (A) and step changes in mean arterial pressure (B and C) and to phasic changes in blood pressure (D). The fiber demonstrated sparse activity at low pressures and irregular discharge activity as pressure was increased that became more constant at higher pressures. Pulse changes in pressure caused synchronous activity (Thoren and Jones, 1977).

arch stretch receptors have a higher threshold, but once activated, they exhibit similar sensitivities.

More information is available on mechanisms of activation of myelinated compared to unmyelinated baroreceptor sensory nerve fibers and most of the earlier comments refer to the former group. Fundamental differences between these two sets of afferents include a higher threshold, reduced responsiveness to changes in intravascular pressure, less regular discharge, and lower maximal frequency of discharge activity of C versus A fibers. Unmyelinated baroreceptor afferents are relatively inactive under basal conditions. Conversely, the threshold for discharge of most myelinated baroreceptor afferents is near resting mean arterial pressure.

Adaptive Responses

Arterial baroreceptors can be reset (Fig. 36.9). *Resetting*, as originally defined by McCubbin, means that afferents respond differently to the same pressure after they have been subjected to an altered pressure for a period of time (Chapleau *et al.*, 1991; Mark and Mancia, 1983; Sagawa, 1983). Thus, resetting is defined as a shift in the relationship between arterial pressure and reflex hemodynamic or autonomic response. The period required to induce this phenomenon ranges from 15 to 30 min (acute resetting), which is promptly reversible, or hours to days (chronic resetting), which is slowly reversible or irreversible. Wall viscoelastic elements (creep/stress relaxation) and possibly changes in myogenic tone underlie acute resetting, whereas altered wall structure is involved in chronic resetting. Alterations in the afferents themselves,

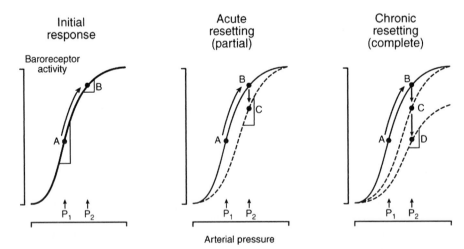

FIGURE 36.9 Acute and chronic resetting of arterial baroreceptors in response to a sustained increase in arterial blood pressure. Note that the immediate response occurring after seconds to minutes (C in middle panel) during an increase in blood pressure partially restores the increase in discharge frequency, whereas the chronic response occurring after days to weeks (D in right panel) completely restores the activity (A in all panels) (Chapleau *et al.*, 1989).

perhaps caused by changes in biophysical properties of sensory nerve membranes, such as ionic mechanisms or local release of substances from the endothelium and even resetting in the central nervous system, may underlie this phenomenon. Central resetting occurs when neural–humoral responses are altered during direct stimulation of afferent nerves. Although inhibition of central processing is observed most commonly, paradoxical central facilitation can occur and may briefly oppose central inhibition. Other than localized vascular changes that lead to resetting of afferent endings, mechanisms underlying peripheral and central neuronal resetting remain ill defined.

Influence of Chemical Stimuli on Mechanoreflexes

Although arterial baroreceptors primarily act as mechanoreceptors, their function can be modified by changes in the regional chemical environment. For instance, circulating or locally released catecholamines constrict vascular smooth muscle and hence unload baroreceptor nerve endings (Chapleau et al., 1991; Sagawa, 1983). Elevated catecholamines also directly excite myelinated and unmyelinated baroreceptor afferents. Other substances, like atrial natriuretic factor, substance P, and vasopressin, influence baroreceptor activity largely through their action on vascular smooth muscle. Paracrine factors, including NO, prostacyclin, reactive oxygen species, and various platelet-derived substances, modulate baroreceptor sensitivity by altering neural excitability (Chapleau, 2001). Stretch, shear stress, and chemical substances influence discharge activity of myelinated nerve endings by stimulating the release of NO or prostacyclin from vascular endothelium. Baroreceptors are sensitized by cyclooxygenase products, independent of changes in mechanical deformation. Group III mechanosensitive somatic sensory endings similarly are sensitized by paracrine factors during muscle contraction (Box 36.2).

Several hormonal systems, including angiotensin, arginine vasopressin, and atrial natriuretic factor (ANF), strongly influence the vascular system and end organ function, for example the kidney's regulation of body fluid balance. These hormones also influence blood pressure by regulating sympathetic tone, in part, through their baroreflex effects. Angiotensin and vasopressin are potent vasoconstrictors. Angiotensin also has many actions in the peripheral and central nervous systems, for example by facilitating activation of sympathetic ganglia and stimulating the adrenal medulla to release norepinephrine and epinephrine. Angiotensin decreases baroreflex sensitivity and facilitates its ability to increase blood pressure through a central

action on several circumventricular organs, including the area postrema in the hindbrain, the subfornical organ (SFO), anteroventral third ventricle (AV3V), organum vasculosum of the lamina terminalis, and median preoptic nucleus (MNO) in the forebrain. Through an action on V_1 receptors in the area postrema, vasopressin acts on V_1 receptors inhibit sympathetic outflow and facilitate baroreceptor reflex function. Likewise, ANF reduces sympathetic activity. Although controversial, ANF may stimulate cardiopulmonary afferents to suppress baroreflex function and promote renal sodium excretion. The AV3V region is critical for fluid and electrolyte balance and likely participates as part of a long-loop supramedullary pathway in the baroreflex (Dampney, 1994a).

Baroreflexes in Health and Disease

The primary function of the arterial baroreflex system is to oppose changes in blood pressure (Longhurst, 1982), for example during changes in orthostatic position when an upright posture is assumed. Under normal circumstances, baroreceptor endings sense orthostatic-induced changes in arterial blood pressure and prevent syncope by increasing heart rate, cardiac output, and peripheral vascular resistance. Individuals who have a defect in the baroreflex arc (e.g., in multiple systems atrophy) have difficulty withstanding orthostatic challenges. In response to extreme changes in blood pressure, such as during shock, baroreceptors restore total peripheral resistance by 50 to 70% and increase heart rate sufficiently to increase cardiac output by 20 to 25%. Carotid baroreceptors likely contribute more to restoring blood pressure than aortic baroreceptors, as the threshold for afferent activation and reflex regulation of regional vascular resistance is higher for mechanoreceptors in the arch than in the sinus (100 vs. 60 mm Hg). As hemorrhage progresses, cardiopulmonary stretch receptors and both central and peripheral chemoreceptors, responding to changes in central filling pressures, blood gases and pH, contribute to increased sympathetic activity. Ultimately, in severe shock when blood pressure and acid–base balance are profoundly disordered, baroreflexes fail and may even contribute to terminal bradycardia and demise when stimulated paradoxically.

Abnormal baroreceptor function contributes to the pathogenesis of certain disorders like hypertension (Mark and Mancia, 1983; Oparil et al., 1991). Reduced baroreflex function does not initiate hypertension, since complete denervation of both carotid and aortic afferent pathways does not cause sustained hypertension. However, baroreceptors adapt in chronic hypertension, documented by the downward resetting and rightward shift of the blood pressure–afferent nerve activity relationship. Resetting to a higher set point prevents baroreceptor opposition of elevated blood pressure. Decreased elastin and increased collagen in the media of the walls of vessels in hypertensive patients reduces vascular compliance and ultimately the stretch of mechanosensitive nerve endings when arterial pressure is increased chronically. Similarly, as stiffening of vessels occurs in advancing age, baroreflex function is reduced (Mark and Mancia, 1983).

Aortic baroreflexes are more impaired in hypertension than carotid baroreflexes. Also, the heart rate component is affected more profoundly by hypertension than the blood pressure response, indicating that baroreflex control of parasympathetic activity is influenced more than sympathetic activity. Central resetting also occurs in hypertensive subjects although the underlying mechanisms are unknown.

Patients with *congestive heart failure* (CHF) experience baroreflex dysfunction (Zucker, 1991). Retained salt and water that decrease vascular compliance (increase stiffness) may reduce baroreceptor afferent activity. Additionally, physiological and pathological alterations of the heart, particularly in the atria, may influence arterial and cardiopulmonary baroreflex interactions and ultimately reflex responses to changes in arterial pressure.

Carotid sinus hypersensitivity results from local and systemic disease near the carotid bifurcation (Longhurst, 1982). Thus, tumors and atherosclerosis, digitalis intoxication, and Takayasu's arteritis, among other conditions, can cause profound bradycardia or arteriolar vasodilation, or a combination of the two, and recurrent presyncope or syncope.

PERIPHERAL ARTERIAL CHEMORECEPTORS

Anatomy

Arterial chemoreceptors are located principally in two regions along the great vessels: the *carotid bodies* at the bifurcation of the common carotid arteries and the *aortic bodies* in the aortic arch (Biscoe, 1971; Gonzalez *et al.*, 1994). Chemoreceptors are composed of clumps of *glomus* tissue perfused with a high blood flow by branches of the carotid artery or aorta. Glomera contain extensive capillary networks. A characteristic feature of carotid (and presumably aortic) bodies is their very high oxygen consumption, which is four times greater than brain. Glomus tissue includes type I (glomus or chemoreceptor) and type II (sustentacular, satellite, or supporting) cells. Type I cells, comprising more than 75% of glomus cells, are secretory, contain catecholamines, and are heavily innervated by sensory nerves. Type II cells, closely associated with type I cells, are not innervated directly; rather, they encircle unmyelinated nerve fibers. Sympathetic and parasympathetic efferent fibers innervate blood vessels in chemoreceptor regions as well as type I and II cells. Carotid body parenchymal cells are essential for chemotransduction, although only afferent nerve endings are chemosensitive.

Sensory fibers from the carotid body course through the carotid sinus (Hering's) nerve, a branch of the glossopharyngeal (IX cranial) nerve; cell bodies of carotid body sensory afferents are in the petrosal and, to a smaller extent, superior cervical ganglia (Eyzaguirre *et al.*, 1983). Sensory pathways of aortic bodies travel in the aortic depressor (Cyan's) nerve, a branch of the vagus (X cranial) nerve, with perikarya in the nodose ganglia. Both A and C fiber chemoreceptor afferents exist. Unmyelinated afferents project to the lateral NTS, rostral to the obex, or to the medial and commissural NTS, caudal to the obex. Most chemoreceptor afferents terminate caudal to the obex, whereas baroreceptor afferents terminate mainly rostral to the obex. Chemoreceptor NTS inputs excite inspiratory neurons in the ventrolateral subnucleus and converge with baroreceptor inputs on nonrespiratory neurons in the ventrolateral subnucleus and neighboring reticular formation. The NTS projects to the rVLM where, along with cVLM-routed signals from baroreceptors, axons containing chemoreceptor information converge on common sympathoexcitatory premotor neurons (Spyer, 1990, 1994; Dampney, 1994b).

Effective Stimuli

Carotid and aortic body chemoreceptors are stimulated by decreases in the partial pressure of oxygen (Pa_{O_2}), with a threshold beginning at about 85 mm Hg, and increases in partial pressure of carbon dioxide (Pa_{CO_2}) (Eyzaguirre *et al.*, 1983; Gonzalez *et al.*, 1994). Although CO_2 stimulation of chemosensitive nerve endings occurs largely through carbonic anhydrase-mediated changes in H^+, CO_2 also exerts a small direct effect on these sensory endings. Aortic body afferents respond more vigorously to CO_2 at low Pa_{O_2} (Fig. 36.10). Aortic chemoreceptors are stimulated less than carotid body chemoreceptors by changes in Pa_{O_2} (Fig. 36.11). Afferent activity is increased with maneuvers that influence local Pa_{O_2}, including changes in arterial O_2 content in response to carboxy-hemoglobinemia (aortic chemoreceptors), arterial hypotension (aortic chemoreceptors), and sympathetic vasoconstriction, which reduces local blood flow. Stimulation of para-

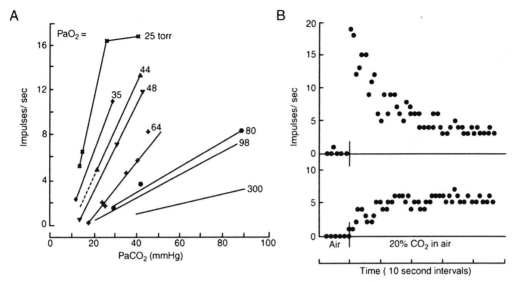

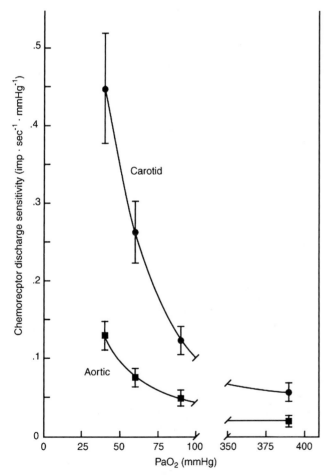

FIGURE 36.10   Response of single carotid body chemoreceptor fibers to increases in arterial Pco_2 (A). The response is diminished as Po_2 is increased. Inhibition of carbonic anhydrase, preventing production of H^+, reduces the immediate but not the more delayed response to a step increase in Pco_2 (B) (McCloskey, 1968).

FIGURE 36.11   Sensitivity of aortic and carotid body chemoreceptors to arterial hypoxia. Compared to discharge activity of carotid chemoreceptors, aortic chemoreceptors manifest a smaller response of the arterial Po_2 levels (Lahiri et al., 1981).

sympathetic nerves to the carotid body reduces afferent discharge activity, either through an increase in blood flow or release of a local neurotransmitter, like dopamine from type I cells.

Peripheral chemoreceptors respond not only to changes in Pao_2, $Paco_2$, and pH, but also are strongly stimulated by chemicals like lobeline, nicotine, and cyanide (Eyzaguirre et al., 1983) that have been used extensively in experimental studies of chemoreceptor function. Chemoreceptors also respond to changes in blood temperature and plasma osmolality. Their *thermosensitivity* may be explained by their high Q_{10} (75) and a high mean energy of activation (81.5 kcal/mol). The high Q_{10} value may represent multiplication of a series of reactions, each with a low individual Q_{10}. Their *osmoreceptor* response causes vasopressin release. Changes in blood temperature and plasma osmolality during prolonged exercise increase plasma ADH through a chemoreflex, which assists in reducing renal blood flow.

Receptor Mechanisms

Oxygen sensing of the carotid body occurs as an instantaneous acute response and as a more chronic cellular change over days to weeks (Lahiri, 1996). Oxygen chemoreception is linked mainly to vesicular neurotransmitter release and generated neural discharge, although nonvesicular neurotransmitters like nitric oxide may play an important role. Functional neurotransmitters responsible for chemoreceptor stimulation have not been identified.

Reflex Cardiovascular Responses

Although the best known effect of these receptors is on respiration, stimulation of peripheral chemoreceptors both directly and indirectly influences the cardiovascular system (Biscoe, 1971; Eyzaguirre et al., 1983; Marshall, 1994). Ventilation is increased by the stimulation of aortic and carotid body chemoreceptors, although more profound responses result from the stimulation of carotid bodies. In the intact animal, carotid body stimulation evokes reflex hyperventilation, tachycardia, increased myocardial contractility and cardiac output, and decreased systemic vascular resistance (Fig. 36.12). When ventilation is controlled, stimulation of carotid bodies consistently causes reflex bradycardia, decreased cardiac output, and increased total peripheral resistance. In contrast, stimulation of aortic bodies with cyanide or nicotine during controlled respiration leads to tachycardia and hypertension, resulting from increases in cardiac contractility and vasoconstriction of both veins and arteries. Thus, the direct influence of chemoreceptor stimulation on the cardiovascular system is affected profoundly by their influence on respiration. The secondary reflex, caused by stimulation of pulmonary stretch receptors, reverses the direct influence of the carotid chemoreceptors on the cardiovascular system. In addition, increased ventilation during chemoreceptor stimulation (e.g., during hypoxia) reduces arterial P_{CO_2}, which further increases cardiac output and modulates reduced systemic vascular resistance.

Peripheral chemoreceptors are responsible for hypoxic (e.g., high altitude) arterial hypertension and venoconstriction, although CNS chemoreceptors may facilitate these reflexes. The vasoconstrictor response is modified to some extent by the direct effect of low P_{O_2} that relaxes vascular smooth muscle. Carotid and aortic chemoreceptor stimulation generally causes vasoconstriction in skeletal muscle, kidney, and intestine and cutaneous vasodilation (Eyzaguirre et al., 1983; Marshall, 1994). Chemoreceptor stimulation can lead to parasympathetic coronary vasodilation. Regional vasoconstriction observed in most circulations is reduced or reversed during arterial chemoreceptor stimulation as ventilation increases. Peripheral chemoreceptors do not regulate brain blood flow. Since there are species differences in chemoreceptor-related cardiovascular responses, extrapolation of responses across species should be made cautiously.

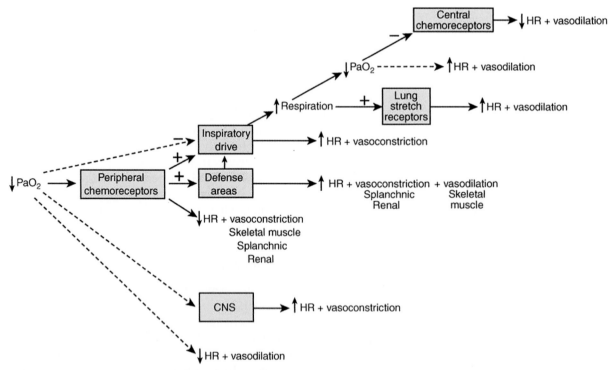

FIGURE 36.12 Cardiovascular and respiratory responses induced by systemic hypoxia. The responses are a composite of the influence of low arterial P_{O_2} on peripheral chemoreceptors and their influence on both respiratory and cardiovascular centers in the brain stem (solid arrows), as well as direct effects of low P_{O_2} on the CNS and effector organs, including the heart and blood vessels (dotted arrows) (Marshall, 1994).

CARDIAC RECEPTORS

Intravenous veratridine reduces heart rate and blood pressure through stimulation of cardiac vagal afferent endings, a response known as the *von Bezold-Jarisch reflex*. The heart is innervated by atrial and ventricular vagal and sympathetic afferent endings that respond to mechanical, chemical, or both types of stimuli. The term *cardiopulmonary baroreceptors* is used frequently to describe reflex vagal events that arise from the four cardiac chambers, great veins, and other vessels in the lungs. Cardiac vagal afferents first synapse in the medial subnucleus of the NTS. Like arterial baroreceptors, second-order cardiac reflex pathways project from the NTS to the cVLM (via glutaminergic synapses), which in turn sends GABAergic projections to sympathetic premotor neurons in the rVLM. Sympathetic afferents project to many regions in the hypothalamus, pons, midbrain, and brain stem, including the rVLM where they influence sympathetic outflow.

Atrial Receptors and Reflexes

Receptive fields of myelinated afferents are located in venoatrial junctions; afferent connections to the CNS principally course through the vagus (Hainsworth, 1991) (Linden and Kappagoda, 1982). These sensory endings are stimulated by stretch and often fire either in late systole just after the upstroke of the aortic pressure wave (type B receptors) or just before the start of the atrial *a* wave (type A receptors). Discharge activity of type B receptors likely reflects atrial volume, whereas type A receptors are influenced more by heart rate. Thus, while the rate of firing of type A receptors generally is maximal for each cardiac cycle, type B receptors operate over a range of discharge frequencies (Fig. 36.13). Other myelinated atrial afferents

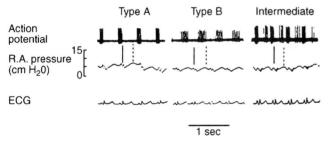

FIGURE 36.13 Type A, B, and intermediate atrial myelinated afferent patterns of discharge activity. Activity of type A receptors correspond to atrial *a* wave, whereas type B receptor activity occur during atrial *v* wave. Receptors with an intermediate response coincide with both *a* and *v* waves (Kappagoda *et al.*, 1976).

display both type A and B patterns of discharge activity.

In 1915, Bainbridge described a reflex increase in the heart rate in dogs in response to infusion of saline, most evident at low basal heart rates. Interestingly, despite the chronotropic response that originates from activation of myelinated afferents, arterial pressure, systemic vascular resistance, and myocardial contractility do not change consistently during stimulation of venoatrial junctions by distending small intravascular balloons. Several studies have failed to document the Bainbridge reflex in humans. Unmyelinated atrial vagal and sympathetic afferents also innervate atria, but their physiological function is uncertain.

Atrial Reflexes in Humans

Mild *lower-body negative pressure* (LBNP), which primarily unloads cardiopulmonary receptors to reduce blood flow to both skin and muscle, has only a small effect on splanchnic resistance vessels and very little influence on capacitance vessels (Fig. 36.14) (Mark and Mancia, 1983). Contrast dye injection during cardiac angiography, in addition to a direct action on cardiac pacemaker cells, activates cardiac chemoreceptors (von Bezold–Jarisch reflex), causing profound reflex bradycardia in humans. In addition, cardiopulmonary baroreflex unloading potentiates arterial depressor baroreflex function. Cardiopulmonary afferents also oppose the somatic exercise pressor reflex.

Cardiopulmonary Regulation of the Kidney

Cardiopulmonary receptor activation is more effective in decreasing renin and plasma vasopressin than arterial baroreceptors (Mark and Mancia, 1983). There is no clear consensus on the relative importance of vagal cardiopulmonary versus ventricular receptors in vasopressin regulation, as there is substantial variability between species. Studies of cardiac transplant recipients, although complicated by the underlying disorder, which frequently includes renal dysfunction, suggest an important role of cardiac reflexes in regulating renal function. Cardiopulmonary and ventricular receptors tonically inhibit cardiovascular sympathetic activity and renin release from the kidney (Goetz *et al.*, 1991). Ventricular receptors in humans primarily influence muscle resistance vessels, whereas arterial baroreceptors exert more control of the splanchnic circulation.

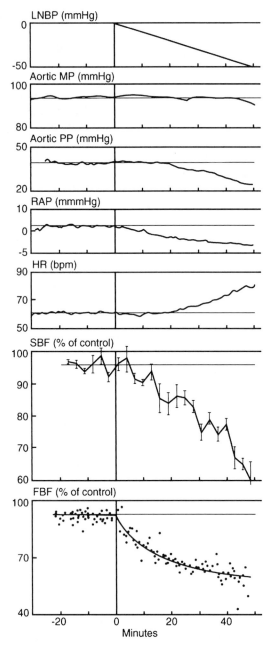

FIGURE 36.14 Effects of LBNP on aortic mean pressure (MP), aortic pulse pressure (PP), right atrial pressure (RAP), HR, splanchnic blood flow (SBF), and forearm blood flow (FBF). Low-level LBNP (0–20 mm Hg) lowered venous return and RAP and, by unloading cardiopulmonary receptors, significantly decreased skin and muscle FBF and slightly reduced SBF. High-level LBNP (20–50 mm Hg), which reduced aortic PP and eventually aortic MP (therefore also unloading high-pressure arterial baroreceptors), increased HR and decreased both FBF and SBF (Johnson *et al.*, 1974).

Ventricular Receptors and Reflexes

Ventricular sensory receptors mostly exist as bare nerve endings situated in the interstitium. Finely myelinated (Aδ) or unmyelinated cardiac afferent fibers (C) project to the CNS through vagal or sympathetic (spinal) pathways (Longhurst *et al.*, 2001). There is little difference in sensitivity between myelinated and unmyelinated cardiac afferents. Mechanosensitive cardiac C fibers respond mainly to changes in ventricular diastolic pressure, suggesting that they may be more responsive to stretch than to compression. Reflex depressor responses to changes in arterial pressure may originate from coronary rather than ventricular mechanoreceptors. Ventricular chemosensitive C fibers respond to exogenous stimuli such as veratridine or contrast dye injected during angiographic procedures and to a number of endogenous chemical stimuli produced during pathologic conditions like ischemia and reperfusion (Box 36.3). Cardiac transplantation and myocardial infarction globally or regionally interrupt ventricular afferent activity.

Reflex cardiovascular responses to stimulation of either mechano- or chemosensitive vagal endings are primarily inhibitory, resulting in reduced sympathetic nerve activity, depressor responses, and bradycardia. Conversely, activation of sympathetic cardiac ventricular afferents reflexly excites the cardiovascular system, leading to an increase in blood pressure and heart rate (Fig. 36.15). Although stimulation of ventricular vagal C fiber afferents causes bradycardia in anesthetized animals, heart rate is minimally altered in conscious animals and humans. Activation of ventricular vagal afferent endings attenuates baroreflex sensitivity and baroreflex control of heart rate.

Cardiac Reflexes in Disease

Adverse cardiovascular reflexes originating from the atria and ventricles are associated with clinical conditions like *aortic stenosis*, where high left ventricular pressures generated during stress stimulates vagal afferent mechanoreceptors causing vasodilation and occasionally syncope (Longhurst, 1984).

Prolonged supraventricular tachyarrhythmias like atrial fibrillation and *paroxysmal supraventricular arrhythmias* are associated occasionally with reflex diuresis and orthostatic intolerance (Longhurst, 1984).

Hypertensive patients demonstrate reduced reflex forearm vasoconstriction during LBNP ranging from −5 to −40 mm Hg, suggesting that both cardiopulmonary and arterial baroreflexes are impaired. Also, by reducing atrial compliance and diminishing responses of stretch sensitive atrial myelinated endings to elevated volume and pressure, CHF impairs the reflex response to increased plasma volume, thus promoting higher levels of circulating ADH and volume retention. As such, reduced cardiac afferent information in response to high atrial pressures contributes to periph-

---

BOX 36.3

CARDIAC RECEPTORS IN MYOCARDIAL ISCHEMIA

Myocardial ischemia is associated with a number of mechanical and chemical changes in myocardium. Chemosensitive receptors in the heart have been studied extensively. For instance, bradykinin is released, particularly during conditions of high sympathetic activity, reactive oxygen species are formed during ischemia and more so during reperfusion, and platelet activation during plaque rupture increases the regional concentration of 5-hydroxytryptamine, histamine and thromboxane A_2 (Longhurst *el al.*, 2001). Prostaglandins also are released during ischemia. Cyclooxygenase products either directly stimulate cardiac vagal afferents to induce a depressor response or sensitize vagal and sympathetic cardiac afferents to the action of other metabolic factors like bradykinin. These chemical mediators lead to profound stimulation of vagal and sympathetic cardiac afferents (Fig. 36.16). Receptive fields of vagal afferents are located transmurally, especially in the inferior–posterior ventricle, whereas sympathetic afferents are located closer to the epicardial surface and more in the anterior left ventricle. Thus, ischemia and reperfusion involving the inferior surface of the heart are associated more commonly with vasodepressor reflex responses, bradyarrhythmias, nausea, and vomiting, whereas excitatory pressor responses, including pain (*angina*), peripheral vasoconstriction, and tachyarrhythmias, are associated more commonly with anterior ischemia and reperfusion. Depressor reflexes reduce myocardial oxygen demand and thus lessen the imbalance between oxygen supply and demand during ischemia. Conversely, in addition to the warning signs of angina, activation of sympathetic afferents improves coronary perfusion pressure and hence maintains myocardial blood flow. It is confusing to imagine, however, that two diametrically opposed reflexes are engendered by myocardial ischemia. Clearly, overactivity of either reflex could also result in untoward cardiac events by exaggerating ischemia, either through marked decreases in perfusion or through exaggerated increases in oxygen demand. However, simultaneous activation of both afferent pathways from the heart lessens the influence of either pathway alone due to neural occlusion that occurs in the NTS and perhaps elsewhere (Fig. 36.15). While stimulation of ventricular sensory endings by endogenous chemical mediators leads to complex but frequently observed clinical responses, the reflex role of cardiac mechanoreceptor stimulation by changes in regional or global wall motion abnormalities is much less clear and deserves further study.

John C. Longhurst

Reference

Longhurst, J. C., Tjen-A-Looi, S., and Fu, L-W. (2001). Cardiac sympathetic afferent activation provoked by myocardial ischemia and reperfusion: Mechanisms and reflexes. *N.Y. Acad. Sci.*

---

eral edema and ascites (Schultz, 2005). Interestingly, digitalis glycosides can reverse the reduced cardiac sensory receptor responsiveness associated with CHF.

ABDOMINAL VISCERAL REFLEXES

Studies of visceral reflexes have yielded substantial information on the mechanisms by which visceral spinal sensory nerve endings are activated as well as their central integrative mechanisms.

Reflexes from Visceral Ischemia

Mesenteric ischemia reflexly increases blood pressure through temporal and spatial summation of input to cardiovascular centers in the brain stem. The pattern of response includes increases in heart rate, myocardial contractility, maintenance of cardiac output in the face of increased afterload, and vasoconstriction of splanchnic, hindlimb, and the coronary arterial circulation. Chemical mediators formed during *mesenteric ischemia*, including bradykinin, prostaglandin E_2, hydroxyl radicals, lactic acid, 8R,15S-dihydroxyeicosan tetraenoic acid (diHETE), serotonin, and histamine, stimulate or sensitize predominantly unmyelinated afferent endings. Low Po_2 triggers sensory nerve discharge through the production of other chemical mediators. Conversely, leukotriene B_4 modulates the afferent response during regional ischemia, for example, in patients with splanchnic vascular atherosclerosis. Hypercapnia, prostacyclin, thromboxane A_2, lactate, and mechanical stimulation are not involved in the ischemic reflex. The physiological importance of

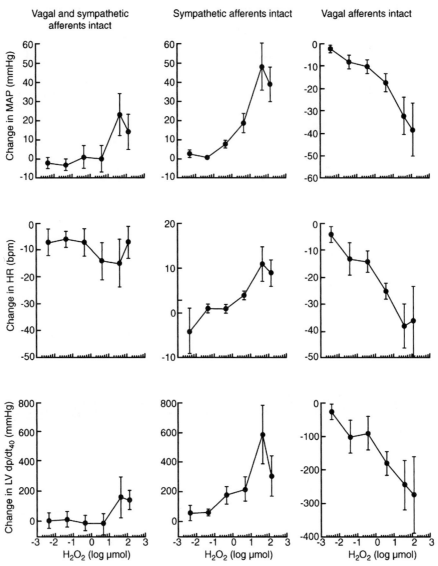

FIGURE 36.15 Changes in mean arterial pressure (MAP), HR, and left ventricular dP/dt at 40 mm Hg developed pressure, as an index of myocardial contractile function, during stimulation of the epicardial surface of the heart with increasing concentrations of the reactive oxygen species, hydrogen peroxide (H_2O_2). Transection of both vagi to interrupt the afferents coursing in this pathway, leaving the sympathetic afferent system intact (middle panels), resulted in a concentration-dependent reflex pressor response in association with increases in HR and myocardial contractility. Denervation of sympathetic (spinal) afferent pathways from the heart, leaving the vagal afferent system intact (right panels), converted the response to reflex cardiovascular inhibition, consisting of graded reductions in MAP, HR, and dP/dt. Stimulation of the ventricle when both afferent pathways were intact (left panels) led to a small pressor response with little change in HR or dP/dt as a result of occlusive interaction in the NTS (Huang *et al.*, 1995).

mechanoreceptor activation in the abdomen in ischemically mediated sympathoexcitation requires further investigation.

Renal Reflexes

The kidney is innervated by myelinated mechanosensitive and chemosensitive afferents. Myelinated afferents respond to increased ureteral pressure and distension of renal pelvis in most species, except nonhuman primates, where these receptors are sparse.

These mechanoreceptors frequently respond to changes in arterial or venous pressure and frequently a number of chemical mediators, including bradykinin, prostaglandins, histamine, adenosine, acetylcholine, and nicotine. The role of these mediators in stimulating these endings during conditions such as ischemia has not been determined. Two types of renal chemoreceptors have been described, including the CR_1 receptor that responds to hypoxia, ischemia, or cyanide infusion, and the CR_2 receptor that responds to ion movement, including back infusion of urine, NaCl, and

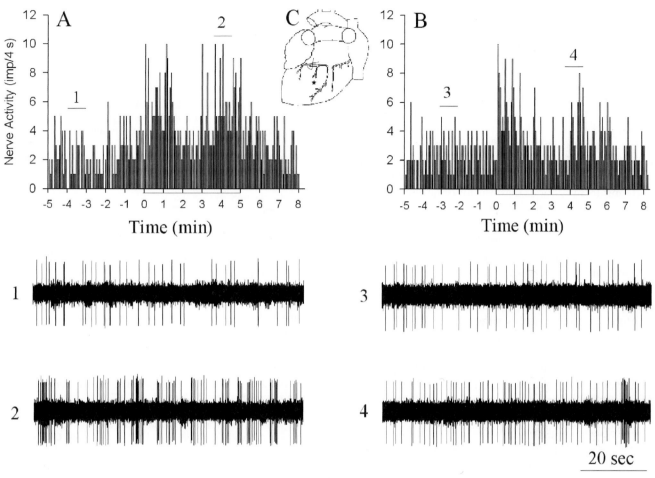

FIGURE 36.16 Responses of a sympathetic afferent fiber to repeated myocardial ischemia before (A) and after (B) blockade of thromboxane A₂ receptors in the heart with BM13177. This C-fiber afferent with a conduction velocity of 1.83 m/s innervating the posterior left ventricle (☆, inset) was excited biphasically during five minutes of regional ischemia (see ligature). Blockade of thromboxane's action reduced the increase in total discharge activity by ~50% from 214 to 103 imp/5 min, as demonstrated by actual responses in panels below histograms at indicated times.

KC1, but not urea or mannitol (Dietz and Gilmore, 1991; Kopp and DiBona, 1991).

Reflex cardiovascular activation commonly is observed during renal afferent stimulation. However, blood pressure is not tonically regulated by renal afferent input. Renorenal reflexes occur during stimulation of renal mechano- or chemoreceptors, leading to diuresis and natriuresis in association with decreased contralateral renal nerve activity. Renal afferents likely play a role in hypertension resulting from renal artery stenosis (Goldblatt kidney model) or coarctation involving the renal circulation (Oparil *et al.*, 1991).

Interactions between Somatic and Visceral Reflexes

There are many examples of interactions between reflex systems. For example, low frequency, low intensity somatic afferent stimulation during electroacu-

puncture produces long-lasting reduction in visceral afferent-induced sympathoexcitatory increases in blood pressure by as much as 40% (Longhurst, 2004). *Acupuncture* stimulates group III and IV somatic afferents that provide input to both spinal and supraspinal centers in the CNS. During acupuncture polysynaptic input to the ventral hypothalamic arcuate nucleus, midbrain ventrolateral PAG and midline medullary (raphe) nuclei, causes prolonged inhibition of rVLM sympathoexcitatory cardiovascular premotor neuronal activity and sympathetic outflow. Visceral afferent-induced activity in rVLM bulbospinal glutaminergic neurons is inhibited by modulatory neuropeptides, including enkephalins and endorphins, nociceptin and GABA released during low level somatic stimulation. Ultimately these somatic-visceral interactions have the capability of lowering elevated blood pressure and reducing demand-induced myocardial ischemia.

Summary

Reflex control of the cardiovascular system provides a unique advantage to mammalian species by allowing them to cope with the environment. The ability to respond rapidly to changes in blood pressure, regional blood flow, chemical composition of the blood and interstitial fluid, and adverse events such as pain, stress, and ischemia is managed by a set of sensory afferent systems that provide input into cardiovascular centers in the hypothalamus, midbrain, and brain stem with the end result of inducing both immediate and long-term changes in the autonomic and hormonal neuroeffector systems. Over the last century, an in-depth understanding of many neural mechanisms underlying these reflex events has been achieved. It is now known, for example, that even though these systems are largely responsible for beneficial short-term adaptations, they can play a role in a number of diseases that contribute to substantial morbidity and mortality. Recent insights suggest that it may be possible to activate these systems to more optimally regulate disease states like hypertension. Future challenges are to further define mechanisms underlying activation of sensory endings and central neural integrative functions, including interactions of multiple systems that add complexity to the behavior of interneurons in multiple regions concerned with conditioning reflex responses. Of particular interest are the new opportunities for investigation that are now present as a result of recently developed molecular and genetic approaches.

References

Barman, S. M. and Gebber, G. L. (2000). "Rapid" rhythmic discharges of sympathetic nerves: Sources, mechanisms of generation, and physiological relevance. *J. Biol. Rhythms* **15**, 365–379.

Biscoe, T. J. (1971). Carotid body: Structure and function. *Physiol. Rev.* **51**, 437–495.

Carrive, P., Bandler, R., and Campney, R. A. L. (1988). Anatomical evidence that hypertension associated with the defence reaction in the cat is mediated by a direct projection from a restricted portion of the midbrain periaqueductal gray to the subretrofacial nucleus of the medulla. *Brain Res.* **460**, 339–345.

Chapleau, M. W. (2001). Neuro-cardiovascular regulation: From molecules to man. *Ann. NY Acad. Sci.* **940**, xiii–xxii.

Chapleau, M. W., Hajduczok, G., and Abboud, F. M. (1989). Peripheral and central mechanisms of baroreflex resetting. *Clin. Exp. Pharmacol. Physiol.* **15** (suppl.), 31.

Chapleau, M. W., Hajduczok, G., and Abboud, F. M. (1991). Resetting of the arterial baroreflex: Peripheral and central mechanisms. In: Zucker, I. H. and Gilmore, J. P. (Eds.), "Reflex Control of the Circulation," CRC Press, Boston, pp. 165–194.

Dampney, R. A. L. (1994a). Functional organization of central pathways regulating the cardiovascular system. *Physiol. Rev.* **74**, 323–364.

Dampney, R. A. L. (1994b). The subretrofacial vasomotor nucleus: Anatomical, chemical and pharmacological properties and role in cardiovascular regulation. *Prog. Neurobiol.* **42**, 197–227.

Dietz, J. R. and Gilmore, J. P. (1991). The role of renal afferent nerves in circulatory control. In: Zucker, I. H. and Gilmore, J. P. (Eds.), "Reflex Control of the Circulation," CRC Press, Boston, pp. 435–449.

Eyzaguirre, C., Fitzgerald, R. S., Lahiri, S., and Zapata, P. (1983). Arterial chemoreceptors. In: Shepherd, J. T., Abboud, F. M., and Geiger, S. R. (Eds.), "Handbook of Physiology, Section 2: The Cardiovascular System, Volume III." Peripheral Circulation and Organ Blood Flow, Part 2, American Physiological Society, Bethesda, pp. 557–621.

Gonzalez, C., Almaraz, L., Obeso, A., and Rigual, R. (1994). Carotid body chemoreceptors: From natural stimuli to sensory discharges. *Physiol. Rev.* **74**, 829–898.

Guyenet, P. (1990). Role of ventral medulla oblongata in blood pressure regulation. In: Loewy, A. D. and Spyer, K. M. (Eds.), "Central Regulation of Autonomic Functions," Oxford University Press, pp. 145–167.

Hainsworth, R. (1991). Atrial receptors. In: Zucker, I. H. and Gilmore, J. P. (Eds.), "Reflex Control of the Circulation," CRC Press, Boston, pp. 273–289.

Huang, H.-S., Stahl, G., and Longhurst, J. (1995). Cardiac-cardiovascular reflexes induced by hydrogen peroxide in cats. *Am. J. Physiol.* **268**, H2114–H2124.

Izzo, P. N., Deuchars, J., and Spyer, K. G. (1993). Localization of cardiac vagal preganglionic motoneurons in the rat: immunocytochemical evidence of synaptic inputs containing 5-hydroxytryptamine. *J. Comp. Neurol.* **327**, 572–583.

Johnson, J. M., Rowell, L. B., Niederberger, M., and Eisman, M. M. (1974). Human splanchnic and forearm vasoconstrictor responses to reductions of right atrial and aortic pressures. *Circ. Res.* **34**, 515–524.

Jordan, D. (1995). CNS integration of cardiovascular regulation. In: Jordan, D. and Marshall, J. J. (Eds.), "Cardiovascular Regulation," Portland Press, London, pp. 1–14.

Kappagoda, C. T., Linden, R. J., and Mary, D. A. S. G. (1976). Atrial receptors in the cat. *J. Physiol.* **262**, 431.

Kopp, U. C. and DiBona, G. F. (1991). Neural control of renal function. In: Zucker, I. H. and Gilmore, J. P. (Eds.), "Reflex Control of the Circulation," CRC Press, Boston, pp. 493–528.

Lahiri, S. (1996). Peripheral chemoreceptors and their sensory neurons in chronic states of hypo- and hyperoxygenation. In: Fregly, M. J. and Blatteis, C. M. (Eds.), "Handbook of Physiology, Section 4, Environmental Physiology, Vol. II," American Physiological Society, Oxford University Press, New York, pp. 1183–1206.

Lahiri, S., Mokashi, A., Mulligan, E., and Nishino, T. (1981). Comparison of aortic and carotid chemoreceptor responses to hypercapnia and hypoxia. *J. Appl. Physiol.* **51**, 55–61.

Linden, R. J. and Kappagoda, C. T. (1982). "Atrial Receptors." Cambridge University Press, Cambridge.

Loewy, A. D. and Neil, J. J. (1981). The role of descending monoaminergic systems in central control of blood pressure. *Fed. Proc.* **40**, 2778–2785.

Loewy, A. D. and Spyer, K. M. (1990). "Central Regulation of Autonomic Functions." Oxford University Press.

Longhurst, J. C. (1982). Arterial baroreceptors in health and disease. *Cardiovasc. Rev. Rep.* **3**, 271–298.

Longhurst, J. C. (1984). Cardiac receptors: Their function in health and disease. *Prog. Cardiovasc. Dis.* **XXVII**, 201–222.

Longhurst, J. (2004). Acupuncture. In: Robertson, D., Low, P., Burnstock, G., and Biaggioni, I. (Eds.), "Primer on the Autonomic Nervous System," Academic Press, New York, pp. 246–429.

Longhurst, J., Tjen-A-Looi, S., and Fu L-W. (2001). Cardiac sympathetic afferent activation provoked by myocardial ischemia and reperfusion: Mechanisms and reflexes. *Ann. Acad. Sci.* **940**, 74–95.

Lovick, T. A. (1987). Cardiovascular control from neurones in the ventrolateral medulla. In: Taylor, E. W. (Ed.), "Neurobiology of the Cardiorespiratory System," Manchester University Press, Manchester, pp. 197–208.

Mark, A. L. and Mancia, G. (1983). Cardiopulmonary baroreflexes in humans. In: Shepherd, J. T., Abboud, F. M., and Geiger, S. R. (Eds.), "Handbook of Physiology, Section 2: The Cardiovascular System, Volume III." Peripheral Circulation and Organ Blood Flow, Part 2, American Physiological Society, Bethesda, pp. 795–813.

Marshall, J. M. (1994). Peripheral chemoreceptors and cardiovascular regulation. *Physiol. Rev.* **74**, 543–594.

McCloskey, D. I. (1968). Carbon dioxide and the carotid body. In: Torrance, R. W. (Ed.), "Arterial Chemoreceptors," Oxford, UK: Blackwell, pp. 279–295.

Oparil, S., Wyss, M. J., and Sripairojthikoon, W. (1991). The role of the renal nerves in the pathogenesis of hypertension. In: Zucker, I. H. and Gilmore, J. P. (Eds.), "Reflex Control of the Circulation," CRC Press, Boston, pp. 451–491.

Sagawa, K. (1983). Baroreflex control of systemic arterial pressure and vascular bed. In: Shepherd, J. T., Abboud, F. M., and Geiger, S. R. (Eds.), "Handbook of Physiology, Section 2: The Cardiovascular System, Vol. III." Peripheral Circulation and Organ Blood Flow, Part 2, American Physiological Society, Bethesda, pp. 453–496.

Schultz, H. D. (2005). Cardiac vagal afferent nerves. In: Undem, B. J. and Weinreich, D. (Eds.), "Advances in Vagal Afferent Neurobiology," Taylor & Francis Group, New York, pp. 351–375.

Spyer, K. M. (1990). The central nervous organization of reflex circulatory control. In: Loewy, A. D. and Spyer, K. M. (Eds.), "Central Regulation of Autonomic Functions," Oxford University Press, New York, pp. 168–188.

Spyer, K. M. (1994). Annual review prize lecture. Central nervous mechanisms contributing to cardiovascular control. *J. Physiol.* **474**, 1–19.

Thoren, P. N. and Jones, J. (1977). Characteristics of aortic baroreceptor C-fibers in the rabbit. *Acta Physiol. Scand.* **99**, 448.

Tjen-A-Looi, S., Li, P., and Longhurst, J. (2006). Midbrain vlPAG inhibits rVLM cardiovascular sympathoexcitatory. *Am. J. Physiol.* **209**, H2543–H2553.

Undem, B. J. and Weinreich, D. (2005). "Advances in Vagal Afferent Neurobiology." Taylor and Francis Group.

Zucker, I. H. (1991). Baro and cardiac reflex abnormalities in chronic heart failure. In: Zucker, I. H. and Gilmore, J. P. (Eds.), "Reflex Control of the Circulation," CRC Press, Boston, pp. 849–873.

John Longhurst

37

Neural Control of Breathing

Organisms must constantly exchange various substances with the environment for homeostasis and ultimately, survival. Different physiological systems handle solid, liquid, and gaseous metabolic precursors and by-products. Aerobic metabolism consumes O_2 and produces CO_2, the so-called blood gases. Organisms with large surface-to-volume ratios, such as bacteria, yeast, plants, and insects, rely on passive diffusion for gas exchange with the environment. However, vertebrates, with low external surface-to-volume ratios, actively pump gas using respiratory muscles to exchange air between the environment and the lungs (which in mammals have an exceptionally high surface-to-volume ratio) and the heart moves blood through and between the pulmonary and systemic circulations. The role of the brain in breathing is preeminent, as it generates the motor outflow driving the contraction and relaxation of skeletal muscles that pump the lung and modulates the tone of skeletal and smooth muscles in the upper airways and bronchi to control resistance to air flow.

Mammals that are neither diving nor hibernating, including humans, need a constant supply of O_2 to support their high metabolic rate, and so must breathe continuously from birth until death without lapses of more than a few minutes. Breathing appropriate for the regulation of blood O_2 and CO_2 is adaptable over a 20-fold range in metabolism (in humans from 0.25 liter $\times$ min^{-1} O_2 consumed at rest to ~3–6 liter $\times$ min^{-1} O_2 consumed during maximal exercise). Respiratory muscles account for ~5% of the body's metabolism at rest, and need to be used efficiently because the metabolic cost adds up over time. Moreover, serious respiratory muscle fatigue must be avoided. During exercise, energy costs of breathing must be kept low when possible (the respiratory pump moves from 5 (at rest) up to 200 (maximal performance in world-class athletes) liters $\times$ min^{-1} of air) to maximize the amount of O_2 available to other working muscles.

The drive for continuous ventilation in humans is strong. Typically, breathing is the last consequential movement to disappear following generalized depression of higher function, such as during surgical anesthesia or following insults to the brain (e.g., hypoglycemic coma). Automatic, homeostatic breathing can remain in people who have lost cortical function, resulting in ethical and emotional dilemmas over people who are "brain dead."

Neuroscientists take advantage of this robustness of breathing to study synaptic, neural, and network mechanisms in anesthetized or decerebrate mammals that continue to breathe. These mechanisms are so robust and sufficiently self-contained that an *in vitro* tissue slice at an appropriate level in neonatal or late fetal rodent brain stem continues to generate respiratory-related patterns of motor nerve activity.

EARLY NEUROSCIENCE AND THE BRAIN STEM

The persistent absence of breathing is the surest sign of death. Many ancient cultures associated life with breathing itself. Animation and animal are derived from the Latin *anima*, which means to breathe (and also soul). Taoism holds breathing sacred and a key to enlightenment: *immortality to the person who holds their breath for the time of 1000 respirations.* Buddhists believe that special states of enlightenment can be attained through effective modulation of breathing.

Although the purpose of breathing—to move air into and out of the lungs—was not established until the late eighteenth century, the underlying sites have been a constant source of speculation in Western culture. Perhaps the earliest suggestion that the brain was involved came from Galen (ca. 131–201 AD), who observed that gladiators and animals injured below the neck continued to breathe, but those injured in the neck stopped. Lorry (1760), who showed that cerebellectomized rabbits continued to breathe, concluded that the critical circuits lay within the brainstem and upper spinal cord, essentially the contemporary view. Legallois (1813) "extracted" brain tissue to determine what was necessary for breathing; he localized sites to the rostral ventrolateral medulla close to key sites, including the preBötzinger Complex (preBötC).

Ramon y Cajal (1909) provided an anatomic rationale for the critical role of the medulla in breathing. Examining the afferent and efferent projections of respiratory-related nerves, he suggested that three brainstem nuclei are important: the solitary tract (NTS) and commissural nuclei, primary targets of pulmonary afferents, and the nucleus ambiguus, which contains cranial motoneurons that innervate upper airway muscles. His network model for breathing is prescient (Fig. 37.1A): afferent containing information about lung volume and blood gases combined with intrinsic properties of brainstem neurons to produce a rhythmic outflow to spinal and cranial respiratory motoneurons. Adrian and Buytendijk (1931) recorded slow, rhythmic potentials in *in vitro* goldfish brainstems that were similar in timing to that of gill movements in intact goldfish (Fig. 37.1B), physiological evidence that the brainstem contained the critical circuits. Gesell and coworkers (1936) subsequently recorded individual medullary neurons discharging bursts of activity in phase with the breathing rhythm. This finding initiated a concerted effort to identify neurons that generate respiratory rhythm.

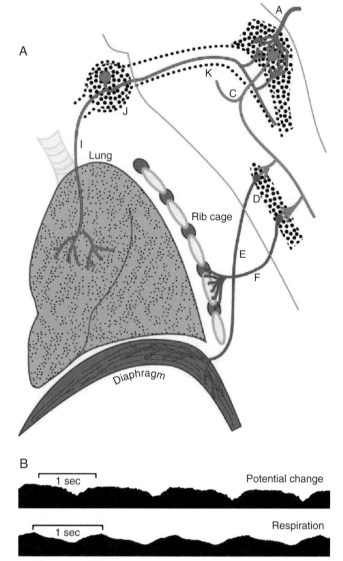

FIGURE 37.1   **A)** Ramon Y Cajal's model for respiratory control. Respiratory neurons in solitary tract (C) process signals from pulmonary afferents (K; cell bodies in nodose ganglion (J)) and some blood factor present in local capillaries (A). Descending control signals go to spinal motoneurons (D), innervating the diaphragm (E) or intercostal muscles (F). From Ramon y Cajal (1909). **B)** Oscillograph recordings of the gross electrical potential of an *in vitro* goldfish brain stem (top) compared to gill movement of an intact goldfish (bottom). From Adrian and Buytendijk (1931).

CENTRAL NERVOUS SYSTEM AND BREATHING

Breathing, chewing, and locomotion are important behaviors that require rhythmic movements with periods ~seconds. The underlying neural networks are programmed genetically and develop *in utero* so that newborn mammals breathe, chew, and, in some cases, locomote. How does the brain generate the rhythmic movements underlying breathing?

Although functionally straightforward (air in, air out), breathing from the brain's point of view is much

more than rhythm generation. Ventilation requires a patterned motor output with appropriate timing and magnitude of muscle contraction and relaxation with precise coordination among synergists and antagonists. A complex and mutable regulatory network composed of many distinct components, including forebrain, brainstem, and spinal cord circuitry, interacts to adjust breathing as necessary during life (Fig. 37.2).

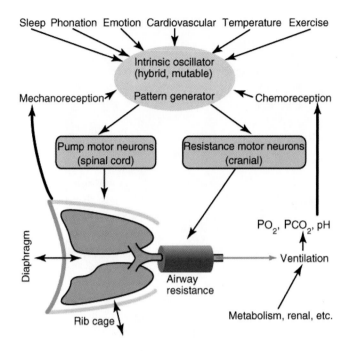

Sleep Phonation Emotion Cardiovascular Temperature Exercise

Intrinsic oscillator
(hybrid, mutable)

Pattern generator

Mechanoreception

Chemoreception

Pump motor neurons
(spinal cord)

Resistance motor neurons
(cranial)

Diaphragm

PO_2, PCO_2, pH

Airway
resistance

Ventilation

Metabolism, renal, etc.

Rib cage

FIGURE 37.2 Functional organization of central nervous system control of breathing (see text for details).

Appropriate adjustment of breathing to meet changing metabolic needs requires information on blood and brain chemistry and the mechanical state of the lung, airways, and respiratory muscles. Nevertheless, even if deprived of sensory information, the central nervous system can generate a basic rhythmic respiratory pattern.

WHERE ARE THE NEURONS GENERATING RESPIRATORY PATTERN?

Three distinct functional classes of rhythmically active (respiratory) neurons underlie the generation of respiratory pattern: motoneurons innervating skeletal muscles of the respiratory pump and airways; rhythm generating neurons; and premotoneurons that link and coordinate rhythm generation to motoneurons. Nonrhythmic neurons that modulate the behavior of these respiratory neurons are also important. Brainstem respiratory neurons are concentrated within three regions (Fig. 37.3).

Dorsal respiratory group (DRG). The DRG is in the dorsomedial medulla caudal to mid-rostral portions of the NTS. The DRG is a source of input to spinal cord respiratory motoneurons in some species, but less so in others.

Ventral respiratory column (VRC). The VRC forms a continuous field in the ventrolateral medulla with five or more physiologically distinct subregions extending from around the facial nucleus to the spinomedullary junction. Distinct anatomical compartments within the VRC are inferred from clustering of similar types of respiratory neurons (e.g., inspiratory or expiratory neurons), presence or absence of spinal cord projections, and responses to chemical stimulation. Specific respiratory neuron types are not restricted to individual compartments, but overlap with the peak concentrations of particular neuron types centered within given compartments.

The caudal half of the VRC, termed the ventral respiratory group (VRG), contains bulbospinal respiratory premotoneurons that receive converging inputs from VRC rhythm generating neurons and neurons outside the VRC, sculpting the activity pattern distributed to various pools of respiratory motoneurons. The VRG is subdivided into rostral (rVRG) and caudal (cVRG) based on the peak concentrations of inspiratory (rVRG) versus expiratory (cVRG) bulbospinal neurons. Inspiratory bulbospinal neurons provide mainly excitatory input to spinal inspiratory motor pools, including monosynaptic input to phrenic motoneurons. Expiratory bulbospinal neurons in the cVRG are mainly excitatory and project to interneurons innervating thoracic and upper lumbar motoneurons that drive expiratory pump muscles (e.g., internal intercostal and abdominal muscles). Inhibitory expiratory neurons in the vicinity of the Bötzinger complex (BötC) and cVRG inhibit spinal inspiratory motor pools. Spinal cord injury that damages the axons of these bulbospinal neurons can cause life-threatening interference with the ability to breathe (Box 37.2)

Respiratory rhythm generation occurs mainly in the rostral VRC, which contains at least three functionally distinct compartments. The BötC is caudal to the facial nucleus, and contains mainly inhibitory glycinergic expiratory neurons. Further caudal is the preBötC an essential structure for rhythm generation (see later). The retrotrapezoid nucleus (RTN) is located at the rostral end of the VRC, ventral to the BötC and continues rostrally beneath the facial nucleus near the surface of the brain to just caudal to the trapezoid body. RTN neurons project to the pons and caudal VRC. The RTN is (part of) a region containing potential central chemosensory cells, an important source of respiratory drive and chemoreflexes. RTN

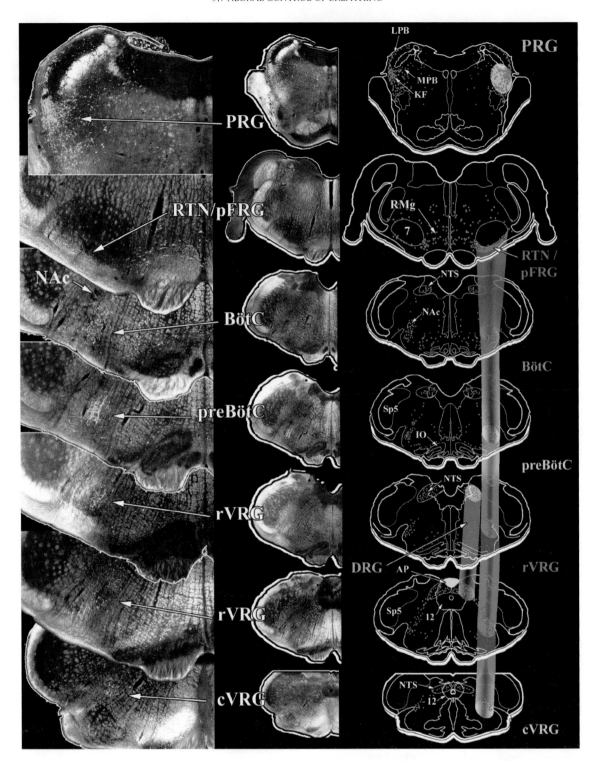

neuronal activity monotonically increases with extracellular acidity (associated with increased CO_2). The parafacial respiratory group (pFRG) is ventral to the facial nucleus, at least in neonatal rats. The RTN/pFRG may play an important role in rhythm generation (see later).

Pontine respiratory group (PRG). The PRG is in the rostral dorsolateral pons and encompasses portions of the parabrachial complex and the ventrally adjacent Kölliker-Fuse nucleus.

The PRG and VRC are at either end of an essentially continuous column of respiratory and

FIGURE 37.3 Respiratory neuroanatomy of the rat brain stem. The three major brain stem clusters of neurons controlling respiration are depicted. On the left are medium- and low-magnification views of labeled neurons in representative transverse sections of the pons and medulla after a retrograde tracer injection (not shown) within the rostral part of the contralateral VRG. The sections were photographed using reflected light so that myelinated axons appear white, retrogradely labeled neurons (immunostained with diaminobenzidine) appear gold, whereas the remaining cellular areas appear gray. On the right, these retrogradely labeled cells are diagrammed (red dots) within outlines of the adjacent low-magnification photomicrographs. On the right side of the diagram, the columns of respiratory neurons are identified: cVRG, pink; rVRG, green; preBötC, yellow; BötC, blue, PRG, orange; RTN/pFRG, burgundy; DRG, purple. Not all the (red) retrogradely labeled neurons in the medulla are included within the respiratory regions; some of these presumably represent extrinsic neuronal relays that can modulate respiration in relation to other behavioral or metabolic demands. 4V, fourth ventricle; 7, facial nucleus; 12, hypoglossal nucleus; AP, area postrema; BötC, Bötzinger Complex; cVRG, caudal VRG; DRG, dorsal respiratory group; IO, inferior olive; KF, Kölliker–Fuse nucleus; LPB, lateral parabrachial nucleus; MPB, medial parabrachial nucleus; NAc, ambiguus nucleus, compact division; NTS, nucleus of the solitary tract; preBötC, pre-Bötzinger complex; PRG, pontine respiratory group; RMg, raphe magnus; RTN, retrotrapezoid nucleus; rVRG, rostral VRG Sp5, spinal trigeminal nucleus; VRG, ventral respiratory group. Courtesy of George F. Alheid.

respiratory-related neurons in the lateral pons and medulla that extends rostrally from the medulla, essentially surrounds the facial nucleus, traverses the ventrolateral pons and regions adjacent to both the motor trigeminal and principal sensory trigeminal nuclei before merging dorsally with the PRG.

DISCHARGE PATTERNS OF RESPIRATORY NEURONS

One consideration that has impacted the classification of respiratory neurons is the number of phases that constitute the mammalian respiratory cycle, and whether the parcellation of motor output into multiple phases requires that rhythm generation *per se* be equivalently multiphasic. Inspiration and expiration constitute two obvious phases. However, a close examination of the discharge patterns of inspiratory (e.g., diaphragm) and expiratory (e.g., internal intercostal) muscle activity shows that there can be three (or more) phases of motor output (Fig. 37.4). Expiration can

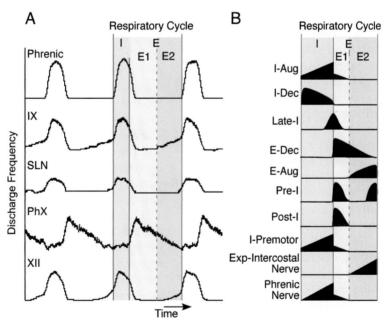

FIGURE 37.4 A) Diverse pattern of respiratory motor activity is exhibited on spinal (phrenic) and cranial nerves innervating the diaphragm and airways in an anesthetized rat. (A) Low-pass filtered trace of activity showing overall pattern of discharge. Red vertical lines indicate phase transitions between inspiration (I) and expiration (E). Note that the onset of inspiratory activity on cranial nerves precedes the onset of activity on the phrenic nerve. IX, glossopharyngeal nerve; PhX, pharyngeal branch of vagus nerve; SLN, superior laryngeal nerve XII, hypoglossal nerve. Adapted from Hayashi and McCrimmon (1996). Traces depict activity in a rat, but activity is similar in other mammals although respiratory frequency varies from greater than $100 \times min^{-1}$ in mice to about $3 \times min^{-1}$ in giraffes. **B)** Discharge patterns of respiratory neurons. For this classification, expiration is divided into E1, early expiration (or post-inspiration), and E2 or late expiration. The transition from inspiration to expiration is marked by a rapid decline in phrenic nerve activity illustrated in the bottom trace. Residual, declining phrenic nerve activity (when present) marks the E1 phase. Augmenting (Aug) activity occurs on the internal intercostal nerve (second trace from bottom) and Aug-E neurons largely during the E2 phase. The ordinate for each trace is discharge frequency with higher levels representing higher frequencies. Aug, augmenting neurons; Dec, decrementing neurons; E, expiratory; I, inspiratory.

consist of at least two phases. The onset of exhalation marks the onset of the first expiratory (E1), or post-inspiratory, phase. During this phase, the diaphragm often exhibits a low level of "postinspiratory" activity that brakes diaphragmatic relaxation and slows exhalation that declines rapidly, frequently ending by mid-expiration (Fig. 37.4B). At this point, an augmenting pattern of expiratory muscle activity begins, marking the onset of the second (E2) phase, characterized by active expiratory muscle contraction. Are these two phases separate components of rhythm generation, or a premotor division of a single expiratory phase? The rationale that the bursts reflect motor output patterns rather than rhythm is that the transition between the two expiratory phases is often ambiguous. For example, in Figure 37.4A, a declining postinspiratory output of the pharyngeal branch of the vagus nerve, which controls upper airway caliber, overlaps with the E2 pattern of activity in other nerves that affect the diameter of the upper airway, such as the glossopharyngeal and hypoglossal nerves.

The naming of respiratory neurons follows from the phase(s) in which their bursting activity occurs. Thus, neurons are designated inspiratory (I), postinspiratory (post-I), or expiratory (E). Neurons that have activity in more than one phase are termed phase-spanning. In addition, neurons that discharge only during inspiration or only during expiration can have distinctly different patterns of discharge. Investigators have interpreted these diverse patterns as evidence of different roles for each group of neurons in respiratory control and have used the discharge patterns to define subgroups of inspiratory and expiratory neurons. Although the total number of subgroups is relatively small, investigators have measured neuronal activities under various experimental conditions—for example, different species, different anesthesia, or decerebration—and have used slightly different criteria to name them. The result is a confusing array of population names, even to experts.

Representative patterns of discharge are shown in Figure 37.4B with the most common names of the predominant neuronal subgroups. Given this array of neuron types, the quandary is how they form a network. In particular, how do rhythm generating neurons in the preBötC, RTN/pFRG and perhaps elsewhere, interact with the rest of the network, including premotor and motoneurons, to produce the appropriate motor pattern? Investigators have produced working models that begin to address plausible mechanisms, however considerably more information on intrinsic membrane properties and synaptic interactions is required before this goal can be fully realized.

Synaptic Excitation and Inhibition Shape the Respiratory Motor Pattern

Two factors determine the firing patterns of motoneurons and premotoneurons: rhythmic synaptic drive (Box 37.1) and intrinsic excitability. Typically, respiratory neurons (other than pacemaker neurons) fire when they receive phasic excitatory drive, and they are silent during the pauses in phasic excitatory drive; during these pauses they typically receive phasic inhibitory drive. Fast excitatory drive is probably similar throughout the respiratory control system. Application of antagonists of excitatory amino acid receptors reduces the phasic activity of brainstem respiratory neurons, including those in the preBötC and in the VRC (Feldman et al., 2003), indicating that most of their excitatory synaptic input is via glutamatergic neurotransmission. The inhibitory drive is a bit more complex. Three types of synaptically mediated inhibition must be considered (Fig. 37.5).

Reciprocal Inhibition. Between phases of excitatory inputs (e.g., during expiration in inspiratory neurons), most respiratory motoneurons and premotoneurons are actively inhibited. Because these neurons are not spontaneously active, why is it necessary to inhibit them when they are not receiving excitatory input? Most likely, reciprocal inhibition prevents their spurious activation from unexpected afferent inputs that would cause inappropriate muscle contraction. Except for motoneurons that control jaw opening, which must be activated rapidly when a hard object is unexpectedly bitten during chewing, all motor and premotoneurons involved in periodic movements appear subject to reciprocal inhibition.

Recurrent Inhibition. There is a push–pull control of neuronal firing during a burst, that is, excitation and inhibition are concurrent, with the net balance determining the firing pattern. Reduction of inhibition (e.g., by the administration of inhibitory receptor blockers to motoneurons) results in increased burst amplitudes.

Phase-Transition Inhibition. Inhibition is postulated in many network models for breathing to arrest the activity of neurons producing one phase, thereby allowing transition into the next phase, as the neurons of the next phase become active.

Fast synaptic inhibition in brainstem respiratory networks is associated with an increase in permeability to Cl^-, mediated by GABA and glycine. Blocking $GABA_A$ receptors increases discharge of all VRG neurons. GABA appears to participate in reciprocal and phase-transition inhibitions to help maintain

---

BOX 37.1

EXPERIMENTAL ANALYSIS OF VENTILATION

Neuroscientists are constrained by what they can measure. Noninvasive experiments in humans reveal the phenomenology of breathing; for example, the relationship between ventilation and blood CO_2 levels. Brain imaging may reveal which regions are involved, but at present, spatiotemporal resolution is too coarse to show networks, much less neurons or synapses. Respiration lends itself to study because mammals suitable for experimentation breathe much like humans. Breathing continues following anesthesia or decerebration, and the neural patterns remain following paralysis (and mechanical ventilation), making possible experimental procedures requiring highly invasive techniques, such as single neuron recording. Compared to studies of neurons in culture or in tissue slices, experiments in whole mammals have serious limitations. For example, movements of the heart and lungs make certain types of single neuron recording difficult in the brain stem and spinal cord. Moreover, in live animals, the blood–brain barrier is intact, making precise control of brain extracellular fluid impossible. Finally, experiments are expensive and can last 36 h or more, requiring the dedicated and concurrent efforts of several investigators. Fortunately, the robust features of breathing that allow it to persist when other behaviors are suppressed by anesthesia or decerebration permit further reduction in the experimental preparation that removes these limitations. For example, a respiratory-related rhythm persists in the brainstem and spinal cord removed from a neonatal rodent. In addition, the rhythm persists in a particular slice of brain stem isolated from this preparation (Fig. 37.7). Using such slices, one can correlate exquisite measures of a neuron's intrinsic and synaptic properties and projections with measures of endogenous behavior, such as its firing pattern. With many common and otherwise useful *in vitro* preparations, such as the hippocampal slice, such correlations cannot be made because no endogenous behavior *in vitro* relates to known behavior *in vivo*. In general, *en bloc in vitro*, or slice, preparations play an important role in the study of basic cellular, synaptic, and integrative mechanisms underlying motor behavior.

Jack L. Feldman and Donald R. McCrimmon

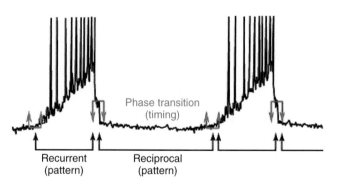

FIGURE 37.5 Membrane potential and discharge pattern of an augmenting inspiratory (or expiratory) neuron. The timing of three distinct types of inhibition is indicated.

appropriate timing of discharges in (pre)motor networks. GABA also appears to play a key role in regulating the baseline discharge frequency of VRG respiratory neurons, as antagonism of $GABA_A$ receptors with bicuculline amplifies the discharge pattern.

In inspiratory neurons, glycinergic transmission appears to mediate the rapid inhibition at the begin-

ning of expiration. Despite the clear role of synaptic inhibition in maintaining appropriate patterns of activity within the premotor circuitry, respiratory rhythms can be generated in the absence of phasic synaptic inhibition, at least in *in vitro* slices (see later).

Neurotransmitters Contribute to the Modulation of Breathing

The waxing and waning membrane potential underlying the periodic bursting of motoneurons can be ascribed simply to the alternating release of glutamate, GABA, and glycine. Given the straightforward task of these motoneurons, especially for muscle as specialized as the diaphragm, one might presume that little more is required to control their excitability. Yet the picture is much richer and provides some insight into possible roles for the diversity of transmitters.

For all respiratory neurons including motoneurons, neuromodulators affect their discharge pattern. Modulators that alter respiratory rhythm and pattern include serotonin, catecholamines (norepinephrine, epinephrine, and dopamine), acetylcholine, adenosine, and numerous peptides (typically colocalized with the amines), including substance P, neuropeptide Y, galanin, met-enkephalin, cholecystokinin, and thyrotropin-releasing hormone (Fig. 37.6). Many of these modulators are present in the phrenic nucleus. Why are so many present at such a simple relay to a muscle with a limited repertoire of functions? Multiple transmitters and receptors may be necessary to control neuronal excitability over the broad time scales important in breathing, from the millisecond scale of synaptic currents, to the seconds required to alter ventilation in response to changes in blood gases, to the hours or days for acclimatization to altitude or adjustment to disease, development, pregnancy, and aging. Multiple transmitters may allow coarse and fine control of P_{O_2}, P_{CO_2}, and pH and may also ensure that breathing is not compromised by changes in state, especially during sleep when global changes in amine levels can affect motoneuronal excitability. During rapid eye movement (REM) sleep, motoneuronal excitability produces a widespread muscle atonia, yet respiratory motoneurons must continue to drive respiratory muscles. Some of these modulators, particularly 5-HT and NE, may underlie persistent changes in excitability (see later). Finally, multiple neurotransmitters may be needed to prevent respiratory muscle fatigue. A strong inspiratory drive will sufficiently elevate glutamate in the synaptic cleft to act via a presynaptic metabotropic receptor to reduce further release. This governor of maximal phrenic motoneuronal activity would limit the likelihood of diaphragmatic fatigue during periods of sustained high ventilatory demand (e.g., fleeing a predator).

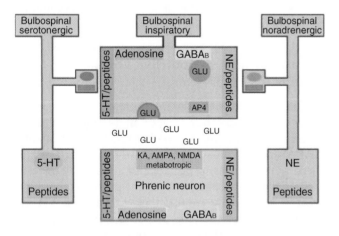

FIGURE 37.6 The diversity of synaptic control of phrenic motoneurons is essential for their proper function. The challenge is to understand how these mechanisms act in an integrated manner. Isolating and studying them one by one or out of context, e.g., in culture, may fail to reveal key properties. AMPA (see text), GLU, glutamate; 5-HT, serotonin; KA, kainate; NE, norepinephrine.

WHERE ARE THE NEURONS THAT GENERATE THE BREATHING RHYTHM?

Although it has been known since the late eighteenth century that the brainstem is responsible for respiratory rhythm generation, precise localization of neurons involved required the development of novel *in vitro* preparations (Box 37.1). In the 1980s, Suzue and colleagues (Ballanyi *et al.*, 1999) noted that the isolated neonatal rat brain stem and spinal cord, termed the *en bloc* preparation, generated a respiratory rhythmic motor output in spinal nerves that innervate muscles of the respiratory pump, as well as in cranial nerves innervating the airways, which control airway resistance. This *in vitro* preparation permitted a refinement of the "blunt spatula" brainstem transection approach used almost 200 years earlier by Legallois (1813) to identify structures critical for respiratory rhythm generation. In the contemporary case, removing thin (50–75 µm) sections from either rostral or caudal ends of the *en bloc* preparation while recording respiratory activity from spinal or cranial nerves resulted in the isolation of a small, ventral medullary region as an essential component in the rhythm generating circuitry (Smith *et al.*, 1991). Lacking an anatomical designation, this region was named the preBötzinger Complex. Confirmation that neurons within this region provide a minimum essential circuitry for respiratory rhythm generation follows from the observations that a thin transverse slice that contains the preBötC can generate a respiratory-related rhythm (Fig. 37.7). In addition, the necessary role of preBötC neurons for rhythm generation is suggested by the observation that local blockade of excitatory, glutamatergic neurotransmission abolishes respiratory rhythm.

Further demonstration of the essential role of the preBöC in respiratory rhythm generation follows from the observation that local activation of receptors for substance P or for opioids, which are expressed on restricted subsets of neurons within the preBötC, change respiratory frequency. Targeted destruction of preBötC neurokinin-1 (NK1) receptor-expressing neurons with the toxin saporin (a ribosomal toxin that interferes with the synthesis of new proteins) conjugated to substance P disrupts breathing. By itself, saporin is not taken up appreciably by neurons. However, when substance P binds to the NK1 receptor, the ligand–receptor is internalized, and when substance P is conjugated to saporin, the toxin is dragged along into the neuron. The result is that neurons expressing the NK1 receptor in the region of the injection die after several days. This delayed effect permits injected rats to fully recover from surgery

WHERE ARE THE NEURONS THAT GENERATE THE BREATHING RHYTHM?

BOX 37.2

SPINAL CORD INJURY CAN DIMINISH VENTILATION

Spinal injuries that affect the long axons of bulbospinal premotoneurons, which transmit respiratory drive to spinal motoneurons, can be life-threatening. With respect to the control of breathing, the severity of the injury depends on the spinal cord level at which the injury occurs. Damage between T1 and S1, in addition to causing sensory and motor deficits in the legs (see Chapters 25 and 29, this volume), can produce loss of control of intercostal muscles, leading to a diminished ability to generate inspiratory or expiratory movements. Injury between C4 and C8 can cause quadriplegia and loss of control of all intercostal musculature. Although innervation of the diaphragm may remain (with low cervical damage), the lack of chest wall control means the diaphragm must do more work; these patients frequently experience an alarming sense of difficulty breathing (dyspnea). Injuries between the lower brain stem and C4 result in pentaplegia, in which there are motor and sensory losses to the legs, arms, diaphragm, and neck. Such injuries require immediate artificial support of ventilation to maintain life. This type of cervical injury is common following dives into shallow pools.

Jack L. Feldman and Donald R. McCrimmon

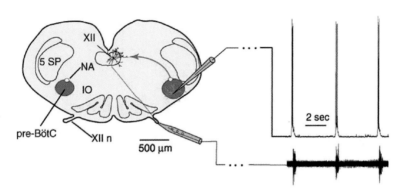

FIGURE 37.7 A respiratory rhythm generating slice preparation from a rodent brain stem exhibits endogenous respiratory rhythm in cranial nerve XII with patch-clamp recordings from a preBötC neuron. Diagram of a transverse slice. Patch-clamp recordings can be made from XII or preBötC cells while drugs (or different solutions) are supplied in a bath or applied locally. The rhythm is generated in preBötC and transmitted to the XII nucleus so it appears in motor axons of the XII nerve. (Right) A recording from a preBötC inspiratory neuron that shows periodic, inspiration modulated depolarizations and associated action potentials (current-clamp) (upper tracing). Lower tracing is a XII nerve recording; bursts of activity correspond to inspiration.

prior to determining the impact of the toxin on breathing. During the initial postinjection period, rats breathe normally, even during sleep. However, after several days, rats continue to breathe normally during wakefulness but have sleep disordered breathing. After a few more days, with substantial bilateral destruction of the NK1 preBötC neurons, rats develop a highly abnormal breathing pattern during wakefulness and are essentially unable to breathe during sleep; they are unable to maintain normal levels of blood O_2 and CO_2, and have pathological responses to hypoxia and hyperoxia. Nevertheless, the rats maintain sufficient ventilation to survive. These results suggest that preBötC neurons are required for normal breathing, but in their absence, other neurons can generate an ataxic breathing pattern sufficient to maintain life. The identity of the neurons generating this remaining rhythm is unknown but could include preBötC neurons lacking NK1 receptors or even neurons that normally generate volitional breathing movements.

Taken together, these experiments underlie the now widely accepted view that the preBötC is an essential component of the circuitry generating breathing rhythm.

A Second Respiratory Rhythm Generator Is Unveiled

Another collection of neurons, near the ventral medullary surface, rostral to the preBötC, may also be essential for rhythmogenesis. They are proximal to the facial nucleus in the pFRG; they may also be coextensive with the RTN, and we will refer to them as the RTN/pFRG (Fig. 37.3). Several factors point to their involvement in generating breathing pattern:

1. RTN/pFRG neurons are rhythmically active neurons in *en bloc* preparations.
2. When isolated from the preBötC by intramedullary transection in these preparations,

both the block containing the preBötC and that containing the RTN/FRG can generate a rhythm.

3. When exposed to μ-opiates, inspiratory activity in slices (which contain the preBötC but not the RTN/pFRG) slow down continuously, whereas inspiratory activity in *en bloc* preparations and *in vivo* slows discontinuously, a phenomena referred to as quantal slowing (Fig. 37.8); in similar conditions *in vivo*, rhythmic expiratory activity remains unaffected (Fig. 37.8).

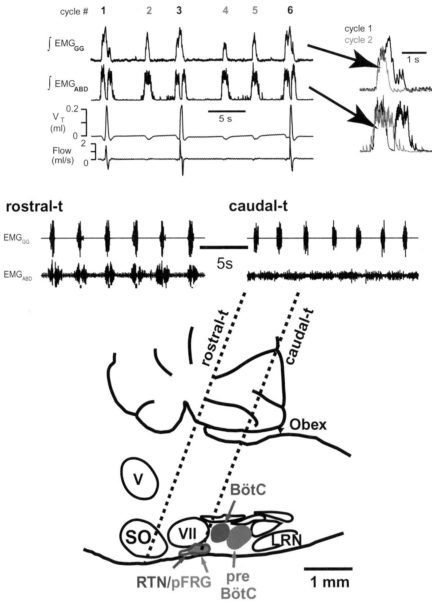

FIGURE 37.8 Inspiratory and expiratory rhythm generators in juvenile rat *in vivo*. **Top.** Left: Typical recording with normal cycles (1, 3, 6) interspersed with cycles without inspiratory activity (2,4,5) Traces from top: integrated genioglossus muscle EMG; integrated abdominal muscle EMG, tidal volume (V_T); and air flow (inspiratory air flow is up). Right: Superposition of ∫EMG$_{GG}$ and ∫EMG$_{ABD}$ in a normal cycle (black) compared to an inspiratory skipped cycle (red). Note the marked differences, especially in the ∫EMG$_{ABD}$ burst. **Bottom.** Alternating inspiratory and active expiratory activity is evident in EMG GG and EMGABD following complete brainstem transection in the plane marked rostral-t. Following a more caudal transection (caudal-t), effectively removing the RTN/pFRG, only active inspiratory activity is present. From Janczewski and Feldman, *J. Physiol.* (2006).

So, it appears that the preBötC and RTN/pFRG each have the intrinsic capacity to generate a respiratory rhythm, but normally are entrained to produce a single rhythm.

Thus, although many structures can modulate respiratory pattern, the working hypothesis of many in the field is that there are (at least) two coupled oscillators, the preBötC and the RTN/pFRG. An additional hypothesis is that inspiratory rhythm is generated by the preBötC and active expiration by the RTN/pFRG (Feldman and Del Negro, 2006).

Why two oscillators? Breathing in mammals is the end result of an evolutionary lineage of rhythmic movements that are closely driving rhythmic activity in gills of fish and adaptations related to the transformation of the swim bladder into the lung. Tadpoles have two distinct breathing patterns that can act separately or in tandem, with the "lung" rhythm opiate-sensitive whereas the "buccal" rhythm is not. The relative role of the two oscillators may differ between mammals and nonmammals, since the power stroke for breathing is primarily inspiratory and diaphragm-driven in mammals, whereas in most reptiles and amphibians, the power stroke is dominated more by expiratory effort.

Opiate-sensitivity clearly has consequences at birth, when there is a significant surge of maternally-derived opiates in the newborn mammal. This opiate surge should depress the opiate-sensitive preBötC; the unaffected RTN-pFRG may play an essential role in providing the necessary respiratory drive during the early postnatal period.

Do Pacemaker Neurons Contribute to Rhythm Generation?

Almost every successful explanation of brain processes underlying complex behavior is based on a combination of cellular, synaptic, and network properties. It is easy to build computational models of simple oscillatory behavior using only network (e.g., reciprocal inhibition) or only cellular (e.g., pacemaker) properties. For several decades neuroscientists speculated that respiratory rhythm was driven by the same kinds of cellular properties that drive heart rhythm, which is generated by the pacemaker activity of specialized cells. The simplest explanation for generation of respiratory rhythm is that it is driven by pacemaker neurons, presumably located in the preBötC and/or RTN/pFRG. The compelling, but now intensely debated, evidence that pacemaker neurons were critical came from the observation that rhythm in slices continues after blockade of postsynaptic inhibition. *In vitro*, neurons with bursting-pacemaker properties repre-

sent a subpopulation (5–25%) of preBötC neurons. There are two types: voltage-dependent and driven by a persistent Na conductance (I_{NaP}), present at birth; and those dependent on Ca^{2+} and/or a Ca^{2+}-activated nonspecific cation current (I_{CAN}), sparse before postnatal day 5.

Since the previous edition of this book, the pacemaker hypothesis has been challenged. Now, there are data that such activity is not essential (Feldman and Del Negro, 2006). While an obligatory role of pacemaker activity is disputed, there is a critical role in rhythmogenesis for intrinsic I_{NaP} and I_{CAN} currents that appear to be present in all preBötC neurons, pacemaker or otherwise, and contribute to burst generation in the context of amplifying inspiratory-related synaptic input.

Summary

At the heart of breathing is rhythm generation. The site for automatic generation of this rhythm is within the brainstem. The consensus hypothesis is that the critical circuits lie within the preBötC and may also include the RTN/pFRG.

SENSORY INPUTS AND ALTERED BREATHING

O_2 and CO_2 Are Measured by Chemoreceptors

The principal homeostatic goal of breathing is the maintenance, within narrow limits, of arterial blood O_2 and CO_2 levels; that is, arterial Po_2 = 80–100 mm Hg and Pco_2 = 35–45 mm Hg. Changes in Po_2, Pco_2, and pH alter chemoreceptor activity, inducing reflex alterations in breathing that restore them toward regulated levels. Decreases in either Po_2 or pH or an increase in Pco_2 stimulates breathing, whereas increases in Po_2 or pH or a decrease in Pco_2 depress breathing.

The consequences of anoxia are catastrophic (particularly for the brain), but the effects of modestly depressed or elevated Po_2 are benign. In contrast, small changes in Pco_2 can profoundly affect cellular metabolism (due in large part to consequent changes in pH). Thus, at rest and in health, ventilation is controlled mostly to regulate CO_2; however, in severe hypoxia, everything else is ignored. For example, the exceptionally low ambient Po_2 at the top of Mount Everest (~40 mm Hg) results in such a strong ventilatory drive that Pco_2 in climbers without supplemental O_2 is exceptionally low (~7 mm Hg compared with normal ~40 mm Hg; see Box 37.3).

BOX 37.3

DISORDERS OF THE CHEMICAL CONTROL OF BREATHING ARE WIDESPREAD AND SERIOUS

Breathing is normally tightly controlled to maintain arterial P_{CO_2} within a narrow range close to 40 mm Hg. However, cardiopulmonary pathology can increase P_{CO_2} markedly. When P_{CO_2} exceeds 90–120 mm Hg, respiratory depression ensues. With further increases in P_{CO_2}, central nervous system function can be impaired severely; a life-threatening positive feedback loop can develop in which depression of breathing elevates P_{CO_2}, which further depresses breathing. In some patients with advanced chronic lung disease, and consequently elevated P_{CO_2}, any acute lung disease, such as bronchitis or pneumonia, can cause a further increase in P_{CO_2} and exacerbate respiratory depression. In these patients, the drive to breathe comes from hypoxemia (low blood P_{O_2}) sensed by peripheral chemoreceptors. If these patients breathe O_2 without medical monitoring, the hypoxemia may disappear, and without this stimulus these patients' breathing may stop.

Drug abuse is another cause of respiratory depression sufficient to elevate P_{CO_2} to produce further respiratory depression. Narcotics, barbiturates, and most general anesthetics depress breathing and reduce the sensitivity of the chemoreceptors to elevations in P_{CO_2} or reductions in P_{O_2}. Therefore, at high doses these drugs can cause death by respiratory failure.

Sleep apnea. Sleep is associated with a modest decrease in ventilation and a reduced responsiveness to deviations in P_{CO_2} and P_{O_2}. As a result, during sleep, arterial P_{CO_2} increases a few mm Hg and P_{O_2} decreases. This change in respiratory control is associated with the loss of a wakefulness-related excitatory drive. The source of this wakefulness stimulus is not known, but it may derive from corollary activity originating in the brain stem reticular-activating system.

With the onset of sleep, the breathing pattern can become unstable, and apnea (defined as at least 10 s without breathing) can occur. Two major types of apnea have been defined (White, 1990). The most common, *obstructive* sleep apnea occurs when inspiratory airflow reduces airway pressure (via the Bernoulli effect), pulling in on the walls of the upper airway and causing airway obstruction. During wakefulness, the activity of airway muscles counteracts this collapsing force. However, during sleep the reduction of airway muscle tone can result in vibration of the walls of the oropharynx (i.e., snoring). In more severe cases, the loss of activity in the glossopharyngeal nerve, which innervates the tongue, can cause the airway to become obstructed. Obstruction

reduces or abolishes ventilation, raising CO_2 and lowering O_2. This in turn can cause arousal, or waking, and restoration of airway muscle tone and airway patency. In serious cases, the cycle of sleep, airway obstruction, and hypoxia-induced waking repeats hundreds of times every night. The marked disruption of sleep can cause debilitating hypersomnolesence, and severe obstructive sleep apnea can also have other sequelae, such as hypertension.

In the less common *central* sleep apnea, pauses in breathing result from the failure of the central pattern generator for breathing to generate a rhythmic motor command. Mechanisms underlying central apneas are not understood, but the problem is associated with a variety of neurodegenerative diseases, including amyotrophic lateral sclerosis, Parkinson's disease, and multiple systems atrophy (where there is evidence of loss of preBötC neurons).

Sudden infant death syndrome (SIDS; Martin, 1990). SIDS is defined as the unexpected sudden death of an infant or young child that cannot be explained by a postmortem examination. SIDS is the leading cause of death of infants between 1 month and 1 year of age in the United States. Although a number of etiologies are likely to contribute to SIDS, current hypotheses tend to focus on abnormalities of cardiorespiratory control. The apnea hypothesis of SIDS attributes apnea to disorders of both the chemical control of breathing and the arousal response to insufficient ventilation. Infants who later succumb to SIDS have been described as exhibiting irregular breathing patterns, including periods of apnea, and depressed arousal responses to hypoxia or hypercapnia. Epidemiologically, SIDS has been associated with infants sleeping face down, where pillows, blankets, and mattress can limit the diffusion of expired air, resulting in elevated CO_2 and lower O_2 in inspired air and, consequently, in arterial blood. This in turn could depress breathing sufficiently to cause respiratory arrest and death.

Jack L. Feldman and Donald R. McCrimmon

References

Martin, R. J. (ed.) (1990). Respiratory disorders during sleep in pediatrics. *In* "Cardiorespiratory Disorders during Sleep," (R. J. Martin, ed.), pp. 283–322. Futura, Mount Kisco, NY.

White, D. P. (1990). Ventilation and the control of respiration during sleep: Normal mechanisms, pathologic nocturnal hypoventilation, and central sleep apnea. *In* "Cardiorespiratory Disorders during Sleep" (R. J. Martin, ed.), pp. 53–108. Futura, Mount Kisco, NY.

O_2

The mammalian brain is extremely sensitive to O_2 deprivation. Several minutes of anoxia can initiate a cascade leading to neuronal (and, if sufficiently widespread, brain) death. Brain hypoxia can cause loss of consciousness. Accordingly, the principal O_2 sensors for breathing for the entire body, the carotid bodies, are in the bifurcation of the common carotid artery, the main portal through which O_2 enters the brain. The transduction of O_2 levels has been intensely studied since Corneille Heyman's Nobel Prize winning discovery (1938) of the physiological function of the carotid bodies. O_2-related signals enter the brain via the glossopharyngeal nerve and terminate in the NTS.

O_2 sensors account for only a small part of the chemical drive to breathe during wakefulness, as breathing 100% O_2 reduces chemoreceptor discharge to near zero but only decreases ventilation by about 15% in awake mammals (Bisgard and Neubauer, 1995). Decreasing Po_2 has relatively little influence on chemoreceptor activity and ventilation until it falls at or below 60 mm Hg. With further reductions in Po_2, chemoreceptor discharge and ventilation increase exponentially.

CO_2

Ventilation is very sensitive to small changes in Pco_2. In an adult human, a 1 mm Hg increase in Pco_2 leads to a 2 liter $\times$ min^{-1} increase in ventilation at rest (which is ~5 liter $\times$ min^{-1}). In other words, a 2.5% increase in Pco_2 at rest leads to a 40% increase in ventilation!

The sites and mechanisms of CO_2 chemoreception are the subject of intense scrutiny (Feldman *et al.*, 2003). Although the carotid bodies are sensitive to changes in Pco_2, a robust CO_2 response remains in peripherally chemodenervated, vagotomized mammals, indicating the existence of intracranial sensors for CO_2 or related variables (pH, HCO_3^-). A number of structures are implicated in intracranial chemotransduction (Feldman *et al.*, 2003), including:

- RTN
- Serotonergic neurons in the medullary raphe nuclei
- NTS
- Noradrenergic neurons in the locus coeruleus
- Cerebellar fastigial nucleus
- PreBötC

For example, neurons within the RTN and raphe robustly respond to changes in CO_2/pH within the physiological range and project to respiratory nuclei (Mulkey *et al.*, 2004; Richerson, 2004).

MECHANORECEPTORS IN THE LUNGS ADJUST BREATHING PATTERN AND INITIATE PROTECTIVE REFLEXES

Many different patterns of muscle activity can produce appropriate ventilation; however, the specific pattern should minimize energy expenditure. Mechanical efficiency, in turn, depends on factors such as posture, and lung and chest wall mechanics; for example, a fibrotic (stiff) lung is harder to inflate to a given tidal volume than a normal lung, and thus for this condition, rapid, shallow breathing is more efficient. Feedback about the mechanical status of the lungs and chest wall is provided by mechanoreceptors.

Because the airways must be patent for airflow, airway receptors produce three of the most powerful and compelling of *all* reflexes: gagging, coughing, and sneezing. Another powerful reflex that occurs at birth is stimulated by pulmonary receptors, when the transition from the liquid uterine to an air environment requires powerful inspiratory efforts (sighs) to overcome surface forces resisting the initial lung inflation. Additional nonspecific reflex stimuli, typically mechanical or thermal, further facilitate the first inspiratory sighs. Special respiratory reflexes, such as sneezing and yawning, are also generated by pulmonary receptors.

Mechanosensory signals are critical in adapting, adjusting, and integrating breathing with other acts of brain and body. Many movements impact directly on breathing, either by using the same muscles—for example, phonation, posture, defecation, and emesis—or through mechanical linkages, such as locomotion.

Lung Afferents

There are three distinct groups of lung afferents, all in the vagus nerve and terminating in the NTS (see Table 37.1; Kubin *et al.*, 2006). Two groups consist of large, myelinated A fibers, with receptors in the airways, and include slowly adapting and rapidly adapting pulmonary stretch receptors. The third group gives rise to small, unmyelinated C fibers and responds to inhaled irritants such as smoke. Most natural stimuli (e.g., pulmonary edema or alveolar collapse) probably activate multiple receptor types.

Slowly Adapting Pulmonary Stretch Receptors (SARs)

Located in airway smooth muscle, SARs are activated when the airways stretch during lung inflation (Kubin *et al.*, 2006). Their activation drives reflexes by

TABLE 37.1   Pulmonary Vagal Afferents and Their Associated Reflexes

Receptor (fiber type)	Location	Stimulus	Reflex Response
Slowly adapting pulmonary stretch receptors (myelinated, A-β)	Airway smooth muscle	Lung inflation (distension of airways)	Breuer–Hering (a) inspiratory termination (b) expiratory prolongation, airway dilation
Rapidly adapting pulmonary stretch receptors (myelinated, A-β)	Airway epithelium and subepithelial layers of mucosa	Rapid lung inflation or deflation, edema in walls of large airways, chemical irritants	Augmented inspiration (sigh), shortened expiration, airway constriction, may contribute to cough
Cough receptors (myelinated, A-∂)	Mucosa of extra-pulmonary airways	Punctate mechanical stimuli, acid	Cough
Broncho-pulmonary C fibers (unmyelinated, C)	Throughout airways and alveolar wall	Chemical irritants, edema	Apnea (cessation of breathing), rapid shallow breathing, airway constriction, mucus secretion, may contribute to cough

which lung inflation shortens inspiration and prolongs expiration (Breuer–Hering reflex), which is an important determinant of inspiratory duration during normal relaxed breathing in most mammalian species, with the notable exception of humans. The activation of these receptors shortens inspiration and increases breathing frequency. In humans, activation of SARs by normal breaths only modestly alters inspiratory timing. Their activation also relaxes airway smooth muscle, dilating the airways to make breathing easier.

Higher Order Neurons in the Breuer–Hering Reflex

The Breuer–Hering reflex is one of the most intensely studied vagal reflexes. Because its activation changes respiratory frequency, the underlying mechanisms illuminate aspects of the central rhythm generator for breathing. SAR afferents monosynaptically activate "pump" neurons in the NTS. Many pump cells are GABAergic and, in some cases, glycinergic and project widely within the VRC as well as to the PRG.

Rapidly Adapting Pulmonary Stretch Receptors (RARs)

Located in airways, RARs initiate protective reflexes, including coughs, in response to a variety of stimuli, including large or rapid lung inflation or deflation, inhaled irritants, and, possibly, airway edema (see Table 37.1; Kubin *et al.*, 2006). Smoke activates RARs and elicits a cough to rid the airway of the offending material. RARs elicit a larger than normal inspiration (sigh) in response to alveolar collapse (atelectasis), inflating the lung and popping open the alveoli. Also contributing to cough is a recently identified group of airway receptors, termed cough receptors, with small myelinated fibers (A-∂ fibers; Canning, 2006).

Bronchopulmonary C Fibers

C fiber afferent activation elicits apnea followed by rapid shallow breathing. Like RARs, C fibers are polymodal, activated by chemical and mechanical stimuli. Their activation also enhances airway mucus secretion. Inhaled particles are trapped in the mucus and removed from the airways by the action of cilia, which continuously (in the nonsmoker) move the mucus and trapped particles toward the mouth to be swallowed or expectorated.

Summary

Chemoreceptors in the carotid bodies and in the brain provide sensory input to the central circuits controlling breathing to tightly regulate Pco_2 and Po_2 within narrow limits, matching a wide range of metabolic demands. Carotid body chemoreceptors are especially sensitive to decreases in arterial Po_2, while also sensing increases in arterial Pco_2 and decreases in arterial pH. Central chemoreceptors are exquisitely sensitive to increases in Pco_2 and decreases in pH, but respond little to changes in Po_2. Mechanoreceptors, especially those in the lungs, are essential for the precise regulation of the timing and amplitude of breathing, and also participate in reflexes, such as coughing, that protect the airways and lungs from compromises in airflow.

MODULATION AND PLASTICITY OF RESPIRATORY MOTOR OUTPUT

The spatiotemporal pattern of motor outputs for breathing must constantly adjust over short-time scales (e.g., due to changes in posture), medium-time scales (e.g., during exercise), and much longer time scales (e.g., to adapt to changes in lung and chest wall mechanics that accompany postnatal development, normal aging, and pulmonary disease). A good example is adaptation to hypoxia, which increases ventilation by multiple mechanisms acting in time domains lasting seconds, minutes, and hours to days (Mitchell *et al.*, 2001). Thus, a single 10–20 min episode of hypoxia increases ventilation and it remains elevated for several minutes after return to breathing room air. This slow posthypoxic decline back to normal ventilation is referred to as *short-term potentiation* (Fig. 37.9) that may be important in smoothly changing respiratory responses during rapid or large changes in afferent input, since abrupt changes can result in an unstable breathing (e.g., Cheyne–Stokes respiration; Fig. 37.10).

A longer-lasting mechanism is revealed following repetitive brief episodes of hypoxia (three episodes of 5 min duration). As illustrated in Figure 37.9, following the decline in short-term potentiation after the third hypoxic episode, ventilation transiently returns toward control levels, but is followed by a slow, progressive augmentation that lasts for several hours; this postepisodic hypoxic response is known as *respiratory long-term facilitation*.

The mechanism of respiratory long-term facilitation has been investigated extensively (Feldman *et al.*, 2003). This form of plasticity requires episodic rather than continuous stimulation. The episodic and repetitive nature of activation of serotonergic and noradrenergic receptors appears necessary for its initiation but not its maintenance. Activation of serotonergic and noradrenergic receptors on respiratory motoneurons is hypothesized to initiate an intracellular signaling cascade, resulting in the synthesis of new proteins, including BDNF, that amplify the synaptic currents associated with descending, glutamatergic synaptic inputs from respiratory premotoneurons (Fig. 37.6). Thus, in many respects, respiratory long-term facilitation has similarities with other forms of synaptic plasticity in its stimulus pattern sensitivity and requirement for protein synthesis. The requirement for episodic serotonin receptor activation is similar to serotonin-dependent plasticity at the sensorimotor synapse of *Aplysia*.

Long-term facilitation could underlie some of the adaptive changes to pathophysiological conditions associated with repeating episodes of hypoxia. For example, it may stablize respiration during repeated periods of sleep apnea that consequently result in episodic hypoxia. During sleep, the activity of

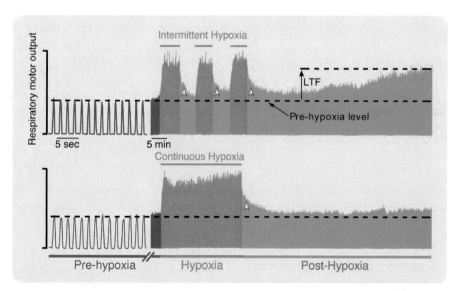

FIGURE 37.9 Respiration-related neural activity in an anesthetized and artificially ventilated rat exhibits *respiratory long-term facilitation*. On the left in each trace, electrical activity in the phrenic nerve (the main nerve innervating the diaphragm) is integrated such that each peak represents the magnitude of a "fictive inspiration" that the artificially ventilated rat had intended to make. Under normal conditions of arterial oxygen and carbon dioxide (baseline), the phrenic bursts are rhythmic and consistent. When the rats are exposed to different patterns of decreased arterial oxygen (intermittent, upper trace; continuous, lower trace), phrenic bursts increase in amplitude, reflecting the drive to take deeper breaths when hypoxia-sensitive chemoreceptors are activated. In both types of low oxygen exposure, phrenic nerve activity slowly returns nearly to baseline over a period of several minutes in a process termed short-term potentiation (STP; open arrowheads). Following STP, in the case of intermittent but not continuous hypoxia, phrenic burst amplitude increases slowly and progressively over an hour, even though arterial oxygen and carbon dioxide are at baseline levels. This slow increase reflects a sort of "respiratory memory" that is elicited by intermittent hypoxia, or respiratory long-term facilitation (LTF). Respiratory LTF appears to result from serotonin-dependent plasticity (see text). Courtesy of T. L. Baker and G. S. Mitchell.

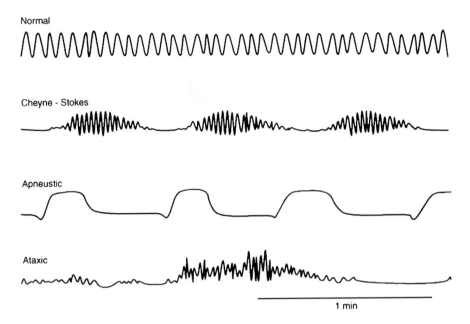

FIGURE 37.10 Abnormal breathing patterns resulting from CNS disorders. The ordinate is lung volume. Adapted from Plum and Posner (1980).

serotonergic raphe and noradrenergic neurons is reduced dramatically. Repeated activation of these neurons secondary to apnea-induced chemoreceptor activation could increase ventilatory drive and reduce the incidence and/or severity of subsequent apneas.

An example of unstable breathing is seen in a subset of patients in whom the central network controlling breathing responds to a change in activity of the chemoreceptors by causing too large or too rapid a change in breathing, as is the case in many individuals with congestive heart failure, who often exhibit Cheyne–Stokes respiration, especially during sleep (Fig. 37.10). Marked derangements of breathing can also occur in response to pathology within a variety of other brain regions that contain projections to respiratory neurons (Box 37.4).

Summary

As metabolism, posture, and sleep–wake state change, the breathing pattern must adjust rapidly to ensure appropriate and efficient ventilation. Development and disease, as well as adaptation to changes in altitude, are processes that require slower adaptations in breathing. Amino acids, amines, peptides, and other neurotransmitters coordinate circuits generating and modulating respiratory pattern to ensure that adaptations proceed smoothly and precisely by mechanisms that remain to be determined.

SUPRAPONTINE STRUCTURES AND BREATHING

We have focused on the roles of the pons, medulla, and spinal cord in the control of breathing. With respect to breathing, emotional and overt behavioral responses serve respiratory homeostasis. For example, although lung irritants such as smoke can elicit a cough as a protective reflex, mammals will avoid approaching environments where lung irritants are present, based on learned or unlearned sensory cues. Escape from environments inhospitable to respiratory homeostasis is also readily evoked and engages a broad range of higher order brain functions. It is interesting to note, however, that many organisms may voluntarily temporarily suspend breathing altogether to serve competing behavioral demands, such as mammals diving to catch prey under water.

Breathing also serves other behaviors. Speech, for example, requires the coordination of jaw and facial muscles, precise tongue, upper airway, and laryngeal control, and coordinated activity of breathing muscles to produce controlled modulation of expiratory airflow. The anterior limbic cortex (including the rostral anterior cingulate gyrus), midbrain periaqueductal gray, parabrachial region, ventrolateral pons, brain stem respiratory nuclei, and cranial and spinal motoneurons contribute to speech.

Sustained or repetitive movement of large muscles involved in such basic behaviors as fleeing predators,

BOX 37.4

CNS LESIONS PRODUCE ABNORMAL BREATHING PATTERNS

Many brain injuries and diseases produce abnormal breathing patterns (Plum and Posner, 1980). Because many brain regions provide afferent input to the neurons generating the breathing rhythm, pathology in regions not normally associated with the generation of breathing can produce abnormal breathing patterns. Despite the diffuse nature of many pathologies that give rise to abnormal breathing patterns, Fred Plum and colleagues systematically characterized several breathing disorders (Fig. 37.10) arising from specific CNS pathologies.

Apneustic breathing is marked by prolonged inspiratory periods. In humans, the most frequent pattern is inspirations lasting 2–3 s alternating with prolonged expiratory pauses. In cats, the prolonged inspiratory periods are associated with plateaus in inspiratory drive that can last minutes (leading to death in the absence of mechanical ventilation). Apneusis is observed in people with lesions of the pons, including, or just ventral to, the pontine respiratory group. In experimental animals, apneusis requires not only lesions of the pontine respiratory group, but also interruption of vagal afferent input.

Lesions of the corticobulbar or corticospinal tracts can lead to *Cheyne–Stokes respiration*, a rhythmic waxing and waning of the depth of breathing. Periods of no breathing (apnea) follow each period of waning inspiratory depth. Lesions of the corticobulbar and corticospinal tracts can

also result in loss of voluntary control of breathing. In pseudobulbar palsy, for example, voluntary control of breathing and of cranial motoneuron function is lost secondary to a lesion often located dorsomedially in the base of the pons.

Extensive bilateral damage to the medullary respiratory groups can severely disrupt or abolish respiratory rhythm, resulting in death unless artificial ventilation is initiated immediately. Fortunately, unilateral damage does not appear to be sufficient to cause severe disruption of respiratory rhythm- and pattern-generating mechanisms. Because two vertebral arteries supply blood to the medulla, bilateral damage from an infarct or embolism is unlikely.

Less extensive damage to medullary respiratory structures can produce *ataxic breathing*, an irregular pattern of breathing with apparent randomly occurring large and small breaths and periods of apnea. Breathing frequency tends to be low.

Jack L. Feldman and Donald R. McCrimmon

Reference

Plum, F., and Posner, J. B. (1980). "The Diagnosis of Stupor and Coma," 3rd ed., *Contemp. Neurol. Sci.* Vol. 19. Davis, Philadelphia.

chasing prey, and exercising also involves higher CNS functions. During exercise the metabolic production of CO_2 increases. However, arterial Pco_2 and Po_2 change little during moderate exercise because of an increase in breathing proportional to the demand for gas exchange. This increase may be produced by a feed-forward descending command from higher brain systems and sensory input from joint and muscle receptors. The latter inputs provide feedback about limb and muscle mechanics, but can be considered feed-forward with respect to the control of arterial blood gases because they may provide cues for adjusting the magnitude of the descending command. Any inadequacy in the feed-forward mechanisms to produce appropriate ventilation is immediately met by chemoreceptor feedback (mostly relevant to Pco_2). Thus arterial blood gases are tightly regulated during moderate exercise.

The volitional initiation of exercise, signaled by suprapontine structures, likely involves parallel activation of spinal locomotor pathways and brainstem mechanisms controlling breathing and cardiovascular function so that pulmonary ventilation and gas transport change appropriately. Commands for the activation of pulmonary and locomotor muscles presumably arise in similar ways in the motor cortex, thalamus, and basal ganglia. Once the motor command has been initiated, respiratory and locomotor activities are likely coordinated at subcortical levels, including the hypothalamus. Electrical or chemical activation of the hypothalamus of an anesthetized, paralyzed cat elicits fictive locomotion paired with proportional increases in respiratory motor output (suggestive of exercise-related increases in ventilation or exercise hyperpnea) and redistribution of blood flow consistent with the induction of locomotion (Waldrop *et al.*, 1996). Because the animals are paralyzed, muscle activation does not occur, and thus the changes in breathing are the result of feed-forward control rather than sensory feedback from joint or muscle receptors.

Summary

Breathing is a basic physiologic function that must be controlled vigilantly by the brain. This chapter has explored basic mechanisms underlying rhythm generation, sensory processing, and motor output. Given the broad range of experimental systems available to study these mechanisms, from *in vitro* rodent brain stem slices to intact awake and sleeping humans, all levels of neurobiological analysis have considerable potential to unravel integrative mechanisms of brain function that may be of general importance.

References

Adrian, E. D., and Buytendijk, F. J. J. (1931). Potential changes in the isolated brainstem of goldfish. *J. Physiol. (Lond.)* **71**, 121–135.

Ballanyi, K., Onimaru, H., and Homma, K. (1999). Respiratory network function in the isolated brainstem-spinal cord of newborn rats. *Prog. Neurobiol.* **59**, 583–634.

Bisgard, G. E., and Neubauer, J. A. (1995). Peripheral and central effects of hypoxia. *In* "Lung Biology in Health and Disease: Regulation of Breathing" (J. A. Dempsey and A. Pack, Eds.), pp. 617–668. Dekker, New York.

Canning, B. J. (2006). Anatomy and neurophysiology of the cough reflex: ACCP evidence-based clinical practice guidelines. *Chest* **129**, 33S-47S.

Feldman, J. L., and Del Negro, C. A. (2006). Looking for inspiration: New perspectives on the generation of respiratory rhythm. *Nat. Rev. Neurosci.* **7**, 232–242.

Feldman, J. L., Mitchell G. S., and Nattie, E. E. (2003). Breathing: Rhythmicity, plasticity, chemosensitivity. *Annu. Rev. Neurosci.* **26**, 239–266.

Galen. (1968). "Usefulness of the Parts of the Body" (M. T. May, ed.). Cornell Univ. Press, Ithaca, NY.

Gesell, R., Bricker, J., and Magee, C. (1936). Structural and functional organization of the central mechanism controlling breathing. *Am. J. Physiol.* **117**, 423–452.

Hayashi, F., and McCrimmon, D. R. (1996). Respiratory motor responses to cranial afferent stimulation in rats. *Am. J. Physiol.* **271**, R1054–R1062.

Koshiya, N., and Smith, J. C. (1999). Neuronal pacemaker for breathing visualized *in vitro*. *Nature* **400**, 360–363.

Kubin, L., Alheid, G. F., Zuperku, E. J., and McCrimmon, D. R. (2006). Central pathways of pulmonary and lower airway vagal afferents. *J. Appl. Physiol.* **101**, 618–627.

Legallois, M. (1813). "Experiments on the Principle of Life." Thomas, Philadelphia.

Lorry, M. (1760). *Les mouvements du cerveau. Mem. Math. Phys. Pres. Acad. Roy. Sci. Div. Sav. Paris* **3**, 344–377.

Lumsden, T. (1923). Observations on the respiratory centres in the cat. *J. Physiol. (Lond.)* **57**, 153–160.

Martin, R. J. (ed.) (1990). Respiratory disorders during sleep in pediatrics. *In* "Cardiorespiratory Disorders during Sleep," pp. 283–322. Futura, Mount Kisco, NY.

Mitchell, G. S., Baker, T. L., Nanda, S. A., Fuller, D. D., Zabka, A. G., Hodgeman, B. A., Bavis, R. W., Mack, K. J., and Olson, E. B. Jr. (2001). Intermittent hypoxia and respiratory plasticity. *J. Appl. Physiol.* **90**, 2466–2475.

Mulkey, D. K., Stornetta, R. L., Weston, M. C., Simmons, J. R., Parker, A., Bayliss, D. A., and Guyenet, P. G. (2004). Respiratory control by ventral surface chemoreceptor neurons in rats. *Nat. Neurosci.* **7**, 1360–1369.

Plum, F. and Posner, J. B. (1980). "The Diagnosis of Stupor and Coma," 3rd ed., Contemp. Neurol. Ser. Vol. 19. Davis, Philadelphia.

Ramon y Cajal, S. (1909). "*Histologie du Systeme Nerveux de l'homme et des Vertebres.*" Maloine, Paris.

Smith J. C., Ellenberger H. H., Ballanyi K., Richter D., and Feldman J. L. (1991). Pre-Bötzinger complex: A brainstem region that may generate respiratory rhythm in mammals. *Science* **254**, 726–729.

Turner, D. L., Bach, K. B., Martin, P. A., Olson, E. B., Brownfield, M., Foley, K. T., and Mitchell, G. S. (1997). Modulation of ventilatory control during exercise. *Respir. Physiol.* **110**, 277–285.

Waldrop, T. G., Eldridge, F. L., Iwamoto, G. A., and Mitchell, J. H. (1996). Central neural control of respiration and circulation during exercise. *In* "Handbook of Physiology. Sec 12. Exercise: Regulation and Integration of Multiple Systems" (L. B. Rowell, and J. T. Shepherd, Eds.), pp. 333–380. Oxford University Press, New York.

White, D. P. (1990). Ventilation and the control of respiration during sleep: Normal mechanisms, pathologic nocturnal hypoventilation, and central sleep apnea. *In* "Cardiorespiratory Disorders during Sleep" (R. J. Martin, ed.), pp. 53–108. Futura, Mount Kisco, NY.

Jack L. Feldman and Donald R. McCrimmon

38

Food Intake and Metabolism

Eating is a familiar behavior. In humans, it is strongly influenced by cultural, social, and experiential factors. Consequently, eating has been a subject of considerable interest to social and behavioral scientists. In mammals, including humans, food intake is also a regulatory behavior with the primary function of supporting the continuous energy demands of body tissues. When considered from this biological perspective, food intake is influenced by hunger, satiety, and the physiological mechanisms that couple eating with internal caloric supplies and a stable body weight. This chapter discusses the signals important for the central nervous system to control food intake and also describes mechanisms thought to integrate those signals.

Despite decades of investigation, considerable differences of opinion still exist about how food intake is controlled. The multiple views follow one of two principles. The first considers eating to be a consequence of depleted energy stores in adipose tissue, reduced use of energy-rich *metabolic fuel* (glucose or lipid) in some critical tissue, or both. In this schema, the purpose of eating is to restore energy reserves in adipose tissue or to increase fuel utilization to some desired level, thereby eliminating the signal to eat. This traditional "depletion–repletion" model is not the approach taken in this chapter. Rather, we subscribe to a second principle, in which hunger is thought to be initiated in part by the gradual disappearance of inhibitory satiety stimuli that were generated by the previous meal. In addition, meal initiation may occur based on learning, habits, and various social and other situations, which allow animals to take maximal advantage of their environment. According to this view, *caloric homeostasis* influences meals, albeit indirectly, and eating contrib-

utes to caloric homeostasis by providing nutrients. The storage and use of those nutrients are regulated independently by physiological mechanisms described later. We describe the properties of caloric homeostasis before we consider the control of food intake.

CALORIC HOMEOSTASIS

Homeostatic Mechanisms Provide a Continuous Supply of Metabolic Fuels to Cells

The purpose of caloric homeostasis is to preserve cellular metabolism. Cells oxidize metabolic fuels to drive all cellular processes; thus, the higher the cellular activity, the greater the demand for energy. Because most cells have limited amounts of stored energy, they rely on a steady supply of calories and oxygen from the blood stream. Oxygen is dependably and instantaneously available via the respiratory system and is not stored in the body. In contrast, food calories can be scarce and require time after ingestion to become available to cells in significant quantities. One consequence of this functional organization is the capacity of animals to store sufficient energy to bridge long intervals during which no food is eaten.

Three categories of macronutrients—carbohydrates, lipids, and proteins—provide usable energy, but the use of specific macronutrients by the body varies depending on the tissue. Most tissues can oxidize carbohydrates in the form of *glucose* or lipids in the form of *free fatty acids*, depending on the availability of these nutrients and the levels of certain hormones in the blood. Notable exceptions are the liver, which requires lipids for proper functioning, and the brain, which has a large, continuous need for glucose despite the ability to oxidize lipids in the form of *ketone bodies*. When the supply of glucose to the brain is compromised, neurons cease to function and in a few minutes consciousness is lost. Death will ensue unless glucose delivery to the brain is restored. Therefore, maintenance of circulating glucose in sufficient amounts to support normal brain function is a critical goal of caloric homeostasis.

Two distinct metabolic states are defined by the availability of recently consumed food to cells. The *prandial*, or fed, state is characterized by an abundance of newly ingested and absorbed nutrients in the blood. These molecules are sequestered rapidly in tissues to prevent them from being excreted wastefully in urine. The *postabsorptive*, or fasted, state is characterized by the absence of calories entering the circulation from the gastrointestinal tract and a consequent reliance on energy from metabolic fuels less recently consumed and stored. These stores are released gradually into the blood during a fast. In both states, tissues take nutrients from the blood as needed for cellular metabolism.

Many tissues store carbohydrates in the form of *glycogen*, a polymer of glucose; the liver and skeletal muscles have the largest depots. Energy is stored more efficiently, mainly in adipose tissue, as *triglyceride*, each molecule of which consists of glycerol with three attached fatty acids. During the prandial period, newly ingested food is used immediately by the body or is stored as glycogen or triglyceride. Excess carbohydrate is largely converted to lipid (*lipogenesis*) because glycogen storage capacity is limited, and triglyceride is a more efficient form of stored energy. During the fasting period, liver glycogen is converted back to glucose (*glycogenolysis*), which enters the blood and is available to all tissues. Similarly, stored triglycerides are mobilized from adipose tissue (*lipolysis*) and enter the circulation as fatty acids and glycerol. The fatty acids are used by tissues as needed or are converted to ketone bodies (*ketogenesis*), whereas glycerol is converted to glucose (*gluconeogenesis*) (Fig. 38.1).

The liver is the key organ in the traffic of energy. Lipogenesis (which also occurs in adipose tissue) and glycogen formation occur in the liver during the prandial period, and glycogenolysis, ketogenesis, and gluconeogenesis occur during the fasting period. These processes are regulated (discussed later), as are delivery of metabolic fuels (from the small intestine into the circulation), storage of excess fuels, and mobilization of stored energy. The control system involves interplay among several hormones and the sympathetic and parasympathetic divisions of the autonomic nervous system.

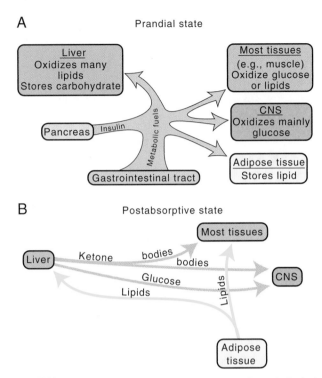

FIGURE 38.1 Schematic diagram of the fluxes of metabolic fuels in the (A) prandial and (B) postabsorptive states. Note the absence of insulin in the postabsorptive state, which greatly facilitates the mobilization of energy stores from the liver and adipose tissue. The adult human liver stores sufficient fuel to support metabolism during fasting for about 7 h, whereas the adipose mass has a far greater capacity.

Insulin Is the Key Hormone Affecting Caloric Homeostasis

Secretion of the peptide hormone *insulin* from B cells of the pancreatic islets is influenced by several factors, among which interstitial glucose is critical. Insulin secretion increases in direct proportion to the concentration of glucose in the blood and does not occur in the absence of glucose. Other substrates, such as amino acids and ketone bodies, also stimulate insulin secretion. In addition, autonomic nerves innervate the pancreatic islets: Cholinergic parasympathetic activity stimulates secretion of insulin, and α-adrenergic sympathetic activity inhibits it.

When a hungry person anticipates a meal, the aroma and, subsequently, the taste of food initiate insulin secretion via neural activity. Neural signals descend from the forebrain (where the smell and taste of food are recognized) through the hypothalamus to the dorsal motor nucleus of the vagus in the caudal brain stem and then to the pancreas by way of cholinergic fibers of the vagus nerve. This *cephalic phase* of insulin secretion helps reverse the mobilization of fuels that occurs during fasting and prepares the body for the entry of fuels from the gut. As ingested food enters the stomach and duodenum, several gastrointestinal hormones, which also are important in digestion, stimulate B cells to secrete more insulin. This *gastrointestinal phase* of insulin secretion ensures that the level of insulin in the circulation is high by the time digested nutrients first appear in the bloodstream. Finally, nutrients absorbed from the small intestine cause even greater stimulation of insulin secretion by their direct effect on the pancreas. This *substrate phase* of insulin secretion increases insulin levels further, and this effect lasts well beyond the cessation of eating. As a result of these coordinated meal-related events, prandial insulin secretion is rapid and appropriate to the caloric load, and ingested fuels are efficiently used and stored (Fig. 38.2).

Neural and endocrine factors control the ebb and flow of metabolic fuels from body stores. Circulating insulin is the most important factor that promotes storage. Insulin enables most tissues to take up glucose from the blood for immediate oxidation or for storage

during the prandial period. Conversely, the most important factor for fuel mobilization is the near disappearance of insulin from the circulation during the postprandial, postabsorptive period, when fuel delivery from the small intestine has ended and parasympathetic activity to the pancreas is no longer elevated. Insulin secretion at this time is reduced greatly but is not inhibited and therefore can be stimulated immediately if necessary. For example, when more than the needed amount of stored fuels is mobilized, substrates directly stimulate the pancreas to secrete insulin, thereby slowing substrate mobilization to a rate appropriate to need. Thus, insulin is pivotal for storing calories in the fed state and for allowing stored calories to be mobilized in measured amounts during the postabsorptive state.

The amount of body fat (*adiposity*) is superimposed on other factors that influence insulin secretion. People with low adiposity have a relatively large number of active insulin receptors on adipose tissue and skeletal muscle. Obese individuals have fewer active insulin receptors. Consequently, in response to a given stimulus, secretion of insulin is lower in people who are lean than in those who are obese, but its net effects on adipose tissue are comparable. In healthy, nondiabetic people, this reciprocal relationship between the secretion of insulin and the sensitivity of most tissues to insulin ensures efficient storage and use of fuels independent of body weight. In these individuals, plasma insulin levels in the prandial and the postprandial periods are reliable correlates of adiposity.

Summary

All cells require continuous supplies of metabolic fuels to support ongoing activity. The fuels enter the circulation from the small intestine during the prandial period and from storage depots during fasting. The availability of fuels to tissues is controlled primarily by the liver and by the hormone insulin. Hepatic function and insulin secretion are in turn controlled in large part by the autonomic nervous system.

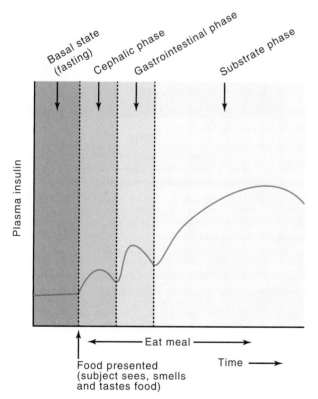

FIGURE 38.2 The three phases of meal-related insulin secretion. The cephalic phase, during which parasympathetic vagal activity stimulates the pancreas, is initiated when food is seen, smelled, and tasted. The gastrointestinal phase is mediated by the direct action of digestive hormones on the insulin-secreting B cells. The prolonged substrate phase is caused by metabolic fuels (mainly glucose) directly stimulating pancreatic B cells. When meals are prolonged, the three stimuli operate simultaneously and have additive effects.

ROLE OF CALORIC HOMEOSTASIS IN CONTROL OF FOOD INTAKE

Meals Generate Biological Satiety Signals

The traditional view of caloric homeostasis is that animals eat when they need calories. However, as discussed earlier, the delivery of metabolic fuels to cells is continuous, and animals experience urgent needs for calories to support cellular metabolism only infrequently. Instead, caloric homeostasis can be related to the control of food intake in ways that are unrelated to acute cellular needs.

Animals consume food in distinct bouts (i.e., meals); therefore, daily food intake reflects the cumulative intake of multiple meals. Control factors influence the time when each meal is initiated and the amount of food consumed before the meal is terminated. The first comprehensive study of meals was reported in 1966 by Le Magnen and Tallon. They maintained laboratory rats in cages and allowed them to eat *ad libitum* for weeks, during which time photosensors and electronic relays recorded intake. They obtained a full record of when meals were initiated, how much was eaten during each meal, how much time passed before the next meal, and so on. No relationship was found between how much a rat ate in a meal and how much time had passed since it had last eaten; in other words, meal sizes were unpredictable. However, the larger the meal, the longer the interval before the next meal, and the relationship was strongest during the night, when rats ate the largest meals. Thus, eating appeared to be inhibited by a satiety signal generated in proportion to the size of a meal, and eating resumed when that signal disappeared. Similar observations have since been made in many laboratories.

Gastric Distension and Cholecystokinin Provide Satiation Signals

Several meal-related factors might plausibly provide satiation signals. These include factors based on the smell, taste, and texture of food. Factors arising from the stomach could also play a role, as could intestinal and postabsorptive factors that arise after ingested food has left the stomach. In studies of rats, pregastric factors have been ruled out because little satiation is achieved when ingested food is immediately drained out through an esophageal or gastric fistula; these animals eat continuously, as if they had no satiation despite the passage of very large amounts of food through the oropharynx (Fig. 38.3). Because meals end long before significant digestion and absorption occur, gastric factors have been considered a likely source of satiation signals. Gastric volume is an obvious possibility, and, in fact, meals are known to end when substantial *gastric distension* occurs. The stomach wall is richly endowed with stretch receptors whose activity increases in proportion to the volume of the stomach. Those signals are communicated by way of the vagus nerve to the *nucleus of the solitary tract* (NST) and adjacent *area postrema* in the brain stem. The signals travel to the hypothalamus and, ultimately, to the cortex, where gastric distension is perceived.

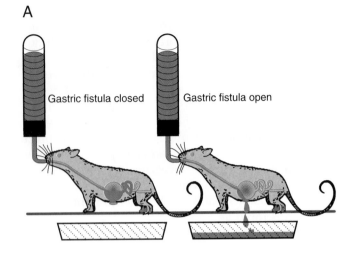

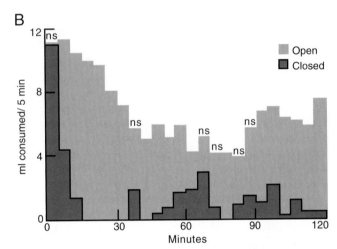

FIGURE 38.3 (A) Rats consuming liquid diet, with gastric fistula closed (real eating) or open ("sham eating"). When the gastric fistula is closed, food passes through the stomach and into the small intestine normally. When the gastric fistula is open, ingested food drains out the open fistula instead of accumulating in the stomach to distend it, and no food enters the small intestine. (B) Mean consumption of liquid diet (ml/5 min) by rats with a gastric fistula after 17 h of food deprivation. The rats consumed food in discrete meals when the fistula was closed. When the fistula was open, they ate continuously throughout the test period and never displayed satiety. Thus, rats consumed significantly more food during the 2-h test period when their fistulas were open than when they were closed. ns, intervals in which the amount of food consumed did not differ significantly between open and closed fistulas. From Smith *et al.* (1974).

Gastric stretch that accompanies meals presumably interacts with other signals to produce satiation. For example, aside from directly affecting digestion, the intestinal peptide *cholecystokinin* (CCK), which is secreted during meals, acts on receptors located on vagal afferent fibers that carry gastric stretch signals from the pyloric region of the stomach to the brain stem. Thus, relatively small amounts of CCK can inhibit feeding in rats by acting synergistically with gastric distension. When given in larger doses, CCK can inhibit food intake even when the stomach is empty (Gibbs *et al.*, 1973). As might be expected, total or selective gastric *vagotomy* (or ablation of brain stem areas to which the gastric vagus projects) decreases the ability of gastric volume and peripherally administered CCK to reduce meal size. Conversely, systemic administration of CCK receptor antagonists increases the size of meals (Box 38.1).

Gastric distension is one of the factors that control suckling in neonatal rats. The dam controls the timing of meals, and at each meal pups consume as much milk as their stomachs allow; gastric volume, not caloric content, provides the signal to stop. Consistent with the importance of gastric distension to suckling, pups are particularly sensitive to the inhibitory effects of CCK on food intake, suggesting that a combination of gastric distension and CCK may be the principal signals that inhibit the initiation of spontaneous meals in suckling rats.

As rat pups mature, signals related to the caloric content of food begin to participate in the control of

---

BOX 38.1

SATIATION FACTORS

Several metabolically important peptides, including CCK, insulin, leptin, and *glucagon*, reduce food intake when administered systemically to laboratory animals. By themselves, these findings do not prove that the peptides normally function as endogenous satiation agents; blood levels of an administered agent must be within the physiological range. In addition, the behavioral effects of the agent must be specific to the inhibition of food intake and cannot merely reflect a secondary consequence of illness, behavioral depression, or motor incapacitation. Because investigators cannot be certain what animals sense, they must infer whether animals experience satiation in association with an observed reduction in food intake.

One common approach is to determine whether the agent can cause a *learned flavor aversion* in animals. People learn readily to avoid food or drink that, when ingested, produces nausea due to the unsuspected presence of some toxic contaminant. Rats and other animals seem to respond in the same way; moreover, when a toxin is administered systemically soon after consumption of an uncontaminated drink with a novel flavor, the animals subsequently avoid fluids of that flavor and behave as if the drink had contained the toxic agent that made them sick. In addition, electrophysiological recording from the first gustatory relay nucleus in the brain stem, the nucleus of the solitary tract, shows that the response elicited by a taste that has been associated with a toxin is similar to the pattern of activity typical of naturally aversive flavor. Nausea is critical to the process; damage to the "emetic center" in the area postrema eliminates the sensation of nausea and prevents the formation of the learned taste aversion.

Another approach to distinguishing nausea from satiation is to monitor biological variables that occur in association with nausea but not satiation, or vice versa. One such variable is neurohypophyseal hormone secretion. For example, administration of lithium chloride causes nausea in humans and vomiting in monkeys, and stimulates vasopressin secretion in both species. In this example, because elevated plasma vasopressin does not itself cause nausea or vomiting, it can be considered a biological marker of nausea. Curiously, rats secrete the other neurohypophyseal peptide, oxytocin (and not vasopressin), in response to large doses of lithium chloride and other nauseants. Control experiments have shown that following an ordinary meal, vasopressin is not secreted in primates nor is oxytocin secreted in rats. Thus, when rats are administered a chemical agent that stimulates pituitary oxytocin secretion, nausea, rather than satiation, is the suspected basis of the reduction in food intake.

Such experiments are helpful for interpreting the anorexia that occurs when a hormone or neurotransmitter is administered exogenously, but insight into the normal effects of endogenously secreted chemical signals requires an alternative experimental approach. For example, a drug that blocks the effect of an endogenous hormone or neurotransmitter on its receptors could be administered. If the hormone normally functions to reduce food intake, then the blocking agent should increase the size of meals. Such drugs are not yet available for all compounds suspected of being endogenous satiation factors, but experiments using drugs to reduce the activity of CCK, insulin, leptin, and glucagon at their respective receptors found increases in meal size. These observations strongly suggest that these compounds normally function as endogenous satiation factors.

Stephen C. Woods and Edward M. Stricker

meal size and become integrated with CCK, gastric distension, and other factors as satiation signals. This developmental change allows adults to ingest, assess, and respond to foods whose caloric density is not as constant as that of milk. For example, when liquid food is diluted with water, adult rats compensate by increasing the volume consumed in each meal, allowing daily caloric intake to remain stable. This observation reveals that gastric distension is but one of several possible satiation signals in adult rats and that greater distension can be accommodated in circumstances in which the calorie-related satiation signals have diminished.

The mechanism by which caloric signals are monitored remains unknown. Although early reports suggested that the stomach monitored caloric nutrients, more recent findings have not supported this hypothesis. For example, when hungry rats were equipped with closed pyloric cuffs to prevent ingested food from entering their small intestine, the rats decreased their food intake in proportion to the volume, and not the caloric content, of intragastric infusions (Phillips and Powley, 1996). In contrast, when the cuff was open so that gastric contents could empty into the duodenum, caloric nutrients were much more effective in reducing food intake than calorie-free loads of equal volume. These observations suggest that the stomach does not detect caloric nutrients but instead contributes primarily inhibitory signals related to gastric distension.

Postgastric Effects of Meals Provide Additional Satiation Signals

Normally, some ingested food enters the small intestine and is absorbed during the course of a meal, thus allowing postgastric signals to contribute to satiation. In this regard, small amounts of specific nutrients infused directly into the duodenum produce a robust suppression of eating that cannot be attributed to gastric factors. It is not clear whether the upper small intestine or a postabsorptive site, or both, is responsible for detecting the administered calories and generating the signals that limit meal size. As discussed earlier, the liver plays a critical role in caloric homeostasis, and the delivery of nutrients absorbed from the small intestine to the liver may be monitored to provide an important postgastric signal for the inhibition of food intake. Consistent with this hypothesis are findings that infusion of glucose or lipids into the hepatic portal vein reduces food intake. Furthermore, such infusions have been found to increase the activity in hepatic vagal afferent fibers, whereas infusion of fructose into the general circulation no longer reduces

food intake in rats after hepatic vagotomy. Collectively, these findings suggest that the liver provides a satiation signal. That signal would disappear as absorption slows, which is when satiety diminishes as well.

Intravenously administered glucose reduces food intake but not by as much as the caloric content of the glucose. In contrast, when glucose solution is fed through a tube directly into the stomach, the compensatory reduction in food (calorie) intake is equivalent to the caloric load. In this case, gastric or intestinal signals might contribute to the increased satiety. Note that more insulin is secreted in response to infusion of glucose into the stomach than into a vein, which might contribute to the observed reduction in appetite. In fact, intravenously infused glucose reduces food intake by an equivalent caloric amount when insulin is added to the infusate. Together, these observations suggest that insulin contributes to the satiety effect of glucose whether given by gavage or ingested during a meal. Consistent with this possibility, rats made diabetic by destruction of their pancreatic B cells eat discrete meals more frequently as if they experienced less postprandial satiety (Box 38.2).

In addition to providing useful nutrients, consumption of food increases *plasma osmolality*, yet another factor influencing the size of a meal. Increased plasma osmolality stimulates thirst, and for this reason water intake usually accompanies eating. The important point, however, is that food intake is lower during meals when drinking water is not available, a phenomenon known as "dehydration anorexia." Similarly, increased plasma osmolality caused by systemic injection of hypertonic saline reduces food intake in proportion to the administered osmotic load. The well-described osmoreceptors that cause thirst and influence vasopressin secretion (Chapter 39) apparently do not mediate this inhibitory effect on food intake, which persists after those forebrain osmoreceptors have been destroyed surgically. Instead, osmoreceptors in the caudal brainstem might mediate the inhibitory effect of increased plasma osmolality on food intake.

Body Weight Influences Food Intake

The control of food intake also is associated with the maintenance of body weight. For example, after a period of food deprivation and forced loss of body weight, animals (including humans) eat larger meals than normal until adiposity returns to pretreatment levels. Conversely, after a period of force feeding, during which the increased daily intake of calories causes weight gain, animals eat smaller meals than

---

BOX 38.2

THE GLUCOSTATIC HYPOTHESIS

Almost half a century ago, Mayer (1955) proposed that the onset and termination of meals are determined by changes in the amounts of glucose used by certain areas of the brain. Critical to Mayer's hypothesis was the observation that food intake was reliably increased by systemic injections of insulin, which reduced delivery of glucose to the brain. Thus, Mayer hypothesized that food intake was controlled by a region of the ventral hypothalamus whose glucose utilization, unlike the rest of the brain, was sensitive to insulin. In diabetic rats, the presence of hyperphagia despite elevated blood glucose emphasized the point that diminished glucose utilization in these insulin-dependent cells (as opposed to a reduction of blood glucose levels *per se*) was the critical factor in removing satiety and stimulating food intake.

At first widely accepted, the glucostatic theory fell into disfavor for several reasons. For one, although eating can be induced by low availability of glucose to the brain, no evidence supports suppression of eating by high glucose availability. In addition, concentrations of blood glucose after administration of large doses of insulin are now known to be far lower than what occurs during a fast or at the normal onset of a meal. Thus, eating in response to insulin-induced hypoglycemia is now considered an "emergency" response that is not relevant to normal feeding. (A similar emergency response occurs after a systemic injection of 2-deoxyglucose induces an acute decrease in cellular glycolysis.) Finally, the prevailing view of the ventromedial hypothalamus as a satiety center has changed considerably since the original formulation of the glucostatic hypothesis. This region of the brain is better regarded as a regulatory control center that influences many aspects of hormonal and autonomic function. Diabetic hyperphagia may therefore represent the loss of satiety promoted by insulin or leptin that activates neural systems to inhibit food intake. In contrast, the earlier glucostatic hypothesis would have explained diabetic hyperphagia as a consequence of the loss of insulin's effect on special hypothalamic neurons that mediate satiety.

In recent years, evidence has emerged that neurons in the hypothalamic arcuate nucleus are sensitive to fluctuations of glucose as well as to certain fatty acids (especially oleic acid) and amino acids (especially leucine). Neurons in the same area are also sensitive to the adiposity signals leptin and insulin, and manipulations of either nutrients or adiposity signals locally influence both food intake and the release of glucose from the liver (Obici *et al.*, 2002). Hence, the view is emerging that the arcuate nucleus integrates long-term stores (body fat) with acute availability of nutrients to control several aspects of energy homeostasis.

Stephen C. Woods and Edward M. Stricker

References

Mayer, J. (1955). Regulation of energy intake and body weight: The glucostatic theory and the lipostatic hypothesis. *Ann. N.Y. Acad. Sci.* **411**, 221–235.

Obici, S., Feng, Z., Morgan, K., Stein, D., Karkanias, G., and Rossetti, L. (2002). Central administration of oleic acid inhibits glucose production and food intake. *Diabetes* **51**, 271–275.

---

normal (or no meals at all) until normal adiposity is restored. Long-term stability of adiposity in adult animals is attributed in part to these compensatory responses to weight fluctuations that are caused by acute changes in food intake or energy expenditure.

Adiposity may indirectly influence food intake by modulating how quickly food passes through the gastrointestinal tract, how long nutrients remain in the circulation, or how nutrients interact with the liver. For example, after fasting, less prolonged gastric distension or more rapid absorption and storage of ingested calories would diminish the duration of satiety signals, thereby increasing the frequency of meals. These effects would promote the *hyperphagia* (overeating) associated with the restoration of body weight in animals that have fasted. Similarly, the satiating properties of ingested food would be reduced by diversions of ingested food calories to the fetus during pregnancy, to milk during lactation, or to skeletal muscle during exposure to cold temperatures, thereby contributing to the hyperphagia associated with each of these conditions.

Alternatively, adipose tissue may directly signal the brain to modulate food intake. Although afferent nerves from adipose tissue to brain have been described, most communication from adipose tissue is thought to be mediated by a humoral (circulating) factor. *Parabionts*, created by surgically joining two animals at the flank muscles and skin, support this hypothesis. Vascular interconnection within a parabiont is demonstrated when a dye injected into one animal appears in the blood of its partner. Because the nervous systems of the two animals remain independent, any physiologic communication between the animals must occur through the vascular system. When two lean rats were joined parabiotically, each consumed its normal amount of food and maintained its usual body weight. Similar results were obtained when two obese rats with lesions in the *ventromedial hypothalamus* (see later) were joined together. However, when a lean animal was joined to an obese animal, the result was striking: The lean animal reduced its food intake markedly and consequently lost a considerable amount of weight. These experiments have been interpreted to mean that a circulating factor, secreted in large amounts (proportional to adiposity) by the obese animal, enters the blood of the lean animal and reduces food intake, as if the lean animal was receiving a signal that it was too fat (Coleman and Hummel, 1969).

Investigations have identified *leptin*, the hormone synthesized by adipose tissue, as a key circulating adiposity signal that regulates food intake and body weight. Leptin concentration in plasma is directly proportional to adiposity, and a receptor-mediated transport process passes it into the brain. Animals that lack the gene for synthesizing leptin (ob/ob mice) are hyperphagic and obese, and administering leptin to ob/ob mice, or to normal animals, causes them to eat less and lose weight. Because lower doses of leptin have this action when administered into the ventricles of the brain than when administered systemically, leptin is believed to reduce food intake and body weight via a central site of action. Consistent with this hypothesis, the receptor for leptin has been identified and found to be located within the hypothalamus. Like ob/ob mice, animals with mutated leptin receptors (db/db mice and fatty Zucker rats) are obese (see Box 38.3), but unlike ob/ob mice, they do not respond to exogenous leptin. Hence, leptin appears to function as a negative feedback signal to the brain. When fat stores increase in adipose tissue, more leptin is secreted and enters the brain, causing a greater inhibition of food intake and loss of body fat. When circulating leptin levels are low, feeding and other anabolic responses are disinhibited. In this way, leptin acts to promote the maintenance of a relatively stable body weight over long intervals. The neural circuitry through which leptin inhibits food intake and influences caloric homeostasis is discussed later.

Insulin provides a second circulating signal that informs the brain of body fat levels. Plasma levels of insulin are directly correlated with adiposity, and, like leptin, insulin is transported through the blood–brain barrier where it can act on insulin receptors in the hypothalamus and other sites. Insulin administered into the brain reduces food intake and body weight, whereas animals that do not secrete insulin are hyperphagic. Of course, severe diabetics do not become obese because their adipocytes cannot store fat in the absence of insulin. Mice lacking insulin receptors on brain cells are both hyperphagic and obese. In short, both leptin and insulin provide blood-borne signals that enter the brain and act on their respective receptors to reduce food intake and body fat.

Obesity, the accumulation of greater than normal fat in the body, is increasing throughout the world and creating a major public health problem, in part because it increases the risk for developing cardiovascular disorders, several cancers, type-2 diabetes mellitus, and other metabolic problems, and increases mortality. Obese individuals have both leptin resistance and insulin resistance; that is, although they secrete large amounts of both leptin and insulin in proportion to their adiposity, these signals are relatively ineffective at controlling metabolic parameters. Consequently, considerably more insulin is required to maintain plasma glucose within normal limits, and the brain is insensitive to the catabolic action of both leptin and insulin. The latter is thought to be due to a combination of reduced passage of these hormones through the blood–brain barrier as well as fewer leptin and insulin receptors within the brain itself.

Obesity often is defined in terms of the *body mass index* (BMI), which is weight in kilograms divided by height in meters squared. A BMI of 18.5 to 24.9 is considered normal, with <18.5 being underweight, 25 to 29.9 being overweight, and >30 being obese. A BMI over 40 is morbid obesity. However, the calculation of BMI does not consider the distribution of fat within the body, yet this is recognized as a major determinant of health risk. Abdominal fat (mainly around the viscera) is more metabolically active, more highly vascularized and innervated, and more likely to be a site of inflammation in the body than subcutaneous fat; abdominal fat (typified by the "pot belly" or "apple" stereotype) therefore carries a much higher health risk than subcutaneous fat (characterized as a "pear" shape). Because of this, assessment of the circumference of the waist (abdomen) at the top of the iliac crest is thought to be a better predictor of health risk than the BMI. By this determination, obesity is defined as a waist circumference of >102 cm in men and >88 cm in women.

Hypotheses as to why the incidence of obesity has been increasing so dramatically in recent years abound, and often presume some combination of increased availability of energy-dense palatable foods

BOX 38.3

VARIATIONS OF STRUCTURE AND FUNCTION AMONG LEPTIN RECEPTORS

Leptin, the peptide hormone secreted by adipose cells, enters the brain through a receptor-mediated transport system in the brain capillary endothelium. Within the brain, it binds to specific leptin receptors that are expressed in discrete populations of neurons. The leptin receptor, termed Ob-R, is a member of the class I cytokine receptor family, a group of structurally similar receptors that act through JAK and STAT intracellular signaling proteins. There is only a single gene for Ob-R, and it is processed differentially by different cells such that many isoforms of the receptor are known. As depicted in Figure 38.4, Ob-R has an extracellular segment that is over 800 amino acids long, which binds selectively to leptin. Variations in function of the different isoforms are conferred by the length of the intracellular segment, with longer segments having more sites where intracellular signaling molecules can interact. Hence, longer isoforms have a greater potential to interact with the greatest number of such signaling molecules. The longest isoform, termed Ob-Rb, has an intracellular component of over 300 amino acids and is the only form capable of initiating the full complement of JAK and STAT proteins and of stimulating transcription at the nucleus of the cell. For these reasons, Ob-Rb is known as the signal-transducing form of the leptin receptor.

Within the brain, Ob-Rb is synthesized in high quantities in various nuclei in the ventral hypothalamus, including the arcuate, ventromedial, and paraventricular nuclei. Ob-Rb is recognized as the receptor that stimulates the synthesis and secretion of α-MSH and inhibits the synthesis and secretion of neuropeptide Y and agouti-related peptide in the arcuate nucleus. Other isoforms, with shorter intracellular segments, are localized in other brain areas and in the choroid plexus. They are thought to mediate activities locally at the cell membrane such as the movement of molecules through the membrane.

Many spontaneously occurring mutations of the leptin receptor have been identified (Fig. 38.4), and all result in an animal that is obese and hyperphagic. In the mouse, where the leptin receptor gene is called DB, at least four mutations are known. All have shortened intracellular segments and are called db/db mice. All are also extremely obese. The fatty Zucker rat (fa/fa) has a mutation resulting in a single amino acid substitution in the extracellular segment. As its name implies, this animal is also extremely obese. The Koletsky rat has a truncated extracellular segment and lacks all leptin signaling and it is also obese. A small number of hyperphagic, obese humans recently have been identified with mutations of the leptin receptor.

Stephen C. Woods and Edward M. Stricker

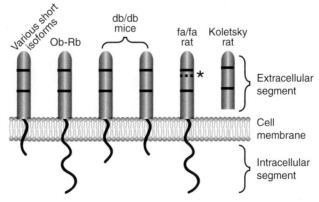

FIGURE 38.4 There are many isoforms of the leptin receptor (Ob-R). All have the same extracellular segment that binds with leptin, such that variations are in the length of the intracellular segment. Most tissues synthesize one or another "short" isoform that binds leptin and initiates cellular events locally near the cell membrane. Only the long form (Ob-Rb) is capable of triggering intracellular signaling cascades that can alter nuclear transcription. Several mutations that shorten the Ob-R occur in mice. These mice, called db/db, are all hyperphagic and obese. Two mutations of the Ob-R also have been identified in rats, and both also result in obesity. Fatty Zucker (fa/fa) rats have single amino acid substitution in the extracellular segment of Ob-R, whereas Koletsky rats have a truncated receptor.

and/or decreased average energy expenditure. One intriguing possibility is that the bacteria in the gastrointestinal (GI) system, termed the *microbiota*, are an important factor. Among other actions, these bacteria determine the proportion of ingested nutrients that actually are absorbed into the body. Recent reports indicate that the GI tracts of some obese humans and obese mice have a higher proportion of bacteria of the phylum *Firmucites* than the GI tracts of lean individuals; those bacteria express enzymes that break down certain complex carbohydrates, enabling the obese individuals to extract and absorb more calories from what they have eaten. Importantly, inoculation of the guts of lean mice with the microbiota from obese mice resulted in the lean mice losing fewer calories in their feces and gaining slightly more weight than controls inoculated with microbiota from lean mice (Turnbaugh *et al.*, 2006), and mice lacking gut bacteria have been found to be resistant to diet-induced obesity (Backhed *et al.*, 2007). Although these observations are preliminary, they suggest that manipulation of the micobiota might provide a novel therapeutic approach to obesity.

Summary

Three effects of eating limit the size of an ongoing meal: gastric distension (potentiated by CCK), postgastric detection of calories (via satiety signals such as CCK, perhaps potentiated in the liver by insulin), and increased plasma osmolality. These signals reach the brain through visceral afferent fibers (especially those traveling in the vagus nerve) and the circulatory system. Meal size invariably increases after the inhibitory effects are experimentally removed by blocking the detection of gastric stretch or of CCK, diluting the food with noncaloric material, or rehydrating the animal. Inhibition appears to be integrated so that as one effect diminishes (e.g., the signal associated with gastric distension), another increases (e.g., the signal associated with the postabsorptive delivery of calories and insulin to the liver), thereby maintaining satiety and prolonging the interval between meals. When the satiety signals disappear, hunger emerges and stimulates initiation of another meal.

Food intake is also normally linked closely with body weight. In experimental animals, periods of starvation or forced feeding are followed by self-regulated compensatory changes in eating patterns until prior body size is regained. Links between body weight and food intake can be mediated indirectly by altered gastrointestinal motility and absorption as well as by the blood-borne, centrally active hormones, leptin and insulin.

CENTRAL CONTROL OF FOOD INTAKE

Early theories of the control of food intake focused on signals derived from stomach and blood. Peripheral factors, such as the concentration of glucose in the blood, were plausibly associated with food intake and were accessible to experimentation. Theories involving the brain emerged after development of a stereotaxic instrument that enabled discrete lesions of the cerebrum to be made in laboratory animals, permitting evaluation of ideas about brain function (Fig. 38.5). The first findings from such experiments were striking: Bilateral electrolytic lesions of the ventromedial hypothalamus (VMH) caused marked hyperphagia and obesity in rats. This observation was interpreted to mean that the animals had become less sensitive to incoming signals of satiety and that they therefore overate and became fat. Bilateral lesions in the adjacent *ventrolateral hypothalamus* (VLH) were later found to cause *aphagia* (absence of eating), which led to death by starvation. This finding was interpreted to mean that these rats no longer detected hunger signals, and thus they starved to death while unaware of their internal state. Collectively, these results provided the foundation of a *dual center hypothesis*, in which a satiety center in the VMH was thought to suppress activity in a hunger center in the VLH. Both the syndrome of hyperphagia and obesity and the syndrome of aphagia and weight loss that result from focal hypothalamic

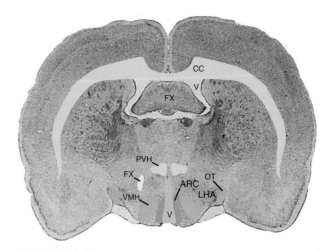

FIGURE 38.5 Coronal section through an adult rat brain at the level where the optic tract enters the two hemispheres. The lateral ventricle and third cerebral ventricle (V) are shaded, and selected fiber tracts are outlined for reference (OT, optic tract; CC, corpus callosum; FX, fornix). Key hypothalamic sites important in caloric homeostasis, present in each hemisphere, are paraventricular nuclei (PVN), ventromedial nuclei (VMN), arcuate nuclei (ARC), and the lateral hypothalamic area (LHA).

lesions have been reproduced in multiple laboratories and in multiple species. However, additional observations have caused reinterpretation of these familiar and reliable findings, and they now have a very different meaning.

Food Intake Is Not Controlled by Hypothalamic Hunger and Satiety Centers

The strength of the traditional dual center model, in which one center mediates hunger and the other satiety, was that it provided a simple answer to the question of how the brain controls intake of food. The principal limitation of the model, however, was that it could not answer the next level of questions: What signals control individual meals? How are these signals integrated with physiological aspects of caloric homeostasis, including long-term maintenance of body weight? What is the influence of other neural sites and systems in the brain? How are excitatory and inhibitory influences on eating integrated? How are behaviors, such as motivation and reinforcement, explained? Investigators are addressing these and related questions, but they now usually take a perspective that does not presume the existence of brain centers mediating hunger and satiety.

In addition to causing hyperphagia, VMH lesions profoundly reduce sympathetic tone and increase parasympathetic tone and vagal reflexes. For example, the high concentration of insulin in animals with VMH lesions is not simply a secondary consequence of hyperphagia but occurs even when food intake is limited. Because of the change in autonomic tone, caloric equilibrium in adipose tissue shifts away from lipolysis and toward lipogenesis, thereby allowing more rapid storage of ingested calories after a meal. Increased parasympathetic activity also allows faster gastric emptying. Importantly, VMH lesions do not cause loss of satiety. Meal sizes continue to correlate with postmeal intervals, but the interval after a meal of any size is shorter in animals with VMH lesions than in control animals.

Because the duration of their postprandial satiety signals is shorter than normal, rats with VMH lesions eat more frequent meals. They remain hyperphagic until the equilibrium between lipogenesis and lipolysis is reestablished in adipocytes. For this to happen, triglyceride stores must accumulate in the cells to levels sufficient to create insulin resistance, whereupon lipogenesis subsides. Meals are eaten less frequently, and daily food intake is reduced. The animal maintains its obese state as if it had a new "set point" for body weight. Thus, after obese rats with VMH lesions have been deprived of food and forced to lose weight, hyperphagia reoccurs until high adiposity is reattained. Similarly, when neurologically normal rats are overfed and become obese prior to receiving a lesion of the VMH, they are not hyperphagic after the lesion but rather eat to maintain their elevated weight.

In sum, rats gain body fat after VMH lesions, and as fat accumulates the duration of satiety after meals increases. This explanation contrasts with early hypotheses that body weight increased after VMH lesions because animals became obligatorily hyperphagic. The primary phenomenon is now recognized as a change in autonomic tone to promote vagal reflexes and fat storage. Because the change in autonomic tone is observed even when increased food intake is prevented, hyperphagia cannot be the primary phenomenon. The key element appears to be the chronic, neurally stimulated increase in insulin secretion. When the increased insulin secretion is prevented by denervating the pancreas, obesity does not develop and hyperphagia is attenuated greatly.

In contrast to VMH lesions, rats with large VLH lesions do not eat (to the point of starvation). However, they also do not drink water when dehydrated, move about in their cages, or respond to diverse stimulation. In other words, the most prominent abnormalities in these rats are akinesia and sensory neglect. In this respect, rats with VLH lesions resemble human patients with Parkinson's disease, a neurological disorder that has been attributed to the degeneration of dopamine-containing neurons of the nigrostriatal bundle (see Chapters 30 and 31). Dopaminergic fibers course through the internal capsule just lateral to the VLH as they ascend from the ventral mesencephalon to the striatum along the medial forebrain bundle (Chapter 43). Large electrolytic lesions of the VLH area interrupt these dopaminergic fibers. More selective damage to the dopaminergic neurons by intracerebral administration of the neurotoxin 6-hydroxydopamine also produces akinesia and sensory neglect in association with loss of food intake, and it does so without disturbing parasympathetic reflexes. Thus, the aphagia induced by large VLH lesions does not result from damage to a putative hunger center in the brain but instead reflects a more general disruption of movement and sensorimotor integration (Ungerstedt, 1971), at least in part (see later).

Summary

The ability of brain circuitry to control food intake initially was based on the consequences of stereotaxic lesions of ventromedial and ventrolateral hypothalamic nuclei leading to hyperphagia or aphagia, respectively. Subsequent studies revealed that each lesion also induces changes in gastrointestinal function and in the equilibrium between lipolysis and lipogenesis in adipocytes. In addition, these nuclei help regulate body weight through influences on spontaneous mobility and sensory responsivity to food-derived stimuli. The nucleus of the solitary tract also relays important visceral sensory cues on gastrointestinal fill that can further influence food intake.

NEUROPEPTIDES AND THE CONTROL OF FOOD INTAKE

Although the details of the central control of food intake are incompletely understood, many neuropeptides have been implicated. To simplify the discussion,

we focus on two broad categories of neuropeptides: those that are mainly anabolic and those that are mainly catabolic. Both of these subgroups in turn are modified by multiple influences, including satiety signals, adiposity signals, experience, habit, emotional state, the social situation, and so on. Anabolic peptides cause increased eating, decreased energy expenditure, and, when chronically elevated, increased body fat accumulation. Catabolic peptides reduce food intake, increase energy expenditure, and lead to loss of body fat. Figure 38.6 presents a simplified model of some of the better-known components of this control system. An interesting feature of the model is that major efferent outputs from the hypothalamus to autonomic nuclei and to motor nuclei controlling ingestion *per se* emanate from the VMH and VLH, the central neural sites that were the main focus of attention over a half-century ago.

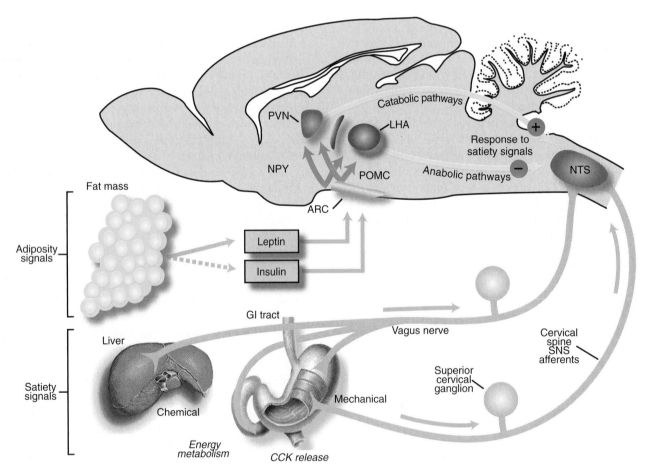

FIGURE 38.6 Schematic diagram of the signals that control caloric homeostasis. Satiety signals arising in the periphery such as gastric distension and CCK are relayed to the nucleus of the solitary tract (NTS) in the brain stem. Leptin and insulin, the two circulating adiposity signals, enter the brain and interact with receptors in the arcuate nucleus (ARC) and other brain areas. These adiposity signals inhibit ARC neurons that synthesize NPY and AgRP (NPY cells in the diagram) and stimulate neurons that synthesize proopiomelanocortin (POMC), the precursor of α-MSH, and CART. These ARC neurons in turn project to other hypothalamic areas, including paraventricular nuclei (PVN) and the lateral hypothalamic area (LHA). Catabolic signals from the PVN and anabolic signals from the LHA are thought to interact with the satiety signals in the brain stem to determine when meals will end. From Schwartz *et al.* (2000).

The Hypothalamic Arcuate Nucleus Mediates the Effect of Adiposity on Food Intake and Body Weight

The amount of body fat is signaled to the brain by leptin and insulin. Receptors for both peptides are located (among other sites) in the hypothalamic arcuate nucleus on two distinct groups of neurons. The first group of arcuate neurons synthesizes the neuropeptides, *α-melanocyte-stimulating hormone* (α-MSH) and *cocaine–amphetamine-related transcript* (CART); these neurons are activated by leptin and insulin. α-MSH and CART are potent catabolic peptides, and when either is administered locally into the third cerebral ventricle, animals eat less food, have increased energy expenditure, and lose weight (Cone, 1999). α-MSH is in the *melanocortin* family of peptides, a group that also includes *adrenocorticotrophic hormone* (ACTH). These peptides act on melanocortin (MC) receptors, and two of these receptors, termed MC3 and MC4, are expressed in several nuclei of the hypothalamus, including the *paraventricular nucleus* (PVN), VMH, and VLH, where α-MSH-containing axons project from the arcuate nucleus and where α-MSH acts as an agonist. Animals that lack either MC3 or MC4 receptors become obese; animals with no MC3 receptors gain weight gradually over their lifetime without overeating, whereas animals without MC4 receptors are hyperphagic and become much fatter. The administration of selective MC4 antagonists causes normal animals to overeat and, if prolonged, to become obese. Hence, hypothalamic melanocortins, exemplified by α-MSH, exert a tonic catabolic effect to keep body weight from increasing (Cone, 1999). Importantly, the ability of exogenous leptin to reduce food intake and body weight is completely attenuated in the presence of an antagonist to MC3 and MC4 receptors, suggesting that the "downstream" pathway by which adiposity influences caloric homeostasis is via melanocortin signaling (Seeley *et al.*, 1997).

The other type of arcuate neuron influenced by adiposity signals synthesizes the neuropeptides, *neuropeptide Y* (NPY) and *agouti-related protein* (AgRP). Both peptides are potent anabolic compounds in that the administration of either into the third ventricle results in hyperphagia, reduced energy expenditure, and weight gain. Although NPY is synthesized in many areas of the brain, AgRP synthesis is limited to the arcuate nucleus. Neurons containing NPY and AgRP express receptors for both leptin and insulin, and the local administration of either insulin or leptin near the arcuate nucleus reduces the synthesis of both NPY and AgRP. Hence, adiposity signals exert net catabolic effects through the joint mechanisms of stimulating α-MSH/CART neurons and simultaneously inhibiting NPY/AgRP neurons. For all these reasons, the arcuate nucleus can be considered to be the brain's sense organ that detects body adiposity by monitoring the levels of leptin and insulin. In turn, axons from these two groups of arcuate neurons innervate many other hypothalamic nuclei as they modulate aspects of caloric homeostasis.

A particularly important tract includes NPY-containing axons that project from the arcuate nucleus to the PVN. The PVN expresses receptors for NPY, and careful mapping studies have identified the PVN as the most sensitive site to the orexigenic effect of exogenously administered NPY. Consistent with this finding, when antisense oligonucleotides to NPY receptors are given locally within the PVN, rats that have been deprived of food eat less than they usually would. Among the numerous sites that synthesize NPY within the brain, only those in the arcuate nucleus are sensitive to changes of food intake as signaled by leptin and insulin (Schwartz *et al.*, 2000). NPY mRNA is increased in the arcuate nucleus when animals are fasted and returns to baseline upon refeeding. Fasting also increases the levels of NPY in axon terminals in the PVN and increases secretion of NPY within the PVN. Hence, the fasted animal is primed to eat more food due in part to the action of elevated NPY in the area of the PVN. Circulating insulin and leptin are decreased during fasting, and local administration of either into the brain of a fasted rat lowers the elevation of NPY mRNA in the arcuate nucleus and lowers the amount of food eaten by food-deprived rats (Schwartz *et al.*, 2000). Hence, the metabolic state, as reflected by leptin and insulin, has a marked influence on arcuate NPY-containing neurons and consequently on neural pathways that influence energy intake and expenditure. Endogenous levels of NPY in the arcuate–PVN system normally peak when daylight ends and nocturnal activity begins (Leibowitz, 1990), which is also the time when rats typically eat their largest meal of the day. Hence, NPY activity in the PVN appears to be a key mediator of anabolic activity.

The other neuropeptide synthesized in NPY neurons in the arcuate nucleus, AgRP, is an endogenous antagonist of MC3 and MC4 receptors (Cone, 1999). AgRP therefore promotes food intake by acting at MC receptors in the PVN and other areas to block the action of α-MSH. When exogenous AgRP is administered into the third cerebral ventricle, the resultant hyperphagia lasts for up to six days (Hagan *et al.*, 2000). In contrast, the hyperphagia elicited by NPY lasts only a few hours. Hence, activation of the arcuate NPY/AgRP neurons results in an acute but robust increase of food intake caused by the actions of NPY and a steady and more prolonged increase caused by the actions of AgRP. Furthermore, the two peptides act in different ways. NPY acts through its receptors to stimulate anabolic pathways directly, whereas AgRP acts through a different receptor by antagonizing tonically active catabolic peptides. It is not known whether NPY and AgRP from the arcuate neurons act on the same or different target cells in the PVN, VLH, and elsewhere.

In sum, the arcuate nucleus appears to be a chemosensor in the brain that detects body adiposity by means of insulin and leptin receptors. One type of arcuate neuron synthesizes peptide transmitters (α-MSH and CART) having a net catabolic effect, whereas another type synthesizes peptide transmitters (NPY and AgRP) with a net anabolic effect. Both types of neuron project to other hypothalamic areas where they influence specific aspects of caloric homeostasis.

Other Hypothalamic Peptides Also Influence Food Intake

Several newly discovered neuropeptides important in the control of caloric homeostasis are synthesized in the VLH, including *orexin A* and *melanin-concentrating hormone* (MCH) (Shimada *et al.*, 1998). Administration of either of these peptides into the brain stimulates food intake, whereas mice that lack the gene for MCH are lean and obesity resistant. It appears likely that the aphagia exhibited by animals with VLH lesions is due in part to reduced levels of these peptides and that the hyperphagia caused by electrical stimulation of the VLH is due in part to release of these peptides.

Oxytocin is a peptide synthesized in the PVN and supraoptic nuclei of the hypothalamus and is secreted from the posterior lobe of the pituitary. In the systemic circulation of female mammals, this hormone is well known for stimulating uterine contractions for parturition and stimulating milk let-down in mammary glands during lactation. However, the hypothalamus and pituitary of males contain as much oxytocin as those of females; therefore, the peptide has long been suspected of being secreted in other circumstances and having additional functions. In fact, research has revealed that gastric distension, CCK administration, and plasma hyperosmolality, treatments that decrease food intake in rats, elicit pituitary oxytocin secretion in male and female rats. In addition, increases in plasma concentrations of oxytocin correlate highly with observed decreases in food intake.

Although observations were consistent with oxytocin being an appetite suppressant, later experiments showed that this was not the case: Systemic administration of oxytocin does not affect food intake. Instead, the reduced food intake seen after treatment with CCK or hypertonic saline appears to be mediated by oxytocin secreted from neurons whose cell bodies are located in the hypothalamic PVN but whose axons project within the central nervous system rather than to the pituitary gland. Consistent with this hypothesis, central injection of oxytocin into the cerebral ventricles (icv) decreases food intake in rats, an effect that is blocked by icv pretreatment of the animals with an oxytocin receptor antagonist. More to the point, the effects of exogenous CCK and hypertonic saline on food intake are blunted by icv pretreatment with an oxytocin receptor antagonist. A summary of some of the hypothalamic peptides that modulate food intake is depicted in Figure 38.7.

Stress is also well known to decrease food intake in animals, and various stressors activate the hypothalamic–pituitary–adrenal axis. A key mediator of this effect is *corticotropin-releasing hormone* (CRH), which is synthesized and secreted from the PVN. CRH stimulates the secretion of ACTH from the anterior pituitary, and ACTH in turn elicits steroid hormone secretion from the adrenal cortex (see Chapter 40). In rats, icv injection of CRH decreases food intake in the absence of stress and in association with oxytocin secretion from the posterior pituitary. In addition, icv pretreatment with an oxytocin receptor antagonist eliminates CRH-induced inhibition of food intake. These findings are consistent with central oxytocinergic neurons mediating the inhibition of food intake that occurs during stress. Such inhibitory effects on eating would complement the known inhibitory effects of central oxytocin on gastric motility and emptying.

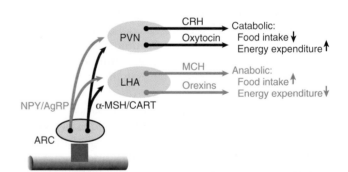

FIGURE 38.7 Hypothalamic neuropeptides that influence caloric homeostasis. The adiposity hormones, leptin and insulin, are transported through the blood–brain barrier and influence neurons in the arcuate nucleus (ARC). ARC neurons that synthesize and release NPY and AgRP are inhibited by adiposity signals, whereas ARC neurons that synthesize and release α-MSH and CART are stimulated by adiposity signals. NPY/AgRP neurons are inhibitory to the PVN and stimulatory to the LHA, whereas α-MSH/CART neurons are stimulatory of the PVN and inhibitory of the LHA. The PVN in turn has a net catabolic action, whereas the LHA has a net anabolic action.

Two additional features of these findings deserve note. First, despite prominent pituitary oxytocin secretion, central oxytocinergic pathways are not stimulated by suckling. Therefore, nursing does not inevitably reduce food intake in dams; in fact, lactating rats are hyperphagic in association with the substantial loss of calories through milk. Second, because the various stimuli for oxytocin release are not related to caloric homeostasis, oxytocin is not a factor in the termination of normal meals; in fact, oxytocin is not secreted by rats during meals taken *ad libitum*.

The Brain Stem Plays an Important Role in the Control of Food Intake

The NST and the adjacent area postrema in the brain stem receive sensory fibers from gustatory receptors in the mouth and throat (see Chapter 24), as well as afferent information from the stomach, small intestine, pancreas, and liver. In the central control of food intake, these linked brain stem sites likely are where sensory input from the viscera is first integrated with input from taste buds. This sensory information is relayed along prominent neural projections from the brain stem to the hypothalamus, amygdala, and other portions of the limbic system, as well as to the thalamus and gustatory cortex. Reciprocal neural connections to the brain stem from these rostral sites allow emotion and cognitive function to influence the control of eating. Complementary and linked efferent elements participate in the central control of the gastrointestinal tract and liver, and of autonomic function generally.

Evidence of the control of the brain stem over food intake comes from investigations of the chronic decerebrate rat in which all axonal connections between the caudal brain stem and the forebrain are severed at the midcollicular level of the midbrain (Grigson *et al.*, 1997). This animal does not seek food or initiate spontaneous meals, but, like an intact animal, will reflexively swallow liquid food put directly into its mouth via an oral fistula. The decerebrate rat will swallow a sucrose solution when its stomach is empty, such as after a period of food deprivation; however, it will not swallow water or saline when food deprived, nor will it swallow sucrose solution when its stomach is full or when it has received a systemic injection of CCK or hypertonic saline (but instead lets the administered fluid passively drip from its mouth). These experiments suggest that considerable control of food intake exists entirely within the caudal brain stem, which receives signals that arise when ingested food contacts peripheral sensory organs (i.e., taste buds, gastric stretch receptors, intestinal CCK-secreting cells). Nonetheless, decerebrate animals are incapable of receiving input from forebrain sites that could modify intake based on learning or other experience, on body adiposity, or on time of day.

Other observations provide additional support for an important role of the brain stem in the control of meal size. When the area postrema is ablated, rats eat larger meals than normal, although total daily food intake is normal; that is, they compensate for their large meals by eating less frequently, indicating that long-term controls of food intake are not impaired. A similar effect is seen in rats pretreated with capsaicin, a neurotoxin that destroys most gastric afferent vagal axons, eliminating the food intake-reducing effects of CCK and presumably damaging the signal of gastric distension.

Summary

To provide a biological context within which to consider the purpose of eating and the fate of ingested food, we began this chapter with a description of the physiology of caloric homeostasis. Two principles were emphasized. First, meal size is determined by the integrated effects of several acute stimuli. Delivery of calories to the stomach and small intestine upon food consumption and postabsorptive delivery of nutrients to the liver elicit sensory neural signals from the stomach and the liver, respectively, to the brain stem. Second, food intake is also influenced by chronic signals associated with body adiposity. Adiposity indirectly affects gastric and hepatic signals of satiety. In addition, humoral signals proportional to adiposity enter the brain and affect food intake. One signal emanates from adipose tissue itself (i.e., leptin), and the other comes from the pancreas (i.e., insulin). These two peptides appear to be important in the central control of food intake; they modulate the response of the brain to the acute neural signals that affect food intake.

These neural and humoral signals are related to caloric homeostasis and are inhibitory in nature. During meals, they combine to stop ongoing ingestion; later, when these signals disappear in the postabsorptive state, new meals begin. Other inhibitory signals are unrelated to caloric homeostasis yet can have a pronounced effect on the cessation of eating. These signals include gastric distension, toxins (in the food or body), and dehydration.

The central nervous system exerts control over eating at many levels. The spinal cord and brain stem influence all aspects of caloric homeostasis via the autonomic nervous system. The hypothalamus and limbic forebrain receive signals about ingested food and body adiposity and integrate them with information about the taste of the food, the memory of that food, the experience of previous meals, the competition of other desires, and aspects of the environment. Thus, the central control of eating involves many areas of the central nervous system in the collective maintenance of caloric homeostasis. In short, food intake is a simple behavior influenced by a complex array of stimuli and situational variables, the details of which remain to be fully understood.

References

Backhed, F., Manchester, J. K., Semenkovich, C. F., and Gordon, J. I. (2007). Mechanisms underlying the resistance to diet-induced obesity in germ-free mice. *Proc. Natl. Acad. Sci. USA* **104**, 979–984.

Coleman, D. L. and Hummel, K. P. (1969). Effects of parabiosis of normal with genetically diabetic mice. *Am. J. Physiol.* **217**, 1298–1304.

Cone, R. D. (1999). The central melanocortin system and energy homeostasis. *Trends Endocrinol. Metab.* **10**, 211–216.

Gibbs, J., Young, R. C., and Smith, G. P. (1973). Cholecystokinin decreases food intake in rats. *J. Comp. Physiol. Psychol.* **84**, 488–495.

Grigson, P. S., Kaplan, J. M., Roitman, M. F., Norgen, R., and Grill, H. J. (1997). Reward comparison in chronic decerebrate rats. *Am. J. Physiol.* **273**, R479–R486.

Hagan, M. M., Rushing, P. A., Pritchard, L. M., Schwartz, M. W., Strack, A. M., Van Der Ploeg, L. H., Woods, S. C., and Seeley, R. J. (2000). Long-term orexigenic effects of AgRP-(83–132) involve mechanisms other than melanocortin receptor blockade. *Am. J. Physiol.* **279**, R47–R52.

Le Magnen, J. and Tallon, S. (1966). *La periodicite spontanee de la prise d'aliments ad libitum du rat blanc. J. Physiol. (Paris)* **58**, 323–349.

Leibowitz, S. F. (1990). Hypothalamic neuropeptide Y, galanin, and amines: Concepts of coexistence in relation to feeding behavior. *Ann. N.Y. Acad. Sci.* **611**, 221–235.

Mayer, J. (1955). Regulation of energy intake and the body weight: The glucostatic theory and the lipostatic hypothesis. *Ann. N.Y. Acad. Sci.* **63**, 15–43.

Phillips, R. J. and Powley, T. L. (1996). Gastric volume rather than nutrient content inhibits food intake. *Am. J. Physiol.* **271**, R766–R779.

Schwartz, M. W., Woods, S. C., Porte, D., Jr., Seeley, R. J., and Baskin, D. G. (2000). Central nervous system control of food intake. *Nature* **404**, 661–671.

Seeley, R. J., Yagaloff, K. A., Fisher, S. L., Burn, P., Thiele, T. E., van Dijk, G., Baskin, D. G., and Schwartz, M. W. (1997). Melanocortin receptors in leptin effects. *Nature* **390**, 349.

Shimada, M., Tritos, N. A., Lowell, B. B., Flier, J. S., and Maratos-Flier, E. (1998). Mice lacking melanin-concentrating hormone are hypophagic and lean. *Nature* **396**, 670–674.

Smith, G. P., Gibbs, J., and Young, R. C. (1974). Cholecystokinin and intestinal satiety in the rat. *Fed. Proc.* **33**, 1146–1150.

Turnbaugh, P. J., Ley, R. E., Mahowald, M. A., Magrini, V., Mardis, E. R., and Gordon, J. I. (2006). An obesity-associated gut microbiome with increased capacity for energy harvest. *Nature* **444**, 1027–1031.

Ungerstedt, U. (1971). Adipsia and aphagia after 6-hydroxydopamine induced degeneration of the nigro-striatal dopamine system. *Acta Physiol. Scandin. Suppl.* **367**, 95–122.

Woods, S. C. (1995). Insulin and the brain: A mutual dependency. *Prog. Psychobiol. Physiol. Psychol.* **16**, 53–81.

Suggested Readings

Barsh, G. S., Farooqi, I. S., and O'Rahilly, S. (2000). Genetics of body-weight regulation. *Nature* **404**, 644–651.

Elmquist, J. K., Elias, C. F., and Saper, C. B. (1999). From lesions to leptin: Hypothalamic control of food intake and body weight. *Neuron* **22**, 221–232.

Friedman, M. I. and Stricker, E. M. (1976). The physiological psychology of hunger: A physiological perspective. *Psychol. Rev.* **83**, 409–431.

Langhans, W. (1996). Metabolic and glucostatic control of feeding. *Proc. Nutr. Soc.* **55**, 497–515.

Le Magnen, J. (1985). "Hunger." Cambridge Univ. Press, London.

Newsholme, E. A. and Start, C. (1973). "Regulation in Metabolism." Wiley, London.

Obici, S., Feng, Z., Morgan, K., Stein, D., Karkanias, G., and Rossetti, L. (2002). Central administration of oleic acid inhibits glucose production and food intake. *Diabetes* **51**, 271–275.

Schwartz, M. W., Figlewicz, D. P., Baskin, D. G., Woods, S. C., and Porte, D., Jr. (1992). Insulin in the brain: A hormonal regulator of energy balance. *Endocr. Rev.* **13**, 387–414.

Smith, G. P. and Gibbs, J. (1992). The development and proof of the CCK hypothesis of satiety. *In* "Multiple Cholecystokinin Receptors in the CNS" (C. T. Dourish, S. J. Cooper, S. D. Iversen, and L. L. Iversen, eds.), pp. 166–182. Oxford Univ. Press, London.

Stricker, E. M. (ed.) (1990). "Handbook of Behavioral Neurobiology," Vol. 10. Plenum, New York.

Woods, S. C., Schwartz, M. W., Baskin, D. G., and Seeley, R. S. (2000). Food intake and the regulation of body weight. *Ann. Rev. Psychol.* **51**, 255–277.

Stephen C. Woods and Edward M. Stricker

39

Water Intake and Body Fluids

Body fluids are the watery matrix in which the biochemical reactions of cellular metabolism occur. The concentration of substrates in cellular fluid is a key factor in determining the rate at which those reactions take place. All tissues depend on circulating blood to deliver the nutrients needed to support cellular metabolism and to carry away unwanted metabolites for excretion. Thus, the maintenance of solute concentrations or osmolalities—*osmotic homeostasis*—and the regulation of plasma volume—*volume homeostasis*—are essential functions in the physiology of animals.

When normal body fluid osmolality or plasma volume is threatened, various physiological and behavioral responses are stimulated to maintain or restore the basal state adaptively. For example, during water deprivation, animals decrease water lost in urine to prevent dehydration from worsening and consume water to replace the fluid they have lost. Similarly, hemorrhage stimulates the urinary conservation of water and sodium, as well as the ingestion of water and NaCl. Water retention and sodium retention are accomplished through actions of the antidiuretic hormone *arginine vasopressin* (AVP) and the antinatriuretic hormone *aldosterone*, whereas water ingestion and NaCl ingestion are motivated by *thirst* and *salt appetite*. These complementary responses are mediated and coordinated by the brain.

This chapter describes the various mechanisms by which the signals for fluid homeostasis are detected and integrated by the central nervous system. However, we first present a brief overview of body fluid physiology to provide a context in which to consider the regulated functions.

BODY FLUID PHYSIOLOGY

Water is the largest constituent of the body. It contributes 55 to 65% of the body weight of animals, including humans, varying mostly in relation to the amount of body fat. Total body water is distributed between *intracellular fluid* (ICF) and *extracellular fluid* (ECF) *compartments,* with approximately two-thirds in the former and one-third in the latter. The ECF can be further subdivided into the *interstitial fluid* surrounding the cells and the *intravascular fluid* within blood vessels. The intravascular fluid, the plasma (or serum) of blood, averages 7 to 8% of total body water, or 20 to 25% of the ECF.

Fluid compartments differ not only in their volumes, but also in the solutes they contain. Specifically, membrane-bound Na^+-K^+ pumps move Na^+ outside the cells and K^+ inside. Despite the differences in solute composition, the *osmotic pressure*, which reflects the concentrations of all solutes in a fluid compartment, is equivalent between ECF and ICF compartments. This equilibrium occurs because water is driven across cell membranes by osmosis from a relatively dilute compartment into one with a higher solute concentration until the osmotic pressures are equivalent on both sides of the cell membrane.

Multiple Mechanisms Help Maintain Blood Volume and Pressure

The distribution of fluid between intravascular and interstitial fluid compartments is determined by a balance between the *hydrostatic pressure* of the blood, which is maintained by cardiac output and arteriolar vasoconstriction, and the opposing osmotic pressure contributed by plasma proteins. Although those proteins contribute only 1 to 2% to overall plasma osmolality, the permeability of capillary membranes to such large molecules is low; therefore, proteins exert a pressure differential (of approximately 15–20 mm Hg), called the *colloid osmotic (oncotic) pressure*, that tends to pull interstitial fluid into the circulation. Figure 39.1 summarizes the forces governing transcapillary fluid transfer between the two extracellular compartments, a phenomenon first described by the English physiologist Starling (1896). Note that according to this arrangement, interstitial fluid acts as a reservoir for plasma. The Starling equilibrium is such that when blood pressure falls, as after hemorrhage, interstitial fluid moves into the circulation, thereby helping restore plasma volume. The reverse occurs when saline is added to the blood because plasma proteins are then diluted and the oncotic pressure they provide diminishes, allowing fluid to flow into the interstitial space.

Blood pressure also is maintained by two other mechanisms intrinsic to the cardiovascular system.

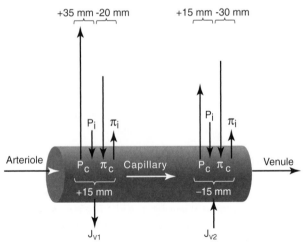

FIGURE 39.1 Starling forces governing transcapillary fluid transfer. At the arteriolar end of the capillary, the difference between the intravascular hydrostatic pressure (P_c) and the interstitial hydrostatic pressure (P_i) exceeds the oppositely oriented difference between the intravascular oncotic pressure (π_c) and the interstitial oncotic pressure (π_i); the resultant pressure gradient drives capillary fluid into the interstitial space (J_{v1}). As fluid leaves the capillary, P_c decreases due to fluid loss and π_c increases due to hemoconcentration. Consequently, at the venous end of the capillary, interstitial fluid is pulled back into the vascular space (J_{v2}). Numerical values indicate approximate net pressure differences (in mm Hg) between intravascular and interstitial spaces. Relative sizes of P and π are indicated by arrow length. Note that fluid accumulating in the interstitial space ultimately returns to the blood via the lymphatic system (not shown).

First, although the arteries that receive the cardiac output of blood have thick walls to preserve blood pressure, the veins are thin-walled, distensible vessels. After a moderate hemorrhage, veins collapse, redistributing blood to arteries that cannot collapse and thus remain full. Consequently, arterial blood pressure is not compromised, and the blood deficit occurs primarily on the venous side of the circulation. Conversely, fluid accumulates in the veins when blood volume is expanded, again without much effect on arterial blood pressure. This property is called the *capacitance* or *compliance* of the vascular system. Second, the filtration of blood in the glomeruli of the kidneys is determined in large part by blood pressure in the renal arteries. A drop in blood pressure reduces the *glomerular filtration rate* (GFR) and decreases urine formation, whereas a rise in blood pressure elevates GFR and promotes urinary fluid loss. This normal function of the kidneys is so efficient that the development of hypertension always implicates renal dysfunction as a contributing factor because the kidneys failed to adjust to the elevated blood pressure by increasing fluid excretion in urine.

Summary

Body fluid homeostasis is directed at achieving stability in the osmolality of body fluids and the volume of plasma. Such homeostatic regulation is promoted by several mechanisms intrinsic to the physiology of body fluids and the cardiovascular system. Nevertheless, changes in body fluid osmolality and plasma volume may be so large that additional mechanisms must be recruited to maintain homeostasis. These other responses involve the central control of water and sodium excretion in urine through the actions of specific hormones, and the central control of water and NaCl consumption motivated by thirst and salt appetite.

OSMOTIC HOMEOSTASIS

Osmolality is an expression of concentration—that is, a ratio of the total amount of solute dissolved in a given weight of water:

$$\frac{\text{solute (osmoles)}}{\text{water (kilograms)}}.$$

Dehydration and the consequent need for water occur whenever this ratio is elevated, whether by a decrease in its denominator or by an increase in its numerator. In fact, both changes occur naturally and often: Body water decreases as a result of water deprivation or the loss of dilute fluids to accomplish evaporative cooling (e.g., sweating) and increases as a result of solute load (e.g., the consumption of sodium as NaCl in foods).

The water loss associated with dehydration is borne by the ECF and the ICF in proportion to their sizes because the osmolality of fluid in the two compartments remains in equilibrium. In contrast, the increase in plasma osmolality that results from a solute load results only in cellular dehydration because the water leaving cells by osmosis expands ECF volume. Thus, a solute load is a more abrupt, less complex treatment than water deprivation for stimulating AVP secretion and thirst, the two main osmoregulatory responses of the brain.

Arginine Vasopressin Is the Antidiuretic Hormone

AVP is a nine amino acid peptide that is synthesized in the magnocellular neurons in the supraoptic nucleus (SON) and the paraventricular nucleus (PVN) of the hypothalamus. The peptide is cleaved enzymatically from its prohormone and transported along axons projecting from the hypothalamus to the nearby posterior lobe of the pituitary gland (*neurohypophysis*). There, AVP is stored within neurosecretory granules until specific stimuli, such as an increase in the effective osmolality of body fluids, activate its secretion into the bloodstream. The importance of AVP in maintaining water balance is underscored by the fact that its stores in the pituitary contain sufficient hormone to enable maximal antidiuresis when dehydration is sustained for more than a week.

The circulating hormone acts on a subset of AVP receptors (the V_2 subtype) in the kidney to increase water permeability of the collecting ducts of nephrons through insertion of water channels, called *aquaporins*, into the apical membranes of tubular epithelial cells. Antidiuresis occurs when water moves out of the distal convoluted tubule and collecting duct by osmosis. As secondary responses to the increased net water reabsorption, urine flow decreases and urine osmolality increases. See Box 39.1 for more on renal water channels (Knepper, 1997).

BOX 39.1

WATER CHANNELS

Water conservation by the kidney is dependent on its ability to concentrate the urine. Urine is concentrated through the combined actions of the loop of Henle and the collecting duct of the kidney. The loop of Henle generates a high osmolality in the renal medulla via a mechanism known as the countercurrent multiplier system. AVP acts in the collecting duct to increase water permeability, thereby allowing osmotic equilibration between the urine and the hypertonic interstitium of the renal medulla. The net effect of this process is to extract water from the urine into the medullary interstitial blood vessels, resulting in increased urine concentration and decreased urine volume.

AVP produces antidiuresis by virtue of its effects on the epithelial principal cells of the collecting duct in the kidney, which are endowed with AVP receptors of the V_2 type. It has long been known that binding of AVP to G protein-coupled V_2 receptors causes cAMP generation via activation of adenylate cyclase. However, the intracellular mechanisms responsible for the subsequent increased water reabsorption across the collecting duct cells have been elucidated only recently following the discovery of

aquaporins, which are a widely expressed family of water channels that facilitate rapid water transport across cell membranes.

Many different aquaporin water channels are expressed in various body tissues, including the brain. Several of these channels have been localized in the kidney. Aquaporin-1 is expressed constitutively in the proximal tubule and thin descending limb of the loop of Henle and is believed to be responsible for reabsorption of a large fraction of the water filtered by the glomerulus. The other three water channels, aquaporin-2, -3, and -4, are expressed in collecting duct principal cells, target cells for the action of AVP to regulate collecting duct water permeability. Aquaporin-2 is the only water channel known to be expressed in the apical membrane (i.e., the side bordering the tubular lumen, through which urine flows) of collecting duct principal cells and is also abundant in intracellular vesicles located below the apical membrane. Aquaporin-2 is the major AVP-regulated water channel and mediates water transport across the apical plasma membrane of the principal cells of the collecting ducts. In contrast, aquaporins-3 and -4 are

(continues)

BOX 39.1   *(cont'd)*

expressed at high levels in the basolateral plasma membranes (i.e., the side bordering the blood) of principal cells and are responsible for the constitutively high water permeability of the basolateral plasma membrane.

There are at least two ways by which AVP regulates osmotic water permeability in the collecting duct: short-term and long-term regulation (Knepper, 1997). Short-term regulation is associated with increases in water permeability within a few minutes of AVP exposure, an effect that is rapidly reversible. AVP triggers this response by binding to the V_2 receptor and increasing intracellular cyclic AMP levels by activating adenylate cyclase. Studies have demonstrated that the increase in collecting duct water permeability is a consequence of fusion of aquaporin-2-containing intracytoplasmic vesicles with the apical plasma membranes of the principal cells, a process that increases apical water permeability by markedly increasing the number of water-conducting pores in the apical plasma membrane. Dissociation of AVP from the V_2 receptor allows intracellular cAMP levels to decrease, and the water channels are reinternalized into the intracytoplasmic vesicles, thereby terminating the increased water permeability. By virtue of the subapical membrane

localization of the aquaporin-containing vesicles, they can be quickly shuttled into and out of the membrane in response to changes in intracellular cAMP levels. This mechanism therefore allows minute-to-minute regulation of renal water excretion through changes in ambient plasma AVP levels. Long-term regulation of collecting duct water permeability represents a sustained increase in collecting duct water permeability in response to prolonged high levels of circulating AVP. This response requires at least 24h to elicit and is not as rapidly reversible. Studies have demonstrated that this conditioning effect is due largely to the ability of AVP to induce large increases in the abundance of aquaporin-2 and -3 water channels in the collecting duct principal cells. Greater total expression of the number of aquaporin-2 and -3 water channels, when combined with the short-term effect of AVP to shift aquaporin-2 into the apical plasma membrane, allows the collecting ducts to achieve extremely high water permeabilities during conditions of prolonged dehydration, thereby further enhancing the urine concentrating capacity in response to elevated levels of circulating AVP.

Edward M. Stricker and Joseph G. Verbalis

With refinement of radioimmunoassays for AVP, the unique sensitivity of the hormone to small changes in osmolality has become apparent, as has the remarkable sensitivity of the kidney to small changes in plasma AVP levels. Circulating AVP is linearly related to plasma osmolality above a threshold of 1 to 2% pg ml^{-1} (Fig. 39.2). Urine osmolality, in turn, is related linearly to AVP levels from 0.5 to 5–6 pg ml^{-1}, in association with increases in plasma osmolality to only 4% above the threshold for AVP secretion. However, because urine volume is related inversely to urine osmolality, small increases in plasma AVP concentration (e.g., from 0.5 to 2 pg ml^{-1}) have the effect of decreasing urine flow much more than subsequent larger increases (e.g., from 2 to 5 pg ml^{-1}; Fig. 39.3). This relationship emphasizes the marked physiological effects of small initial changes in plasma AVP levels. The net result of these relations among plasma osmolality, AVP secretion, urine volume, and urine osmolality is a finely tuned regulatory system that adjusts the rate of free water excretion according to plasma osmolality via changes in pituitary hormone secretion.

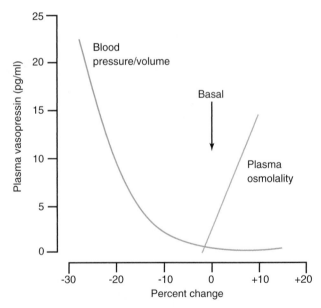

FIGURE 39.2 Plasma concentrations of AVP as a function of changes in plasma osmolality, blood volume, or blood pressure in humans. The arrow indicates the plasma AVP concentration at basal plasma osmolality, volume, and blood pressure. Modified with permission from Robertson (1986).

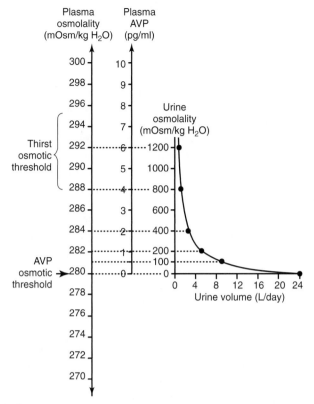

FIGURE 39.3 Relationship of plasma osmolality, plasma AVP concentrations, urine osmolality, and urine volume in humans. Note that small changes in plasma AVP concentrations have larger effects on urine volume at low plasma AVP concentrations than at high plasma AVP concentrations. Modified with permission from Robinson (1985).

Thirst Is Another Effective Osmoregulatory Response to Dehydration

Accompanying the excretion of concentrated urine is reabsorption of conserved water, which can considerably dilute remaining body fluids. Thirst, and the water intake it provokes, is a much more rapid and less limited response to dehydration than antidiuresis. Thirst may be defined as a strong motivation to seek, to obtain, and to consume water in response to deficits in body fluids. Like AVP secretion, thirst can be stimulated by cellular dehydration caused by increases in the effective osmolality of ECF (Gilman, 1937). Studies in animals consistently indicate that drinking behavior is elicited by 1 to 3% increases in plasma osmolality above basal levels, and analogous research in humans has revealed that similar thresholds must be reached to produce thirst (Robertson, 1986). This arrangement, in which the threshold for drinking is slightly higher than that for secretion of AVP (Fig. 39.3), ensures that ongoing behavior is not disrupted by thirst unless the buffering effects of osmosis and antidiuresis are insuf-

ficient to maintain osmoregulation. It also ensures that dehydration does not become severe before thirst is stimulated. See Box 39.2 for more on how the neurohypophysis controls the volume of water in the body.

Also like AVP secretion, water intake increases linearly in proportion to increases in the effective osmolality of ECF. The dilution of body fluids by ingested water complements the retention of water that occurs during antidiuresis, and both responses occur when drinking water is available. However, there may be marked individual differences in whether dehydrated subjects respond to their need for water promptly by drinking, or more slowly by increasing renal water conservation (Kanter, 1953). AVP secretion and urine osmolality are more elevated when an induced increase in plasma osmolality is not compensated for by water intake (e.g., because drinking water is not available). Conversely, water intake in response to a solute load increases in animals after their kidneys are removed or their capability to secrete AVP is compromised, thereby precluding rapid excretion of the load in concentrated urine. See Box 39.3 on the rapid inhibitory feedback control of drinking.

Osmoreceptor Cells Stimulate AVP Secretion and Thirst

All body cells lose water by osmosis when the effective osmolality of ECF is increased. Thus, cells that provoke AVP secretion and thirst do not have unique osmosensitive properties (unlike retinal photoreceptor cells, e.g., which are uniquely responsive to light). Instead, the exceptional feature of osmoreceptor cells is thought to be their neural circuitry, which activates the central systems for AVP secretion and thirst when the cells are dehydrated.

Destruction of osmoreceptor neurons should eliminate detection of increased plasma osmolality and thus the AVP secretion and thirst responses that are elicited by dehydration. In fact, ample research has confirmed that certain brain lesions eliminate AVP secretion and thirst responses. Such studies also have revealed that osmoreceptor cells (Johnson and Buggy, 1978; Thrasher et al., 1982) appear to be located in the vascular organ of the lamina terminalis (OVLT) and areas of the adjacent anterior hypothalamus, near the anterior wall of the third cerebral ventricle (Fig. 39.4). Surgical destruction of that area of the brain in animals abolishes the AVP secretion and thirst responses to hyperosmolality but not their responses to other stimuli. The same conclusion was drawn after a study of people who were unable to osmoregulate when water deprived or when given a NaCl load; these patients were found to have

---

BOX 39.2

DIABETES INSIPIDUS

The disease *diabetes insipidus* (DI), in which the secretion of AVP is impaired or absent, illustrates the crucial role of AVP in controlling the volume of water in the body. An eighteenth century English physician, Thomas Willis, named the disease diabetes insipidus, or insipid urine, to distinguish it from *diabetes mellitus*, or sweet urine, in which abnormally high concentrations of glucose in the blood result in glucose appearing in the urine.

AVP is the only known antidiuretic substance in the body. In its absence, the kidney is unable to concentrate urine maximally to conserve water. The result is a continued excretion of copious amounts of very dilute urine. Patients with severe cases of DI, in whom the ability to excrete AVP is completely lost, can excrete up to 25 liters of urine each day. Such patients urinate almost hourly, which renders the completion of even simple tasks and activities of daily living, including sleeping, exceedingly difficult.

Patients with DI can quickly become dehydrated if their urinary fluid losses are not replaced by drinking water. Fortunately, thirst remains intact in most patients with DI because lesions that destroy the magnocellular neurons in the SON and PVN that synthesize AVP generally leave intact the osmoreceptors in the anterior hypothalamus as well as the higher brain centers that control thirst. Consequently, extreme thirst is one of the hallmarks of this disease, leading to the characteristic symptoms of *polydipsia* (excessive drinking) and *polyuria* (excessive urination). If drinking water is unavailable, or if a person with DI is unable to drink, then the unreplaced urinary water loss leads to dehydration and death in the absence of medical intervention.

DI can be caused when tumors and infiltrative diseases of the hypothalamus destroy the magnocellular neurons that produce AVP. Because four nuclei contain magnocellular neurons, and 10 to 20% of AVP-producing neurons are sufficient for normal urine concentration, brain lesions that cause DI are generally large. Less commonly, DI is idiopathic (of unknown cause), most likely from an autoimmune basis. DI can also be genetic, transmitted as an autosomal-dominant trait. Some patients with DI do not have any defect in AVP secretion but rather have defects in the V_2 AVP receptors in the kidney that respond to circulating AVP. These cases are called *nephrogenic* (of kidney origin) DI.

The treatment for DI, like that of other endocrine deficiency disorders, is replacement of the deficient hormone, in this case AVP. The short half-life of AVP in the circulation allows mammals to have minute-to-minute control of their urine output. However, longer acting synthetic analogs of AVP are more convenient for treatment because they need not be taken as frequently as short-acting drugs (Robinson, 1985). These agents can restore urinary concentration and allow a person with DI to lead a more normal life.

Edward M. Stricker and Joseph G. Verbalis

---

focal brain tumors that destroyed the region around the OVLT.

The location of osmoreceptor cells in the OVLT is consistent with the results of pioneering investigations in which hyperosmotic solutions injected into blood vessels perfusing the anterior hypothalamus stimulated AVP secretion in dogs (Verney, 1947). The OVLT and surrounding areas of the anterior hypothalamus were subsequently implicated by studies in rats. After systemic injections of hypertonic NaCl solution, modern immunocytochemical techniques were used to detect early gene products in cells associated with the production of new protein. Dense staining in the OVLT (and in AVP-secreting cells in the hypothalamus) indicated that it had been strongly stimulated by the experimentally induced dehydration. See Box 39.4 on the use of cFos immunocytochemistry in studies of brain function.

The neural pathways connecting the OVLT with magnocellular AVP-secreting cells in the hypothalamic SON and PVN have been identified, whereas the neural circuits in the forebrain that control thirst are less well known. Early reports identified the lateral hypothalamus as a "thirst center" because its destruction in rats eliminated the drinking response, but not the associated AVP secretion, to increased osmolality of body fluids. However, later investigations showed that the critical damage was not to hypothalamic cells but to dopamine-containing fibers that coursed through the area. The induced disruption of behavior was not

BOX 39.4

USE OF cFOS IMMUNOCYTOCHEMISTRY IN STUDIES OF BRAIN FUNCTION

cFos was named for its homology to an FBR osteogenic sarcoma virus protein, v-fos. The cFos protein is the product of the *c-fos* gene, which belongs to a family of genes whose products are involved in the regulation of gene transcription. The protein has a "leucine zipper" motif that promotes dimerization (pairing) with other regulatory proteins, most commonly with members of the Jun family. Genes encoding Jun and Fos are *immediate-early genes* (IEGs), the first genes expressed in response to activation of cells. Jun and Fos protein dimers bind specific DNA sequences (e.g., APS sites) to modulate the expression of genes expressed later.

The precise relationship between cFos expression and transcription of later gene products remains uncertain for most neural systems. However, many neurons are known to express cFos only on synaptic activation; therefore, the mRNA or protein product (cFos) of the *c-fos* gene can be used as a marker of neuronal activation in the brain. The protein product is especially useful because it is detected easily by antibodies directed against cFos. This immunohistochemical staining allows investigators to determine which neurons are activated in the brain in response to specific stimuli given to experimental animals.

Many of the initial studies simply localized the IEG expression to magnocellular neurons in the SON and PVN, but later studies used antibodies against proteins located in cytoplasm, along with the antibody against IEG products, and showed that the IEG products were located in the nucleus (Hoffman *et al.*, 1993). Such studies have now confirmed that in response to these treatments, IEG expression in AVP- and oxytocin-secreting neurons closely parallels pituitary AVP and oxytocin secretion. Thus, in rats, hyperosmolality, hypovolemia, and administration of AII stimulate secretion of AVP and oxytocin and activate cFos expression in AVP and OT neurons; CCK, which stimulates secretion of oxytocin but not AVP, causes expression of cFos only in neurons that secrete oxytocin. Similarly, although severe hemorrhage stimulates secretion of AVP and oxytocin and activates cFos expression in both types of neurons, lower intensity hypovolemia predominantly causes peptide secretion and cFos expression only in AVP neurons. During CCK administration and during hemorrhage, plasma AVP and oxytocin correlate well with cFos expression. Consequently, IEG expression is now considered a specific and sensitive marker of magnocellular secretory activity in response to most acute stimuli.

Immunocytochemical detection of cFos expression, phenotypic characterization of activated cells, and standard techniques of tract tracing together have enabled investigators to map brain circuits activated by specific stimuli.

Edward M. Stricker and Joseph G. Verbalis

---

is the hormone *atrial natriuretic peptide* (ANP), which is synthesized in the atria of the heart and released when increased intravascular volume distends the atria. Another is the hormone *oxytocin*. Like AVP, oxytocin is synthesized in magnocellular neurons in the PVN and SON and secreted from the posterior pituitary in proportion to induced hyperosmolality (Stricker and Verbalis, 1986). In rats, oxytocin is as potent in stimulating natriuresis as AVP is in stimulating antidiuresis (Verbalis *et al.*, 1991).

Salt loads are also known to decrease the intake of osmolytes, whether in the form of NaCl solution or food, complementing the stimulation of thirst and the secretion of AVP and oxytocin. However, because destruction of the OVLT eliminates the two latter effects but not the dehydration-induced reduction in NaCl intake, osmo- or Na^+ receptors located outside

the basal forebrain must mediate the inhibition of NaCl and food intake. Possible sites for such cells include the *hepatic portal vein* and the *area postrema*, both of which have been suspected of having Na^+ receptor functions. In addition, cells in the hepatic portal vein are well situated to detect the sodium content of ingested food and to modulate its intake accordingly.

Diuresis and Inhibition of Water Intake Promote Osmotic Homeostasis during Overhydration

Osmoregulation is required not only under conditions of dehydration, but also during periods of acute overhydration and hypoosmolality, as may result when beverages are consumed in excess of water needs. Such consumption occurs not because of thirst but, for example, because of the palatability of or chemical substances in the beverages (e.g., caffeine, alcohol). Unlike excess food, which is stored as triglycerides in adipose tissue, excess water is not stored for later use but instead is excreted in urine. When plasma osmolality is below normal, circulating levels of AVP are reduced, and in consequence, the kidneys excrete dilute urine and thereby raise plasma osmolality. The major behavioral contribution of osmotic dilution to osmoregulation is inhibition of free water intake; the ingestion of osmolytes in food and NaCl is not stimulated by osmotic dilution.

Summary

Osmoregulation in animals and humans is accomplished by a combination of physiological responses to dehydration, resulting in antidiuresis and natriuresis, and the behavioral response of increased water intake. Osmoreceptor cells critical for mediating these functions have been identified in the basal forebrain. These neurons respond to very small increases in plasma osmolality, and the effector systems they control correct any increase in plasma osmolality. In addition, an early stimulus must exist, generated by peripheral osmo- or Na^+ receptor cells, which signals the brain in anticipation of subsequent rehydration. Other osmo- or Na^+ receptor cells appear to mediate the inhibition of NaCl and food intake, which also contributes to osmoregulation during dehydration. Conversely, diuresis and inhibition of water intake promote osmoregulation during overhydration, whereas excitation of NaCl or food intake does not.

VOLUME HOMEOSTASIS

Like osmotic dehydration, loss of blood volume (*hypovolemia*) stimulates several adaptive compensatory responses appropriate for restoring circulatory volume. The physiological contributions to volume regulation have been studied extensively in laboratory animals subjected to a controlled loss of blood. Because behavior is compromised by the anemia and hypotension that result from extensive hemorrhage, researchers studying thirst in rats may instead produce hypovolemia by subcutaneous injection of a colloidal solution. Such treatment disrupts the Starling equilibrium in capillaries near the injection site because the extravascular colloid opposes the oncotic effect of plasma proteins. Consequently, fluid leaches out of capillaries and remains in the interstitial space. Ingested

fluid is also drawn into the interstitial space by the colloid so the total fluid volume required to correct the plasma volume deficit may be substantial. Water does not move from the cells by osmosis, however, because the osmolality of the injected colloidal solutions actually is similar to that of cells.

After colloid injections, rats conserve water and Na^+ in urine and increase their consumption of water and saline solution (Stricker, 1981). When given an isotonic NaCl solution to drink, these rats ingest volumes appropriate to their needs. When given separate bottles of water and concentrated NaCl solution, remarkably the rats drink appropriate amounts of each to create an isotonic fluid mixture. These observations raise several questions: How do the hypovolemic rats detect their plasma volume depletion? How are their thirst and salt appetite coordinated so that they consume the desired isotonic mixture of fluid? How are these two behavioral responses integrated with the complementary physiological responses of water and Na^+ conservation in urine, as well as with the vasoconstrictor responses needed to support blood pressure? Research has provided answers to these and related questions, which are discussed in the following sections.

Neural and Endocrine Signals of Hypovolemia Stimulate AVP Secretion

An appropriate physiological response to volume depletion should include water conservation and urine concentration. In fact, like plasma hyperosmolality, hypovolemia is an effective stimulus for AVP secretion (Robertson, 1986; Stricker and Verbalis, 1986). However, AVP secretion does not occur until blood loss exceeds 10% of total blood volume, meaning AVP secretion is much less sensitive to hypovolemia than to increases in ECF osmolality (Fig. 39.2). The effects of hypovolemia and osmotic dehydration are additive; that is, a given increase in osmolality causes greater secretion of AVP when animals are hypovolemic than when they are euvolemic (Fig. 39.5).

Loss of blood volume is first detected by stretch receptors in the great veins entering the right atrium of the heart. These stretch receptors provide an afferent vagal signal to the *nucleus of the solitary tract* (NST) in the brain stem (Chapter 36). Still larger decreases in blood volume may also lower arterial blood pressure and reduce the stretch of receptors in the walls of the carotid sinus and aortic arch. That information is integrated in the NST with neural messages from the low-pressure, venous side of the circulation. Note that these sensory neurons are not actually baroreceptors (literally, pressure receptors), although commonly they are referred to as such.

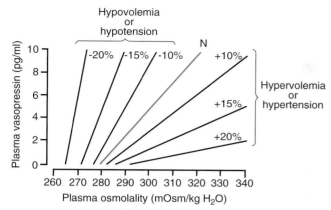

FIGURE 39.5 The relationship between the osmolality of plasma and the concentration of vasopressin (AVP) in plasma depends on blood volume and pressure. The line labeled N shows plasma vasopressin concentration across a range of plasma osmolality in an adult with normal intravascular volume (euvolemic) and normal blood pressure (normotensive). Lines to the left of N show the relationship between plasma vasopressin concentration and plasma osmolality in adults whose low intravascular volume (hypovolemia) or blood pressure (hypotension) is 10, 15, and 20% below normal. Lines to the right of N are for volumes and blood pressures 10, 15, and 20% above normal. Modified with permission from Robertson (1986).

The ascending pathway between the NST in the brain stem and the SON and PVN in the hypothalamus includes noradrenergic fibers arising from A1 cells in the ventrolateral medulla. Volume depletion stimulates AVP secretion via other pathways as well. In addition, the kidneys secrete *renin* during hypovolemia, a response mediated in part by sympathetic neural input to β-adrenergic receptors on cells that secrete renin. Renin is an enzyme that initiates a cascade of biochemical steps that result in the formation of *angiotensin II* (AII) (Fig. 39.6), an extremely potent vasoconstrictor. AII stimulates AVP secretion by acting in the brain at the *subfornical organ* (SFO) (Ferguson and Renaud, 1986), which is located in the dorsal portion of the third cerebral ventricle. Because this circumventricular organ lacks a blood–brain barrier, the AII receptors can detect very small increases in the blood levels of AII. Neural pathways from the SFO to the SON and PVN in the hypothalamus mediate AVP secretion and may use AII as a neurotransmitter. A branch of this pathway goes first to the OVLT, perhaps providing an opportunity for the integration of information about volume states and osmotic concentration.

Neural and Endocrine Signals of Hypovolemia Also Stimulate Thirst

In addition to the antidiuresis and vasoconstriction produced by the secretion of AVP and activation of the renin–angiotensin system, colloid treatment increases water intake in rats (Fitzsimons, 1961). Once 5 to 10% of the normal plasma volume has been lost, the water intake elicited by hypovolemia increases linearly in relation to further deficits in plasma volume. This stimulus for water intake and the effect of a NaCl load on thirst are additive. The stimulus of thirst during hypovolemia appears to be the same as the signal for AVP secretion—that is, a combination of neural afferents from cardiovascular baroreceptors and endocrine stimulation by AII (Fitzsimons, 1969). Each signal can stimulate water intake in the absence of the other. For example, water intake by hypovolemic rats is not diminished by destruction of the NST and loss of neural input from baroreceptors, nor is it eliminated by the loss of AII resulting from bilateral removal of the kidneys (Fitzsimons, 1961). Presumably, colloid treatment would not elicit thirst in rats subjected concurrently to bilateral nephrectomy and NST lesions.

Intravenous infusion of AII strongly stimulates thirst in rats and most other animals studied in the laboratory (including humans). AII adds to the thirst stimulated by an osmotic load when the two treatments are combined. There had been considerable controversy about whether AII functions as a normal physiological stimulus of thirst because the doses of AII required to stimulate significant water intake produce blood levels of AII well above the physiological range. However, such doses also increase arterial blood pressure, and it has been shown that this acute hypertension inhibits thirst and limits the induced water intake. For example, drinking was enhanced considerably when the hypertensive effects of AII were blunted by a simultaneous administration of vasodilating agents (Robinson and Evered, 1987).

Osmotic Dilution Inhibits Thirst and AVP Secretion during Hypovolemia

Hypovolemic rats need isotonic saline, not water alone, to repair their plasma volume deficits. When the rats drink only water, about two-thirds of the ingested volume moves into cells by osmosis, and much of what remains extracellular is captured by the colloid. Thus, most of the water that is consumed does not remain in the vascular compartment, and the volume deficit within the vascular compartment persists. This situation may be contrasted with the negative feedback control of osmoregulatory thirst in which the dehydration of osmoreceptor cells causes thirst and ingested water repairs dehydration, eliminating the stimulus for thirst.

Rather than repairing the volume deficit, the water consumed by hypovolemic rats causes a second serious

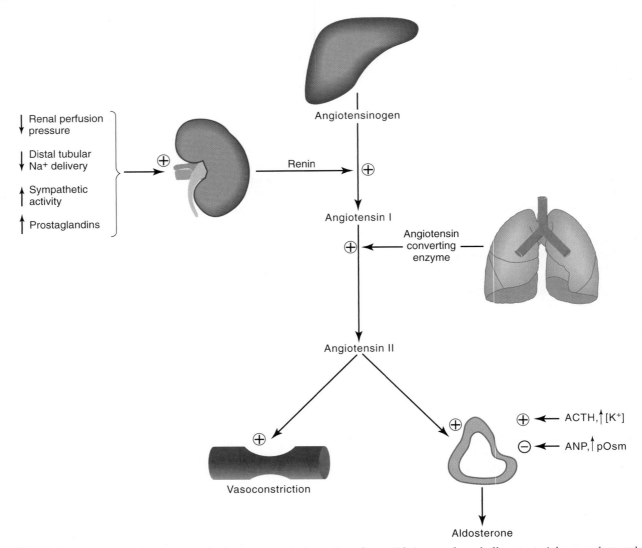

FIGURE 39.6   The renin–angiotensin cascade. Baroreceptors in the aortic arch, carotid sinus, and renal afferent arterioles sense hypovolemia and then cause the kidneys to secrete the enzyme renin. Renin cleaves angiotensinogen, which is synthesized by the liver, to produce angiotensin I. The angiotensin-converting enzyme, primarily in the lungs, cleaves angiotensin I to produce angiotensin II (AII), a peptide made of eight amino acid residues. AII is a potent vasoconstrictor and one of several stimulants of aldosterone secretion. Stimulation of aldosterone secretion from the adrenal cortex, along with possible direct intrarenal effects of AII, promotes renal conservation of sodium ions, complementing the pressor effect of AII in stabilizing arterial pressure and volume. ACTH, adrenocorticotropic hormone; ANP, atrial natriuretic peptide; pOsm, plasma osmolality. Modified with permission from Stricker and Verbalis (1992).

challenge to body fluid homeostasis: osmotic dilution. The animals cannot readily eliminate this self-administered water load because hypovolemia reduces GFR and thereby diminishes urinary excretion independent of AVP. Therefore, when water is the only drinking fluid available, an appropriate response of colloid-treated rats would be to stop drinking and thereby limit the secondary problem of osmotic dilution. In experiments, colloid-treated rats actually do stop drinking water despite persistent hypovolemia, and

the stimulus for the inhibition of thirst has been found to be osmotic dilution of body fluids (Stricker, 1969). A mere 4 to 7% decrease in osmolality is sufficient to stop drinking that has been motivated by a 30 to 40% loss of plasma volume. Thus, the osmoregulatory system appears dominant in the control of thirst. Comparable data demonstrate the same to be true of AVP secretion in colloid-treated rats: osmotic dilution inhibits AVP secretion even in severely hypovolemic rats (Stricker and Verbalis, 1986).

Hypovolemia Also Stimulates Aldosterone Secretion and Salt Appetite

As mentioned earlier, hypovolemic animals need to consume and retain water and NaCl, not just water. Appropriately, colloid-treated rats drink NaCl solution and conserve Na^+ in urine. Renal Na^+ retention is mediated largely by aldosterone secreted from the adrenal cortex, although Na^+ conservation also occurs in association with the decrease in GFR. The central nervous system does not innervate the adrenal cortex, as it does the adrenal medulla. However, a neural influence on aldosterone secretion is provided indirectly because AII is a very potent stimulus of aldosterone secretion, and renin secretion from the kidneys during hypovolemia is stimulated in part by the sympathetic nervous system (Fig. 39.6). The secretion of aldosterone is also stimulated by another peptide hormone, *adrenocorticotrophic hormone* (ACTH), which is secreted from the anterior lobe of the pituitary gland in response to *corticotrophic-releasing hormone* (CRH). The release of CRH from the PVN is triggered by various stressors, including activated cardiovascular baroreceptors. Yet another stimulus of aldosterone secretion is increased plasma levels of K^+, which can develop as a consequence of reduced GFR and an associated decrease in urinary K^+ excretion. The effects of these three independent stimuli of aldosterone secretion are additive. Aldosterone has also been found to stimulate salt appetite in animals, which therefore complements its renal effects to cause sodium conservation. (See Box 39.5 on salt appetite after adrenocortical dysfunction.)

The onset of salt appetite induced in rats by colloid treatment is curiously delayed relative to the appearance of thirst. That is, water intake increases within 1 to 2 h after colloid treatment, whereas the intake of NaCl solution does not increase until at least 5 h later (Stricker, 1981). Investigations showed that the delay is caused both by the gradual appearance of an excitatory stimulus of salt appetite (i.e., AII) and by the gradual disappearance of an inhibitory stimulus for NaCl intake. With regard to the latter, recall that during the delay, hypovolemic rats drank water and thereby caused an osmotic dilution of body fluids, which inhibits secretion of pituitary AVP and oxytocin. Either of these neurohypophyseal hormones might have provided an inhibitory stimulus for salt appetite that

---

BOX 39.5

SALT APPETITE AFTER ADRENOCORTICAL DYSFUNCTION

Fifty years ago, a 4-year-old boy was brought to the Johns Hopkins University Hospital in Baltimore with the peculiar behavior of eating large amounts of table salt. The child heavily salted all his food, even juice, and often consumed salt directly from its container. His parents tried unsuccessfully to keep him from consuming salt and brought him to the hospital in the hope that his strange behavior could be understood and stopped. The physicians were able to prevent the boy from having access to NaCl, but they were shocked when he died a few days later. An autopsy revealed that the child had bilateral tumors in his adrenal glands. In retrospect, it seems clear that he could not secrete aldosterone and therefore lost Na^+ in urine uncontrollably. Like patients with diabetes insipidus, who excrete a copious amount of dilute urine and drink comparable volumes of water in compensation (see Box 39.2 on diabetes insipidus), this boy evidently was ingesting salt as an adaptive compensation to his recurrent need for sodium (Wilkins and Richter, 1940). He died when that response was prevented.

After surgical removal of their adrenal glands, rats increase their consumption of NaCl by amounts proportional to the sodium loss in urine (Richter, 1936). The stimulus for this salt appetite appears to be a combination of elevated circulating levels of AII and reduced activity in a central oxytocinergic inhibitory system. In humans, however, a comparable salt appetite is not as common in patients with Addison disease, whose adrenocortical dysfunction causes them to lose Na^+ in urine excessively. Nor is salt appetite usually stimulated in people who lose excessive amounts of Na^+ in perspiration. No explanation is available for why an adaptive salt appetite appears readily in laboratory animals but not in humans, but this appears to be related, in part, to factors associated with herbivorous versus carnivorous eating behaviors.

Edward M. Stricker and Joseph G. Verbalis

disappeared as the animals drank water. In other studies, however, rats showed a strong inverse relationship between intake of NaCl and plasma levels of oxytocin (but not AVP), suggesting that circulating oxytocin was an inhibitory stimulus of salt appetite.

Central Oxytocin Inhibits Salt Appetite

However, in tests of the hypothesis that circulating oxytocin inhibits salt appetite, an intravenous infusion of physiological doses of oxytocin did not decrease NaCl intake in hypovolemic rats, nor did infusion of an oxytocin receptor blocker increase NaCl intake. These unexpected findings were clarified by the observation that, coincident with the secretion of oxytocin from magnocellular neurons, oxytocin was released from parvicellular neurons projecting centrally from the PVN. Thus, plasma oxytocin may have been a peripheral marker of the centrally acting oxytocin that mediated the inhibition of salt appetite in rats (Blackburn et al., 1992). This hypothesis has been strongly supported by the results of a series of investigations. For example, salt appetite in hypovolemic rats was eliminated by a systemic injection of naloxone, an opioid receptor antagonist that disinhibits oxytocin secretion, and this effect was blocked by the prior injection of an oxytocin receptor antagonist directly into the cerebrospinal fluid. Conversely, NaCl ingestion in response to hypovolemia or AII was potentiated by diverse treatments that inhibit the secretion of oxytocin. In addition to osmotic dilution, these treatments included systemic injection of ethanol, maintenance on a sodium-deficient diet (instead of the standard laboratory diet rich in Na$^+$), and peripheral administration of mineralocorticoid hormones such as aldosterone. Thus, excitatory and inhibitory components together regulate NaCl intake in a dual-control system in the brain.

Central oxytocinergic neurons can inhibit salt intake and food intake (Chapter 38). Thus, when osmotic dehydration activates oxytocinergic neurons, the intake of food and salt is inhibited and hyperosmolality is thereby prevented. From this perspective, food is a source of osmolytes (not just of calories) and NaCl solution provides osmolytes without calories. Inhibition of intake of osmolytes complements the natriuretic effect of pituitary oxytocin in supporting osmoregulation.

Salt Appetite Is Also Inhibited by Atrial Natriuretic Peptide

The role of oxytocin in osmoregulation has become apparent only during the last decade, and the identification of other peptide hormones as important factors in the homeostasis of body fluids may be anticipated. One likely candidate is atrial natriuretic peptide. Like oxytocin, ANP promotes Na$^+$ loss in urine; in fact, ANP may mediate the natriuretric effects of OT. Also, like oxytocin, ANP is secreted by neurons within the brain, and when administered directly into the cerebrospinal fluid, it inhibits an experimentally induced salt appetite in rats. Moreover, destruction of ANP receptors in the brain eliminates the inhibition of salt appetite caused by a NaCl load (Blackburn et al., 1995). Further work will be needed to understand the role of ANP-containing neurons in the central control of salt intake and the relationship of the neurons with other systems that participate in the regulation of water and NaCl intake.

Summary

Regulation of blood volume, like osmoregulation, is accomplished by a combination of physiological responses to hypovolemia, resulting in antidiuresis and antinatriuresis, and the complementary behavioral responses of increased water and NaCl intake. Cardiovascular baroreceptor cells detect hypovolemia and send neural signals to the brain stem, which communicates to the hypothalamus and forebrain structures mediating neurohypophyseal AVP and oxytocin secretion, as well as thirst and salt appetite. AII also appears to stimulate these responses while additionally supporting blood pressure as a potent vasoconstrictor. However, AII and hypovolemia both increase central oxytocin secretion, which mediates inhibition of NaCl intake, so for salt appetite to emerge this effect must be inhibited (as by osmotic dilution of body fluids, resulting from the renal retention of ingested water). The integration of these stimuli ensures that behavioral and physiological responses occur simultaneously, and their redundancy allows these vital regulatory processes to occur even when one stimulus is lost due to injury or disease. More generally, studies of water and NaCl ingestion provide insights into how the brain controls motivation, and how peptide and steroid hormones interact with neural signals in the control of behavior.

References

Blackburn, R. E., Samson, W. K., Fulton, R. J., Stricker, E. M., and Verbalis, J. G. (1995). Central oxytocin and atrial natriuretic peptide receptors mediate osmotic inhibition of salt appetite in rats. Am. J. Physiol. **269**, R245–R251.

Blackburn, R. E., Verbalis, J. G., and Stricker, E. M. (1992). Central oxytocin mediates inhibition of sodium appetite by naloxone in hypovolemic rats. Neuroendocrinology **56**, 255–263.

Curtis, K. S., Verbalis, J. G., and Stricker, E. M. (1996). Area postrema lesions in rats appear to disrupt rapid feedback inhibition of fluid intake. *Brain Res.* **726**, 31–38.

Egan, G., Silk, T., Zamarripa, F., Williams, J., Federico, P., Cunnington, R., Carabott, L., Blair-West, J., Shade, R., McKinley, M., Farrell, M., Lancaster, J., Jackson, G., Fox, P., and Denton, D. (2003). Neural correlates of the emergence of consciousness of thirst. *Proc. Natl. Acad. Sci. USA* **100**, 15241–15246.

Ferguson, A. V. and Renaud, L. P. (1986). Systemic angiotensin acts at subfornical organ to facilitate activity of neurohypophysial neurons. *Am. J. Physiol.* **251**, R712–R717.

Fitzsimons, J. T. (1961). Drinking by rats depleted of body fluid without increase in osmotic pressure. *J. Physiol. (Lond.)* **159**, 297–309.

Fitzsimons, J. T. (1969). The role of a renal thirst factor in drinking induced by extracellular stimuli. *J. Physiol. (Lond.)* **155**, 563–579.

Gilman, A. (1937). The relation between blood osmotic pressure, fluid distribution and voluntary water intake. *Am. J. Physiol.* **120**, 323–328.

Hoffman, G. E., Smith, M. S., and Verbalis, J. G. (1993). c-Fos and related immediate early gene products as markers of activity in neuroendocrine systems. *Front. Neuroendocrinol.* **14**, 173–213.

Huang, W., Sved, A. F., and Stricker, E. M. (2000). Water ingestion provides an early signal inhibiting osmotically stimulated vasopressin secretion in rats. *Am. J. Physiol.* **279**, R756–R760.

Johnson, A. K. and Buggy, J. (1978). Periventricular preoptic-hypothalamus is vital for thirst and normal water economy. *Am. J. Physiol.* **234**, R122–R125.

Kanter, G. S. (1953). Excretion and drinking after salt loading in dogs. *Am. J. Physiol.* **174**, 87–94.

Knepper, M. A. (1997). Molecular physiology of urinary concentrating mechanism: Regulation of aquaporin water channels by vasopressin. *Am. J. Physiol.* **272**, F3–F12.

Richter, C. P. (1936). Increased salt appetite in adrenalectomized rats. *Am. J. Physiol.* **115**, 155–161.

Robertson, G. L. (1986). Posterior pituitary. *In* "Endocrinology and Metabolism" (P. Felig, J. Baxter, and L. A. Frohman, eds.), pp. 338–385. McGraw-Hill, New York.

Robinson, A. G. (1985). Disorders of antidiuretic hormone secretion. *Clin. Endocrinol. Metab.* **14**, 55–88.

Robinson, M. M. and Evered, M. D. (1987). Pressor action of intravenous angiotensin II reduces drinking response in rats. *Am. J. Physiol.* **252**, R754–R759.

Simpson, J. B., Epstein, A. N., and Camardo, J. S., Jr. (1978). Localization of receptors for the dipsogenic action of angiotensin II in the subfornical organ of rat. *J. Comp. Physiol. Psychol.* **92**, 581–601.

Starling, E. H. (1896). On the absorption of fluids from the connective tissue spaces. *J. Physiol. (Lond.)* **19**, 312–326.

Stricker, E. M. (1969). Osmoregulation and volume regulation in rats: Inhibition of hypovolemic thirst by water. *Am. J. Physiol.* **217**, 98–105.

Stricker, E. M. (1976). Drinking by rats after lateral hypothalamic lesions: A new look at the lateral hypothalamic syndrome. *J. Comp. Physiol. Psychol.* **90**, 127–143.

Stricker, E. M. (1981). Thirst and sodium appetite after colloid treatment in rats. *J. Comp. Physiol. Psychol.* **95**, 1–25.

Stricker, E. M. and Verbalis, J. G. (1986). Interaction of osmotic and volume stimuli in regulation of neurohypophyseal secretion in rats. *Am. J. Physiol.* **250**, R267–R275.

Stricker, E. M. and Verbalis, J. G. (1988). Hormones and behavior: The biology of thirst and sodium appetite. *Am. Sci.* **76**, 261–267.

Stricker, E. M. and Verbalis, J. G. (1992). Ingestive behaviors. *In* "Behavioral Endocrinology" (J. B. Becker, S. M. Breedlove, and D. Crews, eds.), pp. 451–472. MIT Press, Cambridge, MA.

Thrasher, T. N., Keil, L. C., and Ramsay, D. J. (1982). Lesions of the laminar terminalis (OVLT) attenuate osmotically-induced drinking and vasopressin secretion in the dog. *Endocrinology (Baltimore)* **110**, 1837–1839.

Thrasher, T. N., Nistal-Herrera, J. F., Keil, L. C., and Ramsay, D. J. (1981). Satiety and inhibition of vasopressin secretion in dogs. *Am. J. Physiol.* **240**, E394–E401.

Verbalis, J. G., Mangione, M. P., and Stricker, E. M. (1991). Oxytocin produces natriuresis in rats at physiological plasma concentrations. *Endocrinology (Baltimore)* **128**, 1317–1322.

Verney, E. B. (1947). The antidiuretic hormone and the factors which determine its release. *Proc. Soc. (Lond.)* **B135**, 25–105.

Wilkins, L. and Richter, C. P. (1940). A great craving for salt by a child with cortico-adrenal insufficiency. *J. Am. Med. Assoc.* **114**, 866–868.

Suggested Readings

Denton, D. (1982). "The Hunger for Salt: An Anthropological, Physiological and Medical Analysis." Springer-Verlag, Berlin.

Fitzsimons, J. T. (1979). "The Physiology of Thirst and Sodium Appetite." Cambridge Univ. Press, Cambridge.

Gauer, O. H. and Henry, J. P. (1963). Circulatory basis of fluid volume control. *Physiol. Rev.* **43**, 423–481.

Guyton, A. C., Hall, J. E., Lohmeier, T. E., Jackson, T. E., and Manning, R. D., Jr. (1981). The many roles of the kidney in arterial pressure control and hypertension. *Can. J. Physiol. Pharmacol.* **59**, 513–519.

Ramsay, D. J. and Thrasher, T. N. (1990). Thirst and water balance. *In* "Handbook of Behavioral Neurobiology" (E. M. Stricker, ed.), Vol. 10, pp. 353–386. Plenum, New York.

Stricker, E. M. (2004). Thirst. *In* "Handbook of Behavioral Neurobiology" (E. M. Stricker and S. C. Woods, eds.), Vol. 14, pp. 505–543. Kluwer Academic/Plenum, New York.

Stricker, E. M. and Verbalis, J. G. (1990). Sodium appetite. *In* "Handbook of Behavioral Neurobiology" (E. M. Stricker, ed.), Vol. 10, pp. 387–419. Plenum, New York.

Verbalis, J. G. (1990). Clinical aspects of body fluid homeostasis in humans. *In* "Handbook of Behavioral Neurobiology" (E. M. Stricker, ed.), Vol. 10, pp. 421–462. Plenum, New York.

Wolf, A. V. (1958). "Thirst: Physiology of the Urge to Drink and Problems of Water Lack." Thomas, Springfield, IL.

Edward M. Stricker and Joseph G. Verbalis

Neuroendocrine Systems

THE HYPOTHALAMUS IS A NEUROENDOCRINE ORGAN

The hypothalamus, located at the most rostral region of the brain stem in the diencephalon, is a key center regulating numerous and diverse physiological functions, including growth, metabolism, stress responses, reproduction, osmoregulation, and circadian rhythms. All these functions are critically involved in maintaining the homeostasis of the organism and in coordinating the occurrence and timing of physiological functions with the appropriate environmental conditions. In order to accomplish this, the hypothalamus acts as an integrator, receiving converging inputs from virtually every sensory and autonomic system related to the internal and external environment of the organism. The hypothalamus responds rapidly to this convergent information by changing its output of neurotransmitters and neuropeptides, thereby affecting change at its targets within the central nervous system (CNS), the pituitary gland, and the bloodstream.

The hypothalamus contains several groups of cells that are defined as neuroendocrine, meaning that they have both neuronal and endocrine features. Neuroendocrine cells in the central nervous system have a neuronal phenotype, and release a peptide or neurotransmitter in response to a depolarizing stimulus. In this way, they are similar to other nonneuroendocrine neurons in the brain. However, neuroendocrine cells differ from other "traditional" neurons in their targets and modes of release. Instead of releasing their neuropeptide or neurotransmitter at another neuron, for example, via synaptic connections, these hypothalamic neuroendocrine cells release their neurotransmitter into a blood system. An endocrine gland is characterized by its direct release of a substance into a blood system. Thus, these neuroendocrine cells in the brain are true endocrine organs in the sense that their targets are a vascular system. Although the substances released by most neuroendocrine cells are peptides and neurotransmitters, they are also appropriately referred to as hormones because they are released into blood vessels to act remotely at a target organ. Thus, the molecules released by these hypothalamic neuroendocrine cells, to be discussed later in this chapter, often are referred to as neurohormones.

The primary target of hypothalamic neuroendocrine cells is the pituitary gland (Fig. 40.1). The pituitary gland comprises the anterior pituitary (adenohypophysis) and the posterior pituitary (neurohypophysis). Some species also possess an intermediate pituitary lobe. The hypothalamic cells that regulate the anterior pituitary are referred to as hypothalamic releasing or inhibiting cells, and are the major focus of this current chapter. These neurons have their cell somata in the hypothalamus, and they have a nerve process that extends to the median eminence at the base of the hypothalamus. The median eminence is highly vascularized by a blood system called the portal capillary plexus. Hypothalamic releasing/inhibiting hormones are released into the portal capillary vasculature, which then transports these neurohormones to the anterior pituitary gland.

The posterior pituitary gland also is regulated by hypothalamic cells, but the anatomical connectivity between the hypothalamus and the neurohypophysis differs from the hypothalamic-adenohypophysial system. Hypothalamic cells that regulate the neurohypophysis have their cell bodies in the hypothalamus, but their nerve terminals project directly into the posterior pituitary gland (Fig. 40.1). From there, the

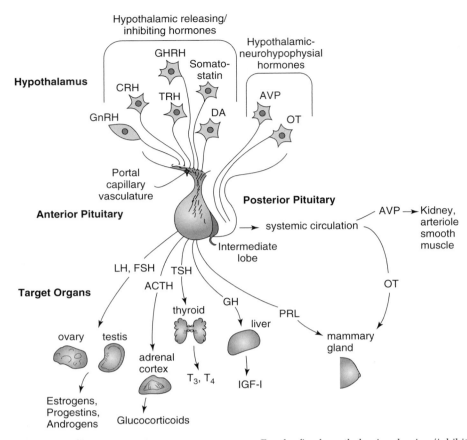

FIGURE 40.1 General features of brain-pituitary-target organ systems. For the five hypothalamic releasing/inhibiting hormonal systems that control reproduction, stress, metabolism, growth, and lactation, each system has a hypothalamic neurosecretory cell, a corresponding pituitary hormone, and a target organ and/or hormone. The hypothalamic-neurohypophysial hormones, AVP and OT, are shown at the right. Abbreviations: ACTH, adrenocorticotrophic hormone; AVP, arginine vasopressin; CRH, corticotropin-releasing hormone; DA, dopamine; FSH, follicle-stimulating hormone; GHRH, growth hormone-releasing hormone; GnRH, gonadotropin-releasing hormone; IGF-I, insulin-like growth factor I; LH, luteinizing hormone; OT, oxytocin; PRL, prolactin; TRH, thyrotropin-releasing hormone; TSH, thyroid-stimulating hormone; T_3, triiodothyronine; T_4, thyroxine.

neurohypophyseal hormones are released into the general circulatory system, and are transported to reach their targets elsewhere in the body.

The concept of a link between the hypothalamus and the pituitary gland was suggested and later demonstrated by Geoffrey Harris, who proved the existence of a tiny series of blood vessels leading from the median eminence at the base of the hypothalamus to the anterior pituitary gland (Harris, 1971). Harris proposed that the hypothalamus controls the release of anterior pituitary hormones via this vascular connection. Subsequent studies have shown that the hypothalamic–anterior pituitary connection is critical for maintaining numerous physiological functions involved in metabolism, growth, reproduction, stress responses, and lactation. The isolation and purification of the hypothalamic releasing hormones in the late 1960s by the laboratories of Roger Guillemin, Andrew Schally, and others led to an enormous increase in

knowledge of these molecules. Guillemin and Schally were awarded the Nobel prize in Physiology and Medicine in 1977 for these fundamental discoveries in neuroendocrinology.

HYPOTHALAMIC RELEASING/ INHIBITING HORMONES AND THEIR TARGETS

Overview of Hypothalamic–Anterior Pituitary–Target Organ Axis Systems

To date, five neuroendocrine systems have been identified involving hypothalamic–adenohypophysial–target organ regulation. These systems each are characterized by a primary hypothalamic hormone that either stimulates ("releasing" hormone) or inhibits ("inhibiting" hormone) and its corresponding anterior pituitary hormone(s) (Fig. 40.1). The hypothalamic releasing hormones are gonadotropin-releasing hormone (GnRH), corticotropin-releasing hormone (CRH), thyrotropin-releasing hormone (TRH), and growth-hormone releasing hormone (GHRH), which stimulate the reproductive, stress, thyroid, and growth axes, respectively. The growth axis also has a hypothalamic inhibiting hormone, somatostatin, and the lactational axis has only a known inhibiting hormone, dopamine, as no unique releasing hormone has been identified for this latter axis. Each of the neuroendocrine axes also has in common a secondary anterior pituitary hormone, or hormones, that is/are released into the general circulatory system. These hormones are luteinizing hormone (LH) and follicle-stimulating hormone (FSH) for the reproductive axis; adrenocorticotropic hormone (ACTH) for the stress axis; thyroid-stimulating hormone (TSH) for the thyroid axis; growth hormone (GH; also called somatotropin) for the growth axis; and prolactin (PRL) for the lactational axis. Finally, four of the neuroendocrine systems (reproduction, stress, thyroid, growth) have a tertiary hormone or hormones, produced by a target organ in the body that expresses receptors for the specific anterior pituitary hormone (Fig. 40.1). The fifth system, lactation, has as its target the mammary gland, whose output is milk, rather than secretion of a hormone into the blood system.

At the hypothalamic level, neuronal perikarya are localized in hypothalamic or preoptic nuclei that vary depending upon the system. The preoptic areas of the brain, just rostral to the hypothalamus, are functionally and neurochemically related to the hypothalamus. Although the precise anatomical localization of the somata of neuroendocrine cells within the hypothalamus and preoptic area may differ, each system projects neuroterminals to a common target, the median eminence, located at the floor of the hypothalamus just above the anterior pituitary gland (Fig. 40.2). The external zone of the median eminence contains the portal blood vessels into which the hypothalamic hormone is released. These blood vessels then carry the hypothalamic hormone to the anterior pituitary gland.

The anterior pituitary contains a heterogeneous group of cells that respond to a specific releasing or inhibiting hormone. The anterior pituitary cells are specialized, and each has specific receptors that bind a specific hypothalamic hormone. Thus, although all anterior pituitary cells are exposed to the same "cocktail" of hypothalamic releasing hormones via the portal capillary vasculature, the only cells that will respond to a hypothalamic hormone are those that express the appropriate receptors. The binding of the hypothalamic hormone to its receptor triggers a cascade of biosynthetic and secretory events, stimulating (or inhibiting) the synthesis and release of the pituitary hormone into the general circulatory system. The anterior pituitary gland, like the hypothalamus, is an endocrine gland, in that it releases its hormones into the circulatory system. Anterior pituitary hormones are transported throughout the general circulation to act remotely at organs in the body that are specific to each neuroendocrine axis.

The target organs of the neuroendocrine axes have specific receptors that bind the corresponding pituitary hormone. These target organs respond to the binding of these pituitary hormones to their receptors by increasing (or decreasing) synthesis and secretion of tertiary hormone, such as sex steroid hormones

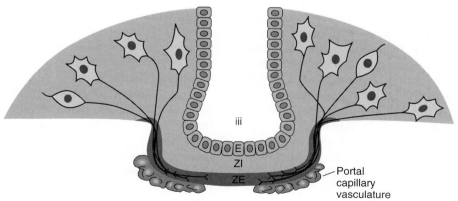

FIGURE 40.2 Depiction of a coronal section through the median eminence showing its three major zones and the typical projections of neuroendocrine cells to the external zone containing the portal capillaries leading to the anterior pituitary gland. The portal capillary system is shown. Abbreviations: iii, third ventricle; E, ependymal zone; ZI, internal zone; ZE, external zone.

(from the gonads), thyroxine and triiodothyronine (from the thyroid), insulin-like growth factor-I (from the liver), and glucocorticoids (from the adrenal cortex). It is the actions of these hormones at somatic targets that cause the ultimate physiological effect on the body such as growth, metabolism, mediation of stress, and reproductive function. In addition, each of these tertiary hormones feeds back at the level of the brain, hypothalamus, and/or pituitary gland to regulate its neuroendocrine axis. In this way, the delicate hormonal balance is maintained for each system.

The hypothalamic–pituitary target organ axes are not an independent series of parallel regulatory systems. These axes have considerable crosstalk with one another and can interact at multiple levels. For example, reproductive and stress axes regulate each other at specific hypothalamic neurons, both through direct actions at the CNS, as well as through feedback from the target steroid hormones. During stressful situations, elevated adrenal glucocorticoids will feedback to the brain and have actions not only at the CRH neurons, but at the GnRH neurons as well, thereby suppressing reproduction at a time when it is not optimal. Therefore these hypothalamic neurons serve as integrators of multiple, complex pieces of information about the homeostasis and environment of the organism.

The Hypothalamus Is Regulated by Inputs from the Central Nervous System and Feedback of Hormones from Target Organs

Hypothalamic-adenohypophysial axes often are referred to as hypothalamic–pituitary–target axes (the targets being the gonad, adrenal gland, thyroid gland, liver, mammary gland, and other somatic organs). However, although this description is accurate in that hypothalamic hormones are the primary regulators of each of these axes, the central nervous system plays a more complex role in the regulation of neuroendocrine function than simply the release of a hypothalamic hormone. For example, each hypothalamic releasing or inhibiting hormone neuron receives afferent inputs from other elements of the brain, such as neurons or glia, which regulate their functions. Additionally, feedback of hormones released by target organs may not occur directly or exclusively at the respective hypothalamic cells, and therefore feedback regulation of hypothalamic hormones may be mediated by other cells in the brain.

A good example is provided by the hypothalamic–pituitary–gonadal system that controls reproduction. Hypothalamic GnRH neurons are the primary activators of this axis. These cells stimulate the release of the gonadotropins, LH and FSH, from the anterior pituitary, and these molecules in turn target the gonads to stimulate the synthesis and release of sex steroid hormones, particularly estrogens, progestins, and androgens. GnRH neurons express one type of sex steroid hormone receptor, the estrogen receptor β (ERβ), but they are not believed to possess other types of sex steroid hormone receptors (Wintermantel et al., 2006). The presence of ERβ within GnRH neurons is not sufficient to explain the ability of GnRH neurons to respond to positive and negative feedback from estrogens. Therefore at least some of the feedback regulation of GnRH cells by these steroid hormones is probably mediated indirectly by other central nervous system afferents to GnRH neurons. Some of these afferents may be localized in the hypothalamus, whereas others may lie in extrahypothalamic regions, that express sex steroid hormone receptors. Thus, many researchers have begun to refer to the hypothalamic–pituitary–gonadal axis as the brain–pituitary–gonadal axis. This concept can be extended to the other neuroendocrine axes as well, as each is regulated by extrahypothalamic inputs, and steroid hormone feedback may occur directly on the hypothalamic cells, but may also be mediated by indirect inputs to these cells.

Characteristics of Neurohormone Release: Ultradian, Circadian/Diurnal, and Seasonal Rhythms

In considering the release of a hypothalamic hormone and its resulting effect on downstream targets, it is important to note the rhythmicity of this release. Hypothalamic hormones are released in a pulsatile manner, characterized by brief bursts of release, usually at intervals of 1–2 h. This rhythmic pattern also is referred to as ultradian (i.e., shorter than daily rhythms) or circhoral (approximately hourly intervals of release). The corresponding anterior pituitary cells respond to each pulse of a hypothalamic hormone with a corresponding pulse of its pituitary hormone shortly thereafter (Fig. 40.3). It has been speculated that the pulsatile release of hypothalamic hormones is necessary to prevent the desensitization of their receptors in the anterior pituitary gland. Indeed, infusion of exogenous hypothalamic hormone analogs in a continuous manner may result in an initial activation of pituitary hormone release, but subsequent refractoriness of the pituitary cell and overall decreases in release. In contrast, infusion of exogenous hypothalamic hormones in a pulsatile manner maintains the maximal pituitary sensitivity to its corresponding hypothalamic hormone.

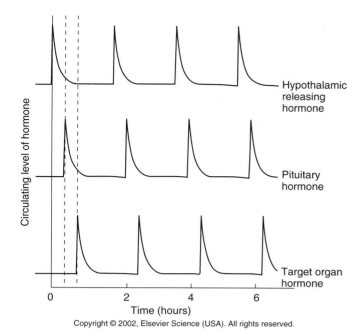

FIGURE 40.3 Representation of pulsatile hormone release in a neuroendocrine system. A representative hypothalamic-releasing hormone is released in discrete pulses at 1- to 2-h intervals. This causes pulses of the corresponding anterior pituitary hormone to occur shortly thereafter. The release of the pituitary hormone stimulates the target organ to release its hormones.

Along with these ultradian pulses of hypothalamic pulses are the circadian or diurnal rhythms of hormonal release that are characteristic of the neuroendocrine system. These 24-h rhythms of hormone release, which can occur independently of the light/dark cycle (circadian) or are entrained to the light/dark cycle (diurnal), are characteristic of neuroendocrine systems (Gore, 1998). Circadian rhythms are driven by a part of the hypothalamus called the suprachiasmatic nucleus (SCN), and the SCN makes projections within the hypothalamus to regulate neuroendocrine release in a 24-h cycle.

Cycles of hormonal release longer than 24h can be observed for the reproductive axis. This is discussed in more detail later, but in brief, female vertebrates often exhibit reproductive cycles, such as the four-day estrous cycles of rats or the 28-day menstrual cycles of many primates, including humans. Reproductive hormones undergo substantial fluctuations across these cycles. Even longer cycles are observed in the cases of seasonal breeders that may be sexually active during certain periods of the year, and as infrequently as once per year or even less. This seasonality enables the females of the species to coordinate the energetically demanding process of pregnancy, lactation, and rearing of the offspring with the appropriate environmental conditions.

THE HYPOTHALAMIC–ADENOHYPOPHYSIAL NEUROENDOCRINE SYSTEMS

Five neuroendocrine axes are involved in the control of metabolism, growth, reproduction, lactation, and stress. These systems are summarized in Figure 40.1, and each is discussed in detail next.

Metabolism: The Brain–Pituitary–Thyroid Axis

Overview and Characteristics

The proper regulation of metabolism is critical for the development and survival of the organism. The brain–pituitary–thyroid axis plays a critical role in the control of metabolic activity, stimulating an increase in metabolic processes in target cells. The hypothalamic hormone involved in this function is thyrotropin-releasing hormone (TRH), produced by a group of cells in a specific nucleus in the hypothalamus, the paraventricular nucleus (Fig. 40.4). As is the case for all hypothalamic releasing hormones, TRH neurons have long axons that terminate in the external zone of the median eminence (Fig. 40.2). This region is highly vascularized by the portal capillary system, which transports TRH to the anterior pituitary gland. There, TRH targets receptors on thyrotropes, which are cells that, when stimulated, produce the thyroid-stimulating hormone (TSH; Fig. 40.4). Thyrotropes constitute approximately 10% of the cells in the anterior pituitary gland. TSH is released from the anterior pituitary gland into the general circulation and acts at receptors in the thyroid gland, located in the neck. These cells then produce the thyroid hormones, thyroxine (T_4) and triiodothyronine (T_3), the latter being the more potent thyroid hormone, which are also released into the general circulation (Fig. 40.4).

Thyroid hormones act at target organs throughout the body to regulate cellular metabolism. T_3 and T_4 also feed back at the brain and anterior pituitary gland to regulate levels of TRH and TSH, respectively. The thyroid hormone axis is a classical negative feedback loop, as thyroid hormone feedback inhibits the secretion of these hypothalamic and pituitary hormones, thereby maintaining appropriate levels of free thyroid hormone in the bloodstream. Such precise regulation of thyroid hormone levels is critically important because elevated or depressed levels of this hormone can have substantial adverse consequences on the body, particularly during sensitive periods of development.

It is important to note the terminology of *free* versus *bound* hormone. When the thyroid hormones are

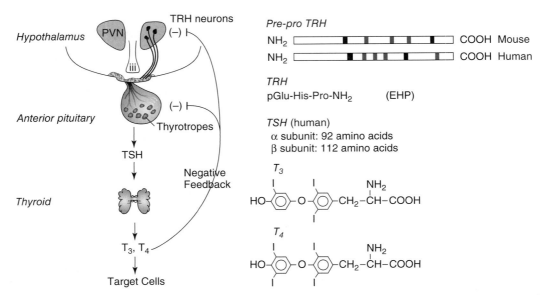

FIGURE 40.4 Left: Schematic representation of the brain–pituitary–thyroid axis showing TRH neurons in the paraventricular nucleus projecting their neuroterminals to the external zone of the median eminence. This projection occurs bilaterally, although only a unilateral projection is shown. TRH travels through the portal vasculature to cause the synthesis and release of TSH from thyrotropes. This molecule acts at the thyroid gland to cause the biosynthesis and release of the thyroid hormones, T_3 and T_4. These are released into the general circulation to affect metabolism. Negative feedback of T_3 and T_4 at the thyrotropes of the pituitary and at the TRH neurons of the PVN is indicated by the long arrows. Right: the pre-proTRH molecule of two representative species, mouse and human, is shown on top, with shaded bars indicating sites where multiple copies of the TRH tripeptide are cleaved from this precursor. Structural information about the TRH tripeptide sequence, the TSH protein, and T_3 and T_4 are provided. Abbreviations: PVN, paraventricular nucleus; iii, third ventricle; TRH, thyrotropin-releasing hormone; TSH, thyroid-stimulating hormone; T_3, triiodothyronine; T_4, thyroxine.

released into the circulation from the thyroid gland, a large portion bind to binding globulins. The free (or not protein-bound) thyroid hormone constitutes only <1% of all thyroid hormone circulating in the bloodstream. It is the free thyroid hormone that is responsible for the negative feedback regulation of TRH neurons and thyrotropes, as well as for actions at target tissues. Bound hormone is not immediately bioavailable but may provide a reservoir for thyroid hormones when they are needed.

Physiological Functions

Thyroid Hormone Is Critical for Cellular Metabolism. The brain–pituitary–thyroid axis is involved in metabolism, protein synthesis, and development of the brain. Thyroid hormone receptors are located throughout the brain, highlighting their importance in central nervous system development and function. They are also found in peripheral and autonomic nervous systems, pituitary gland, bone, muscle, liver, heart, testis, lung, placenta, and other tissues. The major function of the thyroid hormone is the regulation of basal metabolic rate and calorigenic responses. In the target cells, thyroid hormones regulate ATP levels produced by mitochondria and alter protein

synthesis. The maintenance of thyroid hormone levels is extremely important for adult metabolic activity, and thyroid hormone abnormalities in the developing organism can have catastrophic consequences. Thyroid hormone deprivation in the fetus causes permanent nervous system abnormalities and substantial effects on growth and bone development that may be reversed or mitigated only if the thyroid hormone is replaced.

Pulsatile and Circadian Release of TRH and TSH. The pituitary hormone TSH, and presumably the hypothalamic hormone TRH (which is much more difficult to measure due to the inaccessibility of the portal capillary system), is released in a pulsatile manner, with pulses occurring at approximately 90- to 180-min intervals in humans (Gore, 1998). Studies on the pulsatile release of TSH in humans demonstrate that its levels also change across the 24-h period, peaking from the evening through 0400h, decreasing through noon, and remaining low until the beginning of the next cycle in the evening. The increase in TSH release during the evening is referred to as the nocturnal TSH surge. This appears to be a true circadian rhythm, and it is driven by the suprachiasmatic nucleus of the hypothalamus (Chapter 41).

Structures and Properties of TRH, TSH, Thyroid Hormone, and Their Receptors

TRH and the TRH Receptor. TRH is a tripeptide with the sequence pGlu-His-Pro-NH₂ (Fig. 40.4). This tiny peptide, when bound to its receptor on the pituitary thyrotrope, rapidly and potently stimulates the pulsatile release of TSH. As is true for other neurohormones (Chapter 7), TRH is cleaved and processed from a prohormone in which its sequence is expressed in multiple copies (Yamada *et al.*, 1990; Fig. 40.4). The resulting TRH tripeptide is released from the cell to act on its target, the TRH receptor.

The TRH receptor, localized in the anterior pituitary on thyrotropes, is a G-protein-coupled membrane receptor (Gershengorn, 1993). Binding of TRH to its receptor causes the activation of membrane-bound phospholipase C and the hydrolysis of phosphatidylinositol 4,5-biphosphate (PIP_2) to inositol 1,4,5-triphosphate (IP_3) and diacylglycerol (DAG). This results in calcium mobilization and the phosphorylation or changes in the concentrations of nuclear proteins that interact with the genes involved in the synthesis of TSH and causes an increase in TSH gene transcription.

TSH and the TSH Receptor. TSH is an ~30-kDa glycoprotein hormone, comprising an α and β subunit, that are synthesized from two different genes (Fig. 40.4). The α subunit of TSH is identical to that of the gonadotropins, LH and FSH (see later). The β subunit confers the unique identity and biological specificity to TSH, LH, and FSH. The common glycoprotein α subunit is 92 amino acids in humans. The unique TSH β subunit gene encodes a 112 amino acid glycoprotein in humans (Szkudlinski *et al.*, 2000).

The TSH receptor is a seven-transmembrane domain G-protein coupled receptor located predominately in the thyroid gland (Graves and Davies, 2000). Binding of TSH to its receptor activates cAMP generation and stimulates the production and release of the thyroid hormone in the thyroid gland.

Thyroid Hormones. The two major thyroid hormones are T_4 and T_3, of which T_3 has greater biological activity (Fig. 40.4). Thyroid hormones contain two phenolic rings, which make these molecules lipophilic, facilitating their transport through the lipophilic membranes of target cells. Some plasma membrane transport molecules assist with this process because the thyroid hormones possess a polar alanine sidechain that appears to impede the process of passive diffusion. Within the cell, thyroid hormones interact with the thyroid hormone receptor. The thyroid hormones contain three (T_3) or four (T_4) iodide molecules. T_3 binds to the thyroid hormone receptors in target

tissues and also mediates negative feedback effects on TRH cells in the hypothalamus and pituitary. Although T_4 is released from the thyroid gland in greater quantities than T_3, most T_4 is converted to T_3, as catalyzed by two enzymes, type I and type II deiodinases. Type I deiodinase is found mostly in peripheral tissues and the pituitary gland, and type II deiodinase is found in the brain, pituitary gland, and brown fat. Because an important component in the synthesis of thyroid hormones is iodine, an iodine-deficient diet results in the impairment of thyroid hormone metabolism, leading to developmental abnormalities.

With respect to the negative feedback regulation by thyroid hormone (Fig. 40.4), thyroid hormone receptors have been localized within TRH neurons in the paraventricular nucleus (Segerson *et al.*, 1987). Thus, these actions can be mediated directly on TRH neurons in the brain. In addition, the expression of the thyroid hormone receptor is abundant in the anterior pituitary gland, particularly thyrotrope cells, indicating direct feedback action of the thyroid hormone on this organ as well.

Along with its regulation by TSH, the thyroid gland also receives direct neuronal inputs from the autonomic nervous system. Fibers from the sympathetic and parasympathetic nervous systems penetrate the thyroid gland and terminate around arterioles, capillaries, and, less frequently, venules and near follicular cells, providing additional mechanisms for regulation of thyroid hormone output.

The Thyroid Hormone Receptor Family. The thyroid hormone receptor is a member of the nuclear hormone superfamily. These are nuclear transcription factors that bind to a regulatory element on the DNA to activate or inhibit gene transcription. Several thyroid hormone receptor isoforms have been identified. These are termed thyroid hormone receptors (TR) α and β, which are derived from different genes and can be alternatively spliced. Thus, each TR subtype has a splice variant, termed $\alpha1$, $\alpha2$, $\beta1$, and $\beta2$. $\alpha1$ and $\beta2$ are most abundant in the hypothalamus and are found in more than 80% of TRH neurons in the PVN (Lechan *et al.*, 1994). As is the case for most nuclear hormone receptors, binding of the thyroid hormone to its receptor causes the heterodimerization of the receptors with another member of the nuclear receptor superfamily, RXR, and facilitates the binding of the dimerized receptor-ligand complex to the promoter of the target gene, thereby modulating gene transcription.

Neuroanatomy of the TRH System

TRH neurons are concentrated in the paraventricular nucleus (PVN) of the hypothalamus (Fig. 40.4). This hypothalamic nucleus is localized at the dorsal region of the third ventricle and can be divided into a magnocellular division (large neurons) and a parvocellular division (small- to medium-sized neurons). TRH neurons exist exclusively in the parvocellular region of the PVN and are concentrated particularly along the medial and periventricular parts of the parvocellular PVN. Other non-TRH neurons also are localized in this region, including corticotropin-releasing hormone, neurotensin, galanin, enkephalin, and vasoactive intestinal peptide. It has been reported that there are also nonneuroendocrine TRH cells localized outside the PVN that do not project to the median eminence. Because these play nonneuroendocrine roles in nervous system function they are beyond the scope of discussion of this chapter.

Disruptions in the Thyrotropic Axis

Thyroid hormone excess or insufficiency can have profound effects on the central nervous system, particularly during development. Thus, a normal or euthyroid tone is necessary for the health and survival of the organism. Excessive thyroid hormone, or hyperthyroidism, can result in symptoms such as tremor, nervousness, tremulousness, insomnia and sleep disorders, and impairments in memory and concentration. More profound effects include neuropsychiatric syndromes, seizure, chorea, and even coma. Deficiencies of the thyroid hormone, called hypothyroidism, can cause problems such as sleep apnea, hypothermia, hypoventilation, neuropsychiatric syndromes, peripheral neuropathy, cerebellar ataxia, and coma. During chronic thyroid hormone deficiency, fluids can accumulate and result in puffiness of the hands, feet, face, and tongue and cause speech impairment. This is called myxedema. Hypothyroidism generally is treated by thyroid hormone replacement.

When an organism is exposed to insufficient thyroid hormone during fetal or early neonatal development, it can develop cretinism, characterized by severe deficits in mental, neurological, and physical development. Most of these deficiencies can be mitigated or even reversed by newborn infant screening and thyroid hormone replacement.

Growth: The Somatotropic Axis

Overview and Characteristics

Somatic growth is an obvious characteristic of the development of vertebrate organisms. The somatotropic axis has three levels: the primary hypothalamic hormones, of which there are two, growth hormone-releasing hormone (GHRH) and a growth hormone-

inhibiting hormone, called somatostatin; the secondary pituitary hormone, called growth hormone (GH) or somatotropin; and the tertiary target organ, in this case the liver, which produces insulin-like growth factor I (IGF-I) in response to the GH stimulus. A special characteristic of the somatotropic axis is that its pituitary hormone, GH, as well as its tertiary hormone, IGF-I, can exert growth-promoting effects alone or in combination. In fact, GH and IGF-I act synergistically with each other to affect somatic growth. However, both of these hormones play important roles, as mutations resulting in the absence of either GH or IGF-I result in dwarfism.

Unlike most other neuroendocrine systems, the somatotropic axis has two known hypothalamic hormones that play important but opposing roles (Fig. 40.5). One hormone, growth hormone-releasing hormone (GHRH), is stimulatory to the release of the pituitary GH. The other hypothalamic hormone, somatostatin, inhibits GH release. These hormones are both released into the portal capillary vasculature to affect the synthesis and release of GH, from specialized cells in the anterior pituitary gland called somatotropes, constituting approximately 30% of pituitary cells. The GH binds to its receptors in the liver to stimulate the release of the tertiary hormone, IGF-I. Both GH and IGF-I have somatotropic actions on target organs and tissues undergoing growth, regeneration, or metabolic homeostasis, such as bone and muscle.

Growth hormone and IGF-I both appear to be involved in the feedback regulation of the somatotropic axis (Fig. 40.5). In the case of GHRH, GH acts as a classical negative feedback molecule, downregulating the synthesis and secretion of GHRH. By contrast, GH *stimulates* gene expression of the inhibitory hypothalamic hormone somatostatin. In the absence of GH, somatostatin mRNA levels are downregulated. IGF-I also acts as a feedback regulator of the somatotropic axis, negatively regulating GHRH levels and stimulating somatostatin gene expression. As shown in Figure 40.5, this is a highly complicated and tightly regulated feedback loop.

Physiological Functions

The Somatotropic Axis Controls Somatic Growth and Development. GH plays its major role in neonatal and postnatal growth and is particularly important in the pubertal growth spurt. Its anabolic activities are critical for metabolic homeostasis. The growth hormone stimulates protein synthesis, increases lipolysis, influences carbohydrate metabolism, enhances glucose uptake and utilization, and has lipolytic effects. At the systems level, it stimulates the growth of organs, such as the heart and kidney, and plays a role in hormone

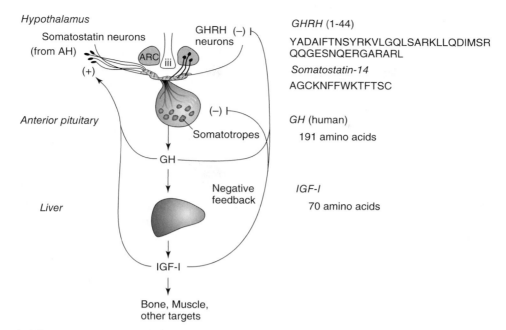

FIGURE 40.5 Left: Schematic representation of the somatotropic axis showing GHRH neurons in the arcuate nucleus and somatostatin neurons in the periventricular region of the anterior hypothalamus. Both sets of neurons project their neuroterminals to the external zone of the median eminence. This occurs bilaterally for both hypothalamic hormones, although only a unilateral projection is shown. GHRH and somatostatin travel through the portal vasculature to stimulate or inhibit, respectively, the synthesis and release of GH from somatotropes. This molecule acts at the liver to cause the biosynthesis and release of IGF-I. Both GH and IGF-I are released into the general circulation to affect somatic growth, and they both feed back in a negative manner at the somatotropes of the pituitary and the GHRH neurons of the arcuate nucleus. IGF-I and growth hormone also feed back to stimulate somatostatin, which in turn negatively regulates the somatotropic axis. Right: The amino acid sequences for GHRH and somatostatin are shown, and structural information about GH and IGF-I is provided. Abbreviations: AH, anterior hypothalamus; ARC, arcuate nucleus; GH, growth hormone; GHRH, growth hormone-releasing hormone; IGF-I, insulin-like growth factor I; iii, third ventricle.

production, skeletal growth and maturation, and immune function. Many of its effects are sexually dimorphic, as are the distribution of its receptors and other proteins involved in this axis.

Pulsatile and Diurnal Release of GHRH and Somatostatin. Both GHRH and somatostatin are released from hypothalamic nerve terminals into the portal circulation in a pulsatile manner, similar to other neurohormones. It might be predicted that these neurohormones might be released out of phase, but in fact, their release often is coordinated. Pulses of GHRH appear to drive pulses of growth hormone, which occur at intervals of about 1–3 h.

The growth hormone is released in a diurnal pattern. The 24-h cycle of GH release is strongly associated with sleep, with highest levels occurring shortly after the onset of slow-wave sleep, and most GH secretion occurring at night in humans. Thus, this is not a true circadian rhythm but rather a diurnal pattern of release (Gore, 1998).

Structures and Properties of GHRH, Somatostatin, Growth Hormone, IGF-I, and Their Receptors

Growth Hormone-Releasing Hormone and Somatostatin. GHRH is derived from a 108 amino acid precursor protein, called preproGHRH. Following a multistep processing procedure involving cleavage by endopeptidases, mature GHRH molecules are produced (Fig. 40.5). In some species, there are two GHRHs with similar biological activity; other species have a single GHRH peptide. In humans, the active GHRH is 44 amino acids (Clynen *et al.*, 2004).

GHRH binds to its receptor on somatotropes in the anterior pituitary gland to affect the release of growth hormone. The GHRH receptor is a member of the G-protein-coupled receptor family (Mayo *et al.*, 1995). Binding of GHRH to its receptor triggers the G protein, in this case G_s, to trigger a cascade of intracellular events involving cyclic AMP (cAMP) as the second messenger. This causes calcium mobilization, the phosphorylation of proteins that are involved in the synthesis of growth hormone, and the release of growth hormone from secretory granules. The synthesis of the growth hormone that is triggered by GHRH probably involves the cAMP-induced phosphorylation of the cAMP response element-binding protein (CREB), which binds to an enhancer element in the promoter of the somatotropin gene to stimulate its transcription.

The somatostatin gene contains two exons that are transcribed into a 0.85-kb mRNA. This mRNA is translated into a 116 amino acid precursor protein, preprosomatostatin. This precursor is cleaved by endopeptidases and is a tissue-specific process. In the hypothalamus, somatostatin is cleaved into two biologically active forms, somatostatin-14 and somatostatin-28, the latter having an N-terminal extension of the somatostatin-14 molecule (Fig. 40.5).

The binding of somatostatin to its receptor causes a cascade of events involved in the inhibition of growth hormone release. To date, there are five known somatostatin receptors (SS-Rs), with different properties and affinities. All these receptors involve an inhibitory G protein (G_i), which reduces cAMP and causes cell hyperpolarization and a diminution of Ca^{2+} influx into cells.

Growth Hormone. Growth hormone, or somatotropin, is a 191 amino acid protein that contains two disulfide bonds that give it a three-dimensional conformation that is necessary for its biological activity. Release of the growth hormone is triggered by GHRH and is inhibited by somatostatin. The growth hormone plays a negative feedback role in the somatotropic axis, feeding back at the level of the GHRH neuron in the hypothalamus to inhibit GHRH mRNA levels and synthesis and stimulating somatostatin gene expression and release. Not only does the regulation of growth hormone involve the balance between GHRH and somatostatin, but it also involves other neurotransmitters and neuroactive substances, such as dopamine, vasoactive intestinal peptide, pituitary adenylate cyclase-activating peptide, TRH, galanin, neuropeptide Y, motilin, and interleukins. In addition, other hormones, such as the thyroid hormone, glucocorticoids, sex steroid hormones, and metabolic fuels, can influence the release of growth hormone, highlighting the interwoven nature of the different neuroendocrine systems.

The growth hormone is released into the general circulation, where approximately 50% of this molecule is transported in association with a growth hormone-binding protein that is the same as the extracellular domain of the growth hormone receptor. This molecule and the growth hormone receptor (GH-R) are alternative splice products of the same precursor RNA

into two distinct mRNAs: one for the growth hormone-binding protein and one for the GH-R.

The GH-R is a member of the cytokine receptor family (Behncken and Waters, 1999). Binding of growth hormone to its plasma membrane receptor causes receptor dimerization and triggers a cascade of protein phosphorylations mediated by the JAK-STAT system, the activation of nuclear transcription factors, and the stimulation of somatotropin-regulated genes.

IGF-I. A major target for the growth hormone is the IGF-I system. IGF-I is a growth factor that exerts effects on neurite outgrowth and nervous system development. In addition, it is an important mediator of the somatotropic axis and, in fact, was originally referred to as somatomedin. Both GH and IGF-I individually have similar effects on somatic growth and development; however, the two hormones act much more efficiently in combination. In addition, it appears that the growth-promoting effects of IGF-I are particularly important in the central nervous system. This is consistent with the observation that along with its production by the liver, IGF-I is also produced in the brain, and it is probably this central IGF-I that is responsible for many of its neurotrophic effects.

Like the growth hormone, IGF-I plays a feedback role in the inhibition of GHRH biosynthesis (Fig. 40.5). This may involve central and not peripheral IGF-I, as most animal studies have demonstrated that only centrally but not peripherally administered IGF-I has this effect. At the level of the pituitary, IGF-I appears to be the major regulator of negative feedback of growth hormone synthesis and release. Somatotropes contain IGF-I receptors, and IGF-I inhibits growth hormone gene expression and release under basal and GHRH-stimulated conditions. IGF-I can also act as a negative feedback regulator of GHRH neurons, as it can cross the blood–brain barrier to a limited extent. Moreover, IGF-I is synthesized in the central nervous system, including the hypothalamus, and therefore local IGF-I can exert an influence on GHRH neurons. It also has been shown that IGF-I can stimulate somatostatin release, albeit at higher doses than those necessary to inhibit GHRH release, suggesting that IGF-I may be more involved in the suppression of GHRH than the enhancement of somatostatin levels.

The IGF-I gene has been cloned in a number of species and encodes a precursor molecule of 105 amino acids that is cleaved into the mature IGF-I of 70 amino acids. IGF-I in the circulation binds to several binding proteins, the major one being IGF-I binding protein 3 (IGFBP-3). Free IGF-I is more biologically active than bound IGF-I, and therefore IGFBP-3 levels are an important regulator of the actions of IGF-I. At its target tissues, IGF-I binds to its receptor, a member of the receptor tyrosine kinase family, which utilizes the MAP kinase system as its second messenger system.

Neuroanatomy of GHRH/Somatostatin Systems

Most GHRH-containing neurons are located in the basomedial hypothalamus in the arcuate nucleus, lateral to the floor of the third ventricle (Fig. 40.5). These neurons project axons to the external zone of the median eminence, from which GHRH is released into the portal capillary system (Fig. 40.2). GHRH neurons can also send axons to other hypothalamic regions, and this may be involved in the regulation of feeding behavior by GHRH. GHRH neurons receive inputs from other neuronal systems in the central nervous system, including α-adrenergic, dopaminergic, serotonergic, and enkephalinergic neurons.

Somatostatin cell bodies have a very different localization from the GHRH system. Unlike GHRH neurons, whose perikarya are concentrated in a discrete nucleus, somatostatin somata are more dispersed throughout the central nervous system. Perikarya of the somatostatin neurons involved in the regulation of growth hormone release are found primarily in the periventricular region of the anterior hypothalamus (Fig. 40.5). As is the case for other hypothalamic hormones, their axon terminals project to the external zone of the median eminence. Somatostatin neurons also make projections to GHRH neurons, and it is believed that this is one mechanism for the inhibition of GHRH neurons by somatostatin cells.

Inputs from metabolically active substances affect the somatostatin system. Amino acids are thought to stimulate, and fatty acids and glucose to inhibit, somatostatin secretion. There are also a number of neurotransmitters involved in the regulation of somatostatin levels. Catecholamines such as dopamine and norepinephrine are implicated in this process, as are acetylcholine and GABA. Neuropeptides, including substance P, neurotensin, and bombesin, can alter somatostatin levels, and factors involved in glucoregulation, such as glucose and glucoregulatory enzymes and hormones, modify activity of the somatostatin system.

Disruptions of the Somatotropic Axis

Somatic growth and development are affected by disruptions of the somatotropic axis, and the consequences are dependent on when during development the system is affected. In early development, growth hormone deficiency can result in an extremely short stature, with adults sometimes reaching a maximal height of less than 1 m. Growth hormone deficiency is generally treated with growth hormone replacement using recombinant growth hormone. In some cases, GHRH, rather than growth hormone, is replaced in a pulsatile manner. Dwarfism is sometimes caused by mutations in the IGF-I system, or in the GH receptor. Laron dwarfism, characterized by mutations in the GH receptor gene, results in very low IGF-I levels, but GH levels are high, probably due to reduced negative feedback from the low IGF-I concentrations.

If GH is oversecreted or dysregulated during development, the consequence can be gigantism, resulting in an extremely large stature. When abnormal GH stimulation occurs after adulthood, this results in acromegaly. In adulthood, only certain tissues remain responsive to growth hormone stimulation, including the jaws, nose, orbital ridges, fingers, elbows, and knees, and these tissues are stimulated to grow while other tissues are not, causing deformations in proportions and appearance.

Stress: The Brain–Pituitary–Adrenal Axis

Overview and Characteristics

In order to survive, an animal must be able to adapt to stress, which can be manifested in a variety of forms. These stressors can include the acute stress of escaping predation or the chronic stress of a disease, the social situation, or social status. Thus, a stress is an internal or external cue that disrupts the homeostatic status of the animal.

The neuroendocrine system involved in mediating the stress response is the brain–pituitary–adrenal axis. The hypothalamic releasing hormone controlling this axis is corticotropin-releasing hormone (CRH), produced in highest levels in the medial parvocellular part of the paraventricular nucleus (mpPVN; Fig. 40.6). CRH somata send neural projections to the external layer of the median eminence (Fig. 40.2), typical of all hypothalamic neuroendocrine cells. From here, CRH is released from the external zone of the median eminence and targets the corticotropes of the anterior pituitary gland. Additionally, arginine vasopressin (AVP), another hypothalamic hormone produced in the magnocellular elements of the PVN, acts at pituitary corticotropes to exert synergistic effects. The corticotropes have receptors for and respond to both CRH and AVP with the synthesis and release of adrenocorticotropic hormone (ACTH). ACTH is released from the pituitary gland into the general circulation, where it acts at receptors on the adrenal cortex to stimulate the production and release of glucocorticoids. The

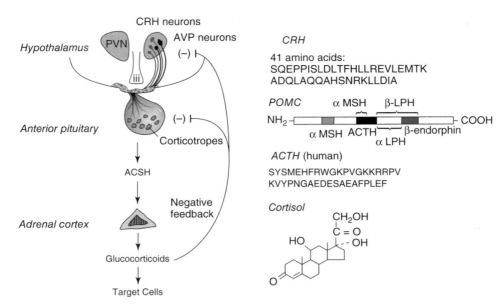

FIGURE 40.6 Left: Schematic representation of the brain–pituitary–adrenal axis showing CRH neurons in the paraventricular nucleus projecting their neuroterminals to the external zone of the median eminence. This occurs bilaterally, although only a unilateral projection is shown. CRH travels through the portal vasculature to cause the synthesis and release of ACTH from corticotropes. This molecule acts at the adrenal cortex to cause the biosynthesis and release of glucocorticoids. These are released into general circulation to mediate the stress response, and they feed back in a negative manner at the corticotropes of the pituitary and the CRH neurons of the PVN (indicated by long arrows). The AVP neurons of the PVN are also shown, as they act together with CRH to regulate the stress axis. Right: The amino acid sequences of CRH and ACTH are shown, as is the structure of the glucocorticoid, cortisol. The precursor molecular from which ACTH is cleaved is shown, with many of its hormone products (including ACTH, β-endorphin, β-LPH, γ-LPH, α-MSH, γ-MSH) indicated. Abbreviations: ACTH, adrenocorticotropic hormone; AVP, arginine vasopressin; CRH, corticotropin-releasing hormone; LPH, lipotropic hormone; MSH, melanocyte-stimulating hormone; POMC, proopiomelanocortin; PVN, paraventricular nucleus; iii, third ventricle.

feedback of glucocorticoids onto the brain and pituitary negatively regulates the synthesis of CRH and ACTH: therefore, this neuroendocrine system is a classical negative feedback loop (Fig. 40.6).

Physiological Functions

Glucocorticoids Mediate Stress Responses and Affect Cellular Metabolism. Corticosteroids are necessary for normal fetal development. Enzymes in the liver, gastrointestinal system, lungs, and adrenal medulla are stimulated by glucocorticoids. There, glucocorticoids play important roles in the regulation of glycogen utilization and storage, with glucocorticoids promoting the production of glycogen synthase. The stress axis is subject to developmental and age-related regulation due to changes in afferent connections to the hypothalamus, glucocorticoid and mineralocorticoid receptor-binding changes, and alterations in the negative feedback of glucocorticoids on CRH neurons.

Glucocorticoids exert effects on the cardiovascular system, causing elevations in heart rate and blood pressure and alterations in blood flow. They act in close coordination with the autonomic nervous system to exert their effects. In general, their role is to mobilize energy stores and to improve cardiovascular tone. Glucocorticoids may also modulate immune responses and inhibit cytokine production, thereby playing a role in the inhibition of the inflammatory response. Glucocorticoids can interact with hormones produced by other neuroendocrine axes, such as those involved in thyroid function, reproductive function, and growth. Chronic stress can suppress the growth axis and reproductive function and cause hyperthyroidism (O'Connor *et al.*, 2000).

Pulsatile and Circadian Release of Stress Hormones. Stress hormones exhibit both pulsatile and circadian cycles (Gore, 1998). When the hypothalamic nucleus responsible for circadian rhythms, the SCN, is lesioned, the 24-h pattern of CRH release is obliterated. During the 24-h day, glucocorticoid levels peak during the early morning (~0700h) and reach a nadir during the afternoon in humans. A similar pattern of release is seen in rats, with highest levels when the animals are awakening (in the case of rats, which are a nocturnal species, this is the evening) and lowest levels in the morning. The adrenal hormone synthesized is species-dependent; in humans, the major glucocorticoid is cortisol, and in rats and mice, it is corticosterone. However, these glucocorticoids serve similar functions and both hormones target glucocorticoid receptors in their respective species.

Structures and Properties of CRH, ACTH, Glucocorticoids, and Their Receptors

CRH. Corticotropin-releasing hormone is a 41 amino acid peptide expressed in the parvocellular region of the PVN in the hypothalamus (Fig. 40.6). It is also found in other CNS regions and outside of the brain in adrenal gland, placenta, ovary, immune tissues, and at inflammatory sites. The CRH receptor is linked to the activation of adenylate cyclase. The CRH receptor is also found in the central nervous system, with highest concentrations in the olfactory bulb, cerebellum, cerebral cortex, and striatum and lower concentrations in the spinal cord, hypothalamus, medulla, midbrain, thalamus, pons, and hippocampus (DeSouza, 1987).

ACTH. ACTH derives from the cleavage of a precursor protein, proopiomelanocortin (POMC), in the corticotropes (Fig. 40.6). POMC is cleaved to a number of smaller peptides, including β-endorphin, β-lipotropin (LPH), γ-LPH, a-melanocyte-stimulating hormone (MSH), β-MSH, and γ-MSH. ACTH is the primary effector of the stress axis, and its release into the general circulation from the pituitary corticotrope has effects on the target organ, the adrenal cortex, to affect the production of glucocorticoids.

The ACTH receptor is a member of the melanocortin family of receptors (Naville *et al.*, 1999). It is a seven-transmembrane domain G-protein-coupled receptor expressed primarily in adrenal cortex. The binding of ACTH to its receptor activates adenylate cyclase and cAMP production.

Glucocorticoids. Glucocorticoids are members of the steroid hormone family. As is the case for sex steroid hormones (discussed later), they derive from a cholesterol precursor. Following a series of enzymatic reactions, the major corticosteroids, such as cortisol (in humans; Fig. 40.6) or corticosterone (in rats), are synthesized.

The glucocorticoid receptor (GR) is widely distributed in the central nervous system, including the hypothalamus and, more specifically, the CRH cells of the PVN. Thus, glucocorticoid negative feedback can occur directly on the hypothalamic cells driving the stress axis. GRs are also abundant in the hippocampus, in pyramidal cells, and in dentate granule cells. The hippocampus appears to play a role in mediating the stress response, and there are multiple redundant pathways between the hippocampus and the PVN.

GRs are found in many other brain regions such as the olfactory cortex, cerebral cortex, amygdala, septum, thalamus, midbrain raphe nuclei, cerebellum, and brain stem (Morimoto *et al.*, 1996). The widespread distribution of this receptor suggests important and diverse effects of glucocorticoids in the central nervous system.

Glucocorticoids can also bind to the mineralocorticoid receptor (MR), the primary ligand of which is aldosterone. Moreover, the brain MR has a higher (~10-fold higher) affinity for glucocorticoids than the GR. Both of these receptors are members of the steroid hormone superfamily. They are thought to reside in the cytoplasm of target cells in the unliganded state. When bound to their ligands, GRs or MRs translocate into the nucleus, where they bind to specific sites on the promoter of the responsive gene. Thus, GR and MR are transcription factors. Their efficacy in stimulating or inhibiting gene transcription is modulated by their binding to other molecules, such as heat shock proteins, which can alter the translocation of the receptor–ligand complex to the nucleus, or transcriptional coactivators, which modulate the complexes associated with the gene (Chapters 3 and 4).

Chronic exposure to glucocorticoids can cause damage to the brain, which is particularly apparent in the hippocampus, a brain region abundant in GRs. It has been observed that chronic corticosteroids changes the morphology of CA3 dendritic fields and may lead to neurodegeneration and eventually cell death. Nevertheless, whereas chronic high levels of glucocorticoids may be toxic, a lack of glucocorticoids can also be detrimental. In the dentate gyrus, maintenance of the dentate granule cell number appears to require glucocorticoids, and adrenalectomy results in the loss of granule cells. Moreover, neurogenesis in the adult dentate gyrus also appears to be glucocorticoid dependent. There are functional correlates to the effects of glucocorticoids on the hippocampus, with these hormones playing roles in learning and memory. Acute stress can interfere with the ability of an organism to learn, and chronic stress is correlated with deficits in spatial learning.

Neuroanatomy of the CRH System

CRH neuronal perikarya are localized primarily in the medial parvocellular part of the paraventricular nucleus (mpPVN; Fig. 40.6) and project neuroterminals to the external zone of the median eminence (Fig. 40.2). The PVN is an important integrative region of the hypothalamus, as it summates inputs from other regions throughout the brain including the brain stem, the midbrain/pons groups, the limbic system, the circumventricular organs, and other hypothalamic nuclei.

Inputs from the brain stem are catecholaminergic pathways involved in the transmission of visceral information. The catecholamine system plays an important role as a stress mediator, possibly as an autonomic nervous system neurotransmitter. Noradrenergic projections to the PVN arise from the nucleus of the solitary tract (A2 cell group), the A1 cell group of the ventrolateral medulla, and the locus coeruleus. Inputs from the midbrain may involve the midbrain raphe cell groups, which are serotonergic in nature. The midbrain periaqueductal gray and pontine central gray regions relay sensory information to the PVN, as do caudal thalamus and cholinergic brain stem regions (laterodorsal tegmental nucleus and the pedunculopontine nucleus). These inputs appear to play a role in relaying somatic and special sensory information. Inputs to the PVN from the limbic system arise from the prefrontal cortex, septum, amygdala, and hippocampus, and a notable projection to the PVN is from the bed nucleus of the stria terminalis to the PVN. These inputs relay information on cognition and emotion. The circumventricular organs (subfornical organ, OVLT) convey information to the PVN from blood-borne chemosensory signals. The major neurotransmitters involved in this function are GABA and angiotensin II. Finally, the hypothalamus itself signals to the PVN, and these connections may provide signals about the motivational state of the animal or relay other stress-specific signals. However, the functions of these interhypothalamic projections are not well understood.

Disruptions in the Stress Axis

The negative feedback loop of CRH–ACTH–glucocorticoids is kept in a delicate balance. In response to stress, there is a large increase in the activity of the stress axis, but the system is downregulated rapidly by negative feedback from the glucocorticoids to the brain and pituitary gland, causing it to return the output of the stress axis to basal low levels.

Overactivity of the stress axis can increase susceptibility to infection and tumors and reduce inflammatory and autoimmune disease (O'Connor et al., 2000). For example, Cushing's disease is characterized by hypercortisolemia and abnormal secretion and/or bioactivity of ACTH. The high levels of cortisol suppress CRH release and cause alterations in immune function. People with Cushing's disease have an increase in appetite and can experience changes in fat distribution. However, the brain–pituitary–adrenal axis can adapt to overactivity, and thus the immune system suppression that occurs in Cushing's disease is milder than that which would be predicted by the high levels of circulating cortisol. Elevations in activity of the stress axis can also impair memory, probably through actions of glucocorticoids on the hippocampus.

Underactivity of the stress axis can cause the opposite effects of hyperactivity, namely resistance to infection and tumors but increased susceptibility to inflammatory and autoimmune disease. In the case of adrenocortical insufficiency, there is generally a failure of the adrenal cortex to respond to ACTH. Addison's disease is characterized by an insufficient release of glucocorticoid secretion, and its symptoms include muscle weakness and fatigue, psychological symptoms, gastrointestinal symptoms, and changes in skin pigmentation. Fortunately, glucocorticoid and mineralocorticoid replacement can help mitigate these symptoms.

Reproduction: The Brain–Pituitary–Gonadal Axis

Overview and Characteristics

Reproduction is critical to the survival of the species. In vertebrate organisms, the reproductive axis comprises (1) the gonadotropin-releasing hormone (GnRH) neurons, also called luteinizing hormone-releasing hormone (LHRH) neurons, of the hypothalamus and preoptic area; (2) the gonadotropes of the anterior pituitary gland; and (3) the gonads (ovary in females and testis in males) (Fig. 40.7). GnRH neurons of the hypothalamus and preoptic area project their neuroterminals to the median eminence, where the GnRH peptide is released in a pulsatile manner (Gore, 2002). GnRH acts at the gonadotropes of the anterior pituitary to cause the production of two hormones collectively called the gonadotropins: luteinizing hormone (LH) and follicle-stimulating hormone (FSH) (Fig. 40.7). These hormones in turn are released in a pulsatile manner into the general circulation and bind to receptors on specialized cells in the ovaries or testes to stimulate steroido genesis, spermatogenesis in males, and oogenesis in females. The gonads of both sexes respond to this activation by synthesizing and secreting the sex steroid hormones, estrogens, progestins, and androgens. Although the dominant steroid hormones produced by the ovaries are estrogens and progestins, ovaries also produce appreciable amounts of androgens. The testes produce androgens at higher levels than females, but they also secrete estrogens and progestins. Thus, there is no such thing as a "male" or "female" sex steroid hormone. Rather, what differs between the sexes are the concentrations of these hormones.

The sex steroid hormones, once released into the circulatory system, act at receptors that are located throughout the body to exert their classical effects on

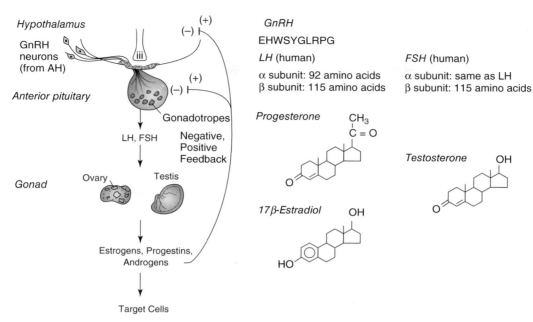

FIGURE 40.7 Left: Schematic representation of the brain–pituitary–gonadal axis showing GnRH neurons in the preoptic area/anterior hypothalamus projecting their neuroterminals to the external zone of the median eminence. This occurs bilaterally, although only a unilateral projection is shown. GnRH travels through the portal vasculature to cause the synthesis and release of the gonadotropins, LH and FSH, from gonadotropes. These molecules act at the gonads (ovary and testis) to cause the biosynthesis and release of sex steroid hormones. These are released into general circulation to affect secondary sex characteristics and act at other somatic tissues. Sex steroid hormones feed back negatively at the brain and pituitary to regulate biosynthesis of GnRH and the gonadotropins in males and females. In addition, prior to ovulation, these steroids exert positive feedback in females. Right: The decapeptide structure of GnRH, general structural features of the gonadotropins, and the structures of progesterone, 17β-estradiol, and testosterone are shown. Abbreviations: AH, anterior hypothalamus; FSH, follicle-stimulating hormone; GnRH, gonadotropin-releasing hormone; LH, luteinizing hormone; iii, third ventricle.

secondary sex characteristic development and on adult reproductive function. In addition, sex steroid hormones act on receptors in the brain, including the hypothalamus, to exert diverse effects on the CNS. One of the major effects of sex steroid hormones is as a feedback modulator of the hypothalamic GnRH neurons. However, along with the traditional negative feedback exerted in neuroendocrine systems, in females, ovulation is controlled by *positive* feedback from sex steroid hormones. This concept of negative/positive feedback regulation is discussed later, and is unique to the reproductive neuroendocrine system of females.

Physiological Functions

Activity of the Reproductive Axis Varies during the Life Cycle. Not surprisingly, functions of the reproductive axis change during the life cycle of the organism, as reproductive function varies with developmental stage (Plant, 1988). During the late embryonic/early postnatal period, as the GnRH system finishes its neuroanatomical development, there is increased activity of the brain–pituitary–testicular axis in males and, to a lesser extent, the brain–pituitary–ovarian axis in females. Shortly after birth, the reproductive system becomes quiescent, possibly due to the development of inhibitory inputs to GnRH neurons or the lack of stimulatory inputs to this system. This quiescent period, sometimes called the "prepubertal hiatus," is characterized by low GnRH, gonadotropin, and steroid hormone release.

At the onset of puberty, the GnRH system becomes activated, probably due to the removal of inhibitory afferents to GnRH neurons, dominated by GABAergic neurons, and the ability of stimulatory afferents to GnRH neurons to exert a stimulatory tone on GnRH regulation. To date, numerous substances have been implicated in the activation of GnRH neurons at the onset of puberty, including glutamate (through the NMDA receptor), kisspeptin, norepinephrine, neuropeptide Y, and growth factors such as transforming growth factor-α and IGF-I. Once mature reproductive function has been attained (ovulation in females, viable sperm in ejaculates of males), the pulsatile release of GnRH and its regulation of the reproductive axis continues for much of the life cycle. In many species, females exhibit reproductive cycles (estrous cycles, menstrual cycles) of varying lengths, such as the four-day estrous cycles of small rodents and the 28-day menstrual cycles of humans and many nonhuman primates.

In human females but not most other species, menopause, the cessation of menstrual cycles and lack of ovulation, signals the end of the reproductive life cycle. The most obvious characteristic of menopause is a loss of ovarian follicles, called follicular atresia. Although many other mammals undergo reproductive senescence, characterized by a lack of spontaneous reproductive cycles, this is not always due to follicular atresia. Instead, in these species, such as rodents, reproductive senescence is probably due to the loss of drive from hypothalamic GnRH neurons and/or changes in pituitary gonadotropin responses to GnRH rather than overt ovarian changes. In humans, it is quite possible that there are also hypothalamic and pituitary changes that may play a role in the menopausal process that are only just beginning to be understood.

Pulsatile, Diurnal, and Seasonal Release of GnRH and LH/FSH. GnRH is released in a pulsatile manner from neuroterminals in the external zone of the median eminence (Fig. 40.2). Virtually every pulse of GnRH causes a pulse of LH (e.g., Fig. 40.3) and, to a slightly lesser extent, FSH, although some "silent" pulses of GnRH (that occur in the absence of the corresponding release of the gonadotropins) can be observed. Although the pulsatile release of LH appears to be driven entirely by pulses of GnRH, as is evidenced by the concordance of these pulses, the regulation of pulsatile FSH also involves the gonadal peptide inhibin. Inhibin is a glycoprotein produced in the gonad that selectively acts at the pituitary gland to suppress FSH release.

The mechanisms by which GnRH pulses are generated are not entirely understood, as these neurons are widely dispersed through the hypothalamus and preoptic area; although contacts between GnRH somata have been observed, these are not abundant. Recent evidence suggests that GnRH dendrites may make anatomical contacts. It is also possible that GnRH neurons coordinate their pulses via some signals at the level of the median eminence where the GnRH neuroterminals converge. Finally, other afferent systems contacting and regulating GnRH neurons may release their neurotransmitters in a pulsatile manner, and this may be a way of coordinating the widely scattered GnRH perikarya to release pulses of GnRH in a coordinated manner.

A pulsatile manner of GnRH release is necessary for the maintenance of pituitary gonadotropin function. When GnRH is infused continuously, the GnRH receptor in the pituitary is downregulated and gonadotropin secretion is inhibited. If exogenous GnRH is infused in a pulsatile manner, the GnRH receptor is upregulated on the gonadotropes. The pulsatile release of GnRH is more or less constant in males, with pulses occurring approximately every 1–2h. However, in females, the nature of GnRH pulses varies across the reproductive cycle. In primates that have a 28-day menstrual cycle, LH pulses (measured as proxy for GnRH pulses) occur approximately once per hour during the early follicular phase, but slow to a frequency of once every 90min later in the follicular phase as estrogen negative feedback increases. Just prior to ovulation, a large increase in GnRH release can be measured, although it is still controversial whether discrete pulses continue to be superimposed on this "GnRH surge." Nevertheless, it is likely that there is a change in the mode of GnRH release during the preovulatory period.

The pulsatile release pattern of GnRH, LH, and FSH is superimposed upon a diurnal pattern of the release of these hormones (Gore, 1998). Thus, the timing of the preovulatory LH surge in rodents is coordinated with the 24-h light/dark cycle (i.e., it is a diurnal but not a circadian rhythm). Although in women these diurnal rhythms are not as obvious as those in rats, it appears that the release of gonadotropins has a weak diurnal rhythm in humans. Men have also been reported to have nocturnal increases in reproductive hormone release. Interestingly, the diurnal rhythm of reproductive hormone release is regulated developmentally. Prior to the onset of puberty, there is little day/night difference in GnRH, LH, or FSH release. At the onset of puberty, a diurnal rhythm develops, characterized by nocturnal increases in the release of these hormones. This nocturnal increase is most pronounced late in puberty, and once adult reproductive function is attained, the rhythm diminishes. The mechanism for, and the function of, the pubertal diurnal rhythm of GnRH and gonadotropins currently is not understood but is one of the most robust diurnal rhythms observed in all the neuroendocrine systems.

The reproductive axis can exhibit rhythms even longer than those entrained to the 24-h clock. Many species are seasonal breeders, in which reproductive processes are coordinated to a time of the year when the likelihood of the survival of the offspring is highest. Seasonal breeding animals such as sheep have an almost completely inactive reproductive axis during the nonbreeding season, with very low GnRH release, low basal and GnRH-stimulated gonadotropin release, and a lack of ovulation. In many of these species, day length is what drives the seasonal breeding cycles, although this varies among the species. For hamsters, short nights stimulate the reproductive system, whereas short nights inhibit the reproductive axes of sheep.

Structures and Properties of GnRH, Gonadotropins, Sex Steroid Hormones, and Their Receptors

GnRH. GnRH is a 10 amino acid peptide that is synthesized from a 92 amino acid precursor molecule (Fig. 40.7). During cleavage of the prohormone, GnRH is produced together with a 56 amino acid peptide referred to as GnRH-associated peptide (GAP). Both GnRH and GAP are released from neuroterminals of these cells into the portal vasculature; however, whereas GnRH is a strong stimulator of LH and FSH release from pituitary gonadotropes, GAP has only a weak ability in this regard. GAP also is able to inhibit prolactin secretion; nevertheless, a specific physiological role for GAP is still unknown.

The GnRH peptide is evolutionarily conserved. All mammals studied to date, with the exception of the guinea pig, have an identical GnRH decapeptide in the hypothalamus-preoptic area. Nonmammalian vertebrates have decapeptides similar to the mammalian GnRH molecule, and the conservation is maintained all the way across the evolutionary spectrum to the jawless fish, which have a GnRH molecule that is 60 to 70% structurally homologous with the mammalian GnRH decapeptide (Morgan and Millar, 2004).

Unlike other neuroendocrine cells, GnRH neurons of most vertebrates have an unusual embryonic origin, deriving from cells that are first detected in the olfactory placode (Wray, 2002). During embryonic development, GnRH neurons migrate into the brain, reaching their final destinations in the preoptic area and hypothalamic regions during mid- to late embryogenesis. Also, unlike most other neuroendocrine peptide systems, GnRH neurons are relatively scattered through these regions and are not tightly concentrated in a single discrete nucleus.

GnRH binds to its receptor in pituitary gonadotropes to activate the synthesis and release of the gonadotropins. The GnRH receptor is a member of the G-protein coupled receptor family, containing seven transmembrane domains. The G protein involved in this function is probably G_q and/or G_{11}. The binding of GnRH to its receptor appears to activate several second messenger systems, including the protein kinase A and protein kinase C pathways, as well as the phospholipase C pathway (Sealfon *et al.*, 1997).

Gonadotropins. The gonadotropins, LH and FSH, originally were named after their functions in the female ovary, namely their ability to cause luteinization of the follicle (hence LH) and to promote follicular development (hence FSH). However, these same gonadotropins are synthesized and released from the anterior pituitary gland of males and play a critical role in the regulation of testicular function. Thus, in the testis, LH is primarily responsible for steroidogenesis, and FSH for spermatogenesis.

LH and FSH are glycoprotein hormones (Fig. 40.7) and are in the same family as TSH. These three molecules have a common α-subunit, but vary in the β-subunit. They are glycosylated polypeptides, and the α and β-subunits are linked covalently. Both LH and FSH are approximately 28 kDa. The common α-subunit is a 92 amino acid molecule, the LH-β molecule is 115 amino acids, and the FSH-β molecule is 115 amino acids in humans. These hormones bind to their respective LH or FSH receptors, which are G-protein coupled receptors found in target tissues of the ovary or testis.

Sex Steroid Hormones. Sex steroid hormones are synthesized from the precursor cholesterol in a multistep process. Cholesterol first is converted to pregnenolone, which is converted to progesterone (Fig. 40.7). This hormone is a major end point for biosynthesis in gonadal tissues and exerts potent effects on the body and brain. Progesterone serves as a precursor for the synthesis of testosterone (Fig. 40.7) following conversion first to 17α-hydroxyprogesterone and then androstenedione. Testosterone is the dominant steroid hormone in males, although it is also found in females. It serves as a precursor via the enzyme 5α-reductase for 5α-dihydrotestosterone, which is responsible for many of the masculinizing effects of steroids in males. Testosterone is also the direct precursor via the enzyme p450 aromatase for 17β-estradiol (Fig. 40.7), which is the primary estrogen responsible for sexual behavior and physiology in females, and which also plays roles in masculine physiology and behavior. The sex steroid hormones bind to their receptors, members of the nuclear hormone superfamily. The binding of a steroid hormone to its receptor, and subsequent receptor dimerization, allows these transcription factors to bind to specific sites on the promoter of the target gene, called androgen-response elements (AREs), estrogen-response elements (EREs), and progesterone-response elements (PREs). This enables the receptor-ligand complexes to induce or repress gene transcription in their target tissues.

Sex steroid hormones are synthesized and secreted from the ovaries or testes to regulate the expression of secondary sex characteristics and to maintain adult reproductive function. In the Leydig cells of the testis, LH released from the pituitary gland is responsible for testosterone biosynthesis. Together with FSH, testosterone acts to coordinate and stimulate spermatogenesis in testicular seminiferous tubules. The testosterone that is released from the testis acts at secondary sex targets such as the internal and external genitalia (prostate, seminal vesicles, penis) to cause their development and maintain their functions. In addition, testosterone acts at muscle to increase muscle mass, causes the growth of facial and body hair, and enlarges the epiglottis, which is responsible for the deepening of the voice. In the ovary, FSH stimulates follicular development during each ovarian cycle, and LH is the primary stimulator of ovulation. Steroidogenesis in the ovary is also stimulated by FSH and LH. The resulting ovarian hormones, primarily 17β-estradiol and progesterone, are responsible for the development of female secondary sex characteristics such as the genitalia and mammary tissue.

Sex steroid hormone feedback in the male is accomplished by the usual neuroendocrine negative feedback mechanism (Fig. 40.7). Pulses of GnRH cause pulses of the gonadotropins, which in turn are responsible for pulses of testosterone. Feedback of testosterone onto the brain causes an inhibition of hypothalamic GnRH neurons, and testosterone also feeds back negatively onto the pituitary gonadotropes. At the pituitary gland, sex steroid hormones appear to act primarily to decrease the sensitivity of pituitary gonadotropes to GnRH, resulting in a decrease in LH pulse amplitude.

In the female, the ovarian steroid hormones estradiol and progesterone can act at the hypothalamus to provide negative feedback. During most of the reproductive cycle, these hormones feed back at the brain and the pituitary gland to suppress the release of GnRH and the gonadotropins, respectively. However, during the mid- to late follicular stage in primates, or on the day of proestrus in rats with a four- to five-day estrous cycle, feedback effects of estrogens on GnRH release become positive. The stimulatory effects of estradiol on the hypothalamus result in the occurrence of the "preovulatory GnRH surge," which in turn triggers a preovulatory LH surge, that triggers the final maturation of the ovum in the follicle and results in ovulation and the subsequent development of the corpus luteum. After ovulation, feedback of ovarian steroid hormones upon the hypothalamus and pituitary reverts to the negative mode until the next cycle.

Unlike the other neuroendocrine systems, the primary feedback site of gonadal steroid hormones does not appear to be the GnRH neurons. Although GnRH neurons express the ERβ, they do not express appreciable levels of ERα, androgen receptors (AR), or progesterone receptors (PR). Therefore, it is thought that other neural cells that make afferent inputs to GnRH neurons, and which express nuclear sex steroid receptors, may mediate effects of sex steroid feedback onto GnRH cells. The mediation of gonadal steroid feedback by intermediary cells in the CNS, primarily those expressing the ERα, may also help to explain the "switch" between negative and positive feedback onto the GnRH neurons in females (Wintermantel et al., 2006). For example, some afferent inputs to GnRH neurons that inhibit GnRH release may exert a dominant tone during most of the reproductive cycle and be responsible for negative feedback. Other afferent inputs that stimulate GnRH release may be regulated by higher levels of estrogens and underlie the positive feedback response to ovarian steroids that results in the preovulatory GnRH surge.

Neuroanatomy of the GnRH System

GnRH neurons are not localized in a discrete hypothalamic nucleus, but are relatively widely dispersed across a continuum of the hypothalamus and preoptic area (Fig. 40.7). Although the rostral-caudal extent of the localization of GnRH perikarya can vary among species, in general, most GnRH cells are found in the preoptic area, septal nuclei, diagonal band of Broca, and anterior hypothalamus (particularly basal regions). All the neuroendocrine GnRH neurons project a process to the median eminence, where the GnRH decapeptide is released in a pulsatile manner. It should be noted that other groups of GnRH neurons have been detected in other, nonneuroendocrine regions such as olfactory regions and midbrain, but their functions are not well understood. It is speculated that they may play roles in the links between pheromones and reproduction, or between reproductive physiology and behavior. The GnRH neurons involved in neuroendocrine control that are described in this chapter are also referred to as GnRH-1 cells, to differentiate them from the GnRH cells in nonneuroendocrine regions, deemed GnRH-2 and GnRH-3 neurons.

The morphology of the hypophysiotropic GnRH-1 neurons is somewhat different from other neuroendocrine neurons, possibly due to differences in the cells' embryonic origin in the olfactory placode. GnRH neurons are oval or fusiform in shape and extend a single axon to the median eminence. They also receive relatively little synaptic input on their perikarya, although recent evidence suggests they may receive afferent input on their dendrites. Nevertheless, GnRH neurons are functionally regulated by numerous neurotransmitters and neurotrophic factors, and this is probably related to the integrative nature of the GnRH neuron. In order for reproduction to be successful, the organism must be able to assess the conditions of internal and external environment to determine that conditions are optimal. The numerous inputs to GnRH neurons from other sensory and neuroendocrine cells enables these key reproductive neurons to evaluate the organism's health and fitness and the external conditions (e.g., season, food availability, the presence of predators or a conspecific mate) and to coordinate the timing of reproduction, which is quite costly, with the optimal external and internal circumstances. Thus, GnRH cells are regulated by inputs from neurotransmitters such as glutamate, GABA, catecholamines and monoamines, neuropeptides such as kisspeptin, CRH, β-endorphin, vasopressin, neurotensin, neurotrophic factors, and many other substances. In this way, GnRH neurons receive cues regarding nutritional status, thermogenesis and exercise, stress, and social cues and can integrate these inputs to coordinate the timing of reproduction with the appropriate stimuli.

Disruptions in the Reproductive Axis

There are two natural genetic mutations lacking normal GnRH systems. In humans, Kallmann's syndrome is characterized by the failure of GnRH neurons to migrate properly during embryonic development. Although these patients develop essentially normally in most other ways (except that they are anosmic due to abnormalities of the olfactory placode from which both GnRH neurons and olfactory cells derive), they never enter puberty and are infertile. Replacement of pulsatile GnRH to these individuals can stimulate the onset of puberty and maturation of the secondary sex characteristics and fertility. The hypogonadal (hpg) mouse has a defect in the GnRH gene. Similar to their human Kallmann's syndrome counterparts, these animals never undergo reproductive development and can be "rescued" by GnRH replacement.

Other defects of the reproductive system involve the timing of the onset of puberty. In idiopathic central precocious puberty, GnRH is released far earlier than is normal, for example, in patients as young as 1 year of age. Although the mechanism by which the GnRH release is stimulated precocially is unknown, this results in activation of the release of pituitary gonadotropins, stimulation of the gonads, and the early development of secondary sexual characteristics. Treatment with long-lasting GnRH agonists, which initially stimulate but rapidly down-regulate pituitary GnRH receptors, cause a downregulation of the reproductive axis and can halt pubertal development until the normal timing of puberty occurs. In the case of delayed puberty, the normal pubertal release of GnRH does not occur, and exogenous GnRH can be infused in a pulsatile manner to cause the stimulation of puberty in these individuals.

Lactation and Maternal Behavior

Overview and Characteristics of the Lactotrophic Axis

The pituitary hormone prolactin is a major regulator of lactation in mammals. However, unlike the other pituitary hormones, a unique releasing neuropeptide hormone for prolactin has not been identified in the hypothalamus, although there is a prolactin-inhibiting factor, dopamine (Fig. 40.8). Prolactin is synthesized in pituitary cells called lactotropes. Its release into the circulatory system from the anterior pituitary gland causes milk synthesis and secretion into the alveoli of the mammary glands in response to the suckling stimulus. Prolactin, together with oxytocin, which is involved in milk ejection (see later), is responsible for the maintenance of lactation.

It is generally accepted that hypothalamic dopamine plays a major role in the inhibition of prolactin release, and dopamine is probably the prolactin-inhibiting factor (Ben-Jonathan et al., 1989; Fig. 40.8). Dopamine is abundant in the hypothalamus and is found in high concentrations in the median eminence, where it can be released to influence the function of the anterior pituitary gland via the portal capillary system. Evidence for a negative influence of the hypothalamus on prolactin secretion was provided by studies demonstrating that experimental isolation of the pituitary gland from the hypothalamus resulted in a large increase in basal prolactin secretion. Moreover, drugs that decrease dopamine levels in the pituitary portal circulation increase prolactin secretion, and application of exogenous dopamine can inhibit endogenous prolactin release. It was also reported that dopamine receptors are present on pituitary lactotropes, and finally, that changes in dopamine concentrations in the portal capillary system correlate with changes in prolactin release under physiological conditions.

Regarding the stimulation of prolactin release by a hypothalamic factor(s), though many molecules have been proposed, none appears to be a unique prolactin-releasing hormone. TRH, vasoactive intestinal polypeptide, and oxytocin can stimulate prolactin release but they do not appear to have the one-to-one relationship between a true hypothalamic releasing factor and its adenohypophysial hormone. Thus, the search for a unique hypothalamic-releasing hormone for prolactin is ongoing.

Physiological Functions of the Lactotrophic Axis

Role of Prolactin in Reproductive Physiology and Lactation. The release of prolactin varies by age, sex, and reproductive status. Levels of prolactin are higher in females than males, probably due to the higher

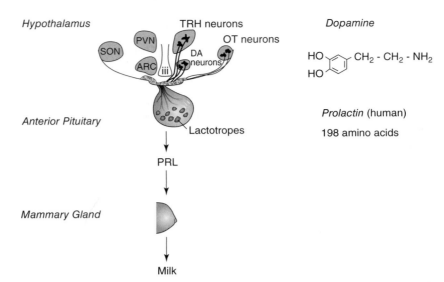

FIGURE 40.8 Left: Major regulatory pathway of prolactin secretion from anterior pituitary lactotropes. Prolactin release is inhibited by tuberinfundibular dopaminergic neurons in the arcuate nucleus. The release of prolactin from the pituitary stimulates the mammary gland to produce and release milk. Other hypothalamic factors shown, TRH and OT, are thought to regulate the prolactin system, although their roles are less well defined than that of dopamine. In all cases a unilateral projection is shown, although this regulation occurs bilaterally. Right: The structure of dopamine is shown, and general structural information about the PRL sequence provided. Abbreviations: ARC, arcuate nucleus; DA, dopamine; OT, oxytocin; PRL, prolactin; PVN, paraventricular nucleus; SON, supraoptic nucleus; TRH, thyrotropin-releasing hormone; iii, third ventricle.

estrogen levels in females, which stimulate prolactin release. During reproductive cycles in females, prolactin levels vary, with levels being higher during the luteal than the follicular phase in humans. In rats, prolactin levels peak shortly before the preovulatory LH surge that causes ovulation, a time when estrogen levels are high. Prolactin levels also increase during pregnancy, decrease just before birth, and increase again as a response to suckling.

The role of prolactin in lactation is its best-studied and probably most important function. In all mammals, prolactin acts on the mammary gland to prepare and maintain its secretory activity. It causes the growth of ductal and lobuloalveolar tissues and causes increases in mRNA levels and translation of the milk protein, casein. Prolactin release is stimulated directly by suckling and results in the synthesis of milk. In addition, there is a relationship between suckling and the inhibition of reproductive function, as suckling can delay implantation of fetuses conceived during a postpartum estrus or cause long periods of anovulation. Prolactin may also play a role in maternal behavior, although this is not well understood.

Prolactin can influence the reproductive axis through its actions on the ovary. There, prolactin causes increases in LH receptors in the corpus luteum, progesterone synthesis by the corpus luteum, and high- and low-density lipoprotein binding in membranes of the corpus luteum. These functions may be related to uterine endometrial development and implantation of the blastocyst in rats, although it is not clear if this is the case in humans.

Other functions of prolactin include osmoregulation, primarily in fish and amphibians. The expression of prolactin receptors in kidney, gut, skin, and urinary bladder is consistent with this role. Prolactin receptors have a widespread distribution throughout the body of many species. Predictably, they are abundant in mammary gland, but they are also found in other tissues involved in reproduction, as well as the liver, kidney, gastrointestinal tract, and brain. Thus, it is likely that prolactin may play roles in these tissues that are independent of lactation or reproductive function.

Pulsatile and Circadian Release of Prolactin. Like the other pituitary hormones, prolactin is released from the anterior pituitary in a pulsatile manner. Its release is entrained to the 24-h clock, similar to other hormones. This appears to be a diurnal rhythm, as it is driven by sleep cues rather than the circadian clock, similar to its related hormone, growth hormone (Gore, 1998).

Structures and Properties of Prolactin and Its Receptors

Prolactin is a member of the somatotropin (growth hormone) family of molecules. Its gene shares considerable homology with the growth hormone gene. The prolactin protein is a 23-to 24-kDa protein containing three disulfide bridges, giving it a unique structure. There are multiple forms of the prolactin receptor, and its expression is widespread throughout the body. Isoforms of the prolactin receptor range in size from 329 to 591 amino acids and they have structural overlap with cytokine receptor family members. Prolactin receptors are found in the mammary gland, liver, kidney, adrenals, gonads, uterus, placenta, prostate, seminal vesicles, lymphoid tissues, intestine, choroid plexus, hypothalamus, and other tissues.

Neuroanatomy of the Hypothalamic Dopamine System

The major hypothalamic population of dopamine neurons believed to regulate prolactin release is called the tuberoinfundibular dopaminergic system, also referred to as the A12 cell group. Their perikarya lie in arcuate nuclei (Fig. 40.8), and their neuroterminals project to the median eminence in the external zone, the latter projection being similar to that of other hypothalamic releasing and inhibiting hormones.

At the anterior pituitary gland, dopamine binds to the D2 subtype of dopamine receptor on lactotropes to cause its inhibitory effects on the synthesis and release of prolactin. The binding of dopamine to the D2 receptor causes a decrease in cAMP concentrations in lactotropes and causes an inhibition of prolactin gene transcription. Another second messenger system, the Ca^{2+}/phosphokinase C system, may also play a role in this process, although this latter mechanism is not well understood.

As discussed earlier, prolactin release is regulated by neurotransmitters other than dopamine. The hypothalamic hormone thyrotropin-releasing hormone (TRH), the primary function of which is the regulation of the thyroid hormone axis, also has been shown to stimulate prolactin release. Another neuropeptide, vasoactive intestinal peptide (VIP), can stimulate prolactin release, and blockade of the VIP system with

antisera inhibits basal prolactin secretion. In addition, oxytocin neurons regulate the prolactin system, possibly through direct actions on lactotrope cells (Fig. 40.8).

Disruptions in the Prolactin System

Hyperprolactinemia is the elevated secretion of prolactin. This is a common human pituitary disorder and can be a result of disrupted dopamine transmission, which normally would suppress prolactin release. Hyperprolactinemia can result from pituitary adenomas that secrete prolactin. As a result of hyperprolactinemia, inappropriate lactation can occur, as can be the case with drugs that affect the dopamine system (neuroleptics, dopamine receptor blockers, antidepressants, antihypertensives, oral contraceptives).

THE HYPOTHALAMIC-NEUROHYPOPHYSIAL SYSTEMS

Although the major focus of this chapter is the regulation of anterior pituitary function by hypothalamic releasing/inhibiting factors, following is a brief discussion of the hypothalamic neurons that innervate the posterior gland, specifically, arginine vasopressin (AVP) and oxytocin (OT) neurons. Further discussion of these hormones is found in Chapters 39 and 7, respectively.

Oxytocin and AVP

The OT and AVP Genes and Peptide Structures

The oxytocin and AVP neuropeptides are synthesized from genes localized on the same chromosome, but they are in a "head to head" organization separated by a domain called the intergenic region. Therefore, these genes are transcribed in opposite directions (Fig. 40.9; Young and Gainer, 2003). The AVP and OT

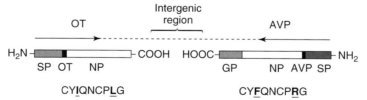

FIGURE 40.9 Structural organization of the AVP and OT genes. The RNA sequences are transcribed off of a single gene but in opposite directions, indicated by arrows. Each gene is made up of three exons that encode a signal peptide, the neurohormone (either AVP or OT) and a neurophysin that may be involved in chaperoning the neuropeptides. AVP also contains an additional sequence for a glycoprotein. The regions are separated by an intergenic sequence. The nonapeptide sequence of each neurohormone is shown below their respective genes, with differences indicated by underlines. Abbreviations: AVP, arginine vasopressin; GP, glycoprotein; NP, neurophysin; OT, oxytocin; SP, signal peptide. Modified from Young and Gainer (2003).

genes are structurally similar, comprising three exons that encode a signal peptide, the neuropeptide (either AVP or OT), and a neurophysin whose function is not entirely known but may be involved in chaperoning the neuropeptides as they are processed in secretory vesicles (Fig. 40.9; Burbach *et al.*, 2001). AVP and OT are highly homologous nonapeptides (Fig. 40.9).

Physiological Functions

AVP and OT have numerous functions in the brain and body, and a brief discussion of their diverse neuro-endocrine-neurohypophysial functions is provided here. For further details, refer to Burbach *et al.* (2001). One of the most important roles of AVP, also called antidiuretic hormone, is salt and water balance. In mammalian systems, changes in osmolality of extra-cellular fluids alter secretion of AVP, and to a lesser extent OT, which can regulate salt and water intake through actions on kidney tubles and by exerting behavioral effects on drinking and salt appetite. AVP plays a further role as a vasoconstrictor in response to hemorrhage. As discussed earlier, AVP also is involved in the regulation of the stress response. Some of these effects may be mediated through transport of AVP to pituitary corticotropes via the portal capillary vasculature leading to the anterior pituitary gland. In addition, neurohypophysial actions of AVP on osmolality may indirectly modulate the brain–pituitary–adrenal axis.

Oxytocin is best known for its roles related to parturition and milk ejection in mammals. OT plays roles in uterine contraction during delivery. With regard to lactation, OT responds to the suckling stimulus by stimulating contraction of the cells around the alveoli of the mammary gland to stimulate milk ejection. OT, together with AVP, plays key roles in social behaviors related to successful reproduction, such maternal and paternal behaviors and pair bonding (Lim and Young, 2006).

Properties of the OT and AVP Receptors

Effects of OT and AVP are mediated by their specific receptors, localized in various target organs. To date, one OT receptor and three AVP receptors (V1a, V1b, V2) have been identified (Young and Gainer, 2003). All are G-protein coupled receptors. The V2 receptor is not found in the CNS, but is highly expressed in kidney. The V1a and V1b receptors are found in brain, and the V1b receptor is highly expressed in pituitary corticotropes. Knockouts of the mouse V1b receptor cause deficits in social aggression (Young and Gainer, 2003).

Neuroanatomy of the Oxytocin and AVP System

The hypothalamic-neurohypophysial AVP and OT neurons have large cell bodies that are 20–40 μm in diameter, and are referred to as magnocellular neurons (as opposed to their parvocellular counterparts of 10–15 μm). These cells are localized in two major hypothalamic regions: the supraoptic nucleus (SON), comprised entirely of magnocellular neurons; and the paraventricular nucleus (PVN), which has both magnocellular and parvocellular divisions (Burbach *et al.*, 2001). The hypothalamic-neurohypophysial neurons in the SON and PVN project processes that enter the pituitary stalk and terminate in the posterior pituitary gland, where the hormones can gain access to the general circulatory system (Fig. 40.1). In addition, some of neurohypophysial axons may branch into other regions, including the median eminence, whereby OT and AVP may influence anterior pituitary gland functions. For example, OT contributes to prolactin release through actions on the lactotropes, and AVP interacts with CRH on the corticotrope to modulate stress responses.

HORMONES AND THE BRAIN

Steroid Hormone Receptors in the Brain

In order for steroid hormones to be able to exert effects on brain and behavior, they must have specific steroid hormone receptor-positive targets. The hypothalamus, the site of steroid hormone regulatory feedback, is a predictable target for steroid hormones. Indeed, the hypothalamus and related regions such as preoptic area are abundant in estrogen receptors (ERs), androgen receptors (ARs), and progesterone receptors (PRs), which can be detected even during fetal development (Roselli *et al.*, 2006). Neuroscientists have begun to look beyond the hypothalamus as a CNS target for steroid hormones, in order to understand how these molecules may mediate or mitigate nonreproductive CNS functions including neuroprotection, synaptic plasticity, neurogenesis, cognition, and neurodegeneration. The findings that sex steroid hormone receptors are widespread in the CNS help explain some of these observations. In addition, along with the nuclear sex steroid hormone family, nonnuclear membrane steroid hormone receptors have been identified that may explain actions of steroid hormones, particularly those that occur more rapidly than those mediated by nuclear receptors, in the CNS.

Sexual Differentiation of the Brain

The brain is sexually differentiated, meaning that there are morphological, anatomical, and functional differences between males and females. In mammals and many other vertebrates, these differences appear to be established, or "organized," in late gestation and early postnatal life, due to early hormonal exposures. The developing fetus is exposed not only to its own gonadal hormones, but also to maternal and environmental factors, exposures to all of which have permanent effects on brain development, and ultimately, on adult neural and behavioral functions.

The early effects of sex steroid hormones are mediated by specific receptors that are abundant in the hypothalamus and other brain regions. Two examples of brain regions that are particularly sensitive to perinatal sex steroid exposure and are sexually differentiated include (but are not limited to) specific hypothalamic and preoptic areas, as well as the spinal nucleus of the bulbocavernosus (SNB), the latter which innervates the base of the penis or clitoris. Perinatal differences in sex steroid hormones act to alter the size of these brain regions such that they differ between males and females; in general, because the fetal testis is much more active than the ovary, testicular hormones are responsible for these neuroanatomical changes.

Although it might be predicted that testosterone is responsible for these effects in males, depending upon the brain region, it is often testosterone's metabolite, 17β-estradiol, that is much more potent in many of these actions in mammalian species. Testosterone is a direct precursor to estradiol, metabolized by the P450 aromatase enzyme. This enzyme is found in the hypothalamus, as well as many other brain regions that are steroid sensitive. The perinatal *organizational* effects of testosterone and estradiol are crucial in order for normal sex-specific reproductive physiology and behavior to occur in adulthood. These behaviors also require further actions of gonadal steroid hormones, which are secreted in increasing amounts during the pubertal process. These so-called *activational* effects of steroids at puberty require that normal organization has occurred early in life. It is also becoming clear that the pubertal period is characterized by continued development of the brain, including neural plasticity that is modulated by steroid hormones. Thus, puberty may be considered as a second organizational period, as well as one during which activation of neonatally-organized circuits occurs (Sisk and Foster, 2004).

The issue of sexual dimorphism in the human brain has been quite controversial, but there is no question

that there is strong evidence for morphological and functional differences in specific brain nuclei in humans (Cahill, 2006). Homologues to the sexually dimorphic nuclei of rats also appear to be dimorphic in humans. For example, Onuf's nucleus, which is homologous to the rat SNB, has more motor neurons in men than in women. In the human anterior hypothalamus, roughly equivalent to the rat sexually dimorphic nucleus of the POA (SDN-POA), there are sex differences in the size of some of the interstitial nuclei, called INAH (for interstitial nuclei of the anterior hypothalamus), with men having larger volumes than women of some of these nuclei. This issue is particularly controversial, as it has been reported that the INAH-3 of homosexual men is intermediate in size between that of heterosexual men (having the largest INAH-3) and that of women (having the smallest INAH-3). Further research is needed to confirm these results and to determine the implications of different-sized INAHs on sexual behavior and identity. Other brain regions have been shown to be sexually dimorphic in humans. For example, hippocampus, amygdala and neocortex have robust sex differences in size and neurochemistry (Cahill, 2006; Fig. 40.10).

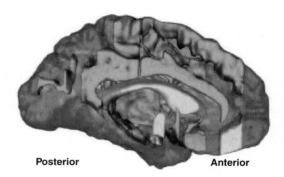

Posterior　　　　　　　**Anterior**

▨ Structures that are larger in the healthy female brain, relative to cerebrum size

▨ Structures that are larger in the healthy male brain, relative to cerebrum size

FIGURE 40.10 (This legend is quoted from the review by Cahill, 2006.) An illustration of sex differences in the size of various human brain regions. Goldstein *et al.* measured the volume of 45 brain structures taken from MRI scans in a sample of male (n = 27) and female (n = 21) subjects. For both sexes, the size of each region was determined relative to volume of the cerebrum. As shown here, significant differences between the sexes were detected in widespread brain regions. The authors also found that the size of the sex differences were related to the presence of sex steroid receptors in homologous brain regions during critical developmental periods, as determined in animal studies, suggesting that sex differences in the adult stem from sex hormone influences on brain development. Data from Goldstein, J. M. *et al.* (2001). Normal sexual dimorphism of the adult human brain assessed by *in vivo* magnetic resonance imaging. *Cereb. Cortex* **11**, 490–497.

Although the mechanisms for these differences cannot be studied experimentally in humans, Cahill (2006) has speculated that some but not all differences are attributable to sex steroid hormones.

Hormones and Sexual Behavior

There are clearly differences in the sexual behavior of males and females, which are dependent upon organizational and activational effects of gonadal steroid hormones. For example, copulatory behavior in male rodents comprises mounts of the female by the male, penile intromissions, and ejaculations. A receptive female rodent will respond to male copulatory behaviors with a lordosis response, characterized by a rigid stance and an arched back. These behaviors are driven in large part by the responses of hypothalamic nuclei to hormones, and the nature of the adult behavior is determined by the prenatal/neonatal exposure to hormones. Thus, the hypothalamus of adult males, which has been masculinized neonatally by testicular testosterone (aromatized to estradiol), will respond to the testosterone produced by the adult gonad (metabolized to DHT) by causing the typical male pattern of sexual behavior. In contrast, the female hypothalamus, which is not exposed neonatally to sufficient levels of ovarian hormones to cause masculinization of her brain, responds to estrogen produced by her adult ovary with the appropriate female lordotic response. Moreover, a normal adult male will not express lordosis behavior in response to exogenous estrogen, nor will a normal adult female exhibit male-typical sexual behaviors in response to exogenous testosterone, as their brains have not been organized in such as way as to produce a gender-inappropriate response in adulthood. However, a neonatally castrated male can respond to estradiol in adulthood with the expression of lordosis behavior, as his brain was not masculinized neonatally. Similarly, a female who was exposed to exogenous testosterone or estradiol neonatally can exhibit male-typical sexual behaviors as an adult if she is treated with testosterone.

It is necessary to distinguish between masculinizing and defeminizing effects of hormones in females, and the feminizing and demasculinizing effects of hormones in males. As mentioned earlier, female sexual behavior in rats is characterized most overtly by lordosis. The prevention of this behavior by exposure to sex steroid hormones neonatally would be an example of a defeminizing effect. In contrast, a female rat whose brain has been exposed to steroid hormones at birth, and in adulthood is injected with testosterone, may exhibit male-typical sexual behavior such as mounting another female. This is an example of the masculiniz-

ing effects of hormones. For males, the feminizing effects of hormones could be seen in a neonatally castrated male who, as an adult, is exposed to female sex steroids and responds with lordosis behavior in response to the mounting by another male. Demasculinization would occur in this male if, as an adult, he fails to perform normal male-typical sexual behavior (mounts, intromissions, ejaculations) in response to a female. Thus, hormones can cause both of these effects in males or females, depending on the time and duration of exposure.

The regions of the brain responsible for sex behavior differ between males and females. Sexual behavior in male rodents is largely driven by the preoptic area (POA) of the brain, just rostral to the hypothalamus, although other regions contribute to this process. Lesions of the POA in males cause the loss of sexual behavior, and testosterone implanted into the POA can induce sex behavior. In female rodents, the primary hypothalamic site regulating receptive sexual behavior is the ventromedial nucleus of the hypothalamus (VMH), although as in males, other regions also play contributing roles. Estradiol, followed by progesterone, facilitates the expression of the lordosis reflex and coordinates the timing of female receptivity with the preovulatory GnRH/LH surge that causes ovulation. Thus, the timing of mating behavior and ovulation is coordinated by the release of ovarian steroid hormones. During the reproductive cycle, circulating estrogen levels increase, and estrogen binds to its receptor in the VMH to cause the transcription of several genes, including that of the progesterone receptor. Then, progesterone levels rise and can bind to its newly induced receptor in the VMH. This causes proceptive behaviors in females and facilitates the lordosis response. Females that have been androgenized neonatally have a large reduction in the lordosis response to exogenous estrogen and progesterone replacement.

Pheromones, the Vomeronasal Organ, and Sexual Behavior

Reproductive behavior in many species involves signals from pheromones, which are chemical signals used for communication between animals (Meredith, 1998). Pheromones are chemical messengers that typically are produced by specialized glands, such as the sebaceous glands in the flanks of animals, or released in urine or vaginal fluid. They are released not into the body but rather into the external environment to convey information about the physiological status of the animal to other individuals.

The ability to detect pheromones involves the vomeronasal organ, an auxiliary olfactory organ found near the nasal cavity, and its projections to the accessory olfactory bulbs. Cells in these latter regions contact a number of brain regions, including the medial amygdala, cortical regions, and the bed nucleus of the stria terminalis, all of which are implicated in sex behavior. These regions in turn project to the hypothalamus and preoptic area. Thus, information from pheromones about mating status can be conveyed to parts of the brain coordinating reproductive physiology and behavior.

In most species, information about reproductive status, dominance status, or other social cues can be conveyed by pheromones, although it is still controversial whether humans have a vomeronasal organ and/or can sense pheromones. Those organisms using pheromones as chemosensory cues have developed mechanisms for depositing or releasing pheromones to "advertise" sexual competency or receptivity in order to attract a mate or to mark territory, to protect food supplies, mates, or offspring. Pheromones and the systems that mediate their effects play broader roles than those involved in reproductive physiology and behavior. Pheromones can dramatically influence the timing of sexual development and function. In female mice, exposure to male urinary pheromones accelerates the timing of pubertal development. Males respond to female urinary pheromones with sharp increases in luteinizing hormone release. It has also been reported that pheromones can coordinate the timing of reproductive cycles, even in humans, although the mechanisms and functions of these pheromones are still controversial and poorly understood. Nevertheless, it is clear that pheromones play important roles in coordinating physiology and behavior.

References

Behncken, S. N. and Waters, M. J. (1999). Molecular recognition events involved in the activation of the growth hormone receptor by growth hormone. *J. Mol. Recogn.* **12**, 355–362.

Ben-Jonathan, N., Arbogast, L. A., and Hyde, J. F. (1989). Neuroendocrine regulation of prolactin release. *Prog. Neurobiol.* **33**, 399–447.

Burbach, J. P. H., Luckman, S. M., Murphy, D., and Gainer, H. (2001). Gene regulation in the magnocellular hypothalamo-neurohypophysial system. *Physiol. Rev.* **81**, 1197–1267.

Cahill, L. (2006) Why sex matters for neuroscience. *Nature Rev. Neurosci.* **7**, 477–484.

Clynen, E., De Loof, A., and Schoofs, L. (2004). New insights into the evolution of the GRF superfamily based on sequence similarity between the locust APRPs and human GRF. *Gen. Comp. Endocrinol.* **139**, 173–178.

DeSouza, E. B. (1987). Corticotropin-releasing factor receptors in the rat central nervous system: Characterization and regional distribution. *J. Neurosci.* **7**, 88–100.

Gershengorn, M. C. (1993). Thyrotropin-releasing hormone receptor: Cloning and regulation of its expression. *Recent Progr. Horm. Res.* **48**, 341–363.

Gore, A. C. (1998). Circadian rhythms during aging. *In* "Functional Endocrinology of Aging" (C. V. Mobbs and P. R. Hof, eds.), pp. 127–165. Karger, Basel.

Gore, A. C. (2002). "GnRH: The Master Molecule of Reproduction." Kluwer Academic Publishers.

Graves, P. N. and Davies, T. F. (2000). New insights into the thyroid-stimulating hormone receptor: The major antigen of Graves' disease. *Endocrinol. Metabol. Clin. North Am.* **29**, 267–286.

Harris, G. W. (1971). Humours and hormones. *J. Endocrinol.* **53**, ii–xxiii.

Lechan, R. M., Qi, Y., Jackson, I. M. D., and Mahdavi, V. (1994). Identification of thyroid hormone receptor isoforms in thyrotropin-releasing hormone neurons of the hypothalamic paraventricular nucleus. *Endocrinology.* **135**, 92–100.

Lim, M. M. and Young, L. J. (2006). Neuropeptidergic regulation of affiliative behavior and social bonding in animals. *Horm. Behav.* **50**, 506–517.

Mayo, K. E., Godfrey, P. A., Suhr, S. T., Kulik, D. J., and Rahal, J. O. (1995). Growth hormone-releasing hormone: Synthesis and signaling. *Recent Progr. Horm. Res.* **50**, 35–73.

Meredith, M. (1998). Vomeronasal, olfactory, hormonal convergence in the brain: Cooperation or coincidence? *Ann. N.Y. Acad. Sci.* **855**, 349–361.

Morgan, K. and Millar, R. P. (2004). Evolution of GnRH ligand precursors and GnRH receptors in protochordate and vertebrate species. *Gen. Comp. Endocrinol.* **139**, 191–197.

Morimoto, M., Morita, N., Ozawa, H., Yokoyama, K., and Kawata, M. (1996). Distribution of glucocorticoid receptor immunoreactivity and mRNA in the rat brain: An immunohistochemical and in situ hybridization study. *Neurosci. Res.* **26**, 235–269.

Naville, D., Penhoat, A., Durant, P., and Begeot, M. (1999). Three steroidogenic factor-1 binding elements are required for constitutive and cAMP-regulated expression of the human adrenocorticotropin receptor gene. *Biochem. Biophys. Res. Commun.* **255**, 28–33.

O'Connor, T. M., O'Halloran, D. J., and Shanahan, F. (2000). The stress response and the hypothalamic–pituitary–adrenal axis: From molecule to melancholia. *Q. J. Med.* **93**, 323–333.

Plant, T. M. (1988). Puberty in primates. *In* "The Physiology of Reproduction" (E. Knobil and J. Neill, eds.), pp. 1763–1788. Raven Press, New York.

Roselli, C. E., Resko, J. A., and Stormshak, F. (2006). Expression of steroid hormone receptors in the fetal sheep brain during the critical period for sexual differentiation. *Brain Res.* **1110**, 76–80.

Sealfon, S. C., Weinstein, H., and Millar, R. P. (1997). Molecular mechanisms of ligand interaction with the gonadotropin-releasing hormone receptor. *Endocr. Rev.* **18**, 180–205.

Segerson, T. P., Kauer, J., Wolfe, H., Mobitker, H., Wu, P., Jackson, I. M. D., and Lechan, R. M. (1987). Thyroid hormone regulates TRH biosynthesis in the paraventricular nucleus of the rat hypothalamus. *Science* **238**, 78–80.

Sisk, C. L. and Foster, D. L. (2004). The neural basis of puberty and adolescence. *Nature Neurosci.* **7**, 1040–1047.

Szkudlinski, M. W., Grossmann, M., Leitolf, H., and Weintraub, B. D. (2000). Human thyroid-stimulating hormone: Structure-function analysis. *Methods* **21**, 67–81.

Wintermantel, T. M., Campbell, R. E., Porteous, R., Bock, D., Grone, H. J., Todman, M. G., Korach, K. S., Greiner, E., Perez, C. A., Schutz, G., and Herbison, A. E. (2006). Definition of estrogen receptor pathway critical for estrogen positive feedback to gonadotropin-releasing hormone neurons and fertility. *Neuron* **52**, 271–280.

Wray, S. (2002) Development of gonadotropin-releasing hormone-1 neurons. *Front. Neuroendocrinol.* **23**, 292–316.

Yamada, M., Radovick, S., Wondisford, F. E., Nakayama, Y., Weintraub, B. D., and Wilber, J. F. (1990). Cloning and structure of human genomic DNA and hypothalamic cDNA encoding human preprothyrotropin-releasing hormone. *Mol. Endocrinol.* **4**, 551–556.

Young, W. S. III, Gainer, H. (2003). Transgenesis and the study of expression, cellular targeting and function of oxytocin, vasopressin and their receptors. *Neuroendocrinol.* **78**, 185–203.

Andrea C. Gore

41

Circadian Timekeeping

Life on Earth has adapted to the cyclic variation in conditions imposed by our planet's daily rotation on its axis and its annual revolution around the sun. Almost all organisms studied, from cyanobacteria to *Homo sapiens*, have intrinsic time-keeping systems that adjust physiology and behavior to anticipate the daily cycles of light and darkness. Physiological processes that vary over the 24-hour day include activity, alertness, hormone secretion, organ physiology, and gene expression. These variations are not merely passive responses to a rhythmic environment, but instead reflect an underlying biological mechanism that can measure time on a 24-hour time scale. This mechanism orchestrates physiology to achieve predictive, rather than reactive, homeostasis.

Biological time measurement on the 24-hour time scale is achieved by a circadian timing mechanism. The term *circadian* is derived from the Latin words *circa* and *diem*, meaning "about a day," referring to the cycle length of these rhythms. Circadian rhythms are defined as rhythms that persist with a cycle length of approximately 24 hours in constant environmental conditions (Figs. 41.1, 41.2). These rhythms are generated by a biological timing mechanism that normally is synchronized (entrained) to the 24-hour day by environmental cues. Light is the most widely used signal for entrainment of circadian clocks, but temperature, hormone levels, nutrient availability, and other cues can also affect oscillations.

In this chapter, we will present the current state of knowledge regarding the mammalian circadian timing system, including discussion of the molecular basis for oscillations within individual cells, and the anatomical and neurochemical substrates regulating circadian rhythmicity.

OVERVIEW OF THE MAMMALIAN CIRCADIAN TIMING SYSTEM

A biological timing system in its simplest form consists of an intrinsic clock mechanism that measures time, an input mechanism that allows the clock to become synchronized or reset by changes in the environment, and output pathways that lead to generation of overt rhythms such as daily changes in locomotor activity, sleep, and hormone levels. The mammalian circadian timing system is considerably more complicated than this simple linear scheme, because it is composed of a hierarchy of circadian oscillators (Fig. 41.3). At the pinnacle of this oscillatory system is a small brain area, the suprachiasmatic nuclei (SCN) of the anterior hypothalamus. The SCN often are called the master circadian clock, because these nuclei (one on each side of the brain) play a key role in coordinating oscillations in other tissues, and in regulating behavior. Many other cells and tissues also have the capacity to oscillate with a cycle length of approximately 24 hours. The molecular mechanism underlying these cell-autonomous circadian oscillations is a transcriptional feedback loop described in detail later.

The primary input pathway to the SCN circadian clock is through retinal detection of light. Remarkably, the retinal photoreceptors that lead to visual image formation are not needed for circadian photoreception. Instead, a specialized population of intrinsically photosensitive retinal ganglion cells detect light and also project directly to the SCN. In rodents, the retinohypothalamic tract (RHT) is composed primarily of the axons of these specialized ganglion cells, and the RHT is necessary for synchronization of the SCN clock to light.

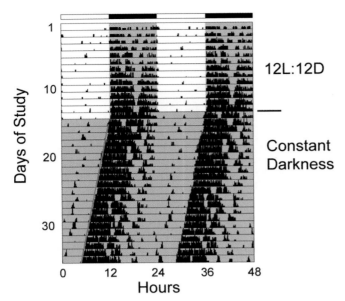

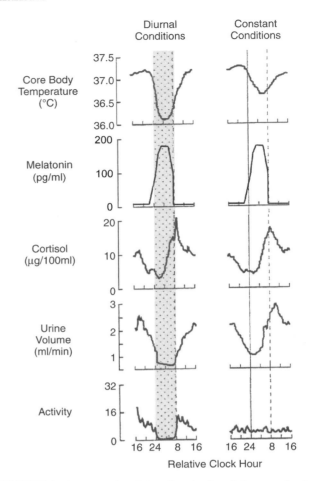

FIGURE 41.1   Locomotor activity rhythm of a mouse. This figure is a "double-plotted actogram" that shows the activity pattern of an adult mouse housed individually in a cage with a running wheel. When the mouse rotates the wheel by running within it, a computerized system detects and records the wheel revolutions and then can be used to plot the wheel revolutions (as bars coming up from the baseline, with their height proportional to the number of revolutions) versus time, for weeks of data. Each horizontal line represents 48 hours of data. On the first line of the record, the first 48 hours of data are shown. In the second line of the record, the data shown are from hours 24–72 of data collection. Thus, each 24-hour period of data after the first 24 hr is shown both to the right of and below the previous day's data; this artificial reproduction of the data (double-plotting) helps to see the rhythmicity when the cycle length of rhythmicity differs from 24 hours (see for example Fig. 41.7 lower). When housed in a 12L:12D lighting cycle (LD), the mouse is active almost exclusively at night. When the lighting cycle is disabled so the animal is in constant darkness (DD), the animal continues to show rhythmicity but with a cycle length slightly less than 24 hours. This record reveals two defining principles of circadian rhythms: circadian rhythmicity is intrinsic, rather than being generated in response to environmental variation, and the circadian clock normally is synchronized to the 24-hour day length by light. Note also the precision of the rhythmicity when the mouse is in DD. Activity onset is defined as a spontaneous, sustained increase in wheel running that occurs after a sustained period of inactivity. It is relatively easy to draw a "best-fit" line through the activity onsets (yellow line). The actual time of activity onset each day does not deviate far from this "best-fit" line, indicating the cycle length between successive activity onsets is measured with great accuracy. Spend a few minutes to think about how you would design a biological timekeeping system that can measure a day while having a variation between successive cycles that is less than 1% (15 minutes per 24 hours; note that the actual variation is often considerably less than this amount). Actogram provided by Jason DeBruyne, UMass Medical School.

FIGURE 41.2   Diurnal and circadian rhythms in human physiology. The panels on the left show rhythms in several functions in human subjects maintained in a lighting cycle, with a period of sleep in the dark indicated by the shading. These rhythms are true circadian rhythms, as they persist in constant conditions. The panels on the right show the same endpoints studied in subjects maintained for 40 hours on a constant routine, with very low levels of illumination and with wakefulness maintained, while the subjects remain constantly in a semi-recumbent posture and receive small meals at regular intervals throughout the study period. The apparent loss of the activity rhythm in constant conditions is due to the protocol-enforced inactivity of the subject. (Adapted from Czeisler and Khalsa, 2000.)

The main output of the SCN is encoded by the neuronal firing rate of SCN neurons. In addition to direct synaptic interactions, rhythmic neuropeptide secretion into the cerebrospinal fluid may be an important mechanism for regulation of downstream targets. Anatomically, the outputs of the SCN are focused within the hypothalamus, but their influence is widespread, consistent with the pervasive influence of the SCN on physiological functions ranging from the modulation of cognitive function to neuroendocrine

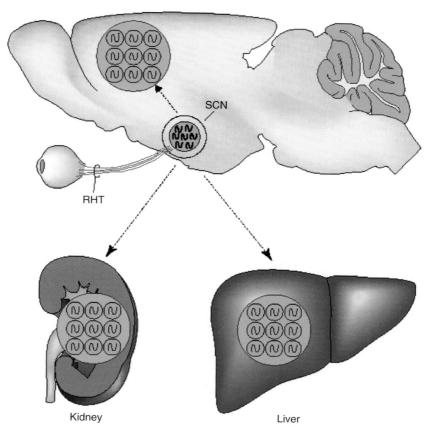

FIGURE 41.3 The mammalian circadian timing system consists of a hierarchy of oscillators. Oscillatory neurons in the SCN interact with each other to produce a set of coherent outputs. These outputs, which include behavioral and physiological rhythms, synchronize cell-autonomous oscillations in other brain regions and in peripheral tissues. (From Reppert and Weaver, 2002.)

responses. Output from the SCN serves to synchronize autonomous cellular oscillators in other tissues. Local circadian oscillators regulate gene expression in a tissue-specific manner, likely contributing to local organ physiology. Microarray studies conducted on brain, liver, heart, and kidney tissue in mammals reveal that as much as 10% of the genome is rhythmically transcribed, and many of the genes that are rhythmically expressed represent key, tissue-specific metabolic control points (Miller *et al.*, 2007). Thus, the circadian timing system has a pervasive impact on physiological processes via its output pathways.

THE SUPRACHIASMATIC NUCLEI ARE THE SITE OF THE PRIMARY CIRCADIAN PACEMAKER IN MAMMALS

The SCN have been recognized as the master circadian clock in rodents since the early 1970s (Weaver, 1998). Lighting conditions (especially constant light)

were known to affect neuroendocrine function in rats, and the eyes were known to be necessary for the effects of light, but the pathways by which light had these effects were not known. At that time, "new" techniques for anterograde tract-tracing were used to assess retinal projections, and revealed a direct retina-to-SCN projection, the retinohypothalamic tract (RHT). To determine if the SCN were an important node on the anatomical pathway for detection of light, Moore and Eichler (1972) destroyed the rat SCN and observed loss of rhythmic serum corticosterone in a population of rats. Using a similar approach, Stephan and Zucker (1972) discovered that the rhythm of drinking in individual rats was lost following destruction of the SCN. Together, these two studies demonstrated the necessity of the SCN for circadian rhythmicity. Hundreds of subsequent studies have confirmed the necessity of the SCN for rhythmicity in physiology and behavior.

SCN: Oscillator and Pacemaker

Lesion studies leading to loss of rhythmicity do not necessarily show that the SCN function as the primary circadian pacemaker. An alternative possibility is that the SCN could be a necessary element on a key output pathway leading to physiological rhythmicity, rather than a pacemaker. The presence of rhythms in SCN metabolic activity, electrical activity, and gene expression profiles *in vivo* do not distinguish between these possibilities. Studies demonstrating rhythms in neuropeptide secretion, metabolic activity, and electrical firing rate in SCN tissue maintained *in vitro* do show that the SCN contain a functional oscillator. The SCN do not simply oscillate, however; through their outputs, the SCN regulate rhythms in physiology and behavior, and thus serve as a circadian pacemaker. This pacemaker role of the SCN is revealed most clearly by studies showing that locomotor activity rhythms can be restored in hamsters by transplanting fetal SCN tissue into the hypothalamus of animals previously made arrhythmic by SCN lesion. Furthermore, transplants between hamsters with different circadian periods reveal that the cycle length (period) of restored rhythmicity is dictated by the genotype of the SCN tissue donor and not by the recipient (Ralph *et al.*, 1990). Thus, the SCN serve as a pacemaker that communicates rhythmicity to tissues regulating behavioral rhythms.

Summary

The SCN are the master circadian pacemakers in mammalian brain. The SCN oscillate *in vivo*, and also when placed *in vitro*. More importantly, however, the SCN generate output signals that lead to physiological and behavioral rhythms. Critical to the function of the SCN as the master pacemaker is its position at the interface between the outside world (detected primarily by retinal photoreception) and the light-insensitive tissues that comprise the rest of the body.

A HIERARCHY OF CELL-AUTONOMOUS CIRCADIAN OSCILLATORS

Single-Cell SCN Oscillators Are Coupled to Form a Coherent Clock

The rodent SCN are composed of approximately 10,000 neurons per side of the brain, with heterogeneity of neuropeptide phenotype, connectivity, and other properties (Silver and Schwartz, 2005; Morin, 2007). Despite this heterogeneity, most cells within the SCN are believed to have the capacity for cell-autonomous rhythmicity. Rhythmicity was first detected in individual SCN neurons using extracellular recording of SCN neuronal firing (Welsh *et al.*, 1995; Liu *et al.*, 1997;

Fig. 41.4A). More recent studies reveal rhythmicity in single SCN cells using imaging methods designed to detect temporal variations in reporter gene expression.

The cycle lengths of individual oscillating SCN cells are much more variable than the periods of animal behavior or the period observed for rhythms (e.g., in neuropeptide secretion) from slices of SCN tissue maintained *in vitro*. Thus, some mechanisms for interaction among the heterogeneous SCN neurons must occur for SCN neurons to synchronize their diverse periods to an intermediate value, to coordinate their phase (timing), and thus to provide a coherent output (Aton and Herzog, 2005). GABA and vasoactive intestinal peptide (VIP) have both been proposed as neurochemicals that may coordinate SCN neurons. VIP is expressed in a subset of neurons in the central "core" region of the SCN, whereas the vast majority of SCN neurons express the inhibitory neurotransmitter,

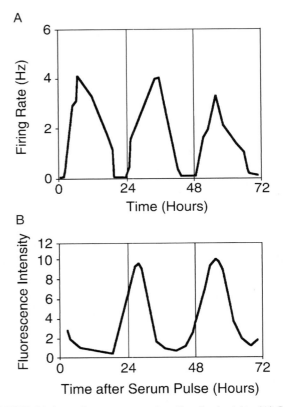

FIGURE 41.4 Cell-autonomous circadian rhythmicity. (A) Circadian rhythmicity in an individual SCN neuron. SCN neurons in dispersed cell culture express rhythms of spontaneous firing rate, as detected by electrodes embedded in the floor of the culture dish on which the neurons were plated. (Adapted from Welsh *et al.*, 1995; Liu *et al.*, 1997). (B) Circadian rhythmicity in an individual fibroblast. Imaging of reporter gene expression was used to detect rhythmicity in culture. (Adapted from Nagoshi *et al.*, 2004.)

GABA (Moore *et al.*, 2002). Mice lacking VIP (or the VPAC2 receptor for VIP) have very weak and unstable rhythms in locomotor activity, yet individual SCN neurons from these mice appear to maintain rhythmicity (Maywood *et al.*, 2006). These findings strongly suggest that a "coupling deficit" leads to altered circadian rhythmicity in these lines of mice.

The entire population of SCN neurons usually functions in concert as an orchestrating oscillator. Under specific conditions, however, SCN cells can be induced to become coupled into two functionally relevant subpopulations that can independently influence circadian behavior. For example, by long-term exposure to constant light, the left and right hamster SCN can be induced to oscillate in opposite phases, with each side regulating locomotor activity, leading to activity profiles with two peaks per 24 hours (de la Iglesia and Schwartz, 2000). In rats, abnormal lighting cycles can be used to dissociate ventrolateral and dorsomedial subdivisions of the SCN from each other, resulting in simultaneous expression of rhythms with two different periods; each activity component apparently is driven by a different region of the SCN. These experiments emphasize that the multioscillatory nature of the circadian system can be appreciated even within the SCN, and that coupling among SCN neurons normally prevents conflicting output signals.

Peripheral Oscillators

For many years, the widely accepted dogma in the circadian field was that only the SCN possessed the capacity for self-sustaining, autonomous rhythmicity. The identification of a family of mammalian *Period* genes (homologues of *period*, a gene essential for circadian rhythmicity in *Drosophila*) helped to change this perception. The mammalian *Period* genes are widely expressed (Zylka *et al.*, 1998), suggesting that rhythmicity could occur in additional tissues. This was clearly shown to be the case in a landmark study by Balsalobre *et al.* (1998). Their demonstration that brief exposure to a high concentration of serum could lead to rhythmic gene expression in fibroblasts immediately transformed thinking about circadian oscillators. More recent single-cell analysis reveals that fibroblasts have ongoing (but uncoordinated) rhythms, and a serum pulse merely synchronizes these rhythms so that it can be detected in the population (Nagoshi *et al.*, 2004; Fig. 41.4B). The serum pulse is not a unique stimulus; acute activation of each of several distinct signaling pathways can induce a rapid molecular response and subsequent rhythmicity in fibroblasts, suggesting that multiple pathways converge upon the oscillator.

It is now generally believed that most mammalian cell types are capable of circadian rhythmicity. In normal circumstances, the rhythmicity in tissues outside the SCN is synchronized by SCN-dependent output signals. These signals include daily fluctuations in hormone levels (e.g., melatonin and glucocorticoids) and other physiological and behavioral rhythms (e.g., body temperature and food intake). Thus, there is a hierarchy of oscillators within the body. Rhythmic outputs controlled by the SCN serve as inputs to peripheral oscillators, synchronizing molecular rhythmicity among cells within a peripheral tissue (Fig. 41.3). In the absence of rhythmic input from the SCN, oscillators in some tissues become desynchronized and thus the organ as a whole loses detectable rhythmicity. In some other tissues, it appears that interactions among oscillating cells are sufficiently robust to maintain rhythmicity of the organ as a whole (Granados-Fuentes *et al.*, 2004). Although this suggests that there are circadian pacemakers in additional tissues besides the SCN, it is worth noting that only the SCN pacemaker appears capable of performing all roles expected of a circadian clock. Even peripheral organs that maintain rhythmicity appear insensitive to environmental lighting conditions and thus different tissues would oscillate with different periods and phases, so that their normal phase relationships to each other would be lost. It is thought that coordination of rhythms within a tissue may optimize organ-level physiological processes, whereas a circadian timing system in general may achieve "internal temporal order" among organs so that organs are prepared to function most efficiently.

The SCN generally function as master circadian pacemakers, but there is now compelling evidence that, under specific circumstances, oscillators in other brain areas including the dorsomedial hypothalamus can regulate circadian behavior in the absence of the SCN (Stephan, 2002; Gooley *et al.*, 2006; Taratoglu *et al.*, 2006). SCN-independent oscillators in the olfactory bulbs can also influence olfactory response rhythms (Granados-Fuentes *et al.*, 2006). Although these oscillatory mechanisms are likely normally overshadowed or coordinated in the presence of a functioning SCN, these examples reveal the importance of recognizing the potential physiological impact of tissue-autonomous oscillators outside the SCN. Our understanding of the implications of the hierarchical nature of the mammalian circadian timing system is advancing rapidly.

Summary

Many cell types possess the capacity for circadian oscillation. Neurons within the SCN oscillate, and also communicate with each other, such that the SCN are able to provide a coherent output resulting in rhythmicity of other functions. Oscillating cells in other tissues generally require occasional SCN-dependent input to remain synchronized. In some cases, however, cell–cell communication is adequate to maintain coordinated cellular rhythmicity in a tissue even in the absence of SCN input.

THE MOLECULAR BASIS FOR CIRCADIAN OSCILLATION IS A TRANSCRIPTIONAL FEEDBACK LOOP

The preceding discussion emphasizes that circadian rhythmicity occurs within single, isolated cells. How do individual cells measure 24-hour time intervals? Studies conducted in many labs over the past 10 years have revealed a great deal about the molecular mecha-

nisms for mammalian circadian rhythmicity (Reppert and Weaver, 2002; Fig. 41.5).

Progress made in understanding the circadian clock mechanisms in the fruit fly, *Drosophila melanogaster*, was instrumental in developing our current understanding of mammalian clock mechanisms. Most of the genes involved in the fly circadian mechanism have mammalian homologs, and their roles in circadian function have now been examined in mammals. Although there have been some alterations in gene products performing specific functions, and gene duplication in mammals has led to greater complexity, the general features of the core clock mechanism are similar between flies and mammals (Box 41.1).

The molecular mechanism for measuring time on a 24-hour time-scale is based on a self-sustaining transcriptional/translational feedback loop present within individual cells (Fig. 41.5). According to the current molecular model for this feedback loop (Reppert and Weaver, 2002; DeBruyne *et al.*, 2007), heterodimers of

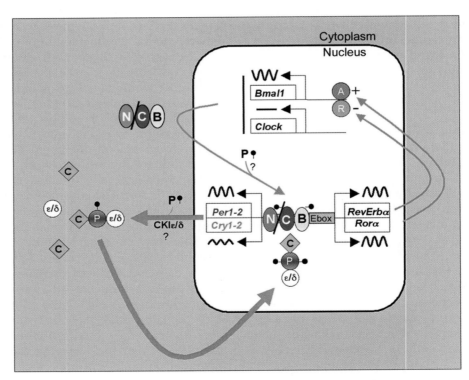

FIGURE 41.5 The molecular mechanism for circadian rhythms in mammals is a transcriptional feedback loop. Transcriptional activation leads to expression of *Per* and *Cry* genes. Posttranslational modifications, including phosphorylation (black P lollipop) mediated in part by casein kinases (CKIε and δ), affect the interactions of PER (P in blue circle) and CRY (C in gold diamond) proteins with each other and regulate nuclear entry of the inhibitory complex. CRY proteins are the major effector of negative feedback, whereas the high-amplitude rhythm of PER production controls the timing of negative feedback. Collectively, these events lead to a delayed, negative feedback to shut off the transcriptional activation, resulting in a negative feedback loop (red arrows). The positive drive to the system comes from the transcription factors CLOCK (C in red oval) and BMAL1 (B in yellow oval), or NPAS2 (N in green oval) and BMAL1. An interlocking loop involving the orphan nuclear receptors RORA (A) and REVERB-alpha (R) controls the expression of *Bmal1*. The antiphase rhythmicity in this positive loop (green arrows) is not essential for rhythm generation, but BMAL1 itself is necessary. (Adapted from Reppert and Weaver, 2002.)

BOX 41.1

CIRCADIAN TIMING MECHANISMS IN *DROSOPHILA*

Virtually nothing was known about the genetic and molecular mechanisms controlling circadian rhythms until the late 1960s, when Seymour Benzer and his graduate student, Ron Konopka, identified the first mutations affecting circadian behavior. Most scientists at the time were skeptical that individual genes could affect complex behaviors. Benzer was nevertheless determined to identify the genetic underpinnings of behavior, and embarked on an ambitious program using the animal model of choice for genetic studies, the fruit fly *Drosophila melanogaster*. Benzer and colleagues exposed large numbers of flies to a chemical mutagen to isolate mutant lines with abnormal behavior in various assays. To identify circadian genes, Konopka screened for mutant fly lines that had lost the ability to predict dawn, as assessed by observing the time of day at which the young flies emerged from their pupal cases. Using this approach, Konopka and Benzer isolated three mutant lines with abnormal circadian rhythms. In one line, the flies emerged from their pupal cases too early, in a second line they were too late, and in a third line the flies emerged at random times. All three mutations mapped to the same

genetic locus, named "*period*" (*per*). Importantly, the three *period* mutants had similar circadian defects in their locomotor activity rhythms, which provided strong evidence that Konopka and Benzer had identified a gene central to the circadian pacemaker.

Benzer's pioneering idea of using unbiased genetic screens to identify genes involved in circadian behavior has led to identification of additional "clock genes" by several other labs. The subsequent genetic, molecular, and biochemical studies have led to a remarkably detailed understanding of the *Drosophila* circadian pacemaker (Figure 1). CLOCK (CLK) and CYCLE (CYC) are two basic helix-loop-helix transcription factors that heterodimerize and bind to E-boxes within the *per* and *timeless* (*tim*) promoters. PER and TIM dimerize in the cytoplasm, with TIM stabilizing PER. Several kinases (DOUBLETIME, CASEIN KINASE 2, and SHAGGY [the *Drosophila* homolog of glycogen synthase kinase 3]) and PROTEIN PHOSPHATASE 2A regulate the stability and thus the rate of PER and TIM accumulation, and the timing of their nuclear entry. Once in the nucleus, the PER/TIM heterodimer (or possibly PER alone) binds to

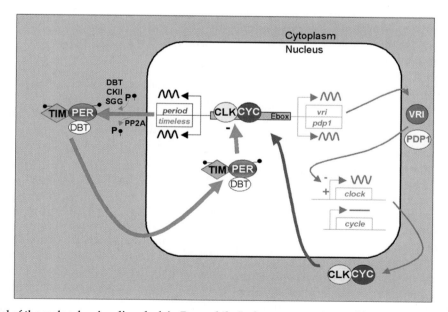

FIGURE 1   **Model of the molecular circadian clock in *Drosophila*.** In the core transcriptional feedback loop (left side of the figure), CLOCK (CLK) and CYCLE (CYC) positively regulate *period* (*per*) and *timeless* (*tim*) transcription. PER is phosphorylated by DOUBLETIME (DBT) and CASEIN KINASE II (CKII), while TIM is phosphorylated by SHAGGY (SGG). PER phosphorylation is also negatively regulated by PROTEIN PHOSPHATASE 2A. PER and TIM enter into the nucleus with DBT to inhibit the activity of CLK/CYC. In a secondary feedback loop shown in grey, VRILLE (VRI) and PAR DOMAIN PROTEIN 1ε (PDP1) regulate *clk* transcription.

(continues)

BOX 41.1   (cont'd)

the CLK/CYC dimer and disrupts its interaction with E-box sequences. The inhibition of *per* and *tim* gene expression is maintained for several hours, until the hyperphosphorylated and unstable PER and TIM proteins decay. CLK/CYC can once again bind to E-Box sequences and restart the circadian cycle. The kinase DOUBLETIME, which translocates into the nucleus with PER and TIM, may also play an important role for repression by regulating CLK phosphorylation and degradation. The CLK/CYC, PER/TIM feedback loop is essential for maintaining rhythms in *Drosophila*. There is a secondary feedback loop that appears to contribute to the precision and robustness of circadian oscillations through the regulation of *clk* mRNA levels.

The *Drosophila* molecular clock is synchronized primarily to environmental light/dark cycles by an intracellular photoreceptor, CRYPTOCHROME (CRY). Light activates CRY, and CRY then binds to and triggers the rapid degradation of TIM. Since TIM stabilizes PER, the degradation of TIM also results in a reduction in PER levels that resets the circadian pacemaker.

Despite the ability of cells throughout the fly to respond directly to light, there is increasing evidence for important interactions between oscillatory neurons in the regulation of behavior. About 150 neurons in the *Drosophila* brain have been identified as putative clock neurons due to their expression of circadian proteins, including PER (Figure 2). To dissect the roles of different groups of neurons, sophisticated genetic approaches have been used to target PER expression to specific neurons of mutant flies otherwise lacking PER, and to genetically eliminate specific cells by targeted expression of pro-apoptotic genes. These studies show that the ventral subpopulation of lateral neurons (the LN_vs) is necessary and sufficient for locomotor activity rhythms to persist under constant darkness. The LN_vs secrete a neuropeptide, pigment dispersing factor (PDF), that synchronizes other circadian neurons and regulates behavior. In addition, the LN_vs play an important role in the control of morning behavioral activity, whereas a dorsal group of lateral neurons, the LN_ds, is more important for the control of evening activity. A subpopulation of dorsal neurons also influences circadian behavior, particularly its responses to light. Future studies using more anatomically restricted manipulations will further elucidate the function of smaller subsets of neurons. Additionally, advances in live imaging techniques and *in vivo* electrophysiological recordings will provide information about how the activ-

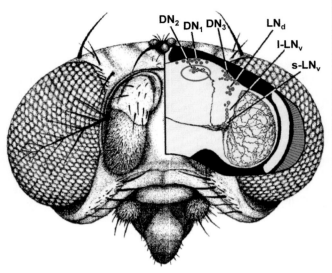

FIGURE 2   **Circadian neurons in the brain of *Drosophila melanogaster*.** Neurons expressing circadian proteins are shown in blue (dorsal neurons 1–3 [DN_1, DN_2, DN_3] and dorsal lateral neurons [LN_d]). Neurons expressing circadian proteins and the circadian neuropeptide pigment dispersing factor (PDF) are shown in red (small and large ventral lateral neurons [$s-LN_v$ and $l-LN_v$]). The PDF-positive neural projections are drawn in black. The most important projections for the synchronization of the circadian network are the $s-LN_v$ projections directed toward the DN_1 and DN_2 groups. Indeed, PDF level oscillates in the termini of these projections (circled in red), presumably because PDF is rhythmically released. (Adapted from Helfrich-Förster, 2005.)

ity of neurons within the circadian network is integrated to control circadian behavior.

As the first components of the mammalian molecular clock were identified in the late 1990s, it quickly became clear that important concepts learned from the *Drosophila* circadian clock also apply to mammals. At the molecular level, the circadian pacemaker of both *Drosophila* and mammals involve homologous proteins in analogous transcriptional feedback loops. Interestingly, the first identified genetic cause of a human circadian rhythm disorder, familial advanced sleep phase syndrome, is due to a mutation in human *Per2*, a human homolog of the *Drosophila per* gene (Box 41.2). Furthermore, recent studies suggest that the neural mechanisms controlling *Drosophila* circadian behavior may also provide useful insights on mammalian circadian rhythms. In mammals, vasoactive intestinal polypeptide (VIP) plays a similar role as PDF in the fly. In both systems, a small subset of peptide-expressing pacemaker cells remain

---

BOX 41.1 *(cont'd)*

rhythmic under constant conditions, and are believed to synchronize each other and maintain rhythmicity in other clock cells. Moreover, mutations affecting the *Drosophila* and mammalian ligands (PDF or VIP) or their receptors lead to a loss of coordinated rhythms among the clock neurons, resulting in arrhythmic behavior after several days in constant conditions. Thus, even the mechanisms that coordinate neurons of the circadian network appear to have been preserved, despite gross differences in the anatomy of insect and mammalian brains.

Ania Busza and Patrick Emery

Further Readings

Hardin, P. E. (2005). The circadian timekeeping system of *Drosophila*. *Curr Biol* **15**, R714–722.

Helfrich-Förster, C. (2005). PDF has found its receptor. *Neuron* **48**, 161–163.

Konopka, R. J. and Benzer, S. (1971). Clock mutants of *Drosophila melanogaster*. **68**, 2112–2116.

Taghert, P. H. and Shafer, O. T. (2006). Mechanisms of clock output in the *Drosophila* circadian pacemaker system. *J Biol Rhythms* **21**, 445–457.

Weiner, J. (2000). "Time, Love, Memory: A Great Biologist and His Quest for the Origins of Behavior." Alfred A. Knopf, New York.

---

several basic helix-loop-helix, PAS-domain containing (bHLH-PAS) transcription factors (CLOCK and BMAL1, or NPAS2 and BMAL1) activate transcription of several genes, including *Period* and *Cryptochrome* genes, through E-box elements. High-amplitude rhythmicity in the rate of transcription of two *Period* genes (*Per1* and *Per2*; Fig. 41.6) is believed to be rate-limiting for formation of complexes between the PERIOD proteins and products of *Cry1* and *Cry2*. The CRY proteins are the primary transcriptional inhibitor proteins that disrupt the activity of the CLOCK:BMAL1 complex (Kume *et al.*, 1999; Sato *et al.*, 2006). The temporally regulated nuclear accumulation of PER/CRY complexes is believed to be critical for closing the feedback loop, disrupting the transcriptional activation of the *Per* and *Cry* genes (Lee *et al.*, 2001). This molecular oscillation, with alternation between a phase of transcriptional activation and a phase of PER/CRY-mediated repression, occurs with a cycle length of approximately 24 hours and results in rhythmic expression of a large number of clock-controlled and output genes.

Posttranslational modifications build a time delay into the oscillatory mechanism. Without this delay step, transcription of *Per* genes rapidly would lead to nuclear accumulation of PER:CRY complexes and a premature inhibition of their transcriptional activation, and the clock would either cycle with a period substantially less than 24 hours, or (if the feedback was very quick) the system would equilibrate to a nonrhythmic steady state. Phosphorylation of the PER proteins alters their intracellular localization, protein:protein interactions, and stability, and thereby contributes to regulating circadian cycle length. Rhythmic phosphorylation of PER proteins is due at least in part to the activity of casein kinase 1 epsilon and casein kinase 1 delta, and mutations of these kinase genes alter circadian cycle length in flies, hamsters, mice, and humans. A circadian-based sleep disorder in humans is due to alterations in casein kinase-mediated phosphorylation of PER2 proteins, leading to rhythms having an aberrantly early phase (Box 41.2). Other important posttranslational processes likely further regulate protein stability; enzymes involved in proteasomal degradation of clock proteins likely influence circadian cycle length (Gatfield and Schibler, 2007).

An interlocked, second feedback loop is important for the rhythmic regulation of BMAL1. Among the E-box driven genes regulated by CLOCK and BMAL1 are two members of the orphan nuclear receptor gene family, *Rev-erb-alpha* and *Rora*. The protein products of these genes compete for binding to a orphan nuclear receptor binding site in the promoter of the *Bmal1* gene, and have antagonistic effects on *Bmal1* gene expression. Overexpression of the repressor, REV-ERB-α, in the liver greatly reduces *Bmal1* expression in this tissue, leading to loss of tissue-autonomous rhythmicity (Kornmann *et al.*, 2007). Thus, this loop is essential for regulating the production of BMAL1. Although rhythmicity of BMAL1 production is of relatively minor importance for maintaining rhythmicity, there is an absolute requirement for BMAL1 in the circadian oscillator (see later).

Molecular Redundancies Revealed by Studies of Knockout Mice

The importance of the genes just discussed in the circadian mechanism has been revealed in part through studies of mice with targeted gene disruption. These

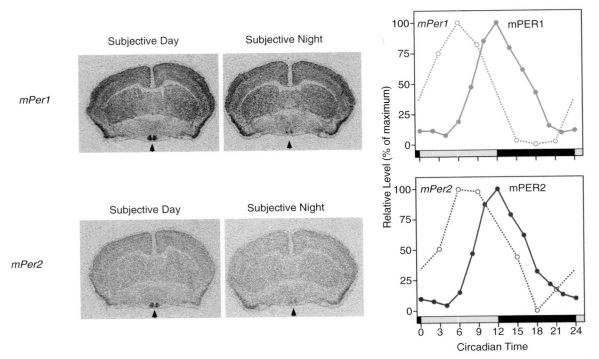

FIGURE 41.6 High-amplitude circadian rhythms of *mPer* gene expression in the mouse SCN. **Left:** Adjacent sections through the SCN from brains collected at the time of peak Per transcript levels during the subjective day (left column) or during the middle of the subjective night (right column) were processed for *in situ* hybridization to detect *mPer1* and *mPer2* mRNAs. "Subjective" refers to the fact that the animals were in constant darkness on the day of tissue collection, rather than in a lighting cycle. The SCN are indicated by an arrowhead in each panel. **Right:** *Per* RNA (open circles and dashed lines) and PER protein (filled circles and solid lines) rhythms in mouse SCN. *mPer1* (upper panel) and *mPer2* (lower panel) RNA levels were assessed by *in situ* hybridization from samples collected on the first day in constant darkness. Protein levels were determined by counting nuclei within the SCN from sections processed for immunohistochemical detection of mPER1 and mPER2. The bar below the panel represents the lighting cycle the animals were exposed to on days prior to the day of tissue collection. Circadian time 0–12 corresponds to subjective day, when the lights would have been on had the animals remained in a lighting cycle.

studies reveal that, in most cases, there is at least partial redundancy of function among members of these gene families (Table 41.1). Mice homozygous for mutation of either *mPer1* or *mPer2* alone have period defects and/or a gradual loss of rhythmicity, whereas mice homozygous for disruption of both *mPer1* and *mPer2* become arrhythmic immediately in constant darkness. Similarly, mice with targeted disruption of either *mCry1* or *mCry2* have altered circadian period length, whereas the double-mutant mice lose circadian rhythmicity (e.g., become arrhythmic) immediately upon placement in constant darkness. Whereas mice homozygous for a semidominant mutation in the *Clock* gene (*Clock^{Δ19}*) have severe behavioral and molecular phenotypes, mice with targeted disruption of *Clock* (and thus lacking CLOCK protein) unexpectedly retain behavioral rhythmicity. The closely related bHLH-PAS transcription factor NPAS2 (also called MOP4) apparently maintains the circadian clock in the absence

of CLOCK (Fig. 41.7). CLOCK appears to be much more abundant than NPAS2 in the SCN, and under normal circumstances it appears that transcriptional regulation occurs primarily through CLOCK:BMAL1 heterodimers. It remains possible that specific genes, or specific tissues, may be controlled to a greater extent by NPAS2:BMAL1 heterodimers.

An exception to the general rule of redundancy within gene families is BMAL1, which is necessary for maintenance of behavioral rhythmicity. A closely related protein, BMAL2, is not able to substitute for BMAL1 in maintaining behavioral rhythmicity.

Clock-Controlled Genes: From the Core Clockwork to Cellular Rhythms

Genes that participate in the core negative feedback loop and are necessary for its function are called "clock genes." The impact of the molecular feedback loop

---

BOX 41.2

CIRCADIAN-BASED SLEEP DISORDERS

The normal timing of sleep in young adult humans is from approximately 11 P.M. to 7 A.M. Most people will voluntarily move sleep to a later time when free of work constraints, and readily readjust their schedule back to an earlier time when the work week begins again. Several disorders affecting the timing of sleep and wake have been described, however, that affect the timing of sleep.

Based on information discussed elsewhere in this chapter, the most obvious disorder would be "Non-24-hour sleep wake syndrome," in which individuals fail to entrain to the 24-hour day. This is relatively frequent in blind individuals, and is likely due to inadequate input to the SCN via the retinohypothalamic tract. These individuals have intervals where their biological day and night coincide with the solar day and night, but then the difference between their free-running period (usually ~ 24.5 hours) and the 24-hour day causes them to drift out of phase with the environment. This results in periods of insomnia and daytime drowsiness (when the individuals are out-of-phase) alternating with periods of coordination with the environment. Melatonin, administered at 24-hour intervals, is useful in synchronizing some blind individuals to a 24-hour cycle length.

Two additional disorders are delayed sleep phase syndrome (DSPS) and advanced sleep phase insomnia (ASPS). The former is characterized by late awakening and late bedtimes (typical wake interval from noon to 0400), and an apparent inability to reset the clock to earlier times of day. This disorder frequently begins in adolescence or in college, but is frequently noticed when individuals enter the working world as young adults; it is not considered unusual for young adults to have late-shifted schedules. ASPS, in contrast, is characterized by early morning awakening and an inability to maintain wakefulness into the evening (typical wake interval is from 0400 to 2000).

Studies of familial cases of ASPS have identified two genes which, when mutated, contribute to ASPS. Remarkably, these genes were already discussed in the context of the core circadian feedback loop defined in mice. The mutations defined in humans are a Serine → Glycine mutation at position 662 of human PER2 (within the casein kinase interacting domain of PER2), and a Threonine → Alanine mutation at position 44 of casein kinase I delta (a mutation that destroys the ability of the kinase to phosphorylate substrates, including PER2). Many cases of ASPS are sporadic, and there is a natural tendency for normal, healthy elderly to live in an advanced phase (for reasons that are unclear). Nevertheless, finding that ASPS can be linked to mutations in human genes that are part of the core circadian oscillator strongly suggests that clock mechanisms studied in mice provide important information on the regulation of human rhythmicity.

Environmentally induced sleep disruption is perhaps more familiar. Jet lag is induced by flying across several time zones, and represents the lack of coordination of the flyer's body to the new local time. Individuals usually entrain to a phase-shifted environment at the rate of about 1 hour per day, although careful control over when light exposure occurs may improve this rate.

Another environmentally imposed disruption of circadian rhythms and sleep occurs with shift work. Most people working the night shift remain entrained to the solar day, so that their biological rhythms (e.g., alertness, temperature, and hormonal rhythms) are in opposition to the demands of the work schedule. Some of the negative health effects of shift work may be due to the stress caused by trying to sleep when the biological clock dictates it is time to be awake, and trying to be alert and productive when our body expects rest. Clear performance deficits occur when individuals attempt to work for extended periods, ignoring the need for sleep and the deterimental effects of performing complex job tasks at an adverse circadian phase.

Further Readings

Jones, C. R., Campbell, S. S., Zone, S. E. *et al.* (1999). Familial advanced sleep-phase syndrome: A short-period circadian rhythm variant in humans. *Nat Med* **5**, 1062–1065.

Lockley, S. W., Landrigan, C. P., Barger, L. K. *et al.* (2006). When policy meets physiology: The challenge of reducing resident work hours. *Clin Orthop Relat Res* **449**, 116–127.

Lu, B. S. and Zee, P. C. (2006). Circadian rhythm sleep disorders. *Chest* **130**, 1915–1923.

Toh, K. L., Jones, C. R., He, Y. *et al.* (2001). An *hPer2* phosphorylation site mutation in familial advanced sleep phase syndrome. *Science* **291**, 1040–1043.

Xu, Y., Padiath, Q. S., Shapiro, R. E. *et al.* (2005). Functional consequences of a CKIdelta mutation causing familial advanced sleep phase syndrome. *Nature* **434**, 640–644.

TABLE 41.1   Gene Families Regulating Mammalian Circadian Rhythms

Gene family	Gene	Mutation	Allele type	Effect on behavioral rhythmicity
bHLH/PAS	Clock	Clock$^{\Delta 19}$ (exon skipped)	Dominant negative	Long period then arrhythmic
	Clock	Targeted disruption	Null	Slightly short period
	Npas2 (Mop4)	Targeted disruption	Deletion	Slightly short period
	Clock + Npas2			Immediately arrhythmic
	Bmal1 (Mop3)	Targeted disruption	Null	Immediately arrhythmic
Period	Per1	Targeted disruption	Null	Short period, gradual arrhythmicity
	Per2	Targeted disruption	Deletion	Short period, gradual arrhythmicity
	Per2*	Missense (S662G)	Dominant negative	Short period, advanced phase
	Per3	Targeted disruption	Null	Slightly short period
	Per1 + Per2			Immediately arrhythmic
	Per1 + Per2 + Per3			Immediately arrhythmic
Cryptochrome	Cry1	Targeted disruption	Null	Short period
	Cry2	Targeted disruption	Null	Long period
	Cry1 + Cry2			Immediately arrhythmic
Casein kinase	CKIδ*	Missense (T44A)	Dominant negative	Short period, advanced phase
	CKIδ	Targeted disruption	Null	Not known; lethal when homozygous
	CKIε**	Missense (R178C)	Dominant negative	Short period, advanced phase
Orphan nuclear receptor	Rev-erb-alpha	Targeted disruption	Null	Short period
	RorA	Deletion (Staggerer)	Null	Rhythm instability

Mutations and mutant phenotypes listed were studied in mice, except as noted by asterisks: * indicates human, ** indicates Syrian hamster. Missense mutations are those that lead to a change in a single amino acid residue; the residue and the change are indicated in single letter format: S662G indicates that a Serine residue at position 662 of human PER2 is mutated to glycine.

Null alleles are those without detectable protein products. Alleles designated as "Deletion" produce some protein products, but the deletion of genomic material results in deletion-mutant protein products that appear to lack functional activity. From a functional point of view, these are viewed as null alleles. "Null" alleles are those where there is either no product or the product is so truncated that it is irrelevant. "Dominant negative" alleles result in protein products that are altered in activity, such that the mutant product interferes with the activity of the product from the normal allele, thus producing a mutant phenotype in heterozygotes and a more severe phenotype in homozygotes (semidominant inheritance).

Additional genes not listed may modify the circadian feedback loop in more subtle ways. Only mutations that disrupt gene function are listed; some of these proteins also produce a circadian phenotype when their protein product is overexpressed (e.g., in transgenic animals).

occurs through the rhythmic regulation of additional genes, so-called "clock-controlled genes," which are distinguished from clock genes by the fact that they do not feed back on the core oscillatory mechanism in a manner that makes them necessary for rhythmicity. Some clock-controlled genes with E-box elements in their regulatory regions are controlled in the same manner as Pers and Crys: their transcription is activated directly by CLOCK:BMAL1 and inhibited by the activity of the PER/CRY complexes disrupting this transcriptional activation. The regulation of arginine vasopressin (AVP) in the SCN is a prototype for this mechanism (Jin et al., 1999). The rhythmic expression of the AVP gene leads to a high-amplitude circadian rhythm of AVP in brain and in the cerebrospinal fluid that is insulated from the osmotic regulation of AVP in blood. The rhythmic synthesis and secretion of AVP from SCN neurons may modulate electrical activity of nearby SCN neurons, increasing the amplitude of the SCN neuronal firing rhythm and thus affecting many rhythms. AVP of SCN origin also appears to play a role in regulation of the hypothalamo–pituitary–adrenal axis.

Notably, products of some genes regulated by the CLOCK:BMAL1 complex are themselves able to regulate transcription through additional regulatory elements. For example, a family of D-element binding proteins (DBP/TEF/TBP and E4BP4) are controlled via E-box elements, and these proteins compete for occupation of elements in responsive genes. Thus, the

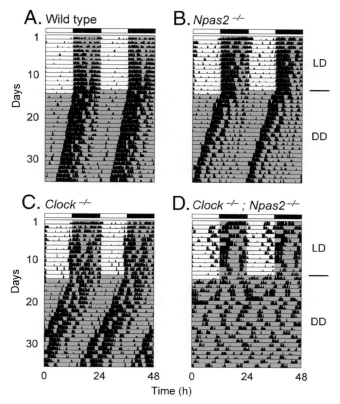

FIGURE 41.7 Molecular redundancy in the maintenance of mouse circadian rhythms. Panels A–D show double-plotted actograms from a wild-type mouse (A), an NPAS2-deficient mouse (B), a CLOCK-deficient mouse (C), and double-mutant CLOCK-deficient, NPAS2-deficient (D) mouse. Note that the double-mutant mouse becomes arrhythmic when placed in constant darkness (DD), whereas the mice of the other genotypes "free run" with a cycle length of less than 24 hours. The apparent rhythmicity when the double-mutant mouse is housed in a light-dark cycle (LD) is likely due to partial suppression of locomotor activity by light, a phenomenon called negative masking. Plotting conventions as in Fig. 41.1. Actograms provided by Jason DeBruyne, UMass Medical School (DeBruyne *et al.*, 2007).

rhythmic regulation of these transcription factors, and the multiple regulatory element combinations that can occur in responsive genes, results in a cascade leading to rhythmic gene expression with many possible peak times of expression and with tissue-specific regulation (Ueda *et al.*, 2005).

Summary

The molecular basis for circadian oscillation is a transcriptional feedback loop. For rhythmic gene expression, a phase of transcriptional activation alternates with a phase of transcriptional repression, and the cycle length of this rhythm is approximately 24 hours. The bHLH-PAS proteins CLOCK and NPAS2

heterodimerize with BMAL1 to produce E-box dependent transcriptional activation. Among the genes controlled primarily by E-box regulation are *Per1* and *Per2*; their high-amplitude temporal regulation is critical to determining the timing of the phase of transcriptional repression, through regulating the formation of inhibitory complexes containing PER, CRY, and casein kinase proteins. Casein kinases phosphorylate the PER proteins, influencing their abundance and activity. This is likely important for building a posttranslational time delay into the feedback loop. The CRY proteins are the primary inhibitory components of the complex. In parallel with the regulation of "clock genes," the feedback loop regulates the expression of effector genes and also of transcription factors that widen the influence of the circadian clock to affect key (often rate-limiting) metabolic steps in a tissue-specific manner.

CIRCADIAN PHOTORECEPTION

In the following sections, we will discuss the mechanisms by which the circadian clock in the mammalian SCN is synchronized to the daily light:dark cycle.

Entrainment of the SCN Pacemaker Occurs through Daily Phase Shifts

The intrinsic period of the circadian pacemaker is revealed by measuring it in constant environmental conditions, or in animals made insensitive to light (e.g., by removal of the eyes). Rhythms that are not entrained (e.g., not synchronized by a rhythmic environmental cue) are referred to as "free-running" rhythms The period length of rhythms that are free-running is usually not exactly 24 hours (Fig. 41.1). Instead, the circadian oscillator has a species-typical intrinsic period. Rats housed in constant darkness have a free-running period of approximately 24.3 hr, whereas mice under the same conditions free-run with a period of approximately 23.6 hr. There are strain variations within these species as well. For an animal to entrain to a light:dark cycle that has a 24.0-hour cycle length, light must shift the animal's clock each day by an amount equal to the difference between their intrinsic, free-running period and the 24-hour period of the lighting cycle. Although it might seem simpler for the clock to have a cycle length of exactly 24 hours (and forget about daily readjustments), there is seasonal variation in lighting conditions in natural environments, most notably a seasonal change in the length of the light period each day, and a timekeeping system that tracks dusk and dawn each day is needed for accurate entrainment under these circumstances.

In assessing the response of the circadian system to light, it is often more informative to assess the response of animals to brief light pulses rather than to full light-dark cycles (Fig. 41.8). A brief pulse of light is provided when an animal is free-running in constant conditions, allowing control over the timing, intensity, and even wavelength of the stimulus. Experiments of this type reveal that the response of the circadian system to light is very dependent upon the time of the day at which the light exposure occurs (Fig. 41.8). In a night-active (nocturnal) animal like a mouse or hamster studied in constant conditions, the time when the animal "thinks" it is night (as reported by its locomotor activity rhythm) is referred to as "subjective night," and the time that corresponds to internal daytime (the inactive period) is "subjective day." Light falling soon after the animal becomes active (early in the subjective night) leads to a phase delay, for example, activity on subsequent

days begins at a later time, consistent with a light-induced shift in the timing of the oscillator. Conversely, light exposure late in the night will cause a phase advance. During large portions of the animal's subjective day, light exposure does not lead to a phase shift. Plotting the response to brief pulses of light occurring at different times in the animal's rest/activity cycle leads to generation of a phase response curve (Fig. 41.8). In general, the shape of the phase response curve to light is very similar across species, whether nocturnal or diurnal and regardless of whether their intrinsic period is less than or greater than 24 hours. Many pharmacological stimuli also cause phase-dependent phase shifts of the SCN oscillator, and the shape of the phase response curve to nonphotic stimuli differs from the phase-response curve to light.

A Specific Retinal Photoreceptive System Mediates Circadian Entrainment

In mammals, detection of light by the circadian system requires the presence of the retina. Remark-

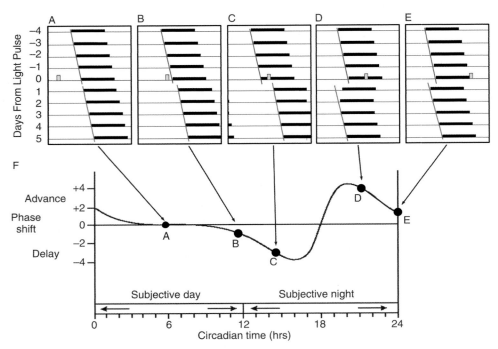

FIGURE 41.8 Schematic representation of the phase-dependent phase shifts that occur following exposure of free-running animals to brief pulses of light. (A–E) Each panel represents an actogram of one animal housed in constant darkness and exposed to one, 1-hr light pulse (yellow box). The phase shift is the difference, on the day of the light pulse, between a line drawn through activity onsets before the light pulse (blue lines) with a line through activity onsets after the light pulse (yellow line). In some cases, the behavioral shift is not complete on the first cycle after the light exposure, and the shift seems to occur over several days. The intermediate activity onsets are called "transients" and are ignored when determining the phase after the shift. (F) Phase response curve. The phase shift induced by a brief light pulse is highly dependent upon the time at which the light exposure occurred. Circadian time (CT) 12 is defined as the time of activity onset in a nocturnal animal. Light exposure occurring during the biological daytime (CT 0–12) does not cause a shift. Responses occur only during the animals' biological night-time. Early in the subjective night, light exposure delays the clock, so activity begins at a later time on subsequent cycles. (C) Late in the subjective night, light exposure causes a phase advance as detected by earlier activity onsets after the light pulse. (D,E) Phase-dependent phase shifts allow the circadian oscillator to be entrained to environmental cycles.

ably, however, the retinal photoreceptors that are required for visual image formation are not needed for circadian photoreception. Studies of mutant mice show that circadian sensitivity to light persists even in mice with virtually complete loss of the rods and cones. Other nonvisual responses to light (inhibition of locomotor activity, suppression of nocturnal melatonin secretion, and pupillary constriction) also occur in the absence of rods and cones. These findings suggest that a photoreceptive protein expressed in the inner retina mediates nonvisual responses to light.

In mammals, a small percentage of retinal ganglion cells express an opsin-like pigment, melanopsin, that initially was identified as a putative photopigment in the light-responsive dermal melanophores of frog skin (Provencio et al., 1998, 2000). More recently, melanopsin has been shown to contribute to circadian photoreception in the mammalian retina (for review see Peirson and Foster, 2006). Melanopsin-containing retinal ganglion cells respond directly to light, even when isolated from the rods and cones that normally provide their input, and their responsiveness requires the presence of melanopsin. These intrinsically photosensitive retinal ganglion cells (ipRGCs) have extremely large dendritic trees (spanning several hundred micrometers) in the retina, making them ill-suited for spatial localization (Fig. 41.9A). Remarkably, the phototransduction mechanism in melanopsin-containing cells appears to be more similar to the invertebrate visual signaling cascade (in which phosphoinositide signaling activates cation channels of the transient receptor potential channel type), rather than through the typical vertebrate photoreceptor cascade involving transducin, cGMP, and cyclic nucleotide gated ion channels (Peirson and Foster, 2006).

A direct anatomical connection from the retina to the SCN, the RHT, was first identified in the early 1970s. The axons of melanopsin-containing ipRGCs comprise the majority (ca. 80%) of the RHT fibers (Fig. 41.9B). The RHT is necessary for entrainment of circadian rhythms. The RHT also appears sufficient for entrainment, as destruction of virtually all other retinal pathways does not disturb photic entrainment of circadian rhythms. Thus, the melanopsin-expressing ganglion cells are ideally suited for a role as luminance detectors mediating circadian photoreception. Axons from the melanopsin-containing ganglion cells also project to the intergeniculate leaflet, modulating entrainment, and to the olivary pretectal nucleus, a site critical for pupillary constriction in response to light (Fig. 41.9C, D). Additional projections of the melanopsin-expressing retinal ganglion cells have been described that contribute to other, nonvisual responses to illumination (Fig. 41.9E). These include

projections to the ventrolateral preoptic area, important in the regulation of sleep, and to the ventral subparaventricular zone (SPZ) implicated in the acute inhibition of locomotor activity by light (called "negative masking").

Despite being well-positioned to serve as *the* circadian photoreceptive molecule, melanopsin is in fact not the only photopigment involved in circadian photoreception. Melanopsin-deficient mice have a reduction in the effects of light on phase shifting, negative masking of locomotor activity, and pupillary responses at high light intensities, but these responses do persist so long as rods and cones are intact. Importantly, mice that lack rod and cone function and also lack melanopsin appear completely blind with respect to both image formation and nonimage forming responses to light (Hattar et al., 2003; Fig. 41.10; see also Panda et al., 2003). Rods and cones thus appear able to signal through other retinal ganglion cells or through the melanopsin-expressing RGCs, even if the latter cells are melanopsin-deficient. It remains to be determined whether destruction of the melanopsin-expressing cells prevents circadian photoreception. If this is the case, it will indicate that the melanopsin-expressing ipRGCs represent a specific conduit or "labeled line" through which information for nonvisual photoreception reaches the brain. The distinct pattern of anatomical connections formed by the melanopsin-expressing RGCs suggests that this is likely (Fig. 41.9E).

Retinal Input to the SCN: Anatomy and Neurochemistry

The melanopsin-expressing retinal ganglion cells that project to the SCN in the retinohypothalamic tract use glutamate as their primary neurotransmitter. A large percentage of these retinal neurons also contain the neuropeptides Substance P and/or Pituitary Adenylate Cyclase Activating Peptide (PACAP). Application of these agents to SCN slices *in vitro* causes phase-dependent phase shifts of intrinsic SCN rhythms. Furthermore, glutamate receptor antagonists block light-induced phase shifts *in vivo*, revealing the functional importance of glutamatergic input to the SCN.

Axon collaterals of the melanopsin-positive retinal ganglion cells innervate neurons in the intergeniculate leaflet that express neuropeptide Y (NPY). These NPY-containing neurons project to the SCN as the geniculohypothalamic tract. The geniculohypothalamic tract is thought to play a major role in nonphotic phase shifting, a phenomenon in which behavioral arousal (usually accompanied by intense locomotor activity) during the circadian day leads to increased NPY release

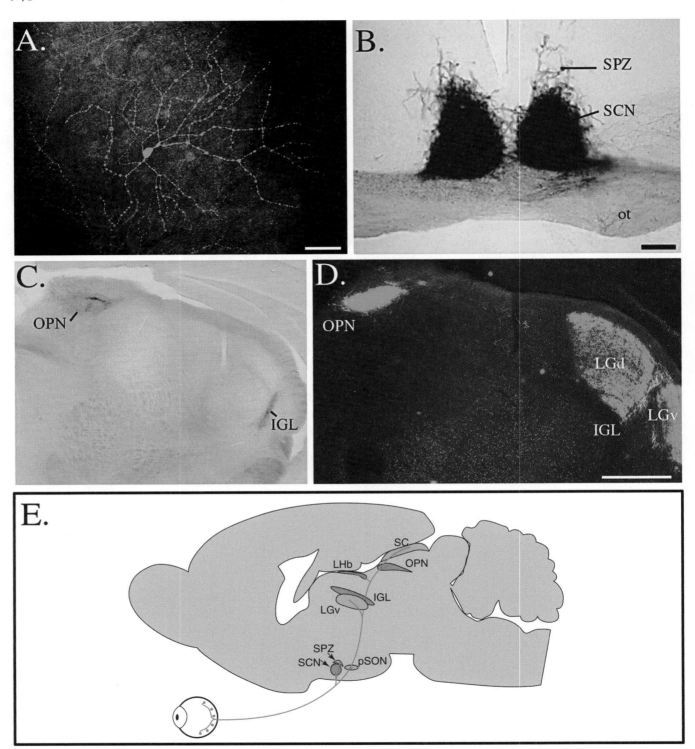

FIGURE 41.9 Circadian photoreception occurs primarily through retinal ganglion cells containing melanopsin. (A) Melanopsin-expressing retinal ganglion cells in the mammalian retina have large dendritic trees. These cells are ill-suited for spatial localization, but are ideally suited for luminance detection. This image shows an ipRGC's immunolabeled with an anti-melanopsin antiserum, in a flat-mounted human retina. Melanopsin immunoreactivity appears green. Note the localization of melanopsin in the cell body and throughout the dendritic arbor rendering the entire cell photosensitive. Also note the characteristic varicosities decorating the dendrites. Image provided by Ignacio Provencio, University of Virginia. Scale bar = 100 microns. (B) The SCN receives a strong, bilateral input from melanopsin-containing RGCs. Central projections of melanopsin-expressing retinal ganglion cells were detected using a beta-galactosidase reporter gene inserted into the melanopsin locus. Tissue was processed for histochemical detection of beta-galactosidase activity, resulting in deposition of blue reaction product in cells and processes containing melanopsin. Melanopsin-containing cell bodies are present only in the retina, so the blue staining seen in the brain represents axons of these retinal neurons. The right eye of the animal had been removed over 2 weeks prior to staining. Labeled, retinal fibers in the optic tract (ot) have degenerated on the apparent left side of the image, leaving weak, granular labeling. The staining in the SCN remains strong on both sides, indicating the extensive, bilateral projections to the SCN. (In contrast, other projections of these cells are ~95% crossed.) Melanopsin-containing processes also extend into the subparaventricular zone (SPZ). Scale bar = 100 microns. (C) Central projections of melanopsin-containing retinal ganglion cells, detected using beta-galactosidase activity as discussed in panel B. The olivary pretectal nucleus (OPN) and intergeniculate leaflet (IGL) receive a crossed input from melanopsin-expressing retinal ganglion cells. Image provided by Samer Hattar, Johns Hopkins University School of Medicine (Hattar *et al.*, 2006). (D) Melanopsin is localized within a subset of retinal projections. The anatomical level is comparable to Panel C. The projections of the entire population of retinal ganglion cells were revealed by intraocular injection of cholera toxin B (CTB) subunit three days prior to tissue collection, allowing anterograde labeling of the fibers of all retinal ganglion, followed by detection of the toxin (green). Note the intense labeling of the dorsal subdivision of the lateral geniculate nucleus (LGd). Melanopsin expression in retinal terminals was detected by red immunofluorescence; in this merged image, the overlap of red and green channels in the melanopsin-containing retinal ganglion cells yields yellow labeling in the IGL and OPN. Scale bar in D (applies to C and D) = 500 microns. Image provided by David Berson, Brown University (Hattar *et al.*, 2006). (E) Schematic illustration of the axonal projections of melanopsin-expressing retinal ganglion cell. Principle targets (SCN, OPN, IGL, and lateral habenula [LHb]) are indicated by dark red shading. Secondary targets indicated by light red shading are the SPZ, superior colliculus, peri-supraoptic nucleus (pSON), and ventral subdivision of the lateral geniculate nucleus (LGv). Minor targets (not shown) include the preoptic area, bed nucleus of the stria terminalis, medial amygdaloid nucleus, anterior hypothalamic area, lateral hypothalamus, LGd, and periaqueductal gray. Adapted from Hattar *et al.*, 2006.

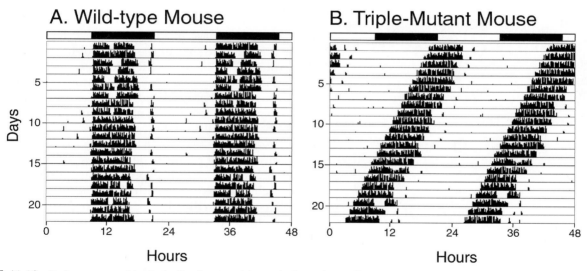

FIGURE 41.10 Rods, cones, and intrinsically photosensitive retinal ganglion cells expressing melanopsin mediate circadian entrainment. Panel A is a double-plotted actogram of a control mouse showing normal entrainment to the light : dark cycle. Panel B shows a double-plotted actogram from a triple-mutant mouse, lacking rod and cone phototransduction mechanisms and also lacking melanopsin. Triple-mutant animals free-run as though there is no lighting cycle. Triple-mutant animals also lack pupillary constriction responses and acute inhibition of locomotor activity by light (Hattar *et al.*, 2003). The disruption of rod and cone signaling is accomplished by breeding to generate animals homozygous for disruption of the alpha subunit of rod transducin and the cone cyclic nucleotide gated channel A3 subunit. Actograms provided by Samer Hattar, Johns Hopkins University School of Medicine.

within the SCN and leads to a phase shift of activity rhythms. The indirect retinal pathway through the intergeniculate leaflet also appears to modulate photic entrainment.

The direct RHT and the indirect retino-geniculohypothalamic pathway converge in the SCN, in the "core" region defined by the dense retinal input and by a subpopulation of neurons expressing vasoactive intestinal peptide (VIP). As noted earlier, VIP appears to play a key role in synchronizing the activity of SCN neurons. Another major input to the SCN core is a serotonergic input from the raphe nuclei, which modulates the impact of light input to the SCN through presynaptic serotonin receptors on RHT terminals.

Neurons of the core region project heavily within the SCN, to both the core and the surrounding "shell" region, and also beyond the SCN borders, primarily to the subparaventricular zone (SPZ) of the hypothalamus, a region along the third ventricle lying just caudal and dorsal of the SCN (Fig. 41.11; see later). The SCN shell receives inputs from the SCN core, and also from other hypothalamic and thalamic regions, limbic cortex, and brain stem. The core/shell organization of the SCN is most readily apparent in the Syrian hamster. There is increasing evidence for functional, as well as phenotypic, differences among neuronal subpopulations within the SCN (Silver and Schwartz, 2005). Indeed, molecular, phenotypic, and functional heterogeneity of the SCN strongly suggests that a simple subdivision of the SCN into core and shell is inadequate for describing its complexity (Morin, 2007).

Molecular Mechanisms of Entrainment: Light-Induced *Per* Expression

Per1 and *Per2* gene expression are rhythmic in the SCN, with high levels during the daytime (Fig. 41.6). Peak nuclear PER protein levels are achieved at the time of the light-to-dark transition. These molecular rhythms persist in constant darkness. Exposure to light at night causes a transient increase in both *Per1* and *Per2* transcript levels. Light-induced increases in PER1 and PER2 protein also occur, but are difficult to detect early in the night due to the high levels of PER already present. As with many other signaling responses, *Per* gene expression in the SCN is unresponsive to light exposure during the subjective daytime. (A third *Period* gene, *Per3*, is unresponsive to light even at night. *Per3* appears to be an output gene, rather than being part of the core circadian oscillatory mechanism or the resetting mechanism.)

An increase in PER1 and PER2 levels early in the night would extend the period of transcriptional inhibition by the PER/CRY complex. This may be the mechanism by which light exposure early in the night causes a phase delay. In contrast, light-induced PER expression late in the night would lead to an earlier-than-normal initiation of PER/CRY-mediated inhibition, resulting in an advance of SCN molecular rhythmicity in subsequent cycles. Thus, light-induced alterations of PER expression and interaction of the light-induced PER with the molecular feedback loop may explain the phase-dependent phase shifting effect of light shown in Figure 41.8 (earlier).

Although light does not influence *Per* levels during the animal's subjective daytime and does not cause phase shifts at this time, nonphotic manipulations can cause phase shifts when administered during the daytime (e.g., novelty-induced voluntary locomotor activity, arousal, serotonin receptor activation). These manipulations lead to reduced *Per* gene expression in the SCN. Thus, acute regulation of *Per* gene expression may be the mechanism by which a variety of manipulations affect circadian phase.

Summary

The circadian clock in the SCN is entrained to the daily light-dark cycle by light causing phase-dependent phase shifts each day. The retinae are

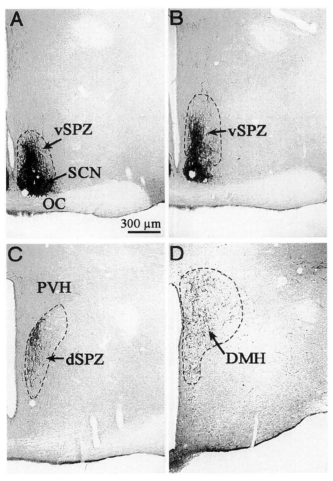

FIGURE 41.11 Distribution of vasoactive intestinal peptide (VIP) immunoreactivity in the rat brain reveals the major output pathways from the SCN. Staining reveals VIP-immunoreactive cell bodies in the SCN and their projections. The major projection of the SCN is to a cell column dorsal and caudal of the SCN, including the ventral (v) and dorsal (d) parts of the subparaventricular zone (SPZ) and the dorsomedial nucleus of the hypothalamus (DMH). (From Lu *et al.*, 2001.)

needed for entrainment, but the retinal photoreceptors necessary for visual image formation, the rods and cones, are not needed. Instead, a population of intrinsically photoreceptive retinal ganglion cells conveys information about the level of light to the SCN, via the RHT, and to other brain regions that regulate nonvisual photoreception. The photopigment in these retinal ganglion cells is melanopsin, and they utilize glutamate and neuropeptides as their transmitters. In the absence of melanopsin, rods and cones appear capable of signaling, perhaps through the ipRGCs, to result in nonvisual responses to light, including entrainment. The impact of nocturnal light exposure on SCN physiology includes numerous signaling responses and induction of *Per* gene expression. Light-induced alterations in *Per* gene expression likely provide a key role in resetting of the molecular feedback loop within SCN neurons.

CIRCADIAN OUTPUT MECHANISMS

From Molecular Feedback Loops to Firing Rate Rhythms

The molecular feedback loop within individual SCN neurons ultimately has its impact on physiological processes through regulation of SCN neuronal firing rate. Rhythmic regulation of potassium channels in the membrane of SCN neurons appears to be critical for producing circadian rhythms in membrane potential and firing rate. Large-conductance, calcium-activated potassium channels (BK channels) are rhythmically expressed in the SCN, with high levels of transcript and of BK channel activity during the night, contributing to the reduction in firing rate at night (Meredith *et al.*, 2006). Disruption of the BK channel (*Kcnma1–/–*) results in higher SCN firing rate selectively at night, and the mutant mice have greatly reduced amplitude of several circadian output rhythms. In addition, fast delayed rectifier potassium currents (Kv3.1b and Kv3.2) are larger in the SCN during the circadian daytime, and blocking activity of these currents leads to loss of rhythmicity in SCN neurons (Itri *et al.*, 2005). It thus appears that multiple ionic mechanisms contribute to rhythmicity in SCN neuronal firing rate, and regulation of channel activity is probably accomplished through both transcriptional and posttranscriptional mechanisms.

Rhythmic firing of SCN neurons not only serves as its mechanism of communication with downstream target regions, but also plays an important role in synchronizing the cellular oscillators within the SCN to each other. When sodium-dependent action potentials are blocked for several days, SCN neurons become desynchronized from one another. The mechanism for this may be as simple as the loss of rhythmic, synaptic release of neurotransmitters including VIP. There is some evidence for gap junction-mediated synchronization of SCN neuronal firing, but little evidence for gap junction- mediated coordination of circadian phase (Aton and Herzog, 2005).

Humoral Outputs from the SCN also Influence Locomotor Activity Rhythms

The rhythmic electrical activity of SCN neurons likely leads to rhythmic neurotransmitter release and modulation of downstream, synaptic targets along defined anatomical pathways as discussed in greater detail later. Several lines of evidence indicate that the SCN also influences locomotor activity through secretion of humoral output molecules. Among these lines of evidence are the findings that (1) transplants of fetal SCN tissue that restore locomotor activity rhythms do not restore other rhythms, including endocrine rhythms, and (2) transplanted SCN tissue encapsulated so as to allow diffusion of small molecules while preventing axonal outgrowth nevertheless can restore locomotor activity rhythms in SCN-lesioned hamsters (Silver *et al.*, 1996). The latter finding suggests that substances secreted by the SCN act in regions near the SCN (but not in the SCN itself) to regulate locomotor activity. Three substances have been identified that are rhythmically expressed in the SCN and inhibit locomotor activity when infused near the SCN, and thus are proposed to play a role in regulating locomotor activity rhythms. These substances are transforming growth factor alpha, cardiotropin-like cytokine 1, and prokineticin 2.

Summary

For oscillations in the abundance of circadian clock gene products in SCN neurons to affect physiology and behavior, the molecular feedback loop must impact on more overt aspects of cellular rhythmicity. One mechanism for this to occur is through circadian regulation of transcripts encoding ion channels, and indeed, two types of potassium channels have been identified that contribute to circadian rhythmicity. Through regulation of potassium channel levels, rhythmic transcription factor activity is translated into rhythmicity in neuronal firing rate and neurotransmitter release. Though much of the output of the SCN comes via synaptic interactions to downstream targets, substances released from the SCN by a humoral mechanism can influence locomotor activity.

BOX 41.3

CHRONOPHARMACOLOGY

Many drugs, toxins, and stressors have effects that vary over the 24-hour day. Many of these variations represent true circadian rhythms in responses (e.g., they persist in constant environmental conditions). Chronopharmacology is the subdiscipline of chronobiology that seeks to identify and understand these rhythms, and is an important field as altering the time of drug administration can have a profound impact on its efficacy, toxicity, and the ratio between these two measures (the therapeutic index). For this reason, there is often a specific time for optimal response to a chemotherapeutic agent, and this optimal timing varies between agents depending on their mechanism of action. Rhythmicity in drug effects could be due to alterations in bioavailability (absorption, metabolism, or excretion) or in end organ sensitivity. Rhythmic variation in therapeutic index has been well documented for several chemotherapeutic agents and anesthetics. There are also robust rhythms in responses to toxic agents, including infectious agents such as *E. coli* and endotoxin derived from bacterial cell walls. Recent studies have begun to unravel the molecular mechanisms underlying rhythms in detoxification and drug sensitivity, and focus attention on rhythmicity within target cell populations.

Further Readings

Gachon, F., Olela, F. F., Schaad, O., *et al.* (2006). The circadian PAR-domain basic leucine zipper transcription factors DBP, TEF, and HLF modulate basal and inducible xenobiotic detoxification. *Cell Metab* **4**, 25–36.

Gorbacheva, V. Y., Kondratov, R. V., Zhang, R., *et al.* (2005). Circadian sensitivity to the chemotherapeutic agent cyclophosphamide depends on the functional status of the CLOCK/BMAL1 transactivation complex. *Proc Natl Acad Sci USA* **102**, 3407–3412.

Levi, F. and Schibler, U. (2007) Circadian rhythms: Mechanisms and therapeutic implications. *Annu Rev Pharmacol Toxicol* **47**, 593–628.

Mormont, M. C. and Levi, F. (2003). Cancer chronotherapy: Principles, applications, and perspectives. *Cancer* **97**, 155–169.

Ohdo, S. (2007). Chronopharmacology focused on biological clock. *Drug Metab Pharmacokinet* **22**, 3–14.

DIVERSITY OF OUTPUT PATHWAYS LEADING TO PHYSIOLOGICAL RHYTHMS

Animals that lack circadian rhythms (those with destruction of the SCN or genetic disruption of key circadian clock genes) survive and these animals generally can reproduce. Nevertheless, the circadian timing system is thought to optimize "internal temporal order" by coordinating rhythms in different tissues, and to allow anticipation of physiological demands imposed by the predictably changing environment. The impact of the circadian oscillator is exerted through the rhythmicity it imposes on physiological functions (Box 41.3).

The following sections will address the regulation of several physiological rhythms. In most cases, we emphasize how these rhythms are "wired" to the SCN through neuroanatomical pathways and emphasize rhythms for which the SCN can be viewed as a driving pacemaker. It is important to note that, to date, there has been relatively little effort to address the role of tissue-autonomous rhythmicity in the neural and peripheral tissues regulated by the SCN. (For important exceptions to this statement, see Kornmann *et al.*, 2006; Granados-Fuentes *et al.*, 2006.)

Amplification of SCN Output and Divergence of Output Pathways

Considering the pervasive influence of the SCN on rhythmic physiology, the direct anatomical output of the SCN is surprisingly limited. SCN efferents project heavily to a column of cells dorsal and caudal of the SCN that includes the subparaventricular zone (SPZ) and the dorsomedial nucleus of the hypothalamus (DMH) (Figs. 41.11, 41.12). Neurons in the SPZ and DMH in turn project to brain regions that ultimately regulate rhythmic activities (Fig. 41.12). Many effector regions receive a small, direct innervation from the SCN that converges with the larger, indirect output via the SPZ. The SPZ thus appears to relay and reinforce the direct output of the SCN, and allows convergence of influences from other brain areas. Lesions that kill cell bodies in the ventral SPZ (while leaving axons passing through this region intact) disrupt the circadian rhythms of sleep/wakefulness and locomotor activity (Lu *et al.*, 2001). Amplification of SCN output in the SPZ is thus necessary for these rhythms; direct

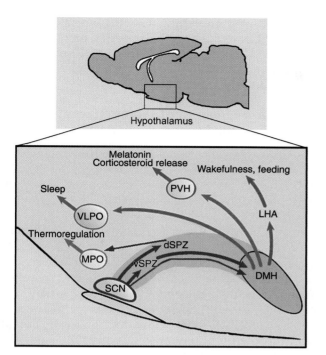

FIGURE 41.12 Schematic representation of neuroanatomical outputs from the rodent SCN. Location of target structures is not accurate to scale. Abbreviations: DMH, dorsomedial hypothalamic nucleus; dSPZ, dorsal subparaventricular zone; LHA, lateral hypothalamic area; MPO, medial preoptic area; PVN, paraventricular nucleus of the hypothalamus; VLPO, ventrolateral preoptic area; vSPZ, ventral subparaventricular zone. (Adapted from Saper *et al.*, 2005.)

SCN output to target sites is insufficient for rhythmicity. In a similar manner, the dorsal part of the SPZ is a key relay for amplifying output from the SCN to the medial preoptic area that generates the rhythm in core body temperature. Fiber-sparing lesions of the dorsal SPZ selectively disrupt the body temperature rhythm while leaving sleep and activity rhythms intact (Fig. 41.12).

Another level of amplification and integration occurs in the DMH, which is heavily innervated by the ventral SPZ as well as receiving direct innervation from the SCN (Figs. 41.11, 41.12). Lesions of the DMH specifically disrupt circadian rhythms of sleep, locomotor activity, corticosterone secretion and feeding, but not body temperature (Chou *et al.*, 2003).

Circadian Regulation of Sleep and Wake

Among the most widely recognized roles of the circadian timing system is in the regulation of behavioral state. Our level of alertness is strongly influenced by two biological factors, the time spent awake and the circadian time, and also by environmental factors including lighting level and the degree of monotony in the task. Studies conducted with human subjects reveal a strong and opposing influence of circadian time and the time spent awake, a phenomenon described as the "opponent-process" model for behavioral state control. With increasing time awake, subjects experience (see Fig. 42.6) a greater homeostatic drive to sleep. The circadian timing system opposes this drive, by providing an arousing signal that reaches peak levels late in the afternoon/early evening in humans (Chapter 42).

The influence of the circadian timing system on sleep is mediated by SCN efferents that only indirectly impact upon the brain regions that regulate sleep (see Chapter 42). As noted earlier, a pathway from the SCN via the vSPZ and DMH is necessary for circadian regulation of sleep. In contrast to the sparse connections of the SPZ to sleep-regulatory areas, the DMH is a primary source of input to the ventrolateral preoptic area (VLPO) and to the neurons of the lateral hypothalamus that secrete the neuropeptide, orexin, that are critical for regulation of sleep. The output from the DMH to the VLPO is GABAergic, therefore promoting wakefulness by inhibiting the "sleep switch" neurons. The output from the DMH to the orexin neurons in the lateral hypothalamus contains glutamate and thyrotropin-releasing hormone, and is presumably excitatory, thus promoting wakefulness. The DMH has few direct projections to the brain stem areas that comprise the ascending arousal system, but the orexin neurons have extensive, excitatory projections to these regions (Saper *et al.*, 2005). Thus, a multistage pathway from the SCN to sleep regulatory centers allows integration of other inputs (emotional and cognitive inputs as well as the homeostatic drive for sleep) and sharp transitions between states, rather than simple and direct command control of behavioral state. The importance of orexin neurons in coordinating the activity of this system is revealed by narcolepsy, a sleep disorder caused by an apparently autoimmune destruction of the orexin-containing neurons in the lateral hypothalamus, and mimicked in animal models by disruption of orexin or orexin receptors.

One relatively small subclass of sleep disorders are the circadian-based sleep disorders. As discussed in Box 41.2, in circadian-based sleep disorders, the problem is with the relative timing of sleep, rather than its initiation or maintenance (as in more classical types of insomnia).

Circadian Regulation of Neuroendocrine Functions

The SCN output pathway leading to rhythms in adrenal corticosteroid secretion starts with a projection from the SCN to the paraventricular nucleus of the hypothalamus (PVN). A subset of neurons in the PVN contain corticotrophin releasing hormone (CRH), and project to the external zone of the median eminence. CRH, released into the pituitary portal system, regulates the release of adrenocorticotropic hormone (ACTH) from the anterior pituitary. ACTH, in turn, is carried in the bloodstream to the adrenal gland, where it stimulates secretion of adrenal steroids. Adrenal steroids feed back to inhibit CRH and ACTH release in a typical neuroendocrine feedback loop. One would expect that a feedback loop of this type has a set point, where a steady state is reached that balances the releasing factor drive to increase secretion with the end-products inhibiting the releasing factors. This feedback loop is modulated by the circadian timing system, however: the set point for feedback apparently varies over the course of the day, such that higher corticosteroid levels are required to reduce ACTH release at some times of day than at others. As a result, corticosteroid levels vary over the course of the day with a sinusoidal rhythm. Light exposure at night can directly activate the sympathetic innervation of the adrenal gland, leading to increased corticosteroid production (Ishida et al., 2005).

The SCN also regulates circadian rhythms in other neuroendocrine activities. In rodents, the timing of ovulation is strictly regulated and occurs early in the active period on the day of behavioral estrus (see de la Iglesia and Schwartz, 2006). In nocturnal species, ovulation is blocked by injection of the anesthetic, pentobarbital, early in the afternoon on the day in which ovulation is expected. Remarkably, after pentobarbital blockage of ovulation, ovulation occurs the following day—exactly 24 hours after it should have occurred. Ovariectomized, estradiol-treated female rats also have daily surges of luteinizing hormone in the late afternoon, indicating the neuroendocrine equivalent of ovulation even in the absence of the ovaries. These luteinizing hormone surges persist when animals are housed in constant darkness, and are not present in animals with SCN lesions. There is an SCN-dependent output that leads to daily activation of gonadotropin releasing hormone-containing neurons. This SCN output is an ipsilateral neuronal connection (rather than a humoral one) in hamsters (de la Iglesia et al., 2003).

Twice-daily prolactin surges can be induced in female rats by mating or by vagino-cervical stimulation that mimics mating and induces pseudopregnancy. The prolactin surges are timed by the SCN. These prolactin surges are ultimately regulated by dopamine released from the median eminence. Feedback of prolactin on the hypothalamic dopamine neurons is also important. The anatomical pathway from the SCN to the hypothalamic dopamine neurons has not been described in detail, but VIP neurons in the SCN and oxytocin neurons in the PVN appear to be involved (Egli et al., 2004).

Circadian Regulation of Autonomic and Pineal Function

A major class of physiological rhythms is in autonomic function. These pathways originate in the SCN, utilize GABA as a neurotransmitter, and generally project to the paraventricular nucleus of the hypothalamus (PVN). Neurons in the PVN (many containing oxytocin) have direct projections to the spinal cord, where they regulate the activity of autonomic preganglionic neurons. Additional pathways include modulation of vagal activity. These pathways regulate autonomic responses leading to circadian variation in blood pressure and heart rate. These pathways also influence metabolism through direct innervation of fat pads and the liver, the latter of which regulates the circadian rhythm in plasma glucose levels (Kalsbeek et al., 2006; Cailotto et al., 2005).

One very well-characterized pathway affecting an output rhythm through autonomic pathways is the pathway from the SCN to the pineal gland, critical for the rhythmic production of melatonin (Fig. 41.13A). GABA released from SCN neurons inhibits neurons in the PVN during the day, and the reduction in SCN GABA release at night apparently allows the PVN neurons to be more active, driving melatonin synthesis (Kalsbeek et al., 2000). The PVN neurons project to autonomic preganglionic neurons in the intermediolateral cell column of the spinal cord. The preganglionic neurons project to the superior cervical ganglion, whose fibers innervate the pineal gland. Norepinephrine from these fibers increases the activity of the penultimate enzyme in melatonin biosynthesis, arylalkylamine N-acetyltransferase (AANAT), leading to melatonin synthesis (Fig. 41.13B). AANAT activity is controlled by a combination of transcriptional mechanisms and switch-like regulation of proteasomal degradation of AANAT protein.

Circadian Regulation of Melatonin Is Critical for Seasonal Reproduction

Several strategies exist to deal with the varying environmental conditions and varying levels of resources that occur over the course of the year. These strategies include seasonal migration to allow highly mobile species to escape challenging conditions (Box 41.4), and seasonal regulation of physiology for those species that remain.

A. SCN-Pineal Pathway

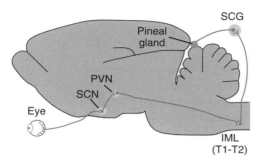

B. Melatonin Biosynthesis

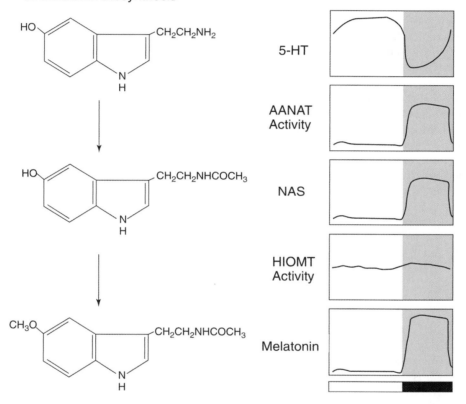

FIGURE 41.13 (A) Schematic illustration of the neuroanatomical circuit regulating pineal melatonin secretion. Photic input detected in the retina is relayed to the suprachiasmatic nucleus (SCN), and from there to the paraventricular nucleus of the hypothalamus (PVN). A subset of PVN neurons projects to the intermediolateral cell column (IML) of the spinal cord. Preganglionic sympathetic cell bodies at thoracic levels 1 and 2 (T1 and T2) project to the superior cervical ganglion (SCG), which innervates the pineal gland. Norepinephrine released from the terminals of SCG neurons is the primary input to the pineal responsible for stimulating melatonin production at night. Light entrains the SCN clock controlling melatonin, but light at night also has an acute inhibitory effect on melatonin synthesis, by disrupting the sympathetic input to the gland. Blood melatonin levels drop within minutes after exposure to light at night. (B) Biosynthesis of melatonin from serotonin (5-hydroxytryptamine, 5-HT). The rate-limiting and highly rhythmic step is the regulation of arylalkylamine N-acetyltransferase (AANAT) activity, which modestly depletes the pineal gland of 5-HT at night as it converts 5HT to N-acetyl-serotonin (NAS). Hydroxy-indole-O-methyltransferase (HIOMT) converts NAS to melatonin (N-acetyl, 5-methoxytryptamine). There is no storage mechanism for melatonin; melatonin diffuses from the pineal into the cerebrospinal fluid and bloodstream as it is produced, making blood melatonin levels a good reflection of the sympathetic input to the gland.

BOX 41.4

LOST WITHOUT A CLOCK

An intriguing function of circadian clocks is their use in time-compensated sun compass orientation, a phenomenon that was first described by Karl von Frisch in foraging honeybees and by Gustav Kramer in migratory birds. Time-compensated sun compass orientation also is used by eastern North American monarch butterflies (*Danaus plexippus*) during their spectacular fall migration. During the fall journey, the butterflies use their navigation skills to fly south from the United States and southern Canada to overwinter in a few restricted sites atop the Transvolcanic Mountains in central Mexico, some having traveled distances approaching 3000 kilometers.

The amazing navigational abilities of monarch butterflies are part of a genetic program initiated in the migrants, as those that make the trip south are at least two generations removed from the previous generation of migrants. Time-compensated sun compass orientation is believed to be an essential component of navigation required for successful migration. Time compensation is provided by the butterfly's circadian clock. The clock allows the butterflies to continually correct their flight direction, relative to changing skylight parameters, to maintain a fixed flight bearing as the sun moves across the sky during the day (Figure 1A). The importance of the circadian clock in regulating the time-compensated component of flight orientation in monarch butterflies has been shown in two ways. First, clock-shift experiments, in which the light-dark cycle in either advanced or delayed, cause predictable alterations in the direction the butterflies fly. Second, constant light, which disrupts the molecular clock mechanism, abolishes the time-compensated component of flight orientation. It is thus clear that time compensation is an absolute requirement for successful migration—without it the butterflies would simply follow the sun (Figure 1B).

Steven M. Reppert

Further Reading

Reppert, S. M. (2006). A colorful model of the circadian clock. *Cell* **124**, 233–236.

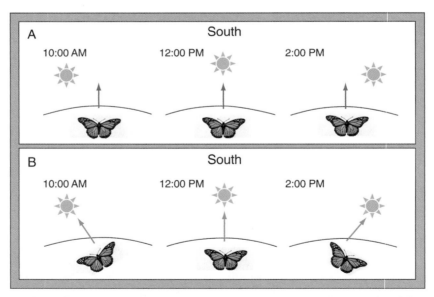

FIGURE 1 A. Monarch butterflies use a time-compensated sun compass to orient south during their fall migration. The butterfly circadian clock allows the butterflies to compensate for the movement of the sun. They are thereby able to maintain a constant bearing in the southerly direction over the course of the day. B. Monarch butterflies follow the sun without a functioning circadian clock. A broken circadian clock would disrupt the migration south, and the butterflies would not be able to travel successfully to their overwintering grounds.

Many mammalian species restrict breeding activity to specific times of year, resulting in birth of young in the spring and summer months when food is abundant and other conditions are most favorable for their survival. In many mammalian species, seasonal reproduction is accomplished through regulation of pineal melatonin secretion by the circadian timing system. Responses to seasonal changes in day length (so-called "photoperiodic responses") include seasonal reproduction, but also extend to a much wider array of physiological responses including seasonal molting of the coat (usually mediated by alterations in prolactin secretion), thermoregulatory adjustments (daily torpor and hibernation), alterations in metabolism and body weight, and alterations in behavior. These adjustments likely improve the probability of survival through challenging times. Like circadian rhythms, these seasonal responses are driven by a system that anticipates predictable environmental changes and prepares for them in a predictive, rather than reactive, manner.

The SCN-controlled circadian rhythm of pineal melatonin production controls photoperiodic responses in mammals. Melatonin is secreted into the bloodstream at night by the pineal gland. The nocturnal duration of melatonin secretion changes in parallel with the length of the night, making the melatonin signal a reliable internal cue encoding season. Removal of the pineal gland prevents seasonal responses to changes in day length, while administration of melatonin in patterns that mimic the profiles occurring at different times of year can predictably drive reproduction and other photoperiodic responses.

The actions of melatonin are mediated by specific, G-protein coupled receptors. Two receptors subtypes with high affinity for melatonin have been identified in mammals. The Mel_{1a} (MT1) receptor subtype is responsible for the majority of melatonin receptor binding found in SCN and elsewhere in brain, while the Mel_{1b} (MT2) receptor is expressed at considerably lower levels. Several photoperiodic species lack a functional Mel_{1b} receptor gene, suggesting that the Mel_{1a} receptor may be key for regulating seasonal responses.

The duration of melatonin receptor occupation is "measured" by responsive tissues, such that varying melatonin duration (as occurs over the course of the year) has profound effects on neuroendocrine function. The anatomical sites of melatonin receptor expression and melatonin action vary by species. In many species, Mel_{1a} melatonin receptors in the pars tuberalis of the pituitary contribute to the regulation of prolactin secretion from the anterior pituitary. Melatonin receptors in the mediobasal hypothalamus contribute to gonadotropic responses important for the seasonal regulation of reproduction.

Summary

The proximate output of the SCN is a rhythm of neuronal firing, which results in rhythmic synaptic and perhaps humoral activation of additional brain regions. The major target site for SCN axons is a region dorsal and caudal of the SCN, including the subparaventricular zone (SPZ) and extending into the dorsomedial hypothalamic nucleus (DMH). This extensive projection appears to amplify SCN neuronal output and provides a mechanism for the SCN to influence a wide variety of downstream targets in a manner that can be site-specific. The anatomical pathways leading to specific behavioral and physiological rhythms have been defined to varying extents. Already within the subparaventricular zone, there is divergence of the output pathway controlling rhythms in sleep/wake and locomotor activity (which requires the ventral SPZ) from the pathway regulating body temperature (which requires the dorsal SPZ).

One of the most extensively studied outputs is the pathway from the SCN, via the PVN and the sympathetic nervous system, to the pineal, regulating melatonin production. The circadian regulation of melatonin secretory duration regulates reproduction and other photoperiodic responses in seasonally breeding species. The regulation of photoperiodic responses is the most obviously adaptive role for circadian rhythmicity in mammals.

GENERAL SUMMARY

Behavioral and physiological rhythmicity is widespread throughout the animal kingdom. A capacity for cell-autonomous rhythmicity exists in many cell types. Circadian rhythmicity in vertebrates and flies is based on an intracellular molecular feedback loop. Many mechanistic details of this molecular feedback loop have been elucidated in the past 10 years. There is frequently redundancy of function between closely related genes in mammals. The SCN serve as the coordinating pacemaker in the mammalian circadian timing system. In the absence of the SCN, rhythms in other tissues generally degrade because the synchronization between oscillators is lost. The SCN clock is entrained primarily by light. Lighting information is detected by a specific luminance-coding retinal system that involves intrinsically photosensitive retinal ganglion cells, that also serve as a conduit for information originating in rod and cone photoreceptors. SCN axonal output is focused dorsocaudally to the subparaventricular zone, and then diverges along specific anatomical pathways to control a diverse array of rhythmic functions. Among the most physiologically important output rhythms in mammals is the regulation of pineal melatonin, as this hormone is key to regulating photo-periodic adjustments in response to changing day length. Other circadian outputs are of less obvious survival value, but an increasing understanding of the importance of the circadian influence on sleep/wake, metabolism, cardiovascular and neuroendocrine function, and resistance to toxicological challenges indicates that the circadian system dictates an "internal temporal order" that improves physiological and mental performance.

References

Aton, S. J. and Herzog, E. D. (2005). Come together, right . . . now: Synchronization of rhythms in a mammalian circadian clock. *Neuron* **48**, 531–534.

Balsalobre, A., Damiola, F., and Schibler, U. (1998). A serum shock induces circadian gene expression in mammalian tissue culture cells. *Cell* **93**, 929–937.

Cailotto, C., La Fleur, S. E., Van Heijningen, C., *et al.* (2005). The suprachiasmatic nucleus controls the daily variation of plasma glucose via the autonomic output to the liver: are the clock genes involved? *Eur J Neurosci* **22**, 2531–2540.

Chou, T. C., Scammell, T. E., Gooley, J. J., *et al.* (2003). Critical role of dorsomedial hypothalamic nucleus in a wide range of behavioral circadian rhythms. *J Neurosci* **23**, 10691–10702.

Czeisler, C. A. and Khalsa, S. B. S. (2000). The human circadian timing system and sleep-wake regulation. *In* "Principles and Practice of Sleep Medicine."

DeBruyne, J. P., Weaver, D. R., and Reppert, S. M. (2007). CLOCK and NPAS2 have overlapping roles in the suprachiasmatic circadian clock. *Nat Neurosci* **10**, 543–545.

de la Iglesia, H. O. and Schwartz, W. J. (2006). Timely ovulation: Circadian regulation of the female hypothalamo-pituitary-gonadal axis. *Endocrinology* **147**, 1148–1153.

de la Iglesia, H. O., Meyer, J., Carpino, A., Jr, and Schwartz, W. J. (2000). Antiphase oscillation of the left and right suprachiasmatic nuclei. *Science* **290**, 799–801.

de la Iglesia, H. O., Meyer, J., and Schwartz, W. J. (2003). Lateralization of circadian pacemaker output: activation of left- and right-sided luteinizing hormone-releasing hormone neurons involves a neural rather than a humoral pathway. *J Neurosci* **23**, 7412–7414.

Egli, M., Bertram, R., Sellix, M. T., and Freeman, M. E. (2004). Rhythmic secretion of prolactin in rats: action of oxytocin coordinated by vasoactive intestinal polypeptide of suprachiasmatic nucleus origin. *Endocrinology* **145**, 3386–3394.

Gatfield D, Schibler U (2007). Proteasomes keep the circadian clock ticking. *Science* **316**, 1135–1136.

Gooley, J. J., Schomer, A., and Saper, C. B. (2006). The dorsomedial hypothalamic nucleus is critical for the expression of food-entrainable circadian rhythms. *Nat Neurosci* **9**, 398–407.

Granados-Fuentes, D., Prolo, L. M., Abraham, U., and Herzog, E. D. (2004). The suprachiasmatic nucleus entrains, but does not sustain, circadian rhythmicity in the olfactory bulb. *J Neurosci* **24**, 615–619.

Granados-Fuentes, D., Tseng, A., and Herzog, E. D. (2006). A circadian clock in the olfactory bulb controls olfactory responsivity. *J Neurosci* **26**, 12219–12225.

Hattar, S., Lucas, R. J., Mrosovsky, N., *et al.* (2006). Melanopsin and rod-cone photoreceptive systems account for all major accessory visual functions in mice. *Nature* **424**, 76–81.

Hattar, S., Kumar, M., Park, A., *et al.* (2006). Central projections of melanopsin-expressing retinal ganglion cells in the mouse. *J Comp Neurol* **497**, 326–349.

Ishida, A., Mutoh, T., Ueyama, T., Bando, H., *et al.* (2005) Light activates the adrenal gland: Timing of gene expression and glucocorticoid release. *Cell Metab* **2**, 297–307.

Itri, J. N., Michel, S., Vansteensel, M. J., *et al.* (2005). Fast delayed rectifier potassium current is required for circadian neural activity. *Nat Neurosci* **8**, 650–656.

Jin, X., Shearman, L. P., Weaver, D. R., *et al.* (1999). A molecular mechanism regulating rhythmic output from the suprachiasmatic circadian clock. *Cell* **96**, 57–68.

Kalsbeek, A., Garidou, M. L., Palm, I. F., *et al.* (2000). Melatonin sees the light: Blocking GABA-ergic transmission in the paraventricular nucleus induces daytime secretion of melatonin. *Eur J Neurosci* **12**, 3146–3154.

Kalsbeek, A., Ruiter, M., La Fleur, S. E., *et al.* (2006). The hypothalamic clock and its control of glucose homeostasis. *Prog Brain Res* **153**, 283–307.

Kornmann, B., Schaad, O., Bujard, H., *et al.* (2007). System-driven and oscillator-dependent circadian transcription in mice with a conditionally active liver clock. *PLoS Biol* **5**, e34.

Kume, K., Zylka, M. J., Sriram, S., *et al.* (1999). mCRY1 and mCRY2 are essential components of the negative limb of the circadian clock feedback loop. *Cell* **98**, 193–205.

Lee, C., Etchegaray, J. P., Cagampang, F. R., *et al.* (2001). Posttranslational mechanisms regulate the mammalian circadian clock. *Cell* **107**, 855–867.

Liu, C., Weaver, D. R., Strogatz, S. H., Reppert, S. M. (1997). Cellular construction of a circadian clock: Period determination in the suprachiasmatic nuclei. *Cell* **91**, 855–860.

Lu, J., Zhang, Y. H., Chou, T. C., *et al.* (2001). Contrasting effects of ibotenate lesions of the paraventricular nucleus and subparaventricular zone on sleep-wake cycle and temperature regulation. *J Neurosci* **21**, 4864–4874.

Maywood, E. S., Reddy, A. B., Wong, G. K., *et al.* (2006). Synchronization and maintenance of timekeeping in suprachiasmatic circadian clock cells by neuropeptidergic signaling. *Curr Biol* **16**, 599–605.

Meredith, A. L., Wiler, S. W., Miller, B. H., *et al.* (2006). BK calcium-activated potassium channels regulate circadian behavioral rhythms and pacemaker output. *Nat Neurosci* **9**, 1041–1049, 1193.

Miller, B. H., McDearmon, E. L., Panda, S., *et al.* (2007). Circadian and CLOCK-controlled regulation of the mouse transcriptome and cell proliferation. *Proc Natl Acad Sci USA* **104**, 3342–3347.

Moore, R. Y. and Eichler, V. B. (1972). Loss of a circadian adrenal corticosterone rhythm following suprachiasmatic lesions in the rat. *Brain Res* **42**, 201–206.

Moore, R. Y., Speh, J. C., and Leak, R. K. (2002). Suprachiasmatic nucleus organization. *Cell Tissue Res* **309**, 89–98.

Morin, L. P. (2007). SCN organization reconsidered. *J Biol Rhythms* **22**, 3–13.

Nagoshi, E., Saini, C., Bauer, C., *et al.* (2004). Circadian gene expression in individual fibroblasts: Cell-autonomous and self-sustained oscillators pass time to daughter cells. *Cell* **119**, 693–705.

Panda, S., Provencio, I., Tu, D. C., *et al.* (2003). Melanopsin is required for non-image-forming photic responses in blind mice. *Science* **301**, 525–527.

Peirson, S. and Foster, R. G. (2006). Melanopsin: Another way of signaling light. *Neuron* **49**, 331–339.

Provencio, I., Jiang, G., De Grip, W. J., *et al.* (1998). Melanopsin: An opsin in melanophores, brain, and eye. *Proc Natl Acad Sci USA* **95**, 340–345.

Provencio, I., Rodriguez, I. R., Jiang, G., *et al.* (2000). A novel human opsin in the inner retina. *J Neurosci* **20**, 600–605.

Ralph, M. R., Foster, R. G., Davis, F. C., Menaker, M. (1990). Transplanted suprachiasmatic nucleus determines circadian period. *Science* **247**, 975–978.

Reppert, S. M. and Weaver, D. R. (2002). Coordination of circadian timing in mammals. *Nature* **418**, 935–941.

Saper, C. B., Lu, J., Chou, T. C., Gooley, J. (2005). The hypothalamic integrator for circadian rhythms. *Trends Neurosci* **28**, 152–157.

Saper, C. B., Scammell, T. E., and Lu, J. (2005). Hypothalamic regulation of sleep and circadian rhythms. *Nature* **437**, 1257–1263.

Sato, T. K., Yamada, R. G., Ukai, H., *et al.* (2006). Feedback repression is required for mammalian circadian clock function. *Nat Genet* **38**, 312–319.

Silver, R., LeSauter, J., Tresco, P. A., Lehman, M. N. (1996). A diffusible coupling signal from the transplanted suprachiasmatic nucleus controlling circadian locomotor rhythms. *Nature* **382**, 810–813.

Silver, R. and Schwartz, W. J. (2005). The suprachiasmatic nucleus is a functionally heterogeneous timekeeping organ. *Methods Enzymol* **393**, 451–465.

Stephan, F. K. and Zucker, I. (1972). Circadian rhythms in drinking behavior and locomotor activity of rats are eliminated by hypothalamic lesions. *Proc Natl Acad Sci USA* **69**, 1583–1586.

Stephan, F. K. (2002). The "other" circadian system: Food as a Zeitgeber. *J Biol Rhythms* **17**, 284–292.

Tataroglu, O., Davidson, A. J., Benvenuto, L. J., Menaker, M. (2006). The methamphetamine-sensitive circadian oscillator (MASCO) in mice. *J Biol Rhythms* **21**, 185–194.

Ueda, H. R., Hayashi, S., Chen, W., *et al.* (2005). System-level identification of transcriptional circuits underlying mammalian circadian clocks. *Nat Genet* **37**, 187–192.

Weaver, D. R. (1998). The suprachiasmatic nucleus: a 25-year retrospective. *J Biol Rhythms* **13**, 100–112.

Welsh, D. K., Logothetis, D. E., Meister, M., Reppert, S. M. (1995). Individual neurons dissociated from rat suprachiasmatic nucleus express independently phased circadian firing rhythms. *Neuron* **14**, 697–706.

Welsh, D. K., Yoo, S. H., Liu, A. C., *et al.* (2004). Bioluminescence imaging of individual fibroblasts reveals persistent, independently phased circadian rhythms of clock gene expression. *Curr Biol* **14**, 2289–2295.

Zylka, M. J., Shearman, L. P., Weaver, D. R., and Reppert, S. M. (1998). Three period homologs in mammals: Differential light responses in the suprachiasmatic circadian clock and oscillating transcripts outside of brain. *Neuron* **20**, 1103–1110.

Suggested Readings

Arendt, J. (1995). "Melatonin and the Mammalian Pineal Gland." Chapman & Hall, London.

Dunlap, J. C., Loros J., and DeCoursey, P. J. (eds.) (2004). "Chronobiology: Biological Timekeeping." Sinauer Associates, Sunderland, MA. 406 pp.

Klein, D. C., Moore, R. Y., and Reppert, S. M. (eds.). (1991). "Suprachiasmatic Nucleus: The Mind's Clock." Oxford University Press, New York.

Ko, C. H. and Takahashi, J. S. (2006). Molecular components of the mammalian circadian clock. *Hum Mol Genet* **15**, R271–R277.

Moore-Ede, M. C., Sulzman F. M., and Fuller, C. A. (1982). "The Clocks That Time Us." Harvard University Press, Cambridge, MA. 448 pp.

Schibler, U. and Naef, F. (2005). Cellular oscillators: Rhythmic gene expression and metabolism. *Curr Opin Cell Biol* **17**, 223–229.

von Gall, C., Stehle, J. H., and Weaver, D. R. (2002). Mammalian melatonin receptors: Molecular biology and signal transduction. *Cell Tissue Res* **309**, 151–162.

David R. Weaver and Steven M. Reppert

42

Sleep, Dreaming, and Wakefulness

Sleep is a behavioral state that alternates with waking. It is characterized by a recumbent posture, a raised threshold to sensory stimulation, a low level of motor output, and a unique behavior—dreaming. The complex neurobiology of these behavioral features of sleep has been explored at the systemic, cellular, and molecular levels since the mid-1960s. Although these studies have provided substantial insight into the physiology and pathology of sleep, only recently have we begun to obtain definitive answers to the adaptive significance of sleep, the behavioral state that takes up one-third of our lives.

In contrast to sleep, the conscious behavior of waking is characterized by an active and deliberate sensorimotor discourse with the environment. For maintenance of waking behavior, neural gates must remain open for sensory input and motor output, the brain must be tuned and activated, and the chemical microclimate must be appropriate for processing and recording of information. Wakefulness is accompanied by conscious experience that reaches its highest level of complexity in adult humans. Waking behavior includes a number of cognitive aspects, such as sensation, perception, attention, memory, instinct, emotion, volition, cognition, and language, that make up our awareness of the world and self and form the basis of an adaptive interaction with our environment. This chapter focuses on the mechanisms of behavioral state control that make such interactions possible, with an emphasis on sleep as a component of adaptive behavior.

THE TWO STATES OF SLEEP: RAPID EYE MOVEMENT AND NONRAPID EYE MOVEMENT

The behavioral signs of sleep vary regularly during every period of sleep; posture, arousal, threshold, and motor output change in a stereotyped, cyclic manner. Although these changes can be observed directly, their study is facilitated by placing electrodes on the scalp to measure the electrical activity of the brain. From these electroencephalograms (EEGs) (Fig. 42.1), two distinct substates of sleep—rapid eye movement (REM) and non-REM (NREM)—can be discerned. Both show interesting contrasts with waking, and these contrasts improve our understanding of states of consciousness.

At the onset of sleep, the brain begins to deactivate, and awareness of the outside world is lost. A well-described sequence of thalamocortical events causes the progressive slowing of brain waves, seen on EEG during this phase of NREM sleep. The threshold for arousal rises in proportion to the degree of EEG slowing, and at the greatest depth of sleep, awakenings are often difficult, incomplete, and brief. This sleep state is designated the slow wave or delta phase of NREM sleep. People have little or no recall of conscious experience in deep NREM sleep. Many autonomic and regulatory functions, such as heart rate, blood pressure, and respiration rate, diminish in NREM sleep, but some neuroendocrine activity

959

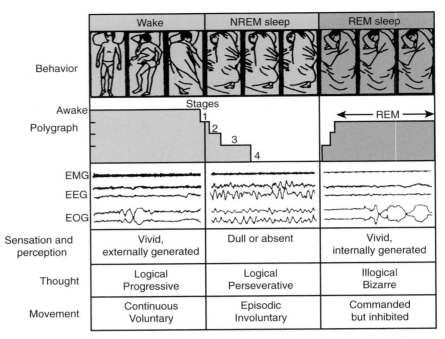

FIGURE 42.1 Behavioral states in humans. Body position changes during waking and at the time of phase changes in the sleep cycle. Removal of facilitation (during stages 1–4 of NREM sleep) and addition of inhibition (during REM sleep) account for immobility during sleep. In dreams, we imagine that we move, but no movement occurs. Tracings of electrical activity are shown in ~20-s sample records. The amplitude of the electromyogram (EMG) is highest in waking, intermediate in NREM sleep, and lowest in REM sleep. The electroencephalogram (EEG) and electrooculogram (EOG) are activated in waking and REM sleep and inactivated in NREM sleep. Reprinted with permission from Hobson and Steriade (1986).

increases. The pulsatile release of growth hormone and sexual maturation hormones from the pituitary is maximal during sleep, with over 95% of the daily output occurring in NREM sleep.

At regular intervals during sleep, the brain reactivates into a state characterized by fast, low-voltage activity in the EEG, along with muscle atonia and rapid eye movements. This sleep state, termed rapid eye movement sleep because of prominent eye movements, differs from waking in several ways, but most notably by an inhibition of sensory input and motor output. Postural shifts precede and follow REM sleep; eye movements, intermittent small muscle twitching, and penile erection occur during REM sleep. The presence (or absence) of REM sleep erection is used clinically to distinguish between psychological and physiological male impotence. In REM sleep, many autonomic functions change; the fine control of temperature and cardiopulmonary function is lost.

People aroused from NREM sleep, especially early in the night, are confused, have difficulty reporting conscious experience, and return to sleep rapidly. Subjects aroused from REM sleep, especially during periods with frequent eye movements, often give detailed reports of dreams characterized by vivid hallucinations, bizarre thinking, and intense emotion.

Although dreaming also occurs during NREM sleep, it is much more strongly associated with the activated brain of REM sleep.

NREM and REM Phases Alternate throughout Sleep

REM and NREM sleep states alternate, with a cycle of about 90 min in adult humans. NREM phases are deeper and longer early in the night (Fig. 42.2A). Together with the regularity of the periods (Fig. 42.2B), this property has suggested that a damped oscillator is the underlying neural mechanism for NREM–REM cycles. The portion of the 24-h day devoted to NREM and REM sleep varies across the lifespan and across species (Fig. 42.3; Table 42.1).

Sleep Appears to Have Multiple Functions

As a quiescent and ecologically protected behavior, sleep fosters conservation of energy, defense from predation, and an opportunity for repair of injury. The anabolic character of NREM sleep (e.g., brain and body inactivity and hormone release) suggests a rest and restoration function. In contrast, the high proportion of REM sleep in the developing brain, the high level

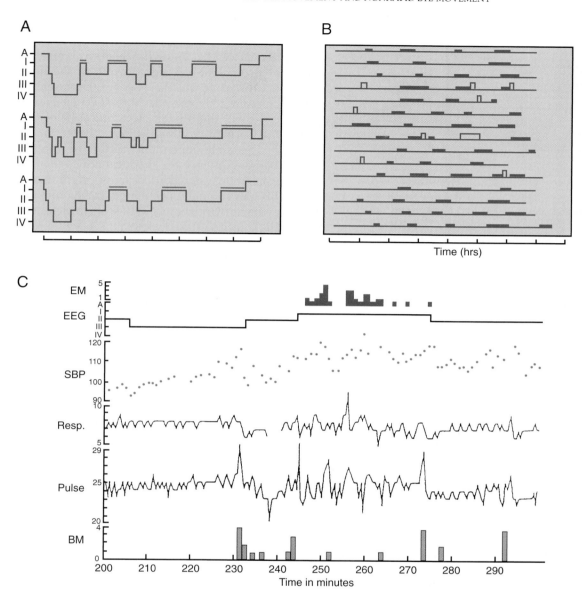

FIGURE 42.2 Periodic activation in sleep cycles. (A) The sleep stages of three people are graphed. The first two or three cycles of the night are dominated by deep stages (3 and 4) of NREM sleep, and REM sleep (indicated by red bars) is brief or nonexistent. During the last two cycles of the night, NREM sleep is lighter (stage 2), and REM episodes are longer, sometimes more than an hour. (B) Fifteen nights of sleep. Each line represents one night of sleep, with REM periods shown as solid bars and periods of wake as taller, open bars. Each record began at the onset of sleep. The amount of time before the first episode of REM varies, but once REM has begun, the interval between episodes is fairly constant. (C) Eye movements (EM), EEG, systolic blood pressure (SBP), respiration (Resp.), pulse, and body movement (BM) over 100 min of uninterrupted sleep. The interval from 242 to 273 min is considered the REM period, although eye movements are not continuous during that interval.

of forebrain activity, and the stereotyped movements in REM sleep suggest a role of REM sleep in brain development and plasticity. Sleep is now known to be important for immune and hormone regulations and memory consolidation. Restriction of sleep to even four hours per night is sufficient to produce impairment of these functions. The crucial role of sleep is illustrated by studies showing that prolonged sleep deprivation results in the disruption of metabolic and caloric homeostasis and eventually death.

Summary

Sleep and waking are distinct, alternate behavioral states that are mutually exclusive, but interrelated. Within sleep, two substates of behavior are readily

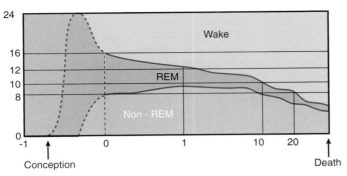

FIGURE 42.3 Portions of a 24-h day that are devoted to waking, REM sleep, and non-REM (NREM) sleep change over a lifetime. Although the timing of these changes *in utero* is not known with certainty (dotted lines), data from premature infants are consistent with REM sleep occupying most of life at a gestational age of 26 weeks. After 26 weeks, the time spent in waking increases until death.

TABLE 42.1 Phylogeny of Rest and Sleep

Organism	Rest	Sleep	REM Sleep
Mammals			
Adults	+	+	+
Neonates	+	+	+
Birds			
Adults	+	+	+, −
Neonates	+	+	+, −
Reptiles	+	+	−
Amphibians	+	+, −	−
Fish	+	+, −	−

REM, rapid eye movement; +, present; +, −, ambiguously or inconsistently present; −, absent.

distinguishable by electroencephalographic recordings, termed rapid eye movement (REM) and non-REM (NREM) sleep. NREM sleep in humans is subdivided further into four levels of increasingly deep sleep (stages I–IV). The characteristic behaviors of sleep include relaxed posture, elevated thresholds for sensory arousal, and diminished motor activity. The NREM and REM sleep phases alternate throughout the sleep period in cycles. Although all the functions of sleep are not yet known, much scientific evidence favors roles for sleep in the restoration, maintenance, and development of a range of physiological and cognitive functions.

SLEEP IN THE MODERN ERA OF NEUROSCIENCE

Philosophical speculation about the nature of sleep and conscious behavior is as old as recorded history, and many philosophers, including the Ionian Greeks,

anticipated the physicalistic models that only recently have been articulated in modern neuroscience (for details of the following historical material, refer to Hobson, 1988). A signal event of the modern era of neuroscience was the discovery of the electrical nature of brain activity and, more specifically, the 1928 discovery of the human electroencephalogram (EEG) by the German psychiatrist Adolf Berger.

The state-dependent nature of the EEG helped Berger convince his skeptical critics that the rhythmic oscillations he recorded across the human scalp with his galvanometer originated in the brain and were not artifacts of movement or of scalp muscle activity. When his subjects relaxed, closed their eyes, or dozed off, the low-voltage brain waves associated with alertness gave way to higher voltage, lower frequency patterns (Fig. 42.4). These patterns stopped immediately when the subjects were aroused.

Waking and Sleep Initially Were Viewed as Activated and Nonactivated States

Following Berger's discovery, a flurry of descriptive and experimental studies aimed at understanding the EEG itself, the full range of its state-dependent variability, and the control of that variability by the brain, were performed. Loomis and Harvey were the first to describe the tendency of the EEG to show systematic changes in activity as subjects fell asleep at night (tracings 3–5 of Fig. 42.4).

Because other mammals shared these correlations between arousal level and EEG, the Belgian physiologist Frederick Bremer made experimental transections of the mammalian brain to determine the nature and source of EEG activation (the low-voltage, fast pattern of waking) and deactivation (the high-voltage, slow pattern of sleep). Thinking the activity level probably depended on sensory input, Bremer transected the brain of the cat at the level of the first cervical spinal cord segment, producing a condition called *encephale isole*. He was surprised to find that the isolated forebrain was activated and alert despite this transection of a major portion of its sensory input. When he then transected the midbrain at the intercollicular level, creating the *cerveau isole* condition, he observed a sleep-like state with persistent EEG slowing and unresponsiveness. Bremer incorrectly inferred that removal of the trigeminal nerve afferents (which entered the brain stem between the level of the two cuts) accounted for the sleep-like state of the *cerveau isole*.

Sleep Can Be Induced Electrically

That sleep might be an active brain process—and not simply the absence of waking as Bremer sup-

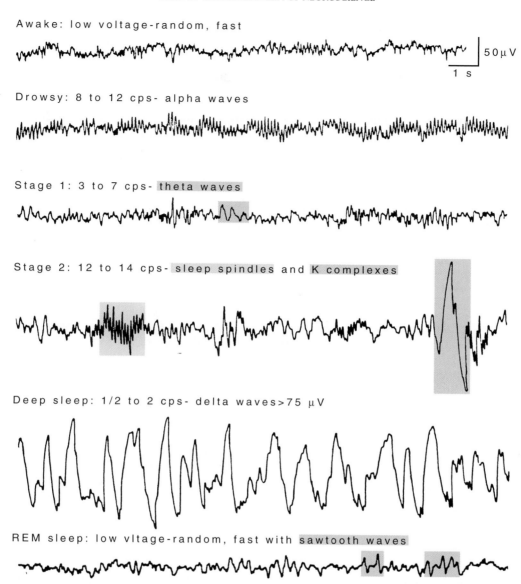

Awake: low voltage-random, fast

50 μV

1 s

Drowsy: 8 to 12 cps- alpha waves

Stage 1: 3 to 7 cps- theta waves

Stage 2: 12 to 14 cps- sleep spindles and K complexes

Deep sleep: 1/2 to 2 cps- delta waves>75 μV

REM sleep: low vltage-random, fast with sawtooth waves

FIGURE 42.4 Electroencephalograms showing electrical activity of the human brain during different stages of sleep.

posed—was first experimentally suggested by the work of the Swiss Nobel laureate W. R. Hess. Hess was involved in a broad program investigating the effects of electrical stimulation of the subcortical regions mediating autonomic control (especially the hypothalamus). He discovered that by stimulating the thalamocortical system at the frequencies of intrinsic EEG spindles (short series of waves) and slow waves, he could induce the behavioral and electrographic signs of sleeping in unanesthetized cats. This discovery opened the door to the idea that sleep and waking might be active processes, each with its own specific cellular and metabolic mechanisms and functional consequences. This paradigm has since born abundant scientific fruit. For

example, the precise details of spindle and slow wave elaboration have been worked out (Steriade, 2000). In addition, the paradigm correctly anticipated the finding that central homologues of sympathetic neurons mediate waking, whereas central cholinergic neurons mediate REM sleep (Hobson and Steriade, 1986; Steriade and McCarley, 1990).

The intrinsic nature of brain activation was clearly demonstrated in 1949 when Giuseppe Moruzzi and Horace Magoun discovered that EEG desynchronization and behavioral arousal could be produced by high-frequency electrical stimulation of the midbrain. To explain their observation, Moruzzi and Magoun proposed that the nonspecific (i.e., nonsensory) reticular activating system operates both in series and in parallel with the ascending sensory pathways. This concept allowed for the translation of afferent stimuli into central activation (Bremer's idea) and opened the door to the more radical idea that not only waking but also adaptive behavior during sleep could be activated by brain stem mechanisms without a need for sensory stimulation.

Now that we accept the idea that the spontaneous activity of neurons determines the substages of sleep, it is difficult for us to appreciate the strength and persistence of its predecessor, the reflex concept of sleep. Scientific giants such as Ivan Pavlov and Charles Sherrington were so inspired by the reflex doctrine that they were convinced that brain activity simply ceased in the absence of sensory input (Box 42.1).

The Brain–Mind Is Activated during REM Sleep

In 1953, Eugene Aserinsky and Nathaniel Kleitman, working in Chicago, discovered that the brain–mind did indeed self-activate during sleep. They observed regularly timed, spontaneous desynchronization of the EEG, accompanied by clusters of rapid, saccadic eye movements and acute accelerations of heart and respiration rates (see Figs. 42.1 and 42.2C). Working with Kleitman, William Dement then showed that these periods of spontaneous autoactivation of the brain–mind, namely REM sleep, were associated with dreaming and that this autoactivation process was also found in another mammal, the cat.

In adult humans, the intrinsic cycle of inactivation (NREM sleep) and activation (REM sleep) recurs with a period length of 90–100 min (Fig. 42.2). REM sleep occupies 20 to 25% of the recording time and NREM the remainder (75–80%) (Hobson, 1989). The NREM phases of the first two cycles are deep and long, whereas REM phases, during which sleep lightens, occupy more of the last two or three cycles.

Input–Output Gating Occurs during Sleep

The paradoxical preservation of sleep in the face of dreaming in REM sleep began to be explained when Francois Michel and Michel Jouvet, working in Lyon in 1959, demonstrated that inhibition of muscle activity was a component of REM sleep in cats. Using transection, lesion, and stimulation techniques, the Jouvet team also discovered that the control system for REM sleep was located in the pontine brain stem. The pons is the source of EEG activation and REMs. Muscle inhibition is also mediated by pontine signals, but these are relayed via the inhibitory bulbar reticular formation to the spinal cord. Synchronous with each flurry of REMs, periodic activation signals, or pontogeniculo-occipital (PGO) waves, are sent from the pons up to the forebrain (and down to the spinal cord). The PGO waves trigger bursts of firing by geniculate and cortical neurons which, during waking, process sensory inputs and motor outputs, whereas other signals originating in the brain stem damp sensory input (via presynaptic inhibition) and motor output (via postsynaptic inhibition) (Calloway et al., 1987).

BOX 42.1

DEAFFERENTATION VERSUS INTERNAL MODULATION: A PARADIGM SHIFT FOR SLEEP NEUROBIOLOGY

Before the discovery of the neuromodulatory systems of the brain by the Swedish team led by Kjell Fuxe in the early 1960s, scientists who studied sleep could not understand that the brain could control its own state and that the stages of sleep were actively generated. Thus such neurobiological giants as Charles Sherrington were convinced that because sleep occurred *when* environmental stimulus levels were low (which is normally true), it also occurred *because* stimulus levels were low (which is false). In light of the discovery of the reticular activating system by Giuseppe Moruzzi and Horace Magoun in 1949 and REM sleep in 1953 by Eugene Aserinsky and Nathaniel Kleitman, this theory of deafferentation gradually

was replaced by a model of active internal control of the brain by the regular variations in balance between different brain regions and neuromodulators, especially NE, 5-HT, and ACh. This paradigm shift has not only led to the specification of increasingly detailed mechanisms for regulation of the brain state, but has opened the door to an appreciation of sleep as functionally significant in ways that are actively complementary to waking. This understanding of brain activity in sleep inevitably makes the emerging panoply of sleep disorders explicable and potentially treatable via a deeper understanding of sleep neurobiology.

Edward F. Pace-Schott, J. Allan Hobson and Robert Stickgold

Thus, in REM sleep the brain–mind is effectively offline with respect to external inputs and outputs and receives internally generated signals. Sensory input is also blocked during NREM sleep, but neither PGO waves nor the active inhibition of motor output is seen outside of REM sleep.

The cellular and molecular bases of these dramatic changes in input–output gating have been detailed using Sherrington's reflex paradigm together with extracellular and intracellular recording techniques (Hobson and Steriade, 1986). For example, such techniques revealed that during REM sleep, each motor neuron was subject to a 10-mV hyperpolarization, which blocked all but a few of the activation signals generated by the REM–PGO system. Furthermore, studies showed that this inhibition was mediated by glycine (Chase et al., 1989). When this motor inhibition was disrupted experimentally by lesions in the pons, the cats, still in a REM-like sleep state but without the atonia, showed stereotyped behaviors (such as defense and attack postures). These behaviors reflected the activation, in REM, of the generators that produce the motor patterns for these instinctual, fixed acts (see Jouvet, 1999). In humans with REM sleep behavior disorder (see later), a similar acting out of dreams is observed.

These studies indicated that in normal REM sleep, motor inhibition prevents the motor commands of the generators of instinctual behavior patterns from being acted out. As a result, during REM these patterns are unexpressed in the outside world but occur in a fictive manner within the internal world of the brain–mind. One strong significance of these findings for a theory of dream consciousness lies in their ability to explain the ubiquity of imagined movement in dreams (Hobson et al., 2000).

Summary

The scientific history of sleep began with the application of electroencephalography in 1928, demonstrating that the electrical activity of the brain changed but did not cease during sleep. Subsequent refinements revealed the necessary role of the reticular formation for cortical arousal during sleep and the association of this arousal with desynchronized cortical activity. In addition, emulating the oscillatory EEG spindles of thalamic projections to cortex allowed sleep to be induced by direct thalamic electrical stimulation in experimental animals. Sleep studies in the 1950s showed REM sleep to be a "paradoxical" state in which sleep was at its deepest, yet accompanied by rapid eye movements and sympathetic signs of arousal, interpreted as dreaming, and accompanied by still further reductions of postural muscle tone. During REM sleep,

pontine waves of activity arise with each flurry of eye movements, during which motor activity is deeply depressed.

ANATOMY AND PHYSIOLOGY OF BRAIN STEM REGULATORY SYSTEMS

Brain Stem Reticular Formation Contains Specific Neuronal Groups Involved in Behavioral State Regulation

Moruzzi and Magoun's original concept of a nonspecific reticular activating system has been elaborated and greatly modified by subsequent anatomical and physiological studies. Two general principles have emerged. One is that most of the classic reticular core neurons have very specific afferent inputs and highly organized outputs. The other is that the reticular formation contains small groups of neurons that send widely branching axons to distant parts of the brain, where their neurotransmitters modulate brain function.

The input–output characteristics of each of the reticular formation's multiple subsystems reflect specific sensorimotor function, modulatory function, or both. For example, many neurons of the reticular formation are involved in the integration of eye, head, and trunk positions as these change to accommodate specific behavioral challenges and tasks. These reticular neurons receive inputs from skin, muscle, bone, and joint receptors in the periphery, which they integrate and link to the vestibular and cerebellar circuits that determine posture and movement. This information must also be integrated by higher brain structures in the visual, somatosensory, and motor systems to develop the complex motor patterns of adaptive behavior.

Several chemically specified neuronal groups also lie in the reticular formation. They have patterns of connections that differ from those of the previously discussed neurons of the reticular formation. Dahlstrom and Fuxe (1964) were the first to identify neuronal populations in the brain that produced norepinephrine (NE) and serotonin (5-hydroxytryptamine, 5-HT). (For additional discussion of these and other neurotransmitter systems discussed in this chapter, see Chapter 7 and Fig. 43.2.)

The groups of NE neurons (designated A1–A7 by Dahlstrom and Fuxe) are located in the pons and medulla in two major groups. One group consists of scattered neurons largely located in the ventral and lateral reticular formation. The most caudal neurons (A1–A3) project rostrally to the brain stem, hypothalamus, and basal forebrain, whereas the rostral groups (A5 and A7) project caudally to the brain stem and spinal cord. These neuronal groups, together called the lateral tegmental neuron group (Moore and Card, 1984) appear to be involved in hypothalamic regulation and motor control.

The major NE cell group is the locus coeruleus (A4 and A6). This compact cell group is located in the rostral pontine reticular formation and central gray matter. The neurons of the locus coeruleus project widely but in a highly specific pattern. One group of locus coeruleus neurons appears to project largely caudally to sensory regions of the brain stem and spinal cord. Other neurons of the locus coeruleus project widely to the cerebellar cortex, dorsal thalamus, and cerebral cortex (Moore and Card, 1984). Thus, the projection patterns of the locus coeruleus appear to be primarily to sensory structures and to cortical structures involved in integration. From this we could expect the locus coeruleus to be involved in regulating sensory input and cortical activation. As we shall see, this expectation is in accord with the available information on function.

Two additional subsets of chemically identified neurons appear critically involved in behavioral state regulation. The first subset to be discovered contained the 5-HT neurons of the brain stem raphé (Dahlstrom and Fuxe, 1964) (B1–B9 in their nomenclature). These neurons extend from caudal medulla to midbrain and are located predominantly in the raphé nuclei, a set of neuronal groups located in the midline of the brain stem reticular formation. The largest numbers of 5-HT neurons are found in the midbrain nuclei, the dorsal raphé nuclei, and the median raphé nuclei (B8 and B9). These groups project rostrally, innervating nearly the entire forebrain in a pattern that also suggests a role in regulation of behavioral state.

The other important set of reticular formation nuclei involved in behavioral state control produces acetylcholine (ACh) as its neurotransmitter. ACh was well known as the transmitter of motor neurons, but early work indicated that ACh is found in nonmotor brain areas. Two sets of cholinergic neurons are involved in control of the behavioral state. The first set is two pontine nuclei: the laterodorsal tegmental nucleus and the pedunculopontine nucleus. Cholinergic neurons in these nuclei project to the brain stem reticular formation, hypothalamus, thalamus, and basal forebrain. The projection to the forebrain involves the second set of cholinergic neurons, those in the medial septum, the nucleus of the diagonal band, and the substantia innominata–nucleus basalis complex. This set projects to limbic forebrain, including the hippocampus, and to the neocortex.

In contrast to the modulatory neurons, which are characterized by their production of norepinephrine, 5-HT, and ACh, neurons in reticular formation nuclei that are involved in sensorimotor integration typically produce either the excitatory transmitter glutamate or the inhibitory transmitter GABA.

Sensorimotor and Modulatory Reticular Neurons Differ Functionally

Sensorimotor and modulatory neurons of the reticular formation differ markedly in their firing properties. Sensorimotor neurons, approximately 50 to 75 μm in diameter, can fire continuously at high rates of up to 50 Hz and can generate bursts of up to 500 Hz. Their larger axons, especially those projecting to the spinal cord, have conduction velocities in excess of 100 m/s (Hobson and Steriade, 1986), making these neurons well suited to rapid posture adjustment and motor control.

Modulatory neurons are smaller (10–25 μm in diameter) than sensorimotor neurons, and even those with long axons conduct their signals very slowly (1 m/s). They fire wider spikes (2 ms in duration) at much slower tonic frequencies (1–10 Hz) than sensorimotor neurons. In addition, they often show a very regular, metronome-like firing pattern, a reflection of the pacemaker properties they share with the Purkinje cells of the heart. As their leaky membranes spontaneously and slowly depolarize, they reach threshold, fire, and then self-inhibit, becoming refractory even to exogenous excitatory inputs. Modulatory neurons are much less likely than sensorimotor neurons to fire in clusters or bursts unless powerfully excited; even then, they adapt rapidly. Thus, modulatory neurons are well suited to detect novel input. Because of their vast postsynaptic domain, they can also help set behavioral and mental states.

The contrasting features of sensorimotor and modulatory neurons confer functionally important distinctions on their neuronal populations. The high rate of discharge of the sensorimotor neurons, acting through extensive interconnections that tie together sensorimotor neurons, causes exponential recruitment. This amplification occurs in response to novel stimuli requiring analysis and directed action (in the wake state) and in the generation of REM sleep, a state of sustained activation (when the system is offline with respect to its inputs and outputs).

In contrast, the feedback inhibition and pacemaker potentials of modulatory neurons allow the production of highly synchronized output during waking and sleeping. These features also somehow cause an exponential decline in the firing rate of these neurons during REM sleep, until firing stops. Studies suggest that this REM sleep inhibition may involve a GABAergic action that arises in the hypothalamus at sleep onset and spreads through the brain stem as NREM sleep progresses to REM (Fig. 42.5). These contrasting properties of sensorimotor and modulatory neurons of the reticular formation are the physical basis for the reciprocal interaction that causes NREM–REM sleep cycles.

Waking Requires Active Maintenance

Moruzzi and Magoun (1949) demonstrated that reticular formation is necessary to maintain the waking conscious state. Lesions in the midbrain reticular formation, sparing the lemniscal sensory pathways, result in a state that resembles NREM sleep. Although maintenance of the waking state often seems effortless, it requires brain mechanisms that compete with other active mechanisms that promote and mediate sleep.

Later, Dann et al. (1984) proposed that two mechanisms interact to regulate sleep–wake cycles. One of

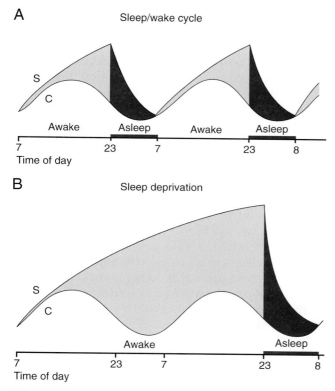

FIGURE 42.5 The Borbely and Daan model of sleep regulation. Sleep is assumed to result from the actions of process C and process S. Process C follows a circadian rhythm and is independent of sleeping and waking. Process S, on the other hand, depends on sleep–wake behavior; S declines during sleep and rises continuously during sleep deprivation. The period of recovery sleep that follows sleep deprivation is more intensive but only slightly longer than normal. If curve C represents the threshold for waking up, then at any time, "sleep pressure" is the (vertical) distance between the S and C curves. The greater the distance, the greater the pressure to fall asleep. Reprinted with permission from Daan et al. (1984).

these, outlined in detail in Chapter 41, is a circadian rhythm in the propensity to fall asleep (curve C, Fig. 42.5). The tendency to fall asleep is normally lowest early in the day, peaks at about late afternoon, and then declines in the evening. Thus sleepiness cycles independent of sleeping and waking. That is, if a subject is deprived of sleep, sleepiness will continue to follow a circadian rhythm. The second mechanism is a homeostatic property, designated S in Fig. 42.5. It increases as a function of the amount of time since the last sleep episode. S can be conceptualized as a sleep-promoting substance that accumulates during waking and dissipates during sleep. Although a large amount of work has been devoted to identifying a "sleep substance," whether one or more such substances normally is involved in sleep regulation remains unclear (Strecker et al., 2000; Krueger and Fang, in Lydic and Baghdoyan, 1999). Many factors likely contribute to the homeostatic regulation of sleep.

Waking is a complex state. Its fundamental features are the maintenance of sensory input from multiple receptors, the capacity for directing attention and accessing memory, the constant readjustments of posture, the maintenance of forebrain activation, and an array of motor output. It is worth noting here, however, that reticular formation plays a crucial role in the initiation and maintenance of both wakefulness and sleep. We now focus on the active control of sleep.

The onset of sleep occurs when the combined S and C factors near their peak and the environmental milieu is conducive to sleep. At this point, the brain stem and cortical mechanisms of waking are relaxed: vigilance lapses, muscle tone declines, eyelids close, and the EEG slows. Because the cortex is still active, removal of ascending influences often results in sudden unresponsiveness, which may be associated with dreamlike imagery. This sleep onset mentation usually is abolished quickly by the thalamocortical oscillation of deeper NREM sleep. As NREM sleep begins, the incomplete relaxation of the mechanisms maintaining muscle tone may result in paroxysmal twitching, a form of myoclonus often affecting the legs.

Hypothalamic Control of Sleep–Wake Transitions

Clinical and experimental findings have shown that lesions to anterior portions of the hypothalamus cause insomnia, and stimulation of this area promotes sleep, whereas lesions of the posterior hypothalamus cause hypersomnolence and neurons in this region decrease firing during sleep (Saper et al. 2001, 2005a). During NREM sleep, neurons in the ventrolateral preoptic area (VLPO) of the anterior hypothalamus show an increase in metabolic activity (Saper et al., 2001, 2005). These VLPO neurons produce both the inhibitory amino acid gamma-aminobutyric acid (GABA) and the inhibitory neuropeptide galanin and project to histaminergic neurons in the tuberomammillary nucleus (TMN) of the posterior hypothalamus as well as to brain stem noradrenaline (NE) and serotonin (5-HT) producing neurons of the locus coeruleus (LC) and dorsal raphé (DR) nuclei, respectively, all of which project widely to the thalamus and cortex where they promote waking (Saper et al. 2001, 2005, Fig. 42.6). The LC, DRN and TMN, in turn, innervate and inhibit the GABA and galaninergic neurons of the VLPO (Saper *et al.* 2001).

Saper and colleagues (2001, 2005) have suggested that this mutually inhibitory arrangement constitutes a biological analog of a bistable electrical "flip-flop" switch in which either the sleep (VLPO dominant) or wake (LC, DRN, TMN dominant) conditions are self-reinforcing stable states whereas intermediate states are highly transient (Fig. 42.7). The neuropeptide orexin (also called hypocretin) produced by cells in the lateral

hypothalamus plays a key modulatory role in this flip-flop mechanism by providing excitatory drive on aminergic neurons, thereby further stabilizing the waking end of the bipolar sleep–wake switch (Borbely, 1982; Borbely and Achermann, 2005; Daan et al. 1984, Fig. 42.5). In the sleep disorder narcolepsy, these orexin-producing neurons are lost, most likely due to an autoimmune mechanism (Black, 2005), and wake to sleep transitions can occur at inappropriate times.

Homeostatic and Circadian Control of Sleep Onset

The two-process model of sleep propensity suggests that homeostatic mechanisms (Process S) interact with circadian factors (Process C) to regulate behaviorally expressed sleep and waking (Borbely, 1982; Borbely and Achermann, 2005 Daan *et al.* 1984, Fig. 42.5). The purine nucleoside adenosine is currently the most widely suggested endogenous somnogen—a substance whose accumulation in the brain over prolonged wakefulness constitutes the physiological basis of Process S (Basheer, *et al.* 2004). Although adenosine may directly inhibit basal forebrain cholinergic neurons that project to the cortex and promote waking (Basheer *et al.*, 2004) recent findings suggest that adenosine's effects may occur elsewhere (Blanco-Centurion *et al.*, 2006) such as at A2A adenosine receptors in ventral striatal areas (Satoh *et al.*, 1999) or on GABAergic anterior hypothalamic and basal forebrain neurons projecting to the VLPO (Shiromani *et al.*, 1999). Evidence also exists for endogenous somnogenic actions of interleukin-1, prostaglandin D1, growth hormone releasing hormone (McGinty and Szymusiak, 2005) and the fatty acid oleomide (Mendelson and Basile, 2001).

Details of the molecular biological and cellular basis of Process C are in Chapter 41 (Lowrey and Takahashi, 2004). Briefly, molecular clocks in cells of the suprachiasmatic nucleus (SCN) of the anterior hypothalamus maintain a near-perfect 24-hour periodicity (Czeisler *et al.*, 1999) via interlocking positive and negative feedback control of transcription and translation of circadian genes (Lowrey and Takahashi, 2004). This endogenous clock is entrained to the ambient photoperiod via photoreceptive pigments like melanopsin in retinal ganglion cells (Peirson and Foster, 2006) that send glutamatergic signals to the SCN via the retinohypothalamic tract (Antle and Silver, 2005). The SCN conveys circadian information via the subparaventricular and dorsomedial nuclei of the hypothalamus to the VLPO to influence sleep–wake transitions (Saper *et al.*, 2005). Secretion of the pineal sleep hormone melatonin is also controlled by the SCN and provides an internal indicator of circadian time as well as feedback influence on the SCN via its melatonin receptors (Arendt, 2006).

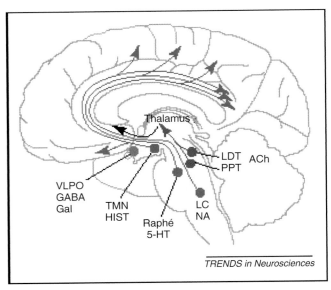

TRENDS in Neurosciences

FIGURE 42.6 Integrated model of sleep onset and REM–NREM oscillation proposed by Shiromani *et al.* I. During prolonged wakefulness, accumulating adenosine inhibits specific GABAergic anterior hypothalamic and basal forebrain neurons, which have been inhibiting the sleep–active VLPO neurons during waking. II. Disinhibited sleep–active GABAergic neurons of the VLPO and adjacent structures then inhibit the wake–active histaminergic neurons of the TMN, as well as those of the pontine aminergic (DR and LC) and cholinergic (LDT/PPT) ascending arousal systems, thereby initiating NREM sleep. III. Forebrain activation by ascending aminergic and orexinergic arousal systems is disfacilitated. IV. Once NREM sleep is thus established, the executive networks of the pons initiate and maintain the ultradian REM/NREM cycle. ACh, acetylcholine; DRN, dorsal raphe nucleus; GABA, γ-amino butyric acid; LC, locus coeruleus; 5-HT, serotonin; LDT, laterodorsal tegmental nucleus; NE, norepinephrine; PRF, pontine reticular formation; TMN, tuberomammillary nucleus. From Pace-Schott and Hobson (2001).

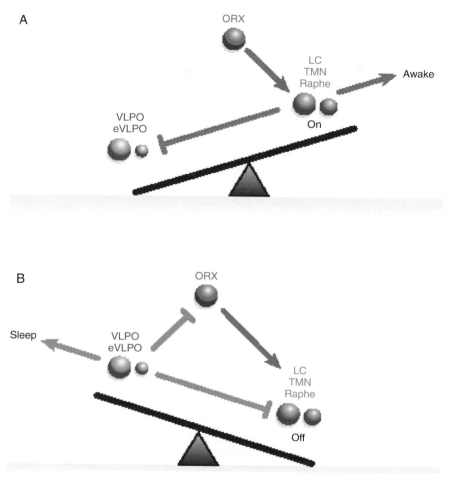

FIGURE 42.7 Thalamocortical oscillations in vivo. Sleep spindle oscillations are generated by synapses in the thalamus. (A, left) Potentials recorded through a microelectrode inserted in the deafferented reticular thalamic nucleus of a cat. The arrow points to one spindle sequence. (A, right) Spindle oscillations recorded from the thalamus of a cat with an upper brainstem transection that created an isolated forebrain preparation. The figure shows two spindle sequences (the second marked by an arrow) and, between them, lower frequency (delta) waves. (B) Neuronal connections involved in the generation of spindle oscillations. (C) Intracellular recordings of one spindle sequence (see A) in three types of neurons (cortical, reticular thalamic, and thalamocortical). Ca2+, calcium ions; IPSP, inhibitory postsynaptic potential.

NREM Sleep Rhythms Require Thalamo-cortical Interaction

Reciprocally interconnected thalamic and cortical neurons form the circuits that are the physiological basis of the NREM oscillations including delta and slow (<1 Hz) oscillatory rhythms and the spindle and K-complex wave forms ((Steriade, 2000, 2006, see Fig. 42.4). In NREM, thalamic and cortical neurons shift into this oscillatory mode when the influence of ascending arousal systems of the brain stem, hypothalamus, and basal forebrain decrease. During waking, these ascending systems modulate thalamo-cortical circuits with the arousal-promoting neuro-modulators acetylcholine (Ach), NE, 5-HT, and histamine. Similarly, during REM sleep, thalamo-

cortical rhythms are prevented by cholinergic thalamic activation originating from brain stem cholinergic nuclei (Steriade, 2000, 2006).

Intrinsic oscillations of NREM represent the combined influence of two neuronal mechanisms: (1) inhibitory influence by neurons of the thalamic reticular nucleus—a thin sheet of GABAergic neurons surrounding the thalamic periphery—on thalamic neurons projecting to the cortex; and (2) an oscillation within cortical neurons involving alternation between a prolonged hyperpolarization phase and a shorter, rapidly spiking, depolarized phase (Steriade, 2000, 2006, Fig. 42.8). This latter rhythm is generated entirely within cortical networks, is the neuronal basis of the <1 Hz slow EEG oscillation, and exerts an organizing or grouping influence on other intrinsic NREM sleep rhythms (Steriade, 2000, 2006).

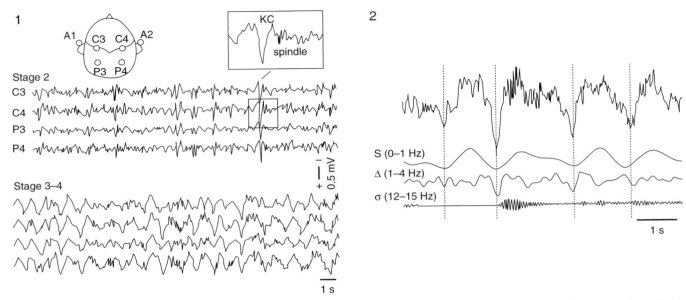

FIGURE 42.8 `Schematic representation of the REM sleep generation process. A distributed network involves cells at many brain levels (left). The network is represented as comprising 3 neuronal systems (center) that mediate REM sleep electrographic phenomena (right). Postulated inhibitory connections are shown as red circles; postulated excitatory connections as green circles; and cholinergic pontine nuclei are shown as blue circles. It should be noted that the actual synaptic signs of many of the aminergic and reticular pathways remain to be demonstrated and, in many cases, the neuronal architecture is known to be far more complex than indicated here (e.g., the thalamus and cortex). During REM, additive facilitatory effects on pontine REM-on cells are postulated to occur via disinhibition (resulting from the marked reduction in firing rate by aminergic neurons at REM sleep onset) and through excitation (resulting from mutually excitatory cholinergic–noncholinergic cell interactions within the pontine tegmentum). The net result is strong tonic and phasic activation of reticular and sensorimotor neurons in REM sleep. REM sleep phenomena are postulated to be mediated as follows: EEG desynchronization results from a net tonic increase in reticular, thalamocortical, and cortical neuronal firing rates. PGO waves are the result of tonic disinhibition and phasic excitation of burst cells in the lateral pontomesencephalic tegmentum. Rapid eye movements are the consequence of phasic firing by reticular and vestibular cells; the latter (not shown) excite oculomotor neurons directly. Muscular atonia is the consequence of tonic postsynaptic inhibition of spinal anterior horn cells by the pontomedullary reticular formation. Muscle twitches occur when excitation by reticular and pyramidal tract motorneurons phasically overcomes the tonic inhibition of the anterior horn cells. RN, raphé nuclei; LC, locus coeruleus; P, peribrachial region; PPT, pedunculopontine tegmental nucleus; LDT, laterodorsal tegmental nucleus; mPRF, meso- and mediopontine tegmentum (e.g., gigantocellular tegmental field, parvocellular tegmental field); RAS, midbrain reticular activating system; BIRF, bulbospinal inhibitory reticular formation (e.g., gigantocellular tegmental field, parvocellular tegmental field, magnocellular tegmental field); TC, thalamocortical; CT, cortical; PT cell, pyramidal cell; III, oculomotor; IV, trochlear; V, trigmenial motor nuclei; AHC, anterior horn cell. From Hobson et al. (2000).

The cortical slow oscillation initiates sleep spindles during its depolarized phase when cortico-thalamic neurons excite thalamic reticular neurons (Steriade, 2000, 2006). The resulting periodic inhibition by thalamic reticular neurons on thalamo-cortical neurons changes their firing from a transmission mode in which firing frequency is proportional to the strength of a sensory input to a bursting mode in which transmission of sensory information to the cortex is blocked (McCormick and Bal, 1997). This bursting mode is initiated when thalamo-cortical neurons are sufficiently hyperpolarized by thalamic reticular inhibition to activate a unique ion channel that opens only when their membrane potential falls well below its normal resting potential (McCormick and Bal, 1997). Opening of this channel begins a series of membrane events ("H" currents followed by "T" or "low-threshold calcium spike" currents) that culminates in thalamo-cortical neurons emitting a series of spikes that impinge on the cortex at spindle frequency (Steriade, 2000, 2006, Fig. 42.9).

Delta frequency EEG oscillations may have both thalamic and cortico-thalamic origins (Steriade, 2000, 2006). Delta-frequency oscillations can arise via thalamic neurons' intrinsic delta-frequency pacemaker property or, alternatively, when cortico-thalamic neurons excite thalamic reticular neurons that, in turn, hyperpolarize the thalamo-cortical neurons causing them to stimulate the cortex at delta frequency (Steriade, 2000, 2006).

In the transition from NREM to REM, activation of the thalamus by brain stem cholinergic neurons blocks the intrinsic oscillations resulting in a cortical "desynchronization." This appears as a speeding and diversification of EEG frequencies along with a decrease in the amplitude of oscillations in the EEG (Steriade,

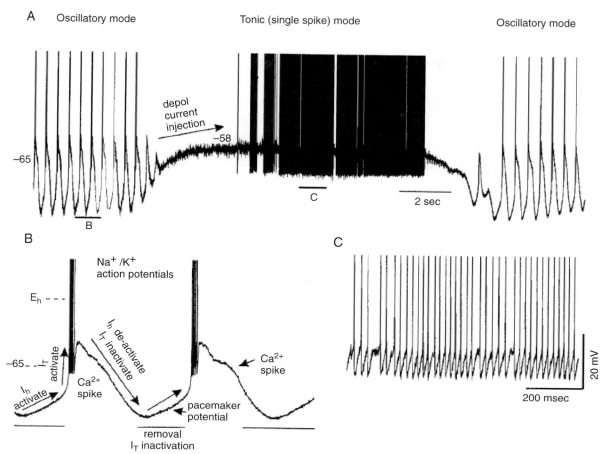

FIGURE 42.9 Sites of REM induction. (A, left) Filled circle indicates site of injection of carbachol into peribrachial pons. Small dots indicate location of cholinergic cells and crosses location of PGO burst cells. (A, right) Injection of carbachol, an acetylcholine agonist, in the perimedian pons (magenta dot) causes short-term induction of REM sleep. Injection of the peribrachial pons (yellow dot) causes long-term induction of REM sleep. After injection of a cholinergic agonist into the peribrachial pons, ponto-geniculo-occipital (PGO) burst cell activity and PGO waves can be measured immediately in the lateral geniculate body (LGB) on the same side (ipsilateral) of the brain as the injections (B, left). Hypothesized connections among the PGO trigger zone (PB), modulatory aminergic raphé (R), locus coeruleus (LC), and paramedian pontine reticular cells (FTC) are shown. (C, left) Injections of carbachol at the PB site shown in A produce immediate PGO waves in the ipsilateral LGB. After 24 h, REM sleep intensifies (C, middle) and remains so for 6 days (C, right). FTC, FTG, and FTP are reticular tegmented nuclei. ACh are cholinergic nuclei. RN, red nucleus; IC, inferior colliculus; SC, superior colliculus; BC, brachium conjunctiuum.

2000, 2006). Cholinergic desynchronization of EEG in REM and waking occurs via (1) an inhibitory effect on GABAergic thalamic reticular neurons that disinhibits their thalamo-cortical targets, and (2) by cholinergic excitatory influence on thalamo-cortical neurons that facilitates their glutamatergic excitation of the cortex (Steriade, 2000, 2006).

REM Sleep Is Initiated in the Brain Stem

The cellular and molecular events that produce the regular alternation of REM and NREM sleep emerge under brain stem control once sleep has been achieved (Hobson *et al.*, 2000; Jones, 2005). As described in the previous section, activation of the cortex in REM sleep occurs when the intrinsic oscillations of NREM sleep are blocked by the brain stem ascending reticular activating system's (ARAS) cholinergic reactivation of the thalamus (Steriade, 2000, 2006). A second, more ventral ascending activation pathway by which the ARAS can produce cortical activation and desynchrony involves excitation of the cholinergic neurons of the basal forebrain that project widely to the cortex and promote arousal in waking (Jones, 2005; Lu *et al.*, 2006a). The cardinal physiological features of REM include:

- EEG desynchronization
- Phasic potentials recorded sequentially in the pons, thalamic lateral geniculate body and the occipital cortex of the cat, termed ponto-geniculo-occipital (PGO) waves or P-waves in the rat (Datta, 1998)

- Rapid eye movements
- A change in hippocampal EEG from irregular rhythms to a regular theta rhythm
- Skeletal muscle atonia

The earliest cellular model of the brain stem REM sleep generator, the reciprocal interaction (RI) model, suggested that regular alternation of REM and NREM sleep results from the interaction of aminergic REM-off and cholinergic REM-on neuronal populations in the pontine brain stem (Hobson et al., 1975; McCarley and Hobson, 1975, Figs. 42.10 and 42.11).

The aminergic and cholinergic neurons described in the RI hypothesis are components of the brain stem ARAS, a system that also includes large and varied populations of glutamatergic and GABAergic neurons (Hobson *et al.*, 2000; Lydic and Baghdoyan, 2005; Jones, 2005). The key cholinergic neurons supporting thalamic activation in REM reside in the laterodorsal teg-

mental (LDT) and pedunculopontine (PPT) nuclei of the pons-midbrain junction (mesopontine tegmentum). Aminergic REM-off neurons of the LC and DR also occur in the mesopontine region. Many glutamatergic and GABAergic ARAS neurons are located ventral to the mesopontine tegmentum in the medial pontine reticular formation (mPRF). Many of these neurons bear Ach receptors (are "cholinoceptive").

The RI hypothesis is supported by extensive experimental and clinical findings showing powerful effects of Ach and the monoamines NE and 5-HT on the REM-NREM oscillator. For example, in many regions of the pontine reticular activating system of cats, micro-injection of either drugs like carbachol that mimic the action of acetylcholine (cholinergic agonists) or drugs like neostigmine that block the enzymatic degradation of naturally produced acetylcholine (acetylcholinesterase inhibitors) produces a state indistinguishable from naturally occurring REM sleep (Hobson *et al.*, 2000; Lydic and Baghdoyan, 2005). Moreover, REM sleep is reduced or blocked by microinjection of noradrenergic or serotonergic agonists into the pontine brain stem of animals as well as by systemic administration, in humans, of drugs that block the synaptic reuptake or enzymatic degradation of naturally produced monoamines such as many antidepressants (Hobson *et al.*, 2000). PGO waves result from tonic disinhibition and phasic excitation of bursting cells in the peribrachial region of the mesopontine tegmentum and, like REM sleep, they can be triggered and augmented by cholinergic stimulation (Datta, 1997).

Major roles for acetylcholine and the monoamines in REM-NREM control continue to be reported (reviewed in Lydic and Baghdoyan, 2005; Pace-Schott and Hobson, 2002), and many intermediate synaptic steps for both the activation of REM-on neurons in the mesopontine tegmentum and the suppression of REM-off aminergic nuclei also have been described. For example, mutually excitatory interactions between cholinergic and glutamatergic neurons may underlie the rapidly escalating firing of pontine reticular REM-on neurons during REM sleep (see Pace-Schott and Hobson, 2002). Similarly, during REM, inhibition of REM-off aminergic neurons in the LC and DR by brain stem GABAergic neurons may remove their inhibitory influence on REM-on cell populations (see Pace-Schott and Hobson, 2002). GABAergic inhibition by neurons of periaqueductal gray area (PAG), substantia nigra pars reticulata, dorsal paragigantocellular nucleus, or local pontine interneurons may be responsible for shutting down REM-off serotonergic and noradrenergic cells (Gervasoni *et al.*, 2000; reviewed in Jones, 2005).

Other models on the control of REM onset and offset have emphasized interactions between brain

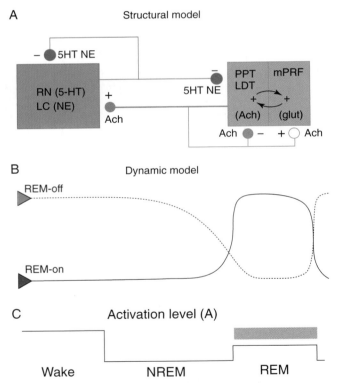

FIGURE 42.10 Structural and dynamic models of activation in REM sleep. (A) In the structural model, REM-on cells of the pontine reticular formation are excited by acetylcholine (ACh) or they produce excitatory signals using ACh in their synaptic terminals (or they do both). REM-off cells respond to or use NE or 5-HT as inhibitory neurotransmitters (or they do both). (B) In the dynamic model, during waking the pontine aminergic system (dashed line) is active continuously and inhibits the pontine cholinergic system (solid line). During NREM sleep, aminergic activity decreases, allowing cholinergic activity to rise. By the time REM sleep begins, the aminergic system is off and cholinergic excitation has peaked.

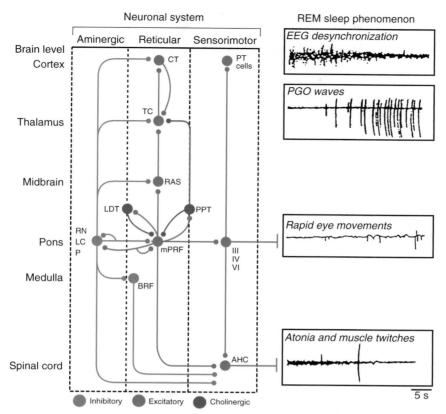

FIGURE 42.11 Convergent findings on relative regional brain activation and deactivation in REM compared to waking. A schematic sagittal view of the human brain showing those areas of relative activation and deactivation in REM sleep compared to waking and/or NREM sleep, which were reported in two or more of the three PET studies published to date (Braun et al., 1997; Maquet et al., 1996; Nofzinger et al., 1997; see Hobson et al., 2000 for citations). Only those areas that could be matched easily between two or more studies are illustrated here and a realistic morphology of the depicted areas is not implied. Note that considerably more extensive areas of activation and deactivation are reported in the individual studies; these more detailed findings are reviewed in Hobson et al. (2000). The depicted areas are thus viewed most realistically as representative portions of larger CNS areas subserving similar functions (e.g., limbic-related cortex, ascending activation pathways, and multimodal association cortex).

stem GABAergic cell populations in which GABAergic inhibition plays both REM facilitory and REM inhibitory roles. For example, Boissard *et al.* (2002) have suggested that the onset of REM in the rat is triggered by the release of neurons of the pontine sublaterodorsal (SLD) nucleus from tonic GABAergic inhibition. In addition, a GABAergic "switch" between waking and REM has been reported in the nucleus pontine oralis (within the mPRF) of cats (Xi *et al.*, 2001). Most recently, Lu *et al.* (2006b) have proposed that a GABAergic flip-flop switch, analogous to the earlier hypothalamic flip-flop switch, controls the alternation of REM and NREM sleep states in the rat via REM-off regions in the ventrolateral periaqueductal gray and lateral pontine tegmentum that project to REM-on regions in the SLD, precoeruleus, and periventricular grey mesopontine areas. Discrepancies between models proposing that REM results from brain stem GABAergic interactions and those suggest-

ing release of mesopontine cholinergic neurons from inhibition by aminergic nuclei may result from: (1) GABAergic models being derived mainly from work in the rat whereas aminergic-cholinergic models were derived in cats, and (2) hierarchical, horizontal, or redundant control by overlapping GABAergic and aminergic/cholinergic brain stem circuits paralleling the multiple overlapping ascending arousal circuits whose cholinergic, aminergic, and orexinergic components promote waking (Saper *et al.*, 2005).

Other Brain Stem and Diencephalic Neurotransmitter Systems Participating in Behavioral State Control

Despite the prominence of cholinergic, aminergic, and amino acid (GABAergic and glutamatergic) neuronal populations in the control of the sleep–wake and REM–NREM cycles, many additional neurotransmitter systems participate in the modulation sleep states. Such neurochemicals may exert their effects on behavioral state by interacting with aminergic, cholinergic, and amino acid neurons that exert a more direct "executive" control on sleep and waking states. In addition, such neurochemicals may have roles in mediating specific physiological signs of REM and NREM sleep.

Other Monoamines

In addition to the prominent roles of norepinephrine, serotonin, and histamine, recent findings have begun to identify roles for a fourth monoamine, dopamine, in sleep wake regulation. Until just recently, little attention was paid to potential roles of dopamine in sleep–wake control because, unlike the other monoamines, firing rate of dopaminergic cells in the substantia nigra pars compacta (SNpc) and ventral tegmental area (VTA) did not appear to vary with behavioral state (see Hobson et al., 2000). However, Lu et al. (2006b) have identified wake–active dopaminergic neurons in the ventral PAG that project to many of the traditional neuronal components of sleep–wake regulation. These investigators hypothesize that these neurons constitute a dopaminergic component of monoaminergic ascending wake–promoting arousal systems analogous to serotonergic, noradrenergic, and histaminergic systems in the DR, LC, and TMN, respectively (Saper et al., 2005). Additional emerging evidence for a dopaminergic component to behavioral state control includes the findings that (1) dopamine concentrations in the medial prefrontal cortex (mPFC) and nucleus accumbens (NAc) are greater during REM and waking than during NREM sleep (Lena et al., 2005), (2) VTA dopaminergic neurons switch to a burst firing mode in REM (Dahan et al., 2007), (3) metabolism increases in the dopaminergic ventral tegmental area during REM rebound (Maloney et al., 2002), and (4) mice with abnormally high brain dopamine levels display an abnormal REM-like waking that is reversible with a D2 antagonist (Dzirasa et al., 2006).

Histaminergic projections of the TMN diffusely innervate the entire forebrain and brain stem (Shiromani et al., 1999). Histaminergic TMN neurons show a behavioral state-dependent pattern of firing similar to LC and DR neurons (Saper et al., 2001). In addition, like serotonergic and noradrenergic neurons, histaminergic neurons are tonically inhibited during sleep by GABAergic and galaninergic projections from the VLPO (Saper et al., 2001).

Amino Acid Neurotransmitters

The brain's most ubiquitous inhibitory neurotransmitter, GABA, is involved at many stages in the control of the sleep–wake and the REM–NREM cycle (Jones, 2005). In addition to its roles in sleep onset and maintenance, thalamo-cortical oscillations and REM facilitation/inhibition described earlier, GABAergic inhibition also controls the neurons that generate PGO waves (Datta, 1997). Another inhibitory neurotransmitter, glycine, is responsible for the postsynaptic inhibition of somatic motoneurons that results in REM atonia (Chase and Morales, 2005). In addition to its important roles in ascending reticular activation and thalamo-cortical oscillations described earlier, the brain's most ubiquitous excitatory neurotransmitter, glutamate, transmits signals from pontine REM-control regions to inhibitory glycinergic and GABAergic cells of the medulla that, in turn, suppress somatic motoneurons to produce REM atonia (Boissard et al., 2002; Jones, 2005).

Intercellularly Diffusible Gaseous Neurotransmitter: Nitric Oxide (NO)

NO has recently been widely implicated in sleep cycle modulation and functions primarily as an intercellular messenger, which can enhance the synaptic release of neurotransmitters such as ACh and enhance capillary vasodilation (Leonard and Lydic, 1999). NO is coproduced by cholinergic mesopontine neurons and may play a role in maintaining the cholinergically mediated REM sleep state in both the pons and the thalamus (Leonard and Lydic, 1999).

Neuropeptides

Neuropeptides are increasingly seen to play diverse roles in regulation of the sleep–wake and REM–NREM cycles. As described earlier, the hypothalamic peptide galanin participates in the inhibition of wake-promoting centers during sleep onset, the hypothalamic peptide orexin promotes waking (Saper et al., 2001) and cytokine peptides may play roles as endogenous somnogens (McGinty and Szmusiak, 2005). Other peptides implicated in the REM–NREM and sleep–wake cycles include vasoactive intestinal polypeptide (Steiger and Holsboer, 1997), as well as numerous hormones (Obal and Krueger, 1999) such as corticotropin releasing hormone (Chang and Opp, 2001).

Second Messengers and Intranuclear Events

As in other areas of neuroscience, research on behavioral state control is now beginning to extend its inquiry beyond the neurotransmitter and its receptors to the roles of intracellular second messengers and gene transcription. For example, Bandyopadhya et al. (2006) have shown variation with prior amount of REM sleep in the intracellular concentrations of the catalytic subunit of protein kinase A (PKA), the component of PKA that enters the nucleus to phosphorylate CRE-binding protein. Similarly, manipulation of the cAMP-PKA pathway has been shown to modify the REM sleep effects of cholinergic activation of muscarinic receptors in the mPRF (Capece et al., 1997). The specific profile of genes selectively activated in different sleep–wake states is also a current area of intense investigation (Wisor and Kilduff, 2005; Tononi and Cirelli, 2001).

Summary

Circadian and homeostatic factors together regulate sleep–wake propensity. Circadian information is provided by reliable molecular clocks in cells of the SCN of the anterior hypothalamus. SCN neurons transmit circadian information via several hypothalamic nuclei to sites where its influence is combined with information on homeostatic sleep pressure indexed by endogenous somnogenic substances such as adenosine that accumulate during waking.

A key structure in the initiation of NREM sleep is the VLPO in the anterior hypothalamus whose inhibitory GABA and galaninergic neurons inhibit wake-promoting arousal systems originating in cholinergic, noradrenergic, serotonergic, and dopaminergic brain stem nuclei, the histaminergic TMN in the posterior hypothalamus, and cholinergic arousal systems of the basal forebrain and brain stem. During waking, the peptide neuromodulator orexin from cells in the lateral hypothalamus further excites these aminergic and cholinergic arousal systems and stabilizes the waking pole of a bistable sleep–wake mechanism analogous to an electrical flip-flop circuit.

The ascending arousal systems of waking and REM suppress the endogenous oscillatory rhythms generated in thalamo-cortical circuits that manifest as the slow oscillation, spindle, and delta brain waves of the EEG in NREM sleep. The slow oscillatory rhythm is generated by cortical neurons and reflects a pattern of prolonged hyperpolarization followed by burst firing that controls the timing of spindles, delta waves, and K-complexes in the NREM sleep EEG.

Aminergic, cholinergic, and orexinergic neuromodulation promotes the activated brain states of waking but cholinergic and perhaps dopaminergic arousal systems promote the activated state of REM sleep. After sleep is initiated, an ultradian oscillator located near the junction of the pons and midbrain controls the regular alternation of NREM and REM sleep stages. This alternation of states involves a reciprocal interaction between aminergic REM-on and cholinergic REM-off cell groups, the interactions of which are mediated by interposed excitatory, inhibitory, and autoregulatory circuits that utilize the amino acid neurotransmitters GABA and glutamate. Substantial species differences may exist in mechanisms of REM-NREM control and mutually inhibitory GABAergic circuits may exert the most proximal control of sleep state alternation in rodents.

Additional neurochemicals such as the amino acid glycine and neuropeptide hormones contribute to the regulation of sleep–wake and REM–NREM cycles as well as to specific physiological signs of particular behavioral states. Contemporary neurophysiological research is elucidating the intracellular events taking place in second messenger systems and the cell nucleus. Emerging details on the molecular expression of specific sleep-related genes will undoubtedly provide an increasingly detailed understanding of causal mechanisms underlying the control of behavioral state.

MODELING THE CONTROL OF BEHAVIORAL STATE

Two linked models—one neurobiological (McCarley and Hobson, 1975) and the other psychophysiological (Hobson and McCarley, 1977; Hobson et al., 2000)—have been advanced to organize experimental findings and their implications for a theory of consciousness. According to the neurobiological model (Fig. 42.10), in waking, brain activation and open input–output gates result from a combination of continual aminergic activity and phasic (waxing and waning) aminergic and cholinergic activities. In contrast, in REM sleep the continually low level of aminergic activity and the periodic increases in cholinergic activity close input–output gates, activate the brain, and periodically stimulate the brain.

The subjective experience of waking, NREM sleep, and REM sleep can tentatively be linked to the accompanying changes in physiology (Table 42.2). Sensation and perception decline progressively during the cortical deactivation that occurs at the onset of sleep; responsiveness declines further as NREM sleep deepens. During REM sleep the brain reactivates, but presynaptic inhibition blocks sensory signals. REMs and their associated PGO waves act as internal stimuli, which take the place of stimuli from external sources. The brain stem sends information about eye movements to the thalamocortical visual system, possibly contributing to the intense visual hallucinations of dreams. Furthermore, these PGO signals also drive the amygdala, perhaps accounting for such emotions as anxiety and surprise in dreams. Interestingly, PET studies show an increase of regional cerebral blood flow in limbic structures, including the amygdala during REM sleep.

In the early days of the EEG, brain waves with frequencies greater than 25 Hz were either ignored or

TABLE 42.2   Physiological Basis of Changes That Occur during Dreaming

Function	Change (compared with waking)	Hypothesized Cause
Sensory input	Blocked	Presynaptic inhibition
Perception (external)	Diminished	Blockade of sensory input
Perception (internal)	Enhanced	Removal of inhibition from networks that store sensory representations
Attention	Lost	Aminergic modulation decreases, causing a decrease in the ratio of signal to noise
Memory (recent)	Diminished	Because of a decrease in aminergic activity, activated representations are not stored in memory
Memory (remote)	Enhanced	Removal of inhibition from networks that store memory representations
Orientation	Unstable	Internally inconsistent signals are generated by cholinergic systems
Thought	Poor reasoning, processing hyper-associative	Loss of attention, memory, and volition leads to failure of sequencing and rule inconstancy; analogy replaces analysis
Insight	Self-reflection lost	Failures of attention, logic, and memory weaken second- (and third-) order representations
Language (internal)	Confabulatory	Aminergic demodulation frees the use of language from the restraint of logic
Emotion	Episodically strong	Cholinergic hyperstimulation of the amygdala and related structures of the temporal lobe trigger emotional storms, which are not modulated by aminergic activity
Instinct	Episodically strong	Cholinergic hyperstimulation of the hypothalamus and limbic forebrain triggers fixed motor programs, which are experienced fictively but not enacted
Volition	Weak	Cortical motor control cannot compete with disinhibited subcortical networks
Output	Blocked	Postsynaptic inhibition

filtered out because of the problem of interference by 50- or 60-Hz artifacts from electrical power sources. Later, researchers were able to focus on the possibility that 40-Hz EEG activity synchronizes processing in cortical areas.

Rodolfo Llinás proposed that as the cortex is scanned by the thalamus, 40-Hz waves propagate from the frontal to the occipital poles (Kahn *et al.*, 1997). Noting that 40-Hz activity is observed in REM sleep as well as in waking, he suggested that intrinsic neuronal oscillations are important in the genesis of both states of consciousness and that the main difference between waking and dreaming arises from differences in input–output gating (Llinás and Pare, 1991). In addition to being without temporal or spatial input, in REM sleep the brain has little background aminergic activity; this lack may contribute to the declines in attention, orientation, memory, and logic that characterize dreaming and make dreams difficult to remember (Hobson *et al.*, 2000).

Neuroimaging and Lesion Studies of Sleep and Dreaming

Functional neuroimaging studies have revealed distinctive differences in regional brain activation between the cardinal behavioral states of waking, NREM sleep

and REM sleep (for a review and original citations, see Hobson *et al.*, 2000). In brief, widespread deactivation of the forebrain and brain stem is seen following sleep onset, and activation further decreases with the increasing depth of NREM sleep from NREM stage 1 to 2 to slow-wave sleep (SWS). Subsequently in REM sleep, however, there is selective reactivation to waking or above-waking levels in brain stem, diencephalic, subcortical limbic, cortico-limbic, and selected cortical association areas. Notably however, during REM, significant deactivation of the dorsolateral prefrontal cortex relative to waking persists. Figure 42.12 illustrates the structures found to be activated or deactivated in two or more of these recent neuroimaging studies.

These findings have led to an updating of the original activation synthesis hypothesis of dreaming (Hobson and McCarley, 1977), which now emphasizes that the distinctive cognitive features of dreams result from simultaneous autoactivation of the limbic brain (with consequent dream emotionality and salience) combined with the deactivation of executive structures subserving logic and working memory (with consequent dream bizarreness and disorientation) (Hobson *et al.*, 2000).

Complementing these findings have been neuropsychological findings on brain-damaged patients by

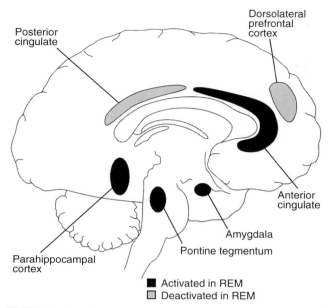

Posterior cingulate

Dorsolateral prefrontal cortex

Anterior cingulate

Amygdala

Pontine tegmentum

Parahippocampal cortex

■ Activated in REM
□ Deactivated in REM

FIGURE 42.12 Central and obstructive sleep apnea. During waking, the respiratory oscillator of the medulla receives tonic drive from other neural structures and can respond to voluntary and metabolic signals to change breathing pattern. Muscle tone keeps the oropharynx open to the flow of air. In NREM sleep, central drive decreases, and the rate and depth of ventilation fall. If the decrease is excessive, central sleep apnea results, including a complete, albeit temporary, cessation of breathing. If the airway collapses, obstructive sleep apnea may result, with a similar cessation of breathing. During REM sleep, activation of pontine generator neurons drives the respiratory oscillator, and desynchronization may lead to breathing efforts that are too frequent or strong (hyperpnea) or that stop. During REM sleep the oscillator also becomes unresponsive to metabolic signals.

Solms (1997). In these studies, dorsolateral prefrontal damage was found to have little effect on dreaming. This is in keeping with its relative inactivity during REM and the apparent diminution of its mnemonic, orientational, and logical functioning in dreaming. In contrast, destructive lesions of multimodal parietal areas resulted in global cessation of dreaming, and lesions of the visual association cortex led to nonvisual dreaming. This is in keeping with the activation, in REM, of the supramarginal gyrus and inferotemporal cortex in some PET studies and the highly visuospatial nature of dreaming. Similarly, disconnective lesions in upper brain stem-limbic-prefrontal areas led to global cessation of dreaming in keeping with their activation in REM and the highly emotional and personally salient quality of dreaming (Box 42.2).

Central Autonomic Control Systems Are State Dependent

Alterations in cognitive and vegetative functions are linked to changes in central modulatory systems during sleep. For example, active hypothalamic temperature control is diminished or abandoned in REM sleep (Hobson, 1989), and the sensitivity of the respiratory control system is likewise diminished. As a result, a neuron that is sensitive to temperature or pO_2 during waking loses that responsiveness during REM sleep. Sleep, then, involves significant changes in the reflex

---

BOX 42.2

REGIONAL ACTIVATION AND DEACTIVATION: IMAGING THE HUMAN BRAIN AWAKE AND ASLEEP

Not only is the brain surprisingly and strongly activated during REM sleep, but its various subregions are activated and deactivated in a pattern quite different from that of waking. The mechanism of this differential activation pattern, revealed in PET, SPECT, and fMRI functional neuroimaging studies, is unknown, but it seems quite likely that the underlying shifts in regional metabolism and blood flow are orchestrated by the neuromodulators, which play this role in the rest of the body.

Compared to waking, deep NREM sleep shows a rather global deactivation pattern in keeping with the observed EEG slowing and synchronization. However, in REM, some areas are hyperactivated (among which the pontine brain stem, the limbic system, and the paralimbic cortex are most notable), whereas others are deactivated

(among which the posterior cingulate and the dorsolateral prefrontal cortex are most notable).

These shifts in regional blood flow, and by inference regional neuronal activity, map nicely onto the observed changes in consciousness. Thus activation of the amygdala and other sub-cortical and cortical limbic structures is relevant to the domination of dream consciousness by emotion, the selective activation of only certain of the multimodal cortices is relevant to the hyper-associative quality of dreaming, whereas the inactivation of the dorsolateral prefrontal cortex seems relevant to the defects in cognition and memory. At last, the human brain and its wondrous array of conscious states is amenable to neurobiological analysis, in real time, in real people, and in patients as well as in healthy individuals.

Edward F. Pace-Schott, J. Allan Hobson and Robert Stickgold

control of autonomic function. Our understanding of changes in the control of vegetative and sensorimotor functions during waking and sleep can be applied to understand sleep-related dysfunctions such as insomnia (inability to sleep), hypersomnia (excessive sleepiness), and parasomnia (sleep with reflex responses to stimuli) (Hobson, 1989).

Sleep Disorders Result from Disrupted Control of Behavioral State

When central nervous system (CNS) aminergic activity increases, the brain is shifted in the direction of hyperarousal, and stress and insomnia can result. If stress and insomnia are prolonged, the chronic sympathetic overdrive can harm cardiovascular, cognitive, and behavioral functions of an animal. Behavioral and pharmacologic interventions that decrease sympathetic output can be used to reduce this hyperarousal, allowing the restoration of more normal sleep. But pharmacologic interventions invariably additionally modify the structure of sleep itself, reducing its quality. (For reviews of sleep disorders discussed later, see Kryger *et al.*, 2005, and Hobson, 1989.)

The decrease in CNS aminergic activity (as in depression or narcolepsy) has opposite effects on cardiovascular, cognitive, and behavioral systems. As sleepiness appears, attention and cognition decline, and motor activity is impaired. Under these conditions, the cholinergic system is disinhibited, and the REM generator is triggered abnormally easily. The result is unwanted sleepiness or frank REM sleep attacks. These symptoms are decreased by aminergic agonists, such as blockers of amine reuptake, and psychostimulants, both of which boost sympathetic activity. Notably, many such drugs have both proaminergic and anticholinergic actions.

Respiratory Dysfunction Is Common in Sleep

One of the more dramatic problems associated with sleep is an exaggeration, at sleep onset, of the normal decline in respiratory drive (Chapter 37). Also, respiratory irregularity may be provoked by the altered brain stem activity of REM sleep. This sleep-dependent disruption of respiratory function, known as central sleep apnea, involves a decrease in respiratory drive during sleep (Fig. 42.13). More commonly, especially in obese people, mechanical collapse or compression of the airway can result in peripheral (or obstructive) sleep apnea, in which ventilation decreases because of a mechanical problem in the oropharynx. In peripheral sleep apnea, central respiratory drive causes continued respiratory effort. Eventually, the increase in CO_2 in the brain leads to an arousal which in turn produces

increased muscle tone in the oropharynx, and the airway opens. As the person falls into deeper sleep again, the oropharynx relaxes and the airway can become obstructed again until arousal recurs. In severe cases, arousals every one to two minutes all night are common.

People with sleep apnea usually feel excessively sleepy during daytime because frequent arousal prevents deep, sustained sleep. They are often unaware of their sleep-dependent breathing pattern because they are never fully aroused from sleep. Instead, observant friends or family often describe seeing these signs of respiratory dysfunction. In a sleep laboratory, EEG, respiratory effort, and cardiovascular parameters (e.g., heart rate, blood pressure, and oxygen saturation) are measured to quantify the problem. Sleep apnea can be treated by having the person wear a mask through which air is forced gently, helping to keep the airways open. The pressure of the air fluctuates to allow regular exhalation. People with sleep apnea are treated not only to alleviate their constant drowsiness, but to prevent long-term effects of sleep apnea, such as cardiopulmonary complications.

Motor Disturbances Can Occur during Sleep

Parasomnias that affect the skeletal motor system can occur in either REM or NREM sleep. In NREM sleep these parasomnias (e.g., sleep walking, sleep talking, tooth grinding, and night terrors) are seen mainly in young people, and reflect motor responses to the outputs of central motor pattern generators. Normally, responses to these generators are disinhibited at sleep onset and inhibited during REM sleep, when no motor commands normally are enacted. Extensive motor output in REM sleep requires the overpowering of the normal muscle atonia of REM sleep, and REM sleep behavior disorder can be a sign of degenerative brain disease in the elderly. Because of a failure to inhibit the expression of motor acts in REM, people with this disorder may literally enact their dream scenarios in their bedrooms. Given the normal lack of this active motor suppression during NREM sleep, it remains a mystery why dreams arising during NREM sleep fail to produce robust motor output.

A syndrome strikingly similar to REM sleep behavior disorder can be induced experimentally by bilateral lesions of the pontine tegmentum in cats. When in REM sleep, these animals perform stereotyped behavior sequences, such as hissing, piloerection, pouncing, and jumping. Jouvet called these "hallucinatory behaviors," and Morrison called the phenomenon "REM sleep without atonia" (Jouvet, 1999). Because the ability of a cat to inhibit motor pattern commands during

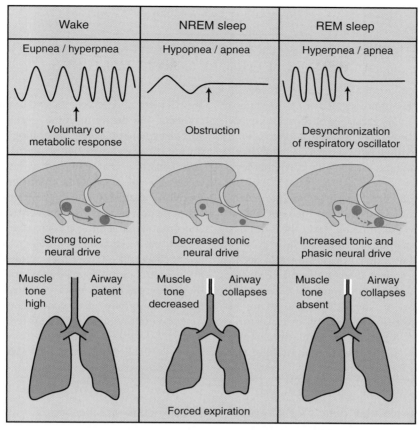

FIGURE 42.13 Central and obstructive sleep apnea. During waking, the respiratory oscillator of the medulla receives tonic drive from other neural structures and can respond to voluntary and metabolic signals to change breathing pattern. Muscle tone keeps the oropharynx open to the flow of air. In NREM sleep, central drive decreases, and the rate and depth of ventilation fall. If the decrease is excessive, central sleep apnea results, including a complete, albeit temporary, cessation of breathing. If the airway collapses, obstructive sleep apnea may result, with a similar cessation of breathing. During REM sleep, activation of pontine generator neurons drives the respiratory oscillator, and desynchronization may lead to breathing efforts that are too frequent or strong (hyperpnea) or that stop. During REM sleep the oscillator also becomes unresponsive to metabolic signals.

REM sleep is impaired by the lesions, they act out commands that are normally initiated during REM sleep. The instinctual aspect of these released behaviors is relevant to the psychophysiology of dreaming and has important implications for the hypothesis that REM sleep serves an active maintenance function.

Summary

Two models attempt to explain sleep and wakefulness. In the neurobiological model, sleep and waking are the result of a hypothalamic sleep-wake switch followed by reciprocal, phasic interactions between aminergic and cholinergic neurons and related circuitry. In the psychophysiological model, the influences of external and internal sensations are emphasized. During slow-wave sleep, thalamocortical synchronous activity diminishes responsiveness to the external world. During REM sleep, under the influence of cholinergic projections to the forebrain, internal representations of visual information predominate. In addition to changes in levels of sensory arousal and cognitive performance during sleeping, vegetative functions such as circadian rhythm recognition, body temperature, and autonomic activity also are regulated coordinately. Disorders of sleep influence emotional and cognitive functions adversely. Prolonged sleep deprivation is lethal, ending in a septic state of immunodeficiency, strongly suggesting that sleep serves important but uncertainly mediated maintenance functions.

The brain undergoes daily, complex, systematic changes in state that profoundly alter the nature of our consciousness, behavior, autonomic control, and physiologic homeostasis. At the root of these changes is the circadian clock. Located in the hypothalamus, this clock programs rest–activity and body temperature cycles. It also gates the NREM–REM sleep cycle control system in the pons.

BOX 42.3

CONSCIOUSNESS IN WAKING, SLEEPING, AND DREAMING: A MODEL FOR MENTAL ILLNESS

Since the brain is continuously active across states of sleep, so is the mind. Every modality of conscious experience can be shown to be state dependent, and particularly strong contrasts are evident between waking and dreaming, the two conscious states associated with higher levels of brain activation.

These alterations in consciousness are particularly informative to neurologists and psychiatrists because they provide new ways of thinking about the brain basis of mental illness. When brain activation declines in NREM sleep, most aspects of waking consciousness also decline. But when the brain is activated again during REM sleep, it does not reinstate waking consciousness. Instead, consciousness changes in an almost qualitative manner: Perception, instead of being shaped by external forms, becomes hallucinatory. Cognition is similarly deranged. Instead of being oriented, the dreaming mind loses track of time place and person. Instead of thinking actively and critically, the dreaming mind indulges in nonsequiturs, ad-hoc explanations, and other illogical whims. Memory, during dreams, instead of being declaratively faithful to recent history, is fragmented in its disconnection from current events and globally deficient in recalling them. Emotion, instead of being restrained and focused in response to percepts and thoughts, comes to dominate and organize dreaming consciousness often in strikingly salient ways. All these features of dream consciousness are normal and all are closely related to the changes in brain modulation. Because they are formally similar to the symptoms of the major psychoses, they suggest that mental illness may reflect genetic and environmental dysfunction of brain modulatory systems.

Edward F. Pace-Schott, J. Allan Hobson and Robert Stickgold

Some cellular and molecular details of the brain stem diencephalic behavioral state control system are clear: For the awake state, the noradrenergic locus coeruleus and the serotonergic raphé neurons of the pons, histaminergic and orexinergic neurons of the hypothalamus and, possibly, midbrain dopaminergic neurons must fire regularly to support alertness, attentiveness, memory, orientation, logical thought, and emotional stability. When the output of these chemical systems diminishes, drowsiness occurs. When sleep begins, the output of these systems declines further, and the brain–mind enters NREM sleep, a phase of lowered consciousness which, at times, might lapse into true unconsciousness. As NREM sleep continues, cholinergic activity increases gradually. At the NREM–REM transition, the activity of the serotonergic and nonadrenergic systems is at its nadir, and the output of the cholinergic system is increasing exponentially. This chemical switch results in the cholinergic activation and autostimulation of REM sleep. Combined with the lack of aminergic stimulation, this cholinergic overdrive contributes to the characteristic hallucination, delusion, disorientation, memory loss, and emotional intensity of REM sleep dreaming. What replaces these influences during NREM dreaming remains unknown.

Although the details of these daily changes in brain chemistry remain to be specified, consequences of their dysfunction are becoming more clearly understood through the study of insomnias, hypersomnias, and parasomnias. Integration of data from basic neurobiology, cognitive psychology, and clinical science enables the construction of specific, testable models of how waking and sleep affect our conscious experience. These models, in turn, constitute the building blocks of a scientific theory of consciousness.

SLEEP HAS MULTIPLE FUNCTIONS

Although it is easy to conceptualize the function of waking and to understand the adaptive value of consciousness, it has been difficult to move beyond the subjectively compelling but scientifically unsatisfactory notion of sleep as rest. But recent evidence indicates that sleep plays important roles in many physiological and cognitive systems. Evidence strongly suggests that sleep has an anabolic and actively conservative function that is related to the complexity of the mammalian brain (Table 42.1). In fact, Bennington and Heller (1995) propose that an important function of NREM sleep is to restore brain glycogen stores.

The most dramatic evidence of this homeostatic function comes from sleep deprivation studies in rats in which sleep deprivation was fatal when it persisted for four to six weeks. Early in the deprivation period, the rats began to eat more but could not maintain their body weight. Later, they lost their ability to maintain body temperature and developed strong heat-seeking behavior, and, finally, they died. Although the cause of this death remains uncertain, it is suspected to result from overwhelming sepsis because of immunodeficiency. The study implied that the maintenance of metabolic caloric balance, thermal equilibrium, and immune competence require sleep (Rechtschaffen and Bergman, 2002).

Additional evidence links immune function and sleep. Pappenheimer found that NREM sleep in rabbits was enhanced by dimuramyl peptides of bacterial cell wall origin. NREM sleep is also enhanced by the cytokines interleukin-1 and interleukin-2, both of which are released during NREM sleep (Krueger et al., 1999). But perhaps more to the point, immunizing subjects with an anti-influenza vaccine during a period when sleep was restricted to 4 hours per night reduced the normal production of anti-influenza antibodies by 57% (Spiegel et al., 2002).

Contributing to the notion that sleep has an anabolic function is the abundance of sleep in early life. REM sleep predominates *in utero*, where its stereotypic pattern of motor activation and its chemical microenvironment could promote CNS development (Hobson, 1989). Sleep also plays a role in hormonal regulation. Growth and development are most likely enhanced by the release of growth hormone and gonadotropins in slow wave sleep. Such hormone release declines, along with slow wave sleep, after growth and sexual maturation are complete, around 30 years of age. Recent studies have demonstrated the importance of sleep for normal regulation of glucose and prevention of diabetes (Knutson et al., 2006).

Sleep and Memory Are Interdependent

Perhaps the clearest evidence of a function for sleep is for its role in consolidating recently formed memories (Stickgold, 2005). One of the most notable examples is the discovery, in rats, by Carlyle Smith, of "REM windows" (Smith, 1995), which are time periods after an animal is trained on specific tasks when the animal shows enhanced quantities of REM sleep and when retention of learning could be decreased by selective REM sleep deprivation. Thus, not only does REM sleep appear to be critical for the maintenance of this learning, but animals also increase their REM sleep after training, presumably by a homeostatic mechanism that regulates the amounts of different sleep stages based on the needs of the organism. Subsequent studies in humans have led to the theory that REM sleep is especially important for the consolidation of proce-

dural memories, which are defined as memories of how to do things in the absence of explicit rules, such as how to ride a bicycle or read a map. In contrast, SWS has been implicated in the consolidation of explicit or declarative memory, which includes autobiographical memories as well as general knowledge of facts. However, studies on a procedural visual discrimination task have shown a requirement for both REM and SWS on the night following training in order for next day improvement on the task to be observed (Stickgold et al., 2000b). This finding supports an earlier theory based on animal studies in which a two-step process involving both REM and SWS was hypothesized. In humans, the requirement for sleep immediately following learning was shown when subjects who were sleep deprived the night following training, but who then were allowed two nights of unrestricted recovery sleep before testing, showed no significant task improvement (Stickgold et al., 2000a). In addition to procedural learning tasks, REM effects have been reported in humans on learning of complex logic games, foreign language acquisition, intensive studying in general, and both REM and SWS effects have been reported for the consolidation and processing of emotional memories. In contrast, sleep-dependent consolidation of simple motor skill learning, such as for a sequential finger-tapping task, correlates with Stage II NREM sleep (Walker et al., 2002).

There is also evidence that normal sleep-dependent memory consolidation is altered in patients with psychiatric disorders. Patients with schizophrenia show normal learning of the finger-tapping task during training, but no overnight improvement (Fig. 42.14; Manoach et al., 2004).

No Function for Dreaming Has Been Identified

Does dreaming also serve a function? Answering this question is not yet possible, because we do not know how to separate possible functions of the experience of dreaming from functions of the neurobiological processes that produce it. But REM sleep dream content can predict who will become clinically depressed following marital divorce or separation (Cartwright et al., 1998), suggesting that dreaming may serve to modulate emotional states during waking.

Dreaming may also play a role in modifying associative memory networks. Subjects awakened from REM sleep show preferential activation of more distantly related associated than seen either during waking or during NREM sleep (Stickgold et al., 1999). Such preferential activation of weak associations might explain the bizarre content of dreams during REM

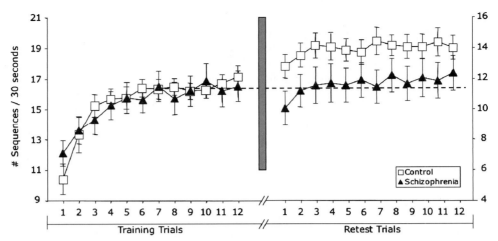

FIGURE 42.14 Sleep-dependent memory consolidation. Subjects were asked to type the sequence "41324" as quickly and accurately as they could for 30 sec and then reset for 30 sec before doing it again. Twelve trials were performed on each of two successive days. The control subjects showed a 16% increase in speed overnight, whereas the schizophrenia patients showed none.

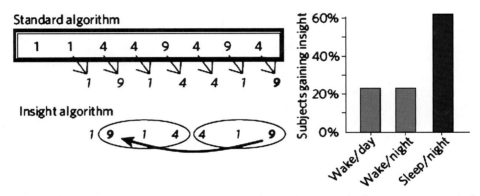

FIGURE 42.15 Sleep facilitates mathematical insight. Subjects are taught to reduce an eight-digit sequence to a single digit (9 at the right), through six intermediate calculations (in italics) with a standard algorithm. Unknown to the subjects, the task is designed so that the last three calculations form a mirror image of the preceding three, and thus the second intermediate calculation matches the final answer. Right: subjects who slept between training and testing were more than twice as likely to discover the insight algorithm than those not allowed sleep.

sleep, and may permit the brain to discover creative solutions to problems during sleep. Indeed, sleep has been shown to increase the likelihood of such discoveries (Fig. 42.15; Wagner *et al.*, 2004). But whether the conscious experience of dreaming is crucial for any of these sleep-dependent processes remains unknown.

Summary

Sleep represents a complex behavioral and neurobiological process that is conserved across much of the animal kingdom. REM sleep is similarly conserved across the class Mammalia. The complex, REM–NREM cycle suggests that these distinct neurophysiological and neurochemical brain states have evolved to subserve specific functions. One hypothesis is that the types of sleep stages and their timing across the night evolved to meet the demands of varying memory systems for

effective offline processing during sleep. Whether dreaming also evolved to permit offline cognitive and affective processing remains an open question.

References

Achermann, P. and Borbely, A. A. (1997). Low frequency (<1 Hz) oscillations in the human sleep electroencephalogram. *Neuroscience* **81**, 213–222.

Antle, M. C. and Silver, R. (2005). Orchestrating time: arrangements of the brain circadian clock. *Trends Neurosci* **28**(3), 145–151.

Arendt, J. (2006). Melatonin and human rhythms. *Chronobiol Int* **23**(1–2), 21–37.

Aston-Jones, G. and Bloom, F. E. (1981). Activity of norepinephrine-containing locus coeruleus neurons in behaving rats anticipates fluctuations in the sleep-waking cycle. *J. Neurosci.* **1**, 876–886.

Baghdoyan, H. A., Monaco, A. P., Rodrigo-Angulo, M. L., Assens, F., McCarley, R. W., and Hobson, J. A. (1984). Microinjection of neostigmine into the pontine reticular formation of cats enhances desynchronized sleep signs. *J. Pharmacol. Exp. Ther.* **231**, 173–180.

Bandyopadhya, R. S., Datta, S., and Saha, S. (2006). Activation of pedunculopontine tegmental protein kinase A: a mechanism for rapid eye movement sleep generation in the freely moving rat. *J Neurosci* **26**(35), 8931–8942.

Basheer, R., Strecker, R. E., Thakkar, M. M., and McCarley, R. W. (2004). Adenosine and sleep-wake regulation. *Prog Neurobiol* **73**(6), 379–396.

Bell-Pedersen, D., Cassone, V. M., Earnest, D. J., Golden, S. S., Hardin, P. E., Thomas, T. L. *et al.* (2005). Circadian rhythms from multiple oscillators: lessons from diverse organisms. *Nat Rev Genet* **6**(7), 544–556.

Bennington, J. H. and Heller, H. C. (1995). Restoration of brain energy metabolism as the function of sleep. *Progress Neurobiol.* **45**, 347–360.

Berson, D. M., Dunn, F. A., and Takao, M. (2002). Phototransduction by retinal ganglion cells that set the circadian clock. *Science* **295**(5557), 1070–1073.

Black, J. L., 3rd. (2005). Narcolepsy: a review of evidence for autoimmune diathesis. *Int Rev Psychiatry* **17**(6), 461–469.

Blanco-Centurion, C., Xu, M., Murillo-Rodriguez, E., Gerashchenko, D., Shiromani, A. M., Salin-Pascual, R. J. *et al.* (2006). Adenosine and sleep homeostasis in the Basal forebrain. *J Neurosci* **26**(31), 8092–8100.

Boissard, R., Gervasoni, D., Schmidt, M. H., Barbagli, B., Fort, P., and Luppi, P. H. (2002). The rat ponto-medullary network responsible for paradoxical sleep onset and maintenance: A combined microinjection and functional neuroanatomical study. *Eur J Neurosci* **16**(10), 1959–1973.

Borbely, A. A. and Achermann, P. (2005). Sleep homeostasis and models of sleep regulation. *In* "Principles and Practice of Sleep Medicine," 4th ed. (M. H. Kryger, T. Roth, and W. C. Dement, eds.), pp. 405–417. Elsevier, Philadelphia.

Callaway, C. W., Lydic, R., Baghdoyan, H. A., and Hobson, J. A. (1987). Ponto-geniculo-occipital waves: Spontaneous visual system activation occurring in REM sleep. *Cell Mol. Neurobiol.* **7**, 105–149.

Calvo, J., Datta, S., Quattrochi, J. J., and Hobson, J. A. (1992). Cholinergic microstimulation of the peribrachial nucleus in the cat. II. Delayed and prolonged increases in REM sleep. *Arch. Ital. Biol.* **130**, 285–301.

Capece, M. L. and Lydic, R. (1997). cAMP and protein kinase A modulate cholinergic rapid eye movement sleep generation. *Am J Physiol* **273**(4 Pt 2), R1430–1440.

Cartwright, R., Young, M. A., Mercer, P., and Bears, M. (1998). Role of REM sleep and dream variables in the prediction of remission from depression. *Psychiatry Research* **80**, 249–255.

Chang, F. C. and Opp, M. R. (2001). Corticotropin-releasing hormone (CRH) as a regulator of waking. *Neurosci Biobehav Rev* **25**(5), 445–453.

Chase, M. H. and Morales, F. R. (2005). Control of motoneurons during sleep. *In* "Principles and Practice of Sleep Medicine," 4th ed. (M. H. Kryger, T. Roth, and W. C. Dement, eds.), pp. 154–168. Elsevier, Philadelphia.

Chase, M. H., Soja, P. J., and Morales, F. R. (1989). Evidence that glycine mediates the post synaptic potentials that inhibit lumbar motorneurons during the atonia of active sleep. *J. Neurosci.* **9**, 743–751.

Cirelli, C. (2005). A molecular window on sleep: changes in gene expression between sleep and wakefulness. *Neuroscientist* **11**(1), 63–74.

Czeisler, C. A., Duffy, J. F., Shanahan, T. L., Brown, E. N., Mitchell, J. F., Rimmer, D. W. *et al.* (1999). Stability, precision, and near-24-hour period of the human circadian pacemaker. *Science* **284**(5423), 2177–2181.

Daan, S., Beersma, D. G. M., and Borbely, A. A. (1984). Timing of human sleep: Recovery process gated by a circadian pacemaker. *Am. J. Physiol.* **246**, R161–R178.

Dahan, L., Astier, B., Vautrelle, N., Urbain, N., Kocsis, B., and Chouvet, G. Prominent burst firing of dopaminergic neurons in the ventral tegmental area during paradoxical sleep. *Neuropsychopharmacology* **32**, 1232–1241.

Dahlstrom, A. and Fuxe, K. (1964). Evidence for the existence of monoamine neurons in the central nervous system. I. Demonstration of monoamines in the cell bodies of brain stem neurons. *Acta Physiol. Scand.* **62**(S232), 1–55.

Datta, S. (1997). Cellular basis of pontine ponto-geniculo-occipital wave generation and modulation. *Cell Mol Neurobiol* **17**(3), 341–365.

Datta, S., Siwek, D. F., Patterson, E. H., and Cipolloni, P. B. (1998). Localization of pontine PGO wave generation sites and their anatomical projections in the rat. *Synapse* **30**(4), 409–423.

Dubocovich, M. L. and Markowska, M. (2005). Functional MT1 and MT2 melatonin receptors in mammals. *Endocrine* **27**(2), 101–110.

Dzirasa, K., Ribeiro, S., Costa, R., Santos, L. M., Lin, S. C., Grosmark, A. *et al.* (2006). Dopaminergic control of sleep-wake states. *J Neurosci* **26**(41), 10577–10589.

Gervasoni, D., Peyron, C., Rampon, C., Barbagli, B., Chouvet, G., Urbain, N. *et al.* (2000). Role and origin of the GABAergic innervation of dorsal raphe serotonergic neurons. *J Neurosci* **20**(11), 4217–4225.

Hobson, J. A. and Pace-Schott, E. F. (2002). The cognitive neuroscience of sleep: Neuronal systems, consciousness and learning. *Nat Rev Neurosci* **3**(9), 679–693.

Hobson, J. A. and Steriade, M. (1986). The neuronal basis of behavioral state control. *In* "Handbook of Physiology" (F. E. Bloom, ed.), Sect. 1, Vol. 4, pp. 701–823. American Physiological Society, Bethesda, MD.

Hobson, J. A., Datta, S., Calvo, J. M., and Quattrochi, J. (1993). Acetylcholine as a brain state modulator: Triggering and long-term regulation of REM sleep. *Prog. Brain Res.* **98**, 389–404.

Hobson, J. A., McCarley, R. W., and Wyzinski, P. W. (1975). Sleep cycle oscillation: Reciprocal discharge by two brainstem neuronal groups. *Science* **189**(4196), 55–58.

Hobson, J. A., McCarley, R. W., and Qyzinki, P. W. (1975). Sleep cycle oscillation: Reciprocal discharge by two brainstem neuronal groups. *Science* **189**, 55–58.

Hobson, J. A., Pace-Schott, E. F. and Stickgold, R. (2000). Dreaming and the brain: toward a cognitive neuroscience of conscious states. *Behav Brain Sci*, 23(6), 793–842; discussion 904–1121.

Jacobs, B. L. and Azmitia, E. C. (1992). Structure and function of the brain serotonin system. *Physiol. Rev.* **72**, 165–229.

Jones, B. E. (2005). Basic mechanisms of sleep-wake states. *In* "Principles and Practice of Sleep Medicine," 4th ed. (M. H. Kryger, T. Roth, and W. C. Dement, eds.), pp. 136–153. Elsevier, Philadelphia.

Kahn, D., Pace-Schott, E. F., and Hobson, J. A. (1997). Consciousness in waking and dreaming: The roles of neuronal oscillation and neuromodulation in determining similarities and differences. *Neuroscience* **78**, 13–38.

Knutson, K. L., Ryden, A. M., Mander, B. A., and Van Cauter, E. (2006). Role of sleep duration and quality in the risk and severity of type 2 diabetes mellitus. *Arch Intern Med* **166**, 1768–1774.

Krueger, J. M., Obal, F., and Fang, J. (1999). Humoral regulation of physiological sleep: Cytokines and GHRH. *Sleep Res.* **8** (Suppl. 1), 53–59.

Lena, I., Parrot, S., Deschaux, O., Muffat-Joly, S., Sauvinet, V., Renaud, B. *et al.* (2005). Variations in extracellular levels of dopamine, noradrenaline, glutamate, and aspartate across the sleep–wake cycle in the medial prefrontal cortex and nucleus accumbens of freely moving rats. *J Neurosci Res* **81**(6), 891–899.

Leonard, T. O. and Lydic, R. (1999). Nitric oxide, a diffusible modulator of physiological traits and behavioral states. *In* "Rapid Eye Movement Sleep" (B. N. Mallick and S. Inoue, eds.), pp. 167–193. Marcel Dekker, New York.

Llinas, R. R. and Pare, D. (1991). Of dreaming and wakefulness. *Neuroscience* **44**, 521–535.

Lowrey, P. L. and Takahashi, J. S. (2004). Mammalian circadian biology: Elucidating genome-wide levels of temporal organization. *Annu Rev Genomics Hum Genet* **5**, 407–441.

Lu, J., Bjorkum, A. A., Xu, M., Gaus, S. E., Shiromani, P. J., and Saper, C. B. (2002). Selective activation of the extended ventrolateral preoptic nucleus during rapid eye movement sleep. *J Neurosci* **22**(11), 4568–4576.

Lu, J., Greco, M. A., Shiromani, P., and Saper, C. B. (2000). Effect of lesions of the ventrolateral preoptic nucleus on NREM and REM sleep. *J Neurosci* **20**(10), 3830–3842.

Lu, J., Jhou, T. C., and Saper, C. B. (2006). Identification of wake-active dopaminergic neurons in the ventral periaqueductal gray matter. *J Neurosci* **26**(1), 193–202.

Lu, J., Sherman, D., Devor, M., and Saper, C. B. (2006). A putative flip-flop switch for control of REM sleep. *Nature* **441**(7093), 589–594.

Luppi, P. H., Peyron, C., Rampon, C., Gervasoni, D., Barbagli, B., Boissard, R. *et al.* (1999). Inhibitory mechanisms in the dorsal raphe nucleus and locus coeruleus during sleep. *In* "Handbook of behavioral state control: Molecular and cellular mechanisms" (R. Lydic and H. A. Baghdoyan, eds.). CRC Press.

Lydic, R. and Baghdoyan, H. A. (2005). Sleep, anesthesiology, and the neurobiology of arousal state control. *Anesthesiology* **103**(6), 1268–1295.

Lydic, R., Baghdoyan, H. A., and Lorinc, Z. (1991). Microdialysis of cat pons reveals enhanced acetylcholine release during statedependent respiratory depression. *Am. J. Physiol.* **261**, 766–770.

Maloney, K. J., Mainville, L., and Jones, B. E. (2002). c-Fos expression in dopaminergic and GABAergic neurons of the ventral mesencephalic tegmentum after paradoxical sleep deprivation and recovery. *Eur J Neurosci* **15**(4), 774–778.

Manoach, D. S., Cain, M. S., Vangel, M. G., Khurana, A., Goff, D. C., and Stickgold, R. (2004). A failure of sleep-dependent procedural learning in chronic, medicated schizophrenia. *Biol Psychiatry* **56**, 951–956.

Maquet, P. (2000). Functional neuroimaging of normal human sleep by positron emission tomography. *J Sleep Res* **9**(3), 207–231.

Maquet, P., Ruby, P., Maudoux, A., Albouy, G., Sterpenich, V., Dang-Vu, T. *et al.* (2005). Human cognition during REM sleep and the activity profile within frontal and parietal cortices: A reappraisal of functional neuroimaging data. *Prog Brain Res* **150**, 219–227.

McCarley, R. W. (2004). Mechanisms and models of REM sleep control. *Arch Ital Biol* **142**(4), 429–467.

McCarley, R. W. and Hobson, J. A. (1975). Neuronal excitability modulation over the sleep cycle: A structural and mathematical model. *Science* **189**(4196), 58–60.

McCormick, D. A. and Bal, T. (1997). Sleep and arousal: Thalamocortical mechanisms. *Annu Rev Neurosci* **20**, 185–215.

McGinty, D. and Szymusiak, R. (2005). Sleep-promoting mechanisms in mammals. *In* "Principles and Practice of Sleep Medicine," 4th ed. (M. H. Kryger, T. Roth, and W. C. Dement, eds.), pp. 169–184. Elsevier, Philadelphia.

Mendelson, W. B. and Basile, A. S. (2001). The hypnotic actions of the fatty acid amide, oleamide. *Neuropsychopharmacology* **25**(5 Suppl), S36–S39.

Miller, J. D., Farber, J., Gatz, P., Roffwarg, H., and German, D. C. (1983). Activity of mesencephalic dopamine and non-dopamine neurons across stages of sleep and walking in the rat. *Brain Res* **273**(1), 133–141.

Moore, R. Y., Abrahamson, E. A., and Van Den Pol, A. (2001). The hypocretin neuron system: An arousal system in the human brain. *Arch. Ital. Biol.* **139**, 2195–2205.

Moore, R. Y. and Card, J. P. (eds.) (1984). "Noradrenaline-Containing Neuron System." Elsevier, Amsterdam.

Moruzzi, G. and Magoun, H. W. (1949). Brainstem reticular formation and activation of the EEG. *Electroencephal. Clin. Neurophysiol.* **1**, 455–473.

Nitz, D. and Siegel, J. M. (1997). GABA release in the dorsal raphe nucleus: Role in the control of REM sleep. *Am. J. Physiol.* **273**, R451–R455.

Obal, F. and Krueger, J. M. (1999). Hormones and REM sleep. *In* "Rapid Eye Movement Sleep" (B. N. Mallick and S. Inoue, eds.), pp. 233–247. Marcel Dekker, New York.

Pace-Schott, E. F. (2005). The neurobiology of dreaming. *In* "Principles and Practice of Sleep Medicine," 4th ed. (M. H. Kryger, T. Roth, and W. C. Dement, eds.), pp. 551–564. Elsevier, Philadelphia.

Pace-Schott, E. F. (2007; in press). The frontal lobes and dreaming. *In* "The New Science of Dreaming," Vol. I: The Biology of Dreaming (D. Barrett and P. McNamara, eds.). Praeger, Greenwood Press, Westport, CT.

Pace-Schott, E. F., and Hobson, J. A. (2002). The neurobiology of sleep: genetics, cellular physiology and subcortical networks. *Nat Rev Neurosci* **3**(8), 591–605.

Pappenheimer, J. R. (1982). Induction of sleep by moramyl peptides. *J. Physiol.* **336**, 1–11.

Peirson, S. and Foster, R. G. (2006). Melanopsin: Another way of signaling light. *Neuron* **49**(3), 331–339.

Rechtschaffen, A. and Bergman, B. M. (2002). Sleep deprivation in the rat: An update of the 1989 paper. *Sleep* **24**, 18–24.

Rye, D. B. (1997). Contributions of the peduculopontine region to normal and altered REM sleep. *Sleep* **20**, 757–788.

Sakai, K., Crochet, S., and Onoe, H. (2001). Pontine structures and mechanisms involved in the generation of paradoxical (REM) sleep. *Arch Ital Biol* **139**(1–2), 93–107.

Saper, C. B., Chou, T. C., and Scammell, T. E. (2001). The sleep switch: Hypothalamic control of sleep and wakefulness. *Trends Neurosci.* **24**, 726–731.

Saper, C. B., Scammell, T. E., and Lu, J. (2005). Hypothalamic regulation of sleep and circadian rhythms. *Nature* **437**(7063), 1257–1263.

Satoh, S., Matsumura, H., Koike, N., Tokunaga, Y., Maeda, T., and Hayaishi, O. (1999). Region-dependent difference in the sleep-promoting potency of an adenosine A2A receptor agonist. *Eur J Neurosci* **11**(5), 1587–1597.

Scheer, F. A., Cajochen, C., Turek, F. W., and Czeisler, C. A. (2005). Melatonin in the regulation of sleep and circadian rhythms. *In* "Principles and Practice of Sleep Medicine," 4th ed. (M. H. Kryger, T. Roth, and W. C. Dement, eds.), pp. 395–404. Elsevier, Philadelphia.

Sherin, J. E., Shiromani, P. J., McCarley, R. W. *et al.* (1996). Activation of ventrolateral preoptic neurons during sleep. *Science* **271**, 216–219.

Siegel, J. M. (2005). REM sleep. *In* "Principles and Practice of Sleep Medicine," 4th ed. (M. H. Kryger, T. Roth, and W. C. Dement, eds.), pp. 120–135. Elsevier, Philadelphia.

Smith, C. (1995). Sleep states and memory processes. *Behav. Brain res.* **69**, 137–145.

Solms, M. (1997). "The Neuropsychology of Dreams: A Clinicoanatomical Study." Lawrence Erlbaum Associates.

Spiegel, K., Sheridan, J. F., and Van Cauter, E. (2002). Effect of sleep deprivation on response to immunization. *Jama* **288**, 1471–1472.

Steiger, A. and Holsboer, F. (1997). Neuropeptides and human sleep. *Sleep* **20**, 1038–1052.

Steriade, M. (2000). Corticothalamic resonance, states of vigilance and mentation. *Neuroscience* **101**, 243–276.

Steriade, M. (2006). Grouping of brain rhythms in corticothalamic systems. *Neuroscience* **137**(4), 1087–1106.

Steriade, M. and McCarley, R. W. (1990). "Brainstem Control of Wakefulness and Sleep." Plenum, New York.

Stickgold, R. (1998). Sleep: Off-line memory reprocessing. *Trends Cog. Sci.* **2**, 484–492.

Stickgold, R. (2005). Sleep-dependent memory consolidation. *Nature* **437**, 1272–1278.

Stickgold, R., Hobson, J. A., Fosse, R., and Fosse, M. (2001). Sleep, learning and dreams: Off-line memory reprocessing. *Science* **294**, 1052–1057.

Stickgold, R., James, L., and Hobson, J. A. (2000a). Visual discrimination learning requires sleep after training. *Nature Neurosci.* **3**, 1237–1238.

Stickgold, R., Scott, L., Rittenhouse, C., and Hobson, J. A. (1999). Sleep-induced changes in associative memory. *J Cogn Neurosci* **11**, 182–193.

Stickgold, R., Whidbee, D., Schirmer, B., Patel, V., and Hobson, J. A. (2000b). Visual discrimination task improvement: A multistep process occurring during sleep. *J. Cog. Neurosci.* **12**, 246–254.

Strecker, R. E., Morairty, S., Thakkar, M. M., Porkka-Heiskanen, T., Basheer, R., Dauphin, L. J., Rainnie, D. G., Portas, C. M., Greene, R. W., and McCarley, R. W. (2000). Adenosinergic modulation of basal forebrain and preoptic/anterior hypothalamic neuronal activity in the control of behavioral state. *Behav. Brain Res.* **115**, 183–204.

Tononi, G. and Cirelli, C. (2001). Modulation of brain gene expression during sleep and wakefulness: A review of recent findings. *Neuropsychopharm.* **25**, S28–S35.

Trulson, M. E., and Preussler, D. W. (1984). Dopamine-containing ventral tegmental area neurons in freely moving cats: activity during the sleep-waking cycle and effects of stress. *Exp Neurol* **83**(2), 367–377.

Wagner, U., Gais, S., Haider, H., Verleger, R., and Born, J. (2004). Sleep inspires insight. *Nature* **427**, 352–355.

Walker, M., Brakefield, T., Morgan, A., Hobson, J. A., and Stickgold, R. (2002). Practice with sleep makes perfect: Sleep dependent motor skill learning. *Neuron* **35**, 205–211.

Wisor, J. P. and Kilduff, T. S. (2005). Molecular genetic advances in sleep research and their relevance to sleep medicine. *Sleep* **28**(3), 357–367.

Xi, M. C., Morales, F. R., and Chase, M. H. (2001). Induction of wakefulness and inhibition of active (REM) sleep by GABAergic processes in the nucleus pontis oralis. *Arch Ital Biol* **139**(1–2), 125–145.

Suggested Readings

Barrett, D. and McNamara, P. (eds.) The New Science of Dreaming. Praeger, Greenwood Press, Westport, CT.

Hobson, J. A. (1988). "The Dreaming Brain." Basic Books, New York.

Hobson, J. A. (1989). "Sleep." Scientific American Library.

Hobson, J. A., Pace-Schott, E. F., and Stickgold, R. (2000). Dreaming and the brain: Toward a cognitive neuroscience of conscious states. *Behav. Brain Sci.* **23**, 793–842.

Jouvet, M. (1999). "The Paradox of Sleep: The Story of Dreaming." MIT Press.

Kryger, M. H., Roth, T., and Dement, W. C. (eds.) (2005). "Principles and Practice of Sleep Medicine, 4th Edition." Elsevier, Philadelphia.

Lydic, R. and Baghdoyan, H. A. (eds.) (1999). "Handbook of Behavioral State Control: Molecular and cellular mechanisms." CRC Press.

Mallick, B. N. and Inoue, S. (eds.) (1999). "Rapid Eye Movement Sleep." Dekker.

Pace-Schott, E. F. and Hobson, J. A. (2002). Basic mechanisms of sleep: New evidence on the neuroanatomy and neuromodulation of the NREM-REM cycle. *In* "American College of Neuropsychopharmacology, Fifth Generation of Progress" (D. Charney, J. Coyle, K. Davis, and C. Nemeroff, eds.). Lippincott, Williams & Wilkins, New York.

Edward F. Pace-Schott, J. Allan Hobson and Robert Stickgold

Reward, Motivation, and Addiction

REWARD AND MOTIVATION

Motivation is the process by which animals adapt to internal or external change. Such adjustments to change may involve the integration of endocrine, autonomic, and behavioral responses. Some adjustments are part of the regulatory process of homeostasis and act through negative feedback loops to correct an internal change. Indeed, motivation used to be explained largely in terms of reductions of needs, or "drives," such as those for sodium (salt) or water. This purely homeostatic explanation of motivation could plausibly account for ingestive and thermoregulatory behaviors, and possibly even the drug-seeking behavior generated by the state produced by drug withdrawal. However, it fails to explain satisfactorily behavior such as aggression, mating, or exploration, for which there are obvious external triggering stimuli but no identifiable deficit state. It also fails to explain what happens when homeostatic mechanisms are apparently overridden by powerful external incentives, as occurs during drug binges and during overingestion of delicious food. Instead, such behavior often is explained by attractions to external stimuli that have appetitive or rewarding properties (incentive–motivation).

In many cases, goals such as food or a sexual partner are not available, and an animal must search or forage for them. Therefore, motivated behavior is more than the simple control of consummatory responses that end sequences of motivated behavior, such as eating, drinking, and sexual mounting or lordosis. Consummatory behavior tends to be stereotyped and reflexive and is acquired early in an animal's life. In sequences of motivated behavior, however, consummatory behavior usually is preceded by adaptive, flexible forms of appetitive behavior (e.g., foraging for food), which enables an animal to come into physical contact with its goal. This appetitive behavior may be simple locomotor approach responses to the goal or may include exploratory behavior and complex response sequences. In addition, appetitive behavior may occur in parallel with endocrine and autonomic responses (e.g., secretion of saliva and insulin) (Chapter 38) that prepare the animal for efficient interaction with the goal.

Conditioning and Learning

In the absence of immediate goals, an animal usually must use past experience to predict the likelihood of an occurrence. This learning may involve classically conditioned (i.e., Pavlovian) reflexes and goal-directed instrumental (or operant) behaviors. With the latter behaviors, an outcome that increases the occurrence of a preceding behavior is a positive reinforcer. This term often is used interchangeably with the more colloquial term *reward*, which connotes a positive emotional component of positive reinforcement or pleasure. However, accurately inferring subjective states in animals is difficult, if not impossible and both reward and positive reinforcer should be used operationally. Identification of neural structures that mediate subjective phenomena, such as pleasure, may need to depend on functional neuroimaging of humans. Incentive, however, generally refers to the attractiveness of a goal, and positive reinforcement strengthens specific responses by presenting stimuli contingent on performance. For example, rats can learn arbitrary instrumental actions, such as a lever press, to gain access to positive reinforcers, such as food or drugs (i.e., self-administration behavior) (Fig. 43.1) or to

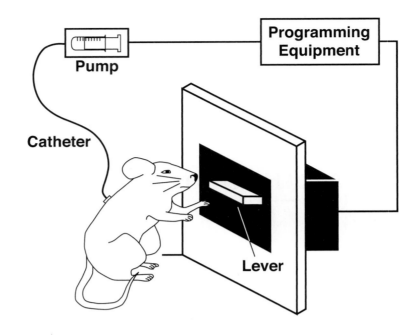

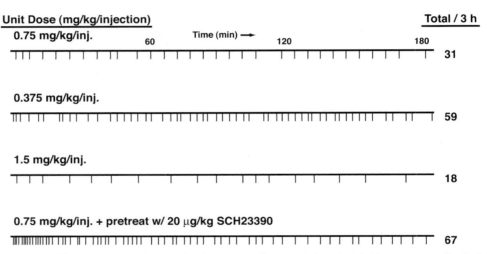

FIGURE 43.1   Intravenous self-administration by rats. (Top) Drawing illustrating the setup for intravenous self-administration of cocaine by rats. (Bottom) Event records for different unit doses of cocaine show a dose-response relationship relating dose of cocaine to number of infusions. Rats implanted with intravenous catheters and trained to self-administer cocaine with limited access (3h per day) show stable and regular drug intake over each daily session. No obvious tolerance or dependence develops. Rats generally are maintained on a fixed-ratio (FR) schedule of drug infusion, such as FR-1 or FR-5. In an FR-1 schedule, one lever press is required to deliver an intravenous infusion of cocaine; in an FR-5 schedule, five lever presses are required to deliver an infusion of cocaine, and so on. With an FR schedule, rats appear to regulate the amount of drug self-administered. Lowering the dose from the training level of 0.75mg/kg per injection increases the number of self-administered infusions. Raising the unit dose decreases the number of infusions. Pretreatment with the dopamine receptor antagonist SCH 23390 also increases the number of self-administered infusions. Reprinted with permission from Caine *et al.* (1993).

stimuli associated with these primary reinforcers–conditioned reinforcers. Schedules of reinforcement vary the relationship between activity and positive reinforcement. Thus, an animal's degree of motivation can be assessed by its capacity to work for a goal.

Another measure of motivation often involves the choice an animal makes (e.g., between actions that produce one stimulus associated with reward over another stimulus that is not associated with reward). Conditioned place preference is a commonly used procedure based on this principle in which choice is made between two environments, only one of which is linked to reward. A distinctive location is associated with a reinforcer such as food or a drug. Following several

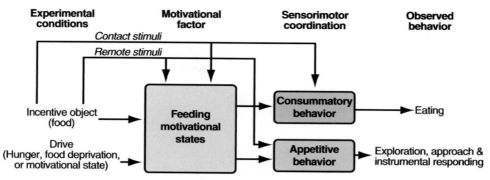

FIGURE 43.2 Theory of incentive-motivation, as applied to feeding. Note that drive (internal stimuli) and food (external stimuli) separately affect central motivation to feed. Also, incentive (i.e., food) and discriminative (i.e., sensorimotor) influences exert their effects separately to control consummatory behavior (when food is contacted) and appetitive behavior (when food remains remote). Modified with permission from Robbins (1986).

learning trials, a test trial is conducted in the absence of the goal itself. In the test, the preference for that place (as measured by the time spent there) relative to another place without that association is taken as an index of the incentive properties of the goal. Negative reinforcement, as distinct from positive reinforcement, is defined by the removal or postponement of an aversive event, the process that increases the probability of a preceding response and that may lead to a positive emotional state sometimes associated with relief. Negative reinforcement should not be confused with punishment. Punishment is defined as the suppression of instrumental actions with which it is associated.

Apparently motivated behavior can sometimes be shown to be resistant to motivational changes (e.g., satiety or other forms of devaluation of the goal such as lithium-induced illness). The assumption then is that it has acquired some habitual properties, based on the association between stimulus and response (S-R), rather than between the instrumental action and goal outcome. Such S-R "habits" may predominate after certain forms of training, particularly if prolonged (Dickinson 1994; Yin *et al.*, 2004).

The complexity of motivated behavior (Fig. 43.2), from selection of voluntary actions based on past experience to reflexive control of consummatory behavior, involves coordination of several levels of neural control, including the neocortex, older areas of the cortex often grouped as the "limbic system" (including the amygdala), the extrapyramidal motor system, integrative centers (e.g., the hypothalamus), and the basic motor mechanisms of the brain stem.

The Hypothalamus Plays a Role in Motivation

Hypothetical motivational centers were based on early concepts of brain stem centers for fundamental processes, such as respiration (Chapter 37). Though the construct of *brain centers* eventually fell out of favor, the original concept of homeostatic centers led many investigators to approach motivation as a search for specific neural or hormonal signals that could trigger neurocircuitry to initiate the appropriate behavioral responses. Such models could explain the efficient integration of many different signals to produce coordinated sequences of behavior. The hypothalamus is an ideal key structure in a circuitry for such integrative functions, given its rich blood supply and exposure to cerebrospinal fluid (and thus access to a variety of chemical signals and the capacity also to deliver them), control over the pituitary, and neural connections (through brain stem and other centers) with visceral afferent information (e.g., taste and olfaction), autonomic and somatic motor outflow. The concept of mutually inhibitory centers for the initiation (hunger) and suppression (satiety) of feeding arose from such a view (Chapter 34) (Box 43.1).

Hypothalamic Contribution to Other Forms of Motivation

On the basis of the criteria just described, the list of hypothalamic motivational centers originally was extended to include behaviors as diverse as drinking, thermoregulation, aggression, and sex (Chapter 34). Thus, for example, thermoregulatory behavior, such as reflexive shivering, panting, and grooming, is impaired by lesions of the preoptic region of the hypothalamus and is elicited by cooling or warming of this region (Satinoff, 1982). Defensive (affective) aggression and accompanying autonomic changes, such as piloerection, are produced by electrical stimulation of the ventromedial hypothalamus, whereas predatory ("quiet") aggression is produced by electrical stimulation of the

---

BOX 43.1

LOGIC OF METHODOLOGY USED IN BEHAVIORAL NEUROSCIENCE

Theorists (e.g., Hoebel, 1974) have defined principles for identifying the involvement of particular brain regions in motivational (and other behavioral) functions. For example, electrophysiological, neurochemical, or metabolic activity may be observed in a structure correlated with a behavioral response of interest, which suggests an involvement of that structure with the function under study. However, this correlative relationship does not establish whether the functioning of that region is causally important for the behavior. This issue may be addressed by an intervention that disrupts (impairs or improves) the function under study. For example, if the function is lost following selective lesioning, this suggests that the brain structure is implicated in that function. Similarly, if the function is enhanced following a boosting of neurotransmission in that region (e.g., via mild levels of electrical or chemical stimulation through implanted electrodes or cannulae), this is similarly converging evidence of a causative role. However, it is important to realize that the use of these techniques is each associated with interpretative problems, and their effects must be critically appraised. For example, the designation of the lateral hypothalamic region as a "feeding center" was based on the following:

1. Electrolytic lesions produced profound aphagia (abstention from eating).

2. Electrical stimulation of the brain through electrodes implanted in the lateral hypothalamus induced eating in sated rats (Hoebel, 1974).
3. Infusions of the neurotransmitter norepinephrine into the hypothalamus also induced eating in sated rats (Leibowitz, 1980).
4. Electrophysiological recording showed that when monkeys were deprived of food, individual neurons in the lateral hypothalamus were sensitive to the sight and taste of food (Rolls, 1985).

Although many of these experimental observations have remained valid, subsequent evidence has suggested that the lateral hypothalamus has a much more specific role in feeding motivation, as a result of using improved techniques (see main text). Moreover, many of the effects associated with diminishing and increasing activity within the lateral hypothalamus have been shown subsequently to depend upon coincidental influences on axons within the medial forebrain bundle that merely passes through the lateral hypothalamus (e.g., between midbrain and striatum or between prefrontal cortex and brainstem) and had little to do functionally with the lateral hypothalamus itself. Only through the use of several neurobiological techniques providing converging evidence can the relationships between brain and behavior become more certain.

---

lateral hypothalamus in a number of species (Flynn *et al.*, 1970).

A particularly well-studied example is that of sexual motivation. Some hypothalamic nuclei contain high concentrations of steroid receptors (e.g., for estradiol and testosterone), especially the medial preoptic area and the ventromedial hypothalamic nucleus. In male rats, copulatory behavior is abolished by medial preoptic area lesions (Heimer and Larsson, 1966–1967), whereas electrical stimulation of this region elicits copulation. In female rats, lordosis is blocked (and aggression toward males increased) by lesions of the ventromedial hypothalamic nucleus. Lordosis also is induced by electrical stimulation of this region (Pfaff, 1982). Most of these responses are hormone dependent. In ovariectomized females, proceptive ("ear wiggling," hopping, and darting) and receptive (lordosis) behaviors can be elicited by implanting

estradiol into the ventromedial hypothalamic nucleus. Similarly, the sexual behavior of castrated males can be restored by implanting testosterone into the medial preoptic area. This hormone dependence underlines the interaction between external (environmental) and internal (plasma steroids) states in inducing motivated responses. Also, sex hormones influence the organization of sexually dimorphic brain development (e.g., development of the sexually dimorphic nuclei in the preoptic region), as well as sexually dimorphic behavior. Maternal behavior also is dependent on hormones and is impaired by lesions of the medial preoptic area. Thus, it seems clear that the hypothalamus plays a central role in the regulation of sexual behavior.

The intermingling of hypothalamic neural mechanisms controlling thermoregulation, maternal behavior, feeding, and mating suggests that these behaviors share controls in many circumstances. In summary, a wide variety of motivational states expressed as easily elicited, sometimes stereotypical and species-specific behaviors are represented within the hypothalamus and related structures.

Hypothalamic Role in Consummatory Behavior

Many studies of the role of the hypothalamus in motivation have concentrated on measuring consummatory rather than learned instrumental behavior, a research strategy that may lead to misleading conclusions about motivational changes in general. For example, rats with preoptic lesions will not respond to thermal stress with reflex mechanisms such as shivering, adjustment of food intake, nest building, or changes in locomotion. However, they will learn to press a lever for hot or cool air and can achieve thermoregulation by this means (Carlisle, 1969). Similarly, male monkeys with preoptic lesions still show motivated behavior (e.g., they will masturbate in the presence of females), but are unable to mount and intromit. Moreover, male rats with preoptic lesions still show displacement behaviors and locomotor excitement in the presence of a receptive female, as well as lever pressing to gain access to the female, but cannot perform the consummatory behavior of copulation.

This dissociation between consummatory and instrumental behaviors is important because damage to a component of the limbic system, the amygdala, is known to produce the opposite pattern of effects—unimpaired mounting but reduced instrumental behavior (Everitt, 1990). These examples show that hypothalamic mechanisms operate mainly at the level of controlling consummatory rather than learned instrumental behavior leading to a goal. Behavioral flexibility illustrated by learning appears to require recruitment of additional neural systems, including the amygdala (see later).

The Lateral Hypothalamic Syndrome

Electrolytic lesions of the lateral hypothalamus cause dramatic deficits in food intake. However, not only do such lesions affect more than simply feeding behavior, they also destroy far more than the lateral hypothalamic region itself, mainly by interrupting ascending and descending fibers in the medial forebrain bundle (Fig. 43.3). Thus, the effects of lateral hypothalamic lesions on eating were unlikely all to be caused by damage to a lateral hypothalamus "feeding" center. For example, the nigrostriatal dopamine pathway courses through the lateral hypothalamus

without synapsing there. Injection of a catecholamine-selective neurotoxin, 6-hydroxydopamine (6-OHDA), into the vicinity of the lateral hypothalamus reproduced most of the classic lateral hypothalamic syndrome (Ungerstedt, 1971a). The injected rats suffered large depletions (>90%) of dopamine from the striatum. They were cataleptic and had difficulty initiating head, limb, axial, and oral movements. Further experiments showed that the syndrome could be produced by injection of 6-OHDA at any of several points along the nigrostriatal dopamine pathway between its origin in the substantia nigra and its termination in the caudate-putamen. Thus, dopamine depletion at sites well away from the lateral hypothalamus reproduced much of the lateral hypothalamic syndrome. 6-OHDA-induced lesions of the mesolimbic rather than the nigrostriatal pathway minimally affected ingestion but did affect other behaviors, such as locomotion and exploration, that may accompany feeding (Koob et al., 1978). Therefore, both consummatory and appetitive behaviors are affected by dopamine depletion, although probably in different regions of the forebrain.

Lateral Hypothalamic Syndrome Is Similar to, but Not the Same As, the Syndrome Produced by 6-Hydroxydopamine Lesions

Recovery from the lateral hypothalamic syndrome and the syndrome produced by 6-OHDA is similar (Marshall and Teitelbaum, 1977). At first, animals show adipsia (absence of drinking), aphagia (absence of eating), sensorimotor neglect (lack of response to stimuli), and akinesia (dearth of movements). Recovery from sensorimotor neglect parallels recovery from aphagia in both syndromes. (Recovered rats do not respond normally to physiological challenges, such as hyperosmolality.) Stricker and Zigmond (1976) explained this recovery as intrinsic recovery of function in dopaminergic systems. They proposed that changes in dopamine turnover and reuptake, as well as increases in the number of dopamine receptors, combine to minimize the impact of a lesion. In fact, in animals recovered from 6-OHDA lesions, aphagia and adipsia can be reinstated by drugs that block the synthesis of catecholamines or that block dopamine receptors. These studies led to the hypothesis that the entire lateral hypothalamic syndrome is caused by a deficit in behavioral activation and results from depletion of brain dopamine rather than damage to lateral hypothalamic cells.

Despite these similarities, subtle differences between the two syndromes cannot be explained by this dopamine hypothesis. For example, 6-OHDA-lesioned rats, unlike animals with lateral hypothalamic lesions, are

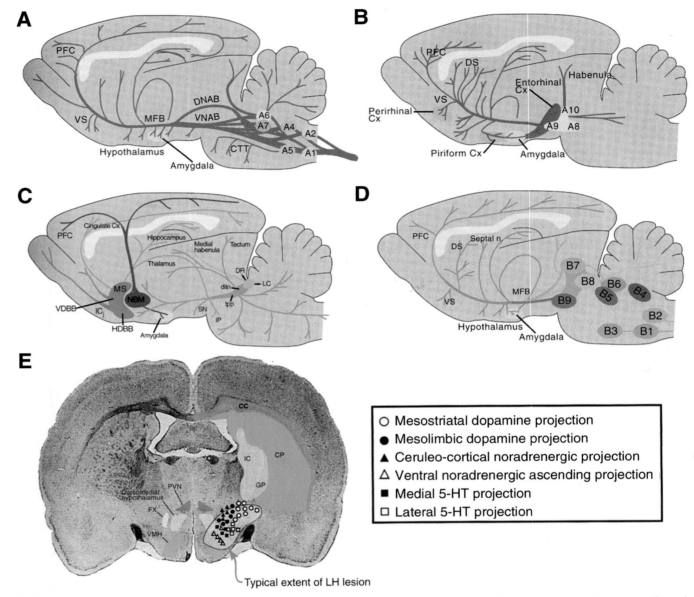

FIGURE 43.3 Ascending monoamine neurotransmitter systems. Figure shows schematic sagittal (A–D) and coronal (E) sections through the lateral hypothalamus of a rat brain. (A) Origin and distribution of central noradrenergic pathways. Note noradrenergic cell groups A1–A7, including the locus coeruleus (A6). DNAB, dorsal noradrenergic ascending bundle; VNAB, ventral noradrenergic ascending bundle. (B) Origin and distribution of central dopamine pathways. Note dopaminergic cell groups A8–A10. (C) Origin and distribution of central cholinergic pathways. Note rostral cell groups. NBM, nucleus basalis magnocellularis (Meynert in primates); MS, medial septum; VDBB, vertical limb nucleus of the diagonal band of Broca; HDBB, horizontal limb nucleus of the diagonal band of Broca. (D) Origin and distribution of central serotonergic pathways. Note cell groups in the raphe nucleus, B4–B9. MFB, medial forebrain bundle; PFC, prefrontal cortex; VS, ventral striatum; DS, dorsal striatum. Based on T. W. Robbins and B. J. Everitt, in *The Cognitive Neurosciences*, MIT Press, Cambridge MA, 1995; reproduced with permission. (E) Schematic coronal section through the rat brain at the level of the ventromedial hypothalamus. LH lesions could disrupt various ascending monoaminergic fibers. CC, corpus callosum; CP, caudate-putamen; DMH, dorsomedial hypothalamus; IC, internal capsule; GP, globus pallidus; PVH, paraventricular hypothalamus; LH, lateral hypothalamus; FX, fornix; VMH, ventromedial hypothalamus. Modified from Robbins (1986), with permission.

not somnolent (drowsy), are less finicky about food, and have fewer deficits in thermoregulation, taste aversion, learning, and oral motor performance. However, the most compelling difference is that rats with lateral hypothalamic lesions and the full syndrome have only 50% depletion of striatal dopamine, whereas 6-OHDA-lesioned rats (even with less than the full syndrome) have greater than 90% depletion of striatal dopamine. Therefore, damage to the lateral hypothalamus itself or to fibers of passage, in addition to those of the ascending dopamine system, apparently contributes to the lateral hypothalamic syndrome.

Cell Body Lesions in the Hypothalamus Produce a Less Severe Lateral Hypothalamic Syndrome

The neurotoxins kainic, ibotenic, and quisqualic acids are structurally related to the excitatory neurotransmitter glutamate. They bind to various subtypes of glutamate receptors and essentially stimulate neurons to death. These excitotoxins can be used to produce lesions of cell bodies while sparing fibers of passage. Electrolytic and mechanical lesions, in contrast, damage soma and fibers. Despite sparing striatal dopamine, such cell body lesions lead to severe problems with eating, followed by long-lasting deficits in homeostasis (Winn *et al.*, 1984). However, there are no signs of akinesia or sensorimotor neglect. Hence, at least two factors are required to explain the lateral hypothalamic syndrome: damage to the lateral hypothalamus disrupts homeostasis and striatal dopamine loss disrupts the initiation and execution of feeding and related behaviors. Neurons in the lateral hypothalamus containing the neuropeptide orexin project directly to the dopamine neurons in the ventral tegmental area and may therefore provide a neuroanatomical and functional link between regulatory and incentive motivational components of ingestive behavior.

Animals with Dopamine Denervation Syndrome Show Rotation and Sensorimotor Neglect

A central question for research in this area is what is the precise nature of the contribution by the dopamine system to motivation? Some clues to answering this question were provided by experiments in which rats received unilateral 6-OHDA lesions of the nigrostriatal system. The rats rotated toward the side of the lesion following treatment with a dopamine-releasing drug, such as D-amphetamine (Ungerstedt, 1971b). Presumably, this rotation occurred because more dopamine is released on the intact side than the lesioned side of the brain and because any output pathway from the striatum influences, via crossed pathways, the opposite side of the body. Of even greater interest, the direct dopamine receptor agonist apomorphine caused the rats to turn away from the side of the lesion so long as the lesion was extensive. This rotation could be explained by supersensitive dopamine receptors on the lesioned side. These studies of drug-induced movement have been important in identifying mechanisms of receptor regulation that might contribute to recovery from motivational deficits following dopamine denervation. However, they do not by themselves answer the central question posed earlier of the precise contribution of dopamine to motivation.

More relevant to this question is that unilateral lesions of the lateral hypothalamus or nigrostriatal dopamine system also produce polymodal sensorimotor neglect, measured by responses to contralateral somatic, auditory, and olfactory stimuli—a syndrome initially attributed to sensory inattention (Marshall and Teitelbaum, 1977). A more detailed analysis suggests that the fundamental deficit is a failure in initiating responses rather than attending to stimuli (Box 43.2 and Fig. 43.4). This leads to the conclusion that the nigrostriatal dopamine system contributes to *activation*, energizing the vigor of behavioral output (Chapter 31).

Mesencephalic Dopamine Mediates Activation of Behavior

Animals with lesions of the nigrostriatal system retain the capacity for movement; they do not have an irreversible motor deficit. However, they have great difficulty in starting that movement without an intact dopamine system. It is instructive to compare this with the problems that patients with Parkinson's disease have with movement initiation, since they too have a severe deficit in nigrostriatal dopamine (Chapter 31). To identify the circumstances in which the dopamine system normally becomes active, *in vivo* neurochemical techniques, such as microdialysis and voltammetry (Phillips *et al.*, 1991), have been used to monitor the release of dopamine in animals exhibiting particular behaviors. In addition, researchers have examined the firing of dopamine cells in response to particular stimuli (Schultz *et al.*, 1995). For example, dopamine cells have been identified in the ventral tegmental area that respond to rewards such as food (Fig. 43.5). However, if the food is reliably predicted by a conditioned stimulus, then, as a consequence of learning, the firing is initiated instead in response to the previously neutral stimulus. This and other findings support the theory that the dopamine cells provide an *error prediction signal* that enables reinforcement learning

---

BOX 43.2

SEPARATION OF SENSORY AND RESPONSE FACTORS IN THE DOPAMINERGIC SYNDROME OF NEGLECT

Sensory and motor aspects of the neglect syndrome were separated in the following way (Carli *et al.*, 1985). Rats were trained to detect brief flashes of light presented to either side of the head. In one case, they were trained to respond where the light was. In the alternate case, rats were trained to respond in a location away from the side of the light (Fig. 43.4). Rats with dopamine depleted from one side of the striatum could not respond properly toward

the contralateral side, whether or not the stimulus was presented on that side. Measurement of reaction time showed that the problem was in initiation, rather than completion, of the movement. Consistent with these data, dopamine-depleted rats could detect a contralateral stimulus and respond to it with an unlateralized response. Thus, their deficit was in initiation of movement in response to stimuli rather than in detection of the stimuli.

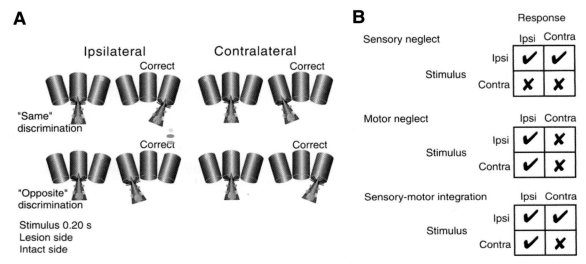

FIGURE 43.4 Distinguishing between sensory and response factors. Rats were trained to move toward the same or opposite side on which a brief (0.2 s) visual stimulus was presented. If a rat did not respond to the stimulus, it remained in a central location. After training, unilateral striatal lesions (oval) were made. Rats then were observed for their ability to respond to stimuli presented ipsilateral or contralateral to their lesions. Based on the responses of lesioned rats to stimuli, conclusions were drawn about the sensory and motor nature of the deficits caused by striatal lesions. (B) The hypothetical pattern of behavioral results obtained in the different test conditions following unilateral dopamine depletion is shown. Data were, in fact, consistent with a deficit in responding to contralateral space (i.e., motor neglect). From Robbins and Everitt (2003), with permission.

by networks in the striatum to influence appetitive behaviors.

Electrical Stimulation of the Brain Can Be a Positive Reinforcer

Olds and Milner (1954) discovered that rats learned new responses when their learning was positively reinforced by electrical stimulation of the brain. That finding affirmed the importance of incentive factors in instrumental behavior. The best natural parallel appeared to be behavior in response to large incentives by animals

in low states of drive (e.g., chocolate milk in undeprived rats). However, any attempt to directly equate rewarding brain stimulation with subjective feelings of pleasure and hedonism is limited because the individuals used as subjects in human intracranial self-stimulation (ICSS) experiments often suffer from schizophrenia or intractable pain, and they frequently had difficulty verbalizing their responses. Measurement of the strength of reinforcers in microamperes (Box 43.3; Fig. 43.6) promised a pragmatic, as well as objective, new approach to the psychophysics of an operational hedonism. However, its greatest contribution was to enable

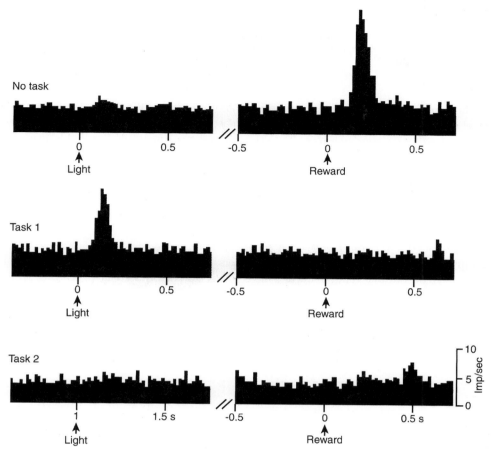

FIGURE 43.5 Responses of dopamine neurons to unpredicted primary reward (Top) and the transfer of this response to progressively earlier reward-predicting conditioned stimuli with training (Middle). (Bottom) the record shows a control baseline task when the reward is predicted by an earlier stimulus and not the light. From Schultz *et al.* (1995), with permission.

the anatomical mapping of the central reward or reinforcement mechanisms of the brain (Phillips, 1984).

Catecholamine Hypothesis of Rewarding Brain Stimulation

Later work showed that animals also can learn behaviors to stimulate other sites through ICSS (e.g., the locus coeruleus, midbrain central gray, cerebellum, trigeminal motor nucleus, substantia nigra, caudate nucleus, and nucleus accumbens) (Phillips, 1984). The diversity of these sites has not been fully elucidated. Many of them are known to lie along the trajectory of catecholamine-containing neurons. Although this includes both norepinephrine and dopamine projections, most pharmacologic data can be interpreted in terms of dopaminergic mechanisms (Box 43.3; Fig. 43.6). In particular, very high rates of responding for ICSS were found in the ventral tegmental area, ventral striatum (including the nucleus accumbens

to which dopamine neurons project), and limbic system. However, of several possibly independent neural systems that may subserve rewarding brain stimulation, only some depend on dopamine. The difficulty in interpreting the role of striatal dopamine in ICSS lies in distinguishing motor from reward effects (Box 43.3). The phenomenon of human drug dependence and the discovery in the early 1960s of drug self-administration in animals opened the way to an analysis of the neurochemical basis of reinforcement.

The Mesolimbic Dopamine System Plays a Key Role in Incentive Motivation

Wise (1982) proposed that natural reinforcers are perceived as rewards because they increase activity of the mesolimbic dopamine system. If the hypothesis of Wise is correct, the mesolimbic dopamine system should mediate natural rewards, such as food and sex.

BOX 43.3

PSYCHOPHYSICS OF HEDONISM AS MEASURED BY BRAIN STIMULATION REWARD

Electrical stimulation of the brain can be used as positive reinforcement of learned behaviors. An apparatus is set up so that when an animal presses a lever, its brain is stimulated. Alternatively, a threshold procedure can be used where a lever or wheel response is required. An advantage of using electrical stimulation to measure motivation is that the strength of the reward can be titrated easily by adjusting the amount of current delivered for each response. If the current is adjusted too low (i.e., below the threshold for reinforcement), then the animals will not press the lever. As current is increased, the relationship between the rate of lever pressing and the strength of current typically follows a sigmoidal curve called rate-intensity function. In the threshold procedure, a series of discrete trials are presented at different current intensities, and the responses within a certain cutoff period are recorded. Current levels are lowered and

raised to capture the threshold current (Fig. 43.2). Pharmacologic manipulation of catecholamine neurotransmitter systems can lower thresholds (e.g., treatment with amphetamines or cocaine) or raise thresholds (e.g., treatment with the dopamine receptor antagonist pimozide). Such changes mean the amount of electrical current required to support lever presses varies, whereas the latency to respond is unchanged. Thus, these drugs affect reward processes rather than motor function. The threshold of reinforcement also can be measured by allowing rats to adjust the strength of the current. In this setup, a process is employed in which rats press lever one to produce ICSS, but each lever-one press also reduces the intensity of the current. To reset current intensity, rats must press lever two. Such "self-titration" experiments yield conclusions similar to those based on threshold and rate-intensity functions.

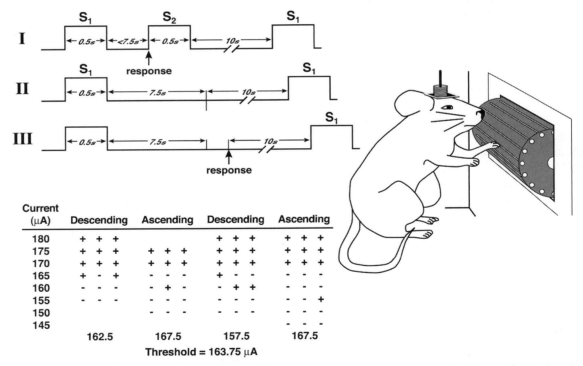

Current (μA)	Descending			Ascending			Descending			Ascending		
180	+	+	+				+	+	+	+	+	+
175	+	+	+	+	+	+	+	+	+	+	+	+
170	+	+	+	+	+	+	+	+	+	+	+	+
165	+	-	+	-	-	-	+	-	-	-	-	-
160	-	-	-	-	+	-	-	+	+	-	-	-
155	-	-	-							-	-	+
150				-	-	-	-	-	-	-	-	-
145										-	-	-
	162.5			167.5			157.5			167.5		

Threshold = 163.75 μA

FIGURE 43.6 Intracranial self-stimulation threshold procedure. Panels I, II, and III illustrate the timing of events during three hypothetical discrete trials. Panel I shows a trial during which the rat responded within the 7.5 s following the delivery of the noncontingent stimulus (positive response). Panel II shows a trial during which the animal did not respond (negative response). Panel III shows a trial during which the animal responded during the intertrial interval (negative response). For demonstration purposes, the intertrial interval was set at 10 s. In reality, the interresponse interval had an average duration of 10 s and ranged from 7.5–12.5 s. The table depicts a hypothetical session and demonstrates how thresholds were defined for the four individual series. The threshold of the session is the mean of the four series' thresholds. Taken with permission from Markou and Koob (1992).

In fact, depletion of dopamine in the nucleus accumbens does not impair consummatory behavior in rats (unlike dopamine depletion from the caudate-putamen), but it does reduce incentive-motivational responses. For example, proceptive behaviors (such as "ear wiggling," hopping, and darting, all of which serve to solicit the attention of the male) are reduced in female rats with dopamine depletions. Also, depletion of dopamine from the nucleus accumbens decreases locomotor excitement of hungry rats in the presence of food (Koob *et al.*, 1978). Intriguingly, consummatory behaviors, including eating of high incentive foods, appears to be mediated by distinct mechanisms within the nucleus accumbens that depend upon opiate receptors (Kelley and Berridge, 2002).

Mesolimbic dopamine also may control behavior motivated by reward. When a neutral light predicts a primary reinforcer (e.g., water), rats will learn a new behavior (e.g., lever-pressing) to turn on the light, which thus has rewarding properties of its own and is called a conditioned reinforcer. Taylor and Robbins showed that microinjections of D-amphetamine into the nucleus accumbens, but not the caudate, increased the frequency of this behavior (Robbins *et al.*, 1989). This increased frequency of response suggested that amphetamine enhanced the motivational properties of this reward-related stimulus (i.e., the light, which was related to the reward, water). Amphetamine itself was well known to suppress eating and drinking; thus, the increased frequency of behavior was not due to increased thirst, for example. Furthermore, amphetamine did not increase the frequency of responding for a randomly paired stimulus. Finally, amphetamine-induced facilitation was blocked completely by dopamine depletion in the nucleus accumbens (but not by dopamine depletion in the caudate). These results indicated that increased activity in the mesolimbic dopamine system could enhance the motivational properties of stimuli predictive of natural rewards (Robbins *et al.*, 1989). Another way of describing this activational role of the dopamine system is to suggest that it mediates "incentive salience" (Berridge and Robinson, 1998).

The Amygdala Also Has a Role in Appetitive Conditioning and Motivation

There is growing evidence that the amygdala subserves positive emotional functions, as well as negative functions such as fear. For example, monkeys with lesions of the temporal pole, including the amygdala, poorly learn new stimulus/reward associations, suggesting an altered emotional response to reward. In rats, excitotoxic, axon-sparing lesions of the amygdala

diminish the capacity of stimuli associated with reward to motivate behavior. Such lesions also impair responses that are rewarded with access to female rats. However, mating *per se* is unaffected, in contrast with the effect of medial preoptic area lesions (Everitt *et al.*, 2000; see above). This separation of control of consummatory and learned aspects of sexual behavior indicates that motivated sequences of behavior are constructed through the coordination of neural systems that are at least partly independent. Lesions of the basolateral amygdala also impair the phenomenon of place preference in which experimental animals will consistently return to locations—generally in a well-defined environmental enclosure—where they were previously given reward. In contrast, selective lesions of the central nucleus of the amygdala impair the learning of Pavlovian conditioned approach responses. Thus, different nuclei of the amygdala clearly function in associative processes that contribute to appetitive behavior.

Motivated Behavior Is Controlled in Part by Learning

The ventral striatum, including the nucleus accumbens, is part of a system that receives afferents from several limbic cortical structures, including the basolateral amygdala, the hippocampal formation, and the prefrontal cortex, and projects to structures such as the lateral hypothalamus and ventral pallidum (Fig. 43.7). Output from the pallidum is routed in several different ways, including to brain stem motor regions and looping back to the prefrontal cortex through the mediodorsal thalamus. This circuit (limbic cortex-ventral striatum-ventral pallidum), in principle, could provide "a neural mechanism by which motivation gets translated into action" (Mogenson *et al.*, 1980). To understand the function of this circuit, Mogenson and colleagues used electrophysiological experiments to demonstrate interactions among its components (e.g., between the amygdala and the nucleus accumbens).

As described earlier, the basal and lateral amygdala convey associative information about stimuli that predict the occurrence of reinforcers. The recipients of this information are systems that select responses. A link between the amygdala and the ventral striatum is important in the translation of emotion (and motivational effects of stimuli) to behavioral output, or action (Everitt *et al.*, 2000). In some circumstances, therefore, reinforcement involves glutamatergic inputs from these limbic afferents to the striatum. These inputs interact with the ascending dopamine system, and together they determine the output of the ventral striatal γ-aminobutyric acid (GABA)-ergic medium spiny

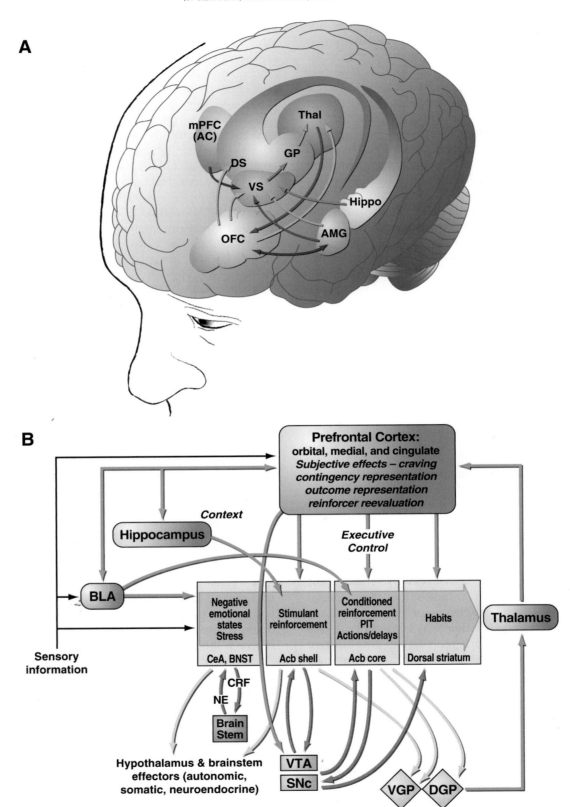

◀ **FIGURE 43.7** Representation of limbic cortical and associated striatal circuitry, with tentative localization of functions involved in drug addiction. (A) Key structures include the medial prefrontal cortex (mPFC), which also includes the anterior cingulate (AC), orbitofrontal cortex (OFC), hippocampus (Hippo), ventral striatum (VS), nucleus accumbens, dorsal striatum (DS), globus pallidus (GP), amygdala (AMYG), and thalamus (Thal). Redrawn from Salloway et al. (2001) with permission. (B) Limbic cortical-striatopallidal circuitry. (i) Processing of conditioned reinforcement and delays by basolateral amygdala and of contextual information by hippocampus. (ii) Goal-directed actions involve interaction of the prefrontal cortex with other structures, possibly including the nucleus accumbens and dorsomedial striatum. (iii) "Habits" depend on interactions between the prefrontal cortex and dorsolateral striatum. (iv) "Executive control" depends on prefrontal cortex and includes representation of contingencies, representation of outcomes, and their value and subjective states (craving and, presumably, feelings) associated with drugs. (v) Drug craving involves activation of the orbital and anterior cingulate cortex, and temporal lobe including the amygdala in functional imaging studies. (vi) Connections between dopaminergic neurons and striatum reflect "spirals"—serial interactions organized in a ventral-to-dorsal cascade. (vii) Reinforcing effects of drugs may engage stimulant, Pavlovian instrumental transfer, and conditioned reinforcement processes in the nucleus accumbens shell and core and then engage stimulus-response habits that depend on the dorsal striatum. (viii) The extended amygdala is composed of several basal forebrain structures including the bed nucleus of the stria terminalis, the centromedial amygdala, and more controversially, the medial portion (or shell) of the nucleus accumbens. A major transmitter in the extended amygdala is corticotropin-releasing factor, which projects to the brain stem, where noradrenergic neurons provide a major projection reciprocally to the extended amygdala. Activation of this system is closely associated with the negative emotional state that occurs during withdrawal. Green/blue arrows, glutamatergic projections; orange arrows, dopaminergic projections; pink arrows, GABAergic projections; Acb, nucleus accumbens; BLA, basolateral amygdala; VTA, ventral tegmental area; SNc, substantia nigra pars compacta; VGP, ventral globus pallidus; DGP, dorsal globus pallidus; BNST, bed nucleus of the stria terminalis; CeA, central nucleus of the amygdala; NE, norepinephrine; CRF, corticotropin-releasing factor; PIT, Pavlovian instrumental transfer.

neurons that project to the globus pallidus (ventral pallidum) (Chapter 31). The nucleus accumbens has turned out to be a heterogeneous structure. Its medial "shell" and lateral "core" can be distinguished anatomically, neurochemically, and functionally. The shell region is considered by some to be part of an "extended amygdala" encompassing the central nucleus of the amygdala and the bed nucleus of the stria terminalis (Heimer et al., 1995; see later). The basolateral amygdala projects predominantly to the nucleus accumbens core subregion and to specific parts of the dorsal striatum. These connections are paralleled by other corticostriatal "loops" (Chapter 31 and 52) (e.g., between different parts of the neocortex that project topographically to different sectors of the dorsal striatum). Moreover, there appears to be the possibility of "cross-talk" between these systems, arising from anatomical connections that potentially allow communication from the amygdala-ventral striatal circuitry to the cortico-dorsal striatal loops (Haber et al., 2000). In functional terms, the ventral striatum has been linked to the initial acquisition of drug-seeking and drug-taking behavior, and the subsequent cascade of loop circuitry involving the dorsal striatum to the gradual acquisition of S-R habits that help to give drug addiction its "compulsive" quality (Everitt and Robbins, 2005).

ADDICTION

Introduction

Drug abuse is one of the world's major public health problems. For example, statistics from the 2005 *National Household Survey on Drug Abuse* of the National Institute on Drug Abuse of the United States reveal that approximately 20 million Americans aged 12 and older were past-month users of illicit drugs 126 million were past-month users of alcohol. Furthermore, 22 million past-month users of illicit drugs or alcohol could be classified with substance dependence or abuse in the past year. In addition, 60 million were current smokers (Substance Abuse and Mental Health Services Administration, 2006). The cost of drug abuse to society is staggering. Total economic costs of alcohol abuse per year in the United State is listed as 166.5 billion dollars, smoking 138 billion dollars, and that of illicit drugs 109.9 billion dollars. This totals over 400 billion dollars in lost productivity due to illness or death, health care expenditures, motor vehicle crashes, social welfare, and so on (Robert Wood Johnson Foundation, Princeton, New Jersey, 2001).

Addiction Cycle

Drug addiction, also known as substance dependence, is a chronically relapsing disorder characterized by (1) compulsion to seek and take the drug, (2) loss of control in limiting intake, and (3) emergence of a negative emotional state (e.g., dysphoria, anxiety, irritability) when access to the drug is prevented (defined here as *dependence*) (Koob and Le Moal, 1997). Addiction and substance dependence (as currently defined by the *Diagnostic and Statistical Manual of Mental Disorders*, 4th edition; American Psychiatric Association, 1994) will be used interchangeably throughout this text and refer to a final stage of a usage process that moves from drug use to addiction. Clinically, the occasional but limited use of a drug with the *potential* for abuse or dependence is distinct from escalated drug use and the emergence of a chronic drug-dependent state (Table 43.1). An important goal of current neurobiological research is to understand the neuropharmacological and neuroadaptive mechanisms within specific neurocircuits that mediate the transition from occasional, controlled drug use and the loss of behavioral control over drug-seeking and drug-taking that defines chronic addiction.

Addiction has been conceptualized as a chronic relapsing disorder with roots both in impulsivity and compulsivity and neurobiological mechanisms that change as an individual moves from one domain to the other. Subjects with impulse control disorders experience an increasing sense of tension or arousal before committing an impulsive act; pleasure, gratification, or

TABLE 43.1　DSM-IV and ICD-10 Diagnostic Criteria for *Alcohol and Drug Dependence*

	DSM-IV	**ICD-10**
Clustering criterion	A. A maladaptive pattern of substance use, leading to clinically significant impairment or distress as manifested by three or more of the following occurring at any time in the same 12-month period:	A. Three or more of the following have been experienced or exhibited at some time during the previous year:
Tolerance	1. Need for markedly increased amounts of a substance to achieve intoxication or desired effect; or markedly diminished effect with continued use of the same amount of the substance.	1. Evidence of tolerance, such that increased doses are required in order to achieve effects originally produced by lower doses.
Withdrawal	2. The characteristic withdrawal syndrome for a substance or use of a substance (or a closely related substance) to relieve or avoid withdrawal symptoms.	2. A physiological withdrawal state when substance use has ceased or been reduced as evidenced by: the characteristic substance withdrawal syndrome, or use of substance (or a closely related substance) to relieve or avoid withdrawal symptoms.
Impaired control	3. Persistent desire or one or more unsuccessful efforts to cut down or control substance use. 4. Substance used in larger amounts or over a longer period than the person intended.	3. Difficulties in controlling substance use in terms of onset, termination, or levels of use.
Neglect of activities	5. Important social, occupational, or recreational activities given up or reduced because of substance use.	4. Progressive neglect of alternative pleasures or interests in favor of substance use; or
Time spent	6. A great deal of time spent in activities necessary to obtain, to use, or to recover from the effects of substance used.	A great deal of time spent in activities necessary to obtain, to use, or to recover from the effects of substance use.
Inability to fulfill roles	None	None
Hazardous use	None	None
Continued use despite problems	7. Continued substance use despite knowledge of having a persistent or recurrent physical or psychological problem that is likely to be caused or exacerbated by use.	5. Continued substance use despite clear evidence of overtly harmful physical or psychological consequences.
Compulsive use	None	6. A strong desire or sense of compulsion to use substance.
Duration criterion	B. No duration criterion separately specified. However, several dependence criteria must occur repeatedly as specified by duration qualifiers associated with criteria (e.g., "often," "persistent," "continued").	B. No duration criterion separately specified.
Criterion for subtyping dependence	*With physiological dependence*: Evidence of tolerance or withdrawal (i.e., any of items A-1 or A-2 above are present). *Without physiological dependence*: No evidence of tolerance or withdrawal (i.e., none of items A-1 or A-2 above are present).	None

relief at the time of committing the act; and finally regret, self-reproach, or guilt following the act. In contrast, individuals with compulsive disorders experience anxiety and stress before committing a compulsive repetitive behavior, and then relief from the stress by performing the compulsive behavior. In addiction, drug-taking behavior progresses from impulsivity to compulsivity in a three-stage cycle: *binge/intoxication*, *withdrawal/negative affect*, and *preoccupation/anticipation*. As individuals move from an impulsive to a compulsive disorder, the drive for the drug-taking behavior shifts from positive to negative reinforcement (Figs. 43.8, 43.9).

Much of the recent progress in understanding the neurobiology of addiction has derived from the study of animal models of addiction on specific drugs such as opiates, stimulants, and alcohol (Shippenberg and Koob, 2002). Although no animal model of addiction fully emulates the human condition, animal models do permit investigation of specific elements of the process of drug addiction. Such elements can be defined by psychological constructs such as positive and negative reinforcement and different stages of the addiction cycle.

Neurobiological Substrates of Drug Reward

Positive Reinforcing Effects of Drugs Are Mediated by Multiple Systems Converging on Common Targets in the Basal Forebrain

The midbrain and forebrain appear to be involved in motivated behavior through connections of the medial forebrain bundle, composed of ascending and descending pathways, including most of the brain monoamine systems (Koob, 1992). The medial forebrain bundle is involved in brain stimulation reward and natural rewards, and work in the neurobiology of addiction has led to an understanding of the neurochemical and neuroanatomical components of this system. The principal components of this system include the ventral tegmental area, the basal forebrain (nucleus accumbens, olfactory tubercle, frontal cortex, and amygdala), and the dopamine connection between the ventral tegmental area and the basal forebrain, called the mesolimbic dopamine system. Additional components are the opioid peptide, GABA, serotonin (5-hydroxytryptamine; 5-HT), and endocannabinoid systems that interact with the ventral tegmental area and the basal forebrain (Koob, 1992)

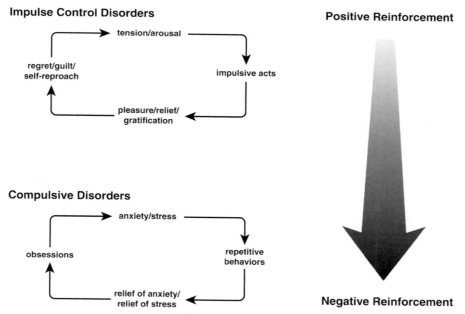

FIGURE 43.8 Diagram showing the stages of impulse control disorder and compulsive disorder cycles related to the sources of reinforcement. In impulse control disorders, an increasing tension and arousal occurs before the impulsive act, with pleasure, gratification, or relief during the act. Following the act there may or may not be regret or guilt. In compulsive disorders, there are recurrent and persistent thoughts (obsessions) that cause marked anxiety and stress followed by repetitive behaviors (compulsions) that are aimed at preventing or reducing distress (American Psychiatric Association, 1994). Positive reinforcement (pleasure/gratification) is more closely associated with impulse control disorders. Negative reinforcement (relief of anxiety or relief of stress) is more closely associated with compulsive disorders. Taken with permission from Koob (2004).

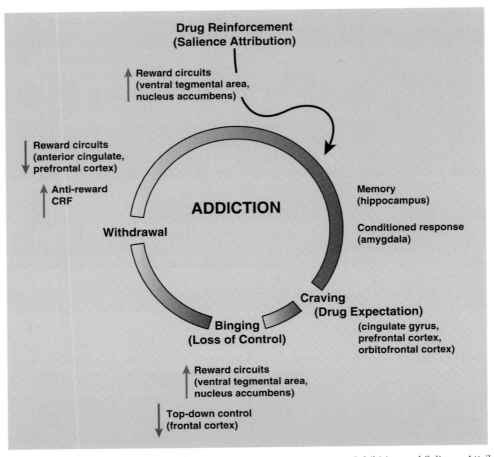

FIGURE 43.9 Integrative model of brain and behavior describing the Impaired Response Inhibition and Salience Attribution (I-RISA) syndrome of drug addiction (Goldstein and Volkow, 2002). The mesolimbic dopamine circuit, which includes the nucleus accumbens, amygdala, and hippocampus, has been traditionally associated with the acute reinforcing effects of a drug and with the memory and conditioned responses that have been linked to craving. It also is likely to be involved in the emotional and motivational changes seen in drug abusers during withdrawal. The mesocortical dopamine circuit, which includes the prefrontal cortex, orbitofrontal cortex, and anterior cingulate, is likely to be involved in the conscious experience of drug intoxication, drug incentive salience, drug expectation/craving, and compulsive drug administration. Note the circular nature of this interaction: the attribution of salience to a given stimulus, which is a function of the orbitofrontal cortex, depends on the relative value of a reinforcer compared to simultaneously available reinforcers, which requires knowledge of the strength of the stimulus as a reinforcer, a function of the hippocampus and amygdala. Consumption of the drug in turn will further activate cortical circuits (orbitofrontal cortex and anterior cingulate) in proportion to the dopamine stimulation by favoring the target response and decreasing nontarget-related background activity. The activation of these interacting circuits may be indispensable for maintaining the compulsive drug administration observed during bingeing and to the vicious circle of addiction. Modified from Goldstein and Volkow (2002), with permission.

(Fig. 43.10). The functional role of each component of the basal forebrain reward system is discussed in the following sections, and the relationship between drug reward and natural reward systems is explored. The actions of several groups of drugs with respect to this system, including indirect sympathomimetics (cocaine and amphetamine), opiates (heroin), nicotine, sedative hypnotics (alcohol, barbiturates, and benzodiazepines), and Δ⁹-tetrahydrocannabinol (THC) also will be explored.

The Positive Reinforcing Effects of Cocaine and Other Indirect Sympathomimetics Depend Critically on the Mesolimbic Dopamine System

Amphetamine and cocaine are psychomotor stimulants that, in humans, have behavioral effects such as suppressing hunger and fatigue and inducing euphoria. In animals, these drugs increase locomotor activity, decrease food intake, stimulate operant behavior, enhance conditioned responding, and produce place

Neurochemical Neurocircuits in Drug Reward

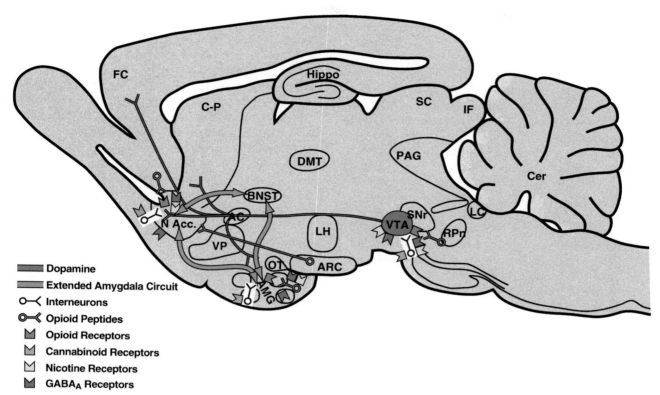

FIGURE 43.10 Sagittal section through a representative rodent brain illustrating the pathways and receptor systems implicated in the acute reinforcing actions of drugs of abuse. Cocaine and amphetamines activate the release of dopamine in the nucleus accumbens and amygdala via direct actions on dopamine terminals. Opioids activate opioid receptors in the ventral tegmental area, nucleus accumbens, and amygdala via direct actions on interneurons. Opioids facilitate the release of dopamine in the nucleus accumbens via an action either in the ventral tegmental area or the nucleus accumbens (but also are hypothesized to activate elements independent of the dopamine system). Alcohol activates γ-aminobutyric acid-A ($GABA_A$) receptors in the ventral tegmental area, nucleus accumbens, and amygdala via either direct actions at the $GABA_A$ receptor or through indirect release of GABA. Alcohol is hypothesized to facilitate the release of opioid peptides in the ventral tegmental area, nucleus accumbens, and central nucleus of the amygdala. Alcohol facilitates the release of dopamine in the nucleus accumbens via an action either in the ventral tegmental area or the nucleus accumbens. Nicotine activates nicotinic acetylcholine receptors in the ventral tegmental area, nucleus accumbens, and amygdala, either directly or indirectly, via actions on interneurons. Nicotine also may activate opioid peptide release in the nucleus accumbens or amygdala independent of the dopamine system. Cannabinoids activate CB_1 receptors in the ventral tegmental area, nucleus accumbens, and amygdala via direct actions on interneurons. Cannabinoids facilitate the release of dopamine in the nucleus accumbens via an action either in the ventral tegmental area or the nucleus accumbens, but also are hypothesized to activate elements independent of the dopamine system. Endogenous cannabinoids may interact with postsynaptic elements in the nucleus accumbens involving dopamine and/or opioid peptide systems. The blue arrows represent the interactions within the extended amygdala system hypothesized to have a key role in psychostimulant reinforcement. AC, anterior commissure; AMG, amygdala; ARC, arcuate nucleus; BNST, bed nucleus of the stria terminalis; Cer, cerebellum; C-P, caudate-putamen; DMT, dorsomedial thalamus; FC, frontal cortex; Hippo, hippocampus; IF, inferior colliculus; LC, locus coeruleus; LH, lateral hypothalamus; N Acc., nucleus accumbens; OT, olfactory tract; PAG, periaqueductal gray; RPn, reticular pontine nucleus; SC, superior colliculus; SNr, substantia nigra pars reticulata; VP, ventral pallidum; VTA, ventral tegmental area. Taken with permission from Koob (2005).

preferences (Koob, 1992). Psychomotor stimulants also decrease thresholds for reinforcing brain stimulation and act as reinforcers for drug self-administration (Koob, 1992) (Fig. 43.10).

Psychomotor stimulants with a high potential for abuse have effects that lead to increases in the availability of monoamine neurotransmitters at synapses. By blocking reuptake or enhancing release, cocaine and amphetamine increase the synaptic availability of dopamine, norepinephrine, and serotonin. However, the acute reinforcing effects of these drugs depend critically on dopamine (Koob, 1992). Studies of intravenous self-administration have provided the most direct evidence implicating dopamine, and more specifically the mesolimbic dopamine system, in the reinforcing actions of cocaine and amphetamines. Low doses of dopamine receptor antagonists injected either systemically or centrally into the nucleus accumbens, amygdala, or bed nucleus of the stria terminalis block cocaine and amphetamine self-administration in rats (Koob, 1992). A specific role for the mesolimbic dopamine system in the reinforcing properties of cocaine and amphetamine was deduced from the observation that neurotoxin-induced lesions of the terminal regions of the mesolimbic dopamine system in the nucleus accumbens produce a significant and long-lasting decrease in self-administration of cocaine and amphetamine over days (Koob, 1992). The neurotoxin 6-OHDA is an analog of dopamine that, when injected into the brain, is taken up by presynaptic dopamine nerve terminals, converted to reactive oxygen species, and destroys the terminals. 6-OHDA-induced lesions of the dopamine terminals in the nucleus accumbens even decrease the amount of work an animal will perform for cocaine without affecting many other reinforcers, such as food.

Neural Substrates of Sensitization Involve Changes in the Mesolimbic Dopamine System

Under certain circumstances, repeated administration of stimulants and opiates results in increasingly larger motor-activating effects, and these effects have been hypothesized to contribute to the development of the neuroadaptations associated with addiction, particularly those involving initial use and/or reinstatement of drug-taking following abstinence. This sensitization to the activating effects of stimulants involves activation of the mesolimbic dopamine system (Robinson and Berridge, 1993). Repeated microinjections of amphetamine into the somatodendritic region of the ventral tegmental dopamine cells, at doses that do not cause behavioral activation, are sufficient to sensitize the dopamine cells to later systemic injections of amphetamine. Thus, changes in the activity of these dopamine cells are sufficient to produce sensitization.

Multiple mechanisms for sensitization have been proposed, and a common theme is a time-dependent chain of adaptations that ultimately leads to long-lasting changes in the function of the mesolimbic dopamine system (Robinson and Berridge, 1993). For example, repeated administration of cocaine produces a decrease in the sensitivity of impulse-regulating somatodendritic dopamine D_2 receptors on dopamine neurons (termed *autoreceptors* because the neuron is itself responding to the transmitter it releases). Because

these autoreceptors normally exert an inhibitory influence on the activity of dopaminergic neurons, this change in sensitivity could translate into enhanced dopamine availability with subsequent injections of cocaine (Robinson and Berridge, 1993). However, the decrease in D_2 receptor sensitivity lasts only four to eight days, and behavioral sensitization can last for weeks; thus, other mechanisms also must be involved. N-methyl-D-aspartate (NMDA) glutamate receptor antagonists selectively block the development of sensitization to psychomotor stimulants, suggesting a role for brain glutamate systems and specific glutamate receptors in sensitization (Koob and Le Moal, 2001).

The hypothalamic–pituitary–adrenal stress axis also may play an important role in sensitization. Stressors can cause sensitization to stimulant drugs, and both the stress axis and the extrahypothalamic corticotropin-releasing factor (CRF) system appear to be important in stress-induced sensitization. Sensitization may be hypothesized to play a key role in initial sensitivity to drugs and contribute to the high incentive salience attributed to drugs during the course of addiction.

Neurobiological Substrates for the Acute Reinforcing Effects of Opiates Involve Opioid Peptide Systems

Much like psychostimulants, opiate drugs such as heroin are readily self-administered intravenously by animals, and with limited access opiates do not produce dependence (Schuster and Thompson, 1969). Advances in opiate pharmacology, such as identification of high specific binding for opiates in the brain and discovery of endogenous opioid peptides, provided insight into the brain systems responsible for the reinforcing effects of these drugs (Koob, 1992) (Box 43.4; Figs. 43.7, 43.10). Both systemic administration and central administration of competitive opiate antagonists into the nucleus accumbens and ventral tegmental area decrease intravenous opiate self-administration. The reinforcing actions of heroin and morphine appear to be mediated largely by the μ opioid receptor subtype. μ receptor antagonists produce dose-dependent decreases in heroin reinforcement, and μ receptor knockout mice show loss of morphine-induced reward and loss of morphine-induced analgesia. Intracerebral opioid receptor antagonists block heroin self-administration in nondependent rats when these drugs are injected into the ventral tegmental area or the region of the nucleus accumbens. Microinjections of opioids into the ventral tegmental area also lower thresholds for ICSS and produce robust place preferences, and these effects appear to be dependent on mesolimbic dopamine function (Di Chiara and North, 1992). Rats will self-administer opioid peptides in the region of the ventral

BOX 43.4

OPIATE RECEPTORS

In 1978, Professors Solomon H. Snyder, John Hughes, and Hans W. Kosterlitz were awarded the Lasker award for their combined discovery of opiate receptors in the brain and for the even more amazing discovery of the existence of endogenous ligands for opiate receptors. The endogenous brain compounds were peptides, called enkephalins and endorphins, which shared all the physiologic properties of opiates. These observations opened a rich field of inquiry into the mechanism of action of opiate drugs, and these findings began a new outlook on neuropeptide neurotransmitters, an area of intense research. First, many new neuropeptides have been discovered with brain receptors potentially amenable to nonpeptidergic agonists and antagonists. Second, because neuropeptides are contained within neurons that also utilize more conventional amino acid or amine transmitters, the possibility of multiple mediators for any specific synapse had to be considered.

---

tegmental area and nucleus accumbens. However, heroin self-administration is not blocked by dopamine receptor antagonists administered at doses that block cocaine self-administration nor by large neurotoxin-induced lesions of the mesolimbic dopamine system. These results demonstrate that neural elements in the regions of the ventral tegmental area and nucleus accumbens are responsible for the reinforcing properties of opiates, suggesting both dopamine-dependent and dopamine-independent opiate actions (Di Chiara and North, 1992; Koob, 1992).

Nicotine Activates Mesolimbic Dopamine

Nicotine has anti-fatigue, stimulant-like, and anti-anxiety effects and appears to improve cognitive performance in animals and humans. It has direct reinforcing actions, as measured by intravenous self-administration in animals and humans. Nicotine is a direct agonist at nicotinic acetylcholine receptors and appears to activate nicotinic acetylcholine receptors in the mesolimbic dopamine system both at the level of the ventral tegmental area and nucleus accumbens (Figs. 43.7, 43.10). Knockout of the $\beta2$ nicotinic acetylcholine receptor subunit blocks nicotine self-administration (Picciotto *et al.*, 1998). Blockade of dopaminergic systems with antagonists or lesions to the mesolimbic dopamine system can block nicotine self-administration, suggesting a key role for dopamine in the acute reinforcing actions of nicotine.

Alcohol Has Multiple Neurochemical Substrates within Brain Reinforcement Systems

Alcohol (i.e., ethanol), barbiturates, and benzodiazepines all have measurable sedative–hypnotic actions, including euphoria, disinhibition, anxiety reduction, sedation, and hypnosis. All of these drugs also have anxiolytic (or antianxiety) effects, reflected in conflict situations as a reduction of behavior suppressed by punishment. The neurobiological basis of the anxiolytic properties of these sedative–hypnotic drugs has provided clues to their reinforcing properties and their abuse potential. For example, GABAergic receptor antagonists reverse many of the behavioral effects of alcohol, an observation that has led to the hypothesis that GABA has a role in its intoxicating effects. Further support for a role of brain GABA in alcohol reinforcement is the observation that potent antagonists of GABA receptor function decrease alcohol reinforcement, and one particularly affected brain site is the central nucleus of the amygdala. Activation of opioid peptide systems has been implicated in alcohol reinforcement by numerous reports that the opiate antagonists naloxone and naltrexone reduce alcohol self-administration in several animal models (Tabakoff and Hoffman, 1992). The brain site most sensitive to opioid receptor antagonists also is the central nucleus of the amygdala (Koob and Le Moal, 2001) (see Figs. 43.7, 43.10).

Dopamine receptor antagonists also reduce lever pressing for alcohol in nondeprived rats (Koob, 1992), and extracellular dopamine levels increase in nondependent rats orally self-administering low doses of alcohol (Koob and Le Moal, 2001). However, virtually complete destruction of dopamine terminals in the nucleus accumbens by 6-OHDA failed to alter voluntary responding for alcohol. Combined with the pharmacologic data discussed earlier, these results suggest that although mesolimbic dopamine transmission may be associated with important aspects of alcohol reinforcement, it is not critical for the reinforcing properties of alcohol (Koob and Le Moal, 2001). Alcohol in a physiologic dose range may antagonize the actions of glutamate (Tabakoff and Hoffman, 1992). Also, increases and decreases in synaptic availability of serotonin (e.g., via blockade of serotonin reuptake or via administration of selective serotonin receptor antagonists) reduce the voluntary intake of alcohol. Thus, multiple neurotransmitters combine to orchestrate the reward profile of this drug (Engel et al., 1987).

Δ^9-Tetrahydrocannabinol Has Effects Similar to Other Drugs of Abuse

THC shares effects in animal models of drug reinforcement similar to those of other drugs of abuse. Upon acute administration, THC decreases ICSS reward thresholds and produces a place preference in rats. It maintains self-administration behavior in squirrel monkeys, and a synthetic THC analog is self-administered intravenously in mice. THC binds to the cannabinoid CB_1 receptor that is distributed widely throughout the brain, particularly in the striatum, and is thought to mediate some of the actions of the endogenous endocannabinoids. THC self-administration is blocked by both CB_1 and opioid receptor antagonists (Justinova et al., 2004, 2005) (Figs. 43.7, 43.10).

Neurobiological Substrates for the Withdrawal/ Negative Affect State of Drug Dependence

Substance dependence is defined as compulsive, uncontrollable drug use (see above); however, the etiology of that compulsive use is controversial. One position is that the positive reinforcing effects of a drug are critical for addiction. Partially as a result of this concept, another position—substance dependence as the alleviation of withdrawal symptoms—fell out of favor as an explanation for compulsive drug use. However, although simple positive reinforcement clearly is necessary for the development of drug use, it falls short in explaining the development of compulsive use. For example, tolerance or apparent tolerance develops to drug reward during chronic use, with positive reinforcement either being absent or subsumed by negative reinforcement. Indeed, this issue raises the question of what factors distinguish drug *use* from *abuse* and *dependence* or addiction. Sensitization theory addresses this issue by invoking a shift to a state of *incentive salience*, defined as a hypersensitive neural state that produces the experience of craving. It has been hypothesized that this state of incentive salience is produced by drug-induced sensitization of the mesolimbic dopamine system (Robinson and Berridge, 1993). According to this model, the pathologically strong craving or yearning encountered during protracted abstinence would be due in large part to an overactive mesolimbic dopamine system (see earlier). Yet other theorists have invoked compensatory changes in the reward system.

Negative Affect Is a Common Result of Withdrawal from Chronic Administration of Drugs of Abuse

Withdrawal signs associated with cessation of chronic drug administration usually are characterized by responses that are opposite to the initial effects of the drug. Thus, all drugs of abuse produce rewarding effects and a withdrawal syndrome associated with subjective symptoms of dysphoria, negative affect, and anxiety (Koob and Le Moal, 2001). However, the overt symptoms of the withdrawal experience can vary depending on the nature of the effects of the drug itself. For example, whereas few physical signs of withdrawal usually are observed during cocaine withdrawal in humans, cessation of cocaine use often is characterized by severe depressive symptoms combined with irritability, anxiety, and anhedonia lasting several hours to several days (i.e., the "crash") and may be a motivating factor in the maintenance of the cocaine dependence cycle (Koob and Le Moal, 2001). However, in the case of opiate withdrawal, subjective symptoms such as dysphoria, anxiety, craving for drugs, and malaise are accompanied by both physical symptoms of extreme discomfort, including pupillary dilation, hot and cold flashes, goose bumps, and a flu-like state. In the case of alcohol withdrawal, humans show tremor and increases in heartbeat rate, blood pressure, and body temperature, as well as subjective signs, including dysphoria, anxiety, insomnia, and malaise. Withdrawal from nicotine is characterized by anxiety, negative affect, fatigue, irritability, sleep disturbances, and an intense craving for cigarettes. Withdrawal from THC resembles a combination of sedative–hypnotic or opiate withdrawal with irritability, restlessness, hot flashes, sweating, disturbances in appetite and sleep, and low-level dysphoria. Withdrawal from benzodiazepines resembles that of alcohol withdrawal but generally is more prolonged in onset and less intense.

Chronic Drug Administration Compromises the Brain Reward Systems

In animal studies, withdrawal from chronic administration of stimulants has been studied using ICSS after repeated administration of cocaine or amphetamine over several weeks or 12–48 h of cocaine self-administration. Withdrawal from prolonged self-administration of cocaine in rats increases ICSS reward thresholds. The increase in threshold dose is time-dependent and is an effect opposite to that of acute cocaine (Koob and Le Moal, 2001). Similar effects have been observed during withdrawal from chronic amphetamine administration. Opiate withdrawal induced by the administration of low doses of the opiate receptor antagonist naloxone also elevates ICSS reward thresholds, as does spontaneous withdrawal from alcohol (Koob and Le Moal, 2001). Withdrawal from chronic nicotine and THC also produces profound increases in ICSS reward thresholds, suggesting that brain reward dysregulation is a common element of acute withdrawal from drugs of abuse (Table 43.2).

Withdrawal/Negative Reinforcement Brain Substrate

A brain system that mediates aversion was identified early on by a combination of anatomical and behavioral experiments. For example, mapping studies identified sites eliciting escape from brain stimulation, sites at which electrical brain stimulation elicits defensive responses, and sites at which fear or pain naturally evokes aggressive responses. Many of these pathways

TABLE 43.2 Drug Effects on Thresholds of Rewarding Brain Stimulation

Drug class	Acute administration	Withdrawal from chronic treatment
Psychostimulants (cocaine amphetamines)	↓	↑
Opiates (morphine, heroin)	↓	↑
Nicotine	↓	↑
Sedative–hypnotics (ethanol)	↓	↑

appear to be central reflex pathways, with both ascending and descending components modulated by influences from the central nucleus of the amygdala.

Some electrical brain stimulation sites (e.g., ventral medial hypothalamus) produced markedly ambivalent reinforcing effects, which were interpreted as an interaction between positive reinforcement and aversion systems. Electrical brain stimulation itself has been suggested to have both appetitive and aversive properties, as have drugs of abuse and, in some situations, electric shock. In fact, animal learning theory has led to the concept of opponent motivational processes (one appetitive, one aversive).

Aversion May Be Mediated by Opponent Motivational Processes

It has been suggested that many reinforcers produce affective and hedonic effects (*a-processes*) that are opposed by *b-processes* of opposite affective sign in a simple, dynamic control system for affect (Fig. 43.11) (Solomon and Corbit, 1974). The *a-process* follows the reinforcing event with short latency and then decays. Its size decreases with repeated presentation of the reinforcer. The *b-process* has a longer latency of onset, reaches its maximum only after repeated trials, and decays very slowly. These components of emotional responses are exhibited in response to many reinforcers, including not only drugs such as heroin, but also natural rewards that also may lead to different forms

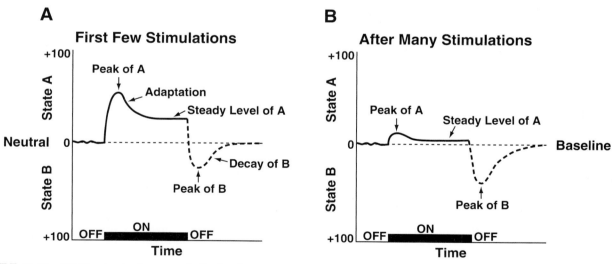

FIGURE 43.11 (A) The standard pattern of affective dynamics produced by a relatively novel unconditioned stimulus (nondependent drug intake). (B) The standard pattern of affective dynamics produced by a familiar, frequently repeated unconditioned stimulus (dependent drug intake). First, an unconditional arousing stimulus triggers a primary affective process, termed the *a-process*. An unconditional reaction that translates the intensity, quality, and duration of the stimulus (for example, the first injection of a drug). Second, as a consequence of the *a-process*, the *b-process* is evoked after a short delay, an opponent process. The two responses are consequently and temporarily linked (*a* triggers *b*) but are hypothesized to depend on different neurobiological mechanisms. The *b-process* has a longer latency, but has more inertia, a slower recruitment, and a more sluggish decay. At a given moment, the pattern of affect will be the algebraic sum of these opposite influences and the dynamics reveal, with the passage of time, the net product of the opponent process. Taken with permission from Solomon (1980).

of "dependence," such as attachment of offspring to a mother. So, for example, either heroin or the mother elicits a hedonic *a-process*, which is followed by a rebound negative or aversive *b-process* when the heroin wears off (or the mother leaves). These effects may get stronger with repeated experiences of heroin (or the mother at a critical period of development), leading to a form of dependence in each case. The opponent process also can work in reverse. For example, if tolerance develops for an initially aversive experience, a rebound hedonic response can occur when this experience stops. Indeed, this has been suggested partly to explain the "pleasure" of jogging.

How might these opponent processes be represented in the brain? The neurobiological basis for neuroadaptive mechanisms that reflect changes in reward function have been hypothesized to be alterations of the same neurotransmitters implicated in the acute reinforcing effects of drugs (Koob and Bloom, 1988). Examples of such homeostatic, "within-system" adaptive neurobiological changes include decreases in dopaminergic and serotonergic transmission in the nucleus accumbens during drug withdrawal as measured by *in vivo* microdialysis (Weiss *et al.*, 1992; Parsons *et al.*, 1995), increased sensitivity of opioid receptor transduction mechanisms in the nucleus accumbens during opiate withdrawal (Shaw-Lutchman *et al.*, 2002), decreased GABAergic and increased NMDA glutamatergic transmission during ethanol withdrawal (Roberts *et al.*, 1996; Weiss *et al.*, 1996; Fitzgerald and Nestler, 1995). However, recruitment of other neurotransmitter systems in the adaptive responses to drugs of abuse may involve neurotransmitter systems not linked to the acute reinforcing effects of the drug, or a "between-system" neuroadaptation (Koob and Bloom, 1988). An example of a between-system neuroadaptation is the activation of brain stress systems such as CRF and dysregulation of other stress-mediating neurotransmitter systems during withdrawal (Koob and Le Moal, 2005) (see below).

Neurochemical Adaptation in Reward Neurotransmitters

Dopamine and Serotonin Systems Show Neuroadaptive Changes with Substance Dependence

Monoamines, including the dopamine and serotonin systems, have been implicated in the reinforcing actions of each of the drugs discussed: psychostimulants, opiates, nicotine, alcohol, and THC. One way of assessing the prolonged effects of drug self-administration is to measure extracellular levels of a neurotransmitter during ongoing behaviors, such as intravenous self-administration of a drug or withdrawal from the drug

after a period of chronic self-administration. Such measurements can be made by using the technique of *in vivo* microdialysis, which involves implanting into a specific site in the brain a guide cannula through which a probe with a semipermeable membrane at the tip perfuses and collects the neurotransmitter in extracellular fluid. Several studies using this technique have shown that extracellular dopamine and serotonin levels in the nucleus accumbens decrease during acute withdrawal from cocaine, opiates, and alcohol, an effect opposite to that of acute drug administration.

Brain Corticotropin-Releasing Factor and Noradrenergic Systems Show Neuroadaptive Changes with Substance Dependence

Anxiety and stress are common elements of substance dependence and acute withdrawal from drugs of abuse in humans and produce stress-like behavior in animals (Sarnyai *et al.*, 2001). Stress can be defined simply as *any challenge to or alteration in psychological homeostatic processes*. Animal studies exploring the role of CRF in the actions of alcohol, opiates, nicotine, THC, and cocaine have provided evidence of a neurochemical basis for the stress associated with abstinence following chronic drug administration. Not only is CRF a major hypothalamic releasing factor controlling the classic stress response, but it also appears to have a neurotropic role in the central nervous system, modulating behavioral responses to stress. CRF itself produces stress-related behaviors, and CRF antagonists reverse a number of behavioral responses to stress. Rats treated repeatedly with drugs such as alcohol, opiates, and cocaine show significant stress-like responses in behavioral tests that follow cessation of drug administration, and these responses are reversed by administration of a CRF antagonist directly into the brain. For example, injection of a CRF antagonist into the central nucleus of the amygdala reverses the stress-like effects of withdrawal from alcohol, cocaine, nicotine, opioid, and THC in rats. Withdrawal from alcohol, cocaine, nicotine, opioids, and THC is associated with an increase in the release of CRF into the amygdala, suggesting that alcohol withdrawal can activate CRF systems previously implicated in behavioral responses to stress (Fig. 43.12). Also, stress-induced reinstatement of drug self-administration following extinction can be reversed by CRF antagonists (Sarnyai *et al.*, 2001).

Norepinephrine in the limbic forebrain also has been associated with the physical and motivational signs of opiate withdrawal. More recent evidence suggests that projections to the bed nucleus of the stria terminalis—a critical part of the extended amygdala—may have a role in the motivational effects of opiate withdrawal (Delfs *et al.*, 2000). In addition, the

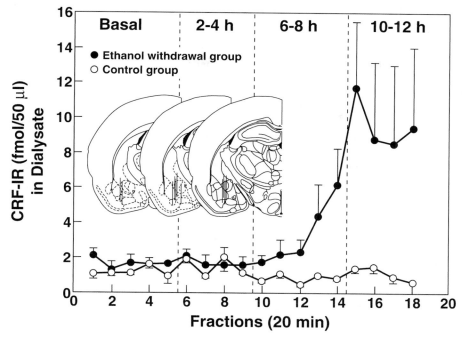

FIGURE 43.12 Extracellular levels of corticotropin-releasing factor immunoreactivity (CRF-IR) in the central nucleus of the amygdala during withdrawal from chronic alcohol administration. Dialysate was collected over four 2-h periods, which alternated with nonsampling 2-h periods. The four sampling periods correspond to the basal collection (before removal of alcohol) and 1–4 h, 6–8 h, and 10–12 h after withdrawal of alcohol. Fractions were collected every 20 min. Data are represented as means ± SEM ($n = 5$ per group). ANOVA confirmed significant differences between the two groups over time ($p < 0.05$). (Inset) An anatomical analysis identifying these locations (vertical lines in inset) as sites where changes in CRF concentration were detected in the alcohol experiment ($n = 10$) and as being within the amygdaloid complex. Reprinted with permission from Merlo Pich *et al.* (1995).

bed nucleus of the stria terminalis is rich in CRF neurons and has been implicated in a feed-forward norepinephrine-CRF-norepinephrine system that may be involved in stress and drug dependence. In this hypothesized feed-forward system, CRF activates brain stem noradrenergic activity, which in turn activates forebrain CRF activity, effectively closing the loop. This mechanism could explain the potentiation of stress responses with repeated exposure that could lead to psychopathology. However, note that behavioral and autonomic responses may be activated in both the brain stem and the forebrain as partial exits from the loop (Koob, 1999).

In contrast to the sensitization theory, neuroadaptation theories such as the opponent process theory postulate that the processes of tolerance to the positive emotional state (hedonic tolerance) and subsequent development of a negative emotional state play important roles in the transition from drug use to drug dependence (Solomon, 1977). Whereas initial drug use may be motivated by the positive emotional state produced by a drug and potentiated by sensitization, continued drug use leads to neuroadaptation to the presence of the drug, to development of a negative emotional state, and ultimately to self-medication of

that negative emotional state. Some theorists have gone so far as to argue that the presence of a negative emotional state is the defining feature of addiction and that dysregulation of the residual reward system represents a state of allostasis, the process of obtaining stability through change (Box 43.5). The *allostasis* hypothesis suggests that *both* sensitization and opponent process mechanisms may contribute to changes in incentive motivation (or incentive salience) that convey the vulnerability to relapse for drug addiction.

Neuroadaptation, Prolonged Abstinence, and Relapse

An explanation for the chronic relapsing nature of drug addiction is that reinforcement can be derived from secondary sources that motivate continued drug use. Both positive and negative emotional states can become associated with stimuli in the drug-taking environment through classical conditioning processes. When the drug taker is reexposed to these conditioned stimuli, compulsive drug use is reestablished (Box 43.6).

Conditioning to the positive emotional states induced by drugs has been demonstrated in animals.

BOX 43.5

ALLOSTASIS

Counteradaptive processes, such as opponent process, that are part of normal homeostatic limitations of reward function can fail to return within the normal range and are hypothesized to contribute to the allostatic state of addiction. Allostasis from the addiction perspective is defined as the process of maintaining apparent reward function stability by changes in brain reward mechanisms. The allostatic state is fueled not only by dysregulation of reward circuits *per se*, but also by the activation of

brain and hormonal stress systems (Koob and Le Moal, 2001). The following definitions apply: (1) *allostasis*, the process of achieving stability through change; (2) *allostatic state*, a state of chronic deviation of the regulatory system from its normal, or homeostatic, operating level; and (3) *allostatic load*, the cost to the brain and body of the deviation, accumulating over time, and reflecting in many cases pathological states and accumulation of damage.

BOX 43.6

CRAVING

Craving is a hypothetical construct that can be defined as the memory of the rewarding effects of a drug, superimposed upon a negative motivational state (Markou *et al.*, 1998). Human studies of craving have been fraught with difficulty in linking it to relapse *per se* in humans with drug addiction, but it is a ubiquitous construct that pervades the fabric of the clinical setting (Tiffany *et al.*, 2000). Two types of craving can be conceptualized and linked to animal models.

Craving Type 1: Craving is induced by stimuli that have been paired with drug self-administration such as environmental cues. Termed conditioned positive reinforcement in experimental psychology, an animal model of Craving Type 1 is cue-induced reinstatement where a cue previously paired with access to drug reinstates

responding for a lever that has been extinguished. Another example is drug-induced reinstatement where a small dose of the drug elicits responding for a lever that has been extinguished.

Craving Type-2: Craving is conceptualized as a state change characterized by anxiety and dysphoria or a residual negative emotional state that combines with Craving Type 1 situations to produce relapse to drug seeking. Animal models of Craving Type 2 include stress-induced reinstatement of drug seeking after extinction, or increased drug taking in animals during or after a prolonged deprivation. Craving Type 2 can be conceptualized as a residual negative emotional state or an acute stress-like state, both of which can elicit drug-seeking behavior.

Stimuli associated with drugs of abuse can maintain responding in rats and monkeys when presented without the drug. It is also possible to demonstrate conditioned withdrawal. For example, when previously neutral stimuli are paired with naloxone-induced withdrawal, the neutral stimuli themselves elicit signs of opiate withdrawal. Conditioned withdrawal has been observed in opiate-dependent animals and humans.

Neurobiological substrates for conditioned drug effects have been investigated in some detail (Everitt *et al.*, 2000; Cardinal *et al.*, 2002). The enhancement of conditioned reinforcement produced by psychomotor stimulants appears to involve the basolateral amygdala and its glutamatergic projection to the nucleus accumbens with dopaminergic modulation. The neu-

robiological substrates for conditioned withdrawal also may depend on the basolateral amygdala. In contrast, drug-induced reinstatement of drug-seeking appears to be mediated via a prefrontal cortex glutamatergic projection to the nucleus accumbens with dopaminergic modulation (Kalivas and McFarland, 2003). Another powerful source of relapse in human subjects is exposure to a stressor or the state of stress. Both of these constructs involve activation of the brain CRF systems in the basal forebrain structures that include the central nucleus of the amygdala and the bed nucleus of the stria terminalis.

Neuroanatomic and functional data have provided the basis for the hypothesis that the central nucleus of the amygdala and the bed nucleus of the stria terminalis form a separate entity within the basal forebrain termed the *extended amygdala* (Heimer and Alheid, 1991) that may mediate some of the emotional dysregulation associated with the stress component of addiction (Fig. 43.7). CRF and norepinephrine systems within the extended amygdala appear to be a critical substrate for not only stress-induced reinstatement (Shaham *et al.*, 2003) but also for the dysregulated hypersensitivity to stress observed in protracted abstinence (Valdez *et al.*, 2004) (see above).

Brain Imaging Circuits Involved in Addiction

Brain imaging studies using positron emission tomography with ligands for measuring oxygen utilization or glucose metabolism or using functional magnetic resonance imaging techniques have provided important insights into the neurocircuitry changes in the human brain associated with the development, maintenance, and even vulnerability to addiction.

These imaging results, overall, have a striking resemblance to the neurocircuitry identified by animal studies. During acute intoxication with drugs such as cocaine, methamphetamine, nicotine, alcohol, and THC there are changes in the orbitofrontal cortex, prefrontal cortex, anterior cingulate cortex, amygdala, and ventral striatum. These changes are reflected in activation as measured by functional magnetic resonance imaging, and changes in oxygen utilization as measured by positron emission tomography. The changes during acute intoxication often are accompanied by an increase in the availability of dopamine, particularly in striatal regions.

In contrast, during acute and chronic withdrawal there is a reversal of these changes, with decreases in metabolic activity in the orbitofrontal cortex, prefrontal cortex, and anterior cingulate cortex, and decreases in basal dopamine activity in the striatum and prefrontal cortex as reflected in decreases in D_2 receptor binding (Fig. 43.13). Cue-induced reinstatement for

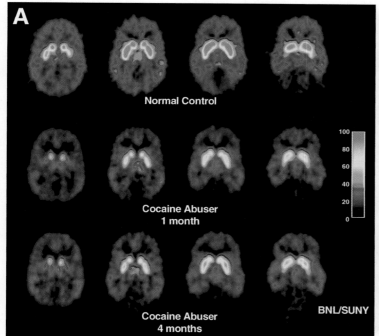

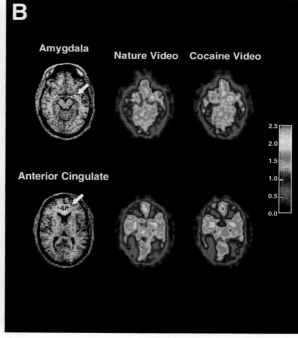

FIGURE 43.13 **(A)** [^{18}F]*N*-methylspiroperidol images in a normal control and in a cocaine abuser tested 1 month and 4 months after the last cocaine use. The images correspond to the four sequential planes where the basal ganglia are located. The color scale has been normalized to the injected dose. Notice the lower uptake of the tracer in the cocaine abuser compared to the normal control. Notice also the persistence of the decreased uptake even after 4 months of cocaine discontinuation. BNL, Brookhaven National Laboratory; SUNY, State University of New York. Taken with permission from Volkow *et al.* (1993b). **(B)** Transaxial images illustrating the differential increase in relative regional cerebral blood flow (CBF) in the amygdala and anterior cingulate of a detoxified cocaine patient during a nondrug-related (nature) video and a cocaine-related video. Anatomical regions of interest first were localized on the patient's magnetic resonance image (the first image in each row). Region templates subsequently were superimposed on [^{15}O] positron emission tomography (PET) images, yielding radioactive count files for conversion to normalized (relative) regional CBF. The middle and final images in each row show relative regional CBF as measured by PET. The range on the arbitrary color scale is from 0.0 to 2.5 times background activity. Whole brain average regional CBF is 1.0 on the scale. Areas with greatest relative regional CBF are shown in red. Activity in the amygdala and in the anterior cingulate differentially increased during the cocaine video. Taken with permission from Childress *et al.* (1999).

cocaine, alcohol, nicotine, and THC appears to involve a reactivation of these same circuits, much like acute intoxication, but with the reactivation of the prefrontal cortex, orbitofrontal cortex, and anterior cingulate cortex and the additional recruitment of activation in the amygdala (Childress *et al.*, 1999) (Fig. 43.13).

Signal Transduction Mechanisms Show Neuroadaptive Changes with Substance Dependence

Not only can the functional activity of neurotransmitters change due to the increased presynaptic activity of chronic drug use, but these changes also can be translated to postsynaptic signal transduction, where changes can be long-lasting and may result in long-term adaptations in receptor function. These changes

take place at the level of receptor binding or in the impact that the association of a drug with its receptor has on a postsynaptic cell. For example, whereas chronic opiate use has minimal effects on the binding of opiate receptors by opioid peptides, it dramatically changes signal transduction in the terminal regions of the mesolimbic dopamine system, such as the nucleus accumbens (Nestler, 2001). This occurs at least in part by increasing adenylate cyclase activity and immediate-early gene expression. Many of these effects presumably occur postsynaptic to dopamine release, but some of the same effects may be observed on dopamine neurons themselves. These changes can result in the expression of novel proteins and, consequently, changes in neuronal function (see above) that could explain clinical phenomena, such as protracted absti-

FIGURE 43.14 Molecular mechanisms of neuroadaptation. *Cocaine and amphetamines*, as indirect sympathomimetics, stimulate the release of dopamine, which acts at G protein-coupled receptors, specifically D_1, D_2, D_3, D_4, and D_5. These receptors modulate the levels of second-messengers like cyclic adenosine monophosphate (cAMP) and Ca^{2+}, which in turn regulate the activity of protein kinase transducers. Such protein kinases affect the functions of proteins located in the cytoplasm, plasma membrane, and nucleus. Among membrane proteins affected are ligand-gated and voltage-gated Ca^{2+} channels (VGCC). G_i and G_o proteins also can regulate K^+ and Ca^{2+} channels directly through their $\beta\gamma$ subunits. Protein kinase transduction pathways also affect the activities of transcription factors. Some of these factors, like cAMP element binding protein (CREB), are regulated posttranslationally by phosphorylation; others, like Fos, are regulated transcriptionally; still others, like Jun, are regulated both posttranslationally and/or transcriptionally. Whereas membrane and cytoplasmic changes may be only local (e.g., dendritic domains or synaptic boutons), changes in the activity of transcription factors may result in long-term functional changes. These may include changes in gene expression of proteins involved in signal transduction and/or neurotransmission, resulting in altered neuronal responses. For example, chronic exposure to psychostimulants has been reported to increase levels of protein kinase A (PKA) and adenylyl cyclase in the nucleus accumbens and to decrease levels of $G_{i\alpha}$. Chronic exposure to psychostimulants also alters the expression of transcription factors themselves. CREB expression, for instance, is depressed in the nucleus accumbens by chronic cocaine treatment. Chronic cocaine induces a transition from Fos induction to the induction of the much longer-lasting Fos-related antigens such as ΔFosB. *Opioids*, by acting on neurotransmitter systems, affect the phenotypic and functional properties of neurons through the general mechanisms outlined in the diagram. Shown are examples of ligand-gated ion channels such as the γ-aminobutyric acid-A (GABA$_A$) and glutamate N-methyl-D-aspartate (NMDA) receptor (NMR) and G protein-coupled receptors such as opioid, dopamine (DA), or cannabinoid CB$_1$ receptors, among others. These receptors modulate the levels of second messengers like cAMP and Ca^{2+}, which in turn regulate the activity of protein kinase transducers. Chronic exposure to opioids has been reported to increase levels of PKA and adenylyl cyclase in the nucleus accumbens and to decrease levels of $G_{i\alpha}$. Chronic exposure to opioids also alters the expression of transcription factors themselves. CREB expression, for instance, is depressed in the nucleus accumbens and increased in the locus coeruleus by chronic morphine treatment, whereas chronic opioid exposure activates Fos-related antigens such as ΔFosB. *Alcohol*, by acting on neurotransmitter systems, affects the phenotypic and functional properties of neurons through the general mechanisms outlined in the diagram. Shown are examples of ligand-gated ion channels such as the GABA$_A$ and the NMDA receptor or G protein-coupled receptors such as opioid, dopamine, or cannabinoid CB$_1$ receptors, among others. The latter also are activated by endogenous cannabinoids such as anandamide. These receptors modulate the levels of second messengers such as cAMP and Ca^{2+}, which in turn regulate the activity of protein kinase transducers. Such protein kinases affect the functions of proteins located in the cytoplasm, plasma membrane, and nucleus. Among membrane proteins affected are ligand-gated and VGCCs. Alcohol, for instance, has been proposed to affect the GABA$_A$ response via protein kinase C (PKC) phosphorylation. G_i and G_o proteins also can regulate K^+ and Ca^{2+} channels directly through their $\beta\gamma$ subunits. Chronic exposure to alcohol has been reported to increase levels of PKA and adenylyl cyclase in the nucleus accumbens and to decrease levels of $G_{i\alpha}$. Moreover, chronic ethanol induces differential changes in subunit composition in the GABA$_A$ and glutamate inotropic receptors and increases expression of VGCCs. Chronic exposure to alcohol also alters the expression of transcription factors themselves. CREB expression, for instance, is increased in the nucleus accumbens and decreased in the amygdala by chronic alcohol treatment. Chronic alcohol induces a transition from Fos induction to the induction of the longer-lasting Fos-related antigens. *Nicotine* acts directly on ligand-gated ion channels. These receptors modulate the levels of Ca^{2+}, which in turn regulate the activity of protein kinase transducers. Chronic exposure to nicotine has been reported to increase levels of PKA in the nucleus accumbens. Chronic exposure to nicotine also alters the expression of transcription factors themselves. CREB expression, for instance, is depressed in the amygdala and prefrontal cortex and increased in the nucleus accumbens and ventral tegmental area. Δ^9-*Tetrahydrocannabinol (THC)*, by acting on neurotransmitter systems, affects the phenotypic and functional properties of neurons through the general mechanisms outlined in the diagram. Cannabinoids act on the cannabinoid CB$_1$ G protein-coupled receptor. The CB$_1$ receptor also is activated by endogenous cannabinoids such as anandamide. This receptor modulates (inhibits) the levels of second messengers like cAMP and Ca^{2+}, which in turn regulate the activity of protein kinase transducers. Chronic exposure to THC also alters the expression of transcription factors themselves. CaMK, Ca^{2+}/calmodulin-dependent protein kinase; ELK-1, E-26-like protein 1; PLCβ, phosphlipase C β; IP3, inositol triphosphate; MAPK, mitogen-activated protein kinase; PI3K, phosphoinositide 3-kinase; R, receptor; Modified with permission from Koob *et al.* (1998).

nence and vulnerability to relapse, in people formerly addicted to drugs (Fig. 43.14).

The persistence of changes in drug reinforcement mechanisms that characterize drug addiction suggests that the underlying molecular mechanisms are long-lasting, and considerable attention has been given to the drug regulation of gene expression. Current research focuses on several types of transcription factors, including cyclic adenosine monophosphate response element binding protein and novel Fos-like proteins termed chronic Fos-related antigens. More recently, FosB, a member of the Fos family of transcription factors that dimerizes with a member of the Jun family to form activator protein-1 transcription factor complexes, has been implicated in the adaptation to chronic exposure to drugs of abuse. Activator protein-

1 transcription factor complexes bind to sites in the regulatory regions of many genes. Acute administration of several drugs of abuse activates c-*fos* and Fos-related antigens, but chronic administration of drugs of abuse eliminates these activations and produces a gradual accumulation of FosB. This highly stable protein is hypothesized to function as a sustained molecular switch contributing to vulnerability to relapse after prolonged periods of abstinence (Nestler, 2001). Thus, these transcription factors may be possible mediators of chronic drug action. However, it has not been possible yet to relate regulation of a specific transcription factor to specific features of drug reinforcement or addiction.

Genetic and molecular genetic animal models have provided a convergence of data to provide hypotheses

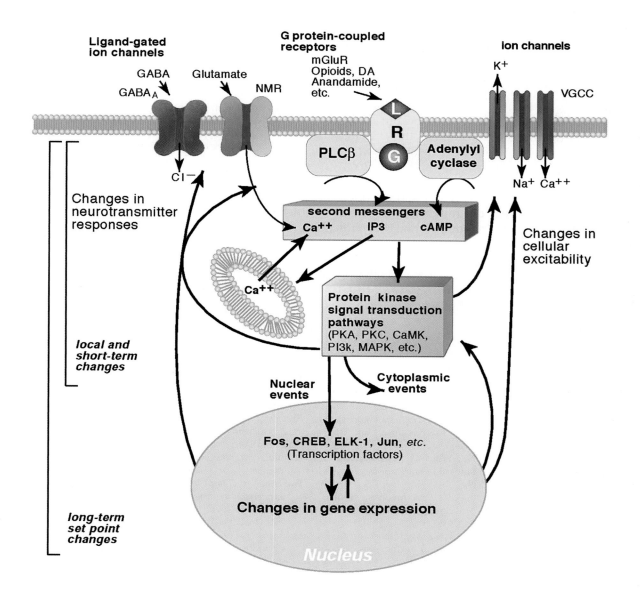

regarding vulnerability to addiction in some of the neuropharmacological substrates identified in neurocircuitry studies. High alcohol-preferring rats have been bred that show high voluntary consumption of alcohol, increased anxiety-like responses, and numerous neuropharmacological phenotypes such as decreased dopaminergic activity and decreased neuropeptide Y activity, both of which have support from genetic studies of chromosomal linkage (McBride *et al.*, 1990; Murphy *et al.*, 2002; Carr *et al.*, 1998).

Advances in molecular biology have led to the ability to systematically inactivate the genes that control the expression of proteins that make up receptors or neurotransmitter/neuromodulators in the central nervous system using the gene knockout approach. Knockout mice have a gene inactivated by homologous recombination. A knockout mouse deficient in both alleles of a gene is homozygous for the deletion and is termed a null mutation ($-/-$). Although such an approach does not guarantee that these genes are the ones that are vulnerable in the human population, they provide viable candidates for exploring the genetic basis of endophenotypes associated with addiction (Koob *et al.*, 2001).

Notable positive results with gene knockout studies in mice have focused on knockout of the μ opioid receptor, which eliminates opioid, nicotine, alcohol and cannabinoid reward in mice (Contet *et al.*, 2004). Selective deletion of the genes for expression of the dopamine D_1 and D_2 receptor subtypes and the dopamine transporter has revealed significant decreases in activation and reward produced by psychomotor stimulants. Although developmental factors must be taken into account for the compensatory effect of deleting any one or a combination of genes, it is clear that D_1 and D_2 receptors and the dopamine transporter play important roles in the actions of psychomotor stimulants (Caine *et al.*, 2002, 2007).

References

American Psychiatric Association (1994). "Diagnostic and Statistical Manual of Mental Disorders," 4th ed. American Psychiatric Press, Washington, DC.

Bergman, J., Kamien, J. B., and Spealman, R. D. (1990). Antagonism of cocaine self-adminstration by selective dopamine D1 and D2 antagonists. *Behav. Pharmacol.* **1**, 355–363.

Berridge, K. C. and Robinson, T. E. (1998). What is the role of dopamine in reward: Hedonic impact. Reward learning, or incentive salience? *Brain Res. Rev.* **28**, 309–369.

Cador, M., Robbins, T. W., and Everitt, B. J. (1989). Involvement of the amygdala in stimulus-reward associations: Interaction with the ventral striatum. *Neuroscience* **30**, 77–86.

Caine, S. B., Gabriel, K. I., Berkowitz, J. S., Zhang, J., and Xu, M. (2002). Decreased cocaine self-administration in dopamine D-1 receptor knockout mice. *Drug Alcohol Depend.* **66**, s25.

Caine, S. B., Thomsen, M., Gabriel, K. I., Berkowitz, J. S., Gold, L. H., Koob, G. F., Tonegawa, S., Zhang, J., and Xu, M. (2007). Lack of self-administration of cocaine in dopamine D_1 receptor knockout mice. *J. Neurosci.* in press.

Caine, S. B., Lintz, R., and Koob, G. F. (1993). Intravenous drug self-administration techniques in animals. In "Behavioral Neuroscience: A Practical Approach" (A. Sahgal, ed.), Vol. 2, pp. 117–143. Oxford Univ. Press, New York.

Cardinal, R. N., Parkinson, J. A., Hall, J., and Everitt, B. J. (2002). Emotion and motivation: The role of the amygdala, ventral striatum and prefrontal cortex. *Neurosci. Biobehav. Rev.* **26**, 321–352.

Carli, M., Evenden, J. L., and Robbins, T. W. (1985). Depletion of unilateral striatal dopamine impairs initiation of contralateral actions and not sensory attention. *Nature (Lond.)* **313**, 679–682.

Carlisle, H. J. (1969). The effects of preoptic and anterior hypothalamic lesions on behavioral thermoregulation in the cold. *J. Comp. Physiol. Psychol.* **69**, 391–402.

Carr, L. G., Foroud, T., Bice, P., Gobbett, T., Ivashina, J., Edenberg, H., Lumeng, L., and Li, T. K. (1998). A quantitative trait locus for alcohol consumption in selectively bred rat lines. *Alcohol. Clin. Exp. Res.* **22**, 884–887.

Childress, A. R., Mozley, P. D., McElgin, W., Fitzgerald, J., Reivich, M., and O'Brien, C. P. (1999). Limbic activation during cue-induced cocaine craving. *Am. J. Psychiatry* **156**, 11–18.

Contet, C., Kieffer, B. L., and Befort, K. (2004). Mu opioid receptor: A gateway to drug addiction. *Curr. Opin. Neurobiol.* **14**, 370–378.

Delfs, J. M., Zhu, Y., Druhan, J. P., and Aston-Jones, G. (2000). Noradrenaline in the ventral forebrain is critical for opiate withdrawal-induced aversion. *Nature* **403**, 430–434.

Di Chiara, G. and North, R. A. (1992). Neurobiology of opiate abuse. *Trends Pharmacol. Sci.* **13**, 185–193.

Dickinson, A. (1994). Contemporary animal learning theories. In "Animal Learning and Cognition" (N. J. Mackintosh, ed.), pp. 45–78. Academic Press, London.

Engel, J., Oreland, L., Ingvar, D. H., Pernow, B., Rossner, S., and Pellborn, L. A. (1987). "Brain Reward Systems and Abuse," Raven Press, New York.

Everitt, B. J. (1990). Sexual motivation: A neural and behavioral analysis of the mechanisms underlying appetitive copulatory responses of male rats. *Neurosci. Biobehav. Rev.* **14**, 217–232.

Everitt, B. J., Cardinal, R. N., Hall, J., Parkinson, J. A., and Robbins, T. W. (2000). Differential involvement of amygdala sub-systems in appetitive conditioning and drug addiction. In "The Amygdala: A Functional Analysis." (J. P. Aggleton, ed.), pp. 353–390. Oxford Univ. Press, Oxford.

Everitt, B. J. and Robbins, T. W. (2005). Neural systems of reinforcement for drug addiction: From actions to habits to compulsion. *Nat. Neurosci.* **8**, 1481–1489.

Fitzgerald, L. W. and Nestler, E. J. (1995). Molecular and cellular adaptations in signal transduction pathways following ethanol exposure. *Clin. Neurosci.* **3**, 165–173.

Flynn, J. P., Vanegas, H., Foote, W., and Edwards, S. (1970). Neural mechanisms involved in a cat's attack on a rat. In "The Neural Control of Behaviour" (R. Whalen, R. F. Thompson, M. Verzeano, and N. M. Weinberger, eds.), pp. 135–173. Academic Press, New York.

Goldstein, R. Z. and Volkow, N. D. (2002). Drug addiction and its underlying neurobiological basis: Neuroimaging evidence for the involvement of the frontal cortex. *Am. J. Psychiatry* **159**, 1642–1652.

Haber, S. N., Fudge, J. L., and McFarland, N. R. (2000). Striatonigro-striatal pathways in primates form an ascending spiral from the shell to the dorsolateral striatum. *J. Neurosci.* **20**, 2369–2382.

Heimer, L. and Alheid, G. (1991). Piecing together the puzzle of basal forebrain anatomy. *In* "Advances in Experimental Medicine and Biology" (T. C. Napier, P. Kalivas, and I. Hanin, eds.), Vol. 295, pp. 1–42. Plenum, New York.

Heimer, L. and Larsson, K. (1966–1967). Impairment of mating behaviour in male rats following lesions in the preoptic-anterior hypothalamic contimuum. *Brain Res.* **3**, 248–263.

Heimer, L., Zahm, D. S., and Alheid, G. F. (1995). Basal ganglia. *In* "The Rat Nervous System" (G. Paxinos, ed.), 2nd ed., pp. 579–614. Academic Press, Sydney.

Hoebel, B. G. (1974). Brain reward and aversion systems in the control of feeding and sexual behavior. *In* "Nebraska Symposium on Motivation" (J. Cole and T. Sonderegger, eds.), pp. 49–112. University of Nebraska Press, Lincoln.

Johnston, J. B. (1923). Further contributions to the study of the evolution of the forebrain. *J. Comp. Neurol.* **35**, 337–481.

Justinova, Z., Solinas, M., Tanda, G., Redhi, G. H., and Goldberg, S. R. (2005). The endogenous cannabinoid anandamide and its synthetic analog R(+)-methanandamide are intravenously self-administered by squirrel monkeys. *J. Neurosci.* **25**, 5645–5650.

Justinova, Z., Tanda, G., Munzar, P., and Goldberg, S. R. (2004). The opioid antagonist naltrexone reduces the reinforcing effects of delta 9 tetrahydrocannabinol (THC) in squirrel monkeys. *Psychopharmacol.* **173**, 186–194.

Kalivas, P. W. and McFarland, K. (2003). Brain circuitry and the reinstatement of cocaine-seeking behavior. *Psychopharmacology* **168**, 44–56.

Kelley, A. E. and Berridge, K. C. (2002). The neuroscience of natural rewards: Relevance to addictive drugs. *J. Neurosci.* **22**, 3309–3311.

Koob, G. F. (2004). Allostatic view of motivation: Implications for psychopathology. *In* "Motivational Factors in the Etiology of Drug Abuse" (series title: Nebraska Symposium on Motivation), (R. A. Bevins and M. T. Bardo, eds.), Vol. 50, pp. 1–18. University of Nebraska Press, Lincoln, NE.

Koob, G. F. (1992). Drugs of abuse: Anatomy, pharmacology, and function of reward pathways. *Trends Pharmacol. Sci.* **13**, 177–184.

Koob, G. F. (1999). Corticotropin-releasing factor, norepinephrine and stress. *Biol. Psychiatry* **46**, 1167–1180.

Koob, G. F. (2005). The neurocircuitry of addiction: Implications for treatment. *Clin. Neurosci. Res.* **5**, 89–101.

Koob, G. F., Bartfai, T., and Roberts, A. J. (2001). The use of molecular genetic approaches in the neuropharmacology of corticotropin-releasing factor. *Int. J. Comp. Psychol.* **14**, 90–110.

Koob, G. F. and Bloom, F. E. (1988). Cellular and molecular mechanisms of drug dependence. *Science* **242**, 715–723.

Koob, G. F. and Le Moal, M. (1997). Drug abuse: Hedonic homeostatic dysregulation. *Science* **278**, 52–58.

Koob, G. F. and Le Moal, M. (2001). Drug addiction, dysregulation of reward, and allostasis. *Neuropsychopharmacology* **24**, 97–129.

Koob, G. F. and Le Moal, M. (2005). Plasticity of reward neurocircuitry and the 'dark side' of drug addiction. *Nat. Neurosci.* **8**, 1442–1444.

Koob, G. F., Riley, S., Smith, S. C., and Robbins, T. W. (1978). Effects of 6-hydroxydopamine lesions to nucleus accumbens septi and olfactory tubercle on food intake, locomotor activity and amphetamine anorexia in the rat. *J. Comp. Physiol. Psychol.* **92**, 917–927.

Koob, G. F., Sanna, P. P., and Bloom, F. E. (1998). Neuroscience of addiction. *Neuron* **21**, 467–476.

Leibowitz, S. F. (1980). Neurochemical systems of the hypothalamus: Control of feeding and drinking behaviour and water-electrolyte excretion. *In* "Handbook of the Hypothalamus" (P. J. Morgan and J. Panksepp, eds.), Vol. 3, Part A, pp. 299–437. Raven Press, New York.

Markou, A. and Koob, G. F. (1992). Construct validity of a self-stimulation threshold paradigm: Effects of reward and performance manipulations. *Physiol. Behav.* **51**, 111–119.

Markou, A., Kosten, T. R., and Koob, G. F. (1998). Neurobiological similarities in depression and drug dependence: A self-medication hypothesis. *Neuropsychopharmacology* **18**, 135–174.

Marshall, J. F. and Teitelbaum, P. (1977). New considerations in the neuropsychology of motivated behaviors. *In* "Handbook of Psychopharmacology" (L. L. Iversen, S. D. Iversen, and S. H. Snyder, eds.), Vol. 7, pp. 201–229. Plenum, New York.

McBride, W. J., Murphy, J. M., Lumeng, L., and Li, T. K. (1990). Serotonin, dopamine and GABA involvement in alcohol drinking of selectively bred rats. *Alcohol* **7**, 199–205.

Merlo Pich, E., Lorang, M., Yeganeh, M., Rodriguez de Fonseca, F., Koob, G. F., and Weiss, F. (1995). Increase of extracellular corticotropin-releasing factor-like immunoreactivity levels in the amygdala of awake rats during restraint stress and alcohol withdrawal as measured by microdialysis. *J. Neurosci.* **15**, 5439–5447.

Mogenson, G. J., Jones, D. L., and Yim, C. Y. (1980). From motivation to action: Functional interface between the limbic system and the motor system. *Prog. Neurobiol.* **14**, 69–97.

Murphy, J. M., Stewart, R. B., Bell, R. L., Badia-Elder, N. E., Carr, L. G., McBride, W. J., Lumeng, L., and Li, T. K. (2002). Phenotypic and genotypic characterization of the Indiana University rat lines selectively bred for high and low alcohol preference. *Behav. Genet.* **32**, 363–388.

Nestler, E. J. (2001). Molecular basis of long-term plasticity underlying addiction. *Nature Rev. Neurosci.* **2**, 119–128.

Olds, J. and Milner, P. (1954). Positive reinforcement produced by electrical stimulation of septal area and other regions of rat brain. *J. Comp. Physiol. Psychol.* **47**, 419–427.

Parsons, L. H., Koob, G. F., and Weiss, F. (1995). Serotonin dysfunction in the nucleus accumbens of rats during withdrawal after unlimited access to intravenous cocaine. *J. Pharmacol. Exp. Ther.* **274**, 1182–1191.

Pfaff, D. W. (1982). "The Physiological Mechanisms of Motivation," pp. 217–251. Springer-Verlag, New York.

Phillips, A. G. (1984). Brain reward circuitry: A case for separate systems. *Brain Res. Bull.* **12**, 195–201.

Phillips, A. G., Pfaus, J. G., and Blaha, D. C. (1991). Dopamine and motivated behavior. *In* "The Mesolimbic Dopamine System: From Motivation to Action" (P. Willner and J. Scheel-Kruger, eds.), pp. 199–224. Wiley, Chichester.

Picciotto, M. R., Zoli, M., Rimondini, R., Lena, C., Marubio, L. M., Pich, E. M., Fuxe, K., and Changeux, J. P. (1998). Acetylcholine receptors containing the beta2 subunit are involved in the reinforcing properties of nicotine. *Nature* **391**, 173–177.

Robbins, T. W. (1986). Hunger. *In* "Neuroendocrinology" (S. L. Lightman and B. J. Everitt, eds.), pp. 252–303. Blackwell, Oxford.

Robbins, T. W. and Everitt, B. J. (2003). Motivation and reward. *In* "Fundamental Neuroscience" (L. R. Squire, F. E. Bloom, S. K. McConnell, J. L. Roberts, N. C. Spitzer and M. J. Zigmond, eds.), 2nd edition, pp. 1109–1126. Academic Press, New York.

Robbins, T. W., Taylor, J. R., Cador, M., and Everitt, B. J. (1989). Limbic-striatal interactions and reward-related processes. *Neurosci. Biobehav. Rev.* **13**, 155–162.

Roberts, A. J., Cole, M., and Koob, G. F. (1996). Intra-amygdala muscimol decreases operant ethanol self-administration in dependent rats. *Alcohol. Clin. Exp. Res.* **20**, 1289–1298.

Robinson, T. E. and Berridge, K. C. (1993). The neural basis of drug craving: An incentive-sensitization theory of addiction. *Brain Res. Rev.* **18**, 247–291.

Rolls, E. T. (1985). The neurophysiology of feeding. *In* "Psychopharmacology and Food" (M. Sandler and T. Silverstone, eds.), pp. 1–16. Oxford Univ. Press, Oxford.

Salloway, S. P., Malloy, P. E., and Duffy, J. D., eds. (2001). "The Frontal Lobes and Neuropsychiatric Illness." America Psychiatric Press, Washington, DC.

Sarnyai, Z., Shaham, Y., and Heinrichs, S. C. (2001). The role of corticotropin-releasing factor in drug addiction. *Pharmacol. Rev.* **53**, 209–243.

Satinoff, E. (1982). Are there similarities between thermoregulation and sexual behaviour? *In* "The Physiological Mechanisms of Motivation" (D. W. Pfaff, ed.), pp. 217–251. Springer-Verlag, New York.

Schultz, W., Romo, R., Ljungberg, T., Mirenowicz, J., Hollerman, J. R., and Dickinson, A. (1995). Reward-related signals carried by dopamine neurons. *In* "Models of Information Processing in the Basal Ganglia" (J. R. Houk, J. L. Davis, and D. Beiser, eds.), pp. 233–248. MIT Press, Cambridge, MA.

Schuster, C. R. and Thompson, T. (1969). Self administration and behavioral dependence on drugs. *Annu. Rev. Pharmacol. Toxicol.* **9**, 483–502.

Shaham, Y., Shalev, U., Lu, L., De Wit, H., and Stewart, J. (2003). The reinstatement model of drug relapse: History, methodology and major findings. *Psychopharmacology* **168**, 3–20.

Shaw-Lutchman, T. Z., Barrot, M., Wallace, T., Gilden, L., Zachariou, V., Impey, S., Duman, R. S., Storm, D., and Nestler, E. J. (2002). Regional and cellular mapping of cAMP response element-mediated transcription during naltrexone-precipitated morphine withdrawal. *J. Neurosci.* **22**, 3663–3672.

Shippenberg, T. S. and Koob, G. F. (2002). Recent advances in animal models of drug addiction and alcoholism. *In* "Neuropsychopharmacology: The Fifth Generation of Progress" (K. L. Davis, D. Charney, J. T. Coyle, and C. Nemeroff, eds.), pp. 1381–1397. Lippincott Williams and Wilkins, Philadelphia.

Solomon, R. L. (1977). The opponent-process theory of acquired motivation: The affective dynamics of addiction. *In* "Psychopathology: Experimental Models" (J. D. Maser and M. E. P. Seligman, eds.), pp. 124–145. Freeman, San Francisco.

Solomon, R. L. (1980). The opponent-process theory of acquired motivation: The costs of pleasure and the benefits of pain. *Am. Psychol.* **35**, 691–712.

Solomon, R. L. and Corbit, J. D. (1974). An opponent-process theory of motivation: I. Temporal dynamics of affect. *Psychol. Rev.* **81**, 119–145.

Stinus, L., Le Moal, M., and Koob, G. F. (1990). Nucleus accumbens and amygdala are possible substrates for the aversive stimulus effects of opiate withdrawal. *Neuroscience* **37**, 767–773.

Stricker, E. M. and Zigmond, M. J. (1976). Recovery of function after damage to central catecholamine-containing neurons: A neuro-chemical model for the lateral hypothalamic syndrome. *Prog. Psychol. Physiol. Psychol.* **6**, 121–188.

Substance Abuse and Mental Health Services Administration. (2006). "Results from the 2005 National Survey on Drug Use and Health: National Findings," Office of Applied Studies, NSDUH Series H-30, DHHS Publication No. SMA 06–4194. Rockville, MD.

Tabakoff, B. and Hoffman, P. L. (1992). Alcohol: Neurobiology. *In* "Substance Abuse: A Comprehensive Textbook" (J. H. Lowenstein, P. Ruiz, and R. B. Millman, eds.), 2nd ed., pp. 152–185. Williams & Wilkins, Baltimore, MD.

Tiffany, S. T., Carter, B. L., and Singleton, E. G. (2000). Challenges in the manipulation, assessment and interpretation of craving relevant variables. *Addiction* **95**, s177–s187.

Ungerstedt, U. (1971a). Adipsia and aphagia after 6-hydroxydopamine induced degeneration of the nigrostriatal dopamine system. *Acta Physiol. Scand. Suppl.* **367**, 95–122.

Ungerstedt, U. (1971b). Striatal dopamine release after amphetamine or nerve degeneration revealed by rotational behavior. *Acta Physiol. Scand. Suppl.* **367**, 49–68.

Valdez, G. R., Sabino, V., and Koob, G. F. (2004). Increased anxiety-like behavior and ethanol self-administration in dependent rats: Reversal via corticotropin-releasing factor-2 receptor activation. *Alcohol. Clin. Exp. Res.* **28**, 865–872.

Volkow, N. D., Fowler, J. S., Wang, G. J., Hitzemann, R., Logan, J., Schlyer, D. J., Dewey, S. L., and Wolf, A. P. (1993b). Decreased dopamine D2 receptor availability is associated with reduced frontal metabolism in cocaine abusers. *Synapse* **14**, 169–177.

Weiss, F., Markou, A., Lorang, M. T., and Koob, G. F. (1992). Basal extracellular dopamine levels in the nucleus accumbens are decreased during cocaine withdrawal after unlimited-access self-administration. *Brain Res.* **593**, 314–318.

Weiss, F., Parsons, L. H., Schulteis, G., Hyytia, P., Lorang, M. T., Bloom, F. E., and Koob, G. F. (1996). Ethanol self-administration restores withdrawal-associated deficiencies in accumbal dopamine and 5-hydroxytryptamine release in dependent rats. *J. Neurosci.* **16**, 3474–3485.

Winn, P., Tarbuck, A., and Dunnett, S. B. (1984). Ibotenic acid lesions of the lateral hypothalamus: Comparison with the electrolytic lesion syndrome. *Neuroscience* **12**, 225–240.

Wise, R. (1982). Neuroleptics and operant behavior: The anhedonia hypothesis. *Behav. Brain Sci.* **5**, 39–87.

Yin, H. H., Knowlton, B. J., and Balleine, B. W. (2004). Lesions of the dorsolateral striatum preserve previous outcome expectancy but disrupt habit formation in instrumental learning. *Eur. J. Neurosci.* **19**, 181–189.

George F. Koob, Barry J. Everitt, and Trevor W. Robbins

BEHAVIORAL AND COGNITIVE NEUROSCIENCE

44

Human Brain Evolution

Much of the allure of the neurosciences stems from the common conviction that there is something unusual about the human brain and its behavioral capacities. Nevertheless, modern neuroscientists have paid rather little attention to the study of brain evolution, and so our understanding of how the human brain differs from that of other animals is very rudimentary. In part, this neglect is due to a widely held belief that mammalian brains are all essentially similar in their internal structure and that species differ mainly in the size of the brain. This chapter reviews the modern evidence concerning brain evolution and shows that brain structure, far from being uniform across species, exhibits some remarkable variations. Because the subject is vast, the discussion is necessarily selective. Thus, after a brief review of evolutionary principles, this chapter describes the evolutionary history of three groups of vertebrates that are of special interest to people: mammals, primates, and humans themselves. The major steps are outlined in the evolution of our large, complex, and extremely useful brains from the smaller, simpler brains of the first mammals, focusing on the neocortex, as this part of the brain has been studied most extensively. The neocortex is disproportionately large in humans and is critically involved in mental activities and processes that are considered to be distinctly human.

EVOLUTIONARY AND COMPARATIVE PRINCIPLES

How Do We Learn about Brain Evolution?

There are three main ways to learn about how different brains have evolved. First, the fossil record can be studied. Much of what we have learned about the evolution of vertebrates in general has come from studying fossils. However, because bones readily fossilize, whereas soft tissues seldom do, we know a lot about the bones of our ancestors, but much less about everything else. Of course, one can infer much about some soft tissues, such as muscles, from their effects on bones, and this is true for brains as well. The brains of mammals fill the skull tightly, and thus the skull cavity of fossils rather closely reflects the size and shape of the brain, and even the locations of major fissures. Much has been learned and written about the changes in brain size from the fossil record (Jerison, 2007), and we could learn more by considering changes in the proportions of brain parts and even the proportions of parts of neocortex. For example, early primates already differed from most early mammals by having more neocortex in proportion to the rest of the brain, and more neocortex devoted to the temporal lobe where visual processing occurs. This implies a greater emphasis on functions mediated by neocortex and a greater emphasis especially on the processing of visual information.

We can even learn something about the functional organization of the neocortex from fossils. For instance, subdivisions of the body representation in the primary somatosensory cortex often are marked by fissures in the brain, and fissure patterns revealed by endocasts (internal casts of the brain case) from fossil skulls have been used to suggest specializations of the somatosensory cortex in extinct mammals. Recently, David Van Essen (1997, 2007) has proposed that an important factor in the development of brain fissures is the pattern of connections within the developing brain. According to this theory, densely interconnected regions tend to resist separation during brain growth and form bulges (gyri) that limit the separation distance, whereas poorly interconnected regions are free to fold and form fissures (sulci) that would increase the separation distance. Thus, it is because the hand and face regions of the somatosensory cortex are poorly interconnected that a fissure may develop between the two. If this explanation of fissures is correct, then the locations of brain fissures seen in fossil endocasts can potentially tell us something about anatomical connections in the brains of extinct mammals.

Unfortunately, little of the brain's great internal complexity is revealed by its size, shape, and fissures. Thus, to learn more about brain evolution, it is necessary to study the brains of extant (present-day) species and use comparative methods to deduce the organization of ancestral brains. There has been great progress since the early 1980s in understanding how to use comparative approaches to study evolution. In the final analysis, even studies of the fossil record involve a comparative approach. It is seldom known, for example, whether any given fossil was an actual ancestor of another fossil, but only that the comparative evidence, together with suitable times of existence, suggests that they could be.

As Darwin (1859) recognized, each living species represents the living tip of a largely dead branch of an extremely bushy tree of life. By examining other living tips of the tree, we can infer much about the organizations of the brains (and other body parts) of ancestors that occupied the branching points of this tree. Theories of brain evolution, including those of the evolution of the human brain, depend on reconstructing the probable features of the brains of ever more distant relatives.

The comparative method depends on (1) examining the brains of suitable ranges of extant species and (2) determining what features they share and whether these features are shared because they were inherited from a common ancestor or because they evolved separately. The 50-year-old field of cladistics provides guidelines for making such judgments. The choice of species for comparison depends on the question being asked. For example, to deduce what the brains of early mammals were like, one should examine brains from each of the major branches of the mammalian evolutionary tree (Fig. 44.1), and thus consider members of the monotreme, marsupial, and placental mammal branches. To know about early primates, members of the major branches of primate evolution should be considered, along with mammals thought to be closely related to primates, such as tree shrews. In principle, the more branches considered, the better, because a broad comparative approach is required to accurately reconstruct ancestral brain organization.

As brain studies can be difficult, time-consuming, and labor intensive, it is not always possible to study a large number of species, and one must concentrate on the most informative species. The brains of all living mammals contain mixtures of ancestral and derived features, and comparative studies are needed to distinguish the two, however one might first consider the brains of those mammals that are likely to have changed the least since the time of their divergence. Because we know from the fossil record that early mammals had small brains with little neocortex, mammals with large brains and much neocortex obviously have changed quite a bit, and it is likely to be more useful to concentrate on present-day mammals with brain proportions similar to those of early mammals. Whereas the very large brains of humans and whales undoubtedly share features inherited from a small-brained common ancestor, it may be difficult to detect the common features among the multitude of changes. As early primates more closely resembled many of the living prosimian primates in brain size and proportions than monkeys, apes, and humans, studies of the brains of prosimians are likely to be especially relevant for theories of early primate brain organization. Also, it is useful to consider the probable impact of obvious specializations on the brain organizations of living mammals. Although monotremes represent a very early major branch of the mammalian tree, the living monotremes, consisting of the duck-billed platypus with a "bill" and a capacity for electroreception and the spiny echidna with spines and a long sticky tongue for eating ants, are quite specialized, as are their somatosensory systems.

Although the brains of some mammals may have retained more primitive characteristics than others, it is dangerous to assume that the brain of any living mammal fully represents an ancestral condition. Brain features need to be evaluated trait by trait in a comparative context, as any particular feature could be primitive (i.e., ancestral) or derived. To reconstruct the course of brain evolution, we need to distinguish ancestral and derived characters rather than ancestral and derived species. The existence of a mixture of ancestral and derived features in a single species is referred to as *mosaic evolution.*

A third source of information about brain evolution is based on understandings of the mechanisms and modes of brain development and the constraints they impose on evolution. For example, Finlay and Darlington (1995) have presented evidence that brains change in orderly ways as they get bigger. In general, larger brains have proportionately more neocortex and less brain stem. Finlay and Darlington suggest this is the case because the late-maturing neocortex of large

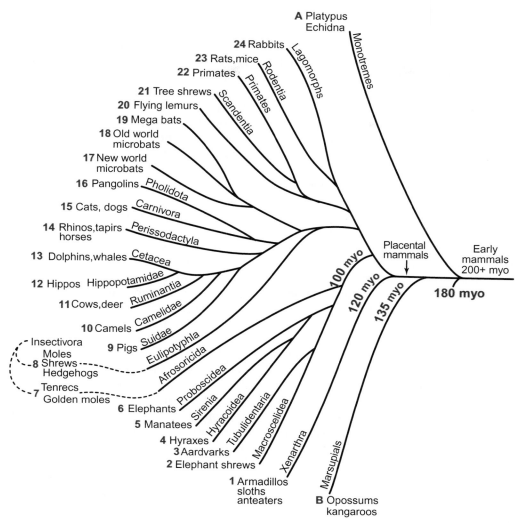

FIGURE 44.1 The probable course of the major branches of mammalian evolution (e.g., mammalian orders). Proposed clades of placental mammals are numbered, whereas monotremes (A) and marsupials (B) are lettered. Each branch of the tree also has branched many times given the great numbers of present-day species. Note how this branching pattern differs from long-standing notions of a scale of nature from simple to complex. Based on Springer and deJong (2001).

brains grows proportionately longer and larger. Another difference between large and small brains is that large brains have more neurons and longer connections. The increase in neurons makes it difficult for each neuron in a large brain to maintain the same proportion of connections with other neurons, as do neurons in a small brain, and to maintain the same transmission times over longer axons. Thus, as larger brains evolve, changes in organization are needed to reduce the commitment to connections, especially connections requiring long, thick axons. A deeper understanding of the genetic, developmental, and structural constraints on brain design could allow us to better postulate how brains are likely to change in organization with changes in brain size.

Cladistics and Phylogenetic Trees

To understand brain evolution, we need to understand the evolutionary relationships among mammals, which are summarized in biological classifications (taxonomies). The science of classification (known as "systematics") has a long history, but an early classification that divided the world into "things that belong to the Emperor and things that don't" is clearly outmoded. Our modern understanding of plant and animal relationships emerged from the efforts of the Swedish naturalist Linnaeus (1707–1778), who grouped species into ever-larger categories (species, genus, family, order, class, phylum) according to degrees of resemblance and dissimilarity. We now understand why life forms exhibit the particular pattern of similarities and differences they do, and we have a logic for extending and refining the Linnaean system. In brief, complex life on Earth appears to have evolved only once. It is based on a molecular template that is passed on from generation to generation and yet is modifiable. Because this molecular material usually is not exchanged between individuals from different species, a phylogenetic classification can be derived that reflects times of divergence from common ancestors. Similarities retained over time reflect the preservation of parts of the code, whereas differences reflect alterations in the code. The many different modifications of the genome in the many diverging lines of descent over billions of years have led to the great diversity of life that we see today. Our current classification scheme is based on our understanding of phylogenetic (ancestor-descendant) relationships. Because this understanding continues to grow, parts of the classification scheme continue to be modified.

Phylogenetic relationships are deduced from comparative evidence. The entomologist Willi Hennig (1966) helped reinvigorate this field of study when he formulated a rigorous comparative method of reconstructing phylogenetic relationships, sometimes known as "cladistics." The term reflects Hennig's emphasis on the correct identification of "clades," groups of organisms that share a common ancestor. A clade is simply a branch of the evolutionary tree, which is connected through a set of ancestors (which are the branching points of the tree) to all the other branches in the tree. In Hennig's method, evolutionary relationships are reconstructed by a process of "character analysis." A character is any observable feature or attribute of an organism. A character could be a feature of the brain, such as the corpus callosum between the two cerebral hemispheres, or a feature of any other part of the body, or (as is often the case today) a molecule or a DNA sequence. By considering the states of as many characters as possible—for example, whether a corpus callosum is present (as it is in placental mammals) or absent (as in other vertebrates)—and by adopting the assumption that closely related species will share more character states than distantly related species, one can arrive at a hypothesis about the relationships of the groups being examined. Because a given character state can evolve independently in different lineages (e.g., forelimbs that function as wings evolved independently in birds and bats), not every character will yield an accurate picture of evolutionary relationships. It is therefore important to base reconstructions of evolutionary relationships on as many characters as possible. Typically, a very large number of possible trees can be generated from a given character analysis; the

tree that requires the smallest number of changes to account for the observed pattern of character states (the maximum-parsimony solution) usually is considered to be the best estimate of the correct tree. The growth of molecular biology has provided a new source of comparative information to supplement character analyses based on anatomical characteristics, helping to improve the resolution of modern mammalian trees (Fig. 44.1). These trees guide our interpretation of the evolutionary history of brain organization.

To communicate precisely, comparative biologists have developed a specialized nomenclature. The concepts of *homology* and *analogy* are central to comparative biology (Box 44.1). A group of species that all share a common evolutionary history is a *natural taxon* or a *monophyletic group*, which is the same thing as a *clade*. "Unnatural" taxa are groups that either exclude one or more of an ancestor's descendants (*paraphyletic groups*) or that combine descendants of multiple ancestors (*polyphyletic groups*). The traditional classification of the great apes (orangutans, gorillas, chimpanzees) in the family Pongidae now is known to be paraphyletic because it excludes humans, which share a recent common ancestor with chimpanzees and gorillas. If a character state is found throughout a monophyletic group, it likely was present in the common ancestor of the group or even earlier. Thus, comparisons are necessary with members of a *sister group*, which is the monophyletic taxon thought to be related most closely to the group under study.

To determine whether a character is derived (new) or ancestral, one examines members of one or more *outgroup*, the more distant relatives of the group under examination. The direction of change is called its *polarity*. For example, mammals include forms that lay eggs, the monotremes, whereas the sister group of monotremes, the marsupial–placental group, gives birth to live offspring. We can determine the polarity of these character states (egg laying, live birth) by examining out groups; that is, by examining other vertebrates like reptiles and amphibians. Because most nonmammalian vertebrates lay eggs, we conclude that egg laying is ancestral for mammals, and live birth derived.

Misconceptions about Brain Evolution

Probably the most serious misconception about brain evolution is that theories of evolutionary change are necessarily highly speculative (Striedter, 1998). As in other historical sciences, direct observation of the process of brain evolution is usually not possible, but objective criteria for evaluating theories and reconstructions of ancestral brains do exist (e.g., character

---

BOX 44.1

HOMOLOGY AND ANALOGY

Homology and analogy are two of the most important concepts in evolutionary biology. Both terms refer to similarity, but to similarity arising from different sources. The terms can be applied to any biological characteristic or feature, including brain structures and even behavior. In our efforts to understand brain evolution, features of brains are compared across species, and it is important to deduce if these features have been inherited from a common ancestor or emerged independently. The two terms reflect conclusions based on the available evidence, and uncertainty is common.

Homology

When different species possess similar characteristics because they inherited them from a common ancestor, the characteristics are said to be homologous. This does not mean they are identical in structure or function. For example, all mammals appear to have a primary somatosensory area, S1, as a subdivision of neocortex. This structure appears to be involved in electroreception in the duck-billed platypus, a monotreme, but not in other mammals. Humans and monotremes have inherited an S1 from a distant ancestor (at least 150 million years ago), but subsequently, ancestors have specialized S1 in quite different ways.

Analogy and Homoplasy

Characteristics that have evolved independently are referred to as analogous or homoplaseous. Some authors prefer to refer to similarities in function as analogous and similarities in appearance as homoplaseous, whereas others use the terms interchangeably or prefer analogous as the more common term. As an example, most primates and some carnivores such as cats divide the primary visual cortex, V1, into alternating bands of tissue activated mostly by one eye or the other, the so-called ocular dominance columns or bands. Because carnivores and primates almost certainly diverged from a common ancestor that did not have ocular dominance columns, we conclude that this way of subdividing visual cortex evolved independently at least two times. When analogous similarities evolve, the process is called *convergent* or *parallel* evolution (parallel if the sequences of changes were similar).

Both homology and analogy can be applied to structures or to specific features of structures. The forelimbs of bats and birds are homologous as forelimbs because both bats and birds inherited forelimbs from their common ancestors. The common ancestor of bats and birds did not fly, however, and the modifications that have transformed the forelimbs of bats and birds into wings evolved independently; these wing-like characteristics are therefore analogous. Correctly identifying homologies is a very important step in deducing the course of evolution of the human brain. Features such as S1, which are homologous in most or nearly all mammals, must have evolved early with the first mammals or before, whereas features common to only primates, such as the middle temporal visual area (MT), would have emerged much later in only the line leading to the first primates. Some brain features related to language production may have arisen quite recently with the emergence of archaic or modern humans. Identifying homologies and their distributions across groups of related mammals (clades) allows us to reconstruct the details of brain evolution in the different lines of descent, including the one leading to modern humans.

The problem of distinguishing homologies from analogies is that both are identified by similarities. However, analogous structures have similarities based on common adaptations for functional roles, and thus they should also vary greatly in details unrelated to function. In contrast, homologous structures likely have retained many details from a common ancestor that may not be functionally necessary. Thus, aspects of structures that appear to be unessential may count more in judging homologies. Much can also be deduced from studying brain development and the development of similarities. Of course, knowledge about the phyletic (cladistic) distribution of similarities is critical, as is accounting for differences in structures by finding species with intermediate types. Ultimately, one hopes to have evidence for so many similarities that the independent evolution of them all seems extremely improbable, or sufficient evidence from differences to indicate that the similarities in structures were acquired independently. Often the evidence is far from compelling, and opinions may change with new evidence.

Jon H. Kaas and Todd M. Preuss

analysis). It is not the case that one opinion is as good as the next, although such a view has allowed poorly founded theories to persist.

Another serious misconception is that evolution has a single goal or direction. This stems from the persistent belief in a "phylogenetic scale" that starts with lowly forms such as sponges; proceeds with insects, fish, amphibians, reptiles, and various mammals at successively higher levels; and reaches its pinnacle with humans. This view that the history of life is like a scale or ladder reflects the popular idea that evolution is primarily a process of progressive improvement, a philosophical and religious viewpoint that reassuringly identifies us as the most perfect species. However, evolutionary biologists have long recognized the great diversity of life and have adopted from Darwin the branching tree as a metaphor for the process whereby parent species divide to form daughter species, with each branch becoming adapted to its particular environment through natural selection. This perspective still leaves a place for progress in evolution, if progress means to become better adapted, but there are so many different ways organisms can become better adapted—most of which do not involve becoming bigger brained or smarter—that there can be no single dimension of progress, as the phylogenetic scale implies. Progress, for example, might mean reducing the size of the visual system to reduce metabolic costs in mammals living underground. Thus, there is also no universal trend toward increased complexity, as evolving brains sometimes simplify by reducing or losing parts.

Overall, the brains of many mammalian groups evolved to get bigger, but this change partly reflects the fact that they started off small, and it was more often adaptive to get bigger than smaller. However, mammals with very small brains, near the theoretical limit of smallness in mammalian brain size, persist today and there are reasons to think that small brain size is not the primitive condition in all these groups, but sometimes resulted from relatively recent reductions in brain size. Interestingly, domestic mammals, with our efforts to improve a wild stock, generally have reduced brain size. Evolutionary biologists make a distinction between traits that are ancestral (also termed primitive or *plesiomorphic*) from those that are recently acquired (also known as derived or *apomorphic*), but this distinction does not imply that primitive is simple and that derived is complex.

A third misconception is that ontogeny recapitulates phylogeny. At one time, evolution was thought to proceed by sequentially adding new parts (*terminal addition*), and that in the development of complex forms, the newest parts were those added last. Thus, the evolutionary history of any organism would be revealed in its sequence of development. Though this is not the case in general, the study of brain development remains relevant to the study of brain evolution because evolution occurs through alterations in the course of development. It is also the case that many changes in the course of development that have led to new adaptations have occurred in the later stages of development, primarily because alterations in early stages often are lethal or produce profoundly different and maladaptive adult forms. Nevertheless, it is important to recognize that the course of development can be altered in many ways and at many stages. Studies of brain development are useful because they can indicate how homologous structures that appear dissimilar in adults arose from forms that were more similar early in development.

The concepts of progress and terminal addition have led some investigators to consider certain features of the brains of such mammals as humans, monkeys, and cats as either relatively old or new on the basis of their histological appearance. Poorly differentiated areas of neocortex—areas with indistinct lamination—were considered to be ancient, and well-differentiated areas were thought to be new. For example, because the primary visual cortex (V1) has very highly differentiated layers in many primates, including humans, V1 has been considered by some to be a recently acquired area of the cortex. However, a comparative analysis indicates that the primary visual cortex (V1) is as old as the first mammals, perhaps much older. The cellular layers of V1 were poorly differentiated in early mammals, and they remain poorly differentiated in many living species. Yet, humans do not have the most highly laminated V1: this distinction belongs to the tarsier, a tiny, nocturnal primate with enormous eyes. As another example, the superior colliculus or optic tectum, an ancient visual structure, has well-differentiated layers in many birds and reptiles, but they are poorly or moderately differentiated in many mammals (including humans), yet well differentiated in other mammals, such as squirrels and tree shrews.

Origin of the Neocortex

The hallmark of the evolution of mammalian brains was the emergence of the neocortex. The cerebral cortex (or pallium) covers the deeper parts of the forebrain (telencephalon). In mammals, the cortex generally is divided into three parts: the lateral *paleocortex* or olfactory piriform cortex, the medial *archicortex* or hippocampus and subiculum, and the *neocortex* or *isocortex* lying in between (Fig. 44.2). The archicortex

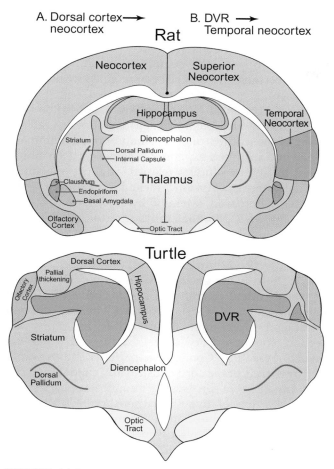

A. Dorsal cortex→ neocortex

B. DVR → Temporal neocortex

Rat

Neocortex

Superior Neocortex

Hippocampus

Temporal Neocortex

Striatum

Diencephalon

Dorsal Pallidum

Internal Capsule

Claustrum

Endopiriform

Basal Amygdala

Thalamus

Olfactory Cortex

Optic Tract

Turtle

Dorsal Cortex

Pallial thickening

Hippocampus

Olfactory Cortex

DVR

Striatum

Diencephalon

Dorsal Pallidum

Optic Tract

FIGURE 44.2 Theories of the origin of neocortex. One view is that (A) only the dorsal cortex of stem amniotes gave rise to the neocortex. A less supported view is that (B) the dorsal cortex gave rise only to the superior part of the neocortex, whereas the dorsal ventricular ridge (DVR) gave rise to the temporal neocortex of extant mammals such as rats (top). See below for the location of dorsal cortex and the DVR in extant reptiles such as turtles. The neocortex and dorsal cortex are much different in size, as well as in cellular organization (see text).

and paleocortex can be recognized in reptiles, and their names reflect the early conclusion that they are phylogenetically old parts of the forebrain. All extant mammals have an obvious neocortex, and the presence of a neocortex is clearly indicated in the endocasts of the skulls of early mammals, but nothing quite like the neocortex exists in sauropsids, reptiles, and birds, which represent the nonmammalian branches of amniote evolution. Hence, the term *neo*cortex was applied to this seemingly new part of the brain. Nevertheless, the neocortex as a structure is not really new, as current evidence indicates that it is homologous to a structure in reptiles called the dorsal cortex. In contrast to the neocortex of mammals, however, which has multiple layers with different cell types and

packing densities, the dorsal cortex of reptiles is a rather small and thin sheet of tissue. Whereas dorsal cortex has little more than a single row of neurons, an imaginary line drawn through the thickness of the neocortex would likely encounter over 100 neurons. Thus, the neocortex is much different in structure than the dorsal cortex. Unfortunately, no species exist today to show us what intermediate states were like in the evolution of the mammalian neocortex. Because the neocortex as a structure did not really originate with mammals and because it is not "new," some investigators refer to the neocortex as the *iso*cortex, using a term that refers to the relatively uniform appearance of the neocortex throughout all regions. However, the changes in the dorsal cortex that produced the neocortex are impressive, and no other vertebrates have a structure that clearly resembles the neocortex. Thus, mammals are characterized as much by their neocortex as a modified structure as they are by their mammary glands.

Although the neocortex has a nearly uniform histological appearance, its considerable variability in size and organization is what allows mammals to differ so much in behavior and abilities. To understand how variations in the neocortex make this possible, it is necessary to identify ancestral features of the neocortex, that is, features that were present in the last common ancestor of living mammals, and then determine how this organization was modified in different lineages of living mammals, such as primates.

The laminar organization of the neocortex appears to be similar in most mammals, suggesting that the ancestral design was so useful that many features have been retained in modern groups. For example, in most mammals and over most of the neocortex, six layers can be recognized (Brodmann, 1909). Of the six layers of neurons, layer 4 receives activating inputs from the thalamus or from other parts of the cortex. Layer 3 communicates with other regions of the cortex, layer 5 projects to subcortical structures, and layer 6 sends feedback to the thalamic nuclei or cortical area providing activating inputs. This ancestral framework for cortical neural circuits has been modified and elaborated in various lines of descent to give us the great variability in brain function we see today. The neocortex has changed by diversifying its neuron types, differentiating (and in some cases simplifying) its laminar structure in various ways, altering connections, changing its overall size and the sizes of individual cortical areas, adding cortical areas, and dividing areas into specialized modular processing units or cortical "columns." What was achieved through these changes ranges from echolocation to language. Of course, changes in the neocortex have been accompanied by modifications occurring in other parts of the brain because cortical and subcortical structures often have integrated functions. For example, the superior colliculus of the midbrain has been variously modified in function, largely by changing and expanding the direct inputs from neocortex, but also through other alterations. To begin to understand how cortical organization has changed to produce the remarkable human brain, we first focus on the neocortex of early mammals.

Brains of Early Mammals

Early mammals had small brains with little neocortex. By comparing the histological appearance of the neocortex in extant mammals of differing lines of descent, we can conclude that the neocortex of early mammals was not as highly differentiated in terms of distinct layers and neuron types as is the cortex of many modern mammals. However, the cortex likely was not homogeneous in appearance either. In present-day mammals, the primary sensory areas typically have a noticeably different layer 4 (the receiving layer) with somewhat smaller and more densely packed neurons than in other areas. From such slight regional differences in appearance, early investigators such as Brodmann (1909) surmised that all mammals have functionally significant subdivisions of the cortex, called *areas*, that some areas are shared by many species (homologous areas), and that mammals differ in numbers of areas. What was difficult for Brodmann and other early investigators was to reliably identify areas by their histological appearance, especially in the poorly differentiated cortex of many small-brained mammals, but sometimes even in the large expanses of the rather homogeneous cortex in large-brained mammals. As a result, areas were often incorrectly delimited, and the sometimes radical changes in the appearance of certain areas that took place across species resulted in mistaken interpretations of homology. To Brodmann's credit, he correctly identified the primary visual cortex in species as different from each other as humans, where the task is easy because of the area's histological distinctiveness, and hedgehogs, which are insectivores with poor cortical differentiation.

Although Brodmann's subdivisions of human and other brains are commonly portrayed in textbooks today, many of his proposed areas and proposed homologies have little validity. Fortunately, modern methods allow us to compare many features of cortical biology in great detail, including connection patterns, neuron-response properties, and cortical histochemistry, which allows us to identify cortical areas and evaluate homologies with a high degree of assurance. From these methods, we can conclude that the neocortex of early mammals was subdivided into a small number of functionally distinct areas, on the order of 10–20, and these areas have been retained in most lines of descent.

The neocortex of North American opossums (Fig. 44.3), which are small-brained marsupials, reflects many of the features of other small-brained mammals. Much of the limited expanse of the neocortex is dominated by sensory inputs relayed from the thalamus. Caudally, the neocortex includes a large primary visual area, V1, bordered laterally by a strip-like second visual area, V2. More laterally, an additional strip of cortex responds to visual stimuli, but the organization of this cortex is not known. Nearly all existing mammals have a V1, V2, and a more lateral zone of visual cortex. This pattern likely emerged with (or before) the first mammals. More rostrally, opossums have a primary somatosensory area, S1, bordered laterally by two additional representations of tactile receptors, the second somatosensory areas, S2, and the parietal ventral area, PV. Connection patterns indicate that narrow bands of the cortex rostral and caudal to S1 are also involved in processing somatosensory inputs. This collection of five somatosensory fields is seen repeatedly in small-brained mammals, although S2 and PV are not always distinct from each other. A region of cortex caudal to S2 and PV responds to auditory stimuli, and much of this region is occupied by the primary auditory area, A1, which is present in all or nearly all living mammals. However, the auditory cortex contains a number of additional areas in most mammals, including many of the studied small-brained mammals, although homologies are often uncertain. Thus, early mammals probably had several auditory fields.

More rostrally, the frontal cortex of opossums is very small and does not contain any obvious motor areas. Instead, motor-related information from the cerebellum is relayed to S1. However, opossums are marsupials (Fig. 44.1), and most or all placental mammals have a primary motor area, M1, just rostral to S1 (and the narrow somatosensory band bordering S1), and possibly a second motor area, M2, also known as the supplementary motor area, SMA. The frontal cortex also includes an orbitofrontal region, which mediates autonomic responses to exteroceptive stimuli. On the medial wall of the cerebral hemisphere, opossums and other mammals share several divisions of the limbic cortex with inputs from the anterior and lateral dorsal nuclei of the thalamus. In addition, narrow strips of the entorhinal cortex are present that connect the neocortex with the hippocampus.

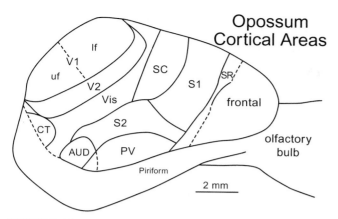

Opossum Cortical Areas

2 mm

FIGURE 44.3 Some of the proposed neocortical areas in North American opossums. Somatosensory areas include the primary area (S1), a secondary area (S2), a parietal ventral area (PV), and caudal (SC) and rostral (SR) somatosensory belts. The auditory cortex (AUD) is limited and likely contains a primary field (A1) and possibly another area or two. The visual cortex includes primary (V1) and secondary (V2) areas and a visual (Vis) belt. The caudotemporal (ct) field is probably visual. In V1, the upper visual field (uf) is represented caudal to the lower visual field (lf). Modified from Beck *et al.* (1996).

From this ancestral pattern, a great variety of brain organizations have evolved through alterations in size and the number of parts and the connections within and between parts (Kaas, 2007a). Consider, for example, the variations of the primary somatosensory cortex. The duck-billed platypus devotes most of S1 to representing tactile receptors of its highly sensitive bill, and it has added inputs from electroreceptors. The star-nosed mole devotes most of S1 to its long, fleshy nose appendages, rats mostly activate S1 with their facial whiskers, and human S1 has a large representation of the hand, lips, and mouth. In addition, the amount of neocortex varies greatly across species of mammals (Jerison, 2007), and some of this variation is due to differences in numbers of cortical areas present in different groups of mammals (Finlay and Brodsky, 2007; Kaas, 2007a).

The number of sensory areas increased independently in several lines of evolution. For example, both cats and monkeys have a large number of visual areas, but the carnivore and primate lines appear to have acquired most of these areas independently rather than from a common ancestor. Thus, many of the visual areas in cats have no homologies in monkeys or humans. A similar situation holds for other regions of the brain. New areas were added, most commonly to sensory and motor regions of cortex rather than to multisensory "association" cortex as once thought. With new cortical areas, new connections between areas and subcortical structures emerged. Thus,

mammals in various lines of descent evolved new cortical areas and connections, as well as other many specializations of previously existing ones. As a long-recognized example of the emergence of a new brain feature, the corpus callosum, the major pathway interconnecting the neocortex of the two cerebral hemispheres, is a derived character of placental mammals, having emerged in the ancestors of placentals after they diverged from marsupials. Whereas connections between the two hemispheres are mediated by the anterior commissure in marsupials and monotremes, most of these connections are carried in the shorter, more direct callosal pathway in placental mammals.

EVOLUTION OF PRIMATE BRAINS

Evolution of Primates

Early primates emerged from small-brained, nocturnal, insect-eating mammals some 60 to 70 million years ago and soon branched into three main lines leading to present-day prosimians, tarsiers, and anthropoids (Fig. 44.4). The prosimian suborder of primates includes lorises, lemurs, and galagos; the anthropoid suborder consists of New World monkeys, Old World monkeys, and the ape–human group. Tarsiers are small, prosimian-like animals. Two main schemes of classification of primates have been in use, mainly because it has not been obvious where to place tarsiers. Many authors use a traditional classification and distinguish prosimian from anthropoid primates and include tarsiers with prosimians. However, this is now thought to be an unnatural paraphyletic grouping because tarsiers, despite their generally prosimian-like appearance, generally are considered to be more closely related to anthropoids than to lemurs, lorises, and galagos. This conclusion is reflected in the cladistic classification preferred by other authors, in which lemurs, lorises, and galagos are placed in the suborder, Strepsirhini, a group that has retained ancestral features, including a naked, moist rhinarum (wet nose). Tarsiers and anthropoids, which have a reduced olfactory system (and thus a dry nose), are placed in the suborder Haplorhini. The anthropoid primates are divided into the infraorders Platyrrhini (New World monkeys) and Catarrhini (Old World monkeys, apes, and humans).

The earliest primates are thought to have been small-bodied, nocturnal visual predators living on insects and small vertebrates, as well as fruit. They adapted to the tropical rainforests by emphasizing vision and visually guided reaching and grasping.

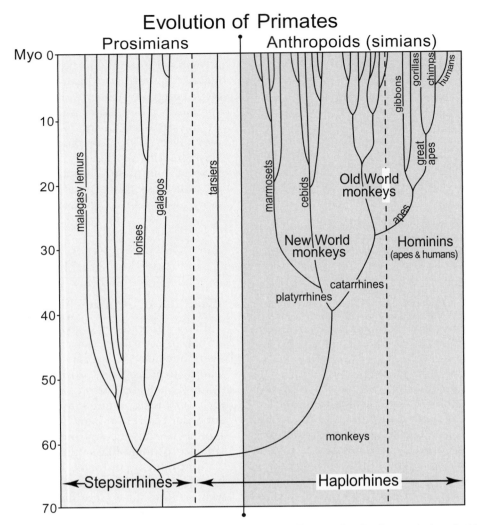

FIGURE 44.4 The evolution and classification of primates. Tarsiers are generally considered to be prosimians, but they are related more closely to anthropoids, so they are recognized as haplorhine primates. Despite the ancient split of prosimian and anthropoid primates, they share many brain features that are unique to primates. Tree shrews and flying lemurs are thought to be close relatives of primates, and together with them constitute the superorder Archonta. See Preuss (2007) for discussion.

This involved having larger, forward-facing eyes, opposable big toes and thumbs, and digits tipped with nails. These primates produced the largely nocturnal strepsirhine radiation with its varied forms, including some species now living in Madagascar that have become diurnal. The haplorine primates emerged about 60 million years ago in association with a shift from nocturnal to diurnal life, together with an increased emphasis on fruit eating (Ross, 1996).

With the shift to diurnality came reduced dependence on olfaction, enhancement of the visual system, enlarged body size, and sometimes a more gregarious mode of life. Specifically, the olfactory apparatus was reduced in size, and the eyes enlarged and brought close together. Early haplorhines evolved a fovea, a specialized region of the retina filled with small photoreceptors and devoid of blood vessels, that enhances central visual acuity. The reflecting surface at the back of the eye (tapetum lucidum), an adaptation to nocturnal vision present in prosimians and many other nocturnal mammals, was lost. Anthropoids underwent further specializations, modifying cone morphology, increasing the proportion cones to rods, and filling the fovea with cones to the near-exclusion of rods. These adaptations enabled anthropoids to achieve a high degree of color discrimination. Nevertheless, whereas humans and other Old World anthropoid primates have three types of retinal cones, ancestral anthropoids probably possess the two types of cones, similar to most modern prosimians and New World monkeys. A third cone type, enabling full trichromatic vision, appears to have evolved independently in the ancestors of Old World anthropoids and in several New World monkey groups. The enhancements of color vision in the anthropoids are plausibly regarded as adaptations for distinguishing ripe, edible fruits. The shift to diurnality was also marked by larger social groupings, which may offer enhanced protection from predation. These changes were accompanied by increased brain size, including increases in the temporal lobe visual region, and in regions of the parietal and frontal cortex mediating motor control and social interactions.

The ancestors of present-day tarsiers evidently abandoned the diurnal niche to become nocturnal visual predators again. Tarsiers retain a fovea, but they have a rod-dominated retina and their enormous eyes are sensitive at low light levels without the aid of a reflecting tapetum. The primary visual cortex became very large relative to the rest of the brain and is extremely well differentiated, with a multiplicity of layers and sublayers (Collins *et al.*, 2005). Tarsiers became such extremely specialized visual predators that they eat no plant food. Other haplorhines (anthropoids) remained diurnal and spread to many niches, including those outside the rainforest and niches based more on eating leaves as well as fruits. Some increased considerably in body size. Later anthropoids were able to process hard, husked fruits with their hands and teeth. Some early anthropoids managed to reach South America from Africa, apparently by rafting, to form the New World monkey radiation. All modern anthropoids are diurnal with the exception of owl monkeys, a New World monkey group, that has (like tarsiers) become secondarily nocturnal.

In Africa, early anthropoids diverged some 25 to 30 million years ago into lines leading to modern Old World monkeys and to apes. At first, apes were the most successful radiation, coming to occupy a range of rainforest and open woodland environments, whereas monkeys were quite rare. Perhaps as many as 30 different types of apes existed at one time. This condition changed radically some 10 million years ago when monkeys became abundant and apes rare. This change may have resulted from the advent of a more seasonably variable and drier climate. Ancestral Old World monkeys acquired specialized teeth suited to a leaf and fruit diet in drier climates. Also, the more rapid reproduction of monkeys may have made them more resistant to extinction than apes.

Some 5 or 6 million years ago, a line of apes diverged into two branches: one that gave rise to modern common chimpanzees and bonobos (pygmy chimpanzees), and a second branch that led to the group of bipedal apes, the "hominins," that includes modern humans (Fig. 44.5). Traditionally members of the human branch were referred to as "hominids," that is, as members of the Family Hominidae. Today, taking into account the close relationship of humans to the great apes, it is customary to extend the term "hominid" to the entire great ape clade, to use the term "hominines" (Subfamily Homininae) to refer to the African ape clade (chimpanzees, bonobos, humans, and gorillas), and to apply the term "hominins" (Tribe Hominini) to members of the human branch of the African ape clade. (A tribe is, taxonomically speaking, a sub-subfamily.)

The oldest known hominins, the so-called australopithecines, date back at least 4 million years. These early hominins were bipedal, but skeletal traits suggest they retained considerable ability for climbing trees. Body size was rather small and males were much bigger than females. The brain was only slightly larger than for apes of similar body size. Hominin traits may have emerged as adaptations to a drier environment with grassland and savanna.

Early australopithecines soon gave rise to a number of species differing in body size and limb proportions, as well as in characteristics of the jaws and teeth and brain size. Primitive members of our own genus began to emerge some 2 million years ago with *Homo habilis* (or "handy man," due to its use of stone tools). *Homo habilis* had a slightly increased cranial capacity compared to australopithecines, reduced face and teeth, and pelvic modifications for improved bipedal locomotion and the birth of neonates with larger heads. About 1.7 million years ago, *H. habilis* appears to have been replaced by *H. erectus*, a larger hominin with a further reduction in face and teeth and a larger brain. Shortly thereafter, *Homo erectus* spread out of Africa to central Asia. *Homo sapiens* emerged from an African population of *H. erectus* some 250,000 to 300,000 years ago. Other members of the genus *Homo* coexisted with *Homo sapiens* until as recently as 35,000 years ago, notably the Neanderthals, who adapted to a colder Europe and southwest Asia over 130,000 years ago. They disappeared and were replaced by modern *H. sapiens* some 35,000 years ago. Neanderthals were shorter and more heavily built than modern humans, but had a comparable or slightly larger cranial capacity. Over the last 15,000 years, some populations of humans have become smaller and have reduced their brain size, possibly as a result of a poorer diet as agriculture replaced hunting and gathering.

Chimpanzees and bonobos (also misleadingly known as "pygmy chimpanzees") appear to be our closest surviving relatives. Humans and chimpanzees are 98 to 99% similar in the coding sequences of genes,

Evolution of Hominins

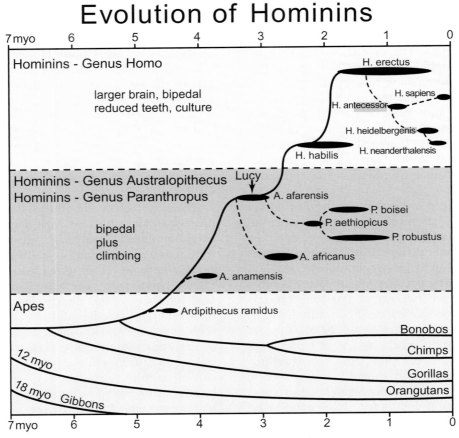

FIGURE 44.5 The evolution of apes and humans. Apes include living apes and a late fossil judged to not be bipedal (*Ardipithecus ramidus*). *Australopithecus* and *Paranthropus* appear to have been bipedal. The age of the famous fossil, Lucy, is indicated. Relationships are somewhat uncertain, and more branches on the tree exist (de Sousa and Wood, 2007).

which is more similar than found among morphologically indistinguishable sister species of some genera. The bases for the great phenotypic differences between humans and chimpanzees, including those in brain size and presumably brain organization, are not well understood, but changes in patterns of gene expression are likely to be important.

Brains of Early Primates

The brains of primates vary greatly in size and fissure patterns. Humans have the largest of primate brains, whereas mouse lemurs have the smallest (Fig. 44.6). Judging from fossil endocasts and other parts of the skeleton, early primates were lemur-like in body form, and their brains were shaped like those of present-day lemurs, although smaller. The modern mouse lemur has moderately expanded temporal and occipital lobes, indicating an emphasis on vision, a calcarine fissure or fold within the primary visual cortex, and a lateral (Sylvian) fissure separating

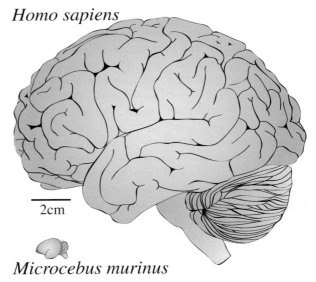

Homo sapiens

2cm

Microcebus murinus

FIGURE 44.6 Lateral views of the brains of a human and a small prosimian primate, the mouse lemur, to illustrate the great range of sizes for present-day primates.

somatosensory areas of the parietal lobe from auditory areas of the temporal lobe. Most modern primates share these two fissures. Many early as well as present-day primates also have an additional prominent fissure in the temporal lobe (the superior temporal sulcus). Other fissures are more variable.

Among living prosimians, evidence about brain organization comes mostly from studies of galagos, nocturnal animals from Africa that are cat-sized or smaller and eat mainly fruit, tree gums, and insects. Their brains have only a few fissures (Fig. 44.7). Their visual system contains several features that are common to all primates that have been examined, including a distinctive type of lamination in the lateral geniculate nucleus of the visual thalamus (with magnocellular and parvocellular layers) and a pulvinar complex with the distinct inferior and lateral pulvinar divisions. As in other primates, the superior colliculus of the midbrain represents only the contralateral half of the visual field, whereas other mammals have a more extensive representation that includes the complete visual field of the contralateral eye. The visual

cortex includes areas V1 and V2, both retained from nonprimate ancestors but with primate modifications including "blobs," which are cytochrome oxidase-rich modules in V1, and band-like modules that span the width of V2.

Other visual areas apparently shared with all other primates include V3, a dorsomedial area (DM), and a middle temporal visual area (MT), all of which receive direct inputs from V1. Other areas, such as the dorsolateral area (DL or V4), also are shared widely among primates, but not enough is known about visual cortex organization in the various primate species to be certain of how many of the more than 30 proposed visual areas are common among primates. Some or most of these areas, such as MT, are not found in nonprimates, or at least not in a primate-like form, and these areas are distinctive features of primate brains (Kaas, 2007b; Preuss, 2007).

Organization of the auditory system of galagos and other prosimians is not well understood, but they appear to share two primary or primary-like areas, A1 and R ("rostral"), with other primates, and they likely share several nonprimary areas as well. A1 and possibly R are likely to have been retained from nonprimate ancestors. The somatosensory cortex appears to be relatively unchanged from nonprimate ancestors, as an S1, S2, and PV have been identified in galagos, and there are additional, narrow belts of somatosensory cortex along the rostral and caudal borders of S1. S2 and PV retain the primitive feature of being activated directly by inputs from the ventroposterior nucleus of the thalamus (Fig. 44.8). Motor cortex organization is surprisingly advanced, with the five or more premotor areas that are also found in anthropoid primates.

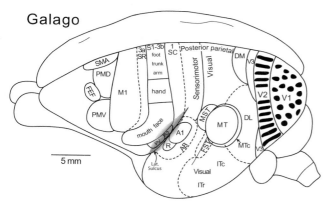

Galago

FIGURE 44.7 A lateral view of the brain of a prosimian primate, *Galago garnetti*, showing some of the proposed visual, somatosensory, auditory, and motor areas. Visual areas include the primary (V1) and secondary (V2) areas, common to most mammals, but with the modular subdivisions (blobs in V1; bands in V2) characteristic of primates. As in other primates, galagos have a third visual area (V3), a dorsomedial area (DM), a middle temporal area (MT), a dorsolateral area (DL), a fundal-sucal-temporal area (FST), an inferior temporal visual region (IT) with subdivisions. Posterior parietal cortex (PP) contains a caudal division with visual inputs and a rostral division with sensorimotor functions. The auditory cortex includes a primary field (A1), a rostral area (R), and an auditory belt (AB), which includes several areas and regions of the parabelt auditory cortex. The somatosensory cortex includes a primary (S1 or 3b), a parietal ventral area (PV), a secondary area (S2), a somatosensory rostral belt (SR or 3a), and a somatosensory caudal belt (SC or possibly area 1 or areas 1 plus 2). Motor areas include a primary area (M1), ventral (PMV) and dorsal (PMD) premotor areas, a supplementary motor area (SMA), a frontal eye field (FEF), and other motor areas on the medial wall of the cerebral hemisphere. Modified from Kaas (2007b).

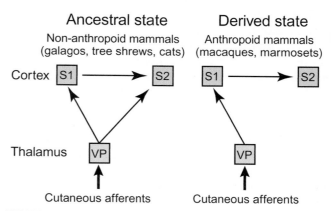

FIGURE 44.8 Somatosensory processing in prosimian primates and anthropoid primates. The processing in anthropoids is serial, rather than parallel and serial. Because the prosimian type also is found in a number of nonprimate mammals, we infer that this is the ancestral state. The ventroposterior nucleus (VP) of the somatosensory thalamus and the first (S1) and second (S2) somatosensory areas of the cortex are shown.

Although the brains of prosimians are not especially large, they share a number of primate-specific areas with anthropoids over and above the complement inherited from their mammalian ancestors. Overall, prosimians and other primates share a large number of brain specializations that likely evolved in early primates or their immediate ancestors (Preuss, 2007). The study of prosimians is critical for understanding that features of brain organization are distinctive of primates.

Brains of Early Anthropoids

The anatomy of early anthropoids suggests that an emphasis on high-acuity, diurnal vision and reduced emphasis on smell were important in their evolution. The endocasts of the skulls of early anthropoids indicate that the visual cortex of the temporal lobe was more expansive than in prosimians.

Collectively, present-day New World and Old World monkeys cover a wide range of body and brain sizes. The extent to which fissures develop on the brain surface of these primates depends largely on brain size. All primates, no matter how small, have a deep calcarine fissure in the occipital lobe. Most also have a deep lateral sulcus (Sylvian fissure). These are the only well-developed sulci in marmosets, the smallest of the New World monkeys. Slightly larger New World monkeys, such as owl monkeys and squirrel monkeys, also have a shallow central sulcus and a shallow superior temporal sulcus. The largest New World monkeys, such as spider monkeys and cebus monkeys (capuchins), have even more sulci and resemble (at least superficially) the well-fissured brains of Old World monkeys, such as macaques and baboons.

Both Old World and New World monkeys have all the areas of the neocortex described for prosimians, as well as some additional areas. Most notably, the somatosensory cortex has been altered so that areas 3a, 3b, 1, and 2 are well differentiated from each other, with each area corresponding to a separate representation of receptors of the contralateral body surface (Kaas, 2007b). Within S1 (area 3b), the proportional representation of body parts varies somewhat across species, so that in some monkeys as much as half of the area is devoted to the face and oral cavity. The hand is also prominently represented, especially in monkeys such as macaques. Some New World monkeys, such as spider monkeys, have a large representation of their highly tactile, prehensile tail. Marmosets appear to have a relatively poorly differentiated somatosensory region, which lacks a definite area 2, a field that is responsive to tactile stimuli and muscle movements in other monkeys. This may be a consequence of brain size reduction in marmosets, which have evolved smaller brains and bodies from larger ancestors.

In all anthropoids, S2 and PV appear to have lost activating inputs from the ventroposterior nucleus of the somatosensory thalamus, and they depend on inputs from areas 3a, 3b, 1, and 2 (Fig. 44.8). S2 and PV receive modulatory inputs from the ventroposterior inferior nucleus. Thus, processing in the somatosensory system became more serial than parallel with the advent of monkeys. Other differences likely exist between anthropoids and prosimians in the somatosensory portions of the lateral sulcus and posterior parietal cortex, but more research is needed. Both of these regions of somatosensory processing have expanded greatly in anthropoids compared to prosimians, and several areas involved in visually guided reaching have been described in macaque monkeys.

Currently, there is much interest in how the visual system of monkeys is subdivided into cortical areas and how these visual areas function in behavior. Elaborate proposals have been presented, but considerable uncertainty remains. In Old World monkeys, over 30 visual areas have been proposed, and it seems likely that anthropoids in general have more visual areas than prosimians, although the full extent of this difference is not yet clear. In the auditory cortex, a core of three primary-like areas, a belt of seven to eight additional fields, and a "parabelt" of several additional areas have been proposed, but not enough is known about auditory cortex of prosimians to know what differences exist. The proposed subdivisions of the motor cortex in prosimians and simians are quite similar, although some of the premotor areas of prosimians have been subdivided into two or three areas in Old World monkeys. Finally, comparisons of cortical connections and architectonics in galagos and macaque monkeys have led to the conclusion that macaques have several areas of dorsolateral prefrontal cortex in addition to those found in galagos.

Evolution of Hominin Brains

In trying to determine the recent course of the evolution of the human brain, we depend more on the fossil record than on a comparative approach, as we are the only surviving hominin species. Human brains are much bigger than those of our closest living relatives, the African apes. From the fossil record (Fig. 44.9), we can see that early australopithecines had brains that were only 10–25% larger than the brains of present day African apes when body size is taken into account. However, brains increased rapidly in size as the various species of *Homo* evolved over the last 2 million years. Early hominins had brains in the 600- to

Brain size (in cm³) plotted against time (Myr) for specimens attributed to Hominidae

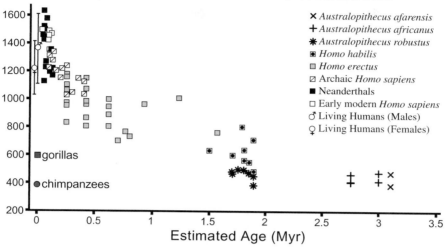

FIGURE 44.9 Evidence for the rapid growth of brains of hominins over the last 2 million years. The brain sizes of modern chimpanzees and gorillas have been added for comparison. Modified from McHenry (1994).

800-cc range; *H. erectus*, about 500,000 years ago, had brain volumes close to 1000 cc; and soon thereafter brains reached the volumes within the range of modern *H. sapiens*, which averages about 1400 cc.

In deducing the changes in internal organization that likely occurred over this remarkable increase in brain size, it would be useful to know more about the organization of the brains of the living apes. However, only limited information is available from early motor and somatosensory mapping experiments on apes, and further noninvasive studies, as in humans, are needed. Much can be learned by studying the histology of tissue from brains of apes that have died natural deaths using modern histochemical techniques (Preuss, 2001), but at the present time, we have only a limited understanding of how the brains of apes differ from those of monkeys. Traditionally, it was postulated that the higher-order "association" regions of the frontal, parietal, and temporal lobes underwent great expansion in human evolution. This view recently has been called into question by the finding that the relative proportions of the frontal, parietal, and temporal lobes are similar in great apes and humans, notwithstanding the much larger absolute size of the human brain. The fact remains, however, that the primary sensory cortical and thalamic structures are of roughly the same absolute size in humans and great apes, whereas the association regions of humans are much larger in absolute terms. This is consistent with classical accounts of human brain evolution that emphasize the expansion of higher-order regions. Changes in human evolution did not only involve an increase in brain size. Recent

comparative studies have documented human specializations at the level of cellular and modular organization as well. At the present time, however, we do not know enough about the specializations of human brain organization to understand how brain changes are related to human specializations of cognitive organization, such as the capacity for conceptualizing the mental states of other individuals ("theory of mind"; Povinelli and Preuss, 1995).

One of the signature specializations of *Homo sapiens* is of course language, and it often has been assumed that the evolution of language involved the evolution of specialized language areas in the brain, such as Wernicke's area in the temporoparietal cortex and Broca's area in the frontal lobe. Human brains are not symmetrical in shape, so that the planum temporale, the sheet of cortex on the lower surface of the lateral sulcus (the upper face of the temporal lobe), is usually larger in the left cerebral hemisphere than on the right. As the left hemisphere usually becomes dominant for language, the larger size of the left planum temporale has long been assumed to be related to language. Interestingly, however, great apes exhibit the same asymmetry of the temporal lobe as humans, despite their lack of language. When and how language emerged in hominins remain issues of much speculation, as does the nature of the changes in brain organization that support language (Deacon, 2007). What seems likely is that previously existing brain regions used for other, nonlinguistic functions in nonhuman primates, such as the ventral premotor area and dorsal-stream auditory areas, became specialized for language in the ancestors of humans, ultimately acquiring the functions of Broca's and Wernicke's areas. Presumably, the evolution of language involved changes in the internal organizations of these areas (see, for example, Buxhoeveden *et al.*, 2001).

WHY BRAIN SIZE IS IMPORTANT

A general assumption is that larger brains are better because they can do more. This does not imply that brains of the same size (or same size in proportion to the body) do the same things, because brain organization is modified for different functions. Nevertheless, larger brains do have obvious design problems that are likely to have been solved in similar ways in the different lines of evolution. To some extent, neurons usually enlarge as brains get bigger, but dendrites and axons cannot enlarge much without compromising their functions. To maintain passive cable conduction in dendrites, Bekkers and Stevens (1970) calculated that dendrites would need to increase four times in diameter when they double in length. Similarly, when axons double in length, axon diameter must also double, to maintain conduction times. As brain size increases, some dendrites and axons do become longer and enlarge disproportionately. Given the problems associated with increasing neuron size, the major way of increasing brain size is to increase the number of neurons. This introduces a related problem as the number of neurons increases: it becomes more and more difficult to maintain each neuron's connections with the same percentage of other neurons, since the required number of connections grows much faster than the number of neurons. Cell body densities typically decrease and cortical thickness increases, but the increase in connections is not nearly enough to maintain ratios of connectivity or to fully compensate for longer connection distances. As a result, larger brains generally devote much more of their mass to connections.

There are two major ways in which larger brains can be modified to reduce the design problems produced by larger distances and more neurons. First, the brain can become more modular, so that most connections of individual neurons are with neighboring, rather than more distant, neurons. This can be done, for example, by increasing the number of processing areas so that areas are usually smaller. In addition, areas can be subdivided into smaller functional divisions (columns or modules), which likewise reduce the need for long connections. Thus, larger brains are likely to have larger numbers of cortical areas, and larger areas

are likely to contain several types of modules. Second, connections that require long, thick axons should be reduced as much as possible. This can be done by grouping functionally related areas together so that necessary connections between areas are shorter. An effective way of reducing the need for long connections, as brains get bigger, is to increase the degree of specialization of each cerebral hemisphere. As regions and areas of each hemisphere become differently specialized, the major cortical connections become more focused within the same hemisphere, rather than between hemispheres. As a result, the size of the corpus callosum does not increase proportionately with the size of the brain. In addition, proportionately fewer of the callosal axons are of the larger diameters that would be needed to preserve fast conduction times. Therefore, large brains should be less symmetrical than small brains. The large human brain appears to be extreme in this respect.

Another issue is that large cortical areas are unlikely to function in the same manner as small cortical areas. Unless neurons compensate with larger dendrites and intrinsic connections as areas get larger, the computational window or scope of neurons will decrease. For example, as a visual area gets bigger, individual neurons would evaluate information from less and less of the total visual field (Fig. 44.10). This implies that as areas get bigger, their neurons become less capable of global center-surround comparisons and more devoted to local center-surround comparisons. Thus, some of the integrative functions of large areas must be displaced to smaller areas. It is also apparent that changes in the sizes of dendritic arbors and the lengths of intrinsic axons in smaller areas would have more impact on the sizes of computational windows of neurons. Comparable changes in dendrites and axons would enlarge or reduce receptive field sizes more in a small than a large visual area (Fig. 44.10). Because their functions are more modifiable by small structural modifications, smaller areas may be specialized more easily for different functions. Recent measurements suggest that neurons in large areas typically do not have longer dendritic arbors and larger intrinsic connections, and indeed, they may have smaller dendritic arbors. In addition, although primary sensory areas are often larger in larger brains, they are not enlarged in proportion to the rest of cortex (Fig. 44.11). Thus, the primary visual cortex is less than three times larger in human brains than in the brains of macaque monkeys, whereas the neocortex as a whole is over 10 times larger (Fig. 44.10). In humans, primary visual cortex is about the same size as in the much smaller chimpanzee brain. The general lack of proportional growth of cortical areas with brain size reduces the

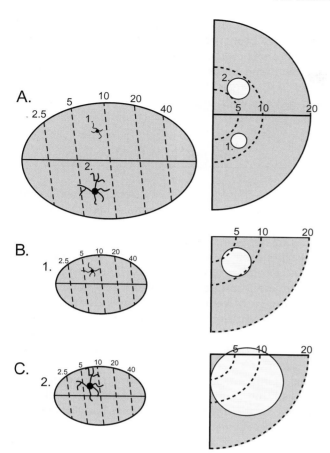

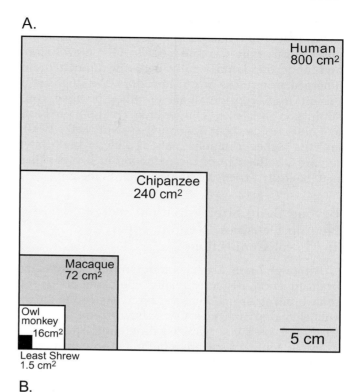

FIGURE 44.10 The effects of varying the horizontal spread of dendritic arbors of neurons in large (A) and small (B and C) visual areas. An increase in arbor size (1 to 2) in a large area (A) produces little change in receptive field size (circles 1 to 2 in the central 20° of the visual hemifield on the right), whereas such a change (B to C) in a small area changes the scope of the receptive field greatly. Thus, the functions of small areas are changed more dramatically by small morphological adjustments. Surface view schematics of retinotopically organized visual areas are on the left, whereas schematics of receptive fields in the visual hemifield and the lower visual quadrant are on the right. From Kaas (2000).

impact of the changing of functions of areas with size, and reflects the addition of other smaller cortical areas.

Other size-related constraints relate to mechanisms of development. We often assume that natural selection can act independently on each brain trait, but this is unlikely to be the case. Instead, selection may operate on a few developmental mechanisms, such as those that control the number of neurons or the extent to which correlated activity levels maintain functional connections. Along this line of reasoning, Finlay and Darlington (1995) have provided evidence that late-developing brain structures such as neocortex have enlarged disproportionately in mammals with larger brains ("late makes large"). As we learn more about

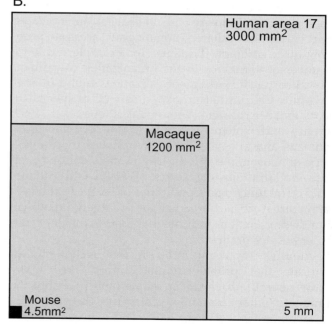

FIGURE 44.11 Species differences in (B) the surface area of the neocortex and (A) of the primary visual cortex in one cerebral hemisphere. The neocortex of humans is over 500 times larger in surface area and over twice as thick as the neocortex in the smallest mammals that resembled those leading to the first primates and over three times the surface area of our closest living relatives, the chimpanzees. Some of the areas of the brain are also larger in humans, but not to the extent expected from the great enlargement of the neocortex. From Kaas (2000).

the genetics and mechanisms of development and how development is modified in evolution, we should be able to form more accurate models of brain evolution, and understand more fully how the human brain emerged from those of our ancestors. We might also benefit from considerations of other possible constraints on brain size. A larger brain creates more heat, and thus needs a better cooling system (Falk, 1990), and the higher metabolic costs of a larger brain may require a better diet or the reduction in size of other metabolically expensive tissues, such as the gut.

Beyond Brain Size: Variations in Neuron Densities, Neuron Types, and Local Connections

Although brain size has important implications for brain function, distantly related species of about the same brain size sometimes appear to have quite different behavioral capacities. One reason, of course, is that the brains may be organized differently; sharing for example, some cortical areas but not others, or specializing shared cortical areas in different ways. Although more comparative studies are needed, it seems unlikely that large rodents with brains as large as those of some monkeys have as many cortical areas as the monkeys. Specializations of shared (homologous) areas are probably quite common. Primates, for example, exhibit a number of variations in the organization of primary visual cortex, including specializations found in apes and humans, but not monkeys, as well as specializations that are restricted to humans (Preuss *et al.*, 1999; Preuss and Coleman, 2002). At the cellular level, humans and apes are the only primates that have a type of neuron, the spindle cell, in the anterior cingulate and frontoinsular cortex (Watson and Allman, 2007). The functional significance of such variations in brain structure and organization is largely unknown, but these variations indicate that there is much to consider besides brain size.

Another important factor is that recent studies indicate that primate brains simply have many more neurons than rodent brains (and probably the brains of other taxanomic groups) of the same size (Herculano-Houzel *et al.*, 2007). This is because as the brains of rodents increase in size from small to large species, the density of neurons in the brain decreases, and so, the gain in neuronal numbers is not as great as one might expect. However, in primates, the neuronal density does not change very much with brain size so that more neurons are gained with each increase in brain size. If the same scaling rules relating numbers of neurons to brain size in rodents applied to primates, a human brain with about 100 billion neurons would have to weigh over 45 kg., and be supported by a body the size of a blue whale. The greater neuronal densities in the brains of larger primates would seem to present an advantage in processing capacity that gives primates with larger brains a considerable advantage over rodents with large brains.

CONCLUSIONS

Based on comparative studies and the fossil record, we conclude that early mammals had small brains with little neocortex and few functional subdivisions (areas or fields) of cortex. Vision was emphasized in the early primates, and the visual cortex in the temporal and occipital lobes enlarged. These primates also had several unique features of the visual system, including new visual areas such as MT, distinctive kinds of modules in V1 (blobs) and V2 (bands), separate magnocellular and parvocellular layers in the lateral geniculate nucleus, and a representation in the superior colliculus restricted to the contralateral visual hemifield. Several premotor areas were present, whereas the somatosensory system was relatively primitive. Later anthropoid primates had larger brains, more neocortex, and more areas of neocortex. The somatosensory cortex had expanded and included the four strip-like areas on the anterior parietal lobe, areas 3a, 3b, 1, and 2. We know little about possible specializations of the brains of apes. However, over the last 6 million years of evolution in the human lineage, brains increased three to four times in size, due mainly to the enlargement of the cortex. Although the relative proportions of the different cortical lobes remain similar in humans and apes, the available evidence suggests that the cortex did not expand uniformly in human evolution: the association regions of frontal, temporal, and parietal lobes are much larger in absolute terms in humans than in apes, whereas human primary sensory areas are about the same absolute size as those of apes. The expansion of nonprimary cortex was probably accompanied by a further increase in the number of cortical areas, modifications leading to functional and anatomical asymmetries in the two cerebral hemispheres, specializations for language and cognition, and larger expanses of prefrontal, parietal, and temporal cortex. The larger brains had many more neurons with a greater proportion of tissue devoted to connections relative to cell bodies, and presumably had a higher ratio of local connections to long-distance connections.

Further progress in understanding the course of the evolution of the human brain can be achieved with current methods of investigation. We have the opportunity to learn much about the similarities and differences among the brains of various primates. Neuroscientists have generally concentrated on studies of brain features that are widely shared, but we need to know more about the differences in brain structure and function that make us distinctively human.

References

Beck, P. D., Pospichal, M. W., and Kaas, J. H. (1996). Topography, architecture, and connections of somatosensory cortex in opossums: Evidence for five somatosensory areas. *J. Comp. Neurol.* **366**, 109–133.

Bekkers, J. M. and Stevens, C. F. (1970). Two different ways evolution makes neurons larger. *Prog. Brain Res.* **83**, 37–45.

Brodmann, K. (1909). *"Vergleichende Lokalisationslehre der Grosshirnrhinde."* Barth, Leipzig. [Reprinted as Brodmann's "Localisation in the Cerebral cortex," translated and edited by L. J. Garey, Smith-Gordon, London, 1994.]

Buxhoeveden, D. P., Switala, A. E., Roy, E., Litaker, M. and Casanova, M. F. (2001). Morphological differences between minicolumns in human and nonhuman primate cortex. *Am. J. Phys. Anthropol.* **115**, 361–371.

Collins, C. E., Hendrickson, A., and Kaas, J. H. (2005). Overview of the visual system of Tarsius. *Anat. Rec.* **287A**, 1013–1025.

Darwin, C. (1859). "On the Origin of Species." John Murray, London. [Facsimile of the first edition: Harvard Univ. Press, Cambridge, MA, 1984].

Deacon. (2007). The Evolution of Language Systems in the Human Brain. *In* "Evolution of Nervous Systems," Vol. 3 (J. H. Kaas and L. H. Krubitzer, eds.), pp. 529–547. Elsevier, London.

De Sousa, A., and Wood, B. (2007). The hominin fossil record and the emergence of the modern human central nervous system. *In* "Evolution of Nervous Systems," Vol. 4 (J. H. Kaas and T. M. Preuss, eds.), pp. 292–336. Elsevier, London.

Falk, D. (1990). Brain evolution in *Homo*: The "radiator" theory. *Behav. Brain Sci.* **13**, 339–381.

Finlay, B. L., and Brodsky, P. (2007). Cortical evolution as the expression of a program for disproportionate growth and the proliferation of areas. *In* "Evolution of Nervous Systems," Vol. 3 (J. H. Kaas and L. A. Krubitzer, eds.), pp. 73–96. Elsevier, London.

Finlay, B. L., and Darlington, R. B. (1995). Linked regularities in the development and evolution of mammalian brains. *Science* **268**, 1578–1584.

Hennig, W. (1966). "Phylogenetic Systematics." University of Illinois Press, Urbana.

Herculano-Houzel, S., Collins, C., Wong P., and Kaas, J. H. (2007). Cellular scaling rules for primate brains. *Proc. Natl. Acad. Sci. USA.* In press.

Jerison, H. J. (2007). What fossils tell us about the evolution of the neocortex. *In* "Evolution of Nervous Systems," Vol. 3 (J. H. Kaas and L. A. Krubitzer, eds.), pp. 1–12. Elsevier, London.

Kaas, J. H. (2000). Why is brain size so important: Design problems and solutions as neocortex gets bigger or smaller. *Brain Mind* **1**, 7–23.

Kaas, J. H. (2007a). Reconstructing the organization of neocortex of the first mammals and subsequent modifications. *In* "Evolution of Nervous Systems," Vol. 3 (J. H. Kaas and L. A. Krubitzer, eds.), pp. 27–48. Elsevier, London.

Kaas, J. H. (2007b). The evolution of sensory and motor systems in primates. *In* "Evolution of Nervous Systems," Vol. 4 (J. H. Kaas and T. M. Preuss, eds.), pp. 35–57. Elsevier, London.

McHenry, H. M. (1994). Tempo and mode in human evolution. *Proc. Natl. Acad. Sci. USA* **91**, 6780–6786.

Povinelli, D. J. and Preuss, T. M. (1995). Theory of mind: Evolutionary history of a cognitive specialization. *Trends Neurosci.* **18**, 418–424.

Preuss, T. M. (2001). The discovery of cerebral diversity: An unwelcome scientific revolution. *In* "Evolutionary Anatomy of the Primate Cerebral Cortex" (D. Falk and K. R. Gibson, eds.), pp. 138–164.

Preuss, T. M. (2007). Primate brain evolution in phylogenetic context. *In* "Evolution of Nervous Systems," Vol. 4 (J. H. Kaas and T. M. Preuss, eds.), pp. 1–34. Elsevier, London.

Preuss, T. M., and Coleman, G. Q. (2002). Human specific organization of primary visual cortex: alternating compartments of dense Cat-301 and calbindin immunoreactivity in layer 4A. *Cereb. Cortex,* **12**, 671–691.

Preuss, T. M., Qi, H., and Kaas, J. H. (1999). Distinctive compartmental organization of human primary visual cortex. *Proc. Natl. Acad. Sci. USA* **96**, 11601–11606.

Ross, C. (1996). Adaptive explanation for the origins of the Anthropoidea (Primates). *Am. J. Primatol.* **40**, 205–230.

Springer, M. S., and deJong, W. W. (2001). Which mammalian supertree to bark up. *Science* **291**, 1709–1711.

Striedter, G. F. (1998). Progress in the study of brain evolution: From speculative theories to testable hypotheses. *Anat. Rec. (New Anat.)* **253**, 105–112.

Van Essen, D. C. (1997). A tension-based theory of morphogenesis and compact wiring in the central nervous system. *Nature* **385**, 313–318.

Van Essen, D. C. (2007). Cerebral cortical folding patterns in primates: Why they vary and what they signify. *In* "Evolution of Nervous Systems," Vol. 4 (J. H. Kaas and T. M. Preuss, eds.), pp. 267–289. Elsevier, London.

Watson, K. K., and Allman, J. M. (2007). Role of spindle cells in the social cognition of apes and humans. *In* "Evolution of Nervous Systems," Vol. 4 (J. H. Kaas and T. M. Preuss, eds.), pp. 479–484. Elsevier, London.

Suggested Readings

Allman, J. M. (1999). "Evolving Brains." Freeman, New York.

Butler, A. B., and Hodos, W. (1996). "Comparative Vertebrate Neuroanatomy." Wiley-Liss, New York.

Deacon, T. W. (1997). "The Symbolic Species: The Co-Evolution of Language and the Brain." Norton, New York.

Falk D., and Gibson, K. R. (2001). "Evolutionary Anatomy of the Primate Cerebral Cortex." Cambridge University Press, Cambridge.

Fleagle, J. G. (1999). "Primate Adaptation and Evolution," 2nd ed. Academic Press, San Diego.

Kaas, J. H. (2007). "Evolution of Nervous Systems," 4 Vols. (J. H. Kaas series ed.). Elsevier, London.

Northcutt, R. G. and Kaas, J. H. (1995). The emergence and evolution of mammalian neocortex. *Trends Neurosci.* **18**, 373–379.

Preuss, T. M., ed. (2000). "The Diversity of Cerebral Cortex." *Brain Behav. Evol.* **55**, 283–347.

Preuss, T. M. (2004). What is it like to be a human? *In* "The Cognitive Neurosciences," 3rd ed. (M. S. Gazzaniga, ed.), pp. 5–22. MIT Press, Cambridge, MA.

Striedter, G. F. (2005). "Principles of Brain Evolution." Sinauer Associates, Sunderland, MA.

Jon H. Kaas and Todd M. Preuss

45

Cognitive Development and Aging

The human brain has evolved over a very extensive period (Chapter 44), but an individual human brain develops rapidly over the first few years of life, allowing the nearly helpless human infant to gain control of locomotion, language, and thought. The story of brain development in the early years of infancy through adolescence occupies the first part of this chapter. This is followed by consideration of the cognitive changes that occur during this period of the life span and how these capacities are altered over the course of normal aging. The final parts of the chapter discuss pathological processes that (1) affect cognitive development in childhood and adolescence and (2) can cause devastating cognitive impairment in the elderly.

BRAIN DEVELOPMENT

Basic Concepts of Brain Development

Brain Maturation Progresses Well into Adolescence

Brain development is an organized, predetermined, and highly dynamic multistep process that continues beyond birth into the postnatal period, well into adolescence in humans. Most of our knowledge about human brain development has been obtained through analyses of postmortem specimens and, more recently, powerful neuroimaging techniques. These approaches have provided extensive information on how the size and shape of different brain regions change from conception through adulthood (Fig. 45.1). The brain grows rapidly in the first two years of life, by which time it has achieved 80% of its adult weight. Thereafter, brain growth is slower, reaching over 90% of average adult weight by the sixth year. The development of the ce-

rebral cortical surface proceeds gradually from the flat (lissencephalic) brain of the fetus to the adult gyral pattern at birth. Primary sulci (Sylvian fissure, and rolandic, parietal, and superior temporal sulci) and gyri (pre- and postcentral gyri, superior temporal and middle temporal gyri, superior and middle frontal gyri, and superior and inferior occipital gyri) become well defined between 26 and 28 weeks of gestation. Development of secondary and tertiary gyri occurs later in gestation, and in the last trimester the sulci become deeply enfolded.

Cortical development in most mammals is broadly divided into two phases. The first is a genetically determined sequence of events occurring *in utero* that can be modified by both the local environment within the fetal nervous system and the maternal environment. The second phase is one that occurs in the perinatal period when the connectivity of the cortex becomes sensitive to patterns of neural activity. Glial processes also critically participate in normal development, supporting cell proliferation, differentiation and guidance, and the formation of myelin sheets around axons. Brain structure size and developmental trajectories are highly variable and sexually dimorphic.

The Cortex Thickens during Development

Development of the cortical mantle in the human brain was documented extensively in the classic studies of Conel (1939–1963). Longitudinal neuroimaging studies in large samples of children (Lenroot and Giedd, 2006) have shown changes in volume of cortical gray matter that are nonlinear and regionally specific. Volume of gray matter increases during preadolescence with a maximum occurring around 10 to 12 years of age in the parietal lobe, 11 to 12 years of age in the frontal lobe, and 16 to 17 years of age in

1039

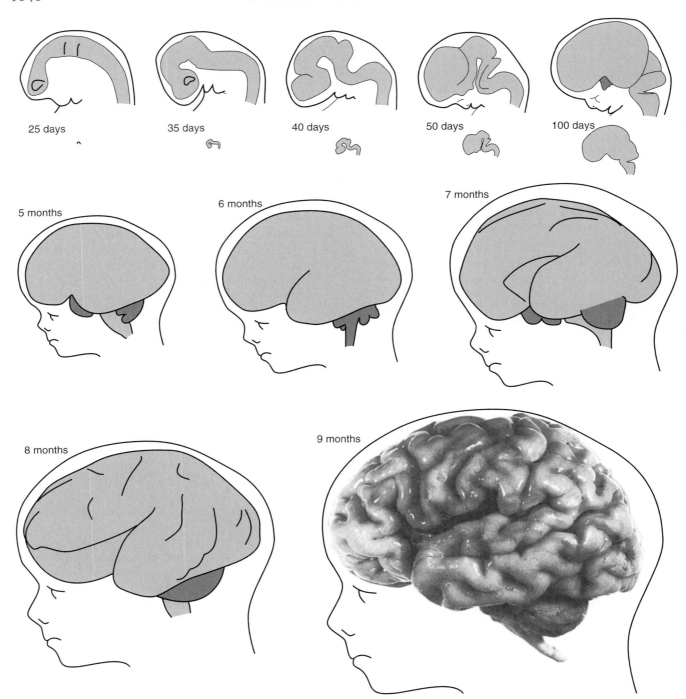

FIGURE 45.1 The size and form of the human brain as it develops through gestation and early infancy.

the temporal lobe, followed by a decline in post-adolescence. Although neuroimaging techniques lack the resolution to identify the specific structural modifications that underlie changes in cortical volume during development (i.e., cell number, synapse density, myelination), they do indicate that development of the human cerebral cortex differs in males and females, is highly dynamic over time, and varies across different cortical areas.

Dendritic Spine Number Increases during Development

The formation of dendritic spines and the time course of changes in the length and branching patterns of dendrites have been described for visual and frontal cortical areas in humans. Within the visual cortex, peak spine density is achieved around 5 months of age; this number then decreases until adult values are obtained around 21 months of age. Progressive elongation of dendrites occurs up to 24 months. Thus, during the period from 5 to 24 months, the major event may be a decrease in spine density rather than an actual decline in total spine number. Dendritic development in the visual system reaches mature levels in deep cortical layers earlier than in superficial layers, displaying the "inside-out" pattern of development that is characteristic of neurogenesis and migration (Chapters 15 and 16).

Development and maturation of the frontal cortex proceed more slowly. Whereas neuronal density in the primary visual cortex (V1) reaches adult levels by 4 to 5 months, neuronal density in the frontal cortex fails to reach adult levels even at 7 years of age. Additionally, by 2 years of age, dendritic length in frontal cortex (which is mature by 18–24 months in V1) is only half that found in adults. Left–right asymmetries exist in the dendritic branching patterns of pyramidal neurons within layer V of the inferior frontal and anterior precentral cortex. During the first year, growth is more advanced on the right side, but by 6 to 8 years of age, the maturation of distal dendrites on the left exceeds that of the right.

Synapses First Increase and Then Decrease in Number

As in the brains of other animals (Chapter 20), the immature human brain contains many more synapses than the mature brain (Huttenlocher, 1990). Within the primary visual cortex, synaptic density increases gradually during late gestation and early postnatal life, and then doubles from 2 to 4 months of age. After 1 year of age, however, there is a decline in synaptic density until adult values (50–60% of the maximum) are attained at about 11 years. The time course of the decrease in synaptic density varies across different cortical layers (Fig. 45.2). The decrease does not display the "inside-out" pattern of development; rather, there is a considerable decrease, over time, in the number of synapses in every layer. A postnatal increase and subsequent decline in synaptic density also occurs in the middle frontal gyrus (layer III), but these frontal cortex changes take place over a longer time course than in the primary visual cortex (Fig. 45.3). The maximum density of synapses occurs around 1 year of age (compared to 4 months in the visual cortex), and adult values are not reached until roughly 16 years of age (compared to 7–11 years for the visual cortex).

In summary, these quantitative anatomical results suggest that, in contrast to other animals (Chapter 19), programmed cell death plays only a limited role in

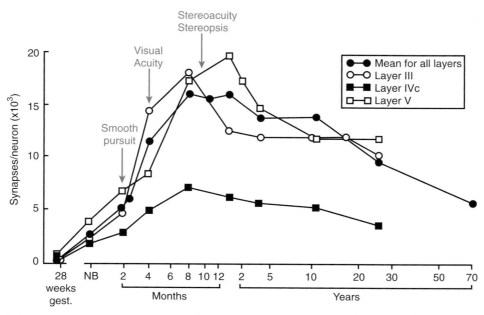

FIGURE 45.2 Variations in the density of synapses of different cortical layers of the human primary visual cortex during development. Arrows point to the emergence of various visual functions in relation to the increase in synapses in the visual cortex. NB, newborn. Adapted with permission from Huttenlocher and De Courten (1987).

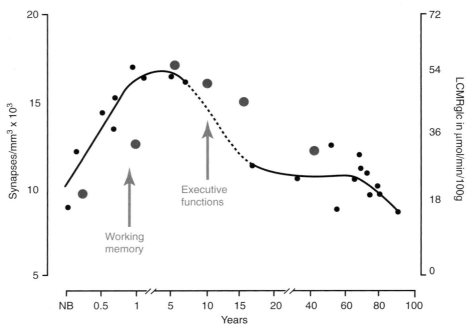

FIGURE 45.3 Variations in the human prefrontal cortex during development: Density of synapses in layer III of the medial frontal gyrus (black circles) and resting glucose uptake (LCMRglc) in the frontal cortex (blue circles). Arrows point to the approximate periods of emergence of various prefrontal cortex functions. NB, newborn. Adapted with permission from Huttenlocker (1990) and Chugani *et al.* (1993).

human brain development. Synapse elimination in humans and other primates exceeds that observed in many species, however, suggesting that modifications in synaptic connectivity may be particularly important for the development of complex nervous systems. Current findings are consistent with a role for incoming afferent input in selectively stabilizing functional synapses and in eliminating or suppressing inactive contacts. The exuberant connectivity observed during development may also support compensatory processes in cases of early brain trauma, allowing surviving synapses to assume functions normally subserved by damaged inputs.

Postnatal Refinement of Synaptic Contacts Occurs in the Neostriatum and Hippocampus as Well

Although the morphological development of certain subcortical structures, such as the thalamus and cerebellum, is nearly complete at birth, other structures continue to mature into early postnatal life. In the neostriatum, synaptic density continues to increase until the end of the first postnatal month, and changes in neuronal and neuropil morphology are observed until 2 to 4 months postnatally. Refinement of synaptic contacts proceeds until the end of the first postnatal year when the striatal neurochemical mosaic attains adult characteristics. In the hippocampus, neurons have reached their destination at birth in all CA fields, but immature cells continue to accrue within the

dentate gyrus throughout the first year of life (Seress, 2001). In addition, continued dendritic remodeling in the hippocampus has been observed as late as the fifth postnatal year in humans (Seress, 2001). Myelination in the subicular and presubicular regions, a key relay zone between the hippocampus and many cortical areas, continues through adolescence and adulthood (Benes *et al.*, 1994) with female subjects showing a greater extent of myelination between the ages of 6 and 29 than males. These changes are accompanied by corresponding changes in the volume of the hippocampus during early development (Giedd *et al.*, 1996).

White Matter Increases throughout Childhood and Adolescence

In contrast to the inverted U shape of gray matter development, white matter maturation in the brain continues long after the gray matter has reached its definitive volume (Yakovlev and Lecours, 1967; Lenroot and Giedd, 2006). In general, myelination progresses from caudal to cephalic. Thus, the brain stem myelinates prior to the cerebellum and basal ganglia. Similarly, the cerebellum and basal ganglia myelinate prior to the cerebral hemispheres. Sensory and motor systems display mature myelination within the first two years of life. However, nonspecific thalamic radiations do not reach mature levels until 5 to 7 years of age, and intracortical fibers continue their myelination processes well into the third decade of life. Numerous morphological studies demonstrate that the corpus callosum is uniformly thin in the first month postnatally. The genu and then the splenium undergo rapid growth spurts during the next five months, resulting in their characteristic "bulbous" appearance.

Cerebral Metabolic Rate Increases and Then Decreases after Birth

Increases in neuronal activity have been linked to increases in cerebral metabolism, as measured by neuroimaging techniques. Data show substantial subcortical activation in newborns, notably in the thalamus and phylogenetically old portions of the cerebellum, but little activation of the cerebral cortex. However, over the first 3 to 4 years of life, cortical metabolic rate increases until it reaches levels twice those observed in adults. After 4 years of age, metabolic activity declines gradually to adult levels at around age 15. The time course of the rise and decline of activity, as revealed by positron emission tomography (PET), varies across cortical regions but parallels the rise and decline in the number of synapses in these areas, suggesting that the exuberant cortical synapses observed during development are metabolically active. As these synapses are eliminated, metabolic activity decreases.

Dorsolateral Prefrontal Cortex Is One of the Last Brain Regions to Develop

Refinement in the circuit organization of the dorsolateral prefrontal cortex, like that of the hippocampal formation (see earlier discussion), continues until quite late in development. In the case of the dorsolateral prefrontal cortex, a gradual decline in synaptic density continues well into late adolescence and suggests that connectivity in this cortical region undergoes substantial change until 15 to 20 years of age. These changes are accompanied by a decrease in metabolic rate during the same period (Fig. 45.3). It is believed that changes in the neuromodulatory effects of dopamine may influence the adolescent refinement of excitatory and inhibitory inputs to layer III pyramidal neurons and that dopamine may have a particularly strong influence on cortical information processing around the time of puberty (for the relevance of this late prefrontal cortex development in the emergence of psychopathologic states, such as schizophrenia, see the section on pathological processes).

Summary

Structural, metabolic, and physiological indices of human brain development all point to a long postnatal time course that displays considerable variability from region to region and from system to system. Thus, whereas postnatal development in the hippocampus proceeds until approximately 4 to 5 years of age, it continues until 7 to 11 years for the visual cortex and until around 16 years of age for the frontal cortical areas. The dorsolateral portion of the prefrontal cortex appears to be one of the last regions of the brain to develop. Interestingly, the development of myelin has been observed as late as the third to fourth decade of life for fibers within the association areas of the neocortex. Finally, brain structure size and developmental trajectories are sexually dimorphic.

COGNITIVE DEVELOPMENT AND AGING: A LIFE SPAN PERSPECTIVE

Investigators are now attempting to link the types of neural changes described in the preceding section to the maturation of sensory, behavioral, and cognitive capacities in infants and children. Such studies are limited in number. Until recently, the predominant view was that postnatal development of the cortex is largely intrinsically determined. However, new studies of both animals and humans have revealed a central role for extrinsic factors in shaping the organization of neural systems and in permitting recovery from brain damage. Indeed, calculations showing that information in the genome is not sufficient to specify the connectivity of the brain, together with evidence for the long-lasting existence of transient, redundant connections in primates, suggest that neural changes under the influence of environmental input play a significant and persistent role in the development of functional specificity in the human brain.

Cognitive Development in the First Years of Life

Early Subcortical Behavioral Responses Are Inhibited by Cortical Development

The increasing metabolic rate observed in basal ganglia and many cortical regions during the second and third postnatal months is correlated with several organizational changes that occur in the infant brain during this period. Thus, the frequency of several subcortical reflexes (e.g., the tonic neck reflex, the grasp reflex, as well as the frequency of endogeneous smiles and spontaneous crying) decreases, presumably as a consequence of increased cortical drive on the brain stem centers that control these reflexes. A similar mechanism may underlie the observation that the risk of sudden infant death syndrome peaks during this postnatal interval. Specifically, this syndrome is thought to be associated with a developmentally regulated increase in cortical inhibition of the brain stem centers that control normal respiration.

Anatomical Development of the Visual System in Infants Is Linked to Functional Development

Several parallels have been noted between anatomical changes and the emergence of visual system function in humans. The visual abilities of the newborn have been linked to subcortical structures that show mature anatomy and high metabolic rates during this period. The limited visual abilities of the newborn are augmented by the appearance of smooth pursuit tracking around 6 to 8 weeks. The emergence of this capacity may be related to the maturation of cortical layers IV, V, and VI in V1 (Fig. 45.2). These layers connect magnocellular afferents from the lateral geniculate nucleus and the middle temporal (MT) region, a pathway important for the detection of visual movement. Visual acuity and visual alertness increase dramatically around 4 months of age, when the volume of visual cortex reaches adult levels and the highest density of synapses is present. Shortly thereafter, cortically mediated binocular interactions become apparent. These interactions, which include stereoacuity, binocular summation of the light reflex, and stereopsis (binocular depth perception), appear in the same time frame as maturation of the middle cortical layers and rapid synaptogenesis in V1. This period of rapid growth appears to be a time of increased vulnerability to altered afferent input. Strabismic amblyopia and amblyopia due to the absence of patterned input (i.e., centrally mediated visual impairments) are reported to occur at this age unless corrected early.

The time period during which visual impairments can be corrected is different for different visual functions. For example, correction for an absent lens (aphakia) due to cataracts may be completely effective only when performed prior to 2 months of age (i.e., just prior to the onset of exuberant synaptogenesis in V1). However, very high synaptic density persists to at least the age of 4 years in V1. The presence of this extensive connectivity may account for the ability to recover from amblyopia (with forced use of the strabismic eye) during this time. If the disruption of binocular convergence by strabismus is not corrected within the first year of life, the ability to see stereoscopically may not develop even though the development of acuity and contrast sensitivity is normal. Thus, there may be separate critical periods (Chapter 22) for the development of resolution acuity and stereopsis. Because the parvocellular system is thought to under-

lie acuity, whereas disparity detection is mediated by the magnocellular system, the magnocellular system may be more modifiable by environmental input than the parvocellular system. This differential sensitivity may be due to the slower maturation of the magnocellular system and/or to differences in the number of exuberant synapses within these systems.

Studies of individuals born deaf and blind suggest that in humans, as in other animals, there is a time period when cortical areas that normally process information from the deprived modality can be reorganized to process information from intact modalities. From the alteration of visual functions seen in individuals born deaf, it appears that areas of the auditory cortex have been recruited for visual function. Deaf subjects, by comparison, show abnormally strong electrical responses to peripheral visual stimuli. These responses are not found in normally hearing individuals, even those exposed to sign language by deaf parents. The prolonged persistence of exuberant cortico-cortico connections may provide the substrate for such cortical reorganization.

Development of the Ventral Visual Stream and Object Representation

A multitude of visual cortical areas located in the striate and extrastriate cortex contribute to different aspects of visual processing (Chapter 46). These visual areas have been divided into two visual processing systems, one coursing ventrally from the striate and extrastriate areas to the ventrolateral temporal cortex and the other coursing dorsally from the striate and extrastriate cortex to the parietal cortex. The dorsal stream mediates spatial processing associated with attention to movement and location, whereas the ventral stream is involved primarily in processing information about patterns and objects. Both streams project rostrally, each reaching common and adjacent areas of the prefrontal cortex.

Major functions of the ventral stream include the ability both to segment a pattern into a set of constituent parts (local level of processing) and to integrate those parts into a coherent whole (global level of processing). In adults, systematic differences exist in the distribution of global and local levels of processing within the brain, such that global levels of processing are associated with the right posterior temporal region and local levels of processing with the left posterior temporal regions. Around 4 months of age, infants show hemispheric differences for global and local processing that are similar to adults. Thus, at this age, children segment out well-formed units and use simple associations to organize these units into a configuration. With further maturation, children show changes in the way they decompose objects into parts and in the relations they use to organize the parts. These changes may reflect the protracted development of the posterior temporal cortical areas, a finding consistent with ERP studies in human infants indicating functional activation within the temporal cortical areas around 6 months of age. The latter observation parallels the emergence, around the same age, of object representation and visual rule learning (Alvarado and Bachevalier, 2000).

Development of the Dorsal Visual Stream and Spatial Attention

A variety of spatial processes have been associated with activation of the dorsal visual pathway, such as the processing of information about spatial attention and spatial location (Chapter 47). Considerable clinical and experimental evidence shows that the posterior parietal lobe plays a crucial role in the ability to shift attention to different spatial locations. By at least 6 months of age, infants, like adults, show facilitated stimulus detection as a function of prior cueing; that is, they take less time to detect an object if it is presented at a location that has been cued previously. They also demonstrate, as do adults, quicker target detection when the interval between the cue and target is short than when this interval is long. Even though this basic attentional response may be robust as early as 4 months of age, the timing parameters that elicit the response change with development, suggesting that further maturational processes occur in the parietal cortex.

Precursors of Declarative Memory Emerge Early in Development

Looking preference has been used to study the emergence of recognition memory ability in both human infants and monkeys (Bachevalier and Vargha-Khadem, 2005). In both species, the most common procedure is to first present the infant with an attractive stimulus for a period of time (generally 20 to 30 s). After an intervening delay that can vary from a few seconds to many hours, the familiar stimulus is shown together with a new item. Human infants as young as 1 day old display a strong preference for looking at the novel object after delays as long as 24 h, demonstrating that they recognize the previously presented item as familiar. The early emergence of recognition memory is also observed in infant monkeys and, in this species, surgical removal of the hippocampal formation and adjacent tissue before the age of 3 weeks disrupts this ability. The capacity for recall over long periods of time, as measured by nonverbal deferred imitation

tasks, also emerges by the end of first year in humans and is impaired after early hippocampal injury (Bauer, 2002; de Haan *et al.*, 2006).

The early emergence and continued improvement of recognition and recall abilities is consistent with what is known about maturation of the hippocampal formation in both humans and monkeys (see earlier discussion). In both species, the hippocampal formation is almost adult-like at birth, although there are a number of postnatal morphological refinements that continue until the end of the first year in monkeys and until about 4 to 5 years of age in humans. These protracted modifications in hippocampal circuitry, together with the further maturation of cortical areas in the temporal and prefrontal lobes, may provide a basis for the relatively late elaboration of certain capacities, including spatial and relational memory (Chapters 50 and 52) around 1 year of age in monkeys and 4 to 5 years of age in humans.

Orientation to Faces, Recognition of Mother's Face, and Imitation of Facial Expressions Are Present at Birth

The human face is a powerful visual source of social information, and the development of face-processing abilities appears to follow a delayed postnatal development. Newborns preferentially orient to faces, recognize the mother and other familiar faces, and imitate facial expressions. These precocious tendencies appear to be mediated by a subcortical retinotectal pathway. For example, newborns display preferences selectively for moving stimuli presented in the peripheral visual field (i.e., under conditions that engage the subcortical systems), but not when they are displayed in the central visual field.

The fact that newborns just hours to days after birth look longer at the mother's face than a stranger's face (even when cues from her smell and voice are eliminated) indicates that, from very early on, there is a mechanism capable of learning about individual faces based on experience. This early learning system may be mediated by the hippocampal formation, as this area is known to be involved in memory and it matures early relative to memory–related neocortical areas (see earlier discussion). Nevertheless, the face-processing abilities of the newborn are qualitatively different than the sophisticated capacities present in the adult, and none of these tendencies are necessarily the direct precursors of the cortically mediated, adult face-processing system (Chapter 46). Instead, these responses might serve the purpose of providing input to developing cortical circuits that will at some later time functionally emerge to mediate face processing.

At 3 Months of Age Infants Are Able to Form Categories of Visual Stimuli

A marked change in infants' visual attention to faces occurs at approximately 8 weeks of age. At this time, infants' preferential following of peripheral moving faces declines, and a preference emerges for fixating faces compared to other patterns presented in the central visual field. This behavioral change is thought to be due to the functional development of visual cortical pathways that inhibit the preferential following response and mediate the new preferential fixation response. At 2 months of age, infants become more sensitive to the internal facial features of static faces. In addition, within the face, the eyes are a more salient feature of the face than the nose or mouth, but where the eyes are located is immaterial to the babies' preference for face-like drawings. Furthermore, at roughly the same age, infants begin to relate information between individual faces and to form categories; that is, they perceive visual stimuli with comparable features as being more similar than visual stimuli with different features. They also exhibit inversion effects (i.e., recognition of upright faces is easier than recognition of inverted faces). The change between 1 and 3 months has been associated with the functional development of temporal cortical areas and their connections with the hippocampus and adjacent structures. However, at this early age, the change may not be specific to faces, as 3-month-old infants can form perceptual categories of a variety of other complex categories, such as tables and trees.

Hemispheric Differences in Face Processing Emerge around 4 Months of Age

By 4 to 9 months, hemispheric differences in face processing emerge, as babies at this age move their eyes more quickly to the mother's face than a stranger's face when the stimuli are presented in the left visual field, but not when they are presented in the right visual field. In contrast, simple geometric shapes are discriminated equally well in either visual field. These results suggest that by this age the right hemisphere may be more proficient than the left in recognizing faces. Finally, during childhood, the child's prototype of the face becomes more tuned to the types of faces that the child sees most often, and with this may emerge race effects and other species effects seen in adult face processing. Other race effects refer to the ability to distinguish among faces of one's own race more quickly than faces of other races. Other species effects refer to the ability to distinguish between human faces more quickly than faces of other species. These last changes may be based on experience due to increased exposure to faces, the number and type of features children attend to and encode, and the maturation of a posterior temporal cortical area (the fusiform area) that is known to be involved in face processing in adults.

Fear of Unfamiliar Faces Emerges between 7 and 9 Months

Although infants are able to discriminate among some facial expressions of emotion by 3 months of age, they do have difficulty discriminating between sadness and surprise. It is only by 7 months of age that infants, like adults, show categorical perception of facial expressions. For example, with pictures of faces in which emotional expressions are progressively degraded from happy to fear, human infants and adults exhibit more accurate discrimination for pictures of faces that cross emotional categories (happy versus fear) than for pictures of faces within the same emotional category, despite equal physical differences in the pictures. With this increased ability to categorize faces, and the increased exposure to familiar faces, two remarkable events occur between 7–9 months of age: the emergence of fear for unfamiliar people and anxiety during temporary separation from the caretaker. Distress under both conditions must involve more than the ability to discriminate strangers from parents because 3 month-olds can make this discrimination, but do not show fear reactions.

One speculation concerning stranger and separation fear is that improvements in working memory, and an enhanced ability to retrieve memories of a past event, are required for the appearance of these reactions. This change appears to be associated with the maturation of the orbital frontal cortex (Schore, 1996), a region of the prefrontal cortex intimately interconnected with limbic areas in the temporal lobe (temporal pole cortex and amygdala), with subcortical drive centers in the hypothalamus, and with dopamine neurons in reward centers in the ventral tegmental area. By the end of the first year, the initial phase of orbital frontal maturation is achieved and allows for developmental advances that enable the individual to react to situations on the basis of stored representations, rather than on information immediately present in the environment. In the case of the orbital frontal cortex, this capacity applies specifically to socioemotional information. Indeed, by 10 months, infants are first able to construct and store abstract prototypes of human visual facial patterns, and can use these prototypes to evaluate novel information.

In an illuminating experiment, a preferential looking task was used to habituate 10-month-old infants to pictures of faces. One condition presented faces of different females, and in the other condition, the same female face was presented in various poses. After infants habituated to the stimuli (i.e., decreased their looking), they were given two test trials: one with a face of a familiar female and another with a face of a novel female. When infants were previously habituated using different female faces, they spent equivalent amounts of time gazing at familiar and novel faces on the test trials. However, when infants were habituated to the same face with different poses, they generalized their looking response to the familiar face but dishabituated (or looked longer) at the novel face. This pattern of results suggests that infants can abstract relevant categorical information by 10 month of age.

Changes in Visual Search Are Related to Development of the Prefrontal Cortex

Infants younger than 7 or 8 months of age will not uncover a hidden object. If a cloth is thrown over a toy while an infant of 5 or 6 months is reaching for it, the infant will withdraw his or her hand and stop reaching. By 7 or 8 months, most infants who watch an object being hidden can retrieve it. However, if the infant then watches as the object is hidden at a second location, most infants of 7 to 8 months search for the object at its first hiding location, which is now empty. Jean Piaget called this the A-not-B error because the infant is correct at the first hiding place (A) but not at the second (B). It is rare to see this error when there is no delay between when the object is hidden and when the infant is allowed to reach. However, a delay as brief as 1, 2, or 3 s is sufficient to produce the error in infants 7 to 8 months of age. As they grow older, the error is still seen, but only if the delay between hiding and retrieval is increased. The A-not-B error suggests that during the second half of the first year the brain has matured sufficiently to enable the infant to hold a representation of the object's location in mind (working memory) for a few seconds (Chapter 52) or to understand the relationship of (a) their previous action of retrieving the object, (b) the subsequent hiding of the object in a different location (even though they observed the hiding), and (c) the object's present whereabouts (Diamond, 1990). The enhanced working memory observed in the 7- to 10-month-old infant parallels the protracted maturation and refinement of dorsolateral prefrontal cortex circuitry (Fig. 45.3). After 12 months of age, it is difficult to elicit the A-not-B error.

Language Undergoes Rapid but Prolonged Development

In humans, the prolonged structural, metabolic, and neurophysiological maturation of "association" corti-

cal areas, which continues well into adolescence, provides the substrate for the panoply of higher cognitive functions that continue to develop during this time. Attempts have been made to link the very rapid development of speech and language skills over the first three years of life to these general changes. This goal has remained elusive, probably because so many aspects of the brain and behavior are changing together. Moreover, key elements of language appear to be processed at cellular and synaptic levels of organization that are difficult to investigate in humans with even the most sophisticated techniques available.

There is wide agreement that language is strongly dependent on structures within the left perisylvian region (Chapter 51). Very early on (by 28 weeks of gestation), structural asymmetries appear between the temporal lobes; these may provide the substrate for the functional asymmetries that appear later. The rapid and early acquisition of phonological information and speech production and comprehension, and the subsequent burst in size of the vocabulary, may be linked to the rapid rise in the number of synapses and the marked increases in cortical metabolism that occur during the second year of life. Also, the persistence of large numbers of exuberant synapses through adolescence may provide the anatomical substrate for prolonged neural plasticity and recovery of language skills following cortical damage in the first decade of life. It is well established that language skills can display considerable recovery following large lesions involving the left hemisphere during the first 7 to 10 years of life. The impressive language recovery in children who have undergone left hemisphere removal is even more striking in light of the enduring impairment seen in specifically language-impaired children or the reading difficulties encountered by dyslexics, in whom macroscopic aspects of brain structure are basically normal. These findings underscore the importance of characterizing microscopic structural aspects of the brain and the functional organization of the brain in relation to processing.

The prolonged time course of development that may confer plasticity on the immature brain appears to be characterized by optimal or critical periods (Chapter 22) for language acquisition. Several studies report that both first and second language acquisitions are impaired, and cerebral organization is altered, when language is acquired after the first decade of life. Moreover, as has been observed for vision (Fig. 45.2), different aspects of language appear to display different critical periods. Vocabulary items can be acquired long past the first decade of life, but the grammatical rules of a language appear to be acquired most readily before the age of 10. Along with other evidence, this pattern suggests that different neural systems, with differing developmental time courses, mediate these various aspects of language.

Summary

Cognitive development parallels some of the most important changes in brain development. Thus, the visual abilities of the newborn are limited and control by subcortical systems predominates. With the maturation of visual cortical areas in the first few postnatal months, visual acuity, visual alertness, and stereopsis develop. With further development of visual cortical areas within the temporal lobe, an infant's ability to recognize faces using internal facial features, mostly the eye region, emerges around 3 months of age. However, it is only by 7 to 9 months of age that infants develop fear of unfamiliar people. It is believed that this behavior is associated with the development, around this age, of the orbital frontal cortex, a cortical area that enables the infant to react to situations on the basis of stored representations (working memory). The enhanced spatial working memory abilities observed by the middle of the first year parallel the protracted maturation and refinement of the dorsolateral prefrontal circuitry. Finally, the maturation of association areas of the cortex, which continues well into adolescence, provides the substrate for the development of higher cognitive functions, such as language. In the future, it will be possible to analyze the neural basis of cognition in increasing detail by using high-resolution methods for imaging brain structure and function in normally developing children and children with specific structural or functional deficits.

Cognitive Aging

Average Human Life Expectancy Has Increased Dramatically

The world's population is growing older. Whereas life expectancy in the United States was approximately 50 years in 1900, infant mortality rates decreased dramatically during the last century, and combined with advances in disease prevention and treatment, current life expectancy is approaching 80 years in most industrialized countries. A significant demographic trend is that the percentage of the population over 65 years of age is projected to double over the next several decades, with particularly rapid growth among those over 85. Neurodegenerative disorders such as Alzheimer's disease are among the most devastating consequences of growing older (see section on Dementia, this chapter). There is increasing recognition, however, that many otherwise healthy aged individuals experi-

ence deficits in memory and other aspects of cognitive function that, although less severe, significantly compromise the quality of life. The following sections outline key concepts that have emerged from the study of normal cognitive aging in the absence of frank disease.

Effects of Normal Aging on Memory Are Variable

Cognition encompasses a variety of dissociable capacities mediated by partially distinct neural systems, including perception, attention, executive function, and memory (Section V). This organization yields a valuable framework for the study of cognitive aging. Through careful analyses aimed at documenting the specific types and severity of cognitive deficits that accompany aging, neuropsychological research has provided important clues about the brain systems that are likely to mediate dysfunction. This strategy has been widely exploited to examine the effects of normal aging on learning and memory, where a rich background of information is available concerning the characteristics of impairment that follows damage to selective brain regions in young adults.

An important concept to emerge from neuropsychological studies is that the effect of aging on the type of memory mediated by the hippocampal formation and related medial temporal lobe structures is highly variable among older individuals (Figure 45.4). In humans and animal models, memory ability among the aged is continuously distributed across a broad range such that some individuals exhibit considerable impairment while others—at the same chronological age—perform as well as young adults. Long considered a complicating factor in gerontological research, this variability suggests the important conclusion that marked deterioration in memory is not an inescapable consequence of growing older and that in a substantial proportion of aged individuals, the functional integrity of the medial temporal lobe system is preserved. The next section considers findings from animal models that support this perspective, and that have guided efforts to identify the neurobiological alterations that account for individual differences in normal cognitive aging.

Animal Models of Cognitive and Neurobiological Aging

As described in Chapter 50, normal adult rats solve the spatial, or "place," version of the Morris water maze by learning and remembering the escape location in relation to the configuration of cues surrounding the testing apparatus, and this capacity requires the functional integrity of the hippocampal formation. Guided by this background, many laboratories

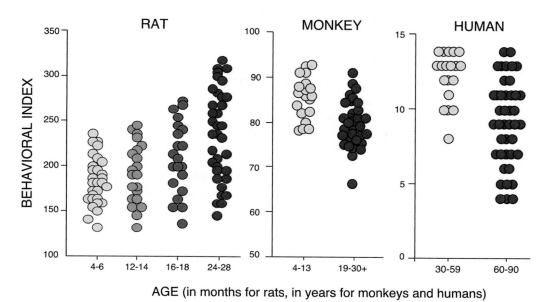

FIGURE 45.4 The status of memory varies considerably across aged individuals. The data comprise a summary measure of spatial memory in rats (courtesy of M. Gallagher, Johns Hopkins University), object recognition memory in monkeys (courtesy of P. Rapp, Mount Sinai School of Medicine), and delayed recall in humans (courtesy of M. Albert, Johns Hopkins School of Medicine). For rats, low scores represent better memory, and high values reflect better performance in monkeys and humans. Symbols signify scores for individual subjects. Note that in all species the status of memory among the aged is distributed across a broad range, from individuals that score on a par with the best young adults to other individuals that exhibit substantial impairment.

have used the Morris water maze and other tests of spatial information processing to examine the effects of aging on learning and memory that are mediated by the hippocampal system (Rosenzweig and Barnes, 2003).

Rats generally are considered "aged" at approximately 24 to 28 months of age, roughly equivalent to humans in their 70s. Relative to young adults, aged rats are significantly slower to learn the hidden platform location in the place version of the water maze. Although these findings are consistent with a possible hippocampal contribution to cognitive aging, an important alternative is that impaired visual acuity, motor function, or other factors might account for the deficits observed in older subjects. This possibility has been tested using control procedures that make many of the same sensory, motor, and motivational demands as the hidden platform task, but that lack an explicit spatial learning component. In a common variant of testing, for example, rats swim to a platform that protrudes above the surface of the water, providing a salient visual goal to guide navigation. Because the position of the cued platform is varied randomly across trials, learning and memory for spatial information are not required. Many aged rats acquire this task normally, demonstrating that they can swim as rapidly as young adults, are motivated to escape, and retain sufficient visual acuity to support cue-directed navigation

(Gallagher *et al.*, 1993). Thus, although certain aspects of motor and visual function are susceptible to age-related decline, these impairments fail to explain the deficits that aged rats exhibit in the spatial version of the water maze. Rather, the overall profile that emerges during aging—impaired learning and memory for spatial information against a background of intact goal-approach learning—is qualitatively similar to the effects of damage to the hippocampal system in young subjects. Results from other testing procedures confirm that learning and memory mediated by the hippocampus and related brain structures are susceptible to an age-related decline (Rosenzweig and Barnes, 2003).

As noted earlier, research in animal models has revealed substantial individual variability in the cognitive outcome of aging. Figure 45.4, for example, illustrates individual learning scores for large numbers of young, middle-aged, and aged rats tested in the spatial version of the water maze. The performance measure in this case reflects the average of each rat's distance from the escape platform, assessed over the course of multiple trials. Accordingly, low scores represent better learning focused on the escape location. Similar to the distribution of performance observed in humans, spatial learning scores for aged rats are distributed continuously across a broad range such that some perform as well as even the most proficient young subjects. By comparison, approximately half the older animals exhibit marked deficits, scoring outside the range of the young adult group.

Variability in the cognitive effects of aging has been documented in other animal models as well. The neuropsychology of aging in nonhuman primates has been examined using testing procedures adopted from studies on the effects of medial temporal lobe damage in young adult monkeys (Chapter 50). In the delayed nonmatching to sample task, subjects are tested for their ability to recognize a visual stimulus across delay intervals ranging from seconds to many minutes or more. Young adult monkeys with sufficiently extensive damage to any component of the medial temporal lobe memory system display significant deficits on this task, particularly when the memory demands are increased by lengthening the retention interval. Recognition memory also declines during aging in the monkey, and as in humans and rats, the degree of impairment varies substantially across individual subjects (Fig. 45.4). As discussed in the next section, this variability makes it possible to ask what neurobiological alterations distinguish the brains of individuals with age-related memory impairment from other, age-matched subjects with preserved function. By this approach the aim is to identify those changes in the structure and physiology of the aged brain that are specifically coupled to individual differences in the cognitive outcome of normal aging.

Neurobiology of Age-Related Memory Decline

Age-Related Learning and Memory Impairment Is Associated with Compromised Hippocampal Physiology

Behavioral studies suggest that the functional integrity of the hippocampal system declines during aging. Electrophysiological investigations extend these observations by directly assessing the computational processing functions of the aged hippocampus. Models of learning-related cellular plasticity, such as hippocampal long-term potentiation (LTP; Chapter 49), establish a useful background against which to explore the effects of aging. A prominent theme is that aging influences hippocampal physiology in a highly selective manner that spares many properties of hippocampal function (Burke and Barnes, 2006; Wilson et al., 2006). Parameters that exhibit little or no change in older subjects include the resting potential, the input resistance, and the amplitude and duration of evoked action potentials of principal hippocampal neurons. By comparison, although the peak magnitude of LTP is comparable in the young adult and aged hippocampus, the intensity and frequency of stimulation necessary to achieve that response increase with advanced age.

Once established, LTP also decays to prepotentiated baseline levels more rapidly in aged subjects than in young animals. This enhanced rate of decay is correlated with the rapid forgetting aged rats exhibit on tests of spatial memory, consistent with the conclusion that a reduced capacity for synaptic enhancement in the hippocampus contributes to this behavioral deficit. Recent studies also reveal significant change in hippocampal long-term depression (i.e., an enduring decrease in synaptic efficacy; LTD), demonstrating that multiple plasticity mechanisms are affected in relation to variability in the cognitive outcome of aging.

Electrophysiological studies of ongoing neuronal activity in awake-behaving rats shed additional light on how normal aging influences hippocampal information processing. As described in Chapter 50, a particularly well-characterized phenomenon is that the firing rate of individual pyramidal neurons in the hippocampus increases dramatically as animals navigate through a restricted area within a testing environment. This "place field" activity is largely independent of specific behaviors exhibited during exploration, and instead seems to be controlled by geometric relations between the multiple sensory cues that define a particular spatial location. Recapitulating a theme from studies of LTP and LTD many aspects of place field firing remain relatively normal in the aged hippocampus, including the percentage of neurons that exhibit location-specific firing, and the spatial selectivity of place fields. The informational content encoded by the aged hippocampus, however, is reliably altered (Burke and Barnes, 2006; Wilson et al., 2006). For example, hippocampal pyramidal cells in old rats are abnormally prone to confuse familiar environments that share certain stimulus elements, unpredictably engaging the wrong distribution, or "map," of place fields during successive bouts of exploration in a given setting.

Related findings demonstrate that hippocampal encoding also becomes increasingly rigid with age such that a relatively narrow subset of available spatial cues comes to control place field activity. These changes are more prevalent in the CA3 field of the hippocampus than in CA1, and among aged individuals that demonstrate impaired spatial learning (Wilson et al., 2006). Research on the structural integrity of the hippocampus, discussed in the following sections, points to a potential link between the behavioral and electrophysiological consequences of aging.

Early Findings Suggested Neuron Loss Is Distributed Diffusely throughout the Aged Brain

One of the most widely held notions about brain aging is that a substantial number of neurons inevitably die as we grow older. Early studies supported this view, suggesting that neuron death occurs throughout life, with the cumulative loss exceeding 50% in many neocortical areas by age 95 (for an historical review, see Brody, 1970). Although not all brain regions seemed affected to the same degree, significant age-related neuron loss was reported for all regions examined, including both primary sensory and association areas of the cortex. On the basis of these early observations, it seemed reasonable to suppose that widely distributed neuron death might account for many of the cognitive deficits associated with normal aging. A significant concern, however, was that the finding of widespread neuronal loss might be attributable to preclinical dementing disease in an unknown proportion of the aged individuals who were examined. This confound is overcome effectively in studies using animal models of cognitive aging because rats and monkeys do not develop dementia spontaneously. In addition, as described next, advanced quantitative tools for examining neuron loss in the aged brain are now widely available.

Memory Impairment during Normal Aging Does Not Require Substantial Hippocampal Neuron Loss

Improved methods for quantifying cell number have led to significant revision in traditional views on age-related neuron loss (Morrison and Hof, 1997). Until recently, most investigators had focused on cell density, defined as the number of neurons present in a fixed area or volume of tissue. A significant limitation of this approach, however, is that density can vary widely in the absence of any actual difference in cell number. Assume, for example, that total neuron number is identical in two brains, but that the overall size of the brains differs due to normal biological variability among individuals, gliosis, white matter abnormalities, or other neuropil alterations. Under these conditions, neuron density will be lower in the larger brain, simply as a consequence of the cells being distributed in a larger volume. In this way, real or processing-related volumetric differences between young and aged brains could substantially influence cell density in the absence of any actual age-dependent difference in neuron number.

The field of stereology has provided standardized tools for directly estimating the total number of neurons in any defined brain region, yielding an unequivocal measure for examining potential neuron loss during normal aging (Box 45.1). In contrast to early studies measuring cell density, the conclusion from recent investigations using modern stereological methods is that the total number of principal neurons

(i.e., the granule cells of the dentate gyrus and pyramidal neurons in the CA3 and CA1 fields) generally is preserved in the aged hippocampus. Substantial preservation also has been documented in adjacent parahippocampal cortical regions that critically participate in memory. Whereas similar results have been observed in humans, data from animal models of cognitive aging are particularly compelling in demonstrating that hippocampal neuron number remains stable even among aged individuals with pronounced learning and memory deficits. As illustrated in Figure 45.5, for example, stereological analyses have revealed that the total number of neurons comprising the hippocampal region is comparable in young adult subjects and aged rats, regardless of their capacity for spatial learning and memory. Prompted by such observations, researchers have looked toward changes in synaptic connectivity and other markers of functional deterioration as a more likely basis for normal cognitive aging.

Hippocampal Connectivity Is Compromised during Normal Aging

The entorhinal cortex originates the major source of cortical input to the hippocampus, giving rise to the so-called perforant path, which synapses on the distal dendrites of dentate gyrus granule cells in the outer portions of the molecular layer. The same entorhinal cortex neurons also innervate other fields of the hippocampus, including the most distal aspects of CA3 pyramidal cell dendrites. Multiple lines of investigation suggest that the integrity of these inputs is compromised with age. Electron microscopic quantification using modern stereology, for example, has documented that aging is associated with significant loss among a morphologically distinct subset of synapses in the outer molecular layer of the dentate gyrus (i.e., the zone innervated by the entorhinal cortex) (Geinisman et al., 1992), together with alterations of synaptic structure (e.g., decreased postsynaptic density length) in other hippocampal subfields. Input originating in the entorhinal cortex conveys much of the neocortically derived information that the hippocampus processes in support of normal learning and memory, and accordingly, it is reasonable to suppose that the disruption of entorhinal–hippocampal connectivity might contribute to the cognitive outcome of aging. Consistent with this proposal, the magnitude of the morphological alterations observed in the hippocampal termination zones of the entorhinal cortex is greatest among aged animals with documented deficits on tasks that require the hippocampus and among older rats that exhibit abnormalities in various physiological measures of hippocampal plasticity.

BOX 45.1

THE OPTICAL FRACTIONATOR STEREOLOGICAL METHOD

OPTICAL FRACTIONATOR

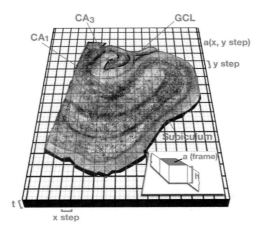

by routine methods for visualizing neurons microscopically. An evenly spaced series of the sections is then chosen for analysis (positions represented schematically in top panel). This first level of sampling, the "section fraction," therefore comprises the fraction of the total number of sections examined. For example, if every tenth section through the hippocampus is analyzed, the section fraction equals 1/10. The appropriate sections are then surveyed according to a systematic sampling scheme, typically carried out using a microscope with a motorized, computer-controlled stage. The lower part of the figure illustrates this design in which the microscope stage is moved in even X and Y intervals, and neurons are counted within the areas defined by the small red squares ("a (frame)" in the inset). The second level of the fractionator sampling scheme is therefore the "area fraction," or the fraction of the XY step from which the cell counts are derived.

The last level of sampling is counting cells only within a known fraction (h) of the total section thickness (t), avoiding a variety of known errors introduced by including the cut surfaces of the histological preparations in the analysis. This is accomplished using a high-magnification microscope objective (usually 100X) with a shallow focal depth. In the illustration provided, the "thickness fraction" is defined as h/t. Neurons are counted as they first come into focus, according to an unbiased counting rule, called the "optical disector," that eliminates the possibility of counting a given cell more than once.

Finally, total neuron number in the region of interest (N) is estimated as the sum of the neurons counted (sumQ), multiplied by the reciprocal of the three sampling fractions; the "section fraction," "area fraction," and the "thickness fraction." For the present example, the total estimated neuron number is given by

$$N = (\text{sumQ}-) \times (10/1) \times (\text{XYstep area}/a(\text{frame})) \times (t/h).$$

The figure shows the key features of a stereological technique designed to provide accurate and efficient estimates of total neuron number in a brain region of interest. The hippocampal formation is used as an example. The method consists of counting the number of neurons in a known and representative fraction of a neuroanatomically defined structure in such a way that each cell has an equal probability of being counted. The sum of the neurons counted, multiplied by the reciprocal of the fraction of the structure that was sampled, provides an estimate of total neuron number.

Serial histological sections are prepared through the rostrocaudal extent of the hippocampus and are stained

Original illustration design by L. E. Mekiou and P. R. Rapp.

Peter R. Rapp

Executive Function Mediated by the Prefrontal Cortex Declines during Normal Aging

Normal aging is sometimes thought of as a generalized process of nonspecific deterioration across many neural systems, implying that age-related cognitive decline might be the cumulative effect of subtle altera-

tions distributed diffusely throughout the brain. There is actually little empirical support for this notion, however, and as we have seen in the case of hippocampal circuitry, the current perspective is that the neurobiological consequences of aging exhibit considerable selectivity. At the same time, it is important to recognize that the effects of aging on memory do not occur

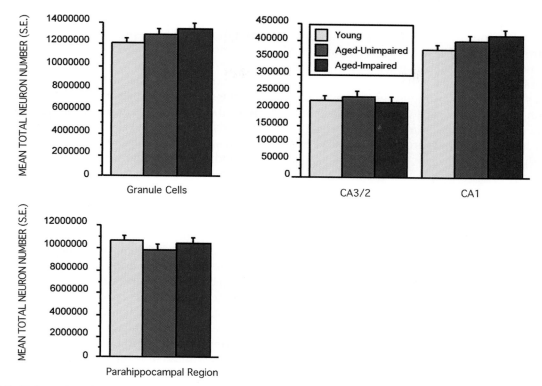

FIGURE 45.5 Estimated total neuron number (plus S.E.) in principal cell layers of the hippocampus and parahippocampal cortices (i.e., entorhinal, perirhinal, and postrhinal) for behaviorally characterized young and aged rats. All values are unilateral. Half of the aged rats exhibited substantial spatial learning deficits in the Morris water maze (aged impaired); the other half performed within the range of learning scores for the young group (aged unimpaired). Neuron number is stable with age and across a broad range of learning capacities. Adapted with permission from Rapp and Gallagher (1996) and Rapp *et al*. (2002).

in isolation and that cognitive capacities mediated by other neural systems are also susceptible to decline.

As discussed in Chapter 52, the functionally and neuroanatomically heterogeneous areas comprising the prefrontal cortex support a constellation of capacities referred to as "executive function." Whereas damage to these regions does not produce the frank amnesia that follows medial temporal lobe lesions, significant deficits are observed in a variety of processing functions that influence memory, including the strategic use and manipulation of remembered information, the ability to identify the source from which information was acquired, and the order in which it was presented. It is therefore of interest that a qualitatively similar profile of deficits emerges during the course of normal aging (Gallagher and Rapp, 1997). Elderly humans, for example, frequently experience difficulty recalling the original source of remembered information, even under conditions in which recollection of a target item is successful. These failures in "source memory" predict performance on other tests of frontal lobe function, including memory for temporal order, suggesting that a common neurobiological basis might underlie these deficits.

Neuropsychological studies in animal models reinforce the conclusion that prefrontal cortex dysfunction contributes to certain features of cognitive aging. One of the best-documented deficits associated with aging in monkeys, for example, is poor performance under testing conditions that emphasize the spatial and temporal, or the "where and when," components of memory (Gallagher and Rapp, 1997). In the classic delayed response task, subjects observe while a food treat is hidden in one of two locations that are then covered with identical plaques (Fig. 52.5). Because each of the reward locations are baited frequently across trials within a single test session, there is substantial opportunity for interference, and accurate performance depends on the ability to keep in mind the spatial location that was baited most recently. Damage involving the dorsolateral prefrontal cortex in young monkeys produces impairment at retention intervals of just a few seconds or more in the delayed response procedure. Impairment on this task following lesions of the hippocampal system, in contrast, is selectively observed at longer delays.

Numerous studies since the late 1970s have documented that aged monkeys exhibit a profile of delayed response deficits qualitatively similar to the effects of prefrontal cortex lesions in young animals. Additional signs of prefrontal decline have also been documented in the aged monkey, including marked perseveration and behavioral rigidity when older subjects are confronted with shifting task contingencies. For example, whereas young adult monkeys quickly cease responding to a previously positive stimulus when it is no longer rewarded, aged subjects are slower to make this behavioral adjustment.

Are the effects of normal aging on capacities mediated by the prefrontal cortex coupled to other signatures of cognitive decline, consistent with the view of aging as a process of generalized deterioration? Available evidence suggests that this is not the case. For example, age-related deficits on tests of prefrontal cortex function tend to emerge earlier during the course of aging than alterations in declarative memory supported by the medial temporal lobe system. In addition, when impairments indicative of both frontal and medial temporal lobe dysfunction cooccur in the same subjects, these effects are statistically unrelated to each other (Gallagher and Rapp, 1997). This issue has been examined by testing healthy elderly people on a neuropsychological battery that includes assessments of both frontal lobe function and declarative memory. Results from this type of experiment indicate that whereas measures of information processing within each domain are correlated with each other (e.g., the severity of impairment on different tests of hippocampal memory are correlated), there is no apparent relationship in performance across tests of frontal and medial temporal lobe function. Thus, different aspects of cognitive function can decline independently during aging and these effects are not the consequence of a global, brain-wide degenerative process.

Neurobiology of Aging in the Prefrontal Cortex

Although the basis of age-related decline in capacities supported by the prefrontal cortex remains to be established, an outline has begun to emerge. Similar to the hippocampal system, neuron death is not a prominent feature of normal aging in frontal lobe areas that mediate information processing functions known to decline with age. Other structural alterations have been documented, however, including a substantial decline in the density of prefrontal cortex synapses, marked changes in dendritic architecture, and a variety of abnormalities in the myelination of axons (Peters et al., 1998). These parameters are correlated with behavioral signatures of cognitive aging in monkeys, raising the possibility that, like the hippocampal system, changes in the organization of critical prefrontal circuitry may play a significant role. Results from

in vivo imaging studies in humans support this interpretation, and moreover, suggest that the functional organization of the aged brain may be more dynamic than previously presumed.

Gerontological research traditionally has focused on the nature and severity of age-related impairment with the implicit assumption that capacities found to be resistant to aging reflect a preservation of the same mechanisms that support these abilities in younger individuals. A number of investigations, however, have demonstrated that, rather than exhibiting a simple pattern of regional decline or maintenance, normal aging is associated with a reliable shift in the relative balance of activations in brain areas important for normal cognitive function, including the prefrontal cortex and medial temporal lobe. Indeed a profile of increased prefrontal and blunted medial temporal activation has been observed even under conditions where aged people score as accurately as young adults. These findings imply that the distribution of neural systems engaged during testing is altered significantly over the course of normal aging, and that successful cognitive outcomes may rely, in part, on adaptive functional reorganization (Hedden and Gabrieli, 2005). Intervention aimed at harnessing this dynamic capacity in support of optimally healthy cognitive aging is an area of active investigation.

Neurochemically Specific Subcortical Systems Are Susceptible to Aging and Positioned to Broadly Influence Behavior

Current evidence challenges the view that normal aging is associated with generalized deterioration distributed diffusely throughout the brain. It should be noted, however, that aging does affect the integrity of neurochemically specific classes of cells whose ascending projections influence widespread brain areas. Cholinergic cell groups in the basal forebrain have been studied particularly intensively because this system exhibits pronounced degeneration in Alzheimer's disease. Less severe effects are also observed during normal aging, involving acetylcholine-containing neurons that project to the hippocampus, amygdala, and neocortex. A significant implication is that cholinergic cell loss might disrupt the information processing functions of these target regions, and consistent with this possibility, reliable correlations have been reported between the magnitude of cholinergic deficits and behavioral impairment in aged individuals. Cholinergic abnormalities alone, however, fail to account for the full profile of cognitive deficits observed during aging since neurotoxin lesions that selectively destroy these neurons fail to reproduce key features of the behavioral decline observed in older individuals. One possibility is that age-related cholinergic impairment comprises an important component of a broader constellation of alterations that leads to cognitive dysfunction. Other neurochemical systems that might contribute include dopamine-containing circuitry that originates in mid-brain cell groups and noradrenergic inputs that arise from the locus coeruleus. These systems innervate a variety of more anterior brain regions, and consistent with this broad distribution, neurochemical alterations have been linked to a diverse array of behavioral deficits ranging from age-related cognitive impairment to disturbances in motor function. These findings encourage the view that appropriately designed pharmacological interventions might have beneficial effects on a similarly broad profile of outcome measures of aging.

Summary

This section considered the cognitive and neurobiological effects of normal aging in humans and animal models. Key themes include the observation that the impact of aging on memory varies widely from one individual to the next. Whereas neuron death was once presumed to be the proximal cause of many cognitive deficits associated with aging, including impairments in learning and memory, it is now clear that age-related cognitive decline does not require marked neuron loss in the hippocampus and related cortical areas. Instead, it now appears that subtle alterations in connectivity, and downstream changes in cellular function, are more likely the causative factors. Building on advances from *in vivo* imaging studies, a comprehensive, neural systems account of normal cognitive aging is beginning to emerge.

PATHOLOGICAL PROCESSES IN COGNITIVE DEVELOPMENT AND AGING

Developmental Psychopathologies

Research in this field focuses on elucidating the interplay among the biological, psychological, and social-contextual aspects of normal and abnormal development across the life span. Recent advances in neuroimaging and gene linkage technology have vastly expanded research possibilities for noninvasive clinical studies. Even with a new arsenal of research technologies, however, it has not been possible to isolate and define characteristics of each developmental disability. Such efforts are limited by several factors:

1. Disagreement regarding the criteria for diagnosis. Because different researchers often use differing behavioral criteria, cross-study comparisons frequently show inconsistencies in results.
2. Heterogeneity of disorders. Even when consistent behavioral criteria are carefully applied, different underlying etiologies can result in the same behavioral profile.
3. Difficulties inherent in the study of children. Although modern brain imaging techniques such as magnetic resonance imaging are considered noninvasive, they are stressful and time-consuming, and parents will not always allow their affected children to participate in such studies. Although studies can be conducted using adults who have been affected since childhood, data obtained from such studies may not accurately reflect anomalies that characterize the early disruption of brain development because brains change over time. Moreover, retrospective diagnoses of childhood disorders frequently rely on memory (e.g., a patient is asked "Did you have difficulty reading as a young child?") and hence can be unreliable.

Despite these challenges, ongoing research has uncovered important neurological and genetic features that seem to be associated with specific developmental disorders (see Boxes 45.2 and 45.3 for examples). The following sections outline what is known about how specific injuries or anomalies in development of the brain may result in specific patterns of cognitive and behavioral impairment. Two examples of developmental psychopathology are described: autism and schizophrenia.

Autism

Autism is a neurodevelopmental syndrome characterized by deficits in a broad domain of social interaction, play, language, and communication, and a restrictive range of interests and activities, including repetitive and stereotypic movements and persistent preoccupation that begin in infancy and become apparent by the end of the third year of life. Although autism was once believed to be a rare disorder, recent surveys report a prevalence of 40 to 50 per 10,000 births. Parents often first become concerned because their child fails to use words to communicate, even though he or she recites passages from videotapes or says the alphabet. Like other kinds of abnormal behavior, autism is a heterogeneous condition that can range widely in severity.

Characteristics of Autism

A multitude of symptoms occur in autism, including social isolation (the autistic child largely ignores other people, shows little attachment to parents or

BOX 45.2

DYSLEXIA

Reading skills occupy a uniquely important position in overall cognitive development, strongly dependent on antecedent skill acquisition and critically supportive of educational success. After accounting for the effects of inadequate instruction, the most common cause of severe reading problems in childhood is *developmental dyslexia*, a disorder typically characterized by impairment in applying the sound–correspondence rules necessary to decode print (Chapter 51). Although reading impairment is common, with an incidence of 5–10% among school-age children, our understanding of its biological roots is relatively recent.

Neurobiological studies of dyslexia are challenging, as the behavioral manifestations of this disorder are complex. Although defined by poor reading performance, dyslexia involves additional deficits, including poor phonological processing, poor verbal working memory, and slow naming ability. Dyslexics frequently show subtle deficits in motor control and early sensory processing, most notably in the visual system. Imaging studies have revealed that brain structure in dyslexics differs from that seen in individuals with normal reading skills. Micro and macroscopic structural abnormalities have been detected in perisylvian language regions (temporal and parietal banks of the sylvian fissure as well as the insula), visual system structures, the thalamus, and the corpus callosum. The planum temporale is an expanse of neocortex on the temporal bank of the sylvian fissure, its anterior border defined by Heschl's gyrus. From the time of birth the left planum temporale is larger than the homologous region in the right hemisphere in 70–80% of individuals. Reductions or reversal of this leftward asymmetry has been documented in individuals with developmental dyslexia. The functional concomitants of these anatomical variations have been explored in functional imaging studies that have revealed differences in patterns of task-related activity between dyslexics and controls during the performance of reading and phonological decoding tasks. Interestingly, the cortical areas identified are similar in English, French, and Italian dyslexics, supporting a biological origin for this reading abnormality, independent of cultural upbringing. Similarly, candidate genes for dyslexia have been identified and replicated in different countries, suggesting the multigenetic contribution to reading disability.

Besides the problems related to reading, dyslexics also exhibit subtle abnormalities in visual processing. Evidence for selective involvement of magnocellular pathways of the visual system (Chapter 27) has accumulated, beginning with the demonstration that cell bodies in the magnocellular layers of the lateral geniculate nucleus are smaller in dyslexics compared to a control group. The fact that both contrast sensitivity and visual persistence are abnormal in reading disabled children is consistent with the notion that these children have disturbances in the magnocellular system, which is specialized for temporally demanding visual processing. These behavioral results are supported by functional imaging studies demonstrating less task-related activity in extrastriate and parietal cortex during visual motion detection by dyslexics.

Future studies will determine the nature of the neural mechanisms that link deficits in early visual processing with more cognitive skills such as phonological awareness. Also remaining to be clarified are issues concerning which of these deficits is a primary cause of dyslexia as opposed to a consequence of having partially compensated for a reading disorder. Such information will aid in identifying better avenues for reading remediation. To date, the best methods for improving the reading abilities of dyslexics focus on structured teaching of the code that allows individuals to sound out words (phonological awareness) in the context of fluency and comprehension exercises.

Guinevere F. Eden

other relatives, and retreats into a world of his or her own), stereotyped behaviors (the autistic child rocks back and forth, stares at neutral stimuli, rotates an object, or engages in other repetitive behaviors for long, uninterrupted time periods), resistance to change in routine, abnormal responses to sensory stimuli (the autistic child may ignore visual stimuli and sounds, especially speech sounds, sometimes to such an extent that it might be thought the child is deaf), inappropriate emotional expressions (the autistic child has sudden bouts of fear and crying for no obvious reason; at other times the child displays utter fearlessness and unprovoked laughter), and poor use of speech. Individuals with autism can also exhibit symptoms that fall along a continuum of disorders. For example, many have mental retardation and show delays in the development of language. Others display neither language delay nor mental retardation, although they are clearly socially inept. Finally, autism is also associated with idiosyncratic interests and restricted and repetitive behavioral repertoire. For a long time, autism was thought to be the result of poor parenting. However, a growing number of studies over the last two decades provide mounting evidence that autism results from an early onset brain dysfunction.

Neuropathological Alterations in Autism

Several studies of autistic individuals provide evidence for megalencephaly (whole brain enlargement) and for increased head circumference. The nature of this enlargement has yet to be studied carefully, but it is specifically observed in the occipital, parietal, and temporal lobes. Postmortem analyses of the brains of autistic individuals have revealed increased cortical thickness and abnormalities in cortical morphology. Furthermore, abnormalities in the size and density of neurons in the medial temporal lobe and limbic system, including the amygdala, hippocampus, and entorhinal cortex, septal nuclei, and cingulate cortex, have also been documented. These neuropathological abnormalities make sense with respect to certain major symptoms of autism because the limbic system is known to play a central role in coordinating social–emotional functioning. Other evidence implicating this neural system in autism includes (1) case studies of patients with temporal lobe lesions and autistic-like behaviors, (2) a significant and specific association between temporal lobe lesions from tuberous sclerosis and autism, (3) a volumetric decrease in the amygdala in autism, (4) reduced functional activity on SPECT scans in the temporal lobe in autism, and (5) animal models of autism based on early lesions to medial temporal lobe structures.

A fMRI study of autism has documented dysfunction localized to the ventral temporal lobe using a face recognition task (Baron-Cohen et al., 1999). In this study, high-functioning individuals with autism and matched controls were presented with a series of photographs of eyes while their brains were being scanned to assess brain regions activated by two different tasks. In the first task (experimental condition), each photograph was presented simultaneously with words (either unconcerned or concerned, or sympathetic or unsympathetic). Participants were asked to indicate by pressing a button which word best describes what the person in the photograph is feeling or thinking. In the second task (control condition), each photograph was presented with the words "male" or "female" and the subjects were asked to make a gender recognition by pressing a button. In both tasks, the performance of autistic individuals and controls was above chance. However, the controls were more accurate in the two

tasks than the autistic individuals. To investigate the brain regions activated by the experimental task, brain maps obtained during the control task were subtracted from brain maps obtained during the experimental task. In the control group, the experimental task activated two main regions of the brain: the frontotemporal neocortical areas, including the left dorsolateral prefrontal cortex and supplementary motor areas, bilateral temporoparietal regions, and subcortical structures, including the left amygdala and parahippocampal gyrus. The autism group activated the frontal components but did not activate the amygdala. The significance of these data is that one critical role of the amygdala is to identify the mental states or emotions of other individuals and that this structure appears to be dysfunctional in persons with autism (see Pelphrey et al., 2004 for additional information on the neuroanatomical substrates of social cognition dysfunction in autism).

Significant atrophy of the vermal lobules in the cerebellum has also been reported in autistic individuals, although this effect does not appear specific to autism and is found widely in neurodevelopmental disorders. By comparison, there is now extensive evidence for the involvement of higher association areas in the pathophysiology of autism, particularly the frontal and parietal cortices.

Autism Is a Complex Genetic Disorder

The most compelling evidence for a genetic contribution to autism comes from studies of monozygotic (identical) and dizygotic (fraternal) twins. When one member of a pair of monozygotic twins is autistic, the other twin has more than a 50% probability of being autistic. However, if one member of a pair of dizygotic twins is autistic, the other twin has only a 3% probability of developing autism. Using conventional nomenclature, we say that monozygotic twins have a 50% or higher concordance for autism and that dizygotic twins have a 3% concordance. The relative risk factor for siblings is one of the highest for a complex genetic disorder.

The most significant genetic finding relevant to autism is the recent identification of genes responsible for Rett's syndrome and fragile X syndrome, two conditions associated with autistic features. Rett's syndrome is caused by mutations in the methyl-CpG binding protein 2 (MECP2) gene. Decreased expression of this gene leads to a failure to suppress expression of genes regulated by methylation. Fragile X syndrome is a disorder caused by mutation of the gene FMR1 (Box 45.3). A better understanding of how reduced expression of the MECP2 and FMR1 genes leads to mental retardation as well as social and lan-

BOX 45.3

TRIPLET REPEAT DISORDERS

Not all neurogenetic diseases follow simple inheritance patterns; not all developmental disorders reveal themselves immediately; and diseases with similar genetic etiologies can have widely different manifestations. Trinucleotide repeat mutations, or dynamic mutations, were discovered only in the early 1990s but are responsible for a number of important neurodegenerative disorders, such as Huntington's Disease (Box 31.5), Fragile X mental retardation, and myotonic dystrophy. To date, 14 neurological disorders have been found to result from such a mutational mechanism, and the list will probably continue to grow.

Triplet repeat mutations are unstable and prone to expand as alleles are passed from one generation to the next, although contractions can occur. Repeats at different loci differ markedly in their rates and extents of expansion, and the instability of repeats at each locus often depends upon the sex of the transmitting parent.

Fragile X syndrome, an X-linked dominant mental retardation syndrome, was the first disease shown to result from a triplet repeat mutation. There is a CGG repeat expansion at a locus (Xq27.3) giving rise to a folate-sensitive fragile site (FRAXA). The CGG repeat ranges in length in the general population from 5 to 50 triplets. Affected individuals carry more than 230 repeats, up to several thousand triplets. CGG tracts between 45 and 200 repeats in length are termed "premutations"; they are not long enough to result in mental retardation, yet are prerequisite for further expansion to the full mutation. Escalating risk of affected offspring is associated with increasing size of premutations in female carriers. Curiously, offspring of male carriers of premuations are not at risk for fragile X syndrome, as these alleles have not been observed to expand to full mutations upon male transmission. The normal function of the affected protein FMR1 remains to be fully determined, but it appears to play a role in RNA metabolism, likely in the areas of transport, stability, and/or translation.

Myotonic dystrophy (DM) remains a particularly enigmatic triplet repeat disorder affecting multiple systems, causing myotonia, muscle weakness, cardiac conduction defects, diabetes, cataracts, premature balding, and sometimes dysmorphyic features and mild mental retardation. There is a tendency for these features to appear in more severe form in subsequent generations of a family. In DM there is a CTG repeat in the 3' untranslated region of the myotonin protein kinase gene (DMPK).

This CTG tract is small in the general population (5–37 triplets) but can expand dramatically to thousands of repeats. Slightly increased lengths (70–100) can result in mild effects, such as late-onset cataracts, and the classic symptoms of early adulthood myotonia are found in individuals with hundreds of repeats. Very large expansions can cause severe disease in newborns.

The mechanism(s) by which the mutation leads to the various clinical symptoms remains uncertain. A variety of effects have been described and the repeat expansions clearly affect both the DMPK gene and nearby genes. It is also likely that the expanded CTG repeat carrying mRNAs can act to diminish the levels of nuclear proteins that bind at these sequences, resulting in misregulation of genes not linked to DMPK. Mutations in equivalent mouse genes have not provided a complete model of the human disease. The phenotype in DM results from multiple effects of the repeat expansion, rather than a simple loss or gain of function mechanism.

Friedreich's ataxia (FA) causes both limb and gait ataxia, loss of position and vibration sense, decreased tendon reflexes, cardiomyopathy, diabetes, and, in some patients, optic atrophy and deafness. FA is a recessive disorder and the only triplet repeat disorder known to involve a GAA sequence. Carriers of FA are rather common (1/85), and, consequently, so is the GAA expansion. Members of the general population carry between 7 and 34 repeats, and pathogenic alleles can number in the hundreds. The affected frataxin gene product appears to be involved in mitochondrial iron homeostasis, and its absence affects postmitotic cells rich in mitochondria. The mechanism by which the GAA repeat affects gene function appears to be inhibition of primary RNA transcripts.

A second class of triplet repeat disorders includes those caused by expansion of translated CAG repeats that encode a polyglutamine tract in each of the respective proteins. These "polyglutamine" disorders share many features, suggesting that a common pathogenetic mechanism is at play in spite of the fact that the mutated genes share no homology outside of the CAG repeats. They are progressive neurologic diseases with onset of symptoms in young to mid-adulthood with only a specific group of neurons vulnerable in each disease. Symptoms develop when the number of uninterrupted repeats exceed about 35 glutamines. The phenotype is caused not by loss of function of the relevant protein but rather by a gain of function conferred by the expanded polyglutamine tract.

BOX 45.3   *(cont'd)*

The **spinocerebellar ataxias (SCA)** are genetically distinct but overlap a great deal in their clinical presentation—between subtypes and even within families affected by the same disease. It is nearly impossible to distinguish one subtype from another on the basis of clinical signs alone. **Spinocerebellar Ataxia Type 1 (SCA1)** was the first of the inherited ataxias to be mapped. Patients with SCA1 suffer from progressive ataxia (loss of balance and coordination), dysarthria, and eventual respiratory failure. SCA1 is characterized pathologically by cerebellar atrophy with severe loss of Purkinje cells and brain stem neurons. Patients with SCA1 have an expanded "perfect" CAG repeat tract encoding ataxin-1, the function of which is largely unknown.

Studies of human patients with deletions of, for example, the HD, or SCA1 genes (and of mice lacking these genes) confirm that disease is not due to loss of function of the respective proteins. More importantly, studies of mouse and fruit fly models that overexpress full-length or truncated forms of either ataxin-1, ataxin-3, huntingtin, AR, atrophin-1, or ataxin-7 demonstrate progressive neuronal dysfunction consistent with a gain of function mechanism. Several illuminating findings have emerged from the study of human patients, cell culture systems, and the various animal models. In transgenic mice, expression of truncated polypeptides containing expanded glutamine tracts caused widespread dysfunc-

tion that extends beyond the specific groups of neurons affected by the full-length protein. The expanded polyglutamine tracts may cause the respective proteins to misfold or resist degradation. Aberrant protein–protein interactions or alterations in levels of proteins essential for the normal functioning of neurons (such as proteins that regulate Ca^{+2} levels and neurotransmitters), in turn, could exacerbate neuronal dysfunction and lead to neuronal death.

The finding that chaperone overexpression mitigates the phenotype in one vertebrate and several invertebrate models of polyglutamine disorders, and the identification of several additional modifiers of the neuronal phenotype, provide a platform for identifying key pathways of pathogenesis and potential therapeutic targets.

Suggested Readings

Adapted by Graham V. Lees with permission from Nelson, D., and Zoghbi, H. (2003). Disorders of trinucleotide repeat expansions. *In* "Encyclopedia of the Human Genome." *Nature* Macmillan, London.

Wells, R. D. and Warren, S. T. (eds.) (1998). "Genetic Instabilities and Hereditary Neurological Diseases." Academic Press, San Diego.

Cummings, C. J. and Zoghbi, H. Y. (2000). Trinucleotide repeats: Mechanisms and pathophysiology. *Ann. Rev. Genom. Hum. Gen.* **1**, 281–328.

---

guage disorders will likely provide a better understanding of the autistic disorder.

Schizophrenia

Schizophrenia is a catastrophic illness with an onset typically in adolescence or early adulthood. It was identified and defined in the early twentieth century as *dementia praecox* (Latin for deterioration of the mind at an early age). In 1911, Eugen Bleuler introduced the term schizophrenia. This term can be confusing because its translation from the Greek means "split mind," which is sometimes interpreted as multiple personality, a condition in which a person alternates between one personality and another. A schizophrenic person has only one personality and the split in the schizophrenic mind is between emotional and intellectual realms. The combination of significant incapacity, onset early in life, and chronicity of illness makes schizophrenia a particularly tragic disorder, occurring in 0.5 to 1% of the population.

Characteristics of Schizophrenia

The clinical symptoms of schizophrenia often are divided into two broad categories: positive and negative. Positive symptoms include delusions, hallucinations, disorganized speech, and disorganized or bizarre behavior. These symptoms are referred to as positive because they represent distortions or exaggerations of normal cognitive or emotional functions. The negative symptoms of schizophrenia reflect a loss or diminution of normal function. Negative symptoms include alogia (marked poverty of speech or speech devoid of coherent content), affective flattening (diminution in the ability to express emotion), anhedonia (inability to experience pleasure), avolition (inability to initiate or persist in goal-directed behavior), and attentional impairment. Although the positive symptoms of schizophrenia are often colorful and draw attention to the patient's illness, the negative symptoms are an important feature of the disease that impair the person's ability to function normally in daily life.

The profound and pervasive cognitive and emotional disturbances that characterize schizophrenia suggest that it is a serious brain disease affecting multiple functions and systems. Auditory hallucinations and disruptions in linguistic expression suggest involvement of the auditory cortex and perisylvian language regions. Other positive symptoms, such as delusions or disorganized behavior, are more difficult to localize to specific regions and suggest dysfunction distributed across multiple neural systems and circuits. Negative symptoms may be related to the prefrontal cortex, which mediates goal-directed behavior and fluency of thought and speech. The affective flattening and asociality suggest involvement of the limbic areas within the temporal lobe.

The Dopamine Hypothesis of Schizophrenia

A great deal of research since the 1950s has made it clear that schizophrenia is associated with abnormalities in the dopaminergic synapses of the brain. Still, the exact nature of this perturbation remains elusive. The strongest link between dopamine synapses and schizophrenia comes from studies of drugs that alleviate the symptoms of schizophrenia. A large number of antischizophrenic, or neuroleptic, drugs have been found, most of which belong to two chemical families, the phenothiazines, which include chlorpromazine, and the butyropherones, which include haloperidol. All these drugs have two properties in common: they block postsynaptic dopamine receptors and they inhibit the release of dopamine from the presynaptic neuron. One interpretation of these results is that schizophrenia is due to excess activity at dopamine synapses. Establishing cause-and-effect relationships in human psychiatric disorders, however, is challenging. For example, although a substantial increase in the number of dopaminergic receptors is observed in the schizophrenic brain, because virtually all patients receive neuroleptic medication, it is not known if increased receptor density is a consequence of the disease itself or a response to an extended use of anti-dopaminergic drugs. An additional complication for the dopamine receptor hypothesis of schizophrenia concerns the time course for the effects of neuroleptic drugs. Although these treatments block dopamine receptors almost at once and reach their full pharmacological effectiveness within a few days, effects on behavior emerge more slowly, over two or three weeks. Clearly the symptomatic relief achieved with neuroleptic intervention involves mechanisms in addition to the blockade of dopamine receptors.

One possibility is that the prolonged use of neuroleptic drugs decreases the number of spontaneously active dopamine neurons in what is known as the mesolimbic system, a set of neurons that projects from the midbrain tegmentum to the limbic system. Another possibility is that the underlying problem in schizophrenia is not an excess of dopamine activity at all, but a deficit of glutamate activity. Glutamate is the predominant excitatory amino acid released by neurons in the cerebral cortex that project widely throughout the limbic system, and dopamine synapses are known to inhibit glutamate release in these target regions. If glutamate release is reduced in the schizophrenic brain, one way to correct this deficiency would be to block activity at dopaminergic receptors, relieving glutamate synapses from inhibition.

Neuropathological Alterations in Schizophrenia

The most consistently described structural alterations in schizophrenia are ventricular enlargement and decreases in the volume of temporal lobe structures, including the hippocampus, amygdala, and entorhinal cortex. Cytoarchitecture and cell densities also reportedly are altered in limbic structures: reduced cell number and size have been observed in the hippocampus, parahippocampal gyrus, and entorhinal cortex, along with disturbed cytoarchitecture involving a cellular disarray in hippocampus, and the entorhinal and cingulate cortices. This distribution of structural abnormalities may help account for the symptoms of schizophrenia in that these limbic structures link neocortical association areas with the septum–hypothalamic complex and are therefore key in sensory information processing and sensory gating. Defects in this neural circuit could lead to a dissociation between neocortically mediated cognitive processing and limbic–hypothalamic emotional reactions and to a disturbed emotional experience of external sensory perceptions. The structural alterations reported in schizophrenia are often subtle and certainly heterogeneous, contributing minimally to the elucidation of the disease if symptoms are to be attributed to defects in single brain regions. Alternatively, it is currently hypothesized that the variety and severity of symptoms observed in schizophrenia can be explained by defects in several brain regions affecting the neural circuits that mediate higher cognitive functions. Such neural circuitry would include neocortical association areas, the limbic system and midline thalamic structures, with possible involvement of the basal ganglia.

The Neurodevelopmental Hypothesis of Schizophrenia

Several clinical observations point to neurodevelopmental abnormalities in schizophrenia: onset in adolescence in most patients and earlier onset among males, pronounced premorbid neurodevelopmental abnormalities, such as asociality and "soft" neurological signs; impaired cognitive and neuromotor functioning; minor physical anomalies; presence of structural abnormalities at the onset of the illness, which may predate the illness; and the absence of neurodegenerative processes.

Postmortem morphological observations in the brains of schizophrenic patients suggest that multiple aspects of neural development may be disrupted. The spatial disorganization of neurons found in the hippocampus and the inappropriate laminar location of cells in the entorhinal cortex, for example, suggest a potential defect in neuronal migration. Other evidence indicates that the organization of synaptic connectivity is affected. The functional circuitry of the frontal cortex undergoes significant remodeling until adolescence, mediated by an NMDA receptor-dependent mechanism. The number of these receptors is reduced in the brain of schizophrenics. Still other abnormalities involve biochemical alterations of membrane phospholipids in the dorsolateral prefrontal region. These changes include a decrease in phosphomonoesters and an increase in phosphodiesters. Such biochemical changes in membrane phospholipids are known to occur during normal synaptic pruning in adolescence, but appear to be exaggerated in the dorsolateral prefrontal cortex of schizophrenics. These enhanced biochemical changes may lead to greater synaptic pruning in schizophrenics than in normal individuals and contribute to the reduction in prefrontal neuropil volume that accompanies the disease.

The precise timing of neurodevelopmental abnormalities in schizophrenia remains controversial. Some implicate a fixed neural lesion from early life interacting with normal neurodevelopmental events that take place much later. Others posit that schizophrenia is caused by a deviation from normal brain maturational processes during late adolescence, involving large-scale synaptic elimination or pruning in brain regions critical for cognitive development. Still others propose that early brain pathology acts as a risk factor rather than a sufficient cause so that its effects can only be understood in the light of the individual's later exposure to other risk and protective factors.

Genetics Play a Role in Schizophrenia

As for autism, a substantial body of evidence indicates that schizophrenia is associated with both genetic and environmental factors. Findings pointing to a significant genetic contribution come from family, twin, and adoption studies. Evidence for environmental factors derives from studies of prenatal intrauterine environment, neonatal obstetric complications, and postnatal brain insults. Currently, it remains an open question whether these environmental factors play a necessary role in the neurodevelopmental abnormalities responsible for psychotic brain disorders or whether their role is additive or interactive with genetic contributions.

People with schizophrenia are more likely than others to have schizophrenic relatives. The most compelling evidence is the 50% concordance rate for monozygotic twins relative to 15% concordance for dizygotic twins. Neuroimaging studies of discordant monozygotic twins consistently indicate more prominent structural brain abnormalities in the ill twin. More importantly, a neuroimaging study demonstrated that schizophrenia is associated with dysfunction of a prefrontal-limbic network (Weinberger et al., 1992). In this study, monozygotic twins discordant for schizophrenia underwent PET blood flow scans while they performed a task known to activate the dorsolateral prefrontal cortex in normal subjects. This task, known as the Wisconsin Card Sorting Task (WCST), measures abstract problem solving requiring attention and working memory. The participant sees four cards bearing designs that differ in color, form, and number of elements. The subject's task is to sort the stack of cards into piles in front of one or another of the stimulus cards. The subject is told whether the sorting choice is correct or incorrect. The test follows this principle: the correct solution is first color; once the subject has figured out this solution, without warning the correct solution then becomes form. Thus, the subject must now inhibit classifying cards on the basis of color and shift to classifying on the basis of form. Once the subject has succeeded at selecting form, the correct solution again changes unexpectedly, now to the number of elements.

Shifting response strategies is difficult for patients with frontal lobe dysfunction. In the affected schizophrenic twin, prefrontal activation correlated strongly with both left and right hippocampal volumes; a relationship that was not found in the unaffected twins. These data suggest that schizophrenia could result from a dysfunction of a distributed neural network involving the prefrontal cortical areas and the limbic structures.

Molecular studies also have demonstrated schizophrenia to be a complex genetic disorder and have implicated several possible susceptibility loci, including regions on chromosomes 1q, 6p, 8p, 13q, and 22q. Of genes mapped, the 22q11 gene (a common functional polymorphism of catechol-O-methyltransferase (COMT), which is a methylation enzyme that metabolizes released dopamine) has been a popular candidate because of the long-hypothesized role of dopamine in schizophrenia.

Sex Hormones May Influence Neurodevelopmental Disorders

There are gender differences in the occurrence of neurodevelopmental disorders. The gender ratio of incidence is skewed toward boys for a variety of developmental disabilities, including severe mental retardation (1.3 boys/1 girl), speech and language disorders (2.6/1), learning difficulties (2.2/1), dyslexia (4.3/1), schizophrenia (1.9/1), and autism (4/1). Why? Although the existence of sex differences in the area of language and learning disorders has been contested on the grounds that it is a result of teacher and clinician bias in identifying these disorders, studies have supported the existence of a significant gender difference in the incidence of these disorders (Flannery et al., 2000). Moreover, gender differences are also seen in clear-cut phenomena such as complications of pregnancy and birth, which are more common in male infants, and these differences cannot be explained as a reflection of investigator or clinician bias. Several theories have attempted to explain these gender differences.

One theory posits that male but not female fetuses invoke a form of "antigenic" response from the mother during pregnancy, being recognized by the immune system as "foreign." The hostile environment thus created for the male fetus in utero would explain not only a higher incidence of developmental disorders among boys, but also the finding that later-born sons are increasingly likely to be affected adversely (consistent with increased immune responsiveness on repeated exposure).

Another theory (the Geschwind theory, as it came to be known, after the neurologist Norman Geschwind) proposed that the uneven gender ratios in developmental disorders are associated with exposure to some "male factor" (possibly androgen) during the last trimester of fetal development. This factor acts to slow cortical maturation in male fetuses, particularly in the left hemisphere, rendering them more susceptible to perturbation during the normal course of development. This exposure to a male hormone would explain the higher incidence of developmental disorders among boys. Conversely, faster central neurons system development in female fetuses would enable them to better withstand insult during late pregnancy and birth. This idea is supported by evidence that female infants appear to show better cognitive recovery than male infants from intracranial hemorrhage associated with prematurity.

Summary

This section examined two forms of developmental abnormalities: autism and schizophrenia. In both cases it is possible to link the behavioral manifestations of the disorder to anomalies in neurobiological systems and to identify a genetic component. In both disorders, too, affected individuals do not show localized damage in circumscribed brain regions, but show evidence of cellular disturbance early in brain development (i.e., during neuromigration), which gives rise to pervasive and dysfunctional reorganization of critical neural systems underlying complex behaviors. Emerging evidence suggests that similar mechanisms may characterize other developmental disabilities, including dyslexia, mental retardation, and attention deficit disorders. Current findings also support the existence of sex differences in brain development and organization, which may in turn differentially affect the response of the brain to early injury.

Pathological Manifestations of Cognitive Aging: Dementia

Dementia is a generic term that refers to debilitating deterioration in more than one domain of cognitive function. The involvement of multiple capacities distinguishes dementia from other disorders, such as amnesia and aphasia, that affect a single functional domain (memory and language, respectively). Approximately 50 disorders are known to cause dementia. Most of these are progressive in nature, increasing in severity over time. The age at onset and the rate of progression of symptoms differ dramatically among the major dementing disorders. Most have an insidious onset and develop slowly, sometimes over a period of many years. These include Alzheimer's disease, Huntington's disease, frontotemporal dementia, and the dementia observed in a subpopulation of patients with Parkinson's disease. The rare neurodegenerative disorder Creutzfeldt–Jakob disease also develops insidiously, but is distinguished by a very rapid rate of progression, often spanning a year or less from the onset of dementia to death. Vascular dementia (or multi-infarct dementia) follows still another pattern of decline. Initial cognitive symptoms develop acutely, but in this case the clinical course typically proceeds in a step-wise fashion over many years, with periods of relative stability punctuated by abrupt deterioration. Age is a major risk factor for dementing illnesses, and the following sections focus on the most prevalent example of age-related dementia, Alzheimer's disease.

Alzheimer's Disease Is the Most Common Cause of Dementia

In 1906, the German scientist Alois Alzheimer reported a case study of a woman in her late 40s who developed memory impairment and deficits in other mental faculties that progressively worsened in severity until her death in her early 50s. Subsequent neuropathological examination of this patient's brain revealed grossly apparent and widespread neocortical atrophy that Alzheimer concluded was inconsistent with any previously described disease. Although the disorder that would come to bear Alzheimer's name originally was thought to be quite rare, it is now recognized as the most common form of dementia, accounting for an estimated 50% of all cases. Current estimates are that more than 4 million people in the United States, and 26 million people worldwide, suffer from Alzheimer's disease. The prevalence of the disease is tightly coupled to age, and increases after the fourth decade of life. The disease affects about 10% of persons over the age of 65, and by 85, at least one in four individuals is afflicted. Demographic trends indicate that the elderly comprise the fastest growing segment of the population, and as a consequence, the number of people suffering the devastating impact of Alzheimer's disease is projected to quadruple by 2050. Not surprisingly, efforts to understand the basis of this and related disorders are among the most active research areas in modern neuroscience.

Memory Impairment Is a Central Feature of Alzheimer's Disease

Alzheimer's original case study provided an accurate description of the clinical course for many patients with this disease. Impairment in establishing and maintaining memories for recent events is frequently the initial sign that brings the patient to the attention of health care professionals. At this early stage, memory for the remote past is relatively preserved, as are other forms of memory that do not require medial temporal lobe structures. Thus, the neuropsychological profile of the Alzheimer's patient provides important clues about the neuropathological progression of this disorder, suggesting that the systems responsible for declarative memory are especially vulnerable. Neuroanatomical studies of the diagnostic hallmarks of Alzheimer's disease confirm this proposal (described in the next section). Progression of the disorder is marked by inexorable decline across multiple cognitive capacities. Although generally alert and responsive in early and middle phases of the disorder, Alzheimer's patients often experience difficulty identifying the meaning of simple words, uses for common household objects, or the meaning of numbers. Other accompanying disturbances can include general confusion, agitation, delusions, social disinhibition, and paranoia. Language abilities can deteriorate progressively, and at the end stage of disease, patients are often mute and densely amnesic for both recent and more remote events. As might be predicted on the basis of this broad profile of impairment, the distribution of neuropathology in individuals dying with late-stage Alzheimer's disease is extensive, involving widespread areas of the limbic system and association cortices, along with many subcortical brain regions, including the basal forebrain cholinergic system, striatum, thalamus, and cerebellum.

Amyloid-Containing Plaques and Neurofibrillary Tangles Are Diagnostic Hallmarks of Alzheimer's Disease That Preferentially Target Memory-Related Brain Regions

There are two types of neuropathological lesions that, when present in sufficiently high numbers, are the diagnostic hallmarks of Alzheimer's disease (Fig. 45.6). Plaques comprise extracellular deposits of amyloid β-protein (Aβ) that, in their mature form, typically are surrounded by a shell of dystrophic neurites (Fig. 45.6B). The neuritic component of plaques consists of morphologically abnormal dendrites and axons of multiple neuronal types, typically associated with substantial numbers of glial cells. The amyloid core of plaques can contain multiple species of Aβ, including a form ending at amino acid 42 that is prone to aggregation (Aβ_{42}), and a slightly shorter species (Aβ_{40}) that is normally produced more abundantly in neurons. Plaques vary substantially in size, and although the time it takes for mature neuritic plaques to form is not known precisely, they are likely to evolve slowly, over many years.

In contrast to extracellular plaques, the second neuropathological hallmark of Alzheimer's disease

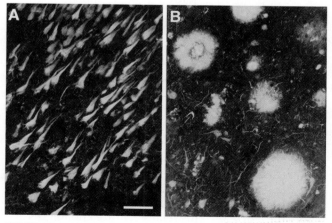

FIGURE 45.6 Photomicrographs of neurofibrillary tangles (A) and amyloid-containing plaques (B) in the hippocampal formation of a patient that died with late-stage Alzheimer's disease. Note that the size and morphological characteristics of plaques in B vary widely. Scale bar: 50 μm. Courtesy of P. Hof, Mount Sinai School of Medicine.

is localized intracellularly. Termed neurofibrillary tangles, these are composed of densely packed, abnormal fibers that occupy much of the cytoplasm and proximal neuronal processes of affected cells (Fig. 45.6A). Indeed the density of these fibers in many tangle-bearing neurons is sufficiently high that the cell body appears swollen and the nucleus displaced. The term paired helical filaments (PHF), sometimes used to describe this type of pathology, derives from electron microscopic evidence that tangles consist of pairs of twisted filaments with a highly regular periodicity. Studies using biochemical and immunocytochemical methods have demonstrated that PHF are composed of a cytoskeletal protein found in normal, healthy neurons; the microtuble-associated protein tau. Cytoskeletal abnormalities can disrupt intracellular trafficking of materials that are critical for cell viability, culminating in neuron death. Defining the mechanisms that initiate the hyperphosphorylation of tau and its aggregation into mature neurofibrillary pathology is a prominent avenue of research on Alzheimer's disease.

Investigations aimed at mapping the neuroanatomical distribution of pathology during early and later stages of Alzheimer's disease help account for the progression of clinical symptoms in patients with this disorder. Early in the course of the disease, when memory impairment is often the predominant complaint, plaques and tangles are particularly abundant in the entorhinal cortex and associated projection targets in the hippocampus. Indeed, in many cases virtually all of the neurons that originate the perforant path projection to the hippocampus exhibit neurofibrillary tangles. Thus, the neuropathology of Alzheimer's disease appears to initially target circuitry critical for normal declarative memory, disrupting a major source of neocortical input that the medial temporal lobe system uses to establish memory for ongoing events. As the severity of cognitive impairment progresses, the distribution of neuropathology becomes more widespread, affecting cortical and subcortical systems that support language, semantic knowledge, abstract reasoning, and other capacities.

Genetic Contributions to Alzheimer's Disease

Important clues about the basis of Alzheimer's disease have come from the study of people afflicted with an "early onset" form of the disorder that emerges before the age of 65. This relatively rare and aggressive condition exhibits an increased prevalence in certain families and therefore is thought to have a significant genetic component. To date, three gene mutations linked to early onset Alzheimer's have been identified: one involving the amyloid precursor protein gene on chromosome 21, another in the presenilin 1 gene on chromosome 14, and the third in the presenilin 2 gene on chromosome 1. Each of these mutations is expressed in an autosomal-dominant fashion such that half the offspring of an affected parent will develop the disease. Importantly, the common thread among gene mutations associated with early onset Alzheimer's disease is that each of them influences processing of the amyloid precursor protein (Selkoe, 2001).

Several additional lines of evidence support the suggestion that the deposition of amyloid is central to the pathogenesis of Alzheimer's disease. By far the most common form of the disorder (accounting for an estimated 98% of cases) has a relatively late onset, after age 65. Among a number of recently identified candidate genes, one risk factor for late-onset Alzheimer's disease is the allelic composition of the apolipoprotein E gene on chromosome 19 (ApoE); that is, a gene that codes for a glycoprotein involved in cholesterol transport and metabolism. The ApoE gene has three alleles (designated ApoE2, ApoE3, and ApoE4), and patients with Alzheimer's disease carry the ApoE4 allele with a frequency greatly exceeding individuals without the disease. Moreover, the risk of disease is coupled to the number of copies of this specific allele: individuals with no copies of ApoE4 are less likely than the general population to develop Alzheimer's disease, the presence of one ApoE4 allele increases the disease risk four times, and people with two copies of ApoE4 are eight times more likely than the population at large to develop Alzheimer's. Although the precise mechanism by which ApoE4 mediates disease susceptibility is the subject of intensive investigation, one hypothesis is that it involves the disruption of processes responsible for scavenging and clearing amyloid peptide from the extracelluar space in brain.

Additional findings relevant to the role of amyloid in Alzheimer's disease come from research on Down's syndrome. Down's syndrome results from the presence of an extra copy of the same chromosome that carries the gene for the amyloid precursor protein; chromosome 21. Thus, it is of considerable interest that people with Down's syndrome who survive until the fourth decade of life invariably develop a distribution of amyloid plaques in the brain similar to that observed in Alzheimer's patients. To the degree that aberrant processing or clearance of amyloid comprises the single common pathogenic mechanism of Alzheimer's disease, current findings encourage the view that interventions targeting this pathway might be developed to prevent or reverse this defect, realizing an effective treatment for the most common cause of age-related dementia.

Prospects for the Treatment of Alzheimer's Disease

Although Alzheimer's disease is presently incurable, several pharmacological treatments have been approved for the clinical management of cognitive impairment associated with the disorder. These interventions all target the function of the neurotransmitter acetylcholine. Although available treatments offer modest, transient benefit for certain Alzheimer's patients, they fall far short of an effective cure or prevention. Encouraging preliminary progress toward the latter goal is based on the proposal that aberrant amyloid deposition represents the central pathogenic event responsible for the disorder. Whereas laboratory animals fail to develop Alzheimer's disease spontaneously, one can take advantage of a transgenic mouse model that accumulates amyloid-containing plaques in an age-dependent manner. These mice also develop spatial learning deficits relatively early in life, in temporal correspondence with the emergence of amyloid deposits. When transgenic animals were vaccinated at multiple time points during the first four months of life with Aβ, and subsequently examined for the effects of immunization on behavior and neuropathology, treatment attenuated the spatial learning deficits substantially (Janus et al., 2000). Nevertheless, performance remained mildly impaired relative to nontransgenic controls. This partial protection from cognitive decline was also accompanied by roughly a 50% reduction in the number and size of mature, Aβ-containing plaques, relative to nonimmunized, age-matched transgenic controls.

Related findings using a similar experimental approach also demonstrate that immunization with Aβ is even effective in reducing the degree and progression of amyloid deposition when provided to older animals at a point in the life span when this pathology is already well established (Selkoe, 2001). The development of similar approaches for clinical use faces significant hurdles, including early indications that parallel treatment in humans can produce life threatening cerebral inflammation. Nonetheless, the recent findings from animal models encourage the view that vaccine therapies might ultimately yield an effective route for the treatment and prevention of Alzheimer's disease.

Summary

This chapter has considered the neurobiological basis of cognitive development and aging. Early cognitive development is marked by the emergence of a remarkably complex set of adaptive capacities, including language skills, reasoning abilities, attention, declarative memory, and executive functions. The vast majority of neurons—the essential building blocks of the central nervous system—are present at birth in humans, and rather than depending on the recruitment of new neurons, cognitive development appears to involve the fine tuning and establishment of appropriate circuitry among existing cells. Given this complex orchestration of processes, it is not surprising that many major disorders affecting cognitive function appear to result from a disruption of normal neurodevelopmental events. Higher order cognitive capacities are also vulnerable to deterioration during normal aging. Whereas widely distributed neuron death was previously presumed to underlie these deficits, it now appears that regionally selective changes in connectivity and cell biological function in critical circuits are the causative factors. Aging is also associated with some of the most devastating pathological conditions affecting cognitive function, including Alzheimer's disease. The dramatic progress realized in recent years, for the first time, has established a basis for the rationale development of strategies for disease prevention and treatment.

References

Alvarado, M. C. and Bachevalier, J. (2000). Revisiting the maturation of medial temporal lobe memory functions in primates. *Learn. Mem.* **7**, 244–256.

Bachevalier, J. and Vargha-Khadem, F. (2005). The primate hippocampus: Ontogeny, early insult and memory. *Curr. Opin. Neurobiol.* **15**, 168–174.

Baron-Cohen, S., Ring, H. A., Wheelwright, S., Bullmore, E. T., Brammer, M. J., Simmons, A., and Williams, S. C. R. (1999). Social intelligence in the normal and autistic brain: An fMRI study. *Eur. J. Neurosc.* **11**, 1891–1898.

Bauer, P. J. (2002). Long-term recall memory: Behavioral and neurondevelopmental changes in the first 2 years of life. *Cur. Direct. Psychol. Sci.* **11**, 137–141.

Benes, F. M., Turtle, M., Khan, Y., and Farol, P. (1994). Myelination of a key relay zone in the hippocampal formation occurs in the human brain during childhood, adolescence, and adulthood. *Arch. Gen. Psychiatr.* **51**, 477–484.

Brody, H. (1970). Structural changes in the aging nervous system. *Interdisc. Top. Gerontol.* **7**, 9–21.

Burke, S. N. (2006). Neural plasticity in the ageing brain. *Nat. Rev. Neurosci.* **7**, 30–40.

Chugani, H. T., Phelps, M. E., and Mazziotta, J. C. (1993). Positron emission tomography study of human brain functional development. *In* "Brain Development and Cognition: A Reader" (M. H. Johnson, ed.), Chap. 8, pp. 125–143. Blackwell, Cambridge, MA.

Conel, J. L. (1939–1963). "The Postnatal Development of the Human Cerebral Cortex," Vols. I–VI. Harvard Univ. Press, Cambridge, MA.

de Haan, M., Mishkin, M., Baldeweg, T., and Vargha-Khadem, F. (2006). Human memory development and its dysfunction after early hippocampal injury. *TINS* **29**, 374–381.

Diamond, A. (1990). The development and neural bases of memory functions as indexed by the AB and delayed response tasks in human infants and infant monkeys. *Ann. N.Y. Acad. Sci.* **608**, 637–676.

Flannery, K. A., Liederman, J., Daly, L., and Schultz, J. (2000). Male prevalence for reading disability is found in a large sample of black and white children free from ascertainment bias. *J. Int. Neuropsychol. Soc.* **6**, 433–442.

Gallagher, M., Burwell, R., and Burchinal, M. (1993). Severity of spatial learning impairment in aging: Development of a learning index for performance in the Morris water maze. *Behav. Neurosci.* **107**, 618–626.

Gallagher, M. and Rapp, P. R. (1997). The use of animal models to study the effects of aging on cognition. *Annu. Rev. Psychol.* **48**, 339–370.

Geinisman, Y., de Toledo-Morrell, L., Morrell, F., Persina, I. S., and Rossi, M. (1992). Age-related loss of axospinous synapses formed by two afferent systems in the rat dentate gyrus as revealed by the unbiased stereological disector technique. *Hippocampus* **2**, 347–444.

Giedd, J. N., Vaituzis, A. C., Hamburger, S. D., Lange, N., Rajapakse, J. C., Kaysen, D., Vauss, Y. C., and Rapoport, J. L. (1996). Quantitative MRI of the temporal lobe, amygdala, and hippocampus in normal human development: Ages 4–18 years. *J. Comp. Neurol.* **366**, 223–230.

Hedden, T. and Gabrieli, J. D. (2005). Healthy and pathological processes in adult development: New evidence from neuroimaging of the aging brain. *Curr. Opin. Neurol.* **18**, 740–747.

Huttenlocher, P. R. (1990). Morphometric study of human cerebral cortex development. *Neuropsychologia* **28**, 517–527.

Huttenlocher, P. R. and De Courten, C. (1987). The development of synapses in striate cortex of man. *Hum. Neurobiol.* **6**, 1–9.

Janus, C., Pearson, J., McLaurin, J., Mathews, P. M., Jiang, Y., Schmidt, S. D., Chishti, M. A., Horne, P., Heslin, D., French, J., Mount H. T., Nixon, R. A., Mercken, M., Bergeron, C., Fraser, P. E., St. George-Hyslop, P., and Westaway, D. (2000). Aβ peptide immunization reduces behavioral impairment and plaques in a model of Alzheimer's disease. *Nature* **408**, 915–916.

Lenroot, R. K. and Giedd, J. N. (2006). Brain development in children and adolescents: Insights from anatomical magnetic resonance imaging. *Neurosci. Biobehav. Rev.* **30**, 718–729.

Morrison, J. H. and Hof, P. R. (1997). Life and death of neurons in the aging brain. *Science* **278**, 412–419.

Pelphrey, K., Adolphs, R., and Morris, J. P. (2004). Neuroanatomical substrates of social cognition dysfunction in autism. *MRDD Res. Rev.* **10**, 259–271.

Peters, A., Morrison, J. H., Rosene, D. L., and Hyman, B. T. (1998). Are neurons lost from the primate cerebral cortex during normal aging? *Cerebr. Cort.* **8**, 295–300.

Rapp, P. R. and Gallagher, M. (1996). Preserved neuron number in the hippocampus of aged rats with spatial learning deficits. *Proc. Natl. Acad. Sci. USA* **93**, 9926–9930.

Rapp, P. R., Deroche, P. S., Mao, Y., and Burwell, R. D. (2002). Neuron number in the parahippocampal region is preserved in aged rats with spatial learning deficits. *Cerebr. Cort.* **12**, 1171–1179.

Rosenzweig, E. S. and Barnes, C. A. (2003). Impact of aging on hippocampal function: Plasticity, network dynamics, and cognition. *Prog. Neurobiol.* **69**, 143–179.

Schore, A. N. (1996). The experience-dependent maturation of a regulatory system in the orbital prefrontal cortex and the origin of developmental psychopathology. *Dev. Psychopathol.* **8**, 59–87.

Selkoe, D. J. (2001). Alzheimer's disease: Genes, proteins, and therapy. *Physiol. Rev.* **81**, 741–766.

Seress, L. (2001). Morphological changes of the human hippocampal formation from midgestation to early childhood. *In* "Handbook of Developmental Cognitive Neuroscience" (C. A. Nelson and M. Luciana, eds) Chap. 4, pp. 45–58. MIT Press, Cambridge, MA.

Weinberger, D. R., Berman, K. F., Suddath, R., and Torrey, E. F. (1992). Evidence of dysfunction of a prefrontal-limbic network in schizophrenia: A magnetic resonance imaging and regional cerebral blood flow study of discordant monozygotic twins. *Am. J. Psychiat.* **149**, 890–897.

Wilson, I. A., Gallagher, M., Eichenbaum, H., and Tanila H. (2006). Neurocognitive aging: Prior memories hinder new hippocampal encoding. *TINS* **29**, 662–670.

Yakovlev, P. I. and Lecours, A. R. (1967). The myelogenetic cycles of regional maturation of the brain. *In* "Regional Development of the Brain in Early Life" (A. Minkowski, ed.), pp. 3–70. Blackwell, Oxford.

Suggested Readings

Barnes, C. A., Suster, M. S., Shen, J., and McNaughton, B. L. (1997). Multistability of cognitive maps in the hippocampus of old rats. *Nature* **388**, 272–275.

Bearden, C. E., Meyer, S. E., Loewy, R. L., Niendam, T. A., and Cannon, T. D. (2006). The neurodevelopmental model of schizophrenia: Updated. *In* "Developmental Psychopathology" (D. Cicchetti and D. J. Cohen, eds.), 2nd ed., Vol. 3, pp. 542–569, John Wiley & Sons, New York.

Dawson G. and Toth, K. (2006). Autism spectrum disorders. *In* "Developmental Psychopathology" (D. Cicchetti, and D. J. Cohen, eds.), 2nd. ed., Vol. 3, pp. 318–357, John Wiley & Sons, New York.

Dean, R. L. and Bartus, R. T. (1988). Behavioral models of aging in nonhuman primates. *In* "Handbook of Psychopharmacology" (L. L. Iversen, S. D. Iversen, and S. H. Snyder, eds.), Vol. 20, pp. 352–392. Plenum, New York.

Deboysson-Bardies, B., de Schonen, S., Jusczyk, P., McNeilage, P., and Morton, J. (1993). "Developmental Neurocognition: Speech and Face Processing in the First Year of Life." Kluwer Academic, Boston.

Foster, T. C. (1999). Involvement of hippocampal synaptic plasticity in age-related memory decline. *Brain Res. Rev.* **30**, 236–249.

Grady, C. L. (2000). Functional brain imaging and age-related changes in cognition. *Biol. Psychol.* **54**, 259–281.

Keshavan, M. S. and Murray, R. M. (1997). "Neurodevelopment and Adult Psychopathology." Cambridge Univ. Press, Cambridge.

Nelson, C. A. and Luciana, M. (2001). "Handbook of Developmental Cognitive Neuroscience." MIT Press, Cambridge, MA.

Stromswold, K. (1995). The cognitive and neural bases of language acquisition. *In* "The Cognitive Neurosciences" (M. Gazzaniga, ed.), pp. 855–870. MIT Press, Cambridge, MA.

Tanila, H., Shapiro, M., Gallagher, M., and Eichenbaum, H. (1997). Brain aging: Changes in the nature of information coding by the hippocampus. *J. Neurosci.* **17**, 5155–5166.

Tanila, H., Sipila, P., Shapiro, M., and Eichenbaum, H. (1997). Brain aging: Impaired coding of novel environmental cues. *J. Neurosci.* **17**, 5167–5174.

Winkler, J., Thal, L. J., Gage, F. H., and Fisher, L. J. (1998). Cholinergic strategies for Alzheimer's disease. *J. Mol. Med.* **76**, 555–567.

Peter R. Rapp and Jocelyne Bachevalier

Visual Perception of Objects

THE PROBLEM OF OBJECT RECOGNITION

Accurate perception and recognition of objects is a crucial ability for visually dependent organisms such as humans—but it is also a complex one. As object and viewer move with respect to one another, the images cast by the object on the retinas of the observer can change radically. Not only will image position and size change, but surfaces may be foreshortened, occluded, or newly revealed. Similarly, changes in ambient illumination can reveal new surfaces, hide others in shadow, or introduce spurious edges where light ends and shadow begins. Despite this, human and nonhuman primates recognize objects quite accurately. This means that the visual system is able to extract a reasonably constant "representation" for the shapes and identities of meaningful objects despite such transformations. Image processing by the retina, visual thalamus, and primary visual cortex begins this process by extracting features and representing them over progressively larger receptive fields. However, in order to understand how the brain represents the identity of an object—its defining, invariant features of shape, relations among parts, and its significance—one must examine how circuits beyond the primary visual cortex encode and retain crucial information of a more global or "high-level" nature.

Visual processing can be subdivided into low-level, intermediate-level, and high-level components. The initial, low-level stages are those in which the brain extracts information about the physical properties of the stimulus, such as the edges, brightness, color, and motion of objects present in scenes (Chapter 27). This processing often is described as *bottom up* because it can proceed from low (input) levels up through the visual system without feedback from cognitive processes such as memory. Intermediate-level vision involves combining information about these physical properties in order to detect global properties of an object, such as object shape and the orientation of the object in depth. High-level processes include those that endow the stimulus with meaning, leading to object recognition and classification. For example, recognition and classification depend on object knowledge and context, and can therefore be considered *top down* because the neural processing is influenced directly and profoundly by cognitive factors. Other top-down mechanisms, such as selective attention, can operate at even early stages of visual processing (Chapter 48).

This chapter is concerned primarily with the neural basis of the high-level components of object vision. First, historical evidence is reviewed that helped localize the crucial circuitry involved to the ventral parts of the temporal lobes. Next, from studies in nonhuman primates, the main pathways are described for object perception and recognition, in addition to neural codes that lead to object vision. Then, neuroimaging and electrophysiological studies are reviewed to show that very similar processing pathways and mechanisms exist in humans. In addition, human studies are reviewed to consider how the brain may treat special categories of objects and the ways in which experience with objects tunes their neural representations. Finally, visual object perception is considered in the broader context of general object knowledge and memory.

SUBSTRATES FOR OBJECT PERCEPTION AND RECOGNITION: EARLY EVIDENCE FROM BRAIN DAMAGE

Early Studies of Human Patients Pointed to the Inferior Temporal Cortex as a Site for Object Agnosia

The earliest clues about the neural bases of object perception and recognition came from the study of brain-damaged humans with *visual agnosia*. Visual agnosia refers to a set of disorders affecting object recognition, in which elementary visual capacities such as acuity and visual fields are preserved or grossly intact (for review, see Farah, 1990). Early attempts to localize the lesions in such cases appeared mainly in the German neurological literature of the early 1900s. Writers as early as Potzl in 1928 noted the importance of inferior regions of the temporal lobes, along with adjacent ventral occipital cortex, in most cases of visual agnosia. Subsequent studies upheld these general findings and confirmed the central role of the lingual and fusiform gyri (Fig. 46.1). Patients with visual agnosia typically have bilateral (or, less commonly, right-sided) damage to these regions.

Two Types of Visual Agnosia Have Been Distinguished Historically

Lissauer is credited with the first detailed discussion of visual agnosia in 1890. He distinguished between two types, which he termed *apperceptive* and *associative*; these terms are still used commonly today. Agnosic patients, whose elementary perception of shape is impaired (despite intact or roughly normal visual acuity), are classified as apperceptive agnosics because their impairment is presumed to be at an early stage of the perceptual processing needed for object recognition. Patients with such deficits are relatively rare. Usually they have suffered diffuse brain damage, often from carbon monoxide poisoning, as opposed to the more localized damage to posterior cortex and inferior temporal regions seen in the majority of agnosics. Agnosics whose elementary shape perception is grossly intact are classified as associative agnosics, following Lissauer's contention that they cannot associate a percept with its meaning (i.e., its visual representation stored in memory).

Although the notion that these latter patients show a deficit in association *per se* has been called into question, the impairments they show reveal critical features of how the object recognition system tends to break down. To be considered an associative visual agnosic,

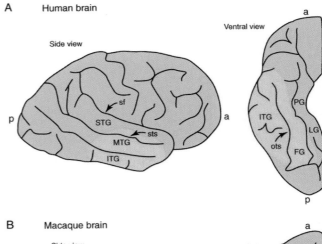

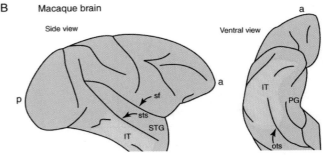

FIGURE 46.1 (Top) Region of the human brain (in shading) within which large bilateral or right-sided lesions cause associative visual agnosia. FG, fusiform gyrus; ITG, inferior temporal gyrus; LG, lingual gyrus; MTG, middle temporal gyrus; PG, parahippocampal gyrus; STG, superior temporal gyrus; ots, occipito-temporal sulcus; sf, Sylvian or lateral fissure; sts, superior temporal sulcus. (Bottom) Region in the macaque monkey (in shading) referred to as the inferior temporal cortex, within which bilateral lesions cause visual discrimination and recognition deficits analogous to agnosic syndromes in humans. a, anterior; p, posterior.

the patient is tested to determine the presence of the following features. First, the patient must show difficulty recognizing visually presented objects, as measured by both naming and nonverbal tests of recognition, such as sorting objects by category (e.g., grouping kitchen utensils together, separate from sports equipment) or pantomiming the uses of the object. Second, the patient must show knowledge of the objects through sensory modalities other than vision; that is, be able to recognize items by sound or touch or through verbal questioning ("What is an egg beater?"). Third, the patient must have sensory and perceptual capacity sufficient to allow him or her to describe the appearance of an unrecognized object, to draw it, or identify whether it is the same as or different from a second item. An illustrative case of associative visual agnosia is summarized in Box 46.1. Figure 46.2 shows some copies drawn by this patient, who

BOX 46.1

CASE STUDY OF ASSOCIATIVE VISUAL AGNOSIA

The subject was a 47-year-old man who had suffered an acute loss of blood pressure with resulting brain damage. His mental status and language abilities were normal, and his visual acuity was 20/30, with a right homonymous hemianopia (blindness in the right visual hemifield). His one severe impairment was an inability to recognize most visual stimuli. For the first three weeks in the hospital, the patient could not identify common objects presented visually and did not know what was on his plate until he tasted it. He identified objects immediately on touching them.

When shown a stethoscope, he described it as "a long cord with a round thing at the end" and asked if it could be a watch. He identified a can opener as a key. Asked to name a cigarette lighter, he said, "I don't know" but named it after the examiner lit it. He said he was "not sure" when shown a toothbrush. He was never able to describe or demonstrate the use of an object if he could not name it. If he misnamed an object, his demonstration of use would correspond to the mistaken identification. Identification improved very slightly when given the category of the object (e.g., something to eat) or when asked to point to a named object instead of being required to give the name. When told the correct name of an object, he usually responded with a quick nod and often said, "Yes, I see it now." Then, often he could point out various parts of the previously unrecognized item as readily as a normal subject (e.g., the stem and bowl of a pipe and the laces, sole, and heel of a shoe). However, if asked by the examiner, "Suppose I told you that the last object was not really a pipe, what would you say?" He would reply, "I would take your word for it. Perhaps it's not a pipe." Similar vacillation never occurred with tactilely or aurally identified objects.

After he had spent three weeks on the ward, his object naming ability improved so that he could name many common objects, but this was variable; he might correctly name an object at one time and misname it later. Performance deteriorated severely when any part of the object was covered by the examiner. He could match identical objects but could not group objects by categories (clothing, food). He could draw the outlines of objects (key, spoon, etc.) that he could not identify.

He was unable to recognize members of his family, the hospital staff, or even his own face in the mirror. Sometimes he had difficulty distinguishing a line drawing of an animal face from a man's face but always recognized it as a face. The ability to recognize pictures of objects was impaired greatly, and after repeated testing he could name only one or two of 10 line drawings. He was always able to name geometrical forms (circle, square, triangle, cube). Remarkably, he could make excellent copies of line drawings and still fail to name the subject. He easily matched drawings of objects that he could not identify and had no difficulty discriminating between complex nonrepresentational patterns, differing from each other only subtly. He occasionally failed in discriminating because he included imperfections in the paper or in the printer's ink. He could never group drawings of objects by class unless he could first name the subject. Reading, both aloud and for comprehension, was limited greatly. He could read, hesitantly, most printed letters but often misread "K" as "R," and "L" as "T," and vice versa. He was able to read words slowly by spelling them aloud.

Luiz Pessoa, Roger B. H. Tootell, and Leslie G. Ungerleider

Excerpted from Rubens, A. B. and Benson, D. F. (1971). Associative visual agnosia. *Arch. Neurol.* 24, 305–316.

nevertheless did not recognize the pictures he copied.

Associative Visual Agnosia Involves Deficits in Invariant Object Representations

Closer examination of the manner in which agnosics copy and match pictures they cannot identify shows that they are often abnormally attentive to the small details of the image, rather than being guided by the overall shape or "gestalt" of the object itself. This copying is done in a slow, line-by-line fashion, and patients may classify two pictures as different because of a small flaw in the printing. These observations suggest that the associative agnosic patients are unable to match what is in front of them to a unified, stored visual representation of the object. Such stored representations likely include knowledge about the *invariant*, distinguishing visual attributes (such as the general body shape and relative placement of features of the pig in Fig. 46.2) and/or stored semantic knowledge (the identity of the object as a pig). The underlying

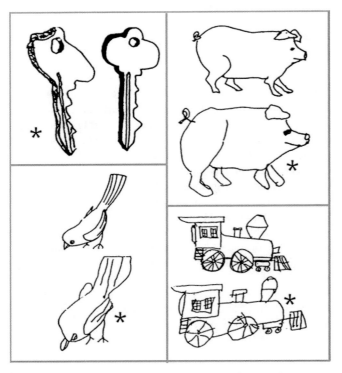

FIGURE 46.2 Pictures that a patient with associative visual agnosia did not recognize but was able to copy. Asterisks indicate the patient's copies. From Rubens and Benson (1971).

problem in visual agnosia could thus result from an inability to generate a high-level representation of the object (a perceptual deficit), a permanent loss of previously stored representations (a memory deficit), or an inability to link the two. In fact, some evidence suggests that different agnosic patients are impaired in distinct stages of visual processing and storage. At the same time, recent work suggests that, at the neural level, "perceptual" and "memory" processes for objects may use the same circuitry (and thus, might be experimentally inseparable), as discussed later in this chapter.

Object Recognition in Monkeys Depends on the Inferior Temporal Cortex

Ablation studies in monkeys have (1) confirmed the crucial role of the temporal lobes in visual object perception and recognition, and (2) clarified the nature of the visual representations used in the object recognition process. One of the earliest experimental studies on the role of the temporal cortex in visual recognition was that of Klüver and Bucy, who described in the 1930s what is now known as the Klüver–Bucy syn-

drome. When these researchers removed the temporal lobes of monkeys bilaterally, the monkeys demonstrated complex changes in visual, social, sexual, and eating behavior. These changes appeared to result from a combination of perceptual, memory, and motivational impairments. Subsequent research eventually identified the region now known as inferior temporal (IT) cortex as the critical area for producing the visual component of the syndrome. Monkeys with bilateral IT lesions are impaired both in learning to distinguish between different visual patterns or objects, and in retaining previously acquired visual discriminations. Figure 46.1 shows the location of this area in the macaque brain, relative to the areas implicated in agnosia in humans.

What type of stimulus representation is impaired after IT lesions? A great deal of research been devoted to this question. Generally, such experiments measure whether monkeys can discriminate different stimuli, before and after experimental lesions in IT cortex—and from any such differences, object-processing mechanisms are inferred. A current summary might be that the IT cortex represents aspects of the intrinsic shape of the stimulus that are useful for recognition (e.g., the salient features in a face and their relative positions), generalizing across idiosyncratic aspects of stimulus appearance that depend on specific viewing conditions (see later).

Position in the visual field (retinotopy) is one of these specific visual properties that is independent of object identity, and normal monkeys will easily generalize a visual discrimination learned in one part of the visual field to the other. For example, if monkeys learn to tell apart stimuli in one half of visual space, they will easily be able to perform the discrimination when the same stimuli are presented to the other half of visual space. However, monkeys with IT lesions are impaired at this generalization; presumably they have lost representations that are invariant across visual field locations. Retinal image size of a stimulus is a related visual property that depends on viewing conditions and not just intrinsic object geometry, and this is another perceptual distinction that is difficult for monkeys with IT lesions. For example, monkeys trained to discriminate between two disks of different absolute sizes are unable to perform the discrimination when those disks are presented at variable distances (hence with variable retinal image sizes). Instead, they respond on the basis of retinal size or distance. This result implies that representations within IT cortex normally encode the absolute size of an object, an intrinsic object property useful for recognition, rather than its distance *per se* or its retinal image size. Finally, variations in illumination prevent monkeys with IT lesions from seeing the equivalence of objects, implying that representations in IT cortex are unaffected by patterns of shadow and light falling on object surfaces.

Summary

Studies of brain damage in humans, and experimental lesions in monkeys, both point to the crucial role of the inferior temporal cortex during object perception and recognition. In both species, analysis of the ensuing deficits suggests that this region participates in creating and storing neural representations of the defining features and invariant aspects that identify a visual object or class of objects.

VISUAL PATHWAYS FOR OBJECT PROCESSING IN NONHUMAN PRIMATES

The Visual Cortex of Monkeys Can Be Divided into Occipitotemporal and Occipitoparietal Pathways

Much of our detailed knowledge of visual cortical organization derives from studies of Old World monkeys of the genus *Macaca*. Similar regions and processing stages are now known to exist in both New World species (Chapter 44) and in humans, as described later in this chapter. The original evidence that the visual cortex is divided into separate processing streams was based on the contrasting effects of inferior temporal and posterior parietal cortex lesions in monkeys. For example, inferior temporal lesions cause severe deficits in visual discrimination tasks, but they do not affect animals' performance on visuospatial tasks (e.g., visually guided reaching and judging which of two objects lies closer to a visual landmark). In contrast, parietal lesions do not affect visual discrimination ability but instead cause severe deficits on visuospatial performance. On the basis of behavioral results such as these, Ungerleider and Mishkin (1982) proposed the existence of two processing streams: an occipitotemporal or ventral processing stream for mediating the visual recognition of objects ("what" an object is) and an occipitoparietal or dorsal processing stream for mediating the appreciation of the spatial relationships among objects and the visual guidance toward them ("where" an object is). Goodale and Milner (1992) have proposed a modification of this model, emphasizing "perception" vs. "action" for ventral and dorsal processing

streams, respectively. The dorsal processing stream and its importance for spatial cognition is the topic of Chapter 47.

As described in Chapter 27, physiological studies of cell properties also support the functional distinction between ventral and dorsal pathways. Neurons within the ventral stream (in distinct modules within V1 and V2, area V4, and inferior temporal areas TEO and TE) respond selectively to visual features relevant for object identification, such as color, shape, and texture. Neurons within the dorsal stream (in other modules within with V1 and V2, areas V3, V3A, middle temporal area MT, medial superior temporal area MST, and additional areas in inferior parietal cortex) instead respond selectively to spatial aspects of stimuli, such as the direction and speed of stimulus motion. Such cells also respond when the animal visually tracks a moving target. The remainder of this section describes the anatomy of the *occipitotemporal pathway* for object recognition.

The Object Recognition Pathway in Monkeys Consists of an Interconnected Set of Cortical Areas

In both monkeys and humans, cortical regions comprising the object recognition pathway lie directly adjacent to the primary visual cortex (V1) in the occipital lobe, extending progressively into more anterior and ventral portions of the temporal lobe. Figure 46.3 illustrates the location of these areas and diagrams the forward flow of information within this occipitotemporal pathway in monkeys (see also Fig. 46.10). The cortical analysis of objects begins in V1, where information about contour, orientation, color composition, and brightness is represented in subsets of neurons coding for each point in the visual field (Chapter 27). Information from V1 projects forward to neural subdivisions—interdigitating thin, thick, and interstripe regions—within V2 (Chapters 27 and 44). From the thin and interstripe regions in V2 (specialized, respectively, for color and form information, roughly speaking), neural signals proceed forward to area V4 on the lateral and ventromedial surfaces of the hemisphere and to a posterior inferior temporal area just in front of V4, area TEO. From both V4 and TEO, signals related to object form, color, and texture proceed forward to area TE, the last exclusively visual area within the ventral stream for object recognition. Together, areas TEO and TE constitute the IT cortex.

Progressing forward along the occipitotemporal pathway from V1 to TE, there is a gradual shift in the nature of connectivity, which is significant for the

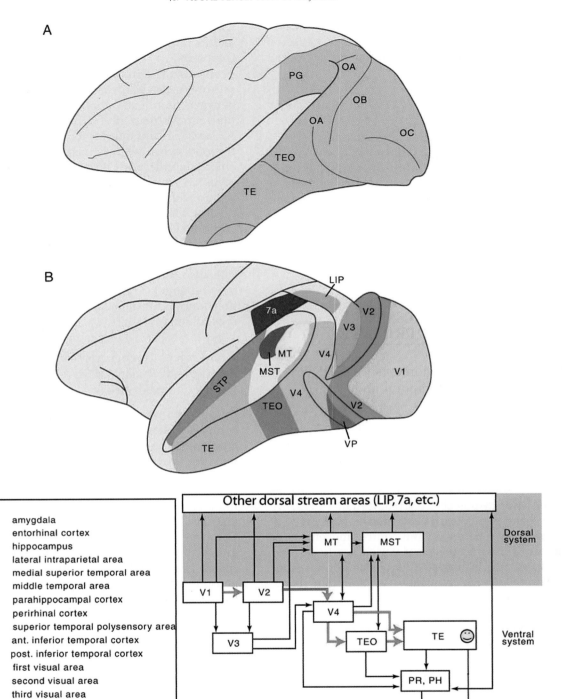

FIGURE 46.3 (A) Schematic diagram of the macaque monkey brain illustrating the two cortical visual system model of Ungerleider and Mishkin (1982). According to this model, there are two major processing pathways, or "streams," in the visual cortex, both originating in the primary visual cortex: a ventral stream, directed into the temporal lobe and crucial for the identification of objects, and a dorsal stream, directed into the parietal lobe and crucial for spatial perception and visuomotor performance. The shaded region indicates the extent of the visual cortex in the monkey, and the labels (OC, OB, OA, TEO, TE, PG) indicate cytoarchitectonic areas according to the nomenclature of von Bonin and Bailey. Adapted from Mishkin *et al.* (1983). (B) A lateral view of the monkey brain illustrating the multiplicity of functional areas within both processing streams. (C) Some of the pertinent connections of the inferior temporal cortex with other cortical areas and medial temporal lobe structures. Red lines indicate the main afferent pathway to area TE, which includes areas V1, V2, V4, and TEO. For simplicity, only projections from lower order to higher order areas are shown, but each of these feed-forward projections is reciprocated by a feedback projection. Faces indicate areas in which neurons selectively responsive to faces have been found. Adapted from Gross *et al.* (1993).

eventual extraction of invariant object features. At successive stages, patterns of cortical projections become successively less topographic or "point to point." While connections from V1 to V2 maintain topographic order, the organization becomes looser between V2 and V4, and inputs to area TE from V4 and TEO retain no obvious retinotopic organization. Likewise, pathways interconnecting areas of the ventral stream on the two sides of the brain via the corpus callosum and/or the anterior commissure tend to be restricted to the representation of the midline of visual space early in the pathway, but their representation becomes more widespread at higher levels in the pathway. This loss of retinotopy within area TE means that single neurons respond to objects anywhere in the visual field. This evidence suggests that the initial explicit information about spatial position in the visual field ("retinotopy") has been converted to information about object identity, in at least some parts of IT cortex.

However, neural processing in the object recognition pathway is not a simple elaboration of information from lower order to higher order areas. At all stages, the neural connections are reciprocal, such that an area receiving feedforward projections from an area earlier in the ventral stream also provides feedback projections to that area. Long-range nonreciprocal connections, such as from area TE back to V1, also exist. On the one hand, feedforward projections provide bottom-up, sensory-driven inputs to subsequent visual areas, and severing these projections disconnects subsequent areas from their visual input. On the other hand, the precise functions of reciprocal,

feedback projections are still unknown. Probably these feedback projections play a top-down role in vision (as in selective attention) by modulating activity in earlier areas of the ventral pathway (Chapter 48). In addition to feedforward and feedback projections, there also appear to be "intermediate-type" projections that link areas at the same level of the visual hierarchy. These are seen most notably between areas of the dorsal and ventral streams. For example, area V4 is interconnected with motion-sensitive areas MT and MST. Thus, an anatomical substrate exists for interactions between the two processing streams, and these might serve to integrate what an object is with where an object is.

All areas within the ventral stream also have heavy interconnections with subcortical structures, notably the pulvinar, claustrum, and basal ganglia. In addition, each ventral stream area also receives subcortical modulatory inputs from ascending cholinergic projections from the basal forebrain, and ascending noradrenergic projections from the locus coeruleus. These projections are thought to play a role in the storage of information in the cortex, and the influence of arousal on information processing, respectively. Finally, visual information is sent *from* IT cortex to the most ventral and anterior reaches of the temporal lobe, notably the perirhinal cortex and parahippocampal areas TF and TH. These regions in turn project, via the entorhinal cortex, to medial temporal lobe structures, such as the hippocampus, which contribute to forming long-term memories of visual objects and their contexts (Chapter 50). Information is also sent from IT cortex to the prefrontal cortex, which plays an important role in working memory, holding an object briefly in mind when it is no longer visible (Chapter 52). Finally, there are direct projections from IT cortex to the amygdala, which is important for attaching emotional valence to a stimulus (Chapter 50).

Summary

Areas along the occipitotemporal pathway of the monkey are organized hierarchically proceeding from V1 > V2 > V4 > TEO > TE. At progressively higher stages of this hierarchy, there is a corresponding loss (or conversion) of retinotopic information, such that information about the spatial location of the object is gradually changed. In addition to this bottom-up information (supplied by feedforward projections from lower order to higher order visual areas within this pathway), there are also reciprocal, feedback projections from higher order to lower order visual areas. These latter projections are thought to play a top-down role in vision.

NEURONAL PROPERTIES WITHIN THE OBJECT RECOGNITION PATHWAY

Cells along the Occipitotemporal Pathway Are Sensitive to Increasingly More Complex Physical Features of Objects

The different cortical areas in the occipitotemporal pathway share a number of physiological characteristics. Consistent with a role in object recognition, all areas in the pathway contain populations of cells sensitive to the shape, color, and/or texture of visual stimuli. However, at progressively higher levels, cells have larger receptive fields and their stimulus selectivity is more complex to characterize. Thus, higher order properties (consistent with invariant representation of objects) usually are attributed to cells in higher tier areas. For instance, many V1 neurons function as local spatial filters, signaling the presence of contours at particular positions and orientations in the visual field. In contrast, an increasingly higher proportion of cells in higher tier areas (e.g., V2 and beyond) apparently respond to *illusory* contours (i.e., contours implied by stimulus context and higher-order properties, not due to simple light-dark contrast), across increasingly larger regions of the visual field.

Additional higher-order, object-related properties are also seen at higher tier cortical levels. One of these properties is increasing stimulus invariance, namely, similar shape selectivity despite variations in the type of lower-order cues used to generate the shape (such as luminance, color, motion, etc). Another of these properties is increasingly strong, silent suppressive zones surrounding the classical, excitatory receptive field of cells. These zones are called silent because visual stimulation of them normally does not cause any change from the baseline activity of the cells—although it can nevertheless powerfully affect responses to a stimulus placed in the classical receptive field. For example, many V4 cells respond maximally to a stimulus only if it stands out from its background based on differences in form—as in the well-known perceptual phenomenon of figure/ground segregation, a computation that remains difficult to achieve in machine-based object recognition.

Farther forward in area TEO of the occipitotemporal pathway, and then into posterior and anterior portions of area TE, there are further increases in the complexity of the critical features needed to activate many neurons, and relative increases in the proportion of cells that are selectively driven by some kind of complex pattern or object (Gross *et al.*, 1972). Thus at the highest levels, it can become very difficult to discover (or prove) the optimal stimulus for a given single cell, by simply testing various visual stimuli and intuiting the response constraints, because the number of possible test stimuli is literally infinite.

This experimental problem is related to the idea, developed sometime in the late 1960s, that every object would be coded by maximal firing in a single cell in IT cortex; this was termed the "grandmother cell" hypothesis because each "grandmother" was thought to cause maximal firing in a single cell (Gross, 2002). This idea, which was controversial from its inception, is considered by most researchers to be overly simplistic. Indeed, there simply are not enough neurons to represent all the different objects we can recognize. Instead, a "population code" is more likely, in which each specific object (and perhaps each viewpoint) causes a correspondingly unique pattern of firing in multiple cells that share connections and overlapping functional selectivity.

This latter idea is supported by evidence suggesting that IT cells are anatomically grouped into functionally similar patches, columns, or areas (see later). Based on findings from single-unit and optical recording, Tanaka (1996) has suggested that IT cells that respond to common visual features are grouped together into cortical columns that run perpendicular to the cortical surface (Fig. 46.4). This object-column model is similar to that for other well-known columns in early visual areas, which are selective either for orientation, ocular dominance, spatial frequency, or direction of motion. However in Tanaka's model, there is a continuous mapping of complex feature space within a larger region that contains several partially overlapping columns (Fig. 46.4), rather than just one periodic feature, such as orientation, mapped repeatedly. Newer evidence from functional magnetic resonance imaging in awake monkeys confirms that cells with common functional properties are grouped together in IT cortex (see later)—although details of this organization remain incompletely understood.

Thus, a given object is apparently not represented by the activity of a single cell, but by the activity of many cells across different columns or regions. This unique-but-overlapping population code satisfies two otherwise conflicting requirements in visual recognition: robustness to subtle changes in input, and precision of representation. Although the image of an object projected on the retina changes in response to variations in illumination and viewing angle, the pattern of activity in IT cortex will be largely maintained, insofar as the clustering of cells with overlapping but slightly different selectivity serve as a buffer to absorb such changes.

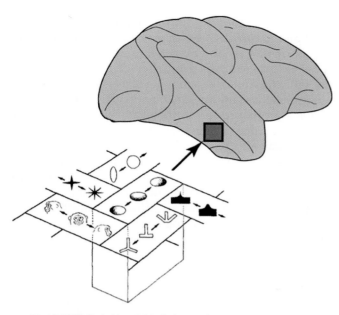

FIGURE 46.4 Schematic diagram illustrating the columnar organization in area TE of the monkey. Cells with similar but slightly different selectivity tend to cluster in elongated vertical columns perpendicular to the cortical surface. Stimuli shown are examples of the critical features for the activation of single cells in TE. Most cells, as shown, require moderately complex features for their activation, although some can be driven by simpler stimuli, such as color, orientation, and texture. Adapted from Tanaka (1996).

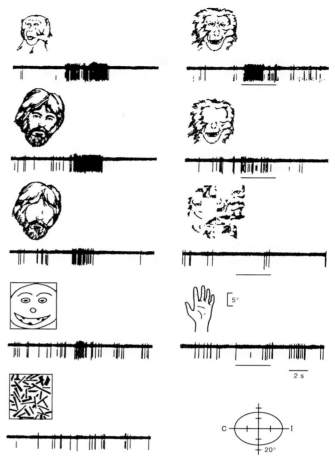

FIGURE 46.5 Activity of an STP neuron that responded better to faces than to all other stimuli tested. Removing eyes on a picture or representing the face as a caricature reduced the response. Cutting the picture into 16 pieces and rearranging the pieces eliminated the response. All the unit records are representative ones chosen from a larger number of trials. The receptive field of the neuron is illustrated on the lower right. C, contralateral; I, ipsilateral visual field. From Bruce *et al.* (1981).

Some Cells in the Occipitotemporal Pathway Are Face Selective

One of the more striking characteristics of monkey temporal cortex is the presence of neurons with responses selective for face stimuli, first described by Bruce *et al.* (1981) and Perrett *et al.* (1982). For several decades, such cells were found by recording blindly, but recently it has been shown that such cells are organized into several column-sized patches in macaque IT cortex, in which up to 97% of the cells are selective for faces (Tsao *et al.*, 2003, 2006). These "face patches," located largely within the superior temporal sulcus, are embedded within a large swath of object-selective cortex extending from V4 to anterior TE. As described later, the macaque face patches appear homologous to face-selective regions in human cortex.

Face-selective cells vary in the degree and nature of their preference for face stimuli. Many respond well to both real faces or face pictures, but give little or no response to any other stimuli tested, including other complex objects, texture patterns, and images in which the features making up the face are rearranged or scrambled (Fig. 46.5). Thus, at least some face-selective neurons appear to be sensitive to very general or global aspects of a face, such as the prototypical arrangement of the eyes, nose, and mouth into a face-like configuration. Other neurons respond to specific face components, such as the presence of eyes *per se*, the distance between the eyes, or the extent of the forehead. Although such neurons may merely be sensitive to facial features, they are nonetheless likely to participate in circuits responsible for recognizing a particular individual. Indeed, recent work has demonstrated that electrical microstimulation of face-selective cells biases object categorization in the direction of face perception, thus establishing a causal relationship between face-selective neurons and face perception (Afraz *et al.*, 2006).

One subset of face-selective cells appears especially sensitive to the direction of gaze of the eyes (looking back directly or turned to the side), which is an important social signal for both monkeys and humans. Other neurons apparently respond selectively to *biological motion*; that is, motion patterns consistent with what one sees when one observes the gait of another animal. Such cells respond when viewing a monkey or human walking, or to equivalent motion patterns generated by so-called Johanssen point-light stimuli (i.e., stimuli generated by attaching lights to the joints of a moving animal, which is otherwise in the dark).

Together, these observations have led to the proposal that the primate temporal lobe has evolved specialized mechanisms for the encoding and recognition of biologically significant stimuli, especially faces. Are such regions innate, or recruited and trained as a result of experience? This is another form of the classical nature-versus-nurture question. Some evidence (e.g., preferential looking at faces in newborns) suggests that face-selectivity may indeed be innate. Alternative interpretations of such *category-specific* phenomena also are discussed later, which take into account the learning history of the individual.

The Occipitotemporal Pathway Extracts Some Invariant Aspects of Objects

As suggested at the start of the chapter, visual information impinging on the retina is a two-dimensional projection of a three-dimensional world—a projection that varies dramatically due to changes in the position, distance, illumination, and orientation of an object relative to the viewer. Thus, the function of the occipitotemporal pathway can be defined in large measure as determining and encoding those *invariant* features of objects that are useful for recognition across myriad circumstances. Exactly how this might be accomplished is unclear even at a theoretical level. However, several findings suggest how this process might at least begin.

As described earlier, as one progresses forward from V1 to TE, receptive fields become larger and the retinotopic maps less precise. By area TE, receptive fields are large and bilateral, typically encompassing the central 20° or more—up to the entire visual field in far anterior TE. As we have seen, the large receptive fields of TE neurons signal the presence of a given feature, collection of features, or object, regardless of where they fall in the visual field. In other words, they exhibit invariance of stimulus coding across changes in retinal position.

In addition to invariance over retinal position, many neurons in TE show maintained selectivity for their "preferred" stimuli over more complex kinds of transformations, such as size, distance from the subject in depth, or degree of ambient illumination; thus, they show invariance over a number of the changes that normally take place as a moving object is viewed under naturalistic conditions. Also, many TE neurons show *form-cue invariance*: for example, a neuron selective for a particular shape defined by luminance borders (e.g., a white star on a black background) will also respond selectively to the same form when it is instead defined by relative texture and motion cues (e.g., a star-shaped region of speckles of the same average luminance as the background, moving in a direction different from the background). Taken together, these various types of response invariance or *generalization* can explain the various deficits seen in experimental monkeys when IT cortex is removed, as described earlier.

As discussed in Box 46.2, one central question in object recognition has been the degree to which neural representations of objects are object-centered (view independent) or viewer-centered (view dependent). The question of *view dependence* has been addressed by a number of single-unit studies of neurons in TE. For example, in one study, monkeys were trained to recognize specific novel three-dimensional objects from any viewpoint after experience with a few prototypical views in a discrimination task (Logothetis *et al.*, 1995). After months of such training, the monkeys were found to have neurons in TE that exhibited strong selectivity for precisely those views on which the animal had been trained. These results support the view-dependent account of object recognition. Studies of face-selective neurons in area TE also initially provided strong support for view dependence by finding separate populations of cells tuned for front and profile views. However, other studies have shown that at least small subsets of TE neurons respond to faces or objects independent of view, providing evidence for view-independent representations as well. Overall, the issue of view dependence remains an open question, with some evidence for both accounts to be found in the neurophysiological literature as well as in psychophysical studies (Fig. 46.6).

Response Properties of Cells in the Object Recognition Pathway Are Affected by Experience

The previous sections illustrate how the object recognition pathway, especially area TE, represents features that are useful for recognition. For example, some neurons respond selectively to the salient features in a face, or their relative positions, while ignoring aspects of stimulus appearance that depend on viewing

BOX 46.2

THE ISSUE OF VIEWPOINT DEPENDENCE IN OBJECT RECOGNITION

Although we know something of the underlying anatomy and physiology, the mechanisms by which object recognition actually is accomplished by the brain have been a matter of heated debate. The pioneering theoretical work of Marr in the late 1970s set the stage for current theories, which can be divided into two main groups: object-centered (view independent), and viewer-centered (view dependent) theories.

According to *object-centered* accounts, objects are represented in the brain as hierarchies of parts that are, to a large degree, independent of any particular viewpoint. One of the best known examples is Biederman's (1990) recognition by components (RBC), which postulates that objects can be defined as a set of 30 or so primitive shapes, so-called *geons*. In this theory, geons are three-dimensional component building blocks, which, when combined, form the objects that we know (Fig. 45.6). Thus, any arbitrary object may be described as a collection of elementary geons. For example, the telephone in Figure 45.6 can be constructed from geons 1, 3, and 5. One compelling feature of geons for object recognition is that particular ones can be inferred from the presence of *nonaccidental relations* between image contours or parts. Nonaccidental relations are those unlikely to occur by chance, such as parallel or colinear edges, closure, and symmetry. Also, because the same nonaccidental features can be derived across a wide range of image variations, detection of geons and their spatial relations would be preserved over changes in position, distance, illumination, and orientation; the telephone, for example, looks pretty much the same (in terms of its constituent geons and how they sit relative to one another) from a variety of angles. Therefore, according to RBC, recognizing an object amounts to determining the set of geons that compose the object and determining the spatial relations between them.

Viewer-centered theories, however, suggest that recognition is linked to the appearance of objects as originally seen. In other words, the brain is not thought of as storing a central, invariant representation of an object, but instead as storing information from multiple independent views (e.g., a view of the telephone as in the figure, but also an end-on or side view, a view from above, etc.). A potential problem with this idea is that it might seem that for reliable object recognition, stored representations for *all* views of objects would need to exist. The solution is to

suggest that a small set of characteristic or *prototypical* views would suffice. Proponents of this approach have offered a variety of different theoretical normalization mechanisms for generalizing from unfamiliar views to the familiar, prototypical stored views (Bulthoff *et al.*, 1995). For example, the visual system might mentally rotate the object as currently seen to bring it into line with a prototypical view or do an interpolation or comparison of the current view with multiple stored views. Therefore, according to viewer-centered theories, recognizing an object amounts to explaining the appearance of an incoming stimulus in terms of transformations or combinations of previously stored representative views.

At first glance, object-centered theories appear more compelling, as they attempt at the outset to solve the central problem associated with object recognition, namely reliable recognition despite variation in the precise retinal image. However, one serious challenge is how the brain chooses the correct structural interpretation among the possible ones, as a given set of geons may be compatible with several different objects. Psychophysical testing also indicates that object-centered encoding is not the whole story; studies have shown substantial and robust viewpoint effects even for extremely simple three-dimensional volumes resembling proposed geon primitives.

However, viewer-centered approaches also face challenges. Viewer-centered theories have been most successful in accounting for *exemplar-level* recognition; that is, discriminating between individual visually similar objects within a relatively homogeneous class of objects such as faces, cars, or birds. Though recognition at the exemplar level is often important, there is little question that so-called *class-level* recognition actually constitutes much of the job of recognition. For example, for objects such as chairs, telephones, and toasters, we rarely need to discriminate two instances of the class and instead we learn to recognize such objects at a basic category or class level. Viewer-centered models suggest that the brain represents the appearance of a specific individual example of an object from specific viewpoints, and thus seem unsuitable or unnecessary for class-level recognition. Moreover, even for a single exemplar, it is uncertain how many prototypical views are enough for robust recognition. One solution would be to store as many views of an object as possible so that new views would always be similar

(continues)

BOX 46.2   *(cont'd)*

---

enough to some previously stored view. However, it is unclear that there can ever be sufficient memory to do so for all the objects that the observer becomes proficient at recognizing under varied conditions.

Luiz Pessoa, Roger B. H. Tootell, and Leslie G. Ungerleider

References

Biederman, I. (1990). Higher-level vision. *In* "Visual Cognition and Action" (D. H. Osherson, S. M. Kozzlyn, and Z. M. Hollerbach, eds.), Vol. 2, pp. 41–72. MIT Press, Cambridge, MA.

Bulthoff, H. H., Edelman, S. Y., and Tan, M. J. (1995). How are 3-dimensional objects represented in the brain? *Cereb. Cortex* **5**, 247–260.

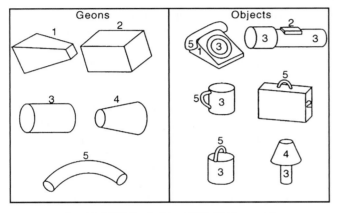

FIGURE 46.6 Illustration of geons and how they are arranged to form objects. (Left) A given view of an object can be represented by an arrangement of simple primitive volumes, or geons, five of which are shown here. (Right) Only two or three geons are required to uniquely specify an object. The relations among the geons matter, as illustrated with the pail and cup. From Biederman (1990).

conditions, such as absolute size and luminance. However, in ecological terms, what is useful and salient within the image of an object or scene changes with experience and context, and these changes are reflected in the way the brain represents stimuli over time. Neurons in IT cortex and the adjacent perirhinal cortex show effects of both recent viewing history and long-term experience with stimuli.

Experiments by Miller and Desimone (1994) have shown how short-term changes in response properties of temporal neurons encode aspects of both stimulus recency and short-term stimulus significance. For example, for many of these cells, responses to an object that is presented repeatedly *decrease* in a systematic fashion, potentially signaling that the object has been seen recently. Such reduced signals may be useful for redirecting attention to novel objects. This kind of response decrement, termed *repetition suppression*, has

been found both under anesthesia and for stimuli that an awake animal is trained to ignore. This habituation-like response is therefore a passive, largely automatic mechanism by which the visual system edits out recently seen objects. In contrast, if a monkey is trained to do a matching task in which it must hold a particular visual pattern in memory over a short delay, some temporal neurons *increase* their firing rates specifically when the to-be-remembered stimulus is recognized. This phenomenon, termed *mnemonic enhancement*, can be thought of as a mechanism in which bottom-up processing of the stimulus is affected by top-down, cognitive factors, namely, expectation of the relevant stimulus (Fig. 46.7). Stimuli that are repeated but are not behaviorally relevant in the context of the task do not show this enhancement effect.

Experiments such as these demonstrate that the magnitude of neuronal responses can be affected by the experience of the subject with the stimulus and by its significance, at least in the short term. Other studies suggest that the kind of familiarity gained over longer periods of time—akin to the development of "expertise" as discussed later in the chapter—can also be reflected in the responses of TE cells. For example, in monkeys that have been trained for as long as a year to discriminate between objects, more than 30% of the cells show selectivity for those objects compared to untrained monkeys in which roughly 10% of the cells show such selectivity (Kobatake *et al.*, 1998).

Intriguingly, some TE and perirhinal cells can represent the *history of association between objects* in the activity patterns they exhibit in response to members of a set of stimuli. Some research groups have examined mechanisms of visual memory by using *paired-associate* tasks, in which monkeys are trained to associate arbitrary patterns with one another for reward. In one experiment, monkeys were first shown one member of a learned pair and then were required to choose the associated "match" from a display of two stimuli while single-unit activity was recorded in ante-

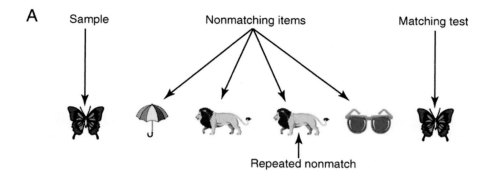

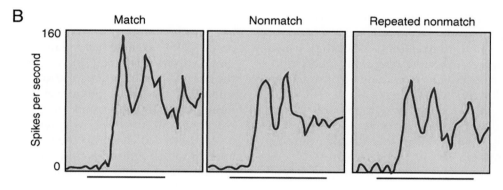

FIGURE 46.7 Experiment demonstrating mnemonic enhancement. (A) On each trial, the monkey was shown a sample stimulus, and then between zero and four intervening nonmatching items appeared before the matching stimulus was presented. Repeated nonmatches appeared after zero (as shown) or one intervening stimulus between them. The task of the monkey was to respond to the matching test stimulus and ignore repeated nonmatches. (B) Responses of a neuron to a single stimulus appearing as a match, nonmatch, and repeated nonmatch. Horizontal bars under the histograms indicated when the stimuli were on, which was 500 ms for nonmatch and repeated nonmatch stimuli. The match stimulus was terminated when the animal made its response. Results show mnemonic enhancement to the match stimulus but not to a stimulus that repeats but is not held in mind (i.e., repeated nonmatch). Adapted from Miller and Desimone (1994).

rior IT cortex (Sakai and Miyashita, 1991). Across all cells recorded, the authors found neurons that responded best to stimuli that had been paired together during training. In other words, some cells in temporal cortex developed conjoint selectivity (*pair coding*) for the members of associated pairs. It was shown that pair coding by TE cells was decreased by surgically interrupting the feedback projections to them from the perirhinal and entorhinal cortices. Thus, return projections appear to play a specific role in associating objects because the pair-coding property depends on them.

Although associative learning effects typically have been seen in studies after weeks or months of training, there is evidence that small but detectable shifts in neuronal population selectivity occur after only a day or so of experience; this suggests that TE cells tend to develop pair-coding properties more rapidly than previously thought (Ericksen *et al.*, 2000; Messinger *et al.*, 2001). Thus, the pattern of selectivity shown by TE cells at any given moment in time may partly reflect a record of their exposure to stimuli seen together. Overall, TE circuits appear to represent not only complex visual features, but also the history of their meanings and relationships in a behavioral context.

Summary

At successive levels of processing along the occipitotemporal pathway, increasingly complex kinds of information about objects are extracted. In area TE of IT cortex, some neurons are preferentially responsive to faces and other biologically significant stimuli. In addition, TE neurons apparently extract some invariant aspects of objects. For example, TE neurons show similar selectivity for object shape, regardless of large changes in their retinal position, size, distance in depth, and degree of ambient illumination. Finally, as the result of experience, some TE neurons come to represent novel objects and/or the associative relations between objects.

FUNCTIONAL NEUROIMAGING AND ELECTROPHYSIOLOGY OF OBJECT RECOGNITION IN HUMANS

The Human Brain Also Has Ventral and Dorsal Processing Streams

The differential visual impairments produced by focal lesions in clinical cases suggest that the human visual cortex, like that of the monkey, contains two anatomically distinct and functionally specialized pathways: the ventral and dorsal streams. For example, one study demonstrated a double dissociation of visual recognition (face perception camouflaged by shadows) and visuospatial performance (maze learning) in two men with lesions of the occipitotemporal and occipitoparietal cortex, respectively, confirmed by postmortem examination (Newcombe *et al.*, 1987). The specific clinical syndromes produced by occipitotemporal lesions include visual object agnosia, as noted earlier, as well as prosopagnosia, an inability to recognize familiar faces, and achromatopsia, or cortical color blindness. In contrast, syndromes produced by occipitoparietal lesions include optic ataxia (misreaching), visuospatial neglect, constructional apraxia, gaze apraxia, akinetopsia (an inability to perceive movement), and disorders of spatial cognition. Interestingly, imagery disorders involving descriptions of either objects (especially faces, animals, and colors of objects) or spatial relations (geographic directions) are also dissociable following temporal and parietal lesions, respectively.

The development of functional brain imaging, including PET (positron emission tomography) and fMRI (functional magnetic resonance imaging), has made it possible to map the organization of the human visual cortex with far greater precision than is possible with human lesion studies, and without the confounding influence of compensatory responses to brain injury. For instance Haxby and colleagues (Haxby *et al.*, 1994) were able to confirm a functional difference between ventral (object) and dorsal (spatial) pathways in humans, by measuring PET activity while subjects performed object identity and spatial location tasks (Fig. 46.8). The results identified occipitotemporal and occipitoparietal regions associated with face and location matching, respectively. In addition, regions in the ventral and dorsal frontal cortex were also selectively activated by face and location matching, respectively. These findings thus indicate the existence in humans, as in monkeys, of two functionally specialized and anatomically segregated visual processing pathways. The findings further suggest the extensions of each into the frontal lobe, which plays a role in working memory for objects and their spatial locations (Chapter 52).

The "Building Blocks" of Object Recognition in Human Cortex Include Early Visual Areas and the Lateral Occipital Complex (LOC)

Functional neuroimaging studies have also revealed a number of distinct visual areas within the ventral and dorsal streams in humans. Many of these areas appear to be equivalent (and perhaps homologous) to specific monkey visual areas, including V1, V2, V3, V3A, V4v, and the middle temporal area (MT) (see also Fig. 46.10). As in monkeys, most of these areas have been defined on the basis of retinotopic maps (Sereno *et al.*, 1995). Area MT, a small area with somewhat disorganized retinotopy, instead has been defined on the basis of its distinctive functional properties (strong selectivity for visual motion and direction) in both human neuroimaging and single-unit recordings in macaque.

As described earlier, receptive field size increases, and stimulus requirements become more stringent, as one progresses from V1 through various stages of the ventral stream. For instance, single units in macaque IT cortex do not respond to "scrambled" versions of object images (e.g., Fig. 46.5). To test for analogous regions in human visual cortex, Malach and colleagues (Malach *et al.*, 1995) compared responses evoked by a range of images to activations evoked by "scrambled" versions of those same images, while attempting to equate several image properties (such as size and contrast). Regions responding more to objects than to scrambled stimuli were interpreted as being involved in object-related processing; collectively these regions were named the lateral occipital complex (LOC) because of their anatomical location (Fig. 46.9). Later neuroimaging studies revealed that the cortical regions activated by such stimuli includes classical IT cortex in macaques (Tootell *et al.*, 2003; Denys *et al.*, 2004); see Figure 46.10.

A number of fMRI findings suggest that the LOC is an important stage in human object processing. First, whereas early visual areas V1–V3 respond similarly to intact, recognizable stimuli and highly scrambled, unrecognizable pictures (scrambled pictures tend to elicit even stronger responses), the LOC *exhibits a high sensitivity to image scrambling* (in fact, the LOC often is defined by contrasting objects to scrambled objects). Second, a fourfold change in visual object size does not affect the LOC activation, although it changes some stimulus properties greatly (e.g., local contrast); responses in the LOC also exhibit some invariance to changes in image position. Third, the LOC responds

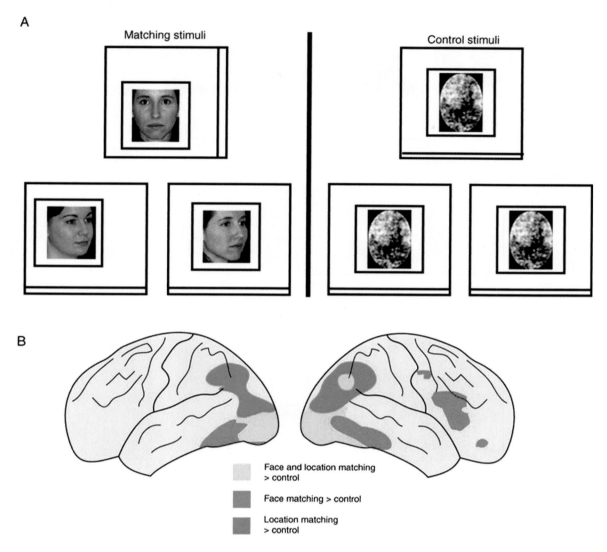

FIGURE 46.8 Ventral and dorsal processing streams in the human cortex, as demonstrated in a PET imaging study of face and location perception. (A) Sample stimuli used during the experiment. (Left) Stimuli for both face and location matching tasks. During face matching trials, subjects had to indicate with a button press which of the two faces at the bottom was the same person shown at the top. In this example, the correct choice is the stimulus on the right. During location matching trials, subjects had to indicate with a button press which of the two small squares at the bottom was in the same location relative to the double line as the small square shown at the top. In this example, the correct choice is the stimulus on the left. (Right) Control stimuli. During control trials, subjects saw an array of three stimuli in which the small squares contained a complex visual image. In these trials, which controlled for both visual stimulation and finger movements, they alternated left and right button presses. (B) Areas shown in red had significantly increased activity during the face matching but not during the location matching task, as compared with activity during the control task. Areas shown in green had significantly increased activity during the location matching but not during the face matching task, as compared with activity during the control task. Areas shown in yellow had significantly increased activity during both face and location matching tasks. Adapted from Haxby *et al.* (1994).

to multiple visual cues. For example, the shape of an object can be defined by luminance, by texture, or by motion cues alone; objects defined by any one or more of these cues reliably evoke responses in the LOC. Fourth, objects that vary widely in their recognizability (e.g., famous faces and unfamiliar three-dimensional abstract sculptures) produce similar activations in the LOC. Together, these findings indicate that the LOC is an intermediate link in the processing of objects; that is, a stage following low-level processing and preceding object processing stages that involve memory.

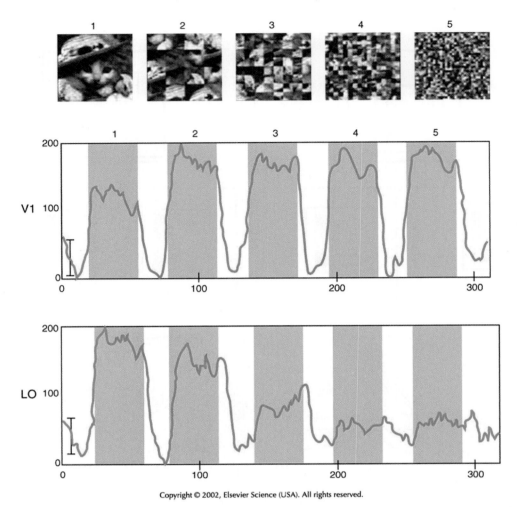

FIGURE 46.9 Shape processing in the lateral occipital cortex (LOC). (Top) Visual stimuli that were used to investigate the effects of image scrambling on activity in the LOC, as measured by fMRI. Subjects viewed the stimuli and were instructed to covertly name them, even the scrambled images. Stimulus 1 is the original image, and stimuli 2–5 are increasingly scrambled. (Bottom) fMRI time series illustrating the averaged activations in V1 and the LOC to each of the images shown at the top. Note that LOC is very sensitive to image scrambling, whereas V1 is not. Adapted from Grill-Spector *et al.* (1998).

Although the processing of general object shape observed in the LOC appears to be important for object perception and recognition, the LOC probably is not responsible for object recognition *per se*. Whereas the LOC responds to essentially any three-dimensional shape, areas in the ventral stream anterior to the LOC, most notably on the fusiform gyrus in the ventral temporal cortex, respond preferentially to recognizable objects. It therefore is thought that object recognition is critically dependent on the ventral temporal cortex. Indeed, as later sections describe, it is likely that the storage of object representations takes place in the ventral temporal cortex as well. The next section considers the processing of specific categories of objects and the functional architecture of the ventral temporal cortex.

Summary

Evidence from lesion studies and neuroimaging studies confirms the existence in humans, as in macaques, of two visual processing pathways: a ventral stream for object vision and a dorsal stream for spatial vision. Within the ventral stream, early visual areas, including V1–V4, are organized retinotopically and appear to be homologous to areas studied in monkeys. The LOC appears to play an important role at the next stage of processing, where general object shape is processed. Activation of the LOC is highly sensitive to image scrambling, but less sensitive to variations in image size or the visual cues that define the shape of an object. Activation of the LOC, however, is not sensitive to object recognizability, sug-

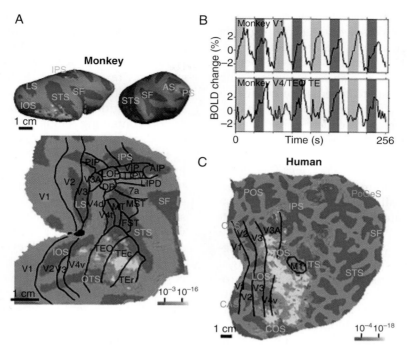

FIGURE 46.10 Object-selective areas in the macaque and human as revealed by fMRI. (A) Areas in the macaque brain significantly more activated by intact objects than scrambled counterparts (Fig. 45.9, top). The activation is shown on two different lateral views of the "inflated" right hemisphere (top), as well as on a "flat" map of the right visual cortex (bottom). Significant activations occurred in ventral stream areas V4, TEO, and TE. (B) Time courses of activation from macaque V1 (top) and V4/TEO/TE (bottom) to alternating blocks of intact (dark gray) and scrambled (light gray) objects, separated by epochs of a blank screen (white). (C) Activity in the right hemisphere of a human subject to the same stimulus comparison. Sulcal abbreviations: LS, lunate; IOS, inferior occipital; OTS, occipito-temporal; IPS, intraparietal; STS; superior temporal; SF, Sylvian fissure; AS, arcuate; ITS, inferior temporal; POS, parieto-occipital; CAS, calcarine; COS, collateral; TOS, transoccipital; LOS, lateral occipital; PoCeS, postcentral. Adapted from Tsao *et al.* (2003).

gesting that it precedes processing stages that involve memory.

PERCEPTION AND RECOGNITION OF SPECIFIC CLASSES OF OBJECTS

Prosopagnosia Is a Specific Deficit of Face Recognition

Evidence from neuropsychology and neurophysiology suggests that certain kinds of objects may have special significance to primates. *Prosopagnosia*, a selective deficit in recognizing familiar faces, has been known since at least the end of the nineteenth century. Although patients with this syndrome are aware that faces are faces (i.e., they know the basic stimulus category), they fail to reliably identify (or even achieve a sense of familiarity from) the faces of coworkers, family members, famous persons, and other individuals previously well known to them. Typically, prosopagnosics also have trouble forming memories of new faces, even

if other new objects can be learned. Because the voice of a visually unrecognized person usually enables the patient to identify and feel familiar with that person, prosopagnosia (face agnosia) appears to be a specifically visual impairment.

A number of explanations have been offered for the apparent selectivity of the prosopagnosic deficit. One explanation is that face perception and recognition are indeed unique behavioral capacities and reflect unique, dedicated neural circuits that can be selectively damaged. Such dedicated neural circuits may have developed through evolution due to the biological significance of faces and facial expressions for social communication. A second explanation suggests that face processing reflects subtle and difficult discriminations between highly similar exemplars within a category and that it is this general capacity, not the processing of the facial configuration *per se*, that is disrupted in prosopagnosia. This second view is consistent with the observation that prosopagnosia often is accompanied by varying levels of object agnosia. A third, and related, suggestion (see later) is that face processing represents the acquisition of a type of expertise derived from very protracted experience with a category of complex visual stimuli. Reports of persons with prosopagnosia who have associated deficits in types of object recognition in which they had previously acquired expertise over long periods of time (e.g., a show dog expert who lost the ability to differentiate breeds) are consistent with this idea. Note that the second view may be considered as a specific form of the expertise hypothesis.

Since the 1980s, a number of cases with prosopagnosia have come to autopsy. The damage common to these lay within the *lingual and fusiform gyri* (Fig. 46.1), ventrally and medially within the cortex at the occipitotemporal junction. In all such cases, this area (or at least the underlying white matter) was damaged bilaterally, and consequently bilateral damage was thought by many to be a necessary precondition for prosopagnosia. Additional cases with prosopagnosia and right cortical damage alone, along with the results of neuroimaging studies (see later), have reopened the debate over the locus of the critical lesion in prosopagnosia.

Several recent reports of developmental prosopagnosia have provided invaluable information concerning the perceptual deficit observed in these patients. Developmental prosopagnosia is characterized by severely impaired face recognition in individuals with no detectable brain damage (i.e., presumably intact visual systems). Many developmental prosopagnosics have normal or relatively good object recognition, indicating that their impairments are not the result of deficits to a unitary visual recognition mechanism. Indeed, extensive testing of particular individuals has revealed especially pure face processing deficits (Duchaine and Nakayama, 2006). In such individuals, the results strongly suggest that faces are recognized by a content-specific, face-processing mechanism. However, it should be noted that neuroimaging studies have demonstrated a variety of neural profiles in developmental prosopagnosia, which is consistent with behavioral studies indicating that, in general, developmental prosopagnosia is a heterogeneous disorder (i.e., in some cases deficits are not restricted to faces).

Neuroimaging and Electrophysiological Studies Also Suggest Specialized Processing Regions

The existence of prosopagnosia suggests that certain regions of the ventral stream should be very responsive to faces. This is exactly what has been observed in neuroimaging studies. For example, a number of early neuroimaging studies revealed that a region on the fusiform gyrus of the right hemisphere is significantly more active when subjects view faces, compared to when they view common nonface objects. This region was called the "fusiform face area" (FFA) by Kanwisher and colleagues (Kanwisher *et al.*, 1997). In addition, the FFA is more active during face matching than location matching, and during the viewing of faces than during the viewing of scrambled faces, letter strings, or textures. Thus, in humans, both lesion and neuroimaging evidence converges to suggest a face-selective region in/near the fusiform gyrus. A similar region containing single units selective for faces has been demonstrated in macaques.

Furthermore, the face selectivity of the FFA has been linked closely to conscious face perception in humans. In one neuroimaging study, the differential sensitivity of the ventral cortex to different visual stimuli was explored in a *binocular rivalry* paradigm in which different stimuli (faces and houses) were presented to the right and left eyes (Tong *et al.*, 1998; Fig. 46.11). During binocular rivalry, the resulting percept is not a combination of the two stimuli. Instead, subjects report that the actual percept alternates between the two stimuli, every few seconds—even though the dual retinal stimulation is kept constant. As expected, subjects' perceptions fluctuated between faces and houses. Critically, while changes from house to face percepts were accompanied by increasing activation in the right FFA, perceived changes from a face to a house led to decreases in FFA responses. This suggests that the right fusiform gyrus plays an important and specialized role in conscious face perception.

Evidence from electrophysiological recordings in patients with implanted electrodes (as part of the preparation for neurosurgery to treat epilepsy) also reveals face-selective processing in the ventral stream (Allison *et al.*, 1999). A large evoked potential, called the *N200* because it occurs roughly 200 ms after stimulus onset, is generated in small regions (i.e., typically at only one electrode within an electrode array) in the ventral occipitotemporal cortex. The selectivity of the N200 has been studied most thoroughly for face stimuli (Fig. 46.12). In this case, it has been found that the N200 amplitude does not vary substantially whether evoked by colored or gray-scale faces, by normal, blurred, or line-drawn faces, or by faces of different sizes. The N200 amplitude is largest to full faces and decreases progressively to eyes, face contours, lips, and noses viewed in isolation. Interestingly, a region just lateral to face-selective N200 sites is more responsive to internal face parts than to faces, and some sites in ventral occipitotemporal cortex appear to be face-part specific.

In other small, circumscribed locations in the ventral occipitotemporal cortex, N200 potentials are elicited selectively by visual stimuli other than faces, such as letter strings (including words and nonwords), numbers, complex objects, and gratings. Importantly,

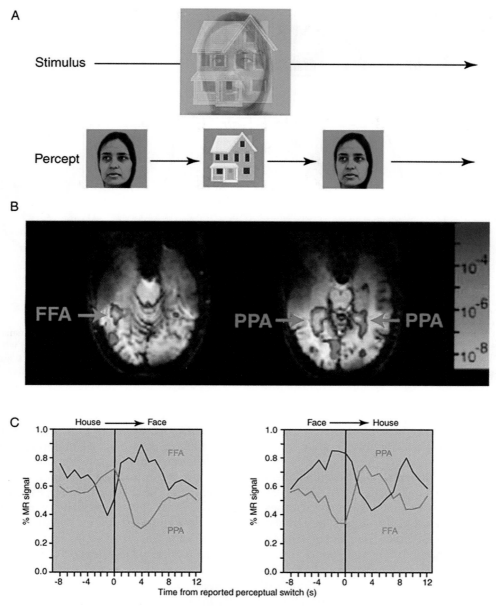

FIGURE 46.11 Design and results from an fMRI experiment on binocular rivalry. (A) The top illustration shows the ambiguous face/house stimulus used to produce rivalry. When viewed through red and green filter glasses, only the face could be seen through one eye and only the house through the other eye. This arrangement led to binocular rivalry, with a face percept alternating with a house percept every few seconds. (B) Two adjacent axial slices through the brain of a single subject showing, in the left slice, the area on the fusiform gyrus that responded more to faces than houses (FFA) and, in the right slice, the area on the parahippocampal gyrus that responded more to houses than to faces (PPA). In the slices shown, left is the right hemisphere and right is the left hemisphere. (C) fMRI time series showing FFA and PPA activity for a single subject illustrating that the activity in these two regions correlated with the subject's visual percept. PPA, parahippocampal place area. Adapted from Tong *et al.* (1998).

the responses are not elicited by control stimuli, such as scrambled versions of the stimuli. The small regions, which vary in their exact location among individuals, are reminiscent of clumps of object-selective neurons found in area TE of monkeys.

Specialized Processing Likely Involves Other Categories

In addition to faces, other visually complex and behaviorally important classes of stimuli may also

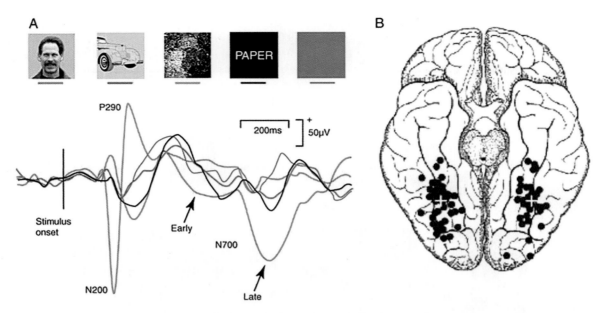

FIGURE 46.12 (A) Event-related potential specific to faces (red line) elicited from the occipitotemporal cortex of humans. Potentials elicited by other stimuli are also shown. (B) Summary of locations (black dots) on the ventral surface of the human brain from which a surface-negative potential (N200) was recorded when patients were shown faces but not when other categories of objects were tested. Locations are primarily along the fusiform gyrus. Adapted from Allison *et al.* (1999).

have specialized substrates in the ventral stream. These include "places," words, and, possibly, human body parts.

Neuroimaging studies have revealed a region within the human parahippocampal cortex that is significantly more active when subjects view complex scenes (such as rooms, landscapes, and city streets) than when they view objects, faces, or other kinds of visual stimuli. This region has been named the parahippocampal place area (PPA) because it responds strongly whenever subjects view an image of a place (Epstein and Kanwisher, 1998). Consistent with the neuroimaging data, damage to the lingual/parahippocampal region may result in landmark agnosia.

Recent findings also suggest that some extrastriate visual areas may respond selectively when viewing human body parts (such as an isolated foot or arm) relative to viewing objects (including man-made objects and food). The region that is activated is located in the lateral occipitotemporal cortex (near the middle occipital gyrus) in an area that does not appear to have evident retinotopy. Continuing with the tradition of naming regions, this region has been called the extrastriate body area (EBA; Downing *et al.*, 2001). Interestingly, it also appears that the EBA is strongly modulated by limb (arm, foot) movements, even in the absence of visual feedback from the movement.

Finally, studies of brain lesions in patients suggest that there is a specialized region for the processing of *written words* in the inferior temporal lobe. Patients with *alexia* cannot read normally despite intact visual capabilities, and despite their ability to understand spoken language and to write. Accordingly, alexia has been conceptualized as essentially a visual agnosia for verbal material. Studies investigating the neural substrates of the processing of words and letter strings have compared activations elicited by (unpronounceable, nonword) letter strings with activations elicited by face and texture stimuli. Regions responding more strongly to letter strings primarily involved the left occipitotemporal and inferior occipital sulci. The suggestion is that these regions comprise a prelexical, presemantic stage of word-form processing in the reading system (Cohen *et al.*, 2002).

Because learning to read is a very recent event in evolution, regions of the cortex specialized for words and letter strings are presumably not present at birth. Thus, if such regions exist, word-selective regions must develop through protracted experience with written words, demonstrating that the environment helps shape the architecture of the visual system. Stimuli within a category, such as letters, tend to occur together and would therefore be encoded in close temporal proximity. In this view, many years of repeated experience with letters could lead to regions in the ventral temporal cortex specialized to process letters, similar to the development of conjoint selectivity by monkey TE cells for visual images that are repeatedly shown together. Consistent with this proposal, brain damaged patients can be selectively impaired at letter recognition compared to, for example, number recognition and vice versa.

Collectively, the evidence suggests that the visual system is organized such that some object categories are represented in a relatively specialized fashion: faces, places, body parts, and words appear to have dedicated neural mechanisms for their representation. Of these, the evidence is perhaps strongest for faces, at least in some individuals with relatively pure forms of prosopagnosia. Not surprisingly, alternative proposals have been advanced concerning the organization of ventral visual cortex. We briefly discuss two of them now.

Alternative Proposals to Category-Specific Representations

It has been proposed that the FFA is specialized not for the processing of faces *per se* but rather for the processing of objects with which one has acquired expertise (Gauthier *et al.*, 1998, 1999). The idea that expertise underlies responses in ventral temporal cortex raises an important distinction between two interpretations. One interpretation suggests that activation is based on visual features (e.g., the visual features composing a face, such as eyes, nose, and mouth). Another interpretation emphasizes the *processes* associated with perceiving or recognizing objects. In this latter view, the ventral stream contains areas that are best suited for different computations, which may be recruited as one acquires expertise with specific types of visual objects. This might involve encoding subtle differences between visually similar objects or, conversely, finding underlying similarities between visually dissimilar objects. In particular, the expertise view may help explain why some patients with prosopagnosia may also have associated deficits in areas of acquired expertise, such as the show dog expert mentioned earlier who lost the ability to differentiate breeds.

According to another alternative framework, the functional architecture of the ventral stream is viewed not as being composed of category-specific regions but instead as containing a distributed representation of information about object form (Ishai *et al.*, 1999; Haxby *et al.*, 2001). Each category is associated with its own pattern of response across the ventral temporal cortex with a highly consistent spatial arrangement across subjects. In this view, the representation of an object is not restricted to a region that responds maximally to that object, but rather is distributed across a broader expanse of cortex. Within this distributed system, however, the representation of faces would be less extensive as compared to the representation of nonface objects, helping explain why it is possible for lesions to produce prosopagnosia while leaving the recognition of nonface objects largely intact.

Object Knowledge Is Stored in a Distributed Network of Cortical Areas

Although the distinction between perception and memory might seem sharp, in reality these two processes are intimately connected. Indeed, each time one identifies an object, one implicitly retrieves stored knowledge about that object. In other words, identifying an object necessarily implies knowing what that object is. Further, it is generally believed that the same neural tissue devoted to object processing also serves to store the long-term representations of objects. For example, when a subject mentally imagines faces or houses (requiring the retrieval of stored object representations from long-term memory), the same patterns of activation across ventral temporal cortex are seen as those observed when actually viewing faces or houses.

Object knowledge implies not only familiarity with the shape of an object, but also other attributes, such as its typical color and function. For example, how does one know that bulldozers are usually yellow and are used for digging? One neuroimaging study that examined this question asked subjects to name either the associated color or the associated action in response to the presentation of black and white line drawings of objects (Martin *et al.*, 1995). By associated action the investigators meant the typical motion associated with an object, such as the motion produced by the arm of the bulldozer digging. Results showed that the generation of color names associated with objects selectively activated a region in the ventral occipital cortex just in front of the area involved in the perception of color, whereas the generation of action names activated a region dorsally in the middle temporal gyrus just in front of the area involved in the perception of motion. These findings suggest that information about the different attributes of an object is not stored in a unified fashion in any single area of the cortex. Rather, object knowledge seems to be stored in a distributed fashion such that information about specific attributes is stored close to the specific regions of the cortex that mediate the perception of those attributes.

These neuroimaging results provide a basis for understanding category-specific disorders of knowledge resulting from patterns of focal brain damage.

This type of deficit, seen most commonly for animals and tools, involves a selective difficulty in naming and retrieving information about objects from a single category. If the distinction between different categories of objects, such as animals and tools, is dependent on access to information about different types of attributes, and if such attributes (e.g., form, color, motion and object-use associated motor movements) are stored in separate regions of the brain, then one may find a category-specific deficit depending on the site of the lesion. Indeed, consistent with neuroimaging data, it has been found that the locations of lesions in patients with impaired recognition and impaired naming of drawings of animals included the medial aspects of the occipital lobe, nearly the same region active during animal naming. In contrast, the locations of lesions in patients with impaired recognition and impaired naming of drawings of tools all included the middle temporal gyrus, nearly the same region active during tool naming (Tranel *et al.*, 1997). It is important to note that the impairment in patients with category-specific deficits is not limited to visual recognition. These deficits occur when knowledge is probed visually or verbally, and therefore they reflect disorders of stored knowledge and not simply disorders of perception.

Summary

Prosopagnosia, a selective deficit in recognizing faces, is associated with damage to the occipitotemporal cortex, including the fusiform gyrus. Consistent with this impairment, neuroimaging and electrophysiological studies have demonstrated that the fusiform gyrus responds selectively to faces and that this activation is linked to face perception. Although these findings suggest the existence of a specialized region for processing faces, an alternative interpretation is that the fusiform gyrus performs specialized computations associated with perceiving certain objects, including but not limited to faces. Such computations would presumably apply to many objects with which one has expertise. Moreover, the existence of a region in the left occipitotemporal cortex selectively activated by letter strings and words demonstrates the role of experience in shaping the functional architecture of the object recognition pathway.

OVERALL SUMMARY

Early studies of brain damage in humans and lesions in monkeys pointed to the crucial role of tissue in the ventral parts of the temporal cortex for object perception and recognition. After many decades of intensive research, one can now say that the ventral portions of the temporal cortex form the last stations of an occipitotemporal ventral pathway, beginning in the primary visual cortex and progressing through multiple visual areas beyond it. In macaque monkeys, important regions for object perception and recognition include areas TEO and TE of the inferior temporal cortex. In humans, homologous circuits appear to exist in both occipital and temporal regions, including the lateral occipital complex, the fusiform gyrus, and nearby regions in the ventral temporal cortex.

Single-unit recording studies in monkeys show a progressive increase in the complexity of object features needed to "trigger" cells at progressively more anterior stations in the occipitotemporal pathway. Many TE neurons show preservation of their selectivity over transformations that change the physical stimulus, such as retinal position, size, distance in depth, and degree of ambient illumination. At the highest levels of the pathway, within area TE, some neurons are preferentially responsive to faces and other biologically significant stimuli.

More recently, neuroimaging and evoked potential studies in humans with implanted electrodes have also shed light on the mechanisms by which the primate brain represents objects. Descriptions of prosopagnosia, a selective deficit of face recognition that is associated with damage to ventral occipitotemporal cortex, foreshadowed discoveries of cortical zones that are activated preferentially by face stimuli. At the same time, selectivity for letter strings and even for learned arbitrary stimuli has suggested that experience also influences activity in the ventral pathway and, in the long term, may tune representations. Finally, more general knowledge about objects (what they are used for and how to manipulate them) appears to be stored across a distributed network of sensory and motor cortical areas.

References

Afraz, S. R., Kiani, R., and Esteky, H. (2006). Microstimulation of inferotemporal cortex influences face categorization. *Nature* **442**, 692–695.

Allison, T., Puce, A., Spencer, D. D., and McCarthy, G. (1999). Electrophysiological studies of human face perception. I: Potential generated in occipitotemporal cortex by face and non-face stimuli. *Cereb. Cortex* **9**, 415–430.

Biederman, I. (1990). Higher-level vision. *In* "Visual Cognition and Action" (D. H. Osherson, S. M. Kosslyn, and J. M. Hollerbach, eds.), Vol. 2, pp. 41–72. MIT Press, Cambridge, MA.

Bruce, C., Desimone, R., and Gross, C. G. (1981). Visual properties of neurons in a polysensory area in superior temporal sulcus of the macaque. *J. Neurophysiol.* **46**, 369–384.

Bulthoff, H. H., Edelman, S. Y., and Tarr, M. J. (1995). How are 3-dimensional objects represented in the brain? *Cereb. Cortex* **5**, 247–260.

Cohen, L., Lehericy, S., Chochon, F., Lemer, C., Rivaud, S., and Dehaene, S. (2002). Language-specific tuning of visual cortex? Functional properties of the Visual Word Form Area. *Brain* **125**, 1054–1069.

Denys, K., Vanduffel, W., Fize, D., Nelissen, K., Peuskens, H., Van Essen, D., and Orban, G. A. (2004). The processing of visual shape in the cerebral cortex of human and nonhuman primate: A functional magnetic resonance imaging study. *J. Neurosci.* **24**, 2551–2565.

Downing, P. E., Jiang, Y., Shuman, M., and Kanwisher, N. (2001). A cortical area selective for the visual processing of the human body. *Science* **293**, 2470–2473.

Duchaine, B. C. and Nakayama, K. (2006). Developmental prosopagnosia: a window to content-specific face processing. *Curr. Opin. Neurobiol.* **16**, 166–173.

Epstein, R. and Kanwisher, N. (1998). A cortical representation of the local visual environment. *Nature* **392**, 598–601.

Erickson, C. A., Jagadeesh, B., and Desimone, R. (2000). Clustering of perirhinal neurons with similar properties following visual experience in adult monkeys. *Nature Neurosci.* **3**, 1143–1148.

Farah, M. J. (1990). "Visual Agnosia: Disorders of Object Recognition and What They Tell Us About Normal Vision." MIT Press, Cambridge, MA.

Gauthier, I., Tarr, M. J., Anderson A. W., Skudlarski P., and Gore, J. C. (1999). Activation of the middle fusiform "face area" increases with expertise in recognizing novel objects. *Nature Neurosci.* **2**, 568–573.

Gauthier, I., Williams, P., Tarr, M. J., and Tanaka, J. (1998). Training "greeble" experts: A framework for studying expert object recognition processes. *Vis. Res.* **38**, 2401–2428.

Goodale, M. A. and Milner, A. D. (1992). Separate visual pathways for perception and action. *Trends Neurosci.* **15**, 20–25.

Grill-Spector, K., Kushnir, T., Hendler, T., Edelman, S., Itzchank, Y., and Malach, R. (1998). A sequence of early object processing stages revealed by fMRI in human occipital lobe. *Hum. Brain Mapp.* **6**, 316–328.

Gross, C. G. (2002). Genealogy of the "grandmother cell." *Neuroscientist* **8**, 512–518.

Gross, C. G., Rocha-Miranda, C. E., and Bender, D. B. (1972). Visual properties of neurons in inferotemporal cortex of the Macaque. *J. Neurophysiol.* **35**, 96–111.

Gross, C. G., Rodman, H. R., Gochin, P. M., and Colombo, M. W. (1993). Inferior temporal cortex as a pattern recognition device. *In* "Computational Learning and Cognition" (E. Baum, ed.), pp. 44–73. Society for Industrial and Applied Mathematics, Philadelphia.

Haxby, J. V., Horwitz, B., Ungerleider, L. C., Maisog, J. M., Pietrini, P., and Grady, C. L. (1994). The functional organization of human extrastriate cortex: A PET-rCBF study of selective attention to faces and locations. *J. Neurosci.* **14**, 6336–6353.

Haxby, J. V., Gobbini, M. I., Furey, M. L., Ishai, A., Schouten, J. L., and Pietrini, P. (2001). Distributed and overlapping representations of faces and objects in ventral temporal cortex. *Science* **293**, 2425–2430.

Ishai, A., Ungerleider, L. G., Martin, A., Schouten, J. L., and Haxby, J. V. (1999). Distributed representation of objects in the human ventral visual pathway. *Proc. Natl. Acad. Sci. USA* **3**, 9379–9384.

Kanwisher, N., McDermott, J., and Chun, M. M. (1997). The fusiform face area: A module in human extrastriate cortex specialized for face perception. *J. Neurosci.* **17**, 4302–4311.

Kobatake, E., Wang, G., and Tanaka, K. (1998). Effects of shape-discrimination training on the selectivity of inferotemporal cells in adult monkeys. *J. Neurophysiol.* **80**, 324–330.

Logothetis, N. K., Pauls, J., and Poggio, T. (1995). Shape representation in the inferior temporal cortex of monkeys. *Curr. Biol.* **5**, 552–563.

Malach, R., Reppas, J. B., Benson, R. R., Kwong, K. K., Jiang. H., Kennedy, W. A., Leden, P. J., Brady, Rosen, B. R., and Tootell, R. B. (1995). Object-related activity revealed by functional magnetic resonance imaging in human occipital cortex. *Proc. Natl. Acad. Sci. USA* **92**, 8135–8139.

Martin, A. J., Haxby, J. V., Lalonde, F. M., Wiggs, C. L., and Ungerleider, L. G. (1995). Discrete cortical regions associated with knowledge of color and knowledge of action. *Science* **270**, 102–105.

Messinger, A., Squire, L. R., Zola, S. M., and Albright, T. O. (2001). Neuronal representatives of visual stimulus associations develop in the temporal lobe during learning. *Proc. Natl. Acad. Sci. USA* **98**, 12239–12244.

Miller, E. K. and Desimone, R. (1994). Parallel neuronal mechanisms for short-term memory. *Science* **263**, 520–522.

Mishkin, M., Ungerleider, L. G., and Macko, K. A. (1983). Object vision and spatial vision: Two cortical pathways. *Trends Neurosci.* **6**, 415–417.

Newcombe, F., Ratcliff, G., and Damasio, H. (1987). Dissociable visual and spatial impairments following right posterior cerebral lesions: Clinical, neuropsychological and anatomical evidence. *Neuropsychologia* **25**, 149–161.

Perrett, D. I., Rolls, E. T., and Caan, W. (1982). Visual neurones responsive to faces in the monkey temporal cortex. *Exp. Brain Res.* **47**, 329–342.

Rubens, A. B. and Benson, D. F. (1971). Associative visual agnosia. *Arch. Neurol.* **24**, 305–316.

Sakai, K. and Miyashita, Y. (1991). Neural organization for the long-term memory of paired associates. *Nature* **354**, 152–155.

Sereno, M. I., Dale, A. M., Reppas, J. B., Kwong, K. K., Belliveau, J. W., Brady, T. J., Rosen, B. R., and Tootell, R. B. (1995). Borders of multiple visual areas in humans revealed by functional magnetic resonance imaging. *Science* **268**, 803–804.

Tanaka, K. (1996). Inferotemporal cortex and object vision. *Annu. Rev. Neurosci.* **19**, 109–139.

Tong, F., Nakayama, K., Vaughan, J. T., and Kanwisher, N. (1998). Binocular rivalry and visual awareness in human extrastriate cortex. *Neuron* **21**, 753–759.

Tootell, R. B., Tsao, D., and Vanduffel, W. (1993). Neuroimaging weighs: Humans meet macques in "primate" visual cortex. *J. Neurosci.* **23**, 3981–3989.

Tranel, D., Damasio, H., and Damasio, A. R. (1997). A neural basis for the retrieval of conceptual knowledge. *Neuropsychologia* **35**, 1319–1327.

Tsao, D. Y., Freiwald, W. A., Tootell, R. B. H., and Livingstone, M. S. (2006). A cortical region consisting entirely of face-selective cells. *Science* **311**, 670–674.

Tsao, D. Y., Freiwald, W. A., Knutsen, T. A., Mandeville, J. B., and Tootell, R. B. H. (2003). Faces and objects on macaque cerebral cortex. *Nature Neurosci.* **6**, 989–995.

Ungerleider, L. G. and Mishkin, M. (1982). Two cortical visual systems. *In* "Analysis of Visual Behavior" (D. J. Ingle, M. A. Goodale, and R. J. W. Mansfield, eds.), pp. 549–586. MIT Press, Cambridge, MA.

Suggested Readings

Bar, M. (2004). Visual objects in context. *Nature Rev. Neurosci.* **5**, 617–629.

Barlow, H. B. (1972). Single units and sensation: A neuron doctrine for perceptual psychology? *Perception* **1**, 371–394.

Damasio, A. R. (1990). Category-related recognition deficits as a clue to the neural substrates of knowledge. *Trends Neurosci.* **13**, 95–98.

Desimone, R. (1991). Face-selective cells in the temporal cortex of monkeys. *J. Cogn. Neurosci.* **3**, 1–8.

Desimone, R. and Ungerleider, L. G. (1989). Neural mechanisms of visual perception in monkeys. *In* "Handbook of Neuropsychology" (F. Boiler and J. Grafman, eds.), Vol. 2, pp. 267–299. Elsevier, Amsterdam, The Netherlands.

Forde, E. M. E. and Humphreys, G. W. (1999). Category-specific recognition impairments: A review of important case studies and influential theories. *Aphasiology* **13**, 169–193.

Logothetis, N. K. and Sheinberg, D. L. (1996). Visual object recognition. *Annu. Rev. Neurosci.* **19**, 577–621.

Marr, D. (1982). "Vision: A Computational Investigation into the Human Representation and Processing of Visual Information." Freeman, San Francisco, CA.

Martin, A., Ungerleider, L. G., and Haxby, J. V. (2000). Category specificity and the brain: The sensory/motor model of semantic representations of objects. *In* "The New Cognitive Neurosci-ences" (M. S. Gazzaniga, ed.), pp. 1023–1036. MIT Press, Cambridge, MA.

Sereno, M. I. and Tootell, B. H. (2005). From monkeys to humans: What do we now know about brain homologies? *Curr. Opin. Neurobiol.* **15**, 135–144.

Tanaka, K. (1993). Neuronal mechanisms of object recognition. *Science* **262**, 685–688.

Ullman S. (1996). "High-Level Vision." MIT Press, Cambridge, MA.

Ungerleider, L. G. and Haxby, J. V. (1994). "What" and "where" in the human brain. *Curr. Opin. Neurobiol.* **4**, 157–165.

Ungerleider, L. G. and Pasternak, T. (2004). Ventral and dorsal cortical processing streams. *In* "The Visual Neurosciences" (L.M. Chalupa and J. S. Werner, eds.), pp. 541–562. MIT Press, Cambridge, MA.

Luiz Pessoa, Roger B. H. Tootell,
and Leslie G. Ungerleider

Spatial Cognition

As we move through the world, new visual, auditory, vestibular, and somatosensory inputs are continuously presented to the brain. Given such constantly changing input, it is remarkable how easily we are able to keep track of where things are. We can reach for an object, look at it, or even kick it without making a conscious effort to assess its location in space, but how do we construct a representation of space that allows us to act so effortlessly? This chapter outlines the contributions of several brain areas in the parietal, frontal, and hippocampal cortices to spatial representation, spatial memory, and the generation of actions in space.

NEURAL SYSTEMS FOR SPATIAL COGNITION

Dorsal Stream Areas Process Visuospatial Information

Within the cerebral hemispheres, visual information is processed serially by a succession of areas beginning with the primary visual cortex of the occipital lobe. Within this hierarchical system, there are two parallel chains, or streams, of areas: the ventral stream, leading downward into the temporal lobe, and the dorsal stream, leading forward into the parietal lobe (Ungerlieder and Mishkin, 1982) (Chapter 46). Although the two streams are interconnected to some degree, it is a fair first approximation to describe them as separate parallel systems. Areas of the ventral stream play a critical role in the recognition of visual patterns, including faces, whereas areas within the dorsal stream contribute selectively to conscious spatial

awareness and to the spatial guidance of actions, such as reaching and grasping. Dorsal stream areas have at least two distinctive functional characteristics: (1) they contain a comparatively extensive representation of the peripheral visual field and (2) they appear to be specialized for the detection and analysis of moving visual images. Both traits would be expected in any system processing visual information for use in spatial awareness and in the visual guidance of behavior.

Although visual input is important for spatial operations, awareness of space is more than just a visual function. It is possible to apprehend the shape of an object and know where it is, regardless of whether it is seen or sensed through touch. Accordingly, spatial awareness, considered as a general phenomenon, depends not simply on visual areas of the dorsal stream but rather on the higher order cortical areas to which they send their output. The transition from areas serving purely visual functions to those mediating generalized spatial awareness is gradual, but has been accomplished by the time the dorsal stream reaches its termination in the association cortex of the posterior parietal lobe.

Association Areas Responsible for Spatial Cognition Form a Tightly Interconnected System

The posterior parietal cortex is considered to be preeminent among cortical areas responsible for spatial awareness because lesions of the posterior parietal cortex lead to the most devastating and specific impairments of spatial cognition. However, numerous other association areas in the cerebral hemisphere mediate cognitive functions that depend in some way on the use of spatial information. These areas occupy a continuous region of the cerebral hemisphere, encompassing large parts of the frontal, cingulate, temporal, parahippocampal, and insular cortices. They are connected to each other anatomically and to the parietal cortex by a parallel distributed pathway through which signals shuttle back and forth (Goldman-Rakic, 1988). The functions of some parts of this system are especially well understood. For example, the frontal cortex is involved in the generation of voluntary behavior, and the medial temporal lobe, including the hippocampus, is important for memory. In most individuals, the association areas of the right hemisphere are particularly important for spatial cognition. In nonhuman species, lateralization is much less pronounced. In general, spatial functions are represented in the same cortical areas in humans, monkeys, and rats. The following sections describe the spatial functions of the parietal cortex; how frontal areas program voluntary movements in spatial terms; and, finally, how the hippocampus and associated cortical areas mediate the formation of memories with a spatial component.

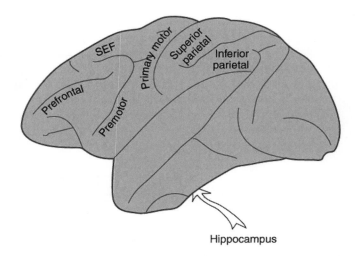

FIGURE 47.1 Lateral view of the left cerebral hemisphere of a rhesus monkey. Areas in the parietal and frontal cortex responsible for motor and cognitive processes of a spatial nature. SEF, supplementary eye field. Brain drawing, courtesy of Laboratory of Neuropsychology, NIMH.

PARIETAL CORTEX

The Parietal Cortex Contributes to Spatial Perception and Attention

The parietal lobe is divided into superior and inferior parietal lobules. Within each lobule, several subdivisions are distinguished by anatomical and functional properties (Colby and Duhamel, 1991). In both humans and monkeys, the superior parietal lobule serves functions related primarily to somesthesis, or tactile perception, whereas the inferior parietal lobule serves functions related primarily to visuospatial cognition (Fig. 47.1). The intraparietal sulcus runs in between the superior and the inferior lobules. Within the sulcus there are several functionally distinct areas with independent representations of space. In monkeys, these include the lateral intraparietal area (LIP), which is important for spatial attention, and the ventral intraparietal area (VIP), where visual and somatosensory representations of space are brought together. Recent functional imaging studies suggest that there are corresponding cortical areas in humans.

Injury to the Human Posterior Parietal Cortex Causes Impairments of Spatial Function

The neurologist Balint in 1909 first described a group of behavioral impairments specifically associated with damage to the parietal lobes. Balint's syn-

drome includes difficulty in executing eye movements to engage visual targets (ocular apraxia), an inability to perceive multiple objects at the same time (simultanagnosia), and inaccuracy in reaching for visual targets (optic ataxia). This collection of symptoms reflects an impairment in the visual guidance of movement and, more generally, in spatial cognition. Another major class of spatial deficits is known as neglect, the failure to attend to portions of space. Many other deficits in spatial behavior and cognition have been observed in patients with parietal damage, such as the inability to judge distances between objects. The following sections describe three specific impairments of spatial behavior known to arise from injury to the parietal lobe.

Simultanagnosia: An Inability to See Multiple Objects Simultaneously

After injury to the posterior parietal cortex, some patients experience difficulty in the visual perception of spatial relations. They can see objects, but cannot judge whether two objects are the same size or which of two objects is closer. When asked to copy simple line drawings, they may omit or transpose parts, as if unable to judge accurately the spatial arrangement of the object's components. In some patients, failure on visuospatial tests arises from a specific inability to see more than one object at a time (Coslett and Saffran, 1991). This condition, called simultanagnosia, commonly is observed after bilateral damage to the parietal cortex. When looking at the flame of a match, a patient with simultanagnosia may be unable to see the hand holding it. When tested in controlled situations, simultanagnosic patients demonstrate a profound inability to perceive more than one object simultaneously (see Box 47.1).

A particularly illuminating test carried out on simultanagnosic patients is shown in Figure 47.2 (Humphreys and Riddoch, 1993). When patients were presented with a set of randomly intermingled red and green circles (random condition), they were unable to say whether the circles were of different colors. Presumably this was because they could see only one circle at a time. When the circles were joined by line segments into red or green barbells (single color condition), these patients still could not say that more than one color was present. Only when pairs of circles of different colors were joined by line segments, unifying them into a single object, were patients able to report that there were different colors present (mixed condition). Note that the average distance between circles of different colors was the same in all three conditions. This indicates that the inability of patients to see and compare pairs of circles in the random and single conditions was the result of the circles belonging to sepa-

rate objects and not simply a result of their being separated by a certain distance. Simultanagnosia, by preventing simultaneous vision of two objects, gives rise to poor performance on tests requiring a comparison between objects. However, when tests are confined to a single object, this condition appears to leave spatial perception intact. Simultanagnosic patients are able to make accurate judgments of spatial relations so long as the judgments pertain to a single object.

Optic Ataxia: An Impairment of Visually Guided Reaching

After damage to the parietal cortex, some patients experience difficulty in making visually guided arm movements. This condition is referred to as misreaching, or optic ataxia (Perenin and Vighetto, 1988). Patients with optic ataxia experience difficulty in real-life situations requiring them to reach accurately under visual guidance. For example, a patient cutting food with a knife and fork may miss the plate altogether and hit the table when attempting to move the knife toward the food. Neuropsychologists have been able to characterize the reaching deficits of such patients in considerable detail by testing them in controlled situations.

BOX 47.1

SIMULTANAGNOSIA

Disorders of spatial awareness arising from parietal lobe injury are not merely a laboratory phenomenon, but can have a profound impact on the patient's daily life, as described in the following case study of simultanagnosia (from Coslett and Saffran, 1991).

When first examined by the authors four months after a right hemisphere infarction, the patient's major complaint was that her environment appeared fragmented; although she saw individual items clearly, they appeared to be isolated and she could not discern any meaningful relationship among them. She stated, for example, that she could find her way in her home (in which she had lived for 25 years) with her eyes closed, but she became confused with her eyes open. On one occasion, for example, she attempted to find her way to her bedroom by using a large lamp as a landmark; while walking toward the lamp, she fell over her dining room table. Although she enjoyed listening to the radio, television programs bewildered her because she could only "see" one person or object at a time

and therefore could not determine who was speaking or being spoken to; she reported watching a movie in which, after hearing a heated argument, she noted to her surprise and consternation that the character she had been watching was suddenly sent reeling across the room, apparently as a consequence of a punch thrown by a character she had never seen. Although she was able to read single words effortlessly, she stopped reading because the "competing words" confused her. She was unable to write, as she claimed to be able to see only a single letter; thus when creating a letter, she saw only the tip of the pencil and the letter under construction and "lost" the previously constructed letters.

Carol L. Colby and Carl R. Olson

Reference

Coslett, H. B. and Saffran, E. (1991). Simultanagnosia: To see but not to see. *Brain* **114**, 1523–1545.

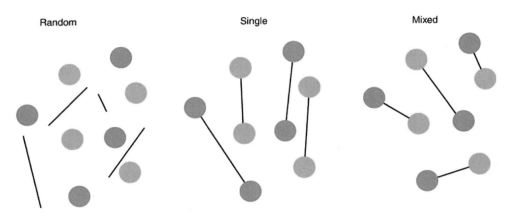

FIGURE 47.2 Three stimuli used for assessing simultanagnosia in human patients. Patients were consistently able to report that circles of different colors were present only in the mixed condition, in which red and green circles were unified into a single object by connecting line segments. From Humphreys and Riddoch (1993).

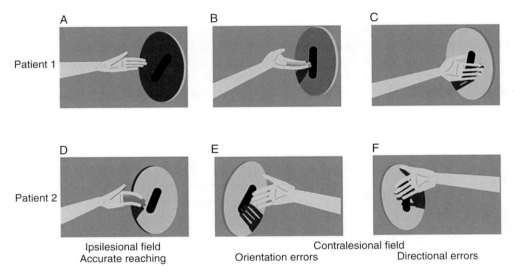

FIGURE 47.3 A test for optic ataxia. Patients can reach accurately (A and D) when using the ipsilesional hand in the ipsilesional visual field, but they make errors of hand orientation (B and E) and of direction (C and F) when reaching into the contralesional visual field. The top and bottom rows show the performance of two different patients. From Perenin and Vighetto (1988).

In the test illustrated in Figure 47.3, patients were required to reach out and insert one hand into a slot in the center of a disk held by the experimenter. The disk could be held in either the right or the left visual hemifield, and the patient could be asked to reach with either the right or the left hand. To perform the task, the patient had to direct the hand toward the center of the slot and orient the hand correctly so that it would pass into it. Pictures in the top row of Figure 47.3 show a patient with right parietal lobe damage reaching with the left hand. When the disk was positioned in the right (ipsilesional) visual field, the patient was able

to reach accurately (Fig. 47.3A). And yet, when the disk was in the left (contralesional) visual field, the patient committed errors of hand orientation (Fig. 47.3B) and of reaching direction (Fig. 47.3C). The lower row of Figure 47.3 shows reaching by a patient with damage to the left parietal lobe. When using the left hand in the left (ipsilesional) visual field, this patient was also able to reach accurately (Fig. 47.3D). However, when using the right hand in the right (contralesional) visual field, the patient committed errors of hand orientation (Fig. 47.3E) and of reaching direction (Fig. 47.3F).

These results demonstrate that optic ataxia may be lateralized, occurring only when the patient is required to point to targets in one visual hemifield or only when one hand is used for pointing. In cases where the deficit is restricted to a single hemifield or a single arm, the affected hemifield or arm is usually opposite the injured parietal lobe. This is what would be expected if the parietal lobe of each hemisphere mediates communication between more posterior visual areas (which represent the contralateral half of visual space) and more anterior motor areas (which represent the contralateral arm). There are, however, cases in which the problem is bilateral or occurs for specific combinations of arm and hemifield. Optic ataxia is not simply a problem with visuospatial perception, as indicated by the fact that performance with one arm may be perfectly normal. Nor is it simply a motor problem, as indicated by the fact that patients unable to reach accurately for visual targets can commonly touch points on their own bodies accurately under proprioceptive guidance. Optic ataxia is best characterized as an inability to use visuospatial information to guide arm movements.

Hemispatial Neglect: Unawareness of the Contralesional Half of Space

Hemispatial neglect is a classic symptom of injury to the posterior parietal cortex. It is a condition in which a lateralized failure of spatial awareness is present. The most common form of neglect arises from damage to the right parietal lobe and is manifested as a failure to detect objects in the left half of space. Patients with neglect experience problems in daily life, such as colliding with obstacles on the contralesional side of the body or mistakenly identifying letters on the contralesional side of a written word. When asked to copy pictures or to draw simple objects from memory, they leave out details on the affected side. When asked to say whether two objects are the same or different, they tend to indicate that two dissimilar objects are the same if the differentiating details are on the contralesional side. Whether the problem underlying neglect is a failure of attention to one half of space or a failure of the ability to form a mental representation of that half of space is not clear.

Neglect involves more than just a defect of attention to contralesional sensory events. For example, patients fail to report detail on the left half of an object even when they are asked to form a mental representation of the object by viewing it one part at a time as it passes slowly behind a vertical slit. In a famous set of experiments, patients were asked to imagine themselves in a familiar public setting and then to report what they would be able to see around them (Bisiach and Luzzatti, 1978). When they imagined standing at one end of the town square, patients described the buildings on the right side and failed to report buildings on the left. However, when instructed to imagine themselves standing at the other end of the square, facing the opposite direction, they now failed to mention buildings on the left that they had described just moments before and were able to describe buildings on the right that they had previously omitted.

Given that neglect affects the half of space opposite an injured parietal lobe, each parietal lobe must represent the opposite half of space. On the face of it, this proposition does not sound particularly surprising. Most visual areas in each hemisphere represent the opposite half of the visual field, most somatosensory areas represent the opposite half of the skin surface, and most motor areas represent muscles on the opposite half of the body. Injury to these areas leads to a sensory loss or motor impairment that affects the opposite half of a functional space defined with respect to some anatomical reference frame (the retina, the skin surface, or the muscles). In contrast, injury to the posterior parietal cortex gives rise to a neglect that may be defined with respect to any of several spatial reference frames. These fall into two broad classes: egocentric reference frames, in which objects and locations are represented relative to the observer; and allocentric reference frames, in which locations are represented in coordinates extrinsic to the observer. Examples of egocentric reference frames include those centered on the hand or the body. Allocentric reference frames include those in environmental coordinates (for example, room centered) and those centered on an object of interest. The full range of reference frames used by the brain is just beginning to be explored. Patients with neglect often show deficits with respect to more than one reference frame. The following sections describe evidence for impairments of multiple spatial reference frames in neglect.

A patient with left hemispatial neglect, looking straight ahead at the center of some object, will tend not to see detail on its left side. This observation is open to several interpretations. The simplest interpretation is that stimuli presented in the left visual field tend not to be registered. However, the fundamental problem might actually be with registering stimuli that are to the left of the head, or to the left of the torso, or the object itself. These possibilities cannot be distinguished without disentangling the various reference frames. Experiments aimed at identifying the spatial reference frame in neglect indicate that a patient's failure to detect a stimulus is affected not only by its location relative to the retina but also by its location relative to the object or array within which it is contained, by its location relative to the body, and even by its location relative to a gravitationally defined reference frame. Patients span a continuum from exhibiting predominantly egocentric forms of neglect to predominantly allocentric neglect.

In many patients, a component of neglect is object centered. If these patients are presented with an image anywhere in the visual field, they tend to ignore its left side. In one experiment, patients with left hemispatial neglect were required to maintain fixation on a small spot while chimeric faces (images formed by joining at the midline half-images representing the faces of two people) were presented at various visual field locations. Their reports of what they saw were based predominantly on the right halves of the chimeric images. This was true even when the entire composite face was presented within the right visual hemifield. In a second experiment, patients with left hemifield neglect were required to maintain fixation on a small spot while four stimuli in a horizontal row were presented. One of the stimuli was a letter that was to be named aloud. When the letter was in the leftmost location, patients were slower to name it, even when the entire array was in the right visual field.

A particularly dramatic way of demonstrating the object-centered nature of neglect is to ask patients to make copies of simple line drawings (Marshall and Halligan, 1993). When asked to copy a picture of two flowers, a patient with right parietal lobe damage saw and copied each flower but omitted the petals on the left half of each, thus showing evidence of left object-centered neglect (Fig. 47.4, top). When the same two flowers were joined by a common stem, the patient saw and copied the plant, but omitted details on its left side, including the entire leftmost flower, thus showing neglect for the left half of the larger composite object (Fig. 47.4, bottom).

Patients with parietal lobe damage can exhibit deficits with respect to other allocentric reference frames in addition to the object-centered neglect described earlier. The role of a gravitational reference frame in neglect has emerged from studies in which patients face a display screen either while sitting upright or while lying on their side (Ladavas, 1987). When the patient sits upright, the right and left halves of the

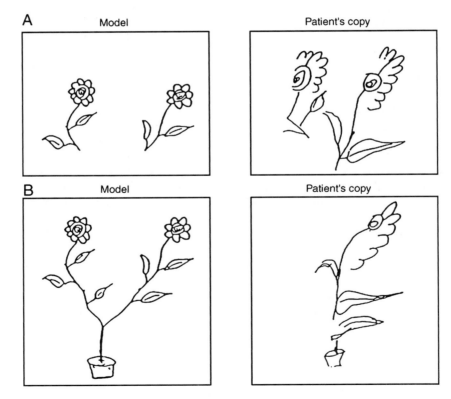

FIGURE 47.4   A test for object-centered neglect. When asked to copy the two drawings on the left, a patient made the two copies on the right. Detail is omitted from the left half of each object rather than from the left half of the drawing as a whole. From Marshall and Halligan (1993).

screen coincide with the right and left retinal visual fields. However, when the patient lies on one side while facing the screen, the situation is changed. Reclining on the right side, the patient sees the right and left halves of the screen as being in the upper and lower visual fields, respectively. Applying this procedure to patients with left hemispatial neglect enables the investigator to pose the question: Is the neglect specific for the left retinal visual field or is it specific for the left half of the screen? Neglect turns out to depend in part on each of these factors.

Further evidence that the parietal cortex represents the locations of objects relative to an external framework comes from the observation that neglect, in some patients, is restricted to stimuli within a certain range of distances from the body. The form of neglect termed peripersonal or proximal is specific for stimuli in the immediate vicinity of the body. Another form of neglect, termed extrapersonal, is specific for more distant stimuli. The implication of these findings is that there is a localization of function within the parietal cortex and that neurons in discrete areas process sensory input from objects at different distances. Discrete areas of the parietal cortex are known to be specialized for relaying sensory information to different motor systems. Sensory input from peripersonal space is uniquely significant for guiding reaching movements. In contrast, sensory input from greater distances (extrapersonal space) is linked primarily to the control of eye movements.

Lesions of the Parietal Cortex in Monkeys Produce Spatial and Attentional Problems

The original distinction between dorsal and ventral processing streams in the primate visual system was based in part on differences in the effects of lesions of the posterior parietal cortex and the inferior temporal cortex (Ungerleider and Mishkin, 1982). Monkeys with posterior parietal lesions are selectively impaired in visuospatial performance, such as judging which of two identical objects is located closer to a visual landmark. In contrast, inferior temporal cortex lesions produce deficits in visual discrimination (for example, shape or pattern recognition; see Chapter 46).

In addition to spatial perceptual deficits, parietal lesions also produce spatial motor deficits. After parietal lesions, monkeys have difficulty directing eye movements toward targets in the hemispace opposite the side of the lesion (Lynch and McLaren, 1989). Further, when two targets are presented simultaneously in the ipsilesional and contralesional fields, lesioned monkeys tend to ignore the contralesional stimulus, an effect termed visual extinction. Parietal

lesions also have profound effects on the ability of a monkey to reach toward an object. Lesions confined to the inferior parietal cortex impair reaching with the contralesional limb toward a target in contralesional space. Lesions extending across the intraparietal sulcus to include the superior parietal cortex produce impairments in reaching with the contralateral arm into either half of space. In sum, monkeys with parietal lesions are unable to assess spatial relations between objects or to judge locations of objects relative to themselves.

Parietal Neurons Have Response Properties Related to Spatial Information Processing

To understand more precisely how the parietal cortex contributes to spatial cognition, several groups of investigators have measured the electrical activity of single neurons during the performance of spatial tasks. These studies were done in alert monkeys trained to make eye movements to visual targets. Because brain tissue itself has no sensory receptors, microelectrodes can be introduced into the brain without disturbing the performance of an animal on a task. By recording single neuron activity during specially designed tasks, neural activity can be related directly to the sensory, cognitive, and motor processes that underlie spatial behavior. The next four sections describe how neurons in specific areas within the parietal cortex are selectively activated during spatial tasks and how they contribute to spatial representation and cognition.

Area VIP

The ventral intraparietal area (area VIP) is a functionally and anatomically defined area that lies at the junction of the visual and somatosensory districts within parietal cortex. Individual neurons respond strongly to moving stimuli and are selective for both the speed and the direction of the stimulus (Colby *et al.*, 1993). In this respect, VIP neurons are similar to those in other dorsal stream visual areas that process stimulus motion, especially areas MT and MST (Chapter 27). An important finding in area VIP is that most of these visually responsive neurons also respond to somatosensory stimuli (Duhamel *et al.*, 1991). These neurons are truly bimodal in the sense that they can be driven equally well by either a visual or a somatosensory stimulus. Most of these bimodal neurons have somatosensory receptive fields on the head or face that match the location of their visual receptive fields. This correspondence is illustrated in Figure 47.5, which shows visual and somatosensory receptive fields for several individual VIP neurons.

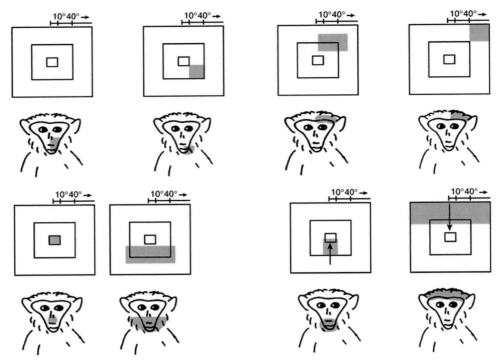

FIGURE 47.5 Matching visual and somatosensory receptive fields of neurons from area VIP. On each outline of the head of the monkey, the patch of color corresponds to the somatosensory receptive field of one neuron. On the square above the head, the patch of texture corresponds to the visual receptive field of the same neuron. The square can be thought of as a screen placed in front of the monkey so that its center is directly ahead of the monkey's eyes. From Duhamel *et al.* (1991) with permission of Oxford University Press.

Three kinds of correspondence are found in VIP neurons. First, for each neuron, receptive fields in each modality match in location. For example, a neuron that responds to a visual stimulus in the upper left visual field also responds when the left brow is touched. Second, receptive fields match in size. A neuron with a restricted visual receptive field also has a small somatosensory receptive field. The smallest visual receptive fields are those located at the fovea. It is striking that the corresponding somatosensory receptive fields for these neurons are not centered around the eye but are found around the mouth. In terms of sensitivity, the mouth can be thought of as the fovea of the facial somatosensory system. Hence it is logical that the two regions of highest spatial acuity should be linked. The third type of correspondence is in the preferred directions of movement. For example, a neuron responsive to a visual stimulus moving toward the right also responds when a small probe is brushed lightly across the face of the monkey moving to the right. In sum, visual and somatosensory receptive fields for individual VIP neurons match in location, size, and directional preference.

The correspondence in receptive field locations across modalities immediately raises a question: What

happens to the relative locations of the visual and somatosensory receptive fields when the eyes move? If the visual receptive field were simply retinotopic, it would move when the eyes do; if the somatosensory receptive field were purely somatotopic, it would be unchanged by eye movements. There could not be a consistent correspondence in location if visual receptive fields were defined only with respect to the retina while somatosensory receptive fields were defined only with respect to the skin surface.

The answer is that visual receptive fields are linked to the skin surface and not to the retina. A neuron that responds best to a visual stimulus approaching the forehead and has a matching somatosensory receptive field on the brow will continue to respond best to a visual stimulus moving toward the brow regardless of where the monkey is looking. Thus, both visual and somatosensory receptive fields are defined with respect to the skin surface. In this sense, the receptive fields are head centered: a given VIP neuron responds to stimulation of a certain portion of the skin surface and to the visual stimulus aligned with it no matter which part of the retina is activated.

In sum, neurons in area VIP encode bimodal sensory information in a head-centered representation of space.

This kind of representation would be most useful for guiding head movements. Anatomical studies indicate that area VIP sends information to premotor cortex including the specific region that is involved in generating head movements.

Area LIP

The lateral intraparietal area (area LIP) is a functionally defined area within the visual district of parietal cortex. Neurons in LIP exhibit several different kinds of activity during spatial tasks (Colby *et al.*, 1996). First, LIP neurons, like neurons in striate and extrastriate visual cortex (Chapter 27), respond to the onset of a visual stimulus in the receptive field of the neuron. Second, these visual responses are enhanced when the monkey attends to the stimulus: the amplitude of the visual response is increased when the stimulus or stimulus location becomes the focus of attention. This enhancement occurs no matter what kind of movement the animal will use to respond to the stimulus. Regardless of whether the task requires a hand movement or an eye movement or requires that the monkey refrain from moving toward the stimulus, the visual response of an LIP neuron becomes larger when the stimulus is made behaviorally relevant. This means that the same physical stimulus arriving at the retina can evoke very different responses in cortex as a result of spatial attention.

A third important feature of LIP neuron activity is the sustained responses observed when the monkey must remember the location at which the stimulus appeared. In this task, a stimulus is flashed only briefly but the neuron continues to fire for several seconds after the stimulus is gone, as though it were holding a memory trace of the target location. A particularly intriguing question in understanding spatial representation concerns the fate of this memory-related activity in area LIP following an eye movement, as will be described later. A fourth kind of activation commonly observed in LIP neurons is related to performance of a saccade—a rapid eye movement—toward the receptive field. Some LIP neurons fire just before the monkey initiates a saccade that will move the fovea onto a target presented in the receptive field. LIP neurons have overlapping sensory and motor fields, just like neurons in the superior colliculus (Chapter 33). Finally, LIP neuron activity can be modulated by the position of the eye in the orbit (Andersen *et al.*, 1990). For instance, the visual response of a given cell may become larger when the monkey is looking toward the left part of the screen than when it is looking toward the right. This property is interesting because it suggests that neurons in area LIP may contribute to spatial repre-

sentations that go beyond a simple replication of the retinal map. This idea is discussed in more detail in the next section.

In sum, individual LIP neurons have receptive fields at particular retinal locations and carry visual, memory, and saccade-related signals that can be modulated by attention and by eye position. Activity in area LIP cannot be characterized as a simple visual or motor signal. Rather, the level of activation in a given LIP neuron reflects the degree to which attention has been allocated to a location within the receptive field.

Spatial Representation in Area LIP

Every time we move our eyes, each object in our surroundings activates a new set of retinal neurons. Despite this ever-changing input, we experience a stable visual world. How is this possible? More than a century ago, Helmholtz proposed that the reason the world appears to stay still when we move our eyes is that the "effort of will" involved in making a saccade simultaneously adjusts our perception to take that specific eye movement into account. He suggested that when a motor command to move the eyes is issued, a copy of that command, or corollary discharge, is sent to brain areas responsible for generating our internal image of the world. This image is then updated so as to be aligned with the new visual information that will arrive in the cortex after the eye movement. A simple experiment shows that Helmholtz's account must be essentially true. When the retina is displaced by pressing on the eye, the world does seem to move, presumably because there is no corollary discharge. Without that internal knowledge of the intended eye movement, there is no way to update the spatial representation of the world around us.

Neurons in area LIP contribute to this updating of the internal image (Duhamel *et al.*, 1992). The experiment illustrated in Figure 47.6 shows that the memory trace of a stimulus is updated when the eyes move. The activity of a single LIP neuron was recorded under three different conditions. In the first set of trials, the monkey looked at a fixed point on the screen while a stimulus was presented in the receptive field (Fig. 47.6A). In the diagram at the top of Figure 47.6A, the dot is the fixation point, the dashed circle shows the location of the receptive field, and the asterisk represents the visual stimulus. The time lines just below the diagram indicate that the vertical and horizontal eye positions remained steady throughout the trial, demonstrating that the monkey maintained fixation. The stimulus time line shows that the stimulus appeared 400 ms after the beginning of the trial and continued for the entire trial. The raster display below plots the

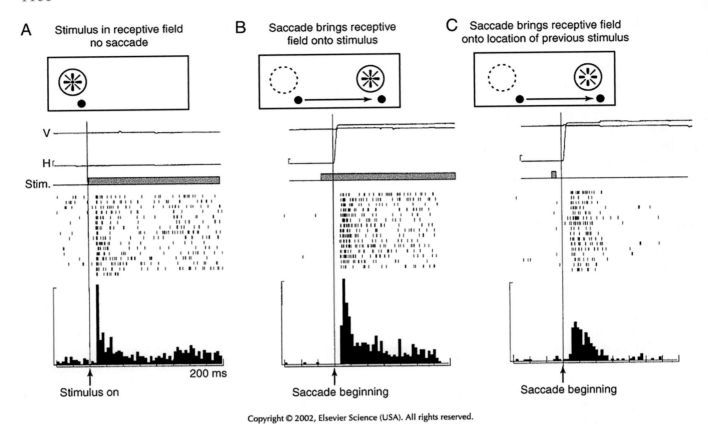

A Stimulus in receptive field
no saccade

B Saccade brings receptive
field onto stimulus

C Saccade brings receptive field
onto location of previous stimulus

200 ms

Stimulus on

Saccade beginning

Saccade beginning

FIGURE 47.6 Remapping of visual memory trace activity in area LIP. The activity of a single neuron was recorded under three different conditions. (A) Simple visual response to a constant stimulus in the receptive field, presented while the monkey is fixating. The rasters and histogram are aligned on the time of stimulus onset. (B) Response following a saccade that brings the receptive field onto the location of a constant visual stimulus. (C) Response following a saccade that brings the receptive field onto the location where a stimulus was presented previously. The stimulus is extinguished before the saccade begins so it is never physically present in the receptive field. The neuron responds to the updated memory trace of the stimulus. V, vertical eye position; H, horizontal eye position. From Duhamel *et al.* (1992).

electrical activity of a single LIP neuron in 16 successive trials. In these rasters, each small vertical line indicates the time at which an action potential occurred, and each horizontal line of dots represents activity in a single trial. In each trial there was a brief initial burst of action potentials shortly after the stimulus appeared, followed by continuing neural activity at a lower rate. The histogram at the bottom of Figure 47.6 shows the average firing rate as a function of time. The visual response in Figure 47.6A is typical of that observed in neurons in many visual areas: the neuron fired when a stimulus appeared in the receptive field of the neuron.

In the second set of trials (Fig. 47.6B), a visual response occurred when the monkey made an eye movement that brought a stimulus into the receptive field. At the beginning of the trial, the monkey looked at the fixation point on the left, and the rest of the screen was blank. Then, simultaneously, a new fixation point appeared on the right and a visual stimulus

appeared above it. The monkey made a saccade from the old fixation point to the new one, indicated by the arrow. The eye movement was straight to the right so only the horizontal eye position trace shows a change. At the end of this saccade, the receptive field had been moved to the screen location containing the visual stimulus. The rasters and histogram in Figure 47.6B are aligned on the time that the saccade began. In each trial, the neuron began to respond after the receptive field had landed on the stimulus. This result is just what would be expected for neurons in any visual area with retinotopic receptive fields.

The novel finding is shown in Figure 47.6C. In this third set of trials, the monkey made a saccade that would bring a stimulus into the receptive field, just as in the second set of trials. The only difference was the duration of the stimulus, which lasted for a mere 50 ms instead of staying on for the entire trial. As can be seen on the stimulus time line, the stimulus actually disappeared before the saccade began. This means that the stimulus was never physically present in the receptive field. Nevertheless, the neuron fired as though there were a stimulus in its receptive field. This result indicates that LIP neurons respond to the memory trace of a previous stimulus. Moreover, the representation of the memory trace is updated at the time of a saccade. The general idea of how a memory trace can be updated is as follows. At the beginning of the trial, while the monkey is looking at the initial fixation point, the onset of the stimulus activates neurons whose receptive fields encompass the stimulated location. Some of these neurons will have tomic activity and continue to respond after stimulus offset, encoding a memory of the location at which the stimulus occurred. When the monkey moves its eyes toward the new fixation point, a copy of the eye movement command is sent to parietal cortex. This corollary discharge causes the active LIP neurons to transmit their activity to a new set of neurons whose receptive fields will encompass the stimulated screen location after the saccade. By means of this transfer, LIP neurons encode the spatially updated memory trace of a previous stimulus.

The significance of this finding lies in what it illustrates about spatial representation in area LIP. It indicates that the internal image is dynamic rather than static. Tonic, memory-related activity in area LIP not only allows neurons to encode a salient spatial location after the stimulus is gone, but also allows for dynamic remapping of visual information in conjunction with eye movements. This updating of the internal visual image has specific consequences for spatial representation in the parietal cortex. Instead of spatial information being encoded in purely retinotopic coordinates, tied to the specific neurons initially activated by the stimulus, the information is encoded in eye-centered coordinates. This is a subtle distinction but a very important one in generating accurate spatial behavior. Maintaining visual information in eye-centered coordinates tells the monkey not just where the stimulus was on the retina when it first appeared, but where it would be on the retina following an intervening eye movement. The result is that the monkey always has accurate information with which to program an eye movement toward a real or a remembered target. Further results from this series of experiments indicate that humans also depend on this kind of remapping for accurate spatial representation.

In sum, neurons in area LIP encode sensory information in an eye-centered representation of space, the format most useful for guiding eye movements. Area LIP sends projections to both the superior colliculus and the frontal eye fields, regions that are involved in generating eye movements. In addition to area LIP, the parietal cortex contains a number of functionally distinct areas, each of which may be specialized for particular types of stimuli and particular regions of space. The general point is that the problem of spatial representation may be solved in several ways, and each solution may contribute to the generation of a different kind of action.

Parietal Cortex and Decision Processes

Spatially selective neuronal activity in area LIP is graded in strength. If the monkey's attention to a given location or intention to make a saccade in a given direction increases in strength, then the firing of LIP neurons with response fields in that direction also increases. This has been demonstrated in decision tasks—tasks requiring the monkey to decide which of several possible saccades to make. A decision can be either sensory-based or value-based. In a sensory decision task, the monkey must judge whether a visual display contains dots moving to the right or left and must report the judgment by executing a rightward or leftward saccade (Roitman and Shadlen, 2002). If the motion stimulus is weak, requiring the monkey to integrate sensory information over time, the neuronal signal reflecting the decision develops slowly. If the motion stimulus is strong, allowing the monkey to make an instantaneous decision, the neuronal signal rises sharply. Thus the evolution of the monkey's decision is reflected in the activity of spatially selective LIP neurons. In a value-based decision task, the monkey chooses between a saccade to the right and a saccade to the left on the basis of prior experience indicating that one of the saccades is more likely than the other to yield a valued juice reward (Sugrue *et al.*, 2004).

During the preparatory period of the trial, LIP neurons fire strongly if the target in their response field is associated with a valuable reward. In this task the activity of spatially selective LIP neurons reflects the value of the target. These two findings do not necessarily indicate that LIP plays a direct role in making sensory- or value-based decisions. They can also be explained on the assumption that LIP neurons mediate attention or motor preparation and that they fire more strongly when attention or motor preparation is more intense (Maunsell, 2004).

Summary

The posterior parietal cortex plays a critical role in spatial awareness. Injury to the parietal cortex in humans and monkeys leads to deficits in spatial perception and action. Physiological studies in monkeys have shown that parietal neurons construct a representation of space by combining signals from multiple sensory modalities with motor signals. An important physiological finding is that parietal neurons represent spatial locations relative to multiple reference frames, including ones centered on the eye and the head. In accord with these physiological findings, human neuropsychological studies have shown that neglect resulting from parietal lobe injury can be expressed with respect to several different reference frames.

FRONTAL CORTEX

The Frontal Cortex Contributes to Voluntary Movement and the Control of Behavior

The frontal lobe is involved in spatial functions as a natural result of its being involved in behavioral control. The three main divisions of the frontal lobe are the primary motor cortex located on the precentral gyrus; the premotor cortex, including the supplementary eye field, located in front of the primary motor cortex; and the prefrontal cortex (Fig. 47.1). The motor, premotor, and prefrontal areas all contribute to behavioral control, but they differ with respect to the quality of their contribution. This difference is seen in the effects of brain injury. Injury to the primary motor cortex leads to weakness and paralysis of the contralateral muscles. In contrast, injury to the premotor cortex leads to difficulty in producing movements in certain circumstances, for example, when the patient is asked to mime the use of a tool or to learn arbitrary associations between stimuli and responses. Finally, injury to the prefrontal cortex results in a classic syndrome characterized by a lack of drive and an impaired ability to execute plans (Chapter 52). These effects indicate that progressively more anterior parts of the frontal cortex contribute to progressively more abstract aspects of behavioral control. Each of these divisions participates in spatial processes insofar as the kind of behavioral control to which it is dedicated has a spatial component. The next four sections describe aspects of spatial representation and function in the primary motor cortex, in the premotor cortex, in the supplementary eye field, and in the prefrontal cortex. The final section examines the relation between motivational modulation of neural activity and the representation of value.

Neurons in the Primary Motor Cortex Represent Movement Direction Relative to a Spatial Frame

The primary motor cortex contains a map of the muscles of the body in which the leg is represented medially, the head laterally, and other body parts at intermediate locations. Within this map are patches of neurons that represent different muscles. Neurons within a given patch receive proprioceptive input from a muscle or small group of synergistic muscles and send their output back to that muscle or group of synergists by way of a multisynaptic pathway through the brain stem and spinal cord. There have been many studies in which the electrical activity of neurons in the primary motor cortex is monitored while animals move (Chapter 30). The general conclusion of these studies is that neurons in the primary motor cortex are active when the corresponding muscles are undergoing active contraction. It is important to note, however, that neurons in the primary motor cortex probably do more than simply encode the levels of activation of individual muscles. One proposal is that they encode movement trajectories. Every voluntary movement can be described in two quite different but perfectly complementary ways: in terms of the lengths of the muscles or in terms of the position of the part of the body being moved. For example, during an arm movement, changes take place both in the lengths of muscles acting on the arm and in the position of the hand. Could it be that neurons in the primary motor cortex encode a spatial variable, such as the direction of movement of the hand, rather than a muscle variable?

Evidence for the idea that neurons of the primary motor cortex encode movement direction has come from studies in which monkeys make reaching movements in various directions. Individual neurons are selective for a specific direction of movement: a given neuron may fire most strongly during movements up and to the right and progressively less strongly for movements that deviate from the preferred direction (Schwartz et al., 1988). The patterns of selectivity are well defined and the preferred directions of different neurons cover the range of possible movements fairly evenly. By recording the activity of the entire population of neurons one could, in principle, quite accurately describe the movement. The question remains whether these neurons are selective for the direction of movement in space or for the specific patterns of muscle activation associated with particular movements. In an elegant series of studies, Kakei et al. (1999) recorded from neurons in the primary motor cortex while the monkey moved its arm in a single direction using different combinations of muscles. They discovered that the primary motor cortex contains both neurons selective for movement direction and neurons selective for patterns of muscle activation.

Neurons in the Premotor Cortex Have Head and Hand-Centered Visual Receptive Fields

One of the functions of premotor cortical areas is to act as a conduit through which sensory signals are relayed to the motor system. The sensory information that reaches these areas has been highly processed already in the posterior cerebral hemispheres. Thus it is not surprising that the sensory receptive fields of some neurons in the premotor cortex are defined with respect to an external spatial framework in a form suitable for use by the motor system.

Both head and hand-centered visual receptive fields have been described in the premotor cortex. First, in portions of the premotor cortex representing orofacial movements, neurons respond to visual stimuli at a certain location relative to the head (Fogassi *et al.*,

1992). These neurons have been characterized by recording from them while objects approach the head of the monkey along various trajectories. A typical experiment is illustrated in Figure 47.7. In the first phase of the experiment, shown in the left column, the monkey looked straight ahead at a fixation target (F) while an object approached the face and then receded. The stimulus moved along a trajectory that brought it either to the right side of the head (trajectory 1) or to the left side (trajectory 2). Records of neuronal activity show that the neuron fired when the object approached along trajectory 1 but not when it approached along trajectory 2. Accordingly, it appears that the neuron had a visual receptive field located in the right visual field (shaded area in Fig. 47.7A).

To determine whether the receptive field was head or retina-centered required a second phase of testing

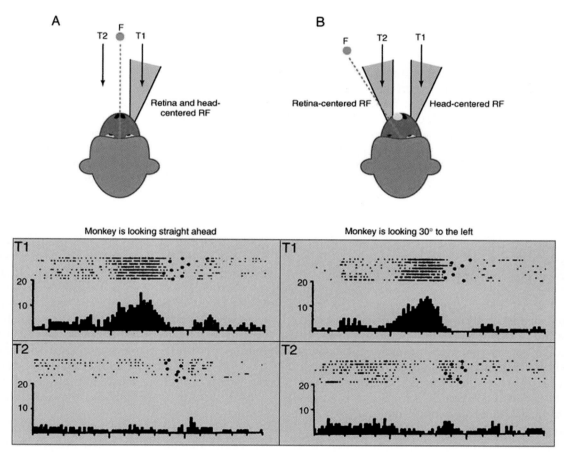

FIGURE 47.7 Data from a neuron in the premotor cortex with a head-centered visual receptive field. (A) While the monkey was looking straight ahead at a fixation point (F), an object approached and receded, traveling either along trajectory 1 (T1, to the right of the head) or along trajectory 2 (T2, to the left of the head). The neuron fired strongly only when the object approached along trajectory 1. (B) When the monkey looked at a new fixation point on the far left, the neuron still responded only when the object approached along trajectory 1, indicating that the receptive field is tied to the location of the head and not to the location of the retina. The time lines above each raster and histogram show the horizontal (H) and vertical (V) eye position during one trial. The time line below shows the position of the visual object as it comes toward and then moves away from the monkey. From Fogassi *et al.* (1992).

(right column). In this phase, the monkey looked far to the left at a fixation point (F). While he maintained a leftward gaze, objects again approached the face, following trajectories 1 and 2. If the receptive field of this neuron were head centered, one would predict that it should continue to respond as the object approached along trajectory 1. However, if the receptive field were fixed to the retina, then the neuron should respond as the object approached along trajectory 2. Records of the activity of the neuron clearly indicate that the neuron had a head-centered visual receptive field. It was responsive to stimuli presented to the right of the midline of the head, but not to stimuli presented to the right of the midline of the retina. This head-centered spatial selectivity of neurons in the premotor cortex is similar to that found in area VIP. That neurons in these two areas should exhibit consistent patterns of spatial selectivity is not surprising because the parietal and premotor cortices are strongly interconnected.

A second type of body-centered spatial representation in the premotor cortex has been observed in areas representing arm movements. Here, neurons respond to visual stimuli if they are presented in the vicinity of the hand (Graziano *et al.*, 1994). When the hand moves to a new location, the visual receptive field moves with it. The visual receptive field remains fixed to the position of the hand regardless of where the monkey is looking and, thus, regardless of the part of the retina on which the image is cast. The two distinct forms of body-centered visual responsiveness seen in the premotor cortex presumably reflect the involvement of these neurons in the visual guidance of orofacial and arm movements.

Neurons in the Supplementary Eye Field Encode the Object-Centered Direction of Eye Movements

A particularly interesting form of allocentric spatial representation is found in the supplementary eye field (SEF). The SEF is a division of the premotor cortex with attentional and oculomotor functions. Neurons here fire before and during the execution of saccadic eye movements. Two characteristics of the SEF set it apart from subcortical oculomotor centers and suggest that its contribution to eye movement control occurs at a comparatively abstract level. First, neurons here fire while the monkey is waiting to make an eye movement in the preferred direction, as well as during the eye movement itself. Second, some SEF neurons become especially active when the monkey is learning to associate arbitrary visual cues with particular directions of eye movements.

SEF neurons exhibit a unique form of spatial selectivity in monkeys trained to make eye movements to particular locations on an object. In this context, SEF neurons encode the direction of the impending eye movement as defined relative to an object-centered reference frame (Olson and Gettner, 1995). Regardless of where in space an object appears, these neurons respond when the animal is planning to make an eye movement to a specific location on that object. An experiment demonstrating this point is shown in Figure 47.8. This figure shows the activity of a single SEF neuron while the monkey performed an eye movement task. At the beginning of each trial, the monkey fixated a dot at the center of a screen. While the monkey fixated, a sample and cue were presented in the right visual field. The cue flashed on either the right or the left end of the short horizontal sample bar. Following extinction of the sample–cue display, the monkey maintained fixation for a brief time. Then, simultaneously, the central spot was extinguished and a target bar appeared at an unpredictable location in the upper visual field of the monkey. The monkey had to make an eye movement to the end of the target bar corresponding to the cued end of the sample bar. Across the set of eight conditions, the object-centered direction of the eye movement was completely independent of its physical direction.

This situation made it possible to ask whether the activity of the neuron was related to the object-centered direction of the movement or to its physical direction. To the right of each panel in Figure 47.8 is shown the average firing rate of a single neuron as a function of time during the trial. Regardless of the direction of the physical movement of the eye, firing was clearly stronger on trials in which the monkey made an eye movement to the left end of the target bar than on trials in which the target was the right end of the bar. For example, in panels 1 and 2, the physical direction of the eye movement was exactly the same, and yet firing was much stronger when the left end of the bar was the target (panel 1) than when the right end of the bar was the target (panel 2). This neuron exhibited object-centered direction selectivity in the sense that it fired most strongly before and during movements to a certain location on an object.

Object-centered direction selectivity serves an important function in natural settings. In scanning the environment, we sometimes look toward locations where things are expected to appear, but which, at the time, contain no detail, for example, the center of a blank screen or the center of an empty doorway. Our eyes are guided to these featureless locations by surrounding features that define them indirectly. It is spe-

A Left with respect to bar

B Right with respect to bar

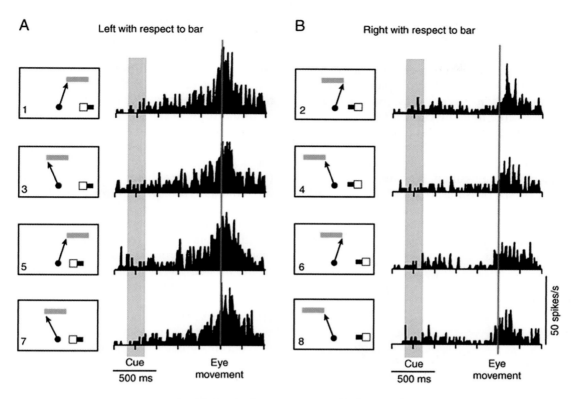

FIGURE 47.8 Data from a neuron in the SEF selective for the object-centered direction of eye movements. The monkey was trained to make eye movements to the right or left end of a horizontal bar. A cue appeared on the small sample bar shown in the lower right of each panel (1–8) to tell the monkey which end of the target bar (top in each panel) was relevant on a given trial. The arrow in the panel next to each histogram indicates the direction of the eye movement. The neuron fired strongly when the eye movement was directed to the left end of the target bar (left column) regardless of whether the physical movement of the eye was up and to the right (panels 1 and 5) or up and to the left (panels 3 and 7). From Olson and Gettner (1995).

cifically in these cases that the SEF may contribute to the selection of the target for an eye movement.

The Prefrontal Cortex Mediates Working Memory for Spatial Information

Working memory is required to hold a plan in mind and carry it out step by step (Chapter 52). The fact that this ability is severely impaired in some patients following prefrontal injury indicates that this region plays a crucial role in working memory. Insofar as plans and working memory have a spatial component, operations carried out by the prefrontal cortex are also spatial in nature. Single neuron-recording experiments in monkeys performing delayed-response tasks have demonstrated the importance of the dorsolateral prefrontal cortex for both spatial and nonspatial forms of working memory (Funahashi *et al.*, 1991). A delayed-response task consists of several trials, each several seconds long. At the beginning of each trial, a cue is presented briefly, instructing the monkey which

response to make, but the response must be withheld until the end of the trial. Delayed-response tasks can be designed so that both the cues (for example, spots flashed to the right or left of fixation) and the responses (for example, eye movements to the right or left) are spatial. In the context of these spatial tasks, prefrontal neurons are active during the period between the cue and the signal to respond, when the monkey is holding spatial information in working memory. Just as for visual responses, some neurons are selective for the direction of the motor response. The simple interpretation of this pattern of activity is that it is a neural correlate of the monkey actively remembering the cue and holding in mind the intended response. This interpretation is consistent with results from brain imaging experiments in humans during working memory tasks.

The idea that the prefrontal cortex mediates spatial working memory has received further support from lesion experiments in monkeys. After injury to or inactivation of specific locations in the prefrontal cortex, monkeys are impaired at remembering locations in the opposite half of space. Their delayed responses are spatially inaccurate and the inaccuracy is exacerbated by longer delays. An experiment demonstrating the importance of the prefrontal cortex for spatial working memory is illustrated in Figure 47.9, which shows behavioral data from a monkey trained to perform an oculomotor delayed-response task (Sawaguchi and Goldman-Rakic, 1994). At the beginning of each trial, the monkey fixated a spot at the center of the screen. While the monkey maintained fixation, a cue was flashed at one of six possible locations at positions. Following presentation of the cue, a delay of 1.5 to 6 sec ensued before the monkey was permitted to make an eye movement to the cued location. Data

from a normal monkey are shown in the left column of Fig. 47.9. In each panel, the six bundles of radiating rays represent the eye trajectories on trials when the cue was at the six different locations. Even when required to remember the cue for 6 s, the monkey made accurate eye movements. Data in the right column are from the same monkey after a dopamine antagonist was injected into the left prefrontal cortex. The ability of the monkey to remember the location of the cue remained intact when the cue was in the left (ipsilesional) visual field, as shown by the accurate eye movements. However, performance deteriorated on trials when the cue was in the right (contralesional) visual field. Especially after long delays (top panel), the direction of the eye movement began to deviate from the location of the cue, as if the working memory of the monkey were fading. The fact that this deficit was specific to long delays is noteworthy because it rules out any simple explanation based on interference with visual or motor processes as opposed to working memory itself.

Neural signals related to working memory are found in object memory tasks as well as spatial tasks (Chapter 52). In a task in which the monkey had to remember the identity of an object as well as its location, some neurons were selectively activated by remembering the object (Rao *et al.*, 1997). Prefrontal cortex is important for both kinds of working memory.

Value Representation and Motivational Modulation in Frontal Cortex

In many areas of the frontal lobe, neurons fire more strongly when a monkey anticipates receiving a larger reward for making a saccade (Roesch and Olson, 2003). The firing of these neurons could signify either of two things. First, it might represent, in an economically meaningful sense, the value of the reward that the monkey expects to receive. The monkey might commit more or less firmly to making the saccade on the basis of how strongly the neurons are firing. Second, it might simply be related to motor planning, attention or arousal—internal states enhanced when the monkey is more motivated. That these interpretations are different can be illustrated by a simple example. The neck muscles of some monkeys grow more tense when they are planning a saccade in expectation of getting a larger reward (Roesch and Olson, 2004). The tension of the neck muscles obviously reflects motivation and does not represent value in an economically meaningful sense. Whether the activity of frontal neurons represents value or reflects motivation cannot be determined by experiments manipulating solely the size of

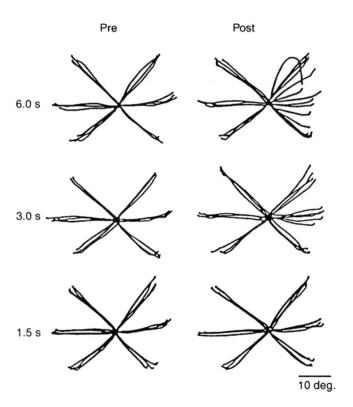

Pre   Post

6.0 s

3.0 s

1.5 s

10 deg.

FIGURE 47.9 Inactivation of the left prefrontal cortex disrupts spatial working memory in the right half of space. A monkey was trained to fixate a central spot while a peripheral cue was presented briefly at one of six locations. After an interval of 6, 3, or 1.5 s, the fixation spot was extinguished and the monkey made an eye movement to the remembered location. Each set of diverging rays shows the trajectories of eye movements executed under a certain set of conditions. The left column shows eye movement performance before inactivation by a dopamine antagonist. The right column shows performance after inactivation. When the cue was in the right hemifield and the monkey was required to remember its location over long delays, the movements became highly inaccurate. From Sawaguchi and Goldman-Rakic (1994).

the expected reward because, as value increases, so does motivation.

This issue has been resolved by manipulating independently the size of the reward promised in the event of the monkey's successfully completing the trial (one or three drops of juice) and the size of the penalty threatened in the event of failure (a one-second or eight-second time-out). A large promised reward and a large threatened penalty have opposite value but both act to increase motivation. Spatially selective neurons in premotor cortex fire more strongly when either a large reward has been promised or a large penalty has been threatened (Fig. 47.10). In other words, activity in these neurons is modulated by the animal's level of motivation. In contrast, neurons in orbitofrontal cortex fire most strongly when a large reward has been promised and least strongly when a large penalty has been threatened. In other words, they represent anticipated value. The orbitofrontal cortex is a part of the limbic system known to be involved in emotional processes. Presumably, when the monkey considers an action that would lead to an outcome, neurons in orbitofrontal cortex signal the emotion that would be elicited by the outcome, thus representing its value. Value representations in orbitofrontal cortex presumably give rise to motivational modulation in premotor cortex.

Summary

Neurons in the frontal cortex represent spatial information as a natural consequence of their role in controlling behavior. Neurons in the primary motor cortex encode the directions of movements. Neurons in the premotor cortex encode locations relative to the body, even when these locations are not the immediate targets of actions. SEF neurons encode locations in an allocentric, object-centered representation. Neurons in the prefrontal cortex encode the locations of and identity of objects being held in short-term memory. Finally, the strength of neural activity in premotor cortex is modulated by the monkey's level of motivation while neurons in orbitofrontal cortex encode the anticipated value of the expected reward.

HIPPOCAMPUS AND ADJACENT CORTEX

The Hippocampal System Is Associated with Memory Formation

Spatial functions of the hippocampus center on its role in memory formation (Chapter 50). The hippo-

campus is an area of primitive cortex, or allocortex, hidden on the underside of the temporal lobe. It is connected to a set of immediately adjacent cortical areas, including the perirhinal, entorhinal, and parahippocampal cortices (Zola-Morgan and Squire, 1993).

Amnesia Resulting from Hippocampal Injury in Humans Includes a Spatial Component

Extensive evidence implicates the hippocampus and related medial temporal structures in the formation of declarative memories in humans. Memories dependent on the hippocampus include, although they are by no means restricted to, memories for spatial material. In addition to many other impairments, the amnesia of the noted patient H.M. was evident in his inability to learn to find his way through new neighborhoods. Patients with damage to the hippocampus, especially the right hippocampus, are impaired on tests requiring them to inspect a scene with many objects and then to recall the locations of individual objects.

The hippocampus has a role in both spatial and nonspatial memory. In rats, as described in Chapter 50, the hippocampus is essential not only for remembering locations but also for remembering odors and for learning associations between stimuli and rewards (Bunsey and Eichenbaum, 1996). Likewise, in humans, memories of many kinds depend on the hippocampus. Patient EP, profoundly amnesic after extensive medial temporal damage resulting from a viral encephalitis, suffers from problems with spatial memory. He cannot learn the layout of the new neighborhood where he now lives (Teng and Squire, 1999). However, he is not impaired at spatial navigation in itself. He is able to find his way around the neighborhood where he grew up and can describe it in detail. Moreover, his deficit extends beyond spatial memory. Like patient HM, he fails to recognize someone to whom he was introduced only minutes before. How the hippocampus uses spatial information, along with nonspatial information, to form new memories is taken up in Chapter 50.

Neurons of the Hippocampal System Have Place Fields and Are Sensitive to Directional Orientation of the Head

Recordings from hippocampal neurons in rats running mazes or exploring open areas have revealed a remarkable degree of spatial selectivity. Many neurons in the hippocampus have place fields: a given neuron will fire most strongly when the rat is within

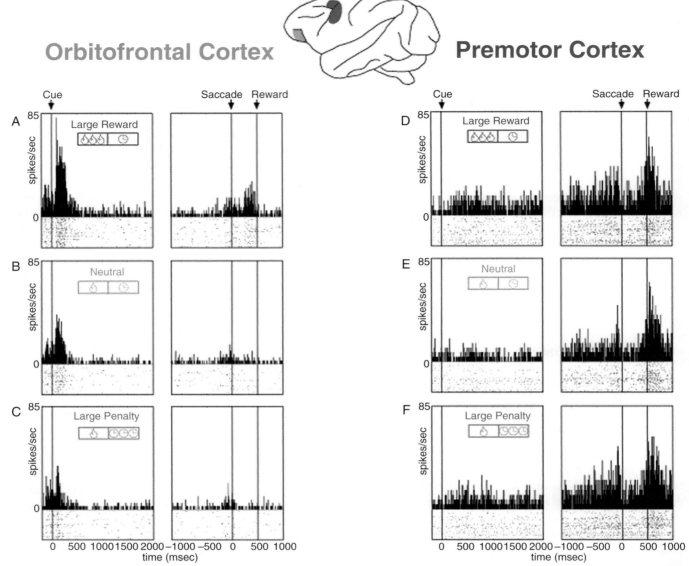

FIGURE 47.10 Evidence that orbitofrontal neurons encode expected value whereas premotor neurons are subject to motivational modulation. A cue presented early in each trial instructed the monkey whether to make a rightward or leftward saccade at the trial's end. It also indicated whether making the instructed saccade would lead to a large or small reward (three drops or one drop of juice). Finally, it indicated whether failure to make the saccade would lead to a large or small penalty (a one-second or eight-second time-out). After a delay of several seconds, the monkey was permitted to respond. For a saccade in the instructed direction, the monkey received the promised reward. For any other response, the monkey incurred the threatened penalty. The orbitofrontal neuron encoded the value of the outcomes predicted by the cue. It responded most strongly when the combined predicted outcome was best (A: large reward promised and small penalty threatened), at an intermediate level when it was middling (B: small reward promised and small penalty threatened) and least strongly when it was worst (C: small reward promised and large penalty threatened). The premotor neuron fired during the delay period at a rate that reflected how motivated the monkey was. It fired most strongly when the monkey had something significant to work for, either earning a large reward (D) or avoiding a large penalty (F) and least strongly when little was at stake (E). In each panel, time is on the horizontal axis. The histogram represents the average firing rate of the neuron as a function of time during the trial. In the underlying raster display, each horizontal line corresponds to a single trial and each dot represents an action potential. From Roesch & Olson (2004).

a certain sector of the workspace (Chapter 50). Different neurons have distinct place fields so that, collectively, they cover the workspace. Place fields are defined relative to prominent environmental landmarks. When a radially symmetric eight-arm maze is rotated relative to a surrounding room containing salient landmarks the place fields remain fixed with respect to the room. Even in darkness, neurons continue to fire when the rat is in their place field, indicating that these responses are not simply visual. If the rat is placed in a new environment, hippocampal neurons develop new place fields rapidly. The location of the place field of a neuron relative to the new environment is not predictable from its location relative to the old environment. This is true even in cases where the workspace is changed without any change in the surrounding room, for example, through replacement of an eight-arm radial maze by an open field.

Neurons in other regions of the hippocampal system are sensitive to the direction in which the head is pointing. Sensitivity to heading is common in the postsubiculum, a cortical area adjacent to and closely linked to the hippocampus. Each postsubicular neuron fires most strongly when the head of the rat is pointing in its preferred direction and fires progressively less strongly as the orientation of the head of the rat deviates farther from that direction. In a room that is suddenly darkened, postsubicular neurons remain sensitive to the heading of the rat so long as the rat retains a sense of spatial orientation, as reflected by error-free performance on spatial tests. When the sense of spatial orientation drifts away from veridicality during prolonged darkness, as evidenced by the occurrence of systematic errors on spatial tests, then the preferred headings of postsubicular neurons exhibit a commensurate drift. These observations establish that neurons of the hippocampal system are sensitive to the spatial orientation of the rat. The place cells in the hippocampus and the head-direction cells encode complementary forms of spatial information.

As an animal moves through space it can keep track of the route it has taken by combining information from place cells and head-direction cells. This path integration produces a sense of position that is independent of the immediate visual environment. The path integration system is thought to depend on grid cells in the medial entorhinal cortex, an area of high-order neocortex adjacent to the hippocampus (Hafting *et al.*, 2005). In contrast to place cells, which generally have single firing fields, grid cells have a multitude of firing fields arranged in a strikingly regular hexagonal grid pattern. Moreover, the grid patterns exist at multiple scales. This is what allows the rat to derive unique location codes for every point in a large environment.

The map in medial entorhinal cortex, like that in other regions, is anchored to external landmarks but, unlike the map in other areas, it persists in the absence of landmarks. The grid cells constitute a map of relative spatial location that is independent of external sensory cues and the specific environment.

Summary

The hippocampal system mediates the formation of memories, including memories with spatial content. Injury to the hippocampal system, both in humans and in other animals, leads to profound deficits of memory, especially navigational memory. In accord with these findings, single neuron recording in rats has revealed that neurons of the hippocampal system have place fields, encoding the location of the rat in its environment, and directional sensitivity, encoding head direction relative to the environment. Information from these neurons is combined so that animals can keep track of their position in space over time.

SPATIAL COGNITION AND SPATIAL ACTION

The preceding sections described how spatial information is processed in cortical areas that serve distinct functions, such as motor control, attention, and working memory. Even within the motor system, there appears to be a fractionation of spatial functions in that neurons controlling movements of the eyes, head, and arm represent the locations of visible targets relative to eye, head, and hand-centered reference frames, respectively. In addition to these distinctions, there may be another fundamental distinction within brain systems mediating spatial functions. Areas mediating conscious awareness of spatial information may be partially separate from those mediating the spatial guidance of motor behavior. These functions may seem inseparable insofar as one must be aware of the location of a thing in order to look at it or reach for it. However, this is not necessarily the case. An indication that spatial awareness and spatially programmed behavior are distinguishable has come from studies of patients with "blindsight" (Weiskrantz, 1996). This condition arises as a result of injury to the primary visual cortex. Patients experience a scotoma, an area of blindness, in the part of the visual field represented by the injured cortex. The blindness is total in the sense that patients do not report seeing visual stimuli when stimuli are presented within the confines of the scotoma. Nevertheless, if patients are asked to look toward or to reach for an unseen stimulus, choosing the target by guesswork, their responses are directed to the correct location.

Similar findings have been reported in patients with diffuse pathology affecting widespread areas, including the prestriate visual cortex. One such patient, DF, has been extensively studied in spatial tasks (Goodale *et al.*, 1994). These studies reveal an important distinction between spatial performance and spatial cognition. When asked to express spatial judgments (for example, to indicate the size of a visible object by spreading the thumb and forefinger), DF performs poorly. Nevertheless, when asked to make visually guided movements (for example, to reach for an object), she accurately adjusts her grip size and hand orientation under visual control to grasp the object efficiently. The fact that intact visuomotor performance coexists with profoundly impaired visuospatial perception in some patients seems to argue for the existence of distinct brain systems specialized for conscious spatial awareness and motor guidance. Many lines of evidence indicate that conscious spatial awareness depends on relatively lateral regions of parietal cortex, whereas motor guidance depends on more medial regions of parietal cortex. Conscious spatial vision is distinct from conscious visual recognition, which depends on occipitotemporal cortex (Chapter 46).

Summary

Spatial cognition is a function of many brain areas. No one area is uniquely responsible for the ability to carry out spatial tasks. Nevertheless, some generalizations can be made about the part of the problem that is solved in each brain region. The parietal lobe plays a crucial role in many aspects of spatial awareness, including spatially focused attention. The representation of space in the parietal cortex takes several forms. Each subdivision within the parietal cortex contributes to different kinds of spatial representations, designed to help guide different kinds of actions.

In contrast to the parietal cortex, the frontal lobe transforms spatial awareness into actions. The motor cortex uses a spatial framework to encode intended actions. The premotor cortex contains a set of separate spatial representations to generate eye, hand, and arm movements. The SEF contains neurons with very high-order, abstract spatial representations. Prefrontal cortex mediates spatial working memory. Finally, the hippocampus mediates spatial declarative memories, including those that underlie spatial navigation.

Beneath the unity of our spatial perception lies a diversity of specific representations. The distributed nature of spatial cognition and the many purposes it serves means that we construct internal representations of space not once but many times in parallel. A challenge for the future is to understand how these many representations function together so seamlessly.

References

Andersen, R. A., Bracewell, R. M., Barash, S., Gnadt, J. W., and Fogassi, L. (1990). Eye position effects on visual, memory, and saccade-related activity in areas LIP and 7a of macaque. *J. Neurosci.* **10**, 1176–1196.

Bisiach, E. and Luzzatti, C. (1978). Unilateral neglect of representational space. *Cortex* **14**, 129–133.

Bunsey, M., Eichenbaum, H. (1996). Conservation of hippocampal memory function in rats and humans. *Nature* **379(6562)**, 255–257.

Colby, C. L. and Duhamel, J.-R. (1991). Heterogeneity of extra-striate visual areas and multiple parietal areas in the macaque monkey. *Neuropsychologia* **29**, 517–537.

Colby, C. L., Duhamel, J.-R., and Goldberg, M. E. (1993). Ventral intraparietal area of the macaque: Anatomic location and visual response properties. *J. Neurophysiol.* **69**, 902–914.

Colby, C. L., Duhamel, J.-R., and Goldberg, M. E. (1996). Visual, pre-saccadic and cognitive activation of single neurons in monkey lateral intraparietal area. *J. Neurophysiol.* **76**, 2841–2852.

Coslett, H. B. and Saffran, E. (1991). Simultanagnosia: To see but not two see. *Brain* **114**, 1523–1545.

Duhamel, J.-R., Colby, C. L., and Goldberg, M. E. (1991). Congruent representations of visual and somatosensory space in single neurons of monkey ventral intraparietal cortex (area VIP). *In* "Brain and Space" (J. Paillard, ed.), pp. 223–236. Oxford Press, Oxford.

Duhamel, J.-R., Colby, C. L., and Goldberg, M. E. (1992). The updating of the representation of visual space in parietal cortex by intended eye movements. *Science* **255**, 90–92.

Fogassi, L., Gallese, V., di Pellegrino, G., Fadiga, L., Gentilucci, M., Luppino, G., Matelli, M., Pedotti, A., and Rizzolatti, G. (1992). Space coding by premotor cortex. *Exp. Brain Res.* **89**, 686–690.

Funahashi, S., Bruce, C. J., and Goldman-Rakic, P. S. (1991). Neuronal activity related to saccadic eye movements in the monkey's dorsolateral prefrontal cortex. *J. Neurophysiol.* **65**, 1464–1483.

Goldman-Rakic, P. S. (1988). Topography of cognition: Parallel distributed networks in primate association cortex. *Annu. Rev. Neurosci.* **11**, 137–156.

Goodale, M. A., Meenan, J. P., Bulthoff, H. H., Nicolle, D. A., Murphy, K. J., and Racicot, C. I. (1994). Separate neural pathways for the visual analysis of object shape in perception and prehension. *Curr. Biol.* **4**, 604–610.

Graziano, M. S., Yap, G. S., and Gross, C. G. (1994). Coding of visual space by premotor neurons. *Science* **266**, 1054–1057.

Hafting, T., Fyhn, M., Molden, S., Moser, M. B., and Moser, E. I. (2005). Microstructure of a spatial map in the entorhinal cortex. *Nature* **436**, 801–806.

Humphreys, G. W. and Riddoch, M. J. (1993). Interactions between object and space systems revealed through neuropsychology. *In* "Attention and Performance" (D. E. Meyer and S. Kornblum, eds.), Vol. XIV, pp. 143–162. MIT Press, Cambridge, MA.

Kakei, S., Hoffman, D. S., and Strick, P. L. (1999). Muscle and movement representations in the primary motor cortex. *Science* **285**, 2136–2139.

Ladavas, E. (1987). Is the hemispatial deficit produced by right parietal lobe damage associated with retinal or gravitational coordinates? *Brain* **110**, 167–180.

Lynch, J. C. and McLaren, J. W. (1989). Deficits of visual attention and saccadic eye movements after lesions of parietooccipital cortex in monkeys. *J. Neurophysiol.* **61**, 74–90.

Marshall, J. C. and Halligan, P. W. (1993). Visuo-spatial neglect: A new copying text to assess perceptual parsing. *J. Neurol.* **240**, 37–40.

Maunsell, J. H. (2004). Neuronal representations of cognitive state: Reward or attention? *Trends Cogn. Sci.* **8**, 261–265.

Olson, C. R., and Gettner, S. N. (1995). Object-centered direction selectivity in the macaque supplementary eye field. *Science* **269**, 985–988.

Perenin, M.-T. and Vighetto, A. (1988). Optic ataxia: A specific disruption in visuomotor mechanisms. I. Different aspects of the deficit in reaching for objects. *Brain* **111**, 643–674.

Rao, S. C., Rainer, G., and Miller, E. K. (1997). Integration of what and where in the primate prefrontal cortex. *Science* **276**, 821–824.

Roesch, M. R. and Olson, C. R. (2003). Impact of expected reward on neuronal activity in prefrontal cortex, frontal and supplementary eye fields and premotor cortex. *J. Neurophysiol.* **90**, 1766–1789.

Roesch, M. R. and Olson, C. R. (2004). Neuronal activity related to reward value and motivation in primate frontal cortex. *Science* **304**, 307–310.

Roitman, J. D. and Shadlen, M. N. (2002). Response of neurons in the lateral intraparietal area during a combined visual discrimination reaction time task. *J. Neurosci.* **22**, 9475–9489.

Sawaguchi, T. and Goldman-Rakic, P. S. (1994). The role of D1-dopamine receptor in working memory: Local injections of dopamine antagonists into the prefrontal cortex of rhesus monkeys performing an oculomotor delayed-response task. *J. Neurophysiol.* **71**, 515–528.

Schwartz, A. B., Kettner, R. E., and Georgopoulos, A. P. (1988). Primate motor cortex and free arm movements to visual targets in three-dimensional space. I. Relations between single cell discharge and direction of movement. *J. Neurosci.* **8**, 2913–2927.

Sugrue, L. P., Corrado, G. S., and Newsome W. T. (2004). Matching behavior and the representation of value in the parietal cortex. *Science* **304**, 1753–1754.

Teng, E., Squire, L. R. (1999). Memory for places learned long ago is intact after hippocampal damage. *Nature* **400(6745)**, 675–677.

Ungerleider, L. G. and Mishkin, M. (1982). Two cortical visual systems. *In* "Analysis of Visual Behavior" (D. J. Ingle, M. A. Goodale, and R. J. W. Mansfield, eds.), pp. 549–586. MIT Press, Cambridge, MA.

Weiskrantz, L. (1996). Blindsight revisited. *Curr. Opin. Neurobiol.* **6**, 215–220.

Zola-Morgan, S. and Squire, L. R. (1993). Neuroanatomy of memory. *Annu. Rev. Neurosci.* **16**, 547–563.

Suggested Readings

Behrmann, M., Geng, J. J., and Shomstein, S. (2004). Parietal cortex and attention. *Curr. Opin. Neurobiol.* **14**, 212–217.

Bisiach, E., Luzzatti, C., and Perani, D. (1979). Unilateral neglect, representational schema and consciousness. *Brain* **102**, 609–618.

Colby, C. L., Goldberg, M. E. (1999). Space and attention in parietal cortex. *Annu. Rev. Neurosci.* **22**, 319–349.

DeRenzi, E. (1985). Disorders of spatial orientation. *In* "Handbook of Clinical Neurology" (J. A. M. Frederiks, ed.), Vol. 1, pp. 405–422. Elsevier, Amsterdam.

Hyvarinen J. (1982). Posterior parietal lobe of the primate brain. *Physiol. Rev.* **62**, 1060–1129.

Kakei, S., Hoffman, D. S., and Strick, P. L. (2001). Direction of action is represented in the ventral premotor cortex. *Nature Neurosci.* **4**, 1020–1025.

O'Keefe, J. and Nadel, L. (1978). "The Hippocampus as a Cognitive Map." Oxford Univ. Press, Oxford.

Olson, C. R. (2003). Brain representation of object-centered space in monkeys and humans. *Annu. Rev. Neurosci.* **26**, 331–354.

Rizzolatti, G., Fogassi, L., and Gallese, V. (1997). Parietal cortex: From sight to action. *Curr. Opin. Neurobiol.* **7**, 562–567.

Snyder, L. H. (2000). Coordinate transformations for eye and arm movements in the brain. *Curr. Opin. Neurobiol.* **10**, 747–754.

Carol L. Colby and Carl R. Olson

48

Attention

INTRODUCTION

Intuition, together with cognitive and psychophysical experiments, shows that the brain is limited in the amount of information that it can process at any moment in time. For instance, when people are asked to identify the objects of a briefly presented scene, they become less accurate as the number of objects increases. Similarly, when people concentrate on one demanding task (e.g., mental arithmetic) this typically comes at the expense of performance on other simultaneous activities (e.g., recalling a familiar tune). Limitations on the ability to simultaneously carry out multiple cognitive or perceptual tasks reflect the limited capacity of some stage or stages of sensory processing, decision-making, or behavioral control. As a result of such computational bottlenecks, it is necessary to have neural mechanisms in place to ensure the selection of stimuli, or tasks, that are immediately relevant to behavior. "Attention" is a broad term denoting the mechanisms that mediate this selection.

Over the past several decades, research has concentrated on the relation between attention and sensory perception. Clearly, it is possible to focus on selected stimuli in any sensory modality—sights, sounds, smells, and touch. Most studies, however, have examined vision, the dominant sensory modality for humans and nonhuman primates. These experiments have investigated visual attention at multiple levels, ranging from behavior to the single neuron. This chapter outlines the view of attention that has emerged from these studies.

A key observation is that attending to an object greatly enhances the ability to perceive and report the object's visual attributes; conversely, withdrawing attention, either by force of the behavioral context or following certain brain lesions, can render observers practically blind to certain visual events. A second key point is that the attentional processes that underlie perceptual selection may also guide the voluntary eye movements with which foveate animals, such as monkeys and humans, scan the environment. In terms of neural organization, although some neural centers have been closely linked with the control of attention, attention appears to affect the activity of neurons at almost all levels of the visual system.

VARIETIES OF ATTENTION

In natural behavior, individuals have great flexibility as to how and when they attend and what they attend to. Attending to an object often is accompanied by overt orienting toward that object using the eyes and possibly also the head and body. When a person enters a room, it is natural to turn in that person's direction. However, one can also attend covertly to objects that are not in the center of gaze, without looking directly at them. When driving, one can monitor a passing car while continuing to look straight ahead. Covert attention can improve peripheral visual acuity, thereby extending the functional field of view. Although one can direct attention without moving the eyes, the converse does not appear to be true: experiments show that covert attention must be deployed to an object before that object can be targeted with an eye movement. Indeed, saccades—rapid eye movements used most commonly for scanning the environment—

may be considered a motor manifestation of visual attention, a relationship that is discussed in more detail later.

Certain external stimuli can summon attention in and of themselves. These can be physically salient objects, such as especially large, bright, or loud objects, or stimuli with special learned significance, such as one's own name or a mother's face. This mode of attentional engagement, known as *exogenous* or *stimulus driven*, ensures that salient and potentially important external events do not pass unnoticed even if they are not being actively sought out. However, purposeful behavior often requires that attention be directed voluntarily, or *endogenously*, based on internally defined goals and against potential external distractions. When reading, one purposefully directs one's attention from one word to the next, tuning out noise and other distractions. In natural behavior, endogenous and exogenous factors interact continuously to control the allocation of sensory processing (Egeth and Yantis, 1997).

Another important issue is *what* is attended. Attention can be directed to a location in space, regardless of what happens to be at that location. It can also be directed to a feature, as when one searches within a complex scene for an object of a particular shape or color. Finally, attention can select whole objects, as demonstrated by studies showing that when attention is directed to one feature of an object, the other features

of the same object are selected automatically for visual processing, and by studies finding that when attention is directed to one of two semitransparent objects, attending to a feature of one object impairs processing of the features of the other object. In each of these cases, attention has been found to facilitate processing of the attended location, feature, or object, as assessed at the behavioral and neural levels (Reynolds and Chelazzi, 2004).

Attention is thus highly flexible and can be deployed in a manner that best serves the organism's momentary behavioral goals: to locations, features, or objects, based on internal goals or the external environment, with or without accompanying orienting movements. It is important to keep in mind that although these phenomena can be isolated in laboratory experiments, all varieties of attention operate seamlessly during natural behavior.

NEGLECT SYNDROME: A DEFICIT OF SPATIAL ATTENTION

Studies of patients with brain lesions have identified regions of the brain that are involved in attention (see Fig. 48.1, which shows the major divisions of the human cortex). Unilateral lesions in the parietal lobe,

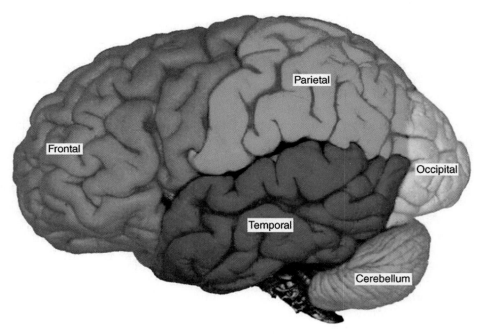

FIGURE 48.1 Lateral view of a human brain. Frontal (purple), parietal (orange), temporal (blue), and occipital (yellow) lobes are outlined.

the frontal lobe, and the anterior cingulate cortex (Heilman, 1979; Vallar, 1993) in humans may cause a profound inability to attend to certain spatial regions, a syndrome known as spatial neglect. At the subcortical level, lesions of the basal ganglia or of the pulvinar thalamic nucleus, which is heavily connected with the parietal cortex, can also cause neglect. In severe cases, patients with neglect behave as if the world contralateral to their lesioned hemisphere (the contralesional world) has ceased to exist. For example, a patient with neglect following a right hemisphere lesion may fail to read from the left side of a book, may ignore the food on the left side of the plate, or may remain unaware of the numerals on the left side of a clock (Fig. 48.2). Neglect patients may also be reluctant to initiate movement in contralesional space, with or without external sensory stimulation. Because the critical feature causing an object to be ignored is its location, this type of neglect is thought of as primarily a disorder of spatial attention.

The behavioral impairments in neglect cannot be explained by simple sensory or motor deficits (Mesulam, 1999). Neglect patients have normal vision in the contralesional visual field once their attention has been directed there, and they have no hemiparesis that could account for their reluctance to move. In addition, the extent of their difficulty is not immutable, as a sensory or motor deficit would be, but is strongly affected by the goals, expectations, and motivational state of the patient. Thus, neglect represents a failure to select, in some circumstances, the appropriate portions of a sensory representation. Neglect operates on high-level representations that differ considerably from the raw sensory input. For example, visual neglect is not confined only to retinal visual coordinates (affecting only objects in the contralesional visual field), but can affect objects in the contralesional space relative to the patient's body, relative to the patient's gaze, or relative to an external object or scene. Thus, neglect affects a visual representation that incorporates information about the position of the body with high-level visual information. Neglect affects not only the perception of the present sensory environment, but also the processing of memorized or imagined scenes.

A related but milder attentional deficit, known as extinction, can follow more limited brain lesions or can occur during recovery from the acute phase of neglect. In visual extinction, patients are able to orient toward a contralesional stimulus presented in isolation, but fail to notice it if the same stimulus is presented simultaneously with an ipsilateral distracter. Extinction-like deficits are demonstrated readily with the spatial cuing task (Posner *et al.*, 1984). In these tasks, subjects are cued to attend to the left or right visual field. Unimpaired observers are better able to process information at the cued location, but can also detect stimuli at an uncued location. Compared to unimpaired observers, neglect patients are slowed only mildly in their ability to detect targets in the contralesional visual field following valid symbolic or peripheral cues in the same field. However, they have enormously prolonged latencies for detecting contralateral targets if their attention had been misdirected to the ipsilesional field by means of an invalid cue. No impairment is seen

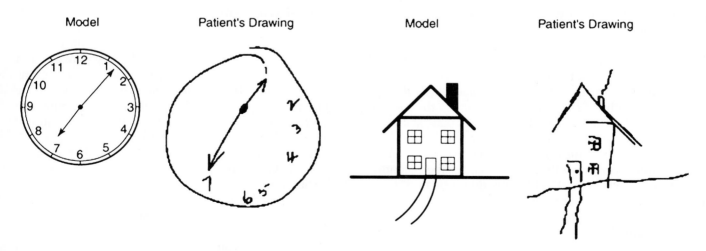

FIGURE 48.2 Two drawings that were made by a patient with spatial neglect. The patient was asked to copy the two models (clock, house). In each case, the copies exclude important elements that appeared on the left side of the model, indicating that the patient was unable to process information about the left side of the model.

when patients are cued to the contralesional field but asked to detect an ipsilesional target. Extinction-like phenomena also have been demonstrated in search tasks in which patients are asked to find a target object located at various locations in a complex scene. The time needed to find a target in contralesional space increases in proportion with the number and salience of ipsilesional distracters.

SINGLE UNIT RECORDING STUDIES IN NONHUMAN PRIMATES PROVIDE CONVERGENT EVIDENCE FOR A FRONTO-PARIETAL ATTENTIONAL CONTROL SYSTEM

Studies in monkeys have supported the conclusion that frontal and parietal cortices play a role in controlling the allocation of attention. These studies point to two interconnected areas, one in the parietal cortex, the lateral intraparietal area (LIP), and one in the frontal lobe, the frontal eye field (FEF), as being important for spatial attention. Many neurons in FEF and LIP have visual receptive fields, meaning that they respond to visual stimuli that activate a limited region of the retina. These receptive fields are mostly contralateral, so that the FEF and LIP in one side of the brain represent the locations of objects in the contralateral portion of the visual field. Receptive fields are retinotopic, that is they are linked to the retina and move relative to the external world every time the eye moves. Thus, during natural behavior LIP and FEF neurons are stimulated by numerous objects that enter or exit their receptive field by virtue of the observers' own movement.

Despite this continuous visual stimulation, LIP neurons respond little to most objects during natural viewing. For instance, the LIP neuron in Figure 48.3 responded robustly whenever an object was flashed in its receptive field (A, left). However, the neuron responded little if the same object was stable in the environment and entered the receptive field by virtue of an eye movement (A, center). The neuron responded very reliably if the object in the receptive field was rendered salient by blinking it on and off on every trial (A, right) or if it was rendered task relevant by being designated as the target for the next eye movement (B). Thus, LIP neurons respond very selectively to objects that attract attention—either by virtue of their intrinsic conspicuity or by becoming relevant to the behavioral task. These findings have given rise to the proposal that LIP encodes a "salience representation," or a "priority map" that specifies only a small

FIGURE 48.3 LIP neurons encode a spatial salience representation. (A) LIP neurons are sensitive to the physical conspicuity of objects in the receptive field. An LIP neuron responds strongly when a visual stimulus is flashed in its receptive field (left). The same neuron does not respond if the physically identical stimulus is stable in the world and enters the receptive field by virtue of the monkeys' eye movement (center). The neuron again responds if the stimulus entering the receptive field is rendered salient by flashing it on and off on each trial (before it enters the receptive field). In each panel, rasters show the times of individual action potentials relative to stimulus onset (left) or the eye movement that brings the stimulus in the receptive field (center and right). Traces underneath the rasters show average firing rate. Bottom traces show the horizontal and vertical eye position. The rapid eye movement (saccade) that brings the stimulus in the receptive field is seen as the abrupt deflection in eye position traces. (B) LIP neurons are sensitive to the behavioral relevance of objects in the receptive field. Traces show average firing rate of an LIP neuron when the monkey is cued to make a saccade to a stimulus in the receptive field. In the left panels the monkey is fixating and a stable stimulus is in the receptive field. The neuron has low firing rate. In the center panel the monkey receives a cue instructing him to make a saccade to the receptive field. The neuron gradually increases its response. In the right panel the monkey executes the instructed saccade. The neural response remains high until after the eye movement.

number of objects and locations worthy of immediate attention. A similar proposal has been advanced for the FEF, and may also hold for area 7a, a parietal area adjacent to LIP.

In the example shown in Figure 48.3B the object in the receptive field became behaviorally relevant by being designated as the target of the next eye movement. Thus, the selective LIP response in this case may have contributed to the attentional selection of an eye movement target. However, LIP neurons, like visually responsive neurons in the FEF also respond strongly to task-relevant objects even when monkeys covertly attend to a stimulus without making eye movements or when they purposefully move their eyes away from an attended cue. Taken together these observations suggest that LIP neurons and visually responsive neurons in the FEF are not limited to guiding eye movements, but play a key role in attentional control, regardless of the ensuing behavior.

A number of studies have shown that directly manipulating activity in the LIP and FEF can affect the distribution of attention. Reversible inactivation of the LIP or FEF using microinjections of the GABA agonist, muscimol, produces deficits in target selection during either saccade-based or covert visual search tasks. In addition, subthreshold microstimulation of FEF can bias the covert distribution of attention. These findings support the conclusion that LIP and FEF are parts of the network of areas that mediate the allocation of attention.

A   Stimulus appears in RF   Stimulus enters RF by virtue of an eye movement

Stable stimulus   Recently appeared stimulus

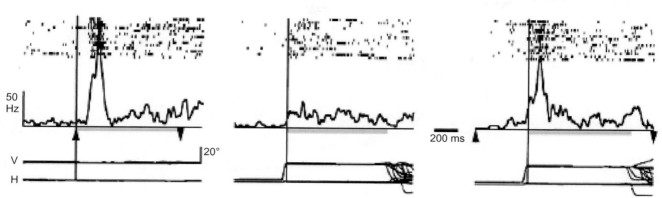

B   Stimulus in RF is designated as saccade target

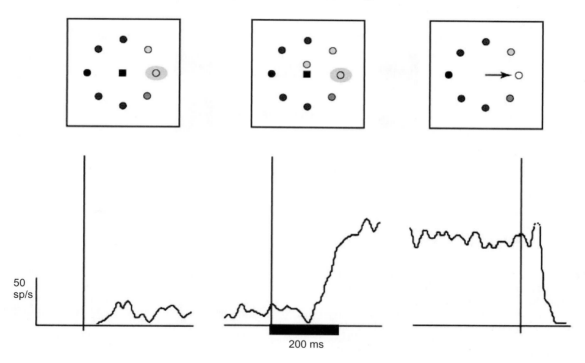

Converging evidence shows that spatially-selective responses in LIP and FEF are modulated by reward expectation, the memory—or expectation—of specific events, by contextual variables (i.e., relationship between task events), and the similarity between target and distracters. In the experiment illustrated in Figure 48.4A monkeys were free to choose between two saccade directions (either toward or away from the receptive field), and the probability that any one direction would be rewarded was systematically varied. Monkeys apportioned their choices in relation to reward probability, tending to choose the target with

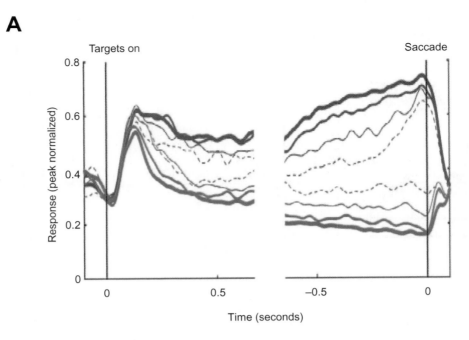

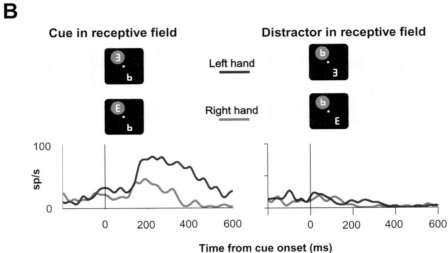

FIGURE 48.4 (A) LIP responses are modulated by reward probability. Monkeys performed a dynamic foraging task in which they tracked the changing reward values of each of two saccade targets. Traces represent population responses for neurons with significant effects of reward probability. Blue traces represent saccades toward the receptive field, and green traces represent saccades opposite the receptive field. Traces are further subdivided according to local fractional income (reward probability during the past few trials): solid thick lines, 0.75–1.0; solid medium lines, 0.5–0.75; solid thin line, 0.25–0.5; dotted thin lines, 0–0.25. Activity for saccades toward the receptive field increased, and that for saccades away decreased as function of local fractional income, resulting in more reliable spatial selectivity (difference between the two saccade directions) with increasing reward probability. (B) LIP neurons are modulated by limb motor planning. Monkeys viewed a display containing a cue (a letter "E") and several distracters. Without moving gaze from straight ahead, they reported the orientation of the cue by releasing one of two bars grasped with their hands. The neuron illustrated here responded much more strongly when the cue rather than a distracter appeared in its receptive field (left versus right panels). This cue-related response was modulated by the manual release: the neuron responded much more when the monkey released the left than when she released the right bar (blue versus red traces). Limb modulations did not always accompany a bar release, but were found only when the attended cue was in the receptive field. Bar-release latencies were on the order of 400–500 ms.

the higher probability of reward. LIP responses to the saccade target (blue traces) increased as a function of reward probability. Note that reward probability modulated the LIP response but did not fundamentally change the neurons' spatial selectivity: neurons always responded more before a saccade to the receptive field than before a saccade away even if the reward expected for the former was less than for the latter.

Behavior related modulations of LIP also include effects related to limb rather than eye movements. In one experiment monkeys were trained to discriminate the orientation of a cue presented in the peripheral visual field and, without making a saccade to the cue, to report the *orientation* of the cue by releasing a bar grasped with either the right or left hands. LIP neurons generated the expected spatial salience signal, responding much more strongly if the cue than if a distracter

appeared in their receptive fields (Figure 48.4B, left versus right panels). However, this response was modulated by the monkey's choice of limb. The neuron illustrated in Figure 48.4B responded more strongly to the cue in the receptive field if the monkey released the left than if he released the right bar, while other neurons showed the opposite preference, responding more for right than for left bar release.

These findings suggest that salience maps integrate information about a wide variety of behavioral cues relevant for directing attention, whether these cues are from the visual, cognitive, motor or motivational domains (Fig. 48.5). Although the output of these maps has an invariant meaning—a selected location—this output is constructed from a wide variety of sources of information that are relevant to a current behavior. By virtue of these integrative properties salience maps

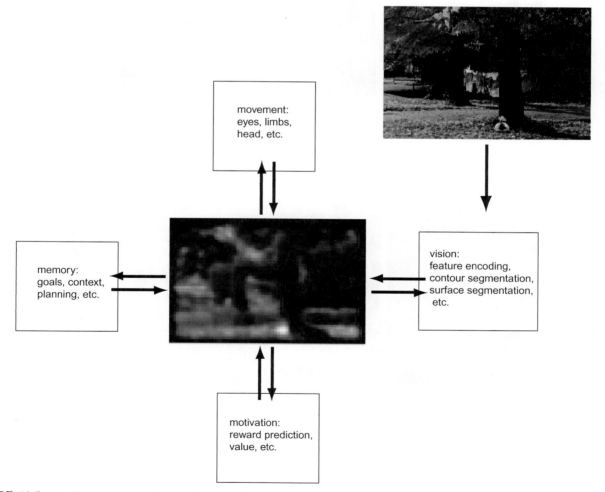

FIGURE 48.5 A salience representation can be viewed as an intermediate representation that interacts with multiple behavioral systems (visual, motor, cognitive, and motivational) and fashions a unified signal of "salience" based on multiple task demands. By virtue of feedback connections to these systems the salience map can help coordinate output processing in multiple "task-relevant areas".

have the potential to coordinate perceptual selection with ongoing action for a wide variety of behaviors.

Human functional brain imaging studies are in agreement with both single-cell physiology and neuropsychological studies in identifying areas in the parietal, frontal, and cingulate cortices as being especially active in relation to spatially directed attention. When observers attend to a location in space in anticipation of the appearance of a stimulus, this is accompanied by elevated levels of activity in a fronto-parietal network consisting of the superior parietal lobule (SPL), the frontal eye field (FEF), and the supplementary eye field (SEF) extending into the anterior cingulate cortex (Fig. 48.6; for review, see Kastner and Ungerleider, 2000). This elevation in activity persists throughout the expectation period and when stimuli later appear at the attended location (Fig. 48.7B). It is important to note that the fronto-parietal attentional control network is not limited to controlling spatial attention. It is also activated when subjects select nonspatial information. Object- or feature-based attention engages the fronto-parietal attention network as well

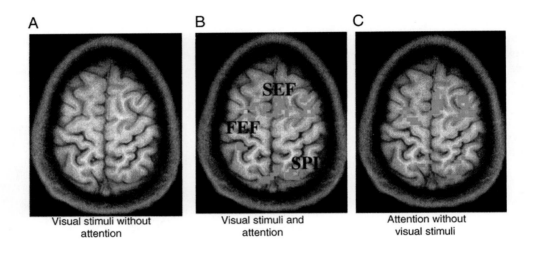

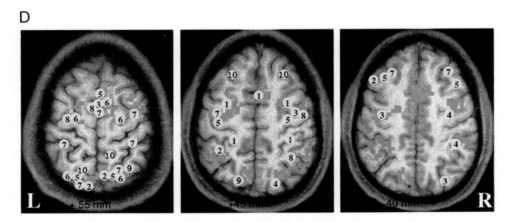

FIGURE 48.6 Regions in human brain activated by attention and regions associated with neglect. (A) Visual stimulation did not activate the frontal or parietal cortex reliably when attention was directed elsewhere in the visual field. (B) When the subject directed attention to a peripheral target location and performed an object discrimination task, a distributed fronto-parietal network was activated, including the SEF, the FEF, and the SPL. (C) The same network of frontal and parietal areas was activated when the subject directed attention to the peripheral target location in the expectation of the stimulus onset—in the absence of any visual input whatsoever. This activity therefore may not reflect attentional modulation of visually evoked responses, but rather attentional control operations themselves. (D) Meta-analysis of studies investigating the spatial attention network. Axial slices at different Talairach planes are indicated. Talairach (peak) coordinates of activated areas in the parietal and frontal cortex from several studies are indicated (for references, see Kastner and Ungerleider, 2000). R, right hemisphere; L, left hemisphere.

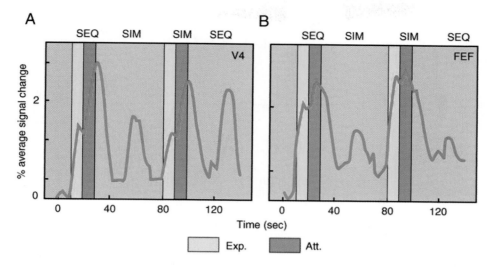

FIGURE 48.7 Directed attention in humans with and without visual stimulation. (A) Time series of fMRI signals in V4. Directing attention to a peripheral target location in the absence of visual stimulation led to an increase of baseline activity (textured blocks), which was followed by a further increase after the onset of stimuli (gray-shaded blocks). Baseline increases were found in both the striate and the extrastriate visual cortex. (B) Time series of fMRI signals in FEF. Directing attention to the peripheral target location in the absence of visual stimulation led to a stronger increase in baseline activity than in the visual cortex; the further increase of activity after the onset of stimuli was not significant. Sustained activity was seen in a distributed network of areas outside the visual cortex, including SPL, FEF, and SEF, suggesting that these areas may provide the source for the attentional top-down signals seen in visual cortex. Adapted from Kastner *et al.* (1999).

indicating a general domain-independent mechanism of attentional control (Yantis and Serences, 2003).

ATTENTION AFFECTS NEURAL ACTIVITY IN THE HUMAN VISUAL CORTEX IN THE PRESENCE AND ABSENCE OF VISUAL STIMULATION

Frontal and parietal areas have been associated with the control of spatial attention, but they are not usually considered crucial for visual processing or object recognition. There is converging evidence from event-related potential (ERP) and functional imaging studies that feedback signals from the fronto-parietal areas that control the allocation of attention can affect the neural processing of visual information in the human visual cortex. In a typical experiment, identical visual stimuli are presented simultaneously to corresponding peripheral field locations to the right and to the left of a central fixation point. Subjects are instructed to direct attention covertly to the right or the left by a symbolic cue presented at the fixation point and to detect the occurrence of a visual stimulus or, in some cases, discriminate a feature of the attended stimulus. Directing attention to the left hemifield increases stimulus-evoked activity in the extrastriate visual cortex of

the right hemisphere, whereas directing attention to the right hemifield increases activity in the extrastriate visual cortex of the left hemisphere (Heinze *et al.*, 1994) (Fig. 48.8). Thus, responses to stimuli are enhanced on the side of the extrastriate cortex that contains representations of the attended hemifield. As is the case in the fronto-parietal areas, this increase in response is observed during the period prior to appearance of a stimulus, reflecting the effects of feedback signals from the attentional control areas.

In one imaging study (Kastner *et al.*, 1999), subjects were cued to direct attention covertly to a target location in the periphery of the visual field and to anticipate the onset of visual stimuli. Visual stimuli occurred with a delay of several seconds. Neural activity associated with spatially directed attention in the absence and in the presence of visual stimulation could therefore be distinguished. As was the case in the fronto-parietal control network, fMRI signals increased when attention was directed to the target location and before any visual stimulus was present on the screen. This increase in baseline activity was followed by an additional increase when stimuli appeared at the attended location (Fig. 48.7A), reflecting attention-dependent increases in the stimulus evoked response itself. These studies and related single-unit studies in the monkey show that spatial attention effects are topographically organized and retinotopically specific, resulting in increased

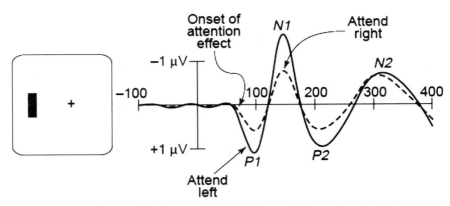

FIGURE 48.8 Attentional modulation of event related potentials (ERPs). In a typical experiment, subjects fixate a central cross and attend either to the left or right visual field. Stimuli are then presented to the left and right visual fields in a rapid sequence. This idealized example illustrated the common finding that the average ERP elicited by an attended visual field stimulus contains larger P1 and N1 components.

activity among neurons coding the attended location. Taken together, these findings suggest that spatial attention modulates visual processing by amplifying the neural signals for stimuli at an attended location.

Earlier we noted that the attentional control network is activated during nonspatial attention. Functional brain imaging studies and single-unit recording studies in nonhuman primates have found evidence that non-spatial attention also modulates sensory cortices. One such study found that selective attention to shape, color, or speed enhanced activity in the regions of the extrastriate visual cortex that selectively process these same attributes. Attention to shape and color led to response enhancement in regions of the posterior portion of the fusiform gyrus, including area V4. Attention to speed led to response enhancement in areas MT/MST (Corbetta *et al.*, 1991). In other studies, attention to faces or houses led to response enhancement in areas of the midanterior portion of the fusiform gyrus, areas responsive to the processing of faces and objects. Taken together, these results support the idea that selective attention to a particular stimulus attribute modulates neural activity in those extrastriate areas that preferentially process the selected attribute.

ATTENTION INCREASES SENSITIVITY AND BOOSTS THE CLARITY OF SIGNALS GENERATED BY NEURONS IN PARTS OF THE VISUAL SYSTEM DEVOTED TO PROCESSING INFORMATION ABOUT OBJECTS

Consistent with the previously mentioned functional imaging studies, single-unit recording studies in the visual cortices have found that when attention is

directed to a single stimulus in the receptive field, there is often an increase in the firing rates of neurons that respond to the attended stimulus (Reynolds and Chelazzi, 2004). How could such a change in firing rate improve an observer's ability to identify properties of stimuli that appear at an attended location? A likely answer is that by increasing neuronal responses, attention enables neurons to send signals that better differentiate between stimuli with different physical characteristics (McAdams and Maunsell, 1999). Why would an increase in firing rate enable the neuron to more reliably signal the identity of the attended stimulus? The reason is that neuronal signals are noisy. Consider a neuron that responds poorly to a grating oriented 45° from vertical, and responds strongly to a 90° grating. Response variability will likely cause some presentations of the poorer 45° orientation stimulus to elicit responses that are higher than responses observed on some presentations of the 90° orientation stimulus. This is illustrated in Figure 48.9, which shows the distribution of responses of a hypothetical neuron, elicited by two different stimuli. The two curves in the upper part of Figure 48.9 indicate the distributions of responses elicited by the two stimuli, without attention. Although the distributions are not identical, they overlap a great deal. Responses within this area of overlap cannot be uniquely associated with either stimulus. The lower part of Figure 48.9 illustrates the effect of attention: to shift response distributions of both stimuli to the right. This shift is larger for the 90° orientation stimulus, and so with attention, the two distributions overlap less. As a result, the neuron can more reliably signal which of the two stimuli is present.

Microstimulation studies support the conclusion that attention-dependent changes in firing rate within the visual cortices result from feedback signals that are

Attention-dependent increases in firing rate improve the quality of visual information

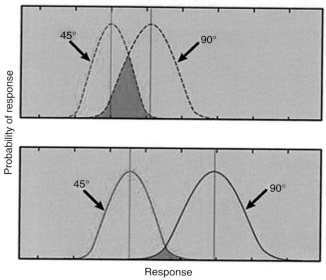

FIGURE 48.9   Attention-dependent increases in firing rate can improve the ability of a neuron to signal stimulus properties. Response distributions of a hypothetical neuron to a 45 and 90° oriented grating. With attention directed away from the receptive field (top), there is substantial overlap between the distributions; consequently, the response on a given trial does not reliably indicate which stimulus is present. Attention boosts responses elicited by the preferred stimulus by more than it boosts responses to the poor stimulus, reducing the ambiguity in the signal.

generated in the fronto-parietal attentional control network. Moore and Armstrong (2003) measured the locations in FEF where stimulating current would move the eyes, the "movement field" of the stimulation site. They then reduced the stimulating current to a level too low to evoke an eye movement, and observed the effect of this stimulation on the responses of neurons in visual area V4. They found that FEF stimulation caused V4 neurons to respond more robustly to a stimulus that appeared at the movement field within the V4 neuron's receptive field (Fig. 48.10A). Figure 48.10B shows responses of a single V4 neuron with and without stimulation. The time course of stimulus presentation (RF stim) and current injection (FEF stim) are illustrated at the top of the panel. The neuronal responses in the two conditions are indicated at the bottom of the figure. Stimulation was injected 500 ms after the appearance of the stimulus, at which point the stimulus-evoked response had begun to diminish in strength. The average response on microstimulation trials (gray) was elevated following electrical stimulation, relative to nonstimulation trials.

ATTENTION MODULATES NEURAL RESPONSES IN THE HUMAN LATERAL GENICULATE NUCLEUS

The lateral geniculate nucleus (LGN) is the thalamic station in the retinocortical projection and traditionally has been viewed as the gateway to the visual cortex. In addition to retinal afferents, the LGN receives input from multiple sources including striate cortex, the thalamic reticular nucleus (TRN), and the brain stem. The LGN therefore represents the first stage in the visual pathway at which cortical top-down feedback signals could affect information processing.

It has proven difficult to study attentional response modulation in the LGN using single-cell physiology due to the small RF sizes of LGN neurons and the possible confound of small eye movements. Several single-cell physiology studies have failed to demonstrate attentional modulation in the LGN, supporting a notion that selective attention affects neural processing only at the cortical level. This notion was recently revisited using fMRI in humans (O'Connor *et al.*, 2002).

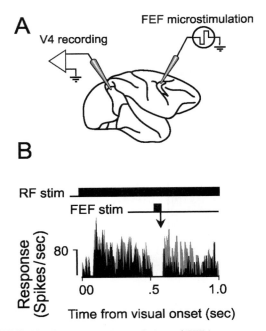

FIGURE 48.10 Electrical stimulation of FEF increases neuronal responsiveness in V4. (A) Current was injected into FEF while neuronal activity was recorded from V4. (B) The visual stimulus appeared in the receptive field for one second (RF stim). 500 ms after the onset of the visual stimulus, a low level of current was applied to a site in FEF for 50 ms (FEF stim). The response of a single V4 neuron with (gray) and without (black) FEF microstimulation appears below. The apparent gap in response reflects the brief period during which V4 recording was paused while current was injected into FEF, when, for technical reasons, the stimulating current interfered with the ability to record action potentials. Following stimulation, the V4 neuron had elevated responses on trials when FEF was electrically stimulated, as compared to trials without FEF stimulation.

As in visual cortex, different modulatory effects of attention were found. Neural responses to attended visual stimuli were enhanced relative to the same stimuli when unattended. As in cortex, this effect of attentional response enhancement was shown to be spatially specific in the LGN. And directing attention to a location in the absence of visual stimulation and in anticipation of the stimulus onset increased neural baseline activity. Together, these studies indicate that the LGN appears to be the first stage in the processing of visual information that is modulated by attentional top-down signals.

These studies were suited to compare the magnitude of the attention effects across the visual system. The magnitude of all attention effects increased from early to more advanced processing levels along both the ventral and dorsal pathways of visual cortex. This is consistent with the idea that attention operates through top-down signals that are transmitted via cortico-cortical feedback connections in a hierarchical fashion. This idea is supported by single-cell recording studies, which have shown that attention effects in area TE of inferior temporal cortex have a latency of approximately 150 ms, whereas attention effects in V1 have a longer latency of approximately 230 ms. According to this account, one would predict smaller attention effects in the LGN than in striate cortex. Surprisingly, it was found that all attention effects tended to be larger in the LGN than in striate cortex. This finding suggests that attentional response modulation in the LGN is unlikely to be due solely to corticothalamic feedback from striate cortex, but may be further influenced by additional sources of input. In addition to corticothalamic feedback projections from V1, which comprise about 30% of its modulatory input, the LGN receives another 30% of modulatory inputs each from the TRN and the parabrachial nucleus of the brain stem. Given its afferent input the LGN may be in an ideal strategic position to serve as an early gatekeeper in attentional gain control.

THE VISUAL SEARCH PARADIGM HAS BEEN USED TO STUDY THE ROLE OF ATTENTION IN SELECTING RELEVANT STIMULI FROM WITHIN A CLUTTERED VISUAL ENVIRONMENT

So far this chapter has considered the situation when attention is cued to a location (or a feature) in the visual field. However, one is rarely told in advance to attend to a particular location. Normally, one needs to find a particular object in a complex world that is composed of a large number of stimuli. Psychologists have used a visual search paradigm to understand how attention selects behaviorally relevant stimuli out of a group of other stimuli.

In a typical task, observers are asked to search among an array of stimuli and indicate whether a particular target is present in the array. This task is easier under some conditions than others. For example, it is easy to determine whether a horizontal green bar is present in Figure 48.11A. It takes longer to make this judgment for stimuli in Figure 48.11B. Treisman and Gelade (1980) found that searches for a target with a unique feature, like the one illustrated in Figure 48.11A, can be completed quickly regardless of the number of elements in the search array—the target seems to "pop out" of the search array. Such searches often are referred to as efficient. In contrast, Treisman and Gelade (1980) found that the amount of time required to find a target that is defined by conjunctions of

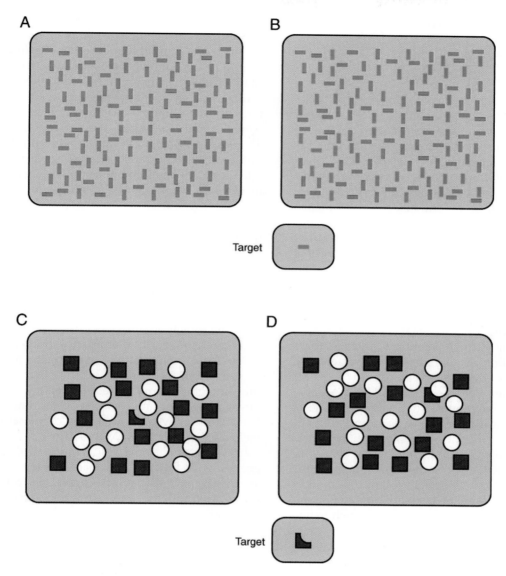

FIGURE 48.11   Efficient and inefficient visual searches. (A) Pop-out search. The horizontal green bar differs from all other array elements in a single feature, color. As a result, it pops out immediately, regardless of how many elements are in the array. (B) The same target is difficult to find when it is embedded among elements that cannot be differentiated from it on the basis of a single feature (color or orientation). (C) The dark target pops out among squares and circles. (D) When the same target is perceived to be a square occluded behind a circle, it is difficult to find among squares and circles. Because this arrangement of the array interferes with visual search, this suggests that integration of objects into wholes occurs prior to the activation of attentional mechanisms used in visual search. Thus, these mechanisms are evidently not necessary for completion. Adapted from Rensink and Enns (1998).

elementary features (e.g., a horizontal green among horizontal reds and vertical greens, as in Fig. 48.11B) increases linearly with the number of elements in the array. Such searches often are referred to as inefficient. The amount of time added per item depends on stimulus conditions, but a rule of thumb is that each addi-

tional item in the array adds about 50 ms to the amount of time taken to locate the target. The fact that it takes time to locate the target in Figure 48.11B implies that the visual system is limited in capacity. If it were not, the target could be identified immediately by evaluating every element in the array simultaneously.

WHERE IS THE COMPUTATIONAL BOTTLENECK AS REVEALED BY SEARCH TASKS?

Considerable debate exists regarding the locus and nature of the computational bottleneck that causes the capacity limitations that are revealed by visual search tasks. One influential theory, feature integration theory (FIT, Treisman and Gelade, 1980), proposes that capacity limitations occur at very early stages of cortical processing where "elementary" features are bound into coherent objects. This theory proposes that in the "pop-out" case, where the target differs from distracters in a single elemental feature, a search can be completed efficiently because there is no need to bind together different elementary features. Thus, a green target among red distracters (Fig. 48.11A) can be found in parallel by V1 neurons that respond to green but not red stimuli, without the need for visual attention. However, in a conjunction search, where each of the elementary target features (color and orientation) is shared with distracters, the observer must bind elementary features into coherent object representations before search can occur. According to this account, one cannot process multiple objects simultaneously because in order to avoid misconjoining features from different objects, attention selects features from one location at a time. Consistent with this idea, patient RM, who suffered damage to his parietal attentional control system, often misconjoins the shape and color of letters, even after viewing them for up to 10 seconds; that is, he mixes up which letters are presented in which colors.

However, the idea that limited capacity occurs at early stages of processing has been challenged. For example, some studies suggest that parts of objects are integrated into wholes prior to the application of attention. Consider the task of searching for the oddly shaped object appearing at the bottom of Figure 48.11 (adapted from Rensink and Enns, 1998). This odd shape pops out from among the squares and circles in Figure 48.11C, but it is harder to find the same shape in Figure 48.11D. The difference is that in Figure 48.11D, the oddly shaped part appears to be a square that is occluded by a superimposed circle; that is, the visual system completes the shape of the occluded square, hiding it among the other squares. Because this completion stops the square from popping out, the completion operation must occur prior to visual search.

Further evidence that features can be integrated preattentively into objects comes from the finding that when an observer attends to one feature of an object (e.g., its orientation), other features of the object are selected automatically as well. In a now classic study, Duncan (1984) found that observers could easily make simultaneous judgments about two features of the same object (e.g., its orientation and whether or not it contained a gap). Making the second judgment did not interfere with the first judgment. However, observers were severely impaired when they simultaneously judged the same two features, one on each of two different stimuli. This finding suggests that when attention is directed to one feature of an object, all the features that make up the object can automatically be selected together because the features are already linked together into objects. These sorts of observations suggest that limited capacity results from information processing bottlenecks that occur relatively late in processing, after features are already integrated into wholes. One possibility is that the bottleneck occurs when objects enter a limited capacity working memory. Under this view, attention plays the role of selecting which objects pass through the bottleneck.

These two perspectives are not mutually exclusive. Some feature integration may occur preattentively (e.g., Rensink and Enns, 1998), but other types of stimulus integration might require spatial selection. As described earlier, attention has been found to modulate neural activity at all stages of visual processing including early stages (Motter, 1993; O'Connor et al., 2002), where neuronal responses first show signs of grouping the features and parts of objects into wholes, as well as late stages, which are likely involved in selecting stimuli for storage in working memory. Thus, both early and late stages of processing may act as limited capacity bottlenecks.

NEURONAL RECEPTIVE FIELDS ARE A POSSIBLE NEURAL CORRELATE OF LIMITED CAPACITY

Extracellular recording studies of attention in awake, behaving monkeys have revealed that the receptive fields of individual neurons might contribute to the limitation in visual capacity. As visual information traverses the successive cortical areas of the ventral stream that underlie object recognition (Ungerleider and Mishkin, 1982), the sizes of receptive fields increase from less than a degree of visual arc in primary visual cortex (area V1) to about 20° of visual arc in area TE, the last purely visual area in the ventral visual processing stream (Chapter 46). Similar increases in receptive field size are observed in the dorsal visual processing stream. In a typical scene, such large receptive fields will contain many different objects. Therefore, a likely explanation for why one cannot process many different objects in a scene simultaneously is that neurons, whose signals are limited in bandwidth, cannot simultaneously send signals about all the stimuli inside their receptive fields. This idea implies that processing limitations exist at all levels of processing but that they become more pronounced in higher order visual areas, where receptive field sizes are larger. How then do neurons transmit signals about behaviorally relevant stimuli in their receptive fields?

One proposal is that the selective processing of behaviorally relevant stimuli is accomplished by two interacting mechanisms: competition among potentially relevant stimuli and biases that determine the outcome of this competition. For example, when an observer is asked to detect the appearance of a target at a particular location, this task is thought to activate spatially selective feedback signals in cortical areas, such as parts of the parietal and frontal cortex discussed earlier. These cortical areas provide a task-appropriate spatial reference frame and transmit signals to the extrastriate cortex, where they bias competition in favor of stimuli that appear at the attended location. Similarly, when an observer searches for an object in a cluttered scene, feature-selective feedback is thought to bias competition in favor of stimuli that share features in common with the searched-for object. When competition is resolved, the winning stimulus controls neuronal responses, and other stimuli are effectively filtered out of the visual stream.

Studies in monkeys have compared the effect of attention on neuronal responses with single and multiple stimuli in the receptive field. Consistent with the proposal just outlined, larger changes in firing rate occur when multiple stimuli appear within the receptive field (and one stimulus must be attended), as compared to when the attended stimulus appears alone and there is no competition to be resolved (Moran and Desimone, 1985). According to the biased competition model just described, neurons that respond to a stimulus should be suppressed when a second stimulus is added that activates a competing population of neurons. To test this idea, pairs of stimuli were presented that activated two nearby populations of neurons (Reynolds et al., 1999). Monkeys attended away from the receptive fields of the neurons. The first stimulus was chosen to be of a color and orientation that would strongly activate the recorded neuron. Then, a second stimulus, which elicited a response in a nearby population of neurons, was added. As predicted, the addition of this second stimulus partially suppressed the responses of many neurons to the first stimulus. To test whether attentional feedback would

resolve competition in favor of the neurons that responded to the attended stimulus, attention was then directed to the poor stimulus. As predicted, this often magnified the suppressive influence of the poor stimulus, further reducing the firing rates of many neurons (Fig. 48.12A). In a final experiment, attention was directed to the preferred stimulus. Consistent with the resolution of competition in favor of the neurons that responded to the preferred stimulus, directing attention to the preferred stimulus strongly reduced the suppressive effect of the poor stimulus (Fig. 48.12B).

Similar results have been reported in the dorsal stream of visual processing (Recanzone and Wurtz, 2000; Treue and Maunsell, 1999). Thus, when multiple stimuli appear together within the receptive fields of neurons in the later stages of the ventral visual processing stream and in motion-sensitive areas MT and MST of the dorsal stream, the neuronal response is driven preferentially by the attended stimulus. In this way, attention helps determine which stimuli are selected for processing in these areas.

COMPETITION CAN BE BIASED BY NONSPATIAL FEEDBACK

The findings just summarized support the proposal that competitive neural circuits can be biased by spatially selective feedback signals. As illustrated in Figure 48.13, feature-selective feedback can also bias competition. Responses of neurons in area TE were recorded as monkeys searched for a target (Chelazzi et al., 1998). On each trial, the monkey viewed a cue stimulus that appeared at the fixation point. After a delay, an array of stimuli (the search array) appeared within the receptive fields of neurons in area TE. The monkey's task was to indicate whether a target matching the cue stimulus was present in the array, by making a saccade to the target. Locations of the stimuli were selected at random so that the monkey could not know where the target would appear, if it appeared at all. During the delay period, many TE neurons had a higher baseline firing rate when the cue stimulus was a preferred stimulus for the cell than when it was a poor stimulus; that is, neurons that were tuned to respond to the target were activated while the monkey prepared to search for the target. An interesting possibility is that this activation might bias the visual system toward detecting the target object, just as spatially specific feedback is thought to bias the system toward a particular location. Consistent with this, within 150 to 200ms after the search array appeared, the response of the neuron

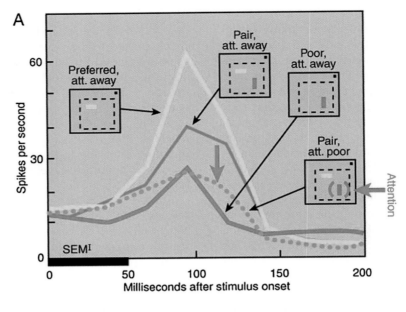

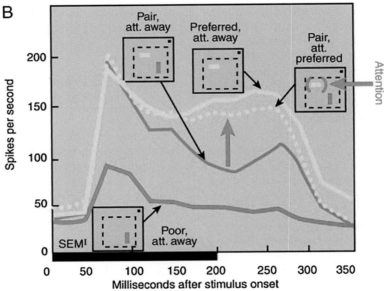

FIGURE 48.12   Attention to one stimulus of a pair filters out the effect of the ignored stimulus. (A) The x-axis shows time (in milliseconds) from stimulus onset, and the thick horizontal bar indicates stimulus duration. Small iconic figures illustrate sensory conditions. Within each icon, the dotted line indicates the receptive field, and the small dot represents the fixation point. The location of attention inside the receptive field is indicated in red. Attention was directed away from the receptive field in all but one condition. The preferred stimulus is indicated by a horizontal yellow bar and the poor stimulus by a vertical blue bar. In fact, the identity of both stimuli varied from cell to cell. The yellow line shows the response of a V2 neuron to the preferred stimulus. The solid blue line shows the response to the poor stimulus. The green line shows the response to the pair. The addition of the poor stimulus suppressed the response to the preferred stimulus. Attention to the poor stimulus (red arrow) magnified its suppressive effect and drove the response down to a level (dotted blue line) that was similar to the response elicited by the poor stimulus alone. (B) Response of a second V2 neuron. The format is the same as in A. As in the neuron above, the response to the preferred stimulus was suppressed by the addition of the poor stimulus. Attention directed to the preferred stimulus filtered out this suppression, returning the neuron to a response (dotted yellow line) similar to the response that was elicited when the preferred stimulus appeared alone inside the RF. Adapted from Reynolds *et al.* (1999).

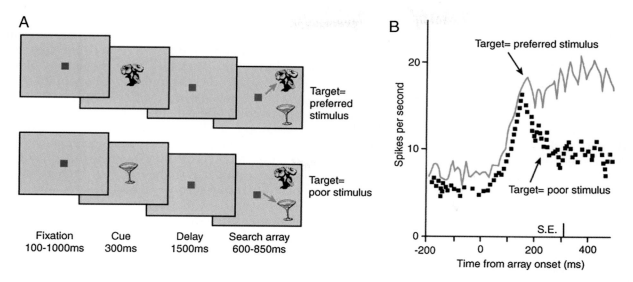

FIGURE 48.13 Responses of inferior temporal (IT) neurons during a memory-guided visual search. (A) Task. Monkeys fixated a spot on a computer screen and were shown a central cue (here, either the flower or the cup). After a delay, two (or more) stimuli appeared within the receptive field and the monkey had to saccade to the stimulus that had appeared earlier as the cue. Sometimes (top four images), the cue was a preferred stimulus for the cell (the flower) and the monkey had to saccade to the preferred stimulus. On separate trials (lower four images), the cue was a poor stimulus (the cup) and the monkey had to saccade to the poor stimulus. (B) Neuronal responses. During the delay period, IT neurons showed an elevated baseline activity that reflected the cue stored in memory. The spontaneous firing rate was higher on trials in which the cue was a preferred stimulus for the cell, relative to trials when the cue was a poor stimulus. After the search array appeared, the responses separated, increasing or decreasing depending on whether the cue was, respectively, a preferred or poor stimulus for the cell. This separation occurred well before the onset of the saccade, which is indicated by the vertical bar on the horizontal axis.

increased or decreased, depending on whether the target was, respectively, a preferred or a poor stimulus for the cell. That is, shortly after the appearance of the search array, the activity of the cell reflected the identity of the target stimulus and was no longer influenced by the nontarget stimulus, even though the nontarget stimulus was still present physically within its receptive field.

FILTERING OF UNWANTED INFORMATION IN HUMANS

Functional brain imaging studies suggest that in the human visual cortex, just as in the monkey cortex, mechanisms exist such that multiple stimuli compete for neural representation and unwanted information can be filtered out (Kastner *et al.*, 1998, 2001). In one fMRI study, subjects maintained fixation while they were presented with colorful visual stimuli in four nearby locations in the periphery of the visual field (Kastner *et al.*, 1998). Stimuli were presented under two different conditions. In a sequential condition, a single stimulus appeared in one of the four peripheral locations, then another appeared in a different location, and so on, until each of the four stimuli had been presented in the four different locations. In a simultaneous condition, the same four stimuli appeared in the same four locations, but they were presented at the same time. Thus, physical stimuli were identical in the two conditions except that in the second case, all four stimuli appeared simultaneously, which should encourage competitive interactions among the stimuli. The subjects' task was to count letters at fixation, thereby ignoring the peripheral stimulus presentations. Activation of V1 and ventral stream extrastriate areas V2 to TEO was found under both presentation conditions. Although the fMRI signal was similar in the two presentation conditions in V1, activation was reduced in the simultaneous condition compared to the sequential condition in V2. This reduction in activity was especially pronounced in V4 and TEO. The most straightforward interpretation of this result is that simultaneously presented stimuli interacted in a mutually suppressive way. This sensory suppression effect may be a neural correlate of competition in the human visual cortex. Importantly, the suppression effects appeared to be scaled to the receptive field size of neurons within visual cortical areas; that is, the effects were larger in areas with larger receptive fields. This observation suggests that, as in the monkey visual cortex, competitive interactions occur most strongly at the level of the receptive field (Kastner et al., 1998).

The effects of spatially directed attention on multiple, competing visual stimuli have been studied in a variation of the same paradigm (Kastner et al., 1998). In addition to the two different visual presentation conditions, sequential and simultaneous, two different attentional conditions were tested, unattended and attended. During the unattended condition, attention was directed away from the visual display by having subjects count the occurrences of letters at the fixation point. In the attended condition, subjects were instructed to attend covertly to the stimulus location that was closest to the fixation point and to count the occurrences of one of the four stimuli. The attended condition led to greater increases in fMRI signals to simultaneously presented stimuli than to sequentially presented stimuli. Thus, in this case, attention partially cancelled out the suppressive interactions among competing stimuli. The magnitude of the attentional effect was greater where the suppressive interactions among stimuli had been strongest, such that the strongest reduction of suppression occurred in areas V4 and TEO. These findings support the idea that directed attention enhances information processing of stimuli at attended locations by counteracting the suppression induced by nearby stimuli that are competing for limited processing resources. In this way, unwanted distracting information is filtered out effectively.

Unwanted information can not only be filtered out effectively by top-down mechanisms that resolve competitive interactions among multiple objects, but also by bottom-up mechanisms that are defined by the context of the stimuli. As we have seen earlier, if a salient stimulus is present in a cluttered scene, it will be effortlessly and quickly detected regardless of the number of distracters, suggesting that competition is biased in favor of the salient stimulus. Bottom-up influences related to stimulus context of a visual display, in which a single, salient stimulus pops out from a homogeneous background were examined using the same design of sequentially and simultaneously presented stimuli described earlier (Beck and Kastner, 2005). The pop-out display condition, in which a single item differed from the others, was compared to a heterogeneous display, in which all four stimuli were different. Subjects were engaged in a difficult letter detection task at fixation and never attended to the pop-out or heterogeneous displays that were shown in the periphery of the visual field. Competitive interactions among multiple stimuli were eliminated in area V4 when the stimuli were presented in the context of pop-out displays relative to heterogeneous displays, indicating that competition was biased in favor of the salient stimulus in a bottom-up stimulus context-dependent fashion. These pop-out effects appeared to originate from area V1, where the strongest responses to pop-out displays were found. The effects of pop-out on competitive interactions among multiple stimuli appeared to operate in an automatic fashion, thus independently of attentional top-down control. Taken together, the studies reviewed in this section indicate that both top-down attention and bottom-up stimulus-driven mechanisms are critical in resolving competitive interactions among multiple stimuli at intermediate processing levels of visual cortex, thereby filtering out unwanted information from cluttered visual scenes.

CONCLUSIONS

"Everyone knows what attention is. It is the taking possession of the mind in clear and vivid form of one out of what seem several simultaneous objects or trains of thought." Thus did William James define the elusive entity commonly known as "attention." A century later, experimental neuroscience and cognitive psychology afford a closer look at the neural mechanisms enabling this "taking possession" to occur.

The behavioral manifestations of attention are clearly not the result of operations in any one dedicated neural center. Selective attention affects neural activity at all levels of the visual system and other sensory systems, with the possible exception of neurons in the sensory periphery. A few structures located at the top of the visual system hierarchy in the parietal, frontal, and cingulate cortices are thought to be relatively more specialized for generating the signals that guide selective attention, and these are among the areas that are thought to provide feedback signals that influence sensory processing in areas that are specialized for analyzing the physical properties of the sensory environment.

Answers are beginning to be available to the question of why attention is necessary in the first place. A possible computational bottleneck may be built into all levels of the visual system in the form of spatial receptive fields. The fact that several objects within a certain spatial region can activate the same neuron reduces the ability of such neurons to transmit accurate information about each one of these objects. Mechanisms therefore are required to shrink the functional receptive field around the object(s) that needs greater scrutiny in a particular situation. Other bottlenecks will doubtlessly be revealed in future research.

References

Beck, D. M. and Kastner, S. (2005). Stimulus context modulates competition in human extrastriate cortex. *Nature Neurosci.* **8**, 1110–1116.

Bichot, N. P., Thompson, K. G., Chenchal Rao, S., and Schall, J. D. (2001). Reliability of macaque frontal eye field neurons signaling saccade targets during visual search. *J. Neurosci.* **21**(2), 713–725.

Bushnell, M. C., Goldberg, M. E., and Robinson, D. L. (1981). Behavioral enhancement of visual responses in monkey cerebral cortex. I. Modulation in posterior parietal cortex related to selective visual attention. *J. Neurophysiol.* **46**, 755–772.

Chelazzi, L., Duncan, J., Miller, E. K., and Desimone, R. (1998). Responses of neurons in inferior temporal cortex during memory-guided visual search. *J. Neurophysiol.* **80**(6), 2918–2940.

Constantinidis, C. and Steinmetz, M. (2001). Neuronal responses in area 7a to multiple stimulus displays: II responses are suppressed at the cued location. *Cereb. Cortex* **11**(7), 592–597.

Corbetta, M., Miezin, F. M., Dobmeyer, S., Shulman, G. L., and Petersen, S. E. (1991). Attentional modulation of neural processing of shape, color, and velocity in humans. *Science* **248**, 1556–1559.

Duncan, J. (1984). Selective attention and the organization of visual information. *J. Exp. Psychol. Gen.* **113**(4), 501–517.

Egeth, H. E. and Yantis, S. (1997). Visual attention: Control, representation, and time course. *Annu. Rev. Psychol.* **48**, 269–297.

Gottlieb, J. M., Kusunoki, M., and Goldberg, M. E. (1998). The representation of visual salience in monkey parietal cortex. *Nature* **391**, 481–484.

Heilman, K. M. (1979). Neglect and related disorders. *In* "Clinical Neuropsychology" (E. Valenstein, ed.), pp. 268–307. Oxford Univ. Press, New York.

Heinze, H. J., Mangun, G. R., Burchert, W., Hinrichs, H., Scholz, M. *et al.* (1994). Combined spatial and temporal imaging of brain activity during visual selective attention in humans. *Nature* **372**, 543–546.

Hillyard, S. A. and Vento, L. A. (1998). Event-related brain potentials in the study of visual selective attention. *Proc. Natl. Acad. Sci. USA* **95**, 781–787.

Karnath, H. O., Ferber, S., and Himmelbach, M. (2001). Spatial awareness is a function of the temporal not the posterior parietal lobe. *Nature* **411**, 950–953.

Kastner, S., De Weerd, P., Desimone, R., and Ungerleider, L. G. (1998). Mechanisms of directed attention in the human extrastriate cortex as revealed by functional MRI. *Science* **282**, 108–111.

Kastner, S., Pinsk, M. A., De Weerd, P., Desimone, R., and Ungerleider, L. G. (1999). Increased activity in human visual cortex during directed attention in the absence of visual stimulation. *Neuron* **22**, 751–761.

Kastner, S. and Ungerleider, L. G. (2000). Mechanisms of visual attention in the human cortex. *Annu. Rev. Neurosci.* **23**, 315–342.

Kowler, E., Anderson, E. *et al.* (1995). The role of attention in the programming of saccades. *Vision Res.* **35**(13), 1897–1916.

McAdams, C. J. and Maunsell, J. H. (1999). Effects of attention on the reliability of individual neurons in monkey visual cortex. *Neuron* **23**(4), 765–773.

Mesulam, M. M. (1999). Spatial attention and neglect: Parietal, frontal and cingulate contributions to the mental representation and attentional targeting of salient extrapersonal events. *Philos. Trans. R. Soc. Lond. B Biol. Sci.* **354**(1387), 1325–1346.

Moran, J. and Desimone, R. (1985). Selective attention gates visual processing in the extrastriate cortex. *Science* **229**(4715), 782–784.

Motter, B. C. (1993). Focal attention produces spatially selective processing in visual cortical areas V1, V2, and V4 in the presence of competing stimuli.

Neisser, U. and Becklen, R. (1975). Selective looking: Attending to visually significant events. *Cog. Psychol.* **7**, 480–494.

O'Connor, D. H., Fukui, M. M., Pinsk, M. A., and Kastner, S. (2002). Attention modulates responses in the human lateral geniculate nucleus. *Nature Neurosci.* **5**, 1203–1209.

Posner, M. I. and Petersen, S. E. (1990). The attention system of the human brain. *Annu. Rev. Neurosci.* **13**, 25–42.

Posner, M. I., Walker, J. A., Friedrich, F. J., and Rafal, R. D. (1984). Effects of parietal lobe injury on covert orienting of visual attention. *J. Neurosci.* **4**, 1863–1874.

Recanzone, G. H. and Wurtz, R. H. (2000). Effects of attention on MT and MST neuronal activity during pursuit initiation. *J. Neurophysiol.* **83**(2), 777–790.

Rensink, R. A. and Enns, J. T. (1998). Early completion of occluded objects. *Vision Res.* **38**(15–16), 2489–2505.

Reynolds, J. H., Chelazzi, L., and Desimone, R. (1999). Competitive mechanisms subserve attention in macaque areas V2 and V4. *J. Neurosci.* **19**(5), 1736–1753.

Reynolds, J. H. and Chelazzi, L. (2004). Attentional Modulation of Visual Processing. *Annu. Rev. Neurosci.* **27**, 611–647.

Rock, L. and Gutman, D. (1981). The effect of inattention on form perception. *J. Exp. Psychol. Hum. Percep. Perform.* **7**(2), 275–285.

Treisman, A. M. and Gelade, G. (1980). A feature-integration theory of attention. *Cognit. Psychol.* **12**(1), 97–136.

Treue, S. and Maunsell, J. H. R. (1999). Effects of attention on the processing of motion in macaque middle temporal and medial superior temporal visual cortical areas. *J. Neurosci.* **19**(17), 7591–7602.

Ungerleider, L. G. and Mishkin, M. (1982). Two cortical visual systems. *In* "The Analysis of Visual Behavior" (D. J. Ingle, R. J. W. Mansfield, and M. A. Goodale, eds.), pp. 549–586. MIT Press, Cambridge, MA.

Vallar, G. (1993). The anatomical basis of spatial neglect in humans. *In* "Unilateral Neglect: Clinical and Experimental Studies" (I. H. Robertson and J. C. Marshall Eds.), pp. 27–62. Lawrence Erlbaum, Hillsdale, NJ.

Yantis, S. and Jonides, J. (1984). Abrupt visual onsets and selective attention: Evidence from visual search. *J. Exp. Psychol. Hum. Percept. Perform.* **10**(5), 601–621.

Suggested Readings

Colby, C. L. and Goldberg, M. E. (1999). Space and attention in parietal cortex. *Annu. Rev. Neurosci.* **22**, 319–349.

Connor, C. E., Preddie, D. C., Gallant, J. L., and Van Essen, D. C. (1997). Spatial attention effects in macaque area V4. *J. Neurosci.* **17**(9), 3201–3214.

Corbetta, M. (1998). Frontoparietal cortical networks for directing attention and the eye to visual locations: Identical, independent, or overlapping neural systems? *Proc. Natl. Acad. Sci. USA* **95**(3), 831–838.

Desimone, R. and Duncan, J. (1995). Neural mechanisms of selective visual attention. *Annu. Rev. Neurosci.* **18**, 193–222.

Driver, J., Davis, G., Russell, C., Turatto, M., and Freeman, E. (2001). Segmentation, attention and phenomenal visual objects. *Cognition* **80**, 61–95.

Heeger, D. J. (1999). Linking visual perception with human brain activity. *Curr. Opin. Neurobiol.* **9**(4), 474–479.

Itti, L. and Koch, C. (2001). Computational modelling of visual attention. *Nature Rev. Neurosci.* **2**(3), 194–203.

Mangun, G. R. (1995). Neural mechanisms of visual selective attention. *Psychophysiology* **32**(1), 4–18.

Posner, M. I. (1994). Attention: The mechanisms of consciousness. *Proc. Natl. Acad. Sci. USA* 91(16), 7398–7403.

Treue, S. (2001). Neural correlates of attention in primate visual cortex. *Trends Neurosci.* **24**(5), 295–300.

John H. Reynolds, Jacqueline P. Gottlieb, and Sabine Kastner

Learning and Memory: Basic Mechanisms

Significant advances have been made in understanding the ways in which the nervous system encodes and retrieves information. One focus of current research is on the understanding of learning and memory at the cellular level, where the information-encoding process can be traced to changes in neuronal properties such as membrane excitability and synaptic strength. One emerging principle is that no single universal mechanism for learning and memory exists. Instead, different memory systems (Chapter 50) can use different mechanisms. Therefore, a comprehensive understanding of memory mechanisms requires an understanding of the general ways in which neurons are changed by learning and the ways in which those changes are maintained and expressed at the cellular level.

Research during the past several decades on several vertebrate and invertebrate model systems has led to the development of several general principles. A list of these principles (Byrne, 1987) might include the following:

1. Multiple memory systems are present in the brain (Chapter 50).
2. Short-term and long-term forms of learning and memory involve changes in existing neural circuits.
3. These changes may involve multiple cellular mechanisms within individual neurons.
4. Second-messenger systems play a role in mediating cellular changes.
5. Changes in membrane channels often are correlated with learning and memory.
6. Long-term memory requires new protein synthesis, whereas short-term memory does not.

This chapter describes several of the types of neural and molecular mechanisms implicated in learning

and memory. The chapter first considers some of the paradigms that have been used to study simple forms of nonassociative and associative learning and provides an example of mechanistic analyses that have been performed in a selected invertebrate model system. The later sections of the chapter describe the mechanisms of two phenomena, which are known as long-term potentiation (LTP) and long-term depression (LTD). Both LTP and LTD occur in forebrain structures, and LTP and LTD are thought to be mechanisms for memory storage in the central nervous system.

PARADIGMS HAVE BEEN DEVELOPED TO STUDY ASSOCIATIVE AND NONASSOCIATIVE LEARNING

Associative Learning

Associative learning is a broad category that includes many of our daily learning activities that involve formation of associations among stimuli and/or responses. It usually is subdivided into classical conditioning and instrumental conditioning. Classical (or Pavlovian) conditioning is induced by a procedure in which a generally neutral stimulus, termed a *conditioned stimulus* (CS), is paired with a stimulus that generally elicits a response, termed an *unconditioned stimulus* (US). Two examples of unconditioned stimuli are food, which elicits salivation, or a shock to the foot, which elicits limb withdrawal. Instrumental (or operant) conditioning is a process by which an organism learns to associate consequences with its own behavior. In an operant conditioning paradigm, the delivery of a reinforcing stimulus is contingent upon the expression of a designated behavior. The probability that this behavior will be expressed is then altered. This chapter focuses on classical conditioning, as it is mechanistically the best understood type of associative conditioning.

An astute observation by Ivan Pavlov, a Russian physiologist who had been studying digestion in dogs, led to his discovery of classical conditioning. He first noticed that the mere sight of the food dish caused dogs to salivate. He continued the experiments to see if dogs would also salivate in response to a bell rung at feeding time. Pavlov trained dogs to stand in a harness and, after the sound of a bell, fed them meat powder (Fig. 49.1). At first, the bell by itself did not elicit any response, but the meat powder elicited reflex salivation, which was termed the *unconditioned response* (UR). He noted that after a few pairings of the bell and meat powder the dogs began to salivate when the bell rang, before they received the meat powder. This response is termed the *conditioned response* (CR). This type of conditioning came to be called reward or appetitive *classical conditioning*. If the bell or another stimulus was followed by an unpleasant event, such as an electric shock, then a variety of autonomic

responses became conditioned. This type of conditioning is termed aversive conditioning.

Nonassociative Learning

Three examples of nonassociative learning have received the most experimental attention: habituation, dishabituation, and sensitization. *Habituation* is defined as a reduction in the response to a stimulus that is delivered repeatedly. *Dishabituation* refers to the restoration or recovery of a habituated response due to the presentation of another, typically strong, stimulus to the animal. *Sensitization* is an enhancement or augmentation of a response produced by the presentation of a strong stimulus. The following sections introduce the *Aplysia* and focus on the neural and molecular mechanisms of sensitization.

INVERTEBRATE STUDIES: KEY INSIGHTS FROM *APLYSIA* INTO BASIC MECHANISMS OF LEARNING

Since the mid-1960s, the marine mollusk *Aplysia* has proven to be an extremely useful model system to gain insights into the neural and molecular mechanisms of simple forms of memory. Indeed, the pioneering discoveries of Eric Kandel using this animal were recognized by his receipt of the Nobel Prize in Physiology or Medicine in 2000. A number of characteristics make *Aplysia* well suited for the examination of the molecular, cellular, morphological, and network mechanisms underlying neuronal modifications (plasticity) and learning and memory. The animal has a relatively simple nervous system with large, individually identifiable neurons that are accessible for detailed anatomical, biophysical, biochemical, and molecular studies. Neurons and neural circuits that mediate many behaviors in *Aplysia* have been identified. In several cases, these behaviors have been shown to be modifiable by learning. Moreover, specific loci within neural circuits at which modifications occur during learning have been identified, and aspects of the cellular mechanisms underlying these modifications have been analyzed (Byrne *et al.*, 1993; Byrne and Kandel, 1996; Hawkins *et al.*, 1993).

A single sensitizing stimulus, such as a brief several second-duration electric shock, can produce a reflex enhancement that lasts minutes (short-term sensitization), whereas prolonged training (e.g., multiple stimuli over an hour or more) produces an enhancement that lasts from days to weeks (long-term sensitization).

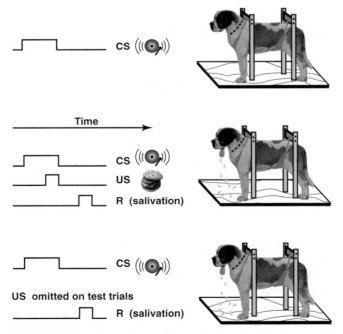

FIGURE 49.1 Classical conditioning. In the procedure introduced by Pavlov, the production of saliva is monitored continuously. Presentation of meat powder reliably leads to salivation, whereas some "neutral" stimulus such as a bell initially does not. With repeated pairings of the bell and meat powder, the animal learns that the bell predicts the food and salivates in response to the bell alone. Modified from Rachlin (1991).

Multiple Cellular Processes Mediate Short- and Long-Term Sensitization in Aplysia

A discussion of memory mechanisms can be divided into three parts: induction, expression, and maintenance. *Induction* refers to the initial events that trigger or initiate the modification process; *expression* concerns how the modification process ultimately is expressed; and *maintenance* addresses the manner in which the modification is made to endure over time.

Short-Term Sensitization

Short-term (minutes) sensitization is induced when a single brief train of shocks to the body wall results in the release of modulatory transmitters, such as serotonin (5-HT), from a separate class of interneurons (IN) referred to as facilitatory neurons (Fig. 49.2A1). These facilitatory neurons regulate the properties of the sensory neurons and the strength of their connections with postsynaptic interneurons and motor neurons (MN) through a process called *heterosynaptic facilitation* (Byrne and Kandel, 1996; Figs. 49.2A2 and 49.2A3). The molecular mechanisms contributing to short-term heterosynaptic facilitation are illustrated in Figure 49.2B. The first step is the binding of 5-HT to one class of receptors on the outer surface of the membrane of the sensory neurons. This leads to the activation of adenylyl cyclase, which in turn leads to an elevation of the intracellular level of the second messenger cyclic adenosine 3′,5′-monophosphate [cyclic AMP (cAMP)] in sensory neurons. When cAMP binds to the regulatory subunit of cAMP-dependent protein kinase [protein kinase A(PKA)], the catalytic subunit is released and can now add phosphate groups to specific substrate proteins and therefore alter their functional properties.

One consequence of this protein phosphorylation is an alteration in the properties of membrane channels. Specifically, the increased levels of cAMP lead to a decrease in the serotonin-sensitive potassium current [S-K$^+$ current ($I_{K,S}$)], a component of the calcium-activated K$^+$ current ($I_{K,Ca}$) and the delayed K$^+$ current ($I_{K,V}$). (See Chapter 6 for more information on these channel types.) These changes in membrane currents lead to depolarization of the membrane potential, enhanced excitability, and an increase in the duration of the action potential (i.e., spike broadening). Reflections of enhanced excitability include an increase in the number of action potentials elicited in a sensory neuron by a fixed extrinsic current injected into the cell or by a fixed stimulus to the skin (Fig. 49.2A3).

Cyclic AMP also appears to activate a facilitatory process that is independent of membrane potential and spike duration. This process is represented in

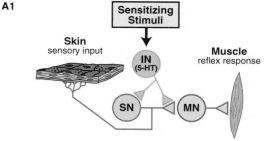

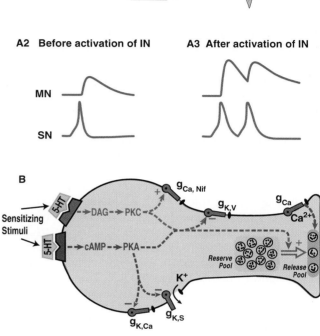

FIGURE 49.2 Model of short-term heterosynaptic facilitation of the sensorimotor connection that contributes to short- and long-term sensitization in *Aplysia*. (A1) Sensitizing stimuli activate facilitatory interneurons (IN) that release modulatory transmitters, one of which is 5-HT. The modulator leads to an alteration of the properties of the sensory neuron (SN). (A2 and A3) An action potential in SN after the sensitizing stimulus results in greater transmitter release and hence a larger postsynaptic potential in the motor neuron (MN, A3) than an action potential before the sensitizing stimulus (A2). For short-term sensitization, the enhancement of transmitter release is due, at least in part, to broadening of the action potential and an enhanced flow of Ca^{2+} into the sensory neuron. (B) Model of a sensory neuron that depicts the multiple processes for short-term facilitation that contribute to short-term sensitization. 5-HT released from facilitatory neurons binds to at least two distinct classes of receptors on the outer surface of the membrane and leads to the transient activation of two intracellular second messengers, DAG and cAMP, and their respective kinases (PKC and PKA). These kinases affect multiple cellular processes, the combined effects of which lead to enhanced transmitter release when subsequent action potentials are fired in the sensory neuron (see text for additional details). Modified from Byrne and Kandel (1996).

Figure 49.2B (large open arrow) as the translocation or mobilization of transmitter vesicles from a reserve pool to a releasable pool. The translocation makes more transmitter-containing vesicles available for release with subsequent action potentials in the sensory neuron. The overall effect is a short-term, cAMP-dependent enhancement of transmitter release. One substrate for cAMP/PKA-dependent phosphorylation is synapsin (Angers *et al.*, 2002). Synapsin is a synaptic vesicle-associated protein that tethers synaptic vesicles to cytoskeletal elements and thus helps control the reserve pool of vesicles in synaptic terminals (Chapter 8). Phosphorylation of synapsin would allow vesicles in the reserve pool to migrate to the releasable pool and thus contribute to enhanced transmitter release.

Serotonin also acts through another class of receptors to increase the level of the second messenger diacylglycerol (DAG). DAG activates protein kinase C (PKC), which, like PKA, contributes to facilitation that is independent of spike duration (e.g., mobilization of vesicles). In addition, PKC regulates a nifedipine-sensitive Ca^{2+} channel ($I_{Ca,Nif}$) and the delayed K^+ channel ($I_{K,V}$). Thus, the delayed K^+ channel ($I_{K,V}$) is dually regulated by PKC and PKA. The modulation of $I_{K,V}$ contributes importantly to the increase in duration of the action potential (Fig. 49.2A3). Due to its small magnitude, the modulation of $I_{Ca,Nif}$ appears to play a minor role in the facilitatory process.

Prolonged treatments of 5-HT (1.5h) also activates mitogen-activated protein kinase (MAPK) (Martin *et al.*, 1997). However, this pathway appears to be important only for the induction of longer-term processes (see later). Of general significance is the observation that a single modulatory transmitter (i.e., 5-HT) activates multiple kinase systems. The involvement of multiple second messenger systems in synaptic plasticity also appears to be a theme emerging from mammalian studies. For example, as discussed in a later section of the chapter, the induction of LTP in the CA1 area of the hippocampus appears to involve MAPK, PKC, calcium/calmodulin-dependent protein kinase (CaM kinase II), and tyrosine kinase (reviewed in Dineley *et al.*, 2001).

The consequences of activating these multiple second-messenger systems and modulating these various cellular processes are expressed when test stimuli elicit action potentials in the sensory neuron at various times after the presentation of the sensitizing stimuli (Fig. 49.2A3). More transmitter is available for release as a result of the mobilization process and each action potential is broader, allowing a larger influx of Ca^{2+} to trigger release of the available transmitter. The combined effects of mobilization and spike broadening lead to the facilitation of transmitter release from the sensory neuron and consequently a larger postsynaptic potential in the motor neuron. Larger postsynaptic potentials lead to enhanced activation of interneurons and motor neurons and thus to an enhanced behavioral response.

The maintenance of short-term sensitization is dependent on the persistence of the PKA- and PKC-induced phosphorylations of the various substrate proteins.

Long-Term Sensitization

Sensitization also exists in a long-term form, which persists for at least 24 h. Whereas short-term sensitization can be produced by a single brief stimulus, the induction of long-term sensitization requires a more extensive training period over an hour or more.

Both short- and long-term sensitization share some common cellular pathways during their *induction*. For example, both forms activate the cAMP/PKA cascade (Fig. 49.3). However, in the long-term form, unlike the short-term form, activation of the cAMP/PKA cascade induces gene transcription and new protein synthesis (Byrne *et al.*, 1993; Hawkins *et al.*, 1993). Repeated training leads to a translocation of PKA to the nucleus where it phosphorylates the transcriptional activator CREB1 (cAMP responsive element binding protein). CREB1 binds to a regulatory region of genes known as CRE (cAMP responsive element). Next, this bound and phosphorylated form of CREB1 leads to increased transcription. Serotonin, either by acting through cAMP or through activation of a tyrosine receptor kinase-like molecule (ApTrk) also leads to the activation of MAPK, which phosphorylates the transcriptional repressor CREB2. Phosphorylation of CREB2 by MAPK leads to a derepression of CREB2 and therefore promotes CREB1-mediated transcriptional activation (Bartsch *et al.*, 1995).

The role of transcription factors in long-term memory formation is not limited to the induction phase but may also extend to the consolidation phase, where consolidation is defined as the time window during which RNA and protein synthesis are required for converting short- to long-term memory. For example, treatment of ganglia with 5 pulses of 5-HT over a 1.5 h period to mimic sensitization training leads to the binding of CREB1 to the promoter of its own gene and induces CREB1 synthesis (Mohamed et al., 2005). This observation agrees well with earlier findings that the requirement for gene expression is not limited to the induction phase. The necessity of prolonged transcription and translation for LTF observed at 24 h persists for at least 7-9 h after induction. These results suggest that CREB1 can regulate its own level of expression, giving rise to a

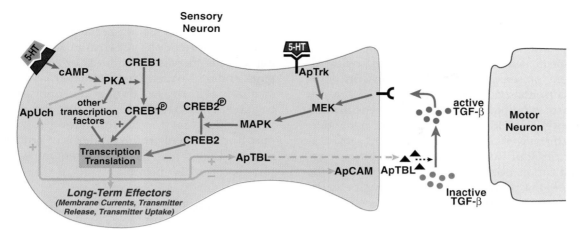

FIGURE 49.3   Simplified scheme of the mechanisms in sensory neurons that contribute to long-term sensitization. Sensitization training leads to the release of 5-HT and the cAMP-dependent phosphorylation of CREB1. 5-HT, either by acting through cAMP or though activation of a tyrosine receptor kinase-like molecule (ApTrk), also leads to the activation of MAPK, which regulates the repressor CREB2. The combined effects of activation of CREB1 and derepression of CREB2 lead to regulation of the synthesis of at least 10 proteins, only 3 of which (ApTBL, ApCAM, and ApUch) are shown. Two of these proteins (ApTBL and ApUch) appear to be components of positive feedback cycles. ApTBL is believed to activate latent forms of TGF-β, which can then bind to receptors on the sensory neuron. TGF-β activates MAPK, which may act by initiating a second round of gene regulation by affecting CREB2-dependent pathways. Increased synthesis of ApUch leads to enhanced degradation of the regulatory subunit of PKA, leading to enhanced activation of the catalytic subunit and increased phosphorylation of CREB1. The third protein ApCAM is downregulated. Downregulation of ApCAM is involved in regulating growth processes associated with long-term facilitation.

CREB1 positive feedback loop that supports memory consolidation.

The combined effects of activation of CREB1 and derepression of CREB2 lead to changes in the synthesis of specific proteins. So far, more than 10 gene products that are regulated by sensitization training have been identified, and others are likely to be found in the future. These results indicate that there is not a single memory gene or protein, but that multiple genes are regulated, and they act in a coordinated way to alter neuronal properties and synaptic strength. The following section discusses three regulated proteins of particular significance.

The down-regulation of a homologue of a neuronal cell adhesion molecule (NCAM), ApCAM plays a key role in long-term facilitation. This down-regulation has two components. First, the synthesis of ApCAM is reduced. Second, preexisting ApCAM is internalized via increased endocytosis. The internalization and degradation of ApCAM allow for the restructuring of the axon arbor (Bailey *et al.*, 1992). The sensory neuron can now form additional connections with the same postsynaptic target or make new connections with other cells. Another protein whose synthesis is regulated by long-term facilitation is *Aplysia* tolloid/BMP-like protein (ApTBL-1). Tolloid and the related molecule BMP-1 appear to function as secreted Zn^{2+} proteases. In some preparations, they activate members

of the transforming growth factor β (TGF-β) family. Indeed, in sensory neurons, TGF-β mimics the effects of 5-HT in that it produces long-term increases in the synaptic strength of the sensory neurons (Zhang *et al.*, 1997). Interestingly, TGF-β activates MAPK in the sensory neurons and induces its translocation to the nucleus. Thus, TGF-β could be part of an *extracellular* positive feedback loop, possibly leading to another round of protein synthesis (Fig. 49.3) to further consolidate the memory (Zhang *et al.*, 1997). A third important protein, *Aplysia* ubiquitin hydrolase (ApUch), appears to be involved in an *intracellular* positive feedback loop. During the induction of long-term facilitation, ApUch levels in sensory neurons are increased, possibly via CREB phosphorylation and a consequent increase in *ApUch* transcription. The increased levels of ApUch increase the rate of degradation, via the ubiquitin-proteosome pathway, of proteins including the regulatory subunit of PKA (Chain *et al.*, 1999). The catalytic subunit of PKA, when freed from the regulatory subunit, is highly active. Thus, increased ApUch will lead to an increase in PKA activity and a more protracted phosphorylation of CREB1. This phosphorylated CREB may act to further prolong *ApUch* expression, thus closing a positive feedback loop.

One simplifying hypothesis is that the mechanisms underlying the *expression* of short- and long-term sensitization are the same, but extended in time for long-term sensitization. Some evidence supports this hypothesis. For example, long-term sensitization, like short-term sensitization, is associated with an enhancement of sensorimotor connections. In addition, K$^+$ currents and the excitability of sensory neurons are modified by long-term sensitization (Cleary *et al.*, 1998). However, the mechanisms underlying the *expression* of short- and long-term sensitization differ in four fundamental ways. First, long-term sensitization training leads to a *decrease* in the duration of the sensory neuron action potential (Antzoulatos and Byrne, 2007). Second, structural changes such as neurite outgrowth and active zone remodeling are associated with the expression of long-term but not with short-term sensitization (Bailey and Kandel, 1993). Interestingly, the structural changes take time to develop. Long-term synaptic facilitation measured one day after four days of training is associated with structural changes, whereas long-term facilitation measured one day after a single training session is not (Wainwright *et al.*, 2004). Third, long-term sensitization is associated with an increase in high-affinity glutamate uptake (Levenson *et al.*, 2000). A change in glutamate uptake could potentially exert a significant effect on synaptic efficacy by regulating the amount of transmitter available for release, the rate of clearance from the cleft, and thereby the duration of the EPSP and the degree of receptor desensitization. Finally, long-term sensitization has been correlated with changes in the postsynaptic cell (i.e., the motor neuron; Cleary *et al.*, 1998). Thus, as with other examples of memory, multiple sites of plasticity exist even within this simple reflex system.

Other Temporal Domains for the Memory of Sensitization

Historically, memory has been divided into two temporal domains, short term and long term. It has become increasing clear from studies of a number of memory systems that this distinction is overly simplistic. For example, in *Aplysia*, Carew and colleagues (Sutton *et al.*, 2001) and Kandel and colleagues (Ghirardi *et al.*, 1995) have discovered an intermediate phase of memory that has distinctive temporal characteristics and a unique molecular signature. The intermediate-phase memory for sensitization is expressed at times approximately 30 min to 3 h after the beginning of training. Like long-term sensitization, its induction requires protein synthesis, but unlike long-term memory it does not require mRNA synthesis. The expression of the intermediate-phase memory requires the persistent activation of PKA.

In addition to intermediate-phase memory, it is likely that *Aplysia* has different phases of long-term memory. For example, at 24 h after sensitization training there is increased synthesis of a number of proteins, some of which are different from those whose synthesis is increased during and immediately after training. These results suggest that the memory for sensitization that persists for times greater than 24 h may be dependent on the synthesis of proteins occurring at 24 h and may have a different molecular signature than the 24-h memory.

Mechanisms Underlying Associative Learning of Withdrawal Reflexes in *Aplysia*

The withdrawal reflexes of *Aplysia* are subject to classical conditioning (Byrne *et al.*, 1993; Hawkins *et al.*, 1993). The short-term classical conditioning observed at the behavioral level reflects, at least in part, a cellular mechanism called *activity-dependent neuromodulation*. A diagram of the general scheme is presented in Figure 49.4. The US pathway is activated by a shock to the animal, which elicits a withdrawal response (the UR). When a CS is paired consistently with the US, the animal will develop a withdrawal response (CR) to the CS. Activity-dependent neuromodulation is proposed as the mechanism for this pairing-specific effect. The US activates both a motor neuron (UR) and a modulatory system. The modulatory system delivers the neurotransmitter serotonin to all the sensory neurons (parts of the various CS pathways), which leads to a nonspecific enhancement of transmitter release from the sensory neurons. This nonspecific enhancement contributes to short-term sensitization (see prior discussion). Sensory neurons whose activity is temporally contiguous with the US-mediated reinforcement are additionally modulated. Spiking in a sensory neuron during the presence of 5-HT leads to changes in that cell relative to other sensory neurons whose activity was not paired with the US. Thus, a subsequent CS will lead to an enhanced activation of the reflex (Fig. 49.4B). Figure 49.5 illustrates a more detailed model of the proposed cellular mechanisms responsible for this example of classical conditioning. The modulator (US) acts by increasing the activity of adenylyl cyclase (AC), which in turn increases the levels of cAMP. Spiking in the sensory neurons (CS) leads to increased levels of intracellular calcium, which greatly enhances the action of the modulator to increase the cAMP cascade. This system determines CS–US contiguity by a method of coincidence detection at the presynaptic terminal.

Now, consider the postsynaptic side of the synapse. The postsynaptic region contains NMDA-type

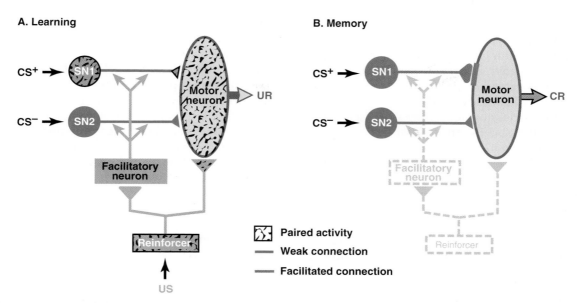

FIGURE 49.4 Model of classical conditioning of a withdrawal reflex in *Aplysia*. (A) Activity in a sensory neuron (SN1) along the CS+ (paired) pathway is coincident with activity in neurons along the reinforcement pathway (US). However, activity in the sensory neuron (SN2) along the CS–(unpaired) pathway is not coincident with activity in neurons along the US pathway. The US directly activates the motor neuron, producing the UR. The US also activates a modulatory system in the form of the facilitatory neuron, resulting in the delivery of a neuromodulatory transmitter to the two sensory neurons. The pairing of activity in SN1 with the delivery of the neuromodulator yields the associative modifications. (B) After the paired activity in A, the synapse from SN1 to the motor neuron is selectively enhanced. Thus, it is more likely to activate the motor neuron and produce the conditioned response (CR) in the absence of US input. Modified from Lechner and Byrne (1998).

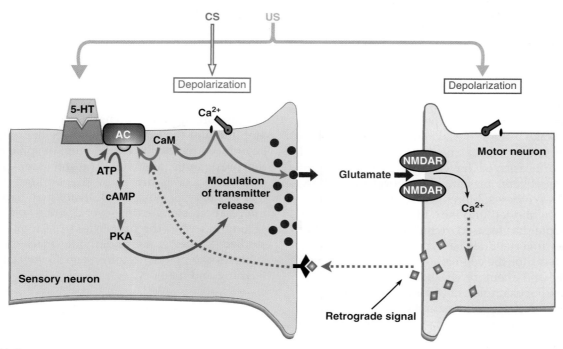

FIGURE 49.5 Model of associative facilitation at the *Aplysia* sensorimotor synapse. This model has both a presynaptic and a postsynaptic detector for the coincidence of the CS and the US. Furthermore, a putative retrograde signal allows for the integration of these two detection systems at the presynaptic level. The CS leads to activity in the sensory neuron, yielding presynaptic calcium influx, which enhances the US-induced cAMP cascade. The CS also induces glutamate release, which results in postsynaptic calcium influx through NMDA receptors if paired with the US-induced depolarization of the postsynaptic neuron. The postsynaptic calcium influx putatively induces a retrograde signal, which further enhances the presynaptic cAMP cascade. The end result of the cAMP cascade is to modulate transmitter release and enhance the strength of the synapse. Modified from Lechner and Byrne (1998).

receptors (Chapters 9 and 11). These receptors need concurrent delivery of glutamate and depolarization in order to allow calcium to enter. The glutamate is provided by the activated sensory neuron (CS), and the depolarization is provided by the US (Antonov *et al.*, 2003; Roberts and Glanzman, 2003). Thus, the postsynaptic neuron provides another example of coincidence detection. The increase in intracellular calcium putatively causes a retrograde signal to be released from the postsynaptic to the presynaptic terminal, ultimately acting to further enhance the cAMP cascade in the sensory neuron. The overall amplification of the cAMP cascade acts to raise the level of PKA, which in turn leads to the modulation of transmitter release. These activity-dependent changes enhance synaptic efficacy between the specific sensory neuron of the CS pathway and the motor neuron. Thus, the sensory neuron along the CS pathway will be better able to activate the motor neuron and produce the CR.

Summary

Certain invertebrates display an enormous capacity for learning and offer particular experimental advantages for analyzing the cellular and molecular mechanisms of learning. For example, behaviors in *Aplysia* are mediated by relatively simple neural circuits, which can be analyzed with conventional anatomical and electrophysiological approaches. Once the circuit is specified, the neural locus for the particular example of learning can be found, and biophysical, biochemical, and molecular approaches can then be used to identify mechanisms underlying the change. The relatively large size of some of these cells allows these analyses to take place at the level of individually identified neurons. Individual neurons can be removed surgically and assayed for changes in the levels of second messengers, protein phosphorylation, RNA, and protein syntheses. Moreover, peptides and nucleotides can be injected into individual neurons.

This chapter has focused exclusively on *Aplysia,* but many other invertebrates have proven to be valuable model systems for the cellular and molecular analysis of learning and memory. Each has its own unique advantages. For example, *Aplysia* is excellent for applying cell biological approaches to the analysis of learning and memory mechanisms. Other invertebrate model systems such as *Drosophila*, **although** not well suited for cell biological approaches because of their small neurons, offer tremendous advantages for obtaining insights into mechanisms of learning and memory through the application of genetic approaches. A frequently used protocol in *Drosophila* employs a two-stage differential odor–shock avoidance proce-

dure, in which animals learn to avoid odors paired (CS+) with shock but not odors explicitly unpaired (CS–). This learning is typically retained for 4–6 h, but retention for 24h to 1 week can be produced by a spaced training procedure. Several mutants deficient in learning have been identified. Analysis of the affected genes has revealed elements of the cAMP signaling pathway as key in learning and memory. It is now known that the formation of long-term memory requires activation of the cAMP signaling pathway, which, in turn, activates members of the CREB transcription family. These proteins appear to be the key molecular switch that enables expression of genes necessary for the formation of long-term memories. These transcription factors are also important for long-term memory in *Aplysia* as described above and in vertebrates (see below). See Byrne (1987) and Carew (2000) for a review of other selected invertebrate model systems that have contributed importantly to the understanding of memory mechanisms.

VERTEBRATE STUDIES: LONG-TERM POTENTIATION

In contrast to the studies on invertebrates like *Aplysia* described earlier, in vertebrates it has been more difficult to link synaptic plasticity with specific examples of learning. However, one exciting and extensively studied candidate memory mechanism is the synaptic phenomenon termed long-term potentiation (LTP). This phenomenon is defined as a persistent increase in synaptic strength (as measured by the amplitude of the EPSP in a follower neuron) that can be induced rapidly by a brief burst of spike activity in the presynaptic afferents. The intense experimental interest in LTP is driven by the working hypothesis that this form of synaptic plasticity may participate in information storage in several brain regions. This section describes the properties of LTP and how it is studied, reviews its underlying mechanisms, and explores the possibility of linkages between LTP and learning and memory.

Long-Term Potentiation Occurs in a Variety of Neural Synapses

The first evidence that long-term modification of mammalian synapses could be induced by experimental means appeared in 1973, when Timothy Bliss and Terje Lomo demonstrated LTP in the hippocampus of the anesthetized rabbit. Brief, high-frequency stimulation of the perforant-pathway input to the dentate gyrus produced a long-lasting enhancement of the extracellularly recorded field potential. Subsequent studies of nonanesthetized animals have shown that LTP can last for weeks or months. Originally thought to be unique to the mammalian hippocampal formation, LTP is now known to occur in the cerebellum, neocortical regions, subcortical regions such as the amygdala, mammalian peripheral nervous system, the arthropod neuromuscular junction, and the *Aplysia* sensorimotor synapse. It is important to bear in mind that no universal mechanism exists for inducing LTP. Indeed, different mechanisms may be used at the same synapse, depending on the experimental conditions. This chapter focuses primarily on the LTP at the synapse made by a pyramidal neuron in the CA3 region of the hippocampus to a pyramidal neuron in the CA1 region of the hippocampus.

Long-Term Potentiation at the CA3–CA1 Synapse

Within the hippocampus proper, by far the best studied synapse is that from the Schaffer collateral/commissural (Sch/com) fibers of the CA3 pyramidal cells to the CA1 pyramidal cells (Fig. 49.6). In fact, this is probably the most commonly studied synapse in the mammalian brain, due in part to its relatively simple circuitry and its laminar organization. These features make it possible to extract useful data from extracellular recordings, which are easier to perform than intracellular recordings and preferable for some purposes. Examples of LTP induction are illustrated in Figure 49.7B. The lower waveforms are extracellularly recorded field EPSPs recorded in the CA1 region in response to a single weak stimulation of the Sch/com pathway. Brief electric stimuli delivered to this pathway lead to the initiation of action potentials in the individual axons in the pathway. These action potentials then propagate to the synaptic terminals. The release of transmitter from the multiple afferent terminals produces a summated EPSP in the postsynaptic cell, which can be detected with an extracellular electrode. Test stimuli are delivered repeatedly at a low rate that produces stable EPSPs in the postsynaptic cell (Figs. 49.7A and B1). After a baseline period, a brief high-frequency tetanus is delivered. Subsequent test stimuli produce enhanced EPSPs (Fig. 49.7B2). The enhancement persists for many hours. Although the synaptic enhancement is stable after the tetanus, LTP, like heterosynaptic facilitation of the sensori-motor

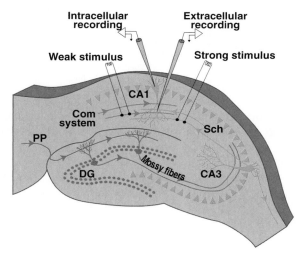

FIGURE 49.6 Schematic of a transverse hippocampal brain slice preparation from the rat. Two extracellular stimulating electrodes are used to activate two nonoverlapping inputs to pyramidal neurons of the CA1 region of the hippocampus. By suitably adjusting the current intensity delivered to the stimulating electrodes, different numbers of Schaffer collateral/commissural (Sch/com) axons can be activated. In this way, one stimulating electrode was made to produce a weak postsynaptic response and the other to produce a strong postsynaptic response. Also illustrated is an extracellular recording electrode placed in the stratum radiatum (the projection zone of the Sch/com inputs) and an intracellular recording electrode in the stratum pyramidale (the cell body layer). Also indicated is the mossy fiber projection from granule cells of the dentate gyrus (DG) to the pyramidal neurons of the CA3 region. Adapted from Barrionuevo and Brown (1983).

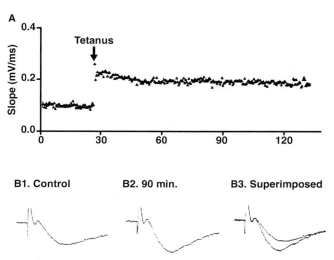

FIGURE 49.7 LTP at the CA3–CA1 synapse in the hippocampus. (A) Test stimuli are delivered repeatedly once every 10 s while the strength of the synaptic connection is monitored. Strength can be assessed by the *amplitude* of the extracellularly recorded EPSP or, as was done in this example, as the *slope* of the rising phase of the EPSP, which provides an accurate reflection of its strength. To induce LTP, two 1 s, 100 Hz tetani were delivered with a 20 interval. Subsequent test stimuli produce enhanced EPSPs. The enhancement is stable and persists for at least 2 h. Examples of extracellulary recorded field EPSPs before (B1) and 90 min after the induction of LTP (B2). In B3 the traces from B1 and B2 are superimposed. Modified from Nicoll *et al.* (1998).

synapse in *Aplysia,* has multiple temporal domains. One domain is associated with an enhancement of the EPSP that persists for about 90 min. This form of LTP is referred to as early LTP (E-LTP). A second domain referred to as late LTP (L-LTP) is associated with synaptic enhancement that persists for periods of time greater than about 90 min. As described later, different mechanisms underlie the induction and maintenance of E- and L-LTP.

Properties of Long-Term Potentiation at the CA3–CA1 Synapse Include Cooperativity, Associativity, and Input Specificity

The CA3-CA1 synapses exhibit a form of LTP characterized by "classical" properties that have been termed "cooperativity," "associativity," and "input specificity" (Fig. 49.8) (Brown *et al.*, 1990; Bliss and Collingridge, 1993). These "classical properties" are actually different manifestations of the same underlying mechanism that is responsible for this type of LTP. Other less commonly studied types of LTP have different signatures.

Cooperativity refers to the fact that the probability of inducing LTP, or the magnitude of the resulting change, increases with the number of stimulated afferents. Weak high-frequency stimulation, which activates fewer afferents, often fails to induce LTP (Fig. 49.8A). In contrast, strong stimulation, which activates more afferents, produces LTP more reliably (Fig. 49.8B). Thus, the additional axons recruited by higher stimulation intensities "cooperate" to trigger LTP.

Associativity was shown in preparations in which two distinct axonal inputs converged onto the same postsynaptic target. Consider the interactions between two stimulus pathways, one termed the *weak* pathway with a small number of stimulated afferents, and the other, termed the *strong* pathway, with a large number of stimulated afferents (Fig. 49.8C). Tetanic (high-frequency) stimulation of the *weak* input by itself failed to produce LTP in that pathway unless this stimulation was paired with tetanic stimulation of the *strong* input. Thus, LTP was induced in a *weak* input only when its activity was associated with activity in the *strong* input.

Input specificity means that LTP is restricted to only the inputs that received the tetanic stimulation. The unstimulated *weak* pathway was not facilitated after the tetanus to the *strong* pathway (Fig. 49.8B).

A Hebbian Mechanism Explains the Properties of Long-Term Potentiation at the CA3-CA1 Synapse in the Hippocampus

How can these classical properties of LTP in the CA1 region of the hippocampus be explained? In the late 1940s, the Canadian psychologist Donald Hebb (1949) formulated a postulate regarding the conditions that cause synapses to change. His thinking proved to be influential and guided later experiments that probed the mechanisms behind LTP. According to Hebb's postulate:

> When an axon of cell A is near enough to excite a cell B and repeatedly or persistently takes part in firing it, some growth process or metabolic change takes place in one or both cells such that A's efficiency, as one of the cells firing B, is increased. (Hebb, 1949, p. 62)

In short, coincident activity in two synaptically coupled neurons was proposed to cause increases in

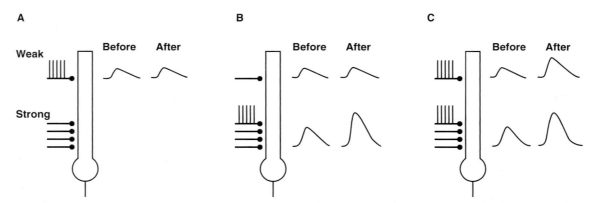

FIGURE 49.8 Features of LTP at CA3–CA1 synapses in the hippocampus. A single hippocampal pyramidal cell is shown receiving a weak and strong synaptic input. (A) Tetanic stimulation of the weak input alone does not cause LTP in that pathway (compare the EPSP before and after the tetanus). (B) Tetanic stimulus of the strong input alone causes LTP in the strong pathway, but not in the weak pathway. (C) Tetanic stimulation of both the weak and the strong pathway together causes LTP in both the weak and the strong pathway. Modified from Nicoll *et al.* (1998).

the synaptic strength between them. Numerous modern interpretations of Hebb's postulate exist, but most are captured by the mnemonic: "Cells that fire together, wire together."

Could the classical properties of LTP all be consequences of synapses that obey a Hebbian rule? Possibly so if a critical amount of postsynaptic depolarization were a necessary condition for inducing LTP in active synapses. In this case, cooperativity would result when enough input fibers were stimulated to produce the critical amount of postsynaptic depolarization. Associativity would emerge from the fact that the strong input caused sufficient depolarization of the postsynaptic membrane during the presynaptic activity in the weak input. Input specificity would occur because LTP was induced only in those inputs to a neuron that were active at the same time that the cell was sufficiently depolarized by the strong input to that neuron. In other words, these classical phenomena could all be manifestations of a single underlying Hebbian mechanism at the CA3–CA1 synapse.

Not all forms of LTP are Hebbian, however. Examples of non-Hebbian LTP can be found at the mossy fiber–CA3 synapse in the hippocampus and at the parallel fiber–Purkinje cell synapse in the cerebellum. These results indicate that the classical properties of cooperativity, associativity, and input specificity are not universal.

Mechanisms for Induction, Expression, and Maintenance of Long-Term Potentiation

LTP Induction

It is currently thought that there are multiple mechanisms or at least multiple second-messenger pathways that can lead to persistent synaptic enhancement. Multiple mechanisms may also contribute to expression and maintenance.

Calcium Ions and LTP. It is generally agreed that the induction of LTP depends on an increase in the intracellular concentration of calcium ions ($[Ca^{2+}]_i$) in some key compartment of pre- and/or postsynaptic cells (Bliss and Collingridge, 1993; Johnston *et al.*, 1992; Nicoll and Malenka, 1995). The exact role of calcium in the induction process depends on the particular form of LTP and the synaptic system. In the CA1 region of the hippocampus, LTP induction in the Sch/com synapse depends on changes in postsynaptic $[Ca^{2+}]_i$.

It is currently thought that many different pathways modulate or control $[Ca^{2+}]_i$ in the critical subcellular

compartment(s) (Bliss and Collingridge, 1993; Nicoll and Malenka, 1995; Teyler *et al.*, 1994) (Fig. 49.9). Two major pathways that have been studied extensively are implicated in some aspect of LTP induction: calcium influx through ionotropic GluRs, especially the *N*-methyl-*D*-aspartate receptor (NMDAR); and calcium influx through voltage-gated calcium channels (VGCCs).

NMDAR-dependent LTP. Recall that the classical form of LTP in the CA1 region of the hippocampus has properties that can be explained in terms of a Hebbian mechanism. For this form of LTP, considerable evidence shows a role for the NMDAR (Bliss and Collingridge, 1993). Numerous pharmacological studies have shown that competitive antagonists of NMDA, such as D-2-amino-5-phosphonopentenoic acid (D-AP5, also termed AP5 or APV) or NMDA ion channel blockers, such as the noncompetitive antagonist (+)–5-methyl-10, 11-dihydro-*5H*-dibenzo[*a,d*] cyclohepten-5,10-imine (MK-801), can prevent the induction of LTP.

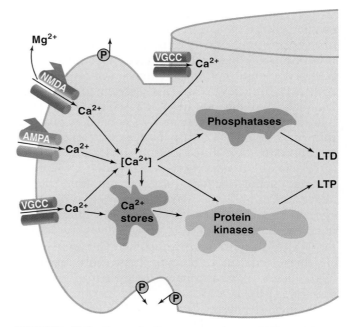

FIGURE 49.9 Events leading to some forms of LTP and LTD. The schematic depicts a postsynaptic spine with various sources of Ca^{2+}. The NMDA receptor channel complex admits Ca^{2+} only after depolarization removes the Mg^{2+} block in the presence of bound glutamate. Calcium may also enter through the ligand-gated AMPA receptor channel or voltage-gated calcium channels (VGCC), which may be located on the spine head or dendritic shaft. Calcium pumps (P), located on the spine head, neck, and dendritic shaft, are hypothesized to help isolate Ca^{2+} concentration changes in the spine head from those in the dendritic shaft.

The NMDAR has two properties that immediately suggest the nature of its role in LTP induction at Hebbian synapses (Gustafsson and Wigstrom, 1988; Brown *et al.*, 1990; Bliss and Collingridge, 1993). First, NMDARs are permeable to Ca^{2+} (in addition to Na^+ and K^+). This property is significant because postsynaptic $[Ca^{2+}]_i$ plays a critical role in inducing NMDAR-dependent LTP. Second, the channel permeability is a function of both pre- and postsynaptic factors. Channel opening requires the neurotransmitter glutamate (or some related agonist) to bind to the NMDA site. This in turn requires presynaptic activity for glutamate release. At the usual resting membrane potential, the ionic channels of NMDARs normally are blocked by magnesium ions (Mg^{2+}), but this channel block is relieved by sufficient depolarization of the postsynaptic membrane. Thus, the NMDAR-mediated conductance is voltage dependent, allowing Ca^{2+} entry only when presynaptic release is combined with postsynaptic depolarization.

At this point, you should recall the distinction between the properties of the NMDAR and those of the AMPAR (α-amino-3-hydroxy-5-methyl-4-isoxazole propionic acid receptor) (see Chapters 9 and 11 for details), which also are found in the postsynaptic membrane. The AMPAR does not exhibit voltage-dependent Mg^{2+} block, has relatively lower Ca^{2+} permeability, and the AMPAR-mediated conductance is essentially voltage independent. Released glutamate can potentially act on both the AMPARs and the NMDARs associated with the membrane on the dendritic spine (Fig. 49.9). With this knowledge, one can easily envision a possible role for the AMPAR and NMDAR in a Hebbian modification. Nearly concurrent presynaptic activity (producing glutamate release and binding) and postsynaptic depolarization (relieving the Mg^{2+} block) allow Ca^{2+} influx into the dendritic spine of the postsynaptic neuron. The increased $[Ca^{2+}]_i$ in some critical region of the dendritic spine, presumably very close to the NMDAR, is thought to activate Ca^{2+}-dependent enzymes, such as calmodulin-dependent kinase II (CaM kinase II), which play a key role in LTP induction (Fig. 49.9).

In qualitative terms, one can understand how these molecular events could help account for the properties of cooperativity, associativity, and spatiotemporal specificity. Active synapses release glutamate, which can bind to the NMDAR, causing Ca^{2+} influx into dendritic spines on the postsynaptic cell. This Ca^{2+} influx acts locally and results in input-specific LTP. However, the Ca^{2+} influx occurs only when the synaptic input is strong enough to depolarize the postsynaptic membrane sufficiently to relieve the Mg^{2+} block, giving rise to cooperativity (see Fig. 49.8). The depolarization itself is mediated in large part by the (voltage-independent) AMPARs, which are also colocalized on the dendritic spine (Fig. 49.9). Note that activity in a weak input by itself would not depolarize the postsynaptic cell sufficiently to relieve the Mg^{2+} block unless this activity were properly timed in relationship to activity in a strong input to the same cell. The combined depolarization of the two inputs gives rise to associativity and input specificity (Fig. 49.8).

NMDAR-independent LTP. Most of the preceding accounts of the Hebbian, NMDAR-dependent form of LTP apply only to certain synapses under some conditions, and even then it may be only one aspect of the story (Johnston *et al.*, 1992; Nicoll and Malenka, 1995). In many synapses, LTP induction does not appear to require the NMDA receptor (Johnston *et al.*, 1992; Teyler *et al.*, 1994). Even within the hippocampus, some synapses exhibit NMDAR-independent forms of LTP. For example, in the presence of the competitive antagonist APV, even the Sch/com inputs to CA1 pyramidal neurons, which are known to exhibit the classical Hebbian form of LTP that relies on the NMDAR, can exhibit an NMDAR-independent type of LTP. The onset of NMDAR-independent LTP is relatively slow (20–30 min), exhibits input specificity, and is prevented by a blocker of voltage-gated calcium channels (VGCCs). Thus, a distinction exists between "NMDA LTP" and "VGCC LTP." Other work has suggested that VGCCs are likely to be responsible for certain types of LTP in the CA3 region of the hippocampus and in the visual cortex.

The general case may be that both NMDAR-dependent and NMDAR-independent forms of LTP may coexist in the same brain region, among different classes of synaptic inputs onto the same postsynaptic neuron, and even among the same class of synaptic inputs to the same postsynaptic neuron.

Induction of late LTP. A common requirement for the induction of both early LTP and late LTP at the CA3–CA1 synapse is the elevation of levels of intracellular calcium in the postsynaptic (i.e., CA1) neuron. Additional steps are involved in the induction of L-LTP, however. As was the case for the induction of long-term facilitation of the sensorimotor synapse in *Aplysia*, activation of PKA appears to be necessary for the induction of L-LTP (Huang *et al.*, 1996). However, in the CA1 neuron the cAMP pathway is engaged directly by the activation of a calcium-sensitive adenylyl cyclase rather than being activated by 5-HT as in *Aplysia*. (Transmitters that regulate cAMP in CA1 neurons can profoundly modulate LTP, however.) Thus, at the CA1 neuron elevated levels of calcium lead to activation of adenylyl cyclase, increased synthesis of cAMP, and activation of PKA. Also, like long-term facilitation in *Aplysia*, MAPK kinase is necessary for the induction of L-LTP (Dineley *et al.*, 2001). The mechanisms of activation of MAPK in CA1 neurons have not been elucidated in detail, but may involve PKA, PKC, or both acting together. The final steps in the induction of L-LTP involve a PKA- and MAPK-dependent phosphorylation of CREB and induction of CREB responsive genes. These genes would encode the proteins underlying changes at the synapse (but see Box 49.1).

The mechanisms underlying the induction of long-term neuronal plasticity and memory seem to be highly conserved across species. NMDAR-dependent LTP has been characterized at the sensorimotor synapse of *Aplysia* (Lin and Glanzman, 1994). Formation of some long-term memories in *Drosophila* depends on both an increase in the activity of a transcriptional activator of the bZip family, dCREB2a, and a decrease in the activity of a related repressor, dCREB2b. Establishment of LTM in *Drosophila* therefore appears analogous to induction of *Aplysia* LTF, involving transcriptional activation by one bZip family member (CREB1 in *Aplysia*) and relief of repression by another member

(CREB2 in *Aplysia*). CREB activation also correlates with mammalian long-term memories. Interference with CREB function can inhibit long-term memories whereas overexpression of CREB can facilitate them.

LTP Expression

Up to this point, evidence related to early events in the causal chain that triggers the induction process has been emphasized. Another question follows naturally: What biochemical and biophysical changes incorporate this modification once it has been triggered? Most of the ideas about enhanced synaptic transmission concern either increased transmitter release or increased receptivity to released transmitter. The former entails presynaptic changes, and the latter, postsynaptic. Although some of the induction mechanisms discussed previously implicated a postsynaptic increase in $[Ca^{2+}]_i$, this does not necessarily imply that expression must also be postsynaptic. Ample evidence is available for ongoing two-way communication across the synaptic cleft, so a postsynaptic trigger could, in principle, give rise to a pre- and/or postsynaptic modification (see also Fig. 49.5). There are extensive and seemingly conflicting accounts in the literature regarding the nature of the changes responsible for the observed increase in synaptic efficacy following LTP induction. Some of the possibilities are illustrated schematically in Figure 49.10. However, regarding LTP at the CA3–CA1 synapse, expression of LTP is

---

BOX 49.1

IS MEMORY MORE THAN CHANGES IN SYNAPTIC STRENGTH?

The search for the biological basis of learning and memory has led many of the leading neuroscientists of the twentieth century to direct their efforts to investigating the synapse. The focus of most recent work on LTP (and indeed the bulk of this chapter) has been to elucidate the mechanisms underlying changes in synaptic strength. Although changes in synaptic strength are certainly ubiquitous, they are not the exclusive means for the expression of neuronal plasticity associated with learning and memory. Both short-term and long-term sensitization and classical conditioning of defensive reflexes in *Aplysia* are associated with an enhancement of excitability of the sensory neurons in addition to changes in synaptic strength. Classical and operant conditioning of feeding behavior in *Aplysia* also leads to changes in excitability of

neuron B51. Interestingly, in this case there is a bidirectional control. The excitability of B51 is increased by operant conditioning, whereas it is decreased by classical conditioning. Changes in excitability of sensory neurons in the mollusc *Hermissenda* are produced by classical conditioning. In vertebrates, classical conditioning of eyeblink reflexes produces changes in the excitability of cortical neurons. Eye-blink conditioning also produces changes in the spike afterpotential of hippocampal pyramidal neurons. Finally, as described in their original report on LTP, Bliss and Lomo found that the expression of LTP was also associated with an apparent enhanced excitability.

John H. Byrne

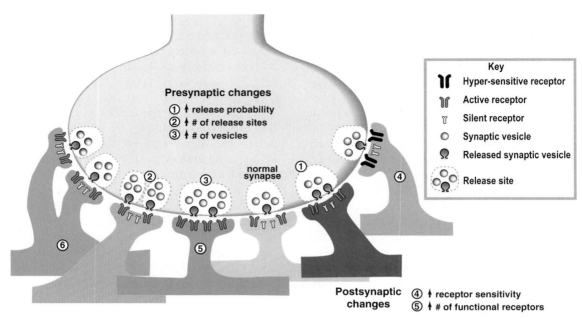

FIGURE 49.10 Schematic representation of possible loci for cellular changes involved in the enhancement of synaptic efficacy. The efficacy of a synapse can be potentiated through at least six mechanisms. First, there could be an increase in the fraction (release probability) of available presynaptic vesicles that undergo exocytosis. For example, in *mechanism 1*, two out of four available vesicles are released (i.e., 50% release probability), in contrast to the *normal synapse*, where only one out of four vesicles is released (i.e., 25% release probability). Second, there could be an increase in the number of release sites at the presynaptic neuron (*mechanism 2*). Third, the synapse could be potentiated through an increase in the number of vesicles available for release. For example, at a release site with eight vesicles, two of them will be exocytosed (instead of one) (*mechanism 3*), even if the release probability (25%) is the same as at the normal synapse. Fourth, there could be an increase in the sensitivity of the preexisting receptors to presynaptically released neurotransmitter or a greater conductance of the channel (*mechanism 4*). Fifth, there could be an increase in the number of functional receptors (illustrated as an increase from two active receptors at the normal synapse to four at the potentiated synapse; *mechanism 5*). Finally, the synapse could also be potentiated through coordinated presynaptic and postsynaptic morphological changes, such as the growth of new synaptic contacts between the same pair of neurons (*mechanism 6*). Adapted from Wang *et al.* (1997).

associated with an increase in the number of functional AMPA receptors in the postsynaptic neuron (*mechanism No. 5* in Fig. 49.10; for review see Malinow and Malenka, 2002). LTP at this synapse also involves phosphorylation of the AMPA receptor, which leads to an increased conductance that contributes to enhanced synaptic efficacy (*mechanism No. 4* in Fig. 49.10). Recent results indicate that a family of small transmembrane AMPA receptor regulatory proteins (TARPs) controls both AMPA receptor trafficking and channel gating (for review see Nicoll *et al.*, 2006). In contrast to the postsynaptic expression of LTP at the CA3-CA1 synapse, LTP at the mossy fiber-CA3 synapse appears to involve an increase in transmitter release.

LTP Maintenance

Regardless of the ultimate nature and locus of the modification that gives rise to LTP expression, the more general problem remains of how a synaptic change can endure over long periods of time in the face of constant molecular turnover.

Maintenance of Early LTP. The maintenance of E-LTP is due to the persistence of phosphorylation of substrate proteins involved in expression mechanisms (see earlier discussion). This phosphorylation, in turn, is regulated by the engagement of protein kinases and phosphatases. For E-LTP, autonomously active forms of PKC and CaM kinase II appear to be particularly important.

Maintenance of Late LTP. The maintenance of persistent forms of LTP ultimately involves alterations in gene expression and changes in protein synthesis. High-frequency electrical stimulation in the rat hippocampus raises levels of specific mRNAs that encode transcription factors (e.g., fos, zif268), cytoskeletal proteins (e.g., arc), and signal transduction molecules such as CaM kinase II. Moreover, protein synthesis inhibitors block the late phase of LTP but not earlier phases. One interesting observation is that brain slices that received protein synthesis inhibitors just 2h after high-frequency stimulation showed no decline in late LTP, indicating that there is a critical time window during which protein synthesis might be necessary to maintain long-term plasticity. One new protein that is synthesized is proten kinase Mzeta (PKMζ). PKMζ is necessary for maintenance of L-LTP and seems necessary for maintaining the increase in postsynaptic AMPA receptors (Ling *et al.*, 2006; Pastalkova *et al.*, 2006). Other work conducted *in vivo* has pointed to a still later stage of LTP that lasts for weeks and that is blocked by treatment with protein synthesis inhibitors during the tetanus (Krug *et al.*, 1984). As indicated earlier, similar multiple stages of long-term synaptic plasticity have been identified in *Aplysia*. They also are found in bees, *Drosophila*, and *Hermissenda*.

Although the involvement of new protein synthesis is consistent with the data, it immediately raises the problem of synaptic input specificity. If neural activity ultimately affects gene expression in the nucleus, then the mRNAs and proteins produced in the somatic compartment could, in principle, travel to any synapse within the cell. The problem is how to modify only the appropriate synapses and maintain synapse specificity. One solution is for the specific kinase(s) activated locally during a tetanus to phosphorylate one or more substrates and that these substrates act to "tag" the synapse as one that has been potentiated. For example, a stimulus that activates one CA1 synaptic input sufficiently to induce L-LTP can lead to the induction of L-LTP by a weaker stimulus to a second synaptic input even when that second input is by itself sufficient to only induce E-LTP. This effect is limited to a time window where the second weaker stimulus occurs within an hour or so of the stronger one. These results indicate that the second stimulus can lead to the transient (~1 h) activation of a "tag" that can specifically "capture" "plasticity factors" induced by the strong stimulus (Frey and Morris, 1997; Martin *et al.*, 1997). A major goal of current LTP research is to identiy the nature of the tag and the plasticity factors.

Support of a more permanent synaptic modification might require a self-perpetuating tag and/or a biochemical positive feedback loop that maintains synaptic strength (Hayer and Bhalla, 2005). Positive feedback based on network activity may also be important. Periodic reactivation of strengthened synapses may be necessary to re-activate mechanisms of L-LTP, maintaining synaptic strength for months or longer (Wang *et al.*, 2006; Smolen, 2007).

Links between Long-Term Potentiation and Learning

Long-term potentiation has properties that have long been considered necessary for the encoding and retrieval of information. Hebbian forms of LTP exhibit associativity, which appears to be a desirable property, and all forms of LTP appear to be well suited to rapid learning. One of the more important challenges entails linking LTP (and LTD, see later) to learning and memory (Martin *et al.*, 2000). This task has proven rather difficult to achieve, although evidence is accumulating for a causal link between LTP and memory.

Several different experimental strategies have been used to search for links between LTP and learning. These include correlating changes in synaptic efficacy with learning and blocking or saturating LTP mechanisms to show that learning has been blocked. Correlating synaptic enhancement with memory has been possible in several systems such as in amygdala circuits that mediate fear conditioning (see Chapter 50). However, there has been only limited success so far. The first obstacle to surmount is the "needle in the haystack" problem. The synaptic changes underlying learning are presumably rather specific and localized. Consequently, it is difficult to place a recording electrode in the correct anatomical location. Moreover, as described later, learning may involve decreases as well as increases in synaptic strength. Coarse recording techniques might see no net change. Saturation of LTP by the massive stimulation of afferents has blocked learning, but the concern here is that the saturation may affect the basic operation of the circuit. The most widely used approach has been to block LTP with pharmacological or gene knockout techniques. This approach has generated many successes but also some striking dissociations, as blocking experiments have two problems. First, it is difficult to be sure that the agent used to block LTP does not cause some secondary effect. Thus, if memory is blocked successfully, it may be because of a block of some other process critical for memory and not the block of LTP. Second, if, for example, an agent that blocks NMDAR-*dependent* LTP fails to block learning, it may not necessarily imply that LTP does not mediate learning, but rather that the animal simply has used another non-NMDAR-*independent* memory mechanism to solve the problem.

Some of the more promising work in this area has come from studies of genetically engineered knockout mice. The first generation of knockouts prevented the expression of some factor that was thought to be necessary for LTP, such as CaM kinase-II. Although these studies were intriguing, they suffered from the fact that the consequences of a gene knockout might alter brain development and might not be confined to a specific part of the brain under study. Some of these problems have been overcome in a second generation of knockouts that have temporal as well as spatial specificity. For example, temporal and spatial control of gene expression can now be achieved using binary transgene systems such as tetracycline transactivating systems and Cre/LoxP recombination systems. The NMDAR gene in only CAI pyramidal cells of the hippocampus has been knocked out using the Crc/LoxP system (Tsien *et al.*, 1996). The results provide strong evidence in favor of the notion that NMDA receptor-dependent synaptic plasticity at CAI synapses is required for the acquisition of spatial memory. This general approach holds tremendous promise for testing hypotheses about the functional role of various forms of LTP in specific brain regions.

Recently, learning in rats was verified to induce hippocampal LTP (Whitlock *et al.*, 2006). Also, inhibition of PKMz in the hippocampus *in vivo* was demonstrated to disrupt LTP maintenance and eliminate spatial learning (Pastalkova *et al.*, 2006). These results have strengthened the link between LTP and learning.

LONG-TERM DEPRESSION

LTD is believed by many to be the mechanism by which learning is encoded in the cerebellum (Ito, 2001), as well as a process whereby LTP could be reversed in the hippocampus and neocortex (Bear and Malenka, 1994). In the hippocampus, brief, high-frequency stimulation (HFS) (e.g., four trains of 10 shocks at 100 Hz) can induce classical LTP, whereas low-frequency stimulation (LFS) over longer periods (1 Hz for 10 min) can induce LTD.

Some forms of LTD appear to be mediated by the NMDAR and these forms appear to be due to dephosphorylation of AMPAR subunits, which decreases their conductance and hence the synaptic efficacy, as well as endocytosis of AMPARs, which decreases their surface number and hence the synaptic efficacy. In addition, NMDAR-independent forms of LTD exist. These often depend on activation of metabotropic glutamate receptors. For example, NMDAR-independent LTD in the parallel fiber input to Purkinje cells in the cerebellum is induced by the combined increase in intracellular levels of Ca^{2+} (by activity) in postsynaptic Purkinje cells and activation of metabotropic glutamate receptors produced by the release of glutamate from parallel fibers.

The form of LTD may vary in different brain regions and sometimes among different inputs to the same brain region. Even within the CA1 region, the Sch/com input may exhibit both NMDAR-dependent and NMDAR-independent forms of LTD.

Calcium levels in the dendritic spines appear to be a common locus for the induction on NMDAR-dependent LTP and LTD. For example, at the Sch/com input to CA1, both LTP and LTD can be blocked by injecting Ca^{2+} chelators into the postsynaptic cell. If both LTP and LTD are triggered by Ca^{2+} entry, then how are their induction processes different? Presumably, more Ca^{2+} influx occurs during an LTP-inducing HFS than during an LTD-inducing LFS. One formal molecular model developed by John Lisman incorporates this Ca^{2+}-dependent, bidirectional control of synaptic strength. In this model, high $[Ca^{2+}]_i$ activates a protein kinase that phosphorylates a protein causing LTP induction, whereas low $[Ca^{2+}]_i$ activates a protein phosphatase that dephosphorylates this protein and causes LTD. The synaptic strength thus depends on which of these competing processes is most active, which in turn will be a function of the pattern of activity experienced by the cell (Fig. 49.9).

LTP and LTD can be readily reversed. If low frequency stimulation (LFS) is applied after a synapse is potentiated by high frequency stimulation (HFS), the synaptic strength decreases. This phenomenon is called depotentiation. In contrast, if HFS is applied after a synapse is depressed by LFS, the synaptic strength increases. This phenomenon is called dedepression. Accumulating evidence indicates that the mechanisms of depotentiation differ from that of LTD, and the mechanisms of dedepression differ from that of LTP.

Summary

The understanding of LTP and LTD is evolving rapidly (Malenka and Nicoll, 1999). For example, whereas the *N*-methyl-D-aspartate receptor was once the pivotal focus of long-term potentiation research, it is now clear that other mechanisms should be considered. It has also become evident that high-frequency stimulation is but one end of a spectrum of stimulations that can induce synaptic changes. Lower stimulation frequencies can induce long-term depression, which may share some common molecular mechanisms with LTP. Finally, several stages in the maintenance of LTP have been identified, and probably more will be found. A remaining challenge is to clarify the varieties of LTP and LTD mechanisms and to demonstrate their functional significance by establishing convincing links to the encoding and retrieval of information.

HOW DOES A CHANGE IN SYNAPTIC STRENGTH STORE A COMPLEX MEMORY?

The relationship between the synaptic changes and the behavior of conditioned reflexes can be straightforward because the locus for the plastic change is part of the mediating circuit. Thus, the change in the strength of the sensorimotor synapse in *Aplysia* can be related to the memory for sensitization (e.g., Fig. 49.2A1). However, the idea that an increase in synaptic strength leads to an enhanced behavioral response, and a decrease in synaptic strength leads to a decreased behavioral response, can be misleading. For example, a decrease in synaptic strength in a postsynaptic neuron that exerts an inhibitory action can be translated into an enhanced behavioral response. Indeed, in

the parallel fiber-to-Purkinje cell connection in the cerebellum, such a disinhibition is precisely the mechanism that has been proposed to contribute to classical conditioning of the eye-blink reflex (see Chapter 50 for a discussion of this form of conditioning). Nevertheless, for relatively simple reflex systems in which the circuit is well understood, it is possible to directly relate a change in synaptic strength to learning. However, in most other examples of memory, it is considerably less clear how the synaptic changes are induced and, once induced, how the information is retrieved. This is especially true in memory systems that involve the storage of information for patterns, facts, and events. Neurobiologists have turned to artificial neural circuits to gain insights into these issues.

A simple network that can store and "recognize" patterns is illustrated in Figure 49.11. The network is artificial, but nevertheless is inspired by actual cir-

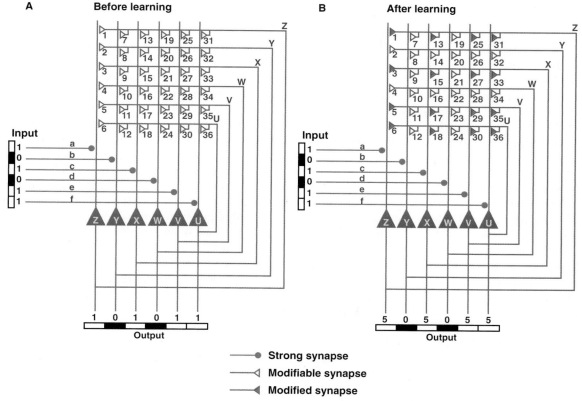

FIGURE 49.11 Autoassociation network for recognition memory. The artificial circuit consists of six input pathways that make strong connections to each of six output neurons. The output neurons have axon collaterals that make synaptic connections (numbered 1–36) with each of the output cells. (A) A pattern represented by activity in the input lines or axons (a, b, c, d, e, f) is presented to the network. A 1 represents an active axon (e.g., a spike), whereas a 0 represents an inactive axon. The input pathways make strong synapses (•) with the postsynaptic output cells. Thus, output cells (u, v, w, x, y, z) generate a pattern that is a replica of the input pattern. The collateral synapses were initially weak and do not contribute to the output. Nevertheless, the activity in the collaterals that occurred in conjunction (assume minimal delays within the circuit) with the input pattern led to a strengthening of a subset of the 36 synapses. (B) A second presentation of the input produces an output pattern that is an amplified, but an otherwise intact, replica of the input. An incomplete input pattern can be used as a cue to retrieve the complete pattern.

cuitry in the CA3 region of the hippocampus. In this example, six different input projections make synaptic connections with the dendrites of each of six postsynaptic neurons (Fig. 49.11A). The postsynaptic neurons serve as the output of the network. Input projections can carry multiple types of patterned information, and these patterns can be complex. In order to simplify the present discussion, consider that the particular input pathway in Figure 49.11A carries information regarding the pattern of neural activity induced by a single brief flash of a spatial pattern of light. For example, activity in the top pathway (line a) might represent light falling on the temporal region of the retina, whereas activity in the pathway on the bottom (line f) might represent light falling on the nasal region of the retina. Thus, the spatial pattern of an image falling upon the retina could be reconstructed from the pattern of neuronal activity over the n (in this case 6) input projections to the network.

Three aspects of the circuit endow it with the ability to store and retrieve patterns. First, each of the input lines makes a sufficiently strong connection with its corresponding postsynaptic cell to activate it reliably. Second, each output cell (z to u) sends an axon collateral that makes an excitatory connection with itself as well as the other five output cells. This pattern of synaptic connectivity leads to a network of 36 synapses (42, including the 6 input synapses). Third, each of the 36 synaptic connections are modifiable through an LTP-like mechanism (see earlier discussion). Specifically, the strength of a particular synaptic connection is initially weak, but it will increase if the presynaptic *and* postsynaptic neurons are active at the same time. The circuit configuration with the embedded synaptic "learning rule" leads to an autoassociation or autocorrelation matrix. The autoassociation is derived from the fact that the output is fed back to the input where it associates with itself.

Now consider the consequences of presenting the patterned input to the network of Figure 49.11A. The input pattern will activate the six postsynaptic cells in such a way as to produce an output pattern that will be a replica of the input pattern. In addition, the pattern will induce changes in the synaptic strength of the active synapses in the network. For example, synapse 3 will be strengthened because the postsynaptic cell, cell z, and the presynaptic cell, cell x, will be active at the same time. Note also that synapses 1, 5, and 6 will be strengthened as well. This occurs because these input pathways to cell z are also active. Thus, all synapses that are active at the same time as cell z will be strengthened. When the pattern is presented again as in Figure 49.11B, the output of the cell will not only be governed by the input, but also by the feedback con-

nections, a subset of which were strengthened (Fig. 49.11B, filled synapses) by the initial presentation of the stimulus. Thus, for output cell z, a component of its activity will be derived from input a, but components will also come from synapses 1, 3, 5, and 6. If each of the initially strong and newly modified synapses is assumed to contribute equally to the firing of output cell z, the activity would be five times greater than the activity produced by input a before the learning. After learning, the output is an amplified version of the input but the basic features of the pattern are preserved.

Note that the "memory" for the pattern does not reside in any one synapse or in any one cell. Rather, it is *distributed* throughout the network at multiple sites. The properties of these types of autoassociation networks have been examined by James Anderson, Teuvo Kohonen, David Marr, Edmond Rolls, David Wilshaw, and their colleagues and found to exhibit a number of phenomena that would be desirable for a biological recognition memory system. For example, such networks exhibit pattern completion. If a partial input pattern is presented, the autoassociation network can complete the pattern in the sense that it can produce an output that is approximately what is expected for the full input pattern. Thus, any part of the stored pattern can be used as a cue to retrieve the complete pattern. For the example of Figure 49.11, the input pattern was {101011}. This pattern led to an output pattern of {505055}. If the input pattern were degraded to {101000}, the output pattern would be {303022}. Some change in the strength of firing occurs, but the basic pattern is preserved. This preservation is a simple example of the phenomenon known as graceful degradation, which is exhibited by autoassociative networks. With graceful degradation the network can still function if some of the input connections or postsynaptic cells are lost. This property arises from the distributed representation of the memory within the circuit.

Summary

The concept of distributed representation of memory crosses multiple levels of organization of memory systems. Multiple brain systems are involved in memory, and memory is distributed among synapses in a particular memory circuit (Fig. 49.11). Also, memory at any one synapse is represented by multiple cellular changes (Figs. 49.2, 49.3, and 49.9). The reductionist approaches described in this chapter have provided key insights into cellular memory mechanisms. In the near future, a major experimental question to be answered is the extent to which the mechanisms for learning are common both within any one animal and between different species. Although many common features are emerging, there seem to be some differences. Thus, it will be important to understand the extent to which specific mechanisms are used selectively for one type of learning and not another. Irrespective of the particular example of learning and memory that is analyzed, whether it be simple or complex, it will be important to pay attention to three major details: details of the circuit interactions, details of the learning rule, and details of the intrinsic biophysical properties of the neurons within the circuit.

References

Antonov, I., Antonova, I., Kandel, E. R. and Hawkins, R. D. (2003) Activity-dependent presynaptic facilitation and Hebbian LTP are both required and interact during classical conditioning in *Aplysia. Neuron* **37**, 135–147.

Agners, A., Fioravante, D., Chin, J., Cleary, L. J., Bean, A. J., and Byrne, J. H. (2002). Serotonin stimulates phosphorylation of *Aplysia* synapsin and alters its subcellular distribution in sensory neurons. *J. Neurosci.* **22**, 5412–5422.

Antzoulatos, E. G. and Byrne, J. H. (2007). Long-term sensitization training produces spike narrowing in *Aplysia* sensory neurons. *J. Neuroscience,* **27**, 676–683.

Bailey, C. H., Chen, M., Keller, F., and Kandel, E. R. (1992). Serotonin-mediated endocytosis of apCAM: An early step of learning-related synaptic growth in *Aplysia. Science* **256**, 645–649.

Bailey, C. H. and Kandel, E. R. (1993). Structural changes accompanying memory storage. *Annu. Rev. Physiol.* **55**, 397–426.

Barrionuevo, G. and Brown, T. H. (1983). Associative long-term potentiation in hippocampal slices. *Proc. Natl. Acad. Sci. USA* **80**, 7347–7351.

Bartsch, D., Ghirardi, M., Skehel, P. A., Karl, K. A., Herder, S. P., Chen, M., Bailey, C. H., and Kandel, E. R. (1995). Aplysia CREB2 represses long-term facilitation: Relief of repression converts transient facilitation into long-term functional and structural change. *Cell* **83**, 979–992.

Bliss, T. V. P. and Collingridge, G. L. (1993) A synaptic model of memory: Long-term potentiation in the hippocampus. *Nature (Lond.).* **361**, 31–39.

Brown, T. H., Ganong, A. H., Kairiss, E. W., and Keenan, C. L. (1990). Hebbian synapses: Biophysical mechanisms and algorithms. *Annu. Rev. Neurosci.* **13**, 475–512.

Byrne, J. H. and Kandel, E. R. (1996). Presynaptic facilitation revisited: State- and time-dependence. *J. Neurosci.* **16**(2), 425–435.

Byrne, J. H., Zwartjes, R., Homayouni, R., Critz, S., and Eskin, A. (1993). Roles of second messenger pathways in neuronal plasticity and in learning and memory: Insights gained from *Aplysia. Adv. Second Messenger Phosphoprotein Res.* **27**, 47–108.

Chain, D. G., Casadio, A., Schacher, S., Hegde, A. N., Valbrun, M., Yamamoto, N., Goldberg, A. L., Bartsch, D., Kandel, E. R., and Schwartz, J. H. (1999). Mechanisms for generating the autonomous cAMP-dependent protein kinase required for long-term facilitation in *Aplysia. Neuron.* **22**, 147–156.

Cleary, L. J., Lee, W. L., and Byrne, J. H. (1998). Cellular correlates of long-term sensitization in *Aplysia. J. Neurosci.* **18**, 5988–5998.

Frey, U. and Morris, R. G. M. (1997). Synaptic tagging and long-term potentiation. *Nature (Lond.).* **385**, 533–536.

Ghirardi, M., Montarolo, P. G., and Kandel, E. R. (1995). A novel intermediate stage in the transition between short- and long-term

facilitation in the sensory to motor neuron synapse of *Aplysia. Neuron* **14**, 413–420.

Gustafsson, B. and Wigstrom, H. (1988). Physiological mechanisms of long-term potentiation. *Trends in Neuroscience.* **11**, 156–162.

Hawkins, R. D., Kandel, E. R., and Siegelbaum, S. (1993). Learning to modulate transmitter release: Themes and variations in synaptic plasticity. *Annu. Rev. Neurosci.* **16**, 625–665.

Hayer, A. and Bhalla, U. S. (2005). Molecular switches at the synapse emerge from receptor and kinase traffic. *PLoS Comput. Biol.* **1**, 137–154.

Hebb, D. O. (1949). "The Organization of Behavior." Wiley (Interscience), New York.

Huang, Y.-Y., Nguyen, P. V., Abel, T., and Kandel, E. R. (1996). Long-lasting forms of synaptic potentiation in the mammalian hippocampus. *Learn. Mem.* **3**, 74–85.

Johnston, D., Williams, D., Jaffe, D., and Gray, R. (1992). NMDA-receptor-independent long-term potentiation. *Annu. Rev. Physiol.* **54**, 489–505.

Krug, M., Loessner, B., and Ott, T. (1984). Anisomycin blocks the late phase of long-term potentiation in the dentate gyrus of freely moving rats. *Brain Res. Bull.* **13**, 39–42.

Lechner, H. A. and Byrne, J. H. (1998). New perspectives on classical conditioning: A synthesis of Hebbian and Non-Hebbian mechanisms. *Neuron* **20**, 355–385.

Levenson, J., Endo, S., Kategaya, L. S., Fernandez, R. I., Brabham, D. G., Chin, J., Byrne, J. H., and Eskin, A. (2000). Long-term regulation of neuronal high-affinity glutamate and glutamate uptake in *Aplysia. Proc. Nat. Acad. Sci.* **97**, 12858–12863.

Lin, X. Y. and Glanzman, D. L. (1994). Hebbian induction of long-term potentiation of *Aplysia* sensorimotor synapses: partial requirement for activation of an NMDA-related receptor. *Proc. Biol. Sci.* **22**, 215–221.

Ling, D. S. F., Benardo, L. S. and Sacktor, T. C. (2006). Protein kinase Mζ enhances excitatory synaptic transmission by increasing the number of active postsynaptic AMPA receptors. *Hippocampus* **16**, 443–452.

Malinow, R. and Malenka, R. C. (2002). AMPA receptor trafficking and synaptic plasticity. *Annu Rev Neurosci.* **25**, 103–126.

Martin, K. C., Michael, D., Rose, J. C., Barad, M., Casadio, A., Zhu, H., and Kandel, E. R. (1997). MAP kinase translocates into the nucleus of the presynaptic cell and is required for long-term facilitation in *Aplysia. Neuron* **18**, 899–912.

Martin, K. C., Casadeo, A., Zhu, H. E. Y., Rose, J. C., Chen, M., Bailey, C. H., and Kandel, E. R. (1997). Synapse-specific, long-term facilitation of *Aplysia* sensory to motor synapses: A function for local protein synthesis in memory storage. *Cell* **91**, 927–938.

Mohamed, H. A., Yao, W., Fioravante, D., Smolen, P. D., and Byrne, J. H. (2005). cAMP-response elements in *Aplysia* creb1, creb2, and Ap-uch promoters: implications for feedback loops modulating long term memory. *J Biol Chem* **280**, 27035–27043.

Nicoll, R. A., Kauer, J. A., and Malenka, R. C. (1988). The curent excitement of long-term potentiation. *Neuron* **1**, 97–103.

Nicoll, R. A. and Malenka, R. C. (1995). Contrasting properties of two forms of long-term potentiation in the hippocampus. *Nature (Lond.)* **377**, 115–118.

Pastalkova, E., Serrano, P., Pinkhasova, D., Wallace, E., Fenton, A. A., and Sacktor, T. C. (2006). Storage of spatial information by the maintenance mechanism of LTP. *Science* **313**, 1141–1144.

Rachlin, H. (1991). "Introduction to Modern Behaviourism," 3rd ed. Freesman, New York.

Roberts, A. C. and Glanzman, D. L. (2003). Learning in *Aplysia*: looking at synaptic plasticity from both sides. *Trends in Neurosci.* **26**, 662–670.

Smolen, P. (2007). A model of late long-term potentiation simulates aspects of memory maintenance. *PLoS ONE* **2**, e445.

Sutton, M. A., Masters, S. E., Bagnall, M. W., and Carew, T. J. (2001). Molecular mechanisms underlying a unique intermediate phase of memory in *Aplysia. Neuron* **31**, 143–154.

Teyler, T. J., Cavus, I., Coussens, C., DiScenna, P., Grover, L., Lee, Y. P., and Little, Z. (1994). Multideterminant role of calcium in hippocampal synaptic plasticity. *Hippocampus* **4**(6), 623–634.

Tsien, J. Z., Huerta, P. T., and Tonegawa, S. (1996). The essential role of hippocampal CA1 NMDA receptor-dependent synaptic plasticity in spatial memory. *Cell* **87**, 1317–1326.

Wainwright, M. L., Byrne, J. H., and Cleary, L. J. (2004). Dissociation of morphological and physiological changes associated with long-term memory in *Aplysia. J. Neurophysiol.*, **92**, 2628–2632.

Wang, J. H., Ko, G. Y., and Kelly, P. T. (1997). Cellular and molecular basis of memory: Synaptic and neuronal plasticity. *J. Chin. NeuroPhysiol.* **14**, 264–293.

Wang, H., Hu, Y., and Tsien, J. Z. (2006). Molecular and systems mechanisms of memory consolidation and storage. *Prog. Neurobiol.* **79**, 123–135.

Whitlock, J. R., Heynen, A. J., Shuler, M. G., and Bear, M. F. (2006). Learning induces long-term potentiation in the hippocampus. *Science* **313**, 1093–1097.

Zhang, F., Endo, S., Cleary, L. J., Eskin, A., and Byrne, J. H. (1997). Role of transforming growth factor-β in long-term facilitation in *Aplysia. Science* **275**, 1318–1320.

Suggested Readings

Bear, M. F. and Malenka, R. C. (1994). Synaptic plasticity: LTP and LTD. *Curr. Opin. Neurobiol.* **4**, 389–399.

Byrne, J. H. (1987). Cellular analysis of associative learning. *Physiol. Rev.* **67**, 329–439.

Bliss, T. V. P. and Collingridge, G. L. (1993). A synaptic model of memory: Long-term potentiation in the hippocampus. *Nature (Lond.)* **361**, 31–39.

Carew, T. J. (2000). *Behavioral Neurobiology.* Sinauer, Sunderland, MA.

Dineley, K. T., Weeber, E. J., Atkins, C., Adams, J. P., Anderson, A. E., and Sweatt, J. D. (2001). Leitmotifs in the biochemistry of LTP induction: Amplification, integration and coordination. *J. Neurochem.* **77**, 961–971.

Ito, M. (2001). Cerebellar long-term depression: Characterization, signal transduction, and functional roles. *Physiol. Rev.* **81**(3), 1143–1195.

Malenka, R. C. and Nicoll, R. A. (1999). Long-term potentiation: A decade of progress? *Science* **285**, 1870–1874.

Martin, S. J., Grimwood, P. D., and Morris, R. G. (2000). Synaptic plasticity and memory: An evaluation of the hypothesis. *Annu. Rev. Neurosci.* **23**, 649–711.

John H. Byrne

Learning and Memory: Brain Systems

INTRODUCTION

The previous chapter summarized the cellular and molecular mechanisms of memory and explained that individual neurons contain complex machinery capable of altering membrane excitability and synaptic strength. These neuronal building blocks are the fundamental basis of plasticity, and this plasticity supports behavior through memory systems, complex circuits involving a large number of interconnected neurons. The present chapter focuses on the major memory systems of the brain and considers how each system supports learning. The chapter begins with a brief history of the concept of multiple forms of memory, followed by a summary of each of the major memory systems of the mammalian brain (Fig. 50.1). The final section considers how these systems might collaborate, compete, or operate in parallel to support learned behaviors.

HISTORY OF MEMORY SYSTEMS

The history of memory systems can be usefully framed by the early question of whether a specific memory could be localized or whether it would be distributed throughout the brain. Even early on there was evidence that the answer might lie somewhere between the extremes. In the 1920s, the psychologist Karl Lashley conducted a series of experiments in which he carefully damaged various areas in the cerebral cortex of rats who had learned a route through a simple maze. He found that the specific location of the brain damage did not relate to memory performance as well as the total amount of cortex damaged and concluded that memory was distributed through-

out the brain. Yet a short time later, the neurosurgeon Wilder Penfield found that when he stimulated various areas of the brains of awake epileptic patients (in order to identify functional areas deemed too important to be removed during surgery to treat the epilepsy), stimulation in some brain regions, particularly in the temporal lobe, led the patient to experience specific memories. He reasoned that, if focal stimulations bring specific memories to mind, then individual brain regions might contain individual memories. The psychologist Donald Hebb offered a reconciliation for these diverse findings by suggesting that thoughts and memories were supported by "cell assemblies," networks of neurons, and that learning experiences resulted in changes in the connections between cells (Hebb, 1949). Thus, to the extent that a change at an individual juncture between cells (i.e., a synapse) recorded experience, memory was localized. Yet to the extent that full memories resulted from a reactivation of an entire cell assembly, memory was distributed.

Although it was not appreciated at this time that different networks, or "cell assemblies," might support different forms of memory, several philosophers and psychologists had already suggested that an important difference existed between the type of everyday memory that one brings to mind in the form of words, pictures, or sounds and the type of memory that accrues with our actions as habits and dispositions. For example, in the early nineteenth century the philosopher Maine de Biran distinguished between *representative memory*, which he imagined as the recollection of ideas and events, *mechanical memory*, which he described as the acquisition of habits and skills, and *sensitive memory*, which he described as the acquisition of affective values for otherwise neutral stimuli. In the

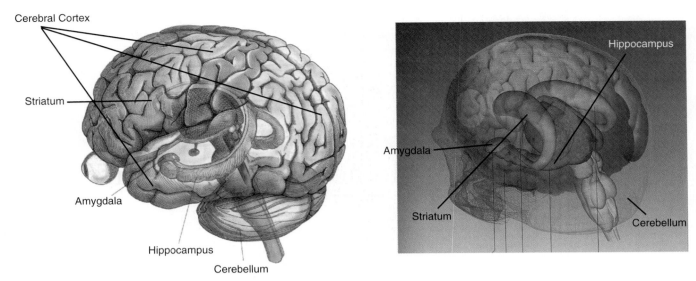

FIGURE 50.1 Drawing of the human brain showing components of the major memory systems.

late nineteenth century, the philosopher and psychologist William James similarly envisioned walking, writing, fencing, singing, and other habitual routines as being fundamentally different from memories produced by conscious recollection.

Thus, by the mid-twentieth century, the work of philosophers, psychologists, and neurosurgeons would have seemed to create an intellectual readiness for the idea of multiple memory systems in the brain. However, at this time the study of memory was guided not by the pursuit of brain systems but by behavioral psychology, a field based on the principle that the opaque details of brain and consciousness were best set aside as more productive research was accomplished by focusing on observable behavior. Despite this principled disregard of the brain, a distinction emerged between a learning based on stimulus-response habit-like learning, such as that described by Clark Hull, and a more cognitive stimulus-stimulus form of learning, such as that described by Edward Tolman. Nevertheless, both ideas were strongly entrenched in a behaviorist tradition and were based on the idea that memory was a single capacity—they disagreed on only how to best depict that capacity. Although behavioral psychology did not actually argue against the idea that particular brain areas might be specialized for particular kinds of memory, its steadfast insistence that the details of the brain be ignored created a vacuum of relevant data. Thus, despite the contributions of earlier work, there was no

clear evidence that brain areas are specialized to support different forms of memory.

Incontrovertible evidence that at least one brain region was specialized to support everyday memory came from a man who underwent brain surgery to treat severe epilepsy and who became perhaps the most famous patient in neuroscience (Scoville and Milner, 1957). This man, known as H.M., began experiencing seizures at the age of 10. It was unclear what caused the seizures, but it was very clear that something drastic had to be done. Heavy doses of anticonvulsant medications did not stop the seizures from worsening to a point at which they were debilitating. Acting on the idea that dysfunctional brain tissue might be causing the seizures, on September 1, 1953 the neurosurgeon William Scoville removed substantial tissue from the medial aspect of H.M.'s temporal lobes (Fig. 50.2).

In one sense, the surgery was a success. The frequency and severity of the seizures were reduced. In another sense, the outcome was tragic. On April 26, 1955, Brenda Milner, a colleague of both Penfield and Hebb, conducted a neuropsychological examination of patient H.M. The profound memory impairment was immediately obvious. H.M. gave the date as March, 1953 and his age as 27 (2 years younger than his actual age). He performed poorly on memory for short stories, word lists, pictures, and a wide range of other materials. Remarkably, it was unclear that he even remembered that he had undergone brain surgery.

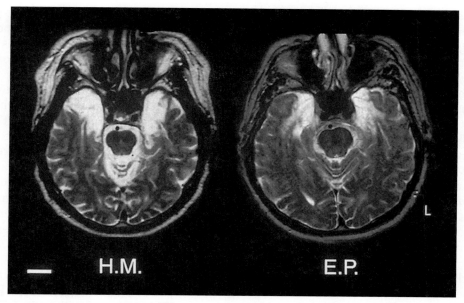

FIGURE 50.2   Magnetic resonance images showing the brains of amnesic patients H.M. and E.P. The images show axial sections through the medial temporal lobes and reveal damaged tissue as a bright signal. H.M.'s damage resulted from surgery, and E.P.'s damage was caused by viral encephalitis. Nevertheless, the resulting lesion was similar for the two patients. Both patients sustained extensive damage to the medial temporal lobes and are profoundly amnesic. From Stefanacci *et al.* (2000).

The severity of H.M.'s amnesia was shocking—he showed almost no capacity for new learning. Also, the fact that the brain damage was known to be confined to a particular region (the medial temporal lobes) added to the intrigue. Yet the observation that a major neurological deficit followed resection of a substantial amount of brain tissue was perhaps not surprising. Indeed, earlier reports already had described brain damaged individuals with memory impairments. Instead, the reason that H.M.'s case ushered in the modern era of research on memory systems came from the four aspects of his mental capacity that remained intact.

First, formal testing identified that cognitive abilities other than memory were intact. H.M.'s IQ score generally was unaffected by the surgery, and a battery of tests found no deficits in perception, abstract thinking, or reasoning ability. These intact abilities indicted that memory could be separated from perception and intelligence. Second, H.M. could hold on to small amounts of information as long as he was actively rehearsing the information. This finding suggested that the ability to maintain information online (now usually referred to as working memory) was distinct from the ability to make a lasting record in the brain (Chapter 51). Third, H.M.'s childhood memories were relatively intact. This finding suggested that, although the medial temporal lobes might be important for

forming new memories, this region was unlikely to be the final storage site for memory. Fourth, H.M. had an intact ability to acquire new motor skill learning (Milner, 1962). Over several days of practice, H.M. gradually improved at tracing the outline of a star when viewing the paper only through a mirror (a task that initially is challenging even for healthy individuals) despite never forming a memory for the testing experience. Thus, the contrast between H.M.'s profound memory loss and intact cognitive abilities outside the ability to form new memory set the stage for the work that has led to our current understanding of memory systems in the brain. H.M.'s case showed that memory is a dissociable cognitive capacity and that day to day memory was supported by brain structures that differed from those that supported the acquisition of motor skills.

H.M. has participated in numerous studies in the decades since his surgery and has provided invaluable insights into the organization of memory systems in the brain. However, studies of other amnesic patients have also been crucial to the emergence of the current understanding of memory and the brain. For example, a later study by Neal Cohen and Larry Squire (1980) suggested that the domain of preserved memory in amnesia was much larger than just motor skill learning. They asked a group of amnesic patients to practice reading words that had been reversed to appear as if they were being viewed in a mirror. Similar to H.M.'s improvement in mirror tracing, the group of amnesic patients steadily improved in the speed with which they could read the mirror-reversed words despite having great trouble remembering the practice sessions. The study helped to show that a great deal of learning can occur outside the scope of conscious recollection and that this nonconscious learning is not limited to motor skill learning. Cohen and Squire borrowed terminology from work in artificial intelligence (Winograd, 1975) and suggested that amnesia was characterized by an impairment in *declarative memory* and that *procedural memory* was left intact.

Another amnesic patient, patient R.B., helped to show that damage limited to the hippocampus could result in substantial memory impairment (Zola-Morgan *et al.*, 1986). Like H.M., R.B. performed poorly on a range of declarative memory tests, although his deficit was not nearly as severe as H.M.'s. Detailed histological analysis of R.B.'s brain indicated that his memory impairment resulted from damage to a specific subregion (region CA1) in his hippocampus. Thus, patient R.B.'s case showed that damage limited to the hippocampus was sufficient to produce amnesia.

Soon after the report on patient H.M., researchers began work in experimental animals, such as rats and monkeys, in an attempt to create a model of human amnesia. Initially, the separate lines of investigation on humans and animals were divergent and suggested to some that the functions of the hippocampus might differ between species. However, subsequent findings from these two areas converged, both in their characterizations of the kind of memory that is dependent on the hippocampal region and in their identification of functional domains and anatomical pathways associated with other types of memory. In early studies, this line of work focused particularly on the hippocampus. In one proposal, O'Keefe and Nadel (1978) summarized a large body of literature on the effects of hippocampal damage on different behavioral tasks and concluded that animals with damage to the hippocampus are severely impaired at many forms of spatial learning. It is now generally believed that the hippocampus is important for nonspatial as well as spatial memory, yet this early proposal added to ongoing work in humans and monkeys by identifying several key properties of hippocampus-dependent memory in experimental animals. In particular, they characterized this kind of memory as rapidly acquired and driven by curiosity rather than by rewards and punishments, properties that are consistent with characterizations of hippocampus-dependent memory in humans.

Our understanding of memory systems has advanced greatly since H.M.'s surgery and it is still

unfolding. The study of amnesia has provided insights not only about the role of the hippocampal memory system in declarative memory but also about the ability of other memory systems to support learning in the absence of conscious recollection. A large body of work, including research on humans and on experimental animals, has allowed researchers to understand better the unique contributions of the striatum, cerebellum, amygdala, and cerebral cortex to different forms of memory. Thus, we now know that the hippocampus anchors only one of several memory systems in the brain and that each of these brain structures are components of other memory systems that contribute to our ability to benefit from experience. The following sections provide an overview of these systems.

MAJOR MEMORY SYSTEMS OF THE MAMMALIAN BRAIN

Overview

The following sections consider the best understood memory systems of the brain. The discussion focuses on individual structures that play central roles in each of the different systems and presents separate sections on the hippocampus, striatum, cerebellum, amygdala, and cerebral cortex (Fig. 50.1). However, each of these regions anchors a much larger network of distributed brain areas that work in concert to support particular forms of memory. For example, the hippocampal memory system includes not only the hippocampus but also areas in the adjacent parahippocampal region as well as many distributed unimodal and polymodal neocortical areas. These areas normally work together to support conscious recollection of facts and events, a capacity termed declarative memory. Even though declarative memory depends on very specialized information processing and plasticity within the hippocampus, a complicated and distributed circuitry outside the hippocampus is also required for these alterations to emerge as the experience of remembering. Most importantly, the cerebral cortex plays a role in many types of memory, and the present chapter will discuss both its specific role as a part of multiple memory systems and the memory functions it supports outside of those systems.

A memory system is most usefully defined in psychological terms as well as anatomical terms. That is, a memory system is most clearly identified by both a distinct circuitry and a unique set of operating characteristics. Memory systems are not separated according to stimulus modality (e.g., auditory vs. visual modality) or between response modalities (e.g., manual vs.

verbal responses). Instead, the critical distinctions involve how each system's anatomy and physiology support a particular type of memory representation in terms of its organization and psychological characteristics. Finally, memory systems collaborate, compete, or operate in parallel to support behavior in the course of one's day-to-day activities, and this issue will be considered in the final section of the chapter.

Hippocampus

Anatomy

The hippocampal memory system includes the hippocampus (defined here as the CA fields, dentate gyrus, and subiculum) and the entorhinal, perirhinal, and postrhinal cortices in the adjacent parahippocampal region (the postrhinal cortex is referred to as the parahippocampal cortex in primates; Fig. 50.3; Burwell *et al.*, 1995; Suzuki, 1996). The anatomy and circuitry of these regions, especially the hippocampus, are largely conserved across mammalian species (Manns and Eichenbaum, 2007). The hippocampal memory system also depends on diverse and widespread higher order regions in the cortex that are both the source of information to the parahippocampal region and hippocampus and the targets of projections originating from these regions. The parahippocampal region serves as a convergence site for input from these cortical association areas and mediates the distribution of cortical afferents to the hippocampus. These parahippocampal cortical areas are interconnected and send major efferents to multiple subdivisions of the hippocampus. Within the hippocampus, an intricate pattern of connectivity mediates a large network of associations (Amaral and Witter, 2004), and these connections support forms of long term potentiation that could participate in the rapid coding of novel conjunctions of information (Chapter 49). The outcomes of hippocampal processing are directed back to the adjacent cortical areas in the parahippocampal region, and the outputs of that region are directed in turn back to the same areas of the cerebral cortex that were the source of its inputs. Additional structures also have been included as components of this system, including midline diencephalic structures.

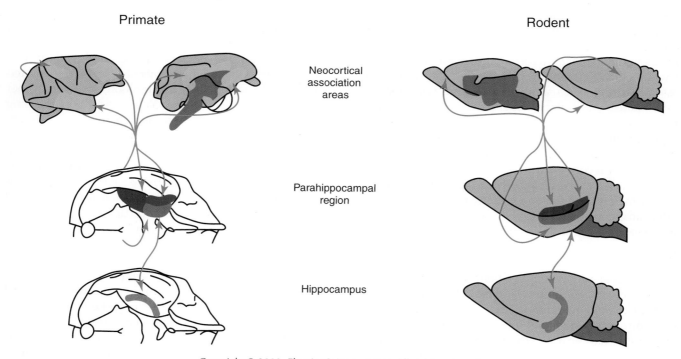

Primate   Rodent

Neocortical association areas

Parahippocampal region

Hippocampus

FIGURE 50.3 The anatomy of the hippocampal memory system in monkeys and rats. The hippocampal memory system includes the hippocampus (CA fields, dentate gyrus, and subiculum) and the parahippocampal region, which includes the entorhinal cortex, the perirhinal cortex, and the parahippocampal cortex. The hippocampal memory system also includes midline diencephalic nuclei. Multiple association areas in the cerebral cortex send outputs that converge on cortical areas in the parahippocampal region, which in turn sends its outputs to the hippocampus. The output path involves return projections from the hippocampus to the surrounding parahippocampal region, which in turn projects back to the same cortical association areas. From Eichenbaum (2001).

Amnesia and Declarative Memory

Damage to the hippocampus and parahippocampal region produces anterograde amnesia, a memory deficit characterized by an inability to make lasting memories of one's daily experiences. That is, the hippocampal memory system ordinarily supports remembering of new facts and events and makes this information subsequently available for conscious recollection. This capacity is called declarative memory. In most amnesic patients, the deficit in declarative memory extends to both remembering specific personal events (*episodic memory*) and learning new facts (*semantic memory*). Their deficit encompasses all stimulus modalities and impacts nonverbal expression as well as verbal report. Moreover, amnesic patients are impaired whenever the memory task requires the explicit expression of memory, as in tests of free recall or recognition memory. Further, damage to the hippocampus and parahippocampal region results in loss of memory acquired for some period before the damage. This memory impairment is referred to as retrograde amnesia.

Amnesia resulting from damage restricted to the hippocampus and parahippocampal region is highly selective in four important ways, the same four ways discussed earlier with regard to the intact abilities of patient H.M. First, perceptual, motor, and intellectual functions are intact. Across a broad range of neuropsychological tests, amnesic patients perform well on assessments of sensory capacities, motor coordination, intelligence, and language performance. Second, memory acquired long before the onset of amnesia is typically intact. The preservation of these remotely acquired memories generally is thought to be the consequence of a consolidation process in which the hippocampus plays a critical role for a prolonged period, after which retrieval of memories can be supported by other areas, including the neocortex. Thus, the impact of retrograde amnesia is typically time limited. The process of consolidation is discussed in more detail later. Third, the capacity for immediate memory is typically intact in amnesic patients, and, just as in the case for healthy individuals, they can extend the contents of immediate memory in time through rehearsal or elaboration, a capacity often referred to as working memory (Chapter 52). Amnesic patients can immediately reproduce a list of six or seven numbers as well as healthy individuals, but the memory deficit becomes evident as soon as immediate memory span is exceeded or after a delay is interposed that includes some distraction to interrupt rehearsal.

Fourth, the various forms of memory that are supported by brain systems outside the hippocampal memory system are intact in amnesic patients. For example, amnesic patients demonstrate normal acquisition of a broad variety of tasks in which memory is expressed through behavioral performance such as in the mirror-guided tracing task and the mirror-reversed word reading task described earlier in the chapter. Other examples include artificial grammar learning and improving reaction times on a repetitive sequence of finger taps. One profoundly amnesic patient, patient E.P., has a memory impairment and brain damage very similar to that of patient H.M. (Fig. 50.2). Patient E.P. has shown normal performance on a broad variety of nonconscious memory tasks, despite the fact that his nearly complete anterograde amnesia has kept him from forming conscious memories of the testing situations themselves. In some of these examples, the learning was simple and reflexive, such as in the case of standard eyeblink classical conditioning (discussed in more detail later in the chapter in the section on the cerebellum).

In other examples, performance on more complicated tasks came to resemble that of healthy individuals despite the fact that the learning was supported by a type of nonconscious and relatively inflexible type of learning. For example, in one study E.P. was shown eight pairs of objects one at a time, each pair containing a correct object and an incorrect object. He slowly learned to pick up the correct object in each pair and after 18 weeks was able to select the correct object almost every time (Bayley *et al.*, 2005). When he was asked how he knew which object to pick up he pointed to his head and replied, "It's here somehow or another and the hand goes for it." Yet, in addition to being unavailable to awareness and being very slowly acquired, E.P.'s memory for the objects also appeared to lack the flexibility of that shown by healthy individuals, who had learned the task after only three sessions. When the task was changed by presenting all 16 objects together and by asking E.P. to sort them into correct and incorrect piles, his performance deteriorated precipitously. In contrast, healthy individuals performed nearly perfectly on the sorting task. In another study, over the course of 12 weeks E.P. came to learn a list of 60 three-word nonsense sentences (e.g., "speech caused laughter"), such that he eventually was able to respond correctly some of the time when asked to fill in the missing third word (e.g., "speech caused ???"; Bayley and Squire, 2002). Yet E.P. continually expressed surprise when informed that he had seen the sentences previously. In addition, his memory also differed from control subjects in that it appeared to be somewhat rigid. When the middle word of each sentence was replaced by a synonym, E.P. was able to answer correctly for only one sentence. His performance contrasted with that of healthy individuals, who learned the sentences after only two weeks and who had no trouble adapting to the new testing format.

These studies with patient E.P. illustrate that although some examples of memory supported by structures outside the hippocampal memory system can be simple and reflexive, other examples can be more complex and can include acquisition of new verbal information and biases for one of the response choices that is readily available. These studies also highlight by contrast the characteristics of memory supported by the hippocampal memory system. Memories formed via the hippocampal system typically are acquired rapidly, flexible, and available to conscious recollection.

Functional Brain Imaging and the Hippocampal Memory System

Functional brain imaging studies in humans, including functional magnetic resonance imaging (fMRI) and positron emission tomography (PET), have shown that the hippocampal memory system is engaged in a variety of declarative memory tasks. For example, in one of the earliest studies, increased activity was observed in the hippocampus and parahippocampal region when participants viewed previously unseen photographs as compared to when participants viewed photographs that they had already seen on multiple occasions (Stern et al., 1996). Subsequent studies have also found that activity in the hippocampal memory system often is related to the success of memory encoding or retrieval. For example, in one study, participants saw a list of words one at a time while being scanned (Wagner et al., 1998). After the scanning session, participants were given a test to assess which words were remembered and which words were forgotten. The results indicated that activity in the hippocampal memory system during encoding (in this case, in the parahippocampal region) was greater for words that were subsequently remembered as compared to words that were forgotten. Many subsequent functional brain imaging studies have also identified memory-related activity in the hippocampus itself. Activity in the hippocampus is particularly strong in tasks that demand memory for associations that compose specific experiences (Eichenbaum et al., 2007). Moreover, functional brain imaging has become a key tool for understanding how the hippocampal memory system, as well as other memory systems, ordinarily operate in the intact human brain.

Models of Anterograde Amnesia in Experimental Animals

Neuroscientists have used a specific, carefully selected set of behavioral tests in developing a nonhu-

man primate model of human amnesia to identify the particular medial temporal lobe structures that support declarative memory. The tasks involve learning about three-dimensional objects or complex pictures. In one task, delayed nonmatching to sample, subjects are shown an object once and then, after a delay, are shown two objects (the original object and a new one). The task of the monkeys is to select the new object (Fig. 50.4A). When the delay is only a few seconds, monkeys with experimental lesions that include the same medial temporal lobe structures damaged in H.M. (including the hippocampus and adjacent cortices) performed as well as normal monkeys (Mishkin, 1978). As the delay was increased, the monkeys became progressively more impaired. Thus, monkeys with medial temporal damage, like humans with amnesia, have intact immediate memory but perform poorly when the memory demand increases over time. Monkeys with medial temporal lobe damage also perform poorly when they must retain rapidly acquired object discriminations for a prolonged period. In contrast to these impairments in object memory, monkeys with medial temporal lobe damage have intact capacities for skill acquisition as measured in a task that involves learning to retrieve a candy by manipulating it along a bent rod. Thus, the pattern of both preserved and impaired memory in human amnesic patients is closely modeled by the performance of monkeys with similar brain damage (Zola-Morgan and Squire, 1985).

Using this animal model, investigators were able to identify the structures of the medial temporal lobe critical to supporting declarative memory. In H.M. the damage included the amygdala, the hippocampus, and the surrounding parahippocampal region. However, studies with monkeys have shown that the amygdala is not a part of the declarative memory system. In addition, the severity of memory impairment depends on the extent and locus of damage within the medial temporal lobe. Damage limited to the hippocampus, or to its major connections through the fornix, produces only a modest impairment. In contrast, damage that includes the adjacent cortices produces severe amnesia. Thus, the perirhinal and parahippocampal cortical regions themselves make major contributions to memory, and the hippocampus itself is a critical component of the system. Particularly compelling evidence about the role of the hippocampus in recognition memory has come from studies using a test known as the visual paired-comparison task (Zola et al., 2000; Fig. 50.4B). In this task, the monkey initially is shown a pair of duplicate pictures. Then, following a variable delay, two pictures again are shown. One picture is identical to the initial sample and the other is a novel picture. Memory for the sample

A

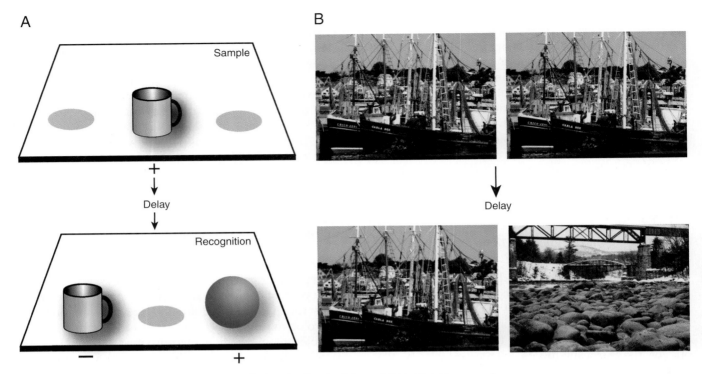

FIGURE 50.4 Recognition memory tasks used for studies of memory in nonhuman primates. Both tasks are easily adapted for human participants as well. (A) The delayed nonmatching-to-sample task using unique objects as stimuli. The subject initially is presented with a single novel object as the sample and must displace the object. This is followed by a variable delay during which the subject cannot see any objects. In the subsequent recognition test, two objects are presented, one of which is the same as the sample and the other of which is novel. Correct performance requires the subject to recognize and avoid the sample object and instead choose the novel one to receive a food reward. (B) Visual paired comparison task. During the sample phase the monkey looks at two identical pictures. In the test phase, one of the sample pictures is represented along with a novel picture. Memory for the repeated picture is inferred by measuring the subject's tendency to look away from the repeated picture and toward the new picture.

picture is measured in normal monkeys (and humans) by a lesser amount of time looking at the familiar picture than the novel one. On this test, damage limited to the hippocampus results in a rapid loss of memory such that, at a brief interval after presentation of the sample picture memory is intact, but after around 10 seconds memory is severely impaired. The overall pattern of findings supports the view that both the hippocampus as well as the adjacent cortical areas support performance on these and other tests of recognition memory.

Studies with rodents have also offered insights into the nature of memory representations in networks of hippocampal neurons. Many studies in rats have demonstrated that damage to the hippocampus results in deficits in a variety of spatial learning and memory tasks. A particularly useful example is place learning in the Morris water maze task (Morris *et al.*, 1982; Fig. 50.5). In this task, rats are trained to find a hidden escape platform submerged just below the surface in

a pool of cloudy water. Because there is no specific cue at the escape site, the rat must learn the location of the platform on the basis of spatial relationships among the cues that are visible in the room. Rats with hippocampal damage are severely impaired at this task. Additional evidence for the importance of the hippocampus in spatial memory comes from recordings of the firing patterns of hippocampal neurons in behaving rats. A major early finding in studies of hippocampal cells was that many of these neurons fire when the rat is in a particular location in its environment. This activity was shown to reflect an encoding of the spatial relationships among physical stimuli in the environment, leading experimenters to call these neurons "place cells" (O'Keefe, 1976).

Although these data indicate an essential role for the hippocampus in spatial learning, all forms of spatial learning are not dependent on the hippocampus. For example, rats with hippocampal damage can learn simple spatial discriminations, such as whether

During learning

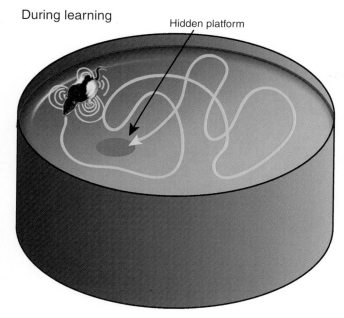

Hidden platform

After learning

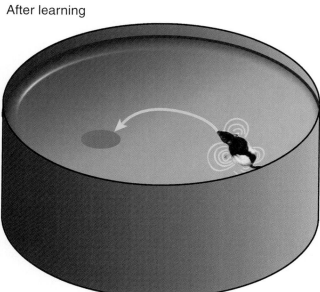

FIGURE 50.5 The Morris water maze task. Early in training, rats search for the submerged platform for extended periods. After training the rat swims directly to the platform.

to turn left rather than right in a T-shaped maze. However they are impaired if they are forced initially to visit either the left or the right arm of the maze, and then asked to remember that experience and choose to visit the opposite arm. Rats with hippocampal damage can also learn to locate the escape platform in the Morris water maze when they are trained to find the platform from a single starting point. However, they

cannot learn if trials from different starting points are intermixed. These studies show that rats with hippocampal damage can sometimes use spatial memory to guide performance, but they cannot organize spatial information gained from different episodes to express memory flexibly according to the demands on a particular trial.

Other experimental evidence indicates that the hippocampus is also critically involved in the organization and flexible expression of nonspatial memories. For example, in one experiment intact rats and rats with damage to the hippocampus were trained to dig through sand mixed with one of a list of odors to obtain a cereal reward (Bunsey and Eichenbaum, 1996; Fig. 50.6). Initially, both healthy rats and rats with hippocampal lesions learned a set of simple associations between pairs of odors. For example, rats learned to dig in odor B when presented with odor A and to dig in odor C when presented with odor B. Subsequently, the rats were given probe tests to determine the extent to which learned representations supported two forms of flexible memory expression. One of these tests, a test for transitivity, measured the ability to infer an association between two odors that shared a common associate. For example, having learned that odor A is associated with odor B and that odor B is associated with odor C, could they infer that A is associated indirectly with C?

The other test, a test for symmetry, measured the ability to recognize associated odors when they were presented in the reverse of their training order. For example, if B is associated with C, is C associated with B? Intact rats showed strong transitivity and successfully inferred an association between A and C that had been learned only indirectly. In contrast, rats with selective hippocampal damage were severely impaired in that they showed no evidence of transitivity. In the symmetry test, healthy rats showed their associations were indeed symmetrical. In contrast, rats with hippocampal damage again were severely impaired, showing no detectable capacity for symmetry. Correspondingly, in humans the hippocampus is activated during retrieval of indirect associations during the same transitive inference judgment (Preston et al., 2004; Fig. 50.7). These findings show that the role of the hippocampus in rats extends beyond spatial memory and suggest that the hippocampus encodes and retrieves nonspatial as well as spatial memories in both animals and humans.

Consistent with this view, numerous recording studies in rats, monkeys, and humans have shown that hippocampal cells also fire in association with visual, auditory, and olfactory stimuli, as well as combinations of these stimuli and the places where they occur,

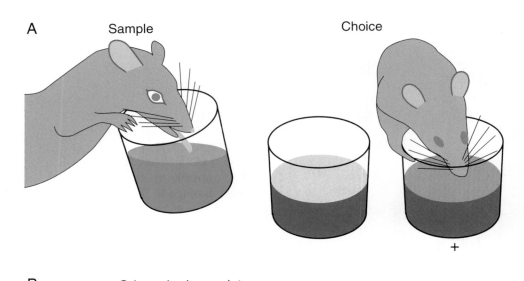

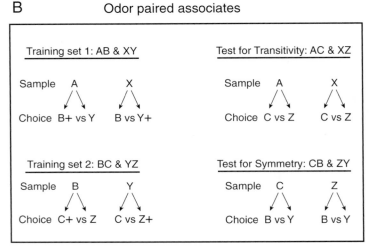

FIGURE 50.6 Associative transitivity and symmetry in paired associate learning. (A) On each training trial, one of two odors is presented as the sample. On the subsequent choice trial the animal must select the assigned associate, indicated by a "+". (B) Outline of odor pairings used in training on two sequential sets of paired associates, plus stimuli used in tests for transitivity (C for A; Z for X) and symmetry (B for C; Y for Z). From Bunsey and Eichenbaum (1996).

suggesting that hippocampal neurons represent important stimuli in the context in which they are remembered (Ekstrom *et al.*, 2003; Wirth *et al.*, 2003; Wood *et al.*, 1999). For example, in one experiment, rats performed a variant of the delayed nonmatching to sample task that was guided by olfactory cues that were presented at several locations in an open field (Fig. 50.8A; Wood *et al.*, 1999). The activity of some hippocampal cells correlated with spatial features of the task (i.e., the rat's location when it encountered odors), and the activity of other cells correlated with nonspatial features of the task (e.g., the identity of odors). However, a substantial number of hippocampal cells fired only when the animal encountered a particular combina-

tion of task features. For example, some cells fired when the rat sniffed a particular odor at a particular place but did so only when the odor represented the correct (i.e., nonmatch) choice. Thus, the hippocampus appeared to represent not only the locations that were common across trials but also the unique combinations of nonspatial and spatial features that may have helped the rat perform the task.

Observations also suggest that hippocampal neurons can represent sequences of events and places that compose episodic memories. For example, evidence of episodic-like coding was found in a study when rats performed a spatial alternation task on a T maze (Wood *et al.*, 2000; Fig. 50.9). Each trial began when the rat

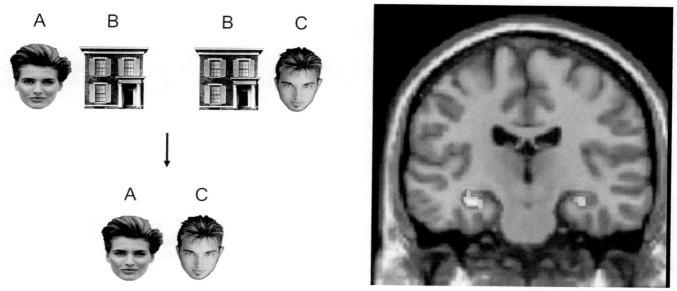

FIGURE 50.7 fMRI activity in the human hippocampus during judgments of associative transitivity. Bilateral regions in the hippocampus (indicated by yellow areas) were more active when participants used memory of face-house pairs (A-B and B-C) to infer face-face pairings (A-C) as compared to when participants simply memorized face-face pairings. Adapted from Preston *et al.* (2004).

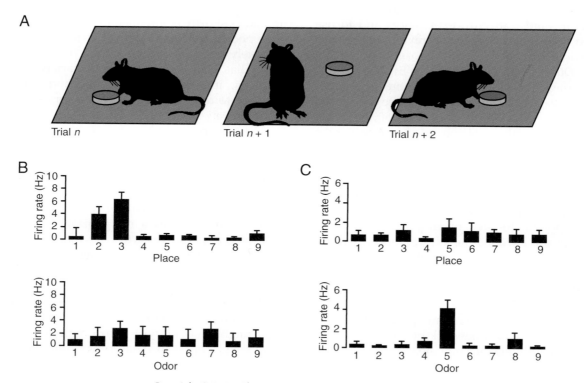

FIGURE 50.8 Hippocampal neuronal firing patterns. (A) A rat performing the delayed nonmatching to sample task with odorized cups as stimuli. The three panels indicate a sequence of trials that vary the position of the odor and whether the presented odor matches the odor used on the previous trial. (B) Example of a cell that fires when trials are performed at two adjacent places (2 and 3), regardless of the odor presented. (C) Example of a cell that does not fire differentially in association with trials at different locations, but fires selectively on trials when odor 5 is presented. From Wood *et al.* (1999).

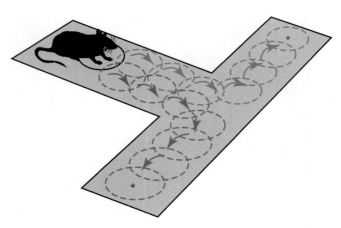

FIGURE 50.9 Selective firing during different types of memory episodes in rats performing a T maze alternation task. On left turn trials, different cells fire as the rat runs forward through a series of locations on the maze. On right turn trials, a different set of cells fire, even when the overt behavior and places are the same. From Eichenbaum (2000).

traversed the stem of the "T" and then selected either the left- or the right-choice arm. To alternate successfully, the rats were required to distinguish between their left-turn and right-turn experiences and to use the memory for their most recent previous experience so they could turn in the opposite direction on the next trial. Different hippocampal cells fired as the rats passed through the sequence of locations within the maze during each trial. These cells could be considered canonical place cells. However, in addition, the firing patterns of many of the cells depended on whether the rat was in the midst of a left- or right-turn episode, even when the rat was in the stem of the T and running similarly on both types of trials. Although the majority of the cells strongly preferred one trial type, most of the cells fired at least to some extent when the rat was at the same point in the stem on either trial type. Thus, the hippocampus encoded the left-turn and right-turn experiences using distinct representations, and these representations included information that could link them by their common features. These results suggest that the representations of event sequences, linked by codings of their common events and places, could constitute the substrate of a network of episodic memories, consistent with the idea that the hippocampus has a broad role in declarative memory.

Retrograde Amnesia and Consolidation

Patients with amnesia due to medial temporal lobe damage suffer not only a deficit in learning new material (anterograde amnesia), but also loss of memories that were acquired before the brain damage (retro-

grade amnesia). Importantly, the retrograde deficit is time-limited, and material acquired shortly before the damage is affected most severely, whereas items learned earlier in life are relatively spared. For example, the profoundly amnesic patient E.P. (Fig. 50.2) has almost no capacity for new declarative learning but shows good memory from his childhood. In one study, E.P. was asked to think back to his childhood neighborhood and describe routes he would take from one location to another (e.g., grade school to the town theater; Teng and Squire, 1999). In some instances, he was asked to imagine that a main street was blocked and to describe an alternate route. E.P. performed as well as a group of healthy individuals who had lived in the neighborhood at the same time but who had moved from the area long before the study, indicating that his memory for the spatial layout of his childhood neighborhood had escaped the damage to his medial temporal lobes. By contrast, E.P. has been unable to learn the layout of the neighborhood he moved to subsequent to the onset of amnesia.

Retrograde amnesia is another aspect of the amnesic syndrome that has been studied extensively in experimental animals. For example, studies have also shown that simple object discriminations that were learned by monkeys shortly before medial temporal lobe damage are poorly retained, but discriminations learned remotely are spared. This pattern of memory impairment has been replicated in a variety of tasks in humans, monkeys, rabbits, rats, and mice and is thought to reflect a process of memory consolidation (Squire et al., 2001).

The term consolidation has been used to characterize two kinds of brain events that affect the stability of memory after learning. One event involves the fixation of plasticity within synapses over a period of minutes or hours through a sequence of protein synthesis and morphological changes at synapses (Chapter 49). The other event involves a reorganization of memories, which occurs over weeks to years following new learning. This prolonged consolidation occurs in the hippocampal memory system and is thought to involve interactions between the hippocampus, parahippocampal region, and the cerebral cortex. Several models have been proposed to account for how the hippocampus might interact with other brain regions over a prolonged period in memory consolidation (Alvarez and Squire, 1995; McClelland et al., 1995). These models assume that widespread areas of the neocortex contain the details of the information that is to be remembered and that areas in the hippocampal memory system support the capacity to retrieve the memory during the period shortly after learning. Over time, areas in the hippocampus are thought to reactivate the cortical representations through repetition, rehearsal, or spontaneous activity. The reactivation is then thought to induce plasticity in cortical–cortical connections, and these connections are viewed as a possible network structure for the permanent storage and organization of the memory.

Recent studies using molecular markers of neuronal activity and synaptic change have reported evidence consistent with this idea. For example, one study of maze learning in mice observed that these markers were found in high numbers in the hippocampus one day after learning but their numbers were much lower 30 days later (Maviel et al., 2004). In contrast, the same markers were found to be at low levels in the cerebral cortex initially after learning but were observed to be high in various cortical areas 30 days after learning, areas including prefrontal, anterior cingulate, and retrosplenial cortices. These results suggested that plasticity in the hippocampus was initially important for the spatial memories but that over time plasticity in neocortical areas became more important.

There has long been interest in the role of sleep and dreaming in memory, and one idea has been that processes might occur during sleep that promotes the consolidation of hippocampus-dependent memories. Although the benefit of sleep to nonconscious examples of learning is generally accepted, the status of hippocampus-dependent memory is less certain (Stickgold, 2005). Nevertheless, there are several results that indicate that processes during sleep might contribute to consolidation of hippocampus-dependent memory. For example, specific patterns of activity observed in the hippocampus while rats were awake were observed to "replay" during both REM sleep and slow wave sleep (Sutherland and McNaughton, 2000). These results suggest that reactivation of memories during sleep may participate in the process of consolidation. In another study, induction of slow-wave like oscillations in the brains of human participants while they slept led to slightly improved recall upon waking (Marshall et al., 2006). Thus, there is accumulating evidence regarding the role of sleep in the consolidation of hippocampus-dependent memory, but further work is needed to fully understand this process.

Summary of Hippocampus

Amnesia associated with damage to the hippocampal memory system in humans is characterized by an inability to retain and consciously recollect memories of facts and events. Studies with animal models of amnesia memory provide a framework for thinking about hippocampal function in terms of memories that are represented as sequences of events remembered in the context in which they occurred and that are organized according to relations among distinct experiences. Furthermore, hippocampus-dependent memory is accessible through a variety of routes and forms of behavioral expression and supports the capacity to make generalizations and inferences from memory.

Striatum

Anatomy

The striatum is a major component of the basal ganglia and is involved with reward and motivation (Chapter 43) as well as with motor control (Chapter 31). The focus of the present section is the role of the striatum, particularly the dorsal striatum, in memory. The striatum receives its cortical inputs from many areas of the cerebral cortex (Fig. 50.10). The projections are organized topographically such that divergent and convergent projections into modules within the striatum might sort and associate somatosensory and motor representations. The striatum projects to other components of the basal ganglia and to the thalamus, which project back to both the premotor and motor cortex and the prefrontal association cortex. Notably, there

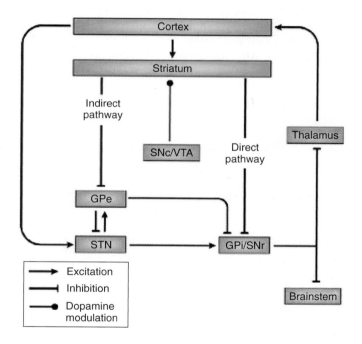

FIGURE 50.10 Anatomy of the striatum and the rest of the basal ganglia. STN, subthalamic nucleus; GPe, external globus pallidus; GPi, internal globus pallidus; SNr, substantia nigra pars reticulata; SNc, substantia nigra pars compacta; VTA, ventral tegmental area. From Yin and Knowlton (2006).

are minimal projections from this system to the brain stem motor nuclei and none to the spinal motor apparatus, which suggests the system operates mainly to modify cortical motor representations rather than control behavior through direct motor outputs.

Memory for Habits, Actions, and Outcomes

The idea that the striatum is critical for many forms of stimulus-response habit learning was introduced in studies in experimental animals that dissociated this system from other memory systems, such as the hippocampal memory system. For example, in one study the role of the striatum in learning specific behavioral responses was demonstrated using a simple T maze apparatus where two possible strategies in solving the maze were compared directly (Packard and McGaugh, 1996; Fig. 50.11). In this study, rats began each trial at the base of the T maze and were rewarded with food at the end of one choice arm (e.g., left). Accordingly, the rats could have acquired the task by learning to make a specific turning response (a "response" strategy, left in this example). Alternatively, rats could have remembered where the reward was located relative to the surrounding stimuli in the test room, independent of any particular behavioral response required to obtain it (a "place" strategy). The critical test to distinguish these two strategies was to rotate the maze by 180° in such a way that the start point was at the opposite end of the room. After training for a week on the T maze task, most rats used a place strategy on the probe test. However, when they were trained for another week with the maze in its original orientation and were then presented with an additional probe trial, most rats adopted a response strategy. These results suggested that under these training circumstances, initial acquisition of the task was guided by memory of the reward locus, but subsequent overtraining led to development of a habitual turning response.

Packard and McGaugh (1996) also examined whether different brain systems supported these different strategies. All animals had been implanted with indwelling cannulae that allowed injection of a local anesthetic or saline directly and locally into one of two brain structures, the hippocampus or the striatum, prior to the probe tests. The effects of the anesthetic were striking. On the first probe trial, when the striatum was inactivated, animals behaved just as controls. That is, they were predominantly place learners and the place strategy did not depend on the striatum. In contrast, when the hippocampus was inactivated, the animals showed no preference on the first probe trial, indicating that the hippocampus was essential for the place strategy and that this was the only strategy available in the early stage of learning. On the second probe test conducted after additional training, control subjects had acquired the response strategy, and inactivation of the hippocampus had no effect. In contrast, animals with the striatum inactivated lost the turning

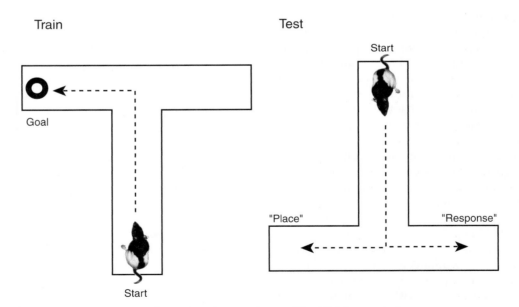

FIGURE 50.11 "Place" versus "response" learning. The rat is trained initially to turn left in order to obtain a reward at a particular location. In a subsequent test, the maze is rotated and the rat is allowed to select whether it will perform the same left turning "response" or remember the "place" of the previous reward.

response and instead showed that they were relying on the place strategy. The results indicated that animals normally develop an initial place strategy that depends on the hippocampus and do so prior to acquiring a habit-like response strategy. With overtraining, a response habit mediated by the striatum is acquired, and it predominates over the hippocampal place strategy. Nevertheless, the place memory can be "uncovered" by inactivating the striatum and suppressing the turning response strategy. These results suggested that there are distinct types of memory that guide place and response strategies, that the response strategy tends to be acquired more slowly than place memory, and that the response strategy is supported by the striatum.

More recent research has suggested that the memory supported by the striatum is not necessarily acquired slowly. For example, one study that recorded from neurons in the striatum and prefrontal cortex in monkeys found that the striatal neurons led prefrontal neurons in tracking the reversal of action-outcome contingencies (Pasupathy and Miller, 2005). On each day of testing, monkeys were trained to move their eyes left after seeing one novel picture and to move their eyes right after seeing a second novel picture. Monkeys were rewarded with juice with each correct response. After the monkeys learned these associations, the picture-direction associations were reversed (e.g., the first picture then indicated that right was the appropriate response). During the trials when the monkeys were learning to reverse their responses to the pictures, neuronal activity related to viewing the pictures in both the striatum and prefrontal cortex mirrored the monkeys' improvement in performance. However, the striatal activity led the prefrontal activity in two ways. First, on each trial the onset of activity in the striatal neurons tended to occur before the onset of the prefrontal neurons. Second, a relationship between the neuronal activity while viewing each picture and the subsequent correct eye movement emerged in fewer trials in the striatum as compared to the prefrontal cortex. The observation that activity in the striatum in this task reflected reversal of a learned association more quickly than the prefrontal cortex suggests that memory supported by the striatum may not necessarily be particularly slow and may be used to instruct the prefrontal cortex in mediating complex stimulus-response associations.

Research with human participants has also indicated that the striatal system may play a critical role in memory even for more complex forms of stimulus-response learning. For example, Knowlton and colleagues (1996) tested patients in the early stages of Parkinson's disease on a probabilistic classification learning task formatted as a weather prediction game (Fig. 50.12). Parkinson's disease is associated with the degeneration of neurons in the substantia nigra and a resulting major loss of input to the striatum. Amnesic patients were also tested, and their results contrasted with that of the Parkinson's patients. The task involved predicting on each trial one of two possible outcomes ("rain" or "sunshine") based on cues that were presented to the subject. On each trial, one, two, or three cards from a deck of four were presented. Each card was associated with the sunshine outcome independently and probabilistically, 75%, 57%, 43%, or 25% of the time, and the outcome with multiple cards was determined by the conjoint probabilities. After the cards were presented on each trial, the subject was asked to choose between rain and shine and was then given feedback (correct or incorrect). The probabilistic nature of the task made it difficult for subjects to improve by recalling specific previous trials because the same configuration of cues could lead to different outcomes. The most useful information was the probability associated with particular cues and combinations of cues, which could be acquired gradually across trials much as habits and skills are acquired. Over a block of 50 trials, normal subjects gradually improved from pure guessing (50% correct) to about 70% correct. However, patients with Parkinson's disease failed to show significant learning, and the failure was particularly evident in those patients with more severe parkinsonian symptoms. In contrast, amnesic patients were successful in learning the task, achieving levels of accuracy similar to the controls by the end of the 50-trial block.

Subsequent to training on the weather prediction task, these subjects were given a set of multiple-choice questions about the nature of the task and the kinds of stimulus materials they had encountered. Normal subjects and patients with Parkinson's disease performed very well in recalling the task events. In contrast, the amnesic patients were severely impaired. Thus this example of probabilistic classification learning was disrupted by striatal damage, whereas declarative memory for the learning events was impaired by hippocampal or diencephalic damage.

Summary of Striatum

The dorsal striatum supports learning that accrues over repeated trials and supports examples of stimulus-response, habit-like learning. However, the dorsal striatum also supports memory that goes beyond simple stimulus-response learning and includes probabilistic classification learning as well as examples in which subjects learn about reversals of outcomes of particular actions.

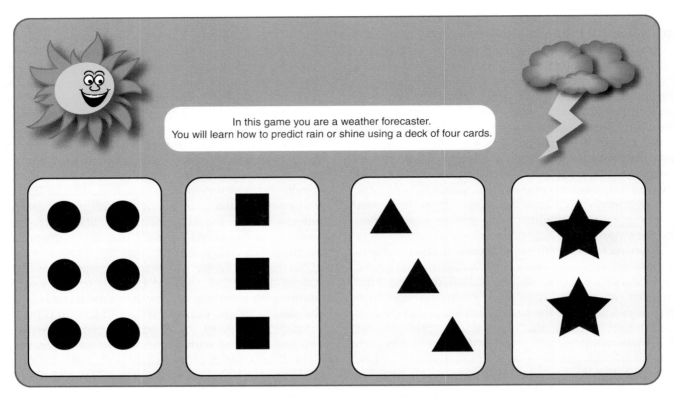

FIGURE 50.12 Weather prediction task. View of the computer screen presented to subjects showing all four stimulus cards and the "sun" or "rain" response choices. From Knowlton *et al.* (1996).

Cerebellum

Anatomy

The cerebellum consists of the cerebellar cortex, which has been divided into lobes and further into lobules, and several deep cerebellar nuclei (see Chapter 32 for more detailed anatomy). The cerebellum receives direct input from the spinal cord and brain stem as well as indirect input from a variety of sensory and motor areas in the cerebral cortex via pontine nuclei in the brain stem. The cerebellum projects directly to the spinal cord, brain stem, hypothalamus, and thalamus, and the thalamic targets in turn project to various motor and nonmotor areas in the cortex, particularly in the frontal lobes (Middleton and Strick, 1998). The internal circuitry of the cerebellum has been described as modular, with a local wiring diagram that is repeated again and again throughout the cerebellar cortex. This modularity raises the possibility that, although different regions of the cerebellar cortex receive different input, each region might perform a similar set of operations on that information before passing the results back out to the rest of the nervous system (Boyden *et al.*, 2004).

The cerebellum has long been associated with aspects of motor learning. For example, the cerebellum supports specific sensory-to-motor adaptations and adjustments of reflexes, such as changing the force that one exerts to compensate for a new load or acquiring conditioned reflexes that involve associating novel motor responses to a new stimulus. More recently, the cerebellum also has been associated with nonmotor learning and other examples of cognition and behavior. Based on the idea that the modular circuitry of the cerebellum offers a generalized capacity for processing of specific inputs, the present section focuses on one example of cerebellar-dependent learning as a framework for highlighting the role of the cerebellum in memory.

Eyeblink Classical Conditioning as a Prototype of Cerebellum-Dependent Memory

Eyeblink classical conditioning has been used frequently as a model of cerebellum-dependent learning, and is perhaps the best understood form of associative memory in vertebrates. In this paradigm, experimental animals (typically rabbits due to their tolerance for physical restraint) are placed in restraining chambers where a well-controlled tone or light is presented as the conditioning stimulus (the CS). In classic "delay" conditioning protocol, the CS lasts 250–1000 ms and coterminates with an air puff or mild electrical shock to the eye (the unconditioned stimulus or US) that produces a reflexive, unconditioned eyeblink (the UR). After many pairings of CS and US, the animal begins to produce the eyeblink after onset of the CS and prior to presentation of the US. With further training, the conditioned response (CR) occurs earlier, and its timing becomes optimized so as to be maximal at the US onset. Amazingly enough, rabbits with no cerebral cortex, basal ganglia, limbic system, thalamus, or hypothalamus showed normal retention of the conditioned eyeblink response (Mauk and Thompson, 1987). Further, in humans, conditioning can proceed without knowledge of the relationships between the conditioning stimuli (Clark *et al.*, 2002).

The essential cerebellar and brain stem circuitry that supports eyeblink conditioning has been carefully delineated by Thompson and colleagues (Fig. 50.13; Thompson, 2005). In their studies, permanent lesions or reversible inactivation of one particular cerebellar nucleus, the interpositus nucleus, resulted in impaired acquisition and retention of classically conditioned eyeblink reflexes without affecting reflexive eyeblinks (URs). Additional compelling data indicating an important role for the interpositus nucleus in this kind of motor memory come from studies using reversible inactivations of particular brain regions during training. These studies showed that drug inactivation of motor nuclei that are essential for production of the CR and UR prevented the elicitation of behavior during training. However, in trials immediately following removal of the inactivation, CRs appeared in full form, showing that the neural circuit that supports UR production is not critical for learning *per se*. A similar pattern of results was obtained with inactivation of the axons leaving the interpositus nucleus or their target in the red nucleus, showing that the final pathway for CR production is also not required to establish the memory trace. These results, in combination with

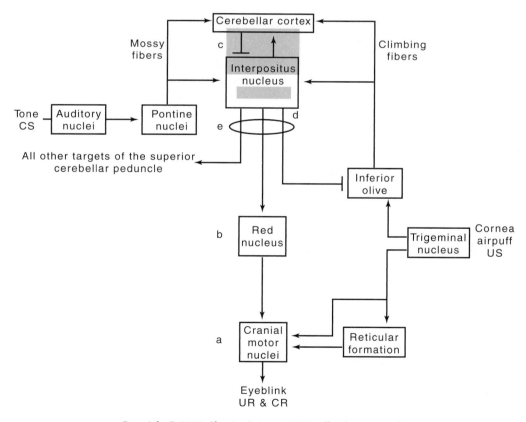

FIGURE 50.13 A schematic diagram of principal pathways involved in classical conditioning of the eyeblink reflex. The role of structures at points a–e has been studied using reversible inactivation with a local anesthetic. Inactivation at point "c" (shaded areas) prevents learning, whereas inactivation at "a," "b," "d," or "e" prevents the behavioral response during inactivation, but does not block learning. From Thompson and Kim (1996).

results from other studies that addressed candidate regions upstream from the interpositus nucleus, point to areas in the interpositus nucleus as an essential locus of plasticity in this form of motor learning. Additional studies have indicated that the cerebellar cortex is important for modulating the learning, such as by influencing the timing of the conditioned response to ensure that it is adaptive. As in animals, damage to the cerebellum in humans also retards the classically conditioned eyeblink reflex.

Summary of Cerebellum

Eyeblink classical conditioning is one example of learning that depends on the cerebellum. Other frequently studied examples include adaptations of the vestibulo-ocular reflex and coordinated motor skill learning. The circuitry essential for eyeblink conditioning has been identified and provides an experimentally tractable opportunity to investigate what may be generalized learning processes supported by the cerebellum. These processes may contribute generally to motor reflex learning or even more generally to mental processing involved in precise timing.

Amygdala

Anatomy

The amygdala is a collection of anatomically and functionally heterogeneous nuclei (Swanson and Petrovich, 1998). The two major areas of the amygdala involved in emotional memory are the basolateral amygdala complex (BLA; includes the lateral, basolateral, and basomedial nuclei) and the central nucleus of the amygdala (CEA; Fig. 50.14B). The BLA receives input from widespread cortical areas as well as from sensory nuclei of the thalamus and thus has access to higher-level information from association areas as well as lower-level sensory information. In addition, the BLA has reciprocal connections with other brain systems, including the hippocampal and striatal memory systems. Areas within the BLA project to the central nucleus, which is the source of outputs to subcortical areas controlling a broad range of fear-related behaviors, including autonomic and motor responses (e.g., changes in heart rate, blood pressure, sweating, hormone release and alterations in startle response).

The amygdala contributes to emotion in several ways, including mediating emotional influences on attention and perception and regulating emotional responses (Phelps and LeDoux, 2005). The topic of the present section is the two ways in which the amygdala contributes to emotional memory. First, the amygdala supports the acquisition of emotional dispositions

toward stimuli. This kind of memory includes preferences and aversions that can be learned unconsciously and independent of declarative memory for the events in which the disposition was acquired. Second, the amygdala mediates the influence of emotion on the consolidation of memory in other memory systems.

Role of the Amygdala in Acquiring Emotional Dispositions to Stimuli

Much of the research directed toward understanding the role of the amygdala in acquiring emotional dispositions to stimuli has focused on the circuitry that supports acquisition of fearful responses to simple auditory or visual stimuli (Davis, 1992; LeDoux, 2000; Fanselow and Gale, 2003). For example, in one frequently used task, rats are initially habituated to a chamber and then are presented with multiple pairings of a tone that terminates with a brief electric shock delivered through the floor of the cage (Fig. 50.14A). Subsequently, conditioned fear elicited by the tone is assessed by measuring autonomic responses, such as changes in arterial pressure, and motor responses, such as stereotypic crouching or freezing behavior. Rats with lesions in the BLA show dramatically reduced conditioned fear responses to the tone in measures of both autonomic and motor responses but still show normal unconditioned fear responses to the shock itself. Rats with lesions to the central nucleus show reduced fear responses to both conditioned and unconditioned stimuli. Thus, the BLA appears to be important for acquiring an aversion to the tone, whereas the central nucleus appears to be important for producing fearful responses in general.

This basic procedure has been adapted for use with human participants, and one study that included patients with damage to the amygdala or hippocampus helped to distinguish the type of memory supported by these two structures (Bechara et al., 1995). In this study, participants were trained in a discriminative fear-conditioning protocol in which, on some trials, a monochrome color stimulus (reinforced

FIGURE 50.14 Fear conditioning. (A) Prior to training, the tone produces a transient orienting response. During training the tone is followed by a brief foot shock. Following training, the rat is reintroduced into the chamber and freezes when the tone is presented. (B) Anatomical pathways that mediate fear conditioning. A hierarchy of sensory inputs converges on the lateral amygdala nucleus, which projects to other amygdala nuclei and then to the central nucleus, which send outputs to several effector systems for emotional responses. BNST, bed nucleus of the stria terminalis; DMV, dorsal nucleus of the vagus; NA, nucleus ambiguus; RPC, nucleus reticularis pontis oralis; RVL, rostral ventral nucleus of the medulla. From LeDoux (1995).

A

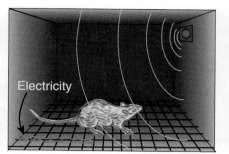
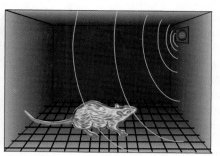

B

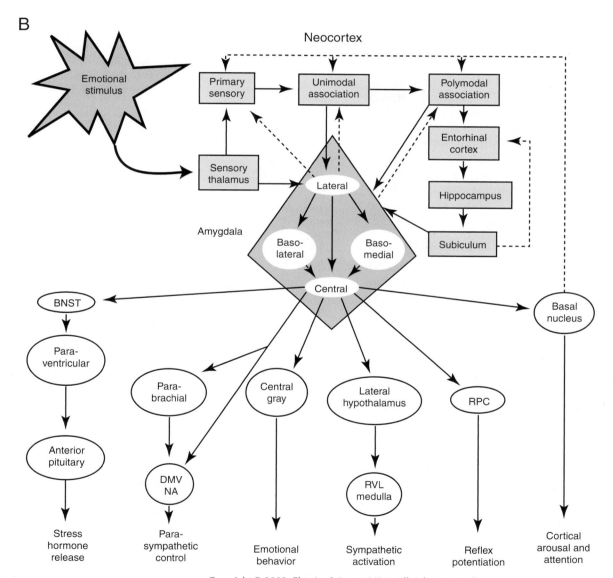

conditioned stimulus, CS+) was presented for 2 s and then was terminated just as an unconditioned stimulus was sounded briefly (US; a loud boat horn). Autonomic responses to these stimuli were measured as skin conductance changes through electrodermal recordings, a measure used often as an indicator of fear or anxiety. On other trials, different colors (unreinforced conditioned stimulus, CS−) were presented without the boat horn. Healthy individuals typically showed strong electrodermal responses to the US and with training came to show conditioned responses to the CS+ but smaller or no responses to CS− stimuli. Thus, the subjects acquired a specific fearful response to the CS+. One patient with selective damage to the amygdala showed normal unconditioned responses to the US but failed to develop conditioned responses to CS+ stimuli. In contrast, a patient with selective hippocampal damage showed robust skin conductance changes to the US and normal conditioning to CS+ stimuli and appropriately smaller responses to CS- stimuli. Control subjects and the patient with selective amygdala damage answered most of these questions correctly, but the patient with hippocampal damage was severely impaired in recollecting the training events. These findings indicated that emotional conditioning was disrupted by amygdala damage and that declarative memory for the learning situation was impaired by hippocampal damage.

Role of the Amygdala in the Modulation of Memory

Memories of emotionally arousing events are often more vivid, more accurate, and longer lasting than memories of more neutral events. Indeed, it is adaptive for organisms to remember important events better than trivial events. Thus, it also makes sense that the brain should have evolved mechanisms for storing information in accordance with how much the information is worth remembering. Evidence from a large number of studies in experimental animals has indicated that the amygdala, in particular the BLA, mediates this memory enhancing effect (McGaugh, 2004).

Emotionally arousing events result in the release of epinephrine and glucocorticoids by the adrenal glands, which in turn result in the release of norepinephrine in the amygdala (Fig. 50.15). The release of norepinephrine leads to an increase in activity in the amygdala, which is thought to influence consolidation of memory in the other parts of the brain, either directly, though connections of the amygdala with the striatum, hippocampus, and cortex, or indirectly, through connections with nucleus basalis (which innervates much of the cortex). Data in support of this model include the findings that damage to the BLA or infusions of drugs in the BLA that interfere with norepinephrine

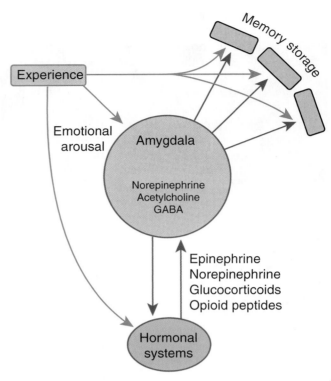

FIGURE 50.15 A schematic representation of how hormonal systems and the amygdala complex can modulate the storage of memory for emotionally arousing events through influences on other brain systems. See text for details. From McGaugh *et al.* (1992).

block the memory enhancing effect of emotional arousal. In addition, damage to pathways out of the amygdala, in particular the stria terminalis, also have a similar effect. Furthermore, the modulatory influence of the BLA seems to be limited to the time of the learning event as well as a short period of time after. Inactivation of the amygdala during subsequent testing of retention does not influence memory performance. Thus, the amygdala is important in modulating storage in emotionally stressful situations, but not in the maintenance or retrieval of the memory that has been modulated.

The amygdala has been strongly implicated in the enhancement of memory for emotional material in humans as well as in experimental animals. In one study, a patient with damage to the amygdala was tested for memory of an emotional story along with a group of healthy individuals (Adolphs *et al.*, 1997). The participants watched a series of slides and listened to an accompanying narrative that told a story about a mother and son. One portion of the story involved a traumatic accident, and all participants rated this portion as being strongly emotional. Compared to the healthy individuals, the patient with damage to the amygdala failed to show enhancement of memory for the emotional part of the story. However, the patient performed as well as others on neutral story material. In brain imaging studies with healthy individuals, the amygdala was activated during the viewing of emotional material that elicited either positive or negative emotional reactions, and this activation was related to the likelihood that participants subsequently were able to remember the emotional material. In these studies, there was no correlation between activation of the amygdala and subsequent memory performance for neutral material.

Summary of Amygdala

The amygdala is well situated anatomically to attach emotional (positive and negative) dispositions to a broad range of stimuli and to mediate the acquisition of such dispositions in the absence of conscious recollection of the circumstances of the emotional experience. In addition, components of the amygdala also mediate the modulation of memory storage during and after emotional events. Memory-modulation mechanisms are an efficient, evolutionarily adaptive method of ensuring that the strength of a memory tends to be proportional to its importance.

Cerebral Cortex

Anatomy

As discussed in Chapter 2, the posterior half of the cerebral cortex contains many functionally distinct areas that are organized into hierarchies of serial and parallel processing for each sensory modality. Additionally, there are cortical areas where information from different modalities converges (so-called "association" areas) in the parietal and temporal lobes. The anterior cerebral cortex contains a similar hierarchy of motor areas, as well as association areas in the prefrontal cortex involved with motor planning, higher order cognition, and working memory (Fuster, 2001; Chapter 52).

Some of the examples provided so far, such as standard eyeblink classical conditioning, illustrate that some simple forms of learning can circumvent the cerebral cortex. Although simplified learning paradigms are indispensable as precise instruments to dissect local circuitry, these examples should not overshadow the fact that the global circuitry of each memory system discussed so far includes the cerebral cortex. The cerebral cortex provides access to incoming information as well as output routes to influence

behavior, access that is important for even the cerebellum. Yet plasticity in the cerebral cortex also directly supports types of memory that do not require components of the other memory systems. For each area of the cortex, the modification of its information processing circuitry through alterations in synaptic connectivity and membrane excitability underlies its direct participation in memory. These alterations can be dramatic, such as in examples of experience-dependent reorganization that occurs in perceptual learning, but can also be more subtle, such as in the phenomenon of repetition priming.

Perceptual Learning

Specific training experiences can result in the modification of specific parts of the cortex. For example, in one study, monkeys were trained over a period of several weeks to discriminate small frequency differences in tone stimuli. In subsequent recordings, increases were observed in the size of the auditory cortex representation of the task-relevant frequencies and the sharpness of tuning to these frequencies (Recanzone et al., 1993). Furthermore, changes in area of the cortical representation were correlated with the improvement in task performance. In other studies using classical conditioning in anesthetized guinea pigs, the tuning curves of single neurons in the auditory cortex were first characterized (Weinberger, 2007). Individual neurons typically showed preferences for a specific tone frequency. Subsequently, the animal was presented with repeated pairings of a nonpreferred tone and foot shocks. This training resulted in long-lasting shifts in the tuning curves of auditory neurons toward the frequency of the training stimulus. A similar expansion of perceptual representations for training stimuli has been observed in the somatosensory cortex of monkeys following acquisition of a tactile discrimination. The alterations of the sensory cortex following experience is thought to contribute to adaptations and tuning of perceptual representation systems and, in coordination with reorganization of cortical motor representations, may also contribute to procedural memories. In both cases, this contribution is derived specifically from alterations in the perceptual and motor processing for the relevant learning tasks.

Functional imaging studies in humans have revealed that perceptual learning can occur for stimuli that are more complex than simple tones, yet similar to the studies in guinea pigs with tones, these studies have also suggested that the cortical area important for perceiving the stimuli is also an important area of plasticity. For example, in one study participants were scanned while viewing pictures of nonsense objects both before and after performing a perceptual discrimination task that included half of the objects (Op de Beeck, 2006). Following the perceptual discrimination task, the pattern of brain activity in object-responsive areas in visual cortex had changed for the objects included in the discrimination task but did not differ for the objects not included in the discrimination task. Thus, it appears that many of the areas in the cerebral cortex that are specialized for specific perceptual or information processing functions—from areas involved in perceiving simple tones to areas involved in perceiving complex objects—are capable of supporting perceptual learning for those stimuli.

Repetition Priming

Another example of memory thought to be supported directly by the cerebral cortex is the phenomenon known as repetition priming (or priming). Priming involves initial presentation of a list of words, pictures of objects, or nonverbal materials and then subsequent reexposure to fragments or very brief presentation of whole items. For example, in one version of priming, words are flashed on a screen so briefly that many of the words are unreadable. The measure of priming is the increased likelihood of identifying previously viewed words as compared to words not encountered recently, and the benefit of repetition has been found to be independent of the ability to recognize which words had been read initially. Studies with brain-damaged patients suggest that priming depends on cortical areas involved in initially processing the stimuli. In one study, the amnesic patient H.M. and a patient with a large lesion in the right visual cortex attempted to read new and repeated words flashed on a computer screen (Keane *et al.*, 1995). Healthy individuals and H.M. showed evidence of priming for the repeated words, but the patient with visual cortex damage did not, suggesting that the priming effect depended on perceptual areas in the visual cortex. In contrast, the patient with visual cortex damage performed normally on conventional memory tests. In functional brain imaging studies with healthy individuals, the visual cortex showed decreased activation to recently presented materials, suggesting that less neural activity is required to identify words recently processed. These findings suggest that the neuronal basis of priming is an increase in the efficiency and bias in the direction of cortical sensory processing associated with perceptual identification.

Summary of Cerebral Cortex

Studies in both animals and humans show that the cerebral cortex is highly plastic in that its representations can be altered after experience. Distinct cortical areas are reorganized by specific training experiences to meet demands for specific types of perceptual representation and memory. Also, modality-relevant cortical circuits are able to support the priming of perceptual representations.

BEHAVIOR SUPPORTED BY MULTIPLE MEMORY SYSTEMS

The discussion of memory systems so far has focused on dissociations between systems in illustrating the unique contributions of the central structure in each system. However, behavior is a product of the entire nervous system and typically is guided by contributions of more than one form of memory. For example, training in a sport normally results in declarative memory of instructions from one's coach but also leads to motor skill learning from repetitive practice. The following section considers emerging data regarding how the memory systems discussed so far might collaborate, compete, or operate in parallel and illustrates some progress in understanding how the distribution of memory throughout the brain ordinarily contributes to our day-to-day activities.

One example of a direct collaboration between two memory systems is the influence of the amygdala on hippocampus-dependent memory for emotion-inducing material. The section earlier in the chapter on the amygdala discussed how emotional arousal leads to the amygdala engaging in modulation of declarative memory, such that stimuli that elicit either positive or negative emotional responses tend to be remembered better than neutral stimuli. This process is thought to result from amygdalar projections modulating synaptic plasticity either in the hippocampus and parahippocampal region or in cortical areas fundamental to the hippocampal memory system. It has been suggested that the amygdala also modulates other memory systems, such as the striatal memory system, and that the amygdala may play a generalized role in memory modulation for emotion inducing stimuli in addition to its role in attaching positive or negative dispositions to stimuli (McGaugh, 2005).

Another example comes from an indirect collaboration between two very dissimilar memory systems, the hippocampal memory system and the cerebellar memory system. The section earlier in the chapter on the cerebellum detailed that the essential circuitry for standard eyeblink classical conditioning involved no forebrain structures and instead centered on deep cerebellar nuclei. Yet the situation changes if the parameters for standard eyeblink classical conditioning, referred to as delay eyeblink conditioning, are altered slightly such that the tone that normally overlaps with the airpuff is moved earlier in time such that it terminates as little as a half second prior to the onset of the airpuff, a version called trace conditioning (Christian and Thompson, 2003). Trace eyeblink conditioning depends on the same cerebellar and hindbrain circuitry as delay eyeblink conditioning, but trace conditioning depends additionally on forebrain structures including the hippocampus and prefrontal cortex. In studies with healthy human participants, trace (but not delay) eyeblink conditioning was found to relate closely to participants' knowledge about the relationship between the tone and the airpuff (Clark et al., 2001). A possible explanation of these findings is that the silent interval that separates the tone and airpuff may outstrip the learning abilities of the cerebellum and may cause it to require inputs from additional forebrain structures, such as the hippocampal system, to bridge that brief interval. The growing understanding of the relationship between hippocampus and cerebellum illustrates the tendency to collaborate between even the most dissimilar memory systems, and exemplifies the progress being made toward understanding how memory systems interact in general.

In other examples, memory systems may operate in parallel or may even compete for control of behavior. For example, the previously discussed studies contrasting the role of the striatum and hippocampus in supporting "response" learning and "place" learning, respectively, illustrate how the memory supported by different brain systems can be at odds with each other. Yet in the example of practicing a sport, the hippocampus-dependent memory for a coach's instructions and the striatal "response" learning would likely support a common behavior. In these instances, it is thought that the striatum and hippocampus contribute to behavior in parallel rather than directly interacting. One can start to get a sense of the often collaborative spirit of brain when one further considers that the emotional arousal that often accompanies athletic competition would likely engage amygdalar modulation of both the hippocampal memory system and the striatal memory systems.

CONCLUSION

Several brain systems support distinct forms of memory characterized by unique operating features and ways in which memory is expressed. The hippocampus supports rapidly acquired memory that is experienced as conscious recollection and that can be expressed flexibly. The striatum supports gradual acquisition of habits. The cerebellum supports learning related to timing and motor reflexes. The amygdala supports acquisition of emotional dispositions toward stimuli and also mediates the role of emotion arousal in modulating other forms of memory. The cerebral cortex participates in each of these types of memory, and mediates perceptual learning and the phenomenon of priming. These systems also often operate in concert to support the various ways that behavior and mental life can be modified by experience.

References

Adolphs, R., Cahill, L., Schul, R., and Babinsky, R. (1997). Impaired declarative memory for emotional material following bilateral amygdala damage in humans. Learn Mem 4, 291–300.

Alvarez, P. and Squire, L. R. (1994). Memory consolidation and the medial temporal lobe: A simple network model. Proc Natl Acad Sci 15, 7041–7045.

Amaral, D. G. and Witter, M. P. (2004). Hippocampal formation. In "The rat nervous system, 3rd ed." (Paxinos, G. ed.) pp. 635–704. Elsevier, San Diego.

Bayley, P. J., Frascino, J. C., and Squire, L. R. (2005). Robust habit learning in the absence of awareness and independent of the medial temporal lobe. Nature 7050, 550–553.

Bayley, P. J. and Squire, L. R. (2002). Medial temporal lobe amnesia: Gradual acquisition of factual information by nondeclarative memory. J Neurosci 13, 5741–5748.

Bechera, A., Tranel, D., Hanna, D., Adolphs, R., Rockland, C., and Damasio, A. R. (1995). Double dissociation of conditioning and declarative knowledge relative to the amygdala and hippocampus in humans. Science 269, 1115–1118.

Boyden, E. S., Katoh, A., and Raymond, J. L. (2004). Cerebellum-dependent learning: The role of multiple plasticity mechanisms. Annu Rev Neurosci 27, 581–609.

Bunsey, M. and Eichenbaum, H. (1996). Conservation of hippocampal memory function in rats and humans. Nature 379, 255–257.

Burwell, R. D., Witter, M. P., and Amaral, D. G. (1995). Perirhinal and postrhinal cortices in the rat: A review of the neuroanatomical literature and comparison with findings from the monkey brain. Hippocampus 5, 390–408.

Christian, K. M. and Thompson, R. F. (2003). Neural substrates of eyeblink conditioning: Acquisition and retention. Learn Mem 10, 427–455.

Clark, R. E., Manns, J. R., and Squire, L. R. (2002). Classical conditioning, awareness, and brain systems. Trends Cogn Sci 6, 524–531.

Cohen, N. J. and Squire, L. R. (1980). Preserved learning and retention of pattern analyzing skill in amnesics: Dissociation of knowing how and knowing that. Science 210, 207–210.

Corkin, S., Amaral, D. G., Gonzalez, R. G., Johnson, K. A., and Hyman, B. T. (1997). H. M.'s medial temporal lobe lesion. Findings from magnetic resonance imaging. J Neurosci 17, 3964–3979.

Davis, M. (1992). The role of the amygdala in fear and anxiety. Annu Rev Neurosci 15, 353–375.

Eichenbaum, H. (2000). A cortical-hippocampal system for declarative memory. Nat Rev Neurosci 1, 41–50.

Eichenbaum, H. (2001). The hippocampus and declarative memory: Cognitive mechanisms and neural codes. Behav Brain Res 127, 199–207.

Eichenbaum, H., Yonelinas, A. R., and Ranganath, C. (2007). The medial temporal lobe and recognition memory. *Annu Rev Neurosci* **30**, 123–152.

Ekstrom, A. D., Kahana, M. J., Caplan, J. B., Fields, T. A., Isham, E. A., Newman, E. L., and Fried, I. (2003). Cellular networks underlying human spatial navigation. *Nature* **425**, 184–188.

Fanselow, M. S. and Gale, G. D. (2003). The amygdala, fear, and memory. *Ann NY Acad Sci* **985**, 125–134.

Fuster, J. M. (2001). The prefrontal cortex—An update: Time is of the essence. *Neuron* **30**, 319–333.

Hebb, D. O. (1949). "The organization of behavior." Wiley, New York.

Keane, M. M., Gabrieli, J. D., Mapstone, H. C., Johnson, K. A., and Corkin, S. (1995). Double dissociation of memory capacities after bilateral occipital-lobe or medial temporal-lobe lesions. *Brain* **118**, 1129–1148.

Knowlton, B. J., Mangels, J. A., and Squire, L. R. (1996). A neostriatal habit learning system in humans. *Science* **273**, 1399–1401.

LeDoux, J. E. (1995). Emotion: Clues from the brain. *Annu Rev Psychol* **46**, 209–235.

LeDoux, J. E. (2000). Emotion circuits in the brain. *Annu Rev Neurosci* **23**, 155–184.

Manns, J. R. and Eichenbaum, H. (2007). Evolution of the hippocampus. *In* "Evolution of Nervous Systems: The Evolution of Nervous Systems in Mammals," Vol. 4 (L. Krubitzer and J. Kaas, eds.). London, Elsevier.

Marshall, L., Helgadottir, H., Molle, M., and Born, J. (2006). Boosting slow oscillations during sleep potentiates memory. *Nature* **444**, 610–613.

Mauk, M. D. and Thompson, R. F. (1987). Retention of classically conditioned eyelid responses following acute decerebration. *Brain Res* **403**, 89–95.

Maviel, T., Durkin, T. P., Menzaghi, F., and Bontempi, B. (2004). Sites of neocortical reorganization critical for remote spatial memory. *Science* **305**, 96–99.

McClelland, J. L., McNaughton, B. L., and O'Reilly, R. C. (1995). Why there are complementary learning systems in the hippocampus and neocortex: Insights from the successes and failures of connectionist models of learning and memory. *Psychol Rev* **102**, 419–457.

McGaugh, J. L. (2004). The amygdala modulates the consolidation of memories of emotionally arousing experiences. *Annu Rev Neurosci* **27**, 1–28.

McGaugh, J. L., Introini-Collison, I. B., Cahill, L., Kim, M., and Liang, K. C. (1992). Involvement of the amygdala in neuromodulatory influences on memory storage. *In* "The Amygdala: Neurobiological Aspects of Emotion, Memory, and Mental Dysfunction" (J. P. Aggleton, ed.), pp. 431–451. Wiley-Liss, New York.

Middleton, F. A. and Strick, P. L. (1998). Cerebellar output: Motor and cognitive channels. *Trend Cog Sci* **2**, 348–354.

Milner, B. (1962). *Les troubles de la memoire accompagnant les lesions hippocampiques bilaterales. In* "Physiologie del'hippocampe, colloques internationaux no. 107" (P. Passouant, ed.), p. 257. Paris: CNRS. [English translation 1965. *In* "Cognitive Processes and the Brain" (B. Milner and S. Glickman, eds.), p. 97. Princeton, NJ, Van Nostrand.]

Mishkin, M. (1978). Memory in monkeys severely impaired by combined but not separate removal of the amygdala and hippocampus. *Nature* **273**, 297–298.

Morris, R. G. M., Garrud, P., Rawlins, J. N. P., and O'Keefe, J. (1982). Place navigation impaired in rats with hippocampal lesions. *Nature* **297**, 681–683.

O'Keefe, J. (1976). Place units in the hippocampus of the freely moving rat. *Exp Neurol* **51**, 78–109.

O'Keefe, J. and Nadel, L. (1978). "The Hippocampus as a Cognitive Map." Oxford Univ. Press (Clarendon), London.

Op de Beeck, H. P., Baker, C. I., DiCarlo, J. J., and Kanwisher, N. G. (2006). Discrimination training alters object representations in human extrastriate cortex. *J Neurosci* **26**, 13025–13036.

Packard, M. G. and McGaugh, J. L. (1996). Inactivation of hippocampus or caudate nucleus with lidocaine differentially affects expression of place and response learning. *Neurobiol Learn Memory* **65**, 65–72.

Pasupathy, A. and Miller, E. K. (2005). Different time courses of learning-related activity in the prefrontal cortex and striatum. *Nature* **433**, 873–876.

Phelps, E. A. and LeDoux, J. E. (2005). Contributions of the amygdala to emotion processing: From animal models to human behavior. *Neuron* **48**, 175–187.

Preston, A. R., Shrager, Y., Dudukovic, N. M., and Gabrieli, J. D. (2004). Hippocampal contribution to the novel use of relational information in declarative memory. *Hippocampus* **14**, 148–152.

Recanzone, G. H., Schreiner, C. E., and Merzenich, M. M. (1993). Plasticity in the frequency representation of primary auditory cortex following discrimination training in adult owl monkeys. *J Neurosci* **13**, 87–103.

Scoville, W. B. and Milner, B. (1957). Loss of recent memory after bilateral hippocampal lesions. *J Neurol Neurosurg Psychiatry* **20**, 11–21.

Squire, L. R., Clark, R. E., and Knowlton, B. J. (2001). Retrograde amnesia. *Hippocampus* **11**, 50–55.

Stefanacci, L., Buffalo, E. A., Schmolck, H., and Squire, L. R. (2000). Profound amnesia after damage to the medial temporal lobe: A neuroanatomical and neuropsychological profile of patient E. P. *J Neurosci* **20**, 7024–7036.

Stern, C. E., Corkin, S., Gonzalez, R. G., Guimaraes, A. R., Baker, J. R., Jennings, P. J., Carr, C. A., Sugiura, R. M., Vedantham, V., and Rosen, B. R. (1996). The hippocampal formation participates in novel picture encoding: Evidence from functional magnetic resonance imaging. *Proc Natl Acad Sci USA* **93**, 8660–8665.

Stickgold, R. (2005). Sleep-dependent memory consolidation. *Nature* **437**, 1272–1278.

Sutherland, G. R. and Mcnaughton, B. (2000). Memory trace reactivation in hippocampal and neocortical neuronal ensembles. *Curr Opin Neurobiol* **10**, 180–186.

Suzuki, W. A. (1996). Neuroanatomy of the monkey entorhinal, perirhinal, and parahippocampal cortices: Organization of cortical inputs and interconnections with amygdala and striatum. *Semin Neurosci* **8**, 3–12.

Swanson, L. W. and Petrovich, G. D. (1998). What is the amygdala? *Trends Neurosci* **21**, 323–331.

Teng, E. and Squire, L. R. (1999). Memory for places learned long ago is intact after hippocampal damage. *Nature* **400**, 675–677.

Thompson, R. F. (2005). In search of memory traces. *Annu Rev Psychol* **56**, 1–23.

Thompson, R. F. and Kim, J. J. (1996). Memory systems in the brain and localization of a memory. *Proc Nat Acad Sci USA* **93**, 13438–13444.

Wagner, A. D., Schacter, D. L., Rotte, M., Koutstaal, W., Maril, A., Dale, A. M., Rosen, B. R., and Buckner, R. L. (1998). Building memories: Remembering and forgetting of verbal experiences as predicted by brain activity. *Science* **281**, 1188–1191.

Wirth, S., Yanike, M., Frank, L. M., Smith, A. C., Brown, E. N., and Suzuki, W. A. (2003). Single neurons in the monkey hippocampus and learning of new associations. *Science* **300**, 1578–1581.

Weinberger, N. M. (2007). Associative representational plasticity in the auditory cortex: A synthesis of two disciplines. *Learn Mem* **14**, 1–16.

Winograd, T. (1975). Frame representations and the declarative-procedural controversy. *In* "Representation and understanding: Studies in cognitive science" (D. Bobrow and A. Collins, eds.), pp. 185–210, Academic Press, New York.

Wood, E. R., Dudchenko, P. A., and Eichenbaum, H. (1999). The global record of memory in hippocampal neuronal activity. *Nature* **397**, 613–616.

Wood, E., Dudchenko, P., Robitsek, J. R., and Eichenbaum, H. (2000). Hippocampal neurons encode information about different types of memory episodes occurring in the same location. *Neuron* **27**, 623–633.

Yin, H. H. and Knowlton, B. J. (2006). The role of the basal ganglia in habit formation. *Nat Rev Neurosci* **7**, 464–476.

Zola-Morgan, S. and Squire, L. R. (1985). Medial temporal lesions in monkeys impair memory on a variety of tasks sensitive to human amnesia. *Behav Neurosci* **99**, 22–34.

Zola-Morgan, S., Squire, L. R., and Amaral, D. G. (1986). Human amnesia and the medial temporal region: Enduring memory impairment following a bilateral lesion limited to field CA1 of the hippocampus. *J Neurosci* **6**, 2950–2967.

Zola, S. M., Squire, L. R., Teng, E., Stefanacci, L., Buffalo, E. A., and Clark, R. E. (2000). Impaired recognition memory in monkeys after damage limited to the hippocampal region. *J Neurosci* **20**, 451–463.

Suggested Readings

Bloedel, J. R., Ebner, T. J., and Wise, S. P. (eds.). (1996). "The Acquisition of Motor Behavior in Vertebrates." MIT Press, Cambridge, MA.

Cahill, L. and McGaugh, J. L. (1998). Mechanisms of emotional arousal and lasting declarative memory. *Trends Neurosci* **21**, 273–313.

Eichenbaum, H. and Cohen, N. J. (2001). "From Conditioning to Conscious Recollection: Memory Systems of the Brain." Oxford Univ. Press.

Fuster, J. M. (1995). "Memory in the Cerebral Cortex: An Empirical Approach to Neural Networks in the Human and Nonhuman Primate." MIT Press, Cambridge, MA.

McClelland, J. L., McNaughton, B. L., and O'Reilly, R. C. (1995). Why there are complementary learning systems in the hippocampus and neocortex: Insights from the successes and failures of connectionist models of learning and memory. *Psychol Rev* **102**, 419–457.

McDonald, R. J. and White, N. M. (1993). A triple dissociation of memory systems: Hippocampus, amygdala, and dorsal striatum. *Behav Neurosci* **107**, 3–22.

Schacter, D. L. (1987). Implicit memory: History and current status. *J Exp Psychol Learn Memory Cogn* **13**, 501–518.

Schacter, D. L. and Tulving, E. (eds.). (1994). "Memory Systems." MIT Press, Cambridge, MA.

Squire, L. R. (1992). Memory and the hippocampus: A synthesis from findings with rats, monkeys, and humans. *Psychol Revi* **99**, 195–231.

Squire, L. R. (2004). Memory systems of the brain: A brief history and current perspective. *Neurobiol Learn Mem* **82**, 171–177.

Suzuki, W. and Clayton, N. S. (2001). The hippocampus and memory: A comparative and ethological perspective. *Curr Opin Neurobiol* **10**, 768–773.

Joseph R. Manns and Howard Eichenbaum

Language and Communication

This chapter discusses the nature of language and the relationship of language to the brain. It begins with a discussion of animal communication, from which human language evolved and which provides a guide to the basic neural elements and processes that underlie our own system of communication. The next section of the chapter provides a brief summary of what human language is and how it is processed. The third section looks at how human language is related to the brain, relying on studies of patients with various types of neurological diseases and on brain imaging in normal subjects.

ANIMAL COMMUNICATION

Communication Is Important for Individual Survival

The ultimate goal of most animal communication, like that of behavior in general, is reproduction. Thus, signals that indicate the sender's species, gender, and degree of reproductive readiness account for the vast majority of natural communication. These are the messages being broadcast by chirping crickets, flashing fireflies, pheromone-releasing moths, and singing birds. Frequently, the mate-attraction call also is used to warn off potential rivals; for example, bird and cricket songs also are used to identify the boundaries of territories or personal space. Differences in signal quality are the usual basis of mate choice in species for which some degree of discrimination is evident. This competition for the attention of members of the opposite sex has led to the development of more conspicuous signals and, in many cases, to the evolution

of displays, signaling morphology (i.e., sounds vs. shapes vs. color vs. movement vs. odors), and messages that go far beyond the basic needs of species and sexual identification (Gould and Gould, 1996). Some researchers argue that sexual competition was an important selective factor in the evolution of human speech.

Most animals are solitary except when mating; they abandon their eggs or larvae before the offspring are born. However, a number of species engage in some degree of parental care. For them, signals between parent and offspring are often very important. Most birds, for instance, have about two dozen innate calls that communicate mundane messages such as the need to eat, defecate, take cover, and so on. In cases in which both parents tend the young—the usual circumstance in birds, for instance—additional signals are required to agree on a nest site, synchronize brooding shifts, and guard the nestlings. Most primates also come equipped with two to three dozen innate signals.

The minority of species that are highly social have the most elaborate communication systems of all. They need messages for a variety of elements of social coordination, including, in many cases, group hunting or foraging, defense, and working out of a social hierarchy.

Animal Communication Strategies and Mechanisms

Attracting a mate is vital to the survival of an individual's genes. To ensure species specificity in mating, most organisms rely on more than one cue to identify a sexual partner. (Exceptions include some of the species that rely on pheromones.) Thus, multiple signals are sent, and a choice must be made between

1179

sending simultaneous messages and sending sequential messages (or a combination of the two). The sequential strategy has the advantage that the individual signals must be correct *and* the order must be appropriate. A female stickleback, for instance, requires the male to have a red ventral stripe, perform a zig-zag dance, poke his nose in a nest, and then vibrate her abdomen; the odds of this concatenation of signals occurring together in this order by chance are remote. A human sentence in a word-order language like English operates roughly in this way. However, sequential signaling is time-consuming. A faster strategy is to provide all the cues in parallel, an approach that accepts the larger chance that these cues can occur together by chance. For example, when a parent herring gull waves its bill in front of chicks to see if they need to be fed, the young simultaneously see a vertical beak, a red spot, and a horizontal motion, each of which is a discrete cue that combine in the mind of the chick to elicit pecking.

Nearly all animal communication is innate: the sender produces the appropriate signal in the correct context even without any opportunity for learning, and it can be recognized for what it is by equally naive conspecifics. The basis of innate recognition appears to lie with feature detectors in the nervous systems—the inborn circuits that automatically isolate iconic visual or acoustic elements (Gould, 1982). The availability of many visual, auditory, tactile, and olfactory feature detectors, which can, in theory, be used in any specific combination or order, accounts for most of the diversity of animal communication, and reliance on these individual feature detectors, rather than pattern detection, accounts for its limitations.

Innate Communication Systems Coupled with Learning

A common misapprehension is that complex behaviors must be learned. In fact, complex behavior in relatively short-lived species is usually innate, reflecting the reality that intricate activities are very difficult to learn and may require more time and risk of errors than an animal can afford (Gould and Gould, 1999). Thus, so far as is known, all bird nests are built on the basis of innate instructions, although some improvement with experience is also evident (Gould and Gould 2007). Therefore, looking at just how complex innate communication can be is a useful calibration for the often-made assumption that something as intricate as human language must be largely learned.

In terms of its ability to communicate information, the most complex system of nonhuman communication known at present is the dance of honey bees (Gould and Gould, 1995). The system has some properties that are reminiscent of human language: it uses arbitrary conventions to describe objects distant in both space and time; that is, it does not reflect a real-time emotive readout, as might be the case when a primate gives an alarm call or grunts at a banana.

In terms of information content, a honey bee dance is second (albeit a very distant second) only to human speech, and so far as is known is far richer than the language of any primate. However the dance language suffers from at least two limitations: It is a closed system—there seems to be no way to introduce new conventions to deal with novel needs. It is also graded; instead of discrete signals for different directions or distances, single components are varied over a range of values (angles and durations). The less complex but more flexible systems of birds and primates are more likely to illustrate the evolutionary precursors of speech.

Some birds have innate songs. In these species, individuals raised in isolation produce songs that are indistinguishable from their socially reared peers and respond to calls appropriately without prior experience. Chickens, doves, gulls, and ducks are familiar examples. Most songbirds, however, illustrate a different pattern—one that shows how potent a combination instinct and innately directed learning can be. Isolated chicks sing a schematic form of the species song, but the richness of a normal song is absent. Adult conspecifics can recognize innate songs as coming from members of their species, but in general these impoverished vocalizations produce lower levels of response. Typically, there is a sensitive period during which exposure to song must occur if it is going to have any impact (Fig. 51.1). In most species, there is a gap between this sensitive period and the process of overt song development—that is, practice and perfection of the adult song are based on the bird's memory of what it heard during its sensitive period (Gould and Marler, 1987).

Given a range of possible song models during isolated rearing, a chick selects an example from its own species and memorizes it. If it hears only songs of other species, the mature song is the unmodified innate song (Fig. 51.1). Thus there is an innate bias in the initial learning. Where this bias has been studied, it appears to depend on acoustic sign stimuli (i.e., species–specific "syllables"). Indeed, chicks are able to extract syllables of their species embedded in foreign songs, or scored in a way never found in their species (e.g., a syllable that is repeated at an accelerating rate presented to a species that sings syllables at a constant rate, or vice versa will be extracted and used in the species-typical manner).

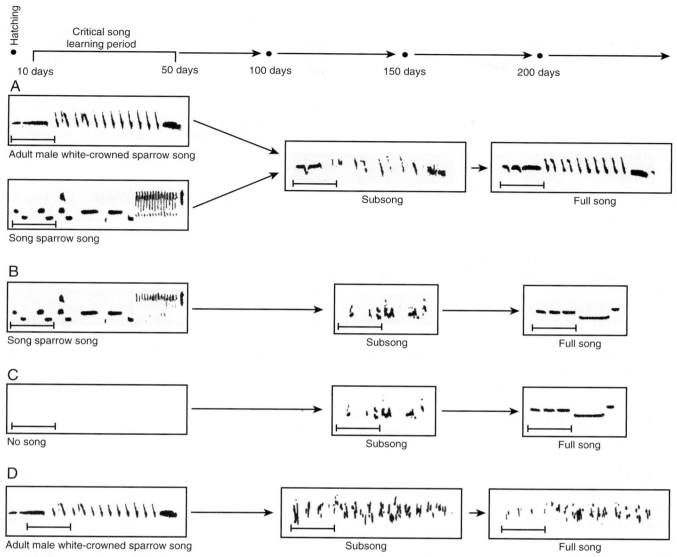

FIGURE 51.1 Birdsong development in most species is characterized by a sensitive period during which a song of the species must be heard. Later, during subsong, the bird practices making notes and assembles them into the correct order and pattern (A). Birds not allowed to hear their species' song sing a schematic version of the song (B and C); birds deafened before subsong cannot sing (D).

Practice is essential in the normal development of birdsong, and part of this practice occurs in a babbling phase known as subsong, which begins at a species-typical age. A bird deafened after its sensitive period, but before it begins producing notes in preparation for singing (subsong), is unable to produce even an innate song (Fig. 51.1). During the earliest parts of subsong, birds try out a number of notes. These notes are typical of the species, but most are absent from the song they eventually sing. The learning process may involve producing each member of an innate repertoire of notes, listening to them, checking to see if they match any element in the memorized song, discarding the unnecessary ones, and rearranging, scoring, and modifying the others to produce a reasonable copy of the original song heard during the sensitive period.

There is some flexibility in song development. For example, when the chick has heard two very different specimens of its own species' song, it will often incorporate elements from each. Similarly, when the chick has been exposed to the sight of a singing conspecific and simultaneously the sound of a heterospecific song, it may pick out elements of the abnormal song and adapt them as best it can into its own species-specific organization.

Birdsong, therefore, depends on two processes that involve an interaction between innate capacities and learning: imprinting the song in memory and then learning to perform it. This system is flexible, but only within clear limits.

Communication Systems in Primates

Primates are the species with the closest evolutionary links to humans, and therefore the communication systems found in these animals are important to study for clues regarding the neural basis of human language.

Vervets, a species of monkey, provide a good case study. Vervets, like all social primates, have a large repertoire of innate calls used for social communication. Among these approximately three dozen signals are four alarm calls (Cheney and Seyfarth, 1990). In some parts of their range, one of the calls is specific for martial eagles; in another the same call is used for certain hawks. In either case, the call causes monkeys to look up; those at the tops of trees drop to the interior, whereas those on the ground move into bushes or under trees. A second call, specific to leopards in one region and to other solitary hunters elsewhere, sends the warned individuals up to the tree tops. A third, specific for snakes, induces the other members of the troop to stand up and look around in the grass. A fourth call is heard in the presence of humans or group-hunting predators.

The development of calling is revealing. Young vervets appear to understand the class of animals each call refers to, but not the particular species that are dangerous. Thus, infants will give the eagle call to harmless vultures, storks, and even falling leaves, but not to snakes or leopards. Consequently, adults generally respond to the alarm calls of infants with a casual look around, followed by their own alarm if there is a genuine danger. Juveniles make fewer mistakes, and adult errors are confined to calls produced when the potential threat is so far away that human observers require binoculars to identify the species. In short, young vervets seem to learn the details of how to apply an innate categorical vocabulary.

Nonhuman primates have virtually no voluntary control over respiration, the vocal tract, or vocalization, so this modality is not available for producing a large number of differentiated signals. Other modalities of communication, such as gesture, however, are also used for relatively restricted purposes, mostly related to items of immediate biological importance and social hierarchy. Recent detailed studies of gesture use in captive chimpanzees indicate that the meaning of particular gestures differ from one group to another.

Researchers have attempted to teach chimpanzees to use a variety of communicative systems that bypass their vocal limitations, such as sign language and manipulation of object tokens, in an effort to determine to what extent these naturally occurring communication systems can be augmented. If anything, this work emphasizes the ways in which nonhuman primates are not capable of full human language. There are dramatic differences between trained chimpanzees and humans in terms of vocabulary size (chimps never get beyond what a 2-year-old can do), conceptual capacities, complex grammar, and so on, despite the fact that chimps have remarkable skills at social cognition.

Neural Structures for Animal Communication Systems

The neural mechanisms that underlie animal communication may give clues as to those that allow for human language. These systems of course differ in different phyla and species. This section briefly reviews aspects of the neural basis for birdsong and primate communication. It would be of great interest to report on the neural basis for the dance of the bees, but nothing is known about this subject.

Birdsong

Birdsong appears to be produced by a small number of nuclei (collections of nerve cells). In the canary, four important nuclei are involved: HVc, RA, DM, and nXIIts. Lesions in any of these nuclei destroy or seriously impair song production. These nuclei also respond when song is presented, and some neurons in some nuclei in some species (e.g., neurons in nXIIts in the zebra finch and the white-crowned sparrow) respond specifically to particular aspects of song. This indicates an overlap in the neural mechanisms involved in song perception and production, a feature that may carry over to humans as well.

Birdsong is produced by a structure between the trachea and the bronchi called the syrinx. There is one syrinx on each side of the trachea. In some species, song is entirely, or mainly, produced by one of these two structures; in other species, both syrinxes participate in song, with each producing different parts of the song. Each syrinx is controlled by structures in its corresponding side of the brain and, when one syrinx is responsible for song, these structures are larger on that side. The perceptual system is also duplicated on both sides of the brain, and, in some species, the two sides of the brain respond to different aspects of song.

In species in which only male birds sing (most species), song nuclei are much larger in males than in females ("sexual dimorphism"). This is not true in those species in which males and females perform duets. In general, there is a correlation between song nucleus size and song complexity in individual birds. Furthermore, song nuclei grow by a factor of almost 100% during the spring mating season (when birds sing) compared to the fall and winter (when they do not).

Changes in the size of song nuclei are related to testosterone production (they can be induced in females by testosterone injection in some circumstances) and by other factors usually related to daylength—part of the birds' innate calendar. They are due to many changes in nerve cells: increase in the number of dendrites, increase in cell size, and, most interestingly, increase in the number of cells. Studies using labeled thymidine, which is taken up by dividing DNA, have shown neurogenesis in neurons in the HVc nucleus in relationship to periods of increased song, indicating that new nerve cells are formed. These neurons project to the RA nucleus, strongly suggesting that they play a role in song.

Primates

The neural basis for primate vocalizations is beginning to be charted. Unlike humans, vocalizations in primates do not appear to begin in a lateral cortical region but rather in the medial portion of the brain (the cingulate), a region involved in connecting basic instinctual to more advanced cognitive functions. Actual motor planning appears to involve mainly brain stem nuclei. On the perception side, there are neurons in the auditory cortex of some species of macaque monkeys that selectively respond to conspecific calls. These regions show some degree of functional asymmetry in some species, with responses from cells in the left hemisphere and not the right.

Possible Lessons from Nonhuman Species Regarding Language and Its Neural Basis

It is dangerous to infer too much about human language, or about how the human brain may support language, from studies of species that are very distant, such as birds. However, these species provide evidence for the possibility of certain aspects of language and its neural basis that could have developed several times in evolution or that could date to a distant common ancestor.

At a behavioral level, human speech and language could have developed from the repertoire of two to three dozen innate calls that are typical of birds and primates. In that case, we would expect to find that language contains elements that are innately recognized and that may be distinguished by the sorts of acoustic features that provide the basis for the innate discrimination of sign stimuli in other species. In fact, babies have an innate ability to discriminate a small number of sounds found in human languages, which function as acoustic sign stimuli. The evolution of human speech could also have retained the sorts of sensitive periods and innate learning biases so evident in song birds and vervet monkeys. In this case, we might expect to see a species-typical babbling phase, and perhaps a sensitive period for easy learning of new languages. If the evolution of speech and language preserves the use of preexisting (innate) forms, we might expect there to be innate aspects of human language, which would presumably be found in all languages and which might surface in situations where we see a "default grammar," not unlike the unlearned and impoverished songs of birds reared in isolation. Creoles potentially exemplify such cases. Creoles are languages developed by the children of people speaking different languages who are drawn together and communicate with very simple pidgin. They have a number of features not present in the languages spoken by the parents of the children who spoke the pidgin, which have been taken to be related to innately specified features of language (Bickerton, 1990).

From the neurological perspective, several features of the neural organization for communication systems appear to be found in humans in a way that is relevant to language. These features are the fact that there are specialized neural nuclei for song production; that the neural basis for song production is often asymmetric, with one hemisphere producing most of song or the two hemispheres producing different aspects of song; that production and perception of song make use of same structures to some degree; and that there is a relationship of size to function. However, other important features of the neurobiology of birdsong, such as the neurogenesis that is important in the seasonal changes in song production, are not known to play a role in human speech and language. As for primates, the neural system responsible for call production is quite different from that in humans, being based in the cortex that is transitional from limbic to association regions (reflecting the limited semantic content and immediate biological relevance of most calls). Asymmetries in the neural basis for call perception, and the existence of neurons in the auditory association cortex that selectively respond to conspecific calls, may be quite direct evolutionary precursors of the human neural substrate for language.

HUMAN LANGUAGE

Although animal communication systems can be quite sophisticated and some of them share attributes with human language, human language is far more complex than animal communication. There are many features of human language that distinguish it from animal communication systems.

One is what messages human language can convey. As we have seen, most animal communication systems serve the purposes of identifying members of a species. These systems do sometimes designate items of immediate biological significance, such as designation of predators by monkeys or food sources by bees. Human language differs from these systems with respect to the number of items that can be designated and the relationship of these items to present biological necessity. Human language allows us to designate an infinitely large number of items, actions, and properties of items and actions, and to do so with respect to items that are not immediately biologically compelling.

Human language also allows us to relate items, actions, and properties to one another. This propositional level of semantic content is beyond the scope of any known animal communication system. At a yet higher level, human language allows us to express relationships between events and states of affairs in the world, such as temporal order and causation. Again, this level of meaning is far beyond anything available to other species as far as we know. It reflects the human ability to conceive of an enormous number of concepts and to relate them to one another.

Language expresses these different semantic values with different types of representations: words, words made from other words, sentences, and discourse. Each of these "levels" of linguistic representation consists of specific forms that are related to specific aspects of meaning. The forms at each level are intricately structured, and there are an infinite number of different structures that can be built at each level of the system.

Words

Simple words are defined and distinguished from each other primarily by their phonemes. A phoneme is a single distinct sound that contrasts with another (Fig. 51.2) and makes it possible to determine the existence of a word in a language. For instance, in English /p/ and /b/ are different phonemes because they determine the separate existence of the members of word pairs such as pat-bat, pale-bale, pull-bull, lap-lab. A major distinction between different types of

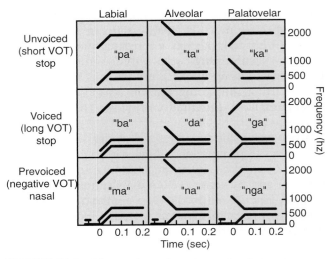

FIGURE 51.2 Schema of sound spectrographs of nine phonemes followed by a short "a" sound. The sound bands are called "formants." The first (bottom) two formants are independent; the third is a harmonic of the second.

phonemes is the difference between vowels and consonants. Consonants involve some degree of constriction of the vocal tract; vowels are produced with a relatively open vocal tract. Each language "selects" its inventory of phonemes from a relatively small set of phonemes found across languages. American English phonemes have 25 consonants and 17 vowels.

At the semantic level, put in very simple terms, words designate concrete objects, abstract concepts, actions, properties, and logical connectives. Many philosophers and psychologists, and most neurologists who have considered the subject, have thought that the meaning of a word consists of a set of features of objects and actions that are associated with the word's sound through one's experience. However, the meaning of a word also includes unobserved properties of the item that the word designates, such as our knowledge that cats can swim even if we have never observed one swimming. This knowledge may be based on inductive generalizations, logical inferences, or innate concepts regarding the structure of items. Tulving (1972), who called the complex set of properties that make up word meaning "semantic memory," emphasized that the hallmark of this type of knowledge is the relations that items have to one another.

Words Formed from Other Words

The word-formation level of language allows words to be formed from other words. Language uses many devices to accomplish this—compounding, affixation, agglutination, and others. English affixation provides a useful illustration. There are two main types of affixes in English: inflection and derivation. Inflection is related to the syntactic structure of a sentence, as in agreement between subjects and objects (I see; he sees). English is relatively poor in overt inflectional features; languages such as Dutch and German have more complete and complex agreement systems that affect adjective–noun agreement, verb inflections, and the case markers on nouns. Derivational processes are those that create new syntactic categories of words (e.g., destroy–destruction; happy–happiness). Derivation allows the meaning associated with a simple lexical item to be used as a different syntactic category without coining a large number of new lexical forms that would have to be learned.

Sentences

The sentential level of language consists of syntactic structures (Chomsky, 1965, 1981, 1986) into which words are inserted. Individual words are marked for syntactic category (e.g., cat is a noun (N); read is a verb (V); of is a preposition (P)). These categories combine in a hierarchy to create nonlexical nodes (or phrasal categories), such as noun phrase (NP), verb phrase (VP), and sentence (S). The way words are inserted into these higher order phrasal categories determines a number of different aspects of sentence meaning. For instance, in the sentence "The dog that scratched the cat killed the mouse," there is a sequence of words—*the cat killed the mouse*—that, in isolation, would mean that the cat killed the mouse. However, "The dog that scratched the cat killed the mouse" does not assert that the cat killed the mouse, but rather that the dog did. This is because *the cat* is not the subject of *killed* and does not play a thematic role around *killed*. The cat is the object of the verb scratched in the relative clause—*that scratched the cat*—and is the theme of *scratched*. *The dog* is the subject of the verb *killed* and is the agent of that verb. The syntactic structure of "The dog that scratched the cat killed the mouse" is shown in Fig. 51.3, which demonstrates these relationships.

At the sentence level the set of semantic values that language can express expands greatly (Pinker, 1994). Sentences convey aspects of the structure of events and states in the world. These semantic values collectively are known as the propositional content of a sentence. These values include thematic roles (information about who did what to whom), attribution of modification (information about which adjectives go with which nouns, such as the fact that in the sentence "The big boy chased the little girl," the boy is big and girl is little), scope of quantification (information about what

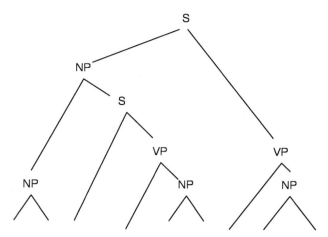

The dog that scratched the cat killed the mouse

FIGURE 51.3 Syntactic structure of the sentence "The dog that scratched the cat killed the mouse" indicating the hierarchical structure responsible for its meaning.

items are included in the scope of quantifiers, such as the fact that in the sentence, "None of the boys wearing hats was cold," the quantifier "none" applies to the boys wearing hats, not just to the boys), the reference of pronouns and other anaphoric elements (information about which words in a set of sentences refer to the same items or actions, such as the fact that in the sentence "The brother of the visitor shaved himself," "himself" refers to "brother" and not to "visitor"), and so on.

Sentences are a crucial level of the language code because the propositions they express make assertions about the world. These assertions can be added to an individual's knowledge of the world (i.e., they can add to semantic memory). Because they can be true or false (something that is not possible with words), they can be entered into logical systems. They are likely critical in certain aspects of thinking, such as decision making and planning actions.

Discourse

The propositional meanings conveyed by sentences are entered into higher order structures that constitute the discourse level of linguistic structure. Discourse includes information about the general topic under discussion, the focus of a speaker's attention, the novelty of the information in a given sentence, the temporal order of events, causation, and so on. Discourse also conveys information about the intentions of a speaker and a listener and the current attentional foci.

The structure of discourse involves establishing relationships between items and propositions. Some part of this structure is derived from the form in which propositions are expressed. In a sequence of sentences such as

> *John and Henry went to Bill's last night. They were really glad they did. The raved about the dessert the whole next day.*

The reader will take "they" as referring to John and Henry, partly because "John and Henry" and "they" are the subjects of successive sentences. However, part of the information conveyed at the discourse level depends upon associations and inferences that listeners make. For instance, in the same passage, the reader will likely infer that John and Henry ate dinner at Bill's—something not stated in the passage but that makes the sentences in the passage coherent.

Language Processing

Current models of language processing subdivide functions such as reading, speaking, auditory comprehension, and writing into many different, semi-independent components, which are sometimes called "modules" or "processors." These components can be further divided into variable numbers of highly specialized operations, such as those involved in mapping features of the acoustic signal onto phonemes or in constructing syntactic structures from words. Each operation accepts only particular types of representations as input and produces only specific types of representations as output.

Models of language processing often are expressed as flow diagrams (or "functional architectures") that indicate the sequence of operations of the different components that perform a language-related task. Figure 51.4 presents a model indicating the sequence of activation of components of the lexical processing system. Figure 51.4 simplifies information flow in four ways. First, it does not specify the nature of the operations in each of the major components of the system. Second, it does not fully convey the extent to which the components of the system operate in parallel. Third, it does not convey the extent of feedback among the components of the system. Lastly, not all components are represented. Despite these simplifications, the model captures enough aspects of information processing in the language system to give an idea of what functional architectures of language processing look like.

Beyond specifying the flow of information through the language processing system, researchers have begun to understand the operating characteristics of its components. Several important features of language processing operations have been demonstrated in experimental laboratories or in studies of patients.

1. They are specialized for language. For instance, recognition of phonemes is probably accomplished by mechanisms that separate very early in the processing stream from those that recognize other auditory stimuli.
2. They are obligatorily activated when their inputs are presented to them. For instance, if we attend to a sound that happens to be the word "elephant," we must hear and understand that word; we cannot hear this sound as just a noise.
3. They generally operate unconsciously. We usually have the subjective impression that we are extracting another person's meaning and producing linguistic forms without being aware of the details of the sounds of words, sentence structure, and so on.
4. They operate quickly and accurately. For instance, it has been estimated on the basis of many different psycholinguistic experimental techniques that spoken words usually are recognized less than 125 ms after their onset—that is, while they are still being uttered. The speed of the language processing system as a whole occurs because of the speed of each of its components, but also is achieved because of the massively parallel functional architecture of the system, which leads to many components of the system being simultaneously active.
5. They require processing resources. Although we do not appreciate it consciously, language processing demands some effort. It is unclear whether there are separate pools of processing resources for each language processing component or for language processing as a whole or whether language processing can "borrow" resources from other systems if it needs to.

We have discussed language processing in the auditory-oral modality. However, humans who are deaf can develop language in the visual and gestural modalities, and the structures of these language are essentially the same as those of spoken language, once allowances are made for the fact that they are signed instead of spoken. Languages can also be represented orthographically. Orthographies range from alphabetical scripts such as English, to syllabic, consonantal, and ideographic orthographies. Orthography is not acquired in the way spoken and signed language is; reading and writing require instruction to master. Many of the features of spoken language processing we have reviewed apply to the processing of spoken, written, and signed language as well. We cannot review

COMPREHENSION

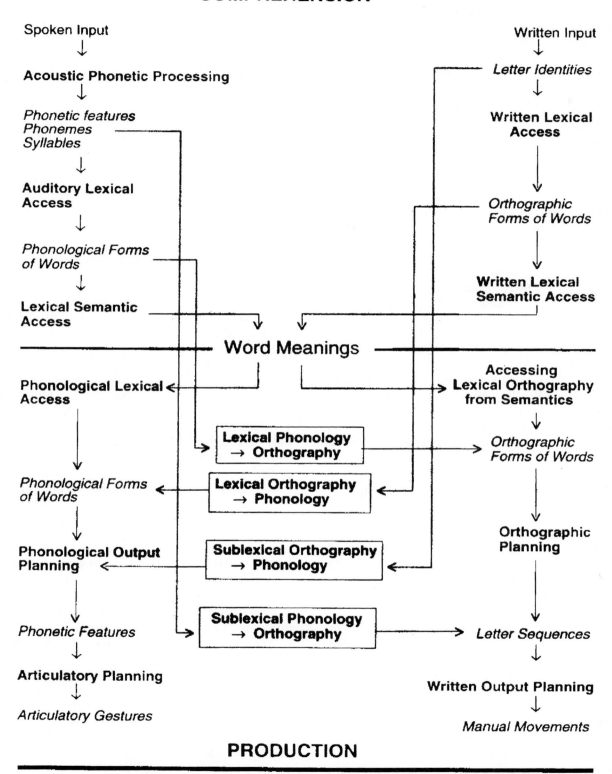

FIGURE 51.4 A model of the major psycholinguistic operations involved in processing simple words.

these modalities of language use here; for an introduction to signed language and its neurology, see Bellugi *et al.* (1990); for a discussion of orthographic representational systems, see Henderson (1982); for models of written language processing, see Plaut *et al.* (1996).

The operations of the language processing system are regulated by a variety of control mechanisms. These control mechanisms include both ones internal to the language processor itself and those that are involved in other aspects of cognition. The first category—language-internal control mechanisms—consists of a large number of operations that schedule psycholinguistic operations on the basis of the ongoing nature of a given psycholinguistic task. The second category of control mechanisms—those that are related to cognitive processing outside the language system—determine what combinations of processors become active in order to accomplish different tasks such as reading, repeating what one has heard, and taking notes on a lecture.

Functional communication involving the language code occurs when people use these processors to inform others, to ask for information, to get things done, and so on. The use of language is thus one type of intentional action. There are conventions (or rules) that regulate the use of language for these purposes. However, it is a mistake to think that language is only used, or even mainly used, for functional communication and that language is, at its core, a communication system. The heart of human language is a system of representations relating forms to meaning. In fact, as Chomsky has pointed out, far and away the most common use of language is not interpersonal communication at all, but talking to oneself.

Language Universals, Language Acquisition and Language Evolution

By some counts, there are over 6500 languages in the world. These languages differ in many ways from one another: they have different vocabularies, use different sounds, and have different ways of forming words and different syntactic rules. However, beyond these surface level differences, there are features that are common to all human language. For instance, the sound systems of all human languages consist of alternations of consonants and vowels; no language forms different words by varying features of sound such as its loudness or whether a word is whispered. These "linguistic universals" form a framework within which the features of individual languages occur.

Linguistic universals can arise because of universal features of motor or sensory systems or, possibly, because of constraints on cognitive computational capacities. They may also arise because language itself is constrained to have certain features. Chomsky's work in the area of syntax is the best known and most controversial example of analyses along these lines. Chomsky and his colleagues and students have argued that very abstract features of syntax are common to all languages. There have been numerous suggestions as to what these features are; one recent proposal is that they include the ability to generate an infinite number of expressions from a finite set of elements (Hauser *et al.*, 2002). Many linguists believe that the structure of language is much less abstract than Chomsky and his colleagues have suggested and, accordingly, that universal "laws" of linguistic structure are also less abstract, mostly taking the form of statements that the presence of one type of element in a language implies that a second type will also be present.

The issues of how abstract linguistic representations are and what universal properties of language exist are bound up with two major issues—how language is acquired and how language evolved—that are in turn relevant to the neural basis for language. Chomsky's view is that language acquisition consists of fixing the parameters in an innate language template that correspond to the constraints on possible forms of human language that are present in the language a child is exposed to. Models of language that see language structures as much less abstract make very different claims about the nature of innate features of language and the balance between innate knowledge and learning in the development of language abilities (e.g., Tomasello, 2003). In the past few years, some theorists have argued that the bulk of language acquisition is the result of a massive pattern-association process, supported by mechanisms that can be simulated in "connectionist" (or "neural net") models (Seidenberg, 1997). Although such models have had important successes in simulating many aspects of language processing at the single word level, they have not been able to account for the most complex combinatorial features of syntax. However, these models tend to be more realistic from a neurological point of view, whereas the neural mechanisms that underlie the representation of the abstract categorical features of language proposed by Chomsky and many linguists remain quite mysterious.

The same disagreements affect ideas about how language evolved. In Chomsky's model, a critical neural development is the emergence of structures and processes in the brain that support recursive properties of language in the human brain. The recent finding that starlings can discriminate recursive from nonrecursive sequences of notes (Gentner *et al.*, 2006) implies that these neural structures are not unique to humans, but if recent studies showing that some species of monkeys do not recognize recursive structures are true of all higher primates, the evolutionary path for these features must have taken a number of divergent paths and reemerged in humans. The major alternative to the view that language evolution consisted of the emergence of neural structures capable of supporting abstract organizational features that are found in all human languages is that language rests upon an evolutionary change that supports much less complex and abstract representations and the complexity of human language is a cultural development, like architectural complexity.

One well-known model of this sort that has a neural basis has been developed by Rizzolati and Arbib (1998), who have argued that the neural structures that are critical for language are modifications of the mirror neurons found in monkeys that fire both when the monkey grasps particular objects and when the animal sees a person grasping those objects. In Arbib and Rizzolati's view, these neurons encode the relation of a number of actors to an action and an object, and thus lay the basis for neural mechanisms that allow an arbitrary number of substitutions of actors and objects in relation to the action described by a verb. In their view, the complexity of human language is a cultural development, like architecture, not due to massive change in neural capacity due to a sudden evolutionary change. Support for this position comes from findings that suggest that culture influences the forms in a language. A much cited recent study (Everett, 2005), for instance, argues that one language spoken by an isolated group of Amazon natives lacks recursive structures because people do not need them to convey messages in that culture. If correct, such analyses would be critical evidence in favor of a minimal evolutionary change and cultural development being the basis for human languages, but the data are very sparse and hard to verify, and the question remains open.

There are very limited empirical data regarding the genetic basis for human language development. Lieberman (2006) argues that the critical evolutionary change that made for human language was not specifically neural, but consisted of the descent of the root of the tongue into the oropharynx, which allows humans to produce the inventory of speech sounds of a language. In his view, two genetic effects—increases in the size of the hominid brain related to the ASPM gene and, more importantly, changes in the ability to exploit the increased channel capacity in humans by changes in the FOXP2 gene—underlie the use of the resulting increase in the vocal repertoire for communicative purposes. Deficits in the FOXP2 gene in a single family have been associated with a variety of language and articulation problems and the FOXP2 gene has been referred to as a (or "the") "language gene" by some researchers.

However, the deficits in affected family members are not specific to syntax or word formation, and, as Lieberman argues, it is more likely that this gene codes for neuronal features of subcortical gray nuclei, in particular the basal ganglia and, to a lesser extent, the cerebellum that make for more elaborate and efficient control of sequencing abilities than for cortical structures that represent and process language *per se*.

Neural Organization for Language: An Inventory of Brain Structures Related to Language

Human language depends on the integrity of the unimodal and multimodal association cortex in the lateral portion of both cerebral hemispheres. This cortex surrounds the sylvian fissure and runs from the pars triangularis and opercularis of the inferior frontal gyrus (Brodmann's areas (BA) 45, 44: Broca's area) through the angular and supramarginal gyri (BA 39 and 40) into the superior temporal gyrus (BA22: Wernicke's area) (Fig. 51.5). For the most part, the connections of these cortical areas are to one another and to the dorsolateral prefrontal cortex and lateral inferior temporal cortex. These regions have only indirect connections to limbic structures. These areas consist of many different types of association cortex, devoted not to sensation or motor function but to a more abstract type of analysis. The nature of this cortex and its patterns of connectivity are thought to combine to give

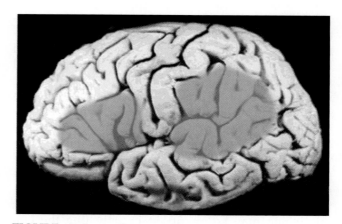

FIGURE 51.5 A depiction of the left hemisphere of the brain showing the main language areas. The area in the inferior frontal lobe is known as Broca's area, and the area in the superior temporal lobe is known as Wernicke's area, named after the nineteenth century physicians who first described their roles in language. Broca's area is adjacent to the motor cortex and is involved in planning speech gestures. It also serves other language functions, such as assigning syntactic structure. Wernicke's area is adjacent to the primary auditory cortex and is involved in representing and recognizing the sound patterns of words.

language its enormous representational power and to allow its use to transcend biological immediacy.

The evidence that language involves these cortical regions originally was derived from deficit-lesion correlations. Patients with lesions in parts of this cortex have been described who have had long-lasting impairments of language ("aphasia"). Disorders affecting language processing after perisylvian lesions have been described in many different types of disease, in all languages that have been studied, in patients of all ages and both sexes, and in both first and subsequent tongues, indicating that this cortical region is involved in language processing independent of these factors. Functional neuroimaging studies have documented increases in regional cerebral blood flow (rCBF) using PET or blood oxygenation level-dependent (BOLD) signal using fMRI in tasks associated with language processing in this region. Event-related potentials whose sources are likely to be in this region have been described in relationship to a variety of language processing operations. Stimulation of this cortex by direct application of electrical current during neurosurgical procedures interrupts language processing. These data all converge on the conclusion that language processing is carried out in the perisylvian cortex.

Regions outside the perisylvian association cortex also appear to support language processing. Working outward from the perisylvian region, evidence shows that the modality of language use affects the location of the neural tissue that supports language, with written language involving the cortex closer to the visual areas of the brain and sign language involving brain regions closer to those involved in movements of the hands than movements of the oral cavity. Some ERP components related to processing improbable or ill-formed language are maximal over high parietal and central scalp electrodes, suggesting that these regions may be involved in language processing. Both lesion studies in stroke patients and functional neuroimaging studies suggest that the inferior and anterior temporal lobe is involved in representing the meanings of nouns. Activation studies also implicate the precentral gyrus and premotor cortex in word meaning. Injury to the supplementary motor cortex along the medial surface of the frontal lobe can lead to speech initiation disturbances; this region may be important in activating the language processing system, at least in production tasks. Activation studies have shown increased rCBF and BOLD signal in the cingulate gyrus in association with many language tasks. This activation, however, appears to be nonspecific, as it occurs in many other, nonlinguistic, tasks as well. It has been suggested that it is due to increased arousal and deployment of attention associated with more complex tasks.

Subcortical structures may also be involved in language processing. Several studies report aphasic disturbances following strokes in deep gray matter nuclei (the caudate, putamen, and parts of the thalamus). It has been suggested that subcortical structures involved in laying down procedural memories for motor functions—in particular, the basal ganglia—are involved in "rule-based" processing in language, such as regular aspects of word formation, as opposed to the long-term maintenance of information in memory, as occurs with simple words and irregularly formed words. The thalamus may play a role in processing the meanings of words. In general, subcortical lesions cause language impairments when the overlying cortex is abnormal (often the abnormality can be seen only with metabolic scanning techniques), and the degree of language impairment is better correlated with measures of cortical than subcortical hypometabolism. It may be that subcortical structures serve to activate a cortically based language processing system but do not themselves process language.

The cerebellum also has increased its rCBF in some activation studies involving both language and other cognitive functions. This may be a result of the role of this part of the brain in processes involved in timing and temporal ordering of events, or in its being directly involved in language and other cognitive functions.

In summary, a large number of brain regions are involved in representing and processing language. Ultimately, they all interact with one another as well as with other brain areas involved in using the products of language processing to accomplish tasks. In this sense, all these regions are part of a "neural system," but this does not imply that every part of this entire system is responsible for computing specific linguistic representations. The most important part of the neural system involved in language is the dominant (usually left-sided) perisylvian cortex.

Lateralization

Most language processing goes on in one hemisphere, called the "dominant" hemisphere. Which hemisphere is dominant shows considerable individual differences and bears a systematic relationship to handedness. In about 98% of right-handed individuals, the left hemisphere is dominant. The extent to which left hemisphere lesions cause language disorders is influenced by the degree to which an individual is right-handed and by the number of nonright-handers in his or her family. About 60 to 65% of nonright-handed individuals are left-hemisphere dominant; about 15 to 20% are right-hemisphere dominant; and the remainder appear to use both hemispheres for lan-

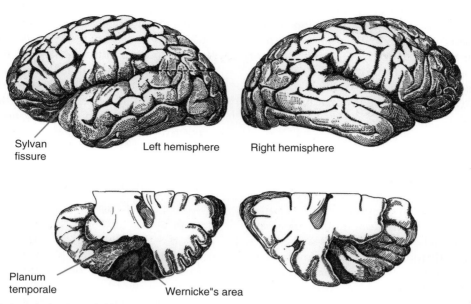

FIGURE 51.6 Depiction of a horizontal slice through the brain showing asymmetry in the size of the planum temporale related to lateralization of language.

guage processing. The relationship of dominance for language to handedness suggests a common determination of both, probably in large part genetic.

The neural basis for lateralization was first suggested by Geschwind and Levitsky (1968), who discovered that part of the language zone (the planum temporale—a portion of the superior temporal lobe; Fig. 51.6) was larger in the left than in the right hemisphere. Subsequent studies have confirmed this finding and identified specific cytoarchitectonically defined regions in this posterior language area that show this asymmetry. Several other asymmetries that may be related to lateralization also have been identified. The exact relationship between size and function is not known, as there are instances of individuals whose dominant hemisphere is not the one with the larger planum temporale. In general, however, relative size is a good predictor of lateralization. This is another example of the "bigger is better" principle that we saw applied to song nuclei in birds.

Although language was the first function known to be lateralized, and is still the best example of a lateralized function, it is not completely lateralized. Although not as important in language functioning as the dominant hemisphere, the nondominant hemisphere is involved in many language operations. Evidence from the effects of lesions and split brain studies, experiments using presentation of stimuli to one or the other hemisphere in normal subjects, and activation studies all indicates that the nondominant hemisphere understands many words, especially concrete nouns,

and suggests that it is involved in other aspects of language processing as well. Some language operations may be carried out primarily in the right hemisphere. The best candidates for these operations are ones that pertain to processing the discourse level of language, interpreting nonliteral language such as metaphors, and appreciating the tone of a discourse; for example, the fact that it is humorous. Some scientists have developed models of the sorts of processing that the right hemisphere carries out. For instance, it has been suggested that the right hemisphere codes information in a more general way compared to the left, representing the overall structure of a stimulus as opposed to its details. This may be true for language as well as for other functions, such as visual perception. This and other suggestions provide the bases for ongoing research programs into the nature of language processing in the right hemisphere.

Organization of the Perisylvian Cortex for Language Processing

The way the perisylvian cortex is organized to support these functions has been the subject of much investigation. Two general classes of theories of the relationship of parts of the perisylvian association cortex to components of the language processing system have been developed. One is based on "holist" or distributed views of neural function and one on localizationist principles. The basic tenet of holist/distributed theories is that linguistic representations are distributed widely and that language processing components rely on broad areas of the association cortex. Localizationist theories maintain that language processing components are localized in specific parts of the cortex. The first author of this chapter has argued that some language operations are localized, but that their localization varies across the population.

Holist Theories

Lashley (1950) identified two functional features of holist/distributed models that determine the effects of lesions upon performance: equipotentiality (every portion of a particular brain region carries out a particular function in every individual) and mass action (the larger the neuronal pool that carries out a function, the more efficiently that function is accomplished). The features of equipotentiality and mass action jointly entail that (1) lesions of similar sizes anywhere in a specified brain region have equivalent effects on function and (2) the magnitude of any functional deficit is directly proportional to the size of a lesion in this specified area. Modern work with parallel distributed processing models provides formal models of holist concepts.

There is some evidence for holist theories. Lesions throughout the language area have been found in some studies to have similar effects on language functions, consistent with the principle of equipotentiality. There is an effect of lesion size on the overall severity of functional impairments in several language spheres, consistent with the principle of mass action. These results are consistent with a distributed, or holistic, neural basis for linguistic representations and processes. However, they do not necessarily show that these models are correct because they also could result from variability in the localization of language processing components across the population.

The functional neuroimaging literature does not support holist models; every study of a language function has yielded neurovascular responses in one area or a small number of areas. However, this may be a result of the way functional neuroimaging data are analyzed: standard analyses are designed to identify small areas in which there are large changes in BOLD signal (relative to the variability of BOLD signal in the entire brain), not large areas in which there are small changes. Thus, if language operations were widely distributed and resulted in small increases in neurophysiological activity throughout the perisylvian cortex, it is not clear that these changes would be detected by present day functional neuroimaging.

Holist models appear to apply to individual areas of the brain. Modern "neural net" (connectionist) computer simulations provide a formal basis for holist models of language processing. As noted earlier, these models have been extremely successful in modeling certain aspects of normal language processing and language deficits. A very reasonable view is that different neural areas support different language functions and that each of these areas works as a whole, as holist models claim. Holism thus may not be wrong; it just needs to be applied on the right scale. As has been said, the brain may be "globally local and locally global."

Localizationist Theories

Evidence against any holist model is the finding that multiple individual language deficits arise in patients with small perisylvian lesions, often in complementary functional spheres. For instance, some patients have trouble producing the small grammatical function words of language (such as *the*, *what*, *is*, *he*), whereas others have trouble producing common nouns. The existence of these two disorders indicates that the tissue involved in producing function words is not involved in producing common nouns in the first set of patients, and vice versa in the second set. Strong evidence against holist models also comes from activation studies that demonstrate vascular responses to particular types of language stimuli in restricted cortical areas. How language is organized at this finer-grained level in the perisylvian association cortex is an ongoing topic of research.

Classical Clinical Localizationist Models

The first localizationist theories emerged from clinical observations in the mid- and late nineteenth century. The pioneers of aphasiology—Paul Broca, Karl Wernicke, John Hughlings Jackson, and other neurologists—described patients with lesions in the left inferior frontal lobe whose speech was hesitant and poorly articulated and other patients with lesions more posteriorly, in the superior temporal lobe, who had disturbances of comprehension and fluent speech with sound and word substitutions. These correlations led to the theory that language comprehension went on in the unimodal auditory association cortex (Wernicke's area, Brodmann area 22) adjacent to the primary auditory cortex (Heschl's gyrus, Brodmann area 41) and that motor speech planning went on in the unimodal motor association cortex in Broca's area (Brodmann areas 44 and 45) adjacent to the primary motor cortex (Brodmann area 4). These theories incorporated the idea that localization of a language operation depends on the way it is related to sensory and motor processes. According to this view, speech planning goes on in Broca's area because Broca's is immediately adjacent to the motor area responsible for movement of the articulators and Wernicke's area is involved in comprehension because it is immediately adjacent to primary auditory cortex.

These ideas and models were extended by Norman Geschwind and colleagues in the 1960s and 1970s. Geschwind (1965) added the hypothesis that word meaning was localized in the inferior parietal lobe (Brodmann areas 39 and 40) because word meanings consist of associations between sounds and properties of objects, and the inferior parietal lobe is an area of multimodal association cortex to which fibers from the unimodal association cortex related to audition, vision, and somasthesis project. Geschwind's model remains the best-known localizationist model of the functional neuroanatomy of language and is widely cited in clinical practice. The model is said to receive support from the existence of about 10 aphasic syndromes that serve the clinical purpose of helping to localize lesions.

However, despite its widespread use, this model has distinct limitations. Although some of the classic aphasic syndromes describe deficits at the level of word formation and sentence production, the model itself deals only with words, not with other levels of the language code. From a linguistic and psycholinguistic point of view, the syndromes all are composed of many processing deficits, which are different in different patients with the same syndrome. The correlations between syndromes and lesions are far from perfect. They are found primarily in the chronic stage of recovery from stroke, not in acute or subacute phases of stroke and not in other diseases, and even in chronic strokes a very large number of patients do not fall into the standard categories or their lesions do not occur where the model predicts they will. Indeed, the correlations between syndromes and lesions may reflect the fact that speech fluency is a critical dimension along which patients are classified, and nonfluent aphasias tend to be due to anterior lesions that affect motor structures whereas fluent aphasias tend to be due to posterior lesions that do not.

In the past 20 years or so, a number of researchers have sought to reinterpret the classic syndromes in terms of more specific components of the language processing system. For instance, the classic syndrome of conduction aphasia, in which patients have difficulty producing the correct sequences of sounds for words despite being able to produce individual sounds clearly and knowing what the words mean, has been divided into two separate deficits—one affecting short-term verbal memory and one affecting speech planning (Shallice and Warrington, 1972). Both of these more fine-grained disorders have been studied in considerable detail, with the consequence that we now have a much better understanding of what aspects of short-term memory and speech planning are affected in these patients. In fact, these studies have led to important new ideas about the nature of both short-term memory and word production. Prior to these studies, it was thought that short-term memory was a gateway to long-term memory, but patients with the short-term memory type of conduction aphasia have excellent long-term memory, making this sequential model untenable; in another vein, additional study of patients with short-term memory disturbances has led to the identification of a short-term semantic memory system, not previously considered to exist on the basis of work with normals (Martin and He, 2004). In the area of word sound planning, study of patients with the second type of conduction aphasia (reproduction conduction aphasia) has led to computer-implemented models of word planning (Dell et al., 1996) that differ in some interesting and important ways from models that emerge from studies of normals (Levelt et al., 1999).

These modern studies of the classic aphasic syndromes have been part of a more radical transition to describing deficits in aphasic patients in terms of specific linguistic representations and psycholinguistic processes. The next sections review studies of location of two language processes—word recognition and comprehension, and syntactic operations—that cover the spectrum of abstractness of linguistic representations.

Lexical Access and Word Meaning

Evidence from normal and impaired human subjects suggests that temporo-spectral acoustic cues to feature identity appear to be integrated in unimodal auditory association cortex lying along the superior temporal sulcus immediately adjacent to the primary auditory koniocortex (Binder et al., 2000). Some researchers have suggested that the unconscious, automatic activation of features and phonemes as a stage in word recognition under normal conditions occurs bilaterally, and that the dominant hemisphere is the sole site only of phonemic processing that is associated with controlled processes such as subvocal rehearsal and conscious processes such as explicit phoneme discrimination and identification, making judgments about rhyme, and other similar functions.

Based on functional neuroimaging results, activation of the long-term representations of the sound patterns of words is thought to occur in the left superior temporal gyrus. Scott and her colleagues have argued that there is a pathway along this gyrus and the corresponding left superior temporal sulcus such that

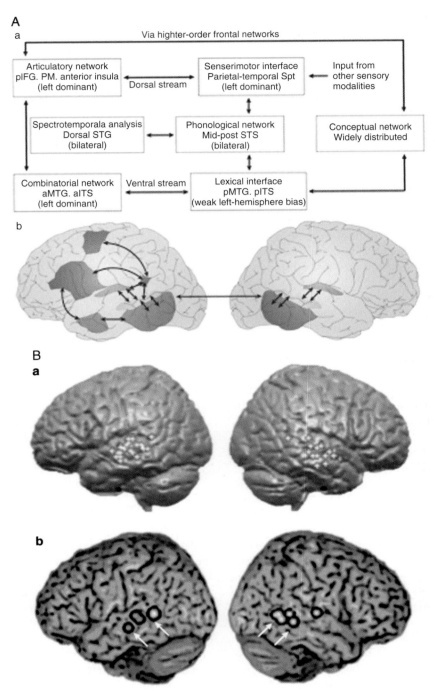

FIGURE 51.7 (A) Depiction of the lateral surface of the brain showing areas involved in the functional neuroanatomy of phonemic processing. H is Heschl's gyrus, the primary auditory cortex. STP is the superior temporal plane, divided into posterior and anterior areas. STG is the superior temporal gyrus. Traditional theories maintain that pSTP and STG are the loci of phonemic processing. Hickok and Poeppel (2000) argue that these areas in both hemispheres are involved in automatic phonemic processing in the process of word recognition. Other research suggests that more anterior structures, aSTP and the area around the superior temporal sulcus (STS), are involved in these processes. The inferior parietal lobe (AG, angular gyrus; SMG, supramarginal gyrus) and Broca's area (areas 44 and 45) are involved in conscious controlled phonological processes such as rehearsal and storage in short-term memory (from Hickok and Poeppel, 2007). (B) Activation of left superior temporal sulcus by the contrast of discrimination of phonemes and discrimination of matched nonlinguistic sounds (from Hickok and Poeppel, 2007).

word recognition occurs in a region anterior and inferior to primary auditory cortex, and that word meanings are activated further along this pathway in anterior inferior temporal lobe bilaterally (Scott and Johnsrude, 2003; Scott and Wise, 2004). This pathway constitutes the auditory counterpart to the visual "what" pathway in the inferior occipital-temporal lobe (Fig. 51.7).

Speech perception is connected to speech production, especially during language acquisition when imitation is crucial for the development of the child's sound inventory and lexicon. On the basis of lesions in patients with reproduction conduction aphasia, the neural substrate for this connection has been thought to consist of the arcuate fibers of the inferior longitudinal fasciculus, which connect auditory association cortex (Wernicke's area in the posterior part of the superior temporal gyrus) to motor association cortex (Broca's area in the posterior part of the inferior frontal gyrus). Recent functional neuroimaging studies and neural models have partially confirmed these ideas, providing evidence that integrated perceptual-motor processing of speech sounds and words makes use of a "dorsal" pathway separate from that involved in word recognition (Hickok and Poeppel, 2007).

As described earlier, traditional neurological models maintained that the meanings of words consist of sets of neural correlates of the physical properties that are associated with a heard word, all converging in the inferior parietal lobe. It is now known that most lesions in the inferior parietal lobe do not affect word meaning and functional neuroimaging studies designed to require word meaning do not tend to activate this region. Evidence is accruing that the associations of words include "retroactivation" of neural patterns

back to unimodal motor and sensory association cortex (Damasio, 1989), and that different types of words activate different cortical regions. Verbs are more likely to activate frontal cortex, and nouns temporal cortex, possibly because verbs refer to actions and nouns refer to static items. More fine-grained relations between words and activation of sensory-motor brain areas have been described. Hauk *et al.* (2004) has found activation in primary motor and premotor cortex at the homuncular level of the leg when subjects recognize the written word "kick," the mouth when the recognize "lick," and the hand when the recognize "pick" (Fig. 51.8).

Both deficits and functional activation studies have suggested that there are unique neural loci for the representation of object categories such as tools (frontal association cortex and middle temporal lobe), animals and foods (inferior temporal lobe and superior temporal sulcus), and faces (fusiform gyrus) (see Caramazza and Mahon, 2006, for review), and some of these localizations are adjacent to areas of the brain that support perception of features that are related to these categories (animals activate an area of lateral temporal adjacent to the area sensitive to biological motion). Debate continues about how these category effects arise, and whether these activations should be taken as reflecting the meanings of these words. At the same time as these specializations receive support, evidence from patients with semantic dementia and from functional neuroimaging indicates that another critical part of the semantic network that represents word meanings and concepts is located in the anterior inferior temporal lobes.

Syntactic Operations

We shall review the use of syntactic structure to determine sentence meaning in the comprehension process. Deficits of this ability typically are established by showing that patients can understand sentences with simple syntactic structures (e.g., *The boy chased the girl.*) and sentences with complex syntactic structures in which the relationships between nouns and verbs can be inferred from a knowledge of real-world events (e.g., *The apple the boy ate was red.*), but not sentences with complex structures in which the relationships between nouns and verbs depend on a syntactic structure that needs to be constructed (e.g., *The boy the girl pushed was tall.*). A variety of models have been proposed regarding the neural basis for sentence comprehension, ranging from localization, though distribution to variable localization.

A very well-known theory—the trace deletion model (Grodzinsky, 2000)—has centered on the finding that Broca's aphasics who produce "agrammatic"

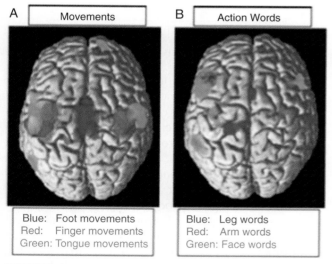

FIGURE 51.8 (A,B) Activation of areas of the motor strip and premotor area by actions and action words (from Hauk *et al.*, 2004).

speech (speech that lacks function words and morphological markers) often have difficulty understanding sentences whose meanings depend the use of these elements to construct syntactic structures. The similarity of comprehension deficits and expressive language problems of these patients—both involving problems with function words, morphology, and syntax—has suggested that these deficits are connected, and that they are linked to lesions in the area of the brain that are associated with expressive agrammatism.

However, the data that are available indicate that deficits in syntactic processing in sentence comprehension occur in all aphasic syndromes and after lesions throughout the perisylvian cortex (Caplan *et al.*, 1985, 1996; Caramazza *et al.*, 2001). Conversely, patients of all types and with all lesion locations have been described with normal syntactic comprehension (Caplan *et al.*, 1985). We have just reported the most detailed study of patients with lesions whose syntactic comprehension has been assessed (Caplan *et al.*, 2007). We studied 42 patients with aphasia secondary to left hemisphere strokes and 25 control subjects for the ability to assign and interpret three syntactic structures in enactment, sentence-picture matching and grammaticality judgment tasks. In regression analyses, lesion measures in both perisylvian and nonperisylvian regions of interest predicted performance after factors such as age, time since stroke, and total lesion volume had been entered into the equations. Patients who performed at similar levels behaviorally had lesions of very different sizes, and patients with equivalent lesion sizes varied greatly in their level of performance. The data are consistent with a model in which the neural tissue that is responsible for the operations underlying sentence comprehension and syntactic processing is localized in different neural regions, possibly varying in different individuals.

Functional neuroimaging studies have led many researchers to articulate models in which one or another aspect of parsing or interpretation is localized in Broca's area, or in portions of this region (Grodzinsky and Freiderici, 2006). However, most neuroimaging studies actually show that multiple cortical areas are activated in tasks that involve syntactic processing (Fig. 51.9). Overall, the data are inconsistent with invariant localization, and suggest multifocal or variable localization of the areas that are sufficient to support syntactic processing within the language area across the adult population. It may be that language operations more closely tied to early perceptual and late motor planning processes are invariantly localized whereas those that compute more abstract aspects of the language code can be supported by a larger number of regions of association cortex.

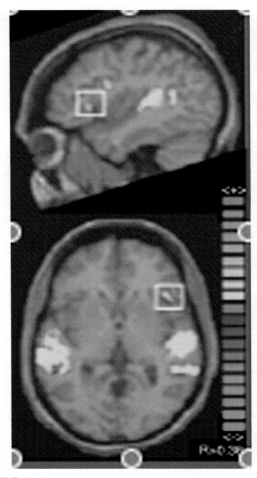

FIGURE 51.9 Activation of Broca's area and superior temporal sulcus when subjects processed sentences with particular syntactic features (questioned noun phrases) compared to matched sentences without those features (from Ben Shachar *et al.*, 2004).

CONCLUSIONS

Human language is a unique representational system that relates aspects of meaning to many types of forms (e.g., phonemes, lexical items, syntax), each with its own complex structure. Animal communication systems are neither as complex nor as powerful as human language. Nonetheless, they have some similarities to language and provide clues as to the neural basis of human language. Some language traits are part of a shared evolutionary ancestry with other species, others represent convergent, independent evolution of analogous processes, and yet others must be unique to our species. Perhaps the most important lessons learned from animal studies are that complex communication systems can be innate, that learning can shape an innately specified range of behavioral possibilities, and that communication systems can rely on specific nuclei within the brain and be lateralized. Other features of the neural basis for animal communication systems, such as seasonal variation in the size of birdsong nuclei and the mechanisms that underlie this phenomenon, are not obviously relevant to human language.

Deficit-lesion correlations and neuroimaging studies are beginning to provide data about the neural structures involved in human language. It appears that one area of the brain—the left perisylvian association cortex—is especially important in representing and processing language (although other regions are involved as well), and that, within this area, particular language operations are localized in specific regions. The exact details of these localizations are just beginning to be understood, as the modern tools of cognitive neuroscience are applied to this problem.

References

Bellugi, U., Poizner, H., *et al.* (1990). Mapping brain function for language: Evidence from sign language. *In* "Signal and Sense: Local and Global Order in Perceptual Maps" (G. Edelman, W. Gall, and W. Cowan, eds.), pp. 521–543. Wiley-Liss, New York.

Ben-Shachar, M., Palti, D., Grodzinsky, Y. (2004). The neural correlates of syntactic movement: Converging evidence from two fMRI experiments. *Neuroimage* 21, 1320–1336.

Bickerton, D. (1990). "Language and Species." University of Chicago Press, Chicago.

Binder, J. (2000). The new neuroanatomy of speech perception. *Brain* 123, 2371–2372.

Caplan, D., Baker, C., and Dehaut, F. (1985). Syntactic determinants of sentence comprehension in aphasia. *Cognition* 21, 117–175.

Caplan, D., Hildebrandt, H., and Makris, N. (1996). Location of lesions in stroke patients with deficits in syntactic processing in sentence comprehension. *Brain* 119, 993–949.

Caplan, D., Waters, G., Kennedy, D., Alpert, N., Makris, N., DeDe, G., Michaud, J., and Reddy, A. (2007). A study of syntactic processing in Aphasia II: Neurological aspects. *Brain and Language.*

Caramazza, A., Capitani, E., Rey, A., and Berndt, R. S. (2001). Agrammatic Broca's aphasia is not associated with a single pattern of comprehension performance. *Brain and Language* 76, 158–184.

Caramazza, A. and Mahon, B. Z. (2006). The organization of conceptual knowledge in the brain: The future's past and some future directions. *Cognitive Neuropsychology* 23, 13–38.

Cheney, D. L. and Seyfarth, R. M. (1990). "How Monkeys See the World." University of Chicago Press, Chicago.

Chomsky, N. (1965). "Aspects of the Theory of Syntax." MIT Press, Cambridge, MA.

Chomsky, N. (1995). "The Minimalist Program." Praeger, New York.

Damasio, A. (1989). Time-locked multiregional retroactivation: A systems-level proposal for the neural substrates of recall and recognition. *Cognition* 33, 25–62.

Dell, G. S., Schwartz, M. F., Martin, N., Saffran, E. M., and Gagnon, D. A. (1997). Lexical access in aphasic and nonaphasic speakers. *Psychological Review* 104, 801–838.

Everett, D. (2005). Cultural constraints on grammar and cognition in pirahã: Another look at the design features of human language. *Current Anthropology* 76, 621–646.

Gentner, T. Q., Fenn, K. M., Margoliash, D., and Nusbaum, H. C. (2006). Recursive syntactic pattern learning by songbirds. *Nature* 440, 1204–1207.

Geschwind, N. (1965). Disconnection syndromes in animals and man. *Brain* 88, 237–294, 585–644.

Geschwind, N. and Levitsky, W. (1968). Human brain: Left-right asymmetries in temporal speech region. *Science* 161, 186–187.

Gould, J. L. (1982). "Ethology." Norton, New York.

Gould, J. L. and Gould, C. G. (1995). "The Honey Bee," 2nd ed. Freeman, New York.

Gould, J. L. and Gould, C. G. (1996). "Sexual Selection," 2nd ed. Freeman, New York.

Gould, J. L. and Gould, C. G. (1999). "The Animal Mind," 2nd ed. Freeman, New York.

Gould, J. L. and Gould, C. G. (2007). "Animal Architects," Basic Books, New York.

Gould, J. L. and Marler, P. (1987). The instinct to learn. *Sci. Am.* 256(1), 74–85.

Grodzinsky, Y. (2000). The neurology of syntax: Language use without Broca'a area. *Behavioral and Brain Sciences,* 23, 47–117.

Grodzinsky, Y. and Freiderici, A. (2006). Neuroimaging of syntax and syntactic processing. *Current Opinion in Neurobiology* 16, 240–246.

Hauk, O., Johnsrude, I., and Pulvermüller, F. (2004). Somatotopic representation of action words in human motor and premotor cortex. *Neuron* 41, 301–307.

Hauser, M., Chomsky, N., and Fitch, W. T. (2002). The faculty of language. *Science* 298, 1569–1579.

Henderson, L. (1982). "Orthography and Word Recognition in Reading." Academic Press, London.

Hickok, G. and Poeppel, D. (2007). The cortical organization of speech processing. *Nature Reviews* 8, 393–402.

Lashley, K. S. (1950). In search of the engram. *Symp. Soc. Exp. Biol.* 4, 454–482.

Levelt, W. J. M., Roelofs, A., and Meyer, A. S. (1999). A theory of lexical access in speech production. *Behavioral and Brain Sciences* 22, 43–44.

Lieberman, P. (2006). "Towards an Evolutionary Biology of Language." Belknap Press of Harvard University Press, Cambridge, MA.

Martin, R. C. and He, T. (2004). Semantic short-term memory deficit and language processing: A replication. *Brain and Language* 89, 76–82.

Plaut, D. C., McClelland, J. L., Seidenberg, M. S., and Patterson, K. (1996). Understanding normal and impaired word reading: Computational principles in quasi-regular domains. *Psychological Review* 103, 56–115.

Pulvermuller, F. (2005). Brain mechanisms linking language and action. *Nature Reviews* 6, 576–582.

Rizzolatti, G. and Arbib, M. A. (1998). Language within our grasp. *Trends in Neurosciences* 21(5), 188–194.

Scott, S. K. and Johnsrude, I. S. (2003). The neuroanatomical and functional organization of speech perception. *Trends in Neuroscience* 26, 100–107.

Scott, S. K. and Wise, R. J. S. (2004). The functional neuroanatomy of prelexical processing of speech. *Cognition* 92, 13–45.

Seidenberg, M. (1997). Language acquisition and use: Learning and applying probabilistic constraints. *Science* 275, 1599–1603.

Shallice, T. and Warrington, E. K. (1970). Independent functioning of verbal memory stores: A neuropsychological study. *Quarterly Journal of Experimental Psychology* 22, 261–273.

Tomasello, M. (2003). "Constructing a Language: A Usage Based Theory of Language Acquisition." Harvard University Press, Cambridge, MA.

Tulving, E. (1972). Episodic and semantic memory. *In* "Organization of Memory" (E. T. A. W. Donaldson, ed.), pp. 381–403. Academic Press, New York.

Suggested Readings

Grodzinsky, Y. and Amunts, K. (eds.) (2006). "Broca's Region." Oxford University Press, Oxford, UK.

Hauser, M. (1996). "The Evaluation of Communication." MIT Press, Cambridge, MA.

Posner, M. and Raichle, M. E. (1997). "Images of Mind." W. H. Freeman, New York.

Pulvermuller, H. (2002). "The Neuroscience of Language." Cambridge University Press, Cambridge, UK.

David N. Caplan and James L. Gould

The Prefrontal Cortex and Executive Brain Functions

INTRODUCTION

What controls your thoughts? How do you decide what to pay attention to? How do you act appropriately while dining in a restaurant or listening to a lecture? How do you plan and execute errands? How do you manage to pursue long-term goals, like obtaining a college degree, in the face of the many distractions that can knock you "off task"? In short, how does the brain manage to orchestrate the activity of millions of neurons to produce behavior that is willful, coordinated, and extends over time? This is called cognitive control, the ability of our thoughts and actions to rise above mere reactions to the immediate environment and be proactive: to anticipate possible futures and coordinate and direct thought and action to them. It is a hallmark of intelligent behavior.

Because cognitive control necessarily involves the coordination of many different brain processes, it no doubt depends on circuits that extend over much of the cerebral cortex. However, one cortical region, the prefrontal cortex (PFC), seems to play a central role. It is the cortical area that reaches the greatest relative size in the human brain and is thus thought to be the neural instantiation of the mental qualities that we think of as "intelligent." Accordingly, the PFC receives converging information from many brain systems processing external and internal information and it is interconnected with motor system structures needed for voluntary action.

Here, we discuss anatomical, neuropsychological, and neurophysiological evidence for, and theories of, the role of the PFC in cognitive control. We will begin with a discussion of the characteristics of volitional, or controlled, processes as they can provide predictions about the neural mechanisms likely to underlie them.

CONTROLLED PROCESSING

Behavior varies along a continuum from mindless, automatic behaviors to willful and purposeful (i.e., controlled) behaviors (see Barsalou, 1992). Automatic processes are reflexive. If a car or a predator is bearing down on us, we leap out of the way before we even have had a chance to form the intention to do so; our mind seems to "catch up" afterward. Automatic processes thus seem to depend on relatively straightforward, hardwired relationships between the brain's input and output systems. In neural terms, it seems that they depend on well-established, neural pathways just waiting to be triggered by the right sensory cue. That is, automatic processes are driven in a "bottom-up" fashion: they are determined largely by the nature of the sensory stimuli and whatever reaction they are most strongly wired to.

These relatively simple, hardwired, cue-response mappings are useful in many situations. Because they can simply be fired off by the environment, they are quick and can be executed without taxing the limited capacity of our conscious thoughts. But a creature that is only at the mercy of its immediate environment is not well equipped to deal with the ambiguity, and exploit the opportunity, of a complex and dynamic world. Sometimes, the course of action is not obvious because cues are ambiguous (i.e., they activate more than one possible internal representation), or multiple behaviors might be triggered and the one that is optimal for a given situation is at a disadvantage relative to better established (more automatic) alternatives. In these situations, ambiguity needs to be resolved by our internal states and intentions, by knowledge of possible and desired future outcomes (goals) and what means have been successful at achieving them in the past as well as their relative costs and benefits. This information is used to activate the same basic sensory, memory, and motor processes that are engaged during automatic behaviors. Only now, they are not triggered by cues from the environment. They are orchestrated in a top-down fashion by our expectations about goals and the means to achieve them. So, this mode of behavior is controlled in the sense that we (our knowledge and intensions) are in charge, not the environment. A tenet of modern studies of behavior is that this knowledge is obtained by learning mechanisms that detect and store associations between cues, internal states, and actions that predict goal attainment (reward) (see Dickinson, 1980). Insofar as primates are capable of navigating situations that involve diverse, arbitrary, relationships across a wide range of informational domains, it follows that a neural system for cognitive control must have access to information from many brain systems and the ability to encode the goal-relevant relationships between them.

The cognitive control system must also have the ability to select which sensory, memory, and motor processes are activated at a given moment. Selection is central because of our very limited capacity to engage in controlled behaviors. This is evident to anyone who has tried to talk on the phone and answer an e-mail at the same time; we can think about only a limited number of things at a time. This is in contrast to our high capacity for automatic processes. Because they can be triggered by the environment, any number of them can be fired off at once, as long as they do not come into direct conflict with one another. It makes sense that brain mechanisms for cognitive control would evolve with this capacity limitation. There is the tradeoff between amount of information and depth of analysis; focusing processing on a narrow band of information relevant to a current goal allows a much more elaborate analysis of a situation and available options. This single-mindedness also allows us to stay on track; processing information not relevant to a current goal increases the chance for distraction. In fact, in many views of cognition, control and attention are virtual synonyms. It follows that a neural system for cognitive control must have the infrastructure and mechanisms for selecting goal-relevant, and suppressing goal-irrelevant, processes throughout the cerebral cortex.

The cognitive control system must also have a way to deal with the gaps in time that are inevitable with goal-directed behaviors. Information about predicted goals and means must be brought online before the behaviors are executed and maintained until the current task is completed. Also, because goal-directed behaviors typically are extended over time, we often must wait for additional information or for a more opportune time to execute an action. The ability to keep goal-relevant information online and available for processing is called working memory (Baddeley, 1986). It is less critical for automatic processes because they are immediate: a stimulus triggers an already established circuit.

Finally, and perhaps most importantly, the cognitive control system must be highly plastic and flexible. Virtually all voluntary, goal-directed behaviors are acquired by experience, and given that humans and other primates are capable of rapid learning of new volitional behaviors, the cognitive control system must have an equal capacity for rapid learning. Primates are also remarkably flexible with their controlled behaviors. One may learn how to navigate a specific circumstance, but one can quickly generalize this to new situations, conjoin them with others, and rearrange them to produce novel behaviors. Goal-directed behaviors can also be temporarily interrupted and resumed at ease. One might temporarily divert oneself from wine shopping, for example, to pick up some cheese. By contrast, automatic behaviors are characterized by their rigidity. They are initiated by specific trigger events and are "ballistic"; once activated they tend to run off in pretty much the same way every time. It stands to reason that the cognitive control system must represent goal-relevant information in a format that is compatible with such flexibility.

Summary

There are two general modes of information processing that underlie thoughts and behaviors, automatic and controlled. Automatic processes are relatively simple, straightforward, reflexive reactions to whatever information happens to be flowing into the senses at a given moment. By contrast, controlled processes are engaged when a task has additional constraints and requires knowledge about goals and means.

From this brief review of some of the main characteristics of controlled processing, it is possible to make some predictions about the nature of any brain system involved in goal-directed (executive) control of cognitive functions:

1. It must have extensive and diverse connections to other regions of the brain in order to have access

to information from, as well as influence, widespread brain systems.

2. It needs to have information about goals and the means to achieve them, and this necessarily involves representing diverse multivariate relationships between a wide variety of sensory information, stored knowledge, actions, and consequences. It needs to be plastic and modifiable by experience in order to acquire this information.
3. There must be short-term storage mechanisms to keep goal-related information online and available until the goal is achieved.
4. It must represent goal-related information in a form that is flexible and easy to modify.

For many reasons, many investigators have argued that the neural instantiation of executive functions are, to a large degree, in the PFC. The remainder of this chapter discusses why.

ANATOMY AND ORGANIZATION OF THE PREFRONTAL CORTEX

The PFC consists of a collection of cortical areas that differ from one another on the basis of size, density, and distribution of their neurons. Its location in the human brain and major divisions in the monkey brain are shown in Figure 52.1, but these can be further subdivided into at least 18 distinct areas. The PFC is well positioned for a central role in cognitive control. Collectively, these areas have interconnections with brain areas processing external information (with all sensory systems and with cortical and subcortical motor system structures) as well as internal information (limbic and midbrain structures involved in affect, memory, and reward). The subdivisions have partly unique, but overlapping, patterns of connections with the rest of the brain, which suggests some regional

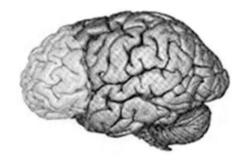

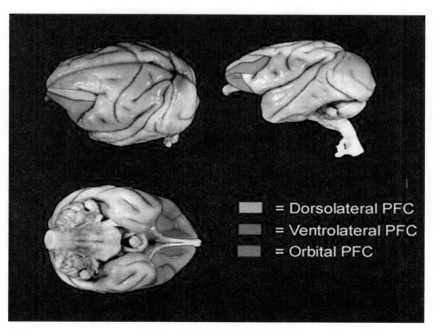

FIGURE 52.1 Location of the prefrontal cortex in the human brain (top) and its major subdivisions in the macaque monkey brain (bottom).

differences in function (Fig. 52.2). As in much of the neocortex, however, there are local connections between different PFC areas that can result in an intermixing and synthesis of disparate information needed for cognitive control. Indeed, the PFC has long been suggested to be the great integrator, a brain region that synthesizes information about the external and internal world for the purpose of producing goal-directed behavior.

The PFC Is Connected with Sensory Systems

The dorsolateral and ventrolateral portions of the PFC receive visual, auditory, and somatosensory information from the occipital, temporal, and parietal cortices, whereas orbitofrontal cortex receives chemo-

sensory and visceral information (Fig. 52.2). Many PFC areas receive converging inputs from at least two sensory modalities. For example, the dorsolateral (areas 8, 9, and 46), and ventrolateral (12 and 45) areas both receive projections from visual, auditory, and somatosensory cortex. Further, the PFC is interconnected with other cortical regions that are themselves sites of multimodal convergence. Many PFC areas (9, 12, 46, and 45) receive inputs from the rostral superior temporal sulcus, which has neurons with bimodal or trimodal (visual, auditory, and somatosensory) responses to external stimulation. The arcuate sulcus region (areas 8 and 45) and area 12 seem to be particularly multimodal. They contain zones that receive overlapping inputs from all three sensory modalities. In all these cases, the PFC is not connected directly

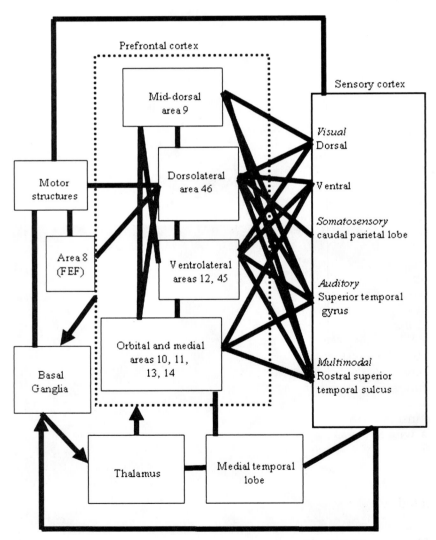

FIGURE 52.2 Schematic diagram of some of the extrinsic and intrinsic connections of the PFC. Most connections are reciprocal; the exceptions are noted by arrows. From Miller and Cohen (2001).

with primary sensory areas, but instead interconnected with secondary or "association" sensory cortex.

The PFC Is Connected with Motor System Structures

The dorsal PFC, particularly dorsolateral area 46, has preferential connections with many motor system structures, and may be the primary region by which the PFC exerts control over behavior. Dorsolateral area 46 is interconnected with motor areas in the medial frontal lobe such as the supplementary motor area (SMA), pre-SMA, and the rostral cingulate, with the premotor cortex on the lateral frontal lobe, and with cerebellum and superior colliculus. Dorsolateral area 46 also sends projections to area 8, which contains the frontal eye fields, a region important for voluntary shifts of gaze. As with sensory cortex, there are no direct connections between the PFC and primary motor cortex; the PFC is instead connected with premotor areas that, in turn, send projections to primary motor cortex and the spinal cord. Also important are the dense interconnections between the PFC and basal ganglia (see later).

The PFC Is Connected with the Basal Ganglia

The basal ganglia (BG) are a set of subcortical structures that are closely associated with both the motor system and the PFC. The anatomy of the BG are similar to that of the PFC in that they receive converging inputs from virtually every cortical system and many subcortical structures. The region of the BG that receives inputs, the striatum (which includes the caudate and putamen) receives a heavy projection from the PFC. Information then travels through loops of subcortical nuclei, which project back to the frontal cortex via the thalamus. Whereas the PFC projects back to the same wide range of cortical areas that give input to the basal ganglia, the basal ganglia project largely to the frontal lobe motor system structures and the PFC. Thus, the BG are well-positioned to influence a wide range of brain functions, albeit with an emphasis on motor output. The anatomical "loops" between the PFC and BG have inspired theories of their close interdependency. This will be discussed later. Basal ganglia anatomy and functions are discussed in greater detail in Chapter 31.

The PFC Is Connected with the Limbic System

The orbitofrontal PFC is more closely associated with medial temporal limbic structures critical for long-term memory and "internal" information, such as affect and motivation, than other PFC regions. It has direct and indirect (via the thalamus) connections with the hippocampus and associated neocortex, the amygdala and hypothalamus. It has strong connections with olfactory and gustatory cortices, but weaker connections than the lateral PFC to parietal and temporal cortices.

Intrinsic PFC Connections May Provide a Substrate for Diverse Information to Interact

Like most of the neocortex, many PF connections are local. There are not only interconnections between the major subdivisions, but also interconnections between their constituent areas. The lateral PFC is particularly well connected. Ventrolateral areas 12 and 45 are interconnected with dorsolateral areas 46 and 8, with dorsal area 9 as well as with orbitofrontal areas 11 and 13. Intrinsic PFC connections presumably allow information from a given PFC afferent or from a given process within a PFC subregion to be distributed to other parts of the PFC.

Summary

The PFC is anatomically well situated to play a role in cognitive control. It receives information from and sends projections to forebrain systems that process information about the external world, motor system structures that produce voluntary movement, systems that consolidate long-term memories, and systems that process information about affect and motivational state. This anatomy long has suggested that the PFC may be important for synthesizing external and internal information needed to produce complex behavior.

EFFECTS OF DAMAGE TO THE PREFRONTAL CORTEX IN HUMANS

PFC Damage in Humans Results in a Dysexecutive Syndrome

The PFC occupies a far greater proportion of the human cerebral cortex than in other animals, suggesting that it might contribute to those elusive cognitive capacities that separate humans from animals. However, unlike electrical stimulation of more posterior regions that produce hallucinations or motoric responses, stimulation of the PFC produces no obvious effect. This led it to be termed *silent cortex*. Similarly, at first glance, PFC damage has remarkably little overt effect; patients can perceive and move, there is little impairment in their memory, and they can appear remarkably normal in casual conversation. The most infamous case is that of Phineas Gage, a railway construction foreman. In 1848 he suffered an accident in which an iron bar exploded and was projected up through his chin and out of the top of his head, destroying much of his medial PFC in the process (Damasio *et al.*, 1994) (Fig. 52.3). He survived but it soon became clear that Gage had changed. He once had been an efficient foreman, but now he was irreverent and profane, rude to his coworkers, and unable to organize and plan the work that needed carrying out.

Despite the superficial appearance of normality, PFC damage seems to devastate a person's life. Take the case of Elliot, a successful, happily married accountant, and by all accounts a responsible person. In his late thirties, he was diagnosed as having a prefrontal meningioma, and during the removal of the tumor his PFC was damaged. Within a few of years of the operation, Elliot had divorced his wife, got married and divorced again, lost touch with his family and friends, acquired a disreputable business partner, and lost his business. All of Elliot's basic mental functions seemed to be intact, his language and memory abilities seemed normal, and on tests of intelligence he consistently performed in the superior range. What Elliot seemed to lack was the ability to coordinate his mental func-

tions in a manner that took long-term concerns (unseen goals) into account. He became impulsive, lurching from one catastrophe to another based on his momentary whims.

Note that this scenario is exactly what would be expected from the loss of a cognitive control system. Damage to the higher-order, executive functions leaves the components of complex behavior intact and able to be elicited by the appropriate external trigger. But the individual would not be able to coordinate these behaviors with respect to goals and task-specific constraints. For example, there was a PFC patient who, when preparing coffee, first stirred and then added cream. Because of this apparent dysfunction in the top-down control over behavior, the pattern of deficits following frontal lobe damage has been thought to reflect a lack of goal-directedness, and the result is called the *dysexecutive syndrome* (see Box 52.2).

Humans with Frontal Damage Are Disinhibited

One of the most striking features of prefrontal patients is their disinhibition and lack of behavioral control. The patients are often impulsive, quick to anger, and prone to making rude and childish remarks. For example, one patient complained of being "much more outspoken" since her injury, and another stated, "I've become impulsive since the accident. If I have something to say I can't wait and have to say it straight away."

An example of disinhibition is utilization behavior. During a medical examination, a patient with PFC damage will pick up and use items that have been left on the doctor's desk. For example, the doctor might leave a comb on the desk, and the patient will pick the comb up and start combing his or her hair. In one extreme case a patient used a urinal that the doctor had left on the desk! More formal tests of disinhibition include the Stroop test (Fig. 52.4A). It requires the patient to name the color of the ink in which a word is written. The catch is that the word is the name of another color. Patients have difficulty inhibiting the saying of the word itself because the word elicits a strong, prepotent tendency to read it.

Such tests require patients to inhibit previously established specific responses, but PFC patients are also impaired at inhibiting more abstract behavioral demands like rules. Consider the Wisconsin Card Sorting Test (WCST) (Fig. 52.4B). The patient is required to sort a deck of cards, on which are printed a number of symbols. The cards differ from one another in the color and shape of the symbols, and the number of symbols per card. The patient's task is to sort the cards

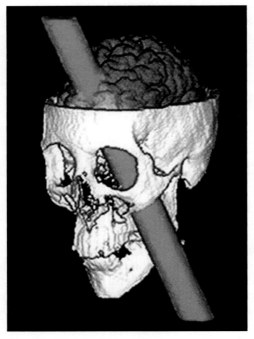

FIGURE 52.3 Reconstruction of damage to Phineas Gage's brain. From Damasio *et al.* (1994).

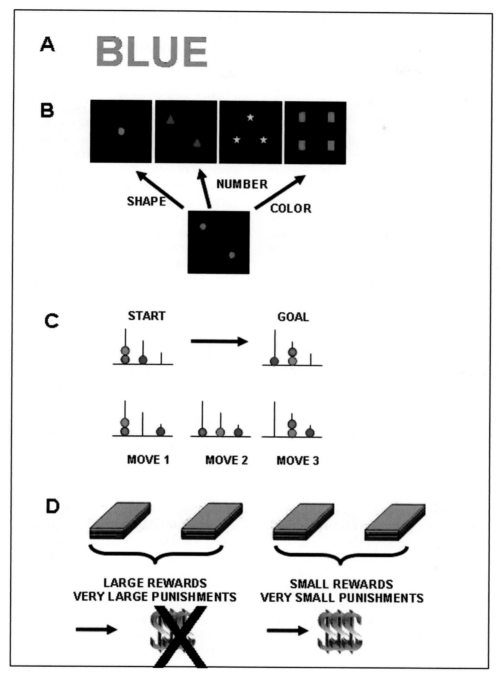

FIGURE 52.4 Tasks used to assess PFC dysfunction in humans. (A) Stroop task. (B) Wisconsin Card Sort. (C) Tower of London task. (D) Gambling task. See text for explanation of tasks.

according to a given criterion; for example, sort by shape. At this point the patients perform the task as well as control subjects. However, when the criterion is changed, for example to sort by number, the patients are impaired; they are unable to inhibit sorting by the previous rule. This behavior, that is continuing to perform a previously reinforced behavior even though

it is no longer rewarded, is called perseveration and is characteristic of a prefrontal patient's performance on a wide variety of tasks.

This inability to ignore irrelevant stimuli and the actions they trigger has led PFC patients to be described as stimulus-bound, and the disruption of the patient's ability to think and plan has been termed goal neglect

(Duncan *et al.*, 1996). Without the influence of the goal-related signals from the executive functions, the systems always respond reflexively to an input with whatever output is strongly connected to it.

Humans with Frontal Lobe Damage Have Emotional Impairments

PFC damage produces a marked alteration of a patient's moods and emotions. In the case of Phineas Gage, these changes were so profound that his friends commented, "Gage is no longer Gage." Patients typically display an emotional shallowness, being indifferent and apathetic both to their own situation and to the needs of others. For example, one patient stated, "If I saw someone cry I'd just laugh—people look silly getting upset." Formal tests show that patients are less able to recognize the emotional expressions of others, though their ability to identify faces remains intact. Patients also become more irritable, aggressive, and prone to angry outbursts. Their humor, termed *Witzelsucht*, becomes compulsive, puerile, and facetious. The patients are aware of these changes even if they are indifferent to them. For example, one patient commented, "Emotion, tears, that's all gone out of the window," and another stated, "I'm much more aggressive, and I feel much less fear. I go fighting for no reason: I don't get anxious."

Humans with Frontal Damage Have Difficulty in Planning

Prefrontal patients have a great deal of difficulty planning and organizing their lives. Wilder Penfield, a neurosurgeon during the mid-twentieth century, described the case of his sister, on whom he operated to remove a large frontal tumor. Prior to the surgery she had been an excellent cook, and though her basic culinary skills remained subsequent to the operation, she seemed unable to organize her behavior so that all the elements of the meal were ready simultaneously. Instead she would move haphazardly from preparing one part of the meal to another, so that some parts of the meal would be burnt while others had hardly been started.

Formal tests have been developed to measure the ability of patients to organize their behavior. For example, the Tower of London Test requires the patient to think several moves in advance (Fig. 52.4C). There are three vertical posts that have three colored rings that can be placed on the posts. The three posts are of different lengths such that the first post can accommodate all three rings, the second post can accommodate two, and the third post just one. From an initial starting position the patient is asked to move the rings one at a time, from one post to another, to attain a specified goal position. This task requires the patient to plan a number of subgoal positions in order to reach the final goal. Patients with prefrontal damage are greatly impaired at this task, making many more moves than necessary to reach the final goal position. The Multiple Errands Test taxes more everyday planning abilities. The test takes place in a shopping center, and the patient is required to perform various errands, such as obtaining a copy of the previous day's newspaper. The patients take much longer than control subjects. They go into inappropriate (irrelevant) shops as they pass them and repeatedly become distracted from their principal goal.

Impairments in planning might be related partly to difficulties evaluating the consequences of actions. Faced with a number of different, competing actions, patients might not choose wisely because they are less sure of the consequences of those actions. Evidence for this arises from a "gambling" task (Fig. 52.4D). The patient has to choose cards from four different decks in an effort to win play money from the experimenter. Each card wins the patient some money, but some cards also lose money and the patient has to pay the experimenter. Unbeknownst to the patient, cards from the first two decks will win large amounts of money, but are also occasionally associated with very large losses, and repeatedly choosing from these two decks will result in a net loss. By contrast, cards in the other two decks win smaller amounts of money, but the losses are also a lot smaller, and overall, choosing from these two decks will result in a net profit. Control subjects initially choose from the two decks associated with the larger rewards, but after a few very large losses, they switch to choosing from the other decks and earn a profit. Patients with PFC damage also initially choose from the decks associated with larger rewards, but then fail to alter their behavior until they lose all their money. This task is appealing because it is closely related to the types of problems that prefrontal patients experience in real life.

Humans with Frontal Lobe Damage Have Impaired Working Memory

Patients also might have difficulties planning due to a problem with holding information in mind for a short while. This is the kind of short-term memory that one relies on when reading a phone number and then dialing that number. This function is called working memory because the information is forgotten as soon as it is no longer relevant. The concept of working memory is influenced by the observation that monkeys with lesions of the PFC have deficits on a spatial delayed-response task that requires them to remember the location of a stimulus or behavioral response over a brief delay of several seconds. Early studies failed to find this deficit in humans with prefrontal damage, possibly because the patients verbalized the location of the stimulus (e.g., "it's on the left"), which presumably makes the task much easier. Later studies, however, in which the prefrontal damage was more extensive and bilateral, did show a spatial delayed response deficit. Other studies used abstract patterns that could not be easily verbally encoded, and found that the patient's nonspatial working memory was also impaired.

Summary

We have seen that the PFC poses a complex problem. Damage to a relatively circumscribed region of cortex in humans produces a bewildering array of symptoms. But one caution in interpreting clinical studies concerns the size of the damage that typifies the prefrontal patient. Not only does the damage often encompass many different PFC regions, each of which may have different functions, but it also often encroaches onto neighboring regions, such as the premotor cortex, the cingulate cortex, and the insular cortex. Plus, it should be remembered that the PFC comprises many different anatomically defined areas, each possessing unique patterns of connections with the rest of the brain; it might be that each component of the dysexecutive syndrome arises from damage to different areas of the PFC. These matters are difficult to explore in humans with PFC damage, because the damage is typically so extensive that it encompasses many brain regions both within and external to the PFC. There are two solutions to this problem. The first is to use neuroimaging techniques, such as functional magnetic resonance imaging (fMRI), in healthy humans. This technique can detect areas of increased neuronal activity in specific brain regions with a resolution of 2–4 mm^3. A second method is to study the effects of experimentally produced focal lesions in monkeys. We will discuss each of these approaches in turn.

NEUROIMAGING STUDIES AND PFC

Neuroimaging studies have shown that many of the tasks with which patients with PFC damage struggle also activate PFC in healthy subjects. However, an advantage of the technique is that specific regions within PFC can be identified that are activated prefer-

entially by specific cognitive processes. For example, fMRI studies show that performing the WCST activates ventrolateral PFC. In general, a number of trends have emerged from this field regarding the functional organization of PFC.

The clearest functional distinction is between orbitofrontal cortex and lateral PFC. Orbitofrontal cortex consistently is activated by tasks requiring the processing of rewarding or emotional information. For example, it shows more activation to pleasant images (e.g., a baby's face) or unpleasant images (e.g., mutilated bodies) as opposed neutral images. Similar results are seen in other modalities such as pleasant or unpleasant smells, tastes, sounds, and touches. Orbitofrontal cortex even responds to abstract rewards such as winning or losing money. Orbitofrontal cortex is not simply responding to the presence of these rewards, but seems to use this information to guide our decisions and choices. For example, orbitofrontal cortex is activated when subjects are choosing a meal from a menu, but not if they are simply reading the menu with no intention of ordering.

In contrast, lateral PFC is activated by a wide assortment of cognitive tasks. These include tests of inhibitory control (such as the Stroop task), rule switching tasks (such as the WCST), and working memory tasks. A distinction that has emerged from these studies concerns the respective roles of dorsolateral and ventrolateral PFC regions. Ventrolateral PFC is activated predominately by tasks that require holding information in working memory whereas dorsolateral PFC is activated when that information must be manipulated in some way. We return to this idea later in the chapter when we discuss the "multiple processing levels" theory of prefrontal function. Recently investigators have begun to focus on the most anterior portions of PFC, referred to as the *frontal pole*, which corresponds to area 10. There is some evidence that this region might be responsible for controlling the transfer of information between dorsolateral and ventrolateral PFC.

In sum, neuroimaging studies have helped to show that we cannot consider the PFC as a single unitary area, but that distinct subregions of PFC are responsible for carrying out specific processes. The clearest distinction is between the emotional and reward-based processing occurring in orbitofrontal cortex, and the more cognitive processes occurring in lateral PFC. A problem with fMRI, however, is that although it can identify which regions are more active during a given task, it does not tell us whether those regions are *necessary* for the performance of the task. To determine whether a given PFC subregion is necessary for a specific cognitive process, findings from monkey neuropsychology are useful. We discuss these findings in the next section.

EFFECTS OF DAMAGE TO THE PREFRONTAL CORTEX IN MONKEYS

Studying the effects of experimentally induced lesions in monkeys has two advantages. First, because the lesions are experimentally induced they can be small and focused on specific subregions of PFC. Second, they are a strong test of whether a specific subregion is necessary for the performance of a given cognitive task. However, we must also bear in mind a number of points when evaluating primate neuropsychological findings.

First, in order to conclude that two different prefrontal areas are performing, or at least emphasize two different functions, it is necessary to demonstrate a double dissociation. A monkey with lesion X must be impaired at process A, but not process B, and a monkey with lesion Y must be impaired at process B, but not process A. A single dissociation, for example showing that lesion X impairs process A, but not process B, is insufficient evidence because this pattern of results could be achieved simply because the test used to assess A is more sensitive than the test used to assess process B. In many studies, monkeys were tested on just one task and any evidence for a double dissociation can only be inferred across different studies and (often) different laboratories. This complicates the interpretation because any differences in training and testing procedures and in the extent of the lesions are potential confounding factors.

Second, early studies produced large lesions. These lesions often produced damage to the underlying white matter, potentially disconnecting regions of cortex outside of the site of the lesion and often damaging underlying structures such as the basal ganglia. In recent years, excitotoxins have been developed. These drugs can be injected into an area and overexcite the cells located there. The cells subsequently die, but the underlying white matter is left intact.

Third, studies in monkeys are useful only if we can show that the results are applicable to understanding the organization of human PFC. The close anatomical correspondence between monkey and human PFC suggested that there might also be close functional correspondence. However, there was a lack of direct evidence that this was indeed the case. Recently, a number of studies have used identical behavioral paradigms in monkeys and humans, and found that the paradigms tax similar PFC regions. For example, a recent fMRI study showed that both monkeys and humans activate ventrolateral PFC when they are performing the WCST. Such findings help validate the monkey as a model for understanding the organization of human PFC.

Monkeys with PFC Damage Are Impaired on Delayed Response Tasks

In 1935, Jacobsen published a seminal paper showing that lesions of the frontal lobe in monkeys produced deficits on a spatial delayed response task. In this task, the monkey sees a reward hidden at one of two locations, and then after a brief delay is allowed to retrieve it (Fig. 52.5A). Monkeys with large PFC lesions behave as if they forgot where the reward was hidden, even after delays of just a few seconds. Subsequent experiments showed that lesions restricted to the dorsolateral PFC alone could produce the same deficit. By contrast, damage to the ventrolateral PFC impaired object delayed response tasks. Monkeys with such damage have difficulty with delayed matching-to-sample (DMS) tasks in which they are shown an object (the sample) and then, after a brief delay of several seconds, must choose the same object (the match) from one or more nonmatching objects (Fig. 52.5B). The most severe deficits are observed when the same two objects are used repeatedly, requiring the monkey to remember which of the two objects it has seen most recently as the sample. Monkeys with orbital (ventral) PFC lesions tend to show the greatest deficit in tasks requiring reversal of previously successful actions. For example, they are impaired at spatial reversal tasks, in which the reward is alternated between two locations (Fig. 52.5C) and object reversal tasks, in which a reward is alternated between each of two objects (Fig. 52.5D). In both tasks the monkeys perseverate, continuing to choose the location or object under which the reward first had been hidden. Such observations led to the suggestion that the dorsal PFC might be responsible for processing spatial information, the ventral PFC for object-related information, and the orbital PFC for inhibitory control in accord with their preferential connections with the parietal, inferior temporal, and limbic system structures, respectively.

The picture, however, is not so simple and straightforward; a clear double dissociation between these impairments has not been demonstrated. Consider the hypothesized spatial versus object dissociation in the PFC. Dorsolateral PFC can impair performance on tasks that do not have a spatial component, such as object self-ordered search tasks. In this task the monkey is presented with three objects covering rewards. They must move each object in turn to get the reward, but

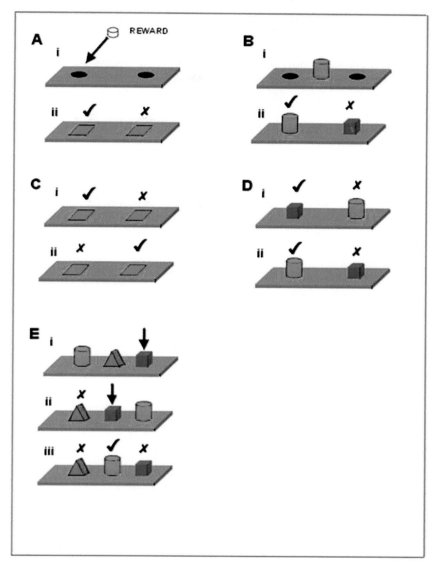

FIGURE 52.5 Some of the tasks on which monkeys with PFC damage are impaired. Check indicates correct choice, ✗ indicates incorrect choice. Arrow in E indicates monkey's choice in self-ordered task. (A) Spatial delayed response. (B) Delayed matching-to-sample. (C) Spatial reversal. (D) Object reversal. (E) Object self-ordered search task.

after each choice a screen is lowered and the object's positions are shuffled (Fig. 52.5E). Thus, the monkey must continually remember which of the three objects it has already checked for a reward. Also, as noted earlier, monkeys with ventral (orbital) PFC lesions are impaired on both object and spatial reversal tasks. Neurophysiological studies paint a similar picture. Although there may be some bias in the distribution of neurons that are sensitive to object identity or location in space (especially for the small population of highly specialized "face cells"), there is also a good deal of intermixing of neurons encoding these attributes, especially when the integration of these attributes is relevant to task performance.

Likewise, the notion that orbital PFC is responsible for inhibitory control is not so straightforward. It has subsequently been shown that inhibitory control is not unique to the orbital PFC. If the appropriate test is used, inhibitory control deficits can also be observed in monkeys with lesions of dorsolateral PFC. Marmosets were trained on a task similar to the WCST. They had to choose complex stimuli on the basis of a rule, such as "choose by shape." When they were required to switch between different rules, monkeys with lesions of the dorsolateral PFC perseverated by continuing to choose on the basis of the previous rule, whereas monkeys with lesions of the orbital PFC were unimpaired. In contrast, when the same stimuli were used in an object reversal task, monkeys with lesions of the orbital PFC perseverated by continuing to choose the stimulus that initially had been rewarded, but the performance of monkeys with lesions of the dorsolateral PFC was normal. This suggests that both the dorsolateral and orbital PFC contribute to inhibitory control, but the dorsolateral PFC may be more engaged by inhibitory control over rules, whereas the orbital PFC may be more engaged by inhibitory control over choices based on objects and rewards. For more discussion of the issue of PFC functional topography, see Box 52.1.

Monkeys with PFC Damage Are Impaired on Conditional Learning Tasks

Conditional learning tasks often are used to test the ability to gain volitional control over behavior. Conditional learning refers to a class of tasks in which associative relationships must be learned that are arbitrary

and extend beyond the simple one-to-one mappings that underlie reflexive reactions. Whether or not a given response is successful depends on additional, contextual, information. For example, reaching for popcorn can be rewarding, but only if one takes other information into account; if the popcorn belongs to another patron, the result could be disastrous. The need to take into account complex relationships in order to decide among alternative actions is presumably why cognitive control has evolved.

Monkeys with PFC damage are impaired at such tasks. For example, when monkeys are trained to learn associations between three visual cues and three directional movements of a joystick, simply moving the joystick is not enough to produce reward. Rather, the specific movement direction is dependent on the specific cue. One means "up," the other "down," and so on. Bilateral damage to the ventrolateral and orbital PFC severely impairs this learning. Depriving the PFC of its inputs from high-order sensory cortex also results

---

BOX 52.1

FUNCTIONAL TOPOGRAPHY

Information about topography of neural properties in the PFC can provide important insights into its function. For example, knowing that primary visual cortex is organized into groups of neurons with similar tuning for orientation provides important clues to how it analyzes visual inputs. The anatomically complex nature of the PFC, coupled with the unique patterns of connections that each prefrontal area possesses, would certainly suggest a degree of functional fractionation. But one immediate problem that investigators face when trying to address this issue is that it is not clear how to subdivide cognitive processing. It is far easier to specify the sensory attributes of an apple, than it is to list the different processes involved in planning or reasoning.

One possibility is that PFC is organized by function, with different regions carrying out qualitatively different operations. One longstanding view is that the orbitofrontal area is associated with behavioral inhibition whereas ventrolateral and dorsolateral regions are associated with memory or attentional functions (Goldman-Rakic, 1987; Fuster, 1989). Other recent suggestions include the idea that ventrolateral regions support maintenance of information (memory) and dorsolateral regions are responsible for the manipulation of such information, as well as schemes based on stimulus dimensions such as segregation of information about object form and color (what)

from information about its location (where). None of these schemes is mutually exclusive.

There is undoubtedly some spatial organization of processing within the PFC, but no clear global scheme has emerged. There is no overwhelming evidence for any one hypothesis and for some there are countervailing data. Perhaps this is not surprising. We do not yet understand the functional topography of the higher-order sensory and motor cortical areas that are interconnected with the PFC. We can, however, make some general predictions. First, PFC functional topography is likely to be based on broad categories or gradients of organization with overlap and intermixing of diverse information. Complex behavior requires that we recognize and respond to relationships across diverse dimensions, and representing these relationships is a prerequisite for cognitive control. Indeed, the majority of outputs of a given prefrontal area are to other prefrontal areas, so it is not surprising that many prefrontal neurons are multimodal. Second, because learning is central to cognitive control and modulates the properties of many PFC neurons (see Neurophysiology of the Prefrontal Cortex), it should have a large influence on PFC topography. Indeed, the flexible nature of controlled behaviors likely requires a dynamic anatomy, with neuronal properties changing to reflect the demands of the current task.

---

in conditional learning impairments. Humans with PFC damage are impaired on conditional learning tasks analogous to those conducted in monkeys. Conditional learning impairments seem to be dependent on which PFC region is damaged; some studies report no conditional learning deficit following dorsolateral PFC damage in monkeys.

Monkeys with PFC Damage Show Emotional Impairments

Like humans, monkeys with large PFC lesions display marked changes in social and emotional behavior. Monkey society is dictated by a dominance hierarchy and each monkey knows its place; a monkey is submissive to monkeys higher in the hierarchy (for example, the monkey grooms them) and is aggressive and threatening to monkeys lower in the hierarchy. Monkeys with large PFC lesions, however, disregard this hierarchy, and also decrease their interactions with troop members. This typically results in their being ostracized from the troop.

Monkeys with PFC Damage Are Impaired at Object Recognition

Monkeys with lesions of the ventromedial PFC show impairments in visual recognition. This ability is tested using a delayed nonmatching-to-sample task (DNMS). The monkey is presented with an object, and then after a brief delay of a few seconds, they are presented with that same object along with a novel object. New objects are used on each trial so the monkey can solve the task by recognizing which of the two objects it has previously seen in order to pick the novel object. However, patients with damage to the ventromedial PFC do not show these recognition deficits. One possibility for this discrepancy is that the deficit does not lie in recognition per se, but rather in the ability to learn the rules of the task. In particular, it has been suggested that the difficulty lies in making the association between the abstract quality of novelty and receiving a reward. This aspect of the task is not required in human patients, because the patients are explicitly told the rules of the task; monkeys have to learn them by trial and error.

Summary

Humans with damage to the PFC present a wide variety of symptoms including a lack of inhibitory control, a shallowness of emotion, difficulty planning and organizing behavior, and impaired working memory. In general, they seem to have difficulty in exerting volitional control and organizing their mental processes and actions in order to direct them toward unseen goals. As a result, their behaviors are simple, reactive, and emitted without concern for future consequences.

Just like humans, monkeys with extensive prefrontal damage can appear remarkably unimpaired to the casual observer. They have no motoric impairments, and their ability to discriminate a variety of sensory stimuli remains intact. They are also capable of simple learning, such as Pavlovian and instrumental conditioning. However, they are impaired when behavioral demands extend beyond this to include additional task-specific constraints and/or involve overcoming a previously established response. Localizing a specific deficit to any one PFC subregion has proven difficult; double dissociations between different PFC regions are rare.

The pattern of deficits observed after PFC damage in humans and monkeys makes sense in light of the cognitive architecture depicted in Figure 52.2. The deficits seem to reflect a selective loss of higher-level functions. Lower-level, well-established routines are intact, but without the information about future goals and means provided by the higher-level functions, processing would be at the mercy of the environment. Behavior would be ruled by whatever sensory inputs happen to flow into the system and by whatever thoughts, emotions, and actions are strongly associated with these inputs. Behavior would seem impulsive and disinhibited, and emotional reactions inappropriate, because they would be emitted reflexively without any consideration of the future. Further, without the influence of future goals to continually drive task-appropriate processes, individuals would be distractible and would easily go "off track" when there are temporal gaps between sensory inputs, or between those inputs and the individual's responses.

NEUROPHYSIOLOGY OF THE PREFRONTAL CORTEX

PFC anatomy suggests it has the means for executive functions, and the effects on behavior after damage to the PFC support this idea. So do neurophysiological studies; they have shown that PFC neurons have the properties one would expect from a region intimately involved in goal-directed behavior.

The PFC Is Multimodal

PFC anatomy indicates that it is interconnected with many different brain systems, a prerequisite for cognitive control. This high degree of connectivity is reflected in the neural activity observed in the PFC. Its neurons are activated by visual, auditory, tactile, and gustatory stimulation (and their memory) as well as voluntary limb and eye movements. Further, these neurons seem capable of synthesizing this information in accord with task demands. For example, after monkeys are trained to make arm movements to visual and auditory targets appearing in different locations, many lateral PFC neurons are activated by both kinds of targets. Interestingly, few bimodal neurons were found in studies in which monkeys passively experienced visual and auditory stimuli. This finding suggests that task demands can exert a strong modulatory influence on PFC activity, which is consistent with its putative role in cognitive control.

PFC Neurons Can Sustain Their Activity

The first observations of PFC neural activity that were related to a cognitive process were made in the 1970s by Fuster, Niki, and their colleagues (Kubota and Niki, 1971; Fuster, 1973). Guided by lesion studies in monkeys, they recorded from the PFC while monkeys performed delayed response tasks. They found that, during the memory delay, many PFC neurons showed sustained activity that seemed to keep information about a spatial or object cue online (Fig. 52.6). In spatial delayed response tasks, cue-related "delay activity," in principle, could reflect either sensory information related to the cue or movement information related to the behavioral response. Both are found in the PFC, each with different temporal characteristics. Sensory-related activity tends to be stronger nearer in time to the cue and then weakens as time passes, whereas movement-related neurons tend to increase their activity over time as the movement draws nearer. Goldman-Rakic and colleagues used a refined oculomotor version of the delayed response task to demonstrate that dorsolateral PFC neurons have precise tuning for memories of particular visual field locations (Funahashi *et al.*, 1989) (Fig.

52.7). They also demonstrated that PFC delay activity depends on dopamine receptors, a neurotransmitter implicated in signaling reinforcements and thus central to acquiring and sustaining goal-directed behaviors.

PFC delay activity has been shown to convey a wide range of behaviorally relevant cues: objects, colors, sounds, the frequency of a vibration to the hand, and forthcoming eye and arm movements. Similar activity is also evident in the sensory cortical areas and motor areas primarily responsible for analyzing inputs and generating behaviors, especially the higher-order areas that are one or more steps removed from primary sensory and motor cortex. This is not surprising; sustained activity is apparent in many brain structures and must play a role in many neural processes. For example, it is likely to underlie iconic memory, the very brief lingering "after image" of recent sensory events. However, what separates more "cognitive" working memory processes from such lower-level processes as iconic memory, is their ability to robustly sustain memories over potential distractions. PFC neurons do have this ability. For example, when monkeys are required to sustain the memory of a sample object across a delay period filled with visual distractors that require attention and processing, sustained activity within the prefrontal cortex can still maintain a memory of the sample object. By contrast, sustained activity in visual cortical areas seems more labile; it is disrupted by the presence of distractors. Following a distracting stimulus, neural activity reflects the distractor rather than the stimulus the monkey is retaining in working memory. This particular ability is not unique to the PFC. Some neurons in the entorhinal cortex, a region critical for memory, can also maintain sample-specific delay activity across intervening stimuli.

How does the PFC "latch" onto goal-relevant information and maintain it without disruption? Several models have been suggested. These typically employ a form of gating signal that instructs the PFC network when to maintain a given activity state. This gating

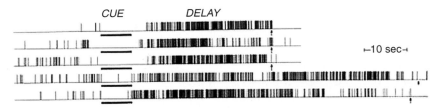

FIGURE 52.6 PFC delay activity. This figure shows the activity of a single PFC neuron during five trials of a spatial delayed response task. The arrow indicates the monkey's behavioral response at the end of the memory delay. Each small vertical line indicates an action potential from the neuron. Note the increased activity in the delay relative to other epochs. From Fuster (1973).

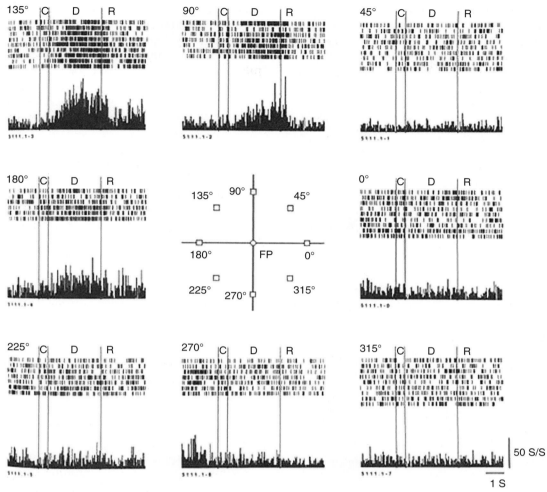

FIGURE 52.7 Activity of a single PFC neuron during performance of an oculomotor spatial delayed response task. Each trial starts with the monkey fixating the central fixation point (FP) indicated in the central diagram. After a delay (D), the monkey was allowed to respond by moving its eyes to the remembered location of the cue. Neural activity from cueing each location is shown by the corresponding histogram. The scale bars in the lower right indicate one second of time (1S) and 50 spikes per second of neural activity (50S/S). From Funahashi *et al.* (1989).

signal may come from dopaminergic (DA) neurons in the midbrain. These neurons respond when monkeys are rewarded or, importantly, when monkeys can first predict that a reward is forthcoming. In many behavioral tasks, presentation of the cue at the start of the trial is a good predictor that reward is soon coming. Such a well-timed DA burst could strengthen current PFC representations, protecting them against interference from disruption by irrelevant, distracting information until another DA influx reinforces another representation. Local injection of DA antagonists in the PFC does block delay activity in the PFC. The anatomical loops between the PFC and basal ganglia have been suggested to play a role here. Neural activity propagating around the loops may help sustain activ-

ity, and the basal ganglia have been suggested to provide a means by which DA signals can influence PFC activity.

PFC Neural Activity Reflects Task Rules

Playing the game requires knowing the rules. To explore how these rules might be represented in the PFC, experimenters manipulate task demands: monkeys are trained on tasks in which the physical cues are held constant while their behavioral meaning changes.

These studies suggest that PFC neurons represent task contingencies, the associative relationships that describe the logic needed to perform a task. Further support comes from investigations that have shown that PFC activity reflects more complex task information such as rules. Monkeys can be taught to rapidly alternate between different rules. For example, following a given cue, they can learn to direct a response to either the cue's location (spatial matching rule) or an alternate location associated with the cue (associative rule), depending on which rule is currently in effect. When tested in this fashion, many PFC neurons show rule-specific activity (Fig. 52.8B). For example, a PFC

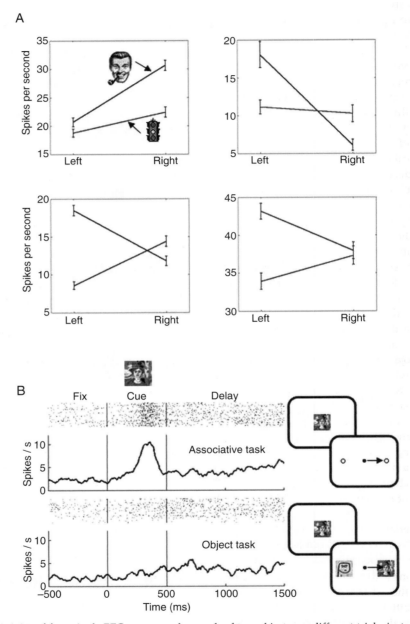

FIGURE 52.8 (A) The activity of four single PFC neurons when each of two objects, on different trials, instructed either a saccade to the right or a saccade to the left. The lines connect the average values obtained when a given object cued on the other saccade. The error bars show the standard error of the mean. Note that in each case, the neuron's activity depends on both the cue object and the saccade direction and that the tuning is nonlinear or conjunctive. That is, the level of activity to a given combination of object and saccade cannot be predicted from the neuron's response to the other combinations. (Adapted from Asaad *et al.*, 1998). (B) A PFC neuron whose neural response to a cue object was highly dependent on the rule the monkey is currently using. The bottom half shows an example of a single PFC neuron's response to the same cue object during an object task (delayed matching to sample) and during an associative task (conditional visual motor). Note that the neuron is responsive to the cue during one task but not the other, even though sensory stimulation is identical across the tasks. (Adapted from Asaad *et al.*, 2000).

BOX 52.2

ATTENTION DEFICIT HYPERACTIVITY DISORDER

Do you pay attention to captivating yet irrelevant stimuli? Do you have trouble acting appropriately, fidgeting while listening to a lecture? Do you have difficulty planning and pursuing long-term goals? Do distractions knock you "off task"? When these not uncommon difficulties occur to a severe degree, they are symptoms of Attention Deficit Hyperactivity Disorder (ADHD), a disorder seen in children and adults. It is estimated that 3 to 5% of children have this disorder, with about half retaining problems into adulthood. The symptoms of ADHD involve dysfunction of the prefrontal cortex (PFC) and its cortical and subcortical connections.

The criteria for diagnosis of ADHD include symptoms of "inattention" and "hyperactivity/impulsivity." Patients can have either the combined type, or predominantly inattentive or predominantly hyperactive/impulsive types. Many of the symptoms for inattention used to diagnose ADHD relate to attentional abilities of the PFC: for example, difficulty sustaining attention or organizing, easily distracted and forgetful. Similarly, many of the symptoms of hyperactivity/impulsivity describe PFC deficits: for example, difficulty awaiting turn. The PFC controls attention via its projections to the parietal and temporal cortices (see Chapter 48), while it controls motor responses via its projections to the motor cortices and striatum. These circuits appear to be impaired in patients with ADHD. ADHD symptoms are especially evident in "boring" settings that require endogenous rather than exogenous regulation of behavior. For example, ADHD children can sit still and play video games for hours, but have trouble attending in school. ADHD patients have difficulty sustaining a behavior or thought over a delay. As described in this chapter, PFC cells exhibit sustained firing over a delay that regulates thought and behavioral output.

Structural and functional imaging studies show consistent alterations in PFC-striatal-cerebellar circuits in ADHD patients. Volumetric measures have detected smaller right-sided PFC regions (the right side is most associated with attention, and the right inferior prefrontal cortex with behavioral inhibition; see Chapter 48). Changes in the size of the striatum also have been reported, and functional imaging studies have shown abnormal activity of both the PFC and striatum in ADHD patients performing tasks that require PFC inhibitory or attentional functions. Structural imaging studies also have shown consistent decreases in the size of the cerebellar vermis, a region that may exert regulatory influences on the noradrenergic

(NE) cells of the locus coeruleus (LC) and the dopaminergic (DA) cells of the ventral tegmental area. These NE and DA cells in turn have profound influences on PFC-striatal circuitry (see later). Thus, a smaller vermis in ADHD patients may lead to impaired catecholamine inputs to PFC and striatum. Some evidence supports this idea: a neuroreceptor imaging study suggests there are reduced numbers of catecholamine terminals in the PFC of adults with ADHD.

NE and DA have a critical influence on PFC-striatal circuits, and thus changes in these inputs can have tremendous import on PFC cognitive functions. The current chapter describes how DA has an important effect on PFC function, and the same is true for NE. NE cells fire when a stimulus is of interest to the animal, releasing NE in the PFC. NE stimulates post-synaptic α_{2A}-adrenoceptors, enhancing delay-related firing and strengthening regulation of behavior and attention. This is accomplished by α_{2A}-adrenoceptor inhibition of cAMP production, resulting in the closure of HCN channels (Hyperpolarization-activated Cyclic Nucleotide-gated cation channels). HCN channels are colocalized with α_{2A}-adrenoceptors on the dendritic spines of PFC neurons in the superficial layers of PFC, especially on spine necks. Thus, α_{2A}-adrenoceptors are ideally localized to control the neurochemical microenvironment around these channels. When cAMP builds up, HCN channels open, and synaptic inputs onto the spines are effectively shunted, thus functionally disconnecting PFC neuronal networks. In contrast, when α_{2A}-adrenoceptor stimulation inhibits the production of cAMP, HCN channels are closed and the networks are able to connect to control attention and behavior. In contrast, very high levels of NE and DA, such as released during stressful conditions, can impair PFC function via excessive cAMP production (e.g., from very high levels of β1 and DA D1 receptor stimulation), and from activation of high levels of protein kinase C signaling initiated by NE α_1-adrenoceptor stimulation. Thus, the PFC has to have just the right neurochemical conditions to function optimally.

What causes ADHD? ADHD is highly heritable, and has been linked with genetic alterations that interfere with NE or DA signaling. These linkages include the DA transporter; the D1, D5, and D4 receptors (the D4 receptor also has very high affinity for NE); the synthetic enzyme for NE (dopamine beta hydroxylase); and the α_{2A}-adrenoceptor. Environmental factors may also lead to disruption of PFC-striatal circuits. For example, some kinds of infection

(continues)

BOX 52.2   (cont'd)

may lead to autoantibodies that attack PFC-striatal circuits, and perinatal injury may impact these circuits.

Medications for ADHD likely reduce symptoms of inattention and impulsivity by optimizing the neurochemical environment in the PFC and in the striatum. All effective treatments for ADHD interact with catecholamines: Stimulants such as Ritalin (methylphenidate) and Adderall (a mixture of amphetamines) increase NE and DA release and/or block catecholamine reuptake. Recent research in rats suggests that low, therapeutic doses of these compounds preferentially increase release in PFC, and have a greater effect on the release of NE than DA. Consistent with these data, NE reuptake blockers are also used to treat ADHD (e.g., desipramine), including highly selective NE agents (atomoxetine, also known as Strattera). Drugs that mimic NE at α_{2A}-adrenoceptors (e.g., guanfacine) are effective, especially in decreasing impulsivity and distractibility. Wellbutrin (also known as Zyban or bupropion) is a DA reuptake blocker that is used to treat ADHD.

How do these medications actually work to ameliorate ADHD symptoms? There have been several ideas proposed: 1) the etiology of ADHD involves insufficient catecholamine innervation of PFC-striatal circuits, and drugs that increase catecholamine receptor stimulation correct

this; 2) ADHD involves excessive catecholamine release and low doses of these medications correct this by promoting negative feedback, decreasing phasic catecholamine release; or 3) catecholamines are normal in ADHD patients (e.g., it is simply that their cortex is immature or inflamed), and amplification of catecholamine mechanisms enhances PFC function and reduces symptoms. Support for this latter idea arises from the fact that normal individuals are improved by small doses of these medications as well, as when college students sometimes take low doses of stimulants to focus on writing a paper.

All of us experience PFC dysfunction as part of daily life, especially when we are tired or stressed. These effects are likely due to neurochemical changes—for example, in NE and DA—that have great impact on the functional integrity of the PFC. But for people with ADHD, dysfunction of the PFC is much more constant and often beyond their control. Perhaps it is this prevalence of PFC dysfunction in otherwise healthy individuals, and its association with stress, that lead to the dismissal of ADHD by some as not a real disorder. It is hoped that as we come to better understand the neurobiological bases of ADHD, we will gain a fuller perspective of ADHD and its symptoms.

Amy Arnsten

---

neuron might respond to a given visual cue when the monkey is using an associative rule but exhibit weak or no activity under identical sensory and attentional conditions that differed only in that the monkey was using a spatial rule instead. Such effects have been found in PFC for associative versus spatial rules, for object matching versus spatial matching versus associative rules, and for shape matching versus object matching rules (Asaad *et al.*, 2000).

Primate behavior is not limited to following literal associations and specific rules. Primates can also abstract general rules or principles, not tied to any particular stimulus or response. This ability allows behavior to extend beyond specific circumstances that have been experienced directly. For example, humans learn the "rules" for restaurant dining from specific experiences and can then generalize and apply them to new restaurants. Wallis and colleagues (Wallis *et al.*, 2001) investigated this process in monkeys by recording from single neurons in the PFC. Monkeys were trained to use two abstract rules; they indicated

whether two successively presented pictures were either the same (match) or different (nonmatch). The monkeys could perform this task with novel pictures, thus demonstrating that they had learned two general principles that could be applied to stimuli that they had not experienced directly. The most prevalent neuronal activity observed in the dorsolateral, ventrolateral, and orbital PFC reflected the coding of the abstract match and nonmatch rules. This finding suggests that rule representation may be a cardinal PFC function. Recent findings using fMRI showed that humans also activate PFC when performing the exact same task.

PFC Neurons Are Flexible

Another hallmark of primates is the ability to quickly learn new behaviors. PFC neurons are capable of a correspondingly rapid modification of their properties to meet task demands. Monkeys can learn new conditional visuomotor associations in just a few trials and as they do, PFC neural activity rapidly modifies to reflect the new contingencies. Striking examples of rapid experience-dependent neural plasticity come from studies of the frontal eye fields, part of Brodmann's area 8 that is important for voluntary shifts of gaze (Bichot and Schall, 1999). Normally, neurons in this area are selectively activated by saccade targets that appear in certain visual field locations. Monkeys also were trained to search for a target defined by a particular visual attribute (e.g., red). In this case, the neurons in the frontal eye fields acquired sensitivity to that attribute. They also trained monkeys to search for a different target every day and found that neurons discriminated not only the current target but also distracting stimuli that had been a target on the previous day relative to stimuli that had been targets even earlier. Monkeys were also more likely to make errors in choosing that distracting stimulus. It was as if the previous day's experience left an impression in the brain that influenced neural activity and task performance.

PFC Neurons Encode Rewards

Feedback about the consequences of actions is key to acquiring new goal-directed behaviors. Consequently, a large proportion of PFC neurons encode information about rewards. For example, some neurons are activated by the delivery of a reward, whereas others respond when an expected reward is not delivered. PFC neurons also encode the type of reward that the monkey expects to receive, for example a piece of cabbage versus a raisin, as well as the quantity of reward the monkey expects. Further, the neurons appear to encode the monkey's relative preference for the rewards that are available, rather than some absolute measure of reward. Thus, a neuron might prefer reward A when the monkey's choice of reward is between A and B, but reward B when the choice is between B and C. The orbital PFC is the area of the PFC that receives gustatory and olfactory information, and indeed neurons responsive to rewards are found there. But many neurons responsive to reward are also found in the dorsolateral PFC.

Summary

As might be expected from a brain region suspected to orchestrate complex behavior, neural activity in the PFC conveys information about a wide range of external and internal information. But its activity is not limited to simple responses to sensory events or actions. Rather, the behavioral context has a pervasive influence on PFC activity. After training, many (up to half) of dorsolateral, ventrolateral, and orbital PFC neurons encode information about the formal demands of behavior: contingencies between a wide range of stimuli, actions, and reward consequences, and behavior-guiding rules. By robustly sustaining their activity, PFC neurons can keep this and other behaviorally relevant information online to guide behaviors that extend over time. Also consistent with its role in guiding goal-directed behavior, PFC neurons process information about the expected results of achieving goals: rewards.

THEORIES OF PREFRONTAL CORTEX FUNCTION

Working Memory

One proposal is that the cardinal function of the PFC is to hold recent sensory inputs and movement-related information in active short-term memory, or working memory (Goldman-Rakic, 1987). In this view, the PFC acts as a temporary scratch pad that sustains the inputs it receives from other cortical areas. This hypothesis has its roots in observations of deficits in performance of delayed response tasks following PFC damage and of PFC neurons that show sustained activity over a memory delay. Other deficits following PFC damage are thought to be sequelae of the working memory deficit. For example, patients might be impaired at planning because they are unable to keep in mind all the different variables that need to be considered. This in turn might make them appear disinhibited as their behavior becomes directed toward more immediate goals, rather than toward longer-term, more delayed goals.

Working memory maintenance functions are central to cognition; complex, goal-directed behaviors typically are drawn out and thus are impossible without mechanisms that extend information over time. There is ample evidence that these mechanisms are abundant in the PFC. Working memory is unlikely to be the whole story, however. Many of the tasks on which prefrontal patients and monkeys show impairment do not have an obvious working memory component, such as the WCST and reversal tasks.

Recent studies have explicitly investigated this issue, by examining the performance of monkeys with lesions of the ventrolateral PFC on a simultaneous color-matching task. In this task the monkey is presented with a sample color and two test colors simultaneously. The monkey must choose the test color that matches the sample color. PFC lesioned monkeys were impaired at this task, despite the fact the task involved contains no delays and therefore no working memory requirements. The task does, however, require conditional rules (e.g., if the sample color is red, then choose the red test stimulus), and the role of the PFC in the representation of such rules (see later) might explain this impairment.

Somatic Marker Hypothesis

It has been suggested that the orbitofrontal PFC is responsible for labeling people, objects or situations with an "affective significance" (Damasio, 1994). This is achieved by associating memories of past, affectively-laden events with a representation of the somatic state which that event evoked. Thus, the memory of the past event is labeled with a "somatic marker" that helps individuals make decisions. Rather than having to evaluate all the pros and cons of prior circumstances to decide on the present, one can simply recall the associated somatic marker, providing the individual with the appropriate affective state and, in effect, a "gut-feeling" of which alternative is correct. Patients with damage to this system might be unable to reach a decision, simply because they become incapacitated by contemplating irrelevant information. None of the outcomes of their actions will be obviously more emotionally preferable than alternative outcomes, and patients might make decisions randomly or impulsively, and as such appear to be disinhibited.

A deficient somatic marker system would explain the impairments that patients with damage to the orbitofrontal PFC exhibit on the gambling task (Fig. 52.4D). The reward contingencies associated with each deck of cards are designed to be too complex to be ascertained purely by logical reasoning. Control subjects succeed at the task because they associate a somatic marker with each deck of cards, warning the subject as to whether the choice he or she is about to make is risky. Further support comes from measures of skin conductance responses (SCRs) that reflect autonomic nervous system processing. Control subjects show SCRs prior to the selection of a card, and choices from risky decks are preceded by a larger SCR than choices from less risky decks. Patients with damage to the ventromedial PFC showed no anticipatory SCRs, although their SCRs when actually receiving the reward or punishment are intact.

This hypothesis is similar to one advanced by William James in the nineteenth century. He proposed that emotional feelings depend upon feedback from the autonomic nervous system. For instance, running from a bear because we are afraid evokes a particular autonomic state (such as increased heart rate and blood pressure) that makes us feel afraid. Thus, as James stated, we do not run because we are afraid, but we are afraid because we run. A problem with simply using autonomic nervous system signals, however, is that autonomic states produced by different emotions can be identical. Later studies showed that manipulations of the autonomic nervous system, for example, by administering adrenaline, could alter the magnitude but not necessarily the type of emotion experienced in human subjects. These problems are avoided if the orbitofrontal cortex does not need to access the autonomic nervous system directly, but rather can use a conditioned association between the autonomic state and various cortical representations. For example, associations with the somatosensory cortex may be involved in emotional interpretations; patients with damage to the somatosensory cortex have difficulty recognizing facial emotions as do patients with orbital frontal damage.

Multiple Processing Levels

It has been proposed that the PFC is organized by level of processing, much as there are primary and higher-order sensory cortices (Petrides, 1996). The ventrolateral PFC is primary. It receives inputs from sensory cortex and is responsible for their simple maintenance in working memory and for retrieval of information from long-term memory. The dorsolateral PFC then takes this information and performs more executive processes, such as monitoring multiple behaviors and manipulating information. For example, simply rehearsing a phone number is a ventrolateral PFC function while saying that same number backwards is a dorsolateral PFC function. It should be noted that such an arrangement would preclude the possibility of demonstrating a functional double dissociation between these two areas, since the dorsolateral PFC is able to perform executive processes only on information after it had been received by the ventrolateral PFC. Thus, removing the dorsolateral PFC would impair the higher-level processes, whereas removing the ventrolateral PFC would impair both.

This hypothesis has been supported by evidence from neuroimaging in humans. Subjects were required to perform five spatial working memory tasks while they were scanned using positron emission tomography. This technique enables the visualization of areas of the brain with the increased blood flow that is related to neural activity. Two of the tasks had simple maintenance requirements; they activated the ventrolateral PFC. The three other tasks also had manipulation requirements and they activated the dorsolateral PFC.

Rule Representation

Several investigators have pointed out that because knowledge about goals, means, and consequences is vital to planning and implementing complex goal-directed behavior, the PFC is likely to play a central role in its acquisition and representation. The neural instantiation of task demands in the PFC was first proposed by Joaquin Fuster (Fuster, 1989). Based on behavioral and neurophysiological experiments in monkeys, he hypothesized that sustained activity in the PFC was not used merely to retain recent sensory events but also to link together behavioral relevant, but temporally separate, events and actions into a representation of task contingencies.

Other investigators have reached similar conclusions. Grafman (1994) suggested that the PFC contains managerial knowledge units (MKUs), sets of task contingencies linked under common themes that can be flexibly combined into extended routines. A MKU may describe the expected events and appropriate responses for dining in a restaurant, for example. Passingham (1993) suggested that many of the deficits following PFC damage can be explained by a loss of the ability to acquire conditional if-then rules. Wise *et al.* (1996) proposed complementary roles in rule learning for the PFC and the basal ganglia. The PFC is thought to act when new rules are learned whereas the basal ganglia is thought to potentiate (select) the previously learned rules that are appropriate for the current situation.

Cognitive control also has been studied by developing simulated neural network models that can perform tasks that depend on controlled processing, such as the Stroop Test. These models invariably include units dedicated to representing task rules and contingencies. Analogies often are drawn between these units and the PFC.

How might the PFC acquire rule information and use it for cognitive control? Miller and Cohen (2001) proposed that reward signals from midbrain dopamine systems act on the multimodal circuitry in the PFC to strengthen connections between neurons that process the information that led to the reward. This action results in a model of the task contingencies, which in neural terms amounts to a representation of the pattern of connections between the sensory, motor, and mnemonic representations needed to solve that task. Maintenance mechanisms in the PFC sustain this representation during task performance. This produces feedback signals to posterior cortex that bias the flow of activity along lines that match the PFC representation, that is, along task-relevant neural pathways. In short, the PFC plots a "map" of which neural pathways in the forebrain are successful at solving a task and then holds it online in the PFC so that the rest of the neocortex can follow it.

Evidence for this proposal comes from the finding that PFC neurons show neural activity that reflects behaviorally relevant associations between disparate and arbitrary events and actions, behavioral context and rules. Behavioral and neurophysiological studies of focal attention also point to the operation of top-down signals on the visual cortex (presumed to come from the PFC) that enhance representations of behaviorally relevant stimuli and visual field locations at the expense of irrelevant ones. Evidence that such signals come from the PFC was supported by a recent observation that disconnecting the PFC from the temporal cortex prevented the activation of stored memories in the latter and caused monkeys to fail at a recall test (Tomita *et al.*, 1999).

The Flexibility and Limited Capacity of Controlled Processing

Sustained activation is central in many views of PFC functions. This is not surprising considering the importance of working memory for cognitive control and that sustained activity is common in the PFC. But as O'Reilly and Munakata (2000) have pointed out, representing information through sustained activity offers more than just the ability to bridge gaps; it has other advantages for cognitive control.

It is universally assumed that long-term storage in the brain depends on changing synaptic weights, by strengthening some neural connections and weakening others. Encoding information in structural changes has obvious advantages for long-term storage, but the resulting memories are relatively inflexible. Also, changing synaptic weights has local effects; only the two neurons sharing the synapse are affected. The information contained in a set of synaptic weights is thus expressed only when the circuit is fired. By contrast, the information needed to guide goal-directed behavior must be expressed in a format that allows it to affect the ongoing processing in other brain systems. Sustained activity is such a format. It is extended over time and information contained in a pattern of activity can be propagated across the brain. Thus, the ability of sustained activity to tonically influence other brain systems is likely important for coordinating diverse processing around a specific goal. It also affords flexibility; if cognitive control stems from a pattern of information maintained in the PFC, changing behavior is as easy as changing the pattern (O'Reilly and Munakata, 2000; Miller and Cohen, 2001).

Finally, the neuronal mechanisms by which information needed for cognitive control is expressed might explain the severely limited capacity of controlled processes. A vast amount of information can be stored by changing synaptic weights of a network that contains many neurons, each with many potential connections. At any given time, however, most of this information is latent; it is expressed only when the relevant circuit is activated. By contrast, if the information for cognitive control is expressed in a unique pattern of ongoing activity distributed across many simultaneously active neurons—a population code—then there will be a natural capacity limitation. Trying to represent more than just a few items at the same time would degrade information because the unique patterns impinging on a given set of neurons would overwrite and interfere with one another.

The PFC versus the Basal Ganglia: Model-Building versus "Snapshots"

The prefrontal cortex (and other frontal cortical areas) forms anatomical loops with the basal ganglia (BG), a set of subcortical structures that have motor and learning functions (see Chapters 31 and 51). Different regions of the frontal lobe (e.g., the lateral PFC, orbital PFC, premotor cortex) have anatomically segregated loops with the BG that are "closed," meaning that the same region of the frontal cortex that gives rise to the loop is also the target of that loop via BG's connections with the thalamus. Closed loops are noteworthy because they may allow recursive processing in which the results of processing are fed back into the loop for further elaboration. It is possible, if not likely, that some form of recursive processing is responsible for the open-ended nature of advanced thought.

Daw *et al.* (2005) suggested functional specializations for the PFC and BG (specifically the striatum, the BG structure that receives input from the cerebral cortex). They suggested that the PFC builds models of an entire behavior—it retains information about the overall structure of the task, following the whole course of action from initial state to ultimate outcome. They liken this to a "tree" structure for a typical operant task: Behaviors begin in an initial state with two or more possible response alternatives. Choosing one response leads to another state with new response alternatives, with this process continuing throughout the task, ultimately leading to a reward. The PFC is able to capture this entire tree structure, essentially providing the animal with an internal model of the task. By contrast, the striatum is thought to learn the task piecemeal, with each state's response alternatives individually captured as separate "frames." This "caching reinforcement learning" system retains infor-

mation about which alternative is "better" in each state but nothing about the overall structure of the task (i.e., the whole tree). In short, the BG learns each fork in the road whereas the PFC puts together the whole route.

This is thought to explain observations of tasks that use reinforcer devaluation. In such tasks, you change the value of the reward by saturating the animal on a given reward (for example, overfeeding on chocolate if chocolate is a reward on that task). This has revealed two classes of behavior. Behaviors that are affected by reinforcer devaluation are considered to be goal-directed because changing the goal changes the behavior. As discussed earlier, goal-directed behaviors depend on the PFC. By contrast, overlearned behaviors whose outcomes remain relatively constant can become habits, impervious to reinforcer devaluation. Because these behaviors are not affected by changing the goal, they seem to reflect control by a frame-based caching system in which the propensity for a given alternative in each situation is stored independently of information about past or future states. Indeed, such a piecemeal caching system could acquire only complex tasks as habits because it can learn them only if they always conducted in the same fashion. Habits have long been thought to be a specialization of the BG.

Support for this model comes from recent observations that during conditional visuomotor rule learning, the dorsal striatum of the BG (the portion that receives direct projections from the lateral PFC and whose output ultimately loops back to the lateral PFC) showed very fast, almost bistable changes in neural activity (Pasupathy and Miller, 2005). By contrast, the lateral PFC showed slower gradual changes in neural activity. This may reflect the much heavier input from midbrain dopaminergic reward-related neurons in the striatum than the PFC. The fast plasticity in the striatum could be ideal for learning the reinforcement-related snapshots that capture the immediate circumstances and which alternative is preferable for a particular state. The slow plasticity in the PFC is better suited for the linking in of additional information about past states that is needed to learn and retain an entire model of the task and thus predict future states.

Summary

The ability to take charge of one's actions and direct them toward future aims is called cognitive control. Virtually all theories of cognition posit that cognition depends on functions specialized for the acquisition of information about goals and means. These functions exert a top-down influence on the lower-level automatic processes that mediate sensory analysis, memory storage, and motor outputs, orchestrating and directing them toward a given goal.

The PFC, a brain structure that reaches its greatest complexity in the primate brain, seems to play a central role in cognitive control. It has the requisite anatomical infrastructure. It has access to, and the means to influence processing in, all major forebrain systems and can provide a means to synthesize the diverse information related to a given goal. Indeed, its damage in humans and monkeys results in a dysexecutive syndrome; individuals are stimulus-bound and seem capable only of reacting to the immediate environment. Information about unseen goals and potential consequences of actions has little influence on behavior. PFC neurons also have properties consistent with their participation in cognitive control: they are multimodal, reflect task demands, and can sustain their activity to keep task-relevant information online during task performance.

Theories of PFC function suggest that it makes several contributions important for cognitive control: maintenance and manipulation of information in working memory, the assignment and recall of the affective tags related to different choices, and the acquisition and representation of behavior-guiding rules that help orchestrate goal-directed behaviors. None of these theories is mutually exclusive, and indeed, different theories may be focusing on different aspects of a complex system that is involved in a wide range of brain processes.

Further Readings

Arnsten, A. F. T. (2006). Stimulants: Therapeutic actions in ADHD. *Neuropsychopharmacology*, **31**, 2376–2383.

Arnsten, A. F. T. and Li, B.-M. (2005). Neurobiology of executive functions: Catecholamine influences on prefrontal cortical function. *Biological Psychiatry* **57**, 1377–1384; epub, Nov. 17, 2004.

Solanto, M., Arnsten, A., and Castellanos, F. X., eds. (2000). The Neuropharmacology of Stimulant Drugs: Implications for AD/HD. Oxford University Press, New York, NY.

Wang, M., Brian, P., Ramos, B. P., Paspalas, C. D., Shu, Y., Simen, A., Duque, A., and Vijayraghavan, S. *et al.* (2007). α2A-Adrenoceptors strengthen working memory networks by inhibiting cAMP-HCN channel signaling in prefrontal cortex. *Cell*, in press.

References

Asaad, W. F., Rainer, G., and Miller, E. K. (1998). Neural activity in the primate prefrontal cortex during associative learning. *Neuron* **21**, 1399–1407.

Asaad, W. F., Rainer, G., and Miller, E. K. (2000). Task-specific activity in the primate prefrontal cortex. *Journal of Neurophysiology* **84**, 451–459.

Bichot, N. P. and Schall, J. D. (1999). Effects of similarity and history on neural mechanisms of visual selection. *Nat Neurosci* **2**, 549–554.

Damasio, A. R. (1994). "Descartes' error: Emotion, reason, and the human brain." Putman, New York.

Damasio, H., Grabowski, T., Frank, R., Galaburda, A. M., and Damasio, A. R. (1994). The return of Phineas Gage: Clues about the brain from the skull of a famous patient. *Science* **264**, 1102–1105.

Duncan, J., Emslie, H., Williams, P., Johnson, R., and Freer, C. (1996). Intelligence and the frontal lobe: The organization of goal-directed behavior. *Cognitive Psychology* **30**, 257–303.

Funahashi, S., Bruce, C. J., and Goldman-Rakic, P. S. (1989). Mnemonic coding of visual space in the monkey's dorsolateral prefrontal cortex. *Journal of Neurophysiology* **61**, 331–349.

Fuster, J. M. (1973). Unit activity in prefrontal cortex during delayed-response performance: Neuronal correlates of transient memory. *Journal of Neurophysiology* **36**, 61–78.

Fuster, J. M. (1989). "The prefrontal cortex." Raven Press, New York.

Goldman-Rakic, P. S. (1987). Circuitry of primate prefrontal cortex and regulation of behavior by representational memory. *In*: "Handbook of Physiology: The Nervous System" (Plum, F., ed.), pp. 373–417. Bethesda: Am.Physio.Soc.

Grafman, J. (1994). Alternative frameworks for the conceptualization of prefrontal functions. *In*: "Handbook of Neuropsychology" (Boller F., Grafman J., eds.), p. 187. Amsterdam: Elsevier.

Kubota, K. and Niki, H. (1971). Prefrontal cortical unit activity and delayed alternation performance in monkeys. *Journal of Neurophysiology* **34**, 337–347.

Miller, E. K. and Cohen, J. D. (2001). An integrative theory of prefrontal function. *Annual Review of Neuroscience* **24**, 167–202.

Norman, D. A. and Shallice, T. (1986). Attention to action: Willed and automatic control of behavior. *In*: "Consciousness and Self-Regulation: Advances in Research and Theory" (Davidson, R. J., Schwartz, G. E., Shapiro, D., eds.), pp. 1–48. New York: Plenum.

O'Reilly, R. C. and Munakata, Y. (2000). "Computational Explorations in Cognitive Neuroscience: Understanding the Mind." MIT Press, Cambridge.

Passingham, R. (1993). "The Frontal Lobes and Voluntary Action." Oxford Univ. Press, Oxford.

Petrides, M. (1996). Specialized systems for the processing of mnemonic information within the primate frontal cortex. *Philosophical Transactions of the Royal Society of London. B: Biological Sciences* **351**, 1455–1461.

Tomita, H., Ohbayashi, M., Nakahara, K., Hasegawa, I., and Miyashita, Y. (1999). Top-down signal from prefrontal cortex in executive control of memory retrieval. *Nature* **401**, 699–703.

Wallis, J. D., Anderson, K. C., and Miller, E. K. (2001). Single neurons in the prefrontal cortex encode abstract rules. *Nature* **411**, 953–956.

Wise, S. P., Murray, E. A., and Gerfen, C. R. (1996). The frontal-basal ganglia system in primates. *Crit. Rev. Neurobiol.* **10**, 317–356.

Suggested Readings

Adolphs, R. (2001). The neurobiology of social cognition. *Curr Opin Neurobiol* **11**, 231–239.

Braver, T. S. and Cohen, J. D. (2000). On the control of control: The role of dopamine in regulating prefrontal function and working memory. *In*: "Attention and Performance 18" (Monsell S., Driver J., eds.), in press. Cambridge: MIT Press.

Dehaene, S., Kerszeberg, M., and Changeux, J. P. (1998). A neuronal model of a global workspace in effortful cognitive tasks. *Proc Nat Acad Sci USA* **95**, 14529–14534.

Fuster, J. M. (1995). "Memory in the cerebral cortex." MIT Press, Cambridge, MA.

Fuster, J. M. (2001). The prefrontal cortex—An update: Time is of the essence. *Neuron* **30**, 319–333.

Goldman-Rakic, P. S. (1994). Working memory dysfunction in schizophrenia. *J Neuropsychiatry Clin Neurosci* **6**, 348–357.

Miller, E. K. (2000). The prefrontal cortex and cognitive control. *Nature Reviews Neuroscience* **1**, 59–65.

Roberts, A. C. and Wallis, J. D. (2000). Inhibitory control and affective processing in the prefrontal cortex: Neuropsychological studies in the common marmoset. *Cereb Cortex* **10**, 252–262.

Shimamura, A. P. (2000). The role of the prefrontal cortex in dynamic filtering. *Psychobiology*; in press.

Earl Miller and Jonathan Wallis

Consciousness

Consciousness is one of the most enigmatic features of the universe. People not only act but feel: they see, hear, smell, recall, plan for the future. These activities are associated with subjective, ineffable, immaterial feelings that are tied in some manner to the material brain. The exact nature of this relationship—the classical mind-body problem—remains elusive and the subject of heated debate. These first-hand, subjective experiences pose a daunting challenge to the scientific method that, in many other areas, has proven so immensely fruitful. Science can describe events microseconds following the Big Bang, offer an increasingly detailed account of matter and how to manipulate it, and uncover the biophysical and neurophysiological nuts and bolts of the brain and its pathologies. However, this same method has as yet failed to provide a satisfactory account of how first-hand, subjective experience fits into the objective, physical universe. The brute fact of consciousness comes as a total surprise; it does not appear to follow from any phenomena in traditional physics or biology. Indeed, some modern philosophers even argue that consciousness is not logically supervenient to physics (Chalmers, 1996). Supervenience is used to describe the relationship between higher-level and lower-level properties such that the property X supervenes on property Y if Y determines X. This implies, for example, that changing Y will, of necessity, change X. In that sense, biology is supervenient to physics. Put differently, two systems that are physically alike will also be biologically alike. Yet it is not at all clear whether two physically identical brains will have the same conscious state.

Note that it is not yet generally accepted that consciousness is an appropriate subject of scientific inquiry. A number of neuroscience textbooks provide extended details about brains over hundreds of pages yet leave out what it feels like to be the owner of such an awake brain, a remarkable omission.

People willingly concede that when it comes to nuclear physics or molecular biology, specialist knowledge is essential; but many assume that there are few relevant facts about consciousness and therefore everybody is entitled to their own theory. Nothing could be further from the truth. There is an immense amount of relevant psychological, clinical, and neuroscientific data and observations that needs to be accounted for. Furthermore, the modern focus on the neuronal basis of consciousness in the brain—rather than on interminable philosophical debates—has given brain scientists tools to greatly increase our knowledge of the conscious mind.

Consciousness is a state-dependent property of certain types of complex, biological, adaptive, and highly interconnected systems. The best example of consciousness is found in a healthy and attentive human brain, for example, the reader of this chapter. In deep sleep, consciousness ceases. Small lesions in the midbrain and thalamus can lead to a complete loss of consciousness, and destruction of circumscribed parts of the cerebral cortex can eliminate very specific aspects of consciousness, such as the ability to be aware of motion or to recognize faces, without a concomitant loss of vision in general.

Brain scientists are exploiting a number of empirical approaches that shed light on the neural basis of consciousness. This chapter reviews these approaches and summarizes what has been learned.

What Phenomena Does Consciousness Encompass?

There are many definitions of consciousness (Koch and Tononi, 2007). A common philosophical one is *"Consciousness is what it is like to be something,"* such as the experience of what it feels like to smell a rose or to be in love. This what-it-feels-like-from-within definition expresses the principal irreducible characteristic of the phenomenal aspect of consciousness: *to experience something*. "What it feels like to be me, to see red, or to be angry" also emphasizes the subjective or *first-person perspective* of consciousness: it is a subject who is having the experience and the experience is inevitably private.

What it feels like to have a particular experience is called the *quale* of that experience: the quale of red is what is common to such disparate conscious states as seeing a red sunset, the red flag of China, arterial blood, or a ruby gemstone. All four subjective states share "redness." There are countless qualia (the plural of quale): the ways things look, sound, and smell, the way it feels to have a pain, the way it feels to have thoughts and desires and so on. To have an experience means to have qualia, and the quale of an experience is what specifies it and makes it different from other experiences.

A science of consciousness must explain the exact relationship between phenomenal, mental states, and brain states. This is the heart of the classical mind–body problem: *What is the nature of the relationship between the immaterial, conscious mind and its physical basis in the electrochemical interactions in the body?* This problem can be divided into several subproblems:

1. Why is there any experience at all? Or, put differently, why does a brain state feel like anything? In philosophy, this is referred to by some as the *Hard Problem* (note the capitalization), or as the explanatory gap between the material, objective world and the subjective, phenomenal world (Chalmers, 1996). Many scholars have argued that the exact nature of this relationship will forever remain a central puzzle of human existence, without an adequate reductionistic, scientific explanation. However, as similar sentiments have been expressed in the past for the problem of seeking to understand life or to determine what material the stars are made of, it is best to put this question aside for the moment and not be taken in by defeatist arguments.
2. Why is the relationship among different experiences the way it is? For instance, red, yellow, green, cyan, blue, magenta are all colors that can be mapped onto the topology of a circle. Why? Furthermore, as a group, these color percepts share certain communalities that make them different from other percepts, such as seeing motion or smelling a rose.
3. Why are feelings private? As expressed by poets and novelists, we cannot communicate an experience to somebody else except by way of example.
4. How do feelings acquire meaning? Subjective states are not abstract states but have an immense amount of associated explicit and implicit feelings. Think of the unmistakable smell of dogs coming in from the rain or the crunchy texture of potato chips.
5. Why are only some behaviors associated with conscious states? Much brain activity and associated behavior occur without any conscious sensation.

The Neurobiology of Free Will

A further aspect of the mind-body problem is the question of free will, a vast topic. Answering this question goes to the heart of the way people think of themselves. The spectrum of views ranges from the traditional and deeply embedded belief that we are free, autonomous, and conscious actors to the view that we are biological machines driven by needs and desires beyond conscious access and without willful control.

Of great relevance are the classical findings by Libet and colleagues (1983) of brain events that precede the conscious initiation of a voluntary action. In this elegant experiment, subjects were sitting in front of an oscilloscope, tracking a spot of light moving every 2.56 sec around a circle. Every now and then, "spontaneously," subjects had to carry out a specific voluntary action, here flexing their wrist. If this action is repeated sufficiently often while electrical activity around the vertex of the head is recorded, a *readiness potential (Bereitschaftspotential)* in the form of a sustained scalp negativity can be recorded long before the muscle starts to move. Libet asked subjects to silently note the position of the spot of light when they first "felt the urge" to flex their wrist and to report this location afterward. This temporal marker for the awareness of willing an action occurs on average 200 msec before initiation of muscular action (with a standard error of about 20 msec), in accordance with commonsense notions of the causal action of free will. However, the readiness potential can be detected at least 350 msec before awareness of the action. In other words, the subject's brain signals the action at least half a second before the subject feels that he or she has initiated it!

This simple result has been replicated but, because of its counterintuitive implication that conscious will has no causal role, continues to be vigorously debated (Haggard and Eimer, 1999).

Psychological work in both normal individuals as well as in patients reveals further dissociations between the conscious perception of a willed action and its actual execution: subjects believe that they perform actions that they did not do while, under different circumstances, subjects feel that they are not responsible for actions that are, demonstrably, their own (Wegner, 2002).

Yet whether volition is illusory or is free in some libertarian sense does not answer the question of how subjective states relate to brain states. The perception of free will, what psychologists call the *feeling of agency* or *authorship* (e.g., "I decided to lift my finger"), is certainly a subjective state with an associated quale no different in kind from the quale of a toothache or seeing marine blue. So even if free will is a complete chimera, the subjective feeling of willing an action must have some neuronal correlate.

Direct electrical brain stimulation during neurosurgery as well as fMRI experiments implicate medial premotor and anterior cingulate cortices in generating the subjective feeling of triggering an action (Lau *et al.*, 2004). In other words, the neural correlate for the feeling of apparent causation involves activity in these regions.

Consciousness in Other Species

Data about subjective states come not only from people that can talk about their subjective experiences but also from nonlinguistic competent individuals— newborn babies or patients with complete paralysis of nearly all voluntary muscles (locked-in syndrome)— and, most importantly, from animals other than humans. There are three reasons to assume that many species, in particular those with complex behaviors such as mammals, share at least some aspects of consciousness with humans:

- **Similar neuronal architectures**. Except for size, there are no large-scale, dramatic differences between the cerebral cortex and thalamus of mice, monkey, humans, and whales. In particular, the macaque monkey is a powerful model organism to study visual perception because it shares with the human visual system three distinct cone photopigments, binocular stereoscopic vision, a foveated retina and similar eye movements.
- **Similar behavior**. Almost all human behaviors have precursors in the animal literature. Take the case of pain. The behaviors seen in humans when

they experience pain and distress—facial contortions, moaning, yelping, or other forms of vocalization; motor activity such as writhing, avoidance behaviors at the prospect of a repetition of the painful stimulus—can be observed in all mammals and in many other species. Likewise for the physiological signals that attend pain— activation of the sympathetic autonomous nervous system resulting in change in blood pressure, dilated pupils, sweating, increased heart rate, release of stress hormones, and so on. The discovery of cortical pain responses in premature babies shows the fallacy of relying on language as sole criteria for consciousness (Slater *et al.*, 2006).

- **Evolutionary continuity**. The first true mammals appeared at the end of the Triassic period, about 220 million years ago, with primates proliferating following the Cretaceous-Tertiary extinction event, about 60 million years ago, whereas humans and macaque monkeys did not diverge until 30 million years ago (Allman, 1999). *Homo sapiens* is part of an evolutionary continuum with its implied structural and behavioral continuity, rather than an independently developed organism.

Although certain aspects of consciousness, in particular those relating to the recursive notion of self and to abstract, culturally transmitted knowledge, are not widespread in nonhuman animals, there is little reason to doubt that other mammals share conscious feelings—sentience—with humans. To believe that humans are special, are singled out by the gift of consciousness above all other species, is a remnant of humanity's atavistic, deeply held belief that *homo sapiens* occupies a privileged place in the universe, a belief with no empirical basis.

The extent to which nonmammalian vertebrates, such as tuna, cichlid and other fish, crows, ravens, magpies, parrots and other birds, or even invertebrates, such as squids, or bees, with complex, non-stereotyped behaviors including delayed-matching, nonmatching-to-sample and other forms of learning (Giurfa *et al.*, 2001) are conscious is difficult to answer at this point in time. Without a sounder understanding of the neuronal architecture necessary to support consciousness, it is unclear where in the animal kingdom to draw the Rubicon that separates species with at least some conscious percepts from those that never experience anything and that are nothing but pure automata (Griffin, 2001).

Arousal and States of Consciousness

There are two common, but quite distinct, usages of the term consciousness, one revolving around *arousal* and *states of consciousness* and another one around *the content of consciousness* and *conscious states*.

To be conscious of anything, the brain must be in a relatively high state of arousal (sometimes also referred to as *vigilance*). This is as true of wakefulness as it is of REM sleep that is vividly, consciously experienced—though usually not remembered—in dreams (see Chapter 42). The level of brain arousal, measured by electrical or metabolic brain activity, fluctuates in a circadian manner, and is influenced by lack of sleep, drugs and alcohol, physical exertion, and so on in a predictable manner. High arousal states are always associated with some conscious state—a percept, thought or memory—that has a specific *content*. We see a face, hear music, remember an incident, plan an experiment, or fantasize about sex. Indeed, it is unlikely that one can be awake without being conscious of something. Referring to such conscious states is conceptually quite distinct from referring to states of consciousness that fluctuate with different levels of arousal. Arousal can be measured behaviorally by the signal amplitude that triggers some criterion reaction (for instance, the sound level necessary to evoke an eye movement or a head turn toward the sound source).

Different levels or states of consciousness are associated with different kinds of conscious experiences. The awake state in a normal functioning individual is quite different from the dreaming state (for instance, the latter has little or no self-reflection) or from the state of deep sleep. In all three cases, the basic physiology of the brain is changed, affecting the space of possible conscious experiences. Physiology is also different in *altered states of consciousness*, for instance after taking psychedelic drugs when events often have a stronger emotional connotation than in normal life. Yet another state of consciousness can occur during certain meditative practices, when conscious perception and insight may be enhanced compared to the normal waking state.

In some obvious but difficult to rigorously define manner, the *richness of conscious experience* increases as an individual transitions from deep sleep to drowsiness to full wakefulness. This richness of possible conscious experience could be quantified using notions from complexity theory that incorporate both the dimensionality as well as the granularity of conscious experience (e.g., Tononi, 2004). For example, inactivating all of visual cortex in an otherwise normal individual would significantly reduce the dimensionality of conscious experience since no color, shape, motion, texture, or depth could be perceived or imagined. As behavioral arousal increases, so does the range and complexity of behaviors of which an individual is capable. A singular exception to this progression is

REM sleep where most motor activity is shut down in the *atonia* that is characteristic of this phase of sleep, and the person is difficult to wake up. Yet this low level of behavioral arousal goes, paradoxically, hand in hand with high metabolic and electrical brain activity and conscious, vivid states.

These observations suggest a two-dimensional graph (Fig. 53.1) in which the richness of conscious experience (its representational capacity) is plotted as a function of levels of behavioral arousal or responsiveness.

Global disorders of consciousness can likewise be mapped onto this plane. Clinicians speak of impaired states of consciousness as in "the comatose state," "the persistent vegetative state" (PVS), and the "minimal conscious state" (MCS). Here, state refers to different levels of consciousness, from a total absence in the case of coma, PVS or (hopefully) general anesthesia, to a fluctuating and limited form of conscious sensation in MCS, sleep walking or during a complex partial epileptic seizure (Schiff, 2004).

The repertoire of distinct conscious states or experiences that are accessible to a patient in MCS is presumably minimal (possibly including pain, discomfort, and sporadic sensory percepts), immeasurably smaller than the possible conscious states that can be experienced by a healthy brain. In the limit of brain death, the origin of this space has been reached with no experience at all (Fig. 53.1). More relevant to clinical practice is the case of global anesthesia, during which the patient should not experience anything so as to avoid traumatic memories and their undesirable sequelae.

Given the absence of any accepted theory for the minimal neuronal criteria necessary for consciousness, the distinction between a PVS patient—who shows regular sleep-wave transitions and who may be able to move eyes or limbs or smile in a reflexive manner as in the widely publicized 2005 case of Terri Schiavo in Florida—and a MCS patient who can communicate (on occasion) in a meaningful manner (for instance, by differential eye movements) and who shows some signs of consciousness, is often difficult in a clinical setting. Functional brain imaging may prove immensely useful here.

Blood-oxygen-level-dependent (BOLD) functional magnetic resonance imaging (fMRI) demonstrated that a patient in a vegetative state following a severe traumatic brain injury exhibited the same pattern of brain activity as normal individuals when asked to imagine playing tennis or to imagine visiting all the rooms in her house (Owen *et al.*, 2006). Differential brain imaging of patients with such global disturbances of consciousness (including akinetic mutism) reveal that dysfunction in a widespread cortical network including medial

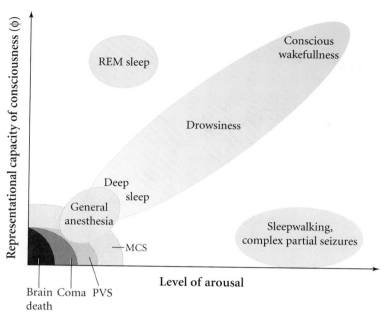

FIGURE 53.1 Normal and pathological brain states can be situated in a two-dimensional graph. Here increasing levels of behaviorally determined arousal is plotted on the x-axis and the "richness" or "representational capacity of consciousness" is plotted on the y-axis. Increasing arousal can be measured by the threshold to obtain some specific behavior (for instance, spatial orientation to a sound). Healthy subjects cycle during a 24-hour period from deep sleep with low arousal and very little conscious experience to increasing levels of arousal and conscious sensation. In REM sleep, low levels of behavioral arousal go hand-in-hand with vivid consciousness. Conversely, various pathologies of clinical relevance are associated with little to no conscious content. Modified from Laureys (2005).

and lateral prefrontal cortex and parietal associative areas is associated with a global loss of consciousness (Laureys, 2005).

In contrast to diffuse cortical damage, relatively discrete bilateral injuries to midline (paramedian) subcortical structures can also cause a complete loss of consciousness. These structures are therefore part of the *enabling* factors that control the level of brain arousal (as determined by metabolic or electrical activity) and that are needed for any form of consciousness to occur. For an example, consider the heterogeneous collection of more than two dozen (on each side) of nuclei in the upper brain stem (pons, midbrain, and in the posterior hypothalamus), collectively referred to as the *reticular activating system*. These nuclei—three-dimensional collections of neurons with their own cytoarchitecture and neurochemical identity—release distinct neuromodulators such as acetylcholine, noradrenaline/norepinephrine, serotonin, histamine, and orexin/hypocretin. Their axons project widely throughout the brain (Fig. 53.2). These neuromodulators control the excitability of thalamus and forebrain, mediate the alternation between wakefulness and sleep, as well as the general level of both behavioral and brain arousal. Acute lesions in the reticular activating system can result in loss of consciousness and coma. However, eventually the excitability of thala-

mus and forebrain can recover and consciousness can return (Villablanca, 2004).

Other enabling factors for consciousness are the five or more *intralaminar nuclei of the thalamus* (ILN). These receive input from many brain stem nuclei and from frontal cortex and project strongly to the basal ganglia and, in a more distributed manner, into layer I of much of neocortex. Comparatively small (1 cm^3 or less), bilateral lesions in the ILN can completely eliminate awareness (Bogen, 1995). Thus, the ILN are necessary for an organism to be conscious at all but do not appear to be responsible for mediating specific conscious percepts or memories.

If a single substance is critical for consciousness, then acetylcholine is the most likely candidate. Two major cholinergic pathways originate in the brain stem and in the basal forebrain (Fig. 53.2). Brain stem cells send an ascending projection to the thalamus, where release of acetylcholine facilitates thalamo-cortical relay cells and suppresses inhibitory interneurons. Cholinergic cells are therefore well positioned to influence all of cortex by controlling the thalamus. In contrast, cholinergic basal forebrain neurons send their axons to a much wider array of target structures. Collectively, brain stem and basal forebrain cholinergic cells innervate the thalamus, hippocampus, amygdala, and neocortex.

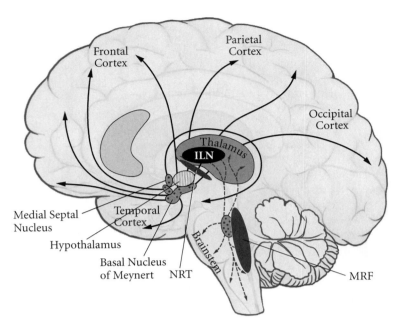

FIGURE 53.2 Midline structures in the brainstem and thalamus necessary to regulate the level of brain arousal include the intralaminar nuclei of the thalamus (ILN), the thalamic reticular nucleus (NRT) encapsulating the dorsal thalamus, and the midbrain reticular formation (MRF) that encompasses the reticular activating system. Small, bilateral lesions in many of these nuclei can cause a global loss of consciousness.

Cholinergic activity fluctuates with the sleep–wake cycle. In general, increasing levels of spiking activity in cholinergic neurons is associated with wakefulness or REM sleep, and decreasing levels occur during non-REM or slow-wave sleep. Lastly, many neurological pathologies whose symptoms include disturbances of consciousness, such as Parkinson's disease, Alzheimer's disease, and other forms of dementia, are associated with a selective loss of cholinergic neurons.

In summary, a plethora of nuclei with distinct chemical signatures in the thalamus, midbrain, and pons must function for the brain of an individual to be sufficiently aroused to experience anything at all. These nuclei belong to the enabling factors for consciousness. It is likely that the specific content of any one conscious sensation is mediated by neurons in cortex and their associated satellite structures, including the amygdala, thalamus, claustrum, and the basal ganglia.

The Neuronal Correlates of Consciousness

Progress in addressing the mind–body problem has come from focusing on empirically accessible questions rather than on eristic philosophical arguments with no clear resolution. One key objective has been to search for the neuronal correlates—and ultimately the causes—of consciousness. As defined by Crick and Koch (2003), the neuronal correlates of consciousness (NCC) are the *minimal neuronal mecha-*

nisms jointly sufficient for any one specific conscious percept (Fig. 53.3).

This definition of the NCC stresses the word "minimal," because the question of interest is which subcomponents of the brain actually are needed. For instance, it is likely that neural activity in the cerebellum does not underlie any conscious perception, and thus is not part of the NCC. That is, trains of spikes in Purkinje cells (or their absence) will not induce a sensory percept although they may ultimately affect some behaviors (such as eye movements).

This definition does not focus on the necessary conditions for consciousness, because of the great redundancy and parallelism found in neurobiological networks. While activity in some population of neurons may underpin a percept in one case, a different population might mediate a related percept if the former population is lost or inactivated.

Every phenomenal, subjective state will have associated NCC: one for seeing a red patch, another one for seeing grandmother, yet a third one for hearing a siren, and so on. Perturbing or inactivating the NCC for any one specific conscious experience will affect the percept or cause it to disappear respectively. If the NCC could be induced artificially, for instance by cortical microstimulation in a prosthetic device or during neurosurgery, the subject will experience the associated percept.

What characterizes the NCC? What are the communalities between the NCC for seeing and for

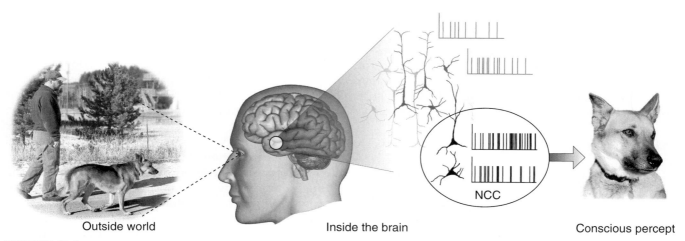

Outside world   Inside the brain   Conscious percept

FIGURE 53.3 The neuronal correlates of consciousness (NCC) are the minimal set of neural events and structures—here synchronized action potentials in neocortical pyramidal neurons—sufficient for a specific conscious percept or memory. From Koch (2004).

hearing? Will the NCC involve all pyramidal neurons in cortex at any given point in time? Or only a subset of long-range projection cells in the frontal lobes that project to the sensory cortices in the back? Only layer 5 cortical cells? Neurons that fire in a rhythmic manner? Neurons that fire in a synchronous manner? These are some of the proposals that have been advanced over the years (Chalmers, 2000).

It is implicitly assumed by most neurobiologists that the relevant variables giving rise to consciousness are to be found at the neuronal level, among the synaptic release or the action potentials in one or more population of cells, rather than at the molecular level. A few scholars have proposed that macroscopic quantum behaviors underlie consciousness. Of particular interest here is *entanglement*, the observation that the quantum states of multiple objects, such as two coupled electrons, may be highly correlated even though they are spatially separated, violating our intuition about locality (entanglement is also the key feature of quantum mechanics that is exploited in quantum computers). The role of quantum mechanics for the photons received by the eye and for the molecules of life is not controversial. But there is no evidence that any components of the nervous system—a 37°C wet and warm tissue strongly coupled to its environment—display quantum entanglement. And even if quantum entanglement were to occur inside individual cells, diffusion and action potential generation and propagation, the principal mechanism for getting information into and out of neurons, would destroy superposition. At the cellular level, the interaction of neurons is governed by classical physics (Koch and Hepp, 2006).

Thus, the vast majority of experiments concerned with the NCC refer to either electrophysiological recordings in monkeys or functional brain imaging data in humans.

The Neuronal Basis of Perceptual Illusions

The possibility of precisely manipulating visual percepts in time and space has made vision a preferred modality for seeking the NCC. Psychologists have perfected a number of techniques—masking, binocular rivalry, continuous flash suppression, motion-induced blindness, change blindness, inattentional blindness—in which the seemingly simple and unambiguous relationship between a physical stimulus in the world and its associated percept in the privacy of the subject's mind is disrupted (Kim and Blake, 2005). With such techniques, a stimulus can be perceptually suppressed for seconds or even minutes at a time: the image is projected into one of the observer's eyes but it is invisible, not seen. In this manner the neural mechanisms that respond to the subjective percept rather than the physical stimulus can be isolated, permitting the footprints of visual consciousness to be tracked in the brain. In a *perceptual illusion*, the physical stimulus remains fixed while the percept fluctuates. The best known example is the *Necker cube*, whose 12 lines can be perceived in one of two different ways in depth (Fig. 53.4).

A perceptual illusion that can be precisely controlled is *binocular rivalry* (Blake and Logothetis, 2002). Here, a small image (e.g., a horizontal grating) is presented to the left eye and another image (e.g., a vertical grating) is shown to the corresponding location in the right eye. In spite of the constant visual stimulus, observers consciously see the horizontal grating alternate every few seconds with the vertical one. The brain does not allow for the simultaneous perception of both images.

Macaque monkeys can be trained to report whether they see the left or the right image. The distribution of the switching times and the way in which changing the contrast in one eye affects the reports, leaves little doubt that monkeys and humans experience the same

basic phenomenon. In a series of elegant experiments, Logothetis and colleagues (Logothetis, 1998) recorded from a variety of visual cortical areas in the awake macaque monkey while the animal performed a binocular rivalry task. In primary visual cortex (V1), only a small fraction of cells weakly modulate their response as a function of the percept of the monkey. The majority of cells responded to one or the other retinal stimulus with little regard to what the animal perceived at the time. In contrast, in a high-level cortical area such as the inferior temporal (IT) cortex along the ventral pathway, almost all neurons responded only to the perceptual dominant stimulus, that is, to the stimulus that was being reported. For example, when a face and a more abstract design were presented, one of these to each eye, a "face" cell fired only when the animal indicated by its performance that it saw the face and not the design presented to the other eye (Fig. 53.5). This result implies that the NCC involves activity in neurons in inferior temporal cortex.

Does this mean that the NCC is local to IT? At this point, this is not clear. However, given known anatomical connections, it is possible that specific reciprocal interactions between IT cells and neurons in parts of the prefrontal cortex are necessary for the NCC. This is compatible with the broadly accepted notion that the NCC must involve positive feedback to insure that neural activity is persistent and strong enough to exceed some threshold and to be broadly distributed to multiple cognitive systems, including working memory, planning, and language.

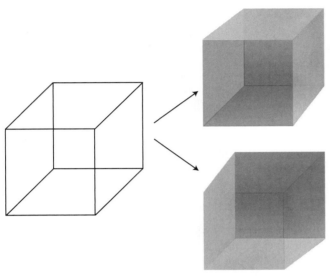

FIGURE 53.4 The left line drawing can be perceived in one of two distinct depth configurations shown on the right. Without any other cue, the visual system flips back and forth between these two interpretations of the Necker cube. From Koch (2004).

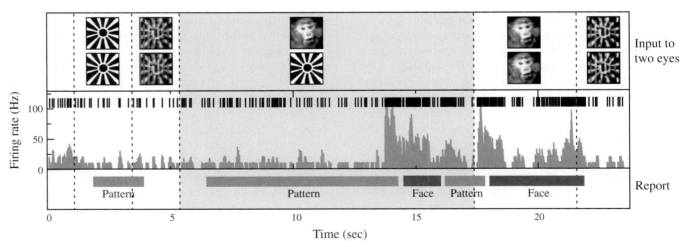

FIGURE 53.5 A fraction of a minute in the life of a typical IT cell while a monkey experiences binocular rivalry. The upper row indicates the visual input, with dotted vertical boundaries marking stimulus transitions. The second row shows the individual spikes, the third the smoothed firing rate, and the bottom row the monkey's behavior. The animal was taught to press a lever when it saw either one or the other image, but not both. The cell responded only weakly to either the sunburst design or to its optical superposition with the image of a monkey's face. During binocular rivalry (gray central zone) between 5 and 18 seconds, the monkey's perception vacillated back and forth between seeing the face and seeing the bursting sun. Perception of the face was consistently accompanied (and preceded) by a strong increase in firing rate. From N. Logothetis (private communication) as modified by Koch (2004).

In a related perceptual phenomena, *flash suppression*, the percept associated with an image projected into one eye is suppressed by flashing another image into the other eye (while the original image remains). Its methodological advantage over binocular rivalry is that the timing of the perceptual transition is determined by an external trigger rather than by an internal event. The majority of responsive cells in inferior temporal cortex and in the superior temporal sulcus follow the monkey's behavior, and therefore its percept. That is, when the animal perceives a cell's preferred stimulus, the neuron fires; when the stimulus is present on the retina but is perceptually suppressed, the cell falls silent, even though legions of V1 neurons fire vigorously to the same stimulus (Sheinberg and Logothetis, 1997). Single neuron recordings in the medial temporal lobe of epileptic patients during flash suppression likewise demonstrate abolition of neuronal responses when the cells preferred stimulus is present on the retina but is not seen (Kreiman, Fried, and Koch, 2002).

A number of fMRI experiments have exploited binocular rivalry and related illusions to identify the hemodynamic activity underlying visual consciousness in humans. They demonstrate quite conclusively that BOLD activity in the upper stages of the ventral pathway (e.g., the fusiform face area and the parahippocampal place area) follow the percept and not simply the retinal stimulus (Rees and Frith, 2007).

There is a lively debate about the extent to which neurons in primary visual cortex simply encode the visual stimulus or are directly responsible for expressing the subject's conscious percept. That is, is V1 part of the NCC (Crick and Koch, 1995)? It is clear that retinal neurons are not part of the NCC for visual experiences. While retinal neurons often correlate with visual experience, the spiking activity of retinal ganglion cells does not accord with visual experience (for example, there are no photoreceptors at the blind spot; yet no hole in the field of view is apparent; in dreams vivid imagery occurs despite closed eyes, and so on). A number of compelling observations link perception with fMRI BOLD activity in human V1 and even LGN (Tong *et al.*, 1998; Lee, Blake, and Heeger, 2005). These data are at odds with single neuron recordings from the monkey. What explains this discrepancy? Most notably, the two methods operate at quite distinct spatial and temporal scales, recording with high granularity and millisecond resolution individual action potentials in monkeys versus measuring hemodynamic activity at the time scale of seconds in people. Clearly, firing rates and fMRI activity are linked but how they are linked is a very active area of research. It is, of course, always possible that the fine structure of visual cortex differs between these two primate species. These questions have yet to

be resolved, particularly in light of more recent experiments that compellingly demonstrate that V1 BOLD activity and conscious perception can be dissociated.

Haynes and Rees (2005) exploited multivariate decoding techniques to read out perceptually suppressed information (the orientation of a masked stimulus) from V1 BOLD activity, even though the stimulus orientation was so efficiently masked that subjects performed at chance levels when trying to guess the orientation. That is, although subjects did not give any behavioral indication that they saw the orientation of the stimulus, its slant could be predicted on a single trial basis with better than chance odds from the V1 (but not from V2 or V3) BOLD signal. This finding supports the hypothesis that information present in V1 may not be accessible to behavior or to consciousness.

In a powerful combination of binocular rivalry and flash suppression, a stationary image in one eye can be suppressed for minutes on end by continuously flashing different images into the other eye (*continuous flash suppression*; Tsuchiya and Koch, 2005). This paradigm lends itself naturally to further investigate the relationship between neural activity—whether assayed at the single neuron or at the brain voxel level—and conscious perception.

Other Perceptual Puzzles of Contemporary Interest

The attributes of even simple percepts seem to vary along a continuum. For instance, a patch of color has a brightness and a hue that are variable, just as a simple tone has an associated loudness and pitch. However, is it possible that each particular, consciously experienced, percept is all-or-none? Might a pure tone of a particular pitch and loudness be experienced as an atom of perception, either heard or not, rather than gradually emerging from the noisy background? The perception of the world around us would then be a superposition of many elementary, binary percepts (Sergent and Dehaene, 2004).

Is perception continuous, like a river, or does it consist of series of discontinuous batches, rather like the discrete frames in a movie (Purves, Paydarfar, and Andrews, 1996; VanRullen and Koch, 2003a). In *cinematographic vision* (Sacks, 2004), a rare form of visual migraine, the subject sees the movement of objects as fractured in time, as a succession of different configurations and positions, without any movement in between. The hypothesis that visual perception is quantized in discrete batches of variable duration, most often related to EEG rhythms in various frequency ranges (from theta to beta), is an old one. This idea is being revisited in light of the discrepancies of timing of perceptual events within and across different sensory modalities. For instance, even though a change in the color of an object occurs simultaneously with a change in its direction of motion, it may not be perceived that way (Zeki, 1998; Bartels and Zeki, 2006; Stetson *et al.*, 2006).

What is the relationship between endogenous, top-down attention and consciousness? Although these are frequently coextensive—subjects are usually conscious of what they attend to—there is a considerable tradition in psychology that argues that these are distinct neurobiological processes (Koch and Tsuchiya, 2007). This question is receiving renewed attention due to the development of ever more refined and powerful visual masking techniques (Kim and Blake, 2005) that independently manipulate attention and consciousness. Indeed, it has been shown that attention can be allocated to a perceptually invisible stimulus (Naccache, Blandin, and Dehaene, 2002) and that subjects can be conscious of a stimulus without attending to it. When exploring the neural basis of these processes, it is therefore critical to not confound attention with consciousness and vice versa.

Forward versus Feedback Projections

Many actions in response to sensory inputs are rapid, transient, stereotyped, and unconscious (Milner and Goodale, 1995). They can be thought of as cortical reflexes and are characterized by rapid and somewhat stereotyped responses sometimes called *zombie behaviors* (Koch and Crick, 2001), in addition to a slower, all-purpose conscious mode. The latter, conscious mode deals more slowly with broader, less stereotyped and more complex aspects of the sensory input (or a reflection of these, as in imagery) and takes time to decide on appropriate thoughts and responses. A consciousness mode is needed because otherwise a vast number of different zombie modes would be required to react to unusual events. The conscious system may interfere somewhat with the concurrent zombie systems (Beilock *et al.*, 2002): focusing consciousness onto the smooth execution of a complex, rapid, and highly trained sensory-motor task—dribbling a soccer ball, to give one example—can interfere with its smooth execution, something well known to athletes and their trainers. Having both a zombie mode that responds in a well-rehearsed and stereotyped manner as well as a slower system that allows time for thinking and planning more complex behavior is a great evolutionary discovery. This latter aspect, planning, may be one of the principal functions of consciousness.

It seems possible that visual zombie modes in the cortex mainly use the dorsal stream in the parietal region (Milner and Goodale, 1995). However, parietal activity can affect consciousness by producing attentional effects on the ventral stream, at least under some circumstances. The basis of this inference are clinical case studies and fMRI experiments in normal subjects (Corbetta and Shulman, 2002). The conscious mode for vision depends largely on intermediate visual areas (beyond V1) and especially on the ventral "what" stream.

Seemingly complex visual processing (such as detecting animals in natural, cluttered images) can be accomplished by cortex within 130–150 msec (Thorpe, Fize, and Marlot, 1996; VanRullen and Koch, 2003b), far too slow for conscious perception to be involved in these tasks. It is quite plausible that such behaviors are mediated by a purely feed-forward moving wave of spiking activity that passes from the retina through V1, into V4, IT, and prefrontal cortex, until it affects motor neurons in the spinal cord that control the finger press (as in a typical laboratory experiment). The hypothesis that the basic processing of information is feed-forward is supported most directly by the short times required for a selective response to appear in IT cells. Indeed, Hung *et al.* (2005) were able to decode from the spiking activity of a couple of hundred neurons in monkey IT (over intervals as short as 12.5 msec and about 100 msec after image onset), the category, identity, and even position of a single image flashed onto the retina of the fixating animal. Such hierarchical, purely feed-forward models of visual perception for object recognition in the ventral stream—an outgrowth of the original Hubel and Wiesel scheme (1968)—has been formalized and performs as well as human observers (for masked natural images) and state-of-the-art machine vision categorization algorithms (Serre *et al.*, 2007). Coupled with a suitable motor output, such a feed-forward network implements a zombie behavior in the Koch and Crick (2001) sense—rapidly and efficiently subserving one task, here distinguishing animal from nonanimal pictures, in the absence of any conscious experience.

Contrariwise, conscious perception is believed to require more sustained, reverberatory neural activity, most likely via global feedback from frontal regions of neocortex back to sensory cortical areas in the back (Crick and Koch, 1995). These feedback loops would explain why in backward masking a second stimulus, flashed up to 80 msec after onset of a first image, can still interfere (mask) with the percept of the first image. The reverberatory activity builds up over time until it exceeds a critical threshold. At this point, the sustained neural activity rapidly propagates to parietal, prefrontal, and anterior cingulate cortical regions, thalamus, claustrum, and related structures that support short-term memory, multimodality integration, planning, speech, and other processes intimately related to consciousness. Competition prevents more than one or a very small number of percepts to be simultaneously and actively represented. This is the hypothesis at the heart of the *global workspace* model of consciousness (Baars, 1977; Dehaene and Changeux, 2005). Sending visual information to more frontal structures would allow the associated visual events to be decoded and placed into context (for instance, by accessing various memory banks) and to have this interpretation feed back to the sensory representation in visual cortex (Jazayeri and Movshon, 2007).

In brief, while rapid but transient neural activity in the thalamo-cortical system can mediate complex behavior without conscious sensation it is surmised that consciousness requires sustained but well-organized neural activity dependent on long-range cortico-cortical feedback.

An Information-Theoretical Theory of Consciousness

At present, it is not known to what extent animals whose nervous systems have an architecture considerably different from the mammalian neocortex are conscious. Furthermore, whether artificial systems, such as computers, robots, or the World Wide Web as a whole, which behave with considerable intelligence, are or can become conscious (as widely assumed in science fiction; e.g., the paranoid computer *HAL* in the film *2001*), remains completely speculative. What is needed is a theory of consciousness, which explains in quantitative terms what type of systems, with what architecture, can possess conscious states.

Progress in the study of the NCC on the one hand, and of the neural correlates of nonconscious zombie behaviors on the other, will hopefully lead to a better understanding of what distinguishes neural structures or processes that are associated with consciousness from those that are not. Yet such an opportunistic, data-driven approach will not lead to an understanding of why certain structures and processes have a privileged relationship with subjective experience. For example, why is it that neurons in corticothalamic circuits are essential for conscious experience, whereas cerebellar neurons, despite their huge numbers, are not? And what aspect of cortical zombie systems make them unsuitable to yield subjective experience? Or why is it that consciousness wanes during slow-wave sleep early in the night, despite levels of neural firing in the thalamocortical system that are comparable to the levels of firing in wakefulness?

Information theory may be such a theoretical approach that establishes at the fundamental level what consciousness is, how it can be measured, and what requisites a physical system must satisfy in order to generate it (Chalmers, 1996; Tononi and Edelman, 1998).

The most promising candidate for such a theoretical framework is the *information integration theory of consciousness* (Tononi, 2004). It posits that the most important property of consciousness is that it is extraordinarily *informative*. Any one particular conscious state rules out a huge number of alternative experiences. Classically, the reduction of uncertainty among a number of alternatives constitutes information. For example, when a subject consciously experiences reading this particular phrase, a huge number of other possible experiences are ruled out (consider all possible written phrases that could have been written in this space, in all possible fonts, ink colors, and sizes, think of the same phrases spoken aloud, or read and spoken, and so on). Thus, every experience represents one particular conscious state out of a huge repertoire of possible conscious states.

Furthermore, information associated with the occurrence of a conscious state is *integrated* information. An experience of a particular conscious state is an integrated whole. It cannot be subdivided into components that are experienced independently (Tononi and Edelman, 1998). For example, the conscious experience of this particular phrase cannot be experienced as subdivided into, say, the conscious experience of how the words look independently of the conscious experience or how they sound in the reader's mind. Similarly, visual shapes cannot be experienced independently of their color, nor can the left half of the visual field of view be experienced independently of the right half.

Based on these and other considerations, the theory claims that *a physical system can generate consciousness to the extent that it can integrate information*. This idea requires that the system has a large repertoire of available states (information) yet cannot be decomposed into a collection of causally independent subsystems (integration).

Importantly, the theory introduces a measure of a system's capacity to integrate information. This measure, called ϕ, is obtained by determining the minimum repertoire of different states that can be produced in one part of the system by perturbations of its other parts (Tononi, 2004). ϕ can loosely be thought of as the representational capacity of the system (as in the y-axis of Fig. 53.1). Although ϕ is not easy to calculate exactly for realistic systems, it can be estimated. Thus, by using simple computer simulations, it is possible to show that ϕ is high for neural architectures that conjoin functional specialization with functional integration, like the mammalian thalamocortical system. Conversely, ϕ is low for systems that are made up of small, quasi-independent modules, like the cerebellum, or for networks of randomly or uniformly connected units (Tononi, 2004).

The notion that consciousness has to do with the brain's ability to integrate information has been tested directly by transcranial magnetic stimulation. In TMS a coil is placed above the skull and a brief and intense magnetic field generates a weak electrical current in the underlying gray matter in a noninvasive manner. Massimini *et al.* (2005) compared multichannel EEG of awake and conscious subjects in response to TMS pulses to the EEG when the same subjects were deeply asleep early in the night—a time during which consciousness is much reduced. During quiet wakefulness, an initial response at the stimulation site was followed by a sequence of waves that moved to connected cortical areas several centimeters away. During slow wave sleep, by contrast, the initial response was stronger but was rapidly extinguished and did not propagate beyond the stimulation site. Thus, the fading of consciousness during certain stages of sleep may be related, as predicted by the theory, to the breakdown of information integration among specialized thalamo-cortical modules.

Conclusion

Ever since the Greeks first considered the mind–body problem more than two millennia ago, it has been the domain of armchair speculations and esoteric debates with no apparent resolution. Yet many aspects of this ancient set of questions now fall squarely within the domain of science.

In order to advance the resolution of these and similar questions, it will be imperative to record from a large number of neurons simultaneously at many locations throughout the cortico-thalamo system and related satellites such as the claustrum (Crick and Koch, 2005) in behaving subjects. Such experiments cannot, of course, be done in humans. Progress in understanding the circuitry of consciousness, therefore, demands a battery of behaviors (akin to but different from the well-known Turing test for intelligence) that the subject—a newborn infant, immobilized patient, or nonhuman animal—has to pass before considering him, her, or it to possess some measure of conscious perception. This is not an insurmountable step for mammals such as the monkey or the mouse that share many behaviors and brain structures with humans. For example, one particular mouse model of contingency awareness (Han *et al.*, 2003) is based on the differential requirement for awareness of trace versus delay associative eyeblink conditioning in humans (Clark and Squire, 1998).

The growing ability of neuroscientists to manipulate in a reversible, transient, deliberate, and delicate manner identified populations of neurons in the mouse, and thereby affecting its behavior, using combined genetic and optical tools (Adamantidis *et al.*, 2007) opens the possibility of moving from correlation—observing that a particular conscious state is associated with some neural or hemodynamic activity—to causation. For example, rather than just noting that motion perception is associated with an elevated firing rate in projection neurons in cortical area MT, we will be able to perturb the system by inactivating genetically distinct subpopulation of cells in a highly targeted manner. Exploiting these increasingly powerful tools depends on the simultaneous development of appropriate behavioral assays and model organisms amenable to large-scale genomic analysis and manipulation, particularly in mice (Lein *et al.*, 2007).

It is the combination of such fine-grained neuronal analysis in mice and monkeys, ever more sensitive psychophysical and brain imaging techniques in patients and healthy individuals, and the development of a robust theoretical framework that lend hope to the belief that human ingenuity can, ultimately, understand in a rational manner one of the central mysteries of life.

References

Adamantidis, A. R., Zhang, F., Aravanis, A. M., Deisseroth, K., and de Lecea, L. (2007). Neural substrates of awakening probed with optogenetic control of hypocretin neurons. *Nature*.

Allman, J. M. (1999). "Evolving Brains." Scientific American Library: New York.

Baars, B. J. (1997). "In the Theater of Consciousness." Oxford University Press: New York.

Bartels, A. and Zeki, S. (2006). The temporal order of binding visual attributes. *Vision Res.* **46**, 2280–2286.

Beilock, S. L., Carr, T. H., MacMahon, C., and Starkes, J. L. (2002). When paying attention becomes counterproductive: Impact of divided versus skill-focused attention on novice and experienced performance of sensorimotor skills. *J. Exp. Psychol. Appl.* **8**, 6–16.

Blake, R. and Logothetis, N. K. (2002). Visual competition. *Nature Reviews Neuroscience* **3**, 13–21.

Bogen, J. E. (1995). On the neurophysiology of consciousness: I. An Overview. *Consciousness & Cognition* **4**, 52–62.

Chalmers, D. J. (1996). "The Conscious Mind: In Search of a Fundamental Theory." Oxford University Press, New York.

Chalmers, D. J. (2000). What is a neural correlate of consciousness? *In* "Neural Correlates of Consciousness: Empirical and Conceptual Questions" (Metzinger, T., ed.), pp. 17–40. MIT Press, Cambridge.

Corbetta, M. and Shulman, G. L. (2002). Control of goal-directed and stimulus-driven attention in the brain. *Nature Rev. Neurosci.* **3**, 201–215.

Crick, F. C. and Koch, C. (1995). Are we aware of neural activity in primary visual cortex? *Nature* **375**, 121–123.

Crick, F. C. and Koch, C. (2003). A framework for consciousness. *Nature Neuroscience* **6**, 119–127.

Dehaene, S. and Changeux, J. P. (2005). Ongoing spontaneous activity controls access to consciousness: A neuronal model for inattentional blindness. *PLoS Biology* **3(5)**:e141.

Edelman, D. B., Baars, J. B., and Seth, A. K. (2005). Identifying hallmarks of consciousness in non-mammalian species. *Conscious Cogn.* **14**, 169–187.

Giurfa, M., Zhang, S., Jenett, A., Menzel, R., and Srinivasan, M. V. (2001). The concepts of "sameness" and "difference" in an insect. *Nature* **410**, 930–933.

Griffin, D. R. (2001). "Animal Minds: Beyond Cognition to Consciousness." University of Chicago Press, Chicago, IL.

Haggard, P. and Eimer, M. (1999). On the relation between brain potentials and conscious awareness. *Exp. Brain Res.* **126**, 128–133.

Han, C. J., O'Tuathaigh, C. M., van Trigt, L., Quinn, J. J., Fanselow, M. S., Mongeau, R., Koch, C., and Anderson, D. J. (2003). Trace but not delay fear conditioning requires attention and the anterior cingulate cortex. *Proc Natl. Acad. Sci. USA* **100**, 13087–13092.

Haynes, J.-D. and Rees, G. (2005). Predicting the orientation of invisible stimuli from activity in human primary visual cortex. *Nature Neurosci.* **8**, 686–691.

Hubel, D. H. and Wiesel, T. N. (1968). Receptive fields and functional architecture of monkey striate cortex. *J. Physiol.* **195**, 215–243.

Hung, C. P., Kreiman, G., Poggio, T. and DiCarlo, J. J. (2005). Fast readout of object identity from macaque inferior temporal cortex. *Science* **310**, 863–866.

Jazayeri, A. and Movshon, J. A. (2007). A new perceptual illusion reveals mechanisms of sensory decoding. *Nature* **446**, 912–915.

Kim, C.-Y. and Blake, R. (2005). Psychophysical magic: Rendering the visible "invisible." *Trends Cogn. Sci.* **9**, 381–388.

Koch, C. (2004). "The Quest for Consciousness: A Neurobiological Approach." Roberts, Denver, CO.

Koch, C. and Crick, F. C. (2001). On the zombie within. *Nature* **411**, 893.

Koch, C. and Hepp, K. (2006). Quantum mechanics and higher brain functions: Lessons from quantum computation and neurobiology. *Nature* **440**, 611–612.

Koch, C. and Tononi, G. (2007). Consciousness. *In* "New Encyclopedia of Neuroscience." Elsevier, in press.

Koch, C. and Tsuchiya, N. (2007). Attention and consciousness: Two distinct brain processes. *Trends Cogn. Sci.* **11**, 16–22.

Kreiman, G., Fried, I., and Koch, C. (2002). Single-neuron correlates of subjective vision in the human medial temporal lobe. *Proc. Natl. Acad. Sci. USA* **99**, 8378–8383.

Lau, H. C., Rogers, R. D., Haggard, P. and Passingham, R. E. (2004). Attention to intention. *Science* **303**, 1208–1210.

Laureys, S. (2005). The neural correlate of (un)awareness: Lessons from the vegetative state. *Trends Cogn. Sci.* **9**, 556–559.

Lee, S. H., Blake, R., and Heeger, D. J. (2005). Traveling waves of activity in primary visual cortex. *Nature Neurosci.* **8**, 22–23.

Lein, E. S., Hawrylycz, M. J., *et al.* (2007). Genome-wide atlas of gene expression in the adult mouse brain. *Nature* **445**, 168–176.

Libet, B., Gleason, C. A., Wright, E. W., and Pearl, D. K. (1983). Time of conscious intention to act in relation to onset of cerebral activity (readiness-potential): The unconscious initiation of a freely voluntary act. *Brain* **106**, 623–642.

Logothetis, N. (1998). Single units and conscious vision. *Philosophical Transactions Royal Society of London B*, **353**, 1801–1818.

Massimini, M., Ferrarelli, F., Huber, R., Esser, S. K., Singh, H., and Tononi, G. (2005). Breakdown of cortical effective connectivity during sleep. *Science* **309**, 2228–2232.

Milner, A. D. and Goodale, M. A. (1995). "The visual brain in action." Oxford University Press, Oxford, UK.

Naccache, L., Blandin, E., and Dehaene, S. (2002). Unconscious masked priming depends on temporal attention. *Psychol. Sciences* **13**, 416–424.

Owen, A. M., Cleman, M. R., Boly, M., Davis, M. H., Laureys, S., and Pickard, J. D. (2006). Detecting awareness in the vegetative state. *Science* **313**, 1402.

Purves, D., Paydarfar, J. A., and Andrews, T. J. (1996). The wagon wheel illusion in movies and reality. *Proc. Natl. Acad. Sci. USA* **93**, 3693–3697.

Rees, G. and Frith, C. (2007). Methodologies for identifying the neural correlates of consciousness. *In* "The Blackwell Companion to Consciousness" (Velmans, M. and Schneider, S., eds.), pp. 553–566. Blackwell, Oxford, UK.

Sacks, O. (2004). In the river of consciousness. *New York Review Books* **51**, 41–44.

Schiff, N. D. (2004). The neurology of impaired consciousness: Challenges for cognitive neuroscience. *In* "The Cognitive Neurosciences III" (Gazzaniga, M. S., ed.), pp. 1121–1132. MIT Press, Cambridge, MA.

Sergent, C. and Dehaene, S. (2004). Is consciousness a gradual phenomenon? Evidence for an all-or-none bifurcation during the attentional blink. *Psychol. Sci.* **15**, 720–728.

Serre, T., Oliva, A., and Poggio, T. (2007). A feedforward architecture accounts for rapid categorization. *Proc. Natl. Acad. Sci. USA* **104**, 6424–6429.

Sheinberg, D. L. and Logothetis, N. K. (1997). The role of temporal cortical areas in perceptual organization. *Proc. Natl. Acad. Sci. USA* **94**, 3408–3413.

Slater, R., Cantarella, A., Gallella, S., Worley, A., Boyd, S., Meek, J., and Fitzgerald, M. (2006). Cortical pain responses in human infants. *J. Neurosci.* **26**, 3662–3666.

Stetson, C., Cui, X., Montague, P. R., and Eagleman, D. M. (2006). Motor-sensory recalibration leads to reversal of action and sensation. *Neuron* **51**, 651–659.

Thorpe, S., Fize, D., and Marlot, C. (1996). Speed of processing in the human visual system. *Nature* **381**, 520–522.

Tong, F., Nakayama, K., Vaughan, J. T., and Kanwisher, N. (1998). Binocular rivalry and visual awareness in human extrastriate cortex. *Neuron* **21**, 753–759.

Tononi, G. (2004). An information integration theory of consciousness. *BMC Neuroscience* **5**, 42–72.

Tononi, G. and Edelman, G. M. (1998). Consciousness and complexity. *Science* **282**, 1846–1851.

Tsuchiya, N. and Koch, C. (2005). Continuous flash suppression reduces negative afterimages. *Nature Neurosci.* **8**, 1096–1101.

VanRullen, R. and Koch, C. (2003a). Is perception discrete or continuous? *Trends Cogn. Sci.* **7**, 207–213.

VanRullen, R. and Koch, C. (2003b). Visual selective behavior can be triggered by a feed-forward process. *Cognitive Neuroscience* **15**, 209–217.

Villablanca, J. R. (2004). Counterpointing the functional role of the forebrain and of the brainstem in the control of the sleep-waking system. *J. Sleep Res.* **13**, 179–208.

Wegner, D. M. (2002). The Illusion of Conscious Will. MIT Press: Cambridge.

Zeki, S. (1998). Parallel processing, asynchronous perception, and a distributed system of consciousness in vision. *Neuroscientist* **4**, 365–372.

Christof Koch

Index